FORMULAS FROM GEOMETRY

Triangle

$h = a \sin \theta$

Area $= \dfrac{1}{2}bh$

Laws of Cosines:
$c^2 = a^2 + b^2 - 2ab \cos \theta$

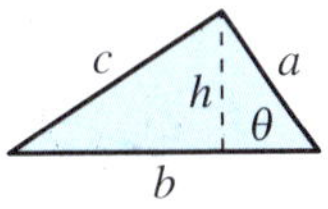

Circular Ring

Area $= \pi(R^2 - r^2)$

$= 2\pi pw$

(p = average radius,
w = width of ring)

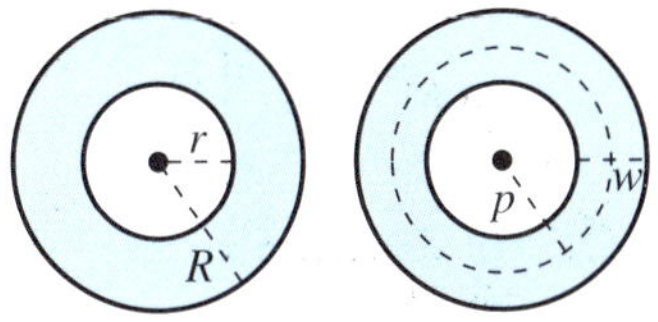

Right Triangle

Pythagorean Theorem:
$c^2 = a^2 + b^2$

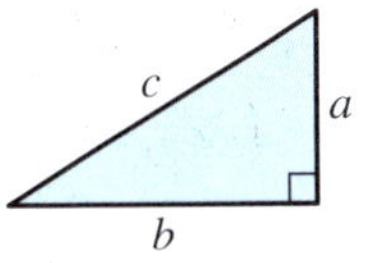

Ellipse

Area $= \pi ab$

Circumference $\approx 2\pi\sqrt{\dfrac{a^2 + b^2}{2}}$

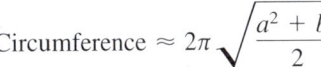

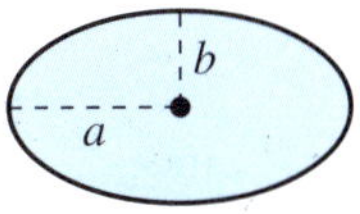

Equilateral Triangle

$h = \dfrac{\sqrt{3}s}{2}$

Area $= \dfrac{\sqrt{3}s^2}{4}$

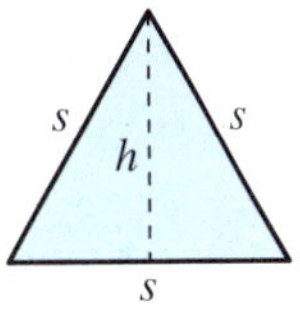

Cone

(A = area of base)

Volume $= \dfrac{Ah}{3}$

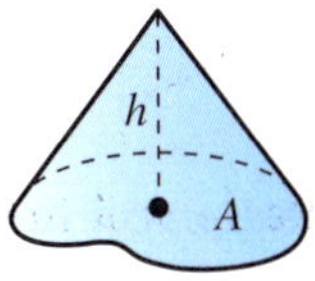

Parallelogram

Area $= bh$

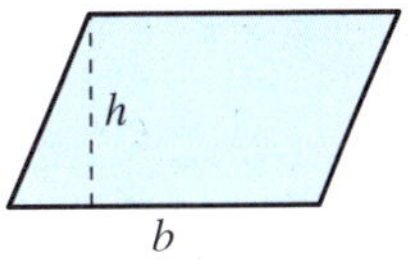

Right Circular Cone

Volume $= \dfrac{\pi r^2 h}{3}$

Lateral Surface Area $= \pi r\sqrt{r^2 + h^2}$

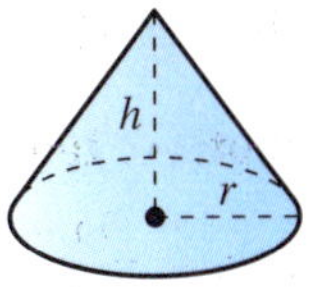

Trapezoid

Area $= \dfrac{h}{2}(a + b)$

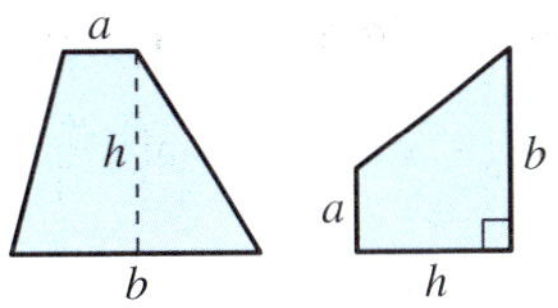

Frustum of Right Circular Cone

Volume $= \dfrac{\pi(r^2 + rR + R^2)h}{3}$

Lateral Surface Area $= \pi s(R + r)$

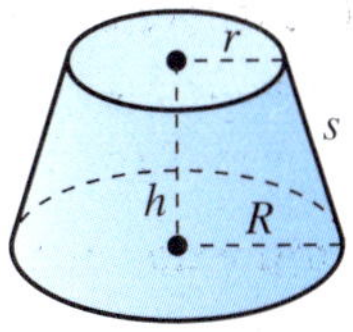

Circle

Area $= \pi r^2$
Circumference $= 2\pi r$

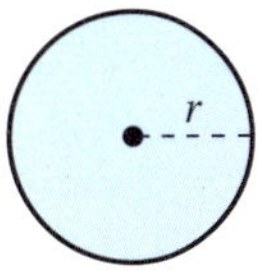

Right Circular Cylinder

Volume $= \pi r^2 h$

Lateral Surface Area $= 2\pi rh$

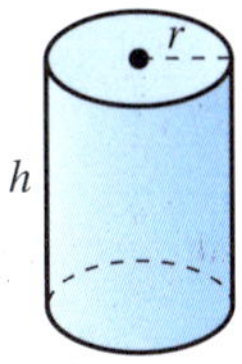

Sector of Circle

Area $= \dfrac{\theta r^2}{2}$

$s = r\theta$

(θ in radians)

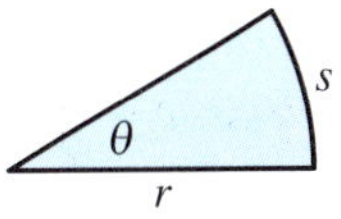

Sphere

Volume $= \dfrac{4}{3}\pi r^3$

Surface Area $= 4\pi r^2$

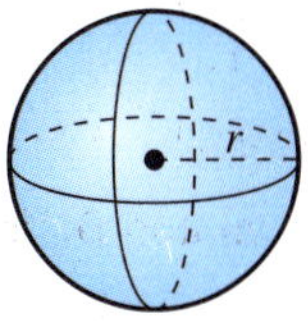

Definition of the Six Trigonometric Functions

Right triangle definitions, where $0 < \theta < \pi/2$

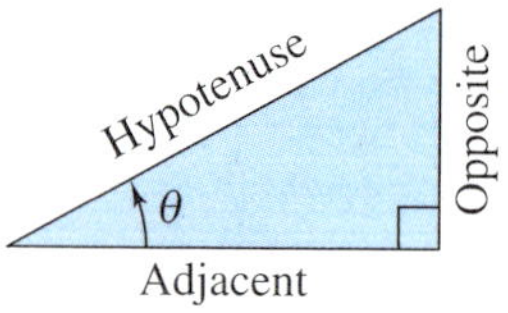

$$\sin\theta = \frac{\text{opp.}}{\text{hyp.}} \qquad \csc\theta = \frac{\text{hyp.}}{\text{opp.}}$$

$$\cos\theta = \frac{\text{adj.}}{\text{hyp.}} \qquad \sec\theta = \frac{\text{hyp.}}{\text{adj.}}$$

$$\tan\theta = \frac{\text{opp.}}{\text{adj.}} \qquad \cot\theta = \frac{\text{adj.}}{\text{opp.}}$$

Circular function definitions, where θ *is any angle*

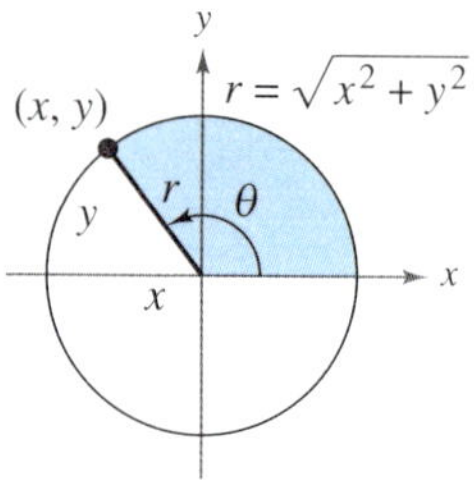

$$\sin\theta = \frac{y}{r} \qquad \csc\theta = \frac{r}{y}$$

$$\cos\theta = \frac{x}{r} \qquad \sec\theta = \frac{r}{x}$$

$$\tan\theta = \frac{y}{x} \qquad \cot\theta = \frac{x}{y}$$

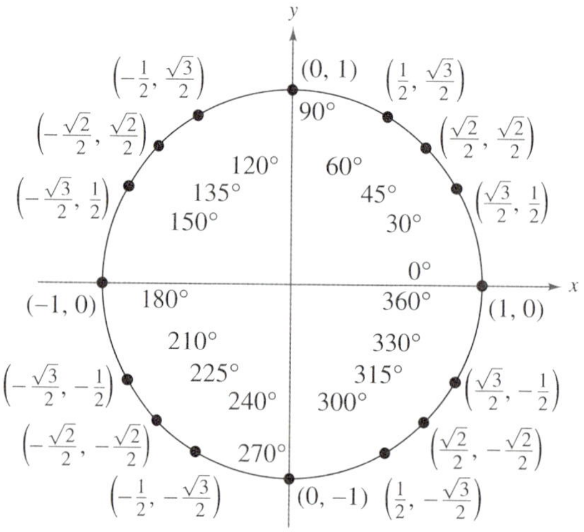

Reciprocal Identities

$$\sin u = \frac{1}{\csc u} \qquad \cos u = \frac{1}{\sec u} \qquad \tan u = \frac{1}{\cot u}$$

$$\csc u = \frac{1}{\sin u} \qquad \sec u = \frac{1}{\cos u} \qquad \cot u = \frac{1}{\tan u}$$

Quotient Identities

$$\tan u = \frac{\sin u}{\cos u} \qquad \cot u = \frac{\cos u}{\sin u}$$

Pythagorean Identities

$$\sin^2 u + \cos^2 u = 1$$

$$1 + \tan^2 u = \sec^2 u \qquad 1 + \cot^2 u = \csc^2 u$$

Cofunction Identities

$$\sin\left(\frac{\pi}{2} - u\right) = \cos u \qquad \cot\left(\frac{\pi}{2} - u\right) = \tan u$$

$$\cos\left(\frac{\pi}{2} - u\right) = \sin u \qquad \sec\left(\frac{\pi}{2} - u\right) = \csc u$$

$$\tan\left(\frac{\pi}{2} - u\right) = \cot u \qquad \csc\left(\frac{\pi}{2} - u\right) = \sec u$$

Even/Odd Identities

$$\sin(-u) = -\sin u \qquad \cot(-u) = -\cot u$$

$$\cos(-u) = \cos u \qquad \sec(-u) = \sec u$$

$$\tan(-u) = -\tan u \qquad \csc(-u) = -\csc u$$

Sum and Difference Formulas

$$\sin(u \pm v) = \sin u \cos v \pm \cos u \sin v$$

$$\cos(u \pm v) = \cos u \cos v \mp \sin u \sin v$$

$$\tan(u \pm v) = \frac{\tan u \pm \tan v}{1 \mp \tan u \tan v}$$

Double-Angle Formulas

$$\sin 2u = 2 \sin u \cos u$$

$$\cos 2u = \cos^2 u - \sin^2 u = 2\cos^2 u - 1 = 1 - 2\sin^2 u$$

$$\tan 2u = \frac{2\tan u}{1 - \tan^2 u}$$

Power-Reducing Formulas

$$\sin^2 u = \frac{1 - \cos 2u}{2}$$

$$\cos^2 u = \frac{1 + \cos 2u}{2}$$

$$\tan^2 u = \frac{1 - \cos 2u}{1 + \cos 2u}$$

Sum-to-Product Formulas

$$\sin u + \sin v = 2\sin\left(\frac{u+v}{2}\right)\cos\left(\frac{u-v}{2}\right)$$

$$\sin u - \sin v = 2\cos\left(\frac{u+v}{2}\right)\sin\left(\frac{u-v}{2}\right)$$

$$\cos u + \cos v = 2\cos\left(\frac{u+v}{2}\right)\cos\left(\frac{u-v}{2}\right)$$

$$\cos u - \cos v = -2\sin\left(\frac{u+v}{2}\right)\sin\left(\frac{u-v}{2}\right)$$

Product-to-Sum Formulas

$$\sin u \sin v = \frac{1}{2}[\cos(u - v) - \cos(u + v)]$$

$$\cos u \cos v = \frac{1}{2}[\cos(u - v) + \cos(u + v)]$$

$$\sin u \cos v = \frac{1}{2}[\sin(u + v) + \sin(u - v)]$$

$$\cos u \sin v = \frac{1}{2}[\sin(u + v) - \sin(u - v)]$$

PRECALCULUS with LIMITS

A Graphing Approach 8e

with CalcChat® and CalcView®

Ron Larson
The Pennsylvania State University
The Behrend College

Paul Battaglia
Brentwood Academy

With the assistance of David C. Falvo
The Pennsylvania State University
The Behrend College

CENGAGE

Australia • Brazil • Mexico • Singapore • United Kingdom • United States

CENGAGE

Precalculus with Limits: A Graphing Approach
with CalcChat and CalcView
Eighth Edition

Ron Larson
Paul Battaglia

Product Director: Daniel Rogers
Product Manager: Raj Desai
Senior Product Marketing Manager: Sebastian Andino
Product Assistant: Emma Collins
Senior Content Project Manager: Jennifer Berry
Senior Content Developer: John Anderson
Manufacturing Planner: Doug Bertke
Senior IP Analyst: Ashley Maynard
Senior Digital Delivery Lead: Mia Bryant
Digital Delivery Lead: Justin Karr
Text and Cover Designer: Larson Texts, Inc.
Cover Image: agsandrew/Shutterstock.com
Compositor: Larson Texts, Inc.

ISBN: 978-1-337-90428-5

Cengage
20 Channel Center Street
Boston, MA 02210
USA

Printed in the United States of America
Print Number: 01 Print Year: 2018

Contents

Preface

Welcome to *Precalculus with Limits: A Graphing Approach,* Eighth Edition. We are excited to present to you this new edition. Although much has changed in this revision, this textbook includes features and resources that continue to make *Precalculus with Limits: A Graphing Approach* a valuable learning tool for students and a trustworthy teaching tool for instructors.

Precalculus with Limits: A Graphing Approach provides the clear instruction, precise mathematics, and thorough coverage that you expect for your course. Additionally, this new edition provides you with **free** access to three companion websites:

- **CalcView.com**—video solutions to selected exercises
- **CalcChat.com**—worked-out solutions to odd-numbered exercises and access to online tutors
- **LarsonPrecalculus.com**—companion website with resources to supplement your learning

These websites will help enhance and reinforce your understanding of the material presented in this text and prepare you for future mathematics courses. CalcView® and CalcChat® are also available as free mobile apps.

Program Features

NEW CalcView®

The website *CalcView.com* contains video solutions of selected exercises. Watch instructors progress step-by-step through solutions, providing guidance to help you solve the exercises. The CalcView mobile app is available for free at the Apple® App Store® or Google Play™ store. The app features an embedded QR Code® reader that can be used to scan the on-page codes and go directly to the videos. You can also access the videos at *CalcView.com.*

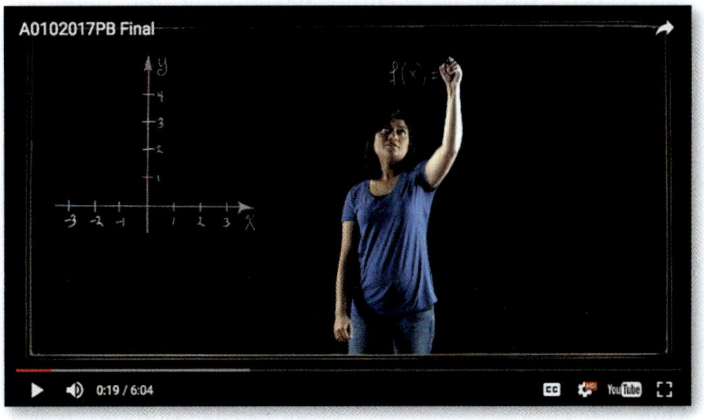

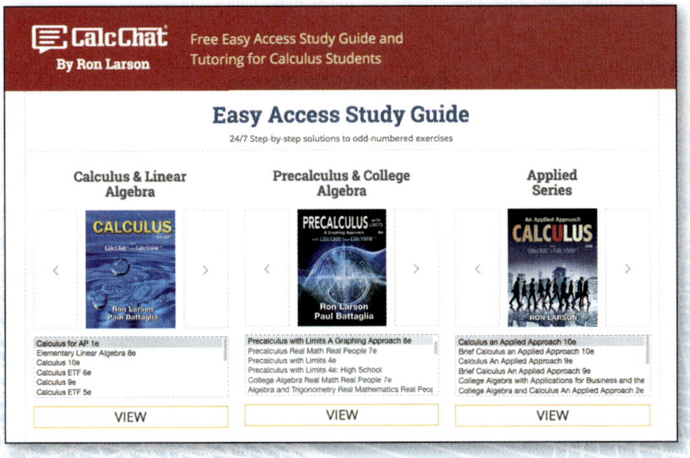

UPDATED CalcChat®

In each exercise set, be sure to notice the reference to *CalcChat.com.* This website provides free step-by-step solutions to all odd-numbered exercises in many of our textbooks. Additionally, you can chat with a tutor, at no charge, during the hours posted at the site. For over 16 years, millions of students have visited this site for help. The CalcChat mobile app is also available as a free download at the Apple® App Store® or Google Play™ store and features an embedded QR Code® reader.

App Store is a service mark of Apple Inc. Google Play is a trademark of Google Inc.
QR Code is a registered trademark of Denso Wave Incorporated.

REVISED LarsonPrecalculus.com

All companion website features have been updated based on this revision, plus we have added a new Collaborative Project feature. Access to these features is free. You can view and listen to worked-out solutions of Checkpoint problems in English or Spanish, explore examples, download data sets, watch lesson videos, and much more.

NEW Collaborative Project

You can find these extended group projects at *LarsonPrecalculus.com.* Check your understanding of the chapter concepts by solving in-depth, real-life problems. These collaborative projects provide an interesting and engaging way for you and other students to work together and investigate ideas.

NEW Insights

Throughout the textbook, numerous Insight notes offer important information to help you plan and prepare for standardized tests and college entrance exams. These helpful notes include tips on how to solve specific questions, details on the different types of questions that could be asked, and ways to be better prepared.

NEW Standardized Test Practice

These questions are modeled after the types of questions you will encounter on standardized tests and college admissions exams and appear after Chapters 3, 6, 9, and 11.

REVISED Exercise Sets

The exercise sets have been carefully and extensively examined to ensure that they are rigorous and relevant, and that they include all topics our users have suggested. The exercises are titled to help you see connections to corresponding examples. Multi-step exercises reinforce problem-solving skills and mastery of concepts by giving you the opportunity to apply the concepts in real-life situations.

REVISED Table of Contents Changes

Section 6.5, The Complex Plane, has been added due to market research and feedback from users. In addition, examples on finding the magnitude of a scalar multiple (Section 6.3), using matrices to transform vectors (Section 7.5), and further applications of 2×2 matrices (Section 7.8) have been added.

Chapter Features

What you should learn/Why you should learn it

These summarize important topics in the section and why they are important in math and in life.

What you should learn
▶ Describe angles.
▶ Use radian measure.
▶ Use degree measure and convert between degrees and radians.
▶ Use angles to model and solve real-life problems.
Why you should learn it
Radian measures of angles are involved in numerous aspects of our daily lives. For instance, in Exercise 104 on page 265, you are asked to determine the measure of the angle generated as a skater performs an axel jump.

Why You Should Learn It Exercise

An engaging real-life application of the concepts in the section. This application exercise is described in the section opener as a motivator for the section.

Dmitry Morgan/Shutterstock.com

Side-By-Side Examples

Throughout the text, we present solutions to examples from multiple perspectives—algebraically, graphically, and numerically. The side-by-side format of this pedagogical feature helps you to see that a problem can be solved in more than one way and to see that different methods yield the same result. The side-by-side format also addresses different learning styles.

What's Wrong?

Each What's Wrong? points out a common error made using graphing utilities.

Library of Parent Functions

To facilitate familiarity with the basic functions, several elementary and nonelementary functions have been compiled as a Library of Parent Functions. Each function is introduced at its first appearance in the text with a definition and description of basic characteristics. The Library of Parent Functions Examples are identified in the title of the example and there is a Review of Library of Parent Functions after Chapter 4. A summary of functions is presented on the inside cover of this text.

Technology Tip

Although a graphing utility can be useful in helping to verify an identity, you must use algebraic techniques to produce a valid proof. For example, graph the two functions $y_1 = \sin 50x$ and $y_2 = \sin 2x$ in a trigonometric viewing window. On some graphing utilities, the graphs appear to be identical. However, $\sin 50x \neq \sin 2x$.

Technology Tip

Technology Tips provide graphing calculator tips or provide alternative methods of solving a problem using a graphing utility.

Algebra of Calculus

Throughout the text, special emphasis is given to the algebraic techniques used in calculus. Algebra of Calculus examples and exercises are integrated throughout the text and are identified by the symbol $\int$.

Algebraic-Graphical-Numerical Exercises

These exercises allow you to solve a problem using multiple approaches—algebraic, graphical, and numerical. This helps you to see that a problem can be solved in more than one way and to see that different methods yield the same result.

Modeling Data Exercises

These multi-part applications that involve real-life data offer you the opportunity to generate and analyze mathematical models.

Error Analysis Exercise

This exercise presents a sample solution that contains a common error which you are asked to identify. We have greatly expanded the number of Error Analysis exercises for this edition.

How Do You See It?

The How Do You See It? feature in each section presents an exercise that you will solve by visual inspection using the concepts learned in the lesson. This exercise is excellent for classroom discussion or test preparation.

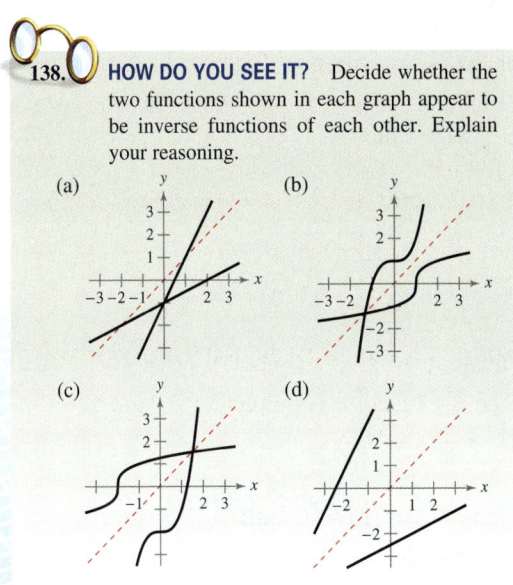

138. HOW DO YOU SEE IT? Decide whether the two functions shown in each graph appear to be inverse functions of each other. Explain your reasoning.

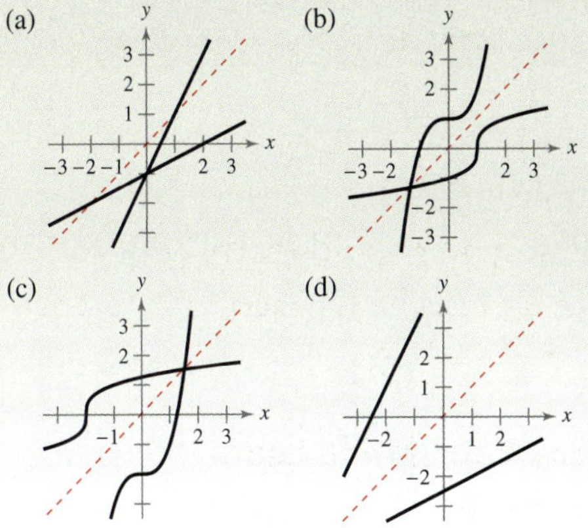

Billion Photos/Shutterstock.com

Checkpoints

Accompanying every example, the Checkpoint problems encourage immediate practice to check your understanding of the concepts presented in the example. View and listen to worked-out solutions of the Checkpoint problems in English or Spanish at *LarsonPrecalculus.com*.

Algebra Help

These hints and tips reinforce or expand upon concepts, help you learn how to study mathematics, address special cases, or show alternative or additional steps to a solution of an example.

Vocabulary and Concept Check

The Vocabulary and Concept Check appears at the beginning of the exercise set for each section. It includes fill-in-the-blank, matching, or non-computational questions designed to help you learn mathematical terminology and to test basic understanding of the concepts of the section.

Explore the Concept

Each Explore the Concept engages you in active discovery of mathematical concepts, strengthens critical thinking skills, and helps build intuition.

> **Explore the Concept**
>
> Find the missing values.
>
> $i^1 = i$ $\quad$ $i^7 = \square$
>
> $i^2 = -1$ $\quad$ $i^8 = \square$
>
> $i^3 = -i$ $\quad$ $i^9 = \square$
>
> $i^4 = 1$ $\quad$ $i^{10} = \square$
>
> $i^5 = \square$ $\quad$ $i^{11} = \square$
>
> $i^6 = \square$ $\quad$ $i^{12} = \square$
>
> What pattern do you see? Write a brief description of how you would find i raised to any positive integer power.

Section Project

The section project at the end of selected sections involves in-depth applied exercises in which you will work with large, real-life data sets, often creating or analyzing models. These exercises are offered online at *LarsonPrecalculus.com*.

Chapter Openers

Each Chapter Opener highlights a real-life modeling problem, showing a graph of the data, a section reference, and a short description of the data.

WEBASSIGN
From Cengage

WebAssign® combines exceptional Precalculus content with the most powerful online homework solution. WebAssign® engages you with immediate feedback, rich tutorial content, and an interactive eBook, MindTap Reader, helping you to develop a deeper conceptual understanding of the subject matter.

Teacher Resources

Teacher's Edition

• ISBN-13: 9780357021996

The Teacher's Edition features facsimile student edition pages with a wrap-around margin containing extensive in-place teacher support material, including Teaching Strategies, Extra Examples, Common Errors, Lesson Closers, and Assignment Guides. Answers to all text exercises, Vocabulary Checks, and Checkpoints are also provided.

Complete Solutions Manual

• ISBN-13: 9780357022009

This manual contains detailed solutions to all exercises from the text, including Chapter Review Exercises and Chapter Tests.

Lesson Plans

• ISBN-13: 9780357022054

Keyed to the text by section, the Lesson Plans provide comprehensive coverage of the course, with section objectives, examples, and tips for the teacher.

Text-Specific DVDs

• ISBN-13: 9781305117143

These text-specific DVDs cover topics from the text—providing explanations of key concepts as well as examples, exercises, and applications in a lecture-based format.

Cengage Learning Testing Powered by Cognero

This flexible online system allows you to author, edit, and manage test bank content; create multiple test versions in an instant; and deliver tests from your LMS, your classroom, or wherever you want. This is available online via *login.cengage.com.*

Instructor Companion Site

Everything you need for your course in one place! This collection of book-specific lecture and class tools is available online via *login.cengage.com.* Access and download PowerPoint Lecture presentations, Note-Taking Guide, Lesson Plans, Solutions Manual, additional test banks, and more.

WEBASSIGN
From Cengage

Instant Access Code: 9780357022078

WebAssign® combines exceptional mathematics content with the most powerful online homework solution. WebAssign® engages your students with immediate feedback, rich tutorial content, and an interactive eBook, MindTap Reader, helping students to develop a deeper conceptual understanding of the subject matter.

Student Resources

Student Solutions Manual

• ISBN-13: 9780357022016

This manual contains step-by-step solutions for all odd-numbered exercises in the text.

Note-Taking Guide, Student Edition

• ISBN-13: 9780357022030

This study aid provides a notebook organizer that helps you develop a section-by-section summary of key concepts.

Student Companion Site

Additional book-specific materials for your course are available online via *login.cengage.com*, including Calculator Keystroke Guides, Note-Taking Guide, and online appendices.

WEBASSIGN From Cengage

Instant Access Code: 9780357022078

WebAssign® combines exceptional mathematics content with the most powerful online homework solution. WebAssign® engages your students with immediate feedback, rich tutorial content, and an interactive eBook, MindTap Reader, helping students to develop a deeper conceptual understanding of the subject matter.

Acknowledgments

We would like to thank the many people who have helped us at various stages of *Precalculus with Limits*. Their encouragement, criticisms, and suggestions have been invaluable to us.

We would particularly like to thank the following reviewers of this and previous editions:

Tony Homayoon Akhlaghi, *Bellevue Community College;* Daniel D. Anderson, *University of Iowa;* Bruce Armbrust, *Lake Tahoe Community College;* Jamie Whitehead Ashby, *Texarkana College;* Teresa Barton, *Western New England College;* Kimberly Bennekin, *Georgia Perimeter College;* Charles M. Biles, *Humboldt State University;* Phyllis Barsch Bolin, *Oklahoma Christian University;* Khristo Boyadzheiv, *Ohio Northern University;* Dave Bregenzer, *Utah State University;* Anne E. Brown, *Indiana University-South Bend;* Diane Burleson, *Central Piedmont Community College;* Beth Burns, *Bowling Green State University;* Alexander Burstein, *University of Rhode Island;* Marilyn Carlson, *University of Kansas;* Hugh Cornell, *University of North Florida;* Victor M. Cornell, *Mesa Community College;* John Dersh, *Grand Rapids Community College;* Jennifer Dollar, *Grand Rapids Community College;* Marcia Drost, *Texas A & M University;* Cameron English, *Rio Hondo College;* Susan E. Enyart, *Otterbein College;* Patricia J. Ernst, *St. Cloud State University;* Eunice Everett, *Seminole Community College;* Kenny Fister, *Murray State University;* Susan C. Fleming, *Virginia Highlands Community College;* Jeff Frost, *Johnson County Community College;* James R. Fryxell, *College of Lake County;* Khadiga H. Gamgoum, *Northern Virginia Community College;* Nicholas E. Geller, *Collin County Community College;* Betty Givan, *Eastern Kentucky University;* Patricia K. Gramling, *Trident Technical College;* Michele Greenfield, *Middlesex County College;* Bernard Greenspan, *University of Akron;* Zenas Hartvigson, *University of Colorado at Denver;* Rodger Hergert, *Rock Valley College;* Allen Hesse, *Rochester Community College;* Rodney Holke-Farnam, *Hawkeye Community College;* Lynda Hollingsworth, *Northwest Missouri State University;* Jean M. Horn, *Northern Virginia Community College;* Spencer Hurd, *The Citadel;* Bill Huston, *Missouri Western State College;* Deborah Johnson, *Cambridge South Dorchester High School;* Francine Winston Johnson, *Howard Community College;* Luella Johnson, *State University of New York, College at Buffalo;* Susan Kellicut, *Seminole Community College;* John Kendall, *Shelby State Community College;* Donna M. Krawczyk, *University of Arizona;* Kewal Krishan, *Hudson County Community College;* Laura Lake, *Center for Advanced Technologies/Lakewood High School;* Peter A. Lappan, *Michigan State University;* Charles G. Laws, *Cleveland State Community College;* JoAnn Lewin, *Edison Community College;* Richard J. Maher, *Loyola University;* Carl Main, *Florida College;* Marilyn McCollum, *North Carolina State University;* Judy McInerney, *Sandhills Community College;* David E. Meel, *Bowling Green University;* Beverly Michael, *University of Pittsburgh;* Wendy Morin, *Dwight D. Eisenhower High School;* Roger B. Nelsen, *Lewis and Clark College;* Stephen Nicoloff, *Paradise Valley Community College;* Jon Odell, *Richland Community College;* Ferdinand Orock, *Hudson County Community College;* Paul Oswood, *Ridgewater College;* Wing M. Park, *College of Lake County;* Rupa M. Patel, *University of Portland;* Robert Pearce, *South Plains College;* David R. Peterson, *University of Central Arkansas;* Marnie Phipps, *North Georgia College and State University;* Sandra Poinsett, *College of Southern Maryland;* James Pommersheim, *Reed College;* Antonio Quesada, *University of Akron;* Laura Reger, *Milwaukee Area Technical College;* Jennifer Rhinehart, *Mars Hill College;* Lila F. Roberts, *Georgia Southern University;* Nancy Schendel, *Iowa Lakes Community College;* Keith Schwingendorf, *Purdue University North Central;* Abdallah Shuaibi, *Truman College;* George W. Shultz, *St. Petersburg Junior College;* Stephen Slack, *Kenyon College;* Judith Smalling, *St. Petersburg Junior College;* Pamela K. M. Smith, *Fort Lewis College;* Cathryn U. Stark, *Collin County Community College;* Craig M. Steenberg, *Lewis-Clark State College;* Mary Jane Sterling, *Bradley University;* G. Bryan Stewart, *Tarrant County Junior College;* Diane Veneziale, *Burlington County College;* Mahbobeh Vezvaei, *Kent State University;* Ellen Vilas, *York Technical College;* Hayat Weiss, *Middlesex Community College;* Rich West, *Francis Marion University;*

Ann Wheeler, *Texas Woman's University;* Vanessa White, *Southern University;* Howard L. Wilson, *Oregon State University;* Joel E. Wilson, *Eastern Kentucky University;* Michelle Wilson, *Franklin University;* Paul Winterbottom, *Montgomery County Community College;* Fred Worth, *Henderson State University;* Karl M. Zilm, *Lewis and Clark Community College;* Cathleen Zucco-Teveloff, *Rowan University*

We hope that you enjoy learning the mathematics presented in this text. More than that, we hope you gain a new appreciation for the relevance of mathematics to careers in science, technology, business, and medicine.

Our thanks to Robert Hostetler, The Behrend College, The Pennsylvania State University, Bruce Edwards, University of Florida, and David Heyd, The Behrend College, The Pennsylvania State University, for their significant contributions to previous editions of this text.

We would also like to thank the staff of Larson Texts, Inc. who assisted in preparing the manuscript, rendering the art package, and typesetting and proofreading the pages and supplements.

On a personal level, we are grateful to our spouses, Deanna Gilbert Larson and Janice Battaglia, for their love, patience, and support. Also, a special thanks goes to R. Scott O'Neil.

If you have suggestions for improving this text, please feel free to write us. Over the years, we have received many useful comments from both instructors and students, and we value these very much.

Ron Larson
Paul Battaglia

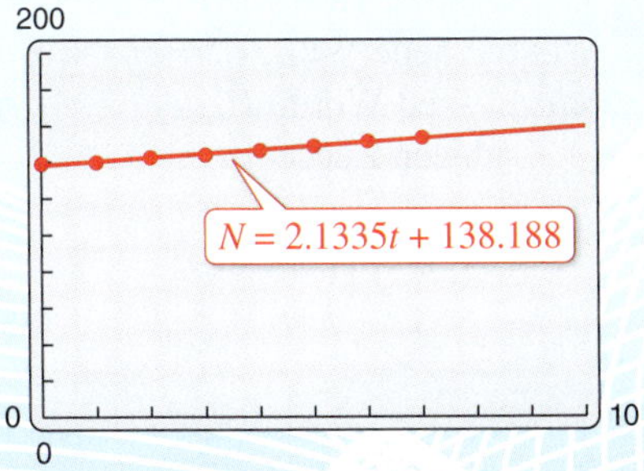

Section 1.7, Example 4
Number of People
Employed in the U.S.

1 Functions and Their Graphs

Student Resources at LarsonPrecalculus.com

- **Videos** explaining the concepts of precalculus
- **Worked-out solution videos** for all *Checkpoint* exercises
- **Editable spreadsheets** of the data sets in the text
- **Group projects** for each chapter applying concepts to real-life problems

kemie/DigitalVision Vectors/Getty Images; DMEPhotography/iStock/Getty Images

Introduction to Library of Parent Functions

In Chapter 1, you will be introduced to the concept of a *function*. As you proceed through the text, you will see that functions play a primary role in modeling real-life situations.

There are three basic types of functions that have proven to be the most important in modeling real-life situations. These functions are algebraic functions, exponential and logarithmic functions, and trigonometric and inverse trigonometric functions. These three types of functions are referred to as the *elementary functions*, though they are often placed in the two categories of *algebraic functions* and *transcendental functions*. Each time a new type of function is studied in detail in this text, it will be highlighted in a box similar to those shown below. The graphs of these functions are shown on the inside cover of this text.

Algebraic Functions

These functions are formed by applying algebraic operations to the linear function $f(x) = x$.

Name	*Function*	*Location*
Linear	$f(x) = x$	Section 1.1
Quadratic	$f(x) = x^2$	Section 2.1
Cubic	$f(x) = x^3$	Section 2.2
Rational	$f(x) = \dfrac{1}{x}$	Section 2.7
Square root	$f(x) = \sqrt{x}$	Section 1.2

Transcendental Functions

These functions cannot be formed from the linear function by using algebraic operations.

Name	*Function*	*Location*
Exponential	$f(x) = a^x, a > 0, a \neq 1$	Section 3.1
Logarithmic	$f(x) = \log_a x, x > 0, a > 0, a \neq 1$	Section 3.2
Trigonometric	$f(x) = \sin x$	Section 4.5
	$f(x) = \cos x$	Section 4.5
	$f(x) = \tan x$	Section 4.6
	$f(x) = \csc x$	Section 4.6
	$f(x) = \sec x$	Section 4.6
	$f(x) = \cot x$	Section 4.6
Inverse trigonometric	$f(x) = \arcsin x$	Section 4.7
	$f(x) = \arccos x$	Section 4.7
	$f(x) = \arctan x$	Section 4.7

Nonelementary Functions

Some useful nonelementary functions include the following.

Name	*Function*	*Location*
Absolute value	$f(x) = \|x\|$	Section 1.2
Greatest integer	$f(x) = [\![x]\!]$	Section 1.3

1.1 Lines in the Plane

The Slope of a Line

In this section, you will study lines and their equations. The **slope** of a nonvertical line represents the number of units the line rises or falls vertically for each unit of horizontal change from left to right. For instance, consider the two points

$$(x_1, y_1) \quad \text{and} \quad (x_2, y_2)$$

on the line shown in Figure 1.1.

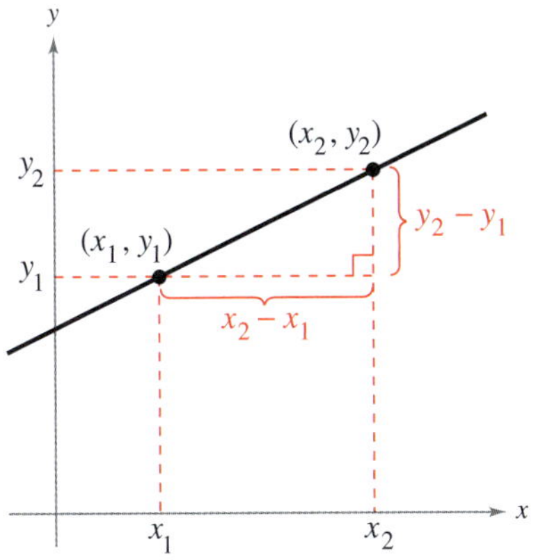

Figure 1.1

As you move from left to right along this line, a change of $(y_2 - y_1)$ units in the vertical direction corresponds to a change of $(x_2 - x_1)$ units in the horizontal direction. That is,

$$y_2 - y_1 = \text{the change in } y$$

and

$$x_2 - x_1 = \text{the change in } x.$$

The slope of the line is given by the ratio of these two changes.

Definition of the Slope of a Line

The **slope** m of the nonvertical line through (x_1, y_1) and (x_2, y_2) is

$$m = \frac{y_2 - y_1}{x_2 - x_1} = \frac{\text{change in } y}{\text{change in } x}$$

where $x_1 \neq x_2$.

When this formula for slope is used, the *order of subtraction* is important. Given two points on a line, you are free to label either one of them as (x_1, y_1) and the other as (x_2, y_2). Once you have done this, however, you must form the numerator and denominator using the same order of subtraction.

$$\underbrace{m = \frac{y_2 - y_1}{x_2 - x_1}}_{\text{Correct}} \qquad \underbrace{m = \frac{y_1 - y_2}{x_1 - x_2}}_{\text{Correct}} \qquad \underbrace{m = \frac{y_2 - y_1}{x_1 - x_2}}_{\text{Incorrect}}$$

Throughout this text, the term *line* always means a *straight* line.

What you should learn

- Find the slopes of lines.
- Write linear equations given points on lines and their slopes.
- Use slope-intercept forms of linear equations to sketch lines.
- Use slope to identify parallel and perpendicular lines.

Why you should learn it

The slope of a line can be used to solve real-life problems. For instance, in Exercise 101 on page 14, you will use a linear equation to model student enrollment at the University of Massachusetts Amherst.

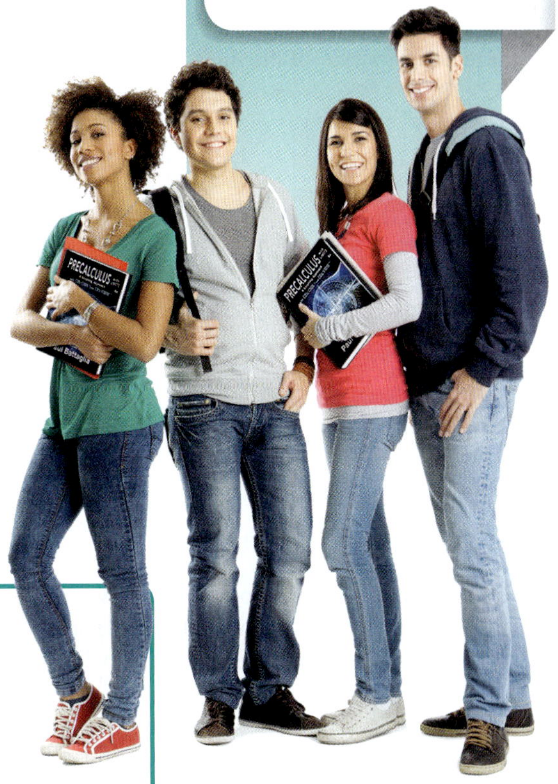

iStock.com/Ridofranz

EXAMPLE 1 Finding the Slope of a Line

Find the slope of the line passing through each pair of points.

a. $(-2, 0)$ and $(3, 1)$ **b.** $(-1, 2)$ and $(2, 2)$ **c.** $(0, 4)$ and $(1, -1)$

Solution

$$\textbf{a. } m = \underbrace{\overbrace{\frac{y_2 - y_1}{x_2 - x_1}}^{\text{Difference in } y\text{-values}}}_{\text{Difference in } x\text{-values}} = \frac{1 - 0}{3 - (-2)} = \frac{1}{3 + 2} = \frac{1}{5}$$

$$\textbf{b. } m = \frac{2 - 2}{2 - (-1)} = \frac{0}{3} = 0$$

$$\textbf{c. } m = \frac{-1 - 4}{1 - 0} = \frac{-5}{1} = -5$$

The graphs of the three lines are shown in Figure 1.2. Note that a square setting gives the correct "steepness" of the lines.

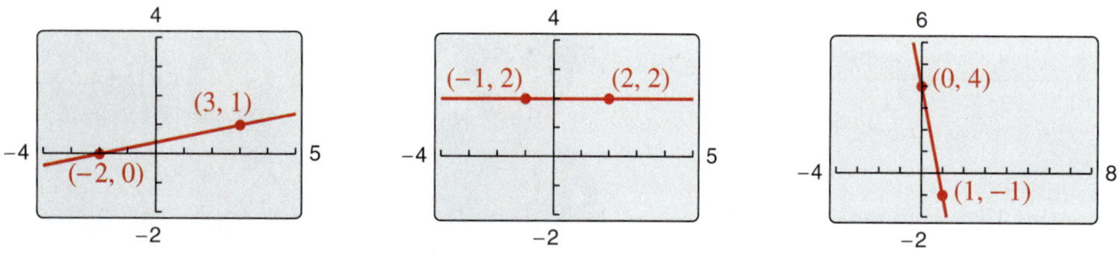

(a) (b) (c)

Figure 1.2

Explore the Concept

Use a graphing utility to compare the slopes of the lines $y = 0.5x$, $y = x$, $y = 2x$, and $y = 4x$. What do you observe about these lines? Compare the slopes of the lines $y = -0.5x$, $y = -x$, $y = -2x$, and $y = -4x$. What do you observe about these lines? (*Hint:* Use a square setting to obtain a true geometric perspective.)

✓ Checkpoint ▶ *Audio-video solution in English & Spanish at LarsonPrecalculus.com*

Find the slope of the line passing through each pair of points.

a. $(-5, -6)$ and $(2, 8)$ **b.** $(4, 2)$ and $(2, 5)$ **c.** $(0, -1)$ and $(3, -1)$

The definition of slope does not apply to vertical lines. For instance, consider the points $(3, 4)$ and $(3, 1)$ on the vertical line shown in Figure 1.3. Applying the formula for slope, you obtain

$$m = \frac{4 - 1}{3 - 3} = \frac{3}{0}. \quad \text{✗} \qquad \text{Undefined}$$

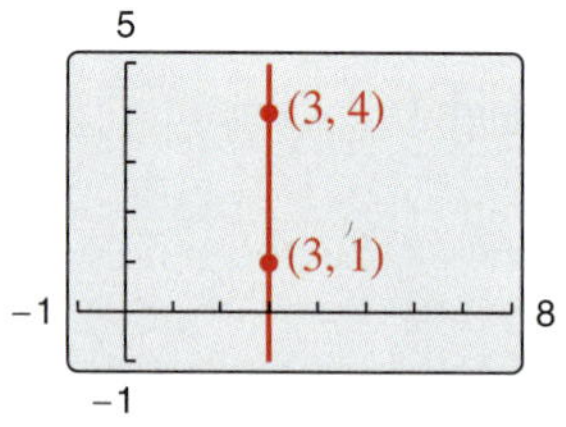

Figure 1.3

Because division by zero is undefined, the slope of a vertical line is undefined.

From the lines shown in Figures 1.2 and 1.3, you can make the following generalizations about the slope of a line.

The Slope of a Line

1. A line with positive slope $(m > 0)$ *rises* from left to right.
2. A line with negative slope $(m < 0)$ *falls* from left to right.
3. A line with zero slope $(m = 0)$ is *horizontal*.
4. A line with undefined slope is *vertical*.

Technology Tip

The standard viewing window on many graphing utilities does not give a true geometric perspective because the screen is rectangular, which distorts the image. That is, perpendicular lines will not appear to be perpendicular, and circles will not appear to be circular. To overcome this, you can use a square setting, in which the horizontal and vertical tick marks have equal spacing, as shown in Figure 1.2. On many graphing utilities, a square setting can be obtained when the ratio of the range of y to the range of x is 2 to 3.

The Point-Slope Form of the Equation of a Line

When you know the slope of a line *and* you also know the coordinates of one point on the line, you can find an equation of the line. For instance, in Figure 1.4, let (x_1, y_1) be a point on the line whose slope is m. When (x, y) is any *other* point on the line, it follows that

$$\frac{y - y_1}{x - x_1} = m.$$

This equation in the variables x and y can be rewritten in the **point-slope form** of the equation of a line.

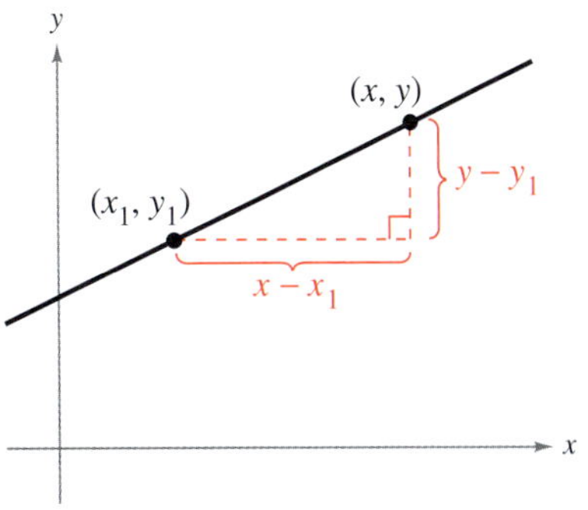

Figure 1.4

Point-Slope Form of the Equation of a Line

The **point-slope form** of the equation of the line that passes through the point (x_1, y_1) and has a slope of m is

$$y - y_1 = m(x - x_1).$$

EXAMPLE 2 The Point-Slope Form of the Equation of a Line

Find an equation of the line that passes through the point

$(1, -2)$

and has a slope of 3.

Solution

$y - y_1 = m(x - x_1)$ Point-slope form

$y - (-2) = 3(x - 1)$ Substitute for y_1, m, and x_1.

$y + 2 = 3x - 3$ Simplify.

$y = 3x - 5$ Solve for y.

The line is shown in Figure 1.5.

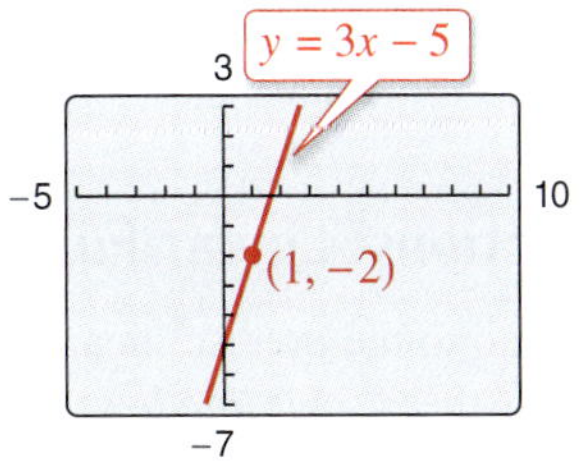

Figure 1.5

✓ ***Checkpoint*** *Audio-video solution in English & Spanish at LarsonPrecalculus.com*

Find an equation of the line that passes through the point $(3, -7)$ and has a slope of 2.

The point-slope form can be used to find an equation of a nonvertical line passing through two points

(x_1, y_1) and (x_2, y_2).

First, find the slope of the line.

$$m = \frac{y_2 - y_1}{x_2 - x_1}, \quad x_1 \neq x_2$$

Then use the point-slope form to obtain the equation

$$y - y_1 = \frac{y_2 - y_1}{x_2 - x_1}(x - x_1).$$

This is sometimes called the **two-point form** of the equation of a line.

Algebra Help

When you find an equation of the line that passes through two given points, you need to substitute the coordinates of only one of the points into the point-slope form. It does not matter which point you choose because both points will yield the same result.

EXAMPLE 3 A Linear Model for Profits Prediction

Comcast had revenues of \$68.775 billion in 2014 and \$74.510 billion in 2015. Write a linear equation giving the revenues y in terms of the year x. Then use the equation to predict the revenues for 2016. (*Source:* Comcast Corp.)

Solution

Let $x = 0$ represent 2010. In Figure 1.6, let $(4, 68.775)$ and $(5, 74.510)$ be two points on the line representing the revenues. The slope of this line is

$$m = \frac{74.510 - 68.775}{5 - 4} = 5.735.$$

Next, use the point-slope form to find the equation of the line.

$$y - 68.775 = 5.735(x - 4)$$

$$y = 5.735x + 45.835$$

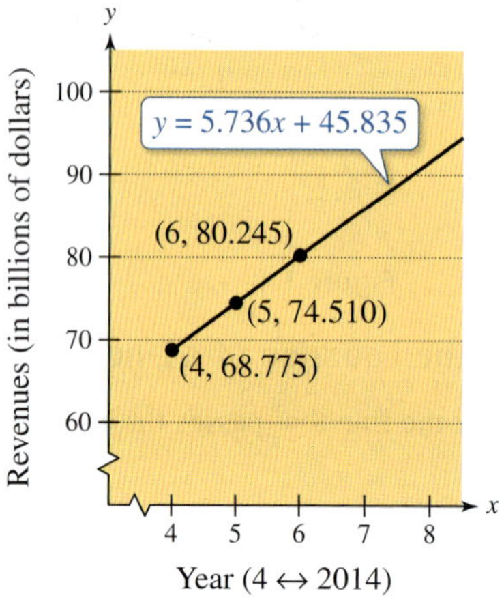

Figure 1.6

Now, using this equation, you can predict the 2016 revenues ($x = 6$) to be

$$y = 5.735(6) + 45.835 = 34.41 + 45.835 = \$80.245 \text{ billion.}$$

(In this case, the prediction is quite good—the actual revenues in 2016 were \$80.403 billion.)

✓ Checkpoint ▶ *Audio-video solution in English & Spanish at LarsonPrecalculus.com*

Costco had revenues of \$112.640 billion in 2014 and \$116.199 billion in 2015. Write a linear equation giving the revenues y in terms of the year x. Then use the equation to predict the revenues for 2016. (*Source:* Costco Wholesale Corp.) ■

Insight

Some standardized tests may asses your ability to interpret linear equations in a real-life context (see Example 3).

Library of Parent Functions: Linear Function

In the next section, you will be introduced to the precise meaning of the term *function*. The simplest type of function is the *parent linear function*

$$f(x) = x.$$

As its name implies, the graph of the parent linear function is a line. The basic characteristics of the parent linear function are summarized below and on the inside cover of this text. (Note that some of the terms below will be defined later in the text.)

Graph of $f(x) = x$
Domain: $(-\infty, \infty)$
Range: $(-\infty, \infty)$
Intercept: $(0, 0)$
Increasing

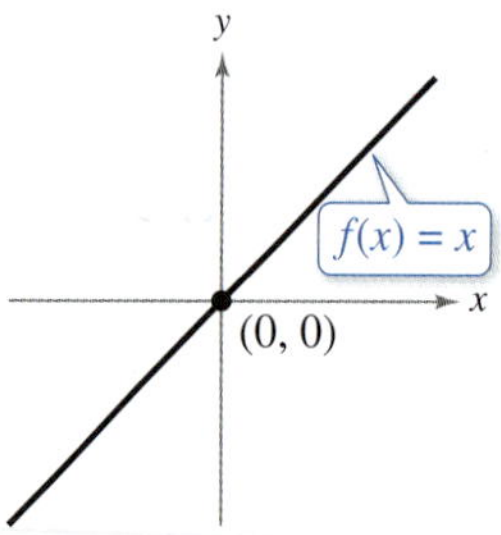

The function $f(x) = x$ is also referred to as the *identity function*. Later in this text, you will learn that the graph of the linear function $f(x) = mx + b$ is a line with slope m and y-intercept $(0, b)$. When $m = 0$, $f(x) = b$ is called a *constant function* and its graph is a horizontal line.

Sketching Graphs of Lines

Many problems in coordinate geometry can be classified in two categories.

1. Given a graph (or parts of it), find its equation.
2. Given an equation, sketch its graph.

For lines, the first problem can be solved by using the point-slope form. This formula, however, is not particularly useful for solving the second type of problem. The form that is better suited to graphing linear equations is the **slope-intercept form** of the equation of a line, $y = mx + b$.

Slope-Intercept Form of the Equation of a Line

The graph of the equation

$$y = mx + b$$

is a line whose slope is m and whose y-intercept is $(0, b)$.

EXAMPLE 4 Using the Slope-Intercept Form

See LarsonPrecalculus.com for an interactive version of this type of example.

Determine the slope and y-intercept of each linear equation. Then describe its graph.

a. $x + y = 2$ **b.** $y = 2$

Algebraic Solution

a. Begin by writing the equation in slope-intercept form.

$x + y = 2$ Write original equation.

$y = 2 - x$ Subtract x from each side.

$y = -x + 2$ Write in slope-intercept form.

From the slope-intercept form of the equation, the slope is -1 and the y-intercept is

$(0, 2)$.

Because the slope is negative, you know that the graph of the equation is a line that falls one unit for every unit it moves to the right.

b. By writing the equation $y = 2$ in slope-intercept form

$y = (0)x + 2$

you can see that the slope is 0 and the y-intercept is

$(0, 2)$.

A zero slope implies that the line is horizontal.

Graphical Solution

a.

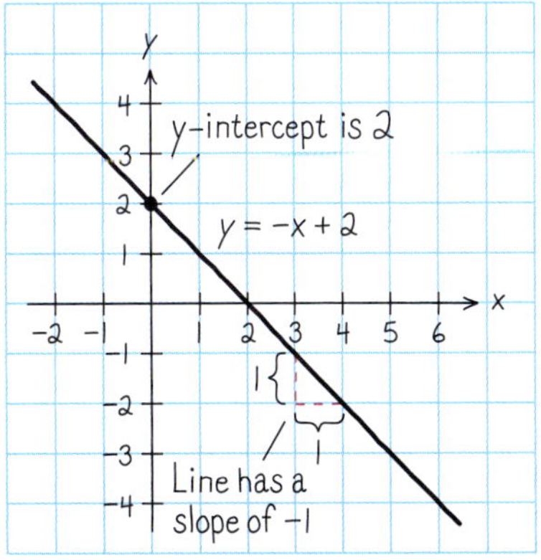

b.

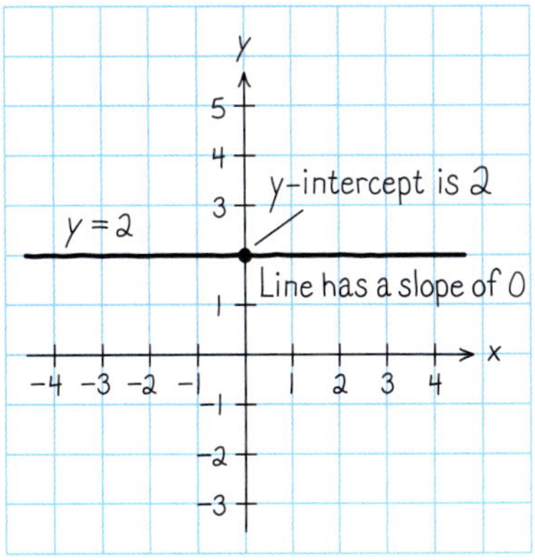

Checkpoint Audio-video solution in English & Spanish at LarsonPrecalculus.com

Determine the slope and y-intercept of $x - 2y = 4$. Then describe its graph.

From the slope-intercept form of the equation of a line, you can see that a horizontal line ($m = 0$) has an equation of the form $y = b$. This is consistent with the fact that each point on a horizontal line through $(0, b)$ has a y-coordinate of b. Similarly, each point on a vertical line through $(a, 0)$ has an x-coordinate of a. So, a vertical line has an equation of the form $x = a$. This equation cannot be written in slope-intercept form because the slope of a vertical line is undefined. However, every line has an equation that can be written in the **general form**

$Ax + By + C = 0$ General form of the equation of a line

where A and B are not *both* zero.

Summary of Equations of Lines

1. General form:	$Ax + By + C = 0$
2. Vertical line:	$x = a$
3. Horizontal line:	$y = b$
4. Slope-intercept form:	$y = mx + b$
5. Point-slope form:	$y - y_1 = m(x - x_1)$

EXAMPLE 5 Different Viewing Windows

When a graphing utility is used to graph a line, it is important to realize that the line may not visually appear to have the slope indicated by its equation. This occurs because of the viewing window used for the graph. For instance, Figure 1.7 shows graphs of $y = 2x + 1$ produced on a graphing utility using three different viewing windows. Notice that the slopes in Figures 1.7(a) and (b) do not visually appear to be equal to 2. When you use a *square setting*, as in Figure 1.7(c), the slope visually appears to be 2.

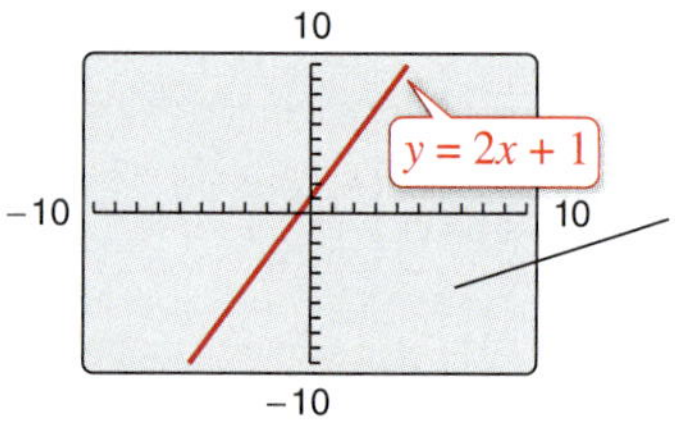

Using a *nonsquare setting*, you do *not* obtain a graph with a true geometric perspective. So, the slope does *not* visually appear to be 2.

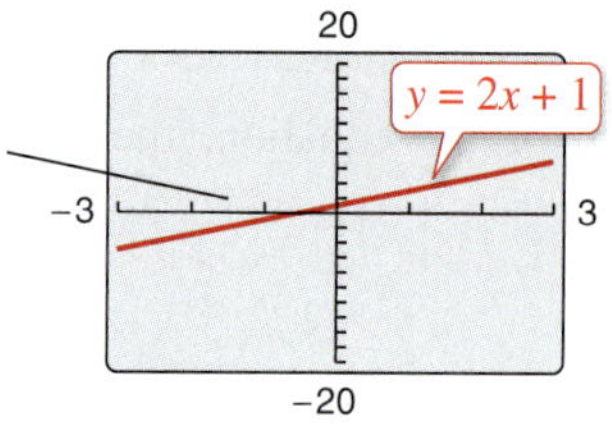

(a) *Nonsquare setting* (b) *Nonsquare setting*

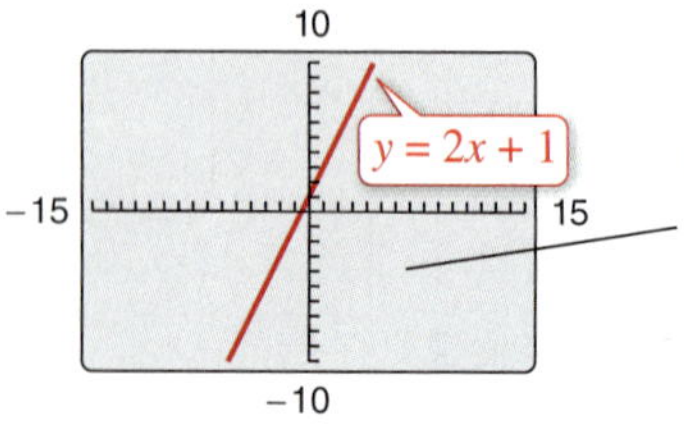

Using a *square setting*, you can obtain a graph with a true geometric perspective. So, the slope visually appears to be 2.

(c) *Square setting*

Figure 1.7

✓ Checkpoint *Audio-video solution in English & Spanish at LarsonPrecalculus.com*

Use a graphing utility to graph $y = 0.5x - 3$ using each viewing window. Describe the difference in the graphs.

a. Xmin = -5, Xmax = 10, Xscl = 1, Ymin = -1, Ymax = 10, Yscl = 1

b. Xmin = -2, Xmax = 10, Xscl = 1, Ymin = -4, Ymax = 1, Yscl = 1

c. Xmin = -5, Xmax = 10, Xscl = 1, Ymin = -7, Ymax = 3, Yscl = 1

Explore the Concept

Graph the lines $y_1 = \frac{1}{2}x + 1$ and $y_2 = -2x + 1$ in the same viewing window using a square setting. What do you observe?

Graph the lines $y_1 = 2x + 1$, $y_2 = 2x$, and $y_3 = 2x - 1$ in the same viewing window using a square setting. What do you observe?

Parallel and Perpendicular Lines

The slope of a line is a convenient tool for determining whether two lines are parallel or perpendicular.

Parallel Lines

Two distinct nonvertical lines are **parallel** if and only if their slopes are equal. That is, $m_1 = m_2$.

Algebra Help

In mathematics, the phrase "if and only if" is a way of stating two propositions in one statement. For instance, the statement at the left could be rewritten as the following two propositions.

a. If two distinct nonvertical lines are parallel, then their slopes are equal.

b. If two distinct nonvertical lines have equal slopes, then they are parallel.

EXAMPLE 6 Equations of Parallel Lines

Find the slope-intercept form of the equation of the line that passes through the point $(2, -1)$ and is parallel to the line $2x - 3y = 5$.

Solution

Begin by writing the equation of the line in slope-intercept form.

$2x - 3y = 5$ Write original equation.

$-2x + 3y = -5$ Multiply by -1.

$3y = 2x - 5$ Add $2x$ to each side.

$y = \frac{2}{3}x - \frac{5}{3}$ Write in slope-intercept form.

Therefore, the given line has a slope of

$$m = \frac{2}{3}.$$

Any line parallel to the given line must also have a slope of $\frac{2}{3}$. So, the line through $(2, -1)$ has the following equation.

$y - y_1 = m(x - x_1)$ Point-slope form

$y - (-1) = \frac{2}{3}(x - 2)$ Substitute for y_1, m, and x_1.

$y + 1 = \frac{2}{3}x - \frac{4}{3}$ Simplify.

$y = \frac{2}{3}x - \frac{7}{3}$ Write in slope-intercept form.

Notice the similarity between the slope-intercept form of the original equation and the slope-intercept form of the parallel equation. The graphs of both equations are shown in Figure 1.8 using a square setting.

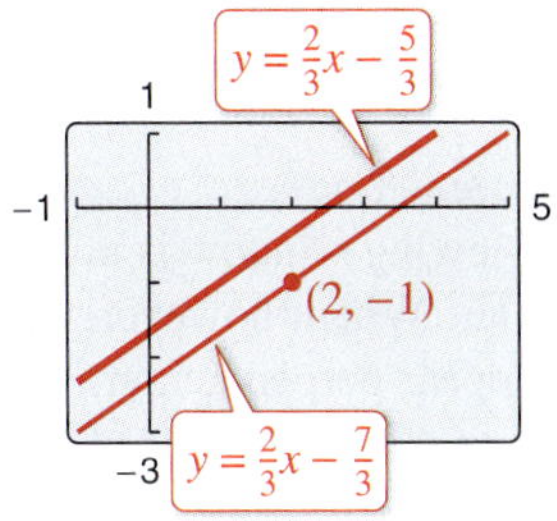

Figure 1.8

✓ ***Checkpoint*** *Audio-video solution in English & Spanish at LarsonPrecalculus.com*

Find the slope-intercept form of the equation of the line that passes through the point $(-4, 1)$ and is parallel to the line $5x - 3y = 8$.

Perpendicular Lines

Two nonvertical lines are **perpendicular** if and only if their slopes are negative reciprocals of each other. That is,

$$m_1 = -\frac{1}{m_2}.$$

EXAMPLE 7 Equations of Perpendicular Lines

Find the slope-intercept form of the equation of the line that passes through the point $(2, -1)$ and is perpendicular to the line $2x - 3y = 5$.

Solution

From Example 6, you know that the equation can be written in the slope-intercept form $y = \frac{2}{3}x - \frac{5}{3}$. You can see that the line has a slope of $\frac{2}{3}$. So, any line perpendicular to this line must have a slope of $-\frac{3}{2}$ (because $-\frac{3}{2}$ is the negative reciprocal of $\frac{2}{3}$). So, the line through the point $(2, -1)$ has the following equation.

$$y - (-1) = -\tfrac{3}{2}(x - 2) \qquad \text{Write in point-slope form.}$$

$$y + 1 = -\tfrac{3}{2}x + 3 \qquad \text{Simplify.}$$

$$y = -\tfrac{3}{2}x + 2 \qquad \text{Write in slope-intercept form.}$$

The graphs of both equations are shown in the figure.

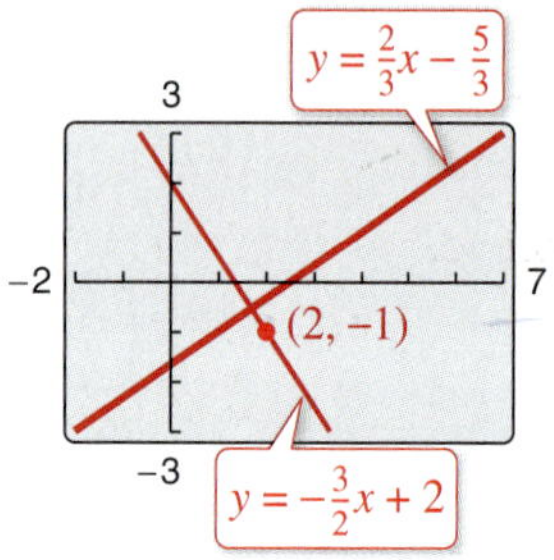

✓ **Checkpoint** ▶ *Audio-video solution in English & Spanish at LarsonPrecalculus.com*

Find the slope-intercept form of the equation of the line that passes through the point $(-4, 1)$ and is perpendicular to the line $5x - 3y = 8$.

EXAMPLE 8 Graphs of Perpendicular Lines

Use a graphing utility to graph the lines $y = x + 1$ and $y = -x + 3$ in the same viewing window. The lines are perpendicular (they have slopes of $m_1 = 1$ and $m_2 = -1$). Do they appear to be perpendicular on the display?

Solution

When the viewing window is nonsquare, as in Figure 1.9, the two lines will not appear perpendicular. When, however, the viewing window is square, as in Figure 1.10, the lines will appear perpendicular.

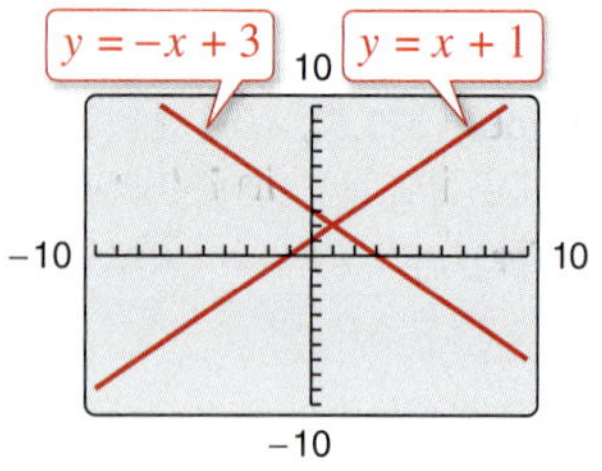

Figure 1.9

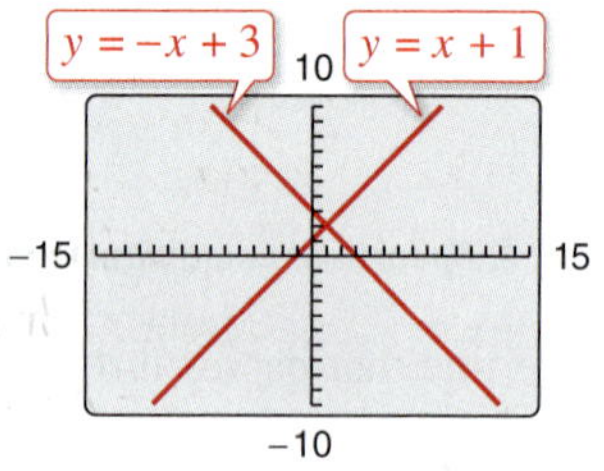

Figure 1.10

✓ **Checkpoint** ▶ *Audio-video solution in English & Spanish at LarsonPrecalculus.com*

Identify any relationships that exist among the lines $y = 2x$, $y = -2x$, and $y = \frac{1}{2}x$. Then use a graphing utility to graph the three equations in the same viewing window. Adjust the viewing window so that each slope appears visually correct. Use the slopes of the lines to verify your results.

What's Wrong?

You use a graphing utility to graph $y_1 = 1.5x$ and $y_2 = -1.5x + 5$, as shown in the figure. You use the graph to conclude that the lines are perpendicular. What's wrong?

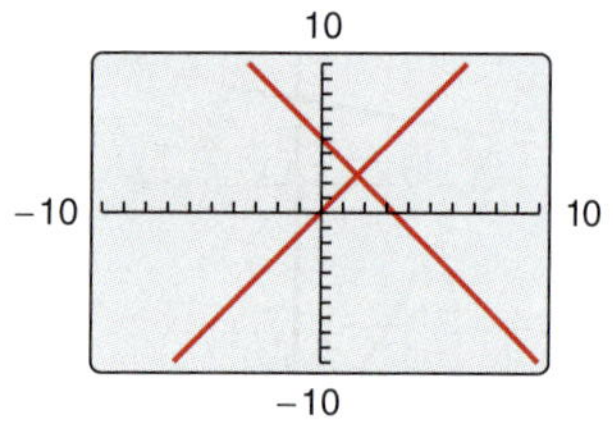

1.1 Exercises

See *CalcChat.com* for tutorial help and worked-out solutions to odd-numbered exercises.
For instructions on how to use a graphing utility, see Appendix A.

Vocabulary and Concept Check

1. Match each equation with its form.

(a) $Ax + By + C = 0$ (i) vertical line

(b) $x = a$ (ii) slope-intercept form

(c) $y = b$ (iii) general form

(d) $y = mx + b$ (iv) point-slope form

(e) $y - y_1 = m(x - x_1)$ (v) horizontal line

In Exercises 2 and 3, fill in the blank.

2. For a line, the ratio of the change in y to the change in x is called the _______ of the line.

3. Two lines are _______ if and only if their slopes are equal.

4. What is the relationship between two lines whose slopes are -3 and $\frac{1}{3}$?

5. What is the slope of a line that is perpendicular to the line represented by $x = 3$?

6. Give the coordinates of a point on the line whose equation in point-slope form is $y - (-1) = \frac{1}{4}(x - 8)$.

Procedures and Problem Solving

Using Slope **In Exercises 7 and 8, identify the line that has the indicated slope.**

7. (a) $m = \frac{2}{3}$ (b) m is undefined. (c) $m = -2$

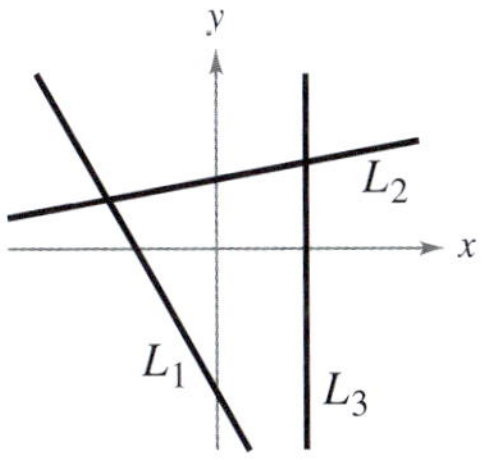

8. (a) $m = 0$ (b) $m = -\frac{3}{4}$ (c) $m = 1$

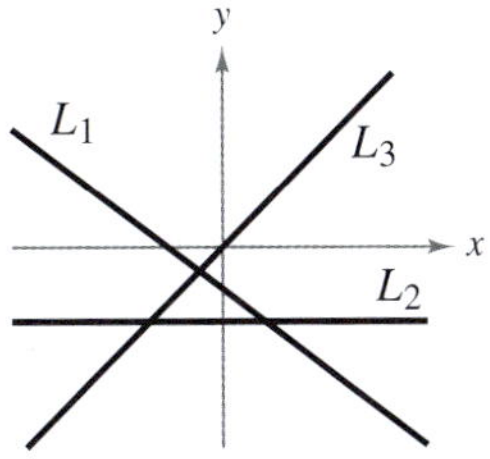

Estimating Slope **In Exercises 9 and 10, estimate the slope of the line.**

9.

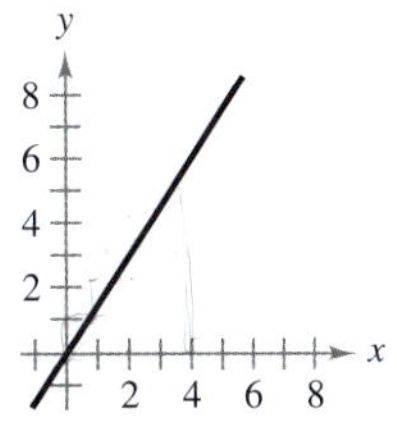

10.

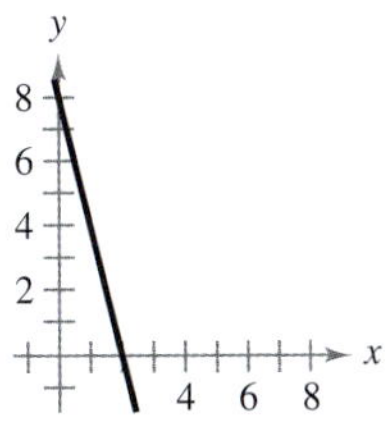

Sketching Lines **In Exercises 11 and 12, sketch the lines through the point with the indicated slopes on the same set of coordinate axes.**

	Point	*Slopes*			
11.	$(2, 3)$	(a) 0	(b) 1	(c) 2	(d) -3
12.	$(-4, 1)$	(a) 4	(b) -2	(c) $\frac{1}{2}$	(d) Undefined

Finding the Slope of a Line **In Exercises 13–22, find the slope of the line passing through the pair of points.**

13. $(0, 9), (6, 0)$

14. $(10, 0), (0, -5)$

15. $(-3, -2), (1, 6)$

16. $(2, -1), (-2, 1)$

17. $(5, -7), (8, -7)$

18. $(-2, 1), (-4, -5)$

19. $(-3, -1), (-3, 4)$

20. $(0, -15), (-6, 0)$

21. $(4.8, 3.1), (-5.2, 1.6)$

22. $\left(\frac{11}{2}, -\frac{4}{3}\right), \left(-\frac{3}{2}, -\frac{1}{3}\right)$

Finding the Slope of a Line **In Exercises 23–26, find the slope of the line passing through the pair of points. Then use a graphing utility to plot the points and use the *draw* feature to graph the line segment connecting the two points. (Use a *square setting*.)**

23. $(0, -10), (-4, 0)$

24. $(2, 4), (4, -4)$

25. $(-6, -1), (-6, 4)$

26. $(4, 9), (6, 12)$

Using Slope **In Exercises 27–32, use the point on the line and the slope of the line to find three additional points through which the line passes. (There are many correct answers.)**

	Point	*Slope*
27.	$(2, 1)$	$m = 0$
28.	$(-4, 1)$	m is undefined.
29.	$(0, -9)$	$m = -2$
30.	$(-4, 5)$	$m = 3$
31.	$(7, -2)$	$m = \frac{1}{2}$
32.	$(-1, -6)$	$m = -\frac{1}{3}$

The symbol and a red exercise number indicate that a video solution can be seen at *CalcView.com*.

The Point-Slope Form of the Equation of a Line **In Exercises 33–40, find an equation of the line that passes through the point with the indicated slope. Sketch the line by hand. Use a graphing utility to verify your sketch, if possible.**

33. $(0, -2)$, $m = 3$
34. $(-3, 6)$, $m = -3$
35. $(2, -3)$, $m = -\frac{1}{2}$
36. $(-2, -5)$, $m = \frac{3}{4}$
37. $\left(-\frac{1}{2}, \frac{3}{2}\right)$, $m = 0$
38. $(2.3, -8.5)$, $m = 0$
39. $(6, -1)$, m is undefined.
40. $(-10, 4)$, m is undefined.

41. **Player Salary** The average player salary in Major League Baseball was \$4.38 million in 2016 and \$4.47 million in 2017. Write a linear equation giving the average salary y in terms of the year x. Then use the equation to predict the average salary in 2021. (*Source:* Statista)

42. **Household Income** The median household income in the United States was \$56,277 in 2015 and \$57,617 in 2016. Write a linear equation giving the median household income y in terms of the year x. Then use the equation to predict the median household income in 2021. (*Source:* U.S. Census Bureau)

Using the Slope-Intercept Form **In Exercises 43–48, determine the slope and y-intercept (if possible) of the linear equation. Then describe its graph.**

43. $2x - 3y = 9$
44. $3x + 4y = 1$
45. $2x - 5y + 10 = 0$
46. $5x - 4y - 12 = 0$
47. $x = -6$
48. $3y + 2 = 0$

Using the Slope-Intercept Form **In Exercises 49–54, (a) find the slope and y-intercept (if possible) of the equation of the line algebraically and (b) sketch the line by hand. Use a graphing utility to verify your answers to parts (a) and (b).**

49. $5x - y + 3 = 0$
50. $2x + 3y - 9 = 0$
51. $5x - 2 = 0$
52. $3x + 7 = 0$
53. $3y + 5 = 0$
54. $-11 - 4y = 0$

Finding the Slope-Intercept Form **In Exercises 55 and 56, find the slope-intercept form of the equation of the line shown.**

55.

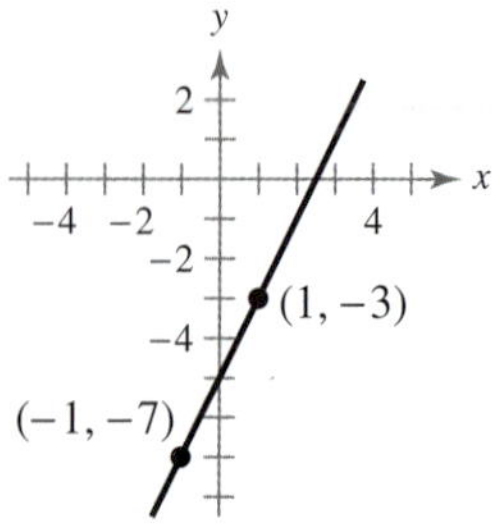

56.

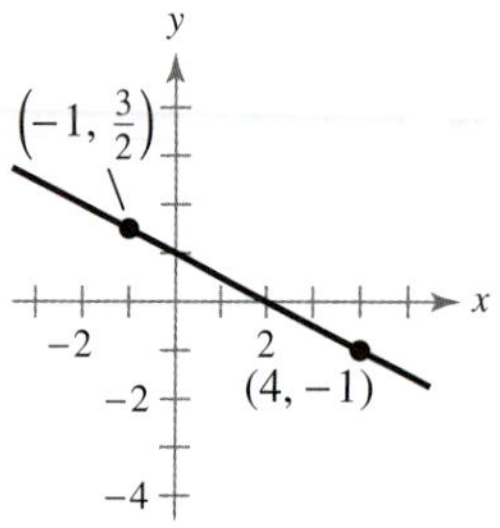

Finding the Slope-Intercept Form **In Exercises 57–64, write an equation of the line that passes through the points. Use the slope-intercept form (if possible). If not possible, explain why and use the general form. Use a graphing utility to graph the line (if possible).**

57. $(5, -1), (-5, 5)$
58. $(4, 3), (-4, -4)$
59. $(-8, 1), (-8, 7)$
60. $(-1, 6), (5, 6)$
61. $\left(2, \frac{1}{2}\right), \left(\frac{1}{2}, \frac{5}{4}\right)$
62. $\left(\frac{3}{4}, \frac{3}{2}\right), \left(-\frac{4}{3}, \frac{7}{4}\right)$
63. $(1, 0.6), (-2, -0.6)$
64. $(-8, 0.6), (2, -2.4)$

Different Viewing Windows **In Exercises 65 and 66, use a graphing utility to graph the equation using each viewing window. Describe the differences in the graphs.**

65. $y = 0.25x - 2$

Xmin = -1	Xmin = -5	Xmin = -5
Xmax = 9	Xmax = 10	Xmax = 10
Xscl = 1	Xscl = 1	Xscl = 1
Ymin = -5	Ymin = -3	Ymin = -5
Ymax = 4	Ymax = 4	Ymax = 5
Yscl = 1	Yscl = 1	Yscl = 1

66. $y = -8x + 5$

Xmin = -5	Xmin = -5	Xmin = -5
Xmax = 5	Xmax = 10	Xmax = 13
Xscl = 1	Xscl = 1	Xscl = 1
Ymin = -10	Ymin = -80	Ymin = -2
Ymax = 10	Ymax = 80	Ymax = 10
Yscl = 1	Yscl = 20	Yscl = 1

Parallel and Perpendicular Lines **In Exercises 67–70, determine whether the lines L_1 and L_2 passing through the pairs of points are parallel, perpendicular, or neither.**

67. L_1: $(0, -1), (5, 9)$
L_2: $(0, 3), (4, 1)$
68. L_1: $(-5, -1)$ and $(1, 2)$
L_2: $(-5, 1)$ and $(1, -2)$
69. L_1: $(3, 6), (-6, 0)$
L_2: $(0, -1), \left(5, \frac{7}{3}\right)$
70. L_1: $(4, 8), (-4, 2)$
L_2: $(3, -5), \left(-1, \frac{1}{3}\right)$

Equations of Parallel and Perpendicular Lines **In Exercises 71–80, write the slope-intercept forms of the equations of the lines through the given point (a) parallel and (b) perpendicular to the given line.**

71. $(2, 1)$, $4x - 2y = 3$
72. $(-3, 2)$, $x + y = 7$
73. $\left(-\frac{2}{3}, \frac{7}{8}\right)$, $3x + 4y = 7$
74. $\left(\frac{2}{5}, -1\right)$, $3x - 2y = 6$
75. $(3, -2)$, $x - 4 = 0$
76. $(3, -1)$, $y - 2 = 0$
77. $(-5, 1)$, $y + 2 = 0$
78. $(-2, 4)$, $x + 5 = 0$
79. $(-3.9, -1.4)$, $6x + 5y = 9$
80. $(-2.4, 1.2)$, $4x + 5y = 1$

Equations of Parallel Lines **In Exercises 81 and 82, the lines are parallel. Find the slope-intercept form of the equation of line y_2.**

81.

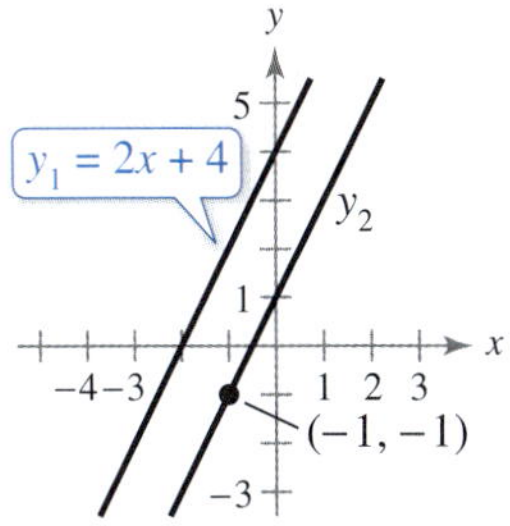

82.

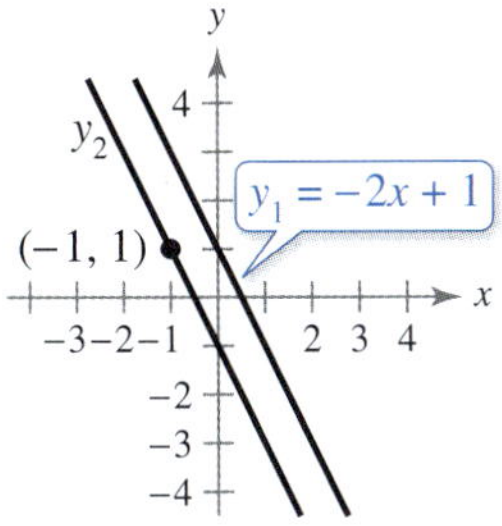

Equations of Perpendicular Lines **In Exercises 83 and 84, the lines are perpendicular. Find the slope-intercept form of the equation of line y_2.**

83.

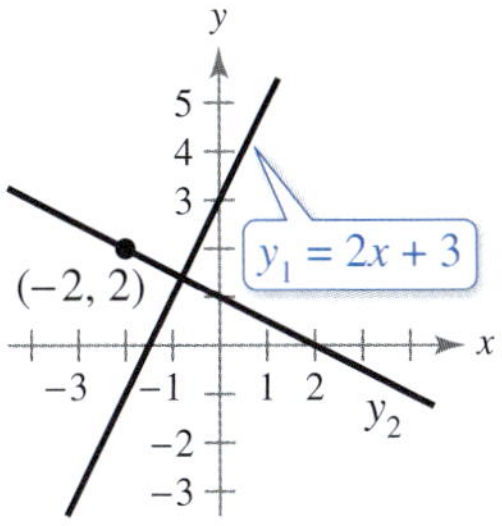

84.

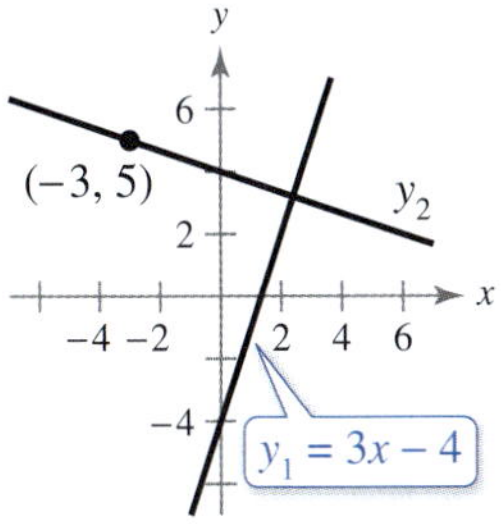

Graphs of Parallel and Perpendicular Lines **In Exercises 85–88, identify any relationships that exist among the lines. Then use a graphing utility to graph the three equations in the same viewing window. Adjust the viewing window so that each slope appears visually correct. Use the slopes of the lines to verify your results.**

85. (a) $y = 4x$ (b) $y = -4x$ (c) $y = \frac{1}{4}x$

86. (a) $y = \frac{2}{3}x$ (b) $y = -\frac{3}{2}x$ (c) $y = \frac{2}{3}x + 2$

87. (a) $y = -\frac{1}{2}x$ (b) $y = -\frac{1}{2}x + 3$ (c) $y = 2x - 4$

88. (a) $y = x - 8$ (b) $y = x + 1$ (c) $y = -x + 3$

89. Architectural Design The rise-to-run ratio of the roof of a house determines the steepness of the roof. The rise-to-run ratio of the roof in the figure is 3 to 4. Determine the maximum height in the attic of the house when the house is 32 feet wide.

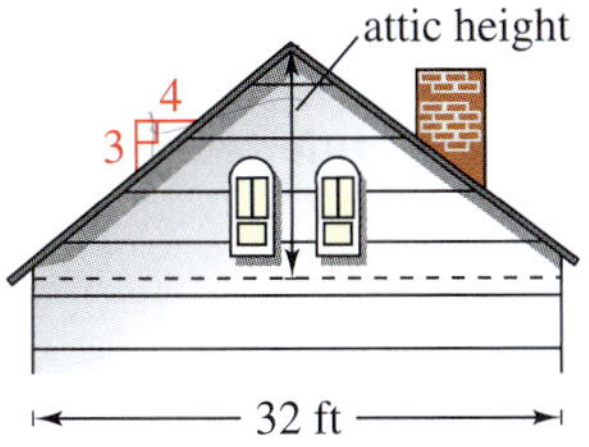

90. Highway Engineering When driving down a mountain road, you notice warning signs indicating that it is a "12% grade." This means that the slope of the road is $-\frac{12}{100}$. Approximate the amount of horizontal change in your position after descending 2000 feet vertically.

91. MODELING DATA

The graph shows the net profits y (in billions of dollars) of Apple, Inc. each year x from 2010 through 2017, where $x = 0$ represents 2010. *(Source: Apple, Inc.)*

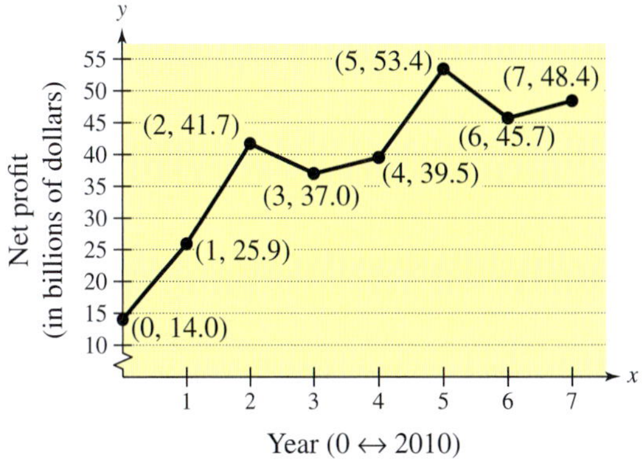

(a) Use the slopes to determine the years in which the net profits showed the greatest increase and greatest decrease.

(b) Find the equation of the line between the years 2010 and 2017.

(c) Interpret the meaning of the slope of the line from part (b) in the context of the problem.

(d) Use the equation from part (b) to estimate the net profits in 2021. Do you think this is an accurate estimate? Explain.

92. MODELING DATA

The table shows the sales y (in billions of dollars) for Nike, Inc. for each year x from 2011 through 2017, where $x = 1$ represents 2011. *(Source: Nike, Inc.)*

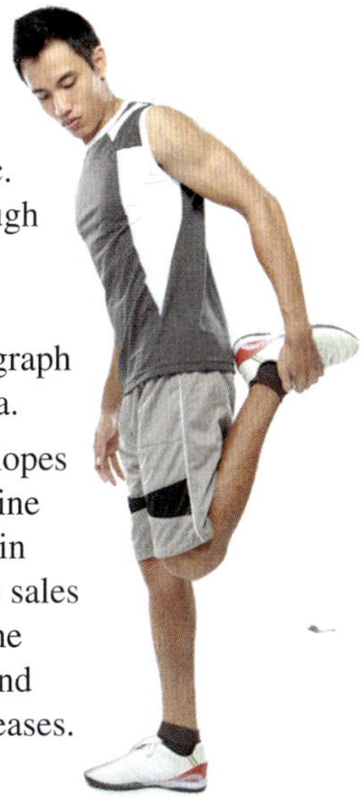

DATA

Year, x	Sales, y
1	20.9
2	24.1
3	25.3
4	27.8
5	30.6
6	32.4
7	34.4

Spreadsheet at LarsonPrecalculus.com

(a) Sketch a graph of the data.

(b) Use the slopes to determine the years in which the sales showed the greatest and least increases.

(c) Find the equation of the line between the years 2011 and 2017.

(d) Interpret the meaning of the slope of the line from part (c) in the context of the problem.

(e) Use the equation from part (c) to estimate the sales in 2021. Do you think this is an accurate estimate? Explain.

Using a Rate of Change to Write an Equation **In Exercises 93–96, you are given the dollar value of a product in 2018 and the rate at which the value of the product is expected to change during the next 5 years. Write a linear equation that gives the dollar value V of the product in terms of the year t. (Let $t = 18$ represent 2018.)**

	2018 Value	*Rate*
93.	\$2540	\$125 increase per year
94.	\$156	\$5.50 increase per year
95.	\$20,400	\$2000 decrease per year
96.	\$245,000	\$5600 decrease per year

97. Accounting An office purchases a high-volume printer, copier, and scanner for \$10,995. After 10 years, the equipment will have to be replaced. Its value at that time is expected to be \$880.

(a) Write a linear equation giving the value V of the equipment for each year t during its 10 years of use.

(b) Use a graphing utility to graph the linear equation representing the depreciation of the equipment and use the *value* or *trace* feature to complete the table. Verify your answers algebraically by using the equation you found in part (a).

t	0	1	2	3	4	5	6	7	8	9	10
V											

98. Meteorology Recall that water freezes at 0°C (32°F) and boils at 100°C (212°F).

(a) Find an equation of the line that shows the relationship between the temperature in degrees Celsius C and degrees Fahrenheit F.

(b) Use the result of part (a) to complete the table.

C		$-10°$	$10°$			$177°$
F	$0°$			$68°$	$90°$	

99. Business A contractor purchases a backhoe for \$84,400. The backhoe requires an average expenditure of \$10 per hour for fuel and maintenance, and \$49 per hour for other owning and operating costs.

(a) Write a linear equation giving the total cost C of operating the backhoe for t hours. (Include the purchase cost of the backhoe.)

(b) Assuming that customers are charged \$150 per hour of backhoe use, write an equation for the revenue R derived from t hours of use.

(c) Use the profit formula $(P = R - C)$ to write an equation for the profit gained from t hours of use.

(d) Use the result of part (c) to find the break-even point (the number of hours the backhoe must be used to gain a profit of 0 dollars).

100. Real Estate A real estate office handles an apartment complex with 50 units. When the rent per unit is \$600 per month, all 50 units are occupied. However, when the rent is \$625 per month, the average number of occupied units drops to 48. Assume that the relationship between the monthly rent p and the demand x is linear.

(a) Write an equation of the line giving the demand x in terms of the rent p.

(b) Use a graphing utility to graph the demand equation and use the *trace* feature to estimate the number of units occupied when the rent is \$675. Verify your answer algebraically.

(c) Use the demand equation to predict the number of units occupied when the rent is \$650. Verify your answer graphically.

101. ***Why you should learn it*** *(p. 3)* In 2000, the University of Massachusetts Amherst had an enrollment of 23,570 students. By 2017, the enrollment had increased to 30,340. (*Source:* University of Massachusetts Amherst)

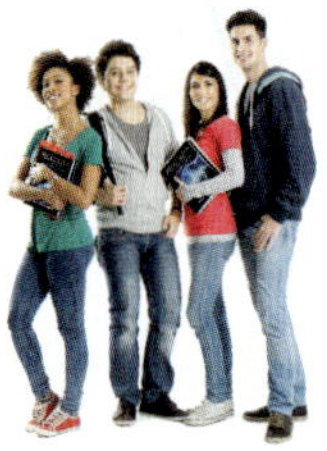

(a) What was the average annual change in enrollment from 2000 to 2017?

(b) Use the average annual change in enrollment to estimate the enrollments in 2005, 2010, and 2015.

(c) Write an equation of a line that represents the given data. What is its slope? Interpret the slope in the context of the problem.

102. Writing Using the results of Exercise 101, write a short paragraph discussing the concepts of *slope* and *average rate of change*.

Focusing on Concepts

True or False? **In Exercises 103 and 104, determine whether the statement is true or false. Justify your answer.**

103. The line through $(-8, 2)$ and $(-1, 4)$ and the line through $(0, -4)$ and $(-7, 7)$ are parallel.

104. If the points $(10, -3)$ and $(2, -9)$ lie on the same line, then the point $\left(-12, -\frac{37}{2}\right)$ also lies on that line.

Exploration **In Exercises 105–108, use a graphing utility to graph the equation of the line in the form**

$$\frac{x}{a} + \frac{y}{b} = 1, \quad a \neq 0, b \neq 0.$$

Use the graphs to make a conjecture about what a and b represent. Verify your conjecture.

105. $\dfrac{x}{7} + \dfrac{y}{-3} = 1$

106. $\dfrac{x}{-6} + \dfrac{y}{2} = 1$

107. $\dfrac{x}{4} + \dfrac{y}{-2/3} = 1$

108. $\dfrac{x}{1/2} + \dfrac{y}{5} = 1$

Using Intercepts **In Exercises 109–112, use the results of Exercises 105–108 to write an equation of the line that passes through the points.**

109. x-intercept: $(2, 0)$
y-intercept: $(0, 9)$

110. x-intercept: $(-5, 0)$
y-intercept: $(0, -4)$

111. x-intercept: $\left(-\frac{1}{6}, 0\right)$
y-intercept: $\left(0, -\frac{2}{3}\right)$

112. x-intercept: $\left(\frac{4}{3}, 0\right)$
y-intercept: $\left(0, \frac{5}{4}\right)$

Think About It **In Exercises 113 and 114, determine whether each equation can be represented by the graph.**

113.

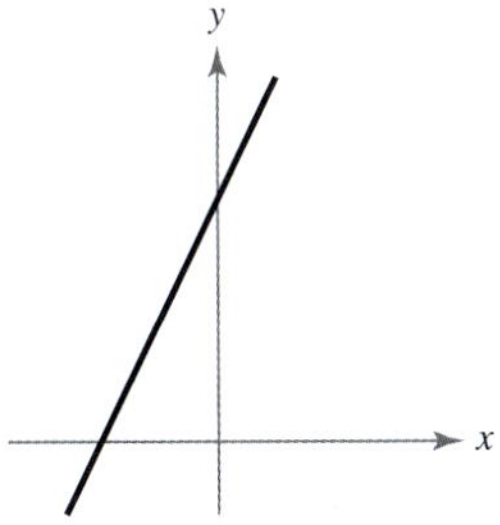

(a) $2x - y = -10$
(b) $2x + y = 10$
(c) $x - 2y = 10$
(d) $x + 2y = 10$

114.

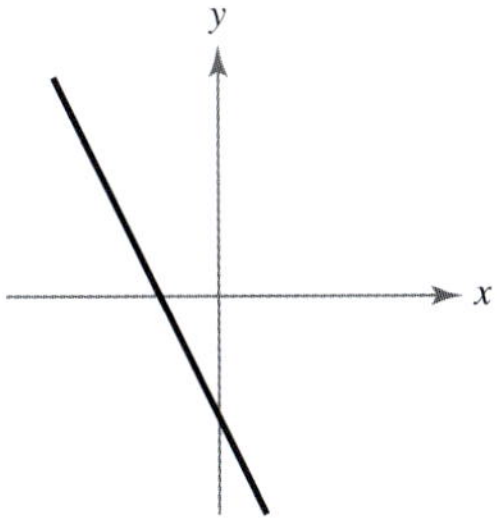

(a) $2x + y = 5$
(b) $2x + y = -5$
(c) $x - 2y = 5$
(d) $x - 2y = -5$

Think About It **In Exercises 115 and 116, determine whether each pair of equations can be represented by the graphs.**

115.

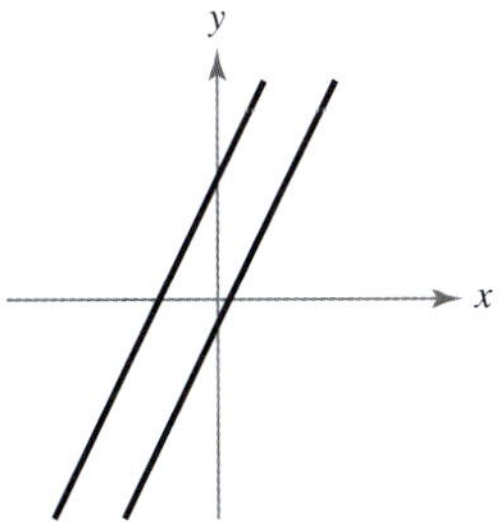

(a) $2x + y = -5$
$2x + y = 1$
(b) $2x - y = -5$
$2x - y = 1$

116.

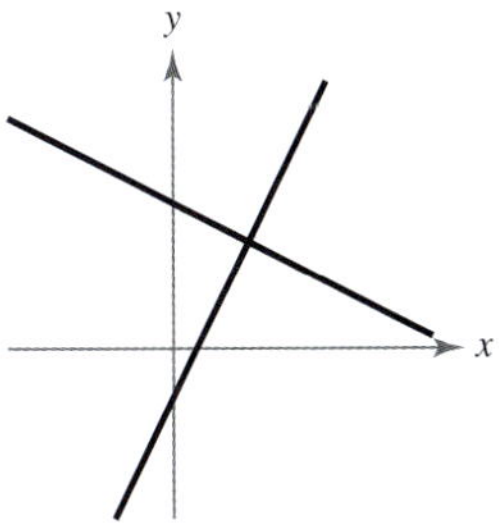

(a) $2x - y = 2$
$x + 2y = 12$
(b) $x - y = 1$
$x + y = 6$

117. Error Analysis Describe the error.

The graph of $y = 2x - 3$ is a line whose slope is -3 and whose y-intercept is $(0, 2)$.

118. Error Analysis Describe the error.

The slope of the line through $(0, 2)$ and $(4, 0)$ is

$$m = \frac{4 - 0}{0 - 2} = -2.$$

119. Think About It Does every line have both an x-intercept and a y-intercept? Explain.

120. HOW DO YOU SEE IT? Match the description with its graph. Determine the slope and y-intercept of each graph and interpret their meaning in the context of the problem. [The graphs are labeled (i), (ii), (iii), and (iv).]

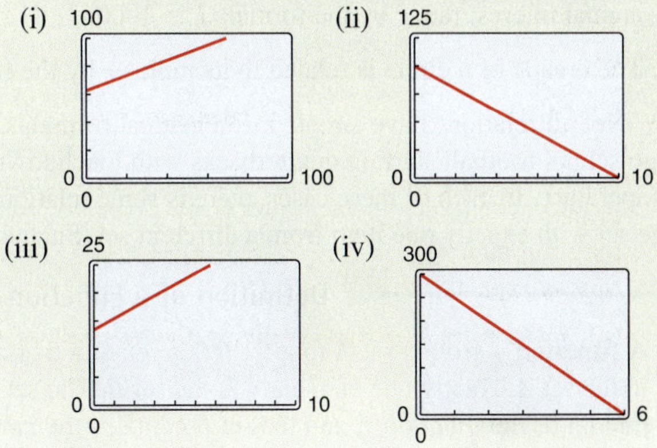

(a) You are paying \$10 per week to repay a \$100 loan.
(b) An employee is paid \$13.50 per hour plus \$2 for each unit produced per hour.
(c) A sales representative receives \$63 per day for expenses plus \$0.535 for each mile traveled.
(d) A tablet computer that was purchased for \$300 depreciates \$50 per year.

Cumulative Mixed Review

Identifying a Polynomial **In Exercises 121–124, determine whether the expression is a polynomial. If it is, write the polynomial in standard form.**

121. $4x^2 + x^{-1} - 3$

122. $3x - 10x^2 + 1$

123. $\dfrac{x^2 + 3x + 4}{x^2 - 9}$

124. $\sqrt{x^2 + 7x + 6}$

Factoring a Trinomial **In Exercises 125–128, factor the trinomial.**

125. $x^2 - 6x - 27$

126. $x^2 + 11x + 28$

127. $2x^2 + 11x - 40$

128. $3x^2 - 16x + 5$

129. ***Project: Bachelor's Degrees Earned by Women*** To work an extended application analyzing the numbers of bachelor's degrees earned by women in the United States from 2005 through 2016, visit this textbook's website at *LarsonPrecalulus.com.* (*Data Source:* National Center for Education Statistics)

The *Project* indicates a multipart exercise using large data sets. Visit this textbook's website at *LarsonPrecalculus.com.*

1.2 Functions

What you should learn

- Determine whether a relation between two variables represents a function.
- Use function notation and evaluate functions.
- Find the domains of functions.
- Use functions to model and solve real-life problems.
- Evaluate difference quotients.

Why you should learn it

Many natural phenomena can be modeled by functions, such as the force of water against the face of a dam, explored in Exercise 82 on page 27.

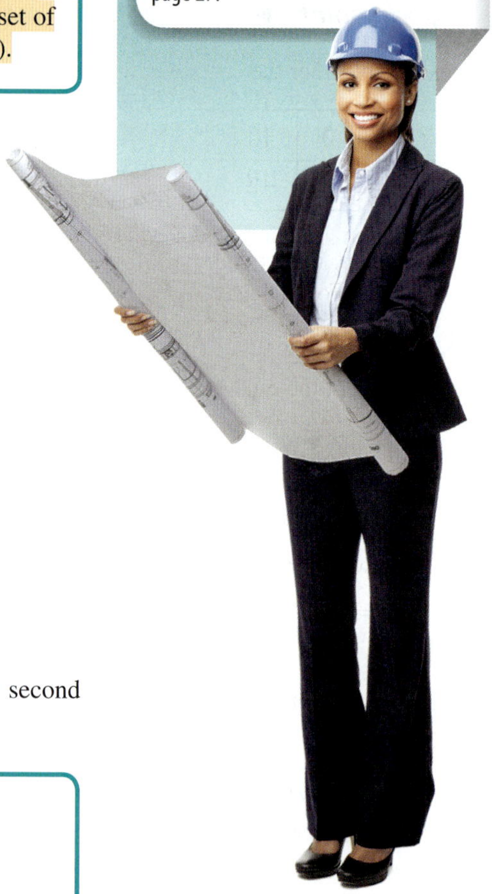

Introduction to Functions

Many everyday phenomena involve two quantities that are related to each other by some rule of correspondence. The mathematical term for such a rule of correspondence is a **relation.** Here are two examples.

1. The simple interest I earned on an investment of \$1000 for 1 year is related to the annual interest rate r by the formula $I = 1000r$.
2. The area A of a circle is related to its radius r by the formula $A = \pi r^2$.

Not all relations have simple mathematical formulas. For instance, you can match high school football starting quarterbacks with touchdown passes and time of day with temperature. In each of these cases, there is some relation that matches each item from one set with exactly one item from a different set. Such a relation is called a **function.**

Definition of a Function

A **function** f from a set A to a set B is a relation that assigns to each element x in the set A exactly one element y in the set B. The set A is the **domain** (or set of inputs) of the function f, and the set B contains the **range** (or set of outputs).

To help understand this definition, look at the function that relates the time of day to the temperature in Figure 1.11.

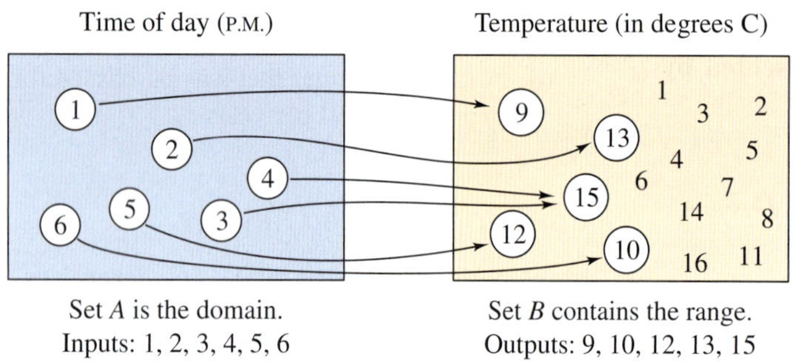

Figure 1.11

This function can be represented by the ordered pairs

$$\{(1, 9°), (2, 13°), (3, 15°), (4, 15°), (5, 12°), (6, 10°)\}.$$

In each ordered pair, the first coordinate (x-value) is the **input** and the second coordinate (y-value) is the **output.**

Characteristics of a Function from Set A to Set B

1. Each element of A must be matched with an element of B.
2. Some elements of B may not be matched with any element of A.
3. Two or more elements of A may be matched with the same element of B.
4. An element of A (the domain) cannot be matched with two different elements of B.

To determine whether a relation is a function, you must decide whether each input value is matched with exactly one output value. When any input value is matched with two or more output values, the relation is not a function.

EXAMPLE 1 Testing for Functions

Determine whether each relation represents y as a function of x.

a.

Input, x	2	2	3	4	5
Output, y	11	10	8	5	1

b.

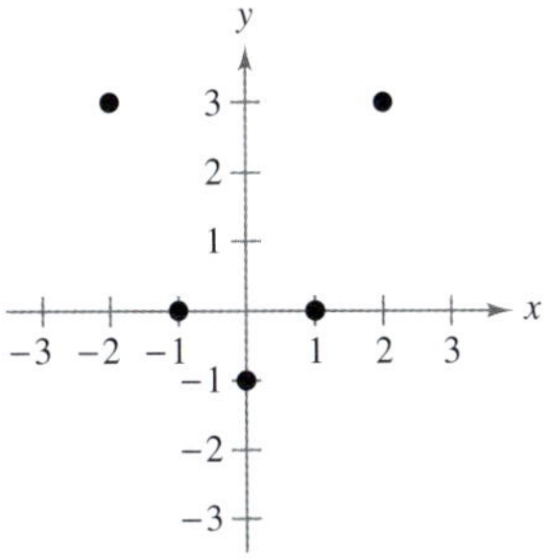

Solution

a. This table *does not* describe y as a function of x. The input value 2 is matched with two different y-values.

b. The graph *does* describe y as a function of x. Each input value is matched with exactly one output value.

Checkpoint Audio-video solution in English & Spanish at LarsonPrecalculus.com

Determine whether the relation represents y as a function of x.

Input, x	0	1	2	3	4
Output, y	−4	−2	0	2	4

In algebra, it is common to represent functions by equations or formulas involving two variables. For instance, $y = x^2$ represents the variable y as a function of the variable x. In this equation, x is the **independent variable** and y is the **dependent variable.** The domain of the function is the set of all values taken on by the independent variable x, and the range of the function is the set of all values taken on by the dependent variable y.

EXAMPLE 2 Testing for Functions Represented Algebraically

See LarsonPrecalculus.com for an interactive version of this type of example.

Determine whether each equation represents y as a function of x.

a. $x^2 + y = 1$ **b.** $-x + y^2 = 1$

Solution

To determine whether y is a function of x, try to solve for y in terms of x.

a. $x^2 + y = 1$ Write original equation.

$y = 1 - x^2$ Solve for y.

Each value of x corresponds to exactly one value of y. So, y is a function of x.

b. $-x + y^2 = 1$ Write original equation.

$y^2 = 1 + x$ Add x to each side.

$y = \pm\sqrt{1 + x}$ Solve for y.

The $\pm$ indicates that for a given value of x, there correspond two values of y. For instance, when $x = 3$, $y = 2$ or $y = -2$. So, y is not a function of x.

Checkpoint Audio-video solution in English & Spanish at LarsonPrecalculus.com

Determine whether each equation represents y as a function of x.

a. $x^2 + y^2 = 8$ **b.** $y - 4x^2 = 36$

Explore the Concept

Use a graphing utility to graph $x^2 + y = 1$. Then use the graph to write a convincing argument that each x-value corresponds to at most one y-value.

Use a graphing utility to graph $-x + y^2 = 1$. (*Hint:* You will need to use two equations.) Does the graph represent y as a function of x? Explain.

Function Notation

When an equation is used to represent a function, it is convenient to name the function so that it can be referenced easily. For instance, you know from Example 2(a) that the equation $y = 1 - x^2$ describes y as a function of x. After giving this function the name "f," you can use **function notation** to write the input, output, and equation as shown.

Input	*Output*	*Equation*
x	$f(x)$	$f(x) = 1 - x^2$

The symbol $f(x)$ is read as the *value of f at x* or simply *f of x*. The symbol $f(x)$ corresponds to the y-value for a given x. So, you can write $y = f(x)$. Keep in mind that f is the *name* of the function, whereas $f(x)$ is the *output value* of the function at the *input value* x. In function notation, the *input* is the independent variable and the *output* is the dependent variable. For instance, the function $f(x) = 3 - 2x$ has *function values* denoted by $f(-1)$, $f(0)$, and so on. To find these values, substitute the specified input values into the given equation.

For $x = -1$, $\quad f(-1) = 3 - 2(-1) = 3 + 2 = 5.$

For $x = 0$, $\quad f(0) = 3 - 2(0) = 3 - 0 = 3.$

Although f is often used as a convenient function name and x is often used as the independent variable, you can use other letters. For instance,

$$f(x) = x^2 - 4x + 7, \quad h(t) = t^2 - 4t + 7, \quad \text{and} \quad g(s) = s^2 - 4s + 7$$

all define the same function. In fact, the role of the independent variable is that of a "placeholder." Consequently, the function can be written as

$$f(\quad) = (\quad)^2 - 4(\quad) + 7.$$

Insight

Your understanding of functional notation will be assessed on standardized tests. For instance, you will use function notation to solve conceptual problems related to *transformations* of functions (see Section 1.4).

EXAMPLE 3 Evaluating a Function

Let $g(x) = -x^2 + 4x + 1$. Find each function value.

a. $g(2)$ **b.** $g(t)$ **c.** $g(x + 2)$

Solution

a. Replace x with 2 in $g(x) = -x^2 + 4x + 1$ and simplify.

$$g(2) = -(2)^2 + 4(2) + 1 = -4 + 8 + 1 = 5$$

b. Replace x with t and simplify.

$$g(t) = -(t)^2 + 4(t) + 1 = -t^2 + 4t + 1$$

c. Replace x with $x + 2$ and simplify.

$g(x+2) = -(x+2)^2 + 4(x+2) + 1$ Substitute $x + 2$ for x.

$= -(x^2 + 4x + 4) + 4x + 8 + 1$ Multiply.

$= -x^2 - 4x - 4 + 4x + 8 + 1$ Distributive Property

$= -x^2 + 5$ Simplify.

✓ ***Checkpoint*** *Audio-video solution in English & Spanish at LarsonPrecalculus.com*

Let $f(x) = 10 - 3x^2$. Find each function value.

a. $f(2)$ **b.** $f(-4)$ **c.** $f(x - 1)$

In part (c) of Example 3, note that $g(x + 2)$ is not equal to $g(x) + g(2)$. In general, $g(u + v) \neq g(u) + g(v)$.

A function defined by two or more equations over a specified domain is called a **piecewise-defined function.**

EXAMPLE 4 Evaluating a Piecewise-Defined Function

Evaluate the function when $x = -1$, 0, and 1.

$$f(x) = \begin{cases} x^2 + 1, & x < 0 \\ x - 1, & x \geq 0 \end{cases}$$

Solution

Because $x = -1$ is less than 0, use $f(x) = x^2 + 1$.

$$f(-1) = (-1)^2 + 1 \qquad \text{Substitute } -1 \text{ for } x.$$
$$= 2 \qquad \text{Simplify.}$$

For $x = 0$, use $f(x) = x - 1$.

$$f(0) = 0 - 1 \qquad \text{Substitute 0 for } x.$$
$$= -1 \qquad \text{Simplify.}$$

For $x = 1$, use $f(x) = x - 1$.

$$f(1) = 1 - 1 \qquad \text{Substitute 1 for } x.$$
$$= 0 \qquad \text{Simplify.}$$

The graph of f is shown in the figure.

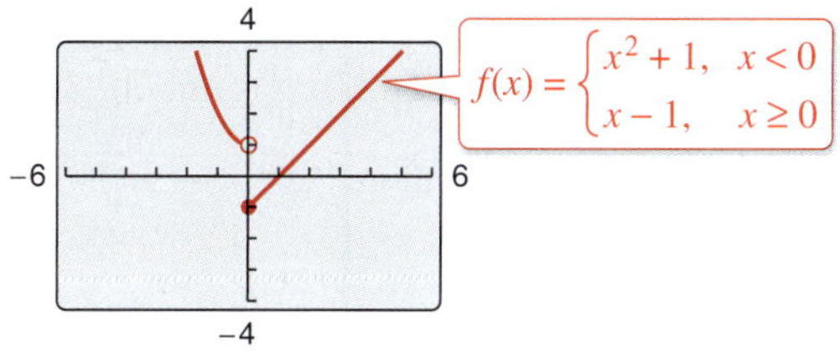

✓ Checkpoint *Audio-video solution in English & Spanish at LarsonPrecalculus.com*

Evaluate the function given in Example 4 when $x = -2$, 2, and 3.

Technology Tip

Most graphing utilities can graph piecewise-defined functions. For instructions on how to enter a piecewise-defined function into your graphing utility, consult your user's manual. You may find it helpful to set your graphing utility to *dot* mode before graphing such functions.

Library of Parent Functions: Absolute Value Function

The *parent absolute value function* given by

$$f(x) = |x|$$

can be written as a piecewise-defined function. The basic characteristics of the parent absolute value function are summarized below and on the inside cover of this text.

Graph of $f(x) = |x| = \begin{cases} x, & x \geq 0 \\ -x, & x < 0 \end{cases}$

Domain: $(-\infty, \infty)$
Range: $[0, \infty)$
Intercept: $(0, 0)$
Decreasing on $(-\infty, 0)$
Increasing on $(0, \infty)$

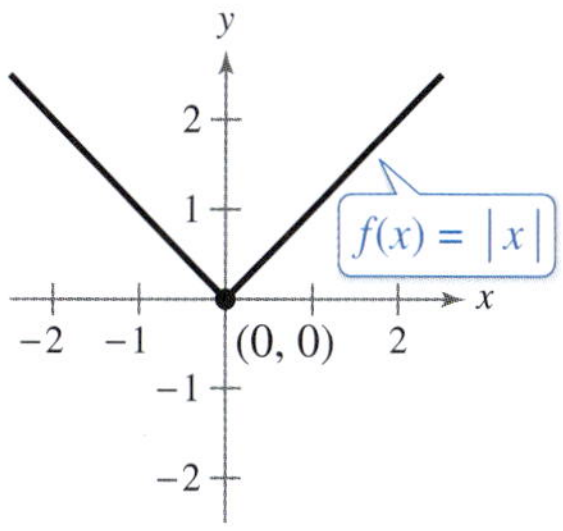

The Domain of a Function

The domain of a function can be described explicitly or it can be *implied* by the expression used to define the function. The **implied domain** is the set of all real numbers for which the expression is defined. For instance, the function

$$f(x) = \frac{1}{x^2 - 4}$$ Domain excludes x-values that result in division by zero.

has an implied domain that consists of all real numbers x other than $x = \pm 2$. These two values are excluded from the domain because division by zero is undefined. Another common type of implied domain is that used to avoid even roots of negative numbers. For example, the function

$$(x) = \sqrt{x}$$ Domain excludes x-values that result in even roots of negative numbers.

is defined only for $x \geq 0$. So, its implied domain is the interval $[0, \infty)$. In general, the domain of a function *excludes* values that cause division by zero *or* result in the even root of a negative number.

Explore the Concept

Use a graphing utility to graph $y = \sqrt{4 - x^2}$. What is the domain of this function? Then graph $y = \sqrt{x^2 - 4}$. What is the domain of this function? Do the domains of these two functions overlap? If so, for what values?

Library of Parent Functions: Square Root Function

Radical functions arise from the use of rational exponents. The most common radical function is the *parent square root* function given by $f(x) = \sqrt{x}$. The basic characteristics of the parent square root function are summarized below and on the inside cover of this text.

Graph of $f(x) = \sqrt{x}$

Domain: $[0, \infty)$
Range: $[0, \infty)$
Intercept: $(0, 0)$
Increasing on $(0, \infty)$

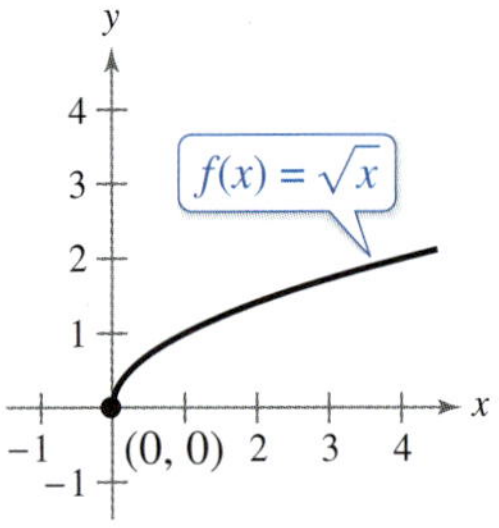

Algebra Help

Because the square root function is not defined for $x < 0$, you must be careful when analyzing the domains of complicated functions involving the square root symbol.

EXAMPLE 5 Finding the Domains of Functions

Find the domain of each function.

a. $f: \{(-3, 0), (-1, 4), (0, 2), (2, 2), (4, -1)\}$

b. $g(x) = -3x^2 + 4x + 5$

c. $h(x) = \dfrac{1}{x + 5}$

Solution

a. The domain of f consists of all first coordinates in the set of ordered pairs.

$$\text{Domain} = \{-3, -1, 0, 2, 4\}$$

b. The domain of g is the set of all *real* numbers.

c. Excluding x-values that yield zero in the denominator, the domain of h is the set of all real numbers x except $x = -5$.

✓ ***Checkpoint*** ▶ *Audio-video solution in English & Spanish at LarsonPrecalculus.com*

Find the domain of each function.

a. $f: \{(-2, 2), (-1, 1), (0, 3), (1, 1), (2, 2)\}$ **b.** $g(x) = \dfrac{1}{3 - x}$

EXAMPLE 6 Finding the Domains of Functions

Find the domain of each function.

a. Volume of a sphere: $V = \frac{4}{3}\pi r^3$ **b.** $k(x) = \sqrt{4 - 3x}$

Solution

a. Because this function represents the volume of a sphere, the values of the radius r must be positive (see Figure 1.12). So, the domain is the set of all real numbers r such that $r > 0$.

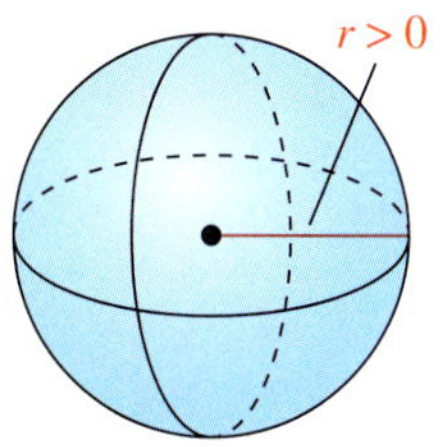

Figure 1.12

b. This function is defined only for x-values for which $4 - 3x \geq 0$. By solving this inequality, you will find that the domain of k is all real numbers that are less than or equal to $\frac{4}{3}$.

Checkpoint *Audio-video solution in English & Spanish at LarsonPrecalculus.com*

Find the domain of each function.

a. Circumference of a circle: $C = 2\pi r$ **b.** $h(x) = \sqrt{x - 16}$

In Example 6(a), note that the *domain of a function may be implied by the physical context.* For instance, from the equation $V = \frac{4}{3}\pi r^3$, you have no reason to restrict r to positive values, but the physical context implies that a sphere cannot have a negative or zero radius.

For some functions, it may be easier to find the domain and range of the function by examining its graph.

EXAMPLE 7 Finding the Domain and Range of a Function

Use a graphing utility to find the domain and range of the function $f(x) = \sqrt{9 - x^2}$.

Solution

Graph the function as $y = \sqrt{9 - x^2}$, as shown in Figure 1.13. Using the *trace* feature of a graphing utility, you can determine that the x-values extend from -3 to 3 and the y-values extend from 0 to 3. So, the domain of the function f is all real numbers such that $-3 \leq x \leq 3$, and the range of f is all real numbers such that $0 \leq y \leq 3$.

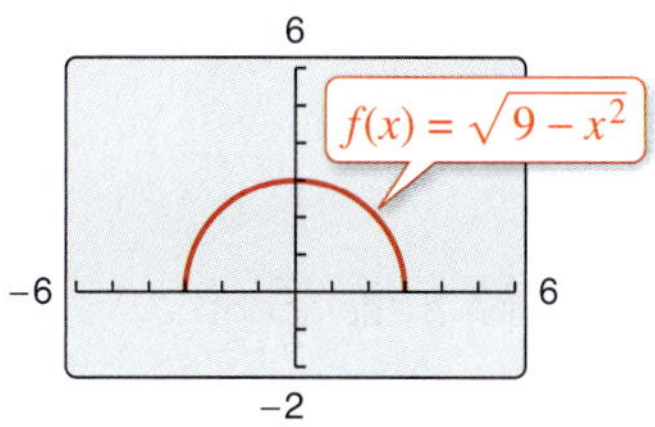

Figure 1.13

Algebra Help

Be sure you see that the *range* of a function is not the same as the use of *range* relating to the viewing window of a graphing utility.

Checkpoint *Audio-video solution in English & Spanish at LarsonPrecalculus.com*

Use a graphing utility to find the domain and range of the function $f(x) = \sqrt{4 - x^2}$.

Applications

EXAMPLE 8 Number of Registered Vessels

The number N (in thousands) of registered vessels in Florida decreased in a linear pattern from 2012 through 2013, as shown in Figure 1.14. In 2014, the number rose and increased through 2016 in a different linear pattern. These two patterns can be approximated by the function

$$N(t) = \begin{cases} -5t + 962, & 12 \le t \le 13 \\ 15.5t + 683, & 14 \le t \le 16 \end{cases}$$

where t represents the year, with $t = 12$ corresponding to 2012. Use this function to approximate the number of registered vessels for the years 2012 and 2015. *(Source: Florida DHSMV)*

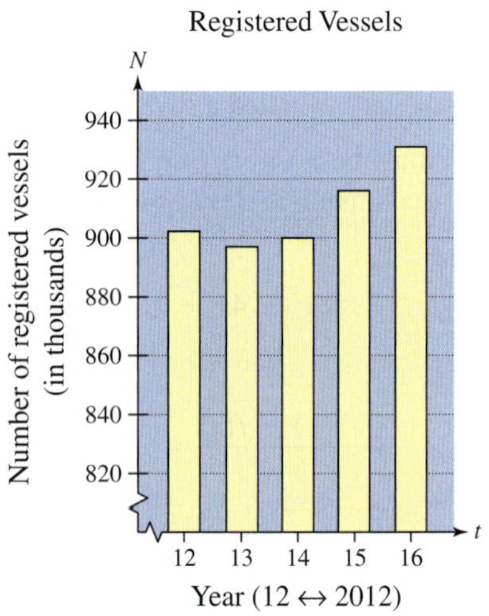

Figure 1.14

Solution

For 2012, $t = 12$, so use $N(t) = -5t + 962$.

$$N(12) = -5(12) + 962 = -60 + 962 = 902 \text{ thousand registered vessels}$$

For 2016, $t = 16$, so use $N(t) = 15.5t + 683$.

$$N(16) = 15.5(16) + 683 = 248 + 683 = 931 \text{ thousand registered vessels}$$

✓ ***Checkpoint*** *Audio-video solution in English & Spanish at LarsonPrecalculus.com*

Use the function in Example 8 to approximate the number of registered vessels for the years 2013 and 2014.

EXAMPLE 9 The Path of a Baseball

A baseball is hit at a point 3 feet above the ground at a velocity of 100 feet per second and an angle of 45°. The path of the baseball is given by $f(x) = -0.0032x^2 + x + 3$, where $f(x)$ is the height of the baseball (in feet) and x is the horizontal distance from home plate (in feet). Will the baseball clear a 10-foot fence located 300 feet from home plate?

Algebraic Solution

The height of the baseball is a function of the horizontal distance from home plate. When $x = 300$, you can find the height of the baseball as follows.

$$f(x) = -0.0032x^2 + x + 3 \quad \text{Write original function.}$$

$$f(300) = -0.0032(300)^2 + 300 + 3 \quad \text{Substitute 300 for } x.$$

$$= 15 \quad \text{Simplify.}$$

When $x = 300$, the height of the baseball is 15 feet. So, the baseball will clear a 10-foot fence.

Graphical Solution

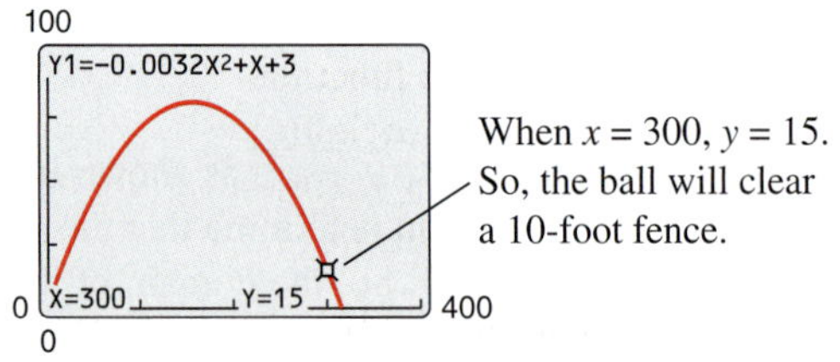

✓ ***Checkpoint*** *Audio-video solution in English & Spanish at LarsonPrecalculus.com*

A second baseman throws a baseball toward the first baseman 60 feet away. The path of the baseball is given by $f(x) = -0.004x^2 + 0.3x + 6$, where $f(x)$ is the height of the baseball (in feet) and x is the horizontal distance from the second baseman (in feet). The first baseman can reach 8 feet high. Can the first baseman catch the baseball without jumping?

Difference Quotients

One of the basic definitions in calculus employs the ratio

$$\frac{f(x+h)-f(x)}{h}, \quad h \neq 0.$$

This ratio is called a **difference quotient,** as illustrated in Example 10.

EXAMPLE 10 Evaluating a Difference Quotient

For $f(x) = x^2 - 4x + 7$, find $\dfrac{f(x+h)-f(x)}{h}$.

Solution

$$\begin{aligned}\frac{f(x+h)-f(x)}{h} &= \frac{[(x+h)^2 - 4(x+h) + 7] - (x^2 - 4x + 7)}{h}\\ &= \frac{x^2 + 2xh + h^2 - 4x - 4h + 7 - x^2 + 4x - 7}{h}\\ &= \frac{2xh + h^2 - 4h}{h}\\ &= \frac{h(2x + h - 4)}{h}\\ &= 2x + h - 4, \quad h \neq 0\end{aligned}$$

Algebra Help

Notice in Example 10 that h cannot be zero in the original expression. Therefore, you must restrict the domain of the simplified expression by listing $h \neq 0$ so that the simplified expression is equivalent to the original expression.

✓ ***Checkpoint*** *Audio-video solution in English & Spanish at LarsonPrecalculus.com*

For $f(x) = x^2 + 2x - 3$, find $\dfrac{f(x+h)-f(x)}{h}$.

Summary of Function Terminology

Function: A **function** is a relationship between two variables such that to each value of the independent variable there corresponds exactly one value of the dependent variable.

Function Notation: $y = f(x)$

f is the *name* of the function.
y is the **dependent variable,** or output value.
x is the **independent variable,** or input value.
$f(x)$ is the *value of the function at* x.

Domain: The **domain** of a function is the set of all values (inputs) of the independent variable for which the function is defined. If x is in the domain of f, then f is said to be *defined* at x. If x is not in the domain of f, then f is said to be *undefined* at x.

Range: The **range** of a function is the set of all values (outputs) assumed by the dependent variable (that is, the set of all function values).

Implied Domain: If f is defined by an algebraic expression and the domain is not specified, then the **implied domain** consists of all real numbers for which the expression is defined.

The symbol $\int$ indicates an example or exercise that highlights algebraic techniques specifically used in calculus.

1.2 Exercises

See *CalcChat.com* for tutorial help and worked-out solutions to odd-numbered exercises.
For instructions on how to use a graphing utility, see Appendix A.

Vocabulary and Concept Check

In Exercises 1 and 2, fill in the blanks.

1. A relation that assigns to each element x from a set of inputs, or _______ , exactly one element y in a set of outputs, or _______ , is called a _______ .

2. For an equation that represents y as a function of x, the _______ variable is the set of all x in the domain, and the _______ variable is the set of all y in the range.

3. Can the ordered pairs (3, 0) and (3, 5) represent a function?

4. To find $g(x + 1)$, what do you substitute for x in the function $g(x) = 3x - 2$?

5. Does the domain of the function $f(x) = \sqrt{1 + x}$ include $x = -2$?

6. Is the domain of a piecewise-defined function *implied* or *explicitly described*?

Procedures and Problem Solving

Testing for Functions **In Exercises 7–10, does the relation describe a function? Explain your reasoning.**

7. *Domain* *Range*

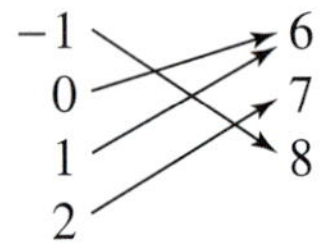

8. *Domain* *Range*

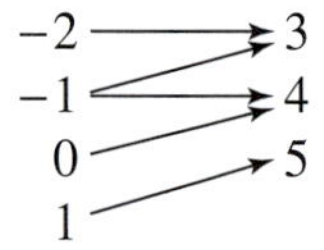

9. *Domain* *Range*

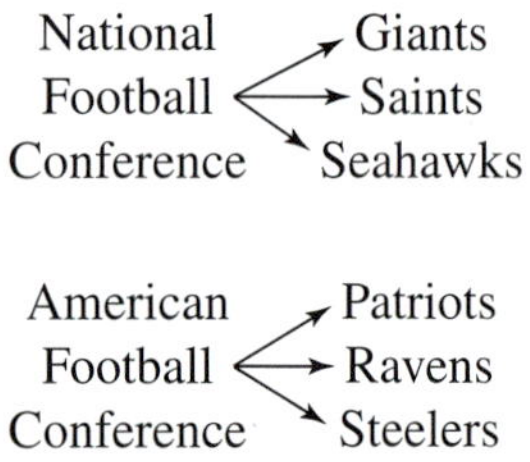

10. *Domain* (State) *Range* (Electoral votes)

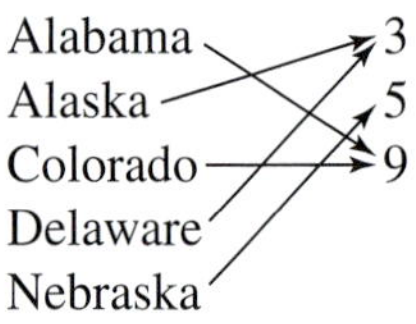

Testing for Functions **In Exercises 11–14, determine whether the relation represents y as a function of x. Explain your reasoning.**

11.

Input, x	10	7	4	7	10
Output, y	3	6	9	12	15

12.

Input, x	−3	−1	0	1	3
Output, y	−9	−1	0	1	−9

13.

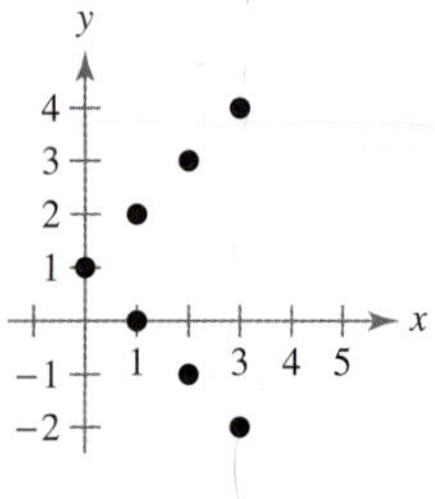

14.

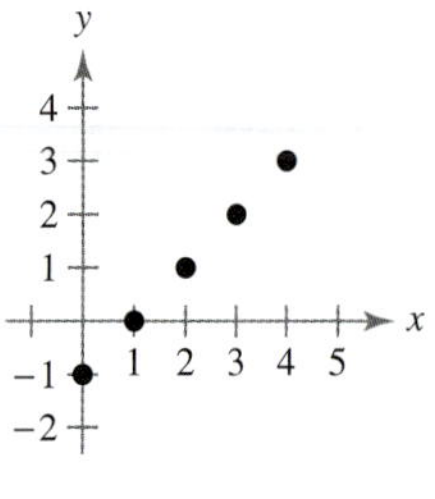

Testing for Functions **In Exercises 15 and 16, which sets of ordered pairs represent functions from A to B? Explain.**

15. $A = \{0, 1, 2, 3\}$ and $B = \{-2, -1, 0, 1, 2\}$

(a) $\{(0, 1), (1, -2), (2, 0), (3, 2)\}$

(b) $\{(0, -1), (2, 2), (1, -2), (3, 0), (1, 1)\}$

(c) $\{(1, 0), (-2, 3), (-1, 3), (0, 0)\}$

16. $A = \{a, b, c\}$ and $B = \{0, 1, 2, 3\}$

(a) $\{(a, 1), (c, 2), (c, 3), (b, 3)\}$

(b) $\{(a, 1), (b, 2), (c, 3)\}$

(c) $\{(1, a), (0, a), (2, c), (3, b)\}$

Pharmaceutical Sales **In Exercises 17 and 18, use the graph, which shows the average prices of a name brand prescription drug and its generic equivalent from 2011 through 2018.**

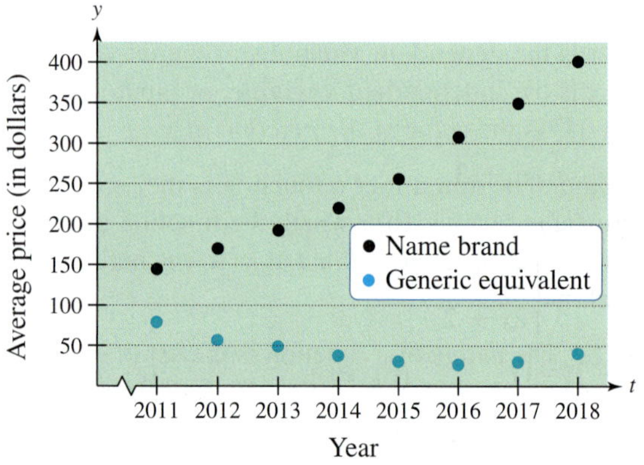

17. Is the average price of the name brand prescription drug a function of the year? Explain.

18. Is the average price of the generic equivalent a function of the year? Explain.

Testing for Functions Represented Algebraically In Exercises 19–30, determine whether the equation represents y as a function of x.

19. $x^2 + y^2 = 4$
20. $x = y^2 + 1$
21. $y = \sqrt{x^2 - 1}$
22. $y = \sqrt{x + 3}$
23. $2x + 3y = 4$
24. $x = -y + 5$
25. $y^2 = x^2 - 1$
26. $x + y^2 = 3$
27. $y = |4 - x|$
28. $|y| = 2 - 5x$
29. $x = -7$
30. $y = 8$

Evaluating a Function In Exercises 31–46, find each function value, if possible.

31. $f(t) = 3t + 1$
 (a) $f(2)$ (b) $f(-4)$ (c) $f(t + 2)$
32. $g(y) = 7 - 3y$
 (a) $g(0)$ (b) $g\left(\frac{7}{3}\right)$ (c) $g(s + 5)$
33. $g(t) = 4t^2 - 3t + 5$
 (a) $g(2)$ (b) $g(t - 2)$ (c) $g(t) - g(2)$
34. $V(r) = \frac{4}{3}\pi r^3$
 (a) $V(3)$ (b) $V\left(\frac{3}{2}\right)$ (c) $V(2r)$
35. $f(y) = 3 - \sqrt{y}$
 (a) $f(4)$ (b) $f(0.25)$ (c) $f(4x^2)$
36. $f(x) = \sqrt{x + 12} + 4$
 (a) $f(-3)$ (b) $f(13)$ (c) $f(x - 12)$
37. $q(x) = \dfrac{1}{x^2 - 9}$
 (a) $q(-3)$ (b) $q(2)$ (c) $q(y + 3)$
38. $q(t) = \dfrac{2t^2 + 3}{t^2}$
 (a) $q(2)$ (b) $q(0)$ (c) $q(-x)$
39. $f(x) = \dfrac{|x|}{x}$
 (a) $f(9)$ (b) $f(-9)$ (c) $f(t)$
40. $f(x) = |x| + 3$
 (a) $f(3)$ (b) $f(-6)$ (c) $f(t)$
41. $f(x) = \begin{cases} 2x + 1, & x < 0 \\ 2x + 2, & x \ge 0 \end{cases}$
 (a) $f(-1)$ (b) $f(0)$ (c) $f(2)$
42. $f(x) = \begin{cases} 2x + 5, & x \le 0 \\ 2 - x, & x > 0 \end{cases}$
 (a) $f(-2)$ (b) $f(0)$ (c) $f(1)$
43. $f(x) = \begin{cases} x^2 + 2, & x \le 1 \\ 2x^2 + 2, & x > 1 \end{cases}$
 (a) $f(-2)$ (b) $f(1)$ (c) $f(2)$
44. $f(x) = \begin{cases} x^2 - 4, & x \le 0 \\ 1 - 2x^2, & x > 0 \end{cases}$
 (a) $f(-2)$ (b) $f(0)$ (c) $f(1)$
45. $f(x) = \begin{cases} x + 2, & x < 0 \\ 4, & 0 \le x < 2 \\ x^2 + 1, & x \ge 2 \end{cases}$
 (a) $f(-2)$ (b) $f(0)$ (c) $f(2)$
46. $f(x) = \begin{cases} 5 - 2x, & x < 0 \\ 5, & 0 \le x < 1 \\ 4x + 1, & x \ge 1 \end{cases}$
 (a) $f(-4)$ (b) $f(0)$ (c) $f(1)$

Evaluating a Function In Exercises 47–50, assume that the domain of f is the set $A = \{-2, -1, 0, 1, 2\}$. Determine the set of ordered pairs representing the function f.

47. $f(x) = (x - 1)^2$
48. $f(x) = x^2 - 3$
49. $f(x) = |x| + 2$
50. $f(x) = |x + 1|$

Evaluating a Function In Exercises 51 and 52, complete the table.

51. $h(t) = \frac{1}{2}|t + 3|$

t	-5	-4	-3	-2	-1
$h(t)$					

52. $f(s) = \dfrac{|s - 2|}{s - 2}$

s	0	1	$\frac{3}{2}$	$\frac{5}{2}$	4
$f(s)$					

Finding the Inputs That Have Outputs of Zero In Exercises 53–56, find all values of x such that $f(x) = 0$.

53. $f(x) = 15 - 3x$
54. $f(x) = 5x + 1$
55. $f(x) = \dfrac{9x - 4}{5}$
56. $f(x) = \dfrac{3 - 2x}{19}$

Finding the Domain of a Function In Exercises 57–66, find the domain of the function.

57. $f(x) = 5x^2 + 2x - 1$
58. $g(x) = 1 - 2x^2$
59. $h(t) = \dfrac{4}{t}$
60. $s(y) = \dfrac{5y}{y + 7}$
61. $g(y) = \sqrt{y + 6}$
62. $f(x) = \sqrt[4]{x^2 + 3x}$
63. $g(x) = \dfrac{1}{x} - \dfrac{3}{x + 2}$
64. $h(x) = \dfrac{10}{x^2 - 2x}$
65. $g(y) = \dfrac{y + 2}{\sqrt{y - 10}}$
66. $f(x) = \dfrac{\sqrt{x + 15}}{15 + x}$

Finding the Domain and Range of a Function **In Exercises 67–72, use a graphing utility to graph the function. Find the domain and range of the function.**

67. $f(x) = \sqrt{16 - x^2}$

68. $f(x) = \sqrt{x^2 + 1}$

69. $g(x) = |2x + 3|$

70. $g(x) = |4x - 21|$

71. $f(x) = \dfrac{x - 4}{\sqrt{x}}$

72. $f(x) = \dfrac{x + 2}{\sqrt[3]{x - 10}}$

73. Geometry Write the area A of a circle as a function of its circumference C.

74. Geometry Write the area A of an equilateral triangle as a function of the length s of its sides.

75. Exploration An open box of maximum volume is to be made from a square piece of material, 24 centimeters on a side, by cutting equal squares from the corners and turning up the sides. (See figure.)

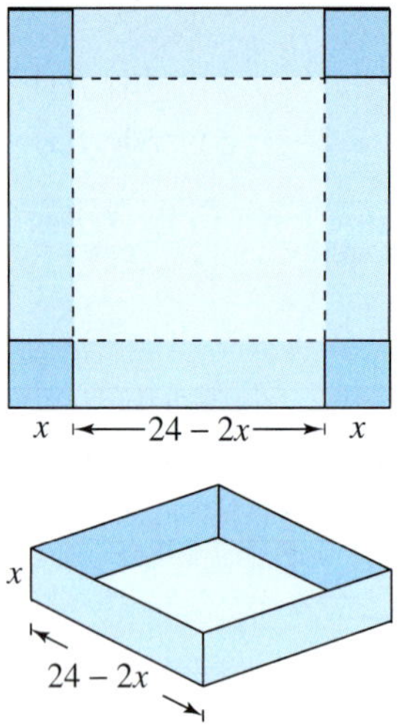

(a) The table shows the volume V (in cubic centimeters) of the box for various heights x (in centimeters). Use the table to estimate the maximum volume.

Height, x	Volume, V
1	484
2	800
3	972
4	1024
5	980
6	864

(b) Does the relation defined by the ordered pairs (x, V) from part (a) represent V as a function of x?

(c) Write an equation that relates V and x. Does the equation represent V as a function of x? If so, determine its domain.

(d) Use a graphing utility to plot the points from the table in part (a) and graph the equation from part (c) in the same viewing window. How closely does the equation represent the data? Explain.

76. Geometry A rectangular package to be sent by the U.S. Postal Service (USPS) using a mail class other than USPS Retail Ground and Parcel Select can have a maximum combined length and girth (perimeter of a cross section) of 108 inches. (See figure.)

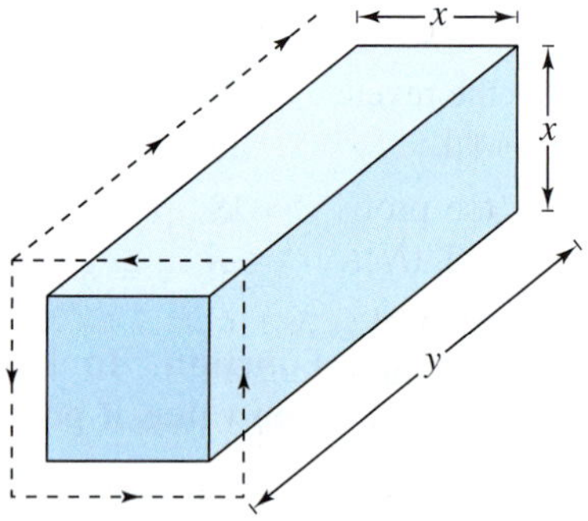

(a) Write the volume V of the package as a function of x. What is the domain of the function?

(b) Use a graphing utility to graph the function. Be sure to use an appropriate viewing window.

(c) What dimensions will maximize the volume of the package? Explain.

77. Geometry A right triangle is formed in the first quadrant by the x- and y-axes and a line through the point (2, 1), as shown in the figure. Write the area A of the triangle as a function of a and determine the domain of the function.

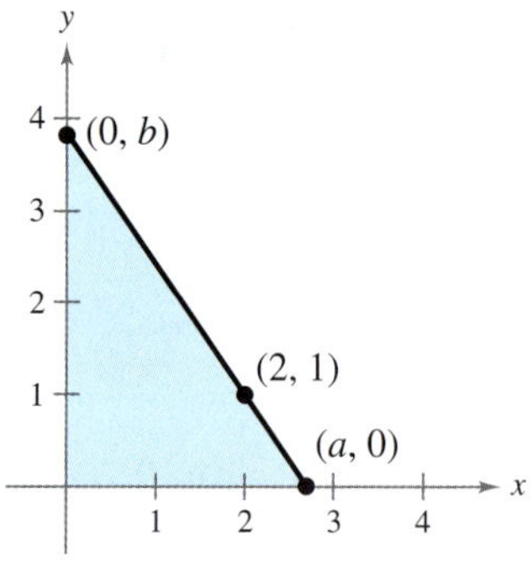

78. Geometry A rectangle is bounded by the x-axis and the semicircle $y = \sqrt{36 - x^2}$, as shown in the figure. Write the area A of the rectangle as a function of x and determine the domain of the function.

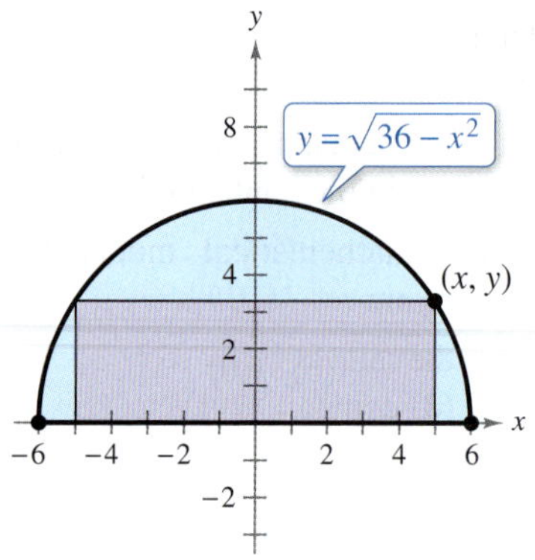

79. Business A company produces a product for which the variable cost is \$68.75 per unit and the fixed costs are \$248,000. The product sells for \$99.99. Let x be the number of units produced and sold.

(a) The total cost for a business is the sum of the variable cost and the fixed costs. Write the total cost C as a function of the number of units produced.

(b) Write the revenue R as a function of the number of units sold.

(c) Write the profit P as a function of the number of units sold. (*Note*: $P = R - C$.)

(d) Use the model in part (c) to find $P(20{,}000)$. Interpret your result in the context of the situation.

(e) Use the model in part (c) to find $P(0)$. Interpret your result in the context of the situation.

80. MODELING DATA

The table shows the profit y (in thousands of dollars) of a property maintenance business for each month of 2018, with $x = 1$ representing January.

Month, x	Profit, y
1	25.2
2	25.6
3	26.6
4	28.3
5	31.5
6	35.8
7	32.8
8	30.1
9	28.6
10	26.9
11	24.5
12	22.7

The mathematical model below represents the data.

$$f(x) = \begin{cases} -1.97x + 46.3 \\ 0.505x^2 - 1.47x + 26.3 \end{cases}$$

(a) Identify the independent and dependent variables and explain what they represent in the context of the problem.

(b) What is the domain of each part of the piecewise-defined function? Explain your reasoning.

(c) Use the mathematical model to find $f(5)$. Interpret your result in the context of the problem.

(d) Use the mathematical model to find $f(11)$. Interpret your result in the context of the problem.

(e) How do the values obtained from the models in parts (c) and (d) compare with the actual data values?

81. Civil Engineering The numbers n (in thousands) of structurally deficient bridges in the United States from 2006 through 2017 can be approximated by the model

$$n(t) = \begin{cases} 0.219t^2 - 4.31t + 93.4, & 6 \le t \le 9 \\ -2.39t + 94.8, & 9 < t \le 17 \end{cases}$$

where t represents the year, with $t = 6$ corresponding to 2006. The actual numbers are shown in the bar graph. (*Source:* U.S. Federal Highway Administration)

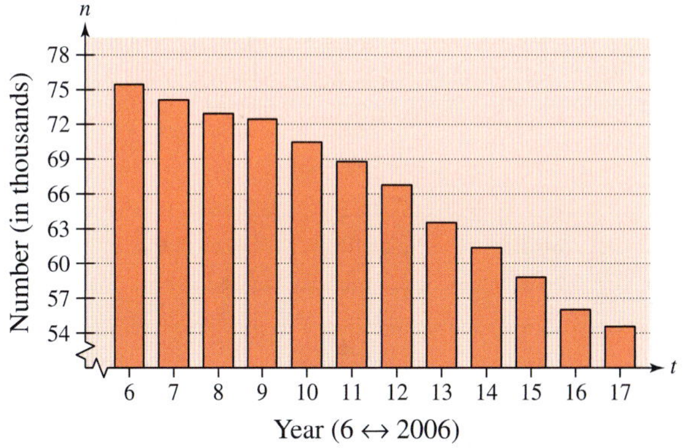

(a) Identify the independent and dependent variables and explain what they represent in the context of the problem.

(b) Use the *table* feature of a graphing utility to approximate the number of structurally deficient bridges each year from 2006 through 2017.

(c) Compare the values in part (b) with the actual values shown in the bar graph. How well does the model fit the data?

(d) Do you think the piecewise-defined function could be used to predict the number of structurally deficient bridges for years outside the domain? Explain your reasoning.

82. *Why you should learn it* *(p. 16)* The force F (in tons) of water against the face of a dam is estimated by the function

$$F(y) = 149.76\sqrt{10}y^{5/2}$$

where y is the depth of the water (in feet).

(a) Complete the table. What can you conclude?

y	5	10	20	30	40
$F(y)$					

(b) Use a graphing utility to graph the function. Describe your viewing window.

(c) Use the table to approximate the depth at which the force against the dam is 1,000,000 tons. Verify your answer graphically. How could you find a better estimate?

83. Projectile Motion The height y (in feet) of a baseball thrown by a child is

$$y = -0.1x^2 + 3x + 6$$

where x is the horizontal distance (in feet) from where the ball was thrown. Will the ball fly over the glove of another child 30 feet away trying to catch the ball? Explain. (Assume that the child who is trying to catch the ball holds a baseball glove at a height of 5 feet.)

84. Business The graph shows the revenues (in billions of dollars) of The Boeing Company from 2010 through 2017. Let $f(x)$ represent the sales in year x. *(Source: The Boeing Company)*

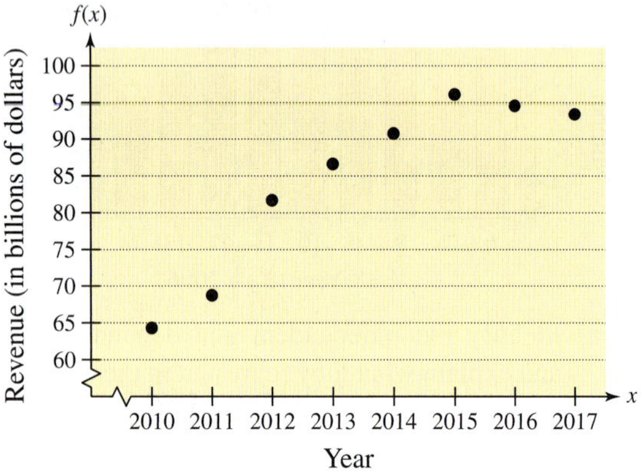

(a) Find $\dfrac{f(2017) - f(2010)}{2017 - 2010}$ and interpret the result in the context of the problem.

(b) An approximate model for the function is

$$R(t) = -0.912t^2 + 10.91t + 62.3, \quad 0 \le t \le 7$$

where R is the revenue (in billions of dollars) and $t = 0$ represents 2010. Complete the table and compare the results with the data in the graph.

t	0	1	2	3	4	5	6	7
$S(t)$								

Evaluating a Difference Quotient **In Exercises 85–88, find the difference quotient and simplify your answer.**

85. $f(x) = 2x, \quad \dfrac{f(x + c) - f(x)}{c}, \quad c \neq 0$

86. $g(x) = 3x - 1, \quad \dfrac{g(x + h) - g(x)}{h}, \quad h \neq 0$

87. $f(x) = x^3 + 3x, \quad \dfrac{f(x + h) - f(x)}{h}, \quad h \neq 0$

88. $f(x) = x^2 - x + 1, \quad \dfrac{f(2 + h) - f(2)}{h}, \quad h \neq 0$

Focusing on Concepts

True or False? **In Exercises 89 and 90, determine whether the statement is true or false. Justify your answer.**

89. The domain of the function $f(x) = x^4 - 1$ is $(-\infty, \infty)$, and the range of f is $(0, \infty)$.

90. The set of ordered pairs $\{(-8, -2), (-6, 0), (-4, 0), (-2, 2), (0, 4), (2, -2)\}$ represents a function.

Think About It **In Exercises 91 and 92, write a square root function for the graph shown. Then identify the domain and range of the function.**

91.

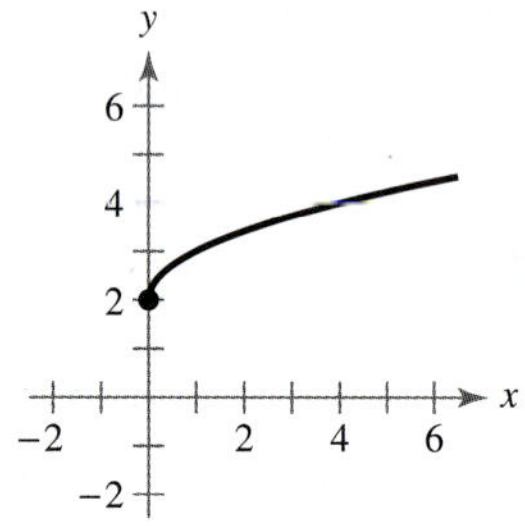

92.

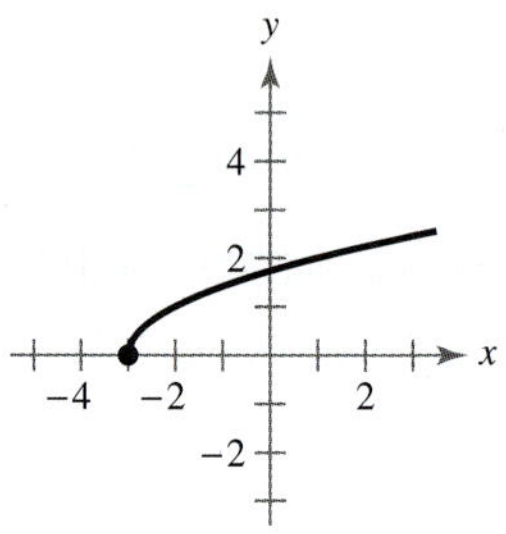

93. Error Analysis Describe the error.

The functions

$$f(x) = \sqrt{x - 1} \quad \text{and} \quad g(x) = \frac{1}{\sqrt{x - 1}}$$

have the same domain, which is the set of all real numbers x such that $x \ge 1$.

94. HOW DO YOU SEE IT? The graph represents the height h of a projectile after t seconds.

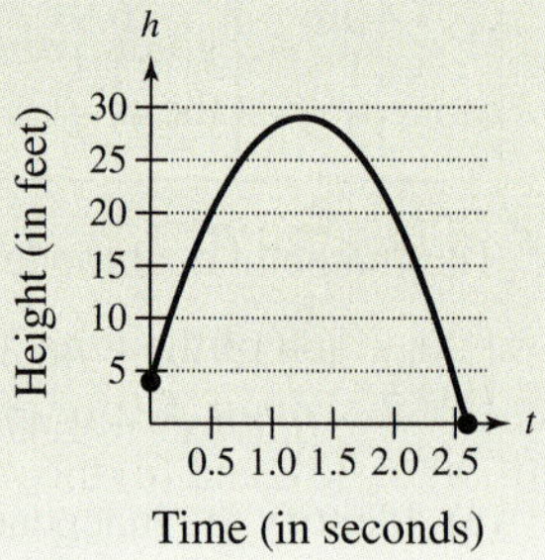

(a) Explain why h is a function of t.

(b) Approximate the height of the projectile after 0.5 second and after 1.25 seconds.

(c) Approximate the domain of h.

(d) Is t a function of h? Explain.

Cumulative Mixed Review

Operations with Rational Expressions **In Exercises 95–98, perform the operation and simplify.**

95. $12 - \dfrac{4}{x + 2}$

96. $\dfrac{3}{x^2 + x - 20} + \dfrac{2x}{x^2 + 4x - 5}$

97. $\dfrac{x^5}{2x^3 + 4x^2} \cdot \dfrac{4x + 8}{3x}$

98. $\dfrac{x + 7}{2(x - 9)} \div \dfrac{x - 7}{2(x - 9)}$

The symbol ∫ indicates an example or exercise that highlights algebraic techniques specifically used in calculus.

1.3 Graphs of Functions

The Graph of a Function

The **graph of a function** f is the collection of ordered pairs $(x, f(x))$ such that x is in the domain of f. As you study this section, remember the geometric interpretations of x and $f(x)$. (See figure at the right.)

x = the directed distance from the y-axis

$f(x)$ = the directed distance from the x-axis

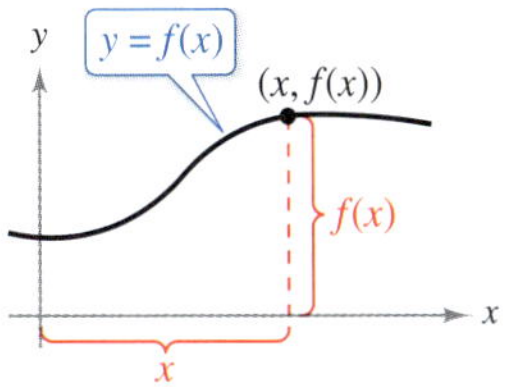

The graph of a function

Example 1 shows how to use the graph of a function to find the domain and range of the function.

EXAMPLE 1 Finding the Domain and Range of a Function

Use the graph of the function f to find (a) the domain of f, (b) the function values $f(-1)$ and $f(2)$, and (c) the range of f.

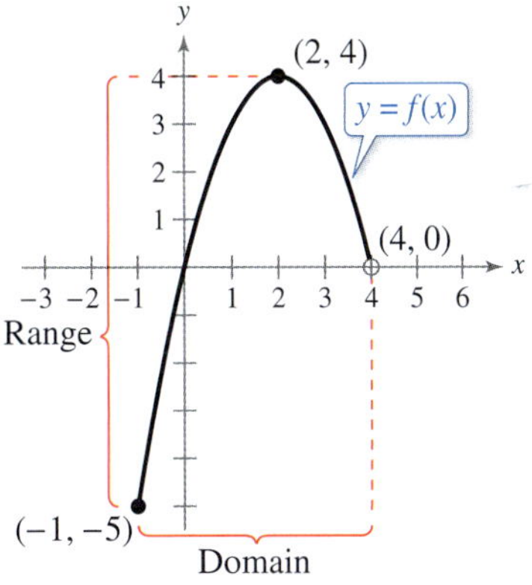

Solution

a. The closed dot at $(-1, -5)$ indicates that $x = -1$ is in the domain of f, whereas the open dot at $(4, 0)$ indicates that $x = 4$ is not in the domain. So, the domain of f is all x in the interval $[-1, 4)$.

b. Because $(-1, -5)$ is a point on the graph of f, you know that

$$f(-1) = -5.$$

Similarly, because $(2, 4)$ is a point on the graph of f, you know that

$$f(2) = 4.$$

c. Because the graph does not extend below $f(-1) = -5$ or above $f(2) = 4$, the range of f is the interval $[-5, 4]$.

✓ **Checkpoint** Audio-video solution in English & Spanish at LarsonPrecalculus.com

Use the graph of the function f to find (a) the domain of f, (b) the function values $f(0)$ and $f(3)$, and (c) the range of f.

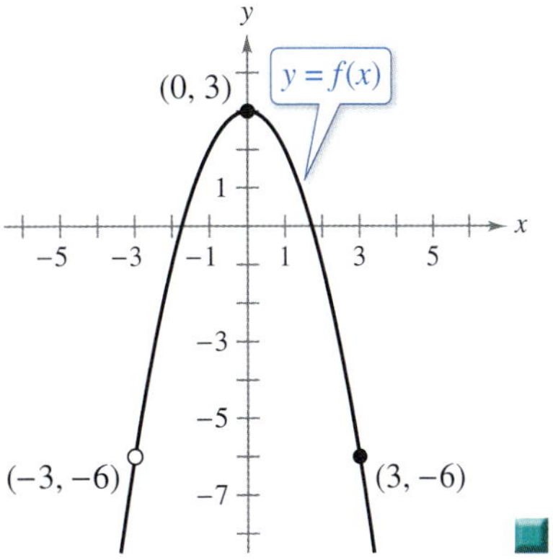

The use of dots (open or closed) at the extreme left and right points of a graph indicates that the graph does not extend beyond these points (for instance, see the figure in Example 1). When no such dots are shown, assume that the graph extends beyond these points.

What you should learn

- Find the domains and ranges of functions and use the Vertical Line Test for functions.
- Determine intervals on which functions are increasing, decreasing, or constant.
- Determine relative minimum and relative maximum values of functions.
- Identify and graph step functions and other piecewise-defined functions.
- Identify even and odd functions.

Why you should learn it

Graphs of functions provide visual relationships between two variables. For example, in Exercise 96 on page 39, you will use the graph of a step function to model the cost of mailing a package.

Franck Boston/Shutterstock.com

EXAMPLE 2 Finding the Domain and Range of a Function

Find the domain and range of $f(x) = \sqrt{x - 4}$.

Algebraic Solution

Because the expression under a radical cannot be negative, the domain of $f(x) = \sqrt{x - 4}$ is the set of all real numbers such that $x - 4 \geq 0$. Solve this linear inequality for x as follows. (For help with solving linear inequalities, see Appendix E at this textbook's *Student Companion Website*.)

$x - 4 \geq 0$ Write original inequality.

$x \geq 4$ Add 4 to each side.

So, the domain is the set of all real numbers greater than or equal to 4. Because the value of a radical expression is never negative, the range of $f(x) = \sqrt{x - 4}$ is the set of all nonnegative real numbers.

Graphical Solution

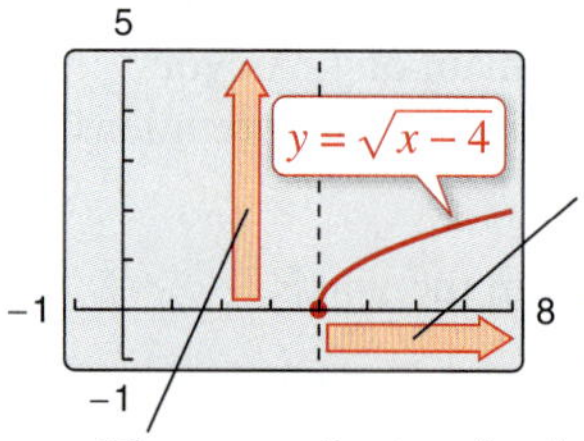

The x-coordinates of points on the graph extend from 4 to the right. So, the domain is the set of all real numbers greater than or equal to 4.

The y-coordinates of points on the graph extend from 0 upwards. So, the range is the set of all nonnegative real numbers.

Checkpoint *Audio-video solution in English & Spanish at LarsonPrecalculus.com*

Find the domain and range of $f(x) = \sqrt{x - 1}$.

By the definition of a function, at most one y-value corresponds to a given x-value. It follows, then, that a vertical line can intersect the graph of a function at most once. This leads to the **Vertical Line Test** for functions.

Vertical Line Test for Functions

A set of points in a coordinate plane is the graph of y as a function of x if and only if no vertical line intersects the graph at more than one point.

EXAMPLE 3 Vertical Line Test for Functions

Use the Vertical Line Test to decide whether each graph represents y as a function of x.

a.

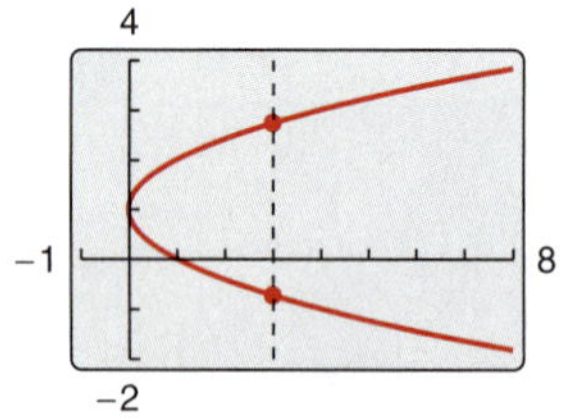

b.

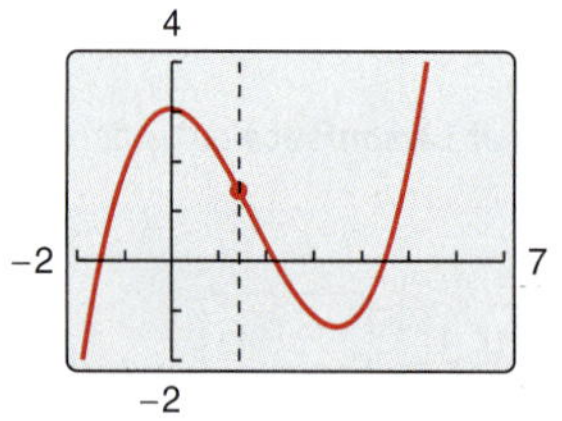

Solution

a. This is *not* a graph of y as a function of x because you can find a vertical line that intersects the graph twice.

b. This *is* a graph of y as a function of x because every vertical line intersects the graph at most once.

Checkpoint *Audio-video solution in English & Spanish at LarsonPrecalculus.com*

Use the Vertical Line Test to decide whether the graph at the right represents y as a function of x.

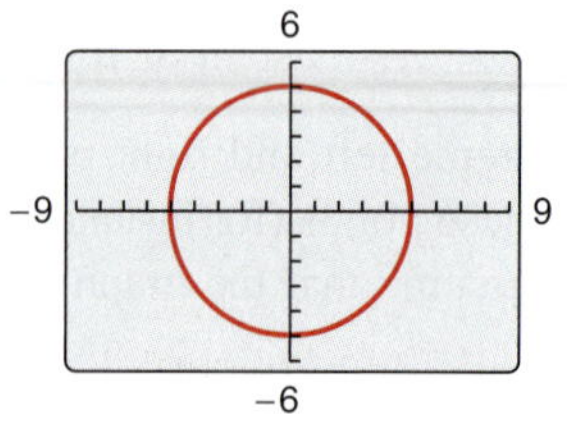

Technology Tip

Most graphing utilities graph functions of x more easily than other types of equations. For instance, the graph shown in Example 3(a) represents the equation $x - (y - 1)^2 = 0$. To use a graphing utility to duplicate this graph, you must first solve the equation for y to obtain $y = 1 \pm \sqrt{x}$, and then graph the two equations $y_1 = 1 + \sqrt{x}$ and $y_2 = 1 - \sqrt{x}$ in the same viewing window.

Increasing and Decreasing Functions

The more you know about the graph of a function, the more you know about the function itself. Consider the graph shown in Figure 1.15. Moving from *left to right,* this graph

- falls from $x = -2$ to $x = 0$.
- is constant from $x = 0$ to $x = 2$.
- rises from $x = 2$ to $x = 4$.

The next definition summarizes when a function is increasing, decreasing, or constant on an open interval.

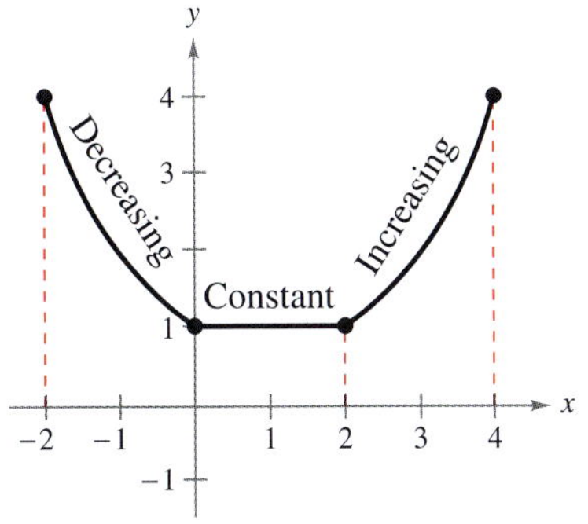

Figure 1.15

Increasing, Decreasing, and Constant Functions

A function f is **increasing** on an interval when, for any x_1 and x_2 in the interval,

$x_1 < x_2$ implies $f(x_1) < f(x_2)$.

A function f is **decreasing** on an interval when, for any x_1 and x_2 in the interval,

$x_1 < x_2$ implies $f(x_1) > f(x_2)$.

A function f is **constant** on an interval when, for any x_1 and x_2 in the interval,

$f(x_1) = f(x_2)$.

EXAMPLE 4 Increasing and Decreasing Functions

In Figure 1.16, determine the open intervals on which each function is increasing, decreasing, or constant.

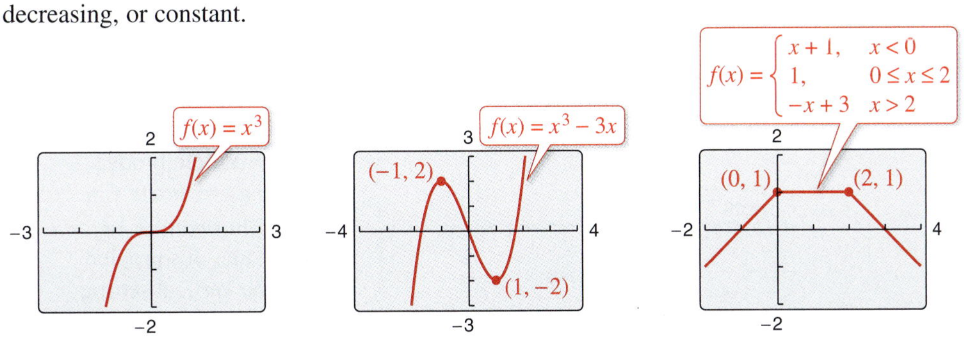

(a) (b) (c)

Figure 1.16

Solution

a. Although it might appear that there is an interval in which this function is constant, you can see that if $x_1 < x_2$, then $(x_1)^3 < (x_2)^3$, which implies that $f(x_1) < f(x_2)$. So, the function is increasing over the entire real number line.

b. This function is increasing on the interval $(-\infty, -1)$, decreasing on the interval $(-1, 1)$, and increasing on the interval $(1, \infty)$.

c. This function is increasing on the interval $(-\infty, 0)$, constant on the interval $(0, 2)$, and decreasing on the interval $(2, \infty)$.

✓ ***Checkpoint*** ▶ *Audio-video solution in English & Spanish at LarsonPrecalculus.com*

Graph the function $f(x) = x^3 + 3x^2 - 1$. Then determine the open intervals on which the function is increasing, decreasing, or constant.

Relative Minimum and Maximum Values

The points at which a function changes its increasing, decreasing, or constant behavior are helpful in determining the relative maximum or relative minimum values of the function.

Definition of Relative Minimum and Relative Maximum

A function value $f(a)$ is a **relative minimum** of f when there exists an interval (x_1, x_2) that contains a such that

$$x_1 < x < x_2 \quad \text{implies} \quad f(a) \le f(x).$$

A function value $f(a)$ is a **relative maximum** of f when there exists an interval (x_1, x_2) that contains a such that

$$x_1 < x < x_2 \quad \text{implies} \quad f(a) \ge f(x).$$

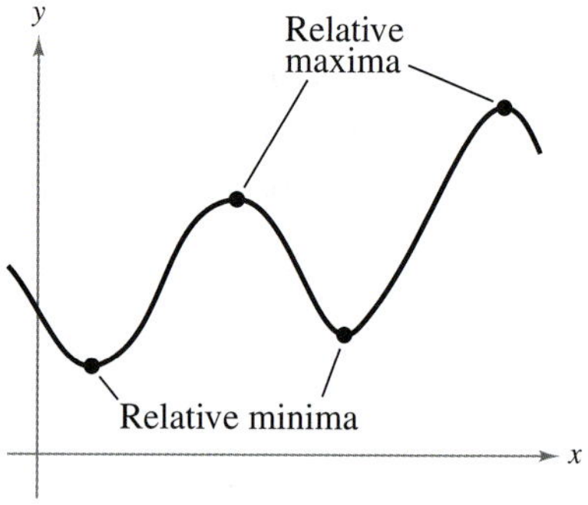

Figure 1.17

Figure 1.17 shows several different examples of relative minima and relative maxima. In Section 2.1, you will study a technique for finding the *exact points* at which a second-degree polynomial function has a relative minimum or relative maximum. For the time being, however, you can use a graphing utility to find reasonable approximations of these points.

EXAMPLE 5 Approximating a Relative Minimum

Use a graphing utility to approximate the relative minimum of the function

$$f(x) = 3x^2 - 4x - 2.$$

Solution

The graph of f is shown in Figure 1.18. By using the *zoom* and *trace* features of a graphing utility, you can estimate that the relative minimum of the function occurs at the point $(0.67, -3.33)$, as shown in Figure 1.19. So, the relative minimum is approximately -3.33. Later, in Section 2.1, you will learn how to determine that the exact point at which the relative minimum occurs is $\left(\frac{2}{3}, -\frac{10}{3}\right)$ and the exact relative minimum is $-\frac{10}{3}$.

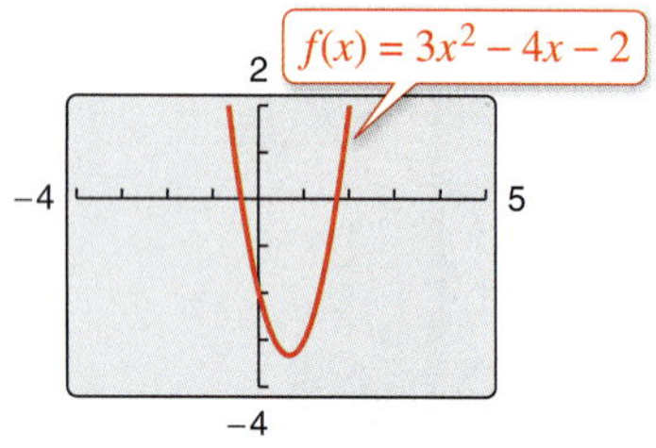

Figure 1.18

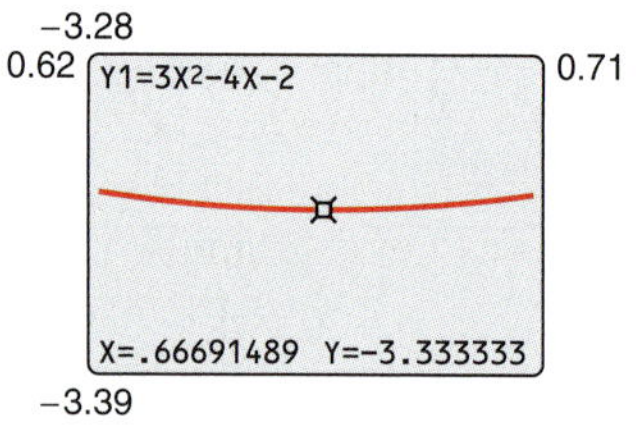

Figure 1.19

Technology Tip

When you use a graphing utility to estimate the x- and y-values of the point at which a relative minimum or relative maximum occurs, the *zoom* feature will often produce graphs that are nearly flat, as shown in Figure 1.19. To overcome this problem, change the vertical setting of the viewing window. The graph will stretch vertically when the values of Ymin and Ymax are closer together.

✓ Checkpoint *Audio-video solution in English & Spanish at LarsonPrecalculus.com*

Use a graphing utility to approximate the relative maximum of the function

$$f(x) = -4x^2 - 7x + 3.$$

You can also use the *table* feature of a graphing utility to numerically approximate the relative minimum of the function in Example 5. Using a table that begins at 0.6 and increments the value of x by 0.01, you can approximate that the minimum of f occurs at the point $(0.67, -3.33)$. Some graphing utilities have built-in programs that will find minimum or maximum values, as shown in Example 6 on the next page.

EXAMPLE 6 Approximating Relative Minima and Maxima

Use a graphing utility to approximate the relative minimum and relative maximum of the function $f(x) = -x^3 + x$.

Solution

By using the *minimum* and *maximum* features of the graphing utility, you can estimate that the relative minimum of the function occurs at the point

$(-0.58, -0.38)$ See Figure 1.20.

and the relative maximum occurs at the point

$(0.58, 0.38)$. See Figure 1.21.

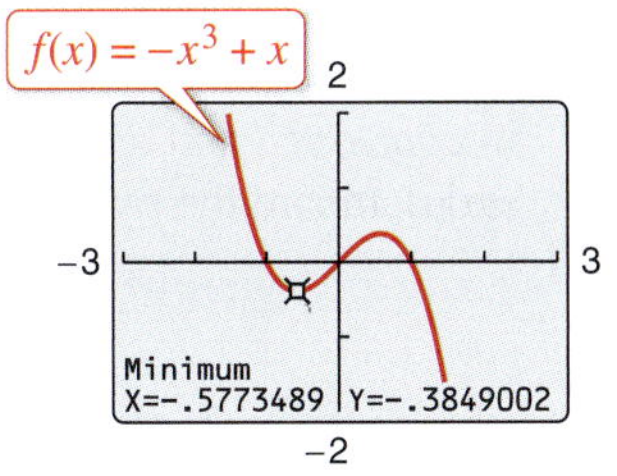

Figure 1.20

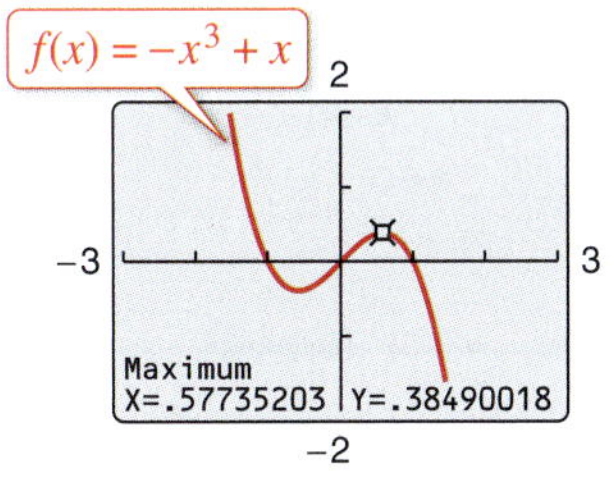

Figure 1.21

If you take a course in calculus, you will learn a technique for finding the exact relative minimum and relative maximum of a function.

Checkpoint *Audio-video solution in English & Spanish at LarsonPrecalculus.com*

Use a graphing utility to approximate the relative minimum and relative maximum of the function $f(x) = 2x^3 + 3x^2 - 12x$.

EXAMPLE 7 Temperature

During a 24-hour period, the temperature y (in degrees Fahrenheit) of a certain city can be approximated by the model

$$y = 0.026x^3 - 1.03x^2 + 10.2x + 34, \quad 0 \le x \le 24$$

where x represents the time of day, with $x = 0$ corresponding to 6 A.M. Approximate the maximum temperature during this 24-hour period.

Solution

Using the *maximum* feature of a graphing utility, you can determine that the maximum temperature during the 24-hour period was approximately 64°F. This temperature occurred at about 12:36 P.M. ($x \approx 6.6$), as shown in Figure 1.22.

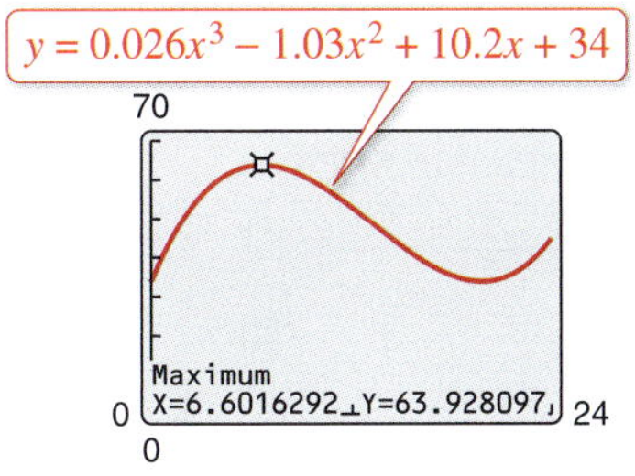

Figure 1.22

Checkpoint *Audio-video solution in English & Spanish at LarsonPrecalculus.com*

In Example 7, approximate the minimum temperature during the 24-hour period.

Step Functions and Piecewise-Defined Functions

Library of Parent Functions: Greatest Integer Function

The *greatest integer function*, denoted by $[\![x]\!]$ and defined as the greatest integer less than or equal to x, has an infinite number of breaks or steps—one at each integer value in its domain. The basic characteristics of the greatest integer function are summarized below.

Graph of $f(x) = [\![x]\!]$

Domain: $(-\infty, \infty)$

Range: the set of integers

x-intercepts: in the interval $[0, 1)$

y-intercept: $(0, 0)$

Constant between each pair of consecutive integers

Jumps vertically one unit at each integer value

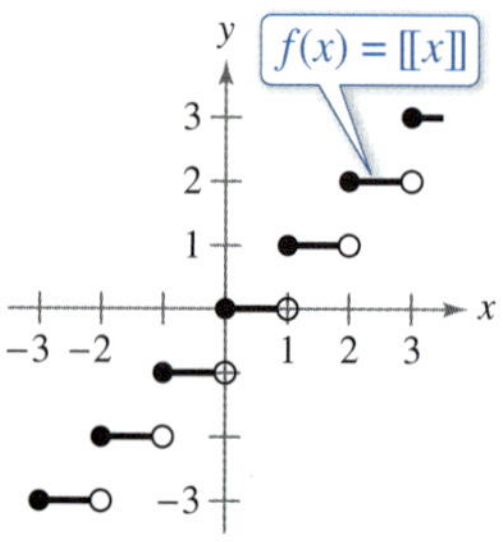

Technology Tip

Most graphing utilities display graphs in *connected* mode, which works well for graphs that do not have breaks. For graphs that do have breaks, such as the greatest integer function, it is better to use *dot* mode. Graph the greatest integer function [often called Int (x)] in *connected* and *dot* modes, and compare the two results.

Because of the vertical jumps described above, the greatest integer function is an example of a **step function** whose graph resembles a set of stairsteps. Here are some values of the greatest integer function.

$$[\![-1]\!] = (\text{greatest integer} \le -1) = -1$$

$$\left[\!\!\left[-\tfrac{1}{2}\right]\!\!\right] = \left(\text{greatest integer} \le -\tfrac{1}{2}\right) = -1$$

$$\left[\!\!\left[\tfrac{1}{10}\right]\!\!\right] = \left(\text{greatest integer} \le \tfrac{1}{10}\right) = 0$$

$$[\![1.5]\!] = (\text{greatest integer} \le 1.5) = 1$$

In Section 1.2, you learned that a piecewise-defined function is a function that is defined by two or more equations over a specified domain. To sketch the graph of a piecewise-defined function, you need to sketch the graph of each equation on the appropriate portion of the domain.

EXAMPLE 8 Sketching a Piecewise-Defined Function

Sketch the graph of

$$f(x) = \begin{cases} 2x + 3, & x \le 1 \\ -x + 4, & x > 1 \end{cases}$$

by hand.

Solution

This piecewise-defined function is composed of two linear functions. At and to the left of $x = 1$, the graph is the line $y = 2x + 3$. To the right of $x = 1$, the graph is the line $y = -x + 4$, as shown in the figure. Notice that the point $(1, 5)$ is a solid dot and the point $(1, 3)$ is an open dot. This is because $f(1) = 5$.

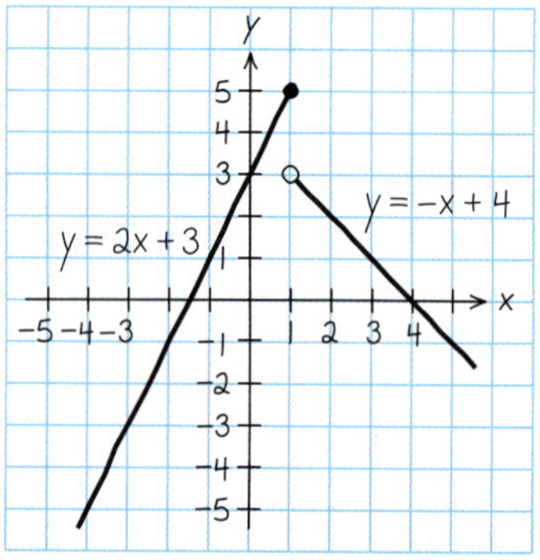

Checkpoint Audio-video solution in English & Spanish at LarsonPrecalculus.com

Sketch the graph of $f(x) = \begin{cases} -\frac{1}{2}x - 6, & x \le -4 \\ x + 5, & x > -4 \end{cases}$ by hand.

What's Wrong?

You use a graphing utility to graph

$$f(x) = \begin{cases} x^2 + 1, & x \le 0 \\ 4 - x, & x > 0 \end{cases}$$

by letting $y_1 = x^2 + 1$ and $y_2 = 4 - x$, as shown in the figure. You conclude that this is the graph of f. What's wrong?

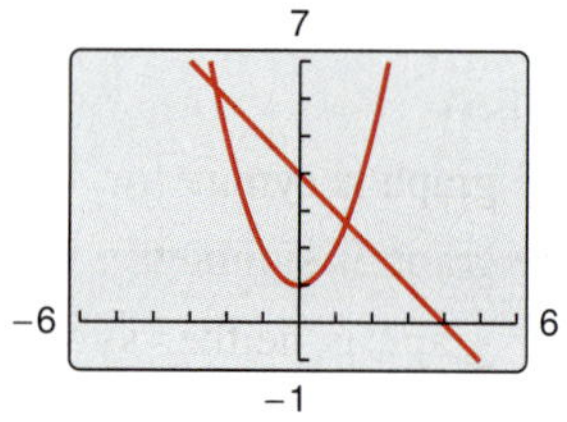

Even and Odd Functions

A graph has *symmetry with respect to the y-axis* if whenever (x, y) is on the graph, then so is the point $(-x, y)$. A graph has *symmetry with respect to the origin* if whenever (x, y) is on the graph, then so is the point $(-x, -y)$. A graph has *symmetry with respect to the x-axis* if whenever (x, y) is on the graph, then so is the point $(x, -y)$. A function whose graph is symmetric with respect to the y-axis is an **even function.** A function whose graph is symmetric with respect to the origin is an **odd function.** A graph that is symmetric with respect to the x-axis is not the graph of a function (except for the graph of $y = 0$). These three types of symmetry are illustrated in Figure 1.23.

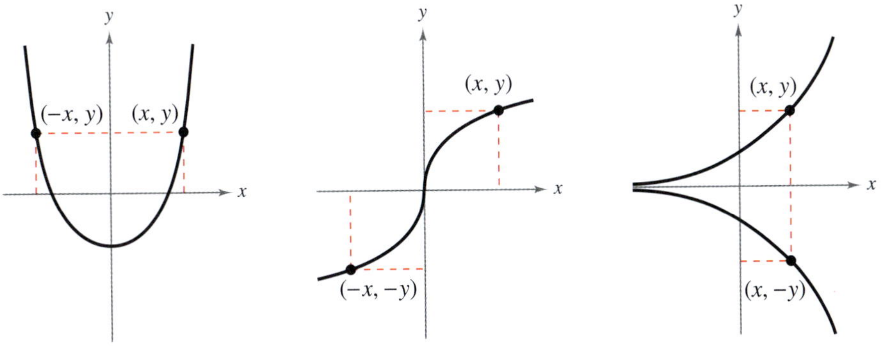

Symmetric to y-axis
Even function
Figure 1.23

Symmetric to origin
Odd function

Symmetric to x-axis
Not a function

Explore the Concept

Graph each function with a graphing utility. Determine whether the function is *even*, *odd*, or *neither*.

$f(x) = x^2 - x^4$

$g(x) = 2x^3 + 1$

$h(x) = x^5 - 2x^3 + x$

$j(x) = 2 - x^6 - x^8$

$k(x) = x^5 - 2x^4 + x - 2$

$p(x) = x^9 + 3x^5 - x^3 + x$

What do you notice about the equations of functions that are (a) odd and (b) even? By inspecting the equation of a function, describe a way to identify the function as (c) odd, (d) even, or (e) neither odd nor even.

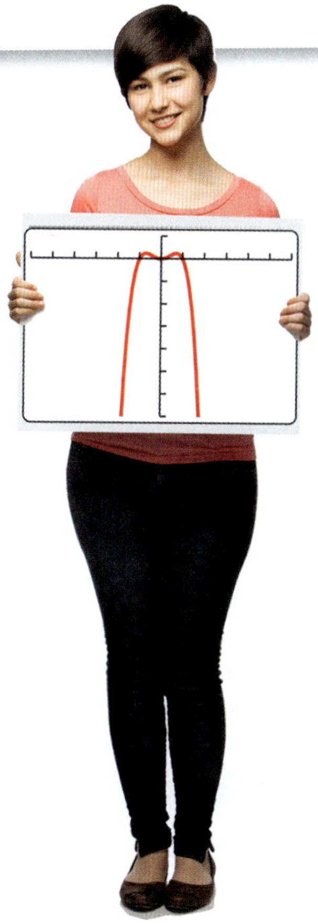

EXAMPLE 9 Even and Odd Functions

See LarsonPrecalculus.com for an interactive version of this type of example.

For each graph, determine whether the function is even, odd, or neither.

a.

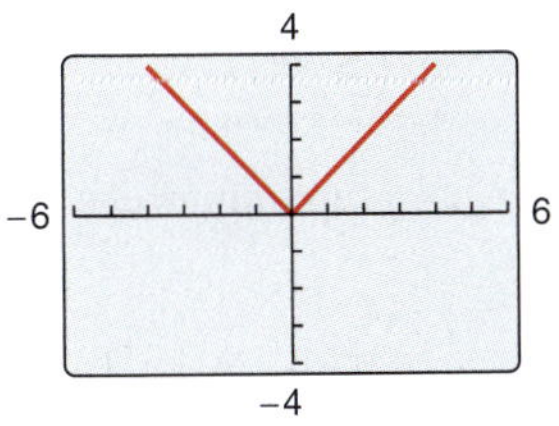

b.

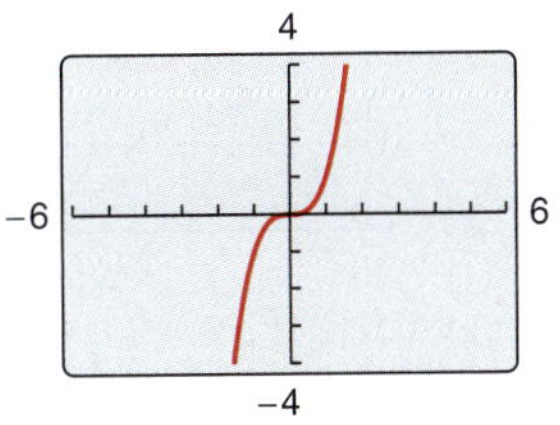

c.

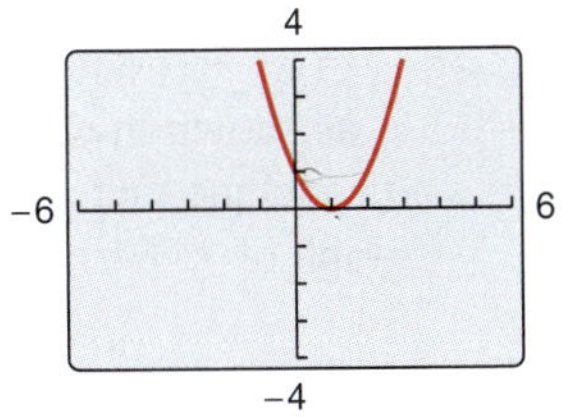

d.

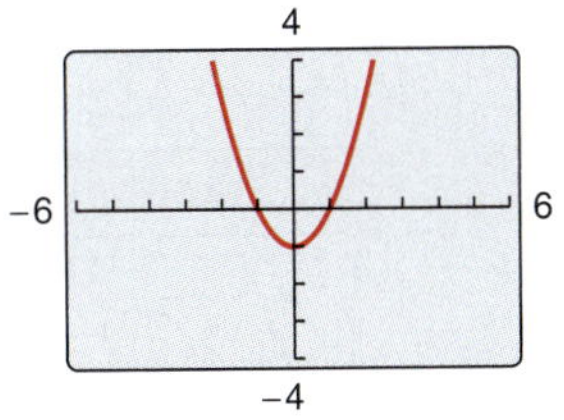

Solution

a. The graph is symmetric with respect to the y-axis. So, the function is even.

b. The graph is symmetric with respect to the origin. So, the function is odd.

c. The graph is neither symmetric with respect to the origin nor with respect to the y-axis. So, the function is neither even nor odd.

d. The graph is symmetric with respect to the y-axis. So, the function is even.

✓ ***Checkpoint*** *Audio-video solution in English & Spanish at LarsonPrecalculus.com*

Use a graphing utility to graph $f(x) = x^2 - 4$ and determine whether it is even, odd, or neither.

Test for Even and Odd Functions

A function f is **even** when, for each x in the domain of f, $f(-x) = f(x)$.
A function f is **odd** when, for each x in the domain of f, $f(-x) = -f(x)$.

EXAMPLE 10 Even and Odd Functions

Determine whether each function is even, odd, or neither.

a. $g(x) = x^3 - x$

b. $h(x) = x^2 + 1$

c. $f(x) = x^3 - 1$

Algebraic Solution

a. This function is odd because

$$\begin{aligned} g(-x) &= (-x)^3 - (-x) \\ &= -x^3 + x \\ &= -(x^3 - x) \\ &= -g(x). \end{aligned}$$

b. This function is even because

$$\begin{aligned} h(-x) &= (-x)^2 + 1 \\ &= x^2 + 1 \\ &= h(x). \end{aligned}$$

c. Substituting $-x$ for x produces

$$\begin{aligned} f(-x) &= (-x)^3 - 1 \\ &= -x^3 - 1. \end{aligned}$$

Because

$$f(x) = x^3 - 1$$

and

$$-f(x) = -x^3 + 1$$

you can conclude that

$$f(-x) \neq f(x)$$

and

$$f(-x) \neq -f(x).$$

So, the function is neither even nor odd.

Graphical Solution

a. The graph is symmetric with respect to the origin. So, this function is odd.

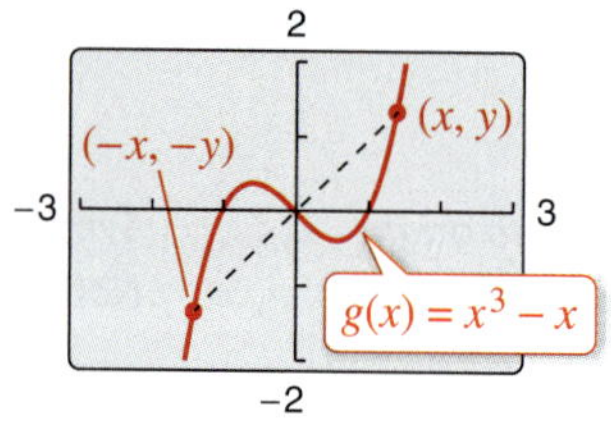

b. The graph is symmetric with respect to the y-axis. So, this function is even.

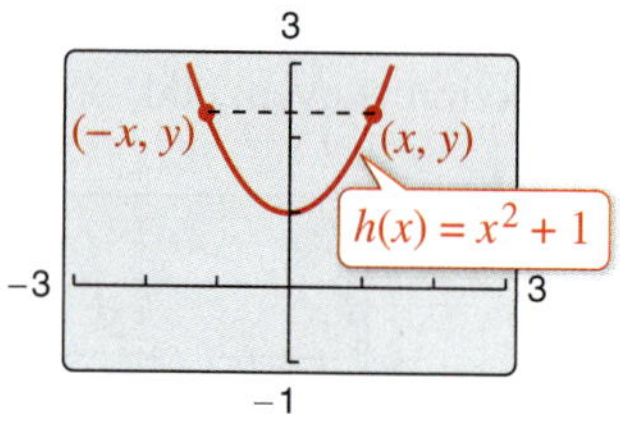

c. The graph is neither symmetric with respect to the origin nor with respect to the y-axis. So, this function is neither even nor odd.

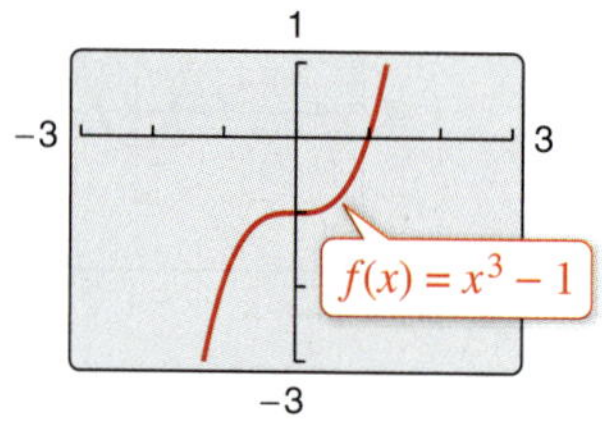

✓ Checkpoint ▶ *Audio-video solution in English & Spanish at LarsonPrecalculus.com*

Determine whether each function is even, odd, or neither. Then describe the symmetry.

a. $f(x) = 5 - 3x$ **b.** $g(x) = x^4 - x^2 - 1$ **c.** $h(x) = 2x^3 + 3x$

To help visualize symmetry with respect to the origin, place a pin at the origin of a graph and rotate the graph 180°. If the result after rotation coincides with the original graph, then the graph is symmetric with respect to the origin.

1.3 Exercises

See *CalcChat.com* for tutorial help and worked-out solutions to odd-numbered exercises.
For instructions on how to use a graphing utility, see Appendix A.

Vocabulary and Concept Check

In Exercises 1 and 2, fill in the blank.

1. A function f is ________ on an interval when, for any x_1 and x_2 in the interval, $x_1 < x_2$ implies $f(x_1) > f(x_2)$.

2. A function f is ________ when, for each x in the domain of f, $f(-x) = f(x)$.

3. The graph of a function f is the segment from $(1, 2)$ to $(4, 5)$, including the endpoints. What is the domain of f?

4. A vertical line intersects a graph twice. Does the graph represent a function?

5. Let f be a function such that $f(2) \geq f(x)$ for all values of x in the interval $(0, 3)$. Does $f(2)$ represent a relative minimum or a relative maximum?

6. For the greatest integer function $f(x) = [\![x]\!]$, in what interval does $f(x) = 5$?

Procedures and Problem Solving

Finding the Domain and Range of a Function **In Exercises 7–10, use the graph of the function to find the domain and range of f. Then find each function value.**

7. (a) $f(-1)$ (b) $f(0)$ (c) $f(1)$ (d) $f(2)$

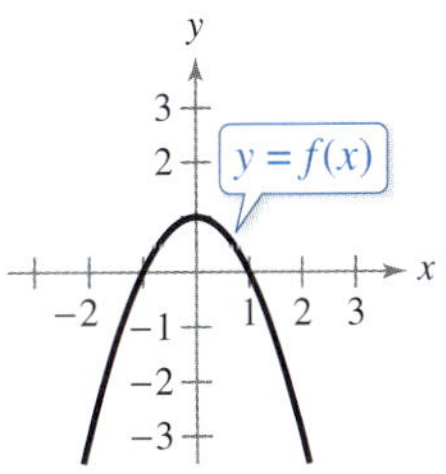

8. (a) $f(-2)$ (b) $f(-1)$ (c) $f(0)$ (d) $f(1)$

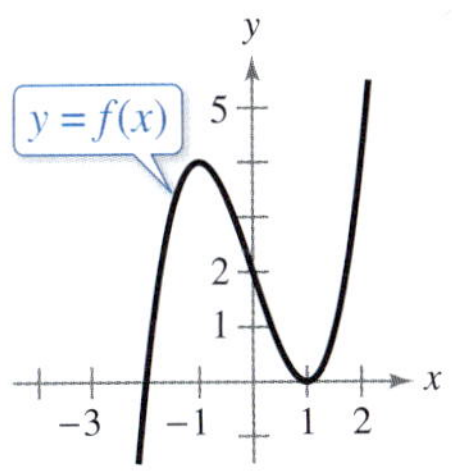

9. (a) $f(2)$ (b) $f(1)$ (c) $f(3)$ (d) $f(-1)$

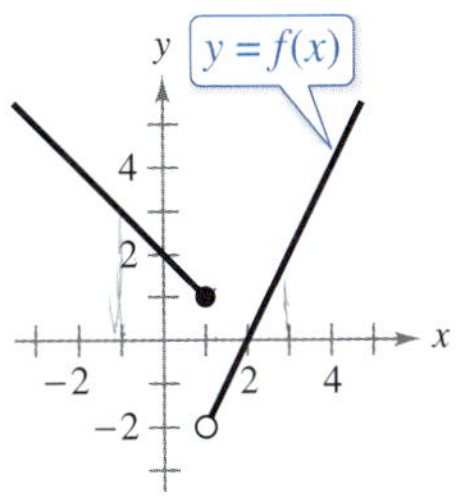

10. (a) $f(-4)$ (b) $f(0)$ (c) $f(2)$ (d) $f(4)$

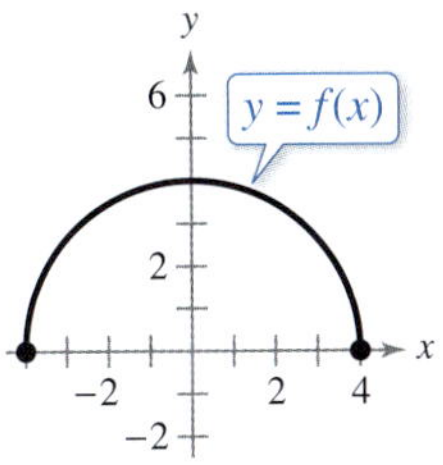

Finding the Domain and Range of a Function **In Exercises 11–16, use a graphing utility to graph the function and estimate its domain and range. Then find the domain and range algebraically.**

11. $f(x) = -2x^2 + 3$

12. $f(x) = x^2 - 1$

13. $f(x) = \sqrt{x + 2}$

14. $h(t) = \sqrt{16 - t^2}$

15. $f(x) = |x + 3|$

16. $f(x) = -\frac{1}{4}|x - 5|$

Analyzing a Graph **In Exercises 17–20, use the graph of the function to answer the questions.**

(a) Determine the domain of the function.

(b) Determine the range of the function.

(c) Find the value(s) of x for which $f(x) = 0$.

(d) Find $f(0)$, if possible.

(e) What is the value of f at $x = 1$? What are the coordinates of the point?

(f) The coordinates of the point on the graph of f at which $x = -3$ can be labeled $(-3, f(-3))$, or $(-3, \quad)$.

17.

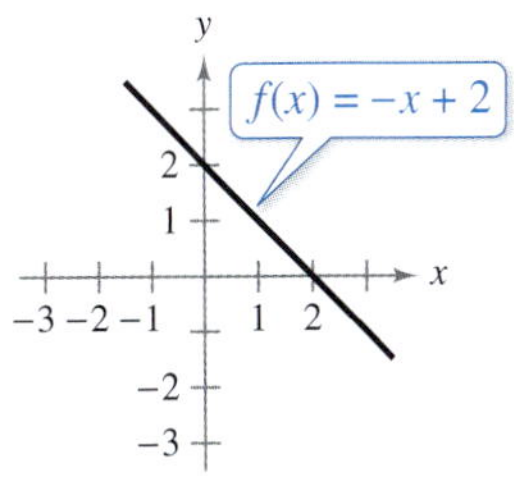

18.

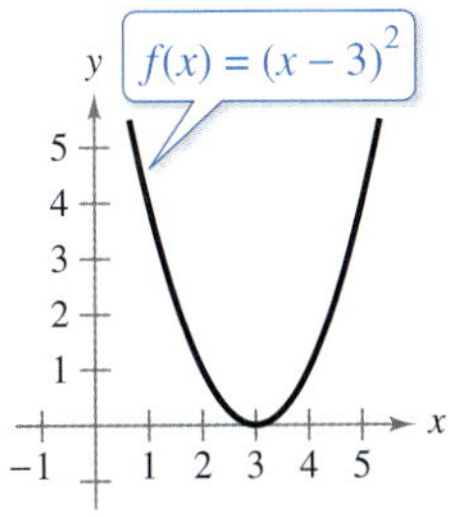

19.

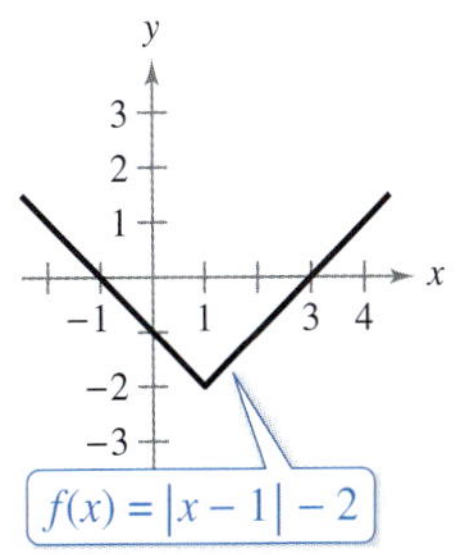

20.

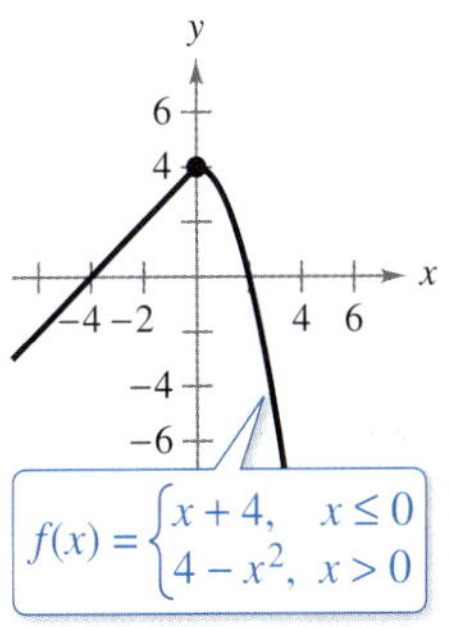

21. **Analyzing a Graph** In Exercises 17–20, what are the values you found in part (c) referred to graphically?

22. **Analyzing a Graph** In Exercises 17–20, what are the values you found in part (d) referred to graphically?

Vertical Line Test for Functions In Exercises 23–26, use the Vertical Line Test to determine whether y is a function of x. Describe how you can use a graphing utility to produce the graph shown.

23. $y = \frac{1}{2}x^2$

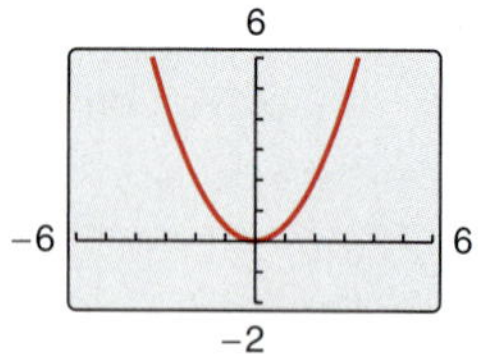

24. $x - y^2 = 1$

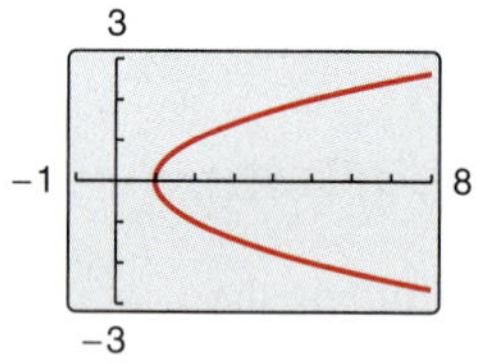

25. $0.25x^2 + y^2 = 1$

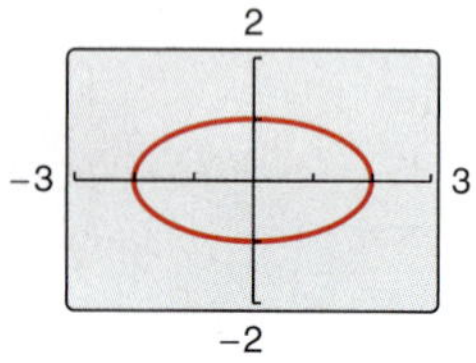

26. $x^2 = 2xy - 1$

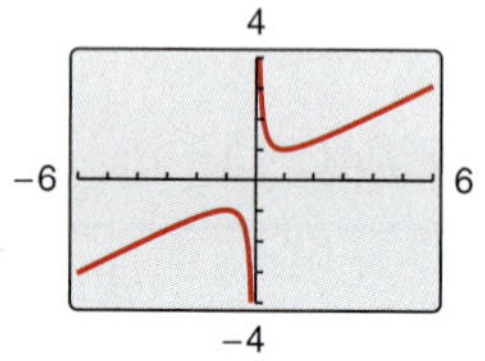

Increasing and Decreasing Functions In Exercises 27–30, determine the open intervals on which the function is increasing, decreasing, or constant.

27. $f(x) = \frac{3}{2}x$

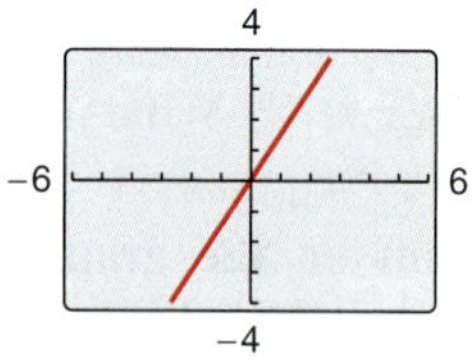

28. $f(x) = x^2 - 4x$

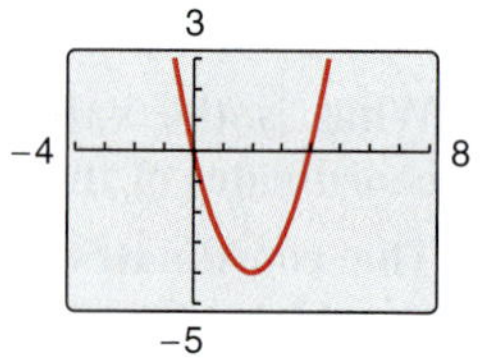

29. $f(x) = \sqrt{x^2 - 1}$

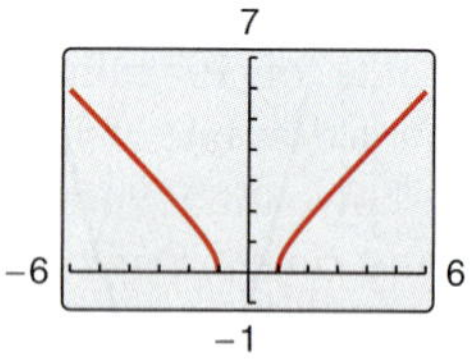

30. $f(x) = x^3 - 3x^2 + 2$

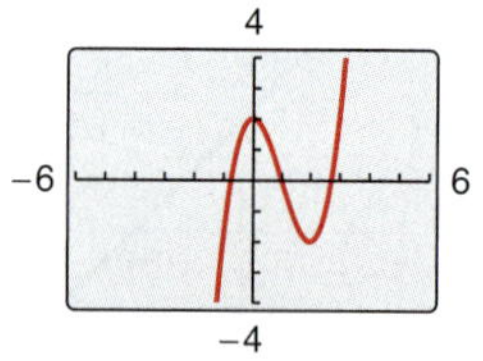

Increasing and Decreasing Functions In Exercises 31–42, (a) use a graphing utility to graph the function and (b) visually determine the open intervals on which the function is increasing, decreasing, or constant. Use a table of values to verify the results of part (b).

31. $f(x) = 3$

32. $f(x) = x$

33. $g(x) = \frac{1}{2}x^2 - 3$

34. $g(x) = 3x^4 - 6x^2$

35. $f(x) = x^{2/3}$

36. $f(x) = -2x^{-1/4}$

37. $f(x) = x\sqrt{x + 3}$

38. $f(x) = x\sqrt{3 - x}$

39. $g(x) = x^3 - 3x + 3$

40. $g(x) = x^3 + 3x^2 - 9x - 27$

41. $f(x) = |x + 1| + |x - 1|$

42. $f(x) = -|x + 4| - |x + 1|$

Approximating Relative Minima and Maxima In Exercises 43–54, use a graphing utility to graph the function and to approximate any relative minimum or relative maximum values of the function.

43. $f(x) = x^2 - 6x$

44. $f(x) = 3x^2 - 2x - 5$

45. $h(x) = x^3 - 6x^2 + 15$

46. $g(x) = -2x^3 - x^2 + 14x$

47. $h(x) = (x - 1)\sqrt{x}$

48. $g(x) = x\sqrt{4 - x}$

49. $f(x) = x^2 - 4x - 5$

50. $f(x) = 3x^2 - 12x$

51. $f(x) = x^3 - 3x$

52. $f(x) = -x^3 + 3x^2$

53. $f(x) = 3x^2 - 6x + 1$

54. $f(x) = 8x - 4x^2$

Library of Parent Functions In Exercises 55–60, sketch the graph of the function by hand. Then use a graphing utility to verify the graph.

55. $f(x) = [\![x]\!] + 2$

56. $f(x) = [\![x]\!] - 3$

57. $f(x) = [\![x - 1]\!] - 2$

58. $f(x) = [\![x + 2]\!] + 1$

59. $f(x) = 2[\![x]\!]$

60. $f(x) = [\![4x]\!]$

Sketching a Piecewise-Defined Function In Exercises 61–66, sketch the graph of the piecewise-defined function by hand.

61. $f(x) = \begin{cases} 2x + 3, & x < 0 \\ 3 - x, & x \ge 0 \end{cases}$

62. $f(x) = \begin{cases} x + 6, & x \le -4 \\ 3x - 4, & x > -4 \end{cases}$

63. $h(x) = \begin{cases} 3 + x, & x < 0 \\ x^2 + 1, & x \ge 0 \end{cases}$

64. $f(x) = \begin{cases} 1 - (x - 1)^2, & x \le 2 \\ \sqrt{x - 2}, & x > 2 \end{cases}$

65. $f(x) = \begin{cases} x + 3, & x \le 0 \\ 3, & 0 < x \le 2 \\ 2x - 1, & x > 2 \end{cases}$

66. $g(x) = \begin{cases} x + 5, & x \le -3 \\ 5, & -3 < x < 1 \\ 5x - 4, & x \ge 1 \end{cases}$

Even and Odd Functions In Exercises 67–78, use a graphing utility to graph the function and determine whether it is even, odd, or neither.

67. $f(x) = 5$

68. $f(x) = -6x$

69. $f(x) = 3x^3 - 2x$

70. $f(x) = 4 - 5x$

71. $h(x) = x^2 + 6$

72. $f(x) = -x^2 - 8$

73. $f(x) = x^6 - 2x^2 + 3$

74. $g(x) = 2x^3 - x - x^5$

75. $f(x) = \sqrt{1 - x}$

76. $g(t) = \sqrt[3]{t - 1}$

77. $f(x) = |x + 2|$

78. $f(x) = -|x^2 - 5|$

Think About It **In Exercises 79–84, find the coordinates of a second point on the graph of a function f when the point provided is on the graph and the function is (a) even and (b) odd.**

79. $\left(\frac{3}{2}, 4\right)$

80. $\left(-\frac{5}{3}, -7\right)$

81. $(-2, -9)$

82. $(5, -1)$

83. $(x, -y)$

84. $(3a, 3b)$

Algebraic-Graphical-Numerical **In Exercises 85–94, determine whether the function is even, odd, or neither (a) algebraically, (b) graphically by using a graphing utility to graph the function, and (c) numerically by using the *table* feature of the graphing utility to compare $f(x)$ and $f(-x)$ for several values of x.**

85. $f(x) = 4 - x$

86. $f(x) = x^2 - 9$

87. $f(x) = x^2 + 2x - 3$

88. $f(x) = x^4 - 4x^2 + 1$

89. $f(x) = x^3 - 5x$

90. $f(x) = x^5 - 4x^3$

91. $f(x) = x\sqrt{1 - x^2}$

92. $f(x) = x\sqrt{3 + x}$

93. $f(x) = 4x^{2/3}$

94. $f(x) = 4x^{3/5}$

95. Cost of Parking The cost of parking in a metered lot is \$1.00 for the first hour and \$0.50 for each additional hour or portion of an hour.

(a) A customer needs a model for the cost C of parking in the metered lot for t hours. Which of the two functions below is an appropriate model?

$C_1(t) = 1 + 0.50[\![t - 1]\!]$

$C_2(t) = 1 - 0.50[\![-(t - 1)]\!]$

(b) Use a graphing utility to graph the appropriate model. Estimate the cost of parking in the metered lot for 7 hours and 10 minutes.

96. *Why you should learn it* *(p. 29)* The cost of mailing a package weighing up to, but not including, 1 pound is \$2.66. Each additional pound or portion of a pound costs \$0.51.

(a) Use the greatest integer function to create a model for the cost C of mailing a package weighing x pounds, where $x > 0$.

(b) Sketch the graph of the function.

Using the Graph of a Function **In Exercises 97 and 98, write the height h of the rectangle as a function of x.**

97.

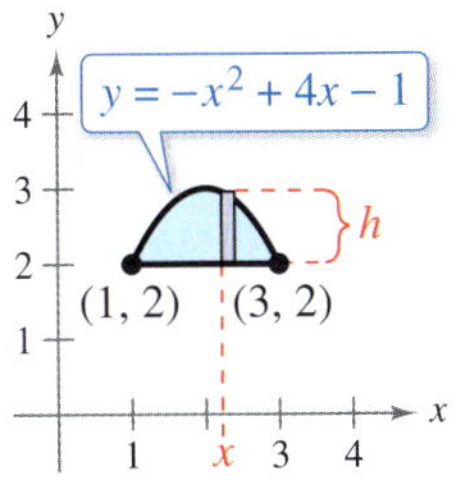

98.

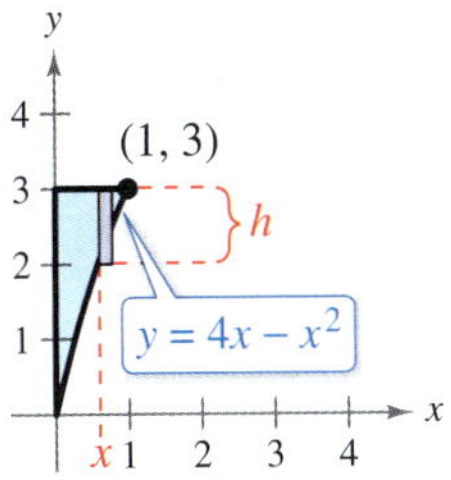

99. MODELING DATA

The number N (in thousands) of new single-family houses sold each year from 2006 through 2017 in the United States is approximated by the model

$N = -2.345t^3 + 97.59t^2 - 1285.0t + 5758$

$6 \le t \le 17$

where t represents the year, with $t = 6$ corresponding to 2006. *(Source:* U.S. Census Bureau)

(a) Use a graphing utility to graph the model over the appropriate domain.

(b) Use the graph from part (a) to determine during which years the number of new single-family houses was increasing. During which years was the number decreasing?

(c) Approximate the minimum number of new single-family houses sold from 2006 through 2017.

100. Mechanical Engineering The intake pipe of a 100-gallon tank has a flow rate of 10 gallons per minute, and two drain pipes have a flow rate of 5 gallons per minute each. The graph shows the volume V of fluid in the tank as a function of time t. Determine in which pipes the fluid is flowing in specific subintervals of the one-hour interval of time shown on the graph. (There are many correct answers.)

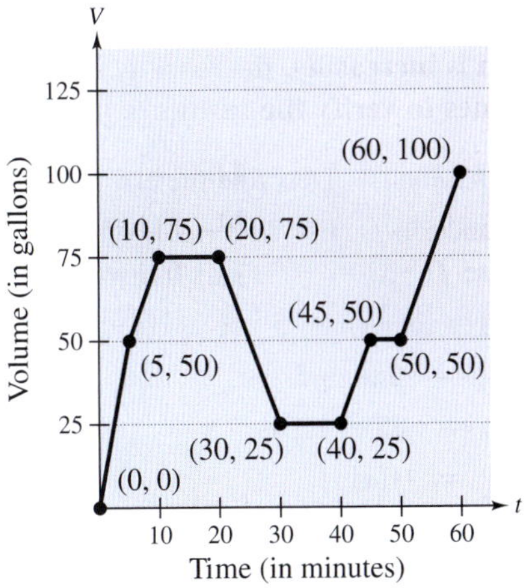

Franck Boston/Shutterstock.com; AVAVA/Shutterstock.com

Focusing on Concepts

True or False? **In Exercises 101 and 102, determine whether the statement is true or false. Justify your answer.**

101. A function with a square root cannot have a domain that is the set of all real numbers.

102. It is possible for an odd function to have the interval $[0, \infty)$ as its domain.

Think About It **In Exercises 103–108, match the graph of the function with the description that best fits the situation.**

(a) The air temperature at a beach on a sunny day

(b) The height of a football kicked in a field goal attempt

(c) The number of children in a family over time

(d) The population of California as a function of time

(e) The depth of the tide at a beach over a 24-hour period

(f) The number of cupcakes on a tray at a party

103.

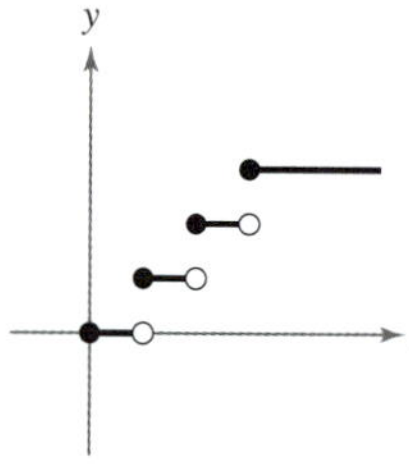

104.

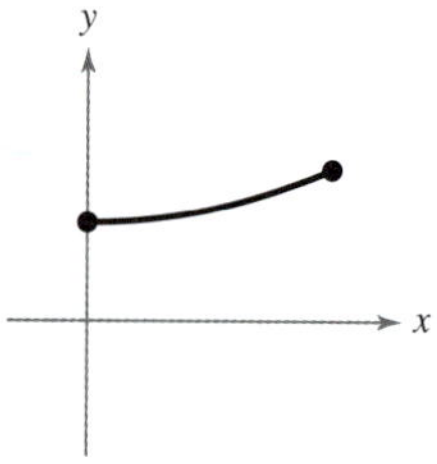

105.

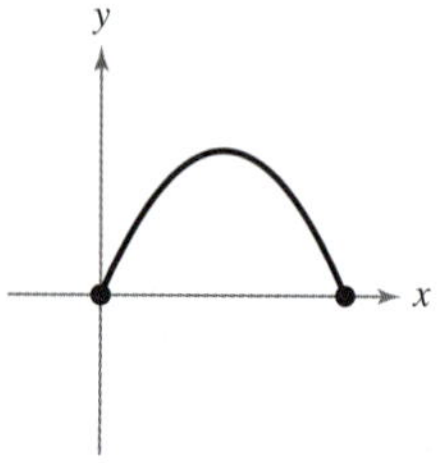

106.

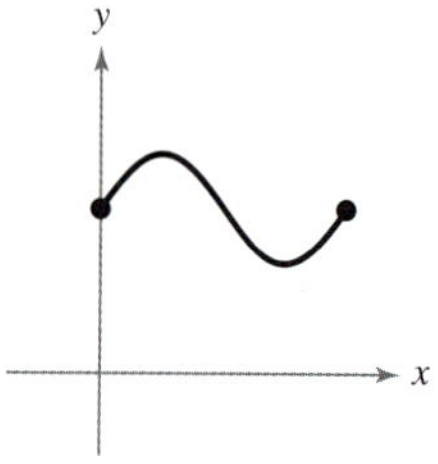

107.

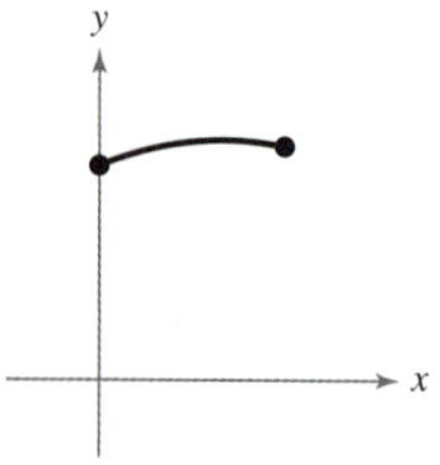

108.

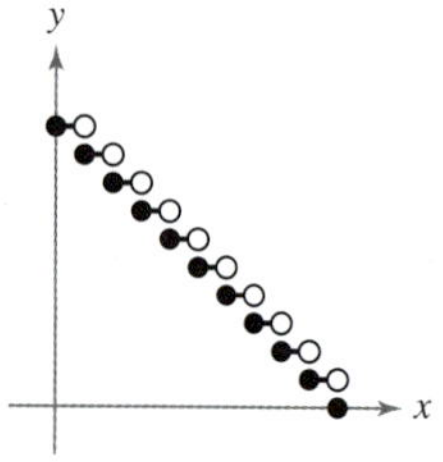

109. Error Analysis Describe the error.

The function $f(x) = 2x^3 - 5$ is odd because $f(-x) = -f(x)$, as follows.

$$\begin{aligned} f(-x) &= 2(-x)^3 - 5 \\ &= -2x^3 - 5 \\ &= -(2x^3 - 5) \\ &= -f(x) \end{aligned}$$

110. Think About It Can you represent the greatest integer function using a piecewise-defined function?

111. Think About It Let f be an even function. Determine whether g is even, odd, or neither. Explain.

(a) $g(x) = -f(x)$ (b) $g(x) = f(-x)$

(c) $g(x) = f(x) - 2$ (d) $g(x) = -f(x + 3)$

112. HOW DO YOU SEE IT? Half of the graph of an odd function is shown.

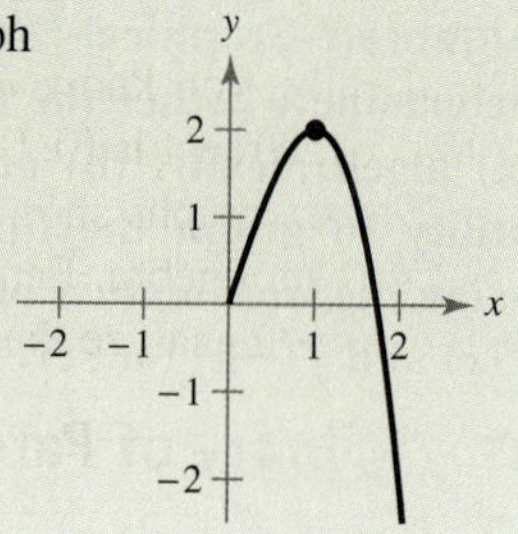

(a) Sketch a complete graph of the function.

(b) Find the domain and range of the function.

(c) Identify the open intervals on which the function is increasing, decreasing, or constant.

(d) Find any relative minimum and relative maximum values of the function.

113. Proof Prove that a function of the form shown below is odd.

$$y = a_{2n+1}x^{2n+1} + a_{2n-1}x^{2n-1} + \cdots + a_3x^3 + a_1x$$

114. Proof Prove that a function of the form shown below is even.

$$y = a_{2n}x^{2n} + a_{2n-2}x^{2n-2} + \cdots + a_2x^2 + a_0$$

Cumulative Mixed Review

Identifying Terms and Coefficients **In Exercises 115–118, identify the terms. Then identify the coefficients of the variable terms of the expression.**

115. $-2x^2 + 11x + 3$ **116.** $10 + 3x$

117. $\dfrac{x}{3} - 5x^2 + x^3$ **118.** $7x^4 + \sqrt{2}x^2 - x$

Evaluating a Function **In Exercises 119 and 120, evaluate the function at each specified value of the independent variable and simplify.**

119. $f(x) = -x^2 - x + 3$

(a) $f(4)$ (b) $f(-5)$ (c) $f(x - 2)$

120. $f(x) = x\sqrt{x - 3}$

(a) $f(3)$ (b) $f(12)$ (c) $f(6)$

Evaluating a Difference Quotient **In Exercises 121 and 122, find the difference quotient and simplify your answer.**

121. $f(x) = x^2 - 2x + 9, \quad \dfrac{f(3 + h) - f(3)}{h}, \quad h \neq 0$

122. $f(x) = 5 + 6x - x^2, \quad \dfrac{f(6 + h) - f(6)}{h}, \quad h \neq 0$

1.4 Shifting, Reflecting, and Stretching Graphs

Summary of Graphs of Parent Functions

One of the goals of this text is to enable you to build your intuition for the basic shapes of the graphs of different types of functions. For instance, from your study of lines in Section 1.1, you can determine the basic shape of the graph of the parent linear function

$f(x) = x.$ Parent linear function

Specifically, you know that the graph of this function is a line whose slope is 1 and whose y-intercept is $(0, 0)$.

The six graphs shown in Figure 1.24 represent the most commonly used types of functions in algebra. Familiarity with the basic characteristics of these parent graphs will help you analyze the shapes of more complicated graphs.

Library of Parent Functions: Commonly Used Functions

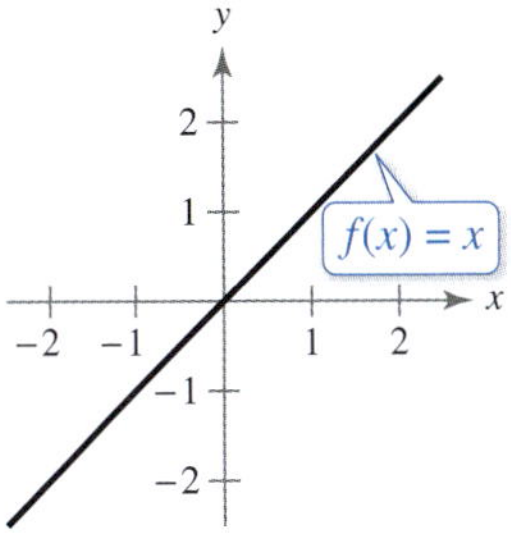

(a) *Linear Function*
(See Section 1.1.)

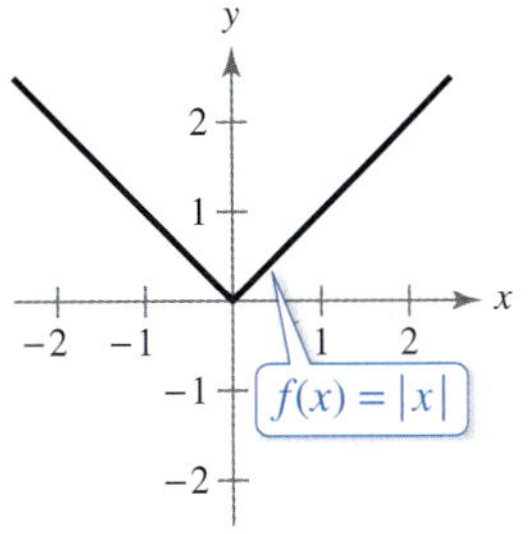

(b) *Absolute Value Function*
(See Section 1.2.)

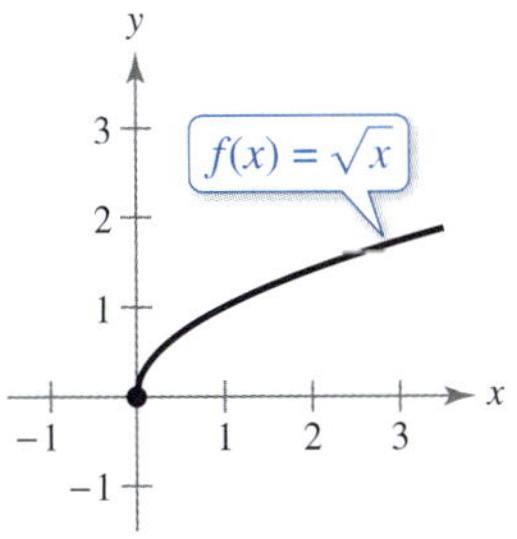

(c) *Square Root Function*
(See Section 1.2.)

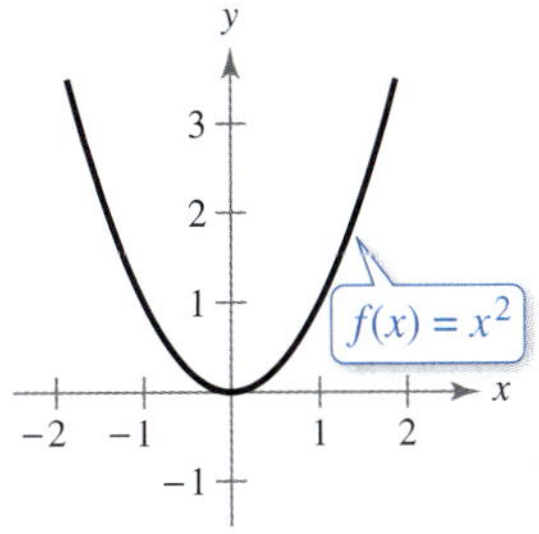

(d) *Quadratic Function*
(See Section 2.1.)

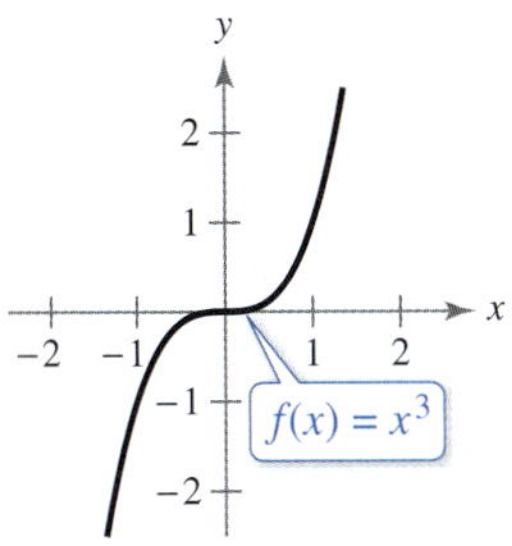

(e) *Cubic Function*
(See Section 2.2.)

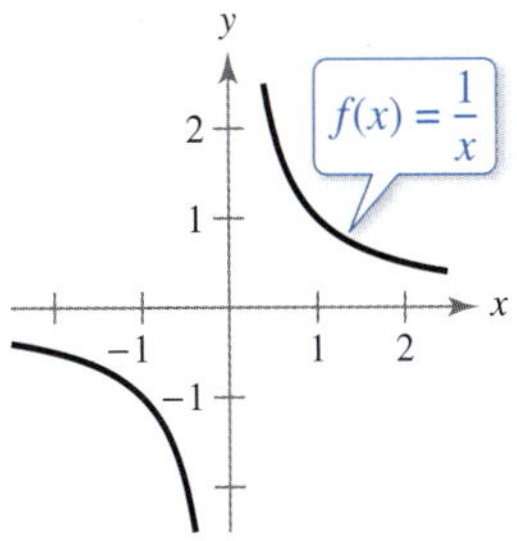

(f) *Rational Function*
(See Section 2.7.)

Figure 1.24

Throughout this section, you will discover how many complicated graphs are derived by shifting, stretching, shrinking, or reflecting the parent graphs shown above. Shifts, stretches, shrinks, and reflections are called *transformations*. Many graphs of functions can be created from combinations of these transformations.

What you should learn

- Recognize graphs of parent functions.
- Use vertical and horizontal shifts to sketch graphs of functions.
- Use reflections to sketch graphs of functions.
- Use nonrigid transformations to sketch graphs of functions.

Why you should learn it

Recognizing the graphs of parent functions and knowing how to shift, reflect, and stretch graphs of functions can help you sketch or describe the graphs of a wide variety of simple functions. For example, in Exercise 66 on page 49, you are asked to describe a transformation that produces the graph of a model for the depreciation of the WD-40 Company's assets.

Vertical and Horizontal Shifts

Many functions have graphs that are simple transformations of the graphs of parent functions summarized in Figure 1.24. For example, to obtain the graph of

$$h(x) = x^2 + 2$$

shift the graph of $f(x) = x^2$ two units *upward*, as shown in Figure 1.25. In function notation, h and f are related as follows.

$$h(x) = x^2 + 2$$
$$= f(x) + 2 \quad \text{Upward shift of two units}$$

Similarly, to obtain the graph of

$$g(x) = (x - 2)^2$$

shift the graph of $f(x) = x^2$ two units to the *right*, as shown in Figure 1.26. In this case, the functions g and f have the following relationship.

$$g(x) = (x - 2)^2$$
$$= f(x - 2) \quad \text{Right shift of two units}$$

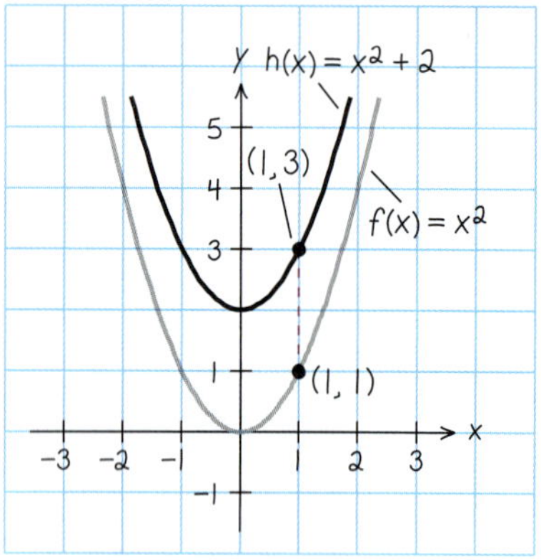

Vertical shift upward: two units
Figure 1.25

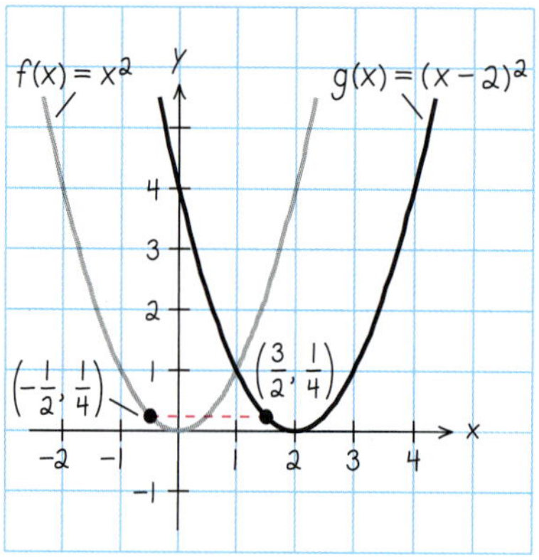

Horizontal shift to the right: two units
Figure 1.26

Explore the Concept

Use a graphing utility to display (in the same viewing window) the graphs of $y = x^2 + c$, where $c = -2, 0, 2$, and 4. Use the results to describe the effect that c has on the graph.

Use a graphing utility to display (in the same viewing window) the graphs of $y = (x + c)^2$, where $c = -2, 0, 2$, and 4. Use the results to describe the effect that c has on the graph.

The list below summarizes vertical and horizontal shifts. In items 3 and 4, be sure you see that $h(x) = f(x - c)$ corresponds to a *right* shift and $h(x) = f(x + c)$ corresponds to a *left* shift for $c > 0$.

Vertical and Horizontal Shifts

Let c be a positive real number. **Vertical and horizontal shifts** in the graph of $y = f(x)$ are represented as follows.

1. Vertical shift c units *upward*: $h(x) = f(x) + c$
2. Vertical shift c units *downward*: $h(x) = f(x) - c$
3. Horizontal shift c units to the *right*: $h(x) = f(x - c)$
4. Horizontal shift c units to the *left*: $h(x) = f(x + c)$

Insight

On a standardized test, you may be asked to determine how a change in an equation will affect the graph of the equation.

Some graphs are obtained from combinations of vertical and horizontal shifts, as shown in Example 1(c) on the next page. Vertical and horizontal shifts generate a *family of functions*, each with a graph that has the same shape but at a different location in the plane.

EXAMPLE 1 Shifts in the Graph of a Function

a. To use the graph of $f(x) = x^3$ to sketch the graph of $g(x) = x^3 - 1$, shift the graph of f one unit downward.

b. To use the graph of $f(x) = x^3$ to sketch the graph of $h(x) = (x - 1)^3$, shift the graph of f one unit to the right.

c. To use the graph of $f(x) = x^3$ to sketch the graph of $k(x) = (x + 2)^3 + 1$, shift the graph of f two units to the left and then one unit upward.

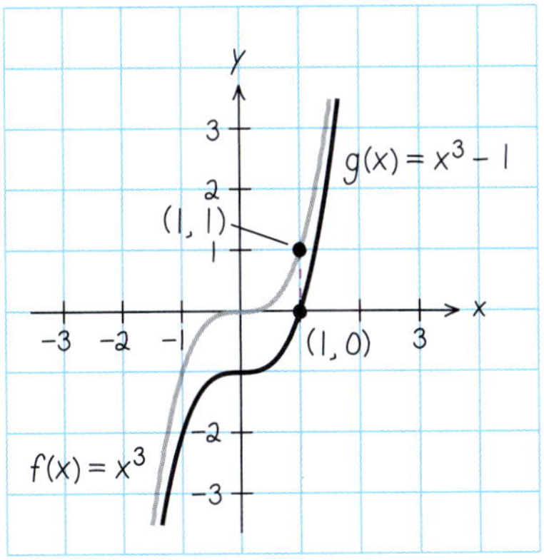

(a) *Vertical shift: one unit downward*

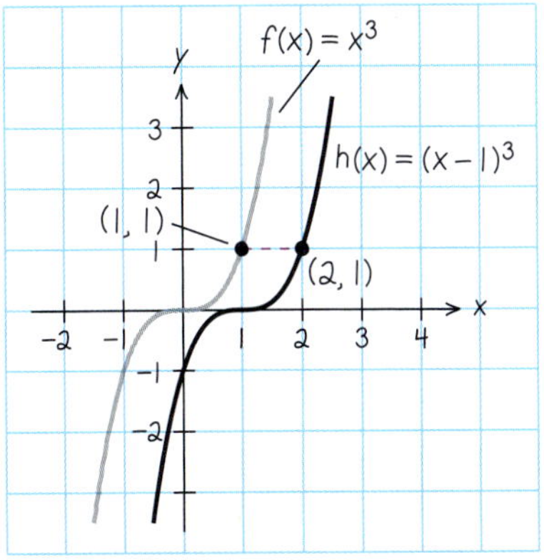

(b) *Horizontal shift: one unit right*

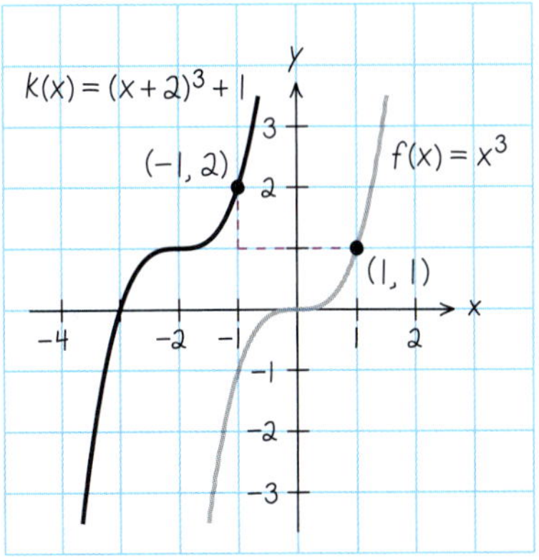

(c) *Two units left and one unit upward*

Checkpoint Audio-video solution in English & Spanish at LarsonPrecalculus.com

Use the graph of $f(x) = x^3$ to sketch the graph of each function.

a. $h(x) = x^3 + 5$ **b.** $g(x) = (x - 3)^3 + 2$

EXAMPLE 2 Writing Equations from Graphs

Each graph is a transformation of the graph of $f(x) = x^2$. Write an equation for the function represented by each graph.

a.

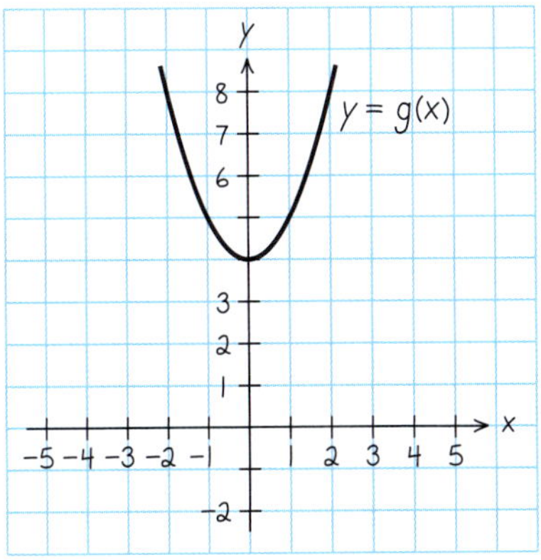

b.

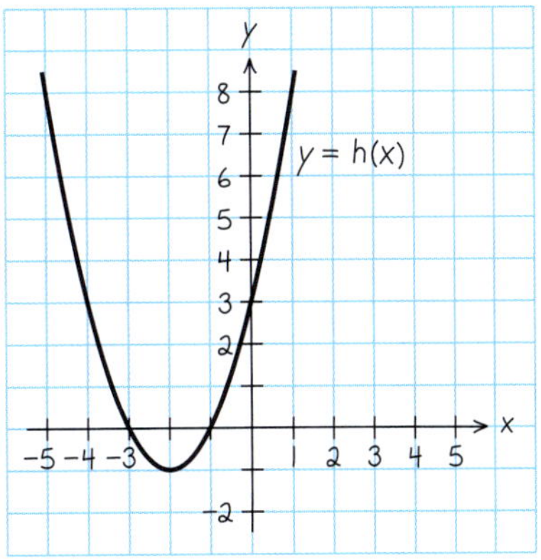

Solution

a. The graph of g is an upward shift of four units of the graph of $f(x) = x^2$. So, an equation for g is $g(x) = x^2 + 4$.

b. The graph of h is a left shift of two units, and a downward shift of one unit, of the graph of $f(x) = x^2$. So, an equation for h is $h(x) = (x + 2)^2 - 1$.

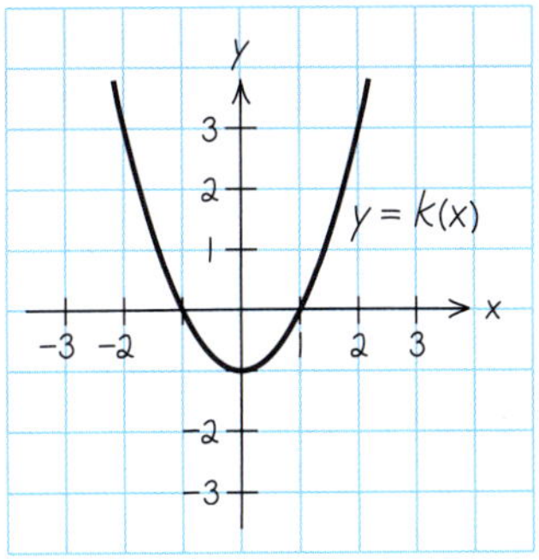

Checkpoint Audio-video solution in English & Spanish at LarsonPrecalculus.com

The graph at the right is a transformation of the graph of $f(x) = x^2$. Write an equation for the function represented by the graph.

Reflecting Graphs

Another common type of transformation is called a **reflection.** For instance, when you consider the x-axis to be a mirror, the graph of $h(x) = -x^2$ is the mirror image (or reflection) of the graph of $f(x) = x^2$, as shown in Figure 1.27.

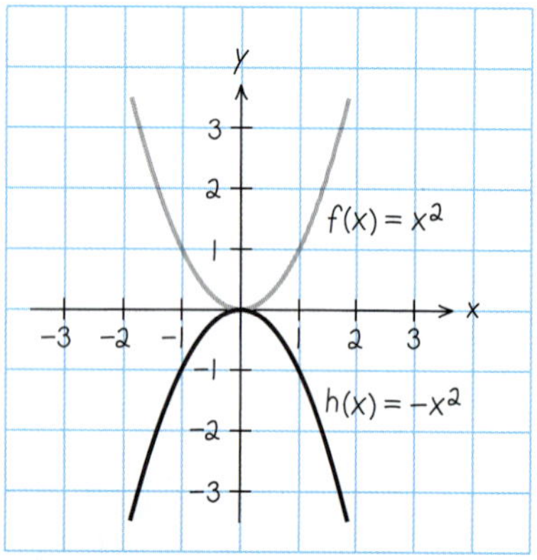

Figure 1.27

Explore the Concept

Compare the graph of each function with the graph of $f(x) = x^2$ by using a graphing utility to graph the function and f in the same viewing window. Describe the transformation.

a. $g(x) = -x^2$

b. $h(x) = (-x)^2$

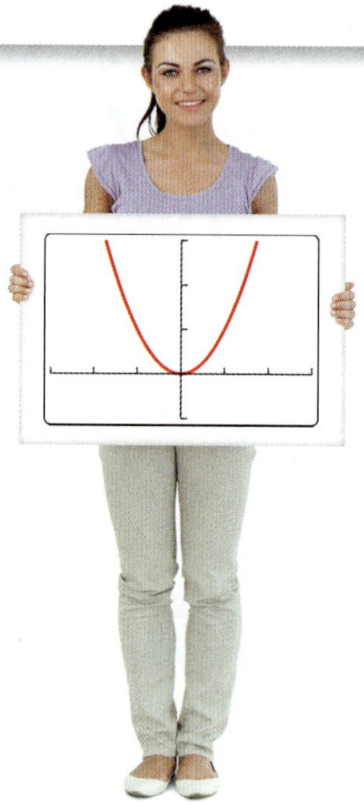

Reflections in the Coordinate Axes

Reflections in the coordinate axes of the graph of $y = f(x)$ are represented as follows.

1. Reflection in the x-axis: $h(x) = -f(x)$
2. Reflection in the y-axis: $h(x) = f(-x)$

EXAMPLE 3 Writing Equations from Graphs

Each graph is a transformation of the graph of $f(x) = x^2$. Write an equation for the function represented by each graph.

a.

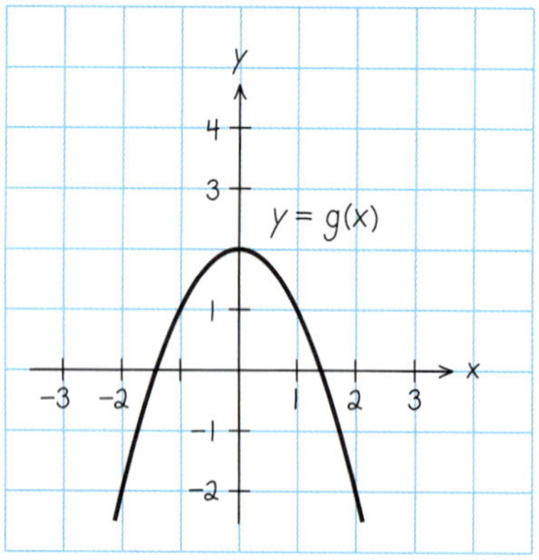

b.

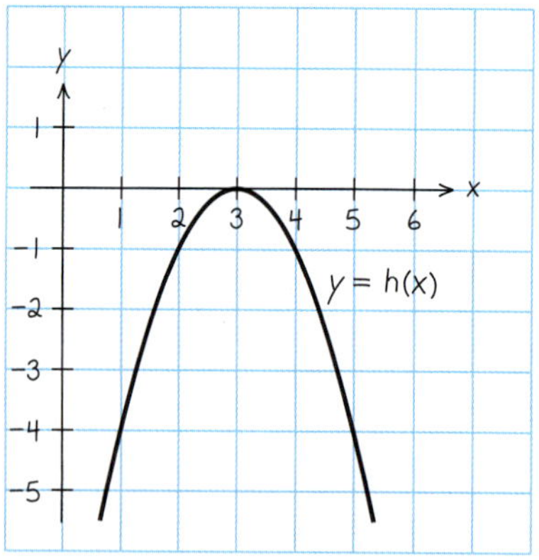

Solution

a. The graph of g is a reflection in the x-axis *followed by* an upward shift of two units of the graph of $f(x) = x^2$. So, an equation for g is $g(x) = -x^2 + 2$.

b. The graph of h is a right shift of three units *followed by* a reflection in the x-axis of the graph of $f(x) = x^2$. So, an equation for h is $h(x) = -(x - 3)^2$.

Checkpoint *Audio-video solution in English & Spanish at LarsonPrecalculus.com*

The graph at the right is a transformation of the graph of $f(x) = x^2$. Write an equation for the function represented by the graph.

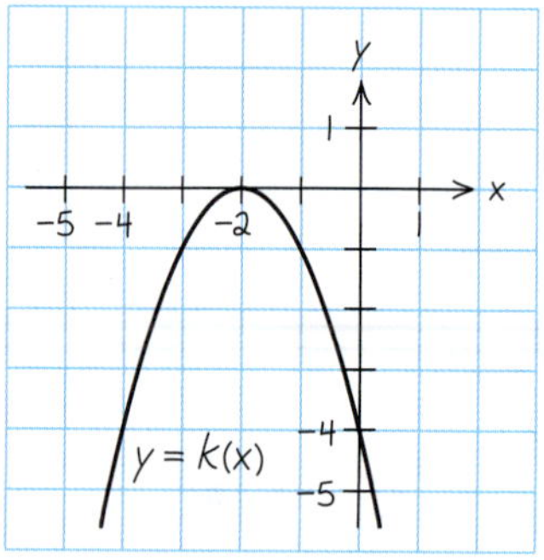

EXAMPLE 4 Reflections and Shifts

Compare the graph of each function with the graph of

$f(x) = \sqrt{x}.$

a. $g(x) = -\sqrt{x}$

b. $h(x) = \sqrt{-x}$

c. $k(x) = -\sqrt{x+2}$

Algebraic Solution

a. Relative to the graph of $f(x) = \sqrt{x}$, the graph of g is a reflection in the x-axis because

$$g(x) = -\sqrt{x}$$
$$= -f(x).$$

b. The graph of h is a reflection of the graph of $f(x) = \sqrt{x}$ in the y-axis because

$$h(x) = \sqrt{-x}$$
$$= f(-x).$$

c. From the equation

$$k(x) = -\sqrt{x+2}$$
$$= -f(x+2)$$

you can conclude that the graph of k is a left shift of two units, followed by a reflection in the x-axis, of the graph of $f(x) = \sqrt{x}$.

Graphical Solution

a. From the graph in Figure 1.28, you can see that the graph of g is a reflection of the graph of f in the x-axis. Note that the domain of g is $x \geq 0$.

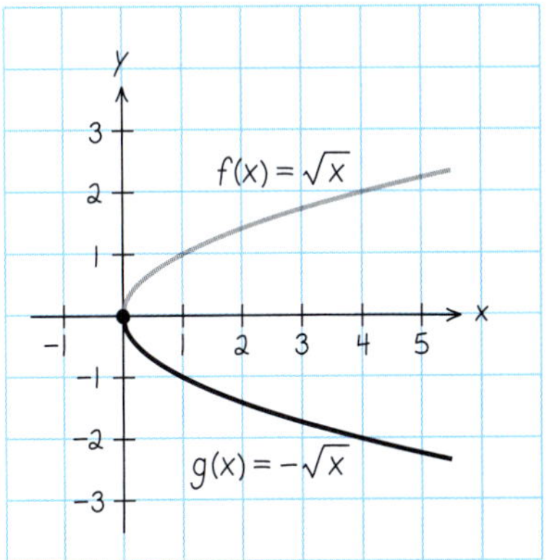

Figure 1.28

b. From the graph in Figure 1.29, you can see that the graph of h is a reflection of the graph of f in the y-axis. Note that the domain of h is $x \leq 0$.

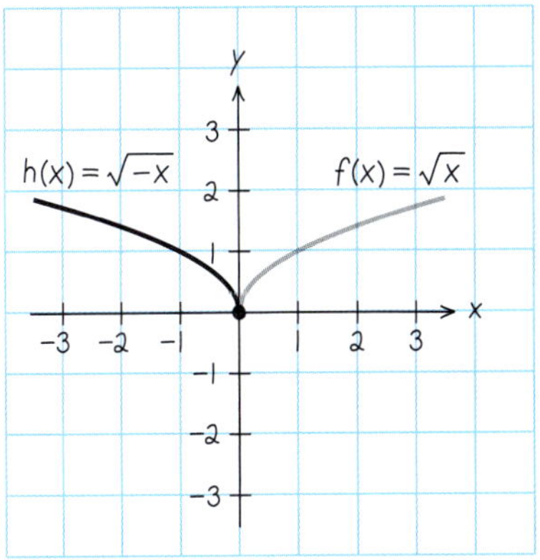

Figure 1.29

c. From the graph in Figure 1.30, you can see that the graph of k is a left shift of two units of the graph of f, followed by a reflection in the x-axis. Note that the domain of k is $x \geq -2$.

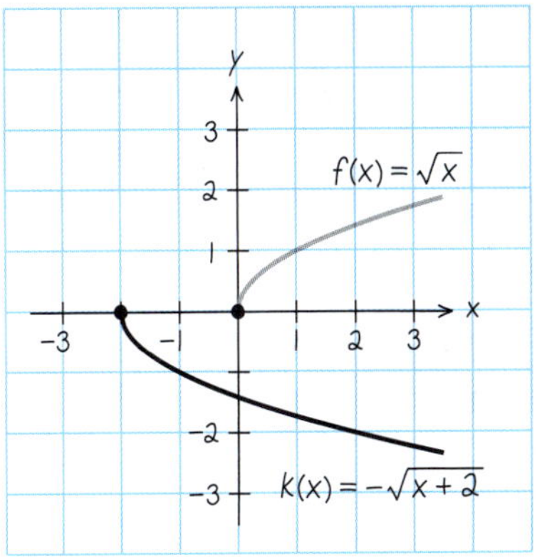

Figure 1.30

✓ ***Checkpoint*** *Audio-video solution in English & Spanish at LarsonPrecalculus.com*

Compare the graph of $g(x) = -|x|$ with the graph of $f(x) = |x|$. ■

Nonrigid Transformations

Horizontal shifts, vertical shifts, and reflections are called **rigid transformations** because the basic shape of the graph is unchanged. These transformations change only the *position* of the graph in the coordinate plane. **Nonrigid transformations** are those that cause a *distortion*—a change in the shape of the original graph. For instance, a nonrigid transformation of the graph of $y = f(x)$ is represented by $g(x) = cf(x)$, where the transformation is a **vertical stretch** when $c > 1$ and a **vertical shrink** when $0 < c < 1$. Another nonrigid transformation of the graph of $y = f(x)$ is represented by $h(x) = f(cx)$, where the transformation is a **horizontal shrink** when $c > 1$ and a **horizontal stretch** when $0 < c < 1$.

EXAMPLE 5 Nonrigid Transformations

See LarsonPrecalculus.com for an interactive version of this type of example.

Compare the graphs of (a) $h(x) = 3|x|$ and (b) $g(x) = \frac{1}{3}|x|$ with the graph of $f(x) = |x|$.

Solution

a. Relative to the graph of $f(x) = |x|$, the graph of $h(x) = 3|x| = 3f(x)$ is a vertical stretch (each y-value is multiplied by 3). (See Figure 1.31.)

b. Similarly, the graph of $g(x) = \frac{1}{3}|x| = \frac{1}{3}f(x)$ is a vertical shrink $\left(\text{each } y\text{-value is multiplied by } \frac{1}{3}\right)$ of the graph of f. (See Figure 1.32.)

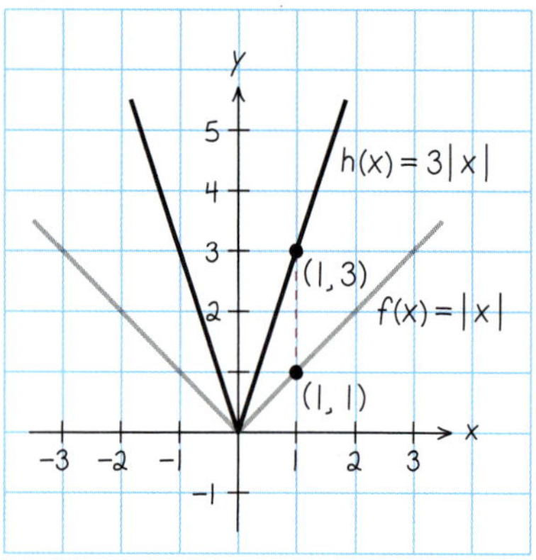

Figure 1.31

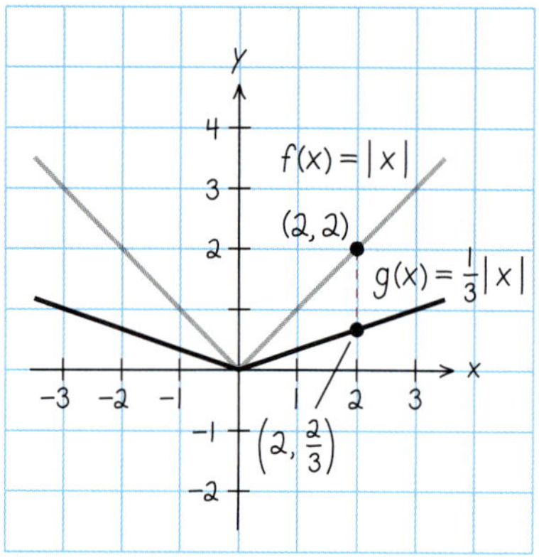

Figure 1.32

✓ ***Checkpoint*** *Audio-video solution in English & Spanish at LarsonPrecalculus.com*

Compare the graphs of (a) $g(x) = 4x^2$ and (b) $h(x) = \frac{1}{4}x^2$ with the graph of $f(x) = x^2$.

EXAMPLE 6 Nonrigid Transformations

Compare the graph of $h(x) = f\left(\frac{1}{2}x\right)$ with the graph of $f(x) = 2 - x^3$.

Solution

Relative to the graph of $f(x) = 2 - x^3$, the graph of

$$h(x) = f\left(\tfrac{1}{2}x\right) = 2 - \left(\tfrac{1}{2}x\right)^3 = 2 - \tfrac{1}{8}x^3$$

is a horizontal stretch (each x-value is multiplied by 2). (See Figure 1.33.)

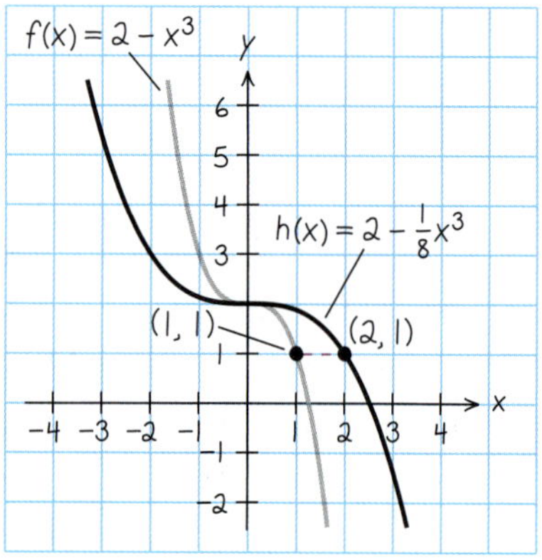

Figure 1.33

✓ ***Checkpoint*** *Audio-video solution in English & Spanish at LarsonPrecalculus.com*

Compare the graphs of (a) $g(x) = f(2x)$ and (b) $h(x) = f\left(\frac{1}{2}x\right)$ with the graph of $f(x) = x^2 + 3$.

1.4 Exercises

See CalcChat.com for tutorial help and worked-out solutions to odd-numbered exercises.
For instructions on how to use a graphing utility, see Appendix A.

Vocabulary and Concept Check

1. Name three types of rigid transformations.

2. Match the rigid transformation of $y = f(x)$ with the correct representation, where $c > 0$.

(a) $h(x) = f(x) + c$ (i) horizontal shift c units to the left

(b) $h(x) = f(x) - c$ (ii) vertical shift c units upward

(c) $h(x) = f(x - c)$ (iii) horizontal shift c units to the right

(d) $h(x) = f(x + c)$ (iv) vertical shift c units downward

In Exercises 3 and 4, fill in the blanks.

3. A reflection in the x-axis of $y = f(x)$ is represented by $h(x) =$ _______ , while a reflection in the y-axis of $y = f(x)$ is represented by $h(x) =$ _______ .

4. A nonrigid transformation of $y = f(x)$ represented by $cf(x)$ is a vertical stretch when _______ and a vertical shrink when _______ .

Procedures and Problem Solving

5. **Shifting the Graph of a Function** For each function, sketch the graphs of the function when $c = -2, -1, 1,$ and 2 on the same rectangular coordinate system.

(a) $f(x) = |x| + c$

(b) $f(x) = |x - c|$

6. **Shifting the Graph of a Function** For each function, sketch the graphs of the function when $c = -3, -2, 2,$ and 3 on the same rectangular coordinate system.

(a) $f(x) = \sqrt{x} + c$

(b) $f(x) = \sqrt{x - c}$

Sketching Transformations **In Exercises 7–20, sketch the graphs of the three functions by hand on the same rectangular coordinate system. Verify your results with a graphing utility.**

7. $f(x) = x$
$g(x) = x - 4$
$h(x) = 3x$

8. $f(x) = \frac{1}{2}x$
$g(x) = \frac{1}{2}x + 2$
$h(x) = 4(x - 2)$

9. $f(x) = x^2$
$g(x) = x^2 + 2$
$h(x) = (x - 2)^2$

10. $f(x) = x^2$
$g(x) = 3x^2$
$h(x) = (x + 2)^2 + 1$

11. $f(x) = -x^2$
$g(x) = -x^2 + 1$
$h(x) = -(x + 3)^2$

12. $f(x) = (x - 2)^2$
$g(x) = (x + 2)^2 + 2$
$h(x) = -(x - 2)^2 - 1$

13. $f(x) = x^2$
$g(x) = \frac{1}{2}x^2$
$h(x) = (2x)^2$

14. $f(x) = x^2$
$g(x) = \frac{1}{4}x^2 + 2$
$h(x) = -\frac{1}{4}x^2$

15. $f(x) = |x|$
$g(x) = |x| - 5$
$h(x) = 4|x - 4|$

16. $f(x) = |x|$
$g(x) = |2x|$
$h(x) = -2|x + 2| - 1$

17. $f(x) = -\sqrt{x}$
$g(x) = \sqrt{x + 1}$
$h(x) = \sqrt{x - 2} + 1$

18. $f(x) = \sqrt{x}$
$g(x) = \frac{1}{2}\sqrt{x}$
$h(x) = -\sqrt{x - 4}$

19. $f(x) = \dfrac{1}{x}$
$g(x) = \dfrac{1}{x} - 2$
$h(x) = \dfrac{1}{x - 1} + 2$

20. $f(x) = \dfrac{1}{x}$
$g(x) = \dfrac{1}{x} - 4$
$h(x) = \dfrac{1}{x + 3} - 1$

Sketching Transformations **In Exercises 21 and 22, use the graph of f to sketch each graph. To print an enlarged copy of the graph, go to *MathGraphs.com*.**

21. (a) $y = f(x) + 2$
(b) $y = f(x) - 3$
(c) $y = -f(x)$
(d) $y = f(x - 2)$
(e) $y = f(x + 3)$
(f) $y = 2f(x)$
(g) $y = f(-x)$
(h) $y = f\left(\frac{1}{2}x\right)$

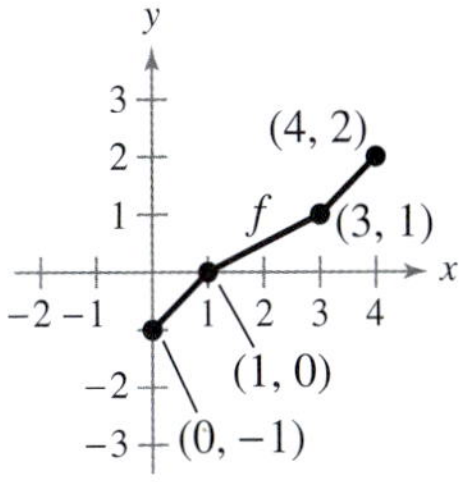

22. (a) $y = f(x) - 1$
(b) $y = f(x) + 3$
(c) $y = f(x + 2)$
(d) $y = f(x - 1)$
(e) $y = -f(x - 2)$
(f) $y = f(-x)$
(g) $y = \frac{1}{2}f(x)$
(h) $y = f(2x)$

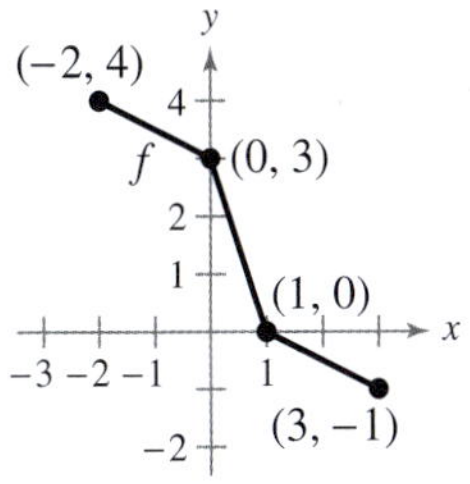

Library of Parent Functions **In Exercises 23–28, compare the graph of the function with the graph of its parent function.**

23. $y = \sqrt{x} + 4$ **24.** $y = \dfrac{1}{x} - 5$

25. $y = (x - 4)^3$ **26.** $y = |x + 5|$

27. $y = x^2 - 12$ **28.** $y = \sqrt{x - 6}$

Library of Parent Functions **In Exercises 29–34, identify the parent function and the transformation shown in the graph. Write an equation for the function represented by the graph.**

29.

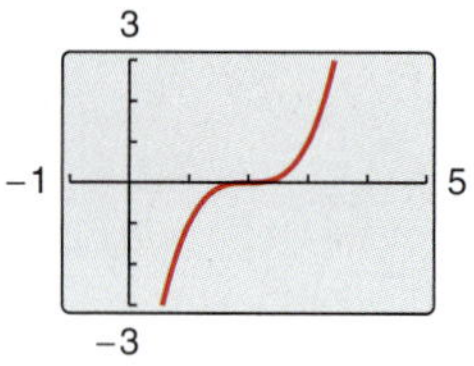

30.

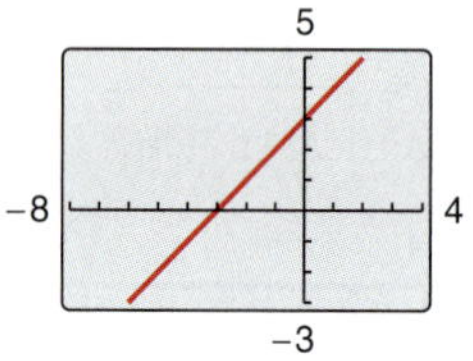

31.

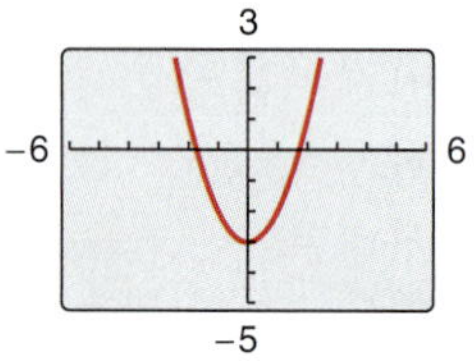

32.

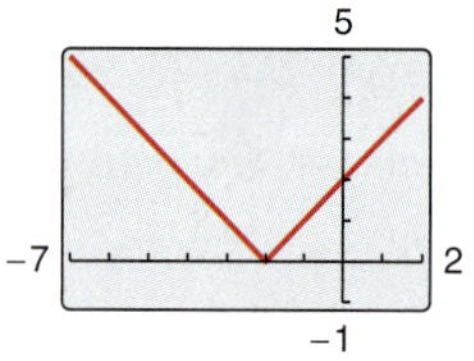

33.

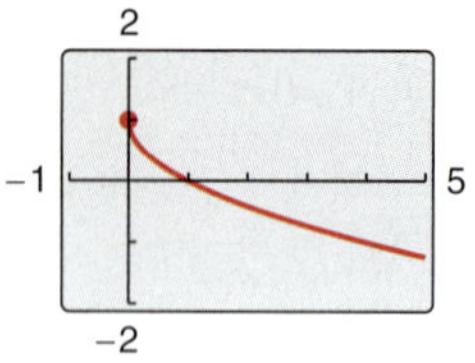

34.

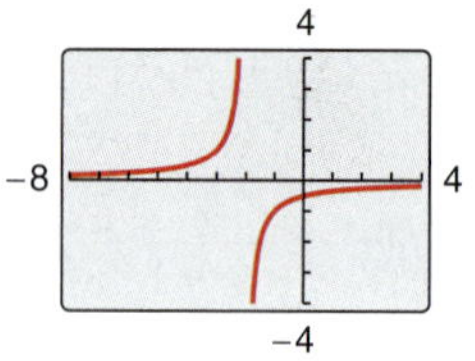

Rigid and Nonrigid Transformations **In Exercises 35–46, compare the graph of the function with the graph of its parent function.**

35. $y = -x$ **36.** $y = |-x|$

37. $y = (-x)^2$ **38.** $y = -x^3$

39. $y = \dfrac{1}{-x}$ **40.** $y = -\dfrac{1}{x}$

41. $g(x) = \frac{1}{4}x^3$ **42.** $p(x) = \frac{1}{2}x^2$

43. $h(x) = 4|x|$ **44.** $y = \left|\frac{1}{2}x\right|$

45. $f(x) = \sqrt{4x}$ **46.** $y = 2\sqrt{x}$

Rigid and Nonrigid Transformations **In Exercises 47 and 48, use a graphing utility to graph the three functions in the same viewing window. Describe the graphs of g and h relative to the graph of f.**

47. $f(x) = x^3 - 3x^2$
$g(x) = f(x + 2)$
$h(x) = \frac{1}{2}f(x)$

48. $f(x) = x^3 - 3x^2 + 2$
$g(x) = -f(x)$
$h(x) = f(2x)$

Describing Transformations **In Exercises 49–60, g is related to one of the six parent functions on page 41. (a) Identify the parent function f. (b) Describe the sequence of transformations from f to g. (c) Sketch the graph of g by hand. (d) Use function notation to write g in terms of the parent function f.**

49. $g(x) = x^2 + 6$ **50.** $g(x) = x^2 - 2$

51. $g(x) = 2 - (x + 5)^2$ **52.** $g(x) = 3 + 2(x - 4)^2$

53. $g(x) = \frac{1}{3}(x - 2)^3$ **54.** $g(x) = -(x + 3)^3 - 1$

55. $g(x) = \dfrac{1}{x + 8} - 9$ **56.** $g(x) = \dfrac{1}{x - 7} + 4$

57. $g(x) = -2|x - 1| - 4$ **58.** $g(x) = \frac{1}{2}|x - 2| - 3$

59. $g(x) = -\frac{1}{2}\sqrt{x + 3} - 1$ **60.** $g(x) = -3\sqrt{x + 1} - 6$

Writing an Equation from a Description **In Exercises 61–64, write an equation for the function whose graph is described.**

61. The shape of $f(x) = x^2$, but shifted three units to the right and seven units down

62. The shape of $f(x) = x^3$, but shifted six units to the left, six units down, and then reflected in the y-axis

63. The shape of $f(x) = |x|$, but shifted 12 units up and then reflected in the x-axis

64. The shape of $f(x) = \sqrt{x}$, but shifted nine units down and then reflected in both the x-axis and y-axis

65. MODELING DATA

The numbers N (in thousands) of new single-family houses sold in the Southern United States from 2010 through 2017 can be expressed by ordered pairs of the form $(t, N(t))$, where $t = 10$ represents 2010. A model for the data is $N(t) = 1.5(t - 4.8)^2 + 123$. *(Spreadsheet at LarsonPrecalculus.com)*
(*Source:* U.S. Census Bureau)

DATA

(10, 173)	(11, 168)
(12, 195)	(13, 233)
(14, 243)	(15, 286)
(16, 318)	(17, 340)

(a) Describe the transformation of the parent function $f(t) = t^2$.

(b) Use a graphing utility to graph the model and the data in the same viewing window.

(c) Rewrite the function so that $t = 0$ represents 2010. Explain how you obtained your answer.

66. ***Why you should learn it*** *(p. 41)* The depreciation D (in millions of dollars) of the WD-40 Company's assets from 2010 through 2017 can be approximated by the function

$$D(t) = 2.11\sqrt{t} + 3.7$$

where $t = 0$ represents 2010. *(Source:* WD-40 Company)

(a) Describe the transformation of the parent function $f(t) = \sqrt{t}$.

(b) Use a graphing utility to graph the model over the interval $0 \le t \le 7$.

(c) According to the model, in what year will the depreciation of WD-40 assets be approximately 7.9 million dollars?

(d) Rewrite the function so that $t = 0$ represents 2017. Explain how you obtained your answer.

Focusing on Concepts

True or False? **In Exercises 67 and 68, determine whether the statement is true or false. Justify your answer.**

67. The graph of $y = f(-x)$ is a reflection of the graph of $y = f(x)$ in the x-axis.

68. The graphs of $f(x) = |x| + 6$ and $f(x) = |-x| + 6$ are identical.

Exploration **In Exercises 69–72, use the fact that the graph of $y = f(x)$ has x-intercepts at $x = 2$ and $x = -3$ to find the x-intercepts of the graph of the transformation, if possible. If not possible, state the reason.**

69. $y = f(-x)$

70. $y = 2f(x)$

71. $y = f(x) + 2$

72. $y = f(x - 3)$

Library of Parent Functions **In Exercises 73–76, determine which equation(s) may be represented by the graph shown. (There may be more than one correct answer.)**

73.

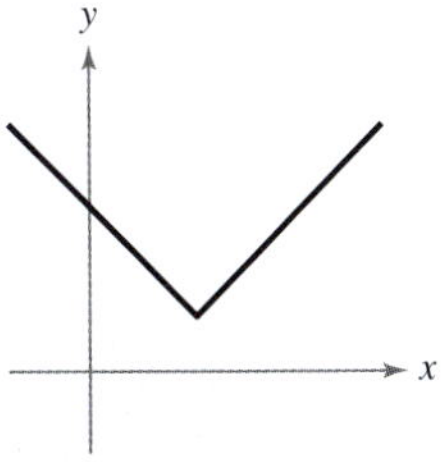

(a) $f(x) = |x + 2| + 1$
(b) $f(x) = |x - 1| + 2$
(c) $f(x) = |x - 2| + 1$
(d) $f(x) = 2 + |x - 2|$
(e) $f(x) = |(x - 2) + 1|$
(f) $f(x) = 1 - |x - 2|$

74.

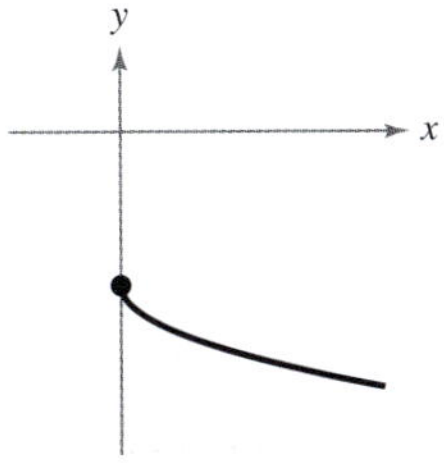

(a) $f(x) = -\sqrt{x} - 4$
(b) $f(x) = -4 - \sqrt{x}$
(c) $f(x) = -4 - \sqrt{-x}$
(d) $f(x) = \sqrt{-x} - 4$
(e) $f(x) = \sqrt{-x} + 4$
(f) $f(x) = \sqrt{x} - 4$

75.

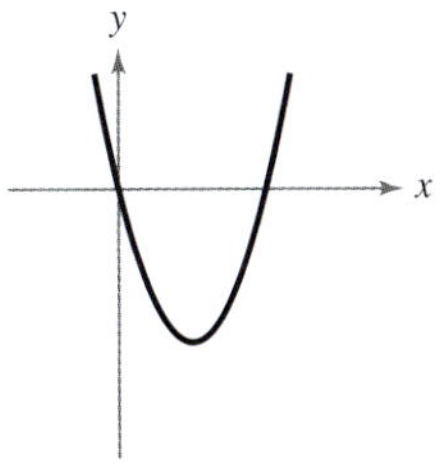

(a) $f(x) = (x - 2)^2 - 2$
(b) $f(x) = (x + 4)^2 - 4$
(c) $f(x) = (x - 2)^2 - 4$
(d) $f(x) = (x + 2)^2 - 4$
(e) $f(x) = 4 - (x - 2)^2$
(f) $f(x) = 4 - (x + 2)^2$

76.

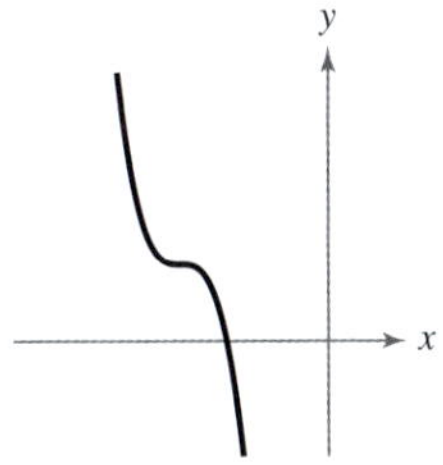

(a) $f(x) = -(x - 4)^3 + 2$
(b) $f(x) = -(x + 4)^3 + 2$
(c) $f(x) = -(x - 2)^3 + 4$
(d) $f(x) = (-x - 4)^3 + 2$
(e) $f(x) = (x + 4)^3 + 2$
(f) $f(x) = (-x + 4)^3 + 2$

77. Error Analysis Describe the error in graphing $f(x) = (x + 1)^2$.

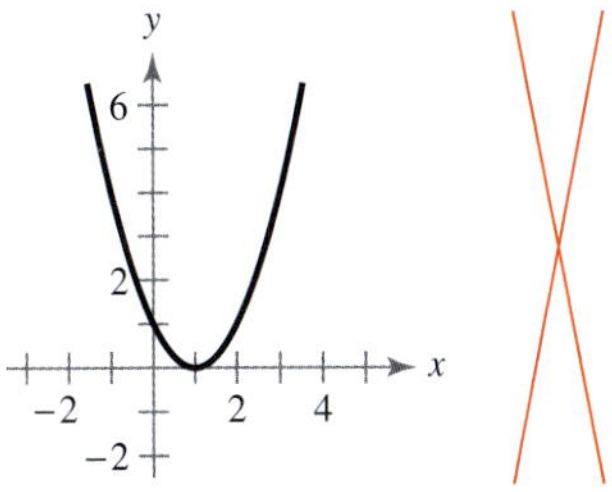

78. HOW DO YOU SEE IT? Use the graph of $y = f(x)$ to find the intervals on which each of the graphs in (a)–(c) is increasing and decreasing. If not possible, state the reason.

(a) $y = -f(x)$
(b) $y = f(x) - 3$
(c) $y = f(x - 1)$

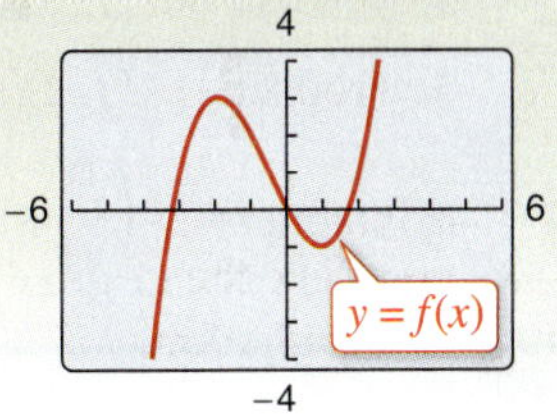

79. Think About It You can use either of two methods to graph a function: plotting points or translating a parent function as shown in this section. Which method do you prefer to use for each function? Explain.

(a) $f(x) = 3x^2 - 4x + 1$ (b) $f(x) = 2(x - 1)^2 - 6$

80. Think About It Compare the graph of $g(x) = ax^2$ with the graph of $f(x) = x^2$ when (a) $0 < a < 1$ and (b) $a > 1$.

Cumulative Mixed Review

Finding the Domain of a Function **In Exercises 81–84, find the domain of the function.**

81. $f(x) = \dfrac{4}{9 - x}$

82. $f(x) = \dfrac{\sqrt{x - 5}}{x - 7}$

83. $f(x) = \sqrt{100 - x^2}$

84. $f(x) = \sqrt[3]{16 - x^2}$

Jacob Wackerhausen/iStock/Getty Images; Tyler Olson/Shutterstock.com

1.5 Combinations of Functions

Arithmetic Combinations of Functions

Just as two real numbers can be combined by the operations of addition, subtraction, multiplication, and division to form other real numbers, two *functions* can be combined to create new functions. When $f(x) = 2x - 3$ and $g(x) = x^2 - 1$, you can form the sum, difference, product, and quotient of f and g as follows.

$$f(x) + g(x) = (2x - 3) + (x^2 - 1) = x^2 + 2x - 4$$ Sum

$$f(x) - g(x) = (2x - 3) - (x^2 - 1) = -x^2 + 2x - 2$$ Difference

$$f(x) \cdot g(x) = (2x - 3)(x^2 - 1) = 2x^3 - 3x^2 - 2x + 3$$ Product

$$\frac{f(x)}{g(x)} = \frac{2x - 3}{x^2 - 1}, \quad x \neq \pm 1$$ Quotient

The domain of an **arithmetic combination** of functions f and g consists of all real numbers that are common to the domains of f and g. In the case of the quotient

$$\frac{f(x)}{g(x)}$$

there is the further restriction that $g(x) \neq 0$.

Sum, Difference, Product, and Quotient of Functions

Let f and g be two functions with overlapping domains. Then, for all x common to both domains, the sum, difference, product, and quotient of f and g are defined as follows.

1. Sum: $(f + g)(x) = f(x) + g(x)$
2. Difference: $(f - g)(x) = f(x) - g(x)$
3. Product: $(fg)(x) = f(x) \cdot g(x)$
4. Quotient: $\left(\frac{f}{g}\right)(x) = \frac{f(x)}{g(x)}, \quad g(x) \neq 0$

What you should learn

- Add, subtract, multiply, and divide functions.
- Find compositions of one function with another function.
- Use combinations of functions to model and solve real-life problems.

Why you should learn it

You can model some situations by combining functions. For instance, in Exercise 79 on page 57, you will model the stopping distance of a car by combining the driver's reaction time with the car's braking distance.

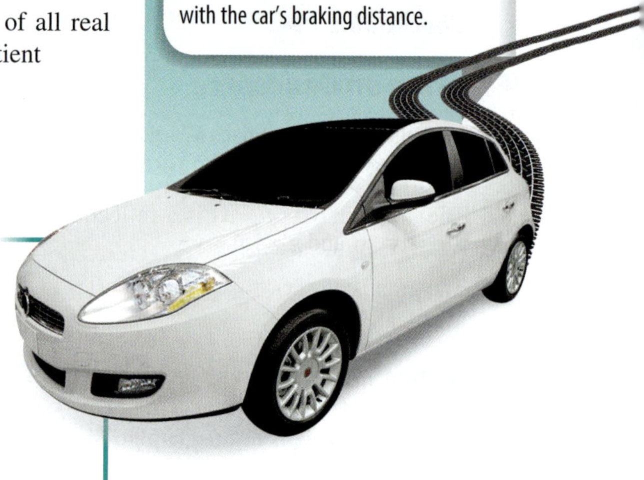

EXAMPLE 1 Finding the Sum of Two Functions

Given

$$f(x) = 2x + 1 \quad \text{and} \quad g(x) = x^2 + 2x - 1$$

find $(f + g)(x)$. Then evaluate the sum when $x = 2$.

Solution

$$(f + g)(x) = f(x) + g(x)$$
$$= (2x + 1) + (x^2 + 2x - 1)$$
$$= x^2 + 4x$$

When $x = 2$, the value of this sum is $(f + g)(2) = 2^2 + 4(2) = 12$.

✓ Checkpoint *Audio-video solution in English & Spanish at LarsonPrecalculus.com*

Given $f(x) = x^2$ and $g(x) = 1 - x$, find $(f + g)(x)$. Then evaluate the sum when $x = 2$.

EXAMPLE 2 Finding the Difference of Two Functions

Given $f(x) = 2x + 1$ and $g(x) = x^2 + 2x - 1$, find $(f - g)(x)$. Then evaluate the difference when $x = 2$.

Algebraic Solution

The difference of the functions f and g is

$$\begin{aligned}(f - g)(x) &= f(x) - g(x) \\ &= (2x + 1) - (x^2 + 2x - 1) \\ &= -x^2 + 2.\end{aligned}$$

When $x = 2$, the value of this difference is

$$\begin{aligned}(f - g)(2) &= -(2)^2 + 2 \\ &= -2.\end{aligned}$$

Graphical Solution

Enter $y_1 = f(x)$, $y_2 = g(x)$, and $y_3 = f(x) - g(x)$. Then graph the difference of the two functions, y_3.

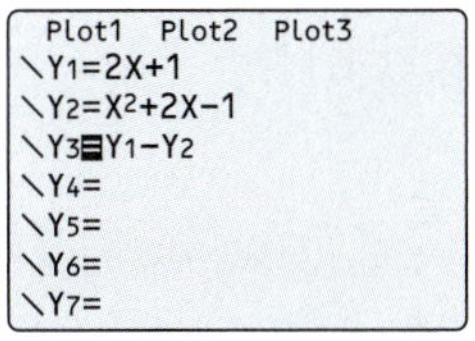

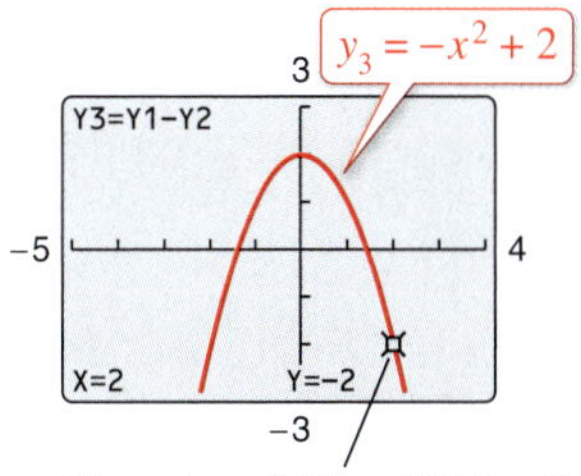

The value of $(f - g)(2)$ is -2.

✓ Checkpoint ▶ *Audio-video solution in English & Spanish at LarsonPrecalculus.com*

Given $f(x) = x^2$ and $g(x) = 1 - x$, find $(f - g)(x)$. Then evaluate the difference when $x = 3$.

EXAMPLE 3 Finding the Product of Two Functions

Given $f(x) = x^2$ and $g(x) = x - 3$, find $(fg)(x)$. Then evaluate the product when $x = 4$.

Solution

$$(fg)(x) = f(x)g(x) = (x^2)(x - 3) = x^3 - 3x^2$$

When $x = 4$, the value of this product is $(fg)(4) = 4^3 - 3(4)^2 = 16$.

✓ Checkpoint ▶ *Audio-video solution in English & Spanish at LarsonPrecalculus.com*

Given $f(x) = x^2$ and $g(x) = 1 - x$, find $(fg)(x)$. Then evaluate the product when $x = 3$.

In Examples 1–3, both f and g have domains that consist of all real numbers. So, the domains of $(f + g)$ and $(f - g)$ are also the set of all real numbers. Remember to consider any restrictions on the domains of f or g when forming the sum, difference, product, or quotient of f and g. For instance, the domain of $f(x) = 1/x$ is all $x \neq 0$, and the domain of $g(x) = \sqrt{x}$ is $[0, \infty)$. This implies that the domain of $(f + g)$ is $(0, \infty)$.

EXAMPLE 4 Finding the Quotient of Two Functions

Given $f(x) = \sqrt{x}$ and $g(x) = \sqrt{4 - x^2}$, find $(f/g)(x)$. Then find the domain of f/g.

Solution

$$\left(\frac{f}{g}\right)(x) = \frac{f(x)}{g(x)} = \frac{\sqrt{x}}{\sqrt{4 - x^2}}$$

The domain of f is $[0, \infty)$ and the domain of g is $[-2, 2]$. The intersection of these domains is $[0, 2]$. So, the domain of f/g is $[0, 2)$.

Algebra Help

Note that the domain of f/g includes $x = 0$ but not $x = 2$. This value is excluded because $x = 2$ yields a zero in the denominator.

✓ Checkpoint ▶ *Audio-video solution in English & Spanish at LarsonPrecalculus.com*

Find $(f/g)(x)$ and $(g/f)(x)$ for the functions $f(x) = \sqrt{x - 3}$ and $g(x) = \sqrt{16 - x^2}$. Then find the domains of f/g and g/f.

Compositions of Functions

Another way of combining two functions is to form the **composition** of one with the other. For instance, when $f(x) = x^2$ and $g(x) = x + 1$, the composition of f with g is

$$f(g(x)) = f(x + 1)$$
$$= (x + 1)^2.$$

This composition is denoted as $f \circ g$ and is read as "f composed with g."

Insight

On standardized tests, the composition of one function with another is tested conceptually and in real-life contexts.

Definition of Composition of Two Functions

The **composition** of the function f with the function g is

$$(f \circ g)(x) = f(g(x)).$$

The domain of $f \circ g$ is the set of all x in the domain of g such that $g(x)$ is in the domain of f. (See Figure 1.34.)

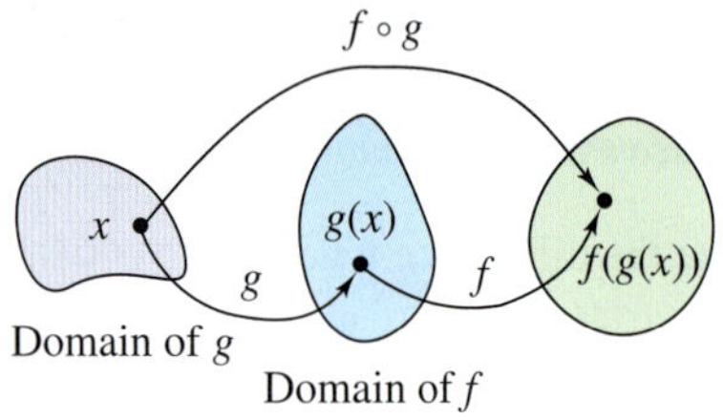

Figure 1.34

EXAMPLE 5 Forming the Composition of f with g

Find $(f \circ g)(x)$ for

$$f(x) = \sqrt{x},\ x \ge 0, \quad \text{and} \quad g(x) = x - 1,\ x \ge 1.$$

If possible, find $(f \circ g)(2)$ and $(f \circ g)(0)$.

Solution

The composition of f with g is

$$(f \circ g)(x) = f(g(x)) \qquad \text{Definition of } f \circ g$$
$$= f(x - 1) \qquad \text{Definition of } g(x)$$
$$= \sqrt{x - 1},\ x \ge 1. \qquad \text{Definition of } f(x)$$

The domain of $f \circ g$ is $[1, \infty)$, as shown in Figure 1.35. So,

$$(f \circ g)(2) = \sqrt{2 - 1} = 1$$

is defined, but $(f \circ g)(0)$ is not defined because 0 is not in the domain of $f \circ g$.

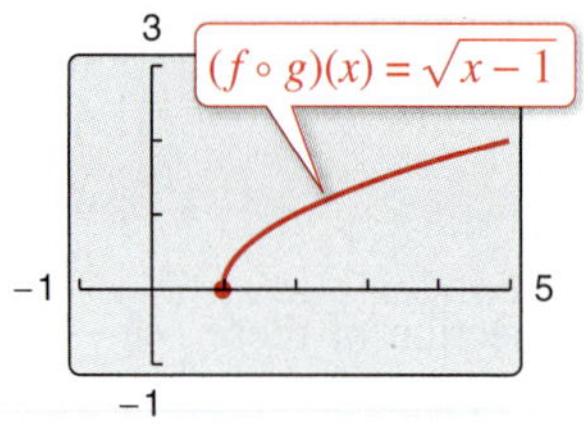

Figure 1.35

✓ **Checkpoint** *Audio-video solution in English & Spanish at LarsonPrecalculus.com*

Find $(f \circ g)(x)$ for $f(x) = x^2$ and $g(x) = x - 1$. If possible, find $(f \circ g)(0)$.

Explore the Concept

Let $f(x) = x + 2$ and $g(x) = 4 - x^2$. Are the compositions $f \circ g$ and $g \circ f$ equal? You can use your graphing utility to answer this question by entering and graphing the following functions.

$$y_1 = (4 - x^2) + 2$$

$$y_2 = 4 - (x + 2)^2$$

What do you observe? Which function represents $f \circ g$ and which represents $g \circ f$?

The composition of f with g is generally not the same as the composition of g with f. This is illustrated in Example 6.

EXAMPLE 6 Compositions of Functions

See LarsonPrecalculus.com for an interactive version of this type of example.

Given $f(x) = x + 2$ and $g(x) = 4 - x^2$, evaluate (a) $(f \circ g)(x)$ and (b) $(g \circ f)(x)$ when $x = 0$ and 1.

Algebraic Solution

a. $(f \circ g)(x) = f(g(x))$ — Definition of $f \circ g$

$= f(4 - x^2)$ — Definition of $g(x)$

$= (4 - x^2) + 2$ — Definition of $f(x)$

$= -x^2 + 6$

$(f \circ g)(0) = -0^2 + 6 = 6$

$(f \circ g)(1) = -1^2 + 6 = 5$

b. $(g \circ f)(x) = g(f(x))$ — Definition of $g \circ f$

$= g(x + 2)$ — Definition of $f(x)$

$= 4 - (x + 2)^2$ — Definition of $g(x)$

$= 4 - (x^2 + 4x + 4)$

$= -x^2 - 4x$

$(g \circ f)(0) = -0^2 - 4(0) = 0$

$(g \circ f)(1) = -1^2 - 4(1) = -5$

Note that $(f \circ g)(x) \neq (g \circ f)(x)$.

Numerical Solution

a. and b. Enter $y_1 = f(x)$, $y_2 = g(x)$, $y_3(f \circ g)(x)$, and $y_4 = (g \circ f)(x)$. Then use the *table* feature to find the desired function values.

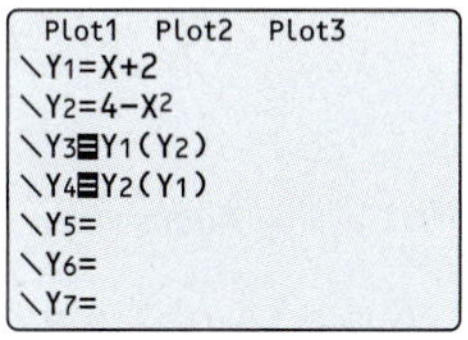

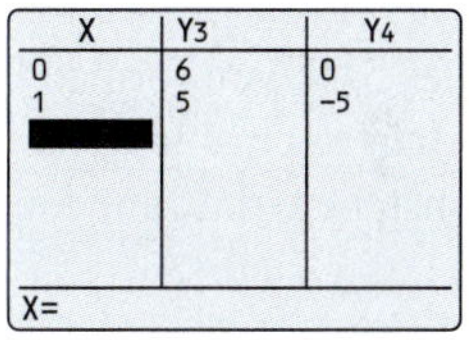

From the table, you can see that $(f \circ g)(x) \neq (g \circ f)(x)$.

✓ Checkpoint ▶ *Audio-video solution in English & Spanish at LarsonPrecalculus.com*

Given $f(x) = 2x + 5$ and $g(x) = 4x^2 + 1$, find the following.

a. $(f \circ g)(x)$ **b.** $(g \circ f)(x)$ **c.** $(f \circ g)\left(-\frac{1}{2}\right)$

EXAMPLE 7 Finding the Domain of a Composite Function

Find the domain of $f \circ g$ for the functions $f(x) = x^2 - 9$ and $g(x) = \sqrt{9 - x^2}$.

Algebraic Solution

The composition of the functions is as follows.

$$(f \circ g)(x) = f(g(x))$$
$$= f\left(\sqrt{9 - x^2}\right)$$
$$= \left(\sqrt{9 - x^2}\right)^2 - 9$$
$$= 9 - x^2 - 9$$
$$= -x^2$$

From this, it might appear that the domain of the composition is the set of all real numbers. This, however, is not true. Because the domain of f is the set of all real numbers and the domain of g is $[-3, 3]$, the domain of $f \circ g$ is $[-3, 3]$.

Graphical Solution

The x-coordinates of points on the graph extend from -3 to 3. So, the domain of $f \circ g$ is $[-3, 3]$.

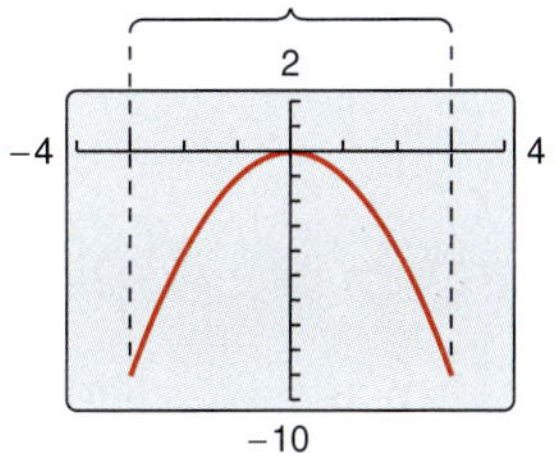

✓ Checkpoint ▶ *Audio-video solution in English & Spanish at LarsonPrecalculus.com*

Find the domain of $f \circ g$ for the functions $f(x) = \sqrt{x}$ and $g(x) = x^2 + 4$.

EXAMPLE 8 A Case in Which $(f \circ g)(x) = (g \circ f)(x)$

Given $f(x) = 2x + 3$ and $g(x) = \frac{1}{2}(x - 3)$, find each composition.

a. $(f \circ g)(x)$ **b.** $(g \circ f)(x)$

Solution

a.
$$\begin{aligned}(f \circ g)(x) &= f(g(x))\\ &= f\left(\frac{1}{2}(x-3)\right)\\ &= 2\left[\frac{1}{2}(x-3)\right] + 3\\ &= x - 3 + 3\\ &= x\end{aligned}$$

b.
$$\begin{aligned}(g \circ f)(x) &= g(f(x))\\ &= g(2x+3)\\ &= \frac{1}{2}[(2x+3) - 3]\\ &= \frac{1}{2}(2x)\\ &= x\end{aligned}$$

Checkpoint Audio-video solution in English & Spanish at LarsonPrecalculus.com

Given $f(x) = x^{1/3}$ and $g(x) = x^6$, find (a) $(f \circ g)(x)$ and (b) $(g \circ f)(x)$.

Algebra Help

In Example 8, note that the two composite functions $f \circ g$ and $g \circ f$ are equal, and both represent the identity function. That is, $(f \circ g)(x) = x$ and $(g \circ f)(x) = x$. You will study this special case in the next section.

In Examples 5–8, you formed the composition of two given functions. In calculus, it is also important to be able to identify two functions that make up a given composite function. For instance, the function $h(x) = (3x - 5)^3$ is the composition of $f(x) = x^3$ and $g(x) = 3x - 5$. That is,

$$h(x) = (3x - 5)^3 = f(3x - 5) = f(g(x)).$$

Basically, to "decompose" a composite function, look for an "inner" function and an "outer" function. In the function h above, $g(x) = 3x - 5$ is the inner function and $f(x) = x^3$ is the outer function.

Explore the Concept

Write each function as a composition of two functions.

a. $h(x) = |x^3 - 2|$

b. $r(x) = |x^3| - 2$

What do you notice about the inner and outer functions?

EXAMPLE 9 Identifying a Composite Function

Write the function

$$h(x) = \frac{1}{(x-2)^2}$$

as a composition of two functions.

Solution

One way to write h as a composition of two functions is to let $g(x) = x - 2$ be the inner function and let

$$f(x) = \frac{1}{x^2} = x^{-2}$$

be the outer function. Then you can write

$$h(x) = \frac{1}{(x-2)^2} = (x-2)^{-2} = f(x-2) = f(g(x)).$$

Checkpoint Audio-video solution in English & Spanish at LarsonPrecalculus.com

Write the function

$$h(x) = \frac{\sqrt[3]{8 - x}}{5}$$

as a composition of two functions.

Application

EXAMPLE 10 Bacteria Count

The number N of bacteria in a refrigerated petri dish is given by

$$N(T) = 20T^2 - 80T + 500, \quad 2 \le T \le 14$$

where T is the temperature of the petri dish (in degrees Celsius). When the petri dish is removed from refrigeration, the temperature of the petri dish is given by

$$T(t) = 4t + 2, \quad 0 \le t \le 3$$

where t is the time (in hours).

a. Find the composition $N(T(t))$ and interpret its meaning in context.

b. Find the number of bacteria in the petri dish when $t = 2$ hours.

c. Find the time when the bacteria count reaches 2000.

Microbiologist

Solution

a.
$$\begin{aligned} N(T(t)) &= 20(4t + 2)^2 - 80(4t + 2) + 500 \\ &= 20(16t^2 + 16t + 4) - 320t - 160 + 500 \\ &= 320t^2 + 320t + 80 - 320t - 160 + 500 \\ &= 320t^2 + 420 \end{aligned}$$

The composite function $N(T(t))$ represents the number of bacteria as a function of the amount of time the petri dish has been out of refrigeration.

b. When $t = 2$, the number of bacteria is

$$N = 320(2)^2 + 420 = 1280 + 420 = 1700.$$

c. The bacteria count will reach $N = 2000$ when $320t^2 + 420 = 2000$. You can solve this equation for t algebraically as follows.

$$320t^2 + 420 = 2000$$

$$320t^2 = 1580$$

$$t^2 = \frac{79}{16}$$

$$t = \frac{\sqrt{79}}{4} \quad \Rightarrow \quad t \approx 2.22 \text{ hours}$$

So, the count will reach 2000 when $t \approx 2.22$ hours. Note that the negative value is rejected because it is not in the domain of the composite function. To confirm your solution, graph the equation $N = 320t^2 + 420$, as shown in Figure 1.36. Then use the *zoom* and *trace* features to approximate $N = 2000$ when $t \approx 2.22$, as shown in Figure 1.37.

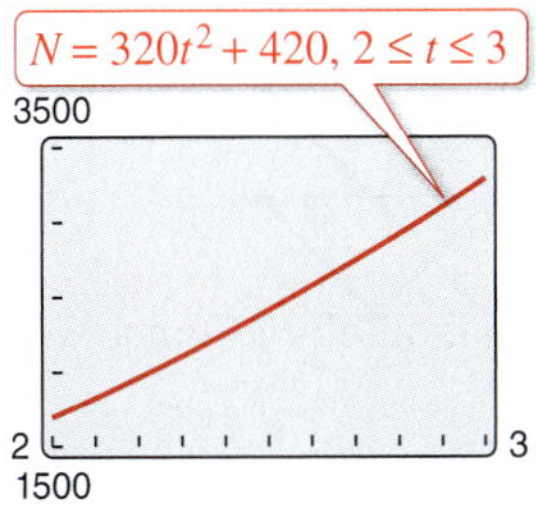

Figure 1.36

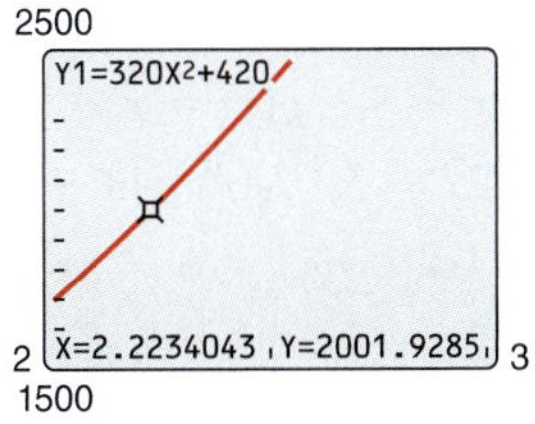

Figure 1.37

Checkpoint *Audio-video solution in English & Spanish at LarsonPrecalculus.com*

The number N of bacteria in a refrigerated food is given by

$$N(T) = 8T^2 - 14T + 200, \quad 2 \le T \le 12$$

where T is the temperature of the food in degrees Celsius. When the food is removed from refrigeration, the temperature of the food is given by

$$T(t) = 2t + 2, \quad 0 \le t \le 5$$

where t is the time in hours. Find (a) $(N \circ T)(t)$ and (b) the time when the bacteria count reaches 1000.

1.5 Exercises

See *CalcChat.com* for tutorial help and worked-out solutions to odd-numbered exercises. For instructions on how to use a graphing utility, see Appendix A.

Vocabulary and Concept Check

In Exercises 1–4, fill in the blank(s).

1. Two functions f and g can be combined by the arithmetic operations of _______, _______, _______, and _______ to create new functions.

2. The _______ of the function f with the function g is $(f \circ g)(x) = f(g(x))$.

3. The domain of $f \circ g$ is the set of all x in the domain of g such that _______ is in the domain of f.

4. To "decompose" a composite function, look for an _______ and an _______ function.

5. If $f(x) = x^2 + 1$ and $(fg)(x) = 2x(x^2 + 1)$, then what is $g(x)$?

6. If $(f \circ g)(x) = f(x^2 + 1)$, then what is $g(x)$?

Procedures and Problem Solving

Graphing the Sum of Two Functions **In Exercises 7–10, use the graphs of f and g to graph $h(x) = (f + g)(x)$. To print an enlarged copy of the graph, go to *MathGraphs.com*.**

7.

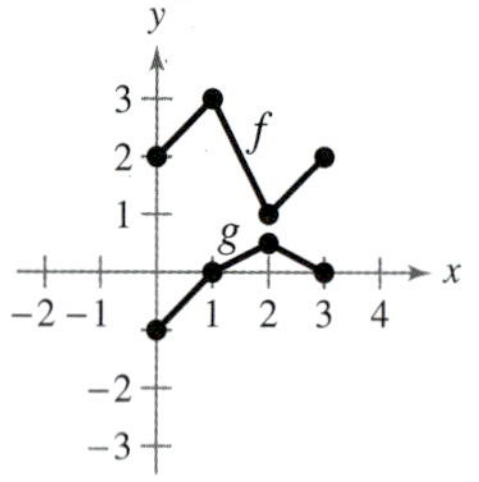

8.

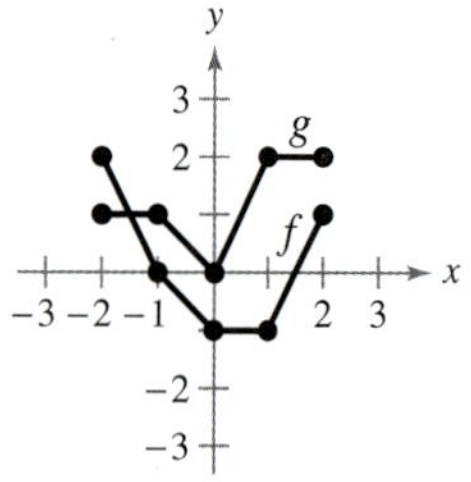

9.

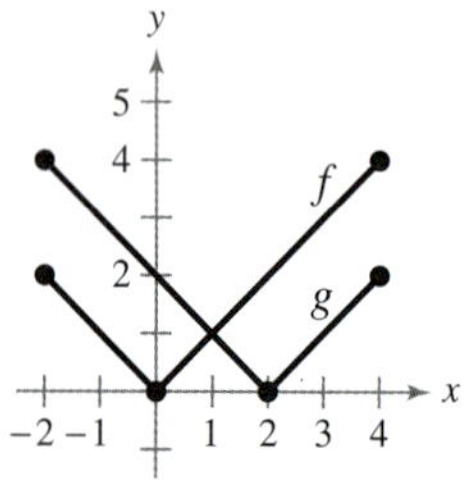

10.

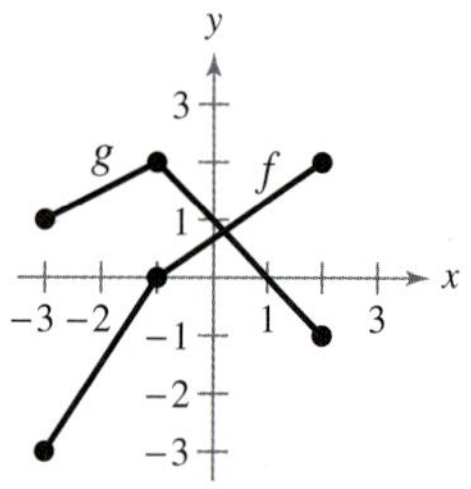

Finding Arithmetic Combinations of Functions **In Exercises 11–18, find (a) $(f + g)(x)$, (b) $(f - g)(x)$, (c) $(fg)(x)$, and (d) $(f/g)(x)$. What is the domain of f/g?**

11. $f(x) = x + 3, \quad g(x) = x - 3$

12. $f(x) = 2x - 5, \quad g(x) = 1 - x$

13. $f(x) = 3x^2, \quad g(x) = 6 - 5x$

14. $f(x) = 2x + 5, \quad g(x) = x^2 - 9$

15. $f(x) = x^2 + 5, \quad g(x) = \sqrt{1 - x}$

16. $f(x) = \sqrt{x^2 - 4}, \quad g(x) = \dfrac{x^2}{x^2 + 1}$

17. $f(x) = \dfrac{x}{x + 1}, \quad g(x) = x^3$

18. $f(x) = \dfrac{1}{x}, \quad g(x) = \dfrac{1}{x^2}$

Evaluating an Arithmetic Combination of Functions **In Exercises 19–32, evaluate the function for $f(x) = x + 3$ and $g(x) = x^2 - 2$. If possible, use a graphing utility to verify your answer.**

19. $(f + g)(2)$

20. $(f + g)(-1)$

21. $(f - g)(0)$

22. $(f - g)(1)$

23. $(fg)(6)$

24. $(fg)(-6)$

25. $(f/g)(5)$

26. $(f/g)(0)$

27. $(f - g)(t + 1)$

28. $(f + g)(t - 3)$

29. $(fg)(-5t)$

30. $(fg)(3t^2)$

31. $(f/g)(t - 4)$

32. $(f/g)(t + 2)$

Graphing an Arithmetic Combination of Functions **In Exercises 33–36, use a graphing utility to graph the functions f, g, and h in the same viewing window.**

33. $f(x) = \frac{1}{2}x, \quad g(x) = x - 1, \quad h(x) = f(x) + g(x)$

34. $f(x) = \frac{1}{3}x, \quad g(x) = -x + 4, \quad h(x) = f(x) - g(x)$

35. $f(x) = x^2, \quad g(x) = -2x + 5, \quad h(x) = f(x) \cdot g(x)$

36. $f(x) = 4 - x^2, \quad g(x) = x, \quad h(x) = f(x)/g(x)$

Graphing a Sum of Functions **In Exercises 37–40, use a graphing utility to graph f, g, and $f + g$ in the same viewing window. Which function contributes most to the magnitude of the sum when $0 \le x \le 2$? Which function contributes most to the magnitude of the sum when $x > 6$?**

37. $f(x) = 3x, \quad g(x) = -\dfrac{x^3}{10}$

38. $f(x) = \frac{1}{2}x, \quad g(x) = \sqrt{x}$

39. $f(x) = 3x + 2, \quad g(x) = -\sqrt{x + 5}$

40. $f(x) = x^2 - \frac{1}{2}, \quad g(x) = -3x^2 - 1$

Compositions of Functions In Exercises 41–44, find (a) $(f \circ g)(x)$, (b) $(g \circ f)(x)$, (c) $(g \circ g)(x)$, and, if possible, (d) $(f \circ g)(0)$.

41. $f(x) = 2x^2, \quad g(x) = x + 4$

42. $f(x) = 3x + 5, \quad g(x) = 5 - x$

43. $f(x) = \sqrt[3]{x - 1}, \quad g(x) = x^3 + 1$

44. $f(x) = x^3, \quad g(x) = \dfrac{1}{x}$

Finding Domains In Exercises 45–54, determine the domains of (a) f, (b) g, and (c) $f \circ g$. Use a graphing utility to verify your results.

45. $f(x) = \sqrt{x - 7}, \quad g(x) = 4x^2$

46. $f(x) = \sqrt{x + 3}, \quad g(x) = \dfrac{x}{2}$

47. $f(x) = x^2 + 1, \quad g(x) = \sqrt{x}$

48. $f(x) = x^{1/4}, \quad g(x) = x^4$

49. $f(x) = \dfrac{1}{x}, \quad g(x) = \dfrac{1}{x + 3}$

50. $f(x) = \dfrac{1}{x}, \quad g(x) = \dfrac{1}{2x}$

51. $f(x) = |x - 4|, \quad g(x) = 3 - x$

52. $f(x) = \dfrac{2}{|x|}, \quad g(x) = x - 5$

53. $f(x) = x + 2, \quad g(x) = \dfrac{1}{x^2 - 4}$

54. $f(x) = \dfrac{3}{x^2 - 1}, \quad g(x) = x + 1$

Determining Whether $(f \circ g)(x) = (g \circ f)(x)$ In Exercises 55–60, (a) find $(f \circ g)(x)$, $(g \circ f)(x)$, and the domain of $f \circ g$. (b) Use a graphing utility to graph $f \circ g$ and $g \circ f$. Determine whether $(f \circ g)(x) = (g \circ f)(x)$.

55. $f(x) = \sqrt{x + 4}, \quad g(x) = x^2$

56. $f(x) = \sqrt[3]{x + 1}, \quad g(x) = x^3 - 1$

57. $f(x) = \frac{1}{3}x - 3, \quad g(x) = 3x + 9$

58. $f(x) = \sqrt{x}, \quad g(x) = \sqrt{x}$

59. $f(x) = x^{2/3}, \quad g(x) = x^6$

60. $f(x) = |x|, \quad g(x) = -x^2 + 1$

Determining Whether $(f \circ g)(x) = (g \circ f)(x)$ In Exercises 61–66, (a) find $(f \circ g)(x)$ and $(g \circ f)(x)$, (b) determine algebraically whether $(f \circ g)(x) = (g \circ f)(x)$, and (c) use a graphing utility to complete a table of values to confirm your answer to part (b).

61. $f(x) = 5x + 4, \quad g(x) = \frac{1}{5}(x - 4)$

62. $f(x) = \frac{1}{4}(x - 1), \quad g(x) = 4x + 1$

63. $f(x) = \sqrt{x + 6}, \quad g(x) = x^2 - 5$

64. $f(x) = x^3 - 4, \quad g(x) = \sqrt[3]{x + 10}$

65. $f(x) = |x|, \quad g(x) = 2x^3$

66. $f(x) = \dfrac{6}{3x - 5}, \quad g(x) = -x$

Evaluating Combinations of Functions In Exercises 67–70, use the graphs of f and g to evaluate the functions.

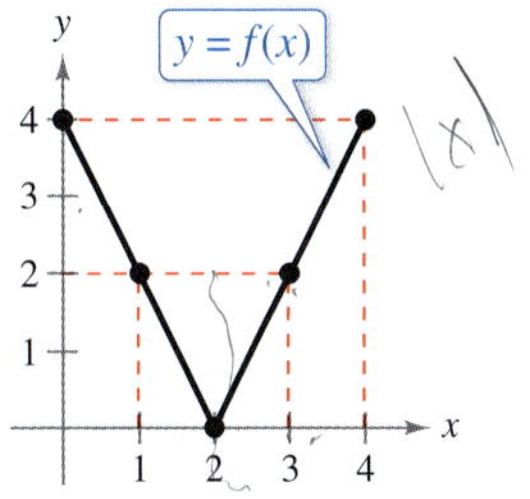

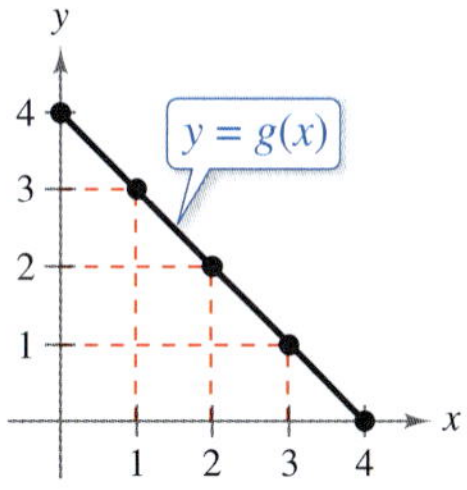

67. (a) $(f + g)(3)$ (b) $(f/g)(2)$

68. (a) $(f - g)(1)$ (b) $(fg)(4)$

69. (a) $(f \circ g)(3)$ (b) $(g \circ f)(2)$

70. (a) $(f \circ g)(1)$ (b) $(g \circ f)(3)$

Identifying a Composite Function In Exercises 71–78, find two functions f and g such that $(f \circ g)(x) = h(x)$. (There are many correct answers.)

71. $h(x) = (2x + 1)^2$

72. $h(x) = (1 - x)^3$

73. $h(x) = \sqrt[3]{x^2 - 4}$

74. $h(x) = \sqrt{9 - x}$

75. $h(x) = \dfrac{1}{x + 2}$

76. $h(x) = \dfrac{4}{(5x + 2)^2}$

77. $h(x) = (x + 4)^2 + 2(x + 4)$

78. $h(x) = (x + 3)^{3/2} + 4(x + 3)^{1/2}$

79. ***Why you should learn it*** *(p. 50)* The research and development department of an automobile manufacturer has determined that the distance (in feet) a car travels during the driver's reaction time, when required to stop quickly, can be modeled by

$$R(x) = \tfrac{3}{4}x$$

where x is the speed of the car in miles per hour. The distance (in feet) traveled while the driver is braking can be modeled by

$$B(x) = \tfrac{1}{15}x^2.$$

(a) Find the function that represents the total stopping distance T.

(b) Use a graphing utility to graph the functions R, B, and T in the same viewing window for $0 \le x \le 60$.

(c) Which function contributes most to the magnitude of the sum at higher speeds? Explain.

80. MODELING DATA

The table shows the total amounts (in billions of dollars) of health consumption expenditures in the United States for the years 2006 through 2016. The variables y_1, y_2, and y_3 represent out-of-pocket payments, insurance premiums, and other types of payments, respectively. (*Source:* U.S. Centers for Medicare and Medicaid Services)

DATA

Year	y_1	y_2	y_3
2006	273	1517	179
2007	290	1609	192
2008	295	1696	188
2009	294	1796	192
2010	300	1877	204
2011	310	1950	205
2012	318	2023	225
2013	325	2088	235
2014	330	2228	239
2015	339	2383	243
2016	353	2487	258

Spreadsheet at LarsonPrecalculus.com

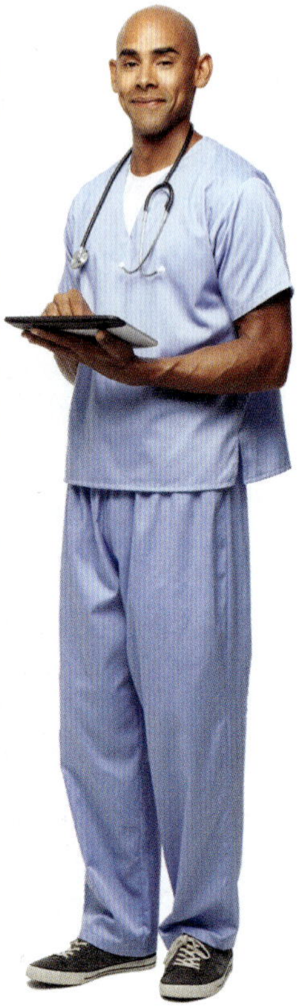

The data are approximated by the three models below, where t represents the year, with $t = 6$ corresponding to 2006.

$$y_1 = 0.19t^2 + 2.9t + 255$$

$$y_2 = 2.55t^2 + 37.4t + 1224$$

$$y_3 = 0.31t^2 + 1.0t + 163$$

(a) Use the models and the *table* feature of a graphing utility to create a table showing the values of y_1, y_2, and y_3 for each year from 2006 through 2016. Compare these models with the original data. Are the models a good fit? Explain.

(b) Use the graphing utility to graph y_1, y_2, y_3, and $y_T = y_1 + y_2 + y_3$ in the same viewing window. What does the function y_T represent?

81. Geometry A square concrete foundation was prepared as a base for a large cylindrical gasoline tank (see figure).

(a) Write the radius r of the tank as a function of the length x of the sides of the square.

(b) Write the area A of the circular base of the tank as a function of the radius r.

(c) Find and interpret $(A \circ r)(x)$.

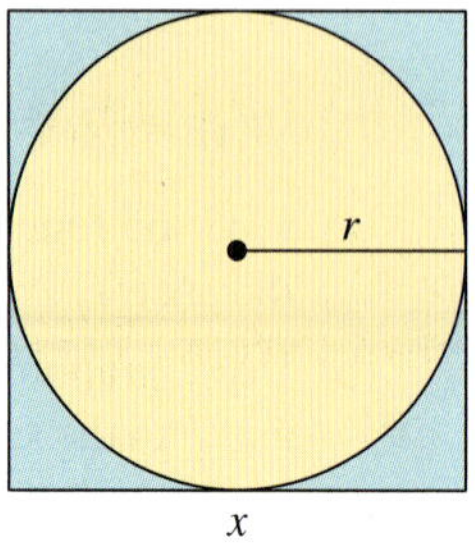

wickedpix/iStock/Getty Images

82. Geometry A pebble is dropped into a calm pond, causing ripples in the form of concentric circles. The function $r(t) = 0.6t$ represents the radius (in feet) of the outermost ripple, where t is the time (in seconds) after the pebble strikes the water. The function $A(r) = \pi r^2$ represents the area of the circle. Find and interpret $(A \circ r)(t)$.

83. Business A company owns two retail stores. The annual sales (in thousands of dollars) of the stores each year from 2012 through 2018 can be approximated by the models

$$S_1 = 973 + 1.3t^2 \quad \text{and} \quad S_2 = 349 + 72.4t$$

where t is the year, with $t = 12$ corresponding to 2012.

(a) Write a function T that represents the total annual sales of the two stores.

(b) Use a graphing utility to graph S_1, S_2, and T in the same viewing window.

84. Business The annual cost C (in thousands of dollars) and revenue R (in thousands of dollars) for a company each year from 2012 through 2018 can be approximated by the models

$$C = 254 - 9t + 1.1t^2 \quad \text{and} \quad R = 341 + 3.2t$$

where t is the year, with $t = 12$ corresponding to 2012.

(a) Write a function P that represents the annual profits of the company.

(b) Use a graphing utility to graph C, R, and P in the same viewing window.

85. Biology The number of bacteria in a refrigerated food product can be modeled by

$$N(T) = 10T^2 - 20T + 600, \quad 1 \le T \le 20$$

where T is the temperature of the food in degrees Celsius. When the food is removed from the refrigerator, the temperature of the food can be modeled by

$$T(t) = 2t + 1$$

where t is the time in hours.

(a) Find the composite function $N(T(t))$ or $(N \circ T)(t)$ and interpret its meaning.

(b) Find $(N \circ T)(12)$ and interpret its meaning.

(c) Find the time when the bacteria count reaches 1200.

86. Environmental Science The spread of a contaminant is increasing in a circular pattern on the surface of a lake. The radius of the contaminant can be modeled by $r(t) = 5.25\sqrt{t}$, where r is the radius in meters and t is time in hours since contamination.

(a) Find a function that gives the contaminated area A in terms of the time t since the spread began.

(b) Find the size of the contaminated area after 36 hours.

(c) Find when the size of the contaminated area is 6250 square meters.

87. Air Traffic Control An air traffic controller spots two planes flying at the same altitude. Their flight paths form a right angle at point P. One plane is 150 miles from point P and is moving at 450 miles per hour. The other plane is 200 miles from point P and is moving at 450 miles per hour. Write the distance s between the planes as a function of time t.

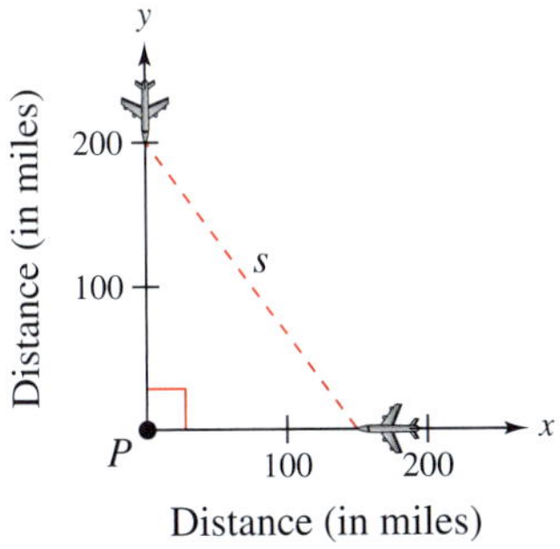

88. Marketing The suggested retail price of a new car is p dollars. The dealership advertised a factory rebate of \$2000 and a 9% discount.

(a) Write a function R in terms of p giving the cost of the car after receiving the rebate from the factory.

(b) Write a function S in terms of p giving the cost of the car after receiving the dealership discount.

(c) Form the composite functions $(R \circ S)(p)$ and $(S \circ R)(p)$ and interpret each.

(d) Find $(R \circ S)(24{,}795)$ and $(S \circ R)(24{,}795)$. Which yields the lower cost for the car? Explain.

Focusing on Concepts

True or False? **In Exercises 89 and 90, determine whether the statement is true or false. Justify your answer.**

89. A function that represents the graph of $f(x) = x^2$ shifted three units to the right is $f(g(x))$, where $g(x) = x + 3$.

90. For two functions f and g, you can calculate $(f \circ g)(x)$ if and only if the range of g is a subset of the domain of f.

91. Proof Use examples to hypothesize whether the product of an odd function and an even function is even or odd. Then prove your hypothesis.

92. Proof Prove that the product of two odd functions is an even function, and that the product of two even functions is an even function.

93. Exploration In Example 9, the function h can be written as the composition of two functions in other ways. For which of the pairs of functions below is $h(x) = 1/(x - 2)^2$ equal to $f(g(x))$?

(a) $g(x) = 1/(x - 2)$ and $f(x) = x^2$

(b) $g(x) = x^2$ and $f(x) = 1/(x - 2)$

(c) $g(x) = (x - 2)^2$ and $f(x) = 1/x$

94. Writing Functions Write two unique functions f and g such that $(f \circ g)(x) = (g \circ f)(x)$.

Exploration **In Exercises 95 and 96, three siblings are of three different ages. The oldest is twice the age of the middle sibling, and the middle sibling is six years older than one-half the age of the youngest.**

95. (a) Write a composite function that gives the oldest sibling's age in terms of the youngest. Explain how you obtained your answer.

(b) The oldest sibling is 16 years old. Find the ages of the other two siblings.

96. (a) Write a composite function that gives the youngest sibling's age in terms of the oldest. Explain how you obtained your answer.

(b) The youngest sibling is two years old. Find the ages of the other two siblings.

97. Error Analysis Describe the error in finding $(f \circ g)(x)$, where $f(x) = 2x + 1$ and $g(x) = 3x + 2$.

$(f \circ g)(x) = 3(2x + 1) + 2 = 6x + 5$ ✗

98. HOW DO YOU SEE IT? The graphs labeled L_1, L_2, L_3, and L_4 represent four different pricing discounts, where p is the original price (in dollars) and S is the sale price (in dollars). Match each function with its graph. Describe the situations in parts (c) and (d).

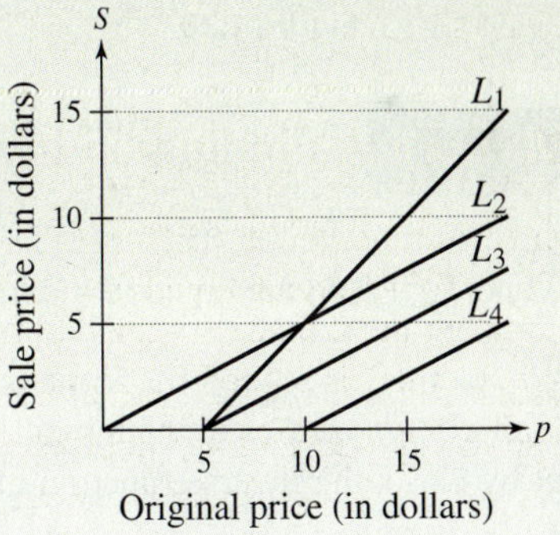

(a) $f(p)$: A 50% discount is applied.

(b) $g(p)$: A \$5 discount is applied.

(c) $(g \circ f)(p)$

(d) $(f \circ g)(p)$

Cumulative Mixed Review

Evaluating an Equation **In Exercises 99–102, find three points that lie on the graph of the equation. (There are many correct answers.)**

99. $y = -x^2 + x - 5$

100. $y = \frac{1}{5}x^3 - 4x^2 + 1$

101. $x^2 + y^2 = 49$

102. $y = \dfrac{x}{x^2 - 5}$

1.6 Inverse Functions

Inverse Functions

Recall from Section 1.2 that a function can be represented by a set of ordered pairs. For instance, the function $f(x) = x + 4$ from the set $A = \{1, 2, 3, 4\}$ to the set $B = \{5, 6, 7, 8\}$ can be written as

$$f(x) = x + 4: \{(1, 5), (2, 6), (3, 7), (4, 8)\}.$$

In this case, by interchanging the first and second coordinates of each ordered pair, you can form the **inverse function** of f, which is denoted by f^{-1}. It is a function from the set B to the set A and can be written as

$$f^{-1}(x) = x - 4: \{(5, 1), (6, 2), (7, 3), (8, 4)\}.$$

Note that the domain of f is equal to the range of f^{-1}, and vice versa, as shown in Figure 1.38. Also note that the functions f and f^{-1} have the effect of "undoing" each other. In other words, when you form the composition of f with f^{-1} or the composition of f^{-1} with f, you obtain the identity function.

$$f(f^{-1}(x)) = f(x - 4) = (x - 4) + 4 = x$$

$$f^{-1}(f(x)) = f^{-1}(x + 4) = (x + 4) - 4 = x$$

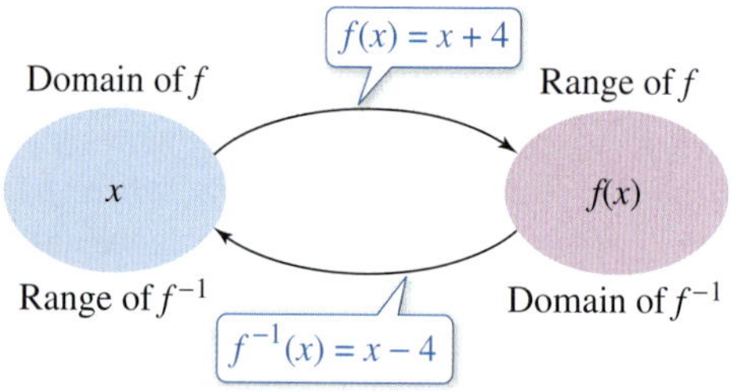

Figure 1.38

What you should learn

- Find inverse functions informally and verify that two functions are inverse functions of each other.
- Use graphs of functions to decide whether functions have inverse functions.
- Determine whether functions are one-to-one.
- Find inverse functions algebraically.

Why you should learn it

Inverse functions can be helpful in further exploring how two variables relate to each other. For example, in Exercise 123 on page 69, you will use inverse functions to find the European shoe sizes from the corresponding U.S. shoe sizes.

EXAMPLE 1 Finding an Inverse Function Informally

Find the inverse function of $f(x) = 4x$. Then verify that both $f(f^{-1}(x))$ and $f^{-1}(f(x))$ are equal to the identity function.

Solution

The function f *multiplies* each input by 4. To "undo" this function, you need to *divide* each input by 4. So, the inverse function of $f(x) = 4x$ is

$$f^{-1}(x) = \frac{x}{4}.$$

To check this result, verify that both $f(f^{-1}(x))$ and $f^{-1}(f(x))$ are equal to the identity function, as shown below.

$$f(f^{-1}(x)) = f\left(\frac{x}{4}\right) = 4\left(\frac{x}{4}\right) = x$$

$$f^{-1}(f(x)) = f^{-1}(4x) = \frac{4x}{4} = x$$

✓ Checkpoint *Audio-video solution in English & Spanish at LarsonPrecalculus.com*

Find the inverse function of $f(x) = \frac{1}{5}x$. Then verify that both $f(f^{-1}(x))$ and $f^{-1}(f(x))$ are equal to the identity function.

talevr/iStock/Getty Images

Do not be confused by the use of -1 to denote the inverse function f^{-1}. In this text, whenever f^{-1} is written, it always refers to the inverse function of the function f and not to the reciprocal of $f(x)$, which is

$$\frac{1}{f(x)}. \qquad \text{Reciprocal of } f(x)$$

EXAMPLE 2 Finding an Inverse Function Informally

Find the inverse function of $f(x) = x - 6$. Then verify that both $f(f^{-1}(x))$ and $f^{-1}(f(x))$ are equal to the identity function.

Solution

The function f *subtracts* 6 from each input. To "undo" this function, you need to *add* 6 to each input. So, the inverse function of $f(x) = x - 6$ is

$$f^{-1}(x) = x + 6.$$

To check this result, verify that both $f(f^{-1}(x))$ and $f^{-1}(f(x))$ are equal to the identity function, as shown below.

$$f(f^{-1}(x)) = f(x + 6) = (x + 6) - 6 = x$$

$$f^{-1}(f(x)) = f^{-1}(x - 6) = (x - 6) + 6 = x$$

Checkpoint *Audio-video solution in English & Spanish at LarsonPrecalculus.com*

Find the inverse function of $f(x) = x + 7$. Then verify that both $f(f^{-1}(x))$ and $f^{-1}(f(x))$ are equal to the identity function.

A table of values can help you understand inverse functions. For instance, the first table below shows several values of the function in Example 2. Interchange the rows of this table to obtain values of the inverse function.

x	-2	-1	0	1	2
$f(x)$	-8	-7	-6	-5	-4

x	-8	-7	-6	-5	-4
$f^{-1}(x)$	-2	-1	0	1	2

In the table at the left, each output is 6 less than the input, and in the table at the right, each output is 6 more than the input.

Definition of Inverse Function

Let f and g be two functions such that

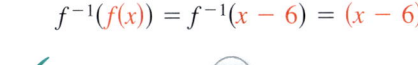

$$f(g(x)) = x \quad \text{for every } x \text{ in the domain of } g$$

and

$$g(f(x)) = x \quad \text{for every } x \text{ in the domain of } f.$$

Under these conditions, the function g is the **inverse function** of the function f. The function g is denoted by f^{-1} (read "f-inverse"). So,

$$f(f^{-1}(x)) = x \quad \text{and} \quad f^{-1}(f(x)) = x.$$

The domain of f must be equal to the range of f^{-1}, and the range of f must be equal to the domain of f^{-1}.

If the function g is the inverse function of the function f, then it must also be true that the function f is the inverse function of the function g. For this reason, you can say that the functions f and g are *inverse functions of each other.*

EXAMPLE 3 Verifying Inverse Functions Algebraically

Show that the functions are inverse functions of each other.

$$f(x) = 2x^3 - 1 \quad \text{and} \quad g(x) = \sqrt[3]{\frac{x+1}{2}}$$

Solution

$$\begin{aligned} f(g(x)) &= f\left(\sqrt[3]{\frac{x+1}{2}}\right) \\ &= 2\left(\sqrt[3]{\frac{x+1}{2}}\right)^3 - 1 \\ &= 2\left(\frac{x+1}{2}\right) - 1 \\ &= x + 1 - 1 \\ &= x \end{aligned}$$

$$\begin{aligned} g(f(x)) &= g(2x^3 - 1) \\ &= \sqrt[3]{\frac{(2x^3 - 1) + 1}{2}} \\ &= \sqrt[3]{\frac{2x^3}{2}} \\ &= \sqrt[3]{x^3} \\ &= x \end{aligned}$$

Checkpoint *Audio-video solution in English & Spanish at LarsonPrecalculus.com*

Show that $f(x) = x^5$ and $g(x) = \sqrt[5]{x}$ are inverse functions of each other.

EXAMPLE 4 Verifying Inverse Functions Algebraically

Which of the functions is the inverse function of $f(x) = \dfrac{5}{x-2}$?

$$g(x) = \frac{x-2}{5} \quad \text{or} \quad h(x) = \frac{5}{x} + 2$$

Solution

By forming the composition of f with g, you have

$$f(g(x)) = f\left(\frac{x-2}{5}\right) = \frac{5}{\left(\frac{x-2}{5}\right) - 2} = \frac{25}{x-12} \neq x.$$

Because this composition is not equal to the identity function x, it follows that g *is not* the inverse function of f. By forming the composition of f with h, you have

$$f(h(x)) = f\left(\frac{5}{x} + 2\right) = \frac{5}{\left(\frac{5}{x} + 2\right) - 2} = \frac{5}{\left(\frac{5}{x}\right)} = x.$$

So, it appears that h is the inverse function of f. You can confirm this by showing that the composition of h with f is also equal to the identity function.

$$h(f(x)) = h\left(\frac{5}{x-2}\right) = \frac{5}{\left(\frac{5}{x-2}\right)} + 2 = x - 2 + 2 = x$$

Checkpoint *Audio-video solution in English & Spanish at LarsonPrecalculus.com*

Which of the functions is the inverse function of $f(x) = \dfrac{x-4}{7}$?

$$g(x) = 7x + 4 \qquad h(x) = \frac{7}{x-4}$$

Technology Tip

Most graphing utilities can graph $y = x^{1/3}$ in two ways:

$y_1 = x \wedge (1/3)$ or

$y_1 = \sqrt[3]{x}$.

On some graphing utilities, you may not be able to obtain the complete graph of $y = x^{2/3}$ by entering $y_1 = x \wedge (2/3)$. If not, you should use

$y_1 = (x \wedge (1/3))^2$ or

$y_1 = \sqrt[3]{x^2}$.

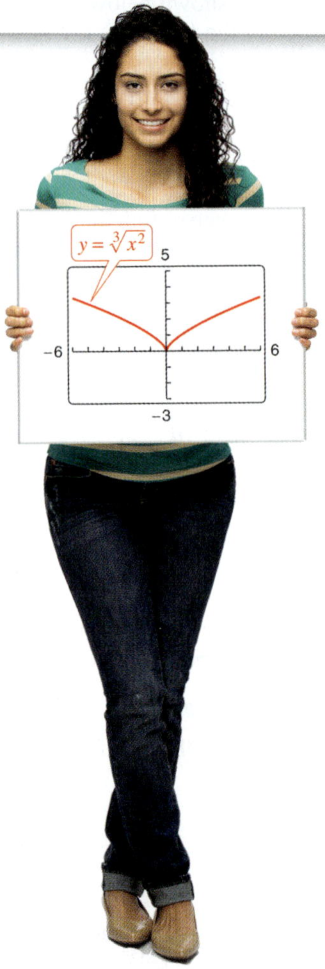

aldomurillo/iStock/Getty Images

The Graph of an Inverse Function

The graphs of a function f and its inverse function f^{-1} are related to each other in the following way. If the point

(a, b)

lies on the graph of f, then the point

(b, a)

must lie on the graph of f^{-1}, and vice versa. This means that the graph of f^{-1} is a reflection of the graph of f in the line $y = x$, as shown in Figure 1.39.

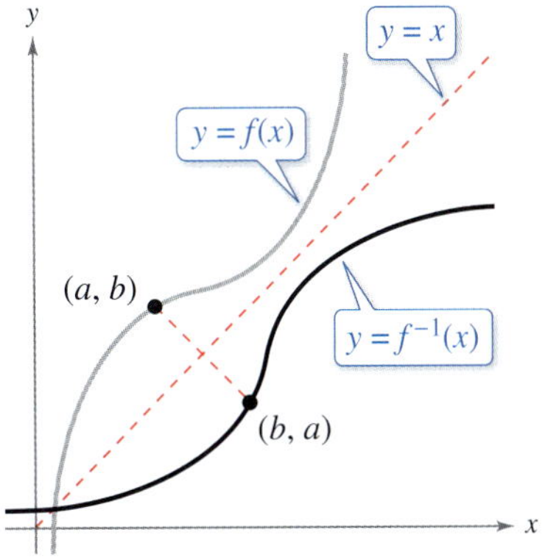

Figure 1.39

Technology Tip

Many graphing utilities have a built-in feature for drawing an inverse function. For instructions on how to use the *draw inverse* feature, see Appendix A; for specific keystrokes, go to this textbook's *Student Companion Website.*

EXAMPLE 5 Verifying Inverse Functions Graphically

Verify that the functions f and g from Example 3 are inverse functions of each other graphically.

Solution

From the figure, it appears that f and g are inverse functions of each other.

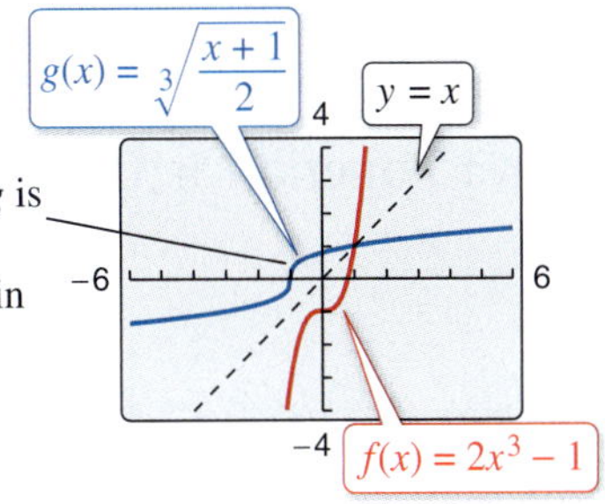

Checkpoint Audio-video solution in English & Spanish at LarsonPrecalculus.com

Verify that $f(x) = x^5$ and $g(x) = \sqrt[5]{x}$ are inverse functions of each other graphically.

EXAMPLE 6 Verifying Inverse Functions Numerically

Verify that the functions $f(x) = \frac{1}{2}(x - 5)$ and $g(x) = 2x + 5$ are inverse functions of each other numerically.

Solution

You can verify that f and g are inverse functions of each other *numerically* by using a graphing utility. Enter $y_1 = f(x)$, $y_2 = g(x)$, $y_3 = f(g(x))$, and $y_4 = g(f(x))$. Then use the *table* feature to create a table.

```
Plot1  Plot2  Plot3
\Y1=1/2*(X-5)
\Y2=2X+5
\Y3=Y1(Y2)
\Y4=Y2(Y1)
\Y5=
\Y6=
\Y7=
```

X	Y3	Y4
-2	-2	-2
-1	-1	-1
0	0	0
1	1	1
2	2	2
3	3	3
4	4	4

X=-2

Note that the entries for x, y_3, and y_4 are the same. So, it appears that $f(g(x)) = x$ and $g(f(x)) = x$, and that f and g are inverse functions of each other.

Checkpoint Audio-video solution in English & Spanish at LarsonPrecalculus.com

Verify that $f(x) = x^5$ and $g(x) = \sqrt[5]{x}$ are inverse functions of each other numerically.

The Existence of an Inverse Function

To have an inverse function, a function must be **one-to-one,** which means that no two elements in the domain of f correspond to the same element in the range of f.

Definition of a One-to-One Function

A function f is **one-to-one** when, for a and b in its domain, $f(a) = f(b)$ implies that $a = b$.

Existence of an Inverse Function

A function f has an inverse function f^{-1} if and only if f is one-to-one.

From its graph, it is easy to tell whether a function of x is one-to-one. Simply check to see that every horizontal line intersects the graph of the function at most once. This is called the **Horizontal Line Test.** For instance, Figure 1.40 shows the graph of $f(x) = x^2$. On the graph, you can find a horizontal line that intersects the graph twice. So, f is not one-to-one and does not have an inverse function.

Two special types of functions that pass the Horizontal Line Test are those that are increasing or decreasing on their entire domains. If f is *increasing* on its entire domain, then f is one-to-one. If f is *decreasing* on its entire domain, then f is one-to-one.

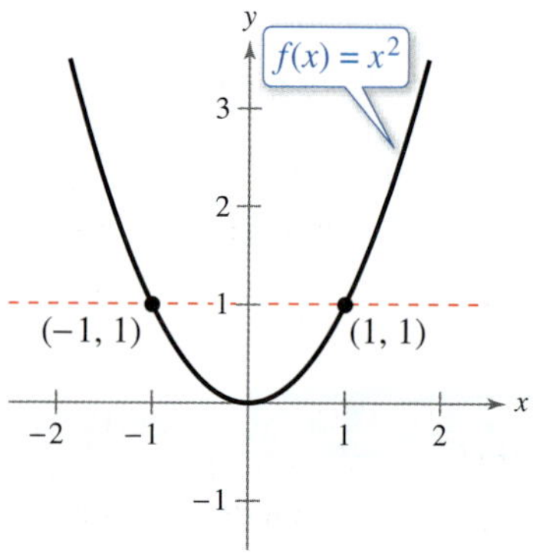

f(x) = x² is not one-to-one.
Figure 1.40

EXAMPLE 7 Testing Whether a Function Is One-to-One

Is the function $f(x) = \sqrt{x} + 1$ one-to-one?

Algebraic Solution

Let a and b be nonnegative real numbers with $f(a) = f(b)$.

$$\sqrt{a} + 1 = \sqrt{b} + 1 \qquad \text{Set } f(a) = f(b).$$

$$\sqrt{a} = \sqrt{b}$$

$$a = b$$

So, $f(a) = f(b)$ implies that $a = b$. You can conclude that f *is* one-to-one and *does* have an inverse function.

Graphical Solution

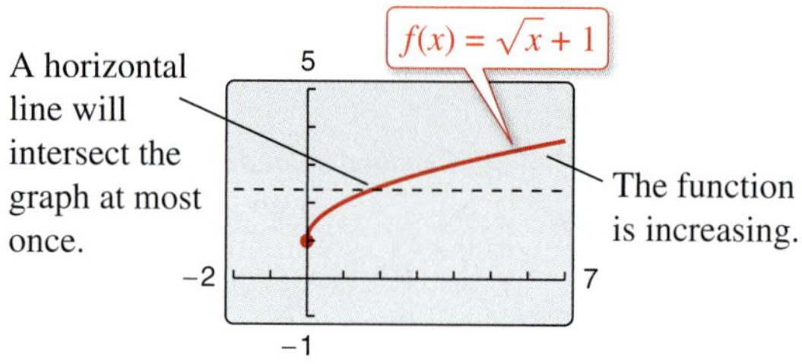

From the figure, you can conclude that f *is* one-to-one and *does* have an inverse function.

✓ **Checkpoint** Audio-video solution in English & Spanish at LarsonPrecalculus.com

Is the function $f(x) = \sqrt[3]{x}$ one-to-one?

EXAMPLE 8 Testing Whether a Function Is One-to-One

See LarsonPrecalculus.com for an interactive version of this type of example.

To determine whether $f(x) = x^2 - x$ is one-to-one, note that $f(-1) = (-1)^2 - (-1) = 2$ and $f(2) = 2^2 - 2 = 2$. Because two inputs correspond to the same output, f *is not* one-to-one and *does not* have an inverse function. You can confirm this graphically by noticing that the horizontal line $y = 2$ intersects the graph of f twice, as shown in Figure 1.41.

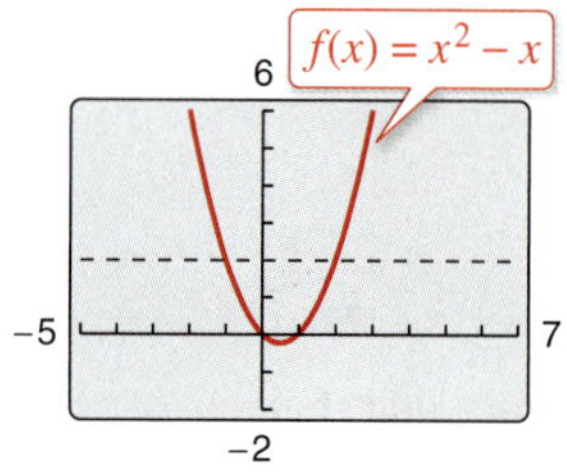

Figure 1.41

✓ **Checkpoint** Audio-video solution in English & Spanish at LarsonPrecalculus.com

Is the function $f(x) = |x|$ one-to-one?

Finding Inverse Functions Algebraically

For relatively simple functions (such as the ones in Examples 1 and 2), you can find inverse functions by inspection. For more complicated functions, however, it is best to use the following guidelines.

Finding an Inverse Function

1. Use the Horizontal Line Test to decide whether f has an inverse function.
2. In the equation for $f(x)$, replace $f(x)$ with y.
3. Interchange the roles of x and y, and solve for y.
4. Replace y with $f^{-1}(x)$ in the new equation.
5. Verify that f and f^{-1} are inverse functions of each other by showing that the domain of f is equal to the range of f^{-1}, the range of f is equal to the domain of f^{-1}, and $f(f^{-1}(x)) = x$ and $f^{-1}(f(x)) = x$.

The key step in these guidelines is Step 3—interchanging the roles of x and y. This step corresponds to the fact that inverse functions have ordered pairs with the coordinates reversed.

What's Wrong?

You use a graphing utility to graph $y_1 = x^2$ and then use the *draw inverse* feature to conclude that $f(x) = x^2$ has an inverse function (see figure). What's wrong?

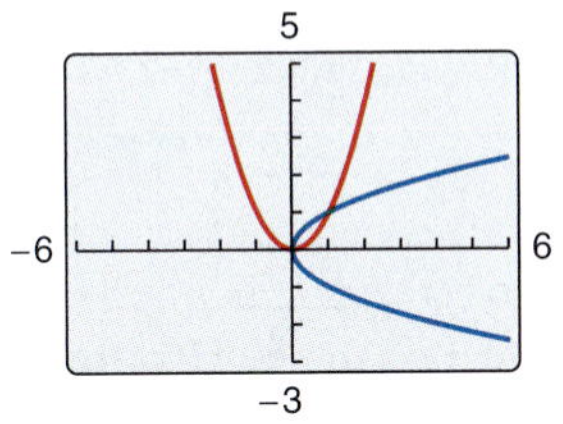

EXAMPLE 9 Finding an Inverse Function Algebraically

Find the inverse function of

$$f(x) = \frac{5 - 3x}{2}.$$

Solution

The graph of f in Figure 1.42 passes the Horizontal Line Test. So, you know that f is one-to-one and has an inverse function.

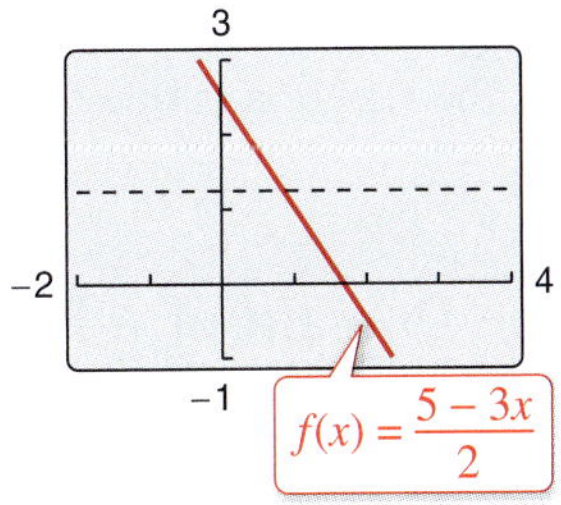

Figure 1.42

$$f(x) = \frac{5 - 3x}{2} \quad \text{Write original function.}$$

$$y = \frac{5 - 3x}{2} \quad \text{Replace } f(x) \text{ with } y.$$

$$x = \frac{5 - 3y}{2} \quad \text{Interchange } x \text{ and } y.$$

$$2x = 5 - 3y \quad \text{Multiply each side by 2.}$$

$$3y = 5 - 2x \quad \text{Isolate the } y\text{-term.}$$

$$y = \frac{5 - 2x}{3} \quad \text{Solve for } y.$$

$$f^{-1}(x) = \frac{5 - 2x}{3} \quad \text{Replace } y \text{ with } f^{-1}(x).$$

The domains and ranges of f and f^{-1} consist of all real numbers. Verify that $f(f^{-1}(x)) = x$ and $f^{-1}(f(x)) = x$.

Checkpoint *Audio-video solution in English & Spanish at LarsonPrecalculus.com*

Find the inverse function of $f(x) = 2x - 3$.

EXAMPLE 10 Finding an Inverse Function Algebraically

Find the inverse function of $f(x) = \sqrt{2x - 3}$.

Solution

The graph of f in Figure 1.43 passes the Horizontal Line Test. So, you know that f is one-to-one and has an inverse function.

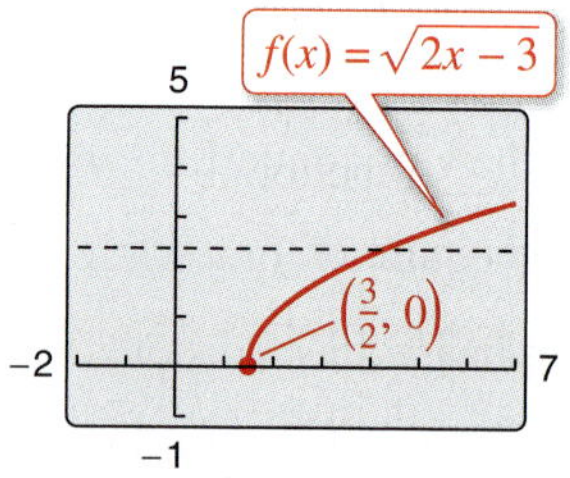

Figure 1.43

$f(x) = \sqrt{2x - 3}$ Write original function.

$y = \sqrt{2x - 3}$ Replace $f(x)$ with y.

$x = \sqrt{2y - 3}$ Interchange x and y.

$x^2 = 2y - 3$ Square each side.

$2y = x^2 + 3$ Isolate y.

$y = \dfrac{x^2 + 3}{2}$ Solve for y.

$f^{-1}(x) = \dfrac{x^2 + 3}{2}, \ x \geq 0$ Replace y with $f^{-1}(x)$.

The range of f is the interval $[0, \infty)$, which implies that the domain of f^{-1} is the interval $[0, \infty)$. Moreover, the domain of f is the interval $\left[\frac{3}{2}, \infty\right)$, which implies that the range of f^{-1} is the interval $\left[\frac{3}{2}, \infty\right)$. Verify that $f(f^{-1}(x)) = x$ and $f^{-1}(f(x)) = x$.

Checkpoint *Audio-video solution in English & Spanish at LarsonPrecalculus.com*

Find the inverse function of $f(x) = \sqrt[3]{x + 10}$.

A function f with an implied domain of all real numbers may not pass the Horizontal Line Test. In this case, the domain of f may be restricted so that f does have an inverse function. Recall from Figure 1.40 that $f(x) = x^2$ is not one-to-one. By restricting the domain of f to $x \geq 0$, the function is one-to-one and does have an inverse function (see Figure 1.44).

EXAMPLE 11 Restricting the Domain of a Function

Find the inverse function of $f(x) = x^2, x \geq 0$.

Solution

The graph of f in Figure 1.44 passes the Horizontal Line Test. So, you know that f is one-to-one and has an inverse function.

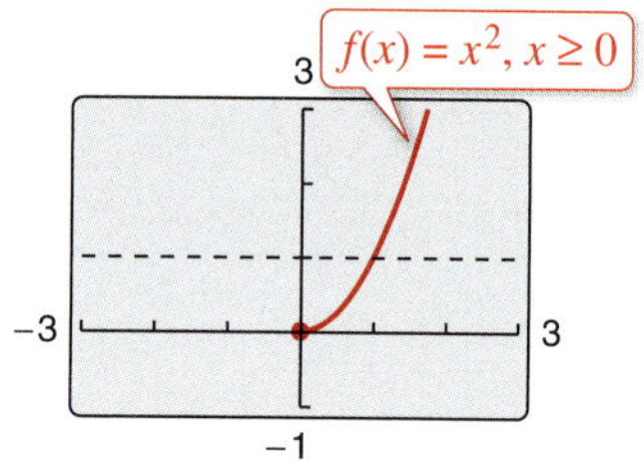

Figure 1.44

$f(x) = x^2, \ x \geq 0$ Write original function with restricted domain.

$y = x^2$ Replace $f(x)$ with y.

$x = y^2, \ y \geq 0$ Interchange x and y.

$y = \sqrt{x}$ Solve for y. Extract positive square root ($y \geq 0$).

$f^{-1}(x) = \sqrt{x}$ Replace y with $f^{-1}(x)$.

The graph of f^{-1} in Figure 1.45 is the reflection of the graph of f in the line $y = x$. The domains and ranges of f and f^{-1} consist of all nonnegative real numbers. Verify that $f(f^{-1}(x)) = x$ and $f^{-1}(f(x)) = x$.

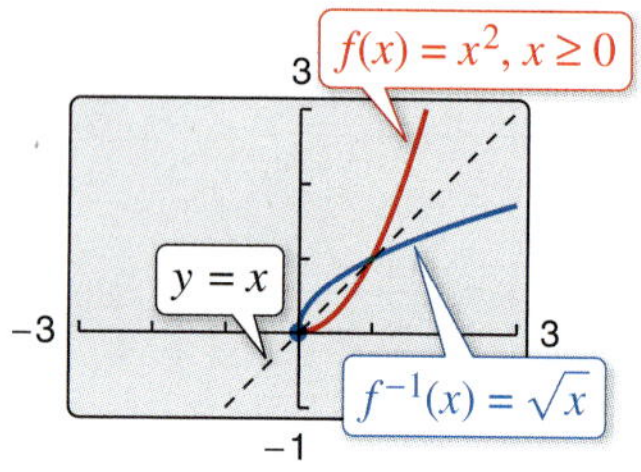

Figure 1.45

Checkpoint *Audio-video solution in English & Spanish at LarsonPrecalculus.com*

Find the inverse function of $f(x) = (x - 1)^2, x \geq 1$.

1.6 Exercises

See *CalcChat.com* for tutorial help and worked-out solutions to odd-numbered exercises.
For instructions on how to use a graphing utility, see Appendix A.

Vocabulary and Concept Check

In Exercises 1–4, fill in the blank(s).

1. If f and g are functions such that $f(g(x)) = x$ and $g(f(x)) = x$, then the function g is the _______ function of f, and is denoted by _______ .

2. The domain of f is the _______ of f^{-1}, and the _______ of f^{-1} is the range of f.

3. The graphs of f and f^{-1} are reflections of each other in the line _______ .

4. To have an inverse function, a function f must be _______ ; that is, $f(a) = f(b)$ implies $a = b$.

5. How many times can a horizontal line intersect the graph of a function that is one-to-one?

6. Can (1, 4) and (2, 4) be two ordered pairs of a one-to-one function?

Procedures and Problem Solving

Finding an Inverse Function Informally **In Exercises 7–12, find the inverse function of f informally. Verify that $f(f^{-1}(x)) = x$ and $f^{-1}(f(x)) = x$.**

7. $f(x) = x + 11$

8. $f(x) = \frac{1}{3}x$

9. $f(x) = 3x + 1$

10. $f(x) = \frac{1}{2}(x - 3)$

11. $f(x) = \sqrt[3]{x}$

12. $f(x) = x^7$

Identifying Graphs of Inverse Functions **In Exercises 13–16, match the graph of the function with the graph of its inverse function. [The graphs of the inverse functions are labeled (a), (b), (c), and (d).]**

(a)
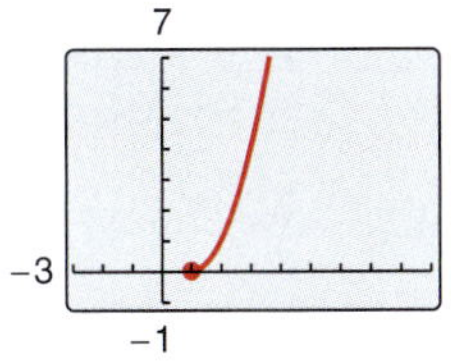

(b)
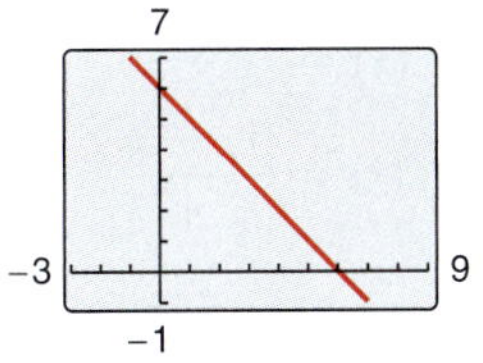

(c)
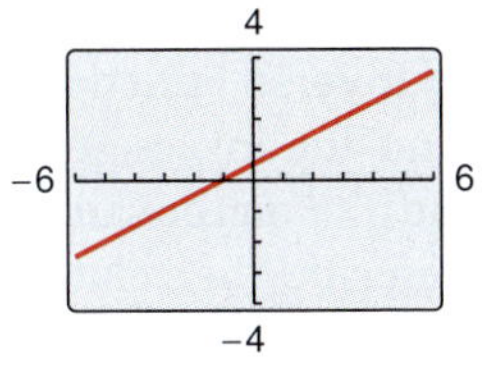

(d)
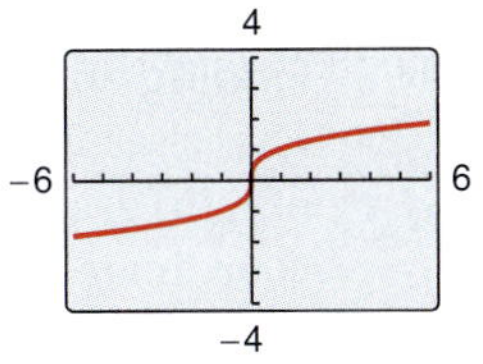

13.
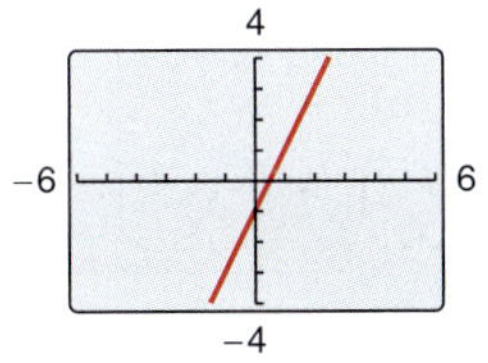

14.
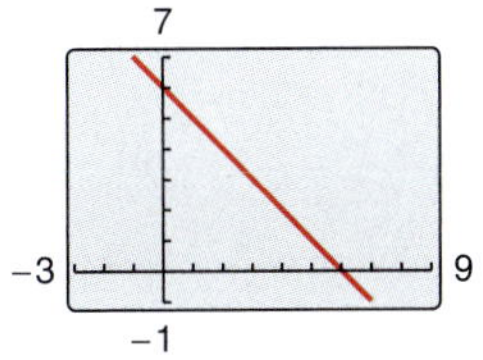

15.
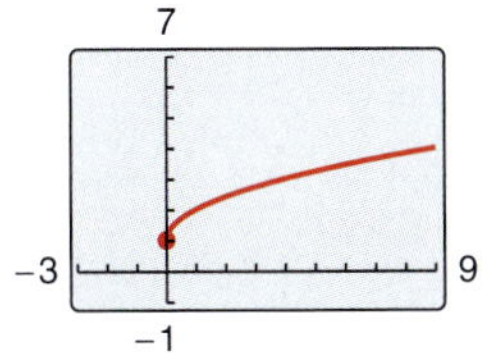

16.
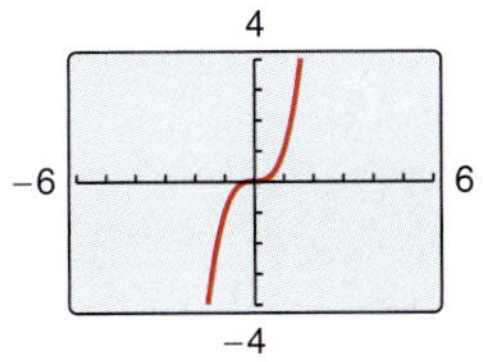

Verifying Inverse Functions Algebraically **In Exercises 17–22, show that f and g are inverse functions algebraically. Use a graphing utility to graph f and g in the same viewing window. Describe the relationship between the graphs.**

17. $f(x) = x^3, \quad g(x) = \sqrt[3]{x}$

18. $f(x) = \dfrac{1}{x}, \quad g(x) = \dfrac{1}{x}$

19. $f(x) = \sqrt{x - 4}; \quad g(x) = x^2 + 4, \quad x \ge 0$

20. $f(x) = 9 - x^2, x \ge 0; \quad g(x) = \sqrt{9 - x}$

21. $f(x) = 1 - x^3, \quad g(x) = \sqrt[3]{1 - x}$

22. $f(x) = \dfrac{1}{1 + x}, x \ge 0; \quad g(x) = \dfrac{1 - x}{x}, 0 < x \le 1$

Algebraic-Graphical-Numerical **In Exercises 23–32, show that f and g are inverse functions (a) algebraically, (b) graphically, and (c) numerically.**

23. $f(x) = -\dfrac{7}{2}x - 3, \quad g(x) = -\dfrac{2x + 6}{7}$

24. $f(x) = \dfrac{x - 9}{4}, \quad g(x) = 4x + 9$

25. $f(x) = x^3 + 5, \quad g(x) = \sqrt[3]{x - 5}$

26. $f(x) = \frac{1}{2}x^3 \quad g(x) = \sqrt[3]{2x}$

27. $f(x) = -\sqrt{x - 8}; \quad g(x) = 8 + x^2, \quad x \le 0$

28. $f(x) = \sqrt[4]{3x - 10}; \quad g(x) = \dfrac{x^4 + 10}{3}, \quad x \ge 0$

29. $f(x) = 2x, \quad g(x) = \frac{1}{2}x$

30. $f(x) = -3x + 5, \quad g(x) = -\dfrac{x - 5}{3}$

31. $f(x) = \dfrac{x - 1}{x + 5}, \quad g(x) = -\dfrac{5x + 1}{x - 1}$

32. $f(x) = \dfrac{x + 3}{x - 2}, \quad g(x) = \dfrac{2x + 3}{x - 1}$

Identifying Whether Functions Have Inverses In Exercises 33–36, does the function have an inverse? Explain.

33. *Domain* → *Range*

1 can → \$1
6 cans → \$5
12 cans → \$9
24 cans → \$16

34. *Domain* → *Range*

1/2 hour → \$40
1 hour → \$40
2 hours → \$70
4 hours → \$120

35. $\{(-3, 6), (-1, 5), (0, 6)\}$

36. $\{(2, 4), (3, 7), (7, 2)\}$

Recognizing One-to-One Functions In Exercises 37–42, determine whether the graph is that of a function. If so, determine whether the function is one-to-one.

37.

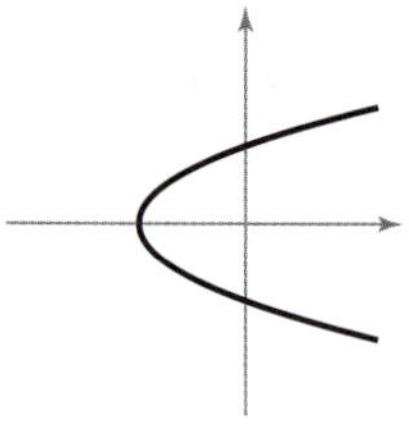

38.

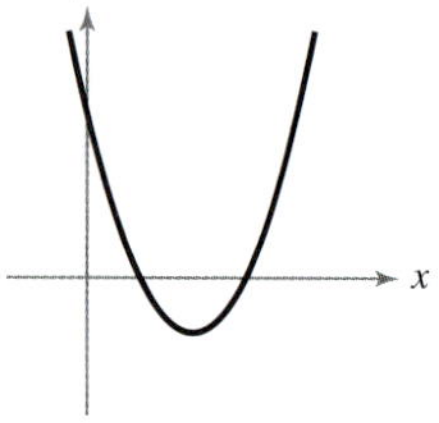

39.

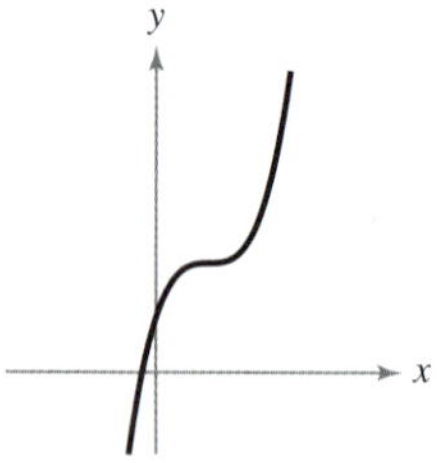

40.

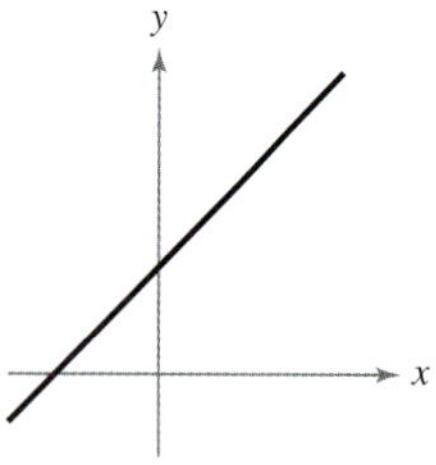

41.

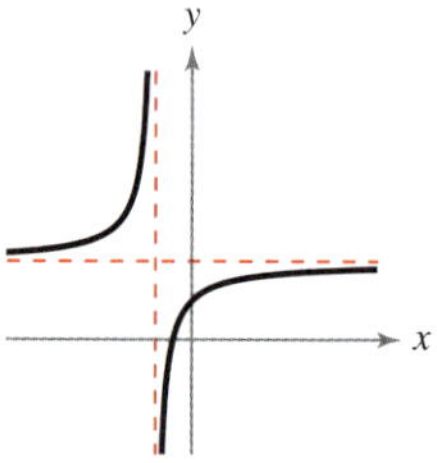

42.

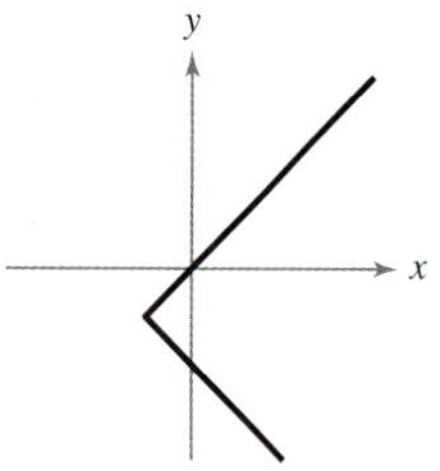

Using the Horizontal Line Test In Exercises 43–54, use a graphing utility to graph the function, and use the Horizontal Line Test to determine whether the function has an inverse function.

43. $f(x) = 10$

44. $f(x) = 10x$

45. $g(x) = (x + 3)^2 + 2$

46. $f(x) = \frac{1}{5}(x + 2)^3$

47. $h(x) = \dfrac{x}{x + 1}$

48. $g(x) = \dfrac{4 - x}{6x^2}$

49. $h(x) = \sqrt[3]{1 - x}$

50. $f(x) = -2x\sqrt{16 - x^2}$

51. $g(x) = (x + 5)^3$

52. $f(x) = x^5 - 7$

53. $h(x) = |x| - |x - 4|$

54. $f(x) = -\dfrac{|x^2 - 9|}{|x^2 + 7|}$

Analyzing a Piecewise-Defined Function In Exercises 55 and 56, sketch the graph of the piecewise-defined function by hand and use the graph to determine whether the function has an inverse function.

55. $f(x) = \begin{cases} x^2, & 0 \le x \le 1 \\ x, & x > 1 \end{cases}$

56. $f(x) = \begin{cases} (x - 2)^3, & x < 3 \\ (x - 4)^2, & x \ge 3 \end{cases}$

Finding an Inverse Function In Exercises 57–72, determine whether the function has an inverse function. If it does, find the inverse function.

57. $f(x) = x^4$

58. $g(x) = x^2 - x^4$

59. $f(x) = \dfrac{3x + 4}{5}$

60. $f(x) = 3x + 5$

61. $p(x) = -4$

62. $f(x) = 0$

63. $f(x) = (x + 3)^2, \quad x \ge -3$

64. $q(x) = (x - 5)^2, \quad x \le 5$

65. $f(x) = \dfrac{1}{x^2}$

66. $h(x) = \dfrac{4}{x^2}$

67. $f(x) = \sqrt{2x + 3}$

68. $f(x) = \sqrt{x - 2}$

69. $f(x) = |x - 2|, \quad x \le 2$

70. $h(x) = |x + 1| - 1$

71. $f(x) = \dfrac{5x - 3}{2x + 5}$

72. $f(x) = \dfrac{x^2}{x^2 + 1}, \quad x \ge 0$

Finding and Analyzing Inverse Functions In Exercises 73–82, (a) find the inverse function of f, (b) graph both f and f^{-1} on the same set of coordinate axes, (c) describe the relationship between the graphs of f and f^{-1}, and (d) state the domains and ranges of f and f^{-1}.

73. $f(x) = 4x - 9$

74. $f(x) = 3x$

75. $f(x) = x^5 - 2$

76. $f(x) = x^3 + 1$

77. $f(x) = x^4, \quad x \le 0$

78. $f(x) = x^2, \quad x \ge 0$

79. $f(x) = \sqrt{4 - x^2}, \quad 0 \le x \le 2$

80. $f(x) = \sqrt{16 - x^2}, \quad -4 \le x \le 0$

81. $f(x) = \dfrac{4}{x^3}$

82. $f(x) = \dfrac{6}{\sqrt{x}}$

Restricting the Domain of a Function In Exercises 83–96, restrict the domain of the function f so that the function is one-to-one and has an inverse function. Then find the inverse function f^{-1}. State the domains and ranges of f and f^{-1}. Explain your results. (There are many correct answers.)

83. $f(x) = (x - 2)^2$ **84.** $f(x) = x^4 + 1$

85. $f(x) = |x + 2|$ **86.** $f(x) = |x - 2|$

87. $f(x) = (x - 1)^4$ **88.** $f(x) = (x + 3)^2$

89. $f(x) = (4x + 3)^2$ **90.** $f(x) = (2x - 5)^4$

91. $f(x) = -2x^2 - 5$ **92.** $f(x) = \frac{1}{2}x^2 + 1$

93. $f(x) = |3 - 3x|$ **94.** $f(x) = |2x|$

95. $f(x) = |x - 4| + 1$

96. $f(x) = -|x - 1| - 2$

Using the Properties of Inverse Functions In Exercises 97 and 98, use the graph of the function f to complete the table and sketch the graph of f^{-1}.

97.

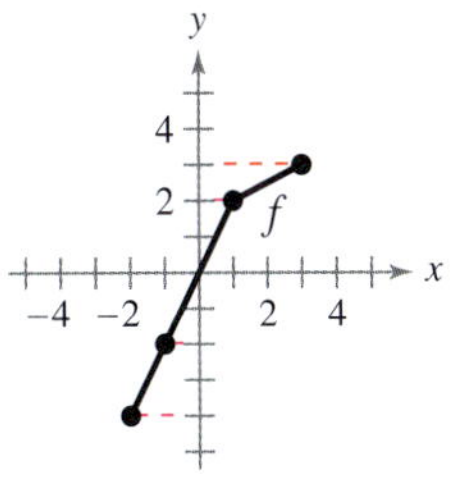

x	$f^{-1}(x)$
−4	
−2	
2	
3	

98.

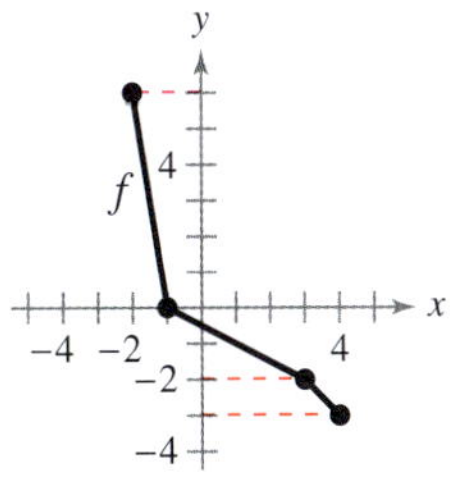

x	$f^{-1}(x)$
−3	
−2	
0	
6	

Using Graphs to Evaluate a Function In Exercises 99–106, use the graphs of $y = f(x)$ and $y = g(x)$ to evaluate the function.

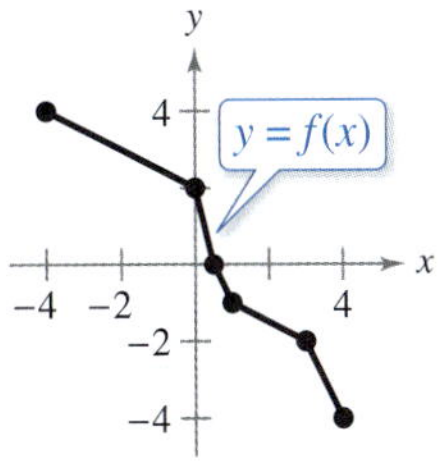

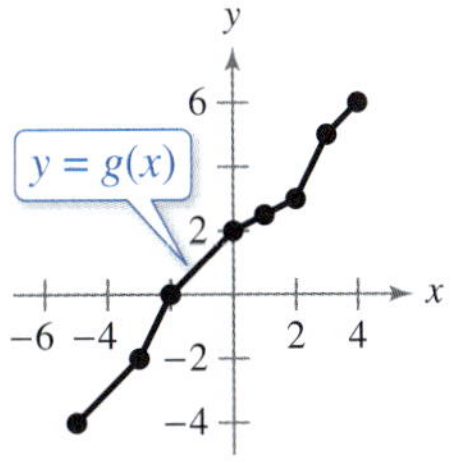

99. $f^{-1}(-4)$ **100.** $g^{-1}(0)$

101. $(f \circ g)(2)$ **102.** $g(f(-4))$

103. $f^{-1}(g(0))$ **104.** $(g^{-1} \circ f)(3)$

105. $(g \circ f^{-1})(2)$ **106.** $(f^{-1} \circ g^{-1})(6)$

Using the *Draw Inverse* Feature In Exercises 107–112, (a) use a graphing utility to graph the function f, (b) use the *draw inverse* feature of the graphing utility to draw the inverse relation of the function, and (c) determine whether the inverse relation is an inverse function. Explain your reasoning.

107. $f(x) = x^3 + x + 1$ **108.** $f(x) = x\sqrt{4 - x^2}$

109. $f(x) = \dfrac{3x^2}{x^2 + 1}$ **110.** $f(x) = \dfrac{4x}{\sqrt{x^2 + 15}}$

111. $f(x) = 2x^3 + 5x^2 + x - 2$

112. $f(x) = \dfrac{1}{x^3 + x^2 + x + 1}$

Evaluating a Composition of Functions In Exercises 113–118, use the functions $f(x) = \frac{1}{8}x - 3$ and $g(x) = x^3$ to find the indicated value or function.

113. $(f^{-1} \circ g^{-1})(1)$ **114.** $(g^{-1} \circ f^{-1})(-3)$

115. $(f^{-1} \circ f^{-1})(-6)$ **116.** $(g^{-1} \circ g^{-1})(4)$

117. $f^{-1} \circ g^{-1}$ **118.** $g^{-1} \circ f^{-1}$

Finding a Composition of Functions In Exercises 119–122, use the functions $f(x) = x + 4$ and $g(x) = 2x - 5$ to find the specified function.

119. $g^{-1} \circ f^{-1}$ **120.** $f^{-1} \circ g^{-1}$

121. $(f \circ g)^{-1}$ **122.** $(g \circ f)^{-1}$

123. ***Why you should learn it*** *(p. 60)* The table shows men's shoe sizes in the United States and the corresponding European shoe sizes. Let $y = f(x)$ represent the function that gives the men's European shoe size in terms of x, the men's U.S. size.

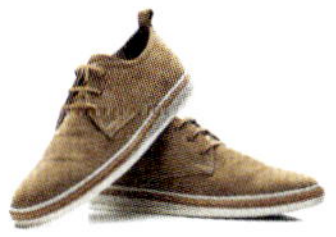

Men's U.S. shoe size	Men's European shoe size
8	41
9	42
10	43
11	44
12	45
13	46

(a) Is f one-to-one? Explain.

(b) Find $f(11)$.

(c) Find $f^{-1}(43)$, if possible.

(d) Find $f(f^{-1}(41))$.

(e) Find $f^{-1}(f(12))$.

124. Shoe Size Let $y = g(x)$ represent the function that gives the women's European shoe size in terms of x, the women's U.S. size. A women's U.S. size 6 shoe corresponds to a European size 37. Find $g^{-1}(g(6))$.

talevr/iStock/Getty Images

125. Encoding and Decoding Messages You can encode and decode messages using functions and their inverses. To code a message, first translate the letters to numbers using 1 for "A," 2 for "B," and so on. Use 0 for a space. So, "A ball" becomes

1 0 2 1 12 12.

Then, use a one-to-one function to convert to coded numbers. Using $f(x) = 2x - 1$, "A ball" becomes

1 −1 3 1 23 23.

(a) Encode "Call me later" using the function $f(x) = 5x + 4$.

(b) Find the inverse function of $f(x) = 5x + 4$ and use it to decode 119 44 9 104 4 104 49 69 29.

126. Hourly Wage Your wage is \$12.00 per hour plus \$0.55 for each unit produced per hour. So, your hourly wage y in terms of the number of units produced x is $y = 12 + 0.55x$.

(a) Find the inverse function. What does each variable in the inverse function represent?

(b) Use a graphing utility to graph the function and its inverse function.

(c) Use the *trace* feature of the graphing utility to find the hourly wage when 9 units are produced per hour.

(d) Use the *trace* feature of the graphing utility to find the number of units produced per hour when your hourly wage is \$21.35.

Focusing on Concepts

True or False? **In Exercises 127 and 128, determine whether the statement is true or false. Justify your answer.**

127. If f is an even function, then f^{-1} exists.

128. If the inverse function of f exists, and the graph of f has a y-intercept, then the y-intercept of f is an x-intercept of f^{-1}.

Think About It **In Exercises 129–132, determine whether the situation could be represented by a one-to-one function. If so, write a statement that describes the inverse function.**

129. The number of miles n a marathon runner has completed in terms of the time t in hours

130. The population p of a town in terms of the year t from 2008 through 2018 given that the population was greatest in 2015

131. The depth of the tide d at a beach in terms of the time t over a 24-hour period

132. The height h in inches of a human child from age 2 to age 14 in terms of his or her age n in years

133. Writing Describe the relationship between the graph of a function f and the graph of its inverse function f^{-1}.

134. Think About It The domain of a one-to-one function f is $[0, 9]$ and the range is $[-3, 3]$. Find the domain and range of f^{-1}.

135. Think About It The function $f(x) = \frac{9}{5}x + 32$ can be used to convert a temperature of x degrees Celsius to its corresponding temperature in degrees Fahrenheit.

(a) Using the expression for f, make a conceptual argument to show that f has an inverse function.

(b) What does $f^{-1}(50)$ represent?

136. Think About It Describe a type of function that is *not* one-to-one on any interval of its domain.

137. Error Analysis Describe the error.

The functions $f(x) = \sqrt[4]{x}$ and $g(x) = x^4$ are inverse functions of each other because

$$f(g(x)) = \sqrt[4]{x^4} = x \text{ and } g(f(x)) = \left(\sqrt[4]{x}\right)^4 = x.$$

138. HOW DO YOU SEE IT? Decide whether the two functions shown in each graph appear to be inverse functions of each other. Explain your reasoning.

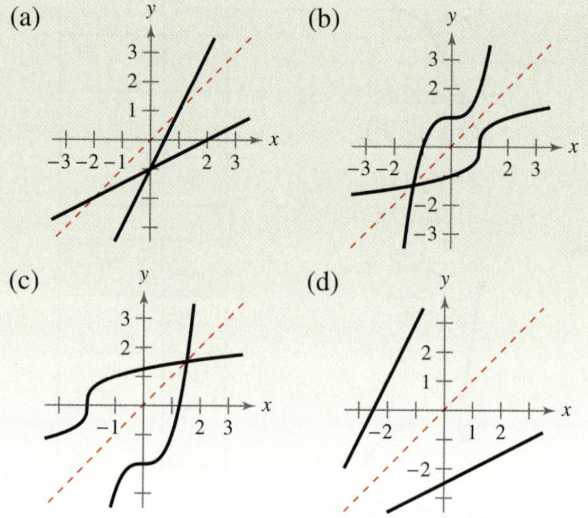

139. Proof Prove that if f and g are one-to-one functions, then $(f \circ g)^{-1}(x) = (g^{-1} \circ f^{-1})(x)$.

140. Proof Prove that if f is a one-to-one odd function, then f^{-1} is an odd function.

Cumulative Mixed Review

Simplifying a Rational Expression **In Exercises 141 and 142, write the rational expression in simplest form.**

141. $\dfrac{5x^2y^2 + 25x^2y}{xy + 5x}$

142. $\dfrac{x^2 + 3x - 40}{10 + 3x - x^2}$

Testing for Functions **In Exercises 143 and 144, determine whether the equation represents y as a function of x.**

143. $y - 7 = -3$

144. $x - y^2 = 0$

1.7 Linear Models and Scatter Plots

Scatter Plots and Correlation

Many real-life situations involve finding relationships between two variables, such as gross domestic product and the population of the United States. In a typical situation, data are collected and written as a set of ordered pairs. The graph of such a set is called a *scatter plot*. (For a brief discussion of scatter plots, see Appendix B.1.)

EXAMPLE 1 Constructing a Scatter Plot

The gross domestic products G (in trillions of dollars) and the populations P (in millions) of the United States from 2012 through 2017 are shown in the table. Construct a scatter plot of the data. (*Source:* U.S. Bureau of Economic Analysis, U.S. Census Bureau)

DATA

Year	Gross domestic product, G (in trillions of dollars)	Population, P (in millions)
2012	16.2	314.0
2013	16.7	316.2
2014	17.4	318.6
2015	18.1	321.0
2016	18.6	323.4
2017	19.4	325.7

Spreadsheet at LarsonPrecalculus.com

Solution

Begin by representing the data with a set of ordered pairs (G, P).

$(16.2, 314.0), (16.7, 316.2), (17.4, 318.6),$
$(18.1, 321.0), (18.6, 323.4), (19.4, 325.7)$

Then plot each point in a coordinate plane, as shown in the figure.

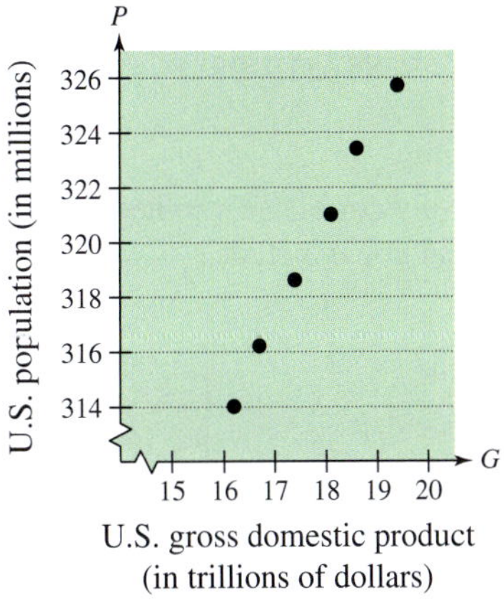

✓ **Checkpoint** *Audio-video solution in English & Spanish at LarsonPrecalculus.com*

The median sales prices (in thousands of dollars) of new homes sold in a neighborhood during the last eight years are given by the ordered pairs shown below. *(Spreadsheet at LarsonPrecalculus.com)* Construct a scatter plot of the data.

DATA (1, 179.4), (2, 185.4), (3, 191.0), (4, 196.7),
(5, 202.6), (6, 208.7), (7, 214.9), (8, 221.4)

From the scatter plot in Example 1, it appears that the points describe a relationship that is nearly linear. To have a relationship that is *exactly* linear, the gross domestic product would have to increase by the same amount each year (for example, by $1 trillion), and the population would have to increase by the same amount each year (for example, by 2 million people).

A mathematical equation that approximates the relationship between G and P is a *mathematical model*. When developing a mathematical model to describe a set of data, you strive for two (often conflicting) goals—accuracy and simplicity. For the data above, a linear model of the form $P = aG + b$, where a and b are constants, appears to be best. It is simple and relatively accurate.

What you should learn

- Construct scatter plots and interpret correlation.
- Use scatter plots and a graphing utility to find linear models for data.

Why you should learn it

Real-life data often follow a linear pattern. For instance, in Exercise 27 on page 79, you will find a linear model for the winning times in the women's 400-meter freestyle Olympic swimming event.

sam100/Shutterstock.com

Consider a collection of ordered pairs of the form (x, y). If y tends to increase as x increases, then the collection is said to have a **positive correlation.** If y tends to decrease as x increases, then the collection is said to have a **negative correlation.** Figure 1.46 shows three examples: one with a positive correlation, one with a negative correlation, and one with no (discernible) correlation.

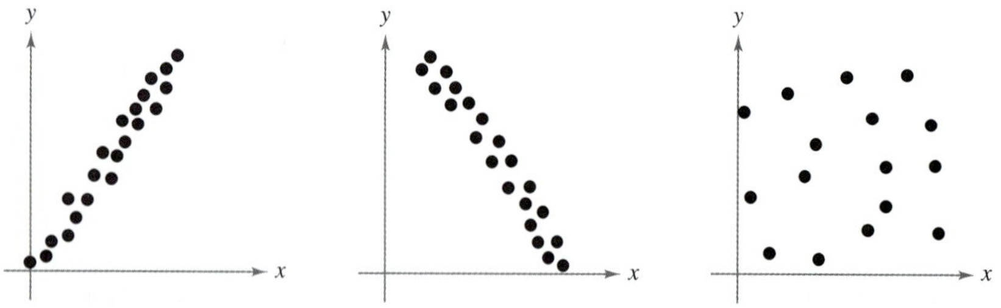

Positive Correlation *Negative Correlation* *No Correlation*

Figure 1.46

EXAMPLE 2 Interpreting Correlation

On a Friday, 22 students in a class were asked to record the numbers of hours they spent studying for a test on Monday. The results are shown below. The first coordinate is the number of hours and the second coordinate is the score obtained on the test. *(Spreadsheet at LarsonPrecalculus.com)*

DATA *Study Hours:* (0, 40), (1, 41), (2, 51), (3, 58), (3, 49), (4, 48), (4, 64), (5, 55), (5, 69), (5, 58), (5, 75), (6, 68), (6, 63), (6, 93), (7, 84), (7, 67), (8, 90), (8, 76), (9, 95), (9, 72), (9, 85), (10, 98)

a. Construct a scatter plot of the data.

b. Determine whether the points are positively correlated, are negatively correlated, or have no discernible correlation. What can you conclude?

Solution

a. The scatter plot is shown in the figure.

b. The scatter plot relating study hours and test scores has a positive correlation. This means that the more a student studied, the higher his or her score tended to be.

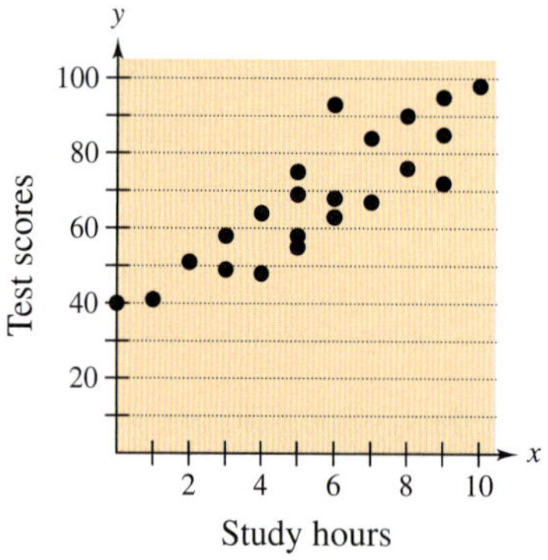

✓ **Checkpoint** *Audio-video solution in English & Spanish at LarsonPrecalculus.com*

The students in Example 2 also recorded the numbers of hours they spent watching television. The results are shown below. The first coordinate is the number of hours and the second coordinate is the score obtained on the test. *(Spreadsheet at LarsonPrecalculus.com)*

DATA *TV Hours:* (0, 98), (1, 85), (2, 72), (2, 90), (3, 67), (3, 93), (3, 95), (4, 68), (4, 84), (5, 76), (7, 75), (7, 58), (9, 63), (9, 69), (11, 55), (12, 58), (14, 64), (16, 48), (17, 51), (18, 41), (19, 49), (20, 40)

a. Construct a scatter plot of the data.

b. Determine whether the points are positively correlated, are negatively correlated, or have no discernible correlation. What can you conclude?

Fitting a Line to Data

Finding a linear model to represent the relationship described by a scatter plot is called **fitting a line to data.** You can do this graphically by simply sketching the line that appears to fit the points, finding two points on the line, and then finding the equation of the line that passes through the two points.

Insight

A standardized test may ask you to find a line of best fit from a scatter plot, as shown in Example 3. You may also be asked to use the line of best fit to make a prediction.

EXAMPLE 3 Fitting a Line to Data

See LarsonPrecalculus.com for an interactive version of this type of example.

Find a linear model that relates the gross domestic product to the population of the United States. (See Example 1.)

DATA

Year	Gross domestic product, G (in trillions of dollars)	Population, P (in millions)
2012	16.2	314.0
2013	16.7	316.2
2014	17.4	318.6
2015	18.1	321.0
2016	18.6	323.4
2017	19.4	325.7

Spreadsheet at LarsonPrecalculus.com

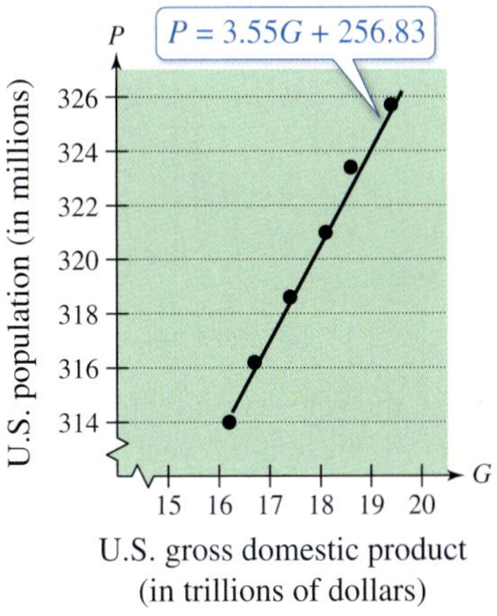

Solution

After plotting the data in the table, draw the line that you think best represents the data, as shown in the figure. Two points that lie on this line are (17.4, 318.6) and (19.4, 325.7). Using the point-slope form, you can find the equation of the line to be

$$P = 3.55(G - 17.4) + 318.6$$

$$= 3.55G + 256.83. \qquad \text{Linear model}$$

Checkpoint Audio-video solution in English & Spanish at LarsonPrecalculus.com

Find a linear model that relates the year to the median sales prices (in thousands of dollars) of new homes sold in a neighborhood during the last eight years given by the ordered pairs shown below. (See Checkpoint for Example 1.) *(Spreadsheet at LarsonPrecalculus.com)*

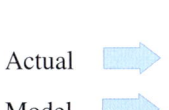

DATA (1, 179.4), (2, 185.4), (3, 191.0), (4, 196.7), (5, 202.6), (6, 208.7), (7, 214.9), (8, 221.4)

After finding a model, you can measure how well the model fits the data by comparing the actual values with the values given by the model, as shown in the table.

	G	16.2	16.7	17.4	18.1	18.6	19.4
Actual	P	314.0	316.2	318.6	321.0	323.4	325.7
Model	P	314.34	316.115	318.6	321.085	322.86	325.7

The sum of the squares of the differences between the actual values and the model values is the **sum of the squared differences.** The model that has the least sum is the **least squares regression line** for the data. For the model in Example 3, the sum of the squared differences is about 0.42. The least squares regression line for the data is

$$P = 3.67G + 254.7. \qquad \text{Best-fitting linear model}$$

Its sum of squared differences is about 0.32. For more on the least squares regression line, see Appendix C.2 at this textbook's *Student Companion Website.*

Algebra Help

The model in Example 3 is based on the two data points chosen. When different points are chosen, the model may change somewhat. For instance, when you choose (16.2, 314.0) and (16.7, 316.2), the new model is

$$P = 4.4(G - 16.2) + 314.0$$

$$= 4.4G + 242.72.$$

Another way to find a linear model to represent the relationship described by a scatter plot is to enter the data points into a graphing utility and use the *linear regression* feature. This method is demonstrated in Example 4.

> **Technology Tip**
>
> For instructions on how to use the *linear regression* feature, see Appendix A; for specific keystrokes, go to this textbook's *Student Companion Website.*

EXAMPLE 4 A Mathematical Model

The table shows the numbers N (in millions) of people employed in the United States from 2010 through 2017. (*Source:* U.S. Bureau of Labor Statistics)

DATA

Year	Number of people employed in the U.S., N (in millions)
2010	139.064
2011	139.869
2012	142.469
2013	143.929
2014	146.305
2015	148.834
2016	151.436
2017	153.337

Spreadsheet at LarsonPrecalculus.com

a. Use the *regression* feature of a graphing utility to find a linear model for the data. Let t represent the year, with $t = 0$ corresponding to 2010.

b. How closely does the model represent the data?

Graphical Solution

a. Use the *linear regression* feature of a graphing utility to obtain the model shown in the figure. You can approximate the model to be $N = 2.1335t + 138.188$.

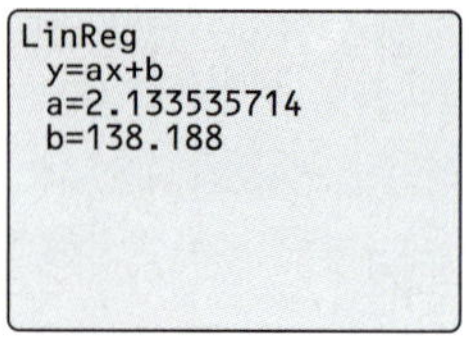

b. Graph the actual data and the model. From the figure, it appears that the model is a good fit for the actual data.

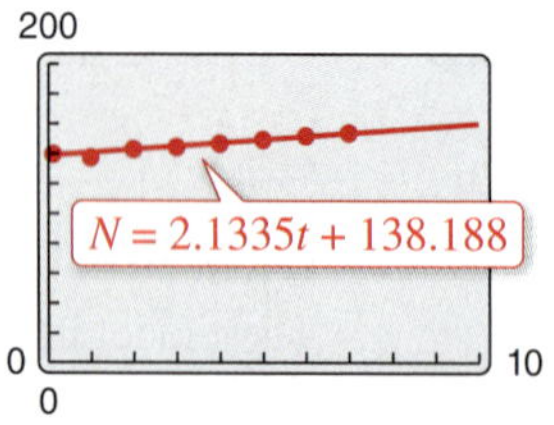

Numerical Solution

a. Using the *linear regression* feature of a graphing utility, you can find that a linear model for the data is $N = 2.1335t + 138.188$.

b. You can see how well the model fits the data by comparing the actual values of N with the values of N given by the model, which are labeled N^* in the table below. From the table, you can see that the model appears to be a good fit for the actual data.

Year	N	N^*
2010	139.064	138.188
2011	139.869	140.322
2012	142.469	142.455
2013	143.929	144.589
2014	146.305	146.722
2015	148.834	148.856
2016	151.436	150.989
2017	153.337	153.123

✓ ***Checkpoint*** *Audio-video solution in English & Spanish at LarsonPrecalculus.com*

The numbers N of alternative fuel stations in the United States from 2010 through 2016 are given by the following ordered pairs. *(Spreadsheet at LarsonPrecalculus.com)* (*Source:* Alternative Fuels Data Center)

DATA (2010, 6,912), (2011, 10,071), (2012, 20,498), (2013, 27,159), (2014, 33,805), (2015, 39,963), (2016, 51,382)

a. Use the *regression* feature of a graphing utility to find a linear model for the data. Let t represent the year, with $t = 0$ corresponding to 2010.

b. How closely does the model represent the data?

When you use the *regression* feature of a graphing utility to find a linear model for data, you will notice that the program may also output an "r-value." For instance, the r-value from Example 4 was $r \approx 0.995$. This r-value is the **correlation coefficient** of the data and gives a measure of how well the model fits the data. The correlation coefficient r varies between -1 and 1. Basically, the closer $|r|$ is to 1, the better the points can be described by a line. Three examples are shown in Figure 1.47.

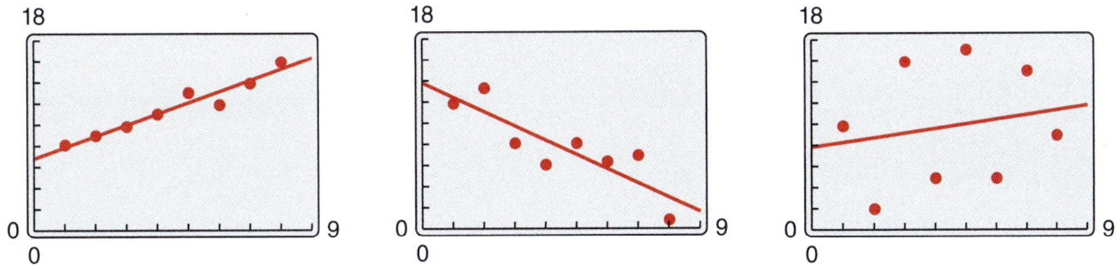

$r = 0.972$ $r = -0.856$ $r = 0.190$

Figure 1.47

> **Technology Tip**
>
> For some graphing utilities, the *diagnostics on* feature must be selected before the *regression* feature is used in order to see the value of the correlation coefficient r. To learn how to use this feature, consult your user's manual.

EXAMPLE 5 Finding a Least Squares Regression Line

The ordered pairs (w, h) represent the shoe sizes w and the heights h (in inches) of 25 men. Use the *regression* feature of a graphing utility to find the least squares regression line for the data.

(10.0, 70.5)	(10.5, 71.0)	(9.5, 69.0)	(11.0, 72.0)	(12.0, 74.0)
(8.5, 67.0)	(9.0, 68.5)	(13.0, 76.0)	(10.5, 71.5)	(10.5, 70.5)
(10.0, 71.0)	(9.5, 70.0)	(10.0, 71.0)	(10.5, 71.0)	(11.0, 71.5)
(12.0, 73.5)	(12.5, 75.0)	(11.0, 72.0)	(9.0, 68.0)	(10.0, 70.0)
(13.0, 75.5)	(10.5, 72.0)	(10.5, 71.0)	(11.0, 73.0)	(8.5, 67.5)

Solution

After entering the data into a graphing utility (see Figure 1.48), you obtain the model shown in Figure 1.49. So, the least squares regression line for the data is

$$h = 1.84w + 51.9.$$

In Figure 1.50, this line is plotted with the data. Note that the plot does not have 25 points because some of the ordered pairs graph as the same point. The correlation coefficient for this model is $r \approx 0.981$, which implies that the model is a good fit for the data.

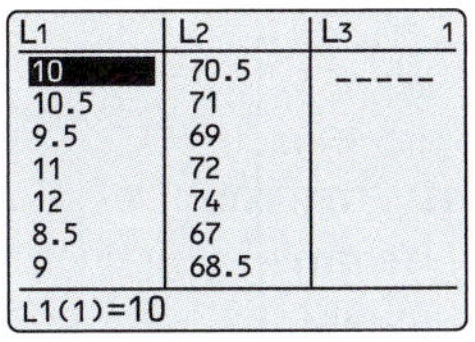

Figure 1.48

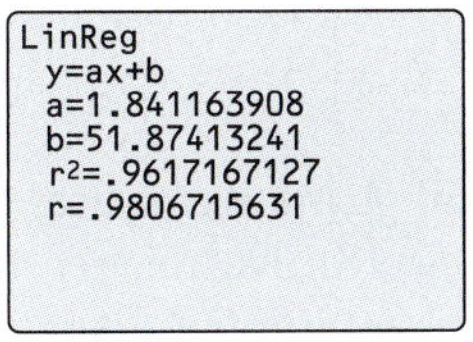

Figure 1.49

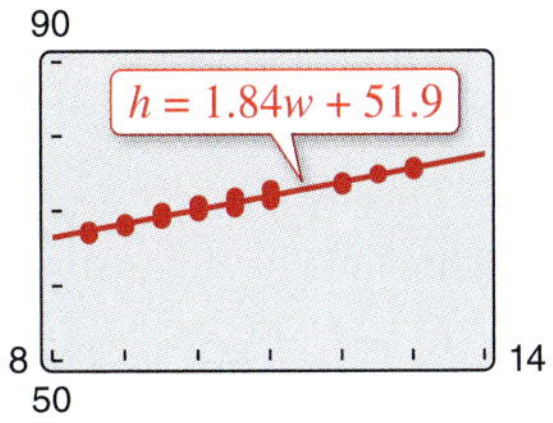

Figure 1.50

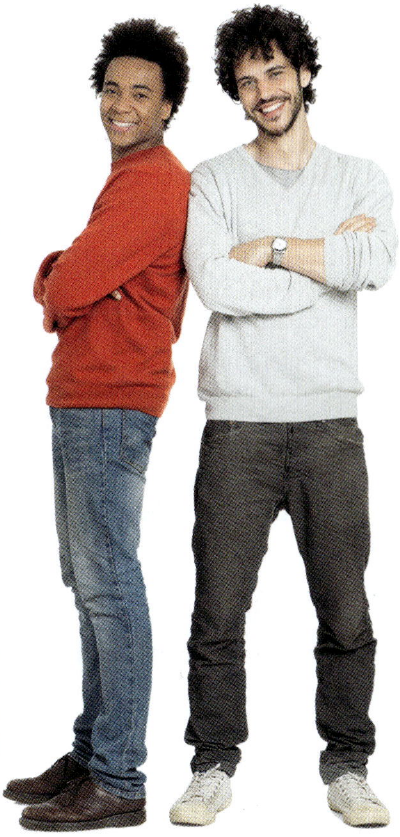

✓ Checkpoint ▶ *Audio-video solution in English & Spanish at LarsonPrecalculus.com*

The ordered pairs (g, e) represent the gross domestic products g (in millions of U.S. dollars) and the carbon dioxide emissions e (in millions of metric tons) for seven countries. *(Spreadsheet at LarsonPrecalculus.com)* Use the *regression* feature of a graphing utility to find the least squares regression line for the data. (*Sources:* World Bank and U.S. Energy Information Administration)

DATA (14.99, 5491), (5.87, 1181), (3.6, 748), (2.77, 374), (2.48, 475), (2.45, 497), (2.19, 401)

1.7 Exercises

See *CalcChat.com* for tutorial help and worked-out solutions to odd-numbered exercises. For instructions on how to use a graphing utility, see Appendix A.

Vocabulary and Concept Check

In Exercises 1 and 2, fill in the blank.

1. Consider a collection of ordered pairs of the form (x, y). If y tends to increase as x increases, then the collection is said to have a _______ correlation.

2. To find the least squares regression line for data, you can use the _______ feature of a graphing utility.

3. In a collection of ordered pairs (x, y), y tends to decrease as x increases. Does the collection have a positive correlation or a negative correlation?

4. You find the least squares regression line for a set of data. The correlation coefficient is 0.114. Is the model a good fit?

Procedures and Problem Solving

5. Constructing a Scatter Plot The ordered pairs below give the years of experience x for 15 sales representatives and the monthly sales y (in thousands of dollars).

(1.5, 41.7), (1.0, 32.4), (0.3, 19.2), (3.0, 48.4), (4.0, 51.2), (0.5, 28.5), (2.5, 50.4), (1.8, 35.5), (2.0, 36.0), (1.5, 40.0), (3.5, 50.3), (4.0, 55.2), (0.5, 29.1), (2.2, 43.2), (2.0, 41.6)

(a) Construct a scatter plot of the data.

(b) Determine whether the points are positively correlated, are negatively correlated, or have no discernible correlation. What can you conclude?

6. Constructing a Scatter Plot The ordered pairs below give the scores on two consecutive 15-point quizzes for a class of 18 students.

(7, 13), (9, 7), (14, 14), (15, 15), (10, 15), (9, 7), (14, 11), (14, 15), (8, 10), (9, 10), (15, 9), (10, 11), (11, 14), (7, 14), (11, 10), (14, 11), (10, 15), (9, 6)

(a) Construct a scatter plot of the data.

(b) Determine whether the points are positively correlated, are negatively correlated, or have no discernible correlation. What can you conclude?

Fitting a Line to Data **In Exercises 7–10, sketch the line that you think best approximates the data in the scatter plot. Then find an equation of the line. To print an enlarged copy of the graph, go to *MathGraphs.com*.**

7.

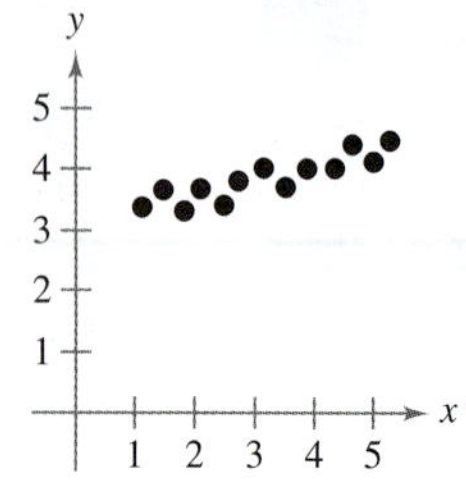

8.

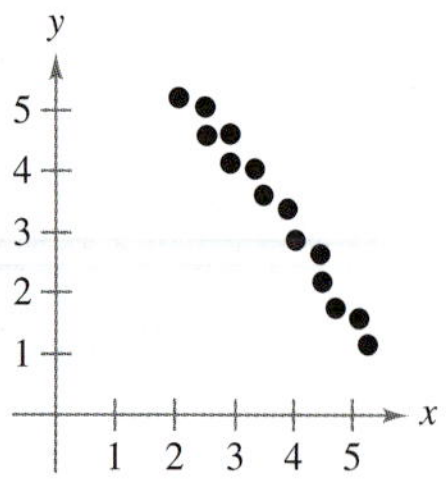

9.

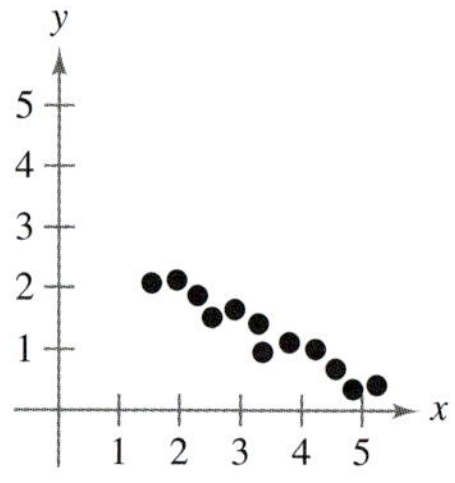

10.

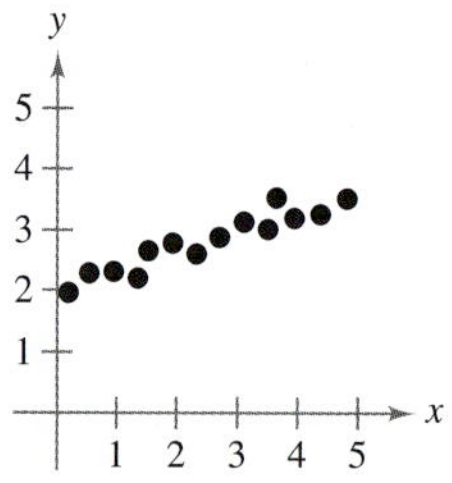

Fitting a Line to Data **In Exercises 11–16, (a) construct a scatter plot of the data, (b) draw a line of fit that passes through two of the points, and (c) use the two points to find an equation of the line.**

11. $(-3, -3), (3, 4), (1, 1), (3, 2), (4, 4), (-1, -1)$

12. $(-2, 3), (-2, 4), (-1, 2), (1, -2), (0, 0), (0, 1)$

13. $(0, 2), (-2, 1), (3, 3), (1, 3), (4, 4)$

14. $(3, 2), (2, 3), (1, 5), (4, 0), (5, 0)$

15. $(0, 7), (3, 2), (6, 0), (4, 3), (2, 5)$

16. $(3, 4), (2, 2), (5, 6), (1, 1), (0, 2)$

A Mathematical Model **In Exercises 17 and 18, use the *regression* feature of a graphing utility to find a linear model for the data. Then use the graphing utility to decide how closely the model fits the data (a) graphically and (b) numerically. To print an enlarged copy of the graph, go to *MathGraphs.com*.**

17.

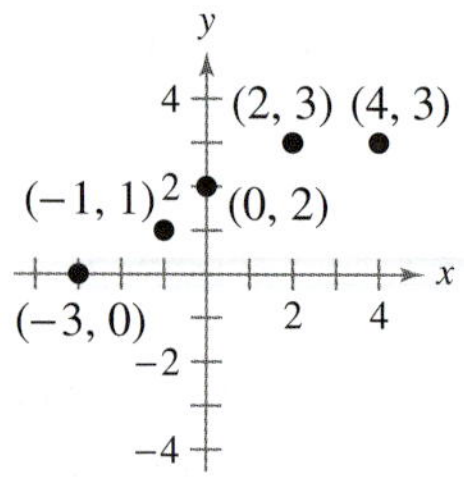

18.

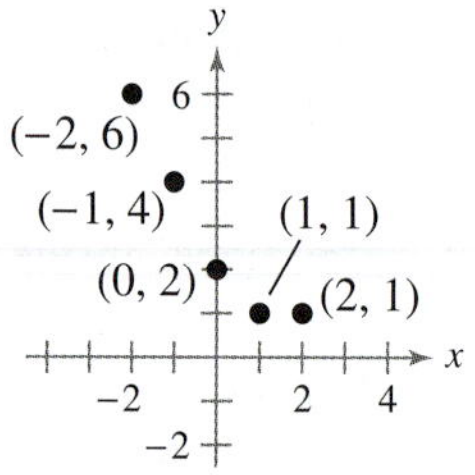

19. MODELING DATA

Hooke's Law states that the force F required to compress or stretch a spring (within its elastic limits) is proportional to the distance d that the spring is compressed or stretched from its original length. That is, $F = kd$, where k is the measure of the stiffness of the spring and is called the *spring constant*. The table shows the elongation d in centimeters of a spring when a force of F kilograms is applied.

Force, F (in kilograms)	Elongation, d (in centimeters)
20	1.4
40	2.5
60	4.0
80	5.3
100	6.6

(a) Construct a scatter plot of the data.

(b) Find the equation of the line that seems to best fit the data.

(c) Use the *regression* feature of a graphing utility to find a linear model for the data.

(d) Use the model from part (c) to estimate the elongation of the spring when a force of 55 kilograms is applied.

20. MODELING DATA

The table shows the numbers of subscribers S (in millions) to wireless networks from 2010 through 2016. (*Source:* CTIA–The Wireless Association)

DATA

Year	Subscribers, S (in millions)
2010	296.3
2011	316.0
2012	326.5
2013	335.7
2014	355.4
2015	377.9
2016	395.9

Spreadsheet at LarsonPrecalculus.com

(a) Use a graphing utility to construct a scatter plot of the data, with $t = 10$ corresponding to 2010.

(b) Use the *regression* feature of the graphing utility to find a linear model for the data.

(c) Use the graphing utility to plot the data and graph the model in the same viewing window. Is the model a good fit? Explain.

(d) Use the model to predict the number of subscribers in 2022. Is your answer reasonable? Explain.

21. MODELING DATA

The table shows the total enterprise values V (in billions of dollars) for the Golden State Warriors from 2013 through 2018. (*Source:* Forbes)

DATA

Year	Total enterprise value, V (in billions of dollars)
2013	0.555
2014	0.750
2015	1.300
2016	1.900
2017	2.600
2018	3.100

Spreadsheet at LarsonPrecalculus.com

(a) Use a graphing utility to construct a scatter plot of the data, with $t = 13$ corresponding to 2013.

(b) Use the *regression* feature of the graphing utility to find a linear model for the data.

(c) Use the graphing utility to plot the data and graph the model in the same viewing window. Is the model a good fit? Explain.

(d) Use the model to predict the values in 2021 and 2026. Do the results seem reasonable? Explain.

(e) What is the slope of your model? What does it tell you about the enterprise value?

22. MODELING DATA

The table shows the average salaries S (in thousands of dollars) of teachers in the United States from 2012 through 2017. (*Source:* National Education Association)

DATA

Year	Average salary, S (in thousands of dollars)
2012	55.5
2013	56.2
2014	56.8
2015	57.6
2016	58.4
2017	59.0

Spreadsheet at LarsonPrecalculus.com

(a) Use a graphing utility to construct a scatter plot of the data, with $t = 12$ corresponding to 2012.

(b) Use the *regression* feature of the graphing utility to find a linear model for the data.

(c) Use the graphing utility to plot the data and graph the model in the same viewing window. Is the model a good fit? Explain.

(d) Use the model to predict the average salaries in 2021 and 2026. Do the results seem reasonable? Explain.

23. MODELING DATA

The table shows the populations P (in millions) of Nebraska from 2013 through 2017. *(Source: U.S. Census Bureau)*

Year	Population, P (in millions)
2013	1.867
2014	1.881
2015	1.894
2016	1.908
2017	1.920

(a) Use a graphing utility to construct a scatter plot of the data, with $t = 13$ corresponding to 2013.

(b) Use the *regression* feature of the graphing utility to find a linear model for the data.

(c) Use the graphing utility to plot the data and graph the model in the same viewing window.

(d) Create a table showing the actual values of P and the values of P obtained from the model. How closely does the model fit the data?

(e) Use the model to predict the population of Nebraska in 2021. Does the result seem reasonable? Explain.

24. MODELING DATA

The table shows the numbers N of Costco Wholesale Corporation stores from 2011 through 2017. *(Source: Costco Wholesale Corporation)*

DATA

Year	Number of stores, N
2011	592
2012	608
2013	634
2014	663
2015	686
2016	715
2017	741

Spreadsheet at LarsonPrecalculus.com

(a) Use a graphing utility to construct a scatter plot of the data, with $t = 11$ corresponding to 2011.

(b) Use the *regression* feature of the graphing utility to find a linear model for the data.

(c) Use the graphing utility to plot the data and graph the model in the same viewing window.

(d) Create a table showing the actual values of N and the values of N obtained from the model. How closely does the model fit the data?

(e) Use the model to predict the number of stores in 2021. Does the result seem reasonable? Explain.

25. MODELING DATA

The table shows the advertising expenditures x and sales volumes y for a company for nine randomly selected months. Both are measured in thousands of dollars.

DATA

Month	Advertising, x (in thousands of dollars)	Sales, y (in thousands of dollars)
1	2.4	215
2	1.6	182
3	1.4	172
4	2.3	214
5	2.0	201
6	2.6	230
7	1.4	176
8	1.6	177

Spreadsheet at LarsonPrecalculus.com

(a) Use the *regression* feature of a graphing utility to find a linear model for the data.

(b) Use the graphing utility to plot the data and graph the model in the same viewing window.

(c) Interpret the slope of the model in the context of the problem.

(d) Use the model to estimate sales for advertising expenditures of \$1500.

26. MODELING DATA

The table shows the sales S (in billions of dollars) of the Campbell Soup Company from 2012 through 2017. *(Source: Campbell Soup Company)*

DATA

Year	Sales, S (in billions of dollars)
2012	7.707
2013	8.052
2014	8.268
2015	8.082
2016	7.961
2017	7.890

Spreadsheet at LarsonPrecalculus.com

(a) Use a graphing utility to construct a scatter plot of the data, with $t = 12$ corresponding to 2012. Identify two sets of points in the scatter plot that are approximately linear.

(b) Use the *regression* feature of the graphing utility to find a linear model for each set of points.

(c) Write a piecewise-defined model for the data. Use the graphing utility to graph the piecewise-defined model.

(d) Describe a scenario that could be the reason that a piecewise-defined function can be used to model the data.

27. ***Why you should learn it*** *(p. 71)* The ordered pairs (t, T) below represent the Olympic year t and the winning time T (in minutes) in the women's 400-meter freestyle swimming event. *(Spreadsheet at LarsonPrecalculus.com)* (*Source:* International Olympic Committee)

DATA

(1956, 4.91)	(1988, 4.06)
(1960, 4.84)	(1992, 4.12)
(1964, 4.72)	(1996, 4.12)
(1968, 4.53)	(2000, 4.10)
(1972, 4.32)	(2004, 4.09)
(1976, 4.16)	(2008, 4.05)
(1980, 4.15)	(2012, 4.02)
(1984, 4.12)	(2016, 3.94)

(a) Use the *regression* feature of a graphing utility to find a linear model for the data and to identify the correlation coefficient. Let t represent the year, with $t = 0$ corresponding to 1950.

(b) Explain what the sign of the slope of the model means in the context of the problem.

(c) Use the graphing utility to plot the data and graph the model in the same viewing window.

(d) Create a table showing the actual values of T and the values of T obtained from the model. How closely does the model fit the data?

(e) How can you use the value of the correlation coefficient to help answer the question in part (d)?

(f) Would you use the model to predict the winning times in the future? Explain.

28. MODELING DATA

In a study, 60 colts were measured every 14 days from birth. The ordered pairs (d, l) represent the average length l (in centimeters) of the 60 colts d days after birth. *(Spreadsheet at LarsonPrecalculus.com)* (*Source:* American Society of Animal Science)

DATA

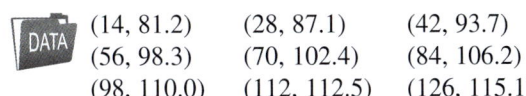

(14, 81.2)	(28, 87.1)	(42, 93.7)
(56, 98.3)	(70, 102.4)	(84, 106.2)
(98, 110.0)	(112, 112.5)	(126, 115.1)

(a) Use the *regression* feature of a graphing utility to find a linear model for the data and to identify the correlation coefficient.

(b) According to the correlation coefficient, does the model represent the data well? Explain.

(c) Use the graphing utility to plot the data and graph the model in the same viewing window. How closely does the model fit the data?

(d) Use the model to predict the average length of a colt 140 days after birth.

Focusing on Concepts

True or False? **In Exercises 29 and 30, determine whether the statement is true or false. Justify your answer.**

29. A linear regression model with a positive correlation will have a slope that is greater than 0.

30. When the correlation coefficient for a linear regression model is close to -1, the regression line is a poor fit for the data.

31. Error Analysis Describe each error in interpreting the graphing calculator display.

(a) An equation of the line of best fit is
$y = 9.18x - 1.02.$

(b) The data have strong positive correlation.

```
LinReg
 y=ax+b
 a=-1.017151515
 b=9.175333333
 r²=.9972867142
 r=.9986424356
```

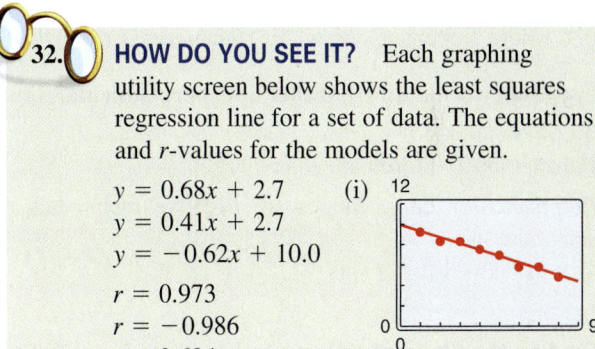

32. HOW DO YOU SEE IT? Each graphing utility screen below shows the least squares regression line for a set of data. The equations and r-values for the models are given.

$y = 0.68x + 2.7$
$y = 0.41x + 2.7$
$y = -0.62x + 10.0$
$r = 0.973$
$r = -0.986$
$r = 0.624$

(i)
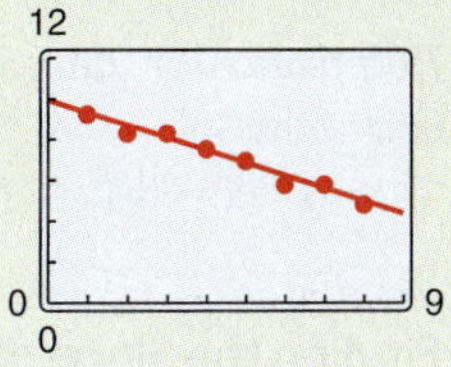

(ii)
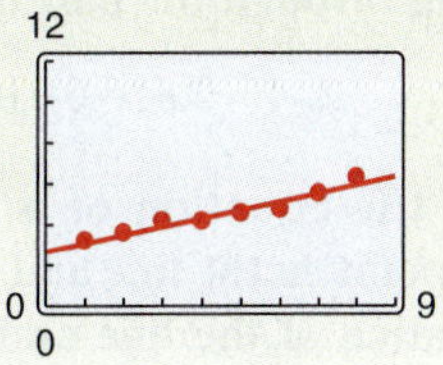

(iii)
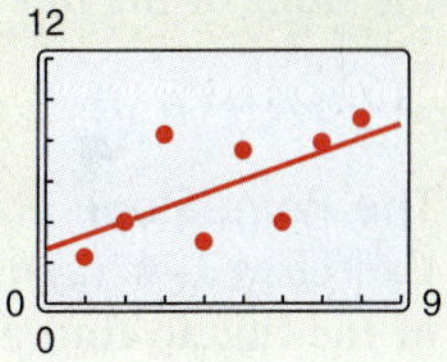

(a) Determine the equation and correlation coefficient (r-value) that represent each graph. Explain how you found your answers.

(b) According to the correlation coefficients, which model is the best fit for its data? Explain.

Cumulative Mixed Review

Evaluating a Function **In Exercises 33 and 34, evaluate the function at each value of the independent variable and simplify.**

33. $f(x) = 2x^2 - 3x + 5$
(a) $f(-1)$
(b) $f(5)$
(c) $f(w + 2)$

34. $g(x) = 5x^2 - 6x + 1$
(a) $g(-2)$
(b) $g(z - 2)$
(c) $g(1 - z)$

sam100/Shutterstock.com

1 Chapter Review

See CalcChat.com for tutorial help and worked-out solutions to odd-numbered exercises.
For instructions on how to use a graphing utility, see Appendix A.

1.1 *What did you learn?*

Find the slopes of lines *(p. 3)*.

The slope m of the nonvertical line through (x_1, y_1) and (x_2, y_2) is

$$m = \frac{y_2 - y_1}{x_2 - x_1}, \quad x_1 \neq x_2.$$

Write linear equations given points on lines and their slopes *(p. 5)*. The point-slope form of the equation of the line that passes through the point (x_1, y_1) and has a slope of m is

$$y - y_1 = m(x - x_1).$$

Use slope-intercept forms of linear equations to sketch lines *(p. 7)*. The graph of the equation $y = mx + b$ is a line whose slope is m and whose y-intercept is $(0, b)$.

Use slope to identify parallel and perpendicular lines *(p. 9)*.

Parallel lines: Slopes are equal.

Perpendicular lines: Slopes are negative reciprocals of each other.

Example To find an equation of the line passing through the points $(2, 0)$ and $(3, -1)$, begin by finding the slope of the line passing through the points.

$$m = \frac{-1 - 0}{3 - 2} = -1$$

In slope-intercept form, the equation of the line is

$$y - 0 = -1(x - 2) \quad \Rightarrow \quad y = -x + 2.$$

Now you can use the slope and y-intercept to sketch the line.

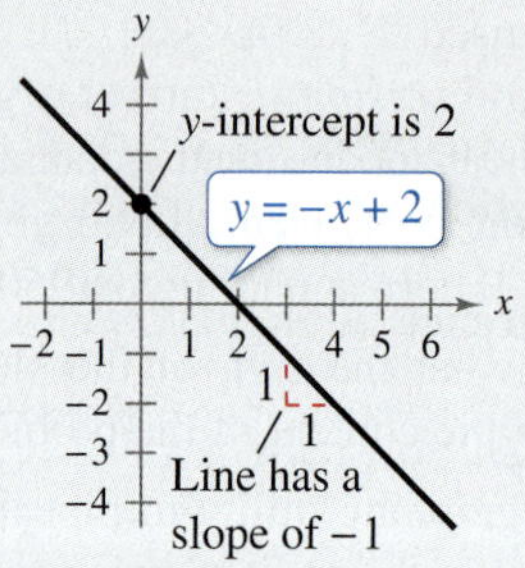

An equation of a line parallel to $y = -x + 2$ is $y = -x - 1$. An equation of a line perpendicular to $y = -x + 2$ is $y = x - 1$.

Finding the Slope of a Line **In Exercises 1 and 2, find the slope of the line passing through the pair of points.**

1. $\left(\frac{3}{2}, 1\right), \left(5, \frac{5}{2}\right)$ **2.** $(-2.4, 6.6), (0, -0.6)$

The Point-Slope Form of the Equation of a Line **In Exercises 3–6, (a) use the point on the line and the slope of the line to find an equation of the line and (b) find three additional points through which the line passes. (There are many correct answers.)**

3. $(-3, 5), m = -\frac{3}{2}$ **4.** $(0, 1), m = \frac{4}{5}$

5. $(10, -6)$, m is undefined. **6.** $(-2, 6), m = 0$

Finding the Slope-Intercept Form **In Exercises 7–10, write an equation of the line that passes through the points. Use the slope-intercept form (if possible). If not possible, explain why and use the general form. Use a graphing utility to graph the line (if possible).**

7. $(2, -1), (4, -1)$ **8.** $\left(\frac{5}{6}, -1\right), \left(\frac{5}{6}, 3\right)$

9. $(-1, 0), (6, 2)$ **10.** $(3, -1), (-3, 2)$

Equations of Parallel and Perpendicular Lines **In Exercises 11 and 12, write the slope-intercept forms of the equations of the lines through the given point (a) parallel and (b) perpendicular to the given line. Verify your results with a graphing utility.**

11. $(3, -2), 5x - 4y = 8$ **12.** $(-8, 3), 2x + 3y = 5$

Using a Rate of Change to Write an Equation **In Exercises 13 and 14, you are given the dollar value of a product in 2018 and the rate at which the value of the product is expected to change during the next 5 years. Write a linear equation that gives the dollar value V of the product in terms of the year t. (Let $t = 18$ represent 2018.)**

13. Value: \$15,800, rate: \$1580 increase per year

14. Value: \$5420, rate: \$725 decrease per year

15. Business During the second and third quarters of the year, an e-commerce business had sales of \$170,000 and \$195,000, respectively. The growth of sales follows a linear pattern. Estimate sales during the fourth quarter.

16. Accounting The dollar value of a game console in 2018 is \$299. The product will decrease in value at an expected rate of \$59.80 per year.

(a) Write a linear equation that gives the dollar value V of the game console in terms of the year t. (Let $t = 0$ represent 2018.)

(b) Use a graphing utility to graph the equation found in part (a). Be sure to choose an appropriate viewing window.

(c) According to the model, when will the game console have no value?

1.2 *What did you learn?*

Determine whether a relation between two variables represents a function *(p. 16)*.

A function f from a set A to a set B is a relation that assigns to each element x in the set A exactly one element y in the set B. The set A is the domain (or set of inputs) of the function f, and the set B contains the range (or set of outputs).

Example The path of a baseball is modeled by

$$f(x) = -0.003x^2 + 0.738x + 3$$

where $f(x)$ is the height of the baseball (in feet) and x is the horizontal distance from home plate (in feet).

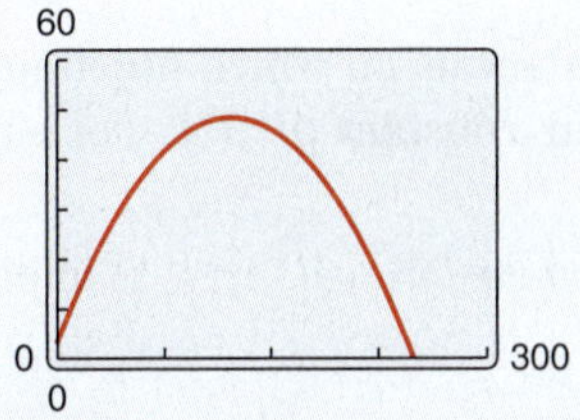

This represents y as a function of x because each value of x corresponds to exactly one value of $y = f(x)$.

$$f(0) = -0.003(0)^2 + 0.738(0) + 3 = 3$$

$$f(250) = -0.003(250)^2 + 0.738(250) + 3 = 0$$

Distance and height cannot be negative. So, the domain of f is $[0, 250]$.

Use function notation and evaluate functions *(p. 18)*. The symbol $f(x)$ is read as the *value of f at x* or *f of x*.

Find the domains of functions *(p. 20)*. The domain of a function can be described explicitly or it can be implied by the expression used to define the function.

Use functions to model and solve real-life problems *(p. 22)*. A function can model the path of a baseball.

Evaluate difference quotients *(p. 23)*.

Different quotient: the ratio $\dfrac{f(x+h) - f(x)}{h}, h \neq 0$

Testing for Functions **In Exercises 17 and 18, which set of ordered pairs represents a function from A to B? Explain.**

17. $A = \{10, 20, 30, 40\}$ and $B = \{0, 2, 4, 6\}$
(a) $\{(20, 4), (40, 0), (20, 6), (30, 2)\}$
(b) $\{(10, 4), (20, 4), (30, 4), (40, 4)\}$

18. $A = \{u, v, w\}$ and $B = \{-2, -1, 0, 1, 2\}$
(a) $\{(u, -2), (v, 2), (w, 1)\}$
(b) $\{(w, -2), (v, 0), (w, 2)\}$

Testing for Functions Represented Algebraically **In Exercises 19–22, determine whether the equation represents y as a function of x.**

19. $16x^2 - y^2 = 0$ **20.** $3x - y - 2 = 0$

21. $y = \sqrt{1 - x}$ **22.** $y = \sqrt{x^2 + 9}$

Evaluating a Function **In Exercises 23–26, find each function value, if possible.**

23. $f(x) = x^2 + 1$
(a) $f(-3)$ (b) $f(b^3)$ (c) $f(x - 1)$

24. $g(x) = \sqrt{x^2 + 1}$
(a) $g(3)$ (b) $g(3x)$ (c) $g(x + 2)$

25. $h(x) = \begin{cases} 2x + 1, & x \leq -1 \\ x^2 + 2, & x > -1 \end{cases}$
(a) $h(-1)$ (b) $h(0)$ (c) $h(2)$

26. $f(x) = \dfrac{5}{3x - 2}$
(a) $f(1)$ (b) $f(-2)$ (c) $f(t)$

Finding the Domain of a Function **In Exercises 27 and 28, find the domain of the function.**

27. $f(x) = \dfrac{x - 1}{x + 2}$ **28.** $f(x) = \sqrt{x^2 - 16}$

29. Industrial Engineering A hand tool manufacturer produces a product for which the variable cost is \$5.25 per unit and the fixed costs are \$17,500. The company sells the product for \$8.43 and can sell all that it produces. Write the total cost C and the profit P as functions of x, the number of units produced.

30. Education The numbers n (in millions) of students enrolled in public schools in the United States from 2009 through 2016 can be approximated by

$$n(t) = \begin{cases} -0.35t^2 + 7.7t + 26, & 9 \leq t \leq 12 \\ -0.187t^3 + 8.22t^2 - 120.4t + 654, & 13 \leq t \leq 16 \end{cases}$$

where t is the year, with $t = 9$ corresponding to 2009. (*Source:* U.S. Census Bureau)

(a) Use the *table* feature of a graphing utility to approximate the enrollment from 2009 through 2016.

(b) Do you think the piecewise-defined function could be used to predict the enrollment for years outside of the domain? Explain your reasoning.

Evaluating a Difference Quotient **In Exercises 31 and 32, find the difference quotient $\dfrac{f(x+h) - f(x)}{h}$ for the function and simplify your answer.**

31. $f(x) = 2x^2 + 3x - 1$ **32.** $f(x) = x^2 - 3x + 5$

1.3 *What did you learn?*

Find the domains and ranges of functions *(p. 29)* and use the Vertical Line Test for functions *(p. 30)*.

Vertical Line Test: A set of points in a coordinate plane is the graph of y as a function of x if and only if no vertical line intersects the graph at more than one point.

Determine intervals on which functions are increasing, decreasing, or constant *(p. 31)*. For any x_1 and x_2 in an interval,

f increasing on the interval: $x_1 < x_2$ implies $f(x_1) < f(x_2)$.

f decreasing on the interval: $x_1 < x_2$ implies $f(x_1) > f(x_2)$.

f constant on the interval: $f(x_1) = f(x_2)$.

Determine relative minimum and relative maximum values of functions *(p. 32)*.

$f(a)$ a relative minimum of f: An interval (x_1, x_2) exists that contains a such that $x_1 < x < x_2$ implies $f(a) \le f(x)$.

$f(a)$ a relative maximum of f: An interval (x_1, x_2) exists that contains a such that $x_1 < x < x_2$ implies $f(a) \ge f(x)$.

Identify and graph step functions and other piecewise-defined functions *(p. 34)*. The greatest integer function $f(x) = [\![x]\!]$ is a step function.

Identify even and odd functions *(p. 35)*.

Even function: For each x in the domain of f, $f(-x) = f(x)$.

Odd function: For each x in the domain of f, $f(-x) = -f(x)$.

Example Consider the graph below.

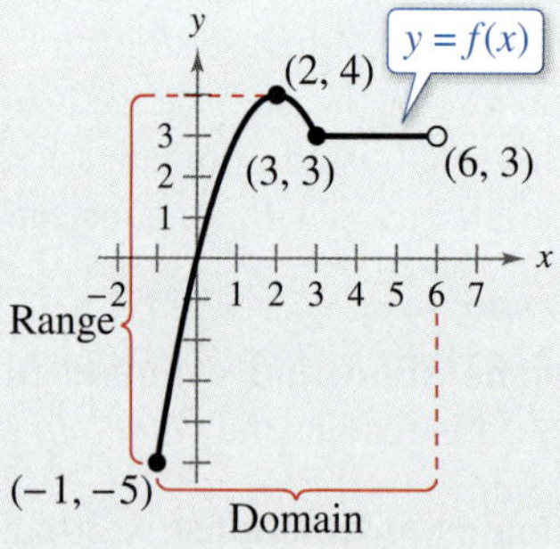

- This is a graph of y as a function of x because every vertical line intersects the graph at most once.
- There is a closed dot at $(-1, -5)$ and an open dot at $(6, 3)$, so the domain of f is all x in the interval $[-1, 6)$.
- The graph does not extend below $f(-1) = -5$ or above $f(2) = 4$, so the range of f is the interval $[-5, 4]$.
- The function is increasing on the interval $(-1, 2)$.
- The function is decreasing on the interval $(2, 3)$.
- The function is constant on the interval $(3, 6)$.
- The relative maximum of the function occurs at the point $(2, 4)$.
- The graph is neither symmetric with respect to the origin nor with respect to the y-axis. So, the function is neither odd nor even.

Finding the Domain and Range of a Function **In Exercises 33–36, use a graphing utility to graph the function and estimate its domain and range. Then find the domain and range algebraically.**

33. $f(x) = 3 - 2x^2$ **34.** $f(x) = \sqrt{x + 3} + 4$

35. $h(x) = \sqrt{36 - x^2}$ **36.** $f(x) = |x - 5| + 2$

Vertical Line Test for Functions **In Exercises 37 and 38, use the Vertical Line Test to determine whether y is a function of x. Describe how you can use a graphing utility to produce the graph shown.**

37. $3x + y^2 - 2 = 0$ **38.** $|x + 5| - 2y = 0$

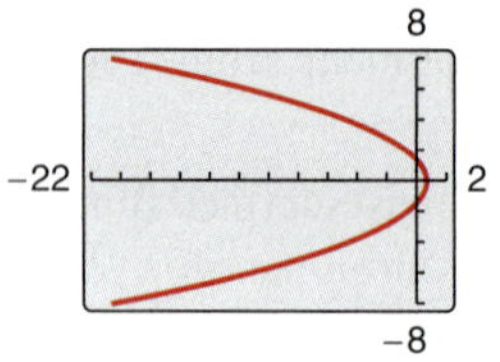

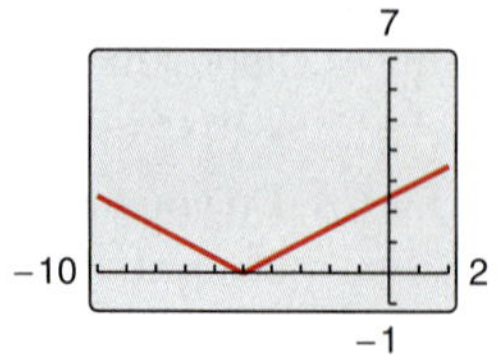

Increasing and Decreasing Functions **In Exercises 39 and 40, determine the open intervals on which the function is increasing, decreasing, or constant.**

39. $f(x) = x^3 - 3x$ **40.** $f(x) = \sqrt{x^2 - 16}$

Approximating Relative Minima and Maxima **In Exercises 41–44, use a graphing utility to graph the function and to approximate any relative minimum or relative maximum values of the function.**

41. $f(x) = (x^2 - 4)^2$ **42.** $f(x) = x^2 + x$

43. $h(x) = x^3 + 4x^2 + 3$ **44.** $f(x) = x^3 - 4x^2 - 1$

Sketching a Piecewise-Defined Function **In Exercises 45–48, sketch the graph of the piecewise-defined function by hand.**

45. $f(x) = \begin{cases} 3x + 5, & x < 0 \\ x - 4, & x \ge 0 \end{cases}$ **46.** $f(x) = \begin{cases} \frac{1}{2}x + 3, & x < 0 \\ 4 - x^2, & x \ge 0 \end{cases}$

47. $f(x) = [\![x]\!] - 4$ **48.** $f(x) = [\![x + 2]\!]$

Even and Odd Functions **In Exercises 49–56, determine algebraically whether the function is even, odd, or neither. Verify your answer using a graphing utility.**

49. $f(x) = x^2 + 5$ **50.** $f(x) = x^2 - x - 1$

51. $f(x) = (x^2 - 8)^2$ **52.** $f(x) = 2x^3 - x$

53. $f(x) = 3x^{5/2}$ **54.** $f(x) = 3x^{2/5}$

55. $f(x) = 2x\sqrt{4 - x^2}$ **56.** $f(x) = x\sqrt{x^2 - 1}$

1.4 What did you learn?

Recognize graphs of parent functions *(p. 41)*.

Linear: $f(x) = x$ | *Absolute value:* $f(x) = |x|$

Square root: $f(x) = \sqrt{x}$ | *Quadratic:* $f(x) = x^2$

Cubic: $f(x) = x^3$ | *Rational:* $f(x) = 1/x$

(See Figure 1.24 on page 41 for graphs.)

Use vertical and horizontal shifts *(p. 42)*, reflections *(p. 44)*, and nonrigid transformations *(p. 46)* to graph functions. Let c be a positive real number.

Vertical shifts: upward: $h(x) = f(x) + c$;
downward: $h(x) = f(x) - c$

Horizontal shifts: right: $h(x) = f(x - c)$;
left: $h(x) = f(x + c)$

Reflection in the x-axis: $h(x) = -f(x)$

Reflection in the y-axis: $h(x) = f(-x)$

Vertical stretch: $h(x) = cf(x), c > 1$

Vertical shrink: $h(x) = cf(x), 0 < c < 1$

Horizontal stretch: $h(x) = f(cx), 0 < c < 1$

Horizontal shrink: $h(x) = f(cx), c > 1$

Example The graph below is a transformation of the graph of

$$f(x) = x^2.$$

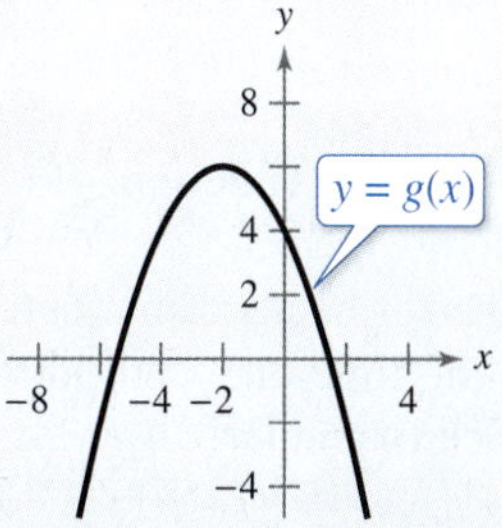

The graph of g is a reflection in the x-axis, a vertical shrink by a factor of $\frac{1}{2}$, a left shift of 2 units, and an upward shift of 6 units of the graph of f. (You can verify this by sketching the graphs of f and g on the same set of coordinate axes.)

So, an equation for g is

$$g(x) = -\frac{1}{2}(x + 2)^2 + 6.$$

Library of Parent Functions **In Exercises 57–64, identify the parent function and the transformation shown in the graph. Write an equation for the function represented by the graph.**

57.

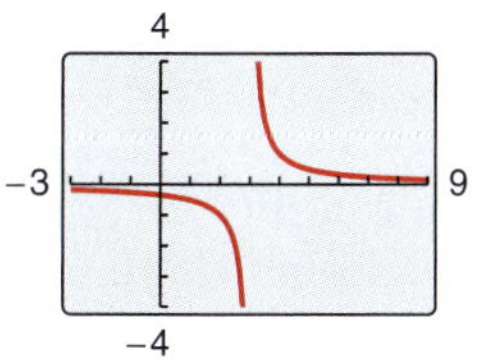

58.

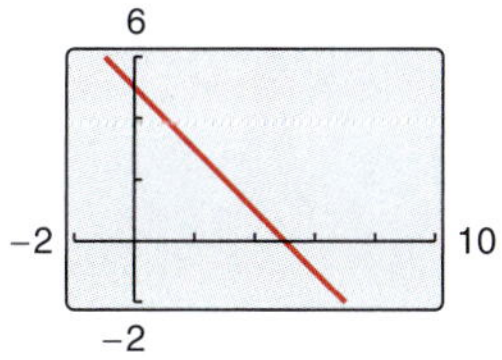

59.

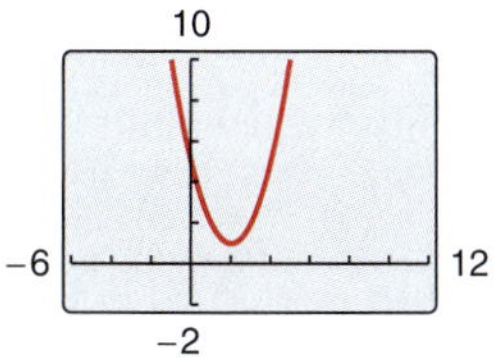

60.

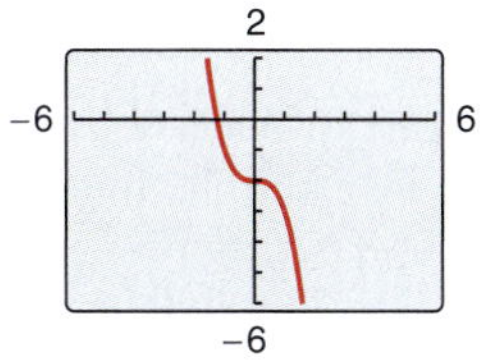

61.

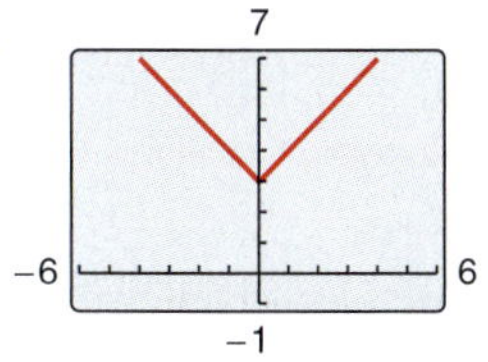

62.

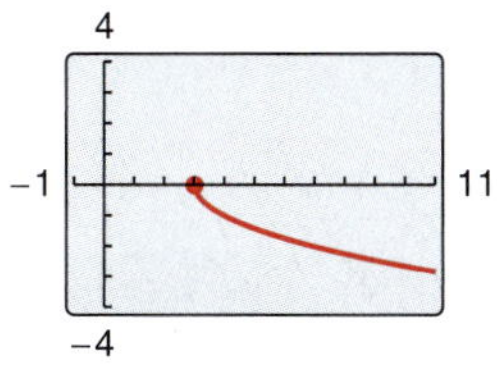

63.

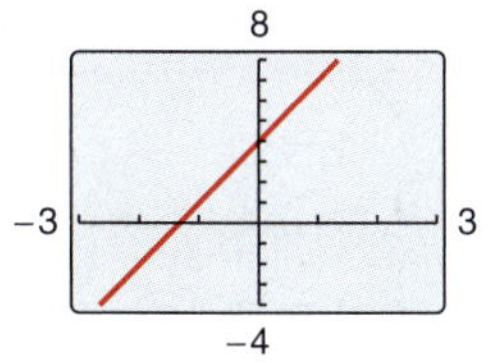

64.

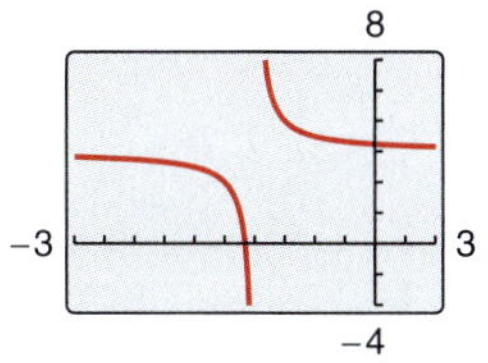

Sketching a Transformation **In Exercises 65–68, use the graph of f to sketch the graph**

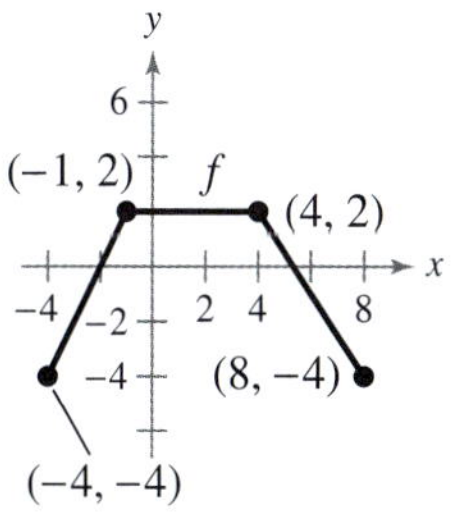

65. $y = f(-x)$

66. $y = -f(x)$

67. $y = f(x) + 2$

68. $y = f(x - 1)$

Describing Transformations **In Exercises 69–78, h is related to one of the six parent functions on page 41. (a) Identify the parent function f. (b) Describe the sequence of transformations from f to h. (c) Sketch the graph of h by hand. (d) Use function notation to write h in terms of the parent function f.**

69. $h(x) = \frac{1}{x} - 6$

70. $h(x) = -\frac{1}{x} + 3$

71. $h(x) = (x - 2)^3 + 5$

72. $h(x) = (-x)^2 - 8$

73. $h(x) = -\sqrt{x} - 6$

74. $h(x) = \sqrt{x - 1} + 4$

75. $h(x) = |3x| + 9$

76. $h(x) = |2x + 8| - 1$

77. $h(x) = \frac{-2}{x + 1} - 3$

78. $h(x) = \frac{5}{x + 2} - 4$

1.5 *What did you learn?*

Add, subtract, multiply, and divide functions *(p. 50)*, and find compositions of functions *(p. 52)*.

$(f + g)(x) = f(x) + g(x)$ $\quad (f - g)(x) = f(x) - g(x)$

$(fg)(x) = f(x) \cdot g(x)$ $\quad (f/g)(x) = f(x)/g(x),\ g(x) \neq 0$

Composition of functions: $(f \circ g)(x) = f(g(x))$

Use combinations of functions to model and solve real-life problems *(p. 55)*. A composite function can model the number of bacteria in a petri dish as a function of the amount of time the petri dish has been out of refrigeration. (See Example 10.)

Example Let $f(x) = x^2 + 1$ and $g(x) = x - 3$.

$(f + g)(x) = (x^2 + 1) + (x - 3) = x^2 + x - 2$

$(f - g)(x) = (x^2 + 1) - (x - 3) = x^2 - x + 4$

$(fg)(x) = (x^2 + 1)(x - 3) = x^3 - 3x^2 + x - 3$

$\left(\dfrac{f}{g}\right)(x) = \dfrac{x^2 + 1}{x - 3},\ x \neq 3$

$(f \circ g)(x) = f(x - 3) = (x - 3)^2 + 1 = x^2 - 6x + 10$

$(g \circ f)(x) = g(x^2 + 1) = (x^2 + 1) - 3 = x^2 - 2$

$(f \circ g)(2) = 2^2 - 6(2) + 10 = 2$

Evaluating an Arithmetic Combination of Functions **In Exercises 79–86, evaluate the function for $f(x) = 3 - 2x$, $g(x) = \sqrt{x}$, and $h(x) = 3x^2 + 2$.**

79. $(f - g)(4)$ **80.** $(f + h)(5)$

81. $(fh)(1)$ **82.** $(g/h)(1)$

83. $(h \circ g)(11)$ **84.** $(g \circ f)(-3)$

85. $(f \circ h)(-4)$ **86.** $(g \circ h)(6)$

Identifying a Composite Function **In Exercises 87–90, find two functions f and g such that $(f \circ g)(x) = h(x)$. (There are many correct answers.)**

87. $h(x) = (x + 3)^2$ **88.** $h(x) = \sqrt[3]{(x + 2)^2}$

89. $h(x) = \dfrac{2}{x + 4}$ **90.** $h(x) = \dfrac{6}{(3x + 1)^3}$

Demography **In Exercises 91 and 92, the populations (in millions) of males (y_1) and females (y_2) in the United States for the years 2010 through 2016 can be modeled by**

$y_1 = -0.018t^2 + 1.66t + 137.3$ **and**

$y_2 = -0.012t^2 + 1.46t + 143.8$

where t represents the year, with $t = 10$ corresponding to 2010. (*Source:* Statista)

91. Use a graphing utility to graph y_1, y_2, and $y_1 + y_2$ in the same viewing window.

92. Use the model $y_1 + y_2$ to estimate (a) the total population of males and females in 2021 and (b) the year in which the total population of males and females reached 320 million.

1.6 *What did you learn?*

Find inverse functions informally and verify that two functions are inverse functions of each other *(p. 60)*.

Let f and g be functions such that $f(g(x)) = x$ for every x in the domain of g and $g(f(x)) = x$ for every x in the domain of f. Under these conditions, the function g is the inverse function of the function f.

Use graphs of functions to decide whether functions have inverse functions *(p. 63)*. If the point (a, b) lies on the graph of f, then the point (b, a) must lie on the graph of f^{-1}, and vice versa. This means that the graph of f^{-1} is a reflection of the graph of f in the line $y = x$.

Determine whether functions are one-to-one *(p. 64)*. A function f is one-to-one when, for a and b in its domain, $f(a) = f(b)$ implies $a = b$. A function is one-to-one if it passes the Horizontal Line Test.

Find inverse functions algebraically *(p. 65)*. To find an inverse function, replace $f(x)$ with y, interchange the roles of x and y, and solve for y. Replace y with $f^{-1}(x)$. Then verify that f and f^{-1} are inverses of each other.

Example The function

$$f(x) = \frac{x + 3}{5}$$

is one-to-one (use a graphing utility to check this), so it has an inverse function.

$y = \dfrac{x + 3}{5}$ Replace $f(x)$ with y.

$x = \dfrac{y + 3}{5}$ Interchange x and y.

$5x = y + 3$ Multiply each side by 5.

$y = 5x - 3$ Solve for y.

$f^{-1}(x) = 5x - 3$ Replace y with $f^{-1}(x)$.

$$f(f^{-1}(x)) = f(5x - 3) = \frac{(5x - 3) + 3}{5} = x$$

$$f^{-1}(f(x)) = f^{-1}\left(\frac{x + 3}{5}\right) = 5\left(\frac{x + 3}{5}\right) - 3 = x$$

Use a graphing utility to verify that the domain of f is equal to the range of f^{-1} and that the range of f is equal to the domain of f^{-1}.

Finding an Inverse Function Informally **In Exercises 93–96, find the inverse function of f informally. Verify that $f(f^{-1}(x)) = x$ and $f^{-1}(f(x)) = x$.**

93. $f(x) = 6x$

94. $f(x) = x + 5$

95. $f(x) = \frac{1}{2}x + 3$

96. $f(x) = \frac{x - 5}{4}$

Algebraic-Graphical-Numerical **In Exercises 97 and 98, show that f and g are inverse functions (a) algebraically, (b) graphically, and (c) numerically.**

97. $f(x) = 3 - 4x, \quad g(x) = \frac{3 - x}{4}$

98. $f(x) = \sqrt{x + 1}; \quad g(x) = x^2 - 1, x \geq 0$

Using the Horizontal Line Test **In Exercises 99–102, use a graphing utility to graph the function and use the Horizontal Line Test to determine whether the function has an inverse function.**

99. $f(x) = \frac{1}{2}x - 3$

100. $f(x) = (x - 1)^2$

101. $h(t) = (t + 5)^{2/3}$

102. $g(x) = \frac{4}{x - 5}$

Finding an Inverse Function Algebraically **In Exercises 103–106, find the inverse function of f algebraically.**

103. $f(x) = \frac{1}{8}(7x + 3)$

104. $f(x) = 5x^3 + 2$

105. $f(x) = 4\sqrt{6 - x}$

106. $f(x) = x^2 + 1, x \geq 0$

1.7 *What did you learn?*

Construct scatter plots and interpret correlation *(p. 71)*. A scatter plot is a graphical representation of data written as a set of ordered pairs. If y tends to increase as x increases, then the set as positive correlation. If y tends to decrease as x increases, then the set has negative correlation.

Example The scatter plot below shows a set of data with positive correlation.

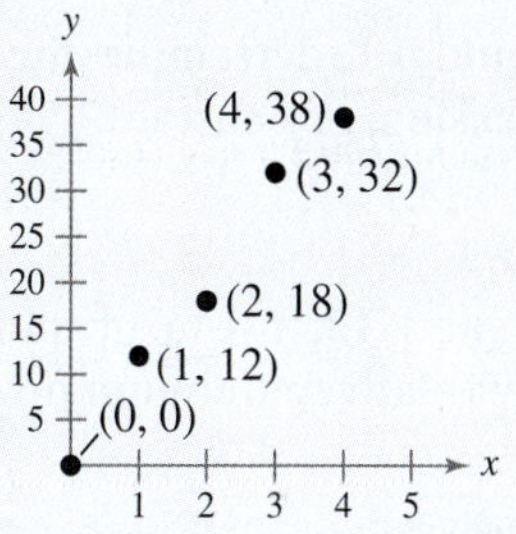

Use scatter plots and a graphing utility to find linear models for data *(p. 73)*. A linear model can be found using the *regression* feature of a graphing utility.

After finding a model, you can measure how well the model fits the data by comparing the actual values with the values given by the model. This can be done numerically, with a table, or graphically.

A linear model for the data, found using the *regression* feature of a graphing utility, is $y = 9.6x + 0.8$.

107. Modeling Data The ordered pairs (x, y) below represent the Olympic year x and the winning time y (in minutes) in the men's 400-meter freestyle swimming event. Use the *regression* feature of a graphing utility to find a linear model for the data. Let x represent the year, with $x = 6$ corresponding to 1996. Is the model a good fit for the data? Explain. *(Spreadsheet at LarsonPrecalculus.com)* (*Source:* International Olympic Committee)

DATA
(1996, 3.80) (2000, 3.68)
(2004, 3.72) (2008, 3.70)
(2012, 3.67) (2016, 3.69)

108. Education The ordered pairs below give the entrance exam scores x and the grade point averages y for 10 students. *(Spreadsheet at LarsonPrecalculus.com)*

DATA
(75, 2.4), (82, 3.0), (90, 3.6), (65, 2.0), (70, 2.1), (88, 3.5), (93, 3.9), (69, 2.0), (80, 2.8), (85, 3.3)

(a) Construct a scatter plot of the data.

(b) Use the *regression* feature of a graphing utility to find a linear model for the data and identify the correlation coefficient.

(c) Use the model from part (b) to predict the grade point average for a student with an entrance exam score of 95.

Focusing on Concepts

True or False? **In Exercises 109 and 110, determine whether the statement is true or false. Justify your answer.**

109. There exists no function f such that $f = f^{-1}$.

110. If the graph of the parent function $f(x) = x^2$ is moved six units to the right, moved three units upward, and reflected in the x-axis, then the point $(-1, 28)$ will lie on the graph of the transformation.

1 Chapter Test

See CalcChat.com for tutorial help and worked-out solutions to odd-numbered exercises.
For instructions on how to use a graphing utility, see Appendix A.

Take this test as you would take a test in class. After you are finished, check your work against the answers given in the back of the book.

1. Find equations of the lines that pass through the point $(0, 4)$ and are (a) parallel and (b) perpendicular to the line $5x + 2y = 3$.

2. Find the slope-intercept form of the equation of the line that passes through the points $(2, -1)$ and $(-3, 4)$.

3. Does the graph at the right represent y as a function of x? Explain.

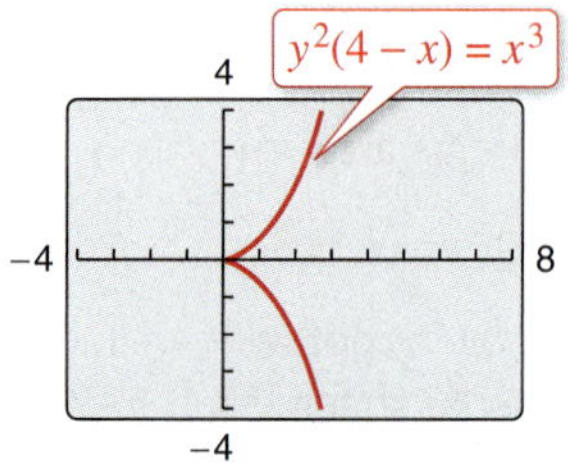

Figure for 3

4. Let $f(x) = |x + 2| - 15$. Find each function value.

(a) $f(-8)$ (b) $f(14)$ (c) $f(t - 6)$

5. Find the domain of $f(x) = 10 - \sqrt{3 - x}$.

6. An electronics company produces a car stereo for which the variable cost is \$48.47 per unit and the fixed costs are \$50,000. The product sells for \$199.99. Write the total cost C as a function of the number of units produced and sold, x. Write the profit P as a function of the number of units produced and sold, x.

In Exercises 7 and 8, determine algebraically whether the function is even, odd, or neither.

7. $f(x) = 2x^3 - 3x$ **8.** $f(x) = 4x^4 + 9x^2$

In Exercises 9 and 10, determine the open intervals on which the function is increasing, decreasing, or constant.

9. $h(x) = \frac{1}{4}x^4 - 2x^2$ **10.** $g(t) = |t + 2| - |t - 2|$

In Exercises 11 and 12, use a graphing utility to graph the function and to approximate any relative minimum or relative maximum values of the function.

11. $f(x) = -x^3 - 5x^2 + 12$ **12.** $f(x) = x^5 - 3x^3 + 4$

In Exercises 13–15, (a) identify the parent function f, (b) describe the sequence of transformations from f to g, and (c) sketch the graph of g by hand.

13. $g(x) = -2(x - 5)^3 + 3$ **14.** $g(x) = \sqrt{-9x - 18}$ **15.** $g(x) = 6|-x| - 8$

16. Use the functions $f(x) = x^2$ and $g(x) = \sqrt{2 - x}$ to find each specified function and its domain.

(a) $(f - g)(x)$ (b) $\left(\dfrac{f}{g}\right)(x)$ (c) $(f \circ g)(x)$ (d) $(g \circ f)(x)$

In Exercises 17–19, determine whether the function has an inverse function. If it does, find the inverse function.

17. $f(x) = x^3 + 8$ **18.** $f(x) = x^2 + 6$ **19.** $f(x) = \dfrac{3x\sqrt{x}}{8}$

20. The table shows the average monthly prices P of expanded basic cable television from 2008 through 2016, where t represents the year, with $t = 8$ corresponding to 2008. Use the *regression* feature of a graphing utility to find a linear model for the data. Use the model to estimate the year in which the average monthly price will reach \$88. *(Source:* Federal Communications Commission)

Collaborative Project

To work a collaborative project involving Functions and Their Graphs, visit this textbook's website at *LarsonPrecalculus.com.*

DATA

Year, t	Average monthly price, P (in dollars)
8	49.65
9	52.37
10	54.44
11	57.46
12	61.63
13	64.41
14	66.61
15	69.03
16	71.37

Spreadsheet at LarsonPrecalculus.com

Table for 20

Proofs in Mathematics

Conditional Statements

Many theorems are written in the **if-then form** "if p, then q," which is denoted by

$p \rightarrow q$ Conditional statement

where p is the **hypothesis** and q is the **conclusion.** Here are some other ways to express the conditional statement $p \rightarrow q$.

p implies q. p, only if q. p is sufficient for q.

Conditional statements can be either true or false. The conditional statement $p \rightarrow q$ is false only when p is true and q is false. To show that a conditional statement is true, you must prove that the conclusion follows for all cases that fulfill the hypothesis. To show that a conditional statement is false, you need to describe only a single **counterexample** that shows that the statement is not always true.

For instance, $x = -4$ is a counterexample that shows that the following statement is false.

If $x^2 = 16$, then $x = 4$.

The hypothesis "$x^2 = 16$" is true because $(-4)^2 = 16$. However, the conclusion "$x = 4$" is false. This implies that the given conditional statement is false.

For the conditional statement $p \rightarrow q$, there are three important associated conditional statements.

1. The **converse** of $p \rightarrow q$: $q \rightarrow p$
2. The **inverse** of $p \rightarrow q$: $\sim p \rightarrow \sim q$
3. The **contrapositive** of $p \rightarrow q$: $\sim q \rightarrow \sim p$

The symbol $\sim$ means the **negation** of a statement. For instance, the negation of "The engine is running" is "The engine is not running."

EXAMPLE 1 Writing the Converse, Inverse, and Contrapositive

Write the converse, inverse, and contrapositive of the conditional statement "If I get a B on my test, then I will pass the course."

Solution

a. *Converse:* If I pass the course, then I got a B on my test.

b. *Inverse:* If I do not get a B on my test, then I will not pass the course.

c. *Contrapositive:* If I do not pass the course, then I did not get a B on my test. ■

In the example above, notice that neither the converse nor the inverse is logically equivalent to the original conditional statement. On the other hand, the contrapositive is logically equivalent to the original conditional statement.

Biconditional Statements

Recall that a conditional statement is a statement of the form "if p, then q." A statement of the form "p if and only if q" is called a **biconditional statement.** A biconditional statement, denoted by

$p \leftrightarrow q$ Biconditional statement

is the conjunction of the conditional statement $p \rightarrow q$ and its converse $q \rightarrow p$.

A biconditional statement can be either true or false. To be true, *both* the conditional statement and its converse must be true.

EXAMPLE 2 Analyzing a Biconditional Statement

Consider the statement $x = 3$ if and only if $x^2 = 9$.

a. Is the statement a biconditional statement?

b. Is the statement true?

Solution

a. The statement is a biconditional statement because it is of the form "p if and only if q."

b. Rewrite the statement as a conditional statement and its converse.

Conditional statement: If $x = 3$, then $x^2 = 9$.

Converse: If $x^2 = 9$, then $x = 3$.

The conditional statement is true, but the converse is false because x could also equal -3. So, the biconditional statement is false. ■

Knowing how to use biconditional statements is an important tool for reasoning in mathematics.

EXAMPLE 3 Analyzing a Biconditional Statement

Determine whether the biconditional statement is true or false. If it is false, provide a counterexample.

A number is divisible by 5 if and only if it ends in 0.

Solution

Rewrite the biconditional statement as a conditional statement and its converse.

Conditional statement: If a number is divisible by 5, then it ends in 0.

Converse: If a number ends in 0, then it is divisible by 5.

The conditional statement is false. A counterexample is the number 15, which is divisible by 5 but does not end in 0. So, the biconditional statement is false. ■

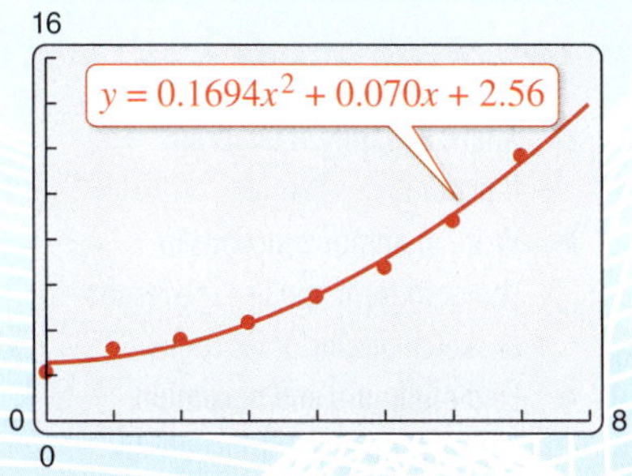

Section 2.8, Example 4
Revenues for Netflix, Inc.

2 Polynomial and Rational Functions

Student Resources at LarsonPrecalculus.com

- **Videos** explaining the concepts of precalculus
- **Worked-out solution videos** for all *Checkpoint* exercises
- **Editable spreadsheets** of the data sets in the text
- **Group projects** for each chapter applying concepts to real-life problems

2.1 Quadratic Functions

The Graph of a Quadratic Function

In this and the next section, you will study the graphs of polynomial functions.

Definition of Polynomial Function

Let n be a nonnegative integer and let $a_n, a_{n-1}, \ldots a_2, a_1, a_0$ be real numbers with $a_n \neq 0$. The function given by

$$f(x) = a_n x^n + a_{n-1}x^{n-1} + \cdots + a_2x^2 + a_1x + a_0$$

is called a **polynomial function of x with degree n.**

Polynomial functions are classified by degree. For instance, the polynomial function

$$f(x) = a, \quad a \neq 0 \qquad \text{Constant function}$$

has degree 0 and is called a **constant function.** In Chapter 1, you learned that the graph of this type of function is a horizontal line. The polynomial function

$$f(x) = mx + b, \quad m \neq 0 \qquad \text{Linear function}$$

has degree 1 and is called a **linear function.** You learned in Chapter 1 that the graph of $f(x) = mx + b$ is a line whose slope is m and whose y-intercept is $(0, b)$. In this section, you will study second-degree polynomial functions, which are called **quadratic functions.**

Definition of Quadratic Function

Let a, b, and c be real numbers with $a \neq 0$. The function given by

$$f(x) = ax^2 + bx + c \qquad \text{Quadratic function}$$

is called a **quadratic function.**

Often, quadratic functions can model real-life data. For instance, the table at the right shows the heights h (in feet) of a projectile fired from a height of 6 feet with an initial velocity of 256 feet per second at selected times t (in seconds). A quadratic model for the data in the table is

$$h(t) = -16t^2 + 256t + 6 \quad \text{for} \quad 0 \leq t \leq 16.$$

The graph of a quadratic function is a special type of U-shaped curve called a **parabola.** For instance, see the graph of h below the table at the right. Parabolas occur in many real-life applications, especially those involving reflective properties, such as satellite dishes or flashlight reflectors. You will study these properties in Section 9.1.

All parabolas are symmetric with respect to a line called the **axis of symmetry,** or simply the **axis** of the parabola. The point where the axis intersects the parabola is called the **vertex** of the parabola. These and other basic characteristics of quadratic functions are summarized on the next page.

Time, t (in seconds)	Height, h (in feet)
0	6
2	454
4	774
6	966
8	1030
10	966
12	774
14	454
16	6

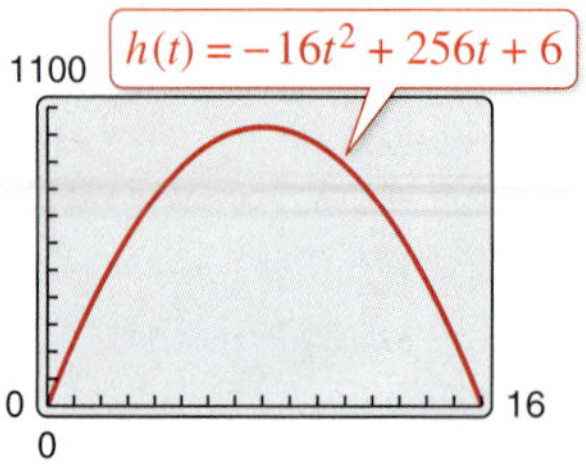

What you should learn

- Analyze graphs of quadratic functions.
- Write quadratic functions in standard form and use the results to sketch graphs of functions.
- Find minimum and maximum values of quadratic functions in real-life applications.

Why you should learn it

Quadratic functions can be used to model the design of a room. For instance, Exercise 59 on page 97 shows how the size of an indoor fitness room with a running track can be modeled.

marpans/iStock/Getty Images

Basic Characteristics of Quadratic Functions

Graph of $f(x) = ax^2, a > 0$
Domain: $(-\infty, \infty)$
Range: $[0, \infty)$
Intercept: $(0, 0)$
Decreasing on $(-\infty, 0)$
Increasing on $(0, \infty)$
Even function
Axis of symmetry: $x = 0$
Relative minimum or vertex: $(0, 0)$

Graph of $f(x) = ax^2, a < 0$
Domain: $(-\infty, \infty)$
Range: $(-\infty, 0]$
Intercept: $(0, 0)$
Increasing on $(-\infty, 0)$
Decreasing on $(0, \infty)$
Even function
Axis of symmetry: $x = 0$
Relative maximum or vertex: $(0, 0)$

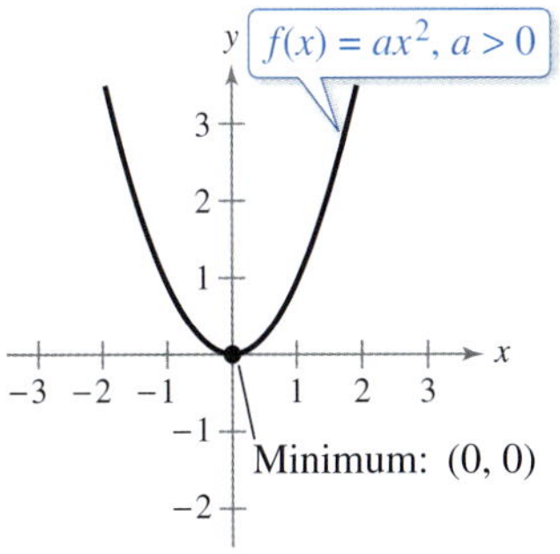

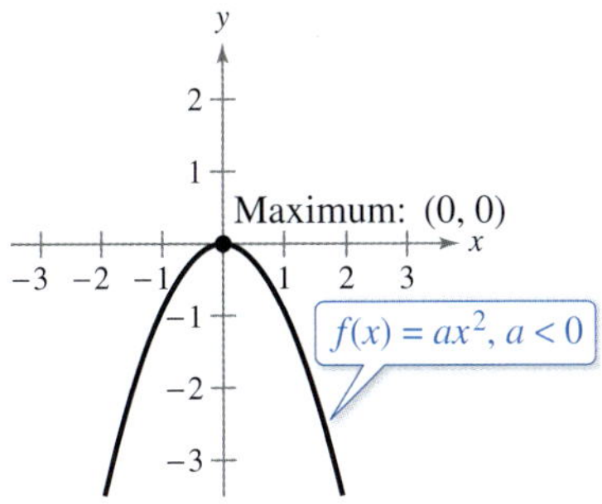

For the general quadratic form $f(x) = ax^2 + bx + c$, when the leading coefficient a is positive, the parabola opens upward, and when the leading coefficient a is negative, the parabola opens downward. Later in this section, you will learn ways to find the coordinates of the vertex of a parabola.

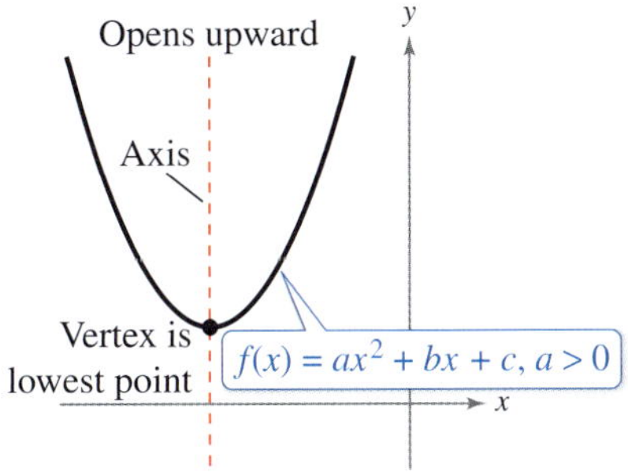

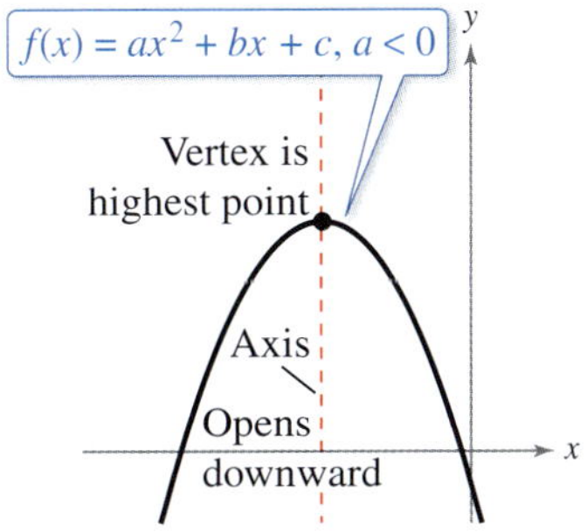

When sketching the graph of $f(x) = ax^2$, it is helpful to use the graph of $y = x^2$ as a reference, as discussed in Section 1.4. There you saw that when $a > 1$, the graph of $y = af(x)$ is a vertical stretch of the graph of $y = f(x)$. When $0 < a < 1$, the graph of $y = af(x)$ is a vertical shrink of the graph of $y = f(x)$. Notice in Figure 2.1 that the coefficient a determines how widely the parabola given by $f(x) = ax^2$ opens. When $|a|$ is small, the parabola opens more widely than when $|a|$ is large.

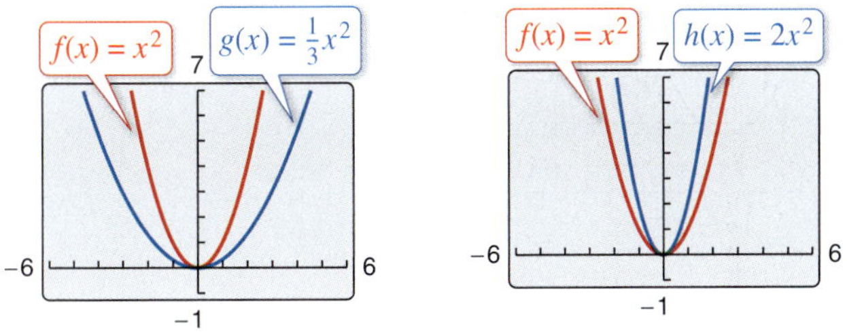

Vertical shrink *Vertical stretch*
Figure 2.1

Explore the Concept

Use a graphing utility to graph $y = ax^2$ with $a = -2, -1, -0.5, 0.5, 1$, and 2. How does changing the value of a affect the graph?

Use a graphing utility to graph $y = (x - h)^2$ with $h = -4, -2, 2$, and 4. How does changing the value of h affect the graph?

Use a graphing utility to graph $y = x^2 + k$ with $k = -4, -2, 2$, and 4. How does changing the value of k affect the graph?

Library of Parent Functions: Quadratic Function

The *parent quadratic function* is $f(x) = x^2$, also known as the *squaring function.* The basic characteristics of the parent quadratic function are summarized below and on the inside cover of this text.

Graph of $f(x) = x^2$

Domain: $(-\infty, \infty)$
Range: $[0, \infty)$
Intercept: $(0, 0)$
Decreasing on $(-\infty, 0)$
Increasing on $(0, \infty)$
Even function
Axis of symmetry: $x = 0$
Relative minimum or vertex: $(0, 0)$

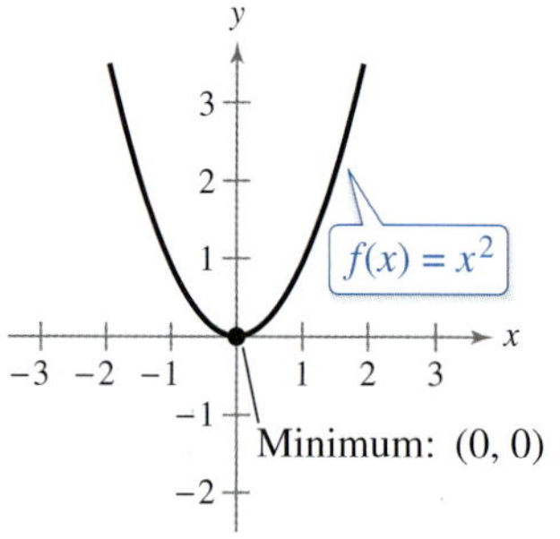

Recall from Section 1.4 that the graphs of $y = f(x \pm c)$, $y = f(x) \pm c$, $y = -f(x)$, and $y = f(-x)$ are rigid transformations of the graph of $y = f(x)$.

$y = f(x \pm c)$ Horizontal shift $\qquad$ $y = -f(x)$ Reflection in x-axis

$y = f(x) \pm c$ Vertical shift $\qquad$ $y = f(-x)$ Reflection in y-axis

EXAMPLE 1 Library of Parent Functions: $f(x) = x^2$

See LarsonPrecalculus.com for an interactive version of this type of example.

Sketch the graph of each function by hand and compare it with the graph of $f(x) = x^2$.

a. $g(x) = -x^2 + 1$ **b.** $h(x) = (x + 2)^2 - 3$

Solution

a. With respect to the graph of $f(x) = x^2$, the graph of g is obtained by a *reflection* in the x-axis and a vertical shift one unit *upward*, as shown in Figure 2.2. Confirm this with a graphing utility.

b. With respect to the graph of $f(x) = x^2$, the graph of h is obtained by a horizontal shift two units *to the left* and a vertical shift three units *downward*, as shown in Figure 2.3. Confirm this with a graphing utility.

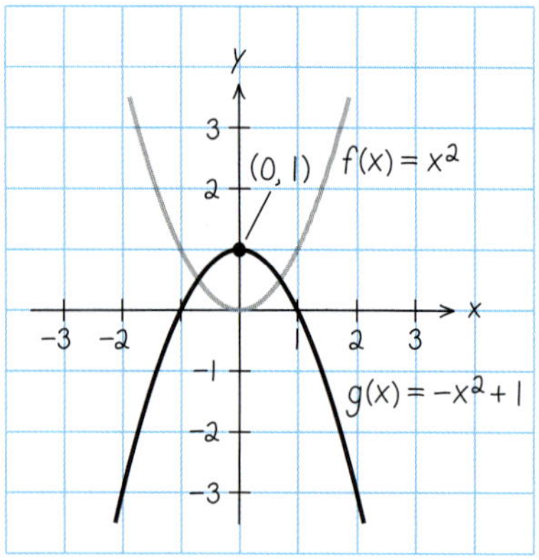

Figure 2.2

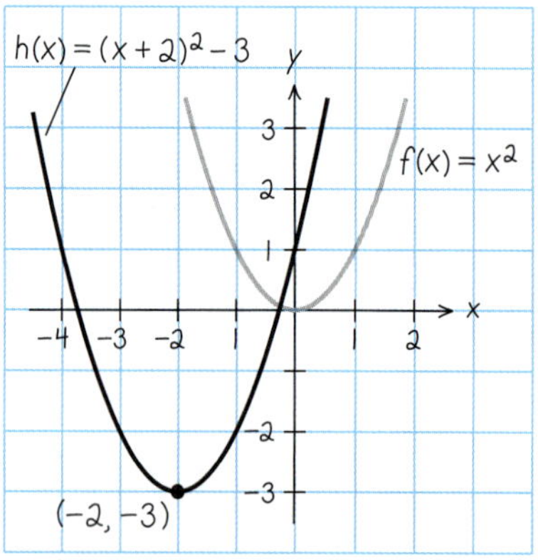

Figure 2.3

✓ **Checkpoint** *Audio-video solution in English & Spanish at LarsonPrecalculus.com*

Sketch the graph of each function by hand and compare it with the graph of $f(x) = x^2$.

a. $g(x) = x^2 - 4$ **b.** $h(x) = (x - 3)^2 + 1$

The Standard Form of a Quadratic Function

The equation in Example 1(b) is written in the **standard form**

$$f(x) = a(x - h)^2 + k.$$

This form is especially convenient for sketching a parabola because it identifies the vertex of the parabola as (h, k).

Standard Form of a Quadratic Function

The quadratic function given by

$$f(x) = a(x - h)^2 + k, \quad a \neq 0$$

is in **standard form.** The graph of f is a parabola whose axis is the vertical line $x = h$ and whose vertex is the point (h, k). When $a > 0$, the parabola opens upward, and when $a < 0$, the parabola opens downward.

Algebra Help

The graph of a quadratic function can have zero, one, or two x-intercepts, but it will always have one y-intercept.

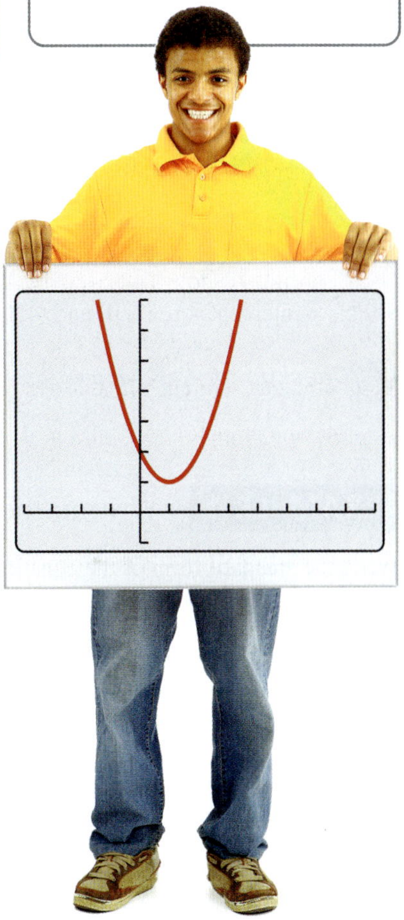

EXAMPLE 2 Identifying the Vertex of a Quadratic Function

Describe the graph of

$$f(x) = 2x^2 + 8x + 7$$

and identify the vertex.

Solution

Write the quadratic function in standard form by completing the square. Recall that the first step is to factor out any coefficient of x^2 that is not 1.

$f(x) = 2x^2 + 8x + 7$	Write original function.
$= (2x^2 + 8x) + 7$	Group x-terms.
$= 2(x^2 + 4x) + 7$	Factor 2 out of x-terms.
$= 2(x^2 + 4x + 4 - 4) + 7$ (where $4 = (4/2)^2$)	Add and subtract $(4/2)^2 = 4$ within parentheses to complete the square.
$= 2(x^2 + 4x + 4) - 2(4) + 7$	Regroup terms.
$= 2(x + 2)^2 - 1$	Write in standard form.

From the standard form, you can see that the graph of f is a parabola that opens upward with vertex

$$(-2, -1)$$

as shown in the figure. This corresponds to a left shift of two units and a downward shift of one unit relative to the graph of

$$y = 2x^2.$$

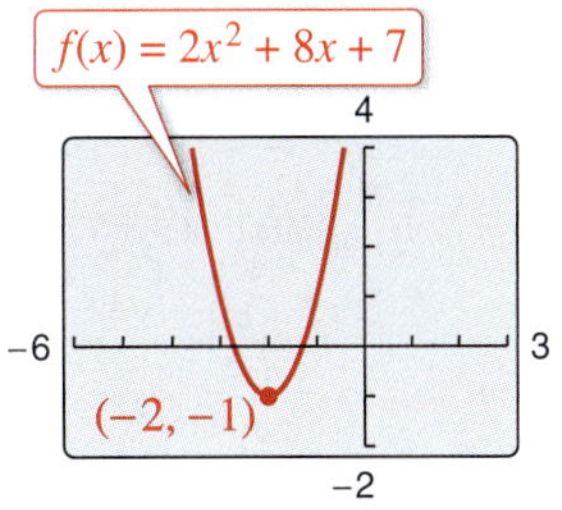

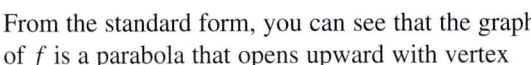

Checkpoint Audio-video solution in English & Spanish at LarsonPrecalculus.com

Describe the graph of $f(x) = 3x^2 - 6x + 4$ and identify the vertex.

To find the x-intercepts of the graph of $f(x) = ax^2 + bx + c$, solve the equation $ax^2 + bx + c = 0$. When $ax^2 + bx + c$ does not factor, you can use the Quadratic Formula to find the x-intercepts or a graphing utility to approximate the x-intercepts. Remember, however, that a parabola may not have x-intercepts.

Insight

When sketching the graph of a quadratic function, write the function in standard form. This will help you identify the vertex, the axis of symmetry, and whether the parabola opens upward or downward.

ASDF_MEDIA/Shutterstock.com

EXAMPLE 3 Identifying *x*-Intercepts of a Quadratic Function

Describe the graph of $f(x) = -x^2 + 6x - 8$ and identify any x-intercepts.

Solution

$f(x) = -x^2 + 6x - 8$ Write original function.

$= -(x^2 - 6x) - 8$ Factor -1 out of x-terms.

$= -(x^2 - 6x + 9 - 9) - 8$ Add and subtract $(-6/2)^2 = 9$ within parentheses.

$(-6/2)^2$

$= -(x^2 - 6x + 9) - (-9) - 8$ Regroup terms.

$= -(x - 3)^2 + 1$ Write in standard form.

The graph of f is a parabola that opens downward with vertex $(3, 1)$, as shown in Figure 2.4. The x-intercepts are determined as follows.

$-(x^2 - 6x + 8) = 0$ Factor out -1.

$-(x - 2)(x - 4) = 0$ Factor.

$x - 2 = 0$ ⇒ $x = 2$ Set 1st factor equal to 0.

$x - 4 = 0$ ⇒ $x = 4$ Set 2nd factor equal to 0.

So, the x-intercepts are $(2, 0)$ and $(4, 0)$, as shown in Figure 2.4.

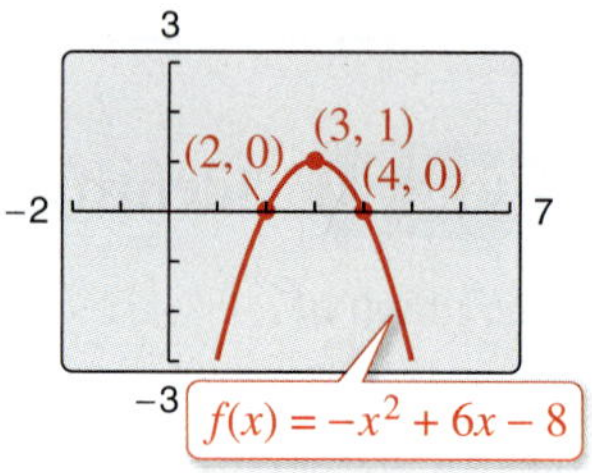

Figure 2.4

✓ ***Checkpoint*** ▶ *Audio-video solution in English & Spanish at LarsonPrecalculus.com*

Describe the graph of $f(x) = x^2 - 4x + 3$ and identify any x-intercepts.

EXAMPLE 4 Writing the Equation of a Parabola

Write the standard form of the equation of the parabola whose vertex is $(1, 2)$ and that passes through the point $(3, -6)$.

Solution

Because the vertex of the parabola is $(h, k) = (1, 2)$, the equation has the form

$f(x) = a(x - 1)^2 + 2.$ Substitute for h and k in standard form.

Because the parabola passes through the point $(3, -6)$, it follows that $f(3) = -6$. So, you obtain

$f(x) = a(x - 1)^2 + 2$ Write in standard form.

$-6 = a(3 - 1)^2 + 2$ Substitute -6 for $f(x)$ and 3 for x.

$-6 = 4a + 2$ Simplify.

$-8 = 4a$ Subtract 2 from each side.

$-2 = a.$ Divide each side by 4.

The equation in standard form is $f(x) = -2(x - 1)^2 + 2$. You can confirm this answer by graphing f with a graphing utility, as shown in Figure 2.5. Use the *zoom* and *trace* features or the *maximum* and *value* features to confirm that its vertex is $(1, 2)$ and that it passes through the point $(3, -6)$.

Algebra Help

In Example 4, there are infinitely many different parabolas that have a vertex at $(1, 2)$. Of these, however, the only one that passes through the point $(3, -6)$ is

$$f(x) = -2(x - 1)^2 + 2.$$

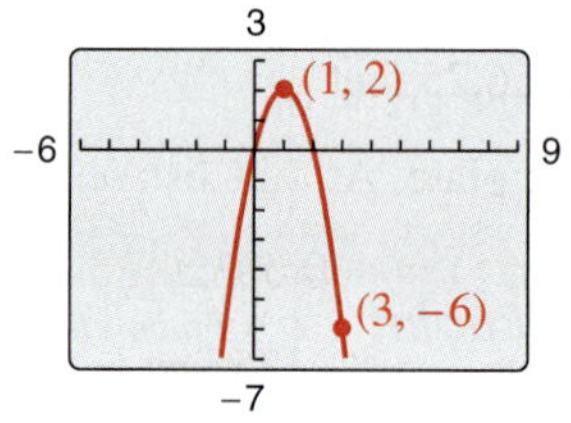

Figure 2.5

✓ ***Checkpoint*** ▶ *Audio-video solution in English & Spanish at LarsonPrecalculus.com*

Write the standard form of the equation of the parabola whose vertex is $(-4, 11)$ and that passes through the point $(-6, 15)$.

Finding Minimum and Maximum Values

Many applications involve finding the maximum or minimum value of a quadratic function. By completing the square of the quadratic function $f(x) = ax^2 + bx + c$, you can rewrite the function in standard form (see Exercise 86).

$$f(x) = a\left(x + \frac{b}{2a}\right)^2 + \left(c - \frac{b^2}{4a}\right) \qquad \text{Standard form}$$

So, the vertex of the graph of f is $\left(-\frac{b}{2a}, f\left(-\frac{b}{2a}\right)\right)$.

Minimum and Maximum Values of Quadratic Functions

Consider the function $f(x) = ax^2 + bx + c$ with vertex $\left(-\frac{b}{2a}, f\left(-\frac{b}{2a}\right)\right)$.

1. If $a > 0$, then f has a *minimum* at $x = -\frac{b}{2a}$.
   The minimum value is $f\left(-\frac{b}{2a}\right)$.
2. If $a < 0$, then f has a *maximum* at $x = -\frac{b}{2a}$.
   The maximum value is $f\left(-\frac{b}{2a}\right)$.

EXAMPLE 5 Finding the Maximum Height of a Projectile

The path of a baseball is given by the function $f(x) = -0.0032x^2 + x + 3$, where $f(x)$ is the height of the baseball (in feet) and x is the horizontal distance from home plate (in feet). What is the maximum height reached by the baseball?

Algebraic Solution

For this quadratic function, you have

$$f(x) = ax^2 + bx + c = -0.0032x^2 + x + 3$$

which implies that $a = -0.0032$ and $b = 1$. Because $a < 0$, the function has a maximum when $x = -b/(2a)$. So, the baseball reaches its maximum height when it is

$$x = -\frac{b}{2a}$$

$$= -\frac{1}{2(-0.0032)}$$

$$= 156.25 \text{ feet}$$

from home plate. At this distance, the maximum height is

$$f(156.25) = -0.0032(156.25)^2 + 156.25 + 3$$

$$= 81.125 \text{ feet.}$$

Graphical Solution

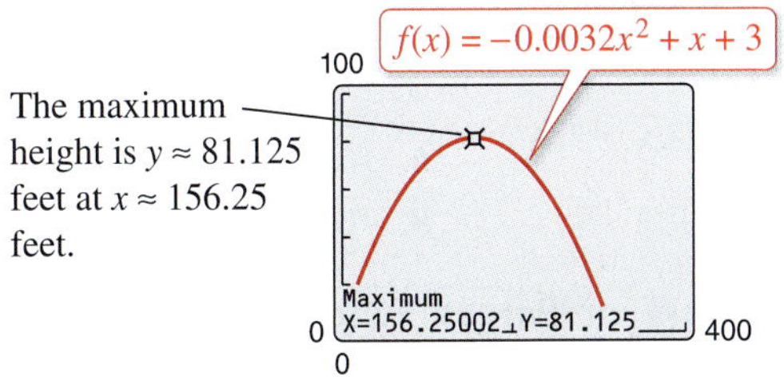

Checkpoint Audio-video solution in English & Spanish at LarsonPrecalculus.com

Rework Example 5 when the path of the baseball is given by the function

$$f(x) = -0.007x^2 + x + 4.$$

2.1 Exercises

See *CalcChat.com* for tutorial help and worked-out solutions to odd-numbered exercises. For instructions on how to use a graphing utility, see Appendix A.

Vocabulary and Concept Check

In Exercises 1 and 2, fill in the blanks.

1. A polynomial function with degree n and leading coefficient a_n is a function of the form $f(x) = a_n x^n + a_{n-1}x^{n-1} + \cdot \cdot \cdot + a_2 x^2 + a_1 x + a_0$, $a_n \neq 0$, where n is a _______ and $a_n, a_{n-1}, . . . , a_2, a_1, a_0$ are _______ numbers.

2. A _______ function is a second-degree polynomial function, and its graph is called a _______ .

3. Is the quadratic function $f(x) = (x - 2)^2 + 3$ written in standard form? Identify the vertex of the graph of f.

4. Does the graph of the quadratic function $f(x) = -3x^2 + 5x + 2$ have a relative minimum value at its vertex?

Procedures and Problem Solving

Graphs of Quadratic Functions **In Exercises 5–8, match the quadratic function with its graph. [The graphs are labeled (a), (b), (c), and (d).]**

(a)

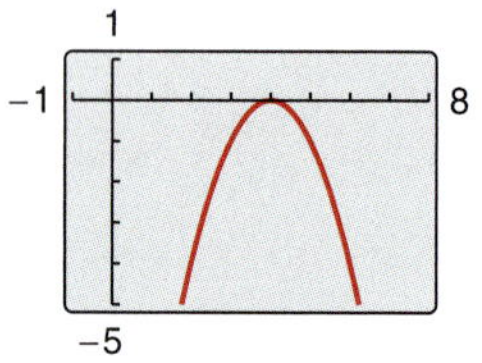

(b)

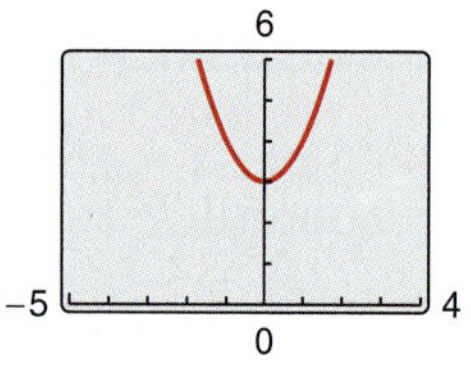

(c)

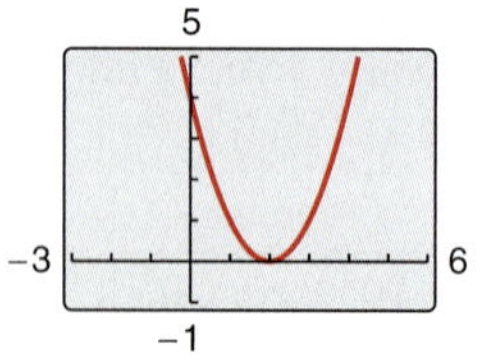

(d)

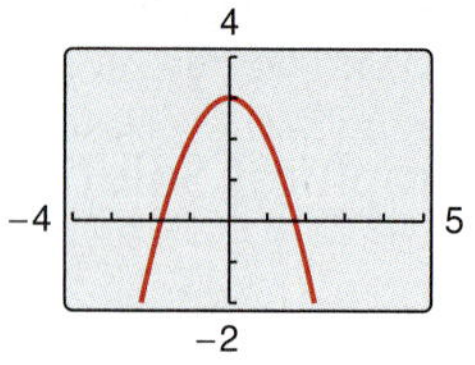

5. $f(x) = (x - 2)^2$ **6.** $f(x) = 3 - x^2$

7. $f(x) = x^2 + 3$ **8.** $f(x) = -(x - 4)^2$

Library of Parent Functions **In Exercises 9–12, sketch the graph of each quadratic function and compare it with the graph of $y = x^2$.**

9. (a) $f(x) = \frac{1}{2}x^2$ (b) $g(x) = -\frac{1}{8}x^2$
(c) $h(x) = \frac{3}{2}x^2$ (d) $k(x) = -3x^2$

10. (a) $f(x) = x^2 + 1$ (b) $g(x) = x^2 - 1$
(c) $h(x) = x^2 + 3$ (d) $k(x) = x^2 - 3$

11. (a) $f(x) = (x - 1)^2$ (b) $g(x) = (3x)^2 + 1$
(c) $h(x) = \left(\frac{1}{3}x\right)^2 - 3$ (d) $k(x) = (x + 3)^2$

12. (a) $f(x) = -\frac{1}{2}(x - 2)^2 + 1$
(b) $g(x) = \left[\frac{1}{2}(x - 1)\right]^2 - 3$
(c) $h(x) = -\frac{1}{2}(x + 2)^2 - 1$
(d) $k(x) = [2(x + 1)]^2 + 4$

Using Standard Form to Graph a Parabola **In Exercises 13–26, write the quadratic function in standard form and sketch its graph. Identify the vertex, axis of symmetry, and x-intercept(s). Use a graphing utility to verify your results.**

13. $f(x) = x^2 - 6x$ **14.** $g(x) = x^2 - 8x$

15. $h(x) = x^2 - 8x + 16$ **16.** $g(x) = x^2 + 2x + 1$

17. $f(x) = x^2 - 6x + 2$ **18.** $f(x) = x^2 + 16x + 61$

19. $f(x) = x^2 - 8x + 21$ **20.** $f(x) = x^2 + 12x + 40$

21. $f(x) = x^2 - x + \frac{5}{4}$ **22.** $f(x) = x^2 + 3x + \frac{1}{4}$

23. $f(x) = -x^2 + 2x + 5$ **24.** $f(x) = -x^2 - 4x + 1$

25. $h(x) = 4x^2 - 4x + 21$ **26.** $f(x) = 2x^2 - x + 1$

Identifying x-Intercepts of a Quadratic Function **In Exercises 27–32, describe the graph of the quadratic function. Identify the vertex and x-intercept(s). Use a graphing utility to verify your results.**

27. $g(x) = x^2 + 8x + 11$ **28.** $f(x) = x^2 + 10x + 14$

29. $f(x) = -(x^2 - 2x - 15)$

30. $f(x) = -(x^2 + 3x - 4)$

31. $f(x) = -2x^2 + 16x - 31$

32. $f(x) = -4x^2 + 24x - 41$

Writing the Equation of a Parabola **In Exercises 33 and 34, write an equation of the parabola in standard form. Use a graphing utility to graph the equation and verify your result.**

33.

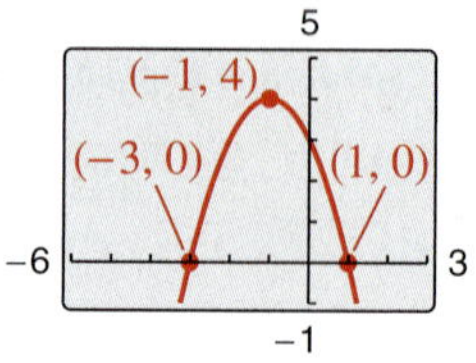

34.

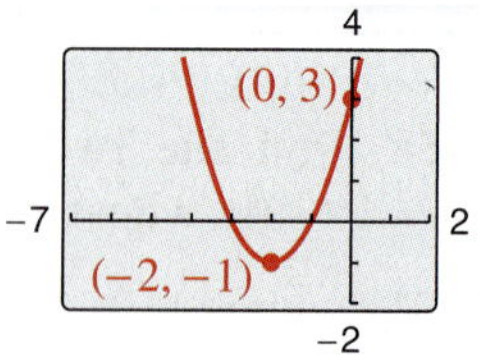

Writing the Equation of a Parabola In Exercises 35–40, write the standard form of the quadratic function whose graph is a parabola with the indicated vertex and that passes through the indicated point. Use a graphing utility to verify your result.

35. Vertex: $(-2, 5)$;	Point: $(0, 9)$
36. Vertex: $(4, 1)$;	Point: $(6, -7)$
37. Vertex: $(1, -2)$;	Point: $(-1, 14)$
38. Vertex: $(-4, -1)$;	Point: $(-2, 4)$
39. Vertex: $\left(\frac{1}{2}, 1\right)$;	Point: $\left(-2, -\frac{21}{5}\right)$
40. Vertex: $\left(-\frac{1}{4}, -1\right)$;	Point: $\left(0, -\frac{17}{16}\right)$

Using a Graph to Identify *x*-Intercepts In Exercises 41–44, determine the x-intercept(s) of the graph visually. Then verify your answer algebraically.

41.

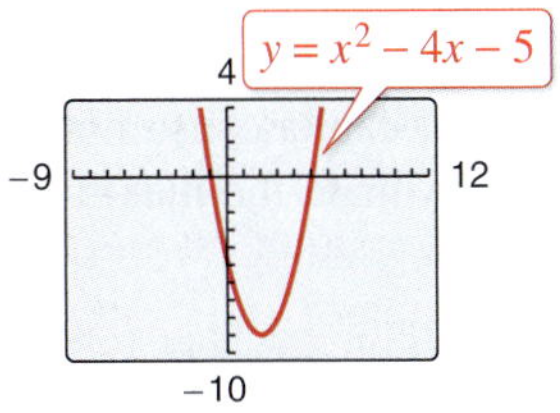

42.

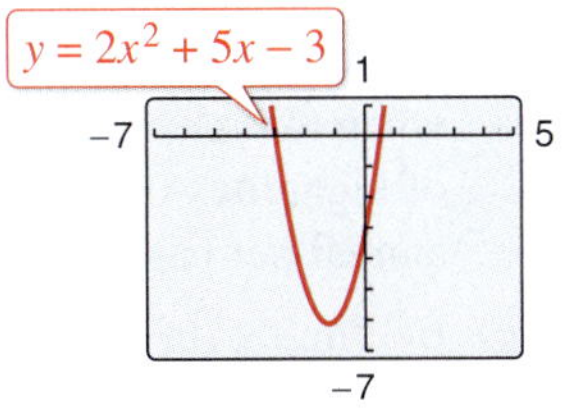

43.

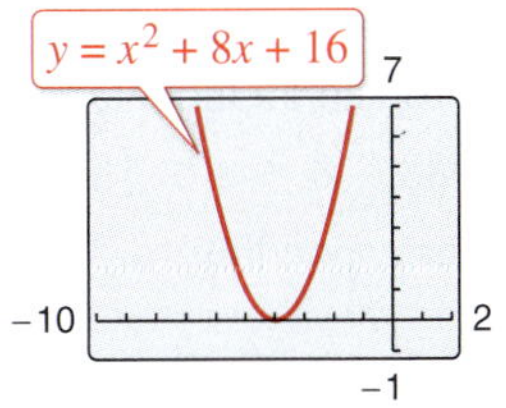

44.

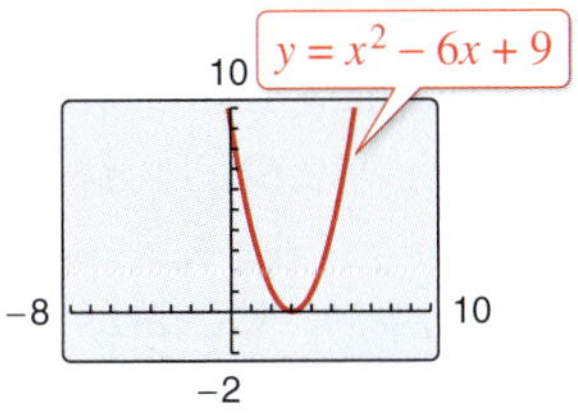

Graphing to Identify *x*-Intercepts In Exercises 45–50, use a graphing utility to graph the quadratic function and find the x-intercepts of the graph. Then find the x-intercepts algebraically to verify your answer.

45. $y = x^2 - 4x$	**46.** $y = -2x^2 + 10x$
47. $y = 2x^2 - 7x - 30$	**48.** $y = 4x^2 + 25x - 21$
49. $y = -\frac{1}{2}(x^2 - 6x - 7)$	**50.** $y = \frac{7}{10}(x^2 + 12x - 45)$

Using the *x*-Intercepts to Write Equations In Exercises 51–54, find two quadratic functions, one that opens upward and one that opens downward, whose graphs have the indicated x-intercepts. (There are many correct answers.)

51. $(-3, 0), (3, 0)$	**52.** $(0, 0), (10, 0)$
53. $(-3, 0), \left(-\frac{1}{2}, 0\right)$	**54.** $\left(-\frac{5}{2}, 0\right), (2, 0)$

Maximizing a Product of Two Numbers In Exercises 55–58, find the two positive real numbers with the indicated sum whose product is a maximum.

55. The sum is 110.

56. The sum is 66.

57. The sum of the first and twice the second is 24.

58. The sum of the first and three times the second is 42.

59. ***Why you should learn it*** *(p. 90)* An indoor physical fitness room consists of a rectangular region with a semicircle on each end. The perimeter of the room is to be a 200-meter single-lane running track.

(a) Draw a diagram that illustrates the situation. Let x and y represent the length and width of the rectangular region, respectively.

(b) Determine the radius of the semicircular ends of the track. Determine the distance, in terms of y, around the inside edge of the two semicircular parts of the track.

(c) Use the result of part (b) to write an equation, in terms of x and y, for the distance traveled in one lap around the track. Solve for y.

(d) Use the result of part (c) to write the area A of the rectangular region as a function of x.

(e) Use a graphing utility to graph the area function from part (d). Use the graph to approximate the dimensions that will produce a rectangle of maximum area.

60. Algebraic-Graphical-Numerical A child care center has 200 feet of fencing to enclose two adjacent rectangular safe play areas (see figure). Use the methods below to determine the dimensions that will produce a maximum enclosed area.

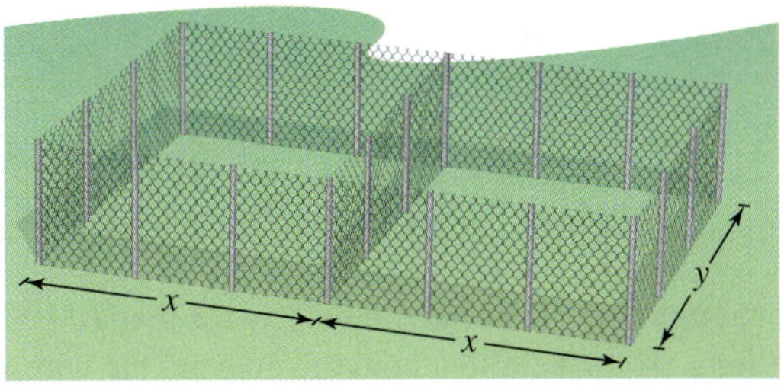

(a) Write the total area A of the play areas as a function of x.

(b) Use the *table* feature of a graphing utility to create a table showing possible values of x and the corresponding total area A of the play areas. Use the table to estimate the dimensions that will produce the maximum enclosed area.

(c) Use the graphing utility to graph the area function. Use the graph to approximate the dimensions that will produce the maximum enclosed area.

(d) Write the area function in standard form to algebraically find the dimensions that will produce the maximum enclosed area.

(e) Compare your results from parts (b), (c), and (d).

61. Height of a Projectile The height y (in feet) of a punted football is approximated by

$$y = -\frac{16}{2025}x^2 + \frac{9}{5}x + \frac{3}{2}$$

where x is the horizontal distance (in feet) from where the football is punted. (See figure.)

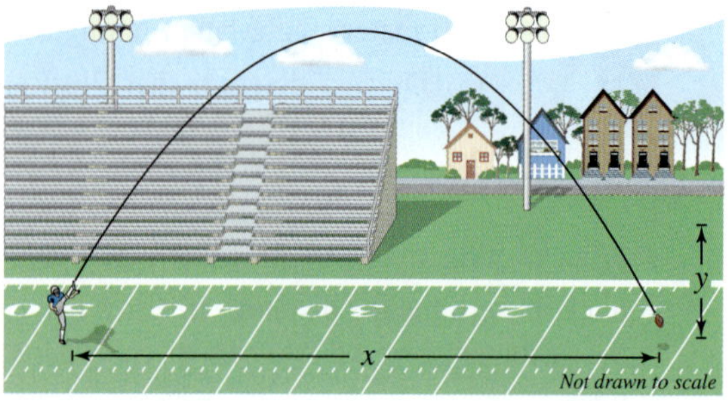

(a) Use a graphing utility to graph the path of the football.

(b) How high is the football when it is punted? (*Hint:* Find y when $x = 0$.)

(c) What is the maximum height of the football?

(d) How far from the punter does the football strike the ground?

62. Physics The path of a diver is approximated by

$$y = -\frac{4}{9}x^2 + \frac{24}{9}x + 12$$

where y is the height (in feet) and x is the horizontal distance (in feet) from the end of the diving board (see figure). What is the maximum height of the diver?

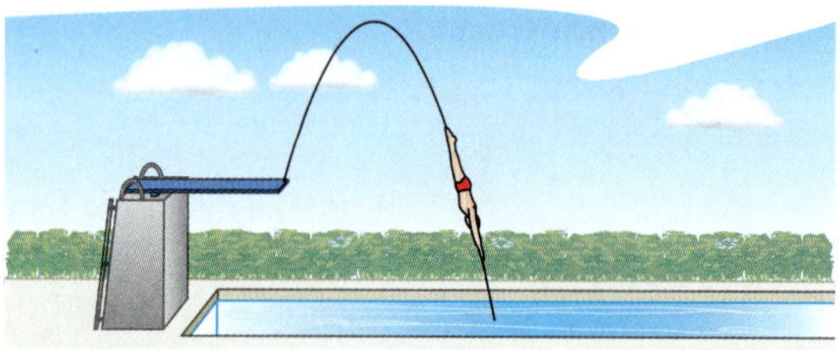

63. Geometry You have a steel wire that is 100 inches long. To make a sign holder, you bend the wire x inches from each end to form two right angles. To use the sign holder, you insert each end 6 inches into the ground. (See figure.)

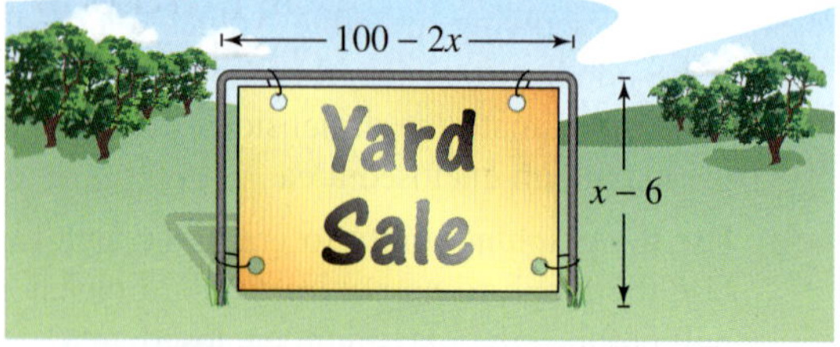

(a) Write a function for the rectangular area A enclosed by the sign holder in terms of x.

(b) Use the *table* feature of a graphing utility to determine the value of x that maximizes the rectangular area enclosed by the sign holder.

64. Economics The monthly revenue R (in thousands of dollars) from the sales of a digital picture frame is approximated by $R(p) = -10p^2 + 1580p$, where p is the price per unit (in dollars).

(a) Find the monthly revenues for unit prices of \$50, \$70, and \$90.

(b) Find the unit price that will yield a maximum monthly revenue.

(c) What is the maximum monthly revenue?

65. Public Health For selected years from 1955 through 2016, the annual per capita consumption C of cigarettes by Americans (ages 18 and older) can be modeled by

$$C(t) = -1.08t^2 + 19.2t + 4048, \quad 5 \le t \le 66$$

where t is the year, with $t = 5$ corresponding to 1955. (*Sources:* Centers for Disease Control and Prevention and Statista)

(a) Use a graphing utility to graph the model.

(b) Use the graph of the model to approximate the year when the maximum annual per capita consumption of cigarettes occurred. Approximate the maximum annual per capita consumption.

(c) Beginning in 1966, all cigarette packages were required by law to carry a health warning. Do you think the warning had any effect? Explain.

66. Demography The population P of Cuba (in thousands) from 2008 through 2018 can be modeled by

$$P(t) = -2.61t^2 + 54.02t + 10{,}988, \quad 8 \le t \le 18$$

where t is the year, with $t = 8$ corresponding to 2008. (*Source:* U.S. Census Bureau)

(a) According to the model, in what year did Cuba have its greatest population? What was the population?

(b) According to the model, what will Cuba's population be in the year 2050? Is this result reasonable? Explain.

Focusing on Concepts

True or False? In Exercises 67 and 68, determine whether the statement is true or false. Justify your answer.

67. The function $f(x) = a(x - 5)^2$ has exactly one x-intercept for any nonzero value of a.

68. The quadratic functions $f(x) = 3x^2 + 6x + 7$ and $g(x) = 3x^2 + 6x - 1$ have the same vertex.

69. Error Analysis Describe the error.

The graph of $y = 2x^2 + 3$ is a vertical shrink by a factor of two and an upward shift of three units relative to the graph of $y = x^2$.

70. Error Analysis Describe the error.

The axis of symmetry of the graph of $y = 4x^2 - 16x + 11$ is $x = -2$.

Library of Parent Functions **In Exercises 71 and 72, determine which equation(s) may be represented by the graph shown. (There may be more than one correct answer.)**

71. (a) $f(x) = -(x - 4)^2 + 2$
(b) $f(x) = -(x + 2)^2 + 4$
(c) $f(x) = -(x + 2)^2 - 4$
(d) $f(x) = -x^2 - 4x - 8$
(e) $f(x) = -(x - 2)^2 - 4$
(f) $f(x) = -x^2 + 4x - 8$

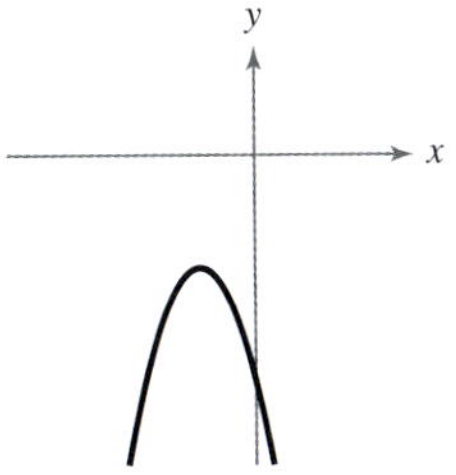

72. (a) $f(x) = (x - 1)^2 + 3$
(b) $f(x) = (x + 1)^2 + 3$
(c) $f(x) = (x - 3)^2 + 1$
(d) $f(x) = x^2 + 2x + 4$
(e) $f(x) = (x + 3)^2 + 1$
(f) $f(x) = x^2 + 6x + 10$

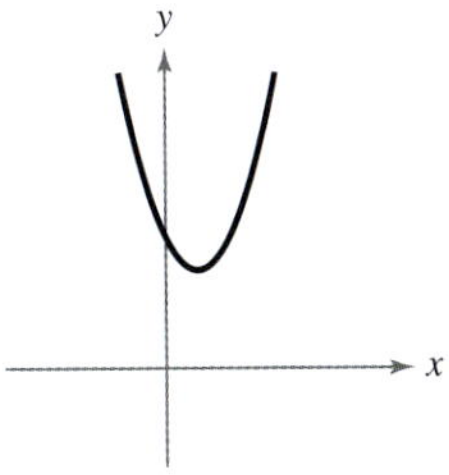

Describing Parabolas **In Exercises 73–76, let z represent a positive real number. Describe how the family of parabolas represented by the function compares with the graph of $g(x) = x^2$.**

73. $f(x) = (x - z)^2$

74. $f(x) = x^2 - z$

75. $f(x) = z(x - 3)^2$

76. $f(x) = zx^2 + 4$

Think About It **In Exercises 77–80, find the value of b such that the function has the indicated maximum or minimum value.**

77. $f(x) = -x^2 + bx - 75$; Maximum value: 25

78. $f(x) = -x^2 + bx - 16$; Maximum value: 48

79. $f(x) = x^2 + bx + 26$; Minimum value: 10

80. $f(x) = x^2 + bx - 25$; Minimum value: -50

81. Proof Let x and y be two positive real numbers whose sum is S. Show that the maximum product of x and y occurs when x and y are both equal to $S/2$.

82. Proof Assume that the function $f(x) = ax^2 + bx + c$, $a \neq 0$, has two real zeros. Show that the x-coordinate of the vertex of the graph is the average of the zeros of f. (*Hint:* Use the Quadratic Formula.)

83. Writing The parabola in the figure has an equation of the form $y = ax^2 + bx - 4$. Find the equation of this parabola two different ways, by hand and with technology. Write a paragraph describing the methods you used and comparing the results.

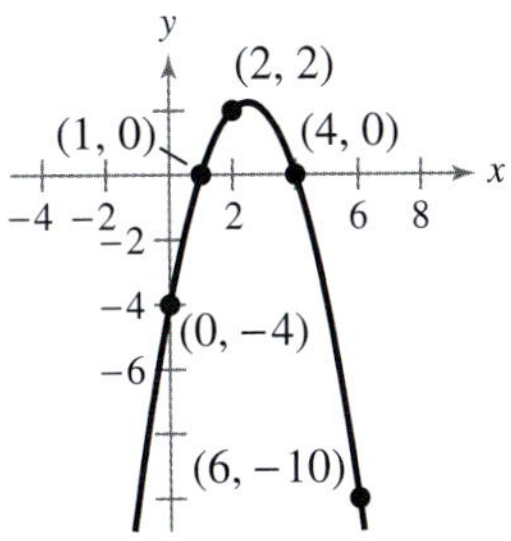

84. HOW DO YOU SEE IT? The graph shows a quadratic function of the form $R(t) = at^2 + bt + c$, which represents the yearly revenues for a company, where $R(t)$ is the revenue in year t.

(a) Is the value of a positive, negative, or zero?

(b) Write an expression in terms of a and b that represents the year t when the company had the least revenue.

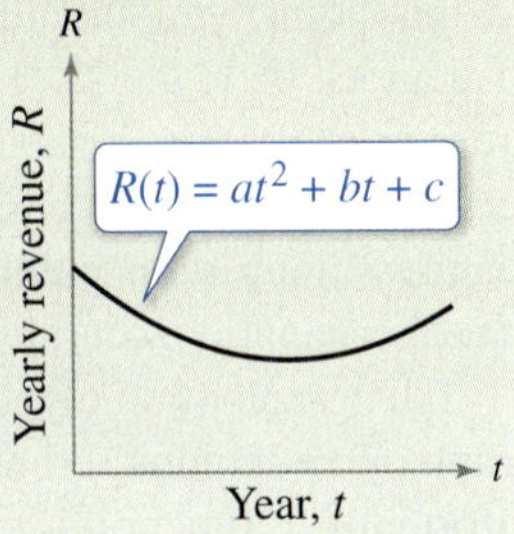

(c) The company made the same yearly revenues in 2008 and 2018. Estimate the year in which the company had the least revenue.

(d) Assume that the model is still valid today. Are the yearly revenues currently increasing, decreasing, or constant? Explain.

85. Think About It The annual profit P (in dollars) of a company is modeled by a function of the form

$$P = at^2 + bt + c$$

where t represents the year. Discuss which of the models below the company might prefer.

(a) a is positive and $t \geq -b/(2a)$.
(b) a is positive and $t \leq -b/(2a)$.
(c) a is negative and $t \geq -b/(2a)$.
(d) a is negative and $t \leq -b/(2a)$.

86. Verifying the Vertex Write the quadratic function

$$f(x) = ax^2 + bx + c$$

in standard form to verify that the vertex occurs at

$$\left(-\frac{b}{2a}, f\left(-\frac{b}{2a}\right)\right).$$

Cumulative Mixed Review

Finding Points of Intersection **In Exercises 87 and 88, algebraically determine any point(s) of intersection of the graphs of the equations. Verify your results using the *intersect* feature of a graphing utility.**

87. $x + y = 8$
$-\frac{2}{3}x + y = 6$

88. $y = 9 - x^2$
$y = x + 3$

89. *Project: Height of a Softball* To work an extended application analyzing the heights of a softball after it has been dropped, visit this textbook's website at *LarsonPrecalculus.com.*

2.2 Polynomial Functions of Higher Degree

Graphs of Polynomial Functions

At this point, you should be able to sketch accurate graphs of polynomial functions of degrees 0, 1, and 2. The graphs of polynomial functions of degree greater than 2 are more difficult to sketch by hand. In this section, however, you will learn how to recognize some of the basic features of the graphs of polynomial functions. Using these features along with point plotting, intercepts, and symmetry, you should be able to make reasonably accurate sketches *by hand*.

The graph of a polynomial function is **continuous.** Essentially, this means that the graph of a polynomial function has no breaks, holes, or gaps, as shown in Figure 2.6(a). Informally, a function is continuous when its graph can be drawn with a pencil without lifting the pencil from the paper. The graph of the piecewise-defined function in Figure 2.6(b) has a gap, so this function is *not* continuous.

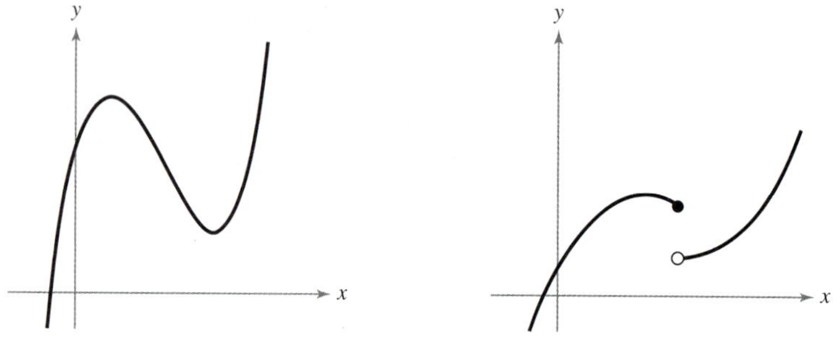

(a) ***Polynomial functions have continuous graphs.***

(b) ***Functions with graphs that are not continuous are not polynomial functions.***

Figure 2.6

Another feature of the graph of a polynomial function is that it has only smooth, rounded turns, as shown in Figure 2.7(a). The graph of a polynomial function cannot have a sharp turn, such as the graph of the function shown in Figure 2.7(b).

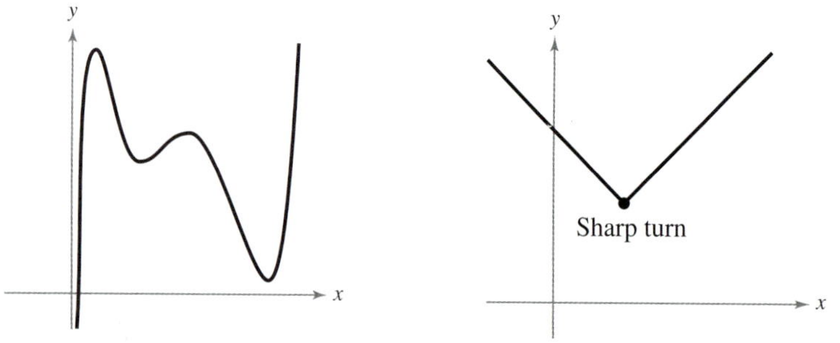

(a) ***Polynomial functions have graphs with smooth, rounded turns.***

(b) ***Functions with graphs that have sharp turns are not polynomial functions.***

Figure 2.7

The graphs of polynomial functions of degree 1 are lines, and those of functions of degree 2 are parabolas. The graphs of all polynomial functions are smooth and continuous. A polynomial function of degree n has the form

$$f(x) = a_n x^n + a_{n-1}x^{n-1} + \cdot \cdot \cdot + a_2x^2 + a_1x + a_0$$

where n is a positive integer and $a_n \neq 0$.

What you should learn

- Use transformations to sketch graphs of polynomial functions.
- Use the Leading Coefficient Test to determine the end behavior of graphs of polynomial functions.
- Find and use zeros of polynomial functions as sketching aids.
- Use the Intermediate Value Theorem to help locate zeros of polynomial functions.

Why you should learn it

You can use polynomial functions to model various aspects of nature, such as the growth of a red oak tree, as shown in Exercise 111 on page 112.

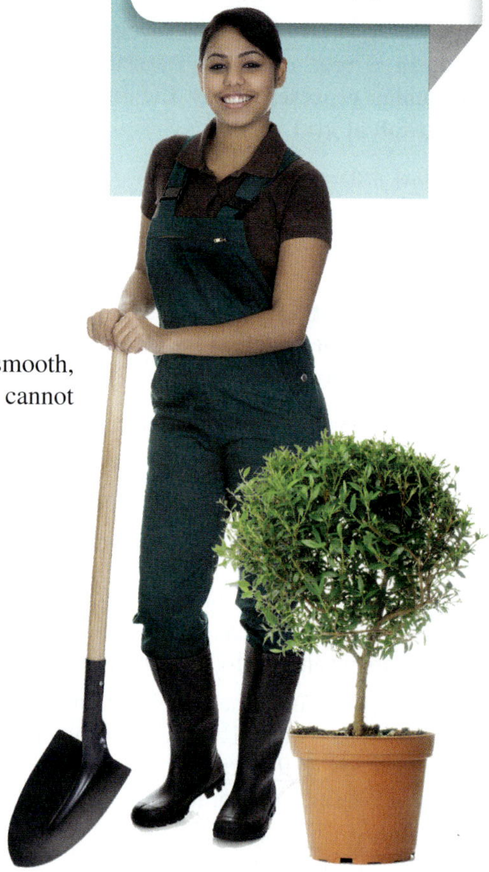

The polynomial functions that have the simplest graphs are monomials of the form $f(x) = x^n$, where n is an integer greater than zero. The greater the value of n, the flatter the graph near the origin. When n is even, the graph is similar to the graph of $f(x) = x^2$, touches the x-axis at the x-intercept, and is symmetric with respect to the y-axis. When n is odd, the graph is similar to the graph of $f(x) = x^3$, crosses the x-axis at the x-intercept, and is symmetric with respect to the origin. Polynomial functions of the form $f(x) = x^n$ are often referred to as **power functions.**

Library of Parent Functions: Cubic Function

The basic characteristics of the *parent cubic function* $f(x) = x^3$ are summarized below and on the inside cover of this text.

Graph of $f(x) = x^3$
Domain: $(-\infty, \infty)$
Range: $(-\infty, \infty)$
Intercept: $(0, 0)$
Increasing on $(-\infty, \infty)$
Odd function
Origin symmetry

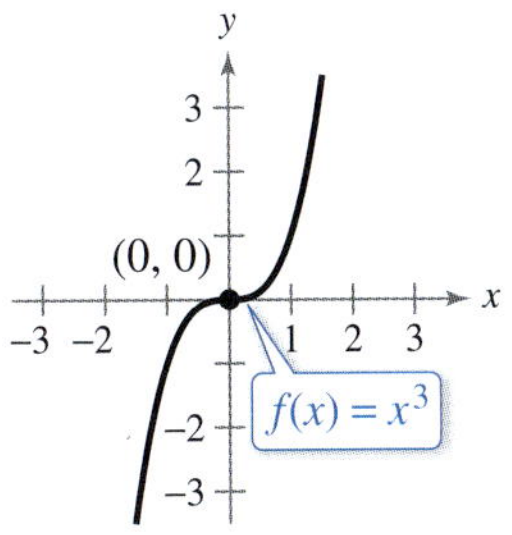

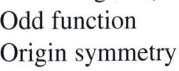

Explore the Concept

Use a graphing utility to graph $y = x^n$ for $n = 2$, 4, and 8. (Use the viewing window $-1.5 \le x \le 1.5$ and $-1 \le y \le 6$.) Compare the graphs. In the interval $(-1, 1)$, which graph is on the bottom? Outside the interval $(-1, 1)$, which graph is on the bottom?

Use a graphing utility to graph $y = x^n$ for $n = 3$, 5, and 7. (Use the viewing window $-1.5 \le x \le 1.5$ and $-4 \le y \le 4$.) Compare the graphs. In the intervals $(-\infty, -1)$ and $(0, 1)$, which graph is on the bottom? In the intervals $(-1, 0)$ and $(1, \infty)$, which graph is on the bottom?

EXAMPLE 1 Library of Parent Functions: $f(x) = x^3$

See LarsonPrecalculus.com for an interactive version of this type of example.

Sketch the graphs of (a) $g(x) = -x^3$, (b) $h(x) = x^3 + 1$, and (c) $k(x) = (x - 1)^3$ by hand.

Solution

a. With respect to the graph of $f(x) = x^3$, the graph of g is obtained by a *reflection* in the x-axis, as shown in Figure 2.8.

b. With respect to the graph of $f(x) = x^3$, the graph of h is obtained by a vertical shift one unit *upward*, as shown in Figure 2.9.

c. With respect to the graph of $f(x) = x^3$, the graph of k is obtained by a horizontal shift one unit *to the right*, as shown in Figure 2.10.

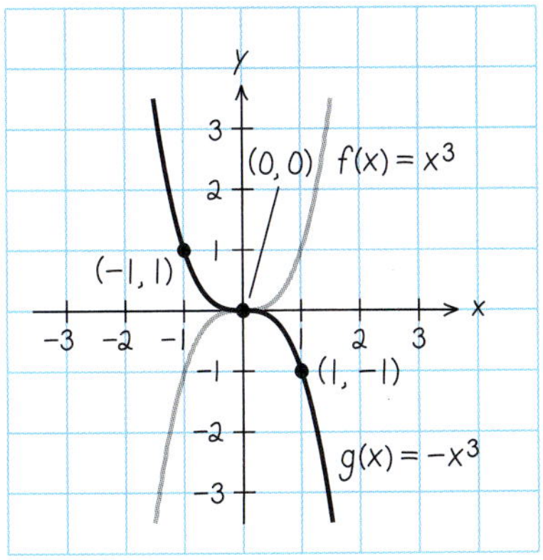

Figure 2.8

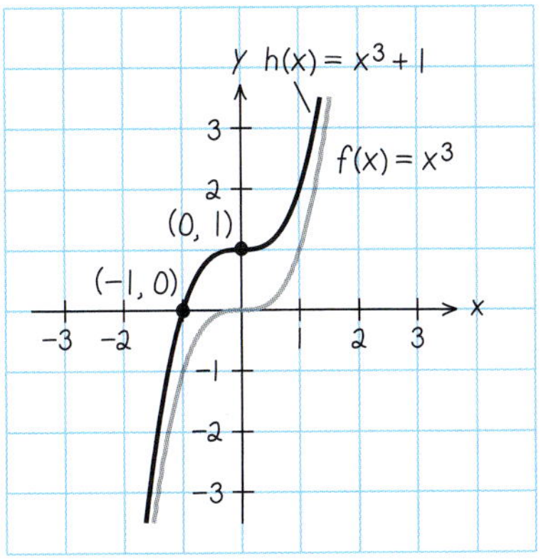

Figure 2.9

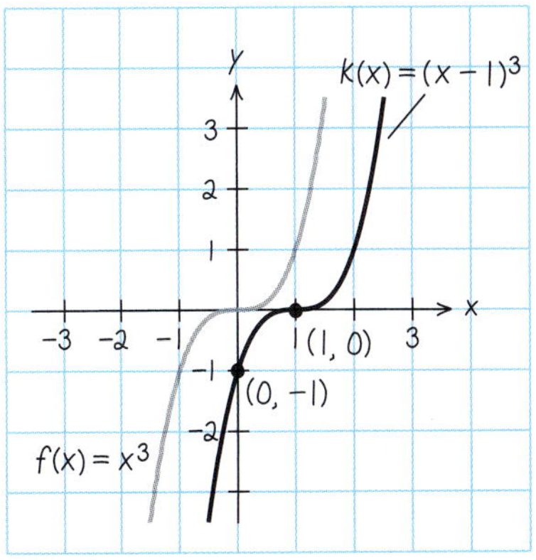

Figure 2.10

✓ ***Checkpoint*** *Audio-video solution in English & Spanish at LarsonPrecalculus.com*

Sketch the graphs of (a) $g(x) = -x^3 + 2$, (b) $h(x) = x^3 - 5$, and (c) $k(x) = (x + 2)^3$ by hand.

The Leading Coefficient Test

In Example 1, note that all three graphs eventually rise or fall without bound as x moves to the right. Whether the graph of a nonconstant polynomial function eventually rises or falls can be determined by the polynomial function's degree (even or odd) and by its leading coefficient, as indicated in the **Leading Coefficient Test.**

Leading Coefficient Test

As x moves without bound to the left or to the right, the graph of the nonconstant polynomial function

$$f(x) = a_n x^n + \cdot \cdot \cdot + a_1 x + a_0, \quad a_n \neq 0$$

eventually rises or falls in the following manner.

1. When n is odd:

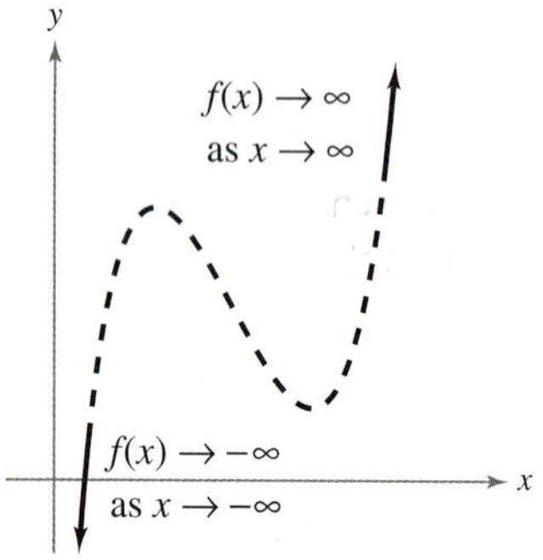

If the leading coefficient is positive ($a_n > 0$), then the graph falls to the left and rises to the right.

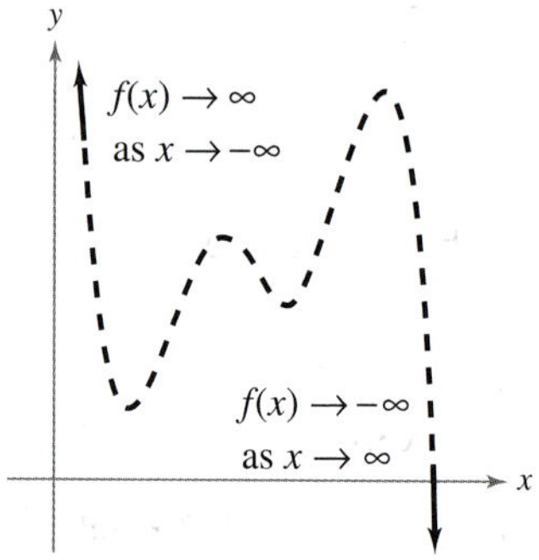

If the leading coefficient is negative ($a_n < 0$), then the graph rises to the left and falls to the right.

2. When n is even:

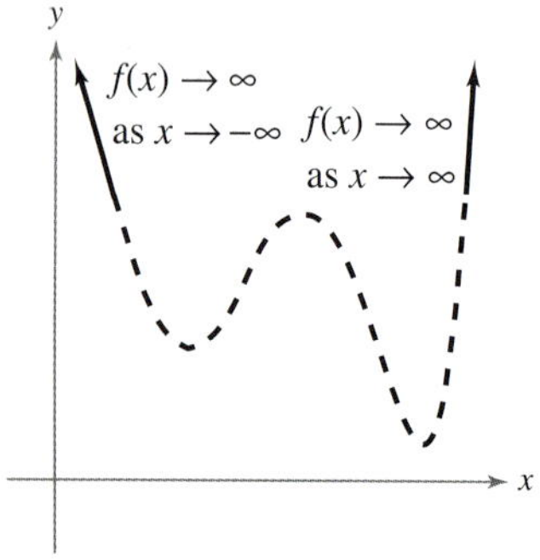

If the leading coefficient is positive ($a_n > 0$), then the graph rises to the left and right.

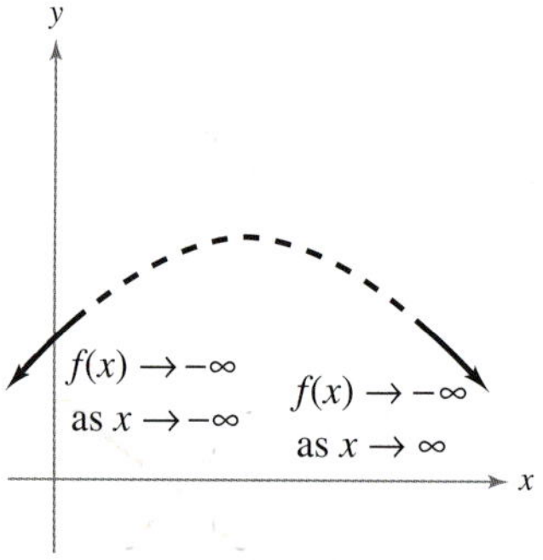

If the leading coefficient is negative ($a_n < 0$), then the graph falls to the left and right.

Note that the dashed portions of the graphs indicate that the test determines only the right-hand and left-hand behavior of the graph.

As you continue to study polynomial functions and their graphs, you will notice that the degree of a polynomial plays an important role in determining other characteristics of the polynomial function and its graph.

Explore the Concept

For each function, identify the degree of the function and whether the degree of the function is even or odd. Identify the leading coefficient and whether the leading coefficient is positive or negative. Use a graphing utility to graph each function. Describe the relationship between the degree and sign of the leading coefficient of the function, and the right- and left-hand behavior of the graph of the function.

a. $y = x^3 - 2x^2 - x + 1$
b. $y = 2x^5 + 2x^2 - 5x + 1$
c. $y = -2x^5 - x^2 + 5x + 3$
d. $y = -x^3 + 5x - 2$
e. $y = 2x^2 + 3x - 4$
f. $y = x^4 - 3x^2 + 2x - 1$
g. $y = -x^2 + 3x + 2$
h. $y = -x^6 - x^2 - 5x + 4$

Algebra Help

The notation "$f(x) \to -\infty$ as $x \to -\infty$" indicates that the graph falls to the left, the notation "$f(x) \to \infty$ as $x \to \infty$" indicates that the graph rises to the right, and so on.

EXAMPLE 2 Applying the Leading Coefficient Test

Use the Leading Coefficient Test to describe the right-hand and left-hand behavior of the graph of $f(x) = -x^3 + 4x$.

Solution

Because the degree is odd and the leading coefficient is negative, the graph rises to the left and falls to the right, as shown in the figure.

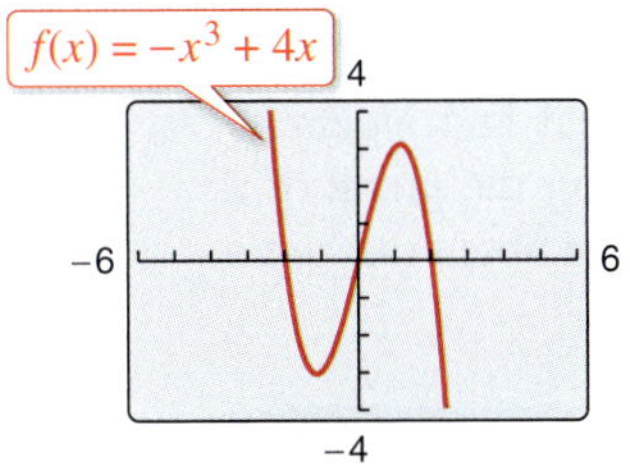

Checkpoint Audio-video solution in English & Spanish at LarsonPrecalculus.com

Use the Leading Coefficient Test to describe the right-hand and left-hand behavior of the graph of $f(x) = 2x^3 - 3x^2 + 5$.

EXAMPLE 3 Applying the Leading Coefficient Test

Use the Leading Coefficient Test to describe the right-hand and left-hand behavior of the graph of each polynomial function.

a. $f(x) = x^4 - 5x^2 + 4$ **b.** $f(x) = x^5 - x$

Solution

a. Because the degree is even and the leading coefficient is positive, the graph rises to the left and right, as shown in the figure.

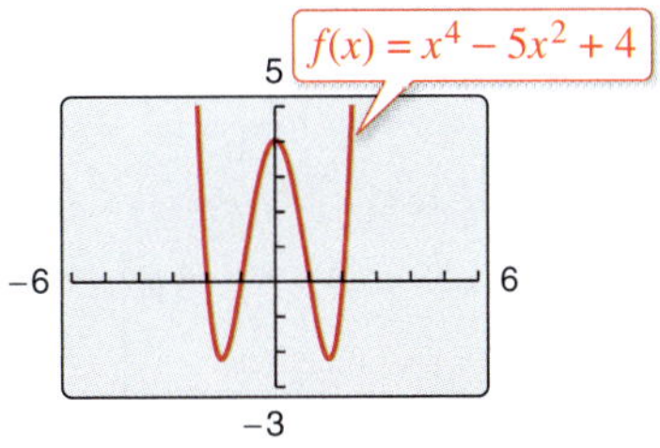

b. Because the degree is odd and the leading coefficient is positive, the graph falls to the left and rises to the right, as shown in the figure.

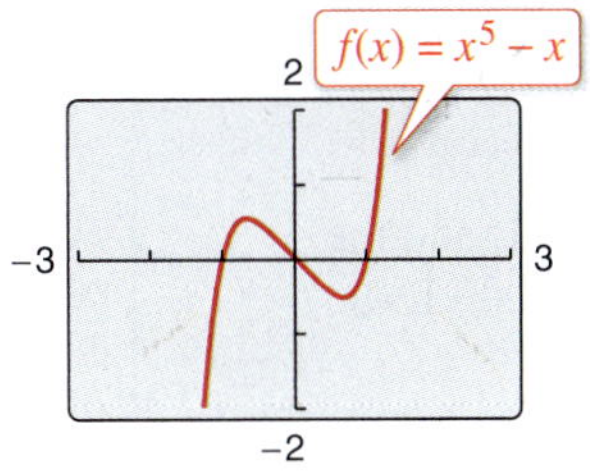

Checkpoint Audio-video solution in English & Spanish at LarsonPrecalculus.com

Use the Leading Coefficient Test to describe the right-hand and left-hand behavior of the graph of each polynomial function.

a. $f(x) = -x^4 + 2x^2 + 6$ **b.** $f(x) = -x^5 + 3x^4 - x$

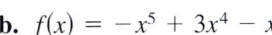

In Examples 2 and 3, note that the Leading Coefficient Test tells you only whether the graph *eventually* rises or falls to the right or left. Other characteristics of the graph, such as intercepts and minimum and maximum points, must be determined by other tests.

Explore the Concept

For each of the graphs in Examples 2 and 3, count the number of zeros of the polynomial function and the number of relative minima and relative maxima, and compare these numbers with the degree of the polynomial. What do you observe?

Zeros of Polynomial Functions

It can be shown that for a polynomial function f of degree $n > 0$, the statements below are true.

1. The function f has at most n real zeros. (You will study this result in detail in Section 2.5 on the Fundamental Theorem of Algebra.)
2. The graph of f has at most $n - 1$ **relative extrema** (relative minima or maxima).

Recall that a zero of a function f is a number x for which $f(x) = 0$. Finding the zeros of polynomial functions is one of the most important problems in algebra. You have already seen that there is a strong interplay between graphical and algebraic approaches to this problem. Sometimes you can use information about the graph of a function to help find its zeros. In other cases, you can use information about the zeros of a function to find a good viewing window.

Real Zeros of Polynomial Functions

When f is a nonconstant polynomial function and a is a real number, the statements below are equivalent.

1. $x = a$ is a *zero* of the function f.
2. $x = a$ is a *solution* of the polynomial equation $f(x) = 0$.
3. $(x - a)$ is a *factor* of the polynomial $f(x)$.
4. $(a, 0)$ is an *x-intercept* of the graph of f.

Finding zeros of polynomial functions is closely related to factoring and finding x-intercepts, as demonstrated in Examples 4, 5, and 6.

EXAMPLE 4 Finding Zeros of a Polynomial Function

Find all real zeros of

$$f(x) = x^3 - x^2 - 2x.$$

Algebraic Solution

$f(x) = x^3 - x^2 - 2x$ — Write original function.

$0 = x^3 - x^2 - 2x$ — Substitute 0 for $f(x)$.

$0 = x(x^2 - x - 2)$ — Remove common monomial factor.

$0 = x(x - 2)(x + 1)$ — Factor completely.

So, the real zeros are $x = 0$, $x = 2$, and $x = -1$, and the corresponding x-intercepts are $(0, 0)$, $(2, 0)$, and $(-1, 0)$.

Check

$(0)^3 - (0)^2 - 2(0) = 0$ — $x = 0$ is a zero. ✓

$(2)^3 - (2)^2 - 2(2) = 0$ — $x = 2$ is a zero. ✓

$(-1)^3 - (-1)^2 - 2(-1) = 0$ — $x = -1$ is a zero. ✓

Graphical Solution

The graph of f has the x-intercepts

$$(0, 0), \quad (2, 0), \quad \text{and} \quad (-1, 0)$$

as shown in the figure. So, the real zeros of f are

$$x = 0, \quad x = 2, \quad \text{and} \quad x = -1.$$

Use the *zero* or *root* feature of a graphing utility to verify these zeros.

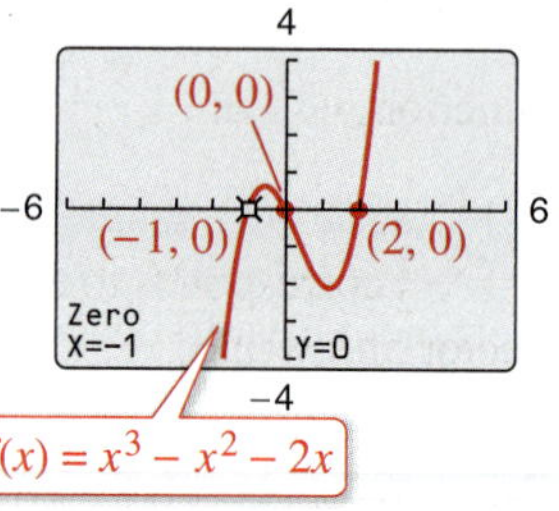

✓ Checkpoint *Audio-video solution in English & Spanish at LarsonPrecalculus.com*

Find all real zeros of $f(x) = x^3 + x^2 - 6x$.

EXAMPLE 5 Analyzing a Polynomial Function

Find all real zeros and relative extrema of $f(x) = -2x^4 + 2x^2$.

Solution

$0 = -2x^4 + 2x^2$ Substitute 0 for $f(x)$.

$0 = -2x^2(x^2 - 1)$ Remove common monomial factor.

$0 = -2x^2(x - 1)(x + 1)$ Factor completely.

So, the real zeros are $x = 0$, $x = 1$, and $x = -1$, and the corresponding x-intercepts are $(0, 0)$, $(1, 0)$, and $(-1, 0)$, as shown in Figure 2.11. Using the *minimum* and *maximum* features of a graphing utility, you can approximate the three relative extrema to be $(-0.71, 0.5)$, $(0, 0)$, and $(0.71, 0.5)$.

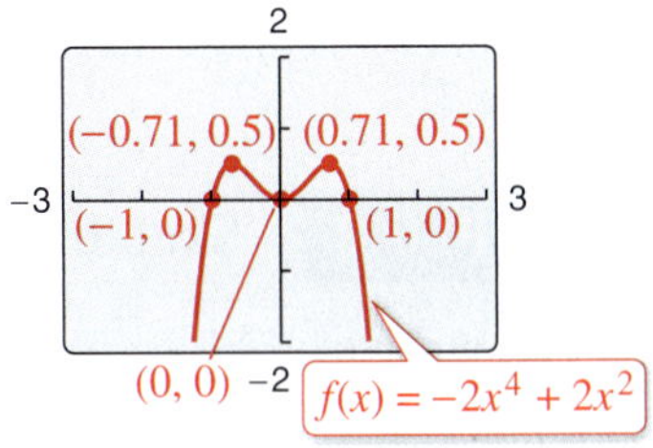

Figure 2.11

Checkpoint Audio-video solution in English & Spanish at LarsonPrecalculus.com

Find all real zeros and relative extrema of $f(x) = x^3 + 2x^2 - 3x$.

Repeated Zeros

If $(x - a)^k$ is a factor of a polynomial function and $k > 1$, then $(x - a)^k$ yields a **repeated zero** $x = a$ of **multiplicity** k.

1. If k is odd, then the graph *crosses* the x-axis at $x = a$.
2. If k is even, then the graph *touches* the x-axis (but does not cross the x-axis) at $x = a$.

Algebra Help

In Example 5, note that because k is even, the factor $-2x^2$ yields the repeated zero $x = 0$. The graph touches (but does not cross) the x-axis at $x = 0$, as shown in Figure 2.11.

EXAMPLE 6 Analyzing a Polynomial Function

To find all real zeros of $f(x) = x^5 - 3x^3 - x^2 - 4x - 1$, use the *zero* feature of a graphing utility. From Figure 2.12, the zeros are $x \approx -1.86$, $x \approx -0.25$, and $x \approx 2.11$. Note that this fifth-degree polynomial factors as

$$f(x) = x^5 - 3x^3 - x^2 - 4x - 1 = (x^2 + 1)(x^3 - 4x - 1).$$

The three zeros obtained above are the zeros of the cubic factor $x^3 - 4x - 1$. The quadratic factor $x^2 + 1$ has no real zeros but does have two *imaginary* zeros. You will learn more about imaginary zeros in Section 2.5.

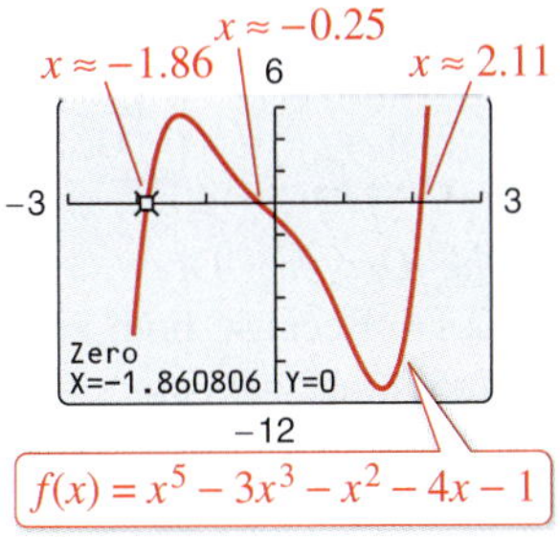

Figure 2.12

Checkpoint Audio-video solution in English & Spanish at LarsonPrecalculus.com

Find all real zeros of $f(x) = x^5 - 2x^4 + x^3 - 2x^2$.

EXAMPLE 7 Finding a Polynomial Function with Given Zeros

Find a polynomial function with zeros $-\frac{1}{2}$, 3, and 3. (There are many correct solutions.)

Solution

Note that the zero $x = -\frac{1}{2}$ corresponds to either $\left(x + \frac{1}{2}\right)$ or $(2x + 1)$. To avoid fractions, choose the second factor and write

$$f(x) = (2x + 1)(x - 3)^2 = (2x + 1)(x^2 - 6x + 9) = 2x^3 - 11x^2 + 12x + 9.$$

Checkpoint Audio-video solution in English & Spanish at LarsonPrecalculus.com

Find a polynomial function with zeros -2, -1, 1, and 2. (There are many correct solutions.)

Note in Example 7 that there are many polynomial functions with the indicated zeros. In fact, multiplying the function by any real number does not change the zeros of the function. For instance, multiply the function from Example 7 by $\frac{1}{2}$ to obtain

$$f(x) = x^3 - \tfrac{11}{2}x^2 + 6x + \tfrac{9}{2}.$$

Then find the zeros of the function. You will obtain the zeros $-\frac{1}{2}$, 3, and 3, as given in Example 7.

EXAMPLE 8 Sketching the Graph of a Polynomial Function

Sketch the graph of

$$f(x) = 3x^4 - 4x^3$$

by hand.

Solution

1. *Apply the Leading Coefficient Test.* Because the leading coefficient is positive and the degree is even, you know that the graph eventually rises to the left and to the right (see Figure 2.13).
2. *Find the Real Zeros of the Polynomial.* By factoring

   $$f(x) = 3x^4 - 4x^3 = x^3(3x - 4)$$

   you can see that the real zeros of f are $x = 0$ (of odd multiplicity 3) and $x = \frac{4}{3}$ (of odd multiplicity 1). So, the x-intercepts occur at $(0, 0)$ and $\left(\frac{4}{3}, 0\right)$. Add these points to your graph, as shown in Figure 2.13.
3. *Plot a Few Additional Points.* To sketch the graph by hand, find a few additional points, as shown in the table. Be sure to choose points between the zeros and to the left and right of the zeros. Then plot the points (see Figure 2.14).

x	-1	0.5	1	1.5
$f(x)$	7	-0.31	-1	1.69

4. *Draw the Graph.* Draw a continuous curve through the points, as shown in Figure 2.14. Because both zeros are of odd multiplicity, you know that the graph should cross the x-axis at $x = 0$ and $x = \frac{4}{3}$. When you are unsure of the shape of a portion of the graph, plot some additional points.

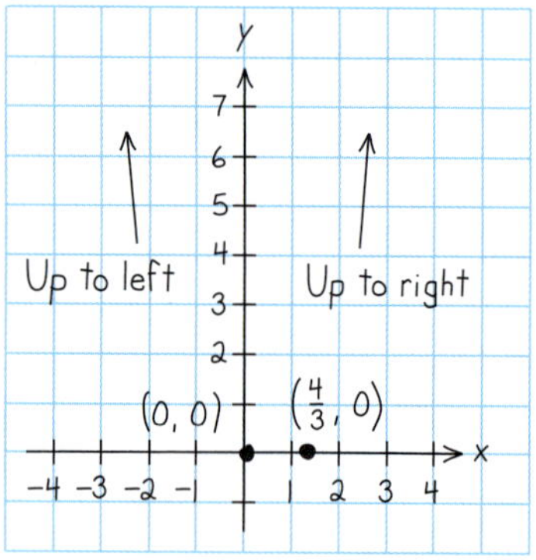

Figure 2.13

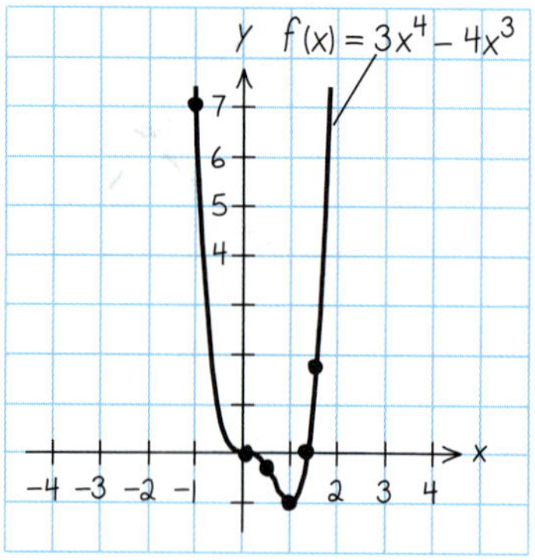

Figure 2.14

Checkpoint *Audio-video solution in English & Spanish at LarsonPrecalculus.com*

Sketch the graph of $f(x) = 2x^3 - 6x^2$ by hand.

Technology Tip

Because it is easy to make mistakes when entering functions into a graphing utility, it is important to understand the basic shapes of graphs and to be able to graph simple polynomials *by hand.* For instance, suppose you had entered the function in Example 8 as $y = 3x^5 - 4x^3$. From the graph, what mathematical principles would alert you to the fact that you had made a mistake?

Insight

Some standardized tests may ask you to sketch the graph of a polynomial function or choose an equation given a graph. Finding the real zeros of the polynomial can help you solve these types of problems.

EXAMPLE 9 Sketching the Graph of a Polynomial Function

Sketch the graph of

$$f(x) = -2x^3 + 6x^2 - \frac{9}{2}x.$$

Solution

1. *Apply the Leading Coefficient Test.* Because the leading coefficient is negative and the degree is odd, you know that the graph eventually rises to the left and falls to the right (see Figure 2.15).

2. *Find the Real Zeros of the Polynomial.* By factoring

$$\begin{aligned} f(x) &= -2x^3 + 6x^2 - \tfrac{9}{2}x \\ &= -\tfrac{1}{2}x(4x^2 - 12x + 9) \\ &= -\tfrac{1}{2}x(2x - 3)^2 \end{aligned}$$

you can see that the real zeros of f are $x = 0$ (of odd multiplicity 1) and $x = \frac{3}{2}$ (of even multiplicity 2). So, the x-intercepts occur at $(0, 0)$ and $\left(\frac{3}{2}, 0\right)$. Add these points to your graph, as shown in Figure 2.15.

3. *Plot a Few Additional Points.* To sketch the graph, find a few additional points, as shown in the table. Then plot the points (see Figure 2.16).

x	-0.5	0.5	1	2
$f(x)$	4	-1	-0.5	-1

4. *Draw the Graph.* Draw a continuous curve through the points, as shown in Figure 2.16. As indicated by the multiplicities of the zeros, the graph crosses the x-axis at $(0, 0)$ and touches (but does not cross) the x-axis at $\left(\frac{3}{2}, 0\right)$.

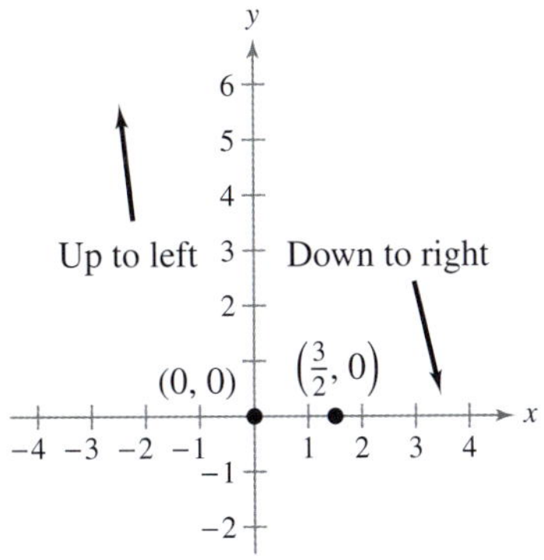

Figure 2.15

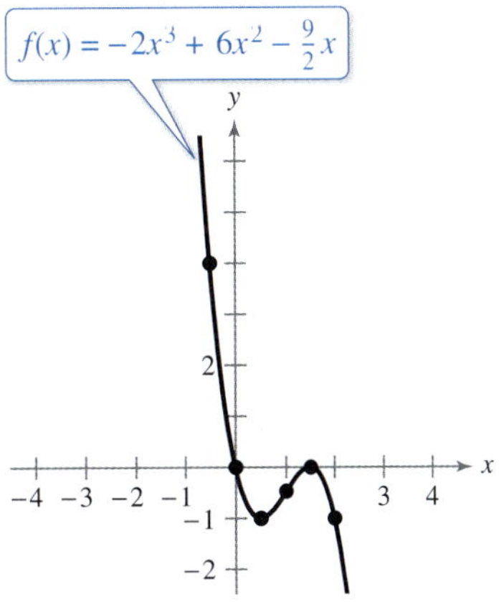

Figure 2.16

Checkpoint *Audio-video solution in English & Spanish at LarsonPrecalculus.com*

Sketch the graph of $f(x) = -\frac{1}{4}x^4 + \frac{3}{2}x^3 - \frac{9}{4}x^2$.

Technology Tip

Remember that when using a graphing utility to verify your graphs, you may need to adjust your viewing window in order to see all the features of the graph.

Algebra Help

Observe in Example 9 that the sign of $f(x)$ is positive to the left of and negative to the right of the zero $x = 0$. Similarly, the sign of $f(x)$ is negative to the left and to the right of the zero $x = \frac{3}{2}$. This suggests that if a zero of a polynomial function is of *odd* multiplicity, then the sign of $f(x)$ changes from one side of the zero to the other side. If a zero is of *even* multiplicity, then the sign of $f(x)$ does not change from one side of the zero to the other side. The following table helps to illustrate this result.

x	-0.5	0	0.5
$f(x)$	4	0	-1
Sign	$+$		$-$

x	1	$\frac{3}{2}$	2
$f(x)$	-0.5	0	-1
Sign			

This sign analysis may be helpful in graphing polynomial functions.

The Intermediate Value Theorem

The **Intermediate Value Theorem** implies that if $(a, f(a))$ and $(b, f(b))$ are two points on the graph of a polynomial function such that $f(a) \neq f(b)$, then for any number d between $f(a)$ and $f(b)$, there must be a number c between a and b such that $f(c) = d$. (See figure shown at the right.)

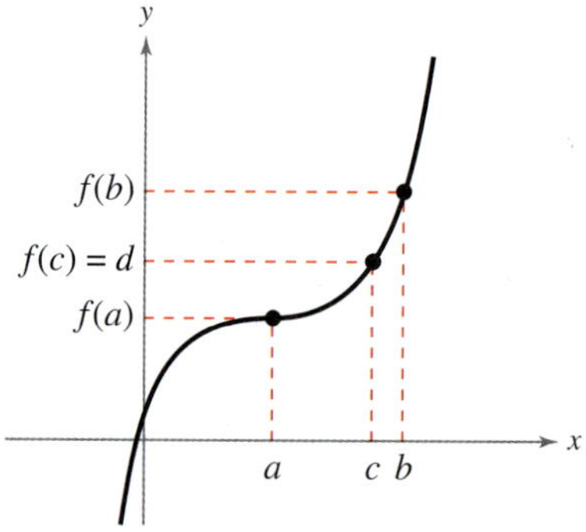

Intermediate Value Theorem

Let a and b be real numbers such that $a < b$. If f is a polynomial function such that $f(a) \neq f(b)$, then in the interval $[a, b]$, f takes on every value between $f(a)$ and $f(b)$.

This theorem helps you locate the real zeros of a polynomial function in the following way. If you can find a value $x = a$ at which a polynomial function is positive, and another value $x = b$ at which it is negative, then you can conclude that the function has at least one real zero between these two values. For example, the function $f(x) = x^3 + x^2 + 1$ is negative when $x = -2$ and positive when $x = -1$. So, it follows from the Intermediate Value Theorem that f must have a real zero somewhere between -2 and -1, as shown in the figure. By continuing this line of reasoning, you can approximate any real zeros of a polynomial function to any desired accuracy.

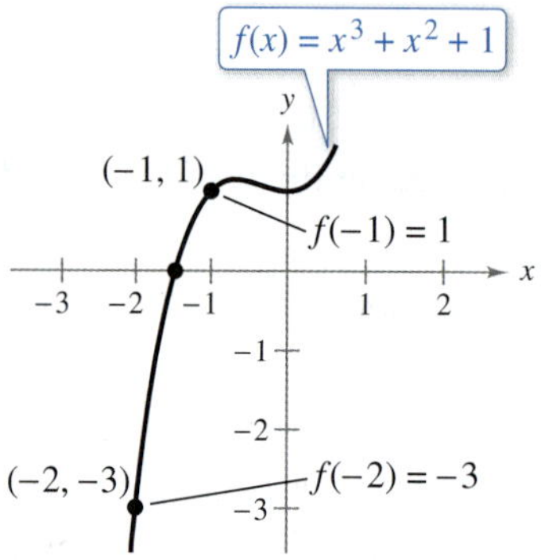

The function f must have a real zero somewhere between −2 and −2.

EXAMPLE 10 Approximating the Zeros of a Function

Find three intervals of length 1 in which the polynomial

$$f(x) = 12x^3 - 32x^2 + 3x + 5$$

is guaranteed to have a zero.

Graphical Solution

From the figure, you can see that the graph of f crosses the x-axis three times—between -1 and 0, between 0 and 1, and between 2 and 3. So, you can conclude that the function has zeros in the intervals $(-1, 0)$, $(0, 1)$, and $(2, 3)$.

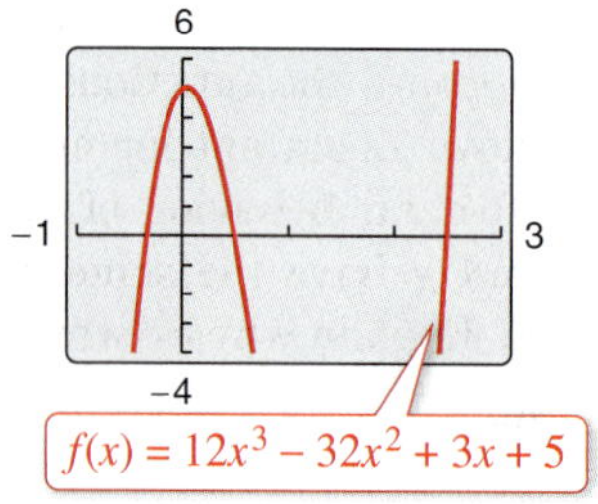

Numerical Solution

From the table, you can see that $f(-1)$ and $f(0)$ differ in sign. So, you can conclude from the Intermediate Value Theorem that the function has a zero between -1 and 0. Similarly, $f(0)$ and $f(1)$ differ in sign, so the function has a zero between 0 and 1. Likewise, $f(2)$ and $f(3)$ differ in sign, so the function has a zero between 2 and 3. So, you can conclude that the function has zeros in the intervals $(-1, 0)$, $(0, 1)$, and $(2, 3)$.

X	Y1
-2	-225
-1	-42
0	5
1	-12
2	-21
3	50
4	273

X=−1

Checkpoint *Audio-video solution in English & Spanish at LarsonPrecalculus.com*

Find three intervals of length 1 in which the polynomial $f(x) = x^3 - 4x^2 + 1$ is guaranteed to have a zero.

2.2 Exercises

See *CalcChat.com* for tutorial help and worked-out solutions to odd-numbered exercises.
For instructions on how to use a graphing utility, see Appendix A.

Vocabulary and Concept Check

In Exercises 1–4, fill in the blank(s).

1. The graph of a polynomial function is _______ , so it has no breaks, holes, or gaps.

2. A polynomial function of degree n has at most _______ real zeros and at most _______ relative extrema.

3. When $x = a$ is a zero of a polynomial function f, the statements below are true.

(a) $x = a$ is a _______ of the polynomial equation $f(x) = 0$.

(b) _______ is a factor of the polynomial $f(x)$.

(c) The point _______ is an x-intercept of the graph of f.

4. If a zero of a polynomial function f is of even multiplicity, then the graph of f _______ the x-axis, and if the zero is of odd multiplicity, then the graph of f _______ the x-axis.

For Exercises 5–8, the graph shows the right-hand and left-hand behavior of a polynomial function f.

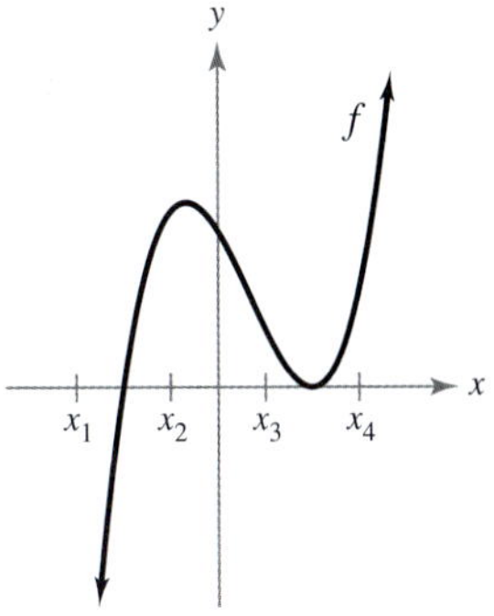

5. Can f be a fourth-degree polynomial function?

6. Can the leading coefficient of f be negative?

7. The graph shows that $f(x_1) < 0$. What other information shown in the graph allows you to apply the Intermediate Value Theorem to guarantee that f has a zero in the interval $[x_1, x_2]$?

8. Is the repeated zero of f in the interval $[x_3, x_4]$ of even or odd multiplicity?

Procedures and Problem Solving

Identifying Graphs of Polynomial Functions **In Exercises 9–16, match the polynomial function with its graph. [The graphs are labeled (a) through (h).]**

(a)

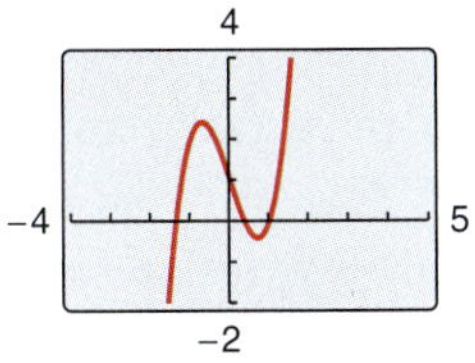

(b)

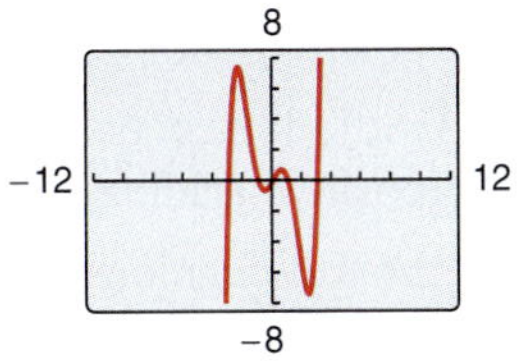

(c)

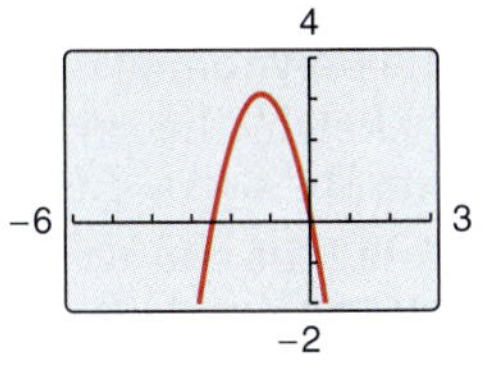

(d)

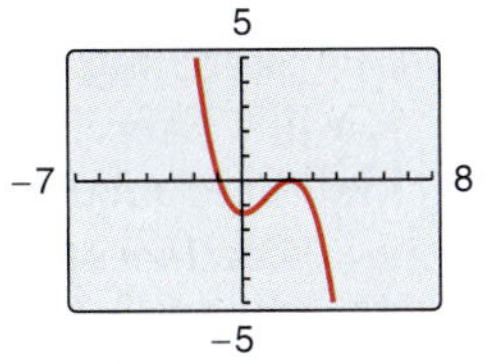

(e)

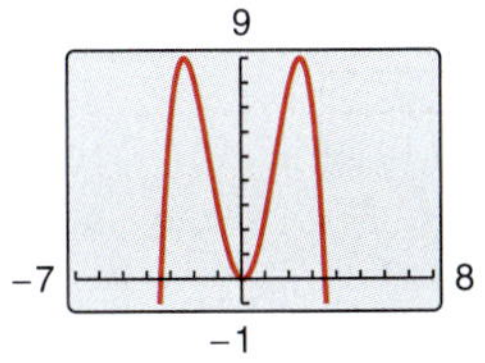

(f)

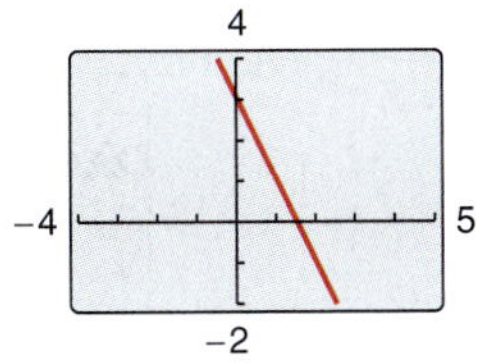

(g)

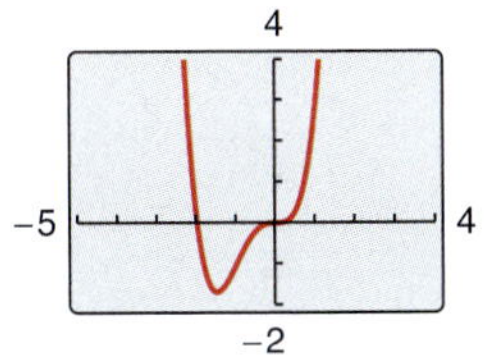

(h)

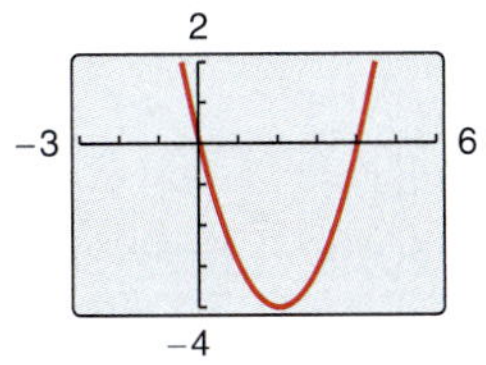

9. $f(x) = -2x + 3$

10. $f(x) = x^2 - 4x$

11. $f(x) = -2x^2 - 5x$

12. $f(x) = 2x^3 - 3x + 1$

13. $f(x) = -\frac{1}{4}x^4 + 3x^2$

14. $f(x) = -\frac{1}{3}x^3 + x^2 - \frac{4}{3}$

15. $f(x) = x^4 + 2x^3$

16. $f(x) = \frac{1}{5}x^5 - 2x^3 + \frac{9}{5}x$

Library of Parent Functions **In Exercises 17–22, sketch the graph of $f(x) = x^3$ and the graph of the function g. Describe the transformation from f to g.**

17. $g(x) = (x - 3)^3$

18. $g(x) = x^3 - 3$

19. $g(x) = -x^3 + 4$

20. $g(x) = -(x - 5)^3$

21. $g(x) = (x - 2)^3 - 3$

22. $g(x) = (x + 4)^3 + 1$

Comparing End Behavior **In Exercises 23–28, use a graphing utility to graph the functions f and g in the same viewing window. Zoom out far enough to see the right-hand and left-hand behavior of each graph. Do the graphs of f and g have the same right-hand and left-hand behavior? Explain why or why not.**

23. $f(x) = 3x^3 - 9x + 1, \quad g(x) = 3x^3$

24. $f(x) = -\frac{1}{3}(x^3 - 2x + 3), \quad g(x) = \frac{1}{3}x^3$

25. $f(x) = -(x^4 - 4x^3 + 16x), \quad g(x) = -x^4$

26. $f(x) = 3x^4 - 6x^2, \quad g(x) = 3x^4$

27. $f(x) = -2x^3 + 4x^2 - 1, \quad g(x) = 2x^3$

28. $f(x) = -(x^4 - 6x^2 - x + 10), \quad g(x) = x^4$

Applying the Leading Coefficient Test In Exercises 29–36, describe the left-hand and right-hand behavior of the graph of the polynomial function. Use a graphing utility to verify your results.

29. $f(x) = 2x^4 - 3x + 1$
30. $h(x) = 1 - x^6$
31. $g(x) = 5 - \frac{7}{2}x - 3x^2$
32. $f(x) = \frac{1}{4}x^3 + 3x$
33. $f(x) = \dfrac{6x^5 - 2x^4 + 4x^2 - 5x}{3}$
34. $f(x) = \dfrac{3x^7 - 2x^5 + 5x^3 + 6x^2}{4}$
35. $h(t) = -\frac{2}{3}(t^2 - 5t + 3)$
36. $f(s) = -\frac{5}{6}(s^3 + 7s^2 - 8s + 1)$

Finding Zeros of a Polynomial Function In Exercises 37–48, (a) find the real zeros algebraically, (b) use a graphing utility to graph the function, and (c) use the graph to approximate any real zeros and compare them with those from part (a).

37. $f(x) = 3x^2 - 12x + 3$
38. $g(x) = 5x^2 - 10x - 5$
39. $g(t) = \frac{1}{2}t^4 - \frac{1}{2}$
40. $y = \frac{1}{4}x^3(x^2 - 9)$
41. $y = 8x^3 - 8x^2 + 2x$
42. $y = x^5 - 5x^3 + 4x$
43. $f(x) = x^5 + x^3 - 6x$
44. $g(t) = t^5 - 6t^3 + 9t$
45. $y = 2x^4 - 2x^2 - 40$
46. $y = 5x^4 + 15x^2 + 10$
47. $f(x) = x^3 - 4x^2 - 25x + 100$
48. $y = 4x^3 + 4x^2 - 9x - 9$

Finding Zeros and Their Multiplicities In Exercises 49–58, find all the real zeros of the polynomial function. Determine the multiplicity of each zero. Use a graphing utility to verify your results.

49. $f(x) = x^2 - 25$
50. $f(x) = 36 - x^2$
51. $h(t) = t^2 - 6t + 9$
52. $f(x) = x^2 + 10x + 25$
53. $f(x) = x^2 + x - 2$
54. $f(x) = 2x^2 - 14x + 24$
55. $f(t) = t^3 - 4t^2 + 4t$
56. $f(x) = x^4 - x^3 - 20x^2$
57. $f(x) = \frac{1}{2}x^2 + \frac{5}{2}x - \frac{3}{2}$
58. $f(x) = \frac{5}{3}x^2 + \frac{8}{3}x - \frac{4}{3}$

Analyzing a Polynomial Function In Exercises 59–64, use a graphing utility to graph the function and approximate (accurate to three decimal places) any real zeros and relative extrema.

59. $f(x) = 2x^4 - 6x^2 + 1$
60. $f(x) = -\frac{3}{8}x^4 - x^3 + 2x^2 + 5$
61. $f(x) = x^5 + 3x^3 - x + 6$
62. $f(x) = -3x^3 - 4x^2 + x - 3$
63. $f(x) = -2x^4 + 5x^2 - x - 1$
64. $f(x) = 4x^5 - 3x^2 - x + 1$

Finding a Polynomial Function In Exercises 65–74, find a polynomial function that has the indicated zeros. (There are many correct answers.)

65. 0, 7
66. −2, 5
67. 0, −2, −4
68. 0, 1, 8
69. 4, −3, 3, 0
70. −2, −1, 0, 1, 2
71. $1 + \sqrt{2}, 1 - \sqrt{2}$
72. $4 + \sqrt{3}, 4 - \sqrt{3}$
73. $2, 2 + \sqrt{5}, 2 - \sqrt{5}$
74. $3, 2 + \sqrt{7}, 2 - \sqrt{7}$

Finding a Polynomial Function In Exercises 75–78, find a polynomial function with the indicated characteristics. (There are many correct answers.)

75. Zero: −4, multiplicity: 2
Zero: 3, multiplicity: 2
Degree: 4

76. Zero: 5, multiplicity: 3
Zero: 0, multiplicity: 2
Degree: 5

77. Zero: −1, multiplicity: 2
Zero: −2, multiplicity: 1
Degree: 3
Rises to the left

78. Zero: 1, multiplicity: 2
Zero: 4, multiplicity: 2
Degree: 4
Falls to the left

Sketching a Polynomial In Exercises 79–82, sketch the graph of a polynomial function that satisfies the conditions. If not possible, explain your reasoning. (There are many correct answers.)

79. Third-degree polynomial with two real zeros and a negative leading coefficient
80. Fourth-degree polynomial with three real zeros and a positive leading coefficient
81. Fifth-degree polynomial with three real zeros and a positive leading coefficient
82. Fourth-degree polynomial with two real zeros and a negative leading coefficient

Sketching the Graph of a Polynomial Function In Exercises 83–92, sketch the graph of the function by (a) applying the Leading Coefficient Test, (b) finding the real zeros of the polynomial, (c) plotting sufficient solution points, and (d) drawing a continuous curve through the points.

83. $f(x) = x^3 - 25x$
84. $g(x) = x^4 - 4x^2$
85. $f(x) = x^3 - 3x^2$
86. $f(x) = 3x^3 - 24x^2$
87. $f(x) = -x^4 + 9x^2 - 20$
88. $f(x) = -x^6 + 7x^3 + 8$
89. $f(x) = x^3 + 3x^2 - 9x - 27$
90. $h(x) = x^5 - 4x^3 + 8x^2 - 32$
91. $g(t) = -\frac{1}{4}t^4 + 2t^2 - 4$
92. $g(x) = \frac{1}{10}(x^4 - 4x^3 + 8x - 32)$

Approximating the Zeros of a Function **In Exercises 93–98, (a) use the Intermediate Value Theorem and the *table* feature of a graphing utility to find intervals one unit in length in which the polynomial function is guaranteed to have a zero. (b) Adjust the table to approximate the zeros of the function to the nearest thousandth.**

93. $f(x) = x^3 - 3x^2 + 3$
94. $f(x) = -2x^3 - 6x^2 + 3$
95. $g(x) = 3x^4 + 4x^3 - 3$
96. $h(x) = x^4 - 10x^2 + 2$
97. $f(x) = x^4 - 3x^3 - 4x - 3$
98. $f(x) = x^3 - 4x^2 - 2x + 10$

Identifying Symmetry and x-Intercepts **In Exercises 99–106, use a graphing utility to graph the function. Identify any symmetry with respect to the x-axis, y-axis, or origin. Determine the number of x-intercepts of the graph.**

99. $f(x) = x^2(x + 5)$
100. $h(x) = x^3(x - 3)^2$
101. $f(x) = x^3 - 4x$
102. $f(x) = x^4 - 2x^2$
103. $g(x) = -\frac{1}{4}(x - 2)^2(x + 2)^2$
104. $g(x) = \frac{1}{8}(x + 1)^2(x - 3)^3$
105. $g(x) = \frac{1}{5}(x + 1)^2(x - 3)(2x - 9)$
106. $h(x) = \frac{1}{3}(x + 4)^2(5x - 4)^2$

107. **Geometry** An open box is made from a square piece of material 36 centimeters on a side by cutting equal squares with sides of length x from the corners and turning up the sides (see figure).

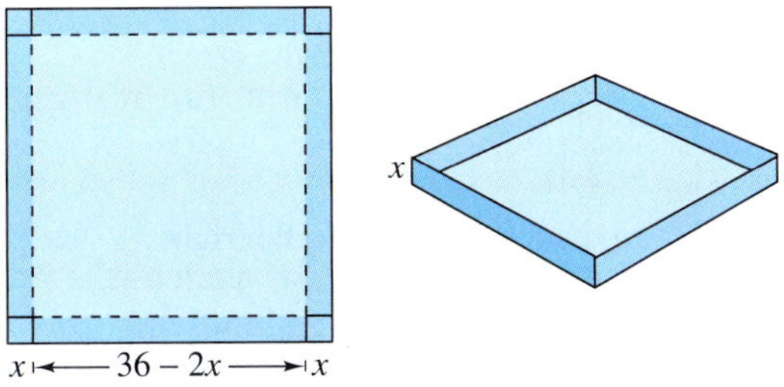

(a) Write a function V that represents the volume of the box.

(b) Determine the domain of the function V.

(c) Use the *table* feature of a graphing utility to create a table that shows various box heights x and the corresponding volumes V. Use the table to estimate the dimensions that produce maximum volume.

(d) Use the graphing utility to graph V and use the graph to estimate the value of x for which $V(x)$ is maximum. Compare your result with that of part (c).

108. **Geometry** An open box with locking tabs is made from a square piece of material 24 inches on a side by cutting equal squares from the corners and folding along the dashed lines, as shown in the figure.

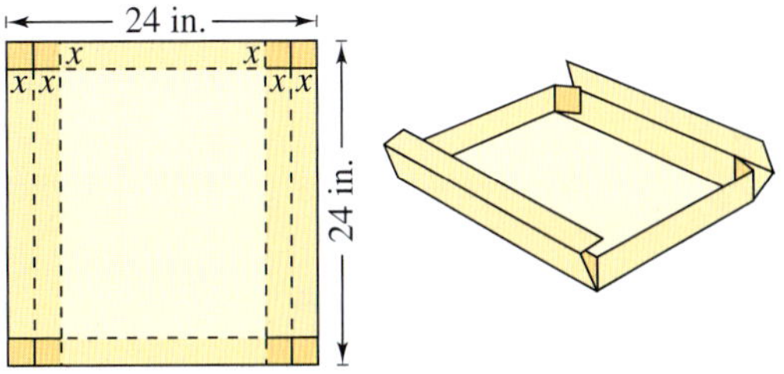

(a) Write a function V that represents the volume of the box.

(b) Determine the domain of the function V.

(c) Sketch a graph of the function and estimate the value of x for which $V(x)$ is maximum.

109. **Marketing** The total revenue R (in millions of dollars) for a company is related to its advertising expense by the function

$$R = 0.00001(-x^3 + 600x^2), \quad 0 \le x \le 400$$

where x is the amount spent on advertising (in tens of thousands of dollars). Use the graph of the function shown in the figure to estimate the point on the graph at which the function is increasing most rapidly. This point is called the **point of diminishing returns** because any expense above this amount will yield less return per dollar invested in advertising.

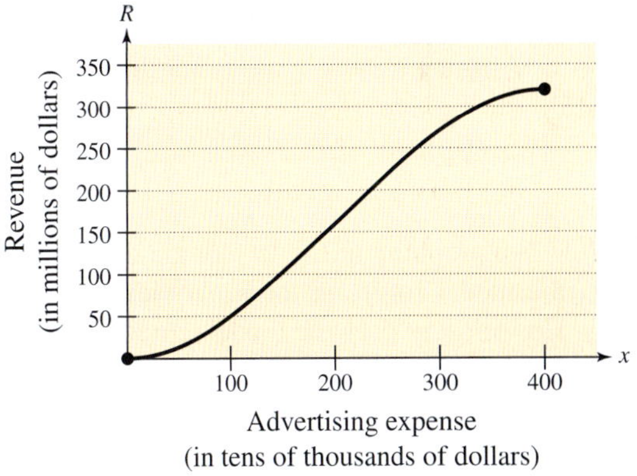

110. **Marketing** The total revenue R (in millions of dollars) for a company is related to its advertising expense by the function

$$R = 0.000005(-x^3 + 750x^2), \quad 0 \le x \le 500$$

where x is the amount spent on advertising (in tens of thousands of dollars).

(a) Sketch the graph of the function.

(b) Use the graph to estimate the point of diminishing returns (described in Exercise 109).

111. ***Why you should learn it*** *(p. 100)* The growth of a red oak tree is approximated by the function

$$G = -0.003t^3 + 0.137t^2 + 0.458t - 0.839, \quad 2 \leq t \leq 32$$

where G is the height of the tree (in feet) and t is its age (in years). Use a graphing utility to graph the function and estimate the age of the tree when it is growing most rapidly.

112. MODELING DATA

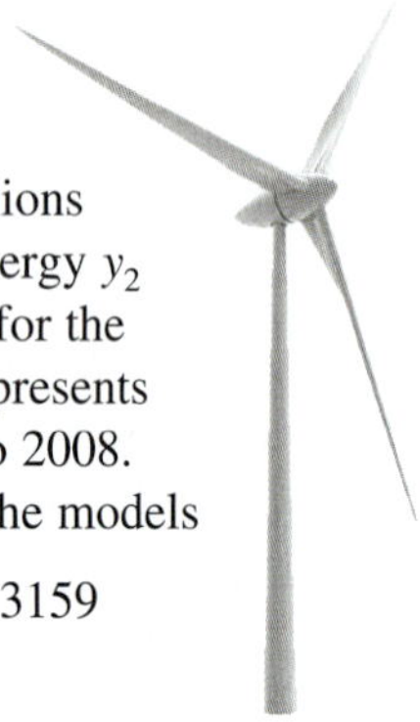

The table shows the estimated U.S. production of wind energy y_1 (in trillions of British thermal units) and solar energy y_2 (in trillions of British thermal units) for the years 2008 through 2017, where t represents the year, with $t = 8$ corresponding to 2008. These data can be approximated by the models

$$y_1 = 1.167t^3 - 44.39t^2 + 739.5t - 3159$$

and

$$y_2 = 0.572t^3 - 10.10t^2 + 49.9t + 29.$$

(*Source:* U.S. Energy Information Administration)

DATA

Year, t	y_1	y_2
8	546	74
9	721	78
10	923	90
11	1168	111
12	1340	157
13	1601	225
14	1728	337
15	1777	426
16	2096	569
17	2347	774

Spreadsheet at LarsonPrecalculus.com

(a) Use a graphing utility to plot the data and graph the model for y_1 in the same viewing window. How closely does the model represent the data?

(b) Extend the viewing window of the graphing utility to show the right-hand behavior of the model y_1. Would you use the model to estimate the production of wind energy in 2021? in 2026? Explain.

(c) Repeat parts (a) and (b) for y_2.

Focusing on Concepts

True or False? **In Exercises 113–116, determine whether the statement is true or false. Justify your answer.**

113. It is possible for a sixth-degree polynomial to have only one real zero.

114. It is possible for a fifth-degree polynomial to have no real zeros.

115. It is possible for a polynomial with an even degree to have a range of $(-\infty, \infty)$.

116. The graph of the function $f(x) = 3x(x + 1)^2(x - 2)^3$ touches, but does not cross, the x-axis.

117. **Exploration** Use a graphing utility to graph

$$y_1 = x + 2 \quad \text{and} \quad y_2 = (x + 2)(x - 1).$$

Predict the shape of the graph of

$$y_3 = (x + 2)(x - 1)(x - 3).$$

Use the graphing utility to verify your answer.

118. **HOW DO YOU SEE IT?** For each graph, describe a polynomial function that could represent the graph. (Indicate the degree of the function and the sign of its leading coefficient.)

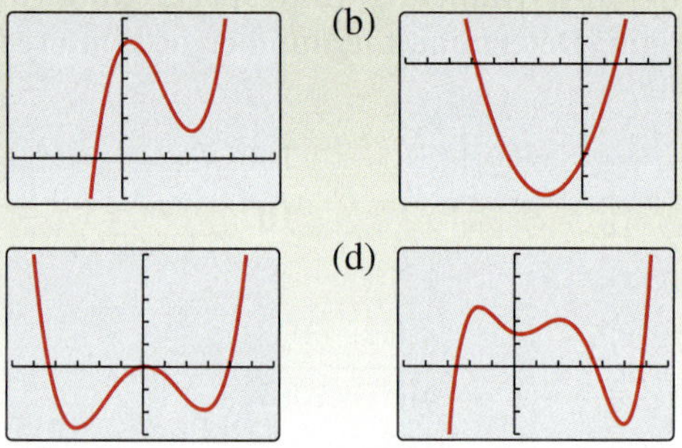

119. **Error Analysis** Describe the error.

The graph of the function

$f(x) = 1 - x^3$

falls to the left and rises to the right.

120. **Error Analysis** Describe the error.

The graph of the function

$f(x) = 2x(x - 1)^2(x + 3)^3$

rises to the left and falls to the right.

Cumulative Mixed Review

Evaluating Combinations of Functions **In Exercises 121–126, let $f(x) = 14x - 3$ and $g(x) = 8x^2$. Find the indicated value.**

121. $(f + g)(-4)$

122. $(g - f)(3)$

123. $(fg)\left(-\frac{4}{7}\right)$

124. $\left(\frac{f}{g}\right)(-1.5)$

125. $(f \circ g)(-1)$

126. $(g \circ f)(0)$

Solving Inequalities **In Exercises 127 and 128, solve the inequality and sketch the solution on the real number line. Use a graphing utility to verify your solution.**

127. $2x^2 - x \geq 1$

128. $|x + 8| - 1 \geq 15$

2.3 Real Zeros of Polynomial Functions

Long Division of Polynomials

In the graph of $f(x) = 6x^3 - 19x^2 + 16x - 4$ shown in Figure 2.17, it appears that $x = 2$ is a zero of f. Because $f(2) = 0$, you know that $x = 2$ is a zero of the polynomial function f and that $(x - 2)$ is a factor of $f(x)$. This means that there exists a second-degree polynomial $q(x)$ such that $f(x) = (x - 2) \cdot q(x)$. One way to find $q(x)$ is to use **long division of polynomials.**

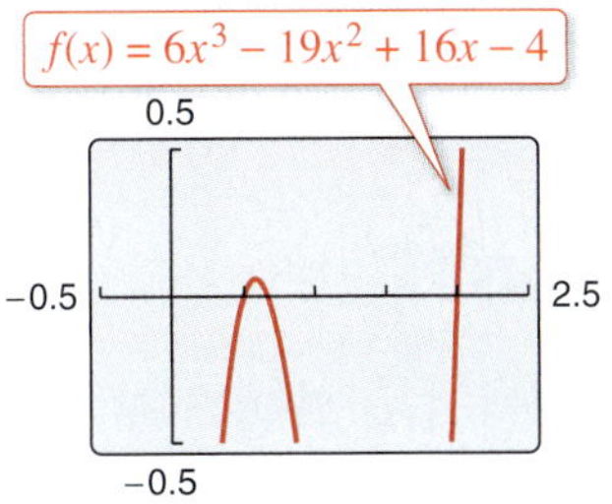

Figure 2.17

What you should learn

- Use long division to divide polynomials by other polynomials.
- Use synthetic division to divide polynomials by binomials of the form $(x - k)$.
- Use the Remainder and Factor Theorems.
- Use the Rational Zero Test to determine possible rational zeros of polynomial functions.
- Use Descartes's Rule of Signs and the Upper and Lower Bound Rules to find real zeros of polynomials.

Why you should learn it

The Remainder Theorem can be used to determine the number of employees in education and health services in the United States in a given year based on a polynomial model, as shown in Exercise 106 on page 127.

EXAMPLE 1 Long Division of Polynomials

Divide the polynomial $6x^3 - 19x^2 + 16x - 4$ by $x - 2$, and use the result to factor the polynomial completely.

Solution

Think $\frac{6x^3}{x} = 6x^2$.

Think $\frac{-7x^2}{x} = -7x$.

Think $\frac{2x}{x} = 2$.

$$
\begin{array}{rrll}
 & 6x^2 - 7x + 2 & & \\
x - 2 \,\big)\!\! & 6x^3 - 19x^2 + 16x - 4 & & \\
 & \underline{6x^3 - 12x^2} & & \text{Multiply: } 6x^2(x - 2). \\
 & -7x^2 + 16x & & \text{Subtract and bring down } 16x. \\
 & \underline{-7x^2 + 14x} & & \text{Multiply: } -7x(x - 2). \\
 & 2x - 4 & & \text{Subtract and bring down } 4. \\
 & \underline{2x - 4} & & \text{Multiply: } 2(x - 2). \\
 & 0 & & \text{Subtract.}
\end{array}
$$

You can see that

$$6x^3 - 19x^2 + 16x - 4 = (x - 2)(6x^2 - 7x + 2)$$
$$= (x - 2)(2x - 1)(3x - 2).$$

Note that this factorization agrees with the graph of f (see Figure 2.17) in that the three x-intercepts occur at $x = 2$, $x = \frac{1}{2}$, and $x = \frac{2}{3}$.

✓ ***Checkpoint*** ▶ *Audio-video solution in English & Spanish at LarsonPrecalculus.com*

Divide the polynomial $3x^2 + 19x + 28$ by $x + 4$, and use the result to factor the polynomial completely.

Note that in Example 1, the division process requires $-7x^2 + 14x$ to be subtracted from $-7x^2 + 16x$. So, it is implied that

$$\begin{array}{r} -7x^2 + 16x \\ \underline{-(-7x^2 + 14x)} \end{array} = \begin{array}{r} -7x^2 + 16x \\ \underline{7x^2 - 14x} \end{array}$$

and instead is written simply as

$$\begin{array}{r} -7x^2 + 16x \\ \underline{-7x^2 + 14x} \\ 2x. \end{array}$$

In Example 1, $x - 2$ is a factor of the polynomial

$$6x^3 - 19x^2 + 16x - 4$$

and the long division process produces a remainder of zero. Often, long division will produce a nonzero remainder. For instance, when you divide $x^2 + 3x + 5$ by $x + 1$, you obtain a remainder of 3.

$$\begin{array}{rl} & x + 2 \quad \leftarrow \text{Quotient} \\ \text{Divisor} \rightarrow \quad x + 1 \overline{)} & x^2 + 3x + 5 \quad \leftarrow \text{Dividend} \\ & \underline{x^2 + x} \\ & 2x + 5 \\ & \underline{2x + 2} \\ & 3 \quad \leftarrow \text{Remainder} \end{array}$$

In fractional form, you can write this result as

$$\underbrace{\overbrace{\frac{x^2 + 3x + 5}{x + 1}}^{\text{Dividend}}}_{\text{Divisor}} = \overbrace{x + 2}^{\text{Quotient}} + \underbrace{\frac{\overset{\text{Remainder}}{3}}{x + 1}}_{\text{Divisor}}.$$

This implies that

$$x^2 + 3x + 5 = (x + 1)(x + 2) + 3 \qquad \text{Multiply each side by } (x + 1).$$

which illustrates the **Division Algorithm.**

The Division Algorithm

If $f(x)$ and $d(x)$ are polynomials such that $d(x) \neq 0$, and the degree of $d(x)$ is less than or equal to the degree of $f(x)$, then there exist unique polynomials $q(x)$ and $r(x)$ such that

$$f(x) = d(x)q(x) + r(x)$$

($f(x)$: Dividend; $d(x)$: Divisor; $q(x)$: Quotient; $r(x)$: Remainder)

where $r(x) = 0$ *or* the degree of $r(x)$ is less than the degree of $d(x)$. If the remainder $r(x)$ is zero, then $d(x)$ *divides evenly* into $f(x)$.

The Division Algorithm can also be written as

$$\frac{f(x)}{d(x)} = q(x) + \frac{r(x)}{d(x)}.$$

In the Division Algorithm, the rational expression $f(x)/d(x)$ is **improper** because the degree of $f(x)$ is greater than or equal to the degree of $d(x)$. On the other hand, the rational expression $r(x)/d(x)$ is **proper** because the degree of $r(x)$ is less than the degree of $d(x)$.

Before you apply the Division Algorithm, follow these steps.

1. Write the dividend and divisor in descending powers of the variable.
2. Insert placeholders with zero coefficients for missing powers of the variable.

Note how these steps are applied in the next two examples.

EXAMPLE 2 Long Division of Polynomials

Divide $8x^3 - 1$ by $2x - 1$.

Solution

Because there is no x^2-term or x-term in the dividend, you need to line up the subtraction by using zero coefficients (or leaving spaces) for the missing terms.

$$\begin{array}{r@{}l}
& 4x^2 + 2x + 1 \\
2x - 1 \,\big)\! & \overline{8x^3 + 0x^2 + 0x - 1} \\
& \underline{8x^3 - 4x^2} \qquad\qquad \text{Multiply: } 4x^2(2x - 1). \\
& \quad 4x^2 + 0x \qquad\qquad \text{Subtract and bring down } 0x. \\
& \quad \underline{4x^2 - 2x} \qquad\qquad \text{Multiply: } 2x(2x - 1). \\
& \qquad\quad 2x - 1 \qquad\qquad \text{Subtract and bring down } -1. \\
& \qquad\quad \underline{2x - 1} \qquad\qquad \text{Multiply: } 1(2x - 1). \\
& \qquad\qquad\quad 0 \qquad\qquad \text{Subtract.}
\end{array}$$

So, $2x - 1$ divides evenly into $8x^3 - 1$, and you can write

$$\frac{8x^3 - 1}{2x - 1} = 4x^2 + 2x + 1, \quad x \neq \frac{1}{2}.$$

You can check this result by multiplying.

$$\begin{aligned}(2x - 1)(4x^2 + 2x + 1) &= 8x^3 + 4x^2 + 2x - 4x^2 - 2x - 1 \\ &= 8x^3 - 1\end{aligned}$$

✓ Checkpoint ▶ *Audio-video solution in English & Spanish at LarsonPrecalculus.com*

Divide $x^3 - 2x^2 - 9$ by $x - 3$.

In each of the long division examples presented so far, the divisor has been a first-degree polynomial. The long division algorithm works just as well with polynomial divisors of degree two or more, as shown in Example 3.

EXAMPLE 3 Long Division of Polynomials

Divide $-2 + 3x - 5x^2 + 4x^3 + 2x^4$ by $x^2 + 2x - 3$.

Solution

Begin by writing the dividend in descending powers of x.

$$\begin{array}{r@{}l}
& 2x^2 \qquad\qquad + 1 \\
x^2 + 2x - 3 \,\big)\! & \overline{2x^4 + 4x^3 - 5x^2 + 3x - 2} \\
& \underline{2x^4 + 4x^3 - 6x^2} \qquad\qquad \text{Multiply: } 2x^2(x^2 + 2x - 3). \\
& \qquad\qquad\quad x^2 + 3x - 2 \qquad\qquad \text{Subtract and bring down } 3x - 2. \\
& \qquad\qquad\quad \underline{x^2 + 2x - 3} \qquad\qquad \text{Multiply: } 1(x^2 + 2x - 3). \\
& \qquad\qquad\qquad\quad x + 1 \qquad\qquad \text{Subtract.}
\end{array}$$

Note that the first subtraction eliminated two terms from the dividend. When this happens, the quotient skips a term. You can write the result as

$$\frac{2x^4 + 4x^3 - 5x^2 + 3x - 2}{x^2 + 2x - 3} = 2x^2 + 1 + \frac{x + 1}{x^2 + 2x - 3}.$$

✓ Checkpoint ▶ *Audio-video solution in English & Spanish at LarsonPrecalculus.com*

Divide $-x^3 + 9x + 6x^4 - x^2 - 3$ by $1 + 3x$.

Synthetic Division

There is a nice shortcut for long division of polynomials when dividing by divisors of the form

$$x - k.$$

The shortcut is called **synthetic division.** The pattern for synthetic division of a cubic polynomial is summarized below. (The pattern for higher-degree polynomials is similar.)

Synthetic Division (of a Cubic Polynomial)

To divide $ax^3 + bx^2 + cx + d$ by $x - k$, use the following pattern.

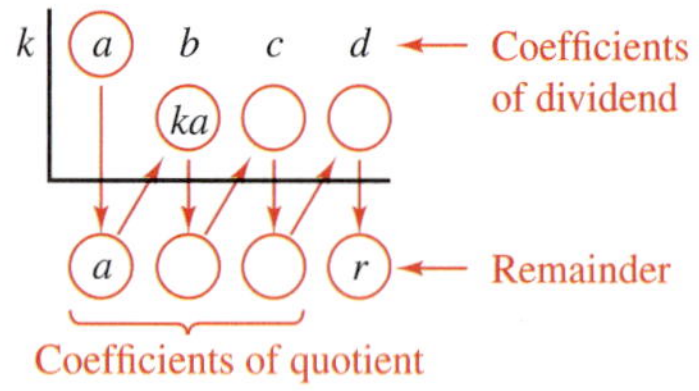

Vertical pattern: Add terms.
Diagonal pattern: Multiply by k.

This algorithm for synthetic division works *only* for divisors of the form $x - k$. Remember that

$$x + k = x - (-k).$$

EXAMPLE 4 Using Synthetic Division

Use synthetic division to divide

$$x^4 - 10x^2 - 2x + 4 \quad \text{by} \quad x + 3.$$

Solution

Begin by setting up an array, as shown below. Include a zero for the missing x^3-term in the dividend.

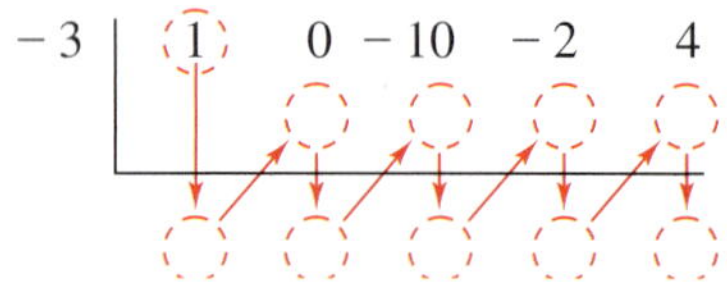

Then, use the synthetic division pattern by adding terms in columns and multiplying the results by -3.

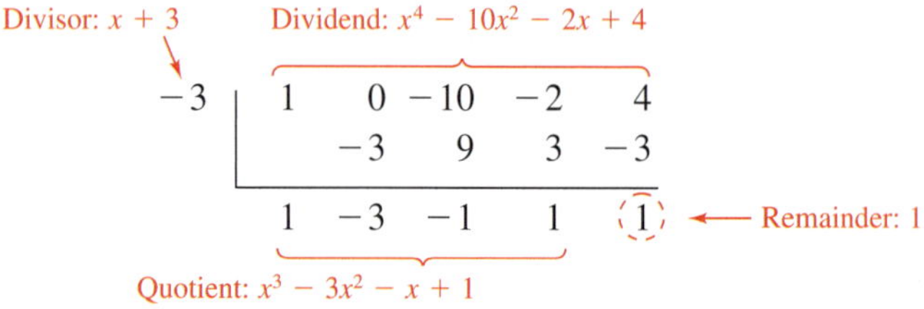

So, you have

$$\frac{x^4 - 10x^2 - 2x + 4}{x + 3} = x^3 - 3x^2 - x + 1 + \frac{1}{x + 3}.$$

Checkpoint *Audio-video solution in English & Spanish at LarsonPrecalculus.com*

Use synthetic division to divide $5x^3 + 8x^2 - x + 6$ by $x + 2$.

Explore the Concept

Evaluate the polynomial $x^4 - 10x^2 - 2x + 4$ when $x = -3$. What do you observe?

pikselstock/Shutterstock.com

The Remainder and Factor Theorems

The remainder obtained in the synthetic division process has an important interpretation, as described in the **Remainder Theorem.**

The Remainder Theorem

If a polynomial $f(x)$ is divided by $x - k$, then the remainder is

$r = f(k)$.

(See the proof on page 178.)

The Remainder Theorem tells you that synthetic division can be used to evaluate a polynomial function. That is, to evaluate a polynomial $f(x)$ when $x = k$, divide $f(x)$ by $x - k$. The remainder will be $f(k)$.

EXAMPLE 5 Using the Remainder Theorem

Use the Remainder Theorem to evaluate

$$f(x) = 3x^3 + 8x^2 + 5x - 7$$

when $x = -2$.

Solution

Using synthetic division, you obtain the following.

$$\begin{array}{r|rrrr} -2 & 3 & 8 & 5 & -7 \\ & & -6 & -4 & -2 \\ \hline & 3 & 2 & 1 & -9 \end{array}$$

Because the remainder is $r = -9$, you can conclude that

$$f(-2) = -9. \qquad r = f(k)$$

This means that $(-2, -9)$ is a point on the graph of f. You can check this by substituting $x = -2$ in the original function.

Check

$$\begin{aligned} f(-2) &= 3(-2)^3 + 8(-2)^2 + 5(-2) - 7 \\ &= 3(-8) + 8(4) - 10 - 7 \\ &= -24 + 32 - 10 - 7 \\ &= -9 \end{aligned}$$

✓ **Checkpoint** ▶ *Audio-video solution in English & Spanish at LarsonPrecalculus.com*

Use the Remainder Theorem to find each function value given

$$f(x) = 4x^3 + 10x^2 - 3x - 8.$$

a. $f(-1)$ **b.** $f(4)$

c. $f(\frac{1}{2})$ **d.** $f(-3)$

Another important theorem is the **Factor Theorem.** This theorem states that you can test whether a polynomial has $(x - k)$ as a factor by evaluating the polynomial at $x = k$. If the result is 0, then $(x - k)$ is a factor.

The Factor Theorem

A polynomial $f(x)$ has a factor $(x - k)$ if and only if $f(k) = 0$.

(See the proof on page 178.)

EXAMPLE 6 Factoring a Polynomial: Repeated Division

Show that $(x - 2)$ and $(x + 3)$ are factors of

$$f(x) = 2x^4 + 7x^3 - 4x^2 - 27x - 18.$$

Then find the remaining factors of $f(x)$.

Algebraic Solution

Using synthetic division with the factor $(x - 2)$, you obtain the following.

2	2	7	−4	−27	−18	
		4	22	36	18	
	2	11	18	9	0	→ 0 remainder; $(x - 2)$ is a factor.

Take the result of this division and perform synthetic division again using the factor $(x + 3)$.

−3	2	11	18	9	
		−6	−15	−9	
	2	5	3	0	→ 0 remainder; $(x + 3)$ is a factor.

$2x^2 + 5x + 3$

Because the resulting quadratic factors as

$$2x^2 + 5x + 3 = (2x + 3)(x + 1)$$

the complete factorization of $f(x)$ is

$$f(x) = (x - 2)(x + 3)(2x + 3)(x + 1).$$

Graphical Solution

The graph of

$$f(x) = 2x^4 + 7x^3 - 4x^2 - 27x - 18$$

has four x-intercepts (see Figure 2.18). These occur at $x = -3$, $x = -\frac{3}{2}$, $x = -1$, and $x = 2$. (Check this algebraically.) This implies that

$$(x + 3),\ \left(x + \tfrac{3}{2}\right),\ (x + 1),\ \text{and}\ (x - 2)$$

are factors of $f(x)$. $\left[\text{Note that } \left(x + \frac{3}{2}\right) \text{ and } (2x + 3) \text{ are equivalent factors because they both yield the same zero, } x = -\frac{3}{2}.\right]$

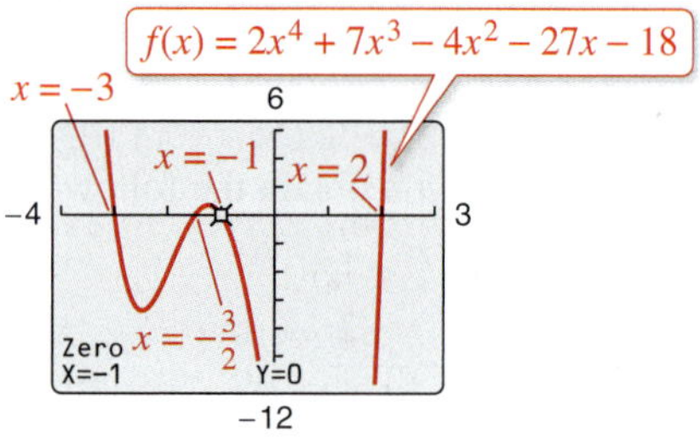

Figure 2.18

Checkpoint *Audio-video solution in English & Spanish at LarsonPrecalculus.com*

Show that $(x + 3)$ is a factor of $f(x) = x^3 - 19x - 30$. Then find the remaining factors of $f(x)$.

Note in Example 6 that the complete factorization of $f(x)$ implies that f has four real zeros:

$$x = 2, \quad x = -3, \quad x = -\tfrac{3}{2}, \quad \text{and} \quad x = -1.$$

This is confirmed by the graph of f, which is shown in Figure 2.18.

Insight

Some standardized tests may ask you to determine whether an expression is a factor of a polynomial (see Example 6).

Using the Remainder in Synthetic Division

In summary, the remainder r, obtained in the synthetic division of a polynomial $f(x)$ by $x - k$, provides the following information.

1. The remainder r gives the *value* of f at $x = k$. That is, $r = f(k)$.
2. If $r = 0$, then $(x - k)$ is a *factor* of $f(x)$.
3. If $r = 0$, then $(k, 0)$ is an *x-intercept* of the graph of f.

This text emphasizes the importance of developing several problem-solving strategies. So, in the exercises for this section, try using more than one strategy to solve several of the exercises. For instance, when you find that $x - k$ divides evenly into $f(x)$, try sketching the graph of f. You should find that $(k, 0)$ is an x-intercept of the graph.

The Rational Zero Test

The **Rational Zero Test** relates the possible rational zeros of a polynomial (having integer coefficients) to the leading coefficient and to the constant term of the polynomial.

The Rational Zero Test

If the polynomial

$$f(x) = a_n x^n + a_{n-1}x^{n-1} + \cdot \cdot \cdot + a_2x^2 + a_1x + a_0 \qquad n \geq 1 \text{ and } a_0 \neq 0$$

has integer coefficients, then every rational zero of f has the form

$$\text{Rational zero} = \frac{p}{q}$$

where p and q have no common factors other than 1, p is a factor of the constant term a_0, and q is a factor of the leading coefficient a_n.

To use the Rational Zero Test, first list all rational numbers whose numerators are factors of the constant term and whose denominators are factors of the leading coefficient.

$$\text{Possible rational zeros} = \frac{\text{factors of constant term}}{\text{factors of leading coefficient}}$$

Now that you have formed this list of *possible rational zeros,* use a trial-and-error method to determine which, if any, are actual zeros of the polynomial. Note that when the leading coefficient is 1, the possible rational zeros are simply the factors of the constant term. This case is illustrated in Example 7.

Algebra Help

Use a graphing utility to graph the polynomial

$$y = x^3 - 53x^2 + 103x - 51$$

in a standard viewing window. From the graph alone, it appears that there is only one zero. From the Leading Coefficient Test, you know that because the degree of the polynomial is odd and the leading coefficient is positive, the graph falls to the left and rises to the right. So, the function must have another zero. From the Rational Zero Test, you know that ± 51 might be zeros of the function. When you zoom out several times, you will see a more complete picture of the graph. Your graph should confirm that $x = 51$ is a zero of f.

EXAMPLE 7 Rational Zero Test with Leading Coefficient of 1

Find the rational zeros of $f(x) = x^3 + x + 1$.

Solution

Because the leading coefficient is 1, the possible rational zeros are simply the factors of the constant term.

Possible rational zeros: ± 1

By testing these possible zeros, you can see that neither works.

$$f(1) = (1)^3 + 1 + 1 = 3$$

$$f(-1) = (-1)^3 + (-1) + 1 = -1$$

So, you can conclude that the polynomial has *no* rational zeros. Note from the graph of f shown below that f does have one real zero between -1 and 0. However, by the Rational Zero Test, you know that this real zero is *not* a rational number.

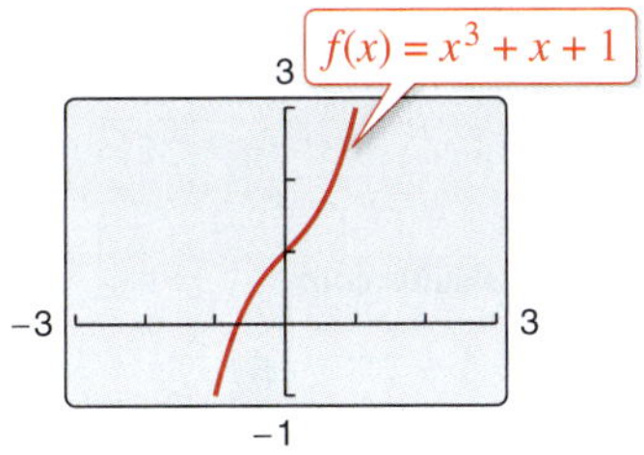

Checkpoint *Audio-video solution in English & Spanish at LarsonPrecalculus.com*

Find the rational zeros of $f(x) = x^3 - 5x^2 + 2x + 8$.

When the leading coefficient of a polynomial is not 1, the list of possible rational zeros can increase dramatically. In such cases, the search can be shortened in several ways.

1. A graphing utility can be used to speed up the calculations.
2. A graph, drawn either by hand or with a graphing utility, can give good estimates of the locations of the zeros.
3. The Intermediate Value Theorem, along with a table generated by a graphing utility, can give approximations of zeros.
4. The Factor Theorem and synthetic division can be used to test the possible rational zeros.

Finding the first zero is often the most difficult part. After that, the search is simplified by working with the lower-degree polynomial obtained in synthetic division, as shown in Example 8.

EXAMPLE 8 Using the Rational Zero Test

Find the rational zeros of $f(x) = 2x^3 + 3x^2 - 8x + 3$.

Solution

The leading coefficient is 2, and the constant term is 3.

Possible rational zeros: $\dfrac{\text{Factors of } 3}{\text{Factors of } 2} = \dfrac{\pm 1, \pm 3}{\pm 1, \pm 2} = \pm 1, \pm 3, \pm \dfrac{1}{2}, \pm \dfrac{3}{2}$

By synthetic division, you can determine that $x = 1$ is a rational zero.

$$\begin{array}{r|rrrr} 1 & 2 & 3 & -8 & 3 \\ & & 2 & 5 & -3 \\ \hline & 2 & 5 & -3 & 0 \end{array}$$

0 remainder; $(x - 1)$ is a factor.

$2x^2 + 5x - 3$

So, $f(x)$ factors as

$$f(x) = (x - 1)(2x^2 + 5x - 3) \quad \text{Factor using result of synthetic division.}$$
$$= (x - 1)(2x - 1)(x + 3) \quad \text{Factor completely.}$$

and you can conclude that the rational zeros of f are $x = 1$, $x = \frac{1}{2}$, and $x = -3$, as shown in the figure.

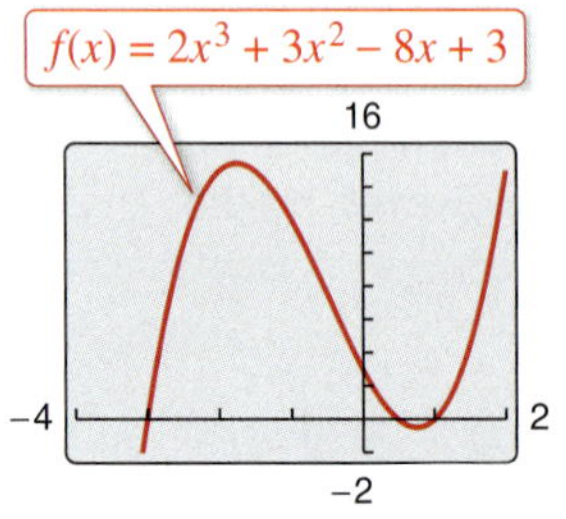

Checkpoint Audio-video solution in English & Spanish at LarsonPrecalculus.com

Find the rational zeros of $f(x) = 2x^4 - 9x^3 - 18x^2 + 71x - 30$.

Remember that when you try to find the rational zeros of a polynomial function with many possible rational zeros, as in Example 8, you must use trial and error. There is no quick algebraic method to determine which of the possibilities is an actual zero; however, sketching a graph may be helpful.

Other Tests for Zeros of Polynomials

An nth-degree polynomial function can have *at most n* real zeros. Of course, many nth-degree polynomials do not have that many real zeros. For instance, $f(x) = x^2 + 1$ has no real zeros, and $f(x) = x^3 + 1$ has only one real zero. The next theorem, called **Descartes's Rule of Signs,** sheds more light on the number of real zeros of a polynomial.

Descartes's Rule of Signs

Let $f(x) = a_n x^n + a_{n-1}x^{n-1} + \cdots + a_2x^2 + a_1x + a_0$ be a polynomial with real coefficients and $a_0 \neq 0$.

1. The number of *positive real zeros* of f is either equal to the number of variations in sign of $f(x)$ or less than that number by an even integer.
2. The number of *negative real zeros* of f is either equal to the number of variations in sign of $f(-x)$ or less than that number by an even integer.

A **variation in sign** means that two consecutive coefficients have opposite signs. Missing terms (those with zero coefficients) can be ignored.

When using Descartes's Rule of Signs, a zero of multiplicity k should be counted as k zeros. For instance, the polynomial $x^3 - 3x + 2$ has two variations in sign and so has either two positive or no positive real zeros. Because

$$x^3 - 3x + 2 = (x - 1)(x - 1)(x + 2)$$

you can see that the two positive real zeros are $x = 1$ of multiplicity 2.

EXAMPLE 9 Using Descartes's Rule of Signs

Determine the possible numbers of positive and negative real zeros of

$$f(x) = 3x^3 - 5x^2 + 6x - 4.$$

Solution

The original polynomial has *three* variations in sign.

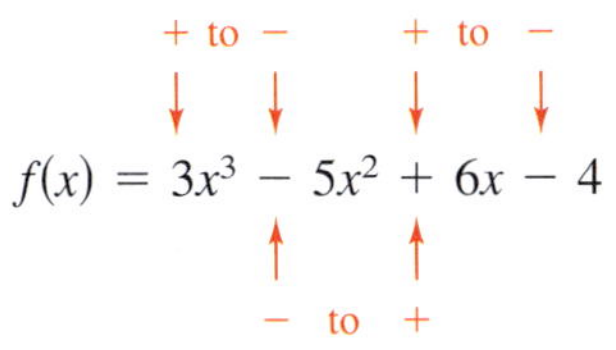

The polynomial

$$f(-x) = 3(-x)^3 - 5(-x)^2 + 6(-x) - 4$$
$$= -3x^3 - 5x^2 - 6x - 4$$

has no variations in sign. So, from Descartes's Rule of Signs, the polynomial $f(x) = 3x^3 - 5x^2 + 6x - 4$ has either three positive real zeros or one positive real zero and has no negative real zeros. By using the *trace* feature of a graphing utility, you can see that the function has only one real zero (it is a positive number near $x = 1$), as shown in the figure.

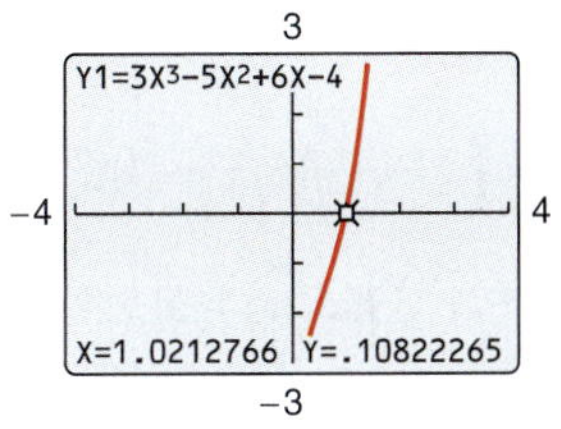

Checkpoint Audio-video solution in English & Spanish at LarsonPrecalculus.com

Determine the possible numbers of positive and negative real zeros of

$$f(x) = -2x^3 + 5x^2 - x + 8.$$

Another test for real zeros of a polynomial function f is related to the sign pattern in the last row of the synthetic division array. This test can give you an upper or lower bound for the real zeros of f, which can help you eliminate possible real zeros. A real number c is an **upper bound** for the real zeros of f when no zeros are greater than c. Similarly, c is a **lower bound** when no real zeros of f are less than c.

Upper and Lower Bound Rules

Let $f(x)$ be a polynomial with real coefficients and a positive leading coefficient. Suppose $f(x)$ is divided by $x - c$, using synthetic division.

1. If $c > 0$ and each number in the last row is either positive or zero, then c is an **upper bound** for the real zeros of f.
2. If $c < 0$ and the numbers in the last row are alternately positive and negative (zero entries count as positive or negative), then c is a **lower bound** for the real zeros of f.

EXAMPLE 10 Finding the Real Zeros of a Polynomial Function

Find all real zeros of $f(x) = 6x^3 - 4x^2 + 3x - 2$.

Solution

The possible real zeros are as follows.

$$\frac{\text{Factors of } -2}{\text{Factors of } 6} = \frac{\pm 1, \pm 2}{\pm 1, \pm 2, \pm 3, \pm 6} = \pm 1, \pm\frac{1}{2}, \pm\frac{1}{3}, \pm\frac{1}{6}, \pm\frac{2}{3}, \pm 2$$

The original polynomial $f(x)$ has three variations in sign. The polynomial

$$f(-x) = 6(-x)^3 - 4(-x)^2 + 3(-x) - 2$$
$$= -6x^3 - 4x^2 - 3x - 2$$

has no variations in sign. So, by Descartes's Rule of Signs, there are either three positive real zeros or one positive real zero, and no negative real zeros. Use synthetic division to test $x = 1$.

1	6	−4	3	−2
		6	2	5
	6	2	5	3

⟶ Nonzero remainder; $(x - 1)$ is not a factor.

So, $x = 1$ is not a zero, but because the last row has all positive entries, you know that $x = 1$ is an upper bound for the real zeros. Therefore, you can restrict the search to zeros between 0 and 1. By trial and error, you can determine that $x = \frac{2}{3}$ is a zero. So, using $\left(x - \frac{2}{3}\right)$ as a factor, you can determine that $f(x)$ factors as

$$f(x) = 6x^3 - 4x^2 + 3x - 2$$
$$= \left(x - \frac{2}{3}\right)(6x^2 + 3).$$

Because $6x^2 + 3$ has no real zeros, it follows that $x = \frac{2}{3}$ is the only real zero of f, as shown in the figure.

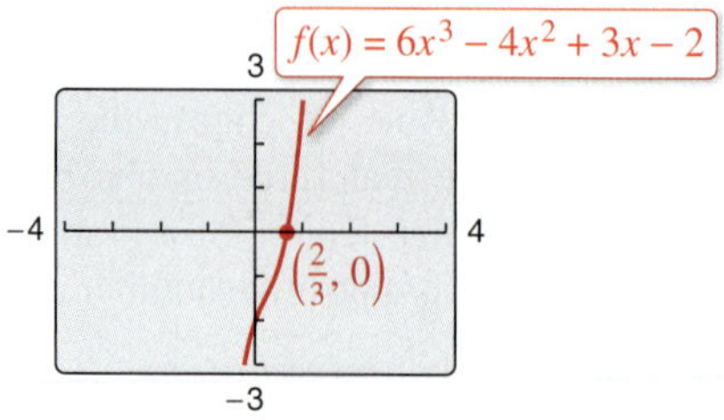

✓ **Checkpoint** Audio-video solution in English & Spanish at LarsonPrecalculus.com

Find all real zeros of

$$f(x) = 8x^3 - 4x^2 + 6x - 3.$$

Explore the Concept

Use a graphing utility to graph the polynomial

$$y_1 = 6x^3 - 4x^2 + 3x - 2.$$

Notice that the graph intersects the x-axis at the point $\left(\frac{2}{3}, 0\right)$. How does this information relate to the real zero found in Example 10? Use the graphing utility to graph

$$y_2 = x^4 - 5x^3 + 3x^2 + x.$$

How many times does the graph intersect the x-axis? How many real zeros does y_2 have?

Here are two additional hints that can help you find the real zeros of a polynomial.

1. When the terms of $f(x)$ have a common monomial factor, it should be factored out before applying the tests in this section. For instance, by writing

$$f(x) = x^4 - 5x^3 + 3x^2 + x = x(x^3 - 5x^2 + 3x + 1)$$

you can see that $x = 0$ is a zero of f and that the remaining zeros can be found by analyzing the cubic factor.

2. When you are able to find all but two zeros of f, you can always use the Quadratic Formula on the remaining quadratic factor. For instance, after writing

$$f(x) = x^4 - 5x^3 + 3x^2 + x = x(x - 1)(x^2 - 4x - 1)$$

you can apply the Quadratic Formula to $x^2 - 4x - 1$ to find that the two remaining zeros are $x = 2 + \sqrt{5}$ and $x = 2 - \sqrt{5}$.

Note how these hints are applied in the next example.

Explore the Concept

Use a graphing utility to graph the polynomial

$$y = x^3 + 4.8x^2 - 127x + 309$$

in a standard viewing window. From the graph, what do the real zeros appear to be? Discuss how the mathematical tools of this section might help you realize that the graph does not show all the important features of the polynomial function. Now use the *zoom* feature to find all the zeros of this function.

EXAMPLE 11 Finding the Zeros of a Polynomial Function

See LarsonPrecalculus.com for an interactive version of this type of example.

Find all real zeros of $f(x) = 10x^4 - 15x^3 - 16x^2 + 12x$.

Solution

Remove the common monomial factor x to write

$$f(x) = 10x^4 - 15x^3 - 16x^2 + 12x = x(10x^3 - 15x^2 - 16x + 12).$$

So, $x = 0$ is a zero of f. You can find the remaining zeros of f by analyzing the cubic factor. Because the leading coefficient is 10 and the constant term is 12, there is a long list of possible rational zeros.

Possible rational zeros:

$$\frac{\text{Factors of } 12}{\text{Factors of } 10} = \frac{\pm 1, \pm 2, \pm 3, \pm 4, \pm 6, \pm 12}{\pm 1, \pm 2, \pm 5, \pm 10}$$

With so many possibilities (32, in fact), it is worth your time to use a graphing utility to focus on just a few. By using the *trace* feature of a graphing utility, it looks like three reasonable choices are $x = -\frac{6}{5}$, $x = \frac{1}{2}$, and $x = 2$ (see figure). Synthetic division shows that only $x = 2$ works. (You could also use the Factor Theorem to test these choices.)

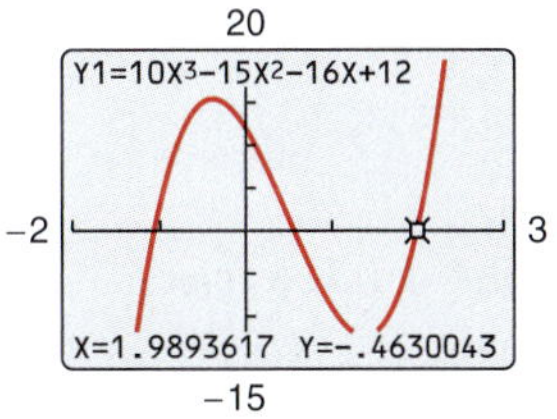

$$\begin{array}{r|rrrr} 2 & 10 & -15 & -16 & 12 \\ & & 20 & 10 & -12 \\ \hline & 10 & 5 & -6 & 0 \end{array}$$

So, $x = 2$ is one zero and you have

$$f(x) = x(x - 2)(10x^2 + 5x - 6).$$

Using the Quadratic Formula, you find that the two additional zeros are irrational numbers.

$$x = \frac{-5 + \sqrt{265}}{20} \approx 0.56 \quad \text{and} \quad x = \frac{-5 - \sqrt{265}}{20} \approx -1.06$$

✓ Checkpoint *Audio-video solution in English & Spanish at LarsonPrecalculus.com*

Find all real zeros of $f(x) = 3x^4 - 14x^2 - 4x$.

2.3 Exercises

See *CalcChat.com* for tutorial help and worked-out solutions to odd-numbered exercises. For instructions on how to use a graphing utility, see Appendix A.

Vocabulary and Concept Check

1. Two forms of the Division Algorithm are shown below. Identify and label each part.

$$f(x) = d(x)q(x) + r(x) \qquad \frac{f(x)}{d(x)} = q(x) + \frac{r(x)}{d(x)}$$

In Exercises 2–5, fill in the blank(s).

2. The rational expression $p(x)/q(x)$ is called _______ when the degree of the numerator is greater than or equal to that of the denominator.

3. Every rational zero of a polynomial function with integer coefficients has the form p/q, where p is a factor of the _______ and q is a factor of the _______ .

4. The theorem that can be used to determine the possible numbers of positive real zeros and negative real zeros of a function is called _______ of _______ .

5. A real number c is a(n) _______ bound for the real zeros of f when no zeros are greater than c, and is a(n) _______ bound when no real zeros of f are less than c.

6. How many negative real zeros are possible for a polynomial function f when $f(-x)$ has 5 variations in sign?

7. You divide the polynomial $f(x)$ by $(x - 2)$ and obtain a remainder of 15. What is $f(2)$?

8. What value should you write in the circle to check whether $(x - 4)$ is a factor of $f(x) = x^3 + 5x^2 - 6x + 2$?

○ | 1 5 −6 2

Procedures and Problem Solving

Long Division of Polynomials **In Exercises 9–12, use long division to divide and use the result to factor the dividend completely.**

9. $(x^2 + 6x + 8) \div (x + 4)$

10. $(5x^2 - 17x - 12) \div (x - 4)$

11. $(x^3 + 5x^2 - 12x - 36) \div (x + 2)$

12. $(2x^3 - 3x^2 - 50x + 75) \div (2x - 3)$

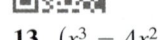

Long Division of Polynomials **In Exercises 13–22, use long division to divide.**

13. $(x^3 - 4x^2 - 17x + 6) \div (x - 3)$

14. $(4x^3 - 7x^2 - 11x + 5) \div (4x + 5)$

15. $(2x^3 + 7) \div (x + 3)$

16. $(8x^4 - 5) \div (2x + 1)$

17. $(5x - 1 + 10x^3 - 2x^2) \div (2x^2 + 1)$

18. $(1 + 3x^2 + x^4) \div (3 - 2x + x^2)$

19. $(x^3 - 9) \div (x^2 + 1)$

20. $\dfrac{x^5 + 7}{x^3 - 1}$

21. $\dfrac{2x^3 - 4x^2 - 15x + 5}{(x - 1)^2}$

22. $\dfrac{x^4}{(x - 1)^3}$

Using Synthetic Division **In Exercises 23–32, use synthetic division to divide.**

23. $(3x^3 - 17x^2 + 15x - 25) \div (x - 5)$

24. $(5x^3 + 18x^2 + 7x - 6) \div (x + 3)$

25. $(6x^3 + 7x^2 - x + 26) \div (x - 3)$

26. $(2x^3 + 14x^2 - 20x + 7) \div (x + 6)$

27. $(9x^3 - 18x^2 - 16x + 32) \div (x - 2)$

28. $(5x^3 + 6x + 8) \div (x + 2)$

29. $(x^3 + 512) \div (x + 8)$ **30.** $(x^3 - 343) \div (x - 7)$

31. $(4x^3 + 16x^2 - 23x - 15) \div \left(x + \frac{1}{2}\right)$

32. $(3x^3 - 4x^2 + 5) \div \left(x - \frac{3}{2}\right)$

Verifying a Quotient **In Exercises 33–36, use a graphing utility to graph the two equations in the same viewing window. Use the graphs to verify that the expressions are equivalent. Verify the results algebraically.**

33. $y_1 = \dfrac{x^2}{x + 2}, \quad y_2 = x - 2 + \dfrac{4}{x + 2}$

34. $y_1 = \dfrac{x^2 + 2x - 2}{x + 4}, \quad y_2 = x - 2 + \dfrac{6}{x + 4}$

35. $y_1 = \dfrac{x^4 - 3x^2 - 1}{x^2 + 5}, \quad y_2 = x^2 - 8 + \dfrac{39}{x^2 + 5}$

36. $y_1 = \dfrac{x^4 + x^2 - 1}{x^2 + 1}, \quad y_2 = x^2 - \dfrac{1}{x^2 + 1}$

Using the Remainder Theorem In Exercises 37–42, write the function in the form $f(x) = (x - k)q(x) + r$ for the indicated value of k and demonstrate that $f(k) = r$.

37. $f(x) = x^3 - x^2 - 18x + 17, \quad k = 2$
38. $f(x) = 15x^4 + 10x^3 - 6x^2 + 14, \quad k = -\frac{2}{3}$
39. $f(x) = x^3 + 3x^2 - 2x - 14, \quad k = \sqrt{2}$
40. $f(x) = x^3 + 2x^2 - 5x - 4, \quad k = -\sqrt{5}$
41. $f(x) = 4x^3 - 6x^2 - 12x - 4, \quad k = 1 - \sqrt{3}$
42. $f(x) = -3x^3 + 8x^2 + 10x - 8, \quad k = 2 + \sqrt{2}$

Using the Remainder Theorem In Exercises 43–46, use the Remainder Theorem and synthetic division to find each function value. Verify your answer using another method.

43. $f(x) = 2x^3 - 7x + 3$
(a) $f(1)$ (b) $f(-2)$ (c) $f(3)$ (d) $f(2)$
44. $g(x) = 2x^6 + 3x^4 - x^2 + 3$
(a) $g(2)$ (b) $g(1)$ (c) $g(3)$ (d) $g(-1)$
45. $h(x) = x^3 - 5x^2 - 7x + 4$
(a) $h(3)$ (b) $h(2)$ (c) $h(-2)$ (d) $h(-5)$
46. $f(x) = 4x^4 - 12x^3 + 11x^2 + 16$
(a) $f(1)$ (b) $f(-2)$ (c) $f(3)$ (d) $f(-6)$

Using the Factor Theorem In Exercises 47–52, use synthetic division to show that x is a solution of the third-degree polynomial equation and use the result to factor the polynomial completely. List all the real solutions of the equation.

47. $x^3 - 21x - 20 = 0, \quad x = 5$
48. $x^3 - 31x + 30 = 0, \quad x = -6$
49. $2x^3 - 17x^2 + 12x + 63 = 0, \quad x = -\frac{3}{2}$
50. $60x^3 - 89x^2 + 41x - 6 = 0, \quad x = \frac{1}{3}$
51. $x^3 + 2x^2 - 3x - 6 = 0, \quad x = \sqrt{3}$
52. $x^3 - x^2 - 13x - 3 = 0, \quad x = 2 - \sqrt{5}$

Factoring a Polynomial In Exercises 53–58, (a) verify the indicated factor(s) of $f(x)$, (b) find the remaining factor(s) of $f(x)$, (c) use your results to write the complete factorization of $f(x)$, (d) list all real zeros of f, and (e) confirm your results by using a graphing utility to graph the function.

	Function	*Factor(s)*
53.	$f(x) = 2x^3 + x^2 - 5x + 2$	$(x + 2), (x - 1)$
54.	$f(x) = x^4 - 4x^3 - 15x^2 + 58x - 40$	$(x - 5), (x + 4)$
55.	$f(x) = 8x^4 - 14x^3 - 71x^2 - 10x + 24$	$(x + 2), (x - 4)$
56.	$f(x) = 3x^3 + 2x^2 - 19x + 6$	$(x + 3)$
57.	$f(x) = 6x^3 + 41x^2 - 9x - 14$	$(2x + 1)$
58.	$f(x) = 3x^3 - x^2 - 15x + 5$	$(3x - 1)$

Using the Rational Zero Test In Exercises 59–62, find (if possible) the rational zeros of the function.

59. $C(x) = 2x^3 + 3x^2 - 1$
60. $g(x) = x^3 + 8x^2 + 12x + 18$
61. $f(x) = 2x^4 - 17x^3 + 35x^2 + 9x - 45$
62. $f(x) = 4x^5 - 8x^4 - 5x^3 + 10x^2 + x - 2$

Using Descartes's Rule of Signs In Exercises 63–66, use Descartes's Rule of Signs to determine the possible numbers of positive and negative real zeros of the function.

63. $f(x) = 2x^4 - x^3 + 6x^2 - x + 5$
64. $f(x) = 3x^4 + 5x^3 - 6x^2 + 8x - 3$
65. $g(x) = 6x^4 + 2x^3 - 3x^2 + 2$
66. $g(x) = 2x^3 - 4x^2 - 5$

Finding the Real Zeros of a Polynomial Function In Exercises 67–72, (a) use Descartes's Rule of Signs to determine the possible numbers of positive and negative real zeros of f, (b) list the possible rational zeros of f, (c) use a graphing utility to graph f so that some of the possible zeros in parts (a) and (b) can be disregarded, and (d) determine all the real zeros of f.

67. $f(x) = x^3 + x^2 - 4x - 4$
68. $f(x) = -3x^3 + 20x^2 - 36x + 16$
69. $f(x) = -2x^4 + 13x^3 - 21x^2 + 2x + 8$
70. $f(x) = 4x^4 - 17x^2 + 4$
71. $f(x) = 32x^3 - 52x^2 + 17x + 3$
72. $f(x) = x^4 - x^3 - 41x^2 - x - 42$

Finding the Real Zeros of a Polynomial Function In Exercises 73–76, use synthetic division to verify the upper and lower bounds of the real zeros of f. Then find all real zeros of the function.

73. $f(x) = x^4 - 4x^3 + 15$
Upper bound: $x = 4$, Lower bound: $x = -1$
74. $f(x) = 2x^4 - 8x + 3$
Upper bound: $x = 3$
Lower bound: $x = -4$
75. $f(x) = 2x^3 - 3x^2 - 12x + 8$
Upper bound: $x = 4$
Lower bound: $x = -3$
76. $f(x) = x^4 - 4x^3 + 16x - 16$
Upper bound: $x = 5$
Lower bound: $x = -3$

Rewriting to Use the Rational Zero Test **In Exercises 77–80, find the rational zeros of the polynomial function.**

77. $P(x) = x^4 - \frac{13}{4}x^2 + \frac{9}{4} = \frac{1}{4}(4x^4 - 13x^2 + 9)$

78. $f(x) = x^3 - \frac{3}{2}x^2 - \frac{23}{2}x + 6 = \frac{1}{2}(2x^3 - 3x^2 - 23x + 12)$

79. $f(z) = z^3 + \frac{11}{6}z^2 - \frac{1}{2}z - \frac{1}{3} = \frac{1}{6}(6z^3 + 11z^2 - 3z - 2)$

80. $f(x) = x^3 - \frac{1}{4}x^2 - 4x + 1 = \frac{1}{4}(4x^3 - x^2 - 16x + 4)$

A Cubic Polynomial with Two Terms **In Exercises 81–84, match the cubic function with the correct number of rational and irrational zeros.**

(a) Rational zeros: 0; Irrational zeros: 1

(b) Rational zeros: 3; Irrational zeros: 0

(c) Rational zeros: 1; Irrational zeros: 2

(d) Rational zeros: 1; Irrational zeros: 0

81. $f(x) = x^3 - 1$

82. $f(x) = x^3 - 2$

83. $f(x) = x^3 - x$

84. $f(x) = x^3 - 3x$

Using a Graph to Help Find Real Zeros **In Exercises 85–88, the graph of $y = f(x)$ is shown. Use the graph as an aid to find all real zeros of the function.**

85. $y = 2x^4 - 9x^3 + 5x^2 + 3x - 1$

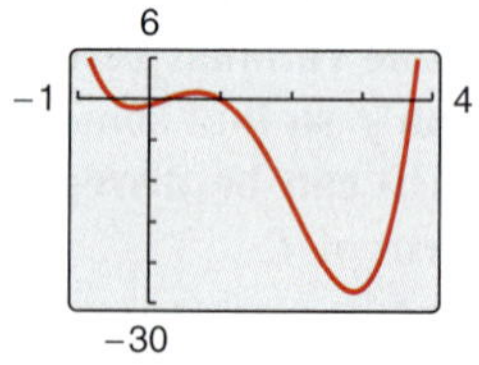

86. $y = x^4 - 5x^3 - 7x^2 + 13x - 2$

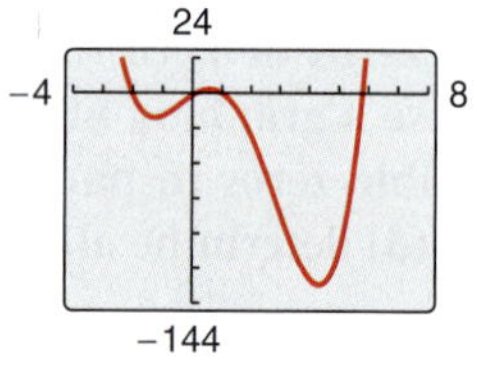

87. $y = -2x^4 + 17x^3 - 3x^2 - 25x - 3$

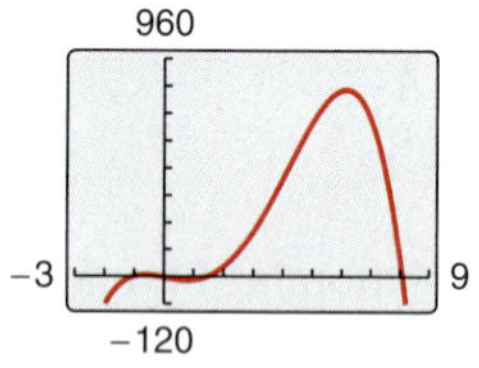

88. $y = -x^4 + 5x^3 - 10x - 4$

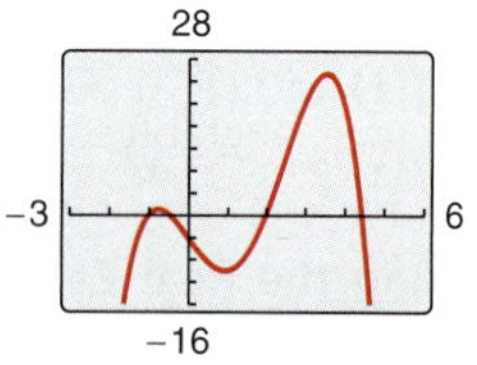

Finding the Real Zeros of a Polynomial Function **In Exercises 89–100, find all real zeros of the polynomial function.**

89. $f(x) = 5x^4 + 9x^3 - 19x^2 - 3x$

90. $g(x) = 4x^4 - 11x^3 - 22x^2 + 8x$

91. $f(z) = z^4 + 2z^3 + 6z - 9$

92. $f(x) = 4x^3 + 7x^2 - 11x - 18$

93. $g(y) = 2y^4 + 7y^3 - 26y^2 + 23y - 6$

94. $h(x) = x^5 - x^4 - 3x^3 + 5x^2 - 2x$

95. $f(x) = 4x^4 - 55x^2 - 45x + 36$

96. $h(x) = 6x^4 + 33x^3 - 69x + 30$

97. $g(x) = 8x^4 + 28x^3 + 9x^2 - 9x$

98. $h(x) = x^5 + 5x^4 - 5x^3 - 15x^2 - 6x$

99. $f(x) = 8x^5 + 6x^4 - 37x^3 - 36x^2 + 29x + 30$

100. $g(x) = 4x^5 + 8x^4 - 15x^3 - 23x^2 + 11x + 15$

Using a Rational Zero **In Exercises 101–104, (a) use the *zero* or *root* feature of a graphing utility to approximate (accurate to the nearest thousandth) the real zeros of the function, (b) determine one of the exact zeros and use synthetic division to verify your result, and (c) factor the polynomial completely.**

101. $h(t) = t^3 - 2t^2 - 7t + 2$

102. $f(s) = s^3 - 12s^2 + 40s - 24$

103. $h(x) = x^5 - 7x^4 + 10x^3 + 14x^2 - 24x$

104. $g(x) = 12x^4 - 11x^3 - 190x^2 + 176x - 32$

105. MODELING DATA

The table shows the numbers S of mobile cellular subscriptions per 100 people in the United States from 2004 through 2016. (*Source:* International Telecommunications Union)

DATA Year	Subscriptions per 100 people, S
2004	63.2
2005	69.0
2006	77.1
2007	82.9
2008	86.1
2009	89.6
2010	92.4
2011	95.6
2012	97.3
2013	98.5
2014	111.9
2015	119.5
2016	122.9

Spreadsheet at LarsonPrecalculus.com

The data can be approximated by the model

$S = 0.0643t^3 - 1.878t^2 + 21.18t + 3.7,$
$4 \le t \le 16$

where t represents the year, with $t = 4$ corresponding to 2004.

(a) Use a graphing utility to plot the data and graph the model in the same viewing window.

(b) How well does the model fit the data?

(c) Use the Remainder Theorem to evaluate the model for the year 2021. Is the value reasonable? Explain.

Occasionally, throughout this text, you will be asked to round to a place value rather than to a number of decimal places.

michaeljung/iStock/Getty Images

106. ***Why you should learn it*** *(p. 113)* The numbers of employees E (in thousands) in education and health services in the United States from 2000 through 2017 are approximated by $E = 0.721t^3 - 19.97t^2 + 595.0t + 15{,}227$, $0 \le t \le 17$, where t is the year, with $t = 0$ corresponding to 2000. *(Source:* U.S. Bureau of Labor Statistics)

(a) Use a graphing utility to graph the model.

(b) Estimate the number of employees in education and health services in 2000. Use the Remainder Theorem to estimate the number in 2015.

(c) Is this a good model for making predictions in future years? Explain.

107. Geometry A rectangular package sent by a delivery service can have a maximum combined length and girth (perimeter of a cross section) of 120 inches (see figure).

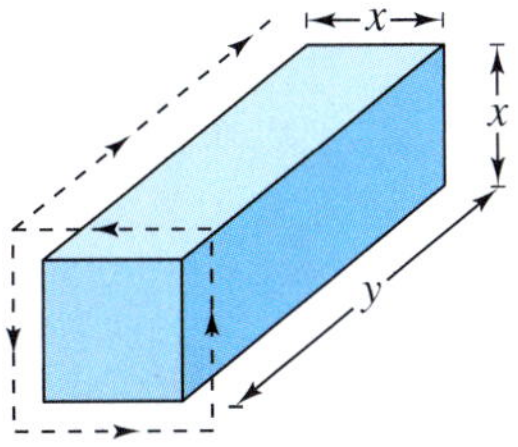

(a) Write a function V that represents the volume of the package.

(b) Use a graphing utility to graph the function and approximate the dimensions of the package that yield a maximum volume.

(c) Find values of x such that $V = 13{,}500$. Which of these values is a physical impossibility in the construction of the package? Explain.

108. Environmental Science The number of parts per million of nitric oxide emissions y from a car engine is approximated by $y = -5.05x^3 + 3857x - 38{,}411.25$, $13 \le x \le 18$, where x is the air-fuel ratio.

(a) Use a graphing utility to graph the model.

(b) There are two air-fuel ratios that produce 2400 parts per million of nitric oxide. One is $x = 15$. Use the graph to approximate the other.

(c) Find the second air-fuel ratio from part (b) algebraically. (*Hint:* Use synthetic division.)

Focusing on Concepts

True or False? **In Exercises 109 and 110, determine whether the statement is true or false. Justify your answer.**

109. If $(4x - 7)$ is a factor of some polynomial function f, then $\frac{7}{4}$ is a zero of f.

110. The value $x = 3$ is a zero of the polynomial function $f(x) = 3x^5 - 2x^4 + x^3 - 16x^2 + 3x - 8$.

Think About It **In Exercises 111 and 112, the graph of a cubic polynomial function $y = f(x)$ with integer zeros is shown. Find the factored form of f.**

111.

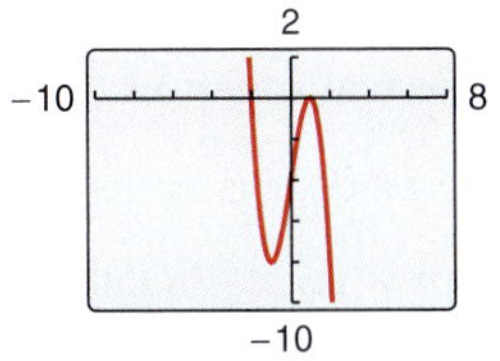

112.

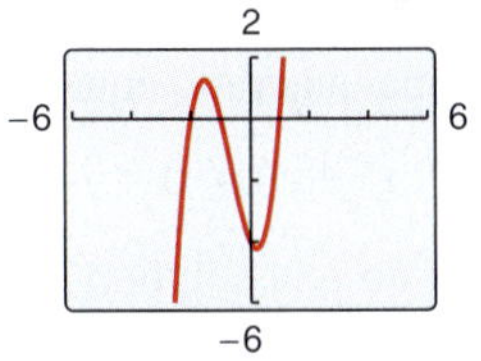

113. Error Analysis Describe the error.

Use synthetic division to find the remainder when $x^2 + 3x - 5$ is divided by $x + 1$.

1	1	3	−5
		1	4
	1	4	−1

← Remainder: −1 ✗

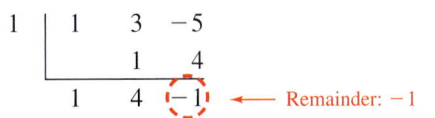

114. HOW DO YOU SEE IT? A graph of $y = f(x)$ is shown, where
$f(x) = 2x^5 - 3x^4 + x^3 - 8x^2 + 5x + 3$ and
$f(-x) = -2x^5 - 3x^4 - x^3 - 8x^2 - 5x + 3$.

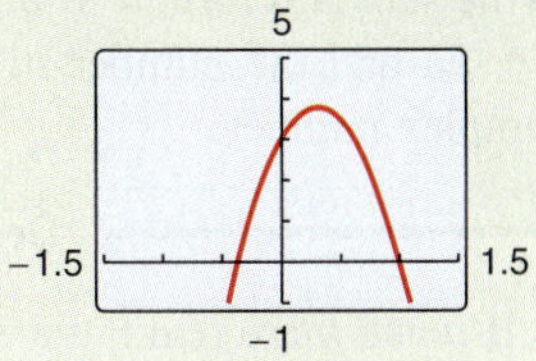

(a) How many negative real zeros does f have? Explain.

(b) How many positive real zeros are *possible* for f? Explain. What does this tell you about the eventual right-hand behavior of the graph?

(c) Is $x = -\frac{1}{3}$ a possible rational zero of f? Explain.

(d) Explain how to check whether $\left(x - \frac{3}{2}\right)$ is a factor of f and whether $x = \frac{3}{2}$ is an upper bound for the real zeros of f.

115. Think About It Find the value of k such that $x - 3$ is a factor of $x^3 - kx^2 + 2kx - 12$.

116. Writing Perform each polynomial division. Write a brief description of the pattern that you obtain, and use your result to find a formula for the polynomial division $(x^n - 1)/(x - 1)$. Create a numerical example to test your formula.

(a) $\dfrac{x^2 - 1}{x - 1}$ (b) $\dfrac{x^3 - 1}{x - 1}$ (c) $\dfrac{x^4 - 1}{x - 1}$

Cumulative Mixed Review

Solving a Quadratic Equation **In Exercises 117–120, use any convenient method to solve the quadratic equation.**

117. $4x^2 - 17 = 0$ **118.** $25x^2 - 1 = 0$

119. $3x^2 - 11x - 20 = 0$ **120.** $6x^2 + 4x - 3 = 0$

2.4 Complex Numbers

What you should learn

- Use the imaginary unit i to write complex numbers.
- Add, subtract, and multiply complex numbers.
- Use complex conjugates to write the quotient of two complex numbers in standard form.
- Find complex solutions of quadratic equations.

Why you should learn it

Complex numbers are used to model numerous aspects of the natural world, such as the impedance of an electrical circuit, as shown in Exercise 87 on page 134.

The Imaginary Unit i

Some quadratic equations have no real solutions. For instance, the quadratic equation

$$x^2 + 1 = 0 \qquad \text{Quadratic equation}$$

has no real solution because there is no real number x that can be squared to produce -1. To overcome this deficiency, mathematicians created an expanded system of numbers using the **imaginary unit *i*,** defined as

$$i = \sqrt{-1} \qquad \text{Imaginary unit}$$

where $i^2 = -1$. By adding real numbers to real multiples of this imaginary unit, you obtain the set of **complex numbers.** Each complex number can be written in the **standard form** $a + bi$. For instance, the standard form of the complex number $\sqrt{-9} - 5$ is $-5 + 3i$ because

$$\begin{aligned}\sqrt{-9} - 5 &= \sqrt{3^2(-1)} - 5\\ &= 3\sqrt{-1} - 5\\ &= 3i - 5\\ &= -5 + 3i.\end{aligned}$$

In the standard form $a + bi$, the real number a is the **real part** of the **complex number** $a + bi$, and the number bi (where b is a real number) is the **imaginary part** of the complex number.

Definition of a Complex Number

If a and b are real numbers, then the number $a + bi$ is a **complex number** written in **standard form.** If $b = 0$, then the number $a + bi = a$ is a real number. If $b \neq 0$, then the number $a + bi$ is called an **imaginary number.** A number of the form bi, where $b \neq 0$, is called a **pure imaginary number.**

The set of real numbers is a subset of the set of complex numbers, as shown in Figure 2.19. This is true because every real number a can be written as a complex number using $b = 0$. That is, for every real number a, you can write $a = a + 0i$.

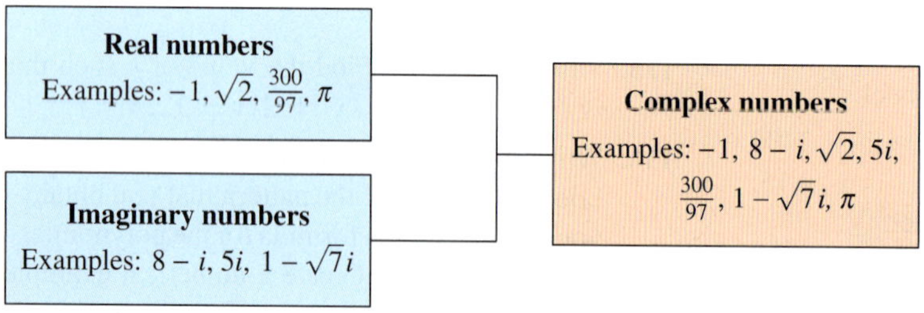

Figure 2.19

Equality of Complex Numbers

Two complex numbers $a + bi$ and $c + di$, written in standard form, are equal to each other

$$a + bi = c + di \qquad \text{Equality of two complex numbers}$$

if and only if $a = c$ and $b = d$.

Operations with Complex Numbers

To add (or subtract) two complex numbers, you add (or subtract) the real and imaginary parts of the numbers separately.

Addition and Subtraction of Complex Numbers

If $a + bi$ and $c + di$ are two complex numbers written in standard form, then their sum is

$$(a + bi) + (c + di) = (a + c) + (b + d)i \qquad \text{Sum}$$

and their difference is

$$(a + bi) - (c + di) = (a - c) + (b - d)i. \qquad \text{Difference}$$

The **additive identity** in the complex number system is zero (the same as in the real number system). Furthermore, the **additive inverse** of the complex number $a + bi$ is

$$-(a + bi) = -a - bi. \qquad \text{Additive inverse}$$

So, you have

$$(a + bi) + (-a - bi) = 0 + 0i = 0.$$

EXAMPLE 1 Adding and Subtracting Complex Numbers

a. $(3 - i) + (2 + 3i) = 3 - i + 2 + 3i$ Remove parentheses.
$= (3 + 2) + (-i + 3i)$ Group like terms.
$= 5 + 2i$ Write in standard form.

b. $(1 + 2i) - (4 + 2i) = 1 + 2i - 4 - 2i$ Remove parentheses.
$= (1 - 4) + (2i - 2i)$ Group like terms.
$= -3 + 0i$ Simplify.
$= -3$ Write in standard form.

c. $3 - (-2 + 3i) + (-5 + i) = 3 + 2 - 3i - 5 + i$
$= (3 + 2 - 5) + (-3i + i)$
$= 0 - 2i$
$= -2i$

d. $(3 + 2i) + (4 - i) - (7 + i) = 3 + 2i + 4 - i - 7 - i$
$= (3 + 4 - 7) + (2i - i - i)$
$= 0 + 0i$
$= 0$

Insight

Be sure you know how to perform operations with complex numbers, as shown in Examples 1 and 2. These operations are usually assessed on standardized tests.

✓ ***Checkpoint*** *Audio-video solution in English & Spanish at LarsonPrecalculus.com*

Perform each operation and write the result in standard form.

a. $(7 + 3i) + (5 - 4i)$

b. $(3 + 4i) - (5 - 3i)$

c. $2i + (-3 - 4i) - (-3 - 3i)$

d. $(5 - 3i) + (3 + 5i) - (8 + 2i)$

In Examples 1(b) and 1(d), note that the sum of complex numbers can be a real number.

Many of the properties of real numbers are valid for complex numbers as well. Here are some examples.

Associative Properties of Addition and Multiplication

Commutative Properties of Addition and Multiplication

Distributive Property of Multiplication over Addition

Notice how these properties are used when two complex numbers are multiplied.

$(a + bi)(c + di) = a(c + di) + bi(c + di)$ Distributive Property

$= ac + (ad)i + (bc)i + (bd)i^2$ Distributive Property

$= ac + (ad)i + (bc)i + (bd)(-1)$ $i^2 = -1$

$= ac - bd + (ad)i + (bc)i$ Commutative Property

$= (ac - bd) + (ad + bc)i$ Associative Property

The procedure above is similar to multiplying two polynomials and combining like terms, as in the FOIL Method.

Explore the Concept

Find the missing values.

$i^1 = i$	$i^7 =$
$i^2 = -1$	$i^8 =$
$i^3 = -i$	$i^9 =$
$i^4 = 1$	$i^{10} =$
$i^5 =$	$i^{11} =$
$i^6 =$	$i^{12} =$

What pattern do you see? Write a brief description of how you would find i raised to any positive integer power.

EXAMPLE 2 Multiplying Complex Numbers

a. $5(-2 + 3i) = 5(-2) + 5(3i)$ Distributive Property

$= -10 + 15i$ Write in standard form.

b. $(2 - i)(4 + 3i) = 2(4 + 3i) - i(4 + 3i)$ Distributive Property

$= 8 + 6i - 4i - 3i^2$ Distributive Property

$= 8 + 6i - 4i - 3(-1)$ $i^2 = -1$

$= 8 + 3 + 6i - 4i$ Group like terms.

$= 11 + 2i$ Write in standard form.

c. $(3 + 2i)(3 - 2i) = 3(3 - 2i) + 2i(3 - 2i)$ Distributive Property

$= 9 - 6i + 6i - 4i^2$ Distributive Property

$= 9 - 4(-1)$ $i^2 = -1$

$= 9 + 4$ Simplify.

$= 13$ Write in standard form.

d. $4i(-1 + 5i) = 4i(-1) + 4i(5i)$ Distributive Property

$= -4i + 20i^2$ Simplify.

$= -4i + 20(-1)$ $i^2 = -1$

$= -20 - 4i$ Write in standard form.

e. $(3 + 2i)^2 = (3 + 2i)(3 + 2i)$ Property of exponents

$= 3(3 + 2i) + 2i(3 + 2i)$ Distributive Property

$= 9 + 6i + 6i + 4i^2$ Distributive Property

$= 9 + 6i + 6i + 4(-1)$ $i^2 = -1$

$= 9 + 12i - 4$ Simplify.

$= 5 + 12i$ Write in standard form.

✓ **Checkpoint** ▶ Audio-video solution in English & Spanish at LarsonPrecalculus.com

Perform each operation and write the result is standard form.

a. $(2 - 4i)(3 + 3i)$

b. $(4 + 5i)(4 - 5i)$

c. $(4 + 2i)^2$

Complex Conjugates

Notice in Example 2(c) that the product of two complex numbers can be a real number. This occurs with pairs of complex numbers of the forms $a + bi$ and $a - bi$, called **complex conjugates.**

$$(a + bi)(a - bi) = a^2 - abi + abi - b^2i^2 = a^2 - b^2(-1) = a^2 + b^2$$

EXAMPLE 3 Multiplying Conjugates

Multiply each complex number by its complex conjugate.

a. $1 + i$ **b.** $4 - 3i$

Solution

a. The complex conjugate of $1 + i$ is $1 - i$.

$$(1 + i)(1 - i) = 1^2 - i^2 = 1 - (-1) = 2$$

b. The complex conjugate of $4 - 3i$ is $4 + 3i$.

$$(4 - 3i)(4 + 3i) = 4^2 - (3i)^2 = 16 - 9i^2 = 16 - 9(-1) = 25$$

✓ ***Checkpoint*** ▶ *Audio-video solution in English & Spanish at LarsonPrecalculus.com*

Multiply each complex number by its complex conjugate.

a. $3 + 6i$ **b.** $2 - 5i$

To write the quotient of $a + bi$ and $c + di$ in standard form, where c and d are not both zero, multiply the numerator and denominator by the complex conjugate of the *denominator* to obtain

$$\frac{a + bi}{c + di} = \frac{a + bi}{c + di}\left(\frac{c - di}{c - di}\right) \quad \text{Multiply numerator and denominator by complex conjugate of denominator.}$$

$$= \frac{(ac + bd) + (bc - ad)i}{c^2 + d^2} \quad \text{Expand and collect like terms.}$$

$$= \frac{ac + bd}{c^2 + d^2} + \left(\frac{bc - ad}{c^2 + d^2}\right)i. \quad \text{Standard form}$$

EXAMPLE 4 Writing a Quotient in Standard Form

$$\frac{2 + 3i}{4 - 2i} = \frac{2 + 3i}{4 - 2i}\left(\frac{4 + 2i}{4 + 2i}\right) \quad \text{Multiply numerator and denominator by complex conjugate of denominator.}$$

$$= \frac{8 + 4i + 12i + 6i^2}{16 - 4i^2} \quad \text{Expand.}$$

$$= \frac{8 - 6 + 16i}{16 + 4} \quad i^2 = -1$$

$$= \frac{2 + 16i}{20} \quad \text{Simplify.}$$

$$= \frac{1}{10} + \frac{4}{5}i \quad \text{Write in standard form.}$$

✓ ***Checkpoint*** ▶ *Audio-video solution in English & Spanish at LarsonPrecalculus.com*

Write $\dfrac{2 + i}{2 - i}$ in standard form.

Technology Tip

Some graphing utilities can perform operations with complex numbers. For instance, on some graphing utilities, to divide $2 + 3i$ by $4 - 2i$, use the following keystrokes.

[(] 2 [+] 3 [i] [)] [÷]

[(] 4 [−] 2 [i] [)] [ENTER]

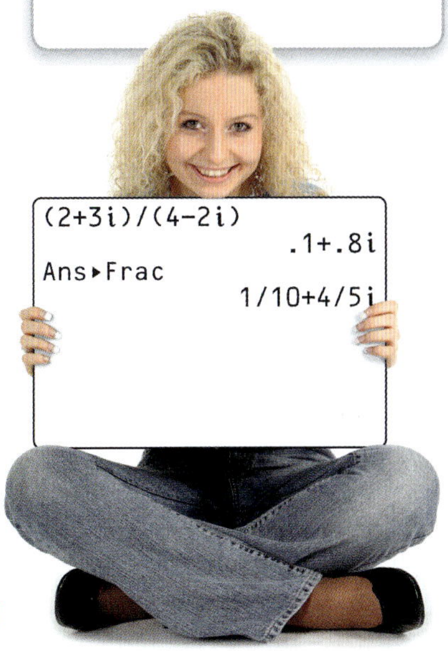

pikselstock/Shutterstock.com

Complex Solutions of Quadratic Equations

When using the Quadratic Formula to solve a quadratic equation, you often obtain a result such as $\sqrt{-3}$, which you know is not a real number. By factoring out $i = \sqrt{-1}$, you can write this number in standard form.

$$\sqrt{-3} = \sqrt{3(-1)} = \sqrt{3}\sqrt{-1} = \sqrt{3}i$$

The number $\sqrt{3}i$ is called the *principal square root* of -3.

Principal Square Root of a Negative Number

If a is a positive real number, then the **principal square root** of the negative number $-a$ is defined as

$$\sqrt{-a} = \sqrt{a}i.$$

Algebra Help

The definition of principal square root uses the rule

$$\sqrt{ab} = \sqrt{a}\sqrt{b}$$

for $a > 0$ and $b < 0$. This rule is not valid when *both* a and b are negative. For example,

$$\begin{aligned}\sqrt{-5}\sqrt{-5} &= \sqrt{5(-1)}\sqrt{5(-1)}\\ &= \sqrt{5}i\sqrt{5}i\\ &= \sqrt{25i^2}\\ &= 5i^2\\ &= -5\end{aligned}$$

whereas

$$\sqrt{(-5)(-5)} = \sqrt{25} = 5.$$

To avoid problems with square roots of negative numbers, be sure to convert complex numbers to standard form *before* performing any operations.

EXAMPLE 5 Writing Complex Numbers in Standard Form

a. $\sqrt{-3}\sqrt{-12} = \sqrt{3}i\sqrt{12}i = \sqrt{36}i^2 = 6(-1) = -6$

b. $$\begin{aligned}\sqrt{-48} - \sqrt{-27} &= \sqrt{48}i - \sqrt{27}i\\ &= 4\sqrt{3}i - 3\sqrt{3}i\\ &= \sqrt{3}i\end{aligned}$$

c. $$\begin{aligned}(-1 + \sqrt{-3})^2 &= (-1 + \sqrt{3}i)^2\\ &= (-1)^2 - 2\sqrt{3}i + (\sqrt{3})^2(i^2)\\ &= 1 - 2\sqrt{3}i + 3(-1)\\ &= -2 - 2\sqrt{3}i\end{aligned}$$

Checkpoint Audio-video solution in English & Spanish at LarsonPrecalculus.com

Write $\sqrt{-14}\sqrt{-2}$ in standard form.

EXAMPLE 6 Complex Solutions of a Quadratic Equation

See LarsonPrecalculus.com for an interactive version of this type of example.

Solve (a) $x^2 + 4 = 0$ and (b) $3x^2 - 2x + 5 = 0$.

Solution

a.
$x^2 + 4 = 0$ — Write original equation.
$x^2 = -4$ — Subtract 4 from each side.
$x = \pm 2i$ — Extract square roots.

b.
$3x^2 - 2x + 5 = 0$ — Write original equation.
$x = \dfrac{-(-2) \pm \sqrt{(-2)^2 - 4(3)(5)}}{2(3)}$ — Quadratic Formula
$= \dfrac{2 \pm \sqrt{-56}}{6}$ — Simplify.
$= \dfrac{2 \pm 2\sqrt{14}i}{6}$ — Write $\sqrt{-56}$ in standard form.
$= \dfrac{1}{3} \pm \dfrac{\sqrt{14}}{3}i$ — Write in standard form.

Checkpoint Audio-video solution in English & Spanish at LarsonPrecalculus.com

Solve $8x^2 + 14x + 9 = 0$.

2.4 Exercises

See *CalcChat.com* for tutorial help and worked-out solutions to odd-numbered exercises. For instructions on how to use a graphing utility, see Appendix A.

Vocabulary and Concept Check

1. Match the type of complex number with its definition.

(a) real number (i) $a + bi, a = 0, b \neq 0$

(b) imaginary number (ii) $a + bi, b = 0$

(c) pure imaginary number (iii) $a + bi, a \neq 0, b \neq 0$

In Exercises 2 and 3, fill in the blanks.

2. The imaginary unit i is defined as $i =$ _______ , where $i^2 =$ _______ .

3. When you add $(6 + 5i)$ and $(9 + 4i)$, the real part of the sum is _______ and the imaginary part of the sum is _______ .

4. What method for multiplying two polynomials can you use when multiplying two complex numbers?

5. What is the additive inverse of the complex number $2 - 4i$?

6. What is the complex conjugate of the complex number $3 - 3i$?

Procedures and Problem Solving

Equality of Complex Numbers **In Exercises 7–10, find real numbers a and b such that the equation is true.**

7. $a + bi = -9 + 4i$ **8.** $a + bi = 16 + 10i$

9. $3a + (b + 3)i = 9 + 8i$ **10.** $(a + 6) + 2bi = 6 - i$

Writing a Complex Number in Standard Form **In Exercises 11–20, write the complex number in standard form.**

11. $2 + \sqrt{-25}$ **12.** $4 + \sqrt{-49}$

13. 12 **14.** -3

15. $-8i - i^2$ **16.** $2i^2 - 6i$

17. $(\sqrt{-16})^2 + 5$ **18.** $-i - (\sqrt{-23})^2$

19. $\sqrt{-0.09}$ **20.** $\sqrt{-0.0036}$

Adding or Subtracting Complex Numbers **In Exercises 21–30, perform the addition or subtraction and write the result in standard form.**

21. $(-1 + 8i) + (8 - 5i)$ **22.** $(7 + 6i) + (3 + 12i)$

23. $(9 - i) - (8 - i)$ **24.** $(11 - 2i) - (-3 + 6i)$

25. $13i - (14 - 7i)$ **26.** $22 + (-5 + 8i) - 9i$

27. $\left(\frac{3}{2} + \frac{5}{2}i\right) + \left(\frac{5}{3} + \frac{11}{3}i\right)$ **28.** $\left(\frac{3}{4} + \frac{7}{5}i\right) - \left(\frac{5}{6} - \frac{1}{6}i\right)$

29. $(1.6 + 3.2i) + (-5.8 + 4.3i)$

30. $-(-2.8 - 13.7i) - (3.6 - 15.4i)$

Multiplying Complex Numbers **In Exercises 31–42, perform the operation(s) and write the result in standard form.**

31. $4(3 + 5i)$ **32.** $-6(5 - 3i)$

33. $(1 + i)(3 - 2i)$ **34.** $(2 - 6i)(4 - 3i)$

35. $4i(8 + 5i)$ **36.** $-3i(6 - i)$

37. $(\sqrt{2} + 3i)(\sqrt{2} - 3i)$ **38.** $(4 + \sqrt{7}i)(4 - \sqrt{7}i)$

39. $(6 + 7i)^2$ **40.** $(5 - 4i)^2$

41. $(4 + 5i)^2 - (4 - 5i)^2$ **42.** $(1 - 2i)^2 - (1 + 2i)^2$

Multiplying Complex Conjugates **In Exercises 43–50, write the complex conjugate of the complex number. Then multiply the number by its complex conjugate.**

43. $6 - 2i$ **44.** $5 + 12i$

45. $-1 + \sqrt{7}i$ **46.** $-4 - \sqrt{3}i$

47. $\sqrt{-29}$ **48.** $\sqrt{-10}$

49. $9 - \sqrt{6}i$ **50.** $-8 + \sqrt{15}i$

Writing a Quotient in Standard Form **In Exercises 51–58, write the quotient in standard form.**

51. $\dfrac{6}{i}$ **52.** $-\dfrac{4}{3i}$

53. $2/(4 - 5i)$ **54.** $3/(1 - i)$

55. $\dfrac{5 + i}{5 - i}$ **56.** $\dfrac{8 - 7i}{1 - 2i}$

57. $i/(4 - 5i)^2$ **58.** $5i/(2 + 3i)^2$

Performing Operations with Complex Numbers **In Exercises 59 and 60, perform the operation and write the result in standard form.**

59. $\dfrac{2}{1 + i} - \dfrac{3}{1 - i}$ **60.** $\dfrac{4i}{4 + i} + \dfrac{3}{4 - i}$

Writing a Complex Number in Standard Form **In Exercises 61–68, write the complex number in standard form.**

61. $\sqrt{-18} - \sqrt{-54}$
62. $\sqrt{-75} + \sqrt{-250}$
63. $\sqrt{-6}\sqrt{-2}$
64. $\sqrt{-5}\sqrt{-10}$
65. $\left(\sqrt{-10}\right)^2$
66. $\left(\sqrt{-75}\right)^2$
67. $\left(2 - \sqrt{-6}\right)^2$
68. $\left(3 + \sqrt{-5}\right)\left(7 - \sqrt{-10}\right)$

Complex Solutions of a Quadratic Equation **In Exercises 69–80, use the Quadratic Formula to solve the quadratic equation.**

69. $x^2 + 25 = 0$
70. $x^2 + 32 = 0$
71. $x^2 - 2x + 2 = 0$
72. $x^2 + 6x + 10 = 0$
73. $4x^2 + 16x + 17 = 0$
74. $9x^2 - 6x + 37 = 0$
75. $16t^2 - 4t + 3 = 0$
76. $4x^2 + 12x + 11 = 0$
77. $\frac{3}{2}x^2 - 6x + 9 = 0$
78. $\frac{7}{8}x^2 - \frac{3}{4}x + \frac{5}{16} = 0$
79. $1.4x^2 - 2x - 10 = 0$
80. $4.5x^2 - 3x + 12 = 0$

Simplifying a Complex Number **In Exercises 81–86, simplify the complex number and write it in standard form.**

81. $-6i^3 + i^2$
82. $2i^2 - 4i^3$
83. $\left(\sqrt{-75}\right)^3$
84. $\left(\sqrt{-2}\right)^6$
85. $\dfrac{1}{i^3}$
86. $\dfrac{1}{(2i)^3}$

87. ***Why you should learn it*** *(p. 128)* The opposition to current in an electrical circuit is called its impedance. The impedance z in a parallel circuit with two pathways satisfies the equation $1/z = 1/z_1 + 1/z_2$, where z_1 is the impedance (in ohms) of pathway 1, and z_2 is the impedance (in ohms) of pathway 2. Use the table to determine the impedance of each parallel circuit. (*Hint:* You can find the impedance of each pathway in a parallel circuit by adding the impedances of all components in the pathway.)

	Resistor	Inductor	Capacitor
Symbol	$a\ \Omega$	$b\ \Omega$	$c\ \Omega$
Impedance	a	bi	$-ci$

(a)

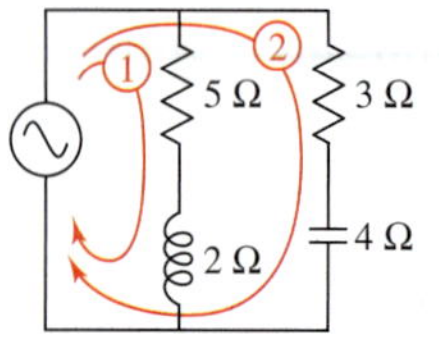

(b)

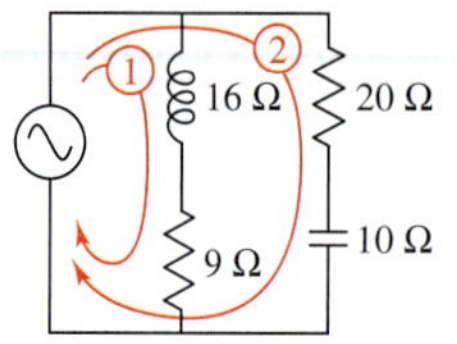

88. **Exploration** Consider the functions

$$f(x) = 2(x - 3)^2 - 4 \quad \text{and} \quad g(x) = -2(x - 3)^2 - 4.$$

(a) Without graphing either function, determine whether the graph of f and the graph of g have x-intercepts. Explain your reasoning.

(b) Solve $f(x) = 0$ and $g(x) = 0$.

(c) Explain how the zeros of f and g are related to whether their graphs have x-intercepts.

(d) For the function $f(x) = a(x - h)^2 + k$, make a general statement about how a, h, and k affect whether the graph of f has x-intercepts and whether the zeros of f are real or complex.

Focusing on Concepts

True or False? **In Exercises 89 and 90, determine whether the statement is true or false. Justify your answer.**

89. No complex number is equal to its complex conjugate.

90. $i^{44} + i^{150} - i^{74} - i^{109} + i^{61} = -1$

91. **Error Analysis** Describe the error.

$$\sqrt{-6}\sqrt{-6} = \sqrt{(-6)(-6)} = \sqrt{36} = 6 \quad \times$$

92. **HOW DO YOU SEE IT?** The coordinate system shown below is called the *complex plane*. In the complex plane, the point (a, b) corresponds to the complex number $a + bi$.

Match each complex number with its corresponding point.

(i) 2

(ii) $2i$

(iii) $-2 + i$

(iv) $1 - 2i$

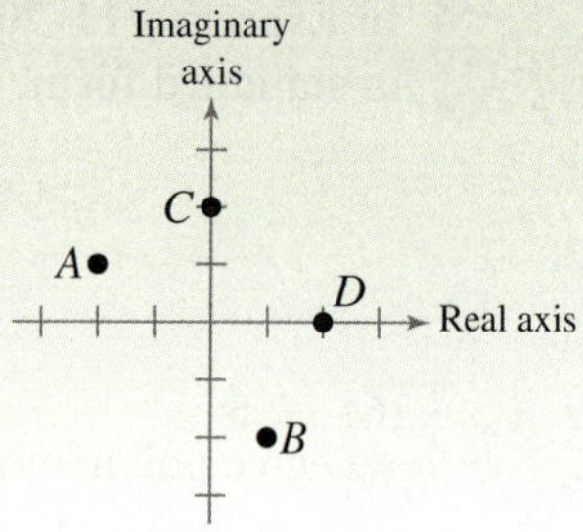

93. **Proof** Prove that the complex conjugate of the product of two complex numbers $a_1 + b_1 i$ and $a_2 + b_2 i$ is the product of their complex conjugates.

94. **Proof** Prove that the complex conjugate of the sum of two complex numbers $a_1 + b_1 i$ and $a_2 + b_2 i$ is the sum of their complex conjugates.

Cumulative Mixed Review

Multiplying Polynomials **In Exercises 95–98, perform the operation and write the result in standard form.**

95. $(4x - 5)(4x + 5)$
96. $(x + 2)^3$
97. $\left(3x - \frac{1}{2}\right)(x + 4)$
98. $(2x - 5)^2$

2.5 The Fundamental Theorem of Algebra

The Fundamental Theorem of Algebra

You know that an nth-degree polynomial can have at most n real zeros. In the complex number system, this statement can be improved. That is, in the complex number system, every nth-degree polynomial function has *precisely n zeros*. This important result is derived from the **Fundamental Theorem of Algebra,** first proved by the German mathematician Carl Friedrich Gauss (1777–1855).

The Fundamental Theorem of Algebra

If $f(x)$ is a polynomial of degree n, where $n > 0$, then f has at least one zero in the complex number system.

Using the Fundamental Theorem of Algebra and the equivalence of zeros and factors, you obtain the **Linear Factorization Theorem.**

Linear Factorization Theorem

If $f(x)$ is a polynomial of degree n, where $n > 0$, then f has precisely n linear factors

$$f(x) = a_n(x - c_1)(x - c_2) \cdot \cdot \cdot (x - c_n)$$

where $c_1, c_2, . . . , c_n$ are complex numbers.

(See the proof on page 179.)

Note that the Fundamental Theorem of Algebra and the Linear Factorization Theorem tell you only that the zeros or factors of a polynomial exist, not how to find them. Such theorems are called *existence theorems*. To find the zeros of a polynomial function, you still must rely on other techniques.

What you should learn

- Use the Fundamental Theorem of Algebra to determine the number of zeros of a polynomial function.
- Find all zeros of polynomial functions.
- Find conjugate pairs of complex zeros.
- Find zeros of polynomials by factoring.

Why you should learn it

Being able to find zeros of polynomial functions is an important part of modeling real-life problems. For instance, Exercise 71 on page 141 shows how to determine whether a football kicked with a given velocity can reach a certain height.

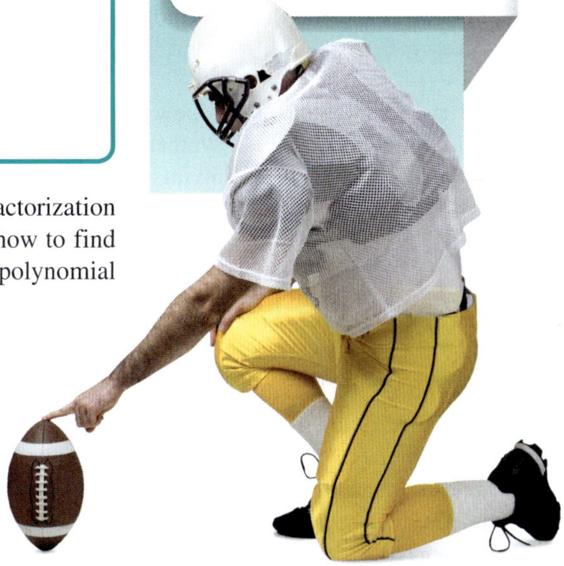

EXAMPLE 1 Zeros of Polynomial Functions

See LarsonPrecalculus.com for an interactive version of this type of example.

a. The first-degree polynomial $f(x) = x - 2$ has exactly *one* zero: $x = 2$.

b. Counting multiplicity, the second-degree polynomial function

$$f(x) = x^2 - 6x + 9$$
$$= (x - 3)(x - 3)$$

has exactly *two* zeros: $x = 3$ and $x = 3$. (This is called a *repeated zero*.)

c. The third-degree polynomial function

$$f(x) = x^3 + 4x = x(x^2 + 4) = x(x - 2i)(x + 2i)$$

has exactly *three* zeros: $x = 0$, $x = 2i$, and $x = -2i$.

d. The fifth-degree polynomial function

$$f(x) = x^5 + 9x^3 = x^3(x^2 + 9) = x^3(x - 3i)(x + 3i)$$

has exactly *five* zeros: $x = 0$, $x = 0$, $x = 0$, $x = 3i$, and $x = -3i$.

✓ Checkpoint *Audio-video solution in English & Spanish at LarsonPrecalculus.com*

Determine the number of zeros of the polynomial function $f(x) = x^4 - 1$.

Finding Zeros of a Polynomial Function

Remember that the n zeros of a polynomial function can be real or imaginary, and they may be repeated. Examples 2 and 3 illustrate several cases.

EXAMPLE 2 Complex Zeros of a Polynomial Function

Confirm that the third-degree polynomial function $f(x) = x^3 + 4x$ has exactly three zeros: $x = 0$, $x = 2i$, and $x = -2i$.

Solution

Factor the polynomial completely as $x(x - 2i)(x + 2i)$. So, the zeros are

$$x(x - 2i)(x + 2i) = 0$$

$x = 0$ — Set 1st factor equal to 0.

$x - 2i = 0 \Rightarrow x = 2i$ — Set 2nd factor equal to 0.

$x + 2i = 0 \Rightarrow x = -2i.$ — Set 3rd factor equal to 0.

In Figure 2.20, only the real zero $x = 0$ appears as an x-intercept.

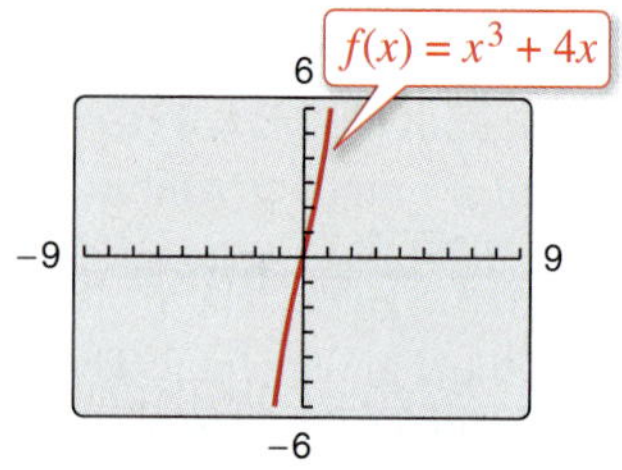

Figure 2.20

✓ Checkpoint *Audio-video solution in English & Spanish at LarsonPrecalculus.com*

Confirm that the fourth-degree polynomial function $f(x) = x^4 - 1$ has exactly four zeros: $x = 1$, $x = -1$, $x = i$, and $x = -i$.

The next example shows how to use the methods described in Sections 2.2 and 2.3 (the Rational Zero Test, synthetic division, and factoring) to find all the zeros of a polynomial function, including imaginary zeros.

EXAMPLE 3 Finding the Zeros of a Polynomial Function

Write $f(x) = x^5 + x^3 + 2x^2 - 12x + 8$ as the product of linear factors, and list all the zeros of f.

Solution

The possible rational zeros are ±1, ±2, ±4, and ±8. The graph shown in Figure 2.21 indicates that 1 and -2 are likely zeros, and that 1 is possibly a repeated zero because it appears that the graph touches (but does not cross) the x-axis at this point. Using synthetic division, you can determine that -2 is a zero and 1 is a repeated zero of f. So, you have

$$f(x) = x^5 + x^3 + 2x^2 - 12x + 8 = (x - 1)(x - 1)(x + 2)(x^2 + 4).$$

By factoring $x^2 + 4$ as

$$x^2 - (-4) = \left(x - \sqrt{-4}\right)\left(x + \sqrt{-4}\right) = (x - 2i)(x + 2i)$$

you obtain

$$f(x) = (x - 1)(x - 1)(x + 2)(x - 2i)(x + 2i)$$

which gives the following five zeros of f.

$$x = 1, x = 1, x = -2, x = 2i, \quad \text{and} \quad x = -2i$$

Note from the graph of f shown in Figure 2.21 that the *real* zeros are the only ones that appear as x-intercepts.

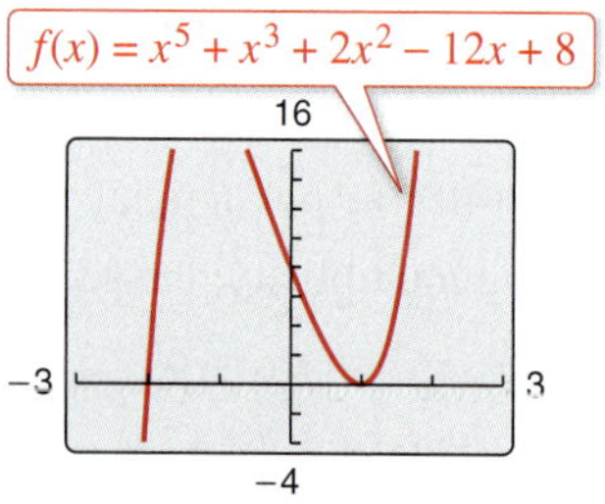

Figure 2.21

✓ Checkpoint *Audio-video solution in English & Spanish at LarsonPrecalculus.com*

Write $f(x) = x^4 - 256$ as the product of linear factors, and list all the zeros of f.

Conjugate Pairs

In Example 3, note that the two complex zeros $2i$ and $-2i$ are **conjugates.** That is, they are of the forms $a + bi$ and $a - bi$.

Complex Zeros Occur in Conjugate Pairs

Let f be a polynomial function that has *real coefficients.* If $a + bi$, where $b \neq 0$, is a zero of the function, then the conjugate $a - bi$ is also a zero of the function.

Be sure you see that this result is true only when the polynomial function has *real coefficients.* For instance, the result applies to the function $f(x) = x^2 + 1$, but not to the function $g(x) = x - i$.

EXAMPLE 4 Finding a Polynomial with Given Zeros

Find a *fourth-degree* polynomial function with real coefficients that has -1, -1, and $3i$ as zeros.

Solution

Because $3i$ is a zero *and* the polynomial is stated to have real coefficients, you know that the conjugate $-3i$ must also be a zero. So, from the Linear Factorization Theorem, $f(x)$ can be written as

$$f(x) = a(x + 1)(x + 1)(x - 3i)(x + 3i).$$

For simplicity, let $a = 1$ to obtain

$$f(x) = (x^2 + 2x + 1)(x^2 + 9) = x^4 + 2x^3 + 10x^2 + 18x + 9.$$

✓ Checkpoint ▶ *Audio-video solution in English & Spanish at LarsonPrecalculus.com*

Find a *fourth-degree* polynomial function with real coefficients that has 2, -2, and $7i$ as zeros.

EXAMPLE 5 Finding a Polynomial with Given Zeros

Find a *cubic* polynomial function f with real coefficients that has 2 and $1-i$ as zeros, and $f(1) = 3$.

Solution

Because $1 - i$ is a zero of f, the conjugate $1 + i$ must also be a zero.

$$\begin{aligned} f(x) &= a(x - 2)[x - (1 - i)][x - (1 + i)] \\ &= a(x - 2)[x^2 - x(1 + i) - x(1 - i) + 1 - i^2] \\ &= a(x - 2)(x^2 - 2x + 2) \\ &= a(x^3 - 4x^2 + 6x - 4) \end{aligned}$$

To find the value of a, use the fact that $f(1) = 3$.

$$a[(1)^3 - 4(1)^2 + 6(1) - 4] = 3 \quad \Rightarrow \quad a(-1) = 3 \quad \Rightarrow \quad a = -3$$

So, $a = -3$ and you can conclude that

$$f(x) = -3(x^3 - 4x^2 + 6x - 4) = -3x^3 + 12x^2 - 18x + 12.$$

✓ Checkpoint ▶ *Audio-video solution in English & Spanish at LarsonPrecalculus.com*

Find a *quartic* polynomial function f with real coefficients that has 1, -2, and $2i$ as zeros, and $f(-1) = 10$.

Factoring a Polynomial

The Linear Factorization Theorem states that you can write any nth-degree polynomial as the product of n linear factors.

$$f(x) = a_n(x - c_1)(x - c_2)(x - c_3) \cdot \cdot \cdot (x - c_n)$$

This result, however, includes the possibility that some of the values of c_i are imaginary. The next theorem states that when you want to avoid "imaginary factors," you can write $f(x)$ as the product of linear and quadratic factors.

Factors of a Polynomial

Every polynomial of degree $n > 0$ with real coefficients can be written as the product of linear and quadratic factors with real coefficients, where the quadratic factors have no real zeros.

(See the proof on page 179.)

Algebra Help

Recall that irrational and rational numbers are subsets of the set of real numbers, and real numbers are a subset of the set of complex numbers.

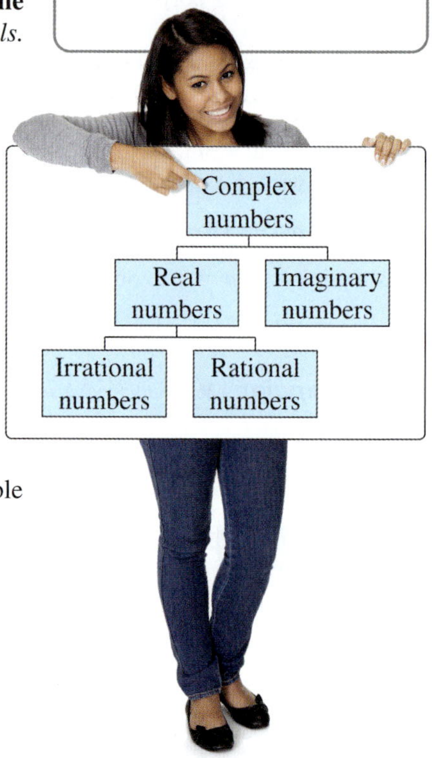

A quadratic factor with no real zeros is said to be **prime** or **irreducible over the reals.** Be sure you see that this is not the same as being *irreducible over the rationals.* For example, the quadratic

$$x^2 + 1 = (x - i)(x + i)$$

is irreducible over the reals (and therefore over the rationals). On the other hand, the quadratic

$$x^2 - 2 = \left(x - \sqrt{2}\right)\left(x + \sqrt{2}\right)$$

is irreducible over the rationals but *reducible* over the reals.

EXAMPLE 6 Factoring a Polynomial

Let $f(x) = x^4 - x^2 - 20$.

a. Write $f(x)$ as the product of factors that are irreducible over the *rationals.*

b. Write $f(x)$ as the product of linear factors and quadratic factors that are irreducible over the *reals.*

c. Write $f(x)$ in completely factored form.

Solution

a. Begin by factoring the polynomial as the product of two quadratic polynomials.

$$x^4 - x^2 - 20 = (x^2 - 5)(x^2 + 4)$$

Both of these factors are irreducible over the rationals.

b. By factoring over the reals, you have

$$x^4 - x^2 - 20 = \left(x + \sqrt{5}\right)\left(x - \sqrt{5}\right)(x^2 + 4)$$

where the quadratic factor is irreducible over the reals.

c. In completely factored form, you have

$$x^4 - x^2 - 20 = \left(x + \sqrt{5}\right)\left(x - \sqrt{5}\right)(x - 2i)(x + 2i).$$

✓ ***Checkpoint*** *Audio-video solution in English & Spanish at LarsonPrecalculus.com*

Write the polynomial $f(x) = x^4 - 2x^2 - 3$ (a) as the product of factors that are irreducible over the *rationals*, (b) as the product of linear factors and quadratic factors that are irreducible over the *reals*, and (c) in completely factored form.

In Example 6, notice from the completely factored form that the *fourth*-degree polynomial has *four* zeros.

Andrey_Popov/Shutterstock.com

Throughout this chapter, the results and theorems have been stated in terms of zeros of polynomial functions. Be sure you see that the same results could have been stated in terms of solutions of polynomial equations. This is true because the zeros of the polynomial function

$$f(x) = a_n x^n + a_{n-1}x^{n-1} + \cdots + a_2 x^2 + a_1 x + a_0$$

are precisely the solutions of the polynomial equation

$$a_n x^n + a_{n-1}x^{n-1} + \cdots + a_2 x^2 + a_1 x + a_0 = 0.$$

EXAMPLE 7 Finding the Zeros of a Polynomial Function

Find all the zeros of

$$f(x) = x^4 - 3x^3 + 6x^2 + 2x - 60$$

given that $1 + 3i$ is a zero of f.

Algebraic Solution

Because complex zeros occur in conjugate pairs, you know that $1 - 3i$ is also a zero of f. This means that both

$$x - (1 + 3i) \quad \text{and} \quad x - (1 - 3i)$$

are factors of f. Multiplying these two factors produces

$$\begin{aligned}[x - (1 + 3i)][x - (1 - 3i)] &= [(x - 1) - 3i][(x - 1) + 3i] \\ &= (x - 1)^2 - 9i^2 \\ &= x^2 - 2x + 10.\end{aligned}$$

Using long division, you can divide $x^2 - 2x + 10$ into f to obtain the following.

$$\begin{array}{r} x^2 - x - 6 \\ x^2 - 2x + 10 \overline{\big)\, x^4 - 3x^3 + 6x^2 + 2x - 60} \\ \underline{x^4 - 2x^3 + 10x^2} \qquad\qquad\quad \\ -x^3 - 4x^2 + 2x \qquad\quad \\ \underline{-x^3 + 2x^2 - 10x} \qquad\quad \\ -6x^2 + 12x - 60 \\ \underline{-6x^2 + 12x - 60} \\ 0 \end{array}$$

So, you have

$$\begin{aligned} f(x) &= (x^2 - 2x + 10)(x^2 - x - 6) \\ &= (x^2 - 2x + 10)(x - 3)(x + 2) \end{aligned}$$

and you can conclude that the zeros of f are

$$x = 1 + 3i,\ x = 1 - 3i,\ x = 3, \text{ and } x = -2.$$

Graphical Solution

Using the *zero* feature of a graphing utility, you can conclude that $x = -2$ and $x = 3$ are zeros of f. (See figure.)

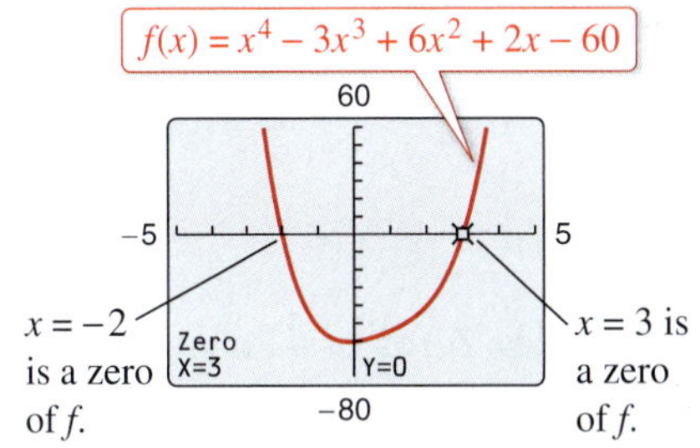

Because $1 + 3i$ is a zero of f, you know that the conjugate $1 - 3i$ must also be a zero. So, you can conclude that the zeros of f are

$$x = 1 + 3i,\ x = 1 - 3i,\ x = 3, \text{ and } x = -2.$$

Checkpoint *Audio-video solution in English & Spanish at LarsonPrecalculus.com*

Find all the zeros of $f(x) = 3x^3 - 2x^2 + 48x - 32$ given that $4i$ is a zero of f.

In Example 7, if you were not told that $1 + 3i$ is a zero of f, you could still find all the zeros of the function by using synthetic division to find the real zeros -2 and 3. Then, you could factor the polynomial as $(x + 2)(x - 3)(x^2 - 2x + 10)$. Finally, by using the Quadratic Formula, you could determine that the zeros are $x = 1 + 3i$, $x = 1 - 3i$, $x = 3$, and $x = -2$.

2.5 Exercises

See *CalcChat.com* for tutorial help and worked-out solutions to odd-numbered exercises. For instructions on how to use a graphing utility, see Appendix A.

Vocabulary and Concept Check

In Exercises 1 and 2, fill in the blanks.

1. The _______ of _______ states that if $f(x)$ is a polynomial of degree n $(n > 0)$, then f has at least one zero in the complex number system.

2. A quadratic factor that cannot be factored as a product of linear factors containing real numbers is said to be _______ over the _______ .

3. How many linear factors does a polynomial function f of degree n have, where $n > 0$?

4. Three of the zeros of a fourth-degree polynomial function f are -1, 3, and $2i$. What is the other zero of f?

Procedures and Problem Solving

Zeros of a Polynomial Function **In Exercises 5–10, determine the number of zeros of the polynomial function.**

5. $f(x) = x^3 + 2x^2 + 1$ **6.** $f(x) = x^4 - 3x$

7. $g(x) = x^4 - x^5$ **8.** $f(x) = x^3 - x^6$

9. $f(x) = (x + 5)^2$

10. $h(t) = (t - 1)^2 - (t + 1)^2$

Complex Zeros of a Polynomial Function **In Exercises 11–14, confirm that the function has the indicated zeros.**

11. $f(x) = x^2 + 6;\ -\sqrt{6}i,\ \sqrt{6}i$

12. $f(x) = x^3 + 9x;\ 0,\ -3i,\ 3i$

13. $f(x) = 3x^4 - 48;\ -2,\ 2,\ -2i,\ 2i$

14. $f(x) = 2x^5 - 2x;\ 0,\ 1,\ -1\ -i,\ i$

Comparing the Zeros and the *x*-Intercepts **In Exercises 15–18, find all the zeros of the function. Is there a relationship between the number of real zeros and the number of *x*-intercepts of the graph? Explain.**

15. $f(x) = x^3 - 4x^2 + x - 4$

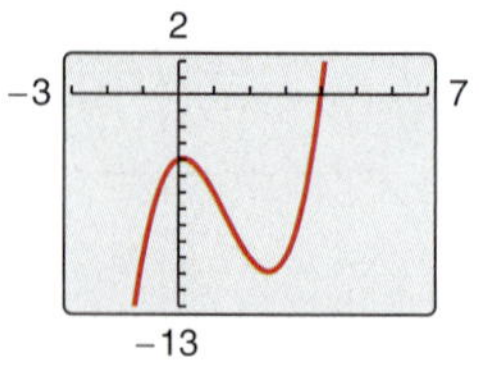

16. $f(x) = x^3 - 4x^2 - 4x + 16$

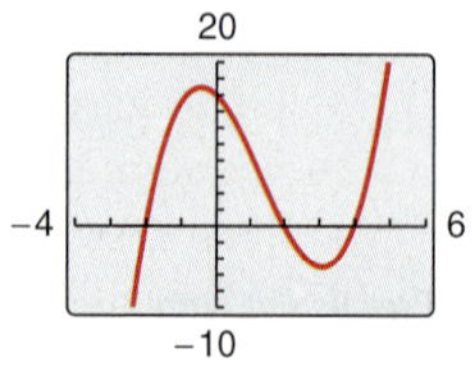

17. $f(x) = x^4 + 4x^2 + 4$

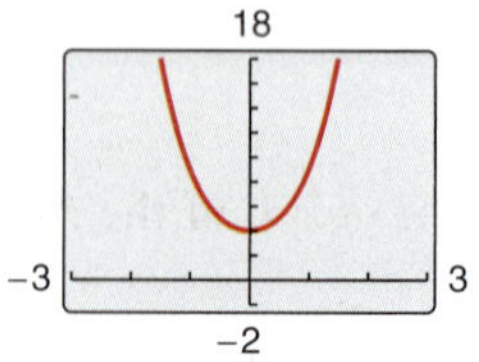

18. $f(x) = x^4 - 3x^2 - 4$

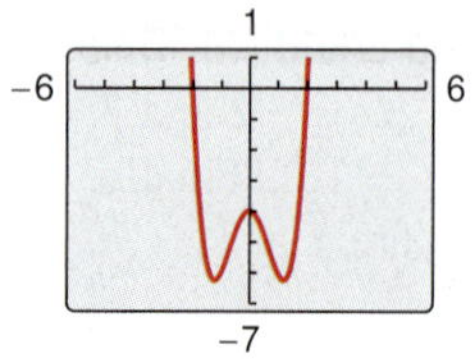

Finding the Zeros of a Polynomial Function **In Exercises 19–38, find all the zeros of the function and write the polynomial as a product of linear factors. Use a graphing utility to verify your results.**

19. $h(x) = x^2 - 4x + 1$ **20.** $g(x) = x^2 + 8x + 3$

21. $f(x) = x^2 - 12x + 26$ **22.** $f(x) = x^2 + 6x - 2$

23. $f(x) = x^2 + 25$ **24.** $f(x) = x^2 + 36$

25. $f(x) = 16x^4 - 81$ **26.** $f(y) = 81y^4 - 625$

27. $f(z) = z^2 - z + 56$ **28.** $h(x) = x^2 - 4x - 3$

29. $f(x) = x^4 + 10x^2 + 9$ **30.** $f(x) = x^4 + 29x^2 + 100$

31. $f(x) = 3x^3 - 5x^2 + 48x - 80$

32. $f(x) = 3x^3 - 2x^2 + 75x - 50$

33. $g(x) = x^3 - 3x^2 + x + 5$

34. $f(x) = x^3 + 11x^2 + 39x + 29$

35. $f(x) = 5x^3 - 9x^2 + 28x + 6$

36. $f(s) = 3s^3 - 4s^2 + 8s + 8$

37. $g(x) = x^4 - 4x^3 + 8x^2 - 16x + 16$

38. $h(x) = x^4 + 6x^3 + 10x^2 + 6x + 9$

Using the Zeros to Find *x*-Intercepts **In Exercises 39–46, (a) find all zeros of the function, (b) write the polynomial as a product of linear factors, and (c) use your factorization to determine the *x*-intercepts of the graph of the function. Use a graphing utility to verify that the real zeros are the only *x*-intercepts.**

39. $f(x) = x^2 - 14x + 46$ **40.** $f(x) = x^2 - 12x + 34$

41. $f(x) = 2x^3 - 3x^2 + 8x - 12$

42. $f(x) = 2x^3 - 5x^2 + 18x - 45$

43. $f(x) = x^3 - 11x + 150$

44. $f(x) = x^3 + 10x^2 + 33x + 34$

45. $f(x) = x^4 + 29x^2 + 100$

46. $f(x) = x^4 - 8x^3 + 17x^2 - 8x + 16$

Finding a Polynomial Function **In Exercises 47–52, find a polynomial function with real coefficients that has the indicated zeros. (There are many correct answers.)**

47. $5, i$ **48.** $3, 4i$

49. $2, 2, 1 + i$ **50.** $-1, 5, 3 - 2i$

51. $0, -5, 1 + \sqrt{2}i$ **52.** $0, 4, 1 + \sqrt{2}i$

Finding a Polynomial Function **In Exercises 53–56, find the polynomial function f with real coefficients that has the indicated degree, zeros, and solution point.**

	Degree	*Zeros*	*Solution Point*
53.	4	$-1, 2, i$	$f(1) = 8$
54.	4	$1, -4, \sqrt{3}i$	$f(0) = -6$
55.	3	$-3, 1 + \sqrt{3}i$	$f(-2) = 12$
56.	3	$-2, 2 + 2\sqrt{2}i$	$f(-1) = -34$

Factoring a Polynomial **In Exercises 57–60, write the polynomial (a) as the product of factors that are irreducible over the *rationals,* (b) as the product of linear and quadratic factors that are irreducible over the *reals,* and (c) in completely factored form.**

57. $f(x) = x^4 - 6x^2 - 7$

58. $f(x) = x^4 + 13x^2 - 48$

59. $f(x) = x^4 - 2x^3 - 3x^2 + 12x - 18$

(*Hint:* One factor is $x^2 - 6$.)

60. $f(x) = x^4 - 3x^3 - x^2 - 12x - 20$

(*Hint:* One factor is $x^2 + 4$.)

Finding the Zeros of a Polynomial Function **In Exercises 61–66, use the specified zero to find all the zeros of the function.**

	Function	*Zero*
61.	$f(x) = x^3 - x^2 + 4x - 4$	$2i$
62.	$f(x) = x^3 + x^2 + 9x + 9$	$3i$
63.	$g(x) = x^3 - 7x^2 - x + 87$	$5 + 2i$
64.	$g(x) = 4x^3 + 23x^2 + 34x - 10$	$-3 + i$
65.	$h(x) = 3x^3 - 4x^2 + 8x + 8$	$1 - \sqrt{3}i$
66.	$f(x) = 25x^3 - 55x^2 - 54x - 18$	$\frac{1}{5}(-2 + \sqrt{2}i)$

Using a Graph to Locate the Real Zeros **In Exercises 67–70, (a) use a graphing utility to find the real zeros of the function, and then (b) use the real zeros to find the exact values of the imaginary zeros.**

67. $f(x) = x^4 + 3x^3 - 5x^2 - 21x + 22$

68. $f(x) = x^3 + 4x^2 + 14x + 20$

69. $h(x) = 8x^3 - 14x^2 + 18x - 9$

70. $f(x) = 25x^3 - 55x^2 - 54x - 18$

71. ***Why you should learn it*** *(p. 135)* A football is kicked off the ground with an initial upward velocity of 48 feet per second. The football's height h (in feet) is modeled by

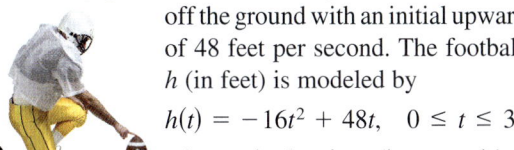

$$h(t) = -16t^2 + 48t, \quad 0 \le t \le 3$$

where t is the time (in seconds). Does the football reach a height of 50 feet? Explain.

72. Marketing The demand equation for a microwave is $p = 140 - 0.001x$, where p is the unit price (in dollars) of the microwave and x is the number of units produced and sold. The cost equation for the microwave is $C = 40x + 150{,}000$, where C is the total cost (in dollars) and x is the number of units produced. The total profit P obtained by producing and selling x units is $P = R - C = xp - C$. Is there a price p that yields a profit of \$3 million? Explain.

Focusing on Concepts

True or False? **In Exercises 73 and 74, decide whether the statement is true or false. Justify your answer.**

73. It is possible for a third-degree polynomial function with integer coefficients to have no real zeros.

74. If $[x + (4 + 3i)]$ is a factor of a polynomial function f with real coefficients, then $[x - (4 + 3i)]$ is also a factor of f.

75. Error Analysis Describe the error.

The graph of a quartic (fourth-degree) polynomial $y = f(x)$ is shown. One of the zeros is i.

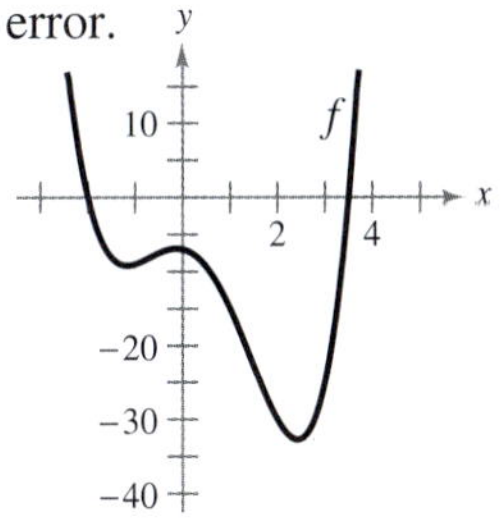

The function is $f(x) = (x + 2)(x - 3.5)(x - i)$. ✗

76. HOW DO YOU SEE IT? Describe a translation of the graph that will result in a function with (a) four distinct real zeros, (b) two real zeros, each of multiplicity 2, (c) two real zeros and two imaginary zeros, and (d) four imaginary zeros.

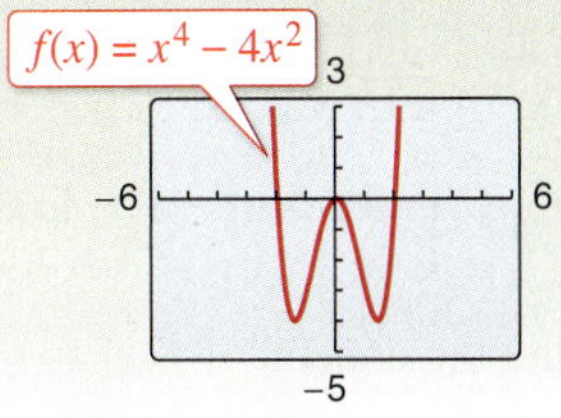

Cumulative Mixed Review

Identifying the Vertex of a Quadratic Function **In Exercises 77 and 78, describe the graph of the function and identify the vertex.**

77. $f(x) = -x^2 + x + 6$ **78.** $f(x) = 4x^2 + 2x - 12$

2.6 Rational Functions and Asymptotes

What you should learn

- Find the domains of rational functions.
- Find vertical and horizontal asymptotes of graphs of rational functions.
- Use rational functions to model and solve real-life problems.

Why you should learn it

Rational functions are convenient in modeling a wide variety of real-life problems, such as environmental scenarios. For instance, Exercise 49 on page 150 shows how to determine the cost of supplying recycling bins to the population of a rural township.

Introduction to Rational Functions

A **rational function** can be written in the form

$$f(x) = \frac{N(x)}{D(x)}$$

where $N(x)$ and $D(x)$ are polynomials and $D(x)$ is not the zero polynomial.

In general, the *domain* of a rational function of x includes all real numbers except x-values that make the denominator zero. Much of the discussion of rational functions will focus on their graphical behavior near these x-values.

EXAMPLE 1 Finding the Domain of a Rational Function

See LarsonPrecalculus.com for an interactive version of this type of example.

Find the domain of $f(x) = 1/x$ and discuss the behavior of f near any excluded x-values.

Solution

Because the denominator is zero when $x = 0$, the domain of f is all real numbers except $x = 0$. To determine the behavior of f near this excluded value, evaluate $f(x)$ to the left and right of $x = 0$, as indicated in the table.

x	-1	-0.5	-0.1	-0.01	-0.001	$\rightarrow 0$
$f(x)$	-1	-2	-10	-100	-1000	$\rightarrow -\infty$

x	$0 \leftarrow$	0.001	0.01	0.1	0.5	1
$f(x)$	$\infty \leftarrow$	1000	100	10	2	1

From the table, note that as x approaches 0 *from the left*, $f(x)$ decreases without bound. In contrast, as x approaches 0 *from the right*, $f(x)$ increases without bound. Because $f(x)$ decreases without bound from the left and increases without bound from the right, you can conclude that f is not continuous. The graph of f is shown in the figure.

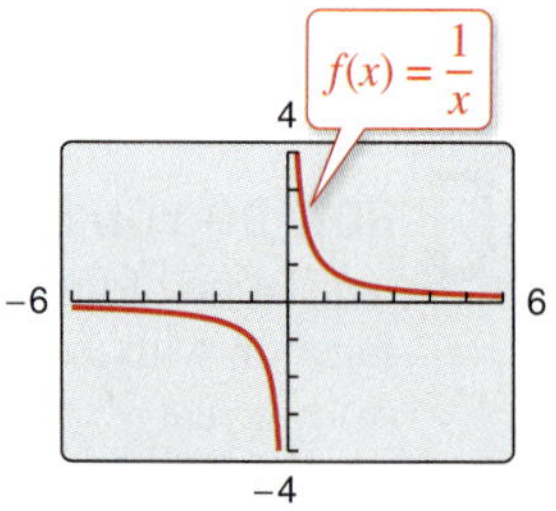

✓ **Checkpoint** ▶ *Audio-video solution in English & Spanish at LarsonPrecalculus.com*

Find the domain of $f(x) = (3x)/(x - 1)$ and discuss the behavior of f near any excluded x-values.

Explore the Concept

Use the *table* and *trace* features of a graphing utility to verify that the function $f(x) = 1/x$ in Example 1 is not continuous.

Vertical and Horizontal Asymptotes

In Example 1, the behavior of the graph of $f(x) = 1/x$ near $x = 0$ is denoted as shown below.

$f(x) \to -\infty$ as $x \to 0^-$ — $f(x)$ decreases without bound as x approaches 0 from the left.

$f(x) \to \infty$ as $x \to 0^+$ — $f(x)$ increases without bound as x approaches 0 from the right.

The line $x = 0$ is a **vertical asymptote** of the graph of f, as shown in Figure 2.22. From this figure, you can see that the graph of f also has a **horizontal asymptote**—the line $y = 0$. This means $f(x)$ approaches 0 as x increases or decreases without bound. This behavior is denoted as shown below.

$f(x) \to 0$ as $x \to -\infty$ — $f(x)$ approaches 0 as x decreases without bound.

$f(x) \to 0$ as $x \to \infty$ — $f(x)$ approaches 0 as x increases without bound.

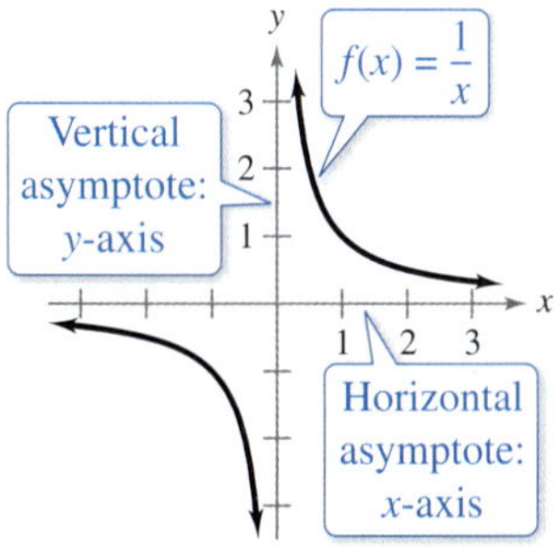

Figure 2.22

Definition of Vertical and Horizontal Asymptotes

1. The line $x = a$ is a **vertical asymptote** of the graph of f when

$$f(x) \to \infty \quad \text{or} \quad f(x) \to -\infty$$

as $x \to a$, either from the right or from the left.

2. The line $y = b$ is a **horizontal asymptote** of the graph of f when

$$f(x) \to b$$

as $x \to \infty$ or $x \to -\infty$.

Figure 2.23 shows the vertical and horizontal asymptotes of the graphs of three rational functions. Note in Figure 2.23 that eventually (as $x \to \infty$ or $x \to -\infty$) the distance between the horizontal asymptote and the points on the graph must approach zero. [The graphs shown in Figures 2.22 and 2.23(a) are **hyperbolas.** You will study hyperbolas in Section 9.3.]

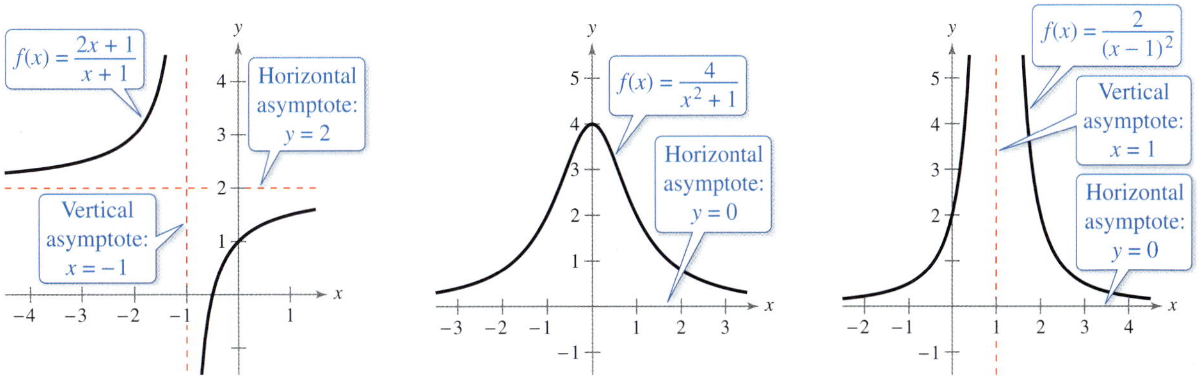

(a) (b) (c)

Figure 2.23

Explore the Concept

Use a table of values to determine whether the functions in Figure 2.23 are continuous. When the graph of a function has an asymptote, can you conclude that the function is not continuous? Explain.

Graphs of Rational Functions: Vertical and Horizontal Asymptotes

Let f be the rational function

$$f(x) = \frac{N(x)}{D(x)} = \frac{a_n x^n + a_{n-1}x^{n-1} + \cdots + a_1 x + a_0}{b_m x^m + b_{m-1}x^{m-1} + \cdots + b_1 x + b_0}$$

where $N(x)$ and $D(x)$ have no common factors.

1. The graph of f has *vertical* asymptotes at the zeros of the denominator, $D(x)$.
2. The graph of f has at most one *horizontal* asymptote determined by comparing the degrees of $N(x)$ and $D(x)$.
   a. If $n < m$, then the graph of f has the line $y = 0$ (the x-axis) as a horizontal asymptote.
   b. If $n = m$, then the graph of f has the line
   $$y = \frac{a_n}{b_m}$$
   as a horizontal asymptote, where a_n is the leading coefficient of the numerator and b_m is the leading coefficient of the denominator.
   c. If $n > m$, then the graph of f has no horizontal asymptote.

What's Wrong?

You use a graphing utility to graph

$$y_1 = \frac{2x^3 + 1000x^2 + x}{x^3 + 1000x^2 + x + 1000}$$

as shown in the figure. You use the graph to conclude that the graph of y_1 has the line $y = 1$ as a horizontal asymptote. What's wrong?

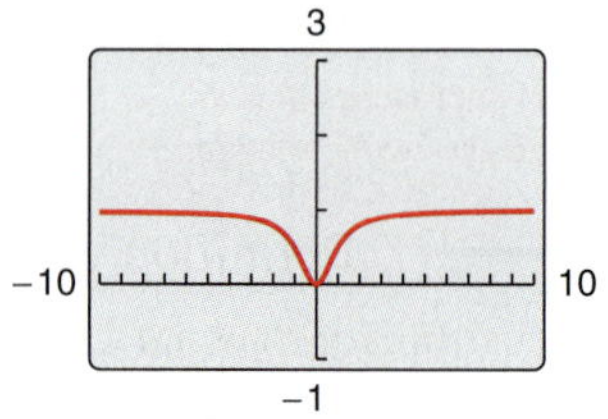

EXAMPLE 2 Finding Vertical and Horizontal Asymptotes

Find all asymptotes of the graph of each rational function.

a. $f(x) = \dfrac{2x}{3x^2 + 1}$ **b.** $f(x) = \dfrac{2x^2}{x^2 - 1}$

Solution

a. For this rational function, the degree of the numerator is *less than* the degree of the denominator, so the graph has the line $y = 0$ as a horizontal asymptote. To find any vertical asymptotes, set the denominator equal to zero and solve the resulting equation for x.

$3x^2 + 1 = 0$ Set denominator equal to zero.

Because this equation has no real solutions, you can conclude that the graph has no vertical asymptote. The graph of the function is shown in Figure 2.24.

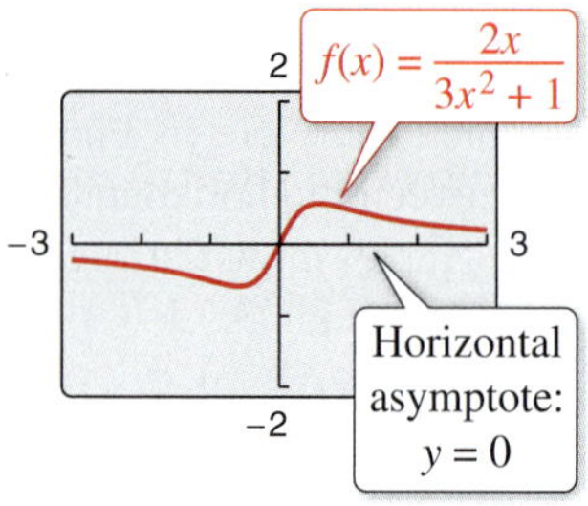

Figure 2.24

b. For this rational function, the degree of the numerator is *equal* to the degree of the denominator. The leading coefficient of the numerator is 2 and the leading coefficient of the denominator is 1, so the graph has the line $y = \frac{2}{1} = 2$ as a horizontal asymptote. To find any vertical asymptotes, set the denominator equal to zero and solve the resulting equation for x.

$x^2 - 1 = 0$ Set denominator equal to zero.

$(x + 1)(x - 1) = 0$ Factor.

$x + 1 = 0 \Rightarrow x = -1$ Set 1st factor equal to 0.

$x - 1 = 0 \Rightarrow x = 1$ Set 2nd factor equal to 0.

This equation has two real solutions, $x = -1$ and $x = 1$, so the graph has the lines $x = -1$ and $x = 1$ as vertical asymptotes, as shown in Figure 2.25.

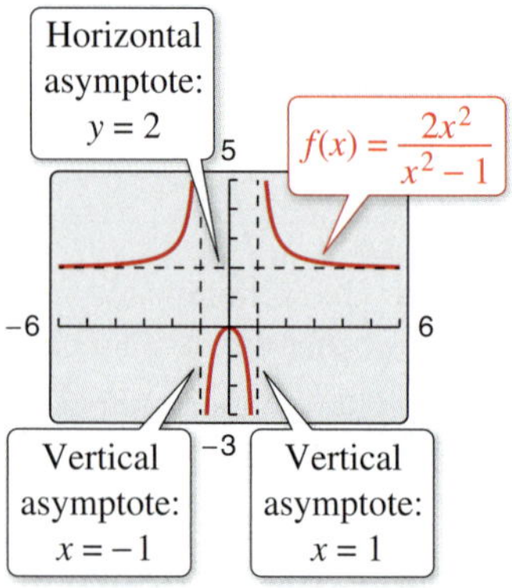

Figure 2.25

✓ Checkpoint Audio-video solution in English & Spanish at LarsonPrecalculus.com

Find all vertical and horizontal asymptotes of the graph of $f(x) = \dfrac{5x^2}{x^2 - 1}$.

Values for which a rational function is undefined (the denominator is zero) result in a vertical asymptote or a hole in the graph, as shown in Example 3.

EXAMPLE 3 Finding Asymptotes and Holes

Find all asymptotes and holes in the graph of

$$f(x) = \frac{x^2 + x - 2}{x^2 - x - 6}.$$

Solution

For this rational function, the degree of the numerator is *equal* to the degree of the denominator. The leading coefficients of the numerator and denominator are both 1, so the graph has the line $y = 1$ as a horizontal asymptote. To find any vertical asymptotes, first factor the numerator and denominator.

$$f(x) = \frac{x^2 + x - 2}{x^2 - x - 6} = \frac{(x - 1)\cancel{(x + 2)}}{\cancel{(x + 2)}(x - 3)} = \frac{x - 1}{x - 3}, \quad x \neq -2$$

By setting the denominator $x - 3$ (of the simplified function) equal to zero, you can determine that the graph has the line $x = 3$ as a vertical asymptote, as shown in the figure. To find any holes in the graph, note that the function is undefined at $x = -2$ and $x = 3$. Because $x = -2$ is not a vertical asymptote of the function, there is a hole in the graph at $x = -2$. To find the y-coordinate of the hole, substitute $x = -2$ into the simplified form of the function.

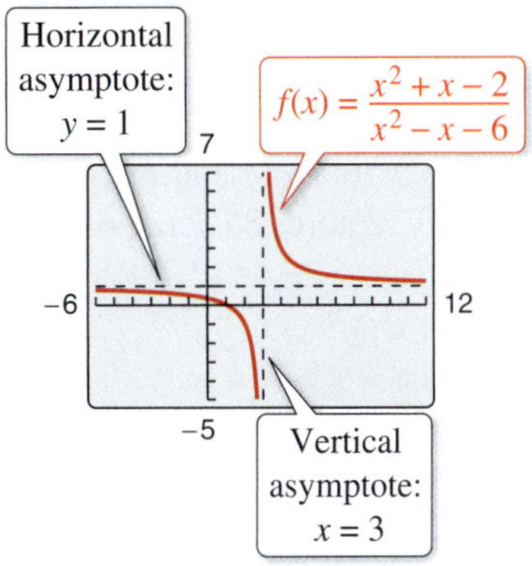

$$y = \frac{x - 1}{x - 3} = \frac{-2 - 1}{-2 - 3} = \frac{3}{5}$$

So, the graph of the rational function has a hole at $\left(-2, \frac{3}{5}\right)$.

✓ **Checkpoint** *Audio-video solution in English & Spanish at LarsonPrecalculus.com*

Find all asymptotes and holes in the graph of $f(x) = \dfrac{x^2 - 25}{x^2 + 5x}$.

EXAMPLE 4 Finding a Function's Domain and Asymptotes

For the function f, find (a) the domain of f, (b) the vertical asymptote of f, and (c) the horizontal asymptote of f.

$$f(x) = \frac{3x^3 + 7x^2 + 2}{-2x^3 + 16}$$

Solution

a. Because the denominator is zero when $-2x^3 + 16 = 0$, solve this equation to determine that the domain of f is all real numbers except $x = 2$.

b. Because the denominator of f has a zero at $x = 2$, and 2 is not a zero of the numerator, the graph of f has the vertical asymptote $x = 2$.

c. Because the degrees of the numerator and denominator are the same, and the leading coefficient of the numerator is 3 and the leading coefficient of the denominator is -2, the horizontal asymptote of f is $y = -\frac{3}{2}$.

✓ **Checkpoint** *Audio-video solution in English & Spanish at LarsonPrecalculus.com*

Repeat Example 4 using the function $f(x) = \dfrac{3x^2 + x - 5}{x^2 + 1}$.

Technology Tip

Graphing utilities are limited in their resolution and therefore may not show a break or hole in the graph. You can use the *table* feature of a graphing utility to verify the values of x at which a function is not defined. Try doing this for the function in Example 3.

Application

There are many examples of asymptotic behavior in real life. For instance, Example 5 shows how a vertical asymptote can be used to analyze the cost of removing pollutants from smokestack emissions.

EXAMPLE 5 Cost-Benefit Model

A utility company burns coal to generate electricity. The cost C (in dollars) of removing $p\%$ of the smokestack pollutants is given by

$$C = \frac{80{,}000p}{100 - p}$$

for $0 \le p < 100$. Use a graphing utility to graph this function. You are a member of a state legislature that is considering a law that would require utility companies to remove 90% of the pollutants from their smokestack emissions. The current law requires 85% removal. How much additional cost would the utility company incur as a result of the new law?

Solution

The graph of this function is shown in Figure 2.26. Note that the graph has a vertical asymptote at $p = 100$. Because the current law requires 85% removal, the current cost to the utility company is

$$C = \frac{80{,}000(85)}{100 - 85} \approx \$453{,}333. \qquad \text{Evaluate } C \text{ when } p = 85.$$

The cost to remove 90% of the pollutants would be

$$C = \frac{80{,}000(90)}{100 - 90} \approx \$720{,}000. \qquad \text{Evaluate } C \text{ when } p = 90.$$

So, the new law would require the utility company to spend an additional

$$720{,}000 - 453{,}333 = \$266{,}667. \qquad \text{Subtract 85\% removal cost from 90\% removal cost.}$$

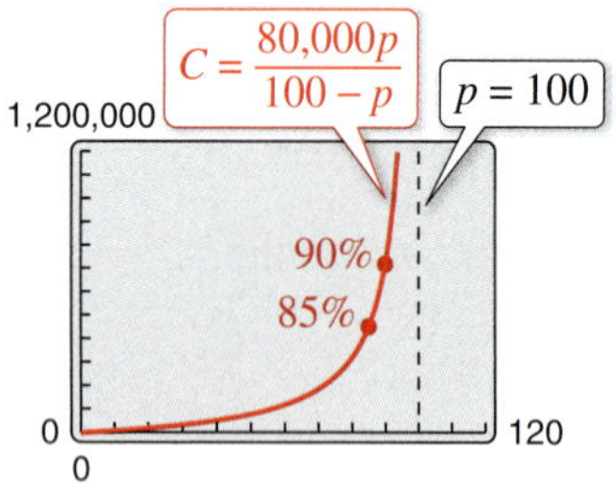

Figure 2.26

Checkpoint *Audio-video solution in English & Spanish at LarsonPrecalculus.com*

The cost C (in millions of dollars) of removing $p\%$ of the industrial and municipal pollutants discharged into a river is given by

$$C = \frac{255p}{100 - p}, \quad 0 \le p < 100.$$

a. Find the costs of removing 20%, 45%, and 80% of the pollutants.

b. According to the model, would it be possible to remove 100% of the pollutants? Explain.

Explore the Concept

The *table* feature of a graphing utility can be used to estimate vertical and horizontal asymptotes of rational functions. Use the *table* feature to find any vertical or horizontal asymptotes of

$$f(x) = \frac{2x}{x + 1}.$$

Write a statement explaining how you found the asymptote(s) using the table.

2.6 Exercises

See *CalcChat.com* for tutorial help and worked-out solutions to odd-numbered exercises.
For instructions on how to use a graphing utility, see Appendix A.

Vocabulary and Concept Check

In Exercises 1 and 2, fill in the blank.

1. Functions of the form $f(x) = N(x)/D(x)$, where $N(x)$ and $D(x)$ are polynomials and $D(x)$ is not the zero polynomial, are called _______ .

2. If $f(x) \to \pm\infty$ as $x \to a$ from the left (or right), then $x = a$ is a _______ of the graph of f.

3. What feature of the graph of $f(x) = \dfrac{9}{x - 3}$ can you find by solving $x - 3 = 0$?

4. Is $y = 4$ a horizontal asymptote of the function $f(x) = \dfrac{4x}{x^2 - 8}$?

Procedures and Problem Solving

Finding the Domain of a Rational Function **In Exercises 5–10, (a) find the domain of the function, (b) complete each table, and (c) discuss the behavior of f near any excluded x-values.**

x	$f(x)$
0.5	
0.9	
0.99	
0.999	

x	$f(x)$
1.5	
1.1	
1.01	
1.001	

x	$f(x)$
5	
10	
100	
1000	

x	$f(x)$
-5	
-10	
-100	
-1000	

5. $f(x) = \dfrac{1}{x - 1}$

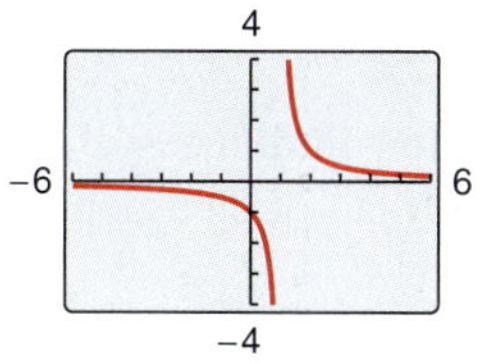

6. $f(x) = \dfrac{5x}{x - 1}$

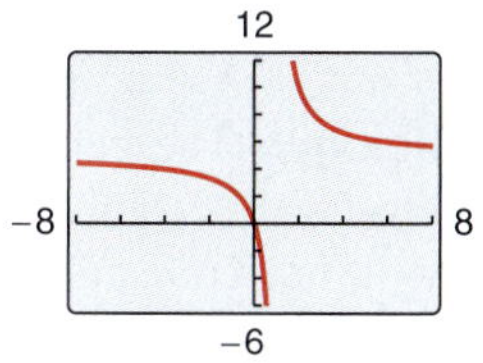

7. $f(x) = \dfrac{3x}{|x - 1|}$

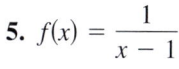

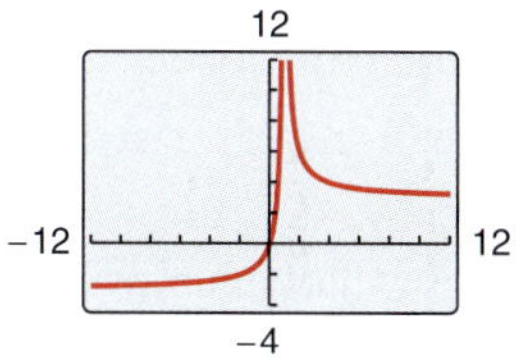

8. $f(x) = \dfrac{3}{|x - 1|}$

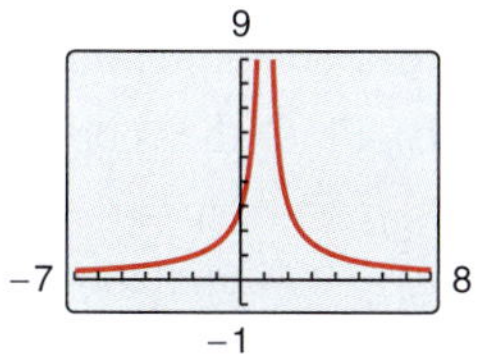

9. $f(x) = \dfrac{3x^2}{x^2 - 1}$

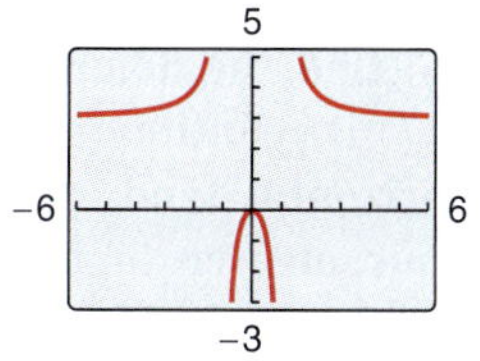

10. $f(x) = \dfrac{4x}{x^2 - 1}$

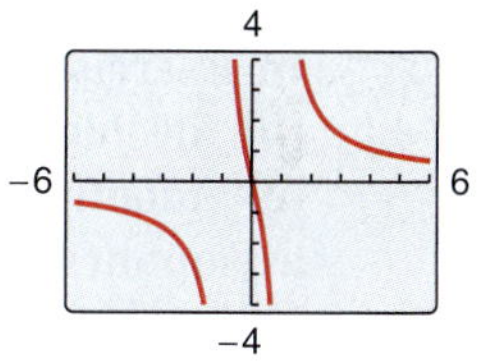

Identifying Graphs of Rational Functions **In Exercises 11–16, match the function with its graph. [The graphs are labeled (a), (b), (c), (d), (e), and (f).]**

(a)

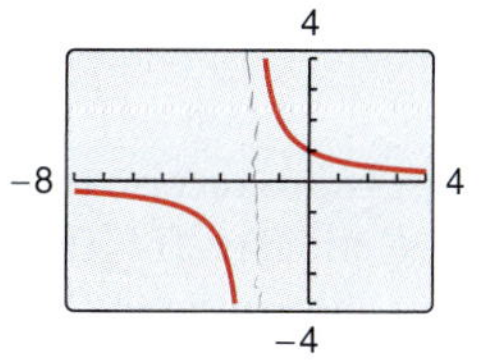

(b)

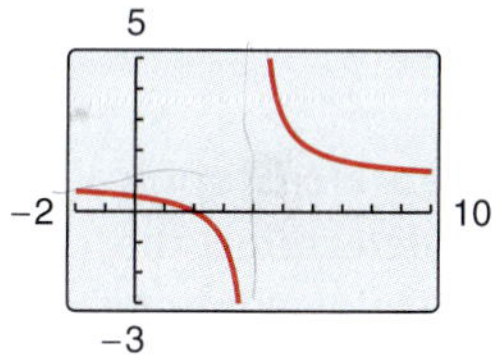

(c)

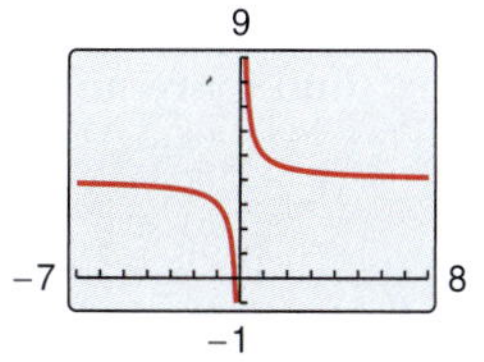

(d)

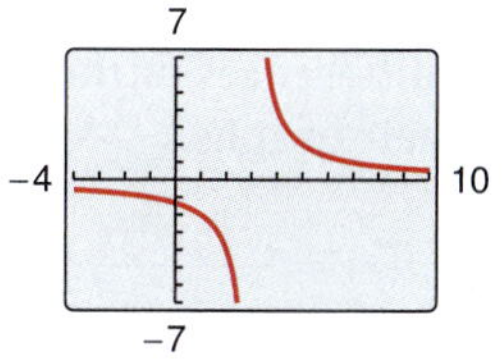

(e)

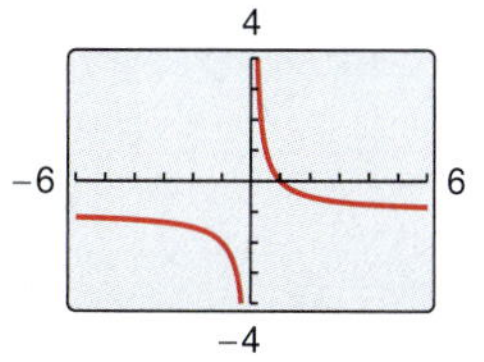

(f)

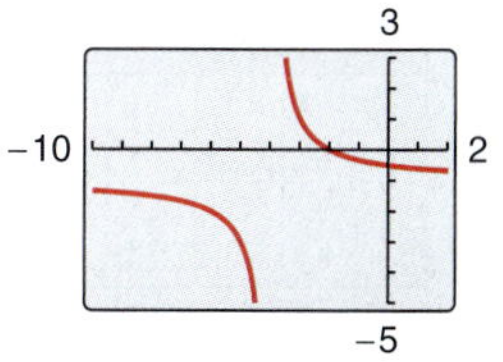

11. $f(x) = \dfrac{2}{x + 2}$

12. $f(x) = \dfrac{4}{x - 3}$

13. $f(x) = \dfrac{4x + 1}{x}$

14. $f(x) = \dfrac{1 - x}{x}$

15. $f(x) = \dfrac{x - 2}{x - 4}$

16. $f(x) = -\dfrac{x + 2}{x + 4}$

Finding Vertical and Horizontal Asymptotes In Exercises 17–20, find all asymptotes of the graph of the rational function. Use a graphing utility to verify your answers.

17. $f(x) = -\dfrac{7}{(x+1)^2}$ **18.** $f(x) = \dfrac{3}{(x-2)^3}$

19. $f(x) = \dfrac{2x^2}{x^2+x-6}$ **20.** $f(x) = \dfrac{x^2-4x}{x^2-4}$

Finding Asymptotes and Holes In Exercises 21–24, find all asymptotes and holes in the graph of the rational function. Use a graphing utility to verify your answers.

21. $f(x) = \dfrac{x(2+x)}{2x-x^2}$ **22.** $f(x) = \dfrac{x^2+2x+1}{2x^2-x-3}$

23. $f(x) = \dfrac{x^2-16}{x^2+4x}$ **24.** $f(x) = \dfrac{3-14x-5x^2}{3+7x+2x^2}$

Finding a Function's Domain and Asymptotes In Exercises 25–28, (a) find the domain of f, (b) decide whether f is continuous, and (c) identify any horizontal and vertical asymptotes. Verify your answer to part (a) both graphically by using a graphing utility and numerically by creating a table of values.

25. $f(x) = \dfrac{5x^2-2x-6}{x^2+4}$ **26.** $f(x) = \dfrac{3x^2+1}{x^2+x+9}$

27. $f(x) = \dfrac{x^2+3x-4}{-x^3+27}$ **28.** $f(x) = \dfrac{4x^3-x^2+3}{3x^3+24}$

Algebraic-Graphical-Numerical In Exercises 29–32, (a) determine the domains of f and g, (b) find any vertical asymptotes and holes in the graphs of f and g, (c) compare f and g by completing the table, (d) use a graphing utility to graph f and g, and (e) explain why the differences in the domains of f and g are not shown in their graphs.

29. $f(x) = \dfrac{x^2-16}{x-4}$, $g(x) = x + 4$

x	1	2	3	4	5	6	7
$f(x)$							
$g(x)$							

30. $f(x) = \dfrac{x^2-9}{x-3}$, $g(x) = x + 3$

x	0	1	2	3	4	5	6
$f(x)$							
$g(x)$							

31. $f(x) = \dfrac{x^2-1}{x^2-2x-3}$, $g(x) = \dfrac{x-1}{x-3}$

x	-2	-1	0	1	2	3	4
$f(x)$							
$g(x)$							

32. $f(x) = \dfrac{x^2-4}{x^2-3x+2}$, $g(x) = \dfrac{x+2}{x-1}$

x	-3	-2	-1	0	1	2	3
$f(x)$							
$g(x)$							

Exploration In Exercises 33–36, determine the value that the function f approaches as the magnitude of x increases. Is $f(x)$ greater than or less than this value when x is positive and large in magnitude? when x is negative and large in magnitude?

33. $f(x) = 4 - \dfrac{1}{x}$ **34.** $f(x) = -5 + \dfrac{1}{x-3}$

35. $f(x) = -\dfrac{7x+6}{7x^2+6}$ **36.** $f(x) = \dfrac{2x-1}{x^2+1}$

Finding the Zeros of a Rational Function In Exercises 37–44, find the zeros (if any) of the rational function. Use a graphing utility to verify your answer.

37. $f(x) = \dfrac{x^2-9}{x^2+5}$ **38.** $h(x) = \dfrac{x^3+8}{x^2-11}$

39. $g(x) = 1 + \dfrac{6}{x-3}$ **40.** $f(x) = 3 - \dfrac{-12}{x^2+2}$

41. $h(x) = \dfrac{x^2-x-20}{x^2+7}$ **42.** $g(x) = \dfrac{x^2-8x+12}{x^2+4}$

43. $f(x) = \dfrac{3x^2-2x-1}{4x^2+7x+3}$

44. $h(x) = \dfrac{2x^2+11x+5}{3x^2+13x-10}$

45. Biology The game commission introduces 100 deer into newly acquired state game lands. The population N of the herd can be modeled by

$$N = \frac{20(5+3t)}{1+0.04t}, \quad t \geq 0$$

where t is the time in years.

(a) Use a graphing utility to graph the model.

(b) Find the populations when $t = 5$, $t = 10$, and $t = 25$.

(c) The population approaches what value over time? Explain.

46. MODELING DATA

The endpoints of the interval over which distinct vision is possible are called the *near point* and *far point* of the eye (see figure). With increasing age, these points normally change. The table shows the approximate near points y (in inches) for various ages x (in years).

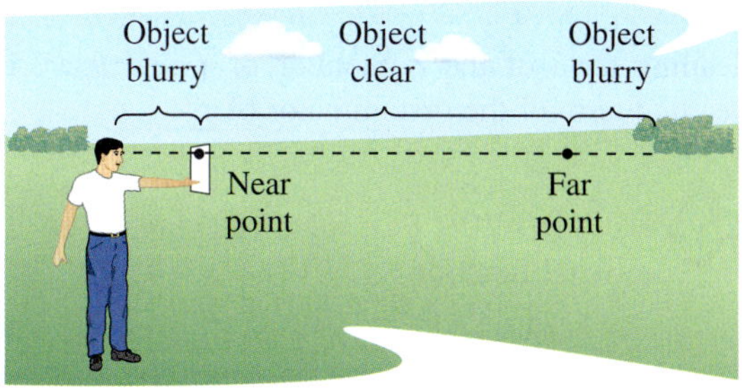

DATA

Age, x	Near point, y
16	3.0
32	4.7
44	9.8
50	19.7
60	39.7

Spreadsheet at LarsonPrecalculus.com

(a) Take the reciprocals of the near points to generate the points

$$\left(x, \frac{1}{y}\right).$$

Use the *regression* feature of a graphing utility to find a linear model for the generated points. The resulting line has the form

$$\frac{1}{y} = ax + b.$$

Solve for y to obtain a rational model for the original data.

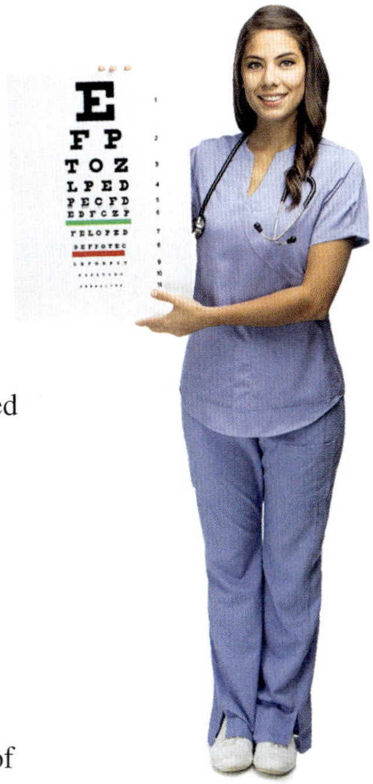

(b) Use the *table* feature of the graphing utility to create a table showing the predicted near point based on the model for each of the ages in the original table.

(c) Is it reasonable to use the model to predict the near point for a person who is 70 years old? Explain.

47. Physics Consider a physics experiment designed to determine an unknown mass. A flexible metal meter stick is clamped to a table with 50 centimeters overhanging the edge (see figure). Known masses M ranging from 200 grams to 2000 grams are attached to the end of the meter stick. For each mass, the meter stick is displaced vertically and then allowed to oscillate. The average time t (in seconds) of one oscillation for each mass is recorded in the table.

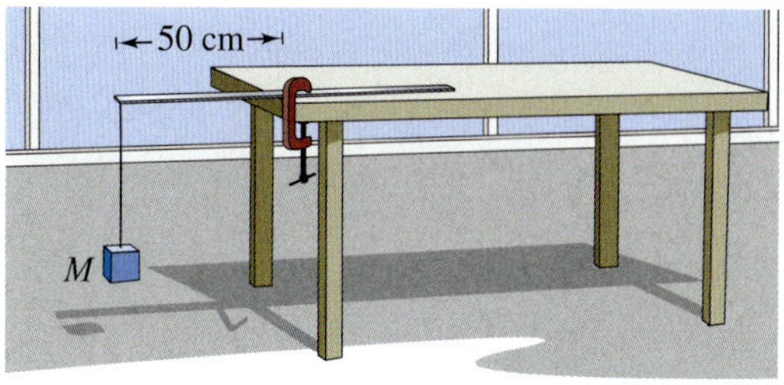

DATA

Mass, M	Time, t
200	0.450
400	0.597
600	0.712
800	0.831
1000	0.906
1200	1.003
1400	1.088
1600	1.126
1800	1.218
2000	1.338

Spreadsheet at LarsonPrecalculus.com

The data can be modeled by

$$t = \frac{38M + 16{,}965}{10(M + 5000)}.$$

(a) Use the *table* feature of a graphing utility to create a table showing the estimated time based on the model for each of the masses shown in the table. What can you conclude?

(b) Use the model to approximate the mass of an object when the average time for one oscillation is 1.056 seconds.

48. Business The sales S (in thousands of units) of a smart speaker during the nth week after the smart speaker is released can be modeled by

$$S = \frac{150n}{n^2 + 100}, \quad n \geq 0.$$

(a) Use a graphing utility to graph the sales function.

(b) Find the sales in week 5, week 10, and week 20.

(c) At least 1000 smart speakers must be sold each week to continue smart speaker production. Use the model to determine whether production will stop. Explain your reasoning.

49. *Why you should learn it* (*p. 142*) The cost C (in dollars) of supplying recycling bins to $p\%$ of the population of a rural township can be modeled by

$$C = \frac{25{,}000p}{100 - p}, \quad 0 \le p < 100.$$

(a) Use a graphing utility to graph the cost function.

(b) Find the costs of supplying bins to 15%, 50%, and 90% of the population.

(c) According to the model, would it be possible to supply bins to 100% of the population? Explain.

Focusing on Concepts

True or False? **In Exercises 50 and 51, determine whether the statement is true or false. Justify your answer.**

50. A rational function can have infinitely many vertical asymptotes.

51. A rational function must have at least one vertical asymptote.

52. **Writing a Rational Function** Write a rational function f that has the specified characteristics.

(a) Vertical asymptotes: $x = -1, x = 2$

Horizontal asymptote: $y = -2$

Zeros: $x = -2, x = 3$

(b) Vertical asymptotes: $x = 0, x = \pm 3$

Horizontal asymptote: $y = 3$

Zeros: $x = -1, x = 1, x = 2$

53. **Error Analysis** Describe the error.

$$f(x) = \frac{x}{x^2 + 2x}$$

The graph of f has a hole when $x = -2$.

54. **Error Analysis** Describe the error.

$$f(x) = \frac{cx}{c - x}, \ c \ne 0$$

The graph of f has a vertical asymptote at $x = -c$ and a horizontal asymptote at $y = c$.

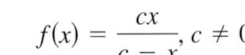

55. **Think About It** A real zero of the numerator of a rational function f is $x = c$. Must $x = c$ also be a zero of f? Explain.

56. **Think About It** When the graph of a rational function f has a vertical asymptote at $x = 4$, can f have a common factor of $(x - 4)$ in the numerator and denominator? Explain.

57. **Exploration** Use a graphing utility to compare the graphs of y_1 and y_2.

$$y_1 = \frac{3x^3 - 5x^2 + 4x - 5}{2x^2 - 6x + 7}, \quad y_2 = \frac{3x^3}{2x^2}$$

Start with a viewing window of $-5 \le x \le 5$ and $-10 \le y \le 10$, and then zoom out. Make a conjecture about how the graph of a rational function f is related to the graph of $y = (a_n x^n)/(b_m x^m)$, where $a_n x^n$ is the leading term of the numerator of f and $b_m x^m$ is the leading term of the denominator of f.

58. **HOW DO YOU SEE IT?** The graph of a rational function

$$f(x) = \frac{N(x)}{D(x)}$$

is shown below. Determine which of the statements about the function is false. Justify your answer.

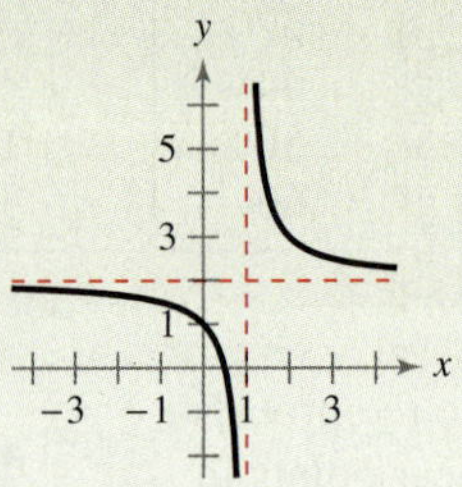

(a) $D(1) = 0$.

(b) The degrees of $N(x)$ and $D(x)$ are equal.

(c) The ratio of the leading coefficients of $N(x)$ and $D(x)$ is 1.

Cumulative Mixed Review

Finding the Slope-Intercept Form **In Exercises 59–62, find an equation of the line that passes through the points. Use the slope-intercept form (if possible). If not possible, explain why and use general form. Use a graphing utility to graph the line.**

59. $(3, 2), (0, -1)$

60. $(-6, 1), (4, -5)$

61. $(2, 7), (2, 10)$

62. $(0, 4), (-9, 4)$

Long Division of Polynomials **In Exercises 63–66, use long division to divide.**

63. $(x^2 + 5x + 6) \div (x - 4)$

64. $(x^2 - 10x + 15) \div (x - 3)$

65. $(2x^4 + x^2 - 11) \div (x^2 + 5)$

66. $(4x^5 + 3x^3 - 10) \div (2x + 3)$

2.7 Graphs of Rational Functions

The Graph of a Rational Function

To sketch the graph of a rational function, use the following guidelines.

Guidelines for Graphing Rational Functions

Let $f(x) = N(x)/D(x)$, where $N(x)$ and $D(x)$ are polynomials and $D(x)$ is not the zero polynomial.

1. Simplify f, if possible. List any restrictions on the domain of f that are not implied by the simplified function.
2. Find and plot the y-intercept (if any) by evaluating $f(0)$.
3. Find the zeros of the numerator (if any) by setting the numerator equal to zero. Then plot the corresponding x-intercepts.
4. Find the zeros of the denominator (if any) by setting the denominator equal to zero. Then sketch the corresponding vertical asymptotes using dashed vertical lines and plot the corresponding holes using open circles.
5. Find and sketch any other asymptotes of the graph using dashed lines.
6. Plot at least one point *between* and one point *beyond* each x-intercept and vertical asymptote.
7. Use smooth curves to complete the graph between and beyond the vertical asymptotes, excluding any points where f is not defined.

When graphing simple rational functions, testing for symmetry can be useful. For instance, the graph of $f(x) = 1/x$ is symmetrical with respect to the origin, and the graph of $g(x) = 1/x^2$ is symmetrical with respect to the y-axis.

Technology Tip

Some graphing utilities have difficulty graphing rational functions that have vertical asymptotes. In *connected* mode, the utility may connect parts of the graph that are not supposed to be connected. Notice that the graph of $f(x) = 1/(x - 2)$ in Figure 2.27(a) should consist of two *unconnected* portions—one to the left of $x = 2$ and the other to the right of $x = 2$. Changing the mode of the graphing utility to *dot* mode eliminates this problem. In *dot* mode, however, the graph is represented as a collection of dots rather than as a smooth curve, as shown in Figure 2.27(b).

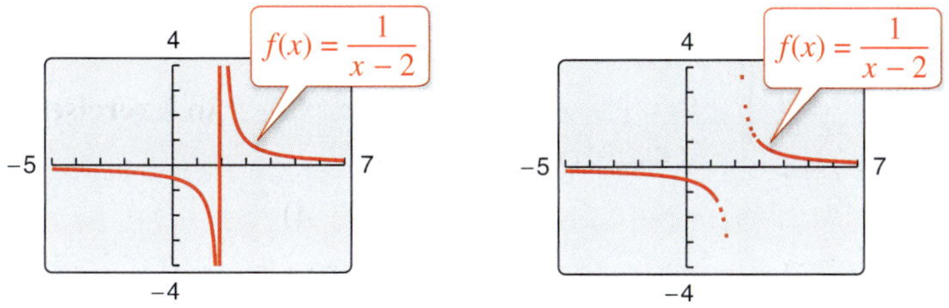

(a) Connected mode (b) Dot mode

Figure 2.27

What you should learn

- Analyze and sketch graphs of rational functions.
- Sketch graphs of rational functions that have slant asymptotes.
- Use graphs of rational functions to model and solve real-life problems.

Why you should learn it

The graph of a rational function provides a good indication of the behavior of a mathematical model. Exercise 85 on page 159 models the concentration of a chemical in the bloodstream after injection.

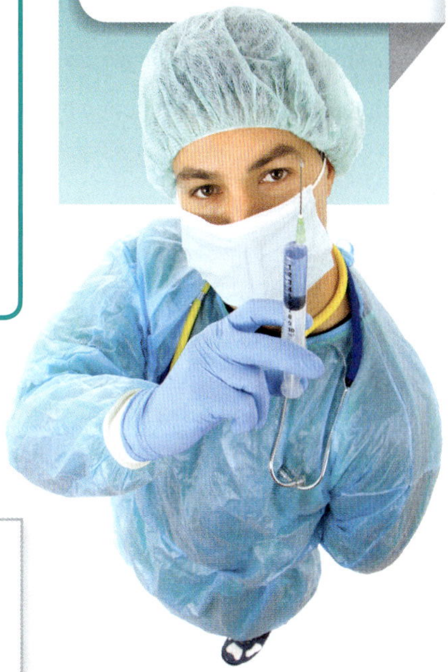

Library of Parent Functions: Rational Function

The simplest type of rational function is the *parent rational function* $f(x) = 1/x$, also known as the *reciprocal function*. The basic characteristics of the parent rational function are summarized below and on the inside cover of this text.

Graph of $f(x) = \dfrac{1}{x}$

Domain: $(-\infty, 0) \cup (0, \infty)$
Range: $(-\infty, 0) \cup (0, \infty)$
No intercepts
Decreasing on $(-\infty, 0)$ and $(0, \infty)$
Odd function
Origin symmetry
Vertical asymptote: y-axis
Horizontal asymptote: x-axis

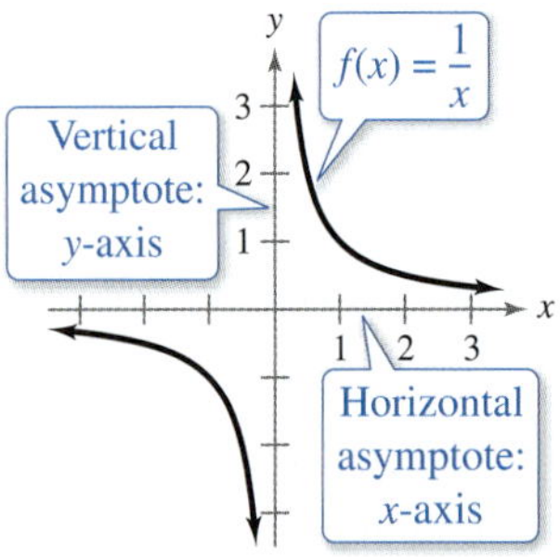

Explore the Concept

Use a graphing utility to graph

$$f(x) = 1 + \frac{1}{x - (1/x)}$$

Set the graphing utility to *dot* mode and use a decimal viewing window. Use the *trace* feature to find three "holes" or "breaks" in the graph. Do all three holes or breaks represent zeros of the denominator $x - (1/x)$? Explain.

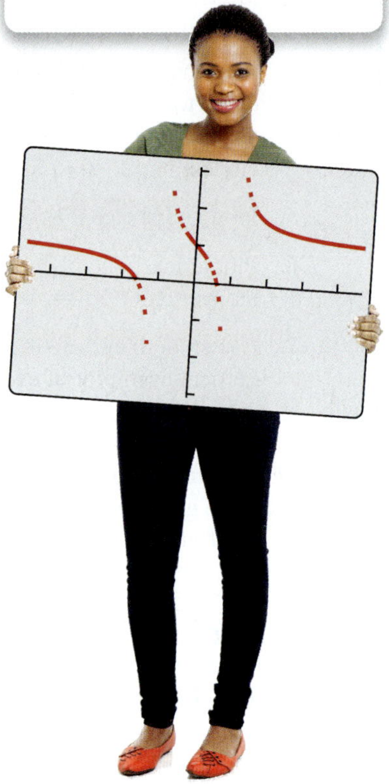

EXAMPLE 1 Library of Parent Functions: $f(x) = 1/x$

See LarsonPrecalculus.com for an interactive version of this type of example.

Sketch the graph of each function by hand and compare it with the graph of $f(x) = 1/x$.

a. $g(x) = \dfrac{-1}{x + 2}$

b. $h(x) = \dfrac{1}{x - 1} + 3$

Solution

a. With respect to the graph of $f(x) = 1/x$, the graph of g is obtained by a *reflection* in the y-axis and a horizontal shift two units *to the left*, as shown in Figure 2.28. Confirm this with a graphing utility.

b. With respect to the graph of $f(x) = 1/x$, the graph of h is obtained by a horizontal shift one unit *to the right* and a vertical shift three units *upward*, as shown in Figure 2.29. Confirm this with a graphing utility.

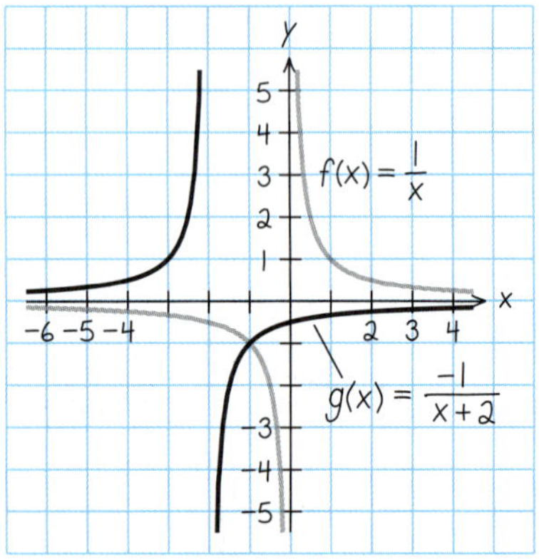

Figure 2.28

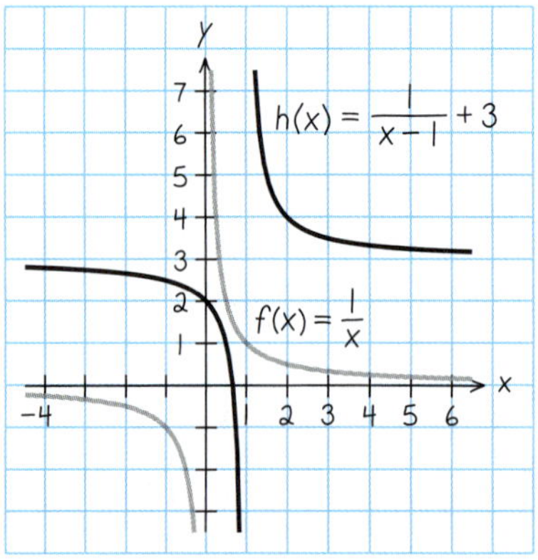

Figure 2.29

✓ ***Checkpoint*** *Audio-video solution in English & Spanish at LarsonPrecalculus.com*

Sketch the graph of each function by hand and compare it with the graph of $f(x) = 1/x$.

a. $g(x) = \dfrac{1}{x - 4}$ **b.** $h(x) = \dfrac{1}{x + 3} - 1$

As you study Examples 2–6, note that the vertical asymptotes are included in the tables of additional points. This is done to emphasize numerically the behavior of the graph of the function.

EXAMPLE 2 Sketching the Graph of a Rational Function

Sketch the graph of $g(x) = \dfrac{3}{x-2}$ by hand.

Solution

y-intercept: $\left(0, -\frac{3}{2}\right)$, because $g(0) = -\frac{3}{2}$

x-intercept: None, because there are no zeros of the numerator $(3 \neq 0)$

Vertical asymptote: $x = 2$, zero of denominator

Horizontal asymptote: $y = 0$, because degree of $N(x) <$ degree of $D(x)$

Additional points:

x	-4	1	2	3	5
$g(x)$	-0.5	-3	Undefined	3	1

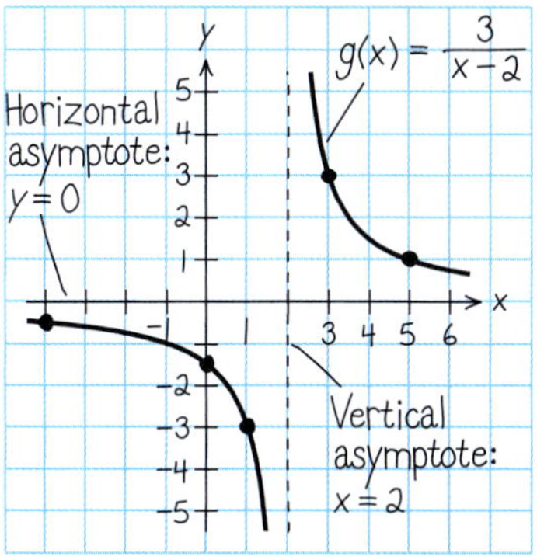

Figure 2.30

By plotting the intercept, asymptotes, and a few additional points, you can obtain the graph shown in Figure 2.30. Confirm this with a graphing utility.

✓ Checkpoint *Audio-video solution in English & Spanish at LarsonPrecalculus.com*

Sketch the graph of $f(x) = \dfrac{1}{x+3}$ by hand.

Note that the graph of g in Example 2 is a vertical stretch and a right shift of the graph of

$$f(x) = \frac{1}{x}$$

because

$$g(x) = \frac{3}{x-2} = 3\left(\frac{1}{x-2}\right) = 3f(x-2).$$

EXAMPLE 3 Sketching the Graph of a Rational Function

Sketch the graph of $f(x) = \dfrac{2x-1}{x}$ by hand.

Solution

y-intercept: None, because $x = 0$ is not in the domain

x-intercept: $\left(\frac{1}{2}, 0\right)$, because $2x - 1 = 0$ when $x = \frac{1}{2}$

Vertical asymptote: $x = 0$, zero of denominator

Horizontal asymptote: $y = 2$, because degree of $N(x) =$ degree of $D(x)$

Additional points:

x	-4	-1	0	0.25	4
$f(x)$	2.25	3	Undefined	-2	1.75

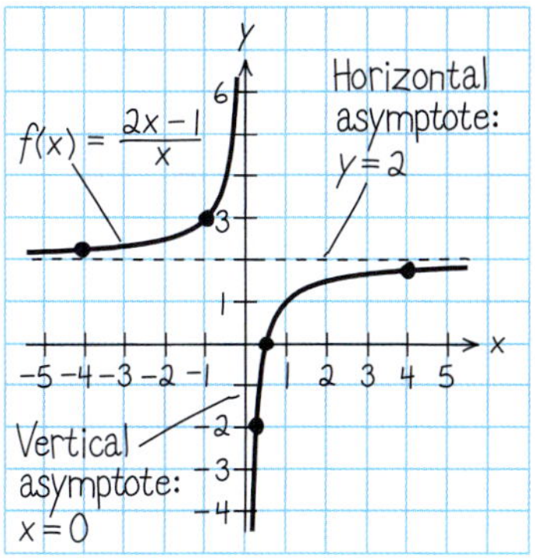

Figure 2.31

By plotting the intercept, asymptotes, and a few additional points, you can obtain the graph shown in Figure 2.31. Confirm this with a graphing utility.

✓ Checkpoint *Audio-video solution in English & Spanish at LarsonPrecalculus.com*

Sketch the graph of $f(x) = \dfrac{2x+3}{x+1}$ by hand.

EXAMPLE 4 Sketching the Graph of a Rational Function

Sketch the graph of $f(x) = \dfrac{x}{x^2 - x - 2}$.

Solution

Factor the denominator to determine the zeros of the denominator.

$$f(x) = \frac{x}{x^2 - x - 2}$$
$$= \frac{x}{(x + 1)(x - 2)}$$

y-intercept: (0, 0), because $f(0) = 0$

x-intercept: (0, 0)

Vertical asymptotes: $x = -1, x = 2$, zeros of denominator

Horizontal asymptote: $y = 0$, because degree of $N(x) <$ degree of $D(x)$

Additional points:

x	-3	-1	-0.5	1	2	3
$f(x)$	-0.3	Undefined	0.4	-0.5	Undefined	0.75

The graph is shown in Figure 2.32.

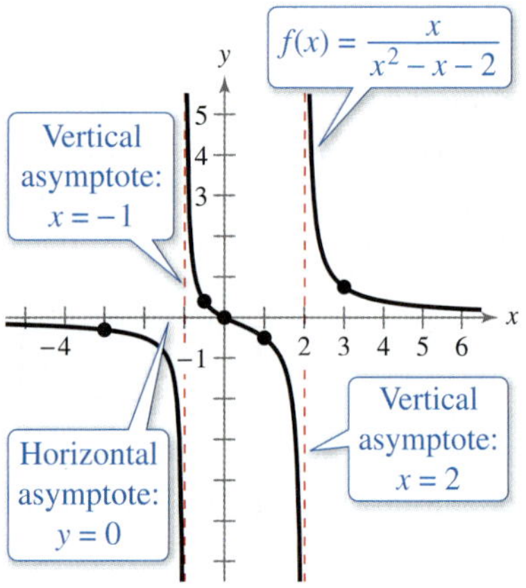

Figure 2.32

Checkpoint *Audio-video solution in English & Spanish at LarsonPrecalculus.com*

Sketch the graph of $f(x) = \dfrac{3x}{x^2 + x - 2}$.

EXAMPLE 5 Sketching the Graph of a Rational Function

Sketch the graph of $f(x) = \dfrac{x^2 - 9}{x^2 - 2x - 3}$.

Solution

By factoring the numerator and denominator, you have

$$f(x) = \frac{x^2 - 9}{x^2 - 2x - 3} = \frac{(x - 3)(x + 3)}{(x - 3)(x + 1)} = \frac{x + 3}{x + 1}, \quad x \neq 3.$$

y-intercept: (0, 3), because $f(0) = 3$

x-intercept: $(-3, 0)$, because $x + 3 = 0$ when $x = -3$

Vertical asymptote: $x = -1$, zero of (simplified) denominator

Hole: $\left(3, \frac{3}{2}\right)$, f is not defined at $x = 3$

Horizontal asymptote: $y = 1$, because degree of $N(x) =$ degree of $D(x)$

Additional points:

x	-5	-2	-1	-0.5	1	3	4
$f(x)$	0.5	-1	Undefined	5	2	Undefined	1.4

The graph is shown in Figure 2.33.

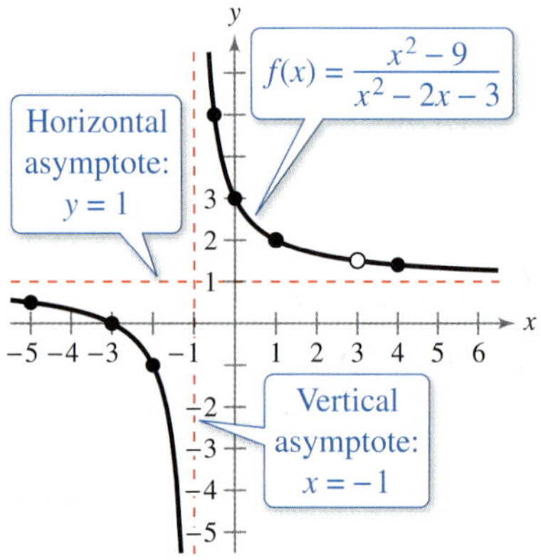

Figure 2.33 *Hole at x = 3*

Checkpoint *Audio-video solution in English & Spanish at LarsonPrecalculus.com*

Sketch the graph of $f(x) = \dfrac{x^2 - 4}{x^2 - x - 6}$.

Slant Asymptotes

Consider a rational function whose denominator is of degree 1 or greater. If the degree of the numerator is exactly *one more* than the degree of the denominator, then the graph of the function has a **slant** (or **oblique**) **asymptote.** For example, the graph of

$$f(x) = \frac{x^2 - x}{x + 1}$$

has a slant asymptote, as shown in Figure 2.34. To find the equation of a slant asymptote, use long division. For instance, by dividing $x + 1$ into $x^2 - x$, you have

$$f(x) = \frac{x^2 - x}{x + 1} = \underbrace{x - 2}_{\text{Slant asymptote } (y = x - 2)} + \frac{2}{x + 1}.$$

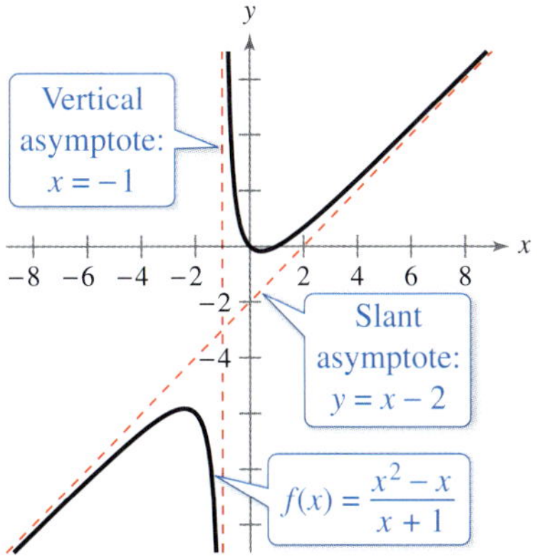

Figure 2.34

As x increases or decreases without bound, the remainder term $2/(x + 1)$ approaches 0, so the graph of f approaches the line $y = x - 2$, as shown in Figure 2.34.

Explore the Concept

Do you think it is possible for the graph of a rational function to cross its horizontal asymptote or its slant asymptote? Use the graphs of the functions below to investigate this question. Write a summary of your conclusion. Explain your reasoning.

$$f(x) = \frac{x}{x^2 + 1}$$

$$g(x) = \frac{2x}{3x^2 - 2x + 1}$$

$$h(x) = \frac{x^3}{x^2 + 1}$$

EXAMPLE 6 A Rational Function with a Slant Asymptote

Sketch the graph of $f(x) = \dfrac{x^2 - x - 2}{x - 1}$.

Solution

First, write $f(x)$ in two different ways. Factoring the numerator

$$f(x) = \frac{x^2 - x - 2}{x - 1} = \frac{(x - 2)(x + 1)}{x - 1}$$

enables you to recognize the x-intercepts. Long division

$$f(x) = \frac{x^2 - x - 2}{x - 1} = x - \frac{2}{x - 1}$$

enables you to recognize that the line $y = x$ is a slant asymptote of the graph.

y-intercept: $(0, 2)$, because $f(0) = 2$

x-intercepts: $(-1, 0)$ and $(2, 0)$

Vertical asymptote: $x = 1$, zero of denominator

Horizontal asymptote: None, because degree of $N(x) >$ degree of $D(x)$

Slant asymptote: $y = x$

Additional points:

x	-2	0.5	1	1.5	3
$f(x)$	-1.33	4.5	Undefined	-2.5	2

The graph is shown in Figure 2.35.

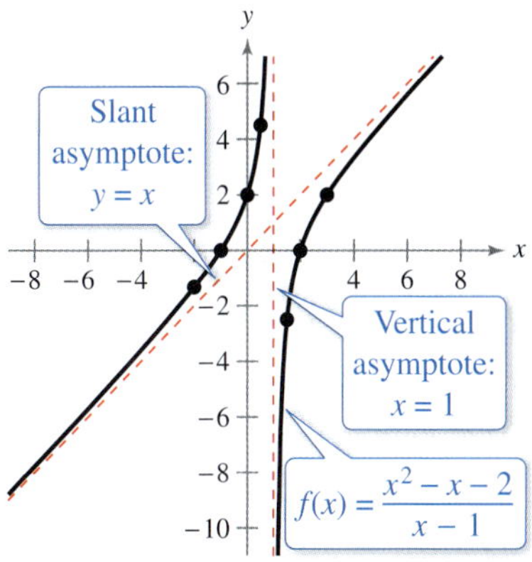

Figure 2.35

Checkpoint Audio-video solution in English & Spanish at LarsonPrecalculus.com

Sketch the graph of $f(x) = \dfrac{3x^2 + 1}{x}$.

Application

EXAMPLE 7 Publishing

A rectangular page is designed to contain 48 square inches of print. The margins on each side of the page are $1\frac{1}{2}$ inches wide. The margins at the top and bottom are each 1 inch deep. What should the dimensions of the page be so that the minimum amount of paper is used?

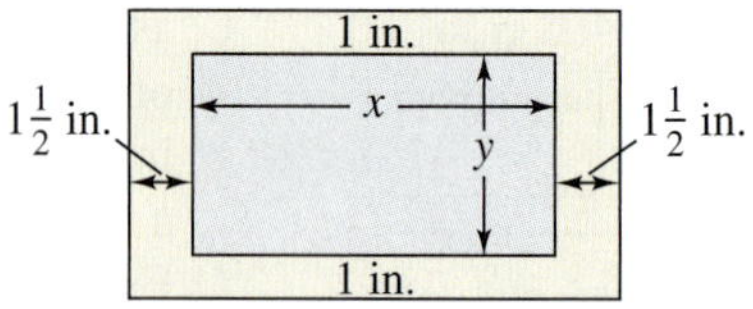

Figure 2.36

Graphical Solution

Let A be the area to be minimized. From Figure 2.36, you can write

$$A = (x + 3)(y + 2).$$

The printed area inside the margins is modeled by $48 = xy$ or $y = 48/x$. To find the minimum area, rewrite the equation for A in terms of just one variable by substituting $48/x$ for y.

$$A = (x + 3)\left(\frac{48}{x} + 2\right) = \frac{(x + 3)(48 + 2x)}{x}, \quad x > 0$$

The graph of this rational function is shown in Figure 2.37. Because x represents the width of the printed area, you need consider only the portion of the graph for which x is positive. Using the *minimum* feature of a graphing utility, you can approximate the minimum value of A to occur when $x \approx 8.5$ inches. The corresponding value of y is $48/8.5 \approx 5.6$ inches. So, the dimensions should be

$$x + 3 \approx 11.5 \text{ inches by } y + 2 \approx 7.6 \text{ inches.}$$

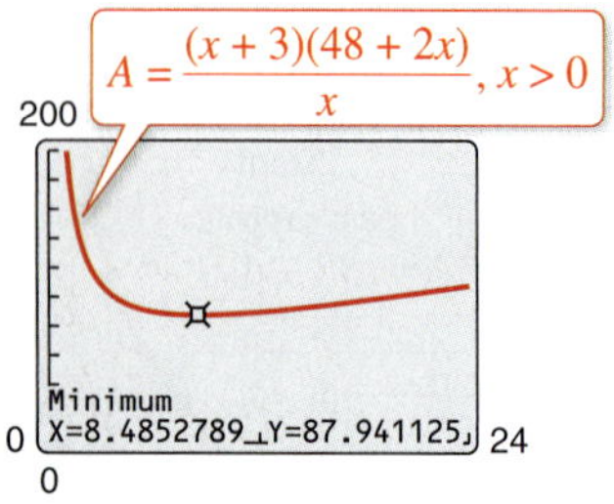

Figure 2.37

Numerical Solution

Let A be the area to be minimized. From Figure 2.36, you can write

$$A = (x + 3)(y + 2).$$

The printed area inside the margins is modeled by $48 = xy$ or $y = 48/x$. To find the minimum area, rewrite the equation for A in terms of just one variable by substituting $48/x$ for y.

$$A = (x + 3)\left(\frac{48}{x} + 2\right) = \frac{(x + 3)(48 + 2x)}{x}, \quad x > 0$$

Use the *table* feature of a graphing utility to create a table of values for the function

$$y_1 = \frac{(x + 3)(48 + 2x)}{x}$$

beginning at $x = 1$. From the table, you can see that the minimum value of y_1 occurs when x is somewhere between 8 and 9, as shown in Figure 2.38. To approximate the minimum value of y_1 to one decimal place, change the table to begin at $x = 8$ and set the table step to 0.1. The minimum value of y_1 occurs when $x \approx 8.5$, as shown in Figure 2.39. The corresponding value of y is $48/8.5 \approx 5.6$ inches. So, the dimensions should be

$$x + 3 \approx 11.5 \text{ inches by } y + 2 \approx 7.6 \text{ inches.}$$

X	Y1
6	90
7	88.571
8	88
9	88
10	88.4
11	89.091
12	90
X=8	

Figure 2.38

X	Y1
8.2	87.961
8.3	87.949
8.4	87.943
8.5	87.941
8.6	87.944
8.7	87.952
8.8	87.964
X=8.5	

Figure 2.39

✓ ***Checkpoint*** *Audio-video solution in English & Spanish at LarsonPrecalculus.com*

Rework Example 7 when the margins on each side are 2 inches wide and the page contains 40 square inches of print.

If you go on to take a course in calculus, you will learn an analytic technique for finding the exact value of x that produces a minimum area in Example 7. In this case, that value is $x = 6\sqrt{2} \approx 8.485$.

2.7 Exercises

See *CalcChat.com* for tutorial help and worked-out solutions to odd-numbered exercises.
For instructions on how to use a graphing utility, see Appendix A.

Vocabulary and Concept Check

In Exercises 1 and 2, fill in the blank(s).

1. For the rational function $f(x) = N(x)/D(x)$, if the degree of $N(x)$ is exactly one more than the degree of $D(x)$, then the graph of f has a _______ (or oblique) _______ .

2. The graph of $f(x) = 9/x^4$ is symmetrical with respect to the _______ -axis.

3. Does the graph of $f(x) = \dfrac{x^3 - 1}{x^2 + 2}$ have a slant asymptote?

4. Using long division, you find that $f(x) = \dfrac{x^2 + 1}{x + 1} = x - 1 + \dfrac{2}{x + 1}$. What is the slant asymptote of the graph of f?

Procedures and Problem Solving

Library of Parent Functions **In Exercises 5–8, sketch the graph of the function g by hand and compare it with the graph of $f(x) = 1/x$.**

5. $g(x) = \dfrac{-1}{x} + 2$

6. $g(x) = \dfrac{1}{x - 6}$

7. $g(x) = \dfrac{1}{x - 3} - 1$

8. $g(x) = \dfrac{-1}{x + 2} - 4$

Describing a Transformation of $f(x) = 2/x$ **In Exercises 9–12, use a graphing utility to graph $f(x) = 2/x$ and the function g in the same viewing window. Describe the relationship between the two graphs.**

9. $g(x) = f(x) - 1$

10. $g(x) = f(x + 3)$

11. $g(x) = 4f(x)$

12. $g(x) = f\left(\frac{1}{2}x\right)$

Describing a Transformation of $f(x) = 3/x^2$ **In Exercises 13–16, use a graphing utility to graph $f(x) = 3/x^2$ and the function g in the same viewing window. Describe the relationship between the two graphs.**

13. $g(x) = f(x - 1)$

14. $g(x) = 4 + f(x)$

15. $g(x) = f(3x)$

16. $g(x) = -f(x)$

Sketching the Graph of a Rational Function **In Exercises 17–32, sketch the graph of the rational function by hand. As sketching aids, check for intercepts, vertical asymptotes, horizontal asymptotes, and holes. Use a graphing utility to verify your graph.**

17. $f(x) = \dfrac{1}{x + 2}$

18. $f(x) = \dfrac{1}{x - 6}$

19. $C(x) = \dfrac{5 + 2x}{1 + x}$

20. $P(x) = \dfrac{1 - 3x}{1 - x}$

21. $f(t) = \dfrac{1 - 2t}{t} - 3$

22. $g(x) = \dfrac{x}{x + 2} + 2$

23. $f(x) = \dfrac{x^2}{x^2 - 4}$

24. $g(t) = \dfrac{t^2}{9 - t^2}$

25. $g(x) = \dfrac{4(x + 1)}{x(x - 4)}$

26. $h(x) = \dfrac{2(x + 2)}{x(x - 3)}$

27. $f(x) = \dfrac{3x}{x^2 - x - 2}$

28. $f(x) = \dfrac{2x}{x^2 + x - 2}$

29. $f(x) = \dfrac{x^2 + 3x}{x^2 + x - 6}$

30. $g(x) = \dfrac{5x(x + 4)}{x^2 + 2x - 8}$

31. $f(x) = \dfrac{x^2 - 1}{x + 1}$

32. $f(x) = \dfrac{x^2 - 16}{x - 4}$

Finding the Domain and Asymptotes **In Exercises 33–40, use a graphing utility to graph the function. Determine its domain and identify any vertical or horizontal asymptotes.**

33. $f(x) = \dfrac{2 + x}{1 - x}$

34. $f(x) = \dfrac{3 - x}{2 - x}$

35. $g(x) = \dfrac{3x - 4}{-x}$

36. $h(x) = \dfrac{2x - 1}{x + 5}$

37. $f(x) = \dfrac{x + 1}{x^2 - x - 6}$

38. $g(x) = -\dfrac{x}{(x - 2)^2}$

39. $f(x) = \dfrac{20x}{x^2 + 1} - \dfrac{1}{x}$

40. $f(x) = 5\left(\dfrac{1}{x - 4} - \dfrac{1}{x + 2}\right)$

Exploration **In Exercises 41–46, use a graphing utility to graph the function. What do you observe about its asymptotes?**

41. $h(x) = \dfrac{6x}{\sqrt{x^2 + 1}}$

42. $f(x) = -\dfrac{x}{\sqrt{9 + x^2}}$

43. $g(x) = \dfrac{4|x - 2|}{x + 1}$

44. $f(x) = -\dfrac{8|3 + x|}{x - 2}$

45. $f(x) = \dfrac{4(x - 1)^2}{x^2 - 4x + 5}$

46. $f(x) = \dfrac{5(1 - x)^2}{x^2 - 3x + 4}$

A Rational Function with a Slant Asymptote **In Exercises 47–54, sketch the graph of the rational function by hand. As sketching aids, check for intercepts, vertical asymptotes, and slant asymptotes.**

47. $f(x) = \dfrac{2x^2 + 1}{x}$ **48.** $g(x) = \dfrac{1 - x^2}{x}$

49. $f(x) = \dfrac{x^3}{6 - x^2}$ **50.** $f(x) = \dfrac{x^3}{x^2 - 1}$

51. $g(x) = \dfrac{x^3}{2x^2 - 8}$ **52.** $f(x) = \dfrac{x^3}{x^2 + 4}$

53. $f(x) = \dfrac{x^3 + 2x^2 + 4}{2x^2 + 1}$ **54.** $f(x) = \dfrac{3 - x^3}{4x^2 + 4}$

Finding the x-Intercepts **In Exercises 55–58, use the graph to estimate any x-intercepts of the rational function. Set $y = 0$ and solve the resulting equation to confirm your result.**

55. $y = \dfrac{x + 1}{x - 3}$

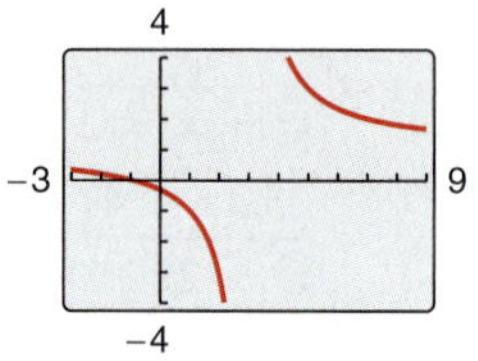

56. $y = \dfrac{2x}{x - 3}$

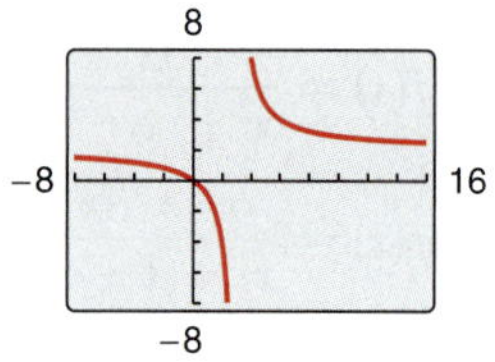

57. $y = \dfrac{1}{x} - x$

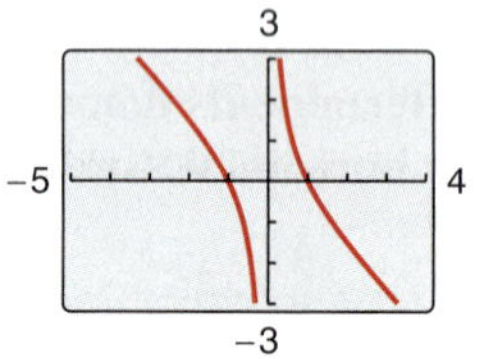

58. $y = x - 3 + \dfrac{2}{x}$

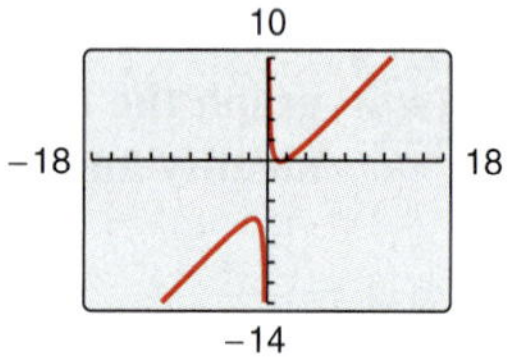

Finding the Domain and Asymptotes **In Exercises 59–62, use a graphing utility to graph the rational function. Determine the domain of the function and identify any asymptotes.**

59. $y = \dfrac{2x^2 + x}{x + 1}$ **60.** $y = \dfrac{x^2 + 5x + 8}{x + 3}$

61. $y = \dfrac{1 + 3x^2 - x^3}{x^2}$ **62.** $y = \dfrac{12 - 2x - x^2}{2(4 + x)}$

Finding Asymptotes and Holes **In Exercises 63–68, find all vertical asymptotes, horizontal asymptotes, slant asymptotes, and holes in the graph of the function. Then use a graphing utility to verify your results.**

63. $f(x) = \dfrac{x^2 - 5x + 4}{x^2 - 4}$ **64.** $f(x) = \dfrac{x^2 - 2x - 8}{x^2 - 9}$

65. $f(x) = \dfrac{2x^2 - 5x + 2}{2x^2 - x - 6}$

66. $f(x) = \dfrac{3x^2 - 8x + 4}{2x^2 - 3x - 2}$

67. $f(x) = \dfrac{2x^3 - x^2 - 2x + 1}{x^2 + 3x + 2}$

68. $f(x) = \dfrac{2x^3 + x^2 - 8x - 4}{x^2 - 3x + 2}$

Finding x-Intercepts Graphically **In Exercises 69–78, use a graphing utility to graph the function and determine any x-intercepts. Set $y = 0$ and solve the resulting equation to confirm your result.**

69. $y = \dfrac{1}{x + 5} + \dfrac{4}{x}$ **70.** $y = \dfrac{1}{x - 2} - \dfrac{5}{x}$

71. $y = \dfrac{2}{x + 2} - \dfrac{3}{x - 1}$ **72.** $y = \dfrac{6}{x + 3} - \dfrac{1}{x + 4}$

73. $y = x - \dfrac{2}{x + 1}$ **74.** $y = 2x - \dfrac{8}{x}$

75. $y = x + 2 - \dfrac{1}{x + 1}$

76. $y = 2x - 1 + \dfrac{1}{x - 2}$

77. $y = x + 1 + \dfrac{2}{x - 1}$

78. $y = x + 2 + \dfrac{2}{x + 2}$

79. **Chemistry** A 1000-liter tank contains 50 liters of a 25% brine solution. You add x liters of a 75% brine solution to the tank.

(a) Show that the concentration C (the ratio of brine to the total solution) of the final mixture can be modeled by

$$C = \frac{3x + 50}{4(x + 50)}.$$

(b) Determine the domain of the function based on the physical constraints of the problem.

(c) Use a graphing utility to graph the function. As the tank is filled, what happens to the rate at which the concentration of brine increases? What percent does the concentration of brine appear to approach?

80. **Geometry** A rectangular region of length x and width y has an area of 500 square meters.

(a) Write the width y as a function of x.

(b) Determine the domain of the function based on the physical constraints of the problem.

(c) Sketch a graph of the function and determine the width of the rectangle when $x = 30$ meters.

81. Publishing A page that is x inches wide and y inches high contains 30 square inches of print (see figure). The margins at the top and bottom are 2 inches deep and the margins on each side are 1 inch wide.

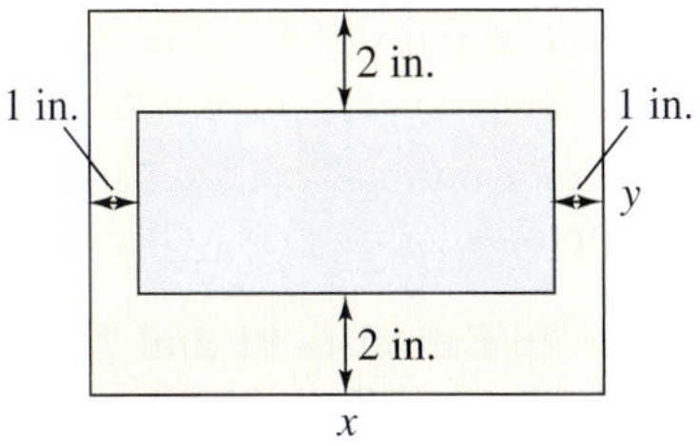

(a) Show that the total area A of the page is given by

$$A = \frac{2x(2x + 11)}{x - 2}.$$

(b) Determine the domain of the function based on the physical constraints of the problem.

(c) Use a graphing utility to graph the area function and approximate the page size such that the minimum amount of paper will be used. Verify your answer numerically using the *table* feature of the graphing utility.

82. Geometry A right triangle is formed in the first quadrant by the x-axis, the y-axis, and a line segment through the point (3, 2). (See figure.)

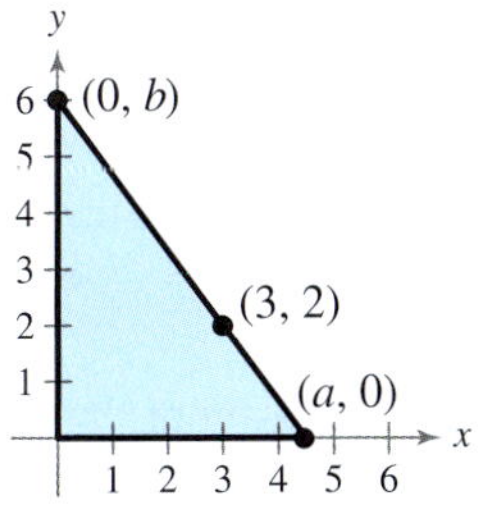

(a) Show that an equation of the line segment is given by

$$y = \frac{2(a - x)}{a - 3}, \quad 0 \le x \le a.$$

(b) Show that the area of the triangle is given by

$$A = \frac{a^2}{a - 3}.$$

(c) Find the area of the triangle when $a = 4.5$.

(d) Use a graphing utility to graph the area function and estimate the value of a that yields a minimum area. Estimate the minimum area. Verify your answer numerically using the *table* feature of the graphing utility.

83. Cost Management The ordering and transportation cost C (in thousands of dollars) for the components used in manufacturing a product can be modeled by

$$C = 100\left(\frac{200}{x^2} + \frac{x}{x + 30}\right), \quad x \ge 1$$

where x is the order size (in hundreds). Use a graphing utility to graph the cost function. From the graph, estimate the order size that minimizes cost.

84. Cost Management The cost C of producing x units of a product is given by $C = 0.2x^2 + 10x + 5$, and the average cost per unit can be modeled by

$$\overline{C} = \frac{C}{x} = \frac{0.2x^2 + 10x + 5}{x}, \quad x > 0.$$

Sketch the graph of the average cost function and estimate the number of units that should be produced to minimize the average cost per unit.

85. *Why you should learn it* *(p. 151)* The concentration C of a chemical in the bloodstream t hours after injection into muscle tissue can be modeled by

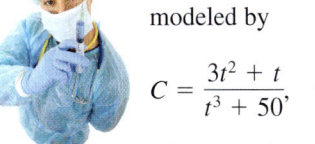

$$C = \frac{3t^2 + t}{t^3 + 50}, \quad t \ge 0.$$

(a) Determine the horizontal asymptote of the function and interpret its meaning in the context of the problem.

(b) Use a graphing utility to graph the function and approximate the time when the bloodstream concentration is greatest.

(c) Use the graphing utility to determine when the concentration is less than 0.345.

86. Algebraic-Graphical-Numerical A driver averaged 50 miles per hour on the round trip between Baltimore, Maryland, and Philadelphia, Pennsylvania, 100 miles away. The average speeds for going and returning were x and y miles per hour, respectively.

(a) Show that $y = \dfrac{25x}{x - 25}$.

(b) Determine the asymptotes of the function.

(c) Use a graphing utility to complete the table. What do you observe?

x	30	35	40	45	50	55	60
y							

(d) Use the graphing utility to graph the function.

(e) Is it possible to average 20 miles per hour in one direction and still average 50 miles per hour on the round trip? Explain.

87. MODELING DATA

The table shows the numbers N of newspaper industry employees in the United States (in hundreds of thousands) each year from 2007 through 2016. (*Source:* Bureau of Labor Statistics)

DATA

Year	Employees, N (in hundreds of thousands)
2007	3.5
2008	3.3
2009	2.9
2010	2.6
2011	2.4
2012	2.3
2013	2.2
2014	2.0
2015	1.9
2016	1.8

Spreadsheet at LarsonPrecalculus.com

(a) Use the *regression* feature of a graphing utility to find a linear model for the data. Let $t = 7$ represent 2007. Use the graphing utility to plot the data and graph the model in the same viewing window.

(b) Take the reciprocals of the numbers of employees to generate the points $(t, 1/N)$. Use the *regression* feature of the graphing utility to find a linear model for the new points. The resulting line has the form $1/N = at + b$. Solve for N to obtain a rational model for the original data. Use the graphing utility to plot the data and graph the rational model in the same viewing window.

(c) Which model do you prefer? Explain.

88. HOW DO YOU SEE IT? A herd of elk is released onto state game lands. The graph shows the expected population P of the herd, where t is the time (in years) since the initial number of elk were released.

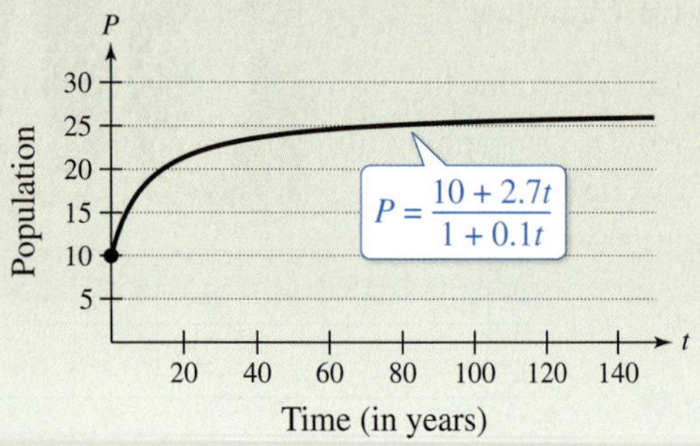

(a) Determine the domain of the function. Explain.

(b) Find the initial number of elk in the herd.

(c) Is there a limit to the size of the herd? If so, what is the expected population?

Focusing on Concepts

True or False? **In Exercises 89 and 90, determine whether the statement is true or false. Justify your answer.**

89. The graph of a rational function is continuous only when the denominator is a constant polynomial.

90. The graph of a rational function can never cross one of its asymptotes.

Think About It **In Exercises 91 and 92, use a graphing utility to graph the function. Explain why there is no vertical asymptote when a superficial examination of the function might indicate that there should be one.**

91. $h(x) = \dfrac{6 - 2x}{3 - x}$ **92.** $g(x) = \dfrac{x^2 + x - 2}{x - 1}$

93. Writing Write a set of guidelines for finding all the asymptotes of a rational function given that the degree of the numerator is not more than 1 greater than the degree of the denominator.

94. Writing a Rational Function Write a rational function with zero $x = -3$ that has the vertical asymptote $x = -2$ and the slant asymptote $y = x + 4$.

Error Analysis **In Exercises 95 and 96, describe the error in graphing the function.**

95. $f(x) = \dfrac{1}{x + 3}$

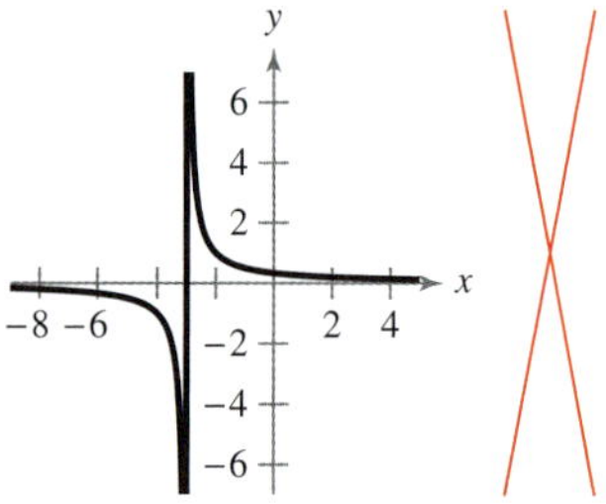

96. $f(x) = -\dfrac{4}{x + 1}$

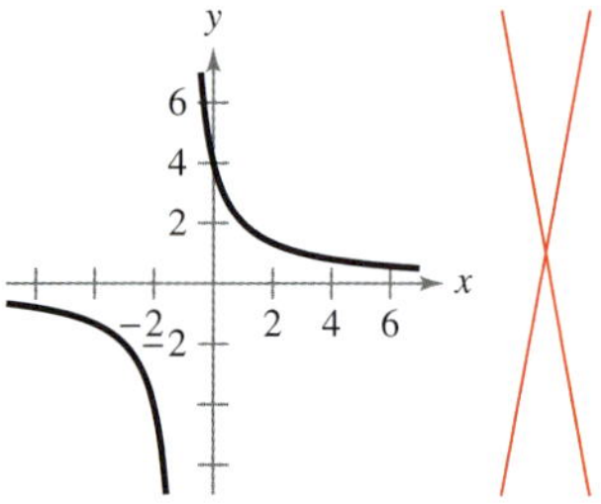

Cumulative Mixed Review

Finding the Domain and Range of a Function **In Exercises 97–100, use a graphing utility to graph the function. Find the domain and range of the function.**

97. $f(x) = \sqrt{6 + x^2}$ **98.** $f(x) = \dfrac{\sqrt{x}}{1 - 2x}$

99. $f(x) = -|x + 9|$ **100.** $f(x) = -x^2 + 9$

101. *Project: Epidemiology* To work an extended application analyzing the numbers of tuberculosis cases in the United States from 1991 through 2016, visit this textbook's website at *LarsonPrecalculus.com.* (*Data Source:* Centers for Disease Control and Prevention)

2.8 Quadratic Models

Classifying Scatter Plots

In real life, many relationships between two variables are parabolic, as in Section 2.1, Example 5. A scatter plot can be used to give you an idea of which type of model will best fit a set of data.

What you should learn

- Classify scatter plots.
- Use scatter plots and a graphing utility to find quadratic models for data.
- Choose a model that best fits a set of data.

Why you should learn it

Many real-life situations can be modeled by quadratic equations. For instance, in Exercise 17 on page 165, a quadratic equation is used to model the monthly precipitation for San Francisco, California.

EXAMPLE 1 Classifying Scatter Plots

See LarsonPrecalculus.com for an interactive version of this type of example.

Decide whether each set of data could be better modeled by a linear model

$$y = ax + b$$

a quadratic model

$$y = ax^2 + bc + c$$

or neither.

a. (0.9, 1.7), (1.2, 2.0), (1.3, 1.9), (1.4, 2.1), (1.6, 2.5), (1.8, 2.8), (2.1, 3.0), (2.5, 3.4), (2.9, 3.7), (3.2, 3.9), (3.3, 4.1), (3.6, 4.4), (4.0, 4.7), (4.2, 4.8), (4.3, 5.0)

b. (0.9, 3.2), (1.2, 4.0), (1.3, 4.1), (1.4, 4.4), (1.6, 5.1), (1.8, 6.0), (2.1, 7.6), (2.5, 9.8), (2.9, 12.4), (3.2, 14.3), (3.3, 15.2), (3.6, 18.1), (4.0, 22.7), (4.2, 24.9), (4.3, 27.2)

c. (0.9, 1.2), (1.2, 6.5), (1.3, 9.3), (1.4, 11.6), (1.6, 15.2), (1.8, 16.9), (2.1, 14.7), (2.5, 8.1), (2.9, 3.7), (3.2, 5.8), (3.3, 7.1), (3.6, 11.5), (4.0, 20.2), (4.2, 23.7), (4.3, 26.9)

Solution

a. Begin by entering the data into a graphing utility. Then display the scatter plot, as shown in Figure 2.40. From the scatter plot, it appears the data follow a linear pattern. So, the data can be better modeled by a linear function.

b. Enter the data into a graphing utility and then display the scatter plot (see Figure 2.41). From the scatter plot, it appears the data follow a parabolic pattern. So, the data can be better modeled by a quadratic function.

c. Enter the data into a graphing utility and then display the scatter plot (see Figure 2.42). From the scatter plot, it appears the data do not follow either a linear or a parabolic pattern. So, the data cannot be modeled by either a linear function or a quadratic function.

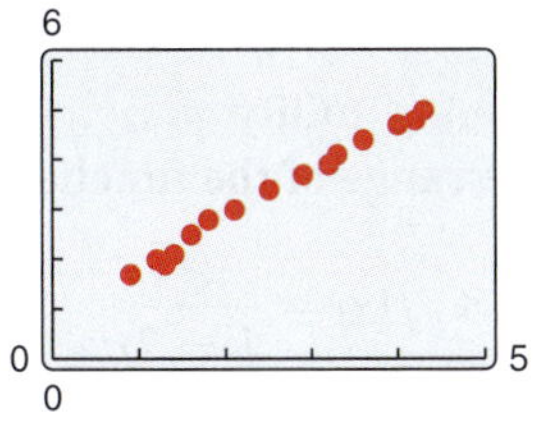

Figure 2.40

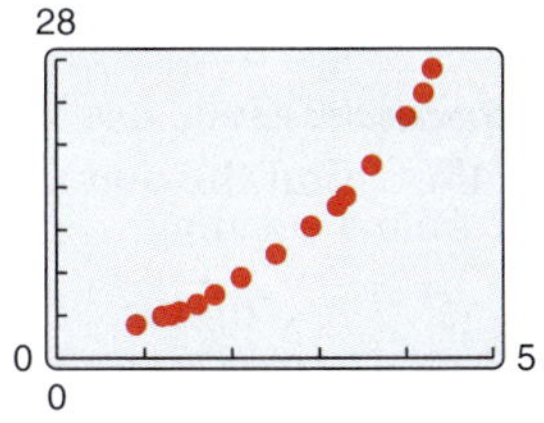

Figure 2.41

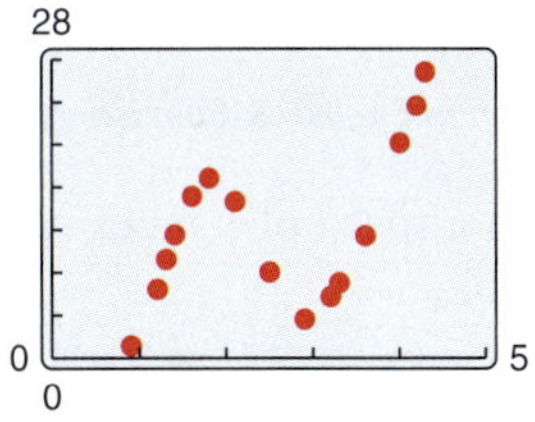

Figure 2.42

✓ ***Checkpoint*** *Audio-video solution in English & Spanish at LarsonPrecalculus.com*

Decide whether the data could be better modeled by a linear model, $y = ax + b$, a quadratic model, $y = ax^2 + bx + c$, or neither.

(0, 3480), (5, 2235), (10, 1250), (15, 565), (20, 150), (25, 12), (30, 145), (35, 575), (40, 1275), (45, 2225), (50, 3500), (55, 5010)

Fitting a Quadratic Model to Data

In Section 1.7, you created scatter plots of data and used a graphing utility to find the least squares regression lines for the data. You can use a similar procedure to find a model for nonlinear data. Once you have used a scatter plot to determine the type of model that would best fit a set of data, there are several ways that you can actually find the model. Each method is best used with a graphing utility, rather than with hand calculations.

EXAMPLE 2 Fitting a Quadratic Model to Data

A study was done to compare the speed x (in miles per hour) with the mileage y (in miles per gallon) of an automobile. The results are shown in the table.

a. Use a graphing utility to create a scatter plot of the data.

b. Use the *regression* feature of the graphing utility to find a model that best fits the data.

c. Approximate the speed at which the mileage is the greatest.

Speed, x	Mileage, y
15	22.3
20	25.5
25	27.5
30	29.0
35	28.7
40	29.9
45	30.4
50	30.2
55	30.0
60	28.8
65	27.4
70	25.3
75	23.3

Spreadsheet at LarsonPrecalculus.com

Solution

a. Begin by entering the data into a graphing utility and displaying the scatter plot, as shown in Figure 2.43. From the scatter plot, you can see that the data appear to follow a parabolic pattern.

b. Using the *regression* feature of the graphing utility, you can find the quadratic model, as shown in Figure 2.44. So, the quadratic equation that best fits the data is given by

$$y = -0.0082x^2 + 0.75x + 13.5. \qquad \text{Quadratic model}$$

c. Graph the data and the model in the same viewing window, as shown in Figure 2.45. Use the *maximum* feature or the *zoom* and *trace* features of the graphing utility to approximate the speed at which the mileage is greatest. You should obtain a maximum of approximately (46, 31), as shown in Figure 2.45. So, the speed at which the mileage is greatest is about 46 miles per hour.

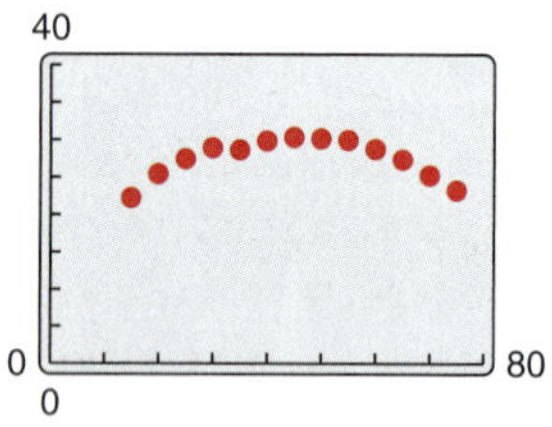

Figure 2.43

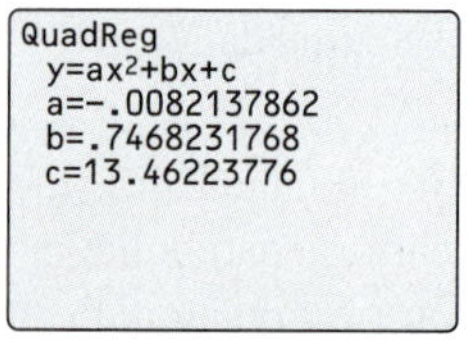

Figure 2.44

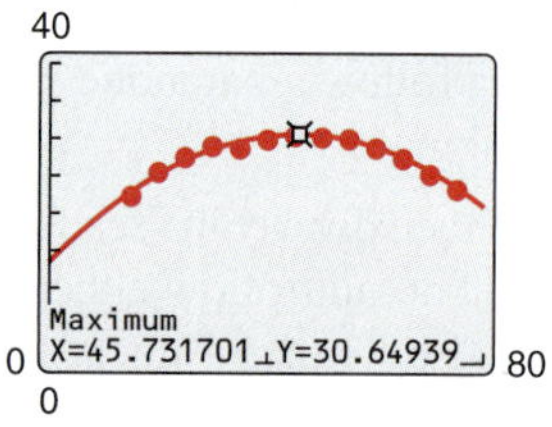

Figure 2.45

Checkpoint Audio-video solution in English & Spanish at LarsonPrecalculus.com

The time y (in seconds) required to attain a speed of x miles per hour from a standing start for an automobile is shown in the table.

Speed, x	0	20	30	40	50	60	70	80
Time, y	0	1.4	2.6	3.8	4.9	6.3	8.0	9.9

Spreadsheet at LarsonPrecalculus.com

a. Use a graphing utility to create a scatter plot of the data.

b. Use the *regression* feature of the graphing utility to find a model that best fits the data.

c. Use the model to estimate how long it takes the automobile, from a standing start, to reach a speed of 55 miles per hour.

EXAMPLE 3 Fitting a Quadratic Model to Data

A basketball is dropped from a height of about 5.25 feet. The height of the basketball is recorded 23 times at intervals of about 0.02 second. The results are shown in the table. Use a graphing utility to find a model that best fits the data. Then use the model to predict the time when the basketball will hit the ground.

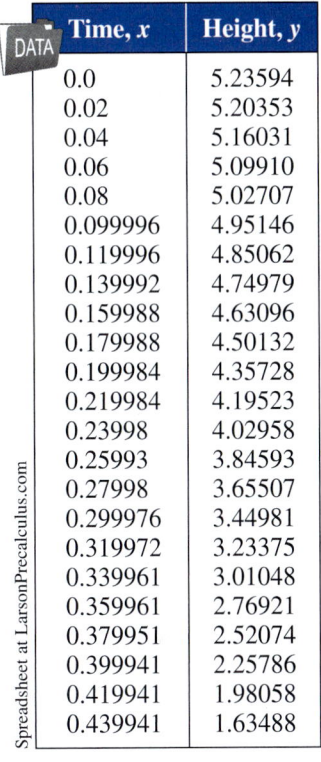

Time, x	Height, y
0.0	5.23594
0.02	5.20353
0.04	5.16031
0.06	5.09910
0.08	5.02707
0.099996	4.95146
0.119996	4.85062
0.139992	4.74979
0.159988	4.63096
0.179988	4.50132
0.199984	4.35728
0.219984	4.19523
0.23998	4.02958
0.25993	3.84593
0.27998	3.65507
0.299976	3.44981
0.319972	3.23375
0.339961	3.01048
0.359961	2.76921
0.379951	2.52074
0.399941	2.25786
0.419941	1.98058
0.439941	1.63488

Spreadsheet at LarsonPrecalculus.com

Solution

Begin by entering the data into a graphing utility and displaying the scatter plot, as shown in Figure 2.46. From the scatter plot, you can see that the data show a parabolic trend. So, using the *regression* feature of the graphing utility, you can find the quadratic model, as shown in Figure 2.47. The quadratic model that best fits the data is given by $y = -15.449x^2 - 1.30x + 5.2$.

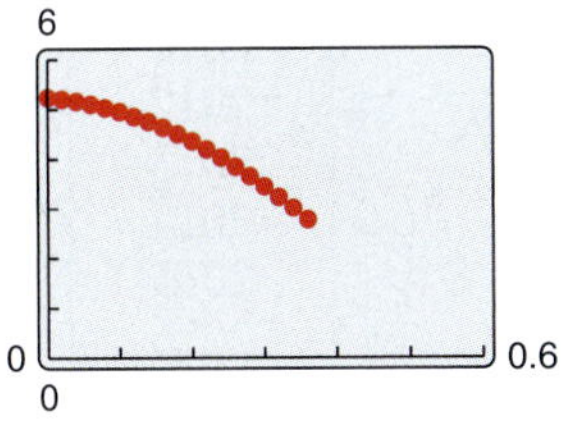

Figure 2.46

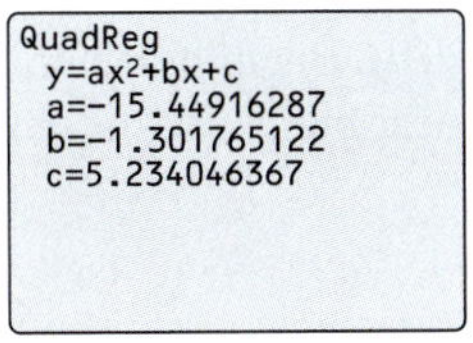

Figure 2.47

You can graph the data and the model in the same viewing window to see that the model fits the data well, as shown in the next figure.

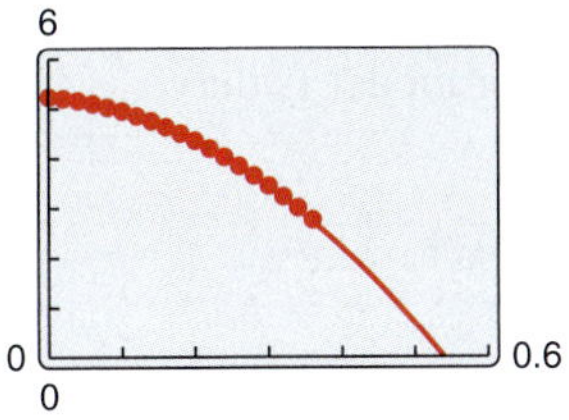

Using this model, you can predict the time when the basketball will hit the ground by substituting 0 for y and solving the resulting equation for x.

$$y = -15.449x^2 - 1.30x + 5.2 \quad \text{Write original model.}$$

$$0 = -15.449x^2 - 1.30x + 5.2 \quad \text{Substitute 0 for } y.$$

$$x = \frac{-b \pm \sqrt{b^2 - 4ac}}{2a} \quad \text{Quadratic Formula}$$

$$= \frac{-(-1.30) \pm \sqrt{(-1.30)^2 - 4(-15.449)(5.2)}}{2(-15.449)} \quad \text{Substitute for } a, b, \text{ and } c.$$

$$\approx 0.54 \quad \text{Choose positive solution.}$$

So, the solution is about 0.54 second. In other words, the basketball will continue to fall for about $0.54 - 0.44 = 0.1$ second more before hitting the ground.

Insight

In addition to identifying which type of model will best fit a set of data (see Example 4), a standardized test may ask you to use the model to make a prediction (see Example 3).

✓ Checkpoint *Audio-video solution in English & Spanish at LarsonPrecalculus.com*

The table shows the annual sales y (in millions of dollars) of a department store chain. Use a graphing utility to find a model that best fits the data. Then use the model to estimate the first year when the annual sales will be less than $190 million.

Year, x	1	2	3	4	5	6
Sales, y	221	222	220	219	216	211

Spreadsheet at LarsonPrecalculus.com

Choosing a Model

Sometimes it is not easy to distinguish from a scatter plot which type of model will best fit the data. You should first find several models for the data, using the *Library of Parent Functions*, and then choose the model that best fits the data by comparing the y-values of each model with the actual y-values.

EXAMPLE 4 Choosing a Model

The table at the right shows the revenues y (in billions of dollars) for Netflix, Inc. from 2010 through 2017. Use the *regression* feature of a graphing utility to find a linear model and a quadratic model for the data. Determine which model better fits the data. (*Source:* Netflix, Inc.)

Year	Revenues, y (in billions of dollars)
2010	2.16
2011	3.20
2012	3.61
2013	4.37
2014	5.50
2015	6.78
2016	8.83
2017	11.69

Spreadsheet at LarsonPrecalculus.com

Solution

Let x represent the year, with $x = 0$ corresponding to 2010. Begin by entering the data into a graphing utility. Using the *regression* feature, a linear model for the data is

$$y = 1.256x + 1.37$$

and a quadratic model for the data is

$$y = 0.1694x^2 + 0.070x + 2.56.$$

Plot the data and the linear model in the same viewing window, as shown in Figure 2.48. Then plot the data and the quadratic model in the same viewing window, as shown in Figure 2.49. To determine which model fits the data better, compare the y-values given by each model with the actual y-values. The model whose y-values are closest to the actual values is the better fit. In this case, the better-fitting model is the quadratic model.

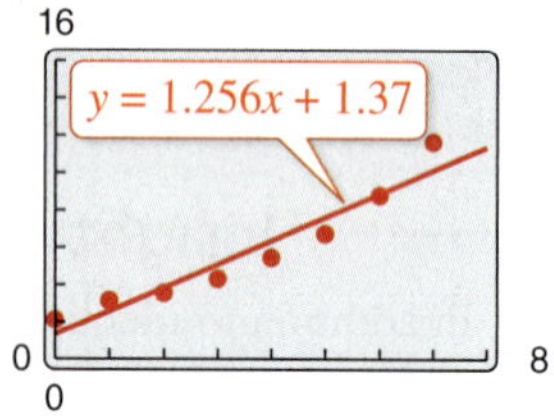

Figure 2.48

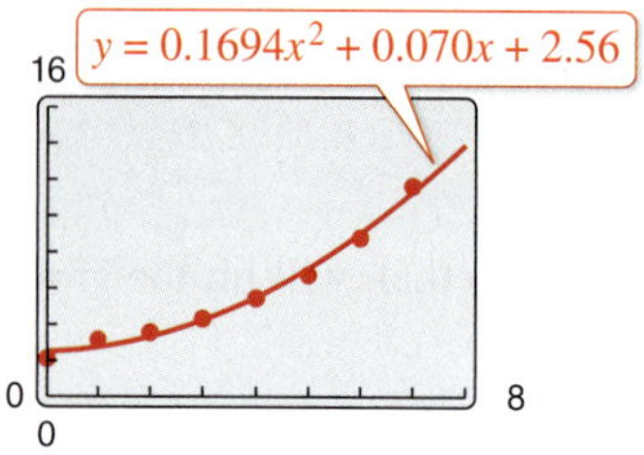

Figure 2.49

✓ **Checkpoint** *Audio-video solution in English & Spanish at LarsonPrecalculus.com*

The table at the right shows the numbers y (in millions) of people ages 16 and older who were not in the U.S. labor force from 2009 through 2017. Use the *regression* feature of a graphing utility to find a linear model and a quadratic model for the data. Determine which model better fits the data. (*Source:* U.S. Department of Labor)

Year	Number of people unemployed, y (in millions)
2009	81.7
2010	83.9
2011	86.0
2012	88.3
2013	90.3
2014	92.0
2015	93.7
2016	94.4
2017	94.8

Spreadsheet at LarsonPrecalculus.com

Technology Tip

When you use the *regression* feature of a graphing utility, the program may output an "r^2-value." This r^2-value is the **coefficient of determination** of the data and gives a measure of how well the model fits the data. The coefficient of determination for the linear model in Example 4 is $r^2 \approx 0.9232$, and the coefficient of determination for the quadratic model is $r^2 \approx 0.9903$. Because the coefficient of determination for the quadratic model is closer to 1, the quadratic model better fits the data.

2.8 Exercises

See *CalcChat.com* for tutorial help and worked-out solutions to odd-numbered exercises.
For instructions on how to use a graphing utility, see Appendix A.

Vocabulary and Concept Check

1. What type of model best represents data that follow a parabolic pattern?

2. Which coefficient of determination indicates a better model for a set of data, $r^2 = 0.0365$ or $r^2 = 0.9688$?

Procedures and Problem Solving

Classifying Scatter Plots **In Exercises 3–8, determine whether the scatter plot could best be modeled by a linear model, a quadratic model, or neither.**

3.

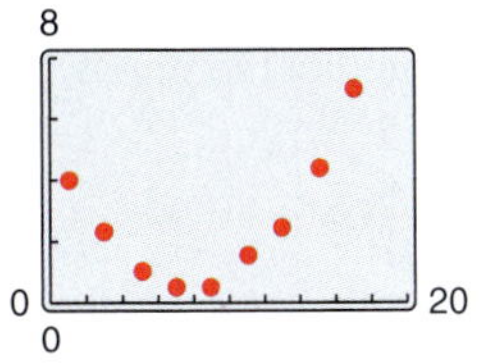

4.

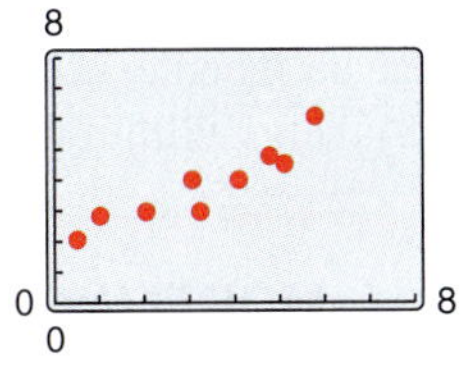

5.

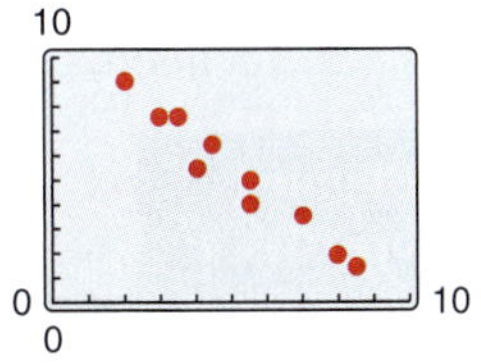

6.

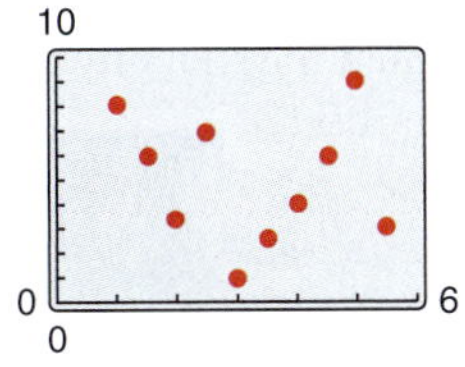

7.

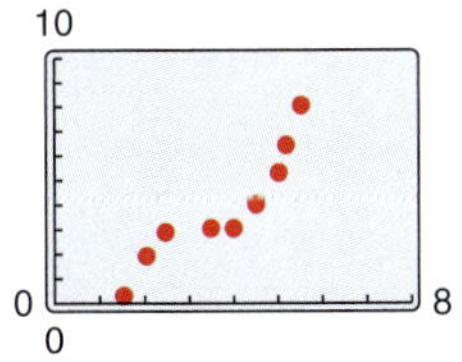

8.

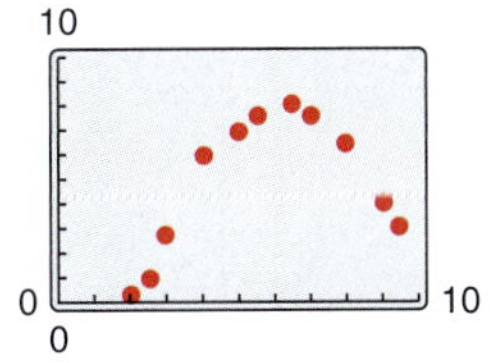

Choosing a Model **In Exercises 9–16, (a) use a graphing utility to create a scatter plot of the data, (b) determine whether the data could be better modeled by a linear model or a quadratic model, (c) use the *regression* feature of the graphing utility to find a model for the data, (d) use the graphing utility to graph the model with the scatter plot from part (a), and (e) create a table comparing the original data with the data given by the model.**

9. (0, 2.1), (1, 2.4), (2, 2.5), (3, 2.8), (4, 2.9), (5, 3.0), (6, 3.0), (7, 3.2), (8, 3.4), (9, 3.5), (10, 3.6)

10. (−2, 11.0), (−1, 10.7), (0, 10.4), (1, 10.3), (2, 10.1), (3, 9.9), (4, 9.6), (5, 9.4), (6, 9.4), (7, 9.2), (8, 9.0)

11. (0, 2795), (5, 1590), (10, 650), (15, −30), (20, −450), (25, −615), (30, −520), (35, −55), (40, 625), (45, 1630), (50, 2845), (55, 4350)

12. (0, 6140), (2, 6815), (4, 7335), (6, 7710), (8, 7915), (10, 7590), (12, 7975), (14, 7700), (16, 7325), (18, 6820), (20, 6125), (22, 5325)

13. (1, 4.0), (2, 6.5), (3, 8.8), (4, 10.6), (5, 13.9), (6, 15.0), (7, 17.5), (8, 20.1), (9, 24.0), (10, 27.1)

14. (−6, 10.7), (−4, 9.0), (−2, 7.0), (0, 5.4), (2, 3.5), (4, 1.7), (6, −0.1), (8, −1.8), (10, −3.6), (12, −5.3)

15. (0, 587), (5, 551), (10, 512), (15, 478), (20, 436), (25, 430), (30, 424), (35, 420), (40, 423), (45, 429), (50, 444)

16. (2, 34.3), (3, 33.8), (4, 32.6), (5, 30.1), (6, 27.8), (7, 22.5), (8, 19.1), (9, 14.8), (10, 9.4), (11, 3.7), (12, −1.6)

17. ***Why you should learn it*** *(p. 161)* The table shows the monthly normal precipitation P (in inches) for San Francisco, California. (*Source:* The Weather Channel)

DATA

Month	Precipitation, P
January	4.50
February	4.61
March	3.26
April	1.46
May	0.70
June	0.16
July	0.00
August	0.06
September	0.21
October	1.12
November	3.16
December	4.56

Spreadsheet at LarsonPrecalculus.com

(a) Use a graphing utility to create a scatter plot of the data. Let t represent the month, with $t = 1$ corresponding to January.

(b) Use the *regression* feature of the graphing utility to find a quadratic model for the data and identify the coefficient of determination.

(c) Use the graphing utility to graph the model with the scatter plot from part (a).

(d) Use the graph from part (c) to determine in which month the normal precipitation in San Francisco is the least.

(e) Use the table to determine the month in which the normal precipitation in San Francisco is the least. Compare your answer with that of part (d).

18. MODELING DATA

The table shows the annual stock S of battery electric cars (in thousands) in the United States from 2011 through 2016. (*Source:* International Energy Agency)

Year	Stock, S (in thousands)
2011	13.52
2012	28.17
2013	75.86
2014	139.28
2015	210.33
2016	297.06

Spreadsheet at LarsonPrecalculus.com

(a) Use a graphing utility to create a scatter plot of the data. Let t represent the year, with $t = 1$ corresponding to 2011.

(b) Use the *regression* feature of the graphing utility to find a quadratic model for the data.

(c) Use the graphing utility to graph the model with the scatter plot from part (a).

(d) Use the model to estimate the first year when the stock of battery electric cars will exceed 530,000. Does the prediction seem reasonable? Explain.

19. MODELING DATA

The table shows the numbers of organ transplants N (in thousands) each year in the United States from 2012 through 2017. (*Source:* United Network for Organ Sharing)

Year	Organ Transplants, N (in thousands)
2012	28.1
2013	29.0
2014	29.5
2015	31.0
2016	33.6
2017	34.8

Spreadsheet at LarsonPrecalculus.com

(a) Use a graphing utility to create a scatter plot of the data. Let t represent the year, with $t = 2$ corresponding to 2012.

(b) Use the *regression* feature of the graphing utility to find a quadratic model for the data.

(c) Use the graphing utility to graph the model with the scatter plot from part (a).

(d) According to the model, in what year will the number of organ transplants be about 47,000? Does the prediction seem reasonable? Explain.

20. MODELING DATA

The height of a thrown javelin is recorded 8 times at intervals of 0.05 second. The results are shown in the table.

Time, x	Height, y
0	5.0422
0.05	6.9758
0.1	8.8329
0.15	10.6102
0.2	12.3111
0.25	13.9347
0.3	15.4692
0.35	16.9306

Spreadsheet at LarsonPrecalculus.com

Use a graphing utility to find a quadratic model for the data. Then use the model to predict the time when the javelin will hit the ground.

21. MODELING DATA

The table shows the numbers of blighted properties P remaining in a city after x weeks of demolition.

Weeks, x	Blighted Properties, P
0	72
1	52
2	37
3	26
4	16
5	5

Spreadsheet at LarsonPrecalculus.com

Use a graphing utility to find a quadratic model for the data. Then use the model to predict whether any blighted properties will remain after 6 weeks.

22. MODELING DATA

The table shows the numbers of industrial robots R (in thousands) sold worldwide from 2011 to 2016. (*Source:* Statista)

Year	Industrial Robots, R (in thousands)
2011	166
2012	159
2013	178
2014	221
2015	254
2016	294

Spreadsheet at LarsonPrecalculus.com

Use the *regression* feature of a graphing utility to find a linear model and a quadratic model for the data. Determine which model better fits the data.

23. MODELING DATA

The table shows the annual revenues R (in millions of dollars) of Under Armour, Inc. from 2007 through 2016. (*Source:* Under Armour, Inc.)

DATA

Year	Revenue, R (in millions of dollars)
2007	606.6
2008	725.2
2009	856.4
2010	1063.9
2011	1472.7
2012	1834.9
2013	2332.1
2014	3084.4
2015	3963.3
2016	4825.3

Spreadsheet at LarsonPrecalculus.com

(a) Use a graphing utility to create a scatter plot of the data. Let t represent the year, with $t = 7$ corresponding to 2007.

(b) Use the *regression* feature of the graphing utility to find a linear model for the data and identify the coefficient of determination.

(c) Use the graphing utility to graph the linear model with the scatter plot from part (a).

(d) Use the *regression* feature of the graphing utility to find a quadratic model for the data and identify the coefficient of determination.

(e) Use the graphing utility to graph the quadratic model with the scatter plot from part (a).

(f) Which model is a better fit for the data? Explain.

(g) Use the better model to approximate the year when the revenue will reach approximately 8185 million dollars.

Focusing on Concepts

True or False? **In Exercises 24 and 25, determine whether the statement is true or false. Justify your answer.**

24. The graph of a quadratic model with a negative leading coefficient will have a maximum value at its vertex.

25. The graph of a quadratic model with a positive leading coefficient will have a minimum value at its vertex.

26. **Error Analysis** Describe the error in fitting a model to the data.

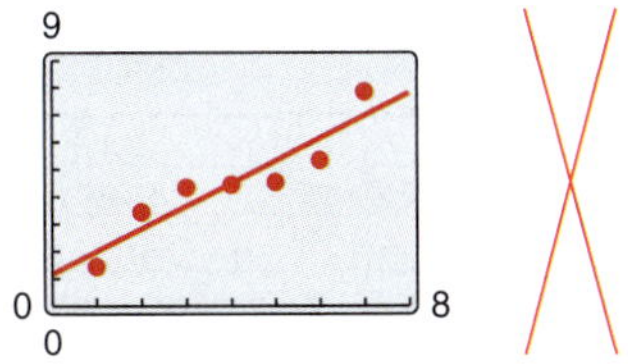

27. **Error Analysis** Describe the error in using a parabola to model the data.

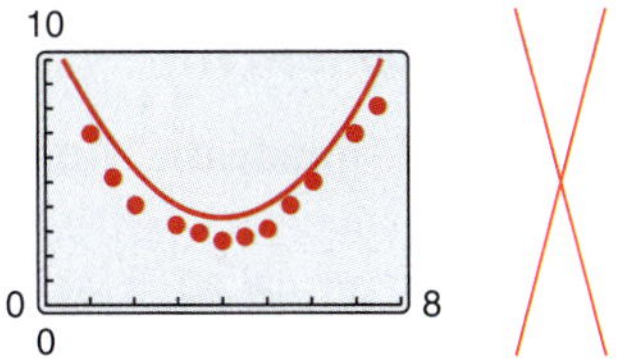

28. **HOW DO YOU SEE IT?** The r^2-values representing the coefficients of determination for the least squares linear model and the least squares quadratic model for the data shown are given below. Which is which? Explain your reasoning.

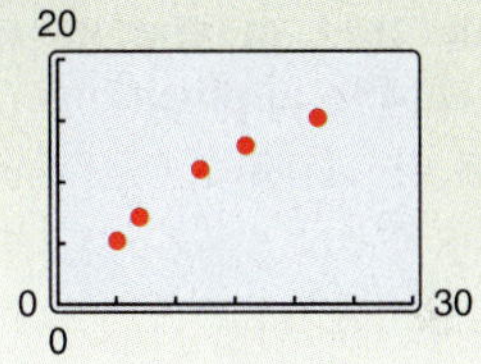

$r^2 \approx 0.9995$

$r^2 \approx 0.9782$

Cumulative Mixed Review

Compositions of Functions **In Exercises 29–32, find (a) $(f \circ g)(x)$, (b) $(g \circ f)(x)$, and (c) $(g \circ g)(x)$.**

29. $f(x) = 2x - 1,\ g(x) = x^2 + 3$

30. $f(x) = 5x + 8,\ g(x) = 2x^2 - 1$

31. $f(x) = x^3 - 1,\ g(x) = \sqrt[3]{x + 1}$

32. $f(x) = \sqrt[3]{x + 5},\ g(x) = x^3 - 5$

Finding an Inverse Function **In Exercises 33–36, determine whether the function has an inverse function. If it does, find the inverse function.**

33. $f(x) = 2x + 5$

34. $f(x) = \dfrac{x - 4}{5}$

35. $f(x) = x^2 + 5,\ x \geq 0$

36. $f(x) = 2x^2 - 3,\ x \geq 0$

Multiplying Complex Conjugates **In Exercises 37–40, write the complex conjugate of the complex number. Then multiply the number by its complex conjugate.**

37. $1 - 3i$

38. $-2 + 4i$

39. $-5i$

40. $8i$

2 Chapter Review

See CalcChat.com for tutorial help and worked-out solutions to odd-numbered exercises. For instructions on how to use a graphing utility, see Appendix A.

2.1 What did you learn?

Analyze graphs of quadratic functions *(p. 90)*. Let a, b, and c be real numbers with $a \neq 0$. The function $f(x) = ax^2 + bx + c$ is called a quadratic function. Its graph is a "U-shaped" curve called a parabola.

Write quadratic functions in standard form and use the results to sketch graphs of functions *(p. 93)*. The quadratic function $f(x) = a(x - h)^2 + k$, $a \neq 0$, is in standard form. The graph of f is a parabola whose axis is the vertical line $x = h$ and whose vertex is the point (h, k). The parabola opens upward when $a > 0$ and opens downward when $a < 0$.

Find minimum and maximum values of quadratic functions in real-life applications *(p. 95)*. Consider the function $f(x) = ax^2 + bx + c$ with vertex

$$\left(-\frac{b}{2a}, f\left(-\frac{b}{2a}\right)\right).$$

If $a > 0$, then f has a *minimum* at $x = -b/(2a)$. The minimum value is $f\left(-\frac{b}{2a}\right)$.

If $a < 0$, then f has a *maximum* at $x = -b/(2a)$. The maximum value is $f\left(-\frac{b}{2a}\right)$.

Example To sketch the graph of the function $g(x) = -(x + 2)^2 - 1$, consider the parent function $f(x) = x^2$.

With respect to the graph of $f(x) = x^2$, the graph of g is obtained by a *reflection* in the x-axis, a vertical shift one unit *downward*, and a horizontal shift two units *to the left*.

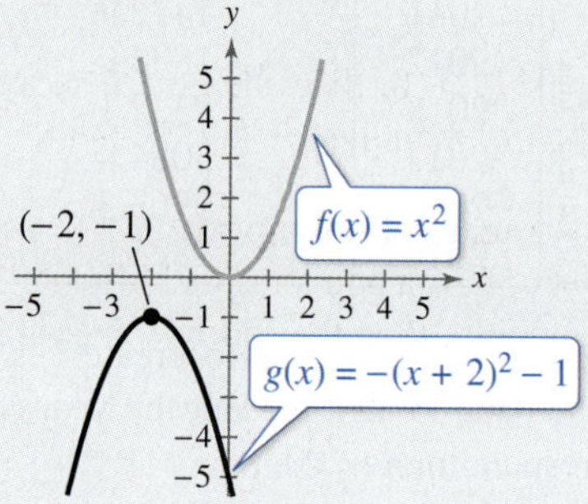

From the graph, the maximum of g occurs at $(-2, -1)$. To verify this algebraically, rewrite g in the form

$$g(x) = ax^2 + bx + c = -x^2 - 4x - 5.$$

The maximum value is

$$g\left(-\frac{b}{2a}\right) = g(-2) = -1.$$

Library of Parent Functions **In Exercises 1–6, sketch the graph of each quadratic function and compare it with the graph of $y = x^2$.**

1. $y = \frac{1}{4}x^2$
2. $y = -\frac{5}{3}x^2$
3. $y = x^2 + 1$
4. $y = -x^2 + 5$
5. $y = (x - 6)^2$
6. $y = -(x + 1)^2$

Using Standard Form to Graph a Parabola **In Exercises 7–10, write the quadratic function in standard form and sketch its graph. Identify the vertex, axis of symmetry, and x-intercept(s). Use a graphing utility to verify your results.**

7. $f(x) = x^2 + 4x - 8$
8. $f(x) = x^2 - 8x + 12$
9. $f(x) = -x^2 - 6x + 10$
10. $f(x) = 3x^2 - 12x + 11$

Writing the Equation of a Parabola **In Exercises 11 and 12, write the standard form of the quadratic function whose graph is a parabola with the indicated vertex and that passes through the indicated point.**

11. Vertex: $(-3, 4)$ Point: $(0, -5)$
12. Vertex: $(2, -1)$ Point: $(5, 2)$

13. **Geometry** A rectangle is inscribed in the region bounded by the x-axis, the y-axis, and the graph of $x + 2y - 8 = 0$.
    (a) Write the area A of the rectangle as a function of x. Determine the domain of the function in the context of the problem.
    (b) Use a graphing utility to graph the area function. Use the graph to approximate the dimensions that produce a maximum area.
    (c) Write the area function in standard form to find algebraically the dimensions that will produce a maximum area. Compare your results with your answer from part (b).

14. **Physical Education** A college has 1500 feet of boards to form three adjacent outdoor ice rinks (see figure). Determine the dimensions that will produce the maximum enclosed area of ice surface.

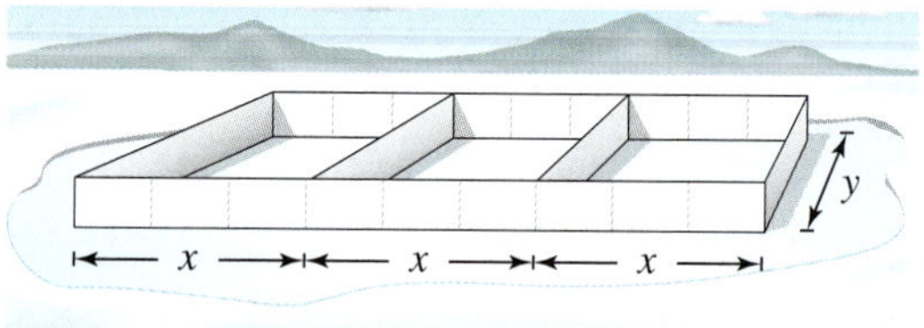

2.2 What did you learn?

Use transformations to sketch graphs of polynomial functions *(p. 100)*. The graph of a polynomial function is continuous (no breaks, holes, or gaps) and has only smooth, rounded turns.

Use the Leading Coefficient Test to determine the end behavior of graphs of polynomial functions *(p. 102)*. Consider $f(x) = a_n x^n + \cdots + a_1 x + a_0,\ a_n \neq 0$.

***n* is odd:** If $a_n > 0$, then the graph falls to the left and rises to the right. If $a_n < 0$, then the graph rises to the left and falls to the right.

***n* is even:** If $a_n > 0$, then the graph rises to the left and right. If $a_n < 0$, then the graph falls to the left and right.

Find and use zeros of polynomial functions as sketching aids *(p. 104)*. If f is a nonconstant polynomial function and a is a real number, then the following are equivalent.

1. $x = a$ is a *zero* of the function f.
2. $x = a$ is a *solution* of the equation $f(x) = 0$.
3. $(x - a)$ is a *factor* of the polynomial $f(x)$.
4. $(a, 0)$ is an x-intercept of the graph of f.

Use the Intermediate Value Theorem to help locate zeros of polynomial functions *(p. 108)*. Let a and b be real numbers such that $a < b$. If f is a polynomial function such that $f(a) \neq f(b)$, then in $[a, b]$, f takes on every value between $f(a)$ and $f(b)$.

Example To sketch the graph of

$$f(x) = -x^3 + 2x^2 - x$$

first determine the end behavior of the function. Because the leading coefficient is negative and the degree is odd, the graph rises to the left and falls to the right.

Factor the function to find the x-intercepts.

$$f(x) = -x^3 + 2x^2 - x = -x(x^2 - 2x + 1) = -x(x - 1)^2$$

The real zeros are $x = 0$ (of odd multiplicity 1) and $x = 1$ (of even multiplicity 2). So, the x-intercepts occur at

$(0, 0)$ and $(1, 0)$.

To complete the sketch, find additional points, such as $(-0.5, 1.125)$, $(0.5, -0.125)$, $(1.5, -0.375)$, and $(2, -2)$.

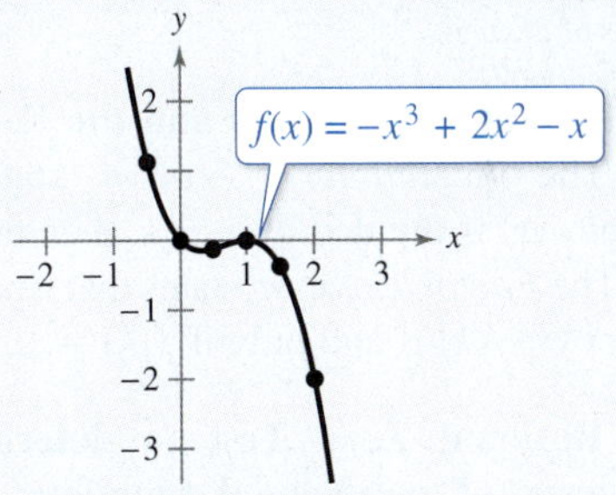

Library of Parent Functions **In Exercises 15–20, sketch the graph of $f(x) = x^3$ and the graph of the function g. Describe the transformation from f to g.**

15. $g(x) = (x - 5)^3$ **16.** $g(x) = x^3 - 2$

17. $g(x) = -(x + 4)^3$ **18.** $g(x) = (x + 4)^3 - 1$

19. $g(x) = -(x - 2)^3 + 6$ **20.** $g(x) = -(x - 1)^3 - 5$

Applying the Leading Coefficient Test **In Exercises 21–24, describe the left-hand and right-hand behavior of the graph of the polynomial function. Use a graphing utility to verify your results.**

21. $f(x) = x^2 - x - 6$ **22.** $g(x) = -3x^3 + 4x^2 - 1$

23. $h(x) = -\frac{2}{3}(x^4 - 7x^3 + 5)$

24. $f(x) = -x^2 + 4x^5 + 5 - 2x^4$

Finding Zeros of a Polynomial Function **In Exercises 25–30, (a) find the real zeros algebraically, (b) use a graphing utility to graph the function, and (c) use the graph to approximate any real zeros and compare them with those from part (a).**

25. $g(x) = x^4 - x^3 - 2x^2$ **26.** $h(x) = -2x^3 - x^2 + x$

27. $f(t) = t^3 - 3t$ **28.** $f(x) = -(x + 6)^3 - 8$

29. $f(x) = x(x + 3)^2$ **30.** $f(t) = t^4 - 4t^2$

Finding a Polynomial Function **In Exercises 31 and 32, find a polynomial function that has the indicated zeros. (There are many correct answers.)**

31. $4, -2, -2$ **32.** $-1, 0, 3, 5$

Sketching the Graph of a Polynomial Function **In Exercises 33 and 34, sketch the graph of the function by (a) applying the Leading Coefficient Test, (b) finding the real zeros of the polynomial, (c) plotting sufficient solution points, and (d) drawing a continuous curve through the points.**

33. $f(x) = x^4 - 2x^3 - 12x^2 + 18x + 27$

34. $f(x) = 18 + 27x - 2x^2 - 3x^3$

Approximating the Zeros of a Function **In Exercises 35–38, (a) use the Intermediate Value Theorem and the *table* feature of a graphing utility to find intervals one unit in length in which the polynomial function is guaranteed to have a zero. (b) Adjust the table to approximate the zeros of the function to the nearest thousandth.**

35. $f(x) = x^3 + 2x^2 - x - 1$ **36.** $f(x) = x^4 - 6x^2 - 4$

37. $f(x) = 0.24x^3 - 2.6x - 1.4$

38. $f(x) = 2x^4 + \frac{7}{2}x^3 - 2$

2.3 *What did you learn?*

Use long division to divide polynomials by other polynomials *(p. 113)*. According to the Division Algorithm, if $f(x)$ and $d(x)$ are polynomials such that $d(x) \neq 0$, and the degree of $d(x)$ is less than or equal to the degree of $f(x)$, then there exist unique polynomials $q(x)$ and $r(x)$ such that $f(x) = d(x)q(x) + r(x)$, where $r(x) = 0$ *or* the degree of $r(x)$ is less than the degree of $d(x)$. If the remainder $r(x)$ is zero, then $d(x)$ *divides evenly* into $f(x)$.

Use synthetic division to divide polynomials by binomials of the form $x - k$ *(p. 116)*.

$(ax^3 + bx^2 + cx + d) \div (x - k)$

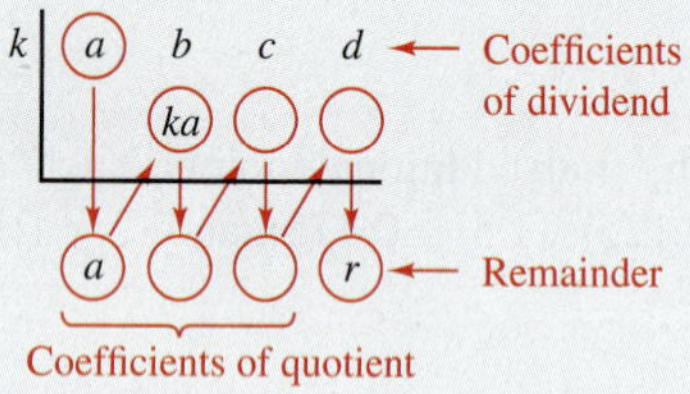

Vertical pattern: Add terms.

Diagonal pattern: Multiply by k.

Use the Remainder Theorem and the Factor Theorem *(p. 117)*. The *Remainder Theorem* states that if a polynomial $f(x)$ is divided by $x - k$, then the remainder is $r = f(k)$. The *Factor Theorem* states that a polynomial $f(x)$ has a factor $(x - k)$ if and only if $f(k) = 0$.

Use the Rational Zero Test to determine possible rational zeros of polynomial functions *(p. 119)*. The Rational Zero Test relates the possible rational zeros of a polynomial to the leading coefficient and to the constant term of the polynomial.

Use Descartes's Rule of Signs *(p. 121)* and the Upper and Lower Bound Rules *(p. 122)* to find real zeros of polynomials. Example 9 shows how to use Descartes's Rule of Signs. Example 10 uses Descartes's Rule of Signs and the Upper and Lower Bound Rules.

Example To find the real zeros of the function $f(x) = 5x^3 - 8x^2 - 10x + 3$, use the Rational Zero Test.

Possible rational zeros:

$$\frac{\text{Factors of } 3}{\text{Factors of } 5} = \frac{\pm 1, \pm 3}{\pm 1, \pm 5} = \pm 1, \pm 3, \pm\frac{1}{5}, \pm\frac{3}{5}$$

Show that $x = -1$ is a rational zero using synthetic division or long division.

$$\begin{array}{r|rrrr} -1 & 5 & -8 & -10 & 3 \\ & & -5 & 13 & -3 \\ \hline & 5 & -13 & 3 & 0 \end{array}$$

or

$$\begin{array}{r} 5x^2 - 13x + 3 \\ x + 1 \overline{)\, 5x^3 - 8x^2 - 10x + 3} \\ \underline{5x^3 + 5x^2 \qquad\qquad} \\ -13x^2 - 10x \quad \\ \underline{-13x^2 - 13x \quad} \\ 3x + 3 \\ \underline{3x + 3} \\ 0 \end{array}$$

So, $f(x)$ factors as $f(x) = (x + 1)(5x^2 - 13x + 3)$. Because the expression $5x^2 - 13x + 3$ does not factor, use the Quadratic Formula to find the remaining real zeros.

$$x = \frac{13 + \sqrt{109}}{10} \approx 2.34$$

$$x = \frac{13 - \sqrt{109}}{10} \approx 0.26$$

Use a graph to verify the intercepts.

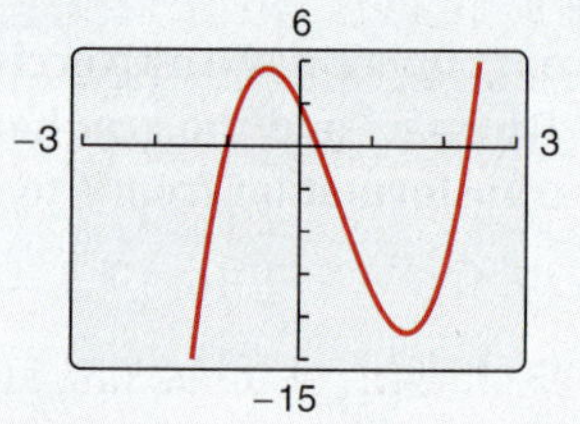

Long Division of Polynomials **In Exercises 39–44, use long division to divide.**

39. $\dfrac{24x^2 - x - 8}{3x - 2}$ **40.** $\dfrac{4x^2 + 7}{3x - 2}$

41. $\dfrac{x^4 - 3x^2 + 2}{x^2 - 1}$ **42.** $\dfrac{3x^4 + x^2 - 1}{x^2 - 1}$

43. $(5x^3 - 13x^2 - x + 2) \div (x^2 - 3x + 1)$

44. $(x^4 + x^3 - x^2 + 2x) \div (x^2 + 2x)$

Using Synthetic Division **In Exercises 45–48, use synthetic division to divide.**

45. $(3x^3 - 10x^2 + 12x - 22) \div (x - 4)$

46. $(2x^3 + 6x^2 - 14x + 9) \div (x - 1)$

47. $(6x^4 - 4x^3 - 27x^2 + 18x) \div \left(x - \frac{2}{3}\right)$

48. $(2x^3 + 2x^2 - x + 2) \div \left(x - \frac{1}{2}\right)$

Using the Remainder Theorem **In Exercises 49 and 50, use the Remainder Theorem and synthetic division to find the function value.**

49. $f(x) = x^4 + 10x^3 - 24x^2 + 20x + 44, \quad f(-3)$

50. $g(t) = 2t^5 - 5t^4 - 8t + 20, \quad g\left(\sqrt{2}\right)$

Factoring a Polynomial **In Exercises 51 and 52, (a) verify the indicated factor(s) of $f(x)$, (b) find the remaining factor(s) of $f(x)$, (c) use your results to write the complete factorization of $f(x)$, (d) list all real zeros of f, and (e) confirm your results by using a graphing utility to graph the function.**

	Function	*Factor(s)*
51.	$f(x) = 2x^3 + 11x^2 - 21x - 90$	$(x + 6)$
52.	$f(x) = x^4 - 4x^3 - 7x^2 + 22x + 24$	$(x + 2), (x - 3)$

Using the Rational Zero Test **In Exercises 53–56, find (if possible) the rational zeros of the function.**

53. $f(x) = x^4 + 5x^3 + 6x^2$ **54.** $f(x) = 2x^5 - 4x + 5$

55. $f(x) = 4x^3 - 11x^2 + 10x - 3$

56. $f(x) = 10x^3 + 21x^2 - x - 6$

Using Descartes's Rule of Signs **In Exercises 57–60, use Descartes's Rule of Signs to determine the possible numbers of positive and negative real zeros of the function.**

57. $g(x) = 5x^3 - 6x + 9$

58. $f(x) = 2x^5 - 3x^2 + 2x - 1$

59. $f(x) = 6x^3 + 9x + 4$

60. $g(x) = -8x^6 + 6x^4 + 4x^3 - 9x + 3$

Finding the Zeros of a Polynomial Function **In Exercises 61 and 62, use synthetic division to verify the upper and lower bounds of the real zeros of *f*. Then find all real zeros of the function.**

61. $f(x) = 4x^3 - 3x^2 + 4x - 3$

Upper bound: $x = 1$; Lower bound: $x = -\frac{1}{4}$

62. $f(x) = 2x^3 - 5x^2 - 14x + 8$

Upper bound: $x = 8$; Lower bound: $x = -4$

Finding the Real Zeros of a Polynomial Function **In Exercises 63 and 64, find all real zeros of the polynomial function.**

63. $f(x) = x^3 - 4x^2 + x + 6$

64. $f(x) = 5x^4 + 126x^2 + 25$

2.4 *What did you learn?*

Use the imaginary unit *i* to write complex numbers *(p. 128)*. The imaginary unit i is defined as $i = \sqrt{-1}$, where $i^2 = -1$. If a and b are real numbers, then the number $a + bi$ is a complex number written in standard form.

Add, subtract, and multiply complex numbers *(p. 129)*.

Sum: $(a + bi) + (c + di) = (a + c) + (b + d)i$

Difference: $(a + bi) - (c + di) = (a - c) + (b - d)i$

Product: $(a + bi)(c + di) = (ac - bd) + (ad + bc)i$

Use complex conjugates to write the quotient of two complex numbers in standard form *(p. 131)*. Complex numbers of the forms $a + bi$ and $a - bi$ are complex conjugates. To write $(a + bi)/(c + di)$ in standard form, multiply by $(c - di)/(c - di)$.

Find complex solutions of quadratic equations *(p. 132)*. If a is a positive real number, then the principal square root of the negative number $-a$ is defined as $\sqrt{-a} = \sqrt{a}i$.

Example Consider $(3 + 5i)$ and $(-1 + 2i)$.

To add (or subtract) the complex numbers, add (or subtract) the real and imaginary parts of the numbers separately.

$$\begin{aligned}(3 + 5i) + (-1 + 2i) &= 3 + 5i + (-1) + 2i \\ &= [3 + (-1)] + (5 + 2)i \\ &= 2 + 7i\end{aligned}$$

To multiply the complex numbers, use properties to distribute and then combine like terms.

$$\begin{aligned}(3 + 5i)(-1 + 2i) &= 3(-1 + 2i) + 5i(-1 + 2i) \\ &= -3 + 6i - 5i + 10i^2 \\ &= -13 + i\end{aligned}$$

To write the quotient of the complex numbers in standard form, multiply the numerator and denominator by the complex conjugate of the denominator.

$$\begin{aligned}\frac{3 + 5i}{-1 + 2i} &= \frac{3 + 5i}{-1 + 2i}\left(\frac{-1 - 2i}{-1 - 2i}\right) = \frac{-3 - 6i - 5i - 10i^2}{1 - 4i^2} \\ &= \frac{-3 - 11i + 10}{1 + 4} = \frac{7 - 11i}{5} = \frac{7}{5} - \frac{11}{5}i\end{aligned}$$

Writing a Complex Number in Standard Form **In Exercises 65–68, write the complex number in standard form.**

65. $3 - \sqrt{-16}$ **66.** $\sqrt{-50} + 8$

67. $-i + 4i^2$ **68.** $7i - 9i^2$

Performing Operations with Complex Numbers **In Exercises 69–76, perform the operation(s) and write the result in standard form.**

69. $(2 + 13i) + (6 - 5i)$ **70.** $(3 - 2i) + 4i$

71. $2 - (3 + 2i)$ **72.** $(7 + 10i) - (1 + 9i)$

73. $(10 - 8i)(2 - 3i)$ **74.** $(1 + 6i)(5 - 2i)$

75. $(3 + 7i)^2 + (3 - 7i)^2$ **76.** $(4 - i)^2 - (4 + i)^2$

Writing a Quotient in Standard Form **In Exercises 77–80, write the quotient in standard form.**

77. $\dfrac{6 + i}{i}$ **78.** $\dfrac{4}{-3i}$

79. $\dfrac{3 + 2i}{5 + i}$ **80.** $\dfrac{1 - 7i}{2 + 3i}$

Complex Solutions of a Quadratic Equation **In Exercises 81–86, use the Quadratic formula to solve the quadratic equation.**

81. $x^2 + 16 = 0$ **82.** $x^2 + 48 = 0$

83. $x^2 + 3x + 6 = 0$ **84.** $x^2 + 4x + 10 = 0$

85. $3x^2 - 5x + 6 = 0$ **86.** $5x^2 - 2x + 4 = 0$

2.5 *What did you learn?*

Use the Fundamental Theorem of Algebra to determine the number of zeros of a polynomial function *(p. 135)*.

The Fundamental Theorem of Algebra: If $f(x)$ is a polynomial of degree n, where $n > 0$, then f has at least one zero in the complex number system.

Example The complex number $2 + i$ is a complex zero of the function $f(x) = x^4 - 4x^3 + x^2 + 16x - 20$. To find the other zeros of f, first recognize that complex zeros occur in conjugate pairs. So, $2 - i$ is also a zero of f.

Find all zeros of polynomial functions *(p. 136)* and find conjugate pairs of complex zeros *(p. 137)*. Let f be a polynomial function that has real coefficients. If $a + bi$ $(b \neq 0)$ is a zero of the function, then the conjugate $a - bi$ is also a zero of the function.

Two factors of f are $x - (2 + i)$ and $x + (2 - i)$. Multiplying these two factors produces

$$[x - (2 + i)][x - (2 - i)] = [(x - 2) - i][(x - 2) + i]$$
$$= (x - 2)^2 - i^2$$
$$= x^2 - 4x + 5.$$

Find zeros of polynomials by factoring *(p. 138)*. Every polynomial of degree $n > 0$ with real coefficients can be written as the product of linear and quadratic factors with real coefficients, where the quadratic factors have no real zeros.

Using long division, divide $x^2 - 4x + 5$ into f to obtain

$f(x) = (x^2 - 4x + 5)(x^2 - 4)$, or

$f(x) = (x^2 - 4x + 5)(x + 2)(x - 2)$.

So, the zeros of f are $x = 1 + 2i$, $x = 1 - 2i$, $x = -2$, and $x = 2$.

Complex Zeros of a Polynomial Function **In Exercises 87–90, confirm that the function has the indicated zeros.**

87. $f(x) = x^2 + 12$; $-2\sqrt{3}i, 2\sqrt{3}i$

88. $f(x) = x^4 - 81$; $-3, 3, -3i, 3i$

89. $f(x) = x^3 + 16x$; $0, -4i, 4i$

90. $f(x) = x^3 + 144x$; $0, -12i, 12i$

Finding the Zeros of a Polynomial Function **In Exercises 91–96, find all the zeros of the function and write the polynomial as a product of linear factors. Use a graphing utility to verify your results.**

91. $h(x) = x^3 - 7x^2 + 18x - 24$

92. $f(x) = 2x^3 - 5x^2 - 9x + 40$

93. $f(x) = 2x^4 - 5x^3 + 10x - 12$

94. $g(x) = 3x^4 - 4x^3 + 7x^2 + 10x - 4$

95. $f(x) = x^5 + x^4 + 5x^3 + 5x^2$

96. $f(x) = x^5 - 5x^3 + 4x$

Using the Zeros to Find the *x*-Intercepts **In Exercises 97–100, (a) find all the zeros of the function, (b) write the polynomial as a product of linear factors, and (c) use your factorization to determine the *x*-intercepts of the graph of the function. Use a graphing utility to verify that the real zeros are the only *x*-intercepts.**

97. $f(x) = x^3 - 5x^2 - 7x + 51$

98. $f(x) = -3x^3 - 19x^2 - 4x + 12$

99. $f(x) = x^4 + 34x^2 + 225$

100. $f(x) = x^4 + 10x^3 + 26x^2 + 10x + 25$

Finding a Polynomial Function **In Exercises 101–104, find a polynomial function with real coefficients that has the indicated zeros. (There are many correct answers.)**

101. $5, 3i$

102. $-6, -i$

103. $-2, -2 - 4i$

104. $1, -5 + \sqrt{2}i$

Factoring a Polynomial **In Exercises 105 and 106, write the polynomial (a) as the product of factors that are irreducible over the *rationals*, (b) as the product of linear and quadratic factors that are irreducible over the *reals*, and (c) in completely factored form.**

105. $f(x) = x^4 - 2x^3 + 8x^2 - 18x - 9$

(*Hint:* One factor is $x^2 + 9$.)

106. $f(x) = x^4 - 4x^3 + 3x^2 + 8x - 16$

(*Hint:* One factor is $x^2 - x - 4$.)

Finding the Zeros of a Polynomial Function **In Exercises 107 and 108, use the specified zero to find all the zeros of the function.**

	Function	*Zero*
107.	$f(x) = x^3 + 3x^2 + 4x + 12$	$-2i$
108.	$f(x) = 2x^3 - 7x^2 + 14x + 9$	$2 + \sqrt{5}i$

Using a Graph to Locate the Real Zeros **In Exercises 109 and 110, (a) use a graphing utility to find the real zeros of the function, and then (b) use the real zeros to find the exact values of the imaginary zeros.**

109. $f(x) = 8x^3 - 12x^2 + 2x - 3$

110. $f(x) = 2x^4 - x^3 + x^2 - x - 1$

2.6 *What did you learn?*

Find the domains of rational functions *(p. 142)*. The domain of a rational function of x includes all real numbers except x-values that make the denominator zero.

Find the vertical and horizontal asymptotes of graphs of rational functions *(p. 143)*.

The line $x = a$ is a *vertical asymptote* of the graph of f when $f(x) \to \infty$ or $f(x) \to -\infty$ as $x \to a$, either from the left or from the right.

The line $y = b$ is a *horizontal asymptote* of the graph of f when $f(x) \to b$ as $x \to \infty$ or $x \to -\infty$.

Use rational functions to model and solve real-life problems *(p. 146)*. A rational function can be used to model the cost of removing a given percent of smokestack pollutants at a utility company that burns coal. (See Example 5.)

Example Consider the function $f(x) = \dfrac{x^2 - x - 12}{3x^2 + 10x + 3}$.

The degree of the numerator is equal to the degree of the denominator, the leading coefficient of the numerator is 1, and the leading coefficient of the denominator is 3. So the graph has the line $y = \frac{1}{3}$ as a horizontal asymptote.

To find any vertical asymptotes, first factor the numerator and denominator.

$$f(x) = \frac{x^2 - x - 12}{3x^2 + 10x + 3} = \frac{\cancel{(x+3)}(x-4)}{(3x+1)\cancel{(x+3)}} = \frac{x-4}{3x+1}, \quad x \neq -3$$

Setting the denominator $3x + 1$ (of the simplified function) equal to zero indicates that the graph has the line $x = -\frac{1}{3}$ as a vertical asymptote. Because $x = -3$ is *not* a vertical asymptote, there is a hole in the graph at $x = -3$.

Finding a Function's Domain and Asymptotes In Exercises 111–122, (a) find the domain of f, (b) decide whether f is continuous, and (c) identify any horizontal and vertical asymptotes.

111. $f(x) = \dfrac{2 - x}{x + 3}$ **112.** $f(x) = \dfrac{4x}{x - 8}$

113. $f(x) = \dfrac{2}{x^2 - 3x - 18}$ **114.** $f(x) = -\dfrac{7}{x^2 + 5x - 6}$

115. $f(x) = \dfrac{7 + x}{7 - x}$ **116.** $f(x) = \dfrac{2x - 9}{2x + 9}$

117. $f(x) = \dfrac{4x^2}{2x^2 - 3}$

118. $f(x) = \dfrac{3x^2 - 11x - 4}{x^2 + 2}$

119. $f(x) = \dfrac{2x - 10}{x^2 - 2x - 15}$

120. $f(x) = \dfrac{4 - x}{x^3 + 6x^2}$

121. $f(x) = \dfrac{3x^2 - 15}{x^3 - 5x^2 - 24x}$

122. $f(x) = \dfrac{x^2 + 3x + 2}{x^3 - 4x^2}$

Exploration In Exercises 123 and 124, determine the value that the function f approaches as the magnitude of x increases. Is $f(x)$ greater than or less than this value when x is positive and large in magnitude? What about when x is negative and large in magnitude?

123. $f(x) = -2 + \dfrac{3}{x}$ **124.** $f(x) = -\dfrac{6x + 1}{x^2 + 2}$

125. Criminology The cost C (in millions of dollars) for the U.S. government to seize $p\%$ of an illegal drug as it enters the country can be modeled by

$$C = \frac{528p}{100 - p}, \quad 0 \le p < 100.$$

(a) Find the costs of seizing 25%, 50%, and 75% of the illegal drug.

(b) Use a graphing utility to graph the function. Be sure to choose an appropriate viewing window. Explain why you chose the values you used in your viewing window.

(c) According to this model, would it be possible to seize 100% of the drug? Explain.

126. Biology Students in a biology class perform an experiment comparing the quantity of food consumed by a moth with the quantity supplied. The experimental data can be modeled by

$$y = \frac{1.568x - 0.001}{6.360x + 1}, \quad x > 0$$

where x is the quantity (in milligrams) of food supplied and y is the quantity (in milligrams) eaten (see figure). At what level of consumption will the moth become satiated?

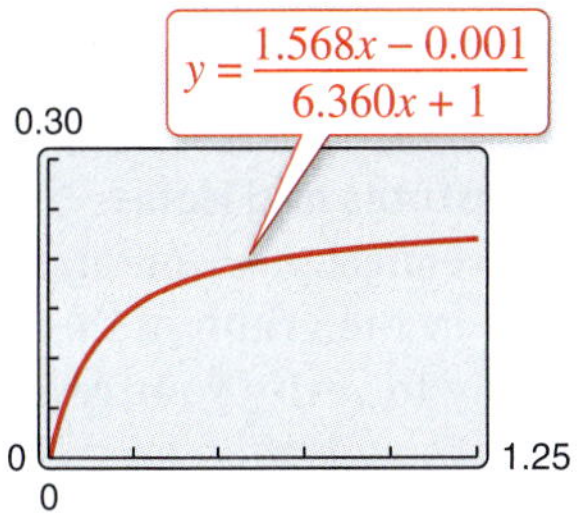

2.7 What did you learn?

Analyze and sketch graphs of rational functions *(p. 151)*. Guidelines for graphing rational functions include simplifying the function algebraically, finding and plotting the intercepts, finding the zeros of the denominator to graph the corresponding vertical asymptotes and holes in the graph, finding and graphing any other asymptotes, and plotting at least one point *between* and one point *beyond* each x-intercept and vertical asymptote. To complete the graph, use smooth curves between and beyond the vertical asymptotes, excluding any points where f is not defined.

Sketch graphs of rational functions that have slant asymptotes *(p. 155)*. Consider a rational function whose denominator is of degree 1 or greater. If the degree of the numerator is exactly *one more* than the degree of the denominator, then the graph of the function has a slant asymptote. Use long division to find the equation of the slant asymptote.

Use rational functions to model and solve real-life problems *(p. 156)*. A rational function can be used to model the area of a page. The model can be used to determine the dimensions of the page that use the minimum amount of paper. (See Example 7.)

Example To sketch the graph of $f(x) = \dfrac{3 - x^2}{x - 1}$, begin by using long division to find the equation of the slant asymptote.

$$f(x) = \frac{3 - x^2}{x - 1} = (-x - 1) + \frac{2}{x - 1}.$$

The slant asymptote is $y = -x - 1$.

y-intercept: $(0, -3)$

x-intercepts: $(\pm\sqrt{3}, 0)$

Vertical asymptote: $x = 1$

Additional points: $(-3, 1.5)$, $(-1, -1)$, $(1.5, 1.5)$, $(2, -1)$, and $(3, -3)$

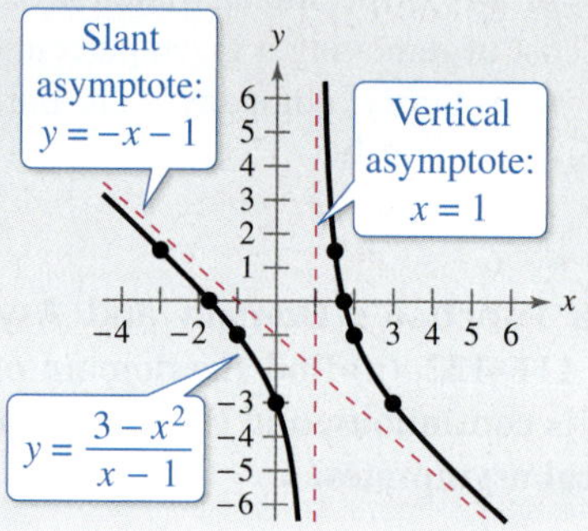

Library of Parent Functions **In Exercises 127 and 128, sketch the graph of the function g by hand and compare it with the graph of $f(x) = 1/x$.**

127. $g(x) = \dfrac{1}{x - 4} + 1$

128. $g(x) = \dfrac{-1}{x + 2} - 2$

Sketching the Graph of a Rational Function **In Exercises 129–134, sketch the graph of the rational function by hand. As sketching aids, check for intercepts, vertical asymptotes, horizontal asymptotes, slant asymptotes, and holes.**

129. $f(x) = \dfrac{2x - 1}{x - 5}$

130. $f(x) = \dfrac{x - 3}{x - 2}$

131. $f(x) = \dfrac{2}{(x + 1)^2}$

132. $f(x) = \dfrac{2x^2}{x^2 - 4}$

133. $f(x) = \dfrac{x^2 - x + 1}{x - 3}$

134. $f(x) = \dfrac{2x^2 + 7x + 3}{x + 1}$

Finding Asymptotes and Holes **In Exercises 135–138, find all of the vertical, horizontal, and slant asymptotes, and any holes in the graph of the function. Then use a graphing utility to verify your results.**

135. $f(x) = \dfrac{x^2 - 5x + 4}{x^2 - 1}$

136. $f(x) = \dfrac{2x^2 - 7x + 3}{2x^2 - 3x - 9}$

137. $f(x) = \dfrac{3x^2 + 5x - 2}{x + 1}$

138. $f(x) = \dfrac{2x^2 + 5x + 3}{x - 2}$

139. **Biology** A parks and wildlife commission releases 80,000 fish into a lake. After t years, the population N of the fish (in thousands) can be modeled by

$$N = \frac{20(4 + 3t)}{1 + 0.05t}, \quad t \geq 0.$$

(a) Use a graphing utility to graph the function and find the populations when $t = 5$, $t = 10$, and $t = 25$.

(b) What is the maximum number of fish in the lake as time passes? Explain your reasoning.

140. **Publishing** A page that is x inches wide and y inches high contains 30 square inches of print. The top and bottom margins are 2 inches deep and the margins on each side are 2 inches wide.

(a) Draw a diagram that illustrates the problem.

(b) Show that the total area A of the page is given by

$$A = \frac{2x(2x + 7)}{x - 4}.$$

(c) Determine the domain of the function based on the physical constraints of the problem.

(d) Use a graphing utility to graph the area function and approximate the page size such that the minimum amount of paper will be used.

2.8 *What did you learn?*

Classify scatter plots ***(p. 161).*** A scatter plot can be used to determine which type of model will best fit a set of data. The scatter plot below represents a linear pattern.

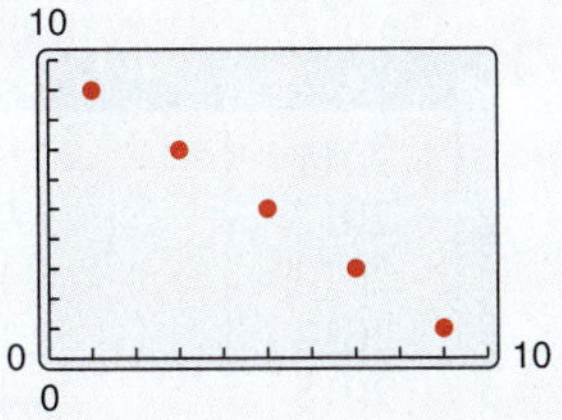

Find quadratic models for data ***(p. 162).*** To fit a quadratic model to data, use a graphing utility to create a scatter plot of the data. Then use the *regression* feature of the graphing utility to find a model of the form $y = ax^2 + bx + c$.

Choose a model that best fits a set of data ***(p. 164).*** To fit a model to data, use a graphing utility to create a scatter plot of the data. Use the scatter plot to determine whether the data fit a pattern. Use the graphing utility to fit the appropriate model to the data.

Example Consider the data set represented by the ordered pairs shown below.

(1, 1), (2, 1.5), (3, 3.1),
(4, 5.6), (5, 8), (6, 9.9)

Use a graphing utility to create a scatter plot to determine whether the data can be best modeled by a linear model, a quadratic model, or neither. The data appear to follow a parabolic pattern, so use the graphing utility to find a quadratic model for the data. The quadratic equation that best fits the data is $y = 0.1821x^2 + 0.625x - 0.1$.

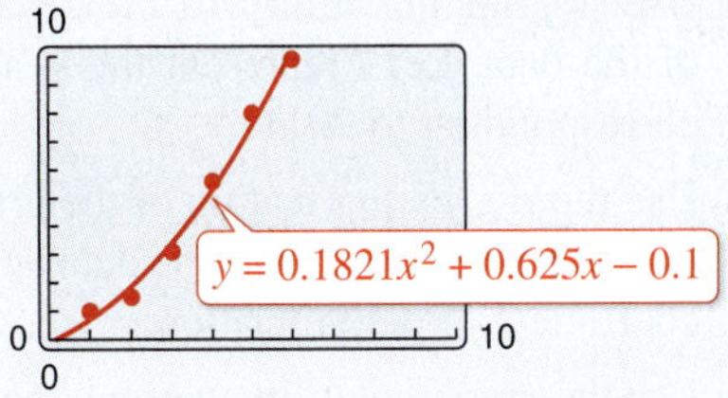

The graph shows the model with the scatter plot. The quadratic model appears to fit the data well.

Classifying Scatter Plots **In Exercises 141–144, determine whether the scatter plot could best be modeled by a linear model, a quadratic model, or neither.**

141.

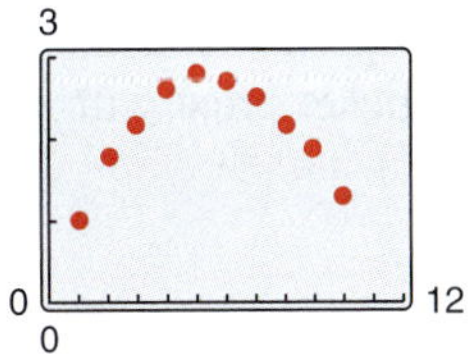

142.

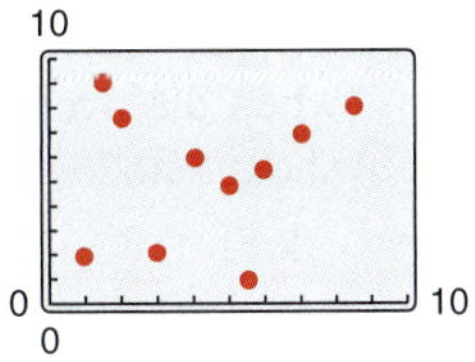

143.

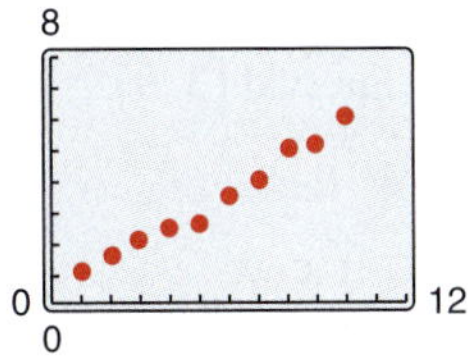

144.

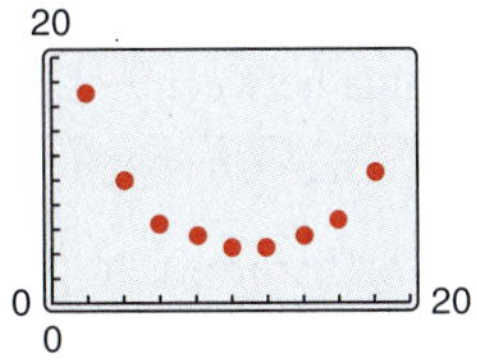

Interpreting Coefficients of Determination **In Exercises 145 and 146, a linear model and a quadratic model were fit to the same data set. Compare the coefficients of determination to determine which model better fits the data.**

145. linear model: $r^2 \approx 0.9247$
quadratic model: $r^2 \approx 0.9876$

146. linear model: $r^2 = 1$
quadratic model: $r^2 \approx 0.9863$

Fitting Linear and Quadratic Models to Data **In Exercises 147–150, use the** ***regression*** **feature of a graphing utility to find a quadratic model for the data. Is the model a good fit for the data? Explain.**

147.

x	y
12	103.32
14	90.25
17	80
21	68.23
23	59.22
27	54.22
30	51.42

148.

x	y
3	34.82
5	49.12
7	59.43
8	64.55
9	82.11
11	104.54
13	134.99

149.

x	y
0.9	7.12
1.2	4.31
1.3	4.41
1.4	4.56
1.8	3.24
1.9	3.01
2.2	8.43

150.

x	y
8.4	1
8.5	4.33
8.6	5.87
9.1	7.55
9.3	7.69
9.4	7.71
9.6	7.85

151. Modeling Data The table shows the Medicare spending M (in billions of dollars) each year from 2011 to 2016. *(Source:* Statista)

DATA

Year	Medicare Spending, M
2011	560.3
2012	550.1
2013	581.7
2014	600.3
2015	638.1
2016	694.6

Spreadsheet at LarsonPrecalculus.com

(a) Use a graphing utility to create a scatter plot of the data. Let t represent the year, with $t = 1$ corresponding to 2011.

(b) Use the *regression* feature of the graphing utility to find a linear model for the data and identify the coefficient of determination.

(c) Use the graphing utility to graph the linear model with the scatter plot from part (a).

(d) Use the *regression* feature of the graphing utility to find a quadratic model for the data and identify the coefficient of determination.

(e) Use the graphing utility to graph the quadratic model with the scatter plot from part (a).

(f) Which model is a better fit for the data? Explain.

(g) Use the better model to approximate the year when the total Medicare spending is about \$830 million.

152. Modeling Data The table shows the average movie ticket price P (in U.S. dollars) in the United States and Canada each year from 2011 to 2017. *(Source:* Statista)

DATA

Year	Average Price, P (in U.S. dollars)
2011	7.93
2012	7.96
2013	8.13
2014	8.17
2015	8.43
2016	8.65
2017	8.93

Spreadsheet at LarsonPrecalculus.com

(a) Use a graphing utility to create a scatter plot of the data. Let t represent the year, with $t = 1$ corresponding to 2011.

(b) Use the *regression* feature of the graphing utility to find a linear model for the data and identify the coefficient of determination.

(c) Use the graphing utility to graph the linear model with the scatter plot from part (a).

(d) Use the *regression* feature of the graphing utility to find a quadratic model for the data and identify the coefficient of determination.

(e) Use the graphing utility to graph the quadratic model with the scatter plot from part (a).

(f) Which model is a better fit for the data? Explain.

(g) Use the better model to approximate the year when the average movie ticket price will reach about \$10.25.

Focusing on Concepts

True or False? **In Exercises 153–160, determine whether the statement is true or false. Justify your answer.**

153. The graph of $f(x) = ax^2$ has a maximum value when $a > 0$.

154. The graph of $f(x) = -x^n$ falls to the left and falls to the right for even values of n.

155. $(x^2 + 4x - 8) \div (x + 4)$ has a remainder of 0.

156. The sum of two complex numbers cannot be a real number.

157. A fourth-degree polynomial with real coefficients can have -5, $-8i$, $4i$, and 5 as its zeros.

158. The graph of $f(x) = (x^2 + 2ax + a^2)/(x + a)$ has no vertical asymptote.

159. The graph of $f(x) = \dfrac{2x^3}{x + 1}$ has a slant asymptote.

160. The coefficient of determination of a quadratic model of a data set can be equal to the coefficient of determination of a linear model of the same data set.

Error Analysis **In Exercises 161–164, describe the error.**

161. The function $f(x) = 2x^2 + 1$ takes on every value between -1 and 2 on the interval $[3, 9]$.

162. $$\sqrt{-8}\sqrt{-8} = \sqrt{(-8)(-8)} = \sqrt{64} = 8$$

163. $$-i(\sqrt{-4} - 1) = -i(4i - 1) = -4i^2 + i = 4 + i$$

164. The graph of the function $f(x) = x^2/(x + 2)$ has a horizontal asymptote at $y = 0$.

165. Think About It What is the value of i^n for each value of n?

(a) $n = 105$ (b) $n = 128$

(c) $n = 214$ (d) $n = 235$

166. Think About It Describe the domain restrictions of a rational function when the denominator divides evenly into the numerator.

2 Chapter Test

See *CalcChat.com* for tutorial help and worked-out solutions to odd-numbered exercises. For instructions on how to use a graphing utility, see Appendix A.

Take this test as you would take a test in class. After you are finished, check your work against the answers given in the back of the book.

1. Identify the vertex and x-intercepts of the graph of $y = x^2 + 5x + 6$.
2. Write an equation in standard form of the parabola shown at the right.
3. Find all the real zeros of $f(x) = 4x^3 - 12x^2 + 9x$. Determine the multiplicity of each zero. Use a graphing utility to verify your results.
4. Sketch the graph of the function $f(x) = -x^3 + 7x + 6$.
5. Divide using long division: $(2x^3 + 5x - 3) \div (x^2 + 2)$.
6. Divide using synthetic division: $(x^4 + 4x^3 + 16x - 35) \div (x + 5)$.
7. Use the Remainder Theorem and synthetic division to evaluate $f(-2)$ for $f(x) = 3x^4 - 6x^2 + 5x - 1$.
8. Find the rational zeros of the function $h(x) = 3x^5 + 2x^4 - 3x - 2$.

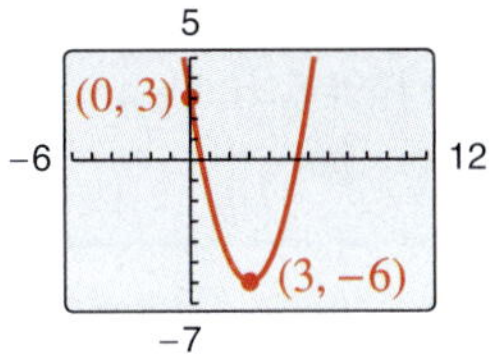

Figure for 2

In Exercises 9–12, perform the operation(s) and write the result in standard form.

9. $(-8 - 3i) + (-1 - 15i)$
10. $(10 + \sqrt{-20}) - (4 - \sqrt{-14})$
11. $(2 + i)(6 - i)$
12. $(4 + 3i)^2 - (5 + i)^2$

In Exercises 13 and 14, write the quotient in standard form.

13. $\dfrac{8 + 5i}{i}$
14. $\dfrac{5i}{2 - i}$

In Exercises 15 and 16, use the Quadratic Formula to solve the quadratic equation.

15. $x^2 + 54 = 0$
16. $x^2 - 2x + 8 = 0$

17. Find all the zeros of the function $f(x) = x^3 - 7x^2 + 11x + 19$ and write the polynomial as a product of linear factors. Use a graphing utility to verify your results.
18. Find all asymptotes and holes in the graph of the rational function $g(x) = (2x + 12)/(x^2 - 36)$. Use a graphing utility to verify your answers.

In Exercises 19 and 20, sketch the graph of the rational function by hand. As sketching aids, check for intercepts, vertical asymptotes, and slant asymptotes.

19. $g(x) = \dfrac{x^2 + 2}{x - 1}$
20. $f(x) = \dfrac{2x^2 + 9}{x + 2}$

21. The table shows the amounts A (in billions of dollars) spent on national defense by the United States for the years 2011 through 2017. (*Source:* U.S. Office of Management and Budget)
    (a) Use a graphing utility to create a scatter plot of the data. Let t represent the year, with $t = 1$ corresponding to 2011.
    (b) Use the *regression* feature of the graphing utility to find a quadratic model for the data.
    (c) Use the graphing utility to graph the quadratic model with the scatter plot from part (a).
    (d) Use the model to approximate the national defense spending in the United States each year from 2011 to 2017.
    (e) Compare the estimated results to the actual data. Is the model a good fit for the data? Explain.

Collaborative Project

To work a collaborative project involving Polynomial and Rational Functions, visit this textbook's website at *LarsonPrecalculus.com.*

DATA

Year	National defense, A (in billions of dollars)
2011	705.6
2012	677.9
2013	633.4
2014	603.5
2015	589.7
2016	593.4
2017	598.7

Spreadsheet at LarsonPrecalculus.com

Table for 21

Proofs in Mathematics

These two pages contain proofs of four important theorems about polynomial functions. The first two theorems are from Section 2.3, and the second two theorems are from Section 2.5.

The Remainder Theorem (p. 117)

If a polynomial $f(x)$ is divided by $x - k$, then the remainder is

$r = f(k)$.

Proof

From the Division Algorithm, you have

$$f(x) = (x - k)q(x) + r(x)$$

and because either $r(x) = 0$ or the degree of $r(x)$ is less than the degree of $x - k$, you know that $r(x)$ must be a constant. That is, $r(x) = r$. Now, by evaluating $f(x)$ at $x = k$, you have

$$\begin{aligned} f(k) &= (k - k)q(k) + r \\ &= (0)q(k) + r \\ &= r. \end{aligned}$$

■

To be successful in algebra, it is important that you understand the connection among the *factors* of a polynomial, the *zeros* of a polynomial function, and the *solutions* or *roots* of a polynomial equation. The Factor Theorem is the basis for this connection.

The Factor Theorem (p. 117)

A polynomial $f(x)$ has a factor $(x - k)$ if and only if $f(k) = 0$.

Proof

Using the Division Algorithm with the factor $(x - k)$, you have

$$f(x) = (x - k)q(x) + r(x).$$

By the Remainder Theorem, $r(x) = r = f(k)$, and you have

$$f(x) = (x - k)q(x) + f(k)$$

where $q(x)$ is a polynomial of lesser degree than $f(x)$. If $f(k) = 0$, then

$$f(x) = (x - k)q(x)$$

and you see that $(x - k)$ is a factor of $f(x)$. Conversely, if $(x - k)$ is a factor of $f(x)$, then division of $f(x)$ by $(x - k)$ yields a remainder of 0. So, by the Remainder Theorem, you have $f(k) = 0$. ■

Linear Factorization Theorem (p. 135)

If $f(x)$ is a polynomial of degree n, where $n > 0$, then f has precisely n linear factors

$$f(x) = a_n(x - c_1)(x - c_2) \cdot \cdot \cdot (x - c_n)$$

where $c_1, c_2, \ldots, c_n$ are complex numbers.

Proof

Using the Fundamental Theorem of Algebra, you know that f must have at least one zero, c_1. Consequently, $(x - c_1)$ is a factor of $f(x)$, and you have

$$f(x) = (x - c_1)f_1(x).$$

If the degree of $f_1(x)$ is greater than zero, then apply the Fundamental Theorem again to conclude that f_1 must have a zero c_2, which implies that

$$f(x) = (x - c_1)(x - c_2)f_2(x).$$

It is clear that the degree of $f_1(x)$ is $n - 1$, that the degree of $f_2(x)$ is $n - 2$, and that you can repeatedly apply the Fundamental Theorem n times until you obtain

$$f(x) = a_n(x - c_1)(x - c_2) \cdot \cdot \cdot (x - c_n)$$

where a_n is the leading coefficient of the polynomial $f(x)$. ■

Factors of a Polynomial (p. 138)

Every polynomial of degree $n > 0$ with real coefficients can be written as the product of linear and quadratic factors with real coefficients, where the quadratic factors have no real zeros.

Proof

To begin, you use the Linear Factorization Theorem to conclude that $f(x)$ can be *completely* factored in the form

$$f(x) = d(x - c_1)(x - c_2)(x - c_3) \ldots (x - c_n).$$

If each c_i is real, then there is nothing more to prove. If any c_i is imaginary ($c_i = a + bi$, $b \neq 0$), then, because the coefficients of $f(x)$ are real, you know that the conjugate $c_j = a - bi$ is also a zero. By multiplying the corresponding factors, you obtain

$$\begin{aligned}(x - c_i)(x - c_j) &= [x - (a + bi)][x - (a - bi)] \\ &= [(x - a) - bi][(x - a) + bi] \\ &= (x - a)^2 + b^2 \\ &= x^2 - 2ax + (a^2 + b^2)\end{aligned}$$

where each coefficient is real. ■

The Fundamental Theorem of Algebra

The Linear Factorization Theorem is closely related to the Fundamental Theorem of Algebra. The Fundamental Theorem of Algebra has a long and interesting history. In the early work with polynomial equations, the Fundamental Theorem of Algebra was thought to have been not true, because imaginary solutions were not considered. In fact, in the very early work by mathematicians such as Abu al-Khwarizmi (c. 800 A.D.), negative solutions were also not considered.

Once imaginary numbers were considered, several mathematicians attempted to give a general proof of the Fundamental Theorem of Algebra. These included Jean Le Rond d'Alembert (1746), Leonhard Euler (1749), Joseph-Louis Lagrange (1772), and Pierre Simon Laplace (1795). The first substantial proof (although not rigorous by today's standards) of the Fundamental Theorem of Algebra is credited to Carl Friedrich Gauss, who published the proof in his doctoral thesis in 1799.

Progressive Summary (Chapters 1–2)

This chart outlines the topics that have been covered so far in this text. Progressive Summary charts appear after Chapters 2, 3, 6, 9, and 11. In each Progressive Summary, new topics encountered for the first time appear in blue.

Algebraic Functions

Polynomial, Rational, Radical

Rewriting

Polynomial form ↔ Factored form
Operations with polynomials
Rationalize denominators
Simplify rational expressions
Operations with complex numbers

Solving

Equation	*Strategy*
Linear	Isolate variable
Quadratic	Factor, set to zero Extract square roots Complete the square Quadratic Formula
Polynomial	Factor, set to zero Rational Zero Test
Rational	Multiply by LCD
Radical	Isolate, raise to power
Absolute value	Isolate, form two equations

Analyzing

Graphically	*Algebraically*
Intercepts	Domain, Range
Symmetry	Transformations
Slope	Composition
Asymptotes	Standard forms of equations
End behavior	Leading Coefficient Test
Minimum values	Synthetic division
Maximum values	Descartes's Rule of Signs

Numerically
Table of values

Transcendental Functions

Rewriting

Solving

Analyzing

Other Topics

Rewriting

Solving

Analyzing

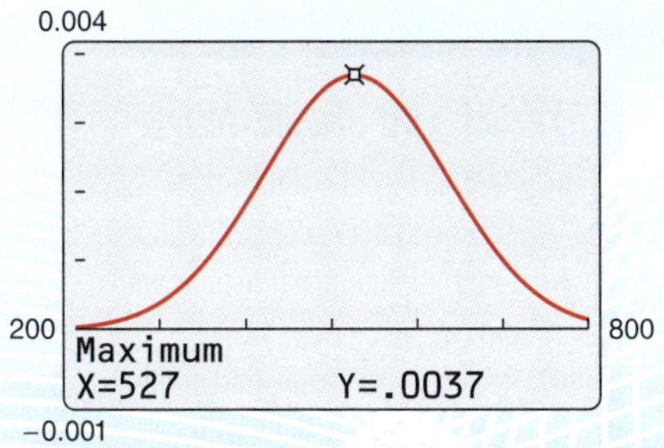

Section 3.5, Example 4
SAT Scores

3 Exponential and Logarithmic Functions

Student Resources at LarsonPrecalculus.com

- **Videos** explaining the concepts of precalculus
- **Worked-out solution videos** for all *Checkpoint* exercises
- **Editable spreadsheets** of the data sets in the text
- **Group projects** for each chapter applying concepts to real-life problems

3.1 Exponential Functions and Their Graphs

Exponential Functions

So far, this text has dealt mainly with **algebraic functions,** which include polynomial functions and rational functions. In this chapter, you will study two types of nonalgebraic functions—*exponential functions* and *logarithmic functions*. These functions are examples of **transcendental functions.**

Definition of Exponential Function

The **exponential function f with base a** is denoted by

$$f(x) = a^x$$

where $a > 0$, $a \neq 1$, and x is any real number.

Note that in the definition of an exponential function, the base $a = 1$ is excluded because it yields

$f(x) = 1^x = 1.$ Constant function

This is a constant function, not an exponential function.

You have already evaluated a^x for integer and rational values of x. For example, you know that

$4^3 = 64$ and $4^{1/2} = 2.$

However, to evaluate 4^x for any real number x, you need to interpret forms with *irrational* exponents. For the purposes of this text, it is sufficient to think of $a^{\sqrt{2}}$, where $\sqrt{2} \approx 1.41421356$, as the number that has the successively closer approximations

$$a^{1.4}, a^{1.41}, a^{1.414}, a^{1.4142}, a^{1.41421}, \ldots .$$

Example 1 shows how to use a calculator to evaluate exponential functions.

What you should learn

- Recognize and evaluate exponential functions with base a.
- Graph exponential functions with base a.
- Recognize, evaluate, and graph exponential functions with base e.
- Use exponential functions to model and solve real-life problems.

Why you should learn it

Exponential functions are useful in modeling data that represent quantities that increase or decrease quickly. For instance, Exercise 78 on page 193 shows how an exponential function is used to model the depreciation of a new vehicle.

EXAMPLE 1 Evaluating Exponential Functions

Use a calculator to evaluate each function at the indicated value of x.

	Function	*Value*
a.	$f(x) = 2^x$	$x = -3.1$
b.	$f(x) = 2^{-x}$	$x = \pi$
c.	$f(x) = 0.6^x$	$x = \frac{3}{2}$
d.	$f(x) = 1.05^{2x}$	$x = 12$

Solution

	Function Value	*Graphing Calculator Keystrokes*	*Display*
a.	$f(-3.1) = 2^{-3.1}$	2 [^] [(-)] 3.1 [ENTER]	0.1166291
b.	$f(\pi) = 2^{-\pi}$	2 [^] [(-)] π [ENTER]	0.1133147
c.	$f(\frac{3}{2}) = (0.6)^{3/2}$	.6 [^] [(] 3 [÷] 2 [)] [ENTER]	0.4647580
d.	$f(12) = (1.05)^{2(12)}$	1.05 [^] [(] 2 [×] 12 [)] [ENTER]	3.2250999

✓ **Checkpoint** Audio-video solution in English & Spanish at LarsonPrecalculus.com

Use a calculator to evaluate $f(x) = 8^{-x}$ at $x = \sqrt{2}$.

Technology Tip

On some graphing utilities, you may need to enclose exponents in parentheses to obtain the correct result.

Graphs of Exponential Functions

The graphs of all exponential functions have similar characteristics, as shown in Examples 2, 3, and 4.

EXAMPLE 2 Graphs of $y = a^x$

In the same coordinate plane, sketch the graphs of $f(x) = 2^x$ and $g(x) = 4^x$.

Solution

The table below lists some values for each function. By plotting these points and connecting them with smooth curves, you obtain the graphs shown in Figure 3.1. Note that both graphs are increasing. Moreover, the graph of $g(x) = 4^x$ is increasing more rapidly than the graph of $f(x) = 2^x$.

x	-3	-2	-1	0	1	2
2^x	$\frac{1}{8}$	$\frac{1}{4}$	$\frac{1}{2}$	1	2	4
4^x	$\frac{1}{64}$	$\frac{1}{16}$	$\frac{1}{4}$	1	4	16

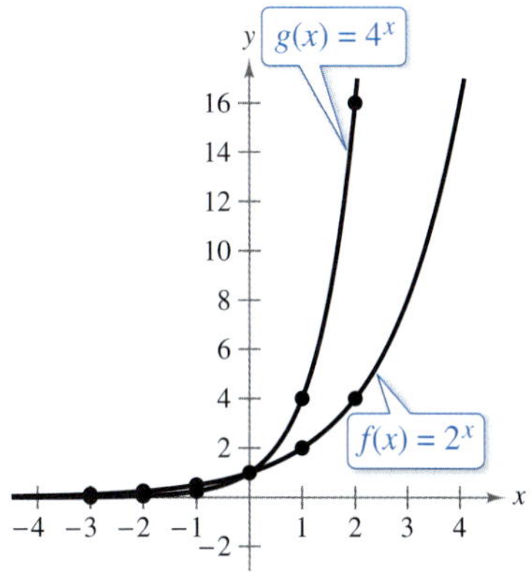

Figure 3.1

Checkpoint Audio-video solution in English & Spanish at LarsonPrecalculus.com

In the same coordinate plane, sketch the graphs of $f(x) = 3^x$ and $g(x) = 9^x$.

EXAMPLE 3 Graphs of $y = a^{-x}$

In the same coordinate plane, sketch the graphs of $F(x) = 2^{-x}$ and $G(x) = 4^{-x}$.

Solution

The table below lists some values for each function. By plotting these points and connecting them with smooth curves, you obtain the graphs shown in Figure 3.2. Note that both graphs are decreasing. Moreover, the graph of $G(x) = 4^{-x}$ is decreasing more rapidly than the graph of $F(x) = 2^{-x}$.

x	-2	-1	0	1	2	3
2^{-x}	4	2	1	$\frac{1}{2}$	$\frac{1}{4}$	$\frac{1}{8}$
4^{-x}	16	4	1	$\frac{1}{4}$	$\frac{1}{16}$	$\frac{1}{64}$

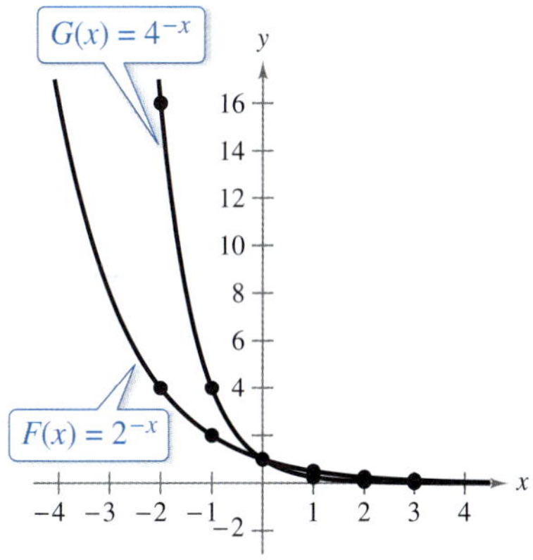

Figure 3.2

Checkpoint Audio-video solution in English & Spanish at LarsonPrecalculus.com

In the same coordinate plane, sketch the graphs of $f(x) = 3^{-x}$ and $g(x) = 9^{-x}$.

The properties of exponents can also be applied to real-number exponents. For review, these properties are listed below.

1. $a^x a^y = a^{x+y}$ **2.** $\dfrac{a^x}{a^y} = a^{x-y}$ **3.** $a^{-x} = \dfrac{1}{a^x} = \left(\dfrac{1}{a}\right)^x$ **4.** $a^0 = 1$

5. $(ab)^x = a^x b^x$ **6.** $(a^x)^y = a^{xy}$ **7.** $\left(\dfrac{a}{b}\right)^x = \dfrac{a^x}{b^x}$ **8.** $|a^2| = |a|^2 = a^2$

In Example 3, note that the functions $F(x) = 2^{-x}$ and $G(x) = 4^{-x}$ can be rewritten with positive exponents as

$$F(x) = 2^{-x} = \left(\frac{1}{2}\right)^x \quad \text{and} \quad G(x) = 4^{-x} = \left(\frac{1}{4}\right)^x.$$

Comparing the functions in Examples 2 and 3, observe that

$$F(x) = 2^{-x} = f(-x) \qquad \text{and} \qquad G(x) = 4^{-x} = g(-x).$$

Consequently, the graph of F is a reflection (in the y-axis) of the graph of f, as shown in Figure 3.3. The graphs of G and g have the same relationship, as shown in Figure 3.4.

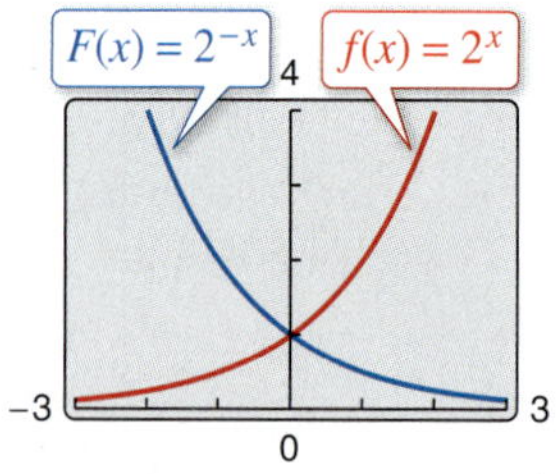

Figure 3.3

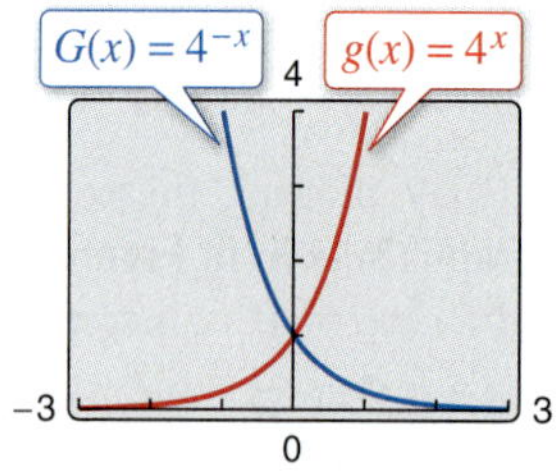

Figure 3.4

The graphs in Figures 3.3 and 3.4 are typical of the graphs of the exponential functions

$$f(x) = a^x \qquad \text{and} \qquad f(x) = a^{-x}.$$

They have one y-intercept and one horizontal asymptote (the x-axis), and they are continuous. The basic characteristics of these exponential functions are summarized below.

Library of Parent Functions: Exponential Function

The *parent exponential function*

$$f(x) = a^x, a > 0, a \neq 1$$

is different from all the functions you have studied so far because the variable x is an *exponent*. A distinguishing characteristic of an exponential function is its rapid increase as x increases (for $a > 1$). Many real-life phenomena with patterns of rapid growth (or decline) can be modeled by exponential functions. The basic characteristics of the exponential function are summarized below and on the inside cover of this text.

Graph of $f(x) = a^x, a > 1$

Domain: $(-\infty, \infty)$

Range: $(0, \infty)$

Intercept: $(0, 1)$

Increasing on $(-\infty, \infty)$

x-axis is a horizontal asymptote

$(a^x \to 0 \text{ as } x \to -\infty)$

Continuous

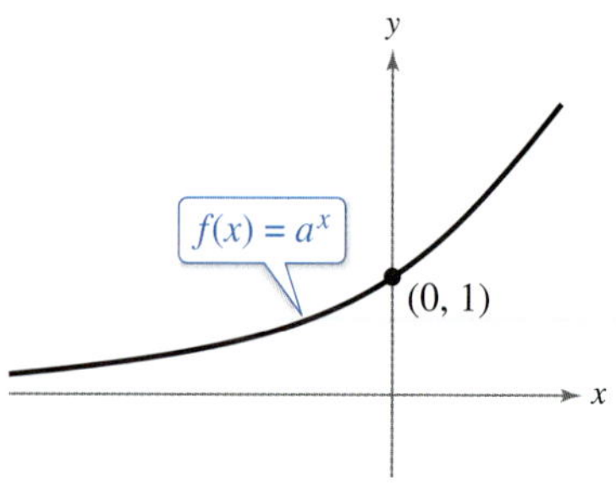

Graph of $f(x) = a^{-x}, a > 1$

Domain: $(-\infty, \infty)$

Range: $(0, \infty)$

Intercept: $(0, 1)$

Decreasing on $(-\infty, \infty)$

x-axis is a horizontal asymptote

$(a^{-x} \to 0 \text{ as } x \to \infty)$

Continuous

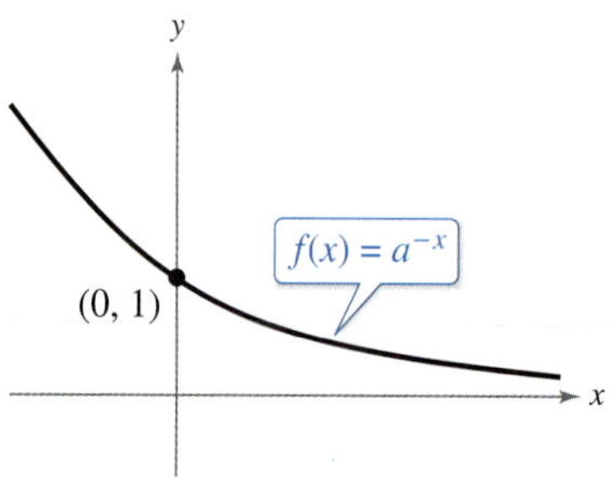

Explore the Concept

Use a graphing utility to graph $y = a^x$ for $a = 3$, 5, and 7 in the same viewing window. (Use a viewing window in which $-2 \leq x \leq 1$ and $0 \leq y \leq 2$.) How do the graphs compare with each other? Which graph is on the top in the interval $(-\infty, 0)$? Which is on the bottom? Which graph is on the top in the interval $(0, \infty)$? Which is on the bottom? Repeat this experiment with the graphs of $y = b^x$ for $b = \frac{1}{3}, \frac{1}{5}$, and $\frac{1}{7}$. (Use a viewing window in which $-1 \leq x \leq 2$ and $0 \leq y \leq 2$.) What can you conclude about the shape of the graph of $y = b^x$ and the value of b?

In the next example, notice how the graph of $y = a^x$ can be used to sketch the graphs of functions of the form $f(x) = b \pm a^{x+c}$.

EXAMPLE 4 Library of Parent Functions: $f(x) = a^x$

See LarsonPrecalculus.com for an interactive version of this type of example.

Each graph shown below is a transformation of the graph of $f(x) = 3^x$.

a. Because $g(x) = 3^{x+1} = f(x + 1)$, the graph of g can be obtained by shifting the graph of f one unit to the *left*, as shown in Figure 3.5.

b. Because $h(x) = 3^x - 2 = f(x) - 2$, the graph of h can be obtained by shifting the graph of f *downward* two units, as shown in Figure 3.6.

c. Because $k(x) = -3^x = -f(x)$, the graph of k can be obtained by *reflecting* the graph of f in the x-axis, as shown in Figure 3.7.

d. Because $j(x) = 3^{-x} = f(-x)$, the graph of j can be obtained by *reflecting* the graph of f in the y-axis, as shown in Figure 3.8.

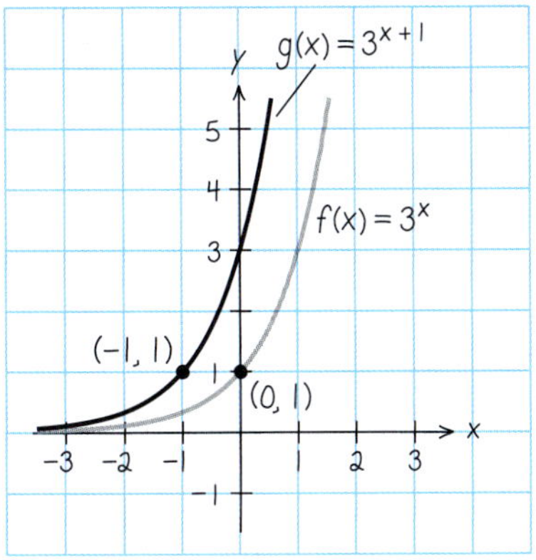

Figure 3.5

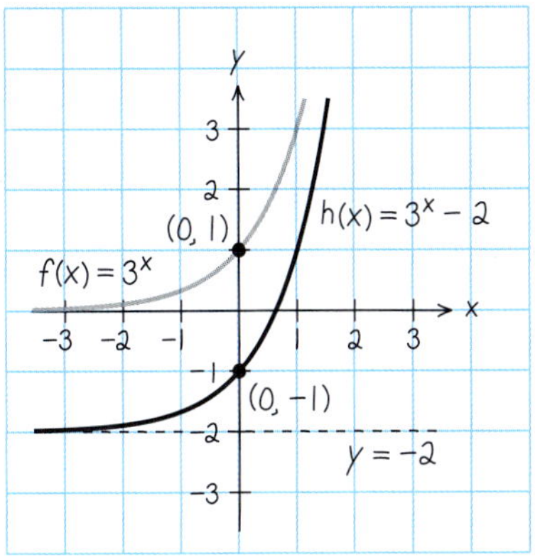

Figure 3.6

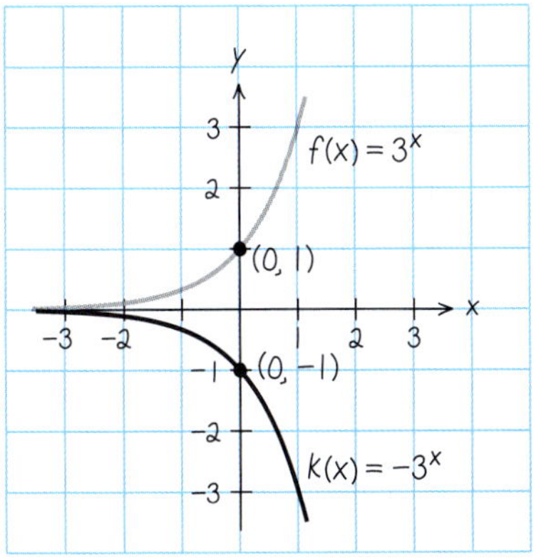

Figure 3.7

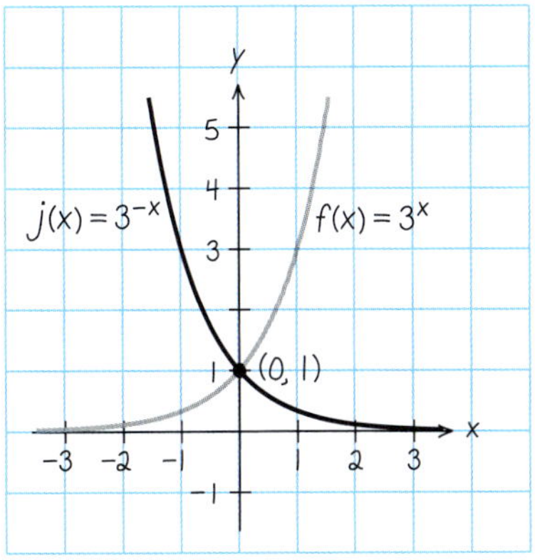

Figure 3.8

Checkpoint Audio-video solution in English & Spanish at LarsonPrecalculus.com

Use the graph of $f(x) = 4^x$ to describe the transformation that yields the graph of each function.

a. $g(x) = 4^{x-2}$ **b.** $h(x) = 4^x + 3$ **c.** $k(x) = 4^{-x} - 3$

Notice that the transformations in Figures 3.5, 3.7, and 3.8 keep the x-axis $(y = 0)$ as a horizontal asymptote, but the transformation in Figure 3.6 yields a new horizontal asymptote of $y = -2$. Also, be sure to note how the y-intercept is affected by each transformation.

What's Wrong?

You use a graphing utility to graph $f(x) = 3^x$ and $g(x) = 3^{x+2}$, as shown in the figure. You use the graph to conclude that the graph of g can be obtained by shifting the graph of f upward two units. What's wrong?

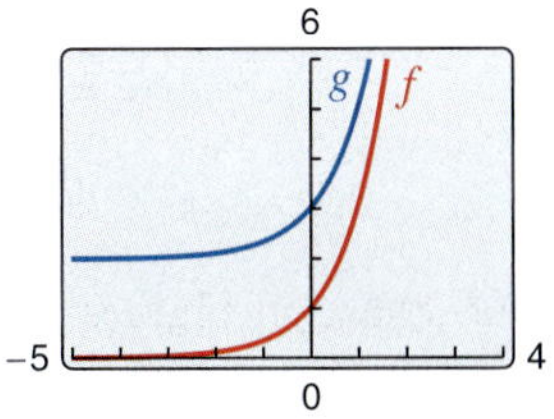

Explore the Concept

The table below shows some points on the graphs in Figure 3.5, where $Y_1 = f(x)$ and $Y_2 = g(x)$. Explain how you can use the table to describe the transformation.

X	Y1	Y2
-3	.03704	.11111
-2	.11111	.33333
-1	.33333	1
0	1	3
1	3	9
2	9	27
3	27	81

X=-3

The Natural Base e

For many applications, the convenient choice for a base is the irrational number

$$e = 2.71828128 \ldots .$$

This number is called the **natural base.** The function $f(x) = e^x$ is called the **natural exponential function,** and its graph is shown in Figure 3.9. The graph of the natural exponential function has the same basic characteristics as the graph of the function $f(x) = a^x$. (See page 184.) Be sure you see that for the natural exponential function $f(x) = e^x$, e is the constant $2.718281828 \ldots$, whereas x is the variable.

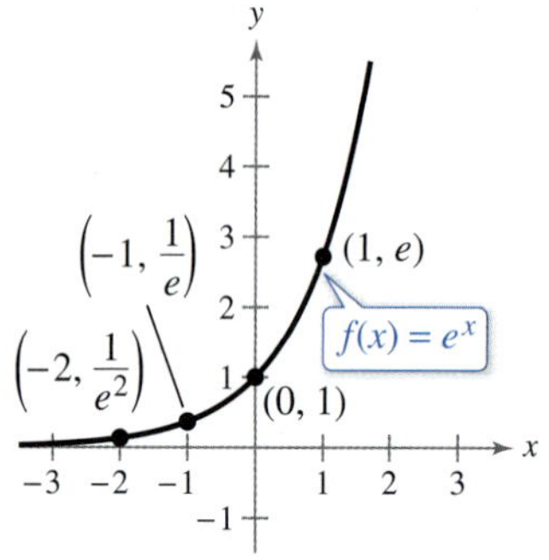

Figure 3.9 ***The Natural Exponential Function***

In Example 5, you will see that, for large values of x, the number e can be approximated by the expression

$$\left(1 + \frac{1}{x}\right)^x.$$

Explore the Concept

Use your graphing utility to graph the functions

$$y_1 = 2^x$$

$$y_2 = e^x$$

$$y_3 = 3^x$$

in the same viewing window. From the relative positions of these graphs, make a guess as to the value of the real number e. Then try to find a number a such that the graphs of $y_2 = e^x$ and $y_4 = a^x$ are as close to each other as possible.

EXAMPLE 5 Approximation of the Number e

Evaluate the expression

$$\left(1 + \frac{1}{x}\right)^x$$

for several large values of x to see that the values approach $e \approx 2.718281828$ as x increases without bound.

Graphical Solution

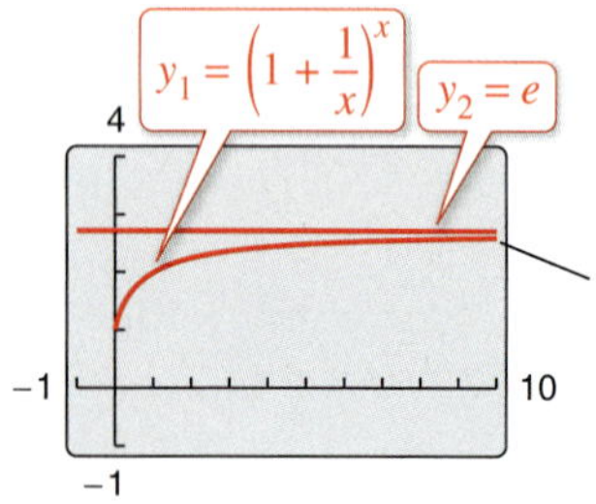

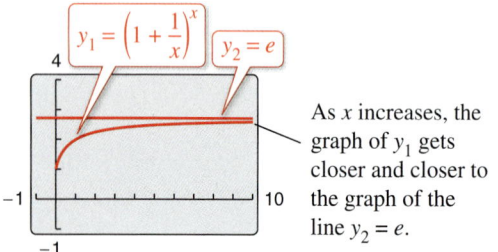

As x increases, the graph of y_1 gets closer and closer to the graph of the line $y_2 = e$.

Numerical Solution

Enter $y_1 = (1 + 1/x)^x$.

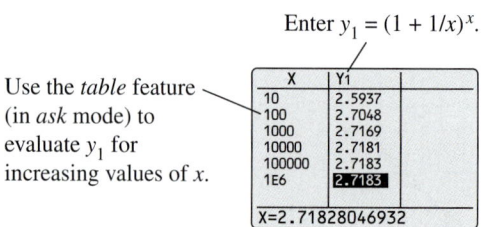

Use the *table* feature (in *ask* mode) to evaluate y_1 for increasing values of x.

X	Y1
10	2.5937
100	2.7048
1000	2.7169
10000	2.7181
100000	2.7183
1E6	2.7183

X=2.71828046932

From the table, it seems reasonable to conclude that

$$\left(1 + \frac{1}{x}\right)^x \to e \text{ as } x \to \infty.$$

✓ ***Checkpoint*** *Audio-video solution in English & Spanish at LarsonPrecalculus.com*

Evaluate the expression $(1 + 2/x)^x$ for several large values of x to see that the values approach $e^2 \approx 7.389056099$ as x increases without bound.

EXAMPLE 6 Evaluating the Natural Exponential Function

Use a calculator to evaluate the function $f(x) = e^x$ at each value of x.

a. $x = -2$

b. $x = 0.25$

c. $x = -0.4$

d. $x = \frac{2}{3}$

Solution

	Function Value	*Graphing Calculator Keystrokes*	*Display*
a.	$f(-2) = e^{-2}$	[e^x] [(−)] 2 [ENTER]	0.1353353
b.	$f(0.25) = e^{0.25}$	[e^x] .25 [ENTER]	1.2840254
c.	$f(-0.4) = e^{-0.4}$	[e^x] [(−)] .4 [ENTER]	0.6703200
d.	$f(\frac{2}{3}) = e^{2/3}$	[e^x] [(] 2 [÷] 3 [)] [ENTER]	1.9477340

✓ ***Checkpoint*** *Audio-video solution in English & Spanish at LarsonPrecalculus.com*

Use a calculator to evaluate the function $f(x) = e^x$ at each value of x.

a. $x = 0.3$ **b.** $x = -1.2$ **c.** $x = 6.2$

EXAMPLE 7 Graphing Natural Exponential Functions

Sketch the graphs of $f(x) = 2e^{0.24x}$ and $g(x) = \frac{1}{2}e^{-0.58x}$.

Solution

To sketch these two graphs, you can use a calculator to construct a table of values, as shown below.

x	-3	-2	-1	0	1	2	3
$f(x)$	0.974	1.238	1.573	2.000	2.542	3.232	4.109
$g(x)$	2.849	1.595	0.893	0.500	0.280	0.157	0.088

After constructing the table, plot the points and connect them with smooth curves. Note that the graph in Figure 3.10 is increasing, whereas the graph in Figure 3.11 is decreasing. Use a graphing utility to verify these graphs.

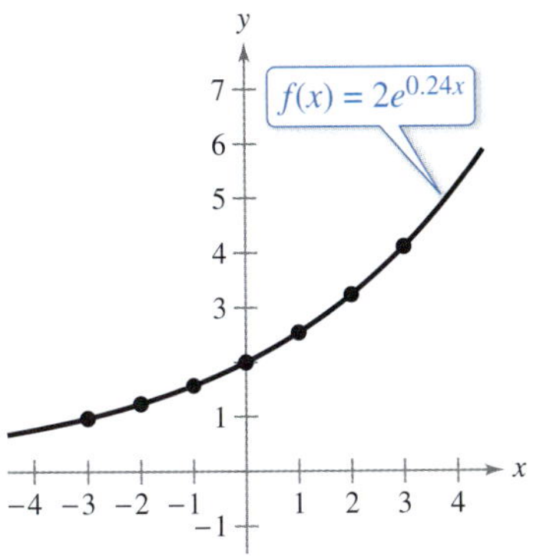

Figure 3.10

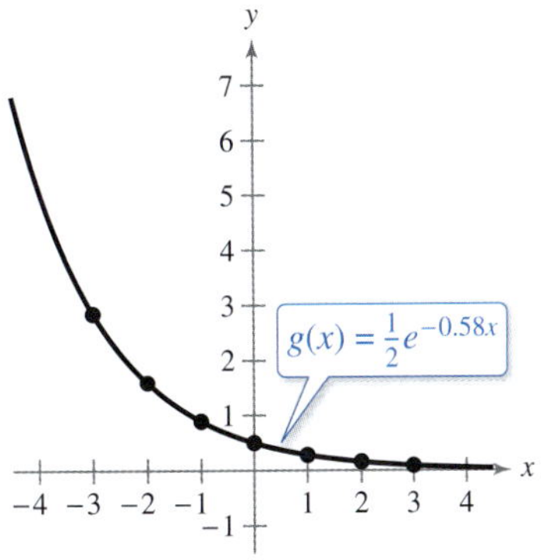

Figure 3.11

✓ ***Checkpoint*** *Audio-video solution in English & Spanish at LarsonPrecalculus.com*

Sketch the graph of $f(x) = 5e^{0.17x}$.

Explore the Concept

Use a graphing utility to graph $y = (1 + x)^{1/x}$. Describe the behavior of the graph near $x = 0$. Is there a y-intercept? How does the behavior of the graph near $x = 0$ relate to the result of Example 5? Use the *table* feature of the graphing utility to create a table that shows values of y for values of x near $x = 0$ to help you describe the behavior of the graph near this point.

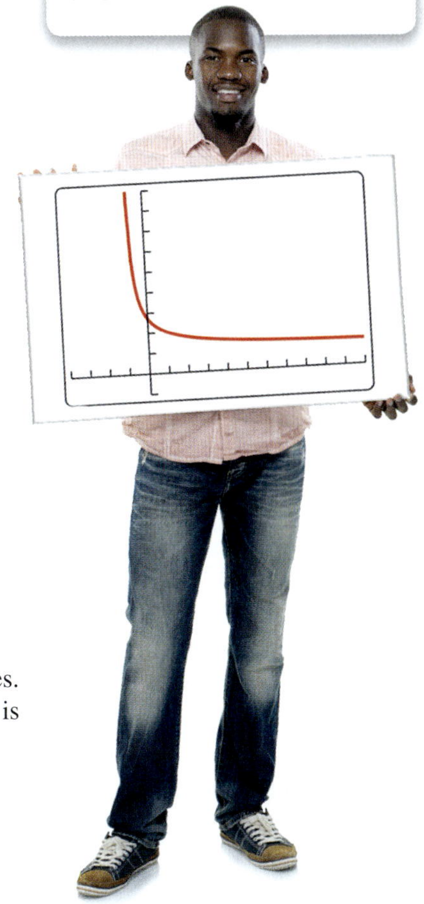

Applications

One of the most familiar examples of exponential growth is an investment earning *continuously compounded interest.* Suppose a principal P is invested at an annual interest rate r, compounded once a year. If the interest is added to the principal at the end of the year, then the new balance P_1 is

$$P_1 = P + Pr = P(1 + r).$$

This pattern of multiplying the previous principal by $1 + r$ is then repeated each successive year, as shown in the table.

Time in years	Balance after each compounding
0	$P = P$
1	$P_1 = P(1 + r)$
2	$P_2 = P_1(1 + r) = P(1 + r)(1 + r) = P(1 + r)^2$
⋮	⋮
t	$P_t = P(1 + r)^t$

To accommodate more frequent (quarterly, monthly, or daily) compounding of interest, let n be the number of compoundings per year and let t be the number of years. (The product nt represents the total number of times the interest will be compounded.) Then the interest rate per compounding period is r/n, and the account balance after t years is

$$A = P\left(1 + \frac{r}{n}\right)^{nt}.$$ Amount (balance) with n compoundings per year

When the number of compoundings n increases without bound, the process approaches what is called **continuous compounding.** In the formula for n compoundings per year, let $m = n/r$. This produces

$$A = P\left(1 + \frac{r}{n}\right)^{nt} = P\left(1 + \frac{1}{m}\right)^{mrt} = P\left[\left(1 + \frac{1}{m}\right)^{m}\right]^{rt}.$$

As m increases without bound, you know from Example 5 that

$$\left(1 + \frac{1}{m}\right)^{m}$$

approaches e. So, for continuous compounding, it follows that

$$P\left[\left(1 + \frac{1}{m}\right)^{m}\right]^{rt} \Rightarrow P[e]^{rt}$$

and you can write $A = Pe^{rt}$. This result is part of the reason that e is the "natural" choice for a base of an exponential function.

Formulas for Compound Interest

After t years, the balance A in an account with principal P and annual interest rate r (in decimal form) is given by the following formulas.

1. For n compoundings per year: $A = P\left(1 + \frac{r}{n}\right)^{nt}$

2. For continuous compounding: $A = Pe^{rt}$

Explore the Concept

Use the formula

$$A = P\left(1 + \frac{r}{n}\right)^{nt}$$

to calculate the amount in an account when $P = \$3000$, $r = 6\%$, $t = 10$ years, and the interest is compounded (a) by the day, (b) by the hour, (c) by the minute, and (d) by the second. Does increasing the number of compoundings per year result in unlimited growth of the amount in the account? Explain.

Algebra Help

The interest rate r in the formulas for compound interest should be written as a decimal. For example, an interest rate of 2.5% would be written as $r = 0.025$.

EXAMPLE 8 Finding the Balance for Compound Interest

A total of \$9000 is invested at an annual interest rate of 2.5%, compounded annually. Find the balance in the account after 5 years.

Algebraic Solution

In this case, $P = 9000$, $r = 2.5\% = 0.025$, $n = 1$, and $t = 5$. Using the formula for compound interest with n compoundings per year, you have

$$A = P\left(1 + \frac{r}{n}\right)^{nt}$$ Formula for compound interest

$$= 9000\left(1 + \frac{0.025}{1}\right)^{1(5)}$$ Substitute for P, r, n, and t.

$$= 9000(1.025)^5$$ Simplify.

$$\approx \$10,182.67.$$ Use a calculator.

So, the balance in the account after 5 years is \$10,182.67.

Graphical Solution

Substitute the values for P, r, and n into the formula for compound interest with n compoundings per year and simplify to obtain $A = 9000(1.025)^t$. Use a graphing utility to graph $A = 9000(1.025)^t$. Then use the *value* feature to approximate the value of A when $t = 5$, as shown in the figure.

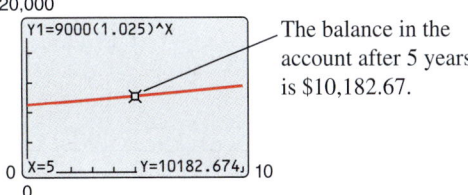

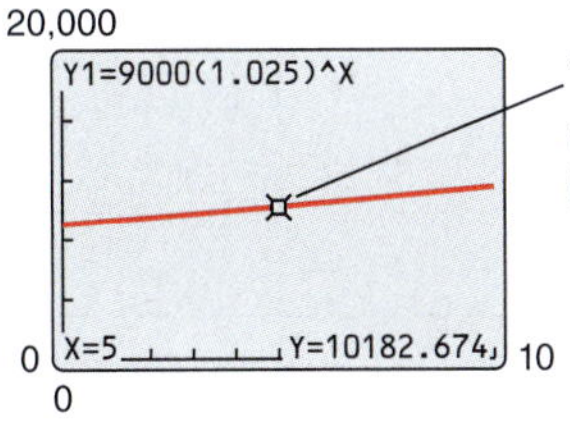

The balance in the account after 5 years is \$10,182.67.

✓ **Checkpoint** ▶ *Audio-video solution in English & Spanish at LarsonPrecalculus.com*

For the account in Example 8, find the balance after 10 years.

EXAMPLE 9 Finding Compound Interest

A total of \$12,000 is invested at an annual interest rate of 3%. Find the balance after 4 years when the interest is compounded (a) quarterly and (b) continuously.

Solution

a. For quarterly compoundings, $n = 4$. So, after 4 years at 3%, the balance is

$$A = P\left(1 + \frac{r}{n}\right)^{nt}$$ Formula for compound interest

$$= 12,000\left(1 + \frac{0.03}{4}\right)^{4(4)}$$ Substitute for P, r, n, and t.

$$\approx \$13,523.91.$$ Use a calculator.

b. For continuous compounding, the balance is

$$A = Pe^{rt}$$ Formula for continuous compounding

$$= 12,000e^{0.03(4)}$$ Substitute for P, r, and t.

$$\approx \$13,529.96.$$ Use a calculator.

Note that a continuous-compounding account yields more than a quarterly-compounding account.

✓ **Checkpoint** ▶ *Audio-video solution in English & Spanish at LarsonPrecalculus.com*

You invest \$6000 at an annual rate of 4%. Find the balance after 7 years when the interest is compounded (a) quarterly, (b) monthly, and (c) continuously.

Financial Analyst

Example 9 illustrates the following general rule. For a given principal, interest rate, and time, the more often the interest is compounded per year, the greater the balance will be. Moreover, the balance obtained by continuous compounding is greater than the balance obtained by compounding n times per year.

EXAMPLE 10 Radioactive Decay

Let y represent a mass, in grams, of radioactive strontium (^{90}Sr), whose half-life is about 29 years. The quantity of strontium present after t years is $y = 10\left(\frac{1}{2}\right)^{t/29}$.

a. What is the initial mass (when $t = 0$)?

b. How much of the initial mass is present after 80 years?

Algebraic Solution

a. $y = 10\left(\dfrac{1}{2}\right)^{t/29}$ Write original equation.

$= 10\left(\dfrac{1}{2}\right)^{0/29}$ Substitute 0 for t.

$= 10$ Simplify.

So, the initial mass is 10 grams.

b. $y = 10\left(\dfrac{1}{2}\right)^{t/29}$ Write original equation.

$= 10\left(\dfrac{1}{2}\right)^{80/29}$ Substitute 80 for t.

$\approx 10\left(\dfrac{1}{2}\right)^{2.759}$ Simplify.

≈ 1.48 Use a calculator.

So, about 1.48 grams are present after 80 years.

Graphical Solution

a.

When $t = 0$, $y = 10$. So, the initial mass is 10 grams.

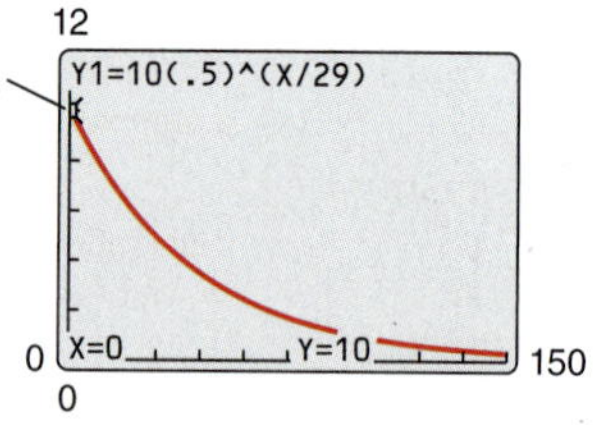

b.

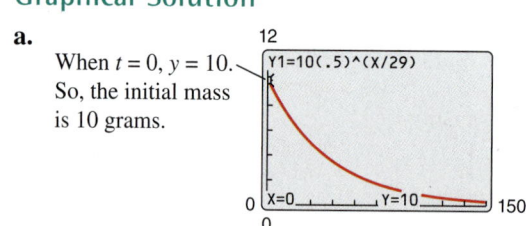

When $t = 80$, $y \approx 1.48$. So, about 1.48 grams are present after 80 years.

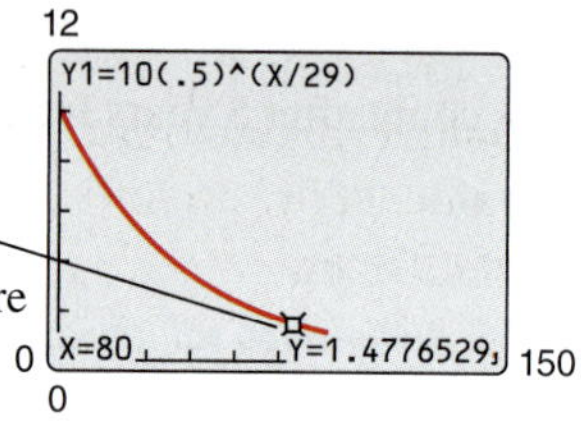

✓ Checkpoint Audio-video solution in English & Spanish at LarsonPrecalculus.com

In Example 10, how much of the initial mass is present after 160 years?

EXAMPLE 11 Population Growth

The approximate number of fruit flies in an experimental population after t hours is given by $Q(t) = 20e^{0.03t}$, where $t \geq 0$.

a. Find the initial number of fruit flies in the population.

b. How large is the population of fruit flies after 72 hours?

c. Graph Q.

Solution

a. To find the initial population, evaluate $Q(t)$ when $t = 0$.

$$Q(0) = 20e^{0.03(0)} = 20e^0 = 20(1) = 20 \text{ flies}$$

b. After 72 hours, the population size is

$$Q(72) = 20e^{0.03(72)} = 20e^{2.16} \approx 173 \text{ flies.}$$

c. The graph of Q is shown in the figure.

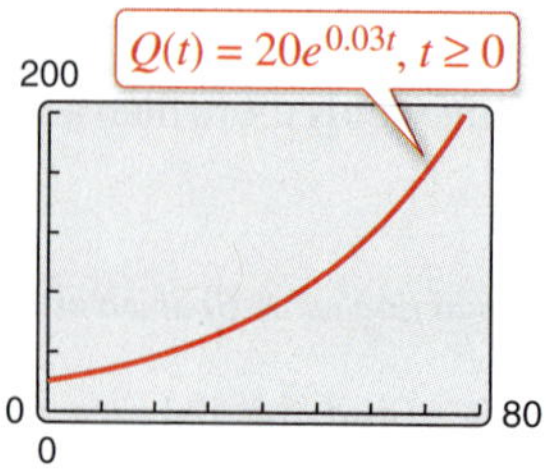

✓ Checkpoint Audio-video solution in English & Spanish at LarsonPrecalculus.com

Rework Example 11 when the approximate number of fruit flies in the experimental population after t hours is given by $Q(t) = 10e^{0.02t}$.

3.1 Exercises

See *CalcChat.com* for tutorial help and worked-out solutions to odd-numbered exercises.
For instructions on how to use a graphing utility, see Appendix A.

Vocabulary and Concept Check

In Exercises 1 and 2, fill in the blank(s).

1. Exponential and logarithmic functions are examples of nonalgebraic functions, also called _______ functions.

2. The exponential function $f(x) = e^x$ is called the _______ function, and the base e is called the _______ base.

3. What type of transformation of the graph of $f(x) = 5^x$ is the graph of $f(x + 1)$?

4. The formula $A = Pe^{rt}$ gives the balance A of an account earning what type of interest?

Procedures and Problem Solving

Evaluating an Exponential Function **In Exercises 5–8, evaluate the function at the indicated value of x. Round your result to three decimal places.**

Function	*Value*
5. $f(x) = 0.9^x$	$x = 1.4$
6. $f(x) = 1.2^x$	$x = \frac{1}{3}$
7. $g(x) = 5^x$	$x = -\pi$
8. $h(x) = 8.6^{-3x}$	$x = -\sqrt{2}$

Graphs of $y = a^x$ and $y = a^{-x}$ **In Exercises 9–12, graph the exponential function by hand. Identify any asymptotes and intercepts and determine whether the graph of the function is increasing or decreasing.**

9. $g(x) = 5^x$

10. $h(x) = 10^{-x}$

11. $g(x) = \left(\frac{5}{4}\right)^{-x}$

12. $g(x) = \left(\frac{3}{4}\right)^x$

Library of Parent Functions **In Exercises 13–16, use the graph of $y = 2^x$ to match the function with its graph. [The graphs are labeled (a), (b), (c), and (d).]**

(a)

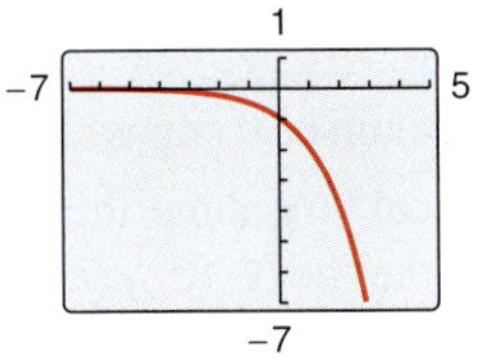

(b)

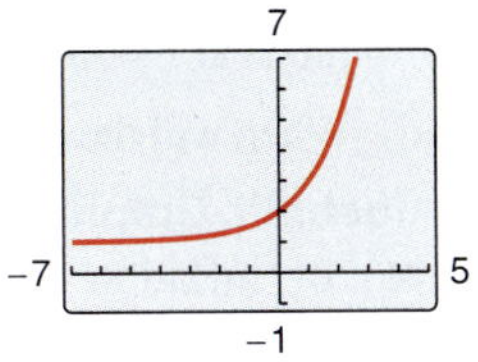

(c)

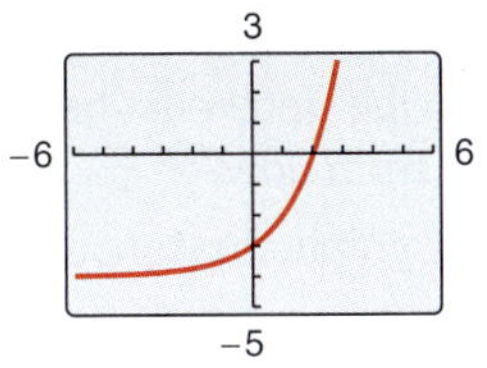

(d)

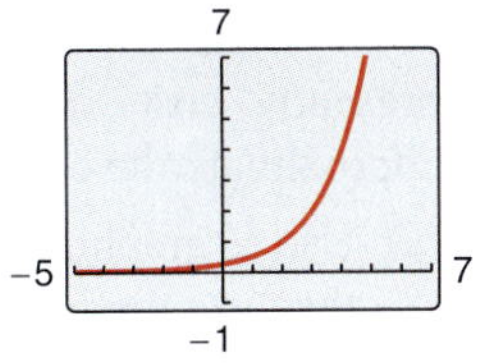

13. $f(x) = 2^{x-2}$

14. $f(x) = -2^x$

15. $f(x) = 2^x - 4$

16. $f(x) = 2^x + 1$

Library of Parent Functions **In Exercises 17–28, use the graph of f to describe the transformation that yields the graph of g. Then sketch the graphs of f and g by hand.**

17. $f(x) = 3^x, \quad g(x) = 3^{x-5}$

18. $f(x) = 10^x, \quad g(x) = 10^{x+4}$

19. $f(x) = 6^x, \quad g(x) = 6^x + 5$

20. $f(x) = -2^x, \quad g(x) = 5 - 2^x$

21. $f(x) = \left(\frac{3}{5}\right)^x, \quad g(x) = -\left(\frac{3}{5}\right)^{x+4}$

22. $f(x) = \left(\frac{7}{2}\right)^x, \quad g(x) = -\left(\frac{7}{2}\right)^{-x}$

23. $f(x) = 0.3^x, \quad g(x) = -0.3^x + 5$

24. $f(x) = 0.4^x, \quad g(x) = 0.4^{x-2} - 3$

25. $f(x) = 10^x, \quad g(x) = 10^{-x+3}$

26. $f(x) = \left(\frac{2}{3}\right)^x, \quad g(x) = \left(\frac{2}{3}\right)^{x+2} - 3$

27. $f(x) = \left(\frac{1}{4}\right)^x, \quad g(x) = \left(\frac{1}{4}\right)^{-x} + 2$

28. $f(x) = \left(\frac{1}{2}\right)^x, \quad g(x) = \left(\frac{1}{2}\right)^{-(x+4)}$

Approximation of a Power with Base e **In Exercises 29 and 30, show that the value of $f(x)$ approaches the value of $g(x)$ as x increases without bound (a) graphically and (b) numerically.**

29. $f(x) = \left(1 + \dfrac{0.5}{x}\right)^x, \quad g(x) = e^{0.5}$

30. $f(x) = \left(1 + \dfrac{3}{x}\right)^x, \quad g(x) = e^3$

Evaluating a Natural Exponential Function **In Exercises 31–34, evaluate the function at the indicated value of x. Round your result to three decimal places.**

31. $f(x) = e^x, \quad x = 9.2$

32. $f(x) = 1.5e^{x/2}, \quad x = 240$

33. $f(x) = 5000e^{0.06x}, \quad x = 6$

34. $h(x) = -5.5e^{-x}, \quad x = 200$

Graphing an Exponential Function In Exercises 35–48, use a graphing utility to construct a table of values for the function. Then sketch the graph of the function. Identify any asymptotes of the graph.

35. $f(x) = \left(\frac{3}{2}\right)^x$
36. $f(x) = 2^{x-1}$
37. $f(x) = 3^{x+2}$
38. $y = 2^{-x^2}$
39. $y = 3^{x-2} + 1$
40. $y = 5^{x+1} - 4$
41. $f(x) = e^{-x}$
42. $s(t) = 3e^{-0.2t}$
43. $f(x) = 3e^{x+4}$
44. $f(x) = 2e^{x-3}$
45. $f(x) = 2 + e^{x-5}$
46. $g(x) = e^{x+1} + 2$
47. $s(t) = 2e^{-0.12t}$
48. $g(x) = e^{0.5x} - 1$

Finding Asymptotes In Exercises 49–52, use a graphing utility to (a) graph the function and (b) find any asymptotes numerically by creating a table of values for the function.

49. $f(x) = \dfrac{8}{1 + e^{-0.5x}}$
50. $g(x) = \dfrac{8}{1 + e^{-0.5/x}}$
51. $f(x) = -\dfrac{6}{2 - e^{0.2x}}$
52. $f(x) = \dfrac{6}{2 - e^{0.2/x}}$

Finding Points of Intersection In Exercises 53–56, use a graphing utility to find the point(s) of intersection, if any, of the graphs of the functions. Round your result to three decimal places.

53. $y = 20e^{0.05x}$
$y = 1500$
54. $y = 100e^{0.01x}$
$y = 12{,}500$
55. $y = 5.2e^{0.5x}$
$y = -0.8$
56. $y = 2.3e^{0.4x}$
$y = 1.9$

Approximating Relative Extrema In Exercises 57–60, (a) use a graphing utility to graph the function, (b) use the graph to find the open intervals on which the function is increasing and decreasing, and (c) approximate any relative maximum or minimum values.

57. $f(x) = x^2e^{-x}$
58. $f(x) = 2x^2e^{x+1}$
59. $f(x) = x^3e^x$
60. $f(x) = x^3e^{-x+2}$

Compound Interest In Exercises 61–64, complete the table by finding the balance A when P dollars is invested at rate r for t years and compounded n times per year.

n	1	2	4	12	365	Continuous
A						

61. $P = \$1500$, $r = 2\%$, $t = 10$ years
62. $P = \$2500$, $r = 6\%$, $t = 10$ years
63. $P = \$2500$, $r = 4\%$, $t = 20$ years
64. $P = \$1500$, $r = 3.5\%$, $t = 40$ years

Compound Interest In Exercises 65–68, complete the table by finding the balance A when \$12,000 is invested at rate r for t years, compounded continuously.

t	1	10	20	30	40	50
A						

65. $r = 4\%$
66. $r = 6\%$
67. $r = 1.5\%$
68. $r = 2.5\%$

Finding the Amount of an Annuity In Exercises 69–72, you build an annuity by investing P dollars every month at interest rate r, compounded monthly. Find the amount A accrued after n months using the formula

$$A = P\left[\frac{(1 + r/12)^n - 1}{r/12}\right], \text{ where } r \text{ is in decimal form.}$$

69. $P = \$25$, $r = 0.12$, $n = 48$ months
70. $P = \$100$, $r = 0.09$, $n = 60$ months
71. $P = \$500$, $r = 0.02$, $n = 36$ months
72. $P = \$75$, $r = 0.03$, $n = 24$ months

73. **Radioactive Decay** Let Q represent a mass, in grams, of carbon 14 (^{14}C), whose half-life is about 5700 years. The quantity present after t years can be found using $Q = 10\left(\frac{1}{2}\right)^{t/5700}$.
(a) Determine the initial quantity (when $t = 0$).
(b) Determine the quantity present after 2000 years.
(c) Sketch the graph of the function over the interval $t = 0$ to $t = 10{,}000$.

74. **Radioactive Decay** Let Q represent a mass, in grams, of radioactive radium (^{226}Ra), whose half-life is about 1600 years. The quantity of radium present after t years can be found using $Q = 25\left(\frac{1}{2}\right)^{t/1600}$.
(a) Determine the initial quantity (when $t = 0$).
(b) Determine the quantity present after 1000 years.
(c) Use a graphing utility to graph the function over the interval $t = 0$ to $t = 5000$.
(d) When will the quantity of radium be 0 grams? Explain.

75. **Algebraic-Graphical-Numerical** Assume the annual rate of inflation is 4% for the next 10 years. The approximate cost C of goods or services during these years is $C(t) = P(1.04)^t$, where t is the time (in years) and P is the present cost. An oil change for your car presently costs \$26.88. Use the methods below to approximate the cost 10 years from now.
(a) Use a graphing utility to graph the function and then use the *value* feature.
(b) Use the *table* feature of the graphing utility to find a numerical approximation.
(c) Evaluate the cost function algebraically.

76. MODELING DATA

There are three options for investing \$1200. The first earns 8% compounded annually, the second earns 8% compounded quarterly, and the third earns 8% compounded continuously.

(a) Find equations that model the growth of each investment and use a graphing utility to graph each model in the same viewing window over a 15-year period.

(b) Use the graph from part (a) to determine which investment yields the highest return after 15 years. What are the differences in earnings among the three investments?

77. Population Growth The projected populations of Americans ages 100 and over for the years 2025 through 2060 can be modeled by $P = 31.752e^{0.0486t}$ where P is the population (in thousands) and t is the year, with $t = 25$ corresponding to 2025. *(Source: U.S. Census Bureau)*

(a) Use a graphing utility to graph the function for the years 2025 through 2060.

(b) Use the *table* feature of the graphing utility to create a table of values for the same time period as in part (a).

(c) According to the model, in what year will the population exceed 300,000?

78. *Why you should learn it* *(p. 182)* In early 2018, a new sedan had a manufacturer's suggested retail price of \$34,950. After t years, the sedan's value V can be found using $V(t) = 34{,}950\left(\frac{4}{5}\right)^t$.

(a) Use a graphing utility to graph the function.

(b) Use the graphing utility to create a table of values that shows the value V for $t = 1$ to $t = 10$ years.

(c) According to the model, when will the sedan have no value?

Focusing on Concepts

True or False? **In Exercises 79 and 80, determine whether the statement is true or false. Justify your answer.**

79. $f(x) = 1^x$ is not an exponential function.

80. $e = \dfrac{271{,}801}{99{,}990}$

81. Library of Parent Functions Determine which equation(s) may be represented by the graph shown. (There may be more than one correct answer.)

(a) $y = e^x + 1$

(b) $y = -e^{-x} + 1$

(c) $y = e^{-x} - 1$

(d) $y = e^{-x} + 1$

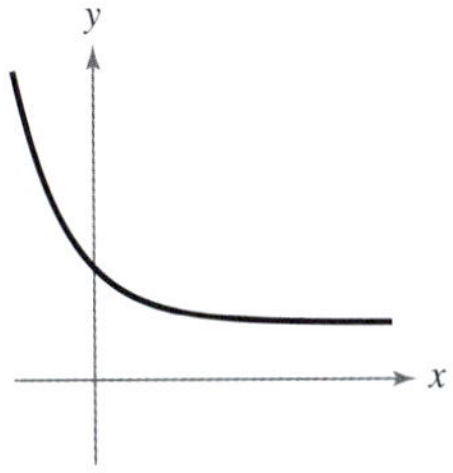

82. Exploration Use a graphing utility to graph $y_1 = e^x$ and each of the functions $y_2 = x^2$, $y_3 = x^3$, $y_4 = \sqrt{x}$, and $y_5 = |x|$ in the same viewing window.

(a) Which function increases at the fastest rate for "large" values of x?

(b) Use the result of part (a) to make a conjecture about the rates of growth of $y_1 = e^x$ and $y = x^n$, where n is a natural number and x is "large."

83. Error Analysis Describe the error.

For the function $f(x) = 2^x - 1$, the domain is $(-\infty, \infty)$ and the range is $(0, \infty)$.

84. HOW DO YOU SEE IT? The figure shows the graphs of $y = 2^x$, $y = e^x$, $y = 10^x$, $y = 2^{-x}$, $y = e^{-x}$, and $y = 10^{-x}$. Match each function with its graph. [The graphs are labeled (a) through (f).] Explain.

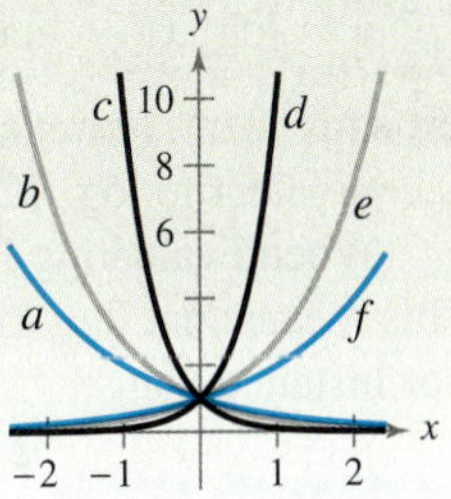

Think About It **In Exercises 85–88, place the correct symbol (< or >) between the two numbers.**

85. e^{π} ▭ π^e

86. 2^{10} ▭ 10^2

87. 6^{-5} ▭ 5^{-6}

88. $4^{1/2}$ ▭ $\left(\frac{1}{2}\right)^4$

Cumulative Mixed Review

Inverse Functions **In Exercises 89–92, determine whether the function has an inverse function. If it does, find f^{-1}.**

89. $f(x) = 5x - 7$

90. $f(x) = -\frac{2}{3}x + \frac{5}{2}$

91. $f(x) = \sqrt[3]{x + 8}$

92. $f(x) = \sqrt{x^2 + 6}$

93. *Project: U.S. Population* To work an extended application analyzing the population per square mile in the United States, visit this textbook's website at *LarsonPrecalculus.com.* *(Data Source: U.S. Census Bureau)*

3.2 Logarithmic Functions and Their Graphs

Logarithmic Functions

In Section 1.6, you studied the concept of an inverse function. There, you learned that when a function is one-to-one—that is, when the function has the property that no horizontal line intersects its graph more than once—the function must have an inverse function. By looking back at the graphs of the exponential functions introduced in Section 3.1, you will see that every function of the form $f(x) = a^x$, where $a > 0$ and $a \neq 1$, passes the Horizontal Line Test and therefore must have an inverse function. This inverse function is called the **logarithmic function with base *a*.**

What you should learn

- Recognize and evaluate logarithmic functions with base *a*.
- Graph logarithmic functions with base *a*.
- Recognize, evaluate, and graph natural logarithmic functions.
- Use logarithmic functions to model and solve real-life problems.

Why you should learn it

Logarithmic functions are useful in modeling data that represent quantities that increase or decrease slowly. For instance, Exercise 114 on page 203 shows how to use a logarithmic function to model the minimum required ventilation rates in public school classrooms.

Definition of Logarithmic Function

For $x > 0$, $a > 0$, and $a \neq 1$,

$$y = \log_a x \quad \text{if and only if} \quad x = a^y.$$

The function given by

$f(x) = \log_a x$ Read as "log base *a* of *x*."

is called the **logarithmic function with base *a*.**

From the definition above, you can see that every logarithmic equation can be written in an equivalent exponential form and every exponential equation can be written in logarithmic form. So, the equations

$$y = \log_a x \quad \text{and} \quad x = a^y$$

are equivalent. For example, $2 = \log_3 9$ is equivalent to $9 = 3^2$, and $5^3 = 125$ is equivalent to $\log_5 125 = 3$.

When evaluating logarithms, remember that *a logarithm is an exponent.* This means that $\log_a x$ is the exponent to which a must be raised to obtain x. For instance, $\log_2 8 = 3$ because 2 must be raised to the third power to get 8.

EXAMPLE 1 Evaluating Logarithms

Evaluate each logarithm at the indicated value of x.

a. $f(x) = \log_2 x, \quad x = 32$

b. $f(x) = \log_3 x, \quad x = 1$

c. $f(x) = \log_4 x, \quad x = 2$

d. $f(x) = \log_{10} x, \quad x = \frac{1}{100}$

Solution

a. $f(32) = \log_2 32 = 5 \quad \text{because} \quad 2^5 = 32.$

b. $f(1) = \log_3 1 = 0 \quad \text{because} \quad 3^0 = 1.$

c. $f(2) = \log_4 2 = \frac{1}{2} \quad \text{because} \quad 4^{1/2} = \sqrt{4} = 2.$

d. $f\left(\frac{1}{100}\right) = \log_{10} \frac{1}{100} = -2 \quad \text{because} \quad 10^{-2} = \frac{1}{10^2} = \frac{1}{100}.$

Algebra Help

In this text, the parentheses in $\log_a(u)$ are sometimes omitted when u is an expression involving exponents, radicals, products, or quotients. For instance, $\log_{10}(2x)$ can be written as $\log_{10} 2x$. To evaluate $\log_{10} 2x$, find the logarithm of the product $2x$.

✓ ***Checkpoint*** *Audio-video solution in English & Spanish at LarsonPrecalculus.com*

Evaluate each logarithm at the indicated value of x.

a. $f(x) = \log_6 x, x = 1$ **b.** $f(x) = \log_5 x, x = \frac{1}{125}$ **c.** $f(x) = \log_{10} x, x = 10{,}000$

The logarithmic function with base 10 is called the **common logarithmic function.** On most calculators, this function is denoted by LOG. Example 2 shows how to use a calculator to evaluate common logarithmic functions. You will learn how to use a calculator to evaluate logarithms with any base in the next section.

EXAMPLE 2 Evaluating Common Logarithms

Use a calculator to evaluate the function $f(x) = \log_{10} x$ at each value of x.

a. $x = 10$ **b.** $x = \frac{1}{3}$ **c.** $x = 2.5$ **d.** $x = -2$

Solution

Function Value	*Graphing Calculator Keystrokes*	*Display*
a. $f(10) = \log_{10} 10$	LOG 10 ENTER	1
b. $f(\frac{1}{3}) = \log_{10} \frac{1}{3}$	LOG (1 ÷ 3) ENTER	−0.4771213
c. $f(2.5) = \log_{10} 2.5$	LOG 2.5 ENTER	0.3979400
d. $f(-2) = \log_{10}(-2)$	LOG (−) 2 ENTER	ERROR

Note that the calculator displays an error message when you try to evaluate $\log_{10}(-2)$. In this case, there is no *real* power to which 10 can be raised to obtain -2.

Checkpoint Audio-video solution in English & Spanish at LarsonPrecalculus.com

Use a calculator to evaluate the function $f(x) = \log_{10} x$ at each value of x.

a. $x = 275$ **b.** $x = 0.275$ **c.** $x = -\frac{1}{2}$ **d.** $x = \frac{1}{2}$

Technology Tip

Some graphing utilities do not give an error message for $\log_{10}(-2)$. Instead, the graphing utility will display an imaginary number. For the purpose of this text, however, the domain of a logarithmic function is the set of positive *real* numbers.

The properties of logarithms listed below follow directly from the definition of the logarithmic function with base a.

Properties of Logarithms

1. $\log_a 1 = 0$ because $a^0 = 1$.

2. $\log_a a = 1$ because $a^1 = a$.

3. $\log_a a^x = x$ and $a^{\log_a x} = x$. Inverse Properties

4. If $\log_a x = \log_a y$, then $x = y$. One-to-One Property

EXAMPLE 3 Using Properties of Logarithms

a. Solve for x: $\log_2 x = \log_2 3$

b. Solve for x: $\log_4 4 = x$

c. Simplify: $\log_5 5^x$

d. Simplify: $7^{\log_7 14}$

Solution

a. Using the One-to-One Property (Property 4), you can conclude that $x = 3$.

b. Using Property 2, you can conclude that $x = 1$.

c. Using the Inverse Property (Property 3), it follows that $\log_5 5^x = x$.

d. Using the Inverse Property (Property 3), it follows that $7^{\log_7 14} = 14$.

Checkpoint Audio-video solution in English & Spanish at LarsonPrecalculus.com

a. Solve for x: $\log_{10} x = \log_{10} 2$ **b.** Simplify: $20^{\log_{20} 3}$

Insight

Be sure you know the properties of logarithms. On standardized tests, you will use them to simplify expressions (see Example 3) and to solve equations (see Section 3.4).

Graphs of Logarithmic Functions

To sketch the graph of $y = \log_a x$, you can use the fact that the graphs of inverse functions are reflections of each other in the line $y = x$.

EXAMPLE 4 Graphs of Exponential and Logarithmic Functions

In the same coordinate plane, sketch the graph of each function by hand.

a. $f(x) = 2^x$

b. $g(x) = \log_2 x$

Solution

a. For $f(x) = 2^x$, construct a table of values. By plotting these points and connecting them with a smooth curve, you obtain the graph of f shown in Figure 3.12.

x	-2	-1	0	1	2	3
$f(x) = 2^x$	$\frac{1}{4}$	$\frac{1}{2}$	1	2	4	8

b. Because $g(x) = \log_2 x$ is the inverse function of $f(x) = 2^x$, the graph of g is obtained by plotting the points $(f(x), x)$ and connecting them with a smooth curve. The graph of g is a reflection of the graph of f in the line $y = x$, as shown in Figure 3.12.

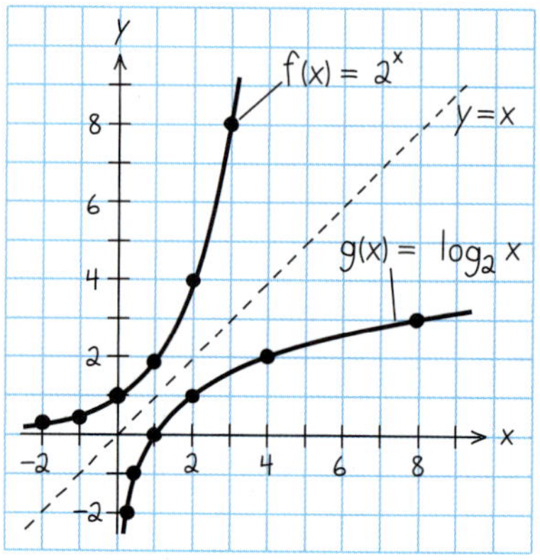

Figure 3.12

✓ Checkpoint Audio-video solution in English & Spanish at LarsonPrecalculus.com

In the same coordinate plane, sketch the graphs of (a) $f(x) = 8^x$ and (b) $g(x) = \log_8 x$.

Before you can confirm the result of Example 4 using a graphing utility, you need to know how to enter $\log_2 x$. You will learn how to do this using the *change-of-base formula* discussed in the next section.

EXAMPLE 5 Sketching the Graph of a Logarithmic Function

Sketch the graph of the common logarithmic function $f(x) = \log_{10} x$ by hand.

Solution

Begin by constructing a table of values. Note that some of the values can be obtained without a calculator by using the Inverse Property of Logarithms. Others require a calculator. Next, plot the points and connect them with a smooth curve, as shown in Figure 3.13. Note that $x = 0$ (the y-axis) is a vertical asymptote of the graph.

	Without calculator				With calculator		
x	$\frac{1}{100}$	$\frac{1}{10}$	1	10	2	5	8
$f(x) = \log_{10} x$	-2	-1	0	1	0.301	0.699	0.903

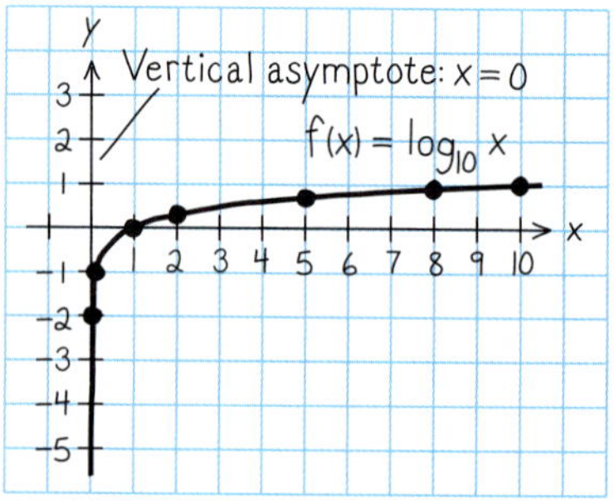

Figure 3.13

✓ Checkpoint Audio-video solution in English & Spanish at LarsonPrecalculus.com

Without using a calculator, sketch the graph of $f(x) = \log_3 x$ by hand.

The nature of the graph in Figure 3.13 is typical of functions of the form $f(x) = \log_a x$, $a > 1$. They have one x-intercept and one vertical asymptote. Notice how slowly the graph rises for $x > 1$.

Library of Parent Functions: Logarithmic Function

The *parent logarithmic function* $f(x) = \log_a x$, $a > 0$, $a \neq 1$ is the inverse function of the exponential function. Its domain is the set of positive real numbers and its range is the set of all real numbers. This is the opposite of the exponential function. Moreover, the logarithmic function has the y-axis as a vertical asymptote, whereas the exponential function has the x-axis as a horizontal asymptote. Many real-life phenomena with slow rates of growth can be modeled by logarithmic functions. The basic characteristics of the logarithmic function are summarized below and on the inside cover of this text.

Graph of $f(x) = \log_a x$, $a > 1$

Domain: $(0, \infty)$

Range: $(-\infty, \infty)$

Intercept: $(1, 0)$

Increasing on $(0, \infty)$

y-axis is a vertical asymptote

($\log_a x \to -\infty$ as $x \to 0^+$)

Continuous

Reflection of graph of $f(x) = a^x$ in the line $y = x$

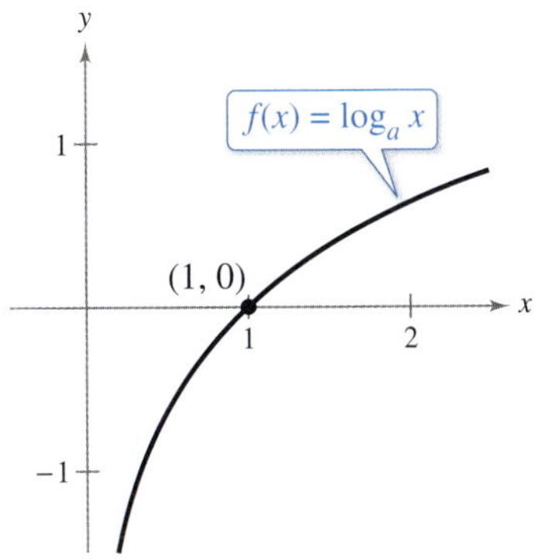

Explore the Concept

Use a graphing utility to graph $y = \log_{10} x$ and $y = 8$ in the same viewing window. Find a viewing window that shows the point of intersection. What is the point of intersection? Use the point of intersection to complete the equation $\log_{10} \boxed{} = 8$.

EXAMPLE 6 Library of Parent Functions $f(x) = \log_a x$

See LarsonPrecalculus.com for an interactive version of this type of example.

Each graph shown below is a transformation of the graph of $f(x) = \log_{10} x$.

a. Because $g(x) = \log_{10}(x - 1) = f(x - 1)$, the graph of g can be obtained by shifting the graph of f one unit to the *right*, as shown in Figure 3.14.

b. Because $h(x) = 2 + \log_{10} x = 2 + f(x)$, the graph of h can be obtained by shifting the graph of f two units *upward*, as shown in Figure 3.15.

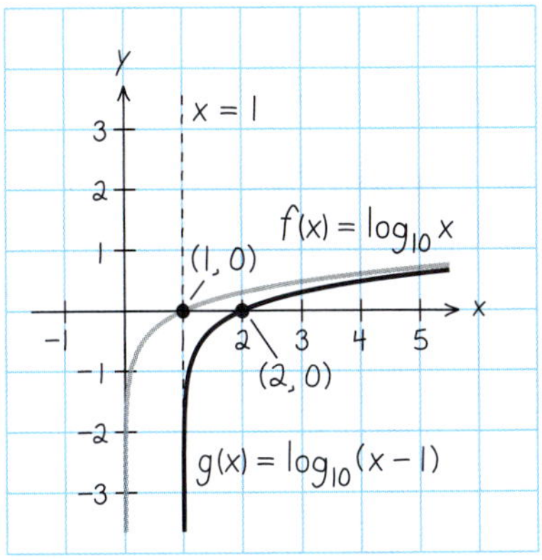

Figure 3.14

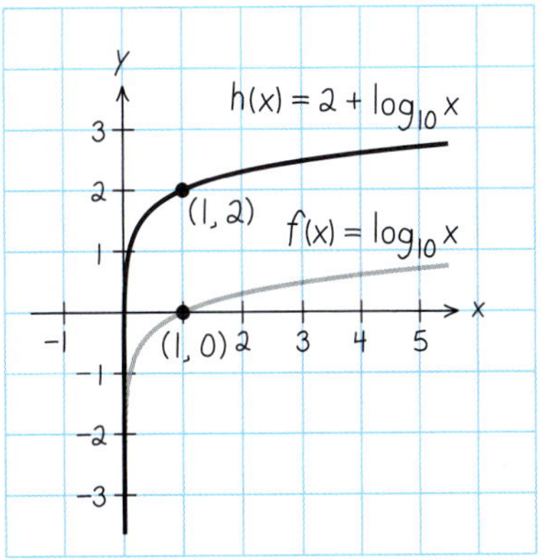

Figure 3.15

Notice that the transformation in Figure 3.15 keeps the y-axis as a vertical asymptote, but the transformation in Figure 3.14 yields the new vertical asymptote $x = 1$.

Checkpoint *Audio-video solution in English & Spanish at LarsonPrecalculus.com*

Use the graph of $f(x) = \log_{10} x$ to sketch the graph of each function by hand.

a. $g(x) = -1 + \log_{10} x$ **b.** $h(x) = \log_{10}(x + 3)$

The Natural Logarithmic Function

By looking back at the graph of the natural exponential function introduced in Section 3.1, you will see that $f(x) = e^x$ is one-to-one and so has an inverse function. This inverse function is called the **natural logarithmic function** and is denoted by the special symbol $\ln x$, read as "the natural log of x" or "el en of x."

The Natural Logarithmic Function

For $x > 0$,

$$y = \ln x \quad \text{if and only if} \quad x = e^y.$$

The function given by

$$f(x) = \log_e x = \ln x$$

is called the **natural logarithmic function.**

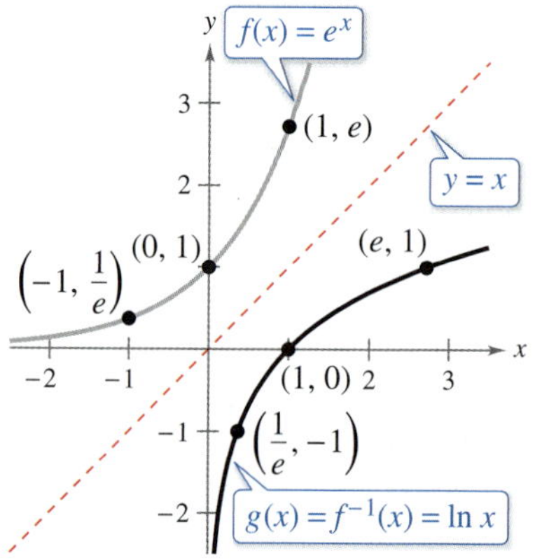

Reflection of graph of $f(x) = e^x$ in the line $y = x$

Figure 3.16

The equations $y = \ln x$ and $x = e^y$ are equivalent. Note that the natural logarithm $\ln x$ is written without a base. The base is understood to be e.

Because the functions $f(x) = e^x$ and $g(x) = \ln x$ are inverse functions of each other, their graphs are reflections of each other in the line $y = x$. This reflective property is illustrated in Figure 3.16.

EXAMPLE 7 Evaluating the Natural Logarithmic Function

Use a calculator to evaluate the function $f(x) = \ln x$ at each value of x.

a. $x = 2$

b. $x = 0.3$

c. $x = -1$

d. $x = 1 + \sqrt{2}$

Solution

	Function Value	*Graphing Calculator Keystrokes*	*Display*
a.	$f(2) = \ln 2$	[LN] 2 [ENTER]	0.6931472
b.	$f(0.3) = \ln 0.3$	[LN] .3 [ENTER]	−1.2039728
c.	$f(-1) = \ln(-1)$	[LN] [(-)] 1 [ENTER]	ERROR
d.	$f(1 + \sqrt{2}) = \ln(1 + \sqrt{2})$	[LN] [(] 1 [+] [√] 2 [)] [ENTER]	0.8813736

✓ *Checkpoint* ▶ ***Audio-video solution in English & Spanish at LarsonPrecalculus.com***

Use a calculator to evaluate the function $f(x) = \ln x$ at each value of x.

a. $x = 0.01$ **b.** $x = 4$ **c.** $x = \sqrt{3} + 2$ **d.** $x = \sqrt{3} - 2$

The four properties of logarithms listed on page 195 are also valid for natural logarithms, as shown below.

Properties of Natural Logarithms

1. $\ln 1 = 0$ because $e^0 = 1$.

2. $\ln e = 1$ because $e^1 = e$.

3. $\ln e^x = x$ and $e^{\ln x} = x$. Inverse Properties

4. If $\ln x = \ln y$, then $x = y$. One-to-One Property

Technology Tip

On most graphing utilities, the natural logarithm is denoted by [LN], as illustrated in Example 7.

Algebra Help

In Example 7(c), be sure you see that $\ln(-1)$ gives an error message on most calculators. This occurs because the domain of $\ln x$ is the set of *positive* real numbers (see Figure 3.16). So, $\ln(-1)$ is undefined.

EXAMPLE 8 Using Properties of Natural Logarithms

Use the properties of natural logarithms to rewrite each expression.

a. $\ln \frac{1}{e}$ **b.** $e^{\ln 5}$ **c.** $4 \ln 1$ **d.** $2 \ln e$

Solution

a. $\ln \frac{1}{e} = \ln e^{-1} = -1$ Inverse Property

b. $e^{\ln 5} = 5$ Inverse Property

c. $4 \ln 1 = 4(0) = 0$ Property 1

d. $2 \ln e = 2(1) = 2$ Property 2

✓ ***Checkpoint*** ▶ *Audio-video solution in English & Spanish at LarsonPrecalculus.com*

Use the properties of natural logarithms to simplify each expression.

a. $\ln e^{1/3}$ **b.** $5 \ln 1$ **c.** $\frac{3}{4} \ln e$ **d.** $e^{\ln 7}$

EXAMPLE 9 Finding the Domains of Logarithmic Functions

Find the domain of each function.

a. $f(x) = \ln(x - 2)$ **b.** $g(x) = \ln(2 - x)$ **c.** $h(x) = \ln x^2$

Algebraic Solution

a. Because $\ln(x - 2)$ is defined only when

$$x - 2 > 0$$

it follows that the domain of f is $(2, \infty)$.

b. Because $\ln(2 - x)$ is defined only when

$$2 - x > 0$$

it follows that the domain of g is $(-\infty, 2)$.

c. Because $\ln x^2$ is defined only when

$$x^2 > 0$$

it follows that the domain of h is all real numbers except $x = 0$.

Graphical Solution

a.

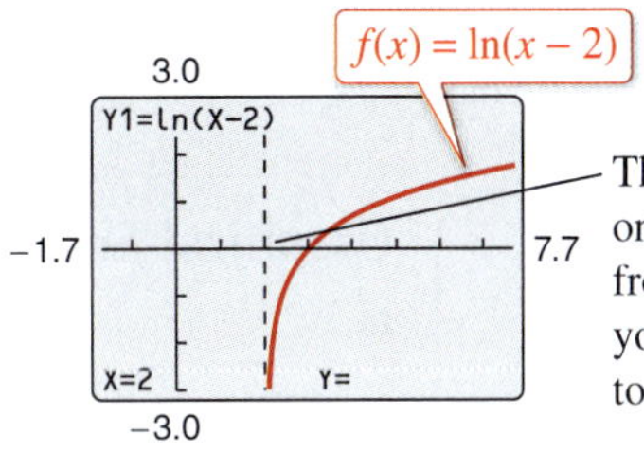

The x-coordinates of points on the graph appear to extend from the right of 2 to ∞. So, you can estimate the domain to be $(2, \infty)$.

b.

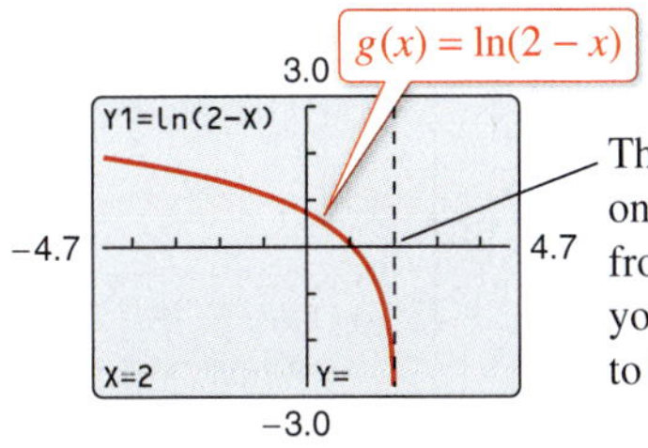

The x-coordinates of points on the graph appear to extend from $-\infty$ to the left of 2. So, you can estimate the domain to be $(-\infty, 2)$.

c.

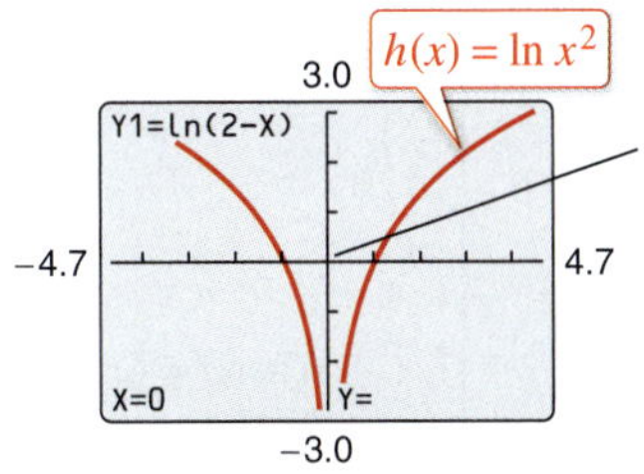

The x-coordinates of points on the graph appear to include all real numbers except 0. So, you can estimate the domain to be all real numbers except $x = 0$.

✓ ***Checkpoint*** ▶ *Audio-video solution in English & Spanish at LarsonPrecalculus.com*

Find the domain of $f(x) = \ln(x + 3)$.

In Example 9, suppose you had been asked to analyze the function $h(x) = \ln|x - 2|$. How would the domain of this function compare with the domains of the functions given in parts (a) and (b) of the example?

Application

EXAMPLE 10 Psychology

Students participating in a psychology experiment attended several lectures on a subject and were given an exam. Every month for a year after the exam, the students were retested to see how much of the material they remembered. The average scores for the group are given by the *human memory model*

$$f(t) = 75 - 6\ln(t + 1), \quad 0 \le t \le 12$$

where t is the time in months. The graph of f is shown in Figure 3.17.

a. What was the average score on the original exam ($t = 0$)?

b. What was the average score at the end of $t = 2$ months?

c. What was the average score at the end of $t = 6$ months?

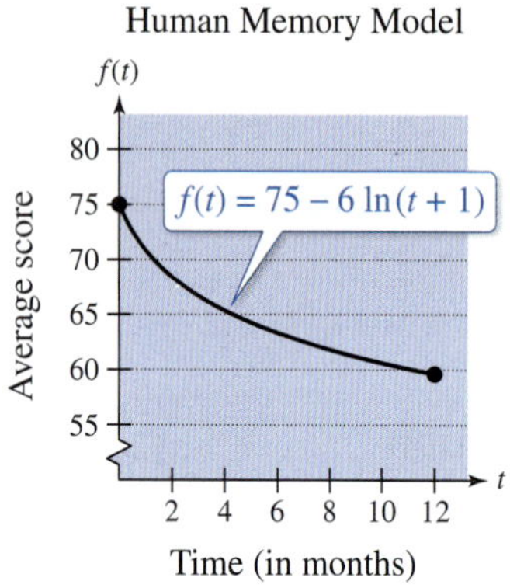

Figure 3.17

Psychologist

Algebraic Solution

a. The original average score was

$$\begin{aligned} f(0) &= 75 - 6\ln(0 + 1) \\ &= 75 - 6\ln 1 \\ &= 75 - 6(0) \\ &= 75. \end{aligned}$$

b. After 2 months, the average score was

$$\begin{aligned} f(2) &= 75 - 6\ln(2 + 1) \\ &= 75 - 6\ln 3 \\ &\approx 75 - 6(1.0986) \\ &\approx 68.41. \end{aligned}$$

c. After 6 months, the average score was

$$\begin{aligned} f(6) &= 75 - 6\ln(6 + 1) \\ &= 75 - 6\ln 7 \\ &\approx 75 - 6(1.9459) \\ &\approx 63.32. \end{aligned}$$

Graphical Solution

a.

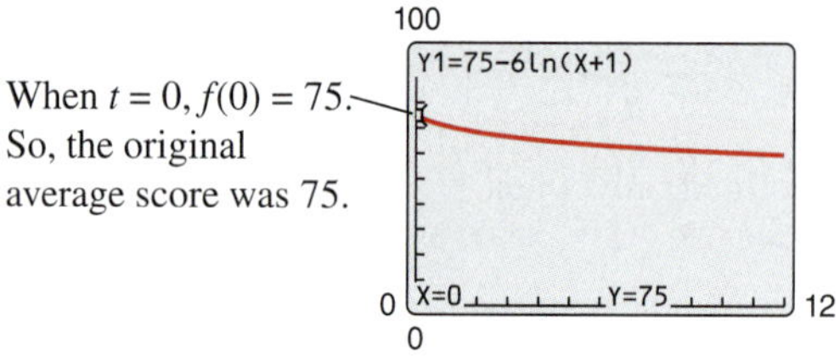

b.

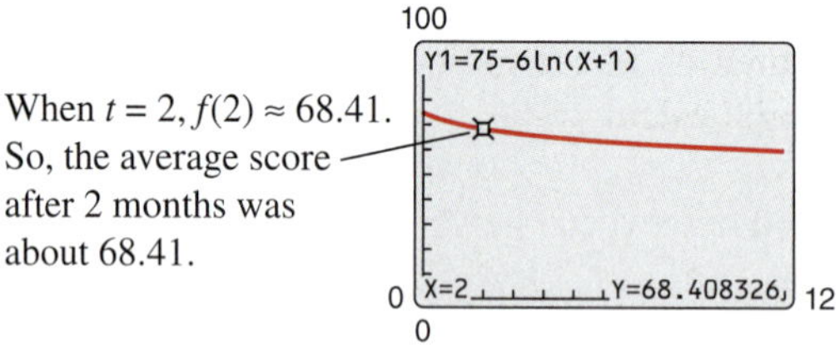

c.

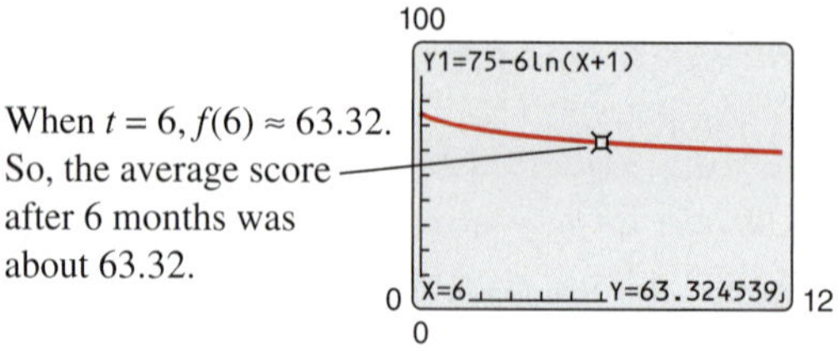

✓ ***Checkpoint*** *Audio-video solution in English & Spanish at LarsonPrecalculus.com*

In Example 10, find the average score at the end of (a) $t = 1$ month, (b) $t = 9$ months, and (c) $t = 12$ months.

3.2 Exercises

See *CalcChat.com* for tutorial help and worked-out solutions to odd-numbered exercises.
For instructions on how to use a graphing utility, see Appendix A.

Vocabulary and Concept Check

In Exercises 1–4, fill in the blank(s).

1. The inverse function of the exponential function $f(x) = a^x$ is called the _______ with base a.

2. The base of the _______ logarithmic function is 10, and the base of the _______ logarithmic function is e.

3. The inverse properties of logarithms are $\log_a a^x = x$ and _______ .

4. If $x = e^y$, then $y =$ _______ .

5. What exponential equation is equivalent to the logarithmic equation $\log_a b = c$?

6. For what value(s) of x is $\ln x = \ln 7$?

Procedures and Problem Solving

Rewriting a Logarithmic Equation **In Exercises 7–14, write the logarithmic equation in exponential form. For example, the exponential form of $\log_5 25 = 2$ is $5^2 = 25$.**

7. $\log_4 16 = 2$
8. $\log_3 81 = 4$
9. $\log_7 \frac{1}{343} = -3$
10. $\log_{10} \frac{1}{100{,}000} = -5$
11. $\log_{32} 4 = \frac{2}{5}$
12. $\log_{16} 8 = \frac{3}{4}$
13. $\log_2 \sqrt{2} = \frac{1}{2}$
14. $\log_5 \sqrt[3]{25} = \frac{2}{3}$

Rewriting an Exponential Equation **In Exercises 15–22, write the exponential equation in logarithmic form. For example, the logarithmic form of $2^3 = 8$ is $\log_2 8 = 3$.**

15. $5^3 = 125$
16. $8^2 = 64$
17. $81^{3/4} = 27$
18. $16^{3/2} = 64$
19. $6^{-2} = \frac{1}{36}$
20. $10^{-3} = 0.001$
21. $g^a = 4$
22. $n^t = 10$

Evaluating a Logarithm **In Exercises 23–26, evaluate the function at the indicated value of x without using a calculator.**

	Function	*Value*
23.	$f(x) = \log_2 x$	$x = 64$
24.	$f(x) = \log_{16} x$	$x = \frac{1}{4}$
25.	$g(x) = \log_{10} x$	$x = \frac{1}{1000}$
26.	$g(x) = \log_{10} x$	$x = 1{,}000{,}000$

Evaluating a Common Logarithm **In Exercises 27–30, use a calculator to evaluate the function at the indicated value of x. Round your result to three decimal places.**

	Function	*Value*
27.	$f(x) = \log_{10} x$	$x = 525$
28.	$f(x) = \log_{10} x$	$x = \frac{4}{5}$
29.	$h(x) = 6 \log_{10} x$	$x = 14.8$
30.	$h(x) = 1.9 \log_{10} x$	$x = 6.7$

Using Properties of Logarithms **In Exercises 31–36, solve the equation for x.**

31. $\log_7 x = \log_7 9$
32. $\log_5 5 = \log_5 x$
33. $\log_8 x = \log_8 10^{-1}$
34. $\log_4 4^3 = \log_4 x$
35. $\log_4 4^2 = \log_4 \sqrt{x}$
36. $\log_3 3^{-5} = \log_3 x^5$

Using Properties of Logarithms **In Exercises 37–40, use the properties of logarithms to simplify the expression.**

37. $\log_4 4^{3x}$
38. $6^{\log_6 36x}$
39. $3 \log_2 \frac{1}{2}$
40. $\frac{1}{15} \log_3 243$

Graphs of Exponential and Logarithmic Functions **In Exercises 41–44, sketch the graphs of f and g in the same coordinate plane.**

41. $f(x) = 7^x$, $g(x) = \log_7 x$
42. $f(x) = 5^x$, $g(x) = \log_5 x$
43. $f(x) = 15^x$, $g(x) = \log_{15} x$
44. $f(x) = 4^x$, $g(x) = \log_4 x$

Library of Parent Functions **In Exercises 45–50, find the domain, x-intercept, and vertical asymptote of the logarithmic function, and sketch its graph by hand.**

45. $f(x) = \log_{10}(x + 4)$
46. $y = \log_4(x - 1)$
47. $y = 1 + \log_{10} x$
48. $f(x) = 2 - \log_{10} x$
49. $f(x) = -\log_6(x + 2)$
50. $y = 1 + \log_8(x - 5)$

Library of Parent Functions **In Exercises 51–54, use the graph of $y = \log_3 x$ to match the function with its graph. [The graphs are labeled (a), (b), (c), and (d).]**

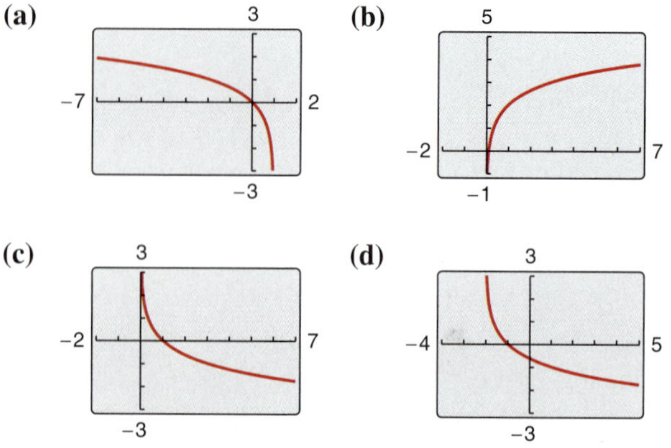

51. $f(x) = \log_3 x + 2$

52. $f(x) = -\log_3 x$

53. $f(x) = -\log_3(x + 2)$

54. $f(x) = \log_3(1 - x)$

Library of Parent Functions **In Exercises 55–60, describe the transformation of the graph of f that yields the graph of g.**

55. $f(x) = \log_{10} x, \quad g(x) = \log_{10}(-x)$

56. $f(x) = \log_{10} x, \quad g(x) = \log_{10}(x + 11)$

57. $f(x) = \log_2 x, \quad g(x) = 5 - \log_2 x$

58. $f(x) = \log_2 x, \quad g(x) = -3 + \log_2 x$

59. $f(x) = \log_8 x, \quad g(x) = -2 + \log_8(x - 3)$

60. $f(x) = \log_8 x, \quad g(x) = 4 + \log_8(x - 1)$

Rewriting a Logarithmic Equation **In Exercises 61–68, write the logarithmic equation in exponential form. For example, the exponential form of $\ln 5 = 1.6094\ldots$ is $e^{1.6094\ldots} = 5$.**

61. $\ln 6 = 1.7917\ldots$ **62.** $\ln 44 = 3.7841\ldots$

63. $\ln e = 1$ **64.** $\ln e^3 = 3$

65. $\ln \sqrt{e} = \dfrac{1}{2}$ **66.** $\ln \dfrac{1}{e^2} = -2$

67. $\ln 9 = 2.1972\ldots$ **68.** $\ln \sqrt[3]{e} = \frac{1}{3}$

Rewriting an Exponential Equation **In Exercises 69–76, write the exponential equation in logarithmic form. For example, the logarithmic form of $e^2 = 7.3890\ldots$ is $\ln 7.3890\ldots = 2$.**

69. $e^3 = 20.0855\ldots$ **70.** $e^0 = 1$

71. $e^{0.9} = 2.4596\ldots$ **72.** $e^{2.5} = 12.1824\ldots$

73. $\sqrt[3]{e} = 1.3956\ldots$ **74.** $\dfrac{1}{e^4} = 0.0183\ldots$

75. $\sqrt{e^3} = 4.4816\ldots$ **76.** $e^{3/4} = 2.1170\ldots$

Evaluating a Natural Logarithm **In Exercises 77–80, use a calculator to evaluate the function at the indicated value of x. Round your result to three decimal places.**

	Function	*Value*
77.	$f(x) = \ln x$	$x = 11$
78.	$f(x) = 3 \ln x$	$x = 0.74$
79.	$g(x) = 8 \ln x$	$x = \sqrt{5}$
80.	$f(x) = -\ln x$	$x = \frac{1}{2}$

Using Properties of Natural Logarithms **In Exercises 81–88, use the properties of natural logarithms to simplify the expression.**

81. $\ln e^2$ **82.** $-\ln e$

83. $e^{\ln 4}$ **84.** $7 \ln e^0$

85. $e \ln 1$ **86.** $e^{\ln 22}$

87. $\ln e^{\ln e}$ **88.** $\ln \dfrac{1}{e^{3\pi}}$

Library of Parent Functions **In Exercises 89–92, find the domain, vertical asymptote, and x-intercept of the logarithmic function, and sketch its graph by hand. Verify using a graphing utility.**

89. $f(x) = \ln(x - 1)$ **90.** $h(x) = \ln(x + 1)$

91. $g(x) = \ln(-x)$ **92.** $f(x) = \ln(3 - x)$

Library of Parent Functions **In Exercises 93–98, use the graph of $f(x) = \ln x$ to describe the transformation that yields the graph of g.**

93. $g(x) = \ln(x + 8)$ **94.** $g(x) = \ln(x - 4)$

95. $g(x) = \ln x - 5$ **96.** $g(x) = \ln x + 4$

97. $g(x) = \ln(x - 1) + 2$

98. $g(x) = \ln(x + 2) - 5$

Analyzing the Graph of a Function **In Exercises 99–108, (a) use a graphing utility to graph the function, (b) find the domain of the function, (c) approximate the open intervals on which the function is increasing and decreasing, and (d) approximate any relative maximum or minimum values of the function. Round your results to three decimal places.**

99. $f(x) = \dfrac{x}{2} - \ln \dfrac{x}{4}$ **100.** $g(x) = 6x \ln x$

101. $h(x) = \dfrac{14 \ln x}{x}$ **102.** $f(x) = \dfrac{x}{\ln x}$

103. $f(x) = \ln \dfrac{x + 2}{x - 1}$ **104.** $f(x) = \ln \dfrac{2x}{x + 2}$

105. $f(x) = \ln \dfrac{x^2}{10}$ **106.** $f(x) = \ln \dfrac{x}{x^2 + 1}$

107. $f(x) = \sqrt{\ln x}$ **108.** $f(x) = (\ln x)^2$

109. Psychology Students in a mathematics class were given an exam and then retested monthly with an equivalent exam. The average scores for the class are approximated by the human memory model

$$f(t) = 80 - 17 \log_{10}(t + 1), \quad 0 \le t \le 12$$

where t is the time in months. Use the model to approximate the average score on the original exam (when $t = 0$), the average score at the end of 2 months, and the average score at the end of 11 months. Verify your answers using a graphing utility.

110. MODELING DATA

The table shows the temperatures T (in degrees Fahrenheit) at which water boils at selected pressures p (in pounds per square inch). (*Source:* Standard Handbook for Mechanical Engineers)

DATA

Pressure, p (in pounds per square inch)	Temperature, T (in degrees Fahrenheit)
5	162.2
10	193.2
14.696 (1 atm)	212.0
20	228.0
30	250.3
40	267.3
60	292.7
80	312.1
100	327.9

Spreadsheet at LarsonPrecalculus.com

A model that approximates the data is

$$T = 87.9 + 35.0 \ln p + 7.9\sqrt{p}.$$

(a) Use a graphing utility to plot the data and graph the model in the same viewing window. How well does the model fit the data?

(b) Use the graph to estimate the pressure at which the boiling point of water is 275°F.

(c) Calculate T when the pressure is 70 pounds per square inch. Verify your answer graphically.

111. Finance A principal P, invested at $3\frac{1}{2}\%$ and compounded continuously, increases to an amount K times the original principal after t years, where

$$T = (\ln K)/0.035.$$

(a) Complete the table and interpret your results.

K	1	2	4	6	8	10	12
t							

(b) Use a graphing utility to graph the function.

112. Audiology The relationship between the number of decibels β and the intensity of a sound I in watts per square meter is

$$\beta = 10 \log_{10}\left(\frac{I}{10^{-12}}\right).$$

(a) Determine the number of decibels of a sound with an intensity of 1 watt per square meter.

(b) Determine the number of decibels of a sound with an intensity of 10^{-2} watt per square meter.

(c) The intensity of the sound in part (a) is 100 times as great as that in part (b). Is the number of decibels 100 times as great? Explain.

113. Real Estate The model

$$t = 22.264 \ln \frac{x}{x - 750}$$

approximates the length of a home mortgage of \$200,000 at 4.5% in terms of the monthly payment. In the model, t is the length of the mortgage in years and x is the monthly payment in dollars.

(a) Use the model to approximate the lengths of a \$200,000 mortgage at 4.5% when the monthly payment is \$1013.37 and when the monthly payment is \$1529.99.

(b) Approximate the total amounts paid over the term of the mortgage with a monthly payment of \$1013.37 and with a monthly payment of \$1529.99. What amount of the total is interest costs for each payment?

114. *Why you should learn it* *(p. 194)* The rate of ventilation required in a public school classroom depends on the volume of air space per child. The model

$$y = 80.4 - 11 \ln x, \quad 100 \le x \le 1500$$

approximates the minimum required rate of ventilation y (in cubic feet per minute per child) in a classroom with x cubic feet of air space per child.

(a) Use a graphing utility to graph the function and approximate the required rate of ventilation in a room with 300 cubic feet of air space per child.

(b) A classroom of 30 students has an air conditioning system that moves 450 cubic feet of air per minute. Determine the rate of ventilation per child.

(c) Use the graph in part (a) to estimate the minimum required air space per child for the classroom in part (b).

(d) The classroom in part (b) has 960 square feet of floor space and a ceiling that is 12 feet high. Is the rate of ventilation for this classroom adequate? Explain.

Kjpargeter/Shutterstock.com

Focusing on Concepts

True or False? **In Exercises 115 and 116, determine whether the statement is true or false. Justify your answer.**

115. You can determine the graph of $f(x) = \log_6 x$ by graphing $g(x) = 6^x$ and reflecting it about the x-axis.

116. The graph of $f(x) = \log_3 x$ contains the point $(27, 3)$.

Think About It **In Exercises 117–120, find the value of the base b so that the graph of $f(x) = \log_b x$ contains the indicated point.**

117. $(32, 5)$ **118.** $(81, 4)$

119. $\left(\frac{1}{81}, 2\right)$ **120.** $\left(\frac{1}{64}, 3\right)$

Library of Parent Functions **In Exercises 121 and 122, determine which equation(s) may be represented by the graph shown. (There may be more than one correct answer.)**

121.

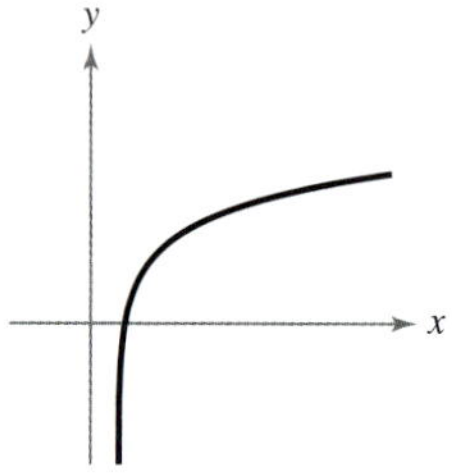

(a) $y = \log_2(x + 1) + 2$
(b) $y = \log_2(x - 1) + 2$
(c) $y = 2 - \log_2(x - 1)$
(d) $y = \log_2(x + 2) + 1$

122.

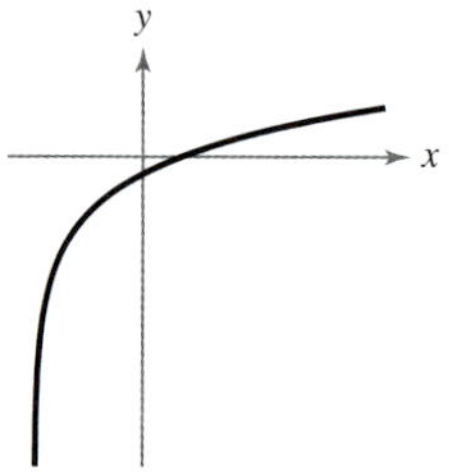

(a) $y = \ln(x - 1) + 2$
(b) $y = \ln(x + 2) - 1$
(c) $y = 2 - \ln(x - 1)$
(d) $y = \ln(x - 2) + 1$

123. Writing Explain why $\log_a x$ is defined only for $0 < a < 1$ and $a > 1$.

124. Exploration Let $f(x) = \ln x$ and $g(x) = x^{1/n}$.

(a) Use a graphing utility to graph g (for $n = 2$) and f in the same viewing window.

(b) Determine which function is increasing at a greater rate as x approaches infinity.

(c) Repeat parts (a) and (b) for $n = 3$, 4, and 5. What do you notice?

125. Exploration

(a) Use a graphing utility to compare the graph of the function $y = \ln x$ with the graph of each function.

$y_1 = x - 1$, $y_2 = (x - 1) - \frac{1}{2}(x - 1)^2$,

$y_3 = (x - 1) - \frac{1}{2}(x - 1)^2 + \frac{1}{3}(x - 1)^3$

(b) Identify the pattern of successive polynomials in part (a). Extend the pattern one more term and compare the graph of the resulting polynomial function with the graph of $y = \ln x$. What do you think the pattern implies?

126. HOW DO YOU SEE IT? The figure shows the graphs of $f(x) = 3^x$ and $g(x) = \log_3 x$. [The graphs are labeled m and n.]

(a) Match each function with its graph.

(b) If $f(a) = b$, what is $g(b)$? Explain.

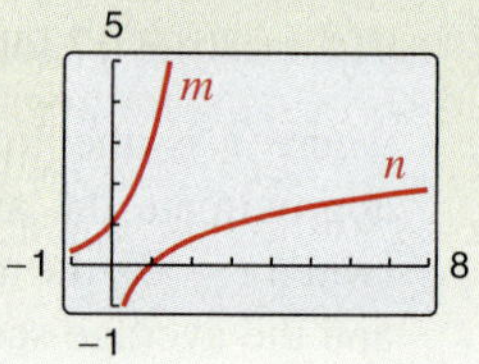

127. Exploration

(a) Use a graphing utility to complete the table for the function $f(x) = (\ln x)/x$.

x	1	5	10	10^2	10^4	10^6
$f(x)$						

(b) Use the table in part (a) to determine what value $f(x)$ approaches as x increases without bound. Use the graphing utility to confirm your result.

128. Writing Use a graphing utility to determine how many months it would take for the average score in Example 10 to decrease to 60. Explain your method of solving the problem. Describe another way that you can use the graphing utility to determine the answer. Also, based on the shape of the graph, does the rate at which a student forgets information *increase* or *decrease* with time? Explain.

Error Analysis **In Exercises 129 and 130, describe the error.**

129.

x	1	2	8
y	0	1	3

From the table, you can conclude that y is an exponential function of x.

130.

x	1	2	5
y	2	4	32

From the table, you can conclude that y is a logarithmic function of x.

Cumulative Mixed Review

Using Graphs **In Exercises 131–134, solve the equation graphically.**

131. $5x - 7 = 7 + 5x$ **132.** $-2x + 3 = 8x$

133. $\sqrt{3x - 2} = 9$ **134.** $\sqrt{x - 11} = x + 2$

3.3 Properties of Logarithms

Change of Base

Most calculators have only two types of log keys, one for common logarithms (base 10) and one for natural logarithms (base e). Although common logs and natural logs are the most frequently used, you may occasionally need to evaluate logarithms with other bases. To do this, you can use the **change-of-base formula.**

Change-of-Base Formula

Let a, b, and x be positive real numbers such that $a \neq 1$ and $b \neq 1$. Then $\log_a x$ can be converted to a different base using any of the following formulas.

Base b	*Base 10*	*Base e*
$\log_a x = \dfrac{\log_b x}{\log_b a}$	$\log_a x = \dfrac{\log_{10} x}{\log_{10} a}$	$\log_a x = \dfrac{\ln x}{\ln a}$

One way to look at the change-of-base formula is that logarithms with base a are *constant multiples* of logarithms with base b. The constant multiplier is $1/(\log_b a)$.

What you should learn

- Rewrite logarithms with different bases.
- Use properties of logarithms to evaluate or rewrite logarithmic expressions.
- Use properties of logarithms to expand or condense logarithmic expressions.
- Use logarithmic functions to model and solve real-life problems.

Why you should learn it

Logarithmic functions can be used to model and solve real-life problems, such as the model for the number of decibels of a sound in Exercise 93 on page 210.

EXAMPLE 1 Changing Bases Using Common Logarithms

a. $\log_4 25 = \dfrac{\log_{10} 25}{\log_{10} 4}$ $\qquad \log_a x = \dfrac{\log_{10} x}{\log_{10} a}$

$\approx \dfrac{1.39794}{0.60206}$ $\qquad$ Use a calculator.

≈ 2.32 $\qquad$ Simplify.

b. $\log_3 17 = \dfrac{\log_{10} 17}{\log_{10} 3} \approx \dfrac{1.23045}{0.47712} \approx 2.58$

✓ **Checkpoint** ▶ Audio-video solution in English & Spanish at LarsonPrecalculus.com

Evaluate $\log_2 12$ using the change-of-base formula and common logarithms.

EXAMPLE 2 Changing Bases Using Natural Logarithms

a. $\log_4 25 = \dfrac{\ln 25}{\ln 4}$ $\qquad \log_a x = \dfrac{\ln x}{\ln a}$

$\approx \dfrac{3.21888}{1.38629}$ $\qquad$ Use a calculator.

≈ 2.32 $\qquad$ Simplify.

b. $\log_3 17 = \dfrac{\ln 17}{\ln 3} \approx \dfrac{2.83321}{1.09861} \approx 2.58$

✓ **Checkpoint** ▶ Audio-video solution in English & Spanish at LarsonPrecalculus.com

Evaluate $\log_2 12$ using the change-of-base formula and natural logarithms.

Notice in Examples 1 and 2 that the result is the same whether common logarithms or natural logarithms are used in the change-of-base formula.

Properties of Logarithms

You know from the previous section that the logarithmic function with base a is the *inverse function* of the exponential function with base a. So, it makes sense that the properties of exponents (see Section 3.1) should have corresponding properties involving logarithms. For instance, the exponential property

$$a^0 = 1$$

has the corresponding logarithmic property

$$\log_a 1 = 0.$$

Properties of Logarithms

Let a be a positive real number such that $a \neq 1$, and let n be a real number. If u and v are positive real numbers, then the following properties are true.

	Logarithm with Base a	*Natural Logarithm*
1. Product Property:	$\log_a(uv) = \log_a u + \log_a v$	$\ln(uv) = \ln u + \ln v$
2. Quotient Property:	$\log_a \frac{u}{v} = \log_a u - \log_a v$	$\ln\frac{u}{v} = \ln u - \ln v$
3. Power Property:	$\log_a u^n = n \log_a u$	$\ln u^n = n \ln u$

(See the proof on page 253.)

Algebra Help

There is no general property that can be used to rewrite $\log_a(u \pm v)$. Specifically, $\log_a(x + y)$ is *not* equal to $\log_a x + \log_a y$.

EXAMPLE 3 Using Properties of Logarithms

Write each logarithm in terms of ln 2 and ln 3.

a. $\ln 6$ **b.** $\ln \frac{2}{27}$

Solution

a. $\ln 6 = \ln(2 \cdot 3)$ Rewrite 6 as $2 \cdot 3$.

$= \ln 2 + \ln 3$ Product Property

b. $\ln \frac{2}{27} = \ln 2 - \ln 27$ Quotient Property

$= \ln 2 - \ln 3^3$ Rewrite 27 as 3^3.

$= \ln 2 - 3 \ln 3$ Power Property

Checkpoint Audio-video solution in English & Spanish at LarsonPrecalculus.com

Write each logarithm in terms of $\log_{10} 3$ and $\log_{10} 5$.

a. $\log_{10} 75$ **b.** $\log_{10} \frac{9}{125}$

Insight

The properties of logarithms on this page (along with the properties in Section 3.2) will help you to solve equations on standardized tests (see Section 3.4).

EXAMPLE 4 Using Properties of Logarithms

Use the properties of logarithms to verify that $-\log_{10} \frac{1}{100} = \log_{10} 100$.

Solution

$-\log_{10} \frac{1}{100} = -\log_{10}(100^{-1})$ Rewrite $\frac{1}{100}$ as 100^{-1}.

$= -(-1) \log_{10} 100$ Power Property

$= \log_{10} 100$ Simplify.

Checkpoint Audio-video solution in English & Spanish at LarsonPrecalculus.com

Use the properties of logarithms to verify that $\ln e^5 = 5$.

Rewriting Logarithmic Expressions

The properties of logarithms are useful for rewriting logarithmic expressions in forms that simplify the operations of algebra. This is true because they convert complicated products, quotients, and exponential forms into simpler sums, differences, and products, respectively.

EXAMPLE 5 Expanding Logarithmic Expressions

Use the properties of logarithms to expand each expression.

a. $\log_4 5x^3y$ **b.** $\ln \dfrac{\sqrt{3x-5}}{7}$

Solution

a. $\log_4 5x^3y = \log_4 5 + \log_4 x^3 + \log_4 y$ — Product Property

$= \log_4 5 + 3\log_4 x + \log_4 y$ — Power Property

b. $\ln \dfrac{\sqrt{3x-5}}{7} = \ln \dfrac{(3x-5)^{1/2}}{7}$ — Rewrite radical using rational exponent.

$= \ln(3x-5)^{1/2} - \ln 7$ — Quotient Property

$= \dfrac{1}{2}\ln(3x-5) - \ln 7$ — Power Property

Checkpoint Audio-video solution in English & Spanish at LarsonPrecalculus.com

Use the properties of logarithms to expand the expression $\log_3 \dfrac{4x^2}{\sqrt{y}}$.

Explore the Concept

Use a graphing utility to graph the functions

$$y = \ln x - \ln(x-3)$$

and

$$y = \ln \frac{x}{x-3}$$

in the same viewing window. Does the graphing utility show the functions with the same domain? Should it? Explain your reasoning.

In Example 5, the properties of logarithms were used to *expand* logarithmic expressions. In Example 6, this procedure is reversed and the properties of logarithms are used to *condense* logarithmic expressions.

EXAMPLE 6 Condensing Logarithmic Expressions

See LarsonPrecalculus.com for an interactive version of this type of example.

Use the properties of logarithms to condense each expression.

a. $\frac{1}{2}\log_{10} x + 3\log_{10}(x+1)$

b. $2\ln(x+2) - \ln x$

c. $\frac{1}{3}[\log_2 x + \log_2(x-4)]$

Solution

a. $\frac{1}{2}\log_{10} x + 3\log_{10}(x+1) = \log_{10} x^{1/2} + \log_{10}(x+1)^3$ — Power Property

$= \log_{10}\left[\sqrt{x}(x+1)^3\right]$ — Product Property

b. $2\ln(x+2) - \ln x = \ln(x+2)^2 - \ln x$ — Power Property

$= \ln \dfrac{(x+2)^2}{x}$ — Quotient Property

c. $\frac{1}{3}[\log_2 x + \log_2(x-4)] = \frac{1}{3}\{\log_2[x(x-4)]\}$ — Product Property

$= \log_2[x(x-4)]^{1/3}$ — Power Property

$= \log_2 \sqrt[3]{x(x-4)}$ — Rewrite with a radical.

Checkpoint Audio-video solution in English & Spanish at LarsonPrecalculus.com

Use the properties of logarithms to condense the expression

$2[\log_{10}(x+3) - 2\log_{10}(x-2)]$.

Application

EXAMPLE 7 Finding a Mathematical Model

The table shows the mean distance x from the sun and the period y (the time it takes a planet to orbit the sun) for each of the six planets that are closest to the sun. In the table, the mean distance is given in astronomical units (where the Earth's mean distance is defined as 1.0), and the period is given in years. The points in the table are plotted in Figure 3.18. Find an equation that relates y and x.

Planet	Mercury	Venus	Earth	Mars	Jupiter	Saturn
Mean distance, x	0.387	0.723	1.000	1.524	5.203	9.537
Period, y	0.241	0.615	1.000	1.881	11.863	29.447

Spreadsheet at LarsonPrecalculus.com

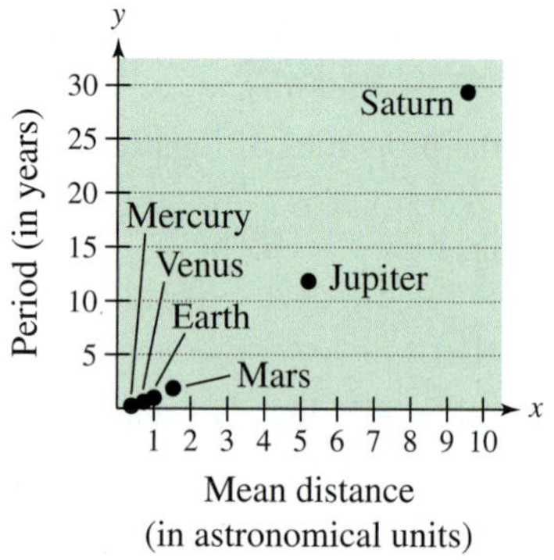

Figure 3.18

Solution

From Figure 3.18, it is not clear how to find an equation that relates y and x. To solve this problem, take the natural log of each of the x- and y-values in the table. This produces the following results.

Planet	Mercury	Venus	Earth	Mars	Jupiter	Saturn
$\ln x = X$	-0.949	-0.324	0.000	0.421	1.649	2.255
$\ln y = Y$	-1.423	-0.486	0.000	0.632	2.473	3.383

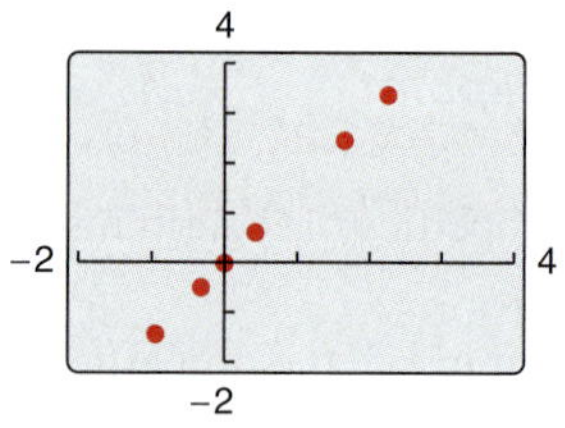

Figure 3.19

Now, by plotting the points in the table, you can see that all six of the points appear to have a linear relationship, as shown in Figure 3.19. To find an equation of the line through these points, you can use one of the following methods.

Method 1: Algebraic

Choose any two points to determine the slope of the line. Using the two points $(0.421, 0.632)$ and $(0, 0)$, you can determine that the slope of the line is

$$m = \frac{0.632 - 0}{0.421 - 0} \approx 1.5 = \frac{3}{2}.$$

By the point-slope form, the equation of the line is

$$Y = \tfrac{3}{2}X$$

where $Y = \ln y$ and $X = \ln x$. You can therefore conclude that

$$\ln y = \frac{3}{2} \ln x.$$

Method 2: Graphical

Using the *linear regression* feature of a graphing utility, you can find a linear model for the data, as shown in Figure 3.20. You can approximate this model to be $Y = 1.5X$, where $Y = \ln y$ and $X = \ln x$. From the model, you can see that the slope of the line is $\frac{3}{2}$. So, you can conclude that $\ln y = \frac{3}{2} \ln x$.

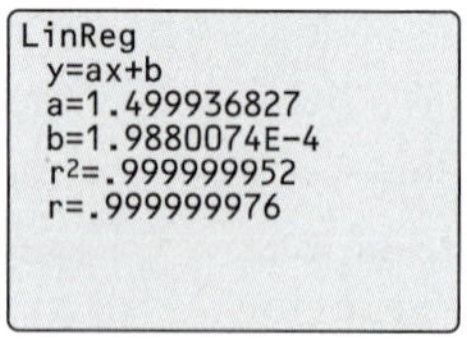

Figure 3.20

✓ ***Checkpoint*** *Audio-video solution in English & Spanish at LarsonPrecalculus.com*

Find a logarithmic equation that relates y and x for the following ordered pairs.

$(0.37, 0.51)$, $(1.00, 1.00)$, $(2.72, 1.95)$, $(7.39, 3.79)$, $(20.09, 7.39)$

In Example 7, try to convert the final equation to $y = f(x)$ form. You will get a function of the form $y = ax^b$, which is called a *power model.*

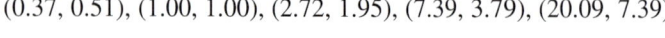

3.3 Exercises

See *CalcChat.com* for tutorial help and worked-out solutions to odd-numbered exercises.
For instructions on how to use a graphing utility, see Appendix A.

Vocabulary and Concept Check

In Exercises 1 and 2, fill in the blank(s).

1. You can evaluate logarithms to any base using the _______ formula.

2. Two properties of logarithms are _______ $= n \log_a u$ and $\ln(uv) =$ _______ .

3. Is $\log_3 24 = \dfrac{\ln 3}{\ln 24}$ or $\log_3 24 = \dfrac{\ln 24}{\ln 3}$ correct?

4. Which property of logarithms can you use to condense the expression $\ln x - \ln 2$?

Procedures and Problem Solving

Rewriting a Logarithm **In Exercises 5–8, rewrite the logarithm as a ratio of (a) common logarithms and (b) natural logarithms.**

5. $\log_5 x$

6. $\log_3 x$

7. $\log_a \frac{5}{6}$

8. $\log_a 4.8$

Changing the Base **In Exercises 9–12, evaluate the logarithm using the change-of-base formula. Round your result to two decimal places.**

9. $\log_5 82$

10. $\log_9 4$

11. $\log_{2/3} 0.9$

12. $\log_{0.3} \frac{1}{8}$

Using Properties of Logarithms **In Exercises 13–16, rewrite the expression in terms of ln 4 and ln 5.**

13. $\ln 20$

14. $\ln 500$

15. $\ln \frac{25}{4}$

16. $\ln \frac{5}{2}$

Using Properties to Evaluate Logarithms **In Exercises 17–20, approximate the logarithm using the properties of logarithms, given the values $\log_b 2 \approx 0.3562$, $\log_b 3 \approx 0.5646$, and $\log_b 5 \approx 0.8271$. Round your result to four decimal places.**

17. $\log_b 8$

18. $\log_b 30$

19. $\log_b \frac{16}{25}$

20. $\log_b \sqrt{3}$

Graphing a Logarithmic Function **In Exercises 21–28, use the change-of-base formula $\log_a x = (\ln x)/(\ln a)$ and a graphing utility to graph the function.**

21. $f(x) = \log_3(x + 1)$

22. $f(x) = \log_2(x - 1)$

23. $f(x) = \log_{1/2}(x - 2)$

24. $f(x) = \log_{1/3}(x + 2)$

25. $f(x) = \log_{1/4} x^2$

26. $f(x) = \log_3 \sqrt{x}$

27. $f(x) = \log_5\left(\dfrac{x}{2}\right)$

28. $f(x) = \log_{1/3}\left(\dfrac{x}{3}\right)$

Simplifying a Logarithm **In Exercises 29–32, use the properties of logarithms to rewrite and simplify the logarithmic expression.**

29. $\log_2(4^2 \cdot 3^4)$

30. $\log_3(9^2 \cdot 2^4)$

31. $\ln \dfrac{6}{e^2}$

32. $\ln \dfrac{e^5}{7}$

Using Properties of Logarithms **In Exercises 33–36, use the properties of logarithms to verify the equation.**

33. $\log_5 \frac{1}{250} = -3 - \log_5 2$

34. $\log_7 50 = 2 \log_7 5 + \log_7 2$

35. $-\ln 24 = -(3 \ln 2 + \ln 3)$

36. $\ln\left(\dfrac{98}{e}\right) = 2 \ln 7 + \ln 2 - 1$

Expanding a Logarithmic Expression **In Exercises 37–50, use the properties of logarithms to expand the expression as a sum, difference, and/or constant multiple of logarithms. (Assume all variables are positive.)**

37. $\log_{10} 9x$

38. $\log_9 23t$

39. $\log_7 \dfrac{t}{8}$

40. $\log_5 \dfrac{7}{z}$

41. $\log_8 x^4$

42. $\log_6 z^{-3}$

43. $\ln \sqrt{z}$

44. $\ln \sqrt[3]{t}$

45. $\log_6 xy^3z^{-2}$

46. $\log_4 xy^{-6} z^4$

47. $\ln \sqrt[3]{\dfrac{x^4}{y^3}}$

48. $\ln \sqrt{\dfrac{x^2}{y^3}}$

49. $\log_b \dfrac{x^4\sqrt{y}}{\sqrt{1 + z}}$

50. $\log_b \dfrac{y^4\sqrt{x + 3}}{z^4}$

Algebraic-Graphical-Numerical **In Exercises 51–54, (a) use a graphing utility to graph the two equations in the same viewing window and (b) use the *table* feature of the graphing utility to create a table of values for each equation. (c) What do the graphs and tables suggest? Verify your conclusion algebraically.**

51. $y_1 = \ln[x^2(x - 4)]$
$y_2 = 2\ln x + \ln(x - 4)$

52. $y_1 = \ln 9x^3$
$y_2 = \ln 9 + 3\ln x$

53. $y_1 = \ln\left(\dfrac{x^4}{x - 2}\right)$
$y_2 = 4\ln x - \ln(x - 2)$

54. $y_1 = \ln\left(\dfrac{\sqrt{x}}{x + 3}\right)$
$y_2 = \dfrac{1}{2}\ln x - \ln(x + 3)$

Condensing a Logarithmic Expression **In Exercises 55–68, use the properties of logarithms to condense the expression.**

55. $\ln x + \ln 4$

56. $\ln y + \ln z$

57. $\log_4 z - \log_4 y$

58. $\log_5 8 - \log_5 t$

59. $4\log_3(x + 2)$

60. $\frac{5}{2}\log_7(z - 4)$

61. $\frac{1}{2}\ln(x^2 + 4) + \ln x$

62. $2\ln x + \ln(x + 1)$

63. $\ln t - 3\ln(t - 5)$

64. $\ln x - 2\ln(x + 2)$

65. $\ln(x - 2) + \ln 2 - 3\ln y$

66. $3\ln x + 2\ln(y + 3) - 4\ln z$

67. $\frac{1}{3}[2\ln(x + 3) + \ln x - \ln(x^2 - 1)]$

68. $\frac{1}{5}[\ln x - \ln(x + 1) - \ln(x - 1)]$

Algebraic-Graphical-Numerical **In Exercises 69–72, (a) use a graphing utility to graph the two equations in the same viewing window and (b) use the *table* feature of the graphing utility to create a table of values for each equation. (c) What do the graphs and tables suggest? Verify your conclusion algebraically.**

69. $y_1 = 2[\ln 8 - \ln(x^2 + 1)]$, $y_2 = \ln\left[\dfrac{64}{(x^2 + 1)^2}\right]$

70. $y_1 = 2[\ln 6 + \ln(x^2 + 1)]$, $y_2 = \ln[36(x^2 + 1)^2]$

71. $y_1 = \ln x + \frac{1}{2}\ln(x + 1)$, $y_2 = \ln\left(x\sqrt{x + 1}\right)$

72. $y_1 = \frac{1}{2}\ln x - \ln(x + 2)$, $y_2 = \ln\left(\dfrac{\sqrt{x}}{x + 2}\right)$

Using Properties to Evaluate Logarithms **In Exercises 73–88, find the exact value of the logarithm without using a calculator. If this is not possible, state the reason.**

73. $\log_3 9$

74. $\log_6 6$

75. $\log_4 16^{3.4}$

76. $\log_5\left(\frac{1}{125}\right)$

77. $\log_2(-4)$

78. $\log_4(-16)$

79. $\log_5 375 - \log_5 3$

80. $\log_4 2 + \log_4 32$

81. $\ln e^{-2} - \ln e^7$

82. $\ln e^6 - 2\ln e^7$

83. $2\ln e^4$

84. $3\ln e^{-4.5}$

85. $\log_9 27$

86. $\log_8 256$

87. $\ln \dfrac{1}{\sqrt{e}}$

88. $\ln \sqrt[5]{e^3}$

Algebraic-Graphical-Numerical **In Exercises 89–92, (a) use a graphing utility to graph the two equations in the same viewing window and (b) use the *table* feature of the graphing utility to create a table of values for each equation. (c) Are the expressions equivalent? Explain. Verify your conclusion algebraically.**

89. $y_1 = \ln x^2$, $y_2 = 2\ln x$

90. $y_1 = 2(\ln 2 + \ln x)$, $y_2 = \ln 4x^2$

91. $y_1 = \ln(x - 2) + \ln(x + 2)$, $y_2 = \ln(x^2 - 4)$

92. $y_1 = \frac{1}{4}\ln[x^4(x^2 + 1)]$, $y_2 = \ln x + \frac{1}{4}\ln(x^2 + 1)$

93. ***Why you should learn it*** *(p. 205)* The relationship between the number of decibels β and the intensity of a sound I in watts per square meter is given by

$$\beta = 10\log_{10}\left(\frac{I}{10^{-12}}\right).$$

(a) Use the properties of logarithms to write the formula in a simpler form.

(b) Use a graphing utility to complete the table. Verify your answers algebraically.

I	10^{-4}	10^{-6}	10^{-8}	10^{-10}	10^{-12}	10^{-14}
β						

94. **Nail Length** The table shows the approximate lengths and diameters (in inches) of common nails. Find a logarithmic equation that relates the diameter y of a common nail to its length x.

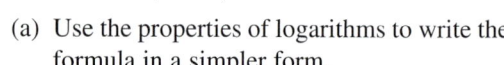

Length, x	Diameter, y
1	0.072
2	0.120
3	0.148
4	0.203
5	0.238
6	0.284

Spreadsheet at LarsonPrecalculus.com

ESB Professional/Shutterstock.com

95. MODELING DATA

A beaker of liquid at an initial temperature of 78°C is placed in a room at a constant temperature of 21°C. The temperature of the liquid is measured every 5 minutes during a half-hour period. The results are recorded as ordered pairs of the form (t, T), where t is the time (in minutes) and T is the temperature (in degrees Celsius). *(Spreadsheet at LarsonPrecalculus.com)*

DATA (0, 78.0°), (5, 66.0°), (10, 57.5°), (15, 51.2°), (20, 46.3°), (25, 42.5°), (30, 39.6°)

(a) The graph of the temperature of the room should be an asymptote of the graph of the model for the data. Subtract the room temperature from each of the temperatures in the ordered pairs. Use a graphing utility to plot the data points (t, T) and $(t, T - 21)$.

(b) An exponential model for the data $(t, T - 21)$ is given by $T - 21 = 54.4(0.964)^t$. Solve for T and graph the model. Compare the result with the plot of the original data.

(c) Take the natural logarithms of the revised temperatures. Use the graphing utility to plot the points $(t, \ln(T - 21))$ and observe that the points appear linear. Use the *regression* feature of the graphing utility to fit a line to the data. The resulting line has the form

$\ln(T - 21) = at + b.$

Use the properties of logarithms to solve for T. Verify that the result is equivalent to the model in part (b).

(d) Take the reciprocals of the y-coordinates of the revised data points to generate the points

$\left(t, \dfrac{1}{T - 21}\right).$

Use the graphing utility to plot these points and observe that they appear linear. Use the *regression* feature of the graphing utility to find a linear model for the points. The resulting line has the form

$\dfrac{1}{T - 21} = at + b.$

Solve for T to obtain a rational model for the original data. Use the graphing utility to graph the rational model and the original data points.

96. Writing Write a short paragraph explaining why the transformations of the data in Exercise 95 were necessary to obtain the models. Why did taking the logarithms of the temperatures lead to a linear scatter plot? Why did taking the reciprocals of the temperatures lead to a linear scatter plot?

Focusing on Concepts

True or False? **In Exercises 97–102, determine whether the statement is true or false given that $f(x) = \ln x$, where $x > 0$. Justify your answer.**

97. $f(ax) = f(a) + f(x),\ a > 0$

98. $f(x - a) = f(x) - f(a),\ x > a$

99. $\sqrt{f(x)} = \frac{1}{2}f(x)$

100. $[f(x)]^n = nf(x)$

101. If $f(x) < 0$, then $0 < x < 1$.

102. If $f(x) > 0$, then $x > e$.

103. Error Analysis Describe the error.

$$\ln\left(\frac{x^2}{\sqrt{x^2 + 4}}\right) = \frac{\ln x^2}{\ln\sqrt{x^2 + 4}} \quad \times$$

104. Think About It Consider the functions below.

$$f(x) = \ln\frac{x}{2},\quad g(x) = \frac{\ln x}{\ln 2},\quad h(x) = \ln x - \ln 2$$

Which two functions have identical graphs? Verify your answer by using a graphing utility to graph all three functions in the same viewing window.

105. Exploration For how many integers between 1 and 20 can the natural logarithms be approximated given that $\ln 2 \approx 0.6931$, $\ln 3 \approx 1.0986$, and $\ln 5 \approx 1.6094$? Approximate these logarithms without a calculator.

106. HOW DO YOU SEE IT? The figure shows the graphs of $y = \ln x$, $y = \ln x^2$, $y = \ln 2x$, and $y = \ln 2$. Match each function with its graph. (The graphs are labeled A through D.) Explain your reasoning.

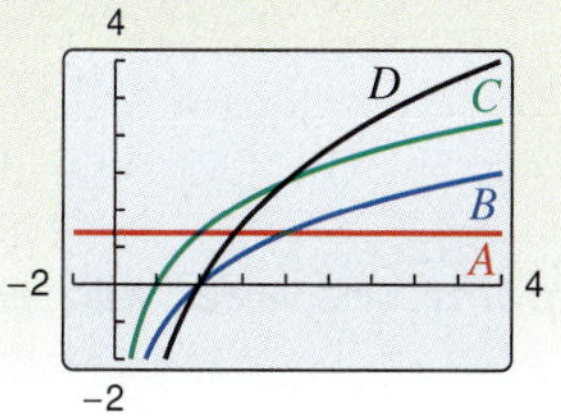

107. Think About It Does $y_1 = \ln[x(x - 2)]$ have the same domain as $y_2 = \ln x + \ln(x - 2)$? Explain.

108. Proof Prove that $\dfrac{\log_a x}{\log_{a/b} x} = 1 + \log_a \dfrac{1}{b}$.

Cumulative Mixed Review

Using Properties of Exponents **In Exercises 109–112, simplify the expression.**

109. $(64x^3y^4)^{-3}(8x^3y^2)^4$

110. $xy(x^{-1} + y^{-1})^{-1}$

111. $\dfrac{24xy^{-2}}{16x^3y}$

112. $\left(\dfrac{2x^3}{3y}\right)^{-3}$

3.4 Solving Exponential and Logarithmic Equations

Introduction

So far in this chapter, you have studied the definitions, graphs, and properties of exponential and logarithmic functions. In this section, you will study procedures for *solving equations* involving exponential and logarithmic functions.

There are two basic strategies for solving exponential or logarithmic equations. The first is based on the One-to-One Properties and the second is based on the Inverse Properties. For $a > 0$ and $a \neq 1$, the properties below are true for all x and y for which $\log_a x$ and $\log_a y$ are defined.

One-to-One Properties

$a^x = a^y$ if and only if $x = y$.

$\log_a x = \log_a y$ if and only if $x = y$.

Inverse Properties

$a^{\log_a x} = x$

$\log_a a^x = x$

What you should learn

- Solve simple exponential and logarithmic equations.
- Solve more complicated exponential equations.
- Solve more complicated logarithmic equations.
- Use exponential and logarithmic equations to model and solve real-life problems.

Why you should learn it

Exponential and logarithmic equations can be used to model and solve real-life problems. For instance, Exercise 132 on page 221 shows how to use an exponential function to model the average heights of men and women.

EXAMPLE 1 Solving Simple Equations

	Original Equation	Rewritten Equation	Solution	Property
a.	$2^x = 32$	$2^x = 2^5$	$x = 5$	One-to-One
b.	$\log_4 x - \log_4 8 = 0$	$\log_4 x = \log_4 8$	$x = 8$	One-to-One
c.	$\ln x - \ln 3 = 0$	$\ln x = \ln 3$	$x = 3$	One-to-One
d.	$\left(\frac{1}{3}\right)^x = 9$	$3^{-x} = 3^2$	$x = -2$	One-to-One
e.	$e^x = 7$	$\ln e^x = \ln 7$	$x = \ln 7$	Inverse
f.	$\ln x = -3$	$e^{\ln x} = e^{-3}$	$x = e^{-3}$	Inverse
g.	$\log_{10} x = -1$	$10^{\log_{10} x} = 10^{-1}$	$x = 10^{-1} = \frac{1}{10}$	Inverse
h.	$\log_3 x = 4$	$3^{\log_3 x} = 3^4$	$x = 81$	Inverse

✓ ***Checkpoint*** ▶ *Audio-video solution in English & Spanish at LarsonPrecalculus.com*

Solve each equation for x.

a. $2^x = 512$ **b.** $\log_6 x = 3$ **c.** $5 - e^x = 0$ **d.** $9^x = \frac{1}{3}$

The strategies used in Example 1 are summarized below.

Strategies for Solving Exponential and Logarithmic Equations

1. Rewrite the original equation in a form that allows the use of the One-to-One Properties of exponential or logarithmic functions.
2. Rewrite an *exponential* equation in logarithmic form and apply the Inverse Property of logarithmic functions.
3. Rewrite a *logarithmic* equation in exponential form and apply the Inverse Property of exponential functions.

Wavebreakmedia Ltd/Wavebreak Media/Getty Images; iStock.com/Allkindza

Solving Exponential Equations

EXAMPLE 2 Solving Exponential Equations

Solve each equation.

a. $e^x = 72$

b. $3(2^x) = 42$

Algebraic Solution

a.
$e^x = 72$ Write original equation.
$\ln e^x = \ln 72$ Take natural log of each side.
$x = \ln 72$ Inverse Property
$x \approx 4.28$ Use a calculator.

The solution is $x = \ln 72 \approx 4.28$. Check this in the original equation.

b.
$3(2^x) = 42$ Write original equation.
$2^x = 14$ Divide each side by 3.
$\log_2 2^x = \log_2 14$ Take log (base 2) of each side.
$x = \log_2 14$ Inverse Property
$x = \dfrac{\ln 14}{\ln 2}$ Change-of-base formula
$x \approx 3.81$ Use a calculator.

The solution is $x = \log_2 14 \approx 3.81$. Check this in the original equation.

Graphical Solution

To solve an equation using a graphing utility, you can graph the left- and right-hand sides of the equation and use the *intersect* feature.

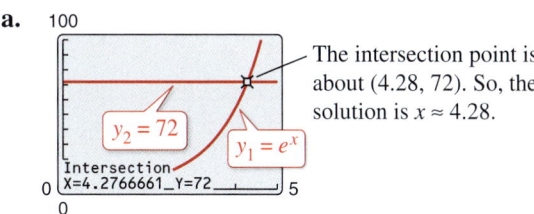

a.

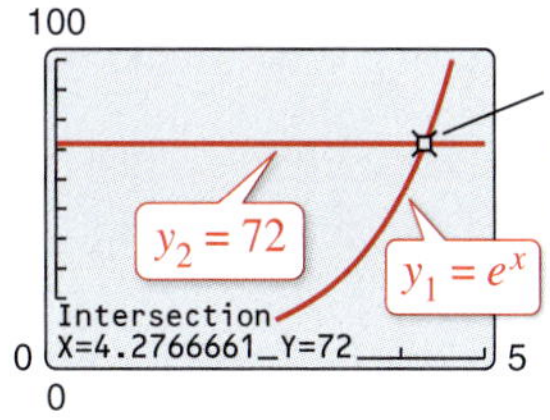

The intersection point is about (4.28, 72). So, the solution is $x \approx 4.28$.

b.

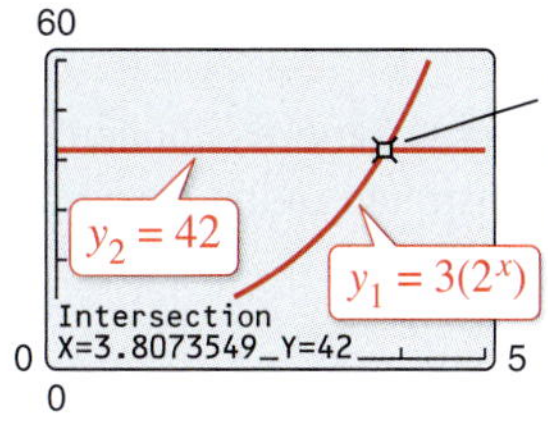

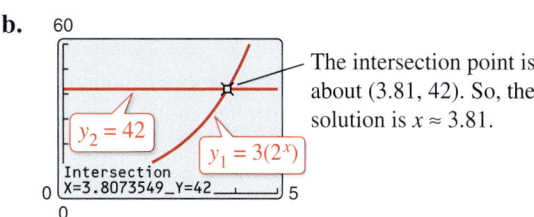

The intersection point is about (3.81, 42). So, the solution is $x \approx 3.81$.

Checkpoint Audio-video solution in English & Spanish at LarsonPrecalculus.com

Solve (a) $e^x = 10$ and (b) $2(5^x) = 32$.

EXAMPLE 3 Solving an Exponential Equation

Solve

$4e^{2x} - 3 = 2.$

Algebraic Solution

$4e^{2x} - 3 = 2$ Write original equation.
$4e^{2x} = 5$ Add 3 to each side.
$e^{2x} = \frac{5}{4}$ Divide each side by 4.
$\ln e^{2x} = \ln \frac{5}{4}$ Take natural log of each side.
$2x = \ln \frac{5}{4}$ Inverse Property
$x = \frac{1}{2} \ln \frac{5}{4}$ Divide each side by 2.
$x \approx 0.11$ Use a calculator.

The solution is $x = \frac{1}{2} \ln \frac{5}{4} \approx 0.11$. Check this in the original equation.

Graphical Solution

Rather than using the procedure in Example 2, another way to solve the equation graphically is first to rewrite the equation as $4e^{2x} - 5 = 0$ and then use a graphing utility to graph $y = 4e^{2x} - 5$. Use the *zero* or *root* feature of the graphing utility to approximate the value of x for which $y = 0$, as shown in the figure.

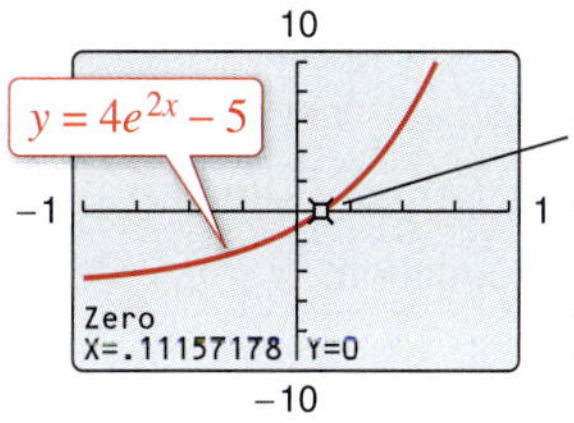

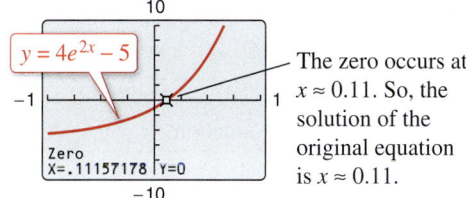

The zero occurs at $x \approx 0.11$. So, the solution of the original equation is $x \approx 0.11$.

Checkpoint Audio-video solution in English & Spanish at LarsonPrecalculus.com

Solve $2e^{3x} + 5 = 29$.

EXAMPLE 4 Solving an Exponential Equation

Solve $2(3^{2t-5}) - 4 = 11$.

Solution

$2(3^{2t-5}) - 4 = 11$	Write original equation.
$2(3^{2t-5}) = 15$	Add 4 to each side.
$3^{2t-5} = \frac{15}{2}$	Divide each side by 2.
$\log_3 3^{2t-5} = \log_3 \frac{15}{2}$	Take log (base 3) of each side.
$2t - 5 = \log_3 \frac{15}{2}$	Inverse Property
$2t = 5 + \log_3 7.5$	Add 5 to each side.
$t = \frac{5}{2} + \frac{1}{2}\log_3 7.5$	Divide each side by 2.
$t \approx 3.42$	Use a calculator.

The solution is $t = \frac{5}{2} + \frac{1}{2}\log_3 7.5 \approx 3.42$. Check this in the original equation.

Checkpoint *Audio-video solution in English & Spanish at LarsonPrecalculus.com*

Solve $6(2^{t+5}) + 4 = 11$.

Algebra Help

Remember that to evaluate a logarithm such as $\log_3 7.5$, you need to use the change-of-base formula.

$$\log_3 7.5 = \frac{\ln 7.5}{\ln 3} \approx 1.834$$

When an equation involves two or more exponential expressions, you can still use a procedure similar to that demonstrated in the previous three examples. However, the algebra is a bit more complicated.

EXAMPLE 5 Solving an Exponential Equation in Quadratic Form

Solve $e^{2x} - 3e^x + 2 = 0$.

Algebraic Solution

$e^{2x} - 3e^x + 2 = 0$	Write original equation.
$(e^x)^2 - 3e^x + 2 = 0$	Write in quadratic form.
$(e^x - 2)(e^x - 1) = 0$	Factor.
$e^x - 2 = 0$	Set 1st factor equal to 0.
$e^x = 2$	Add 2 to each side.
$x = \ln 2$	Solution
$e^x - 1 = 0$	Set 2nd factor equal to 0.
$e^x = 1$	Add 1 to each side.
$x = \ln 1$	Inverse Property
$x = 0$	Solution

The solutions are

$$x = \ln 2 \approx 0.69 \quad \text{and} \quad x = 0.$$

Check these in the original equation.

Graphical Solution

Use a graphing utility to graph $y = e^{2x} - 3e^x + 2$ and then find the zeros.

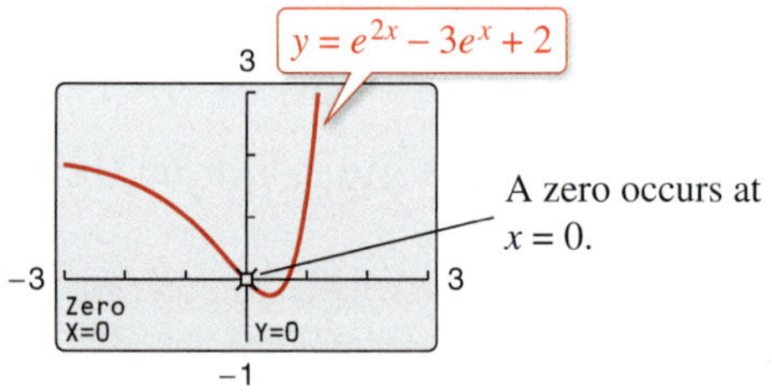

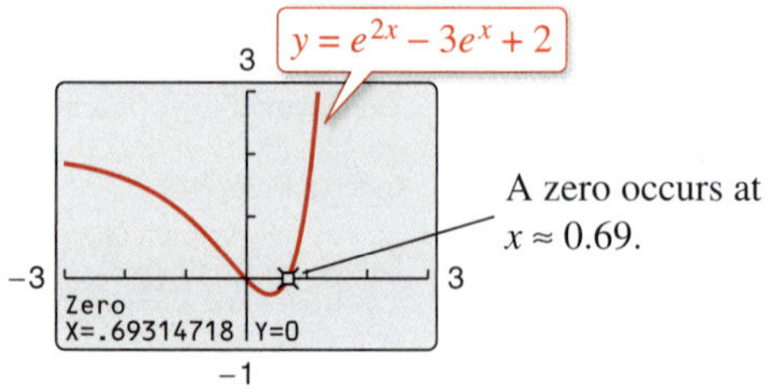

From the figures, you can conclude that the solutions are $x = 0$ and $x \approx 0.69$.

Checkpoint *Audio-video solution in English & Spanish at LarsonPrecalculus.com*

Solve $e^{2x} - 7e^x + 12 = 0$.

Solving Logarithmic Equations

To solve a logarithmic equation, you can write it in exponential form.

$\ln x = 3$ Logarithmic form

$e^{\ln x} = e^3$ Exponentiate each side.

$x = e^3$ Exponential form

This procedure is called *exponentiating* each side of an equation. It is applied after the logarithmic expression has been isolated.

EXAMPLE 6 Solving Logarithmic Equations

Solve each logarithmic equation.

a. $\ln 3x = 2$

b. $\log_3(5x - 1) = \log_3(x + 7)$

Solution

a. $\ln 3x = 2$ Write original equation.

$e^{\ln 3x} = e^2$ Exponentiate each side.

$3x = e^2$ Inverse Property

$x = \frac{1}{3}e^2$ Multiply each side by $\frac{1}{3}$.

$x \approx 2.46$ Use a calculator.

The solution is $x = \frac{1}{3}e^2 \approx 2.46$. Check this in the original equation.

b. $\log_3(5x - 1) = \log_3(x + 7)$ Write original equation.

$5x - 1 = x + 7$ One-to-One Property

$x = 2$ Solve for x.

The solution is $x = 2$. Check this in the original equation.

✓ Checkpoint Audio-video solution in English & Spanish at LarsonPrecalculus.com

Solve each logarithmic equation.

a. $\ln x = \frac{2}{3}$ **b.** $\log_2(2x - 3) = \log_2(x + 4)$

EXAMPLE 7 Solving a Logarithmic Equation

Solve $5 + 2 \ln x = 4$.

Algebraic Solution

$5 + 2 \ln x = 4$ Write original equation.

$2 \ln x = -1$ Subtract 5 from each side.

$\ln x = -\frac{1}{2}$ Divide each side by 2.

$e^{\ln x} = e^{-1/2}$ Exponentiate each side.

$x = e^{-1/2}$ Inverse Property

$x \approx 0.61$ Use a calculator.

The solution is $x = e^{-1/2} \approx 0.61$. Check this in the original equation.

Graphical Solution

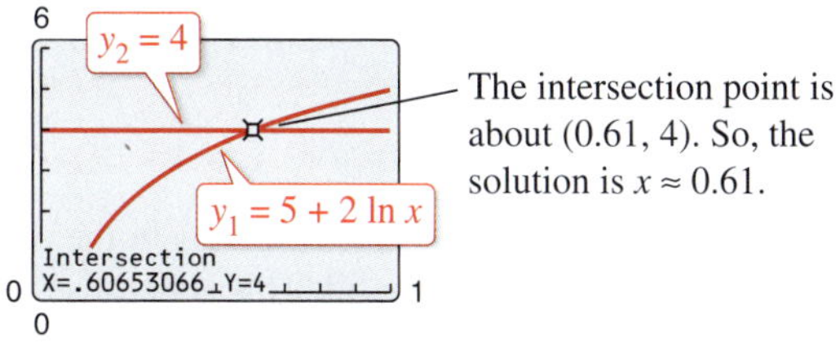

The intersection point is about (0.61, 4). So, the solution is $x \approx 0.61$.

✓ Checkpoint Audio-video solution in English & Spanish at LarsonPrecalculus.com

Solve $7 + 3 \ln x = 5$.

Viorel Sima/Shutterstock.com

EXAMPLE 8 Solving a Logarithmic Equation

Solve $2 \log_5 3x = 4$.

Solution

$$2 \log_5 3x = 4 \quad \text{Write original equation.}$$
$$\log_5 3x = 2 \quad \text{Divide each side by 2.}$$
$$5^{\log_5 3x} = 5^2 \quad \text{Exponentiate each side (base 5).}$$
$$3x = 25 \quad \text{Inverse Property}$$
$$x = \frac{25}{3} \quad \text{Divide each side by 3.}$$

The solution is $x = \frac{25}{3}$. Check this in the original equation. Or, perform a graphical check by graphing

$$y_1 = 2 \log_5 3x = 2\left(\frac{\log_{10} 3x}{\log_{10} 5}\right) \quad \text{and} \quad y_2 = 4$$

in the same viewing window. The two graphs should intersect at $x = \frac{25}{3} \approx 8.33$ and $y = 4$, as shown in Figure 3.21.

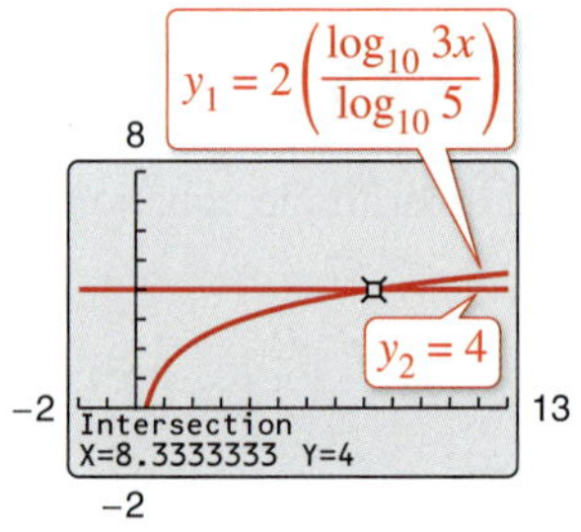

Figure 3.21

Checkpoint *Audio-video solution in English & Spanish at LarsonPrecalculus.com*

Solve $3 \log_4 6x = 9$.

Because the domain of a logarithmic function generally does not include all real numbers, you should be sure to check for extraneous solutions of logarithmic equations, as shown in the next example.

EXAMPLE 9 Checking for Extraneous Solutions

Solve $\ln(x - 2) + \ln(2x - 3) = 2 \ln x$.

Algebraic Solution

$$\ln(x - 2) + \ln(2x - 3) = 2 \ln x \quad \text{Write original equation.}$$
$$\ln[(x - 2)(2x - 3)] = \ln x^2 \quad \text{Use properties of logarithms.}$$
$$\ln(2x^2 - 7x + 6) = \ln x^2 \quad \text{Multiply binomials.}$$
$$2x^2 - 7x + 6 = x^2 \quad \text{One-to-One Property}$$
$$x^2 - 7x + 6 = 0 \quad \text{Write in general form.}$$
$$(x - 6)(x - 1) = 0 \quad \text{Factor.}$$
$$x - 6 = 0 \Rightarrow x = 6 \quad \text{Set 1st factor equal to 0.}$$
$$x - 1 = 0 \Rightarrow x = 1 \quad \text{Set 2nd factor equal to 0.}$$

Finally, by checking these two possible solutions in the original equation, you can conclude that $x = 1$ is not valid. This is because when $x = 1$,

$$\ln(x - 2) + \ln(2x - 3) = \ln(-1) + \ln(-1)$$

which is invalid because -1 is not in the domain of the natural logarithmic function. So, the only solution is $x = 6$.

Graphical Solution

First, rewrite the original equation as

$$\ln(x - 2) + \ln(2x - 3) - 2 \ln x = 0.$$

Then, use a graphing utility to graph the equation

$$y = \ln(x - 2) + \ln(2x - 3) - 2 \ln x$$

and find the zeros, as shown in the figure.

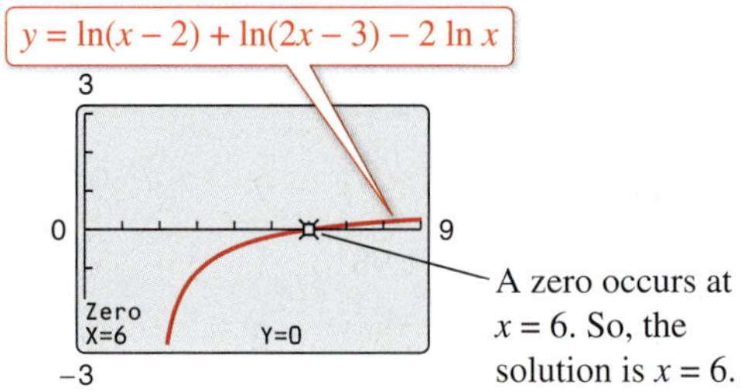

Checkpoint *Audio-video solution in English & Spanish at LarsonPrecalculus.com*

Solve $\log_{10} x + \log_{10}(x - 9) = 1$.

EXAMPLE 10 The Change-of-Base Formula

Prove the change-of-base formula: $\log_a x = \dfrac{\log_b x}{\log_b a}$.

Solution

Begin by letting $y = \log_a x$ and writing the equivalent exponential form

$$a^y = x.$$

Now, taking the logarithms *with base b* of each side produces the following.

$$\log_b a^y = \log_b x$$

$$y \log_b a = \log_b x \qquad \text{Power Property}$$

$$y = \frac{\log_b x}{\log_b a} \qquad \text{Divide each side by } \log_b a.$$

$$\log_a x = \frac{\log_b x}{\log_b a} \qquad \text{Replace } y \text{ with } \log_a x.$$

Checkpoint *Audio-video solution in English & Spanish at LarsonPrecalculus.com*

Prove that $\log_a \dfrac{1}{x} = -\log_a x$.

Equations that involve combinations of algebraic functions, exponential functions, and logarithmic functions can be very difficult to solve by algebraic procedures. Here again, you can take advantage of a graphing utility.

EXAMPLE 11 Approximating the Solution of an Equation

Approximate (to three decimal places) the solution of $\ln x = x^2 - 2$.

Solution

First, rewrite the equation as

$$\ln x - x^2 + 2 = 0.$$

Then, use a graphing utility to graph

$$y = -x^2 + 2 + \ln x$$

as shown in Figure 3.22. From this graph, you can see that the equation has two solutions. Next, using the *zero* or *root* feature, you can approximate the two solutions to be $x \approx 0.138$ and $x \approx 1.564$.

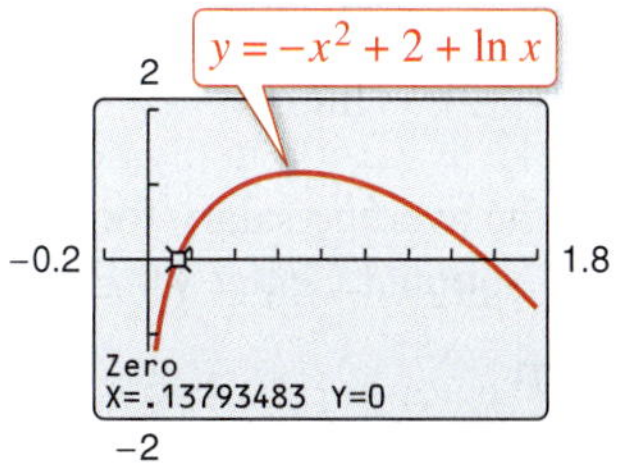

Figure 3.22

Check

$$\ln x = x^2 - 2 \qquad \text{Write original equation.}$$

$$\ln(0.138) \stackrel{?}{\approx} (0.138)^2 - 2 \qquad \text{Substitute 0.138 for } x.$$

$$-1.9805 \approx -1.9810 \qquad \text{Solution checks. ✓}$$

$$\ln(1.564) \stackrel{?}{\approx} (1.564)^2 - 2 \qquad \text{Substitute 1.564 for } x.$$

$$0.4472 \approx 0.4461 \qquad \text{Solution checks. ✓}$$

So, the two solutions $x \approx 0.138$ and $x \approx 1.564$ seem reasonable.

Checkpoint *Audio-video solution in English & Spanish at LarsonPrecalculus.com*

Approximate (to three decimal places) the solution of $5e^{0.3x} = 17$.

Applications

EXAMPLE 12 Doubling an Investment

See LarsonPrecalculus.com for an interactive version of this type of example.

You have deposited \$500 in an account that pays 6.75% interest, compounded continuously. How long will it take your money to double?

Solution

Using the formula for continuous compounding, you can find that the balance in the account is $A = Pe^{rt} = 500e^{0.0675t}$. To find the time required for the balance to double, let $A = 1000$ and solve the resulting equation for t.

$500e^{0.0675t} = 1000$ — Substitute 1000 for A.

$e^{0.0675t} = 2$ — Divide each side by 500.

$\ln e^{0.0675t} = \ln 2$ — Take natural log of each side.

$0.0675t = \ln 2$ — Inverse Property

$t = \dfrac{\ln 2}{0.0675}$ — Divide each side by 0.0675.

$t \approx 10.27$ — Use a calculator.

The balance in the account will double after approximately 10.27 years. This result is demonstrated graphically in Figure 3.23.

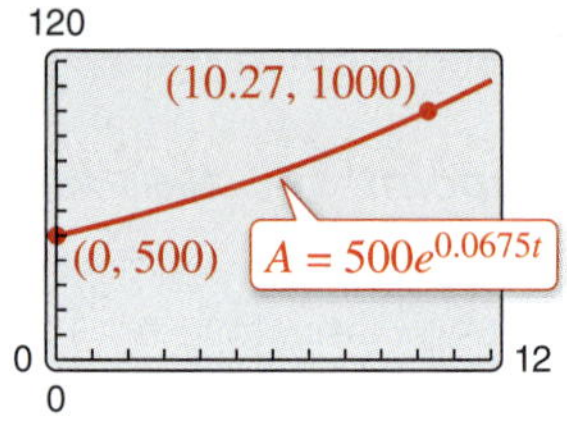

Figure 3.23

Checkpoint *Audio-video solution in English & Spanish at LarsonPrecalculus.com*

You invest \$500 at an annual interest rate of 5.25%, compounded continuously. How long will it take your money to double? Compare your result with that of Example 12.

EXAMPLE 13 Average Salary for Public School Teachers

From 2008 through 2017, the average salary y (in thousands of dollars) for public school teachers for the year t can be modeled by the equation $y = 38.242 + 7.1409 \ln t$, where $t = 8$ represents 2008. During which year did the average salary for public school teachers reach \$56 thousand? (*Source:* National Center for Education Statistics)

Teacher

Solution

$38.242 + 7.1409 \ln t = y$ — Write original equation.

$38.242 + 7.1409 \ln t = 56$ — Substitute 56 for y.

$7.1409 \ln t = 17.758$ — Subtract 38.242 from each side.

$\ln t = \dfrac{17.758}{7.1409}$ — Divide each side by 7.1409.

$e^{\ln t} = e^{17.758/7.1409}$ — Exponentiate each side.

$t = e^{17.758/7.1409}$ — Inverse Property

$t \approx 12.02$ — Use a calculator.

The solution is $t \approx 12.02$ years. Because $t = 8$ represents 2008, it follows that the average salary for public school teachers reached \$56 thousand in 2012.

Checkpoint *Audio-video solution in English & Spanish at LarsonPrecalculus.com*

In Example 13, during which year did the average salary reach \$58.5 thousand?

3.4 Exercises

See *CalcChat.com* for tutorial help and worked-out solutions to odd-numbered exercises. For instructions on how to use a graphing utility, see Appendix A.

Vocabulary and Concept Check

In Exercises 1 and 2, fill in the blank.

1. To solve exponential and logarithmic equations, you can use the following One-to-One and Inverse Properties.

(a) $a^x = a^y$ if and only if _______ .
(b) $\log_a x = \log_a y$ if and only if _______ .
(c) $a^{\log_a x} =$ _______
(d) $\log_a a^x =$ _______

2. A(n) _______ solution does not satisfy the original equation.

3. Is $x = -1$ a solution of $e^{x+1} = 1$?

4. Can you solve $5^x = 125$ using a One-to-One Property?

5. What is the first step in solving the equation $3 + \ln x = 10$?

6. Do you solve $\log_4 x = 2$ by using a One-to-One Property or an Inverse Property?

Procedures and Problem Solving

Solving a Simple Equation **In Exercises 7–16, solve the equation for x. Which property of logarithms or exponents did you use?**

7. $4^x = 16$
8. $\left(\frac{1}{2}\right)^x = 32$
9. $\ln x - \ln 2 = 0$
10. $\log_{10} x - \log_{10} 10 = 0$
11. $e^x = 2$
12. $e^x = \frac{1}{3}$
13. $\ln x = -1$
14. $\log_{10} x = -2$
15. $\log_4 x = 3$
16. $\log_5 x = \frac{1}{2}$

Solving an Exponential Equation Algebraically **In Exercises 17–44, solve the equation algebraically. Approximate the result to three decimal places, if necessary.**

17. $e^x = e^{x^2-2}$
18. $e^{x^2-3} = e^{x-2}$
19. $4(3^x) = 20$
20. $4e^x = 91$
21. $e^x - 8 = 31$
22. $5^x + 8 = 26$
23. $3^{2x} = 80$
24. $4^{-3t} = 0.10$
25. $3^{2-x} = 400$
26. $7^{-3-x} = 242$
27. $8(10^{3x}) = 12$
28. $8(3^{6-x}) = 40$
29. $e^{3x} = 12$
30. $500e^{-2x} = 125$
31. $250e^{0.02x} = 10{,}000$
32. $100e^{0.005x} = 125{,}000$
33. $7 - 2e^x = 5$
34. $-14 + 3e^x = 11$
35. $6(2^{3x-1}) - 7 = 9$
36. $8(4^{6-2x}) + 13 = 41$
37. $3^x = 2^{x-1}$
38. $e^{x+1} + 2^{x+2}$
39. $4^x = 5^{x^2}$
40. $3^{x^2} = 7^{6-x}$
41. $\dfrac{1}{1-e^x} = 5$
42. $\dfrac{100}{1+e^{2x}} = 1$
43. $\left(1 + \dfrac{0.065}{365}\right)^{365t} = 4$
44. $\left(1 + \dfrac{0.10}{12}\right)^{12t} = 2$

Using an Intersection Point **In Exercises 45 and 46, use the graph to approximate the solution of $f(x) = g(x)$ accurate to three decimal places. Then solve the equation $f(x) = g(x)$ algebraically to verify your approximation.**

45. $f(x) = 2^x$, $g(x) = 8$
46. $f(x) = e^{-2x}$, $g(x) = 3$

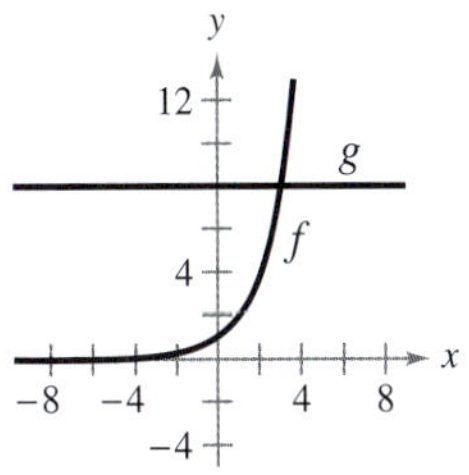

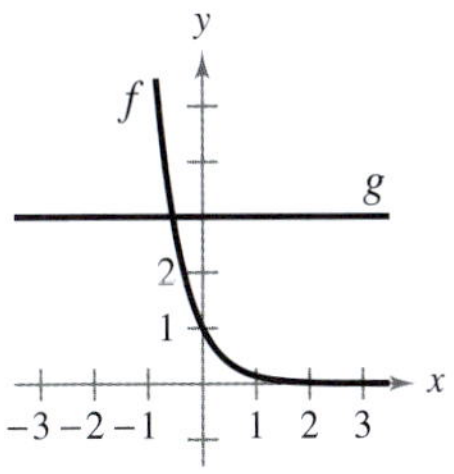

Solving an Exponential Equation Graphically **In Exercises 47–52, use the *zero* or *root* feature or the *zoom* and *trace* features of a graphing utility to approximate the solution of the equation accurate to three decimal places.**

47. $e^{0.2x+3} + 1 = 8$
48. $6^{1-4.7x} = 11$
49. $\left(1 + \dfrac{0.01}{365}\right)^{365t} = 4$
50. $\left(4 - \dfrac{2.471}{40}\right)^{9t} = 21$
51. $\dfrac{7000}{5 + e^{3x}} = 2$
52. $\dfrac{119}{e^{6x} - 14} = 7$

Solving an Exponential Equation in Quadratic Form **In Exercises 53–56, solve the exponential equation.**

53. $e^{2x} - 4e^x - 5 = 0$
54. $e^{2x} - 5e^x + 6 = 0$
55. $6e^{2x} - 11e^x = -4$
56. $5e^{2x} - 7e^x = -2$

Finding the Zero of an Exponential Function In Exercises 57–60, use a graphing utility to graph the function and approximate its zero accurate to three decimal places.

57. $g(x) = 6e^{1-x} - 25$
58. $f(x) = 3e^{3x/2} - 962$
59. $g(t) = e^{0.09t} - 3$
60. $h(t) = e^{-0.125t} - 8$

Solving a Logarithmic Equation Algebraically In Exercises 61–74, solve the equation algebraically. Approximate the result to three decimal places, if necessary.

61. $2.1 = \ln 6x$
62. $\log_{10} 3z = 2$
63. $3 - 4 \ln x = 11$
64. $3 + 8 \ln x = 7$
65. $6 \log_3 0.5x = 11$
66. $4 \log_{10}(x - 6) = 11$
67. $\ln(2x - 1) = 5$
68. $\ln(3x + 5) = 8$
69. $\log_{10}(z^2 + 19) = 2$
70. $\log_{12} x^2 = 6$
71. $\ln\sqrt{x + 2} = 1$
72. $\ln\sqrt{x - 8} = 5$
73. $\ln(x + 1)^2 = 2$
74. $\ln(x - 4)^2 = 5$

Using an Intersection Point In Exercises 75 and 76, use the graph to approximate the solution of $f(x) = g(x)$. Then solve the equation $f(x) = g(x)$ algebraically to verify your approximation.

75. $f(x) = \log_3 x$, $g(x) = 2$

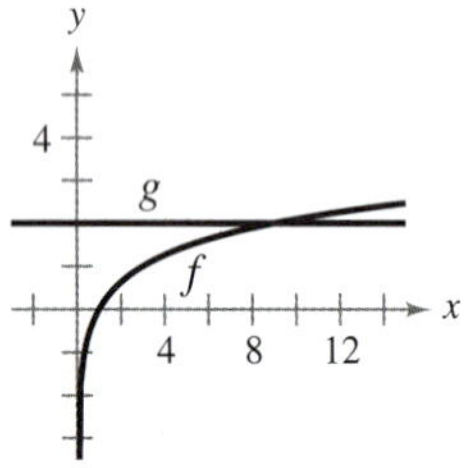

76. $f(x) = \log_5 2x$, $g(x) = -1$

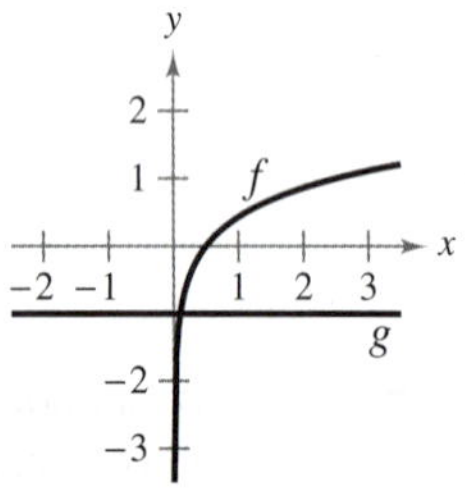

Solving a Logarithmic Equation Graphically In Exercises 77–84, use the *zero* or *root* feature or the *zoom* and *trace* features of a graphing utility to approximate the solution(s) of the equation accurate to three decimal places.

77. $2 \ln(x + 3) = 3$
78. $\ln(x + 1) = 2 - \ln x$
79. $\sqrt{\log_{10}(x + 1)} = 0.4$
80. $\sqrt{\log_{10}(x + 3)} = 0.6$
81. $1 - \log_3 2x = 3$
82. $2 + \log_4 3.5x = 5$
83. $5 \log_8 3x^2 = -2$
84. $2 \log_5 0.5x^2 = -1$

Checking a Logarithmic Equation for Extraneous Solutions In Exercises 85–96, solve the equation (if possible). Identify any extraneous solutions.

85. $\ln x - \ln(x + 1) = 2$
86. $\ln x + \ln(x + 1) = 1$
87. $\ln(x + 5) = \ln(x - 1) - \ln(x + 1)$
88. $\ln(x + 1) - \ln(x - 2) = \ln x$
89. $\log_{10}(3x + 4) = \log_{10}(x - 10)$
90. $\log_2 x + \log_2(x + 2) = \log_2(x + 6)$
91. $\log_9 x + \log_9(x + 2) - \frac{1}{2} = 0$
92. $\log_3 x + \log_3(x - 8) = 2$
93. $\ln(x + 5) = \ln(x - 1) + \ln(x + 1)$
94. $\ln x = \ln(x + 3) - \ln(x - 1)$
95. $\log_{10} 8x - \log_{10}(1 + \sqrt{x}) = 2$
96. $\log_{10} 4x - \log_{10}(12 + \sqrt{x}) = 2$

Verifying a Formula In Exercises 97 and 98, use the properties on page 212 to prove that the equation is true.

97. $\log_a x^n = n \log_a x$
98. $\log_a xy = \log_a x + \log_a y$

Approximating the Solution(s) of an Equation In Exercises 99–106, use the *zero* or *root* feature of a graphing utility to approximate the solution(s) of the equation accurate to three decimal places.

99. $\log_{10} x = x^3 - 3$
100. $\log_{10} x = (x - 3)^2$
101. $\log_{10} x + e^{0.5x} = 6$
102. $e^x \log_{10} x = 7$
103. $\ln(x + 2) - 3^{x-2} + 10 = 5$
104. $\ln x^2 - e^x = -3 - \ln x^2$
105. $x^5 - \log_{10} x = e^{-x}$
106. $\ln x + x^5 = e^{1-2x}$

Algebraic-Graphical-Numerical In Exercises 107–114, (a) complete the table to find an interval containing the solution of the equation, (b) use a graphing utility to graph both sides of the equation to estimate the solution, and (c) solve the equation algebraically. Round your results to three decimal places.

107. $e^{3x} = 12$

x	0.6	0.7	0.8	0.9	1.0
e^{3x}					

108. $4e^{5x} = 24$

x	0.1	0.2	0.3	0.4	0.5
$4e^{5x}$					

109. $20(100 - e^{x/2}) = 500$

x	5	6	7	8	9
$20(100 - e^{x/2})$					

110. $11(77 - e^{x-4}) = 264$

x	5	6	7	8	9
$11(77 - e^{x-4})$					

111. $\ln 2x = 2.4$

x	2	3	4	5	6
$\ln 2x$					

112. $3 \ln 5x = 10$

x	4	5	6	7	8
$3 \ln 5x$					

113. $6 \log_3(0.5x) = 11$

x	12	13	14	15	16
$6 \log_3(0.5x)$					

114. $5 \log_{10}(x - 2) = 11$

x	150	155	160	165	170
$5 \log_{10}(x - 2)$					

Solving Exponential and Logarithmic Equations **In Exercises 115–122, solve the equation algebraically. Round the result to three decimal places. Verify your answer using a graphing utility.**

115. $2x^2e^{2x} + 2xe^{2x} = 0$ **116.** $-x^2e^{-x} + 2xe^{-x} = 0$

117. $-xe^{-x} + e^{-x} = 0$ **118.** $e^{-2x} - 2xe^{-2x} = 0$

119. $2x \ln x + x = 0$ **120.** $\dfrac{1 - \ln x}{x^2} = 0$

121. $\dfrac{1 + \ln x}{2} = 0$ **122.** $3x \ln\left(\dfrac{1}{x}\right) - x = 0$

Solving a Model for x **In Exercises 123–126, the equation represents the given type of model, which you will use in Section 3.5. Solve the equation for x.**

Model Type	*Equation*
123. Exponential growth	$y = ae^{bx}$
124. Exponential decay	$y = ae^{-bx}$
125. Gaussian	$y = ae^{-(x-b)^2/c}$
126. Logarithmic	$y = a + b \ln x$

Doubling and Tripling an Investment **In Exercises 127–130, find the time required for a \$1000 investment to (a) double at interest rate r, compounded continuously, and (b) triple at interest rate r, compounded continuously. Round your results to two decimal places.**

127. $r = 7\%$ **128.** $r = 6\%$

129. $r = 2.5\%$ **130.** $r = 3.75\%$

131. Economics The percent p (in decimal form) of a population who have heard a pop song can be modeled by

$$p = \frac{1}{1 + e^{-(t-93)/22.5}}$$

where t is the number of weeks after the song was released. Find the number of weeks t when the percent of the population who have heard the song is (a) 50% and (b) 80%.

132. ***Why you should learn it*** *(p. 212)* The percent m of American males between the ages of 18 and 24 who are no more than x inches tall is modeled by

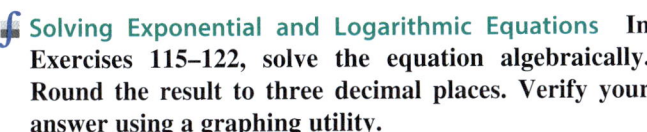

$$m(x) = \frac{100}{1 + e^{-0.6114(x-69.71)}}$$

and the percent f of American females between the ages of 18 and 24 who are no more than x inches tall is modeled by

$$f(x) = \frac{100}{1 + e^{-0.66607(x-64.51)}}.$$

(*Source:* U.S. National Center for Health Statistics)

(a) Use a graphing utility to graph the two functions in the same viewing window.

(b) Use the graphs in part (a) to determine the horizontal asymptotes of the functions. Interpret their meanings in the context of the problem.

(c) What is the average height for each sex?

133. Finance The numbers y (in thousands) of federally insured commercial banks in the United States from 2008 through 2016 can be modeled by

$$y = 13 - 2.8 \ln t$$

where t represents the year, with $t = 8$ corresponding to 2008. In what year were there about 6500 commercial banks? (*Source:* Federal Deposit Insurance Corp.)

134. Forestry The yield V (in millions of cubic feet per acre) for a forest at age t years is given by $V = 6.7e^{-48.1/t}$.

(a) Use a graphing utility to graph the function.

(b) Determine the horizontal asymptote of the function. Interpret its meaning in the context of the problem.

(c) Find the time necessary to obtain a yield of 1.3 million cubic feet.

Wavebreakmedia Ltd/Wavebreak Media/Getty Images; iStock.com/Allkindza

135. Science An object at a temperature of 160°C was removed from a furnace and placed in a room at 20°C. The temperature T (in degrees Celsius) of the object was measured after each hour h and recorded in the table. A model for the data is given by $T = 20[1 + 7(2^{-h})]$.

DATA

Hour, h	Temperature, T
0	160°
1	90°
2	56°
3	38°
4	29°
5	24°

Spreadsheet at LarsonPrecalculus.com

(a) Use a graphing utility to plot the data and graph the model in the same viewing window.

(b) Identify the horizontal asymptote of the graph. Interpret its meaning in the context of the problem.

(c) How long does it take the object to cool to 80°C?

136. MODELING DATA

The table shows the population P (in hundreds) of an endangered species from 2011 through 2016. The data can be modeled by

$P = 7 - 2.72 \ln t$

where t represents the year, with $t = 1$ corresponding to 2011.

DATA

Year	Population, P
2011	6.83
2012	5.12
2013	4.38
2014	3.15
2015	2.8
2016	1.76

Spreadsheet at LarsonPrecalculus.com

(a) Use a graphing utility to plot the data and graph the model in the same viewing window.

(b) Use the model to approximate the population each year from 2011 through 2016.

(c) Compare the estimated values to the actual data. Is the model a good fit for the data? Explain.

(d) According to the model, when will the species become extinct?

Focusing on Concepts

True or False? **In Exercises 137 and 138, determine whether the statement is true or false. Justify your answer.**

137. An exponential equation must have at least one solution.

138. A logarithmic equation can have at most one extraneous solution.

139. Error Analysis Describe the error.

$$2e^x = 10$$
$$\ln(2e^x) = \ln 10$$
$$2x = \ln 10$$
$$x = \tfrac{1}{2}\ln 10$$

140. HOW DO YOU SEE IT? Solving $\log_3 x + \log_3(x - 8) = 2$ algebraically, the solutions appear to be $x = 9$ and $x = -1$. Use the graph of $y = \log_3 x + \log_3(x - 8) - 2$ to determine whether each value is an actual solution of the equation. Explain your reasoning.

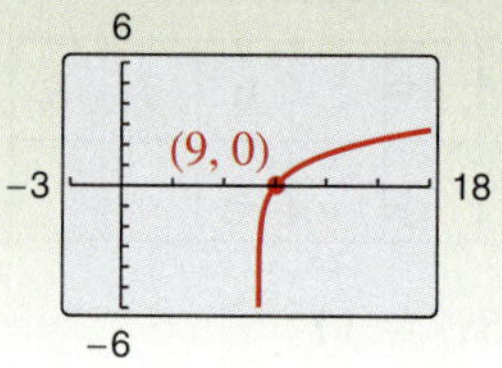

141. Exploration Let

$f(x) = \log_a x$ and $g(x) = a^x$

where $a > 1$.

(a) Let $a = 1.2$ and use a graphing utility to graph the two functions in the same viewing window. What do you observe? Approximate any points of intersection of the two graphs.

(b) Determine the value(s) of a for which the two graphs have one point of intersection.

(c) Determine the value(s) of a for which the two graphs have two points of intersection.

142. Think About It Is the time required for a continuously compounded investment to quadruple twice as long as the time required for it to double? Give a reason for your answer and verify your answer algebraically.

Cumulative Mixed Review

Sketching Graphs **In Exercises 143–146, sketch the graph of the function.**

143. $f(x) = 3x^3 - 4$

144. $f(x) = |x - 2| - 8$

145. $f(x) = \begin{cases} 2x + 1, & x < 0 \\ -x^2, & x \geq 0 \end{cases}$

146. $f(x) = \begin{cases} x - 9, & x \leq 1 \\ x^2 + 1, & x > 1 \end{cases}$

3.5 Exponential and Logarithmic Models

Introduction

There are many examples of exponential and logarithmic models in real life. In Section 3.1, you used the formula

$A = Pe^{rt}$ Exponential model

to find the balance in an account when the interest was compounded continuously. In Section 3.2, Example 10, you used the human memory model

$f(t) = 75 - 6 \ln(t + 1)$. Logarithmic model

The five most common types of mathematical models involving exponential functions or logarithmic functions are listed below.

1. **Exponential growth model:** $y = ae^{bx}, \quad b > 0$
2. **Exponential decay model:** $y = ae^{-bx}, \quad b > 0$
3. **Gaussian model:** $y = ae^{-(x-b)^2/c}$
4. **Logistic growth model:** $y = \dfrac{a}{1 + be^{-rx}}$
5. **Logarithmic models:** $y = a + b \ln x, \quad y = a + b \log_{10} x$

The basic shapes of the graphs of these models are shown in Figure 3.24.

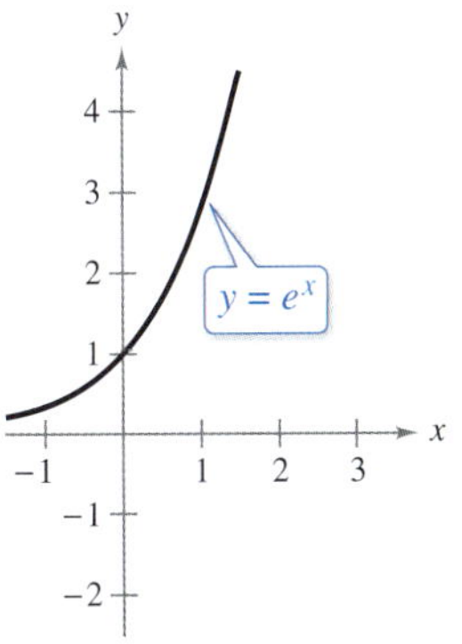

Exponential Growth Model

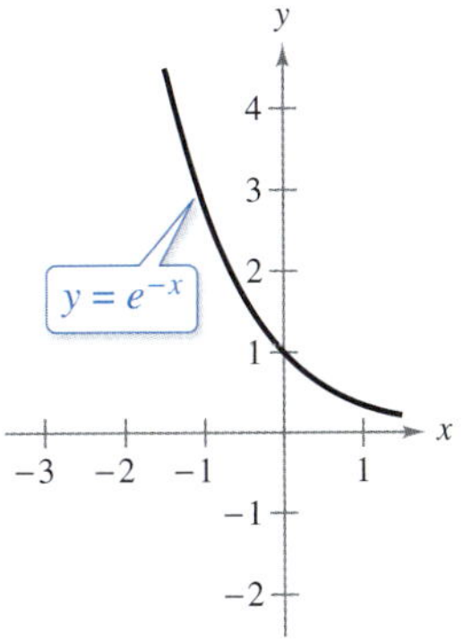

Exponential Decay Model

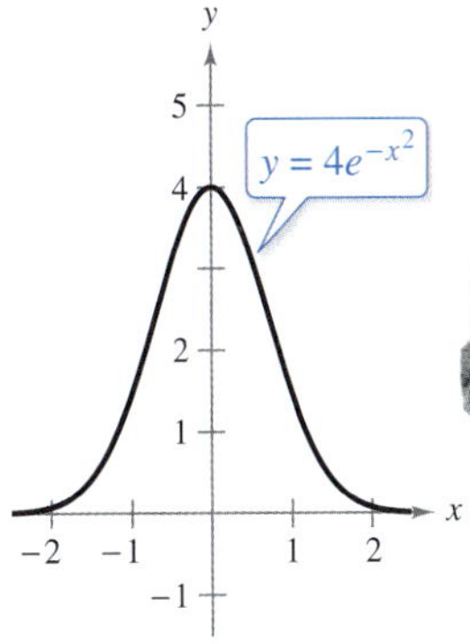

Gaussian Model

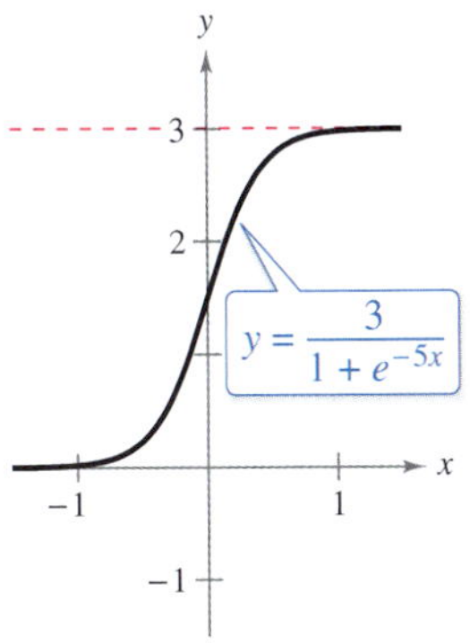

Logistic Growth Model

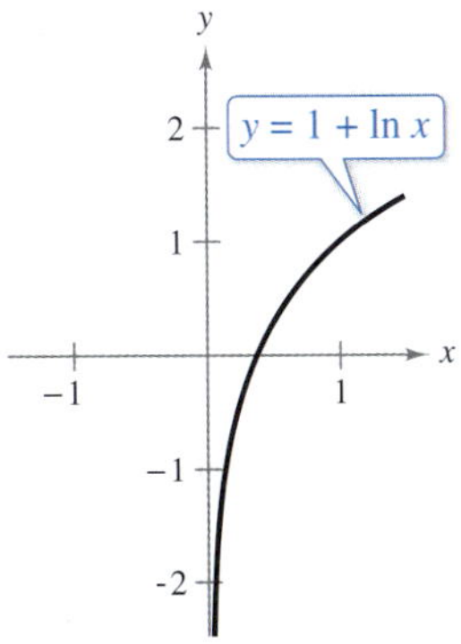

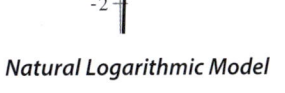

Natural Logarithmic Model

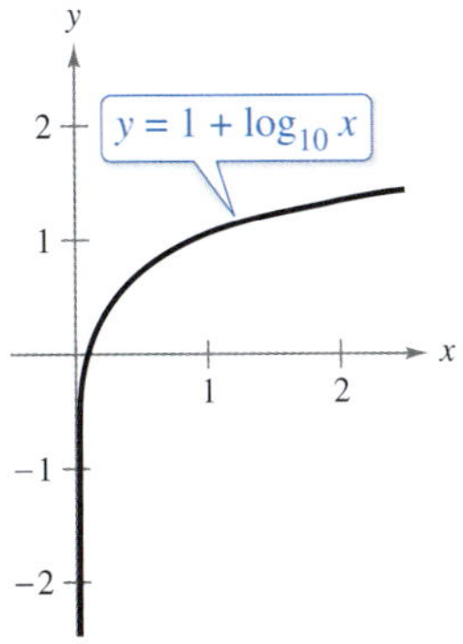

Common Logarithmic Model

Figure 3.24

You can often gain quite a bit of insight into a situation modeled by an exponential or logarithmic function by identifying and interpreting the function's asymptotes.

What you should learn

- Recognize the five most common types of models involving exponential or logarithmic functions.
- Use exponential growth and decay functions to model and solve real-life problems.
- Use Gaussian functions to model and solve real-life problems.
- Use logistic growth functions to model and solve real-life problems.
- Use logarithmic functions to model and solve real-life problems.

Why you should learn it

Exponential decay models are used in carbon dating. For instance, in Exercise 39 on page 232, you will use an exponential decay model to estimate the age of a piece of ancient charcoal.

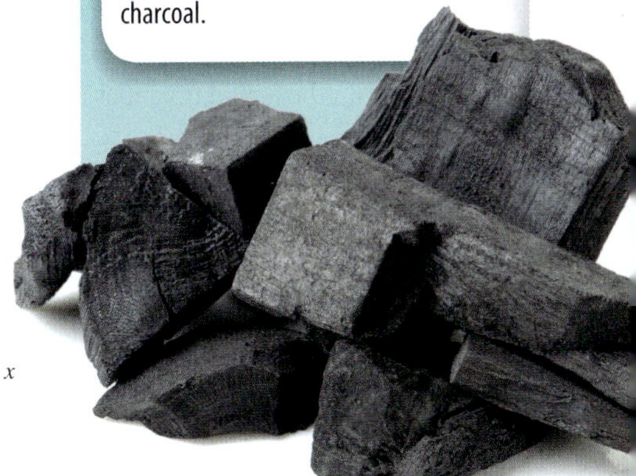

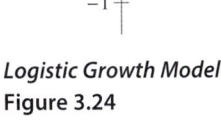

Nor Gal/Shutterstock.com

Exponential Growth and Decay

EXAMPLE 1 Revenues

See LarsonPrecalculus.com for an interactive version of this type of example.

The table shows the revenues (in billions of dollars) for Amazon.com from 2011 through 2016. A scatter plot of the data is shown in the figure at the right. *(Source: Amazon.com)*

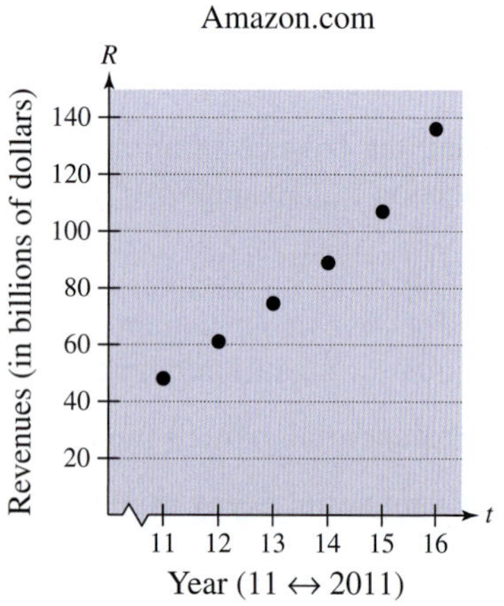

DATA

Year	Revenues, R (in billions of dollars)
2011	48.1
2012	61.1
2013	74.5
2014	89.0
2015	107.0
2016	136.0

Spreadsheet at LarsonPrecalculus.com

An exponential growth model that approximates these data is given by

$$R = 5.3e^{0.2016t}$$

where R is the revenues (in billions of dollars) and $t = 11$ represents 2011. Compare the values given by the model with the revenues shown in the table. According to this model, when will the revenues reach \$300 billion?

Algebraic Solution

The table shows the actual revenues and the values given by the model.

Year	2011	2012	2013	2014	2015	2016
Revenues	48.1	61.1	74.5	89.0	107.0	136.0
Model	48.7	59.6	72.9	89.1	109	133.4

From the table, it appears that the model is a good fit for the data. To find when the revenues will reach \$300 billion, let $R = 300$ in the model and solve for t.

$5.3e^{0.2016t} = R$ — Write original equation.

$5.3e^{0.2016t} = 300$ — Substitute 300 for R.

$e^{0.2016t} = \dfrac{300}{5.3}$ — Divide each side by 5.3.

$\ln e^{0.2016t} = \ln \dfrac{300}{5.3}$ — Take natural log of each side.

$0.2016t = \ln \dfrac{300}{5.3}$ — Inverse Property

$t \approx 20.0$ — Divide each side by 0.2016 and use a calculator.

According to the model, the revenues will reach \$300 billion in 2020.

Graphical Solution

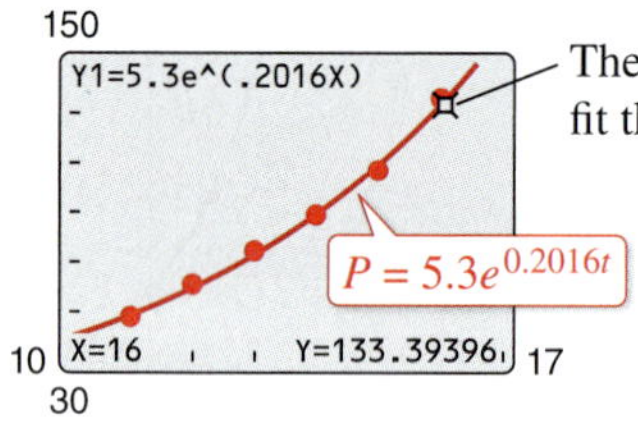

The model appears to fit the data closely.

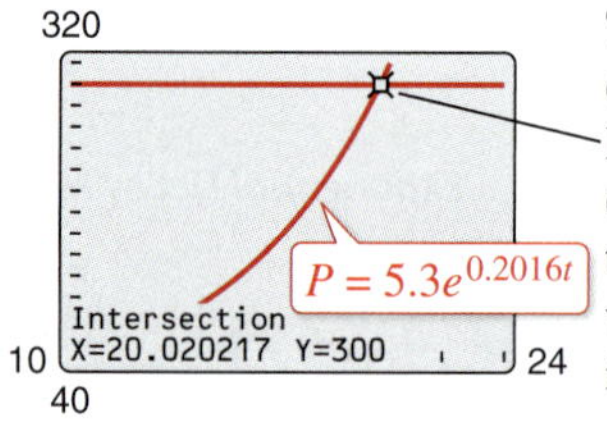

The intersection point of the model and the line $y = 300$ is about (20.0, 300). So, according to the model, the revenues will reach \$300 billion in 2020.

Checkpoint *Audio-video solution in English & Spanish at LarsonPrecalculus.com*

In Example 1, when will the revenues reach \$370 billion?

An exponential model increases (or decreases) by the same percent each year. What is the annual percent increase for the model in Example 1?

In Example 1, you were given the exponential growth model. Sometimes you must find such a model. One technique for doing this is shown in Example 2.

EXAMPLE 2 Modeling Population Growth

In a research experiment, a population of fruit flies is increasing according to the law of exponential growth. After 2 days there are 100 flies, and after 4 days there are 300 flies. How many flies will there be after 5 days?

Entomologist

Solution

Let y be the number of flies at time t (in days). From the given information, you know that $y = 100$ when $t = 2$ and $y = 300$ when $t = 4$. Substituting this information into the model $y = ae^{bt}$ produces

$$100 = ae^{2b} \quad \text{and} \quad 300 = ae^{4b}.$$

To solve for b, solve for a in the first equation.

$$100 = ae^{2b} \quad \Rightarrow \quad a = \frac{100}{e^{2b}} \qquad \text{Solve for } a \text{ in the first equation.}$$

Then substitute the result into the second equation.

$$300 = ae^{4b} \qquad \text{Write second equation.}$$

$$300 = \left(\frac{100}{e^{2b}}\right)e^{4b} \qquad \text{Substitute } \frac{100}{e^{2b}} \text{ for } a.$$

$$300 = 100e^{2b} \qquad \text{Simplify.}$$

$$3 = e^{2b} \qquad \text{Divide each side by 100.}$$

$$\ln 3 = \ln e^{2b} \qquad \text{Take natural log of each side.}$$

$$\ln 3 = 2b \qquad \text{Inverse Property}$$

$$\frac{1}{2}\ln 3 = b \qquad \text{Solve for } b.$$

Using $b = \frac{1}{2}\ln 3$ and the equation you found for a, you can determine that

$$a = \frac{100}{e^{2[(1/2)\ln 3]}} \qquad \text{Substitute } \tfrac{1}{2}\ln 3 \text{ for } b.$$

$$= \frac{100}{e^{\ln 3}} \qquad \text{Simplify.}$$

$$= \frac{100}{3} \qquad \text{Inverse Property}$$

$$\approx 33.33. \qquad \text{Simplify.}$$

So, with $a \approx 33.33$ and $b = \frac{1}{2}\ln 3 \approx 0.5493$, the exponential growth model is

$$y = 33.33e^{0.5493t}$$

as shown in Figure 3.25. After 5 days, the population will be

$$y = 33.33e^{0.5493(5)} \approx 520 \text{ flies.}$$

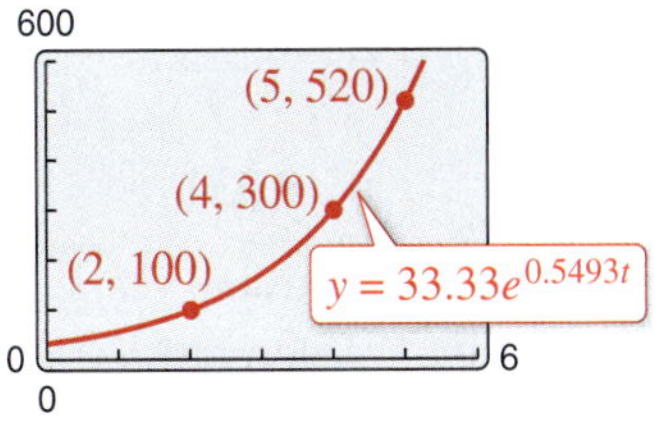

Figure 3.25

✓ **Checkpoint** *Audio-video solution in English & Spanish at LarsonPrecalculus.com*

The number of bacteria in a culture is increasing according to the law of exponential growth. After 1 hour there are 100 bacteria, and after 2 hours there are 200 bacteria. How many bacteria will there be after 3 hours?

Vitalii Hulai/Shutterstock.com

In living organic material, the ratio of the content of radioactive carbon isotopes (carbon-14) to the content of nonradioactive carbon isotopes (carbon-12) is about 1 to 10^{12}. When organic material dies, its carbon-12 content remains fixed, whereas its radioactive carbon-14 begins to decay with a half-life of about 5700 years. To estimate the age of dead organic material, scientists use the carbon dating model

$$R = \frac{1}{10^{12}}e^{-t/8223}$$

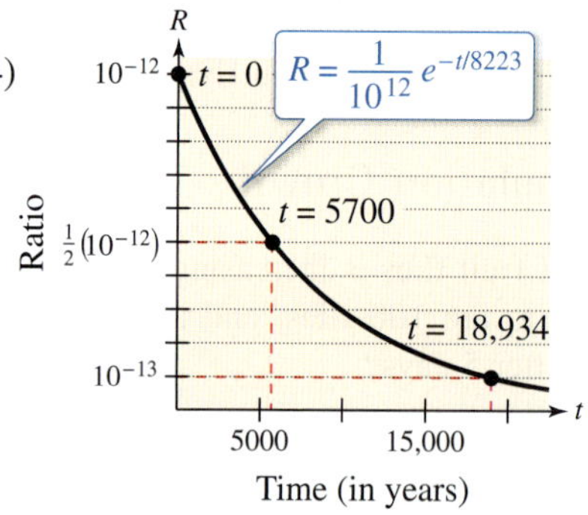

Figure 3.26

where R is the ratio of carbon-14 to carbon-12 at any time t (in years) after death. The graph of R is shown in Figure 3.26. Note that R decreases as t increases.

EXAMPLE 3 Carbon Dating

The ratio of carbon-14 to carbon-12 in a newly discovered fossil is

$$R = \frac{1}{10^{13}}.$$

Estimate the age of the fossil.

Algebraic Solution

In the carbon dating model, substitute the given value of R and solve for t.

$$\frac{1}{10^{12}}e^{-t/8223} = R \qquad \text{Write original model.}$$

$$\frac{e^{-t/8223}}{10^{12}} = \frac{1}{10^{13}} \qquad \text{Substitute } \frac{1}{10^{13}} \text{ for } R.$$

$$e^{-t/8223} = \frac{1}{10} \qquad \text{Multiply each side by } 10^{12}.$$

$$\ln e^{-t/8223} = \ln\frac{1}{10} \qquad \text{Take natural log of each side.}$$

$$-\frac{t}{8223} = \ln\frac{1}{10} \qquad \text{Inverse Property}$$

$$t = -8223 \ln\frac{1}{10} \qquad \text{Multiply each side by } -8223.$$

$$t \approx 18{,}934 \qquad \text{Use a calculator.}$$

So, to the nearest thousand years, you can estimate the age of the fossil to be 19,000 years.

Graphical Solution

Use a graphing utility to graph the carbon dating model

$$y_1 = \frac{1}{10^{12}}e^{-x/8223}.$$

In the same viewing window, graph $y_2 = 1/(10^{13})$, as shown in the figure.

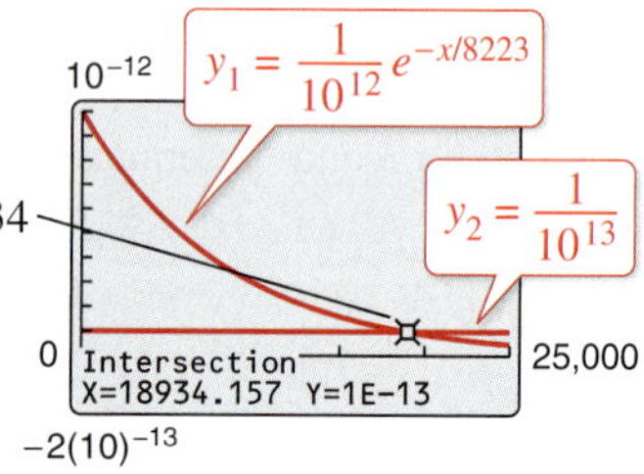

So, to the nearest thousand years, you can estimate the age of the fossil to be 19,000 years.

✓ **Checkpoint** Audio-video solution in English & Spanish at LarsonPrecalculus.com

Estimate the age of a newly discovered fossil for which the ratio of carbon-14 to carbon-12 is $R = 1/10^{14}$.

The carbon dating model in Example 3 assumed that the carbon-14 to carbon-12 ratio of the fossil was one part in 10,000,000,000,000. Suppose an error in measurement occurred and the actual ratio was only one part in 8,000,000,000,000. The fossil age corresponding to the actual ratio would then be approximately 17,000 years. Try checking this result.

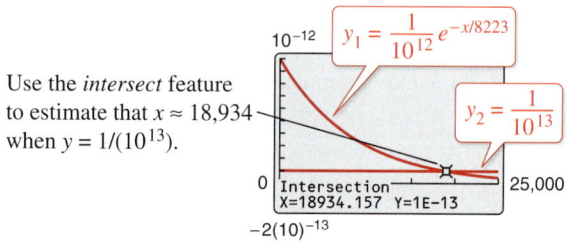

Gaussian Models

As mentioned at the beginning of this section, Gaussian models are of the form

$$y = ae^{-(x-b)^2/c}.$$

This type of model is commonly used in probability and statistics to represent populations that are **normally distributed.** For *standard* normal distributions, the model takes the form

$$y = \frac{1}{\sqrt{2\pi}}e^{-x^2/2}.$$

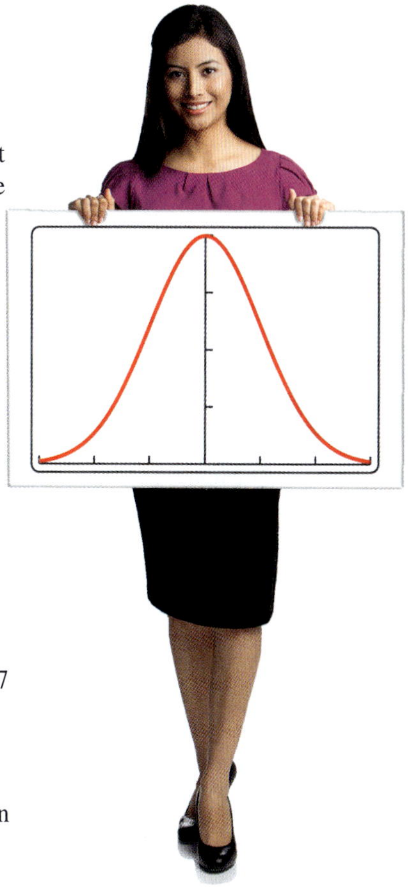

The graph of a Gaussian model is called a **bell-shaped curve.** Use a graphing utility to graph the standard normal distribution curve. Can you see why it is called a bell-shaped curve?

The average value for a population can be found from the bell-shaped curve by observing where the maximum y-value of the function occurs. The x-value corresponding to the maximum y-value of the function represents the average value of the independent variable—in this case, x.

EXAMPLE 4 SAT Scores

The Scholastic Aptitude Test (SAT) mathematics scores for high school seniors in 2017 roughly followed the normal distribution

$$y = 0.0037e^{-(x-527)^2/22,898} \qquad 200 \leq x \leq 800$$

where x is the SAT score for mathematics. Use a graphing utility to graph this function and estimate the average SAT mathematics score. (*Source:* College Board)

Solution

The graph of the function is shown in Figure 3.27. On this bell-shaped curve, the x-value corresponding to the maximum y-value of the curve represents the average score. Using the *maximum* feature of the graphing utility, you can see that the average mathematics score for high school seniors in 2017 was 527.

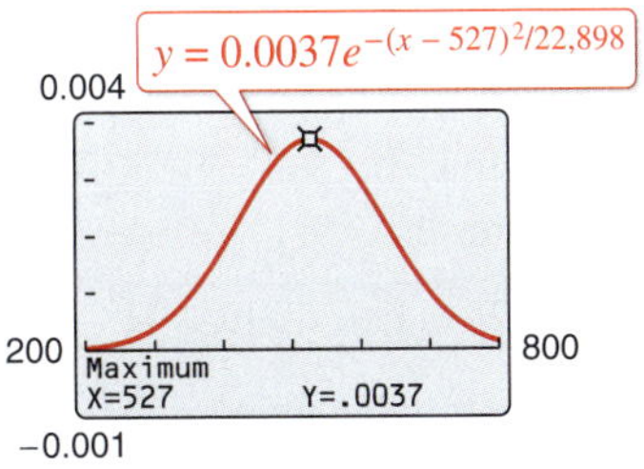

Figure 3.27

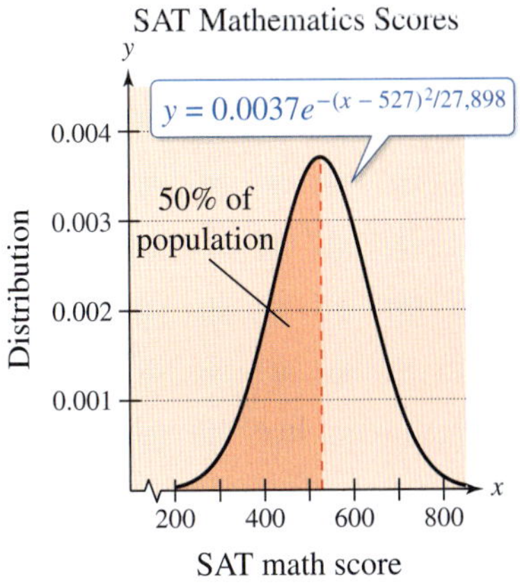

Figure 3.28

Note that 50% of the high school seniors who took the test received scores lower than 527, as shown in Figure 3.28.

Checkpoint *Audio-video solution in English & Spanish at LarsonPrecalculus.com*

The Scholastic Aptitude Test (SAT) reading and writing scores for high school seniors in 2017 roughly followed the normal distribution

$$y = 0.0040e^{-(x-533)^2/20,000}$$

where x is the SAT score for reading and writing. Use a graphing utility to graph this function and estimate the average SAT reading and writing score. (*Source:* College Board)

Logistic Growth Models

Some populations initially have rapid growth, followed by a declining rate of growth, as illustrated by the graph in Figure 3.29. One model for describing this type of growth pattern is the **logistic curve** given by the function

$$y = \frac{a}{1 + be^{-rx}}$$

where y is the population size and x is the time. An example is a bacteria culture that is initially allowed to grow under ideal conditions and then under less favorable conditions that inhibit growth. A logistic growth curve is also called a **sigmoidal curve.**

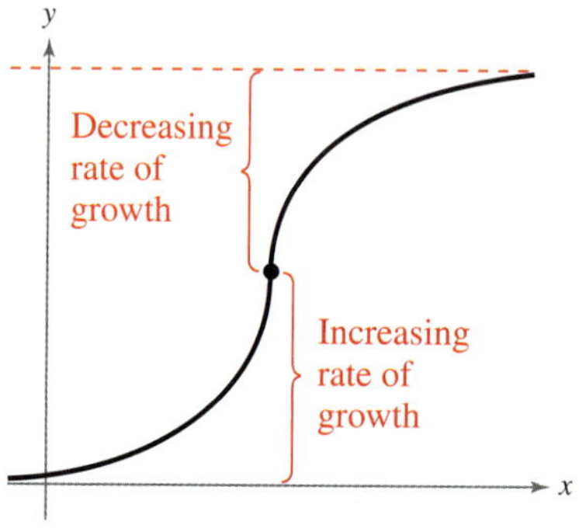

Figure 3.29 *Logistic Curve*

EXAMPLE 5 Spread of a Virus

On a college campus of 5000 students, one student returns from vacation with a contagious flu virus. The spread of the virus is modeled by

$$y = \frac{5000}{1 + 4999e^{-0.8t}}, \quad t \geq 0$$

where y is the total number of students infected after t days. The college will cancel classes when 40% or more of the students are infected.

a. How many students are infected after 5 days?

b. After how many days will the college cancel classes?

Algebraic Solution

a. After 5 days, the number of students infected is

$$y = \frac{5000}{1 + 4999e^{-0.8(5)}} = \frac{5000}{1 + 4999e^{-4}} \approx 54.$$

b. Classes are canceled when the number of infected students is $(0.40)(5000) = 2000$.

$$2000 = \frac{5000}{1 + 4999e^{-0.8t}}$$

$$1 + 4999e^{-0.8t} = 2.5$$

$$e^{-0.8t} = \frac{1.5}{4999}$$

$$\ln e^{-0.8t} = \ln \frac{1.5}{4999}$$

$$-0.8t = \ln \frac{1.5}{4999}$$

$$t = -\frac{1}{0.8} \ln \frac{1.5}{4999}$$

$$t \approx 10.14$$

So, after about 10 days, at least 40% of the students will be infected, and classes will be canceled.

Graphical Solution

a.

Use the *value* feature to estimate that $y \approx 54$ when $x = 5$. So, after 5 days, about 54 students will be infected.

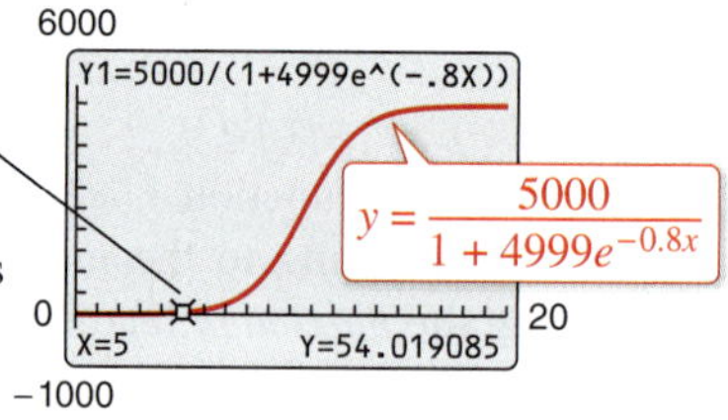

b. Classes are canceled when the number of infected students is $(0.40)(5000) = 2000$. Use a graphing utility to graph

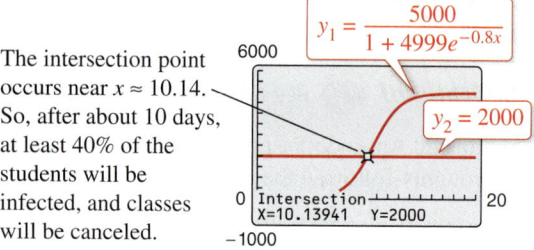

$$y_1 = \frac{5000}{1 + 4999e^{-0.8x}} \quad \text{and} \quad y_2 = 2000$$

in the same viewing window. Use the *intersect* feature of the graphing utility to find the point of intersection of the graphs, as shown in the figure.

The intersection point occurs near $x \approx 10.14$. So, after about 10 days, at least 40% of the students will be infected, and classes will be canceled.

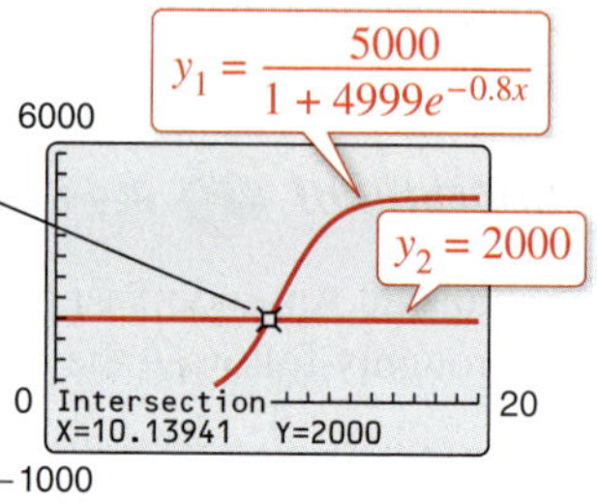

Checkpoint *Audio-video solution in English & Spanish at LarsonPrecalculus.com*

In Example 5, after how many days are 250 students infected?

Logarithmic Models

The level of sound β (in decibels) with an intensity of I (in watts per square meter) is given by

$$\beta = 10 \log_{10} \frac{I}{I_0}$$

where I_0 represents the faintest sound that can be heard by the human ear and is approximately equal to 10^{-12} watt per square meter.

EXAMPLE 6 Sound Intensity

You and your roommate are playing music on your soundbars at the same time and at the same intensity. How much louder is the music when both soundbars are playing than when just one soundbar is playing?

Solution

Let β_1 represent the level of sound when one soundbar is playing and let β_2 represent the level of sound when both soundbars are playing. Using the formula for level of sound, you can express β_1 as

$$\beta_1 = 10 \log_{10} \frac{I}{10^{-12}}.$$

Because β_2 represents the level of sound when *two* soundbars are playing at the same intensity, multiply I by 2 in the formula for level of sound and write β_2 as

$$\beta_2 = 10 \log_{10} \frac{2I}{10^{-12}}.$$

Next, to determine the increase in loudness, subtract β_1 from β_2.

$$\begin{aligned}
\beta_2 - \beta_1 &= 10 \log_{10} \frac{2I}{10^{-12}} - 10 \log_{10} \frac{I}{10^{-12}} \\
&= 10\left(\log_{10} \frac{2I}{10^{-12}} - \log_{10} \frac{I}{10^{-12}}\right) \\
&= 10\left(\log_{10} 2 + \log_{10} \frac{I}{10^{-12}} - \log_{10} \frac{I}{10^{-12}}\right) \\
&= 10 \log_{10} 2 \\
&\approx 3 \text{ decibels}
\end{aligned}$$

So, the music is about 3 decibels louder when both soundbars are playing than when just one soundbar is playing. Notice that the variable I "drops out" of the equation when it is simplified. This means that the loudness increases by about 3 decibels when both soundbars are playing at the same intensity, regardless of the individual intensities of the soundbars.

✓ ***Checkpoint*** ▶ *Audio-video solution in English & Spanish at LarsonPrecalculus.com*

Two sounds have intensities of $I_1 = 10^{-6}$ watt per square meter and $I_2 = 10^{-9}$ watt per square meter. Use the formula for the level of sound to find the difference in loudness between the two sounds. ■

On the Richter scale, the magnitude R of an earthquake of intensity I is given by $R = \log_{10}(I/I_0)$, where $I_0 = 1$ is the minimum intensity used for comparison. Intensity is a measure of the wave energy of an earthquake. The Richter Scale is not commonly used anymore, except for small earthquakes recorded locally.

3.5 Exercises

See *CalcChat.com* for tutorial help and worked-out solutions to odd-numbered exercises.
For instructions on how to use a graphing utility, see Appendix A.

Vocabulary and Concept Check

1. Match each equation with its model.

(a) Exponential growth model	(i) $y = ae^{-bx}, b > 0$
(b) Exponential decay model	(ii) $y = a + b \ln x$
(c) Logistic growth model	(iii) $y = \dfrac{a}{1 + be^{-rx}}$
(d) Gaussian model	(iv) $y = ae^{bx}, b > 0$
(e) Natural logarithmic model	(v) $y = a + b \log_{10} x$
(f) Common logarithmic model	(vi) $y = ae^{-(x-b)^2/c}$

In Exercises 2 and 3, fill in the blank.

2. Gaussian models are commonly used in probability and statistics to represent populations that are _______ distributed.

3. Logistic growth curves are also called _______ curves.

4. Which model in Exercise 1 has a graph called a bell-shaped curve?

5. Does the model $y = 120e^{-0.25x}$ represent exponential growth or exponential decay?

6. Which model in Exercise 1 has a graph with two horizontal asymptotes?

Procedures and Problem Solving

Identifying Graphs of Models **In Exercises 7–12, match the function with its graph. [The graphs are labeled (a), (b), (c), (d), (e), and (f).]**

(a)

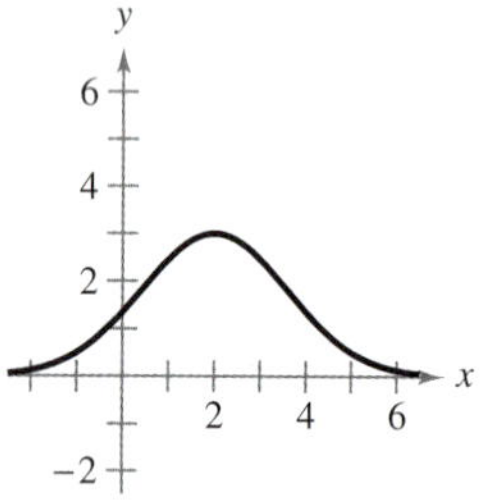

(b)

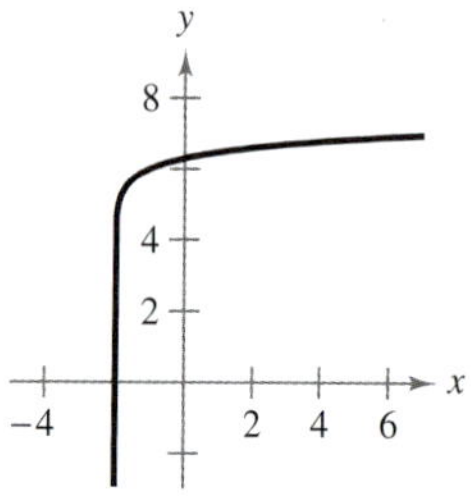

(c)

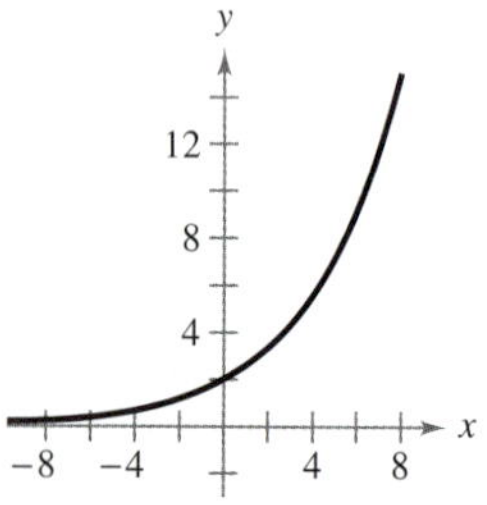

(d)

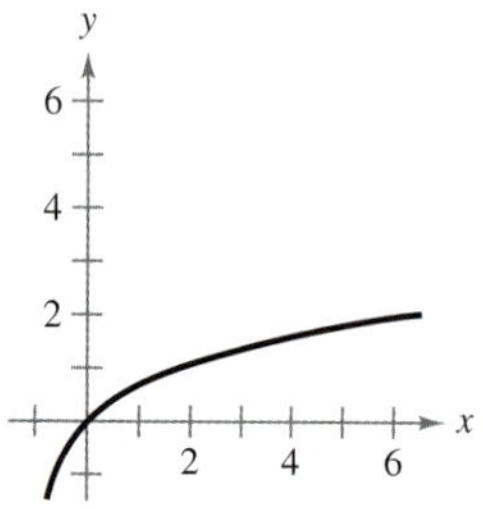

(e)

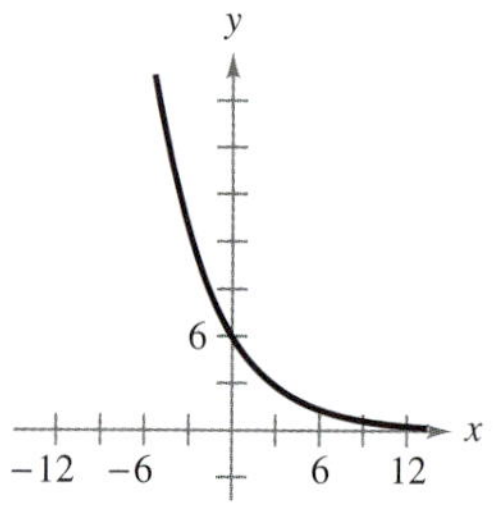

(f)

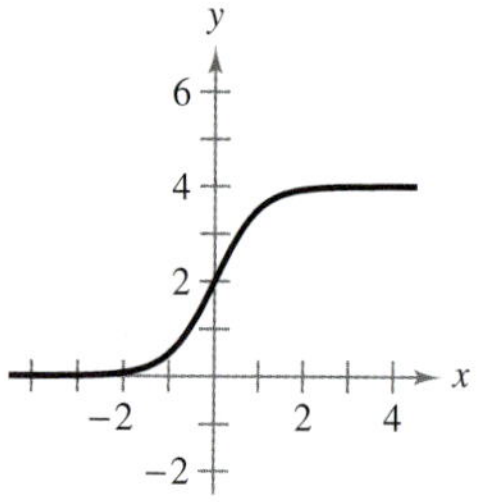

7. $y = 2e^{x/4}$

8. $y = 6e^{-x/4}$

9. $y = 6 + \log_{10}(x + 2)$

10. $y = 3e^{-(x-2)^2/5}$

11. $y = \ln(x + 1)$

12. $y = \dfrac{4}{1 + e^{-2x}}$

Compound Interest **In Exercises 13–20, find the missing values assuming continuously compounded interest.**

	Initial Investment	*Annual % Rate*	*Time to Double*	*Amount After 10 Years*
13.	$1000	3.5%		
14.	$2000	2%		
15.	$7500		21 years	
16.	$10,000		12 years	
17.	$5000			$5665.74
18.	$300			$385.21
19.		4.5%		$100,000.00
20.		4%		$2500.00

21. Tripling an Investment Complete the table for the time t (in years) necessary for P dollars to triple when interest is compounded continuously at rate r. Construct a scatter plot of the data.

r	2%	4%	6%	8%	10%	12%
t						

22. Tripling an Investment Complete the table for the time t (in years) necessary for P dollars to triple when interest is compounded annually at rate r. Construct a scatter plot of the data.

r	2%	4%	6%	8%	10%	12%
t						

23. Finance When \$1 is invested in a fund over a 10-year period, the amount A in the fund after t years can be found using

$$A = 1 + 0.075t \quad \text{or} \quad A = e^{0.07t}$$

depending on whether the fund pays simple interest at $7\frac{1}{2}\%$ or continuous compound interest at 7%. Use a graphing utility to graph each function in the same viewing window. Which grows at a greater rate?

24. Finance When \$1 is invested in a fund over a 10-year period, the amount A in the fund after t years can be found using

$$A = 1 + 0.06t \quad \text{or} \quad A = \left(1 + \frac{0.055}{365}\right)^{365t}$$

depending on whether the fund pays simple interest at 6% or compound interest at $5\frac{1}{2}\%$ compounded daily. Use a graphing utility to graph each function in the same viewing window. Which grows at a greater rate?

Radioactive Decay **In Exercises 25–30, find the missing value for the radioactive isotope.**

	Isotope	Half-Life (years)	Initial Quantity	Amount After 1000 Years
25.	^{14}C	5715		2 g
26.	^{226}Ra	1600	10 g	
27.	^{231}Pa	32,800	1.5 g	
28.	^{239}Pu	24,100		0.4 g
29.	^{241}Am		26.4 g	5.3 g
30.	^{243}Am		100 g	91 g

Writing a Model **In Exercises 31–34, find the exponential model $y = ae^{bx}$ that fits the points shown in the graph or table.**

31.

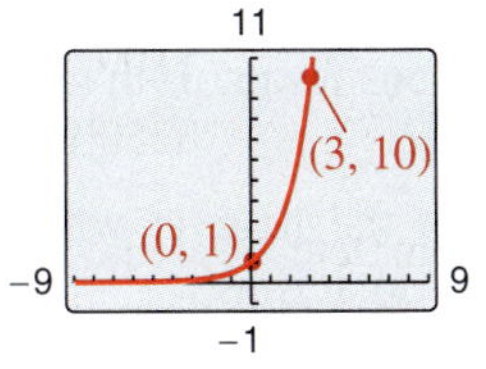

32.

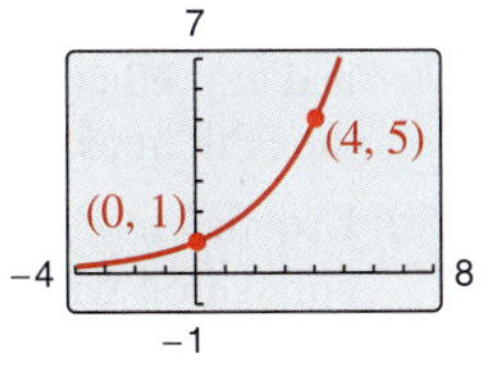

33.

x	0	5
y	4	1

34.

x	0	3
y	5	$\frac{1}{2}$

35. Demography The population P (in millions) of Québec from 2010 through 2017 can be modeled by

$$P = 7.346e^{0.0079t}$$

where t is the year, with $t = 10$ corresponding to 2010. (*Source:* Statistics Canada)

(a) According to the model, was the population of Québec increasing or decreasing from 2010 through 2017? Explain your reasoning.

(b) What were the populations of Québec in 2010, 2013, and 2017?

(c) According to the model, when will the population of Québec be approximately 9 million?

36. MODELING DATA

The table shows the populations (in millions) of five countries in 2018 and the projected populations (in millions) for 2030. (*Source:* U.S. Census Bureau)

Country	2018	2030
Japan	126.2	120.8
Canada	35.9	38.6
Russia	142.1	138.2
Philippines	105.9	125.7
Turkey	81.3	86.7

(a) Find the exponential growth or decay model,

$$y = ae^{bt} \quad \text{or} \quad y = ae^{-bt}$$

for the population of each country, where t is the year, with $t = 18$ corresponding to 2018. Use the model to predict the population of each country in 2045.

(b) You can see that the population of each country is changing at a different rate. What constant in the equation $y = ae^{bt}$ or $y = ae^{-bt}$ is determined by these different rates? Discuss the relationship between the different rates and the magnitude of the constant.

37. Demography The population P (in millions) of Colorado from 2010 through 2017 can be modeled by

$$P = 4.324e^{kt}$$

where t is the year, with $t = 10$ corresponding to 2010. In 2015, the population was 5,440,445. (*Source:* U.S. Census Bureau)

(a) Find the value of k for the model. Round your result to four decimal places.

(b) Use the model to predict the population in 2022.

38. Demography The population P (in millions) of Puerto Rico from 2010 through 2017 can be modeled by $P = 4.369e^{-kt}$, where t is the year, with $t = 10$ corresponding to 2010. In 2015, the population was 3,473,177. (*Source:* U.S. Census Bureau)

(a) Find the value of k for the model. Round your result to four decimal places.

(b) Use your model to predict the population in 2022.

39. *Why you should learn it* *(p. 223)* Carbon-14 (^{14}C) dating assumes that the carbon dioxide on Earth today has the same radioactive content as it did centuries ago. If this is true, then the amount of ^{14}C absorbed by a tree that grew several centuries ago should be the same as the amount of ^{14}C absorbed by a tree growing today. A piece of ancient charcoal contains only 15% as much radioactive carbon as a piece of modern charcoal. How long ago was the tree burned to make the ancient charcoal, assuming that the half-life of ^{14}C is about 5700 years?

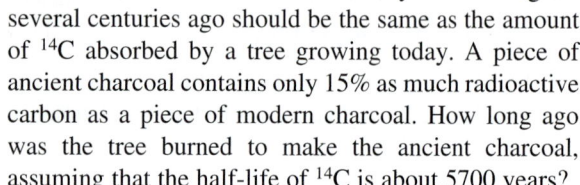

40. Radioactive Decay The half-life of radioactive radium (^{226}Ra) is about 1600 years. What percent of a present amount of radioactive radium will remain after 100 years?

41. MODELING DATA

A new 2018 luxury sedan that sold for \$41,820 has a book value V of \$26,765 after 2 years.

(a) Find a linear model for the value V of the sedan.

(b) Find an exponential model for the value V of the sedan. Round the numbers in the model to four decimal places.

(c) Use a graphing utility to graph the two models in the same viewing window.

(d) Which model represents a greater depreciation rate in the first year?

(e) For what years is the value of the sedan greater using the linear model? the exponential model?

42. MODELING DATA

A new tablet computer that sold for \$300 in 2018 has a book value V of \$165 after 2 years.

(a) Find a linear model for the value V of the tablet.

(b) Find an exponential model for the value V of the tablet. Round the numbers in the model to four decimal places.

(c) Use a graphing utility to graph the two models in the same viewing window.

(d) Which model represents a greater depreciation rate in the first year?

(e) For what years is the value of the tablet greater using the linear model? the exponential model?

43. Psychology The IQ scores for a group of adults roughly follow the normal distribution

$$y = 0.0266e^{-(x-100)^2/450}, \quad 70 \le x \le 115$$

where x is the IQ score.

(a) Use a graphing utility to graph the function.

(b) Use the graph in part (a) to estimate the average IQ score.

44. Education The amount of time (in hours per week) a student utilizes a math-tutoring center roughly follows the normal distribution

$$y = 0.7979e^{-(x-5.4)^2/0.5}, \quad 4 \le x \le 7$$

where x is the number in hours.

(a) Use a graphing utility to graph the function.

(b) From the graph in part (a), estimate the average number of hours per week a student uses the tutoring center.

45. Population A conservation organization releases 100 animals of an endangered species into a game preserve. The organization believes that the preserve has a carrying capacity of 1000 animals and that the growth of the herd will follow the logistic curve

$$p(t) = \frac{1000}{1 + 9e^{-0.1656t}}$$

where t is measured in months.

(a) What is the population after 5 months?

(b) After how many months will the population reach 500?

(c) Use a graphing utility to graph the function.

(d) Use the graph to determine the values of p at which the horizontal asymptotes occur.

(e) Identify the asymptote that is most relevant in the context of the problem and interpret its meaning.

46. Biology The number Y of yeast organisms in a culture is modeled by

$$Y = \frac{663}{1 + 72e^{-0.547t}}$$

where t represents the time (in hours).

(a) Use a graphing utility to graph the model.

(b) Use the model to approximate the initial population and to predict the populations for the 19th hour and the 30th hour.

(c) Use the model to predict the hour at which the population will reach 180 organisms.

(d) According to this model, what is the limiting value of the population?

(e) Why do you think this population of yeast follows a logistic growth model and not an exponential growth model?

Geology **In Exercises 47 and 48, use the Richter scale, in which the magnitude R of an earthquake is related to the intensity I by the formula**

$$R = \log_{10} \frac{I}{I_0}$$

where $I_0 = 1$ is the minimum intensity used for comparison.

47. Find the magnitude of each earthquake.

(a) $I = 39{,}811{,}000$ (b) $I = 12{,}589{,}000$

(c) $I = 251{,}200$ (d) $I = 199{,}500{,}000$

48. Find the intensity of each earthquake.

(a) $R = 7.0$ (b) $R = 8.3$

(c) $R = 6.3$ (d) $R = 5.6$

Audiology **In Exercises 49 and 50, find the level of each sound β using the formula $\beta = 10 \log_{10}(I/I_0)$, where $I_0 = 10^{-12}$ watt per square meter.**

49. (a) $I = 10^{-10}$ watt per m^2 (whisper at 1 meter)

(b) $I = 10^{-5}$ watt per m^2 (busy traffic)

(c) $I \approx 10^0$ watt per m^2 (threshold of pain)

(d) $I = 10^{-11}$ watt per m^2 (rustle of leaves)

50. (a) $I = 10^{-4}$ watt per m^2 (loud radio)

(b) $I = 10^{-3}$ watt per m^2 (inside a heavy truck)

(c) $I = 10^{-2}$ watt per m^2 (siren at 30 meters)

(d) $I = 10^{2}$ watt per m^2 (jet at 30 meters)

51. Audiology As a result of the installation of a muffler, the noise level of an engine was reduced from 88 to 72 decibels. Find the percent decrease in the intensity level of the noise due to the installation of the muffler.

52. Audiology As a result of the installation of noise suppression materials, the noise level in an auditorium was reduced from 93 to 80 decibels. Find the percent decrease in the intensity level of the noise due to the installation of these materials.

Chemistry **In Exercises 53–58, use the acidity model $\text{pH} = -\log_{10}[H^+]$, where acidity (pH) is a measure of the hydrogen ion concentration $[H^+]$ (in moles of hydrogen per liter) of a solution.**

53. Find the pH when $[H^+] = 2.3 \times 10^{-5}$.

54. Find the pH when $[H^+] = 1.13 \times 10^{-5}$.

55. Find $[H^+]$ for a solution in which $\text{pH} = 3.2$.

56. Find $[H^+]$ for a solution for which $\text{pH} = 5.8$.

57. A glass of grape juice has a pH of 3.5, and a baking soda solution has a pH of 8.4. The hydrogen ion concentration of the glass of grape juice is how many times that of the baking soda solution?

58. The pH of a solution is decreased by one unit. The hydrogen ion concentration is increased by what factor?

59. Finance The total interest u paid on a mortgage of P dollars at interest rate r for t years can be found using

$$u = P\left[\frac{rt}{1 - (1 + r/12)^{-12t}} - 1\right].$$

Consider a \$230,000 home mortgage at 4.75%.

(a) Use a graphing utility to graph $u(t)$.

(b) Approximate the length of the mortgage when the total interest paid is the same as the initial amount of the mortgage. Is it possible to be charged twice as much in interest as the initial amount of the mortgage?

60. Finance A \$200,000 mortgage for 30 years at 4.25% has a monthly payment of \$983.88. Part of the monthly payment covers the interest on the unpaid balance, and the rest reduces the principal. The amount paid toward the interest is

$$u = M - \left(M - \frac{Pr}{12}\right)\left(1 + \frac{r}{12}\right)^{12t}$$

and the amount paid toward principal reduction is

$$v = \left(M - \frac{Pr}{12}\right)\left(1 + \frac{r}{12}\right)^{12t}.$$

In these formulas, P is the amount of the mortgage, r is the interest rate, M is the monthly payment, and t is the time (in years).

(a) Use a graphing utility to graph each function in the same viewing window. (The viewing window should show all 30 years of mortgage payments.)

(b) In the early years of the mortgage, is the greater part of the monthly payment paid toward the interest or the principal? Approximate the time when the monthly payment is evenly divided between interest and principal reduction.

(c) Repeat parts (a) and (b) for a repayment period of 20 years ($M = \$1238.47$). What can you conclude?

61. Forensics At 8:30 A.M., a coroner went to the home of a person who had died during the night. To estimate the time of death, the coroner took the person's temperature twice. At 9:00 A.M. the temperature was 85.7°F, and at 11:00 A.M. the temperature was 82.8°F. From these two temperatures, the coroner was able to determine that the time elapsed since death and the body temperature were related by the formula

$$t = -10 \ln \frac{T - 70}{98.6 - 70}$$

where t is the time in hours elapsed since the person died and T is the temperature (in degrees Fahrenheit) of the person's body. (This formula comes from a general cooling principle called *Newton's Law of Cooling*. It uses the assumptions that the person had a normal body temperature of 98.6°F at death and that the room temperature was a constant 70°F.) Use the formula to estimate the time of death of the person.

62. Thawing Time You take a five-pound package of steaks out of a freezer at 11 A.M. and place it in a refrigerator. Assuming that the refrigerator temperature is 40°F and the freezer temperature is 0°F, you can use Newton's Law of Cooling to derive the formula

$$t = -5.05 \ln \frac{T - 40}{0 - 40}$$

where t is the time in hours (with $t = 0$ corresponding to 11 A.M.) and T is the temperature of the package of steaks (in degrees Fahrenheit). Will the steaks be thawed in time to be grilled at 6 P.M.?

Focusing on Concepts

True or False? **In Exercises 63 and 64, determine whether the statement is true or false. Justify your answer.**

63. The domain of a logistic growth function cannot be the set of real numbers.

64. The graph of a logistic growth function will always have an x-intercept.

65. Think About It Can the graph of a Gaussian model ever have an x-intercept? Explain.

66. HOW DO YOU SEE IT? For each graph, state whether an exponential, Gaussian, logarithmic, logistic, or quadratic model will fit the data best. Explain your reasoning. Then describe a real-life situation that could be represented by the data.

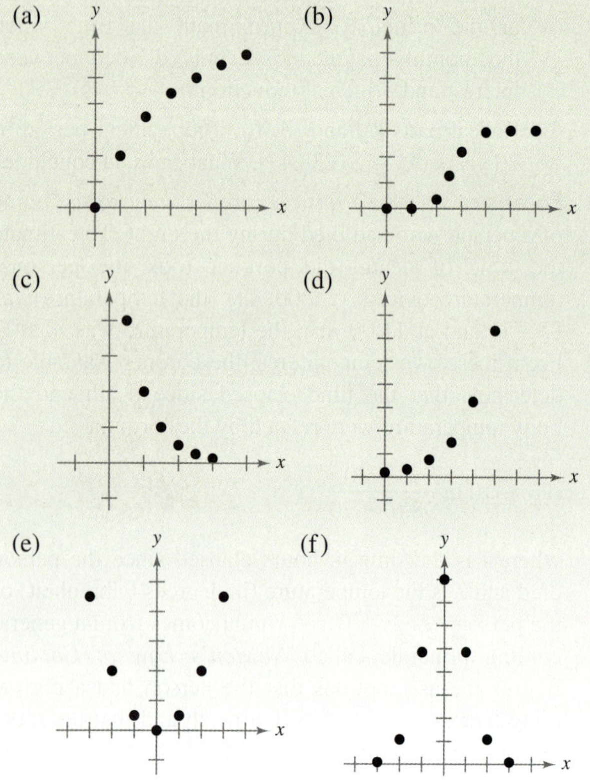

Identifying Models **In Exercises 67–70, identify the type of model you studied in this section that has the indicated characteristic.**

67. The maximum value of the function occurs at the average value of the independent variable.

68. A horizontal asymptote of its graph represents the limiting value of a population.

69. Its graph shows a steadily increasing rate of growth.

70. The only asymptote of its graph is a vertical asymptote.

71. Error Analysis Describe the error.

The graph of the exponential decay model

$y = ae^{-bx}, \quad b > 0$

contains the point $(a, 0)$. ✗

72. Error Analysis Describe the error.

The intensity of an 80-decibel sound is twice the intensity of a 40-decibel sound. ✗

Cumulative Mixed Review

Identifying Graphs of Linear Equations **In Exercises 73–76, match the equation with its graph and identify any intercepts. [The graphs are labeled (a), (b), (c), and (d).]**

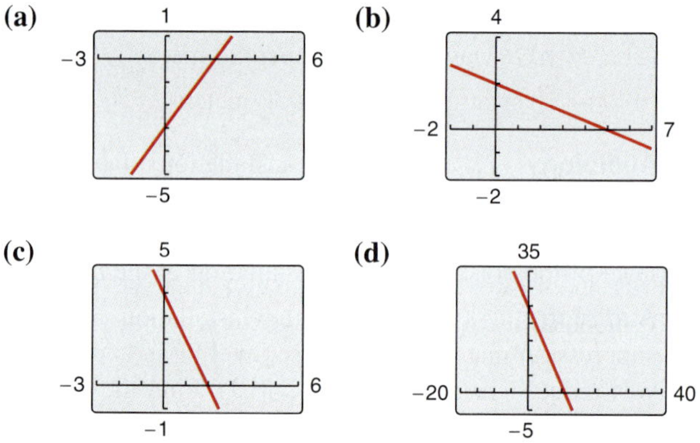

73. $4x - 3y - 9 = 0$

74. $2x + 5y - 10 = 0$

75. $y = 25 - 2.25x$

76. $\dfrac{x}{2} + \dfrac{y}{4} = 1$

Using Synthetic Division **In Exercises 77 and 78, divide using synthetic division.**

77. $(2x^3 - 8x^2 + 3x - 9) \div (x - 4)$

78. $(x^4 - 3x + 1) \div (x + 5)$

79. *Project: Sales Per Share* To work an extended application analyzing the sales per share for Kohl's Corporation from 2002 through 2017, visit this textbook's website at *LarsonPrecalculus.com.* (*Data Source:* Kohl's Corp.)

3.6 Nonlinear Models

Classifying Scatter Plots

In Section 1.7, you saw how to fit linear models to data, and in Section 2.8, you saw how to fit quadratic models to data. In real life, many relationships between two variables are represented by different types of growth patterns. You can use a scatter plot to get an idea of which type of model will best fit a set of data.

What you should learn

- Classify scatter plots.
- Use scatter plots and a graphing utility to find models for data and choose the model that best fits a set of data.
- Use a graphing utility to find exponential and logistic models for data.

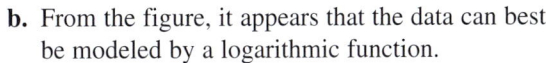

Why you should learn it

Many real-life applications can be modeled by nonlinear equations. For instance, in Exercise 32 on page 242, you are asked to find a nonlinear model that relates air pressure to altitude.

EXAMPLE 1 Classifying Scatter Plots

See LarsonPrecalculus.com for an interactive version of this type of example.

Decide whether each set of data could best be modeled by a linear model, $y = ax + b$, an exponential model, $y = ab^x$, or a logarithmic model, $y = a + b \ln x$.

a. (2, 1), (2.5, 1.2), (3, 1.3), (3.5, 1.5), (4, 1.8), (4.5, 2), (5, 2.4), (5.5, 2.5), (6, 3.1), (6.5, 3.8), (7, 4.5), (7.5, 5), (8, 6.5), (8.5, 7.8), (9, 9), (9.5, 10)

b. (2, 2), (2.5, 3.1), (3, 3.8), (3.5, 4.3), (4, 4.6), (4.5, 5.3), (5, 5.6), (5.5, 5.9), (6, 6.2), (6.5, 6.4), (7, 6.9), (7.5, 7.2), (8, 7.6), (8.5, 7.9), (9, 8), (9.5, 8.2)

c. (2, 1.9), (2.5, 2.5), (3, 3.2), (3.5, 3.6), (4, 4.3), (4.5, 4.7), (5, 5.2), (5.5, 5.7), (6, 6.4), (6.5, 6.8), (7, 7.2), (7.5, 7.9), (8, 8.6), (8.5, 8.9), (9, 9.5), (9.5, 9.9)

Solution

a. From the figure, it appears that the data can best be modeled by an exponential function.

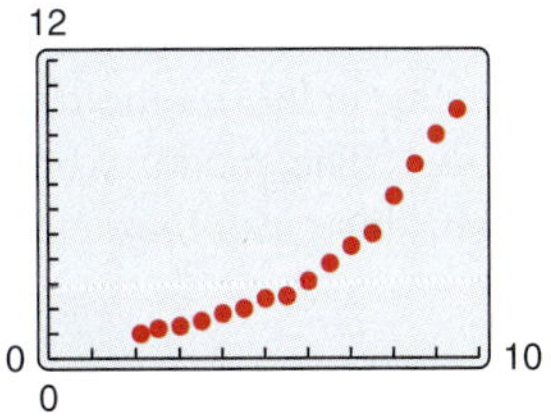

b. From the figure, it appears that the data can best be modeled by a logarithmic function.

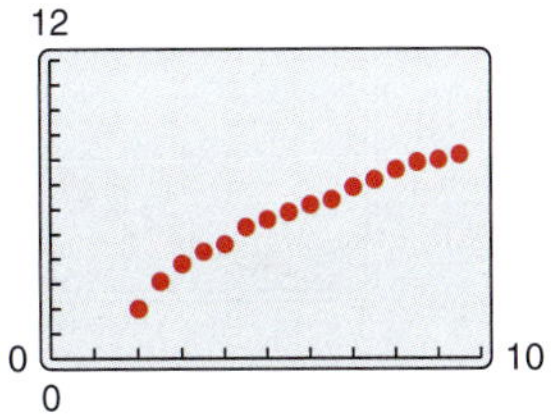

c. From the figure, it appears that the data can best be modeled by a linear function.

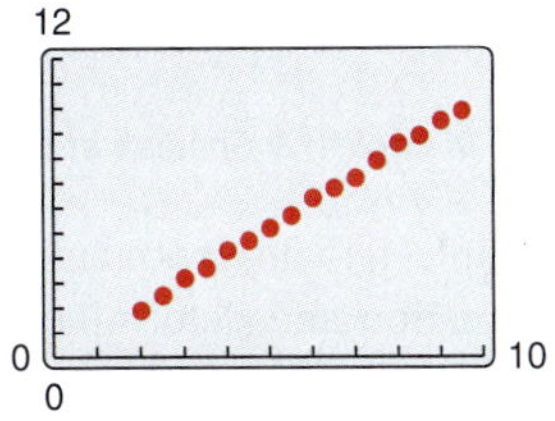

✓ Checkpoint *Audio-video solution in English & Spanish at LarsonPrecalculus.com*

Decide whether the set of data could best be modeled by a linear model, $y = ax + b$, an exponential model, $y = ab^x$, or a logarithmic model, $y = a + b \ln x$.

(0.5, 12), (1, 8), (1.5, 5.7), (2, 4), (2.5, 3.1), (3, 2), (3.5, 1.4), (4, 1), (4.5, 0.8), (5, 0.6), (5.5, 0.4), (6, 0.25), (6.5, 0.18), (7, 0.13), (7.5, 0.09), (8, 0.06)

elwynn/Shutterstock.com

Fitting Nonlinear Models to Data

Once you have used a scatter plot to determine the type of model that would best fit a set of data, there are several ways that you can find the model. Each method is best used with a graphing utility, rather than with hand calculations.

EXAMPLE 2 Fitting a Model to Data

Fit the data from Example 1(a) to an exponential model and a power model. Identify the coefficient of determination and determine which model fits the data better.

(2, 1), (2.5, 1.2), (3, 1.3), (3.5, 1.5), (4, 1.8), (4.5, 2), (5, 2.4), (5.5, 2.5), (6, 3.1), (6.5, 3.8), (7, 4.5), (7.5, 5), (8, 6.5), (8.5, 7.8), (9, 9), (9.5, 10)

Solution

Begin by entering the data into a graphing utility. Then use the *regression* feature of the graphing utility to find exponential and power models for the data, as shown in Figure 3.30.

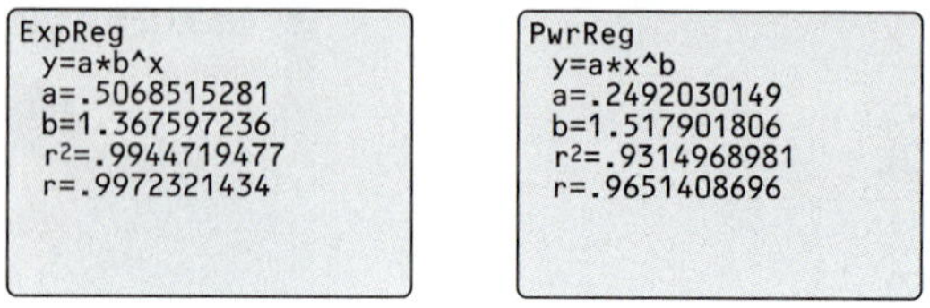

Exponential Model Power Model
Figure 3.30

So, an exponential model for the data is $y = 0.507(1.368)^x$, and a power model for the data is $y = 0.249x^{1.518}$. Plot the data and each model in the same viewing window, as shown in Figure 3.31. To determine which model fits the data better, compare the coefficients of determination for each model. The model whose r^2-value is closest to 1 is the model that better fits the data. In this case, the better-fitting model is the exponential model.

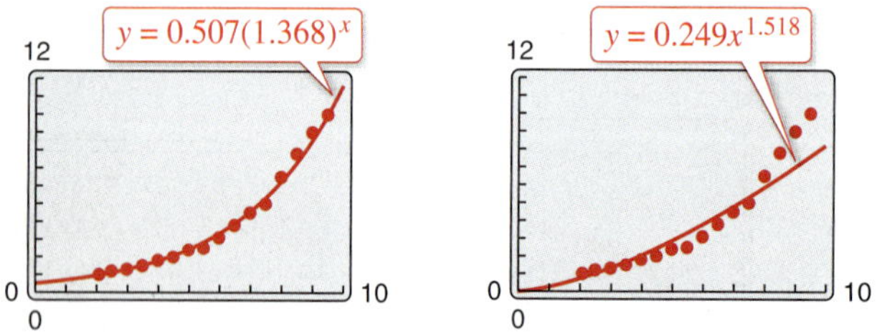

Exponential Model Power Model
Figure 3.31

Insight

Typically, a standardized test will not ask you to fit a nonlinear model to a set of data. However, you may be asked to choose a graph that corresponds to a given model or set of data.

✓ ***Checkpoint*** *Audio-video solution in English & Spanish at LarsonPrecalculus.com*

Fit the data from the Checkpoint for Example 1 to an exponential model and a power model. Identify the coefficient of determination and determine which model fits the data better.

(0.5, 12), (1, 8), (1.5, 5.7), (2, 4), (2.5, 3.1), (3, 2), (3.5, 1.4), (4, 1), (4.5, 0.8), (5, 0.6), (5.5, 0.4), (6, 0.25), (6.5, 0.18), (7, 0.13), (7.5, 0.09), (8, 0.06)

Deciding which model best fits a set of data is a question that is studied in detail in statistics. Recall from Section 1.7 that the model that best fits a set of data is the one whose *sum of the squared differences* is the least. In Example 2, the sums of the squared differences are 0.90 for the exponential model and 14.30 for the power model.

EXAMPLE 3 Fitting a Model to Data

The table shows the yield y (in milligrams) of a chemical reaction after x minutes. Use a graphing utility to find a logarithmic model and a linear model for the data and identify the coefficient of determination for each model. Determine which model fits the data better.

DATA

Minutes, x	Yield, y
1	1.5
2	7.4
3	10.2
4	13.4
5	15.8
6	16.3
7	18.2
8	18.3

Spreadsheet at LarsonPrecalculus.com

Solution

Begin by entering the data into a graphing utility. Then use the *regression* feature of the graphing utility to find logarithmic and linear models for the data, as shown in Figure 3.32.

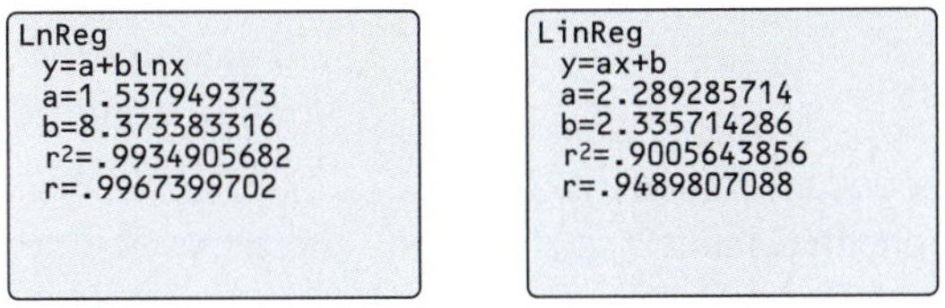

Logarithmic Model **Linear Model**
Figure 3.32

So, a logarithmic model for the data is $y = 1.538 + 8.373 \ln x$, and a linear model for the data is $y = 2.29x + 2.3$. Plot the data and each model in the same viewing window, as shown in Figure 3.33. To determine which model fits the data better, compare the coefficients of determination for each model. The model whose coefficient of determination is closer to 1 is the model that better fits the data. In this case, the better-fitting model is the logarithmic model.

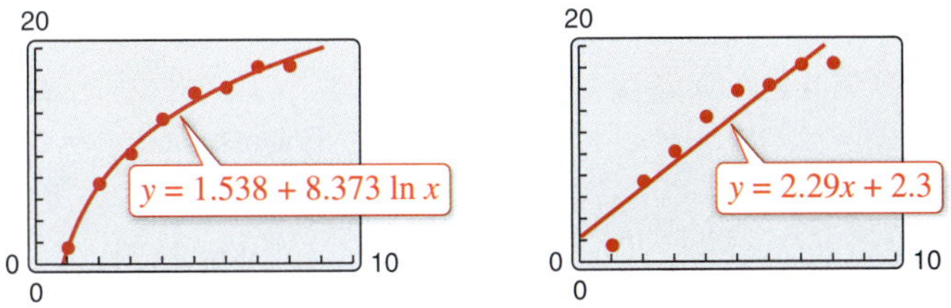

Logarithmic Model **Linear Model**
Figure 3.33

Explore the Concept

Use a graphing utility to find a quadratic model for the data in Example 3. Do you think this model fits the data better than the logarithmic model in Example 3? Explain your reasoning.

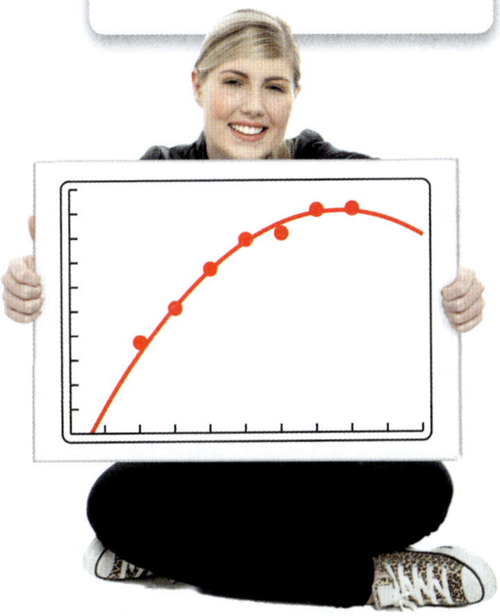

✓ ***Checkpoint*** *Audio-video solution in English & Spanish at LarsonPrecalculus.com*

Use a graphing utility to find a logarithmic model and a linear model for the set of data. Identify the coefficient of determination for each model and determine which model fits the data better.

(0.5, 1.7), (1, 2.5), (1.5, 3), (2, 3.2), (2.5, 3.5), (3, 3.7),
(3.5, 3.8), (4, 3.9), (4.5, 3.9), (5, 4.1), (5.5, 4.2), (6, 4.3),
(6.5, 4.5), (7, 4.6), (7.5, 4.8), (8, 5) ■

In Example 3, the sum of the squared differences for the logarithmic model is 1.59 and the sum of the squared differences for the linear model is 24.31.

Modeling With Exponential and Logistic Functions

EXAMPLE 4 Fitting an Exponential Model to Data

The table shows the amount y (in grams) of a radioactive substance remaining after x days. Use a graphing utility to find a model for the data. How much of the substance remains after 15 days?

DATA

Day, x	Amount, y
0	10.0
1	8.9
2	7.8
3	6.7
4	6.0
5	5.3
6	4.7
7	4.0
8	3.5

Spreadsheet at LarsonPrecalculus.com

Solution

Begin by entering the data into a graphing utility and displaying the scatter plot, as shown in Figure 3.34.

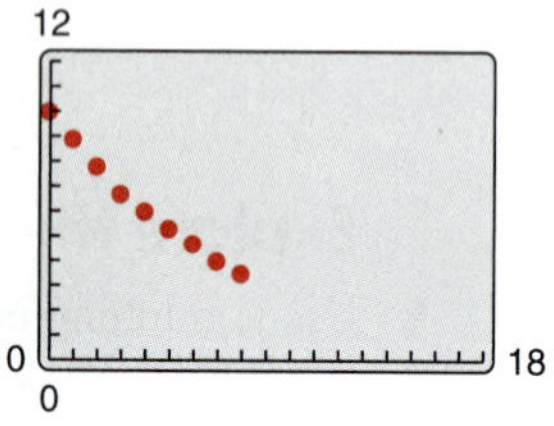

Figure 3.34

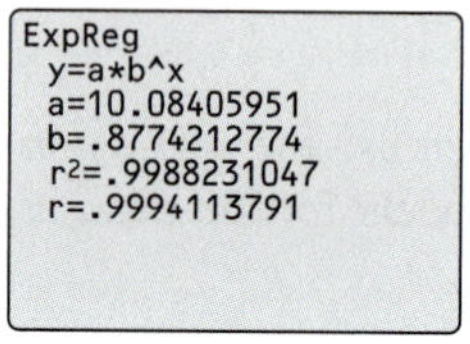

Figure 3.35

From the scatter plot, it appears that an exponential model is a good fit. Use the *regression* feature of the graphing utility to find the exponential model, as shown in Figure 3.35. Change the model to a natural exponential model, as follows.

$$y = 10.08(0.877)^x \qquad \text{Write original model.}$$

$$= 10.08e^{(\ln 0.877)x} \qquad b = e^{\ln b}$$

$$\approx 10.08e^{-0.131x} \qquad \text{Simplify.}$$

Graph the data and the natural exponential model

$$y = 10.08e^{-0.131x}$$

in the same viewing window, as shown in Figure 3.36. From the model, you can see that the amount of the substance decreases by about 13% each day. Also, the amount of the substance remaining after 15 days is

$$y = 10.08e^{-0.131x} \qquad \text{Write natural exponential model.}$$

$$= 10.08e^{-0.131(15)} \qquad \text{Substitute 15 for } x.$$

$$\approx 1.4 \text{ grams.} \qquad \text{Use a calculator.}$$

You can also use the *value* feature of the graphing utility to approximate the amount remaining after 15 days, as shown in Figure 3.37.

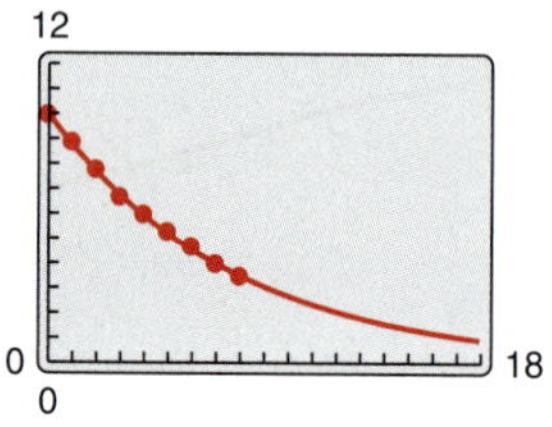

Figure 3.36

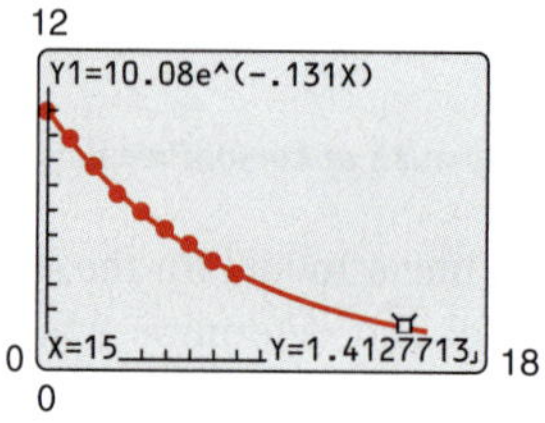

Figure 3.37

Algebra Help

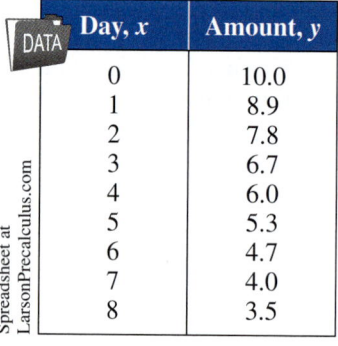

You can change an exponential model of the form

$$y = ab^x$$

to one of the form

$$y = ae^{cx}$$

by rewriting b in the form

$$b = e^{\ln b}.$$

For instance, $y = 3(2^x)$ can be written as

$$y = 3(2^x)$$

$$= 3e^{(\ln 2)x}$$

$$\approx 3e^{0.693x}.$$

✓ ***Checkpoint*** Audio-video solution in English & Spanish at LarsonPrecalculus.com

In Example 4, the initial amount of the substance is 10 grams. According to the natural exponential model, after how many days is the amount of the substance 5 grams? Interpret this value in the context of the problem.

The next example demonstrates how to use a graphing utility to fit a logistic model to data.

Forester

EXAMPLE 5 Fitting a Logistic Model to Data

To estimate the amount of defoliation caused by the gypsy moth during a given year, a forester counts the number x of egg masses on $\frac{1}{40}$ of an acre (circle of radius 18.6 feet) in the fall. The percent of defoliation y the next spring is shown in the table. *(Source: USDA, Forest Service)*

Egg masses, x	Percent of defoliation, y
0	12
25	44
50	81
75	96
100	99

a. Use the *regression* feature of a graphing utility to find a logistic model for the data.

b. How closely does the model represent the data?

Graphical Solution

a. Enter the data into a graphing utility. Using the *regression* feature of the graphing utility, you can find the logistic model, as shown in the figure.

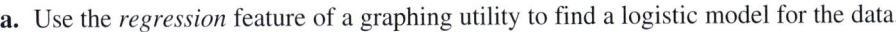

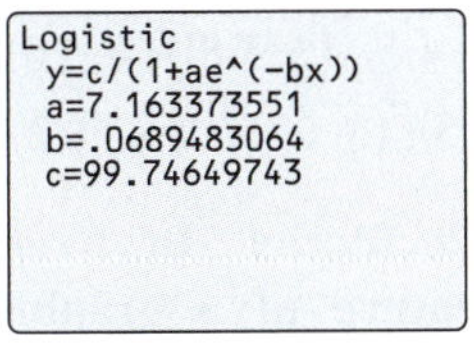
Logistic
y=c/(1+ae^(-bx))
a=7.163373551
b=.0689483064
c=99.74649743

You can approximate this model to be

$$y = \frac{100}{1 + 7e^{-0.069x}}.$$

b. You can use the graphing utility to graph the actual data and the model in the same viewing window. In the figure, it appears that the model is a good fit for the actual data.

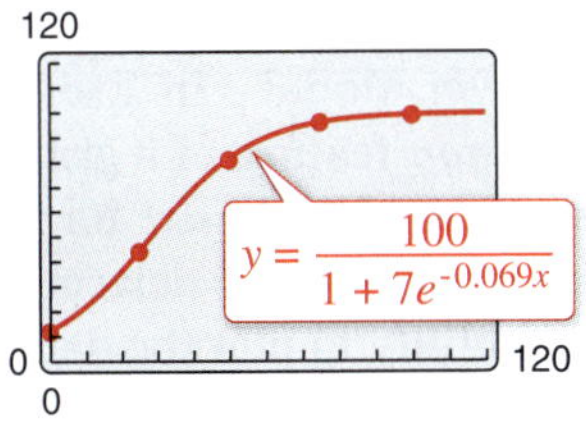

Numerical Solution

a. Enter the data into a graphing utility. Using the *regression* feature of the graphing utility, you can approximate the logistic model to be

$$y = \frac{100}{1 + 7e^{-0.069x}}.$$

b. You can see how well the model fits the data by comparing the actual values of y with the values of y given by the model, which are labeled y^* in the table below.

x	0	25	50	75	100
y	12	44	81	96	99
y^*	12.5	44.5	81.8	96.2	99.3

In the table, you can see that the model appears to be a good fit for the actual data.

✓ Checkpoint *Audio-video solution in English & Spanish at LarsonPrecalculus.com*

Use a graphing utility to graph the logistic model in Example 5. Use the graph to determine the horizontal asymptotes. Interpret the asymptotes in the context of the real-life situation.

3.6 Exercises

See *CalcChat.com* for tutorial help and worked-out solutions to odd-numbered exercises.
For instructions on how to use a graphing utility, see Appendix A.

Vocabulary and Concept Check

In Exercises 1 and 2, fill in the blank.

1. A power model has the form ________ .

2. An exponential model of the form $y = ab^x$ can be rewritten as a natural exponential model of the form ________ .

3. What type of visual display can you create to get an idea of which type of model will best fit the data set?

4. A power model for a set of data has a coefficient of determination of $r^2 \approx 0.901$ and an exponential model for the data has a coefficient of determination of $r^2 \approx 0.967$. Which model fits the data better?

Procedures and Problem Solving

Classifying Scatter Plots **In Exercises 5–10, determine whether the scatter plot could best be modeled by a linear model, a quadratic model, an exponential model, a logarithmic model, or a logistic model.**

5.

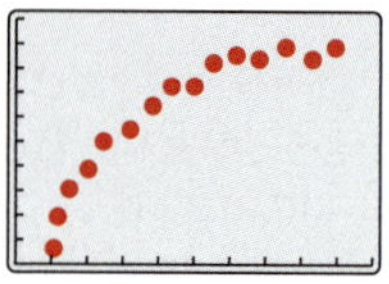

6.

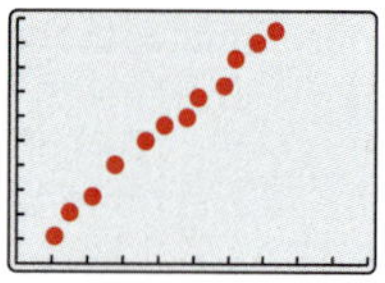

7.

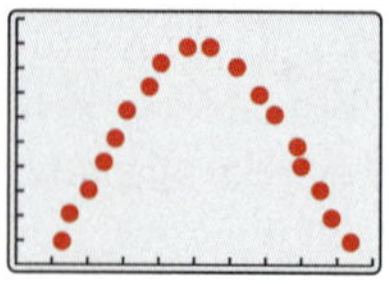

8.

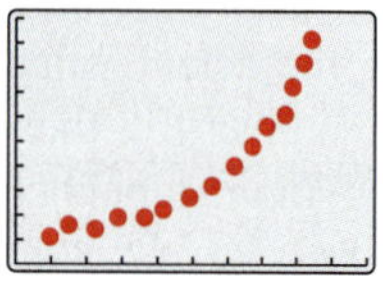

9.

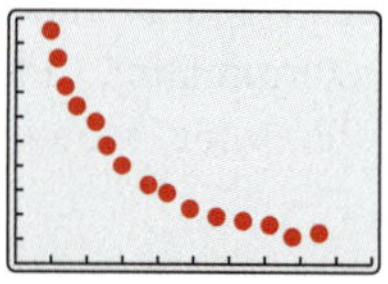

10.

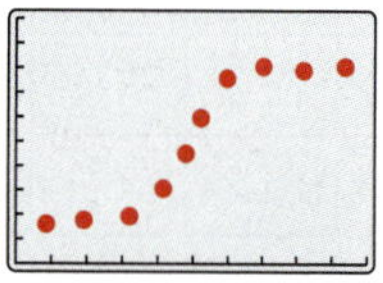

Classifying Scatter Plots **In Exercises 11–16, use a graphing utility to create a scatter plot of the data. Decide whether the data could best be modeled by a linear model $y = ax + b$, an exponential model $y = ab^x$, or a logarithmic model $y = a + b \ln x$.**

11. (1, 2.0), (1.5, 3.5), (2, 4.0), (4, 5.8), (6, 7.0), (8, 7.8)

12. (1, 5.8), (1.5, 6.0), (2, 6.5), (4, 7.6), (6, 8.9), (8, 10.0)

13. (1, 4.4), (1.5, 4.7), (2, 5.5), (4, 9.9), (6, 18.1), (8, 33.0)

14. (1, 11.0), (1.5, 9.6), (2, 8.2), (4, 4.5), (6, 2.5), (8, 1.4)

15. (1, 7.5), (1.5, 7.0), (2, 6.8), (4, 5.0), (6, 3.5), (8, 2.0)

16. (1, 5.0), (1.5, 6.0), (2, 6.4), (4, 7.8), (6, 8.6), (8, 9.0)

Finding an Exponential Model **In Exercises 17–20, use the *regression* feature of a graphing utility to find an exponential model $y = ab^x$ for the data and identify the coefficient of determination. Use the graphing utility to plot the data and graph the model in the same viewing window.**

17. (0, 5), (1, 6), (2, 7), (3, 9), (4, 13), (5, 15)

18. (−2, 2.5), (0, 3.6), (2, 5.7), (4, 9.1)

19. (0, 8.3), (1, 6.1), (2, 4.6), (3, 3.8), (4, 3.6)

20. (−3, 102.2), (0, 80.5), (3, 67.8), (6, 58.2), (10, 55.0)

Finding a Logarithmic Model **In Exercises 21–24, use the *regression* feature of a graphing utility to find a logarithmic model $y = a + b \ln x$ for the data and identify the coefficient of determination. Use the graphing utility to plot the data and graph the model in the same viewing window.**

21. (1, 2.0), (2, 3.0), (3, 3.5), (4, 4.0), (5, 4.1), (6, 4.2), (7, 4.5)

22. (1, 8.5), (2, 11.4), (4, 12.8), (6, 13.6), (8, 14.2), (10, 14.6)

23. (1, 11), (2, 6), (3, 5), (4, 4), (5, 3), (6, 2)

24. (3, 14.6), (6, 11.0), (9, 9.0), (12, 7.6), (15, 6.5)

Finding a Power Model **In Exercises 25–28, use the *regression* feature of a graphing utility to find a power model $y = ax^b$ for the data and identify the coefficient of determination. Use the graphing utility to plot the data and graph the model in the same viewing window.**

25. (1, 2.0), (2, 3.4), (3, 4.6), (5, 6.7), (6, 7.3), (10, 12.0)

26. (0.5, 1.0), (2, 12.5), (4, 33.2), (6, 65.7), (8, 98.5)

27. (1, 10.0), (2, 4.0), (3, 0.7), (4, 0.1)

28. (2, 450), (4, 385), (6, 345), (8, 332), (10, 312)

29. MODELING DATA

The table shows the yearly sales estimates S (in billions of dollars) for the online grocery shopping market for the years 2012 through 2018. (*Source:* Statista)

Year	Sales, S (in billions)
2012	6.0
2013	6.9
2014	8.4
2015	10.1
2016	12.0
2017	14.2
2018	17.5

Spreadsheet at LarsonPrecalculus.com

(a) Use the *regression* feature of a graphing utility to find an exponential model and a power model for the data and identify the coefficient of determination for each model. Let t represent the year, with $t = 2$ corresponding to 2012.

(b) Use the graphing utility to graph each model with the data.

(c) Determine which model fits the data better. Explain.

30. MODELING DATA

The table shows the numbers of third-party skills S available for Amazon's Alexa in six consecutive three-month periods beginning with the first quarter of 2016. (*Source:* Statista)

t	Skills, S
1	135
2	1,000
3	3,000
4	5,000
5	10,000
6	15,000

Spreadsheet at LarsonPrecalculus.com

(a) Use the *regression* feature of a graphing utility to find an exponential model and a power model for the data and identify the coefficient of determination for each model. Let t represent the time, with $t = 1$ corresponding to the first period of 2016.

(b) Use a graphing utility to graph each model with the data.

(c) Determine which model fits the data better. Explain.

31. MODELING DATA

The populations P (in millions) of Jamaica for the years 2001 through 2016 are shown in the table, where t represents the year, with $t = 1$ corresponding to 2001. (*Source:* World Bank Group)

Year	Population, P (in millions)
2001	2.677
2002	2.695
2003	2.713
2004	2.729
2005	2.745
2006	2.760
2007	2.775
2008	2.790
2009	2.804
2010	2.817
2011	2.829
2012	2.841
2013	2.852
2014	2.862
2015	2.872
2016	2.881

Spreadsheet at LarsonPrecalculus.com

(a) Use the *regression* feature of a graphing utility to find a linear model for the data and to identify the coefficient of determination. Plot the model and the data in the same viewing window.

(b) Use the *regression* feature of the graphing utility to find a power model for the data and to identify the coefficient of determination. Plot the model and the data in the same viewing window.

(c) Use the *regression* feature of the graphing utility to find an exponential model for the data and to identify the coefficient of determination. Plot the model and the data in the same viewing window.

(d) Use the *regression* feature of the graphing utility to find a logarithmic model for the data and to identify the coefficient of determination. Plot the model and the data in the same viewing window.

(e) Which model from parts (a)–(d) is the best fit for the data? Explain.

(f) Use each model from parts (a)–(d) to predict the population of Jamaica for the year 2025.

(g) Which model is the best choice for predicting the future population of Jamaica? Explain.

(h) Are your choices of models the same for parts (e) and (g)? If not, explain why your choices are different.

32. ***Why you should learn it*** *(p. 235)* The atmospheric pressure decreases with increasing altitude. At sea level, the average air pressure is approximately 1.03323 kilograms per square centimeter, and this pressure is called one atmosphere. Variations in weather conditions cause changes in the atmospheric pressure of up to ± 5 percent. The ordered pairs (h, p) give the pressures p (in atmospheres) for various altitudes h (in kilometers). *(Spreadsheet at LarsonPrecalculus.com)*

DATA $(0, 1), (10, 0.25), (20, 0.06), (5, 0.55),$
$(15, 0.12), (25, 0.02)$

(a) Use the *regression* feature of a graphing utility to attempt to find the logarithmic model $p = a + b \ln h$ for the data. Explain why the result is an error message.

(b) Use the *regression* feature of the graphing utility to find the logarithmic model $h = a + b \ln p$ for the data.

(c) Use the graphing utility to plot the data and graph the logarithmic model in the same viewing window.

(d) Use the model to estimate the altitude at which the pressure is 0.75 atmosphere.

(e) Use the graph in part (c) to estimate the pressure at an altitude of 13 kilometers.

33. MODELING DATA

The table shows the annual sales S (in billions of dollars) of Lowe's for the years from 2013 through 2017. *(Source: Lowe's Companies, Inc.)*

Year	Sales, S (in billions of dollars)
2013	53.42
2014	56.22
2015	59.07
2016	65.02
2017	68.62

Spreadsheet at LarsonPrecalculus.com

(a) Use the *regression* feature of a graphing utility to find an exponential model for the data. Let t represent the year, with $t = 3$ corresponding to 2013.

(b) Rewrite the model from part (a) as a natural exponential model.

(c) Use the natural exponential model to predict the annual sales of Lowe's in 2020. Is the value reasonable? Explain.

34. MODELING DATA

A beaker of liquid at an initial temperature of 78°C is placed in a room at a constant temperature of 21°C. The temperature of the liquid is measured every 5 minutes for a period of $\frac{1}{2}$ hour. The results are recorded in the table, where t is the time (in minutes) and T is the temperature (in degrees Celsius).

Time, t	Temperature, T (in degrees Celsius)
0	78.0°
5	66.0°
10	57.5°
15	51.2°
20	46.3°
25	42.5°
30	39.6°

Spreadsheet at LarsonPrecalculus.com

(a) Use the *regression* feature of a graphing utility to find a linear model for the data. Use the graphing utility to plot the data and graph the model in the same viewing window. Do the data appear linear? Explain.

(b) Use the *regression* feature of the graphing utility to find a quadratic model for the data. Use the graphing utility to plot the data and graph the model in the same viewing window. Do the data appear quadratic? Even though the quadratic model appears to be a good fit, explain why it might not be a good model for predicting the temperature of the liquid when $t = 60$.

(c) The graph of the temperature of the room should be an asymptote of the graph of the model. Subtract the room temperature from each of the temperatures in the table. Use the *regression* feature of the graphing utility to find an exponential model for the revised data. Add the room temperature to this model. Use the graphing utility to plot the original data and graph the model in the same viewing window.

(d) Explain why the procedure used in part (c) was necessary for finding the exponential model.

35. MODELING DATA

The table shows the percents P of women of different ages x (in years) who have been married at least once. (*Source:* U.S. Census Bureau)

DATA

Age, x	Percent, P
20	7
25	32
30	60
35	74
40	81
45	86
50	87
55	89

Spreadsheet at LarsonPrecalculus.com

(a) Use the *regression* feature of a graphing utility to find a logistic model for the data.

(b) Use the graphing utility to graph the model with the data. How closely does the model represent the data?

36. MODELING DATA

The table shows the lengths y (in centimeters) of yellowtail snappers caught off the coast of Brazil for different ages (in years). (*Source:* Brazilian Journal of Oceanography)

DATA

Age, x	Length, y
1	11.21
2	20.77
3	28.94
4	35.92
5	41.87
6	46.96
7	51.30
8	55.01
9	58.17
10	60.87
11	63.18
12	65.15
13	66.84
14	68.27
15	69.50

Spreadsheet at LarsonPrecalculus.com

(a) Use the *regression* feature of a graphing utility to find a logistic model and a power model for the data.

(b) Use the graphing utility to graph each model from part (a) with the data. Use the graphs to determine which model better fits the data.

(c) Use the model from part (b) to predict the length of a 17-year-old yellowtail snapper.

Focusing on Concepts

True or False? **In Exercises 37 and 38, determine whether the statement is true or false. Justify your answer.**

37. The exponential model $y = ae^{bx}$ represents a growth model when $b > 0$.

38. To change an exponential model of the form $y = ab^x$ to one of the form $y = ae^{cx}$, rewrite b as $b = \ln e^b$.

39. Writing In your own words, explain how to fit a model to a set of data using a graphing utility.

40. Error Analysis Describe the error in rewriting the exponential equation as a natural exponential function.

$y = 3.04(6.948)^x$

$y = 6.948e^{(\ln 3.04)x}$

$y \approx 6.948e^{1.112x}$ ✗

41. Error Analysis Describe the error in using the graph to estimate the value of y when $x = 2.5$.

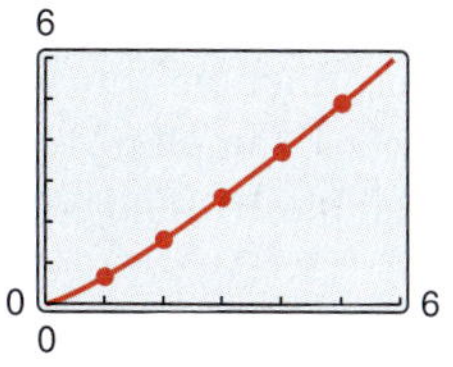

$y \approx 3$ ✗

42. HOW DO YOU SEE IT? Each graphing utility screen below shows a model that fits the set of data. The equations and r^2-values for the models are given. Determine the equation and coefficient of determination that represents each graph. Explain how you found your results.

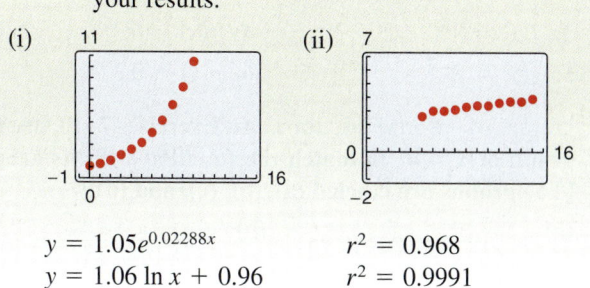

(i)

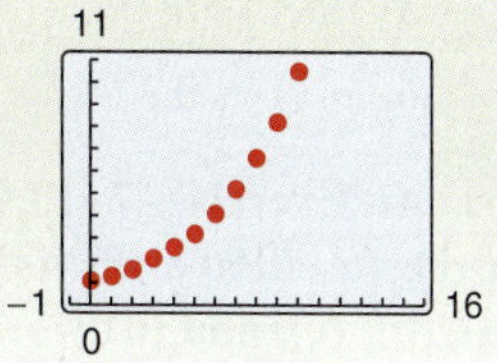

(ii)

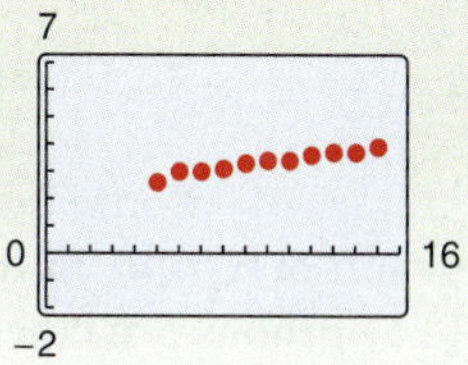

$y = 1.05e^{0.02288x}$ $\qquad r^2 = 0.968$

$y = 1.06 \ln x + 0.96$ $\qquad r^2 = 0.9991$

Cumulative Mixed Review

Using the Slope-Intercept Form **In Exercises 43–46, find the slope and y-intercept of the equation of the line. Then sketch the line by hand.**

43. $2x + 5y = 10$

44. $3x - 2y = 9$

45. $0.4x - 2.5y = 12.5$

46. $1.2x + 3.5y = 10.5$

3 Chapter Review

See CalcChat.com for tutorial help and worked-out solutions to odd-numbered exercises.
For instructions on how to use a graphing utility, see Appendix A.

3.1 *What did you learn?*

Recognize and evaluate exponential functions with base *a* (*p. 182*). The exponential function f with base a is $f(x) = a^x$, where $a > 0$, $a \neq 1$, and x is any real number.

Graph exponential functions with base *a* (*p. 183*). For both $f(x) = a^x$ and $f(x) = a^{-x}$, $a > 1$, the domain is $(-\infty, \infty)$ and the range is $(0, \infty)$. Their graphs are continuous, $(0, 1)$ is the y-intercept, and the x-axis is a horizontal asymptote. $f(x) = a^x$, $a > 1$ is increasing on $(-\infty, \infty)$ and $f(x) = a^{-x}$, $a > 1$ is decreasing on $(-\infty, \infty)$.

Recognize, evaluate, and graph exponential functions with base *e* (*p. 186*). $f(x) = e^x$ is the natural exponential function, where $e = 2.71828128 . . .$ is the natural base.

Use exponential functions to model and solve real-life problems (*p. 188*). Exponential functions are used in compound interest formulas (see Examples 8 and 9), radioactive decay models (see Example 10), and population growth (see Example 11).

Example The graphs of (a) $f(x) = 2^x$, (b) $f(x) = 2^{-x}$, (c) $f(x) = e^x$, and (d) $f(x) = e^{-x}$ are shown below.

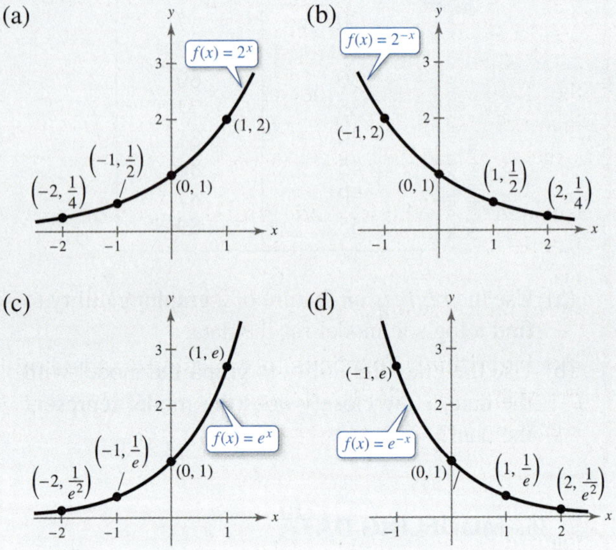

Evaluating an Exponential Function **In Exercises 1 and 2, evaluate the function at the indicated value of x. Round your result to three decimal places.**

1. $g(x) = 30^{2x}$, $x = -0.5$ **2.** $g(x) = 25^{-3x}$, $x = \frac{3}{2}$

Graphs of $y = a^x$ and $y = a^{-x}$ **In Exercises 3–6, graph the exponential function by hand. Identify any asymptotes and intercepts and determine whether the graph of the function is increasing or decreasing.**

3. $f(x) = 8^x$ **4.** $f(x) = 0.3^x$

5. $g(x) = 8^{-x}$ **6.** $g(x) = 0.3^{-x}$

Library of Parent Functions **In Exercises 7–10, use the graph of $y = 4^x$ to match the function with its graph. [The graphs are labeled (a), (b), (c), and (d).]**

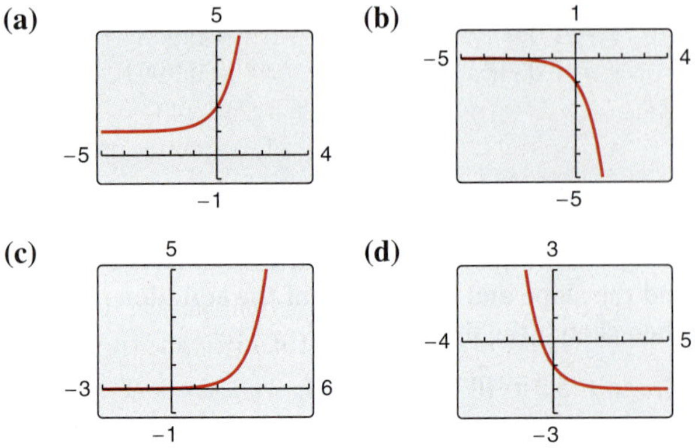

7. $f(x) = 4^{x-3}$ **8.** $f(x) = 4^{-x} - 2$

9. $f(x) = -4^x$ **10.** $f(x) = 4^x + 1$

Graphing a Natural Exponential Function **In Exercises 11–16, use a graphing utility to construct a table of values for the function. Then sketch the graph of the function. Identify any asymptotes of the graph.**

11. $h(x) = e^{x-1}$ **12.** $f(x) = e^{x+2}$

13. $h(x) = -e^x + 3$ **14.** $f(x) = -2e^{-x}$

15. $f(x) = 4e^{-0.5x}$ **16.** $f(x) = 2 + e^{x-3}$

Compound Interest **In Exercises 17 and 18, use a graphing utility to create a table of values that shows the balance A when \$10,000 is invested at rate r, compounded continuously, for $t = 1, 10, 20, 30, 40,$ and 50 years.**

17. $r = 8\%$ **18.** $r = 3\%$

19. Economics A new SUV costs \$28,700. The value V of the SUV after t years is modeled by $V(t) = 28{,}700\left(\frac{4}{5}\right)^t$.

(a) Use a graphing utility to graph the function.

(b) Find the value of the SUV after 3 years.

(c) According to the model, when does the SUV depreciate most rapidly? Is this realistic? Explain.

20. Radioactive Decay Let Q represent a mass, in grams, of radioactive plutonium 241 (^{241}Pu), whose half-life is about 14 years. The quantity of plutonium present after t years can be found using $Q = 50\left(\frac{1}{2}\right)^{t/14}$.

(a) Determine the initial quantity (when $t = 0$).

(b) Determine the quantity present after 10 years.

(c) Use a graphing utility to graph the function over the interval $t = 0$ to $t = 50$.

3.2 What did you learn?

Recognize and evaluate logarithmic functions with base *a* (p. 194). For $x > 0$, $a > 0$, and $a \neq 1$, $y = \log_a x$ if and only if $x = a^y$. The function $g(x) = \log_a x$ is called the logarithmic function with base a.

Graph logarithmic functions with base *a* (p. 196). The graphs of $g(x) = \log_a x$ and $f(x) = a^x$ are reflections of each other in the line $y = x$. For $g(x) = \log_a x$, $a > 1$, the domain is $(0, \infty)$, the range is $(-\infty, \infty)$, and it is increasing on $(0, \infty)$. The graph is continuous, $(1, 0)$ is the x-intercept, and the y-axis is a vertical asymptote.

Recognize, evaluate, and graph natural logarithmic functions (p. 198). The function defined by $g(x) = \log_e x = \ln x$, $x > 0$, is called the natural logarithmic function.

Use logarithmic functions to model and solve real-life problems (p. 200). Human memory models use logarithmic functions (see Example 10).

Example The graphs of

(a) $f(x) = 2^x$ and $g(x) = \log_2 x$

and

(b) $f(x) = e^x$ and $g(x) = \ln x$

are shown below.

(a)

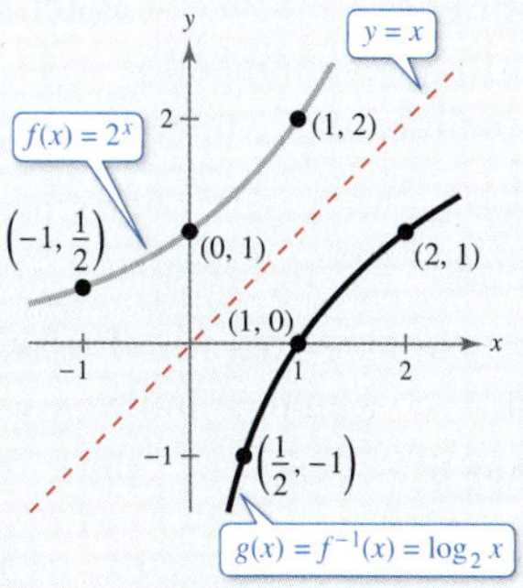

(b)

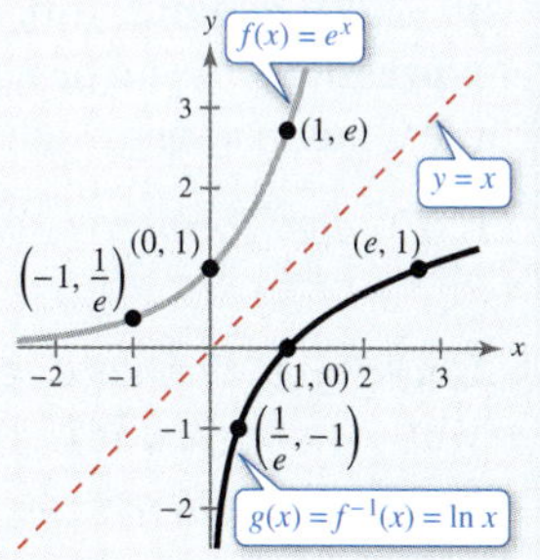

Rewriting an Equation **In Exercises 21–30, write the logarithmic equation in exponential form, or write the exponential equation in logarithmic form.**

21. $\log_5 625 = 4$
22. $\log_9 81 = 2$
23. $\log_{64} 2 = \frac{1}{6}$
24. $\log_{10}\left(\frac{1}{100}\right) = -2$
25. $4^3 = 64$
26. $3^5 = 243$
27. $e^{3.2} = 24.532 \ldots$
28. $\ln 5 = 1.6094 \ldots$
29. $\left(\frac{1}{2}\right)^{-3} = 8$
30. $\left(\frac{2}{3}\right)^{-2} = \frac{9}{4}$

Evaluating a Logarithm **In Exercises 31–34, evaluate the function at the indicated value of x without using a calculator.**

	Function	*Value*
31.	$f(x) = \log_6 x$	$x = 216$
32.	$f(x) = \log_9 x$	$x = 1$
33.	$f(x) = \log_4 x$	$x = \frac{1}{4}$
34.	$f(x) = \log_{10} x$	$x = 0.00001$

Library of Parent Functions **In Exercises 35–38, find the domain, x-intercept, and vertical asymptote of the logarithmic function, and sketch its graph by hand.**

35. $g(x) = -\log_{10} x + 5$
36. $g(x) = \log_{10}(x - 3)$
37. $f(x) = \log_{10}(x - 1) + 6$
38. $f(x) = \log_{10}(x + 2) - 3$

Evaluating a Natural Logarithm **In Exercises 39–42, use a calculator to evaluate the function $f(x) = \ln x$ at the indicated value of x. Round your result to three decimal places.**

39. $x = 52.1$
40. $x = 0.46$
41. $x = \sqrt{13}$
42. $x = \frac{5}{6}$

Analyzing the Graph of a Function **In Exercises 43–46, use a graphing utility to graph the logarithmic function. Determine the domain, x-intercept, and vertical asymptote.**

43. $f(x) = \ln x + 3$
44. $f(x) = \ln(x - 3)$
45. $h(x) = \frac{1}{2} \ln x$
46. $f(x) = 4 \ln x$

47. **Aeronautics** The time t (in minutes) for a small plane to climb to an altitude of h feet is modeled by

$$t = 50 \log_{10}[18{,}000/(18{,}000 - h)]$$

where 18,000 feet is the plane's absolute ceiling.
   (a) Determine the domain of the function.
   (b) Use a graphing utility to graph the function and identify any asymptotes.
   (c) As the plane approaches its absolute ceiling, what can you say about the time required to further increase its altitude?
   (d) Find the amount of time it takes for the plane to climb to an altitude of 4000 feet.

48. **Real Estate** The model $t = 21.094 \ln[x/(x - 855)]$ approximates the length of a home mortgage of \$216,000 at 4.75% in terms of the monthly payment. In the model, t is the length of the mortgage in years and x is the monthly payment in dollars.
   (a) Use the model to approximate the length of a \$216,000 mortgage at 4.75% when the monthly payment is \$1395.84.
   (b) Approximate the total amount paid over the term of the mortgage with a monthly payment of \$1395.84. What amount of the total is interest costs?

3.3 What did you learn?

Rewrite logarithms with different bases (p. 205). Let a, b, and x be positive real numbers such that $a \neq 1$ and $b \neq 1$. Then $\log_a x$ can be converted to a different base.

Base b	*Base 10*	*Base e*
$\log_a x = \dfrac{\log_b x}{\log_b a}$	$\log_a x = \dfrac{\log_{10} x}{\log_{10} a}$	$\log_a x = \dfrac{\ln x}{\ln a}$

Example

(a) $\log_3 15 = \dfrac{\ln 15}{\ln 3}$

≈ 2.46

Use properties of logarithms to evaluate, rewrite, (p. 206) expand, or condense (p. 207) logarithmic expressions. Let a be a positive real number ($a \neq 1$), n be a real number, and u and v be positive real numbers.

1. *Product Property:* $\log_a(uv) = \log_a u + \log_a v$
   $\ln(uv) = \ln u + \ln v$
2. *Quotient Property:* $\log_a(u/v) = \log_a u - \log_a v$
   $\ln(u/v) = \ln u - \ln v$
3. *Power Property:* $\log_a u^n = n \log_a u$
   $\ln u^n = n \ln u$

(b) $\log_5 \dfrac{xy^2}{2\sqrt{x^2+1}} = \log_5 \dfrac{xy^2}{2(x^2+1)^{1/2}}$

$= \log_5 xy^2 - \log_5 \left[2(x^2+1)^{1/2}\right]$

$= \log_5 x + 2\log_5 y - \log_5 2$

$- \frac{1}{2}\log_5(x^2+1)$

(c) $\dfrac{1}{4}[3\log_{10}(x+1) + \log_{10} y - 3\log_{10}(1-y)]$

$= \dfrac{1}{4}[\log_{10}(x+1)^3 y - \log_{10}(1-y)^3]$

$= \dfrac{1}{4}\log_{10}\dfrac{(x+1)^3 y}{(1-y)^3}$

$= \log_{10} \sqrt[4]{\left(\dfrac{x+1}{1-y}\right)^3 y}$

Use logarithmic functions to model and solve real-life problems (p. 208). Logarithmic functions can be used to find an equation that relates the periods of several planets and their distances from the sun (see Example 7).

Changing the Base **In Exercises 49 and 50, evaluate the logarithm using the change-of-base formula with (a) common logarithms and (b) natural logarithms. Round your results to three decimal places.**

49. $\log_3 0.49$
50. $\log_{1/2} 10$

Using Properties to Evaluate Logarithms **In Exercises 51 and 52, approximate the logarithm using the properties of logarithms, given the values $\log_b 2 \approx 0.3562$, $\log_b 3 \approx 0.5646$, and $\log_b 5 \approx 0.8271$.**

51. $\log_b 9$
52. $\log_b \frac{9}{25}$

Graphing a Logarithmic Function **In Exercises 53 and 54, use the change-of-base formula and a graphing utility to graph the function.**

53. $f(x) = \log_2(x - 2)$
54. $f(x) = 2 - \log_3 x$

Simplifying a Logarithm **In Exercises 55 and 56, use the properties of logarithms to rewrite and simplify the logarithmic expression.**

55. $\ln(5e^{-2})$
56. $\log_{10} 200$

Expanding a Logarithmic Expression **In Exercises 57–60, use the properties of logarithms to expand the expression as a sum, difference, and/or constant multiple of logarithms. (Assume all variables are positive.)**

57. $\log_6 6x^2$
58. $\log_4 16xy^2$
59. $\ln \dfrac{x+3}{xy}$
60. $\ln \dfrac{xy^5}{\sqrt{z}}$

Condensing a Logarithmic Expression **In Exercises 61–66, use the properties of logarithms to condense the expression.**

61. $\log_2 9 + \log_2 x$
62. $\log_5 x - 4\log_5 y$
63. $\frac{1}{2}\ln(2x - 1) - 2\ln(x + 1)$
64. $5\ln(x - 2) - \ln(x + 2) + 3\ln x$
65. $\ln 3 + \frac{1}{3}\ln(4 - x^2) - \ln x$
66. $3[\ln x - 2\ln(x^2 + 1)] + 2\ln 5$

67. **Snow Removal** The number of miles s of roads cleared of snow in 1 hour is approximated by the model

$$s = 25 - \frac{13\ln(h/12)}{\ln 3}, \quad 2 \le h \le 15$$

where h is the depth of the snow (in inches).

(a) Use a graphing utility to graph the function.

(b) Using the graph of the function, what conclusion can you make about the number of miles of roads cleared as the depth of the snow increases?

68. **Industrial Engineering** The ordered pairs below represent the number of coats x of sealant applied to a machine part and the corresponding dry thickness y (in millimeters) of the coating. Find a logarithmic equation that relates y and x.

(1, 1.000), (2, 1.189), (3, 1.316), (4, 1.414)

3.4 What did you learn?

Solve simple exponential and logarithmic equations *(p. 212)*. Solve simple exponential or logarithmic equations using the One-to-One Properties and Inverse Properties of exponential and logarithmic functions.

One-to-One Properties	*Inverse Properties*
$a^x = a^y$ if and only if $x = y$.	$a^{\log_a x} = x$
$\log_a x = \log_a y$ if and only if $x = y$.	$\log_a a^x = x$

Solve more complicated exponential *(p. 213)* and logarithmic *(p. 215)* equations. To solve more complicated equations, rewrite the equations so that the One-to-One Properties and Inverse Properties of exponential and logarithmic functions can be used (see Examples 2–9).

Use exponential and logarithmic equations to model and solve real-life problems *(p. 218)*. Exponential and logarithmic equations can be used to find how long it will take to double an investment (see Example 12) and to find the year in which the average salary for public school teachers reached \$54,500 (see Example 13).

Example

(a) $5^x = 125$
$5^x = 5^3$
$x = 3$

(b) $e^x = 15$
$\ln e^x = \ln 15$
$x = \ln 15$
≈ 2.71

(c) $4\log_2 5x = 8$
$\log_2 5x = 2$
$2^{\log_2 5x} = 2^2$
$5x = 4$
$x = \frac{4}{5}$

(d) $\ln(3x - 5) = \ln(x + 13)$
$e^{\ln(3x-5)} = e^{\ln(x+13)}$
$3x - 5 = x + 13$
$2x = 18$
$x = 9$

(e) $e^{2x} - 9e^x + 18 = 0$
$(e^x - 3)(e^x - 6) = 0$

$e^x - 3 = 0$	$e^x - 6 = 0$
$e^x = 3$	$e^x = 6$
$x = \ln 3$	$x = \ln 6$
≈ 1.10	≈ 1.79

Solving a Simple Equation **In Exercises 69–82, solve the equation for x. Which property of logarithms or exponents did you use?**

69. $10^x = 10{,}000$
70. $11^x = 1331$
71. $6^x = \frac{1}{216}$
72. $6^{x-2} = \frac{1}{1296}$
73. $2^{x+1} = \frac{1}{16}$
74. $4^{x/2} = 64$
75. $e^x = 4$
76. $e^{-x} = \frac{1}{6}$
77. $\log_2(x - 1) = \log_2 3$
78. $\ln(2x + 27) = \ln 2$
79. $\log_8 x = 4$
80. $\log_x 729 = 6$
81. $\ln x = 4$
82. $\ln x = e$

Solving an Exponential Equation Algebraically **In Exercises 83–92, solve the equation algebraically. Approximate the result to three decimal places.**

83. $3e^{-5x} = 132$
84. $14e^{3x+2} = 560$
85. $2^x + 13 = 35$
86. $6^x - 28 = -8$
87. $-4(5^x) = -68$
88. $5(12^x) = 240$
89. $2e^{x-3} - 1 = 4$
90. $-e^{x/2} + 1 = \frac{1}{2}$
91. $e^{2x} - 7e^x + 10 = 0$
92. $e^{2x} - 6e^x + 8 = 0$

Solving a Logarithmic Equation Algebraically **In Exercises 93–104, solve the equation algebraically. Approximate the result to three decimal places, if necessary.**

93. $\log_{10}(1 - x) = -1$
94. $\log_{10}(-x - 4) = 2$
95. $\ln 2x = 8.6$
96. $\ln 5x = 4.5$
97. $\ln(x - 1) = 2$
98. $\ln(2x + 1) = -4$
99. $\ln x - \ln 5 = 2$
100. $\ln x - \ln 3 = 4$
101. $\ln\sqrt{x + 1} = 2$
102. $\ln\sqrt{x + 16} = 5$
103. $\log_4(x + 5) = \log_4(13 - x) - \log_4(x - 3)$
104. $\log_5(x + 2) - \log_5 x = \log_5(x + 5)$

Solving an Exponential or Logarithmic Equation **In Exercises 105–108, solve the equation algebraically. Round the result to three decimal places. Verify your answer using a graphing utility.**

105. $xe^x + e^x = 0$
106. $2xe^{2x} + e^{2x} = 0$
107. $x\ln x + x = 0$
108. $\dfrac{1 + \ln x}{x^2} = 0$

109. Psychology Students in a sociology class were given an exam and then retested monthly with an equivalent exam. The average scores for the class are approximated by the human memory model

$$f(t) = 85 - 17\log_{10}(t + 1), \quad 0 \le t \le 12$$

where t is the time in months. When did the average score decrease to 68?

110. Economics The demand x for a television is modeled by

$$p = 5000\left(1 - \frac{4}{4 + e^{-0.0005x}}\right).$$

Find the demands x for prices of (a) $p = \$500$ and (b) $p = \$350$.

3.5 What did you learn?

Recognize the five most common types of models involving exponential or logarithmic functions *(p. 223)*.

1. *Exponential growth model:* $y = ae^{bx}, \quad b > 0$
2. *Exponential decay model:* $y = ae^{-bx}, \quad b > 0$
3. *Gaussian model:* $y = ae^{-(x-b)^2/c}$
4. *Logistic growth model:* $y = \dfrac{a}{1 + be^{-rx}}$
5. *Logarithmic models:* $y = a + b \ln x$, $y = a + b \log_{10} x$

Use exponential growth and decay functions to model and solve real-life problems *(p. 224)*. An exponential growth function can model company revenues (Example 1) or population growth (Example 2). An exponential decay function can be used to estimate the age of a fossil (see Example 3).

Use Gaussian functions *(p. 227)*, logistic growth functions *(p. 228)*, and logarithmic functions *(p. 229)* to model and solve real-life problems. A Gaussian function can model SAT scores (see Example 4). A logistic growth function can model the spread of a virus (see Example 5). A logarithmic function can be used to find sound levels (see Example 6).

Example

(a) The population P (in millions) of Texas from 2010 through 2017 can be modeled by the exponential growth model $P = 21.364e^{0.0166t}$, where t is the year, with $t = 10$ corresponding to 2010. *(Source:* U.S. Census Bureau)

(b) The quantity Q of a mass of plutonium 241 present after t years can be found using $Q = 50\left(\frac{1}{2}\right)^{t/14}$. Another way to write this formula is to use the exponential decay model $Q = 50e^{-0.0495t}$.

(c) The scores for a standardized test are normally distributed, so they can be represented using a Gaussian model such as $y = 0.004e^{-(x-500)^2/25{,}000}$. For this function, the graph is a bell-shaped curve with an average value of 500.

(d) The growth of a population of 500 animals in a state game land with a carrying capacity of 5000 animals can be modeled by the logistic growth model $y = 5000/(1 + 9e^{-0.15t})$, where t is measured in months.

(e) The human memory model $y = 85 - 17 \log_{10}(t + 1)$, $0 \le t \le 12$, where t is the time in months, is a logarithmic model.

Identifying Graphs of Models **In Exercises 111–116, match the function with its graph. [The graphs are labeled (a), (b), (c), (d), (e), and (f).]**

(a)

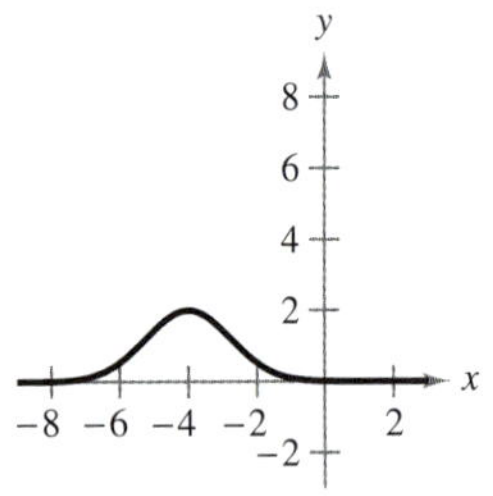

(b)

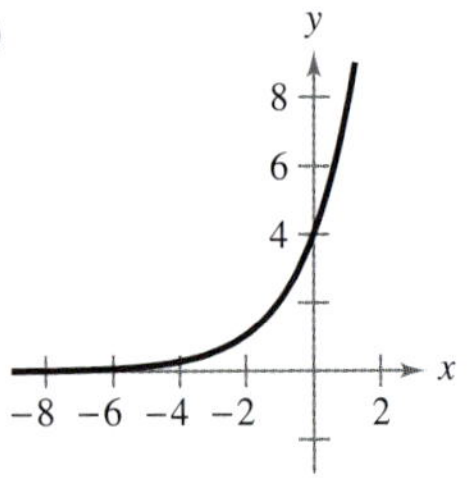

(c)

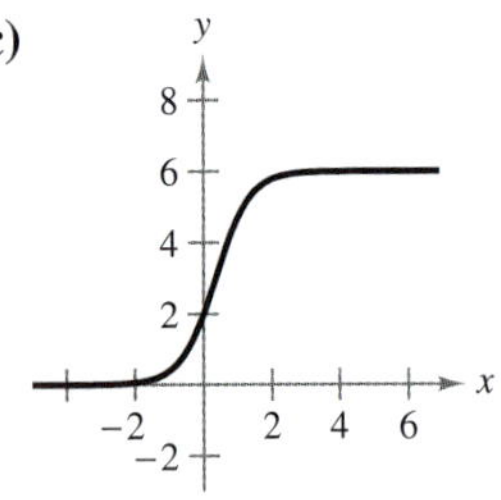

(d)

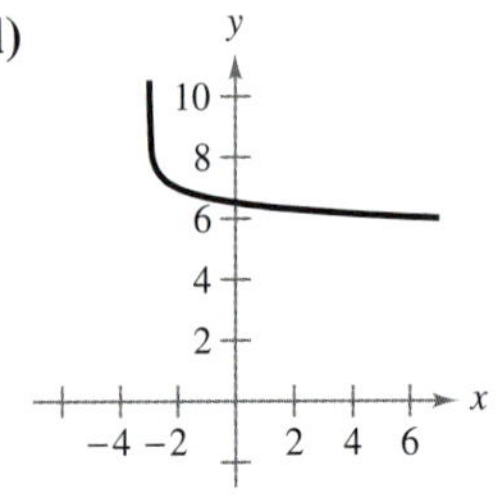

(e)

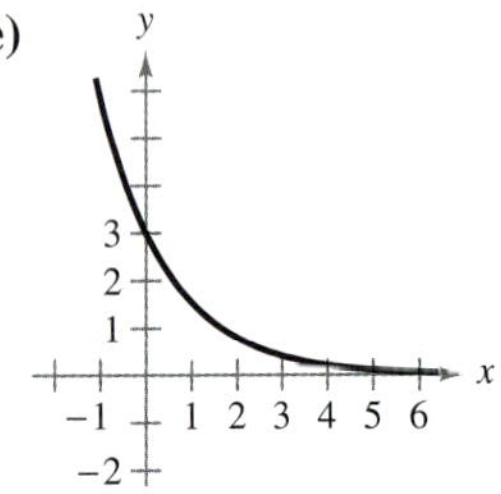

(f)

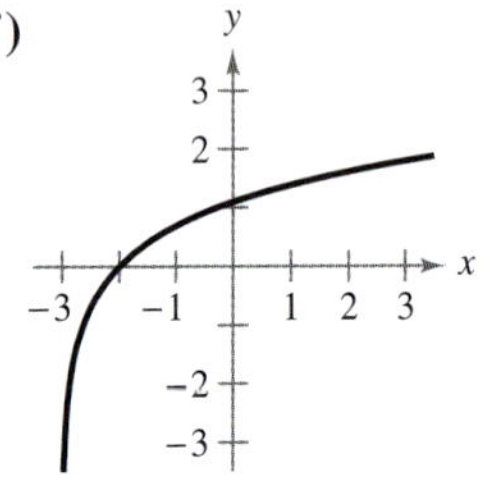

111. $y = 3e^{-2x/3}$

112. $y = 4e^{2x/3}$

113. $y = \dfrac{6}{1 + 2e^{-2x}}$

114. $y = 7 - \log_{10}(x + 3)$

115. $y = 2e^{-(x+4)^2/3}$

116. $y = \ln(x + 3)$

117. Demography The population P (in millions) of Iowa from 2010 through 2017 can be modeled by $P = 2.917e^{kt}$, where t is the year, with $t = 10$ corresponding to 2010. In 2015, the population was about 3,118,473. Find the value of k, to four decimal places, and use the model to predict the population in 2022. *(Source:* U.S. Census Bureau)

118. Education The scores for a test follow the normal distribution $y = 0.0499e^{-(x-74)^2/128}$, $40 \le x \le 100$, where x is the test score. Use a graphing utility to graph the function. Then use the graph to estimate the average test score.

119. Education The average number N of words per minute that the students in a first grade class could read orally after t weeks of school is modeled by $N = 62/(1 + 5.4e^{-0.24t})$. Find the numbers of weeks it took the class to read at average rates of (a) 45 words per minute and (b) 61 words per minute.

120. Audiology Use the formula $\beta = 10 \log_{10}(I/I_0)$, where $I_0 = 10^{-12}$ watt per square meter to find the level of sound β when $I = 10^{-6}$ watt per square meter.

3.6 *What did you learn?*

Classify scatter plots *(p. 235)*, and use scatter plots and a graphing utility to find models for data and choose the model that best fits a set of data *(p. 236)*. You can use a scatter plot and a graphing utility to choose a model that best fits a set of data that represents the yield of a chemical reaction (see Example 3).

Use a graphing utility to find exponential and logistic models for data *(p. 238)*. An exponential model can be used to estimate the amount of a radioactive substance remaining after a number of days (see Example 4), and a logistic model can be used to estimate the percent of defoliation caused by the gypsy moth (see Example 5).

Example The scatter plots below could best be modeled by (a) a linear model, (b) an exponential model, (c) a logarithmic model, and (d) a logistic model.

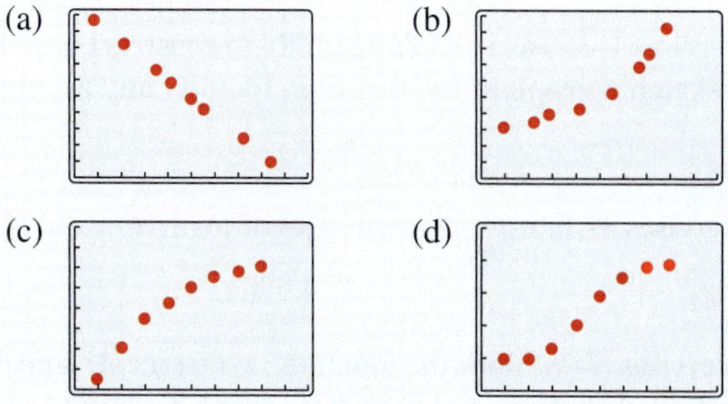

121. Modeling Data You plant a tree when it is 1 meter tall and check its height h (in meters) every 10 years, as shown in the table.

DATA

Year	Height, h
0	1
10	3
20	7.5
30	14.5
40	19
50	20.5
60	21

Spreadsheet at LarsonPrecalculus.com

(a) Use the *regression* feature of a graphing utility to find a logistic model for the data. Let x represent the year.

(b) Use the graphing utility to graph the model with the original data.

(c) How closely does the model represent the data?

(d) What is the limiting height of the tree?

122. Modeling Data Each ordered pair (t, N) represents the year t and the number N (in thousands) of female participants in high school athletic programs during 12 school years, with $t = 6$ corresponding to the 2005–2006 school year. *(Spreadsheet at LarsonPrecalculus.com)* *(Source: National Federation of State High School Associations)*

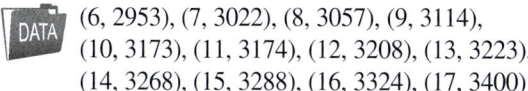

DATA (6, 2953), (7, 3022), (8, 3057), (9, 3114), (10, 3173), (11, 3174), (12, 3208), (13, 3223), (14, 3268), (15, 3288), (16, 3324), (17, 3400)

(a) Use the *regression* feature of a graphing utility to find a linear model, an exponential model, and a power model for the data and identify the coefficient of determination for each model.

(b) Use the graphing utility to graph each model with the original data.

(c) Determine which model best fits the data. Explain.

(d) Use the model you chose in part (c) to predict the school year in which there will be 4 million female participants.

Focusing on Concepts

True or False? **In Exercises 123 and 124, determine whether the statement is true or false. Justify your answer.**

123. The domain of the function

$$f(x) = \log_{10} x$$

is the set of all real numbers.

124. The logarithm of the quotient of two numbers is equal to the difference of the logarithms of the numbers.

125. Think About It Without using a calculator, explain why you know that $3^{\sqrt{3}}$ is greater than 3 and less than 9.

126. Exploration Use a graphing utility to compare the graph of the function $y = e^x$ with the graph of each function below. [$n!$ (read as "n factorial") is defined as $n! = 1 \cdot 2 \cdot 3 \cdot \cdot \cdot (n - 1) \cdot n$.]

$$y_1 = 1 + \frac{x}{1!}$$

$$y_2 = 1 + \frac{x}{1!} + \frac{x^2}{2!}$$

Identify the pattern of successive polynomials. Extend the pattern to five terms and compare the graph of the resulting polynomial function with the graph of $y = e^x$. What do you think this pattern implies?

3 Chapter Test

See *CalcChat.com* for tutorial help and worked-out solutions to odd-numbered exercises. For instructions on how to use a graphing utility, see Appendix A.

Take this test as you would take a test in class. After you are finished, check your work against the answers given in the back of the book.

In Exercises 1–3, use a graphing utility to construct a table of values for the function. Then sketch a graph of the function. Identify any asymptotes of the graph.

1. $f(x) = 10^{-x}$ **2.** $f(x) = -6^{x-2}$ **3.** $f(x) = 1 - e^{2x}$

In Exercises 4–6, use the properties of logarithms to simplify the expression.

4. $\log_7 7^{-0.89}$ **5.** $4.6 \ln e^2$ **6.** $5 - \log_{10} 1000$

In Exercises 7–9, find the domain, x-intercept, and vertical asymptote of the logarithmic function, and sketch its graph by hand.

7. $f(x) = -\log_{10} x - 6$ **8.** $f(x) = \ln(x + 4)$ **9.** $f(x) = 1 + \ln(x - 6)$

In Exercises 10–12, evaluate the logarithm using the change-of-base formula. Round your result to two decimal places.

10. $\log_7 44$ **11.** $\log_{2/5} 1.3$ **12.** $\log_{12} 64$

In Exercises 13–15, use the properties of logarithms to expand the expression as a sum, difference, and/or multiple of logarithms.

13. $\log_2 3a^4$ **14.** $\ln \dfrac{5\sqrt{x}}{6}$ **15.** $\ln \dfrac{x\sqrt{x+1}}{2e^4}$

In Exercises 16–18, use the properties of logarithms to condense the expression.

16. $\log_3 13 + \log_3 y$ **17.** $4 \ln x - 4 \ln y$ **18.** $3 \log_{10} x - \frac{1}{2} \log_{10} x$

Solving an Equation Algebraically **In Exercises 19–24, solve the equation algebraically. Approximate the result to three decimals places, if necessary.**

19. $e^x = e^{x^2-6}$ **20.** $8^{-5t} = 2^5$ **21.** $5^{x^2} = 6^{x+1}$

22. $7 = \ln 2x$ **23.** $2 \log_{10}(x + 2) = 4$ **24.** $\log_3(3x + 1) = 4$

Approximating the Solution(s) of an Equation **In Exercises 25–28, use the *zero* or *root* feature or the *zoom* and *trace* features of a graphing utility to approximate the solution(s) of the equation.**

25. $\log_{10} x = x^4 - 4$ **26.** $e^{-x} \log_{10} 5x = x^3$

27. $4 \ln x = 1 - 3^x$ **28.** $\ln x + e^{x^2} = x^2$

29. The half-life of radioactive actinium (^{227}Ac) is about 22 years. What percent of a present amount of radioactive actinium will remain after 15 years?

30. The table shows the annual revenues R (in millions of dollars) for Zillow Group, Inc. from 2011 to 2017. *(Source:* Zillow Group, Inc.)

(a) Use a graphing utility to construct a scatter plot of the data. Let $t = 1$ correspond to 2011.

(b) Choose a model type for the scatter plot in part (a). Use the *regression* feature of a graphing utility to find a model for the data and identify the coefficient of determination.

(c) Use the graphing utility to graph the model with the data.

(d) Determine whether the model is a good fit for the data. Explain.

Collaborative Project

To work a collaborative project involving Exponential and Logarithmic Functions, visit this textbook's website at *LarsonPrecalculus.com.*

DATA

Year	Revenue, R
2011	66.1
2012	116.9
2013	197.5
2014	325.9
2015	644.7
2016	846.6
2017	1076.8

Spreadsheet at LarsonPrecalculus.com

Table for 30

Standardized Test Practice

See *CalcChat.com* for tutorial help and worked-out solutions to odd-numbered exercises. For instructions on how to use a graphing utility, see Appendix A.

1. $(-1, y)$, $(2, 9)$

Consider the points above. For what value of y is the slope of the line between the points equal to 2?

(A) $y = -3$ (B) $y = 2$

(C) $y = 3$ (D) $y = 7.5$

2. Which function is undefined when $x = 6$?

(A) $f(x) = 6 - x$ (B) $f(x) = \dfrac{1}{6x}$

(C) $f(x) = \dfrac{x-6}{x}$ (D) $f(x) = \dfrac{1}{6-x}$

3.

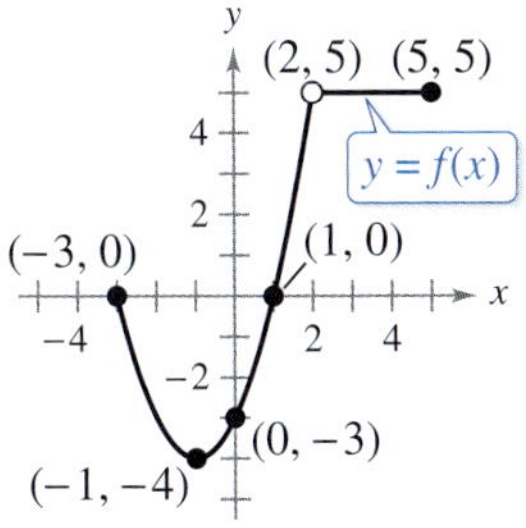

The graph of f is shown above. Which of the following is true?

I. The function is decreasing on the interval $(-3, -1)$.

II. The function is constant on the interval $(-1, 5)$.

III. A relative minimum of the function occurs at the point $(-1, -4)$.

(A) I only (B) II only

(C) I and II only (D) I and III only

4. Which of the following is the parent function of the graph shown below?

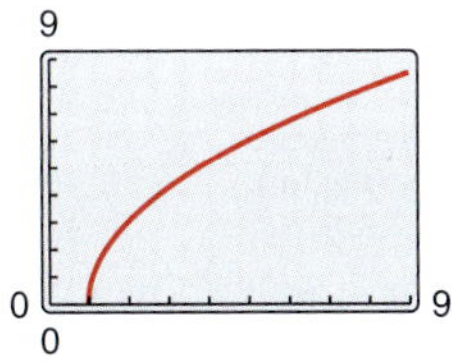

(A) $f(x) = |x|$ (B) $f(x) = \sqrt{x}$

(C) $f(x) = \dfrac{1}{x}$ (D) $f(x) = x^3$

5. Which function represents a vertical translation c units upward of the graph of $f(x)$, where $c > 0$?

(A) $h(x) = f(x + c)$

(B) $h(x) = f\left(\dfrac{x}{c}\right)$

(C) $h(x) = cf(x)$

(D) $h(x) = f(x) + c$

6. A business has two offices. There are 14 workers at the first office and 20 workers at the second office. The first office has a net gain of two workers per year. The second office has a net loss of one worker per year. Which function represents the total number of workers N for the business after x years?

(A) $N = 34 + x$

(B) $N = 6 - x$

(C) $N = 34 + 3x$

(D) $N = 34$

7. If $h(x) = 5x - 2$, what is $h^{-1}(x)$?

(A) $-\dfrac{1}{5}x - 2$ (B) $\dfrac{x+2}{5}$

(C) $5x + 2$ (D) $2 - \dfrac{1}{5}x$

8.

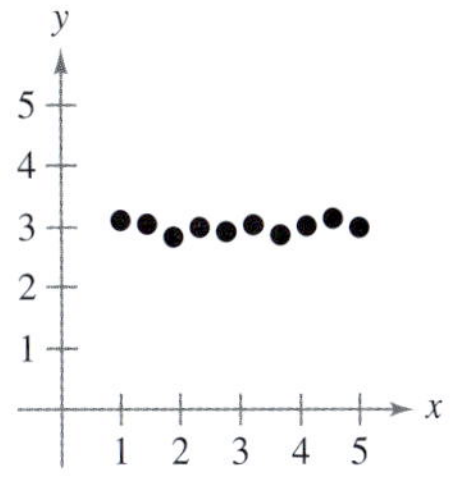

According to the scatter plot above, which of the following statements is true?

(A) y tends to increase as x increases.

(B) y tends to decrease as x increases.

(C) y remains about the same as x increases.

(D) There is no discernible relationship between x and y.

9. $y = x^2 - 10x + 5$

What is the vertex of the parabola represented by the equation above?

(A) $(-10, 5)$ (B) $(-5, 20)$

(C) $(0, 5)$ (D) $(5, -20)$

10. What is the maximum value of $f(x) = -4x^2 + 3$?

(A) -4 (B) $\sqrt{\frac{3}{4}}$

(C) $\frac{4}{3}$ (D) 3

11. The graph of which of the following functions falls to the left and rises to the right?

(A) $y = -\frac{1}{3}x^2 + 4$

(B) $y = 3x^4 + 2x + 3$

(C) $y = -2x^3 + x^2$

(D) $y = \frac{1}{6}x^5 + 17$

12. What is $(3x^3 + 5x^2 - 26x + 8)$ divided by $(x - 2)$?

(A) $3x^2$ (B) $3x^2 - x - 24$

(C) $3x^2 + 11x - 4$ (D) $3x^3 + 7x - 24$

13. What are the zero(s) of $f(x) = x^3 + 2x^2 + 4x + 8$?

(A) $-2i, 2i$ (B) 0

(C) 2 (D) $-2i, 2i, -2$

14. What is the equation of the slant asymptote of the graph of

$$f(x) = \frac{-x^3 + 2x^2 + 5}{x^2 + 1}?$$

(A) $y = x$

(B) $y = -x - 2$

(C) $y = -x + 2$

(D) $y = x^2 + 1$

15.

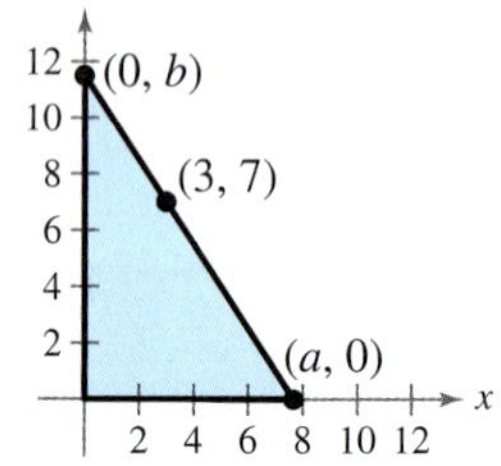

What is the equation of the area A of the shaded triangle above?

(A) $A = \dfrac{7a}{a - 3}$ (B) $A = \dfrac{7a^2}{2(a - 3)}$

(C) $A = \dfrac{3a}{a - 7}$ (D) $A = \dfrac{3a}{2(a - 7)}$

16. Which coefficient of determination represents the best fit of a model to a data set?

(A) $r^2 = 0.98$

(B) $r^2 = 0.13$

(C) $r^2 = 0.04$

(D) $r^2 = 0.95$

17. Which of the following equations is equivalent to $\log_5 5^3 = \log_5 x^2$?

(A) $3 = x^2$ (B) $125 = 2x$

(C) $125 = x^2$ (D) $3 = 2x$

18. Which of the following expressions is equivalent to $\ln(x + 2) + \ln(x - 2) - \ln(x^2 + 1)$?

(A) $\ln(-x^2 + 2x - 1)$

(B) $\ln \dfrac{x^2 - 4}{x^2 + 1}$

(C) $\ln \dfrac{2x}{x^2 + 1}$

(D) $\ln \dfrac{-1 - x^2}{x^2 - 4}$

19. What is the point of intersection of the graphs of

$f(x) = 2^{3x-1}$ and $g(x) = 32$?

(A) $\left(\frac{1}{3}, 0\right)$ (B) $(1, 4)$

(C) $(2, 32)$ (D) $(32, 32)$

20.

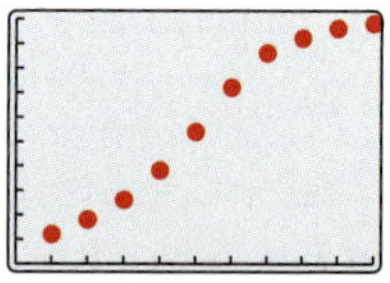

Which of the following model types best fits the scatter plot above?

(A) linear

(B) quadratic

(C) exponential

(D) logistic

21. Which of the following is an exponential growth model?

(A) $y = e^x + a$

(B) $y = ae^{bx}, \quad b < 0$

(C) $y = ae^{bx}, \quad b > 0$

(D) $y = a^x, \quad 0 < a < 1$

22. What is the value of $(fg)(-1)$ when

$f(x) = -x^2 + 3x - 10$ and $g(x) = 4x + 1$?

23. What is the value of $-8 + 3i$ multiplied by its complex conjugate?

24.

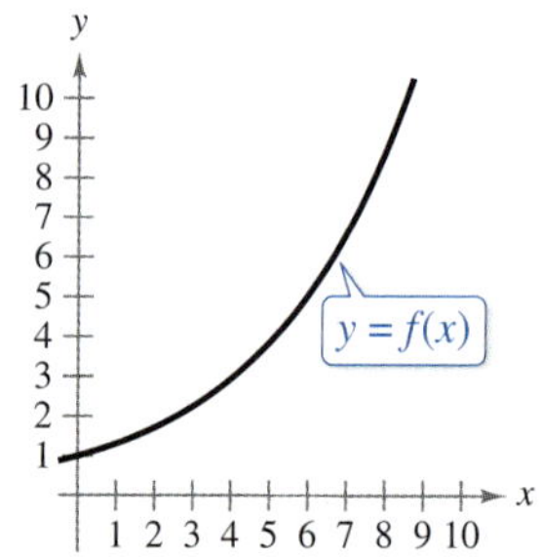

According to the graph above, when $x = 6$, what is the value of $f(x)$ rounded to the nearest whole number?

25. What is the value after 5 years of an initial investment of \$2000 compounded continuously at an annual rate of 6%? Round your answer to the nearest dollar.

26. The rabbit population in a wildlife refuge can be modeled by

$$y = \frac{210}{1 + 6e^{-0.12t}}$$

where y is the population and t is the year, with $t = 11$ corresponding to 2011. What is the projected rabbit population in 2022, rounded to the nearest whole number?

Proofs in Mathematics

Proofs in Mathematics

Each of the following three properties of logarithms can be proved by using properties of exponential functions.

Properties of Logarithms (p. 206)

Let a be a positive real number such that $a \neq 1$, and let n be a real number. If u and v are positive real numbers, then the following properties are true.

	Logarithm with Base a	*Natural Logarithm*
1. Product Property:	$\log_a(uv) = \log_a u + \log_a v$	$\ln(uv) = \ln u + \ln v$
2. Quotient Property:	$\log_a \frac{u}{v} = \log_a u - \log_a v$	$\ln \frac{u}{v} = \ln u - \ln v$
3. Power Property:	$\log_a u^n = n \log_a u$	$\ln u^n = n \ln u$

Slide Rules

The slide rule was invented by William Oughtred (1574–1660) in 1625. The slide rule is a computational device with a sliding portion and a fixed portion. A slide rule enables you to perform multiplication by using the Product Property of logarithms. There are other slide rules that allow for the calculation of roots and trigonometric functions. Slide rules were used by mathematicians and engineers until the invention of the handheld calculator in 1972.

Proof

Let

$$x = \log_a u \quad \text{and} \quad y = \log_a v.$$

The corresponding exponential forms of these two equations are

$$a^x = u \quad \text{and} \quad a^y = v.$$

To prove the Product Property, multiply u and v to obtain

$$uv = a^x a^y = a^{x+y}.$$

The corresponding logarithmic form of $uv = a^{x+y}$ is

$$\log_a(uv) = x + y.$$

So,

$$\log_a(uv) = \log_a u + \log_a v.$$

To prove the Quotient Property, divide u by v to obtain

$$\frac{u}{v} = \frac{a^x}{a^y} = a^{x-y}.$$

The corresponding logarithmic form of $u/v = a^{x-y}$ is

$$\log_a \frac{u}{v} = x - y.$$

So,

$$\log_a \frac{u}{v} = \log_a u - \log_a v.$$

To prove the Power Property, substitute a^x for u in the expression $\log_a u^n$, as follows.

$$\log_a u^n = \log_a(a^x)^n \qquad \text{Substitute } a^x \text{ for } u.$$
$$= \log_a a^{nx} \qquad \text{Property of exponents}$$
$$= nx \qquad \text{Inverse Property of logarithms}$$
$$= n \log_a u \qquad \text{Substitute } \log_a u \text{ for } x.$$

So, $\log_a u^n = n \log_a u$. ■

Progressive Summary (Chapters 1–3)

This chart outlines the topics that have been covered so far in this text. Progressive Summary charts appear after Chapters 2, 3, 6, 9, and 11. In each Progressive Summary, new topics encountered for the first time appear in blue.

Algebraic Functions

Polynomial, Rational, Radical

Rewriting

Polynomial form ↔ Factored form
Operations with polynomials
Rationalize denominators
Simplify rational expressions
Operations with complex numbers

Solving

Equation	*Strategy*
Linear	Isolate variable
Quadratic	Factor, set to zero Extract square roots Complete the square Quadratic Formula
Polynomial	Factor, set to zero Rational Zero Test
Rational	Multiply by LCD
Radical	Isolate, raise to power
Absolute value	Isolate, form two equations

Analyzing

Graphically
Intercepts
Symmetry
Slope
Asymptotes
End behavior
Minimum values
Maximum values

Algebraically
Domain, Range
Transformations
Composition
Standard forms of equations
Leading Coefficient Test
Synthetic division
Descartes's Rule of Signs

Numerically
Table of values

Transcendental Functions

Exponential, Logarithmic

Rewriting

Exponential form ↔ Logarithmic form
Condense/expand logarithmic expressions

Solving

Equation	*Strategy*
Exponential	Take logarithm of each side
Logarithmic	Exponentiate each side

Analyzing

Graphically
Intercepts
Asymptotes

Algebraically
Domain, Range
Transformations
Composition
Inverse Properties

Numerically
Table of values

Other Topics

Rewriting

Solving

Analyzing

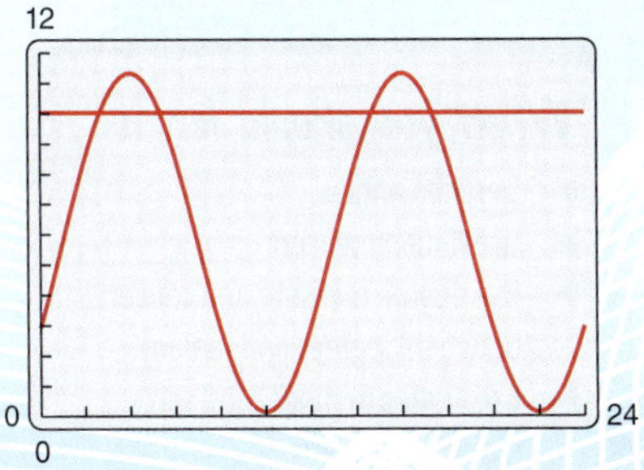

Section 4.5, Example 8
A Depth of at Least 10 Feet

4 Trigonometric Functions

Student Resources at LarsonPrecalculus.com

- **Videos** explaining the concepts of precalculus
- **Worked-out solution videos** for all *Checkpoint* exercises
- **Editable spreadsheets** of the data sets in the text
- **Group projects** for each chapter applying concepts to real-life problems

kemie/DigitalVision Vectors/Getty Images; FatCamera/E+/Getty Images

4.1 Radian and Degree Measure

What you should learn

- Describe angles.
- Use radian measure.
- Use degree measure and convert between degrees and radians.
- Use angles to model and solve real-life problems.

Why you should learn it

Radian measures of angles are involved in numerous aspects of our daily lives. For instance, in Exercise 104 on page 265, you are asked to determine the measure of the angle generated as a skater performs an axel jump.

Angles

As derived from the Greek language, the word **trigonometry** means "measurement of triangles." Initially, trigonometry dealt with relationships among the sides and angles of triangles and was used in the development of astronomy, navigation, and surveying. With the development of calculus and the physical sciences in the seventeenth century, a different perspective arose—one that viewed the classic trigonometric relationships as *functions* with the set of real numbers as their domains. Consequently, the applications of trigonometry expanded to include a vast number of physical phenomena involving rotations and vibrations, including the following.

- sound waves
- light rays
- planetary orbits
- vibrating strings
- pendulums
- orbits of atomic particles

The approach in this text incorporates *both* perspectives, starting with angles and their measure.

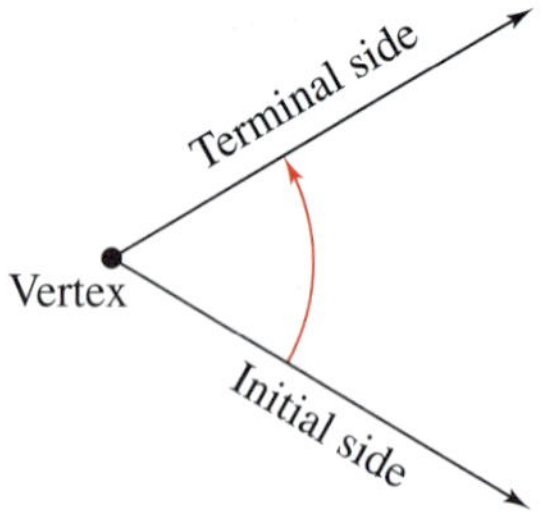

Figure 4.1

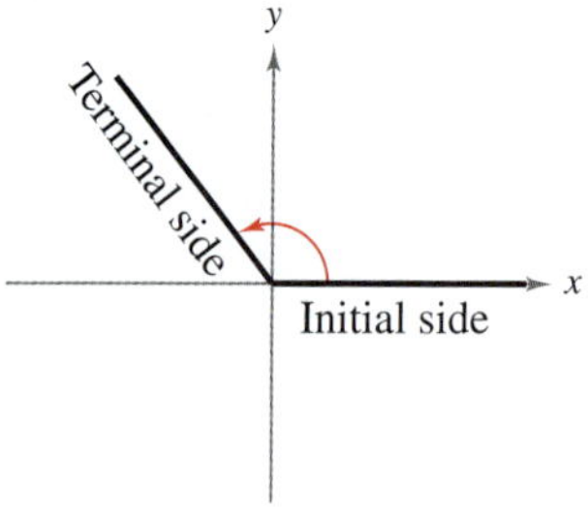

Figure 4.2

An **angle** is determined by rotating a ray (half-line) about its endpoint. The starting position of the ray is the **initial side** of the angle, and the position after rotation is the **terminal side,** as shown in Figure 4.1. The endpoint of the ray is the **vertex** of the angle. This perception of an angle fits a coordinate system in which the origin is the vertex and the initial side coincides with the positive x-axis. Such an angle is in **standard position,** as shown in Figure 4.2. **Positive angles** are generated by counterclockwise rotation and **negative angles** by clockwise rotation, as shown in Figure 4.3. Angles are labeled with Greek letters such as α (alpha), β (beta), and θ (theta), as well as uppercase letters such as A, B, and C. In Figure 4.4, note that angles α and β have the same initial and terminal sides. Such angles are **coterminal.**

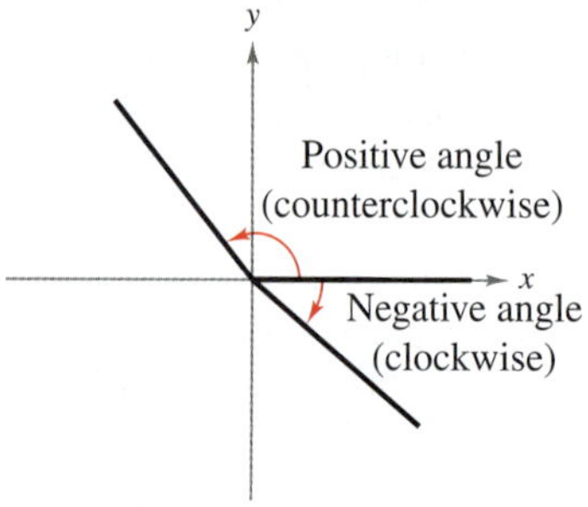

Figure 4.3

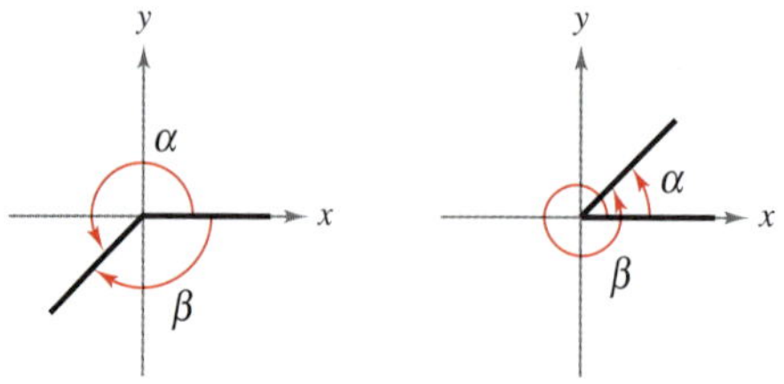

Figure 4.4

Dmitry Morgan/Shutterstock.com

Radian Measure

The **measure of an angle** is determined by the amount of rotation from the initial side to the terminal side. One way to measure angles is in *radians*. This type of measure is especially useful in calculus. To define a radian, you can use a **central angle** of a circle, which is an angle whose vertex is the center of the circle, as shown in Figure 4.5.

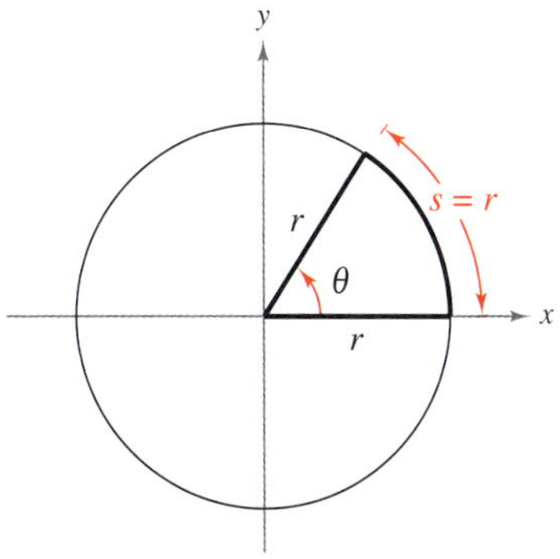

Arc length = radius when θ = 1 radian.
Figure 4.5

Definition of Radian

One **radian** (rad) is the measure of a central angle θ that intercepts an arc s equal in length to the radius r of the circle. (See Figure 4.5.) Algebraically this means that

$$\theta = \frac{s}{r}$$

where θ is measured in radians.

Because the circumference of a circle is $2\pi r$ units, it follows that a central angle of one full revolution (counterclockwise) corresponds to an arc length of $s = 2\pi r$. Moreover, because $2\pi \approx 6.28$, there are just over six radius lengths in a full circle, as shown in Figure 4.6. Because the units of measure for s and r are the same, the ratio s/r has no units—it is simply a real number.

Because the radian measure of an angle of one full revolution is 2π, you can obtain the following.

$$\frac{1}{2}\text{ revolution} = \frac{2\pi}{2} = \pi \text{ radians} \qquad \frac{1}{4}\text{ revolution} = \frac{2\pi}{4} = \frac{\pi}{2}\text{ radians}$$

$$\frac{1}{6}\text{ revolution} = \frac{2\pi}{6} = \frac{\pi}{3}\text{ radians} \qquad \frac{1}{8}\text{ revolution} = \frac{2\pi}{8} = \frac{\pi}{4}\text{ radian}$$

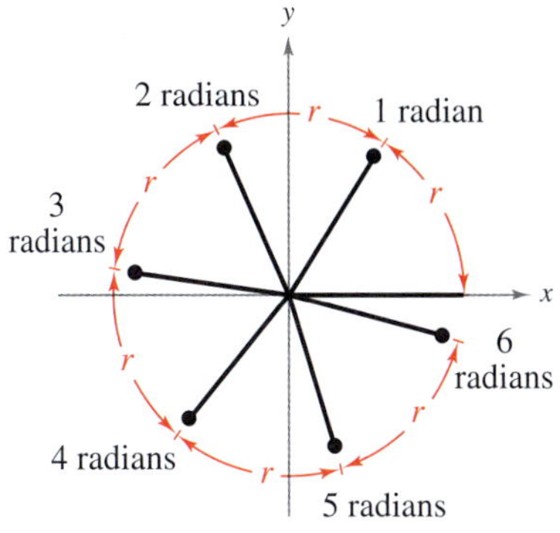

Figure 4.6

These and other common angles are shown below.

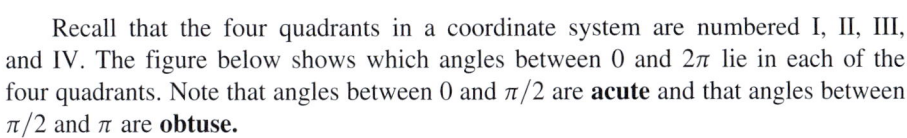

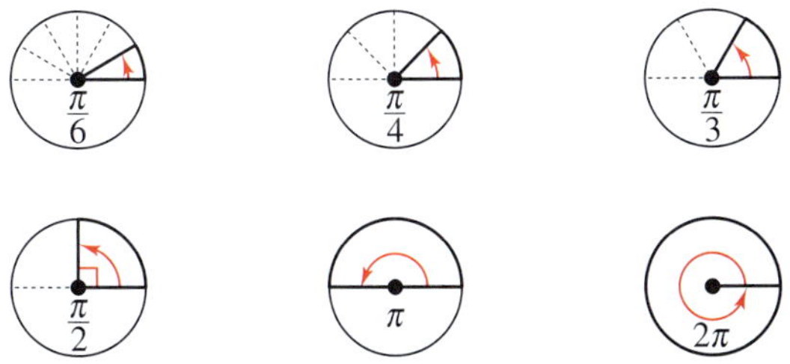

Recall that the four quadrants in a coordinate system are numbered I, II, III, and IV. The figure below shows which angles between 0 and 2π lie in each of the four quadrants. Note that angles between 0 and $\pi/2$ are **acute** and that angles between $\pi/2$ and π are **obtuse.**

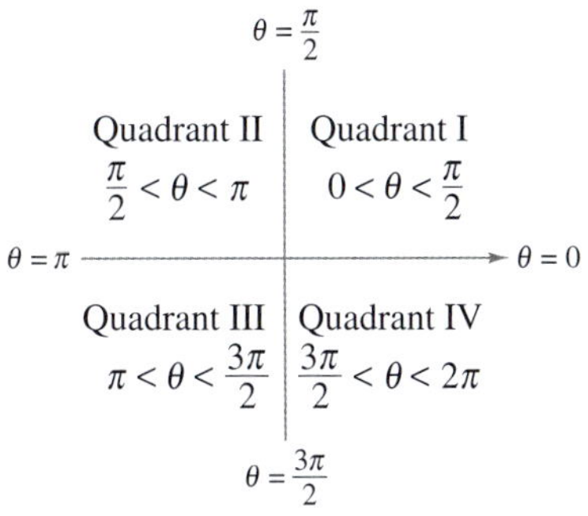

Algebra Help

The phrase "the terminal side of θ lies in a quadrant" is often abbreviated by simply saying that "θ lies in a quadrant." The terminal sides of the quadrantal angles 0, $\pi/2$, π, and $3\pi/2$ do not lie within quadrants.

Two angles are coterminal when they have the same initial and terminal sides. For instance, the angles 0 and 2π are coterminal, as are the angles $\pi/6$ and $13\pi/6$. You can find an angle that is coterminal to a given angle θ by adding or subtracting 2π (one revolution), as demonstrated in Example 1. A given angle θ has infinitely many coterminal angles. For instance, $\theta = \pi/6$ is coterminal with $\pi/6 + 2n\pi$, where n is an integer.

EXAMPLE 1 Sketching and Finding Coterminal Angles

See LarsonPrecalculus.com for an interactive version of this type of example.

a. For the positive angle $\theta = \dfrac{13\pi}{6}$, subtract 2π to obtain a coterminal angle.

$$\frac{13\pi}{6} - 2\pi = \frac{\pi}{6}$$ See figure.

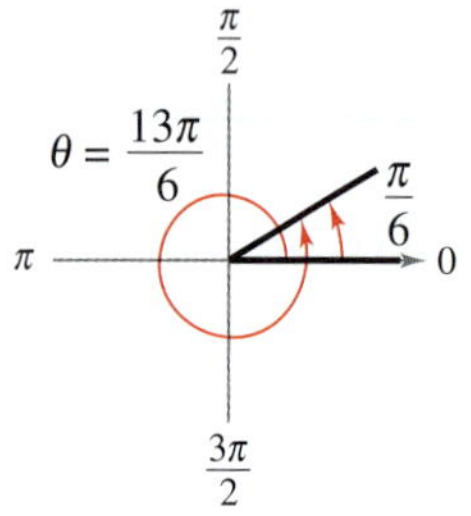

b. For the positive angle $\theta = \dfrac{3\pi}{4}$, subtract 2π to obtain a coterminal angle.

$$\frac{3\pi}{4} - 2\pi = -\frac{5\pi}{4}$$ See figure.

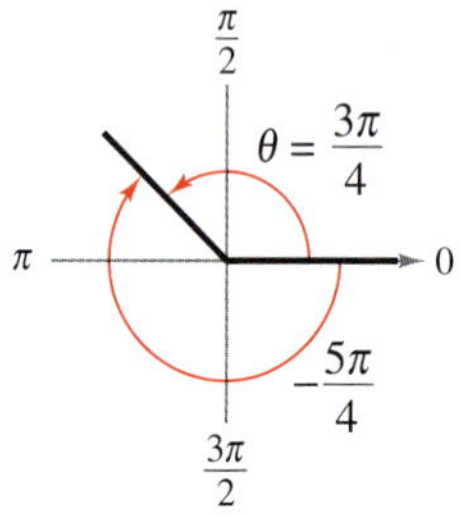

c. For the negative angle $\theta = -\dfrac{2\pi}{3}$, add 2π to obtain a coterminal angle.

$$-\frac{2\pi}{3} + 2\pi = \frac{4\pi}{3}$$ See figure.

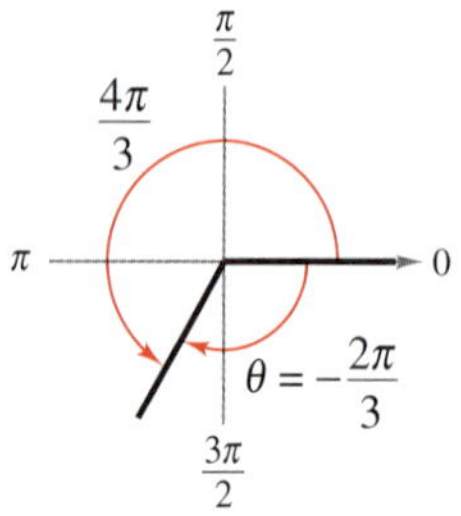

✓ ***Checkpoint*** ▶ *Audio-video solution in English & Spanish at LarsonPrecalculus.com*

Determine two coterminal angles (one positive and one negative) for each angle.

a. $\theta = \dfrac{9\pi}{4}$ **b.** $\theta = -\dfrac{\pi}{3}$

Degree Measure

A second way to measure angles is in terms of **degrees,** denoted by the symbol °. A measure of one degree (1°) is equivalent to a rotation of $\frac{1}{360}$ of a complete revolution about the vertex. To measure angles, it is convenient to mark degrees on the circumference of a circle, as shown in Figure 4.7. So, a full revolution (counterclockwise) corresponds to 360°, a half revolution to 180°, a quarter revolution to 90°, and so on.

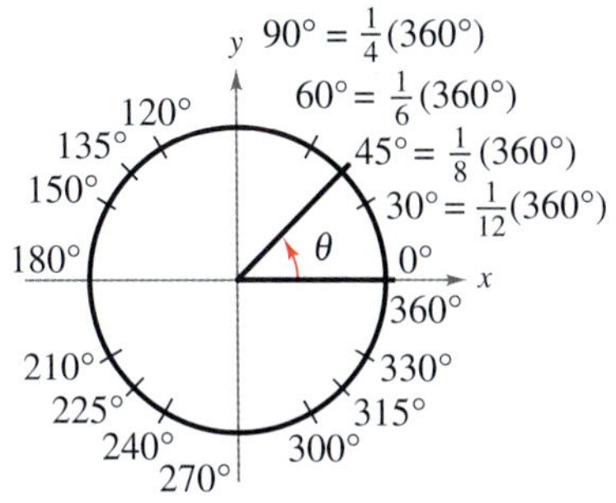

Figure 4.7

One complete revolution corresponds to 2π radians, so degrees and radians are related by the equations

$$360° = 2\pi \text{ rad} \qquad \text{and} \qquad 180° = \pi \text{ rad}.$$

From the second equation, you obtain

$$1° = \frac{\pi}{180} \text{ rad} \qquad \text{and} \qquad 1 \text{ rad} = \frac{180°}{\pi}$$

which lead to the conversion rules below.

Conversions Between Degrees and Radians

1. To convert degrees to radians, multiply degrees by $\dfrac{\pi \text{ rad}}{180°}$.

2. To convert radians to degrees, multiply radians by $\dfrac{180°}{\pi \text{ rad}}$.

To apply these two conversion rules, use the basic relationship π rad = 180°. (See Figure 4.8.)

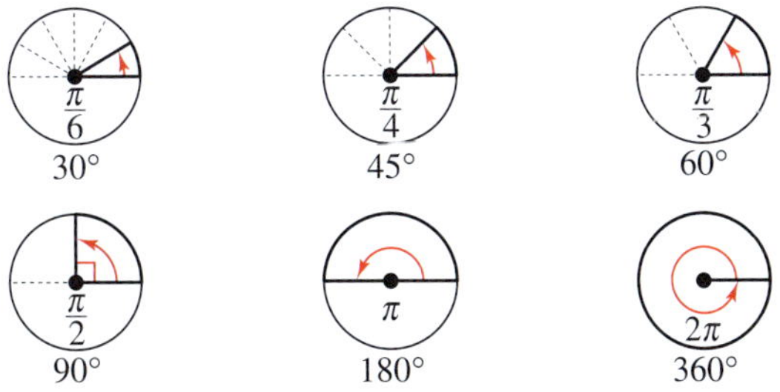

Figure 4.8

When no units of angle measure are specified, *radian measure is implied.* For instance, $\theta = 2$ implies that $\theta = 2$ radians.

EXAMPLE 2 Converting from Degrees to Radians

a. $135° = (135 \text{ deg})\left(\dfrac{\pi \text{ rad}}{180 \text{ deg}}\right) = \dfrac{3\pi}{4}$ radians   Multiply by $\frac{\pi}{180}$.

b. $540° = (540 \text{ deg})\left(\dfrac{\pi \text{ rad}}{180 \text{ deg}}\right) = 3\pi$ radians   Multiply by $\frac{\pi}{180}$.

Checkpoint *Audio-video solution in English & Spanish at LarsonPrecalculus.com*

Rewrite each angle in radian measure as a multiple of π. (Do not use a calculator.)

a. $\theta = 75°$

b. $\theta = 320°$

EXAMPLE 3 Converting from Radians to Degrees

a. $-\frac{\pi}{2} \text{ rad} = \left(-\frac{\pi}{2} \cancel{\text{rad}}\right)\left(\frac{180 \text{ deg}}{\pi \cancel{\text{rad}}}\right) = -90°$ Multiply by $\frac{180}{\pi}$.

b. $2 \text{ rad} = (2 \cancel{\text{rad}})\left(\frac{180 \text{ deg}}{\pi \cancel{\text{rad}}}\right) = \frac{360}{\pi} \approx 114.59°$ Multiply by $\frac{180}{\pi}$.

Checkpoint Audio-video solution in English & Spanish at LarsonPrecalculus.com

Rewrite each angle in degree measure. (Do not use a calculator.)

a. $\frac{\pi}{6}$ **b.** $\frac{5\pi}{3}$

Two positive angles α and β are **complementary** (complements of each other) when their sum is 90° (or $\pi/2$). Two positive angles are **supplementary** (supplements of each other) when their sum is 180° (or π). (See Figure 4.9.)

Complementary angles *Supplementary angles*
Figure 4.9

EXAMPLE 4 Complementary and Supplementary Angles

If possible, find the complement and supplement of each angle.

a. 72° **b.** 148° **c.** $\frac{2\pi}{5}$ **d.** $\frac{4\pi}{5}$

Solution

a. The complement is $90° - 72° = 18°$. The supplement is $180° - 72° = 108°$.

b. Because 148° is greater than 90°, it has no complement. (Remember that complements are *positive* angles.) The supplement is

$$180° - 148° = 32°.$$

c. The complement is

$$\frac{\pi}{2} - \frac{2\pi}{5} = \frac{5\pi}{10} - \frac{4\pi}{10} = \frac{\pi}{10}.$$

The supplement is

$$\pi - \frac{2\pi}{5} = \frac{5\pi}{5} - \frac{2\pi}{5} = \frac{3\pi}{5}.$$

d. Because $4\pi/5$ is greater than $\pi/2$, it has no complement. The supplement is

$$\pi - \frac{4\pi}{5} = \frac{5\pi}{5} - \frac{4\pi}{5} = \frac{\pi}{5}.$$

Checkpoint Audio-video solution in English & Spanish at LarsonPrecalculus.com

If possible, find the complement and supplement of each angle.

a. 36° **b.** 91° **c.** $\frac{\pi}{6}$ **d.** $\frac{5\pi}{6}$

Technology Tip

Historically, fractional parts of degrees were expressed in *minutes* and *seconds*, using the prime (′) and double prime (″) notations, respectively. That is,

$$1' = \text{one minute} = \tfrac{1}{60}(1°)$$

$$1'' = \text{one second} = \tfrac{1}{3600}(1°).$$

Many calculators have special keys for converting angles in degrees, minutes, and seconds (D° M′S″) to decimal degree form, and vice versa.

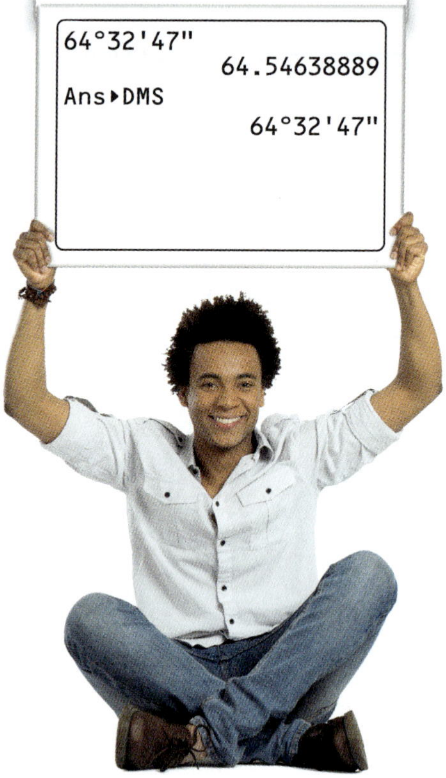

Insight

You will use radian measure and degree measure on most standardized tests. For instance, you may have to find complementary and supplementary angles in degrees or in radians, as shown in Example 4.

Linear and Angular Speed

The *radian measure* formula, $\theta = s/r$, can be used to measure arc length along a circle.

Arc Length

For a circle of radius r, a central angle θ intercepts an arc of length s given by

$s = r\theta$ Length of circular arc

where θ is measured in radians. Note that if $r = 1$, then $s = \theta$, and the radian measure of θ equals the arc length.

EXAMPLE 5 Finding Arc Length

A circle has a radius of 4 inches. Find the length of the arc intercepted by a central angle of 240°, as shown in Figure 4.10.

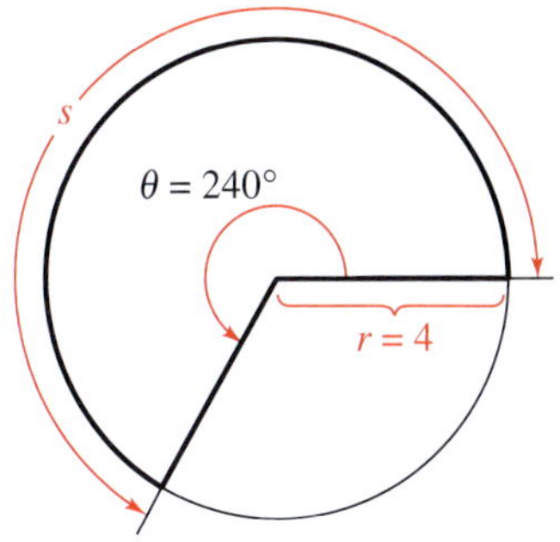

Figure 4.10

Solution

To use the formula

$$s = r\theta$$

first convert 240° to radian measure.

$$240° = (240 \text{ deg})\left(\frac{\pi \text{ rad}}{180 \text{ deg}}\right) = \frac{4\pi}{3} \text{ radians}$$

Then, using a radius of $r = 4$ inches, you can find the arc length to be

$$s = r\theta = 4\left(\frac{4\pi}{3}\right) = \frac{16\pi}{3} \approx 16.76 \text{ inches.}$$

Note that the units for $r\theta$ are determined by the units for r because θ is given in radian measure and therefore has no units.

✓ Checkpoint *Audio-video solution in English & Spanish at LarsonPrecalculus.com*

A circle has a radius of 27 inches. Find the length of the arc intercepted by a central angle of 160°.

Insight

Unless a standardized test specifies otherwise, you can solve a problem using any valid method. For instance, another formula used to find the length of an arc s is

$$s = 2\pi r\left(\frac{\theta}{360°}\right)$$

where θ is measured in degrees. Using this formula in Example 5 yields

$$s = 2\pi(4)\left(\frac{240°}{360°}\right) = \frac{16\pi}{3} \approx 16.76 \text{ inches.}$$

The formula for the length of a circular arc can be used to analyze the motion of a particle moving at a *constant speed* along a circular path.

Linear and Angular Speed

Consider a particle moving at a constant speed along a circular arc of radius r. If s is the length of the arc traveled in time t, then the **linear speed** of the particle is

$$\text{Linear speed} = \frac{\text{arc length}}{\text{time}} = \frac{s}{t}.$$

Moreover, if θ is the angle (in radian measure) corresponding to the arc length s, then the **angular speed** of the particle is

$$\text{Angular speed} = \frac{\text{central angle}}{\text{time}} = \frac{\theta}{t}.$$

Linear speed measures how fast the particle moves, and angular speed measures how fast the angle changes.

EXAMPLE 6 Finding Linear Speed

The second hand of a clock is 10.2 centimeters long, as shown in Figure 4.11. Find the linear speed of the tip of this second hand as it passes around the clock face.

Figure 4.11

Solution

In one revolution, the arc length traveled is

$$s = 2\pi r$$

$$= 2\pi(10.2) \qquad \text{Substitute for } r.$$

$$= 20.4\pi \text{ centimeters.}$$

The time required for the second hand to travel this distance is

$$t = 1 \text{ minute} = 60 \text{ seconds.}$$

So, the linear speed of the tip of the second hand is

$$\text{Linear speed} = \frac{s}{t}$$

$$= \frac{20.4\pi \text{ centimeters}}{60 \text{ seconds}}$$

$$\approx 1.07 \text{ centimeters per second.}$$

Checkpoint *Audio-video solution in English & Spanish at LarsonPrecalculus.com*

The second hand of a clock is 8 centimeters long. Find the linear speed of the tip of this second hand as it passes around the clock face.

EXAMPLE 7 Finding Angular and Linear Speed

The blades of a wind turbine are 116 feet long (see Figure 4.12). The propeller rotates at 15 revolutions per minute.

a. Find the angular speed of the propeller in radians per minute.

b. Find the linear speed of the tips of the blades.

Figure 4.12

Solution

a. Because each revolution generates 2π radians, it follows that the propeller turns

$$(15)(2\pi) = 30\pi \text{ radians per minute.}$$

In other words, the angular speed is

$$\text{Angular speed} = \frac{\theta}{t} = \frac{30\pi \text{ radians}}{1 \text{ minute}} = 30\pi \text{ radians per minute.}$$

b. The linear speed is

$$\text{Linear speed} = \frac{s}{t} = \frac{r\theta}{t} = \frac{(116)(30\pi) \text{ feet}}{1 \text{ minute}} \approx 10{,}933 \text{ feet per minute.}$$

Checkpoint *Audio-video solution in English & Spanish at LarsonPrecalculus.com*

The circular blade on a saw rotates at 2400 revolutions per minute.

a. Find the angular speed of the blade in radians per minute.

b. The blade has a radius of 4 inches. Find the linear speed of a blade tip.

4.1 Exercises

See CalcChat.com for tutorial help and worked-out solutions to odd-numbered exercises.
For instructions on how to use a graphing utility, see Appendix A.

Vocabulary and Concept Check

In Exercises 1–6, fill in the blank.

1. _______ means "measurement of triangles."
2. A(n) _______ is determined by rotating a ray about its endpoint.
3. An angle with its initial side coinciding with the positive x-axis and the origin as its vertex is said to be in _______ .
4. One _______ is the measure of a central angle that intercepts an arc equal in length to the radius of the circle.
5. To convert an angle measure from degrees to radians, multiply the degree measure by _______ .
6. The _______ speed of a particle is a ratio of the change in the central angle to the time.
7. Is one-half revolution of a circle equal to 90° or 180°?
8. What is the sum of two complementary angles?
9. Are the angles 315° and −225° coterminal?
10. Is the angle $\frac{2\pi}{3}$ acute or obtuse?

Procedures and Problem Solving

Estimating an Angle **In Exercises 11 and 12, estimate the angle to the nearest one-half radian.**

11. 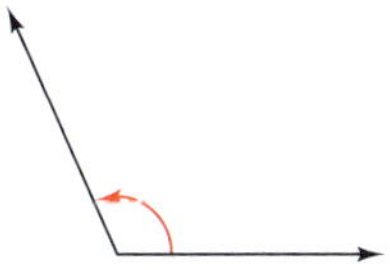

12. 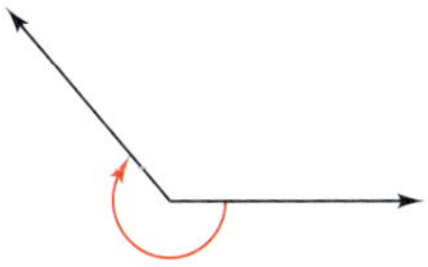

Determining Quadrants **In Exercises 13–18, determine the quadrant in which each angle lies. (The angle measure is given in radians.)**

13. (a) $\frac{\pi}{6}$ (b) $\frac{5\pi}{4}$
14. (a) $\frac{5\pi}{6}$ (b) $-\frac{5\pi}{3}$
15. (a) $\frac{7\pi}{4}$ (b) $\frac{10\pi}{7}$
16. (a) $-\frac{5\pi}{12}$ (b) $-\frac{13\pi}{9}$
17. (a) -1 (b) -2
18. (a) 3.5 (b) 2.25

Sketching Angles **In Exercises 19–24, sketch each angle in standard position.**

19. (a) $\frac{3\pi}{2}$ (b) $-\frac{\pi}{2}$
20. (a) $-\frac{3\pi}{4}$ (b) $\frac{4\pi}{3}$
21. (a) $-\frac{7\pi}{4}$ (b) $-\frac{5\pi}{2}$
22. (a) $\frac{11\pi}{6}$ (b) $\frac{2\pi}{3}$
23. (a) 5π (b) -4
24. (a) 2 (b) -3π

Finding Coterminal Angles **In Exercises 25–28, determine two coterminal angles in radian measure (one positive and one negative) for each angle.**

25. (a) 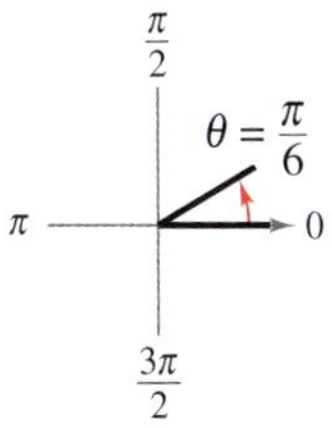

(b) 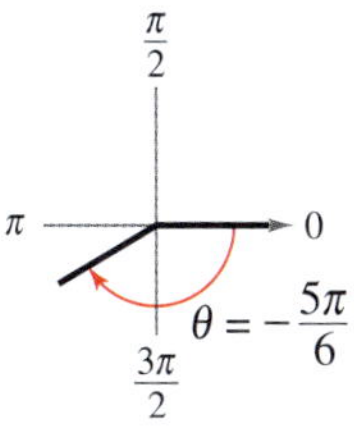

26. (a) 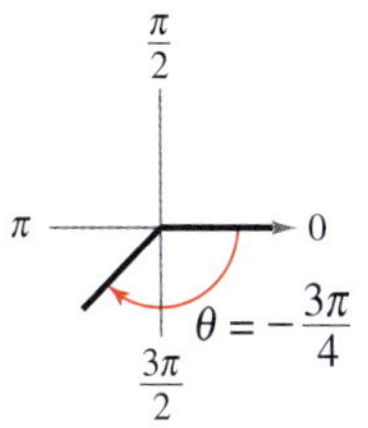

(b) 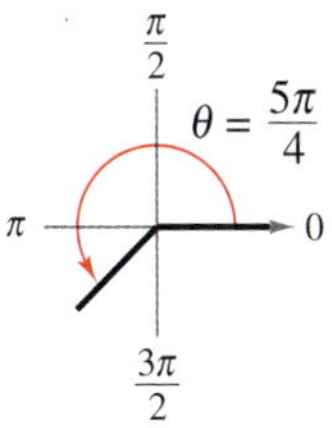

27. (a) $\frac{9\pi}{4}$ (b) $-\frac{2\pi}{15}$
28. (a) $-\frac{7\pi}{8}$ (b) $\frac{\pi}{12}$

Estimating an Angle **In Exercises 29 and 30, estimate the number of degrees in the angle.**

29. 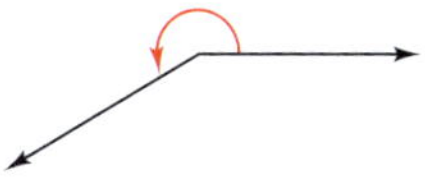

30. 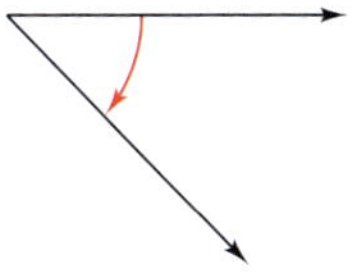

Determining Quadrants **In Exercises 31–36, determine the quadrant in which each angle lies.**

31. (a) 55° (b) 215°

32. (a) 121° (b) 181°

33. (a) -150° (b) 282°

34. (a) 87.9° (b) -8.5°

35. (a) $132^\circ\, 50'$ (b) $-336^\circ\, 30'$

36. (a) -245.25° (b) 12.35°

Sketching Angles **In Exercises 37–42, sketch each angle in standard position.**

37. (a) 45° (b) 90° **38.** (a) 60° (b) 180°

39. (a) -30° (b) 150° **40.** (a) 270° (b) -120°

41. (a) 405° (b) -780° **42.** (a) -450° (b) 600°

Finding Coterminal Angles **In Exercises 43–46, determine two coterminal angles in degree measure (one positive and one negative) for each angle.**

43. (a)
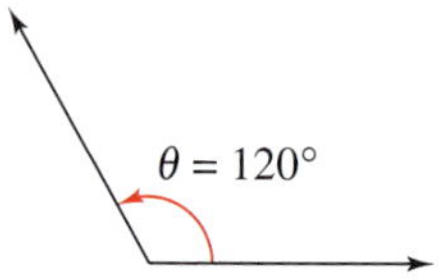

(b)
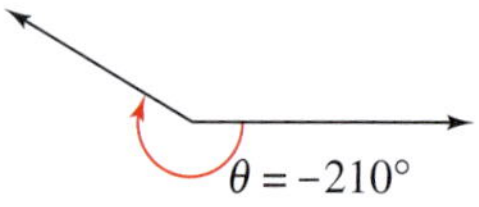

44. (a)
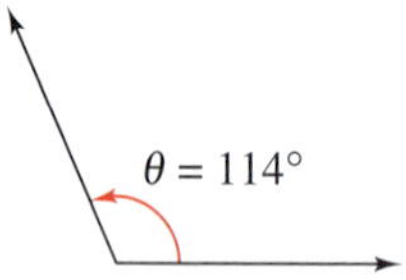

(b)
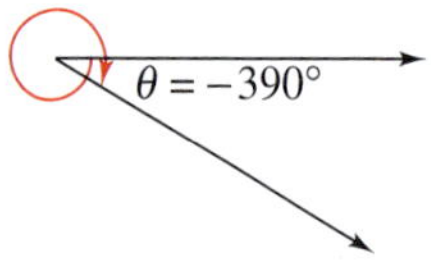

45. (a) -300° (b) 230° **46.** (a) 445° (b) -740°

Converting from Degrees to Radians **In Exercises 47–50, convert each degree measure to radian measure as a multiple of π. (Do not use a calculator.)**

47. (a) 120° (b) -20°

48. (a) -60° (b) 144°

49. (a) 30° (b) -72°

50. (a) -330° (b) 10°

Converting from Radians to Degrees **In Exercises 51–54, convert each radian measure to degree measure. (Do not use a calculator.)**

51. (a) $\frac{3\pi}{2}$ (b) $-\frac{7\pi}{6}$

52. (a) $-\frac{7\pi}{12}$ (b) $\frac{5\pi}{4}$

53. (a) $-\pi$ (b) 3π

54. (a) $\frac{7\pi}{4}$ (b) $-\frac{4\pi}{3}$

Converting from Degrees to Radians **In Exercises 55–60, convert the angle measure from degrees to radians. Round your answer to three decimal places.**

55. 113.8° **56.** 83.7°

57. -216.35° **58.** -46.52°

59. -0.78° **60.** 0.06°

Converting from Radians to Degrees **In Exercises 61–66, convert the angle measure from radians to degrees. Round your answer to three decimal places.**

61. $\frac{\pi}{7}$ **62.** $\frac{5\pi}{22}$

63. 6.5π **64.** -4.2π

65. -2.12 **66.** -0.57

Converting to Decimal Degree Form **In Exercises 67–72, use the angle-conversion capabilities of a graphing utility to convert the angle measure to decimal degree form. Round your answer to three decimal places, if necessary.**

67. $64^\circ\, 45'$ **68.** $-124^\circ\, 30'$

69. $85^\circ\, 18'30''$ **70.** $-408^\circ\, 16'25''$

71. $-125^\circ\, 36''$ **72.** $330^\circ\, 25''$

Finding an Angle Difference **In Exercises 73–76, subtract the second angle measure from the first angle measure. Write your answer in D° M′S″ form.**

73. $51^\circ\, 22'30''$ and $38^\circ\, 17'15''$

74. $120^\circ\, 45'29''$ and $12^\circ\, 36'3''$

75. $9^\circ\, 14'$ and $38^\circ\, 13'13''$

76. $36^\circ\, 8'43''$ and $81^\circ\, 51''$

Converting to D° M′S″ Form **In Exercises 77–82, use the angle-conversion capabilities of a calculator to convert the angle measure to D° M′S″ form.**

77. 280.6° **78.** -115.8°

79. -345.12° **80.** 490.75°

81. -0.355 **82.** 0.7865

Complementary and Supplementary Angles **In Exercises 83–86, find (if possible) the complement and supplement of each angle.**

83. (a) 46° (b) 181° (c) $\frac{\pi}{3}$ (d) $\frac{3\pi}{4}$

84. (a) 35° (b) 164° (c) 4 (d) $\frac{3}{2}$

85. (a) 93° (b) 20° (c) $\frac{\pi}{8}$ (d) 4π

86. (a) 220° (b) 38° (c) $\frac{\pi}{6}$ (d) $\frac{4\pi}{7}$

Finding Arc Length **In Exercises 87–90, find the length of the arc on a circle of radius r intercepted by a central angle θ.**

	Radius, r	*Central Angle, θ*
87.	15 inches	$120°$
88.	3 meters	$150°$
89.	14 inches	π
90.	9 feet	$\frac{\pi}{3}$

Finding the Central Angle **In Exercises 91–94, find the radian measure of the central angle of a circle of radius r that intercepts an arc of length s.**

	Radius, r	*Arc Length, s*
91.	15 inches	8 inches
92.	22 feet	10 feet
93.	14.5 centimeters	35 centimeters
94.	80 kilometers	154 kilometers

Finding the Radius **In Exercises 95–98, find the radius r of a circle with an arc length s and a central angle θ. Round your answer to two decimal places.**

	Arc Length, s	*Central Angle, θ*
95.	36 feet	$\frac{\pi}{2}$
96.	3 meters	$\frac{4\pi}{3}$
97.	82 miles	$135°$
98.	8 inches	$330°$

Earth-Space Science **In Exercises 99 and 100, find the distance between the cities. Assume that Earth is a sphere of radius 4000 miles and the cities are on the same meridian (one city is due north of the other).**

	City	*Latitude*
99.	Bismarck, North Dakota	$46° 49' 33''$
	Pierre, South Dakota	$44° 22' 6''$
100.	San Francisco, California	$37° 47' 36''$N
	Seattle, Washington	$47° 37' 18''$N

101. Earth-Space Science Assuming that Earth is a sphere of radius 6378 kilometers, what is the difference in the latitudes of Syracuse, New York, and Annapolis, Maryland, where Syracuse is 450 kilometers due north of Annapolis?

102. Electrical Engineering A voltmeter's pointer is 6 centimeters in length (see figure). Find the number of degrees through which it rotates when it moves 2.5 centimeters on the scale.

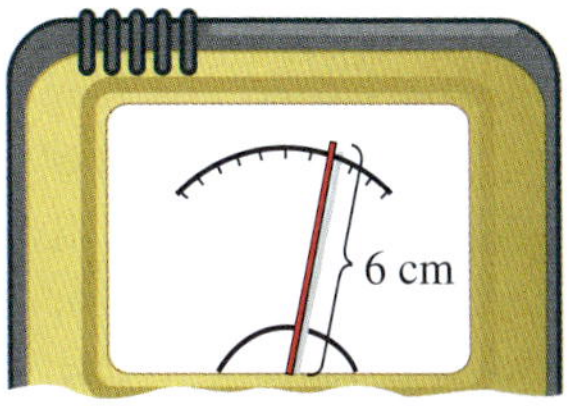

Figure for 102

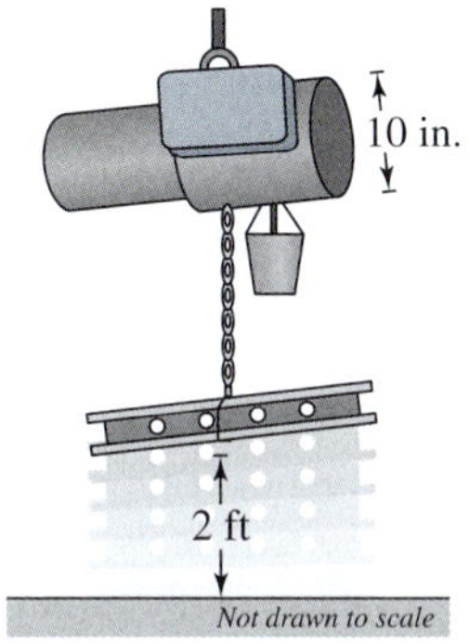

Figure for 103

103. Mechanical Engineering An electric hoist is used to lift a piece of equipment 2 feet. The diameter of the drum on the hoist is 10 inches (see figure). Find the number of degrees through which the drum must rotate.

104. *Why you should learn it* *(p. 256)* The number of revolutions made by a figure skater for each type of axel jump is given. Determine the measure of the angle generated as the skater performs each jump.

(a) Single axel: $1\frac{1}{2}$ revolutions

(b) Double axel: $2\frac{1}{2}$ revolutions

(c) Triple axel: $3\frac{1}{2}$ revolutions

105. Linear Speed A satellite in a circular orbit 1250 kilometers above Earth makes one complete revolution every 110 minutes. What is its linear speed? Assume that Earth is a sphere of radius 6378 kilometers.

106. Mechanical Engineering The circular blade on a saw has a diameter of 7.25 inches and rotates at 4800 revolutions per minute.

(a) Find the angular speed of the blade in radians per minute.

(b) Find the linear speed of the saw teeth (in inches per minute) as they contact the wood being cut.

107. Mechanical Engineering A motorcycle wheel has a diameter of 19.5 inches (see figure) and rotates at 1050 revolutions per minute.

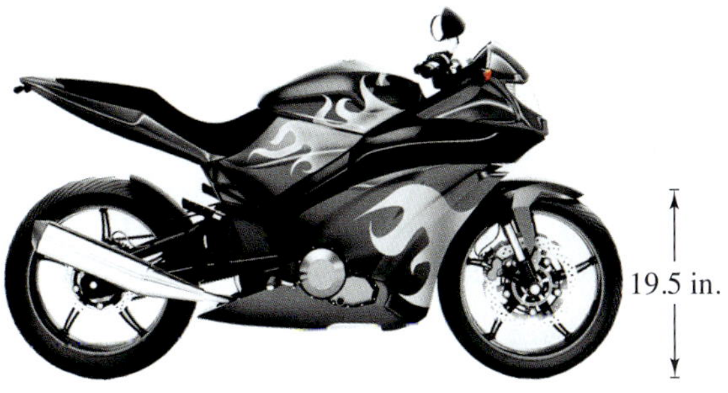

(a) Find the angular speed in radians per minute.

(b) Find the linear speed of the motorcycle (in inches per minute).

108. **Angular Speed** A computerized spin balance machine rotates a 25-inch diameter tire at 480 revolutions per minute.

(a) Find the road speed (in miles per hour) at which the tire is being balanced.

(b) At what rate should the spin balance machine be set so that the tire is being tested for 70 miles per hour?

109. **Mechanical Engineering** A Blu-ray disc is approximately 12 centimeters in diameter. The drive motor of the Blu-ray player is able to rotate up to 10,000 revolutions per minute, depending on what track is being read.

(a) Find the maximum angular speed (in radians per second) of a Blu-ray disc as it rotates.

(b) Find the maximum linear speed (in meters per second) of a point on the outermost track as the disc rotates.

110. **MODELING DATA**

The radii of the pedal sprocket, the wheel sprocket, and the wheel of the bicycle in the figure are 4 inches, 2 inches, and 14 inches, respectively. A cyclist is pedaling at a rate of 1 revolution per second.

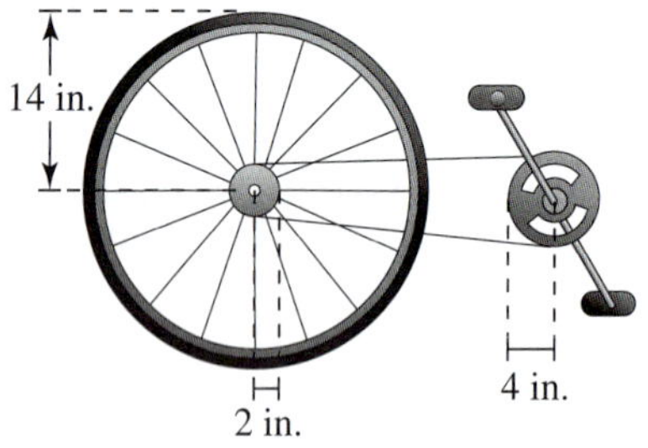

(a) Find the speed of the bicycle in feet per second and miles per hour.

(b) Use your result from part (a) to write a function for the distance d (in miles) a cyclist travels in terms of the number n of revolutions of the pedal sprocket.

(c) Write a function for the distance d (in miles) a cyclist travels in terms of time t (in seconds). Compare this function with the function from part (b).

Focusing on Concepts

True or False? **In Exercises 111 and 112, determine whether the statement is true or false. Justify your answer.**

111. One degree is larger than one radian.

112. An angle that measures $-1260°$ lies in Quadrant III.

113. **Error Analysis** Describe the error.

The interior angle measures of a triangle are $\pi/3$, $3\pi/4$, and $11\pi/12$. ✗

114. **Proof** Prove that the area of a circular sector of radius r with central angle θ is $A = \frac{1}{2}r^2\theta$, where θ is measured in radians.

Geometry **In Exercises 115 and 116, use the result of Exercise 114 to find the area of the sector.**

115.

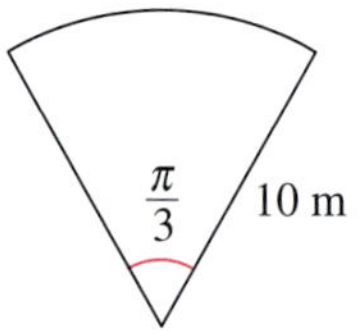

116.

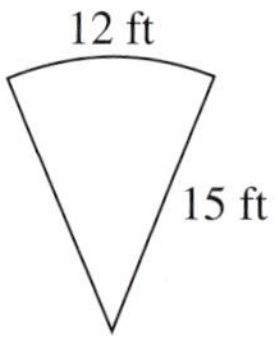

117. **Think About It** The formulas for the area of a circular sector and arc length are $A = \frac{1}{2}r^2\theta$ and $s = r\theta$, respectively. (r is the radius and θ is the angle measured in radians.)

(a) Let $\theta = 0.8$. Write the area and arc length as functions of r and state the domain of each function. Then, use a graphing utility to graph the functions. Use the graphs to determine which function changes more rapidly as r increases. Explain.

(b) Let $r = 10$ centimeters. Write the area and arc length as functions of θ. What is the domain of each function? Use a graphing utility to graph and identify the functions.

118. **HOW DO YOU SEE IT?** Determine which angles in the figure are coterminal angles with angle A. Explain your reasoning.

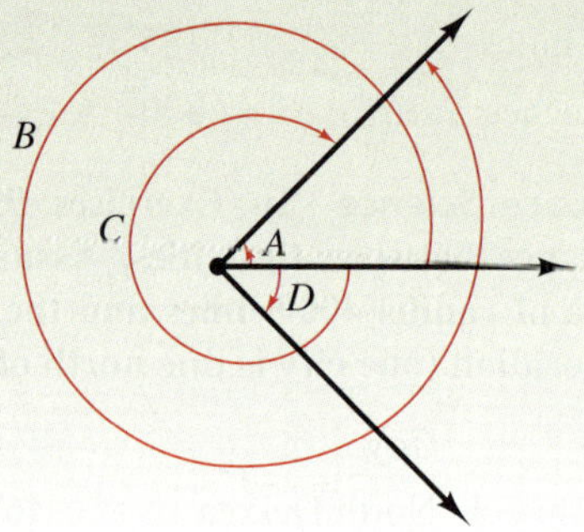

119. **Writing** In your own words, write a definition of 1 radian.

120. **Writing** In your own words, explain the difference between 1 radian and 1 degree.

Cumulative Mixed Review

Library of Parent Functions **In Exercises 121–126, sketch the graph of $f(x) = x^3$ and the graph of the function g. Describe the transformation from f to g.**

121. $g(x) = (x - 1)^3$

122. $g(x) = -(x + 3)^3$

123. $g(x) = 2 - x^3$

124. $g(x) = x^3 - 4$

125. $g(x) = (x + 1)^3 - 3$

126. $g(x) = (x - 5)^3 + 1$

4.2 Trigonometric Functions: The Unit Circle

What you should learn

- Identify the unit circle and describe its relationship to real numbers.
- Evaluate trigonometric functions using the unit circle.
- Use domain and period to evaluate sine and cosine functions and use a calculator to evaluate trigonometric functions.

Why you should learn it

Trigonometric functions are used to model the movement of an oscillating weight. For instance, in Exercise 77 on page 273, the displacement from equilibrium of an oscillating weight suspended by a spring is modeled as a function of time.

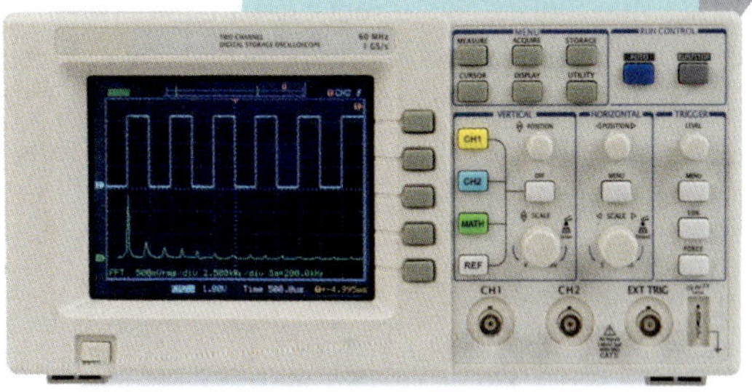

The Unit Circle

The two historical perspectives of trigonometry incorporate different methods of introducing the trigonometric functions. One such perspective is based on the unit circle.

Consider the **unit circle** given by

$x^2 + y^2 = 1$ Unit circle

as shown in the figure.

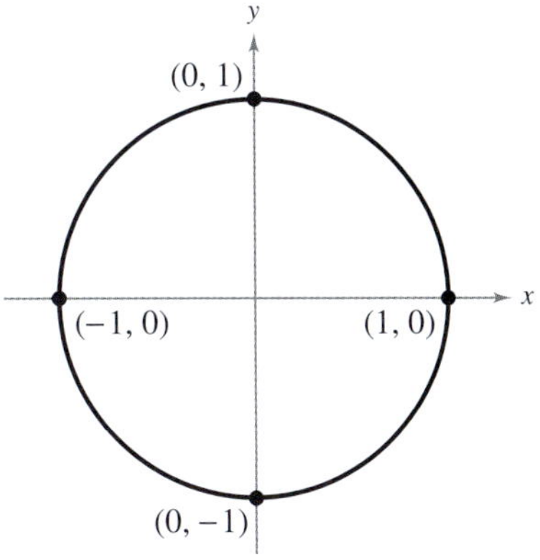

The unit circle

Imagine wrapping the real number line around this circle, with positive numbers corresponding to a counterclockwise wrapping and negative numbers corresponding to a clockwise wrapping, as shown below.

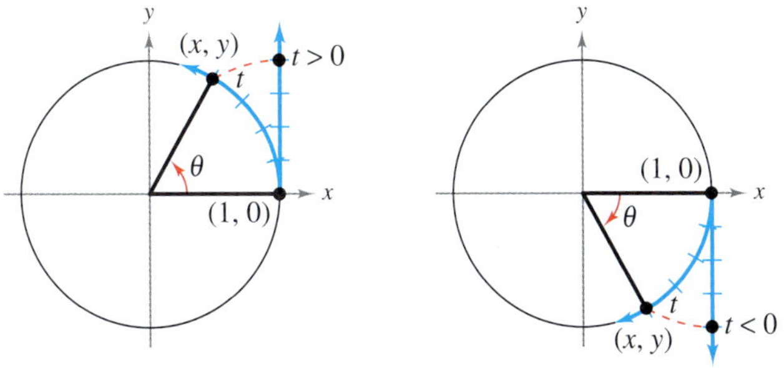

Positive numbers correspond to a counterclockwise wrapping (see figure at the left) and negative numbers correspond to a clockwise wrapping (see figure at the right).

As the real number line wraps around the unit circle, each real number t corresponds to a point (x, y) on the circle. For example, the real number 0 corresponds to the point $(1, 0)$. Moreover, because the unit circle has a circumference of 2π, the real number 2π also corresponds to the point $(1, 0)$.

In general, each real number t also corresponds to a central angle θ (in standard position), whose radian measure is t. With this interpretation of t, the arc length formula

$s = r\theta$

(with $r = 1$) indicates that the real number t is the (directional) length of the arc intercepted by the angle θ, given in radians.

The Trigonometric Functions

From the preceding discussion, it follows that the coordinates x and y are two functions of the real variable t. You can use these coordinates to define the six trigonometric functions of t.

sine	**cosine**	**tangent**
cosecant	**secant**	**cotangent**

These six functions are normally abbreviated sin, cos, tan, csc, sec, and cot, respectively.

Definitions of Trigonometric Functions

Let t be a real number and let (x, y) be the point on the unit circle corresponding to t.

$\sin t = y$ $\qquad\cos t = x$ $\qquad\tan t = \dfrac{y}{x},\quad x \neq 0$

$\csc t = \dfrac{1}{y},\quad y \neq 0$ $\qquad\sec t = \dfrac{1}{x},\quad x \neq 0$ $\qquad\cot t = \dfrac{x}{y},\quad y \neq 0$

Algebra Help

Note in the definitions at the left that the functions in the second row are the *reciprocals* of the corresponding functions in the first row.

In the definitions of the trigonometric functions, note that the tangent and secant are not defined when $x = 0$. For instance, because $t = \pi/2$ corresponds to $(x, y) = (0, 1)$, it follows that $\tan(\pi/2)$ and $\sec(\pi/2)$ are *undefined.* Similarly, the cotangent and cosecant are not defined when $y = 0$. For instance, because $t = 0$ corresponds to $(x, y) = (1, 0)$, cot 0 and csc 0 are *undefined.*

In Figure 4.13, the unit circle is divided into eight equal arcs, corresponding to t-values of

$$0, \frac{\pi}{4}, \frac{\pi}{2}, \frac{3\pi}{4}, \pi, \frac{5\pi}{4}, \frac{3\pi}{2}, \frac{7\pi}{4}, \text{ and } 2\pi.$$

Similarly, in Figure 4.14, the unit circle is divided into 12 equal arcs, corresponding to t-values of

$$0, \frac{\pi}{6}, \frac{\pi}{3}, \frac{\pi}{2}, \frac{2\pi}{3}, \frac{5\pi}{6}, \pi, \frac{7\pi}{6}, \frac{4\pi}{3}, \frac{3\pi}{2}, \frac{5\pi}{3}, \frac{11\pi}{6}, \text{ and } 2\pi.$$

Using the (x, y) coordinates in Figures 4.13 and 4.14, you can evaluate the exact values of trigonometric functions for common t-values. Example 1 demonstrates this procedure. You should study and learn these exact values for common t-values because they will help you to perform calculations in later sections.

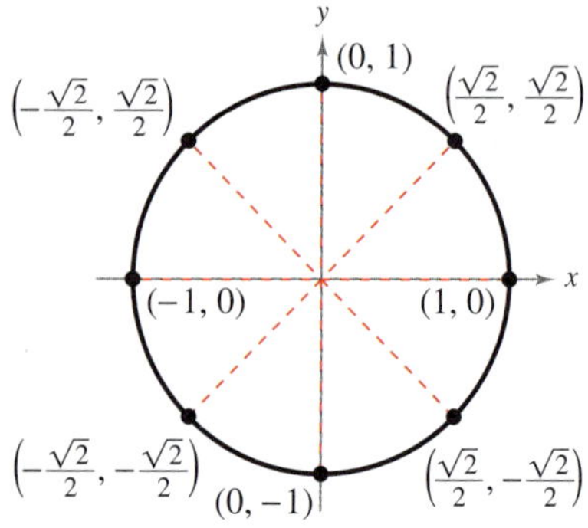

Figure 4.13

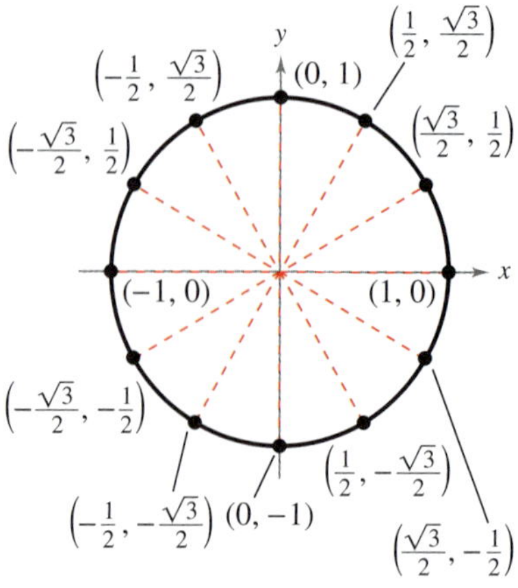

Figure 4.14

Insight

Most standardized tests are timed, so it is important to be efficient. For instance, learning the exact values of trigonometric functions for common t-values (see Figures 4.13 and 4.14, and Example 1) will help you to solve problems more quickly. Also, you may gain a better understanding of the domain and *period* (see page 270) of the sine and cosine functions, which are tested frequently.

EXAMPLE 1 Evaluating Trigonometric Functions

See LarsonPrecalculus.com for an interactive version of this type of example.

Evaluate the six trigonometric functions at each real number.

a. $t = \dfrac{\pi}{6}$ **b.** $t = \dfrac{5\pi}{4}$ **c.** $t = \pi$ **d.** $t = -\dfrac{\pi}{3}$

Solution

For each t-value, begin by finding the corresponding point (x, y) on the unit circle. Then use the definitions of trigonometric functions listed on the preceding page.

a. The real number $t = \pi/6$ corresponds to the point $(x, y) = \left(\sqrt{3}/2, 1/2\right)$.

$$\sin\frac{\pi}{6} = y = \frac{1}{2} \qquad \csc\frac{\pi}{6} = \frac{1}{y} = \frac{1}{1/2} = 2$$

$$\cos\frac{\pi}{6} = x = \frac{\sqrt{3}}{2} \qquad \sec\frac{\pi}{6} = \frac{1}{x} = \frac{2}{\sqrt{3}} = \frac{2\sqrt{3}}{3}$$

$$\tan\frac{\pi}{6} = \frac{y}{x} = \frac{1/2}{\sqrt{3}/2} = \frac{1}{\sqrt{3}} = \frac{\sqrt{3}}{3} \qquad \cot\frac{\pi}{6} = \frac{x}{y} = \frac{\sqrt{3}/2}{1/2} = \sqrt{3}$$

b. The real number $t = 5\pi/4$ corresponds to the point $(x, y) = \left(-\sqrt{2}/2, -\sqrt{2}/2\right)$.

$$\sin\frac{5\pi}{4} = y = -\frac{\sqrt{2}}{2} \qquad \csc\frac{5\pi}{4} = \frac{1}{y} = -\frac{2}{\sqrt{2}} = -\sqrt{2}$$

$$\cos\frac{5\pi}{4} = x = -\frac{\sqrt{2}}{2} \qquad \sec\frac{5\pi}{4} = \frac{1}{x} = -\frac{2}{\sqrt{2}} = -\sqrt{2}$$

$$\tan\frac{5\pi}{4} = \frac{y}{x} = \frac{-\sqrt{2}/2}{-\sqrt{2}/2} = 1 \qquad \cot\frac{5\pi}{4} = \frac{x}{y} = \frac{-\sqrt{2}/2}{-\sqrt{2}/2} = 1$$

c. The real number $t = \pi$ corresponds to the point $(x, y) = (-1, 0)$.

$$\sin\pi = y = 0 \qquad \csc\pi = \frac{1}{y} \text{ is undefined.}$$

$$\cos\pi = x = -1 \qquad \sec\pi = \frac{1}{x} = \frac{1}{-1} = -1$$

$$\tan\pi = \frac{y}{x} = \frac{0}{-1} = 0 \qquad \cot\pi = \frac{x}{y} \text{ is undefined.}$$

d. Moving *clockwise* around the unit circle, it follows that $t = -\pi/3$ corresponds to the point $(x, y) = \left(1/2, -\sqrt{3}/2\right)$.

$$\sin\left(-\frac{\pi}{3}\right) = y = -\frac{\sqrt{3}}{2} \qquad \csc\left(-\frac{\pi}{3}\right) = \frac{1}{y} = -\frac{2}{\sqrt{3}} = -\frac{2\sqrt{3}}{3}$$

$$\cos\left(-\frac{\pi}{3}\right) = x = \frac{1}{2} \qquad \sec\left(-\frac{\pi}{3}\right) = \frac{1}{x} = \frac{1}{1/2} = 2$$

$$\tan\left(-\frac{\pi}{3}\right) = \frac{y}{x} = \frac{-\sqrt{3}/2}{1/2} = -\sqrt{3}$$

$$\cot\left(-\frac{\pi}{3}\right) = \frac{x}{y} = \frac{1/2}{-\sqrt{3}/2} = -\frac{1}{\sqrt{3}} = -\frac{\sqrt{3}}{3}$$

Checkpoint *Audio-video solution in English & Spanish at LarsonPrecalculus.com*

Evaluate the six trigonometric functions at each real number.

a. $t = \pi/2$ **b.** $t = 0$ **c.** $t = -5\pi/6$ **d.** $t = -3\pi/4$

Explore the Concept

With your graphing utility in *radian* and *parametric* modes, enter X1T = cos T and Y1T = sin T and use the settings below.

Tmin = 0, Tmax = 6.3,
Tstep = 0.1
Xmin = −1.5, Xmax = 1.5,
Xscl = 1
Ymin = −1, Ymax = 1,
Yscl = 1

1. Graph the entered equations and describe the graph.
2. Use the *trace* feature to move the cursor around the graph. What do the t-values represent? What do the x- and y-values represent?
3. What are the least and greatest values of x and y?

Domain and Period of Sine and Cosine

The *domain* of the sine and cosine functions is the set of all real numbers. To determine the *range* of these two functions, consider the unit circle shown in the figure. By definition, $\sin t = y$ and $\cos t = x$. Because (x, y) is on the unit circle, you know that $-1 \leq y \leq 1$ and $-1 \leq x \leq 1$. So, the values of sine and cosine also range between -1 and 1.

$$\begin{aligned} -1 \leq \;&y \leq 1 \\ -1 \leq \;&\sin t \leq 1 \end{aligned} \qquad \text{and} \qquad \begin{aligned} -1 \leq \;&x \leq 1 \\ -1 \leq \;&\cos t \leq 1 \end{aligned}$$

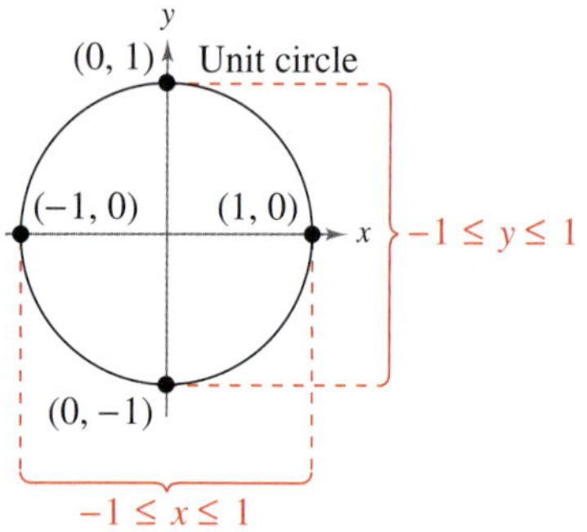

Adding 2π to each value of t in the interval $[0, 2\pi]$ completes a second revolution around the unit circle, as shown in the figure below.

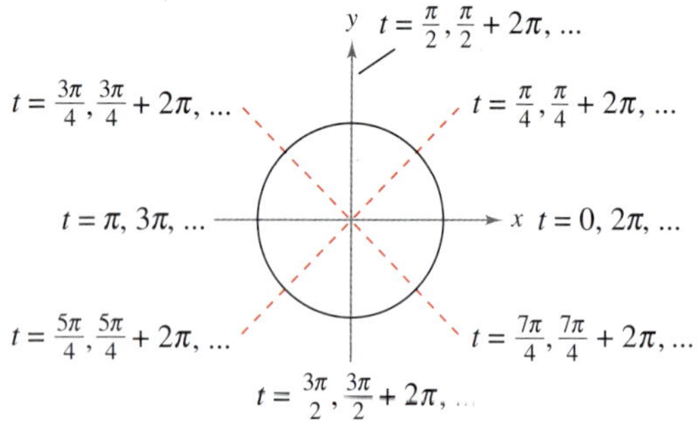

The values of $\sin(t + 2\pi)$ and $\cos(t + 2\pi)$ correspond to those of $\sin t$ and $\cos t$. Similar results can be obtained for repeated revolutions (positive or negative) around the unit circle. This leads to the general result

$$\sin(t + 2\pi n) = \sin t \qquad \text{and} \qquad \cos(t + 2\pi n) = \cos t$$

for any integer n and real number t. Functions that behave in such a repetitive (or cyclic) manner are called **periodic.**

Definition of Periodic Function

A function f is **periodic** when there exists a positive real number c such that

$$f(t + c) = f(t)$$

for all t in the domain of f. The least number c for which f is periodic is called the **period** of f.

From this definition, it follows that the sine and cosine functions are periodic and have a period of 2π. The other four trigonometric functions are also periodic and will be discussed further in Section 4.6.

Dean Drobot/Shutterstock.com

Recall from Section 1.3 that a function f is *even* when $f(-t) = f(t)$ and is *odd* when $f(-t) = -f(t)$.

Even and Odd Trigonometric Functions

The cosine and secant functions are *even*.

$\cos(-t) = \cos t \qquad \sec(-t) = \sec t$

The sine, cosecant, tangent, and cotangent functions are *odd*.

$\sin(-t) = -\sin t \qquad \csc(-t) = -\csc t$

$\tan(-t) = -\tan t \qquad \cot(-t) = -\cot t$

Algebra Help

In this text, the parentheses in $\sin(u)$ are sometimes omitted when u is an expression involving exponents, radicals, products, or quotients. For instance, $\sin(3\pi)$ can be written as $\sin 3\pi$. To evaluate $\sin 3\pi$, find the sine of the product 3π. The other trigonometric functions can be written and evaluated in the same manner.

EXAMPLE 2 Using the Period to Evaluate Sine and Cosine

a. Because $\frac{13\pi}{6} = 2\pi + \frac{\pi}{6}$, you have $\sin\frac{13\pi}{6} = \sin\left(2\pi + \frac{\pi}{6}\right) = \sin\frac{\pi}{6} = \frac{1}{2}$.

b. Because $-\frac{7\pi}{2} = -4\pi + \frac{\pi}{2}$, you have

$$\cos\left(-\frac{7\pi}{2}\right) = \cos\left(-4\pi + \frac{\pi}{2}\right) = \cos\frac{\pi}{2} = 0.$$

c. For $\sin t = \frac{4}{5}$, $\sin(-t) = -\frac{4}{5}$ because the function is odd.

Checkpoint Audio-video solution in English & Spanish at LarsonPrecalculus.com

a. Use the period of the cosine function to evaluate $\cos(9\pi/2)$.

b. Use the period of the sine function to evaluate $\sin(-7\pi/3)$.

c. Evaluate $\cos t$ given that $\cos(-t) = 0.3$.

When evaluating a trigonometric function with a calculator, you need to set the calculator to the desired *mode* of measurement (degrees or radians). Some calculators do not have the cosecant, secant, and cotangent functions. To evaluate these functions, you can use the x^{-1} key with their respective reciprocal functions sine, cosine, and tangent. For example, to evaluate $\csc(\pi/8)$, use the fact that

$$\csc\frac{\pi}{8} = \frac{1}{\sin(\pi/8)}$$

and enter the keystroke sequence below in *radian* mode.

(SIN (π ÷ 8)) x^{-1} ENTER   Display 2.6131259

Technology Tip

When evaluating trigonometric functions with a calculator, remember to enclose all fractional angle measures in parentheses. For instance, to evaluate $\sin t$ for $t = \pi/6$, you should enter

SIN (π ÷ 6) ENTER

to obtain 0.5.

EXAMPLE 3 Using a Calculator

Function	*Mode*	*Graphing Calculator Keystrokes*	*Display*
a. $\sin\frac{2\pi}{3}$	Radian	SIN (2 π ÷ 3) ENTER	0.8660254
b. $\cot 1.5$	Radian	(TAN (1.5)) x^{-1} ENTER	0.0709148

Checkpoint Audio-video solution in English & Spanish at LarsonPrecalculus.com

Use a calculator to evaluate (a) $\sin(5\pi/7)$ and (b) $\csc 2.0$.

4.2 Exercises

See *CalcChat.com* for tutorial help and worked-out solutions to odd-numbered exercises.
For instructions on how to use a graphing utility, see Appendix A.

Vocabulary and Concept Check

In Exercises 1–3, fill in the blank(s).

1. Each real number t corresponds to a point (x, y) on the _______ .

2. A function f is _______ when there exists a positive real number c such that $f(t + c) = f(t)$ for all t in the domain of f.

3. A function f is _______ if $f(-t) = -f(t)$ and _______ if $f(-t) = f(t)$.

4. List the names and abbreviations of the six trigonometric functions.

5. Which trigonometric functions are even? odd?

6. Which trigonometric functions are undefined when $x = 0$?

7. How many equal arcs can you divide the unit circle into using t-values that are multiples of $\pi/4$?

8. How many times is $\tan t = y/x$ undefined on the unit circle?

Procedures and Problem Solving

Determining Values of Trigonometric Functions **In Exercises 9–12, determine the exact values of the six trigonometric functions of the angle θ.**

9.

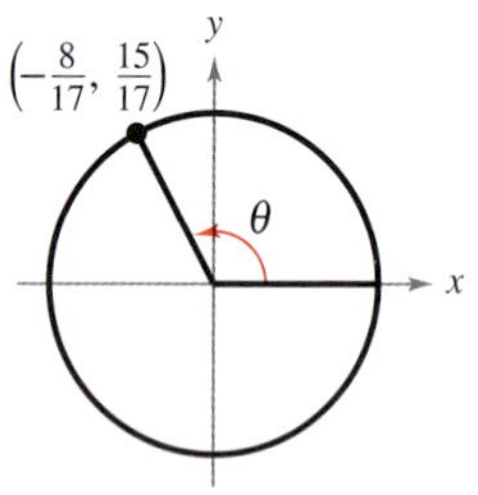

10.

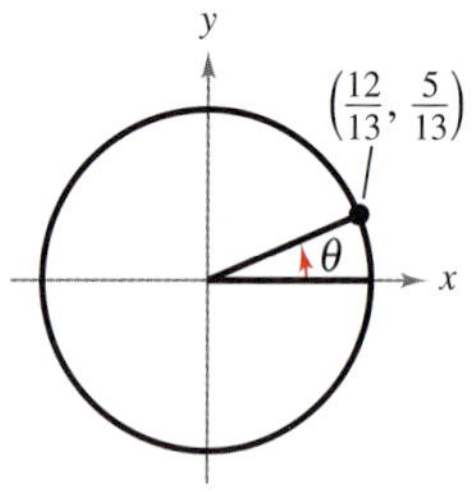

11.

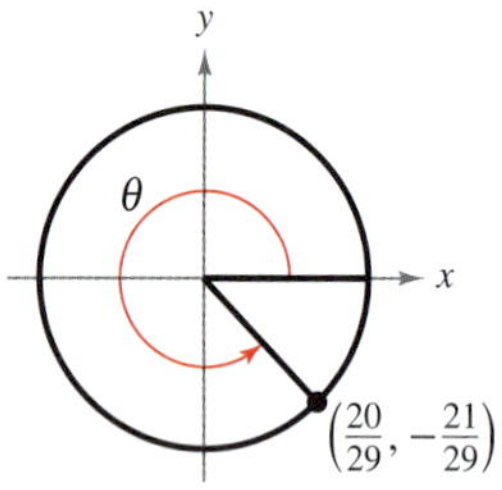

12.

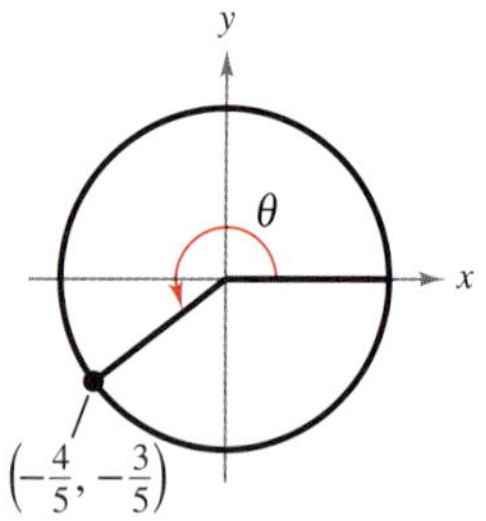

Finding a Point on the Unit Circle **In Exercises 13–22, find the point (x, y) on the unit circle that corresponds to the real number t.**

13. $t = \dfrac{\pi}{6}$ **14.** $t = \dfrac{4\pi}{3}$

15. $t = \dfrac{9\pi}{4}$ **16.** $t = \dfrac{23\pi}{6}$

17. $t = 3\pi/2$ **18.** $t = -3\pi$

19. $t = -\dfrac{\pi}{3}$ **20.** $t = -\dfrac{5\pi}{4}$

21. $t = -\dfrac{17\pi}{6}$ **22.** $t = -\dfrac{7\pi}{4}$

Evaluating Sine, Cosine, and Tangent **In Exercises 23–32, evaluate (if possible) the sine, cosine, and tangent of the real number.**

23. $t = \dfrac{\pi}{4}$ **24.** $t = \dfrac{\pi}{3}$

25. $t = -\dfrac{7\pi}{4}$ **26.** $t = -\dfrac{5\pi}{4}$

27. $t = \dfrac{2\pi}{3}$ **28.** $t = \dfrac{5\pi}{3}$

29. $t = \dfrac{\pi}{2}$ **30.** $t = \dfrac{11\pi}{6}$

31. $t = -\pi/6$ **32.** $t = -2\pi$

Evaluating Trigonometric Functions **In Exercises 33–40, evaluate (if possible) the six trigonometric functions of the real number.**

33. $t = 2\pi/3$ **34.** $t = 5\pi/6$

35. $t = 4\pi/3$ **36.** $t = 7\pi/4$

37. $t = -5\pi/3$ **38.** $t = -3\pi/2$

39. $t = -\pi/2$ **40.** $t = -\pi$

Using the Period to Evaluate Sine and Cosine **In Exercises 41–48, evaluate the trigonometric function using its period as an aid.**

41. $\sin 4\pi$ **42.** $\cos 5\pi$

43. $\cos \dfrac{8\pi}{3}$ **44.** $\sin \dfrac{9\pi}{4}$

45. $\cos\left(-\dfrac{13\pi}{6}\right)$ **46.** $\sin\left(-\dfrac{19\pi}{6}\right)$

47. $\sin\left(-\dfrac{9\pi}{4}\right)$ **48.** $\cos\left(-\dfrac{8\pi}{3}\right)$

Using the Value of a Trigonometric Function In Exercises 49–54, use the value of the trigonometric function to evaluate each function.

49. $\sin t = \frac{1}{3}$
(a) $\sin(-t)$
(b) $\csc(-t)$

50. $\cos t = -\frac{3}{4}$
(a) $\cos(-t)$
(b) $\sec(-t)$

51. $\cos(-t) = -\frac{1}{5}$
(a) $\cos t$
(b) $\sec(-t)$

52. $\sin(-t) = \frac{3}{8}$
(a) $\sin t$
(b) $\csc t$

53. $\tan t = \frac{4}{5}$
(a) $\tan(-t)$
(b) $\cot t$

54. $\cot t = -\frac{1}{4}$
(a) $\cot(-t)$
(b) $\tan(-t)$

Using a Calculator In Exercises 55–70, use a calculator to evaluate the trigonometric expression. Round your answer to four decimal places. (Be sure to use the correct angle mode.)

55. $\sin \frac{2\pi}{9}$ **56.** $\tan \frac{3\pi}{5}$
57. $\cos \frac{11\pi}{5}$ **58.** $\sin \frac{11\pi}{9}$
59. $\tan 7°$ **60.** $\cos 135°$
61. $\cos(-1.7)$ **62.** $\sin(-0.08)$
63. $\tan 433°$ **64.** $\cos(-96°)$
65. $\sec \frac{5\pi}{6}$ **66.** $\cot \frac{4\pi}{3}$
67. $\cot 8°$ **68.** $\csc 75°$
69. $\csc(-5.2)$ **70.** $\sec(-11.3)$

Approximating the Value of a Trigonometric Function In Exercises 71 and 72, use the figure and a straightedge to approximate the value of each trigonometric function. Check your approximation using a graphing utility. To print an enlarged copy of the graph, go to *MathGraphs.com*.

71. (a) $\sin 5$ (b) $\cos 2$
72. (a) $\sin 0.75$ (b) $\cos 2.5$

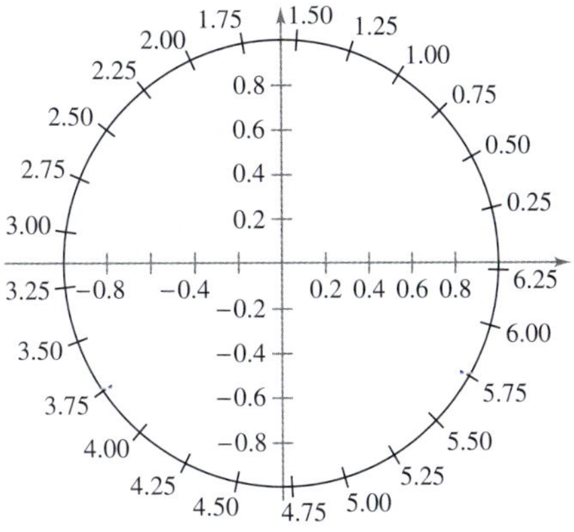

Figure for 71–74

Approximating the Solution of an Equation In Exercises 73 and 74, use the figure and a straightedge to approximate the solution of each equation, where $0 \le t < 2\pi$. Check your approximation using a graphing utility. To print an enlarged copy of the graph, go to *MathGraphs.com*.

73. (a) $\sin t = 0.25$ (b) $\cos t = -0.25$
74. (a) $\sin t = -0.75$ (b) $\cos t = 0.75$

75. Electrical Engineering The initial current and charge of an electrical circuit are zero. The current I (in amperes) t seconds after 100 volts is applied to the circuit is given by $I = 5e^{-2t} \sin t$, where the resistance, inductance, and capacitance are 80 ohms, 20 henrys, and 0.01 farad, respectively. Approximate the current 0.7 second after the voltage is applied.

76. Physics The horizontal distance D (in feet) traveled by a projectile launched at an angle a with level ground is given by

$$D = \frac{100 \sin 2a}{9.8}$$

where the initial velocity is 10 meters per second. Approximate the horizontal distance traveled by a projectile launched at an angle of $\pi/3$ radians.

77. *Why you should learn it* *(p. 267)* The displacement from equilibrium of an oscillating weight suspended by a spring (see figure) is given by

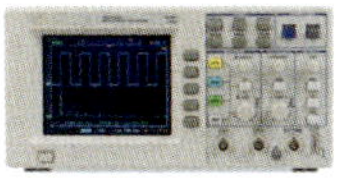

$$y(t) = \frac{1}{4} \cos 6t$$

where y is the displacement (in feet) and t is the time (in seconds). Find the displacement when (a) $t = 0$, (b) $t = \frac{1}{4}$, and (c) $t = \frac{1}{2}$.

Simple Harmonic Motion

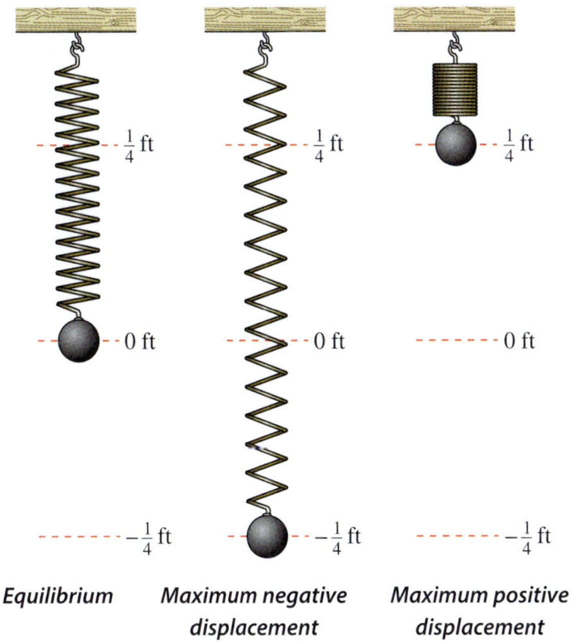

78. MODELING DATA

The displacement from equilibrium of an oscillating weight suspended by a spring and subject to the damping effect of friction can be modeled by

$$y(t) = \frac{1}{4}e^{-t} \cos 6t$$

where y is the displacement (in feet) and t is the time (in seconds).

(a) What is the initial displacement $(t = 0)$?

(b) Use a graphing utility to complete the table.

t	0.50	1.02	1.54	2.07	2.59
y					

(c) The approximate times when the weight is at its maximum distance from equilibrium are shown in the table in part (b). Explain why the magnitude of the maximum displacement is decreasing. What causes this decrease in maximum displacement in the physical system? What factor in the model measures this decrease?

(d) Find the first two times when the weight is at the equilibrium point $(y = 0)$.

Focusing on Concepts

True or False? **In Exercises 79–82, determine whether the statement is true or false. Justify your answer.**

79. Because $\sin(-t) = -\sin t$, you can conclude that the sine of a negative angle is a negative number.

80. $\sin a = \sin(a - 6\pi)$

81. $\tan a = \tan(a - 6\pi)$

82. The real number 0 corresponds to the point (0, 1) on the unit circle.

83. **Think About It** Because $f(t) = \sin t$ is an odd function and $g(t) = \cos t$ is an even function, what can be said about the function $h(t) = f(t)g(t)$?

84. **Think About It** Because $f(t) = \sin t$ and $g(t) = \tan t$ are odd functions, what can be said about the function $h(t) = f(t)g(t)$?

85. **Exploration** Let (x_1, y_1) and (x_2, y_2) be points on the unit circle corresponding to $t = t_1$ and $t = \pi - t_1$, respectively.

(a) Identify the symmetry of the points (x_1, y_1) and (x_2, y_2).

(b) Make a conjecture about any relationship between $\sin t_1$ and $\sin(\pi - t_1)$.

(c) Make a conjecture about any relationship between $\cos t_1$ and $\cos(\pi - t_1)$.

86. **Exploration** Let (x_1, y_1) and (x_2, y_2) be points on the unit circle corresponding to $t = t_1$ and $t = t_1 + \pi$, respectively.

(a) Identify the symmetry of the points (x_1, y_1) and (x_2, y_2).

(b) Make a conjecture about any relationship between $\sin t_1$ and $\sin(t_1 + \pi)$.

(c) Make a conjecture about any relationship between $\cos t_1$ and $\cos(t_1 + \pi)$.

Using the Value of a Trigonometric Function **In Exercises 87 and 88, use the value of the trigonometric function to evaluate the indicated functions.**

87. $\sin t = \frac{2}{3}$

(a) $\sin(\pi - t)$

(b) $\sin(t + \pi)$

88. $\cos t = -\frac{1}{7}$

(a) $\cos(\pi - t)$

(b) $\cos(t + \pi)$

Error Analysis **In Exercises 89 and 90, describe the error.**

89. Using a calculator, $\tan(\pi/7) \approx 0.0078$ ✗

90. $\csc(-\pi) = \frac{1}{\cos(-\pi)} = \frac{1}{\cos \pi} = \frac{1}{-1} = -1$ ✗

91. **Proof** Verify that $\cos 2t = 2 \cos t$ is a false statement by providing a counterexample.

92. **HOW DO YOU SEE IT?** Use the figure below.

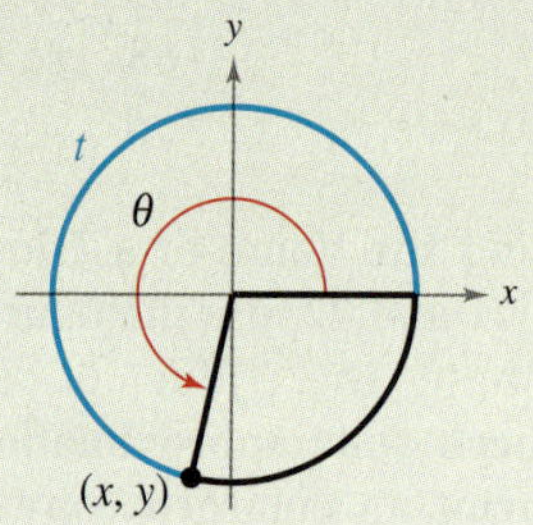

(a) Do all of the trigonometric functions of t exist? Explain your reasoning.

(b) For those trigonometric functions that exist, determine whether the sign of the trigonometric function is positive or negative. Explain your reasoning.

Cumulative Mixed Review

Finding Vertical and Horizontal Asymptotes **In Exercises 93 and 94, find all asymptotes of the graph of the rational function. Use a graphing utility to verify your answer.**

93. $f(x) = \frac{2x}{x - 3}$

94. $f(x) = \frac{5x}{x^2 + x - 6}$

4.3 Right Triangle Trigonometry

The Six Trigonometric Functions

This section introduces the trigonometric functions from a *right triangle* perspective. Consider a right triangle with one acute angle labeled θ, as shown in the figure. Relative to the angle θ, the three sides of the triangle are the **hypotenuse,** the **opposite side** (the side opposite the angle θ), and the **adjacent side** (the side adjacent to the angle θ).

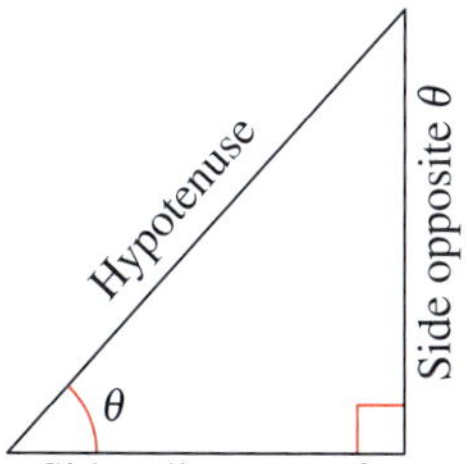

Using the lengths of these three sides, you can form six ratios that define the six trigonometric functions of the acute angle θ.

sine	**cosecant**
cosine	**secant**
tangent	**cotangent**

In the definitions below, it is important to see that

$0° < \theta < 90°$ The angle θ lies in the first quadrant.

and that for such angles, the value of each trigonometric function is *positive*.

Right Triangle Definitions of Trigonometric Functions

Let θ be an *acute* angle of a right triangle. Then the six trigonometric functions *of the angle* θ are defined as shown below. (Note that the functions in the second column are the *reciprocals* of the corresponding functions in the first column.)

$$\sin\theta = \frac{\text{opp}}{\text{hyp}} \qquad \csc\theta = \frac{\text{hyp}}{\text{opp}}$$

$$\cos\theta = \frac{\text{adj}}{\text{hyp}} \qquad \sec\theta = \frac{\text{hyp}}{\text{adj}}$$

$$\tan\theta = \frac{\text{opp}}{\text{adj}} \qquad \cot\theta = \frac{\text{adj}}{\text{opp}}$$

The abbreviations

opp, adj, and hyp

represent the lengths of the three sides of a right triangle.

opp = the length of the side *opposite* θ

adj = the length of the side *adjacent* to θ

hyp = the length of the *hypotenuse*

What you should learn

- Evaluate trigonometric functions of acute angles and use a calculator to evaluate trigonometric functions.
- Use fundamental trigonometric identities.
- Use trigonometric functions to model and solve real-life problems.

Why you should learn it

You can use trigonometry to analyze all aspects of a geometric figure. For instance, Exercise 78 on page 284 shows you how trigonometric functions can be used to approximate the angle of elevation of a zip line.

PHOVOIR/Alamy Stock Photo

EXAMPLE 1 Evaluating Trigonometric Functions

See LarsonPrecalculus.com for an interactive version of this type of example.

Use the triangle in Figure 4.15 to find the exact values of the six trigonometric functions of θ.

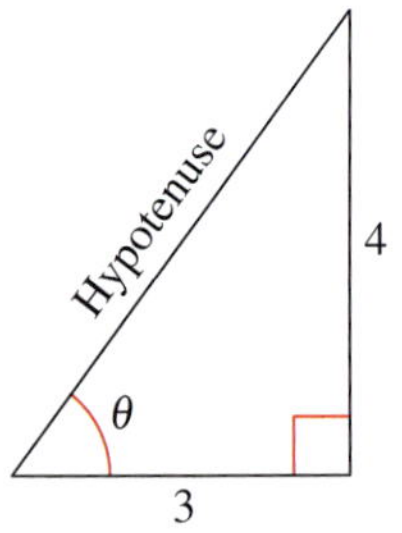

Figure 4.15

Solution

By the Pythagorean Theorem, $(\text{hyp})^2 = (\text{opp})^2 + (\text{adj})^2$, it follows that

$$\text{hyp} = \sqrt{4^2 + 3^2} = \sqrt{25} = 5.$$

So, the six trigonometric functions of θ are

$$\sin\theta = \frac{\text{opp}}{\text{hyp}} = \frac{4}{5} \qquad \csc\theta = \frac{\text{hyp}}{\text{opp}} = \frac{5}{4}$$

$$\cos\theta = \frac{\text{adj}}{\text{hyp}} = \frac{3}{5} \qquad \sec\theta = \frac{\text{hyp}}{\text{adj}} = \frac{5}{3}$$

$$\tan\theta = \frac{\text{opp}}{\text{adj}} = \frac{4}{3} \qquad \cot\theta = \frac{\text{adj}}{\text{opp}} = \frac{3}{4}.$$

✓ ***Checkpoint*** ▶ *Audio-video solution in English & Spanish at LarsonPrecalculus.com*

Use the triangle to find the exact values of the six trigonometric functions of θ.

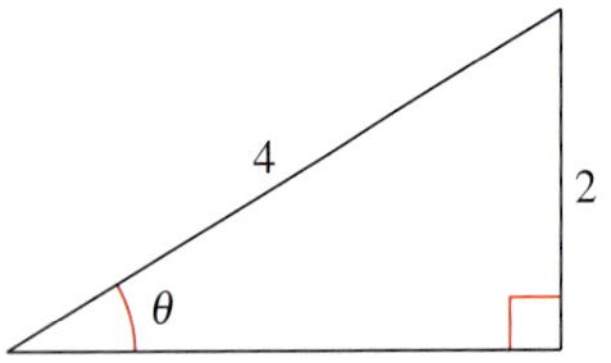

In Example 1, you were given the lengths of two sides of the right triangle but not the angle θ. Often, you will be asked to find the trigonometric functions for a *given* acute angle θ. To do this, you can construct a right triangle having θ as one of its angles.

EXAMPLE 2 Evaluating Trigonometric Functions of 45°

Find the exact values of $\sin 45°$, $\cos 45°$, and $\tan 45°$.

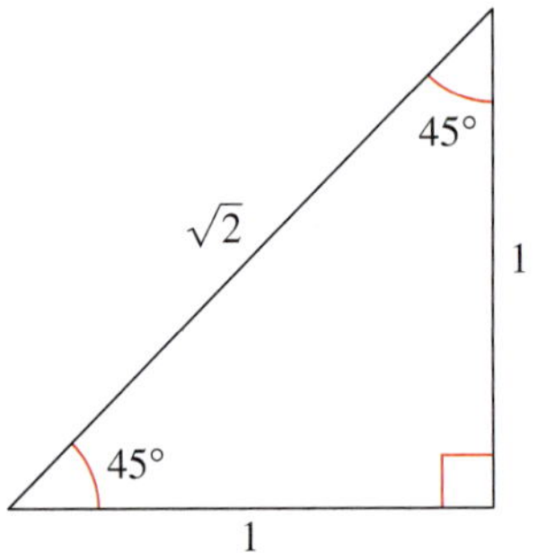

Figure 4.16

Solution

Construct a right triangle having 45° as one of its acute angles, as shown in Figure 4.16. Choose 1 as the length of the adjacent side. From geometry, you know that the other acute angle is also 45°. So, the triangle is isosceles and the length of the opposite side is also 1. By the Pythagorean Theorem, the length of the hypotenuse is $\sqrt{2}$.

$$\sin 45° = \frac{\text{opp}}{\text{hyp}} = \frac{1}{\sqrt{2}} = \frac{\sqrt{2}}{2}$$

$$\cos 45° = \frac{\text{adj}}{\text{hyp}} = \frac{1}{\sqrt{2}} = \frac{\sqrt{2}}{2}$$

$$\tan 45° = \frac{\text{opp}}{\text{adj}} = \frac{1}{1} = 1$$

✓ ***Checkpoint*** ▶ *Audio-video solution in English & Spanish at LarsonPrecalculus.com*

Find the exact values of $\cot 45°$, $\sec 45°$, and $\csc 45°$.

Technology Tip

Use a calculator to convert the answers in Example 2 to decimals. Note that the radical form is the exact value and, in most cases, the exact value is preferred.

EXAMPLE 3 Evaluating Trigonometric Functions of 30° and 60°

Use the equilateral triangle shown in Figure 4.17 to find the exact values of sin 60°, cos 60°, sin 30°, and cos 30°.

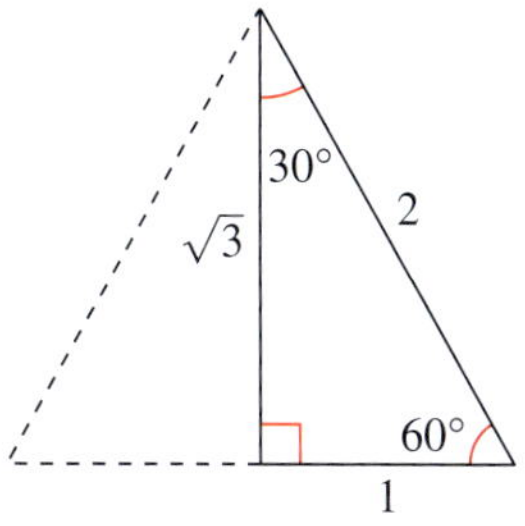

Figure 4.17

Solution

For $\theta = 60°$, you have adj $= 1$, opp $= \sqrt{3}$, and hyp $= 2$. So,

$$\sin 60° = \frac{\text{opp}}{\text{hyp}} = \frac{\sqrt{3}}{2} \quad \text{and} \quad \cos 60° = \frac{\text{adj}}{\text{hyp}} = \frac{1}{2}.$$

For $\theta = 30°$, adj $= \sqrt{3}$, opp $= 1$, and hyp $= 2$. So,

$$\sin 30° = \frac{\text{opp}}{\text{hyp}} = \frac{1}{2} \quad \text{and} \quad \cos 30° = \frac{\text{adj}}{\text{hyp}} = \frac{\sqrt{3}}{2}.$$

✓ **Checkpoint** *Audio-video solution in English & Spanish at LarsonPrecalculus.com*

Use the equilateral triangle shown in Figure 4.17 to find the exact values of tan 60° and tan 30°.

Sines, Cosines, and Tangents of Special Angles

$\sin 30° = \sin \frac{\pi}{6} = \frac{1}{2}$ $\quad$ $\cos 30° = \cos \frac{\pi}{6} = \frac{\sqrt{3}}{2}$ $\quad$ $\tan 30° = \tan \frac{\pi}{6} = \frac{\sqrt{3}}{3}$

$\sin 45° = \sin \frac{\pi}{4} = \frac{\sqrt{2}}{2}$ $\quad$ $\cos 45° = \cos \frac{\pi}{4} = \frac{\sqrt{2}}{2}$ $\quad$ $\tan 45° = \tan \frac{\pi}{4} = 1$

$\sin 60° = \sin \frac{\pi}{3} = \frac{\sqrt{3}}{2}$ $\quad$ $\cos 60° = \cos \frac{\pi}{3} = \frac{1}{2}$ $\quad$ $\tan 60° = \tan \frac{\pi}{3} = \sqrt{3}$

Algebra Help

Because these special angles occur frequently in trigonometry, you should learn to construct the triangles shown in Figures 4.16 and 4.17.

Note that $\sin 30° = \frac{1}{2} = \cos 60°$. This occurs because 30° and 60° are complementary angles. In general, it can be shown from the right triangle definitions that *cofunctions of complementary angles are equal*. That is, if θ is an acute angle, then the relationships below are true.

$$\sin(90° - \theta) = \cos \theta \qquad \cos(90° - \theta) = \sin \theta \qquad \tan(90° - \theta) = \cot \theta$$

$$\cot(90° - \theta) = \tan \theta \qquad \sec(90° - \theta) = \csc \theta \qquad \csc(90° - \theta) = \sec \theta$$

To use a calculator to evaluate trigonometric functions of angles measured in degrees, set the calculator to *degree* mode and then proceed as demonstrated in Section 4.2.

EXAMPLE 4 Using a Calculator

To evaluate $\cos(5° \, 40' 12'')$, begin by converting to decimal degree form.

$$5° \, 40' 12'' = 5° + \left(\frac{40}{60}\right)^{\circ} + \left(\frac{12}{3600}\right)^{\circ} = 5.67°$$

Then use a calculator to evaluate cos 5.67°.

Function	*Graphing Calculator Keystrokes*	*Display*
$\cos(5° \, 40' 12'') = \cos 5.67°$	COS (5.67) ENTER	0.9951074

✓ **Checkpoint** *Audio-video solution in English & Spanish at LarsonPrecalculus.com*

Use a calculator to evaluate $\sin 34° \, 30' 36''$.

Trigonometric Identities

An equation that is true for all values of the variable for which each side of the equation is defined is called an **identity. Trigonometric identities** are identities that involve trigonometric functions.

Fundamental Trigonometric Identities

Reciprocal Identities

$\sin\theta = \dfrac{1}{\csc\theta}$ $\cos\theta = \dfrac{1}{\sec\theta}$ $\tan\theta = \dfrac{1}{\cot\theta}$

$\csc\theta = \dfrac{1}{\sin\theta}$ $\sec\theta = \dfrac{1}{\cos\theta}$ $\cot\theta = \dfrac{1}{\tan\theta}$

Quotient Identities

$\tan\theta = \dfrac{\sin\theta}{\cos\theta}$ $\cot\theta = \dfrac{\cos\theta}{\sin\theta}$

Pythagorean Identities

$\sin^2\theta + \cos^2\theta = 1$

$1 + \tan^2\theta = \sec^2\theta$

$1 + \cot^2\theta = \csc^2\theta$

In the Pythagorean identities listed above, note that $\sin^2\theta$ represents $(\sin\theta)^2$, $\cos^2\theta$ represents $(\cos\theta)^2$, and so on.

Explore the Concept

Select a number t and use your graphing utility to calculate

$$(\sin t)^2 + (\cos t)^2.$$

Repeat this experiment for other values of t and explain why the answer is always the same. Is the result true in both *radian* and *degree* modes?

EXAMPLE 5 Applying Trigonometric Identities

Let θ be an acute angle such that $\cos\theta = 0.8$. Find the values of (a) $\sin\theta$ and (b) $\tan\theta$ using trigonometric identities.

Solution

a. To find the value of $\sin\theta$, use the Pythagorean identity

$$\sin^2\theta + \cos^2\theta = 1.$$

So, you have

$\sin^2\theta + (0.8)^2 = 1$ Substitute 0.8 for $\cos\theta$.

$\sin^2\theta = 1 - (0.8)^2$ Subtract $(0.8)^2$ from each side.

$\sin^2\theta = 0.36$ Simplify.

$\sin\theta = \sqrt{0.36}$ Extract positive square root.

$\sin\theta = 0.6$ Simplify.

b. Now, knowing the sine and cosine of θ, you can find the tangent of θ.

$$\tan\theta = \frac{\sin\theta}{\cos\theta} = \frac{0.6}{0.8} = 0.75$$

Use the definitions of $\sin\theta$ and $\tan\theta$ and the triangle shown in Figure 4.18 to check these results.

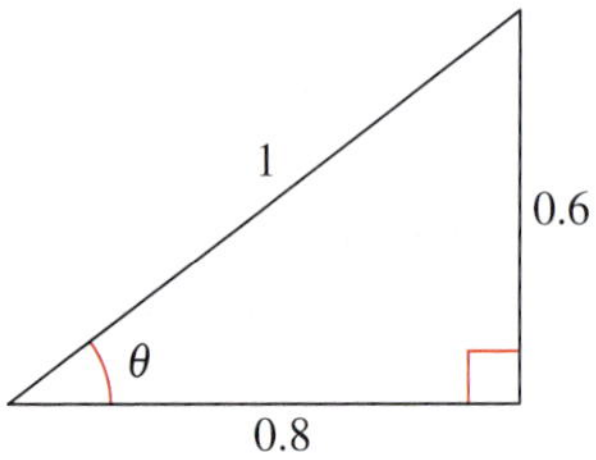

Figure 4.18

Checkpoint Audio-video solution in English & Spanish at LarsonPrecalculus.com

Let θ be an acute angle such that $\sin\theta = 0.28$. Find the values of (a) $\cos\theta$ and (b) $\tan\theta$ using trigonometric identities.

EXAMPLE 6 Applying Trigonometric Identities

Let θ be an acute angle such that $\tan\theta = \frac{1}{3}$. Find the values of (a) $\cot\theta$ and (b) $\sec\theta$ using trigonometric identities.

Solution

a. $\cot\theta = \dfrac{1}{\tan\theta}$ Reciprocal identity

$= \dfrac{1}{1/3}$ Substitute $\frac{1}{3}$ for $\tan\theta$.

$= 3$ Simplify.

b. $\sec^2\theta = 1 + \tan^2\theta$ Pythagorean identity

$\sec^2\theta = 1 + \left(\dfrac{1}{3}\right)^2$ Substitute $\frac{1}{3}$ for $\tan\theta$.

$\sec^2\theta = \dfrac{10}{9}$ Simplify.

$\sec\theta = \dfrac{\sqrt{10}}{3}$ Extract positive square root and simplify.

Use the definitions of $\cot\theta$ and $\sec\theta$ and the triangle to check these results.

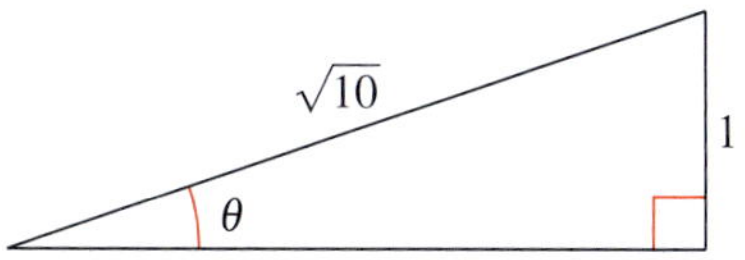

Checkpoint Audio-video solution in English & Spanish at LarsonPrecalculus.com

Let θ be an acute angle such that $\tan\theta = 2$. Find the values of (a) $\cot\theta$ and (b) $\sec\theta$ using trigonometric identities.

EXAMPLE 7 Using Trigonometric Identities

Use trigonometric identities to transform one side of the equation into the other $(0 < \theta < \pi/2)$.

a. $\cos\theta\sec\theta = 1$ **b.** $(\sec\theta + \tan\theta)(\sec\theta - \tan\theta) = 1$

Solution

Simplify the expression on the left-hand side of the equation until you obtain the right-hand side.

a. $\cos\theta\sec\theta = \left(\dfrac{1}{\sec\theta}\right)\sec\theta$ Reciprocal identity

$= 1$ Divide out common factor.

b. $(\sec\theta + \tan\theta)(\sec\theta - \tan\theta)$

$= \sec^2\theta - \sec\theta\tan\theta + \sec\theta\tan\theta - \tan^2\theta$ Distributive Property

$= \sec^2\theta - \tan^2\theta$ Simplify.

$= 1$ Pythagorean identity

Checkpoint Audio-video solution in English & Spanish at LarsonPrecalculus.com

Use trigonometric identities to transform one side of the equation into the other $(0 < \theta < \pi/2)$.

a. $\tan\theta\csc\theta = \sec\theta$ **b.** $(\csc\theta + 1)(\csc\theta - 1) = \cot^2\theta$

Insight

A standardized test may ask you to find the exact value of a trigonometric function. For most angles, a graphing utility provides an *approximate* trigonometric function value because the exact value is an irrational number. To find the exact value, use the appropriate trigonometric definitions and trigonometric identities.

Applications

Many applications of trigonometry involve a process called **solving right triangles.** In this type of application, you are usually given one side of a right triangle and one of the acute angles and are asked to find one of the other sides, *or* you are given two sides and are asked to find one of the acute angles.

In Example 8, you are given the **angle of elevation,** which represents the angle from the horizontal upward to the object. In other applications you may be given the **angle of depression,** which represents the angle from the horizontal downward to the object. (See Figure 4.19.)

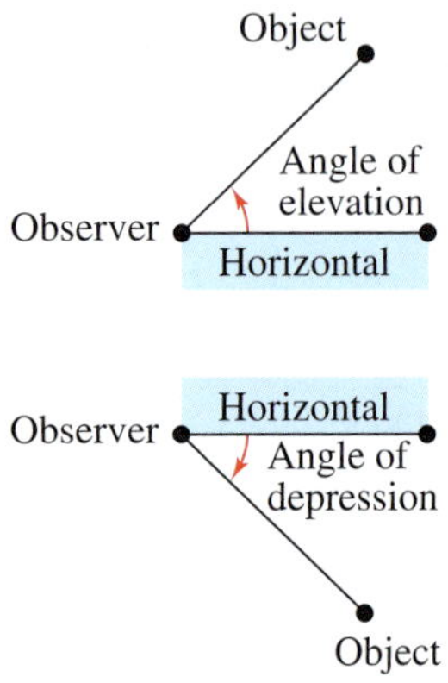

Figure 4.19

EXAMPLE 8 Using Trigonometry to Solve a Right Triangle

A surveyor is standing 50 feet from the base of a large tree, as shown in Figure 4.20. The surveyor measures the angle of elevation to the top of the tree as 71.5°. How tall is the tree?

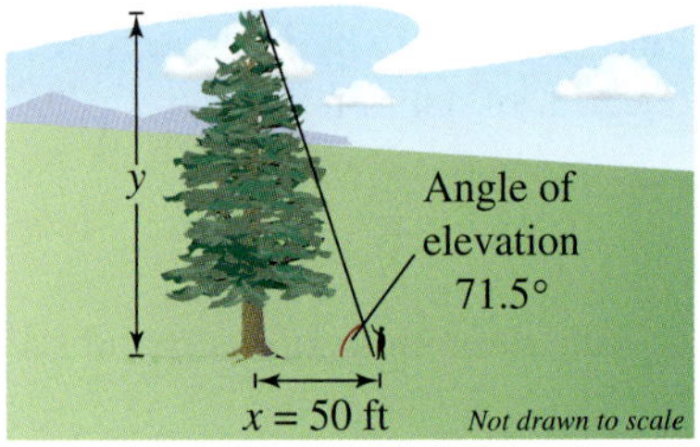

Figure 4.20

Solution

From Figure 4.20, you can see that

$$\tan 71.5^\circ = \frac{\text{opp}}{\text{adj}} = \frac{y}{x}$$

where $x = 50$ and y is the height of the tree. So, the height of the tree is

$$y = x \tan 71.5^\circ = 50 \tan 71.5^\circ \approx 149.43 \text{ feet.}$$

✓ Checkpoint Audio-video solution in English & Spanish at LarsonPrecalculus.com

At a distance of 19 feet from its base, the angle of elevation to the top of a flagpole is 64.6°. (See figure.) How tall is the flagpole?

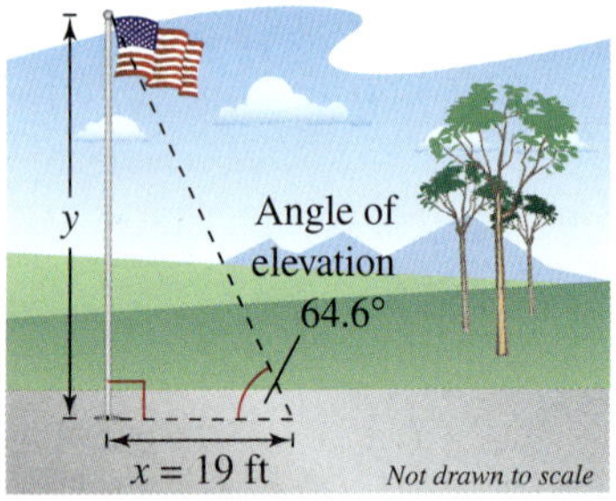

EXAMPLE 9 Using Trigonometry to Solve a Right Triangle

You are 200 yards from a river. Rather than walking directly to the river, you walk 400 yards along a straight path to the river's edge. Find the acute angle θ between this path and the river's edge, as illustrated in Figure 4.21.

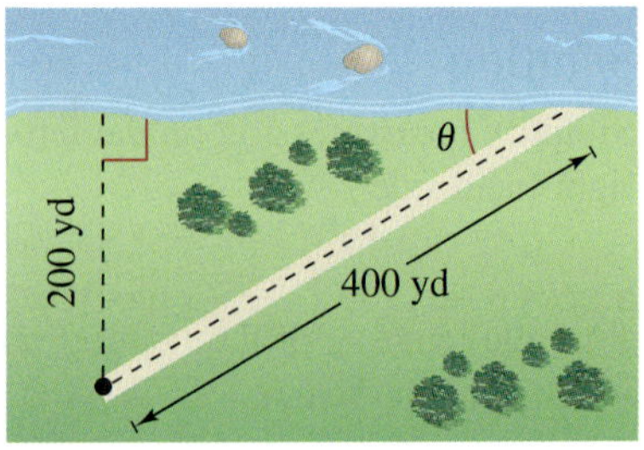

Figure 4.21

Solution

From Figure 4.21, you can see that the sine of the angle θ is

$$\sin \theta = \frac{\text{opp}}{\text{hyp}} = \frac{200}{400} = \frac{1}{2}.$$

Now you should recognize that $\theta = 30^\circ$.

✓ Checkpoint Audio-video solution in English & Spanish at LarsonPrecalculus.com

In Example 9, find the acute angle θ when you are $200\sqrt{3}$ yards from the river and you walk 400 yards along a straight path to the river's edge. ■

In Example 9, you were able to recognize that $\theta = 30°$ is the acute angle that satisfies the equation $\sin \theta = \frac{1}{2}$. Consider, however, the equation $\sin \theta = 0.6$. To find the acute angle θ, note that

$$\sin 30° = \frac{1}{2} = 0.5000$$

and

$$\sin 45° = \frac{1}{\sqrt{2}} \approx 0.7071.$$

So, you might guess that θ lies somewhere between 30° and 45°. In Section 4.7, you will study a method by which a more precise value of θ can be determined.

EXAMPLE 10 Solving a Right Triangle

Find the length c of the skateboard ramp shown in the figure below. Find the horizontal length a of the ramp.

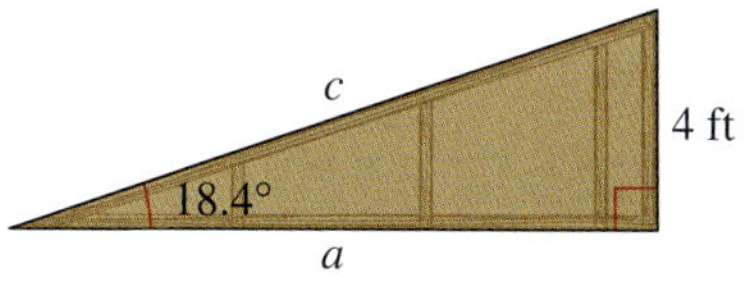

Solution

From the figure, you can see that

$$\sin 18.4° = \frac{\text{opp}}{\text{hyp}} = \frac{4}{c}.$$

So, the length of the skateboard ramp is

$$c = \frac{4}{\sin 18.4°} \approx 12.7 \text{ feet.}$$

Also from the figure, you can see that

$$\tan 18.4° = \frac{\text{opp}}{\text{adj}} = \frac{4}{a}.$$

So, the horizontal length is

$$a = \frac{4}{\tan 18.4°} \approx 12.0 \text{ feet.}$$

Checkpoint *Audio-video solution in English & Spanish at LarsonPrecalculus.com*

Find the length c of the loading ramp shown in the figure below. Find the horizontal length a of the ramp.

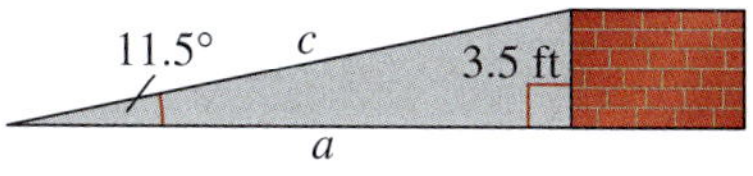

What's Wrong?

You use a graphing utility to check the answers to Example 10 and obtain the screen shown below. What's wrong?

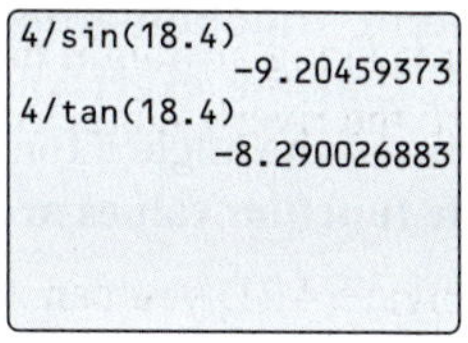

Technology Tip

Graphing utilities have both *degree* and *radian* modes. As you progress through this chapter, be sure you use the correct mode.

4.3 Exercises

See *CalcChat.com* for tutorial help and worked-out solutions to odd-numbered exercises.
For instructions on how to use a graphing utility, see Appendix A.

Vocabulary and Concept Check

1. Match each trigonometric function with its right triangle definition.

(a) sine (b) cosine (c) tangent

(d) cosecant (e) secant (f) cotangent

(i) $\frac{\text{hyp}}{\text{adj}}$ (ii) $\frac{\text{opp}}{\text{adj}}$ (iii) $\frac{\text{opp}}{\text{hyp}}$ (iv) $\frac{\text{adj}}{\text{opp}}$ (v) $\frac{\text{hyp}}{\text{opp}}$ (vi) $\frac{\text{adj}}{\text{hyp}}$

In Exercises 2 and 3, fill in the blanks.

2. Cofunctions of _______ angles are equal.

3. An angle that measures from the horizontal upward to an object is called the angle of _______ , whereas an angle that measures from the horizontal downward to an object is called the angle of _______ .

In Exercises 4–6, use the figure to answer the question.

4. What is the length of the side opposite the angle θ?

5. What is the length of the side adjacent to the angle θ?

6. What is the length of the hypotenuse?

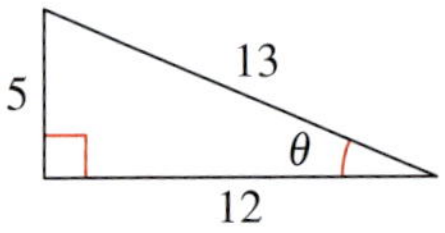

Figure for 4–6

Procedures and Problem Solving

Evaluating Trigonometric Functions **In Exercises 7–10, find the exact values of the six trigonometric functions of the angle θ.**

7.

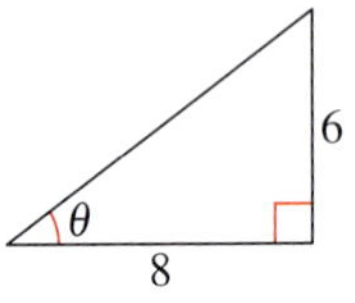

8.

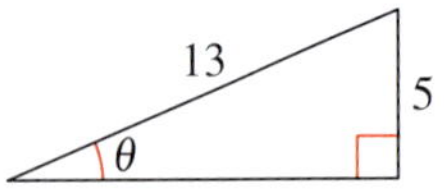

9.

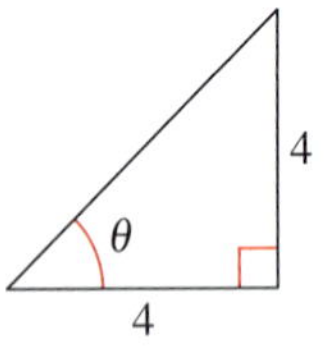

10.

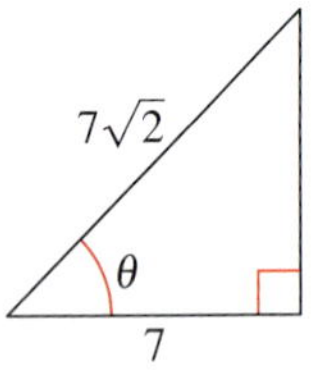

Evaluating Trigonometric Functions **In Exercises 11 and 12, find the exact values of the six trigonometric functions of the angle θ for each of the triangles. Explain why the function values are the same.**

11.

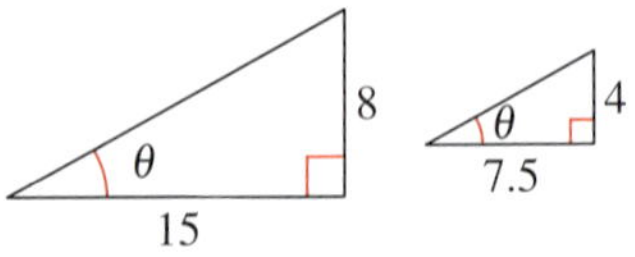

12.

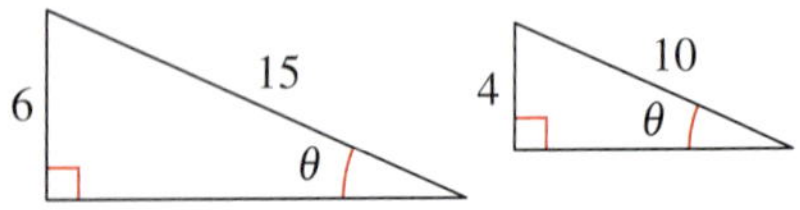

Evaluating Trigonometric Functions **In Exercises 13–20, sketch a right triangle corresponding to the trigonometric function of the acute angle θ. Use the Pythagorean Theorem to determine the third side of the triangle and then find the values of the other five trigonometric functions of θ.**

13. $\sin\theta = \frac{5}{6}$ **14.** $\cot\theta = 5$

15. $\sec\theta = 4$ **16.** $\cos\theta = \frac{3}{7}$

17. $\tan\theta = 3$ **18.** $\csc\theta = \frac{17}{4}$

19. $\cot\theta = \frac{3}{2}$ **20.** $\sin\theta = \frac{3}{8}$

Evaluating Trigonometric Functions of 30°, 45°, and 60° **In Exercises 21–26, construct an appropriate triangle to find the missing values ($0° \le \theta \le 90°, 0 \le \theta \le \pi/2$).**

Function	θ (deg)	θ (rad)	*Function Value*
21. tan	30°		
22. cos	45°		
23. sin		$\frac{\pi}{4}$	
24. tan		$\frac{\pi}{3}$	
25. sec		$\frac{\pi}{4}$	
26. csc		$\frac{\pi}{6}$	

Finding an Angle In Exercises 27 and 28, construct an appropriate triangle to find the acute angle that produces the indicated function value.

	Function	θ (deg)	θ (rad)	*Function Value*
27.	cot			$\frac{\sqrt{3}}{3}$
28.	sec			$\frac{2\sqrt{3}}{3}$

Using a Calculator In Exercises 29–34, use a calculator to evaluate each function. Round your answers to four decimal places. (Be sure to use the correct angle mode.)

29. (a) $\sin 10°$ (b) $\cos 80°$

30. (a) $\tan 18.5°$ (b) $\cot 71.5°$

31. (a) $\sec 42° 12'$ (b) $\csc 48° 7'30''$

32. (a) $\cos 8° 50'25''$ (b) $\sec 56° 10''$

33. (a) $\cot \frac{\pi}{16}$ (b) $\tan \frac{\pi}{8}$

34. (a) $\sec 1.54$ (b) $\cos 1.25$

Fundamental Trigonometric Identities In Exercises 35–50, complete the identity.

35. $\sin \theta = \frac{1}{\square}$

36. $\cos \theta = \frac{1}{\square}$

37. $\tan \theta = \frac{1}{\square}$

38. $\csc \theta = \frac{1}{\square}$

39. $\sec \theta = \frac{1}{\square}$

40. $\cot \theta = \frac{1}{\square}$

41. $\tan \theta = \frac{\square}{\square}$

42. $\cot \theta = \frac{\square}{\square}$

43. $\sin^2 \theta + \cos^2 \theta = \square$

44. $1 + \tan^2 \theta = \square$

45. $\sin(90° - \theta) = \square$

46. $\cos(90° - \theta) = \square$

47. $\tan(90° - \theta) = \square$

48. $\cot(90° - \theta) = \square$

49. $\sec(90° - \theta) = \square$

50. $\csc(90° - \theta) = \square$

Applying Trigonometric Identities In Exercises 51–56, use the given function value(s) and the trigonometric identities to find the exact value of each indicated trigonometric function.

51. $\sin 60° = \frac{\sqrt{3}}{2}$, $\cos 60° = \frac{1}{2}$

(a) $\tan 60°$ (b) $\sin 30°$

(c) $\cos 30°$ (d) $\cot 60°$

52. $\sin 30° = \frac{1}{2}$, $\tan 30° = \frac{\sqrt{3}}{3}$

(a) $\csc 30°$ (b) $\cot 60°$

(c) $\cos 30°$ (d) $\cot 30°$

53. $\cos \theta = \frac{1}{3}$

(a) $\sin \theta$ (b) $\tan \theta$

(c) $\sec \theta$ (d) $\csc(90° - \theta)$

54. $\sec \theta = 5$, $\tan \theta = 2\sqrt{6}$

(a) $\cos \theta$ (b) $\cot \theta$

(c) $\cot(90° - \theta)$ (d) $\sin \theta$

55. $\cot \alpha = 4$

(a) $\tan \alpha$ (b) $\csc \alpha$

(c) $\sec \alpha$ (d) $\tan(90° - \alpha)$

56. $\tan \beta = 3$

(a) $\cot \beta$ (b) $\cos \beta$

(c) $\tan(90° - \beta)$ (d) $\csc \beta$

Using Trigonometric Identities In Exercises 57–66, use trigonometric identities to transform the left side of the equation into the right side ($0 < \theta < \pi/2$).

57. $\tan \theta \cot \theta = 1$

58. $\sin \theta \csc \theta = 1$

59. $\tan \alpha \cos \alpha = \sin \alpha$

60. $\cot \alpha \sin \alpha = \cos \alpha$

61. $(1 + \sin \theta)(1 - \sin \theta) = \cos^2 \theta$

62. $(1 + \cos \theta)(1 - \cos \theta) = \sin^2 \theta$

63. $(\csc \theta + \cot \theta)(\csc \theta - \cot \theta) = 1$

64. $\sin^2 \theta - \cos^2 \theta = 2 \sin^2 \theta - 1$

65. $\frac{\sin \theta}{\cos \theta} + \frac{\cos \theta}{\sin \theta} = \csc \theta \sec \theta$

66. $\frac{\tan \beta + \cot \beta}{\tan \beta} = \csc^2 \beta$

Finding Special Angles of a Triangle In Exercises 67–72, find each value of θ in degrees ($0° < \theta < 90°$) and radians ($0 < \theta < \pi/2$) without using a calculator.

67. (a) $\sin \theta = \frac{1}{2}$ (b) $\csc \theta = 2$

68. (a) $\cos \theta = \frac{\sqrt{2}}{2}$ (b) $\tan \theta = 1$

69. (a) $\sec \theta = 2$ (b) $\cot \theta = 1$

70. (a) $\csc \theta = \frac{2\sqrt{3}}{3}$ (b) $\sin \theta = \frac{1}{2}$

71. (a) $\tan \theta = \sqrt{3}$ (b) $\cos \theta = \frac{\sqrt{2}}{2}$

72. (a) $\cot \theta = \frac{\sqrt{3}}{3}$ (b) $\sec \theta = \sqrt{2}$

Finding Side Lengths of a Triangle **In Exercises 73–76, find the exact values of the indicated variables.**

73. Find y and r.

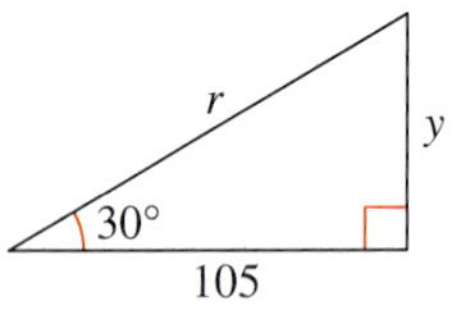

74. Find x and y.

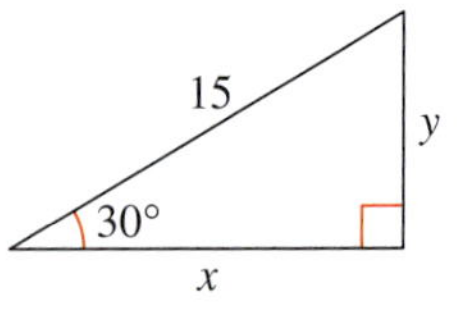

75. Find x and y.

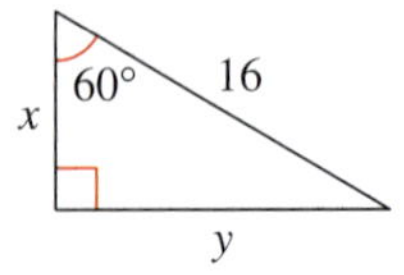

76. Find x and r.

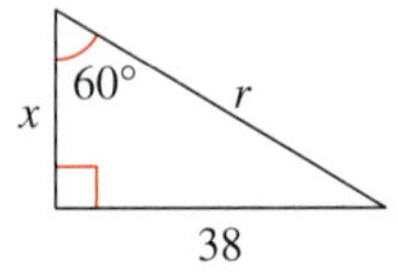

77. Geomatics The Johnstown Inclined Plane in Pennsylvania is one of the longest and steepest hoists in the world. The railway cars travel a distance of 896.5 feet at an angle of approximately 35.4°, rising to a height of 1693.5 feet above sea level, as shown in the figure.

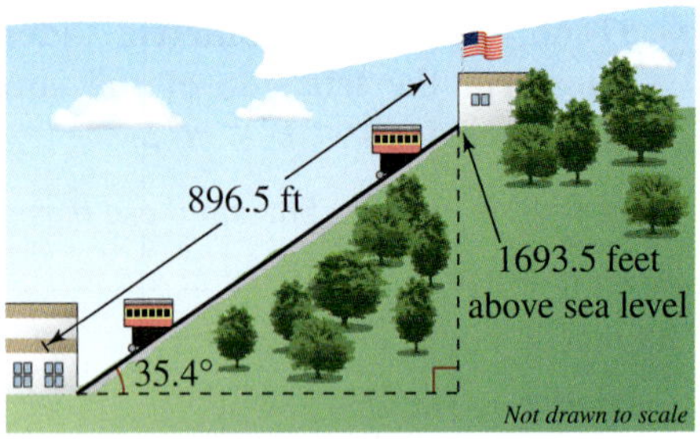

(a) Find the vertical rise of the inclined plane.

(b) Find the elevation of the lower end of the inclined plane.

(c) The cars move up the mountain at a rate of 300 feet per minute. Find the rate at which they rise vertically.

78. ***Why you should learn it*** *(p. 275)* A zip line steel cable is being constructed for a reality television competition show. The high end of the zip line is attached to the top of a 50-foot pole, while the lower end is anchored at ground level to a stake 50 feet from the base of the pole (see figure).

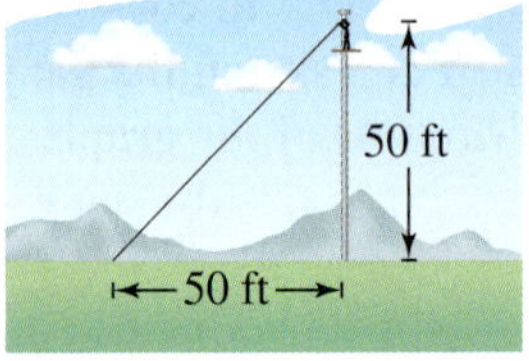

(a) Find the angle of elevation of the zip line.

(b) Find the number of feet of steel cable needed for the zip line.

(c) A contestant takes 6 seconds to reach the ground from the top of the zip line. At what rate is the contestant moving down the line? At what rate is the contestant dropping vertically?

79. Biology A biologist wants to know the width w of a river in order to properly set instruments for studying the pollutants in the water. From point A, the biologist walks upstream 100 feet and sights to point C. It is determined that $\theta = 58°$ (see figure). How wide is the river?

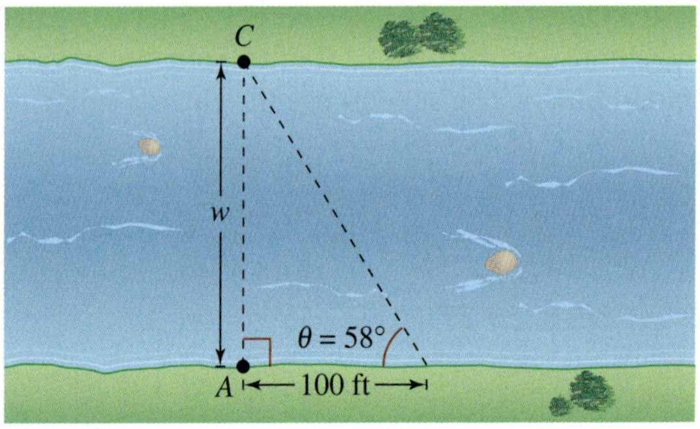

80. Geomatics In traveling across flat land, you notice a mountain directly in front of you. Its angle of elevation (to the peak) is 3.5°. After you drive 13 miles closer to the mountain, the angle of elevation is 9° (see figure). Approximate the height of the mountain.

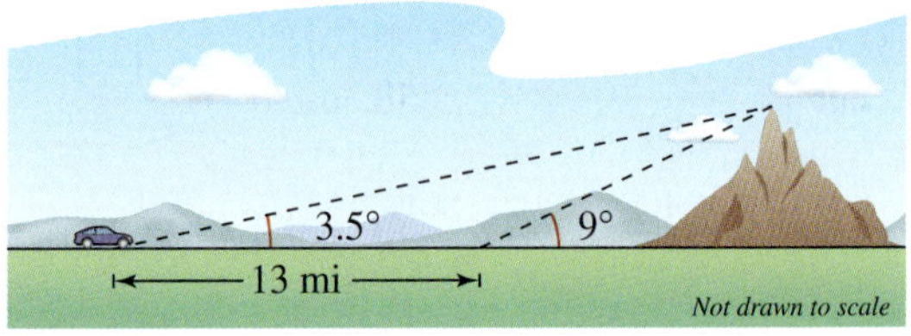

81. Mechanical Engineering A steel plate has the form of one-fourth of a circle with a radius of 60 centimeters. Two 2-centimeter holes are to be drilled in the plate, positioned as shown in the figure. Find the coordinates of the center of each hole.

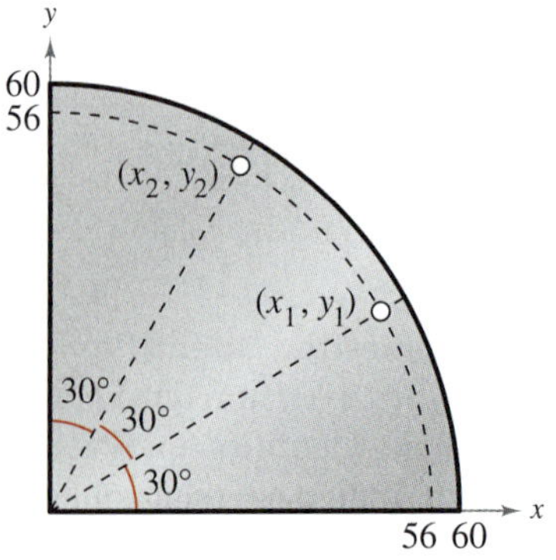

Figure for 81

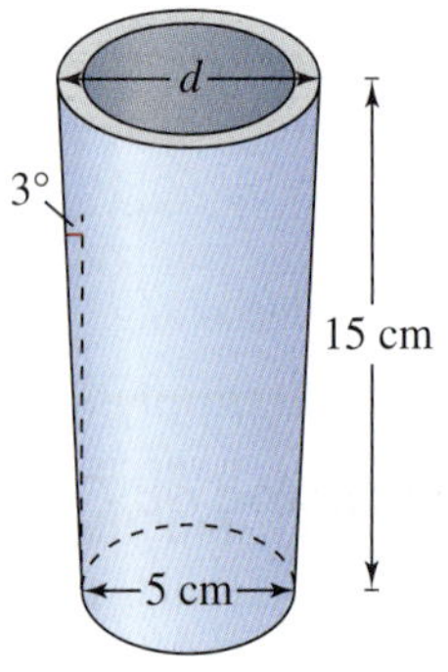

Figure for 82

82. Mechanical Engineering A tapered shaft has a diameter of 5 centimeters at the small end and is 15 centimeters long. The taper is 3° (see figure). Find the diameter d of the large end of the shaft.

83. Geometry A six-foot person walks from the base of a streetlight directly toward the tip of the shadow cast by the streetlight. When the person is 16 feet from the streetlight and 5 feet from the tip of the streetlight's shadow, the person's shadow starts to appear beyond the streetlight's shadow.

(a) Draw a right triangle that gives a visual representation of the problem. Show the known quantities and use a variable to indicate the height of the streetlight.

(b) Use a trigonometric function to write an equation involving the unknown quantity.

(c) What is the height of the streetlight?

84. MODELING DATA

A 20-meter line is used to tether a helium-filled balloon. Because of a breeze, the line makes an angle of approximately 85° with the ground.

(a) Draw a right triangle that gives a visual representation of the problem. Show the known quantities of the triangle and use a variable to indicate the height of the balloon.

(b) Use a trigonometric function to write an equation involving the unknown quantity.

(c) What is the height of the balloon?

(d) The breeze becomes stronger, and the angle the balloon makes with the ground decreases. Complete the table, which shows the heights (in meters) of the balloon for decreasing angle measures θ.

Angle, θ	80°	70°	60°	50°
Height				

Angle, θ	40°	30°	20°	10°
Height				

(e) As the angle the balloon makes with the ground approaches 0°, how does this affect the height of the balloon? Draw a right triangle to explain your reasoning.

Focusing on Concepts

True or False? **In Exercises 85–90, determine whether the statement is true or false. Justify your answer.**

85. $\sin 60° \csc 60° = 1$

86. $\sec 30° = \csc 60°$

87. $\sin 45° + \cos 45° = 1$

88. $\cot^2 10° - \csc^2 10° = -1$

89. $\dfrac{\sin 60°}{\sin 30°} = \sin 2°$

90. $\tan[(5°)^2] = \tan^2 5°$

Kelvin Degree/Shutterstock.com

91. Think About It You are given the value of $\tan \theta$. Is it possible to find the value of $\sec \theta$ without finding the measure of θ? Explain.

92. HOW DO YOU SEE IT? Use the figure to answer each question.

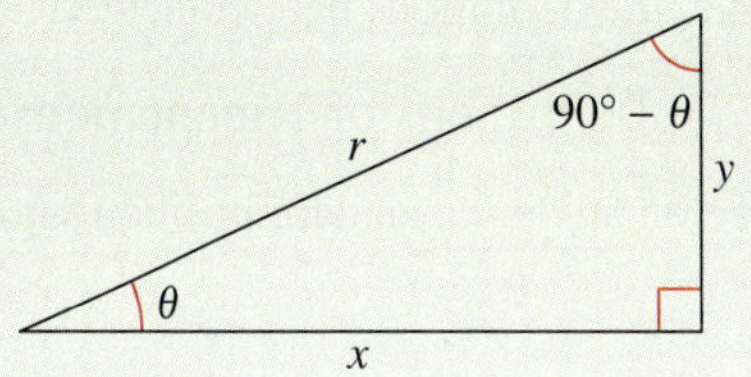

(a) Which side is opposite θ?

(b) Which side is adjacent to $90° - \theta$?

(c) Explain why $\sin \theta = \cos(90° - \theta)$.

93. Exploration

(a) Use a graphing utility to complete the table. Round your results to four decimal places.

θ	0°	20°	40°	60°	80°
$\sin \theta$					
$\cos \theta$					
$\tan \theta$					

(b) Classify each of the three trigonometric functions as increasing or decreasing for the table values.

(c) Using the table, verify that the tangent function is the quotient of the sine and cosine functions.

94. Error Analysis Describe the error.

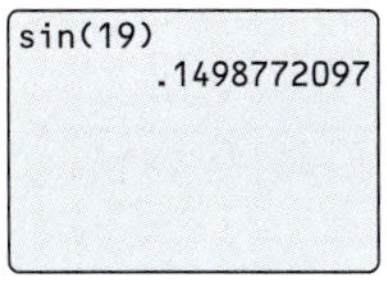

So, $\sin(19°) \approx 0.1498772097$. ✗

Cumulative Mixed Review

Graphing an Exponential Function **In Exercises 95 and 96, use a graphing utility to construct a table of values for the function. Then sketch the graph of the function. Identify any asymptote of the graph.**

95. $f(x) = 2 + e^{3x}$

96. $f(x) = -4 + e^{3x}$

Graphing a Logarithmic Function **In Exercises 97 and 98, use a graphing utility to graph the logarithmic function. Find the domain, vertical asymptote, and x-intercept of the logarithmic function.**

97. $f(x) = \log_3 x + 1$

98. $f(x) = \log_3(x - 4)$

4.4 Trigonometric Functions of Any Angle

What you should learn

- Evaluate trigonometric functions of any angle.
- Find reference angles.
- Evaluate trigonometric functions of real numbers.

Why you should learn it

You can use trigonometric functions to model and solve real-life problems. For instance, Exercise 124 on page 293 shows you how trigonometric functions can be used to model the monthly sales of a seasonal product, such as wakeboards.

Introduction

In Section 4.3, the definitions of trigonometric functions were restricted to acute angles. In this section, the definitions are extended to cover *any* angle. When θ is an *acute* angle, the definitions here coincide with those given in the preceding section.

Definitions of Trigonometric Functions of Any Angle

Let θ be an angle in standard position with (x, y) a point on the terminal side of θ and $r = \sqrt{x^2 + y^2} \neq 0$.

$\sin\theta = \dfrac{y}{r}$ $\qquad$ $\cos\theta = \dfrac{x}{r}$

$\tan\theta = \dfrac{y}{x},\quad x \neq 0$ $\qquad$ $\cot\theta = \dfrac{x}{y},\quad y \neq 0$

$\sec\theta = \dfrac{r}{x},\quad x \neq 0$ $\qquad$ $\csc\theta = \dfrac{r}{y},\quad y \neq 0$

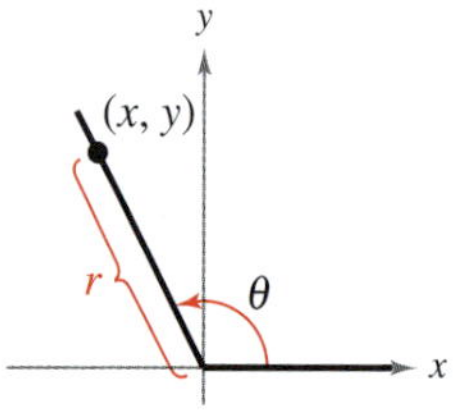

Because $r = \sqrt{x^2 + y^2}$ *cannot* be zero, it follows that the sine and cosine functions are defined for any real value of θ. However, when $x = 0$, the tangent and secant of θ are undefined. For example, the tangent of 90° is undefined. Similarly, when $y = 0$, the cotangent and cosecant of θ are undefined.

EXAMPLE 1 Evaluating Trigonometric Functions

Let $(-3, 4)$ be a point on the terminal side of θ. (See figure.) Find the sine, cosine, and tangent of θ.

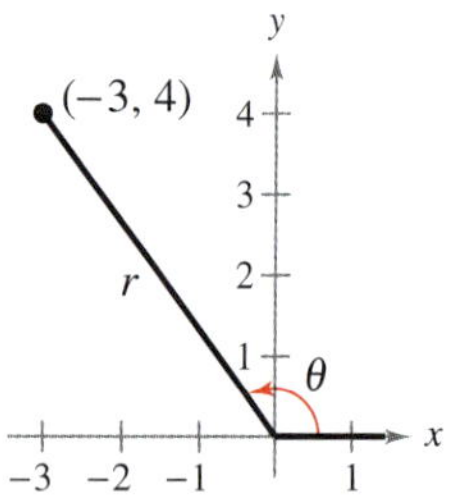

Solution

Referring to the figure, you can see that $x = -3$, $y = 4$, and

$$
\begin{aligned}
r &= \sqrt{x^2 + y^2} \\
&= \sqrt{(-3)^2 + 4^2} \\
&= \sqrt{25} \\
&= 5.
\end{aligned}
$$

So, you have $\sin\theta = \dfrac{y}{r} = \dfrac{4}{5}$, $\cos\theta = \dfrac{x}{r} = -\dfrac{3}{5}$, and $\tan\theta = \dfrac{y}{x} = -\dfrac{4}{3}$.

Checkpoint *Audio-video solution in English & Spanish at LarsonPrecalculus.com*

Let $(-2, 3)$ be a point on the terminal side of θ. Find the sine, cosine, and tangent of θ.

The *signs* of the trigonometric functions in the four quadrants can be determined from the definitions of the functions. For instance, because $\cos\theta = x/r$, it follows that $\cos\theta$ is positive wherever $x > 0$, which is in Quadrants I and IV. (Remember, r is always positive.) In a similar manner, you can verify the other results shown in Figure 4.22.

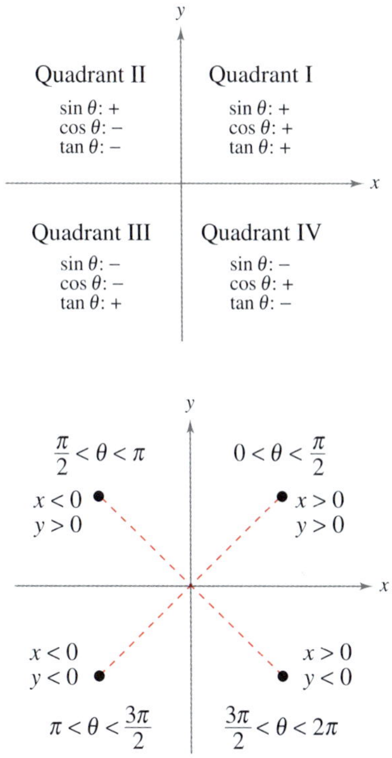

Figure 4.22

EXAMPLE 2 Evaluating Trigonometric Functions

Given $\sin\theta = -\frac{2}{3}$ and $\tan\theta > 0$, find $\cos\theta$ and $\cot\theta$.

Solution

Note that θ lies in Quadrant III because that is the only quadrant in which the sine is negative and the tangent is positive. Moreover, using

$$\sin\theta = \frac{y}{r} = -\frac{2}{3}$$

and the fact that y is negative in Quadrant III, you can let $y = -2$ and $r = 3$. Because x is negative in Quadrant III,

$$x = -\sqrt{9-4} = -\sqrt{5}$$

and you have the following.

$$\cos\theta = \frac{x}{r} = \frac{-\sqrt{5}}{3} \quad \text{Exact value}$$

$$\approx -0.75 \quad \text{Approximate value}$$

$$\cot\theta = \frac{x}{y} = \frac{-\sqrt{5}}{-2} \quad \text{Exact value}$$

$$\approx 1.12 \quad \text{Approximate value}$$

Checkpoint *Audio-video solution in English & Spanish at LarsonPrecalculus.com*

Given $\sin\theta = \frac{4}{5}$ and $\tan\theta < 0$, find $\cos\theta$ and $\tan\theta$.

EXAMPLE 3 Trigonometric Functions of Quadrantal Angles

Evaluate the sine and cosine functions at the angles $0, \frac{\pi}{2}, \pi$, and $\frac{3\pi}{2}$.

Solution

To begin, choose a point on the terminal side of each angle, as shown in Figure 4.23. For each of the four points, $r = 1$. Next, evaluate the sine and cosine functions for each angle.

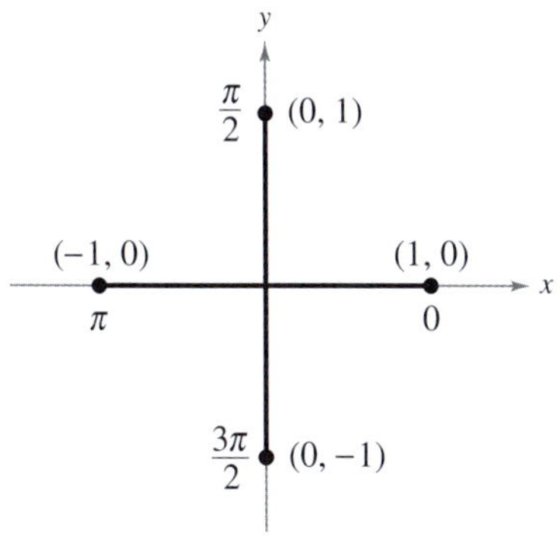

Figure 4.23

$$\sin 0 = \frac{y}{r} = \frac{0}{1} = 0 \qquad \cos 0 = \frac{x}{r} = \frac{1}{1} = 1 \qquad (x, y) = (1, 0)$$

$$\sin\frac{\pi}{2} = \frac{y}{r} = \frac{1}{1} = 1 \qquad \cos\frac{\pi}{2} = \frac{x}{r} = \frac{0}{1} = 0 \qquad (x, y) = (0, 1)$$

$$\sin\pi = \frac{y}{r} = \frac{0}{1} = 0 \qquad \cos\pi = \frac{x}{r} = \frac{-1}{1} = -1 \qquad (x, y) = (-1, 0)$$

$$\sin\frac{3\pi}{2} = \frac{y}{r} = \frac{-1}{1} = -1 \qquad \cos\frac{3\pi}{2} = \frac{x}{r} = \frac{0}{1} = 0 \qquad (x, y) = (0, -1)$$

Checkpoint *Audio-video solution in English & Spanish at LarsonPrecalculus.com*

Evaluate the tangent and cotangent functions at the angle $\frac{3\pi}{2}$.

Reference Angles

The values of the trigonometric functions of angles greater than 90° (or less than 0°) can be determined from their values at corresponding acute angles called **reference angles.**

Definition of Reference Angle

Let θ be an angle in standard position. Its **reference angle** is the acute angle θ' formed by the terminal side of θ and the horizontal axis.

The reference angles for θ in Quadrants II, III, and IV are shown below.

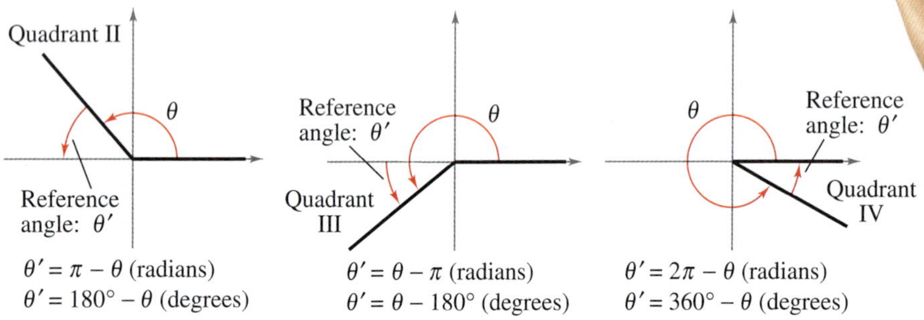

EXAMPLE 4 Finding Reference Angles

Find the reference angle θ' for each value of θ.

a. $\theta = 300°$ **b.** $\theta = 2.3$ **c.** $\theta = -135°$

Solution

a. Because 300° lies in Quadrant IV, the angle it makes with the x-axis is

$$\theta' = 360° - 300° = 60°. \qquad \text{Degrees}$$

b. Because 2.3 lies between $\pi/2 \approx 1.5708$ and $\pi \approx 3.1416$, it follows that it is in Quadrant II and its reference angle is

$$\theta' = \pi - 2.3 \approx 0.8416. \qquad \text{Radians}$$

c. First, determine that $-135°$ is coterminal with 225°, which lies in Quadrant III. So, the reference angle is

$$\theta' = 225° - 180° = 45°. \qquad \text{Degrees}$$

The figure shows each angle θ and its reference angle θ'.

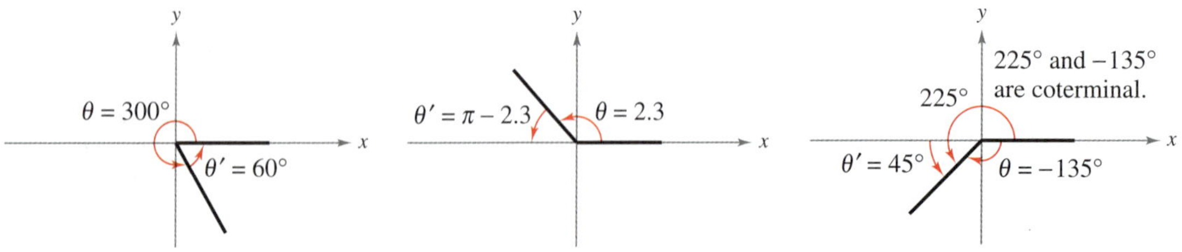

(a) (b) (c)

Checkpoint Audio-video solution in English & Spanish at LarsonPrecalculus.com

Find the reference angle θ'.

a. $\theta = 213°$ **b.** $\theta = \dfrac{14\pi}{9}$ **c.** $\theta = \dfrac{4\pi}{5}$

Trigonometric Functions of Real Numbers

To see how a reference angle is used to evaluate a trigonometric function, consider the point (x, y) on the terminal side of θ, as shown in the figure. By definition, you know that

$$\sin \theta = \frac{y}{r}$$

and

$$\tan \theta = \frac{y}{x}.$$

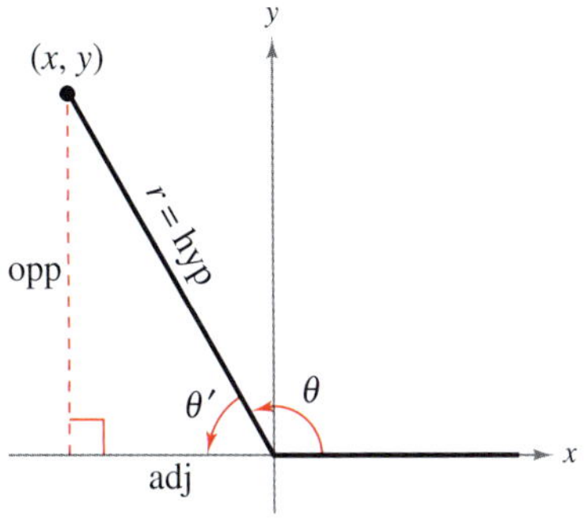

$\text{opp} = |y|, \text{adj} = |x|$

For the right triangle with acute angle θ' and sides of lengths $|x|$ and $|y|$, you have

$$\sin \theta' = \frac{\text{opp}}{\text{hyp}} = \frac{|y|}{r}$$

and

$$\tan \theta' = \frac{\text{opp}}{\text{adj}} = \frac{|y|}{|x|}.$$

So, it follows that $\sin \theta$ and $\sin \theta'$ are equal, *except possibly in sign*. The same is true for $\tan \theta$ and $\tan \theta'$ and for the other four trigonometric functions. In all cases, the sign of the function value can be determined by the quadrant in which θ lies.

Insight

A common error is to find a reference angle using the terminal side of θ and the *vertical* axis. When asked to find a reference angle, be sure you use the terminal side and the *horizontal* axis.

Evaluating Trigonometric Functions of Any Angle

To find the value of a trigonometric function of any angle θ:

1. Determine the function value of the associated reference angle θ'.

2. Depending on the quadrant in which θ lies, affix the appropriate sign to the function value.

By using reference angles and the special angles discussed in the preceding section, you can greatly extend the scope of *exact* trigonometric values. For instance, knowing the function values of 30° means that you know the function values of all angles for which 30° is a reference angle. For convenience, the table below shows the exact values of the sine, cosine, and tangent functions of special angles and quadrantal angles.

Trigonometric Values of Common Angles

θ (degrees)	0°	30°	45°	60°	90°	180°	270°
θ (radians)	0	$\frac{\pi}{6}$	$\frac{\pi}{4}$	$\frac{\pi}{3}$	$\frac{\pi}{2}$	π	$\frac{3\pi}{2}$
$\sin \theta$	0	$\frac{1}{2}$	$\frac{\sqrt{2}}{2}$	$\frac{\sqrt{3}}{2}$	1	0	-1
$\cos \theta$	1	$\frac{\sqrt{3}}{2}$	$\frac{\sqrt{2}}{2}$	$\frac{1}{2}$	0	-1	0
$\tan \theta$	0	$\frac{\sqrt{3}}{3}$	1	$\sqrt{3}$	Undef.	0	Undef.

Algebra Help

Learning the trigonometric values of common angles in the table at the left is worth the effort because doing so will increase both your efficiency and your confidence. Below is a pattern for the sine function that may help you remember the values. Reverse the order to get cosine values of the same angles.

θ	0°	30°	45°	60°	90°
$\sin \theta$	$\frac{\sqrt{0}}{2}$	$\frac{\sqrt{1}}{2}$	$\frac{\sqrt{2}}{2}$	$\frac{\sqrt{3}}{2}$	$\frac{\sqrt{4}}{2}$

EXAMPLE 5 Trigonometric Functions of Nonacute Angles

See LarsonPrecalculus.com for an interactive version of this type of example.

Evaluate (a) $\cos \frac{4\pi}{3}$, (b) $\tan(-210°)$, and (c) $\csc \frac{11\pi}{4}$.

Solution

a. Because $\theta = 4\pi/3$ lies in Quadrant III, the reference angle is $\theta' = (4\pi/3) - \pi = \pi/3$, as shown in Figure 4.24. Moreover, the cosine is negative in Quadrant III, so

$$\cos \frac{4\pi}{3} = (-)\cos \frac{\pi}{3} = -\frac{1}{2}.$$

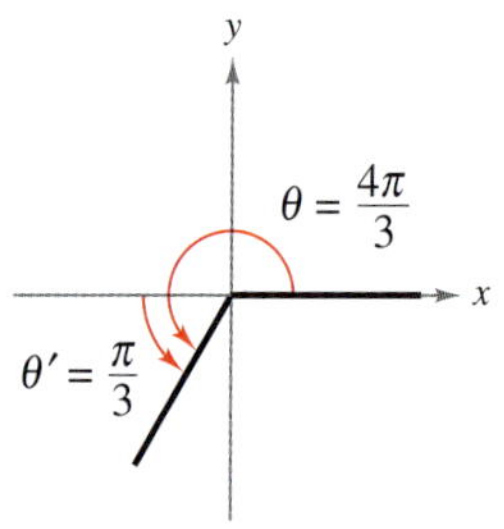

Figure 4.24

b. Because $-210° + 360° = 150°$, it follows that $-210°$ is coterminal with the second-quadrant angle 150°. So, the reference angle is $\theta' = 180° - 150° = 30°$, as shown in Figure 4.25. Finally, because the tangent is negative in Quadrant II, you have

$$\tan(-210°) = (-)\tan 30° = -\frac{\sqrt{3}}{3}.$$

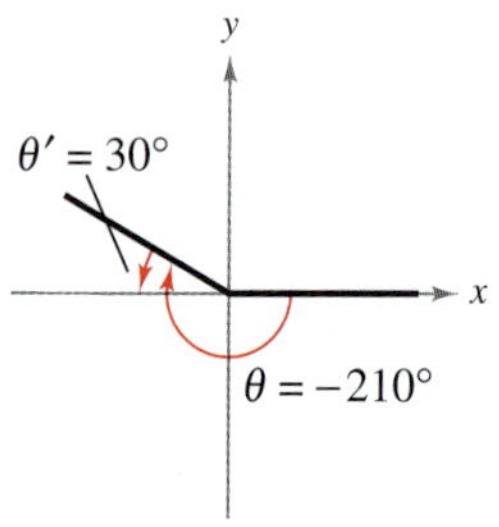

Figure 4.25

c. Because $(11\pi/4) - 2\pi = 3\pi/4$, it follows that $11\pi/4$ is coterminal with the second-quadrant angle $3\pi/4$. So, the reference angle is $\theta' = \pi - (3\pi/4) = \pi/4$, as shown in Figure 4.26. Because the cosecant is positive in Quadrant II, you have

$$\csc \frac{11\pi}{4} = (+)\csc \frac{\pi}{4} = \frac{1}{\sin(\pi/4)} = \sqrt{2}.$$

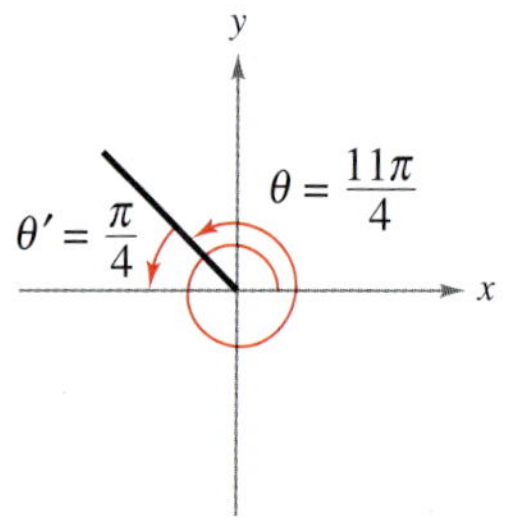

Figure 4.26

✓ ***Checkpoint*** *Audio-video solution in English & Spanish at LarsonPrecalculus.com*

Evaluate (a) $\sin \frac{7\pi}{4}$, (b) $\cos(-120°)$, and (c) $\tan \frac{11\pi}{6}$.

The fundamental trigonometric identities listed in the preceding section (for an acute angle θ) are also valid when θ is any angle in the domain of the function.

EXAMPLE 6 Using Trigonometric Identities

Let θ be an angle in Quadrant II such that $\sin \theta = \frac{1}{3}$. Find $\cos \theta$ by using trigonometric identities.

Solution

Using the Pythagorean identity $\sin^2 \theta + \cos^2 \theta = 1$, you obtain

$$\left(\frac{1}{3}\right)^2 + \cos^2 \theta = 1$$

$$\cos^2 \theta = 1 - \frac{1}{9}$$

$$\cos^2 \theta = \frac{8}{9}.$$

Because $\cos \theta < 0$ in Quadrant II, you can use the negative root to obtain

$$\cos \theta = -\frac{\sqrt{8}}{\sqrt{9}} = -\frac{2\sqrt{2}}{3}.$$

✓ ***Checkpoint*** *Audio-video solution in English & Spanish at LarsonPrecalculus.com*

Let θ be an angle in Quadrant III such that $\sin \theta = -\frac{4}{5}$. Find $\cos \theta$ by using trigonometric identities.

4.4 Exercises

See *CalcChat.com* for tutorial help and worked-out solutions to odd-numbered exercises.
For instructions on how to use a graphing utility, see Appendix A.

Vocabulary and Concept Check

In Exercises 1–6, let θ be an angle in standard position with (x, y) a point on the terminal side of θ and $r = \sqrt{x^2 + y^2} \neq 0$. Fill in the blank.

1. $\sin \theta =$ ________

2. $\dfrac{r}{y} =$ ________

3. $\tan \theta =$ ________

4. $\sec \theta =$ ________

5. $\dfrac{x}{r} =$ ________

6. $\dfrac{x}{y} =$ ________

7. What do you call the acute angle formed by the terminal side of an angle θ in standard position and the horizontal axis?

8. In which quadrants is $\cos \theta$ positive?

9. For which of the quadrantal angles 0, $\pi/2$, π, and $3\pi/2$ is the sine function equal to 0?

10. Is the value of $\cos 170^\circ$ equal to the value of $\cos 10^\circ$?

Procedures and Problem Solving

Evaluating Trigonometric Functions **In Exercises 11–14, find the exact values of the six trigonometric functions of each angle θ.**

11. (a)

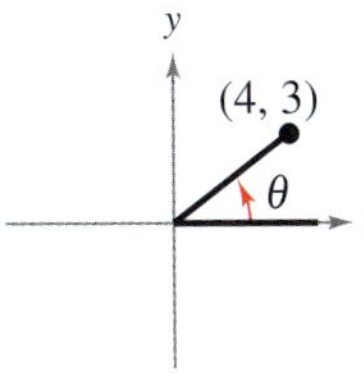

(b)

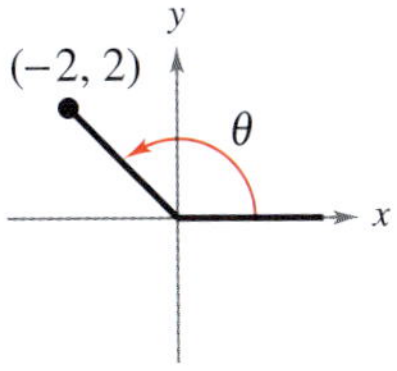

12. (a)

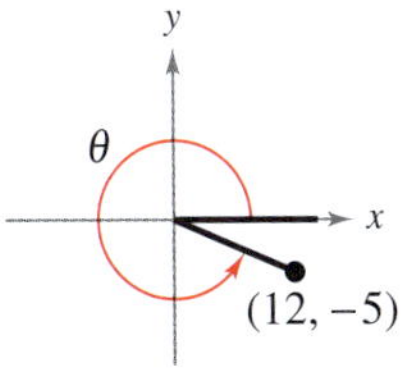

(b)

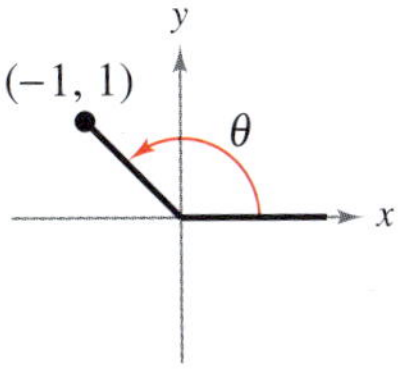

13. (a)

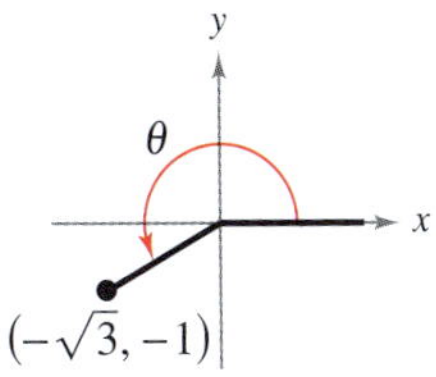

(b)

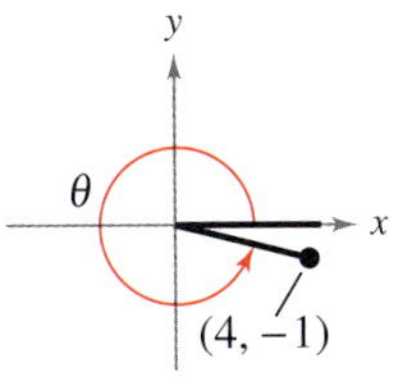

14. (a)

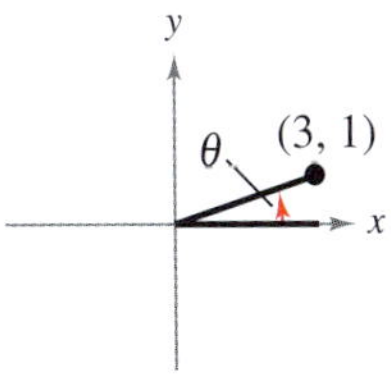

(b)

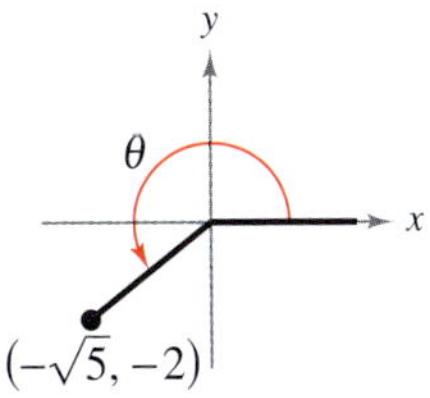

Evaluating Trigonometric Functions **In Exercises 15–22, the point is on the terminal side of an angle θ in standard position. Find the exact values of the six trigonometric functions of θ.**

15. $(7, 24)$

16. $(8, 15)$

17. $(5, -12)$

18. $(-24, 10)$

19. $\left(-2, \sqrt{21}\right)$

20. $\left(-\sqrt{7}, -3\right)$

21. $(-10, -8)$

22. $(5, -15)$

Determining a Quadrant **In Exercises 23–26, determine the quadrant in which θ lies.**

23. $\sin \theta < 0$, $\cos \theta < 0$

24. $\sec \theta > 0$, $\cot \theta < 0$

25. $\cot \theta > 0$, $\cos \theta > 0$

26. $\tan \theta > 0$, $\csc \theta < 0$

Evaluating Trigonometric Functions **In Exercises 27–34, find the exact values of the remaining trigonometric functions of θ satisfying the given condition.**

Function Value	*Condition*
27. $\sin \theta = 0.6$	θ lies in Quadrant II.
28. $\cos \theta = -0.8$	θ lies in Quadrant III.
29. $\tan \theta = -\frac{15}{8}$	$\sin \theta < 0$
30. $\csc \theta = 5$	$\cot \theta < 0$
31. $\sec \theta = -3$	$0 \le \theta \le \pi$
32. $\sin \theta = 0$	$\dfrac{\pi}{2} \le \theta \le \dfrac{3\pi}{2}$
33. $\cot \theta$ is undefined.	$\dfrac{\pi}{2} \le \theta \le \dfrac{3\pi}{2}$
34. $\tan \theta$ is undefined.	$\pi \le \theta \le 2\pi$

An Angle Formed by a Line Through the Origin In Exercises 35–38, the terminal side of θ lies on the indicated line in the specified quadrant. Find the exact values of the six trigonometric functions of θ by finding a point on the line.

	Line	*Quadrant*
35.	$y = -x$	II
36.	$y = \frac{1}{2}x$	I
37.	$3x - y = 0$	III
38.	$2x + 3y = 0$	IV

Trigonometric Function of a Quadrantal Angle In Exercises 39–46, evaluate the trigonometric function of the quadrantal angle, if possible.

39. $\sec \pi$

40. $\tan \dfrac{\pi}{2}$

41. $\sec \dfrac{3\pi}{2}$

42. $\csc 0$

43. $\csc \dfrac{3\pi}{2}$

44. $\sec 0$

45. $\cot \pi$

46. $\csc \dfrac{\pi}{2}$

Finding a Reference Angle In Exercises 47–66, find the reference angle θ'. Sketch θ in standard position and label θ'.

47. $\theta = 120°$

48. $\theta = 225°$

49. $\theta = 150°$

50. $\theta = 315°$

51. $\theta = -45°$

52. $\theta = -330°$

53. $\theta = \dfrac{2\pi}{3}$

54. $\theta = \dfrac{3\pi}{4}$

55. $\theta = -\dfrac{5\pi}{6}$

56. $\theta = \dfrac{7\pi}{6}$

57. $\theta = \dfrac{11\pi}{6}$

58. $\theta = -\dfrac{5\pi}{3}$

59. $\theta = 212°$

60. $\theta = 308°$

61. $\theta = -282°$

62. $\theta = -175°$

63. $\theta = -\dfrac{11\pi}{5}$

64. $\theta = \dfrac{10\pi}{3}$

65. $\theta = 1.7$

66. $\theta = -6.75$

Trigonometric Functions of a Nonacute Angle In Exercises 67–78, evaluate the sine, cosine, and tangent of the angle without using a calculator.

67. $225°$

68. $300°$

69. $-750°$

70. $-495°$

71. $5\pi/3$

72. $3\pi/4$

73. $-5\pi/6$

74. $-4\pi/3$

75. $11\pi/4$

76. $10\pi/3$

77. $-17\pi/6$

78. $-20\pi/3$

Using a Trigonometric Identity In Exercises 79–84, use the function value to find the indicated trigonometric value in the specified quadrant.

	Function Value	*Quadrant*	*Trigonometric Value*
79.	$\sin \theta = -\frac{3}{5}$	IV	$\cos \theta$
80.	$\cot \theta = -2$	II	$\sin \theta$
81.	$\cos \theta = \frac{5}{8}$	I	$\csc \theta$
82.	$\csc \theta = -3$	IV	$\cot \theta$
83.	$\sec \theta = -\frac{9}{4}$	III	$\tan \theta$
84.	$\tan \theta = -\frac{5}{4}$	IV	$\sec \theta$

Using Trigonometric Identities In Exercises 85–90, find the exact values of the remaining trigonometric functions of the angle.

85. $\sin \theta = \frac{2}{5}$, $\cos \theta > 0$

86. $\cos \theta = -\frac{3}{7}$, $\sin \theta < 0$

87. $\tan \theta = -4$, $\cos \theta < 0$

88. $\cot \theta = 5$, $\sin \theta > 0$

89. $\csc \theta = -\frac{3}{2}$, $\tan \theta < 0$

90. $\sec \theta = -\frac{4}{3}$, $\cot \theta > 0$

Using a Calculator In Exercises 91–102, use a calculator to evaluate the trigonometric function. Round your answer to four decimal places. (Be sure to use the correct angle mode.)

91. $\sin 10°$

92. $\tan 245°$

93. $\csc 405°$

94. $\sec 235°$

95. $\cos(-110°)$

96. $\sin(-220°)$

97. $\sec 0.22$

98. $\cot 3.76$

99. $\tan \dfrac{2\pi}{9}$

100. $\cos \dfrac{11\pi}{9}$

101. $\cot\left(-\dfrac{13\pi}{12}\right)$

102. $\csc\left(-\dfrac{7\pi}{9}\right)$

Solving for θ In Exercises 103–108, find two solutions of each equation. Give your answers in degrees $(0° \le \theta < 360°)$ and in radians $(0 \le \theta < 2\pi)$. (Do not use a calculator.)

103. (a) $\sin \theta = \frac{1}{2}$ (b) $\sin \theta = -\frac{1}{2}$

104. (a) $\cos \theta = \dfrac{\sqrt{2}}{2}$ (b) $\cos \theta = -\dfrac{\sqrt{2}}{2}$

105. (a) $\csc \theta = \dfrac{2\sqrt{3}}{3}$ (b) $\cot \theta = -1$

106. (a) $\cot \theta = -\sqrt{3}$ (b) $\csc \theta = 2$

107. (a) $\sec \theta = -\dfrac{2\sqrt{3}}{3}$ (b) $\tan \theta = -\dfrac{\sqrt{3}}{3}$

108. (a) $\tan \theta = 1$ (b) $\sec \theta = \sqrt{2}$

Evaluating Trigonometric Functions **In Exercises 109–122, find the exact value of each function of the angle, where $f(\theta) = \sin\theta$ and $g(\theta) = \cos\theta$. (Do not use a calculator.)**

(a) $(f + g)(\theta)$ (b) $(g - f)(\theta)$ (c) $[g(\theta)]^2$
(d) $(fg)(\theta)$ (e) $f(2\theta)$ (f) $g(-\theta)$

109. $\theta = 30°$ **110.** $\theta = 60°$
111. $\theta = 315°$ **112.** $\theta = 225°$
113. $\theta = -150°$ **114.** $\theta = -300°$
115. $\theta = 7\pi/6$ **116.** $\theta = 5\pi/6$
117. $\theta = 4\pi/3$ **118.** $\theta = -5\pi/3$
119. $\theta = -270°$ **120.** $\theta = 180°$
121. $\theta = 7\pi/2$ **122.** $\theta = 5\pi/2$

123. Meteorology The normal daily mean temperature T (in degrees Fahrenheit) in Bakersfield, California, can be approximated by

$$T = 65.8 + 18\cos\left(\frac{\pi t}{6} - \frac{7\pi}{6}\right)$$

where t is the month, with $t = 1$ corresponding to January. Find the normal daily mean temperature for each month. (*Source:* National Centers for Environmental Information)

(a) January (b) July (c) October

124. *Why you should learn it* *(p. 286)* A company that produces wakeboards forecasts monthly sales S over a two-year period to be

$$S = 3.0 + 0.16t + 2.5\sin\left(\frac{\pi t}{6} - \frac{\pi}{2}\right)$$

where S is measured in hundreds of units and t is the month, with $t = 1$ corresponding to January 2018. Estimate the sales for each month.

(a) January 2018 (b) February 2019
(c) May 2018 (d) June 2019

125. Aeronautics An airplane flying at an altitude of 6 miles is on a flight path that passes directly over an observer (see figure). Let θ be the angle of elevation from the observer to the plane. Find the distance d from the observer to the plane when (a) $\theta = 30°$, (b) $\theta = 90°$, and (c) $\theta = 120°$.

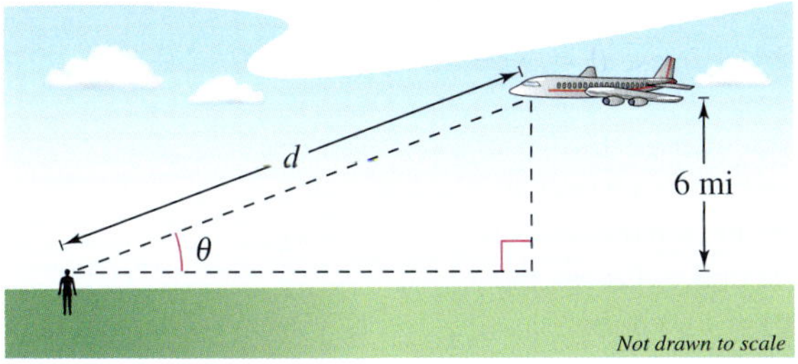

B-D-S Piotr Marcinski/Shutterstock.com

126. HOW DO YOU SEE IT? Consider an angle in standard position with $r = 12$ centimeters, as shown in the figure. Describe the changes in the values of x, y, $\sin\theta$, $\cos\theta$, and $\tan\theta$ as θ increases continuously from 0° to 90°.

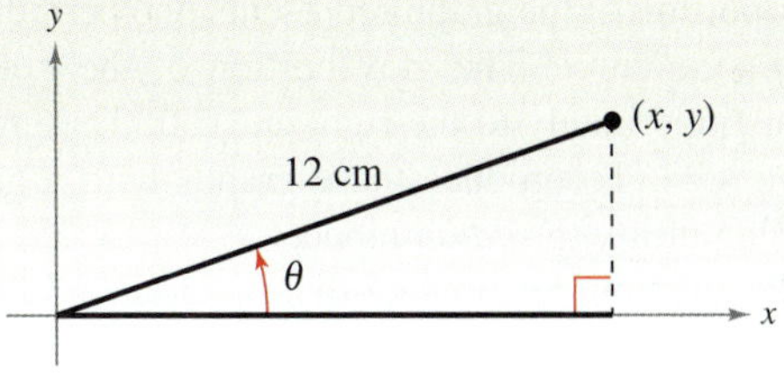

Focusing on Concepts

True or False? **In Exercises 127–130, determine whether the statement is true or false. Justify your answer.**

127. $\sin\theta < \tan\theta$ in Quadrant I

128. $\sin\theta < \cos\theta$ for $0° < \theta < 45°$

129. $\sin\theta = -\sqrt{1 - \cos^2\theta}$ for $90° < \theta < 180°$

130. $\cos\theta = -\sqrt{1 - \sin^2\theta}$ for $90° < \theta < 180°$

131. Exploration Use a graphing utility to complete the table. Then make a conjecture about the relationship between $\sin\theta$ and $\sin(180° - \theta)$.

θ	0°	20°	40°	60°	80°
$\sin\theta$					
$\sin(180° - \theta)$					

132. Exploration Use the procedure in Exercise 131 and a graphing utility to create a table of values and make a conjecture about the relationship between $\cos\theta$ and $\cos(180° - \theta)$ for an acute angle θ.

133. Error Analysis Describe the error in using a graphing utility to evaluate the trigonometric function.

$$\tan\frac{\pi}{2} \approx 0.0274224385$$

134. Writing Write a study sheet that will help you remember how to evaluate the six trigonometric functions of any angle θ in standard position. Include figures and diagrams as needed.

Cumulative Mixed Review

Solving an Equation **In Exercises 135–138, solve the equation. Round your answer to three decimal places, if necessary.**

135. $x^2 - 2x - 5 = 0$ **136.** $2x^2 + x - 4 = 0$

137. $\dfrac{3}{x - 1} = \dfrac{x + 2}{9}$ **138.** $\dfrac{5}{x} = \dfrac{x + 4}{2x}$

4.5 Graphs of Sine and Cosine Functions

Basic Sine and Cosine Curves

In this section, you will study techniques for sketching the graphs of the sine and cosine functions. The graph of the sine function is a **sine curve.** In Figure 4.27, the black portion of the graph represents one period of the function and is called **one cycle** of the sine curve. The gray portion of the graph indicates that the basic sine wave repeats indefinitely to the right and left. The graph of the cosine function is shown in Figure 4.28. To produce these graphs with a graphing utility, make sure you set the graphing utility to *radian* mode.

Recall from Section 4.2 that the domain of the sine and cosine functions is the set of all real numbers. Moreover, the range of each function is the interval

$$[-1, 1]$$

and each function has a period of 2π. Do you see how this information is consistent with the basic graphs shown in Figures 4.27 and 4.28?

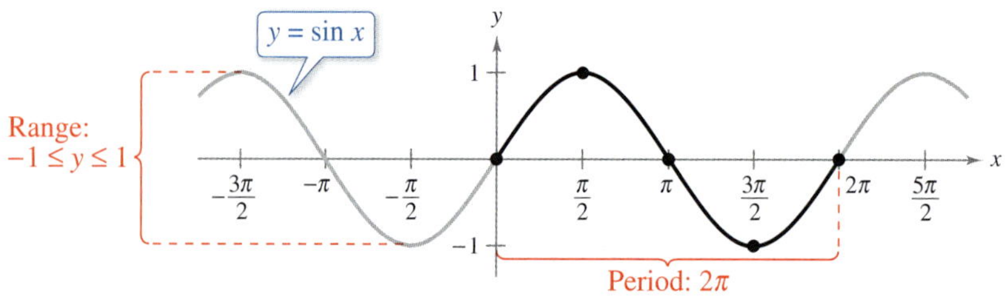

Figure 4.27

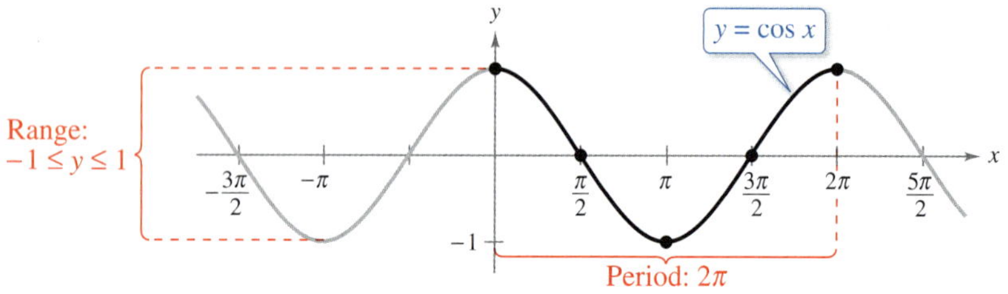

Figure 4.28

To sketch the graphs of the basic sine and cosine functions by hand, it helps to note five *key points* in one period of each graph: the *intercepts*, the *maximum points*, and the *minimum points*. The table below lists the five key points on the graphs of

$$y = \sin x \quad \text{and} \quad y = \cos x.$$

x	0	$\frac{\pi}{2}$	π	$\frac{3\pi}{2}$	2π
$\sin x$	0	1	0	-1	0
$\cos x$	1	0	-1	0	1

Note in Figures 4.27 and 4.28 that the sine curve is symmetric with respect to the *origin*, whereas the cosine curve is symmetric with respect to the *y-axis*. These properties of symmetry follow from the fact that the sine function is odd, whereas the cosine function is even.

What you should learn

- Sketch the graphs of basic sine and cosine functions.
- Use amplitude and period to help sketch the graphs of sine and cosine functions.
- Sketch translations of graphs of sine and cosine functions.
- Use sine and cosine functions to model real-life data.

Why you should learn it

Sine and cosine functions are often used in scientific calculations. For instance, in Exercise 99 on page 303, you can use a trigonometric function to model the percent of the moon's face that is illuminated for any given day in 2020.

Library of Parent Functions: Sine and Cosine Functions

The basic characteristics of the parent sine function and parent cosine function are listed below and summarized on the inside cover of this text.

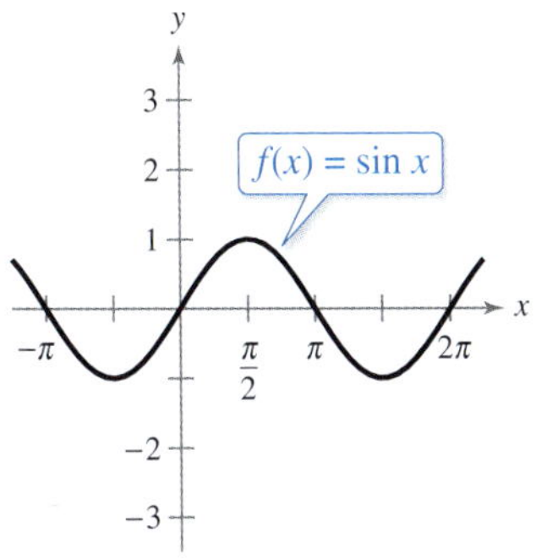

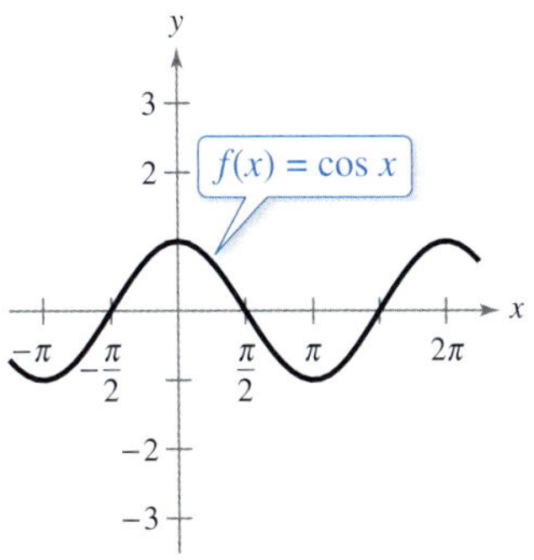

Domain: $(-\infty, \infty)$	Domain: $(-\infty, \infty)$
Range: $[-1, 1]$	Range: $[-1, 1]$
Period: 2π	Period: 2π
x-intercepts: $(n\pi, 0)$	x-intercepts: $(\pi/2 + n\pi, 0)$
y-intercept: $(0, 0)$	y-intercept: $(0, 1)$
Odd function	Even function
Origin symmetry	y-axis symmetry

EXAMPLE 1 Library of Parent Functions: $f(x) = \sin x$

See LarsonPrecalculus.com for an interactive version of this type of example.

Sketch the graph of $g(x) = 2 \sin x$ by hand on the interval $[-\pi, 4\pi]$.

Solution

Note that $g(x) = 2 \sin x = 2(\sin x)$ indicates that the y-values of the key points will have twice the magnitude of those on the graph of $f(x) = \sin x$. Divide the period 2π into four equal parts to get the key points

Intercept	*Maximum*	*Intercept*	*Minimum*		*Intercept*
$(0, 0)$,	$(\pi/2, 2)$,	$(\pi, 0)$,	$(3\pi/2, -2)$,	and	$(2\pi, 0)$.

By connecting these key points with a smooth curve and extending the curve in both directions over the interval $[-\pi, 4\pi]$, you obtain the graph shown below. Use a graphing utility to confirm this graph. Be sure to set the graphing utility to *radian* mode.

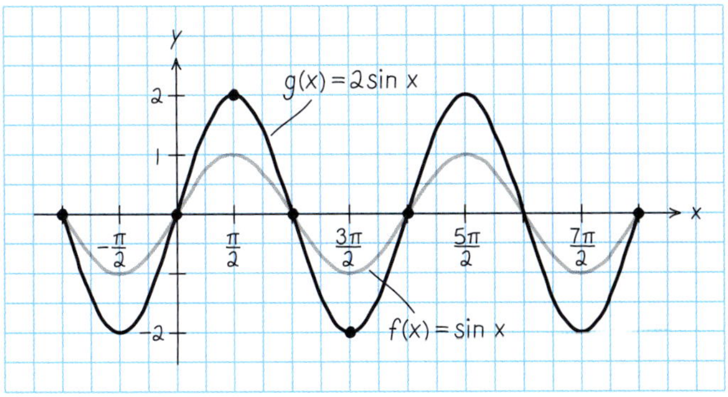

Checkpoint *Audio-video solution in English & Spanish at LarsonPrecalculus.com*

Sketch the graph of $y = 2 \cos x$ by hand on the interval $[-\pi/2, 9\pi/2]$.

Explore the Concept

Enter the Graphing a Sine Function Program, found at this textbook's *Student Companion Website,* into your graphing utility. This program simultaneously draws the unit circle and the corresponding points on the sine curve, as shown below. After the circle and sine curve are drawn, you can connect the points on the unit circle with their corresponding points on the sine curve by pressing ENTER. Discuss the relationship that is illustrated.

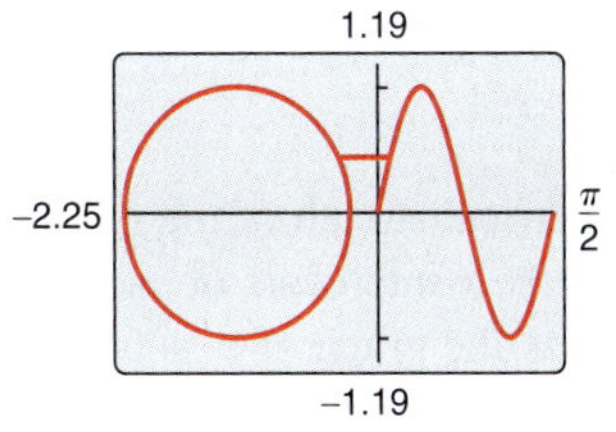

Amplitude and Period of Sine and Cosine Curves

In the rest of this section, you will study the effects of the constants a, b, c, and d on the graphs of equations of the forms

$$y = d + a\sin(bx - c)$$

or

$$y = d + a\cos(bx - c)$$

with $a \neq 0$ and $b \neq 0$.

For the graphs of $y = a\sin x$ and $y = a\cos x$, the constant factor a is a *scaling factor*—a *vertical stretch* when $|a| > 1$ or a *vertical shrink* when $|a| < 1$. For each function, the range is $-|a| \leq y \leq |a|$ and the **amplitude** is $|a|$.

Definition of Amplitude of Sine and Cosine Curves

The **amplitude** of $y = a\sin x$ and $y = a\cos x$ represents half the distance between the maximum and minimum values of the function and is given by

$$\text{Amplitude} = |a|.$$

Insight

On some standardized tests, you may have to find the maximum and minimum values of

$$y = d + a\sin(bx - c)$$

or

$$y = d + a\cos(bx - c)$$

without the use of a graphing utility. Knowing the definition of amplitude will help you to solve this problem by hand.

EXAMPLE 2 Scaling: Vertical Shrinking and Stretching

On the same set of coordinate axes, sketch the graph of each function by hand.

a. $y = \frac{1}{2}\cos x$ **b.** $y = 3\cos x$

Solution

a. Because the amplitude of $y = \frac{1}{2}\cos x$ is $\frac{1}{2}$, the maximum value is $\frac{1}{2}$ and the minimum value is $-\frac{1}{2}$. Divide one cycle, $0 \leq x \leq 2\pi$, into four equal parts to get the key points

Maximum	*Intercept*	*Minimum*	*Intercept*		*Maximum*
$\left(0, \frac{1}{2}\right)$,	$\left(\frac{\pi}{2}, 0\right)$,	$\left(\pi, -\frac{1}{2}\right)$,	$\left(\frac{3\pi}{2}, 0\right)$,	and	$\left(2\pi, \frac{1}{2}\right)$.

b. A similar analysis shows that the amplitude of $y = 3\cos x$ is 3, and the key points are

Maximum	*Intercept*	*Minimum*	*Intercept*		*Maximum*
$(0, 3)$,	$\left(\frac{\pi}{2}, 0\right)$,	$(\pi, -3)$,	$\left(\frac{3\pi}{2}, 0\right)$,	and	$(2\pi, 3)$.

The graphs of these two functions are shown in the figure. Notice that the graph of $y = \frac{1}{2}\cos x$ is a vertical shrink of the graph of $y = \cos x$ and the graph of

$$y = 3\cos x$$

is a vertical stretch of the graph of $y = \cos x$. Also, the x-intercepts of each graph are the same as the graph of $y = \cos x$. Use a graphing utility to confirm these graphs.

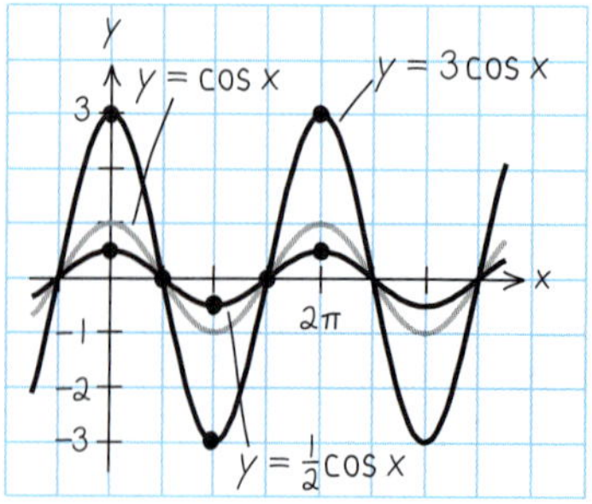

Checkpoint Audio-video solution in English & Spanish at LarsonPrecalculus.com

On the same set of coordinate axes, sketch the graph of each function by hand.

a. $y = \frac{1}{3}\sin x$ **b.** $y = 3\sin x$

What's Wrong?

You use a graphing utility to confirm the graph of $y = \frac{1}{2}\cos x$ in Example 2 and obtain the screen shown below. What's wrong?

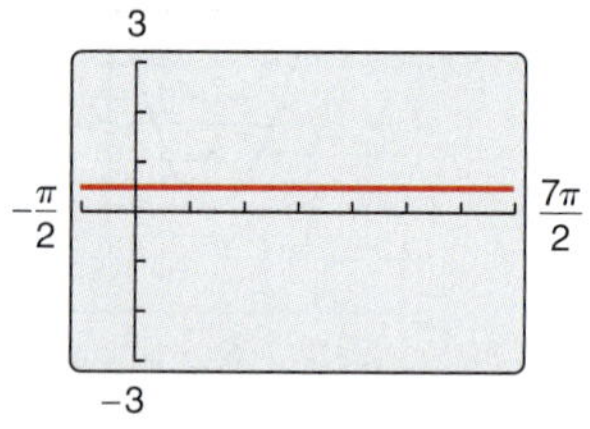

You know from Section 1.4 that the graph of $y = -f(x)$ is a *reflection* in the x-axis of the graph of $y = f(x)$. For instance, the graph of $y = -3 \cos x$ is a reflection of the graph of $y = 3 \cos x$, as shown in Figure 4.29.

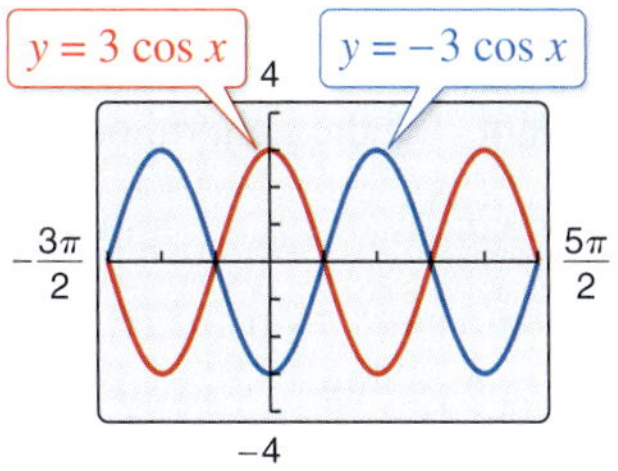

Figure 4.29

Next, consider the effect of the *positive* real number b on the graphs of $y = a \sin bx$ and $y = a \cos bx$. Because $y = a \sin x$ completes one cycle from $x = 0$ to $x = 2\pi$, it follows that $y = a \sin bx$ completes one cycle from $x = 0$ to $x = 2\pi/b$. (The constant b has the same effect on the graph of $y = a \cos bx$.)

Period of Sine and Cosine Functions

Let b be a positive real number. The **period** of

$$y = a \sin bx \quad \text{and} \quad y = a \cos bx$$

is given by

$$\text{Period} = \frac{2\pi}{b}.$$

Note that when $0 < b < 1$, the period of $y = a \sin bx$ is greater than 2π and represents a *horizontal stretch* of the graph of $y = a \sin x$. Similarly, when $b > 1$, the period of $y = a \sin bx$ is less than 2π and represents a *horizontal shrink* of the graph of $y = a \sin x$. (The constant b has the same effect on the graph of $y = a \cos bx$.) When b is negative, use the appropriate identity

$$\sin(-x) = -\sin x \quad \text{or} \quad \cos(-x) = \cos x$$

to rewrite the function.

EXAMPLE 3 Scaling: Horizontal Stretching

Sketch the graph of $y = \sin \frac{x}{2}$ by hand.

Solution

The amplitude is 1. Moreover, because $b = \frac{1}{2}$, the period is

$$\frac{2\pi}{b} = \frac{2\pi}{1/2} = 4\pi. \qquad \text{Substitute for } b.$$

Now, divide the period-interval $[0, 4\pi]$ into four equal parts using the values π, 2π, and 3π to obtain the key points on the graph

Intercept	*Maximum*	*Intercept*	*Minimum*		*Intercept*
$(0, 0)$,	$(\pi, 1)$,	$(2\pi, 0)$,	$(3\pi, -1)$,	and	$(4\pi, 0)$.

The graph is shown below. Use a graphing utility to confirm this graph.

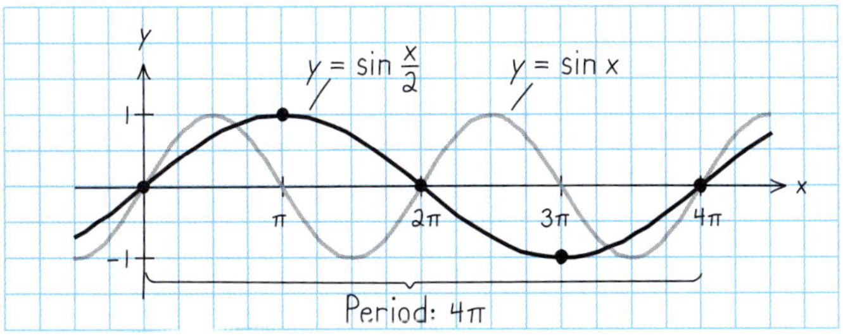

Algebra Help

In general, to divide a period-interval into four equal parts, successively add "period/4," starting with the left endpoint of the interval. For instance, for the period-interval $[-\pi/6, \pi/2]$ of length $2\pi/3$, you would successively add

$$\frac{2\pi/3}{4} = \frac{\pi}{6}$$

to get $-\pi/6, 0, \pi/6, \pi/3$, and $\pi/2$ as the x-values of the key points on the graph.

Checkpoint Audio-video solution in English & Spanish at LarsonPrecalculus.com

Sketch the graph of $y = \cos \frac{x}{3}$ by hand.

Translations of Sine and Cosine Curves

The constant c in the general equations

$$y = a\sin(bx - c) \qquad \text{and} \qquad y = a\cos(bx - c)$$

creates *horizontal translations* (shifts) of the basic sine and cosine curves. For example, comparing $y = a \sin bx$ with $y = a\sin(bx - c)$, you find that the graph of $y = a\sin(bx - c)$ completes one cycle from $bx - c = 0$ to $bx - c = 2\pi$. By solving for x, you can find the interval for one cycle is

$$\underbrace{\frac{c}{b}}_{\text{Left endpoint}} \le x \le \underbrace{\frac{c}{b} + \overbrace{\frac{2\pi}{b}}^{\text{Period}}}_{\text{Right endpoint}}.$$

This implies that the period of $y = a\sin(bx - c)$ is $2\pi/b$, and the graph of $y = a \sin bx$ is shifted by an amount c/b. The number c/b is the **phase shift.**

Technology Tip

When using a graphing utility to graph trigonometric functions, pay special attention to the viewing window you use. For instance, try graphing $y = \frac{1}{10}\sin(10x)$ in the standard viewing window in *radian* mode. What do you observe? Use the *zoom* feature to find a viewing window that displays a good view of the graph.

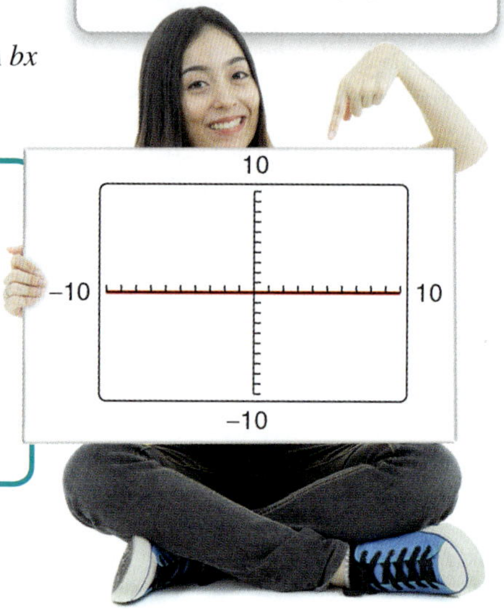

Graphs of Sine and Cosine Functions

The graphs of $y = a\sin(bx - c)$ and $y = a\cos(bx - c)$ have the following characteristics. (Assume $b > 0$.)

$$\text{Amplitude} = |a| \qquad \text{Period} = \frac{2\pi}{b}$$

The left and right endpoints of a one-cycle interval can be determined by solving the equations $bx - c = 0$ and $bx - c = 2\pi$ for x.

EXAMPLE 4 Horizontal Translation

Analyze the graph of $y = \frac{1}{2}\sin\left(x - \frac{\pi}{3}\right)$.

Algebraic Solution

The amplitude is $\frac{1}{2}$ and the period is $2\pi/1 = 2\pi$. By solving the equations

$$x - \frac{\pi}{3} = 0 \qquad \text{and} \qquad x - \frac{\pi}{3} = 2\pi$$

$$x = \frac{\pi}{3} \qquad\qquad x = \frac{7\pi}{3}$$

you see that the interval

$$\left[\frac{\pi}{3}, \frac{7\pi}{3}\right]$$

corresponds to one cycle of the graph. Dividing this interval into four equal parts produces the key points

Intercept	*Maximum*	*Intercept*	*Minimum*		*Intercept*
$\left(\frac{\pi}{3}, 0\right)$,	$\left(\frac{5\pi}{6}, \frac{1}{2}\right)$,	$\left(\frac{4\pi}{3}, 0\right)$,	$\left(\frac{11\pi}{6}, -\frac{1}{2}\right)$,	and	$\left(\frac{7\pi}{3}, 0\right)$.

Graphical Solution

Use a graphing utility set in *radian* mode to graph

$$y = \frac{1}{2}\sin\left(x - \frac{\pi}{3}\right)$$

as shown below. Use the *minimum*, *maximum*, and *zero* or *root* features of the graphing utility to approximate the key points (1.05, 0), (2.62, 0.5), (4.19, 0), (5.76, −0.5), and (7.33, 0).

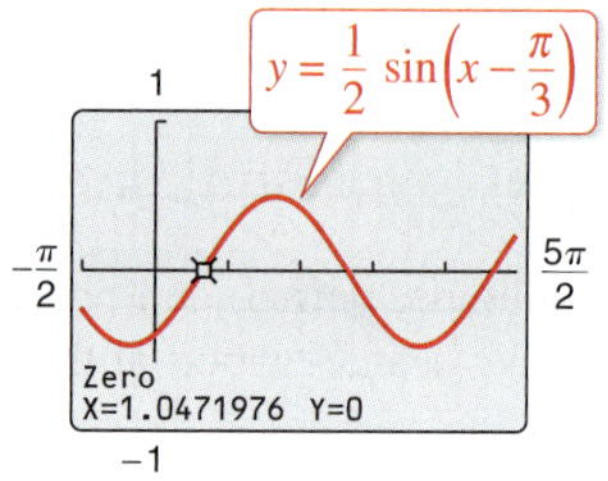

Checkpoint Audio-video solution in English & Spanish at LarsonPrecalculus.com

Analyze the graph of $y = 2\cos\left(x - \frac{\pi}{2}\right)$.

EXAMPLE 5 Horizontal Translation

Analyze the graph of $y = -3\cos(2\pi x + 4\pi)$.

Algebraic Solution

The amplitude is 3 and the period is

$$\frac{2\pi}{2\pi} = 1.$$

By solving the equations

$$2\pi x + 4\pi = 0 \qquad \text{and} \qquad 2\pi x + 4\pi = 2\pi$$

$$2\pi x = -4\pi \qquad\qquad 2\pi x = -2\pi$$

$$x = -2 \qquad\qquad x = -1$$

you see that the interval $[-2, -1]$ corresponds to one cycle of the graph. Dividing this interval into four equal parts produces the key points

Minimum	*Intercept*	*Maximum*	*Intercept*		*Minimum*
$(-2, -3)$,	$(-7/4, 0)$,	$(-3/2, 3)$,	$(-5/4, 0)$,	and	$(-1, -3)$.

Graphical Solution

Use a graphing utility set in *radian* mode to graph $y = -3\cos(2\pi x + 4\pi)$, as shown below. Use the *minimum*, *maximum*, and *zero* or *root* features of the graphing utility to approximate the key points $(-2, -3)$, $(-1.75, 0)$, $(-1.5, 3)$, $(-1.25, 0)$, and $(-1, -3)$.

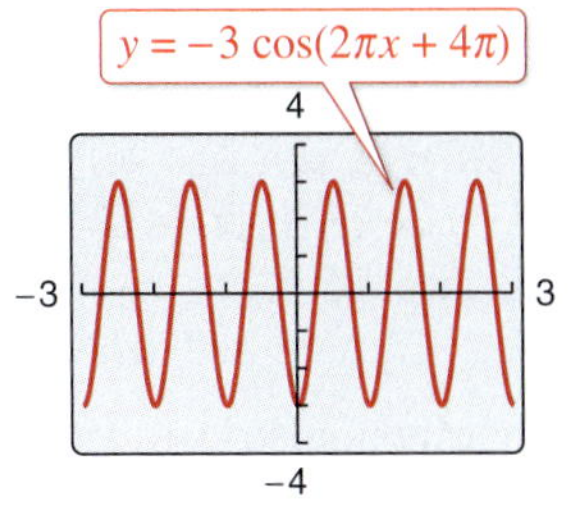

✓ ***Checkpoint*** ▶ *Audio-video solution in English & Spanish at LarsonPrecalculus.com*

Analyze the graph of $y = -\frac{1}{2}\sin(\pi x + \pi)$.

The final type of transformation is the *vertical translation* caused by the constant d in the equations $y = d + a\sin(bx - c)$ and $y = d + a\cos(bx - c)$. The shift is d units upward for $d > 0$ and d units downward for $d < 0$. In other words, the graph oscillates about the horizontal line $y = d$ instead of about the x-axis.

EXAMPLE 6 Vertical Translation

Use a graphing utility to analyze the graph of $y = 2 + 3\cos 2x$.

Solution

The amplitude is 3 and the period is $2\pi/2 = \pi$. The key points over the interval $[0, \pi]$ are

$$(0, 5), \quad (\pi/4, 2), \quad (\pi/2, -1), \quad (3\pi/4, 2), \quad \text{and} \quad (\pi, 5).$$

The graph is shown in Figure 4.30. Compared with the graph of $f(x) = 3\cos 2x$, the graph of $y = 2 + 3\cos 2x$ is shifted upward two units.

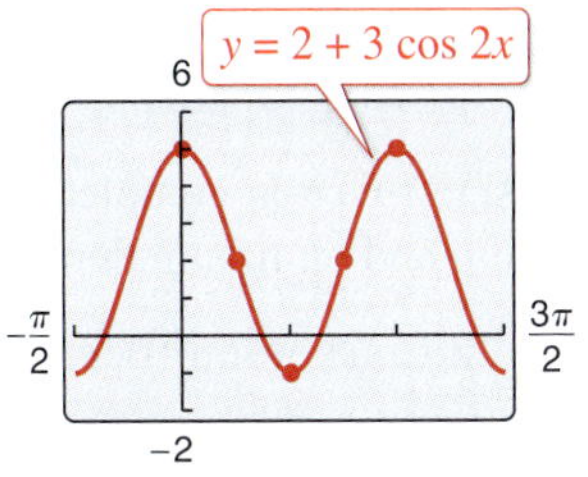

Figure 4.30

✓ ***Checkpoint*** ▶ *Audio-video solution in English & Spanish at LarsonPrecalculus.com*

Use a graphing utility to analyze the graph of $y = 2\cos x - 5$.

EXAMPLE 7 Finding an Equation of a Graph

Find the amplitude, period, and phase shift of the function whose graph is shown in Figure 4.31. Write an equation of this graph in terms of a sine function.

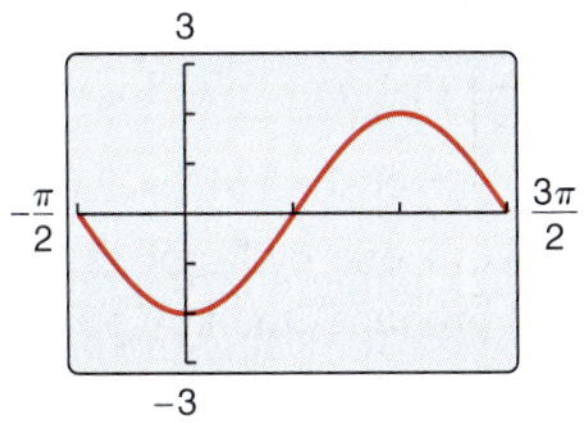

Figure 4.31

Solution

The amplitude of this sine curve is 2. The period is 2π, and there is a right phase shift of $\pi/2$. So, you can write $y = 2\sin(x - \pi/2)$.

✓ ***Checkpoint*** ▶ *Audio-video solution in English & Spanish at LarsonPrecalculus.com*

In Example 7, write an equation of the graph in terms of a cosine function.

Mathematical Modeling

Sine and cosine functions can be used to model many real-life situations, including electric currents, musical tones, radio waves, tides, and weather patterns.

EXAMPLE 8 Finding a Trigonometric Model

Throughout the day, the depth of the water at the end of a dock varies with the tides. The table shows the depths y (in feet) every two hours from midnight to noon.

DATA

Time	Depth, y
Midnight	3.4
2 A.M.	8.7
4 A.M.	11.3
6 A.M.	9.1
8 A.M.	3.8
10 A.M.	0.1
Noon	1.2

Spreadsheet at LarsonPrecalculus.com

a. Use a trigonometric function to model the data. Let t be the time, with $t = 0$ corresponding to midnight.

b. A boat needs at least 10 feet of water to moor at the dock. During what times in the afternoon can it safely dock?

Solution

a. Begin by graphing the data, as shown in Figure 4.32. You can use either a sine or cosine model. So, use a cosine model of the form

$$y = a\cos(bt - c) + d.$$

The difference between the maximum value and minimum value is twice the amplitude of the function. So, the amplitude is

$$a = \tfrac{1}{2}[(\text{maximum depth}) - (\text{minimum depth})] = \tfrac{1}{2}(11.3 - 0.1) = 5.6.$$

The cosine function completes one half of a cycle between the times at which the maximum and minimum depths occur. So, the period p is

$$p = 2[(\text{time of min. depth}) - (\text{time of max. depth})] = 2(10 - 4) = 12$$

which implies that $b = 2\pi/p \approx 0.524$. Because high tide occurs 4 hours after midnight, consider the left endpoint to be $c/b = 4$, so $c \approx 2.094$. Moreover, because the average depth is

$$\tfrac{1}{2}(11.3 + 0.1) = 5.7$$

it follows that $d = 5.7$. So, you can model the depth with the function

$$y = 5.6\cos(0.524t - 2.094) + 5.7.$$

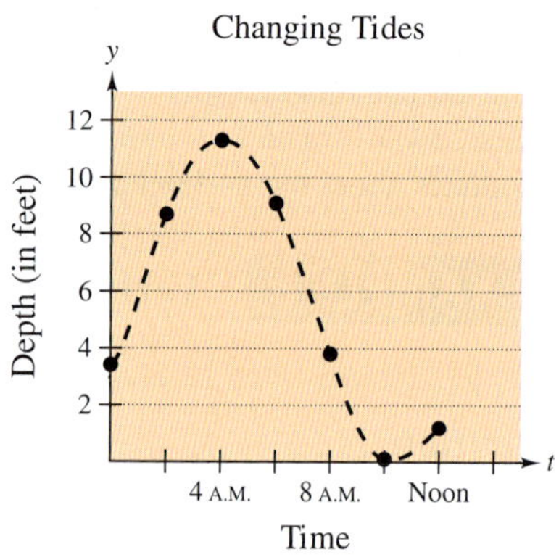

Figure 4.32

b. Using a graphing utility, graph the model with the line $y = 10$. Using the *intersect* feature, you can determine that the depth is at least 10 feet between 2:42 P.M. ($t \approx 14.7$) and 5:18 P.M. ($t \approx 17.3$), as shown at the right.

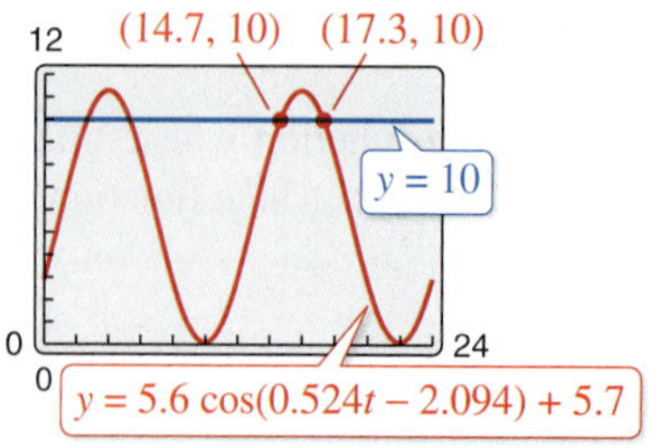

Checkpoint Audio-video solution in English & Spanish at LarsonPrecalculus.com

Find a sine model for the data in Example 8.

4.5 Exercises

See *CalcChat.com* for tutorial help and worked-out solutions to odd-numbered exercises.
For instructions on how to use a graphing utility, see Appendix A.

Vocabulary and Concept Check

In Exercises 1–4, fill in the blank.

1. One period of a sine function is called _______ of the sine curve.

2. The period of $y = d + a\cos(bx - c)$ is _______ .

3. The _______ of a sine or cosine curve represents half the distance between the maximum and minimum values of the function.

4. For the equation $y = a\sin(bx - c)$, $\dfrac{c}{b}$ is the _______ of the graph of the equation.

5. Describe the effect of the constant d on the graph of $y = \sin x + d$.

6. What is the amplitude of $y = -a\sin x$?

Procedures and Problem Solving

Library of Parent Functions **In Exercises 7 and 8, use the graph of the function to answer parts (a)–(d).**

(a) Find the x-intercepts of the graph of $y = f(x)$.

(b) Find the y-intercept of the graph of $y = f(x)$.

(c) Find the intervals on which the graph of $y = f(x)$ is increasing and the intervals on which the graph of $y = f(x)$ is decreasing.

(d) Find the relative extrema of the graph of $y = f(x)$.

7. $f(x) = \sin x$

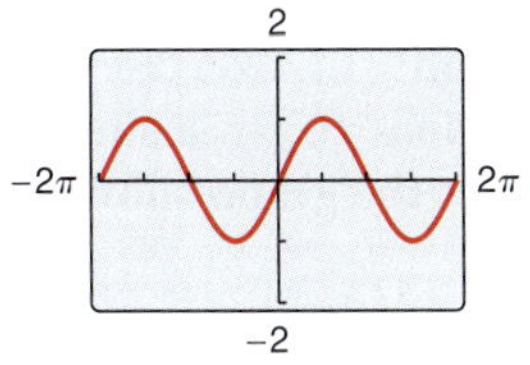

8. $f(x) = \cos x$

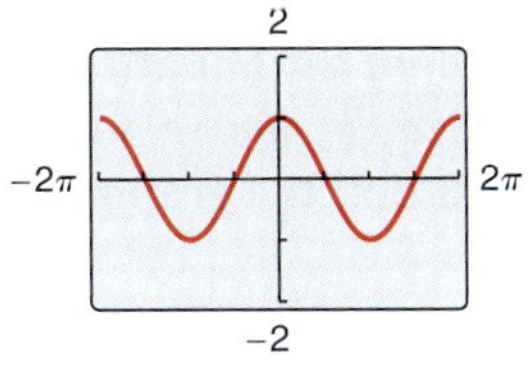

Finding the Period and Amplitude **In Exercises 9–18, find the period and amplitude of the graph of the function.**

9. $y = 4\sin x$

10. $y = -\frac{1}{3}\cos x$

11. $y = \cos\dfrac{x}{4}$

12. $y = \sin 3\pi x$

13. $y = \dfrac{3}{4}\cos\dfrac{\pi x}{2}$

14. $y = -5\sin\dfrac{\pi x}{3}$

15. $y = 1 + \sin\dfrac{x}{2}$

16. $y = 6 - \dfrac{2}{3}\cos 2x$

17. $y = -2\cos(x + 1)$

18. $y = -\sin(\pi x + 2)$

Finding the Maximum and Minimum Values **In Exercises 19–22, find the maximum and minimum values of the function.**

19. $y = 2 + \sin x$

20. $y = -6 + \cos 2x$

21. $y = 4 + 3\cos(\pi x)$

22. $y = 1 - 3\sin(x - 8)$

Determining the Left and Right Endpoints **In Exercises 23–26, determine an interval that corresponds to one cycle of the graph of the function.**

23. $y = \sin(x - 1)$

24. $y = \cos(\pi x + 2)$

25. $y = 1 - \cos(x + \pi)$

26. $y = -2\sin\left(2\pi x - \frac{4}{3}\right)$

Describing the Relationship Between Graphs **In Exercises 27–38, describe the relationship between the graphs of f and g. Consider amplitudes, periods, and shifts.**

27. $f(x) = \sin x$
$g(x) = \sin(x - \pi)$

28. $f(x) = \cos x$
$g(x) = \cos(x + \pi)$

29. $f(x) = \cos 2x$
$g(x) = -\cos 2x$

30. $f(x) = \sin 3x$
$g(x) = \sin(-3x)$

31. $f(x) = \cos 2x$
$g(x) = 3 + \cos 2x$

32. $f(x) = \cos 4x$
$g(x) = -2 + \cos 4x$

33. $f(x) = \sin x$
$g(x) = 5\sin(-x)$

34. $f(x) = \sin x$
$g(x) = -\frac{1}{2}\sin x$

35.

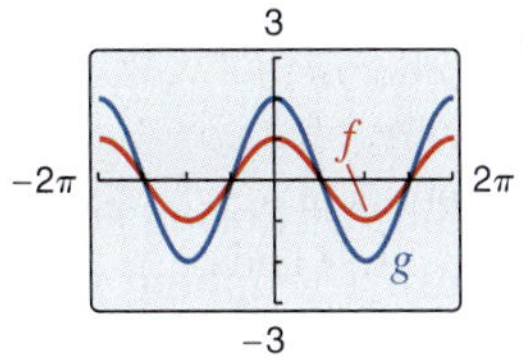

36.

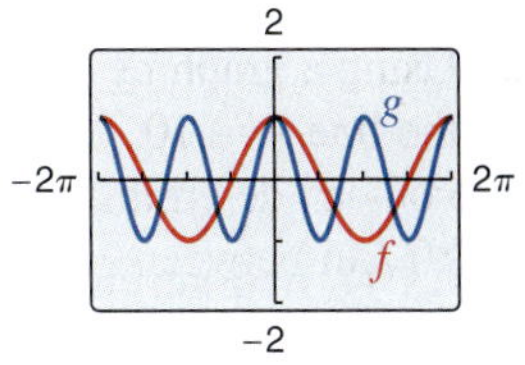

37.

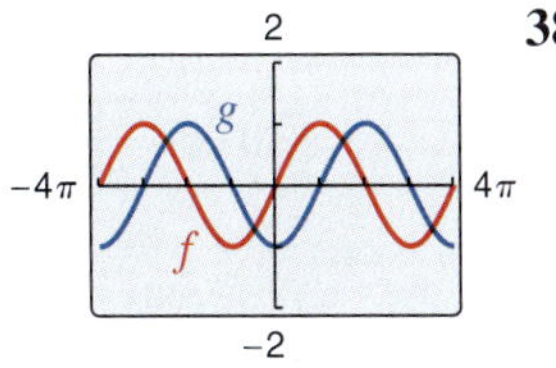

38.

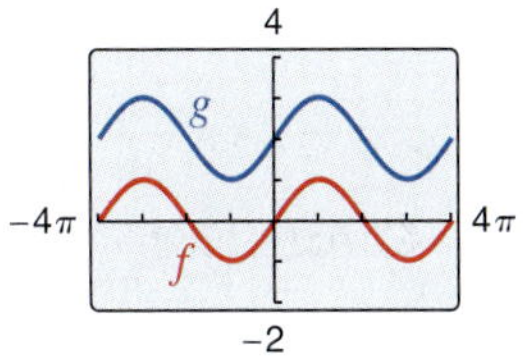

Sketching Graphs of Sine or Cosine Functions

In Exercises 39–46, sketch the graphs of f and g by hand in the same coordinate plane. (Include two full periods.)

39. $f(x) = \sin x$, $g(x) = \sin \dfrac{x}{3}$

40. $f(x) = \sin x$, $g(x) = 4 \sin x$

41. $f(x) = \cos x$, $g(x) = 2 + \cos x$

42. $f(x) = \cos x$, $g(x) = \cos\left(x + \dfrac{\pi}{2}\right)$

43. $f(x) = -\cos x$, $g(x) = -\cos(x - \pi)$

44. $f(x) = -\sin x$, $g(x) = -3 \sin x$

45. $f(x) = \sin x$, $g(x) = \cos\left(x - \dfrac{\pi}{2}\right)$

46. $f(x) = \cos x$, $g(x) = -\sin\left(x - 5\dfrac{\pi}{2}\right)$

Sketching the Graph of a Sine or Cosine Function

In Exercises 47–68, sketch the graph of the function by hand. (Include two full periods.)

47. $y = 5 \sin x$

48. $y = -\frac{1}{4} \sin x$

49. $y = \frac{1}{3} \cos x$

50. $y = -6 \cos x$

51. $y = \cos \dfrac{x}{2}$

52. $y = \sin 4x$

53. $y = \cos 2\pi x$

54. $y = \sin \dfrac{\pi x}{4}$

55. $y = -\sin \dfrac{2\pi x}{3}$

56. $y = -2 \cos \dfrac{\pi x}{6}$

57. $y = \cos\left(x - \dfrac{\pi}{2}\right)$

58. $y = \sin(x - 2\pi)$

59. $y = 3 \sin(x + \pi)$

60. $y = -4 \cos\left(x + \dfrac{\pi}{4}\right)$

61. $y = 2 - \sin \dfrac{\pi x}{8}$

62. $y = -3 + 5 \cos \dfrac{\pi t}{12}$

63. $y = 2 + 5 \cos 6\pi x$

64. $y = 2 \sin 3x + 5$

65. $y = 3 \sin(x + \pi) - 3$

66. $y = -3 \sin(6x + \pi)$

67. $y = \dfrac{2}{3} \cos\left(\dfrac{x}{2} - \dfrac{\pi}{4}\right)$

68. $y = 4 \cos\left(\pi x + \dfrac{\pi}{2}\right) - 1$

Analyzing Graphs of Sine or Cosine Functions

In Exercises 69–82, use a graphing utility to graph the function. (Include two full periods.) Identify the amplitude and period of the graph.

69. $y = -2 \sin \dfrac{2\pi x}{3}$

70. $y = -10 \cos \dfrac{\pi x}{6}$

71. $y = -4 + 5 \cos \dfrac{\pi t}{12}$

72. $y = 2 - 2 \sin \dfrac{2\pi x}{3}$

73. $y = -\dfrac{2}{3} \cos\left(\dfrac{x}{2} - \dfrac{\pi}{4}\right)$

74. $y = \dfrac{5}{2} \cos(6x + \pi)$

75. $y = -2 \sin(4x + \pi)$

76. $y = -4 \sin\left(\dfrac{2}{3}x - \dfrac{\pi}{3}\right)$

77. $y = \cos\left(2\pi x - \dfrac{\pi}{2}\right) + 1$

78. $y = 3 \cos\left(\dfrac{\pi x}{2} + \dfrac{\pi}{2}\right) - 2$

79. $y = 5 \sin(\pi - 2x) + 10$

80. $y = 5 \cos(\pi - 2x) + 6$

81. $y = \frac{1}{100} \sin 120\pi t$

82. $y = \frac{1}{100} \cos 50\pi t$

Finding an Equation of a Graph

In Exercises 83–86, find a and d for the function $f(x) = a \cos x + d$ such that the graph of f matches the figure.

83.

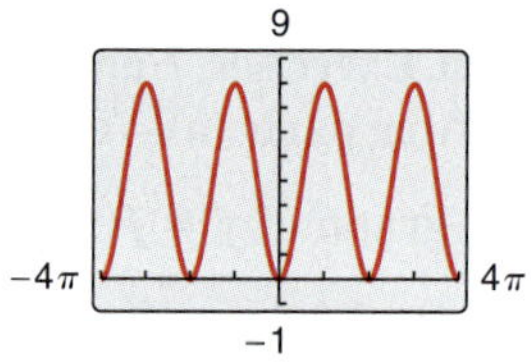

84.

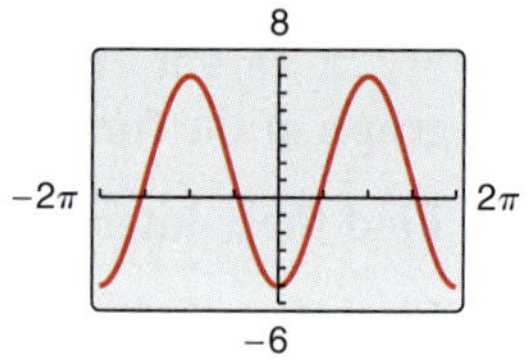

85.

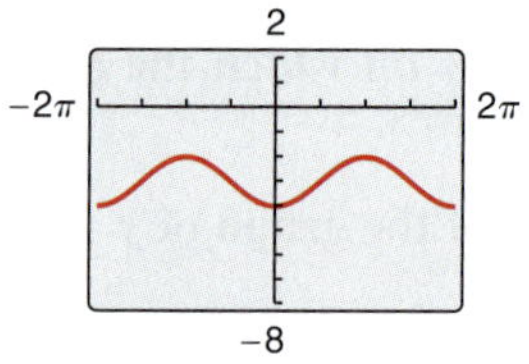

86.

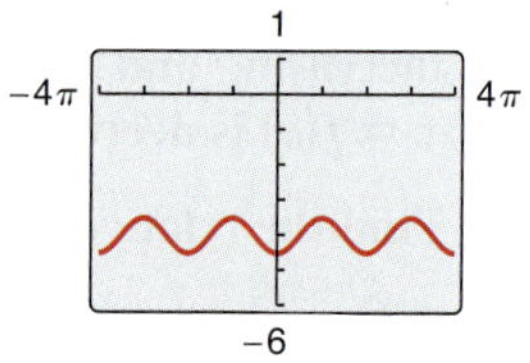

Finding an Equation of a Graph

In Exercises 87–90, find a, b, and c for the function $f(x) = a \sin(bx - c)$ such that the graph of f matches the graph shown.

87.

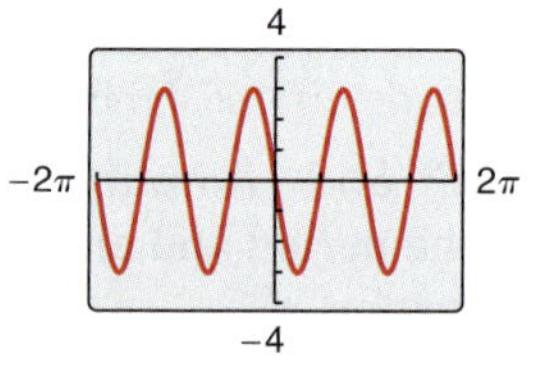

88.

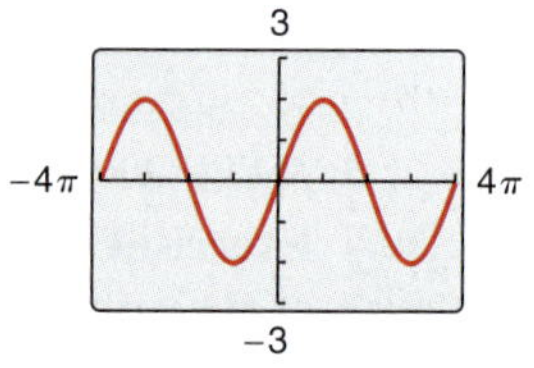

89.

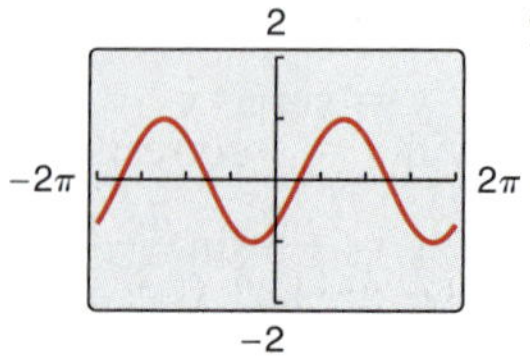

90.

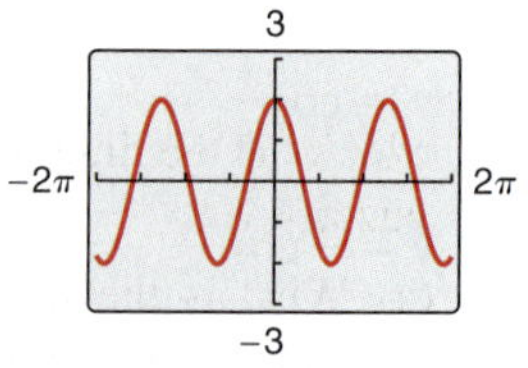

Solving a Trigonometric Equation Graphically

In Exercises 91 and 92, use a graphing utility to graph y_1 and y_2 for all real numbers x in the interval $[-2\pi, 2\pi]$. Use the graphs to find the real numbers x such that $y_1 = y_2$.

91. $y_1 = \sin x$, $y_2 = -\frac{1}{2}$

92. $y_1 = \cos x$, $y_2 = \frac{1}{2}$

93. Health The pressure P (in millimeters of mercury) against the walls of the blood vessels of a person can be modeled by

$$P = 100 - 20\cos\frac{8\pi t}{3}$$

where t is the time (in seconds). Use a graphing utility to graph the model. One cycle is equivalent to one heartbeat. What is the person's pulse rate in heartbeats per minute?

94. Health For a person at rest, the velocity v (in liters per second) of air flow during a respiratory cycle (the time from the beginning of one breath to the beginning of the next) can be modeled by

$$v = 0.85\sin\frac{\pi t}{3}$$

where t is the time (in seconds). (Inhalation occurs when $v > 0$, and exhalation occurs when $v < 0$.)

(a) Use a graphing utility to graph v.

(b) Find the time for one full respiratory cycle.

(c) Find the number of cycles per minute.

(d) The model is for a person at rest. How might the model change for a person who is exercising? Explain.

95. Economics A company that produces skis, which are seasonal products, forecasts monthly sales for one year to be

$$S = 74.50 + 43.75\cos\frac{\pi t}{6}$$

where S is the sales in thousands of units and t is the time in months, with $t = 1$ corresponding to January.

(a) Use a graphing utility to graph the sales function over the one-year period.

(b) Use the graph in part (a) to determine the months of maximum and minimum sales.

96. Agriculture The daily consumption C (in gallons) of diesel fuel on a farm can be modeled by

$$C = 30.3 + 21.6\sin\left(\frac{2\pi t}{365} + 10.9\right)$$

where t is the time in days, with $t = 1$ corresponding to January 1.

(a) What is the period of the model? Is it what you expected? Explain.

(b) What is the average daily fuel consumption? Which term of the model did you use to find your answer? Explain.

(c) Use a graphing utility to graph the model. Use the graph to approximate the time of the year when consumption exceeds 40 gallons per day.

97. Physics You are riding a Ferris wheel. Your height h (in feet) above the ground at any time t (in seconds) can be modeled by

$$h = 25\sin\frac{\pi}{15}(t - 75) + 30.$$

The Ferris wheel turns for 135 seconds before it stops to let the first passengers off.

(a) Use a graphing utility to graph the model.

(b) What are the minimum and maximum heights above the ground?

98. Physics The motion of an oscillating weight suspended from a spring was measured by a motion detector. The data were collected, and the approximate maximum displacements from equilibrium $(y = 2)$ are labeled in the figure. The distance y from the motion detector is measured in centimeters, and the time t is measured in seconds.

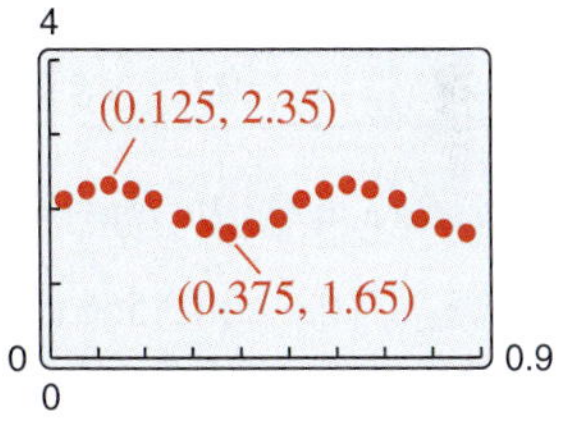

(a) Is y a function of t? Explain.

(b) Approximate the amplitude and period.

(c) Find a model for the data.

(d) Use a graphing utility to graph the model in part (c). Compare the result with the data in the figure.

99. *Why you should learn it* *(p. 294)* The table shows the percent y (in decimal form) of the moon's face that is illuminated on day x of the year 2020, where $x = 1$ represents January 1. (*Source:* U.S. Naval Observatory)

DATA

Day, x	Percent, y
3	0.5
10	1.0
17	0.5
24	0
33	0.5
40	1.0

Spreadsheet at LarsonPrecalculus.com

(a) Create a scatter plot of the data.

(b) Find a trigonometric model for the data.

(c) Add the graph of your model in part (b) to the scatter plot. How well does the model fit the data?

(d) What is the period of the model?

(e) Estimate the percent illumination of the moon on June 28, 2021. (Assume there are 366 days in 2020.)

100. MODELING DATA

The table shows the average daily high temperatures (in degrees Fahrenheit) for Quillayute, Washington, Q and Chicago, Illinois, C for month t, with $t = 1$ corresponding to January. *(Source:* U.S. Weather Bureau and the National Weather Service)

DATA

Month, t	Quillayute, Q	Chicago, C
1	47.1	31.0
2	49.1	35.3
3	51.4	46.6
4	54.8	59.0
5	59.5	70.0
6	63.1	79.7
7	67.4	84.1
8	68.6	81.9
9	66.2	74.8
10	58.2	62.3
11	50.3	48.2
12	46.0	34.8

Spreadsheet at LarsonPrecalculus.com

(a) The temperature in Quillayute can be modeled by

$Q(t) = 57.5 + 10.6 \sin(0.566x - 2.568).$

Find a trigonometric model for Chicago.

(b) Use a graphing utility to graph the data and the model for the temperatures in Quillayute in the same viewing window. How well does the model fit the data?

(c) Use the graphing utility to graph the data and the model for the temperatures in Chicago in the same viewing window. How well does the model fit the data?

(d) Use the models to estimate the average daily high temperature in each city. Which term of the models did you use? Explain.

(e) What is the period of each model? Are the periods what you expected? Explain.

(f) Which city has the greater variability in temperature throughout the year? Which factor of the models determines this variability? Explain.

Focusing on Concepts

True or False? **In Exercises 101–104, determine whether the statement is true or false. Justify your answer.**

101. The graph of the function given by $g(x) = \sin(x + 2\pi)$ translates the graph of $f(x) = \sin x$ one period to the right.

102. The graph of $y = 6 - \frac{3}{4}\sin\frac{3x}{10}$ has a period of $\frac{20\pi}{3}$.

103. The function $y = \frac{1}{2}\cos 2x$ has an amplitude that is twice that of the function $y = \cos x$.

104. The graph of $y = -\cos x$ is a reflection of the graph of $y = \sin\left(x + \frac{\pi}{2}\right)$ in the x-axis.

105. Writing Sketch the graph of $y = \cos bx$ for $b = \frac{1}{2}, 2,$ and 3. How does the value of b affect the graph? How many complete cycles of the graph of y occur between 0 and 2π for each value of b?

106. Writing Use a graphing utility to graph the function given by $y = d + a\sin(bx - c)$ for several different values of a, b, c, and d. Write a paragraph describing how the values of a, b, c, and d affect the graph.

Library of Parent Functions **In Exercises 107 and 108, determine which function is represented by the graph. Do not use a calculator.**

107.

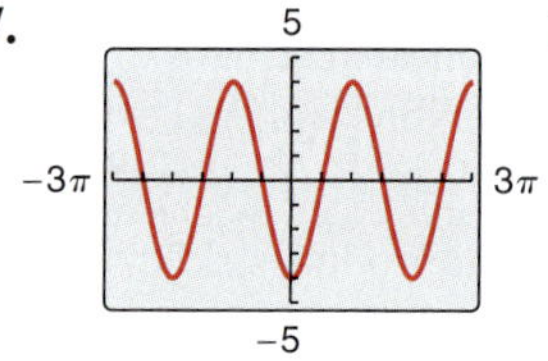

(a) $f(x) = 4\cos(x + \pi)$

(b) $f(x) = 4\cos 4x$

(c) $f(x) = 4\sin(x - \pi)$

(d) $f(x) = -4\cos(x + \pi)$

(e) $f(x) = 1 - \sin\frac{x}{2}$

108.

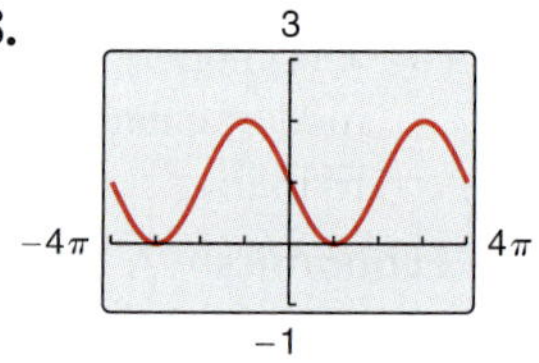

(a) $f(x) = 1 + \sin\frac{x}{2}$

(b) $f(x) = 1 + \cos\frac{x}{2}$

(c) $f(x) = 1 - \sin\frac{x}{2}$

(d) $f(x) = 1 - \cos 2x$

(e) $f(x) = 1 - \sin 2x$

109. Exploration In Section 4.2, it was shown that $f(x) = \cos x$ is an even function and $g(x) = \sin x$ is an odd function. Use a graphing utility to graph h and use the graph to determine whether h is even, odd, or neither.

(a) $h(x) = \cos^2 x$

(b) $h(x) = \sin^2 x$

(c) $h(x) = \sin x \cos x$

110. Conjecture If f is an even function and g is an odd function, use the results of Exercise 109 to make a conjecture about whether h is even, odd, or neither.

(a) $h(x) = [f(x)]^2$

(b) $h(x) = [g(x)]^2$

(c) $h(x) = f(x)g(x)$

Thomas Barrat/Shutterstock.com

111. Exploration Use a graphing utility to explore the ratio $(\sin x)/x$, which appears in calculus.

(a) Complete the table. Round your results to four decimal places.

x	-1	-0.1	-0.01	-0.001
$\frac{\sin x}{x}$				

x	0	0.001	0.01	0.1	1
$\frac{\sin x}{x}$					

(b) Use the graphing utility to graph the function

$$f(x) = \frac{\sin x}{x}.$$

Use the *zoom* and *trace* features to describe the behavior of the graph as x approaches 0.

(c) Write a brief statement regarding the value of the ratio based on your results in parts (a) and (b).

112. Exploration Use a graphing utility to explore the ratio $(1 - \cos x)/x$, which appears in calculus.

(a) Complete the table. Round your results to four decimal places.

x	-1	-0.1	-0.01	-0.001
$\frac{1 - \cos x}{x}$				

x	0	0.001	0.01	0.1	1
$\frac{1 - \cos x}{x}$					

(b) Use the graphing utility to graph the function

$$f(x) = \frac{1 - \cos x}{x}.$$

Use the *zoom* and *trace* features to describe the behavior of the graph as x approaches 0.

(c) Write a brief statement regarding the value of the ratio based on your results in parts (a) and (b).

113. Exploration Using calculus, it can be shown that the sine and cosine functions can be approximated by the polynomials

$$\sin x \approx x - \frac{x^3}{3!} + \frac{x^5}{5!} \quad \text{and} \quad \cos x \approx 1 - \frac{x^2}{2!} + \frac{x^4}{4!}$$

where x is in radians.

(a) Use a graphing utility to graph the sine function and its polynomial approximation in the same viewing window. How do the graphs compare?

(b) Use the graphing utility to graph the cosine function and its polynomial approximation in the same viewing window. How do the graphs compare?

(c) Study the patterns in the polynomial approximations of the sine and cosine functions and predict the next term in each. Then repeat parts (a) and (b). How did the accuracy of the approximations change when an additional term was added?

114. HOW DO YOU SEE IT? The figure below shows the graph of $y = \sin(x - c)$ for

$$c = -\frac{\pi}{4}, \ 0, \ \text{and} \ \frac{\pi}{4}.$$

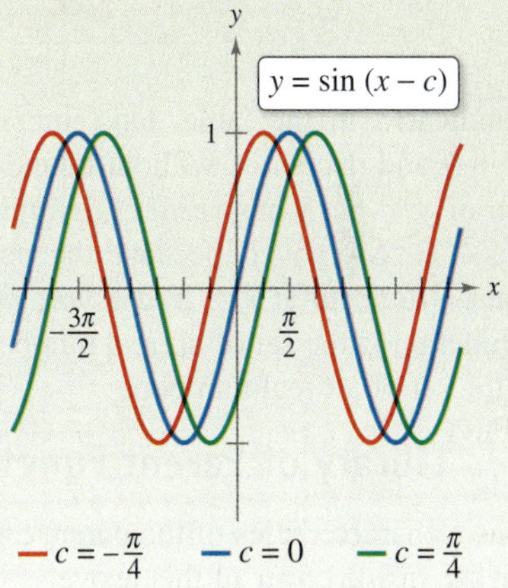

(a) How does the value of c affect the graph?

(b) Which graph is equivalent to that of

$$y = -\cos\left(x + \frac{\pi}{4}\right)?$$

Cumulative Mixed Review

Finding the Slope of a Line **In Exercises 115 and 116, plot the points and find the slope of the line passing through the points.**

115. $(0, 1), (2, 5)$ **116.** $(-1, 4), (3, -2)$

Converting from Radians to Degrees **In Exercises 117 and 118, convert the angle measure from radians to degrees. Round your answer to three decimal places.**

117. 8.5 **118.** -0.48

119. ***Project: Meteorology*** To work an extended application analyzing the mean monthly high temperature and normal precipitation in Cheyenne, Wyoming, visit this textbook's website at *LarsonPrecalculus.com.* (*Data Source:* NOAA)

4.6 Graphs of Other Trigonometric Functions

Graph of the Tangent Function

Recall that the tangent function is odd. That is, $\tan(-x) = -\tan x$. Consequently, the graph of $y = \tan x$ is symmetric with respect to the origin. You also know from the identity $\tan x = \sin x/\cos x$ that the tangent function is undefined when $\cos x = 0$. Two such values are $x = \pm\pi/2 \approx \pm 1.5708$.

x	$-\frac{\pi}{2}$	-1.57	-1.5	$-\frac{\pi}{4}$	0	$\frac{\pi}{4}$	1.5	1.57	$\frac{\pi}{2}$
$\tan x$	Undef.	-1255.8	-14.1	-1	0	1	14.1	1255.8	Undef.

$\tan x$ approaches $-\infty$ as x approaches $-\pi/2$ from the right.

$\tan x$ approaches ∞ as x approaches $\pi/2$ from the left.

As indicated in the table, $\tan x$ increases without bound as x approaches $\pi/2$ from the left and decreases without bound as x approaches $-\pi/2$ from the right. So, the graph of $y = \tan x$ has *vertical asymptotes* at $x = \pi/2$ and $x = -\pi/2$, as shown in Figure 4.33. Moreover, because the period of the tangent function is π, vertical asymptotes also occur at $x = \pi/2 + n\pi$, where n is an integer. The domain of the tangent function is the set of all real numbers other than $x = \pi/2 + n\pi$, and the range is the set of all real numbers.

Library of Parent Functions: Tangent Function

The basic characteristics of the parent tangent function are summarized below and on the inside cover of this text.

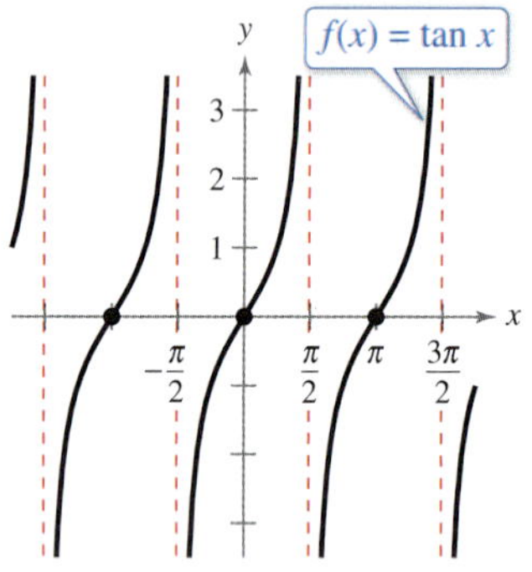

Figure 4.33

Domain: all real numbers x,

$$x \neq \frac{\pi}{2} + n\pi$$

Range: $(-\infty, \infty)$

Period: π

x-intercepts: $(n\pi, 0)$

y-intercept: $(0, 0)$

Vertical asymptotes: $x = \frac{\pi}{2} + n\pi$

Odd function

Origin symmetry

Sketching the graph of $y = a\tan(bx - c)$ is similar to sketching the graph of $y = a\sin(bx - c)$ in that you locate key points that identify the intercepts and asymptotes. Two consecutive asymptotes can be found by solving the equations

$$bx - c = -\pi/2 \quad \text{and} \quad bx - c = \pi/2$$

for x. On the x-axis, the point halfway between two consecutive asymptotes is an x-intercept of the graph. The period of the function $y = a\tan(bx - c)$ is the distance between two consecutive asymptotes. The amplitude of a tangent function is not defined. After plotting the asymptotes and the x-intercept, plot a few additional points between the two asymptotes and sketch one cycle. Finally, sketch one or two additional cycles to the left and right.

What you should learn

- Sketch the graphs of tangent functions.
- Sketch the graphs of cotangent functions.
- Sketch the graphs of secant and cosecant functions.
- Sketch the graphs of damped trigonometric functions.

Why you should learn it

You can use tangent, cotangent, secant, and cosecant functions to model real-life data. For instance, Exercise 64 on page 314 shows you how a tangent function can be used to model and analyze the distance between a television camera and a parade unit.

Believe_In_Me/iStock/Getty Images

EXAMPLE 1 Library of Parent Functions: $f(x) = \tan x$

Sketch the graph of $y = \tan \frac{x}{2}$ by hand.

Solution

By solving the equations $x/2 = -\pi/2$ and $x/2 = \pi/2$, you can see that two consecutive asymptotes occur at $x = -\pi$ and $x = \pi$. Between these two asymptotes, plot a few points, including the x-intercept, as shown in the table. Three cycles of the graph are shown in Figure 4.34. Use a graphing utility to confirm this graph.

x	$-\pi$	$-\frac{\pi}{2}$	0	$\frac{\pi}{2}$	π
$\tan \frac{x}{2}$	Undef.	-1	0	1	Undef.

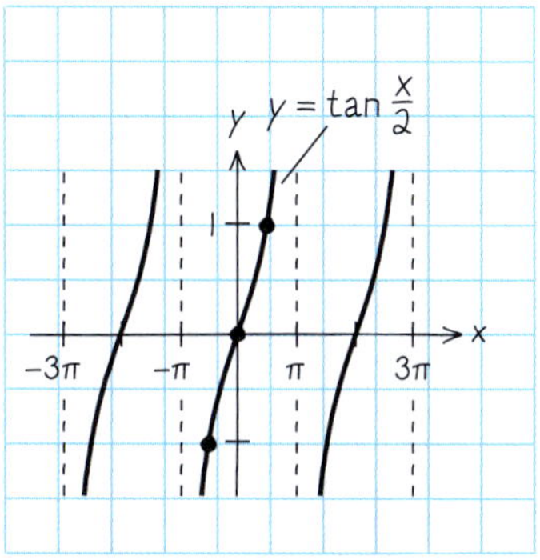

Figure 4.34

Checkpoint *Audio-video solution in English & Spanish at LarsonPrecalculus.com*

Sketch the graph of $y = \tan x/4$ by hand.

EXAMPLE 2 Library of Parent Functions: $f(x) = \tan x$

Sketch the graph of $y = -3 \tan 2x$ by hand.

Solution

By solving the equations $2x = -\pi/2$ and $2x = \pi/2$, you can see that two consecutive asymptotes occur at $x = -\pi/4$ and $x = \pi/4$. Between these two asymptotes, plot a few points, including the x-intercept, as shown in the table. Three cycles of the graph are shown in Figure 4.35. Use a graphing utility to confirm this graph, as shown in Figure 4.36.

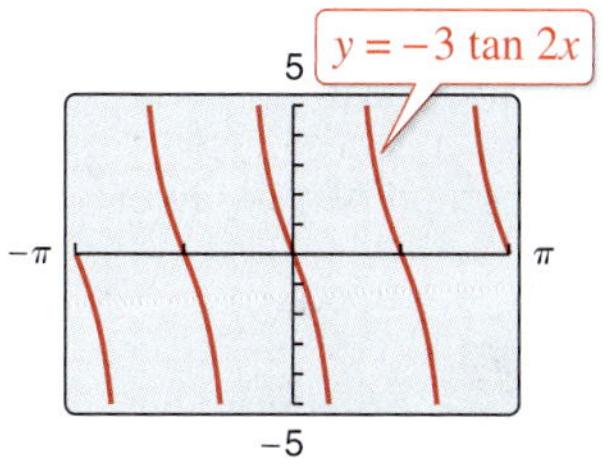

Figure 4.36

x	$-\frac{\pi}{4}$	$-\frac{\pi}{8}$	0	$\frac{\pi}{8}$	$\frac{\pi}{4}$
$-3 \tan 2x$	Undef.	3	0	-3	Undef.

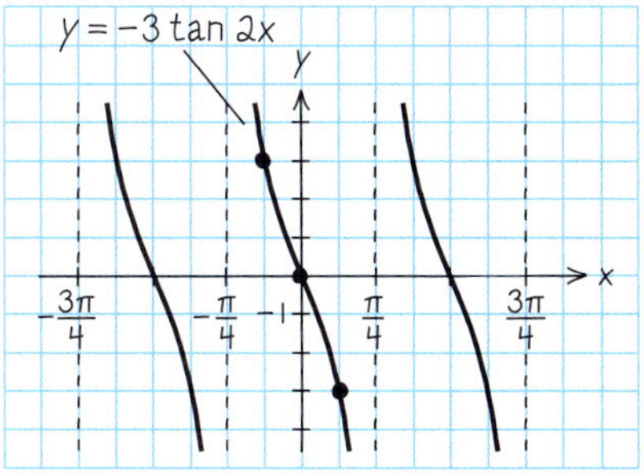

Figure 4.35

Checkpoint *Audio-video solution in English & Spanish at LarsonPrecalculus.com*

Sketch the graph of $y = \tan 2x$ by hand.

By comparing the graphs in Examples 1 and 2, you can see that the graph of $y = a \tan(bx - c)$ increases between consecutive vertical asymptotes when $a > 0$ and decreases between consecutive vertical asymptotes when $a < 0$. In other words, the graph for $a < 0$ is a reflection in the x-axis of the graph for $a > 0$.

Graph of the Cotangent Function

Library of Parent Functions: Cotangent Function

The graph of the parent cotangent function is similar to the graph of the parent tangent function. It also has a period of π. However, from the identity

$$f(x) = \cot x = \frac{\cos x}{\sin x}$$

you can see that the cotangent function has vertical asymptotes when $\sin x$ is zero, which occurs at $x = n\pi$, where n is an integer. The basic characteristics of the parent cotangent function are summarized below and on the inside cover of this text.

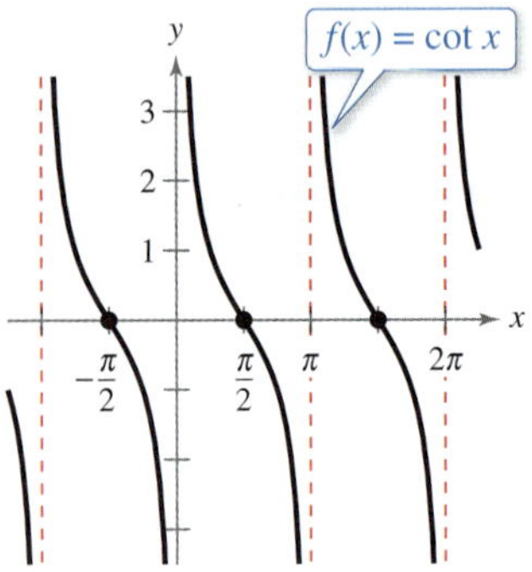

Domain: all real numbers x, $x \neq n\pi$
Range: $(-\infty, \infty)$
Period: π
x-intercepts: $\left(\frac{\pi}{2} + n\pi, 0\right)$
Vertical asymptotes: $x = n\pi$
Odd function
Origin symmetry

EXAMPLE 3 Library of Parent Functions: $f(x) = \cot x$

Sketch the graph of $y = 2 \cot \frac{x}{3}$ by hand.

Solution

To locate two consecutive vertical asymptotes of the graph, solve the equations $x/3 = 0$ and $x/3 = \pi$ to see that two consecutive asymptotes occur at $x = 0$ and $x = 3\pi$. Then, between these two asymptotes, plot a few points, including the x-intercept, as shown in the table. Three cycles of the graph are shown in the figure. Use a graphing utility to confirm this graph. [Enter the function as $y = 2/\tan(x/3)$.] Note that the period is 3π, the distance between consecutive asymptotes.

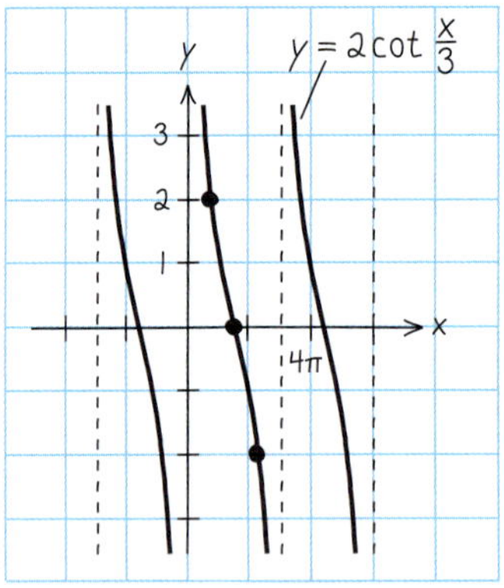

x	0	$\frac{3\pi}{4}$	$\frac{3\pi}{2}$	$\frac{9\pi}{4}$	3π
$2 \cot \frac{x}{3}$	Undef.	2	0	-2	Undef.

✓ **Checkpoint** ▶ *Audio-video solution in English & Spanish at LarsonPrecalculus.com*

Sketch the graph of $y = \cot \frac{x}{4}$ by hand.

Graphs of the Reciprocal Functions

The graphs of the two remaining trigonometric functions can be obtained from the graphs of the sine and cosine functions using the reciprocal identities

$$\csc x = \frac{1}{\sin x} \quad \text{and} \quad \sec x = \frac{1}{\cos x}.$$

For instance, at a given value of x, the y-coordinate of $\sec x$ is the reciprocal of the y-coordinate of $\cos x$. Of course, when $\cos x = 0$, the reciprocal does not exist. Near such values of x, the behavior of the secant function is similar to that of the tangent function. In other words, the graphs of

$$\tan x = \frac{\sin x}{\cos x} \quad \text{and} \quad \sec x = \frac{1}{\cos x}$$

have vertical asymptotes where $\cos x = 0$—that is, at $x = \pi/2 + n\pi$, where n is an integer. Similarly,

$$\cot x = \frac{\cos x}{\sin x} \quad \text{and} \quad \csc x = \frac{1}{\sin x}$$

have vertical asymptotes where $\sin x = 0$—that is, at $x = n\pi$, where n is an integer.

To sketch the graph of a secant or cosecant function, you should first make a sketch of its reciprocal function. For instance, to sketch the graph of $y = \csc x$, first sketch the graph of $y = \sin x$. Then take the reciprocals of the y-coordinates to obtain points on the graph of $y = \csc x$. You can use this procedure to obtain the graphs shown in Figure 4.37.

Explore the Concept

Use a graphing utility to graph the functions $y_1 = \cos x$ and $y_2 = \sec x = 1/\cos x$ in the same viewing window. How are the graphs related? What happens to the graph of the secant function as x approaches the zeros of the cosine function?

Library of Parent Functions: Cosecant and Secant Functions

The basic characteristics of the parent cosecant and secant functions are summarized below and on the inside cover of this text.

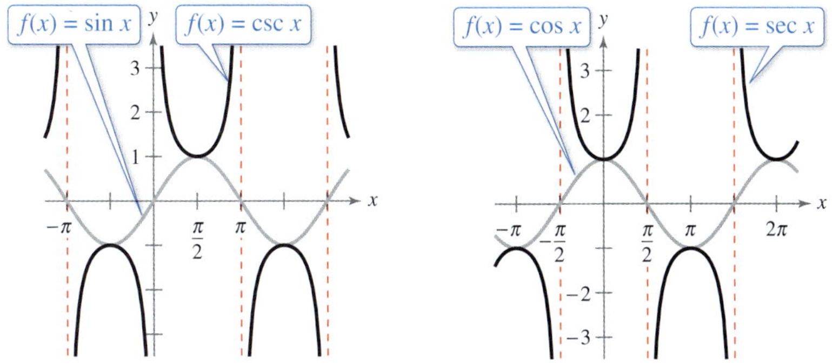

Domain: all real numbers x, $x \neq n\pi$	*Domain: all real numbers* x, $x \neq \frac{\pi}{2} + n\pi$
Range: $(-\infty, -1] \cup [1, \infty)$	*Range:* $(-\infty, -1] \cup [1, \infty)$
Period: 2π	*Period:* 2π
No intercepts	*y-intercept:* $(0, 1)$
Vertical asymptotes: $x = n\pi$	*Vertical asymptotes:* $x = \frac{\pi}{2} + n\pi$
Odd function	*Even function*
Origin symmetry	*y-axis symmetry*

Figure 4.37

In comparing the graphs of the cosecant and secant functions with those of the sine and cosine functions, note that the "hills" and "valleys" are interchanged. For instance, a hill (or maximum point) on the sine curve corresponds to a valley (a relative minimum) on the cosecant curve, and a valley (or minimum point) on the sine curve corresponds to a hill (a relative maximum) on the cosecant curve, as shown in Figure 4.38. Additionally, x-intercepts of the sine and cosine functions become vertical asymptotes of the cosecant and secant functions, respectively (see Figure 4.38).

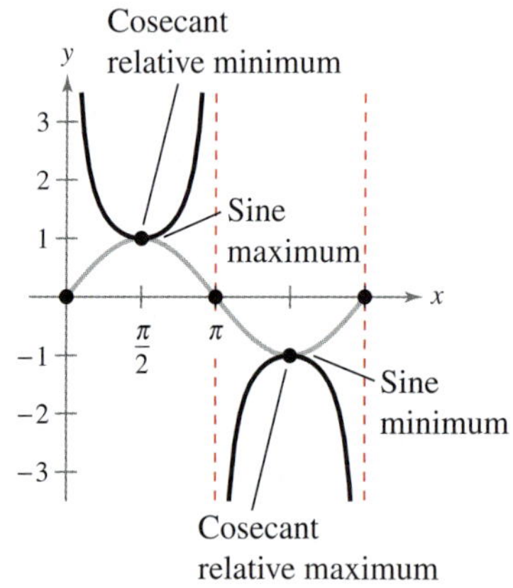

Figure 4.38

EXAMPLE 4 Library of Parent Functions: $f(x) = \csc x$

Sketch the graph of $y = 2\csc\left(x + \dfrac{\pi}{4}\right)$ by hand.

Solution

Begin by sketching the graph of

$$y = 2\sin\left(x + \frac{\pi}{4}\right).$$

For this function, the amplitude is 2 and the period is 2π. By solving the equations

$$x + \frac{\pi}{4} = 0 \qquad \text{and} \qquad x + \frac{\pi}{4} = 2\pi$$

you can see that one cycle of the sine function corresponds to the interval from $x = -\pi/4$ to $x = 7\pi/4$. The graph of this sine function is represented by the gray curve in Figure 4.39. Because the sine function is zero at the midpoint and endpoints of this interval, the corresponding cosecant function

$$y = 2\csc\left(x + \frac{\pi}{4}\right) = 2\left(\frac{1}{\sin[x + (\pi/4)]}\right)$$

has vertical asymptotes at $x = -\pi/4$, $x = 3\pi/4$, $x = 7\pi/4$, and so on. The graph of the cosecant function is represented by the black curve in Figure 4.39.

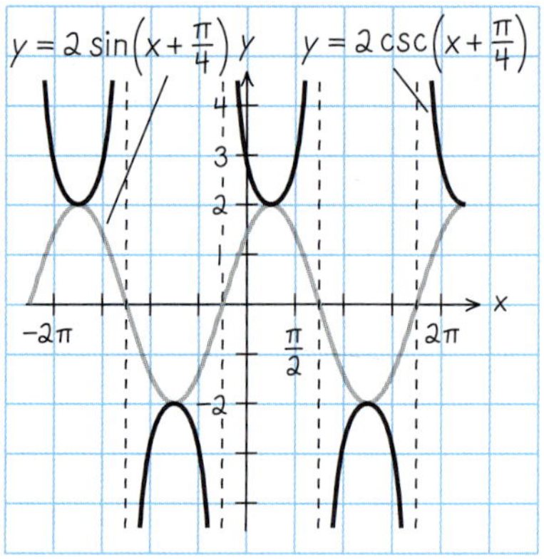

Figure 4.39

Checkpoint *Audio-video solution in English & Spanish at LarsonPrecalculus.com*

Sketch the graph of $y = 2\csc\left(x + \dfrac{\pi}{2}\right)$ by hand.

EXAMPLE 5 Library of Parent Functions: $f(x) = \sec x$

See LarsonPrecalculus.com for an interactive version of this type of example.

To sketch the graph of $y = \sec 2x$ by hand, begin by sketching the graph of $y = \cos 2x$, as indicated by the gray curve in Figure 4.40. Then, form the graph of $y = \sec 2x$, as indicated by the black curve in the figure. Note that the x-intercepts of $y = \cos 2x$

$$\left(-\frac{\pi}{4}, 0\right), \quad \left(\frac{\pi}{4}, 0\right), \quad \left(\frac{3\pi}{4}, 0\right), \ldots$$

correspond to the vertical asymptotes

$$x = -\frac{\pi}{4}, \quad x = \frac{\pi}{4}, \quad x = \frac{3\pi}{4}, \ldots$$

of the graph of $y = \sec 2x$. Moreover, notice that the period of $y = \cos 2x$ and $y = \sec 2x$ is π.

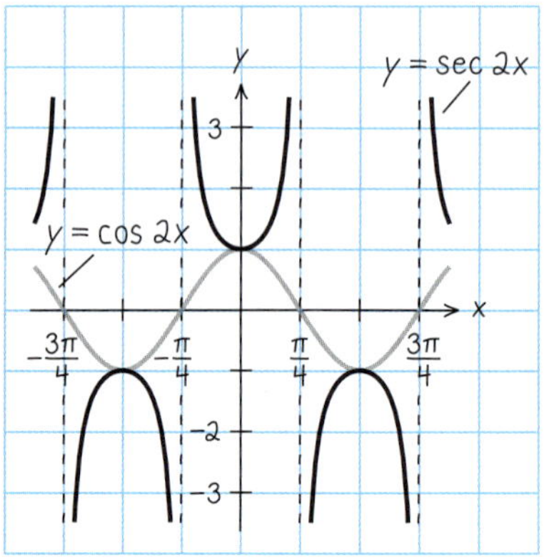

Figure 4.40

Checkpoint *Audio-video solution in English & Spanish at LarsonPrecalculus.com*

Sketch the graph of $y = \sec \dfrac{x}{2}$ by hand.

Damped Trigonometric Graphs

A *product* of two functions can be graphed using properties of the individual functions. For instance, consider the function

$$f(x) = x \sin x$$

as the product of the functions $y = x$ and $y = \sin x$. Using properties of absolute value and the fact that $|\sin x| \le 1$, you have $0 \le |x||\sin x| \le |x|$. Consequently,

$$-|x| \le x \sin x \le |x|$$

which means that the graph of $f(x) = x \sin x$ lies between the lines $y = -x$ and $y = x$. Furthermore, because

$$f(x) = x \sin x = \pm x \quad \text{at} \quad x = \frac{\pi}{2} + n\pi$$

and

$$f(x) = x \sin x = 0 \quad \text{at} \quad x = n\pi$$

where n is an integer, the graph of f touches the line $y = -x$ or the line $y = x$ at $x = \pi/2 + n\pi$ and has x-intercepts at $x = n\pi$. A sketch of f is shown in Figure 4.41. In the function $f(x) = x \sin x$, the factor x is called the **damping factor.**

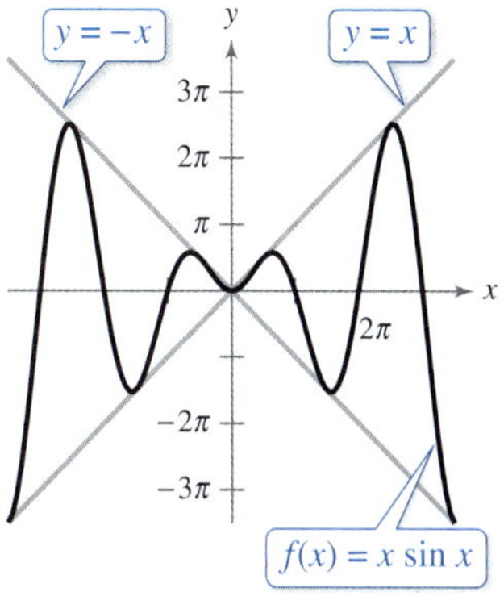

Figure 4.41

Algebra Help

Do you see why the graph of $f(x) = x \sin x$ touches the lines $y = \pm x$ at $x = \pi/2 + n\pi$ and why the graph has x-intercepts at $x = n\pi$? Recall that the sine function is equal to ± 1 at $\pi/2, 3\pi/2, 5\pi/2, . . .$ (odd multiples of $\pi/2$) and is equal to 0 at $\pi, 2\pi, 3\pi, . . .$ (multiples of π).

EXAMPLE 6 Analyzing a Damped Sine Curve

Analyze the graph of $f(x) = e^{-x} \sin 3x$.

Solution

Consider the function $f(x) = e^{-x} \sin 3x$ as the product of the two functions

$$y = e^{-x} \quad \text{and} \quad y = \sin 3x$$

each of which has the set of real numbers as its domain. For any real number x, you know that $e^{-x} > 0$ and $|\sin 3x| \le 1$. So, $e^{-x}|\sin 3x| \le e^{-x}$, which means that

$$-e^{-x} \le e^{-x} \sin 3x \le e^{-x}.$$

Furthermore, because

$$f(x) = e^{-x} \sin 3x = \pm e^{-x} \quad \text{at} \quad x = \frac{\pi}{6} + \frac{n\pi}{3}$$

and

$$f(x) = e^{-x} \sin 3x = 0 \quad \text{at} \quad x = \frac{n\pi}{3}$$

the graph of f touches the curves $y = -e^{-x}$ and $y = e^{-x}$ at $x = \pi/6 + n\pi/3$ and has intercepts at $x = n\pi/3$. The graph is shown below.

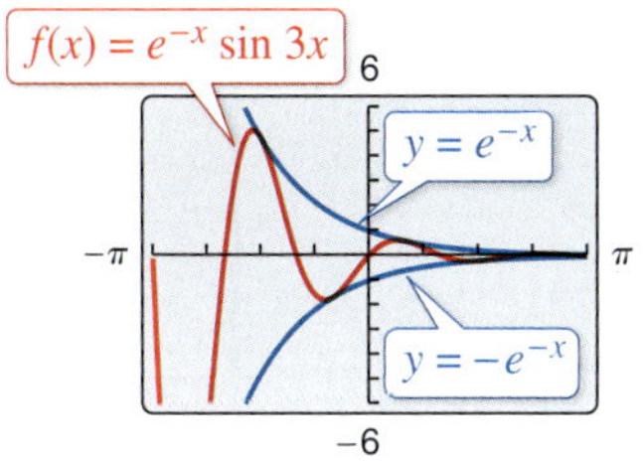

✓ Checkpoint *Audio-video solution in English & Spanish at LarsonPrecalculus.com*

Analyze the graph of $f(x) = e^x \sin 4x$.

Library of Parent Functions: Trigonometric Functions

Figure 4.42 summarizes the six basic trigonometric functions. These functions are also summarized on the inside cover of this text.

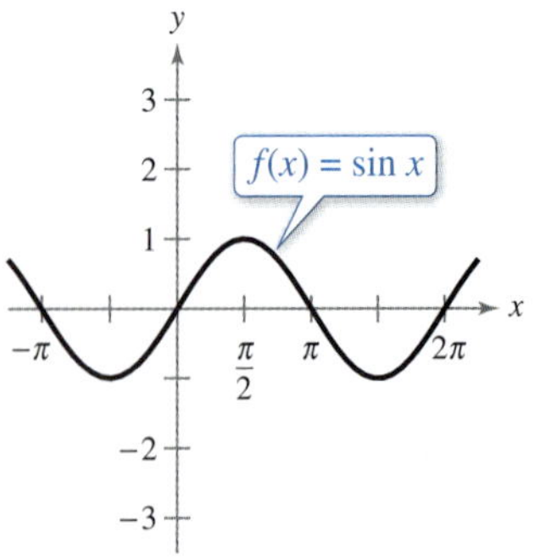

Domain:* all real numbers *x
***Range:* $[-1, 1]$**
***Period:* 2π**

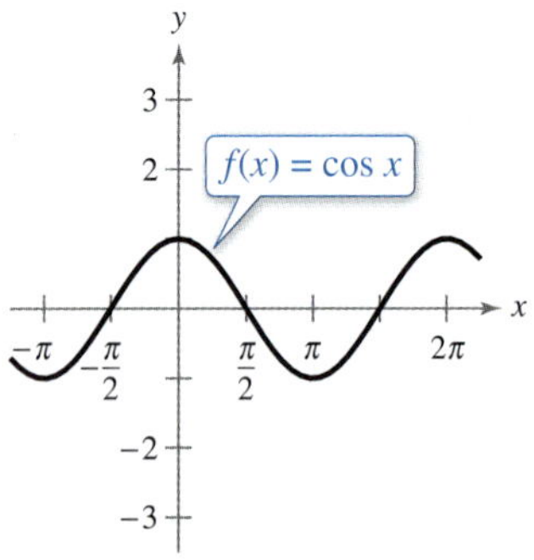

Domain:* all real numbers *x
***Range:* $[-1, 1]$**
***Period:* 2π**

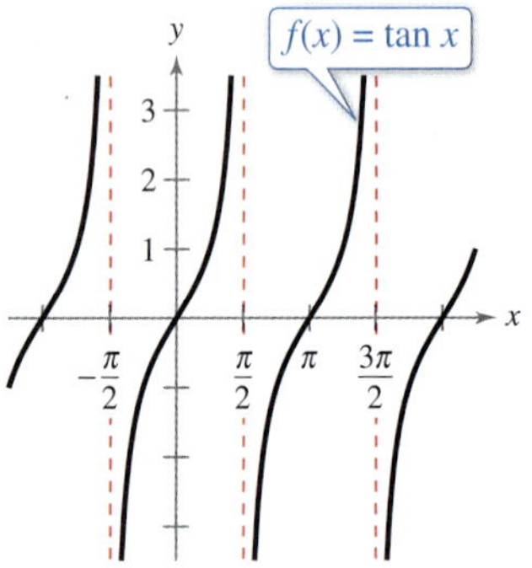

***Domain:* all real numbers *x*,**

$$x \neq \frac{\pi}{2} + n\pi$$

***Range:* $(-\infty, \infty)$**
***Period:* π**

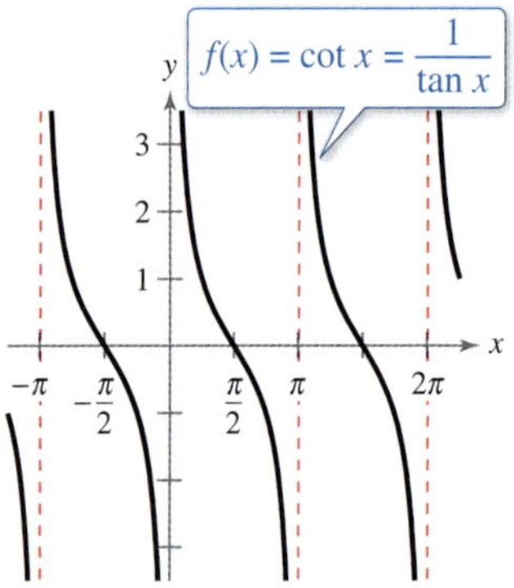

***Domain:* all real numbers *x*,**

$$x \neq n\pi$$

***Range:* $(-\infty, \infty)$**
***Period:* π**

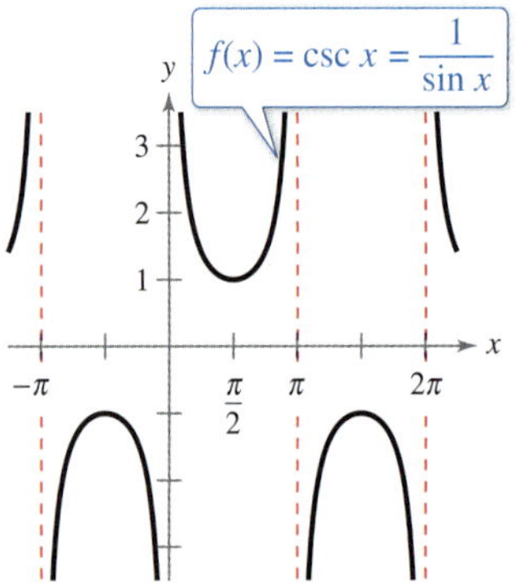

***Domain:* all real numbers *x*,**

$$x \neq n\pi$$

***Range:* $(-\infty, -1] \cup [1, \infty)$**
***Period:* 2π**

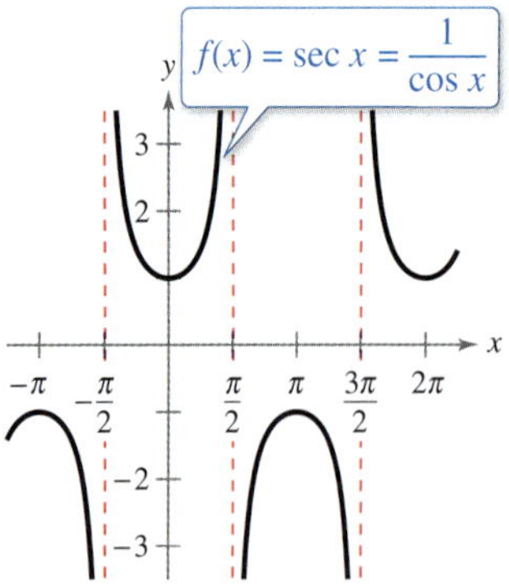

***Domain:* all real numbers *x*,**

$$x \neq \frac{\pi}{2} + n\pi$$

***Range:* $(-\infty, -1] \cup [1, \infty)$**
***Period:* 2π**

Figure 4.42

4.6 Exercises

See *CalcChat.com* for tutorial help and worked-out solutions to odd-numbered exercises. For instructions on how to use a graphing utility, see Appendix A.

Vocabulary and Concept Check

In Exercises 1 and 2, fill in the blank.

1. The graphs of the tangent, cotangent, secant, and cosecant functions have _______ asymptotes.

2. To sketch the graph of a secant or cosecant function, first sketch its _______ function.

3. Which two parent trigonometric functions have a period of π and a range that consists of the set of all real numbers?

4. What is the damping factor of the function $f(x) = e^{2x} \sin x$?

Procedures and Problem Solving

Library of Parent Functions **In Exercises 5–8, use the graph of the function to answer parts (a)–(e).**

(a) Find any x-intercepts.

(b) Find any y-intercepts.

(c) Find any intervals on which the graph is increasing and any intervals on which the graph is decreasing.

(d) Find all relative extrema, if any, of the graph.

(e) Find all vertical asymptotes, if any, of the graph.

5. $f(x) = \tan x$

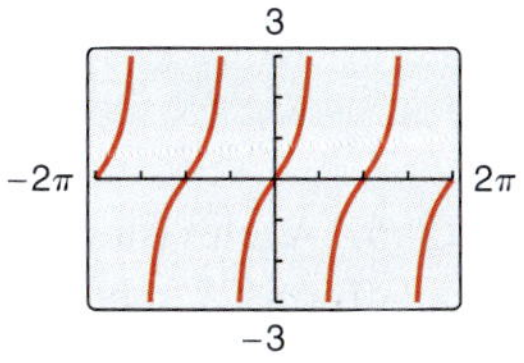

6. $f(x) = \cot x$

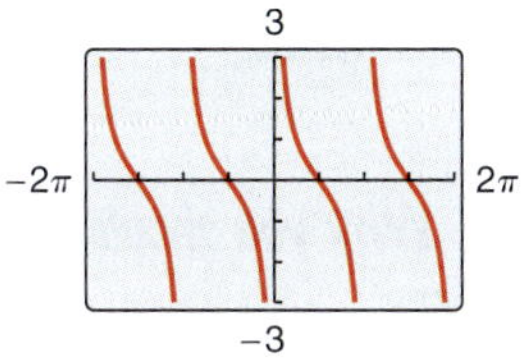

7. $f(x) = \sec x$

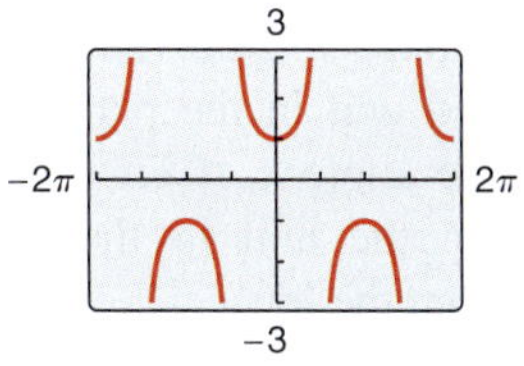

8. $f(x) = \csc x$

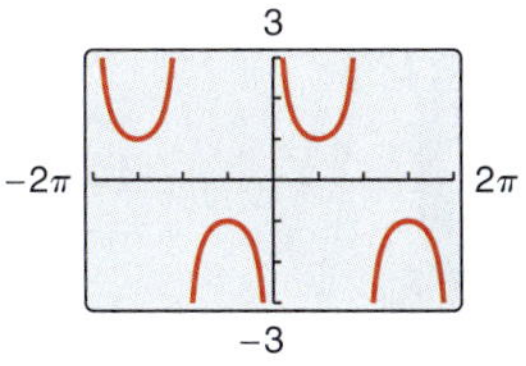

Library of Parent Functions **In Exercises 9–28, sketch the graph of the function by hand. (Include two full periods.) Use a graphing utility to verify your result.**

9. $y = \tan \dfrac{x}{6}$

10. $y = \tan 4x$

11. $y = -2 \tan 3x$

12. $y = -3 \tan \dfrac{x}{4}$

13. $y = \dfrac{1}{2} \cot \dfrac{x}{2}$

14. $y = 3 \cot \pi x$

15. $y = -\frac{1}{2} \sec x$

16. $y = \frac{3}{2} \sec x$

17. $y = 2 \csc \dfrac{x}{3}$

18. $y = -\csc \dfrac{x}{3}$

19. $y = \sec \pi x - 3$

20. $y = 5 \sec 4x + 2$

21. $y = 3 \tan \dfrac{\pi x}{4}$

22. $y = \dfrac{1}{2} \tan \pi x$

23. $y = -\csc(\pi - x)$

24. $y = -\sec(x + \pi)$

25. $y = \frac{1}{2} \sec(2x - \pi)$

26. $y = \csc(3x - \pi)$

27. $y = 4 \cot\left(x + \dfrac{\pi}{2}\right)$

28. $y = \dfrac{1}{2} \cot(x + \pi)$

Comparing Trigonometric Graphs **In Exercises 29–34, use a graphing utility to graph the function (include two full periods). Graph the corresponding reciprocal function in the same viewing window. Describe and compare the graphs.**

29. $y = 2 \csc 3x$

30. $y = -\csc 2x$

31. $y = -2 \sec 4x$

32. $y = \frac{1}{4} \sec \pi x$

33. $y = \dfrac{1}{3} \sec\left(\dfrac{\pi x}{2} + \dfrac{\pi}{2}\right)$

34. $y = \dfrac{1}{2} \csc(4x - \pi)$

Solving a Trigonometric Equation **In Exercises 35–40, find the solutions of the equation on the interval $[-2\pi, 2\pi]$. Use a graphing utility to verify your results.**

35. $\tan x = 1$

36. $\cot x = -1$

37. $\sec x = 2$

38. $\csc x = \sqrt{2}$

39. $\cot x = -\sqrt{3}$

40. $\sec x = -\sqrt{2}$

Even and Odd Trigonometric Functions **In Exercises 41–46, use the graph of the function to determine whether the function is even, odd, or neither. Verify your answer algebraically.**

41. $f(x) = \sec x$

42. $f(x) = \tan x$

43. $f(x) = \csc 2x$

44. $f(x) = \cot 3x$

45. $f(x) = \tan\left(x - \dfrac{\pi}{2}\right)$

46. $f(x) = \sec(x + \pi)$

Using Graphs to Compare Functions **In Exercises 47–50, use a graphing utility to graph the two equations in the same viewing window. Use the graphs to show that the expressions are equivalent. Then show that the expressions are equivalent algebraically.**

47. $y_1 = \sin x \sec x, \quad y_2 = \tan x$

48. $y_1 = \dfrac{\cos x}{\sin x}, \quad y_2 = \cot x$

49. $y_1 = 1 + \cot^2 x, \quad y_2 = \csc^2 x$

50. $y_1 = \sec^2 x - 1, \quad y_2 = \tan^2 x$

Identifying Damped Trigonometric Graphs **In Exercises 51–54, match the function with its graph. Describe the behavior of the function as x approaches zero. [The graphs are labeled (a), (b), (c), and (d).]**

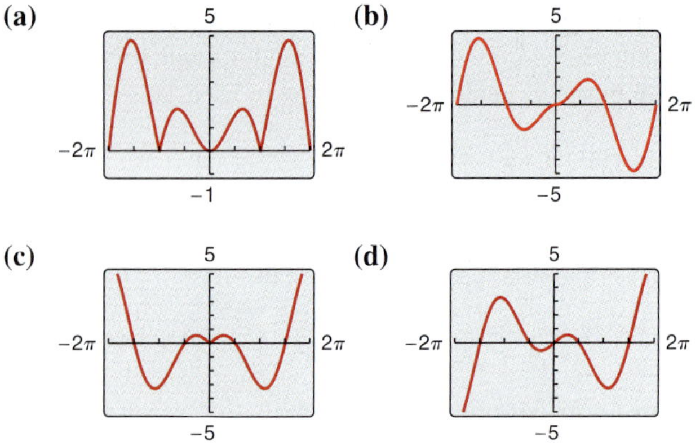

51. $f(x) = x \cos x$ **52.** $f(x) = |x \sin x|$

53. $g(x) = |x| \sin x$ **54.** $g(x) = |x| \cos x$

Analyzing a Damped Trigonometric Graph **In Exercises 55–58, use a graphing utility to graph the function and the damping factor of the function in the same viewing window. Then analyze the graph of the function using the method in Example 6.**

55. $f(x) = e^{-x} \cos x$ **56.** $f(x) = e^{-2x} \sin x$

57. $h(x) = 2^{-x^2/4} \sin x$ **58.** $g(x) = 2^{-x^2/2} \cos x$

Exploration **In Exercises 59 and 60, use a graphing utility to graph the function. Use the graph to determine the behavior of the function**

(a) as $x \to \frac{\pi^+}{2}$ $\left(\text{as } x \text{ approaches } \frac{\pi}{2} \text{ from the right}\right)$,

(b) as $x \to \frac{\pi^-}{2}$ $\left(\text{as } x \text{ approaches } \frac{\pi}{2} \text{ from the left}\right)$,

(c) as $x \to -\frac{\pi^+}{2}$ $\left(\text{as } x \text{ approaches } -\frac{\pi}{2} \text{ from the right}\right)$, and

(d) as $x \to -\frac{\pi^-}{2}$ $\left(\text{as } x \text{ approaches } -\frac{\pi}{2} \text{ from the left}\right)$.

59. $f(x) = \tan x$ **60.** $f(x) = \sec x$

Exploration **In Exercises 61 and 62, use a graphing utility to graph the function. Use the graph to determine the behavior of the function**

(a) as $x \to 0^+$ (as x approaches 0 from the right),

(b) as $x \to 0^-$ (as x approaches 0 from the left),

(c) as $x \to \pi^+$ (as x approaches π from the right), and

(d) as $x \to \pi^-$ (as x approaches π from the left).

61. $f(x) = \csc x$

62. $f(x) = \cot x$

63. Aeronautics A plane flying at an altitude of 7 miles over level ground passes directly over a radar antenna. Let d be the ground distance from the antenna to the point directly under the plane and let x be the angle of elevation to the plane from the antenna (see figure). Write d as a function of x and graph the function over the interval $0 < x < \pi$. (Consider d as positive when the plane approaches the antenna.)

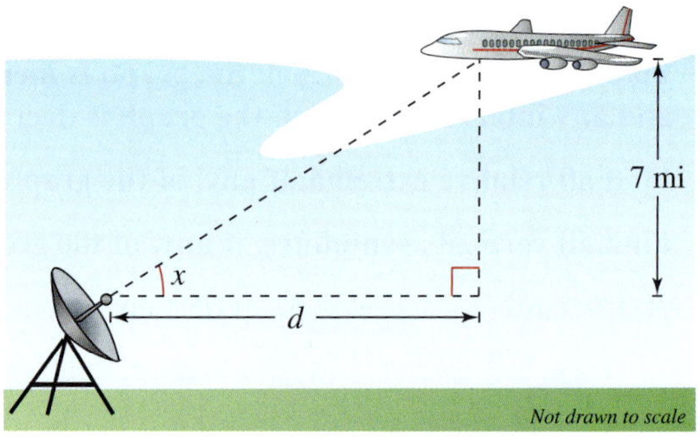

64. ***Why you should learn it*** *(p. 306)* A television camera is on a reviewing platform 27 meters from the street on which a parade will be passing from left to right (see figure).

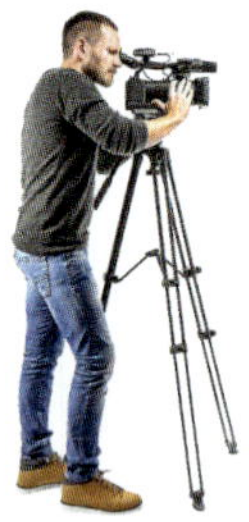

(a) Write the distance d from the camera to a designated unit in the parade as a function of the angle x. (Consider x as negative when the unit in the parade approaches from the left.)

(b) Graph the function over the interval $-\pi/2 < x < \pi/2$.

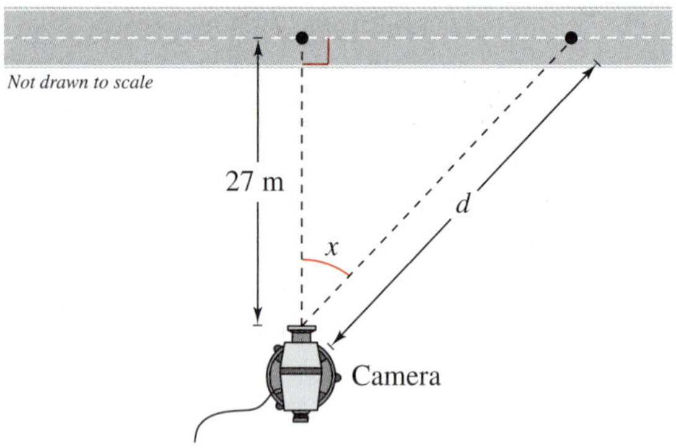

Believe_In_Me/iStock/Getty Images

65. Harmonic Motion An object weighing W pounds is suspended from a ceiling by a steel spring. The weight is pulled downward (which is assumed to be in the positive direction) from its equilibrium position and released (see figure). The resulting motion of the weight is described by the function $y = \frac{1}{2}e^{-t/4} \cos 4t$, where y is the distance in feet and t is the time in seconds ($t > 0$).

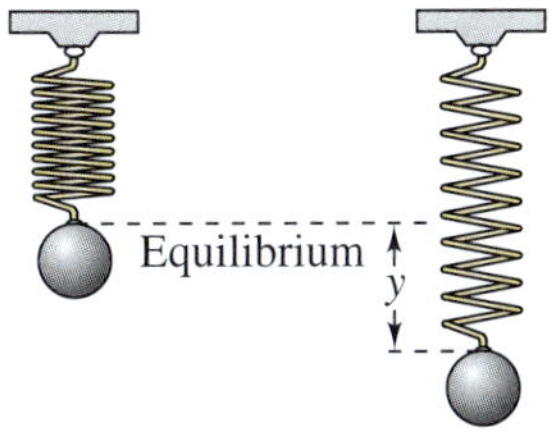

(a) Use a graphing utility to graph the function.

(b) Describe the behavior of the displacement function for increasing values of t.

66. Mechanical Engineering A crossed belt connects a 10-centimeter pulley on an electric motor with a 20-centimeter pulley on a saw arbor (see figure). The electric motor runs at 1700 revolutions per minute.

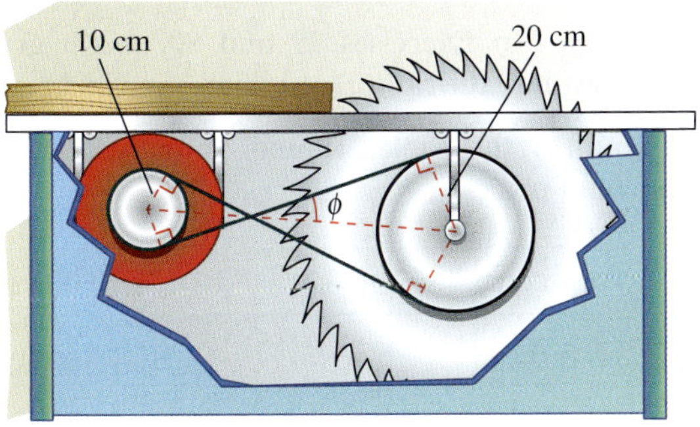

(a) Determine the number of revolutions per minute of the saw.

(b) How does crossing the belt affect the saw in relation to the motor?

(c) Let L be the total length of the belt. Write L as a function of ϕ, where ϕ is measured in radians. What is the domain of the function? (*Hint:* Add the lengths of the straight sections of the belt and the length of the belt around each pulley.)

(d) Use a graphing utility to complete the table.

ϕ	0.3	0.6	0.9	1.2	1.5
L					

(e) As ϕ increases, do the lengths of the straight sections of the belt change faster or slower than the lengths of the belts around each pulley?

(f) Use the graphing utility to graph the function over the appropriate domain.

67. Sales The projected monthly sales S (in thousands of units) of a toy are modeled by

$$S = 2 + e^{-(x-\pi/2)} \sin\left[3\left(x - \frac{\pi}{2}\right)\right], \quad t \geq 0$$

where t is the number of months since December 2017.

(a) Use a graphing utility to graph the sales function over 12 months. Use the *table* feature to determine the sales for each month.

(b) As $t \to 11$, what value does S approach?

68. MODELING DATA

The motion of an oscillating weight suspended by a spring was measured by a motion detector. The data were collected, and the approximate maximum (positive and negative) displacements from equilibrium are shown in the graph. The displacement y is measured in centimeters and the time t is measured in seconds. *Spreadsheet at LarsonPrecalculus.com.*

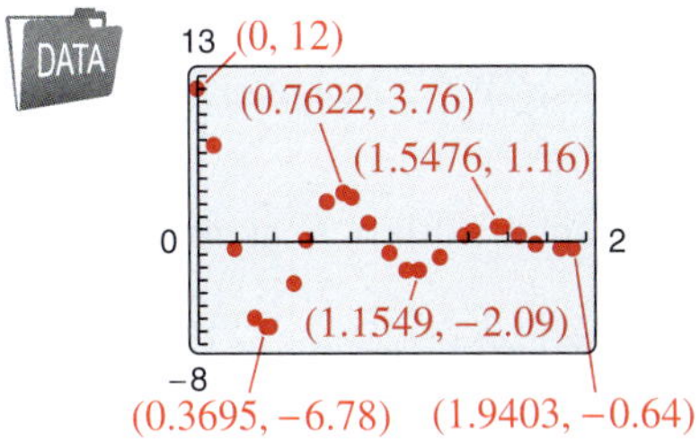

(a) Is y a function of t? Explain.

(b) Approximate the frequency of the oscillations.

(c) Fit a model of the form $y = ab^t \cos ct$ to the data. Use the result of part (b) to approximate c. Use the *regression* feature of a graphing utility to fit an exponential model to the positive maximum displacements of the weight.

(d) Rewrite the model in the form $y = ae^{kt} \cos ct$.

(e) Use the graphing utility to graph the model. Compare the result with the data in the graph above.

Focusing on Concepts

True or False? **In Exercises 69–72, determine whether the statement is true or false. Justify your answer.**

69. The graph of $y = -\frac{1}{6}\tan\left(\frac{x}{2} + \pi\right)$ has an asymptote at $x = -7\pi$.

70. For the graph of $y = 2^x \sin x$, as x approaches $-\infty$, y approaches 0.

71. The graph of $y = \csc x$ can be obtained on a calculator by graphing the reciprocal of $y = \sin x$.

72. The graph of $y = \sec x$ can be obtained on a calculator by graphing a translation of the reciprocal of $y = \sin x$.

73. Exploration Consider the functions

$$f(x) = 2 \sin x \quad \text{and} \quad g(x) = \frac{1}{2} \csc x$$

on the interval $(0, \pi)$.

(a) Use a graphing utility to graph f and g in the same viewing window.

(b) Approximate the interval in which $f(x) > g(x)$.

(c) Describe the behavior of each of the functions as x approaches π. How is the behavior of g related to the behavior of f as x approaches π?

74. Exploration Consider the functions

$$f(x) = \tan \frac{\pi x}{2} \quad \text{and} \quad g(x) = \frac{1}{2} \sec \frac{\pi x}{2}$$

on the interval $(-1, 1)$,

(a) Use a graphing utility to graph f and g in the same viewing window.

(b) Approximate the interval in which $f(x) < g(x)$.

(c) Describe the behavior of each function as x approaches 0. How is the behavior of g related to the behavior of f as x approaches 0?

75. Exploration

(a) Use a graphing utility to graph each function.

$$y_1 = \frac{4}{\pi}\left(\sin \pi x + \frac{1}{3} \sin 3\pi x\right)$$

$$y_2 = \frac{4}{\pi}\left(\sin \pi x + \frac{1}{3} \sin 3\pi x + \frac{1}{5} \sin 5\pi x\right)$$

(b) Identify the pattern in part (a) and find a function y_3 that continues the pattern one more term. Use the graphing utility to graph y_3.

(c) The graphs in parts (a) and (b) approximate the periodic function in the figure. Find a function y_4 that is a better approximation.

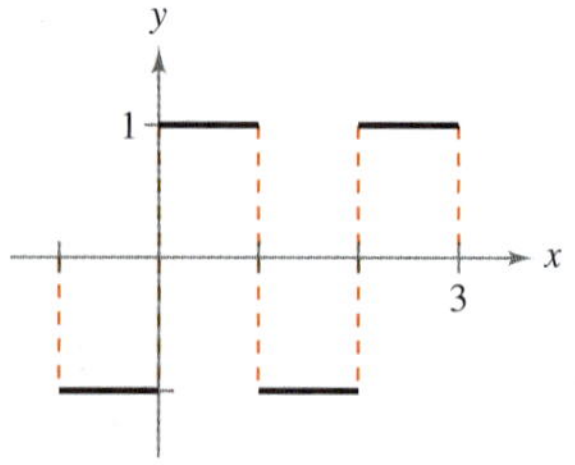

76. Exploration Using calculus, it can be shown that the tangent function can be approximated by the polynomial

$$\tan x \approx x + \frac{2x^3}{3!} + \frac{16x^5}{5!}$$

where x is in radians. Use a graphing utility to graph the tangent function and its polynomial approximation in the same viewing window. How do the graphs compare?

77. Error Analysis Describe the error.

The graph of $g(x) = 3 \tan 4x$ is a vertical stretch by a factor of 4 and a horizontal shrink by a factor of $\frac{1}{3}$ of the graph of $f(x) = \tan x$.

78. HOW DO YOU SEE IT? Determine which function is represented by each graph. (Do not use a calculator.)

(a)

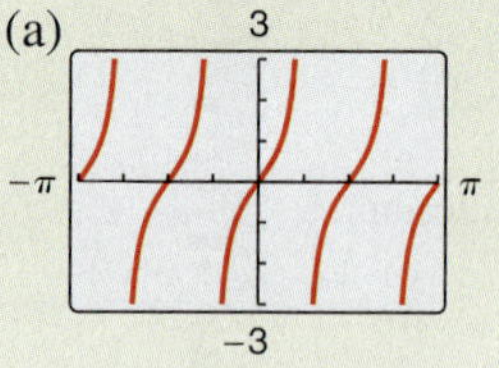

(b)

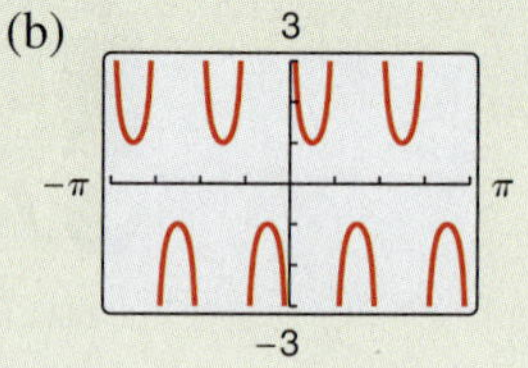

(a)
(i) $f(x) = \tan 2x$
(ii) $f(x) = \tan \frac{x}{2}$
(iii) $f(x) = 2 \tan x$
(iv) $f(x) = -\tan 2x$

(b)
(i) $f(x) = \sec 4x$
(ii) $f(x) = \csc 4x$
(iii) $f(x) = \csc \frac{x}{4}$
(iv) $f(x) = \csc(4x - \pi)$

Exploration In Exercises 79 and 80, use a graphing utility to explore the ratio $f(x)$, which appears in calculus.

(a) Complete the table. Round your results to four decimal places.

x	-1	-0.1	-0.01	-0.001
$f(x)$				

x	0	0.001	0.01	0.1	1
$f(x)$					

(b) Use the graphing utility to graph f. Use the *zoom* and *trace* features to describe the behavior of the graph as x approaches 0.

(c) Write a brief statement regarding the value of the ratio based on your results in parts (a) and (b).

79. $f(x) = \dfrac{\tan x}{x}$

80. $f(x) = \dfrac{\tan 2x}{2x}$

Cumulative Mixed Review

Finding the Domain, Intercepts, and Asymptotes of a Function **In Exercises 81–84, identify the domain, any intercepts, and any asymptotes of the function.**

81. $y = x^2 + 3x - 4$

82. $y = \ln x^4$

83. $f(x) = 3^{x+1} + 2$

84. $f(x) = \dfrac{x - 7}{x^2 + 4x + 4}$

4.7 Inverse Trigonometric Functions

What you should learn

- Evaluate and graph inverse sine functions.
- Evaluate and graph other inverse trigonometric functions.
- Evaluate compositions of trigonometric functions.

Why you should learn it

Inverse trigonometric functions can be useful in exploring how aspects of a real-life problem relate to each other. Exercise 109 on page 326 investigates the relationship between the height of a cone-shaped pile of rock salt, the angle of the cone shape, and the diameter of its base.

Inverse Sine Function

Recall from Section 1.6 that for a function to have an inverse function, it must be one-to-one—that is, it must pass the Horizontal Line Test. Notice from Figure 4.43 that $y = \sin x$ does not pass the test because different values of x yield the same y-value.

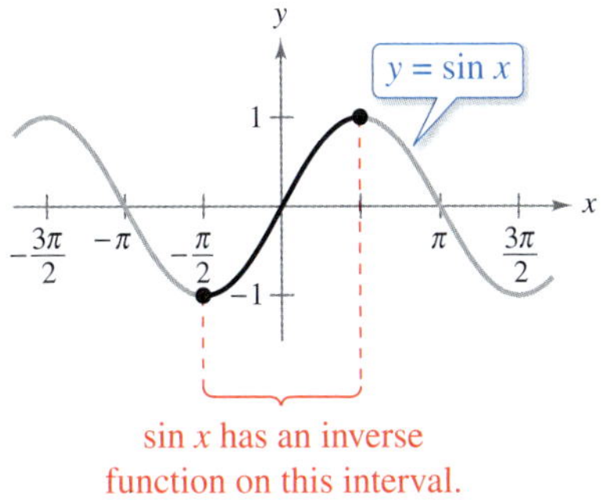

Figure 4.43

However, when you restrict the domain to the interval $-\pi/2 \le x \le \pi/2$ (corresponding to the black portion of the graph in Figure 4.43), the properties listed below hold.

1. On the interval $[-\pi/2, \pi/2]$, the function $y = \sin x$ is increasing.
2. On the interval $[-\pi/2, \pi/2]$, $y = \sin x$ takes on its full range of values, $-1 \le \sin x \le 1$.
3. On the interval $[-\pi/2, \pi/2]$, $y = \sin x$ is one-to-one.

So, on the restricted domain $-\pi/2 \le x \le \pi/2$, $y = \sin x$ has a unique inverse function called the **inverse sine function.** It is denoted by

$$y = \arcsin x \qquad \text{or} \qquad y = \sin^{-1} x.$$

The notation $\sin^{-1} x$ is consistent with the inverse function notation $f^{-1}(x)$. The arcsin x notation (read as "the arcsine of x") comes from the association of a central angle with its intercepted *arc length* on a unit circle. So, arcsin x means the angle (or arc) whose sine is x. Both notations, arcsin x and $\sin^{-1} x$, are commonly used in mathematics, so remember that $\sin^{-1} x$ denotes the *inverse* sine function rather than $1/\sin x$. The values of arcsin x lie in the interval

$$-\frac{\pi}{2} \le \arcsin x \le \frac{\pi}{2}.$$

The graph of $y = \arcsin x$ is shown in Figure 4.44 on the next page.

Definition of Inverse Sine Function

The **inverse sine function** is defined by

$$y = \arcsin x \qquad \text{if and only if} \qquad \sin y = x$$

where $-1 \le x \le 1$ and $-\pi/2 \le y \le \pi/2$. The domain of $y = \arcsin x$ is $[-1, 1]$ and the range is $[-\pi/2, \pi/2]$.

When evaluating the inverse sine function, it helps to remember the phrase "the arcsine of x is the angle (or number) whose sine is x."

EXAMPLE 1 Evaluating the Inverse Sine Function

If possible, find the exact value.

a. $\arcsin\left(-\frac{1}{2}\right)$ **b.** $\sin^{-1}\frac{\sqrt{3}}{2}$ **c.** $\sin^{-1} 2$

Solution

a. Because $\sin\left(-\frac{\pi}{6}\right) = -\frac{1}{2}$, and $-\frac{\pi}{6}$ lies in $\left[-\frac{\pi}{2}, \frac{\pi}{2}\right]$, it follows that

$$\arcsin\left(-\frac{1}{2}\right) = -\frac{\pi}{6}.$$ Angle whose sine is $-\frac{1}{2}$

b. Because $\sin\frac{\pi}{3} = \frac{\sqrt{3}}{2}$, and $\frac{\pi}{3}$ lies in $\left[-\frac{\pi}{2}, \frac{\pi}{2}\right]$, it follows that

$$\sin^{-1}\frac{\sqrt{3}}{2} = \frac{\pi}{3}.$$ Angle whose sine is $\sqrt{3}/2$

c. It is not possible to evaluate $y = \sin^{-1} x$ at $x = 2$ because there is no angle whose sine is 2. Remember that the domain of the inverse sine function is $[-1, 1]$.

Algebra Help

As with the trigonometric functions, much of the work with the inverse trigonometric functions can be done by *exact* calculations rather than by calculator approximations. Exact calculations help to increase your understanding of the inverse functions by relating them to the triangle definitions of the trigonometric functions.

✓ Checkpoint *Audio-video solution in English & Spanish at LarsonPrecalculus.com*

If possible, find the exact value.

a. $\arcsin 1$ **b.** $\sin^{-1}(-2)$

EXAMPLE 2 Graphing the Arcsine Function

Sketch a graph of $y = \arcsin x$ by hand.

Solution

By definition, the equations

$$y = \arcsin x \quad \text{and} \quad \sin y = x$$

are equivalent for $-\pi/2 \le y \le \pi/2$. So, their graphs are the same. For the interval $[-\pi/2, \pi/2]$, you can assign values to y in the equation $\sin y = x$ to make a table of values.

y	$-\frac{\pi}{2}$	$-\frac{\pi}{4}$	$-\frac{\pi}{6}$	0	$\frac{\pi}{6}$	$\frac{\pi}{4}$	$\frac{\pi}{2}$
$x = \sin y$	-1	$-\frac{\sqrt{2}}{2}$	$-\frac{1}{2}$	0	$\frac{1}{2}$	$\frac{\sqrt{2}}{2}$	1

Then plot the points and connect them with a smooth curve. The resulting graph of $y = \arcsin x$ is shown in Figure 4.44. Note that it is the reflection (in the line $y = x$) of the black portion of the graph in Figure 4.43. Be sure you see that Figure 4.44 shows the *entire* graph of the inverse sine function. Remember that the domain of $y = \arcsin x$ is the closed interval $[-1, 1]$ and the range is the closed interval $[-\pi/2, \pi/2]$.

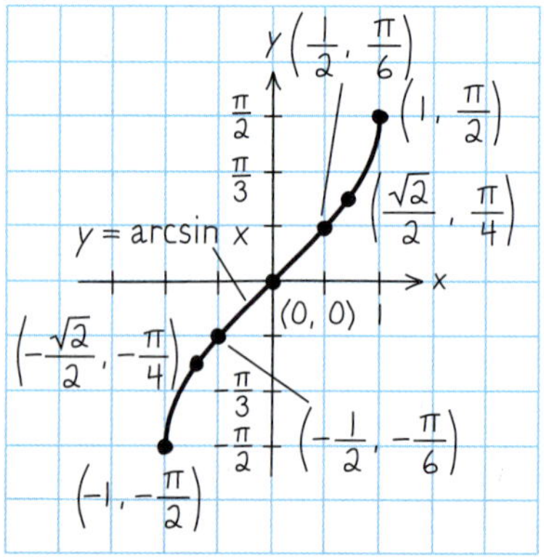

Figure 4.44

✓ Checkpoint *Audio-video solution in English & Spanish at LarsonPrecalculus.com*

Use a graphing utility to graph $f(x) = \sin x$, $g(x) = \arcsin x$, and $y = x$ in the same viewing window to verify geometrically that g is the inverse function of f. (Be sure to restrict the domain of f properly.)

Other Inverse Trigonometric Functions

The cosine function is decreasing and one-to-one on the interval $0 \le x \le \pi$, as shown in the figure.

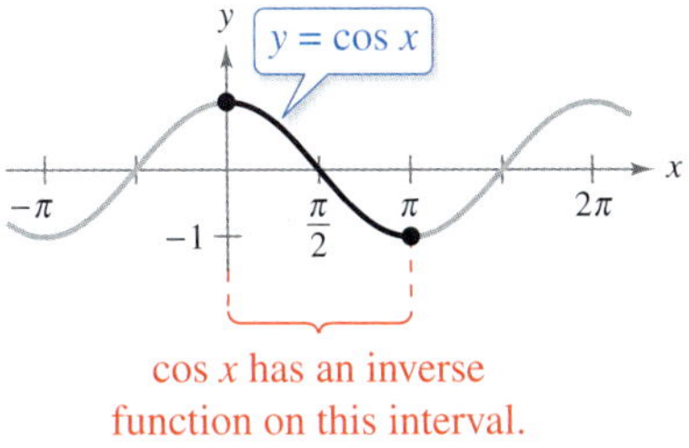
y = cos x
cos x has an inverse function on this interval.

Consequently, on this interval the cosine function has an inverse function—the **inverse cosine function**—denoted by

$y = \arccos x$ or $y = \cos^{-1} x$.

Because $y = \arccos x$ and $x = \cos y$ are equivalent for $0 \le y \le \pi$, their graphs are the same and can be confirmed by the table of values below.

y	0	$\frac{\pi}{6}$	$\frac{\pi}{3}$	$\frac{\pi}{2}$	$\frac{2\pi}{3}$	$\frac{5\pi}{6}$	π
$x = \cos y$	1	$\frac{\sqrt{3}}{2}$	$\frac{1}{2}$	0	$-\frac{1}{2}$	$-\frac{\sqrt{3}}{2}$	-1

Similarly, you can define an **inverse tangent function** by restricting the domain of $y = \tan x$ to the interval $(-\pi/2, \pi/2)$. The inverse tangent function is denoted by

$y = \arctan x$ or $y = \tan^{-1} x$.

Because $y = \arctan x$ and $x = \tan y$ are equivalent for $-\pi/2 < y < \pi/2$, their graphs are the same and can be confirmed by the table of values below.

y	$-\frac{\pi}{4}$	$-\frac{\pi}{6}$	0	$\frac{\pi}{6}$	$\frac{\pi}{4}$
$x = \tan y$	-1	$-\frac{\sqrt{3}}{3}$	0	$\frac{\sqrt{3}}{3}$	1

The list below summarizes the definitions of the three most common inverse trigonometric functions. (Their graphs are shown on page 321.) The remaining three are defined in Exercises 117–119.

Definitions of Inverse Trigonometric Functions

Function	*Domain*	*Range*
$y = \arcsin x$ if and only if $\sin y = x$	$-1 \le x \le 1$	$-\frac{\pi}{2} \le y \le \frac{\pi}{2}$
$y = \arccos x$ if and only if $\cos y = x$	$-1 \le x \le 1$	$0 \le y \le \pi$
$y = \arctan x$ if and only if $\tan y = x$	$-\infty < x < \infty$	$-\frac{\pi}{2} < y < \frac{\pi}{2}$

EXAMPLE 3 Evaluating Inverse Trigonometric Functions

Find the exact value.

a. $\arccos \frac{\sqrt{2}}{2}$

b. $\cos^{-1}(-1)$

c. $\arctan 0$

d. $\tan^{-1}(-1)$

Solution

a. Because $\cos(\pi/4) = \sqrt{2}/2$, and $\pi/4$ lies in $[0, \pi]$, it follows that

$$\arccos \frac{\sqrt{2}}{2} = \frac{\pi}{4}. \qquad \text{Angle whose cosine is } \frac{\sqrt{2}}{2}$$

b. Because $\cos \pi = -1$, and π lies in $[0, \pi]$, it follows that

$$\cos^{-1}(-1) = \pi. \qquad \text{Angle whose cosine is } -1$$

c. Because $\tan 0 = 0$, and 0 lies in $(-\pi/2, \pi/2)$, it follows that

$$\arctan 0 = 0. \qquad \text{Angle whose tangent is } 0$$

d. Because $\tan(-\pi/4) = -1$, and $-\pi/4$ lies in $(-\pi/2, \pi/2)$, it follows that

$$\tan^{-1}(-1) = -\frac{\pi}{4}. \qquad \text{Angle whose tangent is } -1$$

✓ Checkpoint *Audio-video solution in English & Spanish at LarsonPrecalculus.com*

Find the exact value.

a. $\cos^{-1}\left(-\frac{\sqrt{2}}{2}\right)$ **b.** $\arctan \sqrt{3}$

EXAMPLE 4 Calculators and Inverse Trigonometric Functions

Use a calculator to approximate the value, if possible.

a. $\arctan(-8.45)$ **b.** $\sin^{-1} 0.2447$ **c.** $\arccos 2$

Solution

Function	*Mode*	*Graphing Calculator Keystrokes*
a. $\arctan(-8.45)$	Radian	[TAN^{-1}] [(] [(−)] 8.45 [)] [ENTER]

From the display, it follows that $\arctan(-8.45) \approx -1.4530010$.

b. $\sin^{-1} 0.2447$	Radian	[SIN^{-1}] [(] 0.2447 [)] [ENTER]

From the display, it follows that $\sin^{-1} 0.2447 \approx 0.2472103$.

c. $\arccos 2$	Radian	[COS^{-1}] [(] 2 [)] [ENTER]

The calculator should display an *error message* because the domain of the inverse cosine function is $[-1, 1]$.

✓ Checkpoint *Audio-video solution in English & Spanish at LarsonPrecalculus.com*

Use a calculator to approximate the value, if possible.

a. $\arctan 4.84$ **b.** $\arcsin(-1.1)$ **c.** $\arccos(-0.349)$

Remember that the domain of the inverse sine function and the inverse cosine function is $[-1, 1]$, as indicated in Example 4(c).

Algebra Help

In this text, the parentheses in arcsin u are sometimes omitted when u is an expression involving exponents, radicals, products, or quotients. For instance, $\arcsin(2x)$ can be written as $\arcsin 2x$. To evaluate $\arcsin 2x$, find the arcsine of the product $2x$. The other inverse trigonometric functions can be written and evaluated in the same manner.

Technology Tip

You can use the [SIN^{-1}], [COS^{-1}], and [TAN^{-1}] keys on your calculator to approximate values of other inverse trigonometric functions. To evaluate the inverse cosecant function, the inverse secant function, or the inverse cotangent function, you can use the inverse sine, inverse cosine, and inverse tangent functions, respectively. For instance, to evaluate $\sec^{-1} 3.4$, enter the expression as shown below.

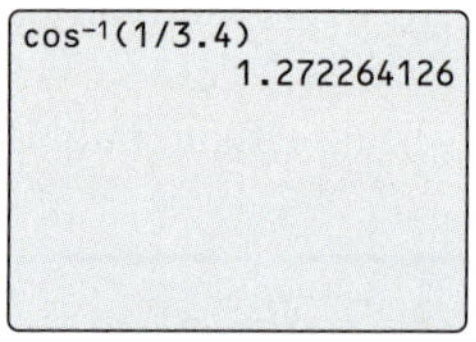

Library of Parent Functions: Inverse Trigonometric Functions

The parent inverse sine function, parent inverse cosine function, and parent inverse tangent function are summarized below and on the inside cover of this text.

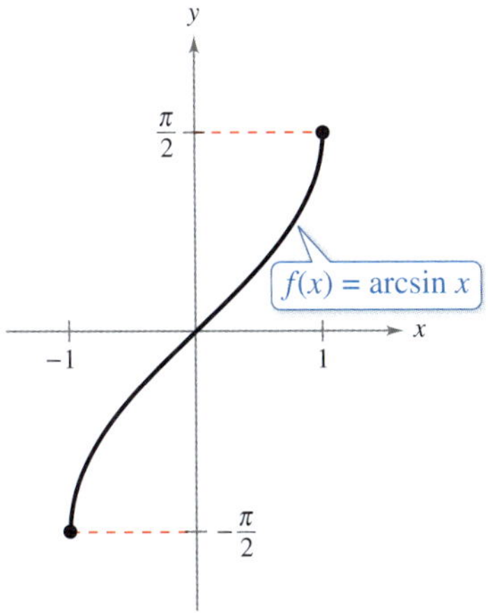

Domain: $[-1, 1]$

Range: $\left[-\frac{\pi}{2}, \frac{\pi}{2}\right]$

Intercept: $(0, 0)$

Odd function

Origin symmetry

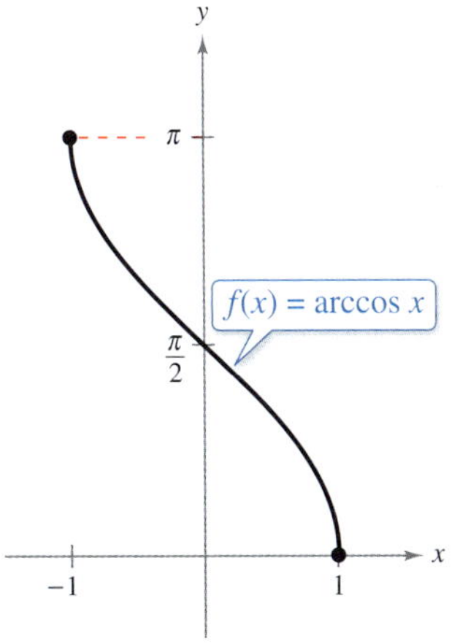

Domain: $[-1, 1]$

Range: $[0, \pi]$

y-intercept: $\left(0, \frac{\pi}{2}\right)$

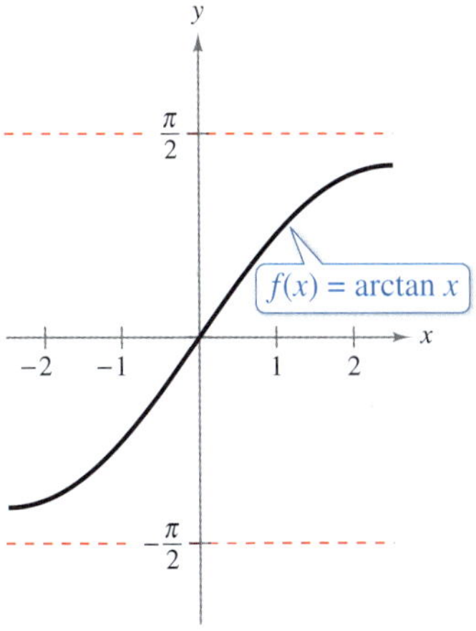

Domain: $(-\infty, \infty)$

Range: $\left(-\frac{\pi}{2}, \frac{\pi}{2}\right)$

Intercept: $(0, 0)$

Horizontal asymptotes: $y = \pm\frac{\pi}{2}$

Odd function

Origin symmetry

EXAMPLE 5 Library of Parent Functions: $f(x) = \arccos x$

See LarsonPrecalculus.com for an interactive version of this type of example.

Note that the graphs of $g(x) = \arccos(x - 2)$ and $h(x) = \arccos(-x)$ are transformations of the graph of $f(x) = \arccos x$.

a. Because $g(x) = \arccos(x - 2) = f(x - 2)$, the graph of g can be obtained by shifting the graph of f two units to the *right*, as shown in Figure 4.45.

b. Because $h(x) = \arccos(-x) = f(-x)$, the graph of h can be obtained by *reflecting* the graph of f in the y-axis, as shown in Figure 4.46.

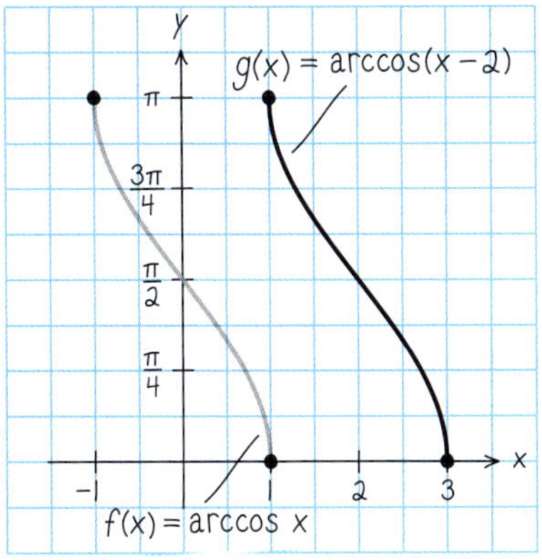

Figure 4.45

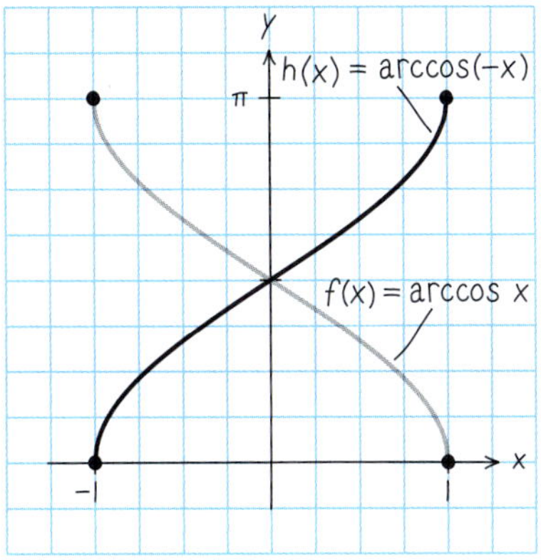

Figure 4.46

✓ ***Checkpoint*** *Audio-video solution in English & Spanish at LarsonPrecalculus.com*

Compare the graph of $g(x) = \arccos(x + 1)$ with the graph of $f(x) = \arccos x$.

Compositions of Functions

Recall from Section 1.6 that for all x in the domains of f and f^{-1}, inverse functions have the properties

$$f(f^{-1}(x)) = x \quad \text{and} \quad f^{-1}(f(x)) = x.$$

Inverse Properties

If $-1 \le x \le 1$ and $-\pi/2 \le y \le \pi/2$, then

$$\sin(\arcsin x) = x \quad \text{and} \quad \arcsin(\sin y) = y.$$

If $-1 \le x \le 1$ and $0 \le y \le \pi$, then

$$\cos(\arccos x) = x \quad \text{and} \quad \arccos(\cos y) = y.$$

If x is a real number and $-\pi/2 < y < \pi/2$, then

$$\tan(\arctan x) = x \quad \text{and} \quad \arctan(\tan y) = y.$$

Keep in mind that these properties do not apply for arbitrary values of x and y. For instance,

$$\arcsin\left(\sin \frac{3\pi}{2}\right) = \arcsin(-1) = -\frac{\pi}{2} \ne \frac{3\pi}{2}.$$

In other words, the property $\arcsin(\sin y) = y$ is not valid for values of y outside the interval $[-\pi/2, \pi/2]$.

EXAMPLE 6 Using Inverse Properties

If possible, find the exact value.

a. $\tan[\arctan(-5)]$ **b.** $\arcsin\left(\sin \frac{5\pi}{3}\right)$ **c.** $\cos(\cos^{-1} \pi)$

Solution

a. Because -5 lies in the domain of the arctangent function, the inverse property applies, and you have

$$\tan[\arctan(-5)] = -5.$$

b. In this case, $5\pi/3$ does not lie within the range of the arcsine function, $-\pi/2 \le y \le \pi/2$. However, $5\pi/3$ is coterminal with

$$\frac{5\pi}{3} - 2\pi = -\frac{\pi}{3}$$

which does lie in the range of the arcsine function, and you have

$$\arcsin\left(\sin \frac{5\pi}{3}\right) = \arcsin\left[\sin\left(-\frac{\pi}{3}\right)\right] = -\frac{\pi}{3}.$$

c. The expression $\cos(\cos^{-1} \pi)$ is not defined because $\cos^{-1} \pi$ is not defined. Remember that the domain of the inverse cosine function is $[-1, 1]$.

Checkpoint Audio-video solution in English & Spanish at LarsonPrecalculus.com

If possible, find the exact value.

a. $\tan[\tan^{-1}(-14)]$ **b.** $\sin^{-1}\left(\sin \frac{7\pi}{4}\right)$ **c.** $\cos(\arccos 0.54)$

Explore the Concept

Use a graphing utility to graph $y = \arcsin(\sin x)$. What are the domain and range of this function? Explain why $\arcsin(\sin 4)$ does not equal 4. Now graph $y = \sin(\arcsin x)$ and determine the domain and range. Explain why $\sin(\arcsin 4)$ is not defined.

elwynn/Shutterstock.com

The algebraic solution of Example 7 shows how to use right triangles to find exact values of compositions of inverse functions.

EXAMPLE 7 Evaluating Compositions of Functions

Find the exact value.

a. $\tan\left(\arccos\frac{2}{3}\right)$ **b.** $\cos\left[\arcsin\left(-\frac{3}{5}\right)\right]$

Algebraic Solution

a. If you let $u = \arccos\frac{2}{3}$, then $\cos u = \frac{2}{3}$. Because the range of the arccosine function is $[0, \pi]$ and $\cos u$ is positive, u is a first-quadrant angle. You can sketch and label angle u as shown in Figure 4.47. Consequently,

$$\tan\left(\arccos\frac{2}{3}\right) = \tan u = \frac{\text{opp}}{\text{adj}} = \frac{\sqrt{5}}{2}.$$

b. If you let $u = \arcsin\left(-\frac{3}{5}\right)$, then $\sin u = -\frac{3}{5}$. Because the range of the arcsine function is $[-\pi/2, \pi/2]$ and $\sin u$ is negative, u is a fourth-quadrant angle. You can sketch and label angle u as shown in Figure 4.48. Consequently,

$$\cos\left[\arcsin\left(-\frac{3}{5}\right)\right] = \cos u = \frac{\text{adj}}{\text{hyp}} = \frac{4}{5}.$$

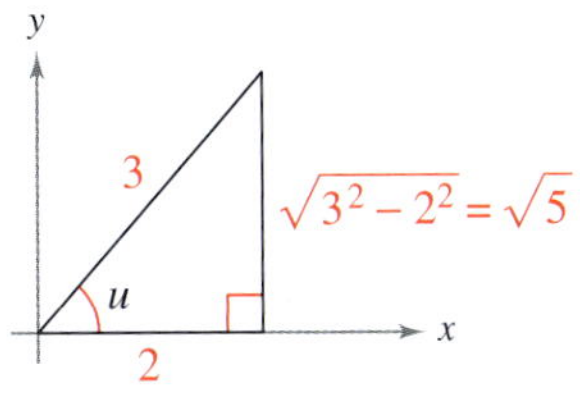

Figure 4.47

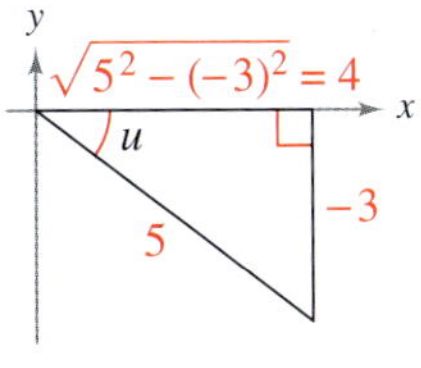

Figure 4.48

Graphical Solution

a. Use a graphing utility set in *radian* mode to graph $y = \tan(\arccos x)$, as shown below.

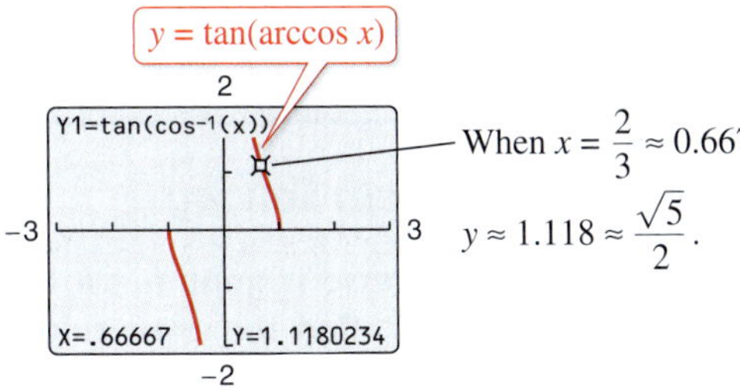

b. Use a graphing utility set in *radian* mode to graph $y = \cos(\arcsin x)$, as shown below.

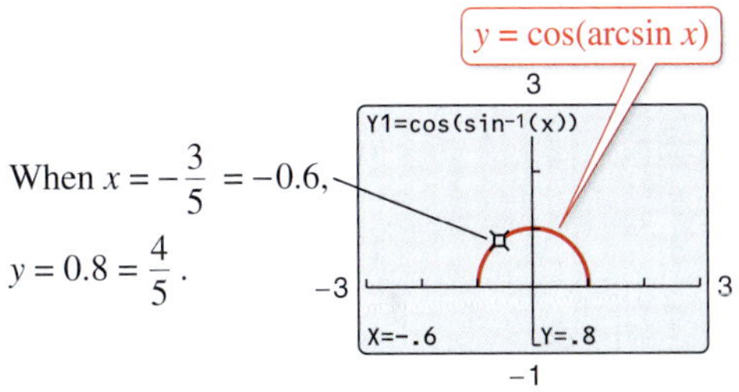

✓ Checkpoint Audio-video solution in English & Spanish at LarsonPrecalculus.com

Find the exact value of $\cos\left[\arctan\left(-\frac{3}{4}\right)\right]$.

EXAMPLE 8 Some Problems from Calculus

Write each trigonometric expression as an algebraic expression in x.

a. $\sin(\arccos 3x),\quad 0 \le x \le \frac{1}{3}$ **b.** $\cot(\arccos 3x),\quad 0 \le x < \frac{1}{3}$

Solution

Let $u = \arccos 3x$. Then $\cos u = 3x$, where $-1 \le 3x \le 1$. Because $\cos u = \text{adj}/\text{hyp} = (3x)/1$, you can sketch a right triangle with acute angle u, as shown in Figure 4.49. From this triangle, you can convert each expression to algebraic form.

a. $\sin(\arccos 3x) = \sin u = \dfrac{\text{opp}}{\text{hyp}} = \sqrt{1 - 9x^2},\quad 0 \le x \le \dfrac{1}{3}$

b. $\cot(\arccos 3x) = \cot u = \dfrac{\text{adj}}{\text{opp}} = \dfrac{3x}{\sqrt{1 - 9x^2}},\quad 0 \le x < \dfrac{1}{3}$

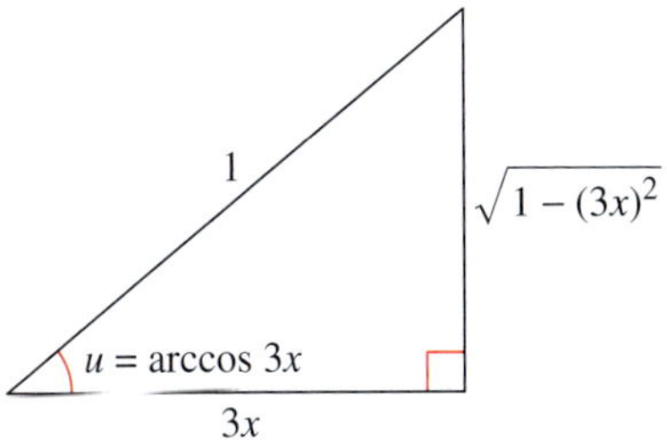

Figure 4.49

✓ Checkpoint Audio-video solution in English & Spanish at LarsonPrecalculus.com

Write $\sec(\arctan x)$ as an algebraic expression in x.

4.7 Exercises

See CalcChat.com for tutorial help and worked-out solutions to odd-numbered exercises. For instructions on how to use a graphing utility, see Appendix A.

Vocabulary and Concept Check

In Exercises 1 and 2, fill in the blanks.

	Function	*Alternative Notation*	*Domain*	*Range*
1.	$y = \arcsin x$	_______	_______	$-\frac{\pi}{2} \leq y \leq \frac{\pi}{2}$
2.	_______	$y = \cos^{-1} x$	$-1 \leq x \leq 1$	_______

3. What notation can you use to represent the inverse tangent function?

4. Does $\arccos x = \dfrac{1}{\cos x}$?

Procedures and Problem Solving

Evaluating an Inverse Trigonometric Function **In Exercises 5–17, find the exact value of the expression, if possible. (Do not use a calculator.)**

5. $\arcsin \frac{1}{2}$

6. $\arcsin 0$

7. $\arccos \frac{\sqrt{3}}{2}$

8. $\arccos(-1)$

9. $\arctan \frac{\sqrt{3}}{3}$

10. $\arctan 1$

11. $\arcsin 3$

12. $\arctan \sqrt{3}$

13. $\arccos\left(-\frac{1}{2}\right)$

14. $\arcsin \frac{\sqrt{2}}{2}$

15. $\tan^{-1}(-\sqrt{3})$

16. $\cos^{-1}(-2)$

17. $\sin^{-1}\left(-\frac{\sqrt{3}}{2}\right)$

18. Graphing the Arccosine Function Consider the function $y = \arccos x$.

(a) Use a graphing utility to complete the table.

x	-1	-0.8	-0.6	-0.4	-0.2
y					

x	0	0.2	0.4	0.6	0.8	1
y						

(b) Plot the points from the table in part (a) and graph the function. (Do not use a graphing utility.)

(c) Use the graphing utility to graph the inverse cosine function and compare the result with your hand-drawn graph in part (b).

(d) Determine any intercepts and symmetry of the graph.

19. Graphing the Arcsine Function Consider the function $y = \arcsin x$.

(a) Use a graphing utility to complete the table.

x	-1	-0.8	-0.6	-0.4	-0.2
y					

x	0	0.2	0.4	0.6	0.8	1
y						

(b) Plot the points from the table in part (a) and graph the function. (Do not use a graphing utility.)

(c) Use the graphing utility to graph the inverse sine function and compare the result with your hand-drawn graph in part (b).

(d) Determine any intercepts and symmetry of the graph.

20. Graphing the Arctangent Function Consider the function $y = \arctan x$.

(a) Use a graphing utility to complete the table.

x	-10	-8	-6	-4	-2
y					

x	0	2	4	6	8	10
y						

(b) Plot the points from the table in part (a) and graph the function. (Do not use a graphing utility.)

(c) Use the graphing utility to graph the inverse tangent function and compare the result with your hand-drawn graph in part (b).

(d) Determine the horizontal asymptotes of the graph.

Finding Missing Coordinates In Exercises 21 and 22, determine the missing coordinates of the points on the graph of the function.

21.

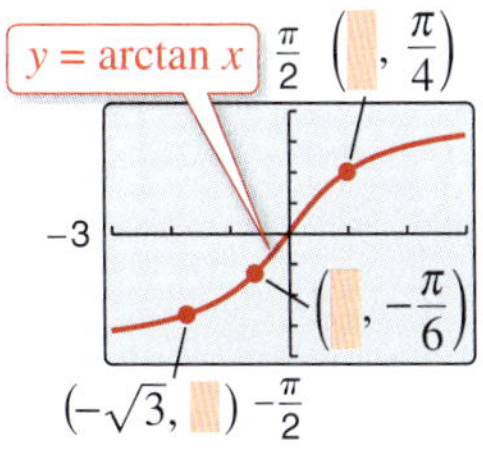

22.

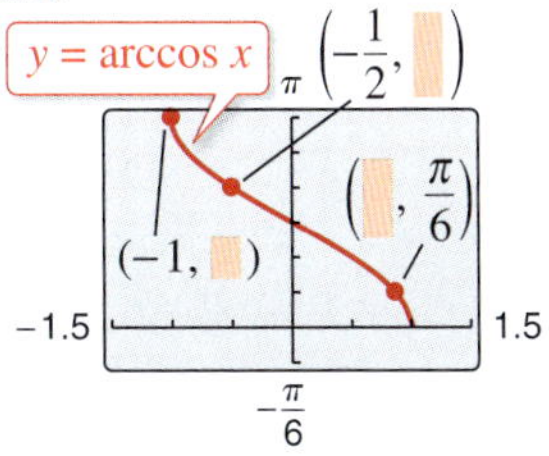

Calculators and Inverse Trigonometric Functions In Exercises 23–38, use a calculator to approximate the value of the expression, if possible. Round your result to two decimal places.

23. arccos 0.37
24. arcsin 0.02
25. $\arcsin(-0.75)$
26. $\arccos(-0.7)$
27. $\arctan(-3)$
28. arctan 25
29. $\sin^{-1} 1.36$
30. $\cos^{-1} 0.26$
31. $\arccos(-0.41)$
32. $\arcsin(-0.125)$
33. arctan 0.92
34. arctan 2.8
35. $\arcsin \frac{7}{8}$
36. $\arccos\left(-\frac{4}{3}\right)$
37. $\tan^{-1}\left(-\frac{95}{7}\right)$
38. $\tan^{-1}\left(-\sqrt{372}\right)$

Library of Parent Functions: $f(x) = \arcsin x$ In Exercises 39–42, compare the graph of the function with the graph of $f(x) = \arcsin x$.

39. $g(x) = \arcsin(x + 3)$
40. $g(x) = \arcsin(x - 4)$
41. $g(x) = \arcsin(-x)$
42. $g(x) = -\arcsin x$

Library of Parent Functions: $f(x) = \arccos x$ In Exercises 43–46, compare the graph of the function with the graph of $f(x) = \arccos x$.

43. $g(x) = \arccos(x + \pi)$
44. $g(x) = \arccos x - 1$
45. $g(x) = \arccos(-x - 2)$
46. $g(x) = \arccos(-x) - 3$

Library of Parent Functions: $f(x) = \arctan x$ In Exercises 47–50, compare the graph of the function with the graph of $f(x) = \arctan x$.

47. $g(x) = \arctan x + 1$
48. $g(x) = \arctan(x - 2)$
49. $g(x) = -\arctan x - 3$
50. $g(x) = \arctan(-x) + 4$

Using an Inverse Trigonometric Function In Exercises 51–54, use an inverse trigonometric function to write θ as a function of x.

51.

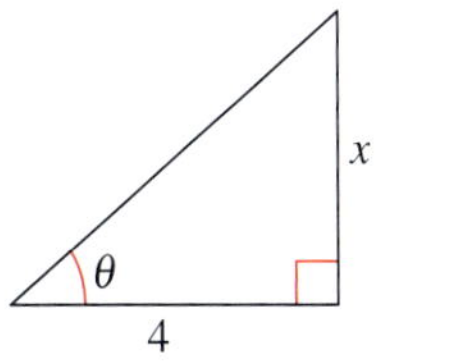

52.

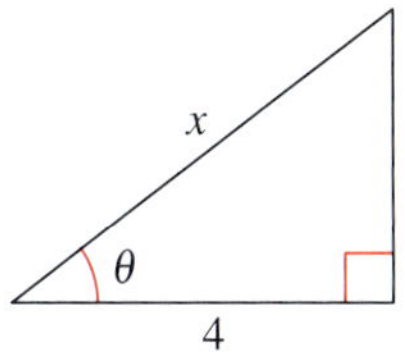

53.

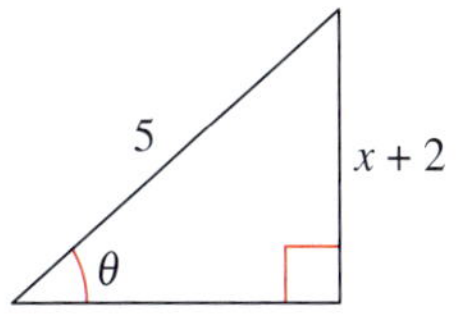

54.

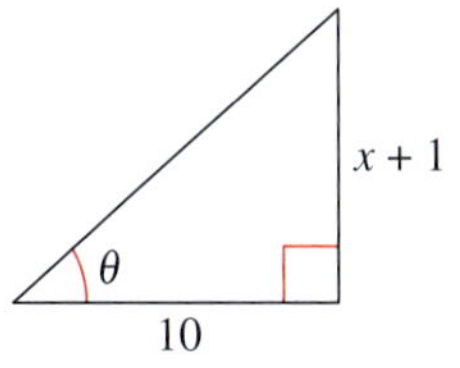

Using an Inverse Trigonometric Function In Exercises 55–58, find the length of the third side of the triangle in terms of x. Then find θ in terms of x for all three inverse trigonometric functions.

55.

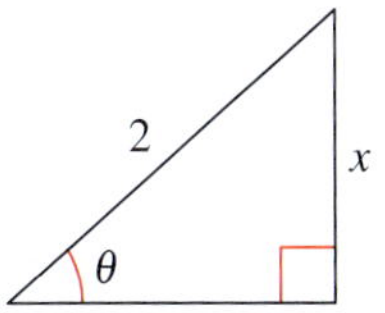

56.

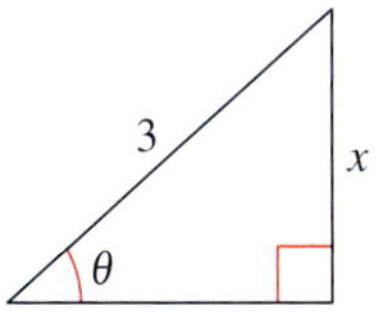

57.

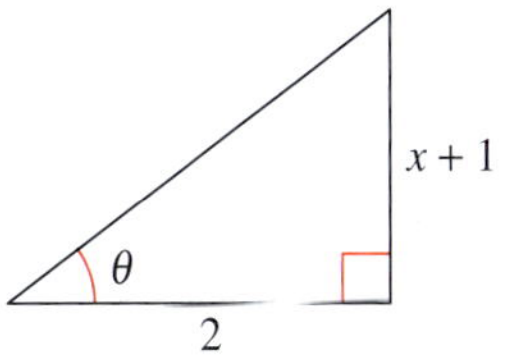

58.

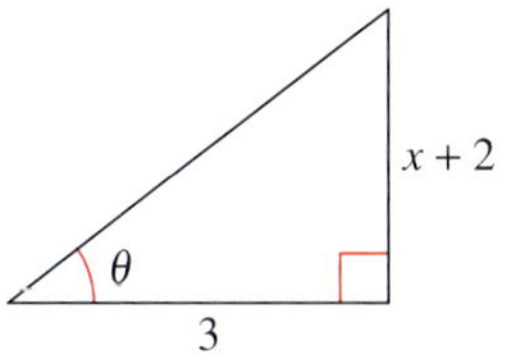

Using Inverse Properties In Exercises 59–68, find the exact value of the expression, if possible.

59. $\sin(\arcsin 0.3)$
60. $\tan(\arctan 45)$
61. $\cos[\arccos(-\sqrt{3})]$
62. $\sin[\arcsin(-0.2)]$
63. $\tan[\tan^{-1}(-0.5)]$
64. $\cos[\cos^{-1}(-2)]$
65. $\arctan\left(\tan \frac{11\pi}{6}\right)$
66. $\arcsin\left(\sin \frac{4\pi}{3}\right)$
67. $\sin^{-1}\left(\tan \frac{5\pi}{4}\right)$
68. $\cos^{-1}\left(\tan \frac{3\pi}{4}\right)$

Evaluating a Composition of Functions In Exercises 69–80, find the exact value of the expression, if possible.

69. $\sin\left(\arctan \frac{3}{4}\right)$
70. $\cos\left(\arcsin \frac{4}{5}\right)$
71. $\cos(\tan^{-1} 2)$
72. $\sin\left(\cos^{-1} \sqrt{5}\right)$
73. $\sec\left(\arcsin \frac{5}{13}\right)$
74. $\csc\left[\arctan\left(-\frac{5}{12}\right)\right]$
75. $\cot\left[\arctan\left(-\frac{3}{5}\right)\right]$
76. $\sec\left[\arccos\left(-\frac{3}{4}\right)\right]$
77. $\tan\left[\arccos\left(-\frac{20}{29}\right)\right]$
78. $\cot\left(\arctan \frac{5}{8}\right)$
79. $\csc\left(\cos^{-1}\frac{\sqrt{3}}{2}\right)$
80. $\tan\left[\sin^{-1}\left(-\frac{\sqrt{2}}{2}\right)\right]$

Writing an Expression In Exercises 81–90, write an algebraic expression that is equivalent to the given expression.

81. $\cos(\arcsin 2x)$
82. $\sin(\arctan 4x)$
83. $\cot(\arctan x)$
84. $\sec(\arctan 3x)$
85. $\sin(\arccos x)$
86. $\csc[\arccos(x - 1)]$
87. $\tan\left(\arccos \dfrac{x}{3}\right)$
88. $\cot\left(\arctan \dfrac{1}{x}\right)$
89. $\csc\left(\arctan \dfrac{x}{a}\right)$
90. $\cos\left(\arcsin \dfrac{x - h}{r}\right)$

Completing an Equation In Exercises 91–94, complete the equation.

91. $\arctan \dfrac{14}{x} = \arcsin(\square), \quad x > 0$

92. $\arcsin \dfrac{\sqrt{36 - x^2}}{6} = \arccos(\square), \quad 0 \le x \le 6$

93. $\arccos \dfrac{3}{\sqrt{x^2 - 2x + 10}} = \arcsin(\square)$

94. $\arccos \dfrac{x - 2}{2} = \arctan(\square), \quad 2 < x < 4$

Graphing an Inverse Trigonometric Function In Exercises 95–100, use a graphing utility to graph the function.

95. $y = 2 \arccos t$
96. $y = \arcsin \dfrac{x}{2}$
97. $f(x) = \arcsin(x - 2)$
98. $g(t) = \arccos(t + 2)$
99. $f(x) = \arctan \dfrac{x}{4}$
100. $f(x) = \arctan 3x$

Using a Trigonometric Identity In Exercises 101 and 102, write the function in terms of the sine function by using the identity

$$A \cos \omega t + B \sin \omega t = \sqrt{A^2 + B^2} \sin\left(\omega t + \arctan \frac{A}{B}\right).$$

Use a graphing utility to graph both forms of the function. What does the graph imply?

101. $f(t) = 3 \cos 2t + 3 \sin 2t$
102. $f(t) = 4 \cos \pi t + 3 \sin \pi t$

Analyzing an Inverse Trigonometric Function In Exercises 103–108, find the value. If not possible, state the reason.

103. As $x \to 1^-$, the value of $\arcsin x \to \square$.
104. As $x \to 1^-$, the value of $\arccos x \to \square$.
105. As $x \to \infty$, the value of $\arctan x \to \square$.
106. As $x \to -1^+$, the value of $\arcsin x \to \square$.
107. As $x \to -1^+$, the value of $\arccos x \to \square$.
108. As $x \to -\infty$, the value of $\arctan x \to \square$.

109. ***Why you should learn it*** *(p. 317)* Different types of granular substances naturally settle at different angles when stored in cone-shaped piles. This angle θ is called the *angle of repose*. When rock salt is stored in a cone-shaped pile 5.5 meters high, the diameter of the pile's base is about 17 meters. (*Source:* Bulk-Store Structures, Inc.)

(a) Draw a diagram that gives a visual representation of the problem. Label all quantities.
(b) Find the angle of repose for rock salt.
(c) How tall is a pile of rock salt that has a base diameter of 25 meters?

110. **MODELING DATA**

A photographer takes a picture of a three-foot painting hanging in an art gallery. The camera lens is 1 foot below the lower edge of the painting (see figure). The angle β subtended by the camera lens x feet from the painting is $\beta = \arctan[3x/(x^2 + 4)]$, $x > 0$.

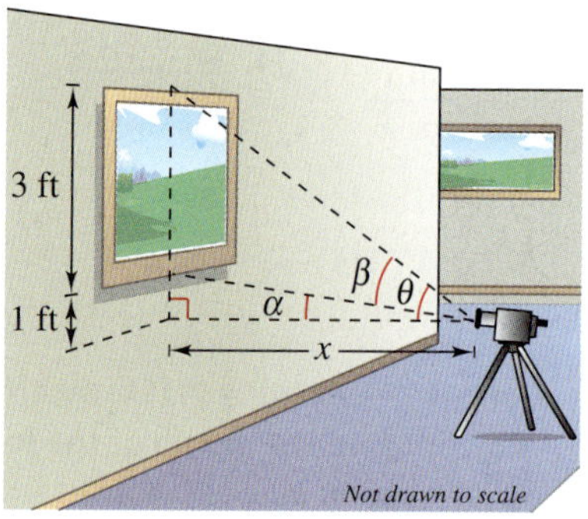

(a) Use a graphing utility to graph β as a function of x.
(b) Use the *trace* feature to approximate the distance from the picture when β is maximum.
(c) Identify the asymptote of the graph and discuss its meaning in the context of the problem.

111. **Angle of Elevation** An airplane flies at an altitude of 6 miles toward a point directly over an observer. Consider θ and x as shown in the figure.

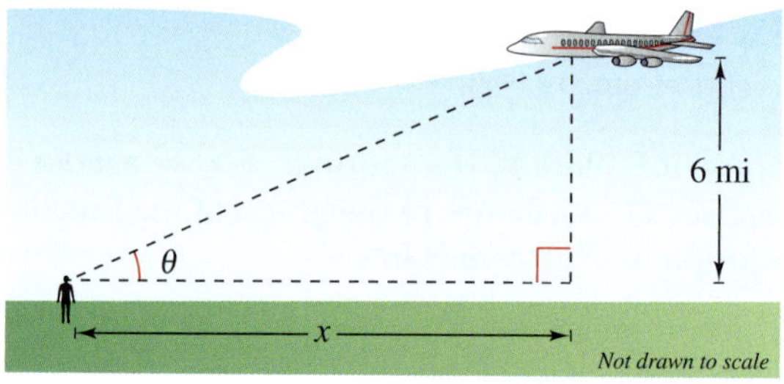

(a) Write θ as a function of x.
(b) Find θ when $x = 10$ miles and $x = 3$ miles.

112. Criminal Justice A police car with its spotlight on is parked 20 meters from a warehouse. Consider θ and x as shown in the figure.

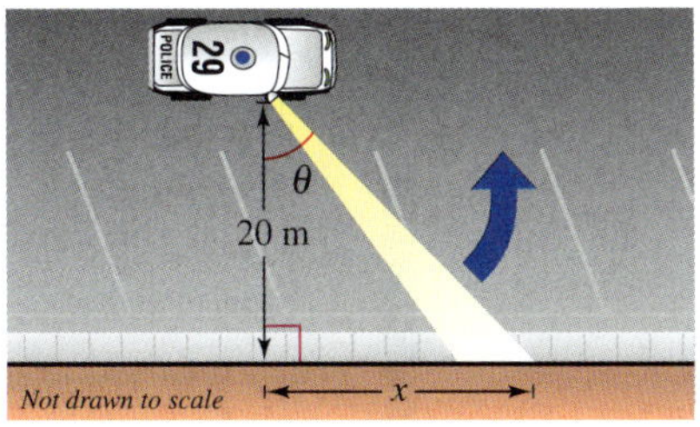

(a) Write θ as a function of x.

(b) Find θ when $x = 5$ meters and $x = 12$ meters.

Focusing on Concepts

True or False? **In Exercises 113–115, determine whether the statement is true or false. Justify your answer.**

113. $\sin \dfrac{5\pi}{6} = \dfrac{1}{2}$ ⇒ $\arcsin \dfrac{1}{2} = \dfrac{5\pi}{6}$

114. $\tan \dfrac{5\pi}{4} = 1$ ⇒ $\arctan 1 = \dfrac{5\pi}{4}$

115. $\arctan x = \dfrac{\arcsin x}{\arccos x}$

116. HOW DO YOU SEE IT? Use the figure below to determine the value(s) of x for which each statement is true.

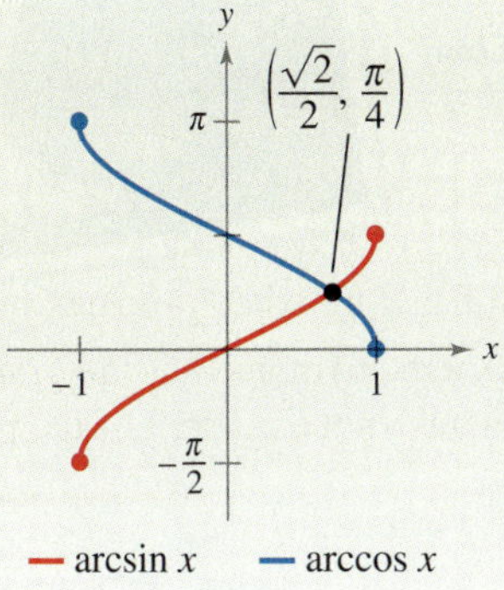

(a) $\arcsin x < \arccos x$

(b) $\arcsin x = \arccos x$

(c) $\arcsin x > \arccos x$

117. Inverse Cotangent Function Define the inverse cotangent function by restricting the domain of the cotangent function to the interval $(0, \pi)$, and sketch the graph of the inverse function.

118. Inverse Secant Function Define the inverse secant function by restricting the domain of the secant function to the intervals $[0, \pi/2)$ and $(\pi/2, \pi]$, and sketch the graph of the inverse function.

119. Inverse Cosecant Function Define the inverse cosecant function by restricting the domain of the cosecant function to the intervals $[-\pi/2, 0)$ and $(0, \pi/2]$, and sketch the graph of the inverse function.

120. Writing Use the results of Exercises 117–119 to explain how to graph (a) the inverse cotangent function, (b) the inverse secant function, and (c) the inverse cosecant function on a graphing utility.

Evaluating an Inverse Trigonometric Function **In Exercises 121–124, use the results of Exercises 117–119 to evaluate the expression without using a calculator.**

121. $\operatorname{arcsec} \sqrt{2}$

122. $\operatorname{arcsec} 1$

123. $\operatorname{arccot}(-\sqrt{3})$

124. $\operatorname{arccsc} 2$

Proof **In Exercises 125–127, prove the identity.**

125. $\arcsin(-x) = -\arcsin x$

126. $\arctan(-x) = -\arctan x$

127. $\arcsin x + \arccos x = \dfrac{\pi}{2}$

128. Finding the Area of a Plane Region In calculus, it is shown that the area of the region bounded by the graphs of $y = 0$, $y = 1/(x^2 + 1)$, $x = a$, and $x = b$ is given by

$$\text{Area} = \arctan b - \arctan a$$

(see figure). Find the area for each value of a and b.

(a) $a = 0, b = 1$

(b) $a = -1, b = 1$

(c) $a = 0, b = 3$

(d) $a = -1, b = 3$

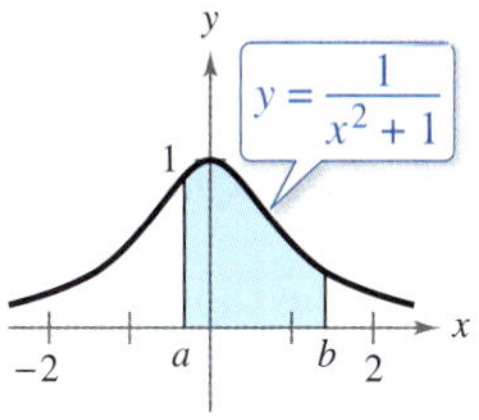

Cumulative Mixed Review

Simplifying a Radical Expression **In Exercises 129–132, simplify the radical expression.**

129. $\dfrac{4}{4\sqrt{2}}$

130. $\dfrac{2}{\sqrt{3}}$

131. $\dfrac{2\sqrt{3}}{6}$

132. $\dfrac{5\sqrt{5}}{2\sqrt{10}}$

Evaluating Trigonometric Functions **In Exercises 133–136, sketch a right triangle corresponding to the trigonometric function of the acute angle θ. Use the Pythagorean Theorem to determine the third side and then find the values of the other five trigonometric functions of θ.**

133. $\sin \theta = \frac{5}{6}$

134. $\tan \theta = 2$

135. $\cos \theta = \frac{3}{4}$

136. $\sec \theta = 3$

4.8 Applications and Models

Applications Involving Right Triangles

In this section, the three angles of a right triangle are denoted by A, B, and C (where C is the right angle), and the lengths of the sides opposite these angles by a, b, and c, respectively (where c is the hypotenuse).

What you should learn

- Solve real-life problems involving right triangles.
- Solve real-life problems involving directional bearings.
- Solve real-life problems involving harmonic motion.

Why you should learn it

You can use trigonometric functions to model and solve real-life problems. For instance, Exercise 37 on page 335 shows you how a trigonometric function can be used to model the length of the shadow of the Sundial Bridge in Redding, California.

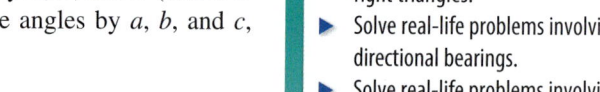

EXAMPLE 1 Solving a Right Triangle

See LarsonPrecalculus.com for an interactive version of this type of example.

Solve the right triangle for all unknown sides and angles.

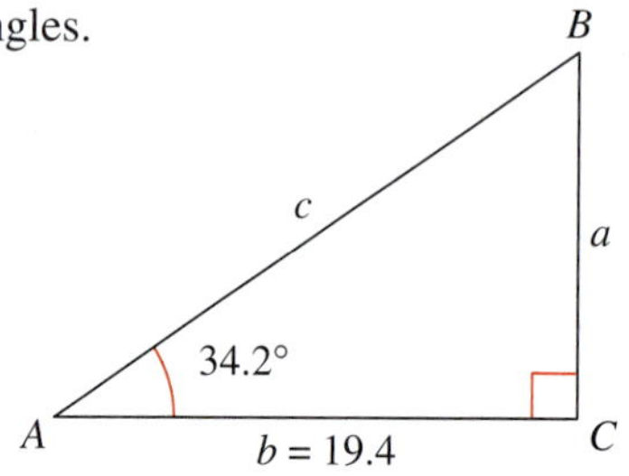

Solution

Because $C = 90°$, it follows that

$$A + B = 90°$$

and

$$B = 90° - 34.2° = 55.8°.$$

To solve for a, use the fact that

$$\tan A = \frac{\text{opp}}{\text{adj}} = \frac{a}{b} \quad \Rightarrow \quad a = b \tan A.$$

So, $a = 19.4 \tan 34.2° \approx 13.18$. Similarly, to solve for c, use the fact that

$$\cos A = \frac{\text{adj}}{\text{hyp}} = \frac{b}{c} \quad \Rightarrow \quad c = \frac{b}{\cos A}.$$

So, $c = \dfrac{19.4}{\cos 34.2°} \approx 23.46.$

✓ **Checkpoint** *Audio-video solution in English & Spanish at LarsonPrecalculus.com*

Solve the right triangle for all unknown sides and angles.

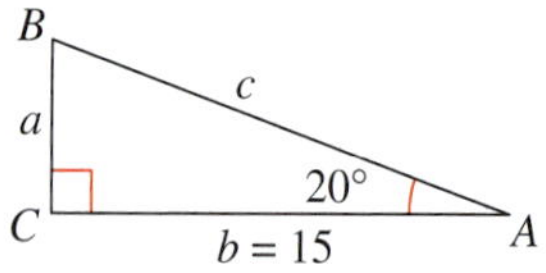

EXAMPLE 2 Finding a Side of a Right Triangle

A safety regulation states that the maximum angle of elevation for a rescue ladder is 72°. A fire department's longest ladder is 110 feet. What is the maximum safe rescue height?

Solution

A sketch is shown in Figure 4.50. From the equation $\sin A = a/c$, it follows that

$$a = c \sin A = 110 \sin 72° \approx 104.6.$$

So, the maximum safe rescue height is about 104.6 feet above the height of the fire truck.

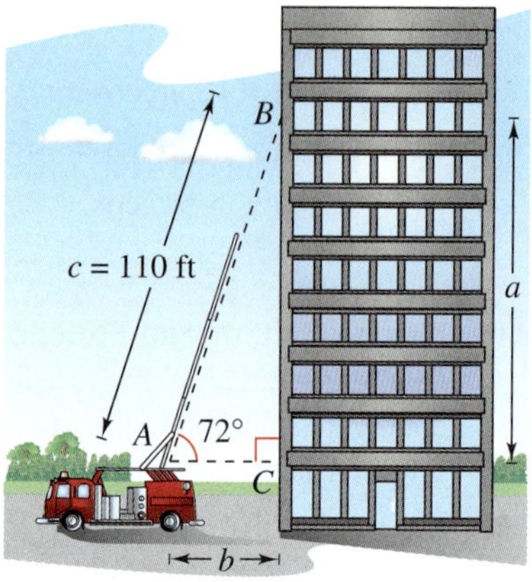

Figure 4.50

✓ **Checkpoint** *Audio-video solution in English & Spanish at LarsonPrecalculus.com*

A ladder that is 16 feet long leans against the side of a house. The angle of elevation of the ladder is 80°. Find the height from the top of the ladder to the ground. ■

Andrew Zarivny/Shutterstock.com

EXAMPLE 3 Finding a Side of a Right Triangle

At a point 200 feet from the base of a building, the angle of elevation to the *bottom* of a smokestack is 35°, and the angle of elevation to the *top* is 53°, as shown in Figure 4.51. Find the height s of the smokestack alone.

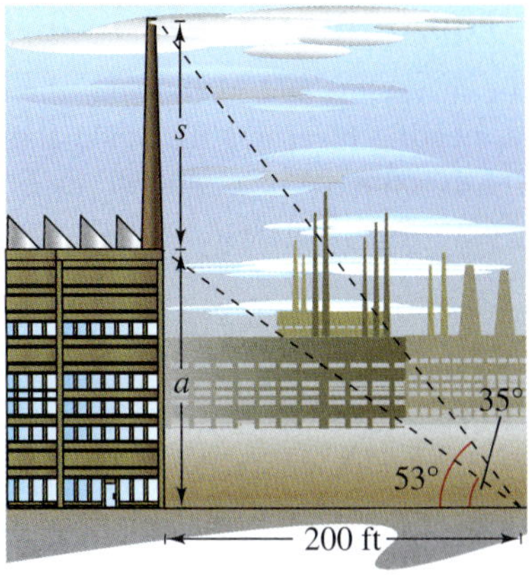

Figure 4.51

Solution

This problem involves two right triangles. For the smaller right triangle, use the fact that

$$\tan 35° = \frac{a}{200}$$

to conclude that the height of the building is

$$a = 200 \tan 35°.$$

For the larger right triangle, use the equation

$$\tan 53° = \frac{a + s}{200}$$

to conclude that

$$a + s = 200 \tan 53°.$$

So, the height of the smokestack is

$$\begin{aligned} s &= 200 \tan 53° - a \\ &= 200 \tan 53° - 200 \tan 35° \\ &\approx 125.4 \text{ feet.} \end{aligned}$$

Checkpoint *Audio-video solution in English & Spanish at LarsonPrecalculus.com*

At a point 65 feet from the base of a church, the angles of elevation to the bottom of the steeple and the top of the steeple are 35° and 43°, respectively. Find the height of the steeple.

EXAMPLE 4 Finding an Angle of Depression

A swimming pool is 20 meters long and 12 meters wide. The bottom of the pool is slanted so that the water depth is 1.3 meters at the shallow end and 4 meters at the deep end, as shown in Figure 4.52. Find the angle of depression (in degrees) of the bottom of the pool.

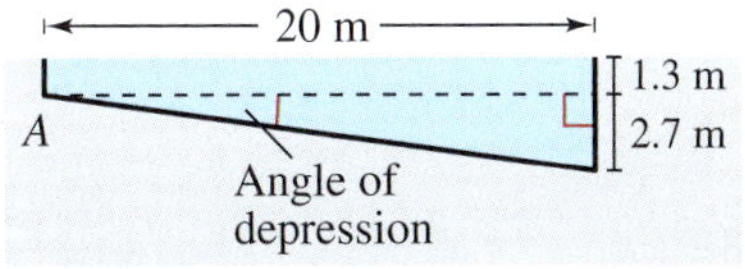

Figure 4.52

Solution

Using the tangent function, you see that

$$\tan A = \frac{\text{opp}}{\text{adj}} = \frac{2.7}{20} = 0.135.$$

So, the angle of depression is

$$A = \arctan 0.135 \approx 0.13419 \text{ radian} \approx 7.69°.$$

Checkpoint *Audio-video solution in English & Spanish at LarsonPrecalculus.com*

From the time a small airplane is 100 feet high and 1600 ground feet from its landing runway, the plane descends in a straight line to the runway. Determine the angle of descent (in degrees) of the plane.

Insight

To solve Examples 2–4, you needed to apply your knowledge of right triangle trigonometric relationships and inverse trigonometric functions. You will need a firm grasp of these concepts to solve similar application problems that may occur on standardized tests.

Trigonometry and Bearings

In surveying and navigation, directions are generally given in terms of **bearings.** A bearing measures the acute angle that a path or line of sight makes with a fixed north-south line, as shown in Figure 4.53. For instance, the bearing S 35° E in Figure 4.53(a) means 35 degrees east of south.

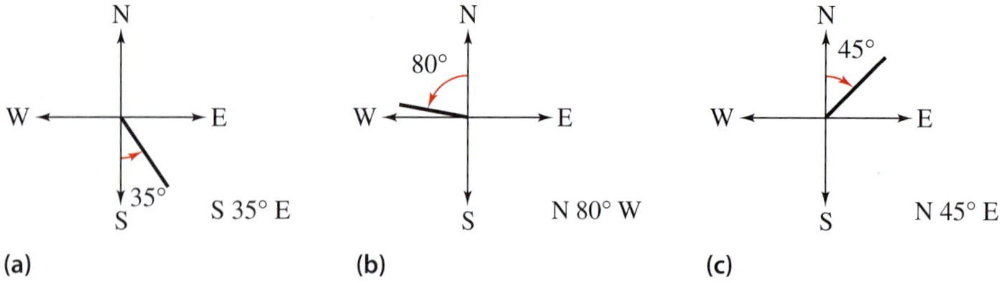

Figure 4.53

> **Algebra Help**
>
> In *air navigation*, bearings are measured in degrees *clockwise* from north. Examples of air navigation bearings are shown below.

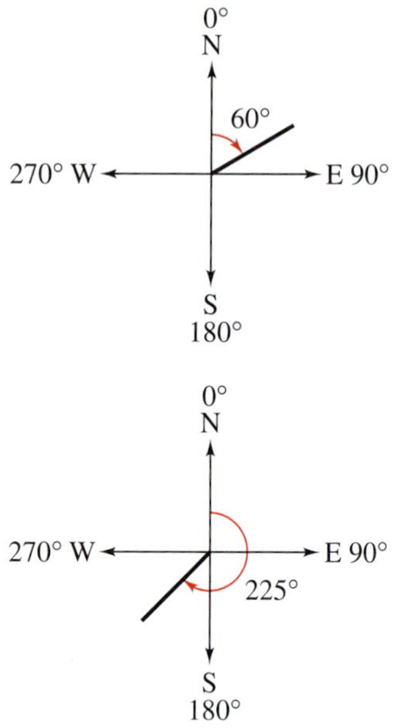

EXAMPLE 5 Finding Directions in Terms of Bearings

A ship leaves port at noon and heads due west at 20 knots, or 20 nautical miles (nm) per hour. At 2 P.M. the ship changes course to N 54° W, as shown in the figure. Find the ship's bearing and distance from the port of departure at 3 P.M.

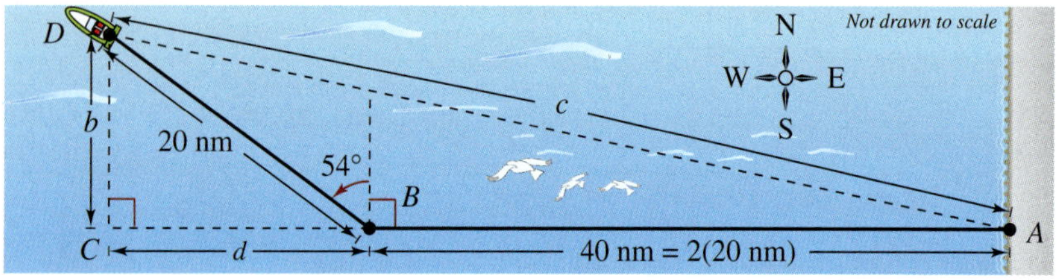

Solution

For triangle BCD, you have $B = 90° - 54° = 36°$. The two sides of this triangle are

$$b = 20 \sin 36° \qquad \text{and} \qquad d = 20 \cos 36°.$$

To find angle A in triangle ACD, note that

$$\tan A = \frac{b}{d + 40} = \frac{20 \sin 36°}{20 \cos 36° + 40}$$

which yields

$$A = \arctan \frac{20 \sin 36°}{20 \cos 36° + 40} \approx 11.82°. \qquad \text{Use a calculator in degree mode.}$$

The angle with the north-south line is $90° - 11.82° = 78.18°$. So, the bearing of the ship is N 78.18° W. Finally, from triangle ACD, you have

$$\sin A = \frac{b}{c}$$

which yields

$$c = \frac{b}{\sin A} = \frac{20 \sin 36°}{\sin 11.82°} \approx 57.39 \text{ nautical miles.} \qquad \text{Distance from port}$$

✓ Checkpoint *Audio-video solution in English & Spanish at LarsonPrecalculus.com*

A sailboat leaves a pier heading due west at 8 knots. After 15 minutes, the sailboat changes its course to N 16° W and changes its speed to 10 knots. Find the sailboat's bearing and distance from the pier after 12 minutes on this course.

Harmonic Motion

The periodic nature of the trigonometric functions is useful for describing the motion of a point on an object that vibrates, oscillates, rotates, or is moved by wave motion.

For example, consider a ball that is bobbing up and down on the end of a spring, as shown in Figure 4.54. Assume that 10 centimeters is the maximum distance the ball moves vertically upward or downward from its equilibrium (at rest) position. Assume further that the time it takes for the ball to move from its maximum displacement above zero to its maximum displacement below zero and back again is

$t = 4$ seconds.

With the ideal conditions of perfect elasticity and no friction or air resistance, the ball would continue to move up and down in a uniform and regular manner.

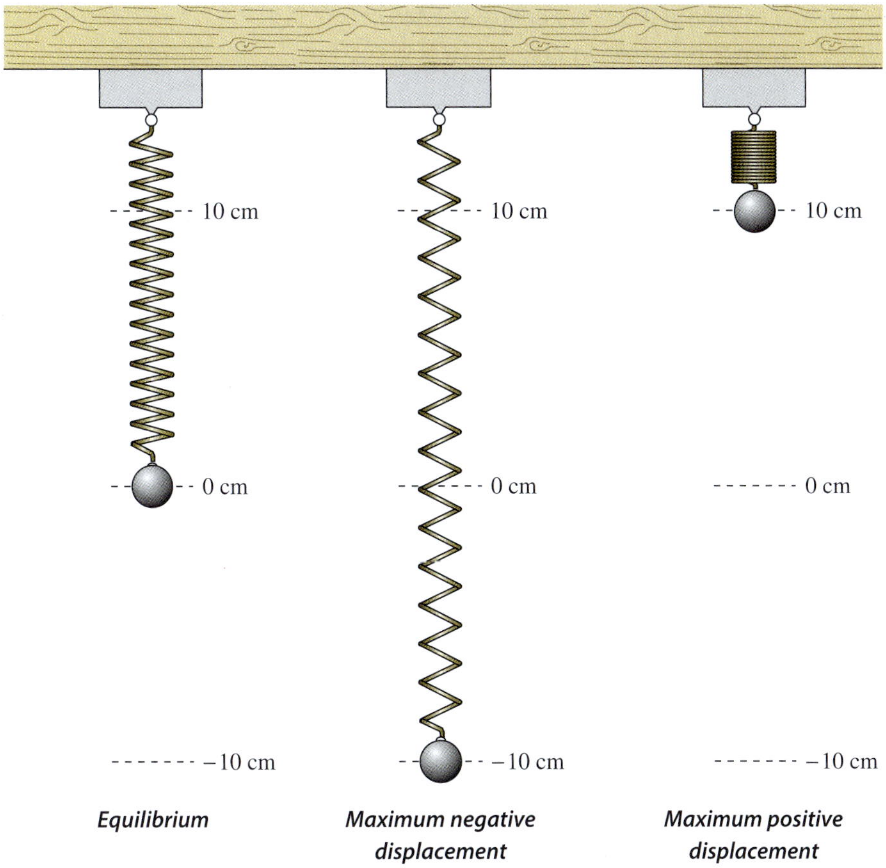

Figure 4.54

Physics Professor

From this spring, you can conclude that the period (time for one complete cycle) of the motion is

Period $= 4$ seconds

its amplitude (maximum displacement from equilibrium) is

Amplitude $= 10$ centimeters

and its **frequency** (number of cycles per second) is

Frequency $= \frac{1}{4}$ cycle per second.

Motion of this nature can be described by a sine or cosine function and is called **simple harmonic motion.**

EHStock/iStock/Getty Images

Definition of Simple Harmonic Motion

A point that moves on a coordinate line is in **simple harmonic motion** when its distance d from the origin at time t is given by either

$$d = a \sin \omega t \qquad \text{or} \qquad d = a \cos \omega t$$

where a and ω are real numbers such that $\omega > 0$. The motion has

$$\text{amplitude } |a|, \qquad \text{period } \frac{2\pi}{\omega}, \qquad \text{and frequency } \frac{\omega}{2\pi}.$$

Algebra Help

Note that the frequency is the reciprocal of the period.

EXAMPLE 6 Simple Harmonic Motion

Write an equation for the simple harmonic motion of the ball described in Figure 4.54, where the period is 4 seconds. Verify that the frequency is $\frac{1}{4}$ cycle per second.

Solution

Because the spring is at equilibrium ($d = 0$) when $t = 0$, use the equation $d = a \sin \omega t$. Moreover, because the maximum displacement from zero is 10, you know that

$$\text{Amplitude} = |a| = 10.$$

Also, because the period is 4 seconds, you find ω as shown.

$$\text{Period} = \frac{2\pi}{\omega} \Rightarrow 4 = \frac{2\pi}{\omega} \Rightarrow \omega = \frac{\pi}{2}$$

Consequently, an equation for the motion of the ball is

$$d = 10 \sin \frac{\pi}{2} t.$$

Note that the choice of $a = 10$ or $a = -10$ depends on whether the ball initially moves up or down. The frequency is

$$\text{Frequency} = \frac{\omega}{2\pi} = \frac{\pi/2}{2\pi} = \frac{1}{4} \text{ cycle per second.}$$

Checkpoint Audio-video solution in English & Spanish at LarsonPrecalculus.com

Find a model for simple harmonic motion that satisfies the following conditions: $d = 0$ when $t = 0$, the amplitude is 6 centimeters, and the period is 3 seconds. Then find the frequency.

One illustration of the relationship between sine waves and harmonic motion is the wave motion that results when a stone is dropped into a calm pool of water. The waves move outward in roughly the shape of sine (or cosine) waves, as shown in Figure 4.55. As an example, assume you are fishing and your fishing bobber is attached so that it does not move horizontally. As the waves move outward from the dropped stone, your fishing bobber will move up and down in simple harmonic motion, as shown in Figure 4.56.

Figure 4.55

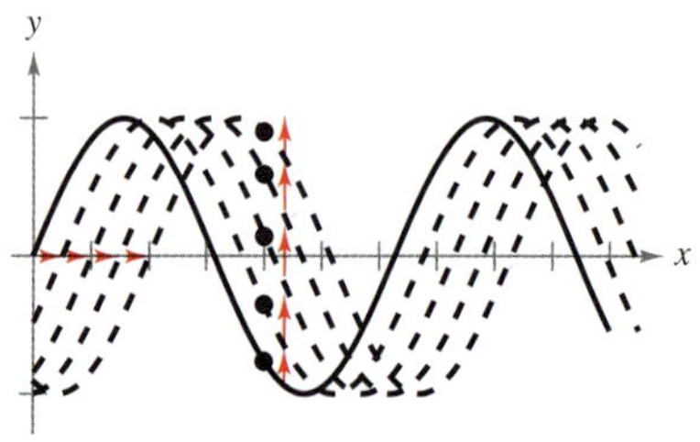

Figure 4.56

EXAMPLE 7 Simple Harmonic Motion

Consider the equation for simple harmonic motion

$$d = 6 \cos \frac{3\pi}{4} t.$$

Find (a) the maximum displacement, (b) the frequency, (c) the value of d when $t = 4$, and (d) the least positive value of t for which $d = 0$.

Algebraic Solution

The equation has the form $d = a \cos \omega t$, with

$$a = 6 \quad \text{and} \quad \omega = \frac{3\pi}{4}.$$

a. The maximum displacement (from the point of equilibrium) is the amplitude. So, the maximum displacement is 6.

b. $\text{Frequency} = \dfrac{\omega}{2\pi}$

$$= \frac{3\pi/4}{2\pi}$$

$$= \frac{3}{8} \text{ cycle per unit of time}$$

c. $d = 6 \cos\left[\dfrac{3\pi}{4}(4)\right]$

$$= 6 \cos 3\pi$$

$$= 6(-1)$$

$$= -6$$

d. To find the least positive value of t for which $d = 0$, solve the equation

$$6 \cos \frac{3\pi}{4} t = 0.$$

First, divide each side by 6 to obtain

$$\cos \frac{3\pi}{4} t = 0.$$

This equation is satisfied when

$$\frac{3\pi}{4} t = \frac{\pi}{2}, \frac{3\pi}{2}, \frac{5\pi}{2}, \ldots.$$

Multiply these values by

$$\frac{4}{3\pi}$$

to obtain

$$t = \tfrac{2}{3}, 2, \tfrac{10}{3}, \ldots.$$

So, the least positive value of t is $t = \frac{2}{3}$.

Graphical Solution

Use a graphing utility in *radian* mode to graph

$$d = 6 \cos \frac{3\pi}{4} t.$$

a.

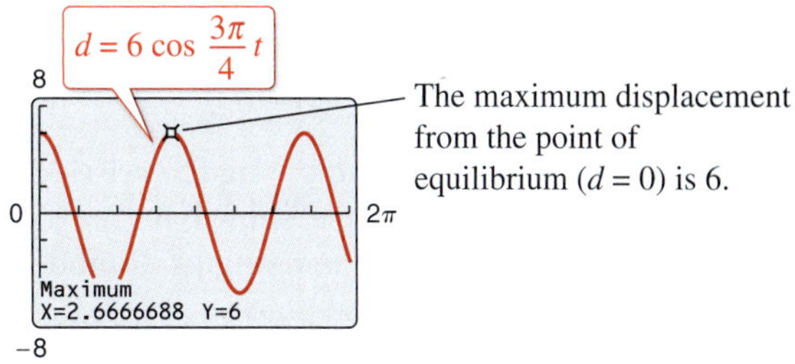
d = 6 cos 3π/4 t
8
0
2π
−8
Maximum
X=2.6666688 Y=6
The maximum displacement from the point of equilibrium (d = 0) is 6.

b.

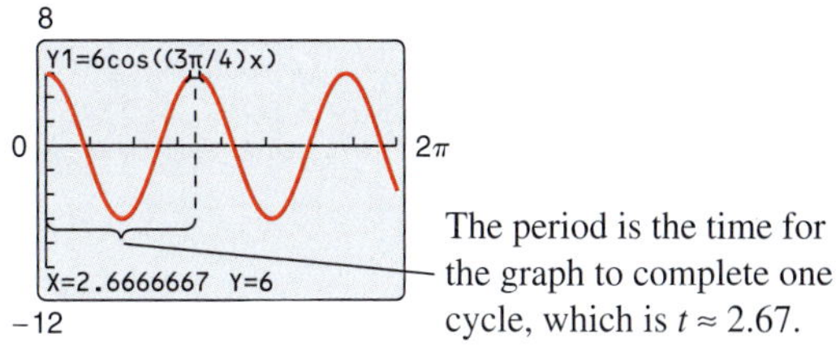
8
0
2π
−12
Y1=6cos((3π/4)x)
X=2.6666667 Y=6
The period is the time for the graph to complete one cycle, which is t ≈ 2.67.

So, the frequency is

$$\text{Frequency} \approx \frac{1}{2.67} \approx 0.37 \text{ cycle per unit of time.}$$

c.

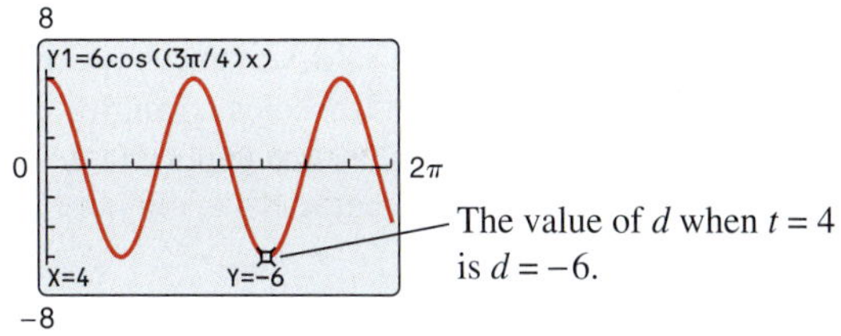
8
0
2π
−8
Y1=6cos((3π/4)x)
X=4 Y=-6
The value of d when t = 4 is d = −6.

d.

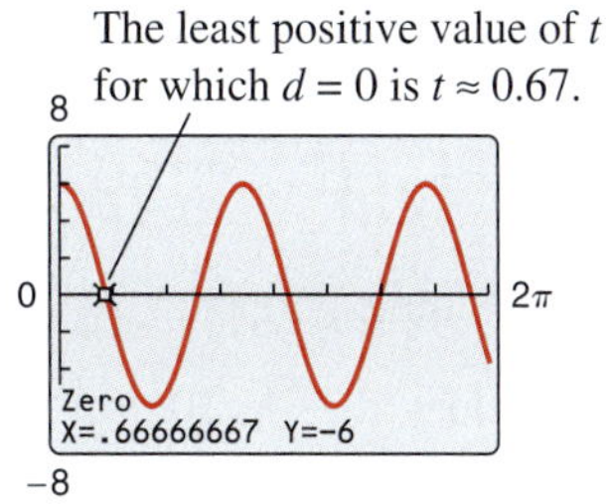
The least positive value of t for which d = 0 is t ≈ 0.67.
8
0
2π
−8
Zero
X=.66666667 Y=-6

✓ **Checkpoint** ▶ *Audio-video solution in English & Spanish at LarsonPrecalculus.com*

Rework Example 7 for the equation $d = 4 \cos 6\pi t$.

4.8 Exercises

See *CalcChat.com* for tutorial help and worked-out solutions to odd-numbered exercises. For instructions on how to use a graphing utility, see Appendix A.

Vocabulary and Concept Check

In Exercises 1 and 2, fill in the blank.

1. A point that moves on a coordinate line is in simple _______ when its distance from the origin at time t is given by either $d = a \sin \omega t$ or $d = a \cos \omega t$.

2. A _______ measures the acute angle a path or line of sight makes with a fixed north-south line.

3. Draw a figure to represent the bearing S 30° W.

4. What is the amplitude of the simple harmonic motion described by $d = 3 \sin \frac{\pi}{2} t$?

Procedures and Problem Solving

Solving a Right Triangle **In Exercises 5–14, solve the right triangle shown in the figure for all unknown sides and angles. Round your answers to two decimal places.**

5. $A = 60°, c = 12$

6. $B = 25°, b = 4$

7. $B = 71°, b = 14$

8. $A = 7.4°, a = 20.5$

9. $a = 3, b = 4$

10. $a = 25, b = 37$

11. $a = 16, c = 54$

12. $b = 1.32, c = 18.9$

13. $A = 12° 15', c = 430.5$

14. $B = 65° 12', a = 145.5$

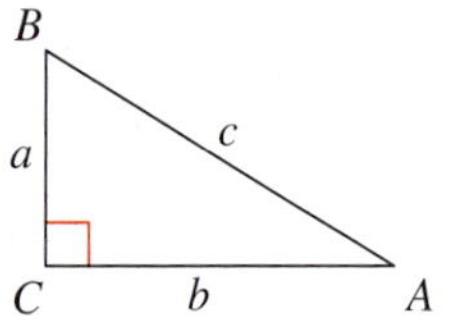

Finding an Altitude **In Exercises 15–20, find the altitude of the isosceles triangle shown in the figure. Round your answers to two decimal places.**

15. $\theta = 32°, b = 8$

16. $\theta = 22°, b = 14$

17. $\theta = 10.5°, b = 5.4$

18. $\theta = 48.5°, b = 13$

19. $\theta = 71° 38', b = 16$

20. $\theta = 62° 23'45'', b = 3.1$

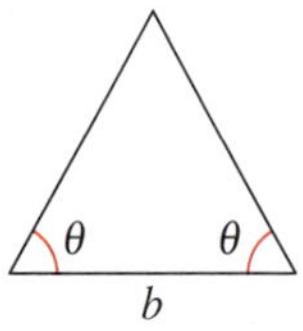

21. Home Maintenance A ladder that is 20 feet long leans against the side of a house. The angle of elevation of the ladder is 75°. Find the height from the top of the ladder to the ground.

22. Electrical Maintenance An electrician is running wire from the electric box on a house to a utility pole 75 feet away. The angle of elevation to the connection on the pole is 16°. How much wire does the electrician need?

23. ROTC A cadet rappelling down a cliff on a rope needs help. A cadet on the ground pulls tight on the end of the rope that hangs down from the rappelling cadet to lock the cadet in place. The length of the rope between the two cadets is 120 feet, and the angle of elevation of the rope is 66°. The cadet on the ground is holding the rope at a height of 4 feet. How high above the ground is the cadet on the rope?

24. Interior Design A display case has the shape of an isosceles triangle. It stands one foot tall and the angle of elevation from its base to its side is 40°. What is the width of the case at its base?

25. Sonar Mapping The sonar of a research vessel detects a coral reef that is 125 feet from the vessel. The angle between the water level and the coral reef is 51° (see figure). How deep is the coral reef?

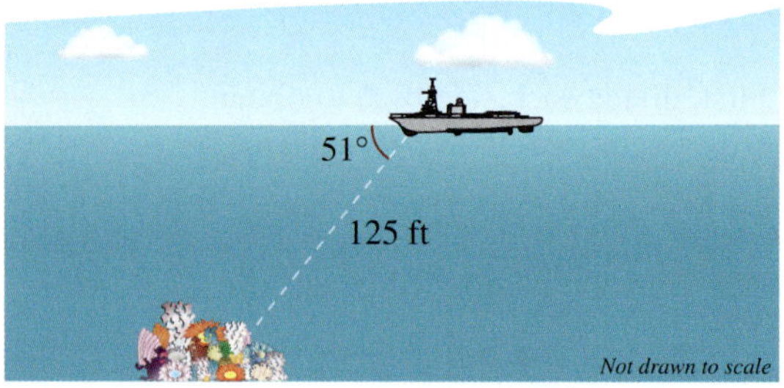

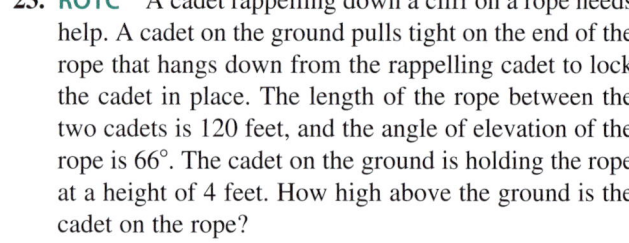

26. Topography A sign on the roadway at the top of a mountain indicates that for the next 4 miles the grade is 9.5° (see figure). Find the change in elevation for a car descending the 4-mile stretch.

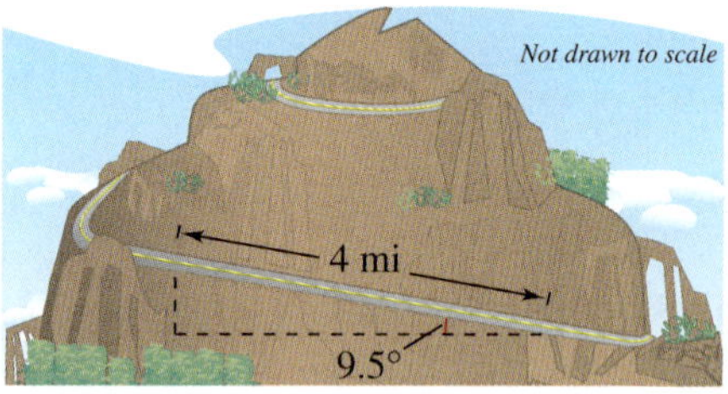

27. Drone Elevation From a point 25 meters in front of an office building, the angle of elevation to the roof is 15° and the angle of elevation to a hovering drone is 21° 30′ (see figure). Use a trigonometric function to write an equation involving the unknown distance. Then find the distance between the roof and the drone.

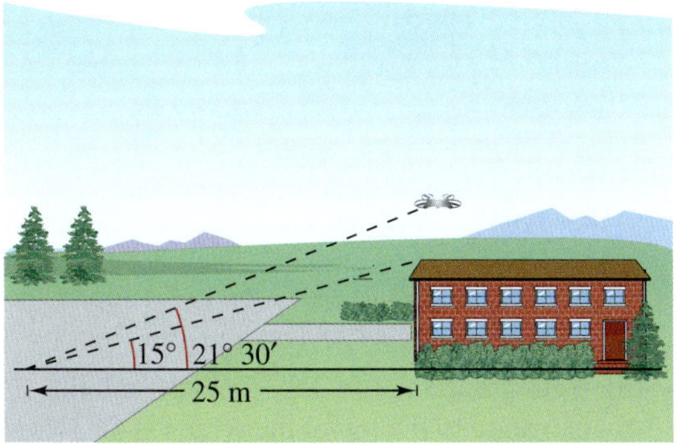

28. Architecture From a point 50 feet in front of a church, the angles of elevation to the base of the steeple and the top of the steeple are 37° and 49° 40′, respectively.

(a) Draw right triangles that give a visual representation of the problem. Label the known and unknown quantities.

(b) Use a trigonometric function to write an equation involving the unknown quantity.

(c) Find the height of the steeple.

29. Marine Transportation An observer in a lighthouse 350 feet above sea level observes two ships in the same vertical plane as the lighthouse. The angles of depression to the ships are 4° and 6.5° (see figure). How far apart are the ships?

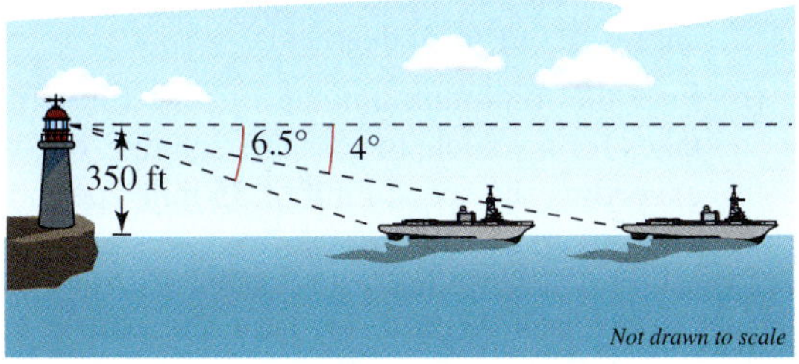

30. Aviation A passenger in an airplane flying at an altitude of 10 kilometers sees two towns due east of the plane. The angles of depression to the towns are 28° and 55° (see figure). How far apart are the towns?

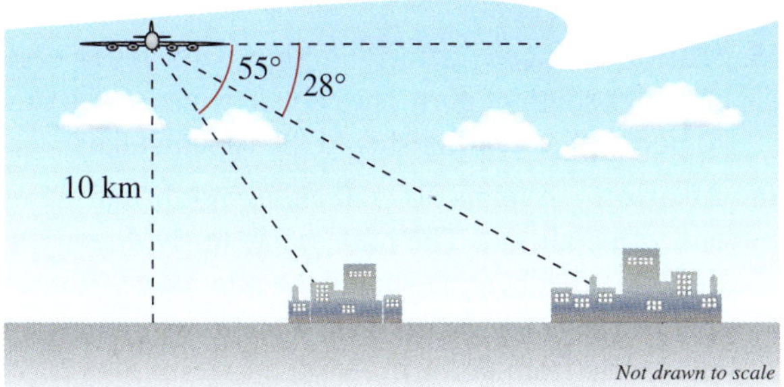

31. Aviation The angle of elevation to a plane approaching your home is 16°. One minute later, it is 57°. You assume that the speed of the plane is 550 miles per hour. Approximate the altitude of the plane.

32. Topography While traveling across flat land, you notice a mountain directly in front of you. The angle of elevation to the peak is 2.5°. After you drive 18 miles closer to the mountain, the angle of elevation is 10°. Approximate the height of the mountain.

33. Communications A Global Positioning System satellite orbits 12,500 miles above Earth's surface (see figure). Find the angle of depression from the satellite to the horizon. Assume the radius of Earth is 4000 miles.

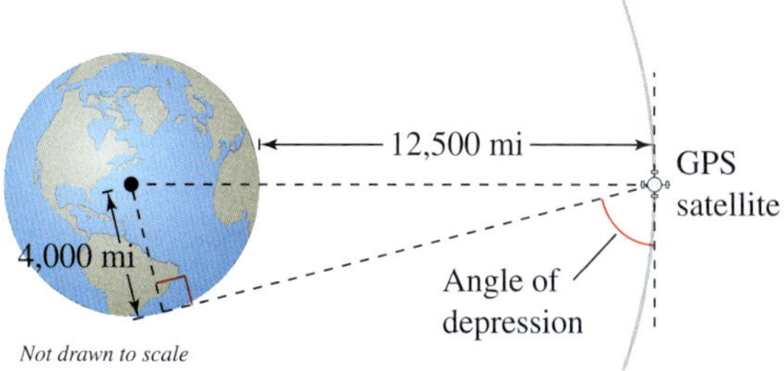

34. Marine Transportation Find the angle of depression from the top of a lighthouse 250 feet above water level to the water line of a ship $2\frac{1}{2}$ miles offshore.

35. Aviation When an airplane leaves the runway, its angle of climb is 18° and its speed is 275 feet per second. Find the plane's altitude after 1 minute.

36. Aviation How long will it take the plane in Exercise 35 to climb to an altitude of 10,000 feet? 16,000 feet?

37. *Why you should learn it* *(p. 328)* The Sundial Bridge in Redding, California, is supported by cables attached to a 217-foot sundial that leans backward at a 42° angle (see figure).

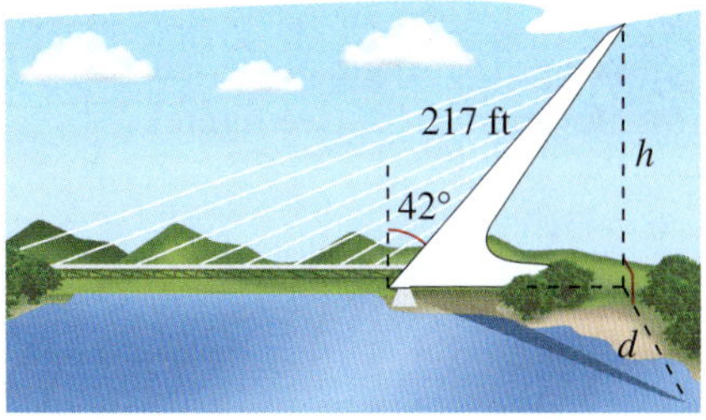

(a) Determine the height h of the sundial.

(b) The distance d shows how far the shadow extends horizontally from the tip of the sundial. Draw the right triangle formed by d and h. Label the angle of elevation of the sun as θ.

(c) Write d as a function of θ.

(d) Find d when $\theta = 10°$, $\theta = 20°$, $\theta = 30°$, $\theta = 40°$, and $\theta = 50°$.

(e) As the angle measure increases in equal increments, does the distance also increase in equal increments? Explain.

38. Basketball The height of an outdoor basketball backboard is $12\frac{1}{2}$ feet, and the backboard casts a shadow $17\frac{1}{3}$ feet long.

(a) Draw a right triangle that gives a visual representation of the problem. Label the known and unknown quantities.

(b) Use a trigonometric function to write an equation involving the unknown angle of elevation.

(c) Find the angle of elevation of the sun.

39. Finding a Height Before a parade, you are holding one of the tethers attached to the top of a character balloon that will be in the parade. The balloon is upright, and the bottom is floating approximately 20 feet above ground level. You are standing approximately 100 feet from the balloon (see figure).

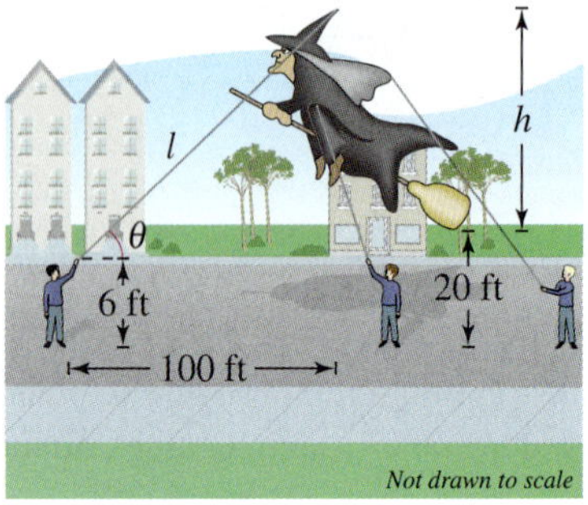

(a) Find an expression for the length l of the tether in terms of h, the height of the balloon from top to bottom.

(b) Find an expression for the angle of elevation θ from you to the top of the balloon.

(c) The angle of elevation to the top of the balloon is 35°. Find the height h.

40. Parks and Recreation The designers of a park are creating a water slide and have sketched a preliminary drawing. The length of the stairs is 30 feet, and its angle of elevation is 45° (see figure).

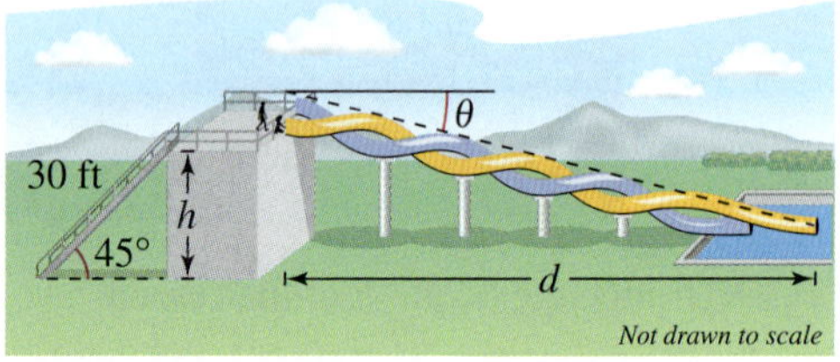

(a) Find the height h of the slide.

(b) Find the angle of depression θ from the top of the slide to the end of the slide at the ground in terms of the horizontal distance d a rider travels.

(c) The designers want the angle of depression of the slide to be at least 25° and at most 30°. Find an interval for how far a rider travels horizontally.

41. Cinema A park is showing a movie on the lawn. The base of the screen is 6 feet off the ground and the screen is 22 feet high (see figure).

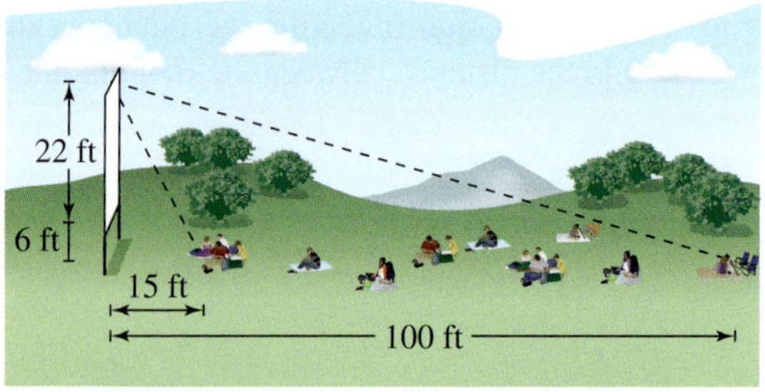

(a) Find the angles of elevation to the top of the screen from distances of 15 feet and 100 feet.

(b) You are lying on the ground and the angle of elevation to the top of the screen is 42°. How far are you from the screen?

42. Speed Enforcement A police department has set up a speed enforcement zone on a straight length of highway. A patrol car is parked parallel to the zone, 200 feet from one end and 150 feet from the other end (see figure).

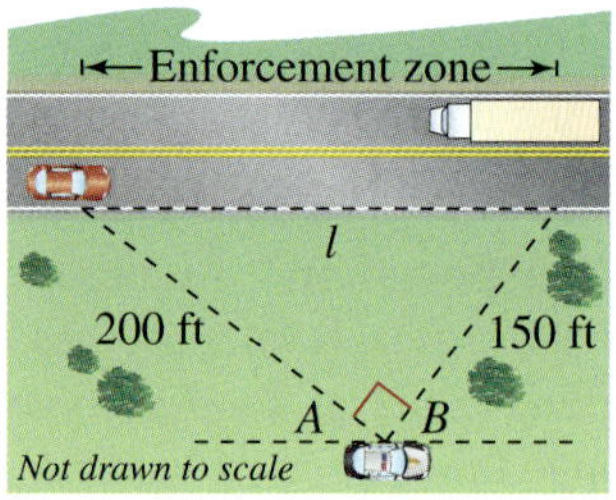

(a) Find the length l of the zone and the measures of the angles A and B (in degrees).

(b) Find the minimum amount of time (in seconds) it takes for a vehicle to pass through the zone without exceeding the speed limit of 35 miles per hour.

Finding Directions in Terms of Bearings In Exercises 43 and 44, find the ship's bearing and distance from the port at 4 P.M.

43. A ship leaves port at noon and heads due east at 10 knots. At 3 P.M., the ship changes course to N 34° E.

44. A ship leaves port at 1 P.M. and heads due north at 8 knots. At 2 P.M., the ship changes course to N 12° W.

45. Marine Transportation A ship is 45 miles east and 30 miles south of port. The captain wants to sail directly to port. What bearing should the captain take?

46. Aviation A plane is 160 miles north and 85 miles east of an airport. The pilot wants to fly directly to the airport. What bearing should the pilot take? (Remember that bearings for air navigation are measured in degrees clockwise from north.)

47. Geomatics A surveyor wants to find the distance across a pond. The bearing from A to B is N 32° W. The surveyor walks 50 meters from A to C. At C, the bearing to B is N 68° W. (See figure.)

(a) Find the bearing from A to C.

(b) Find the distance from A to B.

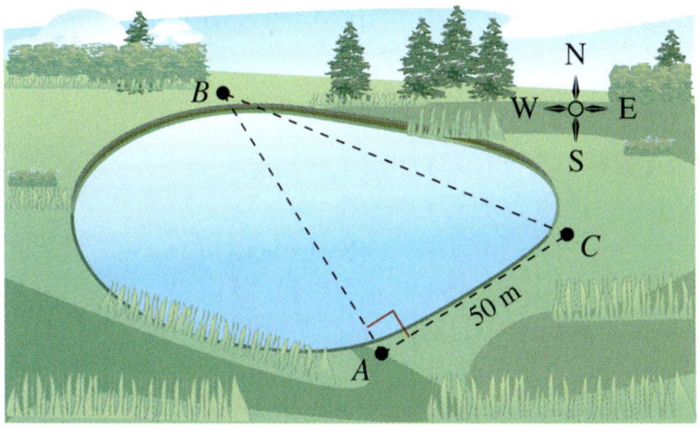

48. Forestry Fire tower A is 30 kilometers due west of fire tower B. A fire is spotted from the towers, and the bearings to the fire from A and B are N 14° E and N 34° W, respectively (see figure). Find the shortest distance d of the fire from the line segment AB.

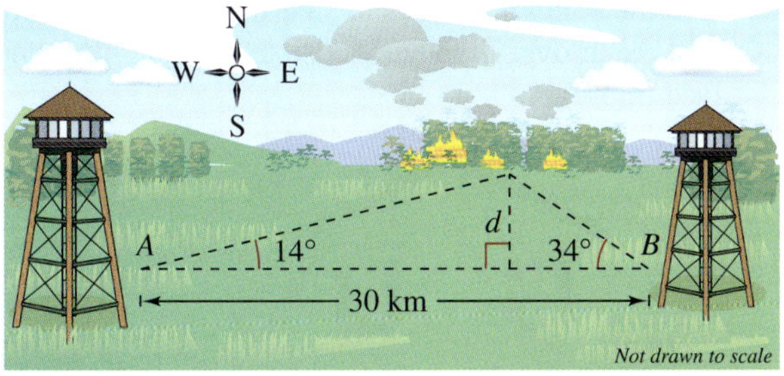

Geometry In Exercises 49 and 50, find the angle α between the two nonvertical lines L_1 and L_2 (assume L_1 and L_2 are not perpendicular). The angle α satisfies the equation

$$\tan \alpha = \left| \frac{m_2 - m_1}{1 + m_2 m_1} \right|$$

where m_1 and m_2 are the slopes of L_1 and L_2, respectively.

49. $L_1: 3x - 2y = 5$
$L_2: x + y = 1$

50. $L_1: 2x + y = 8$
$L_2: x - 5y = -4$

51. Geometry Determine the angle between the diagonal of a cube and the diagonal of its base, as shown in the figure.

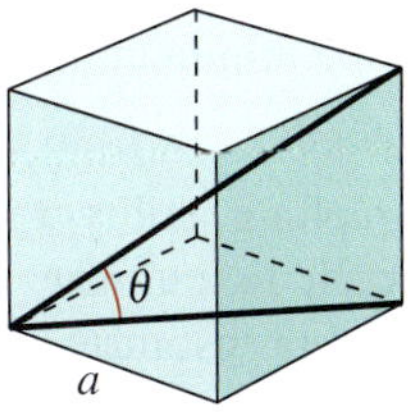

52. Geometry Determine the angle between the diagonal of a cube and its edge, as shown in the figure.

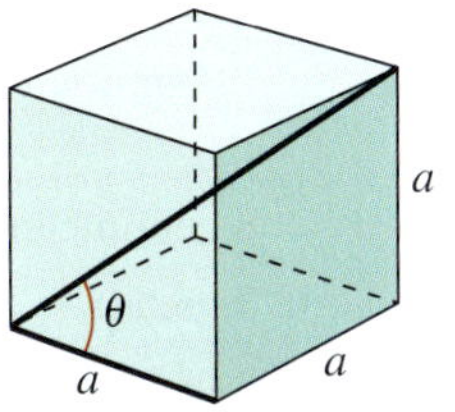

53. Mechanical Engineering Write the distance y from one side of a hexagonal nut to the opposite side as a function of r, as shown in the figure.

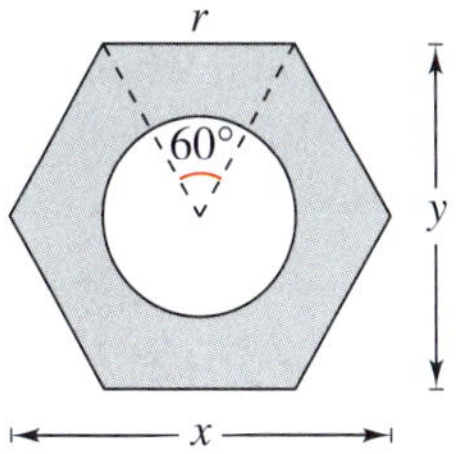

54. Mechanical Engineering The figure shows a circular piece of sheet metal of diameter 40 centimeters. The piece contains 12 equally spaced bolt holes. Determine the straight-line distance between the centers of two consecutive bolt holes.

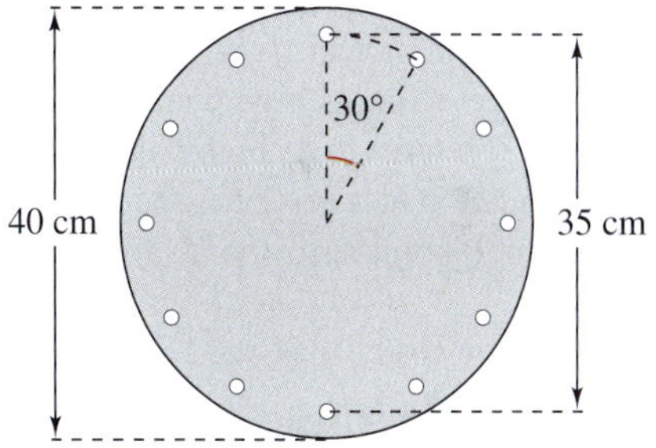

Trusses In Exercises 55 and 56, find the lengths of the indicated members of the truss.

55. Find a and b.

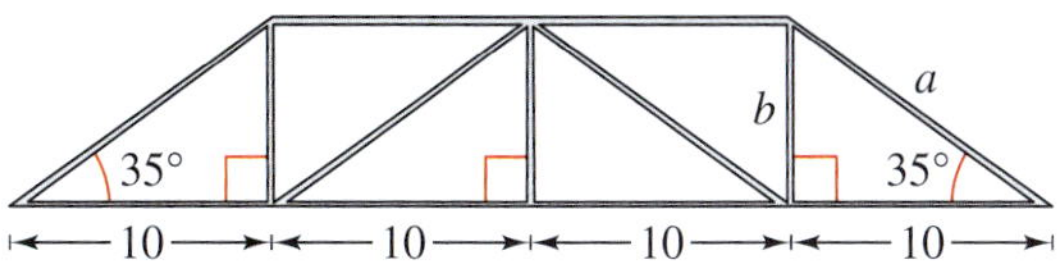

56. Find a, b, and c.

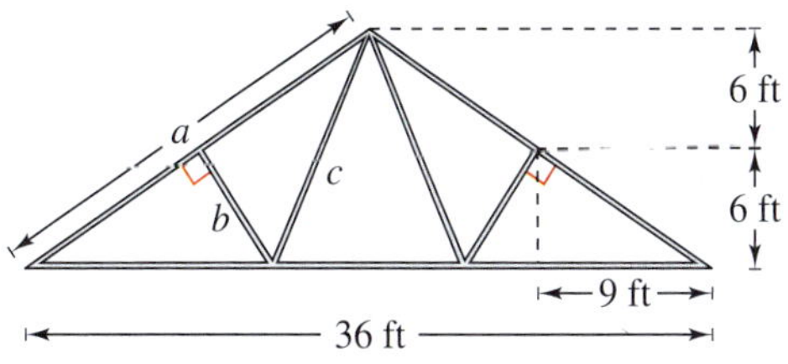

Harmonic Motion In Exercises 57–60, find a model for simple harmonic motion satisfying the specified conditions.

	Displacement $(t = 0)$	*Amplitude*	*Period*
57.	0	8 centimeters	2 seconds
58.	0	3 meters	6 seconds
59.	3 inches	3 inches	1.5 seconds
60.	-2 feet	2 feet	10 seconds

Harmonic Motion In Exercises 61–64, for the simple harmonic motion described by the trigonometric function, find (a) the maximum displacement, (b) the frequency, (c) the value of d when $t = 5$, and (d) the least positive value of t for which $d = 0$. Use a graphing utility to verify your results.

61. $d = 4 \cos 8\pi t$

62. $d = -9.2 \cos 7\pi t$

63. $d = -\frac{1}{16} \sin 140\pi t$

64. $d = \frac{1}{64} \sin 792\pi t$

65. Music A point on the end of a tuning fork moves in the simple harmonic motion described by

$d = a \sin \omega t.$

A tuning fork for middle C has a frequency of 264 vibrations per second. Find ω.

66. Harmonic Motion A buoy oscillates in simple harmonic motion as waves go past. The buoy moves a total of 3.5 feet from its high point to its low point (see figure), and it returns to its high point every 10 seconds. Write an equation that describes the motion of the buoy, where the high point corresponds to the time $t = 0$.

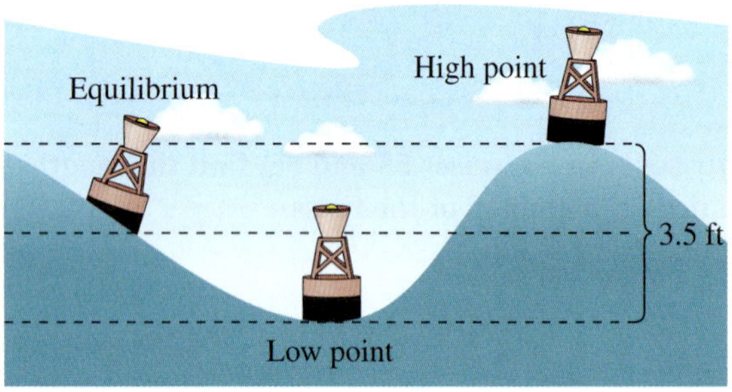

67. Harmonic Motion A ball that is bobbing up and down on the end of a spring has a maximum displacement of 3 inches. Its motion (in ideal conditions) is modeled by

$y = \frac{1}{4} \cos 16t, \quad t > 0$

where y is measured in feet and t is the time in seconds.

(a) Use a graphing utility to graph the function.

(b) What is the period of the oscillations?

(c) Determine the first time the ball passes the point of equilibrium ($y = 0$).

68. Climatology The numbers of hours H of daylight in Denver, Colorado, on the 15th of each month are given by ordered pairs of the form $(t, H(t))$, where $t = 1$ represents January. A model for the data is

$$H(t) = 12.18 + 2.81 \sin\left(\frac{\pi t}{6} - \frac{\pi}{2}\right).$$

(Spreadsheet at LarsonPrecalculus.com) (*Source:* United States Navy)

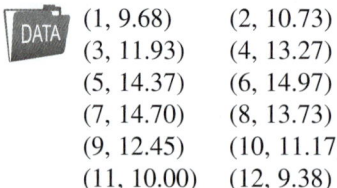

DATA

(1, 9.68)	(2, 10.73)
(3, 11.93)	(4, 13.27)
(5, 14.37)	(6, 14.97)
(7, 14.70)	(8, 13.73)
(9, 12.45)	(10, 11.17)
(11, 10.00)	(12, 9.38)

(a) Use a graphing utility to graph the data points and the model in the same viewing window.

(b) What is the period of the model? Is it what you expected? Explain.

(c) What is the amplitude of the model? What does it represent in the context of the problem? Explain.

69. Home Maintenance You are washing your house. Use the following steps to find the shortest ladder that will reach over the greenhouse to the side of the house (see figure).

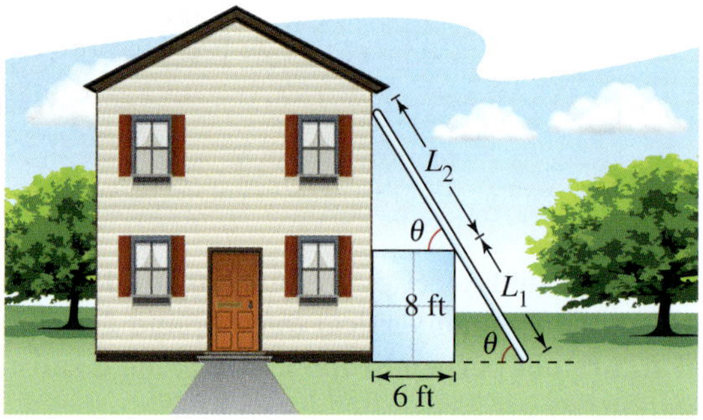

(a) Complete four rows of the table.

θ	L_1	L_2	$L_1 + L_2$
0.1	$\dfrac{8}{\sin 0.1}$	$\dfrac{6}{\cos 0.1}$	86.16
0.2	$\dfrac{8}{\sin 0.2}$	$\dfrac{6}{\cos 0.2}$	46.39

(b) Use the *table* feature of a graphing utility to generate additional rows of the table. Use the table to estimate the minimum length of the ladder.

(c) Write the length $L_1 + L_2$ as a function of θ.

(d) Use the graphing utility to graph the function. Use the graph to estimate the minimum length. How does your estimate compare with that in part (b)?

70. Irrigation Engineering The cross sections of an irrigation canal are isosceles trapezoids, where the lengths of three of the sides are 8 feet (see figure). The objective is to find the angle θ that maximizes the area of the cross sections. [*Hint:* The area of a trapezoid is given by $(h/2)(b_1 + b_2)$.]

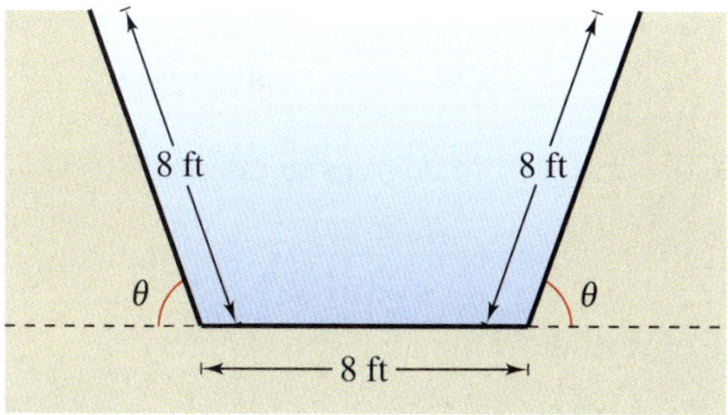

(a) Complete seven rows of the table.

Base 1	*Base 2*	*Altitude*	*Area*
8	$8 + 16 \cos 10°$	$8 \sin 10°$	22.06
8	$8 + 16 \cos 20°$	$8 \sin 20°$	42.46

(b) Use the *table* feature of a graphing utility to generate additional rows of the table. Use the table to estimate the maximum cross-sectional area.

(c) Write the area A as a function of θ.

(d) Use the graphing utility to graph the function. Use the graph to estimate the maximum cross-sectional area. How does your estimate compare with that in part (b)?

71. MODELING DATA

The table shows the average sales S (in millions of dollars) of an outerwear manufacturer for each month t, where $t = 1$ represents January.

DATA

Spreadsheet at LarsonPrecalculus.com

Month, t	Sales, S	Month, t	Sales, S
1	6.73	7	1.27
2	5.58	8	2.43
3	4.00	9	4.00
4	2.43	10	5.58
5	1.27	11	6.73
6	0.85	12	7.15

(a) Create a scatter plot of the data.

(b) Find a trigonometric model that fits the data. Graph the model on your scatter plot. How well does the model fit the data?

(c) What is the period of the model? Do you think it is reasonable given the context? Explain.

(d) Interpret the meaning of the model's amplitude in the context of the problem.

Focusing on Concepts

True or False? **In Exercises 72–74, determine whether the statement is true or false. Justify your answer.**

72. Simple harmonic motion does not involve a damping factor.

73. The tangent function can be used to model harmonic motion.

74. An example of a bearing used in aviation is S 25° W.

75. Writing Find two bearings perpendicular to N 32° E and explain how you found them.

76. HOW DO YOU SEE IT? The figure shows the displacement of an object in simple harmonic motion.

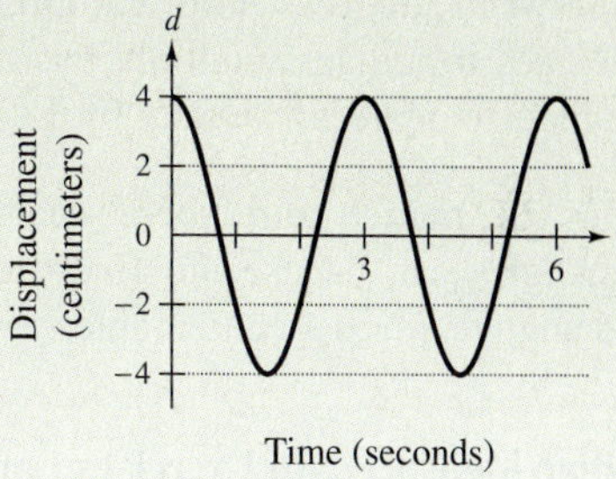

(a) What is the amplitude?

(b) What is the period?

(c) Is the equation of the simple harmonic motion of the form $d = a \sin \omega t$ or $d = a \cos \omega t$?

77. Error Analysis Describe the error in describing the bearing of a ship.

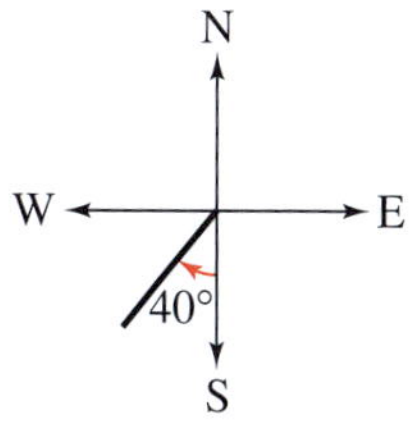

The bearing is N 40° W.

78. Error Analysis Describe the error.

The frequency of the motion given by $d = 5 \sin 3t$ is $2\pi/3$ cycles per second.

Cumulative Mixed Review

Finding the Domain of a Function **In Exercises 79–84, find the domain of the function.**

79. $f(x) = 3x + 8$

80. $f(x) = -x^2 - 1$

81. $g(x) = \sqrt[3]{x + 2}$

82. $g(x) = \sqrt{7 - x}$

83. $h(x) = \dfrac{1}{\sqrt{x - 6}}$

84. $h(x) = \dfrac{x}{x^2 - 9}$

4 Chapter Review

See CalcChat.com for tutorial help and worked-out solutions to odd-numbered exercises.
For instructions on how to use a graphing utility, see Appendix A.

4.1 What did you learn?

Describe angles (*p. 256*).

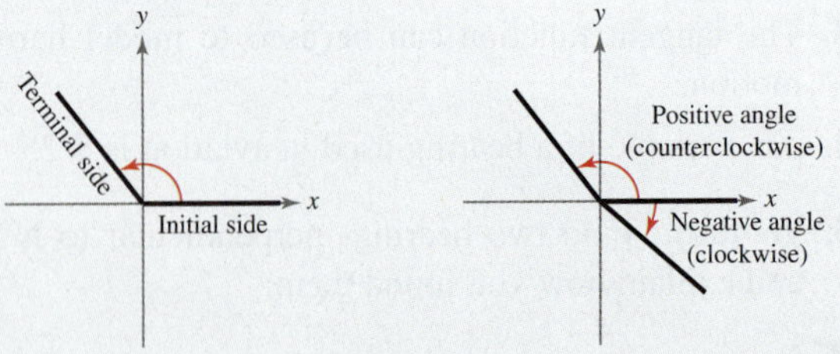

Coterminal angles have the same initial and terminal sides.

Use radian (*p. 257*) and degree (*p. 259*) measure, and convert between degrees and radians (*p. 259*). To convert degrees to radians, multiply by $(\pi \text{ rad})/180°$. To convert radians to degrees, multiply by $180°/(\pi \text{ rad})$.

Use angles to model and solve real-life problems (*p. 261*). Angles can be used to find arc length, linear speed, and angular speed (see Examples 5–7).

Example

(a) Coterminal angles:

$$\frac{2\pi}{3}, \quad \frac{2\pi}{3} - 2\pi = -\frac{4\pi}{3}, \quad \text{and} \quad \frac{2\pi}{3} + 2\pi = \frac{8\pi}{3}$$

(b) Converting from degrees to radians:

$$150° = (150 \text{ deg})\left(\frac{\pi \text{ rad}}{180 \text{ deg}}\right) = \frac{5\pi}{6} \text{ radians}$$

(c) Converting from radians to degrees:

$$\frac{2\pi}{3} \text{ rad} = \left(\frac{2\pi}{3} \text{ rad}\right)\left(\frac{180 \text{ deg}}{\pi \text{ rad}}\right) = 120°$$

(d) The complement of $\pi/5$ is

$$\frac{\pi}{2} - \frac{\pi}{5} = \frac{3\pi}{10}.$$

(e) The supplement of $120°$ is

$$180° - 120° = 60°.$$

Using Radian Measure In Exercises 1 and 2, (a) sketch the angle in standard position, (b) determine the quadrant in which the angle lies, and (c) determine one positive and one negative coterminal angle.

1. $\frac{11\pi}{4}$ **2.** $-\frac{\pi}{6}$

Using Degree Measure In Exercises 3 and 4, (a) sketch the angle in standard position, (b) determine the quadrant in which the angle lies, and (c) determine one positive and one negative coterminal angle.

3. $210°$ **4.** $-405°$

Converting from Degrees to Radians In Exercises 5 and 6, convert the angle measure from degrees to radians. Round your answer to three decimal places.

5. $-88.9°$ **6.** $221.55°$

Converting from Radians to Degrees In Exercises 7 and 8, convert the angle measure from radians to degrees. Round your answer to three decimal places.

7. $\frac{5\pi}{7}$ **8.** -3.3

Converting to Decimal Degree Form In Exercises 9 and 10, use the angle-conversion capabilities of a graphing utility to convert the angle measure to decimal degree form. Round your answer to three decimal places.

9. $153° \, 45' 16''$ **10.** $-243° \, 33''$

Converting to D° M′S″ Form In Exercises 11 and 12, use the angle-conversion capabilities of a calculator to convert the angle measure to D° M′S″ form.

11. $126.93°$ **12.** $-58.63°$

Complementary and Supplementary Angles In Exercises 13 and 14, find (if possible) the complement and supplement of the angle.

13. $\frac{\pi}{7}$ **14.** $108°$

Finding the Central Angle In Exercises 15 and 16, find the central angle θ in radians.

15.

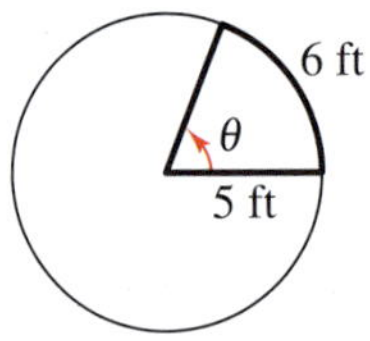

16.

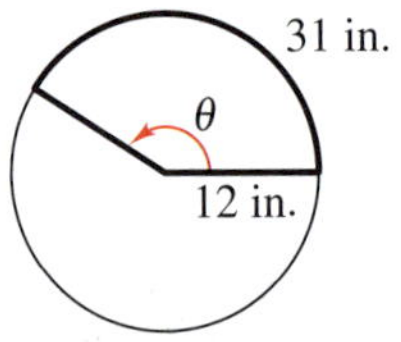

17. Finding Arc Length Find the length of the arc on a circle with (a) a radius of 20 meters intercepted by a central angle of 138°, and (b) a radius of 15 centimeters intercepted by a central angle of 60°.

18. Mechanical Engineering The wheels of a truck have a diameter of 28 inches and rotate at 336 revolutions per minute.

(a) Find the angular speed in radians per minute.

(b) Find the linear speed in inches per minute and miles per hour.

4.2 *What did you learn?*

Identify the unit circle and describe its relationship to real numbers *(p. 267)*.

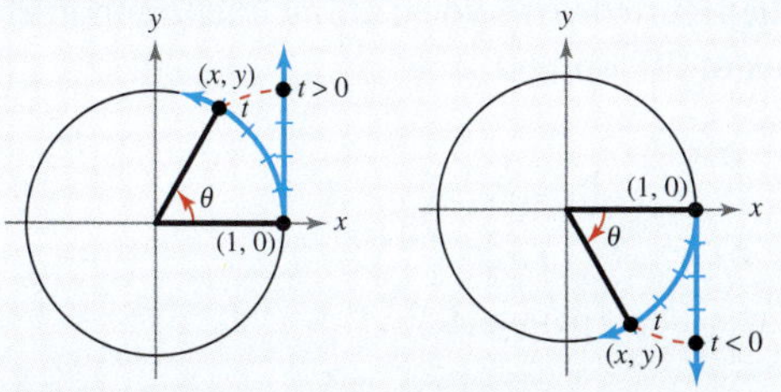

Each real number t on the real number line corresponds to a point (x, y) on the unit circle, and t is the (directional) length of the arc intercepted by the angle θ in radians.

Evaluate trigonometric functions using the unit circle *(p. 268)*. Let t be a real number and let (x, y) be a point on the unit circle corresponding to t.

$\sin t = y$ $\quad\cos t = x$ $\quad\tan t = \dfrac{y}{x}, x \neq 0$

$\csc t = \dfrac{1}{y}, y \neq 0$ $\quad\sec t = \dfrac{1}{x}, x \neq 0$ $\quad\cot t = \dfrac{x}{y}, y \neq 0$

Use domain and period to evaluate sine and cosine functions *(p. 270)*. The domain of the sine and cosine functions is the set of all real numbers. The sine and cosine functions are periodic and have a period of 2π.

The cosine and secant functions are even. The sine, cosecant, tangent, and cotangent functions are odd.

Use a calculator to evaluate trigonometric functions *(p. 271)*. See Example 3.

Example Using the figure below, you can evaluate the exact values of trigonometric functions for common t-values.

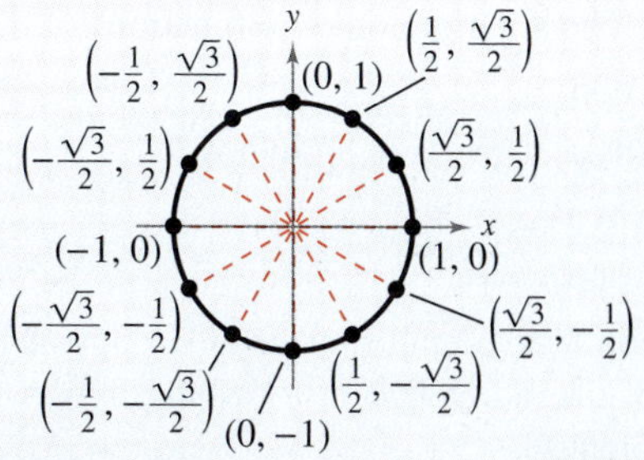

(a) The real number $t = 2\pi/3$ corresponds to

$$(x, y) = \left(-\frac{1}{2}, \frac{\sqrt{3}}{2}\right).$$

So,

$$\sin\frac{2\pi}{3} = \frac{\sqrt{3}}{2},$$

$$\cos\frac{2\pi}{3} = -\frac{1}{2}, \text{ and}$$

$$\tan\frac{2\pi}{3} = \frac{\sqrt{3}/2}{-1/2} = -\sqrt{3}.$$

(b) You can rewrite $17\pi/6$ as

$$2\pi + \frac{5\pi}{6}.$$

So,

$$\cos\frac{17\pi}{6} = \cos\left(2\pi + \frac{5\pi}{6}\right) = \cos\frac{5\pi}{6} = -\frac{\sqrt{3}}{2}.$$

Finding a Point on the Unit Circle **In Exercises 19–22, find the point (x, y) on the unit circle that corresponds to the real number t.**

19. $t = 7\pi/4$ **20.** $t = 3\pi/4$

21. $t = -2\pi/3$ **22.** $t = -7\pi/6$

Evaluating Trigonometric Functions **In Exercises 23–30, evaluate (if possible) the six trigonometric functions at the real number.**

23. $t = 5\pi/4$ **24.** $t = 7\pi/6$

25. $t = 4\pi$ **26.** $t = -2\pi$

27. $t = -11\pi/6$ **28.** $t = -2\pi/3$

29. $t = -7\pi/3$ **30.** $t = -\pi/4$

Using the Period to Evaluate Sine and Cosine **In Exercises 31–34, evaluate the trigonometric function using its period as an aid.**

31. $\cos 6\pi$ **32.** $\sin(11\pi/4)$

33. $\sin(-17\pi/6)$ **34.** $\cos(-13\pi/3)$

Using the Value of a Trigonometric Function **In Exercises 35–38, use the value of the trigonometric function to evaluate each function.**

35. $\sin t = \frac{3}{5}$
(a) $\sin(-t)$
(b) $\csc(-t)$

36. $\cos t = \frac{5}{13}$
(a) $\cos(-t)$
(b) $\sec(-t)$

37. $\sin(-t) = -\frac{2}{3}$
(a) $\sin t$
(b) $\csc t$

38. $\cos(-t) = \frac{4}{9}$
(a) $\cos t$
(b) $\sec(-t)$

Using a Calculator **In Exercises 39–42, use a calculator to evaluate the trigonometric expression. Round your answer to four decimal places. (Be sure to use the correct angle mode.)**

39. $\cos(5\pi/8)$ **40.** $\tan(-11\pi/6)$

41. $\cot 1.9$ **42.** $\sec 5.4$

4.3 *What did you learn?*

Evaluate trigonometric functions of acute angles *(p. 275)* and use a calculator to evaluate trigonometric functions *(p. 277)*. Let θ be an acute angle of a right triangle. The six trigonometric functions of θ are

$$\sin\theta = \frac{\text{opp}}{\text{hyp}},\quad \cos\theta = \frac{\text{adj}}{\text{hyp}},\quad \tan\theta = \frac{\text{opp}}{\text{adj}}$$

$$\csc\theta = \frac{\text{hyp}}{\text{opp}},\quad \sec\theta = \frac{\text{hyp}}{\text{adj}},\quad \cot\theta = \frac{\text{adj}}{\text{opp}}$$

where opp = the length of the side opposite θ, adj = the length of the side adjacent to θ, and hyp = the length of the hypotenuse.

Example

(a) Using the figure below, you can find the trigonometric functions of 30° and 60°.

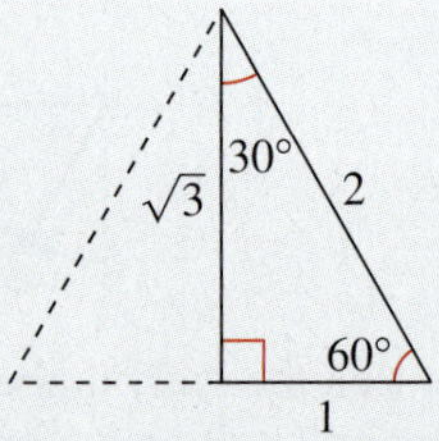

$$\sin 30° = \frac{1}{2},\quad \cos 30° = \frac{\sqrt{3}}{2},\quad \tan 30° = \frac{1}{\sqrt{3}} = \frac{\sqrt{3}}{3},$$

$$\sin 60° = \frac{\sqrt{3}}{2},\quad \cos 60° = \frac{1}{2},\quad \tan 60° = \frac{\sqrt{3}}{1} = \sqrt{3},$$

Use fundamental trigonometric identities *(p. 278)*.

$$\sin\theta = \frac{1}{\csc\theta},\quad \cos\theta = \frac{1}{\sec\theta},\quad \tan\theta = \frac{1}{\cot\theta},$$

$$\tan\theta = \frac{\sin\theta}{\cos\theta},\quad \sin^2\theta + \cos^2\theta = 1,$$

$$1 + \tan^2\theta = \sec^2\theta,\quad 1 + \cot^2\theta = \csc^2\theta$$

(b) Verifying the Pythagorean identity $\sin^2\theta + \cos^2\theta = 1$:

$$\sin^2 45° + \cos^2 45° = \left(\frac{\sqrt{2}}{2}\right)^2 + \left(\frac{\sqrt{2}}{2}\right)^2 = \frac{1}{2} + \frac{1}{2} = 1$$

Use trigonometric functions to model and solve real-life problems *(p. 280)*. Trigonometric functions can be used to find a tree's height, the angle between a path and a river, and the length of a ramp. (See Examples 8–10.)

Evaluating Trigonometric Functions **In Exercises 43–50, find the exact values of the six or five remaining trigonometric functions of the angle θ.**

43.

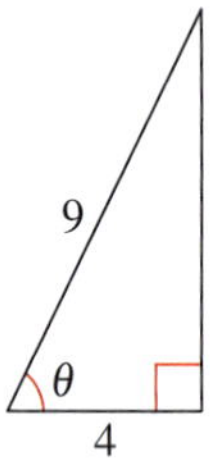

44.

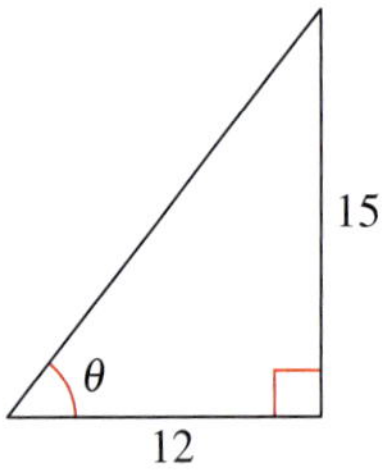

45.

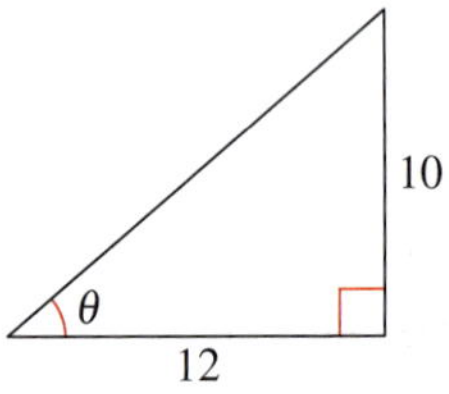

46.

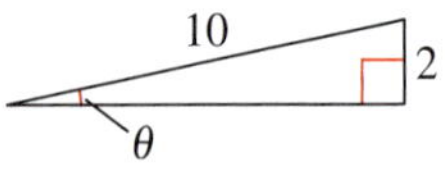

47. $\sin\theta = \frac{7}{12}$

48. $\cos\theta = \frac{2}{3}$

49. $\tan\theta = \frac{1}{5}$

50. $\cot\theta = \frac{9}{40}$

Using a Calculator **In Exercises 51 and 52, use a calculator to evaluate each function. Round your answers to four decimal places. (Be sure to use the correct angle mode.)**

51. (a) $\cos 84°$ (b) $\csc 52° 12'$

52. (a) $\tan(9\pi/20)$ (b) $\cot(9\pi/20)$

Using Trigonometric Identities **In Exercises 53 and 54, use trigonometric identities to transform the left side of the equation into the right side $(0 < \theta < \pi/2)$.**

53. $\sin^2\theta \csc^2\theta = 1$

54. $\dfrac{\cot\theta + \tan\theta}{\cot\theta} = \sec^2\theta$

55. Surveying To determine the width w of a river a surveyor walks downstream 125 feet from point P and sights to point Q (see figure). From this sighting, the surveyor determines that $\theta = 62°$. How wide is the river?

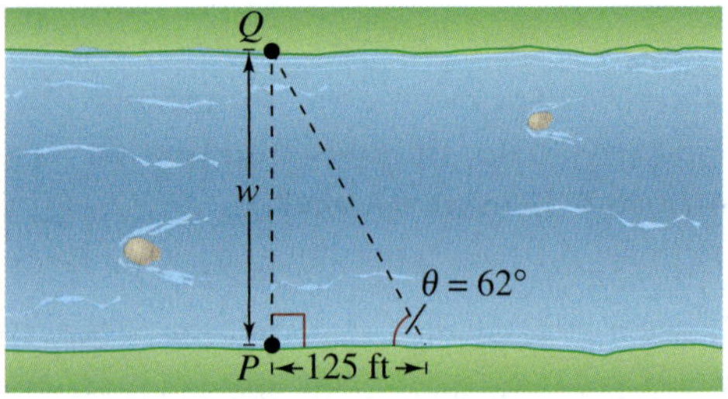

56. Architecture An escalator 152 feet in length rises to a platform and makes a 30° angle with the ground.

(a) Use a trigonometric function to write an equation involving the unknown quantity.

(b) Find the height of the platform above the ground.

4.4 *What did you learn?*

Evaluate trigonometric functions of any angle *(p. 286)*. Let θ be an angle in standard position with (x, y) a point on the terminal side of θ and $r = \sqrt{x^2 + y^2} \neq 0$.

$\sin\theta = \dfrac{y}{r}$ $\quad\cos\theta = \dfrac{x}{r}$

$\tan\theta = \dfrac{y}{x},\ x \neq 0$ $\quad\cot\theta = \dfrac{x}{y},\ y \neq 0$

$\sec\theta = \dfrac{r}{x},\ x \neq 0$ $\quad\csc\theta = \dfrac{r}{y},\ y \neq 0$

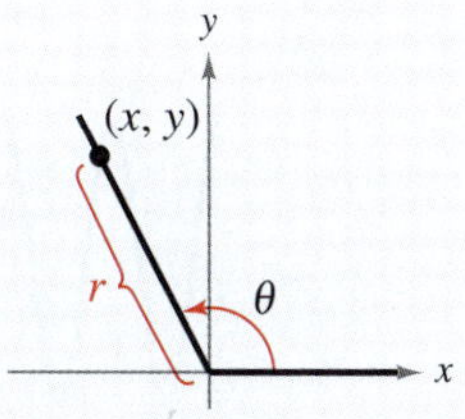

The signs of the trigonometric functions in the four quadrants can be determined from the definitions of the functions. These are summarized for the sine, cosine, and tangent functions in the figure at the right.

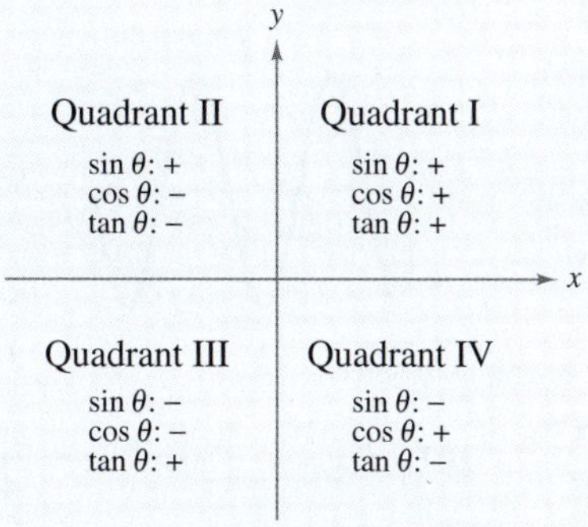

Find reference angles *(p. 288)*. Let θ be an angle in standard position. Its reference angle is the acute angle θ' formed by the terminal side of θ and the horizontal axis.

Evaluate trigonometric functions of real numbers *(p. 289)*. To find the value of a trigonometric function of any angle θ, first determine the function value of the associated reference angle θ'. Then, depending on the quadrant in which θ lies, affix the appropriate sign to the function value.

Example

(a) Let $(-4, 3)$ be a point on the terminal side of θ in standard position. Then $x = -4$, $y = 3$, and $r = \sqrt{(-4)^2 + 3^2} = 5$. So, you have

$$\sin\theta = \frac{3}{5},\quad \cos\theta = -\frac{4}{5},\quad \tan\theta = -\frac{3}{4},$$

$$\cot\theta = -\frac{4}{3},\quad \sec\theta = -\frac{5}{4},\quad \text{and}\quad \csc\theta = \frac{5}{3}.$$

(b) To find $\cos\theta$ given that $\csc\theta = -\frac{9}{5}$ and $\cot\theta < 0$, note that θ lies in Quadrant IV because that is the only quadrant in which the cosecant and the cotangent are both negative. Moreover, using

$$\csc\theta = \frac{r}{y} = -\frac{9}{5}$$

and the fact that y is negative in Quadrant IV, you can let $y = -5$ and $r = 9$. Because x is positive in Quadrant IV,

$$\begin{aligned} x &= (+)\sqrt{r^2 - y^2} \\ &= \sqrt{81 - 25} \\ &= \sqrt{56} \\ &= 2\sqrt{14}. \end{aligned}$$

So, $\cos\theta = \dfrac{x}{r} = \dfrac{2\sqrt{14}}{9}$.

(c) Because $\theta = 229°$ lies in Quadrant III, its reference angle is $\theta' = 229° - 180° = 49°$.

Evaluating Trigonometric Functions In Exercises 57–62, the point is on the terminal side of an angle θ in standard position. Find the exact values of the six trigonometric functions of θ.

57. $(12, 16)$ **58.** $(2, 10)$

59. $(-7, 2)$ **60.** $(3, -7)$

61. $\left(\frac{2}{3}, \frac{5}{8}\right)$ **62.** $\left(-\frac{10}{3}, -\frac{2}{3}\right)$

Evaluating Trigonometric Functions In Exercises 63–66, find the exact values of the remaining trigonometric functions of θ satisfying the given condition.

63. $\sec\theta = \frac{6}{5},\ \tan\theta < 0$ **64.** $\sin\theta = \frac{3}{8},\ \cos\theta < 0$

65. $\cos\theta = -\frac{2}{5},\ \sin\theta > 0$ **66.** $\tan\theta = -\frac{12}{5},\ \sin\theta > 0$

Finding a Reference Angle In Exercises 67–74, find the reference angle θ'. Sketch θ in standard position and label θ'.

67. $\theta = 330°$ **68.** $\theta = 210°$

69. $\theta = \dfrac{5\pi}{4}$ **70.** $\theta = \dfrac{9\pi}{4}$

71. $\theta = -258°$ **72.** $\theta = -645°$

73. $\theta = -\dfrac{6\pi}{5}$ **74.** $\theta = -\dfrac{17\pi}{3}$

Trigonometric Functions of a Nonacute Angle In Exercises 75–82, evaluate the sine, cosine, and tangent of the angle without using a calculator.

75. $240°$ **76.** $315°$

77. $-225°$ **78.** $-315°$

79. 4π **80.** -3π

81. $-\dfrac{9\pi}{4}$ **82.** $\dfrac{7\pi}{3}$

Using a Calculator In Exercises 83–86, use a calculator to evaluate the trigonometric function. Round your answer to four decimal places. (Be sure to use the correct angle mode.)

83. $\tan 33°$ **84.** $\csc 105°$

85. $\sec\dfrac{12\pi}{5}$ **86.** $\sin\left(-\dfrac{\pi}{9}\right)$

4.5 *What did you learn?*

Sketch the graphs of basic sine and cosine functions *(p. 294)*. The basic characteristics of the parent sine and cosine functions are listed on page 295.

Use amplitude and period to help sketch the graphs of sine and cosine functions *(p. 296)*. For $y = a \sin bx$ and $y = a \cos bx$, $b > 0$, the amplitude is $|a|$ and the period is $2\pi/b$. For $b < 0$, use $\sin(-x) = -\sin x$ or $\cos(-x) = \cos x$ to rewrite the function.

Sketch translations of graphs of sine and cosine functions *(p. 298)*. For $y = d + a \sin(bx - c)$ and $y = d + a \cos(bx - c)$, c creates a horizontal translation and d creates a vertical translation.

Use sine and cosine functions to model real-life data *(p. 300)*. A cosine function can model the depth of water at the end of a dock at various times. (See Example 8.)

Example The graphs of (a) $y = 3 \sin 2x$, (b) $y = 2 \cos 3x$, (c) $y = -3 \cos(x + \pi/4)$, and (d) $y = 1 - \sin(x/3 - \pi/3)$ are shown below.

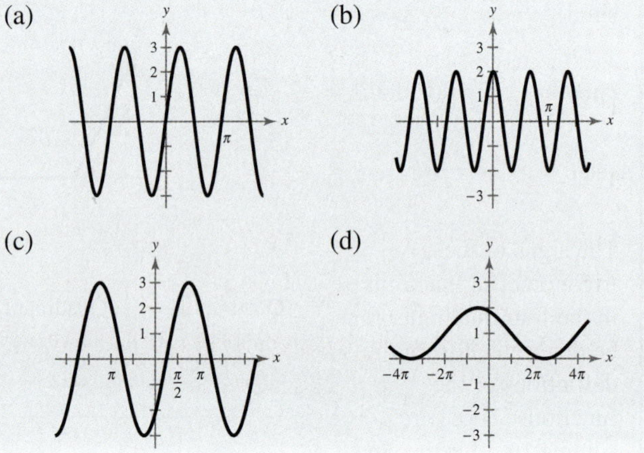

Sketching the Graph of a Sine or Cosine Function **In Exercises 87–102, sketch the graph of the function by hand. (Include two full periods.)**

87. $f(x) = 6 \sin x$

88. $f(x) = 7 \cos x$

89. $f(x) = \frac{1}{5} \cos x$

90. $f(x) = \frac{3}{5} \sin x$

91. $f(x) = 3 \cos 2\pi x$

92. $f(x) = -2 \sin \pi x$

93. $f(x) = 5 \sin(-2x/5)$

94. $f(x) = 8 \cos(-x/4)$

95. $f(x) = -5 \cos(x/4)$

96. $f(x) = -\sin(\pi x/4)$

97. $f(x) = \frac{5}{2} \sin(x - \pi)$

98. $f(x) = 3 \cos(x + \pi)$

99. $f(x) = 2 - \cos(\pi x/2)$

100. $f(x) = \frac{1}{2} \sin \pi x - 3$

101. $f(x) = -3 \sin\left(2x - \dfrac{\pi}{4}\right) - 1$

102. $f(x) = 4 - 2 \cos(4x + \pi)$

Economics **In Exercises 103 and 104, use a graphing utility to graph the sales function over 1 year, where S is the sales (in thousands of units) of a seasonal product, and t is the month, with $t = 1$ corresponding to January. Determine the months of maximum and minimum sales.**

103. $S = 48.4 - 6.1 \cos \dfrac{\pi t}{6}$

104. $S = 66.3 + 8.5 \sin \dfrac{\pi t}{6}$

4.6 *What did you learn?*

Sketch the graphs of tangent *(p. 306)*, cotangent *(p. 308)*, secant *(p. 309)*, and cosecant *(p. 309)* functions. The basic characteristics of the parent tangent, cotangent, secant, and cosecant functions are listed below.

$y = \tan x$: Domain: all real numbers x, $x \neq \pi/2 + n\pi$; range: $(-\infty, \infty)$; period: π; x-intercepts: $(n\pi, 0)$; y-intercept: $(0, 0)$; vertical asymptotes: $x = \pi/2 + n\pi$; odd

$y = \cot x$: Domain: all real numbers x, $x \neq n\pi$; range: $(-\infty, \infty)$; period: π; x-intercepts: $(\pi/2 + n\pi, 0)$; vertical asymptotes: $x = n\pi$; odd

$y = \sec x$: Domain: all real numbers x, $x \neq \pi/2 + n\pi$; range: $(-\infty, -1] \cup [1, \infty)$; period: 2π; y-intercept: $(0, 1)$; vertical asymptotes: $x = \pi/2 + n\pi$; even

$y = \csc x$: Domain: all real numbers x, $x \neq n\pi$; range: $(-\infty, -1] \cup [1, \infty)$; period: 2π; no intercepts; vertical asymptotes: $x = n\pi$; odd

Sketch the graphs of damped trigonometric functions *(p. 311)*. The factor x in $f(x) = x \cos 2x$ and the factor e^{-x} in $f(x) = e^{-x} \sin 4x$ are damping factors.

Example The graphs of (a) $y = \tan x$, (b) $y = \cot x$, (c) $y = \sec x$, and (d) $y = \csc x$ are shown below.

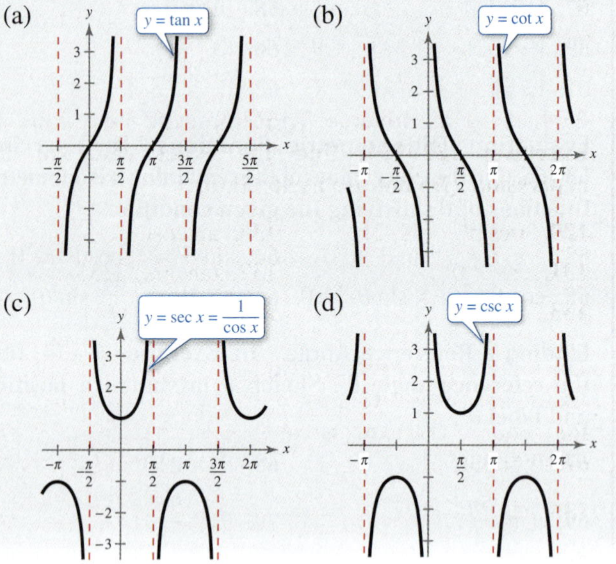

Library of Parent Functions **In Exercises 105–118, sketch the graph of the function by hand. (Include two full periods.) Use a graphing utility to verify your results.**

105. $f(x) = -\tan \frac{\pi x}{4}$ **106.** $f(x) = 4 \tan \pi x$

107. $f(x) = \frac{1}{4} \tan 2x - 1$ **108.** $f(x) = 2 + 2 \tan \frac{x}{3}$

109. $f(x) = 3 \cot \frac{x}{2}$ **110.** $f(x) = \frac{1}{2} \cot \frac{\pi x}{2}$

111. $f(x) = \frac{1}{2} \cot\left(x - \frac{\pi}{2}\right)$ **112.** $f(x) = 4 \cot\left(x + \frac{\pi}{4}\right)$

113. $f(x) = 4 \sec x$ **114.** $f(x) = -\frac{1}{2} \csc x$

115. $f(x) = \frac{1}{4} \csc 2x$ **116.** $f(x) = \frac{1}{2} \sec 2\pi x$

117. $f(x) = \sec\left(x + \frac{\pi}{4}\right)$ **118.** $f(x) = \frac{1}{2} \csc(2x - \pi)$

Comparing Trigonometric Graphs **In Exercises 119–124, use a graphing utility to graph the function (include two full periods). Graph the corresponding reciprocal function in the same viewing window. Describe and compare the graphs.**

119. $f(x) = 4 \csc(2x - \pi)$ **120.** $f(x) = -2 \sec(4x + \pi)$

121. $f(x) = 2 \sec(x - \pi)$ **122.** $f(x) = 2 \csc(x - \pi)$

123. $f(x) = -\csc\left(3x - \frac{\pi}{2}\right)$ **124.** $f(x) = 3 \sec\left(2x - \frac{\pi}{4}\right)$

Analyzing a Damped Trigonometric Graph **In Exercises 125–128, use a graphing utility to graph the function and the damping factor of the function in the same viewing window. Then analyze the graph of the function using the method in Example 6 in Section 4.6.**

125. $f(x) = e^x \sin 2x$ **126.** $f(x) = e^x \cos x$

127. $f(x) = 2x \cos x$ **128.** $f(x) = x \sin \pi x$

4.7 *What did you learn?*

Evaluate and graph inverse trigonometric functions (*p. 317*). See page 321 for graphs of the functions below.

$y = \arcsin x$ if and only if $\sin y = x$, where $-1 \le x \le 1$ and $-\pi/2 \le y \le \pi/2$.

$y = \arccos x$ if and only if $\cos y = x$, where $-1 \le x \le 1$ and $0 \le y \le \pi$.

$y = \arctan x$ if and only if $\tan y = x$, where $-\infty < x < \infty$ and $-\pi/2 < y < \pi/2$.

Example

(a) $\sin^{-1}(1/2) = \pi/6$ because $\sin(\pi/6) = 1/2$ and $\pi/6$ lies in $[-\pi/2, \pi/2]$.

(b) $\cos^{-1}(\sqrt{2}/2) = \pi/4$ because $\cos(\pi/4) = \sqrt{2}/2$ and $\pi/4$ lies in $[0, \pi]$.

(c) $\tan^{-1}(-1) = -\pi/4$ because $\tan(-\pi/4) = -1$ and $-\pi/4$ lies in $(-\pi/2, \pi/2)$.

Evaluate compositions of trigonometric functions (*p. 322*). Use the inverse properties listed below.

If $-1 \le x \le 1$ and $-\pi/2 < y < \pi/2$, then $\sin(\arcsin x) = x$ and $\arcsin(\sin y) = y$.

If $-1 \le x \le 1$ and $0 \le y \le \pi$, then $\cos(\arccos x) = x$ and $\arccos(\cos y) = y$.

If x is a real number and $-\pi/2 < y < \pi/2$, then $\tan(\arctan x) = x$ and $\arctan(\tan y) = y$.

(d) $\sin(\sin^{-1} 0.4) = 0.4$ because 0.4 lies in the domain of the arcsine function, so the inverse property applies.

(e) You can use the figure at the right to determine that

$$\cos\left(\arctan \frac{5}{12}\right) = \frac{12}{13}.$$

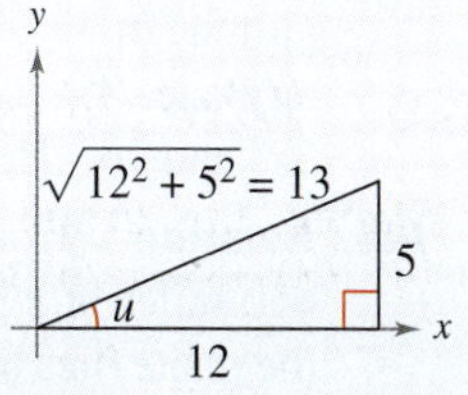

Evaluating an Inverse Trigonometric Function **In Exercises 129–134, find the exact value of the expression. (Do not use a calculator.)**

129. $\arcsin(-\sqrt{2}/2)$ **130.** $\arccos(-\sqrt{3}/2)$

131. $\cos^{-1} 0$ **132.** $\tan^{-1}(\sqrt{3}/3)$

133. $\cos^{-1} 1$ **134.** $\sin^{-1}(-1)$

Calculators and Inverse Trigonometric Functions **In Exercises 135 and 136, use a calculator to approximate the value of the expression. Round your result to two decimal places.**

135. $\cos^{-1} 0.42$

136. $\arctan(-12)$

Using an Inverse Trigonometric Function **In Exercises 137 and 138, use an inverse trigonometric function to write θ as a function of x.**

137.

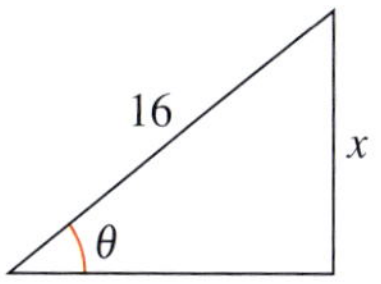

138.

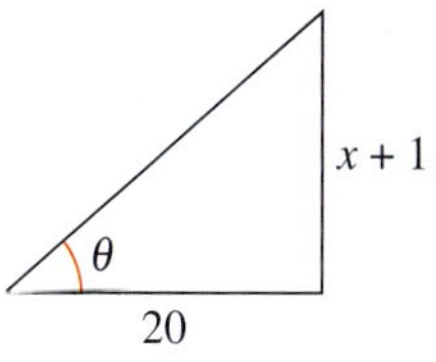

Writing an Expression **In Exercises 139 and 140, write an algebraic expression that is equivalent to the given expression.**

139. $\sec(\arcsin x)$ **140.** $\csc(\arcsin 10x)$

4.8 *What did you learn?*

Solve real-life problems involving right triangles *(p. 328)*. A trigonometric function can be used to find an angle of elevation, height, and an angle of depression. (See Examples 2–4.)

Solve real-life problems involving directional bearings *(p. 330)*. Trigonometric functions can be used to find a ship's bearing and distance from a port at a specified time. (See Example 5.)

Solve real-life problems involving harmonic motion *(p. 331)*. Trigonometric functions can be used to describe the motion of an object that vibrates, oscillates, rotates, or is moved by wave motion. (See Examples 6 and 7.)

Example To solve the right triangle shown, first note that because $C = 90°$, you know that $A + B = 90°$. So,

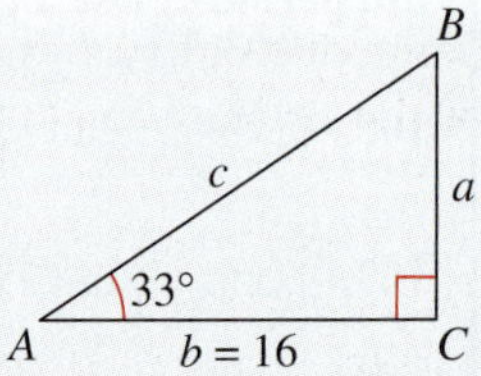

$$B = 90° - 33° = 57°.$$

To solve for a, use the fact that

$$\tan A = \frac{\text{opp}}{\text{adj}} = \frac{a}{b} \quad \Rightarrow \quad a = b \tan A = 16 \tan 33° \approx 10.39.$$

Similarly, to solve for c, use the fact that

$$\cos A = \frac{\text{adj}}{\text{hyp}} = \frac{b}{c} \quad \Rightarrow \quad c = \frac{b}{\cos A} = \frac{16}{\cos 33°} \approx 19.08.$$

141. Topography A train travels 3.5 kilometers on a straight track with a grade of $1° \, 10'$, as shown in the figure. What is the vertical rise of the train in that distance?

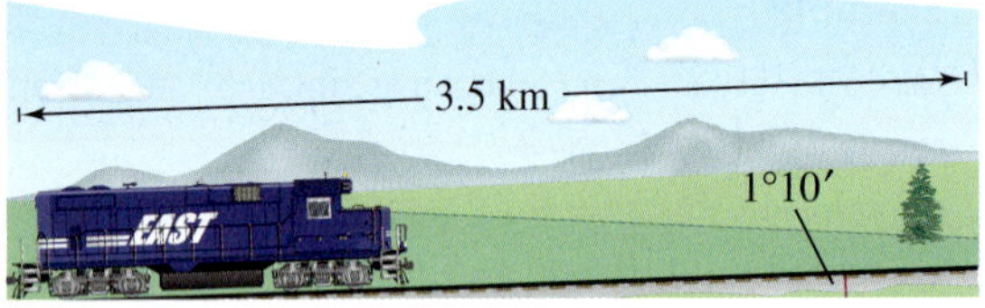

142. Topography A road sign at the top of a mountain indicates that for the next 4 miles the grade is 12%. Find the angle of the grade and the change in elevation for a car descending the 4-mile stretch.

143. Aviation A passenger in an airplane flying at an altitude of 37,000 feet sees two towns due west of the airplane. The angles of depression to the towns are 32° and 76°. How far apart are the towns?

144. Aviation From city A to city B, a plane flies 650 miles at a bearing of 48°. From city B to city C, the plane flies 810 miles at a bearing of 115°. Find the distance from city A to city C and the bearing from city A to city C.

145. Harmonic Motion A buoy oscillates in simple harmonic motion as waves go past. The buoy moves a total of 6 feet from its high point to its low point (see figure), and it returns to its high point every 15 seconds. Write an equation that describes the motion of the buoy, where the high point corresponds to the time $t = 0$.

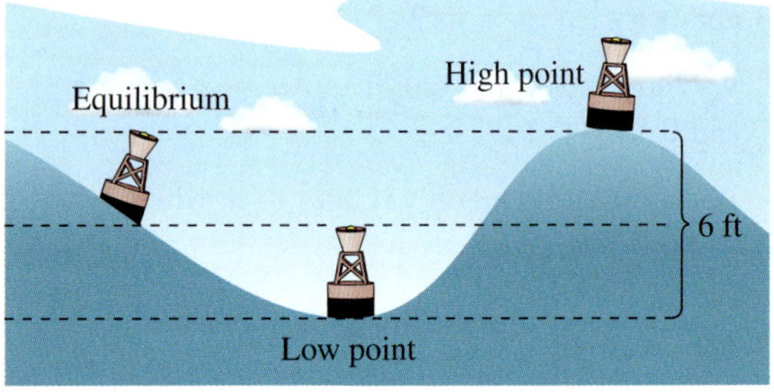

146. Harmonic Motion A fishing bobber is oscillating in simple harmonic motion caused by waves. The bobber moves a total of 7 inches from its high point to its low point and returns to its high point every 2 seconds. Write an equation modeling the motion of the bobber, where the high point corresponds to the time $t = 0$.

Focusing on Concepts

True or False? **In Exercises 147–150, determine whether the statement is true or false. Justify your answer.**

147. $y = \sin \theta$ is not a function because $\sin 30° = \sin 150°$.

148. The equation $y = \cos x$ does not have an inverse function on the interval $-\pi/2 \le x \le \pi/2$.

149. You can use the cotangent function to model simple harmonic motion.

150. The sine of any nonacute angle θ is equal to the sine of the reference angle for θ.

151. Exploration Using calculus, it can be shown that the arctangent function can be approximated by the polynomial

$$\arctan x \approx x - \frac{x^3}{3} + \frac{x^5}{5} - \frac{x^7}{7}$$

where x is in radians.

(a) Use a graphing utility to graph the arctangent function and its polynomial approximation in the same viewing window.

(b) How do the graphs compare?

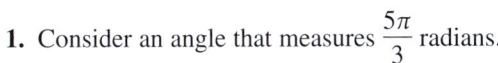

4 Chapter Test

See *CalcChat.com* for tutorial help and worked-out solutions to odd-numbered exercises. For instructions on how to use a graphing utility, see Appendix A.

Take this test as you would take a test in class. After you are finished, check your work against the answers given in the back of the book.

1. Consider an angle that measures $\frac{5\pi}{3}$ radians.

(a) Sketch the angle in standard position.

(b) Determine two coterminal angles (one positive and one negative).

(c) Convert the angle to degree measure.

2. Find the complement and supplement of a 49° angle.

3. A circle has a radius of 6 yards. Find the length of the arc intercepted by a central angle of 130°.

4. A truck is moving at a rate of 100 kilometers per hour, and the diameter of its wheels is 1.25 meters. Find the angular speed of the wheels in radians per second.

5. Find the exact values of the six trigonometric functions of the angle θ shown in the figure.

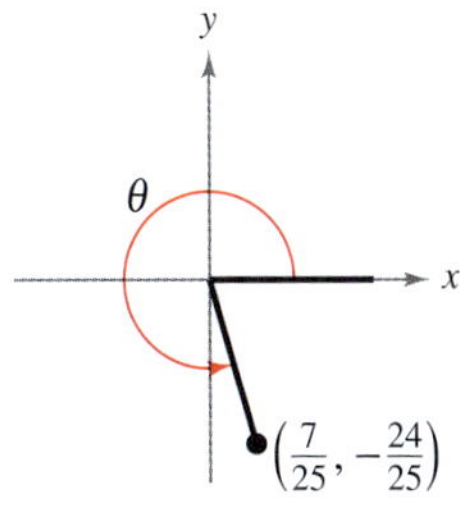

Figure for 5

6. Given that $\tan\theta = \frac{7}{2}$ and θ is an acute angle, find the other five trigonometric functions of θ.

7. Determine the reference angle θ' of the angle $\theta = 255°$.

8. Determine the quadrant in which θ lies when $\sec\theta < 0$ and $\tan\theta > 0$.

9. Find two exact values of θ in degrees $(0 \le \theta < 360°)$ when $\cos\theta = -\sqrt{3}/2$.

10. Use a calculator to evaluate cos 42.6° accurate to four decimal places.

Collaborative Project

To work a collaborative project involving Trigonometric Functions, visit this textbook's website at *LarsonPrecalculus.com*.

In Exercises 11–16, sketch the graph of the function by hand. (Include two full periods.)

11. $g(x) = -2\sin\frac{2x}{3}$

12. $f(x) = \frac{1}{2}\tan 4x$

13. $f(x) = \frac{1}{2}\sec(x - \pi) - 4$

14. $f(x) = 2\cos(-2x) + 3$

15. $f(x) = 2\cot\left(x - \frac{\pi}{2}\right)$

16. $f(x) = 2\csc\left(x + \frac{\pi}{2}\right)$

In Exercises 17 and 18, use a graphing utility to graph the function. If the function is periodic, find its period.

17. $y = \sin 2\pi x + 2\cos\pi x$

18. $y = 6e^{-0.12t}\cos(0.25t), \quad 0 \le t \le 32$

19. Find a, b, and c for the function $f(x) = a\cos(bx - c)$ such that the graph of f matches the graph at the right.

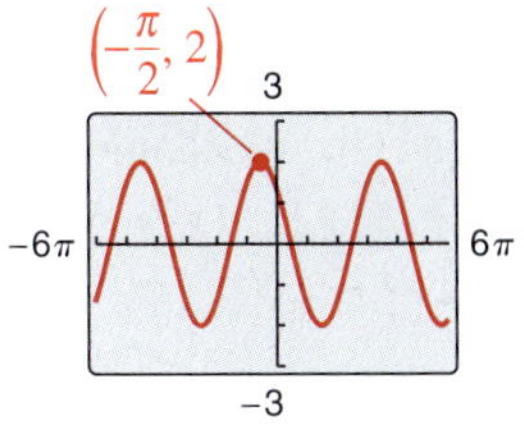

Figure for 19

20. Find the exact value of $\tan\left(\arccos\frac{2}{3}\right)$ without using a calculator.

In Exercises 21–23, use a graphing utility to graph the function.

21. $f(x) = 2\arcsin\frac{x}{2}$

22. $f(x) = 2\arccos x$

23. $f(x) = \arctan\frac{x}{4}$

24. A ship leaves port at noon and heads due south at 5 knots. At 2 P.M., the ship changes course to S 10° E. What is the ship's bearing and distance from port at 6 P.M.?

25. Write an equation for the simple harmonic motion of a ball on a spring that starts at its lowest point of 6 inches below equilibrium, bounces to its maximum height of 6 inches above equilibrium, and returns to its lowest point in a total of 2 seconds.

Library of Parent Functions Review

In Exercises 1–12, fill in the blank to identify the parent function represented by the graph.

1. $f(x) =$ ___________

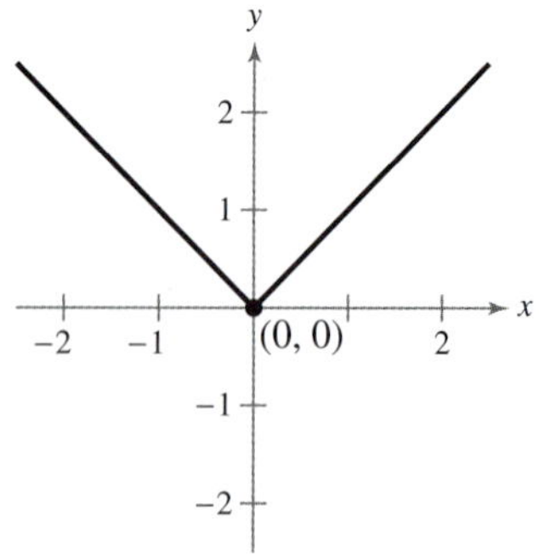

2. $f(x) =$ ___________

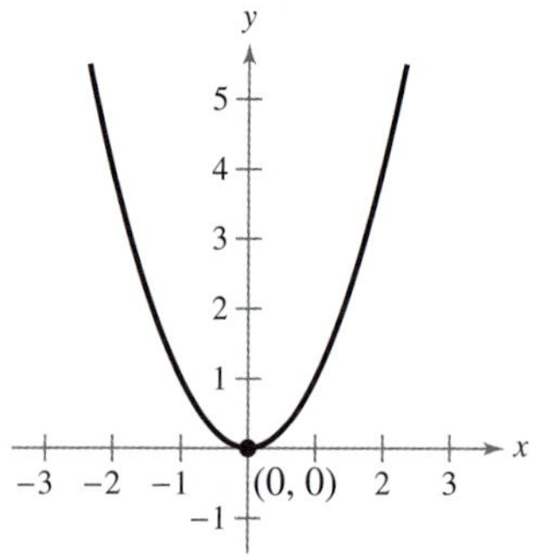

3. $f(x) =$ ___________

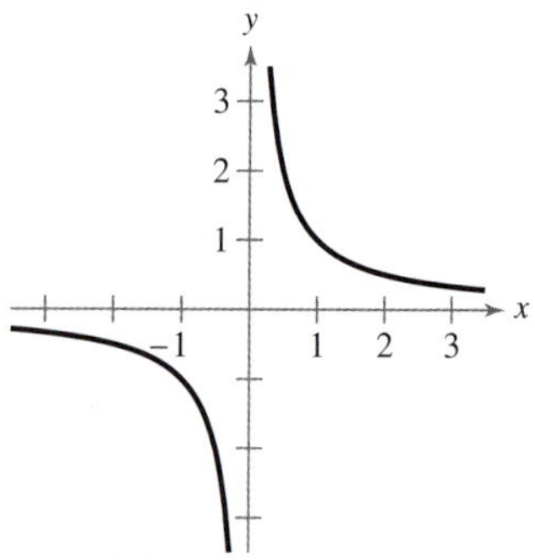

4. $f(x) =$ ___________

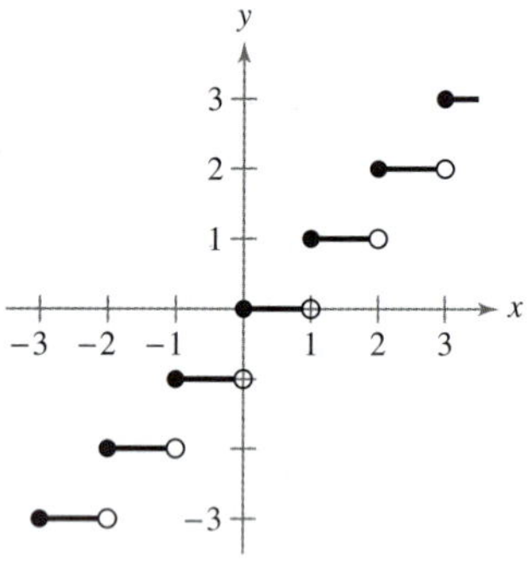

5. $f(x) =$ ___________

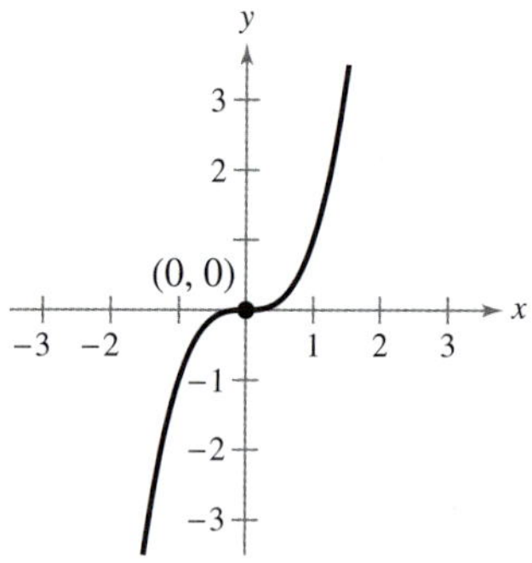

6. $f(x) =$ ___________

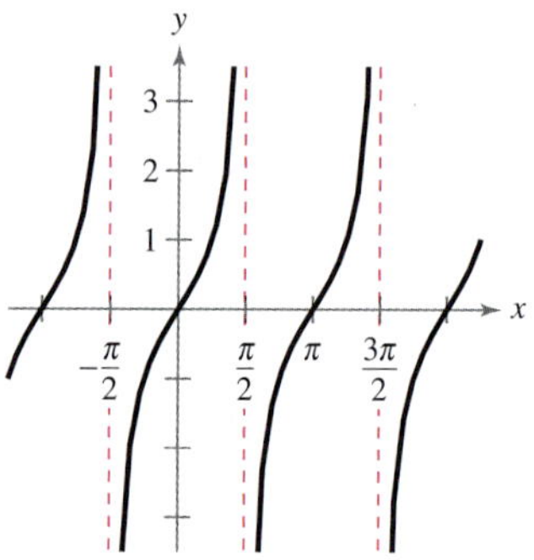

7. $f(x) =$ ___________

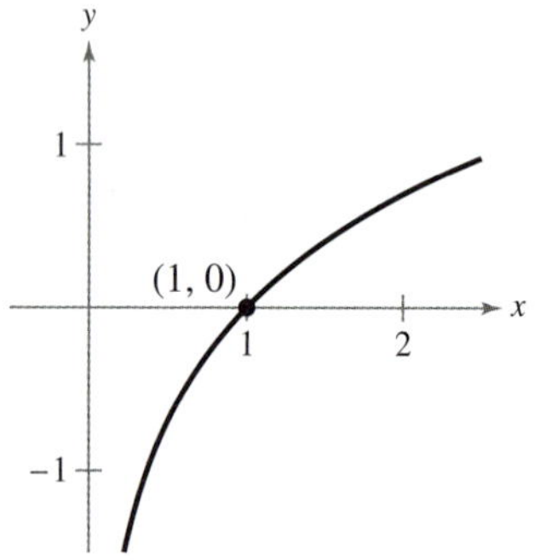

8. $f(x) =$ ___________

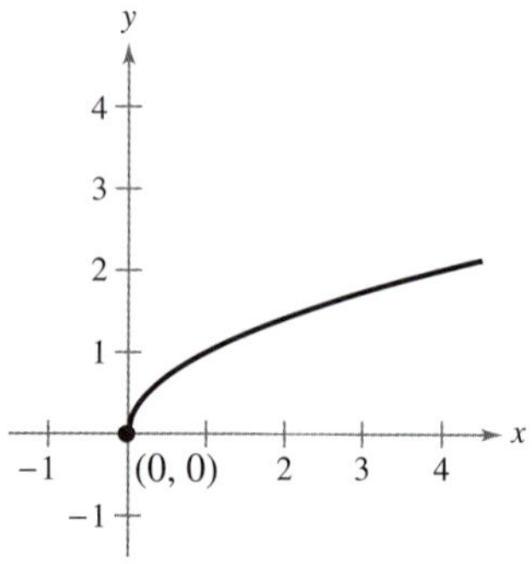

9. $f(x) =$ ___________

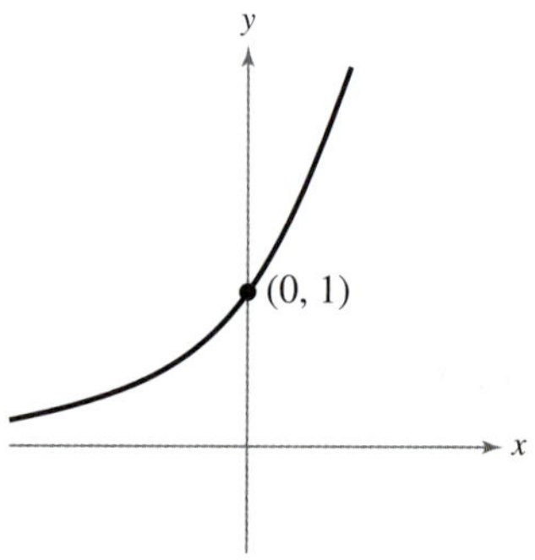

10. $f(x) =$ ____________

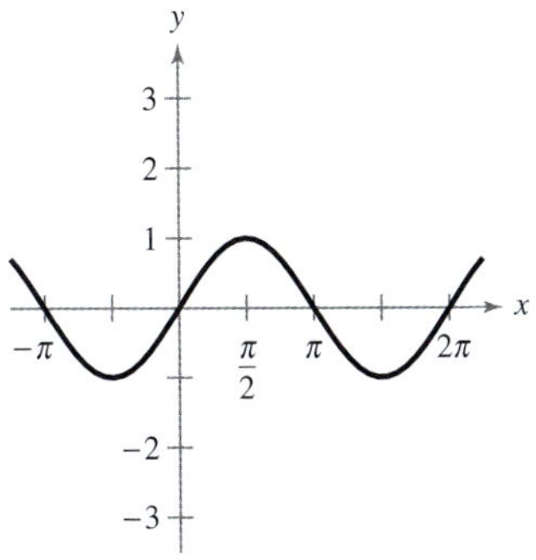

11. $f(x) =$ ____________

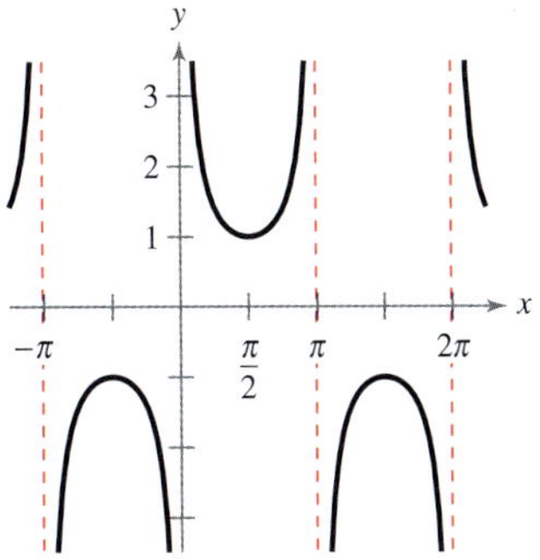

12. $f(x) =$ ____________

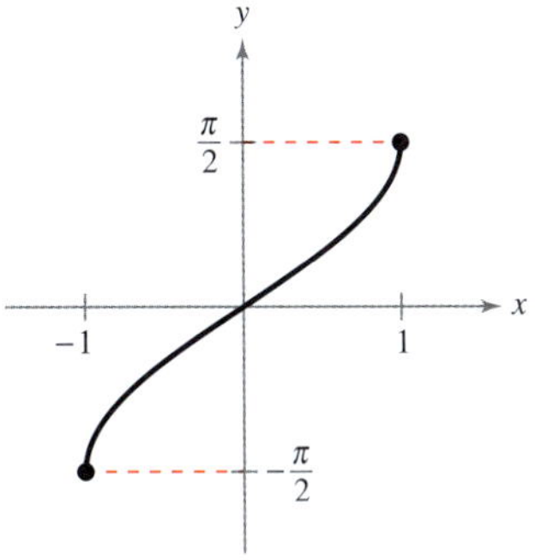

In Exercises 13–18, identify the parent function and describe the transformation shown in the graph. Write an equation for the graphed function. (There may be more than one correct answer.)

13.

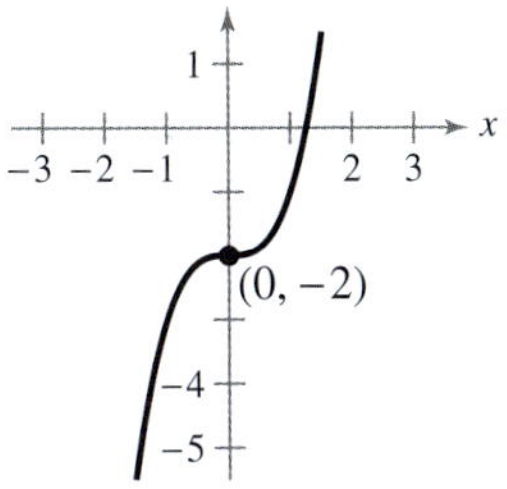

14.

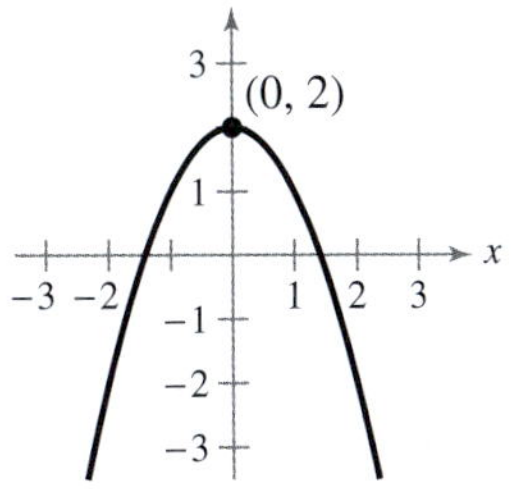

15.

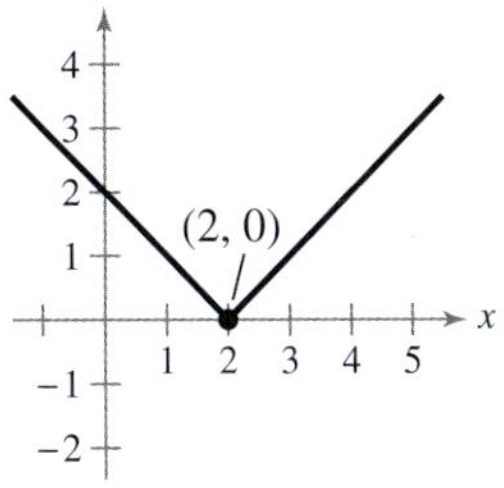

16.

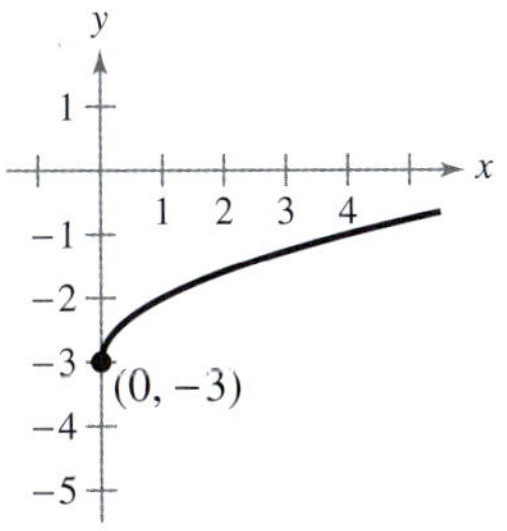

17.

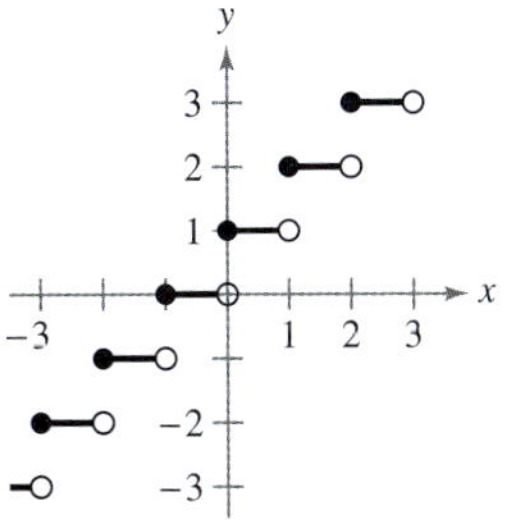

18.

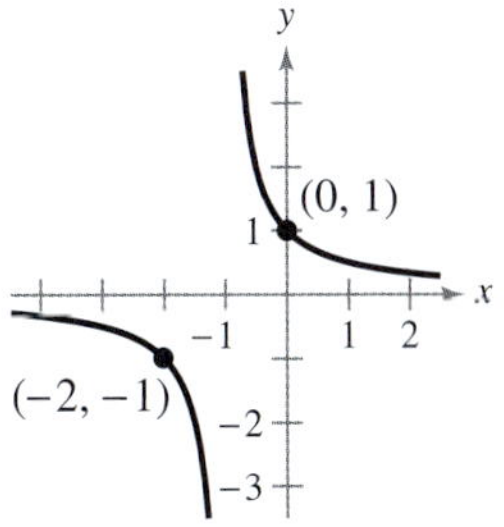

In Exercises 19 and 20, determine which function is represented by the graph. (Do not use a calculator.)

19. (a) $y = -\frac{3}{2}\sin x$

(b) $y = -\frac{3}{2}\cos x$

(c) $y = -\sin\frac{3}{2}x$

(d) $y = \cos\frac{3}{2}x$

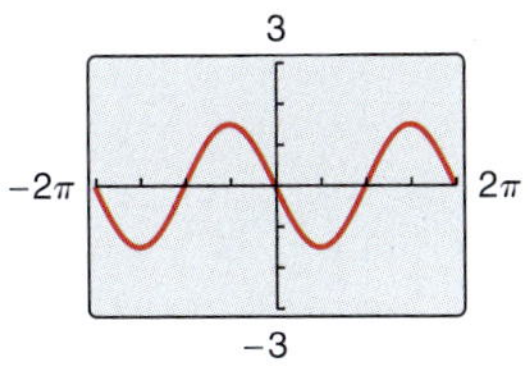

20. (a) $y = \frac{1}{2}\tan x$

(b) $y = \frac{1}{2}\cot x$

(c) $y = \tan\frac{x}{2}$

(d) $y = \cot\frac{x}{2}$

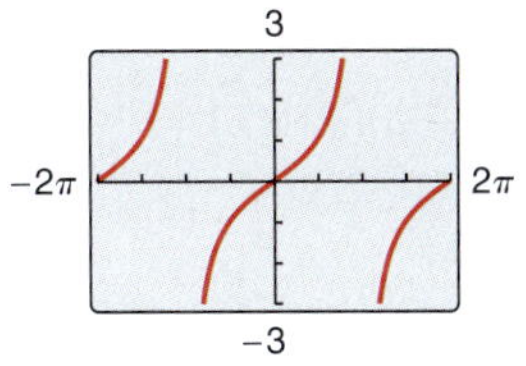

In Exercises 21–24, sketch the graph of the function by hand.

21. $y = (x - 1)^2 - 4$

22. $y = -(x - 4)^3$

23. $y = |x + 1| - 3$

24. $y = -2\cos 2x$

Proofs in Mathematics

The Pythagorean Theorem

The Pythagorean Theorem is one of the most famous theorems in mathematics with hundreds of different proofs. James A. Garfield, the twentieth president of the United States, developed a proof of the Pythagorean Theorem in 1876. His proof, shown below, involves the fact that a trapezoid can be formed from two congruent right triangles and an isosceles right triangle.

The Pythagorean Theorem

In a right triangle, the sum of the squares of the lengths of the legs is equal to the square of the length of the hypotenuse, where a and b are the legs and c is the hypotenuse.

$$a^2 + b^2 = c^2$$

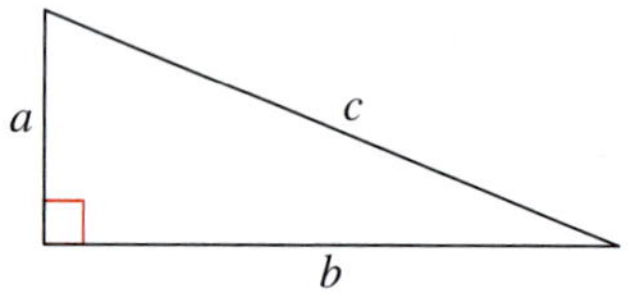

Proof

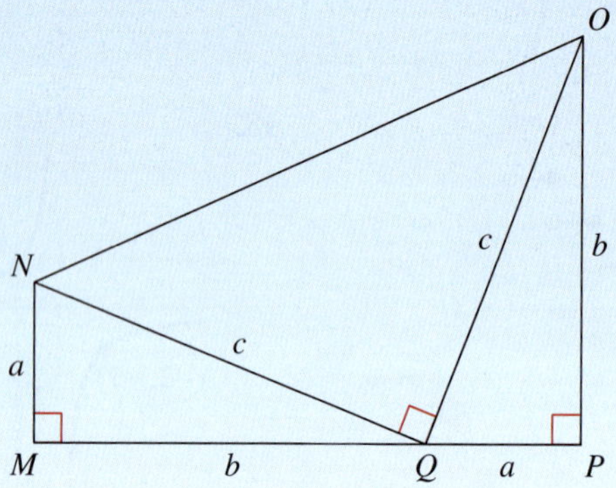

$$\text{Area of trapezoid } MNOP = \text{Area of } \triangle MNQ + \text{Area of } \triangle PQO + \text{Area of } \triangle NOQ$$

$$\frac{1}{2}(a + b)(a + b) = \frac{1}{2}ab + \frac{1}{2}ab + \frac{1}{2}c^2$$

$$\frac{1}{2}(a + b)(a + b) = ab + \frac{1}{2}c^2$$

$$(a + b)(a + b) = 2ab + c^2$$

$$a^2 + 2ab + b^2 = 2ab + c^2$$

$$a^2 + b^2 = c^2$$

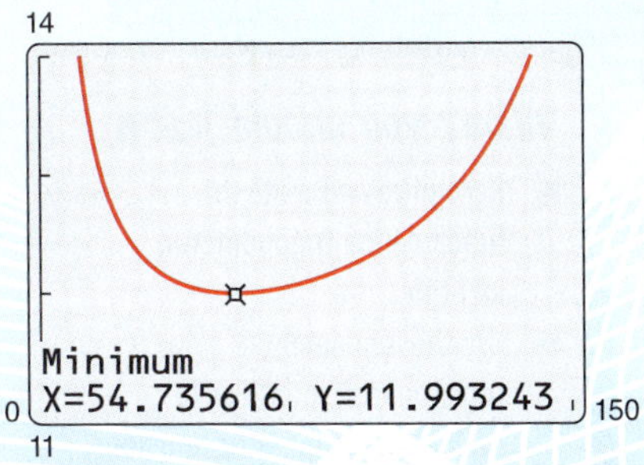

Section 5.3, Example 12
Minimum Surface Area
of a Honeycomb

5 Analytic Trigonometry

Student Resources at LarsonPrecalculus.com

- **Videos** explaining the concepts of precalculus
- **Worked-out solution videos** for all *Checkpoint* exercises
- **Editable spreadsheets** of the data sets in the text
- **Group projects** for each chapter applying concepts to real-life problems

kemie/DigitalVision Vectors/Getty Images, Monkey Business Images/Shutterstock.com

5.1 Using Fundamental Identities

Introduction

In Chapter 4, you studied the basic definitions, properties, graphs, and applications of the individual trigonometric functions. In this chapter, you will learn how to use the fundamental identities to perform four tasks.

1. Evaluate trigonometric functions.
2. Simplify trigonometric expressions.
3. Develop additional trigonometric identities.
4. Solve trigonometric equations.

Fundamental Trigonometric Identities

Reciprocal Identities

$$\sin u = \frac{1}{\csc u} \qquad \cos u = \frac{1}{\sec u} \qquad \tan u = \frac{1}{\cot u}$$

$$\csc u = \frac{1}{\sin u} \qquad \sec u = \frac{1}{\cos u} \qquad \cot u = \frac{1}{\tan u}$$

Quotient Identities

$$\tan u = \frac{\sin u}{\cos u} \qquad \cot u = \frac{\cos u}{\sin u}$$

Pythagorean Identities

$$\sin^2 u + \cos^2 u = 1$$

$$1 + \tan^2 u = \sec^2 u$$

$$1 + \cot^2 u = \csc^2 u$$

Cofunction Identities

$$\sin\left(\frac{\pi}{2} - u\right) = \cos u \qquad \cos\left(\frac{\pi}{2} - u\right) = \sin u$$

$$\tan\left(\frac{\pi}{2} - u\right) = \cot u \qquad \cot\left(\frac{\pi}{2} - u\right) = \tan u$$

$$\sec\left(\frac{\pi}{2} - u\right) = \csc u \qquad \csc\left(\frac{\pi}{2} - u\right) = \sec u$$

Even/Odd Identities

$$\sin(-u) = -\sin u \qquad \csc(-u) = -\csc u$$

$$\cos(-u) = \cos u \qquad \sec(-u) = \sec u$$

$$\tan(-u) = -\tan u \qquad \cot(-u) = -\cot u$$

Pythagorean identities are sometimes used in radical form such as

$$\sin u = \pm\sqrt{1 - \cos^2 u} \qquad \text{Radical form of } \sin^2 u + \cos^2 u = 1$$

or

$$\tan u = \pm\sqrt{\sec^2 u - 1} \qquad \text{Radical form of } 1 + \tan^2 u = \sec^2 u$$

where the sign depends on the choice of u.

What you should learn

- Recognize and write the fundamental trigonometric identities.
- Use the fundamental trigonometric identities to evaluate trigonometric functions, simplify trigonometric expressions, and rewrite trigonometric expressions.

Why you should learn it

The fundamental trigonometric identities can be used to simplify trigonometric expressions. For instance, in Exercise 107 on page 358, you can use trigonometric identities to simplify an expression for the coefficient of friction.

Federico Rostagno/iStock/Getty Images

Using the Fundamental Identities

One common use of trigonometric identities is to use given information about trigonometric functions to evaluate other trigonometric functions.

EXAMPLE 1 Using Identities to Evaluate Functions

Given $\sec u = -\frac{3}{2}$ and $\tan u > 0$, find the values of all six trigonometric functions.

Solution

Using a reciprocal identity, you have

$$\cos u = \frac{1}{\sec u} = \frac{1}{-3/2} = -\frac{2}{3}.$$

Using the Pythagorean identity $\sin^2 u + \cos^2 u = 1$ in the equivalent form $\sin^2 u = 1 - \cos^2 u$, you have

$$\sin^2 u = 1 - \cos^2 u \qquad \text{Pythagorean identity}$$
$$= 1 - \left(-\frac{2}{3}\right)^2 \qquad \text{Substitute } -\tfrac{2}{3} \text{ for } \cos u.$$
$$= 1 - \frac{4}{9} \qquad \text{Evaluate power.}$$
$$= \frac{5}{9}. \qquad \text{Simplify.}$$

Because $\sec u < 0$ and $\tan u > 0$, it follows that u lies in Quadrant III. Moreover, because $\sin u$ is negative when u is in Quadrant III, choose the negative root to obtain $\sin u = -\sqrt{5}/3$. Now, knowing the values of the sine and cosine, you can find the values of all six trigonometric functions.

$$\sin u = -\frac{\sqrt{5}}{3} \qquad \csc u = \frac{1}{\sin u} = -\frac{3}{\sqrt{5}} = -\frac{3\sqrt{5}}{5}$$

$$\cos u = -\frac{2}{3} \qquad \sec u = \frac{1}{\cos u} = -\frac{3}{2}$$

$$\tan u = \frac{\sin u}{\cos u} = \frac{-\sqrt{5}/3}{-2/3} = \frac{\sqrt{5}}{2} \qquad \cot u = \frac{1}{\tan u} = \frac{2}{\sqrt{5}} = \frac{2\sqrt{5}}{5}$$

✓ **Checkpoint** Audio-video solution in English & Spanish at LarsonPrecalculus.com

Given $\tan x = \frac{1}{3}$ and $\cos x < 0$, find the values of all six trigonometric functions.

EXAMPLE 2 Simplifying a Trigonometric Expression

To simplify $\sin x \cos^2 x - \sin x$, factor out the common monomial factor $\sin x$ and then use a fundamental identity.

$$\sin x \cos^2 x - \sin x = \sin x(\cos^2 x - 1) \qquad \text{Factor out monomial factor.}$$
$$= -\sin x(1 - \cos^2 x) \qquad \text{Factor out } -1.$$
$$= -\sin x(\sin^2 x) \qquad \text{Pythagorean identity}$$
$$= -\sin^3 x \qquad \text{Multiply.}$$

✓ **Checkpoint** Audio-video solution in English & Spanish at LarsonPrecalculus.com

Simplify $\cos^2 x \csc x - \csc x$.

Technology Tip

You can use a graphing utility to check the result of Example 2. To do this, let

$$y_1 = \sin x \cos^2 x - \sin x$$

and

$$y_2 = -\sin^3 x.$$

Select the *line* style for y_1 and the *path* style for y_2. Now, graph both equations in the same viewing window. The two graphs *appear* to coincide, so the expressions *appear* to be equivalent. Remember that in order to be certain that two expressions are equivalent, you need to show their equivalence algebraically, as in Example 2.

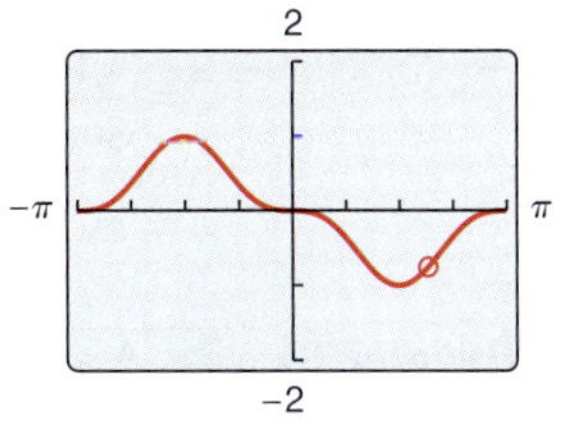

When factoring trigonometric expressions, it is helpful to find a polynomial form that fits the expression, as shown in Example 3.

EXAMPLE 3 Factoring Trigonometric Expressions

Factor (a) $\sec^2 \theta - 1$ and (b) $4 \tan^2 \theta + \tan \theta - 3$.

Solution

a. This expression is a difference of two squares, which factors as

$$\sec^2 \theta - 1 = (\sec \theta - 1)(\sec \theta + 1).$$

b. This expression has the polynomial form $ax^2 + bx + c$, and it factors as

$$4 \tan^2 \theta + \tan \theta - 3 = (4 \tan \theta - 3)(\tan \theta + 1).$$

✓ Checkpoint *Audio-video solution in English & Spanish at LarsonPrecalculus.com*

Factor (a) $1 - \cos^2 \theta$ and (b) $2 \csc^2 \theta - 7 \csc \theta + 6$.

On occasion, factoring or simplifying a trigonometric expression can best be done by first rewriting the expression in terms of just *one* trigonometric function or in terms of *sine or cosine alone*. These strategies are illustrated in Examples 4 and 5.

EXAMPLE 4 Factoring a Trigonometric Expression

Factor $\csc^2 x - \cot x - 3$.

Solution

Use the identity $\csc^2 x = 1 + \cot^2 x$ to rewrite the expression in terms of the cotangent.

$$\csc^2 x - \cot x - 3 = (1 + \cot^2 x) - \cot x - 3 \quad \text{Pythagorean identity}$$

$$= \cot^2 x - \cot x - 2 \quad \text{Combine like terms.}$$

$$= (\cot x - 2)(\cot x + 1) \quad \text{Factor.}$$

✓ Checkpoint *Audio-video solution in English & Spanish at LarsonPrecalculus.com*

Factor $\sec^2 x + 3 \tan x + 1$.

EXAMPLE 5 Simplifying a Trigonometric Expression

See LarsonPrecalculus.com for an interactive version of this type of example.

Simplify $\sin t + \cot t \cos t$.

Solution

Begin by rewriting $\cot t$ in terms of sine and cosine.

$$\sin t + \cot t \cos t = \sin t + \left(\frac{\cos t}{\sin t}\right)\cos t \quad \text{Quotient identity}$$

$$= \frac{\sin^2 t + \cos^2 t}{\sin t} \quad \text{Add fractions.}$$

$$= \frac{1}{\sin t} \quad \text{Pythagorean identity}$$

$$= \csc t \quad \text{Reciprocal identity}$$

✓ Checkpoint *Audio-video solution in English & Spanish at LarsonPrecalculus.com*

Simplify $\csc x - \cos x \cot x$.

Technology Tip

You can use the *table* feature of a graphing utility to check the result of Example 5. To do this, let

$$y_1 = \sin x + \left(\frac{\cos x}{\sin x}\right)\cos x$$

and

$$y_2 = \frac{1}{\sin x}.$$

Now, create a table that shows the values of y_1 and y_2 for different values of x. The values of y_1 and y_2 *appear* to be identical, so the expressions *appear* to be equivalent. To show that the expressions are equivalent, you need to show their equivalence algebraically, as in Example 5.

stockyimages/Shutterstock.com

The next two examples involve techniques for rewriting expressions in forms that are used in calculus.

EXAMPLE 6 Rewriting a Trigonometric Expression

Rewrite $\dfrac{1}{1 + \sin x}$ so that it is *not* in fractional form.

Solution

From the Pythagorean identity

$$\cos^2 x = 1 - \sin^2 x = (1 - \sin x)(1 + \sin x)$$

you can see that multiplying both the numerator and the denominator by $(1 - \sin x)$ will produce a monomial denominator.

$$\frac{1}{1 + \sin x} = \frac{1}{1 + \sin x} \cdot \frac{1 - \sin x}{1 - \sin x}$$ Multiply numerator and denominator by $(1 - \sin x)$.

$$= \frac{1 - \sin x}{1 - \sin^2 x}$$ Multiply.

$$= \frac{1 - \sin x}{\cos^2 x}$$ Pythagorean identity

$$= \frac{1}{\cos^2 x} - \frac{\sin x}{\cos^2 x}$$ Write as separate fractions.

$$= \frac{1}{\cos^2 x} - \frac{\sin x}{\cos x} \cdot \frac{1}{\cos x}$$ Product of fractions

$$= \sec^2 x - \tan x \sec x$$ Reciprocal and quotient identities

✓ **Checkpoint** Audio-video solution in English & Spanish at LarsonPrecalculus.com

Rewrite $\dfrac{\cos^2 \theta}{1 - \sin \theta}$ so that it is *not* in fractional form.

EXAMPLE 7 Trigonometric Substitution

Use the substitution $x = 2 \tan \theta$, $0 < \theta < \pi/2$, to write $\sqrt{4 + x^2}$ as a trigonometric function of θ.

Solution

Begin by letting $x = 2 \tan \theta$. Then you can obtain

$$\sqrt{4 + x^2} = \sqrt{4 + (2 \tan \theta)^2}$$ Substitute $2 \tan \theta$ for x.

$$= \sqrt{4(1 + \tan^2 \theta)}$$ Property of exponents and factor out 4.

$$= \sqrt{4 \sec^2 \theta}$$ Pythagorean identity

$$= 2 \sec \theta.$$ $\sec \theta > 0$ for $0 < \theta < \dfrac{\pi}{2}$

✓ **Checkpoint** Audio-video solution in English & Spanish at LarsonPrecalculus.com

Use the substitution $x = 3 \sin \theta$, $0 < \theta < \pi/2$, to write $\sqrt{9 - x^2}$ as a trigonometric function of θ.

Figure 5.1 shows the right triangle illustration of the trigonometric substitution $x = 2 \tan \theta$ in Example 7. For $0 < \theta < \pi/2$, you have

$$\text{opp} = x, \text{ adj} = 2, \text{ and hyp} = \sqrt{4 + x^2}.$$

Try using these expressions to obtain the result shown in Example 7.

Insight

You will use trigonometric identities on most standardized tests. It is helpful to know the Pythagorean identity

$$\sin^2 \theta + \cos^2 \theta = 1$$

in addition to the reciprocal and quotient identities.

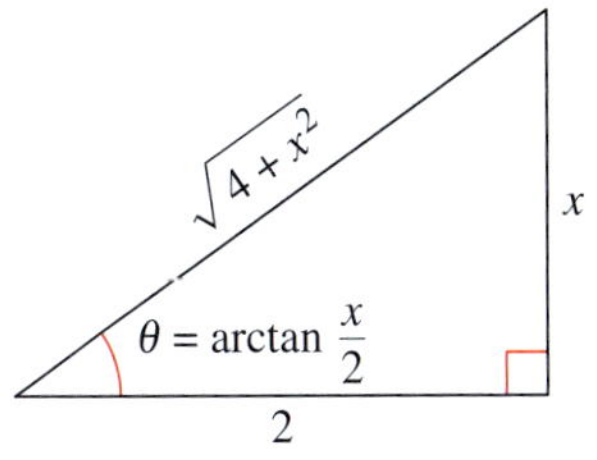

$\tan \theta = \dfrac{x}{2}$

Figure 5.1

5.1 Exercises

See *CalcChat.com* for tutorial help and worked-out solutions to odd-numbered exercises. For instructions on how to use a graphing utility, see Appendix A.

Vocabulary and Concept Check

1. Match each function with an equivalent expression.

(a) $\sin u$ (i) $\dfrac{1}{\sec u}$

(b) $\cos u$ (ii) $\dfrac{1}{\cot u}$

(c) $\tan u$ (iii) $\dfrac{1}{\csc u}$

2. Match each expression with an equivalent expression.

(a) $\sin^2 u$ (i) $1 + \cot^2 u$

(b) $\sec^2 u$ (ii) $1 - \cos^2 u$

(c) $\csc^2 u$ (iii) $1 + \tan^2 u$

In Exercises 3–6, fill in the blank to complete the trigonometric identity.

3. $\cos\left(\dfrac{\pi}{2} - u\right) =$ ______

4. $\csc\left(\dfrac{\pi}{2} - u\right) =$ ______

5. $\sec(-u) =$ ______

6. $\tan(-u) =$ ______

Procedures and Problem Solving

Using Identities to Evaluate Functions **In Exercises 7–18, use the given information to find the values of all six trigonometric functions.**

7. $\sec x = -\dfrac{5}{2},\quad \tan x < 0$

8. $\csc x = -\dfrac{7}{6},\quad \tan x > 0$

9. $\sin \theta = -\dfrac{3}{4},\quad \cos \theta > 0$

10. $\tan \phi = -\dfrac{7}{3},\quad \cos \phi < 0$

11. $\cos \phi = \dfrac{1}{5},\quad \cot \phi > 0$

12. $\cot x = \dfrac{7}{4},\quad \sin x < 0$

13. $\sin x = \dfrac{1}{2},\quad \cos x = \dfrac{\sqrt{3}}{2}$

14. $\tan x = \dfrac{\sqrt{3}}{3},\quad \csc x = -2$

15. $\sec \phi = -\dfrac{17}{15},\quad \tan \phi = -\dfrac{8}{15}$

16. $\cot \phi = -5,\quad \sin \phi = \dfrac{\sqrt{26}}{26}$

17. $\csc \theta$ is undefined, $\cos \theta < 0$

18. $\tan \theta$ is undefined, $\sin \theta > 0$

Matching Trigonometric Expressions **In Exercises 19–24, match the trigonometric expression with its simplified form.**

(a) $\sec x$ **(b)** -1 **(c)** $-\tan x$

(d) 1 **(e)** $-\csc x$ **(f)** $\sin x$

19. $\sec x \cos x$

20. $\tan x \csc x$

21. $\cot^2 x - \csc^2 x$

22. $(1 - \cos^2 x)(\csc x)$

23. $\dfrac{\cos[(5\pi/2) - x]}{\sin[x - (5\pi/2)]}$

24. $\dfrac{\cot(-x)}{\cos x}$

Matching Trigonometric Expressions **In Exercises 25–30, match the trigonometric expression with its simplified form.**

(a) $\csc x$ **(b)** $\tan x$ **(c)** $\sin^2 x$

(d) $\sin x \tan x$ **(e)** $\sec^2 x$ **(f)** $\sec^2 x + \tan^2 x$

25. $\sin x \sec x$

26. $\cos^2 x(\sec^2 x - 1)$

27. $\sec^4 x - \tan^4 x$

28. $\cot x \sec x$

29. $\dfrac{\sec^2 x - 1}{\sin^2 x}$

30. $\dfrac{\cos^2[(\pi/2) - x]}{\cos x}$

Simplifying a Trigonometric Expression **In Exercises 31–46, use the fundamental identities to simplify the expression. Use the *table* feature of a graphing utility to check your result numerically.**

31. $\cot x \sin x$

32. $\cos \beta \tan \beta$

33. $\sin \phi(\csc \phi - \sin \phi)$

34. $\cos x(1 + \tan^2 x)$

35. $\tan^2 x - \tan^2 x \sin^2 x$

36. $\sin^2 x \sec^2 x - \sin^2 x$

37. $\sin \beta \tan \beta + \cos \beta$

38. $\cot u \sin u + \tan u \cos u$

39. $\dfrac{\csc x}{\cot x}$

40. $\dfrac{\sec \theta}{\csc \theta}$

41. $\sec \alpha\left(\dfrac{\sin \alpha}{\tan \alpha}\right)$

42. $\dfrac{1 + \tan^2 \theta}{\sec^2 \theta}$

43. $\sin\left(x - \dfrac{\pi}{2}\right)\csc x$

44. $\cot\left(\dfrac{\pi}{2} - x\right)\cos x$

45. $\dfrac{\cos^2 y}{1 - \sin y}$

46. $\dfrac{1}{\cot^2 x + 1}$

Factoring a Trigonometric Expression **In Exercises 47–56, factor the expression. Use the fundamental identities to simplify, if necessary. (There is more than one correct form of each answer.)**

47. $\dfrac{\sec^2 x - 1}{\sec x - 1}$

48. $\dfrac{\cos x - 2}{\cos^2 x - 4}$

49. $1 - 2\cos^2 x + \cos^4 x$

50. $\sec^4 x - \tan^4 x$

51. $3\sin^2 x - 5\sin x - 2$

52. $6\cos^2 x + 5\cos x - 6$

53. $\cot^3 x + \cot^2 x + \cot x + 1$

54. $\tan^3 x - \tan^2 x - 2\tan x + 2$

55. $\cot^2 x + \csc x - 1$

56. $\sin^2 x + 3\cos x + 3$

Multiplying Trigonometric Expressions **In Exercises 57–60, perform the multiplication and use the fundamental identities to simplify.**

57. $(\sin x + \cos x)^2$

58. $(1 + \tan x)^2$

59. $(\csc x + 1)(\csc x - 1)$

60. $(5 - 5\sin x)(5 + 5\sin x)$

Adding or Subtracting Trigonometric Expressions **In Exercises 61–66, perform the addition or subtraction and use the fundamental identities to simplify.**

61. $\dfrac{1}{1 + \cos x} + \dfrac{1}{1 - \cos x}$

62. $\dfrac{1}{\sec x + 1} - \dfrac{1}{\sec x - 1}$

63. $\tan x - \dfrac{\sec^2 x}{\tan x}$

64. $\tan x + \dfrac{\cos x}{1 + \sin x}$

65. $\dfrac{\cos x}{1 + \sin x} + \dfrac{1 + \sin x}{\cos x}$

66. $\dfrac{\tan x}{1 + \sec x} + \dfrac{1 + \sec x}{\tan x}$

Rewriting a Trigonometric Expression **In Exercises 67–72, rewrite the expression so that it is *not* in fractional form.**

67. $\dfrac{\sin x}{\tan x}$

68. $\dfrac{\csc y}{\cot y}$

69. $\dfrac{\sin^2 y}{1 - \cos y}$

70. $\dfrac{\tan^2 x}{\csc x + 1}$

71. $\dfrac{3}{\sec x - \tan x}$

72. $\dfrac{5}{\tan x + \sec x}$

Graphing Trigonometric Functions **In Exercises 73–76, use a graphing utility to complete the table and graph the functions in the same viewing window. Make a conjecture about y_1 and y_2.**

x	0.2	0.4	0.6	0.8	1.0	1.2	1.4
y_1							
y_2							

73. $y_1 = \cos\left(\dfrac{\pi}{2} - x\right), \quad y_2 = \sin(x + \pi)$

74. $y_1 = \cos x + \sin x \tan x, \quad y_2 = \sec x$

75. $y_1 = \dfrac{\cos x}{1 - \sin x}, \quad y_2 = \dfrac{1 + \sin x}{\cos x}$

76. $y_1 = \sec^4 x - \sec^2 x, \quad y_2 = -\tan^2 x - \tan^4 x$

Graphing Trigonometric Functions **In Exercises 77–80, use a graphing utility to determine which of the six trigonometric functions is equal to the expression. Verify your answer algebraically.**

77. $\cos x \cot x + \sin x$

78. $\sin x(\cot x + \tan x)$

79. $\dfrac{1}{\sin x}\left(\dfrac{1}{\cos x} - \cos x\right)$

80. $\dfrac{1}{2}\left(\dfrac{1 + \sin \theta}{\cos \theta} + \dfrac{\cos \theta}{1 + \sin \theta}\right)$

Trigonometric Substitution **In Exercises 81–92, use the trigonometric substitution to write the algebraic expression as a trigonometric function of θ, where $0 < \theta < \pi/2$.**

81. $\sqrt{9 - x^2}, \quad x = 3\cos \theta$

82. $\sqrt{4 - x^2}, \quad x = 2\cos \theta$

83. $\sqrt{x^2 + 9}, \quad x = 3\tan \theta$

84. $\sqrt{x^2 + 100}, \quad x = 10\tan \theta$

85. $\sqrt{49 - x^2}, \quad x = 7\sin \theta$

86. $\sqrt{64 - x^2}, \quad x = 8\cos \theta$

87. $\sqrt{4x^2 + 9}, \quad 2x = 3\tan \theta$

88. $\sqrt{9x^2 + 4}, \quad 3x = 2\tan \theta$

89. $\sqrt{2 - x^2}, \quad x = \sqrt{2}\sin \theta$

90. $\sqrt{5 - x^2}, \quad x = \sqrt{5}\cos \theta$

91. $\sqrt{(1 + x^2)^3}, \quad x = \tan \theta$

92. $\sqrt{(x^2 - 16)^3}, \quad x = 4\sec \theta$

Solving a Trigonometric Equation **In Exercises 93–96, use a graphing utility to solve the equation for θ, where $0 \le \theta < 2\pi$.**

93. $\sin \theta = \sqrt{1 - \cos^2 \theta}$

94. $\cos \theta = -\sqrt{1 - \sin^2 \theta}$

95. $\sec \theta = \sqrt{1 + \tan^2 \theta}$

96. $\tan \theta = \sqrt{\sec^2 \theta - 1}$

Rewriting an Expression In Exercises 97–102, rewrite the expression as a single logarithm and simplify the result. (*Hint:* Begin by using the properties of logarithms.)

97. $\ln|\cos\theta| - \ln|\sin\theta|$ **98.** $\ln|\cot\theta| + \ln|\sin\theta|$

99. $\ln|\sec x| + \ln|\sin x|$ **100.** $\ln|\tan x| - \ln|\sin x|$

101. $\ln(1 + \sin x) - \ln|\sec x|$

102. $\ln|\cot t| + \ln(1 + \tan^2 t)$

Demonstrating an Identity Numerically In Exercises 103–106, use the *table* feature of a graphing utility to demonstrate the identity for each value of θ.

103. $\csc^2\theta - \cot^2\theta = 1$, (a) $\theta = 132°$ (b) $\theta = \dfrac{2\pi}{7}$

104. $\tan^2\theta + 1 = \sec^2\theta$, (a) $\theta = 346°$ (b) $\theta = 3.1$

105. $\cos\left(\dfrac{\pi}{2} - \theta\right) = \sin\theta$, (a) $\theta = 80°$ (b) $\theta = 0.8$

106. $\sin(-\theta) = -\sin\theta$, (a) $\theta = 250°$ (b) $\theta = \dfrac{\pi}{4}$

107. ***Why you should learn it*** *(p. 352)* The forces acting on an object weighing W units on an inclined plane positioned at an angle of θ with the horizontal are modeled by

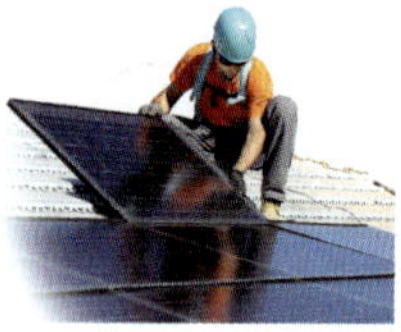

$$\mu W\cos\theta = W\sin\theta$$

where μ is the coefficient of friction (see figure). Solve the equation for μ and simplify the result.

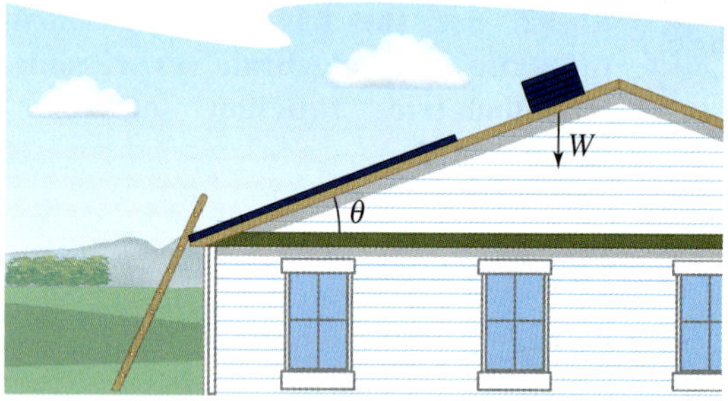

108. Rate of Change The rate of change of the function $f(x) = \tan x + \cot x$ is given by the expression $\sec^2 x - \csc^2 x$. Show that this expression can also be written as $\tan^2 x - \cot^2 x$.

109. Rate of Change The rate of change of the function $f(x) = \sec x + \cos x$ is given by the expression $\sec x \tan x - \sin x$. Show that this expression can also be written as $\sin x \tan^2 x$.

Focusing on Concepts

True or False? In Exercises 110 and 111, determine whether the statement is true or false. Justify your answer.

110. $\sin\theta\csc\theta = 1$ **111.** $\cos\theta\sec\phi = 1$

112. HOW DO YOU SEE IT?

Explain how to use the figure to derive the Pythagorean identities

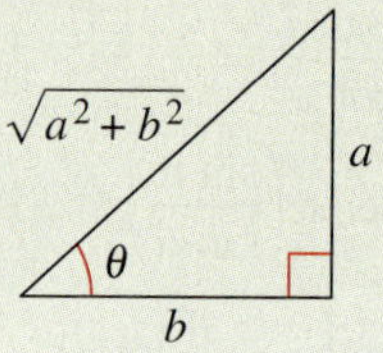

$\sin^2\theta + \cos^2\theta = 1$,

$1 + \tan^2\theta = \sec^2\theta$,

and $1 + \cot^2\theta = \csc^2\theta$.

Discuss how to remember these identities and other fundamental trigonometric identities.

Analyzing Trigonometric Functions In Exercises 113–116, fill in the blanks. (*Note*: $x \to c^+$ indicates that x approaches c from the right, and $x \to c^-$ indicates that x approaches c from the left.)

113. As $x \to \dfrac{\pi}{2}^-$, $\sin x \to$ ____ and $\csc x \to$ ____.

114. As $x \to 0^+$, $\cos x \to$ ____ and $\sec x \to$ ____.

115. As $x \to \dfrac{\pi}{2}^+$, $\tan x \to$ ____ and $\cot x \to$ ____.

116. As $x \to \pi^-$, $\sin x \to$ ____ and $\csc x \to$ ____.

117. Rewriting Trigonometric Functions Rewrite each trigonometric function of θ in terms of $\sin\theta$.

118. Rewriting Trigonometric Functions Rewrite each trigonometric function of θ in terms of $\cos\theta$.

Rewriting a Trigonometric Expression In Exercises 119 and 120, rewrite the expression in terms of $\sin\theta$ and $\cos\theta$.

119. $\dfrac{\sec\theta(1 + \tan\theta)}{\sec\theta + \csc\theta}$ **120.** $\dfrac{\csc\theta(1 + \cot\theta)}{\tan\theta + \cot\theta}$

Cumulative Mixed Review

Finding an Inverse Function In Exercises 121–124, determine whether the function has an inverse function. If it does, find the inverse function.

121. $f(x) = -10$

122. $f(x) = (x - 7)^2 + 3$

123. $f(x) = \sqrt{3x - 14}$

124. $f(x) = \sqrt[4]{x - 5}$

Graphing Trigonometric Functions In Exercises 125–128, sketch the graph of the function. (Include two full periods.)

125. $f(x) = -\sin\pi x - 1$ **126.** $f(x) = -2\tan\dfrac{\pi x}{2}$

127. $f(x) = \dfrac{1}{2}\cot\left(x + \dfrac{\pi}{4}\right)$ **128.** $f(x) = \cos(x - \pi) + 3$

5.2 Verifying Trigonometric Identities

What you should learn

- Verify trigonometric identities.

Why you should learn it

You can use trigonometric identities to rewrite trigonometric expressions. For instance, Exercise 85 on page 366 shows you how trigonometric identities can be used to simplify an equation that models the length of a shadow cast by a gnomon (a device used to tell time).

Verifying Trigonometric Identities

In the preceding section, you used identities to rewrite trigonometric expressions in equivalent forms. In this section, you will study techniques for verifying trigonometric identities. In the next section, you will study techniques for solving trigonometric equations. The key to both verifying identities *and* solving equations is your ability to use the fundamental identities and the rules of algebra to rewrite trigonometric expressions.

Remember that a *conditional equation* is an equation that is true for only some of the values in its domain. For example, the conditional equation

$$\sin x = 0 \qquad \text{Conditional equation}$$

is true only for

$$x = n\pi$$

where n is an integer. When you find these values, you are *solving* the equation.

On the other hand, an equation that is true for all real values in the domain of the variable is an *identity*. For example, the familiar equation

$$\sin^2 x = 1 - \cos^2 x \qquad \text{Identity}$$

is true for all real numbers x. So, it is an identity.

Verifying that a trigonometric equation is an identity is quite different from solving an equation. There is no well-defined set of rules to follow in verifying trigonometric identities, and the process is best learned by practice.

Guidelines for Verifying Trigonometric Identities

1. Work with one side of the equation at a time. It is often better to work with the more complicated side first.
2. Look for opportunities to factor an expression, add fractions, square a binomial, or create a monomial denominator.
3. Look for opportunities to use the fundamental identities. Note which functions are in the final expression you want. Sines and cosines pair up well, as do secants and tangents, and cosecants and cotangents.
4. When the preceding guidelines do not help, try converting all terms to sines and cosines.
5. Always try *something*. Even making an attempt that leads to a dead end provides insight.

Verifying trigonometric identities is a useful process when you need to convert a trigonometric expression into a form that is more useful algebraically. When you verify an identity, you cannot assume that the two sides of the equation are equal because you are trying to verify that they are equal. As a result, when verifying identities, you cannot use operations such as adding the same quantity to each side of the equation or cross multiplication.

Elfstrom/E+/Getty Images; Viorel Sima/Shutterstock.com

EXAMPLE 1 Verifying a Trigonometric Identity

Verify the identity $\dfrac{\sec^2 \theta - 1}{\sec^2 \theta} = \sin^2 \theta$.

Solution

Because the left side is more complicated, start with it.

$$\frac{\sec^2 \theta - 1}{\sec^2 \theta} = \frac{(\tan^2 \theta + 1) - 1}{\sec^2 \theta} \qquad \text{Pythagorean identity}$$

$$= \frac{\tan^2 \theta}{\sec^2 \theta} \qquad \text{Simplify.}$$

$$= \tan^2 \theta(\cos^2 \theta) \qquad \text{Reciprocal identity}$$

$$= \frac{\sin^2 \theta}{\cancel{\cos^2 \theta}}\cancel{(\cos^2 \theta)} \qquad \text{Quotient identity}$$

$$= \sin^2 \theta \qquad \text{Simplify.}$$

Checkpoint *Audio-video solution in English & Spanish at LarsonPrecalculus.com*

Verify the identity $\dfrac{\sin^2 \theta + \cos^2 \theta}{\cos^2 \theta \sec^2 \theta} = 1$.

There can be more than one way to verify an identity. Here is another way to verify the identity in Example 1.

$$\frac{\sec^2 \theta - 1}{\sec^2 \theta} = \frac{\sec^2 \theta}{\sec^2 \theta} - \frac{1}{\sec^2 \theta} \qquad \text{Write as separate fractions.}$$

$$= 1 - \cos^2 \theta \qquad \text{Reciprocal identity}$$

$$= \sin^2 \theta \qquad \text{Pythagorean identity}$$

Remember that an identity is true only for all real values in the domain of the variable. For instance, in Example 1 the identity is not true when $\theta = \pi/2$ because $\sec^2 \theta$ is undefined when $\theta = \pi/2$.

Technology Tip

Although a graphing utility can be useful in helping to verify an identity, you must use algebraic techniques to produce a valid proof. For example, graph the two functions $y_1 = \sin 50x$ and $y_2 = \sin 2x$ in a trigonometric viewing window. On some graphing utilities, the graphs appear to be identical. However, $\sin 50x \neq \sin 2x$.

EXAMPLE 2 Combining Fractions Before Using Identities

Verify the identity $2 \sec^2 \alpha = \dfrac{1}{1 - \sin \alpha} + \dfrac{1}{1 + \sin \alpha}$.

Algebraic Solution

The right side is more complicated, so start with it.

$$\frac{1}{1 - \sin \alpha} + \frac{1}{1 + \sin \alpha} = \frac{1 + \sin \alpha + 1 - \sin \alpha}{(1 - \sin \alpha)(1 + \sin \alpha)} \qquad \text{Add fractions.}$$

$$= \frac{2}{1 - \sin^2 \alpha} \qquad \text{Simplify.}$$

$$= \frac{2}{\cos^2 \alpha} \qquad \text{Pythagorean identity}$$

$$= 2 \sec^2 \alpha \qquad \text{Reciprocal identity}$$

Numerical Solution

Use a graphing utility to create a table that shows the values of $y_1 = 2/\cos^2 x$ and $y_2 = 1/(1 - \sin x) + 1/(1 + \sin x)$ for different values of x, as shown in the figure.

X	Y1	Y2
-.5	2.5969	2.5969
-.25	2.1304	2.1304
0	2	2
.25	2.1304	2.1304
.5	2.5969	2.5969
.75	3.7357	3.7357
1	6.851	6.851
X=-.5		

The values for y_1 and y_2 appear to be identical, so the equation appears to be an identity.

Checkpoint *Audio-video solution in English & Spanish at LarsonPrecalculus.com*

Verify the identity $2 \csc^2 \beta = \dfrac{1}{1 - \cos \beta} + \dfrac{1}{1 + \cos \beta}$.

In Example 2, you needed to write the Pythagorean identity $\sin^2 u + \cos^2 u = 1$ in the equivalent form $\cos^2 u = 1 - \sin^2 u$. When verifying identities, you may find it useful to write Pythagorean identities in one of these equivalent forms.

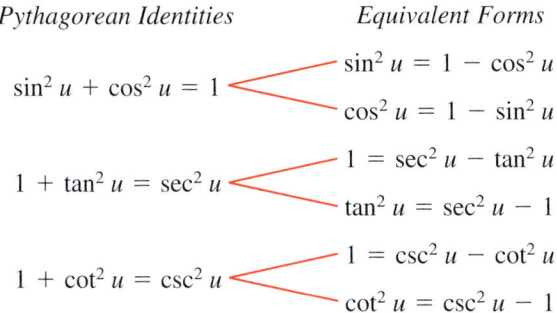

Pythagorean Identities	*Equivalent Forms*
$\sin^2 u + \cos^2 u = 1$	$\sin^2 u = 1 - \cos^2 u$
	$\cos^2 u = 1 - \sin^2 u$
$1 + \tan^2 u = \sec^2 u$	$1 = \sec^2 u - \tan^2 u$
	$\tan^2 u = \sec^2 u - 1$
$1 + \cot^2 u = \csc^2 u$	$1 = \csc^2 u - \cot^2 u$
	$\cot^2 u = \csc^2 u - 1$

EXAMPLE 3 Verifying a Trigonometric Identity

Verify the identity $(\tan^2 x + 1)(\cos^2 x - 1) = -\tan^2 x$.

Algebraic Solution

By applying identities before multiplying, you obtain the following.

$$(\tan^2 x + 1)(\cos^2 x - 1) = (\sec^2 x)(-\sin^2 x) \quad \text{Pythagorean identities}$$

$$= -\frac{\sin^2 x}{\cos^2 x} \quad \text{Reciprocal identity}$$

$$= -\left(\frac{\sin x}{\cos x}\right)^2 \quad \text{Property of exponents}$$

$$= -\tan^2 x \quad \text{Quotient identity}$$

Graphical Solution

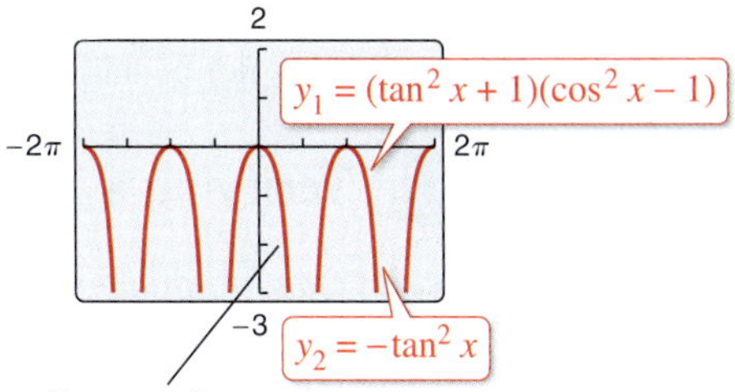

Because the graphs appear to coincide, the given equation appears to be an identity.

Checkpoint *Audio-video solution in English & Spanish at LarsonPrecalculus.com*

Verify the identity $(\sec^2 x - 1)(\sin^2 x - 1) = -\sin^2 x$.

EXAMPLE 4 Converting to Sines and Cosines

Verify the identity $\tan x + \cot x = \sec x \csc x$.

Solution

In this case there appear to be no fractions to add, no products to find, and no opportunities to use the Pythagorean identities. So, try converting the left side to sines and cosines.

$$\tan x + \cot x = \frac{\sin x}{\cos x} + \frac{\cos x}{\sin x} \quad \text{Quotient identities}$$

$$= \frac{\sin^2 x + \cos^2 x}{\cos x \sin x} \quad \text{Add fractions.}$$

$$= \frac{1}{\cos x \sin x} \quad \text{Pythagorean identity}$$

$$= \frac{1}{\cos x} \cdot \frac{1}{\sin x} \quad \text{Product of fractions}$$

$$= \sec x \csc x \quad \text{Reciprocal identities}$$

Checkpoint *Audio-video solution in English & Spanish at LarsonPrecalculus.com*

Verify the identity $\csc x - \sin x = \cos x \cot x$.

Recall from algebra that *rationalizing the denominator* using conjugates is, on occasion, a powerful simplification technique. A related form of this technique works for simplifying trigonometric expressions as well. For instance, to simplify

$$\frac{1}{1 - \cos x}$$

multiply the numerator and the denominator by $(1 + \cos x)$.

$$\begin{aligned}\frac{1}{1 - \cos x} &= \frac{1}{1 - \cos x}\left(\frac{1 + \cos x}{1 + \cos x}\right)\\ &= \frac{1 + \cos x}{1 - \cos^2 x}\\ &= \frac{1 + \cos x}{\sin^2 x}\\ &= \csc^2 x(1 + \cos x)\end{aligned}$$

The expression $\csc^2 x(1 + \cos x)$ is considered a simplified form of

$$\frac{1}{1 - \cos x}$$

because the expression does not contain any fractions.

Insight

Verifying trigonometric identities in the exercises will help you to recognize simplified forms of trigonometric expressions used on standardized tests.

EXAMPLE 5 Verifying a Trigonometric Identity

See LarsonPrecalculus.com for an interactive version of this type of example.

Verify the identity.

$$\sec x + \tan x = \frac{\cos x}{1 - \sin x}$$

Algebraic Solution

Begin with the *right* side because you can create a monomial denominator by multiplying the numerator and denominator by $(1 + \sin x)$.

$$\begin{aligned}\frac{\cos x}{1 - \sin x} &= \frac{\cos x}{1 - \sin x}\left(\frac{1 + \sin x}{1 + \sin x}\right) && \text{Multiply numerator and denominator by } (1 + \sin x).\\ &= \frac{\cos x + \cos x \sin x}{1 - \sin^2 x} && \text{Multiply.}\\ &= \frac{\cos x + \cos x \sin x}{\cos^2 x} && \text{Pythagorean identity}\\ &= \frac{\cos x}{\cos^2 x} + \frac{\cos x \sin x}{\cos^2 x} && \text{Write as separate fractions.}\\ &= \frac{1}{\cos x} + \frac{\sin x}{\cos x} && \text{Simplify.}\\ &= \sec x + \tan x && \text{Identities}\end{aligned}$$

Graphical Solution

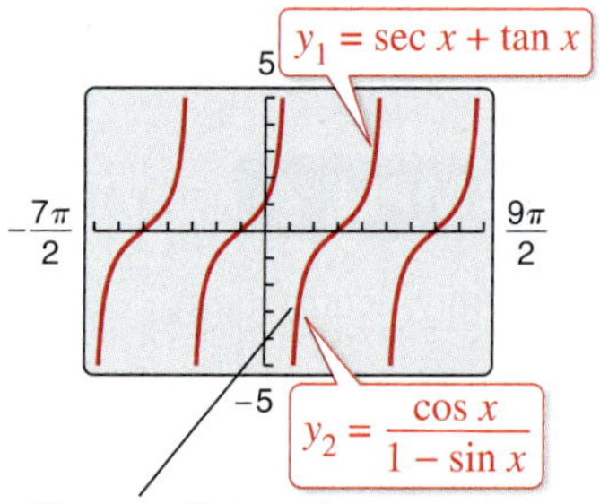

Because the graphs appear to coincide, the given equation appears to be an identity.

Checkpoint *Audio-video solution in English & Spanish at LarsonPrecalculus.com*

Verify the identity $\csc x + \cot x = \dfrac{\sin x}{1 - \cos x}$.

In Examples 1 through 5, you have been verifying trigonometric identities by working with one side of the equation and converting it to the form given on the other side. On occasion, it is practical to work with each side *separately* to obtain one common form equivalent to both sides. This is illustrated in Example 6.

EXAMPLE 6 Working with Each Side Separately

Verify the identity $\dfrac{\cot^2\theta}{1+\csc\theta} = \dfrac{1-\sin\theta}{\sin\theta}$.

Algebraic Solution

Working with the left side, you have

$$\frac{\cot^2\theta}{1+\csc\theta} = \frac{\csc^2\theta - 1}{1+\csc\theta} \qquad \text{Pythagorean identity}$$

$$= \frac{(\csc\theta - 1)(\cancel{\csc\theta + 1})}{\cancel{1+\csc\theta}} \qquad \text{Factor.}$$

$$= \csc\theta - 1. \qquad \text{Simplify.}$$

Now, simplifying the right side, you have

$$\frac{1-\sin\theta}{\sin\theta} = \frac{1}{\sin\theta} - \frac{\sin\theta}{\sin\theta} \qquad \text{Write as separate fractions.}$$

$$= \csc\theta - 1. \qquad \text{Reciprocal identity}$$

The identity is verified because both sides are equal to $\csc\theta - 1$.

Numerical Solution

Use a graphing utility to create a table that shows the values of

$$y_1 = \frac{\cot^2 x}{1+\csc x}$$

and

$$y_2 = \frac{1-\sin x}{\sin x}$$

for different values of x, as shown in the figure.

X	Y1	Y2
-.5	-3.086	-3.086
-.25	-5.042	-5.042
0	ERROR	ERROR
.25	3.042	3.042
.5	1.0858	1.0858
.75	.46705	.46705
1	.1884	.1884

X=1

The values for y_1 and y_2 appear to be identical, so the equation appears to be an identity.

Checkpoint *Audio-video solution in English & Spanish at LarsonPrecalculus.com*

Verify the identity $\dfrac{\tan^2\theta}{1+\sec\theta} = \dfrac{1-\cos\theta}{\cos\theta}$.

In Example 7, powers of trigonometric functions are rewritten as more complicated sums of products of trigonometric functions. This is a common procedure used in calculus.

EXAMPLE 7 Examples from Calculus

Verify each identity.

a. $\tan^4 x = \tan^2 x \sec^2 x - \tan^2 x$

b. $\sin^3 x \cos^4 x = (\cos^4 x - \cos^6 x)\sin x$

c. $\csc^4 x \cot x = \csc^2 x(\cot x + \cot^3 x)$

Solution

a. $\tan^4 x = (\tan^2 x)(\tan^2 x)$ — Write as separate factors.

$= \tan^2 x(\sec^2 x - 1)$ — Pythagorean identity

$= \tan^2 x \sec^2 x - \tan^2 x$ — Multiply.

b. $\sin^3 x \cos^4 x = \sin^2 x \cos^4 x \sin x$ — Write as separate factors.

$= (1 - \cos^2 x)\cos^4 x \sin x$ — Pythagorean identity

$= (\cos^4 x - \cos^6 x)\sin x$ — Multiply.

c. $\csc^4 x \cot x = \csc^2 x \csc^2 x \cot x$ — Write as separate factors.

$= \csc^2 x(1 + \cot^2 x)\cot x$ — Pythagorean identity

$= \csc^2 x(\cot x + \cot^3 x)$ — Multiply.

Checkpoint *Audio-video solution in English & Spanish at LarsonPrecalculus.com*

Verify the identity $\tan^3 x = \tan x \sec^2 x - \tan x$.

What's Wrong?

To determine the validity of the statement $\tan^2 x \sin^2 x \stackrel{?}{=} \frac{5}{6}\tan^2 x$, you use a graphing utility to graph $y_1 = \tan^2 x \sin^2 x$ and $y_2 = \frac{5}{6}\tan^2 x$, as shown in the figure. You use the graph to conclude that the statement is an identity. What's wrong?

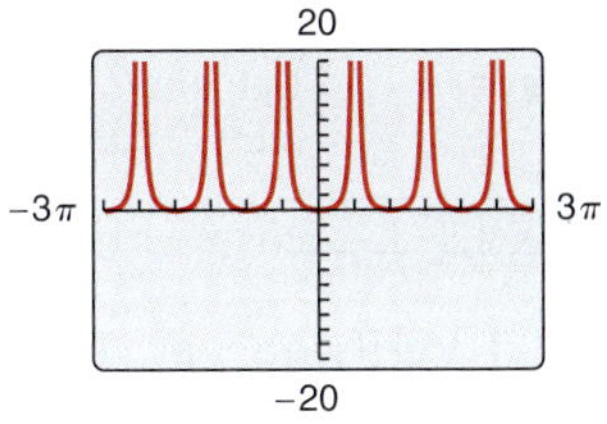

5.2 Exercises

See *CalcChat.com* for tutorial help and worked-out solutions to odd-numbered exercises.
For instructions on how to use a graphing utility, see Appendix A.

Vocabulary and Concept Check

In Exercises 1–8, fill in the blank to complete the trigonometric identity.

1. $\dfrac{1}{\tan u} =$ ________

2. $\dfrac{1}{\csc u} =$ ________

3. $\dfrac{\sin u}{\cos u} =$ ________

4. $\dfrac{1}{\sec u} =$ ________

5. $\sin^2 u +$ ________ $= 1$

6. $\tan\left(\dfrac{\pi}{2} - u\right) =$ ________

7. $\cos(-u) =$ ________

8. $\csc(-u) =$ ________

9. Is a graphical solution sufficient to verify a trigonometric identity?

10. Is a conditional equation true for all real values in its domain?

Procedures and Problem Solving

Verifying a Trigonometric Identity **In Exercises 11–18, verify the identity.**

11. $\tan t \cot t = 1$

12. $\sec y \cos y = 1$

13. $\dfrac{\csc^2 x}{\cot x} = \csc x \sec x$

14. $\dfrac{\sin^2 t}{\tan^2 t} = \cos^2 t$

15. $\dfrac{\csc^2 x - 1}{\tan^2 x} = \cot^4 x$

16. $\dfrac{1 + \sec^2 x}{2 + \tan^2 x} = 1$

17. $\tan^2 \theta + 6 = \sec^2 \theta + 5$

18. $2 + \cot^2 x = \csc^2 x + 1$

Combining Fractions Before Using Identities **In Exercises 19–22, verify the identity algebraically. Use the *table* feature of a graphing utility to check your result numerically.**

19. $\dfrac{1}{\tan x} + \dfrac{1}{\cot x} = \tan x + \cot x$

20. $\dfrac{1}{\sin x} - \dfrac{1}{\csc x} = \csc x - \sin x$

21. $\dfrac{1 + \sin \theta}{\cos \theta} + \dfrac{\cos \theta}{1 + \sin \theta} = 2 \sec \theta$

22. $\dfrac{1}{\cos x + 1} + \dfrac{1}{\cos x - 1} = -2 \csc x \cot x$

Verifying a Trigonometric Identity **In Exercises 23–28, verify the identity algebraically. Use a graphing utility to check your result graphically.**

23. $\sin y \csc y = 1$

24. $\cot^2 y(\sec^2 y - 1) = 1$

25. $\dfrac{\tan^2 \theta}{\sec \theta} = \sin \theta \tan \theta$

26. $\dfrac{\cot^3 t}{\csc t} = \cos t(\csc^2 t - 1)$

27. $\dfrac{1}{\tan \beta} + \tan \beta = \dfrac{\sec^2 \beta}{\tan \beta}$

28. $\dfrac{\sec \theta - 1}{1 - \cos \theta} = \sec \theta$

Converting to Sines and Cosines **In Exercises 29–34, verify the identity by converting the left side into sines and cosines.**

29. $\dfrac{\cot^2 t}{\csc t} = \dfrac{1 - \sin^2 t}{\sin t}$

30. $\cos x + \sin x \tan x = \sec x$

31. $\sec x - \cos x = \sin x \tan x$

32. $\cot x - \tan x = \sec x(\csc x - 2 \sin x)$

33. $\dfrac{\cot x}{\sec x} = \csc x - \sin x$

34. $\dfrac{\csc(-x)}{\sec(-x)} = -\cot x$

Rationalizing a Denominator **In Exercises 35–38, verify the identity.**

35. $\dfrac{\sin \theta}{1 + \sin \theta} = \tan \theta \sec \theta - \tan^2 \theta$

36. $\dfrac{\tan \theta}{1 + \sec \theta} = -\cot \theta + \csc \theta$

37. $\dfrac{\sin \beta}{1 - \cos \beta} = \dfrac{1 + \cos \beta}{\sin \beta}$

38. $\dfrac{\cot \alpha}{\csc \alpha - 1} = \dfrac{\csc \alpha + 1}{\cot \alpha}$

Algebraic-Graphical-Numerical **In Exercises 39–48, use a graphing utility to complete the table and graph the functions in the same viewing window. Use both the table and the graph as evidence that $y_1 = y_2$. Then verify the identity algebraically.**

x	0.2	0.4	0.6	0.8	1.0	1.2	1.4
y_1							
y_2							

39. $y_1 = \dfrac{\sec\theta - 1}{1 - \cos\theta}, \quad y_2 = \sec\theta$

40. $y_1 = \dfrac{\csc x - 1}{1 - \sin x}, \quad y_2 = \csc x$

41. $y_1 = (1 + \cot^2 x)\cos^2 x, \quad y_2 = \cot^2 x$

42. $y_1 = (1 + \tan^2 x)\sin^2 x, \quad y_2 = \tan^2 x$

43. $y_1 = \csc x - \sin x, \quad y_2 = \cos x \cot x$

44. $y_1 = \sec x - \cos x, \quad y_2 = \sin x \tan x$

45. $y_1 = \sin x + \cos x \cot x, \quad y_2 = \csc x$

46. $y_1 = \cos x + \sin x \tan x, \quad y_2 = \sec x$

47. $y_1 = \dfrac{1}{\tan x} + \dfrac{1}{\cot x}, \quad y_2 = \tan x + \cot x$

48. $y_1 = \dfrac{1}{\sin x} - \dfrac{1}{\csc x}, \quad y_2 = \csc x - \sin x$

Working with Each Side Separately **In Exercises 49–52, verify the identity. Use the *table* feature of a graphing utility to check your result numerically.**

49. $\dfrac{\sin^2\theta}{1 - \cos\theta} = \dfrac{2 + 2\cos\theta - \sin^2\theta}{1 + \cos\theta}$

50. $\dfrac{3\tan\theta}{9 - 9\sec^2\theta} = -\dfrac{\cos\theta\sin\theta + \cos\theta}{3(\sin\theta + \sin^2\theta)}$

51. $\dfrac{\sin\alpha\sec\left(\dfrac{\pi}{2} - \alpha\right)}{1 + \tan^2\alpha} = \dfrac{\cos^3\alpha}{\sin\left(\dfrac{\pi}{2} - \alpha\right)}$

52. $\dfrac{1 - \tan^2\alpha}{1 + \cot\left(\alpha - \dfrac{\pi}{2}\right)} = \dfrac{\cos\alpha + \sin\alpha}{\cos\alpha}$

Verifying a Trigonometric Identity **In Exercises 53–66, verify the identity. Use a graphing utility to verify your result.**

53. $\dfrac{\csc(-x)}{\sec(-x)} = -\cot x$

54. $(1 + \sin y)[1 + \sin(-y)] = \cos^2 y$

55. $\sin^{1/2} x \cos x - \sin^{5/2} x \cos x = \cos^3 x\sqrt{\sin x}$

56. $\sec^6 x(\sec x \tan x) - \sec^4 x(\sec x \tan x) = \sec^5 x \tan^3 x$

57. $\dfrac{\cos x - \cos y}{\sin x + \sin y} + \dfrac{\sin x - \sin y}{\cos x + \cos y} = 0$

58. $\dfrac{\tan x + \cot y}{\tan x \cot y} = \tan y + \cot x$

59. $\sin^2\left(\dfrac{\pi}{2} - x\right) + \sin^2 x = 1$

60. $\sec^2 y - \cot^2\left(\dfrac{\pi}{2} - y\right) = 1$

61. $2\sec^2 x - 2\sec^2 x \sin^2 x - \sin^2 x - \cos^2 x = 1$

62. $\csc x(\csc x - \sin x) + \dfrac{\sin x - \cos x}{\sin x} + \cot x = \csc^2 x$

63. $\dfrac{1 + \sin\theta}{\cos\theta} + \dfrac{\cos\theta}{1 + \sin\theta} = 2\sec\theta$

64. $\dfrac{\tan\theta}{1 + \sec\theta} + \dfrac{1 + \sec\theta}{\tan\theta} = 2\csc\theta$

65. $\sqrt{\dfrac{1 + \sin\alpha}{1 - \sin\alpha}} = \dfrac{1 + \sin\alpha}{|\cos\alpha|}$

66. $\sqrt{\dfrac{1 - \cos\beta}{1 + \cos\beta}} = \dfrac{1 - \cos\beta}{|\sin\beta|}$

Graphing a Trigonometric Function **In Exercises 67–70, use a graphing utility to graph the trigonometric function. Use the graph to make a conjecture about a simplification of the expression. Verify the resulting identity algebraically.**

67. $y = \dfrac{1}{\cot x + 1} + \dfrac{1}{\tan x + 1}$

68. $y = \dfrac{\cos x}{1 - \tan x} + \dfrac{\sin x \cos x}{\sin x - \cos x}$

69. $y = \dfrac{1}{\sin x} - \dfrac{\cos^2 x}{\sin x}$

70. $y = \sin t + \dfrac{\cot^2 t}{\csc t}$

Verifying an Identity Involving Logarithms **In Exercises 71 and 72, use the properties of logarithms and trigonometric identities to verify the identity.**

71. $\ln|\cot\theta| = \ln|\cos\theta| - \ln|\sin\theta|$

72. $\ln|\sec\theta| = -\ln|\cos\theta|$

Using Cofunction Identities **In Exercises 73–76, use the cofunction identities to evaluate the expression without using a calculator.**

73. $\sin^2 25^\circ + \sin^2 65^\circ$

74. $\cos^2 18^\circ + \cos^2 72^\circ$

75. $\cos^2 20^\circ + \cos^2 52^\circ + \cos^2 38^\circ + \cos^2 70^\circ$

76. $\sin^2 18^\circ + \sin^2 40^\circ + \sin^2 50^\circ + \sin^2 72^\circ$

Examples from Calculus **In Exercises 77–80, powers of trigonometric functions are rewritten to be useful in calculus. Verify the identity.**

77. $\cot^6 x = \cot^4 x \csc^2 x - \cot^4 x$

78. $\sec^4 x \tan^2 x = (\tan^2 x + \tan^4 x)\sec^2 x$

79. $\cos^3 x \sin^2 x = (\sin^2 x - \sin^4 x)\cos x$

80. $\sin^4 x + \cos^4 x = 1 - 2\cos^2 x + 2\cos^4 x$

Verifying a Trigonometric Identity **In Exercises 81–84, verify the identity.**

81. $\tan(\sin^{-1} x) = \dfrac{x}{\sqrt{1 - x^2}}$

82. $\cos(\sin^{-1} x) = \sqrt{1 - x^2}$

83. $\tan\left(\sin^{-1}\dfrac{x-1}{4}\right) = \dfrac{x-1}{\sqrt{16 - (x-1)^2}}$

84. $\tan\left(\cos^{-1}\dfrac{x+1}{2}\right) = \dfrac{\sqrt{4 - (x+1)^2}}{x+1}$

85. ***Why you should learn it*** *(p. 359)* The length s of a shadow cast by a vertical gnomon (a device used to tell time) of height h when the angle of the sun above the horizon is θ (see figure) can be modeled by

$$s = \frac{h\sin(90° - \theta)}{\sin\theta}.$$

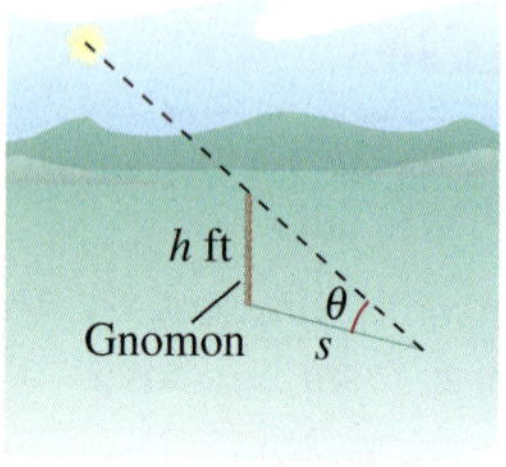

(a) Verify that the expression for s is equivalent to $h\cot\theta$.

(b) Use a graphing utility to complete the table. Let $h = 5$ feet.

θ	15°	30°	45°	60°	75°	90°
s						

(c) Use your table from part (b) to determine the angles of the sun that result in the maximum and minimum lengths of the shadow.

(d) Based on your results from part (c), what time of day do you think it is when the angle of the sun above the horizon is 90°?

86. **Rate of Change** The rate of change of the function $f(x) = \sin x + \csc x$ is given by $\cos x - \csc x \cot x$. Show that the expression for the rate of change can also be given by $-\cos x \cot^2 x$.

Focusing on Concepts

True or False? **In Exercises 87 and 88, determine whether the statement is true or false. Justify your answer.**

87. The equation $\sin^2\theta + \cos^2\theta = 1 + \tan^2\theta$ is an identity, because $\sin^2(0) + \cos^2(0) = 1$ and $1 + \tan^2(0) = 1$.

88. $\sin(x^2) = \sin^2(x)$

Verifying a Trigonometric Identity **In Exercises 89 and 90, (a) verify the identity and (b) determine whether the identity is true for the given value of x. Explain.**

89. $\dfrac{\sin x}{1 + \cos x} = \dfrac{1 - \cos x}{\sin x}, \quad x = 0$

90. $\dfrac{\sec x}{\tan x} = \dfrac{\tan x}{\sec x - \cos x}, \quad x = \pi$

Using Trigonometric Substitution **In Exercises 91–94, use the trigonometric substitution to write the algebraic expression as a trigonometric function of θ, where $0 < \theta < \pi/2$. Assume $a > 0$.**

91. $\sqrt{a^2 - u^2}, \quad u = a\sin\theta$

92. $\sqrt{a^2 - u^2}, \quad u = a\cos\theta$

93. $\sqrt{a^2 + u^2}, \quad u = a\tan\theta$

94. $\sqrt{u^2 - a^2}, \quad u = a\sec\theta$

Think About It **In Exercises 95–98, explain why the equation is not an identity and find one value of the variable for which the equation is not true.**

95. $\sqrt{\sec^2 x - 1} = \tan x$

96. $\sin\theta = \sqrt{1 - \cos^2\theta}$

97. $1 - \cos\theta = \sin\theta$

98. $1 + \tan x = \sec x$

99. **Verifying a Trigonometric Identity** Verify that for all integers n, $\sin[(12n + 1)\pi/6] = 1/2$.

100. **HOW DO YOU SEE IT?** Explain how to use the figure to derive the identity

$$\frac{\sec^2\theta - 1}{\sec^2\theta} = \sin^2\theta$$

given in Example 1.

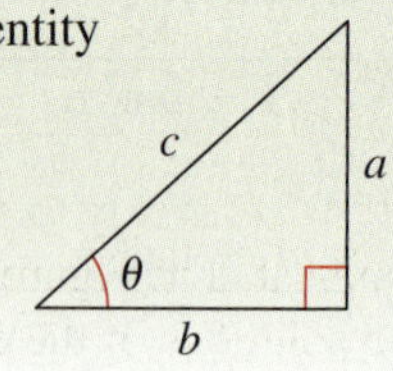

Cumulative Mixed Review

Graphing an Exponential Function **In Exercises 101–104, use a graphing utility to construct a table of values for the function. Then sketch the graph of the function. Identify any asymptotes of the graph.**

101. $f(x) = 2^x + 3$

102. $f(x) = -2^{x-3}$

103. $f(x) = 2^{-x} - 1$

104. $f(x) = 2^{x+1} + 5$

5.3 Solving Trigonometric Equations

Introduction

To solve a trigonometric equation, use standard algebraic techniques such as collecting like terms and factoring. Your preliminary goal is to isolate the trigonometric function involved in the equation.

EXAMPLE 1 Solving a Trigonometric Equation

Solve $2 \sin x - 1 = 0$.

Solution

$2 \sin x - 1 = 0$ Write original equation.

$2 \sin x = 1$ Add 1 to each side.

$\sin x = \frac{1}{2}$ Divide each side by 2.

To solve for x, note in Figure 5.2 that the equation $\sin x = \frac{1}{2}$ has solutions $x = \pi/6$ and $x = 5\pi/6$ in the interval $[0, 2\pi)$. Moreover, because $\sin x$ has a period of 2π, there are infinitely many other solutions, which can be written as

$$x = \frac{\pi}{6} + 2n\pi \quad \text{and} \quad x = \frac{5\pi}{6} + 2n\pi \qquad \text{General solution}$$

where n is an integer. The solutions when $n = \pm 1$ and $n = 0$ are shown in Figure 5.2.

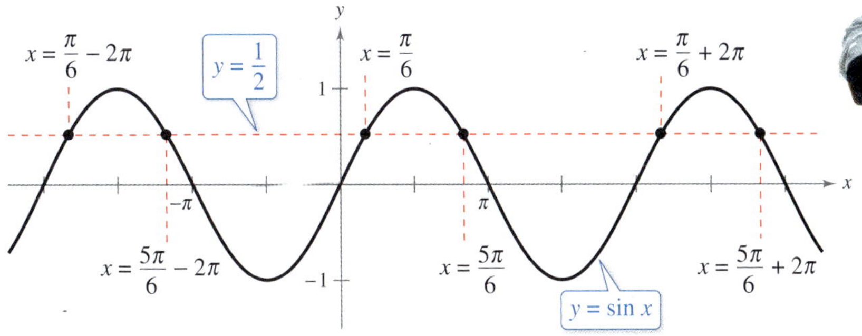

Figure 5.2

✓ ***Checkpoint*** ▶ *Audio-video solution in English & Spanish at LarsonPrecalculus.com*

Solve $\sin x + 1 = 0$.

Another way to show that the equation $\sin x = \frac{1}{2}$ has infinitely many solutions is indicated in Figure 5.3. Any angles that are coterminal with $\pi/6$ or $5\pi/6$ are also solutions of the equation.

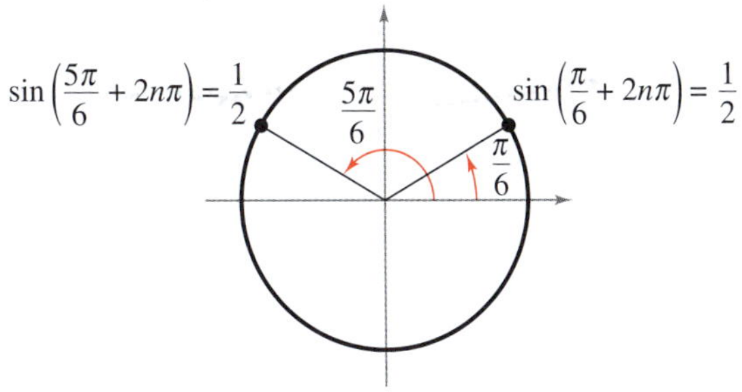

Figure 5.3

What you should learn

- Use standard algebraic techniques to solve trigonometric equations.
- Solve trigonometric equations of quadratic type.
- Solve trigonometric equations involving multiple angles.
- Use inverse trigonometric functions to solve trigonometric equations.

Why you should learn it

You can use trigonometric equations to solve a variety of real-life problems. For instance, in Exercise 96 on page 377, you can solve a trigonometric equation to help answer questions about monthly sales of snowboards.

EXAMPLE 2 Collecting Like Terms

Find all solutions of

$$\sin x + \sqrt{2} = -\sin x$$

in the interval $[0, 2\pi)$.

Algebraic Solution

Rewrite the equation so that $\sin x$ is isolated on one side of the equation.

$\sin x + \sqrt{2} = -\sin x$ Write original equation.

$\sin x + \sin x = -\sqrt{2}$ Add $\sin x$ to and subtract $\sqrt{2}$ from each side.

$2 \sin x = -\sqrt{2}$ Combine like terms.

$\sin x = -\dfrac{\sqrt{2}}{2}$ Divide each side by 2.

The solutions in the interval $[0, 2\pi)$ are

$$x = \frac{5\pi}{4} \quad \text{and} \quad x = \frac{7\pi}{4}.$$

Numerical Solution

Use a graphing utility set in *radian* mode to create a table that shows the values of $y_1 = \sin x + \sqrt{2}$ and $y_2 = -\sin x$ for different values of x. Your table should go from $x = 0$ to $x = 2\pi$ using increments of $\pi/8$, as shown in the figure.

X	Y1	Y2
3.1416	1.4142	0
3.5343	1.0315	.38268
3.927	.70711	.70711
4.3197	.49033	.92388
4.7124	.41421	1
5.1051	.49033	.92388
5.4978	.70711	.70711

X=5.497787143782

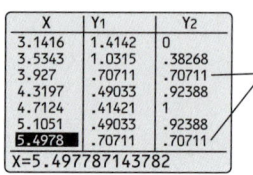

The values of y_1 and y_2 appear to be identical when $x \approx 3.927 \approx 5\pi/4$ and $x \approx 5.4978 \approx 7\pi/4$. These values are the approximate solutions of $\sin x + \sqrt{2} = -\sin x$.

Checkpoint *Audio-video solution in English & Spanish at LarsonPrecalculus.com*

Solve $\sin x - \sqrt{2} = -\sin x$ in the interval $[0, 2\pi)$.

EXAMPLE 3 Extracting Square Roots

Solve $3 \tan^2 x - 1 = 0$.

Solution

Rewrite the equation so that $\tan x$ is isolated on one side of the equation.

$3 \tan^2 x - 1 = 0$ Write original equation.

$3 \tan^2 x = 1$ Add 1 to each side.

$\tan^2 x = \dfrac{1}{3}$ Divide each side by 3.

$\tan x = \pm\dfrac{1}{\sqrt{3}}$ Extract square roots.

$\tan x = \pm\dfrac{\sqrt{3}}{3}$ Rationalize the denominator.

Because $\tan x$ has a period of π, first find all solutions in the interval $[0, \pi)$. These are

$$x = \frac{\pi}{6} \quad \text{and} \quad x = \frac{5\pi}{6}.$$

Finally, add multiples of π to each of these solutions to get the general solution

$$x = \frac{\pi}{6} + n\pi \quad \text{and} \quad x = \frac{5\pi}{6} + n\pi \qquad \text{General solution}$$

where n is an integer. You can support this answer by graphing $y = 3 \tan^2 x - 1$ with a graphing utility, as shown in Figure 5.4. The graph has x-intercepts at $\pi/6, 5\pi/6, 7\pi/6$, and so on. These x-intercepts correspond to the solutions of $3 \tan^2 x - 1 = 0$.

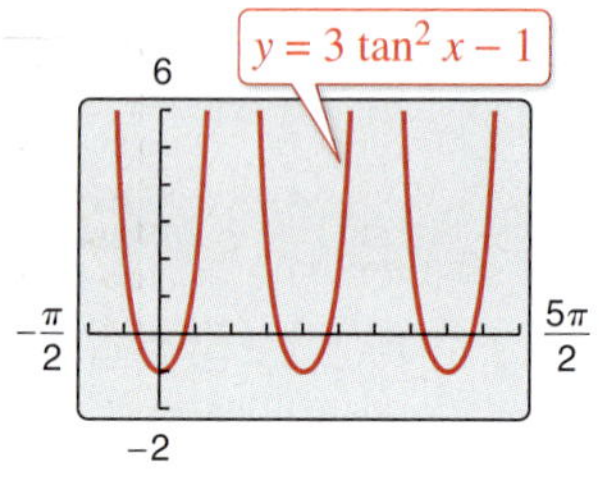

Figure 5.4

Checkpoint *Audio-video solution in English & Spanish at LarsonPrecalculus.com*

Solve $4 \sin^2 x - 3 = 0$.

Insight

You may have to solve trigonometric equations on standardized tests similar to some of the equations in this section. On tests that do not permit calculators, you may see equations like the ones in Examples 1–3 that involve the common angles listed on page 289.

The equations in Examples 1–3 involved only one trigonometric function. When two or more functions occur in the same equation, collect all terms on one side and try to separate the functions by factoring or by using appropriate identities. This may produce factors that yield no solutions, as illustrated in Example 4.

EXAMPLE 4 Factoring

Solve $\cot x \cos^2 x = 2 \cot x$.

Solution

Begin by rewriting the equation so that all terms are collected on one side of the equation.

$$\cot x \cos^2 x = 2 \cot x \quad \text{Write original equation.}$$

$$\cot x \cos^2 x - 2 \cot x = 0 \quad \text{Subtract } 2 \cot x \text{ from each side.}$$

$$\cot x(\cos^2 x - 2) = 0 \quad \text{Factor.}$$

Setting the first factor equal to zero, you obtain the equation

$$\cot x = 0. \quad \text{Set first factor equal to 0.}$$

In the interval $(0, \pi)$, the equation $\cot x = 0$ has the solution $x = \pi/2$. Setting the second factor equal to zero, you obtain

$$\cos^2 x - 2 = 0 \quad \text{Set second factor equal to 0.}$$

$$\cos^2 x = 2 \quad \text{Add 2 to each side.}$$

$$\cos x = \pm\sqrt{2}. \quad \text{Extract square roots.}$$

No solution is obtained for

$$\cos x = \pm\sqrt{2}$$

because $\pm\sqrt{2}$ are outside the range of the cosine function. Because $\cot x$ has a period of π, the general form of the solution is obtained by adding multiples of π to $x = \pi/2$ to get

$$x = \frac{\pi}{2} + n\pi \quad \text{General solution}$$

where n is an integer. The graph of $y = \cot x \cos^2 x - 2 \cot x$, shown in Figure 5.5, supports this result. From the graph, you can see that the x-intercepts occur at

$$-\frac{\pi}{2}, \frac{\pi}{2}, \frac{3\pi}{2}, \frac{5\pi}{2}$$

and so on. These x-intercepts correspond to the solutions of

$$\cot x \cos^2 x = 2 \cot x.$$

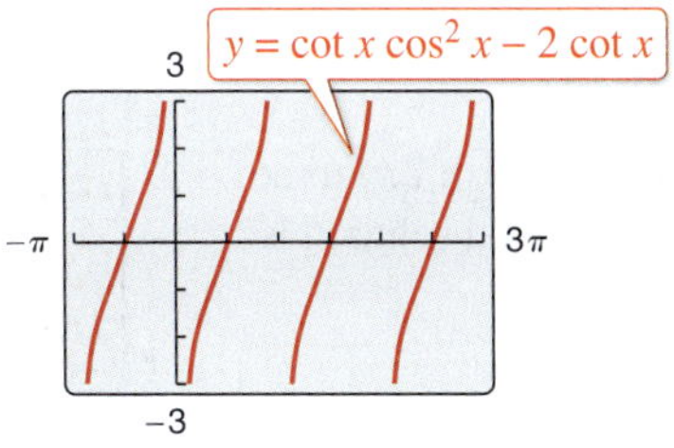

Figure 5.5

Explore the Concept

Using the equation in Example 4, explain what happens when each side of the equation is divided by $\cot x$. Why is this an incorrect method to use when solving an equation?

✓ Checkpoint *Audio-video solution in English & Spanish at LarsonPrecalculus.com*

Solve $\sin^2 x = 2 \sin x$.

Equations of Quadratic Type

Some trigonometric equations are of quadratic type $ax^2 + bx + c = 0$, as shown below. To solve equations of this type, factor the quadratic or, when factoring is not possible, use the Quadratic Formula.

Quadratic in sin x	*Quadratic in sec x*
$2\sin^2 x - \sin x - 1 = 0$	$\sec^2 x - 3\sec x - 2 = 0$
$2(\sin x)^2 - \sin x - 1 = 0$	$(\sec x)^2 - 3(\sec x) - 2 = 0$

EXAMPLE 5 Factoring an Equation of Quadratic Type

Find all solutions of $2\sin^2 x - \sin x - 1 = 0$ in the interval $[0, 2\pi)$.

Algebraic Solution

Treat the equation as a quadratic in $\sin x$ and factor.

$$2\sin^2 x - \sin x - 1 = 0 \quad \text{Write original equation.}$$

$$(2\sin x + 1)(\sin x - 1) = 0 \quad \text{Factor.}$$

Setting each factor equal to zero, you obtain the following solutions in the interval $[0, 2\pi)$.

$$2\sin x + 1 = 0 \quad \text{and} \quad \sin x - 1 = 0$$

$$\sin x = -\frac{1}{2} \qquad \sin x = 1$$

$$x = \frac{7\pi}{6}, \frac{11\pi}{6} \qquad x = \frac{\pi}{2}$$

Graphical Solution

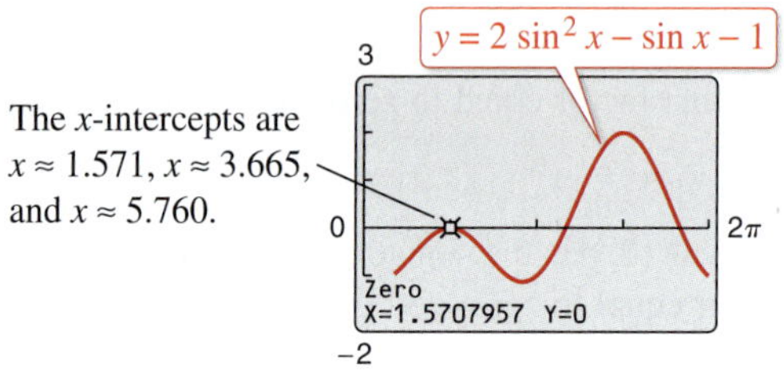

From the figure, you can conclude that the approximate solutions of $2\sin^2 x - \sin x - 1 = 0$ in the interval $[0, 2\pi)$ are

$$x \approx 1.571 \approx \frac{\pi}{2},\ x \approx 3.665 \approx \frac{7\pi}{6}, \text{ and } x \approx 5.760 \approx \frac{11\pi}{6}.$$

✓ Checkpoint *Audio-video solution in English & Spanish at LarsonPrecalculus.com*

Find all solutions of $2\sin^2 x - 3\sin x + 1 = 0$ in the interval $[0, 2\pi)$.

When working with an equation of quadratic type, be sure that the equation involves a *single* trigonometric function, as shown in the next example.

EXAMPLE 6 Rewriting with a Single Trigonometric Function

To solve $2\sin^2 x + 3\cos x - 3 = 0$, begin by rewriting the equation so that it has only cosine functions.

$$2\sin^2 x + 3\cos x - 3 = 0 \quad \text{Write original equation.}$$

$$2(1 - \cos^2 x) + 3\cos x - 3 = 0 \quad \text{Pythagorean identity}$$

$$2\cos^2 x - 3\cos x + 1 = 0 \quad \text{Combine like terms and multiply each side by } -1.$$

$$(2\cos x - 1)(\cos x - 1) = 0 \quad \text{Factor.}$$

By setting each factor equal to zero, you can find the solutions in the interval $[0, 2\pi)$ to be $x = 0$, $x = \pi/3$, and $x = 5\pi/3$. Because $\cos x$ has a period of 2π, the general solution is

$$x = 2n\pi, \quad x = \frac{\pi}{3} + 2n\pi, \quad \text{and} \quad x = \frac{5\pi}{3} + 2n\pi \quad \text{General solution}$$

where n is an integer. The graph of $y = 2\sin^2 x + 3\cos x - 3$, shown in Figure 5.6, supports this result.

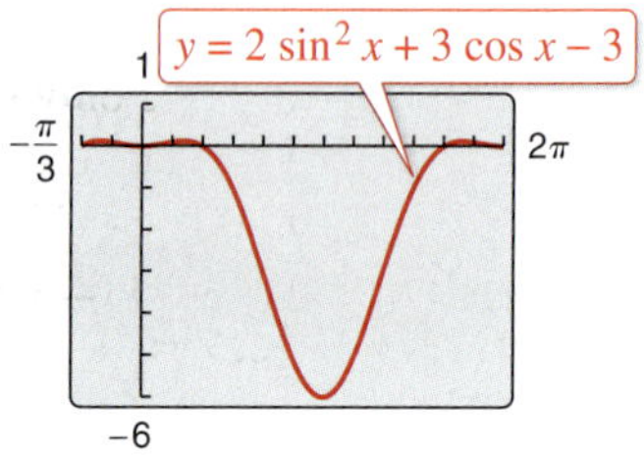

Figure 5.6

✓ Checkpoint *Audio-video solution in English & Spanish at LarsonPrecalculus.com*

Solve $3\sec^2 x - 2\tan^2 x - 4 = 0$.

Sometimes you must square each side of an equation to obtain a quadratic. Because this procedure can introduce extraneous solutions, you should check any solutions in the original equation to see whether they are valid or extraneous.

EXAMPLE 7 Squaring and Converting to Quadratic Type

See LarsonPrecalculus.com for an interactive version of this type of example.

Find all solutions of $\cos x + 1 = \sin x$ in the interval $[0, 2\pi)$.

Solution

It is not clear how to rewrite this equation in terms of a single trigonometric function. Because the squares of the sine and cosine functions are related by a Pythagorean identity, try squaring each side of the equation and then rewriting the equation so that it has only cosine functions.

$$\cos x + 1 = \sin x \qquad \text{Write original equation.}$$

$$\cos^2 x + 2\cos x + 1 = \sin^2 x \qquad \text{Square each side.}$$

$$\cos^2 x + 2\cos x + 1 = 1 - \cos^2 x \qquad \text{Pythagorean identity}$$

$$2\cos^2 x + 2\cos x = 0 \qquad \text{Combine like terms.}$$

$$2\cos x(\cos x + 1) = 0 \qquad \text{Factor.}$$

Setting each factor equal to zero produces the following.

$$2\cos x = 0 \quad \text{and} \quad \cos x + 1 = 0$$

$$\cos x = 0 \qquad \cos x = -1$$

$$x = \frac{\pi}{2}, \frac{3\pi}{2} \qquad x = \pi$$

Because you squared the original equation, check for extraneous solutions.

Check

$$\cos\frac{\pi}{2} + 1 \stackrel{?}{=} \sin\frac{\pi}{2} \qquad \text{Substitute } \pi/2 \text{ for } x.$$

$$0 + 1 = 1 \qquad \text{Solution checks. ✓}$$

$$\cos\frac{3\pi}{2} + 1 \stackrel{?}{=} \sin\frac{3\pi}{2} \qquad \text{Substitute } 3\pi/2 \text{ for } x.$$

$$0 + 1 \neq -1 \qquad \text{Solution does not check.}$$

$$\cos\pi + 1 \stackrel{?}{=} \sin\pi \qquad \text{Substitute } \pi \text{ for } x.$$

$$-1 + 1 = 0 \qquad \text{Solution checks. ✓}$$

Of the three possible solutions, $x = 3\pi/2$ is extraneous. So, in the interval $[0, 2\pi)$, the only solutions are $x = \pi/2$ and $x = \pi$. At the right, the graph of $y = \cos x + 1 - \sin x$ supports this result because the graph has two x-intercepts (at $x = \pi/2$ and $x = \pi$) in the interval $[0, 2\pi)$.

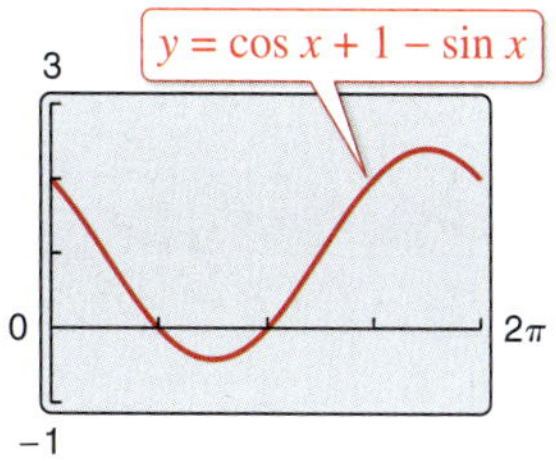

✓ Checkpoint ▶ *Audio-video solution in English & Spanish at LarsonPrecalculus.com*

Find all solutions of $\sin x + 1 = \cos x$ in the interval $[0, 2\pi)$.

Explore the Concept

Use a graphing utility to confirm the solutions found in Example 7 in two different ways. Do both methods produce the same x-values? Which method do you prefer? Why?

1. Graph both sides of the equation and find the x-coordinates of the points at which the graphs intersect.

   Left side: $y = \cos x + 1$

   Right side: $y = \sin x$

2. Graph the equation $y = \cos x + 1 - \sin x$ and find the x-intercepts of the graph.

Functions Involving Multiple Angles

The next two examples involve trigonometric functions of multiple angles of the forms $\sin ku$ and $\cos ku$. To solve equations involving these forms, first solve the equation for ku, then divide your result by k.

EXAMPLE 8 Functions of Multiple Angles

Solve $2 \cos 3t - 1 = 0$.

Solution

$2 \cos 3t - 1 = 0$ Write original equation.

$2 \cos 3t = 1$ Add 1 to each side.

$\cos 3t = \frac{1}{2}$ Divide each side by 2.

In the interval $[0, 2\pi)$, you know that $3t = \pi/3$ and $3t = 5\pi/3$ are the only solutions. So in general, you have $3t = \pi/3 + 2n\pi$ and $3t = 5\pi/3 + 2n\pi$. Dividing this result by 3, you obtain the general solution

$$t = \frac{\pi}{9} + \frac{2n\pi}{3} \quad \text{and} \quad t = \frac{5\pi}{9} + \frac{2n\pi}{3}$$ General solution

where n is an integer. This solution is supported graphically in Figure 5.7.

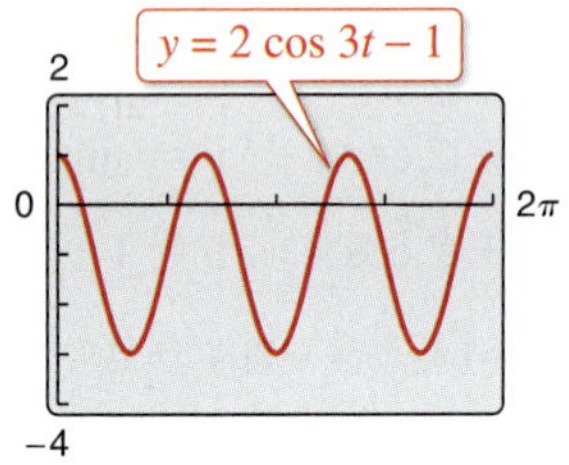

Figure 5.7

✓ **Checkpoint** ▶ *Audio-video solution in English & Spanish at LarsonPrecalculus.com*

Solve $2 \sin 2t - \sqrt{3} = 0$.

EXAMPLE 9 Functions of Multiple Angles

$3 \tan \frac{x}{2} + 3 = 0$ Original equation

$3 \tan \frac{x}{2} = -3$ Subtract 3 from each side.

$\tan \frac{x}{2} = -1$ Divide each side by 3.

In the interval $[0, \pi)$, you know that $x/2 = 3\pi/4$ is the only solution. So in general, you have $x/2 = 3\pi/4 + n\pi$. Multiplying this result by 2, you obtain the general solution

$$x = \frac{3\pi}{2} + 2n\pi$$ General solution

where n is an integer. This solution is supported graphically in Figure 5.8.

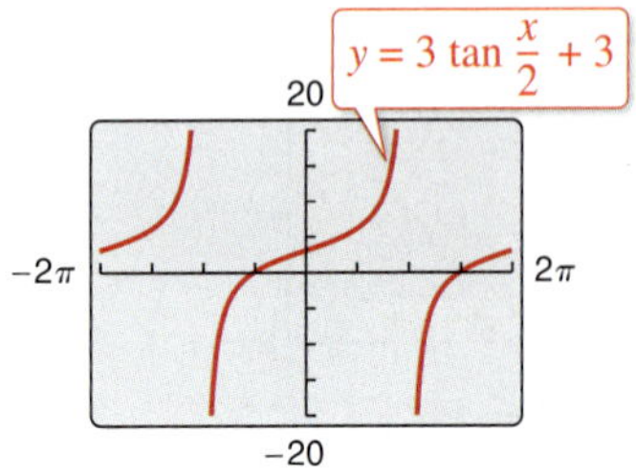

Figure 5.8

✓ **Checkpoint** ▶ *Audio-video solution in English & Spanish at LarsonPrecalculus.com*

Solve $2 \tan \frac{x}{2} - 2 = 0$. ■

Using Inverse Functions

EXAMPLE 10 Using Inverse Functions

Solve $\sec^2 x - 2 \tan x = 4$.

Solution

$$\sec^2 x - 2 \tan x = 4 \quad \text{Write original equation.}$$

$$1 + \tan^2 x - 2 \tan x - 4 = 0 \quad \text{Pythagorean identity}$$

$$\tan^2 x - 2 \tan x - 3 = 0 \quad \text{Combine like terms.}$$

$$(\tan x - 3)(\tan x + 1) = 0 \quad \text{Factor.}$$

Setting each factor equal to zero, you obtain two solutions in the interval $(-\pi/2, \pi/2)$. [Recall that the range of the inverse tangent function is $(-\pi/2, \pi/2)$.]

$$\tan x = 3 \qquad \text{and} \qquad \tan x = -1$$

$$x = \arctan 3 \qquad \qquad x = \arctan(-1) = -\frac{\pi}{4}$$

Finally, because $\tan x$ has a period of π, add multiples of π to obtain

$$x = \arctan 3 + n\pi \qquad \text{and} \qquad x = -\frac{\pi}{4} + n\pi \quad \text{General solution}$$

where n is an integer. This solution is supported graphically in Figure 5.9.

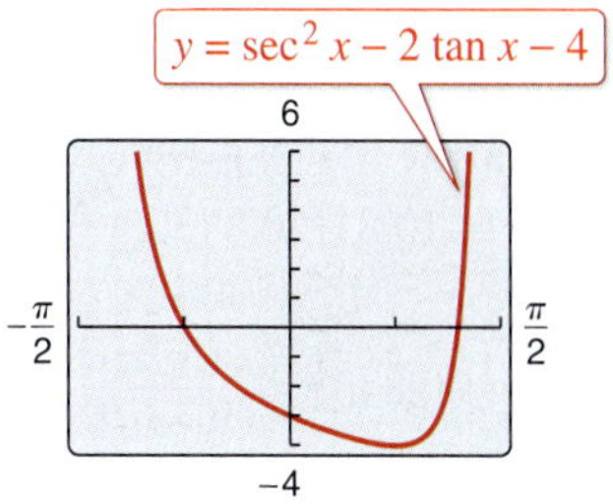

Figure 5.9

✓ ***Checkpoint*** ▶ *Audio-video solution in English & Spanish at LarsonPrecalculus.com*

Solve $4 \tan^2 x + 5 \tan x - 6 = 0$.

With some equations that involve trigonometric functions, there is no reasonable way to find the solutions algebraically. In such cases, you can still use a graphing utility to approximate the solutions.

EXAMPLE 11 Approximating Solutions

Approximate the solutions of

$$x = 2 \sin x$$

in the interval $[-\pi, \pi]$.

Solution

Use a graphing utility to graph $y = x - 2 \sin x$ in the interval $[-\pi, \pi]$. Using the *zero* or *root* feature, you can see that the solutions are

$$x \approx -1.8955, \quad x = 0, \quad \text{and} \quad x \approx 1.8955. \qquad \text{See Figure 5.10.}$$

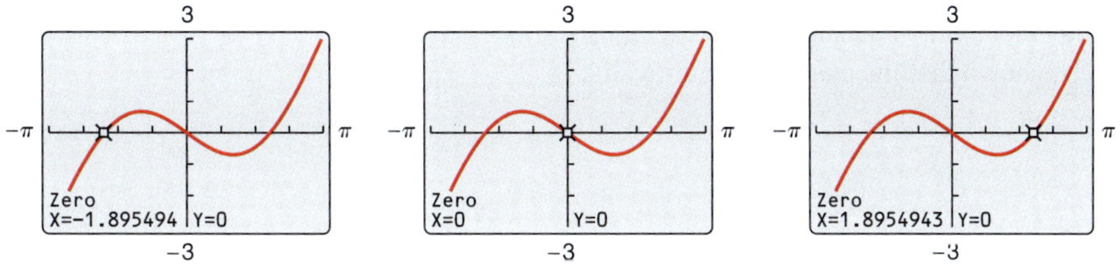

Figure 5.10 $y = x - 2 \sin x$

✓ ***Checkpoint*** ▶ *Audio-video solution in English & Spanish at LarsonPrecalculus.com*

Approximate the solutions of $x = -4 \sin 2x$ in the interval $[-\pi, \pi]$.

EXAMPLE 12 Surface Area of a Honeycomb

The surface area S (in square inches) of a honeycomb is given by

$$S = 6hs + \frac{3}{2}s^2\left(\frac{\sqrt{3} - \cos\theta}{\sin\theta}\right), \qquad 0 < \theta \le 90^\circ$$

where $h = 2.4$ inches, $s = 0.75$ inch, and θ is the angle indicated in Figure 5.11.

a. For what value(s) of θ is the surface area 12 square inches?

b. What value of θ gives the minimum surface area?

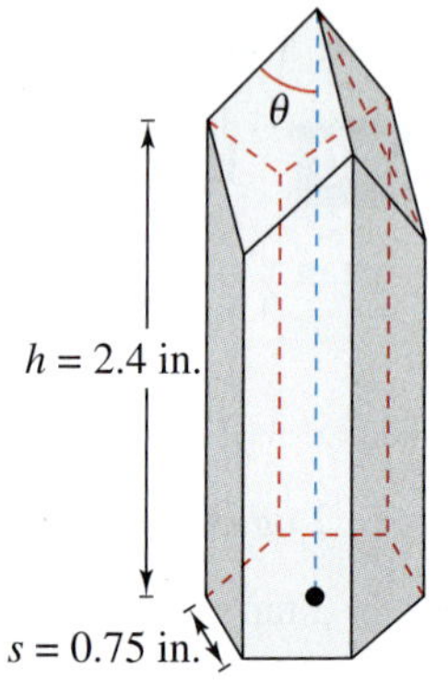

Figure 5.11

Solution

a. Let $h = 2.4$, $s = 0.75$, and $S = 12$.

$$S = 6hs + \frac{3}{2}s^2\left(\frac{\sqrt{3} - \cos\theta}{\sin\theta}\right)$$

$$12 = 6(2.4)(0.75) + \frac{3}{2}(0.75)^2\left(\frac{\sqrt{3} - \cos\theta}{\sin\theta}\right)$$

$$12 = 10.8 + 0.84375\left(\frac{\sqrt{3} - \cos\theta}{\sin\theta}\right)$$

$$0 = 0.84375\left(\frac{\sqrt{3} - \cos\theta}{\sin\theta}\right) - 1.2$$

Using a graphing utility set in *degree* mode, you can graph the function

$$y = 0.84375\left(\frac{\sqrt{3} - \cos x}{\sin x}\right) - 1.2.$$

Using the *zero* or *root* feature, you can determine that

$\theta \approx 49.9^\circ$ and $\theta \approx 59.9^\circ$. See Figure 5.12.

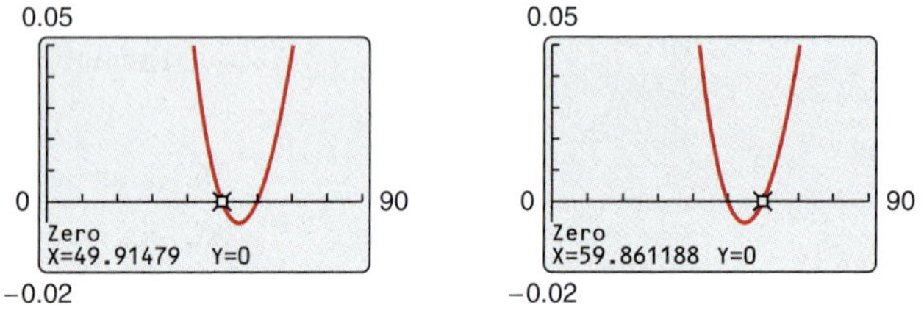

Figure 5.12 $y = 0.84375\left(\frac{\sqrt{3} - \cos x}{\sin x}\right) - 1.2$

b. From part (a), let $h = 2.4$ and $s = 0.75$ to obtain

$$S = 10.8 + 0.84375\left(\frac{\sqrt{3} - \cos\theta}{\sin\theta}\right).$$

Graph this function using a graphing utility set in *degree* mode. Use the *minimum* feature to approximate the minimum point on the graph, which occurs at $\theta \approx 54.7^\circ$, as shown in Figure 5.13. By using calculus, it can be shown that the exact minimum value is

$$\theta = \arccos\left(\frac{1}{\sqrt{3}}\right) \approx 54.7356^\circ.$$

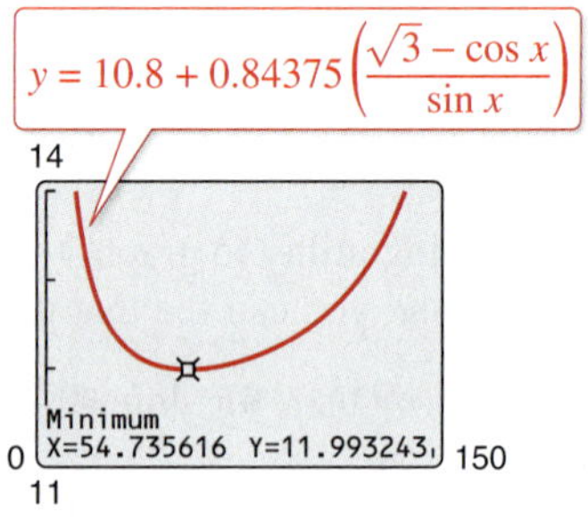

Figure 5.13

Checkpoint *Audio-video solution in English & Spanish at LarsonPrecalculus.com*

Use the equation for the surface area of a honeycomb given in Example 12 with $h = 3.2$ inches and $s = 0.75$ inch to answer each question.

a. For what value(s) of θ is the surface area 15.8 square inches?

b. What value of θ gives the minimum surface area?

5.3 Exercises

See *CalcChat.com* for tutorial help and worked-out solutions to odd-numbered exercises. For instructions on how to use a graphing utility, see Appendix A.

Vocabulary and Concept Check

In Exercises 1 and 2, fill in the blank.

1. The _______ solution of the equation $2 \cos x - 1 = 0$ is given by $x = \frac{\pi}{3} + 2n\pi$ and $x = \frac{5\pi}{3} + 2n\pi$, where n is an integer.

2. The equation $\tan^2 x - 5 \tan x + 6 = 0$ is an equation of _______ type.

3. Is $x = 0$ a solution of the equation $\cos x = 0$?

4. To solve $\sec x \sin^2 x = \sec x$, do you divide each side by $\sec x$?

Procedures and Problem Solving

Verifying a Solution **In Exercises 5–10, verify that the x-value is a solution of the equation.**

5. $\tan x - \sqrt{3} = 0, \quad x = \frac{\pi}{3}$

6. $\sec x - 2 = 0, \quad x = \frac{\pi}{3}$

7. $3 \tan^2 2x - 1 = 0, \quad x = \frac{\pi}{12}$

8. $2 \cos^2 4x - 1 = 0, \quad x = \frac{\pi}{16}$

9. $2 \sin^2 x - \sin x - 1 = 0, \quad x = \frac{\pi}{2}$

10. $\csc^4 x - 4 \csc^2 x = 0, \quad x = \frac{\pi}{6}$

Solving a Trigonometric Equation **In Exercises 11–16, find all solutions of the equation in the interval [0°, 360°).**

11. $\sin x = 0$ **12.** $\cos x = -1$

13. $\sec x = -\frac{2\sqrt{3}}{3}$ **14.** $\csc x = \sqrt{2}$

15. $\tan x = 1$ **16.** $\cot x = -1$

Solving a Trigonometric Equation **In Exercises 17–22, find all solutions of the equation in the interval $[0, 2\pi)$.**

17. $\cos x = -\frac{\sqrt{3}}{2}$ **18.** $\sin x = -\frac{1}{2}$

19. $\cot x = \sqrt{3}$ **20.** $\sec x = 2$

21. $\tan x = -1$ **22.** $\csc x = -\sqrt{2}$

Solving a Trigonometric Equation **In Exercises 23–26, solve the equation.**

23. $2 \sin x + 1 = 0$ **24.** $\sqrt{2} \sin x + 1 = 0$

25. $\sqrt{3} \csc x - 2 = 0$ **26.** $\cot x + 1 = 0$

Collecting Like Terms **In Exercises 27–30, find all solutions of the equation.**

27. $\cos x + 1 = -\cos x$ **28.** $3 \sin x + 1 = \sin x$

29. $1 - \tan x = \tan x + 1$ **30.** $-2 \sec x = 2 - 3 \sec x$

Extracting Square Roots **In Exercises 31–34, solve the equation.**

31. $4 \sin^2 x - 1 = 0$ **32.** $4 - 8 \cos^2 x = 0$

33. $3 - 4 \cos^2 x = 0$ **34.** $4 - 3 \csc^2 x = 0$

Factoring with a Trigonometric Equation **In Exercises 35–38, solve the equation.**

35. $\cos^3 x - \cos x = 0$ **36.** $3 \tan^3 x = \tan x$

37. $2 \cos x \sin x = \sin x$ **38.** $\cos x \cot x = \cot x$

Solving an Equation of Quadratic Type **In Exercises 39–46, find all solutions of the equation in the interval $[0, 2\pi)$.**

39. $2 \sin^2 x + \sin x - 1 = 0$ **40.** $\sec^2 x = 2 + \sec x$

41. $\tan^2 x + 3 \tan x + 1 = 0$ **42.** $\sin^2 x - 6 \sin x + 3 = 0$

43. $2 \sin^2 x = 2 + \cos x$ **44.** $\tan^2 x = \sec x - 1$

45. $\cos x + \sin x = 1$ **46.** $\sec x + \tan x = 1$

Approximating Solutions **In Exercises 47–52, use a graphing utility to approximate the solutions of the equation in the interval $[0, 2\pi)$ by collecting all terms on one side, graphing the new equation, and using the *zero* or *root* feature to approximate the x-intercepts of the graph.**

47. $4 \sin^2 x = 2 \cos x + 1$ **48.** $\csc^2 x - 3 \csc x = 4$

49. $\csc x + \cot x = 1$ **50.** $4 \sin x = \cos x - 2$

51. $\frac{\cos x \cot x}{1 - \sin x} = 3$ **52.** $\frac{1 + \sin x}{\cos^2 x} = 2$

Graphing a Trigonometric Function In Exercises 53–56, (a) use a graphing utility to graph both functions in the interval $[0, 2\pi)$, (b) write an equation whose solutions are the points of intersection of the graphs, and (c) use the *intersect* feature of the graphing utility to find the points of intersection (to four decimal places).

53. $y = \sin 2x, \quad y = x^2 - 2x$

54. $y = \cos x, \quad y = x + x^2$

55. $y = \sin^2 x, \quad y = e^x - 4x$

56. $y = \cos^2 x, \quad y = e^{-x} + x - 1$

Solving a Multiple-Angle Equation In Exercises 57–64, solve the multiple-angle equation.

57. $2 \cos 2x - 1 = 0$

58. $2 \sin 2x + \sqrt{3} = 0$

59. $\tan 3x - 1 = 0$

60. $\sec 4x - 2 = 0$

61. $2 \cos \dfrac{x}{2} - \sqrt{2} = 0$

62. $2 \sin \dfrac{x}{2} + \sqrt{3} = 0$

63. $3 \tan \dfrac{x}{2} - \sqrt{3} = 0$

64. $\tan \dfrac{x}{2} + \sqrt{3} = 0$

Approximating *x*-Intercepts In Exercises 65–68, approximate the x-intercepts of the graph. Use a graphing utility to check your solutions.

65. $y = \sin \dfrac{\pi x}{2} + 1$

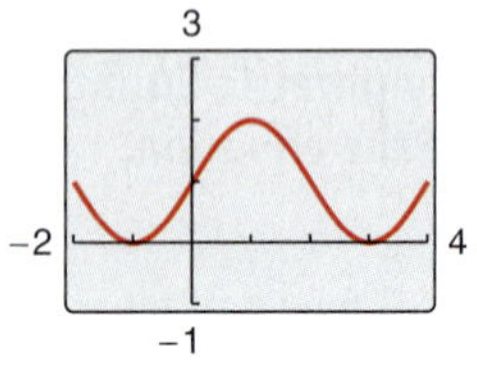

66. $y = \sin \pi x + \cos \pi x$

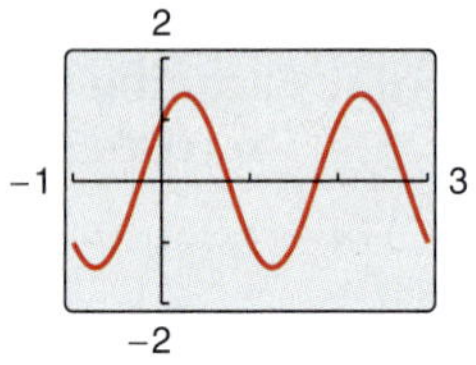

67. $y = \tan^2\left(\dfrac{\pi x}{6}\right) - 3$

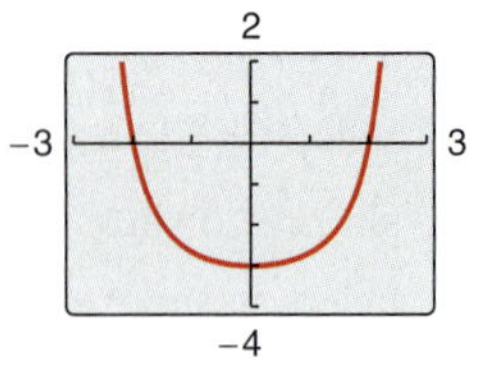

68. $y = \sec^4\left(\dfrac{\pi x}{8}\right) - 4$

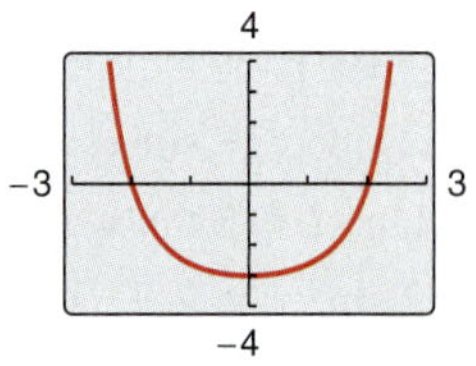

Approximating Solutions In Exercises 69–76, use a graphing utility to approximate the solutions of the equation in the interval $[0, 2\pi)$.

69. $2 \cos x - \sin x = 0$

70. $2 \sin x + \cos x = 0$

71. $x \tan x - 1 = 0$

72. $2x \sin x - 2 = 0$

73. $\sec^2 x + 0.5 \tan x = 1$

74. $\csc^2 x + 0.5 \cot x = 5$

75. $6 \sin^2 x - 7 \sin x + 2 = 0$

76. $2 \tan^2 x + 7 \tan x - 15 = 0$

Using Inverse Functions In Exercises 77–82, use inverse functions where necessary to solve the equation.

77. $\tan^2 x + \tan x - 12 = 0$

78. $\tan^2 x - \tan x - 2 = 0$

79. $\sec^2 x - 6 \tan x = -4$

80. $\sec^2 x + \tan x - 3 = 0$

81. $2 \sin^2 x + 5 \cos x = 4$

82. $2 \cos^2 x + 7 \sin x = 5$

Approximating Solutions In Exercises 83–86, use a graphing utility to approximate the solutions (to three decimal places) of the equation in the given interval.

83. $3 \tan^2 x + 5 \tan x - 4 = 0, \quad \left(-\dfrac{\pi}{2}, \dfrac{\pi}{2}\right)$

84. $\cos^2 x - 2 \cos x - 1 = 0, \quad [0, \pi]$

85. $4 \cos^2 x - 2 \sin x + 1 = 0, \quad \left[-\dfrac{\pi}{2}, \dfrac{\pi}{2}\right]$

86. $2 \sec^2 x + \tan x - 6 = 0, \quad \left(-\dfrac{\pi}{2}, \dfrac{\pi}{2}\right)$

Approximating Maximum and Minimum Points In Exercises 87–92, (a) use a graphing utility to graph the function and approximate the maximum and minimum points (to four decimal places) of the graph in the interval $[0, 2\pi]$ and (b) solve the trigonometric equation and verify that the x-coordinates of the maximum and minimum points of f are among its solutions. (The trigonometric equation is found using calculus.)

	Function	*Trigonometric Equation*
87.	$f(x) = \sin 2x$	$2 \cos 2x = 0$
88.	$f(x) = \cos 2x$	$-2 \sin 2x = 0$
89.	$f(x) = \sin^2 x + \cos x$	$2 \sin x \cos x - \sin x = 0$
90.	$f(x) = \cos^2 x - \sin x$	$-2 \sin x \cos x - \cos x = 0$
91.	$f(x) = \sin x + \cos x$	$\cos x - \sin x = 0$
92.	$f(x) = 2 \sin x + \cos 2x$	$2 \cos x - 4 \sin x \cos x = 0$

Finding a Fixed Point In Exercises 93 and 94, find the smallest positive fixed point of the function f. [A *fixed point* of a function f is a real number c such that $f(c) = c$.]

93. $f(x) = \tan \dfrac{\pi x}{4}$

94. $f(x) = \cos x$

95. Economics The monthly unit sales U (in thousands) of lawn mowers are approximated by

$$U = 74.50 - 43.75 \cos \frac{\pi t}{6}$$

where t is the time (in months), with $t = 1$ corresponding to January. Determine the months in which unit sales exceed 100,000.

96. ***Why you should learn it*** *(p. 367)* The monthly unit sales U (in hundreds) of snowboards for a chain of sports stores can be modeled by

$$U = 58.3 + 32.5 \cos \frac{\pi t}{6}$$

where t is the time (in months), with $t = 1$ corresponding to January. Determine the months in which unit sales exceed 7500.

97. Graphical Reasoning Consider the function

$$f(x) = \cos \frac{1}{x}$$

and its graph shown in the figure.

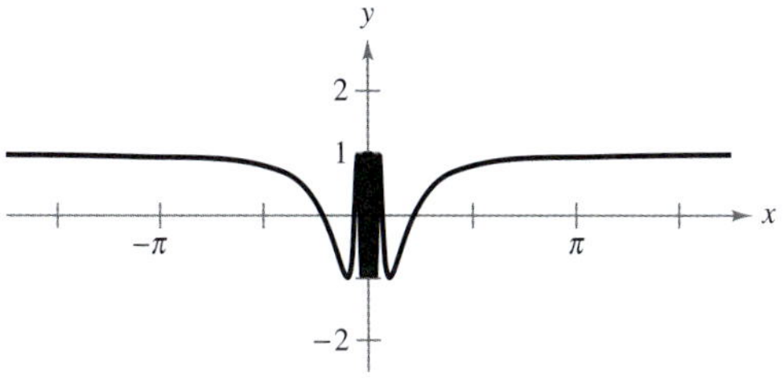

(a) What is the domain of the function?

(b) Identify any symmetry and asymptotes of the graph.

(c) Describe the behavior of the function as $x \to 0$.

(d) How many solutions does the equation $\cos(1/x) = 0$ have in the interval $[-1, 1]$? Find the solutions.

(e) Does the equation $\cos(1/x) = 0$ have a greatest solution? If so, approximate the solution. If not, explain why.

98. Graphical Reasoning Consider the function

$$f(x) = \frac{\sin x}{x}$$

and its graph shown in the figure.

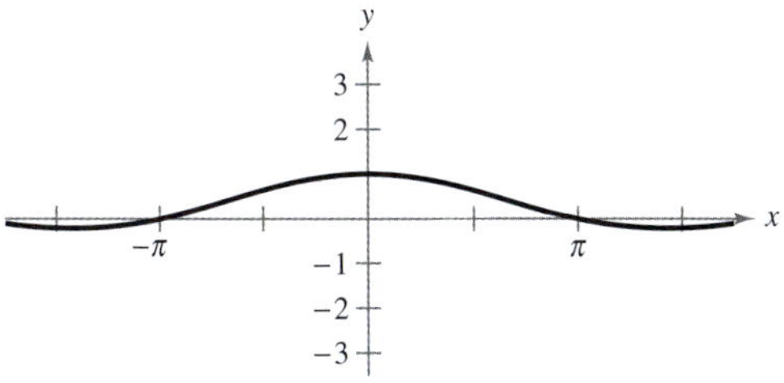

(a) What is the domain of the function?

(b) Identify any symmetry and asymptotes of the graph.

(c) Describe the behavior of the function as $x \to 0$.

(d) How many solutions does the equation $(\sin x)/x = 0$ have in the interval $[-8, 8]$? Find the solutions.

(e) Does the equation $(\sin x)/x = 0$ have a least solution? If so, approximate the solution. If not, explain why.

99. Harmonic Motion A weight is oscillating on the end of a spring (see figure). The position of the weight relative to the point of equilibrium can be modeled by

$$y = \frac{1}{4}(\cos 8t - 3 \sin 8t)$$

where y is the displacement (in meters) and t is the time (in seconds). Find the times at which the weight is at the point of equilibrium ($y = 0$) for $0 \le t \le 1$.

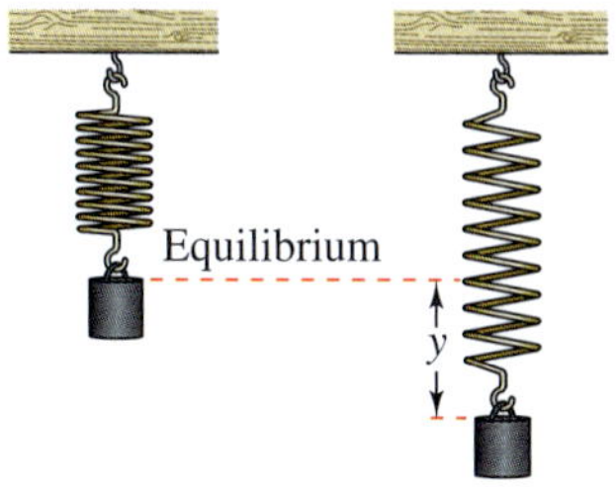

100. Damped Harmonic Motion The displacement from equilibrium of a weight oscillating on the end of a spring is given by $y = 1.56e^{-0.22t} \cos 4.9t$, where y is the displacement (in feet) and t is the time (in seconds). Use a graphing utility to graph the displacement function for $0 \le t \le 10$. Find the time beyond which the displacement does not exceed 1 foot from equilibrium.

101. Projectile Motion A batted baseball leaves the bat at an angle of θ with the horizontal and an initial velocity of $v_0 = 100$ feet per second. The ball is caught by an outfielder 310 feet from home plate (see figure). Find θ where the range r of a projectile is given by

$$r = \frac{1}{32} v_0^2 \sin 2\theta.$$

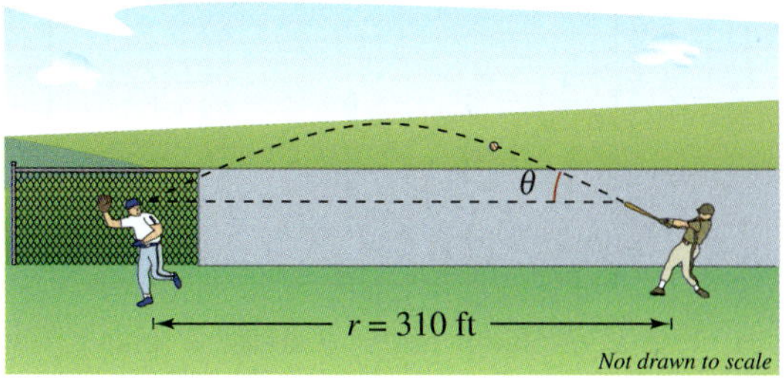

102. Exploration Consider the function

$$f(x) = 3 \sin(0.6x - 2).$$

(a) Find the zero of f in the interval $[0, 6]$.

(b) A quadratic approximation of f near $x = 4$ is

$$g(x) = -0.45x^2 + 5.52x - 13.70.$$

Use a graphing utility to graph f and g in the same viewing window. Describe the result.

(c) Use the Quadratic Formula to approximate the zeros of g. Compare the zero in the interval $[0, 6]$ with the result of part (a).

103. Geometry The area of a rectangle inscribed in one arc of the graph of $y = \cos x$ (see figure) is given by

$$A = 2x \cos x, \quad 0 \le x \le \frac{\pi}{2}.$$

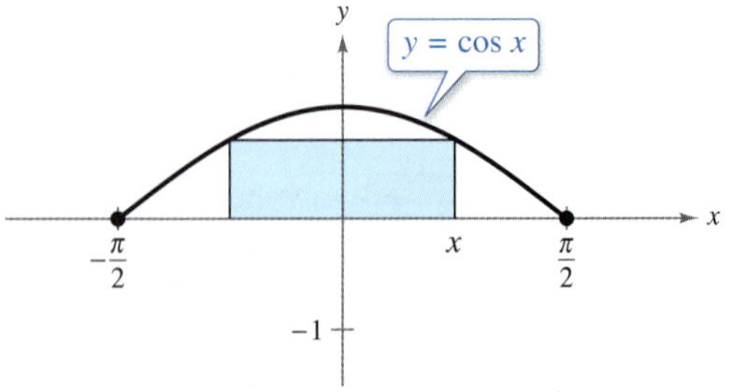

(a) Use a graphing utility to graph the area function and approximate the area of the largest inscribed rectangle.

(b) Determine the values of x for which the area of the inscribed rectangle is greater than 1.

104. MODELING DATA

The table shows the normal daily high temperatures T (in degrees Fahrenheit) in Honolulu, Hawaii, for month t, with $t = 1$ corresponding to January. (*Source:* NOAA)

DATA

Month, t	Temperature, T
1	80.1
2	80.2
3	81.2
4	82.7
5	84.6
6	87.0
7	87.9
8	88.7
9	88.6
10	86.7
11	83.9
12	81.2

Spreadsheet at LarsonPrecalculus.com

(a) Use a graphing utility to create a scatter plot of the data.

(b) Find a sine model for the temperatures.

(c) Use a graphing utility to plot the data and graph the model in the same viewing window. How well does the model fit the data?

(d) What term in the model you found in part (b) gives the average normal daily high temperature? What is the temperature?

(e) Use a graphing utility to identify the months in which the normal daily high temperature is above 86°F.

Points of Intersection **In Exercises 105 and 106, use the graph to approximate the number of points of intersection of the graphs of y_1 and y_2.**

105. $y_1 = 2 \sin x$
$y_2 = 3x + 1$

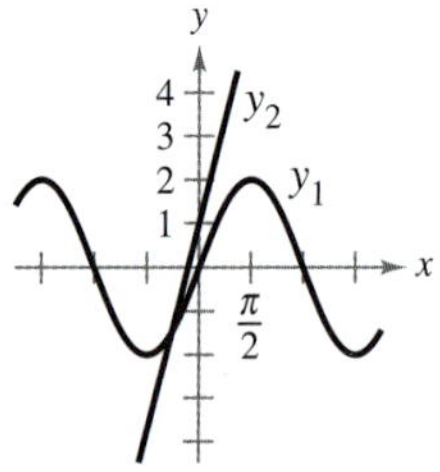

106. $y_1 = 2 \sin x$
$y_2 = \frac{1}{2}x + 1$

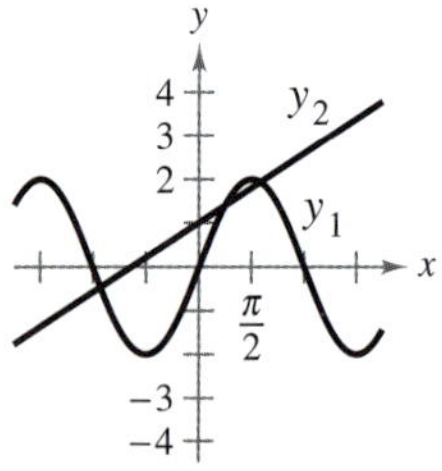

Focusing on Concepts

True or False? **In Exercises 107 and 108, determine whether the statement is true or false. Justify your answer.**

107. If you correctly solve a trigonometric equation down to the statement $\sin x = 3.4$, then you can finish solving the equation by using an inverse trigonometric function.

108. The equation $2 \sin 4t - 1 = 0$ has four times the number of solutions in the interval $[0, 2\pi)$ as the equation $2 \sin t - 1 = 0$.

109. Writing Describe the difference between verifying an identity and solving an equation.

110. HOW DO YOU SEE IT? Explain how to use the figure to solve the equation $2 \cos x - 1 = 0$.

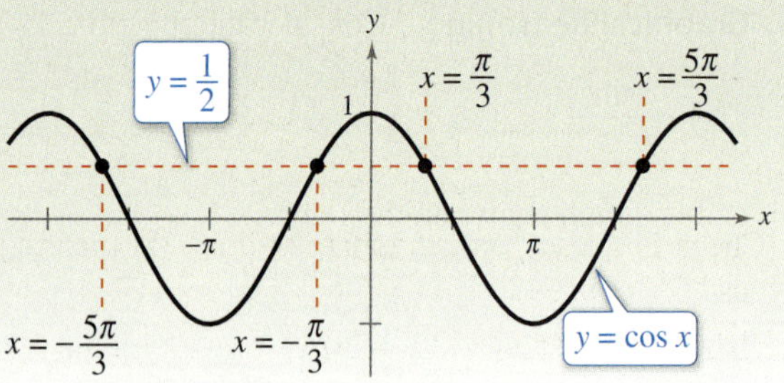

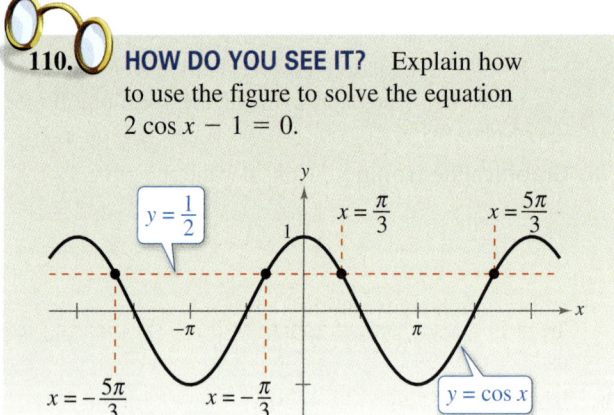

Cumulative Mixed Review

Converting from Degrees to Radians **In Exercises 111–114, convert the angle measure from degrees to radians. Round your answer to three decimal places.**

111. 124°

112. 486°

113. −215.63°

114. −0.46°

115. *Project: Tourism* To work an extended application analyzing the tourism trends of two European islands, visit this textbook's website at *LarsonPrecalulus.com*. (*Data Source:* eurostat)

5.4 Sum and Difference Formulas

Using Sum and Difference Formulas

In this section and the following section, you will study the uses of several trigonometric identities and formulas.

Sum and Difference Formulas

$$\sin(u + v) = \sin u \cos v + \cos u \sin v$$

$$\sin(u - v) = \sin u \cos v - \cos u \sin v$$

$$\cos(u + v) = \cos u \cos v - \sin u \sin v$$

$$\cos(u - v) = \cos u \cos v + \sin u \sin v$$

$$\tan(u + v) = \frac{\tan u + \tan v}{1 - \tan u \tan v}$$

$$\tan(u - v) = \frac{\tan u - \tan v}{1 + \tan u \tan v}$$

(See proofs on page 402.)

What you should learn

- Use sum and difference formulas to evaluate trigonometric functions, verify trigonometric identities, and solve trigonometric equations.

Why you should learn it

You can use sum and difference formulas to rewrite trigonometric expressions. For instance, Exercise 77 on page 384 shows how to use sum and difference formulas to rewrite a trigonometric expression in a form that helps you find the equation of a standing wave.

Example 1 shows how **sum and difference formulas** can be used to find exact values of trigonometric functions involving sums or differences of special angles.

EXAMPLE 1 Evaluating a Trigonometric Function

See LarsonPrecalculus.com for an interactive version of this type of example.

Find the exact value of (a) $\cos 75°$ and (b) $\sin \frac{\pi}{12}$.

Solution

a. Using the fact that $75° = 30° + 45°$ with the formula for $\cos(u + v)$ yields

$$\begin{aligned}\cos 75° &= \cos(30° + 45°)\\ &= \cos 30° \cos 45° - \sin 30° \sin 45°\\ &= \frac{\sqrt{3}}{2}\left(\frac{\sqrt{2}}{2}\right) - \frac{1}{2}\left(\frac{\sqrt{2}}{2}\right)\\ &= \frac{\sqrt{6} - \sqrt{2}}{4}.\end{aligned}$$

Check this result on a graphing utility by comparing its value to $\cos 75° \approx 0.259$.

b. Using the fact that $\pi/12 = \pi/3 - \pi/4$ with the formula for $\sin(u - v)$ yields

$$\begin{aligned}\sin \frac{\pi}{12} &= \sin\left(\frac{\pi}{3} - \frac{\pi}{4}\right)\\ &= \sin \frac{\pi}{3} \cos \frac{\pi}{4} - \cos \frac{\pi}{3} \sin \frac{\pi}{4}\\ &= \frac{\sqrt{3}}{2}\left(\frac{\sqrt{2}}{2}\right) - \frac{1}{2}\left(\frac{\sqrt{2}}{2}\right)\\ &= \frac{\sqrt{6} - \sqrt{2}}{4}.\end{aligned}$$

✓ **Checkpoint** *Audio-video solution in English & Spanish at LarsonPrecalculus.com*

Find the exact value of (a) $\cos \frac{\pi}{12}$ and (b) $\sin 75°$.

Kenneth Man/Shutterstock.com; AlexandrBognat/iStock/Getty Images; 1991 Richard Megna - Fundamental Photographs

EXAMPLE 2 Evaluating a Trigonometric Expression

Find the exact value of $\sin(u + v)$ given $\sin u = 4/5$, where $0 < u < \pi/2$, and $\cos v = -12/13$, where $\pi/2 < v < \pi$.

Solution

Because $\sin u = 4/5$ and u is in Quadrant I, $\cos u = 3/5$, as shown in Figure 5.14. Because $\cos v = -12/13$ and v is in Quadrant II, $\sin v = 5/13$, as shown in Figure 5.15. Next, use the formula for $\sin(u + v)$.

$$\sin(u + v) = \sin u \cos v + \cos u \sin v$$

$$= \left(\frac{4}{5}\right)\left(-\frac{12}{13}\right) + \left(\frac{3}{5}\right)\left(\frac{5}{13}\right) = -\frac{48}{65} + \frac{15}{65} = -\frac{33}{65}$$

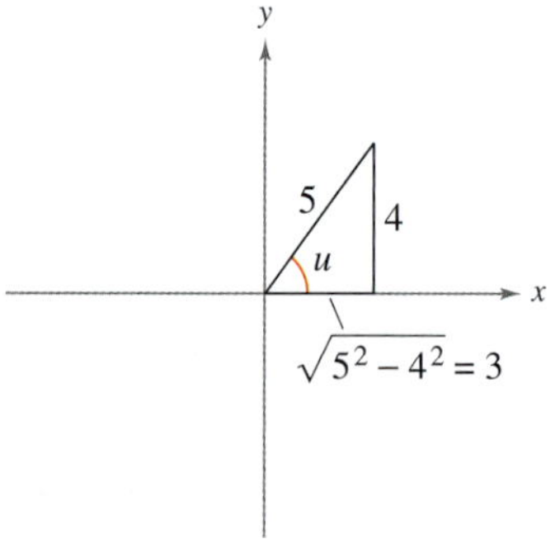

Figure 5.14

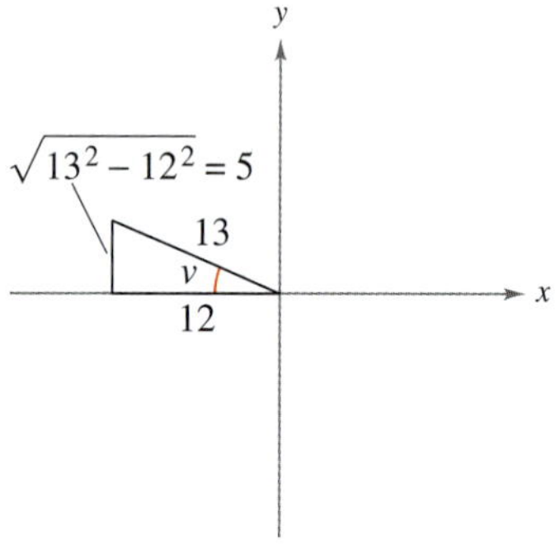

Figure 5.15

Checkpoint *Audio-video solution in English & Spanish at LarsonPrecalculus.com*

Find the exact value of $\cos(u + v)$ given $\sin u = 12/13$, where $0 < u < \pi/2$, and $\cos v = -3/5$, where $\pi/2 < v < \pi$.

EXAMPLE 3 An Application of a Sum Formula

To write $\cos(\arctan 1 + \arccos x)$ as an algebraic expression, note that this expression fits the formula for $\cos(u + v)$. Angles $u = \arctan 1$ and $v = \arccos x$ are shown in Figures 5.16 and 5.17, respectively.

$$\cos(u + v) = \cos(\arctan 1)\cos(\arccos x) - \sin(\arctan 1)\sin(\arccos x)$$

$$= \frac{1}{\sqrt{2}} \cdot x - \frac{1}{\sqrt{2}} \cdot \sqrt{1 - x^2}$$

$$= \frac{x - \sqrt{1 - x^2}}{\sqrt{2}}$$

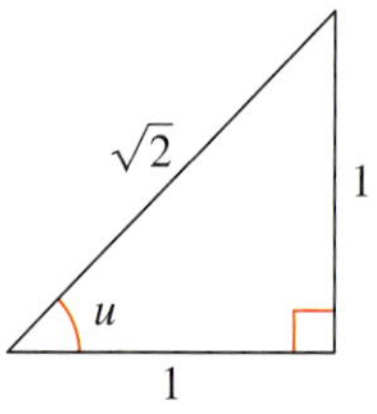

Figure 5.16

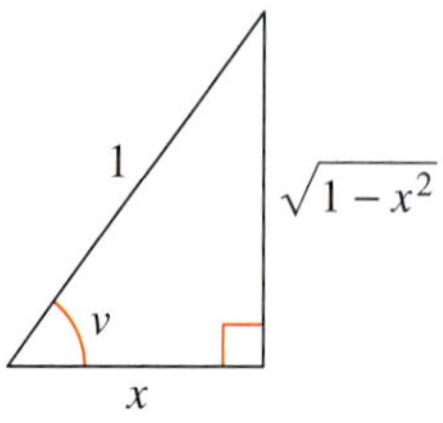

Figure 5.17

Checkpoint *Audio-video solçution in English & Spanish at LarsonPrecalculus.com*

Write $\sin(\arctan 1 + \arccos x)$ as an algebraic expression.

Explore the Concept

Use a graphing utility to graph $y_1 = \cos(x + 2)$ and $y_2 = \cos x + \cos 2$ in the same viewing window. What can you conclude about the graphs? Is it true that

$$\cos(x + 2) = \cos x + \cos 2?$$

Use the graphing utility to graph

$$y_1 = \sin(x + 4)$$

and

$$y_2 = \sin x + \sin 4$$

in the same viewing window. What can you conclude about the graphs? Is it true that

$$\sin(x + 4) = \sin x + \sin 4?$$

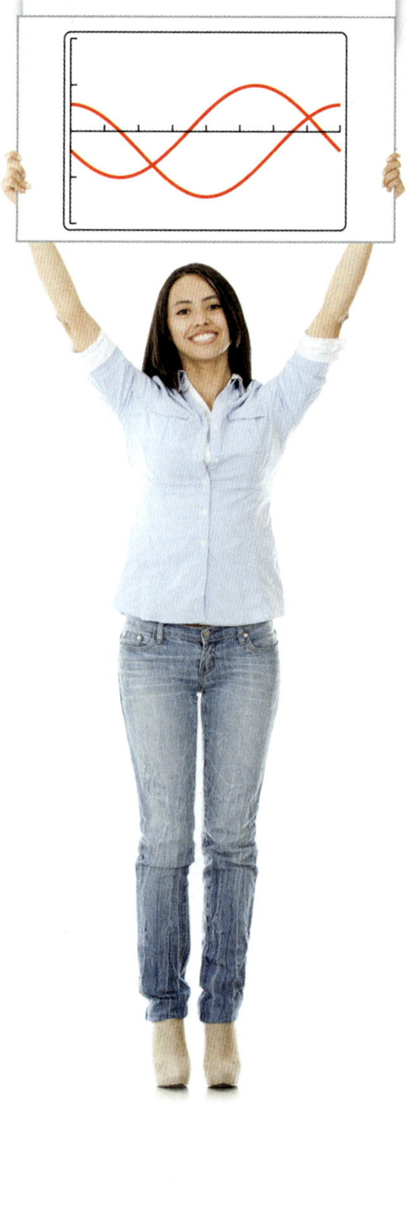

ESB Professional/Shutterstock.com

EXAMPLE 4 Proving a Cofunction Identity

Prove the cofunction identity $\cos\left(\frac{\pi}{2} - x\right) = \sin x$.

Solution

Using the formula for $\cos(u - v)$, you have

$$\cos\left(\frac{\pi}{2} - x\right) = \cos\frac{\pi}{2}\cos x + \sin\frac{\pi}{2}\sin x = (0)(\cos x) + (1)(\sin x) = \sin x.$$

✓ Checkpoint ▶ *Audio-video solution in English & Spanish at LarsonPrecalculus.com*

Use a difference formula to prove the cofunction identity $\sin\left(x - \frac{\pi}{2}\right) = -\cos x$. ■

Sum and difference formulas can be used to simplify an expression involving quadrantal angles such as

$$\sin\left(\theta + \frac{n\pi}{2}\right) \quad \text{or} \quad \cos\left(\theta + \frac{n\pi}{2}\right)$$

where n is an integer. The result will be a single-termed expression involving only $\sin\theta$ or $\cos\theta$. The resulting formula is called a **reduction formula** because it reduces the complexity of the original expression, as shown in Example 5. Similar formulas can be derived for the other trigonometric functions.

EXAMPLE 5 Deriving Reduction Formulas

Simplify each expression.

a. $\cos\left(\theta - \frac{3\pi}{2}\right)$ **b.** $\tan(\theta + 3\pi)$

Solution

a. Using the formula for $\cos(u - v)$, you have

$$\cos\left(\theta - \frac{3\pi}{2}\right) = \cos\theta\cos\frac{3\pi}{2} + \sin\theta\sin\frac{3\pi}{2}$$
$$= (\cos\theta)(0) + (\sin\theta)(-1)$$
$$= -\sin\theta.$$

b. Using the formula for $\tan(u + v)$, you have

$$\tan(\theta + 3\pi) = \frac{\tan\theta + \tan 3\pi}{1 - \tan\theta\tan 3\pi}$$
$$= \frac{\tan\theta + 0}{1 - (\tan\theta)(0)}$$
$$= \tan\theta.$$

Note that the period of $\tan\theta$ is π, so the period of $\tan(\theta + 3\pi)$ is the same as the period of $\tan\theta$.

✓ Checkpoint ▶ *Audio-video solution in English & Spanish at LarsonPrecalculus.com*

Simplify each expression.

a. $\sin\left(\frac{3\pi}{2} - \theta\right)$ **b.** $\tan(2\pi - \theta)$ ■

EXAMPLE 6 Solving a Trigonometric Equation

Find all solutions of

$$\sin\left(x + \frac{\pi}{4}\right) + \sin\left(x - \frac{\pi}{4}\right) = -1$$

in the interval $[0, 2\pi)$.

Algebraic Solution

Using sum and difference formulas, rewrite the equation as

$$\sin x \cos \frac{\pi}{4} + \cos x \sin \frac{\pi}{4} + \sin x \cos \frac{\pi}{4} - \cos x \sin \frac{\pi}{4} = -1$$

$$2 \sin x \cos \frac{\pi}{4} = -1$$

$$2(\sin x)\left(\frac{\sqrt{2}}{2}\right) = -1$$

$$\sin x = -\frac{1}{\sqrt{2}}$$

$$\sin x = -\frac{\sqrt{2}}{2}.$$

So, the only solutions in the interval $[0, 2\pi)$ are

$$x = \frac{5\pi}{4} \quad \text{and} \quad x = \frac{7\pi}{4}.$$

Graphical Solution

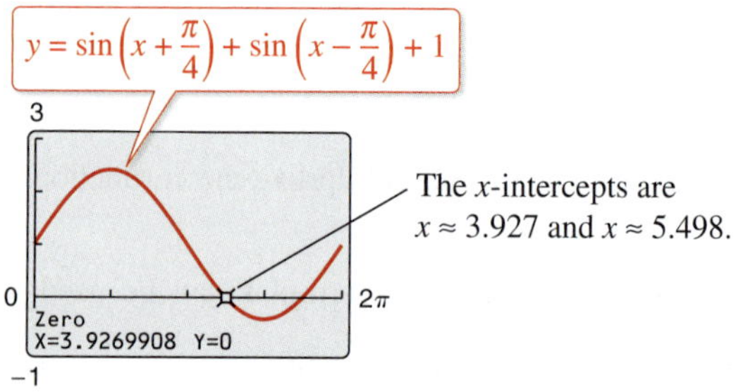

From the figure, you can conclude that the approximate solutions in the interval $[0, 2\pi)$ are

$$x \approx 3.927 \approx \frac{5\pi}{4} \quad \text{and} \quad x \approx 5.498 \approx \frac{7\pi}{4}.$$

✓ Checkpoint ▶ *Audio-video solution in English & Spanish at LarsonPrecalculus.com*

Find all solutions of $\sin\left(x + \frac{\pi}{2}\right) + \sin\left(x - \frac{3\pi}{2}\right) = 1$ in the interval $[0, 2\pi)$.

The next example was taken from calculus. It is used to derive the formula for the derivative of the cosine function.

EXAMPLE 7 An Application from Calculus

Verify that $\dfrac{\sin(x + h) - \sin x}{h} = (\cos x)\left(\dfrac{\sin h}{h}\right) - (\sin x)\left(\dfrac{1 - \cos h}{h}\right)$, where $h \neq 0$.

Solution

Using the formula for $\sin(u + v)$, you have

$$\frac{\sin(x + h) - \sin x}{h} = \frac{\sin x \cos h + \cos x \sin h - \sin x}{h}$$

$$= \frac{\cos x \sin h - \sin x(1 - \cos h)}{h}$$

$$= (\cos x)\left(\frac{\sin h}{h}\right) - (\sin x)\left(\frac{1 - \cos h}{h}\right).$$

✓ Checkpoint ▶ *Audio-video solution in English & Spanish at LarsonPrecalculus.com*

Verify that $\dfrac{\cos(x + h) - \cos x}{h} = (\cos x)\left(\dfrac{\cos h - 1}{h}\right) - (\sin x)\left(\dfrac{\sin h}{h}\right)$, where $h \neq 0$.

5.4 Exercises

See *CalcChat.com* for tutorial help and worked-out solutions to odd-numbered exercises.
For instructions on how to use a graphing utility, see Appendix A.

Vocabulary and Concept Check

In Exercises 1–6, fill in the blank to complete the trigonometric formula.

1. $\sin(u - v) =$ _______ **2.** $\cos(u + v) =$ _______ **3.** $\tan(u + v) =$ _______

4. $\sin(u + v) =$ _______ **5.** $\cos(u - v) =$ _______ **6.** $\tan(u - v) =$ _______

7. Rewrite $\sin 75°$ so that you can use a sum formula.

8. Rewrite $\cos \dfrac{\pi}{12}$ so that you can use a difference formula.

Procedures and Problem Solving

Evaluating Trigonometric Expressions **In Exercises 9–12, find the exact value of each expression.**

9. (a) $\cos(240° - 0°)$ (b) $\cos 240° - \cos 0°$

10. (a) $\sin(405° + 120°)$ (b) $\sin 405° + \sin 120°$

11. (a) $\sin\left(\dfrac{2\pi}{3} + \dfrac{5\pi}{6}\right)$ (b) $\sin \dfrac{2\pi}{3} + \sin \dfrac{5\pi}{6}$

12. (a) $\cos\left(\dfrac{\pi}{4} + \dfrac{\pi}{3}\right)$ (b) $\cos \dfrac{\pi}{4} + \cos \dfrac{\pi}{3}$

Evaluating Trigonometric Functions **In Exercises 13–26, find the exact values of the sine, cosine, and tangent of the angle.**

13. $105° = 60° + 45°$ **14.** $165° = 135° + 30°$

15. $195° = 225° - 30°$ **16.** $255° = 300° - 45°$

17. $-\dfrac{11\pi}{12} = \dfrac{\pi}{4} - \dfrac{7\pi}{6}$ **18.** $-\dfrac{19\pi}{12} = \dfrac{2\pi}{3} - \dfrac{9\pi}{4}$

19. $-\dfrac{5\pi}{12}$ **20.** $-\dfrac{23\pi}{12}$

21. $\dfrac{13\pi}{12}$ **22.** $\dfrac{19\pi}{12}$

23. $285°$ **24.** $15°$

25. $-105°$ **26.** $-165°$

Rewriting a Trigonometric Expression **In Exercises 27–34, write the expression as the sine, cosine, or tangent of an angle.**

27. $\sin 60° \cos 20° - \cos 60° \sin 20°$

28. $\sin 110° \cos 65° + \cos 110° \sin 65°$

29. $\dfrac{\tan 325° - \tan 116°}{1 + \tan 325° \tan 116°}$ **30.** $\dfrac{\tan 154° - \tan 49°}{1 + \tan 154° \tan 49°}$

31. $\cos \dfrac{\pi}{9} \cos \dfrac{\pi}{4} - \sin \dfrac{\pi}{9} \sin \dfrac{\pi}{4}$

32. $\sin \dfrac{4\pi}{9} \cos \dfrac{\pi}{8} + \cos \dfrac{4\pi}{9} \sin \dfrac{\pi}{8}$

33. $\sin 3.5x \cos 1.2y + \cos 3.5x \sin 1.2y$

34. $\cos 0.78y \cos 0.42z - \sin 0.78y \sin 0.42z$

Evaluating a Trigonometric Expression **In Exercises 35–40, find the exact value of the expression.**

35. $\sin \dfrac{\pi}{12} \cos \dfrac{\pi}{4} + \cos \dfrac{\pi}{12} \sin \dfrac{\pi}{4}$

36. $\cos \dfrac{\pi}{16} \cos \dfrac{3\pi}{16} - \sin \dfrac{\pi}{16} \sin \dfrac{3\pi}{16}$

37. $\sin 100° \cos 40° - \cos 100° \sin 40°$

38. $\cos 130° \cos 10° + \sin 130° \sin 10°$

39. $\dfrac{\tan(9\pi/8) - \tan(\pi/8)}{1 + \tan(9\pi/8)\tan(\pi/8)}$ **40.** $\dfrac{\tan 25° + \tan 110°}{1 - \tan 25° \tan 110°}$

Algebraic-Graphical-Numerical **In Exercises 41 and 42, use a graphing utility to complete the table and graph the two functions in the same viewing window. Use both the table and the graph as evidence that $y_1 = y_2$. Then verify the identity algebraically.**

x	0.2	0.4	0.6	0.8	1.0	1.2	1.4
y_1							
y_2							

41. $y_1 = \cos(x + \pi)\cos(x - \pi)$, $y_2 = \cos^2 x$

42. $y_1 = \sin(x + \pi)\sin(x - \pi)$, $y_2 = \sin^2 x$

Evaluating a Trigonometric Expression **In Exercises 43–46, find the exact value of the trigonometric expression given that $\sin u = -\frac{3}{5}$, where $3\pi/2 < u < 2\pi$, and $\cos v = \frac{15}{17}$, where $0 < v < \pi/2$.**

43. $\sin(u + v)$ **44.** $\cos(u - v)$

45. $\tan(u + v)$ **46.** $\sin(u - v)$

Evaluating a Trigonometric Expression **In Exercises 47–50, find the exact value of the trigonometric expression given that $\sin u = -\frac{7}{25}$ and $\cos v = -\frac{4}{5}$. (Both u and v are in Quadrant III.)**

47. $\tan(u + v)$ **48.** $\cos(u + v)$

49. $\sin(v - u)$ **50.** $\cos(v - u)$

An Application of a Sum or Difference Formula **In Exercises 51–54, write the trigonometric expression as an algebraic expression.**

51. $\sin(\arcsin x + \arccos x)$
52. $\cos(\arccos x - \arcsin x)$
53. $\sin(\arctan 2x - \arccos x)$
54. $\cos(\arcsin x - \arctan 2x)$

Evaluating a Trigonometric Expression **In Exercises 55 and 56, use right triangles to evaluate the expression.**

55. $\sin\left(\cos^{-1}\frac{4}{5} + \sin^{-1}\frac{3}{5}\right)$ 56. $\sin\left(\cos^{-1}\frac{3}{5} - \sin^{-1}\frac{5}{13}\right)$

Proving a Trigonometric Identity **In Exercises 57–64, prove the identity.**

57. $\sin(\pi/2 + x) = \cos x$ 58. $\sin(3\pi - x) = \sin x$

59. $\sin\left(\frac{\pi}{6} + x\right) = \frac{1}{2}(\cos x + \sqrt{3}\sin x)$

60. $\cos\left(\frac{5\pi}{4} - x\right) = -\frac{\sqrt{2}}{2}(\cos x + \sin x)$

61. $\tan(x + \pi) - \tan(\pi - x) = 2\tan x$

62. $\tan\left(\frac{\pi}{4} - \theta\right) = \frac{1 - \tan\theta}{1 + \tan\theta}$

63. $\cos(x + y)\cos(x - y) = \cos^2 x - \sin^2 y$

64. $\sin(x + y)\sin(x - y) = \sin^2 x - \sin^2 y$

Deriving a Reduction Formula **In Exercises 65–68, simplify the expression. Use a graphing utility to verify your results.**

65. $\cos(3\pi/2 - \theta)$ 66. $\sin(\pi + \theta)$
67. $\csc(3\pi/2 + \theta)$ 68. $\cot(\theta - \pi)$

Solving a Trigonometric Equation **In Exercises 69–72, find the solution(s) of the equation in the interval $[0, 2\pi)$. Use a graphing utility to verify your results.**

69. $\sin\left(x + \frac{\pi}{3}\right) + \sin\left(x - \frac{\pi}{3}\right) = 1$

70. $\cos\left(x + \frac{\pi}{6}\right) - \cos\left(x - \frac{\pi}{6}\right) = 1$

71. $\tan(x + \pi) + 2\sin(x + \pi) = 0$

72. $2\sin\left(x + \frac{\pi}{2}\right) + 3\tan(\pi - x) = 0$

Approximating Solutions Graphically **In Exercises 73–76, use a graphing utility to approximate the solutions of the equation in the interval $[0, 2\pi)$.**

73. $\cos\left(x + \frac{\pi}{4}\right) + \cos\left(x - \frac{\pi}{4}\right) = 1$

74. $\sin\left(x + \frac{\pi}{2}\right) - \cos\left(x + \frac{3\pi}{2}\right) = 1$

75. $\sin\left(x + \frac{\pi}{2}\right) = -\cos^2 x$

76. $\cos\left(x - \frac{\pi}{2}\right) = \sin^2 x$

77. ***Why you should learn it*** *(p. 379)* The equation of a standing wave is obtained by adding the displacements of two waves traveling in opposite directions (see figure). Assume that each wave has amplitude A, period T, and wavelength λ. The models for two such waves are

$$y_1 = A\cos 2\pi\left(\frac{t}{T} - \frac{x}{\lambda}\right) \text{ and } y_2 = A\cos 2\pi\left(\frac{t}{T} + \frac{x}{\lambda}\right).$$

Show that $y_1 + y_2 = 2A\cos\frac{2\pi t}{T}\cos\frac{2\pi x}{\lambda}$.

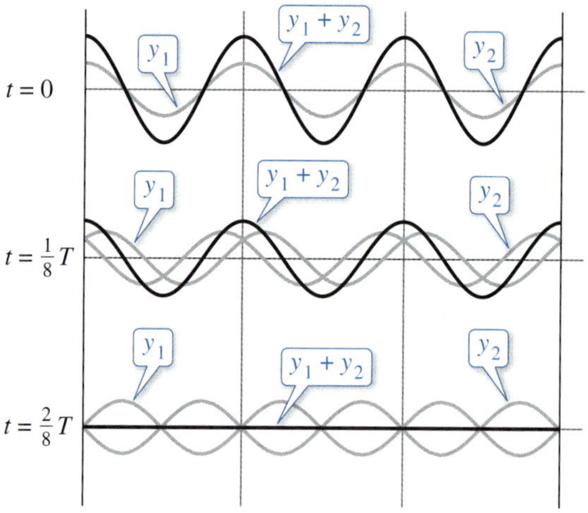

78. **Harmonic Motion** A weight is attached to a spring suspended vertically from a ceiling. When a driving force is applied to the system, the weight moves vertically from its equilibrium position, and this motion is modeled by

$$y = \frac{1}{3}\sin 2t + \frac{1}{4}\cos 2t$$

where y is the displacement (in feet) from equilibrium of the weight and t is the time (in seconds).

(a) Use a graphing utility to graph the model.

(b) Use the identity

$$a\sin B\theta + b\cos B\theta = \sqrt{a^2 + b^2}\sin(B\theta + C)$$

where $C = \arctan(b/a)$, $a > 0$, to write the model in the form $y = \sqrt{a^2 + b^2}\sin(Bt + C)$. Use the graphing utility to verify your result.

(c) Find the amplitude of the oscillations of the weight.

(d) Find the frequency of the oscillations of the weight.

Focusing on Concepts

True or False? **In Exercises 79 and 80, determine whether the statement is true or false. Justify your answer.**

79. $\cos(u \pm v) = \cos u \pm \cos v$

80. $\sin\left(x - \frac{11\pi}{2}\right) = \cos x$

81. An Application from Calculus Let $x = \pi/3$ in the identity in Example 7 and define the functions f and g as follows.

$$f(h) = \frac{\sin[(\pi/3) + h] - \sin(\pi/3)}{h}$$

$$g(h) = \cos\frac{\pi}{3}\left(\frac{\sin h}{h}\right) - \sin\frac{\pi}{3}\left(\frac{1 - \cos h}{h}\right)$$

(a) What are the domains of the functions f and g?

(b) Use a graphing utility to complete the table.

h	0.01	0.02	0.05	0.1	0.2	0.5
$f(h)$						
$g(h)$						

(c) Use the graphing utility to graph the functions f and g.

(d) Use the table and graph to make a conjecture about the values of the functions f and g as $h \to 0^+$.

82. HOW DO YOU SEE IT? Explain how to use the figure to justify each statement.

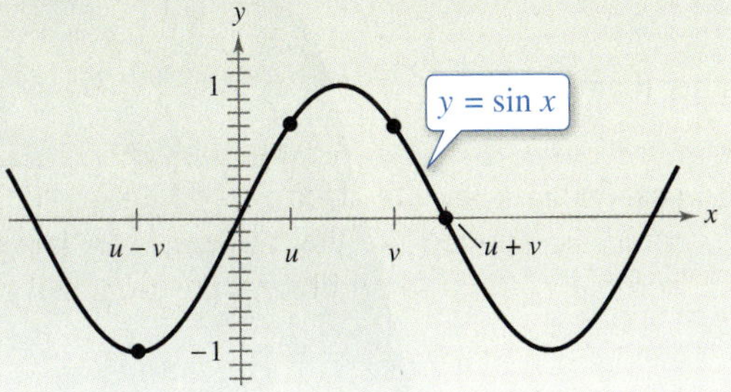

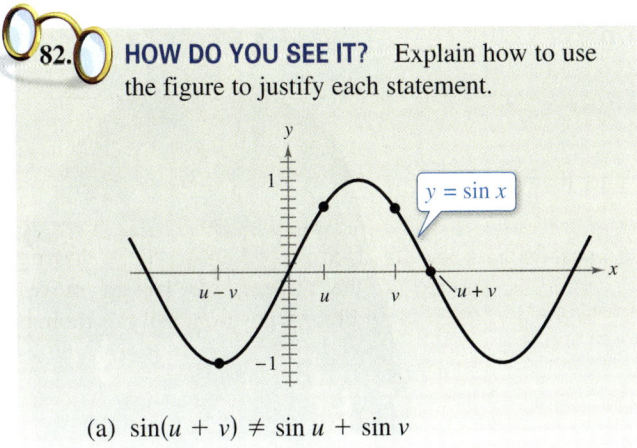

(a) $\sin(u + v) \neq \sin u + \sin v$

(b) $\sin(u - v) \neq \sin u - \sin v$

Verifying an Identity **In Exercises 83–86, verify the identity.**

83. $\cos(n\pi + \theta) = (-1)^n \cos\theta$, n is an integer.

84. $\sin(n\pi + \theta) = (-1)^n \sin\theta$, n is an integer.

85. $a \sin B\theta + b \cos B\theta = \sqrt{a^2 + b^2} \sin(B\theta + C)$, where $C = \arctan(b/a)$ and $a > 0$.

86. $a \sin B\theta + b \cos B\theta = \sqrt{a^2 + b^2} \cos(B\theta - C)$, where $C = \arctan(a/b)$ and $b > 0$.

Rewriting a Trigonometric Expression **In Exercises 87–90, use the formulas given in Exercises 85 and 86 to write the expression in the following forms. Use a graphing utility to verify your results.**

(a) $\sqrt{a^2 + b^2} \sin(B\theta + C)$ **(b)** $\sqrt{a^2 + b^2} \cos(B\theta - C)$

87. $\sin\theta + \cos\theta$

88. $\sin 2\theta + \cos 2\theta$

89. $12 \sin 3\theta + 5 \cos 3\theta$

90. $3 \sin 2\theta + 4 \cos 2\theta$

Rewriting a Trigonometric Expression **In Exercises 91 and 92, use the formulas given in Exercises 85 and 86 to write the trigonometric expression in the form $a \sin B\theta + b \cos B\theta$.**

91. $2 \sin\left(\theta + \frac{\pi}{2}\right)$

92. $5 \cos\left(\theta + \frac{\pi}{4}\right)$

Finding an Angle Between Two Lines **In Exercises 93 and 94, use the figure, which shows two lines whose equations are $y_1 = m_1x + b_1$ and $y_2 = m_2x + b_2$. Assume that both lines have positive slopes. Derive a formula for the angle between the two lines. Then use your formula to find the angle between the given pair of lines.**

93. $y = x$,
$y = \sqrt{3}x$

94. $y = x$,
$y = \frac{1}{\sqrt{3}}x$

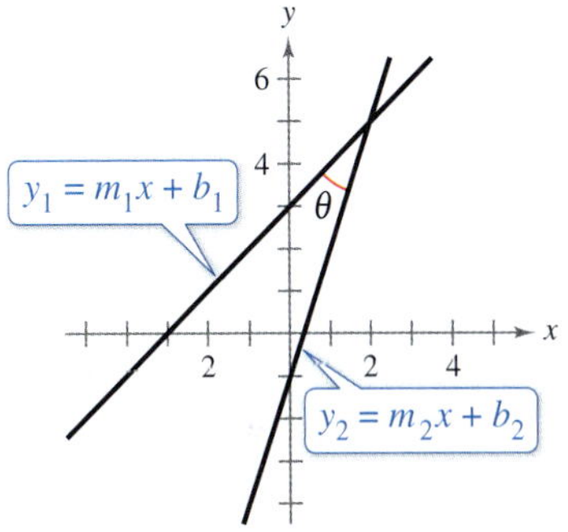

95. Proof Three squares of side s are placed side by side (see figure). Make a conjecture about the relationship between the sum $u + v$ and w. Prove your conjecture by using the formula for $\tan(u + v)$.

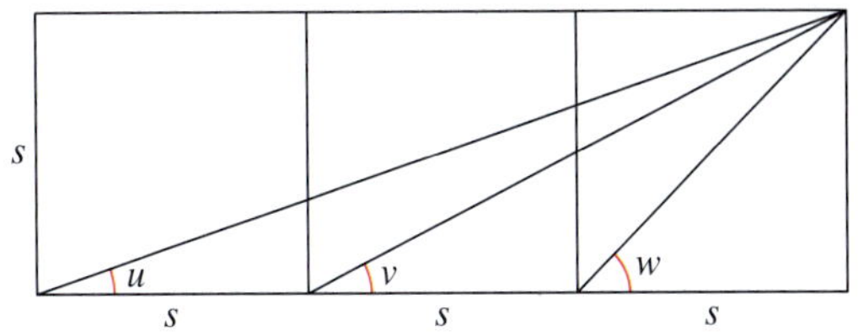

96. Writing a Formula Write a sum formula for $\cos(u + v + w)$.

Cumulative Mixed Review

Finding Intercepts **In Exercises 97–100, find the x- and y-intercepts of the graph of the equation. Use a graphing utility to verify your results.**

97. $y = -\frac{1}{2}(x - 10) + 14$

98. $y = x^2 - 3x - 40$

99. $y = |2x - 9| - 5$

100. $y = 2x\sqrt{x + 7}$

5.5 Multiple-Angle and Product-to-Sum Formulas

Multiple-Angle Formulas

In this section, you will study four additional categories of trigonometric identities.

1. The first category involves *functions of multiple angles* such as

   $\sin ku$ and $\cos ku$.

2. The second category involves *squares of trigonometric functions* such as

   $\sin^2 u$.

3. The third category involves *functions of half-angles* such as

   $\sin \frac{u}{2}$.

4. The fourth category involves *products of trigonometric functions* such as

   $\sin u \cos v$.

You should learn the **double-angle formulas** below because they are used often in trigonometry and calculus.

Double-Angle Formulas

$$\sin 2u = 2 \sin u \cos u$$

$$\tan 2u = \frac{2 \tan u}{1 - \tan^2 u}$$

$$\cos 2u = \cos^2 u - \sin^2 u = 2\cos^2 u - 1 = 1 - 2\sin^2 u$$

(See the proofs on page 403.)

What you should learn

- Use multiple-angle formulas to rewrite and evaluate trigonometric functions.
- Use power-reducing formulas to rewrite and evaluate trigonometric functions.
- Use half-angle formulas to rewrite and evaluate trigonometric functions.
- Use product-to-sum and sum-to-product formulas to rewrite and evaluate trigonometric functions.

Why you should learn it

You can use a variety of trigonometric formulas to rewrite trigonometric functions in more convenient forms. For instance, Exercise 124 on page 395 shows you how to use a half-angle formula to determine the apex angle of a sound wave cone caused by the speed of an airplane.

EXAMPLE 1 Solving a Multiple-Angle Equation

Solve $2 \cos x + \sin 2x = 0$.

Solution

Begin by rewriting the equation so that it involves functions of x (rather than $2x$). Then factor and solve as usual.

$2 \cos x + \sin 2x = 0$ Write original equation.

$2 \cos x + 2 \sin x \cos x = 0$ Double-angle formula

$2 \cos x(1 + \sin x) = 0$ Factor.

$2 \cos x = 0$ $\quad 1 + \sin x = 0$ Set factors equal to zero.

$\cos x = 0$ $\quad \sin x = -1$ Isolate trigonometric functions.

$x = \frac{\pi}{2}, \frac{3\pi}{2}$ $\quad x = \frac{3\pi}{2}$ Solutions in $[0, 2\pi)$

So, the general solution is

$x = \frac{\pi}{2} + 2n\pi$ and $x = \frac{3\pi}{2} + 2n\pi$ General solution

where n is an integer.

Checkpoint *Audio-video solution in English & Spanish at LarsonPrecalculus.com*

Solve $\cos 2x + \cos x = 0$.

EXAMPLE 2 Evaluating Functions Involving Double Angles

Use the following to find $\sin 2\theta$, $\cos 2\theta$, and $\tan 2\theta$.

$$\cos\theta = \frac{5}{13}, \quad \frac{3\pi}{2} < \theta < 2\pi$$

Solution

In Figure 5.18, you can see that

$$\sin\theta = \frac{y}{r} = -\frac{12}{13} \quad \text{and} \quad \tan\theta = \frac{y}{x} = -\frac{12}{5}.$$

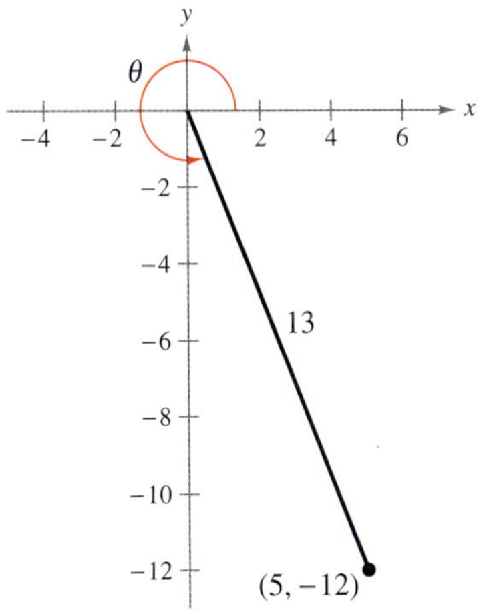

Figure 5.18

Next, use these values with the double-angle formulas, as shown below.

$$\sin 2\theta = 2\sin\theta\cos\theta = 2\left(-\frac{12}{13}\right)\left(\frac{5}{13}\right) = -\frac{120}{169}$$

$$\cos 2\theta = 2\cos^2\theta - 1 = 2\left(\frac{25}{169}\right) - 1 = -\frac{119}{169}$$

$$\tan 2\theta = \frac{2\tan\theta}{1-\tan^2\theta} = \frac{2\left(-\frac{12}{5}\right)}{1-\left(-\frac{12}{5}\right)^2} = \frac{120}{119}$$

✓ **Checkpoint** ▶ *Audio-video solution in English & Spanish at LarsonPrecalculus.com*

Use the following to find $\sin 2\theta$, $\cos 2\theta$, and $\tan 2\theta$.

$$\sin\theta = \frac{3}{5}, \quad 0 < \theta < \frac{\pi}{2}$$

The double-angle formulas are not restricted to the angles 2θ and θ. Other *double* combinations, such as 4θ and 2θ or 6θ and 3θ, are also valid. Here are two examples.

$$\sin 4\theta = 2\sin 2\theta\cos 2\theta \quad \text{and} \quad \cos 6\theta = \cos^2 3\theta - \sin^2 3\theta$$

By using double-angle formulas together with the sum formulas derived in the preceding section, you can form other multiple-angle formulas.

EXAMPLE 3 Deriving a Triple-Angle Formula

Rewrite $\sin 3x$ in terms of $\sin x$.

Solution

$\sin 3x = \sin(2x + x)$	Rewrite as a sum.
$= \sin 2x\cos x + \cos 2x\sin x$	Sum formula
$= 2\sin x\cos x\cos x + (1 - 2\sin^2 x)\sin x$	Double-angle formulas
$= 2\sin x\cos^2 x + \sin x - 2\sin^3 x$	Multiply.
$= 2\sin x(1 - \sin^2 x) + \sin x - 2\sin^3 x$	Pythagorean identity
$= 2\sin x - 2\sin^3 x + \sin x - 2\sin^3 x$	Multiply.
$= 3\sin x - 4\sin^3 x$	Simplify.

✓ **Checkpoint** ▶ *Audio-video solution in English & Spanish at LarsonPrecalculus.com*

Rewrite $\cos 3x$ in terms of $\cos x$.

Power-Reducing Formulas

The double-angle formulas can be used to obtain the following **power-reducing formulas.**

Power-Reducing Formulas

$$\sin^2 u = \frac{1 - \cos 2u}{2}$$

$$\cos^2 u = \frac{1 + \cos 2u}{2}$$

$$\tan^2 u = \frac{1 - \cos 2u}{1 + \cos 2u}$$

(See the proofs on page 403.)

Example 4 shows a typical power reduction that is used in calculus. Note the repeated use of power-reducing formulas.

EXAMPLE 4 Reducing a Power

Rewrite $\sin^4 x$ as a sum of first powers of the cosines of multiple angles.

Solution

$\sin^4 x = (\sin^2 x)^2$ Property of exponents

$= \left(\frac{1 - \cos 2x}{2}\right)^2$ Power-reducing formula

$= \frac{1}{4}(1 - 2\cos 2x + \cos^2 2x)$ Expand binomial.

$= \frac{1}{4}\left(1 - 2\cos 2x + \frac{1 + \cos 4x}{2}\right)$ Power-reducing formula

$= \frac{1}{4} - \frac{1}{2}\cos 2x + \frac{1}{8} + \frac{1}{8}\cos 4x$ Distributive Property

$= \frac{3}{8} - \frac{1}{2}\cos 2x + \frac{1}{8}\cos 4x$ Simplify.

$= \frac{1}{8}(3 - 4\cos 2x + \cos 4x)$ Factor.

You can use a graphing utility to check this result, as shown in Figure 5.19. Notice that the graphs coincide.

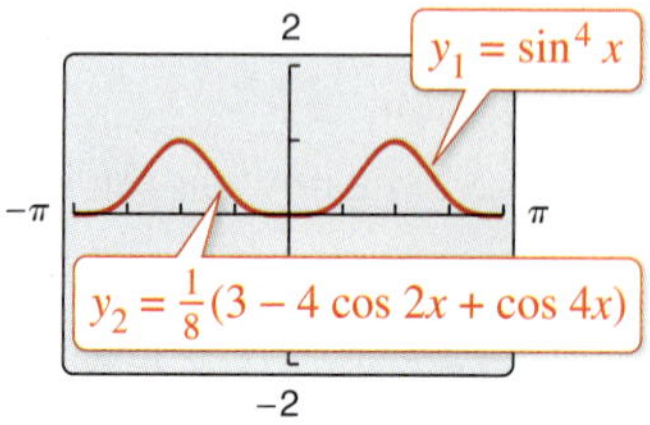

Figure 5.19

Checkpoint *Audio-video solution in English & Spanish at LarsonPrecalculus.com*

Rewrite $\tan^4 x$ in terms of first powers of the cosines of multiple angles.

Half-Angle Formulas

You can derive some useful alternative forms of the power-reducing formulas by replacing u with $u/2$. The results are called **half-angle formulas.**

Half-Angle Formulas

$$\sin\frac{u}{2} = \pm\sqrt{\frac{1-\cos u}{2}} \qquad \cos\frac{u}{2} = \pm\sqrt{\frac{1+\cos u}{2}}$$

$$\tan\frac{u}{2} = \frac{1-\cos u}{\sin u} = \frac{\sin u}{1+\cos u}$$

The signs of $\sin\frac{u}{2}$ and $\cos\frac{u}{2}$ depend on the quadrant in which $\frac{u}{2}$ lies.

Algebra Help

To find the exact value of a trigonometric function with an angle in D° M′S″ form using a half-angle formula, first convert the angle measure to decimal degree form. Then multiply the angle measure by 2.

EXAMPLE 5 Using a Half-Angle Formula

Find the exact value of $\sin 105°$.

Solution

Begin by noting that 105° is half of 210°. Then, using the half-angle formula for $\sin(u/2)$ and the fact that 105° lies in Quadrant II, you have

$$\sin 105° = \sqrt{\frac{1-\cos 210°}{2}} = \sqrt{\frac{1+\left(\sqrt{3}/2\right)}{2}} = \frac{\sqrt{2+\sqrt{3}}}{2}.$$

The positive square root is chosen because $\sin\theta$ is positive in Quadrant II.

Technology Tip

Use your calculator to verify the result obtained in Example 5. That is, evaluate $\sin 105°$ and $\left(\sqrt{2+\sqrt{3}}\right)/2$. You will notice that both expressions yield the same result.

✓ ***Checkpoint*** *Audio-video solution in English & Spanish at LarsonPrecalculus.com*

Find the exact value of $\cos 105°$.

EXAMPLE 6 Solving a Trigonometric Equation

Find all solutions of $1 + \cos^2 x = 2\cos^2\frac{x}{2}$ in the interval $[0, 2\pi)$.

Algebraic Solution

$$1 + \cos^2 x = 2\cos^2\frac{x}{2}$$ Write original equation.

$$1 + \cos^2 x = 2\left(\pm\sqrt{\frac{1+\cos x}{2}}\right)^2$$ Half-angle formula

$$1 + \cos^2 x = 1 + \cos x$$ Simplify.

$$\cos^2 x - \cos x = 0$$ Simplify.

$$\cos x(\cos x - 1) = 0$$ Factor.

By setting the factors $\cos x$ and $\cos x - 1$ equal to zero, you find that the solutions in the interval $[0, 2\pi)$ are $x = \pi/2$, $x = 3\pi/2$, and $x = 0$.

Graphical Solution

The x-intercepts are $x = 0$, $x \approx 1.571$, and $x \approx 4.712$.

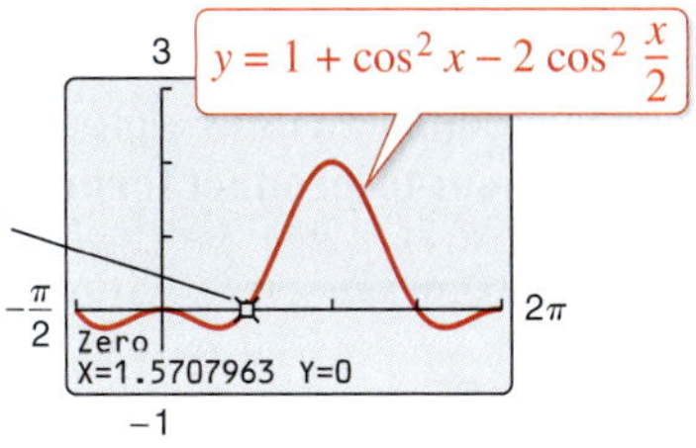

From the figure, you can conclude that the approximate solutions of $1 + \cos^2 x = 2\cos^2\frac{x}{2}$ in the interval $[0, 2\pi)$ are $x = 0$, $x \approx 1.571 \approx \frac{\pi}{2}$, and $x \approx 4.712 \approx \frac{3\pi}{2}$.

✓ ***Checkpoint*** *Audio-video solution in English & Spanish at LarsonPrecalculus.com*

Find all solutions of $\cos^2 x = \sin^2(x/2)$ in the interval $[0, 2\pi)$.

Product-to-Sum and Sum-to-Product Formulas

Each of the following **product-to-sum formulas** is easily verified using the sum and difference formulas discussed in the preceding section.

Product-to-Sum Formulas

$$\sin u \sin v = \frac{1}{2}[\cos(u - v) - \cos(u + v)]$$

$$\cos u \cos v = \frac{1}{2}[\cos(u - v) + \cos(u + v)]$$

$$\sin u \cos v = \frac{1}{2}[\sin(u + v) + \sin(u - v)]$$

$$\cos u \sin v = \frac{1}{2}[\sin(u + v) - \sin(u - v)]$$

Product-to-sum formulas are used in calculus to evaluate integrals involving the products of sines and cosines of two different angles.

EXAMPLE 7 Writing Products as Sums

Rewrite the product as a sum or difference.

$\cos 5x \sin 4x$

Solution

Using the appropriate product-to-sum formula, you obtain

$$\cos 5x \sin 4x = \frac{1}{2}[\sin(5x + 4x) - \sin(5x - 4x)]$$

$$= \frac{1}{2}\sin 9x - \frac{1}{2}\sin x.$$

✓ Checkpoint *Audio-video solution in English & Spanish at LarsonPrecalculus.com*

Rewrite the product $\sin 5x \cos 3x$ as a sum or difference. ■

Occasionally, it is useful to reverse the procedure and write a sum of trigonometric functions as a product. This can be accomplished with the following **sum-to-product formulas.**

Sum-to-Product Formulas

$$\sin u + \sin v = 2\sin\left(\frac{u + v}{2}\right)\cos\left(\frac{u - v}{2}\right)$$

$$\sin u - \sin v = 2\cos\left(\frac{u + v}{2}\right)\sin\left(\frac{u - v}{2}\right)$$

$$\cos u + \cos v = 2\cos\left(\frac{u + v}{2}\right)\cos\left(\frac{u - v}{2}\right)$$

$$\cos u - \cos v = -2\sin\left(\frac{u + v}{2}\right)\sin\left(\frac{u - v}{2}\right)$$

(See proof on page 404.)

Technology Tip

You can use a graphing utility to support the solution to Example 7. Graph $y_1 = \cos 5x \sin 4x$ and $y_2 = \frac{1}{2}\sin 9x - \frac{1}{2}\sin x$ in the same viewing window. Notice that the graphs appear to coincide. So, the two expressions appear to be equivalent.

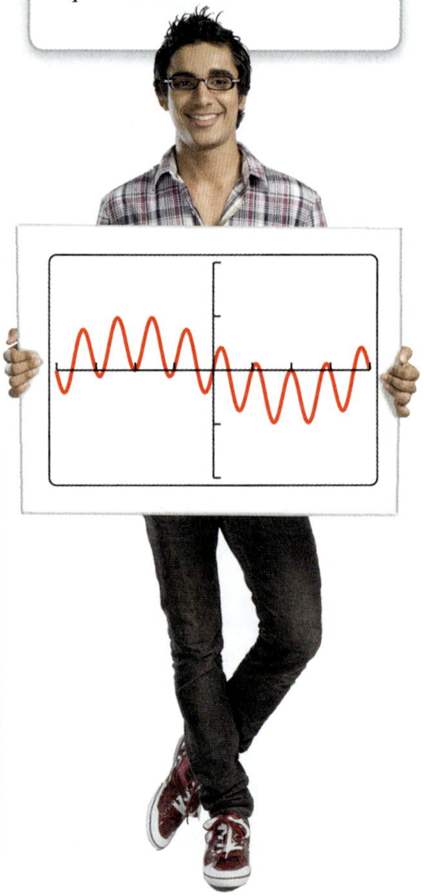

EXAMPLE 8 Using a Sum-to-Product Formula

Find the exact value of $\cos 195° + \cos 105°$.

Solution

Using the appropriate sum-to-product formula, you obtain

$$\cos 195° + \cos 105° = 2\cos\left(\frac{195° + 105°}{2}\right)\cos\left(\frac{195° - 105°}{2}\right)$$
$$= 2\cos 150° \cos 45°$$
$$= 2\left(-\frac{\sqrt{3}}{2}\right)\left(\frac{\sqrt{2}}{2}\right)$$
$$= -\frac{\sqrt{6}}{2}.$$

✓ **Checkpoint** ▶ *Audio-video solution in English & Spanish at LarsonPrecalculus.com*

Find the exact value of $\sin 195° + \sin 105°$.

EXAMPLE 9 Solving a Trigonometric Equation

See LarsonPrecalculus.com for an interactive version of this type of example.

Find all solutions of $\sin 5x + \sin 3x = 0$ in the interval $[0, 2\pi)$.

Solution

$$\sin 5x + \sin 3x = 0 \qquad \text{Write original equation.}$$
$$2\sin\left(\frac{5x + 3x}{2}\right)\cos\left(\frac{5x - 3x}{2}\right) = 0 \qquad \text{Sum-to-product formula}$$
$$2\sin 4x \cos x = 0 \qquad \text{Simplify.}$$

By setting the factor $\sin 4x$ equal to zero, you can find that the solutions in the interval $[0, 2\pi)$ are

$$x = 0, \frac{\pi}{4}, \frac{\pi}{2}, \frac{3\pi}{4}, \pi, \frac{5\pi}{4}, \frac{3\pi}{2}, \frac{7\pi}{4}.$$

The equation $\cos x = 0$ yields no additional solutions. You can use a graphing utility to confirm the solutions, as shown in the figure.

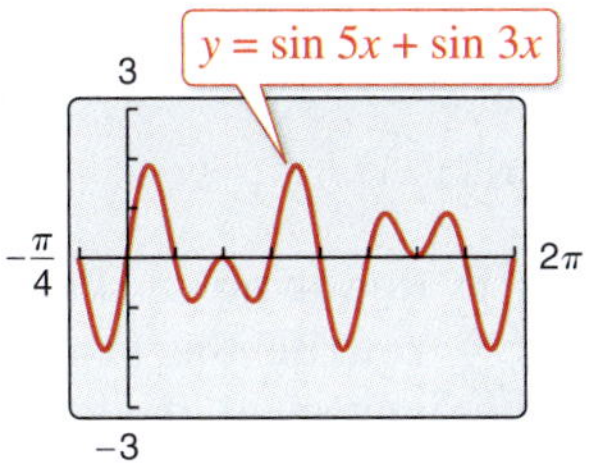

Notice that the general solution is

$$x = \frac{n\pi}{4}$$

where n is an integer.

✓ **Checkpoint** ▶ *Audio-video solution in English & Spanish at LarsonPrecalculus.com*

Solve $\sin 4x - \sin 2x = 0$ in the interval $[0, 2\pi)$.

5.5 Exercises

See *CalcChat.com* for tutorial help and worked-out solutions to odd-numbered exercises. For instructions on how to use a graphing utility, see Appendix A.

Vocabulary and Concept Check

In Exercises 1–6, fill in the blank to complete the trigonometric formula.

1. $\cos^2 u =$ _______ **2.** $\sin u \sin v =$ _______ **3.** $\cos u - \cos v =$ _______

4. $\sin \dfrac{u}{2} =$ _______ **5.** _______ $= \dfrac{\sin u}{1 + \cos u}$ **6.** _______ $= \dfrac{1 - \cos 2u}{1 + \cos 2u}$

7. Match each function with its double-angle formula.

(a) $\sin 2u$ (i) $1 - 2\sin^2 u$

(b) $\cos 2u$ (ii) $2 \sin u \cos u$

(c) $\tan 2u$ (iii) $\dfrac{2 \tan u}{1 - \tan^2 u}$

8. Match each expression with its product-to-sum formula.

(a) $\sin u \cos v$ (i) $\frac{1}{2}[\cos(u - v) + \cos(u + v)]$

(b) $\cos u \sin v$ (ii) $\frac{1}{2}[\sin(u + v) + \sin(u - v)]$

(c) $\cos u \cos v$ (iii) $\frac{1}{2}[\sin(u + v) - \sin(u - v)]$

Procedures and Problem Solving

Finding Exact Values of Trigonometric Functions **In Exercises 9 and 10, use the figure to find the exact value of each trigonometric function.**

(a) $\sin \theta$ **(b)** $\sin 2\theta$

(c) $\cos \theta$ **(d)** $\cos 2\theta$

(e) $\tan \theta$ **(f)** $\tan 2\theta$

(g) $\csc \theta$ **(h)** $\csc 2\theta$

(i) $\sec \theta$ **(j)** $\sec 2\theta$

(k) $\cot \theta$ **(l)** $\cot 2\theta$

9.

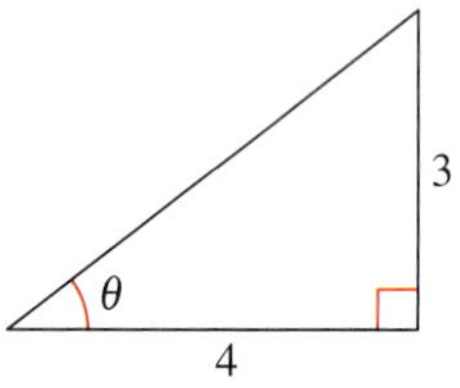

10.

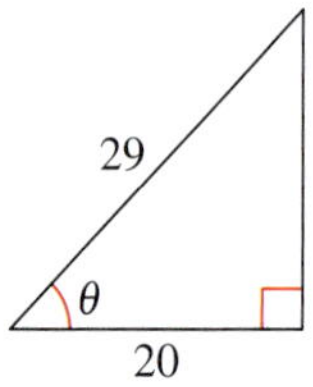

Solving a Multiple-Angle Equation **In Exercises 11–20, solve the equation.**

11. $\sin 2x - \sin x = 0$ **12.** $\sin 2x + \cos x = 0$

13. $4 \sin x \cos x = 1$ **14.** $\sin 2x \sin x = \cos x$

15. $\cos 2x - \cos x = 0$ **16.** $\cos 2x + \sin x = 0$

17. $\sin 4x = -2 \sin 2x$ **18.** $(\sin 2x + \cos 2x)^2 = 1$

19. $\tan 2x - \cot x = 0$ **20.** $\tan 2x - 2 \cos x = 0$

Evaluating Functions Involving Double Angles **In Exercises 21–26, find the exact values of $\sin 2u$, $\cos 2u$, and $\tan 2u$ using the double-angle formulas.**

21. $\sin u = -\frac{3}{5}$, $3\pi/2 < u < 2\pi$

22. $\cos u = -\frac{2}{3}$, $\pi/2 < u < \pi$

23. $\tan u = \frac{1}{2}$, $\pi < u < 3\pi/2$

24. $\cot u = -6$, $3\pi/2 < u < 2\pi$

25. $\sec u = -2$, $\pi/2 < u < \pi$

26. $\csc u = 3$, $0 < u < \pi/2$

27. **Deriving a Triple-Angle Formula** Rewrite $\sec 3x$ in terms of $\cos x$.

28. **Deriving a Triple-Angle Formula** Rewrite $\tan 3x$ in terms of $\tan x$.

Using a Double-Angle Formula **In Exercises 29–32, use a double-angle formula to rewrite the expression. Use a graphing utility to graph both expressions to verify that both forms are the same.**

29. $6 \sin x \cos x$ **30.** $14 \sin x \cos x$

31. $\cos^2 x - \frac{1}{2}$ **32.** $10 \sin^2 x - 5$

Using a Power-Reducing Formula **In Exercises 33–36, use a power-reducing formula to rewrite the expression. Use a graphing utility to graph both expressions to verify that both forms are the same.**

33. $\sin^2 x$ **34.** $\cos^2 x$

35. $\cos^3 x$ **36.** $\sin^3 x$

Reducing a Power **In Exercises 37–46, rewrite the expression in terms of the first power of the cosine. Use a graphing utility to graph both expressions to verify that both forms are the same.**

37. $\cos^4 x$ **38.** $\sin^6 x$

39. $\sin^2 x \cos^4 x$ **40.** $\sin^4 x \cos^2 x$

41. $\sin^2 2x$ **42.** $\cos^2 2x$

43. $\cos^2 \dfrac{x}{2}$ **44.** $\sin^4 \dfrac{x}{2}$

45. $\cos^2 x \tan^2 x$ **46.** $\sin^2 x \tan^2 x$

Finding Exact Values of Trigonometric Functions In Exercises 47 and 48, use the figure to find the exact value of each trigonometric function.

(a) $\sin \frac{\theta}{2}$ (b) $\csc \frac{\theta}{2}$

(c) $\cos \frac{\theta}{2}$ (d) $\sec \frac{\theta}{2}$

(e) $\tan \frac{\theta}{2}$ (f) $\cot \frac{\theta}{2}$

(g) $2 \sin \frac{\theta}{2} \cos \frac{\theta}{2}$ (h) $2 \cos \frac{\theta}{2} \tan \frac{\theta}{2}$

47.

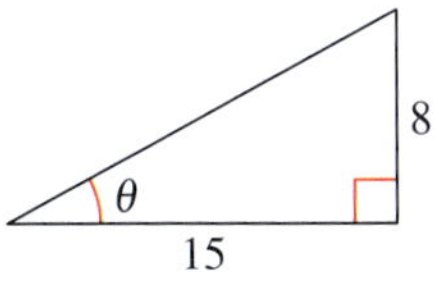

48.

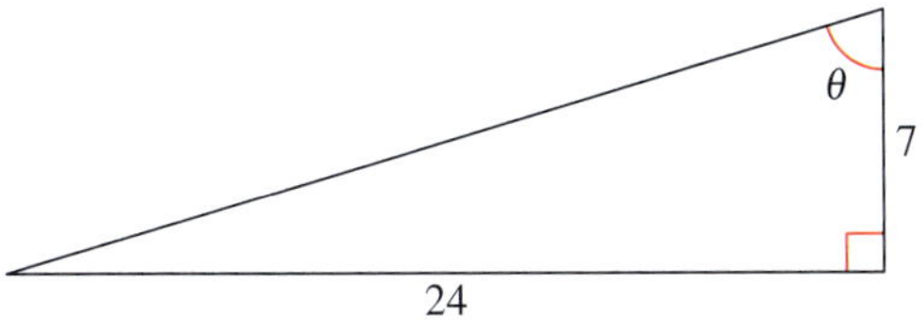

Using a Half-Angle Formula In Exercises 49–56, use the half-angle formulas to determine the exact values of the sine, cosine, and tangent of the angle.

49. $75°$ **50.** $165°$

51. $157° 30'$ **52.** $67° 30'$

53. $\frac{\pi}{8}$ **54.** $\frac{\pi}{12}$

55. $\frac{7\pi}{12}$ **56.** $\frac{5\pi}{8}$

Using a Half-Angle Formula In Exercises 57–62, find the exact values of $\sin(u/2)$, $\cos(u/2)$, and $\tan(u/2)$ using the half-angle formulas.

57. $\cos u = \frac{7}{25}, \quad 0 < u < \pi/2$

58. $\sin u = -\frac{12}{37}, \quad 3\pi/2 < u < 2\pi$

59. $\cot u = 3, \quad \pi < u < 3\pi/2$

60. $\tan u = -\frac{5}{12}, \quad 3\pi/2 < u < 2\pi$

61. $\csc u = -\frac{5}{3}, \quad \pi < u < 3\pi/2$

62. $\sec u = \frac{7}{2}, \quad 0 < u < \pi/2$

Using a Half-Angle Formula In Exercises 63–66, use the half-angle formulas to simplify the expression.

63. $\sqrt{\frac{1 - \cos 6x}{2}}$ **64.** $\sqrt{\frac{1 + \cos 4x}{2}}$

65. $-\sqrt{\frac{1 - \cos 8x}{1 + \cos 8x}}$ **66.** $-\sqrt{\frac{1 - \cos(x - 1)}{2}}$

Solving a Trigonometric Equation In Exercises 67–70, find the solutions of the equation in the interval $[0, 2\pi)$. Use a graphing utility to verify your answers.

67. $\sin \frac{x}{2} + \cos x = 0$ **68.** $\sin \frac{x}{2} + \cos x - 1 = 0$

69. $\cos \frac{x}{2} - \sin x = 0$ **70.** $\tan \frac{x}{2} - \sin x = 0$

Writing Products as Sums In Exercises 71–78, use the product-to-sum formulas to write the product as a sum or difference.

71. $\cos 165° \cos 105°$ **72.** $\sin 45° \cos 15°$

73. $6 \sin \frac{\pi}{3} \cos \frac{\pi}{3}$ **74.** $4 \cos \frac{\pi}{3} \sin \frac{5\pi}{6}$

75. $\sin 5\theta \sin 3\theta$

76. $7 \cos(-5\beta) \sin 3\beta$

77. $\sin(x + y) \cos(x - y)$

78. $\sin(x + y) \sin(x - y)$

Writing Sums as Products In Exercises 79–86, use the sum-to-product formulas to write the sum or difference as a product.

79. $\sin 5\theta - \sin 3\theta$

80. $\sin 3\theta + \sin \theta$

81. $\cos 6x + \cos 2x$

82. $\sin x + \sin 5x$

83. $\sin(\alpha + \beta) - \sin(\alpha - \beta)$

84. $\cos(\phi + \alpha) - \cos(\phi - \alpha)$

85. $\cos\left(\theta + \frac{\pi}{2}\right) - \cos\left(\theta - \frac{\pi}{2}\right)$

86. $\sin\left(x + \frac{\pi}{2}\right) + \sin\left(x - \frac{\pi}{2}\right)$

Using a Sum-to-Product Formula In Exercises 87–90, use the sum-to-product formulas to find the exact value of the expression.

87. $\sin 75° + \sin 15°$ **88.** $\cos 120° + \cos 60°$

89. $\cos \frac{3\pi}{4} - \cos \frac{\pi}{4}$ **90.** $\sin \frac{5\pi}{4} - \sin \frac{3\pi}{4}$

Solving a Trigonometric Equation In Exercises 91–94, find the solutions of the equation in the interval $[0, 2\pi)$. Use a graphing utility to verify your answers.

91. $\sin 6x + \sin 2x = 0$

92. $\cos 2x - \cos 6x = 0$

93. $\frac{\cos 2x}{\sin 3x - \sin x} - 1 = 0$

94. $\sin^2 3x - \sin^2 x = 0$

Using Trigonometric Identities **In Exercises 95–98, use the figure and trigonometric identities to find the exact value of the trigonometric function in two ways.**

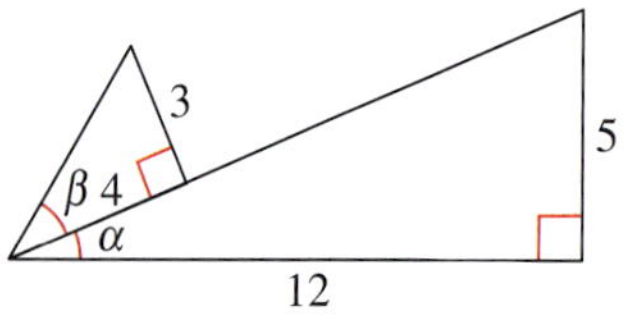

95. $\sin^2 \alpha$

96. $\cos^2 \alpha$

97. $\sin \alpha \cos \beta$

98. $\cos \alpha \sin \beta$

Verifying a Trigonometric Identity **In Exercises 99–112, verify the identity algebraically. Use a graphing utility to check your result graphically.**

99. $\csc 2\theta = \dfrac{\csc \theta}{2 \cos \theta}$

100. $\tan \dfrac{u}{2} = \csc u - \cot u$

101. $\cos^2 2\alpha - \sin^2 2\alpha = \cos 4\alpha$

102. $\cos^4 x - \sin^4 x = \cos 2x$

103. $(\sin x + \cos x)^2 = 1 + \sin 2x$

104. $\sin \dfrac{\alpha}{3} \cos \dfrac{\alpha}{3} = \dfrac{1}{2} \sin \dfrac{2\alpha}{3}$

105. $\dfrac{\sin 3y}{\sin y} = 1 - 2\sin^2 y + 2\cos^2 y$

106. $\dfrac{\cos 3\beta}{\cos \beta} = 1 - 4\sin^2 \beta$

107. $\csc \dfrac{u}{2} = \pm\sqrt{\dfrac{2 \csc u}{\csc u - \cot u}}$

108. $\sec \dfrac{u}{2} = \pm\sqrt{\dfrac{2 \tan u}{\tan u + \sin u}}$

109. $\cos 3\beta = \cos^3 \beta - 3\sin^2 \beta \cos \beta$

110. $\sin 4\beta = 4 \sin \beta \cos \beta(1 - 2\sin^2 \beta)$

111. $\dfrac{\sin x \pm \sin y}{\cos x + \cos y} = \tan \dfrac{x \pm y}{2}$

112. $\dfrac{\sin x + \sin y}{\cos x - \cos y} = -\cot \dfrac{x - y}{2}$

Rewriting a Trigonometric Expression **In Exercises 113–118, write the trigonometric expression as an algebraic expression.**

113. $\sin(2 \arcsin x)$

114. $\cos(2 \arccos x)$

115. $\cos(2 \arcsin x)$

116. $\sin(2 \arccos x)$

117. $\sin(2 \arctan x)$

118. $\cos(2 \arctan x)$

An Application from Calculus **In Exercises 119 and 120, the graph of a function f is shown over the interval $[0, 2\pi]$. (a) Find the x-intercepts of the graph of f algebraically. Verify your solutions by using the *zero* or *root* feature of a graphing utility. (b) The x-coordinates of the extrema of f are solutions of the trigonometric equation. (Calculus is required to find the trigonometric equation.) Find the solutions of the equation algebraically. Verify these solutions using the *maximum* and *minimum* features of the graphing utility.**

119. *Function:* $f(x) = \sin 2x - \sin x$

Trigonometric Equation: $2 \cos 2x - \cos x = 0$

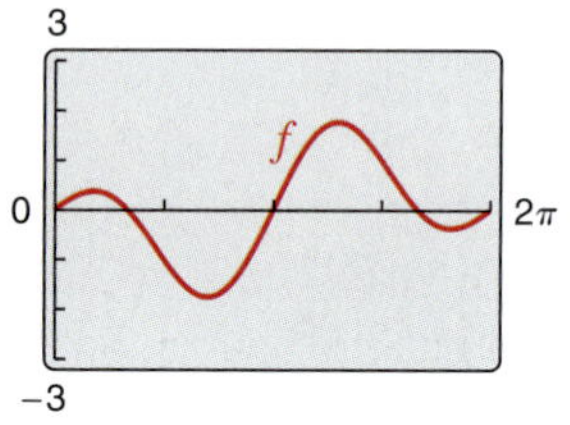

120. *Function:* $f(x) = \cos 2x + \sin x$

Trigonometric Equation: $-2 \sin 2x + \cos x = 0$

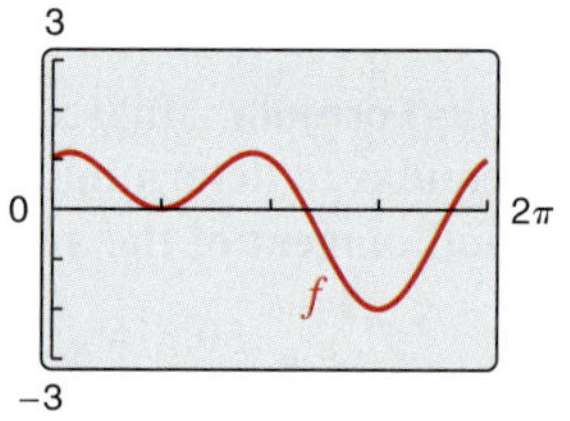

121. Projectile Motion The range of a projectile fired at an angle θ with the horizontal and with an initial velocity of v_0 feet per second is given by

$$r = \frac{1}{32} v_0^2 \sin 2\theta$$

where r is measured in feet. An athlete throws a javelin at 75 feet per second. At what angle must the athlete throw the javelin so that the javelin travels a horizontal distance of 130 feet?

122. Geometry The length of each of the two equal sides of an isosceles triangle is 10 meters (see figure). The angle between the two sides is θ.

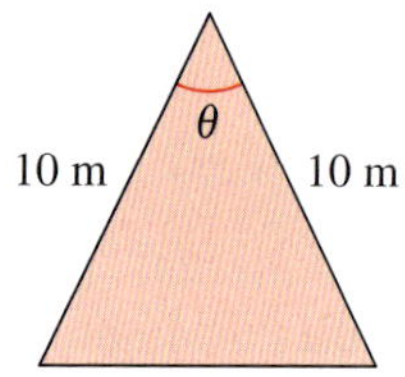

(a) Write the area of the triangle as a function of $\theta/2$.

(b) Write the area of the triangle as a function of θ and determine the value of θ such that the area is a maximum.

123. Mechanical Engineering When two railroad tracks merge, the overlapping portions of the tracks are in the shape of a circular arc (see figure). The radius of each arc r (in feet) and the angle θ are related by

$$\frac{x}{2} = 2r \sin^2 \frac{\theta}{2}.$$

Write a formula for x in terms of $\cos \theta$.

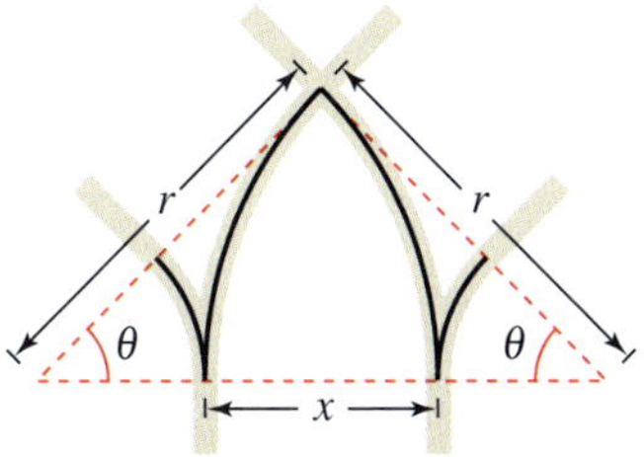

124. *Why you should learn it* *(p. 386)* The Mach number M of an airplane is the ratio of its speed to the speed of sound. When an airplane travels faster than the speed of sound, the sound waves form a cone behind the airplane (see figure). The Mach number is related to the apex angle θ of the cone by

$$\sin \frac{\theta}{2} = \frac{1}{M}.$$

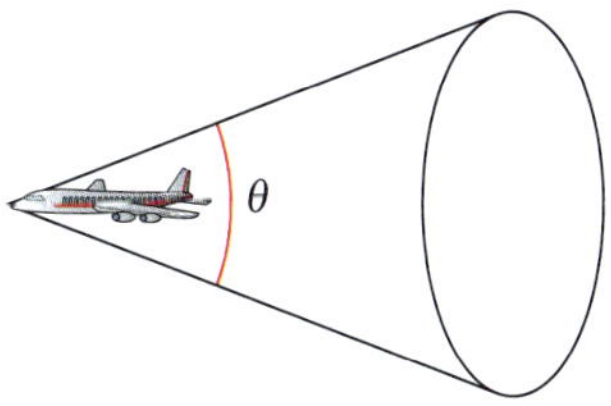

(a) Find the angle θ that corresponds to a Mach number of 1.

(b) Find the angle θ that corresponds to a Mach number of 4.5.

(c) The speed of sound is about 760 miles per hour. Determine the speed of an object having the Mach numbers in parts (a) and (b).

(d) Rewrite the equation as a trigonometric function of θ.

Focusing on Concepts

True or False? **In Exercises 125 and 126, determine whether the statement is true or false. Justify your answer.**

125. $\sin \frac{x}{2} = -\sqrt{\frac{1 - \cos x}{2}}, \quad \pi \le x \le 2\pi$

126. The graph of $y = 4 - 8 \sin^2 x$ has a maximum at $(\pi, 4)$.

127. Exploration Consider $f(x) = \sin^4 x + \cos^4 x$.

(a) Use the power-reducing formulas to write the function in terms of cosine to the first power.

(b) Determine another way of rewriting the function. Use a graphing utility to rule out incorrectly rewritten functions.

(c) Add a trigonometric term to the function so that it becomes a perfect square trinomial. Rewrite the function as a perfect square trinomial minus the term that you added. Use the graphing utility to rule out incorrectly rewritten functions.

(d) Rewrite the result of part (c) in terms of the sine of a double angle. Use the graphing utility to rule out incorrectly rewritten functions.

(e) When you rewrite a trigonometric expression, your result may not be the same as a friend's. Does this mean that one of you is wrong? Explain.

128. HOW DO YOU SEE IT? Explain how to use the figure to verify each double-angle formula.

(a) $\sin 2u = 2 \sin u \cos u$

(b) $\cos 2u = \cos^2 u - \sin^2 u$

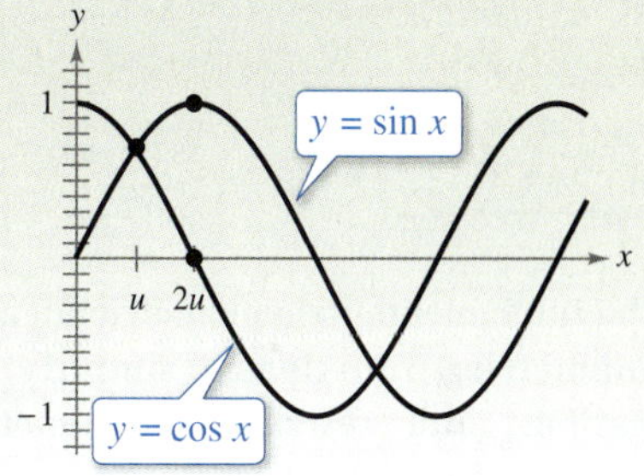

129. Error Analysis Describe the error.

$$\sin 75° = -\sqrt{\frac{1 - \cos 150°}{2}} = -\frac{\sqrt{2 + \sqrt{3}}}{2}$$

130. Error Analysis Describe the error.

$$\cos 4 \sin 3 = \tfrac{1}{2}[\sin 7 - \sin(-1)]$$
$$= \tfrac{1}{2} \sin 7 + \tfrac{1}{2} \sin 1$$

Cumulative Mixed Review

Finding the Midpoint of a Line Segment **In Exercises 131 and 132, (a) plot the points, (b) find the distance between the points, and (c) find the midpoint of the line segment connecting the points.**

131. $(5, 2), (-1, 4)$ **132.** $(3, -1), (7, 4)$

Finding an Arc Length **In Exercises 133 and 134, find the length of the arc on a circle of radius r intercepted by a central angle θ.**

133. $r = 21$ cm, $\theta = 35°$ **134.** $r = 15$ in., $\theta - 110°$

5 Chapter Review

See CalcChat.com for tutorial help and worked-out solutions to odd-numbered exercises.
For instructions on how to use a graphing utility, see Appendix A.

5.1 What did you learn?

Recognize and write the fundamental trigonometric identities *(p. 352)*.

Reciprocal Identities

$\sin u = 1/\csc u$ $\quad \cos u = 1/\sec u$ $\quad \tan u = 1/\cot u$

$\csc u = 1/\sin u$ $\quad \sec u = 1/\cos u$ $\quad \cot u = 1/\tan u$

Quotient Identities

$\tan u = \dfrac{\sin u}{\cos u}$ $\quad \cot u = \dfrac{\cos u}{\sin u}$

Pythagorean Identities

$\sin^2 u + \cos^2 u = 1$ $\quad 1 + \tan^2 u = \sec^2 u$

$1 + \cot^2 u = \csc^2 u$

Cofunction Identities

$\sin[(\pi/2) - u] = \cos u$ $\quad \cos[(\pi/2) - u] = \sin u$

$\tan[(\pi/2) - u] = \cot u$ $\quad \cot[(\pi/2) - u] = \tan u$

$\sec[(\pi/2) - u] = \csc u$ $\quad \csc[(\pi/2) - u] = \sec u$

Even/Odd Identities

$\sin(-u) = -\sin u$ $\quad \cos(-u) = \cos u$

$\tan(-u) = -\tan u$ $\quad \csc(-u) = -\csc u$

$\sec(-u) = \sec u$ $\quad \cot(-u) = -\cot u$

Use the fundamental trigonometric identities to evaluate trigonometric functions, simplify trigonometric expressions, and rewrite trigonometric expressions *(p. 353)*. See Examples 1–7.

Example

(a) Given $\cos\theta = -\frac{4}{5}$ and $\tan\theta > 0$, you can use fundamental trigonometric identities to find the remaining trigonometric functions of θ.

Pythagorean identity:

$$\sin^2\theta = 1 - \cos^2\theta = 1 - \left(-\tfrac{4}{5}\right)^2 = 1 - \tfrac{16}{25} = \tfrac{9}{25}$$

Because $\cos\theta < 0$ and $\tan\theta > 0$, θ lies in Quadrant III, so

$$\sin\theta = -\sqrt{9/25} = -3/5.$$

Reciprocal identities:

$$\tan\theta = \frac{\sin\theta}{\cos\theta} = \frac{-3/5}{-4/5} = \frac{3}{4} \qquad \cot\theta = \frac{1}{\tan\theta} = \frac{4}{3}$$

$$\sec\theta = \frac{1}{\cos\theta} = -\frac{5}{4} \qquad \csc\theta = \frac{1}{\sin\theta} = -\frac{5}{3}$$

(b) $$\tan^4 x + 2\tan^2 x + 1 = (\tan^2 x + 1)^2 = (\sec^2 x)^2 = \sec^4 x$$

(c) $$\frac{\cos t + \tan t \sin t}{\csc t} = \frac{\cos t + (\sin t/\cos t)\sin t}{1/\sin t} = \sin t\left(\frac{\cos^2 t + \sin^2 t}{\cos t}\right) = \sin t(1/\cos t) = \tan t$$

Using Fundamental Identities **In Exercises 1–6, rewrite the expression using the fundamental trigonometric identities.**

1. $1/\tan(-x)$ **2.** $1/\csc(-x)$

3. $\dfrac{1}{\sec[(\pi/2) - x]}$ **4.** $\dfrac{\sin(-x)}{\sin[(\pi/2) - x]}$

5. $\sqrt{1 + \cot^2 x}$ **6.** $\sqrt{1 - \cos^2 x}$

Using Identities to Evaluate a Function **In Exercises 7–14, use the given information to find the values of the remaining trigonometric functions.**

7. $\tan x = \frac{4}{3}$, $\sec x < 0$ **8.** $\csc x = -\frac{13}{5}$, $\cos x > 0$

9. $\sec\theta = -\sqrt{3}$, $\sin\theta > 0$

10. $\cot\theta = \sqrt{2}$, $\sin\theta < 0$

11. $\sin x = \frac{4}{5}$, $\cos x = \frac{3}{5}$

12. $\tan\theta = 2/3$, $\sec\theta = \sqrt{13}/3$

13. $\cos x = 1/\sqrt{2}$, $\sin x = -1/\sqrt{2}$

14. $\sec\theta = -3$, $\sin\theta = 2\sqrt{2}/3$

Simplifying a Trigonometric Expression **In Exercises 15–24, use the fundamental identities to simplify the expression. Use the *table* feature of a graphing utility to check your result numerically.**

15. $1/(\tan^2 x + 1)$ **16.** $(\sin^2\theta + \cos^2\theta)/\sin\theta$

17. $\tan(-\theta)/\sec\theta$ **18.** $\sec^2(-\theta)/\csc^2\theta$

19. $\tan^2\theta(\csc^2\theta - 1)$ **20.** $\cot^2 x(1 - \cos^2 x)$

21. $\tan[(\pi/2) - x)]\sec x$ **22.** $\dfrac{\sin x \cot x}{\sin[(\pi/2) - x]}$

23. $\dfrac{\sec^2 x - 1}{\sec x - 1}$ **24.** $\dfrac{\sin^2\alpha - \cos^2\alpha}{\sin^2\alpha - \sin\alpha\cos\alpha}$

25. Rate of Change The rate of change of the function $f(x) = 2\sqrt{\sin x}$ is given by the expression $\cos x/\sqrt{\sin x}$. Show that this expression can also be written as $\cot x\sqrt{\sin x}$.

26. Rate of Change The rate of change of the function $f(x) = \csc x - \cot x$ is given by the expression $\csc^2 x - \csc x \cot x$. Show that this expression can also be written as $(1 - \cos x)/\sin^2 x$.

5.2 What did you learn?

Verify trigonometric identities *(p. 359).*

Guidelines for Verifying Trigonometric Identities

1. Work with one side of the equation at a time.
2. Look to factor an expression, add fractions, square a binomial, or create a monomial denominator.
3. Look to use the fundamental identities. Note which functions are in the final expression you want. Sines and cosines pair up well, as do secants and tangents, and cosecants and cotangents.
4. When the preceding guidelines do not help, try converting all terms to sines and cosines.
5. Always try *something*.

Example

$$\frac{1-\sin x}{\cos x} = \frac{1}{\sec x(1+\sin x)} \quad \text{Original identity}$$

$$= \frac{\cos x}{1+\sin x} \quad \text{Reciprocal identity}$$

$$= \frac{\cos x(1-\sin x)}{1-\sin^2 x} \quad \text{Multiply numerator and denominator by } (1-\sin x).$$

$$= \frac{\cos x(1-\sin x)}{\cos^2 x} \quad \text{Pythagorean identity}$$

$$= \frac{1-\sin x}{\cos x} \quad \text{Divide numerator and denominator by } \cos x.$$

Verifying a Trigonometric Identity **In Exercises 27–38, verify the identity.**

27. $\cos x(\tan^2 x + 1) = \sec x$

28. $\sec^2 x \cot x - \cot x = \tan x$

29. $\sin^3 \theta + \sin \theta \cos^2 \theta = \sin \theta$

30. $\cot^2 x - \cos^2 x = \cot^2 x \cos^2 x$

31. $\csc^2\left(\frac{\pi}{2} - x\right) - 1 = \tan^2 x$

32. $\tan\left(\frac{\pi}{2} - x\right)\sec x = \csc x$

33. $\sin^4 x \cos^2 x = \cos^2 x - 2\cos^4 x + \cos^6 x$

34. $\cos^4 x \csc^2 x = \csc^2 x + \sin^2 x - 2$

35. $\sqrt{\frac{1-\sin\theta}{1+\sin\theta}} = \frac{1-\sin\theta}{|\cos\theta|}$

36. $\sqrt{1-\cos x} = \frac{|\sin x|}{\sqrt{1+\cos x}}$

37. $\sec(-x)/\csc(-x) = -\tan x$

38. $\frac{1+\sec(-x)}{\sin(-x)+\tan(-x)} = -\csc x$

5.3 What did you learn?

Use standard algebraic techniques to solve trigonometric equations *(p. 367).* Use standard algebraic techniques such as collecting like terms, extracting square roots, and factoring to solve trigonometric equations.

Solve trigonometric equations of quadratic type *(p. 370).* To solve trigonometric equations of quadratic type $ax^2 + bx + c = 0$, factor the quadratic or, when factoring is not possible, use the Quadratic Formula.

Sometimes you must square each side of an equation to obtain a quadratic. This can introduce extraneous solutions, so check any solutions in the original equation.

Solve trigonometric equations involving multiple angles *(p. 372).* To solve equations involving trigonometric functions of multiple angles of the forms $\sin ku$ or $\cos ku$, first solve the equation for ku, then divide the result by k.

Use inverse trigonometric functions to solve equations that involve trigonometric functions *(p. 373).* After factoring an equation and setting the factors equal to 0, you may obtain an equation such as $\tan x - 3 = 0$. In this case, use inverse trigonometric functions to solve. (See Example 10.)

Example

$$\cos\theta - 1 = \sec\theta - \cos\theta$$
$$2\cos\theta - 1 = \sec\theta$$
$$(2\cos\theta - 1)(\cos\theta) = \sec\theta\cos\theta$$
$$2\cos^2\theta - \cos\theta = 1$$
$$2\cos^2\theta - \cos\theta - 1 = 0$$
$$(\cos\theta - 1)(2\cos\theta + 1) = 0$$

By setting each factor equal to zero, you can find the solutions in the interval $[0, 2\pi)$ to be $\theta = 0$, $\theta = 2\pi/3$, and $\theta = 4\pi/3$. Because $\cos\theta$ has a period of 2π, the general solution is

$$\theta = 2n\pi, \quad \theta = \frac{2\pi}{3} + 2n\pi, \quad \text{and} \quad \theta = \frac{4\pi}{3} + 2n\pi$$

where n is an integer. The graph of $y = 2\cos^2\theta - \cos\theta - 1$, shown in the figure, supports this result.

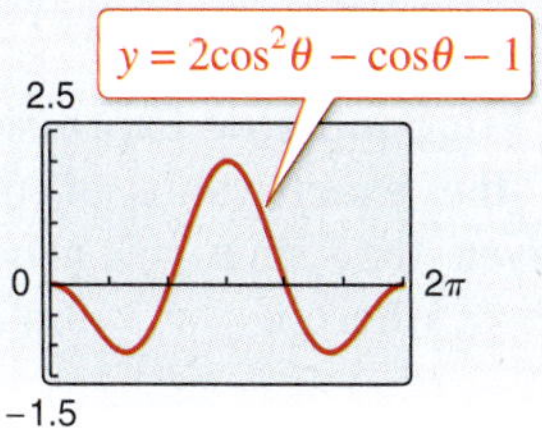

Solving a Trigonometric Equation In Exercises 39–50, solve the equation.

39. $2 \sin x + 2 = 0$ **40.** $\tan x + \sqrt{3} = 0$

41. $\sin x = \sqrt{3} - \sin x$ **42.** $4 \cos x = 1 + 2 \cos x$

43. $3\sqrt{3} \cot x = 3$ **44.** $\frac{1}{2} \sec x - 1 = 0$

45. $3 \csc^2 x = 4$ **46.** $4 \tan^2 x - 1 = \tan^2 x$

47. $2 \cos^2 x - \cos x = 0$ **48.** $\sin^2 x + \sin x = 0$

49. $\sin x - \tan x = 0$ **50.** $\csc x - 2 \cot x = 0$

Solving an Equation of Quadratic Type In Exercises 51–54, find all solutions of the equation in the interval $[0, 2\pi)$.

51. $2 \cos^2 x - \cos x = 1$ **52.** $2 \sin^2 x + 3 \sin x = -1$

53. $\cos^2 x + \sin x = -1$ **54.** $\sin^2 x + 2 \cos x = 2$

Solving a Multiple-Angle Equation In Exercises 55–58, find all solutions of the multiple-angle equation in the interval $[0, 2\pi)$.

55. $2 \sin 2x = \sqrt{2}$ **56.** $\sqrt{3} \tan 3x = 1$

57. $2 \sin(x/2) = 1$ **58.** $\cos(x/3) = 0$

Solving a Multiple-Angle Equation In Exercises 59–62, solve the multiple-angle equation.

59. $2 \sin 2x + 1 = 0$ **60.** $2 \cos 4x + \sqrt{3} = 0$

61. $2 \sin^2 3x - 1 = 0$ **62.** $4 \cos^2 2x - 3 = 0$

Using Inverse Functions In Exercises 63–66, use inverse functions where necessary to solve the equation.

63. $\tan^2 x - 2 \tan x = 0$ **64.** $3 \cos^2 x + 5 \cos x = 0$

65. $3 \cos^2 x - \sin x = 1$ **66.** $\sec^2 x + 6 \tan x + 4 = 0$

5.4 *What did you learn?*

Use sum and difference formulas to evaluate trigonometric functions, verify trigonometric identities, and solve trigonometric equations *(p. 379)*.

Sum and Difference Formulas

$\sin(u + v) = \sin u \cos v + \cos u \sin v$

$\sin(u - v) = \sin u \cos v - \cos u \sin v$

$\cos(u + v) = \cos u \cos v - \sin u \sin v$

$\cos(u - v) = \cos u \cos v + \sin u \sin v$

$$\tan(u + v) = \frac{\tan u + \tan v}{1 - \tan u \tan v}$$

$$\tan(u - v) = \frac{\tan u - \tan v}{1 + \tan u \tan v}$$

Example

(a) $\sin 90° = \sin(60° + 30°)$
$= \sin 60° \cos 30° + \cos 60° \sin 30°$
$= (\sqrt{3}/2)(\sqrt{3}/2) + (1/2)(1/2)$
$= 3/4 + 1/4$
$= 1$

(b) $\cos 90° = \cos(120° - 30°)$
$= \cos 120° \cos 30° + \sin 120° \sin 30°$
$= (-1/2)(\sqrt{3}/2) + (\sqrt{3}/2)(1/2)$
$= -\sqrt{3}/4 + \sqrt{3}/4$
$= 0$

Evaluating Trigonometric Functions In Exercises 67–70, find the exact values of the sine, cosine, and tangent of the angle.

67. $285° = 315° - 30°$ **68.** $345° = 300° + 45°$

69. $\dfrac{31\pi}{12} = \dfrac{11\pi}{6} + \dfrac{3\pi}{4}$ **70.** $\dfrac{17\pi}{12} = \dfrac{7\pi}{4} - \dfrac{\pi}{3}$

Rewriting a Trigonometric Expression In Exercises 71–74, write the expression as the sine, cosine, or tangent of an angle.

71. $\sin 45° \cos 60° - \cos 45° \sin 60°$

72. $\cos 45° \cos 120° - \sin 45° \sin 120°$

73. $\dfrac{\tan 25° + \tan 50°}{1 - \tan 25° \tan 50°}$ **74.** $\dfrac{\tan 63° - \tan 112°}{1 + \tan 63° \tan 112°}$

Evaluating a Trigonometric Expression In Exercises 75–80, find the exact value of the trigonometric expression given that $\sin u = \frac{4}{5}$, where $0 < u < \pi/2$, and $\cos v = -\frac{7}{25}$, where $\pi < v < 3\pi/2$.

75. $\sin(u + v)$ **76.** $\tan(u + v)$

77. $\tan(u - v)$ **78.** $\sin(u - v)$

79. $\cos(u + v)$ **80.** $\cos(u - v)$

Proving a Trigonometric Identity In Exercises 81–86, prove the identity.

81. $\cos\left(x + \dfrac{\pi}{2}\right) = -\sin x$ **82.** $\sin\left(x - \dfrac{3\pi}{2}\right) = \cos x$

83. $\cos(5\pi - x) = -\cos x$ **84.** $\sin(\pi - x) = \sin x$

85. $\cos 3x = 4 \cos^3 x - 3 \cos x$

86. $\dfrac{\sin(\alpha + \beta)}{\cos \alpha \cos \beta} = \tan \alpha + \tan \beta$

Solving a Trigonometric Equation In Exercises 87 and 88, find the solutions of the equation in the interval $[0, 2\pi)$. Use a graphing utility to verify your results.

87. $\sin\left(x + \dfrac{\pi}{2}\right) - \sin\left(x - \dfrac{\pi}{2}\right) = \sqrt{2}$

88. $\cos\left(x + \dfrac{\pi}{4}\right) - \cos\left(x - \dfrac{\pi}{4}\right) = 1$

5.5 What did you learn?

Use multiple-angle formulas to rewrite and evaluate trigonometric functions *(p. 386)*.

$\sin 2u = 2 \sin u \cos u$

$\tan 2u = \dfrac{2 \tan u}{1 - \tan^2 u}$

$$\cos 2u = \cos^2 u - \sin^2 u = 2\cos^2 u - 1 = 1 - 2\sin^2 u$$

Example

(a) $$\cos 60° = \cos^2 30° - \sin^2 30° = \left(\frac{\sqrt{3}}{2}\right)^2 - \left(\frac{1}{2}\right)^2 = \frac{3}{4} - \frac{1}{4} = \frac{1}{2}$$

Use power-reducing formulas to rewrite and evaluate trigonometric functions *(p. 388)*.

$\sin^2 u = \dfrac{1 - \cos 2u}{2}$ $\quad\cos^2 u = \dfrac{1 + \cos 2u}{2}$

$\tan^2 u = \dfrac{1 - \cos 2u}{1 + \cos 2u}$

(b) $$\tan^2 60° = \frac{1 - \cos 120°}{1 + \cos 120°} = \frac{1 - (-1/2)}{1 + (-1/2)} = \frac{3/2}{1/2} = 3$$

Use half-angle formulas to rewrite and evaluate trigonometric functions *(p. 389)*.

$\sin \dfrac{u}{2} = \pm\sqrt{\dfrac{1 - \cos u}{2}}$ $\quad\cos \dfrac{u}{2} = \pm\sqrt{\dfrac{1 + \cos u}{2}}$

$\tan \dfrac{u}{2} = \dfrac{1 - \cos u}{\sin u} = \dfrac{\sin u}{1 + \cos u}$

(c) $$\sin 30° = \sqrt{\frac{1 - \cos 60°}{2}} = \sqrt{\frac{1 - 1/2}{2}} = \sqrt{\frac{1/2}{2}} = \sqrt{\frac{1}{4}} = \frac{1}{2}$$

Use product-to-sum and sum-to-product formulas to rewrite and evaluate trigonometric functions *(p. 390)*.

Product-to-Sum Formulas

$\sin u \sin v = (1/2)[\cos(u - v) - \cos(u + v)]$

$\cos u \cos v = (1/2)[\cos(u - v) + \cos(u + v)]$

$\sin u \cos v = (1/2)[\sin(u + v) + \sin(u - v)]$

$\cos u \sin v = (1/2)[\sin(u + v) - \sin(u - v)]$

Sum-to-Product Formulas

$\sin u + \sin v = 2 \sin\left(\dfrac{u + v}{2}\right)\cos\left(\dfrac{u - v}{2}\right)$

$\sin u - \sin v = 2 \cos\left(\dfrac{u + v}{2}\right)\sin\left(\dfrac{u - v}{2}\right)$

$\cos u + \cos v = 2 \cos\left(\dfrac{u + v}{2}\right)\cos\left(\dfrac{u - v}{2}\right)$

$\cos u - \cos v = -2 \sin\left(\dfrac{u + v}{2}\right)\sin\left(\dfrac{u - v}{2}\right)$

(d) $$\cos 30° \cos 60° = \frac{1}{2}[\cos(-30°) + \cos 90°] = \frac{1}{2}\left(\frac{\sqrt{3}}{2} + 0\right) = \frac{\sqrt{3}}{4}$$

(e) $$\sin 30° - \sin 90° = 2 \cos 60° \sin(-30°) = 2\left(\frac{1}{2}\right)\left(-\frac{1}{2}\right) = -\frac{1}{2}$$

Evaluating Functions Involving Double Angles **In Exercises 89–92, find the exact values of $\sin 2u$, $\cos 2u$, and $\tan 2u$ using the double-angle formulas.**

89. $\sin u = 5/9, \quad 0 < u < \pi/2$

90. $\cos u = 4/5, \quad 3\pi/2 < u < 2\pi$

91. $\tan u = -2/9, \quad \pi/2 < u < \pi$

92. $\cos u = -\dfrac{2}{\sqrt{5}}, \quad \pi < u < \dfrac{3\pi}{2}$

Verifying a Trigonometric Identity **In Exercises 93–96, use double-angle formulas to verify the identity algebraically. Use a graphing utility to check your result graphically.**

93. $8 \sin x \cos x = 4 \sin 2x$

94. $4 \sin x \cos x + 3 = 2 \sin 2x + 3$

95. $1 - 4 \sin^2 x \cos^2 x = \cos^2 2x$

96. $\sin 4x = 8 \cos^3 x \sin x - 4 \cos x \sin x$

Reducing a Power In Exercises 97–100, rewrite the expression in terms of the first power of the cosine. Use a graphing utility to graph both expressions to verify that both forms are the same.

97. $\tan^2 4x$ **98.** $\sin^4 2x$

99. $\cos^4 x \sin^4 x$

100. $\sin^2 2x \tan^2 2x$

Using a Half-Angle Formula In Exercises 101–104, use the half-angle formulas to determine the exact values of the sine, cosine, and tangent of the angle.

101. $15°$ **102.** $202° 30'$

103. $\frac{7\pi}{8}$ **104.** $\frac{11\pi}{12}$

Using a Half-Angle Formula In Exercises 105–108, find the exact values of $\sin(u/2)$, $\cos(u/2)$, and $\tan(u/2)$ using the half-angle formulas.

105. $\sin u = \frac{12}{13}, \quad 0 < u < \pi/2$

106. $\cos u = -\frac{3}{5}, \quad \pi < u < 3\pi/2$

107. $\tan u = -\sqrt{45}/2, \quad \pi/2 < u < \pi$

108. $\sec u = 6, \quad 3\pi/2 < u < 2\pi$

Using a Half-Angle Formula In Exercises 109–112, use the half-angle formulas to simplify the expression.

109. $-\sqrt{\frac{1 + \cos 8x}{2}}$ **110.** $\sqrt{\frac{1 - \cos 4x}{2}}$

111. $\frac{\sin 10x}{1 + \cos 10x}$ **112.** $\frac{1 - \cos 12x}{\sin 12x}$

Writing Products as Sums In Exercises 113–116, use the product-to-sum formulas to write the product as a sum or difference.

113. $4 \sin \frac{\pi}{4} \cos \frac{\pi}{4}$ **114.** $6 \sin 15° \sin 45°$

115. $\sin 5\alpha \sin 4\alpha$ **116.** $\cos 6\theta \sin 8\theta$

Writing Sums as Products In Exercises 117–120, use the sum-to-product formulas to write the sum or difference as a product.

117. $\cos 3\theta + \cos 2\theta$ **118.** $\sin 5\theta + \sin 4\theta$

119. $\sin\left(x + \frac{\pi}{4}\right) - \sin\left(x - \frac{\pi}{4}\right)$

120. $\cos\left(x + \frac{\pi}{6}\right) - \cos\left(x - \frac{\pi}{6}\right)$

Focusing on Concepts

True or False? In Exercises 121–124, determine whether the statement is true or false. Justify your answer.

121. If $\frac{\pi}{2} < \theta < \pi$, then $\cos \frac{\theta}{2} < 0$.

122. $\sin(x + y) = \sin x + \sin y$

123. $4 \sin(-x) \cos(-x) = -2 \sin 2x$

124. $4 \sin 45° \cos 15° = 1 + \sqrt{3}$

Think About It In Exercises 125 and 126, use the graphs of y_1 and y_2 to determine how to change y_2 to a new function y_3 such that $y_1 = y_3$.

125. $y_1 = \sec^2\left(\frac{\pi}{2} - x\right)$
$y_2 = \cot^2 x$

126. $y_1 = \frac{\cos 3x}{\cos x}$
$y_2 = (2 \sin x)^2$

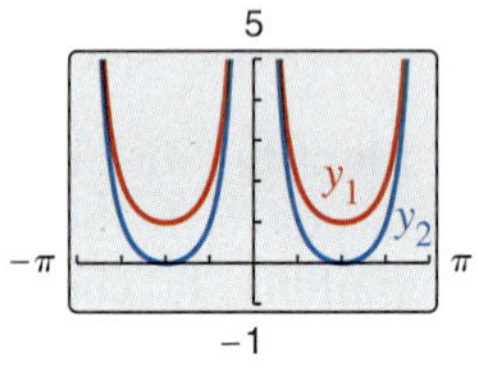

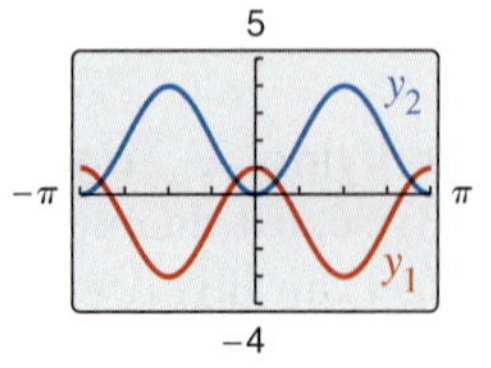

127. **Think About It** List the reciprocal identities, quotient identities, and Pythagorean identities from memory.

128. **Think About It** Is $\sec \theta = \sqrt{1 + \tan^2 \theta}$ an identity? Explain.

129. **Think About It** Is any trigonometric equation with an infinite number of solutions an identity? Explain.

130. **Think About It** Does the equation $a \sin x - b = 0$ have a solution when $|a| < |b|$? Explain.

Harmonic Motion In Exercises 131 and 132, a weight is attached to a spring suspended vertically from a ceiling. When a driving force is applied to the system, the weight moves vertically from its equilibrium position. This motion is modeled by

$$y = \frac{3}{2} \sin 8t - \frac{1}{2} \cos 8t$$

where y is the displacement (in feet) from equilibrium and t is the time (in seconds).

131. Use the identity

$$a \sin B\theta + b \cos B\theta = \sqrt{a^2 + b^2} \sin (B\theta + C)$$

where

$$C = \arctan \frac{b}{a}, \quad a > 0$$

to write the model in the form

$$y = \sqrt{a^2 + b^2} \sin(Bt + C).$$

Use a graphing utility to verify your result.

132. Find the amplitude and frequency of the oscillations of the weight.

5 Chapter Test

See *CalcChat.com* for tutorial help and worked-out solutions to odd-numbered exercises.
For instructions on how to use a graphing utility, see Appendix A.

Take this test as you would take a test in class. After you are finished, check your work against the answers given in the back of the book.

1. Given $\tan\theta = \frac{3}{2}$ and $\cos\theta < 0$, use the fundamental identities to evaluate the other five trigonometric functions of θ.
2. Use the fundamental identities to simplify $\csc^2\beta(1 - \cos^2\beta)$.
3. Factor and simplify $\dfrac{\sec^4 x - \tan^4 x}{\sec^2 x + \tan^2 x}$.
4. Add and simplify $\dfrac{\cos\theta}{\sin\theta} + \dfrac{\sin\theta}{\cos\theta}$.
5. Use a graphing utility to graph the functions

$$y_1 = \tan\left(\frac{3\pi}{2} - x\right) \quad \text{and} \quad y_2 = \cot x$$

in the same viewing window. Make a conjecture about y_1 and y_2.

6. Write $\sqrt{36 - x^2}$ in terms of θ using $x = 6\sin\theta$, where $0 < \theta < \pi/2$.

In Exercises 7–12, verify the identity. Use a graphing utility to verify your result.

7. $\sin\theta\sec\theta = \tan\theta$
8. $\sec^2 x\tan^2 x + \sec^2 x = \sec^4 x$
9. $\dfrac{\csc\alpha + \sec\alpha}{\sin\alpha + \cos\alpha} = \cot\alpha + \tan\alpha$
10. $\sin\left(x - \dfrac{3\pi}{2}\right) + \sin\left(\dfrac{\pi}{2} + x\right) = 2\cos x$
11. $\dfrac{1}{\sin x + 1} + \dfrac{1}{1 - \sin x} = 2\sec^2 x$
12. $(\sin x + \cos x)^2 = 1 + \sin 2x$

Collaborative Project

To work a collaborative project involving Analytic Trigonometry, visit this textbook's website at *LarsonPrecalculus.com.*

In Exercises 13–18, find all solutions of the equation in the interval $[0, 2\pi)$.

13. $3\cos x - 1 = \cos x$
14. $4\cos^2 x - 3 = 0$
15. $\tan^2 x + \tan x = 0$
16. $\csc^2 x - \csc x - 2 = 0$
17. $\cos^2 x - 7\cos x + 4 = 0$
18. $\sin 2\alpha - \cos\alpha = 0$

19. Find the exact value of $\sin(225° + 330°)$.
20. Rewrite $\sin^4 x\tan^2 x$ in terms of the first power of the cosine.
21. Use a half-angle formula to simplify the expression $\dfrac{\sin 4\theta}{1 + \cos 4\theta}$.
22. Write $4\cos 2\theta\sin 4\theta$ as a sum or difference.
23. Write $\cos 3\theta - \cos\theta$ as a product.
24. Use a graphing utility to approximate the solutions of the equation $5\sin x - x = 0$ accurate to three decimal places.
25. The *index of refraction n* of a transparent material is the ratio of the speed of light in a vacuum to the speed of light in the material. For the triangular glass prism in the figure, $n = 1.5$ and $\alpha = 60°$. Find the angle θ for the glass prism given that

$$n = \frac{\sin\left(\frac{\theta}{2} + \frac{\alpha}{2}\right)}{\sin\frac{\theta}{2}}.$$

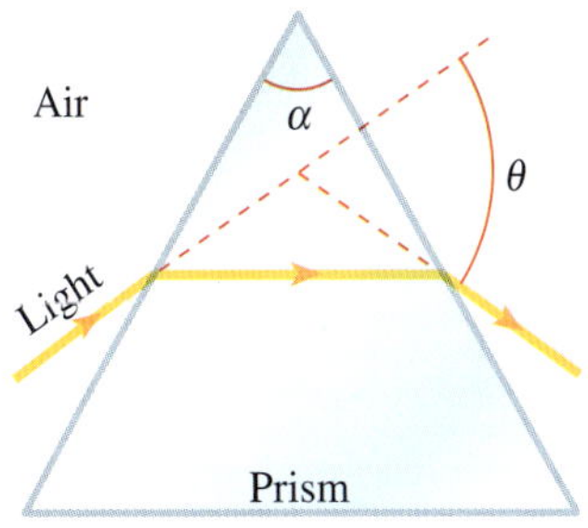

Figure for 25

Proofs in Mathematics

Sum and Difference Formulas (p. 379)

$\sin(u + v) = \sin u \cos v + \cos u \sin v$

$\sin(u - v) = \sin u \cos v - \cos u \sin v$

$\cos(u + v) = \cos u \cos v - \sin u \sin v$

$\cos(u - v) = \cos u \cos v + \sin u \sin v$

$\tan(u + v) = \dfrac{\tan u + \tan v}{1 - \tan u \tan v}$

$\tan(u - v) = \dfrac{\tan u - \tan v}{1 + \tan u \tan v}$

Proof

You can use the figures at the right for the proofs of the formulas for $\cos(u \pm v)$. In the top figure, let A be the point $(1, 0)$ and then use u and v to locate the points $B(x_1, y_1)$, $C(x_2, y_2)$, and $D(x_3, y_3)$ on the unit circle. So, $x_i^2 + y_i^2 = 1$ for $i = 1$, 2, and 3. For convenience, assume that $0 < v < u < 2\pi$. In the bottom figure, note that arcs AC and BD have the same length. So, line segments AC and BD are also equal in length, which implies that

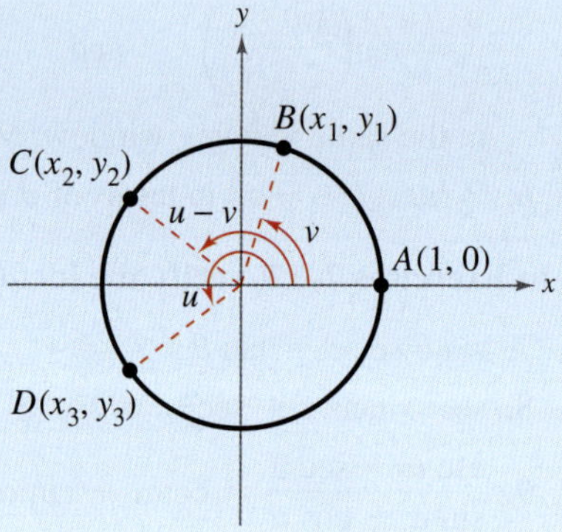

$$\sqrt{(x_2 - 1)^2 + (y_2 - 0)^2} = \sqrt{(x_3 - x_1)^2 + (y_3 - y_1)^2}$$

$$x_2^2 - 2x_2 + 1 + y_2^2 = x_3^2 - 2x_1x_3 + x_1^2 + y_3^2 - 2y_1y_3 + y_1^2$$

$$(x_2^2 + y_2^2) + 1 - 2x_2 = (x_3^2 + y_3^2) + (x_1^2 + y_1^2) - 2x_1x_3 - 2y_1y_3$$

$$1 + 1 - 2x_2 = 1 + 1 - 2x_1x_3 - 2y_1y_3$$

$$x_2 = x_3x_1 + y_3y_1.$$

Finally, by substituting the values $x_2 = \cos(u - v)$, $x_3 = \cos u$, $x_1 = \cos v$, $y_3 = \sin u$, and $y_1 = \sin v$, you obtain $\cos(u - v) = \cos u \cos v + \sin u \sin v$. The formula for $\cos(u + v)$ can be established by considering $u + v = u - (-v)$ and using the formula just derived to obtain

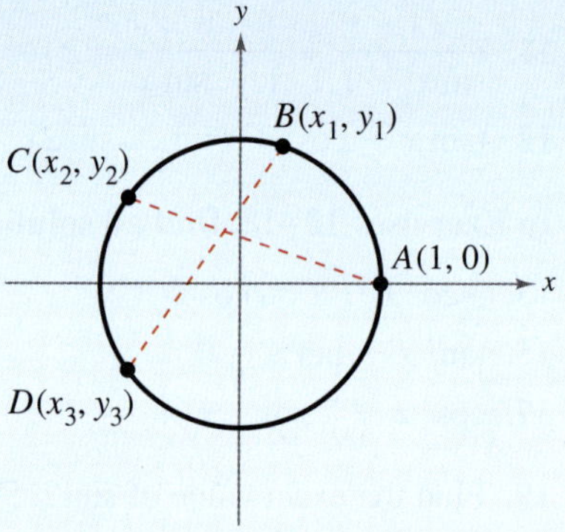

$$\cos(u + v) = \cos[u - (-v)] = \cos u \cos(-v) + \sin u \sin(-v)$$
$$= \cos u \cos v - \sin u \sin v.$$

You can use the sum and difference formulas for sine and cosine to prove the formulas for $\tan(u \pm v)$.

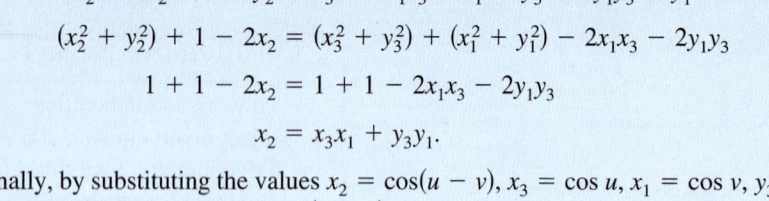

$$\tan(u \pm v) = \frac{\sin(u \pm v)}{\cos(u \pm v)}$$ Quotient identity

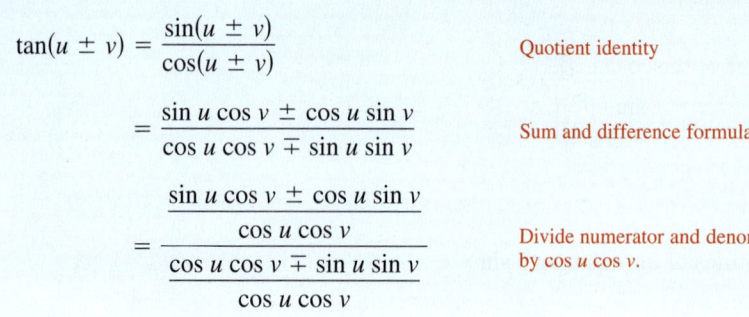

$$= \frac{\sin u \cos v \pm \cos u \sin v}{\cos u \cos v \mp \sin u \sin v}$$ Sum and difference formulas

$$= \frac{\dfrac{\sin u \cos v \pm \cos u \sin v}{\cos u \cos v}}{\dfrac{\cos u \cos v \mp \sin u \sin v}{\cos u \cos v}}$$ Divide numerator and denominator by $\cos u \cos v$.

$$= \frac{\dfrac{\sin u \cos v}{\cos u \cos v} \pm \dfrac{\cos u \sin v}{\cos u \cos v}}{\dfrac{\cos u \cos v}{\cos u \cos v} \mp \dfrac{\sin u \sin v}{\cos u \cos v}}$$ Write as separate fractions.

$$= \frac{\dfrac{\sin u}{\cos u} \pm \dfrac{\sin v}{\cos v}}{1 \mp \dfrac{\sin u}{\cos u} \cdot \dfrac{\sin v}{\cos v}}$$ Product of fractions

$$= \frac{\tan u \pm \tan v}{1 \mp \tan u \tan v}$$ Quotient identity ■

Double-Angle Formulas (p. 386)

$\sin 2u = 2 \sin u \cos u$

$\tan 2u = \dfrac{2 \tan u}{1 - \tan^2 u}$

$\cos 2u = \cos^2 u - \sin^2 u$
$= 2\cos^2 u - 1$
$= 1 - 2\sin^2 u$

Proof

To prove all three formulas, let $v = u$ in the corresponding sum formulas.

$$\sin 2u = \sin(u + u) = \sin u \cos u + \cos u \sin u = 2 \sin u \cos u$$

$$\cos 2u = \cos(u + u) = \cos u \cos u - \sin u \sin u = \cos^2 u - \sin^2 u$$

$$\tan 2u = \tan(u + u) = \frac{\tan u + \tan u}{1 - \tan u \tan u} = \frac{2 \tan u}{1 - \tan^2 u}$$ ■

Power-Reducing Formulas (p. 388)

$\sin^2 u = \dfrac{1 - \cos 2u}{2}$ $\cos^2 u = \dfrac{1 + \cos 2u}{2}$ $\tan^2 u = \dfrac{1 - \cos 2u}{1 + \cos 2u}$

Proof

To prove the first formula, solve for $\sin^2 u$ in the double-angle formula $\cos 2u = 1 - 2\sin^2 u$, as follows.

$$\cos 2u = 1 - 2\sin^2 u$$ Write double-angle formula.

$$2\sin^2 u = 1 - \cos 2u$$ Subtract $\cos 2u$ from and add $2\sin^2 u$ to each side.

$$\sin^2 u = \frac{1 - \cos 2u}{2}$$ Divide each side by 2.

Trigonometry and Astronomy

Trigonometry was used by early astronomers to calculate measurements in the universe. Trigonometry was used to calculate the circumference of Earth and the distance from Earth to the moon. Another major accomplishment in astronomy using trigonometry was computing distances to stars.

In a similar way, you can prove the second formula by solving for $\cos^2 u$ in the double-angle formula

$$\cos 2u = 2\cos^2 u - 1.$$

To prove the third formula, use a quotient identity, as follows.

$$\tan^2 u = \frac{\sin^2 u}{\cos^2 u}$$

$$= \frac{\frac{1 - \cos 2u}{2}}{\frac{1 + \cos 2u}{2}}$$

$$= \frac{1 - \cos 2u}{1 + \cos 2u}$$

Sum-to-Product Formulas (p. 390)

$$\sin u + \sin v = 2\sin\left(\frac{u + v}{2}\right)\cos\left(\frac{u - v}{2}\right)$$

$$\sin u - \sin v = 2\cos\left(\frac{u + v}{2}\right)\sin\left(\frac{u - v}{2}\right)$$

$$\cos u + \cos v = 2\cos\left(\frac{u + v}{2}\right)\cos\left(\frac{u - v}{2}\right)$$

$$\cos u - \cos v = -2\sin\left(\frac{u + v}{2}\right)\sin\left(\frac{u - v}{2}\right)$$

Proof

To prove the first formula, let $x = u + v$ and $y = u - v$. Then substitute $u = (x + y)/2$ and $v = (x - y)/2$ in the product-to-sum formula.

$$\sin u \cos v = \frac{1}{2}[\sin(u + v) + \sin(u - v)]$$

$$\sin\left(\frac{x + y}{2}\right)\cos\left(\frac{x - y}{2}\right) = \frac{1}{2}(\sin x + \sin y)$$

$$2\sin\left(\frac{x + y}{2}\right)\cos\left(\frac{x - y}{2}\right) = \sin x + \sin y$$

The other sum-to-product formulas can be proved in a similar manner.

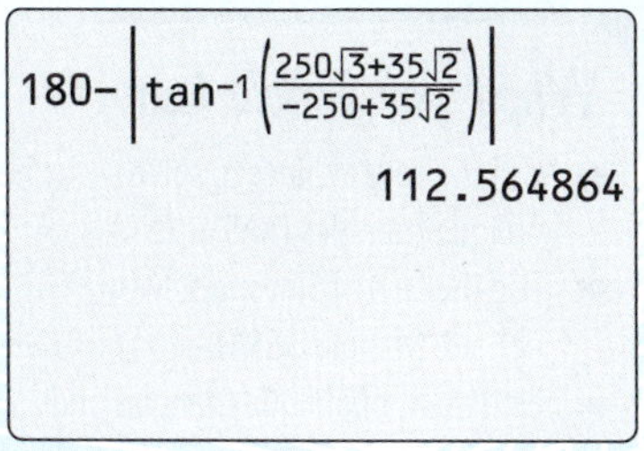

Section 6.3, Example 11
Direction of an Airplane

6 Additional Topics in Trigonometry

Student Resources at LarsonPrecalculus.com

- **Videos** explaining the concepts of precalculus
- **Worked-out solution videos** for all *Checkpoint* exercises
- **Editable spreadsheets** of the data sets in the text
- **Group projects** for each chapter applying concepts to real-life problems

6.1 Law of Sines

Introduction

In Chapter 4, you studied techniques for solving right triangles. In this section and the next, you will solve **oblique triangles**—triangles that have no right angles. As standard notation, the angles of a triangle are labeled A, B, and C, and their opposite sides are labeled a, b, and c, as shown in the figure.

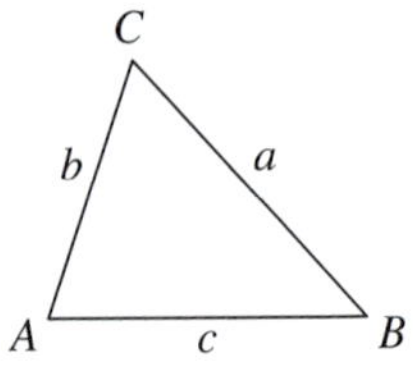

To solve an oblique triangle, you need to know the measure of at least one side and the measures of any two other parts of the triangle—two sides, two angles, or one angle and one side. This breaks down into the following four cases.

1. Two angles and any side (AAS or ASA)
2. Two sides and an angle opposite one of them (SSA)
3. Three sides (SSS)
4. Two sides and their included angle (SAS)

The first two cases can be solved using the **Law of Sines,** whereas the last two cases can be solved using the Law of Cosines (see Section 6.2).

What you should learn

- Use the Law of Sines to solve oblique triangles (AAS or ASA).
- Use the Law of Sines to solve oblique triangles (SSA).
- Find areas of oblique triangles and use the Law of Sines to model and solve real-life problems.

Why you should learn it

You can use the Law of Sines to solve real-life problems involving oblique triangles. For instance, Exercise 48 on page 413 shows how the Law of Sines can be used to help determine the distance from a boat to the shoreline.

Law of Sines

If ABC is a triangle with sides a, b, and c, then

$$\frac{a}{\sin A} = \frac{b}{\sin B} = \frac{c}{\sin C}.$$

Oblique Triangles

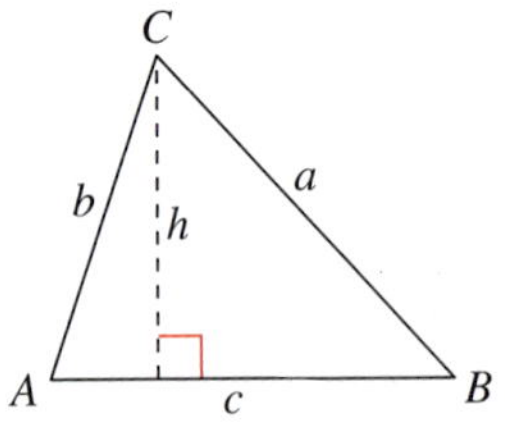

***A* is acute.**

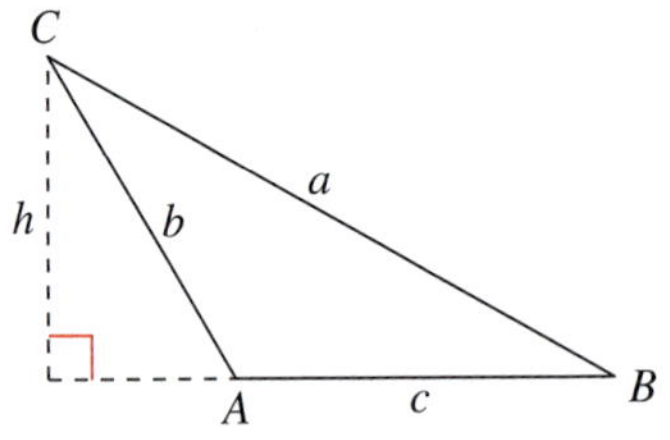

***A* is obtuse.**

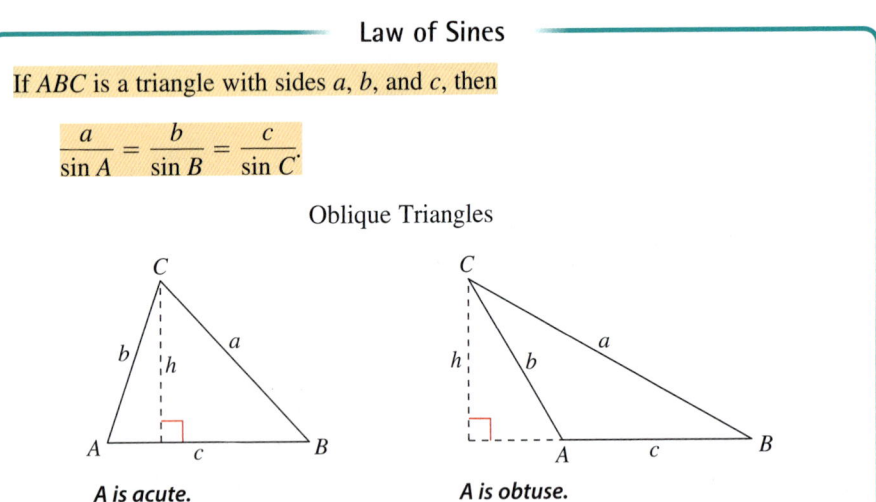

(See the proof on page 472.)

Insight

Some standardized tests may list the Law of Sines and ask you to find an unknown side or angle of a triangle. Also, the Law of Sines is sometimes useful when solving an application problem, in which case the law will likely not be listed.

In the two oblique triangles above, notice that the height h of each triangle can be found using the formula

$$\frac{h}{b} = \sin A \qquad \text{or} \qquad h = b \sin A.$$

Also, note that the Law of Sines can be written in the reciprocal form

$$\frac{\sin A}{a} = \frac{\sin B}{b} = \frac{\sin C}{c}. \qquad \text{Reciprocal form}$$

EXAMPLE 1 Given Two Angles and One Side—AAS

For the triangle shown,

$C = 102.3°, B = 28.7°,$ and $b = 27.4$ feet.

Find the remaining angle and sides.

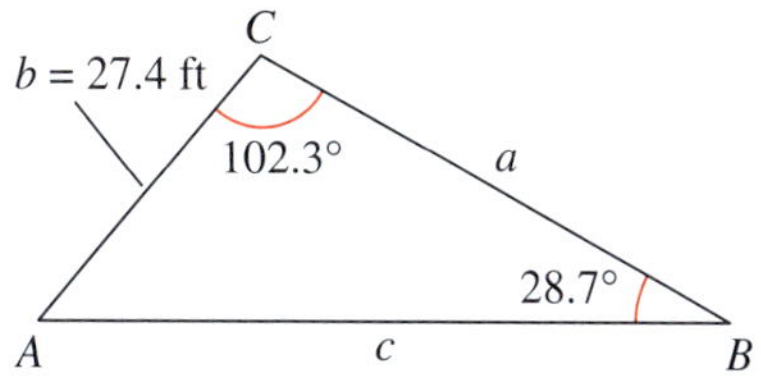

Solution

The third angle of the triangle is

$$A = 180° - B - C$$
$$= 180° - 28.7° - 102.3°$$
$$= 49.0°.$$

By the Law of Sines, you have

$$\frac{a}{\sin A} = \frac{b}{\sin B} = \frac{c}{\sin C}.$$

Using $b = 27.4$ produces

$$a = \frac{b}{\sin B}(\sin A) = \frac{27.4}{\sin 28.7°}(\sin 49.0°) \approx 43.06 \text{ feet}$$

and

$$c = \frac{b}{\sin B}(\sin C) = \frac{27.4}{\sin 28.7°}(\sin 102.3°) \approx 55.75 \text{ feet.}$$

Algebra Help

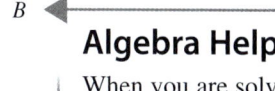

When you are solving a triangle, a careful sketch is useful as a quick test for the feasibility of an answer. Remember that the longest side lies opposite the largest angle, and the shortest side lies opposite the smallest angle.

✓ ***Checkpoint*** *Audio-video solution in English & Spanish at LarsonPrecalculus.com*

For the triangle shown, $A = 30°$, $B = 45°$, and $a = 32$. Find the remaining angle and sides.

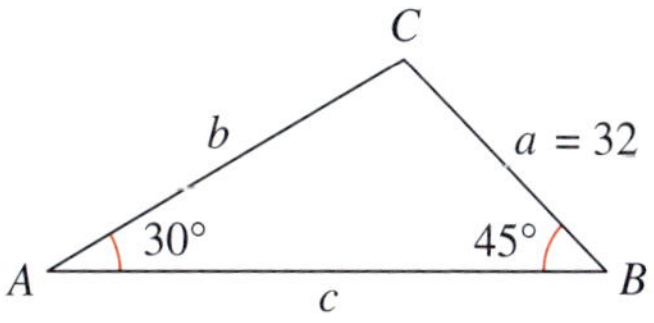

EXAMPLE 2 Given Two Angles and One Side—ASA

A pole tilts *toward* the sun at an 8° angle from the vertical, and it casts a 22-foot shadow. (See Figure 6.1.) The angle of elevation from the tip of the shadow to the top of the pole is 43°. How tall is the pole?

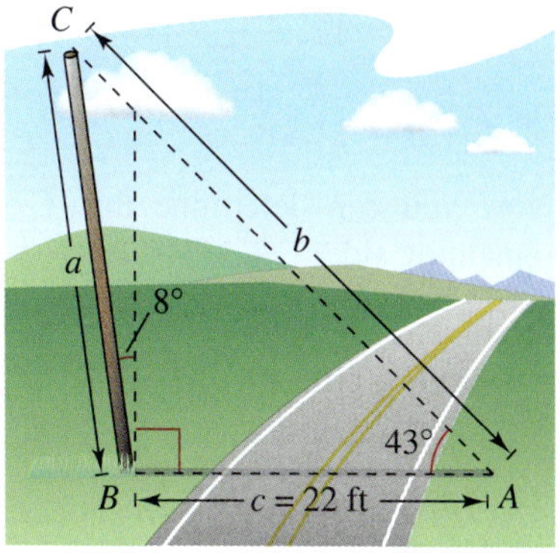

Figure 6.1

Solution

In Figure 6.1, $A = 43°$ and $B = 90° + 8° = 98°$. So, the third angle is

$$C = 180° - A - B = 180° - 43° - 98° = 39°.$$

By the Law of Sines, you have

$$\frac{a}{\sin A} = \frac{c}{\sin C}.$$

Because $c = 22$ feet, the length of the pole is

$$a = \frac{c}{\sin C}(\sin A) = \frac{22}{\sin 39°}(\sin 43°) \approx 23.84 \text{ feet.}$$

✓ ***Checkpoint*** *Audio-video solution in English & Spanish at LarsonPrecalculus.com*

Rework Example 2 for a pole that tilts *away from* the sun under the same conditions.

The Ambiguous Case (SSA)

In Examples 1 and 2, you saw that two angles and one side determine a unique triangle. However, if two sides and one opposite angle are given, then three possible situations can occur: (1) no such triangle exists, (2) one such triangle exists, or (3) two distinct triangles satisfy the conditions.

The Ambiguous Case (SSA)

Consider a triangle in which you are given a, b, and A. ($h = b \sin A$)

	A is acute.	A is acute.	A is acute.	A is acute.	A is obtuse.	A is obtuse.
Sketch	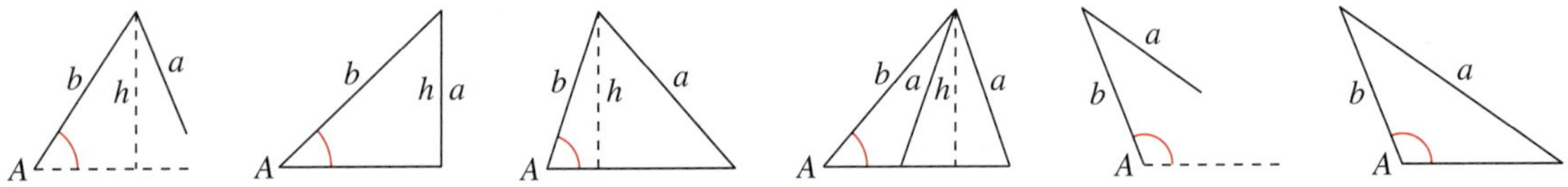					
Necessary condition	$a < h$	$a = h$	$a \ge b$	$h < a < b$	$a \le b$	$a > b$
Possible triangles	None	One	One	Two	None	One

EXAMPLE 3 Single-Solution Case—SSA

See LarsonPrecalculus.com for an interactive version of this type of example.

For the triangle in Figure 6.2, $a = 22$ inches, $b = 12$ inches, and $A = 42°$. Find the remaining side and angles.

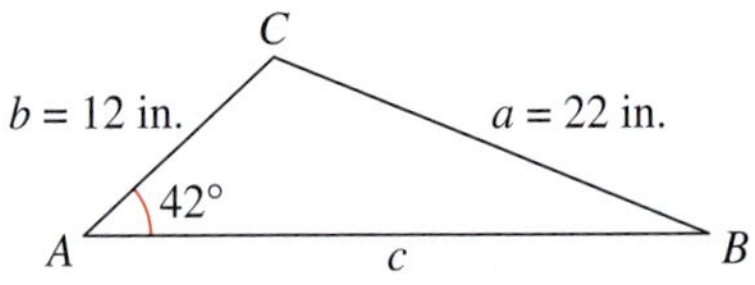

One solution: $a \ge b$
Figure 6.2

Solution

By the Law of Sines, you have

$$\frac{\sin B}{b} = \frac{\sin A}{a}$$ Reciprocal form

$$\sin B = b\left(\frac{\sin A}{a}\right)$$ Multiply each side by b.

$$\sin B = 12\left(\frac{\sin 42°}{22}\right)$$ Substitute for A, a, and b.

$$B \approx 21.41°.$$ Solve for acute angle B.

Now you can determine that $C \approx 180° - 42° - 21.41° = 116.59°$. Then find the remaining side.

$$\frac{c}{\sin C} = \frac{a}{\sin A}$$ Law of Sines

$$c = \frac{a}{\sin A}(\sin C)$$ Multiply each side by $\sin C$.

$$c \approx \frac{22}{\sin 42°}(\sin 116.59°)$$ Substitute for a, A, and C.

$$c \approx 29.40 \text{ inches}$$ Simplify.

✓ ***Checkpoint*** *Audio-video solution in English & Spanish at LarsonPrecalculus.com*

Given $A = 31°$, $a = 12$, and $b = 5$, find the remaining side and angles of the triangle.

EXAMPLE 4 No-Solution Case—SSA

Show that there is no triangle for which $a = 15$, $b = 25$, and $A = 85°$.

Solution Begin by making a sketch, as shown in Figure 6.3. From this figure, it appears that no triangle is formed. You can verify this by using the Law of Sines.

$$\frac{\sin B}{b} = \frac{\sin A}{a} \qquad \text{Reciprocal form}$$

$$\sin B = b\left(\frac{\sin A}{a}\right) = 25\left(\frac{\sin 85°}{15}\right) \approx 1.6603 > 1$$

This contradicts the fact that $|\sin B| \leq 1$. So, no triangle can be formed having sides $a = 15$ and $b = 25$ and an angle of $A = 85°$.

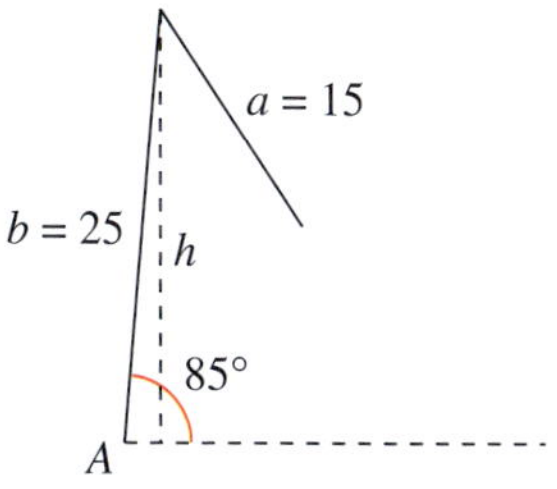

Figure 6.3 *No solution:* $a < h$

✓ **Checkpoint** ▶ *Audio-video solution in English & Spanish at LarsonPrecalculus.com*

Show that there is no triangle for which $a = 4$, $b = 14$, and $A = 60°$.

EXAMPLE 5 Two-Solution Case—SSA

Find two triangles for which $a = 12$ meters, $b = 31$ meters, and $A = 20.5°$.

Solution Because $h = b \sin A = 31(\sin 20.5°) \approx 10.86$ meters and $h < a < b$, you can conclude that there are two possible triangles. Use the Law of Sines to find B.

$$\frac{\sin B}{b} = \frac{\sin A}{a} \qquad \text{Reciprocal form}$$

$$\sin B = b\left(\frac{\sin A}{a}\right) = 31\left(\frac{\sin 20.5°}{12}\right) \quad \Rightarrow \quad B = \arcsin\left(\frac{31 \sin 20.5°}{12}\right)$$

There are two angles between 0° and 180° that satisfy this equation. One angle is

$$B_1 = \arcsin\left(\frac{31 \sin 20.5°}{12}\right) \approx 64.8°.$$

Supplementary angles have the same sine value, so the second angle is

$$B_2 \approx 180° - 64.8° = 115.2°.$$

Now, you can solve each triangle. For $B_1 \approx 64.8°$, you obtain

$$C \approx 180° - 20.5° - 64.8° = 94.7°$$

and

$$c = \frac{a}{\sin A}(\sin C) \approx \frac{12}{\sin 20.5°}(\sin 94.7°) \approx 34.15 \text{ meters.}$$

For $B_2 \approx 115.2°$, you obtain

$$C \approx 180° - 20.5° - 115.2° = 44.3°$$

and

$$c = \frac{a}{\sin A}(\sin C) \approx \frac{12}{\sin 20.5°}(\sin 44.3°) \approx 23.93 \text{ meters.}$$

The resulting triangles are shown in Figure 6.4.

Algebra Help

When using the Law of Sines, choose the form so that the unknown variable is in the numerator.

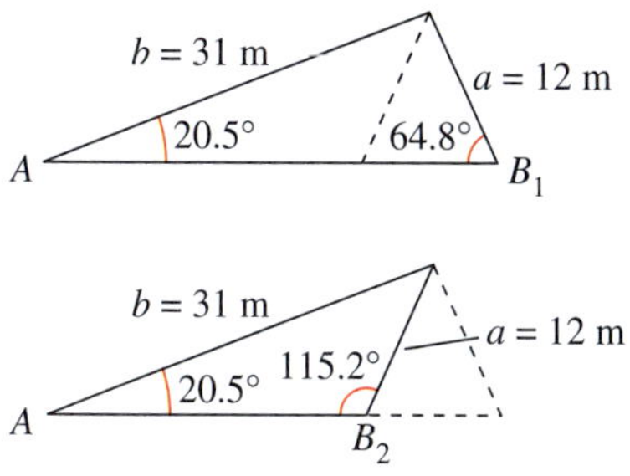

Figure 6.4 *Two solutions:* $h < a < b$

✓ **Checkpoint** ▶ *Audio-video solution in English & Spanish at LarsonPrecalculus.com*

Find two triangles for which $a = 4.5$ feet, $b = 5$ feet, and $A = 58°$.

Area of an Oblique Triangle

The procedure used to prove the Law of Sines leads to a formula for the area of an oblique triangle. Referring to Figure 6.5, note that each triangle has a height of

$$h = b \sin A.$$

To see this when A is obtuse, substitute the reference angle $180° - A$ for A. Now the height of the triangle is given by

$$h = b \sin(180° - A).$$

Using the difference formula for sine, the height is given by

$$\begin{aligned} h &= b(\sin 180° \cos A - \cos 180° \sin A) && \sin(u - v) = \sin u \cos v - \cos u \sin v \\ &= b[0 \cdot \cos A - (-1) \cdot \sin A] \\ &= b \sin A. \end{aligned}$$

Consequently, the area of each triangle is

$$\begin{aligned} \text{Area} &= \frac{1}{2}(\text{base})(\text{height}) \\ &= \frac{1}{2}(c)(b \sin A) \\ &= \frac{1}{2}bc \sin A. \end{aligned}$$

By similar arguments, you can develop the formulas

$$\text{Area} = \frac{1}{2}ab \sin C = \frac{1}{2}ac \sin B.$$

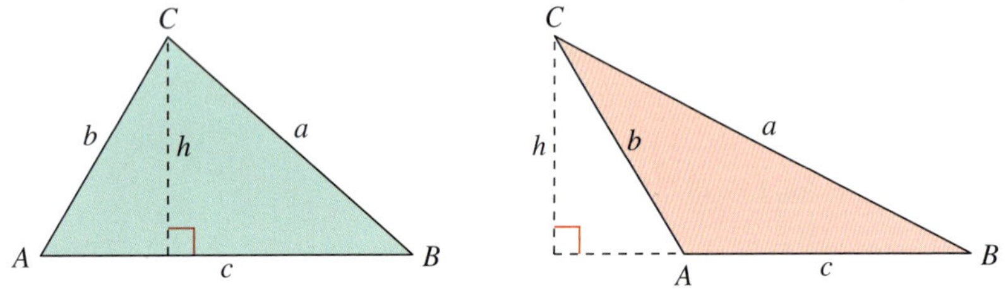

A is acute. *A is obtuse.*

Figure 6.5

Area of an Oblique Triangle

The area of any triangle is one-half the product of the lengths of two sides times the sine of their included angle. That is,

$$\text{Area} = \frac{1}{2}bc \sin A = \frac{1}{2}ab \sin C = \frac{1}{2}ac \sin B.$$

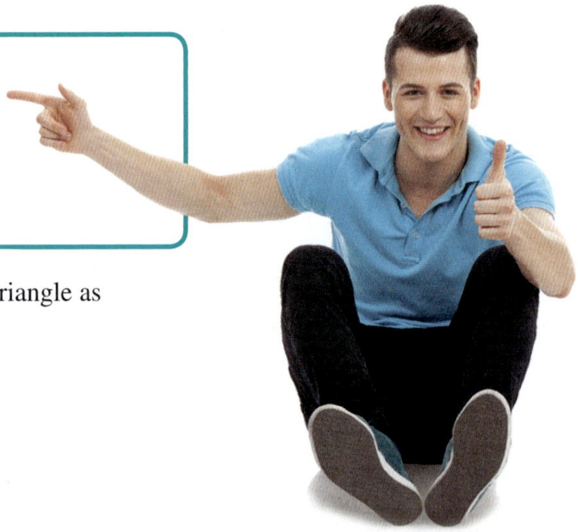

Note that when angle A is 90°, the formula gives the area of a right triangle as

$$\begin{aligned} \text{Area} &= \frac{1}{2}bc \\ &= \frac{1}{2}(\text{base})(\text{height}). \end{aligned}$$

Similar results are obtained for angles C and B equal to 90°.

EXAMPLE 6 Finding the Area of an Oblique Triangle

Find the area of a triangular lot with two sides of lengths 90 meters and 52 meters and an included angle of 102°.

Solution

Consider $a = 90$ meters, $b = 52$ meters, and $C = 102°$, as shown in Figure 6.6. Then the area of the triangle is

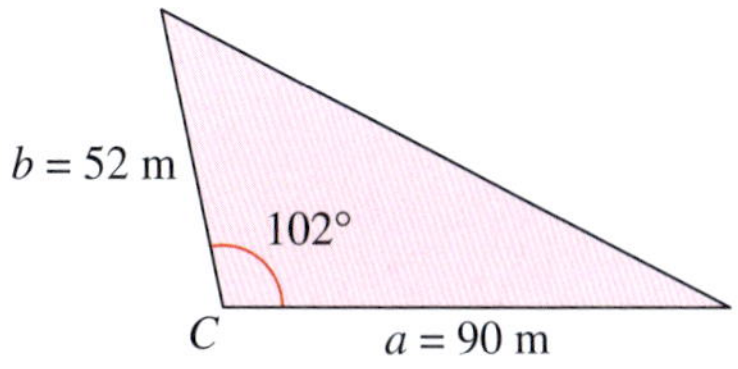

Figure 6.6

$$\text{Area} = \frac{1}{2}ab \sin C$$ Formula for area

$$= \frac{1}{2}(90)(52)(\sin 102°)$$ Substitute for a, b, and C.

$$\approx 2288.87 \text{ square meters.}$$ Simplify.

Checkpoint Audio-video solution in English & Spanish at LarsonPrecalculus.com

Find the area of a triangular lot with two sides of lengths 24 yards and 18 yards and an included angle of 80°.

EXAMPLE 7 An Application of the Law of Sines

The course for a boat race starts at point A and proceeds in the direction S 52° W to point B, then in the direction S 40° E to point C, and finally back to point A, as shown in Figure 6.7. Point C lies 8 kilometers directly south of point A. Approximate the total distance of the race course.

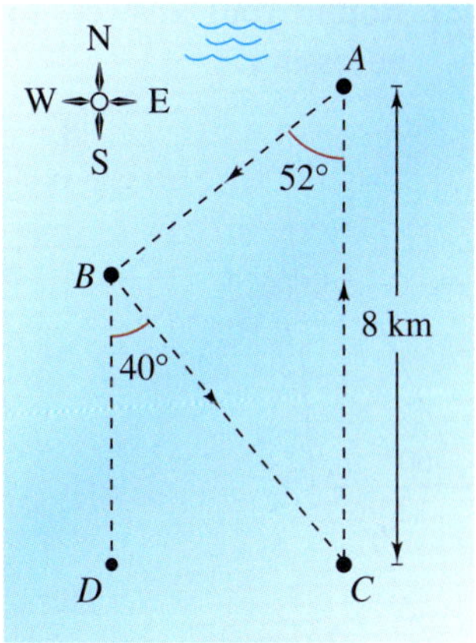

Figure 6.7

Solution

Because lines BD and AC are parallel, it follows that

$$\angle BCA \cong \angle DBC.$$

Consequently, triangle ABC has the measures shown in Figure 6.8. For angle B, you have

$$B = 180° - 52° - 40° = 88°.$$

Using the Law of Sines

$$\frac{a}{\sin 52°} = \frac{b}{\sin 88°} = \frac{c}{\sin 40°}$$

with $b = 8$, you obtain

$$a = \frac{8}{\sin 88°}(\sin 52°) \approx 6.31$$

and

$$c = \frac{8}{\sin 88°}(\sin 40°) \approx 5.15.$$

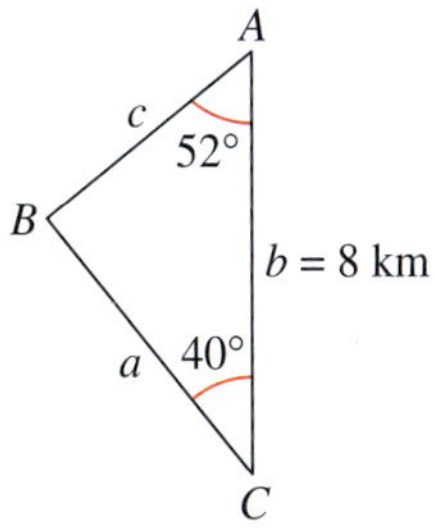

Figure 6.8

The total distance of the course is approximately

$$8 + 6.31 + 5.15 = 19.46 \text{ kilometers.}$$

Checkpoint Audio-video solution in English & Spanish at LarsonPrecalculus.com

On a small lake, you swim from point A to point B at a bearing of N 28° E, then to point C at a bearing of N 58° W, and finally back to point A. Point C lies 800 meters directly north of point A. Draw a diagram to represent this situation. Then approximate the total distance that you swim.

6.1 Exercises

See *CalcChat.com* for tutorial help and worked-out solutions to odd-numbered exercises.
For instructions on how to use a graphing utility, see Appendix A.

Vocabulary and Concept Check

In Exercises 1–4, fill in the blank(s).

1. Law of Sines: $\dfrac{a}{\sin A} = \_\_\_\_\_\_\_ = \dfrac{c}{\sin C}$

2. The ambiguous case occurs when two _______ and one _______ of a triangle are given.

3. To find the area of any triangle, use one of the three formulas: Area = _______, _______, or _______.

4. Two _______ and one _______ determine a unique triangle.

5. Which two cases can be solved using the Law of Sines?

6. Use a formula to represent the height of the triangle at the right.

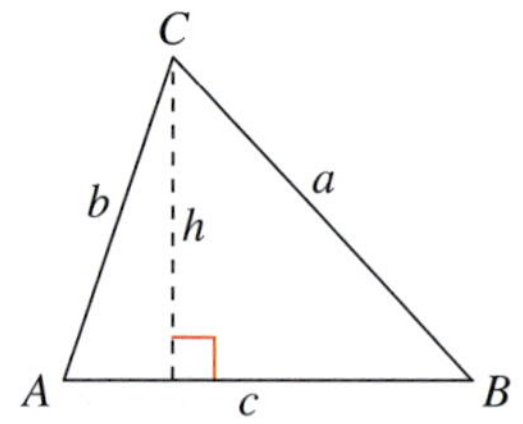

Figure for 6

Procedures and Problem Solving

Using the Law of Sines **In Exercises 7–24, use the Law of Sines to solve the triangle. Round your answers to two decimal places, if necessary.**

7.

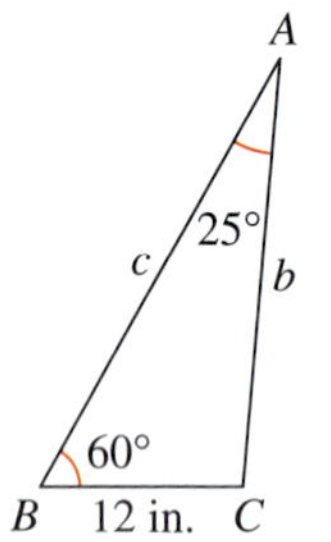

8.

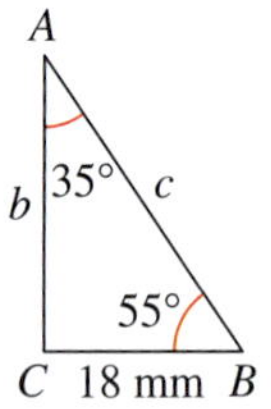

9.

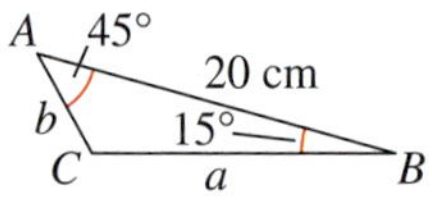

10.

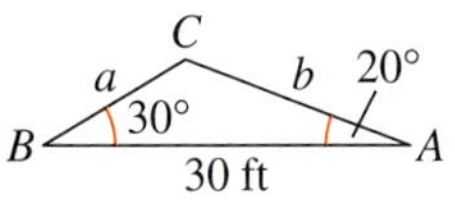

11.

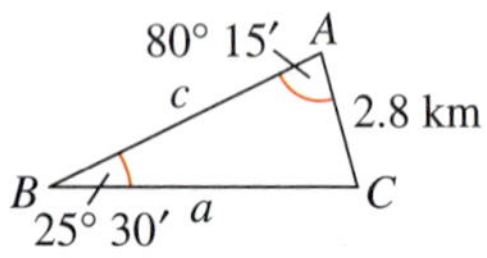

12.

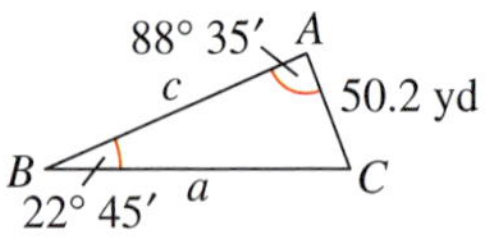

13. $A = 36°, \quad a = 8, \quad b = 5$

14. $A = 76°, \quad a = 34, \quad b = 21$

15. $A = 35°, \quad B = 40°, \quad c = 10$

16. $A = 120°, \quad B = 45°, \quad c = 16$

17. $A = 110°, \quad a = 125, \quad b = 100$

18. $A = 145°, \quad a = 14, \quad b = 4$

19. $A = 83° \, 20', \quad C = 54.6°, \quad c = 18.1$

20. $A = 9° \, 8', \quad C = 142.1°, \quad a = 12.2$

21. $B = 28°, \quad C = 104°, \quad a = 3\frac{5}{8}$

22. $A = 55°, \quad B = 42°, \quad c = \frac{3}{4}$

23. $A = 110° \, 15', \quad a = 48, \quad b = 16$

24. $B = 2° \, 45', \quad b = 6.2, \quad c = 5.8$

Using the Law of Sines **In Exercises 25–30, use the Law of Sines to solve the triangle or show that no triangle with the given characteristics exists. If two solutions exist, find both. Round your answers to two decimal places, if necessary.**

25. $A = 50°, \quad a = 10, \quad b = 40$

26. $A = 110°, \quad a = 125, \quad b = 200$

27. $A = 120°, \quad a = b = 25$

28. $A = 60°, \quad a = 9, \quad c = 10$

29. $A = 58°, \quad a = 11.4, \quad b = 12.8$

30. $A = 58°, \quad a = 4.5, \quad b = 12.8$

Using the Law of Sines **In Exercises 31–34, find the value(s) of b such that the triangle has (a) one solution, (b) two solutions, and (c) no solution.**

31. $A = 36°, \quad a = 5$

32. $A = 60°, \quad a = 10$

33. $A = 10°, \quad a = 10.8$

34. $A = 88°, \quad a = 315.6$

Finding the Area of an Oblique Triangle **In Exercises 35–40, find the area of the triangle with the indicated angle and sides. Round your answer to two decimal places, if necessary.**

35. $C = 110°, \quad a = 6, \quad b = 10$

36. $B = 130°, \quad a = 92, \quad c = 30$

37. $A = 125°, \quad b = 9, \quad c = 6$

38. $C = 120°, \quad a = 4, \quad b = 6$

39. $B = 75° \, 15', \quad a = 103, \quad c = 58$

40. $C = 85° \, 45', \quad a = 16, \quad b = 20$

41. Height A tree leans 4° from the vertical. At a point 40 meters from the tree, the angle of elevation to the top of the tree is 30° (see figure). Find the height h of the tree.

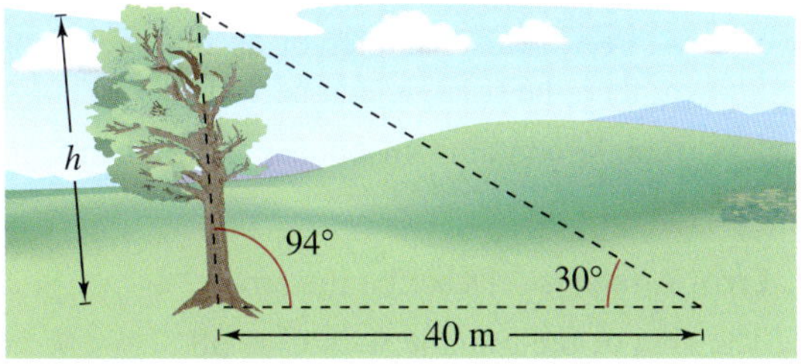

42. Architecture There is a bridge across a lake from a gazebo to a dock. The bearing from the gazebo to the dock is S 41° W. From a tree 100 meters from the gazebo, the bearings to the gazebo and the dock are S 74° E and S 28° E, respectively (see figure). Find the distance from the gazebo to the dock.

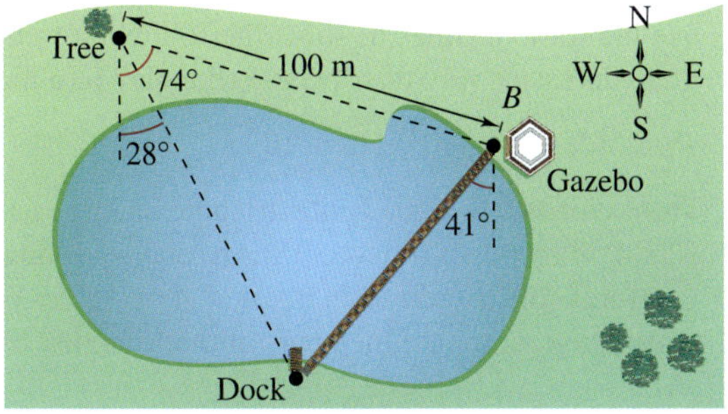

43. Aviation A plane flies 550 kilometers with a bearing of 245° (clockwise from north) from Pittsburgh to Louisville. The plane then flies 1485 kilometers from Louisville to Miami (see figure). Miami is due south of Pittsburgh. Find the bearing of the flight from Louisville to Miami.

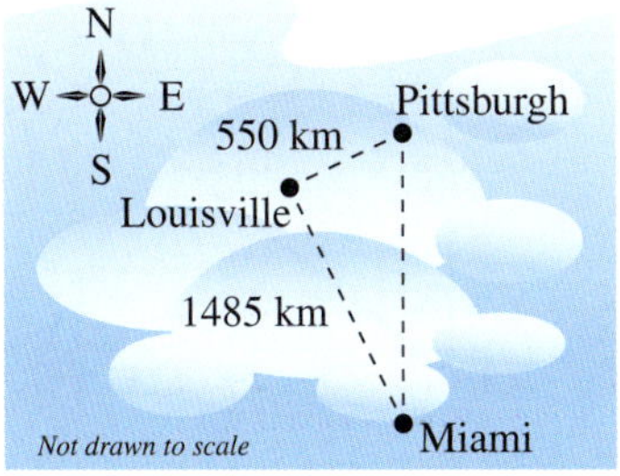

44. Physics A flagpole at a right angle to the horizontal is located on a slope that makes an angle of 12° with the horizontal. The flagpole casts a 16-meter shadow up the slope when the angle of elevation from the tip of the shadow to the sun is 20°.

(a) Draw a diagram that illustrates the problem.

(b) Write an equation involving the unknown quantity.

(c) Find the height of the flagpole.

45. Aviation The angles of elevation θ and ϕ to an airplane are being continuously monitored at two observation points A and B, respectively, which are 2 miles apart, and the airplane is east of both points in the same vertical plane.

(a) Draw a diagram that illustrates the problem.

(b) Write an equation giving the distance d between the plane and point B in terms of θ and ϕ.

(c) Use the equation from part (b) to find the distance between the plane and point B when $\theta = 40°$ and $\phi = 60°$.

46. Environmental Science The bearing from the Pine Knob fire tower to the Colt Station fire tower is N 65° E, and the two towers are 30 kilometers apart. A fire spotted by rangers in each tower has a bearing of N 80° E from Pine Knob and S 70° E from Colt Station (see figure). Find the distance of the fire from each tower.

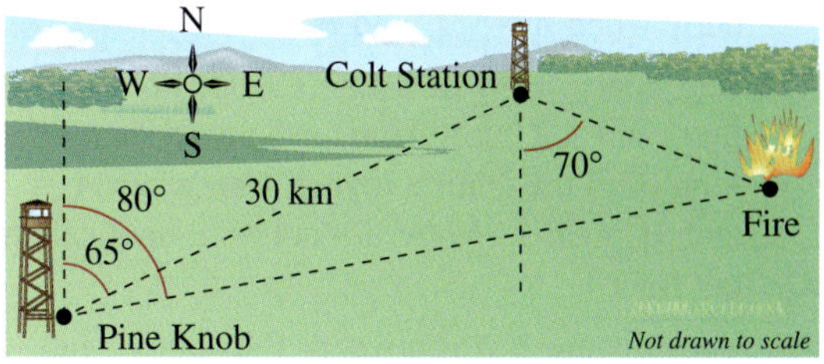

47. Angle of Elevation A 10-meter telephone pole casts a 17-meter shadow directly down a slope when the angle of elevation of the sun is 42° (see figure). Find θ, the angle of elevation of the ground.

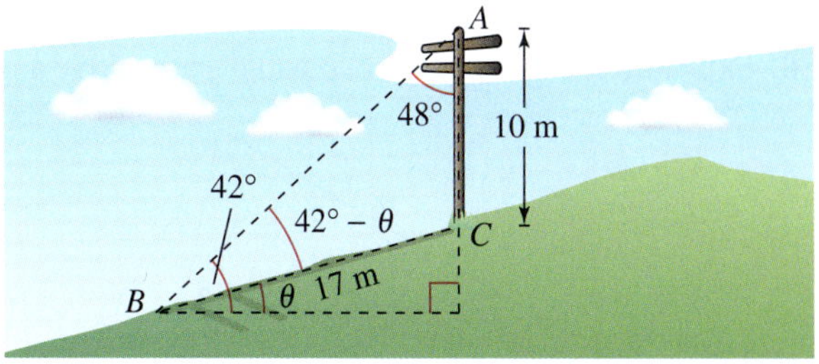

48. *Why you should learn it* *(p. 406)* A boat is sailing due east parallel to the shoreline at a speed of 10 miles per hour. At a given time, the bearing to a lighthouse is S 70° E, and 15 minutes later, the bearing is S 63° E (see figure). The lighthouse is located at the shoreline. Find the distance d from the boat to the shoreline.

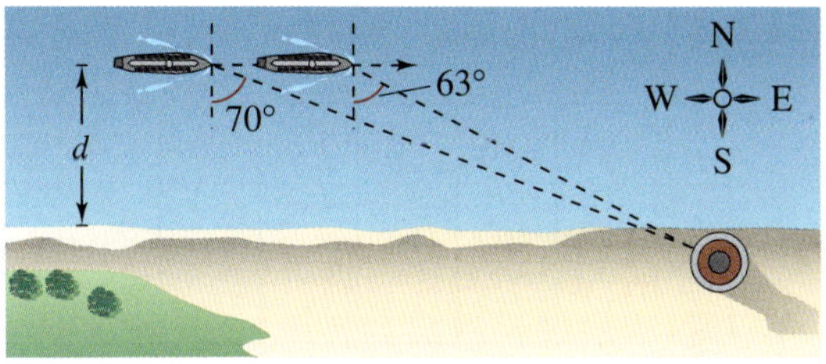

49. MODELING DATA

Port B is due East of Port A. The distance between Port A and Port B is 2.4 nautical miles. A ferry boat leaves Port A and travels to a buoy (see figure). The boat then travels to Port B.

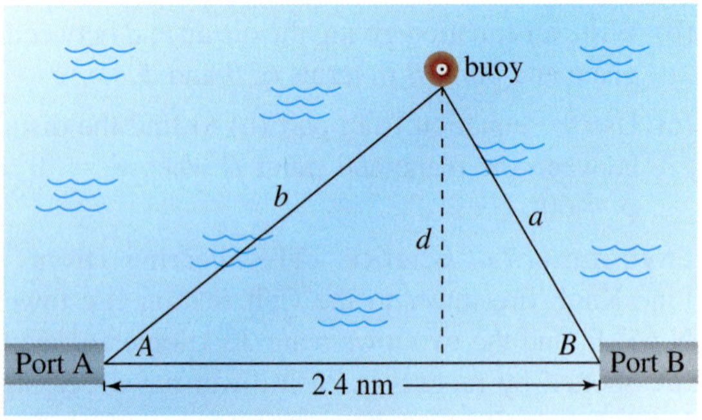

(a) What is the distance b between Port A and the buoy when $A = 22°$ and $B = 36°$?

(b) The boat travels at 2 knots from Port A to the buoy and at 3 knots from the buoy to Port B. How long does this entire voyage take when $A = 22°$ and $B = 36°$? (*Note:* 1 knot = 1 nautical mile per hour.)

(c) The distance d between the buoy and the horizontal between Port A and Port B changes as the buoy drifts. Compare the distance d_1 when $A = 24°$ and $B = 38°$ to the distance d_2 when $A = 28°$ and $B = 40°$.

50. Exploration In the figure, α and β are positive angles.

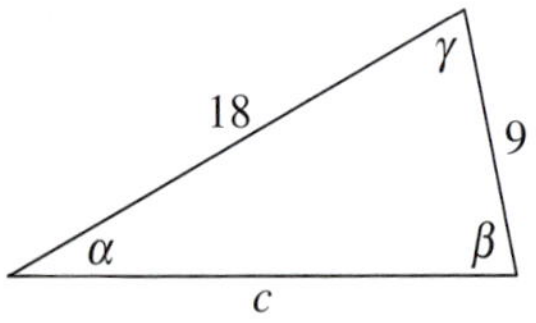

(a) Write α as a function of β.

(b) Use a graphing utility to graph the function. Determine its domain and range.

(c) Use the result of part (a) to write c as a function of β.

(d) Use the graphing utility to graph the function in part (c). Determine its domain and range.

(e) Use the graphing utility to complete the table. What can you conclude?

β	0.4	0.8	1.2	1.6	2.0	2.4	2.8
α							
c							

Focusing on Concepts

True or False? **In Exercises 51 and 52, determine whether the statement is true or false. Justify your answer.**

51. If any three sides or angles of an oblique triangle are known, then the triangle can be solved.

52. If a triangle contains an obtuse angle, then it must be oblique.

53. Error Analysis Describe the error.

The area of the triangle with $C = 58°$, $b = 11$ feet, and $c = 16$ feet is

$$\text{Area} = \frac{1}{2}(11)(16)(\sin 58°)$$
$$= 88(\sin 58°)$$
$$\approx 74.63 \text{ square feet.}$$

54. Writing Write a short paragraph explaining how to use the Law of Sines to solve the following triangle. Is there an easier way to solve the triangle? Explain.

$B = 50°$, $C = 90°$, $a = 10$

55. Think About It An acute triangle has measures of $b = 4$, $c = 6$, and $A = 30°$. An obtuse triangle has measures of $b = 2$, $c = 3$, and $A = 150°$. Without calculating, compare the areas of the triangles.

56. HOW DO YOU SEE IT? In the figure, a triangle is to be formed by drawing a line segment of length a from $(4, 3)$ to the positive x-axis. For what value(s) of a can you form (a) one triangle, (b) two triangles, and (c) no triangles? Explain your reasoning.

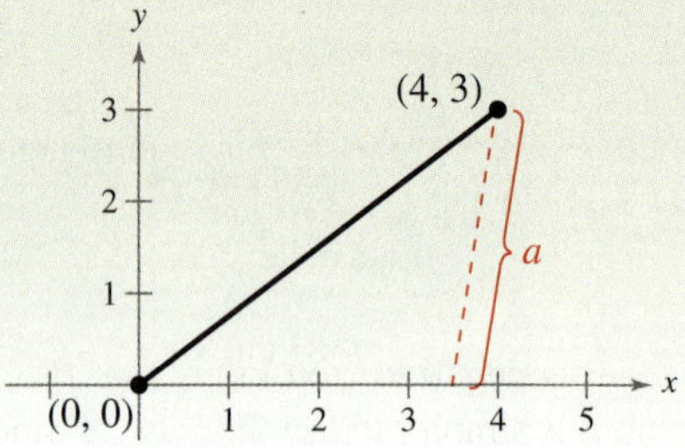

Cumulative Mixed Review

Changing the Base **In Exercises 57–60 evaluate the logarithm using the change-of-base formula. Round your result to two decimal places.**

57. $\log_4 7$

58. $\log_9 1.7$

59. $\log_{0.5} 11$

60. $\log_{1/4} 160$

6.2 Law of Cosines

Introduction

Two cases remain in the list of conditions needed to solve an oblique triangle—SSS and SAS. When you are given three sides (SSS) or two sides and their included angle (SAS), none of the ratios in the Law of Sines is known. In such cases, you can use the **Law of Cosines.**

Law of Cosines

Standard Form	*Alternative Form*
$a^2 = b^2 + c^2 - 2bc\cos A$	$\cos A = \dfrac{b^2 + c^2 - a^2}{2bc}$
$b^2 = a^2 + c^2 - 2ac\cos B$	$\cos B = \dfrac{a^2 + c^2 - b^2}{2ac}$
$c^2 = a^2 + b^2 - 2ab\cos C$	$\cos C = \dfrac{a^2 + b^2 - c^2}{2ab}$

(See the proof on page 473.)

What you should learn

- Use the Law of Cosines to solve oblique triangles (SSS or SAS).
- Use the Law of Cosines to model and solve real-life problems.
- Use Heron's Area Formula to find areas of triangles.

Why you should learn it

You can use the Law of Cosines to solve real-life problems involving oblique triangles. For instance, Exercise 51 on page 420 shows you how the Law of Cosines can be used to determine the lengths of the guy wires that anchor a tower.

Surveyor

EXAMPLE 1 Given Three Sides—SSS

Find the three angles of the triangle shown in the figure.

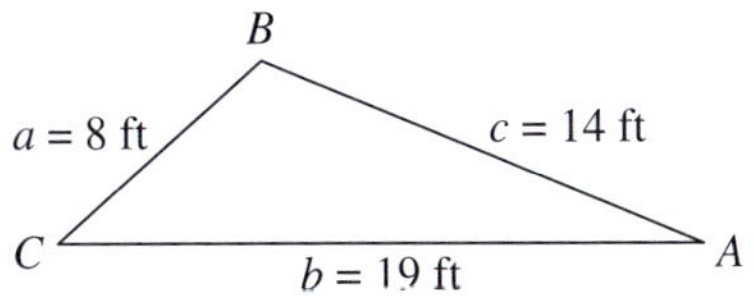

Solution

It is a good idea first to find the angle opposite the longest side—side b in this case. Using the alternative form of the Law of Cosines, you find that

$$\cos B = \frac{a^2 + c^2 - b^2}{2ac} \qquad \text{Alternative form}$$

$$= \frac{8^2 + 14^2 - 19^2}{2(8)(14)} \qquad \text{Substitute for } a, b, \text{ and } c.$$

$$\approx -0.45089. \qquad \text{Simplify.}$$

Because $\cos B$ is negative, you know that B is an *obtuse* angle given by $B \approx 116.80°$. At this point, it is simpler to use the Law of Sines to determine A. Note that because B is obtuse and a triangle can have at most one obtuse angle, you know that A must be acute.

$$\sin A = a\left(\frac{\sin B}{b}\right) \approx 8\left(\frac{\sin 116.80°}{19}\right) \quad \Rightarrow \quad A = \arcsin\left(\frac{8 \sin 116.80°}{19}\right)$$

So, $A \approx 22.08°$ and

$$C \approx 180° - 22.08° - 116.80° = 41.12°.$$

Checkpoint *Audio-video solution in English & Spanish at LarsonPrecalculus.com*

Find the three angles of the triangle whose sides have lengths $a = 6$, $b = 8$, and $c = 12$.

Explore the Concept

What familiar formula do you obtain when you use the third form of the Law of Cosines

$$c^2 = a^2 + b^2 - 2ab\cos C$$

and you let $C = 90°$? What is the relationship between the Law of Cosines and this formula?

Do you see why it was wise to find the largest angle *first* in Example 1? Knowing the cosine of an angle, you can determine whether the angle is acute or obtuse. That is,

$$\cos \theta > 0 \quad \text{for} \quad 0° < \theta < 90°$$ Acute

and

$$\cos \theta < 0 \quad \text{for} \quad 90° < \theta < 180°.$$ Obtuse

So, in Example 1, once you found that angle B was obtuse, you knew that angles A and C were both acute. Furthermore, if the largest angle is acute, then the remaining two angles are also acute.

EXAMPLE 2 Given Two Sides and their Included Angle—SAS

See LarsonPrecalculus.com for an interactive version of this type of example.

Find the remaining angles and side of the triangle shown below.

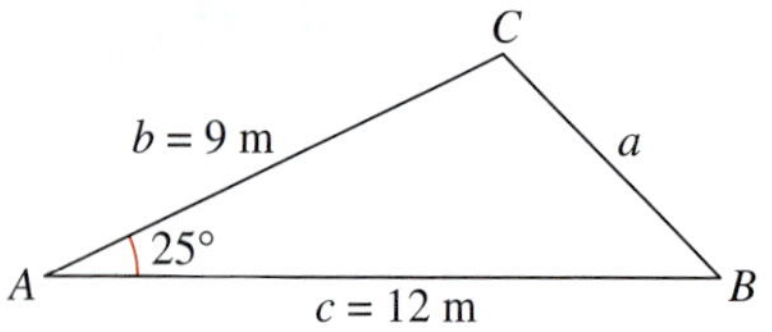

Solution

Use the Law of Cosines to find the unknown side a in the figure.

$$a^2 = b^2 + c^2 - 2bc \cos A$$

$$a^2 = 9^2 + 12^2 - 2(9)(12) \cos 25°$$

$$a^2 = 225 - 216 \cos 25°$$

$$a \approx 5.4072 \text{ meters}$$

Of angles B and C, B must be the smaller angle because it is opposite the shorter side $(b < c)$. So, B cannot be obtuse and must be acute. Because you know a, A, and b, use the reciprocal form of the Law of Sines to solve for acute angle B.

$$\frac{\sin B}{b} = \frac{\sin A}{a}$$ Reciprocal form

$$\sin B = b\left(\frac{\sin A}{a}\right)$$ Multiply each side by b.

$$\sin B \approx 9\left(\frac{\sin 25°}{5.4072}\right)$$ Substitute for A, a, and b.

$$B \approx 44.7°$$ Use the arcsine function and a calculator.

Now, find C.

$$C \approx 180° - 25° - 44.7° = 110.3°$$

✓ **Checkpoint** ▶ ***Audio-video solution in English & Spanish at LarsonPrecalculus.com***

Find the remaining angles and side of the triangle shown below.

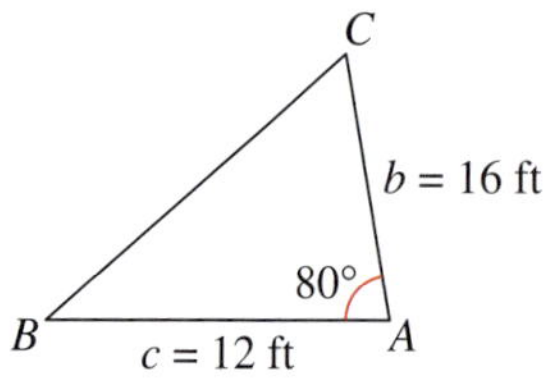

Algebra Help

When solving an oblique triangle given two sides and their included angle, use the standard form of the Law of Cosines to solve for the unknown side. When solving an oblique triangle given three sides, use the alternative form of the Law of Cosines to solve for an angle.

Insight

Some standardized tests may list the Law of Cosines and ask you to find an unknown side or angle of a triangle. Also, the Law of Cosines is sometimes useful when solving an application problem, in which case the law will likely not be listed.

Applications

EXAMPLE 3 An Application of the Law of Cosines

The pitcher's mound on a softball field is 43 feet from home plate. The distance between the bases is 60 feet, as shown in Figure 6.9. (The pitcher's mound is *not* halfway between home plate and second base.) How far is the pitcher's mound from first base?

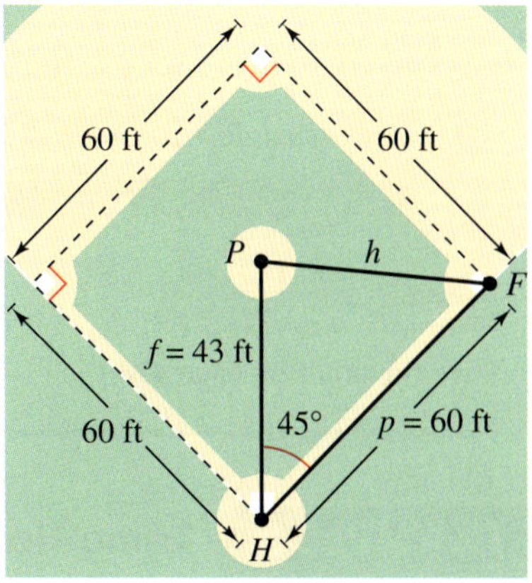

Figure 6.9

Solution

In triangle HPF, $H = 45°$ (line segment HP bisects the right angle at H), $f = 43$, and $p = 60$. Using the Law of Cosines for this SAS case, you have

$$h^2 = f^2 + p^2 - 2fp \cos H \quad \text{Law of Cosines}$$

$$= 43^2 + 60^2 - 2(43)(60) \cos 45° \quad \text{Substitute for } H, f, \text{ and } p.$$

$$= 5449 - 5160 \cos 45°.$$

So, the approximate distance from the pitcher's mound to first base is

$$h = \sqrt{5449 - 5160 \cos 45°} \approx 42.43 \text{ feet.}$$

✓ Checkpoint *Audio-video solution in English & Spanish at LarsonPrecalculus.com*

In a softball game, a batter hits a ball to dead center field, a distance of 240 feet from home plate. The center fielder then throws the ball to third base and gets a runner out. The distance between the bases is 60 feet. How far is the center fielder from third base?

EXAMPLE 4 An Application of the Law of Cosines

A ship travels 60 miles due east and then changes direction, as shown in the figure. After traveling 80 miles in this new direction, the ship is 139 miles from its point of departure. Describe the bearing from point B to point C.

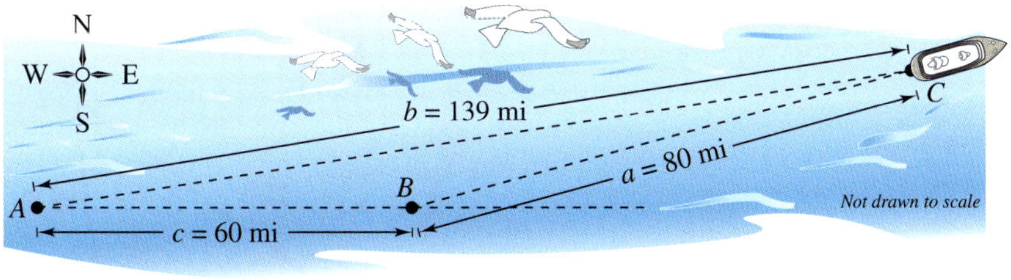

Solution

Using the alternative form of the Law of Cosines, you have

$$\cos B = \frac{a^2 + c^2 - b^2}{2ac} \quad \text{Alternative form}$$

$$= \frac{80^2 + 60^2 - 139^2}{2(80)(60)} \quad \text{Substitute for } a, b, \text{ and } c.$$

$$= -\frac{9321}{9600}.$$

So, $B = \arccos(-9321/9600) \approx 166.15°$, and the bearing measured from due north from point B to point C is approximately $166.15° - 90° = 76.15°$, or N 76.15° E.

✓ Checkpoint *Audio-video solution in English & Spanish at LarsonPrecalculus.com*

A ship travels 40 miles due east and then changes direction, as shown in Figure 6.10. After traveling 30 miles in this new direction, the ship is 56 miles from its point of departure. Describe the bearing from point B to point C.

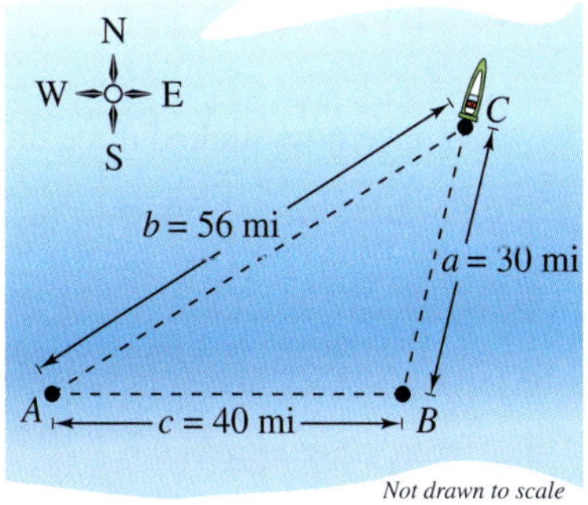

Figure 6.10

Heron's Area Formula

The Law of Cosines can be used to establish the following formula for the area of a triangle. This formula is called **Heron's Area Formula** after the Greek mathematician Heron of Alexandria.

Heron's Area Formula

Given any triangle with sides of lengths a, b, and c, the area of the triangle is

$$\text{Area} = \sqrt{s(s-a)(s-b)(s-c)}$$

where $s = \dfrac{a+b+c}{2}$.

(See the proof on page 474.)

EXAMPLE 5 Using Heron's Area Formula

Find the area of the triangle shown below.

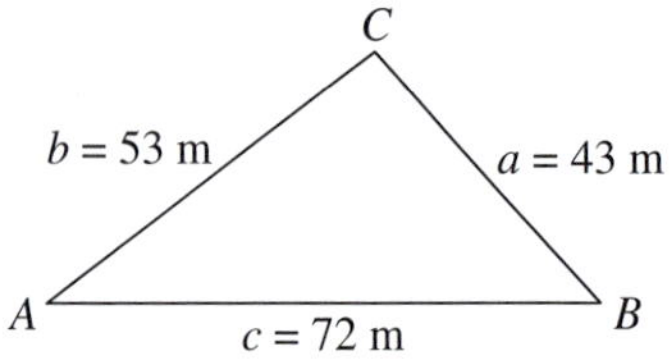

Solution

Because

$$s = \frac{a+b+c}{2} = \frac{168}{2} = 84$$

Heron's Area Formula yields

$$\begin{aligned}\text{Area} &= \sqrt{s(s-a)(s-b)(s-c)} \\ &= \sqrt{84(84-43)(84-53)(84-72)} \\ &= \sqrt{84(41)(31)(12)} \\ &\approx 1131.89 \text{ square meters.}\end{aligned}$$

✓ Checkpoint *Audio-video solution in English & Spanish at LarsonPrecalculus.com*

Find the area of a triangle with sides of lengths $a = 5$ feet, $b = 9$ feet, and $c = 8$ feet.

You have now studied three different formulas for the area of a triangle.

Formulas for Area of a Triangle

1. **Standard Formula:** $\text{Area} = \frac{1}{2}bh$
2. **Oblique Triangle:** $\text{Area} = \frac{1}{2}bc \sin A = \frac{1}{2}ab \sin C = \frac{1}{2}ac \sin B$
3. **Heron's Area Formula:** $\text{Area} = \sqrt{s(s-a)(s-b)(s-c)}$

Explore the Concept

Can the formulas at the bottom of the page be used to find the area of any type of triangle? Explain the advantages and disadvantages of using one formula over another.

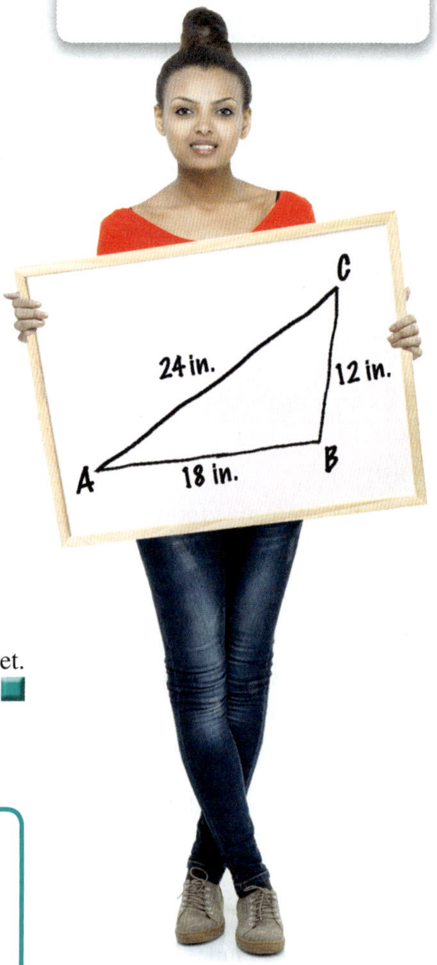

6.2 Exercises

See *CalcChat.com* for tutorial help and worked-out solutions to odd-numbered exercises.
For instructions on how to use a graphing utility, see Appendix A.

Vocabulary and Concept Check

In Exercises 1–4, fill in the blank(s).

1. The standard form of the Law of Cosines for $\cos C = \dfrac{a^2 + b^2 - c^2}{2ab}$ is _______ .

2. When solving an oblique triangle given three sides, use the _______ form of the Law of Cosines to solve for an angle.

3. When solving an oblique triangle given two sides and their included angle, use the _______ form of the Law of Cosines to solve for the remaining side.

4. The Law of Cosines can be used to establish a formula for the area of a triangle called _______ _______ Formula.

Procedures and Problem Solving

Using the Law of Cosines **In Exercises 5–22, use the Law of Cosines to solve the triangle.**

5.

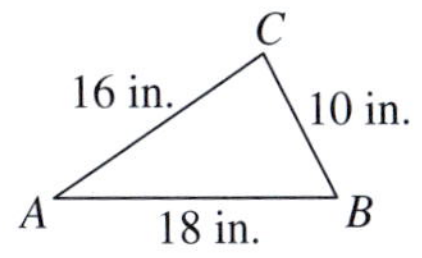

6.

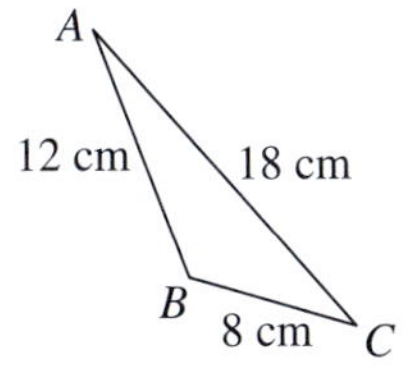

7.

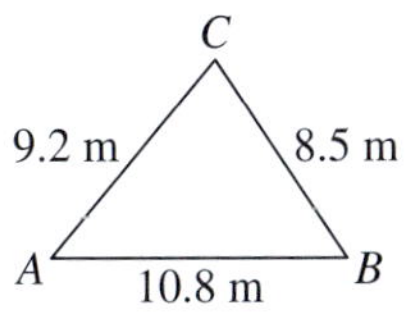

8.

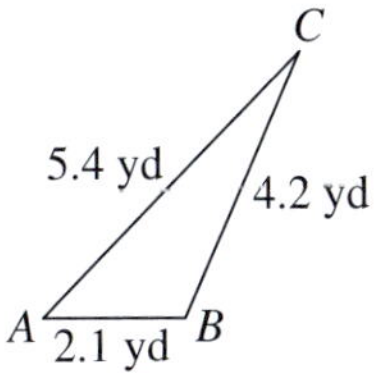

9.

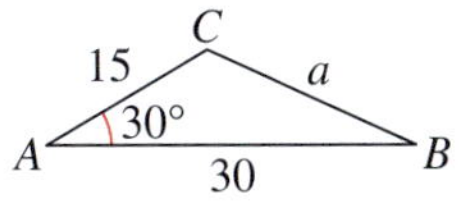

10.

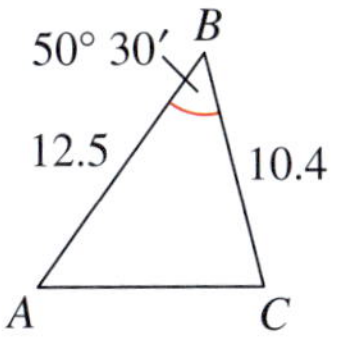

11. $a = 11,\quad b = 15,\quad c = 21$

12. $a = 9,\quad b = 3,\quad c = 11$

13. $A = 50°,\quad b = 15,\quad c = 30$

14. $C = 108°,\quad a = 10,\quad b = 7$

15. $A = 120°,\quad b = 6,\quad c = 7$

16. $A = 48°,\quad b = 3,\quad c = 14$

17. $a = 75.4,\quad b = 48,\quad c = 48$

18. $a = 1.42,\quad b = 0.75,\quad c = 1.25$

19. $B = 8° 15',\quad a = 26,\quad c = 18$

20. $B = 10° 35',\quad a = 40,\quad c = 30$

21. $C = 43°,\quad a = \frac{4}{9},\quad b = \frac{7}{9}$

22. $C = 101°,\quad a = \frac{3}{8},\quad b = \frac{3}{4}$

Finding Measures in a Parallelogram **In Exercises 23–28, solve the parallelogram to find the missing values. (The lengths of the diagonals are represented by *c* and *d*.)**

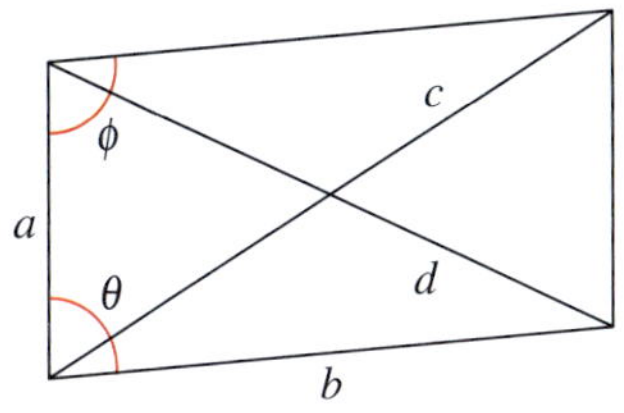

	a	b	c	d	θ	ϕ
23.	5	8			45°	
24.	25	35				120°
25.	10	14	20			
26.	40	60		80		
27.	15		25	20		
28.		25	50	35		

Solving a Triangle **In Exercises 29–36, determine whether the Law of Cosines is needed to solve the triangle. Then solve the triangle (if possible).**

29.

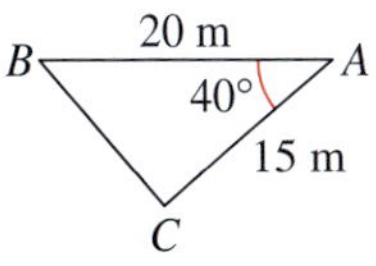

30.

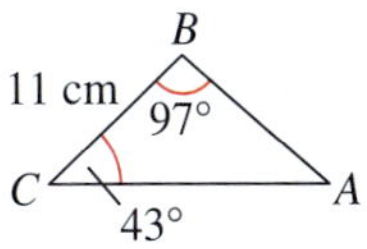

31. $a = 8,\quad c = 5,\quad B = 40°$

32. $a = 10,\quad b = 12,\quad C = 70°$

33. $A = 24°,\quad a = 4,\quad b = 18$

34. $a = 11,\quad b = 13,\quad c = 7$

35. $A = 42°,\quad B = 35°,\quad c = 1.2$

36. $C = 95°,\quad b = 19,\quad c = 25$

Using Heron's Area Formula **In Exercises 37–44, use Heron's Area Formula to find the area of the triangle.**

37. $a = 6, \quad b = 12, \quad c = 17$

38. $a = 25, \quad b = 35, \quad c = 32$

39. $a = 5, \quad b = 8, \quad c = 10$

40. $a = 12, \quad b = 17, \quad c = 8$

41. $a = 1.24, \quad b = 2.45, \quad c = 1.25$

42. $a = 2.4, \quad b = 2.75, \quad c = 2.25$

43. $a = 1, \quad b = \frac{1}{2}, \quad c = \frac{3}{4}$

44. $a = \frac{3}{5}, \quad b = \frac{5}{8}, \quad c = \frac{3}{8}$

45. Surveying To approximate the length of a marsh, a surveyor walks 380 meters from point A to point B. Then the surveyor turns 80° and walks 240 meters to point C (see figure). Approximate the length AC of the marsh.

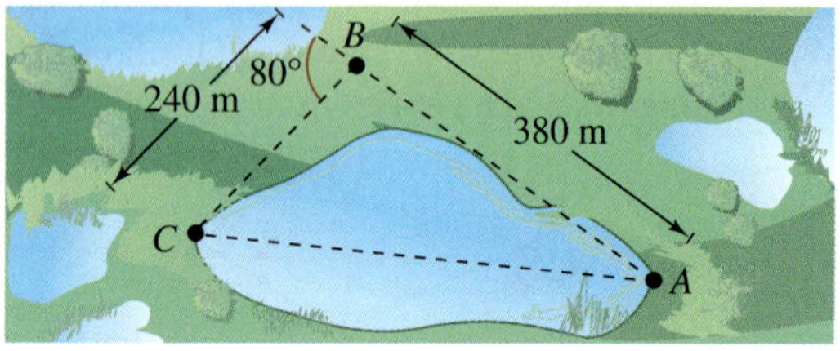

46. Geometry Determine the angle θ in the design of the streetlight shown in the figure.

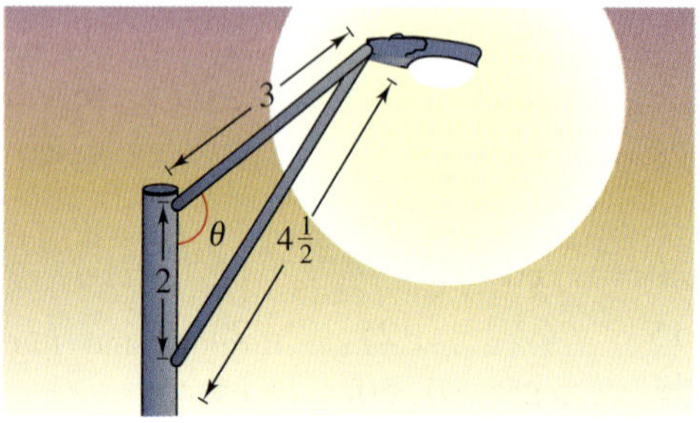

47. Geography On a map, Minneapolis is 165 millimeters due west of Albany, Phoenix is 216 millimeters from Minneapolis, and Phoenix is 368 millimeters from Albany. (See figure.)

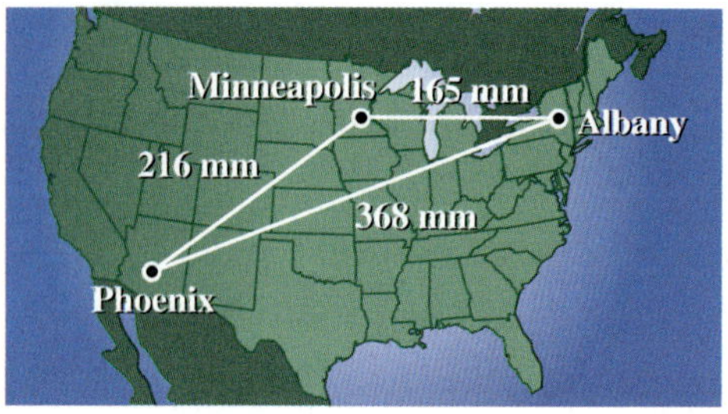

(a) Find the bearing of Minneapolis from Phoenix.

(b) Find the bearing of Albany from Phoenix.

48. Marine Transportation Two ships leave a port at 9 A.M. One travels at a bearing of N 53° W at 12 miles per hour, and the other travels at a bearing of S 67° W at 16 miles per hour. Approximate how far apart the ships are at noon.

49. Surveying A triangular lot has sides of lengths 725 feet, 650 feet, and 575 feet. Find the measure of the largest angle.

50. Structural Engineering Q is the midpoint of the line segment $\overline{PR}$ in the truss rafter shown in the figure. What are the lengths of the line segments $\overline{PQ}$, $\overline{QS}$, and $\overline{RS}$?

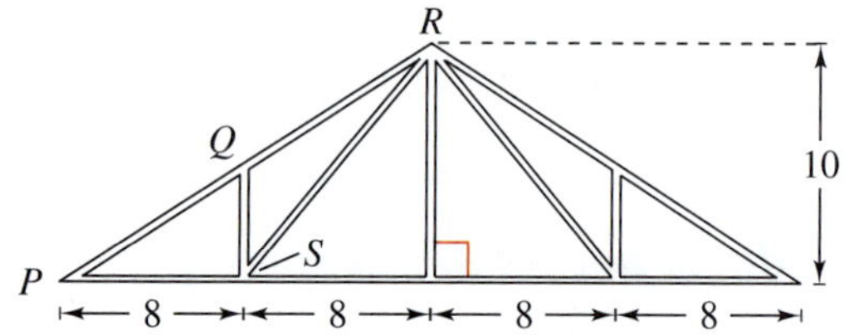

51. *Why you should learn it* *(p. 415)* A 100-foot vertical tower is to be erected on the side of a hill that makes a 6° angle with the horizontal. Find the length of each of the two guy wires that will be anchored 75 feet uphill and downhill from the base of the tower (see figure).

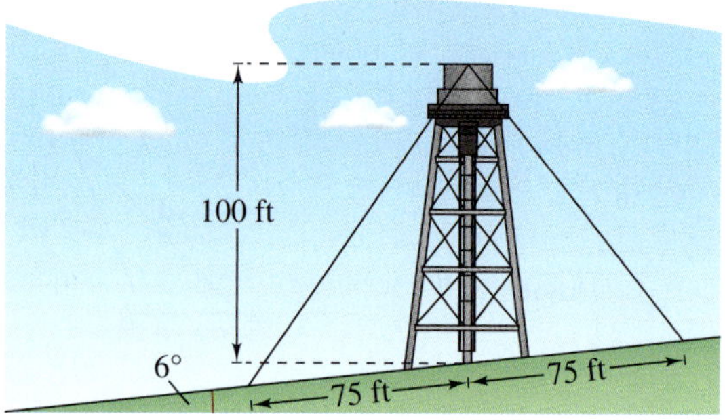

52. Architecture An awning above a patio lowers at an angle of 50° from the horizontal at a height of 10 feet above the ground. Direct sunlight meets the bottom of the door when the angle of elevation of the sun is 70° (see figure). What is the length x of the awning?

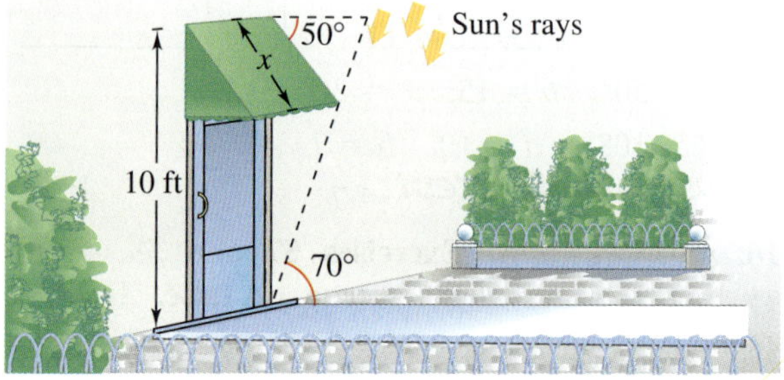

53. Geometry The lengths of the sides of a triangular garden at a university are approximately 160 feet, 150 feet, and 140 feet. Approximate the area of the garden.

54. Geometry A triangular lot has sides of lengths 550 feet, 630 feet, and 720 feet. Find the value of the lot at $3000 per acre. (1 acre = 43,560 square feet)

55. Mechanical Engineering An engine has a seven-inch connecting rod fastened to a crank (see figure).

(a) Use the Law of Cosines to write an equation giving the relationship between x and θ.

(b) Write x as a function of θ. (Select the sign that yields positive values of x.)

(c) Use a graphing utility to graph the function in part (b).

(d) Use the graph in part (c) to determine the total distance the piston moves in one cycle.

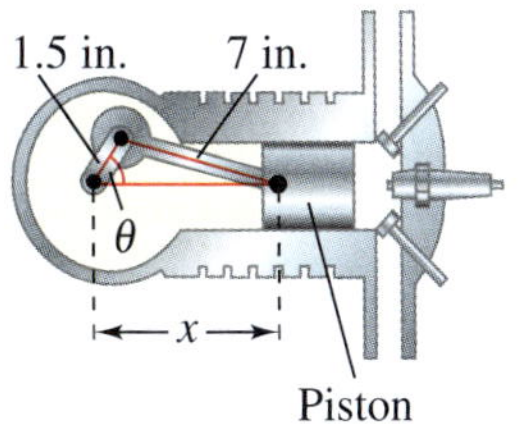

Figure for 55

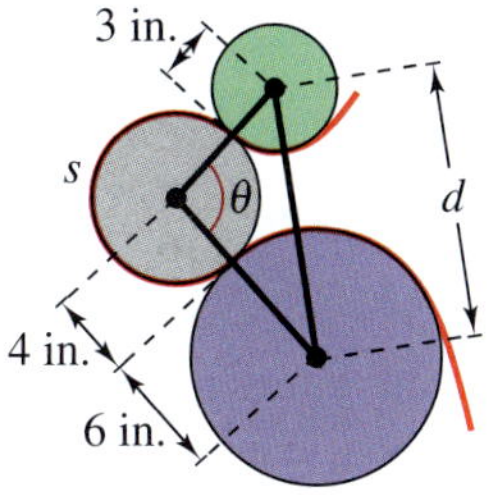

Figure for 56

56. MODELING DATA

In a process with continuous paper, the paper passes across three rollers of radii 3 inches, 4 inches, and 6 inches. The centers of the three-inch and six-inch rollers are d inches apart, and the length of the arc in contact with the paper on the four-inch roller is s inches (see figure).

(a) Use the Law of Cosines to write an equation giving the relationship between d and θ.

(b) Write θ as a function of d.

(c) Write s as a function of θ.

(d) Complete the table.

d (inches)	9	10	12	13	14	15	16
θ (degrees)							
s (inches)							

Focusing on Concepts

True or False? In Exercises 57 and 58, determine whether the statement is true or false. Justify your answer.

57. Two sides and their included angle determine a unique triangle.

58. In Heron's Area Formula, s is the average of the lengths of the three sides of the triangle.

59. Error Analysis Describe the error.

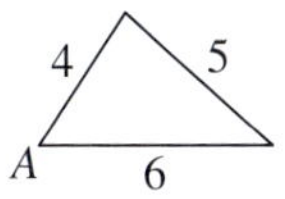

$$\cos A = \frac{5^2 + 6^2 - 4^2}{2(5)(6)} = \frac{45}{60} = 0.75$$

$$A = \arccos(0.75) \approx 41.41°$$

60. HOW DO YOU SEE IT? To solve the triangle, would you begin by using the Law of Sines or the Law of Cosines? Explain.

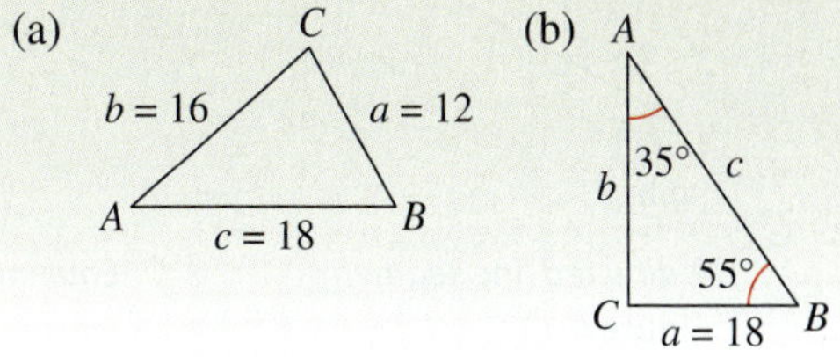

61. Proof Use the Law of Cosines to prove that

$$\frac{1}{2}bc(1 + \cos A) = \left(\frac{a + b + c}{2}\right)\left(\frac{-a + b + c}{2}\right).$$

62. Proof Use the Law of Cosines to prove that

$$\frac{1}{2}bc(1 - \cos A) = \left(\frac{a - b + c}{2}\right)\left(\frac{a + b - c}{2}\right).$$

63. Writing Describe how the Law of Cosines can be used to solve the ambiguous case of the oblique triangle ABC, where $a = 12$ feet, $b = 30$ feet, and $A = 20°$. Is the result the same as when the Law of Sines is used to solve the triangle? Describe the advantages and the disadvantages of each method.

64. Proof Use a half-angle formula and the Law of Cosines to prove that, for any triangle,

(a) $\cos \dfrac{C}{2} = \sqrt{\dfrac{s(s - c)}{ab}}$ and

(b) $\sin \dfrac{C}{2} = \sqrt{\dfrac{(s - a)(s - b)}{ab}}$

where $s = \frac{1}{2}(a + b + c)$.

Cumulative Mixed Review

Evaluating an Inverse Trigonometric Function In Exercises 65–68, find the exact value of the expression. (Do not use a calculator.)

65. $\arcsin(-1)$

66. $\arccos\left(-\dfrac{\sqrt{3}}{2}\right)$

67. $\tan^{-1} \dfrac{\sqrt{3}}{3}$

68. $\tan^{-1} \sqrt{3}$

6.3 Vectors in the Plane

Introduction

Many quantities in geometry and physics, such as area, time, and temperature, can be represented by a single real number. Other quantities, such as force and velocity, involve both *magnitude* and *direction* and cannot be completely characterized by a single real number. To represent such a quantity, you can use a **directed line segment,** as shown in Figure 6.11. The directed line segment $\overrightarrow{PQ}$ has **initial point** P and **terminal point** Q. Its **magnitude,** or **length,** is denoted by $\|\overrightarrow{PQ}\|$ and can be found by using the Distance Formula.

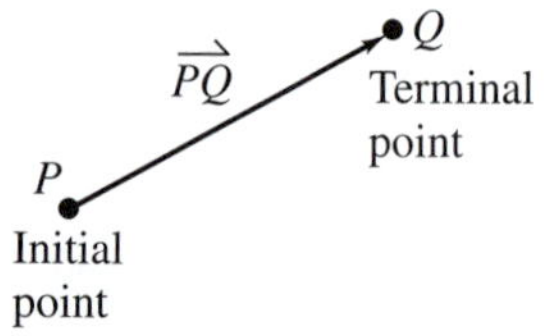

A directed line segment
Figure 6.11

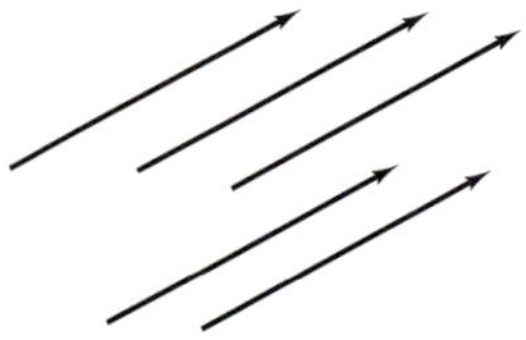

Equivalent directed line segments
Figure 6.12

Two directed line segments that have the same magnitude and direction are *equivalent.* For example, the directed line segments in Figure 6.12 are all equivalent. The set of all directed line segments that are equivalent to a given directed line segment $\overrightarrow{PQ}$ is a **vector v in the plane,** written $\mathbf{v} = \overrightarrow{PQ}$. Vectors are denoted by lowercase, boldface letters such as **u**, **v**, and **w**.

What you should learn

- Represent vectors as directed line segments.
- Write the component forms of vectors.
- Perform basic vector operations and represent vector operations graphically.
- Write vectors as linear combinations of unit vectors.
- Find the direction angles of vectors.
- Use vectors to model and solve real-life problems.

Why you should learn it

Vectors are used to analyze numerous aspects of everyday life. Exercise 103 on page 433 shows you how vectors can be used to determine the tension in the cables of two cranes lifting an object.

EXAMPLE 1 Showing That Two Vectors Are Equivalent

Show that **u** and **v** in the figure at the right are equivalent.

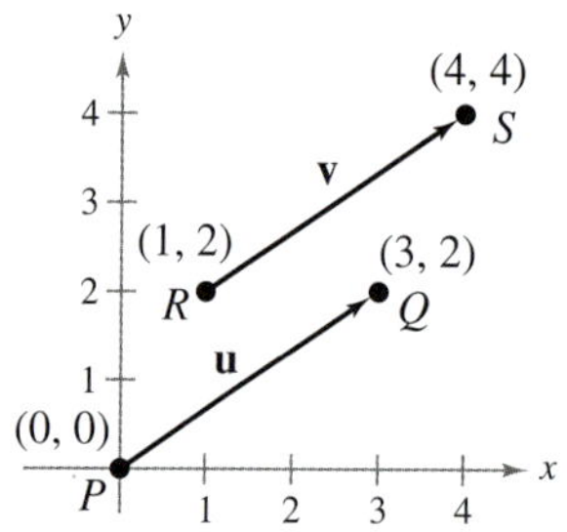

Solution

From the Distance Formula, it follows that $\overrightarrow{PQ}$ and $\overrightarrow{RS}$ have the *same magnitude*.

$$\|\overrightarrow{PQ}\| = \sqrt{(3-0)^2 + (2-0)^2} = \sqrt{13}$$

$$\|\overrightarrow{RS}\| = \sqrt{(4-1)^2 + (4-2)^2} = \sqrt{13}$$

Moreover, both line segments have the *same direction* because they are both directed toward the upper right on lines that have the same slope.

$$\text{Slope of } \overrightarrow{PQ} = \frac{2-0}{3-0} = \frac{2}{3}$$

$$\text{Slope of } \overrightarrow{RS} = \frac{4-2}{4-1} = \frac{2}{3}$$

Because $\overrightarrow{PQ}$ and $\overrightarrow{RS}$ have the same magnitude and direction, **u** and **v** are equivalent.

✓ ***Checkpoint*** *Audio-video solution in English & Spanish at LarsonPrecalculus.com*

Show that **u** and **v** in the figure at the right are equivalent.

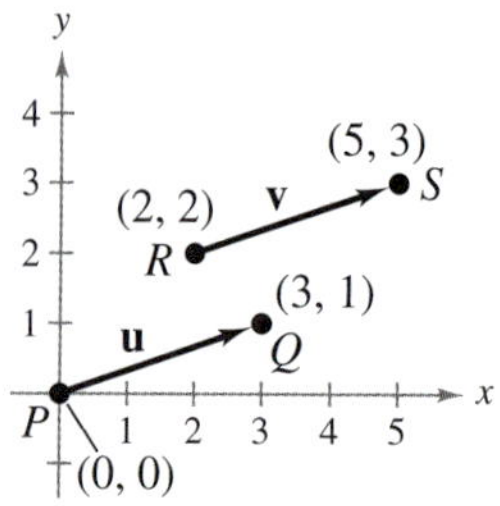

Component Form of a Vector

The directed line segment whose initial point is the origin is often the most convenient representative of a set of equivalent directed line segments. This representative of the vector **v** is in **standard position.**

A vector whose initial point is at the origin (0, 0) can be uniquely represented by the coordinates of its terminal point (v_1, v_2). This is the **component form of a vector v,** written as $\mathbf{v} = \langle v_1, v_2 \rangle$. The coordinates v_1 and v_2 are the *components* of **v**. If both the initial point and the terminal point lie at the origin, then **v** is the **zero vector** and is denoted by $\mathbf{0} = \langle 0, 0 \rangle$.

Component Form of a Vector

The component form of the vector with initial point $P(p_1, p_2)$ and terminal point $Q(q_1, q_2)$ is given by

$$\overrightarrow{PQ} = \langle q_1 - p_1, q_2 - p_2 \rangle = \langle v_1, v_2 \rangle = \mathbf{v}.$$

The **magnitude** (or **length**) of **v** is given by

$$\|\mathbf{v}\| = \sqrt{(q_1 - p_1)^2 + (q_2 - p_2)^2} = \sqrt{v_1^2 + v_2^2}.$$

If $\|\mathbf{v}\| = 1$, then **v** is a **unit vector.** Moreover, $\|\mathbf{v}\| = 0$ if and only if **v** is the zero vector **0**.

Two vectors $\mathbf{u} = \langle u_1, u_2 \rangle$ and $\mathbf{v} = \langle v_1, v_2 \rangle$ are *equal* if and only if $u_1 = v_1$ and $u_2 = v_2$. For instance, in Example 1, the vector **u** from $P(0, 0)$ to $Q(3, 2)$ is

$$\mathbf{u} = \overrightarrow{PQ} = \langle 3 - 0, 2 - 0 \rangle = \langle 3, 2 \rangle$$

and the vector **v** from $R(1, 2)$ to $S(4, 4)$ is

$$\mathbf{v} = \overrightarrow{RS} = \langle 4 - 1, 4 - 2 \rangle = \langle 3, 2 \rangle.$$

So, vectors **u** and **v** are equal.

Technology Tip

You can graph vectors with a graphing utility by graphing directed line segments. Consult the user's guide for your graphing utility for specific instructions.

EXAMPLE 2 Finding the Component Form of a Vector

Find the component form and magnitude of the vector **v** represented by the directed line segment that has initial point $(4, -7)$ and terminal point $(-1, 5)$.

Algebraic Solution

Let

$$P(4, -7) = (p_1, p_2)$$

and

$$Q(-1, 5) = (q_1, q_2).$$

Then, the components of $\mathbf{v} = \langle v_1, v_2 \rangle$ are

$$v_1 = q_1 - p_1 = -1 - 4 = -5$$

$$v_2 = q_2 - p_2 = 5 - (-7) = 12.$$

So, $\mathbf{v} = \langle -5, 12 \rangle$ and the magnitude of **v** is

$$\|\mathbf{v}\| = \sqrt{(-5)^2 + 12^2} = \sqrt{169} = 13.$$

Graphical Solution

Use centimeter graph paper to plot the points $P(4, -7)$ and $Q(-1, 5)$. Carefully sketch the vector **v**. Use the sketch to find the components of $\mathbf{v} = \langle v_1, v_2 \rangle$. Then use a centimeter ruler to find the magnitude of **v**. The figure at the right shows that the components of **v** are $v_1 = -5$ and $v_2 = 12$, so $\mathbf{v} = \langle -5, 12 \rangle$. The figure also shows that the magnitude of **v** is $\|\mathbf{v}\| = 13$.

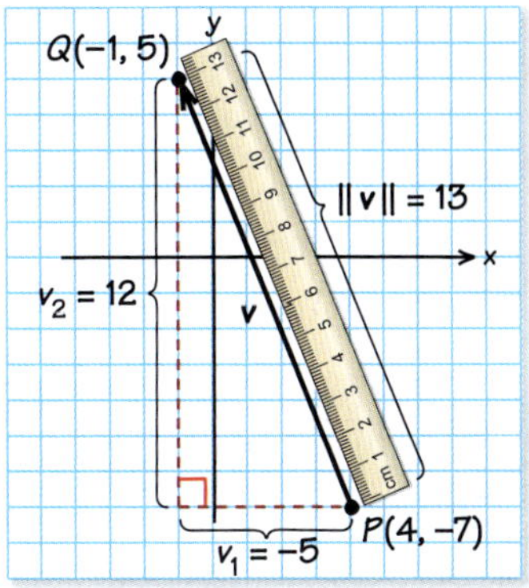

Checkpoint Audio-video solution in English & Spanish at LarsonPrecalculus.com

Find the component form and magnitude of the vector **v** represented by the directed line segment that has initial point $(-2, 3)$ and terminal point $(-7, 9)$.

Vector Operations

The two basic vector operations are **scalar multiplication** and **vector addition.** In operations with vectors, numbers are usually referred to as **scalars.** In this text, scalars will always be real numbers. Geometrically, the product of a vector **v** and a scalar k is the vector that is $|k|$ times as long as **v**. If k is positive, then $k\mathbf{v}$ has the same direction as **v**, and if k is negative, then $k\mathbf{v}$ has the opposite direction of **v**, as shown in Figure 6.13.

To add two vectors **u** and **v** geometrically, first position them (without changing their lengths or directions) so that the initial point of the second vector **v** coincides with the terminal point of the first vector **u**. The sum

$$\mathbf{u} + \mathbf{v}$$

is the vector formed by joining the initial point of the first vector **u** with the terminal point of the second vector **v**, as shown in the figures below. This technique is called the **parallelogram law** for vector addition because the vector $\mathbf{u} + \mathbf{v}$, often called the **resultant** of vector addition, is the diagonal of a parallelogram having adjacent sides **u** and **v**.

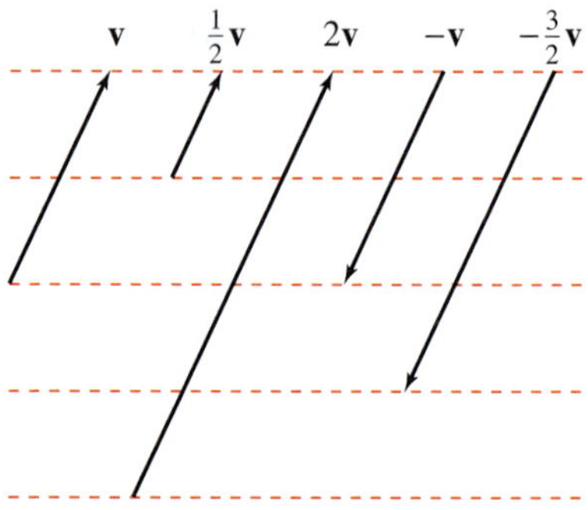

The scalar multiplication of **v**
Figure 6.13

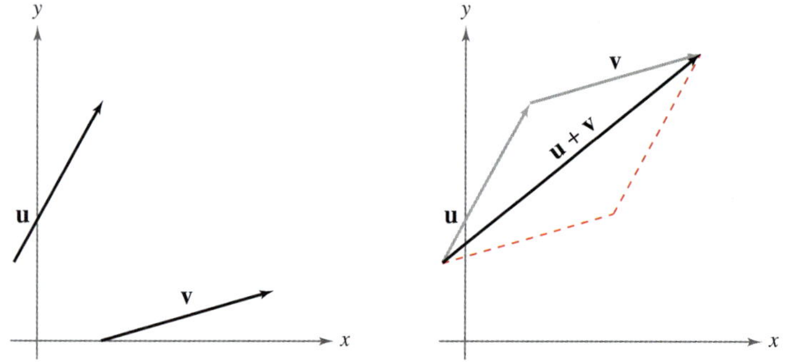

To find **u** + **v***, move the initial point of* **v** *to the terminal point of* **u**.

Definitions of Vector Addition and Scalar Multiplication

Let $\mathbf{u} = \langle u_1, u_2 \rangle$ and $\mathbf{v} = \langle v_1\ v_2 \rangle$ be vectors and let k be a scalar (a real number). Then the **sum** of **u** and **v** is the vector

$$\mathbf{u} + \mathbf{v} = \langle u_1 + v_1, u_2 + v_2 \rangle \qquad \text{Sum}$$

and the **scalar multiple** of k times **u** is the vector

$$k\mathbf{u} = k\langle u_1, u_2 \rangle = \langle ku_1, ku_2 \rangle. \qquad \text{Scalar multiple}$$

The **negative** of $\mathbf{v} = \langle v_1, v_2 \rangle$ is $-\mathbf{v} = (-1)\mathbf{v} = \langle -\mathbf{v}_1, -\mathbf{v}_2 \rangle$, and the **difference** of **u** and **v** is

$$\mathbf{u} - \mathbf{v} = \mathbf{u} + (-\mathbf{v}) \qquad \text{Add } (-\mathbf{v})\text{. See Figure 6.14.}$$
$$= \langle u_1 - v_1, u_2 - v_2 \rangle. \qquad \text{Difference}$$

To represent $\mathbf{u} - \mathbf{v}$ geometrically, you can use directed line segments with the *same* initial point. The difference

$$\mathbf{u} - \mathbf{v}$$

is the vector from the terminal point of **v** to the terminal point of **u**, which is equal to

$$\mathbf{u} + (-\mathbf{v})$$

as shown in Figure 6.14.

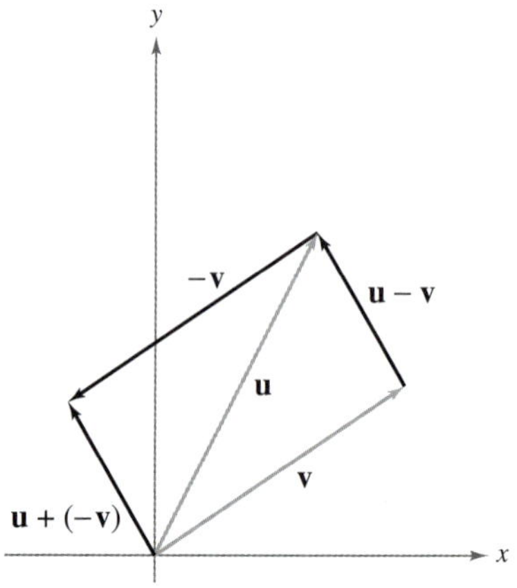

$\mathbf{u} - \mathbf{v} = \mathbf{u} + (-\mathbf{v})$
Figure 6.14

Example 3 illustrates the component definitions of vector addition and scalar multiplication. In this example, notice that each of the vector operations can be interpreted geometrically.

EXAMPLE 3 Vector Operations

See LarsonPrecalculus.com for an interactive version of this type of example.

Let $\mathbf{v} = \langle -2, 5\rangle$ and $\mathbf{w} = \langle 3, 4\rangle$. Find each vector.

a. $2\mathbf{v}$ **b.** $\mathbf{w} - \mathbf{v}$ **c.** $\mathbf{v} + 2\mathbf{w}$

Solution

a. $2\mathbf{v} = 2\langle -2, 5\rangle = \langle 2(-2), 2(5)\rangle = \langle -4, 10\rangle$

A sketch of $2\mathbf{v}$ is shown in Figure 6.15.

b. $\mathbf{w} - \mathbf{v} = \langle 3, 4\rangle - \langle -2, 5\rangle = \langle 3 - (-2), 4 - 5\rangle = \langle 5, -1\rangle$

Note that Figure 6.16 shows the vector difference $\mathbf{w} - \mathbf{v}$ as the sum $\mathbf{w} + (-\mathbf{v})$.

c. $\mathbf{v} + 2\mathbf{w} = \langle -2, 5\rangle + 2\langle 3, 4\rangle = \langle -2, 5\rangle + \langle 6, 8\rangle = \langle 4, 13\rangle$

A sketch of $\mathbf{v} + 2\mathbf{w}$ is shown in Figure 6.17.

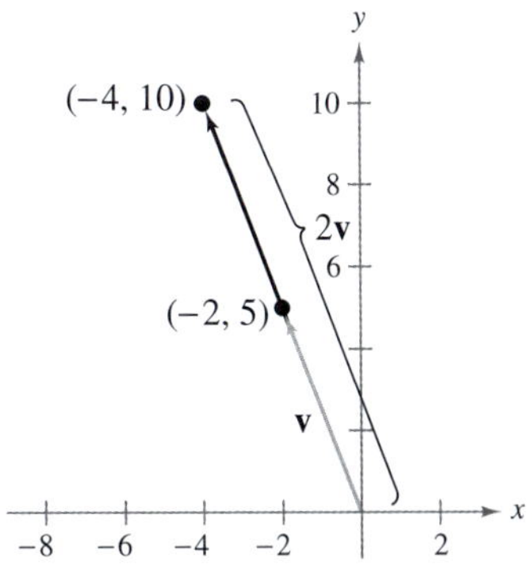

Figure 6.15

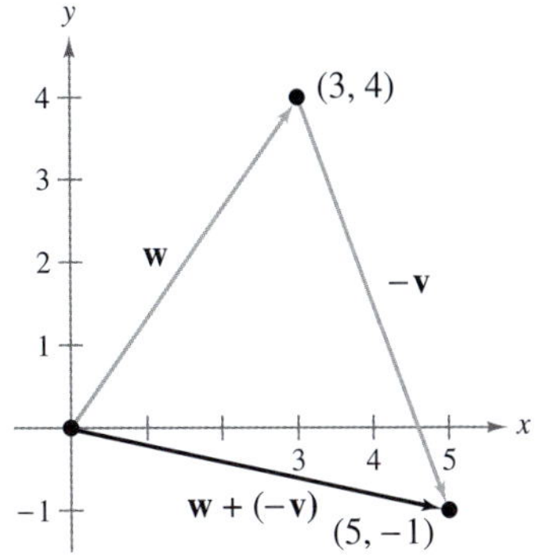

Figure 6.16

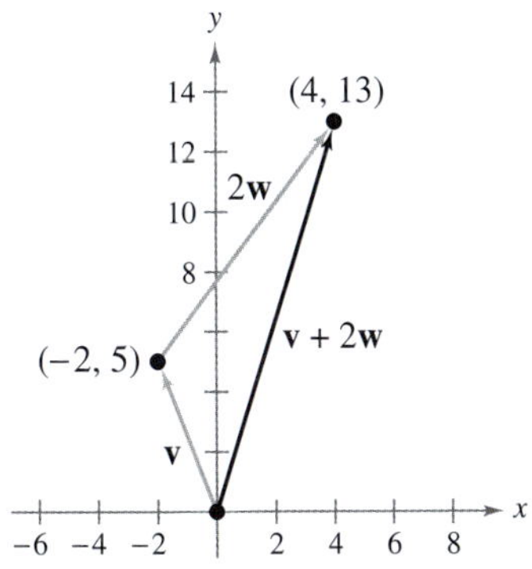

Figure 6.17

Checkpoint *Audio-video solution in English & Spanish at LarsonPrecalculus.com*

Let $\mathbf{u} = \langle 1, 4\rangle$ and $\mathbf{v} = \langle 3, 2\rangle$. Find each vector.

a. $\mathbf{u} + \mathbf{v}$ **b.** $\mathbf{u} - \mathbf{v}$ **c.** $2\mathbf{u} - 3\mathbf{v}$

Vector addition and scalar multiplication share many of the properties of ordinary arithmetic.

Properties of Vector Addition and Scalar Multiplication

Let $\mathbf{u}$, $\mathbf{v}$, and $\mathbf{w}$ be vectors and let c and d be scalars. Then the properties below are true.

1. $\mathbf{u} + \mathbf{v} = \mathbf{v} + \mathbf{u}$

2. $(\mathbf{u} + \mathbf{v}) + \mathbf{w} = \mathbf{u} + (\mathbf{v} + \mathbf{w})$

3. $\mathbf{u} + \mathbf{0} = \mathbf{u}$

4. $\mathbf{u} + (-\mathbf{u}) = \mathbf{0}$

5. $c(d\mathbf{u}) = (cd)\mathbf{u}$

6. $(c + d)\mathbf{u} = c\mathbf{u} + d\mathbf{u}$

7. $c(\mathbf{u} + \mathbf{v}) = c\mathbf{u} + c\mathbf{v}$

8. $1(\mathbf{u}) = \mathbf{u}, 0(\mathbf{u}) = \mathbf{0}$

9. $\|c\mathbf{v}\| = |c|\,\|\mathbf{v}\|$

EXAMPLE 4 Finding the Magnitude of a Scalar Multiple

Let $\mathbf{u} = \langle 1, 3\rangle$ and $\mathbf{v} = \langle -2, 5\rangle$. Find the magnitude of each scalar multiple.

a. $\|2\mathbf{u}\|$ **b.** $\|-5\mathbf{u}\|$ **c.** $\|3\mathbf{v}\|$

Solution

a. $\|2\mathbf{u}\| = |2|\|\mathbf{u}\| = |2|\|\langle 1, 3\rangle\| = |2|\sqrt{1^2 + 3^2} = 2\sqrt{10}$

b. $\|-5\mathbf{u}\| = |-5|\|\mathbf{u}\| = |-5|\|\langle 1, 3\rangle\| = |-5|\sqrt{1^2 + 3^2} = 5\sqrt{10}$

c. $\|3\mathbf{v}\| = |3|\|\mathbf{v}\| = |3|\|\langle -2, 5\rangle\| = |3|\sqrt{(-2)^2 + 5^2} = 3\sqrt{29}$

Checkpoint *Audio-video solution in English & Spanish at LarsonPrecalculus.com*

Let $\mathbf{u} = \langle 4, -1\rangle$ and $\mathbf{v} = \langle 3, 2\rangle$. Find the magnitude of each scalar multiple.

a. $\|3\mathbf{u}\|$ **b.** $\|-2\mathbf{v}\|$ **c.** $\|5\mathbf{v}\|$

Unit Vectors

In many applications of vectors, it is useful to find a unit vector that has the same direction as a given nonzero vector $\mathbf{v}$. To do this, you can divide $\mathbf{v}$ by its magnitude to obtain

$$\mathbf{u} = \text{unit vector} = \frac{\mathbf{v}}{\|\mathbf{v}\|} = \left(\frac{1}{\|\mathbf{v}\|}\right)\mathbf{v}.$$ Unit vector in direction of $\mathbf{v}$

Note that $\mathbf{u}$ is a scalar multiple of $\mathbf{v}$. The vector $\mathbf{u}$ has a magnitude of 1 and the same direction as $\mathbf{v}$. The vector $\mathbf{u}$ is called a **unit vector in the direction of v.**

EXAMPLE 5 Finding a Unit Vector

Find a unit vector in the direction of $\mathbf{v} = \langle -2, 5\rangle$ and verify that the result has a magnitude of 1.

Solution

The unit vector in the direction of $\mathbf{v}$ is

$$\frac{\mathbf{v}}{\|\mathbf{v}\|} = \frac{\langle -2, 5\rangle}{\sqrt{(-2)^2 + 5^2}} = \frac{1}{\sqrt{4 + 25}}\langle -2, 5\rangle = \frac{1}{\sqrt{29}}\langle -2, 5\rangle = \left\langle \frac{-2}{\sqrt{29}}, \frac{5}{\sqrt{29}}\right\rangle.$$

This vector has a magnitude of 1 because

$$\sqrt{\left(\frac{-2}{\sqrt{29}}\right)^2 + \left(\frac{5}{\sqrt{29}}\right)^2} = \sqrt{\frac{4}{29} + \frac{25}{29}} = \sqrt{\frac{29}{29}} = 1.$$

✓ **Checkpoint** ▶ *Audio-video solution in English & Spanish at LarsonPrecalculus.com*

Find a unit vector $\mathbf{u}$ in the direction of $\mathbf{v} = \langle 6, -1\rangle$ and verify that the result has a magnitude of 1.

The unit vectors $\langle 1, 0\rangle$ and $\langle 0, 1\rangle$ are called the **standard unit vectors** and are denoted by $\mathbf{i} = \langle 1, 0\rangle$ and $\mathbf{j} = \langle 0, 1\rangle$, as shown in Figure 6.18. These vectors can be used to represent any vector $\mathbf{v} = \langle v_1, v_2\rangle$ as follows.

$$\mathbf{v} = \langle v_1, v_2\rangle = v_1\langle 1, 0\rangle + v_2\langle 0, 1\rangle = v_1\mathbf{i} + v_2\mathbf{j}$$

The scalars v_1 and v_2 are called the **horizontal** and **vertical components of v,** respectively. The vector sum $v_1\mathbf{i} + v_2\mathbf{j}$ is called a **linear combination** of the vectors $\mathbf{i}$ and $\mathbf{j}$. Any vector in the plane can be written as a linear combination of the standard unit vectors $\mathbf{i}$ and $\mathbf{j}$.

Algebra Help

The unit vector $\mathbf{i} = \langle 1, 0\rangle$ is denoted by a **bold,** nonitalic lowercase **i.** Do not confuse it with the imaginary unit $i = \sqrt{-1}$, which is denoted by an *italic* lowercase *i.*

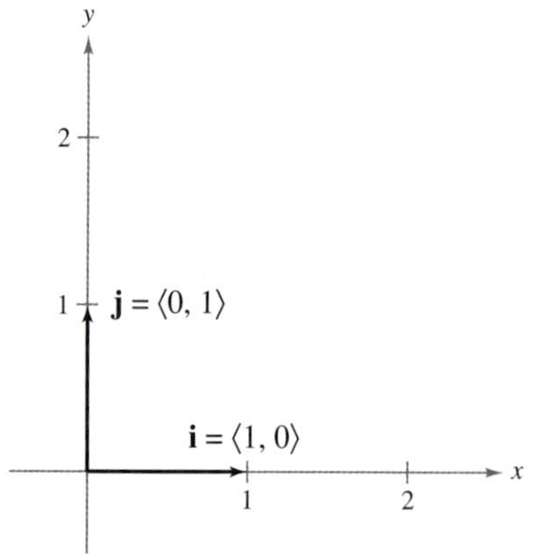

Standard unit vectors i and j
Figure 6.18

EXAMPLE 6 Writing a Linear Combination of Unit Vectors

Let **u** be the vector with initial point $(2, -5)$ and terminal point $(-1, 3)$. Write **u** as a linear combination of the standard unit vectors **i** and **j**.

Solution

Begin by writing the component form of the vector **u**.

$$\mathbf{u} = \langle -1 - 2, 3 - (-5) \rangle = \langle -3, 8 \rangle = -3\mathbf{i} + 8\mathbf{j}$$

This result is shown graphically in the figure.

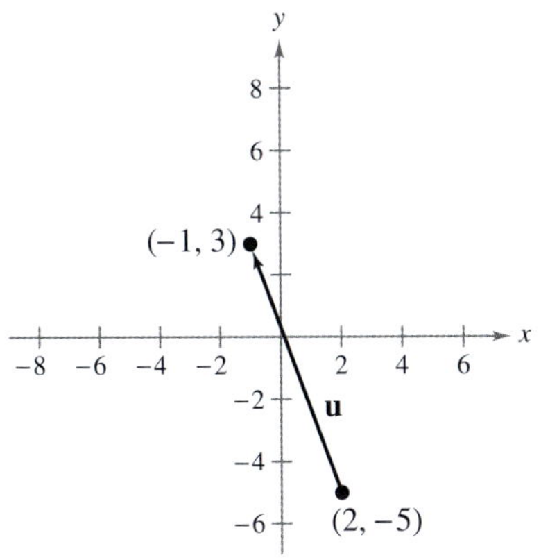

Checkpoint *Audio-video solution in English & Spanish at LarsonPrecalculus.com*

Let **u** be the vector with initial point $(-2, 6)$ and terminal point $(-8, 3)$. Write **u** as a linear combination of the standard unit vectors **i** and **j**.

EXAMPLE 7 Vector Operations

Let $\mathbf{u} = -3\mathbf{i} + 8\mathbf{j}$ and $\mathbf{v} = 2\mathbf{i} - \mathbf{j}$. Find $2\mathbf{u} - 3\mathbf{v}$.

Solution

To find $2\mathbf{u} - 3\mathbf{v}$, note that it is not necessary to convert **u** and **v** to component form. Performing the operations in unit vector form gives the same results.

$$\begin{aligned} 2\mathbf{u} - 3\mathbf{v} &= 2(-3\mathbf{i} + 8\mathbf{j}) - 3(2\mathbf{i} - \mathbf{j}) \\ &= -6\mathbf{i} + 16\mathbf{j} - 6\mathbf{i} + 3\mathbf{j} \\ &= -12\mathbf{i} + 19\mathbf{j} \end{aligned}$$

Checkpoint *Audio-video solution in English & Spanish at LarsonPrecalculus.com*

Let $\mathbf{u} = \mathbf{i} - 2\mathbf{j}$ and $\mathbf{v} = -3\mathbf{i} + 2\mathbf{j}$. Find $5\mathbf{u} - 2\mathbf{v}$.

In Example 7, another way to find $2\mathbf{u} - 3\mathbf{v}$ is to first convert **u** and **v** to component form, as shown below.

$$\mathbf{u} = -3\mathbf{i} + 8\mathbf{j} = \langle -3, 8 \rangle \qquad \text{and} \qquad \mathbf{v} = 2\mathbf{i} - \mathbf{j} = \langle 2, -1 \rangle$$

The difference of $2\mathbf{u}$ and $3\mathbf{v}$ is

$$\begin{aligned} 2\mathbf{u} - 3\mathbf{v} &= 2\langle -3, 8 \rangle - 3\langle 2, -1 \rangle \\ &= \langle -6, 16 \rangle - \langle 6, -3 \rangle \\ &= \langle -6 - 6, 16 - (-3) \rangle \\ &= \langle -12, 19 \rangle. \end{aligned}$$

Compare this result with the solution to Example 7.

Direction Angles

If **u** is a *unit vector* such that θ is the angle (measured counterclockwise) from the positive x-axis to **u**, then the terminal point of **u** lies on the unit circle and you have

$$\mathbf{u} = \langle x, y \rangle = \langle \cos\theta, \sin\theta \rangle = (\cos\theta)\mathbf{i} + (\sin\theta)\mathbf{j}$$

as shown in Figure 6.19. The angle θ is the **direction angle** of the vector **u**.

Let **u** be a unit vector with direction angle θ. If $\mathbf{v} = a\mathbf{i} + b\mathbf{j}$ is any vector that makes an angle θ with the positive x-axis, then it has the same direction as **u** and you can write

$$\mathbf{v} = \|\mathbf{v}\|\langle \cos\theta, \sin\theta \rangle = \|\mathbf{v}\|(\cos\theta)\mathbf{i} + \|\mathbf{v}\|(\sin\theta)\mathbf{j}.$$

For instance, the vector **v** of magnitude 3 that makes an angle of 30° with the positive x-axis is

$$\mathbf{v} = 3(\cos 30°)\mathbf{i} + 3(\sin 30°)\mathbf{j} = \frac{3\sqrt{3}}{2}\mathbf{i} + \frac{3}{2}\mathbf{j}.$$

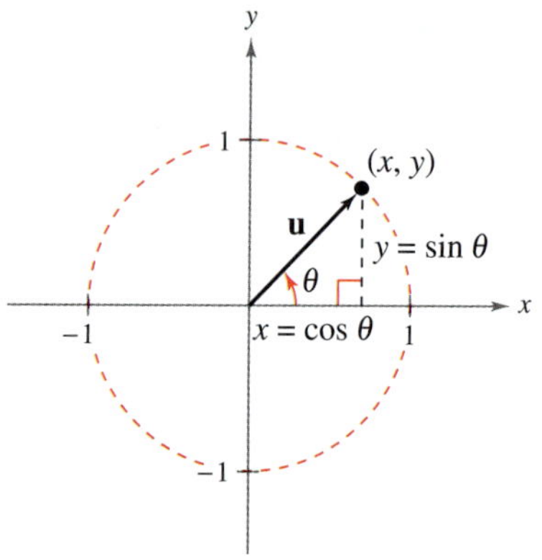

Figure 6.19

Because $\mathbf{v} = a\mathbf{i} + b\mathbf{j} = \|\mathbf{v}\|(\cos\theta)\mathbf{i} + \|\mathbf{v}\|(\sin\theta)\mathbf{j}$, it follows that the direction angle θ for **v** is determined from

$$\tan\theta = \frac{\sin\theta}{\cos\theta} \qquad \text{Quotient identity}$$

$$= \frac{\|\mathbf{v}\|\sin\theta}{\|\mathbf{v}\|\cos\theta} \qquad \text{Multiply numerator and denominator by } \|\mathbf{v}\|.$$

$$= \frac{b}{a}. \qquad \text{Simplify.}$$

EXAMPLE 8 Finding Direction Angles of Vectors

Find the direction angle of each vector.

a. $\mathbf{u} = 3\mathbf{i} + 3\mathbf{j}$ **b.** $\mathbf{v} = 3\mathbf{i} - 4\mathbf{j}$

Solution

a. The direction angle is determined from

$$\tan\theta = \frac{b}{a} = \frac{3}{3} = 1.$$

So, $\theta = 45°$, as shown in Figure 6.20.

b. The direction angle is determined from

$$\tan\theta = \frac{b}{a} = \frac{-4}{3}.$$

Moreover, because $\mathbf{v} = 3\mathbf{i} - 4\mathbf{j}$ lies in Quadrant IV, θ lies in Quadrant IV and its reference angle is

$$\theta' = \left|\arctan\left(-\frac{4}{3}\right)\right| \approx |-53.13°| = 53.13°.$$

So, it follows that $\theta \approx 360° - 53.13° = 306.87°$, as shown in Figure 6.21.

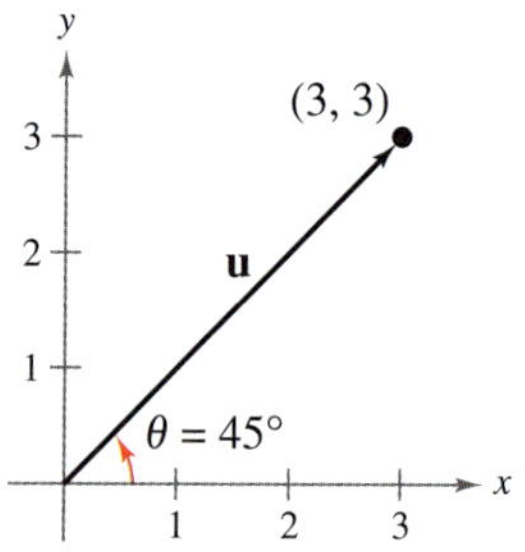

Figure 6.20

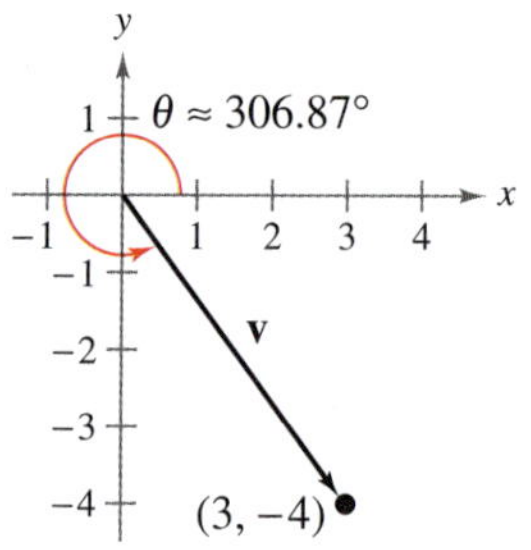

Figure 6.21

✓ Checkpoint *Audio-video solution in English & Spanish at LarsonPrecalculus.com*

Find the direction angle of each vector.

a. $\mathbf{v} = -6\mathbf{i} + 6\mathbf{j}$

b. $\mathbf{v} = -7\mathbf{i} - 4\mathbf{j}$

Applications

EXAMPLE 9 Finding the Component Form of a Vector

Find the component form of the vector that represents the velocity of an airplane descending at a speed of 100 miles per hour at an angle of 30° below the horizontal, as shown in Figure 6.22.

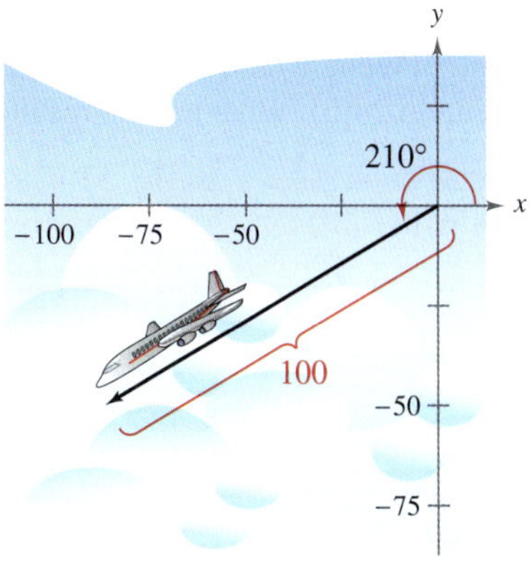

Figure 6.22

Solution

The velocity vector $\mathbf{v}$ has a magnitude of 100 and a direction angle of $\theta = 210°$.

$$\begin{aligned}\mathbf{v} &= \|\mathbf{v}\|(\cos\theta)\mathbf{i} + \|\mathbf{v}\|(\sin\theta)\mathbf{j}\\ &= 100(\cos 210°)\mathbf{i} + 100(\sin 210°)\mathbf{j}\\ &= 100\left(-\frac{\sqrt{3}}{2}\right)\mathbf{i} + 100\left(-\frac{1}{2}\right)\mathbf{j}\\ &= -50\sqrt{3}\mathbf{i} - 50\mathbf{j}\\ &= \langle -50\sqrt{3}, -50\rangle\end{aligned}$$

You can check that $\mathbf{v}$ has a magnitude of 100 as follows.

$$\|\mathbf{v}\| = \sqrt{(-50\sqrt{3})^2 + (-50)^2} = \sqrt{7500 + 2500} = \sqrt{10{,}000} = 100$$ Solution checks. ✓

✓ **Checkpoint** ▶ *Audio-video solution in English & Spanish at LarsonPrecalculus.com*

Rework Example 9 when the airplane descends at an angle 45° below the horizontal under the same conditions.

EXAMPLE 10 Using Vectors to Determine Weight

A force of 600 pounds is required to pull a boat and trailer up a ramp inclined at 15° from the horizontal. Find the combined weight of the boat and trailer.

Solution

Based on Figure 6.23, you can make the following observations.

$\|\overrightarrow{BA}\|$ = force of gravity = combined weight of boat and trailer

$\|\overrightarrow{BC}\|$ = force against ramp

$\|\overrightarrow{AC}\|$ = force required to move boat up ramp = 600 pounds

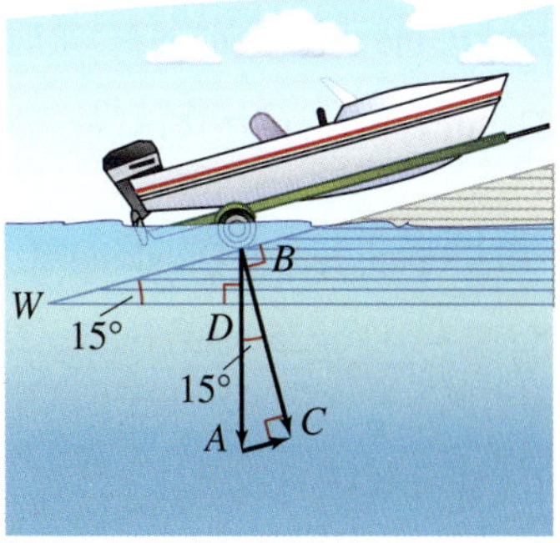

Figure 6.23

By construction, triangles BWD and ABC are similar. So, angle ABC is 15°. In triangle ABC you have

$$\sin 15° = \frac{\|\overrightarrow{AC}\|}{\|\overrightarrow{BA}\|}$$

$$\sin 15° = \frac{600}{\|\overrightarrow{BA}\|}$$

$$\|\overrightarrow{BA}\| = \frac{600}{\sin 15°}$$

$$\|\overrightarrow{BA}\| \approx 2318.$$

So, the combined weight is approximately 2318 pounds. (In Figure 6.23, note that $\overrightarrow{AC}$ is parallel to the ramp.)

✓ **Checkpoint** ▶ *Audio-video solution in English & Spanish at LarsonPrecalculus.com*

A force of 500 pounds is required to pull a boat and trailer up a ramp inclined at 12° from the horizontal. Find the combined weight of the boat and trailer.

EXAMPLE 11 Using Vectors to Find Speed and Direction

An airplane is traveling at a speed of 500 miles per hour with a bearing of 330° at a fixed altitude with a negligible wind velocity, as shown in Figure 6.24(a). (Note that a bearing of 330° corresponds to a direction angle of 120°.) The airplane encounters a wind blowing with a velocity of 70 miles per hour in the direction N 45° E, as shown in Figure 6.24(b). What are the resultant speed and direction of the airplane?

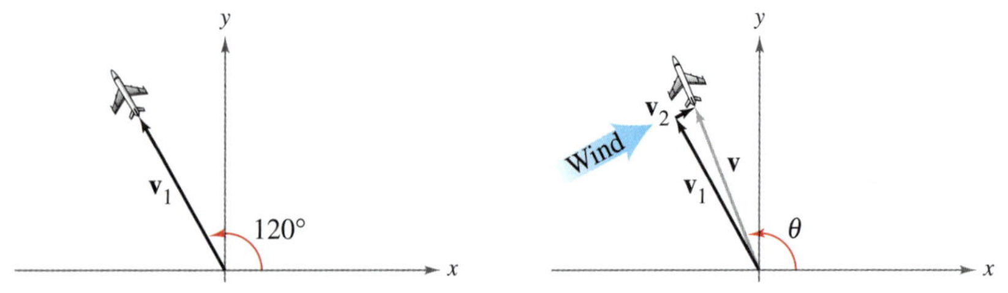

(a) (b)

Figure 6.24

Algebra Help

Recall from Section 4.8 that in air navigation, bearings are measured in degrees clockwise from north. In Figure 6.24, north is in the positive y-direction.

Airplane Pilot

Solution

Using Figure 6.24, the velocity of the airplane (alone) is

$$\mathbf{v}_1 = 500\langle\cos 120°, \sin 120°\rangle = \langle -250, 250\sqrt{3}\rangle$$

and the velocity of the wind is

$$\mathbf{v}_2 = 70\langle\cos 45°, \sin 45°\rangle = \langle 35\sqrt{2}, 35\sqrt{2}\rangle.$$

So, the velocity of the airplane (in the wind) is

$$\begin{aligned}\mathbf{v} &= \mathbf{v}_1 + \mathbf{v}_2\\ &= \langle -250 + 35\sqrt{2}, 250\sqrt{3} + 35\sqrt{2}\rangle\end{aligned}$$

and the resultant speed of the airplane is

$$\|\mathbf{v}\| = \sqrt{(-250 + 35\sqrt{2})^2 + (250\sqrt{3} + 35\sqrt{2})^2} \approx 522.5 \text{ miles per hour.}$$

To find θ, the direction angle of the flight path, note that

$$\tan\theta = \frac{250\sqrt{3} + 35\sqrt{2}}{-250 + 35\sqrt{2}}.$$

The flight path lies in Quadrant II, so θ lies in Quadrant II and its reference angle is

$$\theta = \left|\arctan\left(\frac{250\sqrt{3} + 35\sqrt{2}}{-250 + 35\sqrt{2}}\right)\right| \approx 67.4°.$$

So, the direction angle is

$$\theta \approx 180° - 67.4° = 112.6°.$$

You can use a graphing utility in *degree* mode to check this calculation, as shown in the figure. So, the true direction of the airplane is approximately 337.4°.

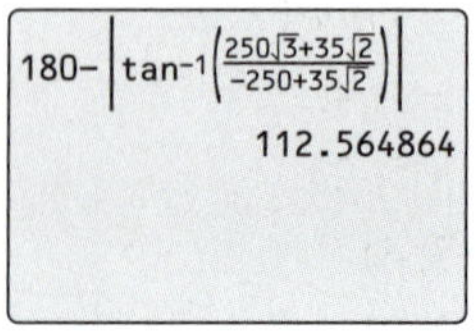

✓ Checkpoint *Audio-video solution in English & Spanish at LarsonPrecalculus.com*

Repeat Example 11 for an airplane traveling at a speed of 450 miles per hour with a bearing of 300° that encounters a wind with a velocity of 40 miles per hour in the direction N 30° E.

6.3 Exercises

See *CalcChat.com* for tutorial help and worked-out solutions to odd-numbered exercises. For instructions on how to use a graphing utility, see Appendix A.

Vocabulary and Concept Check

In Exercises 1–8, fill in the blank(s).

1. A _______ can be used to represent a quantity that involves both magnitude and direction.

2. The directed line segment $\overrightarrow{PQ}$ has _______ point P and _______ point Q.

3. The _______ of the directed line segment $\overrightarrow{PQ}$ is denoted by $\|\overrightarrow{PQ}\|$.

4. The set of all directed line segments that are equivalent to a given directed line segment $\overrightarrow{PQ}$ is a _______ **v** in the plane.

5. The directed line segment whose initial point is the origin is said to be in _______ .

6. The two basic vector operations are scalar _______ and vector _______ .

7. The vector **u** + **v** is called the _______ of vector addition.

8. The vector sum $v_1\mathbf{i} + v_2\mathbf{j}$ is called a _______ of the vectors **i** and **j**, and the scalars v_1 and v_2 are called the _______ and _______ components of **v**, respectively.

9. What two characteristics determine whether two directed line segments are equivalent?

10. What do you call a vector that has a magnitude of 1?

Procedures and Problem Solving

Describing a Quantity **In Exercises 11–16, identify the quantity as a vector or a scalar.**

11. volume of a container

12. velocity of a motorcycle

13. gravitational force

14. area of a cell phone screen

15. amount of money in a bank account

16. force applied while sliding a piece of furniture

Determining Whether Two Vectors Are Equivalent **In Exercises 17–22, determine whether u and v are equivalent. Explain.**

17.

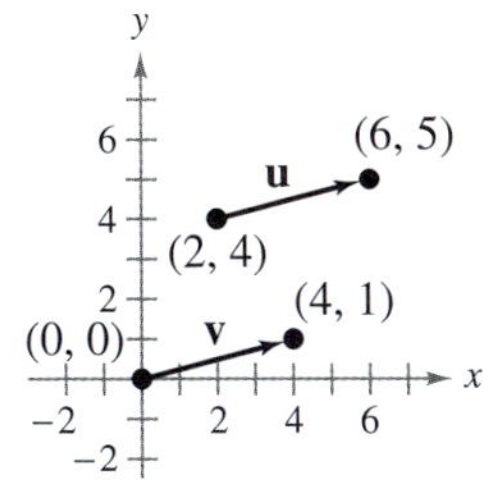

18.

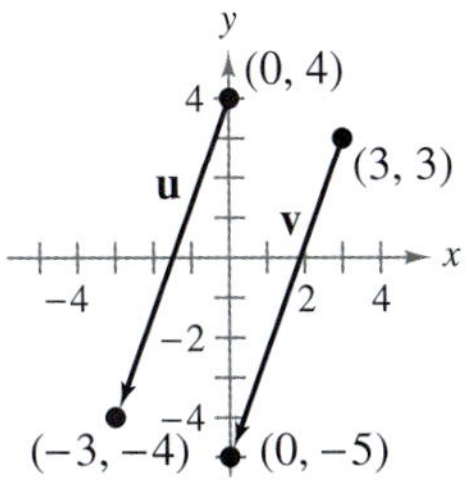

	Vector	Initial Point	Terminal Point
19.	**u**	(2, 2)	(−1, 4)
	v	(−3, −1)	(−5, 2)
20.	**u**	(2, 0)	(7, 4)
	v	(−8, 1)	(2, 9)
21.	**u**	(2, −1)	(5, −10)
	v	(6, 1)	(9, −8)
22.	**u**	(8, 1)	(13, −1)
	v	(−2, 4)	(−7, 6)

Finding the Component Form of a Vector **In Exercises 23–30, find the component form and the magnitude of the vector v.**

23.

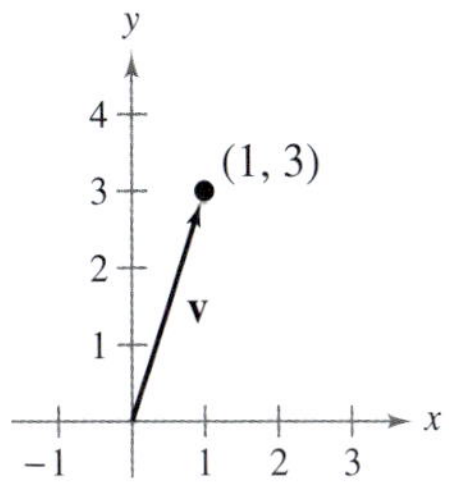

24.

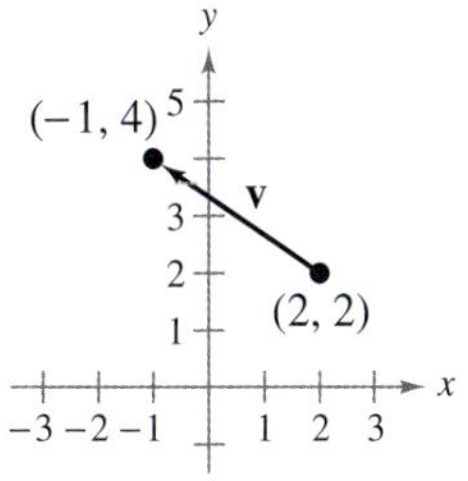

	Initial Point	Terminal Point
25.	(−1, −2)	(4, −2)
26.	(0, −3)	(0, 4)
27.	(−3, −5)	(−11, 1)
28.	(−2, 7)	(5, −17)
29.	(0.6, 3)	(−3, −0.6)
30.	$\left(\frac{3}{8}, -1\right)$	$\left(\frac{1}{4}, -\frac{1}{8}\right)$

Representing a Scalar Multiple **In Exercises 31–34, use the figure to sketch the vector. To print an enlarged copy of the graph, go to *MathGraphs.com*.**

31. $-\mathbf{v}$

32. $3\mathbf{u}$

33. $-\frac{1}{4}\mathbf{u}$

34. $\frac{2}{3}\mathbf{v}$

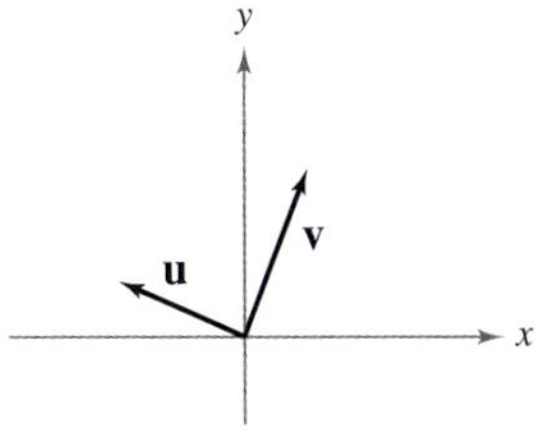

Parallelogram Law In Exercises 35–38, use the figure to sketch the vector. To print an enlarged copy of the graph, go to *MathGraphs.com*.

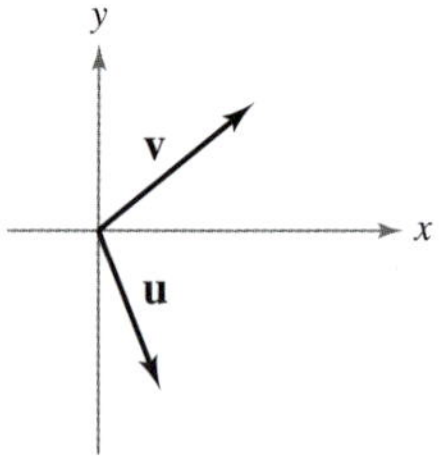

35. $\mathbf{u} + 2\mathbf{v}$

36. $\mathbf{u} + \frac{1}{2}\mathbf{v}$

37. $2\mathbf{v} - \frac{1}{2}\mathbf{u}$

38. $3\mathbf{v} + 2\mathbf{u}$

Vector Operations In Exercises 39–44, find (a) $\mathbf{u} + \mathbf{v}$, (b) $\mathbf{u} - \mathbf{v}$, and (c) $2\mathbf{u} - 3\mathbf{v}$. Then sketch each resultant vector.

39. $\mathbf{u} = \langle 2, 1 \rangle$, $\mathbf{v} = \langle 1, 3 \rangle$

40. $\mathbf{u} = \langle 2, 3 \rangle$, $\mathbf{v} = \langle 4, 0 \rangle$

41. $\mathbf{u} = \langle -5, 3 \rangle$, $\mathbf{v} = \langle 0, 0 \rangle$

42. $\mathbf{u} = \langle 0, 0 \rangle$, $\mathbf{v} = \langle 2, 1 \rangle$

43. $\mathbf{u} = \langle 0, -7 \rangle$, $\mathbf{v} = \langle 1, -2 \rangle$

44. $\mathbf{u} = \langle -3, 1 \rangle$, $\mathbf{v} = \langle 2, -5 \rangle$

Writing a Vector In Exercises 45–48, use the figure to write the vector in terms of the other two vectors.

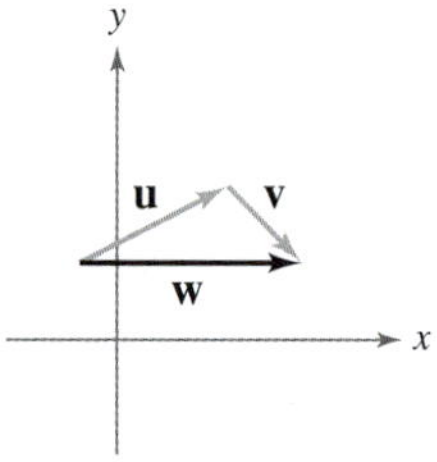

45. $\mathbf{w}$

46. $\mathbf{v}$

47. $\mathbf{u}$

48. $2\mathbf{v}$

Finding the Magnitude of a Scalar Multiple In Exercises 49–52, find the magnitude of the scalar multiple, where $\mathbf{u} = \langle 2, 0 \rangle$ and $\mathbf{v} = \langle -3, 6 \rangle$.

49. $\|5\mathbf{u}\|$

50. $\|4\mathbf{v}\|$

51. $\|-3\mathbf{v}\|$

52. $\left\|-\frac{3}{4}\mathbf{u}\right\|$

Finding a Unit Vector In Exercises 53–62, find a unit vector in the direction of the given vector. Verify that the result has a magnitude of 1.

53. $\mathbf{u} = \langle 6, 0 \rangle$

54. $\mathbf{u} = \langle 0, -2 \rangle$

55. $\mathbf{v} = \langle -2, 2 \rangle$

56. $\mathbf{v} = \langle -1, 1 \rangle$

57. $\mathbf{v} = \langle -24, -7 \rangle$

58. $\mathbf{v} = \langle -9, 12 \rangle$

59. $\mathbf{v} = 4\mathbf{i} - 3\mathbf{j}$

60. $\mathbf{w} = \mathbf{i} - 2\mathbf{j}$

61. $\mathbf{w} = 2\mathbf{j}$

62. $\mathbf{w} = -3\mathbf{i}$

Finding a Vector In Exercises 63–66, find the vector $\mathbf{v}$ with the given magnitude and the same direction as $\mathbf{u}$.

	Magnitude	*Direction*
63.	$\|\mathbf{v}\| = 10$	$\mathbf{u} = \langle -3, 4 \rangle$
64.	$\|\mathbf{v}\| = 3$	$\mathbf{u} = \langle -12, -5 \rangle$
65.	$\|\mathbf{v}\| = 7$	$\mathbf{u} = 3\mathbf{i} + 4\mathbf{j}$
66.	$\|\mathbf{v}\| = 10$	$\mathbf{u} = 2\mathbf{i} - 3\mathbf{j}$

Writing a Linear Combination of Unit Vectors In Exercises 67–70, the initial and terminal points of a vector are given. Write the vector as a linear combination of the standard unit vectors $\mathbf{i}$ and $\mathbf{j}$.

	Initial Point	*Terminal Point*
67.	$(-2, 1)$	$(3, -2)$
68.	$(0, -2)$	$(3, 6)$
69.	$(-6, 4)$	$(0, 1)$
70.	$(-1, -5)$	$(2, 3)$

Vector Operations In Exercises 71–76, find the component form of $\mathbf{v}$ and sketch the specified vector operations geometrically, where $\mathbf{u} = 2\mathbf{i} - \mathbf{j}$ and $\mathbf{w} = \mathbf{i} + 2\mathbf{j}$.

71. $\mathbf{v} = \frac{3}{2}\mathbf{u}$

72. $\mathbf{v} = \frac{3}{4}\mathbf{w}$

73. $\mathbf{v} = \mathbf{u} + 2\mathbf{w}$

74. $\mathbf{v} = -\mathbf{u} + \mathbf{w}$

75. $\mathbf{v} = \frac{1}{2}(3\mathbf{u} + \mathbf{w})$

76. $\mathbf{v} = 2(\mathbf{u} - \mathbf{w})$

Finding the Direction Angle of a Vector In Exercises 77–82, find the magnitude and direction angle of the vector $\mathbf{v}$.

77. $\mathbf{v} = 5(\cos 60^\circ\mathbf{i} + \sin 60^\circ\mathbf{j})$

78. $\mathbf{v} = 8(\cos 135^\circ\mathbf{i} + \sin 135^\circ\mathbf{j})$

79. $\mathbf{v} = 6\mathbf{i} - 6\mathbf{j}$

80. $\mathbf{v} = 8\mathbf{i} - 3\mathbf{j}$

81. $\mathbf{v} = -2\mathbf{i} + 5\mathbf{j}$

82. $\mathbf{v} = -7\mathbf{i} - 6\mathbf{j}$

Finding the Component Form of a Vector In Exercises 83–90, find the component form of $\mathbf{v}$ given its magnitude and the angle it makes with the positive x-axis. Sketch $\mathbf{v}$.

	Magnitude	*Angle*
83.	$\|\mathbf{v}\| = 3$	$\theta = 0^\circ$
84.	$\|\mathbf{v}\| = 1$	$\theta = 45^\circ$
85.	$\|\mathbf{v}\| = \frac{7}{2}$	$\theta = 120^\circ$
86.	$\|\mathbf{v}\| = \frac{5}{2}$	$\theta = 135^\circ$
87.	$\|\mathbf{v}\| = 3\sqrt{2}$	$\theta = 150^\circ$
88.	$\|\mathbf{v}\| = 4\sqrt{3}$	$\theta = 90^\circ$
89.	$\|\mathbf{v}\| = 2$	$\mathbf{v}$ in the direction $\mathbf{i} + 3\mathbf{j}$
90.	$\|\mathbf{v}\| = 3$	$\mathbf{v}$ in the direction $3\mathbf{i} + 4\mathbf{j}$

Finding the Component Form of a Vector In Exercises 91–94, find the component form of the sum of u and v with direction angles $\theta_{\mathbf{u}}$ and $\theta_{\mathbf{v}}$.

	Magnitude	*Angle*
91.	$\|\mathbf{u}\| = 5$	$\theta_{\mathbf{u}} = 0°$
	$\|\mathbf{v}\| = 5$	$\theta_{\mathbf{v}} = 90°$
92.	$\|\mathbf{u}\| = 4$	$\theta_{\mathbf{u}} = 60°$
	$\|\mathbf{v}\| = 4$	$\theta_{\mathbf{v}} = 90°$
93.	$\|\mathbf{u}\| = 20$	$\theta_{\mathbf{u}} = 45°$
	$\|\mathbf{v}\| = 50$	$\theta_{\mathbf{v}} = 160°$
94.	$\|\mathbf{u}\| = 35$	$\theta_{\mathbf{u}} = 25°$
	$\|\mathbf{v}\| = 50$	$\theta_{\mathbf{v}} = 120°$

Using the Law of Cosines In Exercises 95 and 96, use the Law of Cosines to find the angle α between the vectors. (Assume $0° \le \alpha \le 180°$.)

95. $\mathbf{v} = \mathbf{i} + \mathbf{j}$, $\mathbf{w} = 2(\mathbf{i} - \mathbf{j})$

96. $\mathbf{v} = 3\mathbf{i} + \mathbf{j}$, $\mathbf{w} = 2\mathbf{i} - \mathbf{j}$

Graphing Vectors In Exercises 97 and 98, graph the vectors and the resultant of the vectors. Find the magnitude and direction of the resultant.

97.

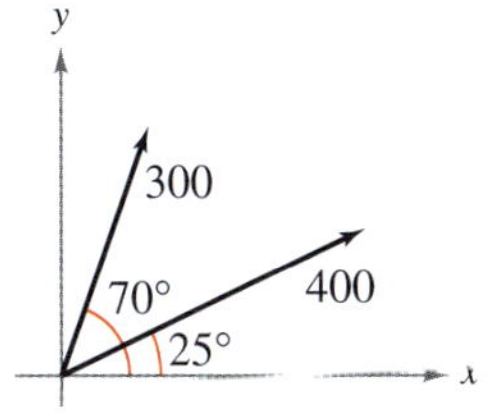

98.

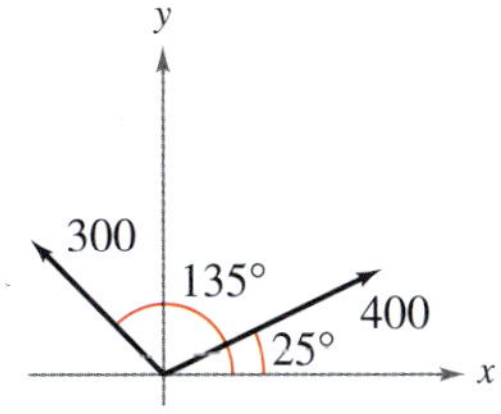

Resultant Force In Exercises 99 and 100, find the angle between the forces given the magnitude of their resultant. (*Hint*: Write force 1 as a vector in the direction of the positive x-axis and force 2 as a vector at an angle θ with the positive x-axis.)

	Force 1	*Force 2*	*Resultant Force*
99.	5 pounds	60 pounds	90 pounds
100.	3000 pounds	1000 pounds	3750 pounds

101. Physical Education A ball is thrown with an initial velocity of 70 feet per second at an angle of 40° with the horizontal (see figure). Find the vertical and horizontal components of the velocity.

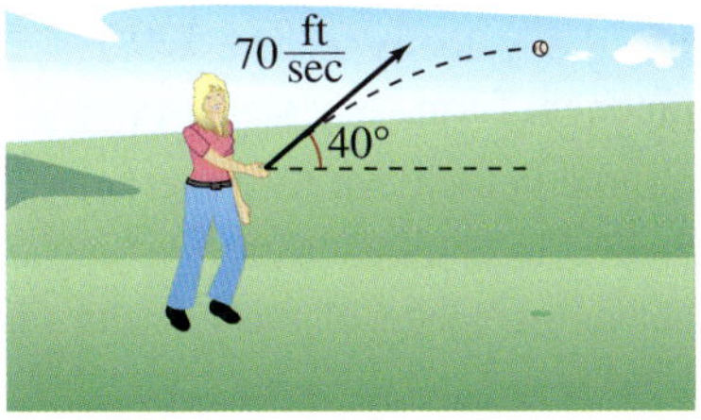

102. Velocity A gun with a muzzle velocity of 1200 feet per second is fired at an angle of 6° above the horizontal. Find the vertical and horizontal components of the velocity.

103. *Why you should learn it* *(p. 422)* The cranes shown in the figure are lifting an object that weighs 20,240 pounds. Find the tension in the cable of each crane.

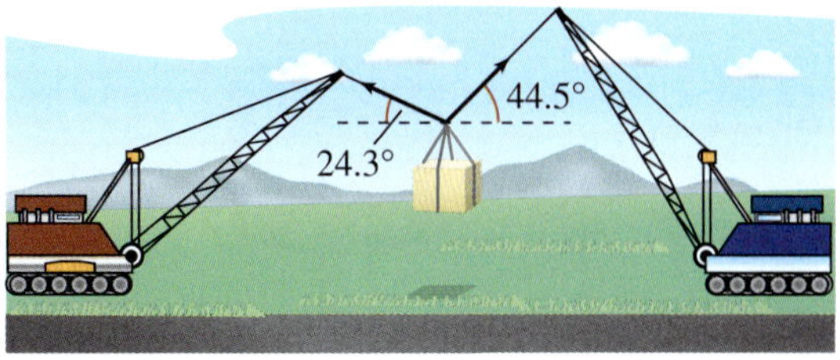

104. Physics Use the figure to determine the tension in each cable supporting the load.

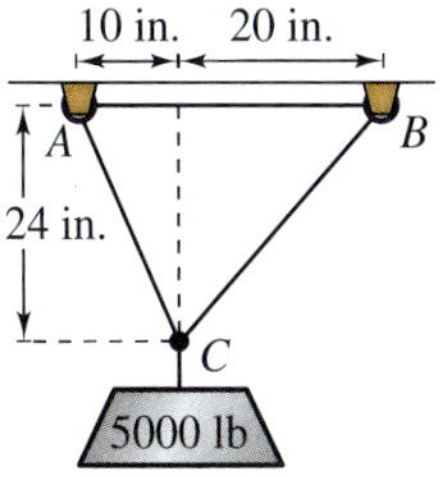

105. MODELING DATA

A loaded barge is being towed by two tugboats, and the magnitude of the resultant is 6000 pounds directed along the axis of the barge. Each towline makes an angle of θ degrees with the axis of the barge (see figure).

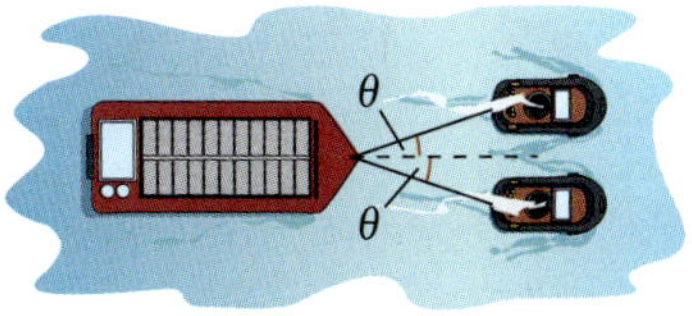

(a) Write the resultant tension T of each towline as a function of θ. Determine the domain of the function.

(b) Use a graphing utility to complete the table.

θ	10°	20°	30°	40°	50°	60°
T						

(c) Use the graphing utility to graph the tension function.

(d) Explain why the tension increases as θ increases.

106. MODELING DATA

To carry a 100-pound cylindrical weight, two people lift on the ends of short ropes that are tied to an eyelet on the top center of the cylinder. Each rope makes an angle of θ degrees with the vertical (see figure).

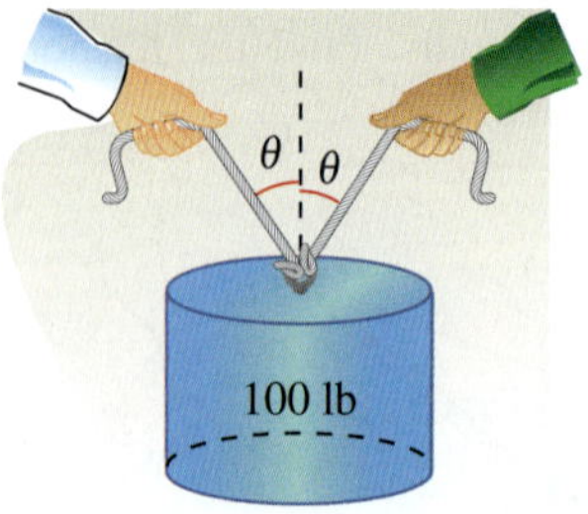

(a) Write the tension T of each rope as a function of θ. Determine the domain of the function.

(b) Use a graphing utility to complete the table.

θ	10°	20°	30°	40°	50°	60°
T						

(c) Use the graphing utility to graph the tension function.

(d) Explain why the tension increases as θ increases.

107. MODELING DATA

Forces with magnitudes of 150 newtons and 220 newtons act on a hook, as shown in the figure.

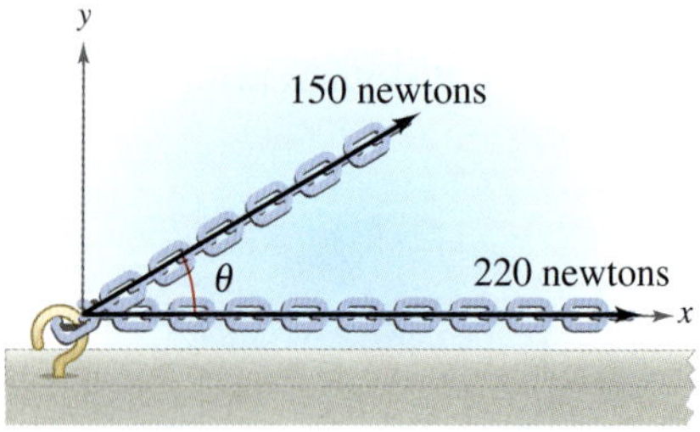

(a) Find the direction and magnitude of the resultant of the forces when $\theta = 30°$.

(b) Write the magnitude M of the resultant and the direction α of the resultant as functions of θ, where $0° \le \theta \le 180°$.

(c) Use a graphing utility to complete the table.

θ	0°	30°	60°	90°	120°	150°	180°
M							
α							

(d) Use the graphing utility to graph the two functions.

(e) Explain why one function decreases for increasing θ, whereas the other does not.

108. MODELING DATA

A tetherball weighing 1 pound is pulled outward from the pole by a horizontal force **u** until the rope makes an angle of θ degrees with the pole (see figure).

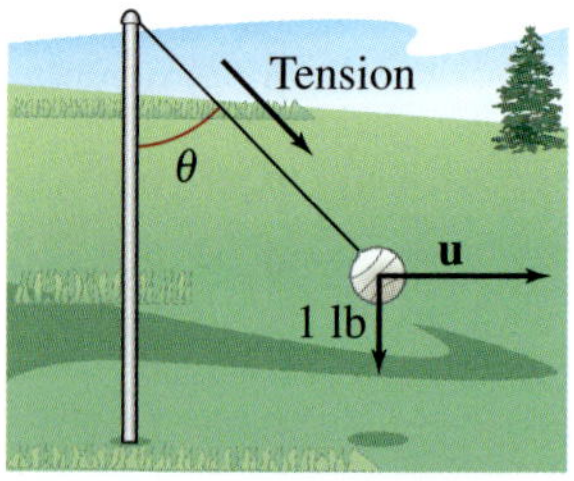

(a) Write the tension T in the rope and the magnitude of **u** as functions of θ. Determine the domain of each function.

(b) Use a graphing utility to graph the two functions for $0° \le \theta \le 60°$.

(c) Compare T and $\|\mathbf{u}\|$ as θ increases.

109. Aviation An airplane is flying in the direction 148° with an airspeed of 860 kilometers per hour. Because of the wind, its groundspeed and direction are 800 kilometers per hour and 140°, respectively (see figure). Find the direction and speed of the wind.

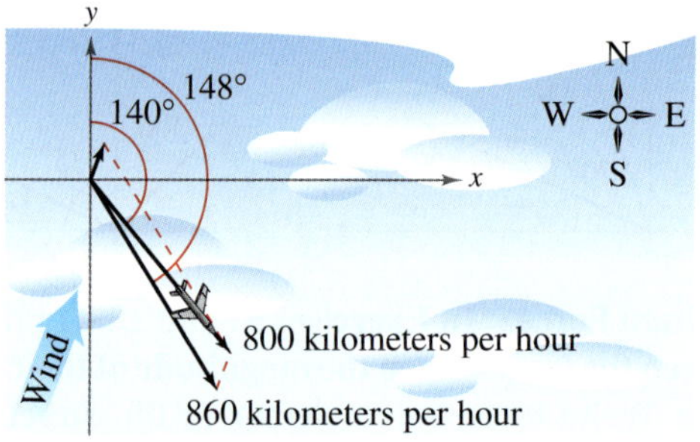

110. MODELING DATA

A commercial jet is flying from Miami to Seattle. The velocity of the jet with respect to the air is 580 miles per hour, and its bearing is 332°. The wind, at the altitude of the jet, is blowing from the southwest with a velocity of 60 miles per hour. (Remember that bearings for air navigation are measured in degrees clockwise from north.)

(a) Draw a figure that gives a visual representation of the problem.

(b) Write the velocity of the wind as a vector in component form.

(c) Write the velocity of the jet relative to the air as a vector in component form.

(d) What is the speed of the jet with respect to the ground?

(e) What is the true direction of the jet?

Focusing on Concepts

True or False? **In Exercises 111–114, determine whether the statement is true or false. Justify your answer.**

111. If $\mathbf{u}$ and $\mathbf{v}$ have the same magnitude and direction, then $\mathbf{u} = \mathbf{v}$.

112. If $\mathbf{u}$ is a unit vector in the direction of $\mathbf{v}$, then $\mathbf{v} = \|\mathbf{v}\|\mathbf{u}$.

113. If $\mathbf{v} = a\mathbf{i} + b\mathbf{j} = \mathbf{0}$, then $a = -b$.

114. If $\mathbf{u} = a\mathbf{i} + b\mathbf{j}$ is a unit vector, then $a^2 + b^2 = 1$.

115. Think About It Consider two forces of equal magnitude acting on a point.

(a) If the magnitude of the resultant is the sum of the magnitudes of the two forces, make a conjecture about the angle between the forces.

(b) If the resultant of the forces is 0, make a conjecture about the angle between the forces.

(c) Can the magnitude of the resultant be greater than the sum of the magnitudes of the two forces? Explain.

116. Exploration Consider two forces

$$\mathbf{F}_1 = \langle 10, 0\rangle \quad \text{and} \quad \mathbf{F}_2 = 5\langle \cos\theta, \sin\theta\rangle.$$

(a) Write $\|\mathbf{F}_1 + \mathbf{F}_2\|$ as a function of θ.

(b) Use a graphing utility to graph the function for $0 \le \theta < 2\pi$.

(c) Use the graph in part (b) to determine the range of the function. What is its maximum, and for what value of θ does it occur? What is its minimum, and for what value of θ does it occur?

(d) Explain why the magnitude of the resultant is never 0.

117. Writing Give geometric descriptions of (a) vector addition and (b) scalar multiplication.

118. HOW DO YOU SEE IT? Use the figure to determine whether each statement is true or false. Justify your answer.

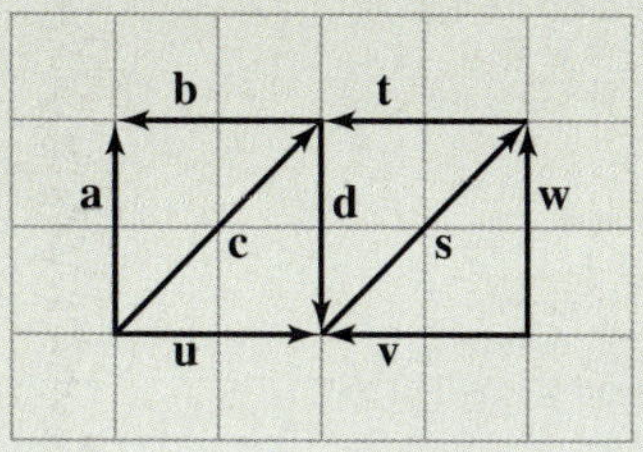

(a) $\mathbf{a} = -\mathbf{d}$ (b) $\mathbf{c} = \mathbf{s}$

(c) $\mathbf{a} + \mathbf{u} = \mathbf{c}$ (d) $\mathbf{v} + \mathbf{w} = -\mathbf{s}$

(e) $\mathbf{a} + \mathbf{w} = -2\mathbf{d}$ (f) $\mathbf{a} + \mathbf{d} = \mathbf{0}$

(g) $\mathbf{u} - \mathbf{v} = -2(\mathbf{b} + \mathbf{t})$ (h) $\mathbf{t} - \mathbf{w} = \mathbf{b} - \mathbf{a}$

119. Proof Prove that $(\cos\theta)\mathbf{i} + (\sin\theta)\mathbf{j}$ is a unit vector for any value of θ.

120. Programming Write a program for a graphing utility that graphs two vectors and their difference given the vectors in component form.

Finding the Difference of Two Vectors **In Exercises 121 and 122, use the program in Exercise 120 to find the difference of the vectors shown in the graph.**

121.

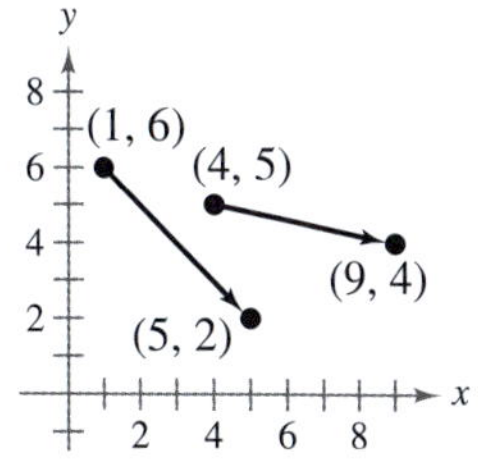

122.

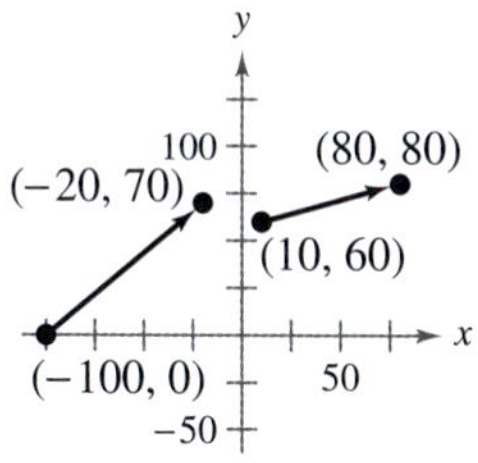

123. Error Analysis Describe the error in finding the component form of the vector $\mathbf{u}$ that has initial point $(-3, 4)$ and terminal point $(6, -1)$.

The components are $u_1 = -3 - 6 = -9$ and $u_2 = 4 - (-1) = 5$. So, $\mathbf{u} = \langle -9, 5\rangle$.

124. Error Analysis Describe the error in finding the direction angle θ of the vector $\mathbf{v} = -5\mathbf{i} + 8\mathbf{j}$.

Because $\tan\theta = \frac{b}{a} = \frac{8}{-5}$, the reference angle is $\theta' = \left|\arctan\left(-\frac{8}{5}\right)\right| \approx |-57.99°| = 57.99°$ and $\theta \approx 360° - 57.99° = 302.01°$.

Cumulative Mixed Review

Simplifying an Expression **In Exercises 125–128, simplify the expression.**

125. $(18x)^0(4xy)^2(3x^{-1})$

126. $(5ab^2)(a^{-3}b^0)(2a^0b)^{-2}$

127. $(2.1 \times 10^9)(3.4 \times 10^{-4})$

128. $(6.5 \times 10^{-6})(3.8 \times 10^4)$

Solving an Equation **In Exercises 129–132, solve the equation.**

129. $\cos x(\cos x + 1) = 0$

130. $\sin x\left(2\sin x + \sqrt{2}\right) = 0$

131. $3\sec x + 4 = 10$

132. $7\csc x + 18 = 4$

Trigonometric Substitution **In Exercises 133 and 134, use the trigonometric substitution to write the algebraic expression as a trigonometric function of θ, where $0 < \theta < \pi/2$.**

133. $\sqrt{x^2 + 100}$, $x = 10\cot\theta$

134. $\sqrt{x^2 - 49}$, $x = 7\sec\theta$

6.4 Vectors and Dot Products

The Dot Product of Two Vectors

So far, you have studied two vector operations—vector addition and multiplication by a scalar—each of which yields another vector. In this section, you will study a third vector operation, the **dot product.** This product yields a scalar, rather than a vector.

Definition of Dot Product

The **dot product** of $\mathbf{u} = \langle u_1, u_2 \rangle$ and $\mathbf{v} = \langle v_1, v_2 \rangle$ is $\mathbf{u} \cdot \mathbf{v} = u_1v_1 + u_2v_2$.

Properties of the Dot Product

Let **u**, **v**, and **w** be vectors and let c be a scalar.

1. $\mathbf{u} \cdot \mathbf{v} = \mathbf{v} \cdot \mathbf{u}$
2. $\mathbf{0} \cdot \mathbf{v} = 0$
3. $\mathbf{u} \cdot (\mathbf{v} + \mathbf{w}) = \mathbf{u} \cdot \mathbf{v} + \mathbf{u} \cdot \mathbf{w}$
4. $\mathbf{v} \cdot \mathbf{v} = \|\mathbf{v}\|^2$
5. $c(\mathbf{u} \cdot \mathbf{v}) = c\mathbf{u} \cdot \mathbf{v} = \mathbf{u} \cdot c\mathbf{v}$

(See the proofs on page 475.)

What you should learn

- Find the dot product of two vectors and use the properties of the dot product.
- Find the angle between two vectors and determine whether two vectors are orthogonal.
- Write vectors as the sums of two vector components.
- Use vectors to find the work done by a force.

Why you should learn it

You can use the dot product of two vectors to solve real-life problems involving two vector quantities. For instance, Exercise 73 on page 443 shows you how the dot product can be used to find the force necessary to prevent a truck from rolling down a hill.

EXAMPLE 1 Finding Dot Products

a. $\langle 4, 5 \rangle \cdot \langle 2, 3 \rangle = 4(2) + 5(3) = 8 + 15 = 23$

b. $\langle 2, -1 \rangle \cdot \langle 1, 2 \rangle = 2(1) + (-1)(2) = 2 - 2 = 0$

c. $\langle 0, 3 \rangle \cdot \langle 4, -2 \rangle = 0(4) + 3(-2) = 0 - 6 = -6$

Checkpoint *Audio-video solution in English & Spanish at LarsonPrecalculus.com*

Find the dot product of $\mathbf{u} = \langle 3, 4 \rangle$ and $\mathbf{v} = \langle 2, -3 \rangle$.

In Example 1, be sure you see that the dot product of two vectors is a scalar (a real number), not a vector. Moreover, notice that the dot product can be positive, zero, or negative.

EXAMPLE 2 Using Properties of the Dot Product

Let $\mathbf{u} = \langle -1, 3 \rangle$, $\mathbf{v} = \langle 2, -4 \rangle$, and $\mathbf{w} = \langle 1, -2 \rangle$. Use the vectors and the properties of the dot product to find (a) $(\mathbf{u} \cdot \mathbf{v})\mathbf{w}$, (b) $\mathbf{u} \cdot 2\mathbf{v}$, and (c) $\|\mathbf{u}\|$.

Solution

Begin by finding the dot product of **u** and **v** and the dot product of **u** and **u**.

$$\mathbf{u} \cdot \mathbf{v} = \langle -1, 3 \rangle \cdot \langle 2, -4 \rangle = -1(2) + 3(-4) = -14$$

$$\mathbf{u} \cdot \mathbf{u} = \langle -1, 3 \rangle \cdot \langle -1, 3 \rangle = -1(-1) + 3(3) = 10$$

a. $(\mathbf{u} \cdot \mathbf{v})\mathbf{w} = -14\langle 1, -2 \rangle = \langle -14, 28 \rangle$

b. $\mathbf{u} \cdot 2\mathbf{v} = 2(\mathbf{u} \cdot \mathbf{v}) = -28$

c. Because $\|\mathbf{u}\|^2 = \mathbf{u} \cdot \mathbf{u} = 10$, it follows that $\|\mathbf{u}\| = \sqrt{\mathbf{u} \cdot \mathbf{u}} = \sqrt{10}$.

Checkpoint *Audio-video solution in English & Spanish at LarsonPrecalculus.com*

Let $\mathbf{u} = \langle 3, 4 \rangle$ and $\mathbf{v} = \langle -2, 6 \rangle$. Use the vectors and the properties of the dot product to find (a) $(\mathbf{u} \cdot \mathbf{v})\mathbf{v}$, (b) $\mathbf{u} \cdot (\mathbf{u} + \mathbf{v})$, and (c) $\|\mathbf{v}\|$.

Algebra Help

In Example 2, notice that the product in part (a) is a vector, whereas the product in part (b) is a scalar. Can you see why?

The Angle Between Two Vectors

Let $\mathbf{u}$ and $\mathbf{v}$ be two nonzero vectors in standard position, as shown in the figure. The **angle between u and v** is the angle θ, where $0 \le \theta \le \pi$ or $0° \le \theta \le 180°$. This angle can be found using the dot product.

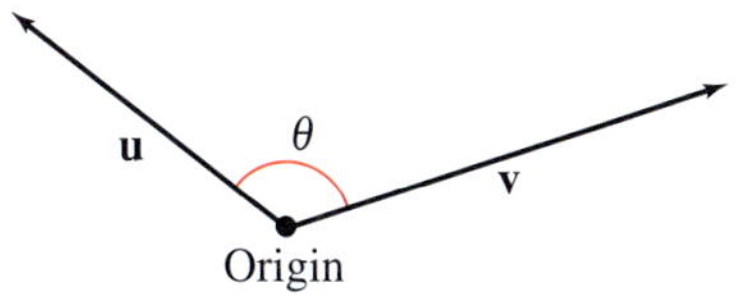

Angle Between Two Vectors

If θ is the angle between two nonzero vectors $\mathbf{u}$ and $\mathbf{v}$, then

$$\cos\theta = \frac{\mathbf{u} \cdot \mathbf{v}}{\|\mathbf{u}\|\,\|\mathbf{v}\|}. \qquad 0 \le \theta \le \pi \text{ or } 0° \le \theta \le 180°$$

(See the proof on page 475.)

Algebra Help

Note that the angle between the zero vector and another vector is not defined.

EXAMPLE 3 Finding the Angle Between Two Vectors

See LarsonPrecalculus.com for an interactive version of this type of example.

Find the angle θ between $\mathbf{u} = \langle 4, 3 \rangle$ and $\mathbf{v} = \langle 3, 5 \rangle$.

Solution

$$\cos\theta = \frac{\mathbf{u} \cdot \mathbf{v}}{\|\mathbf{u}\|\,\|\mathbf{v}\|} \qquad \text{Formula for angle between two vectors}$$

$$= \frac{\langle 4, 3 \rangle \cdot \langle 3, 5 \rangle}{\|\langle 4, 3 \rangle\|\,\|\langle 3, 5 \rangle\|} \qquad \text{Substitute } \mathbf{u} = \langle 4, 3 \rangle \text{ and } \mathbf{v} = \langle 3, 5 \rangle.$$

$$= \frac{4(3) + 3(5)}{\sqrt{4^2 + 3^2}\sqrt{3^2 + 5^2}} \qquad \text{Evaluate.}$$

$$= \frac{27}{5\sqrt{34}} \qquad \text{Simplify.}$$

This implies that the angle between the two vectors is

$$\theta = \arccos \frac{27}{5\sqrt{34}} \approx 22.2° \qquad \text{Use a calculator in degree mode.}$$

as shown below.

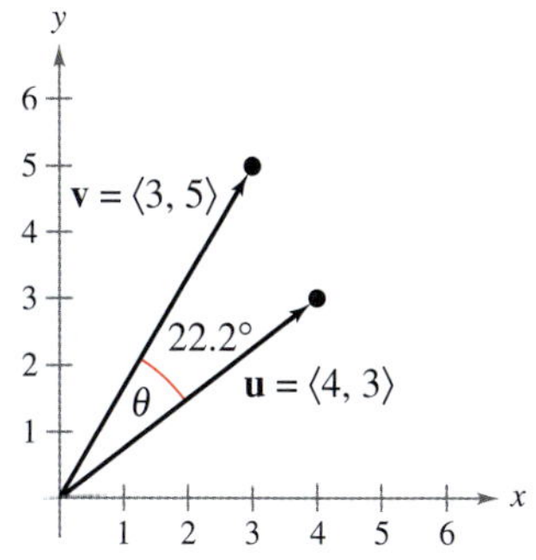

✓ Checkpoint *Audio-video solution in English & Spanish at LarsonPrecalculus.com*

Find the angle θ between $\mathbf{u} = \langle 2, 1 \rangle$ and $\mathbf{v} = \langle 1, 3 \rangle$.

Rewriting the expression for the angle between two vectors in the form

$$\mathbf{u} \cdot \mathbf{v} = \|\mathbf{u}\| \|\mathbf{v}\| \cos \theta \qquad \text{Alternative form of dot product}$$

produces an alternative way to calculate the dot product. From this form, you can see that because $\|\mathbf{u}\|$ and $\|\mathbf{v}\|$ are always positive,

$$\mathbf{u} \cdot \mathbf{v} \text{ and } \cos \theta$$

will always have the same sign. The figures below show the five possible orientations of two vectors.

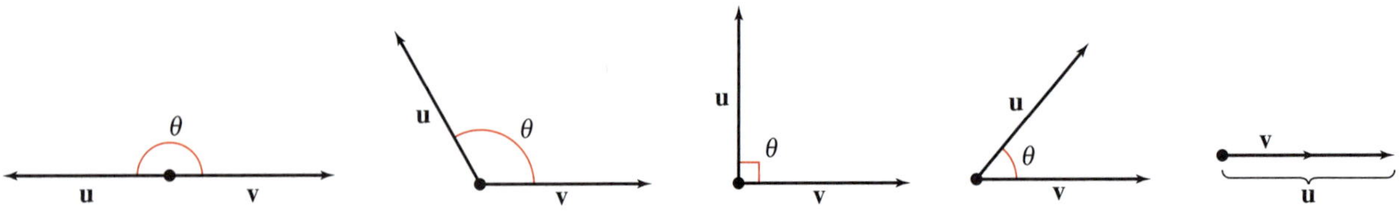

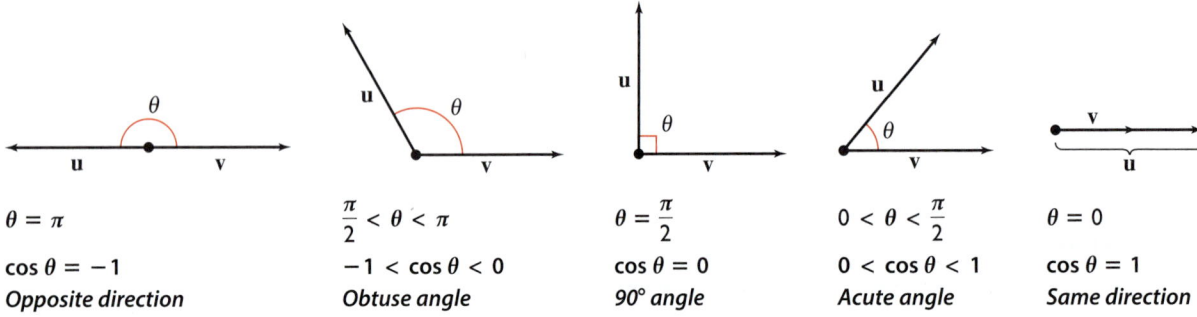

$\theta = \pi$	$\frac{\pi}{2} < \theta < \pi$	$\theta = \frac{\pi}{2}$	$0 < \theta < \frac{\pi}{2}$	$\theta = 0$
$\cos \theta = -1$	$-1 < \cos \theta < 0$	$\cos \theta = 0$	$0 < \cos \theta < 1$	$\cos \theta = 1$
Opposite direction	*Obtuse angle*	*90° angle*	*Acute angle*	*Same direction*

Definition of Orthogonal Vectors

The vectors $\mathbf{u}$ and $\mathbf{v}$ are **orthogonal** when $\mathbf{u} \cdot \mathbf{v} = 0$.

The terms *orthogonal* and *perpendicular* mean essentially the same thing—meeting at right angles. Even though the angle between the zero vector and another vector is not defined, it is convenient to extend the definition of orthogonality to include the zero vector. In other words, the zero vector is orthogonal to every vector $\mathbf{u}$ because $\mathbf{0} \cdot \mathbf{u} = 0$.

Technology Tip

The graphing utility program Finding the Angle Between Two Vectors, found at this textbook's *Student Companion Website*, graphs two vectors $\mathbf{u} = \langle a, b \rangle$ and $\mathbf{v} = \langle c, d \rangle$ in standard position and finds the measure of the angle between them. Use the program to verify the solutions to Examples 3 and 4.

EXAMPLE 4 Determining Orthogonal Vectors

Are the vectors

$$\mathbf{u} = \langle 2, -3 \rangle \text{ and } \mathbf{v} = \langle 6, 4 \rangle$$

orthogonal?

Solution

Find the dot product of the two vectors.

$$\mathbf{u} \cdot \mathbf{v} = \langle 2, -3 \rangle \cdot \langle 6, 4 \rangle = 2(6) + (-3)(4) = 0$$

Because the dot product is 0, the two vectors are orthogonal, as shown below.

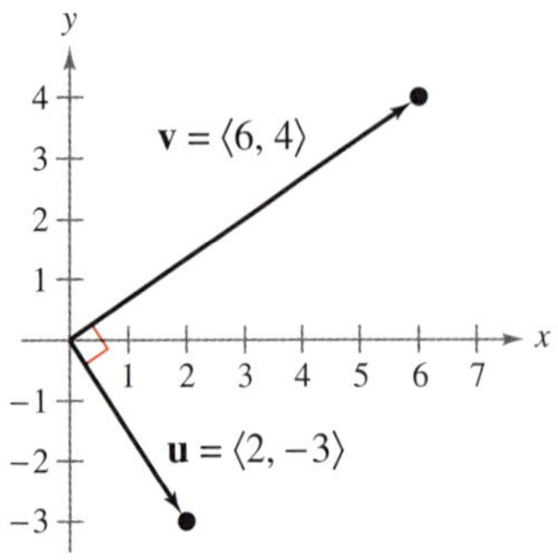

Checkpoint *Audio-video solution in English & Spanish at LarsonPrecalculus.com*

Are the vectors $\mathbf{u} = \langle 6, 10 \rangle$ and $\mathbf{v} = \left\langle -\frac{1}{3}, \frac{1}{5} \right\rangle$ orthogonal?

Finding Vector Components

You have already seen applications in which two vectors are added to produce a resultant vector. Many applications in physics and engineering pose the reverse problem—decomposing a given vector into the sum of two **vector components.**

Consider a boat on an inclined ramp, as shown in Figure 6.25. The force $\mathbf{F}$ due to gravity pulls the boat *down* the ramp and *against* the ramp. These two orthogonal forces, $\mathbf{w}_1$ and $\mathbf{w}_2$, are vector components of $\mathbf{F}$. That is,

$$\mathbf{F} = \mathbf{w}_1 + \mathbf{w}_2. \qquad \text{Vector components of } \mathbf{F}$$

The negative of component $\mathbf{w}_1$ represents the force needed to keep the boat from rolling down the ramp, and $\mathbf{w}_2$ represents the force that the tires must withstand against the ramp. A procedure for finding $\mathbf{w}_1$ and $\mathbf{w}_2$ is shown below.

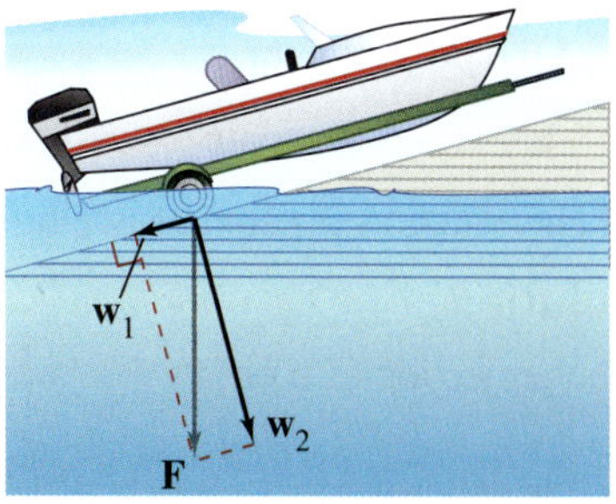

Figure 6.25

Definition of Vector Components

Let $\mathbf{u}$ and $\mathbf{v}$ be nonzero vectors such that

$$\mathbf{u} = \mathbf{w}_1 + \mathbf{w}_2$$

where $\mathbf{w}_1$ and $\mathbf{w}_2$ are orthogonal and $\mathbf{w}_1$ is parallel to (or a scalar multiple of) $\mathbf{v}$, as shown in Figure 6.26. The vectors $\mathbf{w}_1$ and $\mathbf{w}_2$ are called **vector components** of $\mathbf{u}$. The vector $\mathbf{w}_1$ is the **projection** of $\mathbf{u}$ onto $\mathbf{v}$ and is denoted by

$$\mathbf{w}_1 = \text{proj}_{\mathbf{v}}\mathbf{u}.$$

The vector $\mathbf{w}_2$ is given by

$$\mathbf{w}_2 = \mathbf{u} - \mathbf{w}_1.$$

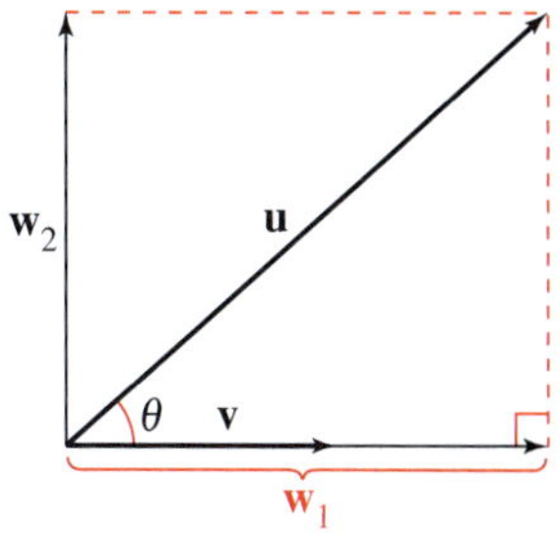

θ is acute.

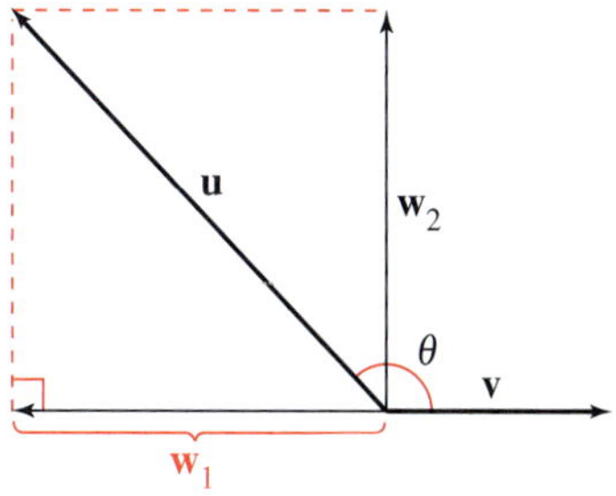

θ is obtuse.

Figure 6.26

From the definition of vector components, notice that you can find the component $\mathbf{w}_2$ once you have found the projection of $\mathbf{u}$ onto $\mathbf{v}$. To find the projection, you can use the dot product, as follows.

$$\mathbf{u} = \mathbf{w}_1 + \mathbf{w}_2$$

$$\mathbf{u} = c\mathbf{v} + \mathbf{w}_2 \qquad \mathbf{w}_1 \text{ is a scalar multiple of } \mathbf{v}.$$

$$\mathbf{u} \cdot \mathbf{v} = (c\mathbf{v} + \mathbf{w}_2) \cdot \mathbf{v} \qquad \text{Take dot product of each side with } \mathbf{v}.$$

$$\mathbf{u} \cdot \mathbf{v} = c\mathbf{v} \cdot \mathbf{v} + \mathbf{w}_2 \cdot \mathbf{v}$$

$$\mathbf{u} \cdot \mathbf{v} = c\|\mathbf{v}\|^2 + 0 \qquad \mathbf{w}_2 \text{ and } \mathbf{v} \text{ are orthogonal.}$$

So,

$$c = \frac{\mathbf{u} \cdot \mathbf{v}}{\|\mathbf{v}\|^2}$$

and

$$\mathbf{w}_1 = \text{proj}_{\mathbf{v}}\mathbf{u} = c\mathbf{v} = \left(\frac{\mathbf{u} \cdot \mathbf{v}}{\|\mathbf{v}\|^2}\right)\mathbf{v}.$$

Projection of u onto v

Let $\mathbf{u}$ and $\mathbf{v}$ be nonzero vectors. The projection of $\mathbf{u}$ onto $\mathbf{v}$ is given by

$$\text{proj}_{\mathbf{v}}\mathbf{u} = \left(\frac{\mathbf{u} \cdot \mathbf{v}}{\|\mathbf{v}\|^2}\right)\mathbf{v}.$$

EXAMPLE 5 Decomposing a Vector into Components

Find the projection of

$\mathbf{u} = \langle 3, -5 \rangle$ onto $\mathbf{v} = \langle 6, 2 \rangle$.

Then write $\mathbf{u}$ as the sum of two orthogonal vectors, one of which is $\text{proj}_{\mathbf{v}}\mathbf{u}$.

Solution

The projection of $\mathbf{u}$ onto $\mathbf{v}$ is

$$\mathbf{w}_1 = \text{proj}_{\mathbf{v}}\mathbf{u} = \left(\frac{\mathbf{u} \cdot \mathbf{v}}{\|\mathbf{v}\|^2}\right)\mathbf{v} = \left(\frac{8}{40}\right)\langle 6, 2 \rangle = \left\langle \frac{6}{5}, \frac{2}{5} \right\rangle$$

as shown in Figure 6.27. The other component, $\mathbf{w}_2$, is

$$\mathbf{w}_2 = \mathbf{u} - \mathbf{w}_1 = \langle 3, -5 \rangle - \left\langle \frac{6}{5}, \frac{2}{5} \right\rangle = \left\langle \frac{9}{5}, -\frac{27}{5} \right\rangle.$$

So,

$$\mathbf{u} = \mathbf{w}_1 + \mathbf{w}_2 = \left\langle \frac{6}{5}, \frac{2}{5} \right\rangle + \left\langle \frac{9}{5}, -\frac{27}{5} \right\rangle = \langle 3, -5 \rangle.$$

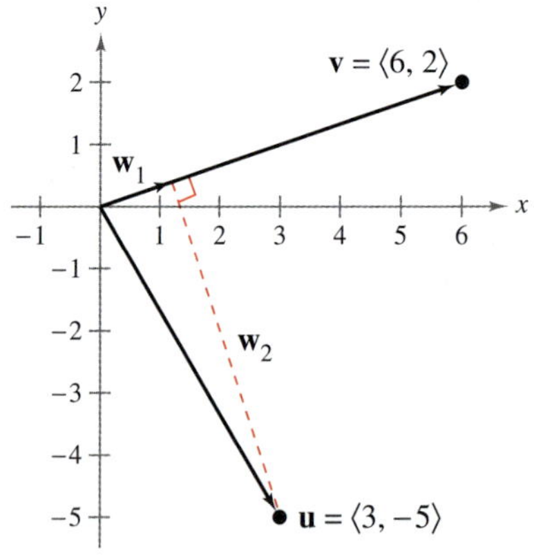

Figure 6.27

Checkpoint *Audio-video solution in English & Spanish at LarsonPrecalculus.com*

Find the projection of $\mathbf{u} = \langle 3, 4 \rangle$ onto $\mathbf{v} = \langle 8, 2 \rangle$. Then write $\mathbf{u}$ as the sum of two orthogonal vectors, one of which is $\text{proj}_{\mathbf{v}}\mathbf{u}$.

EXAMPLE 6 Finding a Force

A 200-pound cart sits on a ramp inclined at 30°, as shown in Figure 6.28. What force is required to keep the cart from rolling down the ramp?

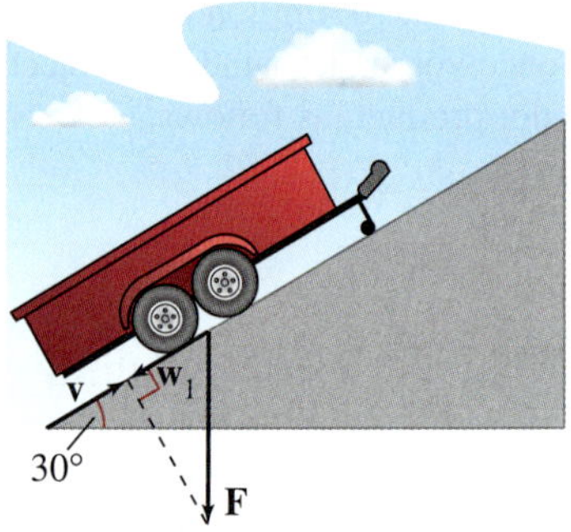

Figure 6.28

Solution

Because the force due to gravity is vertical and downward, you can represent the gravitational force by the vector

$$\mathbf{F} = -200\mathbf{j}. \qquad \text{Force due to gravity}$$

To find the force required to keep the cart from rolling down the ramp, project $\mathbf{F}$ onto a unit vector $\mathbf{v}$ in the direction of the ramp, as follows.

$$\mathbf{v} = (\cos 30°)\mathbf{i} + (\sin 30°)\mathbf{j} = \frac{\sqrt{3}}{2}\mathbf{i} + \frac{1}{2}\mathbf{j} \qquad \text{Unit vector along ramp}$$

So, the projection of $\mathbf{F}$ onto $\mathbf{v}$ is

$$\begin{aligned}\mathbf{w}_1 = \text{proj}_{\mathbf{v}}\mathbf{F} &= \left(\frac{\mathbf{F} \cdot \mathbf{v}}{\|\mathbf{v}\|^2}\right)\mathbf{v} \\ &= (\mathbf{F} \cdot \mathbf{v})\mathbf{v} \\ &= (-200)\left(\frac{1}{2}\right)\mathbf{v} \\ &= -100\left(\frac{\sqrt{3}}{2}\mathbf{i} + \frac{1}{2}\mathbf{j}\right).\end{aligned}$$

The magnitude of this force is 100, so a force of 100 pounds is required to keep the cart from rolling down the ramp.

Checkpoint *Audio-video solution in English & Spanish at LarsonPrecalculus.com*

Rework Example 6 for a 150-pound cart sitting on a ramp inclined at 15°.

Work

The work W done by a constant force $\mathbf{F}$ acting along the line of motion of an object is given by

$$W = \text{(magnitude of force)(distance)}$$
$$= \|\mathbf{F}\| \, \|\overrightarrow{PQ}\|$$

as shown in Figure 6.29. If the constant force $\mathbf{F}$ is not directed along the line of motion (see Figure 6.30), then the work W done by the force is given by

$$W = \|\text{proj}_{\overrightarrow{PQ}}\mathbf{F}\| \, \|\overrightarrow{PQ}\| \qquad \text{Projection form for work}$$
$$= (\cos\theta)\,\|\mathbf{F}\| \, \|\overrightarrow{PQ}\| \qquad \|\text{proj}_{\overrightarrow{PQ}}\mathbf{F}\| = (\cos\theta)\|\mathbf{F}\|$$
$$= \mathbf{F} \cdot \overrightarrow{PQ}. \qquad \text{Dot product form for work}$$

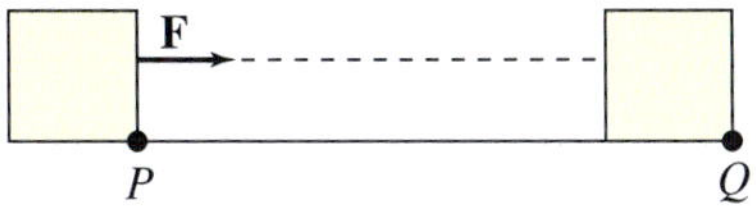

Force acts along the line of motion.
Figure 6.29

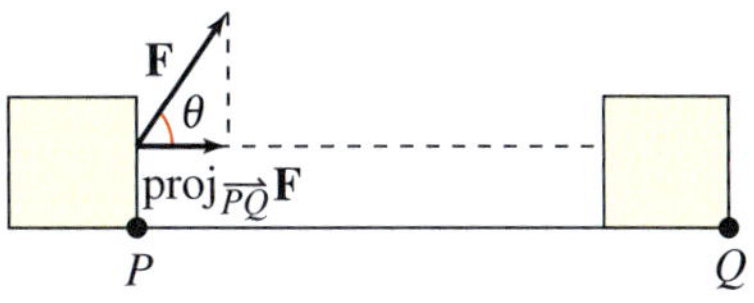

Force acts at angle θ with the line of motion.
Figure 6.30

This notion of work is summarized in the definition below.

Definition of Work

The **work** W done by a constant force $\mathbf{F}$ as its point of application moves along the vector $\overrightarrow{PQ}$ is given by either of the following.

1. $W = \|\text{proj}_{\overrightarrow{PQ}}\mathbf{F}\| \, \|\overrightarrow{PQ}\|$ — Projection form
2. $W = \mathbf{F} \cdot \overrightarrow{PQ}$ — Dot product form

EXAMPLE 7 Finding the Work Done

To close a barn's sliding door, a person pulls on a rope with a constant force of 50 pounds at a constant angle of 60°, as shown in Figure 6.31. Find the work done in moving the door 12 feet to its closed position.

Solution

Using a projection, you can calculate the work as follows.

$$W = \|\text{proj}_{\overrightarrow{PQ}}\mathbf{F}\| \, \|\overrightarrow{PQ}\| \qquad \text{Projection form for work}$$
$$= (\cos 60°)\,\|\mathbf{F}\| \, \|\overrightarrow{PQ}\| \qquad \|\text{proj}_{\overrightarrow{PQ}}\mathbf{F}\| = (\cos\theta)\|\mathbf{F}\|$$
$$= \frac{1}{2}(50)(12)$$
$$= 300 \text{ foot-pounds}$$

So, the work done is 300 foot-pounds. Verify this result by finding the vectors $\mathbf{F}$ and $\overrightarrow{PQ}$ and calculating their dot product.

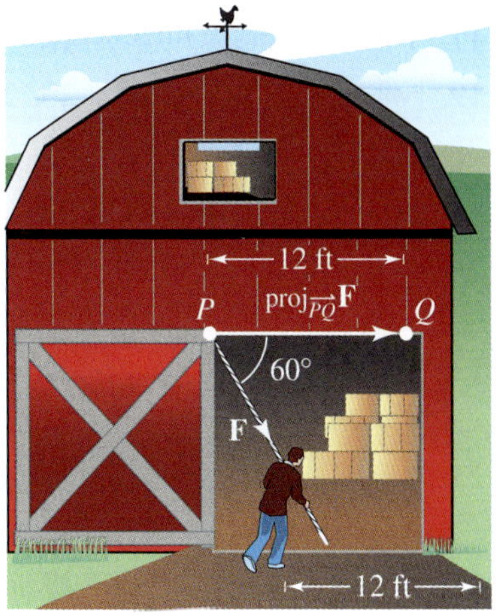

Figure 6.31

✓ ***Checkpoint*** *Audio-video solution in English & Spanish at LarsonPrecalculus.com*

A wagon is pulled by a constant force of 35 pounds on a handle that makes a 30° angle with the horizontal. Find the work done in pulling the wagon 40 feet.

6.4 Exercises

See *CalcChat.com* for tutorial help and worked-out solutions to odd-numbered exercises. For instructions on how to use a graphing utility, see Appendix A.

Vocabulary and Concept Check

1. For two vectors **u** and **v**, does $\mathbf{u} \cdot \mathbf{v} = \mathbf{v} \cdot \mathbf{u}$?

2. What is the dot product of two orthogonal vectors?

3. Is the dot product of two vectors an angle, a vector, or a scalar?

In Exercises 4–6, fill in the blank(s).

4. If θ is the angle between two nonzero vectors **u** and **v**, then $\cos \theta =$ _______ .

5. The projection of **u** onto **v** is given by $\text{proj}_{\mathbf{v}}\mathbf{u} =$ _______ .

6. The work W done by a constant force **F** as its point of application moves along the vector $\overrightarrow{PQ}$ is given by either $W =$ ______ or $W =$ _______ .

Procedures and Problem Solving

Finding the Dot Product **In Exercises 7–10, find the dot product of u and v.**

7. $\mathbf{u} = \langle 7, 1 \rangle$, $\mathbf{v} = \langle -3, 2 \rangle$

8. $\mathbf{u} = \langle -4, 1 \rangle$, $\mathbf{v} = \langle 5, -4 \rangle$

9. $\mathbf{u} = 5\mathbf{i} + \mathbf{j}$, $\mathbf{v} = 3\mathbf{i} - \mathbf{j}$

10. $\mathbf{u} = 2\mathbf{i} + 6\mathbf{j}$, $\mathbf{v} = -3\mathbf{i} + 7\mathbf{j}$

Using Properties of the Dot Product **In Exercises 11–22, use the vectors $\mathbf{u} = \langle 3, 3 \rangle$, $\mathbf{v} = \langle -4, 2 \rangle$, and $\mathbf{w} = \langle 3, -1 \rangle$ to find the quantity. State whether the result is a vector or a scalar.**

11. $\mathbf{v} \cdot \mathbf{v}$

12. $3\mathbf{u} \cdot \mathbf{v}$

13. $(\mathbf{u} \cdot \mathbf{v})\mathbf{v}$

14. $(2\mathbf{v} \cdot \mathbf{u})\mathbf{w}$

15. $(\mathbf{v} \cdot \mathbf{0})\mathbf{w}$

16. $(\mathbf{u} + \mathbf{v}) \cdot \mathbf{0}$

17. $3\|\mathbf{v}\|^2$

18. $-2\|\mathbf{u}\|^2$

19. $\|\mathbf{w}\| - 1$

20. $2 - \|\mathbf{u}\|$

21. $(\mathbf{u} \cdot \mathbf{v}) - (\mathbf{u} \cdot \mathbf{w})$

22. $(\mathbf{v} \cdot \mathbf{u}) - (\mathbf{w} \cdot \mathbf{v})$

Finding the Angle Between Two Vectors **In Exercises 23–30, find the angle θ (in radians) between the vectors.**

23. $\mathbf{u} = \langle -1, 0 \rangle$, $\mathbf{v} = \langle 0, 2 \rangle$

24. $\mathbf{u} = \langle 4, 4 \rangle$, $\mathbf{v} = \langle -2, 0 \rangle$

25. $\mathbf{u} = 3\mathbf{i} + 4\mathbf{j}$, $\mathbf{v} = -2\mathbf{j}$

26. $\mathbf{u} = 7\mathbf{i} - 2\mathbf{j}$, $\mathbf{v} = -8\mathbf{i} + 6\mathbf{j}$

27. $\mathbf{u} = 2\mathbf{i}$, $\mathbf{v} = -3\mathbf{j}$

28. $\mathbf{u} = 4\mathbf{j}$, $\mathbf{v} = -9\mathbf{i}$

29. $\mathbf{u} = \left\langle \cos \frac{\pi}{3}, \sin \frac{\pi}{3} \right\rangle$, $\mathbf{v} = \left\langle \cos \frac{3\pi}{4}, \sin \frac{3\pi}{4} \right\rangle$

30. $\mathbf{u} = \left\langle \cos \frac{\pi}{4}, \sin \frac{\pi}{4} \right\rangle$, $\mathbf{v} = \left\langle \cos \frac{2\pi}{3}, \sin \frac{2\pi}{3} \right\rangle$

Finding the Angle Between Two Vectors **In Exercises 31–34, graph the vectors and find the degree measure of the angle between the vectors.**

31. $\mathbf{u} = 2\mathbf{i} - 4\mathbf{j}$, $\mathbf{v} = 3\mathbf{i} - 5\mathbf{j}$

32. $\mathbf{u} = 6\mathbf{i} - 2\mathbf{j}$, $\mathbf{v} = 8\mathbf{i} - 5\mathbf{j}$

33. $\mathbf{u} = -6\mathbf{i} - 3\mathbf{j}$, $\mathbf{v} = -8\mathbf{i} + 4\mathbf{j}$

34. $\mathbf{u} = -7\mathbf{i} - 4\mathbf{j}$, $\mathbf{v} = -8\mathbf{i} + 2\mathbf{j}$

Finding the Angles in a Triangle **In Exercises 35–38, use vectors to find the interior angles of the triangle with the given vertices.**

35. $(1, 2), (3, 4), (2, 5)$

36. $(-3, -4), (1, 7), (8, 2)$

37. $(-3, 0), (2, 2), (0, 6)$

38. $(-3, 5), (-1, 9), (7, 9)$

Using the Angle Between Two Vectors **In Exercises 39–42, find $\mathbf{u} \cdot \mathbf{v}$, where θ is the angle between u and v.**

39. $\|\mathbf{u}\| = 9, \|\mathbf{v}\| = 36, \theta = 3\pi/4$

40. $\|\mathbf{u}\| = 4, \|\mathbf{v}\| = 12, \theta = \pi/3$

41. $\|\mathbf{u}\| = 4, \|\mathbf{v}\| = 10, \theta = 2\pi/3$

42. $\|\mathbf{u}\| = 100, \|\mathbf{v}\| = 250, \theta = \pi/6$

Determining Orthogonal Vectors **In Exercises 43–46, determine whether u and v are orthogonal.**

43. $\mathbf{u} = \langle 3, 15 \rangle$, $\mathbf{v} = \langle -1, 5 \rangle$

44. $\mathbf{u} = \langle 12, 4 \rangle$, $\mathbf{v} = \left\langle \frac{1}{4}, -\frac{1}{3} \right\rangle$

45. $\mathbf{u} = \mathbf{j}$, $\mathbf{v} = \mathbf{i} - \mathbf{j}$

46. $\mathbf{u} = 2\mathbf{i} - 2\mathbf{j}$, $\mathbf{v} = -\mathbf{i} - \mathbf{j}$

Describing the Relationship Between Two Vectors In Exercises 47–50, determine whether **u** and **v** are orthogonal, parallel, or neither.

47. $\mathbf{u} = \langle 10, 20 \rangle$, $\mathbf{v} = \langle -18, 9 \rangle$

48. $\mathbf{u} = \langle 15, 9 \rangle$, $\mathbf{v} = \langle -5, -3 \rangle$

49. $\mathbf{u} = -\frac{3}{5}\mathbf{i} + \frac{7}{10}\mathbf{j}$, $\mathbf{v} = 12\mathbf{i} - 14\mathbf{j}$

50. $\mathbf{u} = -\frac{9}{10}\mathbf{i} + 3\mathbf{j}$, $\mathbf{v} = -5\mathbf{i} + \frac{3}{2}\mathbf{j}$

Finding an Unknown Vector Component In Exercises 51–56, find the value of k such that the vectors **u** and **v** are orthogonal.

51. $\mathbf{u} = 2\mathbf{i} - k\mathbf{j}$, $\mathbf{v} = 3\mathbf{i} + 2\mathbf{j}$

52. $\mathbf{u} = 8\mathbf{i} + 4\mathbf{j}$, $\mathbf{v} = 2\mathbf{i} - k\mathbf{j}$

53. $\mathbf{u} = \mathbf{i} + 4\mathbf{j}$, $\mathbf{v} = 7k\mathbf{i} - 5\mathbf{j}$

54. $\mathbf{u} = -3k\mathbf{i} + 5\mathbf{j}$, $\mathbf{v} = 2\mathbf{i} - 4\mathbf{j}$

55. $\mathbf{u} = -3k\mathbf{i} + 2\mathbf{j}$, $\mathbf{v} = -6\mathbf{i}$

56. $\mathbf{u} = 4\mathbf{i} - 4k\mathbf{j}$, $\mathbf{v} = 3\mathbf{j}$

Decomposing a Vector into Components In Exercises 57–60, find the projection of **u** onto **v**. Then write **u** as the sum of two orthogonal vectors, one of which is $\text{proj}_{\mathbf{v}}\mathbf{u}$.

57. $\mathbf{u} = \langle 4, 2 \rangle$, $\mathbf{v} = \langle 1, -2 \rangle$

58. $\mathbf{u} = \langle 2, 2 \rangle$, $\mathbf{v} = \langle 6, 1 \rangle$

59. $\mathbf{u} = \langle 0, 3 \rangle$, $\mathbf{v} = \langle 2, 15 \rangle$

60. $\mathbf{u} = \langle -5, -1 \rangle$, $\mathbf{v} = \langle -1, 1 \rangle$

Finding the Projection of u onto v In Exercises 61–64, use the graph to determine mentally the projection of **u** onto **v**. (The coordinates of the terminal points of the vectors in standard position are given.) Use the formula for the projection of **u** onto **v** to verify your result.

61.

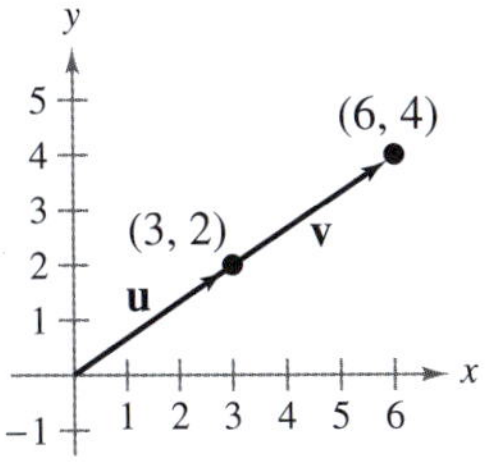

62.

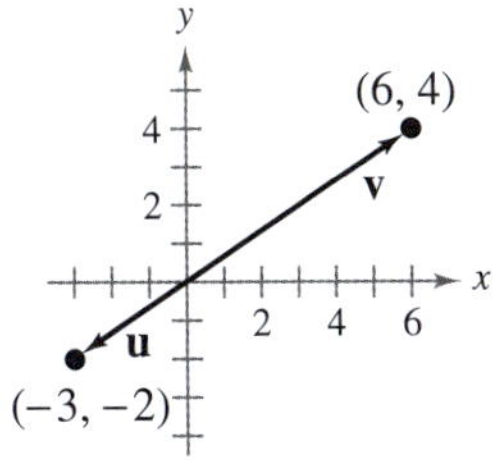

63.

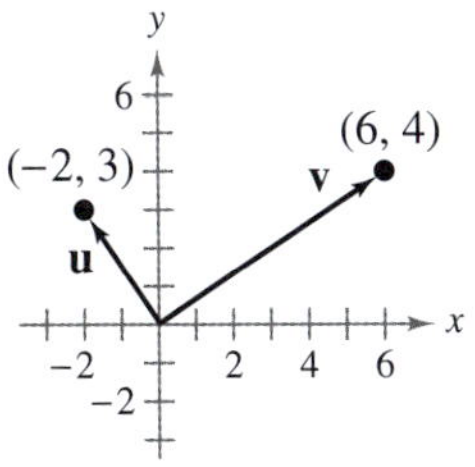

64.

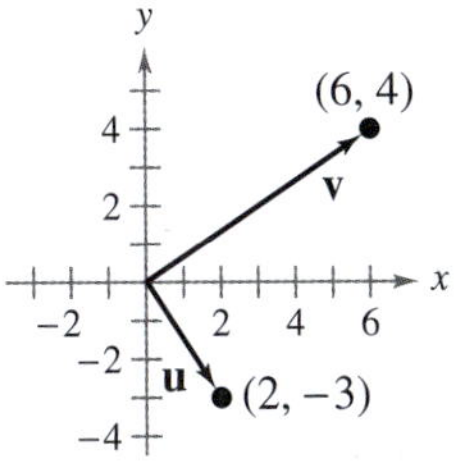

Finding Orthogonal Vectors In Exercises 65–68, find two vectors in opposite directions that are orthogonal to the vector **u**. (There are many correct answers.)

65. $\mathbf{u} = \langle 2, 6 \rangle$

66. $\mathbf{u} = \langle -7, 5 \rangle$

67. $\mathbf{u} = \frac{1}{2}\mathbf{i} - \frac{3}{4}\mathbf{j}$

68. $\mathbf{u} = -\frac{5}{2}\mathbf{i} - 3\mathbf{j}$

Finding the Work Done In Exercises 69 and 70, find the work done in moving a particle from P to Q when the magnitude and direction of the force are given by **v**.

69. $P = (0, 0)$, $Q = (4, 7)$, $\mathbf{v} = \langle 1, 4 \rangle$

70. $P = (1, 3)$, $Q = (-3, 5)$, $\mathbf{v} = -2\mathbf{i} + 3\mathbf{j}$

71. Business The vector $\mathbf{u} = \langle 1225, 2445 \rangle$ gives the numbers of hours worked by employees of a temp agency at two pay levels. The vector $\mathbf{v} = \langle 12.00, 10.25 \rangle$ gives the hourly wage (in dollars) paid at each level, respectively.

(a) Find the dot product $\mathbf{u} \cdot \mathbf{v}$ and explain its meaning in the context of the problem.

(b) Identify the vector operation used to increase wages by 2 percent.

72. Business The vector $\mathbf{u} = \langle 8, 13 \rangle$ gives the numbers of hats and sweatshirts, respectively, sold at a fundraiser. The vector $\mathbf{v} = \langle 13.25, 20.50 \rangle$ gives the prices (in dollars) of the apparel.

(a) Find the dot product $\mathbf{u} \cdot \mathbf{v}$ and explain its meaning in the context of the problem.

(b) Identify the vector operation used to decrease prices by 0.5 percent.

73. *Why you should learn it* *(p. 436)* A truck with a gross weight of 30,000 pounds is parked on a slope of $d°$ (see figure). Assume that the only force to overcome is the force of gravity.

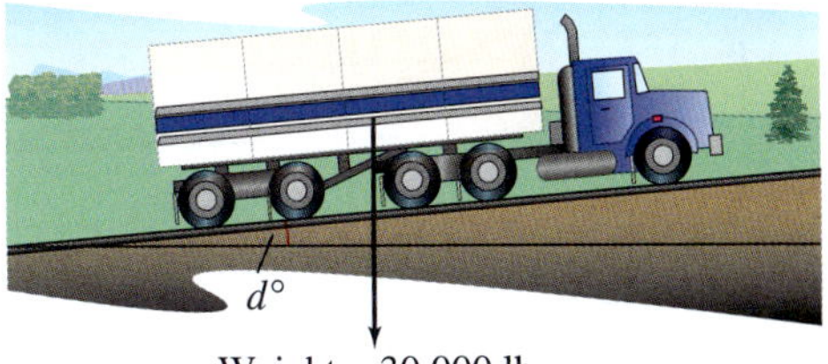

(a) Find the force required to keep the truck from rolling down the hill in terms of the slope d.

(b) Use a graphing utility to complete the table.

d	0°	1°	2°	3°	4°	5°
Force						

d	6°	7°	8°	9°	10°
Force					

(c) Find the force perpendicular to the hill when $d = 5°$.

74. Physics A sport utility vehicle with a gross weight of 5400 pounds is parked on a slope of 10°. Assume that the only force to overcome is the force of gravity. Find the force required to keep the vehicle from rolling down the hill. Find the force perpendicular to the hill.

75. MODELING DATA

One of the events in a local strongman contest is to drag a concrete block. One competitor drags the block with a constant force of 250 pounds at a constant angle of 30° with the horizontal (see figure).

(a) Find the work done in terms of the distance d.

(b) Use a graphing utility to complete the table.

d (in feet)	25	50	100
Work (in foot-pounds)			

76. Public Safety A ski patroller pulls a rescue toboggan across a flat snow surface by exerting a constant force of 35 pounds on a handle that makes a constant angle of 22° with the horizontal (see figure). Find the work done in pulling the toboggan 200 feet.

77. Work A force of 50 pounds, exerted at an angle of 25° with the horizontal, is required to slide a desk across a floor. Determine the work done in sliding the desk 15 feet.

78. Work A mover exerts a horizontal force of 25 pounds on a crate as it is pushed up a ramp that is 12 feet long and inclined at an angle of 20° above the horizontal. Find the work done in pushing the crate up the ramp.

Focusing on Concepts

True or False? **In Exercises 79 and 80, determine whether the statement is true or false. Justify your answer.**

79. The vectors $\mathbf{u} = \langle 0, 0 \rangle$ and $\mathbf{v} = \langle -12, 6 \rangle$ are orthogonal.

80. The work W done by a constant force $\mathbf{F}$ acting along the line of motion of an object is represented by a vector.

81. Think About It If $\mathbf{u} = \langle \cos\theta, \sin\theta \rangle$ and $\mathbf{v} = \langle \sin\theta, -\cos\theta \rangle$, are $\mathbf{u}$ and $\mathbf{v}$ orthogonal, parallel, or neither? Explain.

82. Error Analysis Describe the error.

$$\langle 5, 8 \rangle \cdot \langle -2, 7 \rangle = \langle -10, 56 \rangle \quad \times$$

83. Think About It Let $\mathbf{u}$ be a unit vector. What is the value of $\mathbf{u} \cdot \mathbf{u}$? Explain.

84. HOW DO YOU SEE IT? What is known about θ, the angle between two nonzero vectors $\mathbf{u}$ and $\mathbf{v}$ (see figure), under each condition?

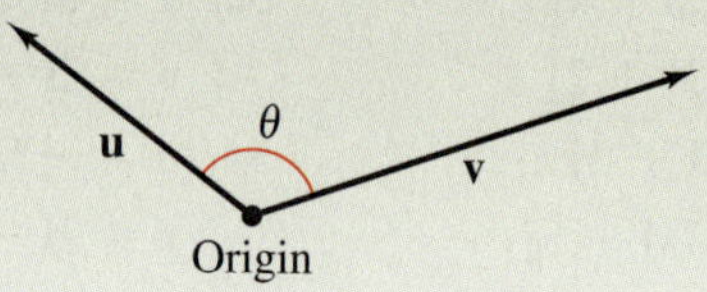

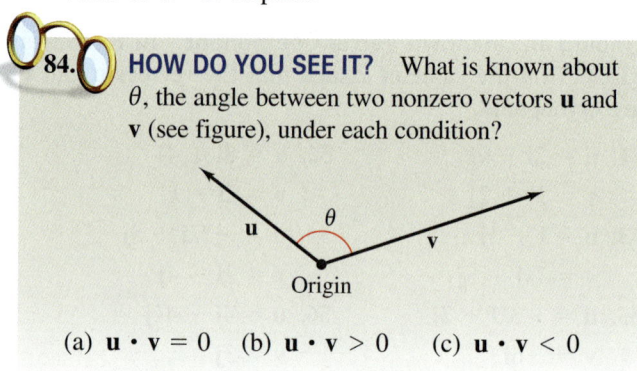

(a) $\mathbf{u} \cdot \mathbf{v} = 0$ (b) $\mathbf{u} \cdot \mathbf{v} > 0$ (c) $\mathbf{u} \cdot \mathbf{v} < 0$

85. Think About It What can be said about the vectors $\mathbf{u}$ and $\mathbf{v}$ under each condition?

(a) The projection of $\mathbf{u}$ onto $\mathbf{v}$ equals $\mathbf{u}$.

(b) The projection of $\mathbf{u}$ onto $\mathbf{v}$ equals $\mathbf{0}$.

86. Proof Use vectors to prove that the diagonals of a rhombus are perpendicular.

87. Proof Prove the following.

$$\|\mathbf{u} - \mathbf{v}\|^2 = \|\mathbf{u}\|^2 + \|\mathbf{v}\|^2 - 2\mathbf{u} \cdot \mathbf{v}$$

88. Proof Prove that if $\mathbf{u}$ is orthogonal to $\mathbf{v}$ and $\mathbf{w}$, then $\mathbf{u}$ is orthogonal to $c\mathbf{v} + d\mathbf{w}$ for any scalars c and d.

89. Proof Prove that if $\mathbf{u}$ is a unit vector that lies in Quadrant I and θ is the angle between $\mathbf{u}$ and $\mathbf{i}$, then $\mathbf{u} = \cos\theta\mathbf{i} + \sin\theta\mathbf{j}$.

90. Proof Prove that if $\mathbf{u}$ is a unit vector that lies in Quadrant I and θ is the angle between $\mathbf{u}$ and $\mathbf{j}$, then

$$\mathbf{u} = \cos\left(\frac{\pi}{2} - \theta\right)\mathbf{i} + \sin\left(\frac{\pi}{2} - \theta\right)\mathbf{j}.$$

Cumulative Mixed Review

Transformation of a Graph **In Exercises 91–94, describe how the graph of g is related to the graph of f.**

91. $g(x) = f(x - 4)$ **92.** $g(x) = -f(x)$

93. $g(x) = f(x) + 6$ **94.** $g(x) = f(2x)$

Performing Operations with Complex Numbers **In Exercises 95–98, perform the operation and write the result in standard form.**

95. $(1 + 3i)(1 - 3i)$ **96.** $(7 - 4i)(7 + 4i)$

97. $\dfrac{3}{1 + i} + \dfrac{2}{2 - 3i}$ **98.** $\dfrac{6}{4 - i} - \dfrac{3}{1 + i}$

6.5 The Complex Plane

The Complex Plane

Just as real numbers can be represented by points on the real number line, you can represent a complex number

$z = a + bi$ The complex number z

as the point (a, b) in a coordinate plane called the **complex plane**. In the complex plane, the horizontal axis is called the **real axis** and the vertical axis is called the **imaginary axis,** as shown in the figure.

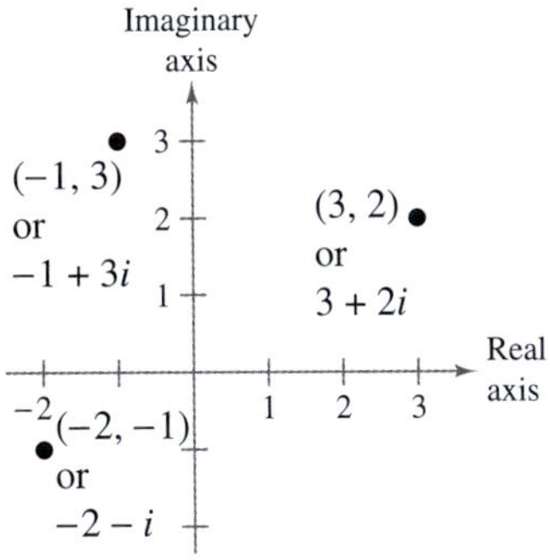

Points plotted in the complex plane

The **absolute value** (or **modulus**) **of a complex number** $a + bi$ is defined as the distance between the origin $(0, 0)$ and the point (a, b).

Definition of the Absolute Value of a Complex Number

The **absolute value** of the complex number $z = a + bi$ is given by

$$|a + bi| = \sqrt{a^2 + b^2}.$$

When the complex number $a + bi$ is a real number (that is, when $b = 0$), this definition agrees with that given for the absolute value of a real number

$$|a + 0i| = \sqrt{a^2 + 0^2} = |a|.$$

EXAMPLE 1 Finding the Absolute Value of a Complex Number

See LarsonPrecalculus.com for an interactive version of this type of example.

Plot $z = -2 + 5i$ and find its absolute value.

Solution

The complex number

$$z = -2 + 5i$$

is plotted in the figure. The absolute value of z is

$$|z| = \sqrt{(-2)^2 + 5^2}$$
$$= \sqrt{29}.$$

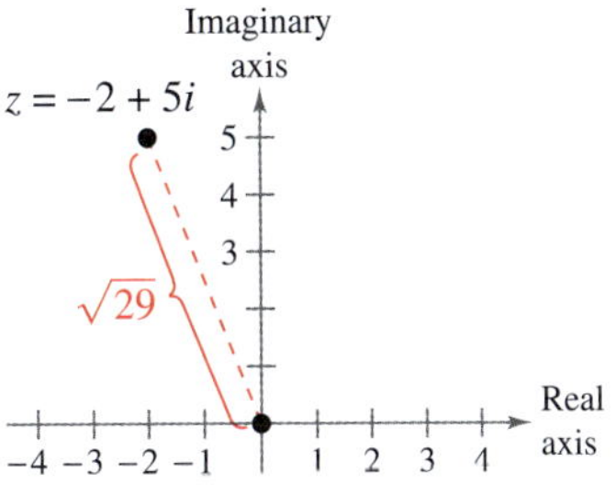

✓ Checkpoint *Audio-video solution in English & Spanish at LarsonPrecalculus.com*

Plot $z = 3 - 4i$ and find its absolute value.

What you should learn

- Plot complex numbers in the complex plane and find absolute values of complex numbers.
- Perform operations with complex numbers in the complex plane.
- Use the Distance and Midpoint Formulas in the complex plane.

Why you should learn it

Digital signals for medical imaging can be represented by equations involving complex numbers. For instance, in Exercise 49 on page 450, you will graph the complex part of an equation of a digital signal.

szefei/Shutterstock.com

Operations with Complex Numbers in the Complex Plane

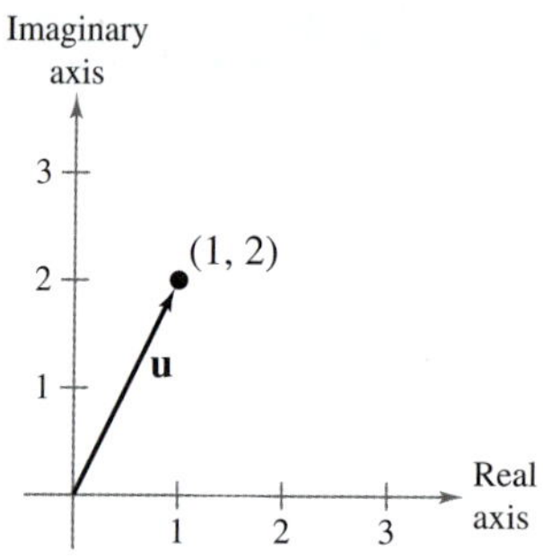

Figure 6.32

In Section 6.3, you learned how to add and subtract vectors geometrically in the coordinate plane. In a similar way, you can add and subtract complex numbers geometrically in the complex plane.

The complex number $z = a + bi$ can be represented by the vector $\mathbf{u} = \langle a, b\rangle$. For instance, in Figure 6.32, the complex number $z = 1 + 2i$ is represented by the vector $\mathbf{u} = \langle 1, 2\rangle$.

To add two complex numbers $z_1 = a + bi$ and $z_2 = c + di$ geometrically, first represent the numbers as vectors $\mathbf{u} = \langle a, b\rangle$ and $\mathbf{v} = \langle c, d\rangle$, respectively (see Figure 6.33). Then use the parallelogram law for vector addition to find $\mathbf{u} + \mathbf{v}$, as shown in Figure 6.34. Because $\mathbf{u} + \mathbf{v}$ represents $z_1 + z_2$, it follows that

$$z_1 + z_2 = (a + c) + (b + d)i.$$

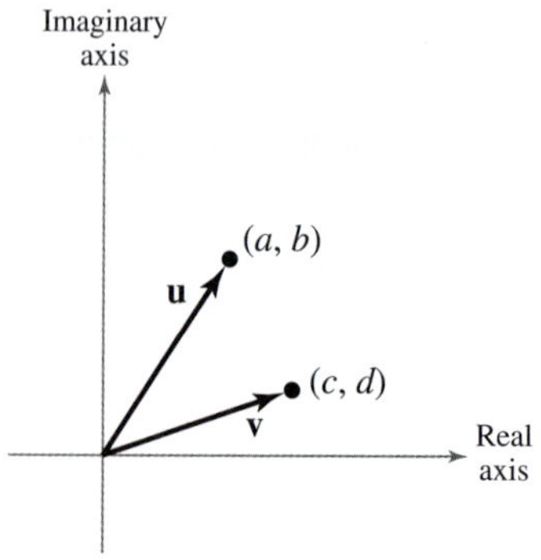

Represent $z_1 = a + bi$ and $z_2 = c + di$ as vectors $\mathbf{u} = \langle a, b\rangle$ and $\mathbf{v} = \langle c, d\rangle$

Figure 6.33

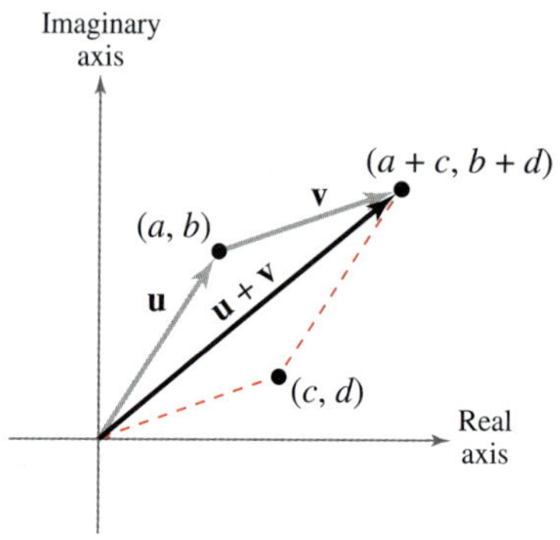

Use parallelogram law for vector addition

Figure 6.34

EXAMPLE 2 Adding Complex Numbers Geometrically

Find $(1 + 3i) + (2 + i)$ geometrically.

Solution

Begin by writing a vector to represent each complex number.

Complex number	*Vector*
$1 + 3i$	$\mathbf{u} = \langle 1, 3\rangle$
$2 + i$	$\mathbf{v} = \langle 2, 1\rangle$

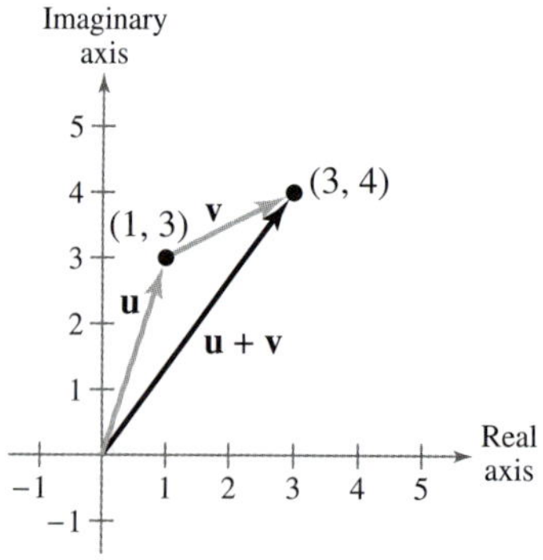

Graph **u** and **v**, where the initial point of **v** coincides with the terminal point of **u**, as shown in the figure. Next, graph $\mathbf{u} + \mathbf{v}$. From the figure, $\mathbf{u} + \mathbf{v} = \langle 3, 4\rangle$, which implies that

$$(1 + 3i) + (2 + i) = 3 + 4i.$$

✓ **Checkpoint** Audio-video solution in English & Spanish at LarsonPrecalculus.com

Find $(3 + i) + (1 + 2i)$ geometrically.

To subtract two complex numbers geometrically, first represent the numbers as vectors **u** and **v**. Then subtract the vectors, as shown in Figure 6.35. The difference of the vectors represents the difference of the complex numbers.

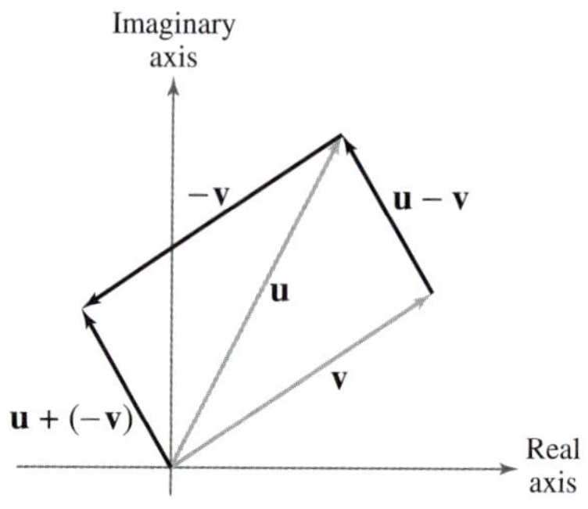

Figure 6.35

EXAMPLE 3 Subtracting Complex Numbers Geometrically

Find $(4 + 2i) - (3 - i)$ geometrically.

Solution

Begin by writing a vector to represent each complex number.

Complex number	*Vector*
$4 + 2i$	$\mathbf{u} = \langle 4, 2\rangle$
$3 - i$	$\mathbf{v} = \langle 3, -1\rangle$

Because $\mathbf{u} - \mathbf{v} = \mathbf{u} + (-\mathbf{v})$, graph $\mathbf{u}$ and $-\mathbf{v} = \langle -3, 1\rangle$, where the initial point of $-\mathbf{v}$ coincides with the terminal point of $\mathbf{u}$, as shown in the figure. Next, graph $\mathbf{u} + (-\mathbf{v})$. From the figure, $\mathbf{u} + (-\mathbf{v}) = \langle 1, 3\rangle$, which implies that

$$(4 + 2i) - (3 - i) = 1 + 3i.$$

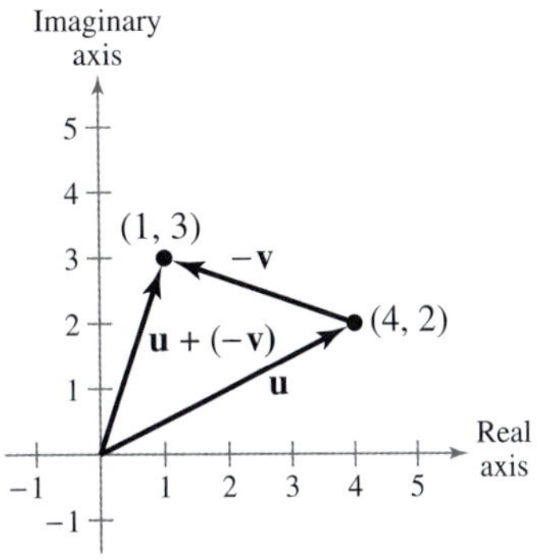

Checkpoint Audio-video solution in English & Spanish at LarsonPrecalculus.com

Find $(2 - 4i) - (1 + i)$ geometrically.

Recall from Section 2.4 that the complex numbers $a + bi$ and $a - bi$ are *complex conjugates*. Plotting each complex number reveals that the points (a, b) and $(a, -b)$ are reflections of each other in the real axis (see Figure 6.36). So, given a complex number, you can find its complex conjugate geometrically, as shown in Example 4.

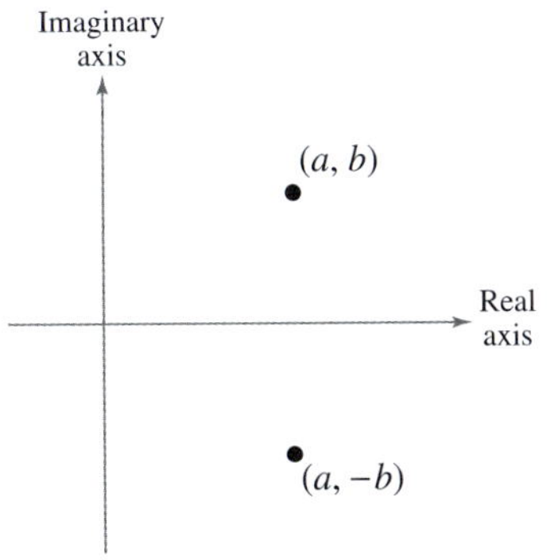

Figure 6.36

EXAMPLE 4 Finding a Complex Conjugate Geometrically

Find the complex conjugate of $-3 + i$ geometrically.

Solution

Begin by plotting $-3 + i$. As shown in Figure 6.37, the reflection of $(-3, 1)$ in the real axis is $(-3, -1)$. So, the complex numbers

$$-3 + i \qquad \text{and} \qquad -3 - i$$

are complex conjugates.

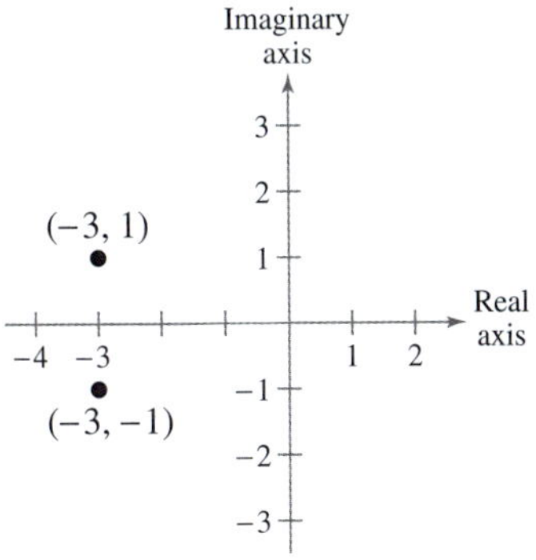

Figure 6.37

Checkpoint Audio-video solution in English & Spanish at LarsonPrecalculus.com

Find the complex conjugate of $2 - 3i$ geometrically.

Distance and Midpoint Formulas in the Complex Plane

Consider the complex numbers $z_1 = a + bi$ and $z_2 = c + di$, and their points in the complex plane. The distance between the points is the modulus (or absolute value) of the difference of the two corresponding complex numbers.

Distance Formula in the Complex Plane

The distance between $z_1 = a + bi$ and $z_2 = c + di$ in the complex plane is given by

$$|z_1 - z_2| = \sqrt{(a - c)^2 + (b - d)^2}.$$

EXAMPLE 5 Finding a Distance

Plot $z_1 = 2 + 3i$ and $z_2 = 5 - 2i$ and find the distance between the points.

Solution

The complex numbers $z_1 = 2 + 3i$ and $z_2 = 5 - 2i$ are plotted in Figure 6.38. The distance between the points is

$$|z_1 - z_2| = \sqrt{(2 - 5)^2 + [3 - (-2)]^2} = \sqrt{(-3)^2 + 5^2} = \sqrt{34} \approx 5.83 \text{ units.}$$

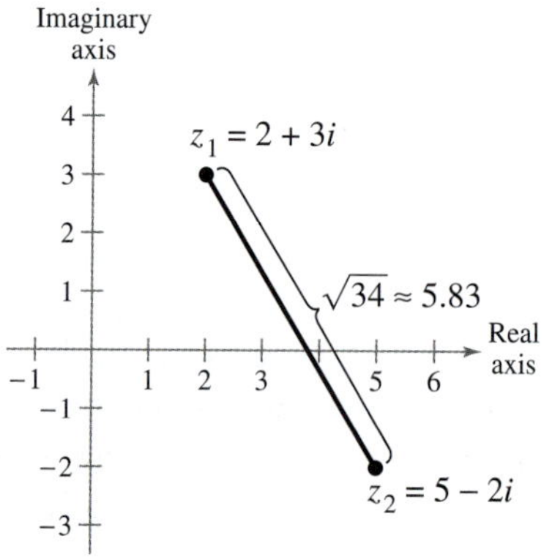

Figure 6.38

Checkpoint Audio-video solution in English & Spanish at LarsonPrecalculus.com

Plot $z_1 = 5 - 4i$ and $z_2 = 6 + 5i$ and find the distance between the points.

To find the midpoint of the line segment joining two points in the complex plane, find the average values of the respective coordinates of the two endpoints.

Midpoint Formula in the Complex Plane

The midpoint of the line segment joining $z_1 = a + bi$ and $z_2 = c + di$ in the complex plane is given by

$$\text{Midpoint} = \left(\frac{a + c}{2}, \frac{b + d}{2}\right).$$

EXAMPLE 6 Finding the Midpoint of a Line Segment

Plot $z_1 = 4 - 3i$ and $z_2 = 2 + 2i$ and find the midpoint of the line segment joining the points.

Solution

The complex numbers $z_1 = 4 - 3i$ and $z_2 = 2 + 2i$ are plotted in Figure 6.39. The midpoint of the line segment joining the points is

$$\text{Midpoint} = \left(\frac{4 + 2}{2}, \frac{-3 + 2}{2}\right) \quad \text{Apply Midpoint Formula.}$$

$$= \left(3, -\frac{1}{2}\right). \quad \text{Simplify.}$$

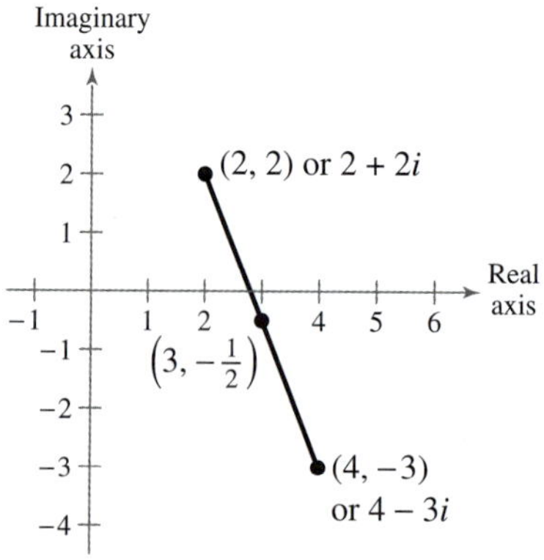

Figure 6.39

Checkpoint Audio-video solution in English & Spanish at LarsonPrecalculus.com

Plot $z_1 = 2 + i$ and $z_2 = 5 - 5i$ and find the midpoint of the line segment joining the points.

6.5 Exercises

See *CalcChat.com* for tutorial help and worked-out solutions to odd-numbered exercises.
For instructions on how to use a graphing utility, see Appendix A.

Vocabulary and Concept Check

In Exercises 1–4, fill in the blank.

1. The complex number $z = a + bi$ is represented by the point _______ in the complex plane.

2. The point $(0, b)$, $b \neq 0$ lies on the _______ axis of the complex plane.

3. The _______ of a complex number $a + bi$ is the distance between $(0, 0)$ and (a, b).

4. To add or subtract complex numbers geometrically, begin by representing each number as a _______ .

Procedures and Problem Solving

Matching In Exercises 5–12, match the complex number with its representation in the complex plane. [The representations are labeled (a)–(h).]

(a)

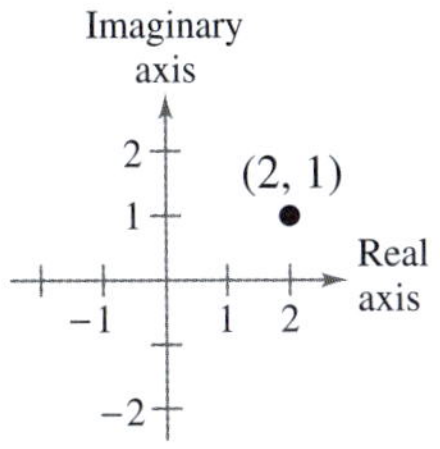

(b)

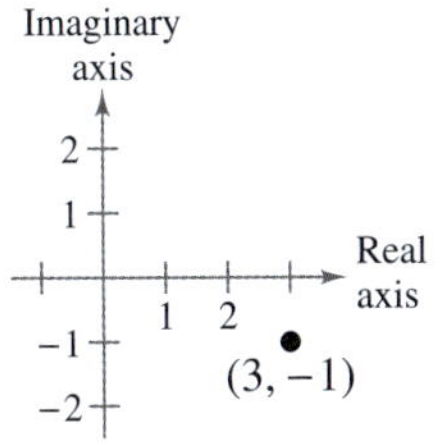

(c)

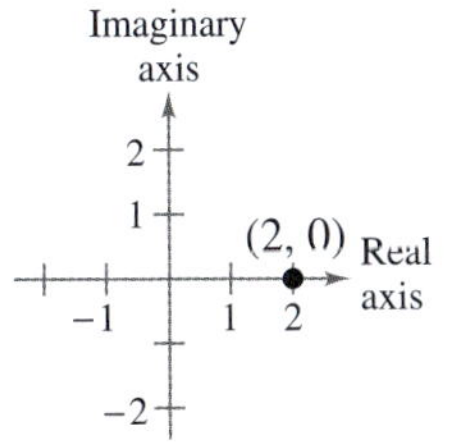

(d)

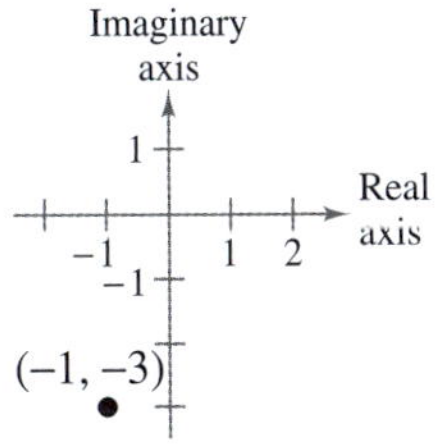

(e)

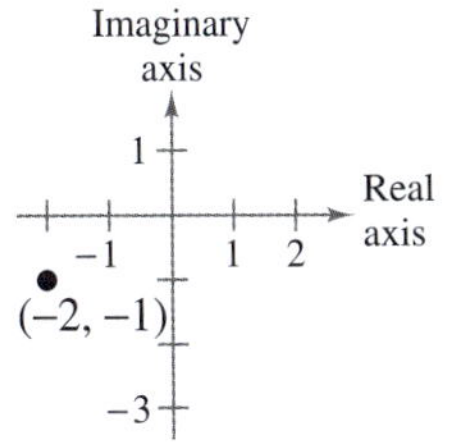

(f)

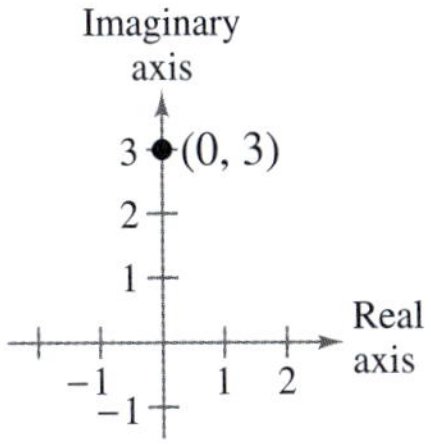

(g)

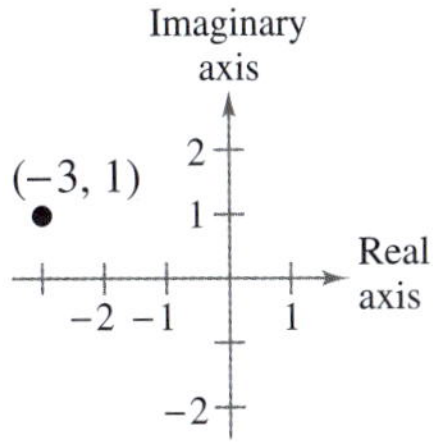

(h)

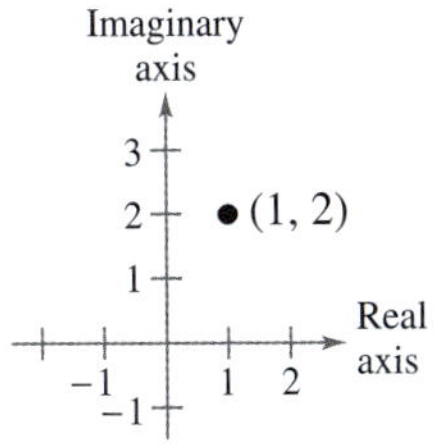

5. 2 **6.** $3i$
7. $1 + 2i$ **8.** $2 + i$
9. $3 - i$ **10.** $-3 + i$
11. $-2 - i$ **12.** $-1 - 3i$

Finding the Absolute Value of a Complex Number In Exercises 13–18, plot the complex number and find its absolute value.

13. $-7i$ **14.** -7
15. $-6 + 8i$ **16.** $5 - 12i$
17. $4 - 6i$ **18.** $-8 + 3i$

Comparing Absolute Values In Exercises 19 and 20, determine which complex number is the greatest distance from the origin.

19. $z_1 = 3 + 2i$
$z_2 = 4 - i$
$z_3 = i$

20. $z_1 = -5 + i$
$z_2 = -7i$
$z_3 = 6 + 4i$

Approximating an Absolute Value In Exercises 21–24, use a graphing utility to approximate the absolute value of the complex number to two decimal places.

21. $z = 4.2 + 0.7i$
22. $z = -8.4 + 3.9i$
23. $z = \sqrt{3} + 0.4i$
24. $z = -3.2 + \sqrt{2}i$

Adding Complex Numbers Geometrically In Exercises 25–30, find the sum of the complex numbers geometrically.

25. $(3 + i) + (2 + 5i)$ **26.** $(5 + 2i) + (3 + 4i)$
27. $(8 - 2i) + (2 + 6i)$ **28.** $(3 - i) + (-1 + 2i)$
29. $3 + (2 - 8i)$ **30.** $2i + (3 - i)$

Subtracting Complex Numbers Geometrically In Exercises 31–36, find the difference of the complex numbers geometrically.

31. $(4 + 2i) - (6 + 4i)$ **32.** $(-3 + i) - (3 + i)$
33. $(5 - i) - (-5 + 2i)$ **34.** $(2 - 3i) - (3 + 2i)$
35. $2 - (2 + 6i)$ **36.** $-3i - (2 + 2i)$

Finding a Complex Conjugate Geometrically **In Exercises 37–40, plot the complex number and its complex conjugate. Write the conjugate as a complex number.**

37. $2 + 3i$ **38.** $5 - 4i$

39. $-1 - 2i$ **40.** $-7 + 3i$

Finding a Distance **In Exercises 41–44, find the distance between the complex numbers in the complex plane.**

41. $1 + 2i, \quad -1 + 4i$

42. $-5 + i, \quad -2 + 5i$

43. $6i, \quad 3 - 4i$

44. $-7 - 3i, \quad 3 + 5i$

Finding the Midpoint of a Line Segment **In Exercises 45–48, find the midpoint of the line segment joining the points corresponding the complex numbers in the complex plane.**

45. $2 + i, \quad 6 + 5i$

46. $-3 + 4i, \quad 1 - 2i$

47. $7i, \quad 9 - 10i$

48. $-1 - \frac{3}{4}i, \quad \frac{1}{2} + \frac{1}{4}i$

49. ***Why you should learn it*** *(p. 445)* A concise expression involving i, such as e^{it}, is an ideal way to represent a digital signal used in medical imaging. Expressions like e^{it} are often easy to manipulate in calculus.

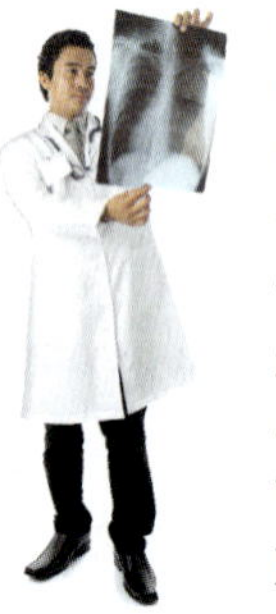

The points $(t, I(t))$ in the complex plane represent the complex part of an equation of a digital signal, where t is the time and $I(t)$ is the output.

Plot the points $(1, 0.54)$, $(2, -0.42)$, $(3, -0.99)$, $(4, -0.65)$, $(5, 0.28)$, and $(6, 0.96)$ in the complex plane and connect them with a smooth curve.

50. Audio Signal The points $(1, 0.84)$, $(2, 0.91)$, $(3, 0.14)$, $(4, -0.76)$, $(5, -0.96)$, and $(6, -0.28)$ in the complex plane represent the complex part of an equation of an audio signal. Plot the points and connect them with a smooth curve.

Force **In Exercises 51 and 52, the horizontal and vertical components of two forces acting on an object are given (in newtons). Use the complex plane to plot the vectors that represent the two forces. Then find the horizontal and vertical components of the resultant force.**

51.	Force A	horizontal: 3 N	vertical: 4 N
	Force B	horizontal: 4 N	vertical: 2 N
52.	Force A	horizontal: 2 N	vertical: 1 N
	Force B	horizontal: 5 N	vertical: 3 N

Focusing on Concepts

True or False? **In Exercises 53–56, determine whether the statement is true or false. Justify your answer.**

53. The modulus of a complex number can be real or imaginary.

54. The distance between two points in the complex plane is always real.

55. The modulus of the sum of two complex numbers is equal to the sum of their moduli.

56. The modulus of the difference of two complex numbers is equal to the difference of their moduli.

The Complex Plane **In Exercises 57–60, consider a real number a, $a > 0$. Describe the process geometrically as a transformation in the complex plane. Then sketch a and the image of a.**

57. Multiply a by i. **58.** Multiply a by $-i$.

59. Multiply a by i^2. **60.** Multiply a by $-i^2$.

61. Think About It What does the set of all points with the same modulus represent in the complex plane?

62. HOW DO YOU SEE IT? Determine which graph represents each expression.

(a) $(a + bi) + (a - bi)$

(b) $(a + bi) - (a - bi)$

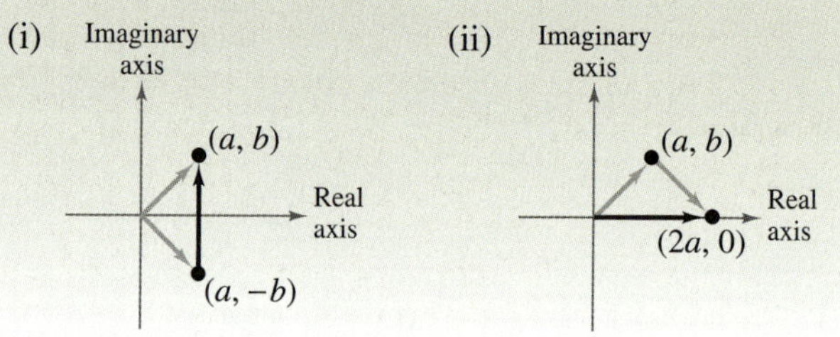

63. Think About It The points corresponding to a complex number and its complex conjugate are plotted in the complex plane. What type of triangle do these points form with the origin?

64. Error Analysis Describe the error.

$z_1 = 2 + 3i$ and $z_2 = 2 - 3i$ are complex conjugates because z_2 is a reflection of z_1 in the imaginary axis.

Cumulative Mixed Review

Verifying a Trigonometric Identity **In Exercises 65–68, verify the identity algebraically. Use a graphing utility to check your result graphically.**

65. $-\sin x \csc x = -1$ **66.** $\tan^2 y(\csc^2 y - 1) = 1$

67. $\dfrac{\cot x}{\cos x} = \csc x$ **68.** $\dfrac{1 - \sin^2 y}{\cos y} = \cos y$

6.6 Trigonometric Form of a Complex Number

Trigonometric Form of a Complex Number

In Section 2.4, you learned how to add, subtract, multiply, and divide complex numbers. To work effectively with *powers* and *roots* of complex numbers, it is helpful to write complex numbers in trigonometric form. Consider the nonzero complex number $a + bi$, plotted at the right. By letting θ be the angle from the positive real axis (measured counterclockwise) to the line segment connecting the origin and the point (a, b), you can write

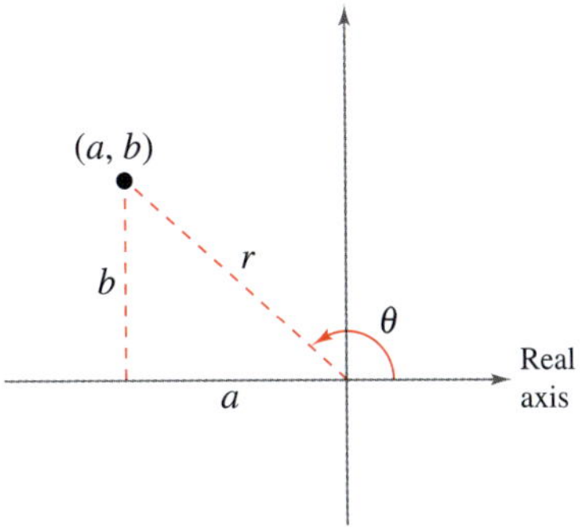

$$a = r\cos\theta \quad \text{and} \quad b = r\sin\theta$$

where $r = \sqrt{a^2 + b^2}$. Consequently,

$$a + bi = (r\cos\theta) + (r\sin\theta)i$$

from which you can obtain the **trigonometric form of a complex number.**

Trigonometric Form of a Complex Number

The **trigonometric form** of the complex number $z = a + bi$ is given by

$$z = r(\cos\theta + i\sin\theta)$$

where $a = r\cos\theta$, $b = r\sin\theta$, $r = \sqrt{a^2 + b^2}$, and $\tan\theta = b/a$. The number r is the **modulus** of z, and θ is called an **argument** of z.

The trigonometric form of a complex number is also called the *polar form*. Because there are infinitely many choices for θ, the trigonometric form of a complex number is not unique. Normally, θ is restricted to the interval $0 \le \theta < 2\pi$, although on occasion it is convenient to use $\theta < 0$.

What you should learn

- Write trigonometric forms of complex numbers.
- Multiply and divide complex numbers written in trigonometric form.
- Use DeMoivre's Theorem to find powers of complex numbers.
- Find *n*th roots of complex numbers.

Why you should learn it

You can use the trigonometric form of a complex number to perform operations with complex numbers. For instance, in Exercises 131–134 on page 461, you will use the trigonometric forms of complex numbers to find voltage, current, or impedance for given conditions.

EXAMPLE 1 Trigonometric Form of a Complex Number

Write the complex number $z = -2i$ in trigonometric form.

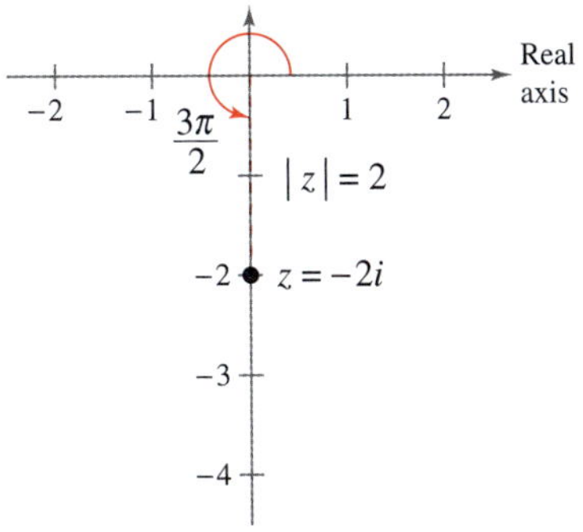

Solution

The modulus of z is

$$r = |-2i| = \sqrt{0^2 + (-2)^2} = \sqrt{4} = 2.$$

With $a = 0$, you cannot use $\tan\theta = b/a$ to find θ. Because $z = -2i$ lies on the negative imaginary axis (see figure), choose $\theta = 3\pi/2$. So, the trigonometric form is

$$z = r(\cos\theta + i\sin\theta)$$

$$= 2\left(\cos\frac{3\pi}{2} + i\sin\frac{3\pi}{2}\right).$$

Checkpoint *Audio-video solution in English & Spanish at LarsonPrecalculus.com*

Write the complex number $z = 3i$ in trigonometric form.

Phovoir/Shutterstock.com

EXAMPLE 2 Trigonometric Form of a Complex Number

Write the complex number $z = -2 - 2\sqrt{3}i$ in trigonometric form.

Solution

The modulus of z is

$$r = |-2 - 2\sqrt{3}i| = \sqrt{(-2)^2 + (-2\sqrt{3})^2} = \sqrt{16} = 4$$

and the angle θ is determined from

$$\tan\theta = \frac{b}{a} = \frac{-2\sqrt{3}}{-2} = \sqrt{3}.$$

The complex number $z = -2 - 2\sqrt{3}i$ lies in Quadrant III, as shown in the figure below. So, the angle θ is

$$\theta = \pi + \arctan\sqrt{3} = \pi + \frac{\pi}{3} = \frac{4\pi}{3}$$

and the trigonometric form of z is

$$z = r(\cos\theta + i\sin\theta) = 4\left(\cos\frac{4\pi}{3} + i\sin\frac{4\pi}{3}\right).$$

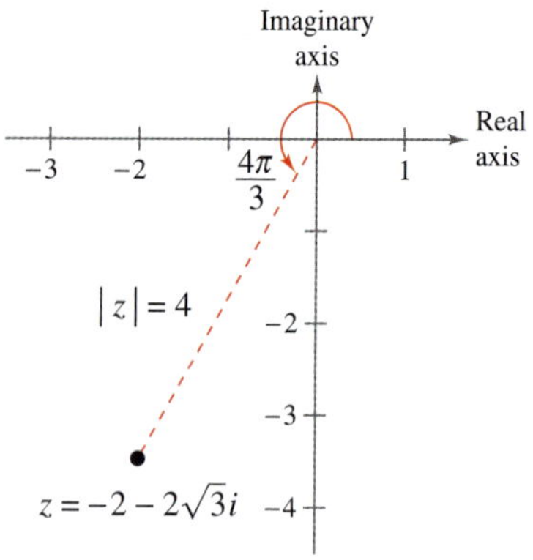

Checkpoint Audio-video solution in English & Spanish at LarsonPrecalculus.com

Write the complex number $z = 6 - 6i$ in trigonometric form.

What's Wrong?

You use a graphing utility to check the answer to Example 2, as shown in the figure. You determine that $r = 4$ and $\theta \approx 1.047 \approx \pi/3$. Your value for θ does not agree with the value found in the example. What's wrong?

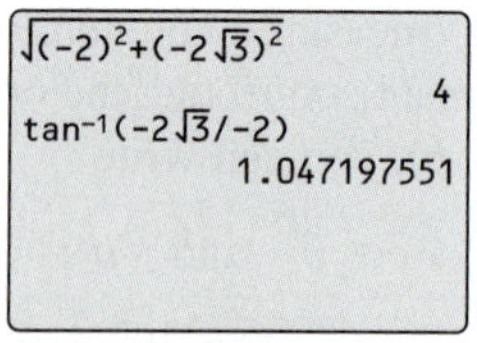

EXAMPLE 3 Writing a Complex Number in Standard Form

Write the complex number in standard form $a + bi$.

$$z = \sqrt{8}\left[\cos\left(-\frac{\pi}{3}\right) + i\sin\left(-\frac{\pi}{3}\right)\right]$$

Solution

Because $\cos(-\pi/3) = 1/2$ and $\sin(-\pi/3) = -\sqrt{3}/2$, you can write

$$z = \sqrt{8}\left[\cos\left(-\frac{\pi}{3}\right) + i\sin\left(-\frac{\pi}{3}\right)\right] \quad \text{Trigonometric form}$$

$$= 2\sqrt{2}\left(\frac{1}{2} - \frac{\sqrt{3}}{2}i\right) \quad \text{Simplify.}$$

$$= \sqrt{2} - \sqrt{6}i. \quad \text{Standard form}$$

Checkpoint Audio-video solution in English & Spanish at LarsonPrecalculus.com

Write $z = 8[\cos(2\pi/3) + i\sin(2\pi/3)]$ in standard form $a + bi$.

Technology Tip

A graphing utility can be used to convert a complex number in trigonometric form to standard form. For instance, enter the complex number $\sqrt{2}(\cos\pi/4 + i\sin\pi/4)$ in your graphing utility and press ENTER. You should obtain the standard form $1 + i$, as shown below.

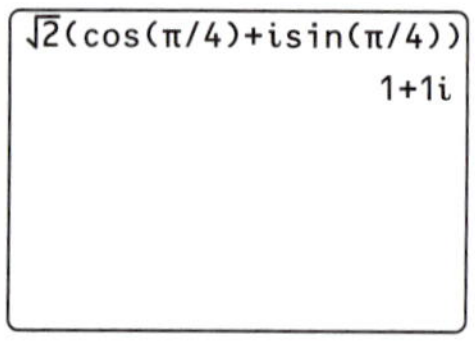

Multiplication and Division of Complex Numbers

The trigonometric form adapts nicely to multiplication and division of complex numbers. For two complex numbers

$$z_1 = r_1(\cos\theta_1 + i\sin\theta_1) \quad \text{and} \quad z_2 = r_2(\cos\theta_2 + i\sin\theta_2)$$

the product of z_1 and z_2 is

$$\begin{aligned} z_1z_2 &= r_1r_2(\cos\theta_1 + i\sin\theta_1)(\cos\theta_2 + i\sin\theta_2) \\ &= r_1r_2[(\cos\theta_1\cos\theta_2 - \sin\theta_1\sin\theta_2) + i(\sin\theta_1\cos\theta_2 + \cos\theta_1\sin\theta_2)] \\ &= r_1r_2[\cos(\theta_1+\theta_2) + i\sin(\theta_1+\theta_2)]. \end{aligned}$$

Sum and difference formulas

This establishes the first part of the following rule. The second part is left for you to verify (see Exercise 139).

Product and Quotient of Two Complex Numbers

Let $z_1 = r_1(\cos\theta_1 + i\sin\theta_1)$ and $z_2 = r_2(\cos\theta_2 + i\sin\theta_2)$ be complex numbers.

$$z_1z_2 = r_1r_2[\cos(\theta_1+\theta_2) + i\sin(\theta_1+\theta_2)]$$ Product

$$\frac{z_1}{z_2} = \frac{r_1}{r_2}[\cos(\theta_1-\theta_2) + i\sin(\theta_1-\theta_2)],\ z_2 \neq 0$$ Quotient

Note that this rule says that to *multiply* two complex numbers you multiply moduli and add arguments, whereas to *divide* two complex numbers you divide moduli and subtract arguments.

EXAMPLE 4 Multiplying Complex Numbers

Find the product z_1z_2 of the complex numbers.

$$z_1 = 3\left(\cos\frac{\pi}{4} + i\sin\frac{\pi}{4}\right)$$

$$z_2 = 2\left(\cos\frac{3\pi}{4} + i\sin\frac{3\pi}{4}\right)$$

Solution

$$\begin{aligned} z_1z_2 &= 3\left(\cos\frac{\pi}{4} + i\sin\frac{\pi}{4}\right) \cdot 2\left(\cos\frac{3\pi}{4} + i\sin\frac{3\pi}{4}\right) \\ &= 6\left[\cos\left(\frac{\pi}{4}+\frac{3\pi}{4}\right) + i\sin\left(\frac{\pi}{4}+\frac{3\pi}{4}\right)\right] \\ &= 6(\cos\pi + i\sin\pi) \\ &= 6[-1 + i(0)] \\ &= -6 \end{aligned}$$

The numbers z_1, z_2, and z_1z_2 are plotted in the figure. Note that multiplying z_1 by z_2 rotates the graph of z_1 about the origin $3\pi/4$ radians and doubles the distance z_1 is from the origin.

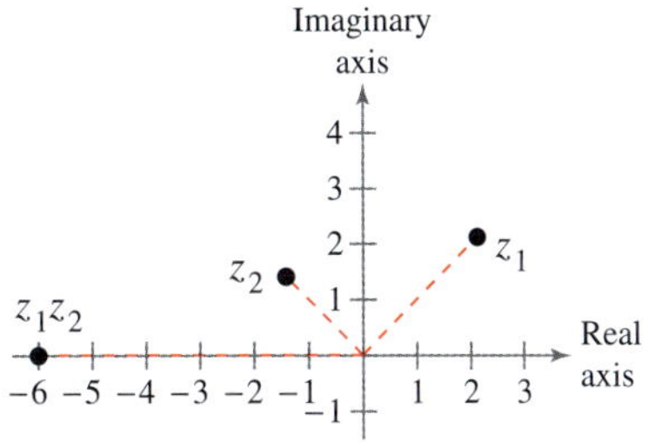

Checkpoint Audio-video solution in English & Spanish at LarsonPrecalculus.com

Find the product z_1z_2 of $z_1 = 2\left(\cos\frac{5\pi}{6} + i\sin\frac{5\pi}{6}\right)$ and $z_2 = 5\left(\cos\frac{7\pi}{6} + i\sin\frac{7\pi}{6}\right)$.

EXAMPLE 5 Multiplying Complex Numbers

Find the product z_1z_2 of the complex numbers.

$$z_1 = 2\left(\cos\frac{2\pi}{3} + i\sin\frac{2\pi}{3}\right) \qquad z_2 = 8\left(\cos\frac{11\pi}{6} + i\sin\frac{11\pi}{6}\right)$$

Solution

$$z_1z_2 = 2\left(\cos\frac{2\pi}{3} + i\sin\frac{2\pi}{3}\right) \cdot 8\left(\cos\frac{11\pi}{6} + i\sin\frac{11\pi}{6}\right)$$

$$= 16\left[\cos\left(\frac{2\pi}{3} + \frac{11\pi}{6}\right) + i\sin\left(\frac{2\pi}{3} + \frac{11\pi}{6}\right)\right]$$

$$= 16\left(\cos\frac{5\pi}{2} + i\sin\frac{5\pi}{2}\right)$$

$$= 16\left(\cos\frac{\pi}{2} + i\sin\frac{\pi}{2}\right) \qquad \frac{5\pi}{2} \text{ and } \frac{\pi}{2} \text{ are coterminal.}$$

$$= 16[0 + i(1)]$$

$$= 16i$$

You can check this result by first converting to the standard forms

$$z_1 = -1 + \sqrt{3}i \quad \text{and} \quad z_2 = 4\sqrt{3} - 4i$$

and then multiplying algebraically.

$$z_1z_2 = (-1 + \sqrt{3}i)(4\sqrt{3} - 4i) = -4\sqrt{3} + 4i + 12i + 4\sqrt{3} = 16i$$

Checkpoint *Audio-video solution in English & Spanish at LarsonPrecalculus.com*

Find the product z_1z_2 of the complex numbers.

$$z_1 = 3\left(\cos\frac{\pi}{3} + i\sin\frac{\pi}{3}\right) \qquad z_2 = 4\left(\cos\frac{\pi}{6} + i\sin\frac{\pi}{6}\right)$$

Technology Tip

Some graphing utilities can multiply and divide complex numbers in trigonometric form. If you have access to such a graphing utility, use it to find z_1z_2 and z_1/z_2 in Examples 5 and 6.

EXAMPLE 6 Dividing Complex Numbers

Find the quotient z_1/z_2 of the complex numbers.

$$z_1 = 24(\cos 300° + i\sin 300°) \qquad z_2 = 8(\cos 75° + i\sin 75°)$$

Solution

$$\frac{z_1}{z_2} = \frac{24(\cos 300° + i\sin 300°)}{8(\cos 75° + i\sin 75°)}$$

$$= \frac{24}{8}[\cos(300° - 75°) + i\sin(300° - 75°)]$$

$$= 3(\cos 225° + i\sin 225°)$$

$$= 3\left[-\frac{\sqrt{2}}{2} + i\left(-\frac{\sqrt{2}}{2}\right)\right]$$

$$= -\frac{3\sqrt{2}}{2} - \frac{3\sqrt{2}}{2}i$$

Checkpoint *Audio-video solution in English & Spanish at LarsonPrecalculus.com*

Find the quotient z_1/z_2 of the complex numbers.

$$z_1 = \cos 40° + i\sin 40° \qquad z_2 = \cos 10° + i\sin 10°$$

Kursad/E+/Getty Images

Powers of Complex Numbers

The trigonometric form of a complex number can be used to raise a complex number to a power. To accomplish this, consider repeated use of the multiplication rule.

$$z = r(\cos\theta + i\sin\theta)$$
$$z^2 = r(\cos\theta + i\sin\theta)r(\cos\theta + i\sin\theta) = r^2(\cos 2\theta + i\sin 2\theta)$$
$$z^3 = r^2(\cos 2\theta + i\sin 2\theta)r(\cos\theta + i\sin\theta) = r^3(\cos 3\theta + i\sin 3\theta)$$
$$z^4 = r^4(\cos 4\theta + i\sin 4\theta)$$
$$z^5 = r^5(\cos 5\theta + i\sin 5\theta)$$
$$\vdots$$

This pattern leads to **DeMoivre's Theorem,** which is named after the French mathematician Abraham DeMoivre (1667–1754).

DeMoivre's Theorem

If $z = r(\cos\theta + i\sin\theta)$ is a complex number and n is a positive integer, then

$$z^n = [r(\cos\theta + i\sin\theta)]^n$$
$$= r^n(\cos n\theta + i\sin n\theta).$$

Explore the Concept

Plot the numbers i, i^2, i^3, i^4, and i^5 in the complex plane. Write each number in trigonometric form and describe what happens to the angle θ as you form higher powers of i.

EXAMPLE 7 Finding a Power of a Complex Number

Use DeMoivre's Theorem to find $(1 + \sqrt{3}i)^{12}$.

Solution

First, convert the complex number to trigonometric form using

$$r = \sqrt{(1)^2 + (\sqrt{3})^2} = 2$$

and

$$\theta = \arctan\frac{\sqrt{3}}{1} = \frac{\pi}{3}.$$

So, the trigonometric form is

$$1 + \sqrt{3}i = 2\left(\cos\frac{\pi}{3} + i\sin\frac{\pi}{3}\right).$$

Then, by DeMoivre's Theorem, you have

$$(1 + \sqrt{3}i)^{12} = \left[2\left(\cos\frac{\pi}{3} + i\sin\frac{\pi}{3}\right)\right]^{12}$$
$$= 2^{12}\left(\cos\frac{12\pi}{3} + i\sin\frac{12\pi}{3}\right)$$
$$= 4096(\cos 4\pi + i\sin 4\pi)$$
$$= 4096(1 + 0)$$
$$= 4096.$$

Checkpoint *Audio-video solution in English & Spanish at LarsonPrecalculus.com*

Use DeMoivre's Theorem to find $(-1 - i)^4$.

Roots of Complex Numbers

Recall that a consequence of the Fundamental Theorem of Algebra is that a polynomial equation of degree n has n solutions in the complex number system. So, an equation such as $x^6 = 1$ has six solutions, and in this particular case you can find the six solutions by factoring and using the Quadratic Formula.

$$x^6 - 1 = 0$$

$$(x^3 - 1)(x^3 + 1) = 0$$

$$(x - 1)(x^2 + x + 1)(x + 1)(x^2 - x + 1) = 0$$

Consequently, the solutions are

$$x = \pm 1, \quad x = \frac{-1 \pm \sqrt{3}i}{2}, \quad \text{and} \quad x = \frac{1 \pm \sqrt{3}i}{2}.$$

Each of these numbers is a sixth root of 1. In general, the ***n*th root of a complex number** is defined as follows.

Definition of an *n*th Root of a Complex Number

The complex number $u = a + bi$ is an ***n*th root** of the complex number z when

$z = u^n = (a + bi)^n$.

To find a formula for an nth root of a complex number, let u be an nth root of z, where $u = s(\cos \beta + i \sin \beta)$ and $z = r(\cos \theta + i \sin \theta)$. By DeMoivre's Theorem and the fact that $u^n = z$, you have

$$s^n(\cos n\beta + i \sin n\beta) = r(\cos \theta + i \sin \theta).$$

Taking the absolute value of each side of this equation, it follows that $s^n = r$. Substituting back into the previous equation and dividing by r, you get

$$\cos n\beta + i \sin n\beta = \cos \theta + i \sin \theta.$$

So, it follows that

$$\cos n\beta = \cos \theta \quad \text{and} \quad \sin n\beta = \sin \theta.$$

Because both sine and cosine have a period of 2π, these last two equations have solutions if and only if the angles differ by a multiple of 2π. Consequently, there must exist an integer k such that

$$n\beta = \theta + 2\pi k$$

$$\beta = \frac{\theta + 2\pi k}{n}.$$

By substituting this value of β and $s = \sqrt[n]{r}$ into the trigonometric form of u, you get the result stated in the theorem on the next page.

Explore the Concept

The nth roots of a complex number are useful for solving some polynomial equations. For instance, explain how you can use DeMoivre's Theorem to solve the polynomial equation $x^4 + 16 = 0$. [*Hint:* Write -16 as $16(\cos \pi + i \sin \pi)$.]

nth Roots of a Complex Number

For a positive integer n, the complex number $z = r(\cos\theta + i\sin\theta)$ has exactly n distinct nth roots given by

$$z_k = \sqrt[n]{r}\left(\cos\frac{\theta + 2\pi k}{n} + i\sin\frac{\theta + 2\pi k}{n}\right)$$

where $k = 0, 1, 2, \ldots, n - 1$.

When $k > n - 1$, the roots begin to repeat. For instance, when $k = n$, the angle

$$\frac{\theta + 2\pi n}{n} = \frac{\theta}{n} + 2\pi$$

is coterminal with θ/n, which is also obtained when $k = 0$.

The formula for the nth roots of a complex number z has a nice geometrical interpretation, as shown in Figure 6.40. Note that because the nth roots of z all have the same magnitude $\sqrt[n]{r}$, they all lie on a circle of radius $\sqrt[n]{r}$ with center at the origin. Furthermore, because successive nth roots have arguments that differ by

$$\frac{2\pi}{n}$$

the nth roots are equally spaced around the circle.

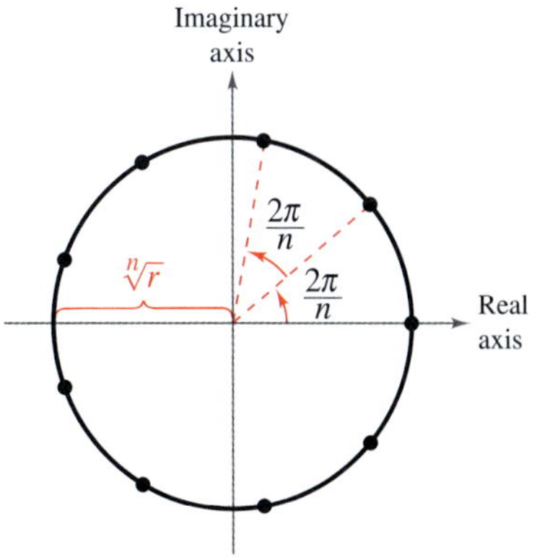

Figure 6.40

You have already found the sixth roots of 1 by factoring and by using the Quadratic Formula. Example 8 shows how you can solve the same problem with the formula for nth roots.

EXAMPLE 8 Finding the nth Roots of a Real Number

Find all the sixth roots of 1.

Solution

First, write 1 in the trigonometric form $1 = 1(\cos 0 + i\sin 0)$. Then, by the nth root formula with $n = 6$, $r = 1$, and $\theta = 0$, the roots have the form

$$z_k = \sqrt[6]{1}\left(\cos\frac{0 + 2\pi k}{6} + i\sin\frac{0 + 2\pi k}{6}\right) = \cos\frac{\pi k}{3} + i\sin\frac{\pi k}{3}.$$

So, for $k = 0, 1, 2, 3, 4,$ and 5, the sixth roots of 1 are shown below. (See Figure 6.41.)

$$z_0 = \cos 0 + i\sin 0 = 1$$

$$z_1 = \cos\frac{\pi}{3} + i\sin\frac{\pi}{3} = \frac{1}{2} + \frac{\sqrt{3}}{2}i \qquad \text{Incremented by } \frac{2\pi}{n} = \frac{2\pi}{6} = \frac{\pi}{3}$$

$$z_2 = \cos\frac{2\pi}{3} + i\sin\frac{2\pi}{3} = -\frac{1}{2} + \frac{\sqrt{3}}{2}i$$

$$z_3 = \cos\pi + i\sin\pi = -1$$

$$z_4 = \cos\frac{4\pi}{3} + i\sin\frac{4\pi}{3} = -\frac{1}{2} - \frac{\sqrt{3}}{2}i$$

$$z_5 = \cos\frac{5\pi}{3} + i\sin\frac{5\pi}{3} = \frac{1}{2} - \frac{\sqrt{3}}{2}i$$

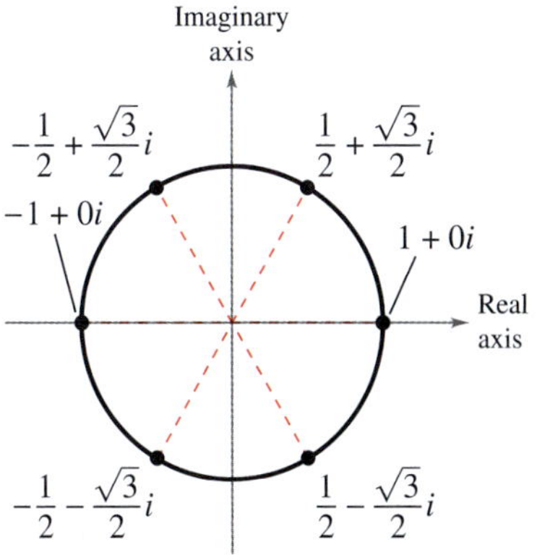

Figure 6.41

Checkpoint *Audio-video solution in English & Spanish at LarsonPrecalculus.com*

Find all fourth roots of 1.

In Figure 6.41, notice that the roots obtained in Example 8 all have a magnitude of 1 and are equally spaced around the unit circle. Also notice that the complex roots occur in conjugate pairs, as discussed in Section 2.5. The n distinct nth roots of 1 are called the ***n*th roots of unity.**

EXAMPLE 9 Finding the nth Roots of a Complex Number

See LarsonPrecalculus.com for an interactive version of this type of example.

Find the three cube roots of $z = -2 + 2i$.

Solution

The modulus of z is $r = |-2 + 2i| = \sqrt{(-2)^2 + 2^2} = \sqrt{8}$, and the angle θ is determined from

$$\tan\theta = \frac{b}{a} = \frac{2}{-2} = -1.$$

Because z lies in Quadrant II, the trigonometric form of z is

$$z = -2 + 2i = \sqrt{8}(\cos 135° + i \sin 135°).$$

By the formula for nth roots, the cube roots have the form

$$z_k = \sqrt[6]{8}\left(\cos\frac{135° + 360°k}{3} + i\sin\frac{135° + 360°k}{3}\right).$$

Finally, for $k = 0$, 1, and 2, you obtain the cube roots of z as shown below.

$$\begin{aligned} z_0 &= \sqrt[6]{8}\left(\cos\frac{135° + 360°(0)}{3} + i\sin\frac{135° + 360°(0)}{3}\right) \\ &= \sqrt{2}(\cos 45° + i\sin 45°) \\ &= 1 + i \end{aligned}$$

$$\begin{aligned} z_1 &= \sqrt[6]{8}\left(\cos\frac{135° + 360°(1)}{3} + i\sin\frac{135° + 360°(1)}{3}\right) \\ &= \sqrt{2}(\cos 165° + i\sin 165°) \\ &\approx -1.3660 + 0.3660i \end{aligned}$$

$$\begin{aligned} z_2 &= \sqrt[6]{8}\left(\cos\frac{135° + 360°(2)}{3} + i\sin\frac{135° + 360°(2)}{3}\right) \\ &= \sqrt{2}(\cos 285° + i\sin 285°) \\ &\approx 0.3660 - 1.3660i \end{aligned}$$

See Figure 6.42.

Algebra Help

In Example 9, because $r = \sqrt{8}$, it follows that

$$\begin{aligned} \sqrt[n]{r} &= \sqrt[3]{\sqrt{8}} \\ &= (8^{1/2})^{1/3} \\ &= 8^{1/6} \\ &= \sqrt[6]{8}. \end{aligned}$$

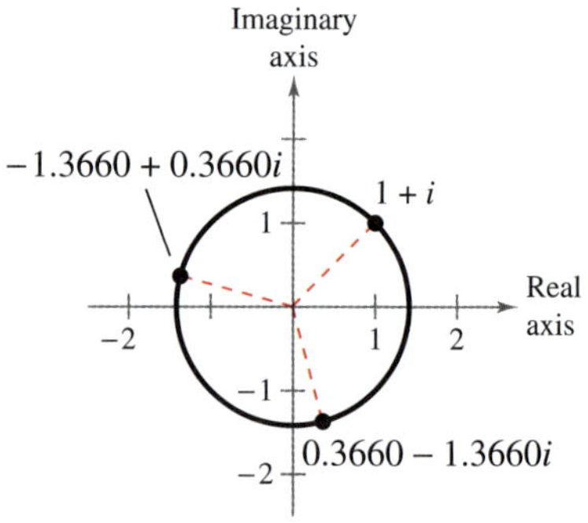

Figure 6.42

✓ ***Checkpoint*** *Audio-video solution in English & Spanish at LarsonPrecalculus.com*

Find the three cube roots of $z = -6 + 6i$.

Explore the Concept

Use a graphing utility set in *parametric* and *radian* modes to display the graphs of X1T = cos T and Y1T = sin T. Set the viewing window so that $-1.5 \le X \le 1.5$ and $-1 \le Y \le 1$. Then, using $0 \le T \le 2\pi$, set the "Tstep" to $2\pi/n$ for various values of n. Explain how the graphing utility can be used to obtain the nth roots of unity.

6.6 Exercises

See *CalcChat.com* for tutorial help and worked-out solutions to odd-numbered exercises.
For instructions on how to use a graphing utility, see Appendix A.

Vocabulary and Concept Check

In Exercises 1–6, fill in the blank(s).

1. The _______ of the complex number $z = a + bi$ is $z = r(\cos\theta + i\sin\theta)$, where r is the _______ of z and θ is an _______ of z.

2. To multiply two complex numbers, you _______ moduli and _______ arguments.

3. To divide two complex numbers, you _______ moduli and _______ arguments.

4. _______ Theorem states that if $z = r(\cos\theta + i\sin\theta)$ is a complex number and n is a positive integer, then $z^n = r^n(\cos n\theta + i\sin n\theta)$.

5. The complex number $u = a + bi$ is an _______ of the complex number z when $z = u^n = (a + bi)^n$.

6. Successive nth roots of a complex number have arguments that differ by _______ .

Procedures and Problem Solving

Trigonometric Form of a Complex Number **In Exercises 7–12, write the complex number in trigonometric form without using a calculator.**

7.

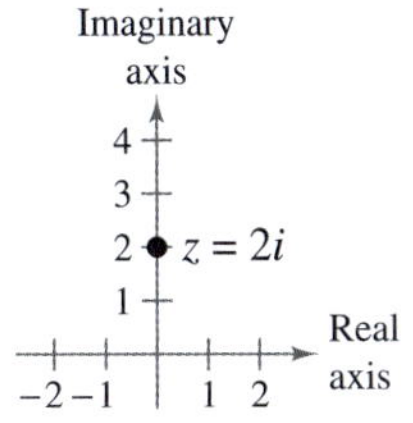

8.

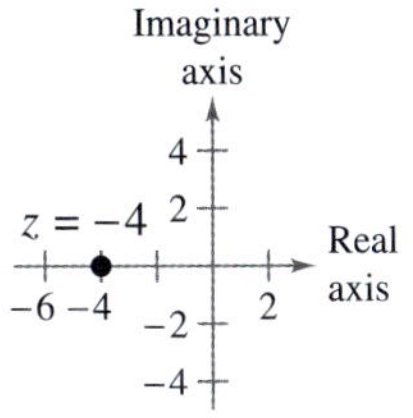

9.

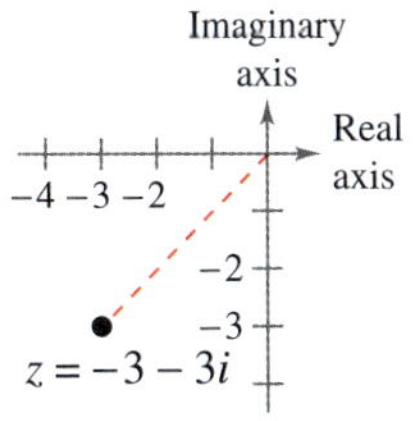

10.

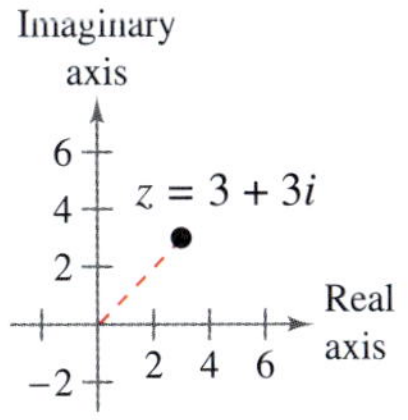

11.

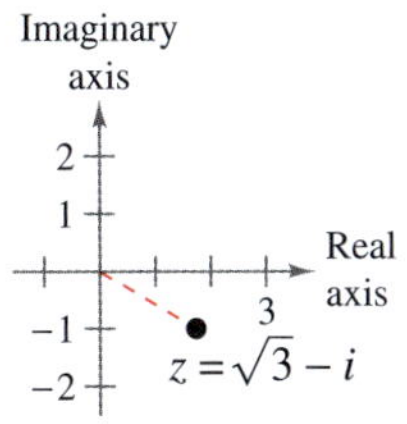

12.

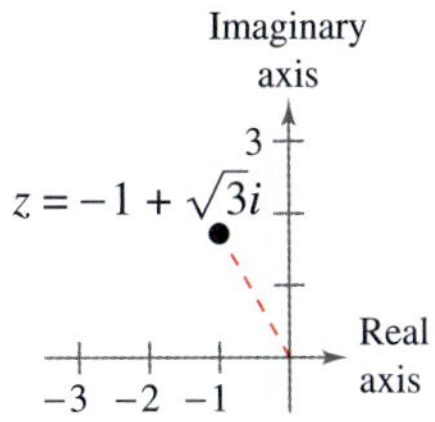

Trigonometric Form of a Complex Number **In Exercises 13–30, plot the complex number. Then write the trigonometric form of the complex number.**

13. $-8i$ **14.** $4i$

15. 9 **16.** 24

17. $7 - 7i$ **18.** $2 + 2i$

19. $1 - \sqrt{3}i$ **20.** $\sqrt{3} + i$

21. $1 + i$ **22.** $10 - 10i$

23. $-2(1 + \sqrt{3}i)$ **24.** $-\frac{5}{2}(\sqrt{3} + i)$

25. $-7 + 4i$ **26.** $3 - 5i$

27. $3 + \sqrt{3}i$ **28.** $-2\sqrt{2} + i$

29. $-1 - 5i$ **30.** $-3 + i$

Writing a Complex Number in Standard Form **In Exercises 31–42, write the standard form of the complex number. Then plot the complex number.**

31. $6\left(\cos\frac{\pi}{3} + i\sin\frac{\pi}{3}\right)$ **32.** $8\left(\cos\frac{5\pi}{6} + i\sin\frac{5\pi}{6}\right)$

33. $5.25\left(\cos\frac{3\pi}{4} + i\sin\frac{3\pi}{4}\right)$

34. $2.3\left(\cos\frac{\pi}{2} + i\sin\frac{\pi}{2}\right)$

35. $2(\cos 120^\circ + i\sin 120^\circ)$

36. $5(\cos 135^\circ + i\sin 135^\circ)$

37. $\sqrt{48}[\cos(-30^\circ) + i\sin(-30^\circ)]$

38. $\sqrt{12}[\cos(-45^\circ) + i\sin(-45^\circ)]$

39. $\frac{5}{2}\left(\cos\frac{3\pi}{2} + i\sin\frac{3\pi}{2}\right)$

40. $\frac{1}{4}(\cos 0 + i\sin 0)$

41. $3[\cos(18^\circ 45') + i\sin(18^\circ 45')]$

42. $6[\cos(230^\circ 30') + i\sin(230^\circ 30')]$

Writing a Complex Number in Standard Form **In Exercises 43–46, use a graphing utility to write the complex number in standard form.**

43. $5\left(\cos\frac{7\pi}{9} + i\sin\frac{7\pi}{9}\right)$ **44.** $12\left(\cos\frac{3\pi}{5} + i\sin\frac{3\pi}{5}\right)$

45. $9(\cos 58^\circ + i\sin 58^\circ)$ **46.** $2(\cos 73^\circ + i\sin 73^\circ)$

Multiplying Complex Numbers **In Exercises 47–52, find the product. Leave the result in trigonometric form.**

47. $\left[2\left(\cos\frac{\pi}{4} + i\sin\frac{\pi}{4}\right)\right]\left[6\left(\cos\frac{\pi}{12} + i\sin\frac{\pi}{12}\right)\right]$

48. $\left[3\left(\cos\frac{\pi}{2} + i\sin\frac{\pi}{2}\right)\right]\left[9\left(\cos\frac{3\pi}{2} + i\sin\frac{3\pi}{2}\right)\right]$

49. $\left[4\left(\cos\frac{\pi}{12} + i\sin\frac{\pi}{12}\right)\right]\left[5\left(\cos\frac{7\pi}{12} + i\sin\frac{7\pi}{12}\right)\right]$

50. $\left[\frac{3}{2}\left(\cos\frac{\pi}{6} + i\sin\frac{\pi}{6}\right)\right]\left[6\left(\cos\frac{\pi}{4} + i\sin\frac{\pi}{4}\right)\right]$

51. $\left[\frac{5}{3}(\cos 140° + i\sin 140°)\right]\left[\frac{2}{3}(\cos 60° + i\sin 60°)\right]$

52. $(\cos 5° + i\sin 5°)(\cos 20° + i\sin 20°)$

Dividing Complex Numbers **In Exercises 53–58, find the quotient. Leave the result in trigonometric form.**

53. $\dfrac{3(\cos 50° + i\sin 50°)}{9(\cos 20° + i\sin 20°)}$

54. $\dfrac{2(\cos 120° + i\sin 120°)}{4(\cos 40° + i\sin 40°)}$

55. $\dfrac{9(\cos 2\pi + i\sin 2\pi)}{8(\cos\pi + i\sin\pi)}$

56. $\dfrac{\cos\left(\frac{7\pi}{4}\right) + i\sin\left(\frac{7\pi}{4}\right)}{\cos\pi + i\sin\pi}$

57. $\dfrac{16(\cos 54° + i\sin 54°)}{4(\cos 102° + i\sin 102°)}$

58. $\dfrac{7(\cos 20° + i\sin 20°)}{4(\cos 75° + i\sin 75°)}$

Multiplying or Dividing Complex Numbers **In Exercises 59–74, (a) write the trigonometric forms of the complex numbers, (b) perform the operation using the trigonometric forms, and (c) perform the operation using the standard forms and check your result with that of part (b).**

59. $(2 - 2i)(1 + i)$

60. $(3 - 3i)(1 - i)$

61. $(2 + 2i)(1 - i)$

62. $(3 + 3i)(1 + i)$

63. $-2i(1 + i)$

64. $3i(1 + \sqrt{2}i)$

65. $-2i(\sqrt{3} - i)$

66. $-i(1 + \sqrt{3}i)$

67. $2(1 - i)$

68. $-4(1 + i)$

69. $\dfrac{3 + 4i}{1 - \sqrt{3}i}$

70. $\dfrac{2 + 2i}{1 + \sqrt{3}i}$

71. $\dfrac{5}{2 + 2i}$

72. $\dfrac{2}{\sqrt{3} - i}$

73. $\dfrac{4i}{-1 + i}$

74. $\dfrac{2i}{1 - \sqrt{3}i}$

Multiplying in the Complex Plane **In Exercises 75 and 76, find the product in the complex plane.**

75. $\left[2\left(\cos\frac{2\pi}{3} + i\sin\frac{2\pi}{3}\right)\right]\left[\frac{1}{2}\left(\cos\frac{\pi}{3} + i\sin\frac{\pi}{3}\right)\right]$

76. $\left[2\left(\cos\frac{\pi}{4} + i\sin\frac{\pi}{4}\right)\right]\left[3\left(\cos\frac{\pi}{4} + i\sin\frac{\pi}{4}\right)\right]$

Sketching the Graph of Complex Numbers **In Exercises 77–80, sketch the graph of all complex numbers z satisfying the specified condition.**

77. $|z| = 1$

78. $|z| = 8$

79. $\theta = \dfrac{\pi}{4}$

80. $\theta = \dfrac{2\pi}{3}$

Finding a Power of a Complex Number **In Exercises 81–98, use DeMoivre's Theorem to find the power of the complex number. Write the result in standard form.**

81. $(1 + i)^5$

82. $(2 + 2i)^6$

83. $(-1 + i)^3$

84. $4(1 - \sqrt{3}i)^3$

85. $[5(\cos 140° + i\sin 140°)]^3$

86. $[3(\cos 150° + i\sin 150°)]^4$

87. $\left(\cos\frac{5\pi}{4} + i\sin\frac{5\pi}{4}\right)^{10}$

88. $\left[2\left(\cos\frac{\pi}{2} + i\sin\frac{\pi}{2}\right)\right]^{12}$

89. $[2(\cos\pi + i\sin\pi)]^{14}$

90. $(\cos 0 + i\sin 0)^{20}$

91. $[4(\cos 10° + i\sin 10°)]^6$

92. $[6(\cos 15° + i\sin 15°)]^4$

93. $\left[3\left(\cos\frac{\pi}{8} + i\sin\frac{\pi}{8}\right)\right]^2$

94. $\left[2\left(\cos\frac{\pi}{10} + i\sin\frac{\pi}{10}\right)\right]^5$

95. $(3 - 2i)^5$

96. $(\sqrt{5} - 4i)^4$

97. $[2(\cos 1.25 + i\sin 1.25)]^4$

98. $[4(\cos 2.8 + i\sin 2.8)]^5$

Representing a Power **In Exercises 99 and 100, represent the powers z, z^2, z^3, and z^4 graphically. Describe the pattern.**

99. $z = \dfrac{\sqrt{2}}{2}(1 + i)$

100. $z = \dfrac{1}{2}(1 + \sqrt{3}i)$

Verifying *n*th Roots **In Exercises 101 and 102, use DeMoivre's Theorem to verify the indicated root of the real number.**

101. $-\frac{1}{2}(1 + \sqrt{3}i)$ is a sixth root of 1.

102. $2^{-1/4}(1 - i)$ is a fourth root of -2.

Finding Square Roots of a Complex Number **In Exercises 103–106, find the square roots of the complex number.**

103. $4i$

104. $-2i$

105. $2 - 2i$

106. $1 + \sqrt{3}i$

Finding the *n*th Roots of a Complex Number

In Exercises 107–122, (a) use the formula on page 457 to find the roots of the complex number, (b) write each of the roots in standard form, and (c) represent each of the roots graphically.

107. Square roots of $5(\cos 120° + i \sin 120°)$

108. Square roots of $16(\cos 60° + i \sin 60°)$

109. Cube roots of $8\left(\cos \frac{2\pi}{3} + i \sin \frac{2\pi}{3}\right)$

110. Fifth roots of $32\left(\cos \frac{5\pi}{6} + i \sin \frac{5\pi}{6}\right)$

111. Cube roots of $-25i$

112. Fourth roots of $625i$

113. Cube roots of $-\frac{125}{2}(1 + \sqrt{3}i)$

114. Cube roots of $-4\sqrt{2}(1 - i)$

115. Fourth roots of $256i$

116. Fourth roots of i

117. Fifth roots of 1

118. Cube roots of 1000

119. Fourth roots of -16

120. Fourth roots of -4

121. Fifth roots of $128(-1 + i)$

122. Fifth roots of $4(1 - i)$

Solving an Equation

In Exercises 123–130, use the formula on page 457 to find all the solutions of the equation, and represent the solutions graphically.

123. $x^4 - i = 0$

124. $x^3 + 1 = 0$

125. $x^5 + 243 = 0$

126. $x^3 - 27 = 0$

127. $x^4 + 16i = 0$

128. $x^6 - 64i = 0$

129. $x^3 - (1 - i) = 0$

130. $x^4 + (1 + i) = 0$

Why you should learn it *(p. 451)* **The formula $E = IZ$, where E represents voltage, I represents current, and Z represents impedance (a measure of opposition to a sinusoidal electric current), is used in electrical engineering. Each variable is a complex number.**

In Exercises 131–134, use the formula to find the missing quantity. Then convert the given quantities to trigonometric form and check your result.

131. $I = 10 + 2i$
$Z = 4 + 3i$

132. $I = 12 + 2i$
$Z = 3 + 5i$

133. $I = 2 + 4i$
$E = 5 + 5i$

134. $E = 12 + 24i$
$Z = 12 + 20i$

Phovoir/Shutterstock.com

Focusing on Concepts

True or False? **In Exercises 135 and 136, determine whether the statement is true or false. Justify your answer.**

135. The product of two complex numbers is 0 only when the modulus of one (or both) of the complex numbers is 0.

136. Geometrically, the nth roots of any complex number z are equally spaced around the unit circle.

137. Complex Conjugates Show that

$$\bar{z} = r[\cos(-\theta) + i \sin(-\theta)]$$

is the complex conjugate of

$$z = r(\cos \theta + i \sin \theta).$$

138. Complex Conjugates Use trigonometric forms of z and $\bar{z}$ in Exercise 137 to find (a) $z\bar{z}$ and (b) $z/\bar{z}$, $\bar{z} \neq 0$.

139. Quotient of Two Complex Numbers Given two complex numbers $z_1 = r_1(\cos \theta_1 + i \sin \theta_1)$ and $z_2 = r_2(\cos \theta_2 + i \sin \theta_2)$, $z_2 \neq 0$, show that

$$\frac{z_1}{z_2} = \frac{r_1}{r_2}[\cos(\theta_1 - \theta_2) + i \sin(\theta_1 - \theta_2)].$$

140. HOW DO YOU SEE IT? The figure shows one of the fourth roots of a complex number z.

(a) How many roots are not shown?

(b) Describe the other roots.

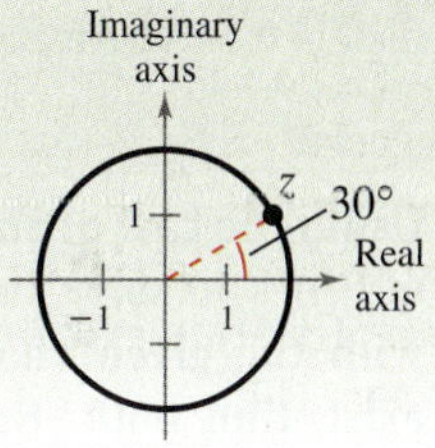

141. Writing The formula $e^{a+bi} = e^a(\cos b + i \sin b)$ is called Euler's Formula, after the Swiss mathematician Leonhard Euler (1707–1783). This formula gives rise to the equation $e^{\pi i} + 1 = 0$. Show how Euler's Formula can be used to derive this equation. Write a short paragraph summarizing your work.

142. Error Analysis Describe the error.

$$\frac{20(\cos 90° + i \sin 90°)}{5(\cos 60° + i \sin 60°)} = 2\sqrt{3} - 2i$$

Cumulative Mixed Review

Harmonic Motion **In Exercises 143 and 144, for the simple harmonic motion described by the trigonometric function, find the maximum displacement from equilibrium and the least possible positive value of t for which $d = 0$.**

143. $d = 16 \cos \frac{\pi}{4}t$

144. $d = \frac{1}{12} \sin 60 \pi t$

6 Chapter Review

See *CalcChat.com* for tutorial help and worked-out solutions to odd-numbered exercises. For instructions on how to use a graphing utility, see Appendix A.

6.1 *What did you learn?*

Use the Law of Sines to solve oblique triangles (AAS or ASA) ***(p. 406).***

Law of Sines

If ABC is a triangle with sides a, b, and c, then

$$\frac{a}{\sin A} = \frac{b}{\sin B} = \frac{c}{\sin C}.$$

Use the Law of Sines to solve oblique triangles (SSA) ***(p. 408).*** If two sides and one opposite angle are given, then three possible situations can occur: (1) no such triangle exists (see Example 4), (2) one such triangle exists (see Example 3), or (3) two distinct triangles satisfy the conditions (see Example 5).

Find areas of oblique triangles ***(p. 410)*** **and use the Law of Sines to model and solve real-life problems** ***(p. 411).*** The area of any triangle is one-half the product of the lengths of two sides times the sine of their included angle. That is,

$$\text{Area} = \tfrac{1}{2}bc \sin A = \tfrac{1}{2}ab \sin C = \tfrac{1}{2}ac \sin B.$$

The Law of Sines can be used to approximate the total distance of a boat race course (see Example 7).

Example

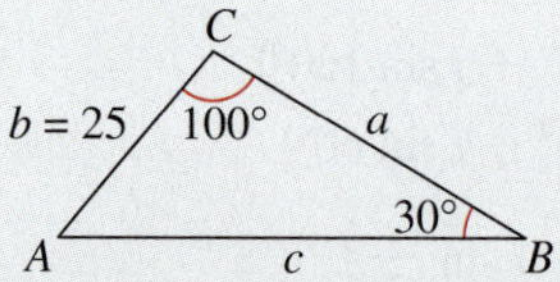

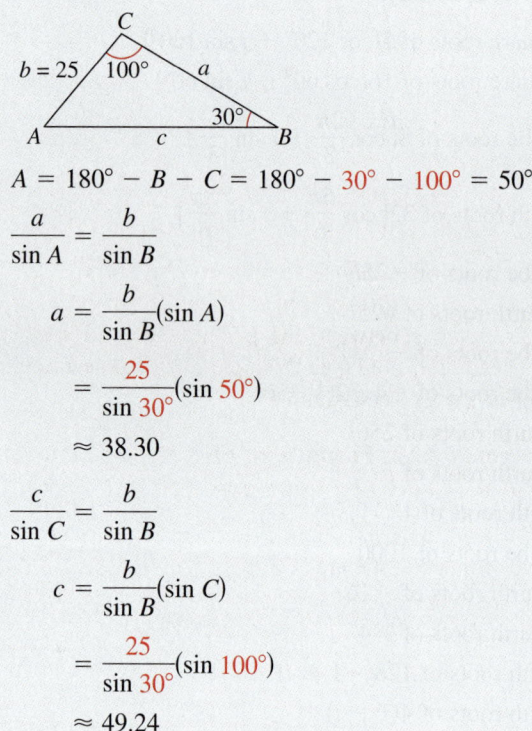

$$A = 180° - B - C = 180° - 30° - 100° = 50°$$

$$\frac{a}{\sin A} = \frac{b}{\sin B}$$

$$a = \frac{b}{\sin B}(\sin A)$$

$$= \frac{25}{\sin 30°}(\sin 50°)$$

$$\approx 38.30$$

$$\frac{c}{\sin C} = \frac{b}{\sin B}$$

$$c = \frac{b}{\sin B}(\sin C)$$

$$= \frac{25}{\sin 30°}(\sin 100°)$$

$$\approx 49.24$$

Using the Law of Sines **In Exercises 1–10, use the Law of Sines to solve the triangle or show that no triangle with the given characteristics exists. If two solutions exist, find both. Round your answers to two decimal places, if necessary.**

1. $A = 32°, \quad B = 50°, \quad a = 16$

2. $A = 38°, \quad B = 58°, \quad a = 12$

3. $B = 16°, \quad C = 98°, \quad a = 8.4$

4. $B = 95°, \quad C = 45°, \quad a = 104.8$

5. $A = 60° 15', \quad B = 45° 30', \quad b = 4.8$

6. $A = 82° 45', \quad B = 28° 45', \quad b = 40.2$

7. $A = 75°, \quad a = 51.2, \quad b = 33.7$

8. $A = 15°, \quad a = 5, \quad b = 10$

9. $B = 25°, \quad a = 6.2, \quad b = 4$

10. $B = 150°, \quad a = 10, \quad b = 3$

Finding the Area of an Oblique Triangle **In Exercises 11–14, find the area of the triangle with the indicated angle and sides. Round your answers to two decimal places, if necessary.**

11. $A = 33°, \quad b = 7, \quad c = 10$

12. $B = 80°, \quad a = 4, \quad c = 8$

13. $C = 122°, \quad b = 18, \quad a = 29$

14. $C = 100°, \quad a = 120, \quad b = 74$

15. Height A tree stands on a hillside of slope 28° from the horizontal. From a point 75 feet down the hill, the angle of elevation to the top of the tree is 45° (see figure). Find the height of the tree.

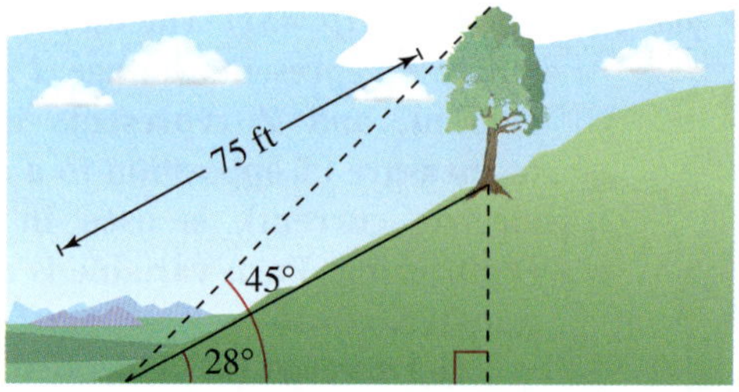

16. Surveying A surveyor finds that a tree on the opposite bank of a river has a bearing of N 22° 30′ E from one point and a bearing of N 15° W from a point 400 feet downstream. Find the width of the river.

6.2 *What did you learn?*

Use the Law of Cosines to solve oblique triangles (SSS or SAS) *(p. 415)*. Consider a triangle ABC with sides a, b, and c.

Standard Form	*Alternative Form*
$a^2 = b^2 + c^2 - 2bc\cos A$	$\cos A = \dfrac{b^2 + c^2 - a^2}{2bc}$
$b^2 = a^2 + c^2 - 2ac\cos B$	$\cos B = \dfrac{a^2 + c^2 - b^2}{2ac}$
$c^2 = a^2 + b^2 - 2ab\cos C$	$\cos C = \dfrac{a^2 + b^2 - c^2}{2ab}$

Use the Law of Cosines to model and solve real-life problems *(p. 417)*. The Law of Cosines can be used to find the distance between the pitcher's mound and first base on a softball field (see Example 3) and to describe the bearing of a ship (see Example 4).

Use Heron's Area Formula to find areas of triangles *(p. 418)*.

Heron's area formula: For any triangle with sides of lengths a, b, and c, the area of the triangle is

$$\text{Area} = \sqrt{s(s-a)(s-b)(c-s)},\ \text{where } s = \frac{a+b+c}{2}.$$

Example

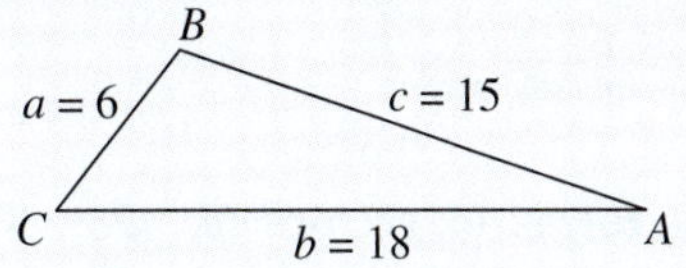

$$\cos B = \frac{a^2 + c^2 - b^2}{2ac}$$

$$\cos B = \frac{6^2 + 15^2 - 18^2}{2(6)(15)}$$

$$\cos B = -0.35$$

$$B \approx 110.49^\circ$$

$$\sin A = a\left(\frac{\sin B}{b}\right)$$

$$\sin A \approx 6\left(\frac{\sin 110.49^\circ}{18}\right)$$

$$A \approx 18.19^\circ$$

$$C \approx 180^\circ - 18.19^\circ - 110.49^\circ = 51.32^\circ$$

Using the Law of Cosines **In Exercises 17–28, use the Law of Cosines to solve the triangle.**

17.

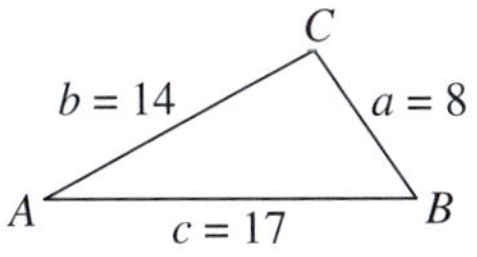

18.

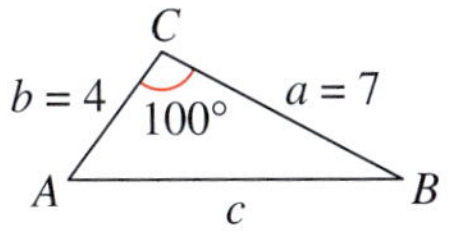

19. $a = 9,\quad b = 12,\quad c = 20$

20. $a = 7,\quad b = 15,\quad c = 19$

21. $A = 150^\circ,\quad b = 10,\quad c = 20$

22. $a = 6.2,\quad b = 6.4,\quad c = 2.1$

23. $C = 65^\circ,\quad a = 25,\quad b = 12$

24. $B = 48^\circ,\quad a = 18,\quad c = 18$

25. $a = 2.5,\quad b = 5.0,\quad c = 4.5$

26. $a = 16.4,\quad b = 8.8,\quad c = 12.2$

27. $C = 43^\circ,\quad a = 22.5,\quad b = 31.4$

28. $A = 62^\circ,\quad b = 11.34,\quad c = 19.52$

Solving a Triangle **In Exercises 29–32, determine whether the Law of Cosines is needed to solve the triangle. Then solve the triangle (if possible).**

29. $b = 9,\quad c = 13,\quad C = 64^\circ$

30. $a = 4,\quad c = 5,\quad B = 52^\circ$

31. $a = 13,\quad b = 15,\quad c = 24$

32. $A = 44^\circ,\quad B = 31^\circ,\quad c = 2.8$

Using Heron's Area Formula **In Exercises 33–38, use Heron's Area Formula to find the area of the triangle.**

33. $a = 4,\quad b = 6,\quad c = 8$

34. $a = 15,\quad b = 8,\quad c = 10$

35. $a = 64.8,\quad b = 49.2,\quad c = 24.1$

36. $a = 8.55,\quad b = 5.14,\quad c = 12.73$

37. $a = 1,\quad b = \frac{3}{5},\quad c = \frac{3}{5}$

38. $a = \frac{4}{5},\quad b = \frac{3}{4},\quad c = \frac{5}{8}$

39. Geometry The lengths of the diagonals of a parallelogram are 10 feet and 16 feet. Find the lengths of the sides of the parallelogram when the diagonals intersect at an angle of 28°.

40. Surveying To approximate the length of a marsh, a surveyor walks 425 meters from point A to point B. Then the surveyor turns 65° and walks 300 meters to point C (see figure). Approximate the length AC of the marsh.

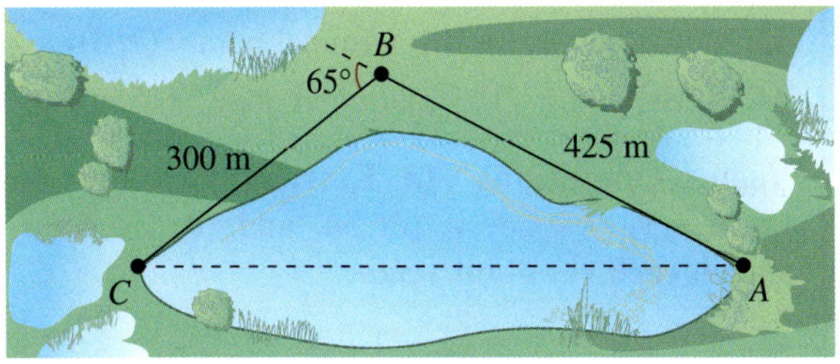

6.3 *What did you learn?*

Represent vectors as directed line segments *(p. 422)*.

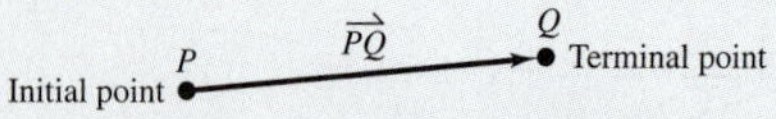

Write the component forms of vectors *(p. 423)*. The component form of the vector with initial point $P(p_1, p_2)$ and terminal point $Q(q_1, q_2)$ is given by

$$\overrightarrow{PQ} = \langle q_1 - p_1, q_2 - p_2 \rangle = \langle v_1, v_2 \rangle = \mathbf{v}.$$

The magnitude (or length) of $\mathbf{v}$ is given by

$$\|\mathbf{v}\| = \sqrt{(q_1 - p_1)^2 + (q_2 - p_2)^2} = \sqrt{v_1^2 + v_2^2}.$$

Perform basic vector operations and represent vector operations graphically *(p. 424)*. Let $\mathbf{u} = \langle u_1, u_2 \rangle$ and $\mathbf{v} = \langle v_1, v_2 \rangle$ be vectors and k be a scalar (a real number).

$\mathbf{u} + \mathbf{v} = \langle u_1 + v_1, u_2 + v_2 \rangle$ $\quad k\mathbf{u} = \langle ku_1, ku_2 \rangle$

$-\mathbf{v} = \langle -v_1, -v_2 \rangle$ $\quad \mathbf{u} - \mathbf{v} = \langle u_1 - v_1, u_2 - v_2 \rangle$

Write vectors as linear combinations of unit vectors *(p. 426)*. Any vector $\mathbf{v} = \langle v_1, v_2 \rangle$ can be represented as $\mathbf{v} = \langle v_1, v_2 \rangle = v_1\langle 1, 0 \rangle + v_2\langle 0, 1 \rangle = v_1\mathbf{i} + v_2\mathbf{j}$. The scalars v_1 and v_2 are called the horizontal and vertical components of $\mathbf{v}$, respectively. The vector sum $v_1\mathbf{i} + v_2\mathbf{j}$ is a linear combination of the standard unit vectors $\mathbf{i}$ and $\mathbf{j}$.

Find the direction angles of vectors *(p. 428)*. The direction angle θ for $\mathbf{v} = a\mathbf{i} + b\mathbf{j}$ is determined from $\tan\theta = b/a$.

Use vectors to model and solve real-life problems *(p. 429)*. Vectors can be used to determine weight (see Example 10) and speed and direction (see Example 11).

Example

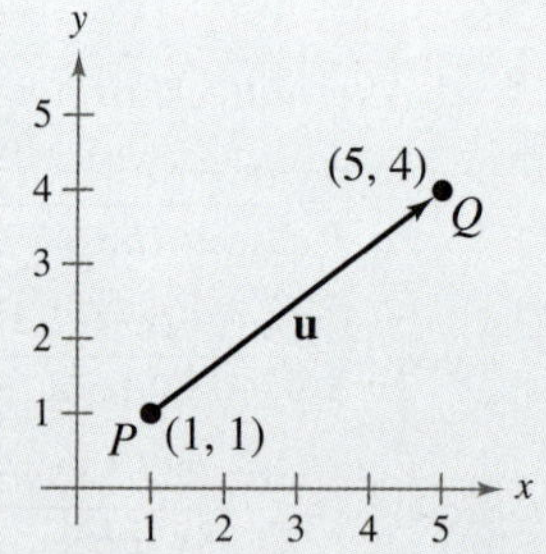

(a) $\|\overrightarrow{PQ}\| = \sqrt{(5 - 1)^2 + (4 - 1)^2} = 5$

slope of $\overrightarrow{PQ} = \dfrac{4 - 1}{5 - 1} = \dfrac{3}{4}$

(b) The component form of $\mathbf{u}$ is $\langle 5 - 1, 4 - 1 \rangle = \langle 4, 3 \rangle$.

(c) Let $\mathbf{v} = \langle -2, 1 \rangle$.

$\mathbf{u} + \mathbf{v} = \langle 4, 3 \rangle + \langle -2, 1 \rangle = \langle 4 + (-2), 3 + 1 \rangle = \langle 2, 4 \rangle$

$\mathbf{u} - \mathbf{v} = \mathbf{u} + (-\mathbf{v}) = \langle 4, 3 \rangle + \langle 2, -1 \rangle = \langle 6, 2 \rangle$

(d) As a linear combination of the standard unit vectors $\mathbf{i}$ and $\mathbf{j}$

$\mathbf{u} = 4\mathbf{i} + 3\mathbf{j}$.

From part (c), $\mathbf{v} = -2\mathbf{i} + \mathbf{j}$ and

$2\mathbf{u} + 3\mathbf{v} = 2(4\mathbf{i} + 3\mathbf{j}) + 3(-2\mathbf{i} + \mathbf{j}) = 2\mathbf{i} + 9\mathbf{j}$.

(e) The direction angle of $\mathbf{u}$ is determined from $\tan\theta = 3/4$, so $\theta \approx 36.87°$.

Determining Whether Two Vectors Are Equivalent **In Exercises 41 and 42, determine whether u and v are equivalent. Explain.**

41. $\mathbf{u}$: Initial point: $(-2, 1)$; terminal point: $(4, 6)$
$\mathbf{v}$: Initial point: $(0, -2)$; terminal point: $(6, 3)$

42. $\mathbf{u}$: Initial point: $(-3, 2)$; terminal point: $(-1, 5)$
$\mathbf{v}$: Initial point: $(-1, -4)$; terminal point: $(3, -2)$

Finding the Component Form **In Exercises 43 and 44, find the component form and the magnitude of the vector v.**

43. Initial point: $(0, 10)$; terminal point: $(7, 3)$

44. Initial point: $(1, 5)$; terminal point: $(15, 9)$

Vector Operations **In Exercises 45 and 46, find (a) u + v, (b) u − v, (c) 3u, and (d) 2v + 5u. Then sketch each resultant vector.**

45. $\mathbf{u} = \langle 4, 5 \rangle$, $\mathbf{v} = \langle 0, -1 \rangle$

46. $\mathbf{u} = -7\mathbf{i} - 3\mathbf{j}$, $\mathbf{v} = 4\mathbf{i} - \mathbf{j}$

Vector Operations **In Exercises 47–50, find the component form of w and sketch the specified vector operations geometrically, where $\mathbf{u} = 6\mathbf{i} - 5\mathbf{j}$ and $\mathbf{v} = 10\mathbf{i} + 3\mathbf{j}$.**

47. $\mathbf{w} = \frac{1}{3}\mathbf{v}$ **48.** $\mathbf{w} = \frac{1}{2}\mathbf{v}$

49. $\mathbf{w} = 4\mathbf{u} + 5\mathbf{v}$ **50.** $\mathbf{w} = 3\mathbf{v} - 2\mathbf{u}$

Finding a Unit Vector **In Exercises 51–54, find a unit vector in the direction of the given vector. Verify that the result has a magnitude of 1.**

51. $\mathbf{u} = \langle 0, -6 \rangle$ **52.** $\mathbf{v} = \langle -12, -5 \rangle$

53. $\mathbf{v} = 8\mathbf{i} - 15\mathbf{j}$ **54.** $\mathbf{w} = -7\mathbf{i}$

Writing a Linear Combination of Unit Vectors **In Exercises 55 and 56, the initial and terminal points of a vector are given. Write the vector as a linear combination of the standard unit vectors i and j.**

55. Initial point: $(-8, 3)$; terminal point: $(1, -5)$

56. Initial point: $(-2, 7)$; terminal point: $(5, -9)$

Finding Direction Angles of Vectors **In Exercises 57–62, find the magnitude and the direction angle of the vector v.**

57. $\mathbf{v} = 2\mathbf{i} - 2\mathbf{j}$ **58.** $\mathbf{v} = \mathbf{i} + \mathbf{j}$

59. $\mathbf{v} = 5\mathbf{i} + 4\mathbf{j}$ **60.** $\mathbf{v} = -3\mathbf{i} - 3\mathbf{j}$

61. $\mathbf{v} = -4\mathbf{i} + 7\mathbf{j}$ **62.** $\mathbf{v} = 8\mathbf{i} - \mathbf{j}$

63. **Resultant Force** Forces with magnitudes of 85 pounds and 50 pounds act on a single point. The angle between the forces is 15°. Describe the resultant force.

64. **Aeronautics** An airplane has an airspeed of 430 miles per hour at a bearing of 135°. The wind velocity is 35 miles per hour in the direction N 30° E. Find the resultant speed and direction of the plane.

65. **Physics** A 180-pound weight is supported by two ropes (see figure). Find the tension in each rope.

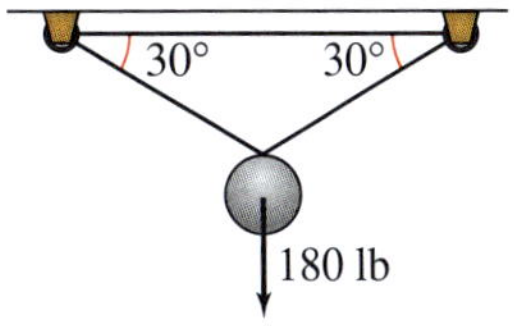

Figure for 65

Figure for 66

66. **Manufacturing** In a manufacturing process, an electric hoist lifts 200-pound ingots (see figure). Find the tension in the supporting cables.

6.4 *What did you learn?*

Find the dot product of two vectors and use the properties of the dot product *(p. 436)*. The dot product of $\mathbf{u} = \langle u_1, u_2\rangle$ and $\mathbf{v} = \langle v_1, v_2\rangle$ is $\mathbf{u} \cdot \mathbf{v} = u_1v_1 + u_2v_2$. Properties of the dot product are listed on page 436.

Find the angle between two vectors and determine whether two vectors are orthogonal *(p. 437)*. If θ is the angle between two nonzero vectors **u** and **v** ($0 \le \theta \le \pi$ or $0° \le \theta \le 180°$), then

$$\cos\theta = \frac{\mathbf{u} \cdot \mathbf{v}}{\|\mathbf{u}\|\|\mathbf{v}\|}.$$

Vectors **u** and **v** are orthogonal when $\mathbf{u} \cdot \mathbf{v} = 0$.

Write vectors as the sums of two vector components *(p. 439)*. Let **u** and **v** be nonzero vectors such that

$$\mathbf{u} = \mathbf{w}_1 + \mathbf{w}_2$$

where $\mathbf{w}_1$ and $\mathbf{w}_2$ are orthogonal and $\mathbf{w}_1$ is parallel to (or a scalar multiple of) **v**. The vectors $\mathbf{w}_1$ and $\mathbf{w}_2$ are called vector components of **u**, where $\mathbf{w}_1$, the projection of **u** onto **v**, is denoted by $\mathbf{w}_1 = \text{proj}_{\mathbf{v}}\mathbf{u}$. The vector $\mathbf{w}_2$ is given by $\mathbf{w}_2 = \mathbf{u} - \mathbf{w}_1$. The projection of **u** onto **v** is given by

$$\text{proj}_{\mathbf{v}}\mathbf{u} = \left(\frac{\mathbf{u} \cdot \mathbf{v}}{\|\mathbf{v}\|^2}\right)\mathbf{v}$$

Use vectors to find the work done by a force *(p. 441)*.

1. $W = \|\text{proj}_{\overrightarrow{PQ}}\mathbf{F}\|\,\|\overrightarrow{PQ}\|$ Projection form
2. $W = \mathbf{F} \cdot \overrightarrow{PQ}$ Dot product form

Example Consider the vectors

$$\mathbf{u} = \langle 3, -2\rangle, \quad \mathbf{v} = \langle 4, 5\rangle, \quad \text{and} \quad \mathbf{w} = \langle 2, 3\rangle.$$

(a) $\mathbf{u} \cdot \mathbf{v} = \langle 3, -2\rangle \cdot \langle 4, 5\rangle$
$= 3(4) + (-2)(5)$
$= 12 - 10$
$= 2$

(b) $(\mathbf{u} \cdot \mathbf{v})\mathbf{w} = 2\langle 2, 3\rangle$
$= \langle 4, 6\rangle$

(c) $\mathbf{0} \cdot \mathbf{v} = \langle 0, 0\rangle \cdot \langle 4, 5\rangle$
$= 0(4) + 0(5)$
$= 0$

(d) $\mathbf{u} \cdot (\mathbf{v} + \mathbf{w}) = \langle 3, -2\rangle \cdot (\langle 4, 5\rangle + \langle 2, 3\rangle)$
$= \langle 3, -2\rangle \cdot \langle 6, 8\rangle$
$= 3(6) + (-2)(8)$
$= 2$

(e) The vectors **u** and **w** are orthogonal because
$\mathbf{u} \cdot \mathbf{w} = \langle 3, -2\rangle \cdot \langle 2, 3\rangle$
$= 6 - 6$
$= 0.$

(f) $\text{proj}_{\mathbf{v}}\mathbf{u} = \left(\frac{2}{41}\right)\langle 4, 5\rangle$
$= \left\langle \frac{8}{41}, \frac{10}{41}\right\rangle$

Finding the Dot Product **In Exercises 67 and 68, find the dot product of u and v.**

67. $\mathbf{u} = \langle -6, 7\rangle$
$\mathbf{v} = \langle -3, -5\rangle$

68. $\mathbf{u} = 8\mathbf{i} - 7\mathbf{j}$
$\mathbf{v} = 3\mathbf{i} - 4\mathbf{j}$

Using Properties of the Dot Product **In Exercises 69 and 70, use the vectors $\mathbf{u} = \langle 6, -4\rangle$ and $\mathbf{v} = \langle 2, 1\rangle$ to find the quantity. State whether the result is a vector or a scalar.**

69. $\mathbf{u} \cdot \mathbf{u}$ **70.** $(\mathbf{u} \cdot \mathbf{v})\mathbf{u}$

Finding the Angle Between Two Vectors In Exercises 71 and 72, find the angle θ (in radians) between the vectors.

71. $\mathbf{u} = \langle 2\sqrt{2}, -4 \rangle$, $\mathbf{v} = \langle -\sqrt{2}, 1 \rangle$

72. $\mathbf{u} = \cos\frac{7\pi}{4}\mathbf{i} + \sin\frac{7\pi}{4}\mathbf{j}$
$\mathbf{v} = \cos\frac{5\pi}{6}\mathbf{i} + \sin\frac{5\pi}{6}\mathbf{j}$

Finding the Angle Between Two Vectors In Exercises 73–76, graph the vectors and find the degree measure of the angle between the vectors.

73. $\mathbf{u} = 4\mathbf{i} + \mathbf{j}$
$\mathbf{v} = \mathbf{i} - 4\mathbf{j}$

74. $\mathbf{u} = 6\mathbf{i} + 2\mathbf{j}$
$\mathbf{v} = -3\mathbf{i} - \mathbf{j}$

75. $\mathbf{u} = 7.2\mathbf{i} - 5.6\mathbf{j}$
$\mathbf{v} = 10\mathbf{i} + 3.1\mathbf{j}$

76. $\mathbf{u} = -5.3\mathbf{i} + 2.8\mathbf{j}$
$\mathbf{v} = -8.1\mathbf{i} - 4\mathbf{j}$

Describing the Relationship Between Two Vectors In Exercises 77–80, determine whether u and v are orthogonal, parallel, or neither.

77. $\mathbf{u} = \langle 12, -8 \rangle$
$\mathbf{v} = \langle 6, 9 \rangle$

78. $\mathbf{u} = \langle 8, -4 \rangle$
$\mathbf{v} = \langle 5, 10 \rangle$.

79. $\mathbf{u} = \langle 8, 5 \rangle$
$\mathbf{v} = \langle -2, 4 \rangle$

80. $\mathbf{u} = \langle -15, 51 \rangle$
$\mathbf{v} = \langle 20, -68 \rangle$

Finding an Unknown Vector Component In Exercises 81 and 82, find the value of k such that the vectors u and v are orthogonal.

81. $\mathbf{u} = \mathbf{i} - k\mathbf{j}$
$\mathbf{v} = \mathbf{i} + 2\mathbf{j}$

82. $\mathbf{u} = -2\mathbf{i} + 4\mathbf{j}$
$\mathbf{v} = 6\mathbf{i} - k\mathbf{j}$

Decomposing a Vector In Exercises 83–86, find the projection of u onto v. Then write u as the sum of two orthogonal vectors, one of which is $\text{proj}_{\mathbf{v}}\mathbf{u}$.

83. $\mathbf{u} = \langle -4, 3 \rangle$, $\mathbf{v} = \langle -8, -2 \rangle$

84. $\mathbf{u} = \langle 5, 6 \rangle$, $\mathbf{v} = \langle 10, 0 \rangle$

85. $\mathbf{u} = \langle 2, 7 \rangle$, $\mathbf{v} = \langle 1, -1 \rangle$

86. $\mathbf{u} = \langle -3, 5 \rangle$, $\mathbf{v} = \langle -5, 2 \rangle$

87. **Work** Determine the work done by a crane lifting an 18,000-pound truck 48 inches.

88. **Physics** A 500-pound motorcycle is stopped on a hill inclined at 12°. What force is required to keep the motorcycle from rolling down the hill?

6.5 *What did you learn?*

Plot complex numbers in the complex plane and find absolute values of complex numbers *(p. 445)*. A complex number $z = a + bi$ can be represented by the point (a, b) in the complex plane. The horizontal axis is the real axis and the vertical axis is the imaginary axis. The absolute value of $z = a + bi$ is $|a + bi| = \sqrt{a^2 + b^2}$.

Perform operations with complex numbers in the complex plane *(p. 446)*. Complex numbers can be added or subtracted geometrically in the complex plane. The points representing complex conjugates $a + bi$ and $a - bi$ are reflections of each other in the real axis.

Use the Distance and Midpoint Formulas in the complex plane. *(p. 448)*. The distance between $z_1 = a + bi$ and $z_2 = c + di$ in the complex plane is

$$|z_1 - z_2| = \sqrt{(a - c)^2 + (b - d)^2}$$

and the midpoint of the line segment joining z_1 and z_2 in the complex plane is

$$\text{Midpoint} = \left(\frac{a + c}{2}, \frac{b + d}{2}\right).$$

Example Consider $z_1 = -3 + 2i$ and $z_2 = 2 - 3i$.

(a) The absolute value of z_1 is

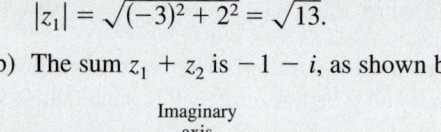

$$|z_1| = \sqrt{(-3)^2 + 2^2} = \sqrt{13}.$$

(b) The sum $z_1 + z_2$ is $-1 - i$, as shown below.

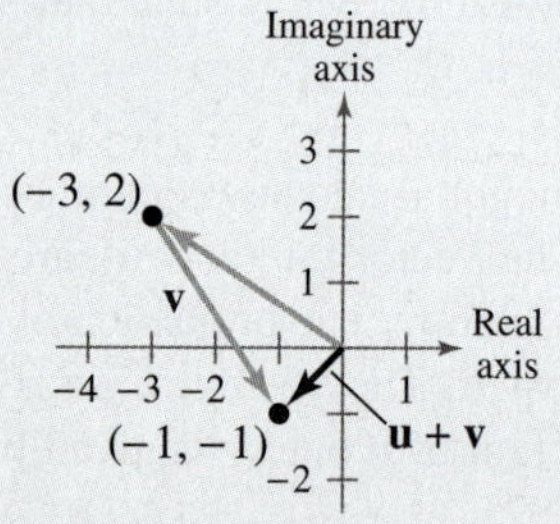

(c) The distance between z_1 and z_2 in the complex plane is

$$|z_1 - z_2| = \sqrt{(-3 - 2)^2 + [2 - (-3)]^2}$$
$$= \sqrt{50}$$
$$\approx 7.07 \text{ units.}$$

Finding the Absolute Value of a Complex Number In Exercises 89–92, plot the complex number and find its absolute value.

89. $5 + 3i$

90. $-10 - 4i$

91. $-12i$

92. $6i$

Adding Complex Numbers Geometrically In Exercises 93 and 94, find the sum of the complex numbers in the complex plane.

93. $(2 + 3i) + (1 - 2i)$

94. $(-4 + 2i) + (2 + i)$

Subtracting Complex Numbers Geometrically In Exercises 95 and 96, find the difference of the complex numbers in the complex plane.

95. $(1 + 2i) - (3 + i)$

96. $(-2 + i) - (1 + 4i)$

Finding a Complex Conjugate Geometrically In Exercises 97 and 98, plot the complex number and its complex conjugate. Write the conjugate as a complex number.

97. $3 + i$

98. $2 - 5i$

Finding a Distance In Exercises 99 and 100, find the distance between the complex numbers in the complex plane.

99. $3 + 2i, \quad 2 - i$

100. $1 + 5i, \quad -1 + 3i$

Finding the Midpoint of a Line Segment In Exercises 101 and 102, find the midpoint of the line segment joining the points corresponding to the complex numbers in the complex plane.

101. $1 + i, \quad 4 + 3i$

102. $2 - i, \quad 1 + 4i$

6.6 *What did you learn?*

Write trigonometric forms of complex numbers *(p. 451)*. The trigonometric form of the complex number $z = a + bi$ is $z = r(\cos\theta + i\sin\theta)$, where $a = r\cos\theta$, $b = r\sin\theta$, $r = \sqrt{a^2 + b^2}$, and $\tan\theta = b/a$. The number r is the modulus of z, and θ is called an argument of z.

Multiply and divide complex numbers written in trigonometric form *(p. 453)*. Let $z_1 = r_1(\cos\theta_1 + i\sin\theta_1)$ and $z_2 = r_2(\cos\theta_2 + i\sin\theta_2)$ be complex numbers.

$$z_1z_2 = r_1r_2[\cos(\theta_1 + \theta_2) + i\sin(\theta_1 + \theta_2)] \quad \text{Product}$$

$$\frac{z_1}{z_2} = \frac{r_1}{r_2}[\cos(\theta_1 - \theta_2) + i\sin(\theta_1 - \theta_2)], \; z_2 \neq 0 \quad \text{Quotient}$$

Use DeMoivre's Theorem to find powers of complex numbers *(p. 455)*.

DeMoivre's Theorem: If $z = r(\cos\theta + i\sin\theta)$ is a complex number and n is a positive integer, then $z^n = [r(\cos\theta + i\sin\theta)]^n = r^n(\cos n\theta + i\sin n\theta)$.

Find *n*th roots of complex numbers *(p. 456)*. The complex number $u = a + bi$ is an nth root of the complex number z when $z = u^n = (a + bi)^n$.

For a positive integer n, the complex number $z = r(\cos\theta + i\sin\theta)$ has exactly n distinct roots given by

$$z_k = \sqrt[n]{r}\left(\cos\frac{\theta + 2\pi k}{n} + i\sin\frac{\theta + 2\pi k}{n}\right)$$

where $k = 0, 1, 2, \ldots, n - 1$.

Example

(a) For the complex number $z = 4\sqrt{3} - 4i$, the modulus is

$$r = \sqrt{(4\sqrt{3})^2 + (-4)^2} = \sqrt{48 + 16} = 8$$

and because z lies in the Quadrant IV, the angle θ is

$$\theta = 360° + \arctan\left[-4/(4\sqrt{3})\right] = 360° - 30° = 330°.$$

So, $z = 8(\cos 330° + i\sin 330°)$ in trigonometric form.

(b) For $z_1 = 8(\cos 330° + i\sin 330°)$ and $z_2 = 4(\cos 30° + i\sin 30°)$,

$$z_1z_2 = 8 \cdot 4[\cos(330° + 30°) + i\sin(330° + 30°)] = 32(\cos 360° + i\sin 360°) = 32$$

$$\frac{z_1}{z_2} = \frac{8}{4}[\cos(330° - 30°) + i\sin(330° - 30°)] = 2(\cos 300° + i\sin 300°) = 1 - \sqrt{3}i.$$

(c) For $z = 4(\cos 30° + i\sin 30°)$,

$$z^3 = 4^3[\cos(3 \cdot 30°) + i\sin(3 \cdot 30°)] = 64i.$$

Trigonometric Form of a Complex Number In Exercises 103–106, write the complex number in trigonometric form without using a calculator.

103. $4i$

104. -9

105. $-\sqrt{3} - i$

106. $-3\sqrt{3} + 3i$

Writing a Complex Number in Standard Form In Exercises 107 and 108, write the standard form of the complex number. Then plot the complex number.

107. $2(\cos 45° + i\sin 45°)$

108. $3(\cos 30° + i\sin 30°)$

Multiplying Complex Numbers **In Exercises 109 and 110, find the product. Leave the result in trigonometric form.**

109. $\left[\frac{5}{2}\left(\cos\frac{\pi}{2} + i\sin\frac{\pi}{2}\right)\right]\left[4\left(\cos\frac{\pi}{4} + i\sin\frac{\pi}{4}\right)\right]$

110. $\left[2\left(\cos\frac{2\pi}{3} + i\sin\frac{2\pi}{3}\right)\right]\left[3\left(\cos\frac{\pi}{6} + i\sin\frac{\pi}{6}\right)\right]$

Dividing Complex Numbers **In Exercises 111 and 112, find the quotient. Leave the result in trigonometric form.**

111. $\dfrac{20(\cos 320° + i\sin 320°)}{5(\cos 80° + i\sin 80°)}$

112. $\dfrac{3(\cos 230° + i\sin 230°)}{9(\cos 95° + i\sin 95°)}$

Multiplying or Dividing Complex Numbers **In Exercises 113–116, (a) write the trigonometric forms of the complex numbers, (b) perform the operation using the trigonometric forms, and (c) perform the operation using the standard forms and check your result with that of part (b).**

113. $(2 - 2i)(3 + 3i)$

114. $(4 + 4i)(-1 - i)$

115. $\dfrac{3 - 3i}{2 + 2i}$

116. $\dfrac{-1 - i}{-2 - 2i}$

Finding a Power of a Complex Number **In Exercises 117 and 118, use DeMoivre's Theorem to find the power of the complex number. Write the result in standard form.**

117. $\left[2\left(\cos\frac{4\pi}{15} + i\sin\frac{4\pi}{15}\right)\right]^5$

118. $(1 - i)^8$

Finding Square Roots of a Complex Number **In Exercises 119–122, find the square roots of the complex number.**

119. $-\sqrt{3} + i$

120. $\sqrt{3} - i$

121. $-4i$

122. $-9i$

Finding the *n*th Roots of a Complex Number **In Exercises 123–126, (a) use the formula on page 457 to find the roots of the complex number, (b) write each of the roots in standard form, and (c) represent each of the roots graphically.**

123. Sixth roots of $-729i$

124. Fourth roots of $1296i$

125. Cube roots of 343

126. Fifth roots of -1024

Solving an Equation **In Exercises 127–130, use the formula on page 457 to find all solutions of the equation, and represent the solutions graphically.**

127. $x^4 + 256 = 0$

128. $x^5 - 32i = 0$

129. $x^3 + 8i = 0$

130. $x^4 + 81 = 0$

Focusing on Concepts

True or False? **In Exercises 131–136, determine whether the statement is true or false. Justify your answer.**

131. The Law of Sines is true when one of the angles in the triangle is a right angle.

132. When the Law of Sines is used, the solution is always unique.

133. Two angles and one side of a triangle do not necessarily determine a unique triangle.

134. In addition to SSS and SAS, the Law of Cosines can be used to solve triangles with AAS conditions.

135. A sliding door moves along the line of vector $\overrightarrow{PQ}$. If a force is applied to the door along a vector that is orthogonal to $\overrightarrow{PQ}$, then no work is done.

136. The complex number $\sqrt{3} + i$ is a solution of the equation $x^2 - 8i = 0$.

137. Law of Sines State the Law of Sines from memory.

138. Law of Cosines State the Law of Cosines from memory.

139. Reasoning What characterizes a vector in the plane?

140. Think About It Identify each quantity as either a scalar or a vector. Explain.

(a) muzzle velocity of a bullet

(b) price of a company's stock

(c) air temperature of a room

(d) weight of an automobile

(e) volume of a fish tank

(f) current of a river

141. Negative of a Complex Number Show that the negative of $z = r(\cos\theta + i\sin\theta)$ is

$$-z = r[\cos(\theta + \pi) + i\sin(\theta + \pi)].$$

142. Graphical Reasoning The figure shows z_1 and z_2. Describe z_1z_2 and z_1/z_2.

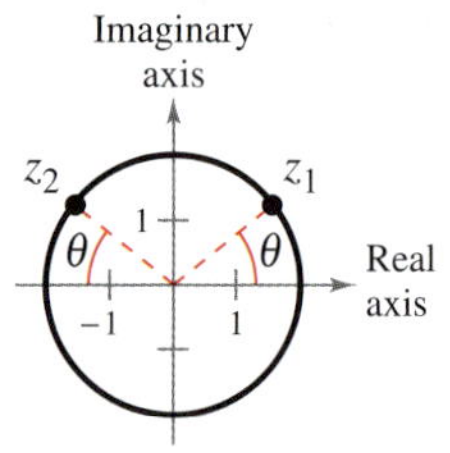

6 Chapter Test

See *CalcChat.com* for tutorial help and worked-out solutions to odd-numbered exercises.
For instructions on how to use a graphing utility, see Appendix A.

Take this test as you would take a test in class. After you are finished, check your work against the answers given in the back of the book.

In Exercises 1–6, use the given information to solve the triangle (if possible). If two solutions exist, find both solutions.

1. $A = 36°, \quad B = 98°, \quad c = 16$

2. $a = 2, \quad b = 4, \quad c = 5$

3. $A = 35°, \quad b = 8, \quad c = 12$

4. $B = 24°, \quad C = 68°, \quad b = 12.2$

5. $A = 25°, \quad b = 28, \quad a = 18$

6. $B = 130°, \quad c = 10.1, \quad b = 5.2$

7. Find the length BC of the pond shown at the right.

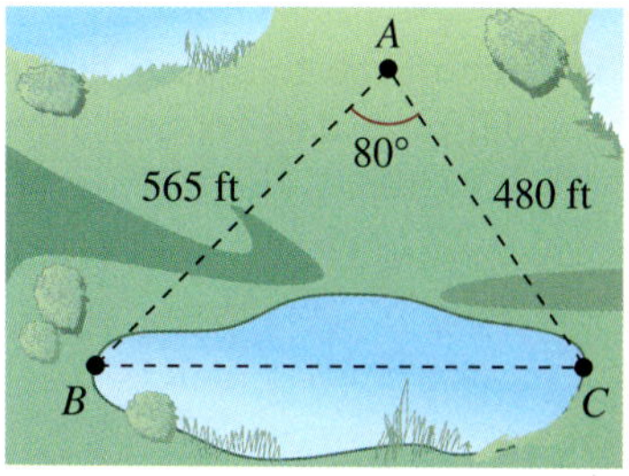

Figure for 7

8. A triangular parcel of land has borders of lengths 55 meters, 85 meters, and 100 meters. Find the area of the parcel of land.

9. Find the component form and magnitude of the vector $\mathbf{w}$ that has initial point $(-8, -12)$ and terminal point $(4, 1)$.

In Exercises 10 and 11, find (a) $2\mathbf{v} + \mathbf{u}$, (b) $\mathbf{u} - 3\mathbf{v}$, and (c) $5\mathbf{u} - 4\mathbf{v}$. Then sketch the resultant vector.

10. $\mathbf{u} = \langle -7, -4 \rangle, \quad \mathbf{v} = \langle 4, 6 \rangle$

11. $\mathbf{u} = 2\mathbf{i} + 3\mathbf{j}, \quad \mathbf{v} = -\mathbf{i} - 2\mathbf{j}$

12. Find a unit vector in the direction of $\mathbf{v} = 24\mathbf{i} - 7\mathbf{j}$.

13. Find the component form of the vector $\mathbf{v}$ with $\|\mathbf{v}\| = 12$, in the same direction as $\mathbf{u} = \langle 3, -5 \rangle$.

14. Forces with magnitudes of 250 pounds and 130 pounds act on an object at angles of 45° and −60°, respectively, with the positive x-axis. Find the direction and magnitude of the resultant of these forces.

15. Find the dot product of $\mathbf{u} = \langle -9, 4 \rangle$ and $\mathbf{v} = \langle 1, 2 \rangle$.

16. Find the angle θ between the vectors $\mathbf{u} = 7\mathbf{i} + 2\mathbf{j}$ and $\mathbf{v} = \quad 4\mathbf{j}$.

17. Are the vectors $\mathbf{u} = \langle 6, -10 \rangle$ and $\mathbf{v} = \langle -5, -3 \rangle$ orthogonal? Explain.

18. Find the projection of $\mathbf{u} = \langle 6, 7 \rangle$ onto $\mathbf{v} = \langle -5, -1 \rangle$. Then write $\mathbf{u}$ as the sum of two orthogonal vectors, one of which is $\text{proj}_{\mathbf{v}}\mathbf{u}$.

19. Plot $z = 3 + 6i$ and find its absolute value.

In Exercises 20 and 21, find the sum or difference of the complex numbers in the complex plane.

20. $(7 - 5i) + (-3 + 2i)$

21. $(6 + 4i) - (5 - 7i)$

22. Find the distance between $4 + 3i$ and $1 - i$ in the complex plane.

23. Plot $z_1 = 5i$ and $z_2 = -6 + i$ and find the midpoint of the segment joining the points.

24. Write the complex number $z = -6 + 6i$ in trigonometric form.

25. Write the complex number $100(\cos 240° + i \sin 240°)$ in standard form. Then plot the complex number.

In Exercises 26 and 27, use DeMoivre's Theorem to find the indicated power of the complex number. Write the result in standard form.

26. $\left[3\left(\cos \frac{5\pi}{6} + i \sin \frac{5\pi}{6}\right)\right]^8$

27. $(3 - 3i)^6$

28. Find the fourth roots of $128(1 + \sqrt{3}i)$. Represent each of the roots graphically.

29. Find all solutions of the equation $x^4 - 625i = 0$ and represent the solutions graphically.

Collaborative Project

To work a collaborative project involving Additional Topics in Trigonometry, visit this textbook's website at *LarsonPrecalculus.com.*

Standardized Test Practice

See *CalcChat.com* for tutorial help and worked-out solutions to odd-numbered exercises.
For instructions on how to use a graphing utility, see Appendix A.

1. Which of the following expressions is equivalent to $(f + g)(x)$, where $f(x) = x + 1$ and $g(x) = x - 1$?

(A) $x^2 - 1$ (B) $x^2 + 1$

(C) 2 (D) $2x$

2. Which of the following equations represents the best model for a data set consisting of the points $(-2, 6)$, $(-1, 4)$, $(0, 3)$, $(1, 1)$, $(2, -1)$?

(A) $y = -2x - 1$ (B) $y = -1.7x + 2.6$

(C) $y = -2x^2 - 1$ (D) $y = -1.7x^2 + 2.6$

3.

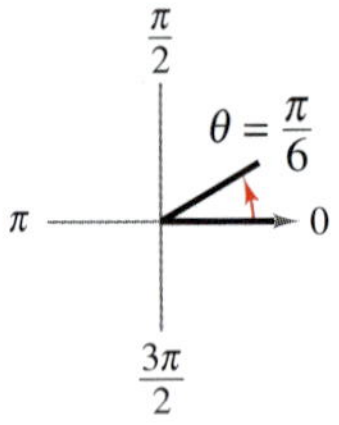

An angle in standard position is shown above. Which of the following is true?

I. The angle in degree measure is 30°.

II. The angle lies in Quadrant IV.

III. The angle is coterminal with $-\dfrac{7\pi}{6}$.

(A) I only (B) II only

(C) I and III only (D) II and III only

4. What are the values of x in the interval $0 \le x \le 2\pi$ such that $-\sin x = \cos x$?

(A) $x = \dfrac{3\pi}{4}, \dfrac{7\pi}{4}$ (B) $x = \dfrac{\pi}{4}, \dfrac{5\pi}{4}$

(C) $x = \dfrac{\pi}{3}, \dfrac{2\pi}{3}$ (D) $x = 0, 2\pi$

5. What is the value of $\cot\theta - \cos\theta$?

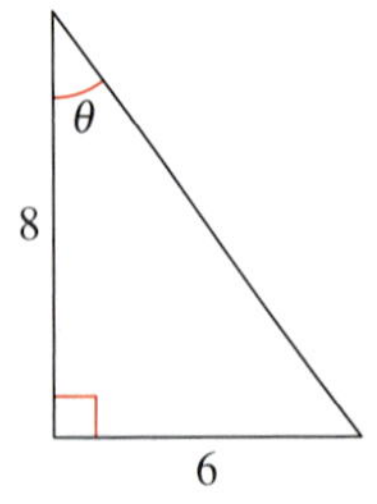

(A) $-\dfrac{1}{20}$

(B) 0

(C) $\dfrac{8}{15}$

(D) $\dfrac{32}{15}$

6. The temperature y (in degrees Fahrenheit) is modeled by $y = 75 + 18\sin 0.2t$, where t is the hours after sunrise. Which of the following is the best approximation of the temperature when $t = 4$?

(A) 62.09°F (B) 75.25°F

(C) 87.54°F (D) 87.91°F

7. What is the period of a graph of $y = 2\cos(3x - 5)$?

(A) $-\dfrac{2\pi}{5}$ (B) $\dfrac{2\pi}{3}$

(C) π (D) 2π

8. Which of the following is a graph of $y = -2\tan x$?

(A)

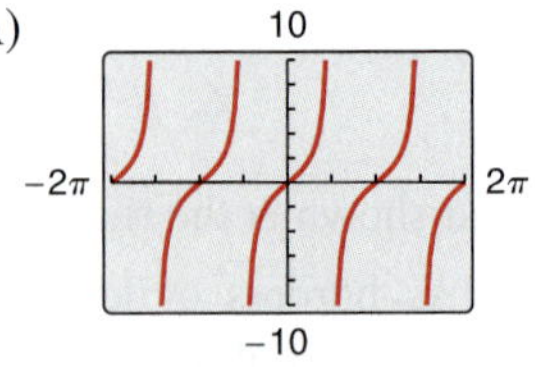

(B)

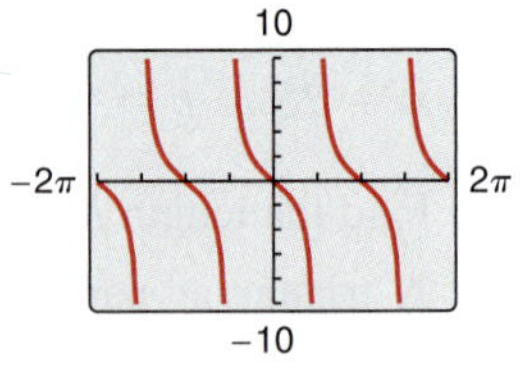

(C)

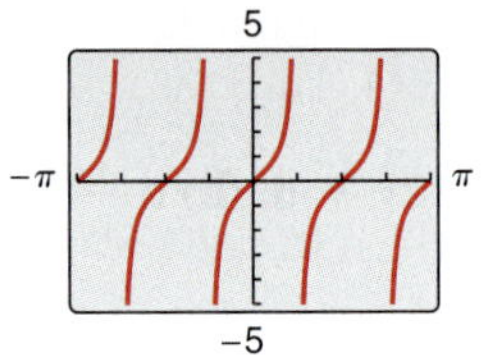

(D)

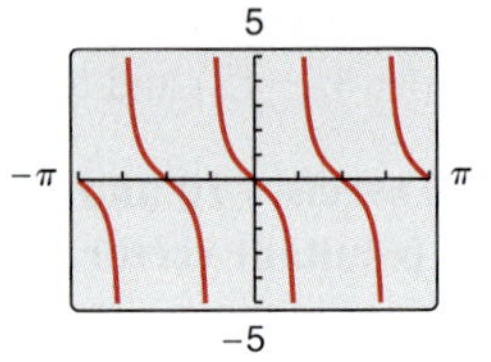

9. $\sin(\arctan 0.75)$

What is the exact value of the expression above?

(A) -0.75

(B) -0.6

(C) 0.6

(D) 0.75

10. Which bearing represents 23° west of south?

(A) W 23° S

(B) S 23° W

(C) N 67° W

(D) E 67° W

11. $\sqrt{100 - x^2}$

Which of the following substitutions allows the expression above to simplify to $10\cos\theta$?

(A) $x = \sqrt{10}\sin\theta$

(B) $x = 10\sin\theta$

(C) $x = \sqrt{10}\cos\theta$

(D) $x = 10\cos\theta$

12. $\tan^2\beta - 4\sec\beta + 5$

Which of the following expressions is equivalent to the expression above?

(A) $(\tan\beta + 2)^2 + 1$

(B) $\dfrac{\sin^2\beta - 9\cos\beta}{\cos\beta}$

(C) $(\sec\beta - 2)^2$

(D) $\dfrac{1}{\cos^2\beta - 4\cos\beta - 4}$

13. What are the solutions of $\tan x + \sqrt{3} = 0$ in the interval $[0, 360°)$?

(A) $150°, 330°$ (B) $120°, 300°$

(C) $120°, 330°$ (D) $150°, 300°$

14. $\cos \frac{5\pi}{3} \cos v + \sin \frac{5\pi}{3} \sin v = -\frac{1}{2}$

What is the measure of v in the equation above, where v is an acute angle?

(A) $\frac{\pi}{12}$ (B) $\frac{\pi}{9}$

(C) $\frac{\pi}{4}$ (D) $\frac{\pi}{3}$

15.

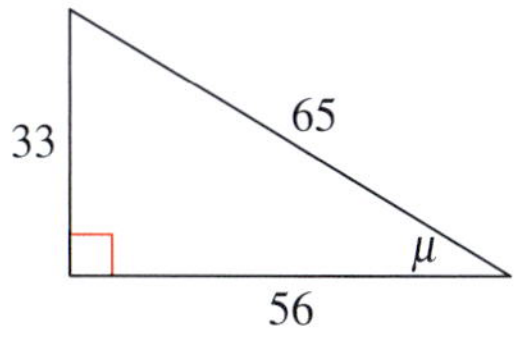

According to the figure above, what is the measure of $\tan(u/2)$?

(A) $\frac{3}{130}$ (B) $\frac{3}{11}$

(C) $\frac{6}{11}$ (D) $\frac{130}{3}$

16. A triangular platform has angle measures of 29°, 51°, and 100°. The side across from the 29° angle is 4.7 meters long. Which of the following best approximates the perimeter of the platform?

(A) 7.1 m (B) 17.1 m

(C) 21.8 m (D) 29.2 m

17. What is the area of a triangle with side lengths of 4 inches, 5 inches, and 7 inches?

(A) $4\sqrt{6}$ in.^{2} (B) 10 in.^{2}

(C) 70 in.^{2} (D) $24\sqrt{33}$ in.^{2}

18.

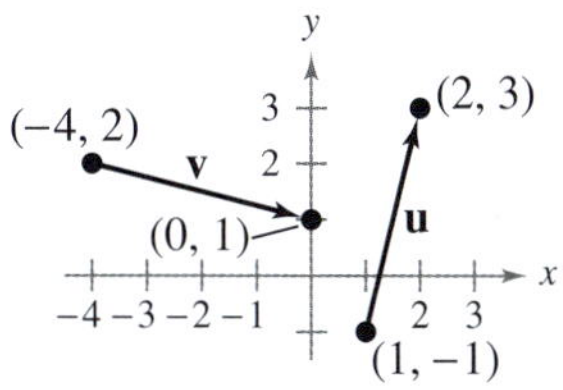

According to the figure above, what is the relationship between the magnitudes of **u** and **v**?

(A) $\|\mathbf{u}\| = \|\mathbf{v}\|$ (B) $\|\mathbf{u}\| = -\|\mathbf{v}\|$

(C) $\|\mathbf{v}\| = 2\|\mathbf{u}\|$ (D) $\|\mathbf{u}\| = 2\|\mathbf{v}\|$

19. What is the dot product of $\mathbf{u} = 2\mathbf{i} - 2\mathbf{j}$ and $\mathbf{v} = 3\mathbf{i} - 5\mathbf{j}$?

(A) 0 (B) 16

(C) $5\mathbf{i} - 7\mathbf{j}$ (D) $6\mathbf{i} + 10\mathbf{j}$

20. Which of the following is the graph of the complex conjugate of $-1 + 2i$ in the complex plane?

(A)

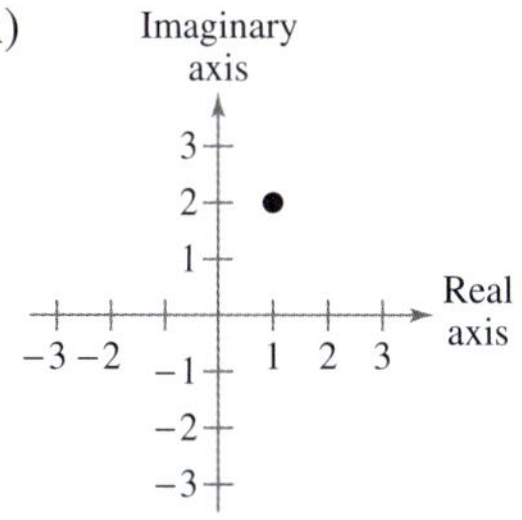

(B)

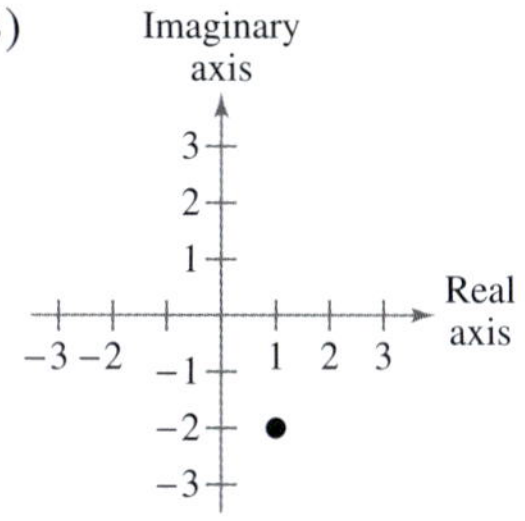

(C)

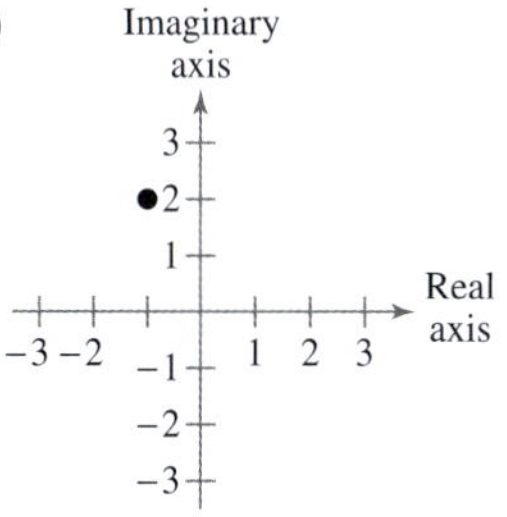

(D)

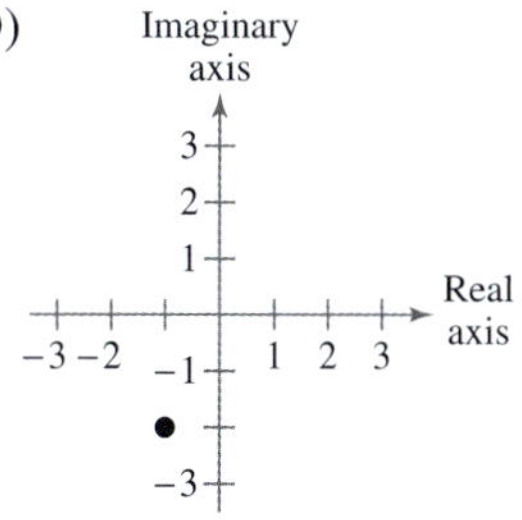

21. Which of the following is the trigonometric form of the complex number $z = 7i$?

(A) $\sqrt{8}\left(\cos \frac{\pi}{2} + i \sin \frac{\pi}{2}\right)$ (B) $7\left(\cos \frac{\pi}{2} + i \sin \frac{\pi}{2}\right)$

(C) $7 \cos \frac{3\pi}{2}$ (D) $\sqrt{8}i \sin \frac{3\pi}{2}$

22. What is the value of $\tan(-219°)$, rounded to the nearest hundredth?

23. What is the amplitude of the graph of

$y = 3.2 \sin(2x + 5.4)$?

24. What is the solution of the equation

$$\sin\left(x + \frac{\pi}{3}\right) + \sin\left(x - \frac{\pi}{4}\right) = 1$$

on the interval $(0, 1]$, rounded to the nearest hundredth?

25. What is the distance between the boat and the lighthouse shown in the figure in miles, rounded to the nearest tenth?

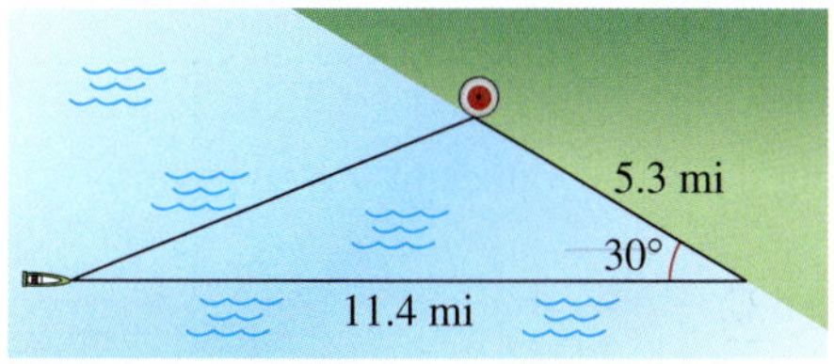

26. $\mathbf{v} = 6\mathbf{i} + 4\mathbf{j}$

The initial point of **v** is (3, 4). The terminal point of **v** is $(x, 8)$. What is the value of x?

27. A 90-pound cart sits on a ramp inclined at 20°. What is the magnitude of the force (in pounds) required to keep the cart from rolling down the ramp, rounded to the nearest hundredth?

28. What is the absolute value of $-4 - 3i$?

Proofs in Mathematics

Law of Sines (p. 406)

If ABC is a triangle with sides a, b, and c, then

$$\frac{a}{\sin A} = \frac{b}{\sin B} = \frac{c}{\sin C}.$$

Oblique Triangles

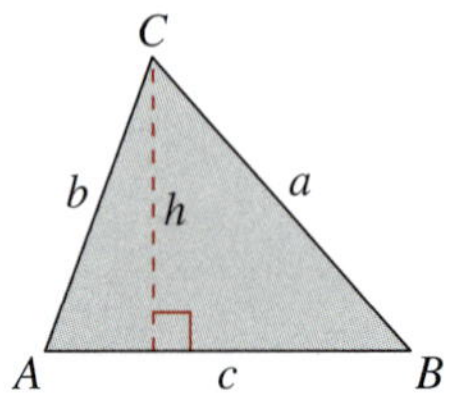

A is acute.

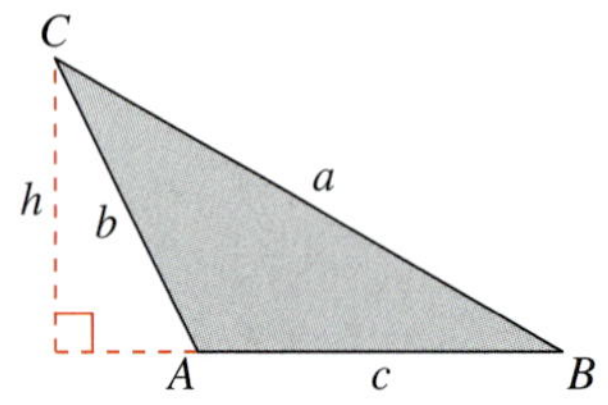

A is obtuse.

Law of Tangents

Besides the Law of Sines and the Law of Cosines, there is also a Law of Tangents, which was developed by Francois Viète (1540–1603). The Law of Tangents follows from the Law of Sines and the sum-to-product formulas for sine and is defined as

$$\frac{a+b}{a-b} = \frac{\tan[(A+B)/2]}{\tan[(A-B)/2]}.$$

The Law of Tangents can be used to solve a triangle when two sides and the included angle are given (SAS). Before calculators were invented, the Law of Tangents was used to solve the SAS case instead of the Law of Cosines, because computation with a table of tangent values was easier.

Proof

Let h be the altitude of either triangle shown in the figure above. Then you have

$$\sin A = \frac{h}{b} \quad \text{or} \quad h = b \sin A$$

and

$$\sin B = \frac{h}{a} \quad \text{or} \quad h = a \sin B.$$

Equating these two values of h, you have

$$a \sin B = b \sin A \quad \text{or} \quad \frac{a}{\sin A} = \frac{b}{\sin B}.$$

Note that $\sin A \neq 0$ and $\sin B \neq 0$ because no angle of a triangle can have a measure of 0° or 180°. In a similar manner, construct an altitude from vertex B to side AC (extended in the obtuse triangle), as shown at the right. Then you have

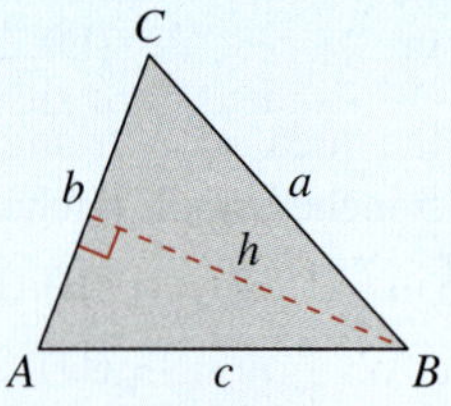

A is acute.

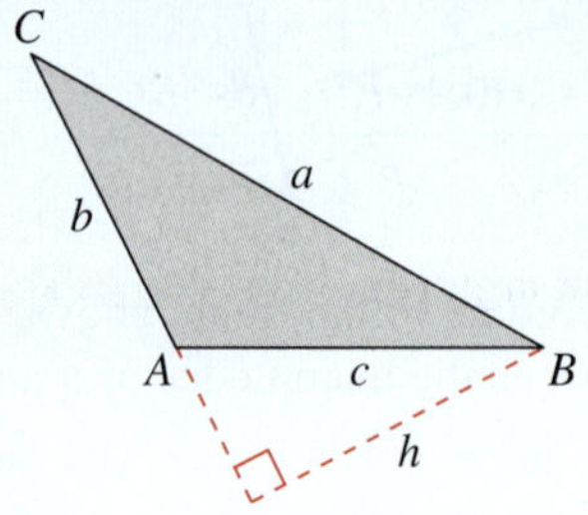

A is obtuse.

$$\sin A = \frac{h}{c} \quad \text{or} \quad h = c \sin A$$

and

$$\sin C = \frac{h}{a} \quad \text{or} \quad h = a \sin C.$$

Equating these two values of h, you have

$$a \sin C = c \sin A \quad \text{or} \quad \frac{a}{\sin A} = \frac{c}{\sin C}.$$

By the Transitive Property of Equality, you know that

$$\frac{a}{\sin A} = \frac{b}{\sin B} = \frac{c}{\sin C}.$$

So, the Law of Sines is established. ∎

Law of Cosines (p. 415)

Standard Form	*Alternative Form*
$a^2 = b^2 + c^2 - 2bc \cos A$	$\cos A = \dfrac{b^2 + c^2 - a^2}{2bc}$
$b^2 = a^2 + c^2 - 2ac \cos B$	$\cos B = \dfrac{a^2 + c^2 - b^2}{2ac}$
$c^2 = a^2 + b^2 - 2ab \cos C$	$\cos C = \dfrac{a^2 + b^2 - c^2}{2ab}$

Proof

To prove the first formula, consider the top triangle at the right, which has three acute angles. Note that vertex B has coordinates $(c, 0)$. Furthermore, C has coordinates (x, y), where

$$x = b \cos A \quad \text{and} \quad y = b \sin A.$$

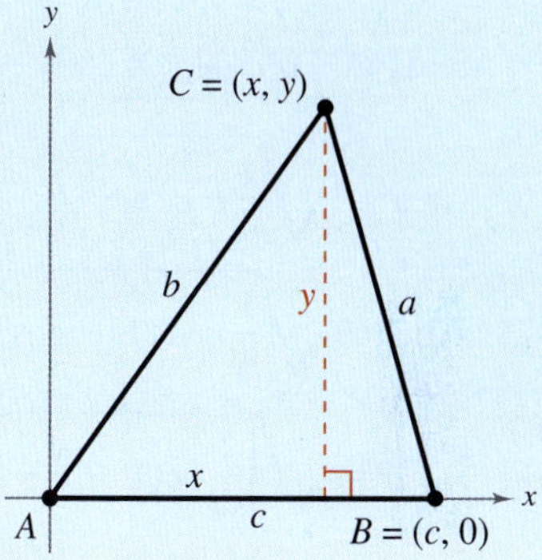

Because a is the distance from vertex C to vertex B, it follows that

$$a = \sqrt{(x - c)^2 + (y - 0)^2} \qquad \text{Distance Formula}$$

$$a^2 = (x - c)^2 + (y - 0)^2 \qquad \text{Square each side.}$$

$$a^2 = (b \cos A - c)^2 + (b \sin A)^2 \qquad \text{Substitute for } x \text{ and } y.$$

$$a^2 = b^2 \cos^2 A - 2bc \cos A + c^2 + b^2 \sin^2 A \qquad \text{Expand.}$$

$$a^2 = b^2(\sin^2 A + \cos^2 A) + c^2 - 2bc \cos A \qquad \text{Factor out } b^2.$$

$$a^2 = b^2 + c^2 - 2bc \cos A. \qquad \sin^2 A + \cos^2 A = 1$$

To prove the second formula, consider the bottom triangle at the right, which also has three acute angles. Note that vertex A has coordinates $(c, 0)$. Furthermore, C has coordinates (x, y), where

$$x = a \cos B \quad \text{and} \quad y = a \sin B.$$

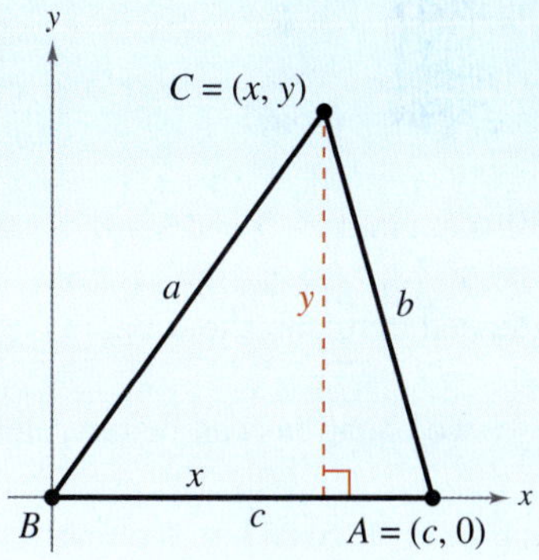

Because b is the distance from vertex C to vertex A, it follows that

$$b = \sqrt{(x - c)^2 + (y - 0)^2} \qquad \text{Distance Formula}$$

$$b^2 = (x - c)^2 + (y - 0)^2 \qquad \text{Square each side.}$$

$$b^2 = (a \cos B - c)^2 + (a \sin B)^2 \qquad \text{Substitute for } x \text{ and } y.$$

$$b^2 = a^2 \cos^2 B - 2ac \cos B + c^2 + a^2 \sin^2 B \qquad \text{Expand.}$$

$$b^2 = a^2(\sin^2 B + \cos^2 B) + c^2 - 2ac \cos B \qquad \text{Factor out } a^2.$$

$$b^2 = a^2 + c^2 - 2ac \cos B. \qquad \sin^2 B + \cos^2 B = 1$$

A similar argument is used to establish the third formula. ■

Heron's Area Formula (p. 418)

For any triangle with sides of lengths a, b, and c, the area of the triangle is

$$\text{Area} = \sqrt{s(s-a)(s-b)(s-c)}$$

where $s = \dfrac{a+b+c}{2}$.

Proof

From Section 6.1, you know that

$$\text{Area} = \frac{1}{2}bc \sin A \qquad \text{Formula for the area of an oblique triangle}$$

$$(\text{Area})^2 = \frac{1}{4}b^2c^2 \sin^2 A \qquad \text{Square each side.}$$

$$\text{Area} = \sqrt{\frac{1}{4}b^2c^2 \sin^2 A} \qquad \text{Take the square root of each side.}$$

$$= \sqrt{\frac{1}{4}b^2c^2(1-\cos^2 A)} \qquad \text{Pythagorean identity}$$

$$= \sqrt{\left[\frac{1}{2}bc(1+\cos A)\right]\left[\frac{1}{2}bc(1-\cos A)\right]}. \qquad \text{Factor.}$$

Using the alternative form of the Law of Cosines,

$$\frac{1}{2}bc(1+\cos A) = \frac{a+b+c}{2} \cdot \frac{-a+b+c}{2}.$$

Similarly,

$$\frac{1}{2}bc(1-\cos A) = \frac{a-b+c}{2} \cdot \frac{a+b-c}{2}.$$

Letting

$$s = \frac{a+b+c}{2}$$

these two equations can be rewritten as

$$\frac{1}{2}bc(1+\cos A) = s(s-a)$$

and

$$\frac{1}{2}bc(1-\cos A) = (s-b)(s-c).$$

By substituting into the last formula for area, you can conclude that

$$\text{Area} = \sqrt{s(s-a)(s-b)(s-c)}.$$

Remark

$$\frac{1}{2}bc(1+\cos A)$$

$$= \frac{1}{2}bc\left(1 + \frac{b^2+c^2-a^2}{2bc}\right)$$

$$= \frac{1}{2}bc\left(\frac{2bc+b^2+c^2-a^2}{2bc}\right)$$

$$= \frac{1}{4}(2bc+b^2+c^2-a^2)$$

$$= \frac{1}{4}(b^2+2bc+c^2-a^2)$$

$$= \frac{1}{4}[(b+c)^2-a^2]$$

$$= \frac{1}{4}(b+c+a)(b+c-a)$$

$$= \frac{a+b+c}{2} \cdot \frac{-a+b+c}{2}$$

Properties of the Dot Product (p. 436)

Let **u**, **v**, and **w** be vectors and let c be a scalar.

1. $\mathbf{u} \cdot \mathbf{v} = \mathbf{v} \cdot \mathbf{u}$
2. $\mathbf{0} \cdot \mathbf{v} = 0$
3. $\mathbf{u} \cdot (\mathbf{v} + \mathbf{w}) = \mathbf{u} \cdot \mathbf{v} + \mathbf{u} \cdot \mathbf{w}$
4. $\mathbf{v} \cdot \mathbf{v} = \|\mathbf{v}\|^2$
5. $c(\mathbf{u} \cdot \mathbf{v}) = c\mathbf{u} \cdot \mathbf{v} = \mathbf{u} \cdot c\mathbf{v}$

Proof

Let $\mathbf{u} = \langle u_1, u_2 \rangle$, $\mathbf{v} = \langle v_1, v_2 \rangle$, $\mathbf{w} = \langle w_1, w_2 \rangle$, and $\mathbf{0} = \langle 0, 0 \rangle$, and let c be a scalar.

1. $\mathbf{u} \cdot \mathbf{v} = u_1v_1 + u_2v_2 = v_1u_1 + v_2u_2 = \mathbf{v} \cdot \mathbf{u}$

2. $\mathbf{0} \cdot \mathbf{v} = 0 \cdot v_1 + 0 \cdot v_2 = 0$

3.
$$\begin{aligned}
\mathbf{u} \cdot (\mathbf{v} + \mathbf{w}) &= \mathbf{u} \cdot \langle v_1 + w_1, v_2 + w_2 \rangle \\
&= u_1(v_1 + w_1) + u_2(v_2 + w_2) \\
&= u_1v_1 + u_1w_1 + u_2v_2 + u_2w_2 \\
&= (u_1v_1 + u_2v_2) + (u_1w_1 + u_2w_2) \\
&= \mathbf{u} \cdot \mathbf{v} + \mathbf{u} \cdot \mathbf{w}
\end{aligned}$$

4. $\mathbf{v} \cdot \mathbf{v} = v_1^2 + v_2^2 = \left(\sqrt{v_1^2 + v_2^2}\right)^2 = \|\mathbf{v}\|^2$

5.
$$\begin{aligned}
c(\mathbf{u} \cdot \mathbf{v}) &= c(\langle u_1, u_2 \rangle \cdot \langle v_1, v_2 \rangle) \\
&= c(u_1v_1 + u_2v_2) \\
&= (cu_1)v_1 + (cu_2)v_2 \\
&= \langle cu_1, cu_2 \rangle \cdot \langle v_1, v_2 \rangle \\
&= c\mathbf{u} \cdot \mathbf{v}
\end{aligned}$$

Angle Between Two Vectors (p. 437)

If θ is the angle between two nonzero vectors **u** and **v**, then

$$\cos\theta = \frac{\mathbf{u} \cdot \mathbf{v}}{\|\mathbf{u}\|\,\|\mathbf{v}\|}.$$

Proof

Consider the triangle determined by vectors **u**, **v**, and **v** − **u**, as shown in the figure. By the Law of Cosines, you can write

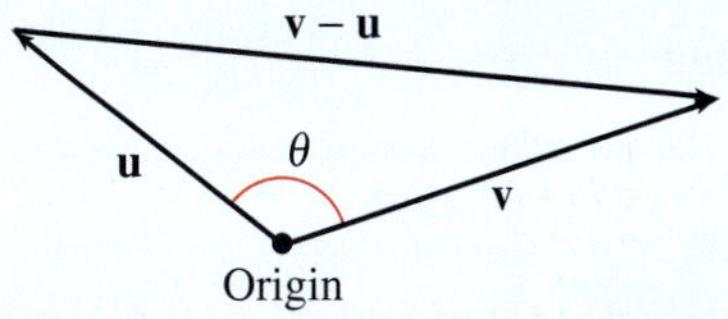

$$\begin{aligned}
\|\mathbf{v} - \mathbf{u}\|^2 &= \|\mathbf{u}\|^2 + \|\mathbf{v}\|^2 - 2\|\mathbf{u}\|\,\|\mathbf{v}\| \cos\theta \\
(\mathbf{v} - \mathbf{u}) \cdot (\mathbf{v} - \mathbf{u}) &= \|\mathbf{u}\|^2 + \|\mathbf{v}\|^2 - 2\|\mathbf{u}\|\,\|\mathbf{v}\| \cos\theta \\
(\mathbf{v} - \mathbf{u}) \cdot \mathbf{v} - (\mathbf{v} - \mathbf{u}) \cdot \mathbf{u} &= \|\mathbf{u}\|^2 + \|\mathbf{v}\|^2 - 2\|\mathbf{u}\|\,\|\mathbf{v}\| \cos\theta \\
\mathbf{v} \cdot \mathbf{v} - \mathbf{u} \cdot \mathbf{v} - \mathbf{v} \cdot \mathbf{u} + \mathbf{u} \cdot \mathbf{u} &= \|\mathbf{u}\|^2 + \|\mathbf{v}\|^2 - 2\|\mathbf{u}\|\,\|\mathbf{v}\| \cos\theta \\
\|\mathbf{v}\|^2 - 2\mathbf{u} \cdot \mathbf{v} + \|\mathbf{u}\|^2 &= \|\mathbf{u}\|^2 + \|\mathbf{v}\|^2 - 2\|\mathbf{u}\|\,\|\mathbf{v}\| \cos\theta
\end{aligned}$$

$$\cos\theta = \frac{\mathbf{u} \cdot \mathbf{v}}{\|\mathbf{u}\|\,\|\mathbf{v}\|}.$$

Progressive Summary (Chapters 1–6)

This chart outlines the topics that have been covered so far in this text. Progressive Summary charts appear after Chapters 2, 3, 6, 9, and 11. In each Progressive Summary, new topics encountered for the first time appear in blue.

Algebraic Functions

Polynomial, Rational, Radical

Rewriting

Polynomial form ↔ Factored form
Operations with polynomials
Rationalize denominators
Simplify rational expressions
Operations with complex numbers

Solving

Equation	*Strategy*
Linear	Isolate variable
Quadratic	Factor, set to zero Extract square roots Complete the square Quadratic Formula
Polynomial	Factor, set to zero Rational Zero Test
Rational	Multiply by LCD
Radical	Isolate, raise to power
Absolute value	Isolate, form two equations

Analyzing

Graphically	*Algebraically*
Intercepts	Domain, Range
Symmetry	Transformations
Slope	Composition
Asymptotes	Standard forms of equations
End behavior	Leading Coefficient Test
Minimum values	Synthetic division
Maximum values	Descartes's Rule of Signs

Numerically
Table of values

Transcendental Functions

Exponential, Logarithmic
Trigonometric, Inverse Trigonometric

Rewriting

Exponential form ↔ Logarithmic form
Condense/expand logarithmic expressions
Simplify trigonometric expressions
Prove trigonometric identities
Use conversion formulas
Operations with vectors
Powers and roots of complex numbers

Solving

Equation	*Strategy*
Exponential	Take logarithm of each side
Logarithmic	Exponentiate each side
Trigonometric	Isolate function Factor, use inverse function
Multiple angle or high powers	Use trigonometric identities

Analyzing

Graphically	*Algebraically*
Intercepts	Domain, Range
Asymptotes	Transformations
Minimum values	Composition
Maximum values	Inverse Properties
	Amplitude, Period
	Reference angles

Numerically
Table of values

Other Topics

Rewriting

Solving

Analyzing

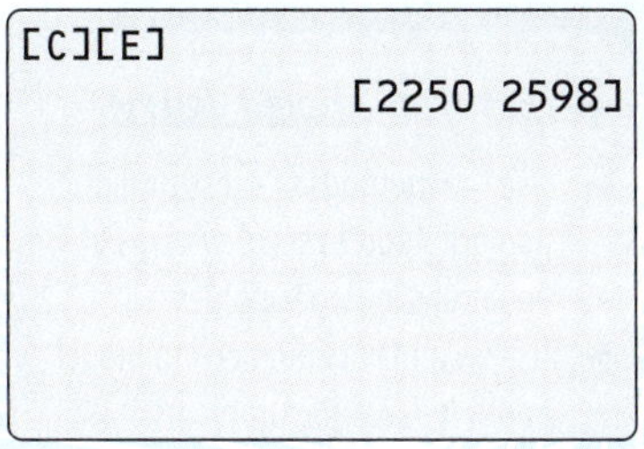

Section 7.5, Example 13
Softball Team Expenses

7 Linear Systems and Matrices

Student Resources at LarsonPrecalculus.com

- **Videos** explaining the concepts of precalculus
- **Worked-out solution videos** for all *Checkpoint* exercises
- **Editable spreadsheets** of the data sets in the text
- **Group projects** for each chapter applying concepts to real-life problems

7.1 Solving Systems of Equations

The Methods of Substitution and Graphing

So far in this text, most problems have involved either a function of one variable or a single equation in two variables. However, many problems in science, business, and engineering involve two or more equations in two or more variables. To solve such problems, you need to find solutions of **systems of equations.** Here is an example of a system of two equations in two unknowns, x and y.

$$\begin{cases} 2x + y = 5 & \text{Equation 1} \\ 3x - 2y = 4 & \text{Equation 2} \end{cases}$$

A **solution** of this system is an ordered pair that satisfies each equation in the system. Finding the set of all such solutions is called **solving the system of equations.** For instance, the ordered pair (2, 1) is a solution of this system. To check this, you can substitute 2 for x and 1 for y in *each* equation.

Check (2, 1) *in Equation 1:*

$$2x + y = 5$$
$$2(2) + 1 \stackrel{?}{=} 5$$
$$5 = 5 \checkmark$$

Check (2, 1) *in Equation 2:*

$$3x - 2y = 4$$
$$3(2) - 2(1) \stackrel{?}{=} 4$$
$$4 = 4 \checkmark$$

In this section, you will study two ways to solve systems of equations, beginning with the **method of substitution.**

The Method of Substitution

To use the **method of substitution** to solve a system of two equations in x and y, perform the following steps.

1. Solve one of the equations for one variable in terms of the other.
2. Substitute the expression found in Step 1 into the other equation to obtain an equation in one variable.
3. Solve the equation obtained in Step 2.
4. Back-substitute the value(s) obtained in Step 3 into the expression obtained in Step 1 to find the value(s) of the other variable.
5. Check that each solution satisfies *both* of the original equations.

Another method is the **method of graphing.** When using this method, note that the solution of the system corresponds to all **points of intersection** of the graphs.

The Method of Graphing

To use the **method of graphing** to solve a system of two equations in x and y, perform the following steps.

1. Solve both equations for y in terms of x.
2. Use a graphing utility to graph both equations in the same viewing window.
3. Use the *intersect* feature or the *zoom* and *trace* features of the graphing utility to approximate all points of intersection of the graphs.
4. Check that each solution satisfies *both* of the original equations.

What you should learn

- Use the methods of substitution and graphing to solve systems of equations in two variables.
- Use systems of equations to model and solve real-life problems.

Why you should learn it

You can use systems of equations in situations in which the variables must satisfy two or more conditions. For instance, Exercise 88 on page 486 shows how to use a system of equations to compare two models for estimating the number of board feet in a 16-foot log.

Algebra Help

When using the method of substitution, it does not matter which variable you choose to solve for first. Whether you solve for y first or x first, you will obtain the same solution. When making your choice, you should choose the variable and equation that are easier to work with.

EXAMPLE 1 Solving a System of Equations

Solve the system of equations.

$$\begin{cases} x + y = 4 & \text{Equation 1} \\ x - y = 2 & \text{Equation 2} \end{cases}$$

Algebraic Solution

Begin by solving for y in Equation 1.

$y = 4 - x$ Solve for y in Equation 1.

Next, substitute this expression for y into Equation 2 and solve the resulting single-variable equation for x.

$x - y = 2$ Write Equation 2.

$x - (4 - x) = 2$ Substitute $4 - x$ for y.

$x - 4 + x = 2$ Distributive Property

$2x - 4 = 2$ Combine like terms.

$2x = 6$ Add 4 to each side.

$x = 3$ Divide each side by 2.

Finally, you can solve for y by *back-substituting* $x = 3$ into the equation $y = 4 - x$ to obtain

$y = 4 - x$ Write revised Equation 1.

$y = 4 - 3$ Substitute 3 for x.

$y = 1.$ Solve for y.

The solution is the ordered pair (3, 1). Check this as follows.

Check (3, 1) *in Equation 1:*

$x + y = 4$ Write Equation 1.

$3 + 1 \stackrel{?}{=} 4$ Substitute for x and y.

$4 = 4$ Solution checks in Equation 1. ✓

Check (3, 1) *in Equation 2:*

$x - y = 2$ Write Equation 2.

$3 - 1 \stackrel{?}{=} 2$ Substitute for x and y.

$2 = 2$ Solution checks in Equation 2. ✓

Because (3, 1) satisfies both equations in the system, it is a solution of the system of equations.

Graphical Solution

Begin by solving both equations for y. Then use a graphing utility to graph the equations

$$y_1 = 4 - x$$

and

$$y_2 = x - 2$$

in the same viewing window. Use the *intersect* feature (see figure) to approximate the point of intersection of the graphs.

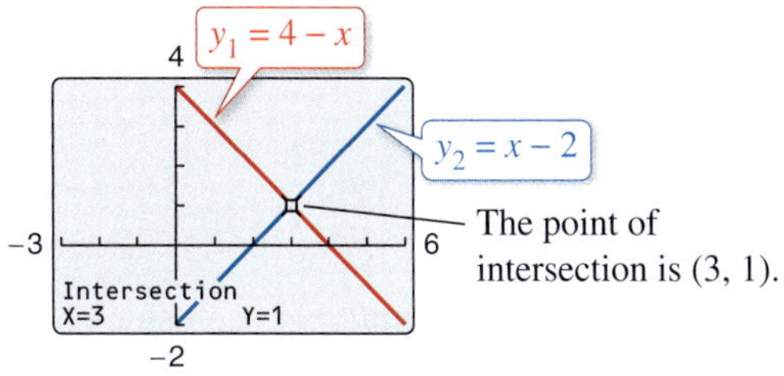

Check that (3, 1) is the exact solution as follows.

Check (3, 1) *in Equation 1:*

$x + y = 4$ Write Equation 1.

$3 + 1 \stackrel{?}{=} 4$ Substitute for x and y.

$4 = 4$ Solution checks in Equation 1. ✓

Check (3, 1) *in Equation 2:*

$x - y = 2$ Write Equation 2.

$3 - 1 \stackrel{?}{=} 2$ Substitute for x and y.

$2 = 2$ Solution checks in Equation 2. ✓

Because (3, 1) satisfies both equations in the system, it is a solution of the system of equations.

✓ **Checkpoint** ▶ *Audio-video solution in English & Spanish at LarsonPrecalculus.com*

Solve the system of equations by the method of substitution.

$$\begin{cases} x - y = 0 \\ 5x - 3y = 6 \end{cases}$$

In the algebraic solution to Example 1, note that the term *back-substitution* implies that you work *backwards*. First you solve for one of the variables, and then you substitute that value *back* into one of the equations in the system to find the value of the other variable.

EXAMPLE 2 Solving a System by Substitution

A total of \$12,000 is invested in two funds paying 5% and 3% simple interest. The yearly interest is \$500. How much is invested at each rate?

Solution

Verbal Model: Amount in 5% fund + Amount in 3% fund = Total investment

Interest for 5% fund + Interest for 3% fund = Total interest

Labels: Amount in 5% fund $= x$ (dollars)

Interest for 5% fund $= 0.05x$ (dollars)

Amount in 3% fund $= y$ (dollars)

Interest for 3% fund $= 0.03y$ (dollars)

Total investment $= 12{,}000$ (dollars)

Total interest $= 500$ (dollars)

System: $\begin{cases} x + y = 12{,}000 & \text{Equation 1} \\ 0.05x + 0.03y = 500 & \text{Equation 2} \end{cases}$

To begin, it is convenient to multiply each side of Equation 2 by 100. This eliminates the need to work with decimals.

$5x + 3y = 50{,}000$ Revised Equation 2

To solve this system, you can solve for x in Equation 1.

$x = 12{,}000 - y$ Revised Equation 1

Next, substitute this expression for x into revised Equation 2 and solve the resulting equation for y.

$5x + 3y = 50{,}000$ Write revised Equation 2.

$5(12{,}000 - y) + 3y = 50{,}000$ Substitute $12{,}000 - y$ for x.

$60{,}000 - 5y + 3y = 50{,}000$ Distributive Property

$-2y = -10{,}000$ Combine like terms.

$y = 5000$ Divide each side by -2.

Finally, back-substitute the value $y = 5000$ to solve for x.

$x = 12{,}000 - y$ Write revised Equation 1.

$x = 12{,}000 - 5000$ Substitute 5000 for y.

$x = 7000$ Simplify.

The solution is (7000, 5000). So, \$7000 is invested at 5% and \$5000 is invested at 3% to yield yearly interest of \$500. Check this in the original system.

✓ Checkpoint *Audio-video solution in English & Spanish at LarsonPrecalculus.com*

A total of \$25,000 is invested in two funds paying 6.5% and 8.5% simple interest. The yearly interest is \$2000. How much is invested at each rate?

The equations in Examples 1 and 2 are linear. Substitution and graphing can also be used to solve systems in which one or both of the equations are nonlinear.

Technology Tip

Remember that a good way to check the answers you obtain in this section is to use a graphing utility. For instance, enter the two equations in Example 2

$$y_1 = 12{,}000 - x$$

$$y_2 = \frac{500 - 0.05x}{0.03}$$

and find an appropriate viewing window that shows where the lines intersect. Then use the *intersect* feature or the *zoom* and *trace* features to find the point of intersection.

Insight

You may see linear systems like those in Examples 1 and 2 on a standardized test. On some tests, you may see a *nonlinear* system. For instance, the nonlinear system in Example 3 (see next page) has a linear equation and a quadratic equation.

photo credits to come

EXAMPLE 3 Substitution: Two-Solution Case

Solve the system of equations.

$$\begin{cases} x^2 + 4x - y = 7 & \text{Equation 1} \\ 2x - y = -1 & \text{Equation 2} \end{cases}$$

Algebraic Solution

Begin by solving for y in Equation 2 to obtain

$y = 2x + 1.$ — Solve for y in Equation 2.

Next, substitute this expression for y into Equation 1 and solve for x.

$x^2 + 4x - y = 7$ — Write Equation 1.

$x^2 + 4x - (2x + 1) = 7$ — Substitute $2x + 1$ for y.

$x^2 + 4x - 2x - 1 = 7$ — Distributive Property

$x^2 + 2x - 8 = 0$ — Write in general form.

$(x + 4)(x - 2) = 0$ — Factor.

$x + 4 = 0 \Rightarrow x = -4$ — Set 1st factor equal to 0.

$x - 2 = 0 \Rightarrow x = 2$ — Set 2nd factor equal to 0.

Back-substituting these values of x into revised Equation 2 produces

$$y = 2(-4) + 1 = -7 \quad \text{and} \quad y = 2(2) + 1 = 5.$$

So, the solutions are $(-4, -7)$ and $(2, 5)$. Check these in the original system.

Graphical Solution

Solve each equation for y and use a graphing utility to graph the equations in the same viewing window. From the figures below, the solutions are $(-4, -7)$ and $(2, 5)$. Check these in the original system.

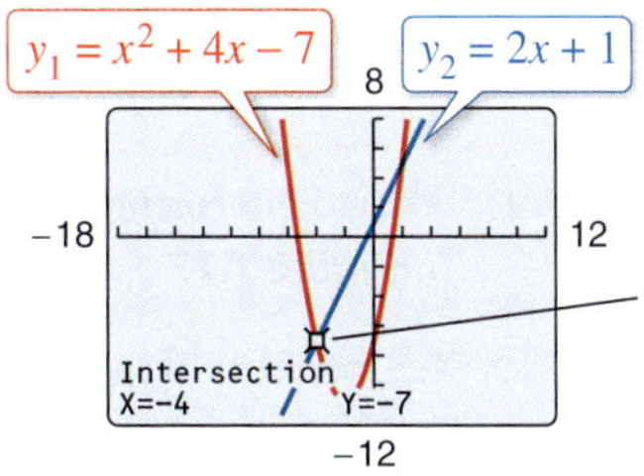

One of the points of intersection is $(-4, -7)$.

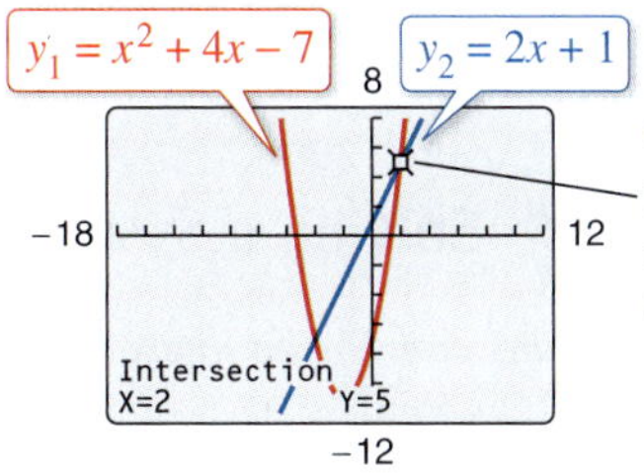

One of the points of intersection is $(2, 5)$.

✓ Checkpoint *Audio-video solution in English & Spanish at LarsonPrecalculus.com*

Solve the system of equations.

$$\begin{cases} -2x + y = 5 \\ x^2 - y + 3x = 1 \end{cases}$$

EXAMPLE 4 Substitution: No-Solution Case

To solve the system of equations

$$\begin{cases} -x + y = 4 & \text{Equation 1} \\ x^2 + y = 3 & \text{Equation 2} \end{cases}$$

begin by solving for y in Equation 1 to obtain $y = x + 4$. Next, substitute this expression for y into Equation 2 and solve for x.

$x^2 + (x + 4) = 3$ — Substitute $x + 4$ for y in Equation 2.

$x^2 + x + 1 = 0$ — Write in general form.

$x = \dfrac{-1 \pm \sqrt{3}i}{2}$ — Quadratic Formula

Because this yields two imaginary values, the equation $x^2 + x + 1 = 0$ has no *real* solution. So, the original system of equations has no *real* solution.

✓ Checkpoint *Audio-video solution in English & Spanish at LarsonPrecalculus.com*

Solve $\begin{cases} 2x - y = -3 \\ 2x^2 + 4x - y^2 = 0 \end{cases}$.

Explore the Concept

Graph the system of equations in Example 4. Do the graphs of the equations intersect? Why or why not?

From Examples 1, 3, and 4, you can see that a system of two equations in two unknowns can have exactly one solution, more than one solution, or no solution. For instance, in Figure 7.1, the two equations in Example 1 graph as two lines with a *single point* of intersection. The two equations in Example 3 graph as a parabola and a line with *two points* of intersection, as shown in Figure 7.2. The two equations in Example 4 graph as a line and a parabola that have *no points* of intersection, as shown in Figure 7.3.

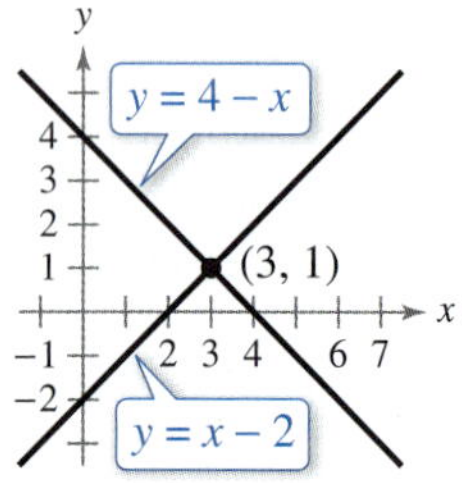

One Intersection Point
Figure 7.1

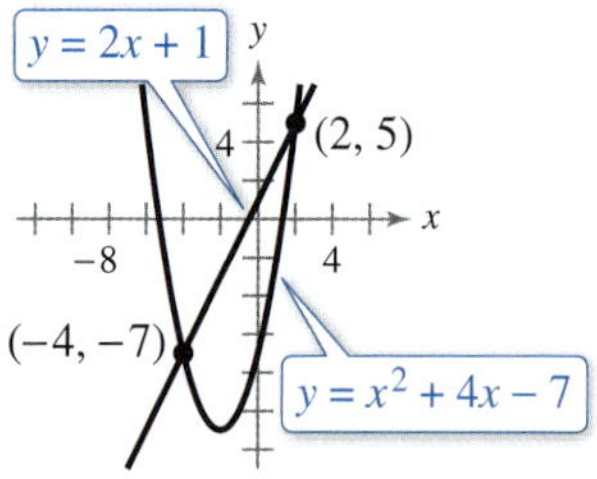

Two Intersection Points
Figure 7.2

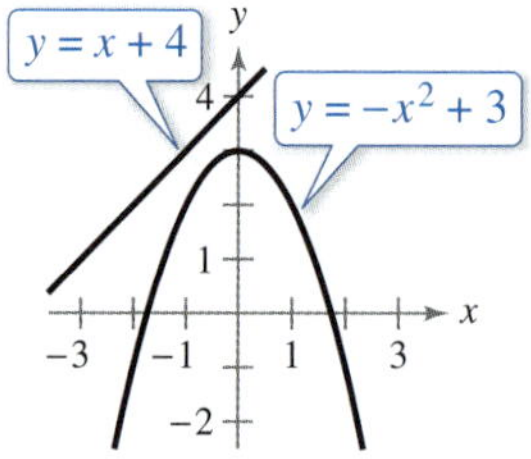

No Intersection Points
Figure 7.3

> ### Insight
> For standardized tests, be prepared to solve systems involving one or more rational coefficients (see Exercise 25). Also, it is possible that a test will have a system that has no solution (see Example 4 and Figure 7.3).

Example 5 shows the benefit of a graphical approach to solving systems of equations in two variables. Notice what happens when you try only the substitution method in Example 5. You obtain the equation $x + \ln x = 1$. It is difficult to solve this equation for x using standard algebraic techniques. In such cases, a graphical approach to solving systems of equations is more convenient.

EXAMPLE 5 Solving a System of Equations Graphically

See LarsonPrecalculus.com for an interactive version of this type of example.

Solve the system of equations.

$$\begin{cases} y = \ln x & \text{Equation 1} \\ x + y = 1 & \text{Equation 2} \end{cases}$$

Solution

From the graphs of these equations, it is clear that there is only one point of intersection. Use the *intersect* feature of a graphing utility to approximate the solution point as $(1, 0)$, as shown in the figure.

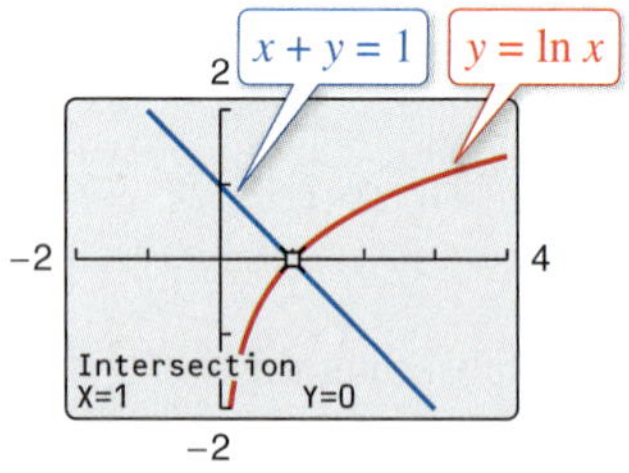

You can confirm this by substituting $(1, 0)$ into *both* equations.

Check $(1, 0)$ *in Equation 1:*

$y = \ln x$ Write Equation 1.

$0 = \ln 1$ Equation 1 checks. ✓

Check $(1, 0)$ *in Equation 2:*

$x + y = 1$ Write Equation 2.

$1 + 0 = 1$ Equation 2 checks. ✓

✓ Checkpoint *Audio-video solution in English & Spanish at LarsonPrecalculus.com*

Solve the system of equations.

$$\begin{cases} y = 3 - \log_{10} x \\ -2x + y = 1 \end{cases}$$

Application

The total cost C of producing x units of a product typically has two components: the initial cost and the cost per unit. When enough units have been sold so that the total revenue R equals the total cost C, the sales are said to have reached the **break-even point.** You will find that the break-even point corresponds to the point of intersection of the cost and revenue curves.

Business Performance Analyst

EXAMPLE 6 Break-Even Analysis

A small business invests \$10,000 in equipment to produce a new soft drink. Each bottle of the soft drink costs \$0.65 to produce and is sold for \$1.20. How many bottles must be sold before the business breaks even?

Solution

The total cost of producing x bottles is

Total cost	=	Cost per bottle	·	Number of bottles	+	Initial cost

$C = 0.65x + 10{,}000.$ Equation 1

The revenue obtained by selling x bottles is

Total revenue	=	Price per bottle	·	Number of bottles

$R = 1.20x.$ Equation 2

Because the break-even point occurs when $R = C$, you have

$$C = 1.20x$$

and the system of equations to solve is

$$\begin{cases} C = 0.65x + 10{,}000 \\ C = 1.20x \end{cases}.$$

Now you can solve by substitution.

$C = 0.65x + 10{,}000$ Write Equation 1.

$1.20x = 0.65x + 10{,}000$ Substitute $1.20x$ for C.

$0.55x = 10{,}000$ Subtract $0.65x$ from each side.

$x = \dfrac{10{,}0000}{0.55}$ Divide each side by 0.55.

$x \approx 18{,}182$ bottles Use a calculator.

So, the business must sell about 18,182 bottles to break even. Note in Figure 7.4 that revenue less than the break-even point corresponds to an overall loss, whereas revenue greater than the break-even point corresponds to a profit. Verify the break-even point using the *intersect* feature or the *zoom* and *trace* features of a graphing utility.

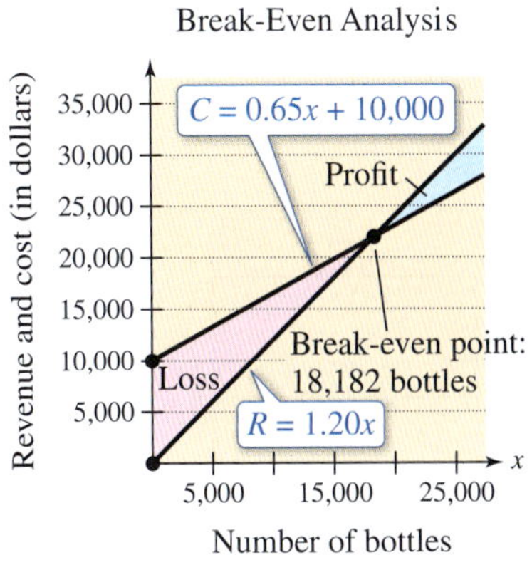

Figure 7.4

Checkpoint Audio-video solution in English & Spanish at LarsonPrecalculus.com

In Example 6, each bottle of the soft drink costs \$0.50 to produce. How many bottles must be sold before the business breaks even?

Another way to view the solution in Example 6 is to consider the profit function $P = R - C$. The break-even point occurs when the profit is 0, which is the same as saying that $R = C$.

7.1 Exercises

See *CalcChat.com* for tutorial help and worked-out solutions to odd-numbered exercises.
For instructions on how to use a graphing utility, see Appendix A.

Vocabulary and Concept Check

In Exercises 1–4, fill in the blank(s).

1. A set of two or more equations in two or more unknowns is called a _______ of _______ .
2. A _______ of a system of equations is an ordered pair that satisfies each equation in the system.
3. The first step in solving a system of two equations in x and y by the method of _______ is to solve one of the equations for one variable in terms of the other.
4. A point of intersection of the graphs of the equations of a system is a _______ of the system.
5. What is the point of intersection of the graphs of the cost and revenue functions called?
6. The graphs of the equations of a system do not intersect. What can you conclude?

Procedures and Problem Solving

Checking Solutions **In Exercises 7–10, determine whether each ordered pair is a solution of the system of equations.**

7. $\begin{cases} 4x - y = 1 \\ 6x + y = -6 \end{cases}$ (a) $(0, -3)$ (b) $(-1, -5)$ (c) $\left(-\frac{3}{2}, 3\right)$ (d) $\left(-\frac{1}{2}, -3\right)$

8. $\begin{cases} 4x^2 + y = 3 \\ -x - y = 11 \end{cases}$ (a) $(2, -13)$ (b) $(-2, -9)$ (c) $\left(-\frac{3}{2}, 6\right)$ (d) $\left(-\frac{7}{4}, -\frac{37}{4}\right)$

9. $\begin{cases} y = -2e^x \\ 3x - y = 2 \end{cases}$ (a) $(-2, 0)$ (b) $(0, -2)$ (c) $(0, -3)$ (d) $(-1, -5)$

10. $\begin{cases} -\log_{10} x + 3 = y \\ \frac{1}{9}x + y = \frac{28}{9} \end{cases}$ (a) $(100, 1)$ (b) $(10, 2)$ (c) $(1, 3)$ (d) $(1, 1)$

Solving a System by Substitution **In Exercises 11–18, solve the system by the method of substitution. Check your solution graphically.**

11. $\begin{cases} 2x + y = 6 \\ -x + y = 0 \end{cases}$

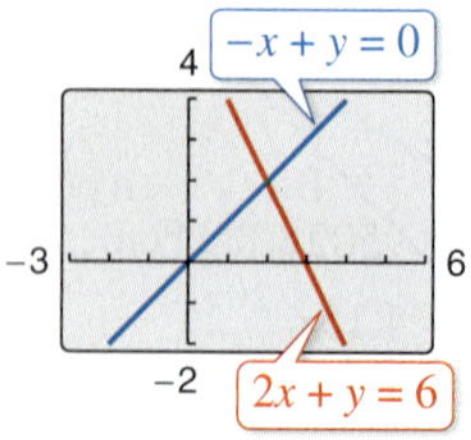

12. $\begin{cases} x - y = -4 \\ x + 2y = 5 \end{cases}$

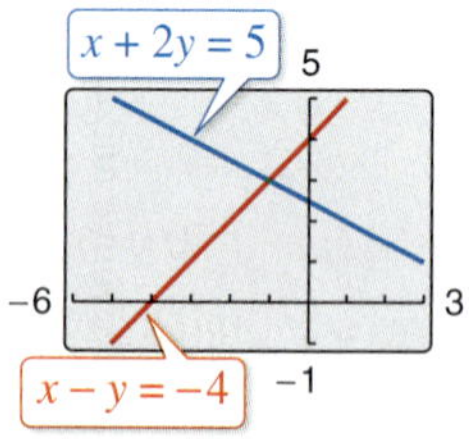

13. $\begin{cases} x - y = -4 \\ x^2 - y = -2 \end{cases}$

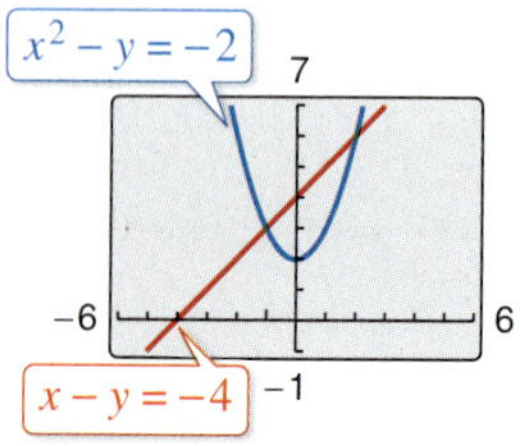

14. $\begin{cases} -2x + y = -5 \\ x^2 + y^2 = 25 \end{cases}$

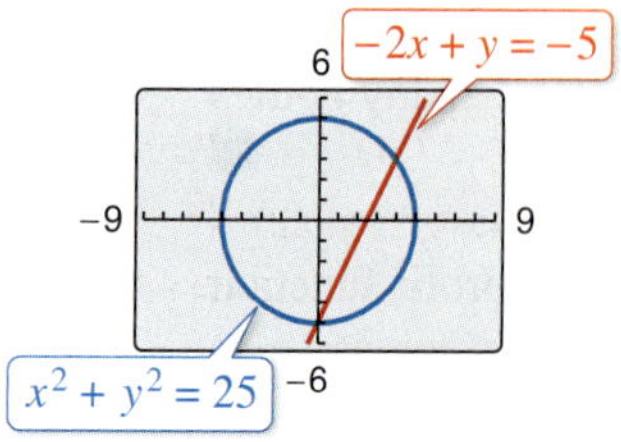

15. $\begin{cases} 3x + y = 2 \\ x^3 - 2 + y = 0 \end{cases}$

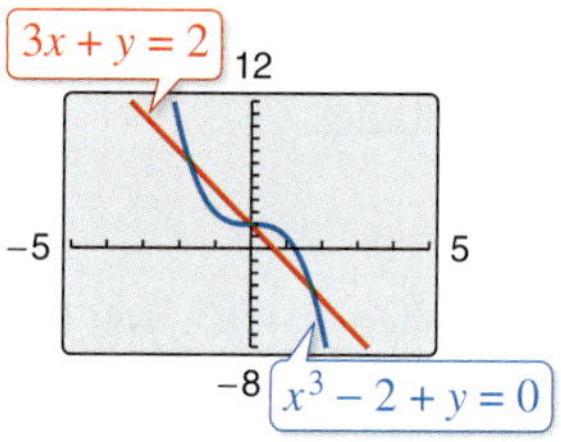

16. $\begin{cases} x + y = 0 \\ x^3 - 5x - y = 0 \end{cases}$

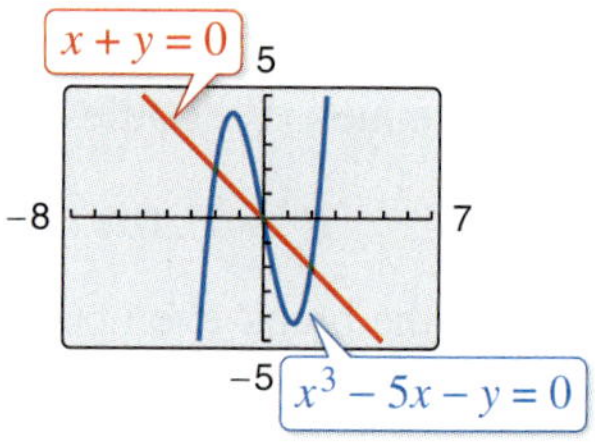

17. $\begin{cases} -\frac{7}{2}x - y = -18 \\ 8x^2 - 2y^3 = 0 \end{cases}$

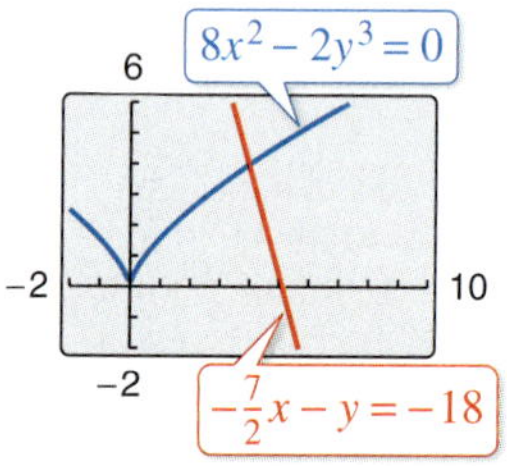

18. $\begin{cases} y = x^3 - 3x^2 + 4 \\ y = -2x + 4 \end{cases}$

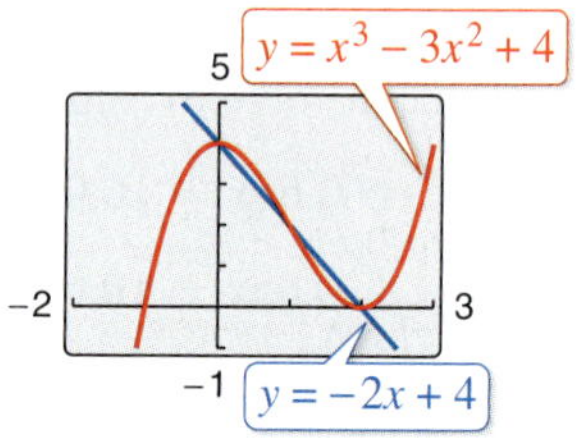

Solving a System of Equations **In Exercises 19–28, solve the system by the method of substitution. Use a graphing utility to verify your results.**

19. $\begin{cases} x + y = 0 \\ 4x + 3y = 10 \end{cases}$

20. $\begin{cases} x + 2y = 1 \\ 5x - 4y = -23 \end{cases}$

21. $\begin{cases} 2x - y + 2 = 0 \\ 4x + y - 5 = 0 \end{cases}$

22. $\begin{cases} 6x - 3y - 4 = 0 \\ x + 2y - 4 = 0 \end{cases}$

23. $\begin{cases} 1.5x + 0.8y = 2.3 \\ 0.3x - 0.2y = 0.1 \end{cases}$

24. $\begin{cases} -0.5x + 4y = 7.8 \\ 0.2x - 1.6y = -3.6 \end{cases}$

25. $\begin{cases} \frac{1}{5}x + \frac{1}{2}y = 8 \\ x + y = 20 \end{cases}$

26. $\begin{cases} \frac{1}{2}x + \frac{3}{4}y = 10 \\ \frac{3}{4}x - y = 4 \end{cases}$

27. $\begin{cases} -\frac{5}{3}x + y = 5 \\ -5x + 3y = 6 \end{cases}$

28. $\begin{cases} -\frac{2}{3}x + y = 2 \\ 3x - \frac{1}{2}y = 4 \end{cases}$

Solving a System by Substitution In Exercises 29–32, the amount of annual interest is earned from a total of \$12,000 invested in two funds paying simple interest. Write and solve a system of equations to find the amount invested at each rate.

	Annual Interest	*Rate 1*	*Rate 2*
29.	\$500	2%	6%
30.	\$630	4%	7%
31.	\$396	2.8%	3.8%
32.	\$254	1.75%	2.25%

Solving a System with a Nonlinear Equation In Exercises 33–38, solve the system by the method of substitution. Use a graphing utility to verify your results.

33. $\begin{cases} x^2 - 2x + y = 8 \\ x - y = -2 \end{cases}$

34. $\begin{cases} 2x^2 - 2x - y = 14 \\ 2x - y = -2 \end{cases}$

35. $\begin{cases} x - y = -1 \\ x^2 - y = -4 \end{cases}$

36. $\begin{cases} 2x^2 + y = 3 \\ x + y = 4 \end{cases}$

37. $\begin{cases} x^3 - y = 0 \\ x - y = 0 \end{cases}$

38. $\begin{cases} y = -x \\ y = x^3 + 3x^2 + 2x \end{cases}$

Solving a System of Equations Graphically In Exercises 39–46, solve the system graphically. Verify your solutions algebraically.

39. $\begin{cases} -2x + y = 7 \\ x + 3y = 0 \end{cases}$

40. $\begin{cases} x + y = 8 \\ 4x + 4y = 0 \end{cases}$

41. $\begin{cases} x - 2y = -3 \\ 5x + 6y = 17 \end{cases}$

42. $\begin{cases} -5x + 2y = -2 \\ x - 2y = 6 \end{cases}$

43. $\begin{cases} x^2 + y = -1 \\ -x + 2y = 5 \end{cases}$

44. $\begin{cases} x^2 - y = 3 \\ x - y = 1 \end{cases}$

45. $\begin{cases} 3x - 2y = 0 \\ x^2 - y^2 = 4 \end{cases}$

46. $\begin{cases} y^2 - x^2 + 9 = 0 \\ -\frac{1}{2}x + y = \frac{3}{2} \end{cases}$

Solving a System of Equations Graphically In Exercises 47–60, use a graphing utility to approximate all points of intersection of the graphs of equations in the system. Round your results to three decimal places. Verify your solutions by checking them in the original system.

47. $\begin{cases} 7x + 8y = 16 \\ x - 8y = 8 \end{cases}$

48. $\begin{cases} x - y = 0 \\ 5x - 2y = 12 \end{cases}$

49. $\begin{cases} x - y^2 = -1 \\ x - y = 5 \end{cases}$

50. $\begin{cases} x - y^2 = -2 \\ x - 2y = 6 \end{cases}$

51. $\begin{cases} x^2 + y^2 = 10 \\ y = x^2 \end{cases}$

52. $\begin{cases} x^2 + y^2 = 25 \\ (x - 8)^2 + y^2 = 41 \end{cases}$

53. $\begin{cases} y = e^x \\ x - y + 1 = 0 \end{cases}$

54. $\begin{cases} y = -4e^{-x} \\ y + 3x + 8 = 0 \end{cases}$

55. $\begin{cases} x + 2y = 10 \\ y = 2 + \ln x \end{cases}$

56. $\begin{cases} 3y + 2x = 9 \\ y = \ln(x - 3) \end{cases}$

57. $\begin{cases} y = \sqrt{x} + 4 \\ y = 2x + 1 \end{cases}$

58. $\begin{cases} x - y = 7 \\ \sqrt{x} - y = 1 \end{cases}$

59. $\begin{cases} x^2 + y^2 = 169 \\ x^2 - 8y = 104 \end{cases}$

60. $\begin{cases} x^2 + y^2 = 8 \\ 2x^2 - y = 2 \end{cases}$

Choosing a Solution Method In Exercises 61–74, solve the system graphically or algebraically. Explain your choice of method.

61. $\begin{cases} 2x - y = 0 \\ x^2 - y = -1 \end{cases}$

62. $\begin{cases} x + y = 4 \\ x^2 + y = 2 \end{cases}$

63. $\begin{cases} 3x - 7y = -6 \\ x^2 - y^2 = 4 \end{cases}$

64. $\begin{cases} x^2 + y^2 = 25 \\ 2x + y = 10 \end{cases}$

65. $\begin{cases} x^2 + y^2 = 1 \\ x + y = 4 \end{cases}$

66. $\begin{cases} x^2 + y^2 = 4 \\ x - y = 5 \end{cases}$

67. $\begin{cases} y = 2x + 1 \\ y = \sqrt{x + 2} \end{cases}$

68. $\begin{cases} y = 2x - 1 \\ y = \sqrt{x + 1} \end{cases}$

69. $\begin{cases} y - e^{-x} = 1 \\ y - \ln x = 3 \end{cases}$

70. $\begin{cases} 2 \ln x + y = 4 \\ e^x - y = 0 \end{cases}$

71. $\begin{cases} y = x^3 - 2x^2 + 1 \\ y = 1 - x^2 \end{cases}$

72. $\begin{cases} y = x^3 - 2x^2 + x - 1 \\ y = -x^2 + 3x - 1 \end{cases}$

73. $\begin{cases} xy - 1 = 0 \\ -5x - 2y + 1 = 0 \end{cases}$

74. $\begin{cases} xy - 2 = 0 \\ 3x - 2y + 4 = 0 \end{cases}$

Break-Even Analysis In Exercises 75–78, use a graphing utility to graph the cost and revenue functions in the same viewing window. Find the sales x necessary to break even $(R = C)$ and the corresponding revenue R obtained by selling x units. (Round to the nearest whole unit.)

	Cost	*Revenue*
75.	$C = 8650x + 250{,}000$	$R = 9502x$
76.	$C = 2.65x + 350{,}000$	$R = 4.15x$
77.	$C = 5.5\sqrt{x} + 10{,}000$	$R = 3.29x$
78.	$C = 7.8\sqrt{x} + 18{,}500$	$R = 12.84x$

Geometry In Exercises 79 and 80, find the dimensions of the rectangle meeting the specified conditions.

79. The perimeter is 30 meters and the length is 3 meters greater than the width.

80. The perimeter is 280 centimeters and the width is 20 centimeters less than the length.

81. Marketing Research The daily DVD rentals of a newly released animated film and a newly released horror film from a movie rental store can be modeled by the equations

$$\begin{cases} N = 360 - 24x & \text{Animated film} \\ N = 24 + 18x & \text{Horror film} \end{cases}$$

where N is the number of DVDs rented and x represents the week, with $x = 1$ corresponding to the first week of release.

(a) Use the *table* feature of a graphing utility to find the numbers of rentals of each movie for each of the first 12 weeks of release.

(b) Use the results of part (a) to determine the solution to the system of equations.

(c) Solve the system of equations algebraically.

(d) Compare your results from parts (b) and (c).

(e) Interpret the results in the context of the situation.

82. Economics You want to buy either a wood pellet stove or an electric furnace. The pellet stove costs \$3660 and produces heat at a cost of \$20.72 per 1 million Btu (British thermal units). The electric furnace costs \$2690 and produces heat at a cost of \$40.26 per 1 million Btu.

(a) Write a function for the total cost y of buying the pellet stove and producing x million Btu of heat.

(b) Write a function for the total cost y of buying the electric furnace and producing x million Btu of heat.

(c) Use a graphing utility to graph and solve the system of equations formed by the two cost functions.

(d) Solve the system of equations algebraically.

(e) Interpret the results in the context of the situation.

83. Break-Even Analysis A small software company invests \$18,000 to produce a software package that will sell for \$55.95. Each unit can be produced for \$9.45.

(a) Write the cost and revenue functions for x units produced and sold.

(b) Use a graphing utility to graph the cost and revenue functions in the same viewing window. Use the graph to approximate the number of units that must be sold to break even and verify the result algebraically.

84. Professional Sales You are offered two jobs selling college textbooks. One company offers an annual salary of \$38,000 plus a year-end bonus of 1% of your total sales. The other company offers an annual salary of \$32,000 plus a year-end bonus of 2.5% of your total sales. How much would you have to sell in a year to make the second offer the better offer?

85. Geometry What are the dimensions of a rectangular tract of land with a perimeter of 40 miles and an area of 96 square miles?

86. Geometry What are the dimensions of a right triangle with a two-inch hypotenuse and an area of 1 square inch?

87. Finance You are deciding how to invest a total of \$20,000 in two funds paying 5.5% and 7.5% simple interest. You want to earn a total of \$1300 in interest from the investments each year.

(a) Write a system of equations in which one equation represents the total amount invested and the other equation represents the \$1300 yearly interest. Let x and y represent the amounts invested at 5.5% and 7.5%, respectively.

(b) Use a graphing utility to graph the two equations in the same viewing window.

(c) How much of the \$20,000 should you invest at 5.5% to earn \$1300 in interest per year? Explain your reasoning.

88. *Why you should learn it* *(p. 478)* You are offered two different rules for estimating the number of board feet in a 16-foot log. (A board foot is a unit of measure for lumber equal to a board 1 foot square and 1 inch thick.) One rule is the Doyle Log Rule modeled by

$$V = (D - 4)^2, \quad 5 \le D \le 40$$

where D is the diameter (in inches) of the log and V is its volume (in board feet). The other rule is the Scribner Log Rule modeled by

$$V = 0.79D^2 - 2D - 4, \quad 5 \le D \le 40.$$

(a) Use a graphing utility to graph the two log rules in the same viewing window.

(b) For what diameter do the two rules agree?

(c) You are selling large logs by the board foot. Which rule would you use? Explain your reasoning.

89. Algebraic-Graphical-Numerical The populations (in thousands) of Arizona A and Massachusetts M from 2010 through 2017 can be modeled by the system

$$\begin{aligned} A &= 87.5t + 6377 && \text{Arizona} \\ M &= 42.5t + 6574 && \text{Massachusetts} \end{aligned}$$

where t is the year, with $t = 0$ corresponding to 2010. (*Source:* U.S. Census Bureau)

(a) Record in a table the populations approximated by the models for the two states in the years 2010 through 2017.

(b) According to the table, in what year(s) was the population of Arizona greater than that of Massachusetts?

(c) Use a graphing utility to graph the models in the same viewing window. Estimate the point of intersection of the models.

(d) Find the point of intersection algebraically.

90. MODELING DATA

The table shows the yearly revenues (in billions of dollars) of the online travel companies Expedia and Booking Holdings (formerly Priceline.com) from 2012 through 2017. (*Sources:* Expedia Group, Inc. and Booking Holdings Inc.)

DATA

Year	Expedia	Booking Holdings
2012	4.03	5.26
2013	4.77	6.79
2014	5.76	8.44
2015	6.67	9.22
2016	8.77	10.74
2017	10.06	12.68

Spreadsheet at LarsonPrecalculus.com

(a) Use the *regression* feature of a graphing utility to find a quadratic model for the yearly revenue E of Expedia and a linear model for the yearly revenue B of Booking Holdings. Let x represent the year, with $x = 2$ corresponding to 2012.

(b) Use the graphing utility to graph the models with the original data in the same viewing window.

(c) In the past, Expedia had greater revenues than Booking Holdings. Use the graph in part (b) to approximate the first year when the revenues of Booking Holdings were greater than the revenues of Expedia.

(d) Algebraically approximate the first year when the revenues of Booking Holdings were greater than the revenues of Expedia.

(e) Compare your results from parts (c) and (d).

Focusing on Concepts

True or False? **In Exercises 91 and 92, determine whether the statement is true or false. Justify your answer.**

91. In order to solve a system of equations by substitution, you must always solve for y in one of the two equations and then back-substitute.

92. If a system consists of a parabola and a circle, then it can have at most two solutions.

93. Think About It Explain how the graphs of a system of two linear equations in two variables are related when the system has (a) one solution, (b) no solution, and (c) more than one solution.

94. Think About It Describe the possible numbers of solutions of each system of equations, where a represents any real number. Explain your reasoning.

a. $\begin{cases} y = x^2 \\ y = a \end{cases}$ **b.** $\begin{cases} y = x^2 \\ x = a \end{cases}$

95. Think About It When solving a system of equations by substitution, how do you recognize that the system has no solution?

96. Exploration Create a system of linear equations in two variables that has the solution $(2, -1)$ as its only solution. (There are many correct answers.)

97. Error Analysis Describe the error in solving the system of equations by graphing.

$$\begin{cases} y = 2x + 1 \\ y = x^2 + 1 \end{cases}$$

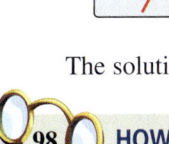

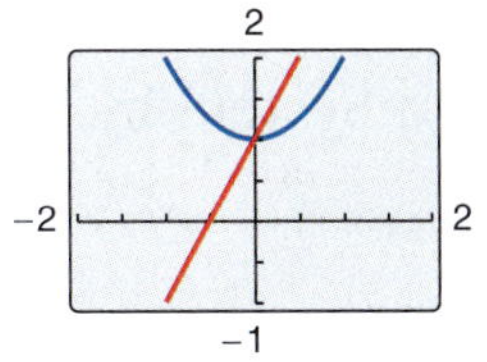

The solution is $(0, 1)$.

98. HOW DO YOU SEE IT? The cost C of producing x units and the revenue R obtained by selling x units are shown in the figure.

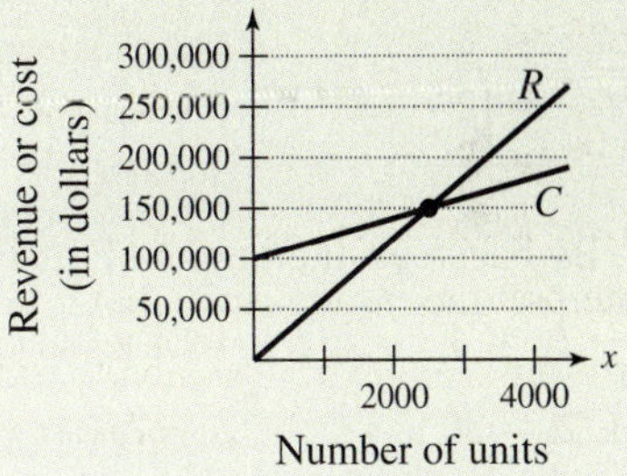

(a) Estimate the point of intersection. What does this point represent?

(b) Use the figure to identify the x-values that correspond to (i) an overall loss and (ii) a profit. Explain your reasoning.

Cumulative Mixed Review

Finding the Slope-Intercept Form **In Exercises 99–102, write an equation of the line passing through the two points. Use the slope-intercept form, if possible. If not possible, explain why.**

99. $(3, 4), (10, 6)$

100. $(6, 3), (10, 3)$

101. $(4, -2), (4, 5)$

102. $\left(\frac{3}{5}, 0\right), (4, 6)$

Lars Christensen/Shutterstock.com

7.2 Systems of Linear Equations in Two Variables

The Method of Elimination

In Section 7.1, you studied two methods for solving a system of equations: substitution and graphing. Now you will study the **method of elimination.** The key step in this method is to obtain, for one of the variables, coefficients that differ only in sign so that *adding* the equations eliminates the variable.

$$\begin{array}{rcll} 3x + 5y &=& 7 & \text{Equation 1} \\ -3x - 2y &=& -1 & \text{Equation 2} \\ \hline 3y &=& 6 & \text{Add equations.} \end{array}$$

Note that by adding the two equations, you eliminate the x-terms and obtain a single equation in y. Solving this equation for y produces

$$y = 2$$

which you can then back-substitute into one of the original equations to solve for x.

What you should learn

- Use the method of elimination to solve systems of linear equations in two variables.
- Graphically interpret the number of solutions of a system of linear equations in two variables.
- Use systems of linear equations in two variables to model and solve real-life problems.

Why you should learn it

You can use systems of linear equations to model many business applications. For instance, Exercise 80 on page 495 shows how to use a system of linear equations to determine the number of running shoes sold.

EXAMPLE 1 Solving a System by Elimination

Solve the system of linear equations.

$$\begin{cases} 3x + 2y = 4 & \text{Equation 1} \\ 5x - 2y = 8 & \text{Equation 2} \end{cases}$$

Solution

Because the coefficients of y differ only in sign, you can eliminate the y-terms by adding the two equations.

$$\begin{array}{rcll} 3x + 2y &=& 4 & \text{Write Equation 1.} \\ 5x - 2y &=& 8 & \text{Write Equation 2.} \\ \hline 8x &=& 12 & \text{Add equations.} \\ x &=& \frac{3}{2} & \text{Solve for } x. \end{array}$$

So, $x = \frac{3}{2}$. By back-substituting into Equation 1, you can solve for y.

$$\begin{array}{rcll} 3x + 2y &=& 4 & \text{Write Equation 1.} \\ 3\left(\frac{3}{2}\right) + 2y &=& 4 & \text{Substitute } \frac{3}{2} \text{ for } x. \\ y &=& -\frac{1}{4} & \text{Solve for } y. \end{array}$$

The solution is $\left(\frac{3}{2}, -\frac{1}{4}\right)$. You can check the solution *algebraically* by substituting into the original system, or graphically, as shown in Section 7.1.

Check

$$\begin{array}{rcll} 3\left(\frac{3}{2}\right) + 2\left(-\frac{1}{4}\right) &\stackrel{?}{=}& 4 & \text{Substitute into Equation 1.} \\ \frac{9}{2} - \frac{1}{2} &=& 4 & \text{Equation 1 checks. ✓} \\ 5\left(\frac{3}{2}\right) - 2\left(-\frac{1}{4}\right) &\stackrel{?}{=}& 8 & \text{Substitute into Equation 2.} \\ \frac{15}{2} + \frac{1}{2} &=& 8 & \text{Equation 2 checks. ✓} \end{array}$$

Explore the Concept

Use the method of substitution to solve the system given in Example 1. Which method is easier?

✓ ***Checkpoint*** *Audio-video solution in English & Spanish at LarsonPrecalculus.com*

Solve the system of linear equations.

$$\begin{cases} 2x + y = 4 \\ 2x - y = -1 \end{cases}$$

The Method of Elimination

To use the **method of elimination** to solve a system of two linear equations in x and y, perform the following steps.

1. Obtain coefficients for x (or y) that differ only in sign by multiplying all terms of one or both equations by suitably chosen constants.
2. Add the equations to eliminate one variable; solve the resulting equation.
3. Back-substitute the value obtained in Step 2 into either of the original equations and solve for the other variable.
4. Check your solution in *both* of the original equations.

EXAMPLE 2 Solving a System by Elimination

Solve the system of linear equations.

$$\begin{cases} 5x + 3y = 9 & \text{Equation 1} \\ 2x - 4y = 14 & \text{Equation 2} \end{cases}$$

Algebraic Solution

You can obtain coefficients of y that differ only in sign by multiplying Equation 1 by 4 and multiplying Equation 2 by 3.

$$\begin{aligned} 5x + 3y &= 9 &\Rightarrow\quad 20x + 12y &= 36 &&\text{Multiply Equation 1 by 4.} \\ 2x - 4y &= 14 &\Rightarrow\quad 6x - 12y &= 42 &&\text{Multiply Equation 2 by 3.} \\ & & 26x \quad &= 78 &&\text{Add equations.} \end{aligned}$$

From this equation, you can see that $x = 3$. By back-substituting this value of x into Equation 2, you can solve for y.

$$\begin{aligned} 2x - 4y &= 14 &&\text{Write Equation 2.} \\ 2(3) - 4y &= 14 &&\text{Substitute 3 for } x. \\ -4y &= 8 &&\text{Combine like terms.} \\ y &= -2 &&\text{Solve for } y. \end{aligned}$$

The solution is $(3, -2)$. You can check the solution algebraically by substituting into the original system.

Graphical Solution

Solve each equation for y and use a graphing utility to graph the equations in the same viewing window. From the figure, the solution is $(3, -2)$. Check this in the original system.

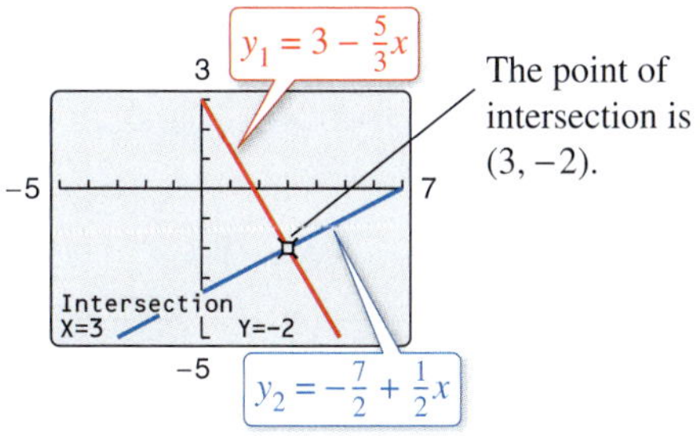

Checkpoint Audio-video solution in English & Spanish at LarsonPrecalculus.com

Solve the system of linear equations.

$$\begin{cases} 3x + 2y = 7 \\ 2x + 5y = 1 \end{cases}$$

In Example 2, the two systems of linear equations (the original system and the system obtained by multiplying by constants)

$$\begin{cases} 5x + 3y = 9 \\ 2x - 4y = 14 \end{cases} \quad \text{and} \quad \begin{cases} 20x + 12y = 36 \\ 6x - 12y = 42 \end{cases}$$

are called **equivalent systems** because they have precisely the same solution set. The operations that can be performed on a system of linear equations to produce an equivalent system are (1) interchanging any two equations, (2) multiplying an equation by a nonzero constant, and (3) adding a multiple of one equation to any other equation in the system.

Graphical Interpretation of Two-Variable Systems

It is possible for any system of equations to have exactly one solution, two or more solutions, or no solution. If a system of *linear* equations has two different solutions, then it must have an *infinite* number of solutions. To see why this is true, consider the following graphical interpretations of a system of two linear equations in two variables.

Graphical Interpretations of Solutions

For a system of two linear equations in two variables, the number of solutions is one of the following.

Number of Solutions	*Graphical Interpretation*
1. Exactly one solution	The two lines intersect at one point.
2. Infinitely many solutions	The two lines are coincident (identical).
3. No solution	The two lines are parallel.

A system of linear equations is **consistent** when it has at least one solution. It is **inconsistent** when it has no solution.

Explore the Concept

Rewrite each system of equations in slope-intercept form and use a graphing utility to graph each system. What is the relationship between the slopes of the two lines and the number of points of intersection?

a. $\begin{cases} y = 5x + 1 \\ y - x = -5 \end{cases}$

b. $\begin{cases} 3y = 4x - 1 \\ -8x + 2 = -6y \end{cases}$

c. $\begin{cases} 2y = -x + 3 \\ -4 = y + \frac{1}{2}x \end{cases}$

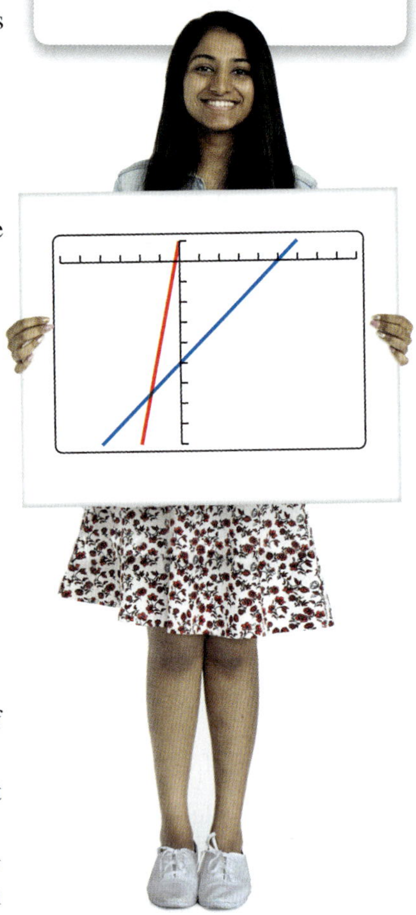

EXAMPLE 3 Recognizing Graphs of Linear Systems

See LarsonPrecalculus.com for an interactive version of this type of example.

Match each system of linear equations (a, b, c) with its graph (i, ii, iii). Describe the number of solutions. Then state whether the system is consistent or inconsistent.

a. $\begin{cases} 2x - 3y = 3 \\ -4x + 6y = 6 \end{cases}$ **b.** $\begin{cases} 2x - 3y = 3 \\ x + 2y = 5 \end{cases}$ **c.** $\begin{cases} 2x - 3y = 3 \\ -4x + 6y = -6 \end{cases}$

i.

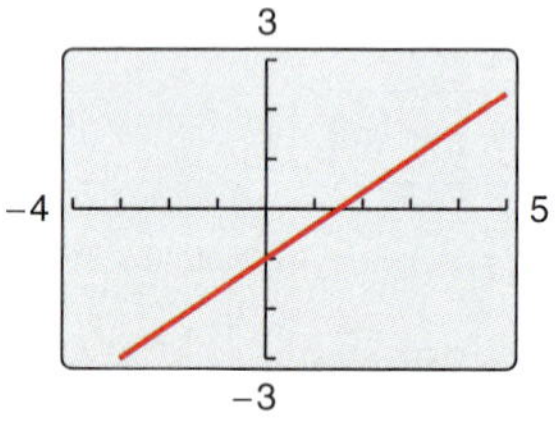

ii.

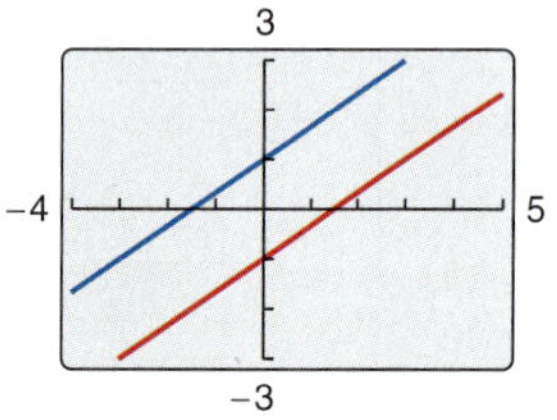

iii.

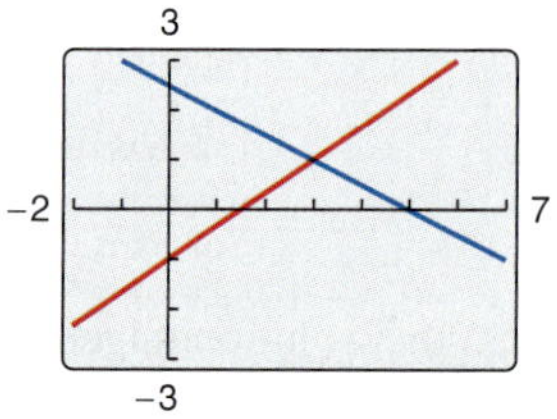

Solution

a. The graph of system (a) is a pair of parallel lines (ii). The lines have no point of intersection, so the system has no solution. The system is inconsistent.

b. The graph of system (b) is a pair of intersecting lines (iii). The lines have one point of intersection, so the system has exactly one solution. The system is consistent.

c. The graph of system (c) is a pair of lines that coincide (i). The lines have infinitely many points of intersection, so the system has infinitely many solutions. The system is consistent.

✓ *Checkpoint* *Audio-video solution in English & Spanish at LarsonPrecalculus.com*

Sketch the graph of the system of linear equations. Then describe the number of solutions and state whether the system is consistent or inconsistent.

$$\begin{cases} -2x + 3y = 6 \\ 4x - 6y = -9 \end{cases}$$

In Examples 4 and 5, note how you can use the method of elimination to determine that a system of linear equations has no solution or infinitely many solutions.

EXAMPLE 4 Method of Elimination: No Solution

Solve the system of linear equations.

$$\begin{cases} x - 2y = 3 & \text{Equation 1} \\ -2x + 4y = 1 & \text{Equation 2} \end{cases}$$

Algebraic Solution

To obtain coefficients that differ only in sign, multiply Equation 1 by 2.

$$x - 2y = 3 \quad \Rightarrow \quad 2x - 4y = 6$$
$$-2x + 4y = 1 \quad \Rightarrow \quad \underline{-2x + 4y = 1}$$
$$0 = 7$$

By adding the equations, you obtain $0 = 7$. Because there are no values of x and y for which

$$0 = 7 \qquad \text{False statement}$$

this is a false statement. So, you can conclude that the system is inconsistent and has no solution.

Graphical Solution

Solve each equation for y and use a graphing utility to graph the equations in the same viewing window. From the figure, it appears that the system has no solution. When you use the *intersect* feature, the graphing utility cannot find a point of intersection and you will get an error message.

The lines have the same slope but different y-intercepts, so they are parallel.

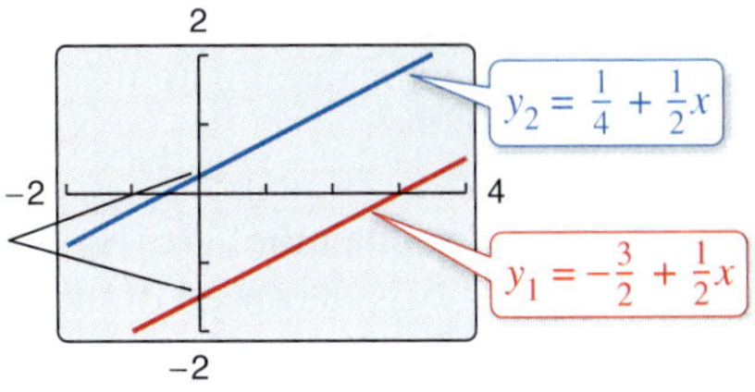

✓ **Checkpoint** ▶ *Audio-video solution in English & Spanish at LarsonPrecalculus.com*

Solve the system of linear equations.

$$\begin{cases} 6x - 5y = 3 \\ -12x + 10y = 5 \end{cases}$$

EXAMPLE 5 Method of Elimination: Infinitely Many Solutions

Solve the system of linear equations.

$$\begin{cases} 2x - y = 1 & \text{Equation 1} \\ 4x - 2y = 2 & \text{Equation 2} \end{cases}$$

Solution

To obtain coefficients that differ only in sign, multiply Equation 1 by -2.

$$2x - y = 1 \quad \Rightarrow \quad -4x + 2y = -2 \qquad \text{Multiply Equation 1 by } -2.$$
$$4x - 2y = 2 \quad \Rightarrow \quad \underline{4x - 2y = 2} \qquad \text{Write Equation 2.}$$
$$0 = 0 \qquad \text{Add equations.}$$

Because $0 = 0$ for all values of x and y, the two equations turn out to be equivalent (have the same solution set). You can conclude that the system has infinitely many solutions. The solution set consists of all points (x, y) lying on the line $2x - y = 1$, as shown in Figure 7.5.

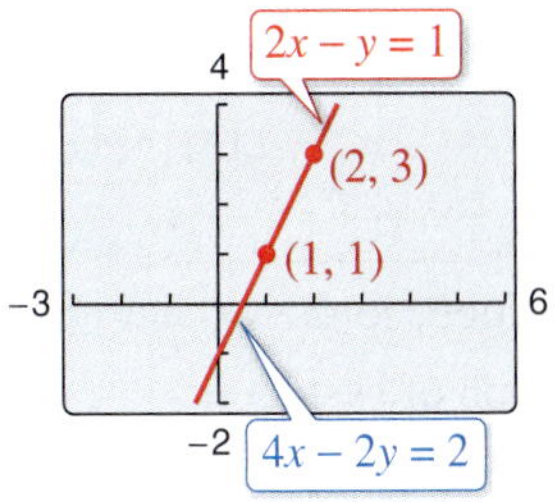

Figure 7.5

✓ **Checkpoint** ▶ *Audio-video solution in English & Spanish at LarsonPrecalculus.com*

Solve the system of linear equations.

$$\begin{cases} \frac{1}{2}x - \frac{1}{8}y = -\frac{3}{8} \\ -4x + y = 3 \end{cases}$$

In Example 4, note that the occurrence of the false statement $0 = 7$ indicates that the system has no solution. In Example 5, note that the occurrence of a statement that is true for all values of the variables—in this case, $0 = 0$—indicates that the system has infinitely many solutions.

Application

At this point, you may be asking the question "How can I tell which application problems can be solved using a system of linear equations?" The answer comes from the following considerations.

1. Does the problem involve more than one unknown quantity?
2. Are there two (or more) equations or conditions to be satisfied?

When one or both of these conditions are met, the appropriate mathematical model for the problem may be a system of linear equations.

EXAMPLE 6 Aviation

An airplane flying into a headwind travels the 2000-mile flying distance between Cleveland, Ohio, and Fresno, California, in 4 hours and 24 minutes. On the return flight, the airplane travels this distance in 4 hours. Find the airspeed of the plane and the speed of the wind, assuming that both remain constant.

Solution

The two unknown quantities are the speeds of the wind and the plane. If r_1 is the speed of the plane and r_2 is the speed of the wind, then

$r_1 - r_2 =$ speed of the plane *against* the wind

and

$r_1 + r_2 =$ speed of the plane *with* the wind

as shown in Figure 7.6. Using the formula Distance = (rate)(time) for these two speeds, you obtain the following equations.

$$2000 = (r_1 - r_2)\left(4 + \frac{24}{60}\right)$$

$$2000 = (r_1 + r_2)(4)$$

These two equations simplify as follows.

$$\begin{cases} 5000 = 11r_1 - 11r_2 \\ 500 = r_1 + r_2 \end{cases}$$ Equation 1; Equation 2

To solve this system by elimination, multiply Equation 2 by 11.

$$5000 = 11r_1 - 11r_2 \quad\Rightarrow\quad 5000 = 11r_1 - 11r_2$$ Write Equation 1.

$$500 = r_1 + r_2 \quad\Rightarrow\quad 5500 = 11r_1 + 11r_2$$ Multiply Equation 2 by 11.

$$10{,}500 = 22r_1$$ Add equations.

So,

$$r_1 = \frac{10{,}500}{22} = \frac{5250}{11} \approx 477.27 \text{ miles per hour}$$ Speed of plane

and

$$r_2 = 500 - \frac{5250}{11} = \frac{250}{11} \approx 22.73 \text{ miles per hour.}$$ Speed of wind

Check this solution in the original statement of the problem.

✓ Checkpoint *Audio-video solution in English & Spanish at LarsonPrecalculus.com*

In Example 6, the return flight takes 4 hours and 6 minutes. Find the airspeed of the plane and the speed of the wind, assuming that both remain constant.

What's Wrong?

You use a graphing utility to graph the system

$$\begin{cases} 100y - x = 200 \\ 99y - x = -198 \end{cases}$$

as shown in the figure. You use the graph to conclude that the system has no solution. What's wrong?

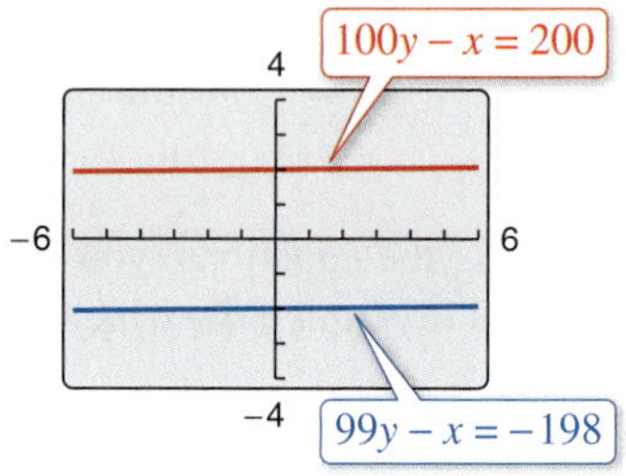

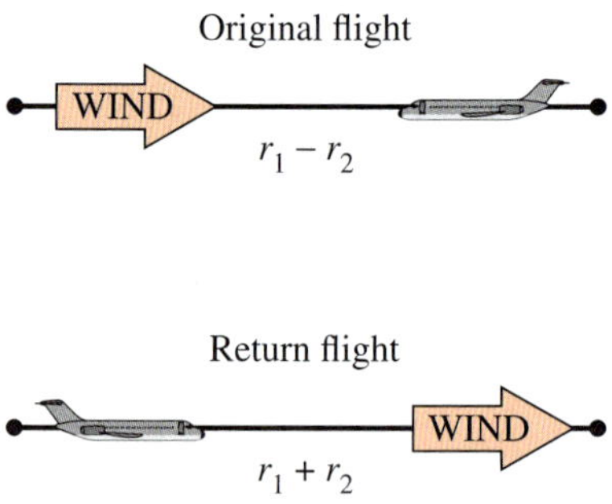

Figure 7.6

Insight

For standardized tests, be prepared to write and solve a system of equations that represents a real-life application, as shown in Example 6.

7.2 Exercises

See *CalcChat.com* for tutorial help and worked-out solutions to odd-numbered exercises.
For instructions on how to use a graphing utility, see Appendix A.

Vocabulary and Concept Check

In Exercises 1 and 2, fill in the blank.

1. The first step in solving a system of equations by the method of _______ is to obtain coefficients for x (or y) that differ only in sign.

2. Two systems of equations that have the same solution set are called _______ systems.

3. Is a system of linear equations with no solution consistent or inconsistent?

4. Is a system of linear equations with at least one solution consistent or inconsistent?

5. Is a system of two linear equations consistent when the lines are coincident?

6. When a system of linear equations has no solution, do the lines intersect?

Procedures and Problem Solving

Solving a System by Elimination **In Exercises 7–12, solve the system by the method of elimination. Label each line with its equation.**

7. $\begin{cases} 2x + y = 5 \\ x - y = 1 \end{cases}$

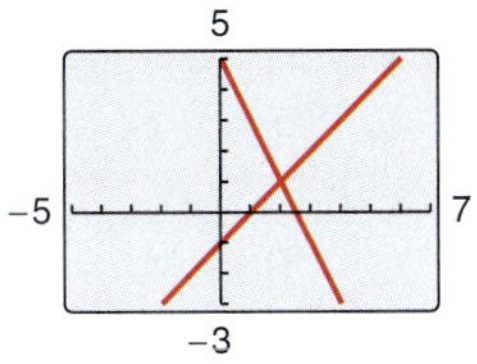

8. $\begin{cases} x + 3y = 1 \\ -x + 2y = 4 \end{cases}$

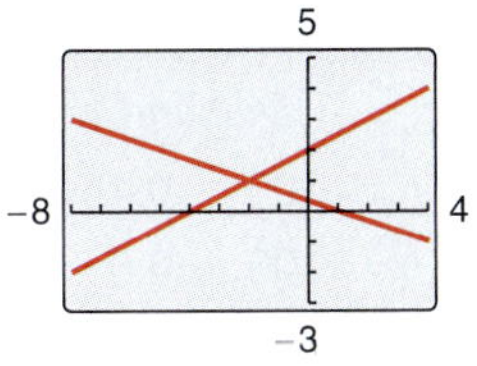

9. $\begin{cases} x + y = 0 \\ 3x + 2y = 1 \end{cases}$

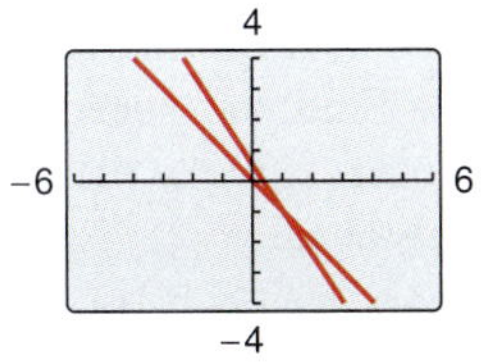

10. $\begin{cases} 2x - y = 3 \\ 4x + 3y = 21 \end{cases}$

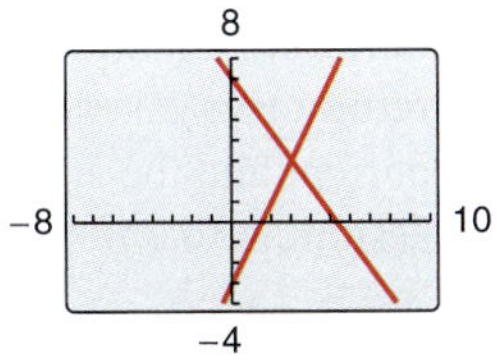

11. $\begin{cases} x - y = 2 \\ -2x + 2y = 5 \end{cases}$

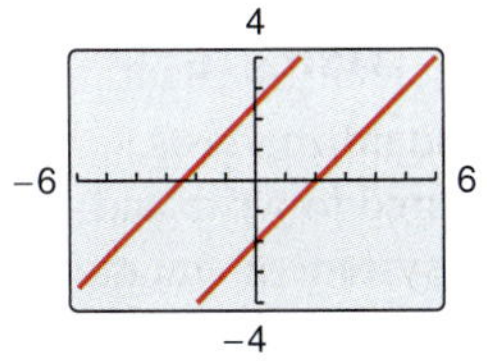

12. $\begin{cases} 3x - 2y = 5 \\ -6x + 4y = -10 \end{cases}$

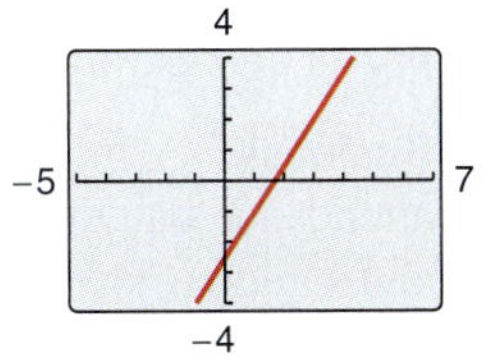

Solving a System by Elimination **In Exercises 13–24, solve the system by the method of elimination and check any solutions algebraically.**

13. $\begin{cases} x + 2y = 3 \\ x - 2y = 1 \end{cases}$

14. $\begin{cases} 3x - 5y = 2 \\ 2x + 5y = 8 \end{cases}$

15. $\begin{cases} 5x + 3y = 6 \\ 3x - y = 5 \end{cases}$

16. $\begin{cases} x + 7y = 12 \\ 3x - 5y = 10 \end{cases}$

17. $\begin{cases} 3r + 2s = -6 \\ 2r + 6s = 3 \end{cases}$

18. $\begin{cases} 8r + 16s = 20 \\ 16r + 50s = 55 \end{cases}$

19. $\begin{cases} -6x + 5y = -15 \\ 4x + 12y = 10 \end{cases}$

20. $\begin{cases} 9x + 3y = 18 \\ 2x - 7y = -19 \end{cases}$

21. $\begin{cases} 1.8x + 1.2y = 4 \\ 9x + 6y = 3 \end{cases}$

22. $\begin{cases} 3.1x - 2.9y = -10.2 \\ 31x - 12y = 34 \end{cases}$

23. $\begin{cases} 3x + \frac{1}{4}y = 1 \\ 2x - \frac{1}{3}y = 0 \end{cases}$

24. $\begin{cases} \frac{1}{2}x - 2y = -\frac{5}{2} \\ -x + 4y = 5 \end{cases}$

Recognizing Graphs of Linear Systems **In Exercises 25–28, match the system of linear equations with its graph. State the number of solutions. Then state whether the system is consistent or inconsistent. [The graphs are labeled (a), (b), (c), and (d).]**

(a)

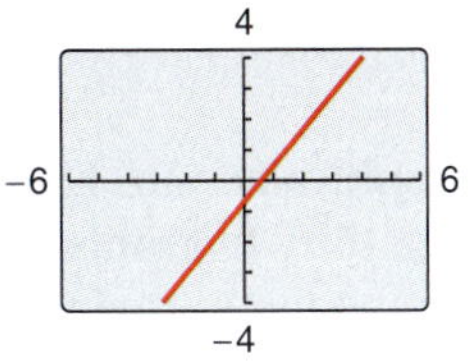

(b)

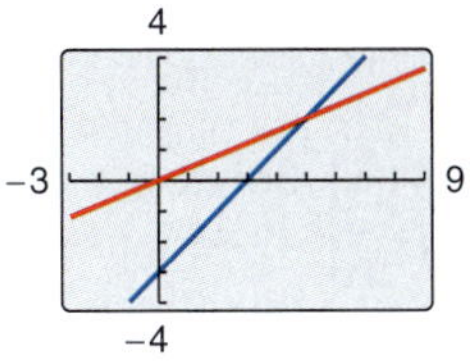

(c)

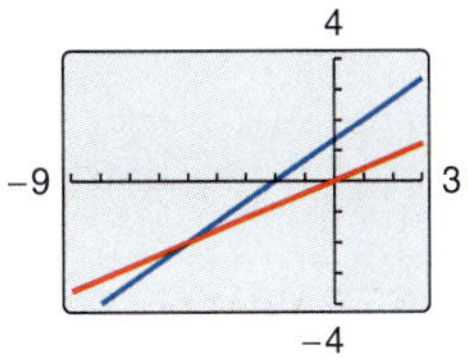

(d)

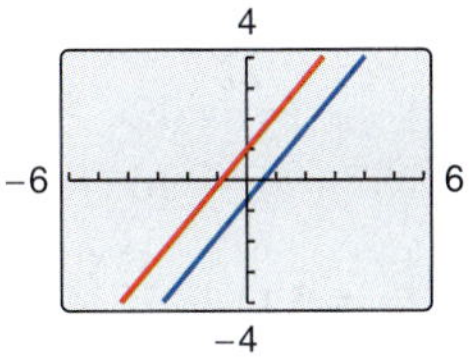

25. $\begin{cases} 2x - 5y = 0 \\ x - y = 3 \end{cases}$

26. $\begin{cases} -7x + 6y = -4 \\ 14x - 12y = 8 \end{cases}$

27. $\begin{cases} 7x - 6y = -6 \\ -7x + 6y = -4 \end{cases}$

28. $\begin{cases} 2x - 5y = 0 \\ 2x - 3y = -4 \end{cases}$

Solving a System by Elimination In Exercises 29–42, solve the system by the method of elimination and check any solutions using a graphing utility.

29. $\begin{cases} 5x + 3y = 6 \\ 3x - y = 5 \end{cases}$

30. $\begin{cases} x + 5y = 10 \\ 3x - 10y = -5 \end{cases}$

31. $\begin{cases} \frac{2}{5}x - \frac{3}{2}y = 4 \\ \frac{1}{5}x - \frac{3}{4}y = -2 \end{cases}$

32. $\begin{cases} \frac{2}{3}x + \frac{1}{6}y = \frac{2}{3} \\ 4x + y = 4 \end{cases}$

33. $\begin{cases} \frac{3}{4}x + y = \frac{1}{8} \\ \frac{9}{4}x + 3y = \frac{3}{8} \end{cases}$

34. $\begin{cases} \frac{1}{4}x + \frac{1}{6}y = 1 \\ -3x - 2y = 0 \end{cases}$

35. $\begin{cases} \dfrac{x+2}{4} + \dfrac{y-1}{4} = 1 \\ x - y = 4 \end{cases}$

36. $\begin{cases} \dfrac{2x+5}{2} + \dfrac{y-1}{3} = -1 \\ 2x - y = 12 \end{cases}$

37. $\begin{cases} -5x + 6y = -3 \\ 20x - 24y = 12 \end{cases}$

38. $\begin{cases} 7x + 8y = 16 \\ -14x - 16y = -4 \end{cases}$

39. $\begin{cases} 2.5x - 3y = 1.5 \\ x - 1.2y = -3.6 \end{cases}$

40. $\begin{cases} 6.3x + 7.2y = 5.4 \\ 5.6x + 6.4y = 4.8 \end{cases}$

41. $\begin{cases} \dfrac{1}{x} + \dfrac{3}{y} = 2 \\ \dfrac{4}{x} - \dfrac{1}{y} = -5 \end{cases}$

42. $\begin{cases} \dfrac{2}{x} - \dfrac{1}{y} = 5 \\ \dfrac{6}{x} + \dfrac{1}{y} = 11 \end{cases}$

Solving a System Graphically In Exercises 43–48, use a graphing utility to graph the lines in the system. Use the graphs to determine whether the system is consistent or inconsistent. If the system is consistent, determine the solution. Verify your results algebraically.

43. $\begin{cases} -x + 3y = 0 \\ 3x - 9y = 14 \end{cases}$

44. $\begin{cases} 3x + 5y = -2 \\ 4x - y = 5 \end{cases}$

45. $\begin{cases} 6x + 3y = -8 \\ -x - \frac{1}{2}y = \frac{4}{3} \end{cases}$

46. $\begin{cases} -\frac{1}{4}x - \frac{1}{2}y = 1 \\ 5x + y = 1 \end{cases}$

47. $\begin{cases} 3.2x - 16y = 7.5 \\ x - 5y = -9 \end{cases}$

48. $\begin{cases} -6x + 4y = -9 \\ 4.5x - 3y = 6.75 \end{cases}$

Solving a System Graphically In Exercises 49–56, use a graphing utility to graph the two equations. Use the graphs to approximate the solution of the system. Round your results to three decimal places.

49. $\begin{cases} 6y = 42 \\ 6x - y = 16 \end{cases}$

50. $\begin{cases} 4y = -8 \\ 7x - 2y = 25 \end{cases}$

51. $\begin{cases} \frac{3}{2}x - \frac{1}{5}y = 8 \\ -2x + 3y = 3 \end{cases}$

52. $\begin{cases} \frac{3}{4}x - \frac{5}{2}y = -9 \\ -x + 6y = 28 \end{cases}$

53. $\begin{cases} \frac{1}{3}x + y = -\frac{1}{3} \\ 5x - 3y = 7 \end{cases}$

54. $\begin{cases} 5x - y = -4 \\ 2x + \frac{3}{5}y = \frac{2}{5} \end{cases}$

55. $\begin{cases} 0.5x + 2.2y = 9 \\ 6x + 0.4y = -22 \end{cases}$

56. $\begin{cases} 2.4x + 3.8y = -17.6 \\ 4x - 0.2y = -3.2 \end{cases}$

Choosing a Solution Method In Exercises 57–64, use any method to solve the system. Explain your choice of method.

57. $\begin{cases} 3x - 5y = 7 \\ 2x + y = 9 \end{cases}$

58. $\begin{cases} -x + 3y = 17 \\ 4x + 3y = 7 \end{cases}$

59. $\begin{cases} y = 2x - 5 \\ y = 5x - 11 \end{cases}$

60. $\begin{cases} 7x + 3y = 16 \\ y = x + 2 \end{cases}$

61. $\begin{cases} x - 5y = 21 \\ 6x + 5y = 21 \end{cases}$

62. $\begin{cases} y = -2x - 17 \\ y = 2 - 3x \end{cases}$

63. $\begin{cases} -5x + 9y = 13 \\ y = x - 4 \end{cases}$

64. $\begin{cases} 4x - 3y = 6 \\ -5x + 7y = -1 \end{cases}$

Finding Slope-Intercept Form In Exercises 65–68, write and solve a system of linear equations to find the values of a and b for which the graph of $y = ax + b$ contains the two points.

65. $(-1, 3), (-3, -1)$

66. $(2, -4), (3, -7)$

67. $(-4, -6), (8, -3)$

68. $(-5, 3), (10 -3)$

Economics In Exercises 69–72, find the *point of equilibrium* of the demand and supply equations. The point of equilibrium is the price p and the number of units x that satisfy both the demand and supply equations.

	Demand	*Supply*
69.	$p = 500 - 0.4x$	$p = 380 + 0.1x$
70.	$p = 100 - 0.05x$	$p = 25 + 0.1x$
71.	$p = 140 - 0.00002x$	$p = 80 + 0.00001x$
72.	$p = 400 - 0.0002x$	$p = 225 + 0.0005x$

73. **Navigation** A motorboat traveling with the current takes 40 minutes to travel 20 miles downstream. The return trip takes 60 minutes. Find the speed of the current and the speed of the boat relative to the current, assuming that both remain constant.

74. **Aviation** An airplane flying into a headwind travels the 1800-mile flying distance between New York City and Albuquerque, New Mexico, in 3 hours and 36 minutes. On the return flight, the same distance is traveled in 3 hours. Find the airspeed of the plane and the speed of the wind, assuming that both remain constant.

75. **Business** A minor league baseball team had a total attendance one evening of 1175. The tickets for adults and children sold for \$15 and \$12, respectively. The ticket revenue that night was \$15,996.
    (a) Create a system of linear equations to find the numbers of adults A and children C at the game.
    (b) Solve your system of equations by elimination or by substitution. Explain your choice.
    (c) Use the *intersect* feature or the *zoom* and *trace* features of a graphing utility to solve your system.

76. Chemistry Thirty liters of a 40% acid solution is obtained by mixing a 25% solution with a 50% solution.

(a) Write a system of equations in which one equation represents the amount of final mixture required and the other represents the percent of acid in the final mixture. Let x and y represent the amounts of the 25% and 50% solutions, respectively.

(b) Use a graphing utility to graph the two equations in part (a) in the same viewing window.

(c) As the amount of the 25% solution increases, how does the amount of the 50% solution change?

(d) How much of each solution is required to obtain the specified concentration of the final mixture?

77. Business A grocer sells oranges for \$0.75 each and grapefruits for \$1.05 each. You purchased a mix of 14 oranges and grapefruits and paid \$12.00. How many of each type of fruit did you buy?

78. Nutrition Two cheeseburgers and one small order of french fries from a fast-food restaurant contain a total of 830 calories. Three cheeseburgers and two small orders of french fries contain a total of 1360 calories. Find the number of calories in each item.

79. Sales The projected sales S (in millions of dollars) of two clothing retailers from 2020 through 2030 can be modeled by

$$\begin{cases} S - 149.9t = 415.5 & \text{Retailer A} \\ S - 224.1t = 117.3 & \text{Retailer B} \end{cases}$$

where t is the year, with $t = 0$ corresponding to 2020.

(a) Solve the system of equations using the method of your choice. Explain why you chose that method.

(b) Interpret the meaning of the solution in the context of the problem.

(c) Interpret the meaning of the coefficient of the t-term in each model.

(d) Suppose the coefficients of t were equal and the models remained the same otherwise. How would this affect your answers in parts (a) and (b)?

80. *Why you should learn it* *(p. 488)* On a Saturday night, the manager of a shoe store evaluates the receipts of the previous week's sales. Two hundred fifty pairs of two different styles of running shoes were sold. One style sold for \$79.50 and the other sold for \$89.95. The receipts totaled \$20,867.75. The cash register that was supposed to record the number of each type of shoe sold malfunctioned. Can you recover the information? If so, how many shoes of each type were sold?

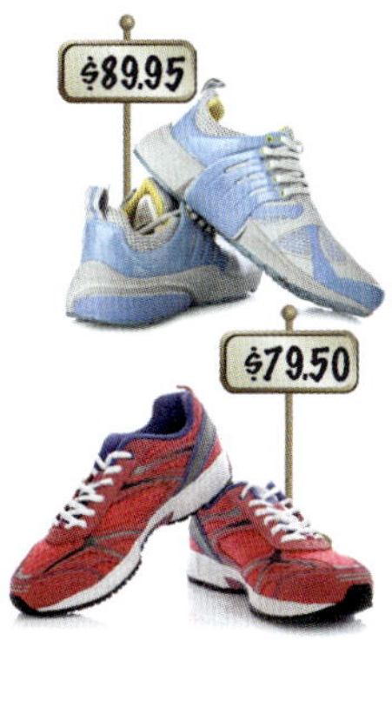

Fitting a Line to Data **To find the least squares regression line $y = ax + b$ for a set of points**

$$(x_1, y_1), (x_2, y_2), \ldots, (x_n, y_n)$$

you can solve the following system for a and b.

$$\begin{cases} nb + \left(\sum_{i=1}^{n} x_i\right)a = \left(\sum_{i=1}^{n} y_i\right) \\ \left(\sum_{i=1}^{n} x_i\right)b + \left(\sum_{i=1}^{n} x_i^2\right)a = \left(\sum_{i=1}^{n} x_i y_i\right) \end{cases}$$

In Exercises 81–84, the sums have been evaluated. Solve the given system for a and b to find the least squares regression line for the points. Use a graphing utility to confirm the results.

81. $\begin{cases} 5b + 10a = 20.2 \\ 10b + 30a = 50.1 \end{cases}$

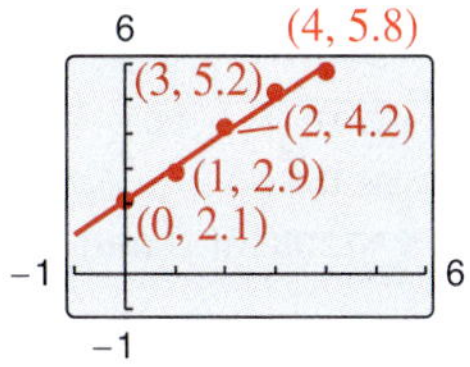

82. $\begin{cases} 5b + 10a = 11.7 \\ 10b + 30a = 25.6 \end{cases}$

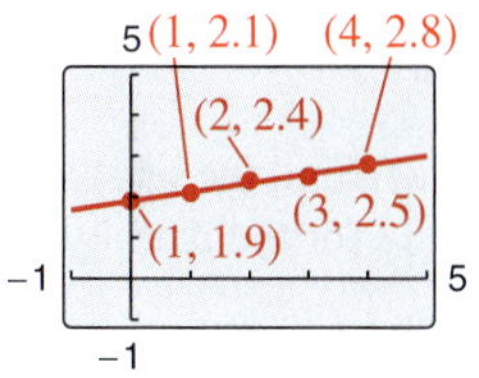

83. $\begin{cases} 6b + 15a = 23.6 \\ 15b + 55a = 48.8 \end{cases}$

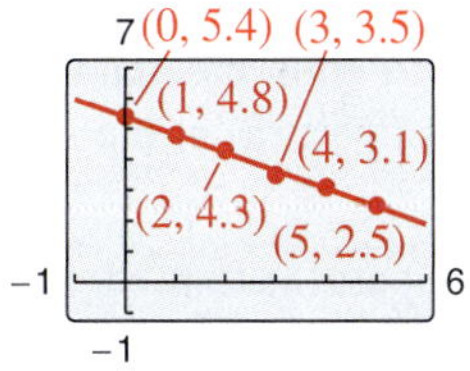

84. $\begin{cases} 7b + 21a = 13.1 \\ 21b + 91a = -2.8 \end{cases}$

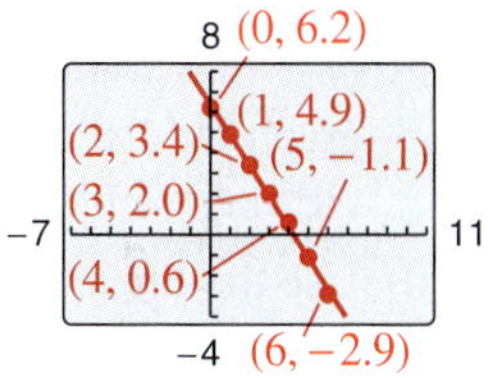

85. MODELING DATA

Four test plots were used to explore the relationship between wheat yield y (in bushels per acre) and amount of fertilizer applied x (in hundreds of pounds per acre). The results are given by the ordered pairs (1.0, 32), (1.5, 41), (2.0, 48), and (2.5, 53).

(a) Find the least squares regression line $y = ax + b$ for the data by solving the system for a and b.

$$\begin{cases} 4b + 7.0a = 174 \\ 7b + 13.5a = 322 \end{cases}$$

(b) Use the *regression* feature of a graphing utility to confirm the result in part (a).

(c) Use the graphing utility to plot the data and graph the linear model from part (a) in the same viewing window.

(d) Use the linear model from part (a) to predict the yield for a fertilizer application of 160 pounds per acre.

86. MODELING DATA

A deli store manager wants to know the demand for a mini sub as a function of the price. The daily sales for different prices of the product are shown in the table.

Price, x	Demand, y
\$3.00	45
\$3.60	37
\$4.50	23

(a) Find the least squares regression line $y = ax + b$ for the data by solving the system for a and b.

$$\begin{cases} 3.00b + 11.10a = 105.00 \\ 11.10b + 42.21a = 371.70 \end{cases}$$

(b) Use the *regression* feature of a graphing utility to confirm the result in part (a).

(c) Use the graphing utility to plot the data and graph the linear model from part (a) in the same viewing window.

(d) Use the linear model from part (a) to predict the demand when the price is \$5.75.

Focusing on Concepts

True or False? **In Exercises 87–89, determine whether the statement is true or false. Justify your answer.**

87. If a system of linear equations has two distinct solutions, then it has an infinite number of solutions.

88. If a system of linear equations has no solution, then the lines must be parallel.

89. Solving a system of equations graphically using a graphing utility always yields an exact solution.

90. Writing The graphs for a system of two linear equations in two variables are *not* parallel lines. Explain whether the system is consistent.

91. Think About It Find all value(s) of k for which the system of linear equations

$$\begin{cases} x + 3y = 9 \\ 2x + 6y = k \end{cases}$$

has (a) infinitely many solutions and (b) no solution.

92. HOW DO YOU SEE IT? Use the graphs of the two equations shown below.

(a) Describe the graphs of the two equations.

(b) Can you conclude that the system of equations shown in the graph is inconsistent? Explain.

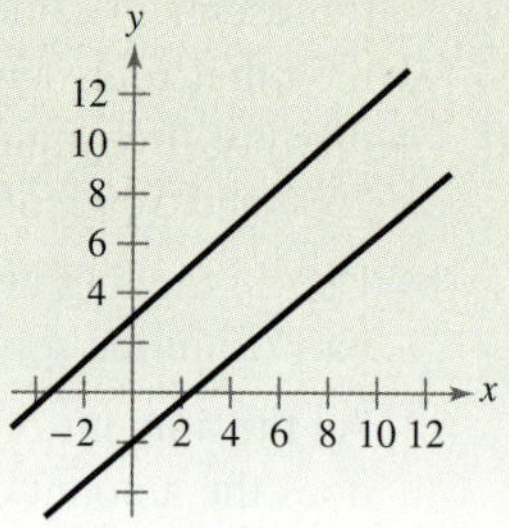

Solving a System **In Exercises 93 and 94, solve the system of equations for u and v. While solving for these variables, consider the transcendental functions as constants. (Systems of this type are found in a course in differential equations.)**

93. $\begin{cases} u \sin x + v \cos x = 0 \\ u \cos x - v \sin x = \sec x \end{cases}$

94. $\begin{cases} u \cos 2x + v \sin 2x = 0 \\ u(-2 \sin 2x) + v(2 \cos 2x) = \csc 2x \end{cases}$

Cumulative Mixed Review

Solving an Inequality **In Exercises 95–100, solve the inequality and graph the solution on a real number line.**

95. $-11 - 6x \geq 33$

96. $-6 \leq 3x - 10 < 6$

97. $|x - 8| < 10$

98. $|x + 10| \geq -3$

99. $2x^2 + 3x - 35 < 0$

100. $3x^2 + 12x > 0$

Rewriting a Logarithmic Expression **In Exercises 101–106, write the expression as the logarithm of a single quantity.**

101. $\ln x + \ln 8$

102. $\ln x - 5 \ln(x + 3)$

103. $\log_9 12 - \log_9 x$

104. $\frac{1}{4} \log_6 3 + \frac{1}{4} \log_6 x$

105. $2 \ln x - \ln(x + 2)$

106. $\frac{1}{2} \ln(x^2 + 4) - \ln x$

107. ***Project: College Expenses*** To work an extended application analyzing the average undergraduate tuition, room, and board charges at private degree-granting institutions in the United States from 1994 through 2016, visit this textbook's website at *LarsonPrecalculus.com*. (*Data Source:* U.S. Department of Education)

7.3 Multivariable Linear Systems

Row-Echelon Form and Back-Substitution

The method of elimination can be applied to a system of linear equations in more than two variables. When elimination is used to solve a system of linear equations, the goal is to rewrite the system in a form to which back-substitution can be applied. To see how this works, consider the following two systems of linear equations.

System of Three Linear Equations in Three Variables (See Example 2):

$$\begin{cases} x - 2y + 3z = 9 \\ -x + 3y + z = -2 \\ 2x - 5y + 5z = 17 \end{cases}$$

Equivalent System in Row-Echelon Form (See Example 1):

$$\begin{cases} x - 2y + 3z = 9 \\ y + 4z = 7 \\ z = 2 \end{cases}$$

The second system is said to be in **row-echelon form,** which means that it has a "stair-step" pattern with leading coefficients of 1. After comparing the two systems, it should be clear that it is easier to solve the system in row-echelon form, using back-substitution.

EXAMPLE 1 Using Back-Substitution in Row-Echelon Form

Solve the system of linear equations.

$$\begin{cases} x - 2y + 3z = 9 & \text{Equation 1} \\ y + 4z = 7 & \text{Equation 2} \\ z = 2 & \text{Equation 3} \end{cases}$$

Solution

From Equation 3, you know the value of z. To solve for y, substitute $z = 2$ into Equation 2 to obtain

$$y + 4(2) = 7 \qquad \text{Substitute 2 for } z.$$

$$y = -1. \qquad \text{Solve for } y.$$

Next, substitute $y = -1$ and $z = 2$ into Equation 1 to obtain

$$x - 2(-1) + 3(2) = 9 \qquad \text{Substitute } -1 \text{ for } y \text{ and 2 for } z.$$

$$x = 1. \qquad \text{Solve for } x.$$

The solution is

$$x = 1,\ y = -1, \text{ and } z = 2$$

which can be written as the **ordered triple**

$$(1, -1, 2).$$

Check this in the original system of equations.

✓ Checkpoint *Audio-video solution in English & Spanish at LarsonPrecalculus.com*

Solve the system of linear equations.

$$\begin{cases} 2x - y + 5z = 22 \\ y + 3z = 6 \\ z = 3 \end{cases}$$

What you should learn

- Use back-substitution to solve linear systems in row-echelon form.
- Use Gaussian elimination to solve systems of linear equations.
- Solve nonsquare systems of linear equations.
- Graphically interpret three-variable linear systems.
- Use systems of linear equations to write partial fraction decompositions of rational expressions.
- Use systems of linear equations in three or more variables to model and solve real-life problems.

Why you should learn it

Systems of linear equations in three or more variables can be used to model and solve real-life problems. For instance, Exercise 88 on page 509 shows how to use a system of linear equations to analyze the numbers of par-3, par-4, and par-5 holes on a golf course.

Gaussian Elimination

Two systems of equations are *equivalent* when they have the same solution set. To solve a system that is not in row-echelon form, first convert it to an *equivalent* system that *is* in row-echelon form by using one or more of the elementary row operations shown below. This process is called **Gaussian elimination,** after the German mathematician Carl Friedrich Gauss (1777–1855).

Elementary Row Operations for Systems of Equations

1. Interchange two equations.

2. Multiply one of the equations by a nonzero constant.

3. Add a multiple of one equation to another equation.

EXAMPLE 2 Using Gaussian Elimination to Solve a System

Solve the system of linear equations.

$$\begin{cases} x - 2y + 3z = 9 & \text{Equation 1} \\ -x + 3y + z = -2 & \text{Equation 2} \\ 2x - 5y + 5z = 17 & \text{Equation 3} \end{cases}$$

Solution

Because the leading coefficient of the first equation is 1, you can begin by saving the x at the upper left and eliminating the other x-terms from the first column.

$$\begin{cases} x - 2y + 3z = 9 \\ y + 4z = 7 \\ 2x - 5y + 5z = 17 \end{cases}$$

Adding the first equation to the second equation produces a new second equation.

$$\begin{cases} x - 2y + 3z = 9 \\ y + 4z = 7 \\ -y - z = -1 \end{cases}$$

Adding -2 times the first equation to the third equation produces a new third equation.

Now that all but the first x have been eliminated from the first column, go to work on the second column. (You need to eliminate y from the third equation.)

$$\begin{cases} x - 2y + 3z = 9 \\ y + 4z = 7 \\ 3z = 6 \end{cases}$$

Adding the second equation to the third equation produces a new third equation.

Finally, you need a coefficient of 1 for z in the third equation.

$$\begin{cases} x - 2y + 3z = 9 \\ y + 4z = 7 \\ z = 2 \end{cases}$$

Multiplying the third equation by $\frac{1}{3}$ produces a new third equation.

This is the same system that was solved in Example 1. So, the solution is $x = 1$, $y = -1$, and $z = 2$.

Algebra Help

Arithmetic errors are often made when elementary row operations are performed. You should note the operation performed in each step so that you can go back and check your work.

✓ Checkpoint *Audio-video solution in English & Spanish at LarsonPrecalculus.com*

Solve the system of linear equations.

$$\begin{cases} x + y + z = 6 \\ 2x - y + z = 3 \\ 3x + y - z = 2 \end{cases}$$

The goal of Gaussian elimination is to use elementary row operations on a system in order to isolate one variable. You can then solve for the value of the variable and use back-substitution to find the values of the remaining variables.

The next example involves an inconsistent system—one that has no solution. The key to recognizing an inconsistent system is that at some stage in the elimination process, you obtain a false statement such as $0 = -2$.

EXAMPLE 3 An Inconsistent System

Solve the system of linear equations.

$$\begin{cases} x - 3y + z = 1 & \text{Equation 1} \\ 2x - y - 2z = 2 & \text{Equation 2} \\ x + 2y - 3z = -1 & \text{Equation 3} \end{cases}$$

Solution

$$\begin{cases} x - 3y + z = 1 \\ 5y - 4z = 0 \\ x + 2y - 3z = -1 \end{cases}$$

Adding -2 times the first equation to the second equation produces a new second equation.

$$\begin{cases} x - 3y + z = 1 \\ 5y - 4z = 0 \\ 5y - 4z = -2 \end{cases}$$

Adding -1 times the first equation to the third equation produces a new third equation.

$$\begin{cases} x - 3y + z = 1 \\ 5y - 4z = 0 \\ 0 = -2 \end{cases}$$

Adding -1 times the second equation to the third equation produces a new third equation.

Because $0 = -2$ is a false statement, you can conclude that this system is inconsistent and has no solution. Moreover, because this system is equivalent to the original system, you can conclude that the original system also has no solution.

✓ Checkpoint *Audio-video solution in English & Spanish at LarsonPrecalculus.com*

Solve the system of linear equations.

$$\begin{cases} x + y - 2z = 3 \\ 3x - 2y + 4z = 1 \\ 2x - 3y + 6z = 8 \end{cases}$$

As with a system of linear equations in two variables, the number of solutions of a system of linear equations in more than two variables must fall into one of three categories.

The Number of Solutions of a Linear System

For a system of linear equations, exactly one of the following is true.

1. There is exactly one solution.
2. There are infinitely many solutions.
3. There is no solution.

A system of linear equations is called *consistent* when it has at least one solution. A consistent system with exactly one solution is **independent.** A consistent system with infinitely many solutions is **dependent.** A system of linear equations is called *inconsistent* when it has no solution.

EXAMPLE 4 A System with Infinitely Many Solutions

Solve the system of linear equations.

$$\begin{cases} x + y - 3z = -1 & \text{Equation 1} \\ y - z = 0 & \text{Equation 2} \\ -x + 2y = 1 & \text{Equation 3} \end{cases}$$

Solution

$$\begin{cases} x + y - 3z = -1 \\ y - z = 0 \\ 3y - 3z = 0 \end{cases}$$

Adding the first equation to the third equation produces a new third equation.

$$\begin{cases} x + y - 3z = -1 \\ y - z = 0 \\ 0 = 0 \end{cases}$$

Adding -3 times the second equation to the third equation produces a new third equation.

This result means that Equation 3 depends on Equations 1 and 2 in the sense that it gives no additional information about the variables. So, the original system is equivalent to

$$\begin{cases} x + y - 3z = -1 \\ y - z = 0 \end{cases}.$$

In the last equation, solve for y in terms of z to obtain $y = z$. Back-substituting for y in the previous equation produces

$$x = 2z - 1.$$

Finally, letting $z = a$, where a is a real number, the solutions of the original system are all of the form

$$x = 2a - 1, \quad y = a, \quad \text{and} \quad z = a.$$

So, every ordered triple of the form

$$(2a - 1, a, a) \qquad a \text{ is a real number.}$$

is a solution of the system.

✓ Checkpoint *Audio-video solution in English & Spanish at LarsonPrecalculus.com*

Solve the system of linear equations.

$$\begin{cases} x + 2y - 7z = -4 \\ 2x + 3y + z = 5 \\ 3x + 7y - 36z = -25 \end{cases}$$

In Example 4, there are other ways to write the same infinite set of solutions. For instance, the solutions could have been written as

$$\left(b, \tfrac{1}{2}(b + 1), \tfrac{1}{2}(b + 1)\right). \qquad b \text{ is a real number.}$$

This description produces the same set of solutions, as shown below.

Substitution	*Solution*	
$a = 0$	$(2(0) - 1, 0, 0) = (-1, 0, 0)$	Same solution
$b = -1$	$\left(-1, \tfrac{1}{2}(-1 + 1), \tfrac{1}{2}(-1 + 1)\right) = (-1, 0, 0)$	
$a = 1$	$(2(1) - 1, 1, 1) = (1, 1, 1)$	Same solution
$b = 1$	$\left(1, \tfrac{1}{2}(1 + 1), \tfrac{1}{2}(1 + 1)\right) = (1, 1, 1)$	
$a = 3$	$(2(3) - 1, 3, 3) = (5, 3, 3)$	Same solution
$b = 5$	$\left(5, \tfrac{1}{2}(5 + 1), \tfrac{1}{2}(5 + 1)\right) = (5, 3, 3)$	

Algebra Help

There are an infinite number of solutions to Example 4, but they are all of a specific form. By selecting, for instance, a-values of 0, 1, and 3, you can verify that $(-1, 0, 0)$, $(1, 1, 1)$, and $(5, 3, 3)$ are specific solutions. It is incorrect to say simply that the solution to Example 4 is "infinite." You must also specify the form of the solutions.

Nonsquare Systems

So far, each system of linear equations you have looked at has been *square*, which means that the number of equations is equal to the number of variables. In a **nonsquare system of equations,** the number of equations differs from the number of variables. A system of linear equations cannot have a unique solution unless there are at least as many equations as there are variables in the system.

EXAMPLE 5 A System with Fewer Equations than Variables

See LarsonPrecalculus.com for an interactive version of this type of example.

Solve the system of linear equations.

$$\begin{cases} x - 2y + z = 2 & \text{Equation 1} \\ 2x - y - z = 1 & \text{Equation 2} \end{cases}$$

Solution

Begin by rewriting the system in row-echelon form.

$$\begin{cases} x - 2y + z = 2 \\ 3y - 3z = -3 \end{cases}$$

Adding -2 times the first equation to the second equation produces a new second equation.

$$\begin{cases} x - 2y + z = 2 \\ y - z = -1 \end{cases}$$

Multiplying the second equation by $\frac{1}{3}$ produces a new second equation.

Solve for y in terms of z to obtain

$$y = z - 1.$$

By back-substituting into Equation 1, you can solve for x as follows.

$$x - 2y + z = 2 \quad \text{Write Equation 1.}$$

$$x - 2(z - 1) + z = 2 \quad \text{Substitute } z - 1 \text{ for } y.$$

$$x - 2z + 2 + z = 2 \quad \text{Distributive Property}$$

$$x = z \quad \text{Solve for } x.$$

Finally, by letting $z = a$, where a is a real number, you have the solution

$$x = a, \quad y = a - 1, \quad \text{and} \quad z = a.$$

So, every ordered triple of the form

$$(a, a - 1, a) \quad a \text{ is a real number.}$$

is a solution of the system.

Checkpoint Audio-video solution in English & Spanish at LarsonPrecalculus.com

Solve the system of linear equations.

$$\begin{cases} x - y + 4z = 3 \\ 4x - z = 0 \end{cases}$$

In Example 5, try choosing some values of a to obtain different solutions of the system, such as

$$(1, 0, 1), \quad (2, 1, 2), \quad \text{and} \quad (3, 2, 3).$$

Then check each ordered triple in the original system to verify that it is a solution of the system.

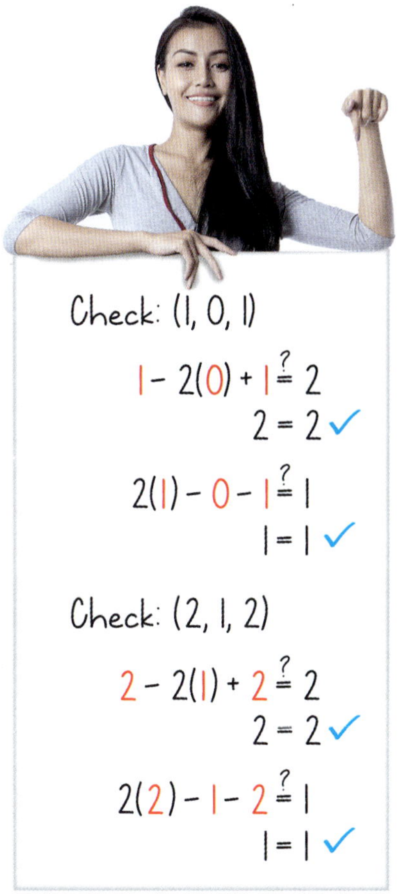

boonchoke/Shutterstock.com

Graphical Interpretation of Three-Variable Systems

Solutions of equations in three variables can be represented graphically using a **three-dimensional coordinate system.** To construct such a system, begin with the xy-coordinate plane in a horizontal position. Then draw the z-axis as a vertical line through the origin.

Every ordered triple (x, y, z) corresponds to a point on the three-dimensional coordinate system. For instance, the points corresponding to $(-2, 5, 4)$, $(2, -5, 3)$, and $(3, 3, -2)$ are shown in Figure 7.7.

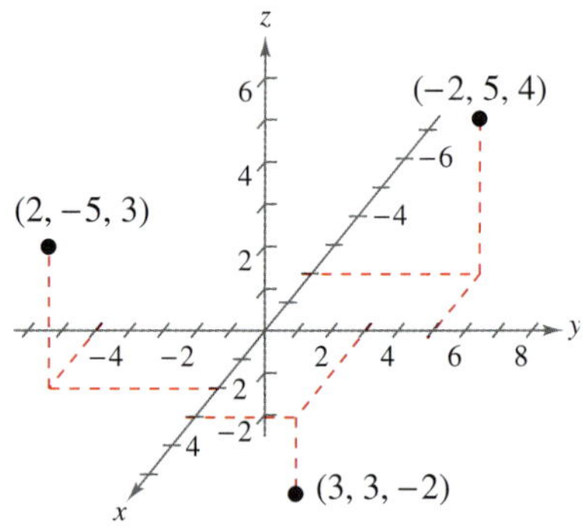

Figure 7.7

The **graph of an equation in three variables** consists of all points (x, y, z) that are solutions of the equation. The graph of a linear equation in three variables is a *plane*. Sketching graphs on a three-dimensional coordinate system is difficult because the sketch itself is only two-dimensional.

One technique for sketching a plane is to find the three points at which the plane intersects the axes. For instance, the plane

$$3x + 2y + 4z = 12$$

intersects the x-axis at the point $(4, 0, 0)$, the y-axis at the point $(0, 6, 0)$, and the z-axis at the point $(0, 0, 3)$. By plotting these three points, connecting them with line segments, and shading the resulting triangular region, you can sketch a portion of the graph, as shown in Figure 7.8.

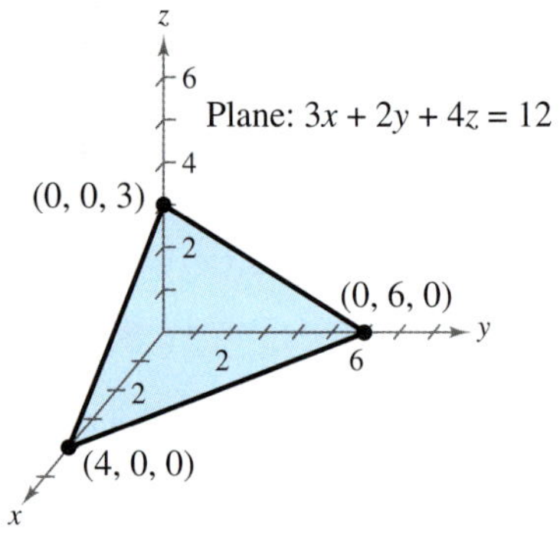

Figure 7.8

The graph of a system of three linear equations in three variables consists of *three* planes. When these planes intersect in a single point, the system has exactly one solution (see Figure 7.9). When the three planes have no point in common, the system has no solution (see Figures 7.10 and 7.11). When the three planes intersect in a line or a plane, the system has infinitely many solutions (see Figures 7.12 and 7.13).

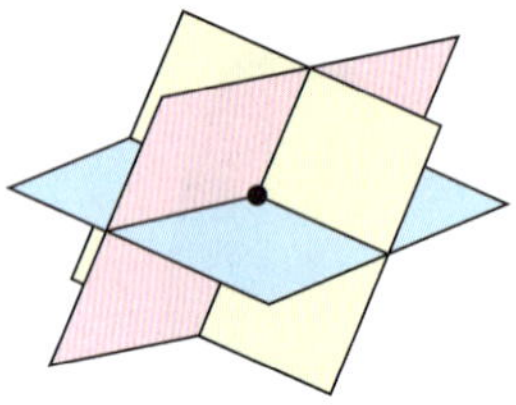

Solution: One point
Figure 7.9

Solution: None
Figure 7.10

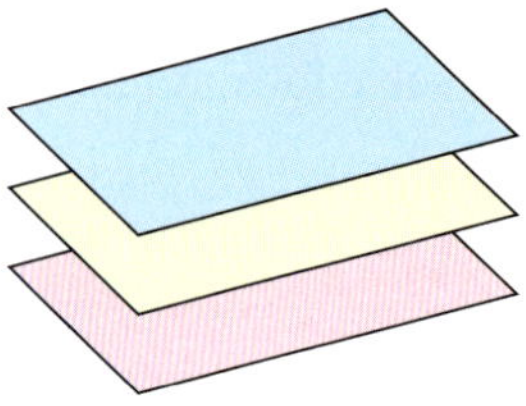

Solution: None
Figure 7.11

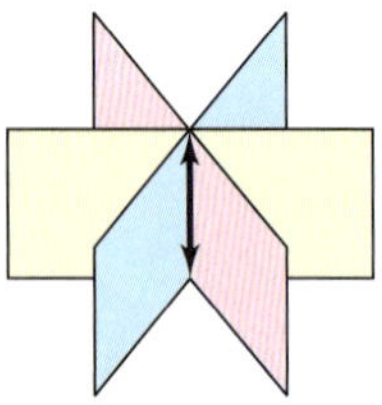

Solution: One line
Figure 7.12

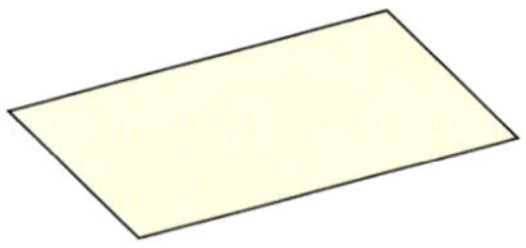

Solution: One plane
Figure 7.13

Technology Tip

Three-dimensional graphing utilities and computer algebra systems, such as *Maple* and *Mathematica*, are very efficient in producing three-dimensional graphs. They are good tools to use while studying calculus. If you have access to such a utility, try reproducing the plane shown in Figure 7.8.

Partial Fraction Decomposition

A rational expression can often be written as the sum of two or more simpler rational expressions. For example, the rational expression

$$\frac{x+7}{x^2-x-6}$$

can be written as the sum of two fractions with linear denominators. That is,

$$\frac{x+7}{x^2-x-6} = \underbrace{\overbrace{\underbrace{\frac{2}{x-3}}_{\text{Partial fraction}} + \underbrace{\frac{-1}{x+2}}_{\text{Partial fraction}}}^{\text{Partial fraction decomposition of } \frac{x+7}{x^2-x-6}}}_{}.$$

Each fraction on the right side of the equation is a **partial fraction,** and together they make up the **partial fraction decomposition** of the left side.

Decomposition of $N(x)/D(x)$ into Partial Fractions

1. *Divide when improper:* When $N(x)/D(x)$ is an improper fraction [degree of $N(x) \geq$ degree of $D(x)$], divide the denominator into the numerator to obtain

$$\frac{N(x)}{D(x)} = (\text{polynomial}) + \frac{N_1(x)}{D(x)}$$

and apply Steps 2, 3, and 4 (below) to the proper rational expression

$$\frac{N_1(x)}{D(x)}.$$

2. *Factor the denominator:* Completely factor the denominator into factors of the form

$$(px+q)^m \quad \text{and} \quad (ax^2+bx+c)^n$$

where (ax^2+bx+c) is irreducible over the reals.

3. *Linear factors:* For *each* factor of the form

$$(px+q)^m$$

the partial fraction decomposition must include the following sum of m fractions.

$$\frac{A_1}{(px+q)} + \frac{A_2}{(px+q)^2} + \cdots + \frac{A_m}{(px+q)^m}$$

4. *Quadratic factors:* For *each* factor of the form

$$(ax^2+bx+c)^n$$

the partial fraction decomposition must include the following sum of n fractions.

$$\frac{B_1x+C_1}{ax^2+bx+c} + \frac{B_2x+C_2}{(ax^2+bx+c)^2} + \cdots + \frac{B_nx+C_n}{(ax^2+bx+c)^n}$$

One of the most important applications of partial fractions is in calculus. Partial fractions can be used in a calculus operation called antidifferentiation.

EXAMPLE 6 Distinct Linear Factors

Write the partial fraction decomposition of

$$\frac{x+7}{x^2-x-6}.$$

Solution

The expression is proper, so factor the denominator. Because

$$x^2 - x - 6 = (x-3)(x+2)$$

you should include one partial fraction with a constant numerator for each linear factor of the denominator and write

$$\frac{x+7}{x^2-x-6} = \frac{A}{x-3} + \frac{B}{x+2}.$$

Multiplying each side of this equation by $(x-3)(x+2)$, the least common denominator (LCD), leads to the basic equation

$$x + 7 = A(x+2) + B(x-3) \qquad \text{Basic equation}$$
$$= Ax + 2A + Bx - 3B \qquad \text{Distributive Property}$$
$$= (A+B)x + 2A - 3B. \qquad \text{Write in polynomial form.}$$

Because two polynomials are equal if and only if the coefficients of like terms are equal, you can equate the coefficients of like terms to opposite sides of the equation.

$$x + 7 = (A+B)x + (2A - 3B) \qquad \text{Equate coefficients of like terms.}$$

You can now write the following system of linear equations.

$$\begin{cases} A + B = 1 & \text{Equation 1} \\ 2A - 3B = 7 & \text{Equation 2} \end{cases}$$

You can solve the system of linear equations as follows.

$$\begin{array}{l} A + B = 1 \\ \underline{2A - 3B = 7} \end{array} \quad \Rightarrow \quad \begin{array}{rl} 3A + 3B = & 3 \\ \underline{2A - 3B =} & \underline{7} \\ 5A \quad = & 10 \end{array} \qquad \begin{array}{l} \text{Multiply Equation 1 by 3.} \\ \text{Write Equation 2.} \\ \text{Add equations.} \end{array}$$

From this equation, you can see that $A = 2$. By back-substituting this value of A into Equation 1, you can solve for B as follows.

$$A + B = 1 \qquad \text{Write Equation 1.}$$
$$2 + B = 1 \qquad \text{Substitute 2 for A.}$$
$$B = -1 \qquad \text{Solve for } B.$$

So, the partial fraction decomposition is

$$\frac{x+7}{x^2-x-6} = \frac{2}{x-3} - \frac{1}{x+2}.$$

Check this result by combining the two partial fractions on the right side of the equation or by using a graphing utility.

✓ Checkpoint *Audio-video solution in English & Spanish at LarsonPrecalculus.com*

Write the partial fraction decomposition of

$$\frac{x+5}{2x^2-x-1}.$$

Technology Tip

You can graphically check the decomposition found in Example 6. To do this, use a graphing utility to graph

$$y_1 = \frac{x+7}{x^2-x-6} \quad \text{and}$$

$$y_2 = \frac{2}{x-3} - \frac{1}{x+2}$$

in the same viewing window. The graphs should be identical.

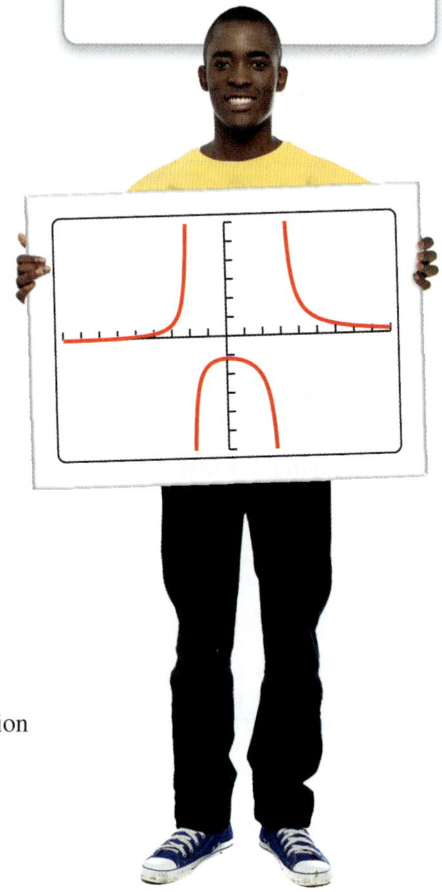

Ishay Botbol/Alamy Stock Photo

The next example shows how to find the partial fraction decomposition for a rational function whose denominator has a repeated linear factor.

EXAMPLE 7 Repeated Linear Factors

Write the partial fraction decomposition of

$$\frac{5x^2 + 20x + 6}{x^3 + 2x^2 + x}.$$

Solution

The expression is proper, so factor the denominator. Because the denominator factors as

$$x^3 + 2x^2 + x = x(x^2 + 2x + 1) = x(x + 1)^2$$

you should include one partial fraction with a constant numerator for each power of x and $(x + 1)$ and write

$$\frac{5x^2 + 20x + 6}{x^3 + 2x^2 + x} = \frac{A}{x} + \frac{B}{x + 1} + \frac{C}{(x + 1)^2}.$$

Multiplying by the LCD $x(x + 1)^2$ leads to the basic equation

$$5x^2 + 20x + 6 = A(x + 1)^2 + Bx(x + 1) + Cx \qquad \text{Basic equation}$$

$$= Ax^2 + 2Ax + A + Bx^2 + Bx + Cx \qquad \text{Expand.}$$

$$= (A + B)x^2 + (2A + B + C)x + A. \qquad \text{Polynomial form}$$

Equating coefficients of like terms on opposite sides of the equation

$$5x^2 + 20x + 6 = (A + B)x^2 + (2A + B + C)x + A$$

produces the following system of linear equations.

$$\begin{cases} A + B = 5 & \text{Equation 1} \\ 2A + B + C = 20 & \text{Equation 2} \\ A = 6 & \text{Equation 3} \end{cases}$$

Substituting 6 for A in Equation 1 yields

$$6 + B = 5$$

$$B = -1.$$

Substituting 6 for A and -1 for B in Equation 2 yields

$$2(6) + (-1) + C = 20$$

$$C = 9.$$

So, the partial fraction decomposition is

$$\frac{5x^2 + 20x + 6}{x^3 + 2x^2 + x} = \frac{6}{x} - \frac{1}{x + 1} + \frac{9}{(x + 1)^2}.$$

Check this result by combining the three partial fractions on the right side of the equation or by using a graphing utility.

Checkpoint *Audio-video solution in English & Spanish at LarsonPrecalculus.com*

Write the partial fraction decomposition of

$$\frac{x + 4}{x^3 + x^2}.$$

Explore the Concept

Partial fraction decomposition is practical only for rational functions whose denominators factor "nicely." For example, the factorization of the expression $x^2 - x - 5$ is

$$\left(x - \frac{1 - \sqrt{21}}{2}\right)\left(x - \frac{1 + \sqrt{21}}{2}\right).$$

Write the basic equation and try to complete the decomposition for

$$\frac{x + 7}{x^2 - x - 5}.$$

What problems do you encounter?

Applications

EXAMPLE 8 Vertical Motion

The height at time t of an object that is moving in a (vertical) line with constant acceleration a is given by the *position equation* $s = \frac{1}{2}at^2 + v_0t + s_0$. The height s is measured in feet, the acceleration a is measured in feet per second squared, t is measured in seconds, v_0 is the initial velocity (in feet per second) at $t = 0$, and s_0 is the initial height (in feet). Find the values of a, v_0, and s_0 when

$$s = 52 \text{ at } t = 1, \quad s = 52 \text{ at } t = 2, \quad \text{and } s = 20 \text{ at } t = 3$$

and interpret the result. (See Figure 7.14.)

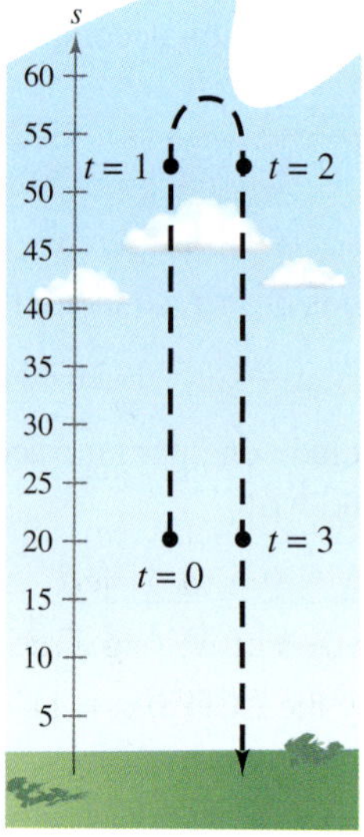

Figure 7.14

Solution

You can obtain three linear equations in a, v_0, and s_0 as follows.

When $t = 1$: $\frac{1}{2}a(1)^2 + v_0(1) + s_0 = 52$ ⟹ $a + 2v_0 + 2s_0 = 104$

When $t = 2$: $\frac{1}{2}a(2)^2 + v_0(2) + s_0 = 52$ ⟹ $2a + 2v_0 + s_0 = 52$

When $t = 3$: $\frac{1}{2}a(3)^2 + v_0(3) + s_0 = 20$ ⟹ $9a + 6v_0 + 2s_0 = 40$

Solving this system yields $a = -32$, $v_0 = 48$, and $s_0 = 20$. This solution results in a position equation of

$$s = -16t^2 + 48t + 20$$

and implies that the object was thrown upward at a velocity of 48 feet per second from a height of 20 feet.

✓ Checkpoint ▶ *Audio-video solution in English & Spanish at LarsonPrecalculus.com*

For the position equation $s = \frac{1}{2}at^2 + v_0t + s_0$ given in Example 8, find the values of a, v_0, and s_0 when $s = 104$ at $t = 1$, $s = 76$ at $t = 2$, and $s = 16$ at $t = 3$, and interpret the result.

EXAMPLE 9 Data Analysis: Curve-Fitting

Find a quadratic equation $y = ax^2 + bx + c$ whose graph passes through the points $(-1, 3)$, $(1, 1)$, and $(2, 6)$.

Solution

Because the graph of $y = ax^2 + bx + c$ passes through the points $(-1, 3)$, $(1, 1)$, and $(2, 6)$, you can write the following.

When $x = -1$, $y = 3$: $a(-1)^2 + b(-1) + c = 3$

When $x = 1$, $y = 1$: $a(1)^2 + b(1) + c = 1$

When $x = 2$, $y = 6$: $a(2)^2 + b(2) + c = 6$

This produces the following system of linear equations.

$$\begin{cases} a - b + c = 3 & \text{Equation 1} \\ a + b + c = 1 & \text{Equation 2} \\ 4a + 2b + c = 6 & \text{Equation 3} \end{cases}$$

The solution of this system is $a = 2$, $b = -1$, and $c = 0$. So, the equation of the parabola is $y = 2x^2 - x$, and its graph is shown in Figure 7.15.

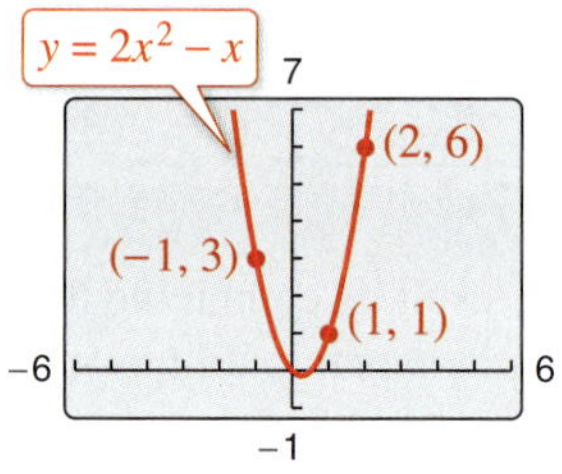

Figure 7.15

✓ Checkpoint ▶ *Audio-video solution in English & Spanish at LarsonPrecalculus.com*

Find a quadratic equation $y = ax^2 + bx + c$ whose graph passes through the points $(0, 0)$, $(3, -3)$, and $(6, 0)$.

7.3 Exercises

See *CalcChat.com* for tutorial help and worked-out solutions to odd-numbered exercises.
For instructions on how to use a graphing utility, see Appendix A.

Vocabulary and Concept Check

In Exercises 1–8, fill in the blank.

1. A system of equations that is in _______ form has a "stair-step" pattern with leading coefficients of 1.

2. A solution of a system of three linear equations in three unknowns can be written as an _______ , which has the form (x, y, z).

3. The process used to write a system of equations in row-echelon form is called _______ elimination.

4. A system of equations is called _______ when the number of equations differs from the number of variables in the system.

5. The equation $s = \frac{1}{2}at^2 + v_0t + s_0$ is called the _______ equation, and it models the height s of an object at time t that is moving in a vertical line with a constant acceleration a.

6. For an object that is thrown upward, the position equation $s(t)$ evaluated when $t = 0$ gives the _______ of the object.

7. The process of writing a rational expression as the sum of two or more simpler rational expressions is called _______ .

8. The partial fraction decomposition of $f(x) = 2x/[(x - 1)(x + 6)]$ involves the sum of _______ linear factors.

9. Is a consistent system with exactly one solution *independent* or *dependent*?

10. Is a consistent system with infinitely many solutions *independent* or *dependent*?

Procedures and Problem Solving

Checking Solutions **In Exercises 11–14, determine whether each ordered triple is a solution of the system of equations.**

11. $\begin{cases} 3x - y + z = 1 \\ 2x - 3z = -14 \\ 5y + 2z = 8 \end{cases}$

(a) $(3, 5, -3)$ (b) $(-1, 0, 4)$
(c) $(0, -1, 3)$ (d) $(1, 0, 4)$

12. $\begin{cases} 3x + 4y - z = 17 \\ 5x - y + 2z = -2 \\ 2x - 3y + 7z = -21 \end{cases}$

(a) $(1, 5, 6)$ (b) $(-2, -4, 2)$
(c) $(1, 3, -2)$ (d) $(0, 7, 0)$

13. $\begin{cases} 4x + y - z = 0 \\ -8x - 6y + z = -\frac{7}{4} \\ 3x - y = -\frac{9}{4} \end{cases}$

(a) $(0, 1, 1)$ (b) $\left(-\frac{3}{2}, \frac{5}{4}, -\frac{5}{4}\right)$
(c) $\left(-\frac{1}{2}, \frac{3}{4}, -\frac{5}{4}\right)$ (d) $\left(-\frac{1}{2}, \frac{1}{6}, -\frac{3}{4}\right)$

14. $\begin{cases} -4x \quad y - 8z = -6 \\ y + z = 0 \\ 4x - 7y = 6 \end{cases}$

(a) $(-2, -2, 2)$ (b) $\left(-\frac{33}{2}, -10, 10\right)$
(c) $\left(\frac{1}{8}, -\frac{1}{2}, \frac{1}{2}\right)$ (d) $\left(-\frac{11}{2}, -4, 4\right)$

Using Back-Substitution **In Exercises 15–20, use back-substitution to solve the system of linear equations.**

15. $\begin{cases} x - y + 5z = 37 \\ y + 2z = 6 \\ z = 8 \end{cases}$

16. $\begin{cases} x - 2y + 2z = 20 \\ y - z = 8 \\ z = -1 \end{cases}$

17. $\begin{cases} 2x + y - 3z = 10 \\ y + z = 12 \\ z = 2 \end{cases}$

18. $\begin{cases} x - y + 2z = 22 \\ 3y - 8z = -9 \\ z = -3 \end{cases}$

19. $\begin{cases} 4x - 2y + z = 8 \\ -y + z = 4 \\ z = 11 \end{cases}$

20. $\begin{cases} 5x - 8z = 22 \\ 3y - 5z = 10 \\ z = -4 \end{cases}$

Performing a Row Operation **In Exercises 21 and 22, perform the row operation and write the equivalent system. What did the operation accomplish?**

21. Add Equation 1 to Equation 2.

$\begin{cases} x - 2y + 3z = 5 & \text{Equation 1} \\ -x + 3y - 5z = 4 & \text{Equation 2} \\ 2x - 3z = 0 & \text{Equation 3} \end{cases}$

22. Add -2 times Equation 1 to Equation 3.

$\begin{cases} x - 2y + 3z = 5 & \text{Equation 1} \\ -x + 3y - 5z = 4 & \text{Equation 2} \\ 2x - 3z = 0 & \text{Equation 3} \end{cases}$

Solving a System of Linear Equations In Exercises 23–34, solve the system of linear equations and check any solution algebraically.

23. $\begin{cases} x + y + z = 5 \\ x - 2y + 4z = -1 \\ 3y + 4z = -1 \end{cases}$

24. $\begin{cases} 2x + 2z = 2 \\ 5x + 3y = 4 \\ 3y - 4z = 4 \end{cases}$

25. $\begin{cases} 4x + y - 3z = 11 \\ 2x - 3y + 2z = 9 \\ x + y + z = -3 \end{cases}$

26. $\begin{cases} x + 2y + z = 1 \\ x - 2y + 3z = -3 \\ 2x + y + z = -1 \end{cases}$

27. $\begin{cases} 2x + y - z = 7 \\ x - 2y + 2z = -9 \\ 3x - y + z = 5 \end{cases}$

28. $\begin{cases} 5x - 3y + 2z = 3 \\ 2x + 4y - z = 7 \\ x - 11y + 4z = 3 \end{cases}$

29. $\begin{cases} 3x - 3y + 6z = 6 \\ x + 2y - z = 5 \\ 5x - 8y + 13z = 7 \end{cases}$

30. $\begin{cases} x + 4z = 13 \\ 4x - 2y + z = 7 \\ 2x - 2y - 7z = -19 \end{cases}$

31. $\begin{cases} 2x + y + 3z = 1 \\ 2x + 6y + 8z = 3 \\ 6x + 8y + 18z = 5 \end{cases}$

32. $\begin{cases} 3x - 2y - 6z = -4 \\ -3x + 2y + 6z = 1 \\ x - y - 5z = -3 \end{cases}$

33. $\begin{cases} x + 4z = 1 \\ x + y + 10z = 10 \\ 2x - y + 2z = -5 \end{cases}$

34. $\begin{cases} x - 2y + z = 2 \\ 2x + 2y - 3z = -4 \\ 5x + z = 1 \end{cases}$

Solving a Nonsquare System In Exercises 35–38, solve the system of linear equations and check any solutions algebraically.

35. $\begin{cases} x - 2y + 5z = 2 \\ 4x - z = 0 \end{cases}$

36. $\begin{cases} 4x - y + z = 1 \\ x - 2y = 2 \end{cases}$

37. $\begin{cases} 2x - 3y + z = -2 \\ -4x + 9y = 7 \end{cases}$

38. $\begin{cases} 2x + 3y + 3z = 7 \\ 4x + 18y + 15z = 44 \end{cases}$

Finding a System In Exercises 39–42, find a system of linear equations that has the given solution. (There are many correct answers.)

39. $(3, -4, 2)$

40. $\left(-\frac{3}{2}, 4, -7\right)$

41. $(a, a + 4, a)$, where a is a real number

42. $(3a, a, a + 2)$, where a is a real number

Sketching a Plane In Exercises 43–46, sketch the plane represented by the linear equation.

43. $x + y + z = 8$

44. $x + 2y + z = 4$

45. $3x + 2y + 2z = 12$

46. $5x + y + 3z = 15$

Graphing a System of Linear Equations In Exercises 47 and 48, describe the solution(s) of the system as a point, line, or plane. Use a three-dimensional graphing utility to check your answer.

47. $\begin{cases} 2x + y - z = 2 \\ 6x + 3y - 3z = 6 \\ 4x + 2y - 2z = 4 \end{cases}$

48. $\begin{cases} x - 2y + 2z = 1 \\ x - 2y - 4z = 1 \\ x - 2y - z = 1 \end{cases}$

Form of a Partial Fraction Decomposition In Exercises 49–56, write the form of the partial fraction decomposition of the rational expression. Do not solve for the constants.

49. $\dfrac{7}{x^2 - 14x}$

50. $\dfrac{x - 2}{x^2 + 4x + 3}$

51. $\dfrac{12}{x^3 - 10x^2}$

52. $\dfrac{x^2 - 3x + 2}{4x^3 + 11x^2}$

53. $\dfrac{4x^2 + 3}{(x - 5)^3}$

54. $\dfrac{6x + 5}{(x + 2)^4}$

55. $\dfrac{8x}{x^2(x^2 + 3)^2}$

56. $\dfrac{x^2 - 9}{x^3(x^2 + 2)^2}$

Partial Fraction Decomposition In Exercises 57–68, write the partial fraction decomposition for the rational expression. Check your result algebraically by combining fractions and check your result graphically by using a graphing utility to graph the rational expression and the partial fractions in the same viewing window.

57. $\dfrac{1}{x^2 + x}$

58. $\dfrac{1}{4x^2 - 9}$

59. $\dfrac{x^2 + 12x + 12}{x^3 - 4x}$

60. $\dfrac{x^2 + 12x - 9}{x^3 - 9x}$

61. $\dfrac{2x - 3}{(x - 1)^2}$

62. $\dfrac{x - 2}{x^2 - 4x + 3}$

63. $\dfrac{x}{16x^4 - 1}$

64. $\dfrac{3}{x^4 + x}$

65. $\dfrac{x^4}{(x - 1)^3}$

66. $\dfrac{4x^4}{(2x - 1)^3}$

67. $\dfrac{x}{x^3 - x^2 - 2x + 2}$

68. $\dfrac{2x^2 + x + 8}{x^4 + 8x^2 + 16}$

Partial Fraction Decomposition In Exercises 69 and 70, write the partial fraction decomposition for the rational function. Identify the graph of the rational function and the graph of each term of its decomposition. State any relationship between the vertical asymptotes of the rational function and the vertical asymptotes of the terms of the decomposition.

69. $y = \dfrac{x - 12}{x(x - 4)}$

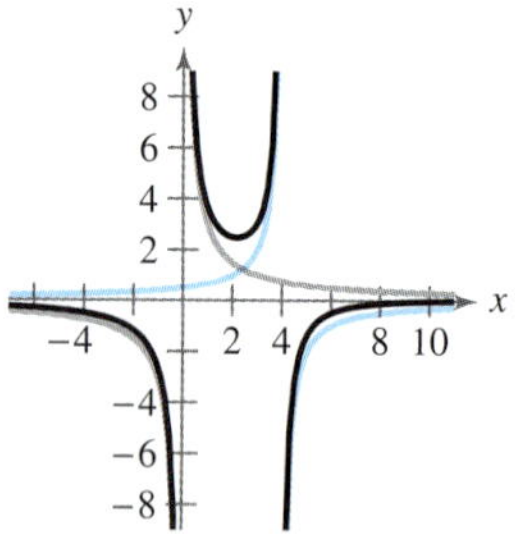

70. $y = \dfrac{2(4x - 3)}{x^2 - 9}$

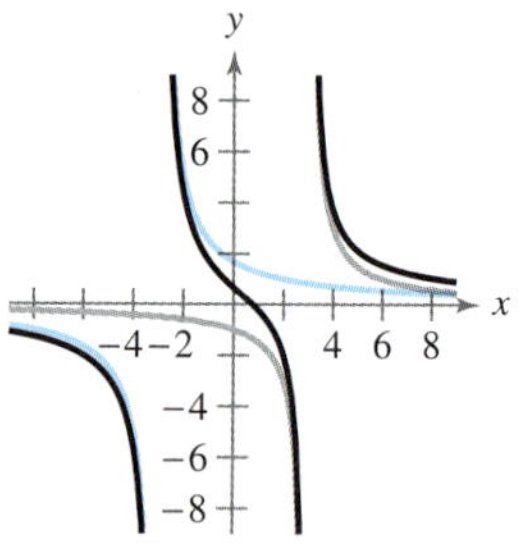

Vertical Motion **In Exercises 71–74, an object moving vertically is at the given heights at the specified times. Find the position equation $s = \frac{1}{2}at^2 + v_0t + s_0$ for the object.**

71. At $t = 1$ second, $s = 128$ feet.
At $t = 2$ seconds, $s = 80$ feet.
At $t = 3$ seconds, $s = 0$ feet.

72. At $t = 1$ second, $s = 32$ feet.
At $t = 2$ seconds, $s = 32$ feet.
At $t = 3$ seconds, $s = 0$ feet.

73. At $t = 1$ second, $s = 352$ feet.
At $t = 2$ seconds, $s = 272$ feet.
At $t = 3$ seconds, $s = 160$ feet.

74. At $t = 1$ second, $s = 132$ feet.
At $t = 2$ seconds, $s = 100$ feet.
At $t = 3$ seconds, $s = 36$ feet.

Data Analysis: Curve-Fitting **In Exercises 75–78, find the equation of the parabola**

$$y = ax^2 + bx + c$$

that passes through the points. To verify your result, use a graphing utility to plot the points and graph the parabola.

75. $(0, 0), (3, 0), (4, 4)$
76. $(0, 5), (1, 6), (2, 5)$
77. $(-1, 1), (0, -4), (1, -13)$
78. $(-2, 9), (-1, 0), (1, 6)$

Finding the Equation of a Circle **In Exercises 79–82, find the equation of the circle**

$$x^2 + y^2 + Dx + Ey + F = 0$$

that passes through the points. To verify your result, use a graphing utility to plot the points and graph the circle.

79. $(0, 0), (5, 5), (10, 0)$
80. $(0, 0), (0, 6), (3, 3)$
81. $(0, 0), (0, -2), (3, 0)$
82. $(3, 6), (7, 2), (3, -2)$

83. **Finance** A college student borrowed \$30,000 to pay for tuition, room, and board. Some of the money was borrowed at 4%, some at 6%, and some at 8%. How much was borrowed at each rate, given that the annual interest was \$1550 and the amount borrowed at 8% was three times the amount borrowed at 6%?

84. **Finance** A small corporation borrowed \$775,000 to expand its software line. Some of the money was borrowed at 8%, some at 9%, and some at 10%. How much was borrowed at each rate, given that the annual interest was \$67,500 and the amount borrowed at 8% was four times the amount borrowed at 10%?

Investment Portfolio **In Exercises 85 and 86, consider an investor with a portfolio totaling \$500,000 that is invested in certificates of deposit, municipal bonds, blue-chip stocks, and growth or speculative stocks. How much is invested in each type of investment?**

85. The certificates of deposit pay 3% annually, and the municipal bonds pay 5% annually. Over a five-year period, the investor expects the blue-chip stocks to return 8% annually and the growth stocks to return 10% annually. The investor wants a combined annual return of 5% and also wants to have only one-fourth of the portfolio invested in stocks.

86. The certificates of deposit pay 2% annually, and the municipal bonds pay 4% annually. Over a five-year period, the investor expects the blue-chip stocks to return 10% annually and the growth stocks to return 14% annually. The investor wants a combined annual return of 6% and also wants to have only one-fourth of the portfolio invested in stocks.

87. **Physical Education** In the 2013 Women's NCAA Championship basketball game, the University of Connecticut defeated Louisville University by a score of 93 to 60. Connecticut won by scoring a combination of two-point baskets, three-point baskets, and one-point free throws. The number of two-point baskets was 12 more than the number of free throws. The number of free throws was three less than the number of three-point baskets. What combination of scoring accounted for Connecticut's 93 points? *(Source: NCAA)*

88. ***Why you should learn it*** *(p. 497)* The Augusta National Golf Club in Augusta, Georgia, is an 18-hole course that consists of par-3 holes, par-4 holes, and par-5 holes. A golfer who shoots par has a total of 72 strokes for the entire course. There are two more par-4 holes than twice the number of par-5 holes, and the number of par-3 holes is equal to the number of par-5 holes. Find the numbers of par-3, par-4, and par-5 holes on the course. *(Source: Augusta National, Inc.)*

89. **Electrical Engineering** When Kirchhoff's Laws are applied to the electrical network in the figure, the currents I_1, I_2, and I_3 (in amperes) are the solution of the system

$$\begin{cases} I_1 - I_2 + I_3 = 0 \\ 3I_1 + 2I_2 = 7. \\ 2I_2 + 4I_3 = 8 \end{cases}$$

Find the currents.

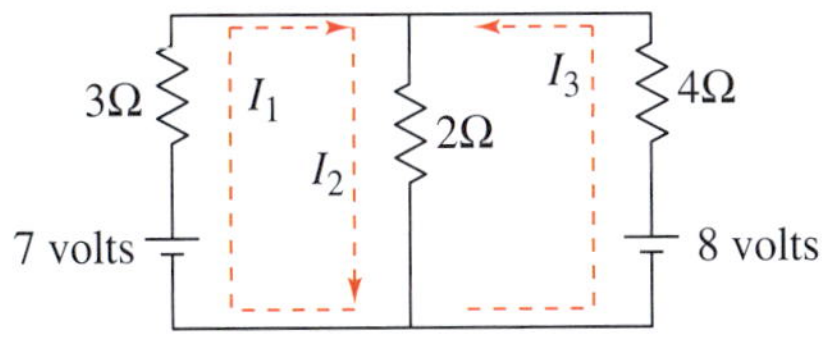

AlbertoChagas/iStock/Getty Images

90. Physics A system of pulleys is loaded with 128-pound and 32-pound weights (see figure). The tensions t_1 and t_2 in the ropes and the acceleration a of the 32-pound weight are modeled by the following system, where t_1 and t_2 are measured in pounds and a is in feet per second squared. Solve the system.

$$\begin{cases} t_1 - 2t_2 = 0 \\ t_1 \quad - 2a = 128 \\ t_2 + a = 32 \end{cases}$$

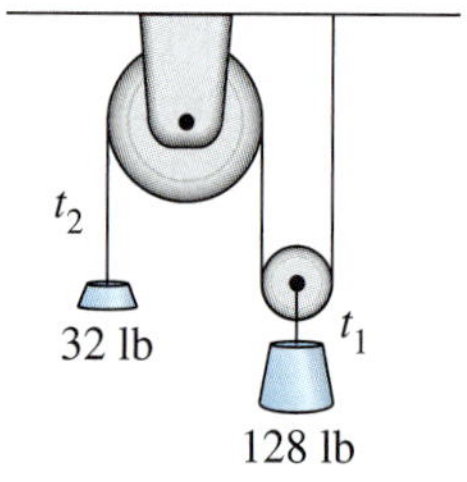

Fitting a Parabola **To find the least squares regression parabola $y = ax^2 + bx + c$ for a set of points (x_1, y_1), $(x_2, y_2), \ldots, (x_n, y_n)$ you can solve the following system of linear equations for a, b, and c.**

$$\begin{cases} nc + \left(\sum_{i=1}^{n} x_i\right)b + \left(\sum_{i=1}^{n} x_i^2\right)a = \sum_{i=1}^{n} y_i \\ \left(\sum_{i=1}^{n} x_i\right)c + \left(\sum_{i=1}^{n} x_i^2\right)b + \left(\sum_{i=1}^{n} x_i^3\right)a = \sum_{i=1}^{n} x_i y_i \\ \left(\sum_{i=1}^{n} x_i^2\right)c + \left(\sum_{i=1}^{n} x_i^3\right)b + \left(\sum_{i=1}^{n} x_i^4\right)a = \sum_{i=1}^{n} x_i^2 y_i \end{cases}$$

In Exercises 91–94, the sums have been evaluated. Solve the given system for a, b, and c to find the least squares regression parabola for the points. Use a graphing utility to confirm the result.

91. $\begin{cases} 4c \quad + 40a = 19 \\ 40b \quad = -12 \\ 40c \quad + 544a = 160 \end{cases}$

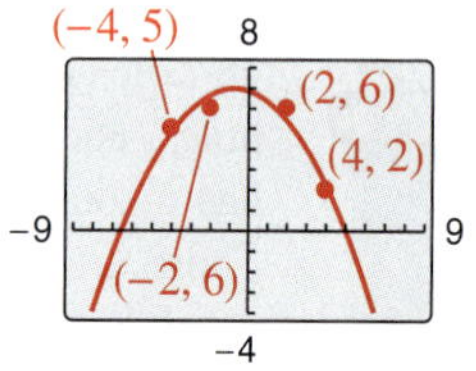

92. $\begin{cases} 5c \quad + 10a = 8 \\ 10b \quad = 12 \\ 10c \quad + 34a = 22 \end{cases}$

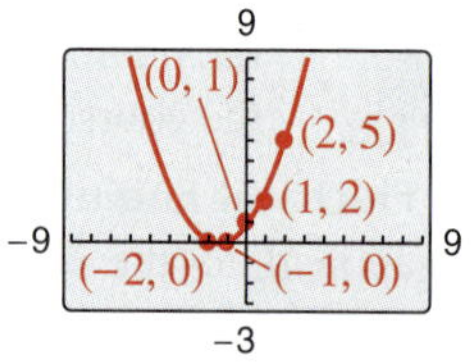

93. $\begin{cases} 4c + 9b + 29a = 20 \\ 9c + 29b + 99a = 70 \\ 29c + 99b + 353a = 254 \end{cases}$

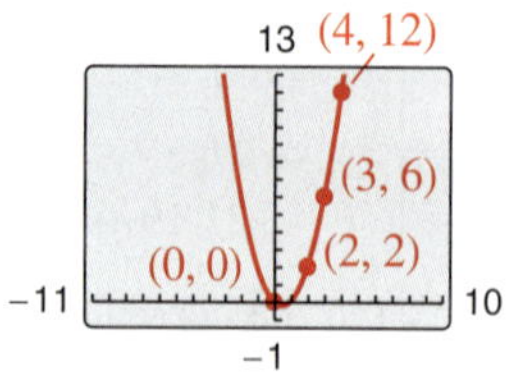

94. $\begin{cases} 4c + 6b + 14a = 25 \\ 6c + 14b + 36a = 21 \\ 14c + 36b + 98a = 33 \end{cases}$

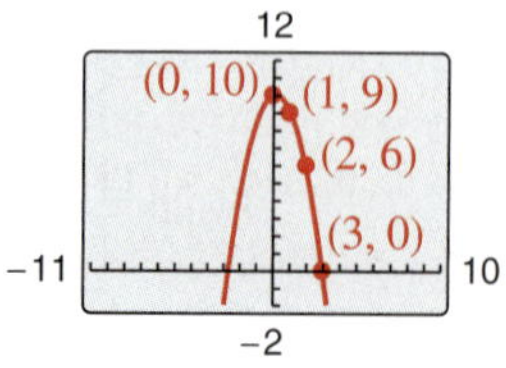

95. MODELING DATA

During the testing of a new automobile braking system, the speeds x (in miles per hour) and the stopping distances y (in feet) were recorded in the table.

Speed, x	Stopping distance, y
30	55
40	105
50	188

(a) Use the data to create a system of linear equations. Then find the least squares regression parabola for the data by solving the system.

(b) Use a graphing utility to graph the parabola and the data in the same viewing window.

(c) Use the model to estimate the stopping distance for a speed of 70 miles per hour.

96. MODELING DATA

A wildlife management team studied the reproduction rates of deer in three five-acre tracts of a wildlife preserve. In each tract, the number of females x and the percent of females y that had offspring the following year were recorded. The results are shown in the table.

Number, x	Percent, y
120	68
140	55
160	30

(a) Use the data to create a system of linear equations. Then find the least squares regression parabola for the data by solving the system.

(b) Use a graphing utility to graph the parabola and the data in the same viewing window.

(c) Use the model to predict the percent of females that had offspring when there were 170 females.

97. Environment The predicted cost C (in thousands of dollars) for a company to remove $p\%$ of a chemical from its wastewater can be modeled by

$$C = \frac{120p}{10{,}000 - p^2}, \quad 0 \le p < 100.$$

Write the partial fraction decomposition of the rational function. Verify your result by using the *table* feature of a graphing utility to create a table comparing the original function with the partial fractions.

98. Thermodynamics The magnitude of the range R of exhaust temperatures (in degrees Fahrenheit) in an experimental diesel engine is approximated by the model

$$R = \frac{2000(4 - 3x)}{(11 - 7x)(7 - 4x)}, \quad 0 \le x \le 1$$

where x is the relative load.

(a) Write the partial fraction decomposition of the rational function.

(b) The decomposition in part (a) is the difference of two fractions. The absolute values of the terms give the expected maximum and minimum temperatures of the exhaust gases. Use a graphing utility to graph each term.

Focusing on Concepts

True or False? **In Exercises 99 and 100, determine whether the statement is true or false. Justify your answer.**

99. If a system consists of three linear equations which represent three parallel planes, then the system has no solution.

100. If a system of three linear equations is inconsistent, then its graph has no points common to all three equations.

101. Error Analysis Describe the error.

The system

$$\begin{cases} x - 2y + 3z = 12 \\ \qquad y + 3z = 5 \\ \qquad\qquad 2z = 4 \end{cases}$$

is in row-echelon form.

102. Error Analysis Describe the error.

A solution of the system

$$\begin{cases} 2x + 3y - z = 7 \\ x - y + z = 2 \\ 3x + y - z = 8 \end{cases}$$

is $(3, 0, -1)$.

103. Error Analysis Describe the error.

$$\frac{4x}{(x-2)^2} = \frac{A}{x-2} + \frac{B}{(x-2)^2}$$

So, $4x = A(x - 2) + B(x - 2)^2$

104. Think About It Find values of a, b, and c (if possible) such that the system of linear equations has (a) a unique solution, (b) no solution, and (c) an infinite number of solutions.

$$\begin{cases} x + y \qquad = 2 \\ \qquad y + z = 2 \\ x \qquad + z = 2 \\ ax + by + cz = 0 \end{cases}$$

105. Think About It Are the two systems of equations equivalent? Give reasons for your answer.

$$\begin{cases} x + 3y - z = 6 \\ 2x - y + 2z = 1 \\ 3x + 2y - z = 2 \end{cases} \qquad \begin{cases} x + 3y - z = 6 \\ -7y + 4z = 1 \\ -7y - 4z = -16 \end{cases}$$

106. HOW DO YOU SEE IT? The number of sides x and the combined number of sides and diagonals y for each of three regular polygons are shown below. Write a system of linear equations to find an equation of the form $y = ax^2 + bx + c$ that represents the relationship between x and y for the three polygons.

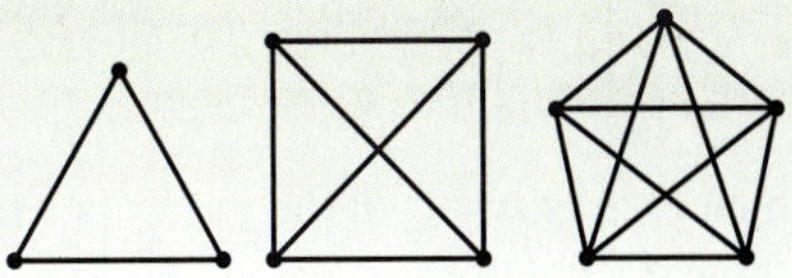

$x = 3, y = 3$ $\quad x = 4, y = 6$ $\quad x = 5, y = 10$

107. Exploration Find a system of equations in three variables that has exactly two equations and no solution.

108. Writing When using Gaussian elimination to solve a system of linear equations, explain how you can recognize that the system has no solution. Give an example that illustrates your answer.

Lagrange Multiplier **In Exercises 109 and 110, find values of x, y, and λ that satisfy the system. These systems arise in certain optimization problems in calculus. (λ is called a *Lagrange multiplier*.)**

109. $\begin{cases} \quad y + \lambda = 0 \\ x \quad + \lambda = 0 \\ x + y - 10 = 0 \end{cases}$

110. $\begin{cases} 2x \quad + \lambda = 0 \\ \quad 2y + \lambda = 0 \\ x + y - 4 = 0 \end{cases}$

Cumulative Mixed Review

Finding a Unit Vector **In Exercises 111–114, find a unit vector in the direction of the given vector.**

111. $\mathbf{v} = \langle 3, 0 \rangle$

112. $\mathbf{u} = \langle 0, -4 \rangle$

113. $\mathbf{u} = \langle 1, 5 \rangle$

114. $\mathbf{v} = \langle -1, -6 \rangle$

Solving a Trigonometric Equation **In Exercises 115 and 116, solve the equation.**

115. $4\sqrt{3} \tan \theta - 3 = 1$

116. $6 \cos x - 2 = 1$

117. ***Project: Fuel Economy*** To work an extended application analyzing the fuel economies of cars from 1998 through 2017, visit this textbook's website at *LarsonPrecalculus.com.* (*Data Source:* United States Environmental Protection Agency)

7.4 Matrices and Systems of Equations

Matrices

In this section, you will study a streamlined technique for solving systems of linear equations. This technique involves the use of a rectangular array of real numbers called a **matrix.** The plural of matrix is *matrices*.

Definition of Matrix

If m and n are positive integers, then an $m \times n$ (read "m by n") matrix is a rectangular array

$$\begin{array}{c} \\ \text{Row 1} \\ \text{Row 2} \\ \text{Row 3} \\ \vdots \\ \text{Row } m \end{array} \begin{array}{c} \begin{array}{ccccc} \text{Column 1} & \text{Column 2} & \text{Column 3} & \dots & \text{Column } n \end{array} \\ \begin{bmatrix} a_{11} & a_{12} & a_{13} & \dots & a_{1n} \\ a_{21} & a_{22} & a_{23} & \dots & a_{2n} \\ a_{31} & a_{32} & a_{33} & \dots & a_{3n} \\ \vdots & \vdots & \vdots & & \vdots \\ a_{m1} & a_{m2} & a_{m3} & \dots & a_{mn} \end{bmatrix} \end{array}$$

in which each **entry** a_{ij} of the matrix is a real number. An $m \times n$ matrix has m rows and n columns.

The entry in the ith row and jth column of a matrix is denoted by the *double subscript* notation a_{ij}. For instance, the entry a_{23} is the entry in the second row and third column. A matrix having m rows and n columns is said to be of **dimension** $m \times n$. If $m = n$, then the matrix is **square** of dimension $m \times m$ (or $n \times n$). For a square matrix, the entries

$$a_{11}, a_{22}, a_{33}, \dots$$

are the **main diagonal** entries.

EXAMPLE 1 Dimension of a Matrix

Determine the dimension of each matrix.

a. $[2]$ **b.** $\begin{bmatrix} 1 & -3 & 0 & \frac{1}{2} \end{bmatrix}$ **c.** $\begin{bmatrix} 0 & 0 \\ 0 & 0 \end{bmatrix}$ **d.** $\begin{bmatrix} 5 & 0 \\ 2 & -2 \\ -7 & 4 \end{bmatrix}$ **e.** $\begin{bmatrix} -2 \\ 0 \\ 1 \end{bmatrix}$

Solution

a. This matrix has *one* row and *one* column. The dimension of the matrix is 1×1.

b. This matrix has *one* row and *four* columns. The dimension of the matrix is 1×4.

c. This matrix has *two* rows and *two* columns. The dimension of the matrix is 2×2.

d. This matrix has *three* rows and *two* columns. The dimension of the matrix is 3×2.

e. This matrix has *three* rows and *one* column. The dimension of the matrix is 3×1.

✓ **Checkpoint** Audio-video solution in English & Spanish at LarsonPrecalculus.com

Determine the dimension of the matrix $\begin{bmatrix} 14 & 7 & 10 \\ -2 & -3 & -8 \end{bmatrix}$.

A matrix that has only one row [such as the matrix in Example 1(b)] is called a **row matrix,** and a matrix that has only one column [such as the matrix in Example 1(e)] is called a **column matrix.**

What you should learn

- Write matrices and determine their dimensions.
- Perform elementary row operations on matrices.
- Use matrices and Gaussian elimination to solve systems of linear equations.
- Use matrices and Gauss-Jordan elimination to solve systems of linear equations.

Why you should learn it

Matrices can be used to solve systems of linear equations in two or more variables. For instance, Exercise 90 on page 524 shows how a matrix can be used to help model an equation for the average annual consumer costs for health insurance.

Physician

Africa Studio/Shutterstock.com

A matrix derived from a system of linear equations (each written in standard form with the constant term on the right) is the **augmented matrix** of the system. Moreover, the matrix derived from the coefficients of the system (but not including the constant terms) is the **coefficient matrix** of the system.

System:
$$\begin{cases} x - 4y + 3z = 5 \\ -x + 3y - z = -3 \\ 2x - 4z = 6 \end{cases}$$

Augmented Matrix:
$$\left[\begin{array}{rrr|r} 1 & -4 & 3 & 5 \\ -1 & 3 & -1 & -3 \\ 2 & 0 & -4 & 6 \end{array}\right]$$

Coefficient Matrix:
$$\begin{bmatrix} 1 & -4 & 3 \\ -1 & 3 & -1 \\ 2 & 0 & -4 \end{bmatrix}$$

Note the use of 0 for the missing coefficient of the y-variable in the third equation, and also note the fourth column (of constant terms) in the augmented matrix.

When forming either the coefficient matrix or the augmented matrix of a system, you should begin by vertically aligning the variables in the equations and using 0's for any missing coefficients of variables.

Algebra Help

The optional dotted line in the augmented matrix helps to separate the coefficients of the linear system from the constant terms.

EXAMPLE 2 Writing an Augmented Matrix

Write the augmented matrix for the system of linear equations. What is the dimension of the augmented matrix?

$$\begin{cases} x + 3y = 9 \\ -y + 4z = -2 \\ x - 5z = 0 \end{cases}$$

Solution

Begin by writing the linear system and aligning the variables.

$$\begin{cases} x + 3y = 9 \\ - y + 4z = -2 \\ x - 5z = 0 \end{cases}$$

Next, use the coefficients and constant terms as the matrix entries. Include zeros for the coefficients of the missing variables.

$$\begin{array}{c} R_1 \\ R_2 \\ R_3 \end{array} \left[\begin{array}{rrr|r} 1 & 3 & 0 & 9 \\ 0 & -1 & 4 & -2 \\ 1 & 0 & -5 & 0 \end{array}\right]$$

The augmented matrix has three rows and four columns, so it is a 3×4 matrix. The notation R_n is used to designate each row in the matrix. For instance, Row 1 is represented by R_1.

Checkpoint *Audio-video solution in English & Spanish at LarsonPrecalculus.com*

Write the augmented matrix for the system of linear equations. What is the dimension of the augmented matrix?

$$\begin{cases} x + y + z = 2 \\ 2x - y + 3z = -1 \\ -x + 2y - z = 4 \end{cases}$$

Elementary Row Operations

In Section 7.3, you studied three operations that can be used on a system of linear equations to produce an equivalent system. Recall that the operations are (1) interchanging any two equations, (2) multiplying one of the equations by a nonzero constant, and (3) adding a multiple of one equation to another equation.

In matrix terminology, these three operations correspond to **elementary row operations.** An elementary row operation on an augmented matrix of a given system of linear equations produces a new augmented matrix corresponding to a new (but equivalent) system of linear equations. Two matrices are **row-equivalent** when one can be obtained from the other by a sequence of elementary row operations.

Elementary Row Operations for Matrices

1. Interchange two rows.
2. Multiply one of the rows by a nonzero constant.
3. Add a multiple of one row to another row.

Although elementary row operations are simple to perform, they involve a lot of arithmetic. Because it is easy to make a mistake, you should get in the habit of noting the elementary row operations performed in each step so that you can go back and check your work. Example 3 demonstrates the elementary row operations described above.

Technology Tip

Most graphing utilities can perform elementary row operations on matrices. For instructions on how to use the *matrix* feature and the elementary row operations features of a graphing utility, see Appendix A; for specific keystrokes, go to this textbook's *Student Companion Website.*

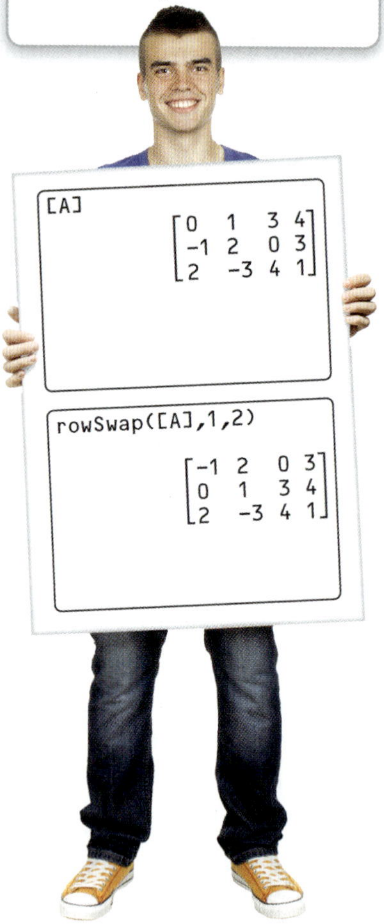

EXAMPLE 3 Elementary Row Operations

a. Interchange the first and second rows of the original matrix.

Original Matrix

$$\begin{bmatrix} 0 & 1 & 3 & 4 \\ -1 & 2 & 0 & 3 \\ 2 & -3 & 4 & 1 \end{bmatrix}$$

New Row-Equivalent Matrix

$$\begin{matrix} \leftarrow R_2 \\ \leftarrow R_1 \\ \\ \end{matrix}\begin{bmatrix} -1 & 2 & 0 & 3 \\ 0 & 1 & 3 & 4 \\ 2 & -3 & 4 & 1 \end{bmatrix}$$

b. Multiply the first row of the original matrix by $\frac{1}{2}$.

Original Matrix

$$\begin{bmatrix} 2 & -4 & 6 & -2 \\ 1 & 3 & -3 & 0 \\ 5 & -2 & 1 & 2 \end{bmatrix}$$

New Row-Equivalent Matrix

$$\begin{matrix} \frac{1}{2}R_1 \rightarrow \\ \\ \\ \end{matrix}\begin{bmatrix} 1 & -2 & 3 & -1 \\ 1 & 3 & -3 & 0 \\ 5 & -2 & 1 & 2 \end{bmatrix}$$

c. Add -2 times the first row of the original matrix to the third row.

Original Matrix

$$\begin{bmatrix} 1 & 2 & -4 & 3 \\ 0 & 3 & -2 & -1 \\ 2 & 1 & 5 & -2 \end{bmatrix}$$

New Row-Equivalent Matrix

$$\begin{matrix} \\ \\ -2R_1 + R_3 \rightarrow \end{matrix}\begin{bmatrix} 1 & 2 & -4 & 3 \\ 0 & 3 & -2 & -1 \\ 0 & -3 & 13 & -8 \end{bmatrix}$$

Note that the elementary row operation is written beside the row that is *changed.*

✓ **Checkpoint** Audio-video solution in English & Spanish at LarsonPrecalculus.com

Fill in the blank using an elementary row operation to form a row-equivalent matrix.

$$\begin{bmatrix} 1 & 2 & 3 \\ 3 & 12 & 5 \end{bmatrix} \qquad \begin{bmatrix} 1 & 2 & 3 \\ 0 & \square & -4 \end{bmatrix}$$

Gaussian Elimination with Back-Substitution

In Example 2 of Section 7.3, you used Gaussian elimination with back-substitution to solve a system of linear equations. The next example demonstrates the matrix version of Gaussian elimination. The basic difference between the two methods is that with matrices you do not need to keep writing the variables.

EXAMPLE 4 Comparing Linear Systems and Matrix Operations

Linear System

$$\begin{cases} x - 2y + 3z = 9 \\ -x + 3y + z = -2 \\ 2x - 5y + 5z = 17 \end{cases}$$

Associated Augmented Matrix

$$\left[\begin{array}{rrr:r} 1 & -2 & 3 & 9 \\ -1 & 3 & 1 & -2 \\ 2 & -5 & 5 & 17 \end{array}\right]$$

Add the first equation to the second equation.

$$\begin{cases} x - 2y + 3z = 9 \\ y + 4z = 7 \\ 2x - 5y + 5z = 17 \end{cases}$$

Add the first row to the second row: $R_1 + R_2$.

$$\begin{array}{r} \\ R_1 + R_2 \rightarrow \\ \\ \end{array}\left[\begin{array}{rrr:r} 1 & -2 & 3 & 9 \\ 0 & 1 & 4 & 7 \\ 2 & -5 & 5 & 17 \end{array}\right]$$

Add -2 times the first equation to the third equation.

$$\begin{cases} x - 2y + 3z = 9 \\ y + 4z = 7 \\ -y - z = -1 \end{cases}$$

Add -2 times the first row to the third row: $-2R_1 + R_3$.

$$\begin{array}{r} \\ \\ -2R_1 + R_3 \rightarrow \end{array}\left[\begin{array}{rrr:r} 1 & -2 & 3 & 9 \\ 0 & 1 & 4 & 7 \\ 0 & -1 & -1 & -1 \end{array}\right]$$

Add the second equation to the third equation.

$$\begin{cases} x - 2y + 3z = 9 \\ y + 4z = 7 \\ 3z = 6 \end{cases}$$

Add the second row to the third row: $R_2 + R_3$.

$$\begin{array}{r} \\ \\ R_2 + R_3 \rightarrow \end{array}\left[\begin{array}{rrr:r} 1 & -2 & 3 & 9 \\ 0 & 1 & 4 & 7 \\ 0 & 0 & 3 & 6 \end{array}\right]$$

Multiply the third equation by $\frac{1}{3}$.

$$\begin{cases} x - 2y + 3z = 9 \\ y + 4z = 7 \\ z = 2 \end{cases}$$

Multiply the third row by $\frac{1}{3}$: $\frac{1}{3}R_3$.

$$\begin{array}{r} \\ \\ \frac{1}{3}R_3 \rightarrow \end{array}\left[\begin{array}{rrr:r} 1 & -2 & 3 & 9 \\ 0 & 1 & 4 & 7 \\ 0 & 0 & 1 & 2 \end{array}\right]$$

At this point, you can use back-substitution to find x and y.

$y + 4(2) = 7$ Substitute 2 for z.

$y = -1$ Solve for y.

$x - 2(-1) + 3(2) = 9$ Substitute -1 for y and 2 for z.

$x = 1$ Solve for x.

The solution is $x = 1$, $y = -1$, and $z = 2$.

Algebra Help

Remember that you should check a solution by substituting the values of x, y, and z into each equation of the original system. For instance, you can check the solution to Example 4 as shown.

Equation 1:
$1 - 2(-1) + 3(2) = 9$ ✓

Equation 2:
$-1 + 3(-1) + 2 = -2$ ✓

Equation 3:
$2(1) - 5(-1) + 5(2) = 17$ ✓

✓ **Checkpoint** Audio-video solution in English & Spanish at LarsonPrecalculus.com

Compare solving the linear system to solving it using its associated augmented matrix.

$$\begin{cases} 2x + y - z = -3 \\ 4x - 2y + 2z = -2 \\ -6x + 5y + 4z = 10 \end{cases}$$

The last matrix in Example 4 is in **row-echelon form.** The term *echelon* refers to the stair-step pattern formed by the nonzero elements of the matrix. To be in this form (or reduced row-echelon form), a matrix must have the following properties.

Row-Echelon Form and Reduced Row-Echelon Form

A matrix in **row-echelon form** has the following properties.

1. Any rows consisting entirely of zeros occur at the bottom of the matrix.
2. For each row that does not consist entirely of zeros, the first nonzero entry is 1 (called a **leading 1**).
3. For two successive (nonzero) rows, the leading 1 in the higher row is farther to the left than the leading 1 in the lower row.

A matrix in *row-echelon form* is in **reduced row-echelon form** when every column that has a leading 1 has zeros in every position above and below its leading 1.

Technology Tip

Some graphing utilities have a feature that transforms a matrix to row-echelon form or to reduced row-echelon form. For instructions on how to use the *row-echelon form* feature and the *reduced row-echelon form* feature of a graphing utility, see Appendix A; for specific keystrokes, go to this textbook's *Student Companion Website.*

It is worth mentioning that the row-echelon form of a matrix is not unique. That is, two different sequences of elementary row operations may yield different row-echelon forms. The *reduced* row-echelon form of a given matrix, however, is unique.

EXAMPLE 5 Row-Echelon Form

Determine whether each matrix is in row-echelon form. If it is, determine whether the matrix is in reduced row-echelon form.

a. $\begin{bmatrix} 1 & 2 & -1 & 4 \\ 0 & 1 & 0 & 3 \\ 0 & 0 & 1 & -2 \end{bmatrix}$ **b.** $\begin{bmatrix} 1 & 2 & -1 & 2 \\ 0 & 0 & 0 & 0 \\ 0 & 1 & 2 & -4 \end{bmatrix}$

c. $\begin{bmatrix} 1 & -5 & 2 & -1 & 3 \\ 0 & 0 & 1 & 3 & -2 \\ 0 & 0 & 0 & 1 & 4 \\ 0 & 0 & 0 & 0 & 1 \end{bmatrix}$ **d.** $\begin{bmatrix} 1 & 0 & 0 & -1 \\ 0 & 1 & 0 & 2 \\ 0 & 0 & 1 & 3 \\ 0 & 0 & 0 & 0 \end{bmatrix}$

e. $\begin{bmatrix} 1 & 2 & -3 & 4 \\ 0 & 2 & 1 & -1 \\ 0 & 0 & 1 & -3 \end{bmatrix}$ **f.** $\begin{bmatrix} 0 & 1 & 0 & 5 \\ 0 & 0 & 1 & 3 \\ 0 & 0 & 0 & 0 \end{bmatrix}$

Solution

The matrices in (a), (c), (d), and (f) are in row-echelon form. The matrices in (d) and (f) are in *reduced* row-echelon form because every column that has a leading 1 has zeros in every position above and below its leading 1. The matrix in (b) is not in row-echelon form because the row of all zeros does not occur at the bottom of the matrix. The matrix in (e) is not in row-echelon form because the first nonzero entry in Row 2 is not a leading 1.

Checkpoint *Audio-video solution in English & Spanish at LarsonPrecalculus.com*

Determine whether the matrix is in row-echelon form. If it is, determine whether it is in reduced row-echelon form.

$$\begin{bmatrix} 1 & 0 & -2 & 4 \\ 0 & 1 & 11 & 3 \\ 0 & 0 & 0 & 0 \end{bmatrix}$$

Every matrix is row-equivalent to a matrix in row-echelon form. For instance, in Example 5, you can change the matrix in part (e) to row-echelon form by multiplying its second row by $\frac{1}{2}$.

Gaussian elimination with back-substitution works well for solving systems of linear equations by hand or with a computer. For this algorithm, the order in which the elementary row operations are performed is important. You should operate *from left to right by columns*, using elementary row operations to obtain zeros in all entries directly below the leading 1's.

EXAMPLE 6 Gaussian Elimination with Back-Substitution

Solve the system of equations.

$$\begin{cases} y + z - 2w = -3 \\ x + 2y - z = 2 \\ 2x + 4y + z - 3w = -2 \\ x - 4y - 7z - w = -19 \end{cases}$$

Solution

$$\begin{bmatrix} 0 & 1 & 1 & -2 & \vdots & -3 \\ 1 & 2 & -1 & 0 & \vdots & 2 \\ 2 & 4 & 1 & -3 & \vdots & -2 \\ 1 & -4 & -7 & -1 & \vdots & -19 \end{bmatrix}$$

Write augmented matrix.

$$\begin{matrix} R_2 \\ R_1 \\ \\ \\ \end{matrix} \begin{bmatrix} 1 & 2 & -1 & 0 & \vdots & 2 \\ 0 & 1 & 1 & -2 & \vdots & -3 \\ 2 & 4 & 1 & -3 & \vdots & -2 \\ 1 & -4 & -7 & -1 & \vdots & -19 \end{bmatrix}$$

Interchange R_1 and R_2 so first column has leading 1 in upper left corner.

$$\begin{matrix} \\ \\ -2R_1 + R_3 \rightarrow \\ -R_1 + R_4 \rightarrow \end{matrix} \begin{bmatrix} 1 & 2 & -1 & 0 & \vdots & 2 \\ 0 & 1 & 1 & -2 & \vdots & -3 \\ 0 & 0 & 3 & -3 & \vdots & -6 \\ 0 & -6 & -6 & -1 & \vdots & -21 \end{bmatrix}$$

Perform operations on R_3 and R_4 so first column has zeros below its leading 1.

$$\begin{matrix} \\ \\ \\ 6R_2 + R_4 \rightarrow \end{matrix} \begin{bmatrix} 1 & 2 & -1 & 0 & \vdots & 2 \\ 0 & 1 & 1 & -2 & \vdots & -3 \\ 0 & 0 & 3 & -3 & \vdots & -6 \\ 0 & 0 & 0 & -13 & \vdots & -39 \end{bmatrix}$$

Perform operations on R_4 so second column has zeros below its leading 1.

$$\begin{matrix} \\ \\ \frac{1}{3}R_3 \rightarrow \\ -\frac{1}{13}R_4 \rightarrow \end{matrix} \begin{bmatrix} 1 & 2 & -1 & 0 & \vdots & 2 \\ 0 & 1 & 1 & -2 & \vdots & -3 \\ 0 & 0 & 1 & -1 & \vdots & -2 \\ 0 & 0 & 0 & 1 & \vdots & 3 \end{bmatrix}$$

Perform operations on R_3 and R_4 so third and fourth columns have leading 1's.

The matrix is now in row-echelon form, and the corresponding system is

$$\begin{cases} x + 2y - z = 2 \\ y + z - 2w = -3 \\ z - w = -2 \\ w = 3 \end{cases}$$

Using back-substitution, you can determine that the solution is

$x = -1, \quad y = 2, \quad z = 1, \quad$ and $\quad w = 3.$

Checkpoint *Audio-video solution in English & Spanish at LarsonPrecalculus.com*

Solve the system of equations.

$$\begin{cases} -3x + 5y + 3z = -19 \\ 3x + 4y + 4z = 8 \\ 4x - 8y - 6z = 26 \end{cases}$$

The following steps summarize the procedure used in Example 6.

Gaussian Elimination with Back-Substitution

1. Write the augmented matrix of the system of linear equations.
2. Use elementary row operations to rewrite the augmented matrix in row-echelon form.
3. Write the system of linear equations corresponding to the matrix in row-echelon form and use back-substitution to find the solution.

Remember that it is possible for a system to have no solution. If, in the elimination process, you obtain a row with zeros except for the last entry, then you can conclude that the system is inconsistent.

EXAMPLE 7 A System with No Solution

Solve the system of equations.

$$\begin{cases} x - y + 2z = 4 \\ x + z = 6 \\ 2x - 3y + 5z = 4 \\ 3x + 2y - z = 1 \end{cases}$$

Solution

$$\begin{bmatrix} 1 & -1 & 2 & \vdots & 4 \\ 1 & 0 & 1 & \vdots & 6 \\ 2 & -3 & 5 & \vdots & 4 \\ 3 & 2 & -1 & \vdots & 1 \end{bmatrix}$$ Write augmented matrix.

$$\begin{matrix} \\ -R_1 + R_2 \rightarrow \\ -2R_1 + R_3 \rightarrow \\ -3R_1 + R_4 \rightarrow \end{matrix} \begin{bmatrix} 1 & -1 & 2 & \vdots & 4 \\ 0 & 1 & -1 & \vdots & 2 \\ 0 & -1 & 1 & \vdots & -4 \\ 0 & 5 & -7 & \vdots & -11 \end{bmatrix}$$ Perform row operations.

$$\begin{matrix} \\ \\ R_2 + R_3 \rightarrow \\ \\ \end{matrix} \begin{bmatrix} 1 & -1 & 2 & \vdots & 4 \\ 0 & 1 & -1 & \vdots & 2 \\ 0 & 0 & 0 & \vdots & -2 \\ 0 & 5 & -7 & \vdots & -11 \end{bmatrix}$$ Perform row operations.

Note that the third row of this matrix consists of zeros except for the last entry. This means that the original system of linear equations is *inconsistent*. You can see why this is true by converting back to a system of linear equations. Because the third equation is not possible, the system has no solution.

$$\begin{cases} x - y + 2z = 4 \\ y - z = 2 \\ 0 = -2 \quad \leftarrow \text{Not possible} \\ 5y - 7z = -11 \end{cases}$$

✓ Checkpoint ▶ *Audio-video solution in English & Spanish at LarsonPrecalculus.com*

Solve the system of equations.

$$\begin{cases} x + y + z = 1 \\ x + 2y + 2z = 2 \\ x - y - z = 1 \end{cases}$$

Gauss–Jordan Elimination

With Gaussian elimination, elementary row operations are applied to a matrix to obtain a (row-equivalent) row-echelon form of the matrix. A second method of elimination, called **Gauss-Jordan elimination** after Carl Friedrich Gauss (1777–1855) and Wilhelm Jordan (1842–1899), continues the reduction process until the *reduced* row-echelon form is obtained. This procedure is demonstrated in Example 8.

EXAMPLE 8 Gauss–Jordan Elimination

See LarsonPrecalculus.com for an interactive version of this type of example.

Use Gauss-Jordan elimination to solve the system.

$$\begin{cases} x - 2y + 3z = 9 \\ -x + 3y + z = -2 \\ 2x - 5y + 5z = 17 \end{cases}$$

Solution

In Example 4, Gaussian elimination was used to obtain the row-echelon form

$$\begin{bmatrix} 1 & -2 & 3 & \vdots & 9 \\ 0 & 1 & 4 & \vdots & 7 \\ 0 & 0 & 1 & \vdots & 2 \end{bmatrix}.$$

Now, rather than using back-substitution, apply additional elementary row operations until you obtain a matrix in *reduced* row-echelon form. To do this, you must produce zeros above each of the leading 1's, as follows.

$$2R_2 + R_1 \rightarrow \begin{bmatrix} 1 & 0 & 11 & \vdots & 23 \\ 0 & 1 & 4 & \vdots & 7 \\ 0 & 0 & 1 & \vdots & 2 \end{bmatrix}$$

Perform operations on R_1 so second column has a zero above its leading 1.

$$\begin{matrix} -11R_3 + R_1 \rightarrow \\ -4R_3 + R_2 \rightarrow \\ \end{matrix} \begin{bmatrix} 1 & 0 & 0 & \vdots & 1 \\ 0 & 1 & 0 & \vdots & -1 \\ 0 & 0 & 1 & \vdots & 2 \end{bmatrix}$$

Perform operations on R_1 and R_2 so third column has zeros above its leading 1.

The matrix is now in reduced row-echelon form. Converting back to a system of linear equations, you have

$$\begin{cases} x = 1 \\ y = -1. \\ z = 2 \end{cases}$$

Now you can simply read the solution, $x = 1$, $y = -1$, and $z = 2$, which can be written as the ordered triple $(1, -1, 2)$. You can check this result using the *reduced row-echelon form* feature of a graphing utility, as shown in Figure 7.16.

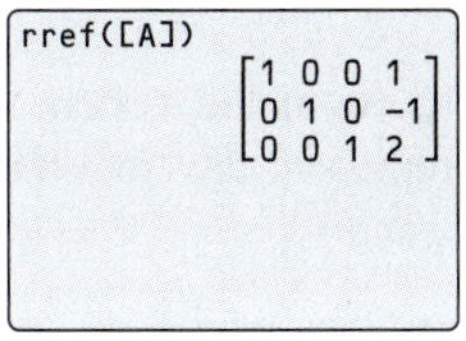

Figure 7.16

✓ Checkpoint ▶ *Audio-video solution in English & Spanish at LarsonPrecalculus.com*

Use Gauss-Jordan elimination to solve the system.

$$\begin{cases} -3x + 7y + 2z = 1 \\ -5x + 3y - 5z = -8 \\ 2x - 2y - 3z = 15 \end{cases}$$

In Example 8, note that the solution is the same as the one obtained using Gaussian elimination in Example 4. The advantage of Gauss-Jordan elimination is that, from the reduced row-echelon form, you can simply read the solution without the need for back-substitution.

Technology Tip

For a demonstration of a graphical approach to Gauss-Jordan elimination on a 2×3 matrix, see the Visualizing Row Operations Program, available for several models of graphing calculators at this textbook's *Student Companion Website.*

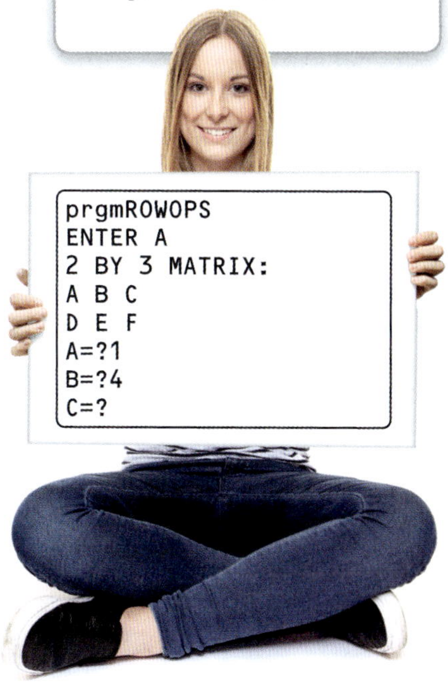

The elimination procedures described in this section employ an algorithmic approach that is easily adapted to computer programs. However, the procedure makes no effort to avoid fractional coefficients. For instance, in the elimination procedure for the system

$$\begin{cases} 2x - 5y + 5z = 17 \\ 3x - 2y + 3z = 11 \\ -3x + 3y = -6 \end{cases}$$

you may be inclined to multiply the first row by $\frac{1}{2}$ to produce a leading 1, which will result in working with fractional coefficients. For hand computations, you can sometimes avoid fractions by judiciously choosing the order in which you apply elementary row operations.

EXAMPLE 9 A System with an Infinite Number of Solutions

Solve the system.

$$\begin{cases} 2x + 4y - 2z = 0 \\ 3x + 5y = 1 \end{cases}$$

Solution

$$\begin{bmatrix} 2 & 4 & -2 & \vdots & 0 \\ 3 & 5 & 0 & \vdots & 1 \end{bmatrix}$$

$$\tfrac{1}{2}R_1 \rightarrow \begin{bmatrix} 1 & 2 & -1 & \vdots & 0 \\ 3 & 5 & 0 & \vdots & 1 \end{bmatrix}$$

$$-3R_1 + R_2 \rightarrow \begin{bmatrix} 1 & 2 & -1 & \vdots & 0 \\ 0 & -1 & 3 & \vdots & 1 \end{bmatrix}$$

$$-R_2 \rightarrow \begin{bmatrix} 1 & 2 & -1 & \vdots & 0 \\ 0 & 1 & -3 & \vdots & -1 \end{bmatrix}$$

$$-2R_2 + R_1 \rightarrow \begin{bmatrix} 1 & 0 & 5 & \vdots & 2 \\ 0 & 1 & -3 & \vdots & -1 \end{bmatrix}$$

The corresponding system of equations is

$$\begin{cases} x + 5z = 2 \\ y - 3z = -1 \end{cases}.$$

Solving for x and y in terms of z, you have

$$x = -5z + 2 \quad \text{and} \quad y = 3z - 1.$$

To write a solution of the system that does not use any of the three variables of the system, let a represent any real number and let $z = a$. Now substitute a for z in the equations for x and y.

$$x = -5z + 2 = -5a + 2 \quad \text{and} \quad y = 3z - 1 = 3a - 1$$

So, the solution set has the form $(-5a + 2, 3a - 1, a)$. Recall from Section 7.3 that a solution set of this form represents an infinite number of solutions. Try substituting values for a to obtain a few solutions. Then check each solution in the original system of equations.

✓ **Checkpoint** ▶ *Audio-video solution in English & Spanish at LarsonPrecalculus.com*

Solve the system.

$$\begin{cases} 2x - 6y + 6z = 46 \\ 2x - 3y = 31 \end{cases}$$

7.4 Exercises

See *CalcChat.com* for tutorial help and worked-out solutions to odd-numbered exercises.
For instructions on how to use a graphing utility, see Appendix A.

Vocabulary and Concept Check

In Exercises 1–3, fill in the blank.

1. A rectangular array of real numbers that can be used to solve a system of linear equations is called a _______ .

2. A matrix in row-echelon form is in _______ when every column that has a leading 1 has zeros in every position above and below its leading 1.

3. The process of using row operations to write a matrix in reduced row-echelon form is called _______ .

In Exercises 4–6, refer to the system of linear equations

$$\begin{cases} -2x + 3y = 5 \\ 6x + 7y = 4 \end{cases}.$$

4. Is the coefficient matrix for the system a *square* matrix?

5. Is the augmented matrix for the system of dimension 3×2?

6. Is the augmented matrix row-equivalent to its reduced row-echelon form?

Procedures and Problem Solving

Double Subscript Notation **In Exercises 7 and 8, represent the verbal statement using double subscript notation.**

7. The entry in the fourth row and sixth column of a matrix is -8.

8. The entry in the second row and first column of a matrix is 0.

Dimension of a Matrix **In Exercises 9–14, determine the dimension of the matrix.**

9. $\begin{bmatrix} 7 & 0 \end{bmatrix}$

10. $\begin{bmatrix} 3 & -1 & 2 & 6 \end{bmatrix}$

11. $\begin{bmatrix} 4 \\ 32 \\ 3 \end{bmatrix}$

12. $\begin{bmatrix} 5 & 4 & 2 \\ 3 & -5 & 1 \\ 7 & -2 & 9 \end{bmatrix}$

13. $\begin{bmatrix} 33 & 45 \\ -9 & 20 \end{bmatrix}$

14. $\begin{bmatrix} 3 & -1 & 6 & 4 \\ -2 & 5 & 7 & 7 \end{bmatrix}$

Writing an Augmented Matrix **In Exercises 15–20, write the augmented matrix for the system of linear equations.**

15. $\begin{cases} 4x - 3y = -5 \\ -x + 3y = 12 \end{cases}$

16. $\begin{cases} 7x + 4y = 22 \\ 5x - 9y = 15 \end{cases}$

17. $\begin{cases} x - y + 2z = 2 \\ 4x - 3y + z = -1 \\ 2x + y = 0 \end{cases}$

18. $\begin{cases} x - 3y + z = 1 \\ 4y = 0 \\ 7z = -5 \end{cases}$

19. $\begin{cases} 7x - 5y + z = 13 \\ 19x - 8z = 10 \end{cases}$

20. $\begin{cases} -2y - 5z = 0 \\ 3x + y = -4 \end{cases}$

Writing a System of Equations **In Exercises 21–24, write the system of linear equations represented by the augmented matrix. (Use the variables x, y, z, and w, if applicable.)**

21. $\left[\begin{array}{cc|c} 3 & 4 & 0 \\ 1 & -1 & 7 \end{array}\right]$

22. $\left[\begin{array}{cc|c} 7 & -5 & 2 \\ 8 & 0 & -2 \end{array}\right]$

23. $\left[\begin{array}{ccc|c} 0 & 12 & 3 & 0 \\ -2 & 18 & 0 & 10 \\ 1 & 7 & -8 & 43 \end{array}\right]$

24. $\left[\begin{array}{cccc|c} 6 & 2 & -1 & -5 & -25 \\ -1 & 0 & 7 & 3 & 7 \\ 4 & -1 & -10 & 6 & 23 \\ 0 & 8 & 1 & -11 & -21 \end{array}\right]$

Identifying an Elementary Row Operation **In Exercises 25–28, identify the elementary row operation(s) performed to obtain the new row-equivalent matrix.**

	Original Matrix	*New Row-Equivalent Matrix*
25.	$\begin{bmatrix} -4 & 8 & -20 \\ -2 & 6 & 7 \end{bmatrix}$	$\begin{bmatrix} 1 & -2 & 5 \\ -2 & 6 & 7 \end{bmatrix}$
26.	$\begin{bmatrix} -3 & 6 & 0 \\ 5 & 2 & -2 \end{bmatrix}$	$\begin{bmatrix} -18 & 0 & 6 \\ 5 & 2 & -2 \end{bmatrix}$
27.	$\begin{bmatrix} 0 & -1 & -5 & 5 \\ -1 & 3 & -7 & 6 \\ 4 & -5 & 1 & 3 \end{bmatrix}$	$\begin{bmatrix} -1 & 3 & -7 & 6 \\ 0 & -1 & -5 & 5 \\ 0 & 7 & -27 & 27 \end{bmatrix}$
28.	$\begin{bmatrix} -1 & -2 & 3 & -2 \\ 2 & -5 & 1 & -7 \\ 5 & 4 & -7 & 6 \end{bmatrix}$	$\begin{bmatrix} -1 & -2 & 3 & -2 \\ 2 & -5 & 1 & -7 \\ 0 & -6 & 8 & -4 \end{bmatrix}$

Elementary Row Operations In Exercises 29–32, fill in the blank(s) using elementary row operations to form a row-equivalent matrix.

29. $\begin{bmatrix} 1 & 4 & 3 \\ 2 & 10 & 5 \end{bmatrix}$

$\begin{bmatrix} 1 & 4 & 3 \\ 0 & \square & -1 \end{bmatrix}$

30. $\begin{bmatrix} 3 & 6 & 8 \\ 4 & -3 & 6 \end{bmatrix}$

$\begin{bmatrix} 1 & \square & \frac{8}{3} \\ 4 & -3 & 6 \end{bmatrix}$

31. $\begin{bmatrix} 1 & 1 & 4 & -1 \\ 3 & 8 & 10 & 3 \\ -2 & 1 & 12 & 6 \end{bmatrix}$

$\begin{bmatrix} 1 & 1 & 4 & -1 \\ 0 & 5 & \square & \square \\ 0 & 3 & \square & \square \end{bmatrix}$

$\begin{bmatrix} 1 & 1 & 4 & -1 \\ 0 & 1 & \square & \square \\ 0 & 3 & \square & \square \end{bmatrix}$

32. $\begin{bmatrix} 2 & 4 & 8 & 3 \\ 1 & -1 & -3 & 2 \\ 2 & 6 & 4 & 9 \end{bmatrix}$

$\begin{bmatrix} 1 & -1 & -3 & 2 \\ 2 & \square & \square & \square \\ 2 & 6 & 4 & 9 \end{bmatrix}$

$\begin{bmatrix} 1 & -1 & -3 & 2 \\ 0 & 6 & \square & \square \\ 0 & 8 & \square & \square \end{bmatrix}$

Performing a Row Operation with Technology In Exercises 33–36, use a graphing utility to perform the operation.

33. Interchange R_2 and R_3.

$\begin{bmatrix} 0 & -2 & 0 & \vdots & 5 \\ 4 & 8 & 4 & \vdots & -4 \\ 0 & 2 & 1 & \vdots & 1 \end{bmatrix}$

34. Add R_1 to R_3.

$\begin{bmatrix} 1 & 6 & -1 & \vdots & 5 \\ 2 & 0 & 9 & \vdots & 12 \\ 7 & -6 & 1 & \vdots & 8 \end{bmatrix}$

35. Add -2 times R_1 to R_2.

$\begin{bmatrix} -1 & 3 & \vdots & 9 \\ 2 & 1 & \vdots & 6 \end{bmatrix}$

36. Multiply R_1 by $\frac{1}{3}$.

$\begin{bmatrix} 3 & -3 & \vdots & 9 \\ 6 & 1 & \vdots & 0 \end{bmatrix}$

Comparing Linear Systems and Matrix Operations In Exercises 37 and 38, (a) perform the row operations to solve the augmented matrix, (b) write and solve the system of linear equations represented by the augmented matrix, and (c) compare the two solution methods. Which do you prefer?

37. $\begin{bmatrix} -3 & 4 & \vdots & 22 \\ 6 & -4 & \vdots & -28 \end{bmatrix}$

(i) Add R_2 to R_1.

(ii) Add -2 times R_1 to R_2.

(iii) Multiply R_2 by $-\frac{1}{4}$.

(iv) Multiply R_1 by $\frac{1}{3}$.

38. $\begin{bmatrix} 7 & 13 & 1 & \vdots & -4 \\ -3 & -5 & -1 & \vdots & -4 \\ 3 & 6 & 1 & \vdots & -2 \end{bmatrix}$

(i) Add R_2 to R_1.

(ii) Multiply R_1 by $\frac{1}{4}$.

(iii) Add R_3 to R_2.

(iv) Add -3 times R_1 to R_3.

(v) Add -2 times R_2 to R_1.

Form of a Matrix In Exercises 39–42, determine whether the matrix is in row-echelon form. If it is, determine whether it is in reduced row-echelon form.

39. $\begin{bmatrix} 1 & 0 & 0 & 0 \\ 0 & 1 & 1 & 5 \\ 0 & 0 & 0 & 0 \end{bmatrix}$

40. $\begin{bmatrix} 1 & 0 & 1 & 0 \\ 0 & 1 & 0 & 2 \\ 0 & 0 & 1 & 0 \end{bmatrix}$

41. $\begin{bmatrix} 1 & 0 & 2 & 1 \\ 0 & 1 & -3 & 10 \\ 0 & 0 & 1 & 0 \end{bmatrix}$

42. $\begin{bmatrix} 1 & 0 & 0 & 1 \\ 0 & 1 & 0 & -1 \\ 0 & 0 & 0 & 2 \end{bmatrix}$

Row-Echelon Form In Exercises 43–46, write the matrix in row-echelon form. Remember that the row-echelon form of a matrix is not unique.

43. $\begin{bmatrix} 1 & -4 & 5 \\ -2 & 6 & -6 \end{bmatrix}$

44. $\begin{bmatrix} 1 & -3 & 2 \\ 5 & 0 & 7 \end{bmatrix}$

45. $\begin{bmatrix} 1 & 2 & -1 & 3 \\ 3 & 7 & -5 & 14 \\ -2 & -1 & -3 & 8 \end{bmatrix}$

46. $\begin{bmatrix} 1 & 4 & 4 & 5 \\ -1 & 8 & 0 & 18 \\ 2 & -4 & 1 & -6 \end{bmatrix}$

Reduced Row-Echelon Form In Exercises 47–50, write the matrix in reduced row-echelon form. Use a graphing utility to verify your result.

47. $\begin{bmatrix} 3 & 3 & 3 \\ -1 & 0 & -4 \\ 2 & 4 & -2 \end{bmatrix}$

48. $\begin{bmatrix} 1 & 3 & 2 \\ 5 & 15 & 9 \\ 2 & 6 & 10 \end{bmatrix}$

49. $\begin{bmatrix} -4 & 1 & 0 & 6 \\ 1 & -2 & 3 & -4 \end{bmatrix}$

50. $\begin{bmatrix} 1 & 0 & -2 & 7 \\ 3 & -4 & 6 & 1 \end{bmatrix}$

Using Back-Substitution In Exercises 51–54, write the system of linear equations represented by the augmented matrix. Then use back-substitution to solve the system. (Use variables x, y, and z, if applicable.)

51. $\begin{bmatrix} 1 & -2 & \vdots & 4 \\ 0 & 1 & \vdots & -1 \end{bmatrix}$

52. $\begin{bmatrix} 1 & 5 & \vdots & 0 \\ 0 & 1 & \vdots & 6 \end{bmatrix}$

53. $\begin{bmatrix} 1 & -1 & 2 & \vdots & 4 \\ 0 & 1 & -1 & \vdots & 2 \\ 0 & 0 & 1 & \vdots & -2 \end{bmatrix}$

54. $\begin{bmatrix} 1 & 2 & -2 & \vdots & -1 \\ 0 & 1 & 1 & \vdots & 9 \\ 0 & 0 & 1 & \vdots & -3 \end{bmatrix}$

Solutions of a System In Exercises 55 and 56, an augmented matrix that represents a system of linear equations (in the variables x and y or x, y, and z) has been reduced using Gauss-Jordan elimination. Write the solution represented by the augmented matrix.

55. $\begin{bmatrix} 1 & 0 & \vdots & 7 \\ 0 & 1 & \vdots & -5 \end{bmatrix}$

56. $\begin{bmatrix} 1 & 0 & 0 & \vdots & 3 \\ 0 & 1 & 0 & \vdots & -1 \\ 0 & 0 & 1 & \vdots & 0 \end{bmatrix}$

Gaussian Elimination with Back-Substitution In Exercises 57–64, use matrices to solve the system of equations, if possible. Use Gaussian elimination with back-substitution.

57. $\begin{cases} x + 2y = 7 \\ 2x + y = 8 \end{cases}$

58. $\begin{cases} 2x + 6y = 16 \\ 2x + 3y = 7 \end{cases}$

59. $\begin{cases} -x + y = -22 \\ 3x + 4y = 4 \\ 4x - 8y = 32 \end{cases}$

60. $\begin{cases} x + 2y = 0 \\ x + y = 6 \\ 3x - 2y = 8 \end{cases}$

61. $\begin{cases} x + 2y - 3z = -28 \\ 4y + 2z = 0 \\ -x + y - z = -5 \end{cases}$

62. $\begin{cases} -x + y - 2z = -4 \\ 4x + 2y - 3z = 8 \\ 2x + 4y - 7z = 1 \end{cases}$

63. $\begin{cases} 3x + 2y - z + w = 0 \\ x - y + 4z + 2w = 25 \\ -2x + y + 2z - w = 2 \\ x + y + z + w = 6 \end{cases}$

64. $\begin{cases} x - 4y + 3z - 2w = 9 \\ 3x - 2y + z - 4w = -13 \\ -4x + 3y - 2z + w = -4 \\ -2x + y - 4z + 3w = -10 \end{cases}$

Gauss-Jordan Elimination In Exercises 65–72, use matrices to solve the system of equations, if possible. Use Gauss-Jordan elimination.

65. $\begin{cases} -2x + 6y = -22 \\ x + 2y = -9 \end{cases}$

66. $\begin{cases} 5x - 5y = -5 \\ -2x - 3y = 7 \end{cases}$

67. $\begin{cases} x - 3z = -2 \\ 3x + y - 2z = 5 \\ 2x + 2y + z = 4 \end{cases}$

68. $\begin{cases} 2x - y + 3z = 24 \\ 2y - z = 14 \\ 7x - 5y = 6 \end{cases}$

69. $\begin{cases} 2x + 3z = 3 \\ 4x - 3y + 7z = 5 \\ 8x - 9y + 15z = 9 \end{cases}$

70. $\begin{cases} x + y - 5z = 3 \\ x - 2z = 1 \\ 2x - y - z = 0 \end{cases}$

71. $\begin{cases} 5x - 5y + z = -3 \\ x + 2y + 2z = 3 \\ -y - z = -8 \end{cases}$

72. $\begin{cases} 2x + 2y - z = 2 \\ x - 3y + z = 28 \\ -x + y = 14 \end{cases}$

Using a Graphing Utility In Exercises 73–76, use the matrix capabilities of a graphing utility to reduce the augmented matrix corresponding to the system of equations and solve the system.

73. $\begin{cases} x + y + 4z = 2 \\ 2x + 5y + 20z = 10 \\ -x + 3y + 8z = -2 \end{cases}$

74. $\begin{cases} x + y + z = 0 \\ 2x + 3y + z = 0 \\ 3x + 5y + z = 0 \end{cases}$

75. $\begin{cases} 2x + 10y + 2z = 6 \\ x + 5y + 2z = 6 \\ x + 5y + z = 3 \\ -3x - 15y - 3z = -9 \end{cases}$

76. $\begin{cases} 2x + y - z + 2w = -6 \\ 3x + 4y + w = 1 \\ x + 5y + 2z + 6w = -3 \\ 5x + 2y - z - w = 3 \end{cases}$

Comparing Solutions of Two Systems In Exercises 77–80, determine whether the two systems of linear equations yield the same solution. If so, find the solution.

77. (a) $\begin{cases} x - 2y + z = -6 \\ y - 5z = 16 \\ z = -3 \end{cases}$ (b) $\begin{cases} x + y - 2z = 6 \\ y + 3z = -8 \\ z = -3 \end{cases}$

78. (a) $\begin{cases} x - 3y + 4z = -11 \\ y - z = -4 \\ z = 2 \end{cases}$ (b) $\begin{cases} x + 4y = -11 \\ y + 3z = 4 \\ z = 2 \end{cases}$

79. (a) $\begin{cases} x - 4y + 5z = 27 \\ y - 7z = -54 \\ z = 8 \end{cases}$ (b) $\begin{cases} x - 6y + z = 15 \\ y + 5z = 42 \\ z = 8 \end{cases}$

80. (a) $\begin{cases} x + 3y - z = 19 \\ y + 6z = -18 \\ z = -4 \end{cases}$ (b) $\begin{cases} x - y + 3z = -21 \\ y - 2z = 14 \\ z = -4 \end{cases}$

Data Analysis: Curve Fitting In Exercises 81–84, use a system of equations to find the equation of the parabola $y = ax^2 + bx + c$ that passes through the points. Solve the system using matrices. Use a graphing utility to verify your result.

81.

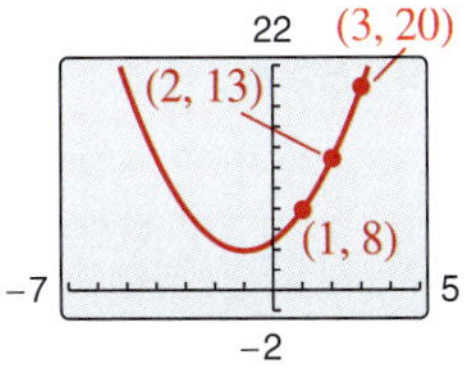

82.

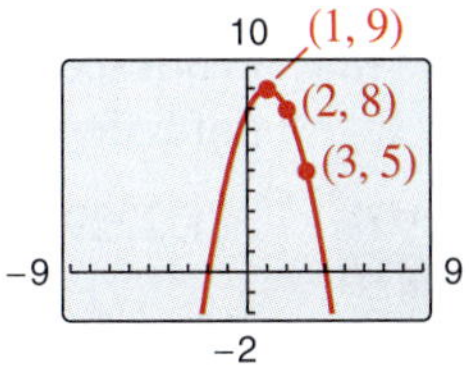

83.

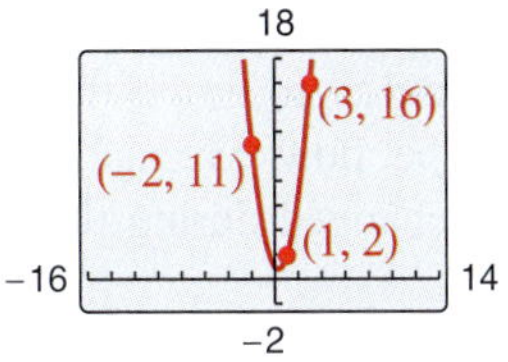

84.

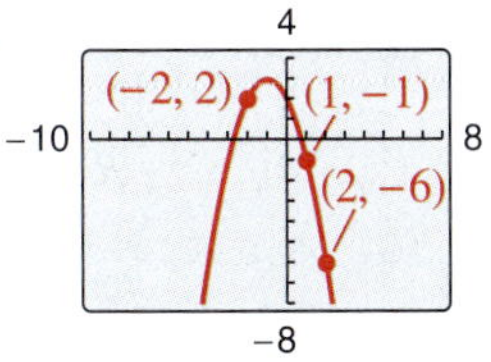

Curve Fitting **In Exercises 85 and 86, use a system of equations to find the quadratic function $f(x) = ax^2 + bx + c$ that satisfies the equations. Solve the system using matrices.**

85. $f(-2) = -15$
$f(-1) = 7$
$f(1) = -3$

86. $f(-2) = -3$
$f(1) = -3$
$f(2) = -11$

Curve Fitting **In Exercises 87 and 88, use a system of equations to find the cubic function $f(x) = ax^3 + bx^2 + cx + d$ that satisfies the equations. Solve the system using matrices.**

87. $f(-2) = -7$
$f(-1) = 2$
$f(1) = -4$
$f(2) = -7$

88. $f(-2) = -17$
$f(-1) = -5$
$f(1) = 1$
$f(2) = 7$

89. MODELING DATA

The table shows the horizontal distances x and vertical distances y of a thrown baseball at three specific times, where x and y are measured in feet.

Horizontal distance, x	Height, y
0	5.0
15	9.6
30	12.4

(a) Use the approach from Exercises 85 and 86 to find the quadratic function whose graph passes through the given points.

(b) Use a graphing utility to graph the quadratic function. Approximate the maximum height of the ball using the graph.

90. ***Why you should learn it*** *(p. 512)* The average annual consumer costs y (in dollars) for health insurance in the United States are given by the ordered pairs $(t, y(t))$, with $t = 3$ corresponding to 2013. (*Source:* U.S. Bureau of Labor Statistics)

(3, 2229), (4, 2868), (5, 2977), (6, 3160)

(a) Use the approach from Exercises 87 and 88 to find the cubic equation whose graph passes through the given points.

(b) Use a graphing utility to graph the cubic function. Describe how average annual consumer costs changed during the four-year period.

Partial Fractions **In Exercises 91 and 92, use a system of equations to write the partial fraction decomposition of the rational expression. Solve the system using matrices.**

91. $$\frac{8x^2}{(x-1)^2(x+1)} = \frac{A}{x+1} + \frac{B}{x-1} + \frac{C}{(x-1)^2}$$

92. $$\frac{x^2+3}{x^2(x-1)} = \frac{A}{x} + \frac{B}{x^2} + \frac{C}{(x-1)}$$

93. **Partial Fractions** Find the values of A, B, C, and D by solving a system using matrices.

$$\frac{x^3 - 2x + 6}{x^2(x^2+2)} = \frac{A}{x} + \frac{B}{x^2} + \frac{Cx + D}{x^2 + 2}$$

94. **Electrical Engineering** The currents in an electrical network are given by the solution of the system

$$\begin{cases} I_1 - I_2 + I_3 = 0 \\ 2I_1 + 2I_2 = 7 \\ 2I_2 + 4I_3 = 8 \end{cases}$$

where I_1, I_2, and I_3 are measured in amperes. Solve the system of equations using matrices.

95. **Finance** The amounts of money invested by a company are given by the solution of the system

$$\begin{cases} A_1 + A_2 + A_3 = 1{,}500{,}000 \\ 0.03A_1 + 0.04A_2 + 0.06A_3 = 74{,}000 \\ 4A_1 - A_3 = 0 \end{cases}$$

where A_1, A_2, and A_3 are the amounts (in dollars) invested in three separate accounts earning 3%, 4%, and 6% interest, respectively. Solve the system using matrices.

96. **Gratuities** A food server examines the amount of money earned in tips after working an 8-hour shift. The server has a total of \$95 in denominations of \$1, \$5, \$10, and \$20 bills. The total number of paper bills is 26. The number of \$5 bills is 4 times the number of \$10 bills, and the number of \$1 bills is 1 less than twice the number of \$5 bills. Write a system of linear equations to represent the situation. Then use matrices to find the number of each denomination.

97. **Marketing** A wholesale paper company sells a 100-pound package of computer paper that consists of three grades, glossy, semi-gloss, and matte, for printing photographs. Glossy costs \$5.50 per pound, semi-gloss costs \$4.25 per pound, and matte costs \$3.75 per pound. One half of the 100-pound package consists of the two less expensive grades. The cost of the 100-pound package is \$480. Set up and solve a system of equations, using matrices, to find the number of pounds of each grade of paper in a 100-pound package.

Africa Studio/Shutterstock.com

98. MODELING DATA

The average school expenditures per student y (in thousands of dollars) in the United States from 2011 through 2018 are given by the ordered pairs of the form $(t, y(t))$, where $t = 1$ corresponds to 2011. (*Source:* National Education Association)

(1, 11.44) (2, 11.36)
(3, 11.42) (4, 11.63)
(5, 11.94) (6, 12.26)
(7, 12.45) (8, 12.76)

The coefficients of the least squares regression parabola

$y = at^2 + bt + c$

can be found by solving the system

$$\begin{cases} 204a + 36b + 8c = 95.26 \\ 1296a + 204b + 36c = 437.43. \\ 8772a + 1296b + 204c = 2512.29 \end{cases}$$

(a) Use matrices to solve the system and write the least squares regression parabola for the data.

(b) Use a graphing utility to graph the parabola with the data.

(c) Use the least squares regression parabola to estimate the expenditure in 2020. Is the estimate reasonable? Explain.

99. Network Analysis Water flowing through a network of pipes (in thousands of cubic meters per hour) is shown below.

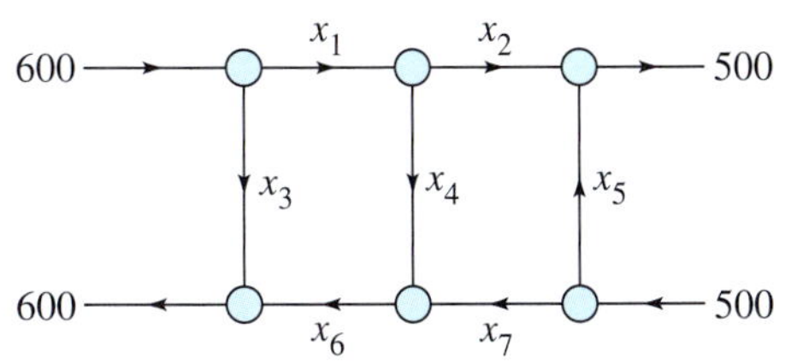

(a) Use matrices to solve this system for the water flow represented by x_i, $i = 1, 2, 3, 4, 5, 6$, and 7.

(b) Find the network flow pattern when $x_6 = 0$ and $x_7 = 0$.

100. Network Analysis The flow of water (in thousands of cubic meters per hour) into and out of the right side of the network of pipes in Exercise 99 is increased from 500 to 700.

(a) Use matrices to solve this system for the water flow represented by x_i, $i = 1, 2, 3, 4, 5, 6$, and 7.

(b) Find the network flow pattern when $x_1 = 600$ and $x_7 = 200$.

Focusing on Concepts

True or False? In Exercises 101 and 102, determine whether the statement is true or false. Justify your answer.

101. When using Gaussian elimination to solve a system of linear equations, you may conclude that the system is inconsistent before you complete the process of rewriting the augmented matrix in row-echelon form.

102. You cannot write an augmented matrix for a dependent system of linear equations in reduced row-echelon form.

103. Think About It The augmented matrix represents a system of linear equations (in the variables x, y, and z) that has been reduced using Gauss-Jordan elimination. Write a system of three equations in three variables with *nonzero* coefficients that is represented by the reduced matrix. (There are many correct answers.)

$$\left[\begin{array}{ccc:c} 1 & 0 & 3 & -2 \\ 0 & 1 & 4 & 1 \\ 0 & 0 & 0 & 0 \end{array}\right]$$

104. HOW DO YOU SEE IT? Determine whether the matrix below is in row-echelon form, reduced row-echelon form, or neither when it satisfies the given conditions.

$$\begin{bmatrix} 1 & b \\ c & 1 \end{bmatrix}$$

(a) $b = 0, c = 0$ (b) $b \neq 0, c = 0$
(c) $b = 0, c \neq 0$ (d) $b \neq 0, c \neq 0$

105. Think About It Can a 2×4 augmented matrix whose entries are all nonzero real numbers represent an independent system of linear equations? Explain.

106. Think About It Determine all values of a and b for which the augmented matrix has each given number of solutions.

$$\left[\begin{array}{cc:c} 1 & 2 & -4 \\ 0 & a & b \end{array}\right]$$

(a) Exactly one solution
(b) Infinitely many solutions
(c) No solution

Cumulative Mixed Review

Graphing a Rational Function In Exercises 107–110, sketch the graph of the function. Identify any asymptotes.

107. $f(x) = \dfrac{7}{-x - 1}$ **108.** $f(x) = \dfrac{4x}{5x^2 + 2}$

109. $f(x) = \dfrac{x^2 - 2x - 3}{x - 4}$ **110.** $f(x) = \dfrac{x^2 - 36}{x + 1}$

7.5 Operations with Matrices

Equality of Matrices

In Section 7.4, you used matrices to solve systems of linear equations. There is a rich mathematical theory of matrices, and its applications are numerous. This section and the next two introduce some fundamentals of matrix theory. It is standard mathematical convention to represent matrices in any of the following three ways.

Representation of Matrices

1. A matrix can be denoted by an uppercase letter such as A, B, or C.
2. A matrix can be denoted by a representative element enclosed in brackets, such as

   $[a_{ij}]$, $[b_{ij}]$, or $[c_{ij}]$.
3. A matrix can be denoted by a rectangular array of numbers such as

$$A = [a_{ij}] = \begin{bmatrix} a_{11} & a_{12} & a_{13} & \cdots & a_{1n} \\ a_{21} & a_{22} & a_{23} & \cdots & a_{2n} \\ a_{31} & a_{32} & a_{33} & \cdots & a_{3n} \\ \vdots & \vdots & \vdots & & \vdots \\ a_{m1} & a_{m2} & a_{m3} & \cdots & a_{mn} \end{bmatrix}.$$

What you should learn

- Decide whether two matrices are equal.
- Add and subtract matrices and multiply matrices by scalars.
- Multiply two matrices.
- Use matrices to perform vector operations and to transform vectors.
- Use matrix operations to model and solve real-life problems.

Why you should learn it

Matrix algebra provides a systematic way of performing mathematical operations on large arrays of numbers. In Exercise 95 on page 539, you will use matrix multiplication to help analyze the labor and wage requirements for a boat manufacturer.

Two matrices

$$A = [a_{ij}] \quad \text{and} \quad B = [b_{ij}]$$

are **equal** when they have the same dimension ($m \times n$) and all of their corresponding entries are equal.

EXAMPLE 1 Equality of Matrices

Solve for a_{11}, a_{12}, a_{21}, and a_{22} in the following matrix equation.

$$\begin{bmatrix} a_{11} & a_{12} \\ a_{21} & a_{22} \end{bmatrix} = \begin{bmatrix} 2 & -1 \\ -3 & 0 \end{bmatrix}$$

Solution

Because two matrices are equal only when their corresponding entries are equal, you can conclude that

$$a_{11} = 2, \quad a_{12} = -1, \quad a_{21} = -3, \quad \text{and} \quad a_{22} = 0.$$

✓ ***Checkpoint*** *Audio-video solution in English & Spanish at LarsonPrecalculus.com*

Solve for a_{11}, a_{12}, a_{21}, and a_{22} in the following matrix equation.

$$\begin{bmatrix} a_{11} & a_{12} \\ a_{21} & a_{22} \end{bmatrix} = \begin{bmatrix} 6 & 3 \\ -2 & 4 \end{bmatrix}$$

Be sure you see that for two matrices to be equal, they must have the same dimension *and* their corresponding entries must be equal. For instance,

$$\begin{bmatrix} 2 & -1 \\ \sqrt{4} & \frac{1}{2} \end{bmatrix} = \begin{bmatrix} 2 & -1 \\ 2 & 0.5 \end{bmatrix} \quad \text{but} \quad \begin{bmatrix} 2 & -1 \\ 3 & 4 \\ 0 & 0 \end{bmatrix} \neq \begin{bmatrix} 2 & -1 \\ 3 & 4 \end{bmatrix}.$$

freestylephoto/iStock/Getty Images

Matrix Addition and Scalar Multiplication

Definition of Matrix Addition

If $A = [a_{ij}]$ and $B = [b_{ij}]$ are matrices of dimension $m \times n$, then their sum is the $m \times n$ matrix given by $A + B = [a_{ij} + b_{ij}]$. The sum of two matrices of different dimensions is undefined.

Algebra Help

With matrix addition, you can add two matrices (of the same dimension) by adding their corresponding entries.

EXAMPLE 2 Addition of Matrices

a. $$\begin{bmatrix} -1 & 2 \\ 0 & 1 \end{bmatrix} + \begin{bmatrix} 1 & 3 \\ -1 & 2 \end{bmatrix} = \begin{bmatrix} -1+1 & 2+3 \\ 0+(-1) & 1+2 \end{bmatrix} = \begin{bmatrix} 0 & 5 \\ -1 & 3 \end{bmatrix}$$

b. $$\begin{bmatrix} 1 \\ -3 \\ -2 \end{bmatrix} + \begin{bmatrix} -1 \\ 3 \\ 2 \end{bmatrix} = \begin{bmatrix} 0 \\ 0 \\ 0 \end{bmatrix}$$

c. The sum of

$$A = \begin{bmatrix} 2 & 1 & 0 \\ 4 & 0 & -1 \end{bmatrix} \quad \text{and} \quad B = \begin{bmatrix} 0 & 1 \\ -1 & 3 \end{bmatrix}$$

is undefined because A is of dimension 2×3 and B is of dimension 2×2.

Checkpoint *Audio-video solution in English & Spanish at LarsonPrecalculus.com*

Evaluate the expression.

$$\begin{bmatrix} 4 & -1 \\ 2 & -3 \end{bmatrix} + \begin{bmatrix} 2 & -1 \\ 0 & 6 \end{bmatrix}$$

Technology Tip

Try using a graphing utility to find the sum of the two matrices in Example 2(c). Your graphing utility should display an error message similar to the one shown below.

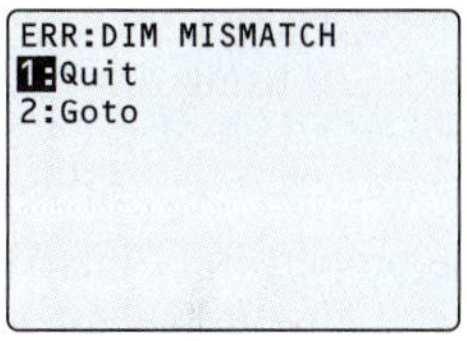

In operations with matrices, numbers are usually referred to as **scalars.** In this text, scalars will always be real numbers. You can multiply a matrix A by a scalar c by multiplying each entry in A by c.

Definition of Scalar Multiplication

If $A = [a_{ij}]$ is an $m \times n$ matrix and c is a scalar, then the **scalar multiple** of A by c is the $m \times n$ matrix given by $cA = [ca_{ij}]$.

EXAMPLE 3 Scalar Multiplication

Find $3A$ using $A = \begin{bmatrix} 2 & 2 & 4 \\ -3 & 0 & -1 \\ 2 & 1 & 2 \end{bmatrix}$.

Solution

$$3A = 3\begin{bmatrix} 2 & 2 & 4 \\ -3 & 0 & -1 \\ 2 & 1 & 2 \end{bmatrix} = \begin{bmatrix} 3(2) & 3(2) & 3(4) \\ 3(-3) & 3(0) & 3(-1) \\ 3(2) & 3(1) & 3(2) \end{bmatrix} = \begin{bmatrix} 6 & 6 & 12 \\ -9 & 0 & -3 \\ 6 & 3 & 6 \end{bmatrix}$$

Checkpoint *Audio-video solution in English & Spanish at LarsonPrecalculus.com*

Find $3A$ using $A = \begin{bmatrix} 4 & -1 \\ 0 & 4 \\ -3 & 8 \end{bmatrix}$.

The symbol $-A$ represents the negation of A, which is the scalar product $(-1)A$. Moreover, if A and B are of the same dimension, then $A - B$ represents the sum of A and $(-1)B$. That is,

$$A - B = A + (-1)B.$$

The order of operations for matrix expressions is similar to that for real numbers. In particular, you perform scalar multiplication before matrix addition and subtraction, as shown in Example 4.

EXAMPLE 4 Scalar Multiplication and Matrix Subtraction

Find $3A - B$ using $A = \begin{bmatrix} 2 & 2 & 4 \\ -3 & 0 & -1 \\ 2 & 1 & 2 \end{bmatrix}$ and $B = \begin{bmatrix} 2 & 0 & 0 \\ 1 & -4 & 3 \\ -1 & 3 & 2 \end{bmatrix}$.

Solution

Note that A is the same matrix from Example 3, where you found $3A$.

$$3A - B = \begin{bmatrix} 6 & 6 & 12 \\ -9 & 0 & -3 \\ 6 & 3 & 6 \end{bmatrix} - \begin{bmatrix} 2 & 0 & 0 \\ 1 & -4 & 3 \\ -1 & 3 & 2 \end{bmatrix}$$ Perform scalar multiplication first.

$$= \begin{bmatrix} 4 & 6 & 12 \\ -10 & 4 & -6 \\ 7 & 0 & 4 \end{bmatrix}$$ Subtract corresponding entries.

Checkpoint Audio-video solution in English & Spanish at LarsonPrecalculus.com

Find $3A - 2B$ using $A = \begin{bmatrix} 4 & -1 \\ 0 & 4 \\ -3 & 8 \end{bmatrix}$ and $B = \begin{bmatrix} 0 & 4 \\ -1 & 3 \\ 1 & 7 \end{bmatrix}$.

The properties of matrix addition and scalar multiplication are similar to those of addition and multiplication of real numbers. One important property of addition of real numbers is that the number 0 is the additive identity. That is, $c + 0 = c$ for any real number c. For matrices, a similar property holds. That is, if A is an $m \times n$ matrix and O is the $m \times n$ **zero matrix** consisting entirely of zeros, then $A + O = A$. In other words, O is the **additive identity** for the set of all $m \times n$ matrices. For example, the following matrices are the additive identities for the sets of all 2×3 and 2×2 matrices.

$$O = \underbrace{\begin{bmatrix} 0 & 0 & 0 \\ 0 & 0 & 0 \end{bmatrix}}_{2 \times 3 \text{ zero matrix}} \quad \text{and} \quad O = \underbrace{\begin{bmatrix} 0 & 0 \\ 0 & 0 \end{bmatrix}}_{2 \times 2 \text{ zero matrix}}$$

Properties of Matrix Addition and Scalar Multiplication

Let A, B, and C be $m \times n$ matrices and let c and d be scalars.

1. $A + B = B + A$ — Commutative Property of Matrix Addition
2. $A + (B + C) = (A + B) + C$ — Associative Property of Matrix Addition
3. $(cd)A = c(dA)$ — Associative Property of Scalar Multiplication
4. $1A = A$ — Scalar Identity
5. $A + O = A$ — Additive Identity
6. $c(A + B) = cA + cB$ — Distributive Property
7. $(c + d)A = cA + dA$ — Distributive Property

Explore the Concept

What do you observe about the relationship between the corresponding entries of A and B below? Use a graphing utility to find $A + B$. What conclusion can you make about the entries of A and B and the sum $A + B$?

$$A = \begin{bmatrix} -1 & 5 \\ 2 & -6 \end{bmatrix}$$

$$B = \begin{bmatrix} 1 & -5 \\ -2 & 6 \end{bmatrix}$$

Algebra Help

Note that the Associative Property of Matrix Addition allows you to write expressions such as $A + B + C$ without ambiguity because the same sum occurs no matter how the matrices are grouped. This same reasoning applies to sums of four or more matrices.

EXAMPLE 5 Using the Distributive Property

$$3\left(\begin{bmatrix} -2 & 0 \\ 4 & 1 \end{bmatrix} + \begin{bmatrix} 4 & -2 \\ 3 & 7 \end{bmatrix}\right) = 3\begin{bmatrix} -2 & 0 \\ 4 & 1 \end{bmatrix} + 3\begin{bmatrix} 4 & -2 \\ 3 & 7 \end{bmatrix}$$

$$= \begin{bmatrix} -6 & 0 \\ 12 & 3 \end{bmatrix} + \begin{bmatrix} 12 & -6 \\ 9 & 21 \end{bmatrix}$$

$$= \begin{bmatrix} 6 & -6 \\ 21 & 24 \end{bmatrix}$$

Checkpoint Audio-video solution in English & Spanish at LarsonPrecalculus.com

Evaluate the expression using the Distributive Property.

$$2\left(\begin{bmatrix} 1 & 3 \\ -2 & 2 \end{bmatrix} + \begin{bmatrix} -4 & 0 \\ -3 & 1 \end{bmatrix}\right)$$

Algebra Help

In Example 5, you could add the two matrices first and then multiply the resulting matrix by 3. The result would be the same.

The algebra of real numbers and the algebra of matrices have many similarities. For example, compare the following solutions.

Real Numbers (*Solve for x.*)	$m \times n$ *Matrices* (*Solve for X.*)
$x + a = b$	$X + A = B$
$x + a + (-a) = b + (-a)$	$X + A + (-A) = B + (-A)$
$x + 0 = b - a$	$X + O = B - A$
$x = b - a$	$X = B - A$

The algebra of real numbers and the algebra of matrices also have important differences, which will be discussed later.

EXAMPLE 6 Solving a Matrix Equation

Solve for X in the equation $3X + A = B$, where

$$A = \begin{bmatrix} 1 & -2 \\ 0 & 3 \end{bmatrix} \quad \text{and} \quad B = \begin{bmatrix} -3 & 4 \\ 2 & 1 \end{bmatrix}.$$

Solution

Begin by solving the matrix equation for X.

$$3X + A = B \quad \Rightarrow \quad 3X = B - A \quad \Rightarrow \quad X = \tfrac{1}{3}(B - A)$$

Now, using the matrices A and B, you have

$$X = \frac{1}{3}\left(\begin{bmatrix} -3 & 4 \\ 2 & 1 \end{bmatrix} - \begin{bmatrix} 1 & -2 \\ 0 & 3 \end{bmatrix}\right)$$ Substitute the matrices.

$$= \frac{1}{3}\begin{bmatrix} -4 & 6 \\ 2 & -2 \end{bmatrix}$$ Subtract matrix A from matrix B.

$$= \begin{bmatrix} -\frac{4}{3} & 2 \\ \frac{2}{3} & -\frac{2}{3} \end{bmatrix}.$$ Multiply the resulting matrix by $\frac{1}{3}$.

Checkpoint Audio-video solution in English & Spanish at LarsonPrecalculus.com

Solve for X in the equation $2X - A = B$, where

$$A = \begin{bmatrix} 6 & 1 \\ 0 & 3 \end{bmatrix} \quad \text{and} \quad B = \begin{bmatrix} 4 & -1 \\ -2 & 5 \end{bmatrix}.$$

Matrix Multiplication

Another basic matrix operation is **matrix multiplication.**

Definition of Matrix Multiplication

If $A = [a_{ij}]$ is an $m \times n$ matrix and $B = [b_{ij}]$ is an $n \times p$ matrix, then the product AB is an $m \times p$ matrix given by

$$AB = [c_{ij}]$$

where

$$c_{ij} = a_{i1}b_{1j} + a_{i2}b_{2j} + a_{i3}b_{3j} + \cdot \cdot \cdot + a_{in}b_{nj}.$$

Algebra Help

At first glance, this definition may seem unusual. You will see later, however, that the definition of the product of two matrices has many practical applications.

The definition of matrix multiplication indicates a *row-by-column* multiplication, where the entry in the ith row and jth column of the product AB is obtained by multiplying the entries in the ith row of A by the corresponding entries in the jth column of B and then adding the results. The general pattern for matrix multiplication is as follows.

$$\begin{bmatrix} a_{11} & a_{12} & a_{13} & \cdots & a_{1n} \\ a_{21} & a_{22} & a_{23} & \cdots & a_{2n} \\ a_{31} & a_{32} & a_{33} & \cdots & a_{3n} \\ \vdots & \vdots & \vdots & & \vdots \\ a_{i1} & a_{i2} & a_{i3} & \cdots & a_{in} \\ \vdots & \vdots & \vdots & & \vdots \\ a_{m1} & a_{m2} & a_{m3} & \cdots & a_{mn} \end{bmatrix} \begin{bmatrix} b_{11} & b_{12} & \cdots & b_{1j} & \cdots & b_{1p} \\ b_{21} & b_{22} & \cdots & b_{2j} & \cdots & b_{2p} \\ b_{31} & b_{32} & \cdots & b_{3j} & \cdots & b_{3p} \\ \vdots & \vdots & & \vdots & & \vdots \\ b_{n1} & b_{n2} & \cdots & b_{nj} & \cdots & b_{np} \end{bmatrix} = \begin{bmatrix} c_{11} & c_{12} & \cdots & c_{1j} & \cdots & c_{1p} \\ c_{21} & c_{22} & \cdots & c_{2j} & \cdots & c_{2p} \\ \vdots & \vdots & & \vdots & & \vdots \\ c_{i1} & c_{i2} & \cdots & c_{ij} & \cdots & c_{ip} \\ \vdots & \vdots & & \vdots & & \vdots \\ c_{m1} & c_{m2} & \cdots & c_{mj} & \cdots & c_{mp} \end{bmatrix}$$

$$a_{i1}b_{1j} + a_{i2}b_{2j} + a_{i3}b_{3j} + \cdot \cdot \cdot + a_{in}b_{nj} = c_{ij}$$

EXAMPLE 7 Finding the Product of Two Matrices

Find the product AB using $A = \begin{bmatrix} -1 & 3 \\ 4 & -2 \\ 5 & 0 \end{bmatrix}$ and $B = \begin{bmatrix} -3 & 2 \\ -4 & 1 \end{bmatrix}$.

Solution

First, note that the product AB is defined because the number of columns of A is equal to the number of rows of B. Moreover, the product AB has dimension 3×2. To find the entries of the product, multiply each row of A by each column of B.

$$\begin{aligned} AB &= \begin{bmatrix} -1 & 3 \\ 4 & -2 \\ 5 & 0 \end{bmatrix} \begin{bmatrix} -3 & 2 \\ -4 & 1 \end{bmatrix} \\ &= \begin{bmatrix} (-1)(-3) + (3)(-4) & (-1)(2) + (3)(1) \\ (4)(-3) + (-2)(-4) & (4)(2) + (-2)(1) \\ (5)(-3) + (0)(-4) & (5)(2) + (0)(1) \end{bmatrix} \\ &= \begin{bmatrix} -9 & 1 \\ -4 & 6 \\ -15 & 10 \end{bmatrix} \end{aligned}$$

✓ Checkpoint *Audio-video solution in English & Spanish at LarsonPrecalculus.com*

Find the product AB using $A = \begin{bmatrix} -1 & 4 \\ 2 & 0 \\ 1 & 2 \end{bmatrix}$ and $B = \begin{bmatrix} 1 & -2 \\ 0 & 7 \end{bmatrix}$.

Be sure you understand that for the product of two matrices to be defined, the number of *columns* of the first matrix must equal the number of *rows* of the second matrix. That is, the middle two indices must be the same. The outside two indices give the dimension of the product, as shown in the following diagram.

$$\underset{m \times n}{A} \times \underset{n \times p}{B} = \underset{m \times p}{AB}$$

Equal (the middle indices n and n)

Dimension of AB (the outside indices m and p)

EXAMPLE 8 Matrix Multiplication

See LarsonPrecalculus.com for an interactive version of this type of example.

a. $\begin{bmatrix} 1 & 0 & 3 \\ 2 & -1 & -2 \end{bmatrix} \begin{bmatrix} -2 & 4 & 2 \\ 1 & 0 & 0 \\ -1 & 1 & -1 \end{bmatrix} = \begin{bmatrix} -5 & 7 & -1 \\ -3 & 6 & 6 \end{bmatrix}$

2×3 $\quad 3 \times 3$ $\quad 2 \times 3$

b. $\begin{bmatrix} 3 & 4 \\ -2 & 5 \end{bmatrix} \begin{bmatrix} 1 & 0 \\ 0 & 1 \end{bmatrix} = \begin{bmatrix} 3 & 4 \\ -2 & 5 \end{bmatrix}$

2×2 $\quad 2 \times 2$ $\quad 2 \times 2$

c. $\begin{bmatrix} 1 & 2 \\ 1 & 1 \end{bmatrix} \begin{bmatrix} -1 & 2 \\ 1 & -1 \end{bmatrix} = \begin{bmatrix} 1 & 0 \\ 0 & 1 \end{bmatrix}$

2×2 $\quad 2 \times 2$ $\quad 2 \times 2$

d. $\begin{bmatrix} 1 & -2 & -3 \end{bmatrix} \begin{bmatrix} 2 \\ -1 \\ 1 \end{bmatrix} = \begin{bmatrix} 1 \end{bmatrix}$

1×3 $\quad 3 \times 1$ $\quad 1 \times 1$

e. $\begin{bmatrix} 2 \\ -1 \\ 1 \end{bmatrix} \begin{bmatrix} 1 & -2 & -3 \end{bmatrix} = \begin{bmatrix} 2 & -4 & -6 \\ -1 & 2 & 3 \\ 1 & -2 & -3 \end{bmatrix}$

3×1 $\quad 1 \times 3$ $\quad 3 \times 3$

f. The product *AB* for the following matrices is not defined.

$$A = \begin{bmatrix} -2 & 1 \\ 1 & -3 \\ 1 & 4 \end{bmatrix} \text{ and } B = \begin{bmatrix} -2 & 3 & 1 & 4 \\ 0 & 1 & -1 & 2 \\ 2 & -1 & 0 & 1 \end{bmatrix}$$

3×2 $\quad 3 \times 4$

✓ Checkpoint *Audio-video solution in English & Spanish at LarsonPrecalculus.com*

Find the product AB using $A = \begin{bmatrix} 0 & 4 & -3 \\ 2 & 1 & 7 \\ 3 & -2 & 1 \end{bmatrix}$ and $B = \begin{bmatrix} -2 & 0 \\ 0 & -4 \\ 1 & 2 \end{bmatrix}$.

In parts (d) and (e) of Example 8, note that the two products are different. Even when both *AB* and *BA* are defined, matrix multiplication is not, in general, commutative. That is, for most matrices,

$$AB \neq BA.$$

This is one way in which the algebra of real numbers and the algebra of matrices differ.

Explore the Concept

Use the following matrices to find AB, BA, $(AB)C$, and $A(BC)$. What do your results tell you about matrix multiplication and commutativity and associativity?

$$A = \begin{bmatrix} 1 & 2 \\ 3 & 4 \end{bmatrix}$$

$$B = \begin{bmatrix} 0 & 1 \\ 2 & 3 \end{bmatrix}$$

$$C = \begin{bmatrix} 3 & 0 \\ 0 & 1 \end{bmatrix}$$

ESB Basic/Shutterstock.com

EXAMPLE 9 Matrix Multiplication

Use a graphing utility to find the product AB using

$$A = \begin{bmatrix} 1 & 2 & 3 \\ 2 & -5 & 1 \end{bmatrix} \quad \text{and} \quad B = \begin{bmatrix} -3 & 2 & 1 \\ 4 & -2 & 0 \\ 1 & 2 & 3 \end{bmatrix}.$$

Solution

Note that the dimension of A is 2×3 and the dimension of B is 3×3. So, the product will have dimension 2×3. Use the *matrix editor* to enter A and B into the graphing utility. Then, find the product, as shown in Figure 7.17.

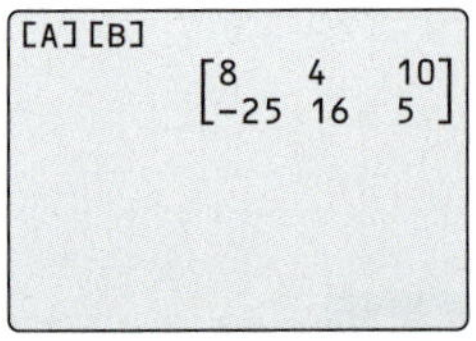

Figure 7.17

✓ Checkpoint *Audio-video solution in English & Spanish at LarsonPrecalculus.com*

Find the product AB using $A = \begin{bmatrix} 7 & 5 & -4 \\ -2 & 5 & 1 \\ 10 & -4 & -7 \end{bmatrix}$ and $B = \begin{bmatrix} 2 & -2 & 3 \\ 8 & 1 & 4 \\ -4 & 2 & -8 \end{bmatrix}$.

Properties of Matrix Multiplication

Let A, B, and C be matrices and let c be a scalar.

1. $A(BC) = (AB)C$ — Associative Property of Matrix Multiplication

2. $A(B + C) = AB + AC$ — Left Distributive Property

3. $(A + B)C = AC + BC$ — Right Distributive Property

4. $c(AB) = (cA)B = A(cB)$ — Associative Property of Scalar Multiplication

Definition of Identity Matrix

The $n \times n$ matrix that consists of 1's on its main diagonal and 0's elsewhere is called the **identity matrix of dimension $n \times n$** and is denoted by

$$I_n = \begin{bmatrix} 1 & 0 & 0 & \cdots & 0 \\ 0 & 1 & 0 & \cdots & 0 \\ 0 & 0 & 1 & \cdots & 0 \\ \vdots & \vdots & \vdots & & \vdots \\ 0 & 0 & 0 & \cdots & 1 \end{bmatrix}. \quad \text{Identity matrix}$$

Note that an identity matrix must be *square*. When the dimension is understood to be $n \times n$, you can denote I_n simply by I.

If A is an $n \times n$ matrix, then the identity matrix has the property that $AI_n = A$ and $I_nA = A$. For example,

$$\begin{bmatrix} 3 & -2 & 5 \\ 1 & 0 & 4 \\ -1 & 2 & -3 \end{bmatrix}\begin{bmatrix} 1 & 0 & 0 \\ 0 & 1 & 0 \\ 0 & 0 & 1 \end{bmatrix} = \begin{bmatrix} 3 & -2 & 5 \\ 1 & 0 & 4 \\ -1 & 2 & -3 \end{bmatrix} \quad AI = A$$

and

$$\begin{bmatrix} 1 & 0 & 0 \\ 0 & 1 & 0 \\ 0 & 0 & 1 \end{bmatrix}\begin{bmatrix} 3 & -2 & 5 \\ 1 & 0 & 4 \\ -1 & 2 & -3 \end{bmatrix} = \begin{bmatrix} 3 & -2 & 5 \\ 1 & 0 & 4 \\ -1 & 2 & -3 \end{bmatrix}. \quad IA = A$$

Matrices and Vectors

In Section 6.3, you represented a vector as $\mathbf{u} = \langle u_1, u_2 \rangle$ or as $\mathbf{u} = u_1\mathbf{i} + u_2\mathbf{j}$. You can also represent a vector as an $n \times 1$ column matrix (or **column vector**). This approach is valid because the matrix operations of addition and scalar multiplication give the same results as the corresponding vector operations. That is, the matrix sum

$$\mathbf{u} + \mathbf{v} = \begin{bmatrix} u_1 \\ u_2 \end{bmatrix} + \begin{bmatrix} v_1 \\ v_2 \end{bmatrix} = \begin{bmatrix} u_1 + v_1 \\ u_2 + v_2 \end{bmatrix}$$

yields the same result as vector addition.

$$\mathbf{u} + \mathbf{v} = \langle u_1, u_2 \rangle + \langle v_1, v_2 \rangle = \langle u_1 + v_1, u_2 + v_2 \rangle$$

The same argument applies to scalar multiplication. The only difference in each set of notations is how the components (entries) are displayed.

EXAMPLE 10 Vector Operations

Let $\mathbf{u} = \begin{bmatrix} 2 \\ 4 \end{bmatrix}$ and $\mathbf{v} = \begin{bmatrix} 6 \\ 2 \end{bmatrix}$. Find (a) $\mathbf{u} + \mathbf{v}$ and (b) $\mathbf{v} - 2\mathbf{u}$.

Solution

a. $\mathbf{u} + \mathbf{v} = \begin{bmatrix} 2 \\ 4 \end{bmatrix} + \begin{bmatrix} 6 \\ 2 \end{bmatrix} = \begin{bmatrix} 8 \\ 6 \end{bmatrix}$

b. $\mathbf{v} - 2\mathbf{u} = \begin{bmatrix} 6 \\ 2 \end{bmatrix} - 2\begin{bmatrix} 2 \\ 4 \end{bmatrix} = \begin{bmatrix} 6 \\ 2 \end{bmatrix} - \begin{bmatrix} 4 \\ 8 \end{bmatrix} = \begin{bmatrix} 2 \\ -6 \end{bmatrix}$

✓ ***Checkpoint*** ▶ *Audio-video solution in English & Spanish at LarsonPrecalculus.com*

Let $\mathbf{u} = \begin{bmatrix} 3 \\ 6 \end{bmatrix}$ and $\mathbf{v} = \begin{bmatrix} 8 \\ 5 \end{bmatrix}$. Find (a) $\mathbf{u} - \mathbf{v}$ and (b) $3\mathbf{u} + \mathbf{v}$.

One way to transform a column vector $\mathbf{v}$ is to multiply $\mathbf{v}$ by a square **transformation matrix** A to produce another vector $A\mathbf{v}$. For a column vector $\mathbf{v}$ with two rows, matrix A must have two columns for the product $A\mathbf{v}$ to be defined.

EXAMPLE 11 Describing a Vector Transformation

Let $A = \begin{bmatrix} 1 & 0 \\ 0 & -1 \end{bmatrix}$ and $\mathbf{v} = \begin{bmatrix} 1 \\ 3 \end{bmatrix}$. Find $A\mathbf{v}$ and describe the transformation.

Solution

$$A\mathbf{v} = \begin{bmatrix} 1 & 0 \\ 0 & -1 \end{bmatrix}\begin{bmatrix} 1 \\ 3 \end{bmatrix} = \begin{bmatrix} 1 \\ -3 \end{bmatrix}$$

Figure 7.18 shows the graphs of the vectors $\mathbf{v}$ and $A\mathbf{v}$. The matrix A transforms $\mathbf{v}$ by reflecting $\mathbf{v}$ in the x-axis.

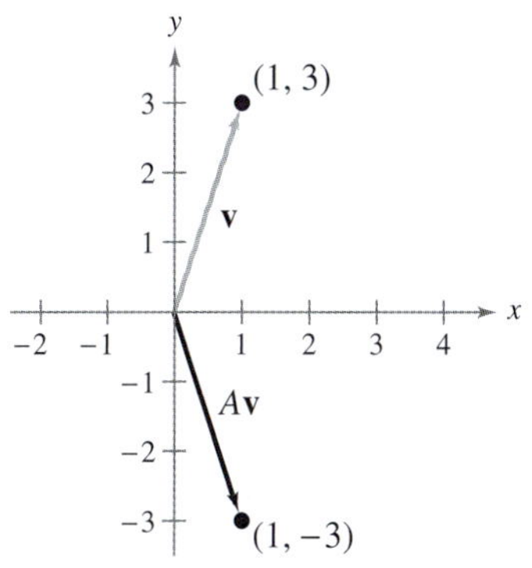

Figure 7.18

✓ ***Checkpoint*** ▶ *Audio-video solution in English & Spanish at LarsonPrecalculus.com*

Let $A = \begin{bmatrix} -1 & 0 \\ 0 & 1 \end{bmatrix}$ and $\mathbf{v} = \begin{bmatrix} 3 \\ 1 \end{bmatrix}$. Find $A\mathbf{v}$ and describe the transformation.

Applications

Matrix multiplication can be used to represent a system of linear equations. Note how the system

$$\begin{cases} a_{11}x_1 + a_{12}x_2 + a_{13}x_3 = b_1 \\ a_{21}x_1 + a_{22}x_2 + a_{23}x_3 = b_2 \\ a_{31}x_1 + a_{32}x_2 + a_{33}x_3 = b_3 \end{cases}$$

can be written as the matrix equation $AX = B$, where A is the *coefficient matrix* of the system, and X and B are column matrices. The column matrix B is also called a *constant matrix*. Its entries are the constant terms in the system of equations.

$$\underbrace{\begin{bmatrix} a_{11} & a_{12} & a_{13} \\ a_{21} & a_{22} & a_{23} \\ a_{31} & a_{32} & a_{33} \end{bmatrix}}_{A} \times \underbrace{\begin{bmatrix} x_1 \\ x_2 \\ x_3 \end{bmatrix}}_{X} = \underbrace{\begin{bmatrix} b_1 \\ b_2 \\ b_3 \end{bmatrix}}_{B}$$

EXAMPLE 12 Solving a System of Linear Equations

For the system of linear equations, (a) write the system as a matrix equation $AX = B$ and (b) use Gauss-Jordan elimination on $[A \vdots B]$ to solve for the matrix X.

$$\begin{cases} x_1 - 2x_2 + x_3 = -4 \\ x_2 + 2x_3 = 4 \\ 2x_1 + 3x_2 - 2x_3 = 2 \end{cases}$$

Solution

a. In matrix form $AX = B$, the system can be written as follows.

$$\begin{bmatrix} 1 & -2 & 1 \\ 0 & 1 & 2 \\ 2 & 3 & -2 \end{bmatrix}\begin{bmatrix} x_1 \\ x_2 \\ x_3 \end{bmatrix} = \begin{bmatrix} -4 \\ 4 \\ 2 \end{bmatrix}$$

b. The augmented matrix is

$$[A \vdots B] = \begin{bmatrix} 1 & -2 & 1 & \vdots & -4 \\ 0 & 1 & 2 & \vdots & 4 \\ 2 & 3 & -2 & \vdots & 2 \end{bmatrix}.$$

Using Gauss-Jordan elimination, you can rewrite this equation as

$$[I \vdots X] = \begin{bmatrix} 1 & 0 & 0 & \vdots & -1 \\ 0 & 1 & 0 & \vdots & 2 \\ 0 & 0 & 1 & \vdots & 1 \end{bmatrix}.$$

So, the solution of the system of linear equations is $x_1 = -1$, $x_2 = 2$, and $x_3 = 1$. The solution of the matrix equation is

$$X = \begin{bmatrix} x_1 \\ x_2 \\ x_3 \end{bmatrix} = \begin{bmatrix} -1 \\ 2 \\ 1 \end{bmatrix}.$$

Technology Tip

Most graphing utilities can be used to obtain the reduced row-echelon form of a matrix. The screen below shows how one graphing utility displays the reduced row-echelon form of the augmented matrix in Example 12.

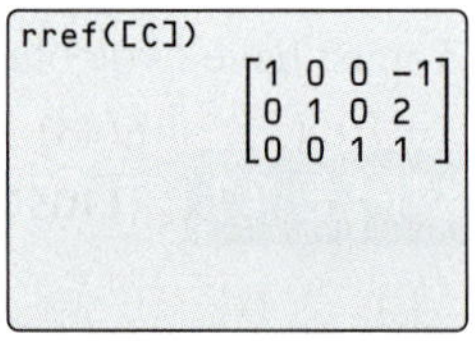

✓ ***Checkpoint*** *Audio-video solution in English & Spanish at LarsonPrecalculus.com*

For the system of linear equations, (a) write the system as a matrix equation $AX = B$ and (b) use Gauss-Jordan elimination on $[A \vdots B]$ to solve for the matrix X.

$$\begin{cases} -2x_1 - 3x_2 = -4 \\ 6x_1 + x_2 = -36 \end{cases}$$

EXAMPLE 13 Softball Team Expenses

Two softball teams submit equipment lists to their sponsors, as shown in the table.

Equipment	Women's team	Men's team
Bats	12	15
Balls	45	38
Gloves	15	17

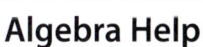

The equipment costs are as follows.

Bats: \$90 per bat

Balls: \$6 per ball

Gloves: \$60 per glove

Use matrices to find the total cost of equipment for each team.

Solution

The equipment list E can be written in matrix form as

$$E = \begin{bmatrix} 12 & 15 \\ 45 & 38 \\ 15 & 17 \end{bmatrix}.$$

The costs per item C can be written in matrix form as

$$C = [90 \quad 6 \quad 60].$$

You can find the total cost of the equipment for each team using the product CE because the number of columns of C (3 columns) equals the number of rows of E (3 rows). Therefore, the total cost of equipment for each team is given by

$$CE = [90 \quad 6 \quad 60]\begin{bmatrix} 12 & 15 \\ 45 & 38 \\ 15 & 17 \end{bmatrix}$$

$$= [90(12) + 6(45) + 60(15) \quad 90(15) + 6(38) + 60(17)]$$

$$= [2250 \quad 2598].$$

Algebra Help

Notice in Example 13 that you cannot find the total cost using the product EC because EC is not defined. That is, the number of columns of E (2 columns) does not equal the number of rows of C (1 row).

So, the total cost of equipment for the women's team is \$2250, and the total cost of equipment for the men's team is \$2598. You can use a graphing utility to check this result, as shown in Figure 7.19.

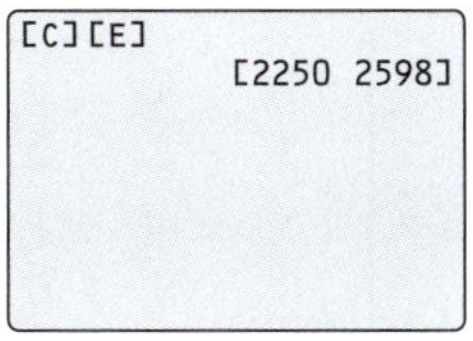

Figure 7.19

Checkpoint *Audio-video solution in English & Spanish at LarsonPrecalculus.com*

Repeat Example 13 when each bat costs \$100, each ball costs \$7, and each glove costs \$65.

7.5 Exercises

See *CalcChat.com* for tutorial help and worked-out solutions to odd-numbered exercises. For instructions on how to use a graphing utility, see Appendix A.

Vocabulary and Concept Check

In Exercises 1–4, fill in the blank(s).

1. Two matrices are _______ when they have the same dimension and all of their corresponding entries are equal.

2. When working with matrices, real numbers are often referred to as _______ .

3. A _______ matrix consists entirely of zeros and is denoted by _______ .

4. The $n \times n$ matrix consisting of 1's on its main diagonal and 0's elsewhere is called the _______ matrix of dimension n.

In Exercises 5 and 6, match the matrix property with the correct form. *A*, *B*, and *C* are matrices, and *c* and *d* are scalars.

5. (a) $(cd)A = c(dA)$ — (i) Commutative Property of Matrix Addition

(b) $A + B = B + A$ — (ii) Associative Property of Matrix Addition

(c) $1A = A$ — (iii) Associative Property of Scalar Multiplication

(d) $c(A + B) = cA + cB$ — (iv) Scalar Identity

(e) $A + (B + C) = (A + B) + C$ — (v) Distributive Property

6. (a) $A(B + C) = AB + AC$ — (i) Associative Property of Matrix Multiplication

(b) $c(AB) = (cA)B = A(cB)$ — (ii) Left Distributive Property

(c) $A(BC) = (AB)C$ — (iii) Right Distributive Property

(d) $(A + B)C = AC + BC$ — (iv) Associative Property of Scalar Multiplication

7. In general, when multiplying matrices A and B, does $AB = BA$?

8. What is the dimension of AB when A is a 2×3 matrix and B is a 3×4 matrix?

Procedures and Problem Solving

Equality of Matrices **In Exercises 9–12, find *x* and *y* or *x*, *y*, and *z*.**

9. $\begin{bmatrix} x & -7 \\ 9 & y \end{bmatrix} = \begin{bmatrix} 5 & -7 \\ 9 & -8 \end{bmatrix}$

10. $\begin{bmatrix} -5 & x \\ y & 8 \end{bmatrix} = \begin{bmatrix} -5 & 13 \\ 12 & 8 \end{bmatrix}$

11. $\begin{bmatrix} 16 & 4 & x & 4 \\ 0 & 2 & 4 & 0 \end{bmatrix} = \begin{bmatrix} 16 & 4 & 2x+1 & 4 \\ 0 & 2 & 3y-5 & 0 \end{bmatrix}$

12. $\begin{bmatrix} x+4 & 8 & -3 \\ 1 & 22 & 2y \\ 7 & -2 & z+2 \end{bmatrix} = \begin{bmatrix} 2x+9 & 8 & -3 \\ 1 & 22 & -8 \\ 7 & -2 & 11 \end{bmatrix}$

Operations with Matrices **In Exercises 13–20, find, if possible, (a) $A + B$, (b) $A - B$, (c) $3A$, and (d) $3A - 2B$. Use the matrix capabilities of a graphing utility to verify your results.**

13. $A = \begin{bmatrix} 5 & -2 \\ 3 & 1 \end{bmatrix}$, $B = \begin{bmatrix} 3 & 1 \\ -2 & 6 \end{bmatrix}$

14. $A = \begin{bmatrix} 1 & 2 \\ 2 & 1 \end{bmatrix}$, $B = \begin{bmatrix} -3 & -2 \\ 4 & 2 \end{bmatrix}$

15. $A = \begin{bmatrix} 8 & -1 \\ 2 & 3 \\ -4 & 5 \end{bmatrix}$, $B = \begin{bmatrix} 1 & 6 \\ -1 & -5 \\ 1 & 10 \end{bmatrix}$

16. $A = \begin{bmatrix} 1 & -1 & 3 \\ 0 & 6 & 9 \end{bmatrix}$, $B = \begin{bmatrix} -2 & 0 & -5 \\ -3 & 4 & -7 \end{bmatrix}$

17. $A = \begin{bmatrix} 4 & 5 & -1 & 3 & 4 \\ 1 & 2 & -2 & -1 & 0 \end{bmatrix}$,

$B = \begin{bmatrix} 1 & 0 & -1 & 1 & 0 \\ -6 & 8 & 2 & -3 & -7 \end{bmatrix}$

18. $A = \begin{bmatrix} -1 & 4 & 0 \\ 3 & -2 & 2 \\ 5 & 4 & -1 \\ 0 & 8 & -6 \\ -4 & -1 & 0 \end{bmatrix}$, $B = \begin{bmatrix} -3 & 5 & 1 \\ 2 & -4 & -7 \\ 10 & -9 & -1 \\ 3 & 2 & -4 \\ 0 & 1 & -2 \end{bmatrix}$

19. $A = \begin{bmatrix} 6 & 0 & 3 \\ -1 & -4 & 0 \end{bmatrix}$, $B = \begin{bmatrix} 8 & -1 \\ 4 & -3 \end{bmatrix}$

20. $A = \begin{bmatrix} 3 \\ 2 \\ -1 \end{bmatrix}$, $B = \begin{bmatrix} -4 & 6 & 2 \end{bmatrix}$

Evaluating an Expression In Exercises 21–24, evaluate the expression.

21. $\begin{bmatrix} -5 & 0 \\ 3 & -6 \end{bmatrix} + \begin{bmatrix} 7 & 1 \\ -2 & -1 \end{bmatrix} + \begin{bmatrix} -10 & -8 \\ 14 & 6 \end{bmatrix}$

22. $\begin{bmatrix} 6 & 9 \\ -1 & 0 \\ 7 & 1 \end{bmatrix} + \begin{bmatrix} 0 & 5 \\ -2 & -1 \\ 3 & -6 \end{bmatrix} + \begin{bmatrix} -13 & -7 \\ 4 & -1 \\ -6 & 0 \end{bmatrix}$

23. $\frac{1}{3}\left(\begin{bmatrix} -4 & 0 & 1 \\ 0 & 2 & -12 \end{bmatrix} - \begin{bmatrix} 5 & 1 & -2 \\ 12 & -6 & 3 \end{bmatrix}\right)$

24. $\frac{1}{2}(\begin{bmatrix} 3 & -2 & 4 & 0 \end{bmatrix} - \begin{bmatrix} 10 & -6 & -18 & 9 \end{bmatrix})$

Operations with Matrices In Exercises 25–28, use the matrix capabilities of a graphing utility to evaluate the expression. Round your results to the nearest thousandths, if necessary.

25. $\frac{3}{7}\begin{bmatrix} 2 & 5 \\ -1 & -4 \end{bmatrix} + 6\begin{bmatrix} -3 & 0 \\ 2 & 2 \end{bmatrix}$

26. $\frac{4}{5}\begin{bmatrix} 14 & -11 \\ -22 & 19 \end{bmatrix} + 7\begin{bmatrix} -22 & 20 \\ 13 & 6 \end{bmatrix}$

27. $-5\begin{bmatrix} 3.211 & 6.829 \\ -1.004 & 4.914 \\ 0.055 & -3.889 \end{bmatrix} - \frac{1}{4}\begin{bmatrix} 1.630 & -3.090 \\ 5.256 & 8.335 \\ -9.768 & 4.251 \end{bmatrix}$

28. $-3\begin{bmatrix} 10 & 15 \\ -20 & 10 \\ 12 & 4 \end{bmatrix} - \frac{1}{8}\left(\begin{bmatrix} 12 & 11 \\ 7 & 0 \\ 6 & 9 \end{bmatrix} + \begin{bmatrix} -3 & 13 \\ -3 & 8 \\ -14 & 15 \end{bmatrix}\right)$

Solving a Matrix Equation In Exercises 29–36, solve for X in the equation, where

$$A = \begin{bmatrix} -2 & 1 & 3 \\ -1 & 0 & 4 \end{bmatrix} \text{ and } B = \begin{bmatrix} 0 & 2 & -4 \\ 3 & 0 & 1 \end{bmatrix}.$$

29. $X = 2A + 2B$

30. $X = 3A - 2B$

31. $2X = 2A - B$

32. $2X = A + B$

33. $2X + 3A = B$

34. $3X - 4A = 2B$

35. $4B = -2X - 2A$

36. $5A = 6B - 3X$

Finding the Product of Two Matrices In Exercises 37–44, find AB, if possible.

37. $A = \begin{bmatrix} -1 & 6 \\ -4 & 5 \\ 0 & 3 \end{bmatrix}, \quad B = \begin{bmatrix} 2 & 3 \\ 0 & 9 \end{bmatrix}$

38. $A = \begin{bmatrix} 3 & -1 \\ 4 & -5 \\ 2 & 6 \end{bmatrix}, \quad B = \begin{bmatrix} 6 & 0 \\ 7 & -1 \end{bmatrix}$

39. $A = \begin{bmatrix} 2 & 1 \\ -3 & 4 \\ -1 & 6 \end{bmatrix}, \quad B = \begin{bmatrix} 0 & -3 & 0 \\ 4 & 0 & 2 \\ 8 & -2 & 7 \end{bmatrix}$

40. $A = \begin{bmatrix} 1 & 0 & 3 & -2 \\ 6 & 13 & 8 & -17 \end{bmatrix}, \quad B = \begin{bmatrix} 1 & 6 \\ 4 & 2 \end{bmatrix}$

41. $A = \begin{bmatrix} 6 & 0 & 0 \\ 0 & 4 & 0 \\ 0 & 0 & -2 \end{bmatrix}, \quad B = \begin{bmatrix} \frac{1}{3} & 0 & 0 \\ 0 & -\frac{1}{4} & 0 \\ 0 & 0 & \frac{1}{6} \end{bmatrix}$

42. $A = \begin{bmatrix} 5 & 0 & 0 \\ 0 & -8 & 0 \\ 0 & 0 & 7 \end{bmatrix}, \quad B = \begin{bmatrix} \frac{1}{5} & 0 & 0 \\ 0 & -\frac{1}{8} & 0 \\ 0 & 0 & \frac{1}{2} \end{bmatrix}$

43. $A = \begin{bmatrix} 5 \\ -3 \\ 4 \end{bmatrix}, \quad B = \begin{bmatrix} 2 & -8 & 4 \end{bmatrix}$

44. $A = \begin{bmatrix} 5 \\ 6 \end{bmatrix}, \quad B = \begin{bmatrix} -3 & -1 & -5 & -9 \end{bmatrix}$

Operations with Matrices In Exercises 45–48, find, if possible, (a) AB, (b) BA, and (c) A^2. (*Note:* $A^2 = AA$.) Use the matrix capabilities of a graphing utility to verify your results.

45. $A = \begin{bmatrix} 1 & 2 \\ 4 & 2 \end{bmatrix}, \quad B = \begin{bmatrix} 2 & -1 \\ -1 & 8 \end{bmatrix}$

46. $A = \begin{bmatrix} 6 & 3 \\ -2 & -4 \end{bmatrix}, \quad B = \begin{bmatrix} -2 & 0 \\ 2 & 4 \end{bmatrix}$

47. $A = \begin{bmatrix} 3 & -1 \\ 1 & 3 \end{bmatrix}, \quad B = \begin{bmatrix} 1 & -3 \\ 3 & 1 \end{bmatrix}$

48. $A = \begin{bmatrix} 1 & -1 \\ 1 & 1 \end{bmatrix}, \quad B = \begin{bmatrix} 1 & 3 \\ -3 & 1 \end{bmatrix}$

Finding the Product of Two Matrices In Exercises 49–52, use the matrix capabilities of a graphing utility to find AB, if possible.

49. $A = \begin{bmatrix} 1 & -12 & 4 \\ 14 & 10 & 12 \\ 6 & -15 & 3 \end{bmatrix}, \quad B = \begin{bmatrix} 12 & 10 \\ -6 & 12 \\ 10 & 16 \end{bmatrix}$

50. $A = \begin{bmatrix} -3 & 8 & -6 & 8 \\ -12 & 15 & 9 & 6 \\ 5 & -1 & 1 & 5 \end{bmatrix}, \quad B = \begin{bmatrix} 3 & 1 & 6 \\ 24 & 15 & 14 \\ 16 & 10 & 21 \\ 8 & -4 & 10 \end{bmatrix}$

51. $A = \begin{bmatrix} 7 & 6 & 9 & -4 \\ 3 & -4 & 11 & -2 \\ -5 & -8 & 1 & 12 \end{bmatrix}, \quad B = \begin{bmatrix} 15 & 8 \\ 23 & -17 \\ 9 & 10 \end{bmatrix}$

52. $A = \begin{bmatrix} -2 & 6 & 12 \\ 21 & -5 & 6 \\ 13 & -2 & 9 \end{bmatrix}, \quad B = \begin{bmatrix} 3 & 0 & 34 & 14 \\ -7 & 18 & 0.5 & 1.4 \end{bmatrix}$

Operations with Matrices **In Exercises 53–56, use the matrix capabilities of a graphing utility to evaluate the expression.**

53. $\begin{bmatrix} 3 & 1 \\ 0 & -2 \end{bmatrix}\begin{bmatrix} 1 & 0 \\ -2 & 2 \end{bmatrix}\begin{bmatrix} 1 & 0 \\ 2 & 4 \end{bmatrix}$

54. $\begin{bmatrix} 6 & 5 & -1 \\ 1 & -2 & 0 \end{bmatrix}\begin{bmatrix} 0 & 3 \\ -1 & -3 \\ 4 & 1 \end{bmatrix}\begin{bmatrix} -2 & 2 \\ 0 & -1 \end{bmatrix}$

55. $\begin{bmatrix} 0 & 2 & -2 \\ 4 & 1 & 2 \end{bmatrix}\left(\begin{bmatrix} 4 & 0 \\ 0 & -1 \\ -1 & 2 \end{bmatrix} + \begin{bmatrix} -2 & 3 \\ -3 & 5 \\ 0 & -3 \end{bmatrix}\right)$

56. $\begin{bmatrix} 3 \\ -1 \\ 5 \\ 7 \end{bmatrix}([5 \quad -6] + [7 \quad -1] + [-8 \quad 9])$

Using Matrix Multiplication **In Exercises 57 and 58, use matrix multiplication to determine whether each matrix is a solution of the system of equations. Use a graphing utility to verify your results.**

57. $\begin{cases} x + 2y = 4 \\ 3x + 2y = 0 \end{cases}$

(a) $\begin{bmatrix} 2 \\ 1 \end{bmatrix}$ (b) $\begin{bmatrix} -2 \\ 3 \end{bmatrix}$

(c) $\begin{bmatrix} -4 \\ 4 \end{bmatrix}$ (d) $\begin{bmatrix} 2 \\ -3 \end{bmatrix}$

58. $\begin{cases} 6x + 2y = 0 \\ -x + 5y = 16 \end{cases}$

(a) $\begin{bmatrix} -1 \\ 3 \end{bmatrix}$ (b) $\begin{bmatrix} 2 \\ -6 \end{bmatrix}$

(c) $\begin{bmatrix} 3 \\ -9 \end{bmatrix}$ (d) $\begin{bmatrix} -3 \\ 9 \end{bmatrix}$

Vector Operations **In Exercises 59–62, use matrices to find (a) u + v, (b) u − v, and (c) 3v − u.**

59. $\mathbf{u} = \langle 1, 5\rangle, \quad \mathbf{v} = \langle 3, 2\rangle$

60. $\mathbf{u} = \langle 4, 2\rangle, \quad \mathbf{v} = \langle 6, -3\rangle$

61. $\mathbf{u} = \langle -2, 2\rangle, \quad \mathbf{v} = \langle 5, 4\rangle$

62. $\mathbf{u} = \langle 7, -4\rangle, \quad \mathbf{v} = \langle 2, 1\rangle$

Describing a Vector Transformation **In Exercises 63–68, find $A\mathbf{v}$, where $\mathbf{v} = \langle 4, 2\rangle$, and describe the transformation.**

63. $A = \begin{bmatrix} 1 & 0 \\ 0 & -1 \end{bmatrix}$ **64.** $A = \begin{bmatrix} -1 & 0 \\ 0 & 1 \end{bmatrix}$

65. $A = \begin{bmatrix} 0 & 1 \\ 1 & 0 \end{bmatrix}$ **66.** $A = \begin{bmatrix} 0 & -1 \\ -1 & 0 \end{bmatrix}$

67. $A = \begin{bmatrix} 2 & 0 \\ 0 & 1 \end{bmatrix}$ **68.** $A = \begin{bmatrix} 1 & 0 \\ 0 & 3 \end{bmatrix}$

Solving a System of Linear Equations **In Exercises 69–74, (a) write the system of equations as a matrix equation $AX = B$ and (b) use Gauss-Jordan elimination on the augmented matrix $[A \,\vdots\, B]$ to solve for X. Use a graphing utility to check your solution.**

69. $\begin{cases} -x_1 + x_2 = 4 \\ -2x_1 + x_2 = 0 \end{cases}$ **70.** $\begin{cases} 2x_1 + 3x_2 = 5 \\ x_1 + 4x_2 = 10 \end{cases}$

71. $\begin{cases} 2x_1 + x_2 = 2 \\ 2x_1 - 3x_2 = 10 \end{cases}$ **72.** $\begin{cases} -4x_1 + 9x_2 = -13 \\ x_1 - 3x_2 = 12 \end{cases}$

73. $\begin{cases} x_1 - 2x_2 + 3x_3 = 9 \\ -x_1 + 3x_2 - x_3 = -6 \\ 2x_1 - 5x_2 + 5x_3 = 17 \end{cases}$

74. $\begin{cases} x_1 + x_2 - 3x_3 = 9 \\ -x_1 + 2x_2 = 6 \\ x_1 - x_2 + x_3 = -5 \end{cases}$

Operations with Matrices **In Exercises 75–78, use a graphing utility to perform the operations. Compare and describe the results of parts (a) and (b).**

$$A = \begin{bmatrix} 1 & 2 & -1 \\ 0 & -2 & 3 \\ 4 & -3 & 2 \end{bmatrix}, \quad B = \begin{bmatrix} 2 & 3 & 0 \\ 4 & 1 & -2 \\ -1 & 2 & 0 \end{bmatrix},$$

$$C = \begin{bmatrix} 3 & -2 & 1 \\ -4 & 0 & 3 \\ -1 & 3 & -2 \end{bmatrix}, \quad \text{and} \quad c = 3$$

75. (a) $A(B + C)$ (b) $AB + AC$

76. (a) $(B + C)A$ (b) $BA + CA$

77. (a) $A(BC)$ (b) $(AB)C$

78. (a) $c(AB)$ (b) $(cA)B$

Operations with Matrices **In Exercises 79–86, perform the operations (a) using a graphing utility and (b) by hand algebraically. If it is not possible to perform the operation(s), state the reason.**

$$A = \begin{bmatrix} 1 & 2 & -2 \\ -1 & 1 & 0 \end{bmatrix}, \quad B = \begin{bmatrix} -1 & 4 & -1 \\ -2 & -1 & 0 \end{bmatrix},$$

$$C = \begin{bmatrix} 1 & 2 \\ -2 & 3 \\ 1 & 0 \end{bmatrix}, \quad c = -2, \quad \text{and} \quad d = -3$$

79. $A + cB$ **80.** $B + dA$

81. $c(AB)$ **82.** $A(B + C)$

83. $CA - BC$ **84.** dAB^2

85. $cdA + B$ **86.** $cA + dB$

Operations with Matrices **In Exercises 87 and 88, use a graphing utility to perform the indicated operations.**

$$A = \begin{bmatrix} 2 & 0 \\ 4 & 5 \end{bmatrix} \quad \text{and} \quad B = \begin{bmatrix} 5 & 3 \\ 1 & 4 \end{bmatrix}$$

87. $A^2 - 5A + 2I_2$ **88.** $B^2 - 7B + 6I_2$

89. Manufacturing A corporation manufactures standard sunglasses and designer sunglasses at four different factories. The production levels are represented by A.

$$A = \overbrace{\begin{bmatrix} 100 & 120 & 60 & 40 \\ 140 & 160 & 200 & 80 \end{bmatrix}}^{\text{Factory: } 1\ \ 2\ \ 3\ \ 4} \begin{matrix} \text{Standard} \\ \text{Designer} \end{matrix} \Big\} \text{Sunglass type}$$

Find the production levels when production is decreased by 15%.

90. Manufacturing A corporation manufactures sport utility vehicles and pickup trucks at four different factories. The production levels are represented by A.

$$A = \overbrace{\begin{bmatrix} 100 & 90 & 70 & 30 \\ 40 & 20 & 60 & 60 \end{bmatrix}}^{\text{Factory: } 1\ \ 2\ \ 3\ \ 4} \begin{matrix} \text{SUV} \\ \text{Pickup} \end{matrix} \Big\} \text{Vehicle type}$$

Find the production levels when production is decreased by 10%.

91. Manufacturing A corporation manufactures acoustic guitars and electric guitars at three different factories. The production levels are represented by A.

$$A = \overbrace{\begin{bmatrix} 70 & 50 & 40 \\ 35 & 100 & 75 \end{bmatrix}}^{\text{Factory: } A\ \ B\ \ C} \begin{matrix} \text{Acoustic} \\ \text{Electric} \end{matrix} \Big\} \text{Guitar type}$$

Find the production levels when production is increased by 20%.

92. Tourism A travel agency offers four-night vacation packages at four different resort hotels, based on double or family (up to 4) occupancy. Off-season rates for the packages are represented by A.

$$A = \overbrace{\begin{bmatrix} 668 & 2666 & 973 & 918 \\ 1214 & 4637 & 1656 & 1699 \end{bmatrix}}^{\text{Hotel: } w\ \ x\ \ y\ \ z} \begin{matrix} \text{Double} \\ \text{Family} \end{matrix} \Big\} \text{Occupancy}$$

During the peak resort season, rates increase by 22%. What are the peak resort season package rates?

93. Agriculture A fruit grower ships bags of apples and peaches to three different outlets. The numbers of bags are represented by A.

$$A = \overbrace{\begin{bmatrix} 125 & 100 & 75 \\ 100 & 175 & 125 \end{bmatrix}}^{\text{Outlet: } 1\ \ 2\ \ 3} \begin{matrix} \text{Apples} \\ \text{Peaches} \end{matrix} \Big\} \text{Crop}$$

(a) The grower receives \$3.50 per bag of apples and \$4.00 per bag of peaches. Organize the amount received per bag of each fruit in a matrix B.

(b) Find the product BA and interpret the result.

94. Physical Education The numbers of calories burned per hour by individuals of different weights performing different types of aerobic exercises are shown in the matrix.

$$B = \overset{\begin{matrix}\text{130-lb person} & \text{155-lb person}\end{matrix}}{\begin{bmatrix} 325 & 387 \\ 236 & 281 \\ 354 & 422 \end{bmatrix}} \begin{matrix} \text{Bicycling} \\ \text{Jogging} \\ \text{Swimming} \end{matrix}$$

(a) A 130-pound person and a 155-pound person bicycle for 30 minutes, jog for 15 minutes, and swim for 60 minutes. Organize the time spent exercising in a matrix A.

(b) Find the product AB and interpret the result.

95. *Why you should learn it* *(p. 526)* A company manufactures boats. Its labor-hour and wage requirements are represented by L and W, respectively.

$$L = \overbrace{\begin{bmatrix} 1.0\text{ h} & 0.5\text{ h} & 0.2\text{ h} \\ 1.6\text{ h} & 1.0\text{ h} & 0.2\text{ h} \\ 2.5\text{ h} & 2.0\text{ h} & 1.4\text{ h} \end{bmatrix}}^{\text{Department: Cutting, Assembly, Packaging}} \begin{matrix} \text{Small} \\ \text{Medium} \\ \text{Large} \end{matrix} \Bigg\} \text{Boat size}$$

$$W = \overbrace{\begin{bmatrix} \$15 & \$13 \\ \$12 & \$11 \\ \$11 & \$10 \end{bmatrix}}^{\text{Plant: } A\ \ B} \begin{matrix} \text{Cutting} \\ \text{Assembly} \\ \text{Packaging} \end{matrix} \Bigg\} \text{Department}$$

(a) Compute LW and interpret the result.

(b) How much more does it cost to build a large boat at plant A than at plant B?

96. Inventory Control A company sells five models of computers through three retail outlets. The inventories are represented by S. The wholesale and retail prices are represented by T. Compute ST and interpret the result.

$$S = \overbrace{\begin{bmatrix} 3 & 2 & 2 & 3 & 0 \\ 0 & 2 & 3 & 4 & 3 \\ 4 & 2 & 1 & 3 & 2 \end{bmatrix}}^{\text{Model: } A\ \ B\ \ C\ \ D\ \ E} \begin{matrix} 1 \\ 2 \\ 3 \end{matrix} \Bigg\} \text{Outlet}$$

$$T = \overbrace{\begin{bmatrix} \$840 & \$1100 \\ \$1200 & \$1350 \\ \$1450 & \$1650 \\ \$2650 & \$3000 \\ \$3050 & \$3200 \end{bmatrix}}^{\text{Price: Wholesale, Retail}} \begin{matrix} A \\ B \\ C \\ D \\ E \end{matrix} \Bigg\} \text{Model}$$

97. Politics The matrix

$$P = \overset{\text{From}}{\begin{array}{ccc} R & D & I \end{array}}$$
$$P = \begin{bmatrix} 0.6 & 0.1 & 0.1 \\ 0.2 & 0.7 & 0.1 \\ 0.2 & 0.2 & 0.8 \end{bmatrix} \begin{matrix} 1 \\ 2 \\ 3 \end{matrix} \Bigg\} \text{To}$$

is called a *stochastic matrix.* Each entry $p_{ij}(i \neq j)$ represents the proportion of the voting population that changes from party i to party j, and p_{ii} represents the proportion that remains loyal to the party from one election to the next. Compute and interpret P^2.

98. Politics Use a graphing utility to find P^3, P^4, P^5, P^6, P^7, and P^8 for the matrix given in Exercise 97. Can you detect a pattern as P is raised to higher powers?

Conclusions

True or False? **In Exercises 99 and 100, determine whether the statement is true or false. Justify your answer.**

99. Two matrices can be added only when they have the same dimension.

100. $\begin{bmatrix} -6 & -2 \\ 2 & -6 \end{bmatrix}\begin{bmatrix} 4 & 0 \\ 0 & -1 \end{bmatrix} = \begin{bmatrix} 4 & 0 \\ 0 & -1 \end{bmatrix}\begin{bmatrix} -6 & -2 \\ 2 & -6 \end{bmatrix}$

Think About It **In Exercises 101–108, let matrices A, B, C, and D be of dimensions 2×3, 2×3, 3×2, and 2×2, respectively. Determine whether the matrices are of proper dimension to perform the operation(s). If so, give the dimension of the answer.**

101. $A + 2C$

102. $B - 3C$

103. CD

104. BC

105. $CA - D$

106. $CB - D$

107. $D(A - 3B)$

108. $C(A + 2B)$

Think About It **In Exercises 109–112, use the matrices**

$$A = \begin{bmatrix} 2 & -1 \\ 1 & 3 \end{bmatrix} \text{ and } B = \begin{bmatrix} -1 & 1 \\ 0 & -2 \end{bmatrix}.$$

109. Show that $(A + B)^2 \neq A^2 + 2AB + B^2$.

110. Show that $(A - B)^2 \neq A^2 - 2AB + B^2$.

111. Show that $(A + B)(A - B) \neq A^2 - B^2$.

112. Show that $(A + B)^2 = A^2 + AB + BA + B^2$.

113. Think About It If a, b, and c are real numbers such that $c \neq 0$ and $ac = bc$, then $a = b$. However, if A, B, and C are nonzero matrices such that $AC = BC$, then A is *not necessarily* equal to B. Illustrate this using the following matrices.

$$A = \begin{bmatrix} 0 & 1 \\ 0 & 1 \end{bmatrix}, \quad B = \begin{bmatrix} 1 & 0 \\ 1 & 0 \end{bmatrix}, \quad C = \begin{bmatrix} 2 & 3 \\ 2 & 3 \end{bmatrix}$$

114. Think About It If a and b are real numbers such that $ab = 0$, then $a = 0$ or $b = 0$. However, if A and B are matrices such that $AB = O$, it is *not necessarily* true that $A = O$ or $B = O$. Illustrate this using the following matrices.

$$A = \begin{bmatrix} 3 & 3 \\ 4 & 4 \end{bmatrix}, \quad B = \begin{bmatrix} 1 & -1 \\ -1 & 1 \end{bmatrix}$$

115. Exploration Let $i = \sqrt{-1}$ and let

$$A = \begin{bmatrix} i & 0 \\ 0 & i \end{bmatrix} \text{ and } B = \begin{bmatrix} 0 & -i \\ i & 0 \end{bmatrix}.$$

(a) Find A^2, A^3, and A^4. Identify any similarities with i^2, i^3, and i^4.

(b) Find and identify B^2.

116. Exploration Let A and B be unequal diagonal matrices of the same dimension. (A **diagonal matrix** is a square matrix in which each entry not on the main diagonal is zero.) Determine the products AB for several pairs of such matrices. Make a conjecture about a quick rule for such products.

117. Error Analysis Describe the error in multiplying the matrices.

$$\begin{bmatrix} 3 & 4 \end{bmatrix}\begin{bmatrix} 1 \\ 2 \end{bmatrix} = \begin{bmatrix} 3(1) \\ 4(2) \end{bmatrix} = \begin{bmatrix} 3 \\ 8 \end{bmatrix} \quad ✗$$

118. HOW DO YOU SEE IT? An electronics manufacturer produces two models of LCD televisions, which are shipped to three warehouses. The shipment levels are represented by A.

$$A = \overset{\text{Warehouse}}{\begin{array}{ccc} 1 & 2 & 3 \end{array}}$$
$$A = \begin{bmatrix} 5{,}000 & 4{,}000 & 8{,}000 \\ 6{,}000 & 10{,}000 & 5{,}000 \end{bmatrix} \begin{matrix} A \\ B \end{matrix} \Big\} \text{Model}$$

(a) Interpret the value of a_{21}.

(b) How could you find the shipment levels when the shipments are increased by 15%?

(c) Each model A television sells for \$700 and each model B television sells for \$900. How could you use matrices to find the cost of the two models of LCD televisions at each warehouse?

Cumulative Mixed Review

Condensing a Logarithmic Expression **In Exercises 119 and 120, condense the expression to the logarithm of a single quantity.**

119. $3 \ln 4 - \frac{1}{3}\ln(x^2 + 3)$

120. $2 \ln 7t^4 - \frac{3}{5}\ln t^5$

7.6 The Inverse of a Square Matrix

The Inverse of a Matrix

This section further develops the algebra of matrices. To begin, consider the real number equation $ax = b$. To solve this equation for x, multiply each side of the equation by a^{-1} (provided that $a \neq 0$).

$$ax = b$$
$$(a^{-1}a)x = a^{-1}b$$
$$(1)x = a^{-1}b$$
$$x = a^{-1}b$$

The number a^{-1} is called the *multiplicative inverse* of a because $a^{-1}a = 1$. The definition of the multiplicative **inverse of a matrix** is similar.

Definition of the Inverse of a Square Matrix

Let A be an $n \times n$ matrix and let I_n be the $n \times n$ identity matrix. If there exists a matrix A^{-1} such that

$$AA^{-1} = I_n = A^{-1}A$$

then A^{-1} is called the **inverse** of A. The symbol A^{-1} is read "A inverse."

What you should learn

- Verify that two matrices are inverses of each other.
- Use Gauss-Jordan elimination to find inverses of matrices.
- Use a formula to find inverses of 2×2 matrices.
- Use inverse matrices to solve systems of linear equations.

Why you should learn it

A system of equations can be solved using the inverse of the coefficient matrix. This method is particularly useful when the coefficients are the same for several systems, but the constants are different. Exercise 64 on page 549 shows how to use an inverse matrix to find a model for the number of international travelers to the United States from Asia.

EXAMPLE 1 The Inverse of a Matrix

Show that B is the inverse of A, where

$$A = \begin{bmatrix} -1 & 2 \\ -1 & 1 \end{bmatrix} \quad \text{and} \quad B = \begin{bmatrix} 1 & -2 \\ 1 & -1 \end{bmatrix}.$$

Solution

To show that B is the inverse of A, show that $AB = I = BA$, as follows.

$$AB = \begin{bmatrix} -1 & 2 \\ -1 & 1 \end{bmatrix}\begin{bmatrix} 1 & -2 \\ 1 & -1 \end{bmatrix} = \begin{bmatrix} -1+2 & 2-2 \\ -1+1 & 2-1 \end{bmatrix} = \begin{bmatrix} 1 & 0 \\ 0 & 1 \end{bmatrix}$$

$$BA = \begin{bmatrix} 1 & -2 \\ 1 & -1 \end{bmatrix}\begin{bmatrix} -1 & 2 \\ -1 & 1 \end{bmatrix} = \begin{bmatrix} -1+2 & 2-2 \\ -1+1 & 2-1 \end{bmatrix} = \begin{bmatrix} 1 & 0 \\ 0 & 1 \end{bmatrix}$$

As you can see,

$$AB = I = BA.$$

This is an example of a square matrix that has an inverse. Note that not all square matrices have inverses.

✓ **Checkpoint** *Audio-video solution in English & Spanish at LarsonPrecalculus.com*

Show that B is the inverse of A, where

$$A = \begin{bmatrix} 2 & -1 \\ -3 & 1 \end{bmatrix} \quad \text{and} \quad B = \begin{bmatrix} -1 & -1 \\ -3 & -2 \end{bmatrix}.$$

Recall from Section 7.5 that it is not always true that $AB = BA$, even when both products are defined. However, if A and B are both square matrices and $AB = I_n$, then it can be shown that $BA = I_n$. So, in Example 1, you need only check that $AB = I_2$.

Finding Inverse Matrices

When a matrix A has an inverse, A is called **invertible** (or **nonsingular**); otherwise, A is called **singular.** A nonsquare matrix cannot have an inverse. To see this, note that if A is of dimension $m \times n$ and B is of dimension $n \times m$ (where $m \neq n$), then the products AB and BA are of different dimensions and so cannot be equal to each other. Not all square matrices have inverses, as you will see later in this section. When a matrix does have an inverse, however, that inverse is unique. Example 2 shows how to use systems of equations to find the inverse of a matrix.

EXAMPLE 2 Finding the Inverse of a Matrix

See LarsonPrecalculus.com for an interactive version of this type of example.

Find the inverse of

$$A = \begin{bmatrix} 1 & 4 \\ -1 & -3 \end{bmatrix}.$$

Solution

To find the inverse of A, try to solve the matrix equation $AX = I$ for X.

$$\overset{A}{\begin{bmatrix} 1 & 4 \\ -1 & -3 \end{bmatrix}} \overset{X}{\begin{bmatrix} x_{11} & x_{12} \\ x_{21} & x_{22} \end{bmatrix}} = \overset{I}{\begin{bmatrix} 1 & 0 \\ 0 & 1 \end{bmatrix}}$$

$$\begin{bmatrix} x_{11} + 4x_{21} & x_{12} + 4x_{22} \\ -x_{11} - 3x_{21} & -x_{12} - 3x_{22} \end{bmatrix} = \begin{bmatrix} 1 & 0 \\ 0 & 1 \end{bmatrix}$$

Equating corresponding entries, you obtain the following two systems of linear equations.

$$\begin{cases} x_{11} + 4x_{21} = 1 \\ -x_{11} - 3x_{21} = 0 \end{cases}$$ Linear system with two variables, x_{11} and x_{21}.

$$\begin{cases} x_{12} + 4x_{22} = 0 \\ -x_{12} - 3x_{22} = 1 \end{cases}$$ Linear system with two variables, x_{12} and x_{22}.

Solve the first system using elementary row operations to determine that $x_{11} = -3$ and $x_{21} = 1$. From the second system you can determine that $x_{12} = -4$ and $x_{22} = 1$. Therefore, the inverse of A is

$$A^{-1} = X$$
$$= \begin{bmatrix} -3 & -4 \\ 1 & 1 \end{bmatrix}.$$

You can use matrix multiplication to check this result.

Check

$$AA^{-1} = \begin{bmatrix} 1 & 4 \\ -1 & -3 \end{bmatrix}\begin{bmatrix} -3 & -4 \\ 1 & 1 \end{bmatrix} = \begin{bmatrix} 1 & 0 \\ 0 & 1 \end{bmatrix} \checkmark$$

$$A^{-1}A = \begin{bmatrix} -3 & -4 \\ 1 & 1 \end{bmatrix}\begin{bmatrix} 1 & 4 \\ -1 & -3 \end{bmatrix} = \begin{bmatrix} 1 & 0 \\ 0 & 1 \end{bmatrix} \checkmark$$

✓ ***Checkpoint*** ▶ ***Audio-video solution in English & Spanish at LarsonPrecalculus.com***

Find the inverse of

$$A = \begin{bmatrix} 1 & -2 \\ -1 & 3 \end{bmatrix}.$$

Explore the Concept

Most graphing utilities are capable of finding the inverse of a square matrix. Try using a graphing utility to find the inverse of the matrix

$$A = \begin{bmatrix} 2 & -3 & 1 \\ -1 & 2 & -1 \\ -2 & 0 & 1 \end{bmatrix}.$$

After you find A^{-1}, store it as $[B]$ and use the graphing utility to find $[A] \times [B]$ and $[B] \times [A]$. What can you conclude?

In Example 2, note that the two systems of linear equations have the *same coefficient matrix* A. Rather than solve the two systems represented by

$$\begin{bmatrix} 1 & 4 & \vdots & 1 \\ -1 & -3 & \vdots & 0 \end{bmatrix}$$

and

$$\begin{bmatrix} 1 & 4 & \vdots & 0 \\ -1 & -3 & \vdots & 1 \end{bmatrix}$$

separately, you can solve them *simultaneously* by *adjoining* the identity matrix to the coefficient matrix to obtain

$$\overset{A}{\begin{bmatrix} 1 & 4 \\ -1 & -3 \end{bmatrix}} \begin{matrix} \vdots \\ \vdots \end{matrix} \overset{I}{\begin{bmatrix} 1 & 0 \\ 0 & 1 \end{bmatrix}}.$$

This "doubly augmented" matrix can be represented as

$[A \,\vdots\, I]$.

By applying Gauss-Jordan elimination to this matrix, you can solve *both* systems with a single elimination process.

$$\begin{bmatrix} 1 & 4 & \vdots & 1 & 0 \\ -1 & -3 & \vdots & 0 & 1 \end{bmatrix}$$

$$R_1 + R_2 \rightarrow \begin{bmatrix} 1 & 4 & \vdots & 1 & 0 \\ 0 & 1 & \vdots & 1 & 1 \end{bmatrix}$$

$$-4R_2 + R_1 \rightarrow \begin{bmatrix} 1 & 0 & \vdots & -3 & -4 \\ 0 & 1 & \vdots & 1 & 1 \end{bmatrix}$$

So, from the "doubly augmented" matrix $[A \,\vdots\, I]$, you obtained the matrix $[I \,\vdots\, A^{-1}]$.

$$\overset{A}{\begin{bmatrix} 1 & 4 \\ -1 & -3 \end{bmatrix}} \begin{matrix} \vdots \\ \vdots \end{matrix} \overset{I}{\begin{bmatrix} 1 & 0 \\ 0 & 1 \end{bmatrix}} \Rightarrow \overset{I}{\begin{bmatrix} 1 & 0 \\ 0 & 1 \end{bmatrix}} \begin{matrix} \vdots \\ \vdots \end{matrix} \overset{A^{-1}}{\begin{bmatrix} -3 & -4 \\ 1 & 1 \end{bmatrix}}$$

This procedure (or algorithm) works for any square matrix that has an inverse.

Finding an Inverse Matrix

Let A be a square matrix of dimension $n \times n$.

1. Write the $n \times 2n$ matrix that consists of the given matrix A on the left and the $n \times n$ identity matrix I on the right to obtain

$[A \,\vdots\, I]$.

2. If possible, row reduce A to I using elementary row operations on the *entire* matrix

$[A \,\vdots\, I]$.

The result will be the matrix

$[I \,\vdots\, A^{-1}]$.

If this is not possible, then A is not invertible.

3. Check your work by multiplying to see that

$AA^{-1} = I = A^{-1}A$.

Explore the Concept

Select two 2×2 matrices A and B that have inverses. Enter them into your graphing utility and calculate $(AB)^{-1}$. Then calculate $B^{-1}A^{-1}$ and $A^{-1}B^{-1}$. Make a conjecture about the inverse of the product of two invertible matrices.

EXAMPLE 3 Finding the Inverse of a Matrix

Find the inverse of $A = \begin{bmatrix} 1 & -1 & 0 \\ 1 & 0 & -1 \\ 6 & -2 & -3 \end{bmatrix}$.

Solution

Begin by adjoining the identity matrix to A to form the matrix

$$[A \;\vdots\; I] = \left[\begin{array}{rrr:rrr} 1 & -1 & 0 & 1 & 0 & 0 \\ 1 & 0 & -1 & 0 & 1 & 0 \\ 6 & -2 & -3 & 0 & 0 & 1 \end{array}\right].$$

Use elementary row operations to obtain the form $[I \;\vdots\; A^{-1}]$, as follows.

$$\left[\begin{array}{rrr:rrr} 1 & 0 & 0 & -2 & -3 & 1 \\ 0 & 1 & 0 & -3 & -3 & 1 \\ 0 & 0 & 1 & -2 & -4 & 1 \end{array}\right]$$

Therefore, the matrix A is invertible and its inverse is

$$A^{-1} = \begin{bmatrix} -2 & -3 & 1 \\ -3 & -3 & 1 \\ -2 & -4 & 1 \end{bmatrix}.$$

Confirm this result by multiplying A by A^{-1} to obtain I as follows.

Check

$$AA^{-1} = \begin{bmatrix} 1 & -1 & 0 \\ 1 & 0 & -1 \\ 6 & -2 & -3 \end{bmatrix}\begin{bmatrix} -2 & -3 & 1 \\ -3 & -3 & 1 \\ -2 & -4 & 1 \end{bmatrix} = \begin{bmatrix} 1 & 0 & 0 \\ 0 & 1 & 0 \\ 0 & 0 & 1 \end{bmatrix} = I$$

✓ Checkpoint *Audio-video solution in English & Spanish at LarsonPrecalculus.com*

Find the inverse of $A = \begin{bmatrix} 1 & -2 & -1 \\ 0 & -1 & 2 \\ 1 & -2 & 0 \end{bmatrix}$.

The algorithm shown in Example 3 applies to any $n \times n$ matrix A. When using this algorithm, if the matrix A does not reduce to the identity matrix, then A does not have an inverse. For instance, the following matrix has no inverse.

$$A = \begin{bmatrix} 1 & 2 & 0 \\ 3 & -1 & 2 \\ -2 & 3 & -2 \end{bmatrix}$$

To see why matrix A above has no inverse, begin by adjoining the identity matrix to A to form

$$[A \;\vdots\; I] = \left[\begin{array}{rrr:rrr} 1 & 2 & 0 & 1 & 0 & 0 \\ 3 & -1 & 2 & 0 & 1 & 0 \\ -2 & 3 & -2 & 0 & 0 & 1 \end{array}\right].$$

Then, use elementary row operations to obtain

$$\left[\begin{array}{rrr:rrr} 1 & 2 & 0 & 1 & 0 & 0 \\ 0 & -7 & 2 & -3 & 1 & 0 \\ 0 & 0 & 0 & -1 & 1 & 1 \end{array}\right].$$

At this point in the elimination process, you can see that it is impossible to obtain the identity matrix I on the left. Therefore, A is not invertible.

Technology Tip

Most graphing utilities can find the inverse of a matrix by using the inverse key x^{-1}. For instructions on how to use the inverse key to find the inverse of a matrix, see Appendix A; for specific keystrokes, go to this textbook's *Student Companion Website.*

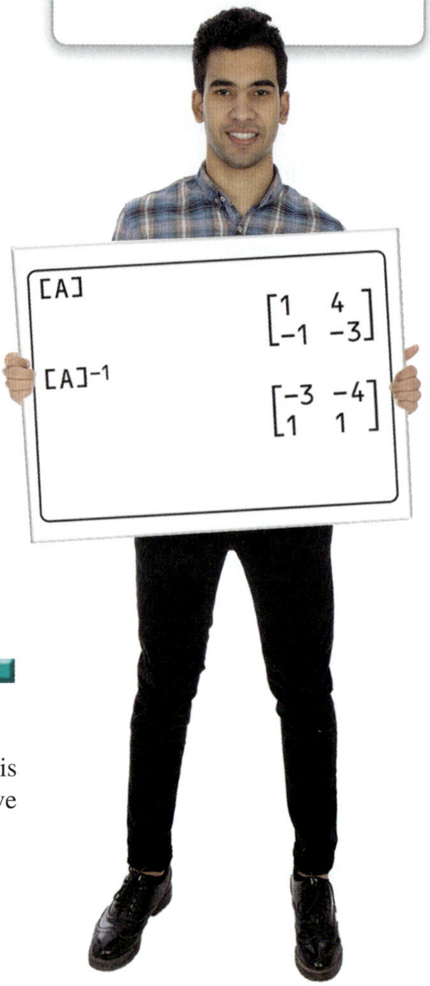

Ebtikar/Shutterstock.com

The Inverse of a 2 × 2 Matrix

Using Gauss-Jordan elimination to find the inverse of a matrix works well (even as a computer technique) for matrices of dimension 3 × 3 or greater. For 2 × 2 matrices, however, many people prefer to use a formula for the inverse rather than Gauss-Jordan elimination. This simple formula, which works *only* for 2 × 2 matrices, is explained as follows. Let A be the 2 × 2 matrix given by

$$A = \begin{bmatrix} a & b \\ c & d \end{bmatrix}.$$

The matrix A is invertible if and only if

$$ad - bc \neq 0.$$

If $ad - bc \neq 0$, then the inverse is given by

$$A^{-1} = \frac{1}{ad - bc}\begin{bmatrix} d & -b \\ -c & a \end{bmatrix}.$$ Formula for the inverse of a 2 × 2 matrix

The denominator $ad - bc$ is called the *determinant* of the 2 × 2 matrix A. You will study determinants in the next section.

EXAMPLE 4 Finding the Inverse of a 2 × 2 Matrix

If possible, find the inverse of each matrix.

a. $A = \begin{bmatrix} 3 & -1 \\ -2 & 2 \end{bmatrix}$

b. $B = \begin{bmatrix} 3 & -1 \\ -6 & 2 \end{bmatrix}$

Solution

a. For matrix A, the determinant is

$$ad - bc = (3)(2) - (-1)(-2)$$
$$= 4.$$

Because this quantity is not zero, the inverse is formed by interchanging the entries on the main diagonal, changing the signs of the other two entries, and multiplying by the scalar $\frac{1}{4}$, as follows.

$$A^{-1} = \frac{1}{4}\begin{bmatrix} 2 & 1 \\ 2 & 3 \end{bmatrix}$$ Substitute for a, b, c, d, and the determinant.

$$= \begin{bmatrix} \frac{1}{2} & \frac{1}{4} \\ \frac{1}{2} & \frac{3}{4} \end{bmatrix}$$ Multiply by the scalar $\frac{1}{4}$ and simplify.

b. For matrix B, you have

$$ad - bc = (3)(2) - (-1)(-6)$$
$$= 0$$

which means that B is not invertible.

Checkpoint Audio-video solution in English & Spanish at LarsonPrecalculus.com

If possible, find the inverse of $A = \begin{bmatrix} 5 & -1 \\ 3 & 4 \end{bmatrix}$.

Explore the Concept

Use a graphing utility to find the inverse of the matrix

$$A = \begin{bmatrix} 1 & -3 \\ -2 & 6 \end{bmatrix}.$$

What message appears on the screen? Why does the graphing utility display this message?

Systems of Linear Equations

You know that a system of linear equations can have exactly one solution, infinitely many solutions, or no solution. If the coefficient matrix A of a *square* system (a system that has the same number of equations as variables) is invertible, then the system has a unique solution, which is defined as follows.

A System of Equations with a Unique Solution

If A is an invertible matrix, then the system of linear equations represented by $AX = B$ has a unique solution given by

$$X = A^{-1}B.$$

The formula $X = A^{-1}B$ is used on most graphing utilities to solve linear systems that have invertible coefficient matrices. That is, you enter the $n \times n$ coefficient matrix $[A]$ and the $n \times 1$ column matrix $[B]$. The solution X is given by $[A]^{-1}[B]$.

EXAMPLE 5 Solving a System of Equations Using an Inverse

Use an inverse matrix to solve the system.

$$\begin{cases} 2x + 3y + z = -1 \\ 3x + 3y + z = 1 \\ 2x + 4y + z = -2 \end{cases}$$

Solution

Begin by writing the system as $AX = B$.

$$\begin{bmatrix} 2 & 3 & 1 \\ 3 & 3 & 1 \\ 2 & 4 & 1 \end{bmatrix}\begin{bmatrix} x \\ y \\ z \end{bmatrix} = \begin{bmatrix} -1 \\ 1 \\ -2 \end{bmatrix}$$

Then, use Gauss-Jordan elimination to find A^{-1}.

$$A^{-1} = \begin{bmatrix} -1 & 1 & 0 \\ -1 & 0 & 1 \\ 6 & -2 & -3 \end{bmatrix}$$

Finally, multiply B by A^{-1} on the left to obtain the solution.

$$\begin{aligned} X &= A^{-1}B \\ &= \begin{bmatrix} -1 & 1 & 0 \\ -1 & 0 & 1 \\ 6 & -2 & -3 \end{bmatrix}\begin{bmatrix} -1 \\ 1 \\ -2 \end{bmatrix} \\ &= \begin{bmatrix} 2 \\ -1 \\ -2 \end{bmatrix} \end{aligned}$$

So, the solution is $x = 2$, $y = -1$, and $z = -2$. Use a graphing utility to verify A^{-1} for the system of equations.

Algebra Help

Remember that matrix multiplication is not commutative. So, you must multiply matrices in the correct order. For instance, in Example 5, you must multiply B by A^{-1} on the left.

✓ Checkpoint *Audio-video solution in English & Spanish at LarsonPrecalculus.com*

Use an inverse matrix to solve the system.

$$\begin{cases} x + y + z = 6 \\ 3x + 5y + 4z = 27 \\ 3x + 6y + 5z = 35 \end{cases}$$

7.6 Exercises

See *CalcChat.com* for tutorial help and worked-out solutions to odd-numbered exercises.
For instructions on how to use a graphing utility, see Appendix A.

Vocabulary and Concept Check

In Exercises 1 and 2, fill in the blank(s).

1. If there exists an $n \times n$ matrix A^{-1} such that $AA^{-1} = I_n = A^{-1}A$, then A^{-1} is called the _______ of A.

2. If a matrix A has an inverse, then it is called invertible or _______ .

3. Do all square matrices have inverses?

4. Given that A and B are square matrices and $AB = I_n$, does $BA = I_n$?

Procedures and Problem Solving

The Inverse of a Matrix **In Exercises 5–10, show that B is the inverse of A.**

5. $A = \begin{bmatrix} 1 & 3 \\ -1 & -2 \end{bmatrix}$, $B = \begin{bmatrix} -2 & -3 \\ 1 & 1 \end{bmatrix}$

6. $A = \begin{bmatrix} -4 & 3 \\ 3 & -2 \end{bmatrix}$, $B = \begin{bmatrix} 2 & 3 \\ 3 & 4 \end{bmatrix}$

7. $A = \begin{bmatrix} 3 & 2 \\ 1 & 4 \end{bmatrix}$, $B = \dfrac{1}{10}\begin{bmatrix} 4 & -2 \\ -1 & 3 \end{bmatrix}$

8. $A = \begin{bmatrix} -\frac{1}{2} & -\frac{5}{4} \\ 1 & 2 \end{bmatrix}$, $B = \begin{bmatrix} 8 & 5 \\ -4 & -2 \end{bmatrix}$

9. $A = \begin{bmatrix} 2 & -17 & 11 \\ -1 & 11 & -7 \\ 0 & 3 & -2 \end{bmatrix}$, $B = \begin{bmatrix} 1 & 1 & 2 \\ 2 & 4 & -3 \\ 3 & 6 & -5 \end{bmatrix}$

10. $A = \begin{bmatrix} -4 & 1 & 5 \\ -1 & 2 & 4 \\ 0 & -1 & -1 \end{bmatrix}$, $B = \dfrac{1}{4}\begin{bmatrix} -2 & 4 & 6 \\ 1 & -4 & -11 \\ -1 & 4 & 7 \end{bmatrix}$

Finding the Inverse of a Matrix **In Exercises 11–20, find the inverse of the matrix (if it exists).**

11. $\begin{bmatrix} 2 & 0 \\ 0 & 3 \end{bmatrix}$

12. $\begin{bmatrix} -9 & -7 \\ -5 & -4 \end{bmatrix}$

13. $\begin{bmatrix} 1 & -2 \\ 2 & -3 \end{bmatrix}$

14. $\begin{bmatrix} -2 & 5 \\ 6 & -15 \\ 0 & 1 \end{bmatrix}$

15. $\begin{bmatrix} 2 & 7 & 1 \\ -3 & -9 & 2 \end{bmatrix}$

16. $\begin{bmatrix} -7 & 33 \\ 4 & -19 \end{bmatrix}$

17. $\begin{bmatrix} 1 & 0 & 0 \\ 3 & 5 & 0 \\ 2 & 5 & 0 \end{bmatrix}$

18. $\begin{bmatrix} 1 & 2 & 2 \\ 3 & 7 & 9 \\ -1 & -4 & -7 \end{bmatrix}$

19. $\begin{bmatrix} 1 & 1 & 1 \\ 3 & 5 & 4 \\ 3 & 6 & 5 \end{bmatrix}$

20. $\begin{bmatrix} 1 & 6 & 10 \\ 3 & 4 & 0 \\ 2 & 5 & 5 \end{bmatrix}$

Finding the Inverse of a Matrix Using Technology **In Exercises 21–28, use the matrix capabilities of a graphing utility to find the inverse of the matrix (if it exists).**

21. $\begin{bmatrix} 1 & 1 & 2 \\ 3 & 1 & 0 \\ -2 & 0 & 3 \end{bmatrix}$

22. $\begin{bmatrix} 1 & 2 & -1 \\ 3 & 7 & -10 \\ -5 & -7 & -15 \end{bmatrix}$

23. $\begin{bmatrix} -\frac{1}{2} & \frac{3}{4} & \frac{1}{4} \\ 1 & 0 & -\frac{3}{2} \\ 0 & -1 & \frac{1}{2} \end{bmatrix}$

24. $\begin{bmatrix} -\frac{5}{6} & \frac{1}{3} & \frac{11}{6} \\ 0 & \frac{2}{3} & 2 \\ 1 & -\frac{1}{2} & -\frac{5}{2} \end{bmatrix}$

25. $\begin{bmatrix} 0.1 & 0.2 & 0.3 \\ -0.3 & 0.2 & 0.2 \\ 0.5 & 0.4 & 0.4 \end{bmatrix}$

26. $\begin{bmatrix} 0.6 & 0 & -0.3 \\ 0.7 & -1 & 0.2 \\ 1 & 0 & -0.9 \end{bmatrix}$

27. $\begin{bmatrix} -1 & 0 & 1 & 0 \\ 0 & 2 & 0 & -2 \\ 2 & 0 & -1 & 0 \\ 0 & -1 & 0 & 1 \end{bmatrix}$

28. $\begin{bmatrix} 1 & -2 & -1 & -2 \\ 3 & -5 & -2 & -3 \\ 2 & -5 & -2 & -5 \\ -1 & 4 & 4 & 11 \end{bmatrix}$

Finding the Inverse of a 2×2 Matrix **In Exercises 29–34, use the formula on page 545 to find the inverse of the 2×2 matrix (if it exists).**

29. $\begin{bmatrix} 4 & -3 \\ 8 & -6 \end{bmatrix}$

30. $\begin{bmatrix} 1 & -2 \\ -3 & 2 \end{bmatrix}$

31. $\begin{bmatrix} \frac{7}{2} & -\frac{3}{4} \\ \frac{1}{5} & \frac{4}{5} \end{bmatrix}$

32. $\begin{bmatrix} -\frac{1}{4} & -\frac{2}{3} \\ \frac{1}{3} & \frac{8}{9} \end{bmatrix}$

33. $\begin{bmatrix} 2 & 3 \\ -1 & 5 \end{bmatrix}$

34. $\begin{bmatrix} 7 & 12 \\ -8 & -5 \end{bmatrix}$

Finding a Matrix Entry **In Exercises 35 and 36, find the value of the constant k such that $B = A^{-1}$.**

35. $A = \begin{bmatrix} 1 & 2 \\ -2 & 0 \end{bmatrix}$, $B = \begin{bmatrix} k & -\frac{1}{2} \\ \frac{1}{2} & \frac{1}{4} \end{bmatrix}$

36. $A = \begin{bmatrix} -1 & 1 \\ 2 & 1 \end{bmatrix}$, $B = \begin{bmatrix} -\frac{1}{3} & \frac{1}{3} \\ k & \frac{1}{3} \end{bmatrix}$

Solving a Two-Variable System In Exercises 37 and 38, use the inverse matrix below to solve the system of linear equations.

$$\begin{bmatrix} 3 & -4 \\ 2 & -3 \end{bmatrix}^{-1} = \begin{bmatrix} 3 & -4 \\ 2 & -3 \end{bmatrix}$$

37. $\begin{cases} 3x - 4y = 3 \\ 2x - 3y = 8 \end{cases}$

38. $\begin{cases} 3x - 4y = -4 \\ 2x - 3y = -6 \end{cases}$

Solving a Three-Variable System In Exercises 39 and 40, use the inverse matrix below to solve the system of linear equations.

$$\begin{bmatrix} 1 & -1 & 1 \\ 0 & -2 & 1 \\ -2 & -3 & 0 \end{bmatrix}^{-1} = \begin{bmatrix} 3 & -3 & 1 \\ -2 & 2 & -1 \\ -4 & 5 & -2 \end{bmatrix}$$

39. $\begin{cases} x - y + z = 4 \\ -2y + z = 1 \\ -2x - 3y = -2 \end{cases}$

40. $\begin{cases} x - y + z = -1 \\ -2y + z = 7 \\ -2x - 3y = 8 \end{cases}$

Solving a System of Equations Using an Inverse In Exercises 41–48, use an inverse matrix to solve (if possible) the system of linear equations.

41. $\begin{cases} 3x + 4y = -2 \\ 5x + 3y = 4 \end{cases}$

42. $\begin{cases} 18x + 12y = 13 \\ 30x + 24y = 23 \end{cases}$

43. $\begin{cases} -0.4x + 0.8y = 1.6 \\ 2x - 4y = 5 \end{cases}$

44. $\begin{cases} 0.2x - 0.6y = 2.4 \\ -x + 1.4y = -8.8 \end{cases}$

45. $\begin{cases} -\frac{1}{4}x + \frac{3}{8}y = -2 \\ \frac{3}{2}x + \frac{3}{4}y = -12 \end{cases}$

46. $\begin{cases} \frac{5}{6}x - y = -10 \\ -\frac{5}{4}x + \frac{3}{2}y = -2 \end{cases}$

47. $\begin{cases} 4x - y + z = -5 \\ 2x + 2y + 3z = 10 \\ 5x - 2y + 6z = 1 \end{cases}$

48. $\begin{cases} 4x - 2y + 3z = -2 \\ 2x + 2y + 5z = 16 \\ 8x - 5y - 2z = 4 \end{cases}$

Solving a System of Equations Using Technology In Exercises 49–54, use the matrix capabilities of a graphing utility to solve the system of linear equations.

49. $\begin{cases} x - 7y = 4 \\ x + y = 1 \end{cases}$

50. $\begin{cases} 3x + 4y = -6 \\ 5x + 9y = 2 \end{cases}$

51. $\begin{cases} 3x + 3y - 2z = -6 \\ -x + 5y - 6z = 3 \\ 4x + y - z = -8 \end{cases}$

52. $\begin{cases} 2x + 3y + 5z = 4 \\ 3x + 5y - 9z = 7 \\ 5x + 9y + 17z = 13 \end{cases}$

53. $\begin{cases} 7x - 3y + 2w = 41 \\ -2x + y - w = -13 \\ 4x + z - 2w = 12 \\ -x + y - w = -8 \end{cases}$

54. $\begin{cases} 2x + 5y + w = 11 \\ x + 4y + 2z - 2w = -7 \\ 2x - 2y + 5z + w = 3 \\ x - 3w = -1 \end{cases}$

Investment Portfolio In Exercises 55–58, consider a person who invests in AAA-rated bonds, A-rated bonds, and B-rated bonds. The average yields are 4.5% on AAA bonds, 5% on A bonds, and 7% on B bonds. The person invests twice as much in B bonds as in A bonds. Let x, y, and z represent the amounts (in dollars) invested in AAA, A, and B bonds, respectively.

$$\begin{cases} x + y + z = \text{(total investment)} \\ 0.045x + 0.05y + 0.07z = \text{(annual return)} \\ 2y - z = 0 \end{cases}$$

Use the inverse of the coefficient matrix of this system to find the amount invested in each type of bond.

	Total Investment	*Annual Return*
55.	$10,000	$560
56.	$20,000	$955
57.	$300,000	$18,450
58.	$500,000	$28,000

Electrical Engineering In Exercises 59 and 60, consider the circuit in the figure. The currents I_1, I_2, and I_3, in amperes, are given by the solution of the system of linear equations

$$\begin{cases} 2I_1 + 4I_3 = E_1 \\ I_2 + 4I_3 = E_2 \\ I_1 + I_2 - I_3 = 0 \end{cases}$$

where E_1 and E_2 are voltages. Use the inverse of the coefficient matrix of this system to find the unknown currents for the given voltages.

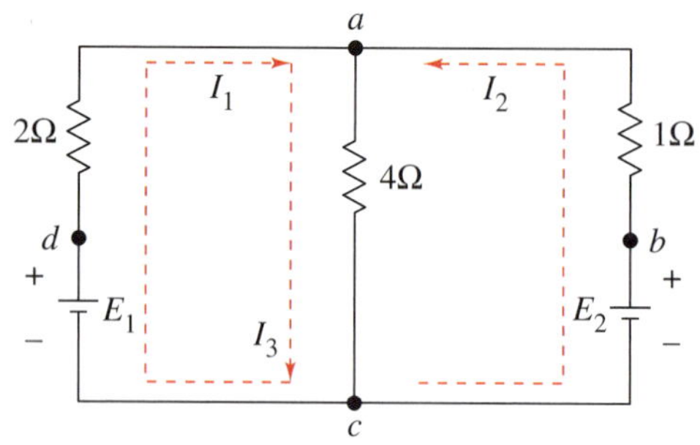

59. $E_1 = 15$ volts, $E_2 = 17$ volts

60. $E_1 = 10$ volts, $E_2 = 10$ volts

Horticulture In Exercises 61 and 62, consider a company that specializes in potting soil. Each bag of potting soil for seedlings requires 2 units of sand, 1 unit of loam, and 1 unit of peat moss. Each bag of potting soil for general potting requires 1 unit of sand, 2 units of loam, and 1 unit of peat moss. Each bag of potting soil for hardwood plants requires 2 units of sand, 2 units of loam, and 2 units of peat moss. Find the numbers of bags of the three types of potting soil that the company can produce with the given amounts of raw materials.

61. 500 units of sand
500 units of loam
400 units of peat moss

62. 500 units of sand
750 units of loam
450 units of peat moss

63. Floral Design A florist is creating 10 centerpieces for the tables at a wedding reception. Roses cost \$2.50 each, lilies cost \$4 each, and irises cost \$2 each. The customer has a budget of \$300 for the centerpieces and wants each centerpiece to contain 12 flowers, with twice as many roses as the number of irises and lilies combined.

(a) Write a linear system that represents the situation.

(b) Write a matrix equation that corresponds to your system.

(c) Solve your linear system using an inverse matrix. Find the number of flowers of each type that the florist can use to create the 10 centerpieces.

64. ***Why you should learn it*** *(p. 541)* The table shows the numbers of international travelers y (in thousands) to the United States from Asia from 2014 through 2016. *(Source:* U.S. Department of Commerce)

Year	Travelers, y (in thousands)
2014	9,697
2015	10,816
2016	11,347

(a) Use a system of equations to find the equation of the parabola $y = at^2 + bt + c$ that passes through the points. Let t represent the year, with $t = 4$ corresponding to 2014. Solve the system using matrices.

(b) Use the graphing utility to graph the parabola with the data points.

(c) Use the result of part (a) to estimate the numbers of international travelers to the United States from Asia in 2017 and 2020.

(d) Are your estimates from part (c) reasonable? Explain.

Focusing on Concepts

True or False? In Exercises 65 and 66, determine whether the statement is true or false. Justify your answer.

65. Multiplication of an invertible matrix and its inverse is commutative.

66. When the product of two square matrices is the identity matrix, the matrices are inverses of one another.

67. Error Analysis Describe the error.

$$A = \begin{bmatrix} 5 & 6 \\ 2 & 3 \end{bmatrix}, \text{ so } A^{-1} = 3\begin{bmatrix} 3 & -6 \\ -2 & 5 \end{bmatrix}. \quad ✗$$

68. Writing Explain how to find the inverse of a 2×2 matrix (assuming that the inverse exists).

69. Writing Explain in your own words how to write a system of three linear equations in three variables as a matrix equation $AX = B$, as well as how to solve the system using an inverse matrix.

70. The Inverse of a 2×2 Matrix If A is a 2×2 matrix given by $A = \begin{bmatrix} a & b \\ c & d \end{bmatrix}$, then A is invertible if and only if $ad - bc \neq 0$. If $ad - bc \neq 0$, verify that the inverse is $A^{-1} = \dfrac{1}{ad - bc}\begin{bmatrix} d & -b \\ -c & a \end{bmatrix}$.

71. Exploration Consider matrices of the form

$$A = \begin{bmatrix} a_{11} & 0 & 0 & 0 & \dots & 0 \\ 0 & a_{22} & 0 & 0 & \dots & 0 \\ 0 & 0 & a_{33} & 0 & \dots & 0 \\ \vdots & \vdots & \vdots & \vdots & \dots & \vdots \\ 0 & 0 & 0 & 0 & \dots & a_{nn} \end{bmatrix}.$$

Write a 2×2 matrix and a 3×3 matrix in the form of A. Find the inverse of each. Then make a conjecture about the inverse of a matrix in the form of A.

72. **HOW DO YOU SEE IT?** Let A be a 2×2 matrix given by

$$A = \begin{bmatrix} x & 0 \\ 0 & y \end{bmatrix}.$$

Use the determinant of A to determine the conditions under which A^{-1} exists.

Cumulative Mixed Review

Solving an Equation In Exercises 73–76, solve the equation algebraically. Approximate the result to three decimal places.

73. $e^{2x} + 2e^x - 15 = 0$

74. $e^{2x} - 10e^x + 24 = 0$

75. $7 \ln 3x = 12$

76. $\log_3 x = -6$

7.7 The Determinant of a Square Matrix

The Determinant of a 2 × 2 Matrix

Every *square* matrix can be associated with a real number called its **determinant.** Determinants have many uses, and several will be discussed in this and the next section. Historically, the use of determinants arose from special number patterns that occur when systems of linear equations are solved. For instance, the system

$$\begin{cases} a_1x + b_1y = c_1 \\ a_2x + b_2y = c_2 \end{cases}$$

has a solution

$$x = \frac{c_1b_2 - c_2b_1}{a_1b_2 - a_2b_1}$$

and

$$y = \frac{a_1c_2 - a_2c_1}{a_1b_2 - a_2b_1}$$

provided that $a_1b_2 - a_2b_1 \neq 0$. Note that each fraction has the same denominator. This denominator is called the *determinant* of the coefficient matrix of the system.

Coefficient Matrix *Determinant*

$$A = \begin{bmatrix} a_1 & b_1 \\ a_2 & b_2 \end{bmatrix} \qquad \det(A) = a_1b_2 - a_2b_1$$

The determinant of matrix A can also be denoted by vertical bars on both sides of the matrix, as indicated in the following definition.

Definition of the Determinant of a 2 × 2 Matrix

The **determinant** of the matrix

$$A = \begin{bmatrix} a_1 & b_1 \\ a_2 & b_2 \end{bmatrix}$$

is given by

$$\det(A) = |A|$$

$$= \begin{vmatrix} a_1 & b_1 \\ a_2 & b_2 \end{vmatrix}$$

$$= a_1b_2 - a_2b_1.$$

In this text, $\det(A)$ and $|A|$ are used interchangeably to represent the determinant of A. Although vertical bars are also used to denote the absolute value of a real number, the context will show which use is intended.

A convenient method for remembering the formula for the determinant of a 2 × 2 matrix is shown in the following diagram.

$$\det(A) = \begin{vmatrix} a_1 & b_1 \\ a_2 & b_2 \end{vmatrix} = a_1b_2 - a_2b_1$$

Note that the determinant is the difference of the products of the two diagonals of the matrix.

What you should learn

- Find the determinants of 2 × 2 matrices.
- Find minors and cofactors of square matrices.
- Find the determinants of square matrices.

Why you should learn it

Determinants and Cramer's Rule can be used to find the least squares regression parabola that models retail sales of family clothing stores, as shown in Exercise 31 of Section 7.8 on page 566.

Neustockimages/iStock/Getty Images

EXAMPLE 1 The Determinant of a 2 × 2 Matrix

Find the determinant of each matrix.

a. $A = \begin{bmatrix} 2 & -3 \\ 1 & 2 \end{bmatrix}$

b. $B = \begin{bmatrix} 2 & 1 \\ 4 & 2 \end{bmatrix}$

c. $C = \begin{bmatrix} 0 & \frac{3}{2} \\ 2 & 4 \end{bmatrix}$

Solution

a. $\det(A) = \begin{vmatrix} 2 & -3 \\ 1 & 2 \end{vmatrix} = 2(2) - 1(-3) = 4 + 3 = 7$

b. $\det(B) = \begin{vmatrix} 2 & 1 \\ 4 & 2 \end{vmatrix} = 2(2) - 4(1) = 4 - 4 = 0$

c. $\det(C) = \begin{vmatrix} 0 & \frac{3}{2} \\ 2 & 4 \end{vmatrix} = 0(4) - 2\left(\frac{3}{2}\right) = 0 - 3 = -3$

✓ ***Checkpoint*** ▶ *Audio-video solution in English & Spanish at LarsonPrecalculus.com*

Find the determinant of each matrix.

a. $A = \begin{bmatrix} 1 & 2 \\ 3 & -1 \end{bmatrix}$ **b.** $B = \begin{bmatrix} 5 & 0 \\ -4 & 2 \end{bmatrix}$ **c.** $C = \begin{bmatrix} 3 & 6 \\ 2 & 4 \end{bmatrix}$

Notice in Example 1 that the determinant of a matrix can be positive, zero, or negative.

The determinant of a matrix of dimension 1 × 1 is defined simply as the entry of the matrix. For instance, if $A = [-2]$, then $\det(A) = -2$.

EXAMPLE 2 Using a Graphing Utility

Use the matrix capabilities of a graphing utility to find the determinant of the matrix.

$$A = \begin{bmatrix} 1.4 & 0.7 \\ -0.3 & -2.5 \end{bmatrix}$$

Solution

Use the *matrix editor* to enter the matrix as $[A]$, as shown in Figure 7.20. Then choose the *determinant* feature. From Figure 7.21, $\det(A) = -3.29$.

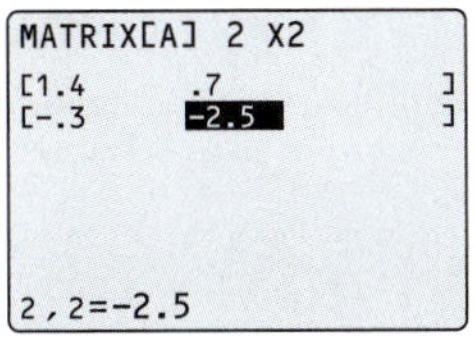

Figure 7.20

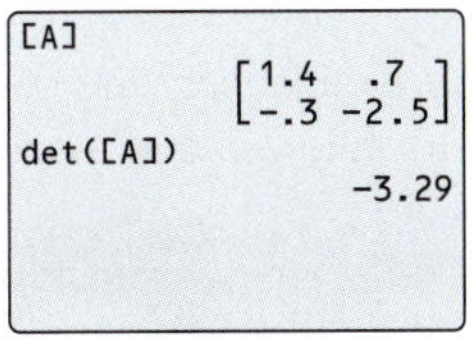

Figure 7.21

✓ ***Checkpoint*** ▶ *Audio-video solution in English & Spanish at LarsonPrecalculus.com*

Use the matrix capabilities of a graphing utility to find the determinant of the matrix.

$$A = \begin{bmatrix} 0.1 & -0.7 \\ 0.8 & 0.8 \end{bmatrix}$$

Explore the Concept

Try using a graphing utility to find the determinant of

$$A = \begin{bmatrix} 3 & -1 & 1 \\ 0 & 2 & 1 \end{bmatrix}.$$

What message appears on the screen? Why does the graphing utility display this message?

photo credits to come

Minors and Cofactors

To define the determinant of a square matrix of dimension 3×3 or greater, it is helpful to introduce the concepts of **minors** and **cofactors.**

Minors and Cofactors of a Square Matrix

If A is a square matrix, then the **minor** M_{ij} of the entry a_{ij} is the determinant of the matrix obtained by deleting the ith row and jth column of A. The **cofactor** C_{ij} of the entry a_{ij} is given by $C_{ij} = (-1)^{i+j}M_{ij}$.

Sign Patterns for Cofactors

$$\begin{bmatrix} + & - & + \\ - & + & - \\ + & - & + \end{bmatrix}$$

3×3 matrix

$$\begin{bmatrix} + & - & + & - \\ - & + & - & + \\ + & - & + & - \\ - & + & - & + \end{bmatrix}$$

4×4 matrix

$$\begin{bmatrix} + & - & + & - & + & \cdots \\ - & + & - & + & - & \cdots \\ + & - & + & - & + & \cdots \\ - & + & - & + & - & \cdots \\ + & - & + & - & + & \cdots \\ \vdots & \vdots & \vdots & \vdots & \vdots & \end{bmatrix}$$

$n \times n$ matrix

In the sign patterns for cofactors at the right, notice that *odd* positions (where $i + j$ is odd) have negative signs and *even* positions (where $i + j$ is even) have positive signs.

EXAMPLE 3 Finding the Minors and Cofactors of a Matrix

Find all the minors and cofactors of

$$A = \begin{bmatrix} 0 & 2 & 1 \\ 3 & -1 & 2 \\ 4 & 0 & 1 \end{bmatrix}.$$

Solution

To find the minor M_{11}, delete the first row and first column of A and evaluate the determinant of the resulting matrix.

$$\begin{bmatrix} 0 & 2 & 1 \\ 3 & -1 & 2 \\ 4 & 0 & 1 \end{bmatrix}, \quad M_{11} = \begin{vmatrix} -1 & 2 \\ 0 & 1 \end{vmatrix} = -1(1) - 0(2) = -1$$

Similarly, to find the minor M_{12}, delete the first row and second column.

$$\begin{bmatrix} 0 & 2 & 1 \\ 3 & -1 & 2 \\ 4 & 0 & 1 \end{bmatrix}, \quad M_{12} = \begin{vmatrix} 3 & 2 \\ 4 & 1 \end{vmatrix} = 3(1) - 4(2) = -5$$

Continuing this pattern, you obtain all the minors.

$M_{11} = -1 \quad M_{12} = -5 \quad M_{13} = 4$

$M_{21} = 2 \quad M_{22} = -4 \quad M_{23} = -8$

$M_{31} = 5 \quad M_{32} = -3 \quad M_{33} = -6$

Now, to find the cofactors, combine these minors with the checkerboard pattern of signs for a 3×3 matrix shown at the upper right.

$C_{11} = -1 \quad C_{12} = 5 \quad C_{13} = 4$

$C_{21} = -2 \quad C_{22} = -4 \quad C_{23} = 8$

$C_{31} = 5 \quad C_{32} = 3 \quad C_{33} = -6$

✓ Checkpoint ▶ *Audio-video solution in English & Spanish at LarsonPrecalculus.com*

Find all the minors and cofactors of

$$A = \begin{bmatrix} 1 & 2 & 3 \\ 0 & -1 & 5 \\ 2 & 1 & 4 \end{bmatrix}.$$

■

The Determinant of a Square Matrix

The following definition is called *inductive* because it uses determinants of matrices of dimension $(n - 1) \times (n - 1)$ to define determinants of matrices of dimension $n \times n$.

Determinant of a Square Matrix

If A is a square matrix (of dimension 2×2 or greater), then the determinant of A is the sum of the entries in any row (or column) of A multiplied by their respective cofactors. For instance, expanding along the first row yields

$$|A| = a_{11}C_{11} + a_{12}C_{12} + \cdot \cdot \cdot + a_{1n}C_{1n}.$$

Applying this definition to find a determinant is called **expanding by cofactors.**

Try checking that for a 2×2 matrix

$$A = \begin{bmatrix} a_1 & b_1 \\ a_2 & b_2 \end{bmatrix}$$

this definition of the determinant yields $|A| = a_1b_2 - a_2b_1$, as previously defined.

EXAMPLE 4 The Determinant of a 3 × 3 Matrix

See LarsonPrecalculus.com for an interactive version of this type of example.

Find the determinant of $A = \begin{bmatrix} 0 & 2 & 1 \\ 3 & -1 & 2 \\ 4 & 0 & 1 \end{bmatrix}$.

Solution

Note that this is the same matrix that was in Example 3. There you found the cofactors of the entries in the first row to be

$$C_{11} = -1, \quad C_{12} = 5, \quad \text{and} \quad C_{13} = 4.$$

So, by the definition of the determinant of a square matrix, you have

$$\begin{aligned} |A| &= a_{11}C_{11} + a_{12}C_{12} + a_{13}C_{13} && \text{First-row expansion} \\ &= 0(-1) + 2(5) + 1(4) \\ &= 14. \end{aligned}$$

Checkpoint *Audio-video solution in English & Spanish at LarsonPrecalculus.com*

Find the determinant of $A = \begin{bmatrix} 3 & 4 & -2 \\ 3 & 5 & 0 \\ -1 & 4 & 1 \end{bmatrix}$.

In Example 4, the determinant was found by expanding by the cofactors in the first row. You could have used any row or column. For instance, you could have expanded along the second row to obtain

$$\begin{aligned} |A| &= a_{21}C_{21} + a_{22}C_{22} + a_{23}C_{23} && \text{Second-row expansion} \\ &= 3(-2) + (-1)(-4) + 2(8) \\ &= 14. \end{aligned}$$

When expanding by cofactors, you do not need to find cofactors of zero entries, because when $a_{ij} = 0$, you have $a_{ij}C_{ij} = (0)C_{ij} = 0$. So, the row (or column) containing the most zeros is usually the best choice for expansion by cofactors.

7.7 Exercises

See *CalcChat.com* for tutorial help and worked-out solutions to odd-numbered exercises. For instructions on how to use a graphing utility, see Appendix A.

Vocabulary and Concept Check

In Exercises 1 and 2, fill in the blank.

1. Both $\det(A)$ and $|A|$ represent the _______ of the matrix A.

2. The determinant of the matrix obtained by deleting the ith row and jth column of a square matrix A is called the _______ of the entry a_{ij}.

3. For a square matrix B, the minor M_{23} is 5. What is the cofactor C_{23} of matrix B?

4. To find the determinant of a matrix using expanding by cofactors, do you need to find all the cofactors?

Procedures and Problem Solving

The Determinant of a Matrix **In Exercises 5–12, find the determinant of the matrix.**

5. $[6]$

6. $[-12]$

7. $\begin{bmatrix} 8 & 4 \\ 2 & 3 \end{bmatrix}$

8. $\begin{bmatrix} -5 & 2 \\ 6 & 3 \end{bmatrix}$

9. $\begin{bmatrix} -7 & 6 \\ \frac{1}{2} & 3 \end{bmatrix}$

10. $\begin{bmatrix} 4 & -3 \\ 0 & 0 \end{bmatrix}$

11. $\begin{bmatrix} \sqrt{3} & 3 \\ 4 & \sqrt{3} \end{bmatrix}$

12. $\begin{bmatrix} 9 & \sqrt{5} \\ \sqrt{5} & 4 \end{bmatrix}$

Using a Graphing Utility **In Exercises 13–16, use a graphing utility to find the determinant of the matrix.**

13. $\begin{bmatrix} 1.9 & -0.3 \\ 5.6 & 3.2 \end{bmatrix}$

14. $\begin{bmatrix} 7.2 & 0.8 \\ 3.6 & -0.4 \end{bmatrix}$

15. $\begin{bmatrix} 1.3 & 0.2 & 3.2 \\ 0.2 & 6.2 & 0.2 \\ -0.4 & 4.4 & 0.3 \end{bmatrix}$

16. $\begin{bmatrix} 5.1 & 0.2 & 7.3 \\ -6.3 & 0.2 & 0.2 \\ 0.5 & 3.4 & 0.4 \end{bmatrix}$

Finding the Minors and Cofactors of a Matrix **In Exercises 17–20, find all the (a) minors and (b) cofactors of the matrix.**

17. $\begin{bmatrix} 4 & 5 \\ 3 & -6 \end{bmatrix}$

18. $\begin{bmatrix} 11 & 6 \\ -3 & 2 \end{bmatrix}$

19. $\begin{bmatrix} -4 & 6 & 3 \\ 7 & -2 & 8 \\ 1 & 0 & -5 \end{bmatrix}$

20. $\begin{bmatrix} -2 & 9 & 4 \\ 7 & -6 & 0 \\ 6 & 7 & -6 \end{bmatrix}$

Finding a Determinant **In Exercises 21–24, find the determinant of the matrix. Expand by cofactors using the indicated row or column.**

21. $\begin{bmatrix} 5 & 0 & -3 \\ 0 & 12 & 4 \\ 1 & 6 & 3 \end{bmatrix}$

(a) Row 2
(b) Column 2

22. $\begin{bmatrix} -3 & 4 & 2 \\ 6 & 3 & 1 \\ 4 & -7 & -8 \end{bmatrix}$

(a) Row 2
(b) Column 3

23. $\begin{bmatrix} 4 & 13 & 6 & -8 \\ 6 & 0 & -3 & 5 \\ -1 & 0 & 7 & 4 \\ 8 & 6 & 0 & 2 \end{bmatrix}$

(a) Row 1
(b) Column 2

24. $\begin{bmatrix} 10 & 8 & 3 & -7 \\ 4 & 0 & 5 & -6 \\ 0 & 3 & 2 & 7 \\ 1 & 0 & -3 & 2 \end{bmatrix}$

(a) Row 3
(b) Column 1

Finding a Determinant **In Exercises 25–34, find the determinant of the matrix. Expand by cofactors using the row or column that appears to make the computations easiest.**

25. $\begin{bmatrix} -1 & 8 & -3 \\ 0 & 3 & -6 \\ 0 & 0 & 3 \end{bmatrix}$

26. $\begin{bmatrix} -3 & 1 & 0 \\ 7 & 11 & 5 \\ 1 & 2 & 2 \end{bmatrix}$

27. $\begin{bmatrix} 6 & 3 & -7 \\ 0 & 0 & 0 \\ 4 & -6 & 3 \end{bmatrix}$

28. $\begin{bmatrix} 1 & 1 & 2 \\ 3 & -5 & 9 \\ 0 & 0 & 0 \end{bmatrix}$

29. $\begin{bmatrix} -1 & 2 & -5 \\ 0 & 3 & -4 \\ 0 & 0 & 3 \end{bmatrix}$

30. $\begin{bmatrix} 1 & 0 & 0 \\ -1 & -1 & 0 \\ 4 & 1 & 5 \end{bmatrix}$

31. $\begin{bmatrix} 2 & 6 & 6 & 2 \\ 2 & 7 & 3 & 6 \\ 1 & 5 & 0 & 1 \\ 3 & 7 & 0 & 7 \end{bmatrix}$

32. $\begin{bmatrix} 3 & 6 & -5 & 4 \\ -2 & 2 & 6 & 0 \\ 1 & 1 & 2 & 0 \\ 0 & 3 & -1 & -1 \end{bmatrix}$

33. $\begin{bmatrix} 3 & 2 & 4 & -1 & 5 \\ -2 & 0 & 1 & 3 & 2 \\ 1 & 0 & 0 & 4 & 0 \\ 6 & 0 & 2 & -1 & 0 \\ 3 & 0 & 5 & 1 & 0 \end{bmatrix}$

34. $\begin{bmatrix} 5 & 2 & 0 & 0 & -2 \\ 0 & 1 & 4 & 3 & 2 \\ 0 & 0 & 2 & 6 & 3 \\ 0 & 0 & 3 & 4 & 1 \\ 0 & 0 & 0 & 0 & 2 \end{bmatrix}$

Using a Graphing Utility In Exercises 35–40, use a graphing utility to find the determinant.

35. $\begin{vmatrix} 1 & -1 & 8 & 4 \\ 2 & 6 & 0 & -4 \\ 2 & 0 & 2 & 6 \\ 0 & 2 & 8 & 0 \end{vmatrix}$ **36.** $\begin{vmatrix} 0 & -3 & 8 & 2 \\ 8 & 1 & -1 & 6 \\ -4 & 6 & 0 & 9 \\ -7 & 0 & 0 & 14 \end{vmatrix}$

37. $\begin{vmatrix} 8 & 5 & 1 & -2 & 0 \\ -1 & 0 & 7 & 1 & 6 \\ 0 & 8 & 6 & 5 & -3 \\ 1 & 2 & 5 & -8 & 4 \\ 2 & 6 & -2 & 0 & 6 \end{vmatrix}$

38. $\begin{vmatrix} 3 & -2 & 4 & 3 & 1 \\ -1 & 0 & 2 & 1 & 0 \\ 5 & -1 & 0 & 3 & 2 \\ 4 & 7 & -8 & 0 & 0 \\ 1 & 2 & 3 & 0 & 2 \end{vmatrix}$

39. $\begin{vmatrix} 3 & -2 & 4 & 3 & 1 & 3 \\ -1 & 0 & 2 & 1 & 0 & 0 \\ 5 & -1 & 0 & 3 & 2 & 1 \\ 4 & 7 & -8 & 0 & 0 & -2 \\ 1 & 2 & 3 & 0 & 2 & 4 \\ 3 & -5 & 1 & 2 & 3 & 1 \end{vmatrix}$

40. $\begin{vmatrix} -2 & 0 & 1 & 4 & 3 & -2 \\ -3 & 3 & 0 & -2 & 1 & -1 \\ 4 & 5 & -1 & 0 & 7 & 3 \\ 2 & 4 & 3 & 2 & 0 & 1 \\ 1 & 3 & 4 & 2 & -4 & 0 \\ -5 & 2 & -1 & 3 & 2 & -3 \end{vmatrix}$

The Determinant of a Matrix Product In Exercises 41 and 42, find (a) $|A|$, (b) $|B|$, (c) AB, and (d) $|AB|$. What do you notice about $|AB|$?

41. $A = \begin{bmatrix} -3 & 1 \\ 5 & -2 \end{bmatrix}, \quad B = \begin{bmatrix} -1 & 2 \\ 0 & 1 \end{bmatrix}$

42. $A = \begin{bmatrix} 2 & 0 & 1 \\ 1 & -4 & 2 \\ 3 & 1 & 0 \end{bmatrix}, \quad B = \begin{bmatrix} 2 & -1 & 4 \\ 0 & 1 & 0 \\ 3 & -2 & 1 \end{bmatrix}$

Using a Graphing Utility In Exercises 43 and 44, use a graphing utility to find (a) $|A|$, (b) $|B|$, (c) AB, and (d) $|AB|$. What do you notice about $|AB|$?

43. $A = \begin{bmatrix} 6 & 4 & 0 & 1 \\ 2 & -3 & -2 & -4 \\ 0 & 1 & 5 & 0 \\ -1 & 0 & -1 & 1 \end{bmatrix}, \quad B = \begin{bmatrix} 0 & -5 & 0 & -2 \\ -2 & 4 & -1 & -4 \\ 3 & 0 & 1 & 0 \\ 1 & -2 & 3 & 0 \end{bmatrix}$

44. $A = \begin{bmatrix} -1 & 5 & 2 & 0 \\ 0 & 0 & 1 & 1 \\ 3 & -3 & -1 & 0 \\ 4 & 2 & 4 & -1 \end{bmatrix}, \quad B = \begin{bmatrix} 1 & 5 & 0 & 0 \\ 10 & -1 & 2 & 4 \\ 2 & 0 & 0 & 1 \\ -3 & 2 & 5 & 0 \end{bmatrix}$

Verifying an Equation In Exercises 45–50, verify the equation.

45. $\begin{vmatrix} w & x \\ y & z \end{vmatrix} = -\begin{vmatrix} y & z \\ w & x \end{vmatrix}$

46. $\begin{vmatrix} w & cx \\ y & cz \end{vmatrix} = c\begin{vmatrix} w & x \\ y & z \end{vmatrix}$

47. $\begin{vmatrix} w & x \\ y & z \end{vmatrix} = \begin{vmatrix} w & x + cw \\ y & z + cy \end{vmatrix}$

48. $\begin{vmatrix} w & x \\ cw & cx \end{vmatrix} = 0$

49. $\begin{vmatrix} 1 & x & x^2 \\ 1 & y & y^2 \\ 1 & z & z^2 \end{vmatrix} = (y - x)(z - x)(z - y)$

50. $\begin{vmatrix} a + b & a & a \\ a & a + b & a \\ a & a & a + b \end{vmatrix} = b^2(3a + b)$

Solving an Equation In Exercises 51–62, solve for x.

51. $\begin{vmatrix} x & 2 \\ 1 & x \end{vmatrix} = 2$ **52.** $\begin{vmatrix} x & 4 \\ -1 & x \end{vmatrix} = 16$

53. $\begin{vmatrix} 2x & -3 \\ -2 & 2x \end{vmatrix} = 19$ **54.** $\begin{vmatrix} x & 2 \\ 4 & 9x \end{vmatrix} = 8$

55. $\begin{vmatrix} x & 2 \\ 2 & x - 2 \end{vmatrix} = -1$ **56.** $\begin{vmatrix} x + 1 & 2 \\ -1 & x \end{vmatrix} = 4$

57. $\begin{vmatrix} x + 3 & 2 \\ 1 & x + 2 \end{vmatrix} = 0$ **58.** $\begin{vmatrix} x - 1 & 2 \\ 3 & x - 2 \end{vmatrix} = 0$

59. $\begin{vmatrix} 2x & 1 \\ -1 & x - 1 \end{vmatrix} = x$ **60.** $\begin{vmatrix} x - 1 & x \\ x + 1 & 2 \end{vmatrix} = -8$

61. $\begin{vmatrix} 1 & 2 & x \\ -1 & 3 & 2 \\ 3 & -2 & 1 \end{vmatrix} = 0$ **62.** $\begin{vmatrix} 1 & x & -2 \\ 1 & 3 & 3 \\ 0 & 2 & -2 \end{vmatrix} = 0$

Entries Involving Expressions In Exercises 63–70, find the determinant in which the entries are functions. Determinants of this type occur in calculus.

63. $\begin{vmatrix} 4u & -1 \\ -1 & 2v \end{vmatrix}$ **64.** $\begin{vmatrix} 3x^2 & -3y^2 \\ 1 & 1 \end{vmatrix}$

65. $\begin{vmatrix} e^{2x} & e^{3x} \\ 2e^{2x} & 3e^{3x} \end{vmatrix}$ **66.** $\begin{vmatrix} e^{-x} & xe^{-x} \\ -e^{-x} & (1 - x)e^{-x} \end{vmatrix}$

67. $\begin{vmatrix} x & \ln x \\ 1 & 1/x \end{vmatrix}$ **68.** $\begin{vmatrix} x & x \ln x \\ 1 & 1 + \ln x \end{vmatrix}$

69. $\begin{vmatrix} \cos \theta & -r \sin \theta & 0 \\ \sin \theta & r \cos \theta & 0 \\ 0 & 0 & 1 \end{vmatrix}$

70. $\begin{vmatrix} 1 - v & -u & 0 \\ v(1 - w) & u(1 - w) & -uv \\ vw & uw & uv \end{vmatrix}$

Focusing on Concepts

True or False? **In Exercises 71 and 72, determine whether the statement is true or false. Justify your answer.**

71. If a square matrix has an entire row of zeros, then the determinant of the matrix is zero.

72. If two columns of a square matrix are the same, then the determinant of the matrix is zero.

73. Exploration Find two 3×3 matrices A and B to demonstrate that $|A + B| \neq |A| + |B|$.

74. Think About It Let A be a 3×3 matrix such that $|A| = 5$. Can you use this information to find $|2A|$? Explain.

Exploration **In Exercises 75–78, (a) find the determinant of A, (b) find A^{-1}, (c) find $\det(A^{-1})$, and (d) compare your results from parts (a) and (c). Make a conjecture based on your results.**

75. $A = \begin{bmatrix} 1 & 3 \\ -2 & 3 \end{bmatrix}$

76. $A = \begin{bmatrix} 2 & -1 \\ -5 & -1 \end{bmatrix}$

77. $A = \begin{bmatrix} 1 & -3 & -2 \\ -1 & 3 & 1 \\ 0 & 2 & -2 \end{bmatrix}$

78. $A = \begin{bmatrix} -1 & 3 & 2 \\ 1 & 3 & -1 \\ 1 & 1 & -2 \end{bmatrix}$

Properties of Determinants **In Exercises 79–81, a property of determinants is given (A and B are square matrices). Explain how each equation uses the property. Then use a graphing utility to verify that each equation is true.**

79. If B is obtained from A by interchanging two rows of A or by interchanging two columns of A, then $|B| = -|A|$.

(a) $\begin{vmatrix} 1 & 3 & 4 \\ -7 & 2 & -5 \\ 6 & 1 & 2 \end{vmatrix} = -\begin{vmatrix} 1 & 4 & 3 \\ -7 & -5 & 2 \\ 6 & 2 & 1 \end{vmatrix}$

(b) $\begin{vmatrix} 1 & 3 & 4 \\ -2 & 2 & 0 \\ 1 & 6 & 2 \end{vmatrix} = -\begin{vmatrix} 1 & 6 & 2 \\ -2 & 2 & 0 \\ 1 & 3 & 4 \end{vmatrix}$

80. If B is obtained from A by adding a multiple of a row of A to another row of A or by adding a multiple of a column of A to another column of A, then $|B| = |A|$.

(a) $\begin{vmatrix} 1 & -3 \\ 5 & 2 \end{vmatrix} = \begin{vmatrix} 1 & -3 \\ 0 & 17 \end{vmatrix}$

(b) $\begin{vmatrix} 5 & 4 & 2 \\ 2 & -3 & 4 \\ 7 & 6 & 3 \end{vmatrix} = \begin{vmatrix} 1 & 10 & -6 \\ 2 & -3 & 4 \\ 7 & 6 & 3 \end{vmatrix}$

81. If B is obtained from A by multiplying a row of A by a nonzero constant c or by multiplying a column of A by a nonzero constant c, then $|B| = c|A|$.

(a) $\begin{vmatrix} 1 & 5 \\ 6 & 9 \end{vmatrix} = 3\begin{vmatrix} 1 & 5 \\ 2 & 3 \end{vmatrix}$ (b) $\begin{vmatrix} 2 & 8 \\ 6 & 8 \end{vmatrix} = 8\begin{vmatrix} 1 & 2 \\ 3 & 2 \end{vmatrix}$

82. HOW DO YOU SEE IT? Explain why the determinant of the matrix is equal to zero.

$$\begin{bmatrix} 2 & -4 & 5 \\ 1 & -2 & 3 \\ 0 & 0 & 0 \end{bmatrix}$$

83. Exploration A **triangular matrix** is a square matrix with all zero entries either above or below its main diagonal. Such a matrix is **upper triangular** when it has all zeros below the main diagonal and **lower triangular** when it has all zeros above the main diagonal. Find the determinant of each triangular matrix. Make a conjecture based on your results.

(a) $\begin{bmatrix} 3 & -2 \\ 0 & 5 \end{bmatrix}$ (b) $\begin{bmatrix} 3 & -7 & 1 \\ 0 & -5 & -9 \\ 0 & 0 & 5 \end{bmatrix}$ (c) $\begin{bmatrix} 4 & 0 & 0 & 0 \\ 3 & -3 & 0 & 0 \\ 3 & 6 & 5 & 0 \\ 2 & -2 & 1 & 2 \end{bmatrix}$

84. Exploration A **diagonal matrix** is a square matrix with all zero entries above and below its main diagonal. (A diagonal matrix is both upper and lower triangular.) Find the determinant of each diagonal matrix. Make a conjecture based on your results.

(a) $\begin{bmatrix} 2 & 0 \\ 0 & 3 \end{bmatrix}$ (b) $\begin{bmatrix} -1 & 0 & 0 \\ 0 & 8 & 0 \\ 0 & 0 & 4 \end{bmatrix}$ (c) $\begin{bmatrix} 2 & 0 & 0 & 0 \\ 0 & -2 & 0 & 0 \\ 0 & 0 & 1 & 0 \\ 0 & 0 & 0 & 6 \end{bmatrix}$

85. Error Analysis Describe the error.

$$\begin{vmatrix} 1 & 1 & 4 \\ 3 & 2 & 0 \\ 2 & 1 & 3 \end{vmatrix} = 3(1)\begin{vmatrix} 1 & 4 \\ 1 & 3 \end{vmatrix} + 2(-1)\begin{vmatrix} 1 & 4 \\ 2 & 3 \end{vmatrix} + 0(1)\begin{vmatrix} 1 & 1 \\ 2 & 1 \end{vmatrix}$$
$$= 3(-1) - 2(-5) + 0$$
$$= 7$$

86. Exploration Consider square matrices in which the entries are consecutive integers. An example of such a matrix is shown below. Use a graphing utility to find the determinants of four matrices of this type. Make a conjecture based on the results. Then verify your conjecture.

$$\begin{bmatrix} 1 & 2 & 3 \\ 4 & 5 & 6 \\ 7 & 8 & 9 \end{bmatrix}$$

Cumulative Mixed Review

Factoring a Quadratic Expression **In Exercises 87 and 88, factor the expression.**

87. $4y^2 - 12y + 9$

88. $4y^2 - 28y + 49$

7.8 Applications of Matrices and Determinants

Area of a Triangle

In this section, you will study some additional applications of matrices and determinants. The first involves a formula for finding the area of a triangle whose vertices are given by three points on a rectangular coordinate system.

Area of a Triangle

The area of a triangle with vertices (x_1, y_1), (x_2, y_2), and (x_3, y_3) is

$$\text{Area} = \pm\frac{1}{2}\begin{vmatrix} x_1 & y_1 & 1 \\ x_2 & y_2 & 1 \\ x_3 & y_3 & 1 \end{vmatrix}$$

where the symbol $(\pm)$ indicates that the appropriate sign should be chosen to yield a positive area.

What you should learn

- Use determinants to find areas of triangles.
- Use determinants to determine whether points are collinear.
- Use matrices to perform transformations in the plane and find areas of parallelograms.
- Use Cramer's Rule to solve systems of linear equations.
- Use matrices to encode and decode messages.

Why you should learn it

Matrices can be used to decode a message, as shown in Exercise 40 on page 567.

EXAMPLE 1 Finding the Area of a Triangle

See LarsonPrecalculus.com for an interactive version of this type of example.

Find the area of the triangle whose vertices are (1, 0), (2, 2), and (4, 3), as shown in the figure.

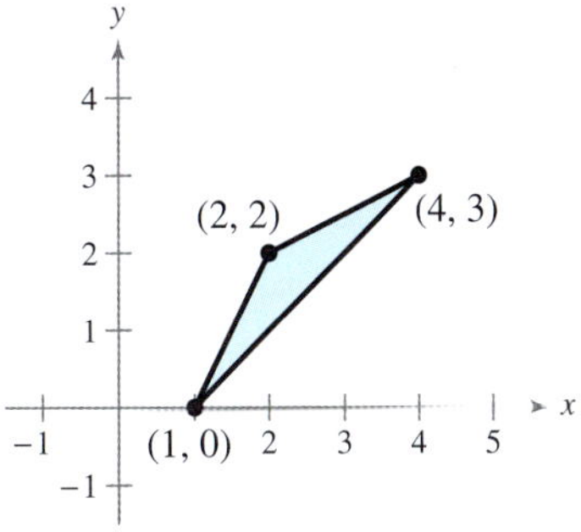

Solution

Begin by letting $(x_1, y_1) = (1, 0)$, $(x_2, y_2) = (2, 2)$, and $(x_3, y_3) = (4, 3)$. Then, to find the area of the triangle, evaluate the determinant by expanding along row 1.

$$\begin{vmatrix} x_1 & y_1 & 1 \\ x_2 & y_2 & 1 \\ x_3 & y_3 & 1 \end{vmatrix} = \begin{vmatrix} 1 & 0 & 1 \\ 2 & 2 & 1 \\ 4 & 3 & 1 \end{vmatrix}$$

$$= 1(-1)^2\begin{vmatrix} 2 & 1 \\ 3 & 1 \end{vmatrix} + 0(-1)^3\begin{vmatrix} 2 & 1 \\ 4 & 1 \end{vmatrix} + 1(-1)^4\begin{vmatrix} 2 & 2 \\ 4 & 3 \end{vmatrix}$$

$$= 1(-1) + 0 + 1(-2)$$

$$= -3$$

Using this value, you can conclude that the area of the triangle is

$$\text{Area} = -\frac{1}{2}\begin{vmatrix} 1 & 0 & 1 \\ 2 & 2 & 1 \\ 4 & 3 & 1 \end{vmatrix} = -\frac{1}{2}(-3) = \frac{3}{2} \text{ square units.}$$

Checkpoint *Audio-video solution in English & Spanish at LarsonPrecalculus.com*

Find the area of the triangle whose vertices are (0, 0), (4, 1), and (2, 5).

nfsphoto/Shutterstock.com

Collinear Points

What if the three points in Example 1 had been on the same line? What would have happened had the area formula been applied to three such points? The answer is that the determinant would have been zero. Consider, for instance, the three collinear points (0, 1), (2, 2), and (4, 3), as shown in Figure 7.22. The area of the "triangle" that has these three points as vertices is

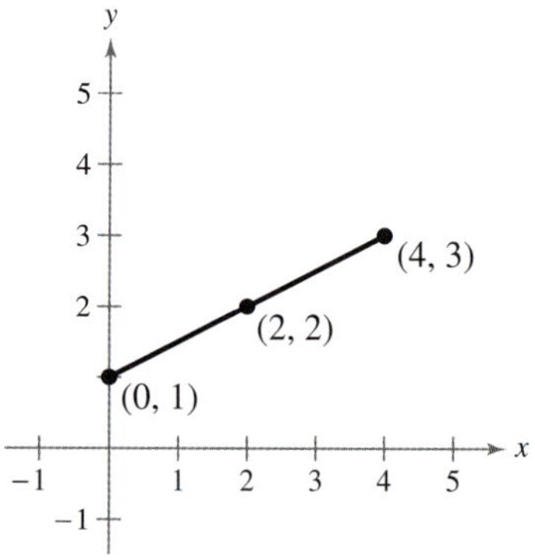

Figure 7.22

$$\frac{1}{2}\begin{vmatrix} 0 & 1 & 1 \\ 2 & 2 & 1 \\ 4 & 3 & 1 \end{vmatrix} = \frac{1}{2}\left[0(-1)^2\begin{vmatrix} 2 & 1 \\ 3 & 1 \end{vmatrix} + 1(-1)^3\begin{vmatrix} 2 & 1 \\ 4 & 1 \end{vmatrix} + 1(-1)^4\begin{vmatrix} 2 & 2 \\ 4 & 3 \end{vmatrix}\right]$$

$$= \frac{1}{2}[0 - 1(-2) + 1(-2)]$$

$$= 0.$$

This result is generalized as follows.

Test for Collinear Points

Three points (x_1, y_1), (x_2, y_2), and (x_3, y_3) are **collinear** (lie on the same line) if and only if

$$\begin{vmatrix} x_1 & y_1 & 1 \\ x_2 & y_2 & 1 \\ x_3 & y_3 & 1 \end{vmatrix} = 0.$$

EXAMPLE 2 Testing for Collinear Points

Determine whether the points

$(-2, -2)$, $(1, 1)$, and $(7, 5)$

are collinear. (See Figure 7.23.)

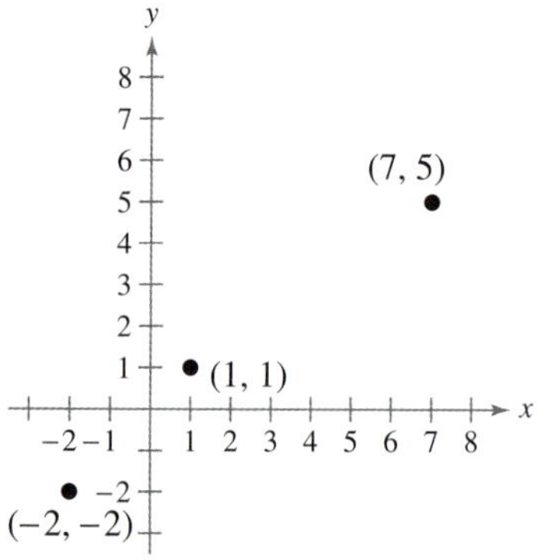

Figure 7.23

Solution

Begin by letting $(x_1, y_1) = (-2, -2)$, $(x_2, y_2) = (1, 1)$, and $(x_3, y_3) = (7, 5)$. Then by expanding along row 1, you have

$$\begin{vmatrix} x_1 & y_1 & 1 \\ x_2 & y_2 & 1 \\ x_3 & y_3 & 1 \end{vmatrix} = \begin{vmatrix} -2 & -2 & 1 \\ 1 & 1 & 1 \\ 7 & 5 & 1 \end{vmatrix}$$

$$= -2(-1)^2\begin{vmatrix} 1 & 1 \\ 5 & 1 \end{vmatrix} + (-2)(-1)^3\begin{vmatrix} 1 & 1 \\ 7 & 1 \end{vmatrix} + 1(-1)^4\begin{vmatrix} 1 & 1 \\ 7 & 5 \end{vmatrix}$$

$$= -2(-4) + 2(-6) + 1(-2)$$

$$= -6.$$

Because the value of this determinant is *not* zero, you can conclude that the three points are not collinear.

✓ **Checkpoint** *Audio-video solution in English & Spanish at LarsonPrecalculus.com*

Determine whether the points

$(-2, 4)$, $(3, -1)$, and $(6, -4)$

are collinear.

Transformation Matrices and Area of a Parallelogram

In addition to transforming vectors (see Section 7.5), you can use transformation matrices to transform graphs (such as squares) in the coordinate plane. Several transformations and their corresponding transformation matrices are listed below.

Transformation Matrices

Reflection in the y-axis	Reflection in the x-axis	Horizontal stretch ($k > 1$) or shrink ($0 < k < 1$)	Vertical stretch ($k > 1$) or shrink ($0 < k < 1$)
$\begin{bmatrix} -1 & 0 \\ 0 & 1 \end{bmatrix}$	$\begin{bmatrix} 1 & 0 \\ 0 & -1 \end{bmatrix}$	$\begin{bmatrix} k & 0 \\ 0 & 1 \end{bmatrix}$	$\begin{bmatrix} 1 & 0 \\ 0 & k \end{bmatrix}$

EXAMPLE 3 Transforming a Square

A square has vertices at (0, 0), (2, 0), (0, 2), and (2, 2). To find the image of the square after a reflection in the y-axis, first write the vertices as column matrices. Then multiply each column matrix by the appropriate transformation matrix on the left.

$$\begin{bmatrix} -1 & 0 \\ 0 & 1 \end{bmatrix}\begin{bmatrix} 0 \\ 0 \end{bmatrix} = \begin{bmatrix} 0 \\ 0 \end{bmatrix} \qquad \begin{bmatrix} -1 & 0 \\ 0 & 1 \end{bmatrix}\begin{bmatrix} 2 \\ 0 \end{bmatrix} = \begin{bmatrix} -2 \\ 0 \end{bmatrix}$$

$$\begin{bmatrix} -1 & 0 \\ 0 & 1 \end{bmatrix}\begin{bmatrix} 0 \\ 2 \end{bmatrix} = \begin{bmatrix} 0 \\ 2 \end{bmatrix} \qquad \begin{bmatrix} -1 & 0 \\ 0 & 1 \end{bmatrix}\begin{bmatrix} 2 \\ 2 \end{bmatrix} = \begin{bmatrix} -2 \\ 2 \end{bmatrix}$$

So, the vertices of the image are (0, 0), (−2, 0), (0, 2), and (−2, 2). Figure 7.24 shows the square and its image.

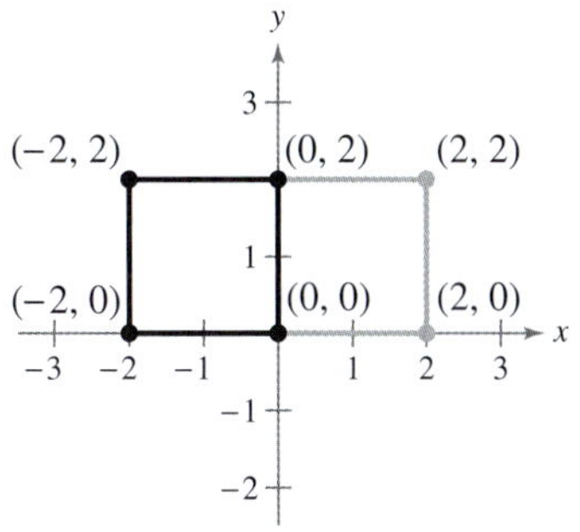

Figure 7.24

✓ Checkpoint ▶ *Audio-video solution in English & Spanish at LarsonPrecalculus.com*

Find the image of the square in Example 3 after a vertical stretch by a factor of $k = 2$. ■

You can find the area of a parallelogram using the determinant of a 2×2 matrix.

Area of a Parallelogram

The area of a parallelogram with vertices $(0, 0)$, (a, b), (c, d), and $(a + c, b + d)$ is

$$\text{Area} = |\det(A)| \qquad |\det(A)| \text{ is the absolute value of the determinant.}$$

where $A = \begin{bmatrix} a & b \\ c & d \end{bmatrix}$.

EXAMPLE 4 Finding the Area of a Parallelogram

To find the area of the parallelogram shown in Figure 7.25 using the formula above, let $(a, b) = (2, 0)$ and $(c, d) = (1, 3)$. Then

$$A = \begin{bmatrix} 2 & 0 \\ 1 & 3 \end{bmatrix} \quad \text{and} \quad \det(A) = 2(3) - 0(1) = 6.$$

So, the area of the parallelogram is

$$\text{Area} = |\det(A)| = |6| = 6 \text{ square units.}$$

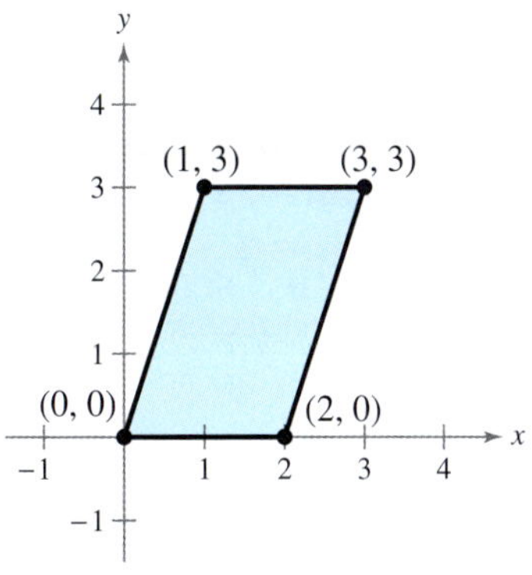

Figure 7.25

✓ Checkpoint ▶ *Audio-video solution in English & Spanish at LarsonPrecalculus.com*

Find the area of the parallelogram with vertices (0, 0), (5, 5), (2, 4), and (7, 9). ■

Cramer's Rule

So far, you have studied several methods for solving a system of linear equations.

- Substitution (Section 7.1)
- Graphing (Section 7.1)
- Elimination (Section 7.2)
- Gaussian elimination (Section 7.3)
- Gauss-Jordan elimination (Section 7.4)
- Inverse matrix (Section 7.6)

Insight

You can use any appropriate method to solve a linear system on a standardized test. On tests that do not permit calculators, solving a system by substitution or elimination is usually the most efficient method.

You will now study one more method, **Cramer's Rule,** named after Gabriel Cramer (1704–1752). This rule uses determinants to write the solution of a system of linear equations. To see how Cramer's Rule works, take another look at the solution described at the beginning of Section 7.7. There, it was pointed out that the system

$$\begin{cases} a_1x + b_1y = c_1 \\ a_2x + b_2y = c_2 \end{cases}$$

has a solution

$$x = \frac{c_1b_2 - c_2b_1}{a_1b_2 - a_2b_1} \quad \text{and} \quad y = \frac{a_1c_2 - a_2c_1}{a_1b_2 - a_2b_1}$$

provided that

$$a_1b_2 - a_2b_1 \neq 0.$$

Each numerator and denominator in this solution can be expressed as a determinant, as follows.

$$x = \frac{c_1b_2 - c_2b_1}{a_1b_2 - a_2b_1} = \frac{\begin{vmatrix} c_1 & b_1 \\ c_2 & b_2 \end{vmatrix}}{\begin{vmatrix} a_1 & b_1 \\ a_2 & b_2 \end{vmatrix}}$$

$$y = \frac{a_1c_2 - a_2c_1}{a_1b_2 - a_2b_1} = \frac{\begin{vmatrix} a_1 & c_1 \\ a_2 & c_2 \end{vmatrix}}{\begin{vmatrix} a_1 & b_1 \\ a_2 & b_2 \end{vmatrix}}$$

Relative to the original system, the denominators of x and y are simply the determinant of the *coefficient* matrix of the system. This determinant is denoted by D. The numerators of x and y are denoted by D_x and D_y, respectively. They are formed by using the column of constants as replacements for the coefficients of x and y, as follows.

Coefficient Matrix	D	D_x	D_y
$\begin{bmatrix} a_1 & b_1 \\ a_2 & b_2 \end{bmatrix}$	$\begin{vmatrix} a_1 & b_1 \\ a_2 & b_2 \end{vmatrix}$	$\begin{vmatrix} c_1 & b_1 \\ c_2 & b_2 \end{vmatrix}$	$\begin{vmatrix} a_1 & c_1 \\ a_2 & c_2 \end{vmatrix}$

For example, given the system

$$\begin{cases} 2x - 5y = 3 \\ -4x + 3y = 8 \end{cases}$$

the coefficient matrix, D, D_x, and D_y are as follows.

Coefficient Matrix	D	D_x	D_y
$\begin{bmatrix} 2 & -5 \\ -4 & 3 \end{bmatrix}$	$\begin{vmatrix} 2 & -5 \\ -4 & 3 \end{vmatrix}$	$\begin{vmatrix} 3 & -5 \\ 8 & 3 \end{vmatrix}$	$\begin{vmatrix} 2 & 3 \\ -4 & 8 \end{vmatrix}$

Cramer's Rule generalizes easily to systems of n equations in n variables. The value of each variable is given as the quotient of two determinants. The denominator is the determinant of the coefficient matrix, and the numerator is the determinant of the matrix formed by replacing the column corresponding to the variable being solved for with the column representing the constants. For instance, the solution for x_3 in the following system is shown.

$$\begin{cases} a_{11}x_1 + a_{12}x_2 + a_{13}x_3 = b_1 \\ a_{21}x_1 + a_{22}x_2 + a_{23}x_3 = b_2 \\ a_{31}x_1 + a_{32}x_2 + a_{33}x_3 = b_3 \end{cases} \qquad x_3 = \frac{|A_3|}{|A|} = \frac{\begin{vmatrix} a_{11} & a_{12} & b_1 \\ a_{21} & a_{22} & b_2 \\ a_{31} & a_{32} & b_3 \end{vmatrix}}{\begin{vmatrix} a_{11} & a_{12} & a_{13} \\ a_{21} & a_{22} & a_{23} \\ a_{31} & a_{32} & a_{33} \end{vmatrix}}$$

Cramer's Rule

If a system of n linear equations in n variables has a coefficient matrix A with a *nonzero* determinant $|A|$, then the solution of the system is

$$x_1 = \frac{|A_1|}{|A|}, \quad x_2 = \frac{|A_2|}{|A|}, \ldots, \quad x_n = \frac{|A_n|}{|A|}$$

where the ith column of A_i is the column of constants in the system of equations. If the determinant of the coefficient matrix is zero, then the system has either no solution or infinitely many solutions.

EXAMPLE 5 Using Cramer's Rule for a 2 × 2 System

Use Cramer's Rule to solve the system.

$$\begin{cases} 4x - 2y = 10 \\ 3x - 5y = 11 \end{cases}$$

Solution

To begin, find the determinant of the coefficient matrix.

$$D = \begin{vmatrix} 4 & -2 \\ 3 & -5 \end{vmatrix} = -20 - (-6) = -14$$

Because this determinant is not zero, apply Cramer's Rule.

$$x = \frac{D_x}{D} = \frac{\begin{vmatrix} 10 & -2 \\ 11 & -5 \end{vmatrix}}{-14} = \frac{(-50) - (-22)}{-14} = \frac{-28}{-14} = 2$$

$$y = \frac{D_y}{D} = \frac{\begin{vmatrix} 4 & 10 \\ 3 & 11 \end{vmatrix}}{-14} = \frac{44 - 30}{-14} = \frac{14}{-14} = -1$$

So, the solution is $x = 2$ and $y = -1$. Check this in the original system.

Checkpoint *Audio-video solution in English & Spanish at LarsonPrecalculus.com*

Use Cramer's Rule to solve the system.

$$\begin{cases} 3x + 4y = 1 \\ 5x + 3y = 9 \end{cases}$$

Nikolas_jkd/iStock/Getty Images

EXAMPLE 6 Using Cramer's Rule for a 3 × 3 System

Use Cramer's Rule, if possible, to solve the system of linear equations.

Coefficient Matrix

$$\begin{cases} -x + 2y - 3z = 1 \\ 2x \qquad + z = 0 \\ 3x - 4y + 4z = 2 \end{cases} \quad \Rightarrow \quad \begin{bmatrix} -1 & 2 & -3 \\ 2 & 0 & 1 \\ 3 & -4 & 4 \end{bmatrix}$$

Solution

The coefficient matrix above can be expanded along the second row, as follows.

$$D = 2(-1)^3\begin{vmatrix} 2 & -3 \\ -4 & 4 \end{vmatrix} + 0(-1)^4\begin{vmatrix} -1 & -3 \\ 3 & 4 \end{vmatrix} + 1(-1)^5\begin{vmatrix} -1 & 2 \\ 3 & -4 \end{vmatrix}$$

$$= -2(-4) + 0 - 1(-2)$$

$$= 10$$

Because this determinant is not zero, you can apply Cramer's Rule.

$$x = \frac{D_x}{D} = \frac{\begin{vmatrix} 1 & 2 & -3 \\ 0 & 0 & 1 \\ 2 & -4 & 4 \end{vmatrix}}{10} = \frac{8}{10} = \frac{4}{5}$$

$$y = \frac{D_y}{D} = \frac{\begin{vmatrix} -1 & 1 & -3 \\ 2 & 0 & 1 \\ 3 & 2 & 4 \end{vmatrix}}{10} = \frac{-15}{10} = -\frac{3}{2}$$

$$z = \frac{D_z}{D} = \frac{\begin{vmatrix} -1 & 2 & 1 \\ 2 & 0 & 0 \\ 3 & -4 & 2 \end{vmatrix}}{10} = \frac{-16}{10} = -\frac{8}{5}$$

The solution is

$$\left(\frac{4}{5}, -\frac{3}{2}, -\frac{8}{5}\right).$$

Check this in the original system.

✓ Checkpoint *Audio-video solution in English & Spanish at LarsonPrecalculus.com*

Use Cramer's Rule to solve the system.

$$\begin{cases} 4x - \ y + \ z = 12 \\ 2x + 2y + 3z = \ 1 \\ 5x - 2y + 6z = 22 \end{cases}$$

Remember that Cramer's Rule does not apply when the determinant of the coefficient matrix is zero. This would create division by zero, which is undefined. For example, consider this system of linear equations.

$$\begin{cases} -x \qquad + \ z = \ \ 4 \\ 2x - y + \ z = -3 \\ \qquad\ \ y - 3z = \ \ 1 \end{cases}$$

Using a graphing utility to evaluate the determinant of the coefficient matrix A, you find that Cramer's Rule cannot be applied because $|A| = 0$.

Technology Tip

Try using a graphing utility to evaluate D_x/D for the system of linear equations shown at the lower left. You should obtain an error message similar to the one shown below.

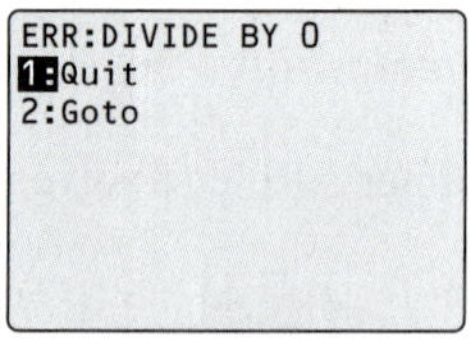

Cryptography

A **cryptogram** is a message written according to a secret code. (The Greek word *kryptos* means "hidden.") Matrix multiplication can be used to encode and decode messages. To begin, you need to assign a number to each letter in the alphabet (with 0 assigned to a blank space), as follows.

0 = _	9 = I	18 = R
1 = A	10 = J	19 = S
2 = B	11 = K	20 = T
3 = C	12 = L	21 = U
4 = D	13 = M	22 = V
5 = E	14 = N	23 = W
6 = F	15 = O	24 = X
7 = G	16 = P	25 = Y
8 = H	17 = Q	26 = Z

Then the message is converted to numbers and partitioned into **uncoded row matrices,** each having n entries, as demonstrated in Example 7.

Applied Cryptography Researcher

EXAMPLE 7 Forming Uncoded Row Matrices

Write the uncoded row matrices of dimension 1×3 for the message

MEET ME MONDAY.

Solution

Partitioning the message (including blank spaces, but ignoring punctuation) into groups of three produces the following uncoded row matrices.

$$[13 \quad 5 \quad 5] \quad [20 \quad 0 \quad 13] \quad [5 \quad 0 \quad 13] \quad [15 \quad 14 \quad 4] \quad [1 \quad 25 \quad 0]$$

M E E T _ M E _ M O N D A Y _

Note that a blank space is used to fill out the last uncoded row matrix.

✓ **Checkpoint** Audio-video solution in English & Spanish at LarsonPrecalculus.com

Write the uncoded row matrices of order 1×3 for the message

OWLS ARE NOCTURNAL.

To encode a message, choose an $n \times n$ invertible matrix (called an **encoding matrix**) such as

$$A = \begin{bmatrix} 1 & -2 & 2 \\ -1 & 1 & 3 \\ 1 & -1 & -4 \end{bmatrix} \qquad \text{Encoding matrix}$$

and multiply the uncoded row matrices by A (on the right) to obtain **coded row matrices.** Here is an example.

Uncoded Matrix *Encoding Matrix A* *Coded Matrix*

$$[13 \quad 5 \quad 5] \begin{bmatrix} 1 & -2 & 2 \\ -1 & 1 & 3 \\ 1 & -1 & -4 \end{bmatrix} = [13 \quad -26 \quad 21]$$

This technique is further illustrated in Example 8.

EXAMPLE 8 Encoding a Message

Use the matrix A to encode the message MEET ME MONDAY.

$$A = \begin{bmatrix} 1 & -2 & 2 \\ -1 & 1 & 3 \\ 1 & -1 & -4 \end{bmatrix}$$

Solution

The coded row matrices are obtained by multiplying each of the uncoded row matrices found in Example 7 by the matrix A, as follows.

Uncoded Matrix *Encoding Matrix A* *Coded Matrix*

$$[13 \quad 5 \quad 5]\begin{bmatrix} 1 & -2 & 2 \\ -1 & 1 & 3 \\ 1 & -1 & -4 \end{bmatrix} = [13 \quad -26 \quad 21]$$

$$[20 \quad 0 \quad 13]\begin{bmatrix} 1 & -2 & 2 \\ -1 & 1 & 3 \\ 1 & -1 & -4 \end{bmatrix} = [33 \quad -53 \quad -12]$$

$$[5 \quad 0 \quad 13]\begin{bmatrix} 1 & -2 & 2 \\ -1 & 1 & 3 \\ 1 & -1 & -4 \end{bmatrix} = [18 \quad -23 \quad -42]$$

$$[15 \quad 14 \quad 4]\begin{bmatrix} 1 & -2 & 2 \\ -1 & 1 & 3 \\ 1 & -1 & -4 \end{bmatrix} = [5 \quad -20 \quad 56]$$

$$[1 \quad 25 \quad 0]\begin{bmatrix} 1 & -2 & 2 \\ -1 & 1 & 3 \\ 1 & -1 & -4 \end{bmatrix} = [-24 \quad 23 \quad 77]$$

So, the sequence of coded row matrices is

$$[13 \quad -26 \quad 21][33 \quad -53 \quad -12][18 \quad -23 \quad -42][5 \quad -20 \quad 56][-24 \quad 23 \quad 77].$$

Finally, removing the matrix notation produces the following cryptogram.

13 −26 21 33 −53 −12 18 −23 −42 5 −20 56 −24 23 77

Checkpoint *Audio-video solution in English & Spanish at LarsonPrecalculus.com*

Use the following invertible matrix to encode the message OWLS ARE NOCTURNAL.

$$A = \begin{bmatrix} 1 & -1 & 0 \\ 1 & 0 & -1 \\ 6 & -2 & -3 \end{bmatrix}$$

For those who do not know the encoding matrix A, decoding the cryptogram found in Example 8 is difficult. But for an authorized receiver who knows the encoding matrix A, decoding is simple. The receiver need only multiply the coded row matrices by A^{-1} (on the right) to retrieve the uncoded row matrices. Here is an example.

$$\underbrace{[13 \quad -26 \quad 21]}_{\text{Coded}}\underbrace{\begin{bmatrix} -1 & -10 & -8 \\ -1 & -6 & -5 \\ 0 & -1 & -1 \end{bmatrix}}_{A^{-1}} = \underbrace{[13 \quad 5 \quad 5]}_{\text{Uncoded}}$$

This technique is further illustrated in Example 9.

Technology Tip

An efficient method for encoding the message at the left with your graphing utility is to enter A as a 3×3 matrix. Let B be the 5×3 matrix whose rows are the uncoded row matrices

$$B = \begin{bmatrix} 13 & 5 & 5 \\ 20 & 0 & 13 \\ 5 & 0 & 13 \\ 15 & 14 & 4 \\ 1 & 25 & 0 \end{bmatrix}.$$

The product BA gives the coded row matrices.

EXAMPLE 9 Decoding a Message

Use the inverse of the matrix A in Example 8 to decode the cryptogram.

13 −26 21 33 −53 −12 18 −23 −42 5 −20 56 −24 23 77

Solution

First, find the decoding matrix A^{-1} by using the techniques demonstrated in Section 7.6. Next partition the message into groups of three to form the coded row matrices. Then multiply each coded row matrix by A^{-1} (on the right).

Coded Matrix Decoding Matrix A^{-1} *Decoded Matrix*

$$[13 \quad -26 \quad 21]\begin{bmatrix} -1 & -10 & -8 \\ -1 & -6 & -5 \\ 0 & -1 & -1 \end{bmatrix} = [13 \quad 5 \quad 5]$$

$$[33 \quad -53 \quad -12]\begin{bmatrix} -1 & -10 & -8 \\ -1 & -6 & -5 \\ 0 & -1 & -1 \end{bmatrix} = [20 \quad 0 \quad 13]$$

$$[18 \quad -23 \quad -42]\begin{bmatrix} -1 & -10 & -8 \\ -1 & -6 & -5 \\ 0 & -1 & -1 \end{bmatrix} = [5 \quad 0 \quad 13]$$

$$[5 \quad -20 \quad 56]\begin{bmatrix} -1 & -10 & -8 \\ -1 & -6 & -5 \\ 0 & -1 & -1 \end{bmatrix} = [15 \quad 14 \quad 4]$$

$$[-24 \quad 23 \quad 77]\begin{bmatrix} -1 & -10 & -8 \\ -1 & -6 & -5 \\ 0 & -1 & -1 \end{bmatrix} = [1 \quad 25 \quad 0]$$

So, the message is as follows.

[13	5	5]	[20	0	13]	[5	0	13]	[15	14	4]	[1	25	0]
M	E	E	T		M	E		M	O	N	D	A	Y	

Checkpoint *Audio-video solution in English & Spanish at LarsonPrecalculus.com*

Use the inverse of the matrix A in the Checkpoint with Example 8 to decode the cryptogram.

110 −39 −59 25 −21 −3 23 −18 −5 47 −20 −24
149 −56 −75 87 −38 −37.

Technology Tip

An efficient method for decoding the cryptogram in Example 9 with your graphing utility is to enter A as a 3×3 matrix and then find A^{-1}. Let B be the 5×3 matrix whose rows are the coded row matrices, as shown below. The product BA^{-1} gives the decoded row matrices.

$$B = \begin{bmatrix} 13 & -26 & 21 \\ 33 & -53 & 12 \\ 18 & -23 & -42 \\ 5 & -20 & 56 \\ -24 & 23 & 77 \end{bmatrix}$$

7.8 Exercises

See *CalcChat.com* for tutorial help and worked-out solutions to odd-numbered exercises.
For instructions on how to use a graphing utility, see Appendix A.

Vocabulary and Concept Check

In Exercises 1 and 2, fill in the blank.

1. _______ is a method for using determinants to solve a system of linear equations.

2. A message written according to a secret code is called a _______ .

Procedures and Problem Solving

Finding the Area of a Triangle **In Exercises 3–6, use a determinant to find the area of the triangle with the given vertices.**

3. $(0, 0), (1, 5), (3, 1)$ **4.** $(-3, 5), (2, 6), (3, -5)$

5. $\left(0, \frac{1}{2}\right), \left(\frac{5}{2}, 0\right), (4, 3)$ **6.** $\left(\frac{9}{2}, 0\right), (2, 6), \left(0, -\frac{3}{2}\right)$

Finding a Coordinate **In Exercises 7 and 8, find x or y such that the triangle has an area of 4 square units.**

7. $(-1, 5), (-2, 0), (x, 2)$ **8.** $(-4, 2), (-3, 5), (-1, y)$

Testing for Collinear Points **In Exercises 9–12, use a determinant to determine whether the points are collinear.**

9. $(3, -1), (0, -3), (12, 5)$

10. $(3, -5), (6, 1), (4, 2)$

11. $\left(2, -\frac{1}{2}\right), (-4, 4), (6, -3)$

12. $\left(0, \frac{1}{2}\right), (2, -1), \left(-4, \frac{7}{2}\right)$

Finding a Coordinate **In Exercises 13 and 14, find x or y such that the points are collinear.**

13. $(1, -2), (x, 2), (5, 6)$ **14.** $(-6, 2), (-5, y), (-3, 5)$

Transforming a Square **In Exercises 15–18, use the matrices to find the vertices of the image of the square with the given vertices after the given transformation. Then sketch the square and its image.**

15. $(0, 0), (0, 3), (3, 0), (3, 3)$: horizontal stretch, $k = 2$

16. $(1, 2), (3, 2), (1, 4), (3, 4)$: reflection in the x-axis

17. $(4, 3), (5, 3), (4, 4), (5, 4)$: reflection in the y-axis

18. $(1, 1), (3, 2), (0, 3), (2, 4)$: vertical shrink, $k = \frac{1}{2}$

Finding the Area of a Parallelogram **In Exercises 19–22, use a determinant to find the area of the parallelogram with the given vertices.**

19. $(0, 0), (1, 0), (2, 2), (3, 2)$

20. $(0, 0), (3, 0), (4, 1), (7, 1)$

21. $(0, 0), (-2, 0), (3, 5), (1, 5)$

22. $(0, 0), (0, 8), (8, -6), (8, 2)$

Using Cramer's Rule for a 2 × 2 System **In Exercises 23–26, use Cramer's Rule to solve (if possible) the system of equations.**

23. $\begin{cases} -5x + 9y = -14 \\ 3x - 7y = 10 \end{cases}$ **24.** $\begin{cases} 3x + 2y = -2 \\ 6x + 4y = 4 \end{cases}$

25. $\begin{cases} 4x - 3y = -10 \\ 6x + 9y = 12 \end{cases}$ **26.** $\begin{cases} 6x - 5y = 17 \\ -13x + 3y = -76 \end{cases}$

Using Cramer's Rule for a 3 × 3 System **In Exercises 27–30, use Cramer's Rule to solve (if possible) the system of equations.**

27. $\begin{cases} 3x + 3y + 5z = 1 \\ 3x + 5y + 9z = 2 \\ 5x + 9y + 17z = 4 \end{cases}$ **28.** $\begin{cases} 2x + 3y - 5z = 1 \\ 3x + 5y + 9z = -16 \\ 5x + 9y + 17z = -30 \end{cases}$

29. $\begin{cases} 4x - y + z = -5 \\ 2x + 2y + 3z = 10 \\ 6x + y + 4z = -5 \end{cases}$ **30.** $\begin{cases} 4x - 2y + 3z = -2 \\ 2x + 2y + 5z = 16 \\ 8x - 5y - 2z = 4 \end{cases}$

31. ***Why you should learn it*** *(p. 550)* The retail sales y (in billions of dollars) of family clothing stores in the United States from 2013 through 2017 can be expressed by ordered pairs of the form $(t, y(t))$, where $t = 3$ represents 2013. *(Source:* U.S. Census Bureau)

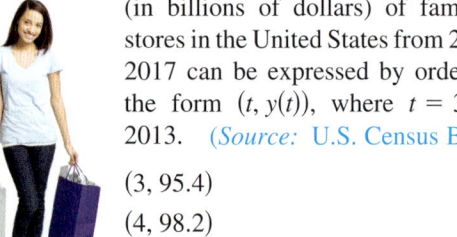

$(3, 95.4)$

$(4, 98.2)$

$(5, 101.4)$

$(6, 103.6)$

$(7, 105.9)$

The coefficients of the least squares regression parabola $y = at^2 + bt + c$ can be found by solving the system

$$\begin{cases} 5c + 25b + 135a = 504.5 \\ 25c + 135b + 775a = 2548.9. \\ 135c + 775b + 4659a = 13{,}883.5 \end{cases}$$

(a) Use Cramer's Rule to solve the system and write the least squares regression parabola for the data.

(b) Use a graphing utility to graph the parabola with the data. How well does the model fit the data?

Neustockimages/iStock/Getty Images

32. MODELING DATA

The retail sales y (in billions of dollars) of automobile and other motor vehicle dealers in the United States from 2013 through 2017 can be expressed by the ordered pairs of the form $(t, y(t))$, where $t = 3$ represents 2013. (*Source:* U.S. Census Bureau)

(3, 874.3), (4, 935.1), (5, 1004.8),
(6, 1053.9), (7, 1103.4)

The coefficients of the least squares regression parabola $y = at^2 + bt + c$ can be found by solving the system

$$\begin{cases} 5c + 25b + 135a = 4971.6 \\ 25c + 135b + 775a = 25{,}434.9 \\ 135c + 775b + 4659a = 139{,}959.8 \end{cases}$$

(a) Use Cramer's Rule to solve the system and write the least squares regression parabola for the data.

(b) Use a graphing utility to graph the parabola with the data. How well does the model fit the data?

Encoding a Message **In Exercises 33 and 34, (a) write the uncoded 1 × 2 row matrices for the message, and then (b) encode the message using the encoding matrix.**

Message	*Encoding Matrix*
33. COME HOME SOON	$\begin{bmatrix} 1 & 2 \\ 3 & 5 \end{bmatrix}$
34. HELP IS ON THE WAY	$\begin{bmatrix} -2 & 3 \\ -1 & 1 \end{bmatrix}$

Encoding a Message **In Exercises 35 and 36, (a) write the uncoded 1 × 3 row matrices for the message, and then (b) encode the message using the encoding matrix.**

Message	*Encoding Matrix*
35. CALL ME TODAY	$\begin{bmatrix} 1 & -1 & 0 \\ 1 & 0 & -1 \\ -6 & 2 & 3 \end{bmatrix}$
36. HAPPY BIRTHDAY	$\begin{bmatrix} 1 & 2 & 2 \\ 3 & 7 & 9 \\ -1 & -4 & -7 \end{bmatrix}$

Decoding a Message **In Exercises 37 and 38, use the inverse of matrix A to decode the cryptogram.**

37. $A = \begin{bmatrix} 1 & 2 \\ 3 & 5 \end{bmatrix}$ 11 21 64 112 25 50 29 53 23 46 40 75 55 92

38. $A = \begin{bmatrix} 3 & -4 & 2 \\ 0 & 2 & 1 \\ 4 & -5 & 3 \end{bmatrix}$ 112 −140 83 19 −25 13 72 −76 61 95 −118 71 20 21 38 35 −23 36 42 −48 32

39. Decoding a Message The following cryptogram was encoded with a 2 × 2 matrix.

5 2 25 11 −2 −7 −15 −15 32 14 −8 −13 38 19 −19 −19 37 16

The last word of the message is _SUE. What is the message?

40. *Why you should learn it* *(p. 557)* The following cryptogram was encoded with a 2 × 2 matrix.

8 21 −15 −10 −13 −13 5 10 5 25 5 19 −1 6 20 40 −18 −18 1 16

The last word of the message is _RON. What is the message?

Focusing on Concepts

True or False? **In Exercises 41 and 42, determine whether the statement is true or false. Justify your answer.**

41. Cramer's Rule cannot be used to solve a system of linear equations when the determinant of the coefficient matrix is zero.

42. In a system of linear equations, when the determinant of the coefficient matrix is zero, the system has no solution.

Error Analysis **In Exercises 43 and 44, describe the error in writing the transformation matrix.**

43. horizontal stretch $(k > 1)$

$\begin{bmatrix} 1 & 0 \\ 0 & k \end{bmatrix}$ ✗

44. reflection in y-axis

$\begin{bmatrix} 1 & -1 \\ 1 & -1 \end{bmatrix}$ ✗

45. Writing Describe a way to use an invertible $n \times n$ matrix to encode a message that is converted to numbers and partitioned into uncoded column matrices.

46. HOW DO YOU SEE IT? Summarize methods you can use to find an equation of the line that passes through the points shown in the figure.

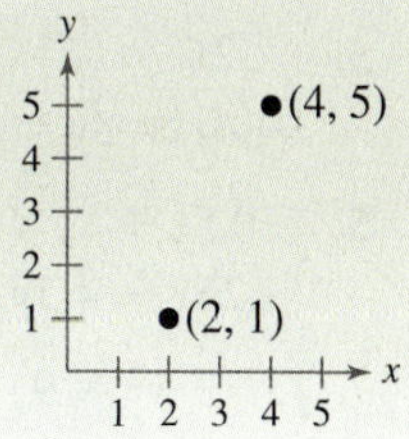

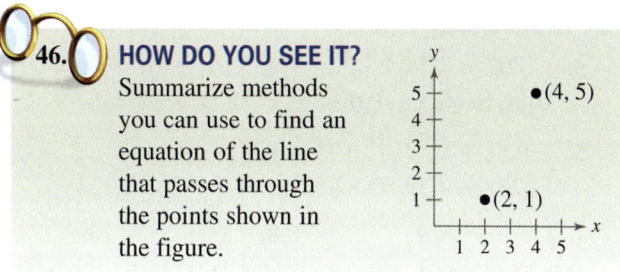

Cumulative Mixed Review

Equation of a Line **In Exercises 47 and 48, find the general form of the equation of the line that passes through the two points.**

47. (−1, 5), (7, 3)

48. (0, −6), (−2, 10)

nfsphoto/Shutterstock.com

7 Chapter Review

See CalcChat.com for tutorial help and worked-out solutions to odd-numbered exercises. For instructions on how to use a graphing utility, see Appendix A.

7.1 *What did you learn?*

Use the methods of substitution and graphing to solve systems of equations in two variables *(p. 478)*.

To use the **method of substitution** to solve a system of equations in x and y, perform these steps.

1. Solve one of the equations for one variable in terms of the other.
2. Substitute the expression found in Step 1 into the other equation to obtain an equation in one variable.
3. Solve the equation obtained in Step 2.
4. Back-substitute the value(s) obtained in Step 3 into the expression obtained in Step 1 to find the value(s) of the other variable.
5. Check each solution in *both* of the original equations.

To use the **method of graphing** to solve a system of equations in x and y, perform these steps.

1. Solve both equations for y in terms of x.
2. Use a graphing utility to graph both equations in the same viewing window.
3. Use the *intersect* feature or the *zoom* and *trace* features of the graphing utility to approximate all points of intersection of the graphs.
4. Check each solution in *both* of the original equations.

Use systems of equations to model and solve real-life problems *(p. 483)*. A system of equations can be used to perform a break-even analysis (see Example 6).

Example You can solve the system of equations

$$\begin{cases} 2x + y = 4 & \text{Equation 1} \\ x + 2y = -1 & \text{Equation 2} \end{cases}$$

by (a) substitution or (b) graphing.

(a)

$$\begin{aligned}
y &= 4 - 2x && \text{Solve for } y \text{ in Equation 1.} \\
x + 2(4 - 2x) &= -1 && \text{Substitute } 4 - 2x \text{ for } y \text{ in Equation 2.} \\
x + 8 - 4x &= -1 && \text{Distributive Property} \\
8 - 3x &= -1 && \text{Combine like terms.} \\
-3x &= -9 && \text{Subtract 8 from each side.} \\
x &= 3 && \text{Divide each side by } -3. \\
y &= 4 - 2(3) && \text{Substitute 3 for } x \text{ in Equation 1.} \\
y &= -2 && \text{Solve for } y.
\end{aligned}$$

The solution is the ordered pair $(3, -2)$. Check this solution by substituting into the original system.

(b)

$$\begin{cases} y = 4 - 2x & \text{Solve Equation 1 for } y. \\ y = -\frac{1}{2}x - \frac{1}{2} & \text{Solve Equation 2 for } y. \end{cases}$$

A graph of both equations is at the right. The solution point is the intersection of the graphs. Note that this is the same result obtained in part (a).

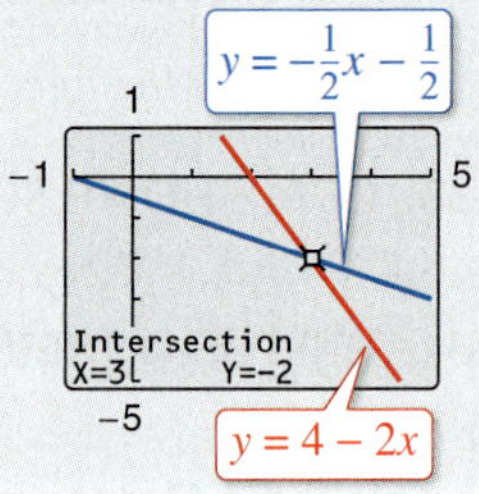

Solving a System by Substitution **In Exercises 1–10, solve the system by the method of substitution.**

1. $\begin{cases} x + y = 6 \\ x - y = 0 \end{cases}$

2. $\begin{cases} 3x - 2y = 2 \\ x - y = 0 \end{cases}$

3. $\begin{cases} 2x - y = 4 \\ -4x + 3y = -10 \end{cases}$

4. $\begin{cases} 10x + 6y = -14 \\ x + 9y = -7 \end{cases}$

5. $\begin{cases} 0.5x + y = 0.75 \\ 1.25x - 4.5y = -2.5 \end{cases}$

6. $\begin{cases} -x + \frac{2}{5}y = \frac{3}{5} \\ -x + \frac{1}{5}y = -\frac{4}{5} \end{cases}$

7. $\begin{cases} x^2 - y^2 = 9 \\ x - y = 1 \end{cases}$

8. $\begin{cases} x^2 + y^2 = 169 \\ 3x + 2y = 39 \end{cases}$

9. $\begin{cases} y = 2x^2 \\ y = x^4 - 2x^2 \end{cases}$

10. $\begin{cases} x = 4y + 5 \\ x = y^2 + 9 \end{cases}$

Solving a System of Equations Graphically **In Exercises 11–16, use a graphing utility to approximate all points of intersection of the graphs of the equations in the system. Check your solutions in the original system.**

11. $\begin{cases} 5x + 6y = 7 \\ -x - 4y = 0 \end{cases}$

12. $\begin{cases} 8x - 3y = -3 \\ 2x + 5y = 28 \end{cases}$

13. $\begin{cases} y^2 - 4x = 0 \\ x + y = 0 \end{cases}$

14. $\begin{cases} y = 2x^2 - 4x + 1 \\ y = x^2 - 4x + 3 \end{cases}$

15. $\begin{cases} y = 2(8 - x) \\ y = 2^{x-1} \end{cases}$

16. $\begin{cases} y = \ln(x + 2) + 1 \\ x + y = 0 \end{cases}$

17. **Finance** You invest \$5000 in a greenhouse. The planter, potting soil, and seed for each plant cost \$6.43, and the selling price of each plant is \$12.68. How many plants must you sell to break even?

18. **Finance** You are offered two sales jobs. One company offers an annual salary of \$55,000 plus a year-end bonus of 1.5% of your total sales. The other company offers an annual salary of \$52,000 plus a year-end bonus of 2% of your total sales. How much would you have to sell in a year to make the second offer the better offer?

19. **Geometry** The perimeter of a rectangle is 560 meters and its length is 2.5 times its width. Find the dimensions of the rectangle.

20. **Geometry** The perimeter of a rectangle is 68 feet and its width is $\frac{8}{9}$ times its length. Find the dimensions of the rectangle.

7.2 What did you learn?

Use the method of elimination to solve systems of linear equations in two variables *(p. 488)*.

To use the **method of elimination** to solve a system of two linear equations in x and y, perform these steps.

1. Obtain coefficients for x (or y) that differ only in sign by multiplying all terms of one or both equations by suitably chosen constants.
2. Add the equations to eliminate one variable. Solve the resulting equation.
3. Back-substitute the value obtained in Step 2 into either of the original equations and solve for the other variable.
4. Check the solution in *both* of the original equations.

Example

$$\begin{cases} 2x + 3y = 1 & \text{Equation 1} \\ 3x - 2y = 21 & \text{Equation 2} \end{cases}$$

You can obtain coefficients of y that differ only in sign by multiplying Equation 1 by 2 and multiplying Equation 2 by 3.

$2x + 3y = 1 \Rightarrow 4x + 6y = 2$ Multiply Equation 1 by 2.

$3x - 2y = 21 \Rightarrow 9x - 6y = 63$ Multiply Equation 2 by 3.

$13x = 65$ Add equations.

From this equation, you can see that $x = 5$. By back-substituting this value into Equation 2, you can solve for y.

$3x - 2y = 21$ Write Equation 2.

$3(5) - 2y = 21$ Substitute 5 for x.

$-2y = 6$ Combine like terms.

$y = -3$ Solve for y.

The solution is $(5, -3)$, Check this solution algebraically by substituting into the original system, as shown below.

Check

$2(5) + 3(-3) \stackrel{?}{=} 1$ Substitute in Equation 1.

$10 - 9 = 1$ Equation 1 checks. ✓

$3(5) - 2(-3) \stackrel{?}{=} 21$ Substitute in Equation 2.

$15 + 6 = 21$ Equation 2 checks. ✓

Graphically interpret the number of solutions of a system of linear equations in two variables *(p. 490)*. For a system of two linear equations in two variables, the number of solutions and corresponding graphical interpretation is one of the following.

1. Exactly one solution (consistent): the two lines intersect at one point.
2. Infinitely many solutions (consistent): the two lines are coincident (identical).
3. No solution (inconsistent): the two lines are parallel.

Use systems of linear equations in two variables to model and solve real-life problems *(p. 492)*. A system of linear equations can be used to find the airspeed of an airplane and the speed of the wind (see Example 6).

Solving a System by Elimination **In Exercises 21–30, solve the system by the method of elimination and check any solutions algebraically.**

21. $\begin{cases} x + 3y = -5 \\ -x - 8y = 0 \end{cases}$ **22.** $\begin{cases} 7x + 2y = 12 \\ x - 2y = -4 \end{cases}$

23. $\begin{cases} 2x - y = 2 \\ 6x + 8y = 39 \end{cases}$ **24.** $\begin{cases} 7x + 12y = 63 \\ 2x + 3y = 15 \end{cases}$

25. $\begin{cases} \frac{1}{5}x + \frac{3}{10}y = 7 \\ \frac{2}{5}x + \frac{3}{5}y = 14 \end{cases}$ **26.** $\begin{cases} 0.2x + 0.3y = 0.14 \\ 0.4x + 0.5y = 0.20 \end{cases}$

27. $\begin{cases} 3x - 2y = 0 \\ 3x + 2(y + 5) = 10 \end{cases}$ **28.** $\begin{cases} 6x - 4(y - 1) = 19 \\ -3x + 2y = 3 \end{cases}$

29. $\begin{cases} 1.5x + 2.5y = 8.5 \\ 6x + 10y = 24 \end{cases}$ **30.** $\begin{cases} 1.25x - 2y = 3.5 \\ 5x - 8y = 14 \end{cases}$

Solving a System Graphically **In Exercises 31–36, use a graphing utility to graph the lines in the system. Use the graphs to determine whether the system is consistent or inconsistent. If the system is consistent, determine the solution. Verify your results algebraically.**

31. $\begin{cases} 3x + 2y = 0 \\ x - y = 4 \end{cases}$ **32.** $\begin{cases} x + y = 6 \\ -2x - 2y = -12 \end{cases}$

33. $\begin{cases} \frac{1}{4}x - \frac{1}{5}y = 2 \\ -5x + 4y = 8 \end{cases}$ **34.** $\begin{cases} \frac{7}{2}x - 7y = -1 \\ -x + 2y = 4 \end{cases}$

35. $\begin{cases} 8x - 2y = 10 \\ -6x + 1.5y = -7.5 \end{cases}$ **36.** $\begin{cases} -x + 3.2y = 10.4 \\ -2x - 9.6y = 6.4 \end{cases}$

Economics **In Exercises 37 and 38, find the point of equilibrium of the demand and supply equations.**

	Demand	*Supply*
37.	$p = 37 - 0.0002x$	$p = 22 + 0.00001x$
38.	$p = 120 - 0.0001x$	$p = 45 + 0.0002x$

39. Aviation Two planes leave Pittsburgh and Philadelphia at the same time, each going to the other city. One plane flies 60 miles per hour faster than the other. Find the airspeed of each plane given that the cities are 257 miles apart and the planes pass each other after 30 minutes of flying time.

40. Economics A total of \$46,000 is invested in two funds that pay 6.75% and 7.25% simple interest. The investor wants an annual interest income of \$3245 from the investments. What is the most that can be invested in the 6.75% funds?

7.3 What did you learn?

Use back-substitution to solve linear systems in row-echelon form *(p. 497)*, and use Gaussian elimination to solve systems of linear equations *(p. 498)*. Use elementary row operations (see page 498) to convert a system of linear equations to an equivalent system in row-echelon form. Then use back-substitution to find the values of the remaining variables.

Example

(a)
$$\begin{cases} x + y + z = 6 & \text{Equation 1} \\ 2x - y + z = 3 & \text{Equation 2} \\ 3x \quad\; - z = 0 & \text{Equation 3} \end{cases}$$

$$\begin{cases} x + y + z = 6 & \\ -3y - z = -9 & -2(\text{Equation 1}) + \text{Equation 2} \\ -3y - 4z = -18 & -3(\text{Equation 1}) + \text{Equation 3} \end{cases}$$

$$\begin{cases} x + y + z = 6 & \\ -3y - z = -9 & \\ -3z = -9 & -1(\text{Equation 2}) + \text{Equation 3} \end{cases}$$

So, $z = \frac{1}{3}(-9) = 3$, $-3y - 3 = -9 \Rightarrow y = 2$, and $x + 2 + 3 = 6 \Rightarrow x = 1$. The solution can be written as $(1, 2, 3)$. Check in the original system.

Solve nonsquare systems of linear equations *(p. 501)*. In a nonsquare system, the number of equations differs from the number of variables. A system of linear equations cannot have a unique solution unless there are at least as many equations as there are variables.

Graphically interpret three-variable linear systems *(p. 502)*. The graph of a system of three linear equations in three variables consists of three planes. When they intersect in a single point, the system has exactly one solution. When they have no point in common, the system has no solution. When they intersect in a line or a plane, the system has infinitely many solutions.

Use systems of linear equations to write partial fraction decompositions of rational expressions *(p. 503)*. See page 503 for finding the decomposition of a rational expression $N(x)/D(x)$ into partial fractions.

(b)
$$\frac{x+6}{x^2+6x+8} = \frac{x+6}{(x+2)(x+4)} = \frac{A}{x+2} + \frac{B}{x+4}$$

$$\begin{aligned} x + 6 &= A(x+4) + B(x+2) && \text{Basic equation} \\ &= Ax + 4A + Bx + 2B && \text{Expand.} \\ &= (A+B)x + 4A + 2B && \text{Polynomial form} \end{aligned}$$

Equating coefficients of like terms leads to the system

$$\begin{cases} A + B = 1 \\ 4a + 2B = 6 \end{cases}.$$

So, $A = 2$, $B = -1$, and

$$\frac{x+6}{x^2+6x+8} = \frac{2}{x+2} - \frac{1}{x+4}.$$

Use systems of linear equations in three or more variables to model and solve real-life problems *(p. 506)*. A system of linear equations in three variables can be used to find a position equation (see Example 8).

Solving a System of Linear Equations **In Exercises 41–46, solve the system of linear equations and check any solution algebraically.**

41. $\begin{cases} x + y + z = 2 \\ -x + 3y + 2z = 8 \\ 4x + y = 4 \end{cases}$

42. $\begin{cases} x - 2y + z = -6 \\ 2x - 3y = -7 \\ -x + 3y - 3z = 11 \end{cases}$

43. $\begin{cases} 2x + 6z = -9 \\ 3x - 2y + 11z = -16 \\ 3x - y + 7z = -11 \end{cases}$

44. $\begin{cases} x - 2y + z = 5 \\ 2x + 3y + z = 5 \\ x + y + 2z = 3 \end{cases}$

45. $\begin{cases} 5x - 12y + 7z = 16 \\ 3x - 7y + 4z = 9 \end{cases}$

46. $\begin{cases} 2x + 5y - 19z = 34 \\ 3x + 8y - 31z = 54 \end{cases}$

Partial Fraction Decomposition **In Exercises 47–50, write the partial fraction decomposition for the rational expression. Check your result algebraically by combining fractions and check your result graphically by using a graphing utility to graph the rational expression and the partial fractions in the same viewing window.**

47. $\dfrac{4-x}{x^2+6x+8}$

48. $\dfrac{16}{x^2-16}$

49. $\dfrac{x^2+2x}{x^3-x^2+x-1}$

50. $\dfrac{2x^2-x+7}{x^4+8x^2+16}$

Data Analysis: Curve-Fitting **In Exercises 51 and 52, find the equation of the parabola $y = ax^2 + bx + c$ that passes through the points. To verify your result, use a graphing utility to plot the points and graph the parabola.**

51. $(-1, -4), (1, -2), (2, 5)$

52. $(-1, 0)\ (1, 4), (2, 3)$

53. **Golf Course** Pebble Beach Golf Links in Pebble Beach, California, is an 18-hole course that consists of par-3 holes, par-4 holes, and par-5 holes. There are two more par-4 holes than twice the number of par-5 holes, and the number of par-3 holes is equal to the number of par-5 holes. Find the number of par-3, par-4, and par-5 holes on the course. (*Source:* Pebble Beach Resorts)

54. **Economics** An inheritance of \$40,000 is divided among three investments yielding \$3500 in interest per year. The interest rates for the three investments are 7%, 9%, and 11%. Find the amount of each investment when the second and third are \$3000 and \$5000 less than the first, respectively.

7.4 What did you learn?

Write matrices and determine their dimensions *(p. 512)*, and perform elementary row operations on matrices *(p. 514)*. Elementary row operations include interchanging two rows, multiplying a row by a nonzero constant, and adding a multiple of a row to another row.

Use matrices and Gaussian elimination to solve systems of linear equations *(p. 515)*. Write the augmented matrix of the system. Use elementary row operations to rewrite the augmented matrix in row-echelon form (see page 516 for definitions of row-echelon form and reduced row-echelon form). Write the system of linear equations corresponding to the matrix in row-echelon form and use back-substitution to find the solution.

Use matrices and Gauss-Jordan elimination to solve systems of linear equations *(p. 519)*. Gauss-Jordan elimination continues the reduction process on a matrix in row-echelon form until the reduced row-echelon form is obtained (see Example 8).

Example

$$\begin{cases} -x + y + 2z = 1 \\ 2x + 3y + z = -2 \\ 5x + 4y + 2z = 4 \end{cases} \Rightarrow \begin{bmatrix} -1 & 1 & 2 & \vdots & 1 \\ 2 & 3 & 1 & \vdots & -2 \\ 5 & 4 & 2 & \vdots & 4 \end{bmatrix}$$

$$\Rightarrow \begin{matrix} -R_1 \to \\ 2R_1 + R_2 \to \\ 5R_1 + R_3 \to \end{matrix} \begin{bmatrix} 1 & -1 & -2 & \vdots & -1 \\ 0 & 5 & 5 & \vdots & 0 \\ 0 & 9 & 12 & \vdots & 9 \end{bmatrix} \Rightarrow \begin{matrix} \\ \frac{1}{5}R_2 \to \\ \\ \end{matrix} \begin{bmatrix} 1 & -1 & -2 & \vdots & -1 \\ 0 & 1 & 1 & \vdots & 0 \\ 0 & 9 & 12 & \vdots & 9 \end{bmatrix}$$

$$\Rightarrow \begin{matrix} \\ \\ R_3 - 9R_2 \to \end{matrix} \begin{bmatrix} 1 & -1 & -2 & \vdots & -1 \\ 0 & 1 & 1 & \vdots & 0 \\ 0 & 0 & 3 & \vdots & 9 \end{bmatrix} \Rightarrow \begin{matrix} \\ \\ \frac{1}{3}R_3 \to \end{matrix} \begin{bmatrix} 1 & -1 & -2 & \vdots & -1 \\ 0 & 1 & 1 & \vdots & 0 \\ 0 & 0 & 1 & \vdots & 3 \end{bmatrix}$$

$$\Rightarrow \begin{matrix} R_1 + R_2 + R_3 \to \\ R_2 - R_3 \to \\ \\ \end{matrix} \begin{bmatrix} 1 & 0 & 0 & \vdots & 2 \\ 0 & 1 & 0 & \vdots & -3 \\ 0 & 0 & 1 & \vdots & 3 \end{bmatrix}$$

So, $x = 2$, $y = -3$, and $z = 3$. Check this solution by substituting into the original system.

Writing an Augmented Matrix **In Exercises 55–58, write the augmented matrix for the system of linear equations.**

55. $\begin{cases} 6x - 7y = 11 \\ -2x + 5y = -1 \end{cases}$

56. $\begin{cases} -x + y = 16 \\ 9x = -18 \end{cases}$

57. $\begin{cases} 3x - 5y + z = 25 \\ -4x - 2z = -14 \end{cases}$

58. $\begin{cases} 8x - 7y + 4z = 12 \\ 3x - 5y + 2z = 20 \\ 5x + 3y - 3z = 26 \end{cases}$

Row-Echelon Form **In Exercises 59–62, write the matrix in row-echelon form. Remember that the row-echelon form of a matrix is not unique.**

59. $\begin{bmatrix} -1 & 4 & 0 \\ 3 & -5 & 9 \end{bmatrix}$

60. $\begin{bmatrix} 0 & -2 & 7 \\ 2 & 6 & -14 \end{bmatrix}$

61. $\begin{bmatrix} 0 & 1 & 1 \\ 1 & 2 & 3 \\ 2 & 2 & 2 \end{bmatrix}$

62. $\begin{bmatrix} 4 & 8 & 16 & -4 \\ 3 & -1 & 2 & 0 \\ -2 & 10 & 12 & 3 \end{bmatrix}$

Reduced Row-Echelon Form **In Exercises 63–66, write the matrix in reduced row-echelon form. Use a graphing utility to verify your result.**

63. $\begin{bmatrix} 4 & -3 \\ 3 & -4 \end{bmatrix}$

64. $\begin{bmatrix} 1 & 1 & 2 & 2 \\ -1 & 0 & 3 & 5 \\ -2 & -2 & -4 & 4 \end{bmatrix}$

65. $\begin{bmatrix} 1.5 & 3.6 & 4.2 \\ 0.2 & 1.4 & 1.8 \\ 2.0 & 4.4 & 6.4 \end{bmatrix}$

66. $\begin{bmatrix} 4.1 & 8.3 & 1.6 \\ 3.2 & -1.7 & 2.4 \\ -2.3 & 1.0 & 1.2 \end{bmatrix}$

Gaussian Elimination with Back-Substitution **In Exercises 67–70, use matrices to solve the system of equations, if possible. Use Gaussian elimination with back-substitution.**

67. $\begin{cases} 5x + 4y = 2 \\ -x + y = -22 \end{cases}$

68. $\begin{cases} 6x - 9y = -3 \\ 4x - 6y = 3 \end{cases}$

69. $\begin{cases} 2x + 3y + 3z = 3 \\ 6x + 6y + 12z = 13 \\ 12x + 9y - z = 2 \end{cases}$

70. $\begin{cases} x + 2y - z = 1 \\ y + z = 0 \end{cases}$

Gauss-Jordan Elimination **In Exercises 71–74, use matrices to solve the system of equations, if possible. Use Gauss-Jordan elimination.**

71. $\begin{cases} x + y + z = -3 \\ 4x + y - 2z = -12 \end{cases}$

72. $\begin{cases} 2x + 3y = 1 \\ -3x - y = -12 \end{cases}$

73. $\begin{cases} x + y + 2z = 4 \\ x - y + 4z = 1 \\ x + 3z = -5 \end{cases}$

74. $\begin{cases} x + 2y - z = 3 \\ x - y - z = -3 \\ 2x + y + 3z = 10 \end{cases}$

Using a Graphing Utility **In Exercises 75 and 76, use the matrix capabilities of a graphing utility to reduce the augmented matrix corresponding to the system of equations, and solve the system.**

75. $\begin{cases} x + 2y - z = 7 \\ -y - z = 4 \\ 4x - z = 16 \end{cases}$

76. $\begin{cases} 3x + 6z = 0 \\ -2x + y = 5 \\ y + 2z = 3 \\ x + 2y + z = 1 \end{cases}$

7.5 What did you learn?

Decide whether two matrices are equal *(p. 526)*. Two matrices are equal when they have the same dimension and all of their corresponding entries are equal.

Add and subtract matrices and multiply matrices by scalars *(p. 527)*. Let $A = [a_{ij}]$ and $B = [b_{ij}]$ be matrices of dimension $m \times n$. Then $A + B = [a_{ij} + b_{ij}]$ and $cA = [ca_{ij}]$, where c is a scalar. Properties of matrix addition and scalar multiplication are listed on page 528.

Multiply two matrices *(p. 530)*. Let $A = [a_{ij}]$ be an $m \times n$ matrix and let $B = [b_{ij}]$ be an $n \times p$ matrix. The product AB is an $m \times p$ matrix given by $AB = [c_{ij}]$, where $c_{ij} = a_{i1}b_{1j} + a_{i2}b_{2j} + a_{i3}b_{3j} + \dots + a_{in}b_{nj}$. Properties of matrix multiplication are listed on page 532.

Use matrices to perform vector operations and to transform vectors *(p. 533)*. You can represent a vector as an $n \times 1$ column matrix (column vector).

One way to transform a column vector **v** is to multiply **v** by a square transformation matrix A to produce another vector $A\mathbf{v}$.

Use matrix operations to model and solve real-life problems *(p. 534)*. A matrix equation can be used to represent and solve a system of linear equations (see Example 12). A matrix product can be used to find total equipment costs (see Example 13).

Example Let $A = \begin{bmatrix} 1 & -2 \\ 3 & -4 \end{bmatrix}$, $B = \begin{bmatrix} -5 & 6 \\ -7 & 8 \end{bmatrix}$, $D = \begin{bmatrix} -1 & 0 \\ 0 & -1 \end{bmatrix}$, and $\mathbf{v} = \langle 3, 4 \rangle$.

(a) $A + B = \begin{bmatrix} 1 & -2 \\ 3 & -4 \end{bmatrix} + \begin{bmatrix} -5 & 6 \\ -7 & 8 \end{bmatrix} = \begin{bmatrix} -4 & 4 \\ -4 & 4 \end{bmatrix}$

(b) $3(A - B) = 3\begin{bmatrix} 6 & -8 \\ 10 & -12 \end{bmatrix} = \begin{bmatrix} 18 & -24 \\ 30 & -36 \end{bmatrix}$

(c) $AB = \begin{bmatrix} 1(-5) + (-2)(-7) & 1(6) + (-2)8 \\ 3(-5) + (-4)(-7) & 3(6) + (-4)8 \end{bmatrix} = \begin{bmatrix} 9 & -10 \\ 13 & -14 \end{bmatrix}$

(d) $BA = \begin{bmatrix} -5(1) + 6(3) & -5(-2) + 6(-4) \\ -7(1) + 8(3) & -7(-2) + 8(-4) \end{bmatrix} = \begin{bmatrix} 13 & -14 \\ 17 & -18 \end{bmatrix}$

(e) $D\mathbf{v} = \begin{bmatrix} -1 & 0 \\ 0 & -1 \end{bmatrix}\begin{bmatrix} 3 \\ 4 \end{bmatrix} = \begin{bmatrix} -3 \\ -4 \end{bmatrix}$

The graph of $D\mathbf{v}$ is a reflection of the graph of **v** in the x-axis and in the y-axis.

Equality of Matrices **In Exercises 77 and 78, find x and y.**

77. $\begin{bmatrix} -1 & x \\ y & 9 \end{bmatrix} = \begin{bmatrix} -1 & 12 \\ -7 & 9 \end{bmatrix}$

78. $\begin{bmatrix} x+3 & 4 & -4y \\ 0 & -3 & 2 \\ -2 & y+5 & 6x \end{bmatrix} = \begin{bmatrix} 5x-1 & 4 & -44 \\ 0 & -3 & 2 \\ -2 & 16 & 6 \end{bmatrix}$

Operations with Matrices **In Exercises 79 and 80, find, if possible, (a) $A + B$, (b) $A - B$, (c) $2A$, and (d) $A + 3B$.**

79. $A = \begin{bmatrix} 7 & 3 \\ -1 & 5 \end{bmatrix}$, $B = \begin{bmatrix} 10 & -20 \\ 14 & -3 \end{bmatrix}$

80. $A = \begin{bmatrix} -11 & 16 & 19 \\ -7 & -2 & 1 \end{bmatrix}$, $B = \begin{bmatrix} 6 & 0 \\ 8 & -4 \\ -2 & 10 \end{bmatrix}$

Evaluating an Expression **In Exercises 81–84, evaluate the expression.**

81. $\begin{bmatrix} 2 & 1 & 0 \\ 0 & 5 & -4 \end{bmatrix} - 3\begin{bmatrix} 5 & 3 & -6 \\ 0 & -2 & 5 \end{bmatrix}$

82. $-1\begin{bmatrix} 8 & -1 \\ -2 & 4 \end{bmatrix} + 5\left(\begin{bmatrix} -2 & 0 \\ 3 & -1 \end{bmatrix} + \begin{bmatrix} 7 & -8 \\ 4 & 3 \end{bmatrix}\right)$

83. $-4\begin{bmatrix} 1 & 2 \\ 5 & -4 \\ 6 & 0 \end{bmatrix} + 8\begin{bmatrix} 7 & 1 \\ 1 & 2 \\ 1 & 4 \end{bmatrix}$

84. $6\left(\begin{bmatrix} -4 & -1 & -3 & 4 \\ 2 & -5 & 7 & -10 \end{bmatrix} - \begin{bmatrix} -1 & 1 & 13 & -7 \\ 14 & -3 & 8 & -1 \end{bmatrix}\right)$

Operations with Matrices **In Exercises 85 and 86, use the matrix capabilities of a graphing utility to evaluate the expression.**

85. $-\frac{3}{8}\begin{bmatrix} 8 & -2 & 5 \\ 1 & 3 & -1 \end{bmatrix} + 6\begin{bmatrix} 4 & -2 & -3 \\ 2 & 7 & 6 \end{bmatrix}$

86. $-5\begin{bmatrix} 2.7 & 0.2 \\ 7.3 & -2.9 \\ 8.6 & 2.1 \end{bmatrix} + \frac{7}{4}\begin{bmatrix} 4.4 & -2.3 \\ 6.6 & 11.6 \\ -1.5 & 3.9 \end{bmatrix}$

Solving a Matrix Equation **In Exercises 87 and 88, solve for X in the equations, where**

$$A = \begin{bmatrix} -4 & 0 \\ 1 & -5 \\ -3 & 2 \end{bmatrix} \text{ and } B = \begin{bmatrix} -2 & 2 \\ -2 & 1 \\ 4 & 4 \end{bmatrix}.$$

87. $6X = 6A + 3B$

88. $3A = 2X + 9B$

Finding the Product of Two Matrices In Exercises 89 and 90, find AB.

89. $A = \begin{bmatrix} 1 & 2 \\ 5 & -4 \\ 6 & 0 \end{bmatrix}, \quad B = \begin{bmatrix} 6 & -2 & 8 \\ 4 & 0 & 0 \end{bmatrix}$

90. $A = \begin{bmatrix} 6 & -5 & 7 \end{bmatrix}, \quad B = \begin{bmatrix} -1 & 2 & 1 \\ 4 & -3 & 0 \\ 8 & 1 & 1 \end{bmatrix}$

Operations with Matrices In Exercises 91 and 92, use the matrix capabilities of a graphing utility to evaluate the expression.

91. $\begin{bmatrix} 4 & 1 \\ 11 & -7 \\ 12 & 3 \end{bmatrix}\begin{bmatrix} 3 & -5 & 6 \\ 2 & -2 & -2 \end{bmatrix}$

92. $\begin{bmatrix} 1 & -1 \\ 5 & 3 \end{bmatrix}\begin{bmatrix} 0 & 3 \\ 2 & 1 \end{bmatrix}\begin{bmatrix} 1 & 0 \\ 4 & -2 \end{bmatrix}$

Describing a Vector Transformation In Exercises 93 and 94, find $A\mathbf{v}$, where $\mathbf{v} = \langle 2, 5 \rangle$, and describe the transformation.

93. $A = \begin{bmatrix} 0 & -1 \\ -1 & 0 \end{bmatrix}$ **94.** $A = \begin{bmatrix} 1 & 0 \\ 0 & 6 \end{bmatrix}$

Manufacturing In Exercises 95 and 96, use the following information. A tire corporation has three factories, each of which manufactures two models of tires. The production levels are represented by A.

Factory (columns 1, 2, 3); Model (rows A, B)

$$A = \begin{bmatrix} 80 & 120 & 140 \\ 40 & 100 & 80 \end{bmatrix}$$

95. The price per unit for a model A tire is \$99 and the price per unit for a model B tire is \$112. Organize the price per unit of each tire in a matrix B.

96. Find the product BA, where B is the matrix you found in Exercise 95, and interpret the result.

7.6 *What did you learn?*

Verify that two matrices are inverses of each other *(p. 541)*. Let A be an $n \times n$ matrix and let I_n be the $n \times n$ identity matrix. If there exists a matrix A^{-1} such that $AA^{-1} = I_n = A^{-1}A$, then A^{-1} is called the inverse of A.

Use Gauss-Jordan elimination to find inverses of matrices *(p. 542)*. Write the $n \times 2n$ matrix $[A \ \vdots \ I]$. If possible, row reduce A to I using elementary row operations. The result will be the matrix $[I \ \vdots \ A^{-1}]$. If this is not possible, then A is singular (not invertible). Check your work by multiplying to see that $AA^{-1} = I_n = A^{-1}A$.

Use a formula to find inverses of 2 × 2 matrices *(p. 545)*.

Let $A = \begin{bmatrix} a & b \\ c & d \end{bmatrix}$.

If $ad - bc \neq 0$, then

$$A^{-1} = \frac{1}{ad - bc}\begin{bmatrix} d & -b \\ -c & a \end{bmatrix}.$$

Use inverse matrices to solve systems of linear equations *(p. 546)*. If A is an invertible matrix, then the system of linear equations represented by $AX = B$ has a unique solution given by $X = A^{-1}B$ (see Example 5).

Example Let $A = \begin{bmatrix} -4 & -1 \\ 7 & 2 \end{bmatrix}$.

(a) $A^{-1} = \dfrac{1}{-4(2) - (-1)(7)}\begin{bmatrix} 2 & 1 \\ -7 & -4 \end{bmatrix} = \begin{bmatrix} -2 & -1 \\ 7 & 4 \end{bmatrix}$.

$AA^{-1} = \begin{bmatrix} -4 & -1 \\ 7 & 2 \end{bmatrix}\begin{bmatrix} -2 & -1 \\ 7 & 4 \end{bmatrix} = \begin{bmatrix} 1 & 0 \\ 0 & 1 \end{bmatrix}$ ✓

$A^{-1}A = \begin{bmatrix} -2 & -1 \\ 7 & 4 \end{bmatrix}\begin{bmatrix} -4 & -1 \\ 7 & 2 \end{bmatrix} = \begin{bmatrix} 1 & 0 \\ 0 & 1 \end{bmatrix}$ ✓

(b) $[A \ \vdots \ I] = \left[\begin{array}{cc:cc} -4 & -1 & 1 & 0 \\ 7 & 2 & 0 & 1 \end{array}\right]$

$\begin{array}{l} 7R_1 \rightarrow \\ 4R_2 \rightarrow \end{array}\left[\begin{array}{cc:cc} -28 & -7 & 7 & 0 \\ 28 & 8 & 0 & 4 \end{array}\right]$

$R_1 + R_2 \rightarrow \left[\begin{array}{cc:cc} -28 & -7 & 7 & 0 \\ 0 & 1 & 7 & 4 \end{array}\right]$

$R_1 + 7R_2 \rightarrow \left[\begin{array}{cc:cc} -28 & 0 & 56 & 28 \\ 0 & 1 & 7 & 4 \end{array}\right]$

$-\frac{1}{28}R_1 \rightarrow \left[\begin{array}{cc:cc} 1 & 0 & -2 & -1 \\ 0 & 1 & 7 & 4 \end{array}\right]$

$= [I \ \vdots \ A^{-1}]$

The Inverse of a Matrix In Exercises 97 and 98, show that B is the inverse of A.

97. $A = \begin{bmatrix} 7 & 2 \\ -4 & -1 \end{bmatrix} \quad B = \begin{bmatrix} -1 & -2 \\ 4 & 7 \end{bmatrix}$

98. $A = \begin{bmatrix} 1 & 1 & 0 \\ 1 & 0 & 1 \\ 6 & 2 & 3 \end{bmatrix}, \quad B = \begin{bmatrix} -2 & -3 & 1 \\ 3 & 3 & -1 \\ 2 & 4 & -1 \end{bmatrix}$

Finding the Inverse of a Matrix In Exercises 99–102, find the inverse of the matrix (if it exists).

99. $\begin{bmatrix} -6 & 5 \\ -5 & 4 \end{bmatrix}$ **100.** $\begin{bmatrix} -4 & -12 \\ 3 & 9 \end{bmatrix}$

101. $\begin{bmatrix} 2 & 0 & 3 \\ -1 & 1 & 1 \\ 2 & -2 & 1 \end{bmatrix}$ **102.** $\begin{bmatrix} 0 & -2 & 1 \\ -5 & -2 & -3 \\ 7 & 3 & 4 \end{bmatrix}$

Finding the Inverse of a Matrix Using Technology **In Exercises 103 and 104, use the matrix capabilities of a graphing utility to find the inverse of the matrix (if it exists).**

103. $\begin{bmatrix} 2 & 6 \\ 3 & -6 \end{bmatrix}$

104. $\begin{bmatrix} 1 & -1 & -2 \\ 0 & 3 & -2 \\ 1 & 2 & -4 \end{bmatrix}$

Finding the Inverse of a 2 × 2 Matrix **In Exercises 105 and 106, use the formula on page 545 to find the inverse of the 2 × 2 matrix (if it exists).**

105. $\begin{bmatrix} -7 & 2 \\ -8 & 2 \end{bmatrix}$

106. $\begin{bmatrix} 10 & 4 \\ 7 & 3 \end{bmatrix}$

Solving a System of Equations Using an Inverse **In Exercises 107–110, use an inverse matrix to solve (if possible) the system of linear equations.**

107. $\begin{cases} -x + 4y = 8 \\ 2x - 7y = -5 \end{cases}$

108. $\begin{cases} 2x + 3y = -10 \\ 4x - y = 1 \end{cases}$

109. $\begin{cases} 3x + 2y - z = 4 \\ x - y + 2z = 3 \\ 5x + y + z = 9 \end{cases}$

110. $\begin{cases} -x + 4y - 2z = 0 \\ 2x - 9y + 5z = -1 \\ -x + 5y - 4z = -2 \end{cases}$

Solving a System of Equations Using Technology **In Exercises 111–114, use the matrix capabilities of a graphing utility to solve the system of linear equations.**

111. $\begin{cases} x + 2y = -1 \\ 3x + 4y = -5 \end{cases}$

112. $\begin{cases} x + 3y = 23 \\ -6x + 2y = -18 \end{cases}$

113. $\begin{cases} -3x - 3y - 4z = 2 \\ y + z = -1 \\ 4x + 3y + 4z = -1 \end{cases}$

114. $\begin{cases} x + y + z + w = 1 \\ x - y + 2z + w = -3 \\ y + w = 2 \\ x + w = 2 \end{cases}$

7.7 *What did you learn?*

Find the determinants of 2 × 2 matrices ***(p. 550).***

$$\det(A) = |A| = \begin{vmatrix} a_1 & b_1 \\ a_2 & b_2 \end{vmatrix} = a_1b_2 - a_2b_1$$

Find minors and cofactors of square matrices ***(p. 552).*** If A is a square matrix, then the minor M_{ij} of the entry a_{ij} is the determinant of the matrix obtained by deleting the ith row and jth column of A. The cofactor C_{ij} of the entry a_{ij} is given by $C_{ij} = (-1)^{i+j}M_{ij}$.

Find the determinants of square matrices ***(p. 553).*** The determinant of a square matrix A (of dimension 2 × 2 or greater) is the sum of the entries in any row (or column) of A multiplied by their respective cofactors.

Example Let $A = \begin{bmatrix} -2 & 4 & 1 \\ -6 & 0 & 2 \\ 5 & 3 & 4 \end{bmatrix}$.

$M_{21} = \begin{vmatrix} 4 & 1 \\ 3 & 4 \end{vmatrix} = 4(4) - 1(3) = 13;$

$C_{21} = (-1)^{2+1}M_{21} = -13;$

$M_{22} = \begin{vmatrix} -2 & 1 \\ 5 & 4 \end{vmatrix} = -13; \; C_{22} = (-1)^{2+2}M_{22} = -13;$

$M_{23} = \begin{vmatrix} -2 & 4 \\ 5 & 3 \end{vmatrix} = -26; \; C_{23} = (-1)^{2+3}M_{23} = 26$

$$|A| = a_{21}C_{21} + a_{22}C_{22} + a_{23}C_{23} = -6(-13) + 2(26) = 130$$

The Determinant of a Matrix **In Exercises 115–118, find the determinant of the matrix.**

115. $[-17]$

116. $[0]$

117. $\begin{bmatrix} 8 & 5 \\ 2 & -4 \end{bmatrix}$

118. $\begin{bmatrix} -9 & -11 \\ 7 & -4 \end{bmatrix}$

Finding the Minors and Cofactors of a Matrix **In Exercises 119–122, find all the (a) minors and (b) cofactors of the matrix.**

119. $\begin{bmatrix} 2 & -1 \\ 7 & 4 \end{bmatrix}$

120. $\begin{bmatrix} 3 & 6 \\ 5 & -4 \end{bmatrix}$

121. $\begin{bmatrix} 3 & 2 & -1 \\ -2 & 5 & 0 \\ 1 & 8 & 6 \end{bmatrix}$

122. $\begin{bmatrix} 8 & 3 & 4 \\ 6 & 5 & -9 \\ -4 & 1 & 2 \end{bmatrix}$

Finding a Determinant **In Exercises 123–128, find the determinant of the matrix. Expand by cofactors using the row or column that appears to make the computations easiest.**

123. $\begin{bmatrix} -2 & 5 & 0 \\ 2 & -1 & 0 \\ -1 & 1 & -3 \end{bmatrix}$

124. $\begin{bmatrix} 4 & 7 & -1 \\ 2 & -3 & 4 \\ -5 & 0 & -1 \end{bmatrix}$

125. $\begin{bmatrix} 0 & 1 & -2 \\ 0 & 1 & 2 \\ -1 & -1 & 3 \end{bmatrix}$

126. $\begin{bmatrix} 0 & 3 & 1 \\ 5 & -2 & 1 \\ 1 & 6 & 1 \end{bmatrix}$

127. $\begin{bmatrix} 3 & 0 & -4 & 0 \\ 0 & 8 & 1 & 2 \\ 6 & 1 & 8 & 2 \\ 0 & 3 & -4 & 1 \end{bmatrix}$

128. $\begin{bmatrix} -5 & 6 & 0 & 0 \\ 0 & 1 & -1 & 2 \\ -3 & 4 & -5 & 1 \\ 1 & 6 & 0 & 3 \end{bmatrix}$

7.8 *What did you learn?*

Use determinants to find areas of triangles *(p. 557)* and determine whether points are collinear *(p. 558)*. The area of a triangle with vertices (x_1, y_1), (x_2, y_2), and (x_3, y_3) is $\pm\frac{1}{2}\det(A)$, where A is a 3×3 matrix with the x-coordinates in column 1, the corresponding y-coordinates in column 2, and 1's in column 3. The symbol $(\pm)$ indicates that the appropriate sign should be chosen to yield a positive area. Three points (x_1, y_1), (x_2, y_2), and (x_3, y_3) are collinear (lie on the same line) if and only if $\det(A) = 0$.

Example

(a) $(-3, 2)$, $(3, 2)$, and $(1, -2)$ are not collinear because

$$\begin{vmatrix} -3 & 2 & 1 \\ 3 & 2 & 1 \\ 1 & -2 & 1 \end{vmatrix} = -24 \neq 0.$$

The area of the triangle with these points as vertices is $-\frac{1}{2}(-24) = 12$ square units.

Use matrices to perform transformations in the plane and find areas of parallelograms *(p. 559)*. See page 559 for several transformations and their corresponding transformation matrices. The area of a parallelogram with vertices $(0, 0)$, (a, b), (c, d), and $(a + c, b + d)$ is $|\det(A)|$, where A is given on page 559.

(b) The area of the parallelogram with vertices $(0, 0)$, $(4, 0)$, $(1, 3)$, and $(5, 3)$ is

$$\begin{vmatrix} 4 & 0 \\ 1 & 3 \end{vmatrix} = 12 \text{ square units.}$$

Use Cramer's Rule to solve systems of linear equations *(p. 560)*. If a system of n linear equations in n variables has a coefficient matrix A with a *nonzero* determinant $|A|$, then the solution of the system is $x_1 = |A_1|/|A|$, $x_2 = |A_2|/|A|, \ldots, x_n = |A_n|/|A|$, where the ith column of A_i is the column of constants in the system of equations. If $|A| = 0$, then the system has either no solution or infinitely many solutions.

(c) $\begin{cases} 2x - y = -10 \\ 3x + 2y = -1 \end{cases}$

The determinant of the coefficient matrix is

$$D = \begin{vmatrix} 2 & -1 \\ 3 & 2 \end{vmatrix} = 7.$$

$$x = \frac{D_x}{D} = \frac{\begin{vmatrix} -10 & -1 \\ -1 & 2 \end{vmatrix}}{7} = \frac{-21}{7} = -3$$

$$y = \frac{D_y}{D} = \frac{\begin{vmatrix} 2 & -10 \\ 3 & -1 \end{vmatrix}}{7} = \frac{28}{7} = 4$$

Use matrices to encode and decode messages *(p. 563)*. See Examples 8 and 9.

Testing for Collinear Points or Finding Area **In Exercises 129 and 130, use a determinant to determine whether the points are collinear. If not, find the area of the triangle with the given points as vertices.**

129. $(2, 4), (5, 6), (4, 1)$ **130.** $(0, -5), (2, 1), (4, 7)$

Finding the Area of a Parallelogram **In Exercises 131 and 132, use a determinant to find the area of the parallelogram with the given vertices.**

131. $(0, 0), (2, 0), (1, 4), (3, 4)$

132. $(0, 0), (-3, 0), (1, 3), (-2, 3)$

Using Cramer's Rule **In Exercises 133–136, use Cramer's Rule to solve (if possible) the system of equations.**

133. $\begin{cases} x + 2y = 5 \\ -x + y = 1 \end{cases}$ **134.** $\begin{cases} 8x - 4y = 3 \\ -12x + 6y = -15 \end{cases}$

135. $\begin{cases} x - 3y + 2z = 3 \\ 2x + 2y - 3z = 3 \\ x - 7y + 8z = -3 \end{cases}$ **136.** $\begin{cases} 14x - 21y - 7z = 14 \\ -4x + 2y - 2z = 4 \\ 56x - 21y + 7z = 14 \end{cases}$

137. Encoding a Message Write the uncoded 1×3 row matrices for the message. Then use the matrix A to encode the message.

LOOK OUT BELOW; $A = \begin{bmatrix} 2 & -2 & 0 \\ 3 & 0 & -3 \\ -6 & 2 & 3 \end{bmatrix}$

138. Decoding a Message Use the inverse of matrix A to decode the cryptogram.

$A = \begin{bmatrix} 1 & 0 & -1 \\ -1 & -2 & 0 \\ 1 & -2 & 2 \end{bmatrix}$

32 −46 37 9 −48
15 3 −14 10 −1
−6 2 −8 −22 −3

Focusing on Concepts

True or False? **In Exercises 139 and 140, determine whether the statement is true or false. Justify your answer.**

139. A nonsingular matrix can have dimension 2×3.

140. Solving a system of equations will always give an exact solution.

141. Think About It Under what conditions does a matrix have an inverse?

7 Chapter Test

See CalcChat.com for tutorial help and worked-out solutions to odd-numbered exercises.
For instructions on how to use a graphing utility, see Appendix A.

Take this test as you would take a test in class. After you are finished, check your work against the answers given in the back of the book.

In Exercises 1 and 2, solve the system by the method of substitution. Use a graphing utility to verify your results.

1. $\begin{cases} x - y = 6 \\ 3x + 5y = 2 \end{cases}$

2. $\begin{cases} y = x - 1 \\ y = \sqrt{x - 1} \end{cases}$

In Exercises 3 and 4, solve the system by the method of elimination and check any solutions algebraically.

3. $\begin{cases} 4x - y = -5 \\ 3x + 2y = -12 \end{cases}$

4. $\begin{cases} 1.5x + 3y = -9 \\ 2x - 1.2y = 5 \end{cases}$

5. Use any method to solve the system below. Explain your choice of method.
$\begin{cases} x + 2y = 10 \\ y = x - 2 \end{cases}$

6. Find the equation of the parabola $y = ax^2 + bx + c$ that passes through the points $(0, 6)$, $(-2, 2)$, and $\left(3, \frac{9}{2}\right)$.

In Exercises 7 and 8, write the partial fraction decomposition for the rational expression.

7. $\dfrac{5x - 2}{(x - 1)^2}$

8. $\dfrac{x^3 + x^2 + x + 2}{x^4 + x^2}$

In Exercises 9 and 10, use matrices to solve the system of equations.

9. $\begin{cases} 2x + y + 2z = 4 \\ 2x + 2y = 5 \\ 2x - y + 6z = 2 \end{cases}$

10. $\begin{cases} 2x + 3y + z = 10 \\ 2x - 3y - 3z = 22 \\ 4x - 2y + 3z = -2 \end{cases}$

11. If possible, find (a) $A - B$, (b) $3A$, (c) $3A - 2B$, and (d) AB.
$A = \begin{bmatrix} 5 & 4 & 4 \\ -4 & -4 & 0 \\ 1 & 2 & 0 \end{bmatrix}$, $B = \begin{bmatrix} 4 & 4 & 0 \\ 3 & 2 & 1 \\ 1 & -2 & 0 \end{bmatrix}$

12. Use an inverse matrix to solve the system below.
$\begin{cases} -2x + 2y + 3z = 7 \\ x - y = -5 \\ y + 4z = -1 \end{cases}$

In Exercises 13 and 14, find the determinant of the matrix.

13. $\begin{bmatrix} -25 & 18 \\ 6 & -7 \end{bmatrix}$

14. $\begin{bmatrix} 4 & 0 & 3 \\ 1 & -8 & 2 \\ 3 & 2 & 2 \end{bmatrix}$

15. Use a determinant to find the area of the triangle whose vertices are $(-2, -3)$, $(0, 4)$, and $(1, 1)$.

16. Use Cramer's Rule to solve $\begin{cases} 2x - 2y = 3 \\ x + 4y = -1 \end{cases}$.

17. The flow of traffic (in vehicles per hour) through a network of streets is shown at the right. Solve the system for the traffic flow represented by x_i, $i = 1, 2, 3, 4$, and 5.

Collaborative Project

To work a collaborative project involving Linear Systems and Matrices, visit this textbook's website at *LarsonPrecalculus.com.*

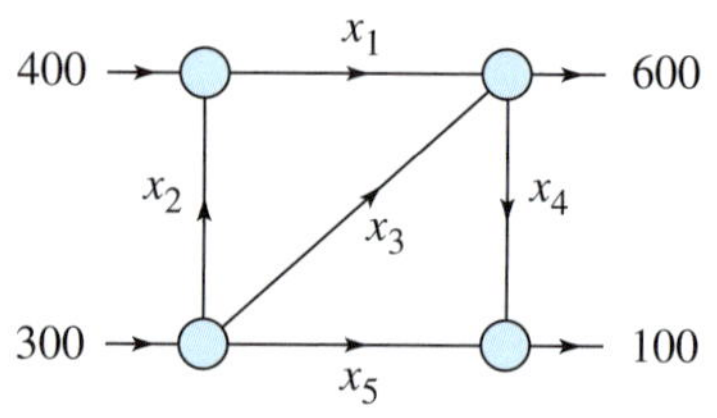

Figure for 17

Proofs in Mathematics

An **indirect proof** can be useful in proving statements of the form "p implies q." Recall that the conditional statement $p \to q$ is false only when p is true and q is false. To prove a conditional statement indirectly, assume that p is true and q is false. If this assumption leads to an impossibility, then you have proved that the conditional statement is true. An indirect proof is also called a **proof by contradiction.**

You can use an indirect proof to prove the conditional statement

"If a is a positive integer and a^2 is divisible by 2, then a is divisible by 2"

as follows. First, assume that p, "a is a positive integer and a^2 is divisible by 2," is true and q, "a is divisible by 2," is false. This means that a is not divisible by 2. If so, then a is odd and can be written as

$$a = 2n + 1$$

where n is an integer.

$$a = 2n + 1 \qquad \text{Definition of an odd integer}$$

$$a^2 = 4n^2 + 4n + 1 \qquad \text{Square each side.}$$

$$a^2 = 2(2n^2 + 2n) + 1 \qquad \text{Distributive Property}$$

So, by the definition of an odd integer, a^2 is odd. This contradicts the assumption, and you can conclude that a is divisible by 2.

EXAMPLE Using an Indirect Proof

Use an indirect proof to prove that $\sqrt{2}$ is an irrational number.

Solution

Begin by assuming that $\sqrt{2}$ is *not* an irrational number. Then $\sqrt{2}$ can be written as the quotient of two integers a and b $(b \neq 0)$ that have no common factors.

$$\sqrt{2} = \frac{a}{b} \qquad \text{Assume that } \sqrt{2} \text{ is a rational number.}$$

$$2 = \frac{a^2}{b^2} \qquad \text{Square each side.}$$

$$2b^2 = a^2 \qquad \text{Multiply each side by } b^2.$$

This implies that 2 is a factor of a^2. So, 2 is also a factor of a, and a can be written as $2c$, where c is an integer.

$$2b^2 = (2c)^2 \qquad \text{Substitute } 2c \text{ for } a.$$

$$2b^2 = 4c^2 \qquad \text{Simplify.}$$

$$b^2 = 2c^2 \qquad \text{Divide each side by 2.}$$

This implies that 2 is a factor of b^2 and also a factor of b. So, 2 is a factor of both a and b. This contradicts the assumption that a and b have no common factors. So, you can conclude that $\sqrt{2}$ is an irrational number. ■

Proofs without words are pictures or diagrams that give a visual understanding of why a theorem or statement is true. They can also provide a starting point for writing a formal proof. The following proof shows that a 2×2 determinant is the area of a parallelogram.

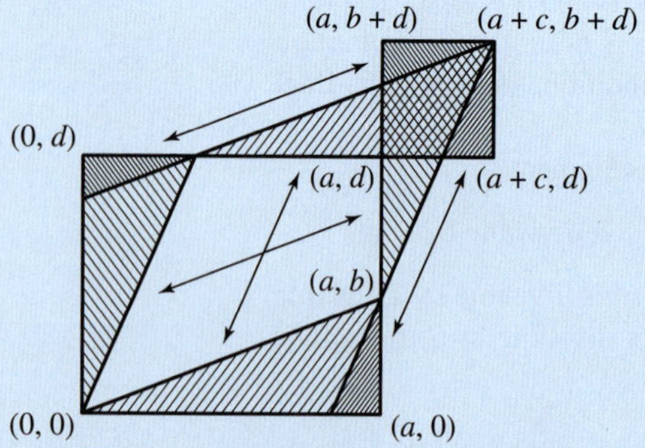

$$\begin{vmatrix} a & b \\ c & d \end{vmatrix} = ad - bc = \|\text{▭}\| - \|\text{□}\| = \|\text{▱}\|$$

The following is a color-coded version of the proof along with a brief explanation of why this proof works.

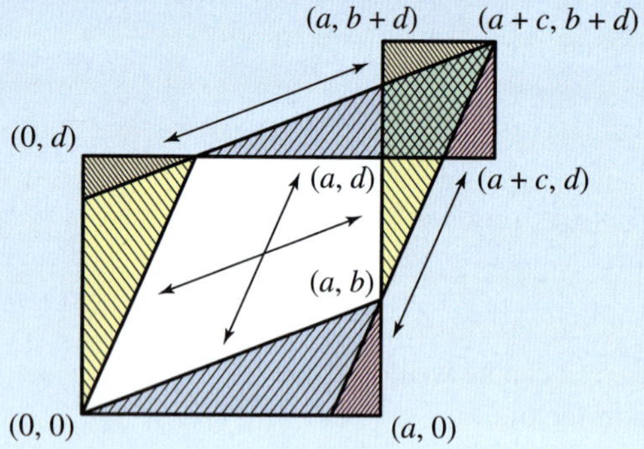

$$\begin{vmatrix} a & b \\ c & d \end{vmatrix} = ad - bc = \|\text{▭}\| - \|\text{□}\| = \|\text{▱}\|$$

Area of ▭ = Area of orange △ + Area of yellow △ + Area of blue △ + Area of pink △ + Area of white quadrilateral

Area of □ = Area of orange △ + Area of pink △ + Area of green quadrilateral

Area of ▱ = Area of white quadrilateral + Area of blue △ + Area of yellow △ − Area of green quadrilateral

= Area of ▭ − Area of □

From "Proof Without Words: A 2×2 Determinant Is the Area of a Parallelogram" by Solomon W. Golomb, *Mathematics Magazine*, March 1985. Vol. 58, No. 2, pg. 107.

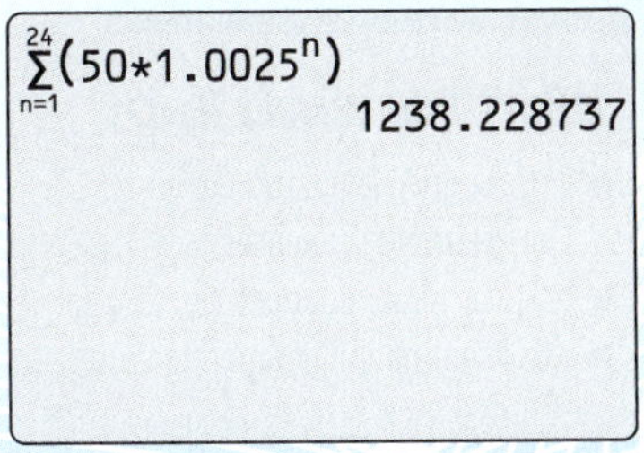

Section 8.3, Example 8
Increasing Annuity

8 Sequences, Series, and Probability

Student Resources at LarsonPrecalculus.com

- **Videos** explaining the concepts of precalculus
- **Worked-out solution videos** for all *Checkpoint* exercises
- **Editable spreadsheets** of the data sets in the text
- **Group projects** for each chapter applying concepts to real-life problems

8.1 Sequences and Series

What you should learn

- Use sequence notation to write the terms of sequences.
- Use factorial notation.
- Use summation notation to write sums.
- Find sums of infinite series.
- Use sequences and series to model and solve real-life problems.

Why you should learn it

Sequences and series are useful in modeling sets of values in order to identify patterns. For instance, Exercise 130 on page 590 shows how a sequence can be used to model the population of trout in a lake.

Sequences

In mathematics, the word *sequence* is used in much the same way as in ordinary English. Saying that a collection is listed in *sequence* means that it is ordered so that it has a first member, a second member, a third member, and so on.

Mathematically, you can think of a sequence as a *function* whose domain is the set of positive integers. Instead of using function notation, sequences are usually written using subscript notation, as shown in the following definition.

Definition of Sequence

An **infinite sequence** is a function whose domain is the set of positive integers. The function values

$$a_1, a_2, a_3, a_4, \ldots, a_n, \ldots$$

are the **terms** of the sequence. If the domain of a function consists of the first n positive integers only, then the sequence is a **finite sequence.**

On occasion, it is convenient to begin subscripting a sequence with 0 instead of 1 so that the terms of the sequence become

$$a_0, a_1, a_2, a_3, \ldots.$$

The domain of the function is the set of nonnegative integers.

EXAMPLE 1 Writing the Terms of a Sequence

Write the first four terms of each sequence.

a. $a_n = 3n - 2$

b. $a_n = 3 + (-1)^n$

Solution

a. The first four terms of the sequence given by $a_n = 3n - 2$ are listed below.

$a_1 = 3(1) - 2 = 1$ 1st term

$a_2 = 3(2) - 2 = 4$ 2nd term

$a_3 = 3(3) - 2 = 7$ 3rd term

$a_4 = 3(4) - 2 = 10$ 4th term

b. The first four terms of the sequence given by $a_n = 3 + (-1)^n$ are listed below.

$a_1 = 3 + (-1)^1 = 3 - 1 = 2$ 1st term

$a_2 = 3 + (-1)^2 = 3 + 1 = 4$ 2nd term

$a_3 = 3 + (-1)^3 = 3 - 1 = 2$ 3rd term

$a_4 = 3 + (-1)^4 = 3 + 1 = 4$ 4th term

✓ Checkpoint ▶ *Audio-video solution in English & Spanish at LarsonPrecalculus.com*

Write the first four terms of the sequence given by $a_n = 2n + 1$. ■

EXAMPLE 2 Writing the Terms of a Sequence

Write the first five terms of the sequence given by $a_n = \dfrac{(-1)^n}{2n - 1}$.

Algebraic Solution

The first five terms of the sequence are as follows.

$$a_1 = \frac{(-1)^1}{2(1) - 1} = \frac{-1}{2 - 1} = -1 \qquad \text{1st term}$$

$$a_2 = \frac{(-1)^2}{2(2) - 1} = \frac{1}{4 - 1} = \frac{1}{3} \qquad \text{2nd term}$$

$$a_3 = \frac{(-1)^3}{2(3) - 1} = \frac{-1}{6 - 1} = -\frac{1}{5} \qquad \text{3rd term}$$

$$a_4 = \frac{(-1)^4}{2(4) - 1} = \frac{1}{8 - 1} = \frac{1}{7} \qquad \text{4th term}$$

$$a_5 = \frac{(-1)^5}{2(5) - 1} = \frac{-1}{10 - 1} = -\frac{1}{9} \qquad \text{5th term}$$

Numerical Solution

Set your graphing utility to *sequence* mode and enter the sequence. Use the *table* feature to create a table showing the terms of the sequence, as shown in the figure.

Use the *table* feature in *ask* mode.

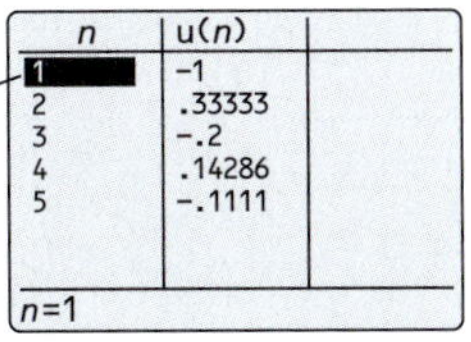

So, you can estimate the first five terms of the sequence as follows.

$$u_1 = -1, \quad u_2 = 0.33333 \approx \tfrac{1}{3}, \quad u_3 = -0.2 = -\tfrac{1}{5},$$

$$u_4 = 0.14286 \approx \tfrac{1}{7}, \quad \text{and} \quad u_5 = -0.1111 \approx -\tfrac{1}{9}$$

✓ **Checkpoint** ▶ *Audio-video solution in English & Spanish at LarsonPrecalculus.com*

Write the first four terms of the sequence given by $a_n = \dfrac{2 + (-1)^n}{n}$.

Simply listing the first few terms is not sufficient to define a unique sequence—the nth term *must be given*. To see this, consider the following sequences, both of which have the same first three terms.

$$\frac{1}{2}, \frac{1}{4}, \frac{1}{8}, \frac{1}{16}, \ldots, \frac{1}{2^n}, \ldots$$

$$\frac{1}{2}, \frac{1}{4}, \frac{1}{8}, \frac{1}{15}, \ldots, \frac{6}{(n + 1)(n^2 - n + 6)}, \ldots$$

EXAMPLE 3 Finding the nth Term of a Sequence

Write an expression for the apparent nth term a_n of each sequence.

a. 1, 3, 5, 7, . . . **b.** 2, 5, 10, 17, . . .

Solution

a. n: 1 2 3 4 . . . n

Terms: 1 3 5 7 . . . a_n

Apparent Pattern: Each term is 1 less than twice n. So, the apparent nth term is $a_n = 2n - 1$.

b. n: 1 2 3 4 . . . n

Terms: 2 5 10 17 . . . a_n

Apparent Pattern: Each term is 1 more than the square of n. So, the apparent nth term is $a_n = n^2 + 1$.

✓ **Checkpoint** ▶ *Audio-video solution in English & Spanish at LarsonPrecalculus.com*

Write an expression for the apparent nth term a_n of each sequence.

a. 1, 5, 9, 13, 17, . . . **b.** 3, 5, 7, 9, 11, . . .

Some sequences are defined **recursively.** To define a sequence recursively, you need to be given one or more of the first few terms. All other terms of the sequence are then defined using previous terms.

EXAMPLE 4 A Recursive Sequence

Write the first five terms of the sequence defined recursively as

$$a_1 = 3, a_k = 2a_{k-1} + 1, \quad \text{where} \quad k \geq 2.$$

Solution

$a_1 = 3$ — 1st term is given.

$a_2 = 2a_{2-1} + 1 = 2a_1 + 1 = 2(3) + 1 = 7$ — Use recursion formula.

$a_3 = 2a_{3-1} + 1 = 2a_2 + 1 = 2(7) + 1 = 15$ — Use recursion formula.

$a_4 = 2a_{4-1} + 1 = 2a_3 + 1 = 2(15) + 1 = 31$ — Use recursion formula.

$a_5 = 2a_{5-1} + 1 = 2a_4 + 1 = 2(31) + 1 = 63$ — Use recursion formula.

✓ Checkpoint ▶ *Audio-video solution in English & Spanish at LarsonPrecalculus.com*

Write the first five terms of the sequence defined recursively as

$$a_1 = 6, \quad a_{k+1} = a_k + 1, \quad \text{where} \quad k \geq 1.$$

In the next example you will study a well-known recursive sequence, the Fibonacci sequence.

EXAMPLE 5 The Fibonacci Sequence: A Recursive Sequence

The Fibonacci sequence is defined recursively as follows.

$$a_0 = 1, a_1 = 1, a_k = a_{k-2} + a_{k-1}, \quad \text{where } k \geq 2$$

Write the first six terms of this sequence.

Solution

$a_0 = 1$ — 0th term is given.

$a_1 = 1$ — 1st term is given.

$a_2 = a_{2-2} + a_{2-1} = a_0 + a_1 = 1 + 1 = 2$ — Use recursion formula.

$a_3 = a_{3-2} + a_{3-1} = a_1 + a_2 = 1 + 2 = 3$ — Use recursion formula.

$a_4 = a_{4-2} + a_{4-1} = a_2 + a_3 = 2 + 3 = 5$ — Use recursion formula.

$a_5 = a_{5-2} + a_{5-1} = a_3 + a_4 = 3 + 5 = 8$ — Use recursion formula.

You can check this result using the *table* feature of a graphing utility, as shown below.

n	u(n)
0	1
1	1
2	2
3	3
4	5
5	8

n=

✓ Checkpoint ▶ *Audio-video solution in English & Spanish at LarsonPrecalculus.com*

Write the first five terms of the sequence defined recursively as

$$a_0 = 1, \quad a_1 = 3, \quad a_k = a_{k-2} + a_{k-1}, \quad \text{where} \quad k \geq 2.$$

Technology Tip

To graph a sequence using a graphing utility, set the mode to *dot* and *sequence* and enter the sequence. Try graphing the sequence in Example 4 and using the *trace* feature to identify the terms. For instructions on how to use the sequence mode, see Appendix A; for specific keystrokes, go to this textbook's *Student Companion Website.*

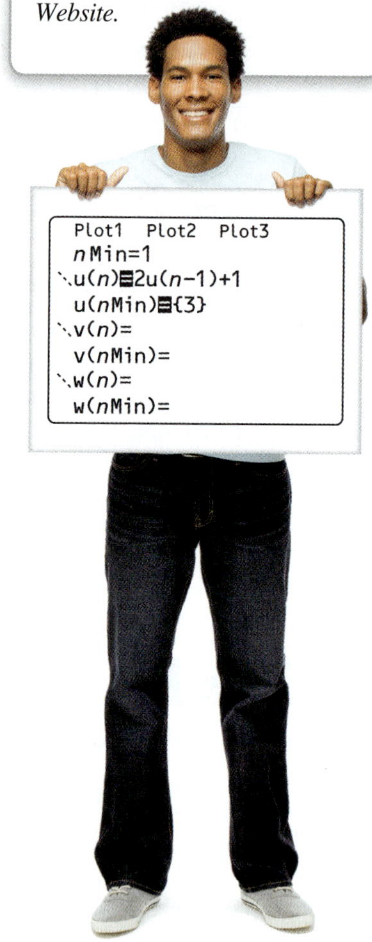

aldomurillo/iStock/Getty Images

Factorial Notation

Some very important sequences in mathematics involve terms that are defined with special types of products called **factorials.**

Definition of Factorial

If n is a positive integer, then ***n* factorial** is defined as

$$n! = 1 \cdot 2 \cdot 3 \cdot 4 \cdot \cdot \cdot (n-1) \cdot n.$$

As a special case, zero factorial is defined as

$$0! = 1.$$

Explore the Concept

Most graphing utilities have the capability to compute $n!$. Use your graphing utility to compare $3 \cdot 5!$ and $(3 \cdot 5)!$. How do they differ? Notice that the value of n does not have to be very large before the value of $n!$ becomes huge. How large a value of $n!$ will your graphing utility allow you to compute?

Notice that $0! = 1$ and $1! = 1$. Here are some other values of $n!$.

$$2! = 1 \cdot 2 = 2, \quad 3! = 1 \cdot 2 \cdot 3 = 6, \quad \text{and} \quad 4! = 1 \cdot 2 \cdot 3 \cdot 4 = 24$$

Factorials follow the same conventions for order of operations as exponents. For instance,

$$2n! = 2(n!) = 2(1 \cdot 2 \cdot 3 \cdot 4 \cdot \cdot \cdot n), \quad \text{whereas} \quad (2n)! = 1 \cdot 2 \cdot 3 \cdot 4 \cdot \cdot \cdot 2n.$$

EXAMPLE 6 Writing the Terms of a Sequence Involving Factorials

Write the first five terms of the sequence given by $a_n = \dfrac{2^n}{n!}$. Begin with $n = 0$.

Algebraic Solution

$a_0 = \dfrac{2^0}{0!} = \dfrac{1}{1} = 1$ 0th term

$a_1 = \dfrac{2^1}{1!} = \dfrac{2}{1} = 2$ 1st term

$a_2 = \dfrac{2^2}{2!} = \dfrac{4}{2} = 2$ 2nd term

$a_3 = \dfrac{2^3}{3!} = \dfrac{8}{6} = \dfrac{4}{3}$ 3rd term

$a_4 = \dfrac{2^4}{4!} = \dfrac{16}{24} = \dfrac{2}{3}$ 4th term

Graphical Solution

Using a graphing utility set to *dot* and *sequence* modes, enter the sequence. Next, graph the sequence, as shown in the figure.

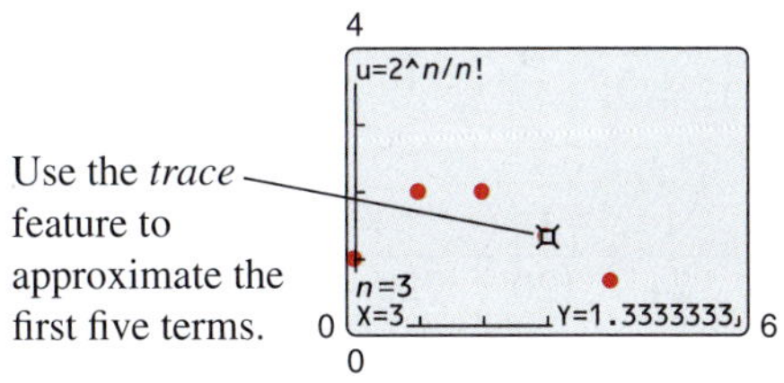

The first five terms are $u_0 = 1$, $u_1 = 2$, $u_2 = 2$, $u_3 \approx 1.333 \approx \frac{4}{3}$, and $u_4 \approx 0.667 \approx \frac{2}{3}$.

Checkpoint Audio-video solution in English & Spanish at LarsonPrecalculus.com

Write the first five terms of the sequence given by $a_n = \dfrac{3^n + 1}{n!}$. Begin with $n = 0$.

EXAMPLE 7 Simplifying Factorial Expressions

a. $\dfrac{8!}{2! \cdot 6!} = \dfrac{\cancel{1 \cdot 2 \cdot 3 \cdot 4 \cdot 5 \cdot 6} \cdot 7 \cdot 8}{1 \cdot 2 \cdot \cancel{1 \cdot 2 \cdot 3 \cdot 4 \cdot 5 \cdot 6}} = \dfrac{7 \cdot 8}{2} = 28$

b. $\dfrac{n!}{(n-1)!} = \dfrac{\cancel{1 \cdot 2 \cdot 3 \cdot \cdot \cdot (n-1)} \cdot n}{\cancel{1 \cdot 2 \cdot 3 \cdot \cdot \cdot (n-1)}} = n$

Checkpoint Audio-video solution in English & Spanish at LarsonPrecalculus.com

Simplify the factorial expression $\dfrac{4!(n+1)!}{3!n!}$.

Algebra Help

In Example 7(a), you can also simplify the computation as

$$\frac{8!}{2! \cdot 6!} = \frac{8 \cdot 7 \cdot \cancel{6!}}{2! \cdot \cancel{6!}} = 28.$$

Summation Notation

There is a convenient notation for the sum of the terms of a finite sequence. It is called **summation notation** or **sigma notation** because it involves the use of the uppercase Greek letter sigma, written as Σ.

Definition of Summation Notation

The sum of the first n terms of a sequence is represented by

$$\sum_{i=1}^{n} a_i = a_1 + a_2 + a_3 + a_4 + \cdot \cdot \cdot + a_n$$

where i is called the **index of summation,** n is the **upper limit of summation,** and 1 is the **lower limit of summation.**

Algebra Help

Summation notation is an instruction to add the terms of a sequence. From the definition at the left, the upper limit of summation tells you where to end the sum. Summation notation helps you generate the appropriate terms of the sequence prior to finding the actual sum.

EXAMPLE 8 Sigma Notation for Sums

a. $$\sum_{i=1}^{5} 3i = 3(1) + 3(2) + 3(3) + 3(4) + 3(5)$$
$$= 3(1 + 2 + 3 + 4 + 5)$$
$$= 3(15)$$
$$= 45$$

b. $$\sum_{k=3}^{6} (1 + k^2) = (1 + 3^2) + (1 + 4^2) + (1 + 5^2) + (1 + 6^2)$$
$$= 10 + 17 + 26 + 37$$
$$= 90$$

c. $$\sum_{n=0}^{8} \frac{1}{n!} = \frac{1}{0!} + \frac{1}{1!} + \frac{1}{2!} + \frac{1}{3!} + \frac{1}{4!} + \frac{1}{5!} + \frac{1}{6!} + \frac{1}{7!} + \frac{1}{8!}$$
$$= 1 + 1 + \frac{1}{2} + \frac{1}{6} + \frac{1}{24} + \frac{1}{120} + \frac{1}{720} + \frac{1}{5040} + \frac{1}{40{,}320}$$
$$\approx 2.71828$$

For the summation in part (c), note that the sum is very close to the irrational number $e \approx 2.718281828$. It can be shown that as more terms of the sequence whose nth term is $1/n!$ are added, the sum becomes closer and closer to e.

Checkpoint *Audio-video solution in English & Spanish at LarsonPrecalculus.com*

Find the sum $\sum_{i=1}^{4} (4i + 1)$.

In Example 8, note that the lower limit of a summation does not have to be 1. Also, the index of summation does not have to be the letter i. For instance, in part (b), the lower limit of summation is 3, and the index of summation is the letter k.

Properties of Sums

1. $\sum_{i=1}^{n} c = cn$, c is a constant.
2. $\sum_{i=1}^{n} ca_i = c\sum_{i=1}^{n} a_i$, c is a constant.
3. $\sum_{i=1}^{n} (a_i + b_i) = \sum_{i=1}^{n} a_i + \sum_{i=1}^{n} b_i$
4. $\sum_{i=1}^{n} (a_i - b_i) = \sum_{i=1}^{n} a_i - \sum_{i=1}^{n} b_i$

(See the proofs on page 642.)

Series

Many applications involve the sum of the terms of a finite or an infinite sequence. Such a sum is called a **series.**

Definition of a Series

Consider the infinite sequence $a_1, a_2, a_3, \ldots, a_i, \ldots$.

1. The sum of the first n terms of the sequence is called a **finite series** or the ***n*th partial sum** of the sequence and is denoted by

$$a_1 + a_2 + a_3 + \cdots + a_n = \sum_{i=1}^{n} a_i.$$

2. The sum of all the terms of the infinite sequence is called an **infinite series** and is denoted by

$$a_1 + a_2 + a_3 + \cdots + a_i + \cdots = \sum_{i=1}^{\infty} a_i.$$

EXAMPLE 9 Finding the Sum of a Series

See LarsonPrecalculus.com for an interactive version of this type of example.

For the series

$$\sum_{i=1}^{\infty} \frac{3}{10^i}$$

find (a) the third partial sum and (b) the sum.

Solution

a. The third partial sum is

$$\sum_{i=1}^{3} \frac{3}{10^i} = \frac{3}{10^1} + \frac{3}{10^2} + \frac{3}{10^3}$$

$$= 0.3 + 0.03 + 0.003$$

$$= 0.333.$$

b. The sum of the series is

$$\sum_{i=1}^{\infty} \frac{3}{10^i} = \frac{3}{10^1} + \frac{3}{10^2} + \frac{3}{10^3} + \frac{3}{10^4} + \frac{3}{10^5} + \cdots$$

$$= 0.3 + 0.03 + 0.003 + 0.0003 + 0.00003 + \cdots$$

$$= 0.33333\ldots$$

$$= \frac{1}{3}.$$

✓ Checkpoint *Audio-video solution in English & Spanish at LarsonPrecalculus.com*

For the series

$$\sum_{i=1}^{\infty} \frac{5}{10^i}$$

find (a) the fourth partial sum and (b) the sum.

Notice in Example 9(b) that the sum of an infinite series can be a finite number.

Technology Tip

Most graphing utilities are able to find the partial sum of a series using the *summation* feature or the *sum sequence* feature. Try using a graphing utility with one of these features to confirm the result in Example 9(a), as shown below.

```
 3
 Σ(3/10^I)
I=1
                    .333
sum(seq(3/10^I,I,1,3))
                    .333
```

For instructions on how to use the *summation* feature or the *sum sequence* feature, see Appendix A; for specific keystrokes, go to this textbook's *Student Companion Website.*

Application

EXAMPLE 10 Compound Interest

An investor deposits $5000 in an account that earns 3% interest compounded quarterly. The balance in the account after n quarters is given by

$$A_n = 5000\left(1 + \frac{0.03}{4}\right)^n, \quad n = 0, 1, 2, \ldots.$$

Use this information to (a) write the first three terms of the sequence and (b) find the balance in the account after 10 years by computing the 40th term of the sequence.

Solution

a. The first three terms of the sequence are as follows.

$$A_0 = 5000\left(1 + \frac{0.03}{4}\right)^0 = \$5000.00 \qquad \text{Original deposit}$$

$$A_1 = 5000\left(1 + \frac{0.03}{4}\right)^1 = \$5037.50 \qquad \text{First-quarter balance}$$

$$A_2 = 5000\left(1 + \frac{0.03}{4}\right)^2 \approx \$5075.28 \qquad \text{Second-quarter balance}$$

b. The 40th term of the sequence is

$$A_{40} = 5000\left(1 + \frac{0.03}{4}\right)^{40} \approx \$6741.74. \qquad \text{Ten-year balance}$$

✓ ***Checkpoint*** ▶ *Audio-video solution in English & Spanish at LarsonPrecalculus.com*

An investor deposits $1000 in an account that earns 3% interest compounded monthly. The balance in the account after n months is given by

$$A_n = 1000\left(1 + \frac{0.03}{12}\right)^n, \quad n = 0, 1, 2, \ldots.$$

Use this information to (a) write the first three terms of the sequence and (b) find the balance in the account after 4 years by computing the 48th term of the sequence.

Insight

You may see applications of sequences on a standardized test. For instance, a sequence is used to solve the compound interest application in Example 10.

Explore the Concept

A $3 \times 3 \times 3$ cube is created using 27 unit cubes (a unit cube has a length, width, and height of 1 unit), and only the faces of the cubes that are visible are painted blue (see the figure at the right). Complete the table below to determine how many unit cubes of the $3 \times 3 \times 3$ cube have no blue faces, one blue face, two blue faces, and three blue faces. Do the same for a $4 \times 4 \times 4$ cube, a $5 \times 5 \times 5$ cube, and a $6 \times 6 \times 6$ cube, and record your results in the table below. What type of pattern do you observe in the table? Write a formula you could use to determine the column values for an $n \times n \times n$ cube.

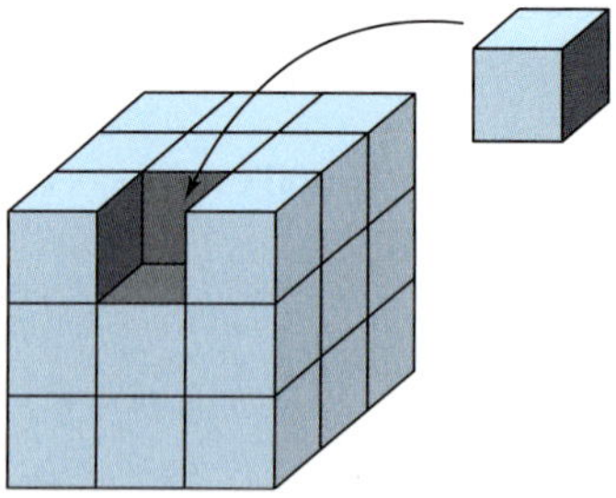

Cube \ Number of blue faces	0	1	2	3
$3 \times 3 \times 3$				

8.1 Exercises

See *CalcChat.com* for tutorial help and worked-out solutions to odd-numbered exercises. For instructions on how to use a graphing utility, see Appendix A.

Vocabulary and Concept Check

In Exercises 1–4, fill in the blank(s).

1. The function values $a_1, a_2, a_3, a_4, \ldots, a_n, \ldots$ are called the _______ of a sequence.
2. When you are given one or more of the first few terms of a sequence, and all other terms of the sequence are defined using previous terms, the sequence is defined _______ .
3. For the sum $\sum_{i=1}^{n} a_i$, i is the _______ of summation, n is the _______ of summation, and 1 is the _______ of summation.
4. The sum of the terms of a finite or infinite sequence is called a _______ .
5. Which describes an infinite sequence? a finite sequence?
   (a) The domain consists of the first n positive integers.
   (b) The domain consists of the set of positive integers.
6. Write $1 \cdot 2 \cdot 3 \cdot 4 \cdot 5$ in factorial notation.

Procedures and Problem Solving

Writing the Terms of a Sequence **In Exercises 7–16, write the first five terms of the sequence. (Assume *n* begins with 1.)**

7. $a_n = 2n - 5$
8. $a_n = 4n - 7$
9. $a_n = 3^n$
10. $a_n = \left(\frac{1}{2}\right)^n$
11. $a_n = (-2)^n$
12. $a_n = \left(-\frac{1}{2}\right)^n$
13. $a_n - \dfrac{n+1}{n}$
14. $a_n = \dfrac{n}{n+1}$
15. $a_n = \dfrac{n}{n^2+1}$
16. $a_n = \dfrac{2n}{n+1}$

Writing the Terms of a Sequence **In Exercises 17–26, write the first five terms of the sequence (a) using the *table* feature of a graphing utility and (b) algebraically. (Assume *n* begins with 1.)**

17. $a_n = \dfrac{1+(-1)^n}{n}$
18. $a_n = \dfrac{1+(-1)^n}{2n}$
19. $a_n = 1 - \dfrac{1}{2^n}$
20. $a_n = \dfrac{3^n}{4^n}$
21. $a_n = \dfrac{1}{n^{3/2}}$
22. $a_n = \dfrac{1}{\sqrt{n}}$
23. $a_n = \dfrac{(-1)^n}{n^2}$
24. $a_n = (-1)^n\left(\dfrac{n}{n+1}\right)$
25. $a_n = (2n-1)(2n+1)$
26. $a_n = n(n-1)(n-2)$

Using a Graphing Utility **In Exercises 27–32, use the *table* feature of a graphing utility to find the first 10 terms of the sequence. (Assume *n* begins with 1.)**

27. $a_n = 2(3n-1) + 5$
28. $a_n = 2n(n+1) - 7$
29. $a_n = 1 + \dfrac{n+1}{n}$
30. $a_n = \dfrac{4n^2}{n+2}$
31. $a_n = (-1)^n + 1$
32. $a_n = (-1)^{n+1} + 8$

Finding a Term of a Sequence **In Exercises 33–38, find the indicated term of the sequence.**

33. $a_n = \dfrac{n^2}{n^2+1}$
    $a_{10} =$
34. $a_n = \dfrac{n^2}{2n+1}$
    $a_5 =$
35. $a_n = (-1)^n(3n-2)$
    $a_{15} =$
36. $a_n = (-1)^{n-1}[n(n-1)]$
    $a_{10} =$
37. $a_n = \dfrac{2^n+1}{2^{n+1}}$
    $a_7 =$
38. $a_n = \dfrac{3^n}{3^n+1}$
    $a_6 =$

Finding the *n*th Term of a Sequence **In Exercises 39–52, write an expression for the apparent *n*th term of the sequence. (Assume *n* begins with 1.)**

39. $3, 8, 13, 18, 23, \ldots$
40. $3, 7, 11, 15, 19, \ldots$
41. $7, 13, 19, 25, 31, \ldots$
42. $9, 11, 13, 15, 17, \ldots$
43. $0, 3, 8, 15, 24, \ldots$
44. $4, 7, 12, 19, 28, \ldots$
45. $\frac{2}{3}, \frac{3}{4}, \frac{4}{5}, \frac{5}{6}, \frac{6}{7}, \ldots$
46. $\frac{2}{1}, \frac{3}{3}, \frac{4}{5}, \frac{5}{7}, \frac{6}{9}, \ldots$
47. $\frac{2}{2}, -\frac{3}{4}, \frac{4}{8}, -\frac{5}{16}, \ldots$
48. $\frac{1}{3}, -\frac{2}{9}, \frac{4}{27}, -\frac{8}{81}, \ldots$
49. $1 + \frac{1}{1}, 1 + \frac{1}{2}, 1 + \frac{1}{3}, 1 + \frac{1}{4}, 1 + \frac{1}{5}, \ldots$
50. $1 + \frac{1}{3}, 1 + \frac{1}{6}, 1 + \frac{1}{11}, 1 + \frac{1}{18}, 1 + \frac{1}{27}, \ldots$
51. $1, 3, 1, 3, 1, \ldots$
52. $1, -1, 1, -1, 1, \ldots$

A Recursive Sequence In Exercises 53–58, write the first five terms of the sequence defined recursively.

53. $a_1 = 28, a_k = a_{k-1} - 4$

54. $a_1 = 15, a_k = a_{k-1} + 3$

55. $a_1 = 3, a_{k+1} = 2(a_k - 1)$

56. $a_1 = 32, a_{k+1} = \frac{1}{2}a_k$

57. $a_0 = 1, a_1 = 3, a_k = a_{k-2} + a_{k-1}$

58. $a_0 = -1, a_1 = 5, a_k = a_{k-2} + a_{k-1}$

Finding the *n*th Term of a Recursive Sequence In Exercises 59–62, write the first five terms of the sequence defined recursively. Use the pattern to write the *n*th term of the sequence as a function of *n*. (Assume *n* begins with 1.)

59. $a_1 = 5, a_{k+1} = a_k + 3$

60. $a_1 = 8, a_{k+1} = a_k - 2$

61. $a_1 = 81, a_{k+1} = \frac{1}{3}a_k$

62. $a_1 = 14, a_{k+1} = -2a_k$

Writing the Terms of a Sequence Involving Factorials In Exercises 63–68, write the first five terms of the sequence (a) using the *table* feature of a graphing utility and (b) algebraically. (Assume *n* begins with 0.)

63. $a_n = \dfrac{1}{n!}$

64. $a_n = \dfrac{1}{(n+1)!}$

65. $a_n = \dfrac{n^2}{(n+1)!}$

66. $a_n = \dfrac{n^3}{(n+2)!}$

67. $a_n = \dfrac{(-1)^n(n+3)!}{n!}$

68. $a_n = \dfrac{(-1)^{2n+1}}{(2n+1)!}$

Simplifying Factorial Expressions In Exercises 69–76, simplify the factorial expression.

69. $\dfrac{2!}{4!}$

70. $\dfrac{5!}{7!}$

71. $\dfrac{12!}{6! \cdot 8!}$

72. $\dfrac{10!}{5! \cdot 5!}$

73. $\dfrac{(n+1)!}{n!}$

74. $\dfrac{(n+2)!}{n!}$

75. $\dfrac{(2n-1)!}{(2n+1)!}$

76. $\dfrac{(2n-4)!}{(2n-2)!}$

Identifying a Graph of a Sequence In Exercises 77–80, match the sequence with its graph. [The graphs are labeled (a), (b), (c), and (d).]

(a)

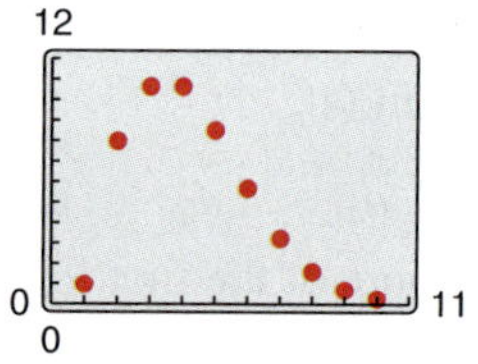

(b)

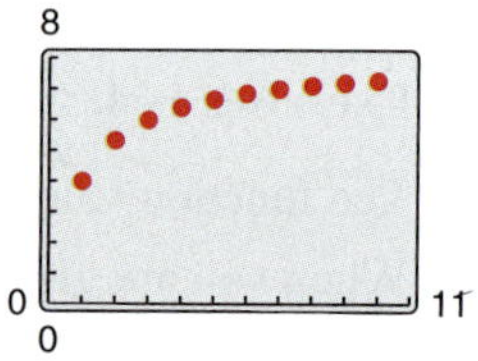

(c)

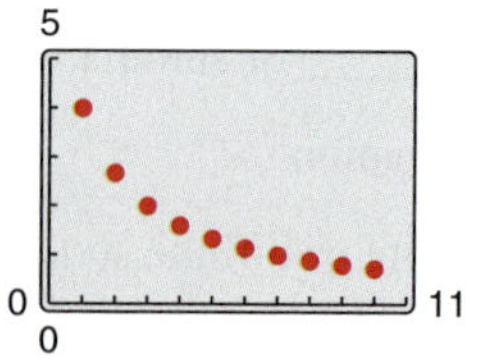

(d)

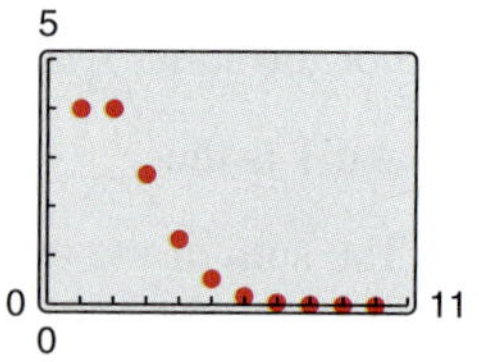

77. $a_n = \dfrac{8}{n+1}$

78. $a_n = \dfrac{8n}{n+1}$

79. $a_n = \dfrac{2^{n+2}}{2n!}$

80. $a_n = \dfrac{4^n}{n!}$

Graphing the Terms of a Sequence In Exercises 81–86, use a graphing utility to graph the first 10 terms of the sequence. (Assume *n* begins with 1.)

81. $a_n = \frac{2}{3}n$

82. $a_n = \frac{1}{2}n + 3$

83. $a_n = 16(-0.5)^{n-1}$

84. $a_n = 8(-0.75)^{n-1}$

85. $a_n = \dfrac{2n}{n+1}$

86. $a_n = \dfrac{3n^2}{n^2+1}$

Sigma Notation for Sums In Exercises 87–96, find the sum.

87. $\displaystyle\sum_{i=1}^{5} (2i + 1)$

88. $\displaystyle\sum_{i=1}^{6} (3i - 1)$

89. $\displaystyle\sum_{i=0}^{6} 4i^2$

90. $\displaystyle\sum_{i=0}^{5} 3i^2$

91. $\displaystyle\sum_{j=3}^{5} \frac{1}{j^2 - 3}$

92. $\displaystyle\sum_{j=3}^{5} \frac{1}{j + 1}$

93. $\displaystyle\sum_{k=1}^{4} 20$

94. $\displaystyle\sum_{k=1}^{5} 6$

95. $\displaystyle\sum_{i=2}^{5} [i^3 + (i+1)^2]$

96. $\displaystyle\sum_{k=2}^{7} [k + (k-3)^2]$

Using a Graphing Utility In Exercises 97–100, use a graphing utility to find the sum.

97. $\displaystyle\sum_{j=1}^{6} (24 - 3j)$

98. $\displaystyle\sum_{j=1}^{10} \frac{6}{3j + 1}$

99. $\displaystyle\sum_{k=0}^{4} \frac{(-1)^k}{(k+1)!}$

100. $\displaystyle\sum_{k=0}^{4} \frac{(-1)^k}{k!}$

Writing a Sum Using Sigma Notation **In Exercises 101–108, use sigma notation to write the sum. Then use a graphing utility to find the sum.**

101. $\frac{1}{3(1)} + \frac{1}{3(2)} + \frac{1}{3(3)} + \cdots + \frac{1}{3(9)}$

102. $\frac{5}{1+1} + \frac{5}{1+2} + \frac{5}{1+3} + \cdots + \frac{5}{1+15}$

103. $\left[2\left(\frac{1}{8}\right) + 3\right] + \left[2\left(\frac{2}{8}\right) + 3\right] + \cdots + \left[2\left(\frac{8}{8}\right) + 3\right]$

104. $\left[1 - \left(\frac{1}{6}\right)^2\right] + \left[1 - \left(\frac{2}{6}\right)^2\right] + \cdots + \left[1 - \left(\frac{6}{6}\right)^2\right]$

105. $-3 + 9 - 27 + 81 - 243 + 729$

106. $1 - \frac{1}{2} + \frac{1}{4} - \frac{1}{8} + \cdots - \frac{1}{128}$

107. $\frac{1}{4} + \frac{3}{8} + \frac{7}{16} + \frac{15}{32} + \frac{31}{64}$

108. $\frac{1}{2} + \frac{2}{4} + \frac{6}{8} + \frac{24}{16} + \frac{120}{32} + \frac{720}{64}$

Finding a Partial Sum of a Series **In Exercises 109–112, find the (a) third, (b) fourth, and (c) fifth partial sums of the series.**

109. $\sum_{i=1}^{\infty} \left(\frac{1}{2}\right)^i$

110. $\sum_{i=1}^{\infty} 2\left(\frac{1}{3}\right)^i$

111. $\sum_{n=1}^{\infty} 4\left(-\frac{1}{2}\right)^n$

112. $\sum_{n=1}^{\infty} 8\left(-\frac{1}{4}\right)^n$

Finding the Sum of a Series **In Exercises 113–116, find the sum of the infinite series.**

113. $\sum_{i=1}^{\infty} \frac{6}{10^i}$

114. $\sum_{k=1}^{\infty} \frac{4}{10^k}$

115. $\sum_{k=1}^{\infty} 7\left(\frac{1}{10}\right)^k$

116. $\sum_{i=1}^{\infty} 2\left(\frac{1}{10}\right)^i$

A Sequence Involving x **In Exercises 117–122, write the first five terms of the sequence.**

117. $a_n = \frac{x^n}{n!}$

118. $a_n = \frac{x^2}{n^2}$

119. $a_n = \frac{(-1)^n x^{2n+1}}{2n+1}$

120. $a_n = \frac{(-1)^n x^{n+1}}{n+1}$

121. $a_n = \frac{(-1)^n x^{2n}}{(2n)!}$

122. $a_n = \frac{(-1)^n x^{2n+1}}{(2n+1)!}$

Writing Partial Sums **In Exercises 123–126, write the first five terms of the sequence. Then find an expression for the nth partial sum.**

123. $a_n = \frac{1}{2n} - \frac{1}{2n+2}$

124. $a_n = \frac{1}{n} - \frac{1}{n+1}$

125. $a_n = \frac{1}{n+1} - \frac{1}{n+2}$

126. $a_n = \frac{1}{n} - \frac{1}{n+2}$

127. Compound Interest You deposit \$5000 in an account that earns 4% interest compounded quarterly. The balance in the account after n quarters is given by

$$A_n = 5000\left(1 + \frac{0.04}{4}\right)^n, \quad n = 1, 2, 3, \ldots .$$

(a) Compute the first eight terms of this sequence.

(b) Find the balance in this account after 10 years by computing the 40th term of the sequence.

128. Compound Interest You deposit \$10,000 in an account that earns 3.5% interest compounded monthly. The balance in the account after n months is given by

$$A_n = 10{,}000\left(1 + \frac{0.035}{12}\right)^n, \quad n = 1, 2, 3, \ldots .$$

(a) Write the first eight terms of the sequence.

(b) Find the balance in the account after 5 years by computing the 60th term of the sequence.

129. MODELING DATA

The revenues R_n (in millions of dollars) of Netflix from 2012 through 2017 are shown in the table. (*Source:* Netflix, Inc.)

Year	Revenue, R_n (in millions of dollars)
2012	3609.3
2013	4374.6
2014	5504.7
2015	6779.5
2016	8830.7
2017	11,692.8

Spreadsheet at LarsonPrecalculus.com

(a) Use the *regression* feature of a graphing utility to find a linear sequence and a quadratic sequence that model the data. Let n represent the year, with $n = 2$ corresponding to 2012. Identify the coefficient of determination for each model.

(b) Graph each model with the data. Decide which model is a better fit for the data. Explain.

(c) Use the model you chose in part (b) to predict the revenue of Netflix in 2021.

(d) Use your model from part (b) to find when the revenue will reach 50 billion dollars.

(e) Use the model you chose in part (b) to approximate the total revenue from 2012 through 2017. Compare this sum with the result of adding the revenues shown in the table.

130. ***Why you should learn it*** *(p. 580)* A landlocked lake has been selected to be stocked in the year 2020 with 5500 trout and to be restocked each year thereafter with 500 trout. Each year the fish population declines 25% due to harvesting as well as natural causes.

(a) Write a recursive sequence that gives the population p_n of trout in the lake in terms of the year n, with $n = 0$ corresponding to 2020.

(b) Use the recursion formula from part (a) to find the numbers of trout in the lake for $n = 1, 2, 3,$ and 4. Interpret these values in the context of the situation.

(c) Use a graphing utility to find the number of trout as time passes infinitely. Explain your result.

Focusing on Concepts

True or False? **In Exercises 131 and 132, determine whether the statement is true or false. Justify your answer.**

131. $\sum_{i=1}^{4}(i^2 + 2i) = \sum_{i=1}^{4} i^2 + 2\sum_{i=1}^{4} i$

132. $\sum_{j=1}^{4} 2^j = \sum_{j=3}^{6} 2^{j-2}$

Fibonacci Sequence **In Exercises 133 and 134, use the Fibonacci sequence. (See Example 5.)**

133. Write the first 12 terms of the Fibonacci sequence a_n and the first 10 terms of the sequence given by

$$b_n = \frac{a_{n+1}}{a_n}, \quad n > 0.$$

134. Using the definition of b_n given in Exercise 133, show that b_n can be defined recursively by

$$b_n = 1 + \frac{1}{b_{n-1}}.$$

Exploration **In Exercises 135–138, let**

$$a_n = \frac{(1 + \sqrt{5})^n - (1 - \sqrt{5})^n}{2^n\sqrt{5}}$$

be a sequence with nth term a_n.

135. Use the *table* feature of a graphing utility to find the first five terms of the sequence.

136. Do you recognize the terms of the sequence in Exercise 135? What sequence is it?

137. Find expressions for a_{n+1} and a_{n+2} in terms of n.

138. Use the result from Exercise 137 to show that $a_{n+2} = a_{n+1} + a_n$. Is this result the same as your answer to Exercise 135? Explain.

139. **Error Analysis** Describe the error in finding the sum.

$$\sum_{k=1}^{4}(3 + 2k^2) = 3(4) + 2(4)\sum_{k=1}^{4} k^2$$
$$= 12 + 8(1 + 4 + 9 + 16)$$
$$= 252$$

140. **HOW DO YOU SEE IT?** The graph represents the first six terms of a sequence.

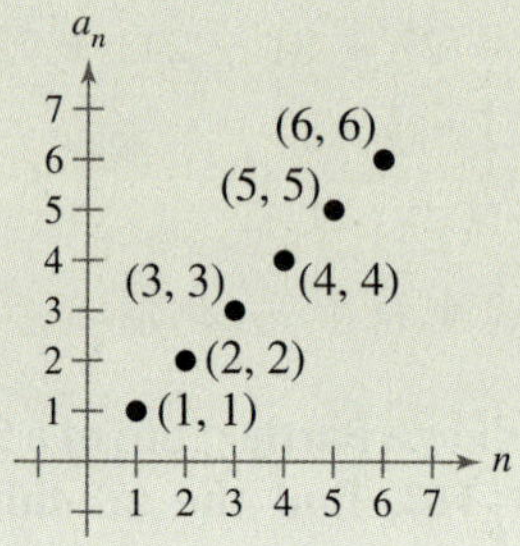

(a) Write the first six terms of the sequence.

(b) Write an expression for the apparent nth term a_n of the sequence.

(c) Use sigma notation to represent the partial sum of the first 50 terms of the sequence.

Arithmetic Mean **In Exercises 141 and 142, use the following definition of the arithmetic mean $\bar{x}$ of a set of n measurements $x_1, x_2, x_3, \ldots, x_n$.**

$$\bar{x} = \frac{1}{n}\sum_{i=1}^{n} x_i$$

141. **Proof** Prove that

$$\sum_{i=1}^{n}(x_i - \bar{x}) = 0.$$

142. **Proof** Prove that

$$\sum_{i=1}^{n}(x_i - \bar{x})^2 = \sum_{i=1}^{n} x_i^2 - \frac{1}{n}\left(\sum_{i=1}^{n} x_i\right)^2.$$

Cumulative Mixed Review

Operations with Matrices **In Exercises 143 and 144, find, if possible, (a) $A - B$, (b) $2B - 3A$, (c) AB, and (d) BA.**

143. $A = \begin{bmatrix} 6 \\ 3 \end{bmatrix}, \quad B = \begin{bmatrix} 4 \\ -3 \end{bmatrix}$

144. $A = \begin{bmatrix} 10 & 7 \\ -4 & 6 \end{bmatrix}, \quad B = \begin{bmatrix} 0 & -12 \\ 8 & 11 \end{bmatrix}$

8.2 Arithmetic Sequences and Partial Sums

What you should learn

- Recognize, write, and find the nth terms of arithmetic sequences.
- Find nth partial sums of arithmetic sequences.
- Use arithmetic sequences to model and solve real-life problems.

Why you should learn it

Arithmetic sequences can reduce the amount of time it takes to find the sum of a sequence of numbers with a common difference. In Exercise 75 on page 597, you will use an arithmetic sequence to find the number of bricks needed to lay a brick patio.

Arithmetic Sequences

A sequence whose consecutive terms have a common difference is called an **arithmetic sequence.**

Definition of Arithmetic Sequence

A sequence is **arithmetic** when the differences between consecutive terms are the same. So, the sequence

$$a_1, a_2, a_3, a_4, \ldots, a_n, \ldots$$

is arithmetic when there is a number d such that

$$a_2 - a_1 = a_3 - a_2 = a_4 - a_3 = \cdots = d.$$

The number d is the **common difference** of the sequence.

EXAMPLE 1 Examples of Arithmetic Sequences

a. The sequence whose nth term is

$$4n + 3$$

is arithmetic. The common difference between consecutive terms is 4.

$$\underbrace{7, 11}_{11 - 7 = 4}, 15, 19, \ldots, 4n + 3, \ldots \qquad \text{Begin with } n = 1.$$

b. The sequence whose nth term is

$$7 - 5n$$

is arithmetic. The common difference between consecutive terms is -5.

$$\underbrace{2, -3}_{-3 - 2 = -5}, -8, -13, \ldots, 7 - 5n, \ldots \qquad \text{Begin with } n = 1.$$

c. The sequence whose nth term is

$$\frac{1}{4}(n + 3)$$

is arithmetic. The common difference between consecutive terms is $\frac{1}{4}$.

$$\underbrace{1, \frac{5}{4}}_{\frac{5}{4} - 1 = \frac{1}{4}}, \frac{3}{2}, \frac{7}{4}, \ldots, \frac{n + 3}{4}, \ldots \qquad \text{Begin with } n = 1.$$

✓ Checkpoint *Audio-video solution in English & Spanish at LarsonPrecalculus.com*

Write the first four terms of the arithmetic sequence whose nth term is $3n - 1$. Then find the common difference between consecutive terms.

The sequence 1, 4, 9, 16, . . . , whose nth term is n^2, is *not* arithmetic. The difference between the first two terms is

$$a_2 - a_1 = 4 - 1 = 3$$

but the difference between the second and third terms is

$$a_3 - a_2 = 9 - 4 = 5.$$

Sylvie Bouchard/Shutterstock.com

The nth term of an arithmetic sequence can be derived from the pattern below.

$a_1 = a_1$ — 1st term

$a_2 = a_1 + d$ — 2nd term

$a_3 = a_1 + 2d$ — 3rd term

$a_4 = a_1 + 3d$ — 4th term

$a_5 = a_1 + 4d$ — 5th term

1 less

⋮

$a_n = a_1 + (n - 1)d$ — nth term

1 less

The result is summarized in the next definition.

The nth Term of an Arithmetic Sequence

The nth term of an arithmetic sequence has the form

$$a_n = a_1 + (n - 1)d$$

where d is the common difference between consecutive terms of the sequence and a_1 is the first term of the sequence.

EXAMPLE 2 Finding the nth Term of an Arithmetic Sequence

Find a formula for the nth term of the arithmetic sequence whose common difference is 3 and whose first term is 2.

Solution

You know that the formula for the nth term is of the form $a_n = a_1 + (n - 1)d$. Moreover, because the common difference is $d = 3$ and the first term is $a_1 = 2$, the formula must have the form

$$a_n = 2 + 3(n - 1) \qquad \text{Substitute 2 for } a_1 \text{ and 3 for } d.$$

or $a_n = 3n - 1$. The sequence therefore has the following form.

$$2, 5, 8, 11, 14, \ldots, 3n - 1, \ldots$$

The figure below shows a graph of the first 15 terms of the sequence. Notice that the points lie on a line. This makes sense because a_n is a linear function of n. In other words, the terms "arithmetic" and "linear" are closely connected.

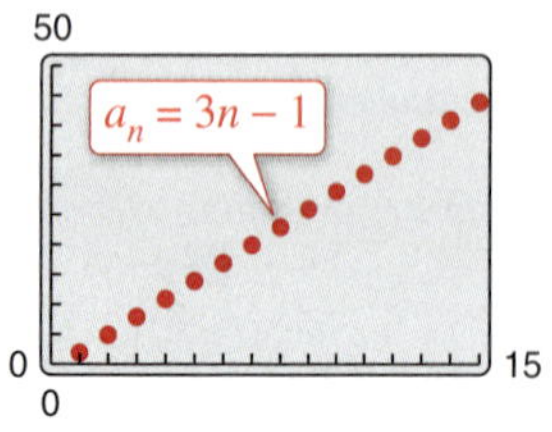

Checkpoint Audio-video solution in English & Spanish at LarsonPrecalculus.com

Find a formula for the nth term of the arithmetic sequence whose common difference is 5 and whose first term is -1.

Explore the Concept

Consider the following sequences.

$1, 4, 7, 10, \ldots, 3n - 2, \ldots$

$-5, 1, 7, 13, \ldots, 6n - 11, \ldots$

$\frac{5}{2}, \frac{3}{2}, \frac{1}{2}, -\frac{1}{2}, \ldots, \frac{7}{2} - n, \ldots$

What relationship do you observe between successive terms of these sequences?

Technology Tip

You can use a graphing utility to generate the arithmetic sequence in Example 2 by using the following steps.

2 ENTER

3 + ANS

Now press the *enter* key repeatedly to generate the terms of the sequence.

EXAMPLE 3 Writing the Terms of an Arithmetic Sequence

The fourth term of an arithmetic sequence is 20, and the 13th term is 65. Write the first 13 terms of this sequence.

Solution

You know that $a_4 = 20$ and $a_{13} = 65$. So, you must add the common difference d nine times to the fourth term to obtain the 13th term. Therefore, the fourth and 13th terms of the sequence are related by

$$a_{13} = a_4 + 9d. \qquad a_4 \text{ and } a_{13} \text{ are nine terms apart.}$$

Using $a_4 = 20$ and $a_{13} = 65$, you have

$$65 = 20 + 9d.$$

So, you can conclude that $d = 5$. Use $d = 5$ and the given terms to write the first 13 terms of the sequence.

a_1	a_2	a_3	a_4	a_5	a_6	a_7	a_8	a_9	a_{10}	a_{11}	a_{12}	a_{13}	. . .
5,	10,	15,	20,	25,	30,	35,	40,	45,	50,	55,	60,	65,	. . .

✓ Checkpoint *Audio-video solution in English & Spanish at LarsonPrecalculus.com*

The eighth term of an arithmetic sequence is 25, and the 12th term is 41. Write the first 11 terms of this sequence.

When you know the nth term of an arithmetic sequence *and* you know the common difference of the sequence, you can find the $(n + 1)$th term by using the *recursion formula*

$$a_{n+1} = a_n + d. \qquad \text{Recursion formula}$$

With this formula, you can find any term of an arithmetic sequence, *provided* that you know the preceding term. For instance, when you know the first term, you can find the second term. Then, knowing the second term, you can find the third term, and so on.

Technology Tip

Most graphing utilities have a built-in function that will display the terms of an arithmetic sequence. For instructions on how to use the *sequence* feature, see Appendix A; for specific keystrokes, go to this textbook's *Student Companion Website.*

EXAMPLE 4 Using a Recursion Formula

Find the ninth term of the arithmetic sequence whose first two terms are 2 and 9.

Solution

You know that the sequence is arithmetic. Also, $a_1 = 2$ and $a_2 = 9$. So, the common difference for this sequence is

$$d = 9 - 2 = 7.$$

There are two ways to find the ninth term. One way is simply to write out the first nine terms (by repeatedly adding 7).

$$2,\ 9,\ 16,\ 23,\ 30,\ 37,\ 44,\ 51,\ 58$$

Another way to find the ninth term is to first find a formula for the nth term. Because the common difference is $d = 7$ and the first term is $a_1 = 2$, the formula must have the form

$$a_n = 2 + 7(n - 1). \qquad \text{Substitute 2 for } a_1 \text{ and 7 for } d.$$

So, a formula for the nth term is $a_n = 7n - 5$, which implies that the ninth term is

$$a_9 = 7(9) - 5 = 58.$$

✓ Checkpoint *Audio-video solution in English & Spanish at LarsonPrecalculus.com*

Find the tenth term of the arithmetic sequence that begins with 7 and 15.

The Sum of a Finite Arithmetic Sequence

There is a simple formula for the *sum* of a finite arithmetic sequence.

The Sum of a Finite Arithmetic Sequence

The sum of a finite arithmetic sequence with n terms is given by $S_n = \frac{n}{2}(a_1 + a_n)$.

(See the proof on page 643.)

Algebra Help

Note that this formula works only for *arithmetic* sequences.

EXAMPLE 5 Sum of a Finite Arithmetic Sequence

Find the sum: $1 + 3 + 5 + 7 + 9 + 11 + 13 + 15 + 17 + 19$.

Solution

To begin, notice that the sequence is arithmetic (with a common difference of 2). Moreover, the sequence has 10 terms. So, the sum of the sequence is

$$S_n = 1 + 3 + 5 + 7 + 9 + 11 + 13 + 15 + 17 + 19$$

$$= \frac{n}{2}(a_1 + a_n) \qquad \text{Formula for sum of an arithmetic sequence}$$

$$= \frac{10}{2}(1 + 19) \qquad \text{Substitute 10 for } n\text{, 1 for } a_1\text{, and 19 for } a_n.$$

$$= 5(20) \qquad \text{Simplify.}$$

$$= 100. \qquad \text{Sum of the sequence}$$

✓ **Checkpoint** ▶ *Audio-video solution in English & Spanish at LarsonPrecalculus.com*

Find the sum: $40 + 37 + 34 + 31 + 28 + 25 + 22$.

Recall from Section 8.1 that the sum of the first n terms of an infinite sequence is the nth partial sum. The ***n*th partial sum of an arithmetic sequence** can be found by using the formula for the sum of a finite arithmetic sequence.

Algebra Help

If you go on to take a course in calculus, you will study sequences and series in detail. You will learn that sequences and series play a major role in the study of calculus.

EXAMPLE 6 Partial Sum of an Arithmetic Sequence

Find the 150th partial sum of the arithmetic sequence $5, 16, 27, 38, 49, \ldots$.

Solution

For this arithmetic sequence, you have $a_1 = 5$ and $d = 16 - 5 = 11$. So,

$$a_n = 5 + 11(n - 1)$$

and the nth term is $a_n = 11n - 6$. Therefore, $a_{150} = 11(150) - 6 = 1644$, and the sum of the first 150 terms is

$$S_n = \frac{n}{2}(a_1 + a_n) \qquad n\text{th partial sum formula}$$

$$= \frac{150}{2}(5 + 1644) \qquad \text{Substitute 150 for } n\text{, 5 for } a_1\text{, and 1644 for } a_n.$$

$$= 75(1649) \qquad \text{Simplify.}$$

$$= 123{,}675. \qquad n\text{th partial sum}$$

✓ **Checkpoint** ▶ *Audio-video solution in English & Spanish at LarsonPrecalculus.com*

Find the 120th partial sum of the arithmetic sequence $6, 12, 18, 24, 30, \ldots$.

Application

EXAMPLE 7 Total Sales

See LarsonPrecalculus.com for an interactive version of this type of example.

A small business sells \$20,000 worth of sports memorabilia during its first year. The owner of the business has set a goal of increasing annual sales by \$15,000 each year for 19 years. Assuming that this goal is met, find the total sales during the first 20 years this business is in operation.

Algebraic Solution

The annual sales form an arithmetic sequence in which $a_1 = 20{,}000$ and $d = 15{,}000$. So,

$$a_n = 20{,}000 + 15{,}000(n - 1)$$

and the nth term of the sequence is

$$a_n = 15{,}000n + 5000.$$

This implies that the 20th term of the sequence is

$$a_{20} = 15{,}000(20) + 5000$$
$$= 300{,}000 + 5000$$
$$= 305{,}000.$$

The sum of the first 20 terms of the sequence is

$$S_n = \frac{n}{2}(a_1 + a_n) \qquad n\text{th partial sum formula}$$

$$= \frac{20}{2}(20{,}000 + 305{,}000) \qquad \text{Substitute 20 for } n, \text{ 20,000 for } a_1, \text{ and 305,000 for } a_n.$$

$$= 10(325{,}000) \qquad \text{Simplify.}$$

$$= 3{,}250{,}000. \qquad \text{Simplify.}$$

So, the total sales for the first 20 years are \$3,250,000.

Numerical Solution

The annual sales form an arithmetic sequence in which $a_1 = 20{,}000$ and $d = 15{,}000$. So,

$$a_n = 20{,}000 + 15{,}000(n - 1)$$

and the nth term of the sequence is

$$a_n = 15{,}000n + 5000.$$

Use the *list editor* of a graphing utility to create a table that shows the sales for each of the first 20 years and the total sales for the first 20 years, as shown in the figure.

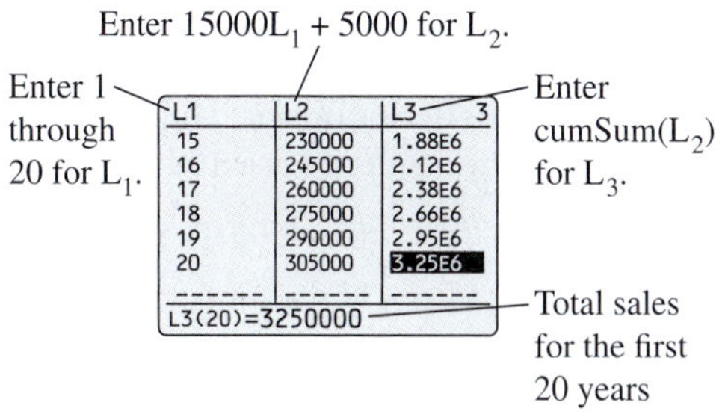

So, the total sales for the first 20 years are \$3,250,000.

✓ Checkpoint *Audio-video solution in English & Spanish at LarsonPrecalculus.com*

A company sells \$160,000 worth of printing paper during its first year. The sales manager has set a goal of increasing annual sales of printing paper by \$20,000 each year for 9 years. Assuming that this goal is met, find the total sales of printing paper during the first 10 years this company is in operation.

The figure below shows the annual sales for the business in Example 7. Notice that the annual sales for the business follow a *linear growth* pattern. In other words, saying that a quantity increases arithmetically is the same as saying that it increases linearly.

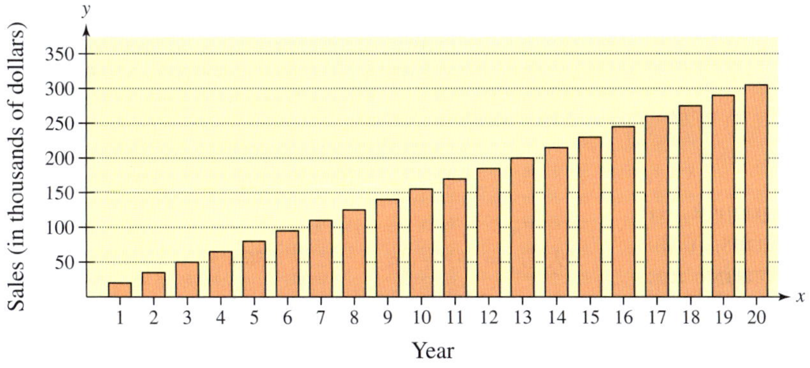

Insight

A standardized test may ask you to use a sequence to make a prediction (see Exercises 79 and 80).

8.2 Exercises

See *CalcChat.com* for tutorial help and worked-out solutions to odd-numbered exercises. For instructions on how to use a graphing utility, see Appendix A.

Vocabulary and Concept Check

In Exercises 1 and 2, fill in the blank.

1. The nth term of an arithmetic sequence has the form _______ .
2. The formula $S_n = \frac{n}{2}(a_1 + a_n)$ can be used to find the sum of the first n terms of an arithmetic sequence, called the _______ .
3. How do you know when a sequence is arithmetic?
4. Is 4 or 1 the common difference of the arithmetic sequence $a_n = 4n + 1$?

Procedures and Problem Solving

Verifying an Arithmetic Sequence In Exercises 5–8, the nth term of a sequence is given. Verify that the sequence is arithmetic by identifying the common difference.

5. $8n - 1$
6. $\frac{1}{2}n + 6$
7. $2 - n$
8. $\frac{3}{4} - 5n$

Identifying an Arithmetic Sequence In Exercises 9–18, determine whether the sequence is arithmetic. Explain.

9. 10, 12, 14, 16, 18, . . .
10. 4, 9, 14, 19, 24, . . .
11. $1^2, 2^2, 3^2, 4^2, 5^2, \ldots$
12. $\frac{1}{3}, \frac{2}{3}, \frac{4}{3}, \frac{8}{3}, \frac{16}{3}, \ldots$
13. $a_n = 8 + 13n$
14. $a_n = 150 - 7n$
15. $a_n = 2^n + n$
16. $a_n = 3 + 2(-1)^n$
17. $a_n = (-1)^{2n}\left(\frac{1}{4}\right)$
18. $a_n = (-1)^{2n+1}$

Finding the nth Term of an Arithmetic Sequence In Exercises 19–28, find a formula for a_n for the arithmetic sequence.

19. $a_1 = 1, d = 6$
20. $a_1 = 15, d = 4$
21. $a_1 = 43, d = -7$
22. $a_1 = 100, d = -8$
23. $a_1 = 5, a_4 = 15$
24. $a_1 = -4, a_5 = 16$
25. $\frac{15}{4}, 3, \frac{9}{4}, \frac{3}{2}, \frac{3}{4}, 0, \ldots$
26. $\frac{13}{2}, 4, \frac{3}{2}, -1, -\frac{7}{2}, \ldots$
27. $a_3 = 94, a_6 = 85$
28. $a_5 = 190, a_{10} = 115$

Writing the Terms of an Arithmetic Sequence In Exercises 29–34, write the first five terms of the arithmetic sequence. Verify your results using a graphing utility.

29. $a_1 = 2, a_{12} = -64$
30. $a_1 = -2.6, d = 0.2$
31. $a_8 = 26, a_{12} = 42$
32. $a_6 = -38, a_{11} = -73$
33. $a_3 = 19, a_{15} = -1.7$
34. $a_5 = 16, a_{14} = 38.5$

Writing the Terms of an Arithmetic Sequence In Exercises 35 and 36, write the first five terms of the arithmetic sequence. Find the common difference and write the nth term of the sequence as a function of n.

35. $a_1 = 15, \; a_{k+1} = a_k + 4$
36. $a_1 = 1.5, \; a_{k+1} = a_k - 2.5$

Using a Graph of a Sequence In Exercises 37 and 38, write the formula for the nth term of the arithmetic sequence whose first 10 terms are represented by the graph.

37.

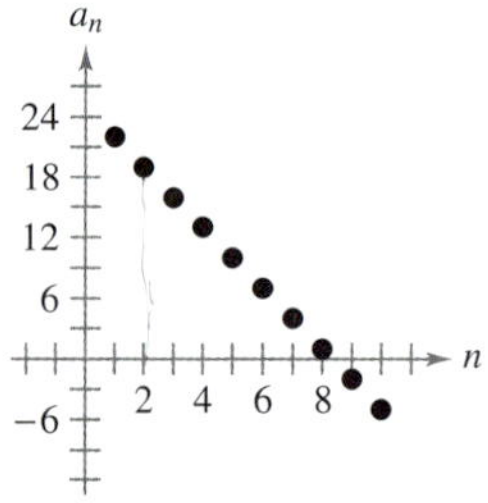

38.

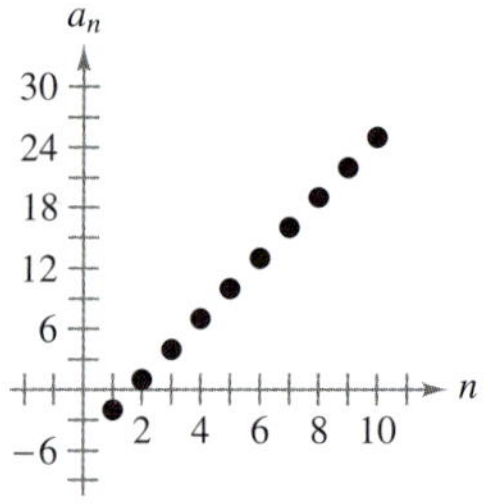

Graphing the Terms of a Sequence In Exercises 39–42, use a graphing utility to graph the first 10 terms of the arithmetic sequence. (Assume n begins with 1.)

39. $a_n = 15 - \frac{3}{2}n$
40. $a_n = -5 + 2n$
41. $a_n = 0.4n - 2$
42. $a_n = -1.3n + 7$

Using a Recursion Formula In Exercises 43–46, the first two terms of the arithmetic sequence are given. Find the missing term.

43. $a_1 = 5, \; a_2 = -1, \; a_{10} = \square$
44. $a_1 = 3, \; a_2 = 13, \; a_9 = \square$
45. $a_1 = 4.2, \; a_2 = 1.8, \; a_7 = \square$
46. $a_1 = -0.7, \; a_2 = -13.8, \; a_8 = \square$

Finding the Terms of a Sequence Using Technology In Exercises 47–52, use the *sequence* feature of a graphing utility to find the first 10 terms of the arithmetic sequence. (Assume n begins with 1.)

47. $a_n = 4n - 5$
48. $a_n = 17 + 3n$
49. $a_n = 20 - \frac{3}{4}n$
50. $a_n = \frac{4}{5}n - 3$
51. $a_n = 1.5 + 0.05n$
52. $a_n = 8 - 12.5n$

Finding the Sum of a Finite Arithmetic Sequence In Exercises 53–60, find the sum of the finite arithmetic sequence.

53. $2 + 4 + 6 + 8 + 10 + 12 + 14 + 16 + 18 + 20$

54. $1 + 4 + 7 + 10 + 13 + 16 + 19$

55. $-1 + (-3) + (-5) + (-7) + (-9)$

56. $-2 + (-5) + (-8) + (-11) + (-14) + (-17)$

57. Sum of the first 100 positive odd integers

58. Sum of the first 50 negative integers

59. Sum of the integers from -100 to 30

60. Sum of the integers from -10 to 50

Finding a Partial Sum of an Arithmetic Sequence In Exercises 61–64, find the indicated nth partial sum of the arithmetic sequence.

61. $0, -9, -18, -27, \ldots; \ n = 40$

62. $4.2, 3.7, 3.2, 2.7, \ldots; \ n = 12$

63. $a_1 = 100, a_{25} = 220, \ n = 25$

64. $a_1 = 15, a_{100} = 307, \ n = 100$

Finding a Partial Sum In Exercises 65–68, find the partial sum without using a graphing utility.

65. $\sum_{n=1}^{500} (n + 8)$

66. $\sum_{n=11}^{30} n - \sum_{n=1}^{10} n$

67. $\sum_{n=1}^{100} 2n$

68. $\sum_{n=1}^{250} (1000 - n)$

Finding a Partial Sum Using a Graphing Utility In Exercises 69–74, use a graphing utility to find the partial sum.

69. $\sum_{n=1}^{20} (2n + 1)$

70. $\sum_{n=1}^{50} (40 - 2n)$

71. $\sum_{n=0}^{100} \frac{n + 5}{2}$

72. $\sum_{n=0}^{100} \frac{4 - n}{4}$

73. $\sum_{i=1}^{60} \left(250 - \frac{2}{5}i\right)$

74. $\sum_{j=1}^{200} (10.5 + 0.025j)$

75. ***Why you should learn it*** *(p. 591)* A brick patio has the approximate shape of a trapezoid, as shown in the figure. The patio has 18 rows of bricks. The first row has 14 bricks and the 18th row has 31 bricks. How many bricks are in the patio?

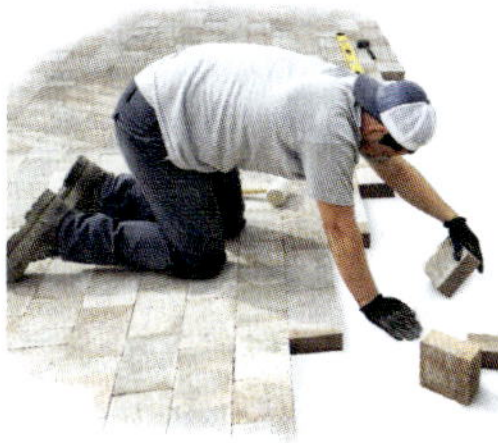

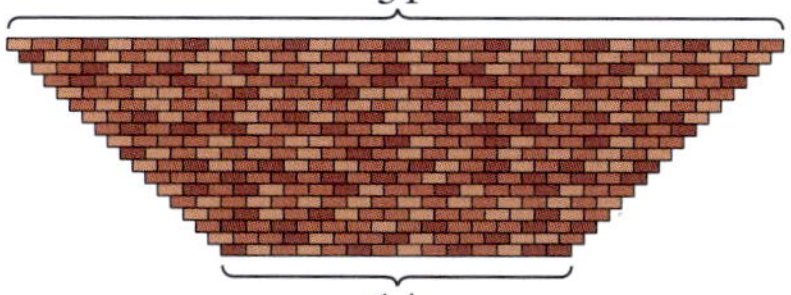

Sylvie Bouchard/Shutterstock.com

76. Performing Arts An auditorium has 20 rows of seats. There are 20 seats in the first row, 21 seats in the second row, 22 seats in the third row, and so on (see figure). How many seats are there in all 20 rows?

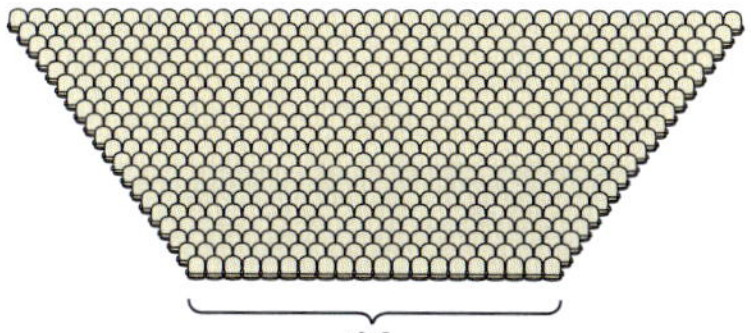

77. Business A hardware store makes a profit of \$30,000 during its first year. The store owner sets a goal of increasing profits by \$5000 each year for 4 years. Assuming that this goal is met, find the total profit during the first 5 years of business.

78. Physics An object with negligible air resistance is dropped from a plane. During the first second of fall, the object falls 16 feet; during the second second, it falls 48 feet; during the third second, it falls 80 feet; and during the fourth second, it falls 112 feet. Assume this pattern continues. How many feet will the object fall in 8 seconds?

79. MODELING DATA

The table shows the sales a_n (in thousands of dollars) for a pet supplies store from 2011 through 2018.

DATA

Year	Sales, a_n (in thousands of dollars)
2011	2.09
2012	2.16
2013	2.24
2014	2.31
2015	2.38
2016	2.44
2017	2.51
2018	2.58

Spreadsheet at LarsonPrecalculus.com

(a) Use the *regression* feature of a graphing utility to find an arithmetic sequence (linear model) for the data. Let n represent the year, with $n = 1$ corresponding to 2011.

(b) Use the sequence from part (a) to approximate the annual sales for the pet supplies store for the years 2011 through 2018.

(c) Use the sequence to find the total sales for the pet supplies store from 2011 through 2018.

(d) Use the sequence to predict the total sales during the period from 2019 through 2025. Is your total reasonable? Explain.

80. MODELING DATA

The table shows the numbers a_n (in thousands) of master's degrees in the biological and biomedical sciences conferred in the United States from 2011 through 2016. (*Source:* U.S. National Center for Education Statistics)

DATA

Year	Master's degrees conferred, a_n (in thousands)
2011	11.32
2012	12.42
2013	13.30
2014	13.96
2015	14.66
2016	15.71

Spreadsheet at LarsonPrecalculus.com

(a) Use the *regression* feature of a graphing utility to find an arithmetic sequence (linear model) for the data. Let n represent the year, with $n = 1$ corresponding to 2011.

(b) Use the sequence from part (a) to approximate the numbers of master's degrees conferred for the years 2011 through 2016.

(c) Use the sequence to find the total number of master's degrees conferred during the period from 2011 through 2016.

(d) Use the sequence to predict the total number of master's degrees conferred during the period from 2017 through 2025. Is your total reasonable? Explain.

Focusing on Concepts

True or False? **In Exercises 81 and 82, determine whether the statement is true or false. Justify your answer.**

81. Given the nth term and the common difference of an arithmetic sequence, it is possible to find the $(n + 1)$th term.

82. If the only known information about a finite arithmetic sequence is its first term and its last term, then it is possible to find the sum of the sequence.

Finding the Terms of a Sequence **In Exercises 83 and 84, find the first 10 terms of the sequence.**

83. $a_1 = x,\ d = 2x$

84. $a_1 = -y,\ d = 5y$

85. Writing Explain how to use the first two terms of an arithmetic sequence to find the nth term.

86. Think About It The sum of the first 20 terms of an arithmetic sequence with a common difference of 3 is 650. Find the first term.

87. Error Analysis Describe the error in finding the sum of the first 50 odd integers.

$$S_n = \frac{n}{2}(a_1 + a_n) = \frac{50}{2}(1 + 101) = 2550$$

88. Error Analysis Describe the error.

S_n is the sum of an arithmetic sequence. When each term is increased by 6, the new sum is equal to $6S_n$.

Think About It **In Exercises 89 and 90, decide whether it is possible to fill in the blanks to form an arithmetic sequence. If so, find a recursion formula for the sequence. Explain how you found your answers.**

89. -7, ▭, ▭, ▭, ▭, ▭, 11

90. 2, 6, ▭, ▭, 162

91. Writing Carl Friedrich Gauss was a famous nineteenth century mathematician. When Gauss was 10, he was asked by his teacher to add the numbers from 1 to 100. Gauss found the answer by mentally finding the summation. Write an explanation of how he arrived at his conclusion, and then find the formula for the sum of the first n natural numbers.

92. HOW DO YOU SEE IT? The graph of a sequence is shown below. Determine which of the statements about the sequence is false. Justify your answer.

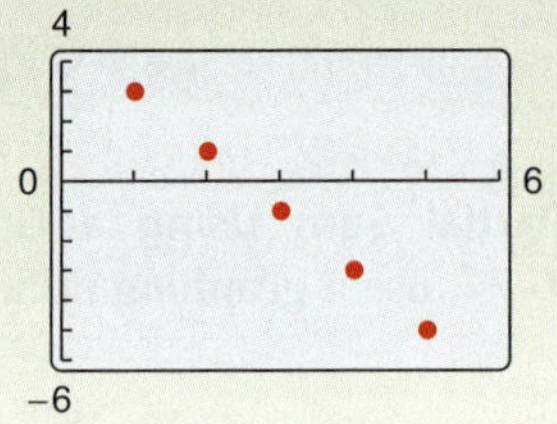

(a) The sequence is arithmetic.

(b) The common difference is 2.

(c) $a_1 = 3$.

Cumulative Mixed Review

Gauss-Jordan Elimination **In Exercises 93 and 94, use Gauss-Jordan elimination to solve the system of equations.**

93. $\begin{cases} 2x - y + 7z = -10 \\ 3x + 2y - 4z = 17 \\ 6x - 5y + z = -20 \end{cases}$

94. $\begin{cases} -x + 4y + 10z = 4 \\ 5x - 3y + z = 31 \\ 8x + 2y - 3z = -5 \end{cases}$

95. *Project: Public Safety* To work an extended application analyzing the numbers of structurally deficient bridges in the United States from 2001 through 2017, visit this textbook's website at *LarsonPrecalculus.com.* (*Data Source:* Bureau of Transportation Statistics)

8.3 Geometric Sequences and Series

What you should learn

- Recognize, write, and find the nth terms of geometric sequences.
- Find nth partial sums of geometric sequences.
- Find sums of infinite geometric series.
- Use geometric sequences to model and solve real-life problems.

Why you should learn it

Geometric sequences can reduce the amount of time it takes to find the sum of a sequence of numbers with a common ratio. For instance, Exercise 111 on page 608 shows how to use a geometric sequence to find the total vertical distance traveled by a bouncing ball.

Geometric Sequences

In Section 8.2, you learned that a sequence whose consecutive terms have a common *difference* is an arithmetic sequence. In this section, you will study another important type of sequence called a **geometric sequence.** Consecutive terms of a geometric sequence have a common *ratio.*

Definition of Geometric Sequence

A sequence is **geometric** when the ratios of consecutive terms are the same. So, the sequence $a_1, a_2, a_3, a_4, \ldots, a_n, \ldots$ is geometric when there is a number r such that

$$\frac{a_2}{a_1} = \frac{a_3}{a_2} = \frac{a_4}{a_3} = \cdots = r, \quad r \neq 0.$$

The number r is the **common ratio** of the sequence.

EXAMPLE 1 Examples of Geometric Sequences

a. The sequence whose nth term is 2^n is geometric. For this sequence, the common ratio between consecutive terms is 2.

$$2, 4, 8, 16, \ldots, 2^n, \ldots \qquad \text{Begin with } n = 1.$$

$$\tfrac{4}{2} = 2$$

b. The sequence whose nth term is $4(3^n)$ is geometric. For this sequence, the common ratio between consecutive terms is 3.

$$12, 36, 108, 324, \ldots, 4(3^n), \ldots \qquad \text{Begin with } n = 1.$$

$$\tfrac{36}{12} = 3$$

c. The sequence whose nth term is $\left(-\frac{1}{3}\right)^n$ is geometric. For this sequence, the common ratio between consecutive terms is $-\frac{1}{3}$.

$$-\frac{1}{3}, \frac{1}{9}, -\frac{1}{27}, \frac{1}{81}, \ldots, \left(-\frac{1}{3}\right)^n, \ldots \qquad \text{Begin with } n = 1.$$

$$\frac{1/9}{-1/3} = -\frac{1}{3}$$

In parts (a), (b), and (c), notice that each of the geometric sequences has an nth term of the form ar^n, where r is the common ratio of the sequence.

✓ Checkpoint *Audio-video solution in English & Spanish at LarsonPrecalculus.com*

Write the first four terms of the geometric sequence whose nth term is $6(-2)^n$. Then find the common ratio of the consecutive terms.

The sequence $1, 4, 9, 16, \ldots$, whose nth term is n^2, is *not* geometric. The ratio of the second term to the first term is

$$\frac{a_2}{a_1} = \frac{4}{1} = 4$$

but the ratio of the third term to the second term is

$$\frac{a_3}{a_2} = \frac{9}{4}.$$

Veronica Louro/Shutterstock.com

The nth Term of a Geometric Sequence

The nth term of a geometric sequence has the form

$$a_n = a_1 r^{n-1}$$

where r is the common ratio of consecutive terms of the sequence. So, every geometric sequence can be written in the following form.

$$a_1,\ a_2,\ a_3,\ a_4,\ a_5,\ \ldots,\ a_n,\ \ldots$$
$$\downarrow$$
$$a_1,\ a_1r,\ a_1r^2,\ a_1r^3,\ a_1r^4,\ \ldots,\ a_1r^{n-1},\ \ldots$$

When you know the nth term of a geometric sequence, you can find the $(n + 1)$th term by multiplying by r. That is, $a_{n+1} = a_n r$.

EXAMPLE 2 Writing the Terms of a Geometric Sequence

Write the first five terms of the geometric sequence whose first term is $a_1 = 3$ and whose common ratio is $r = 2$.

Solution

Starting with 3, repeatedly multiply by 2 to obtain the terms below.

$a_1 = 3$ — 1st term

$a_2 = 3(2^1) = 6$ — 2nd term

$a_3 = 3(2^2) = 12$ — 3rd term

$a_4 = 3(2^3) = 24$ — 4th term

$a_5 = 3(2^4) = 48$ — 5th term

✓ Checkpoint *Audio-video solution in English & Spanish at LarsonPrecalculus.com*

Write the first five terms of the geometric sequence whose first term is $a_1 = 2$ and whose common ratio is $r = 4$.

Technology Tip

You can use a graphing utility to generate the geometric sequence in Example 2 by using the following steps.

3 [ENTER]

2 [×] [ANS]

Now press the *enter* key repeatedly to generate the terms of the sequence.

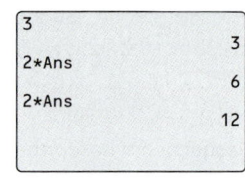

```
3
                3
2*Ans
                6
2*Ans
               12
```

EXAMPLE 3 Finding a Term of a Geometric Sequence

Find the 15th term of the geometric sequence whose first term is 20 and whose common ratio is 1.05.

Algebraic Solution

$a_n = a_1 r^{n-1}$ — Formula for a geometric sequence

$a_{15} = 20(1.05)^{15-1}$ — Substitute 20 for a_1, 1.05 for r, and 15 for n.

≈ 39.60 — Use a calculator.

Numerical Solution

For this sequence, $r = 1.05$ and $a_1 = 20$. So, $a_n = 20(1.05)^{n-1}$. Use the *table* feature of a graphing utility to create a table that shows the terms of the sequence, as shown in the figure.

n	u(n)
9	29.549
10	31.027
11	32.578
12	34.207
13	35.917
14	37.713
15	39.599

u(n)=39.59863199

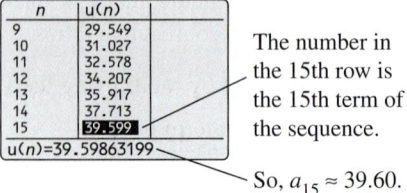

The number in the 15th row is the 15th term of the sequence.

So, $a_{15} \approx 39.60$.

✓ Checkpoint *Audio-video solution in English & Spanish at LarsonPrecalculus.com*

Find the 12th term of the geometric sequence whose first term is 14 and whose common ratio is 1.2.

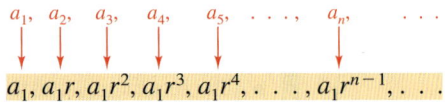

EXAMPLE 4 Finding a Term of a Geometric Sequence

Find a formula for the nth term of the geometric sequence 5, 15, 45, ... What is the ninth term of the sequence?

Solution

The common ratio of this sequence is $r = 15/3 = 3$. Because the first term is $a_1 = 5$, the formula must have the form

$$a_n = a_1 r^{n-1} = 5(3)^{n-1}.$$

You can determine the ninth term ($n = 9$) to be

$$a_9 = 5(3)^{9-1} \qquad \text{Substitute 9 for } n.$$

$$= 5(6561) \qquad \text{Use a calculator.}$$

$$= 32{,}805. \qquad \text{Simplify.}$$

A graph of the first nine terms of the sequence is shown in the figure. Notice that the points lie on an exponential curve. This makes sense because a_n is an exponential function of n.

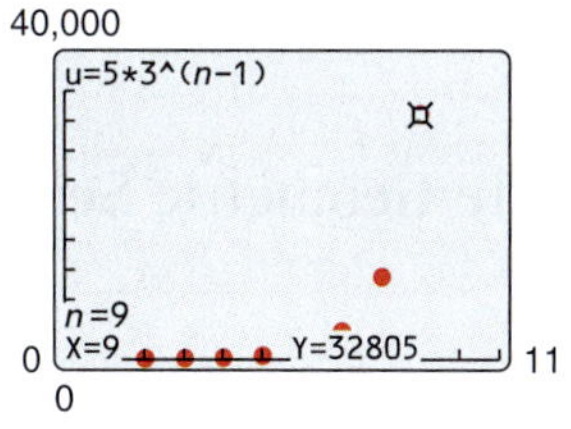

Checkpoint Audio-video solution in English & Spanish at LarsonPrecalculus.com

Find a formula for the nth term of the geometric sequence 4, 20, 100, ... What is the 12th term of the sequence?

When you know *any* two terms of a geometric sequence, you can use that information to find *any other* term of the sequence.

EXAMPLE 5 Finding a Term of a Geometric Sequence

The fourth term of a geometric sequence is 125, and the 10th term is 125/64. Find the 14th term. (Assume that the terms of the sequence are positive.)

Solution

The 10th term is related to the fourth term by the equation

$$a_{10} = a_4 r^6. \qquad \text{Multiply 4th term by } r^{10-4}.$$

Because $a_{10} = 125/64$ and $a_4 = 125$, you can solve for r as follows.

$$\frac{125}{64} = 125r^6 \qquad \text{Substitute } \tfrac{125}{64} \text{ for } a_{10} \text{ and 125 for } a_4.$$

$$\frac{1}{64} = r^6 \qquad \text{Divide each side by 125.}$$

$$\frac{1}{2} = r \qquad \text{Take the sixth root of each side.}$$

You can obtain the 14th term by multiplying the 10th term by $r^{14-10} = r^4$.

$$a_{14} = a_{10}r^4 = \frac{125}{64}\left(\frac{1}{2}\right)^4 = \frac{125}{1024}$$

Checkpoint Audio-video solution in English & Spanish at LarsonPrecalculus.com

The second term of a geometric sequence is 6, and the fifth term is 81/4. Find the eighth term. (Assume that the terms of the sequence are positive.)

Technology Tip

Most graphing utilities have a built-in function that will display the terms of a geometric sequence. For instructions on how to use the *sequence* feature, see Appendix A; for specific keystrokes, go to this textbook's *Student Companion Website.*

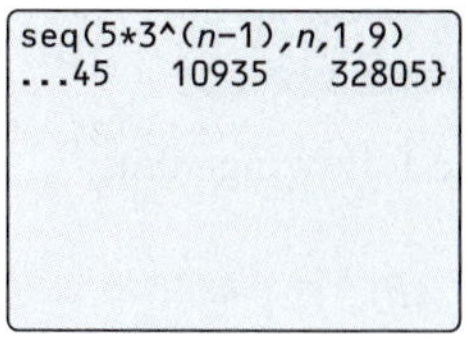

Algebra Help

Remember that r is the common ratio of consecutive terms of a geometric sequence. So, in Example 5,

$$\begin{aligned} a_{10} &= a_1 r^9 \\ &= a_1 \cdot r \cdot r \cdot r \cdot r^6 \\ &= a_1 \cdot \frac{a_2}{a_1} \cdot \frac{a_3}{a_2} \cdot \frac{a_4}{a_3} \cdot r^6 \\ &= a_4 r^6. \end{aligned}$$

The Sum of a Finite Geometric Sequence

The formula for the sum of a *finite* geometric sequence is as follows.

The Sum of a Finite Geometric Sequence

The sum of the finite geometric sequence

$$a_1, a_1r, a_1r^2, a_1r^3, a_1r^4, \ldots, a_1r^{n-1}$$

with common ratio $r \neq 1$ is given by

$$S_n = \sum_{i=1}^{n} a_1r^{i-1} = a_1\left(\frac{1-r^n}{1-r}\right).$$

(See the proof on page 643.)

EXAMPLE 6 Sum of a Finite Geometric Sequence

Find the following sum.

$$\sum_{n=1}^{12} 4(0.3)^n$$

Solution

By writing out a few terms, you have

$$\sum_{n=1}^{12} 4(0.3)^n = 4(0.3)^1 + 4(0.3)^2 + 4(0.3)^3 + \cdots + 4(0.3)^{12}.$$

Now, because

$$a_1 = 4(0.3), \qquad r = 0.3, \qquad \text{and} \qquad n = 12$$

you can apply the formula for the sum of a finite geometric sequence to obtain

$$\sum_{n=1}^{12} 4(0.3)^n = a_1\left(\frac{1-r^n}{1-r}\right)$$ Formula for sum of a finite geometric sequence

$$= 4(0.3)\left[\frac{1-(0.3)^{12}}{1-0.3}\right]$$ Substitute 4(0.3) for a_1, 0.3 for r, and 12 for n.

$$\approx 1.71.$$ Use a calculator.

Technology Tip

Using the *summation* feature or the *sum sequence* feature of a graphing utility, you can calculate the sum of the sequence in Example 6 to be about 1.7142848, as shown below.

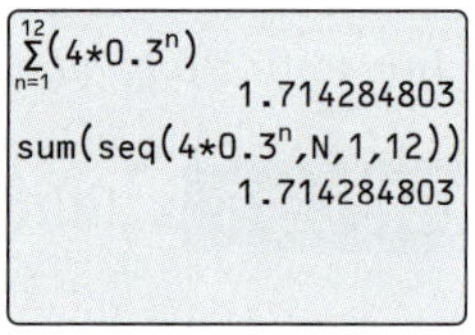

Calculate the sum beginning at $n = 0$. You should obtain a sum of about 5.7142848.

Checkpoint *Audio-video solution in English & Spanish at LarsonPrecalculus.com*

Find the sum $\sum_{i=1}^{10} 2(0.25)^{i-1}$.

When using the formula for the sum of a geometric sequence, be careful to check that the index begins at $i = 1$. For an index that begins at $i = 0$, you must adjust the formula for the nth partial sum. For instance, if the index in Example 6 had begun with $n = 0$, then the sum would have been

$$\sum_{n=0}^{12} 4(0.3)^n = 4(0.3)^0 + \sum_{n=1}^{12} 4(0.3)^n$$

$$= 4 + \sum_{n=1}^{12} 4(0.3)^n$$

$$\approx 4 + 1.71$$

$$= 5.71.$$

What's Wrong?

You use a graphing utility to find the sum

$$\sum_{n=1}^{6} 7(0.5)^n$$

as shown in the figure. What's wrong?

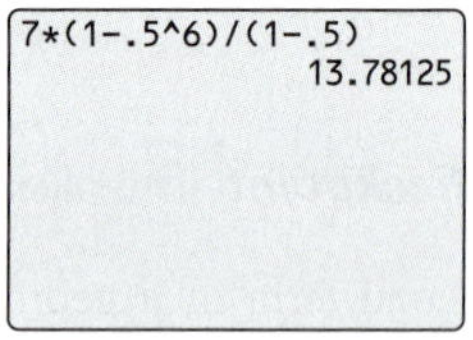

Geometric Series

The sum of the terms of an infinite geometric sequence is called an **infinite geometric series** or simply a **geometric series.**

The formula for the sum of a *finite geometric sequence* can, depending on the value of r, be extended to produce a formula for the sum of an *infinite geometric series*. Specifically, if the common ratio r has the property that $|r| < 1$, then it can be shown that r^n becomes arbitrarily close to zero as n increases without bound. Consequently,

$$a_1\left(\frac{1 - r^n}{1 - r}\right) \longrightarrow a_1\left(\frac{1 - 0}{1 - r}\right) \quad \text{as} \quad n \longrightarrow \infty.$$

This result is summarized as follows.

The Sum of an Infinite Geometric Sequence

If $|r| < 1$, then the infinite geometric series

$$a_1 + a_1r + a_1r^2 + a_1r^3 + \cdots + a_1r^{n-1} + \cdots$$

has the sum

$$S = \sum_{i=0}^{\infty} a_1r^i = \frac{a_1}{1 - r}.$$

Note that when $|r| \geq 1$, the series does not have a sum.

Explore the Concept

Notice that the formula for the sum of an infinite geometric series requires that $|r| < 1$. What happens when $r = 1$ or $r = -1$? Give examples of infinite geometric series for which $|r| > 1$ and convince yourself that they do not have finite sums.

EXAMPLE 7 Finding the Sum of an Infinite Geometric Series

Find each sum.

a. $\displaystyle\sum_{n=0}^{\infty} 4(0.6)^n$

b. $3 + 0.3 + 0.03 + 0.003 + \cdots$

Solution

a. $$\sum_{n=0}^{\infty} 4(0.6)^n = 4 + 4(0.6) + 4(0.6)^2 + 4(0.6)^3 + \cdots + 4(0.6)^n + \cdots$$

$$= \frac{4}{1 - 0.6} \qquad \frac{a_1}{a - r}$$

$$= 10$$

b. $$3 + 0.3 + 0.03 + 0.003 + \cdots = 3 + 3(0.1) + 3(0.1)^2 + 3(0.1)^3 + \cdots$$

$$= \frac{3}{1 - 0.1} \qquad \frac{a_1}{a - r}$$

$$= \frac{10}{3}$$

$$\approx 3.33$$

Checkpoint *Audio-video solution in English & Spanish at LarsonPrecalculus.com*

Find each sum.

a. $\displaystyle\sum_{n=0}^{\infty} 5(0.5)^n$

b. $5 + 1 + 0.2 + 0.04 + \cdots$

Application

EXAMPLE 8 Increasing Annuity

See LarsonPrecalculus.com for an interactive version of this type of example.

A deposit of \$50 is made on the first day of each month in an account that pays 3% interest, compounded monthly. What is the balance at the end of 2 years? (This type of savings plan is called an **increasing annuity.**)

Personal Financial Advisor

Solution

Recall from Section 3.1 that the compound interest formula is

$$A = P\left(1 + \frac{r}{n}\right)^{nt}. \qquad \text{Formula for compound interest}$$

To find the balance in the account after 24 months, consider each of the 24 deposits separately. The first deposit will gain interest for 24 months, and its balance will be

$$A_{24} = 50\left(1 + \frac{0.03}{12}\right)^{24}$$
$$= 50(1.0025)^{24}.$$

The second deposit will gain interest for 23 months, and its balance will be

$$A_{23} = 50\left(1 + \frac{0.03}{12}\right)^{23}$$
$$= 50(1.0025)^{23}.$$

The last deposit will gain interest for only 1 month, and its balance will be

$$A_1 = 50\left(1 + \frac{0.03}{12}\right)^{1}$$
$$= 50(1.0025).$$

The total balance in the annuity will be the sum of the balances of the 24 deposits. Using the formula for the sum of a finite geometric sequence, with $a_1 = 50(1.0025)$, $r = 1.0025$, and $n = 24$, you have

$$S_n = a_1\left(\frac{1 - r^n}{1 - r}\right) \qquad \text{Formula for sum of a finite geometric sequence}$$

$$S_{24} = 50(1.0025)\left[\frac{1 - (1.0025)^{24}}{1 - 1.0025}\right] \qquad \text{Substitute } 50(1.0025) \text{ for } a_1\text{, } 1.0025 \text{ for } r\text{, and 24 for } n.$$

$$\approx \$1238.23. \qquad \text{Use a calculator.}$$

You can use the *summation* feature or the *sum sequence* feature of a graphing utility to verify this result, as shown in the figure.

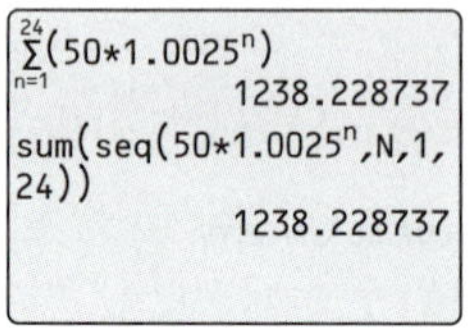

Checkpoint *Audio-video solution in English & Spanish at LarsonPrecalculus.com*

An investor deposits \$70 on the first day of each month in an account that pays 2% interest, compounded monthly. What is the balance at the end of 4 years?

8.3 Exercises

See *CalcChat.com* for tutorial help and worked-out solutions to odd-numbered exercises. For instructions on how to use a graphing utility, see Appendix A.

Vocabulary and Concept Check

In Exercises 1–4, fill in the blank(s).

1. A sequence is _______ when the ratios of consecutive terms are the same. This ratio is called the _______ ratio.

2. The nth term of a geometric sequence has the form $a_n =$ _______ .

3. The sum of the terms of an infinite geometric sequence is called a _______ .

4. The common ratio of a geometric sequence can be any real number except _______ .

5. For what values of the common ratio r is it possible to find the sum of an infinite geometric series?

6. Which formula represents the sum of a *finite geometric sequence*? an *infinite geometric series*?

(a) $S = \dfrac{a_1}{1 - r}, \ |r| < 1$ (b) $S_n = a_1\left(\dfrac{1 - r^n}{1 - r}\right)$

Procedures and Problem Solving

Identifying a Geometric Sequence **In Exercises 7–16, determine whether the sequence is geometric. If it is, find the common ratio.**

7. $3, 6, 12, 24, . . .$

8. $2, 8, 32, 128, . . .$

9. $6, 12, 18, 24, . . .$

10. $4, 16, 28, 40, . . .$

11. $1, -\frac{1}{2}, \frac{1}{4}, -\frac{1}{8}, . . .$

12. $9, -6, 4, -\frac{8}{3}, . . .$

13. $20, 2, 0.2, 0.02, . . .$

14. $5, 1, 0.2, 0.04, . . .$

15. $1, \frac{1}{2}, \frac{1}{3}, \frac{1}{4}, . . .$

16. $\frac{1}{5}, \frac{2}{7}, \frac{3}{9}, \frac{4}{11}, . . .$

Writing the Terms of a Geometric Sequence **In Exercises 17–24, write the first five terms of the geometric sequence.**

17. $a_1 = 6, \ r = 3$

18. $a_1 = 4, \ r = 2$

19. $a_1 = 1, \ r = \frac{1}{2}$

20. $a_1 = 6, \ r = \frac{1}{3}$

21. $a_1 = 5, \ r = -\frac{1}{10}$

22. $a_1 = 12, \ r = -\frac{1}{4}$

23. $a_1 = 1, \ r = e$

24. $a_1 = 5, \ r = \sqrt{5}$

Finding the *n*th Term of a Geometric Sequence **In Exercises 25–30, write the first five terms of the geometric sequence. Find the common ratio and write the *n*th term of the sequence as a function of *n*.**

25. $a_1 = 64, a_{k+1} = \frac{1}{2}a_k$

26. $a_1 = 81, a_{k+1} = \frac{1}{3}a_k$

27. $a_1 = 9, a_{k+1} = 2a_k$

28. $a_1 = 5, a_{k+1} = 3a_k$

29. $a_1 = 6, a_{k+1} = -\frac{3}{2}a_k$

30. $a_1 = 30, a_{k+1} = -\frac{2}{3}a_k$

Finding a Term of a Geometric Sequence **In Exercises 31–38, find the indicated term of the geometric sequence (a) using the *table* feature of a graphing utility and (b) algebraically.**

31. $a_1 = 9, r = 1.12$, 15th term

32. $a_1 = 14, r = 1.4$, 8th term

33. $a_1 = 8, r = -\frac{4}{3}$, 7th term

34. $a_1 = 8, r = -\frac{3}{4}$, 9th term

35. $a_1 = -\frac{1}{16}, r = 8$, 6th term

36. $a_1 = -\frac{1}{256}, r = 2$, 9th term

37. $a_1 = 7, r = \sqrt{2}$, 14th term

38. $a_1 = 2, r = \sqrt{3}$, 11th term

Finding a Term of a Geometric Sequence **In Exercises 39–42, find a formula for the *n*th term of the geometric sequence. Then find the indicated term of the geometric sequence.**

39. $7, 21, 63, . . .$; 9th term

40. $3, 36, 432, . . .$; 7th term

41. $5, 30, 180, . . .$; 10th term

42. $4, 8, 16, . . .$; 22nd term

Finding a Term of a Geometric Sequence **In Exercises 43–46, find the missing term of the geometric sequence.**

43. $a_1 = 16, a_3 = \square, a_4 = \frac{27}{4}$

44. $a_1 = 5, a_3 = \frac{45}{4}, a_8 = \square$

45. $a_2 = -18, a_5 = \frac{2}{3}, a_6 = \square$

46. $a_2 = -8, a_5 = \frac{64}{27}, a_6 = \square$

Graphing the Terms of a Sequence **In Exercises 47–50, use a graphing utility to graph the first 10 terms of the sequence.**

47. $a_n = 12(-0.75)^{n-1}$

48. $a_n = 20(0.85)^{n-1}$

49. $a_n = 2(1.3)^{n-1}$

50. $a_n = 10(-1.2)^{n-1}$

Finding a Sequence of Partial Sums In Exercises 51 and 52, find the sequence of the first five partial sums S_1, S_2, S_3, S_4, and S_5 of the geometric sequence by adding terms.

51. $8, -4, 2, -1, \frac{1}{2}, \ldots$ **52.** $8, 12, 18, 27, \frac{81}{2}, \ldots$

Finding a Sequence of Partial Sums In Exercises 53 and 54, use a graphing utility to create a table showing the sequence of the first 10 partial sums $S_1, S_2, S_3, \ldots,$ and S_{10} for the series.

53. $\sum_{n=1}^{\infty} 16\left(\frac{1}{2}\right)^{n-1}$ **54.** $\sum_{n=1}^{\infty} 4(0.2)^{n-1}$

Finding the Sum of a Finite Geometric Sequence In Exercises 55–64, find the sum. Use a graphing utility to verify your result.

55. $\sum_{n=1}^{9} 2^{n-1}$ **56.** $\sum_{n=1}^{9} (-2)^{n-1}$

57. $\sum_{i=1}^{7} 64\left(-\frac{1}{2}\right)^{i-1}$ **58.** $\sum_{i=1}^{6} 32\left(\frac{1}{4}\right)^{i-1}$

59. $\sum_{n=0}^{20} 3\left(\frac{3}{2}\right)^{n}$ **60.** $\sum_{n=0}^{15} 10\left(\frac{7}{6}\right)^{n}$

61. $\sum_{i=1}^{10} 8\left(-\frac{1}{4}\right)^{i-1}$ **62.** $\sum_{i=1}^{10} 5\left(-\frac{1}{3}\right)^{i-1}$

63. $\sum_{n=0}^{5} 300(1.06)^{n}$ **64.** $\sum_{n=0}^{6} 500(1.04)^{n}$

Using Summation Notation In Exercises 65–68, use summation notation to write the sum.

65. $5 + 20 + 80 + \cdots + 5120$

66. $7 + 21 + 63 + \cdots + 1701$

67. $2 - \frac{1}{2} + \frac{1}{8} - \cdots + \frac{1}{2048}$

68. $15 - 3 + \frac{3}{5} - \cdots - \frac{3}{625}$

Finding the Sum of an Infinite Geometric Series In Exercises 69–84, find the sum of the infinite geometric series, if possible. If not possible, explain why.

69. $\sum_{n=0}^{\infty} \left(\frac{1}{2}\right)^{n}$ **70.** $\sum_{n=0}^{\infty} \left(\frac{2}{3}\right)^{n}$

71. $\sum_{n=0}^{\infty} 4\left(\frac{3}{4}\right)^{n}$ **72.** $\sum_{n=0}^{\infty} 6\left(\frac{1}{5}\right)^{n}$

73. $\sum_{n=0}^{\infty} 5\left(-\frac{1}{2}\right)^{n}$ **74.** $\sum_{n=0}^{\infty} 5\left(-\frac{1}{4}\right)^{n}$

75. $\sum_{n=1}^{\infty} 2\left(\frac{7}{3}\right)^{n-1}$ **76.** $\sum_{n=1}^{\infty} 8\left(\frac{5}{3}\right)^{n-1}$

77. $\sum_{n=0}^{\infty} [-3(-0.9)^{n}]$ **78.** $\sum_{n=0}^{\infty} [-10(-0.2)^{n}]$

79. $9 + 6 + 4 + \frac{8}{3} + \cdots$

80. $8 + 6 + \frac{9}{2} + \frac{27}{8} + \cdots$

81. $3 + \frac{15}{2} + \frac{75}{4} + \frac{375}{8} + \cdots$

82. $2 + \frac{7}{3} + \frac{49}{18} + \frac{343}{108} + \cdots$

83. $-7 + 2 - \frac{4}{7} + \frac{8}{49} - \cdots$

84. $-6 + 5 - \frac{25}{6} + \frac{125}{36} - \cdots$

Writing a Repeating Decimal as a Fraction In Exercises 85–88, write the repeating decimal as a fraction.

85. $0.\overline{36}$ **86.** $0.\overline{297}$

87. $1.2\overline{5}$ **88.** $1.3\overline{8}$

Identifying a Sequence In Exercises 89–96, determine whether the sequence associated with the series is arithmetic or geometric. Find the common difference or ratio and find the sum of the first 15 terms.

89. $8 + 16 + 32 + 64 + \cdots$

90. $17 + 14 + 11 + 8 + \cdots$

91. $90 + 30 + 10 + \frac{10}{3} + \cdots$

92. $\frac{5}{4} + \frac{7}{4} + \frac{9}{4} + \frac{11}{4} + \cdots$

93. $\sum_{n=1}^{\infty} 6n$ **94.** $\sum_{n=1}^{\infty} 3^{n-1}$

95. $\sum_{n=0}^{\infty} 6(0.8)^{n}$ **96.** $\sum_{n=0}^{\infty} \frac{n+8}{8}$

97. Annuity A deposit of \$100 is made at the beginning of each month in an account that pays 3% interest, compounded monthly. The balance A in the account at the end of 5 years is given by

$$A = 100\left(1 + \frac{0.03}{12}\right)^{1} + \cdots + 100\left(1 + \frac{0.03}{12}\right)^{60}.$$

Find A.

98. Annuity A deposit of \$50 is made at the beginning of each month in an account that pays 2% interest, compounded monthly. The balance A in the account at the end of 6 years is given by

$$A = 50\left(1 + \frac{0.02}{12}\right)^{1} + \cdots + 50\left(1 + \frac{0.02}{12}\right)^{72}.$$

Find A.

99. Annuity A deposit of P dollars is made at the beginning of each month in an account earning an annual interest rate r, compounded monthly. The balance A after t years is given by

$$A = P\left(1 + \frac{r}{12}\right) + P\left(1 + \frac{r}{12}\right)^{2} + \cdots + P\left(1 + \frac{r}{12}\right)^{12t}.$$

Show that the balance is given by

$$A = P\left[\left(1 + \frac{r}{12}\right)^{12t} - 1\right]\left(1 + \frac{12}{r}\right).$$

100. Annuity A deposit of P dollars is made at the beginning of each month in an account earning an annual interest rate r, compounded continuously. The balance A after t years is given by $A = Pe^{r/12} + Pe^{2r/12} + \cdot \cdot \cdot + Pe^{12tr/12}$. Show that the balance is given by

$$A = \frac{Pe^{r/12}(e^{rt} - 1)}{e^{r/12} - 1}.$$

Annuities In Exercises 101–104, consider making monthly deposits of P dollars in a savings account earning an annual interest rate r. Use the results of Exercises 99 and 100 to find the balances A after t years when the interest is compounded (a) monthly and (b) continuously.

101. $P = \$200,\ r = 7\%,\ t = 20$ years

102. $P = \$75,\ r = 4\%,\ t = 15$ years

103. $P = \$100,\ r = 5\%,\ t = 40$ years

104. $P = \$20,\ r = 6\%,\ t = 50$ years

105. Geometry The sides of a square are 16 inches in length. A new square is formed by connecting the midpoints of the sides of the original square, and two of the resulting triangles are shaded (see figure). After this process is repeated five more times, determine the total area of the shaded region.

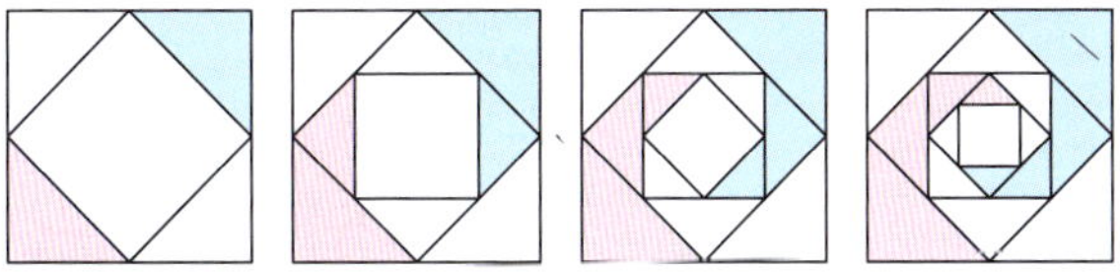

106. Geometry The sides of a square are 27 inches in length. New squares are formed by dividing the original square into nine squares. The center square is then shaded (see figure). After this process is repeated three more times, determine the total area of the shaded region.

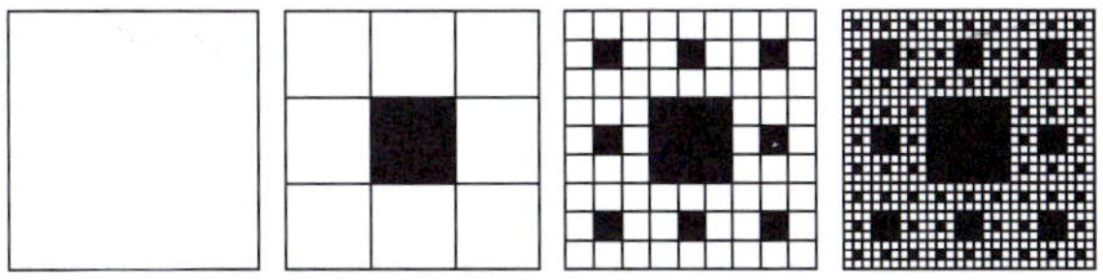

107. Physics The temperature of water in an ice cube tray is 70°F when it is placed in a freezer. Its temperature n hours after being placed in the freezer is 20% less than 1 hour earlier.

(a) Find a formula for the nth term of the geometric sequence that gives the temperature of the water n hours after it is placed in the freezer.

(b) Find the temperatures of the water 6 hours and 12 hours after it is placed in the freezer.

(c) Use a graphing utility to graph the sequence and approximate the time required for the water to freeze.

108. MODELING DATA

The mid-year populations a_n of Peru (in millions) from 2009 through 2017 are given by ordered pairs of the form (n, a_n), with $n = 9$ corresponding to 2009. *(Spreadsheet at LarsonPrecalculus.com)* *(Source: U.S. Census Bureau)*

DATA

(9, 28.65) (10, 28.95) (11, 29.25)
(12, 29.55) (13, 29.85) (14, 30.15)
(15, 30.45) (16, 30.74) (17, 31.04)

(a) Use the *exponential regression* feature of a graphing utility to find a geometric sequence that models the data.

(b) Use the sequence from part (a) to describe the rate at which the population of Peru is growing.

(c) Use the sequence from part (a) to predict the population of Peru in 2025. The U.S. Census Bureau predicts that the population of Peru will be 33.28 million in 2025. How does this value compare with your prediction?

(d) Use the sequence from part (a) to determine when the population of Peru will reach 39 million.

109. Fractals In a *fractal,* a geometric figure is repeated at smaller and smaller scales. The sphereflake shown is a computer-generated fractal that was created by Eric Haines. The radius of the large sphere is 1. Attached to the large sphere are nine spheres of radius $\frac{1}{3}$. Attached to each of the smaller spheres are nine spheres of radius $\frac{1}{9}$. This process is continued infinitely.

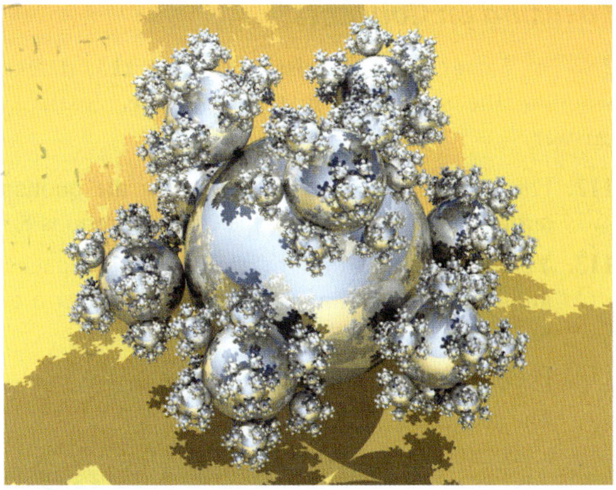

(a) Write a formula in series notation that gives the surface area of the sphereflake.

(b) Write a formula in series notation that gives the volume of the sphereflake.

(c) Is the surface area of the sphereflake finite or infinite? Is the volume finite or infinite? If either is finite, find the value.

Eric Haines

110. Manufacturing The manufacturer of a new food processor plans to produce and sell 8000 units per year. Each year, 10% of all units sold become inoperative. So, 8000 units will be in use after 1 year, $[8000 + 0.9(8000)]$ units will be in use after 2 years, and so on.

(a) Write a formula in series notation for the number of units that will be operative after n years.

(b) Find the numbers of units that will be operative after 10 years, 15 years, and 20 years.

(c) If this trend continues indefinitely, will the number of units that will be operative be finite? If so, how many? If not, explain your reasoning.

111. ***Why you should learn it*** *(p. 599)* A ball is dropped from a height of 6 feet and begins bouncing, as shown in the figure. The height of each bounce is three-fourths the height of the previous bounce. Find the total vertical distance the ball travels before coming to rest.

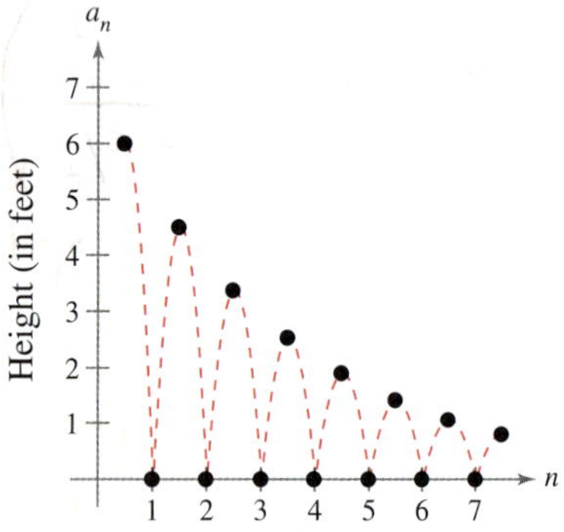

Focusing on Concepts

True or False? **In Exercises 112 and 113, determine whether the statement is true or false. Justify your answer.**

112. For a geometric sequence in which the quotient of term a_7 and term a_5 is 16, the common ratio is 8.

113. You can find the nth term of a geometric sequence by multiplying its common ratio by the first term of the sequence raised to the $(n - 1)$th power.

114. Error Analysis Describe the error.

The sequence 21, 42, 84, 168, . . . is a geometric sequence with a common ratio of $\frac{1}{2}$. ✗

Writing the Terms of a Geometric Sequence **In Exercises 115 and 116, write the first five terms of the geometric sequence.**

115. $a_1 = 3, r = \dfrac{x}{2}$

116. $a_1 = \dfrac{1}{4}, r = 7x$

Veronica Louro/Shutterstock.com

Finding a Term of a Geometric Sequence **In Exercises 117 and 118, find the indicated term of the geometric sequence.**

117. $a_1 = 100, r = e^x$, 9th term

118. $a_1 = 2, r = (4x)/3$, 6th term

119. Exploration Use a graphing utility to graph each function. Identify the horizontal asymptote of the graph and determine its relationship to the sum.

(a) $f(x) = 6\left[\dfrac{1 - (0.5)^x}{1 - (0.5)}\right], \quad \displaystyle\sum_{n=0}^{\infty} 6\left(\frac{1}{2}\right)^n$

(b) $f(x) = 2\left[\dfrac{1 - (0.8)^x}{1 - (0.8)}\right], \quad \displaystyle\sum_{n=0}^{\infty} 2\left(\frac{4}{5}\right)^n$

120. Writing Write a brief paragraph explaining why the terms of a geometric sequence decrease in magnitude when $-1 < r < 1$.

121. Writing Write a brief paragraph explaining how to use the first two terms of a geometric sequence to find the nth term.

122. HOW DO YOU SEE IT? Use the figures shown below.

(i)

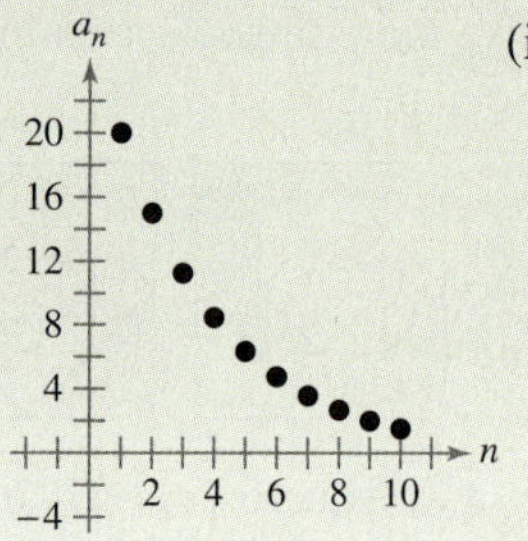

(ii)

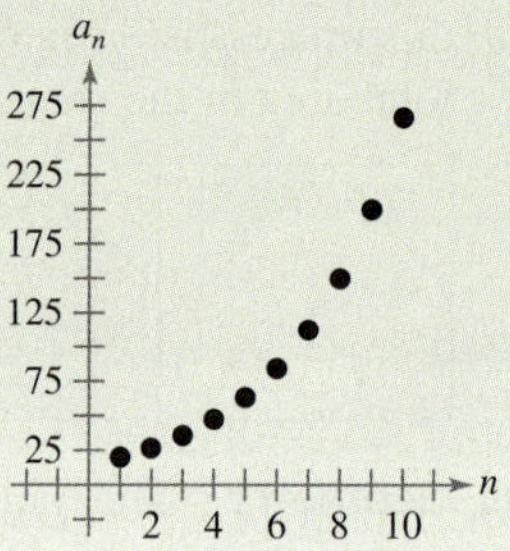

(a) Without performing any calculations, determine which figure shows the terms of the sequence $a_n = 20\left(\frac{4}{3}\right)^{n-1}$, and which shows the terms of $a_n = 20\left(\frac{3}{4}\right)^{n-1}$. Explain your reasoning.

(b) Which sequence has terms that can be summed? Explain your reasoning.

Cumulative Mixed Review

Finding a Determinant **In Exercises 123 and 124, find the determinant of the matrix.**

123. $\begin{bmatrix} -1 & 3 & 4 \\ -2 & 8 & 0 \\ 0 & 5 & -1 \end{bmatrix}$

124. $\begin{bmatrix} -1 & 0 & 4 \\ -4 & 3 & 5 \\ 0 & 2 & -3 \end{bmatrix}$

125. ***Project: Yearly Production*** To work an extended application analyzing the yearly production of an oil well over a period of 36 years, visit the textbook's website at *LarsonPrecalculus.com.*

8.4 The Binomial Theorem

What you should learn

- Use the Binomial Theorem to calculate binomial coefficients.
- Use binomial coefficients to write binomial expansions.
- Use Pascal's Triangle to calculate binomial coefficients.

Why you should learn it

You can use binomial coefficients to model and solve real-life problems. For instance, in Exercise 116 on page 616, you will use binomial coefficients to write the expansion of a model that represents the amount of child support collected in the United States.

Binomial Coefficients

Recall that a *binomial* is a polynomial that has two terms. In this section, you will study a formula that provides a quick method of raising a binomial to a power. To begin, look at the expansion of $(x + y)^n$ for several values of n.

$$(x + y)^0 = 1$$
$$(x + y)^1 = x + y$$
$$(x + y)^2 = x^2 + 2xy + y^2$$
$$(x + y)^3 = x^3 + 3x^2y + 3xy^2 + y^3$$
$$(x + y)^4 = x^4 + 4x^3y + 6x^2y^2 + 4xy^3 + y^4$$
$$(x + y)^5 = x^5 + 5x^4y + 10x^3y^2 + 10x^2y^3 + 5xy^4 + y^5$$
$$(x + y)^6 = x^6 + 6x^5y + 15x^4y^2 + 20x^3y^3 + 15x^2y^4 + 6xy^5 + y^6$$

There are several observations you can make about these expansions.

1. In each expansion, there are $n + 1$ terms.
2. In each expansion, x and y have symmetric roles. The powers of x decrease by 1 in successive terms, whereas the powers of y increase by 1.
3. The sum of the powers of each term is n. For instance, in the expansion of $(x + y)^5$ the sum of the powers of each term is 5.

$$(x + y)^5 = x^5 + \overbrace{5x^4y^1}^{4+1=5} + \overbrace{10x^3y^2}^{3+2=5} + 10x^2y^3 + 5x^1y^4 + y^5$$

4. The coefficients increase and then decrease in a symmetric pattern.

The coefficients of a binomial expansion are called **binomial coefficients.** To find them, you can use the **Binomial Theorem.**

The Binomial Theorem

Let n be a positive integer, and let $r = 0, 1, 2, 3, \ldots, n$. In the expansion of $(x + y)^n$

$$(x + y)^n = x^n + nx^{n-1}y + \cdots + {}_nC_r x^{n-r}y^r + \cdots + nxy^{n-1} + y^n$$

the coefficient of $x^{n-r}y^r$ is

$${}_nC_r = \frac{n!}{(n - r)!r!}.$$

The symbol

$$\binom{n}{r}$$

is sometimes used in place of ${}_nC_r$ to denote binomial coefficients.

(See the proof on page 644.)

Another way to represent the Binomial Theorem is by using summation notation, as shown below.

$$(x + y)^n = \sum_{r=0}^{n} \binom{n}{r} x^{n-r}y^r \qquad \text{Binomial Theorem using summation notation}$$

Bennewitz/iStock/Getty Images

EXAMPLE 1 Finding Binomial Coefficients

Find each binomial coefficient.

a. ${}_8C_2$ **b.** $\binom{10}{3}$ **c.** ${}_7C_0$ **d.** $\binom{8}{8}$

Solution

a. $${}_8C_2 = \frac{8!}{6! \cdot 2!} = \frac{(8 \cdot 7) \cdot \cancel{6!}}{\cancel{6!} \cdot 2!} = \frac{8 \cdot 7}{2 \cdot 1} = 28$$

b. $$\binom{10}{3} = \frac{10!}{7! \cdot 3!} = \frac{(10 \cdot 9 \cdot 8) \cdot \cancel{7!}}{\cancel{7!} \cdot 3!} = \frac{10 \cdot 9 \cdot 8}{3 \cdot 2 \cdot 1} = 120$$

c. $${}_7C_0 = \frac{\cancel{7!}}{\cancel{7!} \cdot 0!} = 1$$

d. $$\binom{8}{8} = \frac{\cancel{8!}}{0! \cdot \cancel{8!}} = 1$$

✓ Checkpoint ▶ Audio-video solution in English & Spanish at LarsonPrecalculus.com

Find each binomial coefficient.

a. $\binom{11}{5}$ **b.** ${}_9C_2$ **c.** $\binom{5}{0}$ **d.** ${}_{15}C_{15}$

When $r \neq 0$ and $r \neq n$, as in parts (a) and (b) of Example 1, there is a simple pattern for evaluating binomial coefficients that works because there will always be factorial terms that divide out from the expression.

$${}_8C_2 = \frac{\overbrace{8 \cdot 7}^{\text{2 factors}}}{\underbrace{2 \cdot 1}_{\text{2 factorial}}} \quad \text{and} \quad \binom{10}{3} = \frac{\overbrace{10 \cdot 9 \cdot 8}^{\text{3 factors}}}{\underbrace{3 \cdot 2 \cdot 1}_{\text{3 factorial}}}$$

EXAMPLE 2 Finding Binomial Coefficients

Notice how the pattern shown above is used in parts (a) and (b).

a. $${}_7C_3 = \frac{7 \cdot \cancel{6} \cdot 5}{\cancel{3} \cdot \cancel{2} \cdot 1} = 35$$

b. $${}_7C_4 = \frac{7 \cdot \cancel{6} \cdot 5 \cdot \cancel{4}}{\cancel{4} \cdot \cancel{3} \cdot \cancel{2} \cdot 1} = 35$$

c. $${}_{12}C_1 = \frac{12}{1} = 12$$

d. $${}_{12}C_{11} = \frac{12!}{1! \cdot 11!} = \frac{(12) \cdot \cancel{11!}}{1! \cdot \cancel{11!}} = \frac{12}{1} = 12$$

✓ Checkpoint ▶ Audio-video solution in English & Spanish at LarsonPrecalculus.com

Find each binomial coefficient.

a. ${}_7C_5$ **b.** $\binom{7}{2}$ **c.** ${}_{14}C_{13}$ **d.** $\binom{14}{1}$

It is not a coincidence that the results in parts (a) and (b) of Example 2 are the same and that the results in parts (c) and (d) are the same. In general, it is true that ${}_nC_r = {}_nC_{n-r}$.

Technology Tip

Most graphing utilities are programmed to evaluate ${}_nC_r$. For instructions on how to use the ${}_nC_r$ feature, see Appendix A; for specific keystrokes, go to this textbook's *Student Companion Website.*

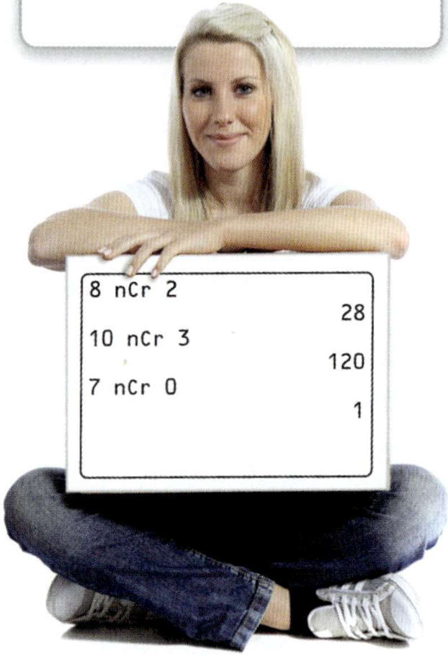

Explore the Concept

Find each pair of binomial coefficients.

a. ${}_7C_0, {}_7C_7$ **d.** ${}_7C_1, {}_7C_6$

b. ${}_8C_0, {}_8C_8$ **e.** ${}_8C_1, {}_8C_7$

c. ${}_{10}C_0, {}_{10}C_{10}$ **f.** ${}_{10}C_1, {}_{10}C_9$

What do you observe about the pairs in (a), (b), and (c)? What do you observe about the pairs in (d), (e), and (f)? Write two conjectures from your observations. Develop a convincing argument for your two conjectures.

Binomial Expansions

As mentioned at the beginning of this section, when you write out the coefficients for a binomial that is raised to a power, you are **expanding a binomial.** The formula for binomial coefficients gives you a way to expand binomials.

EXAMPLE 3 Expanding a Binomial

Write the expansion of the expression $(x + 1)^3$.

Solution

The binomial coefficients are ${}_3C_0 = 1$, ${}_3C_1 = 3$, ${}_3C_2 = 3$, and ${}_3C_3 = 1$. So, the expansion is as follows.

$$(x + 1)^3 = (1)x^3 + (3)x^2(1) + (3)x(1^2) + (1)(1^3)$$
$$= x^3 + 3x^2 + 3x + 1$$

Checkpoint *Audio-video solution in English & Spanish at LarsonPrecalculus.com*

Write the expansion of the expression $(x + 1)^4$.

To expand binomials representing *differences*, rather than sums, you alternate signs, as shown in the next example.

EXAMPLE 4 Expanding a Binomial

Write the expansion of the expression $(x - 1)^3$.

Solution

Expand using the binomial coefficients from Example 3.

$$(x - 1)^3 = [x + (-1)]^3$$
$$= (1)x^3 + (3)x^2(-1) + (3)x(-1)^2 + (1)(-1)^3$$
$$= x^3 - 3x^2 + 3x - 1$$

Checkpoint *Audio-video solution in English & Spanish at LarsonPrecalculus.com*

Write the expansion of the expression $(x - 1)^4$.

EXAMPLE 5 Expanding Binomial Expressions

See LarsonPrecalculus.com for an interactive version of this type of example.

Write the expansion of (a) $(2x - 3)^4$ and (b) $(x - 2y)^4$.

Solution

The binomial coefficients are ${}_4C_0 = 1$, ${}_4C_1 = 4$, ${}_4C_2 = 6$, ${}_4C_3 = 4$, and ${}_4C_4 = 1$. So, the expansions are as follows.

a. $$(2x - 3)^4 = (1)(2x)^4 - (4)(2x)^3(3) + (6)(2x)^2(3^2) - (4)(2x)(3^3) + (1)(3^4)$$
$$= 16x^4 - 96x^3 + 216x^2 - 216x + 81$$

b. $$(x - 2y)^4 = (1)x^4 - (4)x^3(2y) + (6)x^2(2y)^2 - (4)x(2y)^3 + (1)(2y)^4$$
$$= x^4 - 8x^3y + 24x^2y^2 - 32xy^3 + 16y^4$$

Checkpoint *Audio-video solution in English & Spanish at LarsonPrecalculus.com*

Write the expansion of (a) $(3y - 1)^4$ and (b) $(2x - y)^5$.

Technology Tip

You can use a graphing utility to check the expansion in Example 5(a) by graphing the original binomial expression and the expansion in the same viewing window. The graphs should coincide, as shown below.

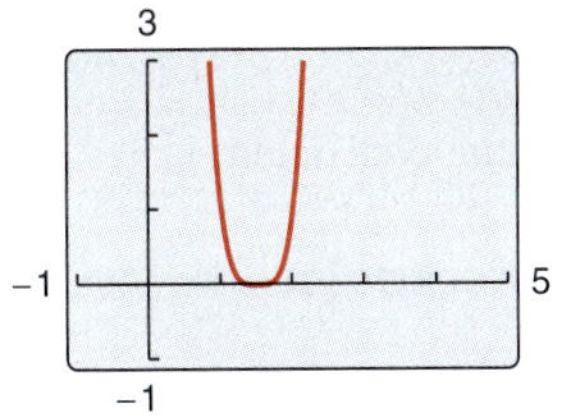

EXAMPLE 6 Expanding a Binomial

Write the expansion of the expression

$$(x^2 + 4)^3.$$

Solution

Expand using the binomial coefficients from Example 3.

$$\begin{aligned}(x^2 + 4)^3 &= (1)(x^2)^3 + (3)(x^2)^2(4) + (3)x^2(4^2) + (1)(4^3)\\ &= x^6 + 12x^4 + 48x^2 + 64\end{aligned}$$

✓ *Checkpoint* ▶ *Audio-video solution in English & Spanish at LarsonPrecalculus.com*

Write the expansion of the expression $(5 + y^2)^3$.

Sometimes you will need to find a specific term in a binomial expansion. Instead of writing out the entire expansion, you can use the fact that, from the Binomial Theorem, the $(r + 1)$th term is

$$_nC_r x^{n-r} y^r.$$

For instance, to find the third term of the expression in Example 6, you could use the formula above with $n = 3$ and $r = 2$ to obtain

$$_3C_2(x^2)^{3-2} \cdot 4^2 = 3(x^2) \cdot 16 = 48x^2.$$

EXAMPLE 7 Finding a Term or Coefficient

a. Find the sixth term of $(a + 2b)^8$.

b. Find the coefficient of the term a^6b^5 in the expansion of $(2a - 5b)^{11}$.

Solution

a. Because the formula is for the $(r + 1)$th term, r is one less than the number of the term you need. So, to find the sixth term in this binomial expansion, use $r = 5$, $n = 8$, $x = a$, and $y = 2b$.

$$\begin{aligned}{}_nC_r x^{n-r} y^r &= {}_8C_5 a^{8-5}(2b)^5\\ &= 56 \cdot a^3 \cdot (2b)^5\\ &= 56(2^5)a^3b^5\\ &= 1792a^3b^5\end{aligned}$$

b. Note that

$$(2a - 5b)^{11} = [2a + (-5b)]^{11}.$$

So, $n = 11$, $r = 5$, $x = 2a$, and $y = -5b$. Substitute these values to obtain

$$\begin{aligned}{}_nC_r x^{n-r} y^r &= {}_{11}C_5(2a)^6(-5b)^5\\ &= (462)(64a^6)(-3125b^5)\\ &= -92{,}400{,}000a^6b^5.\end{aligned}$$

So, the coefficient is $-92{,}400{,}000$.

✓ *Checkpoint* ▶ *Audio-video solution in English & Spanish at LarsonPrecalculus.com*

a. Find the fifth term of $(a + 2b)^8$.

b. Find the coefficient of the term a^4b^7 in the expansion of $(2a - 5b)^{11}$.

Pascal's Triangle

There is a convenient way to remember the pattern for binomial coefficients. By arranging the coefficients in a triangular pattern, you obtain the array shown below, which is called **Pascal's Triangle.** This triangle is named after the famous French mathematician Blaise Pascal (1623–1662).

1
1 1
1 2 1
1 3 3 1
1 4 6 4 1
1 5 10 10 5 1 — $4 + 6 = 10$
1 6 15 20 15 6 1
1 7 21 35 35 21 7 1 — $15 + 6 = 21$

In each row of Pascal's Triangle, the first and last numbers are 1's. Also, each number between the 1's is the sum of the two numbers immediately above that number. Pascal noticed that the numbers in this triangle are precisely the same numbers as the coefficients of binomial expansions, as shown below for $n = 0, 1, 2, \ldots, 7$.

$$(x + y)^0 = 1 \qquad \text{0th row}$$
$$(x + y)^1 = 1x + 1y \qquad \text{1st row}$$
$$(x + y)^2 = 1x^2 + 2xy + 1y^2 \qquad \text{2nd row}$$
$$(x + y)^3 = 1x^3 + 3x^2y + 3xy^2 + 1y^3 \qquad \text{3rd row}$$
$$(x + y)^4 = 1x^4 + 4x^3y + 6x^2y^2 + 4xy^3 + 1y^4 \qquad \vdots$$
$$(x + y)^5 = 1x^5 + 5x^4y + 10x^3y^2 + 10x^2y^3 + 5xy^4 + 1y^5$$
$$(x + y)^6 = 1x^6 + 6x^5y + 15x^4y^2 + 20x^3y^3 + 15x^2y^4 + 6xy^5 + 1y^6$$
$$(x + y)^7 = 1x^7 + 7x^6y + 21x^5y^2 + 35x^4y^3 + 35x^3y^4 + 21x^2y^5 + 7xy^6 + 1y^7$$

The top row of Pascal's Triangle is called the *zeroth row* because it corresponds to the binomial expansion $(x + y)^0 = 1$. Similarly, the next row is called the *first row* because it corresponds to the binomial expansion

$$(x + y)^1 = 1(x) + 1(y).$$

In general, the *nth row* of Pascal's Triangle gives the coefficients of $(x + y)^n$.

EXAMPLE 8 Using Pascal's Triangle

Use the seventh row of Pascal's Triangle to find the binomial coefficients.

$${}_8C_0 \quad {}_8C_1 \quad {}_8C_2 \quad {}_8C_3 \quad {}_8C_4 \quad {}_8C_5 \quad {}_8C_6 \quad {}_8C_7 \quad {}_8C_8$$

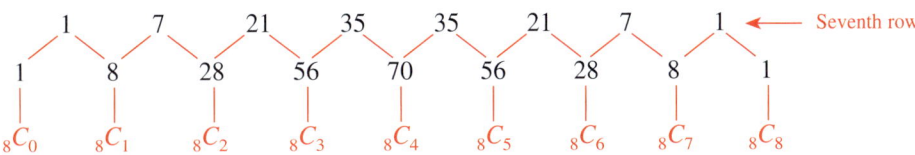

Solution

1 7 21 35 35 21 7 1 ← Seventh row

1 8 28 56 70 56 28 8 1

${}_8C_0$ ${}_8C_1$ ${}_8C_2$ ${}_8C_3$ ${}_8C_4$ ${}_8C_5$ ${}_8C_6$ ${}_8C_7$ ${}_8C_8$

✓ ***Checkpoint*** *Audio-video solution in English & Spanish at LarsonPrecalculus.com*

Use the eighth row of Pascal's Triangle (see Example 8) to find the binomial coefficients.

$${}_9C_0 \quad {}_9C_1 \quad {}_9C_2 \quad {}_9C_3 \quad {}_9C_4 \quad {}_9C_5 \quad {}_9C_6 \quad {}_9C_7 \quad {}_9C_8 \quad {}_9C_9$$

Explore the Concept

Complete the table and describe the result.

n	r	${}_nC_r$	${}_nC_{n-r}$
9	5		
7	1		
12	4		
6	0		
10	7		

What characteristics of Pascal's Triangle are illustrated by this table?

8.4 Exercises

See *CalcChat.com* for tutorial help and worked-out solutions to odd-numbered exercises. For instructions on how to use a graphing utility, see Appendix A.

Vocabulary and Concept Check

In Exercises 1 and 2, fill in the blanks.

1. The notation used to denote a binomial coefficient is ________ or ________ .
2. When you write out the coefficients for a binomial that is raised to a power, you are ________ a ________ .
3. List two ways to find binomial coefficients.
4. In the expression of $(x + y)^3$, what is the sum of the powers of the third term?

Procedures and Problem Solving

Finding Binomial Coefficients **In Exercises 5–16, find the binomial coefficient by hand.**

5. ${}_5C_3$
6. ${}_4C_1$
7. ${}_{12}C_2$
8. ${}_6C_2$
9. ${}_{11}C_8$
10. ${}_{12}C_{10}$
11. $\binom{12}{0}$
12. $\binom{20}{20}$
13. $\binom{10}{4}$
14. $\binom{11}{3}$
15. $\binom{14}{12}$
16. $\binom{11}{10}$

Finding Binomial Coefficients **In Exercises 17–20, use a graphing utility to find ${}_nC_r$.**

17. ${}_{41}C_{36}$
18. ${}_{50}C_{46}$
19. ${}_{500}C_{498}$
20. ${}_{1000}C_2$

Identifying Equivalent Expressions **In Exercises 21–24, without calculating, determine whether the pair of binomial coefficients yields the same value.**

21. ${}_{10}C_1, {}_{10}C_9$
22. ${}_5C_4, {}_{10}C_8$
23. ${}_5C_1, {}_9C_1$
24. ${}_{13}C_5, {}_{13}C_8$

Expanding a Binomial **In Exercises 25–56, use the Binomial Theorem to expand and simplify the expression. Verify your answer using a graphing utility.**

25. $(x + 1)^6$
26. $(x + 1)^5$
27. $(y - 4)^3$
28. $(y - 5)^4$
29. $(x + y)^5$
30. $(x + y)^6$
31. $(2x - y)^5$
32. $(5x - y)^4$
33. $(4y - 3)^3$
34. $(2y - 5)^3$
35. $(2r - 3s)^6$
36. $(7r - 6s)^4$
37. $(x^2 + 2)^4$
38. $(y^2 + 2)^6$
39. $(5 - x^2)^5$
40. $(3 - y^2)^3$
41. $(x^2 + y^2)^4$
42. $(x^2 + y^2)^6$
43. $(3x^3 - y)^6$
44. $(2x^3 - y)^5$
45. $\left(\frac{1}{x} + y\right)^5$
46. $\left(\frac{1}{x} + y\right)^6$
47. $\left(\frac{2}{x} - 2y\right)^4$
48. $\left(\frac{2}{x} - 3y\right)^5$
49. $(4x - 1)^3 - 2(4x - 1)^4$
50. $3(x + 1)^5 + 4(x + 1)^3$
51. $-3(x - 2)^3 - 4(x + 1)^6$
52. $-5(x + 2)^5 - 2(x - 1)^2$
53. $(3\sqrt{x} + 5)^3$
54. $(2\sqrt{t} - 7)^3$
55. $(x^{2/3} - y^{1/3})^3$
56. $(u^{3/5} + v^{1/5})^5$

Finding a Term in a Binomial Expansion **In Exercises 57–64, find the specified nth term in the expansion of the binomial.**

57. $(x + 8)^{10}, \ n = 4$
58. $(x + 6)^6, \ n = 7$
59. $(x - 6y)^5, \ n = 3$
60. $(x - 10z)^7, \ n = 4$
61. $(4x + 3y)^9, \ n = 8$
62. $(5a + 6b)^5, \ n = 5$
63. $(10x - 3y)^{12}, \ n = 10$
64. $(7x - 2y)^{15}, \ n = 7$

Finding a Coefficient in a Binomial Expansion **In Exercises 65–72, find the coefficient a of the given term in the expansion of the binomial.**

	Binomial	*Term*
65.	$(x + 3)^{12}$	ax^5
66.	$(x + 4)^{12}$	ax^4
67.	$(4x - y)^{10}$	ax^2y^8
68.	$(x - 2y)^{10}$	ax^8y^2
69.	$(3x - 2y)^9$	ax^6y^3
70.	$(2x - 3y)^8$	ax^4y^4
71.	$(x^2 + y)^{10}$	ax^8y^6
72.	$(z^2 + y)^{12}$	$az^{14}y^5$

Using Pascal's Triangle **In Exercises 73–76, use Pascal's Triangle to find the binomial coefficient.**

73. ${}_6C_3$
74. ${}_7C_4$
75. ${}_6C_5$
76. ${}_5C_2$

Using Pascal's Triangle **In Exercises 77–82, expand the binomial by using Pascal's Triangle to determine the coefficients.**

77. $(5y + 2)^5$
78. $(2v + 3)^6$
79. $(2x + 3y)^5$
80. $(3x + 4y)^5$
81. $(3t - 2v)^4$
82. $(5v - 2z)^4$

Expanding an Expression **In Exercises 83–86, expand the difference quotient using the given expression for $f(x)$ and simplify.**

$$\frac{f(x + h) - f(x)}{h}, \quad h \neq 0$$

83. $f(x) = x^3$
84. $f(x) = x^4$
85. $f(x) = x^6$
86. $f(x) = x^8$

Expanding a Complex Number **In Exercises 87–100, use the Binomial Theorem to expand the complex number. Simplify your result. (Remember that $i = \sqrt{-1}$.)**

87. $(1 + i)^4$
88. $(1 - i)^6$
89. $(4 + i)^4$
90. $(2 + i)^5$
91. $(2 - 3i)^6$
92. $(3 - 2i)^6$
93. $(5 + \sqrt{-16})^3$
94. $(5 + \sqrt{-9})^3$
95. $(4 + \sqrt{3}i)^4$
96. $(5 - \sqrt{3}i)^4$
97. $\left(-\frac{1}{2} + \frac{\sqrt{3}}{2}i\right)^3$
98. $\left(-\frac{1}{3} + \frac{\sqrt{3}}{3}i\right)^3$
99. $\left(\frac{1}{4} - \frac{\sqrt{3}}{4}i\right)^3$
100. $\left(\frac{1}{2} - \frac{\sqrt{3}}{2}i\right)^3$

Using the Binomial Theorem to Approximate **In Exercises 101–104, use the Binomial Theorem to approximate the quantity accurate to three decimal places. For example, in Exercise 101, use the expansion**

$$\begin{aligned}(1.02)^8 &= (1 + 0.02)^8 \\ &= 1 + 8(0.02) + 28(0.02)^2 + \cdots.\end{aligned}$$

101. $(1.02)^8$
102. $(2.005)^{10}$
103. $(2.99)^{12}$
104. $(1.98)^9$

Using the Binomial Theorem **In Exercises 105–108, use a graphing utility to graph f and g in the same viewing window. What is the relationship between the two graphs? Use the Binomial Theorem to write the polynomial function g in standard form.**

105. $f(x) = x^4 - 5x^2$
$g(x) = f(x - 2)$

106. $f(x) = x^3 - 4x$
$g(x) = f(x + 4)$

107. $f(x) = -x^3 + 3x^2 - 4$
$g(x) = f(x + 5)$

108. $f(x) = -x^4 + 4x^2 - 1$
$g(x) = f(x - 3)$

Comparing Graphs **In Exercises 109 and 110, use a graphing utility to graph the functions in the given order and in the same viewing window. Compare the graphs. Which two functions have identical graphs? Explain.**

109. (a) $f(x) = (1 - x)^3$
(b) $g(x) = 1 - 3x$
(c) $h(x) = 1 - 3x + 3x^2$
(d) $p(x) = 1 - 3x + 3x^2 - x^3$

110. (a) $f(x) = \left(1 - \frac{1}{2}x\right)^4$
(b) $g(x) = 1 - 2x + \frac{3}{2}x^2$
(c) $h(x) = 1 - 2x + \frac{3}{2}x^2 - \frac{1}{2}x^3$
(d) $p(x) = 1 - 2x + \frac{3}{2}x^2 - \frac{1}{2}x^3 + \frac{1}{16}x^4$

Finding a Probability **In Exercises 111–114, consider n independent trials of an experiment in which each trial has two possible outcomes, success or failure. The probability of a success on each trial is p and the probability of a failure is $q = 1 - p$. In this context, the term ${}_nC_k p^k q^{n-k}$ in the expansion of $(p + q)^n$ gives the probability of k successes in the n trials of the experiment.**

111. A fair coin is tossed seven times. To find the probability of obtaining four heads, evaluate the term

$${}_7C_4\left(\tfrac{1}{2}\right)^4\left(\tfrac{1}{2}\right)^3$$

in the expansion of $\left(\frac{1}{2} + \frac{1}{2}\right)^7$.

112. The probability of a baseball player getting a hit during any given time at bat is $\frac{1}{4}$. To find the probability that the player gets three hits during the next 10 times at bat, evaluate the term

$${}_{10}C_3\left(\tfrac{1}{4}\right)^3\left(\tfrac{3}{4}\right)^7$$

in the expansion of $\left(\frac{1}{4} + \frac{3}{4}\right)^{10}$.

113. The probability of a sales representative making a sale to any one customer is $\frac{1}{3}$. The sales representative makes eight contacts a day. To find the probability of making four sales, evaluate the term

$${}_8C_4\left(\tfrac{1}{3}\right)^4\left(\tfrac{2}{3}\right)^4$$

in the expansion of $\left(\frac{1}{3} + \frac{2}{3}\right)^8$.

114. To find the probability that the sales representative in Exercise 113 makes four sales when the probability of a sale to any one customer is $\frac{1}{2}$, evaluate the term

$${}_8C_4\left(\tfrac{1}{2}\right)^4\left(\tfrac{1}{2}\right)^4$$

in the expansion of $\left(\frac{1}{2} + \frac{1}{2}\right)^8$.

115. MODELING DATA

The per capita availability of berries f (in pounds) in the United States from 1970 through 2015 can be approximated by the model

$$f(t) = 0.0055t^2 - 0.027t + 3.46, \quad 0 \le t \le 45$$

where t represents the year, with $t = 0$ corresponding to 1970 (see figure). *(Source:* Economic Research Service, U.S. Department of Agriculture)

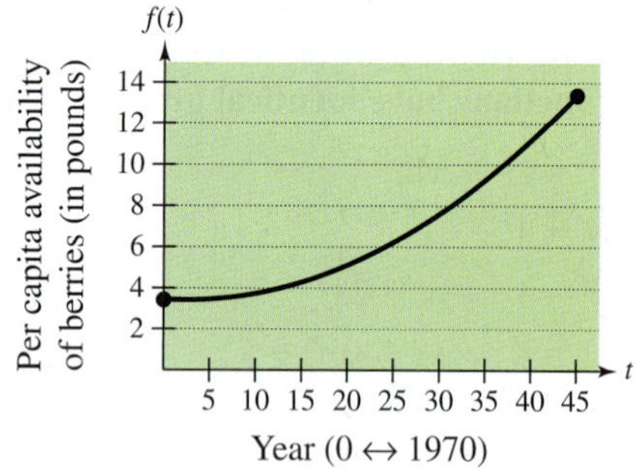

(a) Adjust the model so that $t = 0$ corresponds to 2000 rather than 1970. To do this, shift the graph of f 30 units to the left and obtain $g(t) = f(t + 30)$. Write $g(t)$ in standard form.

(b) Use a graphing utility to graph f and g in the same viewing window.

116. ***Why you should learn it*** *(p. 609)* The amounts f (in billions of dollars) of child support collected in the United States from 2000 through 2017 can be approximated by the model

$$f(t) = -0.038t^2 + 1.30t + 17.7, \quad 0 \le t \le 17$$

where t represents the year, with $t = 0$ corresponding to 2000 (see figure). *(Source:* U.S. Department of Health and Human Services)

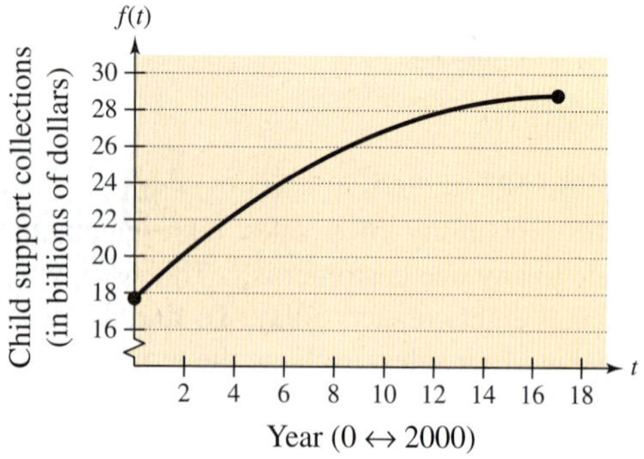

(a) Adjust the model so that $t = 0$ corresponds to 2005 rather than 2000. To do this, shift the graph of f five units to the left and obtain $g(t) = f(t + 5)$. Write $g(t)$ in standard form.

(b) Use a graphing utility to graph f and g in the same viewing window.

Focusing on Concepts

True or False? **In Exercises 117 and 118, determine whether the statement is true or false. Justify your answer.**

117. The Binomial Theorem can be used to produce each row of Pascal's Triangle.

118. A binomial that represents a difference cannot always be accurately expanded using the Binomial Theorem.

119. Writing In your own words, explain how to form the rows of Pascal's Triangle. Then form rows 8–10 of Pascal's Triangle.

120. HOW DO YOU SEE IT? The expansions of $(x + y)^4$, $(x + y)^5$, and $(x + y)^6$ are as follows.

$$(x + y)^4 = 1x^4 + 4x^3y + 6x^2y^2 + 4xy^3 + 1y^4$$

$$(x + y)^5 = 1x^5 + 5x^4y + 10x^3y^2 + 10x^2y^3 + 5xy^4 + 1y^5$$

$$(x + y)^6 = 1x^6 + 6x^5y + 15x^4y^2 + 20x^3y^3 + 15x^2y^4 + 6xy^5 + 1y^6$$

(a) Explain how the exponent of a binomial is related to the number of terms in its expansion.

(b) How many terms are in the expansion of $(x + y)^n$?

Proof **In Exercises 121–124, prove the property for all integers r and n, where $0 \le r \le n$.**

121. ${}_nC_r = {}_nC_{n-r}$

122. ${}_nC_0 - {}_nC_1 + {}_nC_2 - \cdots \pm {}_nC_n = 0$

123. ${}_{n+1}C_r = {}_nC_r + {}_nC_{r-1}$

124. The sum of the numbers in the nth row of Pascal's Triangle is 2^n.

Error Analysis **In Exercises 125 and 126, describe the error in writing the expression using summation notation.**

125. $(3x - 2y)^4 = \sum_{r=0}^{4} \binom{4}{r} (3x)^{4-r}(2y)^r$ ✗

126. $(x + 5y)^7 = \sum_{r=1}^{7} \binom{7}{r} x^{7-r}(5y)^r$ ✗

Cumulative Mixed Review

Finding the Inverse of a Matrix **In Exercises 127 and 128, find the inverse of the matrix.**

127. $\begin{bmatrix} -1 & -4 \\ 1 & 2 \end{bmatrix}$

128. $\begin{bmatrix} 11 & -12 \\ 2 & -2 \end{bmatrix}$

8.5 Counting Principles

What you should learn

- Solve simple counting problems.
- Use the Fundamental Counting Principle to solve more complicated counting problems.
- Use permutations to solve counting problems.
- Use combinations to solve counting problems.

Why you should learn it

You can use counting principles to solve counting problems that occur in real life. For instance, in Exercise 67 on page 625, you are asked to use counting principles to determine in how many ways a player can select six numbers in a Powerball lottery.

Simple Counting Problems

The last two sections of this chapter present a brief introduction to some of the basic counting principles and their application to probability. In the next section, you will see that much of probability has to do with counting the number of ways an event can occur.

EXAMPLE 1 Selecting Pairs of Numbers at Random

You place eight pieces of paper, numbered from 1 to 8, in a box. You draw one piece of paper at random from the box, record its number, and *replace* the paper in the box. Then, you draw a second piece of paper at random from the box and record its number. Finally, you add the two numbers. How many different ways can you obtain a sum of 12?

Solution

To solve this problem, count the number of different ways that a sum of 12 can be obtained using two numbers from 1 to 8. As shown below, a sum of 12 can occur in five different ways.

First number	4	5	6	7	8
Second number	8	7	6	5	4

✓ **Checkpoint** ▶ *Audio-video solution in English & Spanish at LarsonPrecalculus.com*

In Example 1, how many different ways can you obtain a sum of 14?

EXAMPLE 2 Selecting Pairs of Numbers at Random

You place eight pieces of paper, numbered from 1 to 8, in a box. You draw one piece of paper at random from the box, record its number, and *do not* replace the paper in the box. Then, you draw a second piece of paper at random from the box and record its number. Finally, you add the two numbers. How many different ways can you obtain a sum of 12?

Solution

To solve this problem, count the number of different ways that a sum of 12 can be obtained using two *different* numbers from 1 to 8. As shown below, a sum of 12 can occur in four different ways.

First number	4	5	7	8
Second number	8	7	5	4

✓ **Checkpoint** ▶ *Audio-video solution in English & Spanish at LarsonPrecalculus.com*

Repeat Example 2 for drawing *three* pieces of paper. ■

The difference between the counting problems in Examples 1 and 2 can be described by saying that the random selection in Example 1 occurs **with replacement,** whereas the random selection in Example 2 occurs **without replacement,** which eliminates the possibility of choosing two 6's.

The Fundamental Counting Principle

Examples 1 and 2 describe simple counting problems in which you can *list* each possible way that an event can occur. When it is possible, this is always the best way to solve a counting problem. However, some events can occur in so many different ways that it is not feasible to write out the entire list. In such cases, you must rely on formulas and counting principles. The most important of these is the **Fundamental Counting Principle.**

Fundamental Counting Principle

Let E_1 and E_2 be two events. The first event E_1 can occur in m_1 different ways. After E_1 has occurred, E_2 can occur in m_2 different ways. The number of ways that the two events can occur is $m_1 \cdot m_2$.

The Fundamental Counting Principle can be extended to three or more events. For instance, the number of ways that three events E_1, E_2, and E_3 can occur is $m_1 \cdot m_2 \cdot m_3$.

EXAMPLE 3 Using the Fundamental Counting Principle

How many different pairs of letters from the English alphabet are possible?

Solution

There are two events in this situation. The first event is the choice of the first letter, and the second event is the choice of the second letter. Because the English alphabet contains 26 letters, it follows that the number of two-letter pairs is

$$26 \cdot 26 = 676.$$

✓ **Checkpoint** *Audio-video solution in English & Spanish at LarsonPrecalculus.com*

A combination lock will open when you select the right choice of three numbers (from 1 to 30, inclusive). How many different lock combinations are possible?

EXAMPLE 4 Using the Fundamental Counting Principle

Telephone numbers in the United States currently have 10 digits. The first three are the *area code* and the next seven are the *local telephone number*. How many different telephone numbers are possible within each area code? (Note that at this time, a local telephone number cannot begin with 0 or 1.)

Solution

Because the first digit of a local number cannot be 0 or 1, there are only eight choices for the first digit. For each of the other six digits, there are 10 choices.

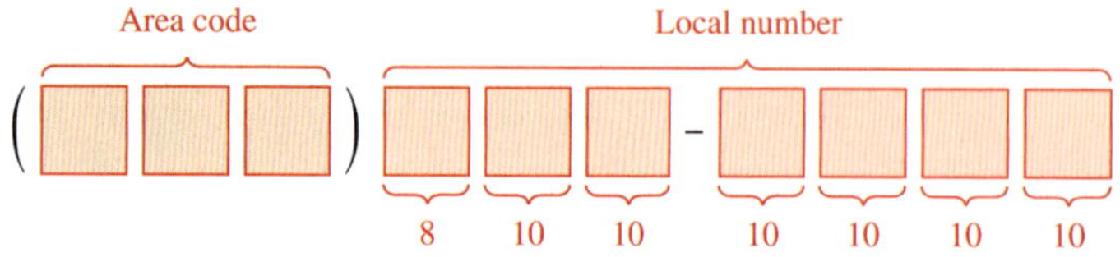

So, the number of local telephone numbers that are possible within each area code is

$$8 \cdot 10 \cdot 10 \cdot 10 \cdot 10 \cdot 10 \cdot 10 = 8{,}000{,}000.$$

✓ **Checkpoint** *Audio-video solution in English & Spanish at LarsonPrecalculus.com*

A product's catalog number is made up of one letter from the English alphabet followed by a five-digit number. How many different catalog numbers are possible? ■

Permutations

One important application of the Fundamental Counting Principle is in determining the number of ways that n elements can be arranged (in order). An ordering of n elements is called a **permutation** of the elements.

Definition of Permutation

A **permutation** of n different elements is an ordering of the elements such that one element is first, one is second, one is third, and so on.

EXAMPLE 5 Finding the Number of Permutations

How many permutations of the letters

A, B, C, D, E, and F

are possible?

Solution

Consider the following reasoning.

First position: Any of the *six* letters
Second position: Any of the remaining *five* letters
Third position: Any of the remaining *four* letters
Fourth position: Any of the remaining *three* letters
Fifth position: Either of the remaining *two* letters
Sixth position: The *one* remaining letter

So, the numbers of choices for the six positions are as follows.

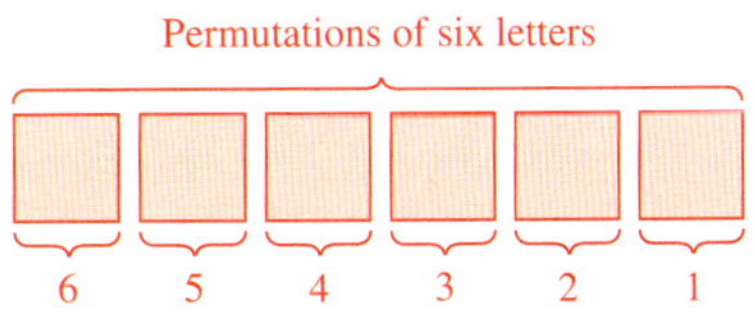

The total number of permutations of the six letters is

$$6! = 6 \cdot 5 \cdot 4 \cdot 3 \cdot 2 \cdot 1$$
$$= 720.$$

✓ Checkpoint Audio-video solution in English & Spanish at LarsonPrecalculus.com

How many permutations of the letters

W, X, Y, and Z

are possible?

The result obtained in Example 5 can be generalized to conclude that the number of permutations of n different elements is $n!$.

Number of Permutations of n Elements

The number of permutations of n elements is given by

$$n \cdot (n-1) \cdot \cdot \cdot 4 \cdot 3 \cdot 2 \cdot 1 = n!.$$

In other words, there are $n!$ different ways that n elements can be ordered.

It is useful, on occasion, to order a *subset* of a collection of elements rather than the entire collection. For instance, you might want to order r elements out of a collection of n elements. Such an ordering is called a **permutation of n elements taken r at a time.** This ordering is demonstrated in the next example.

EXAMPLE 6 Counting Horse Race Finishes

Eight horses are running in a race. In how many different ways can these horses come in first, second, and third? (Assume that there are no ties.)

Solution

Here are the different possibilities.

Win (first position):	*Eight* choices
Place (second position):	*Seven* choices
Show (third position):	*Six* choices

The numbers of choices for the three positions are as follows.

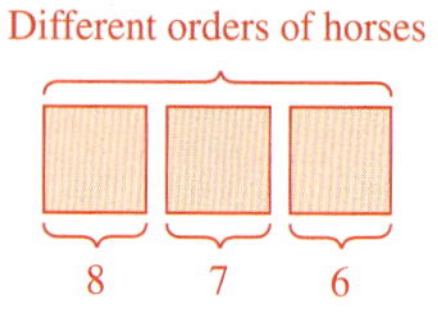

So, using the Fundamental Counting Principle, you can determine that there are

$$8 \cdot 7 \cdot 6 = 336$$

different ways in which the eight horses can come in first, second, and third.

Checkpoint *Audio-video solution in English & Spanish at LarsonPrecalculus.com*

A coin club has five members. In how many different ways can there be a president and a vice-president?

The result obtained in Example 6 can be generalized to the formula in the following definition.

Permutations of n Elements Taken r at a Time

The number of **permutations of n elements taken r at a time** is given by

$$_nP_r = \frac{n!}{(n-r)!} = n(n-1)(n-2) \cdot \cdot \cdot (n-r+1).$$

Using this formula, you can rework Example 6 to find that the number of permutations of eight horses taken three at a time is

$$\begin{aligned} {}_8P_3 &= \frac{8!}{(8-3)!} \\ &= \frac{8!}{5!} \\ &= \frac{8 \cdot 7 \cdot 6 \cdot \cancel{5!}}{\cancel{5!}} \\ &= 336 \end{aligned}$$

which is the same answer obtained in the example.

Technology Tip

Most graphing utilities are programmed to evaluate $_nP_r$. The figure below shows how one graphing utility evaluates the permutation $_8P_3$. For instructions on how to use the $_nP_r$ feature, see Appendix A; for specific keystrokes, go to this textbook's *Student Companion Website.*

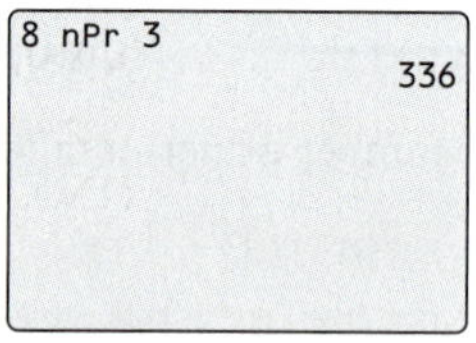

Remember that for permutations, order is important. For instance, to find the possible permutations of the letters A, B, C, and D taken three at a time, you count the permutations (A, B, D) and (B, A, D) as different because the *order* of the elements is different.

Consider, however, the possible permutations of the letters A, A, B, and C. The total number of permutations of the four letters is

$$_4P_4 = 4! = 24.$$

However, not all of these arrangements are *distinguishable* because there are two A's in the list. How many distinguishable permutations are possible? The answer can be found using the formula for the number of distinguishable permutations.

Distinguishable Permutations

Suppose a set of n objects has n_1 of one kind of object, n_2 of a second kind, n_3 of a third kind, and so on, with

$$n = n_1 + n_2 + n_3 + \cdot \cdot \cdot + n_k.$$

The number of **distinguishable permutations** of the n objects is given by

$$\frac{n!}{n_1! \cdot n_2! \cdot n_3! \cdot \cdot \cdot \cdot \cdot n_k!}.$$

EXAMPLE 7 Distinguishable Permutations

See LarsonPrecalculus.com for an interactive version of this type of example.

In how many distinguishable ways can the letters in BANANA be written?

Solution

This word has six letters, of which three are A's, two are N's, and one is a B. So, the number of distinguishable ways in which the letters can be written is

$$\frac{6!}{3! \cdot 2! \cdot 1!} = \frac{6 \cdot 5 \cdot 4 \cdot \cancel{3!}}{\cancel{3!} \cdot 2!} = 60.$$

The 60 different distinguishable permutations are listed below.

AAABNN	AAANBN	AAANNB	AABANN
AABNAN	AABNNA	AANABN	AANANB
AANBAN	AANBNA	AANNAB	AANNBA
ABAANN	ABANAN	ABANNA	ABNAAN
ABNANA	ABNNAA	ANAABN	ANAANB
ANABAN	ANABNA	ANANAB	ANANBA
ANBAAN	ANBANA	ANBNAA	ANNAAB
ANNABA	ANNBAA	BAAANN	BAANAN
BAANNA	BANAAN	BANANA	BANNAA
BNAAAN	BNAANA	BNANAA	BNNAAA
NAAABN	NAAANB	NAABAN	NAABNA
NAANAB	NAANBA	NABAAN	NABANA
NABNAA	NANAAB	NANABA	NANBAA
NBAAAN	NBAANA	NBANAA	NBNAAA
NNAAAB	NNAABA	NNABAA	NNBAAA

✓ ***Checkpoint*** *Audio-video solution in English & Spanish at LarsonPrecalculus.com*

In how many distinguishable ways can the letters in MITOSIS be written? ■

Hugo Felix/Shutterstock.com

Combinations

When you count the number of possible permutations of a set of elements, order is important. As a final topic in this section, you will look at a method for selecting subsets of a larger set in which order *is not* important. Such subsets are called **combinations of n elements taken r at a time.** For instance, the combinations

$$\{A, B, C\} \quad \text{and} \quad \{B, A, C\}$$

are equivalent because both sets contain the same three elements, and the order in which the elements are listed is not important. So, you would count only one of the two sets. A common example of a combination is a card game in which the player is free to reorder the cards after they have been dealt.

Combinations of n Elements Taken r at a Time

The number of **combinations of n elements taken r at a time** is given by

$${}_nC_r = \frac{n!}{(n-r)!r!}.$$

Note that the formula for ${}_nC_r$ is the same one given for binomial coefficients.

EXAMPLE 8 Combinations of n Elements Taken r at a Time

a. In how many different ways can three letters be chosen from the letters A, B, C, D, and E? (The order of the three letters is not important.)

b. A standard poker hand consists of five cards dealt from a deck of 52. How many different poker hands are possible? (After the cards are dealt, the player may reorder them, so order is not important.)

Solution

a. You can find the number of different ways in which the letters can be chosen by using the formula for the number of combinations of five elements taken three at a time, as follows.

$${}_5C_3 = \frac{5!}{2!3!} = \frac{5 \cdot \overset{2}{\cancel{4}} \cdot \cancel{3!}}{\cancel{2} \cdot 1 \cdot \cancel{3!}} = 10$$

b. You can find the number of different poker hands by using the formula for the number of combinations of 52 elements taken five at a time, as follows.

$${}_{52}C_5 = \frac{52!}{47!5!} = \frac{52 \cdot 51 \cdot 50 \cdot 49 \cdot 48 \cdot \cancel{47!}}{5 \cdot 4 \cdot 3 \cdot 2 \cdot 1 \cdot \cancel{47!}} = 2{,}598{,}960$$

Algebra Help

In Example 8(a), you could also make a list of the different combinations of three letters chosen from five letters.

{A, B, C}	{A, B, D}
{A, B, E}	{A, C, D}
{A, C, E}	{A, D, E}
{B, C, D}	{B, C, E}
{B, D, E}	{C, D, E}

From this list, you can conclude that there are 10 different ways in which three letters can be chosen from five letters, which is the same answer obtained in Example 8(a).

✓ ***Checkpoint*** *Audio-video solution in English & Spanish at LarsonPrecalculus.com*

In how many different ways can two letters be chosen from the letters A, B, C, D, E, F, and G? (The order of the two letters is not important.)

When solving a problem involving counting principles, you need to distinguish among the various counting principles to determine which is necessary to solve the problem. To do this, consider the questions below.

1. Is the order of the elements important? *Permutation*
2. Are the chosen elements a subset of a larger set in which order is not important? *Combination*
3. Are there two or more separate events? *Fundamental Counting Principle*

Note that some problems may require you to use more than one counting principle.

8.5 Exercises

See *CalcChat.com* for tutorial help and worked-out solutions to odd-numbered exercises.
For instructions on how to use a graphing utility, see Appendix A.

Vocabulary and Concept Check

In Exercises 1–3, fill in the blank.

1. The _______ states that when there are m_1 ways for one event to occur and m_2 ways for a second event to occur, there are $m_1 \cdot m_2$ ways for both events to occur.

2. The number of _______ of n objects of k different kinds is given by $\dfrac{n!}{n_1! \cdot n_2! \cdot n_3! \cdot \cdots \cdot n_k!}$.

3. The formula ${}_nP_r = \dfrac{n!}{(n-r)!}$ represents the number of permutations of _______ objects taken _______ at a time.

4. Is the ordering of n elements called a *permutation* or a *combination* of the elements?

Procedures and Problem Solving

Selecting Numbers at Random **In Exercises 5–12, determine the number of ways in which a computer can randomly generate one such integer, or pair of integers, from 1 through 15.**

5. An odd integer

6. An even integer

7. A prime integer

8. An integer factor of 24

9. A pair of integers whose sum is 20

10. A pair of integers whose sum is 22

11. A pair of distinct integers whose sum is 8

12. A pair of distinct integers whose sum is 24

13. **Education** To select a course schedule, a student must select one of two mathematics courses, one of three science courses, and one of five courses from the social sciences and humanities. How many different schedules are possible?

14. **Audio Engineering** A customer can choose one of four amplifiers, one of ten stereo receivers, and one of five speaker models for an entertainment system. Determine the number of possible system configurations.

15. **Education** In how many different ways can a 10-question true-false exam be answered? (Assume that no questions are omitted.)

16. **Physiology** In a physiology class, a student must dissect three different specimens. The student will select one of nine earthworms, one of four frogs, and one of seven fetal pigs. In how many different ways can the student select the specimens?

17. **Fundamental Counting Principle** How many different three-digit numbers can be formed under each condition?

(a) The leading digit cannot be 0.

(b) The leading digit cannot be 0 and no repetition of digits is allowed.

18. **Fundamental Counting Principle** How many distinct four-digit numbers can be formed under each condition?

(a) The leading digit cannot be 0 and the number must be less than 5000.

(b) The leading digit cannot be 0 and the number must be even.

19. **Telecommunication** The state of Nevada has three telephone area codes. Using the information about telephone numbers given in Example 4, how many different telephone numbers can Nevada's phone system accommodate?

20. **Transportation** In Pennsylvania, each standard automobile license plate number consists of three letters followed by a four-digit number. How many distinct Pennsylvania license plate numbers can be formed?

21. **Operations Research** In 1963, the United States Postal Service launched the Zoning Improvement Plan (ZIP) Code to streamline the mail-delivery system. A ZIP code consists of a five-digit sequence of numbers.

(a) Find the number of ZIP codes consisting of five digits.

(b) Find the number of ZIP codes consisting of five digits when the first digit is 1 or 2.

22. **Operations Research** In 1983, the U.S. Postal Service began to use an expanded ZIP code called ZIP+4, which is composed of the original five-digit code plus a four-digit add-on code.

(a) Find the total number of ZIP+4 codes possible.

(b) Find the total number of ZIP+4 codes possible when the first number of the five-digit code is 1 or 2.

23. **Banking** ATM personal identification number (PIN) codes consist of four-digit sequences of numbers.

(a) Find the total number of ATM codes possible.

(b) Find the total number of ATM codes possible when the first digit is not 0.

24. Radio Broadcasting Typically, radio stations are identified by four "call letters." Radio stations east of the Mississippi River have call letters that start with the letter W, and radio stations west of the Mississippi River have call letters that start with the letter K.

(a) Find the number of different sets of radio station call letters that are possible in the United States.

(b) Find the number of different sets of radio station call letters that are possible when the call letters must include a Q.

25. Finding Permutations Three couples have reserved seats in a row for a concert. In how many different ways can they be seated when (a) there are no seating restrictions? (b) each couple sits together?

26. Finding Permutations In how many different orders can five girls and three boys walk through a doorway single file when (a) there are no restrictions? (b) the girls walk through before the boys?

27. Photography In how many different ways can five children posing for a photograph line up in a row?

28. Transportation Design In how many different ways can eight people sit in an eight-passenger vehicle?

29. Politics The nine justices of the U.S. Supreme Court pose for a photograph while standing in a straight line, as opposed to the typical pose of two rows. How many different orders of the justices are possible for this photograph?

30. Manufacturing Four processes are involved in assembling a product, and they can be performed in any order. The management wants to test each order to determine which is the least time-consuming. How many different orders will have to be tested?

Evaluating a Permutation **In Exercises 31–36, evaluate $_nP_r$ using the formula from this section.**

31. $_4P_4$ **32.** $_5P_5$

33. $_5P_2$ **34.** $_{20}P_2$

35. $_6P_5$ **36.** $_7P_4$

Evaluating a Permutation Using a Graphing Utility **In Exercises 37–40, evaluate $_nP_r$ using a graphing utility.**

37. $_{15}P_4$ **38.** $_{10}P_8$

39. $_{120}P_4$ **40.** $_{100}P_5$

41. Politics From a pool of 14 candidates, the offices of president, vice-president, secretary, and treasurer will be filled. In how many different ways can the offices be filled?

42. Physical Education How many different batting orders can a baseball coach create from a team of 16 players when there are nine positions to fill?

43. Education A college sponsors a graphic design contest, with awards for the first, second, and third place designs. The college receives entries from 78 students. How many different orders of the top three designs are possible for the 78 entries?

44. Athletics Ten sprinters are running in the finals of the 100-meter dash at a track meet. How many different orders of the top three finishes are possible? (Assume there are no ties.)

Writing Permutations **In Exercises 45 and 46, use the letters A, B, C, and D.**

45. Write all permutations of the letters.

46. Write all permutations of the letters when the letters B and C must remain between the letters A and D.

Distinguishable Permutations **In Exercises 47–50, find the number of distinguishable permutations of the group of letters.**

47. A, A, G, E, E, E, M

48. B, B, B, T, T, T, T, T, Y

49. G, E, O, M, E, T, R, Y

50. M, I, S, S, I, S, S, I, P, P, I

Evaluating a Combination **In Exercises 51–56, evaluate $_nC_r$ using the formula from this section.**

51. $_6C_4$ **52.** $_7C_3$

53. $_{12}C_{10}$ **54.** $_{10}C_7$

55. $_{25}C_0$ **56.** $_{14}C_{14}$

Evaluating a Combination Using a Graphing Utility **In Exercises 57–60, evaluate $_nC_r$ using a graphing utility.**

57. $_{33}C_4$ **58.** $_{10}C_7$

59. $_{42}C_5$ **60.** $_{50}C_6$

Writing Combinations **In Exercises 61 and 62, use the letters A, B, C, D, E, and F.**

61. Write all the possible groups of two of the letters. (The order of the two letters is not important.)

62. Write all the possible groups of three of the letters. (The order of the three letters is not important.)

63. Politics There are 100 U.S. senators. They need to select a committee of 5 senators for a project. How many different committees are possible?

64. Education You can answer any 20 questions from a total of 22 questions on an exam. How many different groups of 20 questions are possible?

65. Jury Selection In how many different ways can a jury of 12 people be randomly selected from a group of 40 people?

66. Geometry Three points that are not collinear determine three lines. How many distinct lines are determined by nine points, of which no three are collinear?

67. *Why you should learn it* *(p. 617)* Powerball is played with 69 white balls, numbered 1 through 69, and 26 red balls, numbered 1 through 26. Five white balls and one red ball, the Powerball, are drawn. In how many different ways can a player select the six numbers?

68. Law A law office interviews paralegals for 10 openings. There are 13 paralegals with two years of experience and 20 paralegals with one year of experience. How many combinations of seven paralegals with two years of experience and three paralegals with one year of experience are possible?

69. Forming a Committee A six-member research committee is to be formed having one administrator, three faculty members, and two students. There are seven administrators, 12 faculty members, and 25 students in contention for the committee. How many different six-member committees are possible?

70. Sociology The number of possible interpersonal relationships increases dramatically as the size of a group increases. Determine the numbers of different two-person relationships that are possible in groups of people of sizes (a) 3, (b) 8, (c) 12, and (d) 20.

71. Game Theory You are dealt five cards from an ordinary deck of 52 playing cards. In how many different ways can you get a full house? (A full house consists of three of one kind and two of another. For example, 8-8-8-5-5 and K-K-K-10-10 are full houses.)

72. Quality Engineering A shipment of 30 flat screen televisions contains three defective units. In how many different ways can a vending company purchase four of these units and receive (a) all good units, (b) two good units, and (c) at least two good units?

Geometry In Exercises 73–76, find the number of diagonals of the polygon. (A line segment connecting any two nonadjacent vertices is called a *diagonal* of a polygon.)

73. Pentagon

74. Hexagon

75. Octagon

76. Decagon (10 sides)

Solving an Equation In Exercises 77–84, solve for n.

77. $14 \cdot {}_nP_3 = {}_{n+2}P_4$

78. ${}_nP_5 = 18 \cdot {}_{n-2}P_4$

79. ${}_nP_4 = 10 \cdot {}_{n-1}P_3$

80. ${}_nP_6 = 12 \cdot {}_{n-1}P_5$

81. ${}_{n+1}P_3 = 4 \cdot {}_nP_2$

82. ${}_{n+2}P_3 = 6 \cdot {}_{n+2}P_1$

83. $4 \cdot {}_{n+1}P_2 = {}_{n+2}P_3$

84. $5 \cdot {}_{n-1}P_1 = {}_nP_2$

Focusing on Concepts

True or False? In Exercises 85 and 86, determine whether the statement is true or false. Justify your answer.

85. The number of permutations of n elements can be derived by using the Fundamental Counting Principle.

86. To find the number of distinguishable permutations of the letters A, A, A, A, B, B, and B, you can use ${}_7C_4$.

Error Analysis In Exercises 87 and 88, describe the error.

87. ${}_5P_3 = 5(5-1)(5-2)(5-3) = 5(4)(3)(2) = 120$ ✗

88. ${}_7C_2 = \dfrac{7!}{5!} = \dfrac{7 \cdot 6 \cdot 5!}{5!} = 42$ ✗

89. Think About It Decide whether each scenario should be counted using permutations or combinations. Explain your reasoning. (Do not calculate.)

(a) Number of ways 10 people can line up in a row for concert tickets

(b) Number of different arrangements of three types of flowers from an array of 20 types

90. HOW DO YOU SEE IT? Without calculating, determine whether the value of ${}_nP_r$ is greater than the value of ${}_nC_r$ for the values of n and r given in the table. Complete the table using yes (Y) or no (N). Is the value of ${}_nP_r$ always greater than the value of ${}_nC_r$? Explain.

n \ r	0	1	2	3	4	5
1						
2						
3						
4						
5						

Proof In Exercises 91–94, prove the identity.

91. ${}_nP_{n-1} = {}_nP_n$

92. ${}_nC_n = {}_nC_0$

93. ${}_nC_{n-1} = {}_nC_1$

94. ${}_nC_r = \dfrac{{}_nP_r}{r!}$

Cumulative Mixed Review

Using Cramer's Rule In Exercises 95 and 96, use Cramer's Rule to solve the system of equations.

95. $\begin{cases} -5x + 3y = -14 \\ 7x - 2y = 2 \end{cases}$

96. $\begin{cases} -3x - 4y = -1 \\ 9x + 5y = -4 \end{cases}$

8.6 Probability

The Probability of an Event

Any happening whose result is uncertain is called an **experiment.** The possible results of the experiment are **outcomes,** the set of all possible outcomes of the experiment is the **sample space** of the experiment, and any subcollection of a sample space is an **event.**

For instance, when a six-sided die is tossed, the sample space can be represented by the numbers 1 through 6. For the experiment to be fair, each of the outcomes must be *equally likely*.

To describe a sample space in such a way that each outcome is equally likely, you must sometimes distinguish between or among various outcomes in ways that appear artificial. Example 1 illustrates such a situation.

What you should learn

- Find probabilities of events.
- Find probabilities of mutually exclusive events.
- Find probabilities of independent events.

Why you should learn it

You can use probability to solve a variety of problems that occur in real life. For instance, in Exercise 54 on page 633, you are asked to use probability to help analyze the age distribution of unemployed workers.

EXAMPLE 1 Finding the Sample Space

Find the sample space for each experiment.

a. One coin is tossed.

b. Two coins are tossed.

c. Three coins are tossed.

Solution

a. Because the coin will land either heads up (denoted by H) or tails up (denoted by T), the sample space is $S = \{H, T\}$.

b. Because either coin can land heads up or tails up, the possible outcomes are as follows.

HH = heads up on both coins

HT = heads up on first coin and tails up on second coin

TH = tails up on first coin and heads up on second coin

TT = tails up on both coins

So, the sample space is $S = \{HH, HT, TH, TT\}$.

Note that this list distinguishes between the two cases

HT and TH

even though these two outcomes appear to be similar.

c. Following the notation in part (b), the sample space is

$$S = \{HHH, HHT, HTH, HTT, THH, THT, TTH, TTT\}.$$

Note that this list distinguishes among the cases

HHT, HTH, and THH

and among the cases

HTT, THT, and TTH.

✓ **Checkpoint** Audio-video solution in English & Spanish at LarsonPrecalculus.com

Find the sample space for the following experiment.

You toss a coin twice and a six-sided die once.

To calculate the probability of an event, count the number of outcomes in the event and in the sample space. The *number of equally likely outcomes* in event E is denoted by $n(E)$, and the number of equally likely outcomes in the sample space S is denoted by $n(S)$. The probability that event E will occur is given by $n(E)/n(S)$.

The Probability of an Event

If an event E has $n(E)$ equally likely outcomes and its sample space S has $n(S)$ equally likely outcomes, then the **probability** of event E is given by

$$P(E) = \frac{n(E)}{n(S)}.$$

Explore the Concept

Toss two coins 40 times and write down the number of heads that occur on each toss (0, 1, or 2). How many times did two heads occur? Without performing the experiment, how many times would you expect two heads to occur when two coins are tossed 1000 times?

Because the number of outcomes in an event must be less than or equal to the number of outcomes in the sample space, the probability of an event must be a number from 0 to 1, inclusive. That is,

$$0 \le P(E) \le 1$$

as shown in the figure. If $P(E) = 0$, then event E *cannot occur*, and E is called an **impossible event.** If $P(E) = 1$, then event E *must occur*, and E is called a **certain event.**

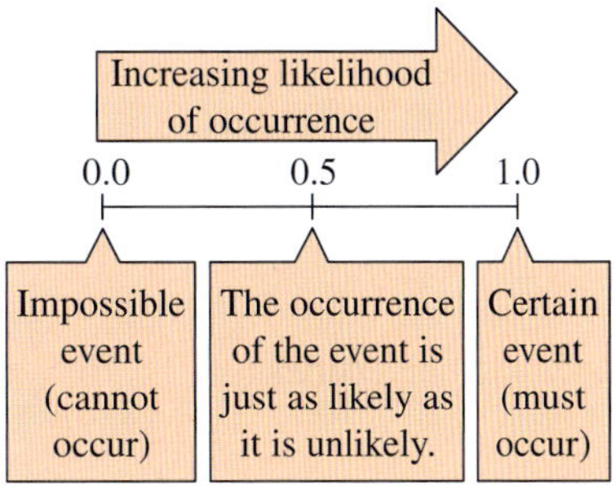

EXAMPLE 2 Finding the Probability of an Event

See LarsonPrecalculus.com for an interactive version of this type of example.

a. Two coins are tossed. What is the probability that both land heads up?

b. A card is drawn at random from a standard deck of playing cards. What is the probability that it is an ace?

Solution

a. Following the procedure in Example 1(b), let

$$E = \{HH\} \qquad \text{and} \qquad S = \{HH, HT, TH, TT\}.$$

The probability of getting two heads is

$$P(E) = \frac{n(E)}{n(S)} = \frac{1}{4}.$$

b. Because there are 52 cards in a standard deck of playing cards and there are four aces (one of each suit), the probability of drawing an ace is

$$P(E) = \frac{n(E)}{n(S)} = \frac{4}{52} = \frac{1}{13}.$$

Checkpoint *Audio-video solution in English & Spanish at LarsonPrecalculus.com*

a. You toss three coins. What is the probability that all three land tails up?

b. You draw one card at random from a standard deck of playing cards. What is the probability that it is a diamond?

Note that a probability can be written as a fraction, a decimal, or a percent. For instance, in Example 2(a), the probability of getting two heads can be written as

$$\frac{1}{4}, \qquad 0.25, \qquad \text{or} \qquad 25\%.$$

EXAMPLE 3 Random Selection

The numbers of colleges and universities in various regions of the United States in 2017 are shown in the figure. One institution is selected at random. What is the probability that the institution is in one of the three southern regions? *(Source:* U.S. National Center for Education Statistics)

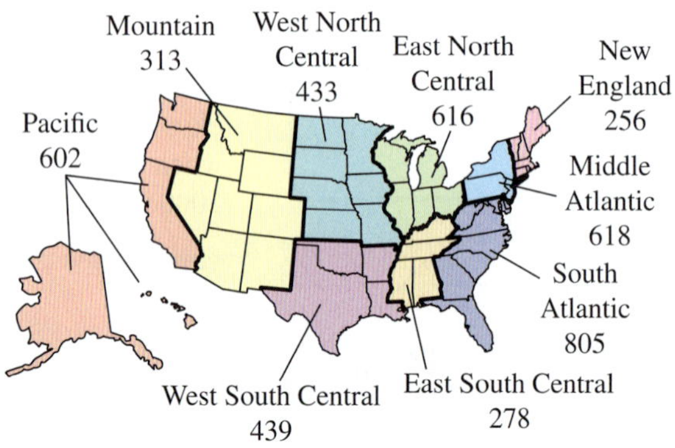

Solution

From the figure, the total number of colleges and universities is 4360. Because there are

$$439 + 278 + 805 = 1522$$

colleges and universities in the three southern regions, the probability that the institution is in one of these regions is

$$P(E) = \frac{n(E)}{n(S)} = \frac{1522}{4360} \approx 0.349.$$

✓ Checkpoint *Audio-video solution in English & Spanish at LarsonPrecalculus.com*

In Example 3, what is the probability that the randomly selected institution is in the Pacific region?

EXAMPLE 4 Finding the Probability of Winning a Lottery

In Arizona's The Pick game, a player chooses six different numbers from 1 to 44. When these six numbers match the six numbers drawn (in any order), the player wins (or shares) the top prize. What is the probability of winning the top prize when the player buys one ticket?

Solution

To find the outcomes in the sample space, use the formula for the number of combinations of 44 numbers taken six at a time.

$$n(S) = {}_{44}C_6 = \frac{44 \cdot 43 \cdot 42 \cdot 41 \cdot 40 \cdot 39}{6 \cdot 5 \cdot 4 \cdot 3 \cdot 2 \cdot 1} = 7{,}059{,}052$$

When a player buys one ticket, the probability of winning is

$$P(E) = \frac{1}{7{,}059{,}052}.$$

✓ Checkpoint *Audio-video solution in English & Spanish at LarsonPrecalculus.com*

In Pennsylvania's Cash 5 game, a player chooses five different numbers from 1 to 43. When these five numbers match the five numbers drawn (in any order), the player wins (or shares) the top prize. What is the probability of winning the top prize when the player buys one ticket?

Algebra Help

In some probability problems, such as Example 2(a), you can simply write out the sample space and count the outcomes in the desired event. For larger sample spaces, such as the one in Example 4, using the counting principles discussed in Section 8.5 should save you time when finding the number of outcomes.

Mutually Exclusive Events

Two events A and B (from the same sample space) are **mutually exclusive** when A and B have no outcomes in common. In the terminology of sets, the intersection of A and B is the empty set, which implies that

$$P(A \cap B) = 0.$$

For instance, when two dice are tossed, the event A of rolling a sum of 6 and the event B of rolling a sum of 9 are mutually exclusive. To find the probability that one or the other of two mutually exclusive events will occur, you can *add* their individual probabilities.

Probability of the Union of Two Events

If A and B are events in the same sample space, then the probability of A or B occurring is given by

$$P(A \cup B) = P(A) + P(B) - P(A \cap B).$$

If A and B are mutually exclusive, then

$$P(A \cup B) = P(A) + P(B).$$

EXAMPLE 5 The Probability of a Union

One card is selected at random from a standard deck of 52 playing cards. What is the probability that the card is either a heart or a face card?

Solution

Because the deck has 13 hearts, the probability of selecting a heart (event A) is

$$P(A) = \frac{13}{52}.$$

Similarly, because the deck has 12 face cards, the probability of selecting a face card (event B) is

$$P(B) = \frac{12}{52}.$$

Because three of the cards are hearts and face cards (see Figure 8.1), it follows that

$$P(A \cap B) = \frac{3}{52}.$$

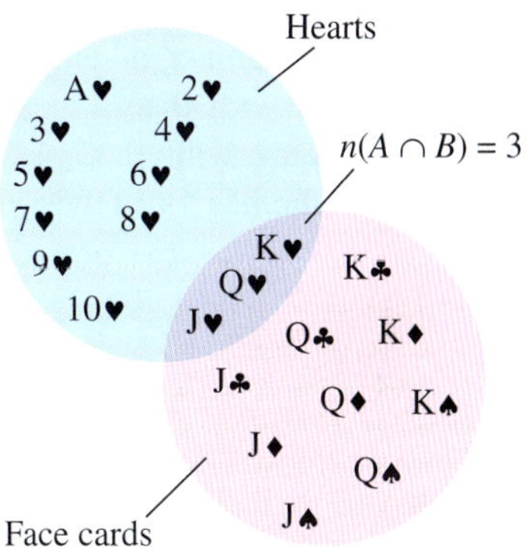

Figure 8.1

Finally, applying the formula for the probability of the union of two events, you can conclude that the probability of selecting either a heart or a face card is

$$\begin{aligned} P(A \cup B) &= P(A) + P(B) - P(A \cap B) \\ &= \frac{13}{52} + \frac{12}{52} - \frac{3}{52} \\ &= \frac{22}{52} \\ &\approx 0.423. \end{aligned}$$

✓ **Checkpoint** ▶ *Audio-video solution in English & Spanish at LarsonPrecalculus.com*

You draw one card at random from a standard deck of 52 playing cards. What is the probability that the card is either an ace or a spade?

EXAMPLE 6 Probability of Mutually Exclusive Events

The human resources department of a company has compiled data on the numbers of employees who have been with the company for various periods of time. The results are shown in the table.

DATA

Years of service	Number of employees
0–4	157
5–9	89
10–14	74
15–19	63
20–24	42
25–29	38
30–34	37
35–39	21
40–44	8

Spreadsheet at LarsonPrecalculus.com

An employee is chosen at random. What is the probability that the employee has

a. 4 or fewer years of service?

b. 9 or fewer years of service?

Solution

a. To begin, add the numbers of employees.

$$157 + 89 + 74 + 63 + 42 + 38 + 37 + 21 + 8 = 529$$

So, the total number of employees is 529. Next, let event A represent choosing an employee with 0 to 4 years of service. From the table, you know there are 157 employees with 0 to 4 years of service. The probability of choosing an employee who has 4 or fewer years of service is

$$P(A) = \frac{157}{529} \approx 0.297.$$

b. Let event B represent choosing an employee with 5 to 9 years of service. From the table, you know that there are 89 employees with 5 to 9 years of service. Then

$$P(B) = \frac{89}{529}.$$

Because event A from part (a) and event B have no outcomes in common, you can conclude that these two events are mutually exclusive and that

$$\begin{aligned} P(A \cup B) &= P(A) + P(B) \\ &= \frac{157}{529} + \frac{89}{529} \\ &= \frac{246}{529} \\ &\approx 0.465. \end{aligned}$$

So, the probability of choosing an employee who has 9 or fewer years of service is about 0.465.

Human Resources Manager

✓ **Checkpoint** *Audio-video solution in English & Spanish at LarsonPrecalculus.com*

In Example 6, what is the probability that an employee chosen at random has 30 or more years of service?

Independent Events

Two events are **independent** when the occurrence of one has no effect on the occurrence of the other. For instance, rolling a total of 12 with two six-sided dice has no effect on the outcome of future rolls of the dice. To find the probability that two independent events will occur, *multiply* the probabilities of each.

Probability of Independent Events

If A and B are **independent events,** then the probability that both A and B will occur is given by

$$P(A \text{ and } B) = P(A) \cdot P(B).$$

This rule can be extended to any number of independent events.

Insight

A standardized test may ask you to find the probability of an event. Be prepared to find sample spaces as well as determining when events are mutually exclusive and when events are independent.

EXAMPLE 7 Probability of Independent Events

A random number generator on a computer selects three integers from 1 to 20. What is the probability that all three numbers are less than or equal to 5?

Solution

Let event A represent selecting a number from 1 to 5. The probability of selecting a number from 1 to 5 is

$$P(A) = \frac{5}{20} = \frac{1}{4}.$$

So, the probability that all three numbers are less than or equal to 5 is

$$P(A) \cdot P(A) \cdot P(A) = \left(\frac{1}{4}\right)\left(\frac{1}{4}\right)\left(\frac{1}{4}\right) = \frac{1}{64}.$$

✓ Checkpoint Audio-video solution in English & Spanish at LarsonPrecalculus.com

A random number generator on a computer selects two integers from 1 to 30. What is the probability that both numbers are less than 12?

EXAMPLE 8 Probability of Independent Events

In 2018, approximately 52% of the adult population in the United States said their cellphone would be very hard to give up. In a survey, 10 people were chosen at random from the adult population. What is the probability that all 10 say their cellphone would be very hard to give up? (*Source:* Pew Research Center)

Solution

Let A represent choosing an adult who says his or her cellphone would be very hard to give up. The probability of choosing an adult who says his or her cellphone would be very hard to give up is 0.52, the probability of choosing a second adult who says his or her cellphone would be very hard to give up is 0.52, and so on. Because these events are independent, the probability that all 10 people say their cellphone would be very hard to give up is

$$[P(A)]^{10} = (0.52)^{10} \approx 0.001446.$$

✓ Checkpoint Audio-video solution in English & Spanish at LarsonPrecalculus.com

In 2018, approximately 31% of the adult population in the United States said their television would be very hard to give up. In a survey, eight people were chosen at random from the adult population. What is the probability that all eight say their television would be very hard to give up? (*Source:* Pew Research Center)

8.6 Exercises

See CalcChat.com for tutorial help and worked-out solutions to odd-numbered exercises.
For instructions on how to use a graphing utility, see Appendix A.

Vocabulary and Concept Check

In Exercises 1–4, fill in the blank(s).

1. The set of all possible outcomes of an experiment is called the _______ .

2. To determine the probability of an event, use the formula $P(E) = \frac{n(E)}{n(S)}$, where $n(E)$ is _______ and $n(S)$ is _______ .

3. If two events from the same sample space have no outcomes in common, then the two events are _______ .

4. If the occurrence of one event has no effect on the occurrence of a second event, then the events are _______ .

5. What is the probability of an impossible event?

6. Match the probability formula with the correct probability name.

(a) Probability of the union of two events (i) $P(A \cup B) = P(A) + P(B)$

(b) Probability of mutually exclusive events (ii) $P(A \cup B) = P(A) + P(B) - P(A \cap B)$

(c) Probability of independent events (iii) $P(A \text{ and } B) = P(A) \cdot P(B)$

Procedures and Problem Solving

Finding the Sample Space **In Exercises 7–10, determine the sample space for the experiment.**

7. A coin and a six-sided die are tossed.

8. A six-sided die is tossed twice and the results of roll 1 and roll 2 are recorded.

9. A taste tester has to rank three varieties of orange juice, A, B, and C, according to preference.

10. Two county supervisors are selected from five supervisors, A, B, C, D, and E, to study a recycling plan.

Tossing a Coin **In Exercises 11–14, find the probability for the experiment of tossing a coin three times. Use the sample space**

$S = \{HHH, HHT, HTH, HTT, THH, THT, TTH, TTT\}$.

11. The probability of getting exactly one tail

12. The probability of getting a head on the first toss

13. The probability of getting at least one head

14. The probability of getting at least two heads

Drawing a Card **In Exercises 15–18, find the probability for the experiment of selecting one card at random from a standard deck of 52 playing cards.**

15. The card is a face card.

16. The card is a black card.

17. The card is a red face card.

18. The card is a numbered card (2–10).

Selecting a Ticket Holder at Random **In Exercises 19–22, one of a team's 2200 season ticket holders is selected at random to win a prize. The circle graph shows the ages of the season ticket holders. Find the probability of the event.**

Ages of Season Ticket Holders

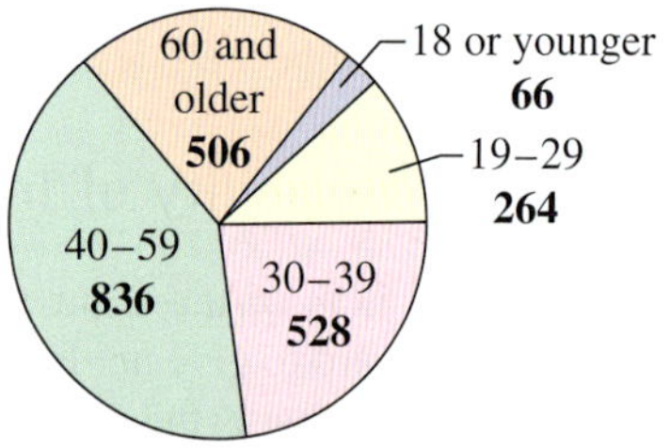

19. The winner is younger than 19 years old.

20. The winner is older than 39 years old.

21. The winner is 19 to 39 years old.

22. The winner is younger than 19 years old or older than 59 years old.

Finding the Probability of Winning a Lottery **In Exercises 23–26, a player chooses n numbers from a to b in a lottery. Find the probability that a player who buys one ticket matches all n numbers drawn (in any order).**

23. $n = 6, a = 1, b = 40$

24. $n = 5, a = 1, b = 53$

25. $n = 3, a = 0, b = 37$

26. $n = 4, a = 0, b = 44$

Drawing Marbles **In Exercises 27 and 28, find the probability for the experiment of drawing two marbles at random (without replacement) from a bag containing one green, two yellow, and three red marbles. (*Hint:* Use combinations to find the numbers of outcomes for the given event and sample space.)**

27. Both marbles are red. **28.** Neither marble is yellow.

Finding the Probability of a Complement **The *complement of an event A* is the collection of all outcomes in the sample space that are *not* in *A*. If the probability of *A* is $P(A)$, then the probability of the complement A' is given by $P(A') = 1 - P(A)$. In Exercises 29 and 30, you are given the probability that an event *will* happen. Find the probability that the event *will not* happen.**

29. $P(E) = 0.75$ **30.** $P(E) = 0.36$

Using the Probability of a Complement **In Exercises 31 and 32, you are given the probability that an event *will not* happen. Find the probability that the event *will* happen.**

31. $P(E') = 0.12$ **32.** $P(E') = 0.84$

Finding the Probability of a Union **In Exercises 33–36, one card is selected at random from a standard deck of 52 playing cards. Use a formula to find the probability of the union of the two events.**

33. The card is a club or a king.

34. The card is a face card or a black card.

35. The card is a 5 or a 2.

36. The card is a heart or a spade.

Identifying Mutually Exclusive Events **In Exercises 37 and 38, determine whether *A* and *B* are mutually exclusive.**

37. $P(A) = 0.4$, $P(B) = 0.7$, $P(A \text{ or } B) = 0.7$

38. $P(A) = 0.3$, $P(B) = 0.35$, $P(A \text{ or } B) = 0.65$

Finding the Probability of Mutually Exclusive Events **In Exercises 39–44, use the table, which shows the age groups of students in a college sociology class.**

Age	Number of students
18–19	11
20–21	18
22–30	2
31–40	1

A student from the class is randomly chosen for a project. Find the probability that the student is the given age.

39. 18 or 19 years old **40.** 22 to 30 years old

41. 20 to 30 years old

42. 18 to 21 years old

43. Older than 21 years old

44. Younger than 31 years old

Identifying Independent Events **In Exercises 45 and 46, determine whether events *A* and *B* are independent.**

45. $P(A) = 0.25$, $P(B) = 0.5$, $P(A \text{ or } B) = 0.75$

46. $P(A) = 0.2$, $P(B) = 0.2$, $P(A \text{ or } B) = 0.04$

Probability of Independent Events **In Exercises 47–52, a random number generator selects three numbers from 1 through 10. Find the probability of the event.**

47. All three numbers are even.

48. All three numbers are a factor of 8.

49. All three numbers are less than or equal to 3.

50. All three numbers are greater than or equal to 9.

51. Two numbers are less than 5 and the other number is 10.

52. One number is 2, 4, or 6, and the other two numbers are odd.

53. **Political Science** The incumbent and two first-time politicians are the only candidates for a public office. The first-time politicians have the same probability of winning. The incumbent has six times the probability of winning as either of the other candidates. Find the probability that the incumbent wins the election.

54. ***Why you should learn it*** *(p. 626)* In 2017, there were approximately 6.98 million unemployed workers in the United States. The circle graph shows the age profile of these unemployed workers. (*Source:* U.S. Bureau of Labor Statistics)

Ages of Unemployed Workers

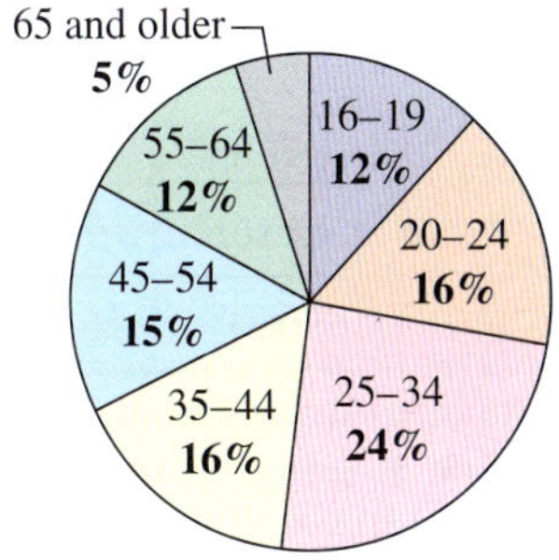

(a) Estimate the number of unemployed workers in the 16–19 age group.

(b) Estimate the number of unemployed workers who are 45 years or older.

(c) What is the probability that a person selected at random from the population of unemployed workers is in the 25–44 age group?

(d) What is the probability that a person selected at random from the population of unemployed workers is 25 to 64 years old?

JohnnyGreig/E+/Getty Images

55. Education The levels of educational attainment of females age 25 years or older in the United States in 2017 are shown in the circle graph. Use the fact that the population of females 25 years or older was 112.6 million in 2017. *(Source:* U.S. Census Bureau)

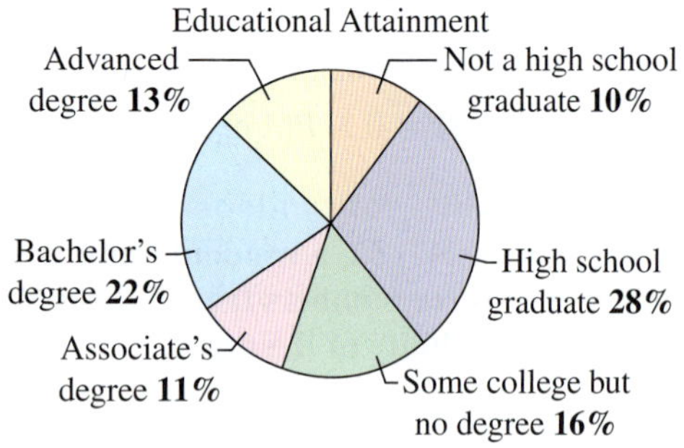

(a) Estimate the number of females 25 years old or older who have advanced degrees.

(b) Find the probability that a female 25 years old or older selected at random has earned a Bachelor's degree or higher.

(c) Find the probability that a female 25 years old or older selected at random has earned a high school diploma or gone on to post-secondary education.

(d) Find the probability that a female 25 years old or older selected at random has earned a degree.

56. Political Science One hundred college students were interviewed to determine their political party affiliations and whether they favored a balanced budget amendment to the U.S. Constitution. The results of the study are listed in the table, where D represents Democrat and R represents Republican.

	D	R	Total
Favor	23	32	55
Oppose	25	9	34
Unsure	7	4	11
Total	55	45	100

A person is selected at random from the sample. Find the probability that the person selected is (a) a person who does not favor the amendment, (b) a Republican, and (c) a Democrat who favors the amendment.

In Exercises 57–60, the sample spaces are large and you should use the counting principles discussed in Section 8.5.

57. Using Counting Principles On a game show, you are given five digits to arrange in the proper order to form the price of a car. If you are correct, you win the car. What is the probability of winning when you (a) randomly guess the position of each digit and (b) know the first digit and randomly guess the others?

58. Using Counting Principles The deck for a card game is made up of 108 cards. Twenty-five each are red, yellow, blue, and green, and eight are wild cards. Each player is randomly dealt a seven-card hand. What is the probability that a hand will contain (a) exactly two wild cards, and (b) two wild cards, two red cards, and three blue cards?

59. Using Counting Principles A shipment of 12 microwave ovens contains three defective units. A vending company has ordered four of these units, and because all are packaged identically, the selection will be random. What is the probability that (a) all four units are good, (b) exactly two units are good, and (c) at least two units are good?

60. Using Counting Principles Two integers from 1 through 40 are chosen by a random number generator. What is the probability that (a) the numbers are both even, (b) one number is even and one is odd, (c) both numbers are less than 30, and (d) the same number is chosen twice?

61. Marketing The circle graph shows the methods used by shoppers to pay for merchandise. Two shoppers are chosen at random. What is the probability that (a) both shoppers paid for their purchases only in cash, (b) at least one shopper paid only in cash, and (c) neither shopper paid only in cash?

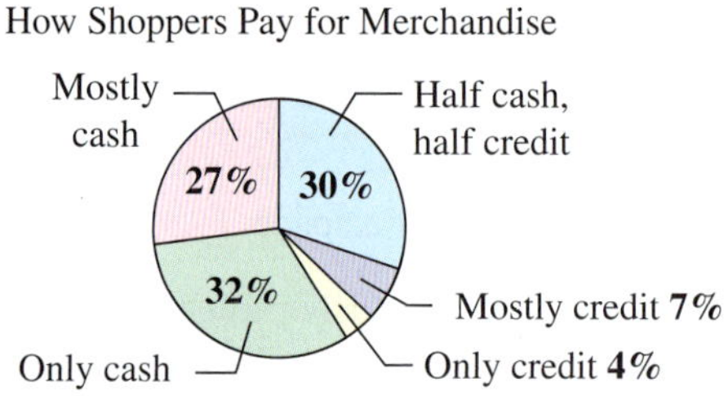

62. Aerospace Engineering A space vehicle has an independent backup system for one of its communication networks. The probability that either system will function satisfactorily for the duration of a flight is 0.985. What is the probability that during a given flight (a) both systems function satisfactorily, (b) at least one system functions satisfactorily, and (c) both systems fail?

63. Marketing A sales representative makes sales on approximately one-fifth of all calls. On a given day, the representative contacts six potential clients. What is the probability that a sale will be made with (a) all six contacts, (b) none of the contacts, and (c) at least one contact?

64. Genetics Assume that the probability of the birth of a child of a particular sex is 50%. In a family with four children, what is the probability that (a) all the children are boys, (b) all the children are the same sex, and (c) there is at least one boy?

65. Estimating π A coin of diameter d is randomly dropped onto a paper that contains a grid of squares d units on a side (see figure).

(a) Find the probability that the coin covers a vertex of one of the squares on the grid.

(b) Perform the experiment 100 times and use the results to approximate π.

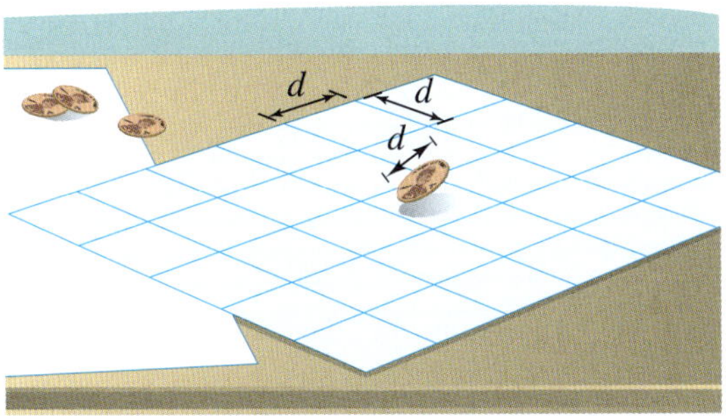

66. Geometry You and a friend agree to meet at your favorite fast food restaurant between 5:00 and 6:00 P.M. The one who arrives first will wait 15 minutes for the other, after which the first person will leave (see figure). What is the probability that the two of you will actually meet, assuming that your arrival times are random within the hour?

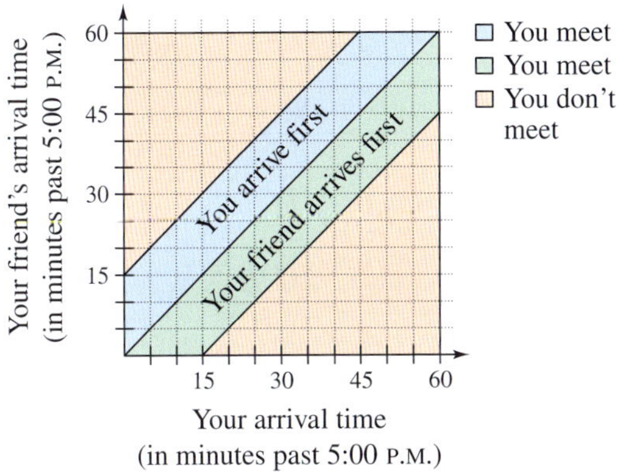

Focusing on Concepts

True or False? **In Exercises 67 and 68, determine whether the statement is true or false. Justify your answer.**

67. If the probability of an outcome in a sample space is 1, then the probability of the other outcomes in the sample space is 0.

68. If A and B are independent events with nonzero probabilities, then A can occur when B occurs.

69. Error Analysis Describe the error.

A random number generator selects two integers from 1 to 15. The generator permits duplicates.

The probability that both numbers are greater than 12 is $\left(\frac{4}{15}\right)\left(\frac{4}{15}\right) = \frac{16}{225}$.

70. Exploration Consider a group of n people.

(a) Explain why the following pattern gives the probability that the n people have distinct birthdays.

$$n = 2: \quad \frac{365}{365} \cdot \frac{364}{365} = \frac{365 \cdot 364}{365^2}$$

$$n = 3: \quad \frac{365}{365} \cdot \frac{364}{365} \cdot \frac{363}{365} = \frac{365 \cdot 364 \cdot 363}{365^3}$$

(b) Use the pattern in part (a) to write an expression for the probability that four people ($n = 4$) have distinct birthdays.

(c) Let P_n be the probability that the n people have distinct birthdays. Verify that this probability can be obtained recursively by

$$P_1 = 1 \quad \text{and} \quad P_n = \frac{365 - (n - 1)}{365} P_{n-1}.$$

(d) Explain why $Q_n = 1 - P_n$ gives the probability that at least two people in a group of n people have the same birthday.

(e) Use the results of parts (c) and (d) to complete the table.

n	10	15	20	23	30	40	50
P_n							
Q_n							

(f) How many people must be in a group so that the probability of at least two of them having the same birthday is greater than $\frac{1}{2}$? Explain.

71. Think About It Let A and B be two events from the same sample space such that $P(A) = 0.76$ and $P(B) = 0.58$.

(a) Is it possible that A and B are mutually exclusive? Explain. Draw a diagram to support your answer.

(b) Determine the possible range of $P(A \cup B)$.

72. HOW DO YOU SEE IT? Use the figure to determine whether events A and B are mutually exclusive. Explain your reasoning.

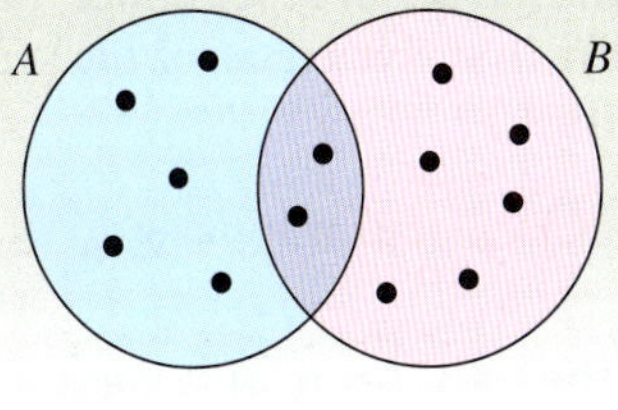

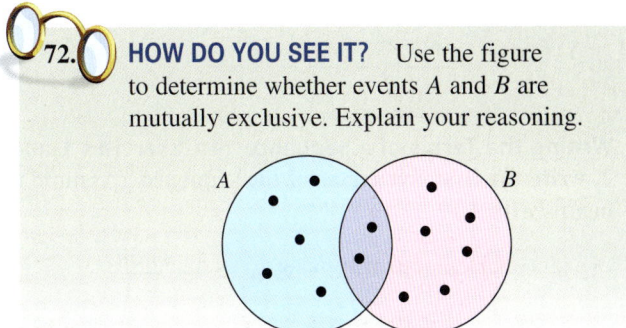

Cumulative Mixed Review

Evaluating a Combination **In Exercises 73 and 74, evaluate ${}_nC_r$. Verify your result using a graphing utility.**

73. ${}_8C_4$

74. ${}_9C_5$

8 Chapter Review

See CalcChat.com for tutorial help and worked-out solutions to odd-numbered exercises. For instructions on how to use a graphing utility, see Appendix A.

8.1 *What did you learn?*

Use sequence notation to write the terms of sequences (*p. 580*). An infinite sequence is a function whose domain is the set of positive integers. The function values $a_1, a_2, a_3, a_4, \ldots, a_n, \ldots$ are the terms of the sequence. If the domain of a function consists of the first n positive integers only, then the sequence is a finite sequence.

Use factorial notation (*p. 583*). If n is a positive integer, then n factorial is $n! = 1 \cdot 2 \cdot 3 \cdot 4 \cdots (n-1) \cdot n$. As a special case, zero factorial is defined as $0! = 1$.

Use summation notation to write sums (*p. 584*). The sum of the first n terms of a sequence is represented by

$$\sum_{i=1}^{n} a_i = a_1 + a_2 + a_3 + a_4 + \cdots + a_n$$

where i is called the index of summation, n is the upper limit of summation, and 1 is the lower limit of summation. Properties of sums are listed on page 584.

Find sums of infinite series (*p. 585*). Consider the infinite sequence $a_1, a_2, a_3, a_4, \ldots, a_i, \ldots$.

1. The sum of the first n terms of the sequence is called a finite series or the nth partial sum of the sequence and is denoted by

$$a_1 + a_2 + a_3 + \cdots + a_n = \sum_{i=1}^{n} a_i.$$

2. The sum of all the terms of the infinite sequence is called an infinite series and is denoted by

$$a_1 + a_2 + a_3 + \cdots + a_i + \cdots = \sum_{i=1}^{\infty} a_i.$$

Use sequences and series to model and solve real-life problems (*p. 586*). A sequence can be used to model the balance in an account (see Example 10).

Example

(a) The first five terms of the sequence $a_n = \dfrac{1}{2^n + 1}$ are

$$a_1 = \frac{1}{2^1 + 1} = \frac{1}{3}, \quad a_2 = \frac{1}{2^2 + 1} = \frac{1}{5},$$

$$a_3 = \frac{1}{2^3 + 1} = \frac{1}{9}, \quad a_4 = \frac{1}{2^4 + 1} = \frac{1}{17},$$

$$\text{and} \quad a_5 = \frac{1}{2^5 + 1} = \frac{1}{33}.$$

(b) An expression for the apparent nth term of the sequence

$$\frac{1}{2}, -\frac{1}{4}, \frac{1}{8}, -\frac{1}{16}, \ldots \text{ is } a_n = \frac{(-1)^{n+1}}{2^n}.$$

(c) The second through fifth terms of the sequence defined recursively as $a_1 = 15$, $a_k = 2a_{k-1} - 5$, where $k \geq 2$, are as follows.

$$a_2 = 2a_{2-1} - 5 = 2a_1 - 5 = 2(15) - 5 = 25$$
$$a_3 = 2a_{3-1} - 5 = 2a_2 - 5 = 2(25) - 5 = 45$$
$$a_4 = 2a_{4-1} - 5 = 2a_3 - 5 = 2(45) - 5 = 85$$
$$a_5 = 2a_{5-1} - 5 = 2a_4 - 5 = 2(85) - 5 = 165$$

(d)
$$\frac{10!n!}{8!(n-2)!} = \frac{10 \cdot 9 \cdot \cancel{8!} \cdot n \cdot (n-1) \cdot \cancel{(n-2)!}}{\cancel{8!}\cancel{(n-2)!}} = 90n(n-1)$$

(e)
$$\sum_{n=1}^{6} 5n = 5(1) + 5(2) + 5(3) + 5(4) + 5(5) + 5(6) = 105$$

(f)
$$\sum_{i=1}^{\infty} \frac{1}{10^i} = \frac{1}{10^1} + \frac{1}{10^2} + \frac{1}{10^3} + \frac{1}{10^4} + \frac{1}{10^5} + \cdots$$
$$= 0.1 + 0.01 + 0.001 + 0.0001 + 0.00001 + \cdots$$
$$= 0.11111\ldots$$
$$= \frac{1}{9}$$

Writing the Terms of a Sequence In Exercises 1 and 2, write the first five terms of the sequence. (Assume n begins with 1.)

1. $a_n = \dfrac{2^n}{2^n + 1}$
2. $a_n = \dfrac{(-1)^n 5n}{2n - 1}$

Finding the nth Term of a Sequence In Exercises 3–6, write an expression for the apparent nth term of the sequence. (Assume n begins with 1.)

3. 7, 12, 17, 22, 27, . . .
4. 50, 48, 46, 44, 42, . . .
5. $2, \frac{2}{3}, \frac{2}{5}, \frac{2}{7}, \frac{2}{9}, \ldots$
6. $\frac{3}{2}, \frac{4}{3}, \frac{5}{4}, \frac{6}{5}, \frac{7}{6}, \ldots$

A Recursive Sequence In Exercises 7 and 8, write the first five terms of the sequence defined recursively.

7. $a_1 = 9$, $a_{k+1} = \frac{1}{3}a_k$
8. $a_1 = 49$, $a_{k+1} = a_k + 6$

Simplifying Factorial Expressions In Exercises 9 and 10, simplify the factorial expression.

9. $\dfrac{10!}{12!}$
10. $\dfrac{(n+1)!}{(n-1)!}$

Finding a Sum In Exercises 11 and 12, find the sum.

11. $\displaystyle\sum_{k=2}^{5} 12k$
12. $\displaystyle\sum_{i=1}^{8} \frac{i}{i+1}$

Writing a Sum Using Sigma Notation **In Exercises 13 and 14, use sigma notation to write the sum. Then use a graphing utility to find the sum.**

13. $3(1^2) + 3(2^2) + 3(3^2) + \cdots + 3(9^2)$

14. $\frac{1}{2} + \frac{2}{3} + \frac{3}{4} + \cdots + \frac{9}{10}$

Finding the Sum of a Series **In Exercises 15 and 16, find the sum of the infinite series.**

15. $\sum_{j=1}^{\infty} \frac{8}{10^j}$

16. $\sum_{k=1}^{\infty} 4(0.01)^k$

Compound Interest **In Exercises 17 and 18, use the following information. You deposit \$3000 in a fund that earns 2% interest compounded quarterly. The balance in the fund after n quarters is given by**

$$A_n = 3000\left(1 + \frac{0.02}{4}\right)^n; n = 1, 2, 3, \ldots.$$

17. Compute the first eight terms of this sequence. What do these values represent?

18. Find the balance in the fund after (a) 10 years and (b) after 20 years. How many times the balance after 10 years is the balance after 20 years?

8.2 *What did you learn?*

Recognize, write, and find the *n*th terms of arithmetic sequences *(p. 591)*. A sequence is arithmetic when the differences between consecutive terms are the same. So, $a_1, a_2, a_3, a_4, \ldots, a_n, \ldots$ is arithmetic when there is a number d such that $a_2 - a_1 = a_3 - a_2 = a_4 - a_3 = \cdots = d$. The number d is the common difference of the sequence.

The nth term of an arithmetic sequence has the form $a_n = a_1 + (n - 1)d$, where d is the common difference between consecutive terms of the sequence.

When you know a_n and d of an arithmetic sequence, you can find the $(n + 1)$th term by using the recursion formula $a_{n+1} = a_n + d$.

Find *n*th partial sums of arithmetic sequences *(p. 594)*. The sum of a finite arithmetic sequence with n terms is given by $S_n = (n/2)(a_1 + a_n)$.

Use arithmetic sequences to model and solve real-life problems *(p. 595)*. An arithmetic sequence can be used to find the total sales of a small business (see Example 7).

Example

(a) To find a formula for the nth term of the arithmetic sequence with common difference $d = 2$ and first term $a_1 = 3$, you have

$a_n = 3 + (n - 1)(2)$

$= 2n + 1.$

(b) To find a_{14} for the arithmetic sequence with $a_1 = 1$ and $a_2 = -9$, you have

$d = -9 - 1 = -10$

and

$a_n = 1 + (n - 1)(-10) = 11 - 10n.$

So, $a_{14} = 11 - 10(14) = -129.$

(c) To find the 100th partial sum of 1, 4, 7, 10, 13, . . . , you have $a_1 = 1$ and $d = 3$,

$a_n = 1(n - 1)(3) = 3n - 2,$

and

$a_{100} = 3(100) - 2 = 298.$

So, $S_{100} = \frac{100}{2}(1 + 298) = 14{,}950.$

Identifying an Arithmetic Sequence **In Exercises 19 and 20, determine whether the sequence is arithmetic. Explain.**

19. 0, 1, 3, 6, 10, . . .

20. $\frac{9}{9}, \frac{8}{9}, \frac{7}{9}, \frac{6}{9}, \frac{5}{9}, \ldots$

Writing the Terms of an Arithmetic Sequence **In Exercises 21–24, write the first five terms of the arithmetic sequence. Verify your results using a graphing utility.**

21. $a_1 = 4, d = 3$

22. $a_1 = 8, d = -2$

23. $a_4 = 10, a_{10} = 28$

24. $a_2 = 14, a_6 = 22$

Writing the Terms of an Arithmetic Sequence **In Exercises 25 and 26, write the first five terms of the arithmetic sequence. Find the common difference and write the *n*th term of the sequence as a function of *n*.**

25. $a_1 = 35, a_{k+1} = a_k - 3$

26. $a_1 = 15, a_{k+1} = a_k + \frac{5}{2}$

Finding the *n*th Term and a Partial Sum **In Exercises 27 and 28, find a formula for a_n for the arithmetic sequence and find the sum of the first 25 terms of the sequence.**

27. $a_1 = 100, d = -3$

28. $a_1 = 10, a_3 = 28$

Finding a Partial Sum **In Exercises 29 and 30, find the sum of the finite arithmetic sequence. Use a graphing utility to verify your result.**

29. $\sum_{j=1}^{11} (2.5j - 5)$

30. $\sum_{k=1}^{11} \left(\frac{2}{3}k + 4\right)$

Accounting **In Exercises 31 and 32, use the following information. The starting salary for an accountant is \$53,300 with a guaranteed increase of \$1066 per year.**

31. Determine the salary during the fifth year.

32. Determine the total compensation for 5 full years.

8.3 What did you learn?

Recognize, write, and find the *n*th terms of geometric sequences *(p. 599)*. The nth term of a geometric sequence has the form $a_n = a_1r^{n-1}$, where r is the common ratio of consecutive terms of the sequence.

Find *n*th partial sums of geometric sequences *(p. 602)*. The sum of the finite geometric sequence $a_1, a_1r, a_1r^2, a_1r^3, a_1r^4, \ldots, a_1r^{n-1}$ with common ratio $r \neq 1$ is given by

$$S_n = \sum_{i=1}^{n} a_1r^{i-1} = a_1\left(\frac{1-r^n}{1-r}\right).$$

Find sums of infinite geometric series *(p. 603)*. If $|r| < 1$, then the infinite geometric series

$$a_1 + a_1r + a_1r^2 + a_1r^3 + \cdots + a_1r^{n-1} + \cdots$$

has the sum

$$S = \sum_{i=0}^{\infty} a_1r^i = \frac{a_1}{1-r}.$$

Use geometric sequences to model and solve real-life problems *(p. 604)*. A finite geometric sequence can be used to find the balance in an annuity (see Example 8).

Example Consider the geometric sequence whose first term is $a_1 = 4$ and whose common ratio is $\frac{1}{5}$.

(a) The first five terms of the sequence are shown below.

$$a_1 = 4$$
$$a_2 = 4\left(\tfrac{1}{5}\right)^1 = \tfrac{4}{5}$$
$$a_3 = 4\left(\tfrac{1}{5}\right)^2 = \tfrac{4}{25}$$
$$a_4 = 4\left(\tfrac{1}{5}\right)^3 = \tfrac{4}{125}$$
$$a_5 = 4\left(\tfrac{1}{5}\right)^4 = \tfrac{4}{625}$$

(b) The sum of the first five terms of the sequence is

$$\sum_{i=1}^{5} 4\left(\frac{1}{5}\right)^{i-1} = 4\left[\frac{1-(1/5)^5}{1-1/5}\right] = 4\left(\frac{1-1/3125}{4/5}\right) = 4.9984.$$

(c) The sum of the corresponding infinite geometric series is

$$\sum_{i=0}^{\infty} 4\left(\frac{1}{5}\right)^i = \frac{4}{1-1/5} = 5.$$

Identifying a Geometric Sequence **In Exercises 33 and 34, determine whether or not the sequence is geometric. If it is, find the common ratio.**

33. $10, 20, 40, 80, \ldots$ **34.** $\frac{1}{2}, \frac{2}{3}, \frac{3}{4}, \frac{4}{5}, \ldots$

Writing Terms of a Geometric Sequence **In Exercises 35 and 36, write the first five terms of the geometric sequence.**

35. $a_1 = 4, \quad r = -\frac{1}{4}$ **36.** $a_1 = 9, \quad a_2 = 6$

Finding the *n*th Term of a Geometric Sequence **In Exercises 37 and 38, write the first five terms of the geometric sequence. Find the common ratio and write the *n*th term of the sequence as a function of *n*.**

37. $a_1 = 120, a_{k+1} = \frac{1}{3}a_k$ **38.** $a_1 = 200, a_{k+1} = 0.1a_k$

Finding a Term of a Geometric Sequence **In Exercises 39 and 40, find the missing term of the geometric sequence (a) using the *table* feature of a graphing utility and (b) algebraically.**

39. $a_1 = 16, a_2 = -8, a_6 = \square$

40. $a_3 = 6, a_4 = 1, a_9 = \square$

Finding the Sum of a Finite Geometric Sequence **In Exercises 41–44, find the sum. Use a graphing utility to verify your result.**

41. $\sum_{i=1}^{7} 5^{i-1}$ **42.** $\sum_{i=1}^{5} 3^{i-1}$

43. $\sum_{n=1}^{7} 9(-4)^{n-1}$ **44.** $\sum_{n=1}^{4} 16\left(-\frac{1}{2}\right)^{n-1}$

Finding the Sum of an Infinite Geometric Series **In Exercises 45–48, find the sum of the infinite geometric series, if possible. If not possible, explain why.**

45. $\sum_{i=1}^{\infty} 3\left(\frac{7}{8}\right)^{i-1}$ **46.** $\sum_{i=1}^{\infty} 4\left(\frac{1}{3}\right)^{i-1}$

47. $\sum_{k=1}^{\infty} 2\left(\frac{3}{2}\right)^{k-1}$ **48.** $\sum_{k=1}^{\infty} 3.1\left(\frac{1}{10}\right)^{k-1}$

49. Finance A company buys a fleet of six vans for \$130,000. During the next 5 years, the fleet will depreciate at a rate of 30% per year.

(a) Find the formula for the nth term of a geometric sequence that gives the value of the fleet n full years after it was purchased.

(b) Find the depreciated value of the fleet at the end of 5 full years.

50. Annuity A deposit of \$80 is made at the beginning of each month in a fund that pays 4% interest, compounded monthly. The balance A in the fund at the end of 4 years is given by

$$A = 80\left(1 + \frac{0.04}{12}\right)^1 + \cdots + 80\left(1 + \frac{0.04}{12}\right)^{48}.$$

Find A.

8.4 What did you learn?

Use the Binomial Theorem to calculate binomial coefficients *(p. 609)*. Let n be a positive integer, and let $r = 0, 1, 2, 3, \ldots, n$. In the expansion of

$$(x + y)^n = x^n + nx^{n-1}y + \cdots + {}_nC_r x^{n-r}y^r + \cdots + y^n$$

the coefficient of $x^{n-r}y^r$ is ${}_nC_r = \dfrac{n!}{(n-r)!r!} = \dbinom{n}{r}$.

Use binomial coefficients to write binomial expansions *(p. 611)*. See Examples 3–6.

Use Pascal's Triangle to calculate binomial coefficients *(p. 613)*. Rows 0 through 7 of Pascal's Triangle are shown on page 613.

Example

Consider the expansion of $(2x - 3y)^9$. The fourth term is

$$\begin{aligned}{}_9C_6(2x)^6(-3y)^3 &= \frac{9!}{(9-6)!6!}(2^6)x^6(-3)^3y^3\\ &= \frac{9 \cdot 8 \cdot 7 \cdot \cancel{6!}}{3!\cancel{6!}}(-1728)x^6y^3\\ &= \frac{9 \cdot 8 \cdot 7}{3 \cdot 2 \cdot 1}(-1728)x^6y^3\\ &= -145{,}152x^6y^3.\end{aligned}$$

Finding Binomial Coefficients **In Exercises 51 and 52, find the binomial coefficient by hand.**

51. ${}_{10}C_8$ **52.** ${}_{12}C_5$

Expanding a Binomial **In Exercises 53–58, use the Binomial Theorem to expand and simplify the expression.**

53. $(x + 5)^4$ **54.** $(y - 3)^3$

55. $(4x - 3y)^4$ **56.** $(x + 3)^5 - 4(x + 3)^4$

57. $(a - 4b)^5$ **58.** $(3x + y)^7$

Using Pascal's Triangle **In Exercises 59–62, use Pascal's Triangle to find the binomial coefficient.**

59. ${}_7C_7$ **60.** ${}_{11}C_7$

61. ${}_{10}C_4$ **62.** ${}_{10}C_5$

8.5 What did you learn?

Solve simple counting problems *(p. 617)*. See Examples 1 and 2.

Use the Fundamental Counting Principle to solve more complicated counting problems *(p. 618)*. Event E_1 can occur in m_1 different ways. After E_1 has occurred, event E_2 can occur in m_2 different ways. The number of ways that the two events can occur is $m_1 \cdot m_2$.

Use permutations to solve counting problems *(p. 619)*. The number of permutations (order is important) of n elements is $n!$. The number of permutations of n elements taken r at a time is ${}_nP_r = n!/(n-r)!$.

The number of distinguishable permutations of n objects is given by $n!/(n_1! \cdot n_2! \cdot n_3! \cdot \cdots \cdot n_k!)$, where $n = n_1 + n_2 + n_3 + \cdots + n_k$.

Use combinations to solve counting problems *(p. 622)*. The number of combinations (order is not important) of n elements taken r at a time is ${}_nC_r$.

Example

(a) To design a home theater system, a customer can choose one of six televisions, one of five video sources, one of three receivers, and one of eight speaker systems. So, there are

$$6 \cdot 5 \cdot 3 \cdot 8 = 720$$

possible system configurations.

(b) There are ten bicyclists entered in a race. The top three places can be decided in

$${}_{10}P_3 = \frac{10!}{(10-3)!} = \frac{10!}{7!} = \frac{10 \cdot 9 \cdot 8 \cdot \cancel{7!}}{\cancel{7!}} = 720$$

different ways.

(c) From a group of 32 potential jurors, a jury of 6 can be randomly selected in

$${}_{32}C_6 = \frac{32!}{(32-6)!6!} = \frac{32!}{26!6!} = 906{,}192$$

different ways.

63. Random Selection Slips of paper numbered 1 through 15 are placed in a hat. In how many different ways can two numbers be drawn so that the sum of the numbers is 12? Assume the random selection is (a) with replacement and (b) without replacement.

64. Education A college student is preparing a course schedule for the next semester. The student can choose from four math courses, three economics courses, six English courses, and three humanities courses. Find the number of possible schedules that the student can create when choosing one course of each subject.

Evaluating a Permutation **In Exercises 65 and 66, evaluate $_nP_r$ using the formula from Section 8.5.**

65. $_{12}P_5$

66. $_6P_4$

Distinguishable Permutations **In Exercises 67 and 68, find the number of distinguishable permutations of the group of letters.**

67. C, A, L, C, U, L, U, S

68. I, N, T, E, G, R, A, T, E

Solving an Equation **In Exercises 69 and 70, solve for n.**

69. $_{n+1}P_2 = 4 \cdot {_nP_1}$

70. $15 \cdot {_nP_2} = {_{n+1}P_3}$

Evaluating a Combination **In Exercises 71 and 72, evaluate $_nC_r$ using the formula from Section 8.5.**

71. $_{12}C_6$

72. $_{50}C_{48}$

73. **Athletics** There are 12 bicyclists entered in a race. In how many different orders could the 12 bicyclists finish? (Assume there are no ties.)

74. **Athletics** From a pool of seven juniors and eleven seniors, four co-captains will be chosen for the football team. How many different combinations are possible if two juniors and two seniors are to be chosen?

8.6 *What did you learn?*

Find probabilities of events *(p. 626)*. If an event E has $n(E)$ equally likely outcomes and its sample space S has $n(S)$ equally likely outcomes, then the probability of event E is

$$P(E) = n(E)/n(S)$$

such that $0 \le P(E) \le 1$.

Find probabilities of mutually exclusive events *(p. 629)*. If A and B are events in the same sample space, then the probability of A or B occurring is

$$P(A \cup B) = P(A) + P(B) - P(A \cap B).$$

If A and B are mutually exclusive (have no outcomes in common), then $P(A \cup B) = P(A) + P(B)$.

Find probabilities of independent events *(p. 631)*. If A and B are independent events, then the probability that both A and B will occur is $P(A \text{ and } B) = P(A) \cdot P(B)$. This can be extended to any number of independent events.

Example A card is drawn at random from a standard deck of playing cards.

(a) The probability that it is a club is

$$P(E) = \frac{13}{52} = \frac{1}{4}.$$

(b) The probability that it is an ace (event A) or a queen (event B) is

$$P(A \cup B) = \frac{4}{52} + \frac{4}{52} = \frac{8}{52} = \frac{2}{13}.$$

(c) The card is replaced, the deck is shuffled, and a second card is drawn. The probability that the first card is a queen (event A) and the second card is also a queen (event B) is

$$P(A \text{ and } B) = \left(\frac{4}{52}\right)\left(\frac{4}{52}\right) = \frac{1}{169}.$$

75. **Random Selection** You have five pairs of socks (no two pairs are the same color). You randomly select two socks from a drawer. What is the probability that you get a matched pair?

76. **Bookshelf** A child returns a five-volume set of books to a bookshelf. The child is not able to read and so cannot distinguish one volume from another. What is the probability that the books are shelved in the correct order?

77. **Rolling a Die** A five-sided die is rolled five times. What is the probability that each side appears exactly once?

78. **Education** A sample of 400 college students, 75 faculty members, and 25 administrators were asked whether they favored a proposed increase in the annual activity fee to enhance student life on campus. Of the students, 237 favor and 163 oppose the proposal. Of the faculty, 37 favor and 38 oppose the proposal. Of the administrators, 18 favor and 7 oppose the proposal. A person is selected at random from the sample. Find the probability that the person selected is (a) a person who does not favor the proposal, (b) a student, and (c) a faculty member who favors the proposal.

Focusing on Concepts

True or False? **In Exercises 79 and 80, determine whether the statement is true or false. Justify your answer.**

79. $\dfrac{(n+2)!}{n!} = (n+2)(n+1)$

80. $\displaystyle\sum_{j=1}^{6} 3^j = \sum_{j=3}^{8} 3^{j-2}$

81. **Writing** In your own words, explain what makes a sequence (a) arithmetic and (b) geometric.

82. **Think About It** Does every finite series whose terms are integers have a finite sum? Explain.

8 Chapter Test

See *CalcChat.com* for tutorial help and worked-out solutions to odd-numbered exercises. For instructions on how to use a graphing utility, see Appendix A.

Take this test as you would take a test in class. After you are finished, check your work against the answers given in the back of the book.

In Exercises 1–3, write the first five terms of the sequence.

1. $a_n = 3\left(\frac{2}{3}\right)^{n-1}$ (Begin with $n = 1$.)

2. $a_1 = 12$ and $a_{k+1} = a_k + 4$

3. $b_n = \dfrac{(-1)^n x^n}{n}$ (Begin with $n = 1$.)

In Exercises 4 and 5, simplify the factorial expression.

4. $\dfrac{11! \cdot 4!}{4! \cdot 7!}$

5. $\dfrac{n!}{(n+1)!}$

6. Write an expression for the apparent nth term of the sequence

2, 5, 10, 17, 26, ... (Assume n begins with 1.)

In Exercises 7 and 8, find a formula for the nth term of the sequence.

7. Arithmetic: $a_1 = 5000$, $d = -100$

8. Geometric: $a_1 = 4$, $a_{k+1} = \frac{1}{2}a_k$

9. Use sigma notation to write $\dfrac{2}{3(1)+1} + \dfrac{2}{3(2)+1} + \cdot \cdot \cdot + \dfrac{2}{3(12)+1}$.

10. Use sigma notation to write $2 + \frac{1}{2} + \frac{1}{8} + \frac{1}{32} + \frac{1}{128} + \cdot \cdot \cdot$.

In Exercises 11–13, determine whether the sequence associated with the series is arithmetic or geometric. Find the common difference or ratio and find the sum.

11. $\sum_{n=1}^{8} 49\left(\frac{1}{7}\right)^{n-1}$

12. $\sum_{n=1}^{15} (n-5)$

13. $5 - 2 + \frac{4}{5} - \frac{8}{25} + \frac{16}{125} - \cdot \cdot \cdot$

14. Use the Binomial Theorem to expand and simplify $(3a - 5b)^4$.

15. Find the fifth term in the expansion of the binomial $(2x - 10)^6$.

In Exercises 16–19, evaluate the expression.

16. ${}_9C_3$ **17.** ${}_{20}C_3$ **18.** ${}_{70}P_3$ **19.** ${}_9P_2$

20. Solve for n in $4 \cdot {}_nP_3 = {}_{n+1}P_4$.

21. How many distinct license plates can be issued consisting of one letter followed by a three-digit number?

22. Four students are randomly selected from a class of 25 to answer questions from a reading assignment. In how many ways can the four be selected?

23. A card is drawn from a standard deck of 52 playing cards. Find the probability that it is a face card in the suit of hearts.

24. In 2018, nine of the fifteen men's basketball teams in the Atlantic Coast Conference participated in the NCAA Men's Basketball Championship Tournament. What is the probability that a random selection of nine teams from the Atlantic Coast Conference will be the same as the nine teams that participated in the tournament?

25. Two integers from 1 to 60 are chosen by a random number generator. What is the probability that (a) both numbers are odd, (b) both numbers are less than 12, and (c) the same number is chosen twice?

Collaborative Project

To work a collaborative project involving Sequences, Series, and Probability, visit this textbook's website at *LarsonPrecalculus.com.*

Proofs in Mathematics

Properties of Sums (p. 584)

1. $\sum_{i=1}^{n} c = cn$, c is a constant.

2. $\sum_{i=1}^{n} ca_i = c\sum_{i=1}^{n} a_i$, c is a constant.

3. $\sum_{i=1}^{n} (a_i + b_i) = \sum_{i=1}^{n} a_i + \sum_{i=1}^{n} b_i$

4. $\sum_{i=1}^{n} (a_i - b_i) = \sum_{i=1}^{n} a_i - \sum_{i=1}^{n} b_i$

Proof

Each of these properties follows directly from the properties of real numbers.

1. $\sum_{i=1}^{n} c = c + c + c + \cdot \cdot \cdot + c = cn$ $\quad$ *n* terms

The Distributive Property is used in the proof of Property 2.

2.
$$\begin{aligned}\sum_{i=1}^{n} ca_i &= ca_1 + ca_2 + ca_3 + \cdot \cdot \cdot + ca_n \\ &= c(a_1 + a_2 + a_3 + \cdot \cdot \cdot + a_n) \\ &= c\sum_{i=1}^{n} a_i\end{aligned}$$

The proof of Property 3 uses the Commutative and Associative Properties of Addition.

3.
$$\begin{aligned}\sum_{i=1}^{n} (a_i + b_i) &= (a_1 + b_1) + (a_2 + b_2) + (a_3 + b_3) + \cdot \cdot \cdot + (a_n + b_n) \\ &= (a_1 + a_2 + a_3 + \cdot \cdot \cdot + a_n) + (b_1 + b_2 + b_3 + \cdot \cdot \cdot + b_n) \\ &= \sum_{i=1}^{n} a_i + \sum_{i=1}^{n} b_i\end{aligned}$$

The proof of Property 4 uses the Commutative and Associative Properties of Addition and the Distributive Property.

4.
$$\begin{aligned}\sum_{i=1}^{n} (a_i - b_i) &= (a_1 - b_1) + (a_2 - b_2) + (a_3 - b_3) + \cdot \cdot \cdot + (a_n - b_n) \\ &= (a_1 + a_2 + a_3 + \cdot \cdot \cdot + a_n) + (-b_1 - b_2 - b_3 - \cdot \cdot \cdot - b_n) \\ &= (a_1 + a_2 + a_3 + \cdot \cdot \cdot + a_n) - (b_1 + b_2 + b_3 + \cdot \cdot \cdot + b_n) \\ &= \sum_{i=1}^{n} a_i - \sum_{i=1}^{n} b_i\end{aligned}$$

Infinite Series

The study of infinite series was considered a novelty in the fourteenth century. Logician Richard Suiseth, whose nickname was Calculator, solved this problem.

> *If throughout the first half of a given time interval a variation continues at a certain intensity; throughout the next quarter of the interval at double the intensity; throughout the following eighth at triple the intensity and so ad infinitum; The average intensity for the whole interval will be the intensity of the variation during the second subinterval (or double the intensity).*

This is the same as saying that the sum of the infinite series

$$\frac{1}{2} + \frac{2}{4} + \frac{3}{8} + \cdot \cdot \cdot + \frac{n}{2^n} + \cdot \cdot \cdot$$

is 2.

The Sum of a Finite Arithmetic Sequence (p. 594)

The sum of a finite arithmetic sequence with n terms is given by

$$S_n = \frac{n}{2}(a_1 + a_n).$$

Proof

Begin by generating the terms of the arithmetic sequence in two ways. In the first way, repeatedly add d to the first term to obtain

$$\begin{aligned} S_n &= a_1 + a_2 + a_3 + \cdot \cdot \cdot + a_{n-2} + a_{n-1} + a_n \\ &= a_1 + [a_1 + d] + [a_1 + 2d] + \cdot \cdot \cdot + [a_1 + (n-1)d]. \end{aligned}$$

In the second way, repeatedly subtract d from the nth term to obtain

$$\begin{aligned} S_n &= a_n + a_{n-1} + a_{n-2} + \cdot \cdot \cdot + a_3 + a_2 + a_1 \\ &= a_n + [a_n - d] + [a_n - 2d] + \cdot \cdot \cdot + [a_n - (n-1)d]. \end{aligned}$$

When you add these two versions of S_n, the multiples of d subtract out and you obtain

$$2S_n = (a_1 + a_n) + (a_1 + a_n) + (a_1 + a_n) + \cdot \cdot \cdot + (a_1 + a_n) \quad n \text{ terms}$$

$$2S_n = n(a_1 + a_n)$$

$$S_n = \frac{n}{2}(a_1 + a_n).$$

■

The Sum of a Finite Geometric Sequence (p. 602)

The sum of the finite geometric sequence

$$a_1, \quad a_1r, \quad a_1r^2, \quad a_1r^3, \quad a_1r^4, \quad . . . , \quad a_1r^{n-1}$$

with common ratio $r \neq 1$ is given by

$$S_n = \sum_{i=1}^{n} a_1 r^{i-1} = a_1\left(\frac{1-r^n}{1-r}\right).$$

Proof

$$S_n = a_1 + a_1r + a_1r^2 + \cdot \cdot \cdot + a_1r^{n-2} + a_1r^{n-1}$$

$$rS_n = a_1r + a_1r^2 + a_1r^3 + \cdot \cdot \cdot + a_1r^{n-1} + a_1r^n \quad \text{Multiply by } r.$$

Subtracting the second equation from the first yields

$$S_n - rS_n = a_1 - a_1r^n.$$

So, $S_n(1 - r) = a_1(1 - r^n)$, and, because $r \neq 1$, you have

$$S_n = a_1\left(\frac{1-r^n}{1-r}\right).$$

■

The Binomial Theorem (p. 609)

Let n be a positive integer, and let $r = 0, 1, 2, 3, \ldots, n$. In the expansion of $(x + y)^n$

$$(x + y)^n = x^n + nx^{n-1}y + \cdots + {}_nC_r x^{n-r}y^r + \cdots + nxy^{n-1} + y^n$$

the coefficient of $x^{n-r}y^r$ is

$${}_nC_r = \frac{n!}{(n-r)!r!}.$$

Proof

The Binomial Theorem can be proved quite nicely using **mathematical induction.** (For information about mathematical induction, see Appendix G at this textbook's *Student Companion Website.*) The steps are straightforward but look a little messy, so only an outline of the proof is presented.

1. For $n = 1$, you have

$$(x + y)^1 = x^1 + y^1 = {}_1C_0x + {}_1C_1y$$

and the formula is valid.

2. Assuming that the formula is true for $n = k$, the coefficient of $x^{k-r}y^r$ is

$${}_kC_r = \frac{k!}{(k-r)!r!} = \frac{k(k-1)(k-2)\cdots(k-r+1)}{r!}.$$

To show that the formula is true for $n = k + 1$, look at the coefficient of $x^{k+1-r}y^r$ in the expansion of

$$(x + y)^{k+1} = (x + y)^k(x + y).$$

From the right-hand side, you can determine that the term involving $x^{k+1-r}y^r$ is the sum of two products.

$$\begin{aligned}
&({}_kC_r x^{k-r}y^r)(x) + ({}_kC_{r-1}x^{k+1-r}y^{r-1})(y)\\
&= \left[\frac{k!}{(k-r)!r!} + \frac{k!}{(k+1-r)!(r-1)!}\right]x^{k+1-r}y^r\\
&= \left[\frac{(k+1-r)k!}{(k+1-r)!r!} + \frac{k!r}{(k+1-r)!r!}\right]x^{k+1-r}y^r\\
&= \left[\frac{k!(k+1-r+r)}{(k+1-r)!r!}\right]x^{k+1-r}y^r\\
&= \left[\frac{(k+1)!}{(k+1-r)!r!}\right]x^{k+1-r}y^r\\
&= {}_{k+1}C_r x^{k+1-r}y^r
\end{aligned}$$

So, by mathematical induction, the Binomial Theorem is valid for all positive integers n. ■

The Principle of Mathematical Induction

Let P_n be a statement involving the positive integer n. If

1. P_1 is true, and
2. the truth of P_k implies the truth of P_{k+1} for every positive integer k,

then P_n must be true for all positive integers n.

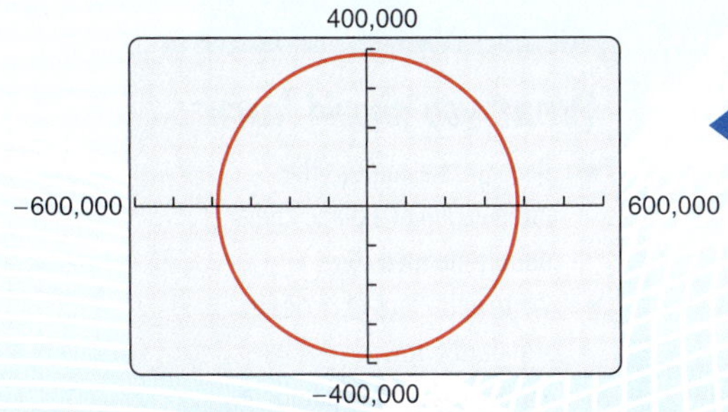

Section 9.2, Example 5
Orbit of the Moon

9 Topics in Analytic Geometry

Student Resources at LarsonPrecalculus.com

- **Videos** explaining the concepts of precalculus
- **Worked-out solution videos** for all *Checkpoint* exercises
- **Editable spreadsheets** of the data sets in the text
- **Group projects** for each chapter applying concepts to real-life problems

9.1 Circles and Parabolas

What you should learn

- Recognize a conic as the intersection of a plane and a double-napped cone.
- Write equations of circles in standard form.
- Write equations of parabolas in standard form.
- Use the reflective property of parabolas to solve real-life problems.

Why you should learn it

Parabolas can be used to model and solve many types of real-life problems. For instance, in Exercise 95 on page 655, a parabola is used to design an entrance ramp for a highway.

Conics

Conic sections were discovered during the classical Greek period. The early Greek studies were largely concerned with the geometric properties of conics. It was not until the early seventeenth century that the broad applicability of conics became apparent and played a prominent role in the early development of calculus.

A **conic section** (or simply **conic**) is the intersection of a plane and a double-napped cone. Notice in Figure 9.1 that in the formation of the four basic conics, the intersecting plane does not pass through the vertex of the cone.

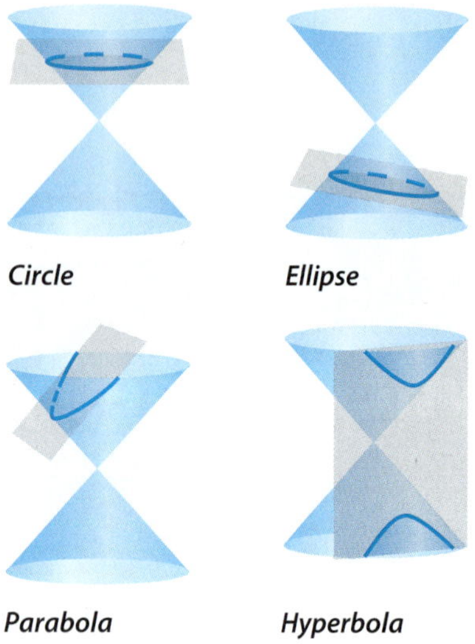

Figure 9.1 *Basic Conics*

When the plane does pass through the vertex, the resulting figure is a **degenerate conic,** as shown in Figure 9.2.

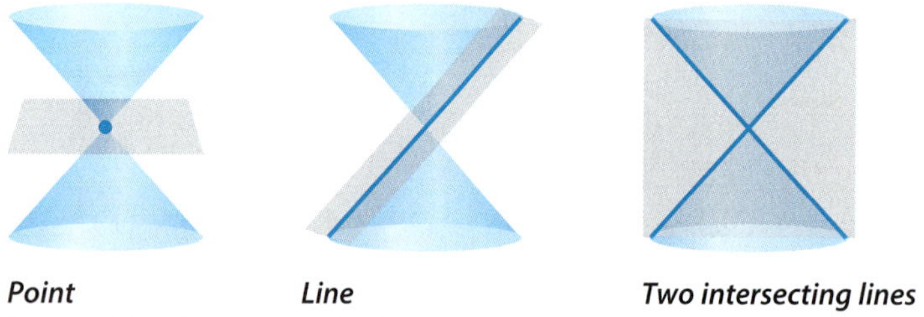

Figure 9.2 *Degenerate Conics*

There are several ways to approach the study of conics. You could begin by defining conics in terms of the intersections of planes and cones, as the Greeks did, or you could define them algebraically, in terms of the general second-degree equation

$$Ax^2 + Bxy + Cy^2 + Dx + Ey + F = 0.$$

However, you will study a third approach, in which each of the conics is defined as a **locus** (collection) of points satisfying a certain geometric property. For example, the definition of a circle as *the collection of all points* (x, y) *that are equidistant from a fixed point* (h, k) leads to the standard equation of a circle

$$(x - h)^2 + (y - k)^2 = r^2. \qquad \text{Equation of circle}$$

Circles

The definition of a circle as a locus of points is a more general definition of a circle as it applies to conics.

Definition of a Circle

A **circle** is the set of all points (x, y) in a plane that are equidistant from a fixed point (h, k), called the **center** of the circle. (See Figure 9.3.) The distance r between the center and any point (x, y) on the circle is the **radius.**

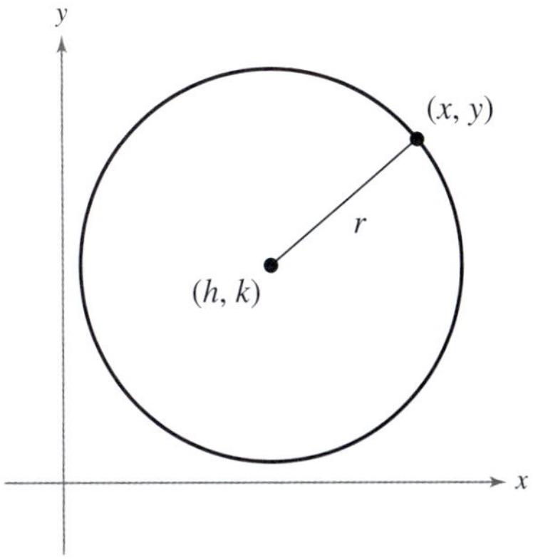

Figure 9.3

The Distance Formula can be used to obtain an equation of a circle whose center is (h, k) and whose radius is r.

$$\sqrt{(x-h)^2+(y-k)^2}=r \quad \text{Distance Formula}$$

$$(x-h)^2+(y-k)^2=r^2 \quad \text{Square each side.}$$

Standard Form of the Equation of a Circle

The **standard form of the equation of a circle** is

$$(x-h)^2+(y-k)^2=r^2.$$

The point (h, k) is the center of the circle, and the positive number r is the radius of the circle. The standard form of the equation of a circle whose center is the origin, $(h, k) = (0, 0)$, is $x^2 + y^2 = r^2$.

EXAMPLE 1 Finding the Standard Equation of a Circle

The point $(1, 4)$ is on a circle whose center is at $(-2, -3)$, as shown in Figure 9.4. Write the standard form of the equation of the circle.

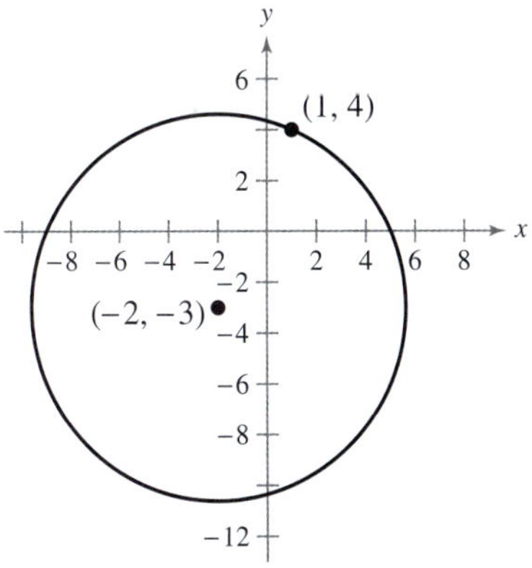

Figure 9.4

Solution

The radius of the circle is the distance between $(-2, -3)$ and $(1, 4)$.

$$r=\sqrt{[1-(-2)]^2+[4-(-3)]^2} \quad \text{Use Distance Formula.}$$

$$=\sqrt{3^2+7^2} \quad \text{Simplify.}$$

$$=\sqrt{58} \quad \text{Radius}$$

The equation of the circle with center $(h, k) = (-2, -3)$ and radius $r = \sqrt{58}$ is

$$(x-h)^2+(y-k)^2=r^2 \quad \text{Equation of a circle}$$

$$[x-(-2)]^2+[y-(-3)]^2=(\sqrt{58})^2 \quad \text{Substitute for } h, k, \text{ and } r.$$

$$(x+2)^2+(y+3)^2=58. \quad \text{Write in standard form.}$$

Checkpoint *Audio-video solution in English & Spanish at LarsonPrecalculus.com*

The point $(3, 1)$ is on a circle whose center is at $(1, -1)$. Write the standard form of the equation of the circle.

Be careful when you are finding h and k from the standard equation of a circle. For instance, to find the correct h and k from the equation of the circle in Example 1, rewrite the quantities $(x + 2)^2$ and $(y + 3)^2$ using subtraction.

$$(x+2)^2=[x-(-2)]^2 \quad \Rightarrow \quad h=-2$$

$$(y+3)^2=[y-(-3)]^2 \quad \Rightarrow \quad k=-3$$

EXAMPLE 2 Sketching a Circle

Sketch the circle $x^2 - 6x + y^2 - 2y + 6 = 0$ by hand and identify its center and radius.

Solution

Begin by writing the equation in standard form. In the third step, note that 9 and 1 are added to *each* side of the equation when completing the squares.

$$x^2 - 6x + y^2 - 2y + 6 = 0$$ Write original equation.

$$(x^2 - 6x + \quad) + (y^2 - 2y + \quad) = -6$$ Group terms.

$$(x^2 - 6x + 9) + (y^2 - 2y + 1) = -6 + 9 + 1$$ Complete the squares.

$$(x - 3)^2 + (y - 1)^2 = 4$$ Write in standard form.

In this form, you can see that the graph is a circle whose center is the point (3, 1) and whose radius is $r = \sqrt{4} = 2$. Plot several points that are two units from the center. The points (5, 1), (3, 3), (1, 1), and (3, −1) are convenient. Draw a circle that passes through the four points, as shown in Figure 9.5.

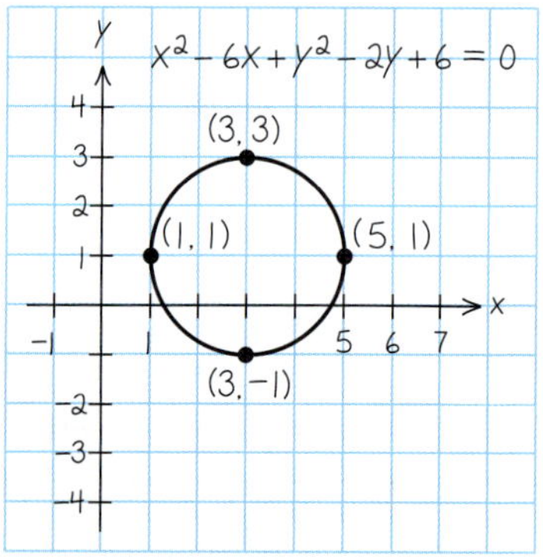

Figure 9.5

Checkpoint Audio-video solution in English & Spanish at LarsonPrecalculus.com

Sketch the circle $x^2 + 4x + y^2 + 4y - 1 = 0$ by hand and identify its center and radius.

EXAMPLE 3 Finding the Intercepts of a Circle

Find the x- and y-intercepts of the graph of the circle $(x - 4)^2 + (y - 2)^2 = 16$.

Solution

To find any x-intercepts, let $y = 0$.

$$(x - 4)^2 + (0 - 2)^2 = 16$$ Substitute 0 for y.

$$(x - 4)^2 = 12$$ Simplify.

$$x - 4 = \pm\sqrt{12}$$ Extract square roots.

$$x = 4 \pm 2\sqrt{3}$$ Add 4 to each side.

To find any y-intercepts, let $x = 0$.

$$(0 - 4)^2 + (y - 2)^2 = 16$$ Substitute 0 for x.

$$(y - 2)^2 = 0$$ Simplify.

$$y - 2 = 0$$ Extract square roots.

$$y = 2$$ Add 2 to each side.

So, the x-intercepts are $(4 + 2\sqrt{3}, 0)$ and $(4 - 2\sqrt{3}, 0)$, and the y-intercept is (0, 2), as shown in the figure.

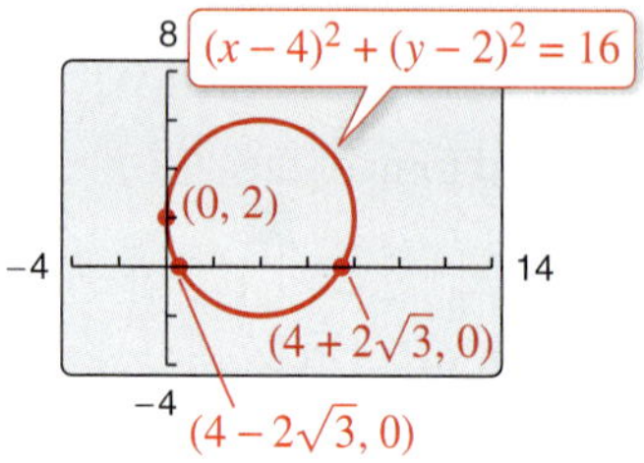

Checkpoint Audio-video solution in English & Spanish at LarsonPrecalculus.com

Find the x- and y-intercepts of the graph of the circle $(x - 2)^2 + (y + 3)^2 = 9$.

Technology Tip

You can use a graphing utility to confirm the result in Example 2 by graphing the upper and lower portions in the same viewing window. First, solve for y to obtain

$$y_1 = 1 + \sqrt{4 - (x - 3)^2}$$

and

$$y_2 = 1 - \sqrt{4 - (x - 3)^2}.$$

Then use a square setting, such as $-1 \le x \le 8$ and $-2 \le y \le 4$, to graph both equations.

Insight

Be sure you are able to recognize the standard form of the equation of a circle and use the equation to identify its center and radius. Also, some standardized tests may ask you to find the equation of a circle given its center and a point on the circle. Additionally, you should always be prepared to find the intercepts of the graph of an equation (see Example 3).

Parabolas

In Section 2.1, you learned that the graph of the quadratic function

$$f(x) = ax^2 + bx + c$$

is a parabola that opens upward or downward. The following definition of a parabola is more general in the sense that it is independent of the orientation of the parabola.

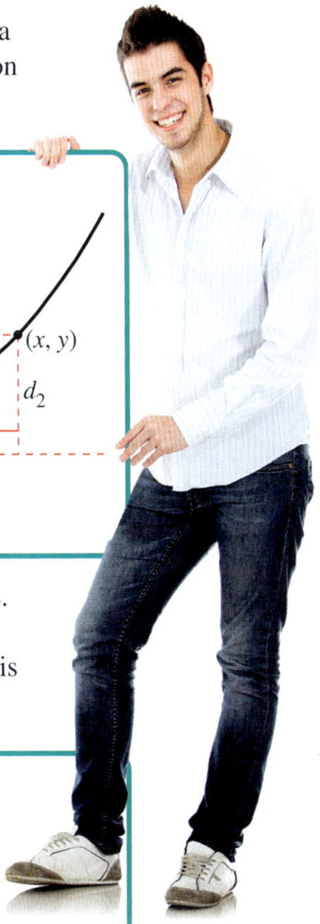

Definition of a Parabola

A **parabola** is the set of all points (x, y) in a plane that are equidistant from a fixed line, the **directrix,** and a fixed point, the **focus,** not on the line. (See Figure 9.6.) The midpoint between the focus and the directrix is the **vertex,** and the line passing through the focus and the vertex is the **axis** of the parabola.

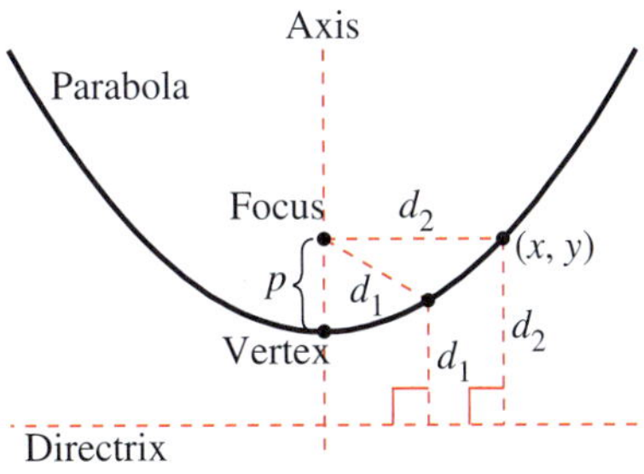

Figure 9.6

Note in Figure 9.6 that a parabola is symmetric with respect to its axis. Using the definition of a parabola, you can derive the following **standard form of the equation of a parabola** whose directrix is parallel to the x-axis or to the y-axis.

Standard Equation of a Parabola

The **standard form of the equation of a parabola** with vertex at (h, k) is

$$(x - h)^2 = 4p(y - k), \quad p \neq 0 \qquad \text{Vertical axis; directrix: } y = k - p$$

or

$$(y - k)^2 = 4p(x - h), \quad p \neq 0. \qquad \text{Horizontal axis; directrix: } x = h - p$$

The focus lies on the axis p units (*directed distance*) from the vertex. If the vertex is at the origin $(0, 0)$, then the equation takes one of the following forms.

$$x^2 = 4py \qquad \text{Vertical axis}$$

$$y^2 = 4px \qquad \text{Horizontal axis}$$

See Figure 9.7.

(See the proof on page 717.)

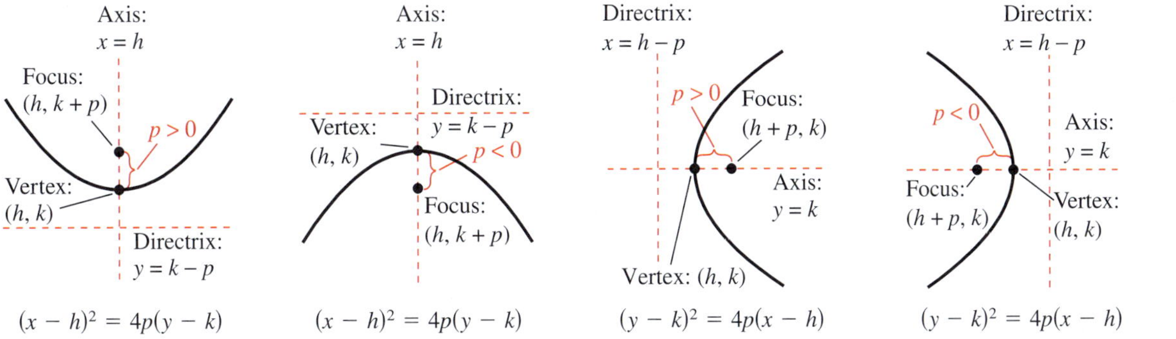

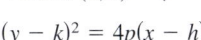

(a) ***Vertical axis:*** $p > 0$ (b) ***Vertical axis:*** $p < 0$ (c) ***Horizontal axis:*** $p > 0$ (d) ***Horizontal axis:*** $p < 0$

Figure 9.7

ESB Professional/Shutterstock.com

EXAMPLE 4 Finding the Standard Equation of a Parabola

Find the standard form of the equation of the parabola with vertex at the origin and focus (0, 4).

Solution

The axis of the parabola is vertical, passing through (0, 0) and (0, 4), as shown in Figure 9.8. The standard form is $x^2 = 4py$, where $p = 4$. So, the equation is $x^2 = 16y$. To check this on a graphing utility, solve the equation for y and graph $y = \frac{1}{16}x^2$.

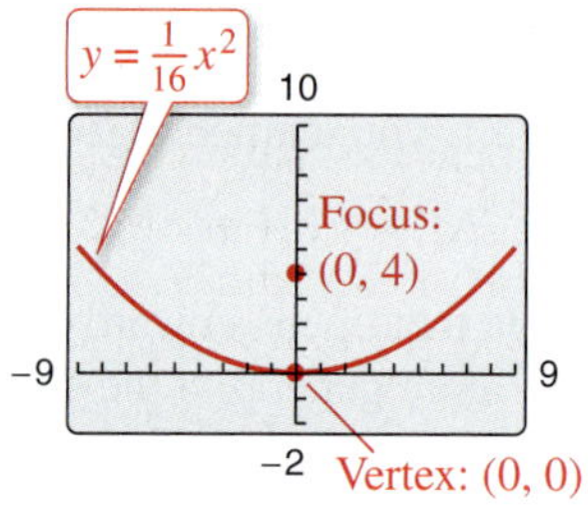

Figure 9.8

Checkpoint *Audio-video solution in English & Spanish at LarsonPrecalculus.com*

Find the standard form of the equation of the parabola with vertex at the origin and focus $\left(0, \frac{3}{8}\right)$.

EXAMPLE 5 Finding the Focus of a Parabola

Find the focus of the parabola given by $y = -\frac{1}{2}x^2 - x + \frac{1}{2}$.

Solution

To find the focus, convert to standard form by completing the square.

$$y = -\frac{1}{2}x^2 - x + \frac{1}{2}$$ Write original equation.

$$-2y = x^2 + 2x - 1$$ Multiply each side by −2.

$$1 - 2y = x^2 + 2x$$ Add 1 to each side.

$$1 + 1 - 2y = x^2 + 2x + 1$$ Complete the square.

$$2 - 2y = x^2 + 2x + 1$$ Combine like terms.

$$-2(y - 1) = (x + 1)^2$$ Write in standard form.

Comparing this equation with $(x - h)^2 = 4p(y - k)$, you can conclude that $h = -1$, $k = 1$, and $p = -\frac{1}{2}$. Because p is negative, the parabola opens downward, as shown in Figure 9.9. So, the focus of the parabola is $(h, k + p) = \left(-1, \frac{1}{2}\right)$.

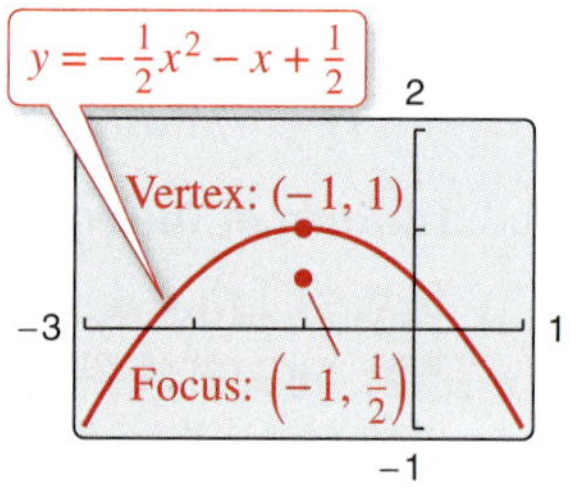

Figure 9.9

Checkpoint *Audio-video solution in English & Spanish at LarsonPrecalculus.com*

Find the focus of the parabola $x = \frac{1}{4}y^2 + \frac{3}{2}y + \frac{13}{4}$.

EXAMPLE 6 Finding the Standard Equation of a Parabola

See LarsonPrecalculus.com for an interactive version of this type of example.

Find the standard form of the equation of the parabola with vertex (1, 0) and focus (2, 0).

Solution

Because the axis of the parabola is horizontal, passing through (1, 0) and (2, 0), consider the equation $(y - k)^2 = 4p(x - h)$, where $h = 1$, $k = 0$, and $p = 2 - 1 = 1$. So, the standard form is

$$(y - 0)^2 = 4(1)(x - 1) \quad \Rightarrow \quad y^2 = 4(x - 1).$$

You can use a graphing utility to confirm this equation. To do this, let $y_1 = \sqrt{4(x - 1)}$ and $y_2 = -\sqrt{4(x - 1)}$, as shown in Figure 9.10.

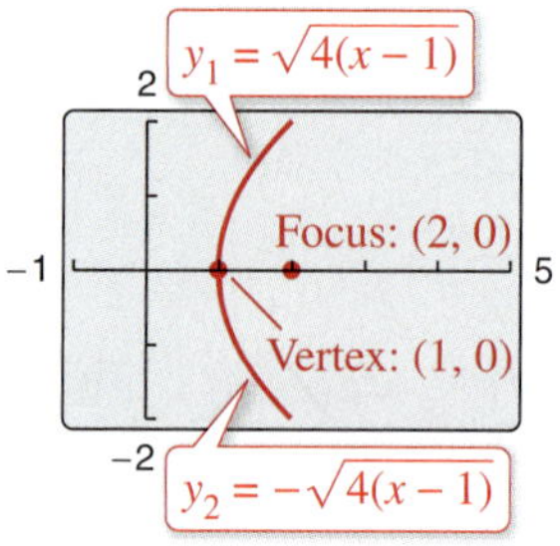

Figure 9.10

Checkpoint *Audio-video solution in English & Spanish at LarsonPrecalculus.com*

Find the standard form of the equation of the parabola with vertex (2, −3) and focus (4, −3).

Reflective Property of Parabolas

A line segment that passes through the focus of a parabola and has endpoints on the parabola is called a **focal chord.** The specific focal chord perpendicular to the axis of the parabola is called the **latus rectum.**

Parabolas occur in a wide variety of applications. For instance, a parabolic reflector can be formed by revolving a parabola about its axis. The resulting surface has the property that all incoming rays parallel to the axis are reflected through the focus of the parabola. This is the principle behind the construction of the parabolic mirrors used in reflecting telescopes. Conversely, the light rays emanating from the focus of a parabolic reflector used in a flashlight are all parallel to one another, as shown in Figure 9.11.

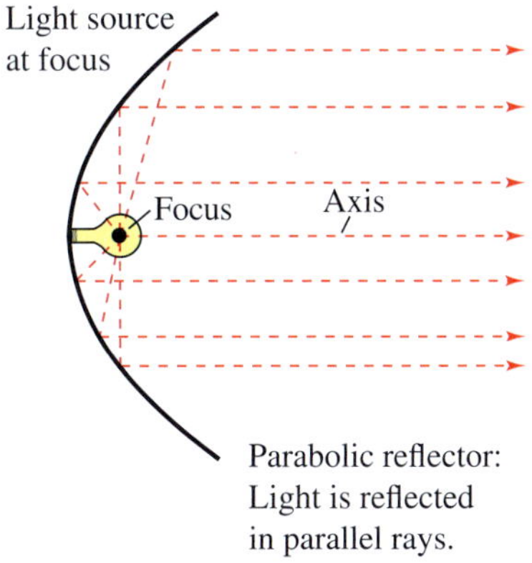

Figure 9.11

A line is **tangent** to a parabola at a point on the parabola when the line intersects, but does not cross, the parabola at the point. Tangent lines to parabolas have special properties related to the use of parabolas in constructing reflective surfaces.

Reflective Property of a Parabola

The tangent line to a parabola at a point P makes equal angles with the following two lines (see Figure 9.12).

1. The line passing through P and the focus
2. The axis of the parabola

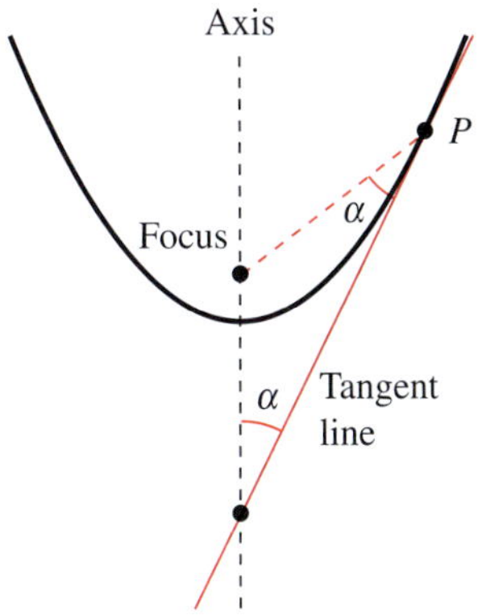

Figure 9.12

EXAMPLE 7 Finding the Tangent Line at a Point on a Parabola

Find an equation of the tangent line to the parabola given by $y = x^2$ at the point $(1, 1)$.

Solution

For this parabola, $p = \frac{1}{4}$ and the focus is $\left(0, \frac{1}{4}\right)$, as shown in Figure 9.13. You can find the y-intercept $(0, b)$ of the tangent line by equating the lengths of the two sides of the isosceles triangle shown in Figure 9.13:

$$d_1 = \frac{1}{4} - b$$

and

$$d_2 = \sqrt{(1 - 0)^2 + \left(1 - \frac{1}{4}\right)^2} = \frac{5}{4}.$$

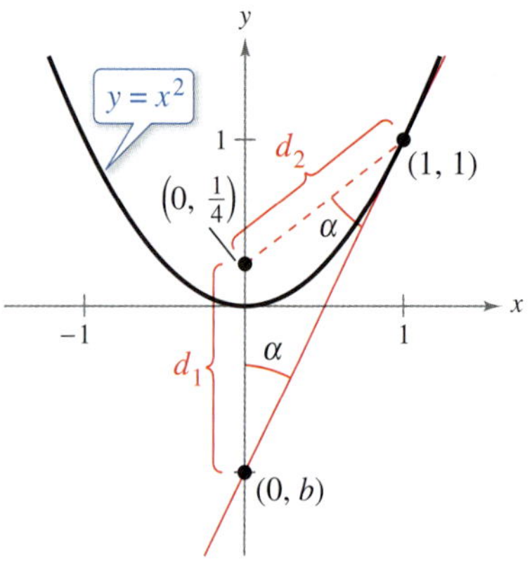

Figure 9.13

Note that $d_1 = \frac{1}{4} - b$ rather than $b - \frac{1}{4}$. The order of subtraction for the distance is important because the distance must be positive. Setting $d_1 = d_2$ produces

$$\frac{1}{4} - b = \frac{5}{4}$$

$$b = -1.$$

So, the slope of the tangent line is

$$m = \frac{1 - (-1)}{1 - 0} = 2$$

and the equation of the tangent line in slope-intercept form is

$$y = 2x - 1.$$

Checkpoint *Audio-video solution in English & Spanish at LarsonPrecalculus.com*

Find an equation of the tangent line to the parabola $y = 3x^2$ at the point $(1, 3)$.

Technology Tip

Use a graphing utility to confirm the result of Example 7. By graphing $y_1 = x^2$ and $y_2 = 2x - 1$ in the same viewing window, you should be able to see that the line touches the parabola at the point $(1, 1)$.

9.1 Exercises

See *CalcChat.com* for tutorial help and worked-out solutions to odd-numbered exercises.
For instructions on how to use a graphing utility, see Appendix A.

Vocabulary and Concept Check

In Exercises 1–4, fill in the blank(s).

1. A _______ is the intersection of a plane and a double-napped cone.

2. A conic can be defined as a _______ (collection) of points satisfying a certain geometric property.

3. A _______ is the set of all points (x, y) in a plane that are equidistant from a fixed point, called the _______ .

4. A _______ is the set of all points (x, y) in a plane that are equidistant from a fixed line, the _______ , and a fixed point, the _______ , not on the line.

Procedures and Problem Solving

Finding the Standard Equation of a Circle **In Exercises 5–10, write the standard form of the equation of the circle with the given characteristics.**

5. Center at origin; radius: 6

6. Center at origin; radius: $\sqrt{5}$

7. Center: $(3, 7)$; point on circle: $(1, 0)$

8. Center: $(6, -3)$; point on circle: $(-2, 4)$

9. Endpoints of a diameter: $(3, 2)$ and $(-9, -8)$

10. Endpoints of a diameter: $(1, -7)$ and $(9, -5)$

Identifying the Center and Radius of a Circle **In Exercises 11–16, identify the center and radius of the circle.**

11. $x^2 + y^2 = 16$

12. $x^2 + y^2 = 49$

13. $(x - 5)^2 + y^2 = 9$

14. $x^2 + (y + 8)^2 = 25$

15. $(x + 1)^2 + (y + 6)^2 = 19$

16. $(x + 7)^2 + (y - 3)^2 = 32$

Writing the Equation of a Circle in Standard Form **In Exercises 17–24, write the equation of the circle in standard form. Then identify its center and radius.**

17. $\frac{1}{4}x^2 + \frac{1}{4}y^2 = 1$

18. $\frac{1}{9}x^2 + \frac{1}{9}y^2 = 1$

19. $\frac{9}{5}x^2 + \frac{9}{5}y^2 = 1$

20. $\frac{25}{2}x^2 + \frac{25}{2}y^2 = 1$

21. $x^2 + y^2 - 2x + 6y + 9 = 0$

22. $x^2 + y^2 - 10x - 6y + 25 = 0$

23. $4x^2 + 4y^2 + 12x - 24y + 41 = 0$

24. $9x^2 + 9y^2 + 54x - 36y + 17 = 0$

Sketching a Circle **In Exercises 25–32, sketch the circle by hand. Identify its center and radius. Use a graphing utility to verify your graph.**

25. $y^2 = 1 - x^2$

26. $x^2 = 100 - y^2$

27. $x^2 + 8x + y^2 + 2y + 8 = 0$

28. $x^2 - 6x + y^2 + 6y + 14 = 0$

29. $x^2 - 14x + y^2 + 8y + 40 = 0$

30. $x^2 + 6x + y^2 - 12y + 41 = 0$

31. $x^2 + 2x + y^2 - 35 = 0$

32. $x^2 + y^2 + 10y + 9 = 0$

Finding the Intercepts of a Circle **In Exercises 33–38, find the x- and y-intercepts of the graph of the circle.**

33. $(x + 5)^2 + (y - 3)^2 = 25$

34. $(x - 1)^2 + (y + 4)^2 = 16$

35. $(x - 6)^2 + (y + 3)^2 = 16$

36. $(x + 7)^2 + (y - 8)^2 = 4$

37. $x^2 - 2x + y^2 - 6y - 27 = 0$

38. $x^2 + 8x + y^2 + 2y + 9 = 0$

Matching an Equation with a Graph **In Exercises 39–42, match the equation with its graph. [The graphs are labeled (a), (b), (c), and (d).]**

(a)

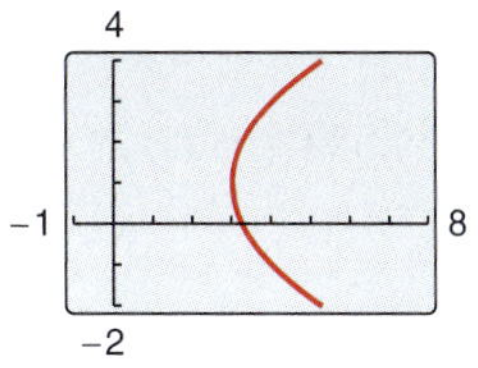

(b)

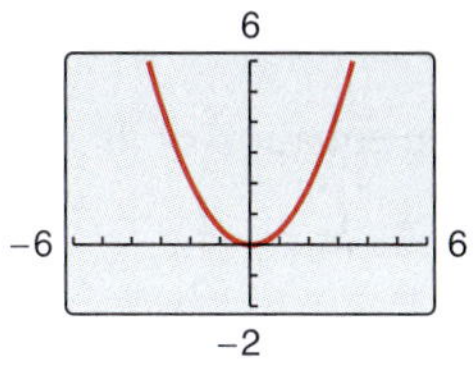

(c)

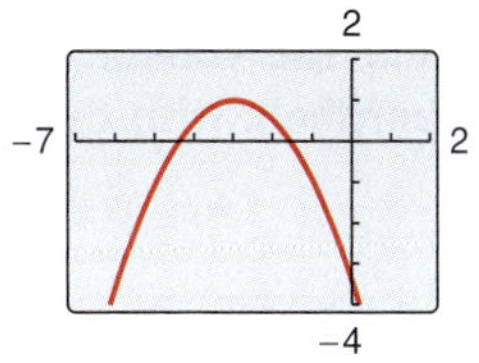

(d)

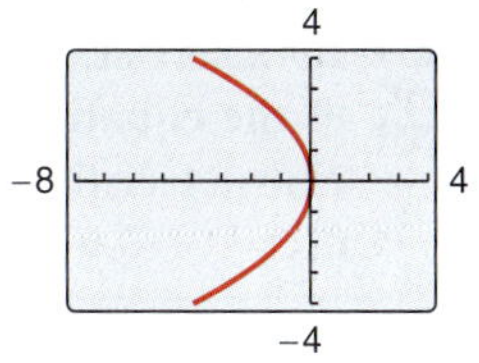

39. $y^2 = -4x$

40. $x^2 = 2y$

41. $(y - 1)^2 = 4(x - 3)$

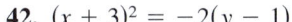

42. $(x + 3)^2 = -2(y - 1)$

Finding the Standard Equation of a Parabola **In Exercises 43–54, find the standard form of the equation of the parabola with the given characteristic(s) and vertex at the origin.**

43.
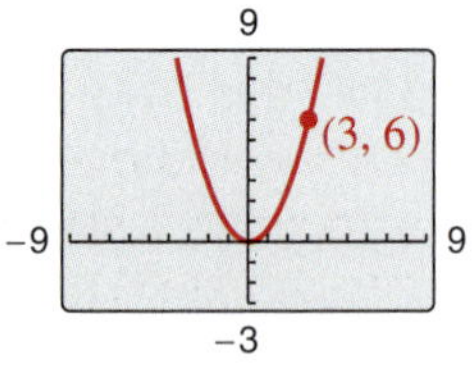

44.
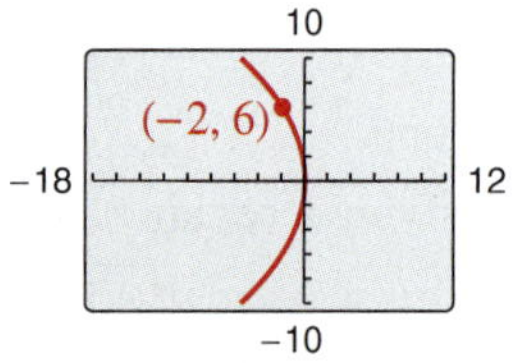

45. Focus: $(0, 2)$

46. Focus: $(0, 1)$

47. Focus: $\left(-\frac{1}{2}, 0\right)$

48. Focus: $\left(-\frac{1}{4}, 0\right)$

49. Directrix: $y = 2$

50. Directrix: $y = -1$

51. Directrix: $x = -2$

52. Directrix: $x = 5$

53. Horizontal axis and passes through the point $(2, 4)$

54. Vertical axis and passes through the point $(-4, 3)$

Finding the Vertex, Focus, and Directrix of a Parabola **In Exercises 55–72, find the vertex, focus, and directrix of the parabola, and sketch its graph. Use a graphing utility to verify your graph.**

55. $y = \frac{1}{2}x^2$

56. $y = -2x^2$

57. $y^2 = -6x$

58. $y^2 = 3x$

59. $x^2 + 6y = 0$

60. $x + y^2 = 0$

61. $\left(x + \frac{3}{2}\right)^2 = 4(y - 2)$

62. $\left(x + \frac{1}{2}\right)^2 = 4(y - 1)$

63. $(y + 7)^2 = 4\left(x - \frac{3}{2}\right)$

64. $(y - 1)^2 = -(x + 5)$

65. $y^2 + 6y + 8x + 25 = 0$

66. $y^2 - 4y - 4x = 0$

67. $x^2 + 4x + 6y - 2 = 0$

68. $x^2 - 2x + 8y + 9 = 0$

69. $y^2 + x + y = 0$

70. $y^2 - 4x - 4 = 0$

71. $y = \frac{1}{4}(x^2 - 2x + 5)$

72. $x = \frac{1}{4}(y^2 + 2y + 33)$

Finding the Standard Equation of a Parabola **In Exercises 73–84, find the standard form of the equation of the parabola with the given characteristics.**

73.
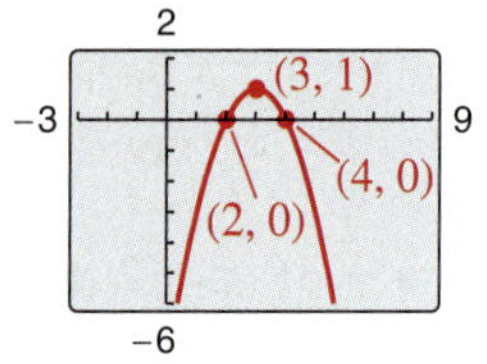

74.
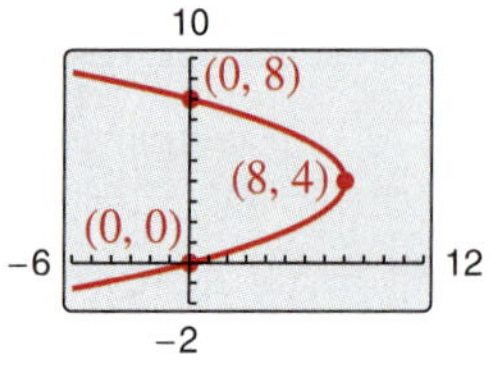

75.
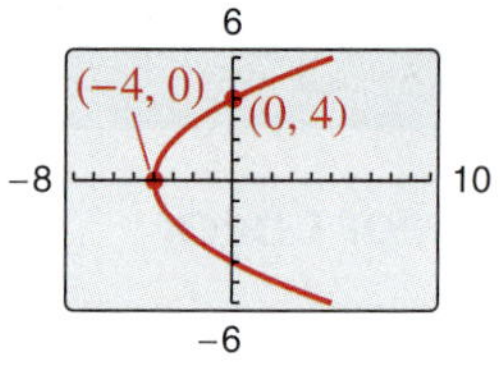

76.
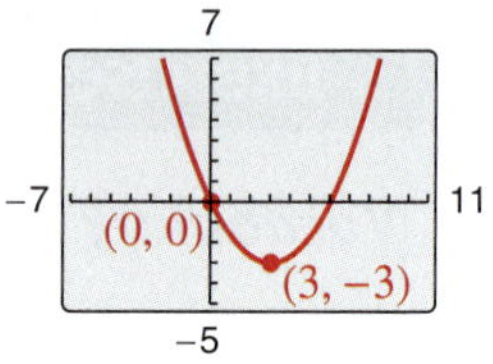

77. Vertex: $(-2, 0)$; focus: $\left(-\frac{3}{2}, 0\right)$

78. Vertex: $(3, -3)$; focus: $\left(3, -\frac{9}{4}\right)$

79. Vertex: $(6, 2)$; focus: $(4, 2)$

80. Vertex: $(-1, 2)$; focus: $(-1, 0)$

81. Vertex: $(0, 5)$; directrix: $y = 3$

82. Vertex: $(-2, 1)$; directrix: $x = 1$

83. Focus: $(2, 2)$; directrix: $x = -2$

84. Focus: $(0, 0)$; directrix: $y = 8$

Finding the Tangent Line at a Point on a Parabola **In Exercises 85–88, find an equation of the tangent line to the parabola at the given point.**

85. $x^2 = 2y,\ (4, 8)$

86. $x^2 = 2y,\ \left(-3, \frac{9}{2}\right)$

87. $y = -2x^2,\ (-1, -2)$

88. $y = -2x^2,\ (2, -8)$

Determining the Point of Tangency **In Exercises 89 and 90, the equations of a parabola and a tangent line to the parabola are given. Use a graphing utility to graph both equations in the same viewing window. Determine the coordinates of the point of tangency.**

	Parabola	*Tangent Line*
89.	$y^2 - 8x = 0$	$x - y + 2 = 0$
90.	$x^2 + 12y = 0$	$x + y - 3 = 0$

91. Seismology An earthquake occurs that is felt up to 52 miles from its epicenter. You are located 40 miles west and 30 miles south of the epicenter.

(a) Let the epicenter be at the point $(0, 0)$. Find the standard equation that describes the outer boundary of the earthquake.

(b) Should you be able to feel the earthquake?

(c) Verify your answer to part (b) by graphing the equation of the outer boundary of the earthquake and plotting your location. How far are you from the outer boundary of the earthquake?

92. Landscape Design A landscaper turns on a sprinkler that covers a circular area of 2000 square feet.

(a) Find the radius of the region covered by the sprinkler. Round your answer to three decimal places.

(b) The landscaper increases the area covered to 2500 square feet by increasing the water pressure. How much longer is the radius?

93. Window Design A church window is bounded above by a parabola (see figure). Find an equation of the parabola.

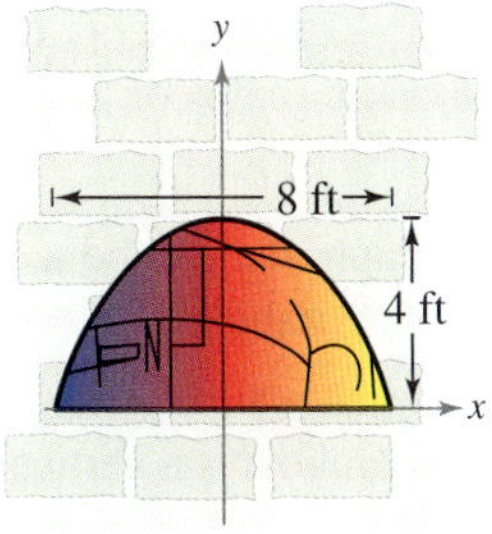

Figure for 93

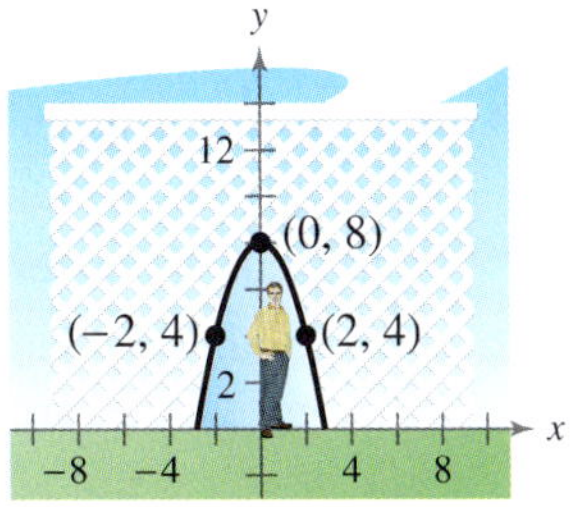

Figure for 94

94. Lattice Arch A parabolic lattice arch is 8 feet high at the vertex. At a height of 4 feet, the width of the lattice arch is 4 feet (see figure). How wide is the lattice arch at ground level?

95. *Why you should learn it* *(p. 646)* Road engineers design a parabolic entrance ramp from a straight street to an interstate highway (see figure). Find an equation of the parabola.

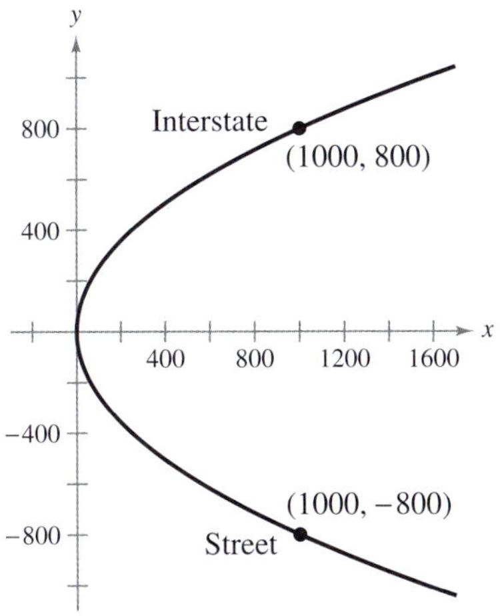

96. Environmental Science Water flows from a horizontal pipe 48 feet above the ocean surface. The stream of water follows the path of a parabola whose vertex (0, 48) is at the end of the pipe (see figure). The stream of water strikes the ocean at the point $(10\sqrt{3}, 0)$. Find an equation of the path taken by the water.

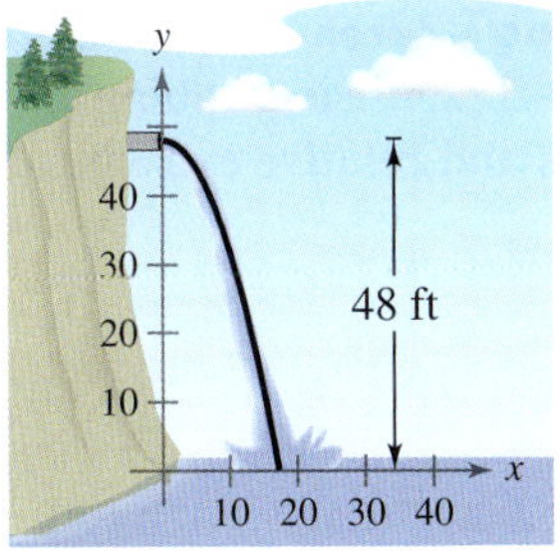

Neustockimages/E+/Getty Images

97. MODELING DATA

A cable of the Golden Gate Bridge is suspended (in the shape of a parabola) between two towers that are 1280 meters apart. The top of each tower is 152 meters above the roadway. The cable touches the roadway midway between the towers.

(a) Draw a sketch of the cable. Locate the origin of a rectangular coordinate system at the center of the roadway. Label the coordinates of the known points.

(b) Write an equation that models the cable.

(c) Complete the table by finding the height y of the suspension cable over the roadway at a distance of x meters from the center of the bridge.

x	0	200	400	500	600
y					

98. Transportation Design Roads are often designed with parabolic surfaces to allow rain water to drain off. The road shown is 32 feet wide and is 0.4 foot higher in the center than it is on the sides. (See figure.)

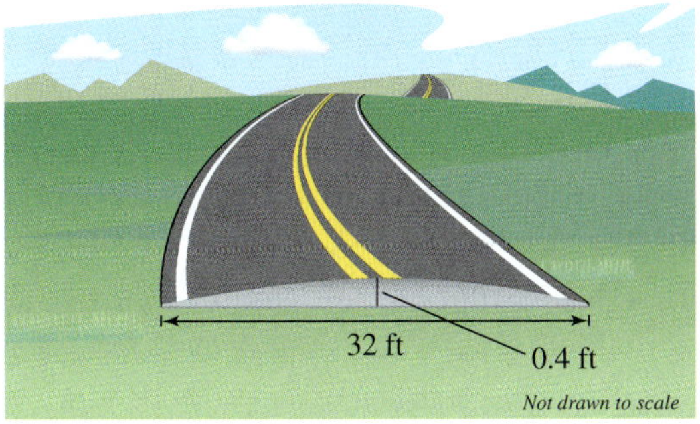

(a) Find an equation of the parabola with its vertex at the origin that models the road surface.

(b) How far from the center of the road is the road surface 0.1 foot lower than in the middle?

99. Mechanical Engineering The filament of an automobile headlight is at the focus of a parabolic reflector, which sends light out in a straight beam. (See figure.)

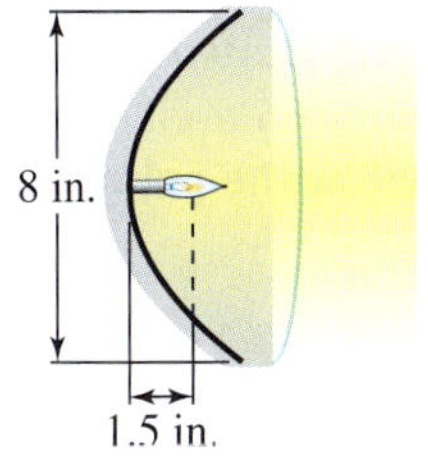

(a) The filament of the headlight is 1.5 inches from the vertex. Find an equation for the cross section of the reflector.

(b) The reflector is 8 inches wide. Find the depth of the reflector.

100. Astronomy A satellite in a 100-mile-high circular orbit around Earth has a velocity of approximately 17,500 miles per hour. When this velocity is multiplied by $\sqrt{2}$, the satellite has the minimum velocity necessary to escape Earth's gravity and follows a parabolic path with the center of Earth as the focus. (See figure.)

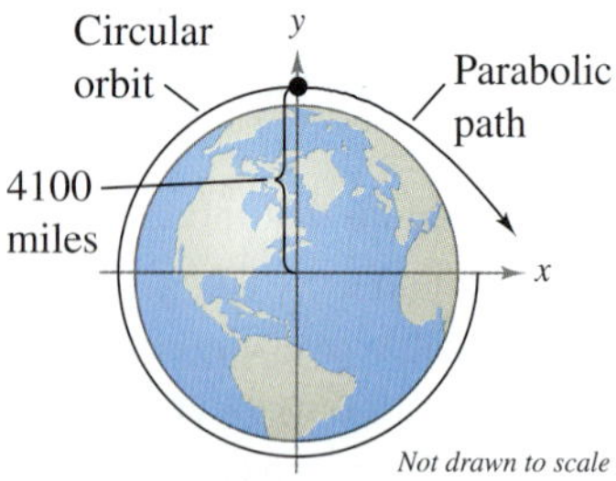

(a) Find the escape velocity of the satellite.

(b) Find an equation of its path (assume the radius of Earth is 4000 miles).

Projectile Motion **In Exercises 101 and 102, consider the path of a projectile projected horizontally with a velocity of v feet per second at a height of s feet, where the model for the path is**

$$x^2 = -\frac{v^2}{16}(y - s).$$

In this model (in which air resistance is disregarded), y is the height (in feet) of the projectile and x is the horizontal distance (in feet) the projectile travels.

101. A ball is thrown horizontally from the top of a 100-foot tower with a velocity of 28 feet per second.

(a) Find the equation of the parabolic path.

(b) How far does the ball travel horizontally before striking the ground?

102. A cargo plane is flying at an altitude of 30,000 feet and a speed of 540 miles per hour. A supply crate is dropped from the plane. How many *feet* will the crate travel horizontally before it hits the ground?

Finding the Tangent Line at a Point on a Circle **In Exercises 103–108, find an equation of the tangent line to the circle at the indicated point. Recall from geometry that the tangent line to a circle is perpendicular to the radius of the circle at the point of tangency.**

	Circle	*Point*
103.	$x^2 + y^2 = 25$	$(3, -4)$
104.	$x^2 + y^2 = 169$	$(-5, 12)$
105.	$x^2 + y^2 = 12$	$(2, -2\sqrt{2})$
106.	$x^2 + y^2 = 24$	$(-2\sqrt{5}, 2)$
107.	$(x + 2)^2 + (y - 1)^2 = 100$	$(4, 9)$
108.	$x^2 + (y - 3)^2 = 16$	$(\sqrt{7}, 0)$

Focusing on Concepts

True or False? **In Exercises 109–112, determine whether the statement is true or false. Justify your answer.**

109. The equation $x^2 + (y + 5)^2 = 25$ represents a circle with its center at the origin and a radius of 5.

110. A circle is a degenerate conic.

111. The point which lies on the graph of a parabola closest to its focus is the vertex of the parabola.

112. If the vertex and focus of a parabola are on a horizontal line, then the directrix of the parabola is a vertical line.

Error Analysis **In Exercises 113 and 114, describe the error in finding the standard form of the equation of the circle or parabola with the given characteristics.**

113. Circle; center: (0, 3), diameter: 4

$x^2 + (y - 3)^2 = 16$ ✗

114. Parabola; vertex: (0, 2), directrix: $y = 3$

$x^2 = 12(y - 2)$ ✗

115. Think About It The graph of $x^2 + y^2 = 0$ is a degenerate conic. Sketch the graph and identify the degenerate conic. Describe the intersection of the plane with the double-napped cone for this conic.

116. HOW DO YOU SEE IT? In parts (a)–(d), describe in words how a plane could intersect the double-napped cone to form the conic section (see figure).

(a) Circle

(b) Ellipse

(c) Parabola

(d) Hyperbola

Think About It **In Exercises 117 and 118, change the equation so that its graph matches the description.**

117. $(y - 3)^2 = 6(x + 1)$; upper half of parabola

118. $(y + 1)^2 = 2(x - 2)$; lower half of parabola

Cumulative Mixed Review

Approximating Zeros and Relative Extrema **In Exercises 119–122, use a graphing utility to approximate any real zeros and relative extrema of the function.**

119. $f(x) = 3x^3 - 4x + 2$

120. $f(x) = 2x^2 + 3x$

121. $f(x) = x^4 + 2x + 2$

122. $f(x) = x^5 - 3x - 1$

9.2 Ellipses

Introduction

The third type of conic is called an **ellipse.**

Definition of an Ellipse

An **ellipse** is the set of all points (x, y) in a plane, the sum of whose distances from two distinct fixed points **(foci)** is constant. (See Figure 9.14.)

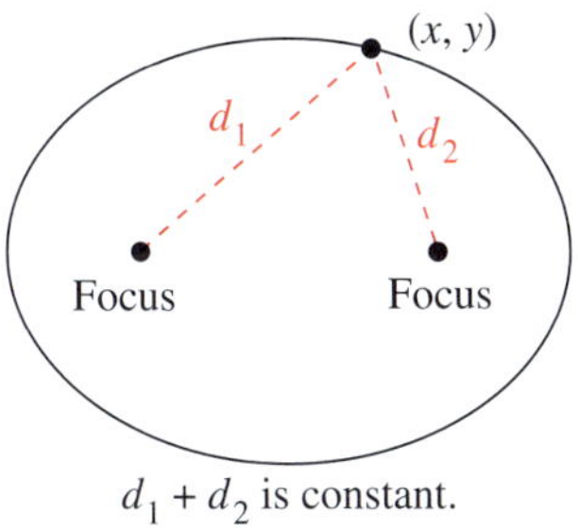

Figure 9.14

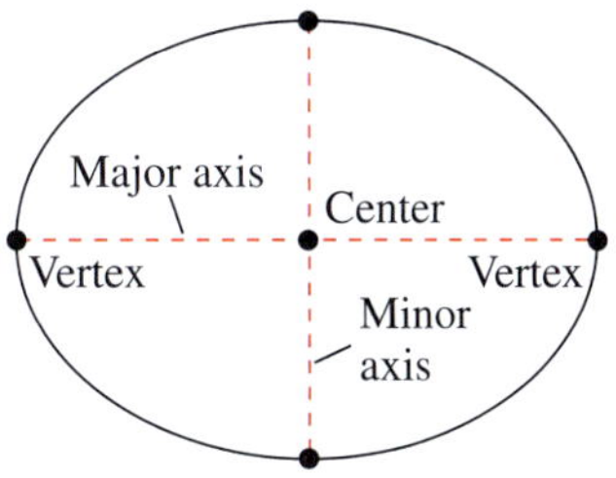

Figure 9.15

The line through the foci intersects the ellipse at two points called **vertices.** The chord joining the vertices is the **major axis,** and its midpoint is the **center** of the ellipse. The chord perpendicular to the major axis at the center is the **minor axis.** (See Figure 9.15.)

You can visualize the definition of an ellipse by imagining two thumbtacks placed at the foci, as shown in Figure 9.16. If the ends of a fixed length of string are fastened to the thumbtacks and the string is drawn taut with a pencil, then the path traced by the pencil will be an ellipse.

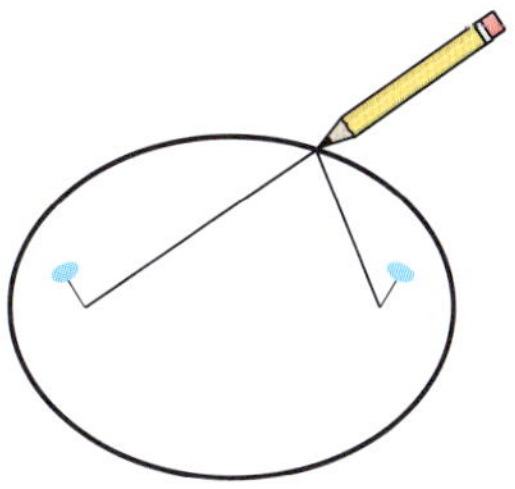

Figure 9.16

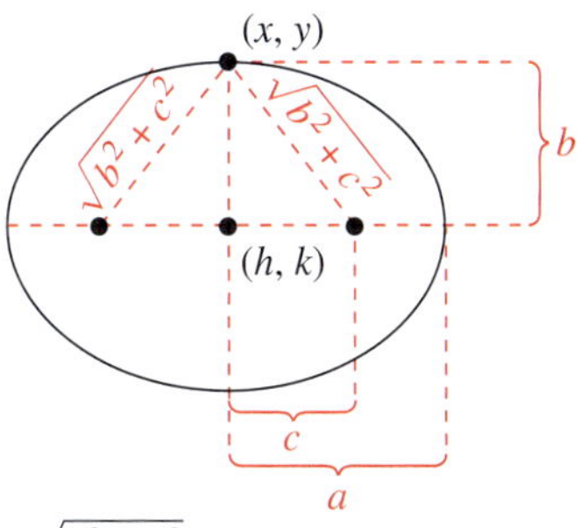

$$2\sqrt{b^2 + c^2} = 2a$$
$$b^2 + c^2 = a^2$$

Figure 9.17

To derive the standard form of the equation of an ellipse, consider the ellipse in Figure 9.17 with the following points.

Center: (h, k) Vertices: $(h \pm a, k)$ Foci: $(h \pm c, k)$

Note that the center is the midpoint of the segment joining the foci.

The sum of the distances from any point on the ellipse to the two foci is constant. Using a vertex point, this constant sum is

$$(a + c) + (a - c) = 2a \quad \text{Length of major axis}$$

or simply the length of the major axis.

What you should learn

- Write equations of ellipses in standard form.
- Use properties of ellipses to model and solve real-life problems.
- Find eccentricities of ellipses.

Why you should learn it

Ellipses can be used to model and solve many types of real-life problems. For instance, Exercise 57 on page 664 shows how the focal properties of an ellipse are used by a lithotripter to break up kidney stones.

Urologist

Now, if you let (x, y) be *any* point on the ellipse, then the sum of the distances between (x, y) and the two foci must also be $2a$. That is,

$$\sqrt{[x-(h-c)]^2+(y-k)^2}+\sqrt{[x-(h+c)]^2+(y-k)^2}=2a$$

which, after expanding and regrouping, reduces to

$$(a^2-c^2)(x-h)^2+a^2(y-k)^2=a^2(a^2-c^2).$$

Finally, in Figure 9.17, you can see that

$$b^2=a^2-c^2$$

which implies that the equation of the ellipse is

$$b^2(x-h)^2+a^2(y-k)^2=a^2b^2$$

$$\frac{(x-h)^2}{a^2}+\frac{(y-k)^2}{b^2}=1.$$

You would obtain a similar equation in the derivation by starting with a vertical major axis. Both results are summarized as follows.

Standard Equation of an Ellipse

The **standard form of the equation of an ellipse** with center (h, k) and major and minor axes of lengths $2a$ and $2b$, respectively, where $0 < b < a$, is

$$\frac{(x-h)^2}{a^2}+\frac{(y-k)^2}{b^2}=1$$ Major axis is horizontal.

or

$$\frac{(x-h)^2}{b^2}+\frac{(y-k)^2}{a^2}=1.$$ Major axis is vertical.

The foci lie on the major axis, c units from the center, with

$$c^2=a^2-b^2.$$

If the center is at the origin $(0, 0)$, then the equation takes one of the following forms.

$$\frac{x^2}{a^2}+\frac{y^2}{b^2}=1$$ Major axis is horizontal.

$$\frac{x^2}{b^2}+\frac{y^2}{a^2}=1$$ Major axis is vertical.

Explore the Concept

On page 657, it was noted that an ellipse can be drawn using two thumbtacks, a string of fixed length (greater than the distance between the two tacks), and a pencil. Try doing this. Vary the length of the string and the distance between the thumbtacks. Explain how to obtain ellipses that are almost circular. Explain how to obtain ellipses that are long and narrow.

Figure 9.18 shows both the vertical and horizontal orientations for an ellipse.

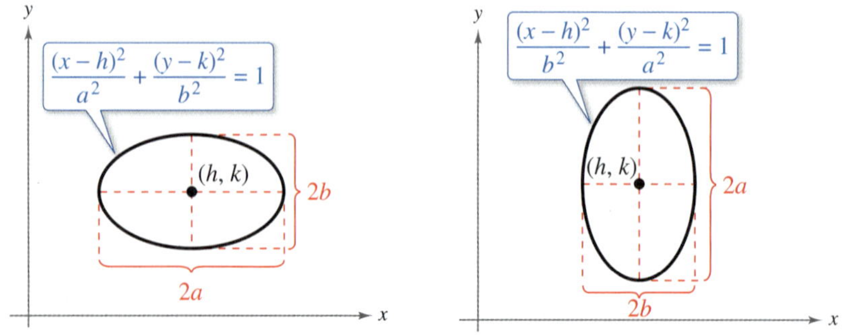

Major axis is horizontal. *Major axis is vertical.*
Figure 9.18

EXAMPLE 1 Finding the Standard Equation of an Ellipse

Find the standard form of the equation of the ellipse having foci at

$(0, 1)$ and $(4, 1)$

and a major axis of length 6, as shown in Figure 9.19.

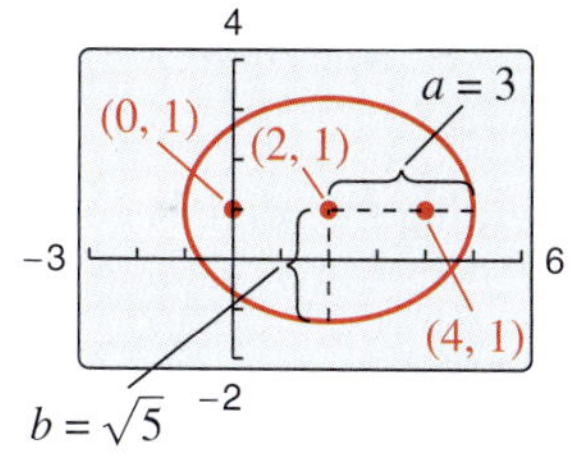

Figure 9.19

Solution

The center of the ellipse is the midpoint of the line segment joining the foci. So, by the Midpoint Formula, the center of the ellipse is $(2, 1)$ and the distance from the center to one of the foci is $c = 2$. Because $2a = 6$, you know that $a = 3$. Now, from $c^2 = a^2 - b^2$, you have

$$b = \sqrt{a^2 - c^2} = \sqrt{9 - 4} = \sqrt{5}.$$

Because the major axis is horizontal, the standard equation is

$$\frac{(x-2)^2}{3^2} + \frac{(y-1)^2}{(\sqrt{5})^2} = 1 \quad \text{or} \quad \frac{(x-2)^2}{9} + \frac{(y-1)^2}{5} = 1.$$

✓ **Checkpoint** ▶ *Audio-video solution in English & Spanish at LarsonPrecalculus.com*

Find the standard form of the equation of the ellipse having foci $(2, 0)$ and $(2, 6)$ and a major axis of length 8.

EXAMPLE 2 Sketching an Ellipse

Sketch the ellipse given by $4x^2 + y^2 = 36$ and identify the center and vertices.

Algebraic Solution

$4x^2 + y^2 = 36$ Write original equation.

$\frac{4x^2}{36} + \frac{y^2}{36} = \frac{36}{36}$ Divide each side by 36.

$\frac{x^2}{3^2} + \frac{y^2}{6^2} = 1$ Write in standard form.

The center of the ellipse is $(0, 0)$. Because the denominator of the y^2-term is greater than the denominator of the x^2-term, you can conclude that the major axis is vertical. Moreover, because $a = 6$, the vertices are $(0, -6)$ and $(0, 6)$. Finally, because $b = 3$, the endpoints of the minor axis are $(-3, 0)$ and $(3, 0)$, as shown in the figure.

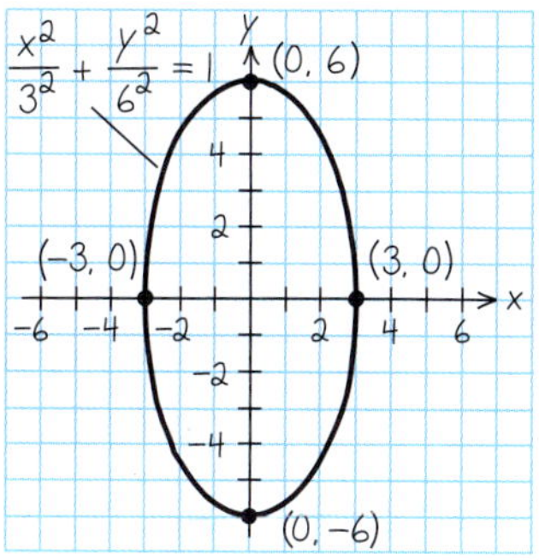

Graphical Solution

Solve the equation of the ellipse for y as follows.

$$4x^2 + y^2 = 36$$

$$y^2 = 36 - 4x^2$$

$$y = \pm\sqrt{36 - 4x^2}$$

Then use a graphing utility to graph

$$y_1 = \sqrt{36 - 4x^2}$$

and

$$y_2 = -\sqrt{36 - 4x^2}$$

in the same viewing window, as shown in the figure. Be sure to use a square setting.

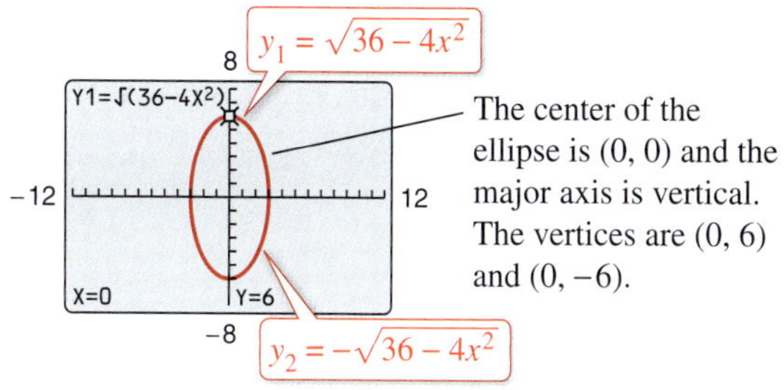

✓ **Checkpoint** ▶ *Audio-video solution in English & Spanish at LarsonPrecalculus.com*

Sketch the ellipse given by $x^2 + 4y^2 = 16$ and identify the center and vertices.

EXAMPLE 3 Graphing an Ellipse

Graph the ellipse given by $x^2 + 4y^2 + 6x - 8y + 9 = 0$.

Solution

Begin by writing the original equation in standard form.

$$x^2 + 4y^2 + 6x - 8y + 9 = 0$$ Write original equation.

$$(x^2 + 6x + \square) + 4(y^2 - 2y + \square) = -9$$ Group terms and factor 4 out of y-terms.

$$(x^2 + 6x + 9) + 4(y^2 - 2y + 1) = -9 + 9 + 4(1)$$ Complete the squares.

$$(x + 3)^2 + 4(y - 1)^2 = 4$$ Write in completed square form.

$$\frac{(x + 3)^2}{2^2} + \frac{(y - 1)^2}{1^2} = 1$$ Write in standard form.

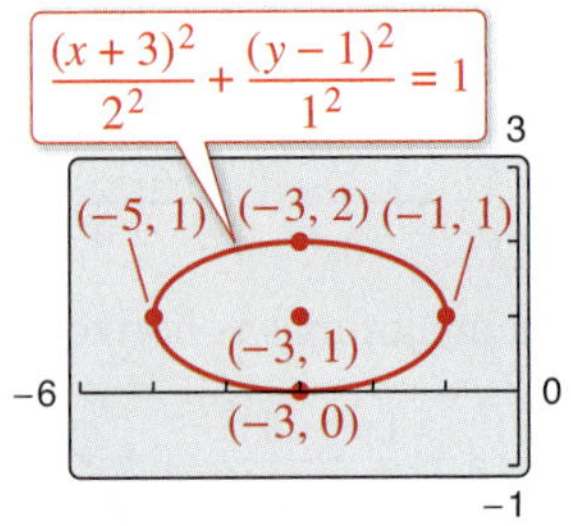

Figure 9.20

Now you see that the center is $(h, k) = (-3, 1)$. Because the denominator of the x-term is $a^2 = 2^2$, the endpoints of the major axis lie two units to the right and left of the center. Similarly, because the denominator of the y-term is $b^2 = 1^2$, the endpoints of the minor axis lie one unit up and down from the center. The ellipse is shown in Figure 9.20.

Checkpoint *Audio-video solution in English & Spanish at LarsonPrecalculus.com*

Graph the ellipse given by $9x^2 + 4y^2 + 36x - 8y + 4 = 0$.

EXAMPLE 4 Analyzing an Ellipse

See LarsonPrecalculus.com for an interactive version of this type of example.

Find the center, vertices, and foci of the ellipse $4x^2 + y^2 - 8x + 4y - 8 = 0$.

Solution

By completing the square, you can write the original equation in standard form.

$$4x^2 + y^2 - 8x + 4y - 8 = 0$$ Write original equation.

$$4(x^2 - 2x + \square) + (y^2 + 4y + \square) = 8$$ Group terms and factor 4 out of x-terms.

$$4(x^2 - 2x + 1) + (y^2 + 4y + 4) = 8 + 4(1) + 4$$ Complete the squares.

$$4(x - 1)^2 + (y + 2)^2 = 16$$ Write in completed square form.

$$\frac{(x - 1)^2}{2^2} + \frac{(y + 2)^2}{4^2} = 1$$ Write in standard form.

So, the major axis is vertical, where $h = 1$, $k = -2$, $a = 4$, $b = 2$, and

$$c = \sqrt{a^2 - b^2} = \sqrt{16 - 4} = \sqrt{12} = 2\sqrt{3}.$$

So, you have the following.

Center: $(1, -2)$

Vertices: $(1, -6)$

$(1, 2)$

Foci: $(1, -2 - 2\sqrt{3})$

$(1, -2 + 2\sqrt{3})$

The ellipse is shown in Figure 9.21.

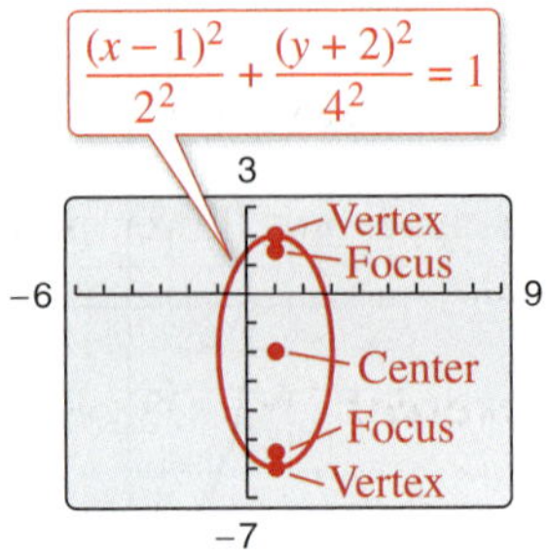

Figure 9.21

Checkpoint *Audio-video solution in English & Spanish at LarsonPrecalculus.com*

Find the center, vertices, and foci of the ellipse $5x^2 + 9y^2 + 10x - 54y + 41 = 0$.

Application

Ellipses have many practical and aesthetic uses. For instance, machine gears, supporting arches, and acoustic designs often involve elliptical shapes. The orbits of satellites and planets are also ellipses. Example 5 investigates the elliptical orbit of the moon about Earth.

EXAMPLE 5 An Application Involving an Elliptical Orbit

The moon travels about Earth in an elliptical orbit with Earth at one focus, as shown in the figure. The major and minor axes of the orbit have lengths of 768,800 kilometers and 767,641 kilometers, respectively. Find the greatest and least distances (the *apogee* and *perigee*) from Earth's center to the moon's center. Then graph the orbit of the moon on a graphing utility.

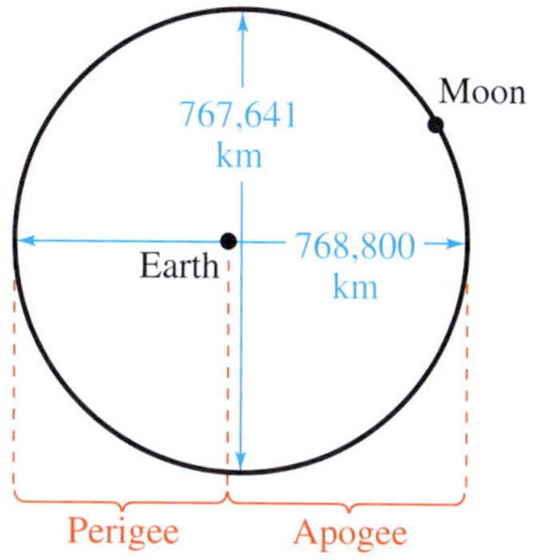

Algebra Help

Note in Example 5 and the figure that Earth *is not* the center of the moon's orbit.

Astronaut

Solution

Because $2a = 768{,}800$ and $2b = 767{,}641$, you have

$$a = 384{,}400 \quad \text{and} \quad b = 383{,}820.5$$

which implies that

$$c = \sqrt{a^2 - b^2} = \sqrt{384{,}400^2 - 383{,}820.5^2} \approx 21{,}099.$$

So, the greatest distance between the center of Earth and the center of the moon is

$$a + c \approx 384{,}400 + 21{,}099 = 405{,}499 \text{ kilometers}$$

and the least distance is

$$a - c \approx 384{,}400 - 21{,}099 = 363{,}301 \text{ kilometers}.$$

To graph the orbit of the moon on a graphing utility, first let $a = 384{,}400$ and $b = 383{,}820.5$ in the standard form of an equation of an ellipse centered at the origin, and then solve for y.

$$\frac{x^2}{384{,}400^2} + \frac{y^2}{383{,}820.5^2} = 1 \quad \Longrightarrow \quad y = \pm 383{,}820.5\sqrt{1 - \frac{x^2}{384{,}400^2}}$$

Graph the upper and lower portions in the same viewing window, as shown in Figure 9.22.

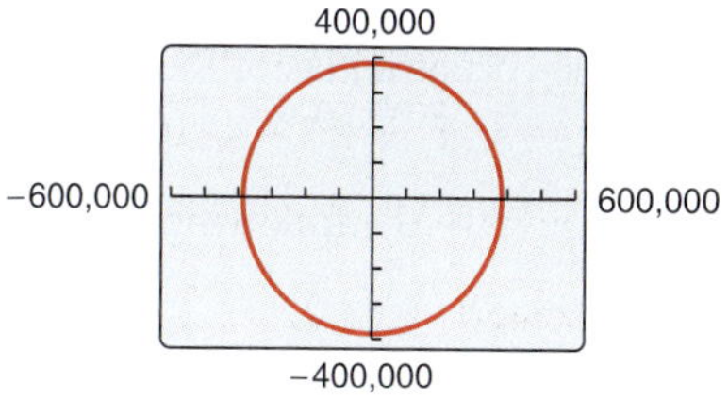

Figure 9.22

Checkpoint *Audio-video solution in English & Spanish at LarsonPrecalculus.com*

Encke's comet travels about the sun in an elliptical orbit with the sun at one focus. The major and minor axes of the orbit have lengths of approximately 4.430 astronomical units and 2.346 astronomical units, respectively. (An astronomical unit is about 93 million miles.) Find the greatest (*aphelion*) and least (*perihelion*) distances from the sun's center to the comet's center.

Eccentricity

One of the reasons it was difficult for early astronomers to detect that the orbits of the planets are ellipses is that the foci of the planetary orbits are relatively close to their centers, so the orbits are nearly circular. To measure the ovalness of an ellipse, you can use the concept of **eccentricity.**

Definition of Eccentricity

The **eccentricity** e of an ellipse is given by the ratio $e = \frac{c}{a}$.

Note that $0 < e < 1$ for *every* ellipse.

To see how this ratio is used to describe the shape of an ellipse, note that because the foci of an ellipse are located along the major axis between the vertices and the center, it follows that

$$0 < c < a.$$

For an ellipse that is nearly circular, the foci are close to the center and the ratio c/a is close to 0, as shown in Figure 9.23. On the other hand, for an elongated ellipse, the foci are close to the vertices and the ratio c/a is close to 1, as shown in Figure 9.24.

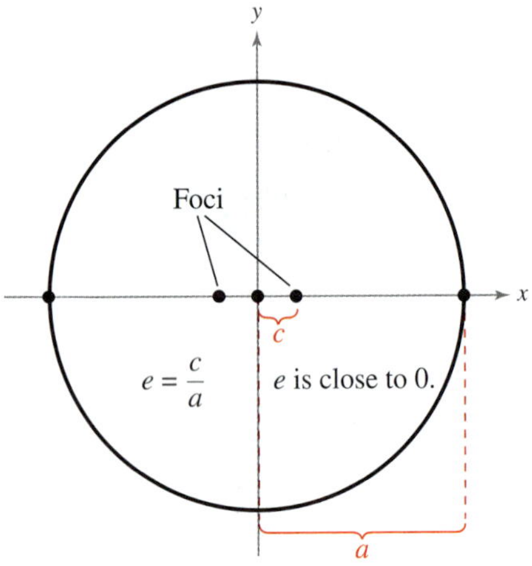

Figure 9.23

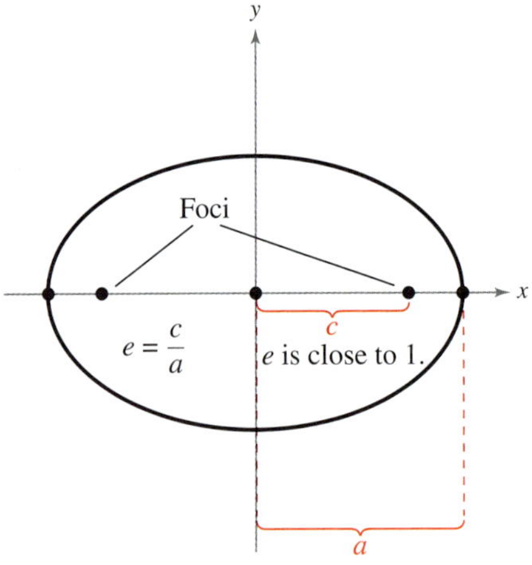

Figure 9.24

The orbit of the moon has an eccentricity of

$e \approx 0.0549$ Eccentricity of the moon

and the eccentricities of the eight planetary orbits are listed in the table.

Planet	Eccentricity, e
Mercury	0.2056
Venus	0.0067
Earth	0.0167
Mars	0.0935
Jupiter	0.0489
Saturn	0.0565
Uranus	0.0457
Neptune	0.0113

9.2 Exercises

See *CalcChat.com* for tutorial help and worked-out solutions to odd-numbered exercises.
For instructions on how to use a graphing utility, see Appendix A.

Vocabulary and Concept Check

In Exercises 1–4, fill in the blank(s).

1. An _______ is the set of all points (x, y) in a plane, the sum of whose distances from two distinct fixed points called ________ is constant.

2. The chord joining the vertices of an ellipse is called the _______ , and its midpoint is the _______ of the ellipse.

3. The chord perpendicular to the major axis at the center of an ellipse is called the _______ of the ellipse.

4. The eccentricity e of an ellipse is given by the ratio $e =$ ________.

In Exercises 5 and 6, consider the ellipse given by $\frac{x^2}{2^2} + \frac{y^2}{8^2} = 1$.

5. Is the major axis horizontal or vertical?

6. What is the length of the major axis?

Procedures and Problem Solving

Matching an Equation with a Graph **In Exercises 7–10, match the equation with its graph. [The graphs are labeled (a), (b), (c), and (d).]**

(a)
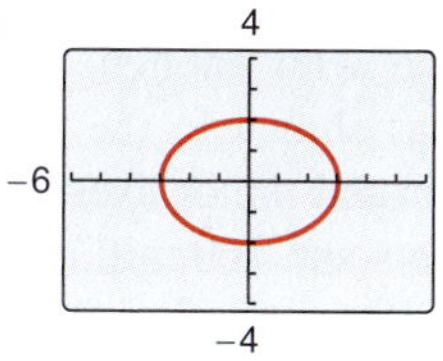

(b)
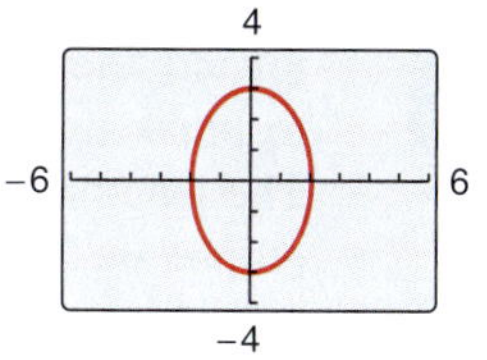

(c)
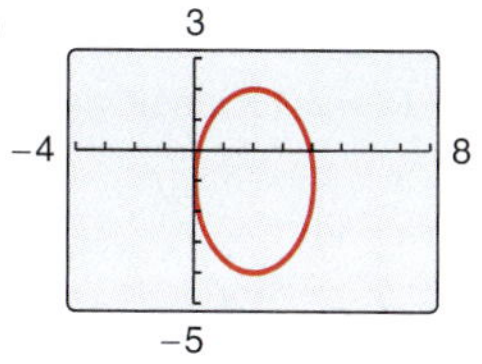

(d)
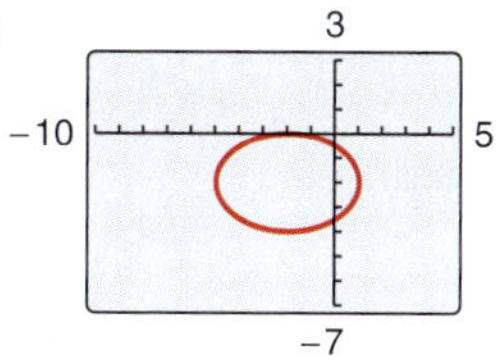

7. $\frac{x^2}{4} + \frac{y^2}{9} = 1$

8. $\frac{x^2}{9} + \frac{y^2}{4} = 1$

9. $\frac{(x-2)^2}{4} + \frac{(y+1)^2}{9} = 1$

10. $\frac{(x+2)^2}{9} + \frac{(y+2)^2}{4} = 1$

An Ellipse Centered at the Origin **In Exercises 11–18, find the standard form of the equation of the ellipse with the given characteristics and center at the origin.**

11.
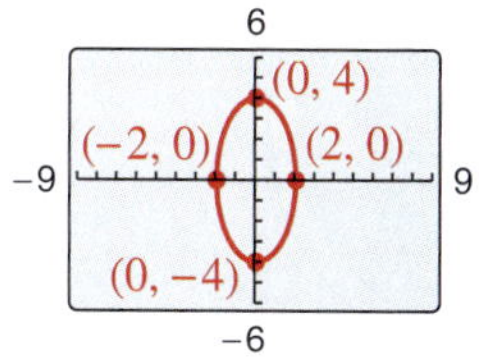

12.
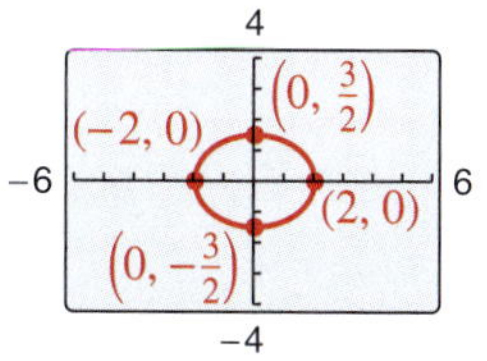

13. Vertices: $(\pm 7, 0)$; foci: $(\pm 2, 0)$

14. Vertices: $(0, \pm 8)$; foci: $(0, \pm 4)$

15. Foci: $(0, \pm 5)$; major axis of length 14

16. Foci: $(\pm 2, 0)$; major axis of length 10

17. Vertices: $(0, \pm 4)$; passes through the point $(3, 1)$

18. Vertices: $(\pm 8, 0)$; passes through the point $(5, -3)$

Finding the Standard Equation of an Ellipse **In Exercises 19–30, find the standard form of the equation of the ellipse with the given characteristics.**

19.
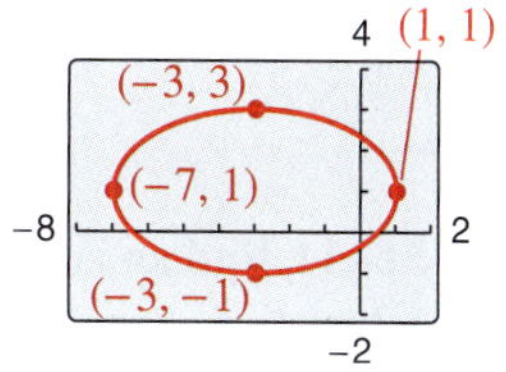

20.
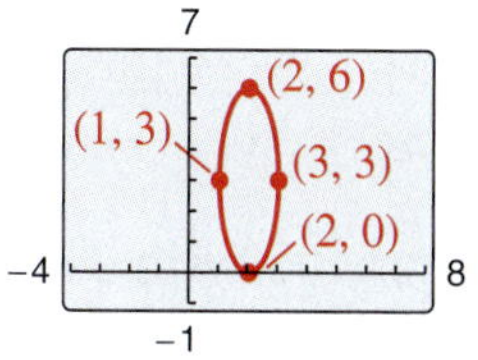

21. Vertices: $(0, 2)$, $(8, 2)$; minor axis of length 2

22. Vertices: $(3, 1)$, $(3, 9)$; minor axis of length 6

23. Foci: $(-1, 1)$, $(3, 1)$; major axis of length 10

24. Foci: $(0, 0)$, $(4, 0)$; major axis of length 6

25. Center: $(1, 3)$; vertex: $(-2, 3)$; major axis of length 4

26. Center: $(2, -1)$; vertex: $\left(2, \frac{1}{2}\right)$; major axis of length 2

27. Center: $(1, 4)$; $a = 2c$; vertices: $(1, 0)$, $(1, 8)$

28. Center: $(1, -5)$; $a = 4c$; vertices: $(-3, -5)$, $(5, -5)$

29. Center: $(3, 2)$; $a = 3c$; foci: $(1, 2)$, $(5, 2)$

30. Center: $(-2, 2)$; $a = 5c$; foci: $(-2, 0)$, $(-2, 4)$

Sketching an Ellipse **In Exercises 31–46, find the center, vertices, foci, and eccentricity of the ellipse, and sketch its graph. Use a graphing utility to verify your graph.**

31. $\dfrac{x^2}{64} + \dfrac{y^2}{9} = 1$

32. $\dfrac{x^2}{16} + \dfrac{y^2}{81} = 1$

33. $\dfrac{(x-4)^2}{16} + \dfrac{y^2}{25} = 1$

34. $\dfrac{x^2}{12} + \dfrac{(y-2)^2}{16} = 1$

35. $\dfrac{(x+5)^2}{9/4} + (y-1)^2 = 1$

36. $(x+2)^2 + \dfrac{(y+4)^2}{1/4} = 1$

37. $x^2 + 9y^2 = 36$

38. $16x^2 + y^2 = 16$

39. $49x^2 + 4y^2 - 196 = 0$

40. $4x^2 + 49y^2 - 196 = 0$

41. $9x^2 + 4y^2 + 36x - 24y + 36 = 0$

42. $9x^2 + 4y^2 - 54x + 40y + 37 = 0$

43. $6x^2 + 2y^2 + 18x - 10y + 2 = 0$

44. $x^2 + 4y^2 - 6x + 20y - 2 = 0$

45. $12x^2 + 20y^2 - 12x + 40y - 37 = 0$

46. $36x^2 + 9y^2 + 48x - 36y + 43 = 0$

Finding Eccentricity **In Exercises 47–50, find the eccentricity of the ellipse.**

47. $\dfrac{x^2}{4} + \dfrac{y^2}{9} = 1$

48. $\dfrac{x^2}{25} + \dfrac{y^2}{49} = 1$

49. $x^2 + 9y^2 - 10x + 36y + 52 = 0$

50. $4x^2 + 3y^2 - 8x + 18y + 19 = 0$

Using Eccentricity **In Exercises 51–54, find an equation of the ellipse with the given characteristics.**

51. Vertices: $(\pm 5, 0)$; eccentricity: $\frac{3}{5}$

52. Vertices: $(0, \pm 8)$; eccentricity: $\frac{1}{2}$

53. Foci: $(1, 1)$, $(1, 13)$; eccentricity: $\frac{2}{3}$

54. Foci: $(-6, 5)$, $(2, 5)$; eccentricity: $\frac{4}{5}$

55. **Astronomy** Halley's comet has an elliptical orbit with the sun at one focus. The eccentricity of the orbit is approximately 0.97. The length of the major axis of the orbit is about 35.67 astronomical units. (An astronomical unit is about 93 million miles.) Find the standard form of the equation of the orbit. Place the center of the orbit at the origin and place the major axis on the x-axis.

56. **Architecture** A fireplace arch is to be constructed in the shape of a semiellipse. The opening is to have a height of 2 feet at the center and a width of 6 feet along the base, as shown in the figure. The contractor draws the outline of the ellipse on the wall by the method discussed on page 657, using tacks, a string, and chalk.

(a) What are the required positions of the tacks?

(b) What is the length of the string?

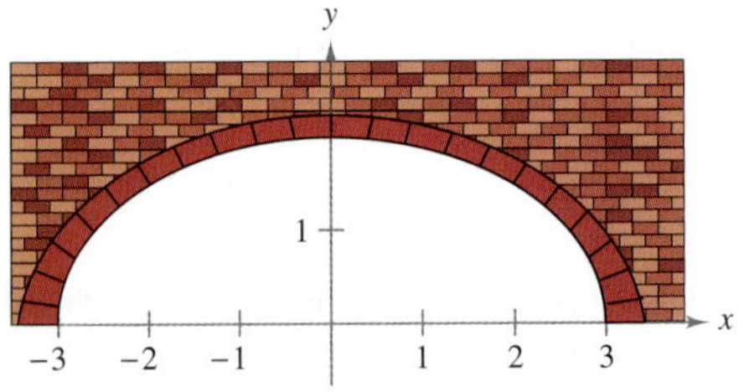

57. ***Why you should learn it*** *(p. 657)* A lithotripter is a machine that uses an elliptical reflector to break up kidney stones nonsurgically. A spark plug in the reflector generates energy waves at one focus of an ellipse. A kidney stone is positioned at the other focus of the ellipse. The reflector directs the energy waves toward the kidney stone with enough energy to break up the stone, as shown in the figure. The lengths of the major and minor axes of the ellipse are 280 millimeters and 160 millimeters, respectively. How far is the spark plug from the kidney stone?

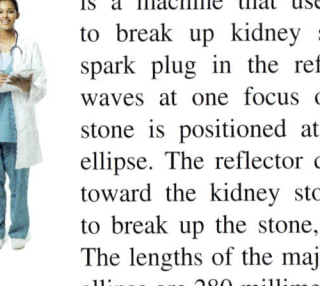

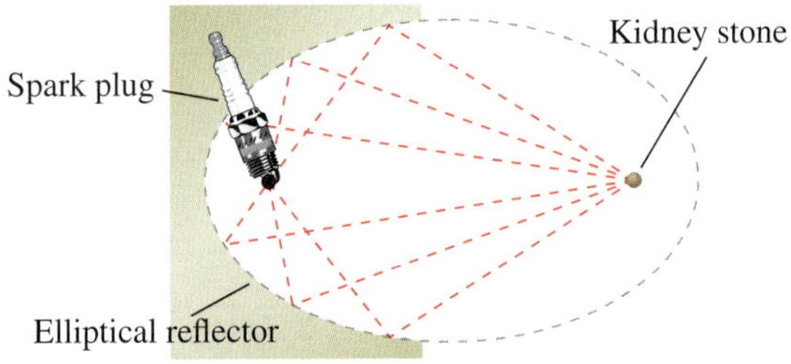

58. **Aeronautics** The first artificial satellite to orbit Earth was Sputnik I (launched by the former Soviet Union in 1957). Its highest point above Earth's surface was 947 kilometers, and its lowest point was 228 kilometers. The center of Earth was a focus of the elliptical orbit, and the radius of Earth is 6378 kilometers (see figure). Find the eccentricity of the orbit.

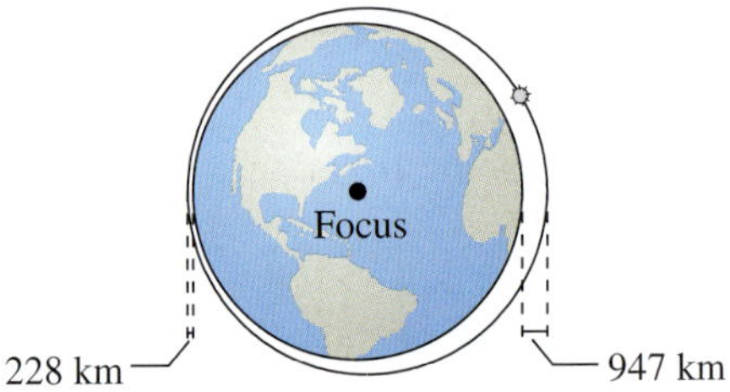

59. Geometry The area of the ellipse in the figure is twice the area of the circle. How long is the major axis? (*Hint:* The area of an ellipse is given by $A = \pi ab$.)

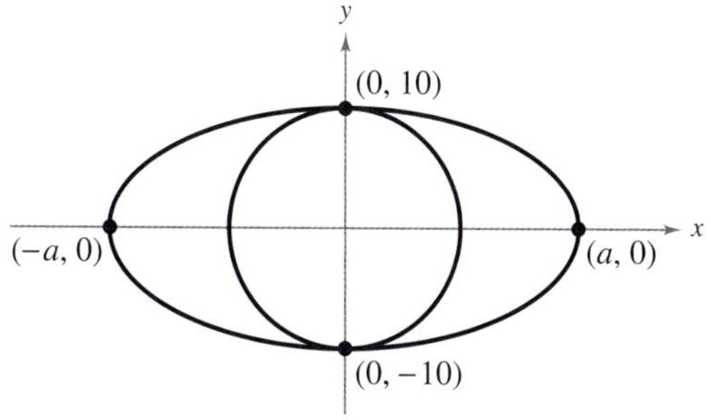

60. Geometry A line segment through a focus with endpoints on an ellipse, perpendicular to the major axis, is called a **latus rectum** of the ellipse. So, an ellipse has two latera recta. Knowing the length of the latera recta is helpful in sketching an ellipse because this information yields other points on the curve (see figure). Show that the length of each latus rectum is $2b^2/a$.

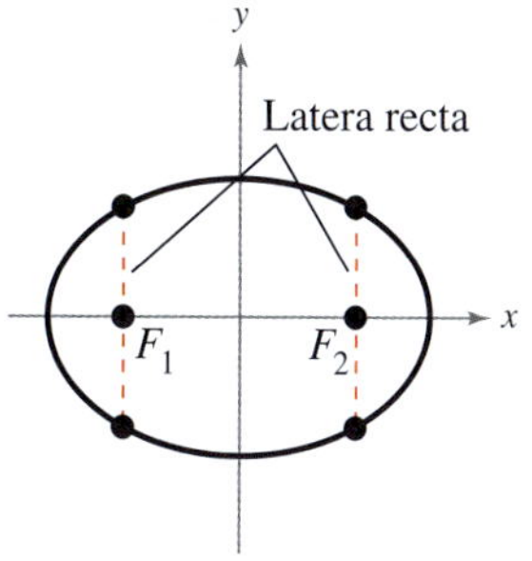

Using Latera Recta **In Exercises 61–64, sketch the ellipse using the latera recta (see Exercise 60).**

61. $\frac{x^2}{9} + \frac{y^2}{1} = 1$ **62.** $\frac{x^2}{4} + \frac{y^2}{25} = 1$

63. $16x^2 + 4y^2 = 64$

64. $6x^2 + 3y^2 = 12$

Focusing on Concepts

True or False? **In Exercises 65 and 66, determine whether the statement is true or false. Justify your answer.**

65. It is easier to distinguish the graph of an ellipse from the graph of a circle when the eccentricity of the ellipse is large (close to 1).

66. The area of a circle with diameter $d = 2r = 8$ is greater than the area of an ellipse with major axis $2a = 8$.

Identifying Degenerate Conics **In Exercises 67 and 68, determine whether the equation represents a degenerate conic. Explain.**

67. $16x^2 + 25y^2 - 32x + 50y + 16 = 0$

68. $9x^2 + 25y^2 - 36x - 50y + 61 = 0$

69. Error Analysis Describe the error in finding the characteristics of the graph of $\frac{(x+7)^2}{6} + \frac{y^2}{10} = 1$.

Center: $(-7, 0)$; Vertices: $(-7, -2)$, $(-7, 2)$;
Foci: $(-7, -\sqrt{10})$, $(7, \sqrt{10})$

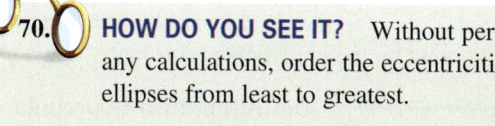

70. HOW DO YOU SEE IT? Without performing any calculations, order the eccentricities of the ellipses from least to greatest.

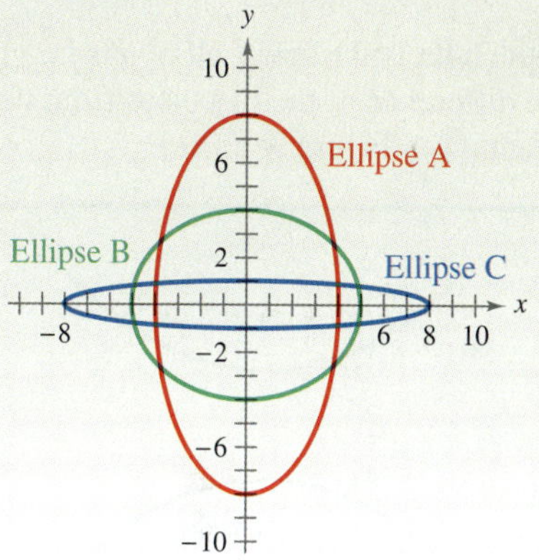

71. Think About It Recall that an ellipse can be drawn using two thumbtacks, a string of fixed length (greater than the distance between the two tacks), and a pencil (see page 657).

(a) What is the length of the string in terms of a? (See Figure 9.17.)

(b) Explain why the path drawn by the pencil is an ellipse.

72. Think About It Find the equation of an ellipse such that for any point on the ellipse, the sum of the distances from the points $(2, 2)$ and $(10, 2)$ is 36.

73. Proof Show that $a^2 = b^2 + c^2$ for the ellipse $x^2/a^2 + y^2/b^2 = 1$ where $a > 0, b > 0$, and the distance from the center of the ellipse $(0, 0)$ to a focus is c.

74. Geometry Write an equation of the circle centered at the origin that has the same area as the ellipse represented by the equation $\frac{(x-4)^2}{100} + \frac{(y-1)^2}{64} = 1$.

Cumulative Mixed Review

Identifying a Sequence **In Exercises 75 and 76, determine whether the sequence is arithmetic, geometric, or neither.**

75. 66, 55, 44, 33, 22, . . . **76.** $-\frac{1}{2}, \frac{1}{2}, \frac{3}{2}, \frac{5}{2}, \frac{7}{2}, \ldots$

Finding the Sum of a Finite Geometric Sequence **In Exercises 77 and 78, find the sum.**

77. $\sum_{n=0}^{6} 3^n$ **78.** $\sum_{n=1}^{10} 4\left(\frac{3}{4}\right)^{n-1}$

9.3 Hyperbolas and Rotation of Conics

Introduction

The definition of a **hyperbola** is similar to that of an ellipse. You know that for an ellipse, the *sum* of the distances between the foci and a point on the ellipse is constant. For a hyperbola, however, the absolute value of the *difference* of the distances between the foci and a point on the hyperbola is constant.

Definition of a Hyperbola

A **hyperbola** is the set of all points (x, y) in a plane for which the absolute value of the difference of the distances from two distinct fixed points, called **foci,** is constant. [See Figure 9.25(a).]

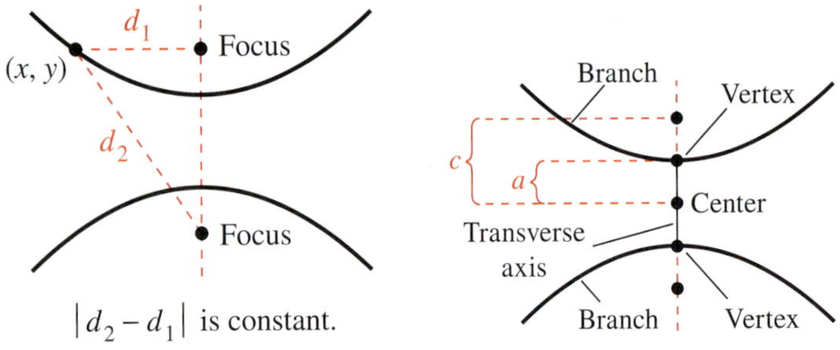

(a) (b)

Figure 9.25

The graph of a hyperbola has two disconnected parts called the **branches.** The line through the two foci intersects the hyperbola at two points called the **vertices.** The line segment connecting the vertices is the **transverse axis,** and the midpoint of the transverse axis is the **center** of the hyperbola [see Figure 9.25(b)]. The development of the **standard form of the equation of a hyperbola** is similar to that of an ellipse. Note, however, that a, b, and c are related differently for hyperbolas than for ellipses. For a hyperbola, the distance between the foci and the center is greater than the distance between the vertices and the center.

Standard Equation of a Hyperbola

The **standard form of the equation of a hyperbola** with center at (h, k) is

$$\frac{(x-h)^2}{a^2} - \frac{(y-k)^2}{b^2} = 1 \qquad \text{Transverse axis is horizontal.}$$

or

$$\frac{(y-k)^2}{a^2} - \frac{(x-h)^2}{b^2} = 1. \qquad \text{Transverse axis is vertical.}$$

The vertices are a units from the center, and the foci are c units from the center. Moreover, $c^2 = a^2 + b^2$. If the center of the hyperbola is at the origin $(0, 0)$, then the equation takes one of the following forms.

$$\frac{x^2}{a^2} - \frac{y^2}{b^2} = 1 \qquad \text{Transverse axis is horizontal.}$$

$$\frac{y^2}{a^2} - \frac{x^2}{b^2} = 1 \qquad \text{Transverse axis is vertical.}$$

What you should learn

- Write equations of hyperbolas in standard form.
- Find asymptotes of and graph hyperbolas.
- Use properties of hyperbolas to solve real-life problems.
- Classify conics from their general equations.
- Rotate the coordinate axes to eliminate the *xy*-term in equations of conics.

Why you should learn it

Hyperbolas can be used to model and solve many types of real-life problems. For instance, in Exercise 52 on page 676, hyperbolas are used to locate the position of an explosion that was recorded by three listening stations.

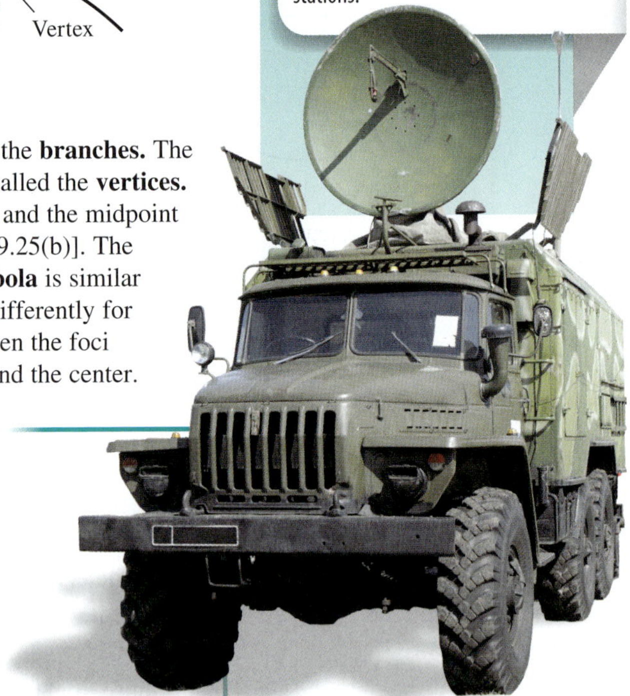

Figure 9.26 shows both the horizontal and vertical orientations for a hyperbola.

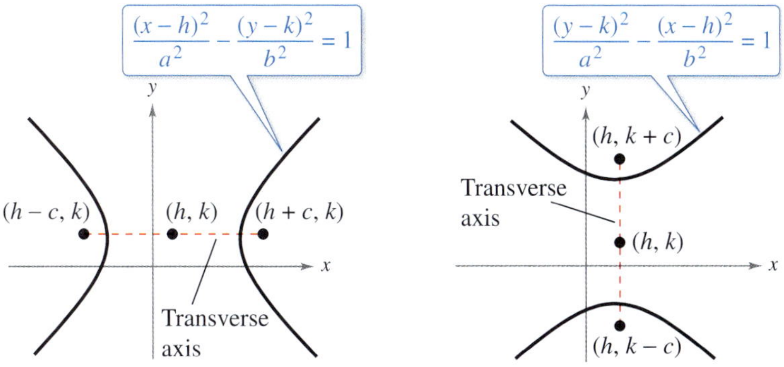

Transverse axis is horizontal. ***Transverse axis is vertical.***

Figure 9.26

EXAMPLE 1 Finding the Standard Equation of a Hyperbola

Find the standard form of the equation of the hyperbola with foci $(-1, 2)$ and $(5, 2)$ and vertices $(0, 2)$ and $(4, 2)$.

Solution

By the Midpoint Formula, the center of the hyperbola occurs at the point $(2, 2)$. Furthermore, $c = 3$ and $a = 2$, and it follows that

$$\begin{aligned} b &= \sqrt{c^2 - a^2} \\ &= \sqrt{3^2 - 2^2} \\ &= \sqrt{9 - 4} \\ &= \sqrt{5}. \end{aligned}$$

So, the hyperbola has a horizontal transverse axis, and the standard form of the equation of the hyperbola is

$$\frac{(x-2)^2}{2^2} - \frac{(y-2)^2}{(\sqrt{5})^2} = 1.$$

This equation simplifies to

$$\frac{(x-2)^2}{4} - \frac{(y-2)^2}{5} = 1.$$

The hyperbola is shown in the figure.

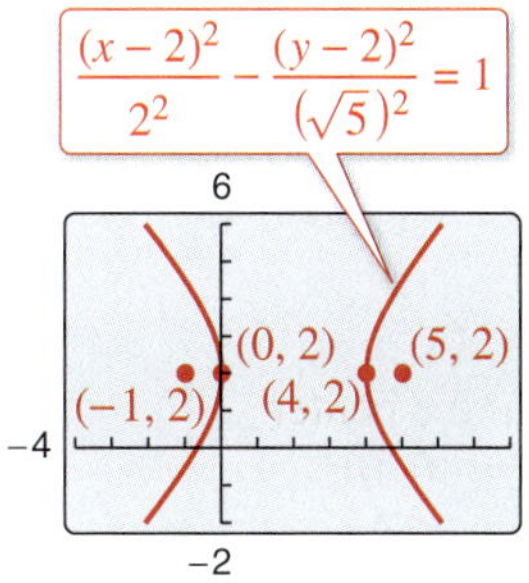

Technology Tip

You can use a graphing utility to graph a hyperbola by graphing the upper and lower portions in the same viewing window. To do this, you must solve the equation for y before entering it into the graphing utility. When graphing equations of conics, it can be difficult to solve for y, which is why it is very important to know the algebra used to solve equations for y.

Checkpoint  *Audio-video solution in English & Spanish at LarsonPrecalculus.com*

Find the standard form of the equation of the hyperbola with foci $(2, -5)$ and $(2, 3)$ and vertices $(2, -4)$ and $(2, 2)$.

Asymptotes of a Hyperbola

Every hyperbola has two **asymptotes** that intersect at the center of the hyperbola. The asymptotes pass through the corners of a rectangle of dimensions $2a$ by $2b$ with its center at (h, k), as shown in Figure 9.27.

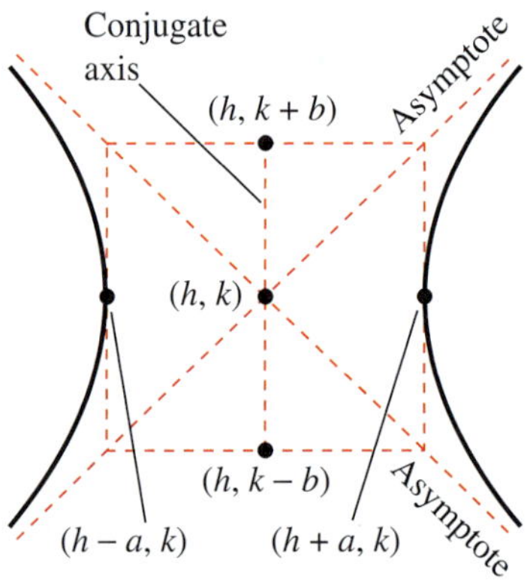

Figure 9.27

Asymptotes of a Hyperbola

$$y = k \pm \frac{b}{a}(x - h)$$ Asymptotes for horizontal transverse axis

$$y = k \pm \frac{a}{b}(x - h)$$ Asymptotes for vertical transverse axis

The **conjugate axis** of a hyperbola is the line segment of length $2b$ joining $(h, k + b)$ and $(h, k - b)$ when the transverse axis is horizontal, and the line segment of length $2b$ joining $(h + b, k)$ and $(h - b, k)$ when the transverse axis is vertical.

EXAMPLE 2 Sketching a Hyperbola

Sketch the hyperbola $4x^2 - y^2 = 16$.

Algebraic Solution

$$4x^2 - y^2 = 16$$ Write original equation.

$$\frac{4x^2}{16} - \frac{y^2}{16} = \frac{16}{16}$$ Divide each side by 16.

$$\frac{x^2}{2^2} - \frac{y^2}{4^2} = 1$$ Write in standard form.

The center of the hyperbola is $(0, 0)$. Because the x^2-term is positive, you can conclude that the transverse axis is horizontal. So, the vertices occur at $(-2, 0)$ and $(2, 0)$, the endpoints of the conjugate axis occur at $(0, -4)$ and $(0, 4)$, and you can sketch the rectangle shown in Figure 9.28. Finally, by drawing the asymptotes $y = 2x$ and $y = -2x$ through the corners of this rectangle, you can complete the sketch, as shown in Figure 9.29.

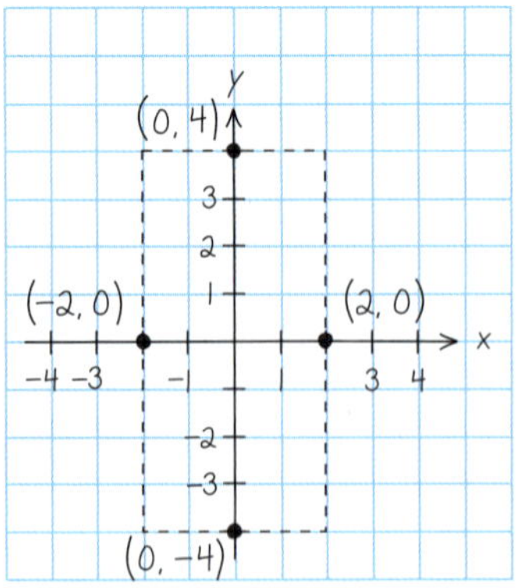

Figure 9.28

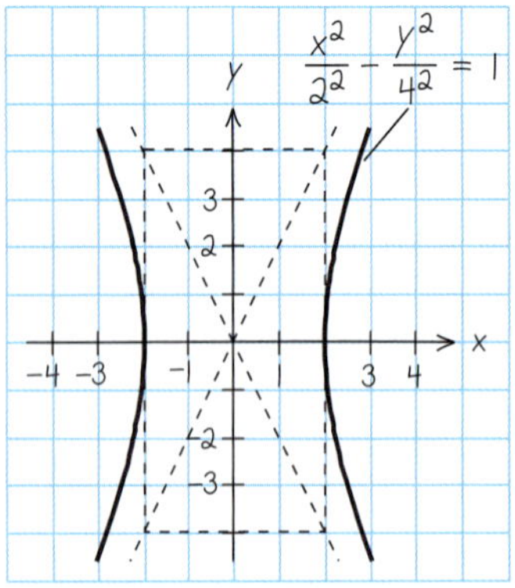

Figure 9.29

Graphical Solution

Solve the equation of the hyperbola for y, as follows.

$$4x^2 - y^2 = 16$$

$$4x^2 - 16 = y^2$$

$$\pm\sqrt{4x^2 - 16} = y$$

Then use a graphing utility to graph

$$y_1 = \sqrt{4x^2 - 16} \quad \text{and} \quad y_2 = -\sqrt{4x^2 - 16}$$

in the same viewing window, as shown in the figure. Be sure to use a square setting.

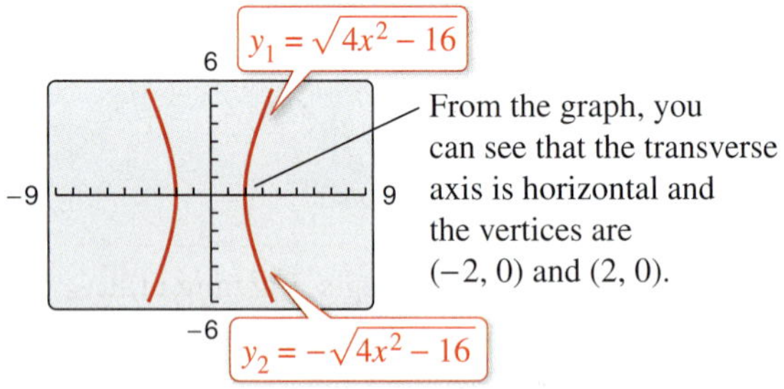

Checkpoint Audio-video solution in English & Spanish at LarsonPrecalculus.com

Sketch the hyperbola $4y^2 - 9x^2 = 36$.

EXAMPLE 3 Finding the Asymptotes of a Hyperbola

Sketch the hyperbola $4x^2 - 3y^2 + 8x + 16 = 0$, find the equations of its asymptotes, and find the foci.

Solution

$$4x^2 - 3y^2 + 8x + 16 = 0 \quad \text{Write original equation.}$$

$$4(x^2 + 2x) - 3y^2 = -16 \quad \text{Subtract 16 from each side and factor.}$$

$$4(x^2 + 2x + 1) - 3y^2 = -16 + 4(1) \quad \text{Complete the square.}$$

$$4(x + 1)^2 - 3y^2 = -12 \quad \text{Write in completed square form.}$$

$$\frac{y^2}{2^2} - \frac{(x+1)^2}{(\sqrt{3})^2} = 1 \quad \text{Write in standard form.}$$

From this equation, you can conclude that the hyperbola has a vertical transverse axis, is centered at $(-1, 0)$, has vertices $(-1, 2)$ and $(-1, -2)$, and has a conjugate axis with endpoints $(-1 - \sqrt{3}, 0)$ and $(-1 + \sqrt{3}, 0)$. To sketch the hyperbola, draw a rectangle through these four points. The asymptotes are the lines passing through the corners of the rectangle. Using $a = 2$ and $b = \sqrt{3}$, you can conclude that the equations of the asymptotes are

$$y = \frac{2}{\sqrt{3}}(x + 1) \quad \text{and} \quad y = -\frac{2}{\sqrt{3}}(x + 1).$$

Finally, you can determine the foci by using the equation $c^2 = a^2 + b^2$. So, you have $c = \sqrt{2^2 + (\sqrt{3})^2} = \sqrt{7}$, and the foci are $(-1, \sqrt{7})$ and $(-1, -\sqrt{7})$. The hyperbola is shown in the figure.

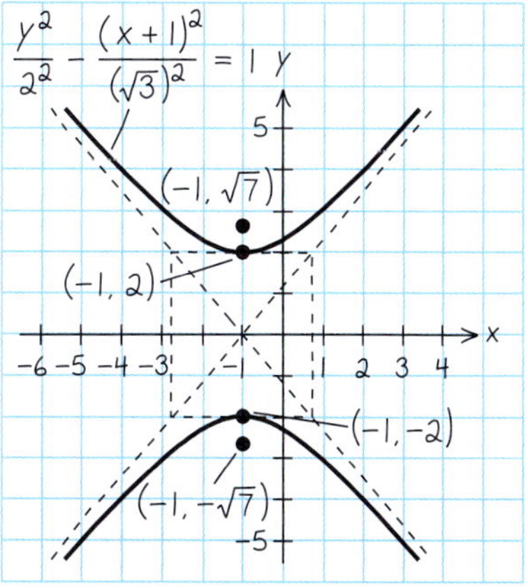

You can check your sketch using a graphing utility, as shown in the figure. Notice that the graphing utility does not draw the asymptotes. When you trace along the branches, however, you will see that the values of the hyperbola approach the asymptotes.

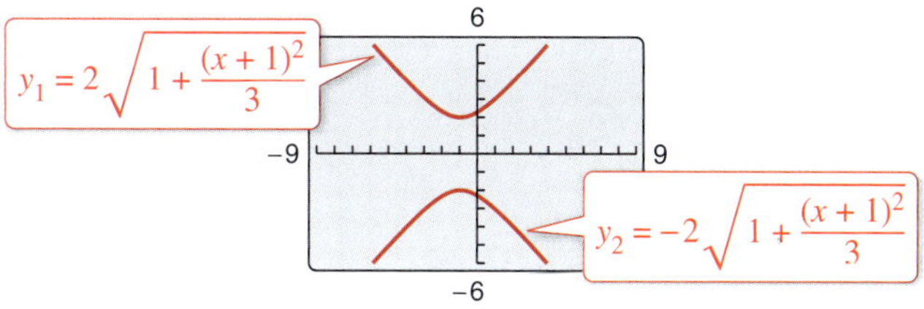

Checkpoint Audio-video solution in English & Spanish at LarsonPrecalculus.com

Sketch the hyperbola $9x^2 - 4y^2 + 8y - 40 = 0$, find the equations of its asymptotes, and find the foci.

EXAMPLE 4 Using Asymptotes to Find the Standard Equation

See LarsonPrecalculus.com for an interactive version of this type of example.

Find the standard form of the equation of the hyperbola having vertices $(3, -5)$ and $(3, 1)$ and having asymptotes

$$y = 2x - 8 \quad \text{and} \quad y = -2x + 4$$

as shown in Figure 9.30.

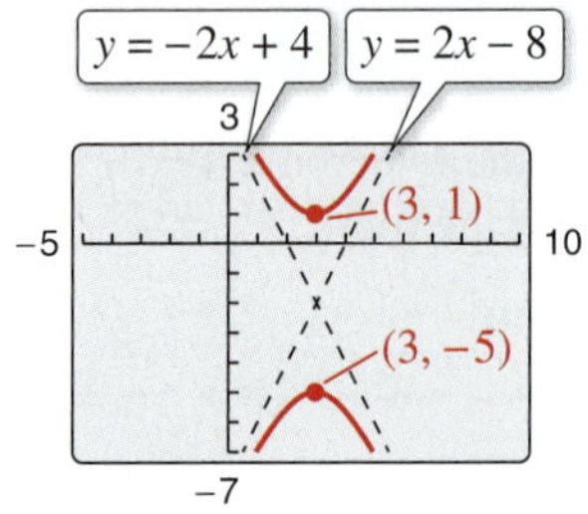

Figure 9.30

Solution

The center of the hyperbola is the midpoint of the line segment joining the vertices. So, by the Midpoint Formula, the center of the hyperbola is $(3, -2)$. Furthermore, the hyperbola has a vertical transverse axis with $a = 3$. From the original equations, you can determine the slopes of the asymptotes to be

$$m_1 = 2 = \frac{a}{b} \quad \text{and} \quad m_2 = -2 = -\frac{a}{b}$$

and because $a = 3$, you can conclude that

$$2 = \frac{a}{b} \quad \Longrightarrow \quad 2 = \frac{3}{b} \quad \Longrightarrow \quad b = \frac{3}{2}.$$

So, the standard form of the equation of the hyperbola is

$$\frac{(y + 2)^2}{3^2} - \frac{(x - 3)^2}{(3/2)^2} = 1.$$

✓ **Checkpoint** *Audio-video solution in English & Spanish at LarsonPrecalculus.com*

Find the standard form of the equation of the hyperbola having vertices $(3, 2)$ and $(9, 2)$ and having asymptotes

$$y = -2 + \frac{2}{3}x \quad \text{and} \quad y = 6 - \frac{2}{3}x.$$

As with ellipses, the *eccentricity* of a hyperbola is

$$e = \frac{c}{a} \qquad \text{Eccentricity}$$

and because $c > a$, it follows that $e > 1$. When the eccentricity is large, the branches of the hyperbola are nearly flat, as shown in Figure 9.31(a). When the eccentricity is close to 1, the branches of the hyperbola are more pointed, as shown in Figure 9.31(b).

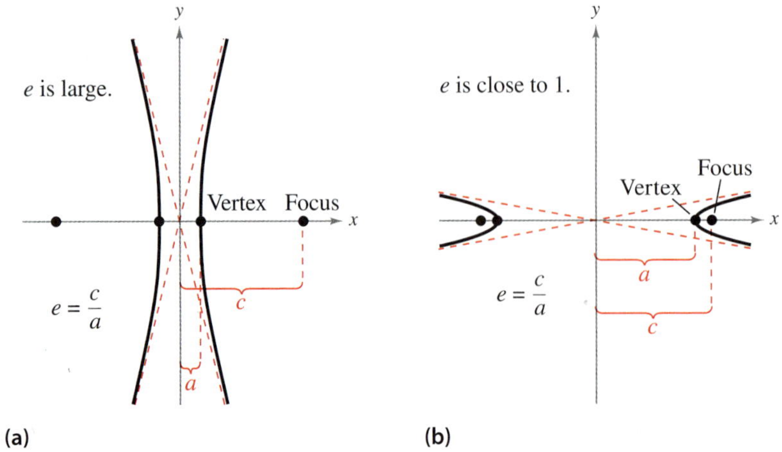

Figure 9.31

Applications

The application in Example 5 was developed during World War II. It shows how the properties of hyperbolas can be used in radar and other detection systems.

EXAMPLE 5 An Application Involving Hyperbolas

Two microphones, 1 mile apart, record an explosion. Microphone A receives the sound 2 seconds before microphone B. Where did the explosion occur?

Solution

Note that 1 mile = 5280 feet. So, place A at (2640, 0) and B at (−2640, 0), as shown in the figure. Because A received the sound 2 seconds before B and assuming sound travels at 1100 feet per second, you know that the explosion took place 2200 feet farther from B than from A. The locus of all points that are 2200 feet closer to A than to B is one branch of the hyperbola with the form

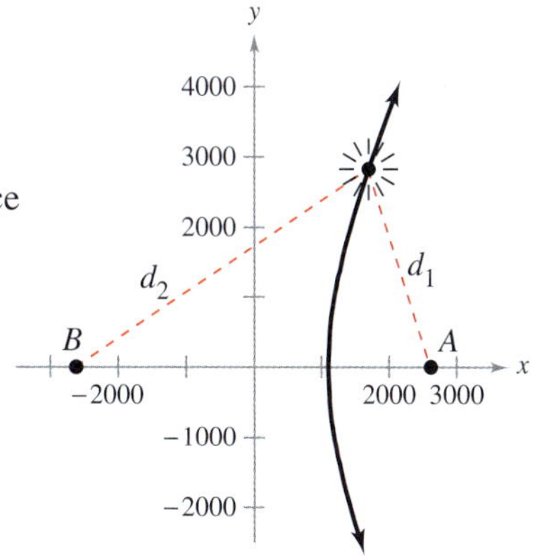

$$\frac{x^2}{a^2} - \frac{y^2}{b^2} = 1.$$

Because the explosion took place 2200 feet farther from B than from A, $d_2 - d_1 = 2a = 2200$, which implies

$$a = \frac{2200}{2} = 1100.$$

The distance from the center of the hyperbola to the focus at A is 2640, so $c = 2640$ and

$$b^2 = c^2 - a^2 = 2640^2 - 1100^2 = 5{,}759{,}600.$$

So, the explosion occurred somewhere on the right branch of the hyperbola

$$\frac{x^2}{1{,}210{,}000} - \frac{y^2}{5{,}759{,}600} = 1.$$

Checkpoint *Audio-video solution in English & Spanish at LarsonPrecalculus.com*

Repeat Example 5 when microphone A receives the sound 4 seconds before microphone B.

Another interesting application of conic sections involves the orbits of comets in our solar system. Comets can have elliptical, parabolic, or hyperbolic orbits. The center of the sun is a focus of each of these orbits, and each orbit has a vertex at the point where the comet is closest to the sun, as shown in Figure 9.32. Undoubtedly, there are many comets with parabolic or hyperbolic orbits that have not been identified. You get to see such comets only *once*. Comets with elliptical orbits, such as Halley's comet, are the only ones that remain in our solar system.

If p is the distance between the vertex and the focus in meters, and v is the velocity of the comet at the vertex in meters per second, then the type of orbit is determined as follows.

1. Ellipse: $v < \sqrt{2GM/p}$
2. Parabola: $v = \sqrt{2GM/p}$
3. Hyperbola: $v > \sqrt{2GM/p}$

Note that $M \approx 1.989 \times 10^{30}$ kilograms (the mass of the sun) and $G \approx 6.67 \times 10^{-11}$ cubic meter per kilogram-second squared (the universal gravitational constant).

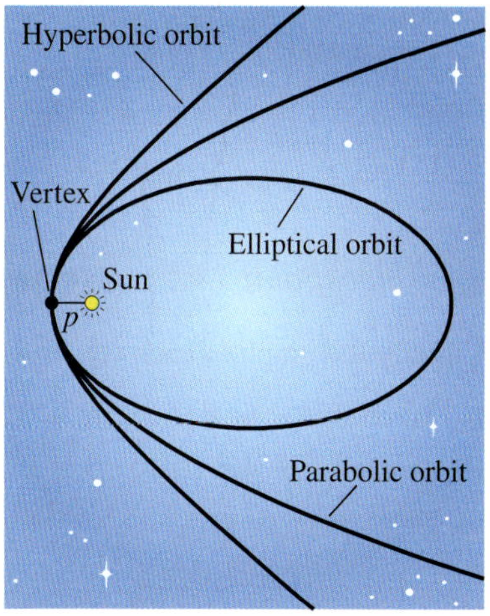

Figure 9.32

General Equations of Conics

Classifying a Conic from Its General Equation

The graph of $Ax^2 + Cy^2 + Dx + Ey + F = 0$ is one of the following.

1. Circle:	$A = C$	$A \neq 0$
2. Parabola:	$AC = 0$	$A = 0$ or $C = 0$, but not both.
3. Ellipse:	$AC > 0$	A and C have like signs ($A \neq C$).
4. Hyperbola:	$AC < 0$	A and C have unlike signs.

The test above is valid when the graph is a *conic*. The test does not apply to equations such as

$$x^2 + y^2 = -1$$

whose graphs are not conics.

EXAMPLE 6 Classifying Conics from General Equations

Classify the graph of each equation.

a. $4x^2 - 9x + y - 5 = 0$
b. $4x^2 - y^2 + 8x - 6y + 4 = 0$
c. $2x^2 + 4y^2 - 4x + 12y = 0$
d. $2x^2 + 2y^2 - 8x + 12y + 2 = 0$

Solution

a. For the equation $4x^2 - 9x + y - 5 = 0$, you have

$AC = 4(0) = 0.$ Parabola

So, the graph is a parabola.

b. For the equation $4x^2 - y^2 + 8x - 6y + 4 = 0$, you have

$AC = 4(-1) < 0.$ Hyperbola

So, the graph is a hyperbola.

c. For the equation $2x^2 + 4y^2 - 4x + 12y = 0$, you have

$AC = 2(4) > 0.$ Ellipse

So, the graph is an ellipse.

d. For the equation $2x^2 + 2y^2 - 8x + 12y + 2 = 0$, you have

$A = C = 2.$ Circle

So, the graph is a circle.

Algebra Help

Notice in Example 6(a) that there is no y^2-term in the equation. So, $C = 0$.

✓ Checkpoint *Audio-video solution in English & Spanish at LarsonPrecalculus.com*

Classify the graph of each equation.

a. $3x^2 + 3y^2 - 6x + 6y + 5 = 0$
b. $2x^2 - 4y^2 + 4x + 8y - 3 = 0$
c. $3x^2 + y^2 + 6x - 2y + 3 = 0$
d. $2x^2 + 4x + y - 2 = 0$

Rotation

The equation of a conic with axes parallel to one of the coordinate axes has a standard form that can be written in the general form

$$Ax^2 + Cy^2 + Dx + Ey + F = 0.$$ Horizontal or vertical axis

You will now study the equations of conics whose axes are rotated so that they are not parallel to either the x-axis or the y-axis. The general equation for such conics contains an xy-term.

$$Ax^2 + Bxy + Cy^2 + Dx + Ey + F = 0$$ Equation in xy-plane

To eliminate this xy-term, you can use a procedure called **rotation of axes.** The objective is to rotate the x- and y-axes until they are parallel to the axes of the conic. The rotated axes are denoted as the x'-axis and the y'-axis, as shown in Figure 9.33.

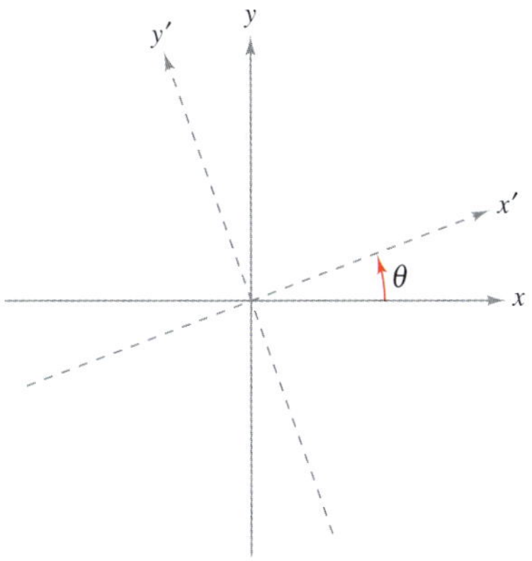

Figure 9.33

After the rotation, the equation of the conic in the new $x'y'$-plane will have the form

$$A'(x')^2 + C'(y')^2 + D'x' + E'y' + F' = 0.$$ Equation in $x'y'$-plane

This equation has no xy-term, so you can obtain a standard form by completing the square. The next theorem identifies how much to rotate the axes to eliminate the xy-term and also the equations for determining the new coefficients A', C', D', E', and F'.

Rotation of Axes to Eliminate an *xy*-Term

The general second-degree equation

$$Ax^2 + Bxy + Cy^2 + Dx + Ey + F = 0$$

where $B \neq 0$, can be rewritten as

$$A'(x')^2 + C'(y')^2 + D'x' + E'y' + F' = 0$$

by rotating the coordinate axes through an angle θ, where

$$\cot 2\theta = \frac{A - C}{B}.$$

The coefficients of the new equation are obtained by making the substitutions

$$x = x' \cos\theta - y' \sin\theta \quad \text{and} \quad y = x' \sin\theta + y' \cos\theta.$$

(See the proof on page 718.)

Note that the substitutions $x = x' \cos\theta - y' \sin\theta$ and $y = x' \sin\theta + y' \cos\theta$ were developed to eliminate the $x'y'$-term in the rotated system. You can use this to check your work. In other words, when your final equation contains an $x'y'$-term, you know that you have made a mistake.

EXAMPLE 7 Rotation of Axes for a Hyperbola

Rotate the axes to eliminate the xy-term in the equation $xy - 1 = 0$. Then write the equation in standard form and sketch its graph.

Solution

Because $A = 0$, $B = 1$, and $C = 0$, you have

$$\cot 2\theta = \frac{A - C}{B} = \frac{0 - 0}{1} = 0.$$

So, $\cot 2\theta = 0$, which implies that $2\theta = \pi/2$, or $\theta = \pi/4$. The equation in the $x'y'$-system is obtained by making the substitutions

$$x = x' \cos \frac{\pi}{4} - y' \sin \frac{\pi}{4}$$
$$= x'\left(\frac{1}{\sqrt{2}}\right) - y'\left(\frac{1}{\sqrt{2}}\right)$$
$$= \frac{x' - y'}{\sqrt{2}}$$

and

$$y = x' \sin \frac{\pi}{4} + y' \cos \frac{\pi}{4}$$
$$= x'\left(\frac{1}{\sqrt{2}}\right) + y'\left(\frac{1}{\sqrt{2}}\right)$$
$$= \frac{x' + y'}{\sqrt{2}}$$

into the equation $xy - 1 = 0$.

$$xy - 1 = 0 \quad \text{Write original equation.}$$
$$\left(\frac{x' - y'}{\sqrt{2}}\right)\left(\frac{x' + y'}{\sqrt{2}}\right) - 1 = 0 \quad \text{Substitute for } x \text{ and } y.$$
$$\frac{(x')^2 - (y')^2}{2} - 1 = 0 \quad \text{Multiply.}$$
$$\frac{(x')^2}{(\sqrt{2})^2} - \frac{(y')^2}{(\sqrt{2})^2} = 1 \quad \text{Write in standard form.}$$

In the $x'y'$-system, this is a hyperbola centered at the origin with vertices at $(\pm\sqrt{2}, 0)$, as shown in Figure 9.34. To find the coordinates of the vertices in the xy-system, substitute the coordinates $(\pm\sqrt{2}, 0)$ into the equations

$$x = \frac{x' - y'}{\sqrt{2}} \quad \text{and} \quad y = \frac{x' + y'}{\sqrt{2}}.$$

This substitution yields the vertices $(1, 1)$ and $(-1, -1)$ in the xy-system. Note also that the asymptotes of the hyperbola have equations

$$y' = \pm x'$$

which correspond to the original x- and y-axes.

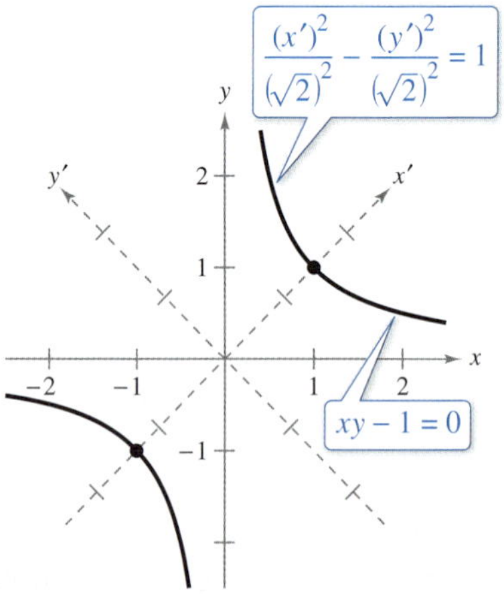

Vertices:
In x'y'-system: $(\sqrt{2}, 0)$, $(-\sqrt{2}, 0)$
In xy-system: $(1, 1)$, $(-1, -1)$
Figure 9.34

✓ Checkpoint ▶ *Audio-video solution in English & Spanish at LarsonPrecalculus.com*

Rotate the axes to eliminate the xy-term in the equation $xy + 6 = 0$. Then write the equation in standard form and sketch its graph.

9.3 Exercises

See *CalcChat.com* for tutorial help and worked-out solutions to odd-numbered exercises.
For instructions on how to use a graphing utility, see Appendix A.

Vocabulary and Concept Check

In Exercises 1–3, fill in the blank(s).

1. A _______ is the set of all points (x, y) in a plane for which the absolute value of the difference of the distances from two distinct fixed points, called foci, is constant.

2. The line segment connecting the vertices of a hyperbola is called the _______ , and the midpoint of the line segment is the _______ of the hyperbola.

3. The procedure used to eliminate the xy-term in a general second-degree equation is called _______ of ______.

4. Which equations represent a hyperbola with a horizontal transverse axis? a vertical transverse axis?

(a) $(x - h)^2/a^2 - (y - k)^2/b^2 = 1$ (b) $(y - k)^2/a^2 - (x - h)^2/b^2 = 1$

(c) $y^2/a^2 - x^2/b^2 = 1$ (d) $x^2/a^2 - y^2/b^2 = 1$

5. How many asymptotes does a hyperbola have? Where do these asymptotes intersect?

6. What type of conic has the general equation $Ax^2 + Cy^2 + Dx + Ey + F = 0$ $(A \neq 0, C \neq 0)$ with $AC < 0$? $AC > 0$? $A = C$?

Procedures and Problem Solving

Matching an Equation with a Graph **In Exercises 7–10, match the equation with its graph. [The graphs are labeled (a), (b), (c), and (d).]**

(a)

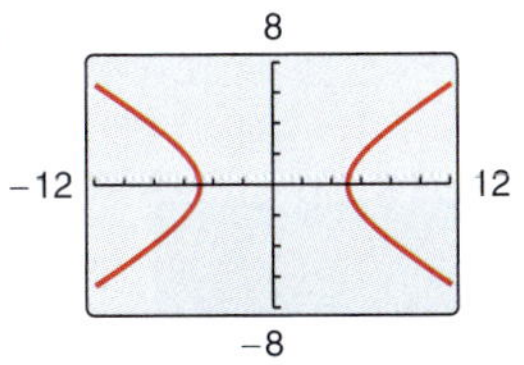

(b)

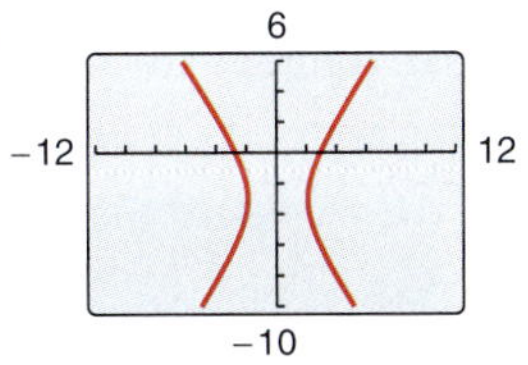

(c)

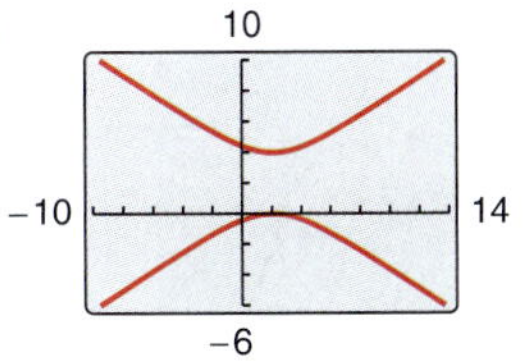

(d)

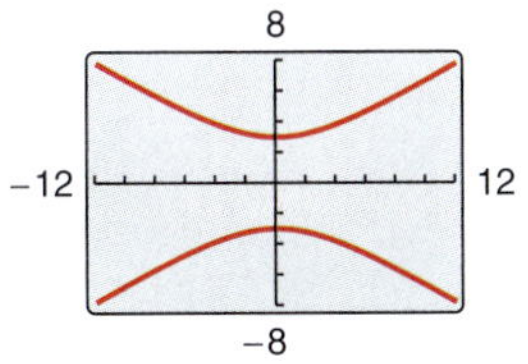

7. $\dfrac{x^2}{25} - \dfrac{y^2}{9} = 1$

8. $\dfrac{y^2}{9} - \dfrac{x^2}{25} = 1$

9. $\dfrac{x^2}{4} - \dfrac{(y + 3)^2}{9} = 1$

10. $\dfrac{(y - 2)^2}{4} - \dfrac{(x - 2)^2}{9} = 1$

Finding the Standard Equation of a Hyperbola **In Exercises 11–20, find the standard form of the equation of the hyperbola with the given characteristics.**

11. Vertices: $(0, \pm 2)$; foci: $(0, \pm 4)$

12. Vertices: $(\pm 3, 0)$; foci: $(\pm 6, 0)$

13. Vertices: $(0, 0)$, $(4, 0)$; foci: $(-2, 0)$, $(6, 0)$

14. Vertices: $(2, 3)$, $(2, -3)$; foci: $(2, 6)$, $(2, -6)$

15. Vertices: $(4, 1)$, $(4, 9)$; foci: $(4, 0)$, $(4, 10)$

16. Vertices: $(0, 1)$, $(4, 1)$; foci: $(-1, 1)$, $(5, 1)$

17. Vertices: $(2, 3)$, $(2, -3)$; passes through the point $(0, 5)$

18. Vertices: $(-2, 1)$, $(2, 1)$; passes through the point $(5, 4)$

19. Vertices: $(0, -3)$, $(4, -3)$; passes through the point $(-4, 5)$

20. Vertices: $(1, -3)$, $(1, -7)$; passes through the point $(5, -11)$

Sketching a Hyperbola **In Exercises 21–30, find the center, vertices, foci, and asymptotes of the hyperbola, and sketch its graph using the asymptotes as an aid. Use a graphing utility to verify your graph.**

21. $x^2 - y^2 = 1$

22. $y^2 - x^2 = 1$

23. $\dfrac{y^2}{1} - \dfrac{x^2}{4} = 1$

24. $\dfrac{x^2}{9} - \dfrac{y^2}{1} = 1$

25. $\dfrac{y^2}{16} - \dfrac{x^2}{4} = 1$

26. $\dfrac{x^2}{25} - \dfrac{y^2}{36} = 1$

27. $\dfrac{(x - 3)^2}{9} - \dfrac{(y - 1)^2}{1} = 1$

28. $\dfrac{(x - 2)^2}{4} - \dfrac{(y + 5)^2}{25} = 1$

29. $\dfrac{(y + 6)^2}{1/9} - \dfrac{(x - 2)^2}{1/4} = 1$

30. $\dfrac{(y - 1)^2}{1/4} - \dfrac{(x + 3)^2}{1/16} = 1$

Analyzing a Hyperbola **In Exercises 31–40, (a) find the standard form of the equation of the hyperbola, (b) find the center, vertices, foci, and asymptotes of the hyperbola, and (c) sketch the hyperbola. Use a graphing utility to verify your graph.**

31. $4x^2 - 9y^2 = 36$ **32.** $25x^2 - 4y^2 = 100$

33. $6y^2 - 3x^2 = 24$ **34.** $3x^2 - 2y^2 = 18$

35. $9x^2 - y^2 - 36x - 6y + 18 = 0$

36. $x^2 - 9y^2 + 36y - 72 = 0$

37. $2x^2 - 7y^2 + 16x + 18 = 0$

38. $3y^2 - 5x^2 + 6y - 60x - 192 = 0$

39. $9y^2 - x^2 + 2x + 54y + 62 = 0$

40. $9x^2 - y^2 + 54x + 10y + 55 = 0$

Using Asymptotes to Find the Standard Equation **In Exercises 41–50, find the standard form of the equation of the hyperbola with the given characteristics.**

41. Vertices: $(\pm 1, 0)$; asymptotes: $y = \pm 5x$

42. Vertices: $(0, \pm 3)$; asymptotes: $y = \pm 3x$

43. Foci: $(0, \pm 8)$; asymptotes: $y = \pm 4x$

44. Foci: $(\pm 10, 0)$; asymptotes: $y = \pm \frac{3}{4}x$

45. Vertices: $(1, 2)$, $(3, 2)$; asymptotes: $y = x$, $y = 4 - x$

46. Vertices: $(3, 0)$, $(3, -6)$; asymptotes: $y = x - 6$, $y = -x$

47. Vertices: $(0, 2)$, $(6, 2)$; asymptotes: $y = \frac{2}{3}x$, $y = 4 - \frac{2}{3}x$

48. Vertices: $(3, 0)$, $(3, 4)$; asymptotes: $y = \frac{2}{3}x$, $y = 4 - \frac{2}{3}x$

49. Foci: $(-1, -1)$, $(9, -1)$; asymptotes: $y = \frac{3}{4}x - 4$, $y = -\frac{3}{4}x + 2$

50. Foci: $(1, 2)$, $(1, 6)$; asymptotes: $y = 2 + 2x$, $y = 6 - 2x$

51. Meteorology You and a friend live 4 miles apart (on the same "east-west" street) and are talking on the phone. You hear a clap of thunder from lightning in a storm, and 18 seconds later your friend hears the thunder. Find an equation that gives the possible places where the lightning could have occurred. (Assume that the coordinate system is measured in feet and that sound travels at 1100 feet per second.)

52. ***Why you should learn it*** *(p. 666)* Three listening stations located at $(3300, 0)$, $(3300, 1100)$, and $(-3300, 0)$ monitor an explosion. The last two stations detect the explosion 1 second and 4 seconds after the first, respectively. Determine the coordinates of the explosion. (Assume that the coordinate system is measured in feet and that sound travels at 1100 feet per second.)

Chubykin Arkady/Shutterstock.com

53. Art and Design The base for the pendulum of a clock has the shape of a hyperbola (see figure).

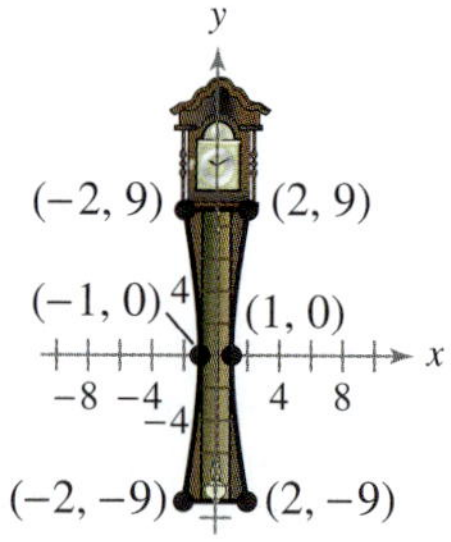

(a) Write an equation of the cross section of the base.

(b) Each unit in the coordinate plane represents $\frac{1}{2}$ foot. Find the width of the base 4 inches from the bottom.

54. MODELING DATA

Long distance radio navigation for aircraft and ships uses synchronized pulses transmitted by widely separated transmitting stations. These pulses travel at the speed of light (186,000 miles per second). The difference in the times of arrival of these pulses at an aircraft or ship is constant on a hyperbola having the transmitting stations as foci. Assume that two stations, 300 miles apart, are positioned on a rectangular coordinate system at coordinates $(-150, 0)$ and $(150, 0)$, and that a ship is traveling on a hyperbolic path with coordinates $(x, 75)$. (See figure.)

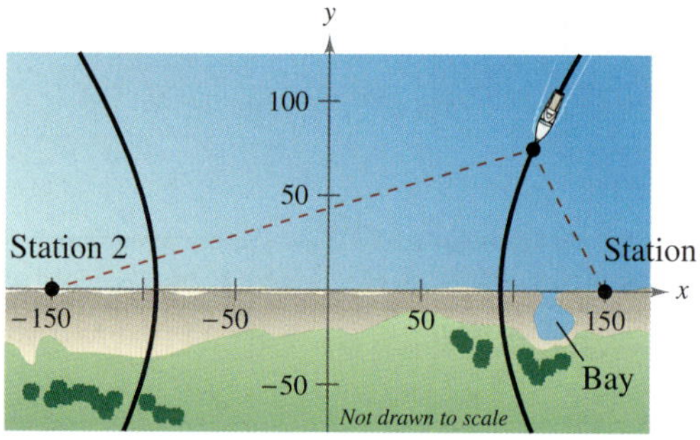

(a) Find the x-coordinate of the position of the ship when the time difference between the pulses from the transmitting stations is 1000 microseconds (0.001 second).

(b) Determine the distance between the ship and station 1 when the ship reaches the shore.

(c) The captain of the ship wants to enter a bay located between the two stations. The bay is 30 miles from station 1. What should be the time difference between the pulses?

(d) The ship is 60 miles offshore when the time difference in part (c) is obtained. What is the position of the ship?

55. Hyperbolic Mirror A hyperbolic mirror (used in some telescopes) has the property that a light ray directed at a focus will be reflected to the other focus. The focus of a hyperbolic mirror (see figure) has coordinates (24, 0). Find the vertex of the mirror given that the mount at the top edge of the mirror has coordinates (24, 24).

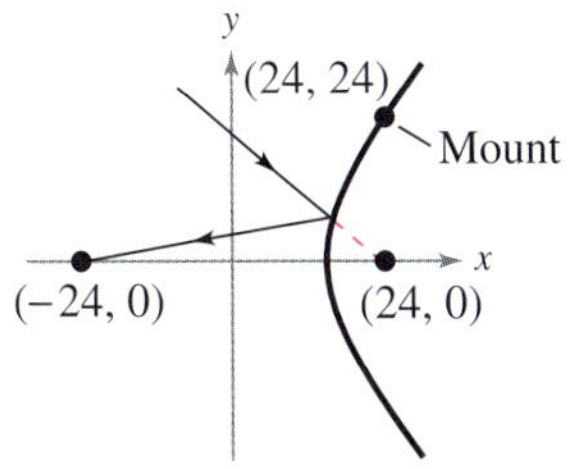

56. Photography A panoramic photo can be taken using a hyperbolic mirror. The camera is pointed toward the vertex of the mirror, and the optical center of the camera is positioned at one focus of the mirror (see figure). An equation for the cross-section of the mirror is

$$\frac{x^2}{25} - \frac{y^2}{16} = 1.$$

Find the distance from the optical center of the camera to the mirror.

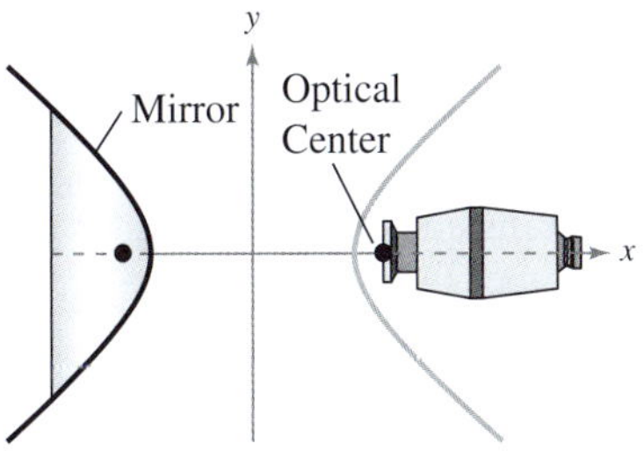

Classifying a Conic from a General Equation **In Exercises 57–62, classify the graph of the equation as a circle, a parabola, an ellipse, or a hyperbola.**

57. $9x^2 + 4y^2 - 18x + 16y - 119 = 0$

58. $x^2 + y^2 - 4x - 6y - 23 = 0$

59. $16x^2 - 9y^2 + 32x + 54y - 209 = 0$

60. $4x^2 + 25y^2 + 16x + 250y + 541 = 0$

61. $x^2 + 4x - 8y + 20 = 0$

62. $y^2 - x^2 + 2x - 6y - 8 = 0$

Finding a Point in a Rotated Coordinate System **In Exercises 63–66, the $x'y'$-coordinate system has been rotated θ degrees from the xy-coordinate system. The coordinates of a point in the xy-coordinate system are given. Find the coordinates of the point in the rotated coordinate system.**

63. $\theta = 90°$, $(0, 3)$

64. $\theta = 45°$, $(3, 3)$

65. $\theta = 30°$, $(1, 3)$

66. $\theta = 60°$, $(1, 2)$

Matching an Equation with a Graph **In Exercises 67–72, match the graph with its equation. [The graphs are labeled (a), (b), (c), (d), (e), and (f).]**

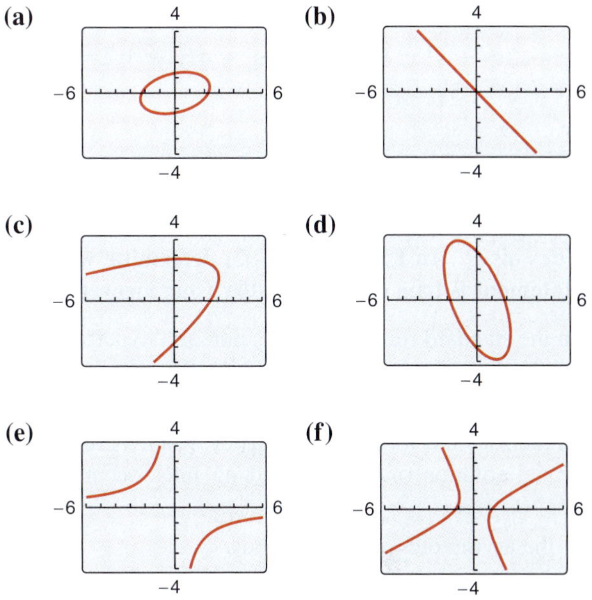

67. $xy + 4 = 0$

68. $x^2 + 2xy + y^2 = 0$

69. $-2x^2 + 3xy + 2y^2 + 3 = 0$

70. $x^2 - xy + 3y^2 - 5 = 0$

71. $3x^2 + 2xy + y^2 - 10 = 0$

72. $x^2 - 4xy + 4y^2 + 10x - 30 = 0$

Rotation of Axes **In Exercises 73–80, rotate the axes to eliminate the xy-term in the equation. Then write the equation in standard form. Sketch the graph of the resulting equation, showing both sets of axes.**

73. $xy + 1 = 0$

74. $xy - 4 = 0$

75. $xy + 2x - y + 4 = 0$

76. $xy - 8x - 4y = 0$

77. $5x^2 - 6xy + 5y^2 - 12 = 0$

78. $2x^2 + xy + 2y^2 - 8 = 0$

79. $16x^2 - 24xy + 9y^2 - 60x - 80y + 100 = 0$

80. $9x^2 + 24xy + 16y^2 + 90x - 130y = 0$

Graphing a Conic **In Exercises 81–84, use a graphing utility to graph the conic. Determine the angle θ through which the axes are rotated. Explain how you used the graphing utility to obtain the graph.**

81. $x^2 - 4xy + 2y^2 = 8$

82. $3x^2 + 5xy - 2y^2 = 10$

83. $24x^2 + 18xy + 12y^2 = 34$

84. $4x^2 - 12xy + 9y^2 + \sqrt{6}x - 29y = 91$

Sketching the Graph of a Degenerate Conic **In Exercises 85–90, sketch (if possible) the graph of the degenerate conic.**

85. $y^2 - 25x^2 = 0$

86. $x^2 - 10xy + y^2 = 0$

87. $x^2 - 2xy + y^2 = 0$

88. $x^2 + 4xy + 4y^2 = 0$

89. $x^2 + y^2 - 2x + 6y + 10 = 0$

90. $x^2 + y^2 + 2x - 4y + 5 = 0$

Focusing on Concepts

True or False? **In Exercises 91–93, determine whether the statement is true or false. Justify your answer.**

91. In the standard form of the equation of a hyperbola, the larger the ratio of b to a, the larger the eccentricity of the hyperbola.

92. In the standard form of the equation of a hyperbola, the trivial solution of two intersecting lines occurs when $b = 0$.

93. If the asymptotes of the hyperbola

$$\frac{x^2}{a^2} - \frac{y^2}{b^2} = 1, \text{ where } a, b > 0$$

intersect at right angles, then $a = b$.

94. Writing Explain how the central rectangle of a hyperbola can be used to sketch its asymptotes.

95. Think About It Consider a hyperbola centered at the origin with a horizontal transverse axis. Use the definition of a hyperbola to derive its standard form.

96. HOW DO YOU SEE IT? Match each equation with its graph.

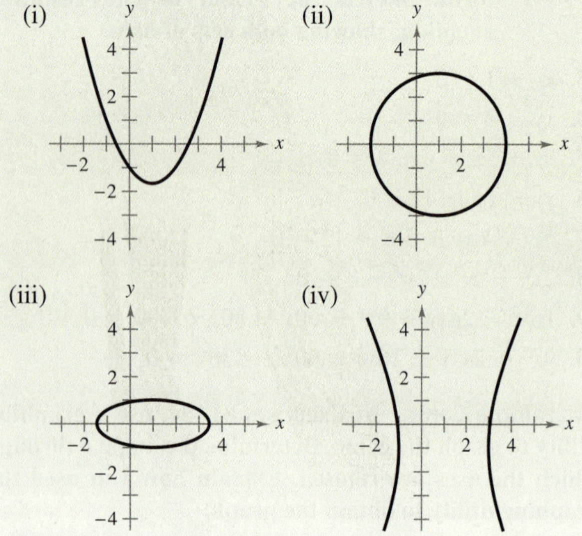

(a) $4x^2 - y^2 - 8x - 2y - 13 = 0$

(b) $x^2 + y^2 - 2x - 8 = 0$

(c) $2x^2 - 4x - 3y - 3 = 0$

(d) $x^2 + 6y^2 - 2x - 5 = 0$

97. Writing Use the figure to explain why $|d_2 - d_1| = 2a$ for a hyperbola.

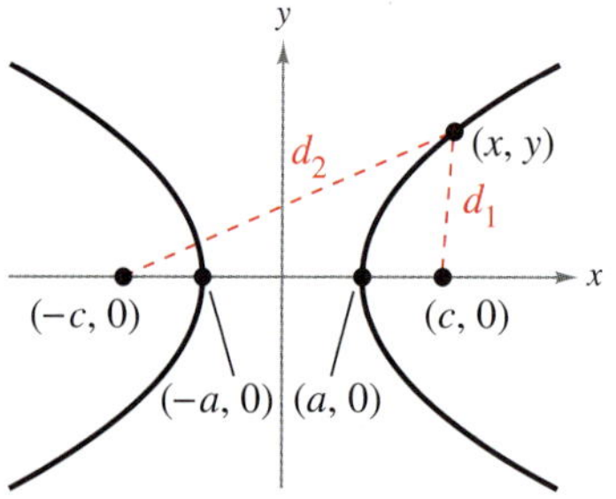

98. Error Analysis Describe the error in finding the asymptotes of the hyperbola

$$\frac{(y+5)^2}{9} - \frac{(x-3)^2}{4} = 1.$$

$$y = k \pm \frac{b}{a}(x - h)$$

$$= -5 \pm \frac{2}{3}(x - 3)$$

The asymptotes are $y = \frac{2}{3}x - 7$ and $y = -\frac{2}{3}x - 3$.

99. Proof Show that $c^2 = a^2 + b^2$ for the equation of the hyperbola

$$\frac{x^2}{a^2} - \frac{y^2}{b^2} = 1$$

where the distance from the center of the hyperbola $(0, 0)$ to a focus is c.

100. Proof Prove that the graph of the equation

$$Ax^2 + Cy^2 + Dx + Ey + F = 0$$

is one of the following (except in degenerate cases).

Conic	*Condition*
(a) Circle	$A = C$
(b) Parabola	$A = 0$ or $C = 0$ (but not both)
(c) Ellipse	$AC > 0$
(d) Hyperbola	$AC < 0$

Cumulative Mixed Review

Factoring a Polynomial **In Exercises 101–106, factor the polynomial completely.**

101. $x^3 - 16x$

102. $x^2 + 14x + 49$

103. $2x^3 - 24x^2 + 72x$

104. $6x^3 - 11x^2 - 10x$

105. $16x^3 + 54$

106. $4 - x + 4x^2 - x^3$

Graphing a Function **In Exercises 107–110, graph the function.**

107. $f(x) = |x + 3|$

108. $f(x) = |x - 4| + 1$

109. $g(x) = \sqrt{4 - x^2}$

110. $g(x) = \sqrt{3x - 2}$

9.4 Parametric Equations

Plane Curves

Up to this point, you have been representing a graph by a single equation involving *two* variables such as x and y. In this section, you will study situations in which it is useful to introduce a *third* variable to represent a curve in the plane.

To see the usefulness of this procedure, consider the path of an object propelled into the air at an angle of 45°. When the initial velocity of the object is 48 feet per second, it can be shown that the object follows the parabolic path

$$y = -\frac{x^2}{72} + x \qquad \text{Rectangular equation}$$

as shown in Figure 9.35. However, this equation does not tell the whole story. Although it does tell you *where* the object has been, it does not tell you *when* the object was at a given point (x, y) on the path. To determine this time, you can introduce a third variable t, called a **parameter.** It is possible to write both x and y as functions of t to obtain the **parametric equations**

$$x = 24\sqrt{2}t \qquad \text{Parametric equation for x}$$

and

$$y = -16t^2 + 24\sqrt{2}t. \qquad \text{Parametric equation for y}$$

From this set of equations, you can determine that at time $t = 0$, the object is at the point $(0, 0)$. Similarly, at time $t = 1$, the object is at the point

$$\left(24\sqrt{2}, 24\sqrt{2} - 16\right)$$

and so on.

Rectangular equation:
$y = -\frac{x^2}{72} + x$

Parametric equations:
$x = 24\sqrt{2}t$
$y = -16t^2 + 24\sqrt{2}t$

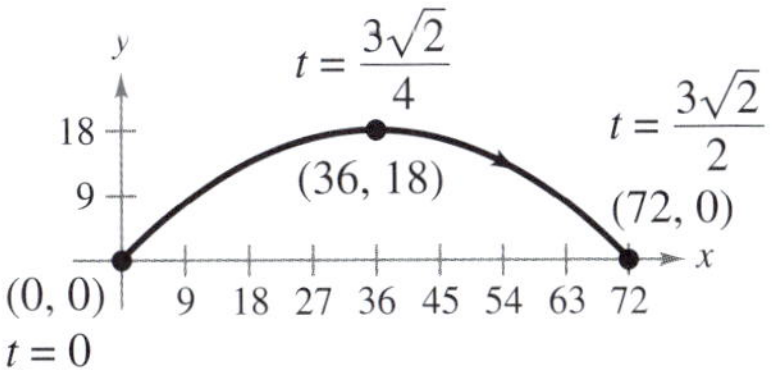

Curvilinear motion: two variables for position, one variable for time
Figure 9.35

For this particular motion problem, x and y are continuous functions of t, and the resulting path is a **plane curve.** (Recall that a *continuous function* is one whose graph has no breaks, holes, or gaps.)

Definition of a Plane Curve

If f and g are continuous functions of t on an interval I, then the set of ordered pairs

$$(f(t), g(t))$$

is a **plane curve** C. The equations

$$x = f(t) \qquad \text{and} \qquad y = g(t)$$

are **parametric equations** for C, and t is the **parameter.**

What you should learn

- Evaluate sets of parametric equations for given values of the parameter.
- Graph curves that are represented by sets of parametric equations.
- Rewrite sets of parametric equations as single rectangular equations by eliminating the parameter.
- Find sets of parametric equations for graphs.

Why you should learn it

Parametric equations are useful for modeling the path of an object. For instance, in Exercise 64 on page 686, a set of parametric equations is used to model the path of an arrow.

Tomasz Trojanowski/Shutterstock.com

Graphs of Plane Curves

One way to sketch a curve represented by a pair of parametric equations is to plot points in the xy-plane. Each set of coordinates (x, y) is determined from a value chosen for the parameter t. By plotting the resulting points in the order of *increasing* values of t, you trace the curve in a specific direction. This is called the **orientation** of the curve.

Technology Tip

Most graphing utilities have a *parametric* mode. So, another way to graph a curve represented by a pair of parametric equations is to use a graphing utility, as shown in Example 2. For instructions on how to use the *parametric* mode, see Appendix A; for specific keystrokes, go to this textbook's *Student Companion Website.*

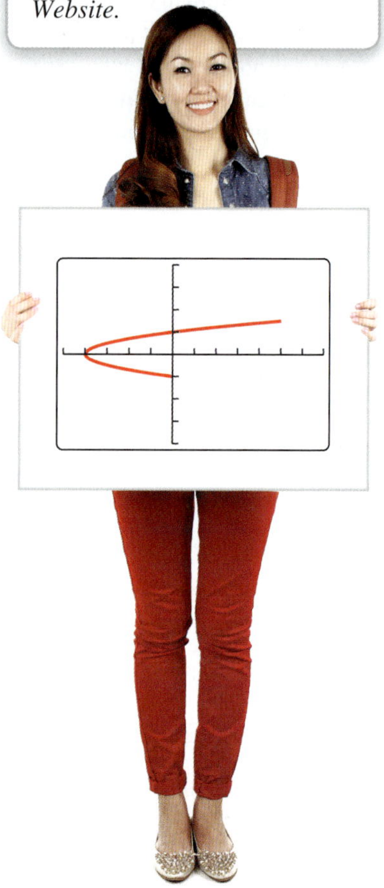

EXAMPLE 1 Sketching a Plane Curve

See LarsonPrecalculus.com for an interactive version of this type of example.

Sketch the curve given by the parametric equations

$$x = t^2 - 4 \quad \text{and} \quad y = \frac{t}{2}, \quad -2 \le t \le 3.$$

Describe the orientation of the curve.

Solution

Using values of t in the interval, the parametric equations yield the points (x, y) shown in the table.

t	-2	-1	0	1	2	3
x	0	-3	-4	-3	0	5
y	-1	$-\frac{1}{2}$	0	$\frac{1}{2}$	1	$\frac{3}{2}$

By plotting these points in the order of increasing t, you obtain the curve shown in Figure 9.36. The arrows on the curve indicate its orientation as t increases from -2 to 3. So, when a particle moves on this curve, it would start at $(0, -1)$ and then move along the curve to the point $\left(5, \frac{3}{2}\right)$.

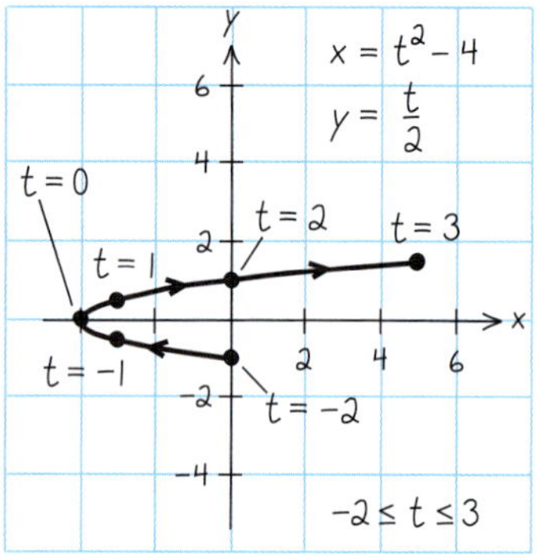

Figure 9.36

Checkpoint *Audio-video solution in English & Spanish at LarsonPrecalculus.com*

Sketch the curve given by the parametric equations

$$x = 2t \quad \text{and} \quad y = 4t^2 + 2, \quad -2 \le t \le 2.$$

Note that the graph shown in Figure 9.36 does not define y as a function of x. This points out one benefit of parametric equations—they can be used to represent graphs that are more general than graphs of functions.

Two different sets of parametric equations can have the same graph. For example, the set of parametric equations

$$x = 4t^2 - 4 \quad \text{and} \quad y = t, \quad -1 \le t \le \tfrac{3}{2}$$

has the same graph as the set given in Example 1. However, by comparing the values of t in Figures 9.36 and 9.37, you can see that this second graph is traced out more *rapidly* (considering t as time) than the first graph. So, in applications, different parametric representations can be used to represent various *speeds* at which objects travel along a given path.

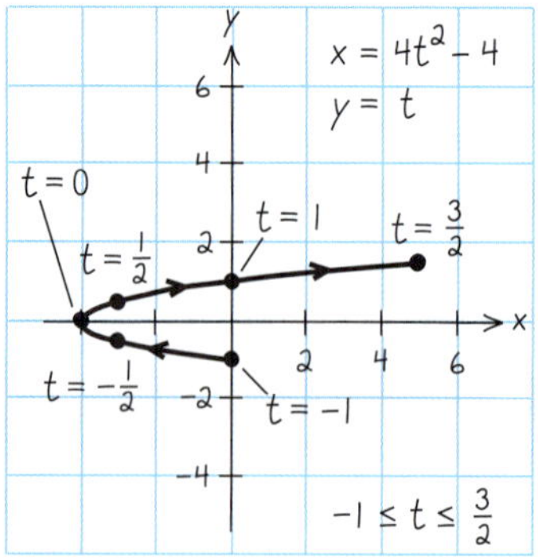

Figure 9.37

EXAMPLE 2 Using a Graphing Utility in Parametric Mode

Use a graphing utility to graph the curves represented by the parametric equations. Use the graph and the Vertical Line Test to determine whether y is a function of x.

a. $x = t^2,\ y = t^3$

b. $x = t,\ y = t^3$

c. $x = t^2,\ y = t$

Solution

Begin by setting the graphing utility to *parametric* mode. When choosing a viewing window, you must set not only minimum and maximum values of x and y but also minimum and maximum values of t.

a. Enter the parametric equations for x and y, as shown in Figure 9.38. Use the viewing window shown in Figure 9.39. The curve is shown in Figure 9.40. From the graph, you can see that y *is not* a function of x.

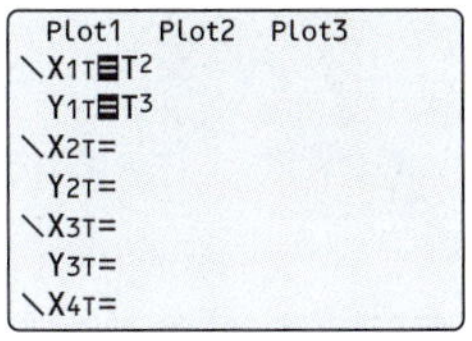

Figure 9.38

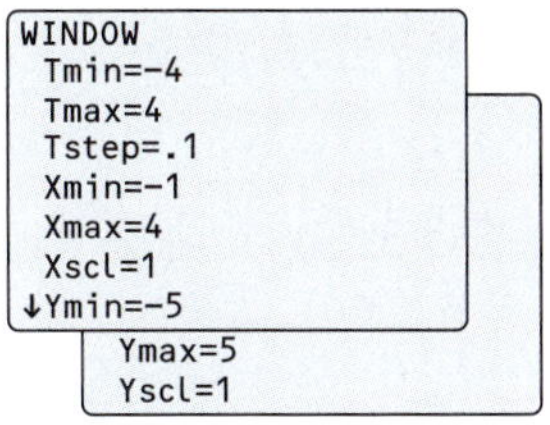

Figure 9.39

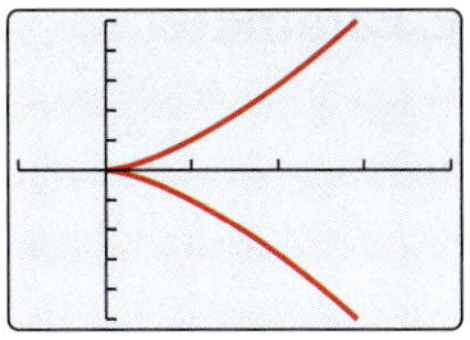

Figure 9.40

b. Enter the parametric equations for x and y, as shown in Figure 9.41. Use the viewing window shown in Figure 9.42. The curve is shown in Figure 9.43. From the graph, you can see that y *is* a function of x.

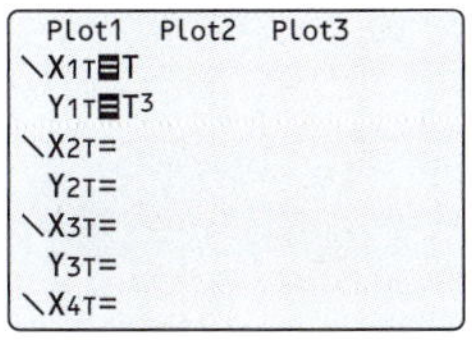

Figure 9.41

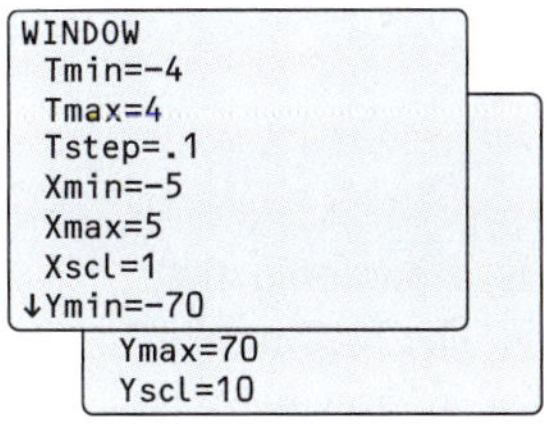

Figure 9.42

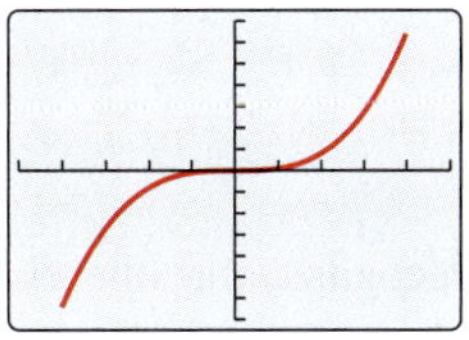

Figure 9.43

c. Enter the parametric equations for x and y, as shown in Figure 9.44. Use the viewing window shown in Figure 9.45. The curve is shown in Figure 9.46. From the graph, you can see that y *is not* a function of x.

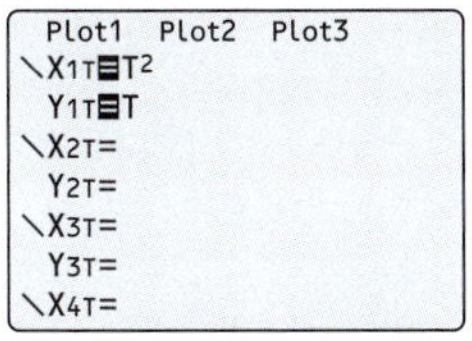

Figure 9.44

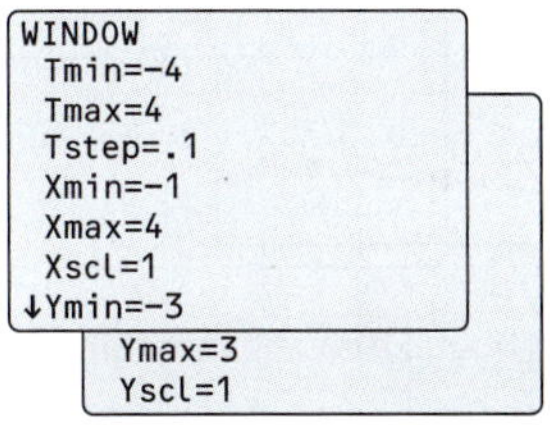

Figure 9.45

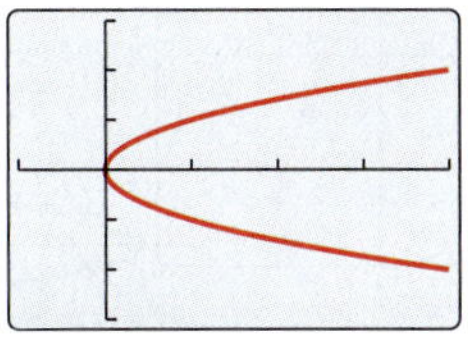

Figure 9.46

Checkpoint *Audio-video solution in English & Spanish at LarsonPrecalculus.com*

Use a graphing utility to graph the curve represented by the equations $x = (t - 2)^2$ and $y = 3t^2$. Use the graph and the Vertical Line Test to determine whether y is a function of x.

Explore the Concept

Use a graphing utility set in *parametric* mode to graph the curve

$$x = t \quad \text{and} \quad y = 1 - t^2.$$

Set the viewing window so that $-4 \le x \le 4$ and $-12 \le y \le 2$. Then graph the curve using each of the following t settings.

a. $0 \le t \le 3$

b. $-3 \le t \le 0$

c. $-3 \le t \le 3$

Compare the curves given by the different t settings. Repeat this experiment using $x = -t$. How does this change the results?

Technology Tip

Notice in Example 2 that to set the viewing windows of parametric graphs, you may have to scroll down to enter the Ymax and Yscl values.

Eliminating the Parameter

Many curves that are represented by sets of parametric equations have graphs that can also be represented by rectangular equations (in x and y). The process of finding the rectangular equation is called **eliminating the parameter.**

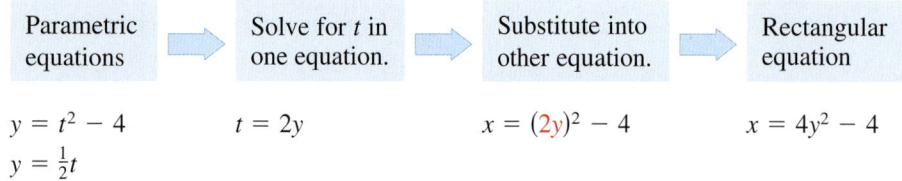

Parametric equations	→	Solve for t in one equation.	→	Substitute into other equation.	→	Rectangular equation
$y = t^2 - 4$ $y = \frac{1}{2}t$		$t = 2y$		$x = (2y)^2 - 4$		$x = 4y^2 - 4$

Now, you can recognize that the equation $x = 4y^2 - 4$ represents a parabola with a horizontal axis and vertex at $(-4, 0)$. When converting equations from parametric to rectangular form, you may need to alter the domain of the rectangular equation so that its graph matches the graph of the parametric equations.

Algebra Help

It is important to realize that eliminating the parameter is primarily an aid to curve sketching. When the parametric equations represent the path of a moving object, the graph alone is not sufficient to describe the object's motion. You still need the parametric equations to determine the *position, direction,* and *speed* at a given time.

EXAMPLE 3 Eliminating the Parameter

Identify the curve represented by the equations $x = \dfrac{1}{\sqrt{t+1}}$ and $y = \dfrac{t}{t+1}$.

Solution

Solving for t in the equation for x produces

$$x^2 = \frac{1}{t+1} \quad \Rightarrow \quad \frac{1}{x^2} = t + 1 \quad \Rightarrow \quad \frac{1}{x^2} - 1 = t.$$

Substituting in the equation for y, you obtain the rectangular equation

$$y = \frac{t}{t+1} = \frac{\dfrac{1}{x^2} - 1}{\dfrac{1}{x^2} - 1 + 1} = \frac{\dfrac{1 - x^2}{x^2}}{\dfrac{1}{x^2}} \cdot \frac{x^2}{x^2} = 1 - x^2.$$

From the rectangular equation, you can recognize that the curve is a parabola that opens downward and has its vertex at $(0, 1)$, as shown in Figure 9.47. The rectangular equation is defined for all values of x. The parametric equation for x, however, is defined only when $t > -1$. From the graph of the parametric equations, you can see that x is always positive, as shown in Figure 9.48. So, you should restrict the domain of x to positive values, as shown in Figure 9.49.

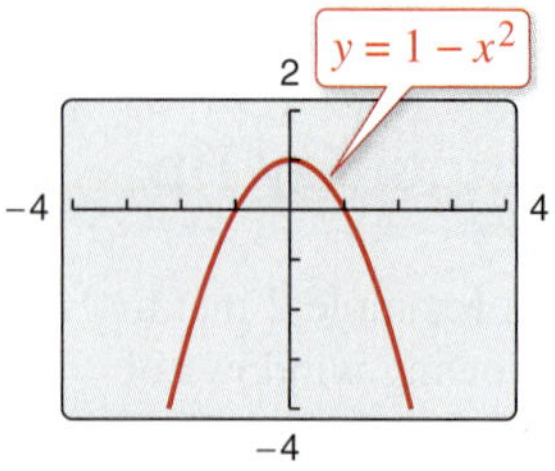

Figure 9.47

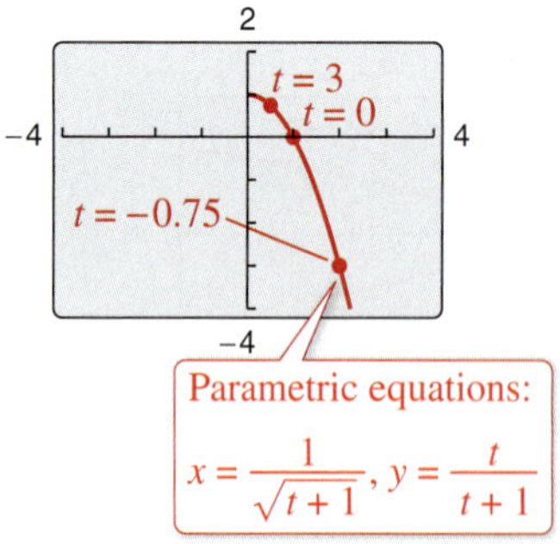

Figure 9.48

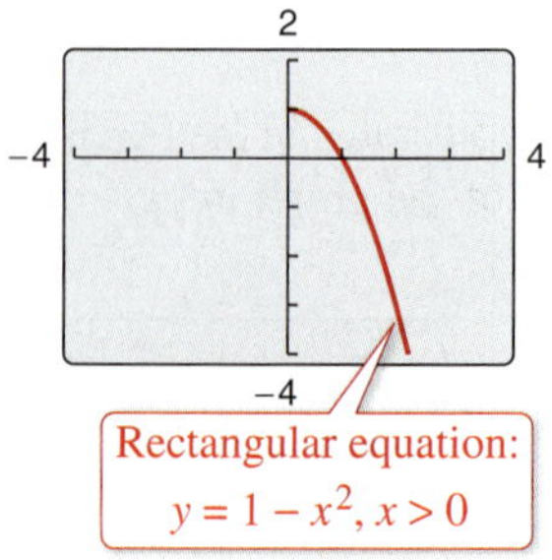

Figure 9.49

Checkpoint Audio-video solution in English & Spanish at LarsonPrecalculus.com

Identify the curve represented by the equations $x = \dfrac{1}{\sqrt{t-1}}$ and $y = \dfrac{t+1}{t-1}$.

Finding Parametric Equations for a Graph

You have been studying techniques for sketching the graph represented by a set of parametric equations. Now consider the *reverse* problem—that is, how can you find a set of parametric equations for a given graph or a given physical description? From the discussion following Example 1, you know that such a representation is not unique. That is, the equations

$$x = 4t^2 - 4 \quad \text{and} \quad y = t, \quad -1 \le t \le \frac{3}{2}$$

produced the same graph as the equations

$$x = t^2 - 4 \quad \text{and} \quad y = \frac{t}{2}, \quad -2 \le t \le 3.$$

This is further demonstrated in Example 4.

EXAMPLE 4 Finding Parametric Equations for a Graph

Find a set of parametric equations to represent the graph of $y = 1 - x^2$ using the parameters (a) $t = x$ and (b) $t = 1 - x$.

Solution

a. Letting $t = x$, you obtain the following parametric equations.

$x = t$ Parametric equation for x

$y = 1 - t^2$ Parametric equation for y

The graph of these equations is shown in the figure.

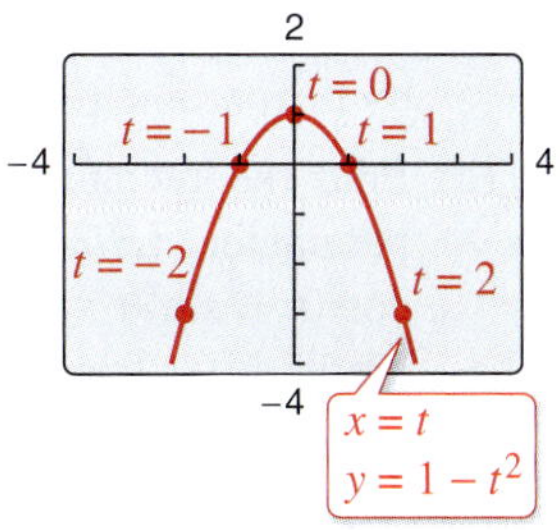

b. Letting $t = 1 - x$, you obtain the following parametric equations.

$x = 1 - t$ Parametric equation for x

$y = 1 - (1 - t)^2 = 2t - t^2$ Parametric equation for y

The graph of these equations is shown in the figure. Note that the graph in this figure has the opposite orientation of the graph in part (a).

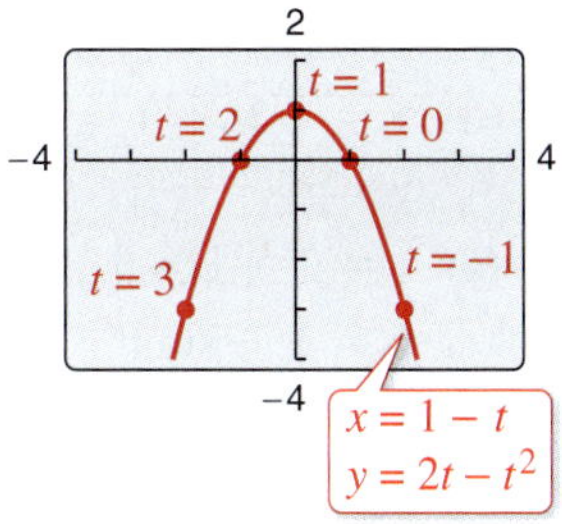

✓ Checkpoint *Audio-video solution in English & Spanish at LarsonPrecalculus.com*

Find a set of parametric equations to represent the graph of $y = x^2 + 2$ using the parameters (a) $t = x$ and (b) $t = 1 - x$.

What's Wrong?

You use a graphing utility in *parametric* mode to graph the parametric equations in Example 4(a). You use a standard viewing window and expect to obtain a parabola similar to the one in Example 4(a), shown at the left. Your result is shown below. What's wrong?

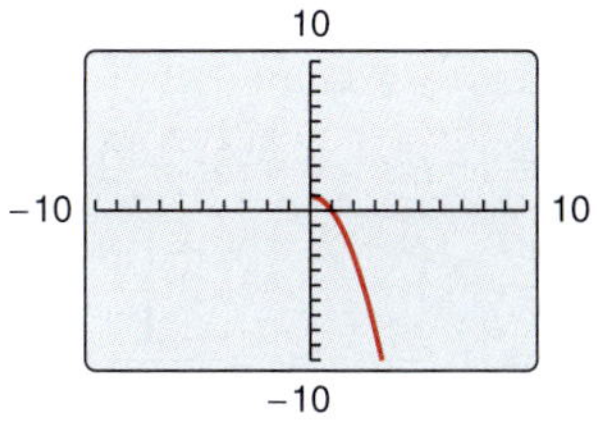

9.4 Exercises

See *CalcChat.com* for tutorial help and worked-out solutions to odd-numbered exercises.
For instructions on how to use a graphing utility, see Appendix A.

Vocabulary and Concept Check

In Exercises 1 and 2, fill in the blank(s).

1. If f and g are continuous functions of t on an interval I, then the set of ordered pairs $(f(t), g(t))$ is a _______ C. The equations given by $x = f(t)$ and $y = g(t)$ are _______ for C, and t is the _______ .

2. The _______ of a curve is the direction in which the curve is traced out for increasing values of the parameter.

3. Given a set of parametric equations, how do you find the corresponding rectangular equation?

4. What point on the plane curve represented by the parametric equations $x = t$ and $y = t$ corresponds to $t = 3$?

Procedures and Problem Solving

Identifying the Graph of Parametric Equations **In Exercises 5–8, match the set of parametric equations with its graph. [The graphs are labeled (a), (b), (c), and (d).]**

(a)

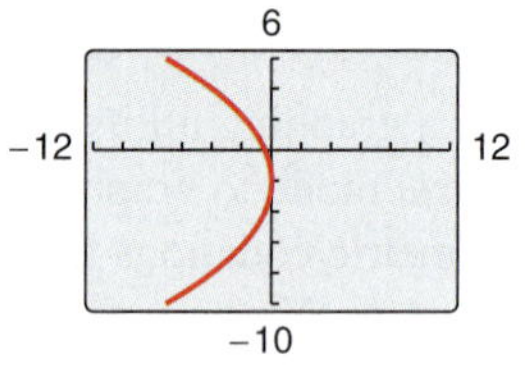

(b)

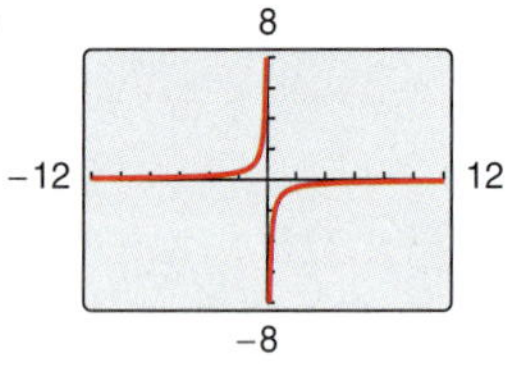

(c)

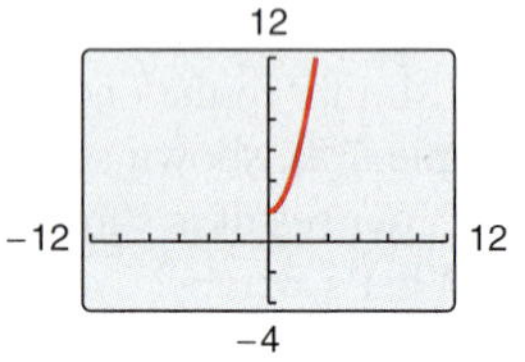

(d)

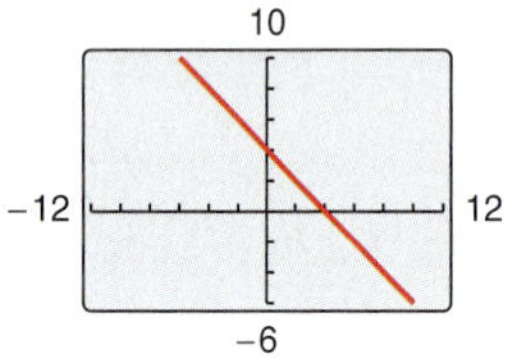

5. $x = t, \quad y = 4 - t$

6. $x = -t^2, \quad y = 3t - 2$

7. $x = \sqrt{t - 2}, \quad y = t$

8. $x = 1 - t, \quad y = \dfrac{1}{t - 1}$

9. Using Parametric Equations Consider the parametric equations $x = \sqrt{t}$ and $y = 2 - t$.

(a) Create a table of x- and y-values using $t = 0, 1, 2, 3$, and 4.

(b) Plot the points (x, y) generated in part (a) and sketch a graph of the parametric equations for $0 \le t \le 4$. Describe the orientation of the curve.

(c) Use a graphing utility to graph the curve represented by the parametric equations.

(d) Find the rectangular equation by eliminating the parameter. Sketch its graph. How does the graph differ from those in parts (b) and (c)?

10. Using Parametric Equations Consider the parametric equations $x = 4\cos^2 t$ and $y = 4 \sin t$.

(a) Create a table of x- and y-values using $t = -\pi/2, -\pi/4, 0, \pi/4$, and $\pi/2$.

(b) Plot the points (x, y) generated in part (a) and sketch a graph of the parametric equations for $-\dfrac{\pi}{2} \le t \le \dfrac{\pi}{2}$. Describe the orientation of the curve.

(c) Use a graphing utility to graph the curve represented by the parametric equations.

(d) Find the rectangular equation by eliminating the parameter. (*Hint:* Use the trigonometric identity $\cos^2 t + \sin^2 t = 1$.) Sketch its graph. How does the graph differ from those in parts (b) and (c)?

Identifying Parametric Equations for a Plane Curve **In Exercises 11 and 12, determine which set of parametric equations represents the graph shown.**

11. (a) $x = t^2$
$y = 2t + 1$
(b) $x = 2t + 1$
$y = t^2$
(c) $x = 2t - 1$
$y = t^2$
(d) $x = -2t + 1$
$y = t^2$

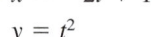

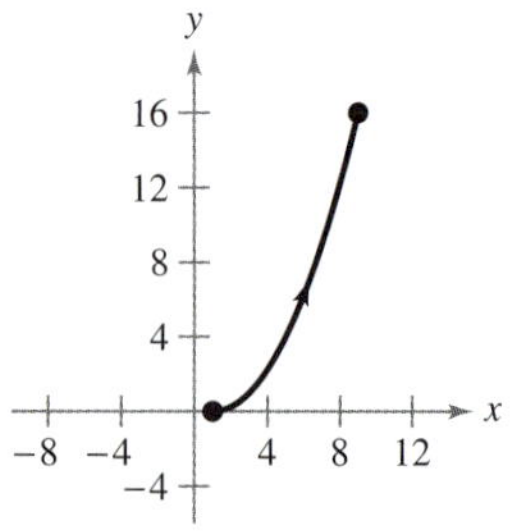

12. (a) $x = 2 - \cos\theta$
$y = 3 - \sin\theta$
(b) $x = 3 - \cos\theta$
$y = 2 + \sin\theta$
(c) $x = 2 + \cos\theta$
$y = 3 - \sin\theta$
(d) $x = 3 + \cos\theta$
$y = 2 - \sin\theta$

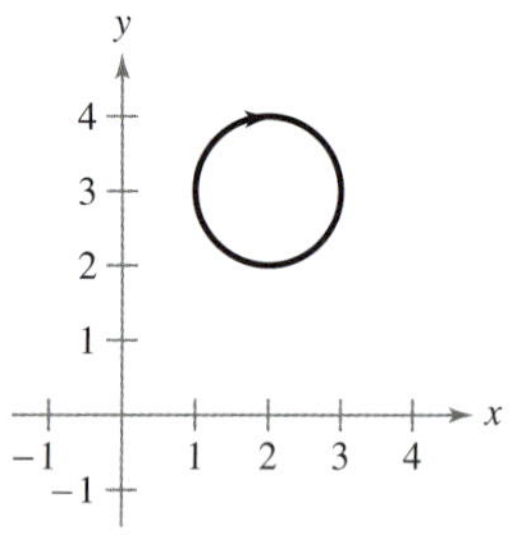

Sketching a Plane Curve and Eliminating the Parameter In Exercises 13–28, (a) sketch the curve represented by the parametric equations (indicate the orientation of the curve). (b) Eliminate the parameter and write the corresponding rectangular equation whose graph represents the curve. Adjust the domain of the resulting rectangular equation, if necessary.

13. $x = t,\ y = -4t$
14. $x = t,\ y = \frac{1}{2}t$
15. $x = 3t - 3,\ y = 2t + 1$
16. $x = 3 - 2t,\ y = 2 + 3t$
17. $x = \frac{1}{4}t,\ y = t^2$
18. $x = t,\ y = t^3$
19. $x = \sqrt{t} - 3,\ y = t^3$
20. $x = \sqrt{t},\ y = 1 - t$
21. $x = 2t,\ y = |t - 2|$
22. $x = |t - 1|,\ y = t + 2$
23. $x = e^{-t},\ y = e^{3t}$
24. $x = 2e^t,\ y = -e^{4t}$
25. $x = 2\cos\theta,\ y = 3\sin\theta$
26. $x = \cos\theta,\ y = 4\sin\theta$
27. $x = t^3,\ y = 3\ln t$
28. $x = \ln 2t,\ y = 2t^2$

Using a Graphing Utility in Parametric Mode In Exercises 29–34, use a graphing utility to graph the curve represented by the parametric equations. Use the Vertical Line Test to determine whether y is a function of x.

29. $x = 4 + 3\cos\theta$, $y = -2 + \sin\theta$
30. $x = 4 + 3\cos\theta$, $y = -2 + 2\sin\theta$
31. $x = 4\sec\theta$, $y = 2\tan\theta$
32. $x = \sec\theta$, $y = \tan\theta$
33. $x = t/2$, $y = \ln(t^2 + 1)$
34. $x = e^t - 4$, $y = t^2$

Comparing Plane Curves In Exercises 35 and 36, determine how the plane curves differ from each other.

35. (a) $x = t$, $y = 2t + 1$
(b) $x = \cos\theta$, $y = 2\cos\theta + 1$
(c) $x = e^{-t}$, $y = 2e^{-t} + 1$
(d) $x = e^t$, $y = 2e^t + 1$

36. (a) $x = 2\sqrt{t}$, $y = 4 - \sqrt{t}$
(b) $x = 2\sqrt[3]{t}$, $y = 4 - \sqrt[3]{t}$
(c) $x = 2(t + 1)$, $y = 3 - t$
(d) $x = -2t^2$, $y = 4 + t^2$

Eliminating the Parameter In Exercises 37–40, eliminate the parameter and obtain the standard form of the rectangular equation.

37. Line through (x_1, y_1) and (x_2, y_2):
$x = x_1 + t(x_2 - x_1),\ y = y_1 + t(y_2 - y_1)$
38. Circle: $x = h + r\cos\theta,\ y = k + r\sin\theta$
39. Ellipse: $x = h + a\cos\theta,\ y = k + b\sin\theta$
40. Hyperbola: $x = h + a\sec\theta,\ y = k + b\tan\theta$

Finding Parametric Equations for a Graph In Exercises 41–44, use the results of Exercises 37–40 to find a set of parametric equations for the line or conic.

41. Line: passes through $(-1, -1)$ and $(6, 9)$
42. Circle: center: $(-2, -5)$; radius: 7
43. Ellipse: vertices: $(\pm 5, 0)$; foci: $(\pm 4, 0)$
44. Hyperbola: vertices: $(\pm 2, 0)$; foci: $(\pm 3, 0)$

Finding Parametric Equations for a Graph In Exercises 45–52, find a set of parametric equations to represent the graph of the given rectangular equation using the parameters (a) $t = x$ and (b) $t = 2 - x$.

45. $y = -2x + 1$
46. $y = 4x - 9$
47. $y = \dfrac{1}{x}$
48. $y = \dfrac{1}{2x}$
49. $y = 4x^2 + 5$
50. $y = x^3 - x^2$
51. $y = e^{-x}$
52. $y = \ln(x + 2)$

Graphing a Curve In Exercises 53–58, use a graphing utility to graph the curve represented by the parametric equations.

53. Hypocycloid: $x = 3\cos^3\theta,\ y = 3\sin^3\theta$
54. Curtate cycloid: $x = 8\theta - 4\sin\theta,\ y = 8 - 4\cos\theta$
55. Witch of Agnesi: $x = 2\cot\theta,\ y = 2\sin^2\theta$
56. Folium of Descartes: $x = \dfrac{3t}{1 + t^3},\ y = \dfrac{3t^2}{1 + t^3}$
57. Cycloid: $x = \theta + \sin\theta,\ y = 1 - \cos\theta$
58. Prolate cycloid: $x = 2\theta - 4\sin\theta,\ y = 2 - 4\cos\theta$

Identifying a Graph In Exercises 59–62, match the parametric equations with the correct graph. [The graphs are labeled (a), (b), (c), and (d).]

(a)
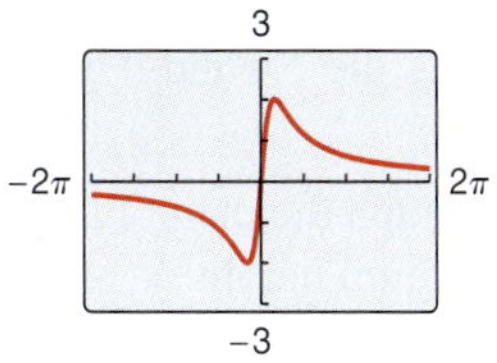

(b)
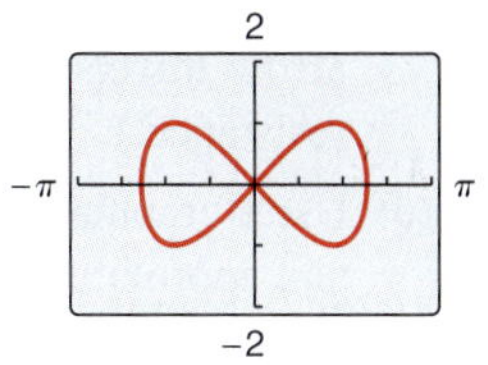

(c)
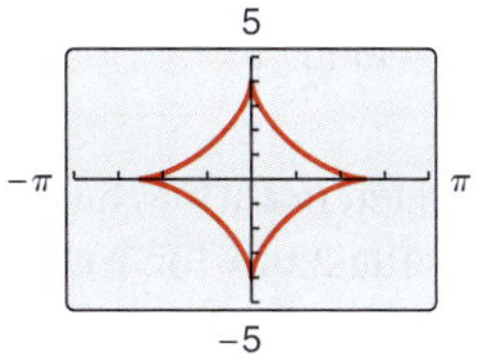

(d)
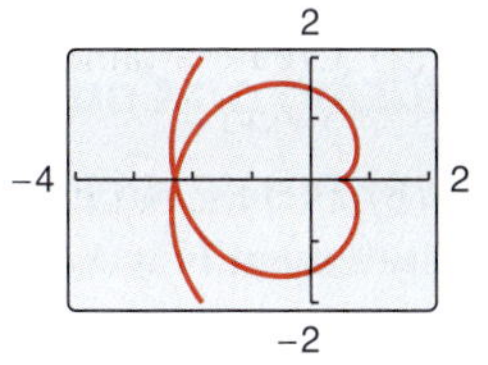

59. Lissajous curve: $x = 2\cos\theta,\ y = \sin 2\theta$
60. Evolute of ellipse: $x = 2\cos^3\theta,\ y = 4\sin^3\theta$
61. Involute of circle: $x = \frac{1}{2}(\cos\theta + \theta\sin\theta)$
$y = \frac{1}{2}(\sin\theta - \theta\cos\theta)$
62. Serpentine curve: $x = \frac{1}{2}\cot\theta,\ y = 4\sin\theta\cos\theta$

Projectile Motion **In Exercises 63 and 64, consider a projectile launched at a height of h feet above the ground at an angle of θ with the horizontal. The initial velocity is v_0 feet per second, and the path of the projectile is modeled by the parametric equations**

$$x = (v_0 \cos \theta)t \text{ and } y = h + (v_0 \sin \theta)t - 16t^2.$$

63. MODELING DATA

The center field fence in a baseball stadium is 7 feet high and 408 feet from home plate. A baseball is hit at a point 3 feet above the ground. It leaves the bat at an angle of θ degrees with the horizontal at a speed of 100 miles per hour (see figure).

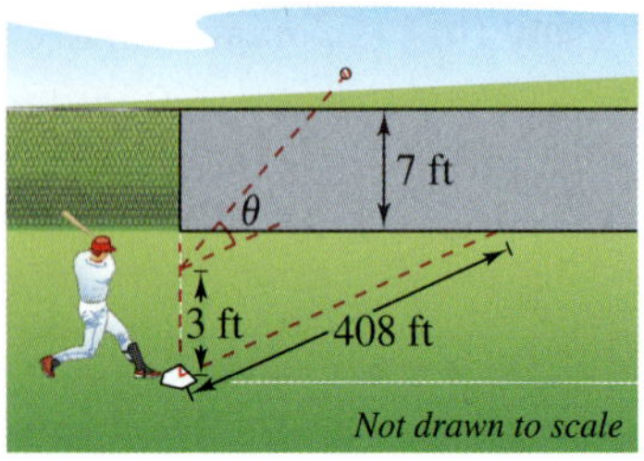

(a) Write a set of parametric equations that model the path of the baseball.

(b) Use a graphing utility to graph the path of the baseball when $\theta = 23°$. Is the hit a home run?

(c) Find the minimum angle required for the hit to be a home run.

64. ***Why you should learn it*** *(p. 679)* An archer releases an arrow from a bow at a point 5 feet above the ground. The arrow leaves the bow at an angle of 15° with the horizontal and at an initial speed of 225 feet per second.

(a) Write a set of parametric equations that model the path of the arrow.

(b) Assuming the ground is level, find the distance the arrow travels before it hits the ground. (Ignore air resistance.)

(c) Use a graphing utility to graph the path of the arrow and approximate its maximum height.

(d) Find the total time the arrow is in the air.

Focusing on Concepts

True or False? **In Exercises 65–68, determine whether the statement is true or false. Justify your answer.**

65. The two sets of parametric equations $x = t$, $y = t^2 + 1$ and $x = 3t$, $y = 9t^2 + 1$ correspond to the same rectangular equation.

66. Because the graphs of the parametric equations $x = t^2$, $y = t^2$ and $x = t$, $y = t$ both represent the line $y = x$, they are the same plane curve.

67. If y is a function of t and x is a function of t, then y must be a function of x.

68. The parametric equations $x = at + h$ and $y = bt + k$, where $a \neq 0$ and $b \neq 0$, represent a circle centered at (h, k) when $a = b$.

Error Analysis **In Exercises 69 and 70, describe the error in graphing the parametric equation, where t is any real number.**

69. $x = -t, y = t^2$

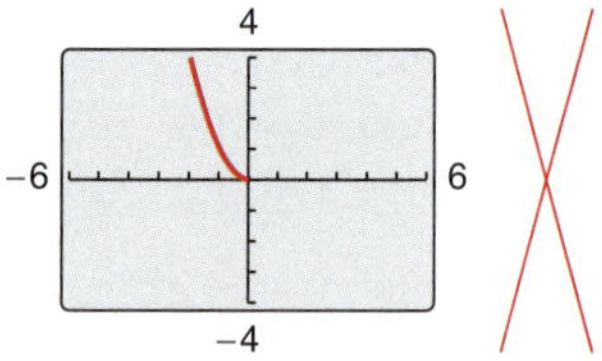

70. $x = -\sin t, y = t$

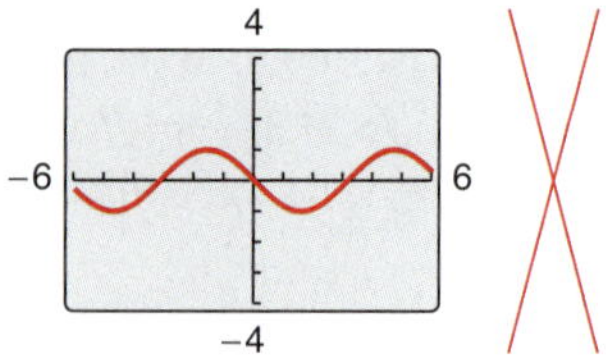

71. Think About It The curve shown is represented by the parametric equations

$$x = 6 \cos \theta \quad \text{and} \quad y = 6 \sin \theta, \quad 0 \leq \theta \leq 6.$$

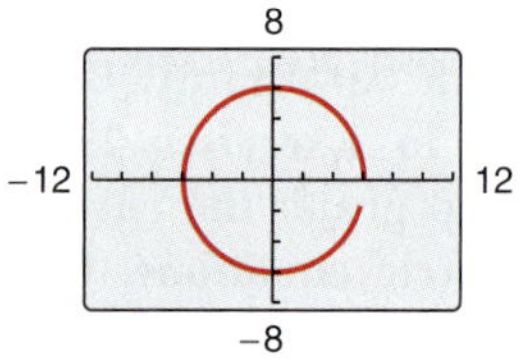

(a) Describe the orientation of the curve.

(b) Determine a range of θ that gives the graph of a circle.

(c) Write a set of parametric equations representing the curve so that the curve traces from the same point as the original curve but in the opposite direction.

(d) How does the original curve change when cosine and sine are interchanged?

72. HOW DO YOU SEE IT? The graph of the parametric equations $x = t$ and $y = t^2$ is shown below. Determine whether the graph would change for each set of parametric equations. If so, how would it change?

(a) $x = -t, y = t^2$

(b) $x = t + 1, y = t^2$

(c) $x = t, y = t^2 + 1$

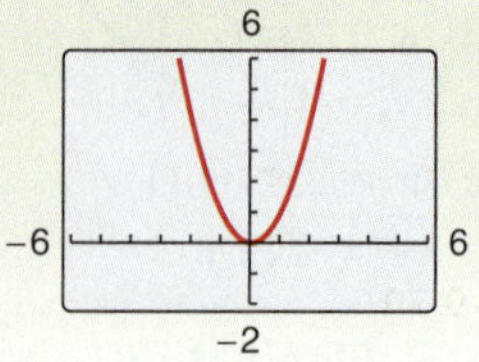

Cumulative Mixed Review

Vector Operations **In Exercises 73 and 74, find the component form of w and sketch the specific vector operations geometrically, where $\mathbf{u} = 4\mathbf{i} - 5\mathbf{j}$ and $\mathbf{v} = 3\mathbf{i}$.**

73. $\mathbf{w} = \frac{1}{6}\mathbf{v}$

74. $\mathbf{w} = \mathbf{u} + \mathbf{v}$

9.5 Polar Coordinates

What you should learn

- Plot points and find multiple representations of points in the polar coordinate system.
- Convert points from rectangular to polar form and vice versa.
- Convert equations from rectangular to polar form and vice versa.

Why you should learn it

You can use polar coordinates in mathematical modeling . For instance, in Exercise 79 on page 692, you will use polar coordinates to model the path of a passenger car on a Ferris wheel.

Introduction

So far, you have been representing graphs of equations as collections of points (x, y) in the rectangular coordinate system, where x and y represent the directed distances from the coordinate axes to the point (x, y). In this section, you will study a second coordinate system called the **polar coordinate system.**

To form the polar coordinate system in the plane, fix a point O, called the **pole** (or **origin**), and construct from O an initial ray called the **polar axis,** as shown in the figure. Then each point P in the plane can be assigned **polar coordinates** (r, θ), as follows.

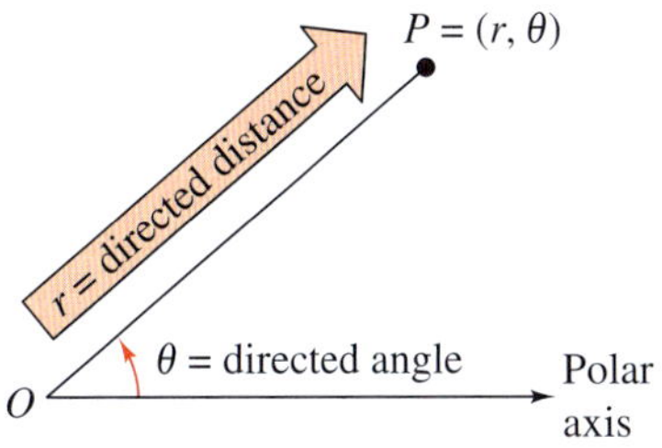

1. r = *directed distance* from O to P
2. θ = *directed angle*, counterclockwise from the polar axis to segment $\overline{OP}$

EXAMPLE 1 Plotting Points in the Polar Coordinate System

Plot each point given in polar coordinates.

a. $\left(2, \frac{\pi}{3}\right)$ **b.** $\left(3, -\frac{\pi}{6}\right)$ **c.** $\left(3, \frac{11\pi}{6}\right)$

Solution

a. The point $(r, \theta) = (2, \pi/3)$ lies two units from the pole on the terminal side of the angle $\theta = \pi/3$, as shown in Figure 9.50.

b. The point $(r, \theta) = (3, -\pi/6)$ lies three units from the pole on the terminal side of the angle $\theta = -\pi/6$, as shown in Figure 9.51.

c. The point $(r, \theta) = (3, 11\pi/6)$ coincides with the point $(3, -\pi/6)$, as shown in Figure 9.52.

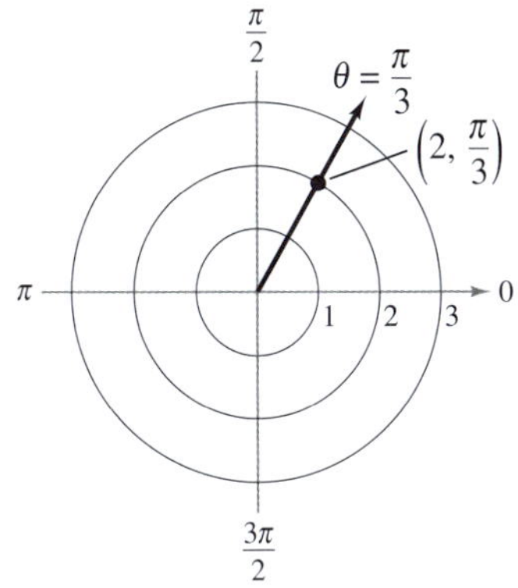

Figure 9.50

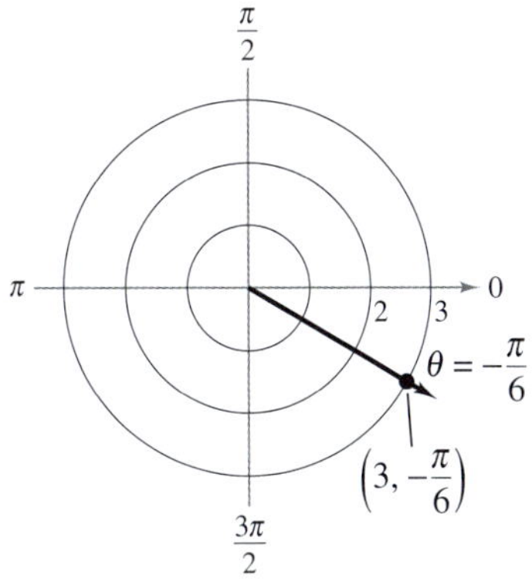

Figure 9.51

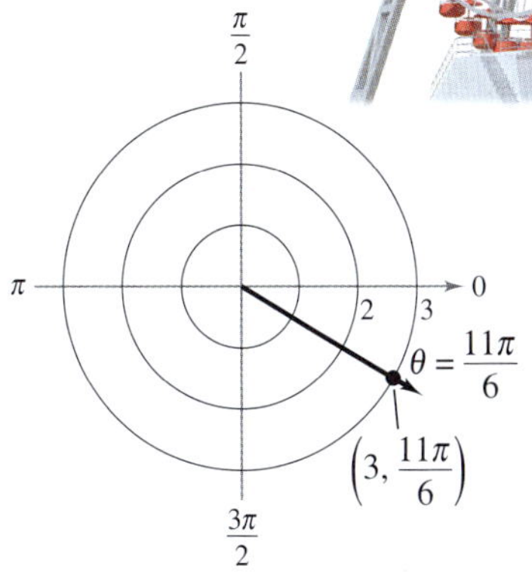

Figure 9.52

Checkpoint Audio-video solution in English & Spanish at LarsonPrecalculus.com

Plot each point given in polar coordinates.

a. $\left(3, \frac{\pi}{4}\right)$ **b.** $\left(2, -\frac{\pi}{3}\right)$ **c.** $\left(2, \frac{5\pi}{3}\right)$

In rectangular coordinates, each point (x, y) has a unique representation. This is not true for polar coordinates. For instance, the coordinates

$$(r, \theta) \quad \text{and} \quad (r, \theta + 2\pi)$$

represent the same point, as illustrated in Example 1. Another way to obtain multiple representations of a point is to use negative values for r. Because r is a *directed distance*, the coordinates

$$(r, \theta) \quad \text{and} \quad (-r, \theta + \pi)$$

represent the same point. In general, the point (r, θ) can be represented as

$$(r, \theta) = (r, \theta \pm 2n\pi) \quad \text{or} \quad (r, \theta) = (-r, \theta \pm (2n + 1)\pi)$$

where n is any integer. Moreover, the pole is represented by $(0, \theta)$, where θ is any angle.

Explore the Concept

Set your graphing utility to *polar* mode. Then graph the equation $r = 3$. (Use a viewing window in which $0 \le \theta \le 2\pi$, $-6 \le x \le 6$, and $-4 \le y \le 4$.) You should obtain a circle of radius 3.

a. Use the *trace* feature to cursor around the circle. Can you locate the point $(3, 5\pi/4)$?

b. Can you locate other representations of the point $(3, 5\pi/4)$? If so, explain how you did it.

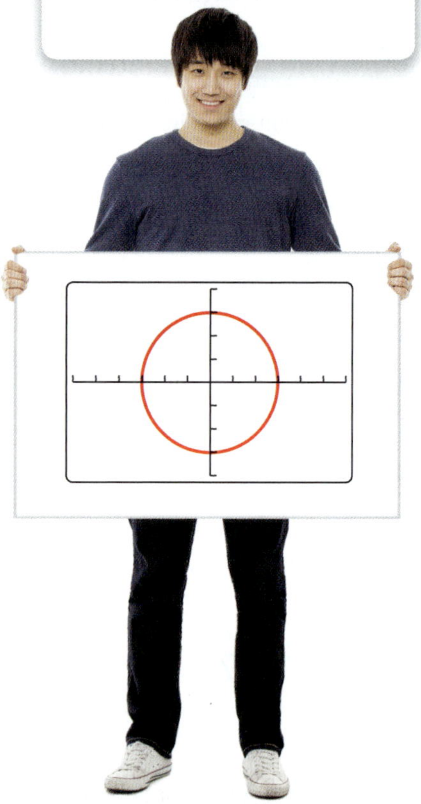

EXAMPLE 2 Multiple Representations of Points

Plot the point

$$\left(3, -\frac{3\pi}{4}\right)$$

and find three additional polar representations of this point, using

$$-2\pi < \theta < 2\pi.$$

Solution

The point is shown in Figure 9.53. Three other representations are as follows.

$$\left(3, -\frac{3\pi}{4} + 2\pi\right) = \left(3, \frac{5\pi}{4}\right) \qquad \text{Add } 2\pi \text{ to } \theta.$$

$$\left(-3, -\frac{3\pi}{4} - \pi\right) = \left(-3, -\frac{7\pi}{4}\right) \qquad \text{Replace } r \text{ with } -r, \text{ and subtract } \pi \text{ from } \theta.$$

$$\left(-3, -\frac{3\pi}{4} + \pi\right) = \left(-3, \frac{\pi}{4}\right) \qquad \text{Replace } r \text{ with } -r, \text{ and add } \pi \text{ to } \theta.$$

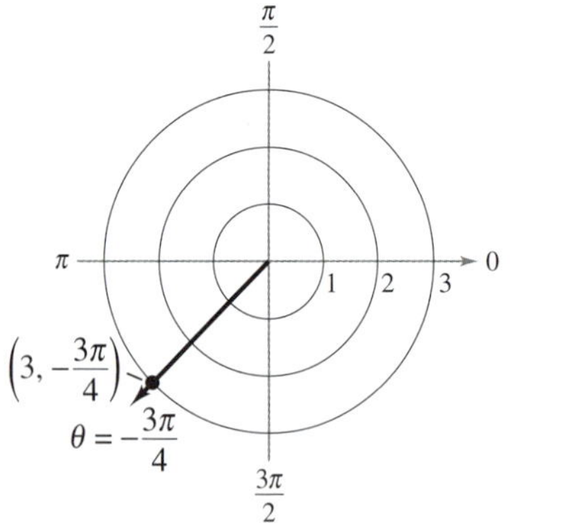

$$\left(3, -\frac{3\pi}{4}\right) = \left(3, \frac{5\pi}{4}\right) = \left(-3, -\frac{7\pi}{4}\right) = \left(-3, \frac{\pi}{4}\right) = \ldots$$

Figure 9.53

✓ ***Checkpoint*** *Audio-video solution in English & Spanish at LarsonPrecalculus.com*

Plot each point and find three additional polar representations of the point, using $-2\pi < \theta < 2\pi$.

a. $\left(3, \frac{4\pi}{3}\right)$ **b.** $\left(2, -\frac{5\pi}{6}\right)$ **c.** $\left(-1, \frac{3\pi}{4}\right)$

Coordinate Conversion

To establish the relationship between polar and rectangular coordinates, let the polar axis coincide with the positive x-axis and the pole with the origin, as shown in Figure 9.54. Because (x, y) lies on a circle of radius r, it follows that $r^2 = x^2 + y^2$. Moreover, for $r > 0$, the definitions of the trigonometric functions imply that

$$\tan\theta = \frac{y}{x}, \qquad \cos\theta = \frac{x}{r}, \qquad \text{and} \qquad \sin\theta = \frac{y}{r}.$$

You can show that the same relationships hold for $r < 0$.

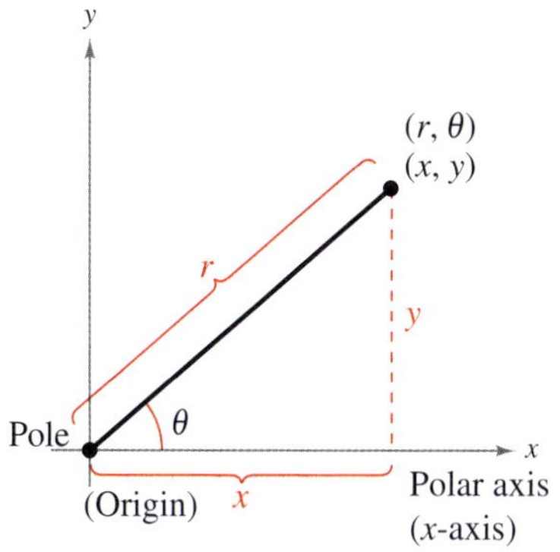

Figure 9.54

Coordinate Conversion

The polar coordinates (r, θ) are related to the rectangular coordinates (x, y), as follows.

Polar-to-Rectangular	*Rectangular-to-Polar*
$x = r\cos\theta$	$\tan\theta = \dfrac{y}{x}$
$y = r\sin\theta$	$r^2 = x^2 + y^2$

EXAMPLE 3 Polar-to-Rectangular Conversion

Convert the point $(2, \pi)$ to rectangular coordinates.

Solution

For the point $(r, \theta) = (2, \pi)$, you have

$$x = r\cos\theta = 2\cos\pi = -2$$

and

$$y = r\sin\theta = 2\sin\pi = 0.$$

The rectangular coordinates are $(x, y) = (-2, 0)$. (See Figure 9.55.)

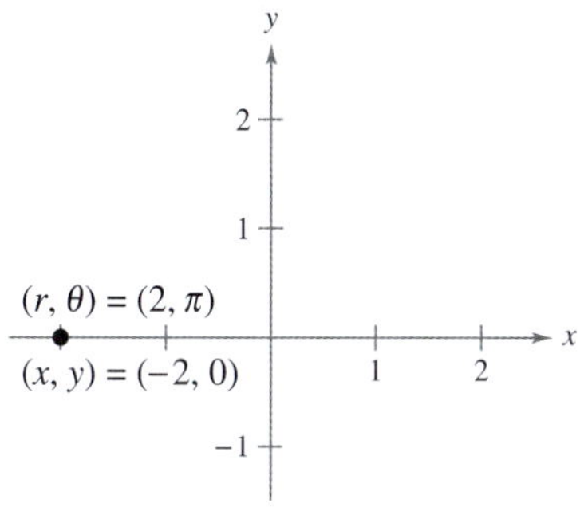

Figure 9.55

Checkpoint *Audio-video solution in English & Spanish at LarsonPrecalculus.com*

Convert the point $\left(4, \dfrac{\pi}{3}\right)$ to rectangular coordinates.

EXAMPLE 4 Rectangular-to-Polar Conversion

Convert the point $(-1, 1)$ to polar coordinates.

Solution

For the second-quadrant point $(x, y) = (-1, 1)$, you have

$$\tan\theta = \frac{y}{x} = \frac{1}{-1} = -1 \quad \Rightarrow \quad \theta = \pi + \arctan(-1) = \frac{3\pi}{4}.$$

Because θ lies in the same quadrant as (x, y), use the positive value of r.

$$r = \sqrt{x^2 + y^2} = \sqrt{(-1)^2 + (1)^2} = \sqrt{2}$$

So, *one* set of polar coordinates is

$$(r, \theta) = \left(\sqrt{2}, \frac{3\pi}{4}\right).$$

See Figure 9.56.

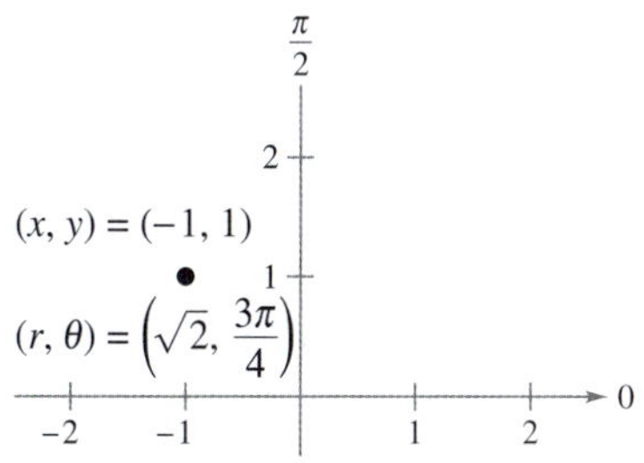

Figure 9.56

Checkpoint *Audio-video solution in English & Spanish at LarsonPrecalculus.com*

Convert the point $(0, 2)$ to polar coordinates.

Equation Conversion

By comparing Examples 3 and 4, you can see that point conversion from the polar to the rectangular system is straightforward, whereas point conversion from the rectangular to the polar system is more involved. For equations, the opposite is true. To convert a rectangular equation to polar form, replace x with $r\cos\theta$ and y with $r\sin\theta$. For instance, the rectangular equation $y = x^2$ can be written in polar form, as follows.

$$y = x^2 \quad \text{Rectangular equation}$$

$$r\sin\theta = (r\cos\theta)^2 \quad \text{Polar equation}$$

$$r = \sec\theta\tan\theta \quad \text{Simplest form}$$

On the other hand, converting a polar equation to rectangular form requires considerable ingenuity. Example 5 demonstrates several polar-to-rectangular conversions that enable you to sketch the graphs of some polar equations.

EXAMPLE 5 Converting Polar Equations to Rectangular Form

See LarsonPrecalculus.com for an interactive version of this type of example.

Describe the graph of each polar equation and find the corresponding rectangular equation.

a. $r = 2$ **b.** $\theta = \dfrac{\pi}{3}$ **c.** $r = \sec\theta$

Solution

a. The graph of the polar equation $r = 2$ consists of all points that are two units from the pole. In other words, this graph is a circle centered at the origin with a radius of 2, as shown in Figure 9.57. You can confirm this by converting to rectangular form, using the relationship $r^2 = x^2 + y^2$.

$$\underbrace{r = 2}_{\text{Polar equation}} \Rightarrow r^2 = 2^2 \Rightarrow \underbrace{x^2 + y^2 = 2^2}_{\text{Rectangular equation}}$$

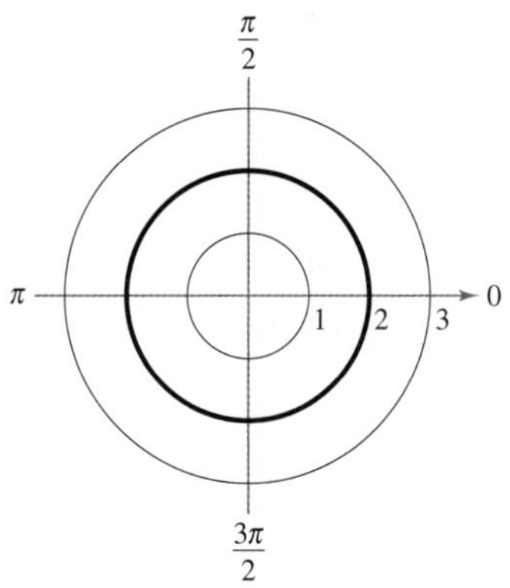

Figure 9.57

b. The graph of the polar equation

$$\theta = \frac{\pi}{3}$$

consists of all points on the line that makes an angle of $\pi/3$ with the positive x-axis, as shown in Figure 9.58. To convert to rectangular form, use the relationship $\tan\theta = y/x$.

$$\underbrace{\theta = \frac{\pi}{3}}_{\text{Polar equation}} \Rightarrow \tan\theta = \sqrt{3} \Rightarrow \underbrace{y = \sqrt{3}x}_{\text{Rectangular equation}}$$

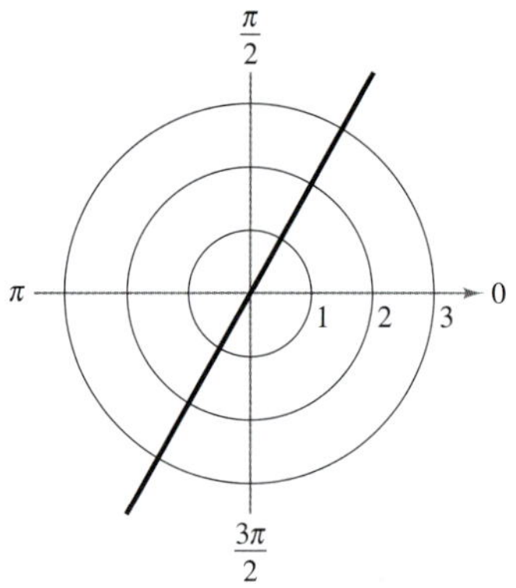

Figure 9.58

c. The graph of the polar equation $r = \sec\theta$ is not evident by inspection, so convert to rectangular form using the relationship $r\cos\theta = x$.

$$\underbrace{r = \sec\theta}_{\text{Polar equation}} \Rightarrow r\cos\theta = 1 \Rightarrow \underbrace{x = 1}_{\text{Rectangular equation}}$$

Now you can see that the graph is a vertical line, as shown in Figure 9.59.

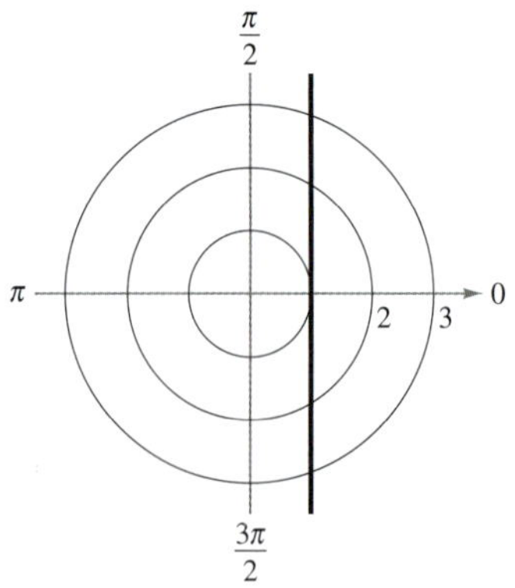

Figure 9.59

✓ **Checkpoint** *Audio-video solution in English & Spanish at LarsonPrecalculus.com*

Describe the graph of $r = 6\sin\theta$ and find the corresponding rectangular equation.

9.5 Exercises

See *CalcChat.com* for tutorial help and worked-out solutions to odd-numbered exercises.
For instructions on how to use a graphing utility, see Appendix A.

Vocabulary and Concept Check

In Exercises 1 and 2, fill in the blank(s).

1. The origin of the polar coordinate system is called the ________ .

2. For the point (r, θ), r is the _______ from O to P and θ is the _______ counterclockwise from the polar axis to segment $\overline{OP}$.

3. How are the rectangular coordinates (x, y) related to the polar coordinates (r, θ)?

4. Do the polar coordinates $(1, \pi)$ and the rectangular coordinates $(-1, 0)$ represent the same point?

Procedures and Problem Solving

Matching Polar Coordinates **In Exercises 5–8, match the polar coordinates with the point on the graph. Then find the rectangular coordinates of the point.**

5. $\left(2, \dfrac{\pi}{2}\right)$

6. $\left(2, \dfrac{3\pi}{2}\right)$

7. $(0, 0)$

8. $\left(-2, \dfrac{5\pi}{4}\right)$

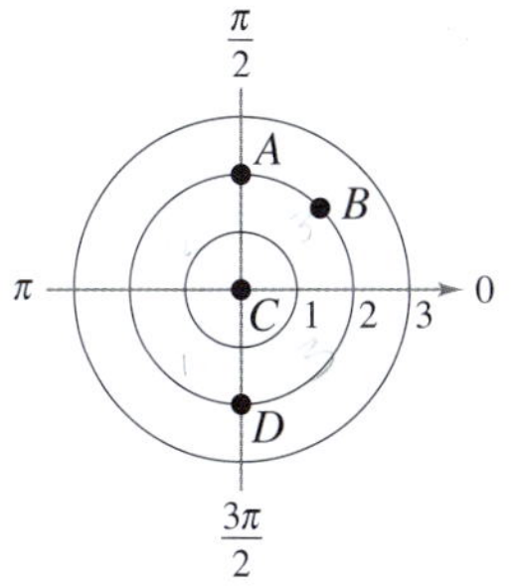

Plotting Points in the Polar Coordinate System **In Exercises 9–16, plot the point given in polar coordinates and find three additional polar representations of the point, using $-2\pi < \theta < 2\pi$.**

9. $\left(3, \dfrac{5\pi}{6}\right)$

10. $\left(2, \dfrac{3\pi}{4}\right)$

11. $\left(-4, \dfrac{5\pi}{6}\right)$

12. $\left(1, -\dfrac{\pi}{3}\right)$

13. $\left(-\dfrac{3}{2}, -\dfrac{7\pi}{4}\right)$

14. $\left(-5\sqrt{2}, \dfrac{2\pi}{3}\right)$

15. $(0, 7\pi/6)$

16. $(-\pi/2, -11\pi/6)$

Polar-to-Rectangular Conversion **In Exercises 17–24, a point is given in polar coordinates. Convert the point to rectangular coordinates.**

17. $(3, \pi)$

18. $(2, 0)$

19. $\left(0, \dfrac{3\pi}{4}\right)$

20. $\left(-16, \dfrac{5\pi}{2}\right)$

21. $\left(-2, \dfrac{7\pi}{6}\right)$

22. $\left(0, \dfrac{5\pi}{4}\right)$

23. $\left(-5, \dfrac{3\pi}{2}\right)$

24. $\left(-3, -\dfrac{\pi}{6}\right)$

Using a Graphing Utility to Find Rectangular Coordinates **In Exercises 25–28, use a graphing utility to find the rectangular coordinates of the point given in polar coordinates. Round your results to two decimal places.**

25. $(2, 2\pi/9)$

26. $(4, 11\pi/9)$

27. $(1.5, -2.82)$

28. $(-5.3, -0.78)$

Rectangular-to-Polar Conversion **In Exercises 29–34, a point is given in rectangular coordinates. Convert the point to polar coordinates. (There are many correct answers.)**

29. $(-7, 0)$

30. $(0, -5)$

31. $(1, 1)$

32. $(-3, -3)$

33. $(-\sqrt{3}, -\sqrt{3})$

34. $(\sqrt{3}, -1)$

Using a Graphing Utility to Find Polar Coordinates **In Exercises 35–38, use a graphing utility to find one set of polar coordinates for the point given in rectangular coordinates. Round your results to two decimal places. (There are many correct answers.)**

35. $(3, -2)$

36. $\left(-\frac{7}{4}, -\frac{3}{2}\right)$

37. $(-\sqrt{3}, 2)$

38. $(3\sqrt{2}, 3\sqrt{2})$

Converting a Rectangular Equation to Polar Form **In Exercises 39–56, convert the rectangular equation to polar form. Assume $a > 0$.**

39. $x^2 + y^2 = 36$

40. $x^2 + y^2 = 169$

41. $y = x$

42. $y = -\sqrt{3}x$

43. $y = a$

44. $x = a$

45. $3x - y + 2 = 0$

46. $2x + 5y - 3 = 0$

47. $xy = 4$

48. $2xy = 1$

49. $(x^2 + y^2)^2 = 9(x^2 - y^2)$

50. $y^2 - 8x - 16 = 0$

51. $x^2 + y^2 - 6x = 0$

52. $x^2 + y^2 - 8y = 0$

53. $x^2 + y^2 - 2ax = 0$

54. $x^2 + y^2 - 2ay = 0$

55. $y^2 = x^3$

56. $x^2 = y^3$

Converting a Polar Equation to Rectangular Form **In Exercises 57–72, convert the polar equation to rectangular form.**

57. $r = 5 \sin \theta$
58. $r = 4 \cos \theta$
59. $\theta = \dfrac{5\pi}{6}$
60. $\theta = \dfrac{11\pi}{6}$
61. $\theta = \pi/2$
62. $\theta = \pi$
63. $r = -2 \csc \theta$
64. $r = 3 \sec \theta$
65. $r^2 = \cos \theta$
66. $r^2 = \sin 2\theta$
67. $r = 3 \sin 3\theta$
68. $r = -2 \cos 2\theta$
69. $r = \dfrac{2}{1 - \cos \theta}$
70. $r = \dfrac{1}{1 + \sin \theta}$
71. $r = \dfrac{6}{2 - 3 \sin \theta}$
72. $r = \dfrac{12}{2 \cos \theta - 3 \sin \theta}$

Converting a Polar Equation to Rectangular Form **In Exercises 73–78, describe the graph of the polar equation and find the corresponding rectangular equation. Sketch its graph.**

73. $r = 7$
74. $r = 10$
75. $\theta = \dfrac{\pi}{4}$
76. $\theta = \dfrac{7\pi}{6}$
77. $r = 5 \sec \theta$
78. $r = 4 \csc \theta$

79. ***Why you should learn it*** *(p. 687)* The center of a Ferris wheel lies at the pole of the polar coordinate system, where the distances are in feet. Passengers enter a car at $(30, -\pi/2)$. It takes 45 seconds for the wheel to complete one clockwise revolution.

(a) Write a polar equation that models the possible positions of a passenger car.

(b) Passengers enter a car. Find and interpret their coordinates after 15 seconds of rotation.

(c) Convert the point in part (b) to rectangular coordinates. Interpret the coordinates.

80. **Ferris Wheel** Repeat Exercise 79 when the distance from a passenger car to the center is 35 feet and it takes 60 seconds to complete one clockwise revolution.

Focusing on Concepts

True or False? **In Exercises 81 and 82, determine whether the statement is true or false. Justify your answer.**

81. If (r_1, θ_1) and (r_2, θ_2) represent the same point in the polar coordinate system, then $|r_1| = |r_2|$.

82. If (r, θ_1) and (r, θ_2) represent the same point in the polar coordinate system, then $\theta_1 = \theta_2 + 2\pi n$ for some integer n.

83. **Think About It**

(a) Show that the distance between the points (r_1, θ_1) and (r_2, θ_2) is $\sqrt{r_1^2 + r_2^2 - 2r_1 r_2 \cos(\theta_1 - \theta_2)}$.

(b) Simplify the Distance Formula for $\theta_1 = \theta_2$. Is the simplification what you expected? Explain.

(c) Simplify the Distance Formula for $\theta_1 - \theta_2 = 90°$. Is the simplification what you expected? Explain.

84. **HOW DO YOU SEE IT?** Use the polar coordinate system shown below.

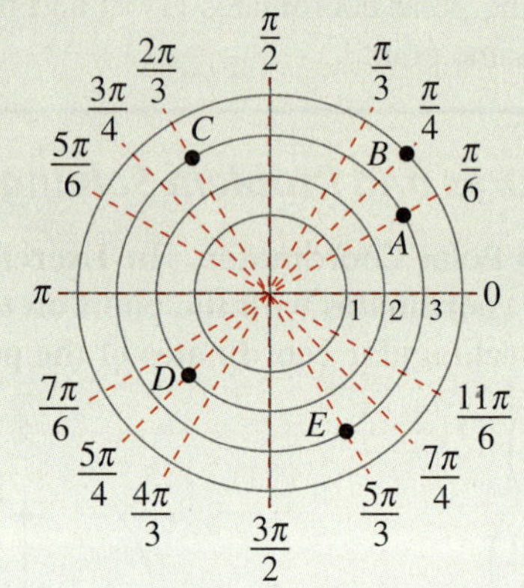

(a) Identify the polar coordinates of the points.

(b) Which points lie on the graph of $r = 3$?

(c) Which points lie on the graph of $\theta = \pi/4$?

85. **Writing** In the rectangular coordinate system, each point (x, y) has a unique representation. Why is this not true for a point (r, θ) in the polar coordinate system?

86. **Error Analysis** Describe the error in converting the rectangular coordinates $(1, -\sqrt{3})$ to polar form.

$\tan \theta = -\sqrt{3}/1 \Rightarrow \theta = 2\pi/3$

$r = \sqrt{1^2 + (-\sqrt{3})^2} = \sqrt{4} = 2$

$(r, \theta) = \left(2, \dfrac{2\pi}{3}\right)$

87. **Think About It** Convert the polar equation $r = 2(h \cos \theta + k \sin \theta)$ to rectangular form and verify that it represents a circle.

88. **Think About It** Use the result of Exercises 87 to convert $r = \cos \theta + 3 \sin \theta$ to rectangular form and find the center and radius of the circle it represents.

Cumulative Mixed Review

Solving a Triangle Using the Law of Sines or Cosines **In Exercises 89 and 90, use the Law of Sines or the Law of Cosines to solve the triangle.**

89. $A = 56°, C = 38°, c = 12$

90. $B = 71°, a = 21, c = 29$

SSSCCC/Shutterstock.com

9.6 Graphs of Polar Equations

What you should learn

- Graph polar equations by point plotting.
- Use symmetry and zeros as sketching aids.
- Recognize special polar graphs.

Why you should learn it

You can use graphs of polar equations in mathematical modeling. For instance, in Exercise 65 on page 700, you will graph the sound pickup pattern of a microphone in the polar coordinates system.

Introduction

In previous chapters, you sketched graphs in rectangular coordinate systems. You began with the basic point-plotting method. Then you used sketching aids such as a graphing utility, symmetry, intercepts, asymptotes, periods, and shifts to further investigate the natures of the graphs. This section approaches curve sketching in the polar coordinate system similarly.

EXAMPLE 1 Graphing a Polar Equation by Point Plotting

Sketch the graph of the polar equation $r = 4 \sin \theta$ by hand.

Solution

The sine function is periodic, so you can get a full range of r-values by considering values of θ in the interval $0 \le \theta \le 2\pi$, as shown in the table.

θ	0	$\frac{\pi}{6}$	$\frac{\pi}{3}$	$\frac{\pi}{2}$	$\frac{2\pi}{3}$	$\frac{5\pi}{6}$	π	$\frac{7\pi}{6}$	$\frac{3\pi}{2}$	$\frac{11\pi}{6}$	2π
r	0	2	$2\sqrt{3}$	4	$2\sqrt{3}$	2	0	-2	-4	-2	0

By plotting these points, as shown in the figure, it appears that the graph is a circle of radius 2, whose center is the point $(x, y) = (0, 2)$.

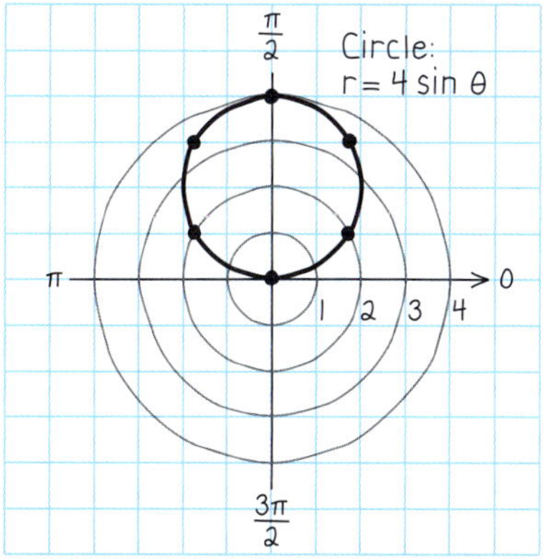

Checkpoint Audio-video solution in English & Spanish at LarsonPrecalculus.com

Sketch the graph of the polar equation $r = 6 \cos \theta$.

Here are three ways to confirm the graph found in Example 1.

1. *Convert to Rectangular Form* Multiply each side of the polar equation by r and convert the result to rectangular form.
2. *Use a Polar Coordinate Mode* Set your graphing utility to *polar* mode and graph the polar equation. (Use $0 \le \theta \le \pi$, $-6 \le x \le 6$, and $-4 \le y \le 4$.)
3. *Use a Parametric Mode* Set your graphing utility to *parametric* mode and graph $x = (4 \sin t) \cos t$ and $y = (4 \sin t) \sin t$.

Most graphing utilities have a *polar* graphing mode. If yours does not, you can rewrite the polar equation $r = f(\theta)$ in parametric form, using t as a parameter with the equations $x = f(t) \cos t$ and $y = f(t) \sin t$.

Di Studio/Shutterstock.com

Symmetry and Zeros

In Example 1, note in the graph of $r = 4 \sin \theta$ that as θ increases from 0 to 2π, the graph is traced out twice. Moreover, note that the graph is *symmetric with respect to the line* $\theta = \pi/2$. Had you known about this symmetry and retracing ahead of time, you could have used fewer points. The three important types of symmetry to consider in polar curve sketching are shown in Figure 9.60.

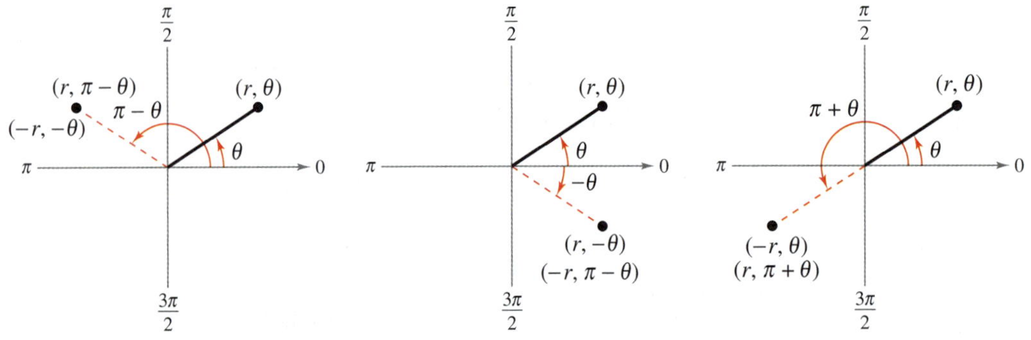

Symmetry with Respect to the Line $\theta = \pi/2$ ***Symmetry with Respect to the Polar Axis*** ***Symmetry with Respect to the Pole***

Figure 9.60

Testing for Symmetry in Polar Coordinates

The graph of a polar equation is symmetric with respect to the following when the given substitution yields an equivalent equation.

1. The line $\theta = \dfrac{\pi}{2}$: Replace (r, θ) with $(r, \pi - \theta)$ or $(-r, -\theta)$.
2. The polar axis: Replace (r, θ) with $(r, -\theta)$ or $(-r, \pi - \theta)$.
3. The pole: Replace (r, θ) with $(r, \pi + \theta)$ or $(-r, \theta)$.

From Example 1, you know that the graph of $r = 4 \sin \theta$ is a circle (see figure below). You can determine the symmetry of this graph as shown below.

1. Replace (r, θ) with $(-r, -\theta)$.

$$-r = 4 \sin(-\theta)$$
$$r = -4 \sin(-\theta)$$
$$r = 4 \sin \theta$$

2. Replace (r, θ) with $(r, -\theta)$.

$$r = 4 \sin(-\theta)$$
$$r = -4 \sin \theta$$

3. Replace (r, θ) with $(-r, \theta)$.

$$-r = 4 \sin \theta$$
$$r = -4 \sin \theta$$

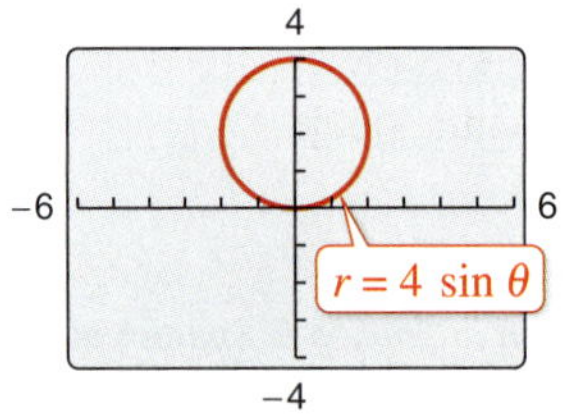

The graph of $r = 4 \sin \theta$ is symmetric with respect to the line $\theta = \pi/2$.

So, the graph of $r = 4 \sin \theta$ is symmetric with respect to the line $\theta = \pi/2$. Recall from Section 4.2 that the sine function is odd. That is,

$$\sin(-\theta) = -\sin \theta.$$

EXAMPLE 2 Using Symmetry to Sketch a Polar Graph

Use symmetry to sketch the graph of $r = 3 + 2\cos\theta$ by hand.

Solution

Replacing (r, θ) with $(r, -\theta)$ produces

$r = 3 + 2\cos(-\theta)$ Replace (r, θ) with $(r, -\theta)$.

$= 3 + 2\cos\theta$. $\cos(-\theta) = \cos\theta$

So, by using the even trigonometric identity, you can conclude that the curve is symmetric with respect to the polar axis. Plotting the points in the table and using polar axis symmetry, you obtain the graph shown in Figure 9.61. This graph is called a **limaçon.**

θ	0	$\frac{\pi}{6}$	$\frac{\pi}{3}$	$\frac{\pi}{2}$	$\frac{2\pi}{3}$	$\frac{5\pi}{6}$	π
r	5	$3 + \sqrt{3}$	4	3	2	$3 - \sqrt{3}$	1

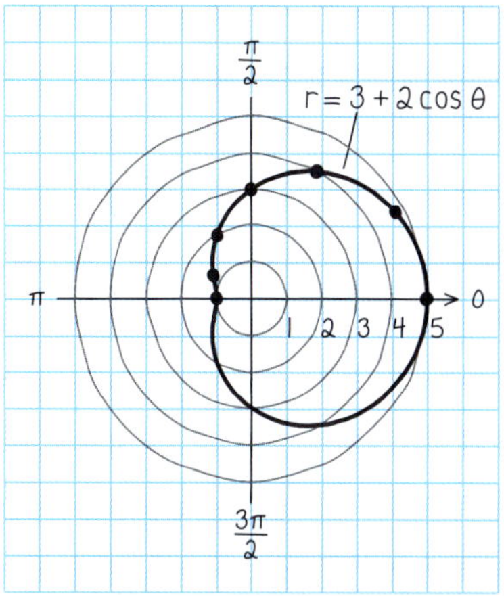

Figure 9.61

Use a graphing utility to confirm this graph.

Checkpoint *Audio-video solution in English & Spanish at LarsonPrecalculus.com*

Use symmetry to sketch the graph of $r = 3 + 2\sin\theta$ by hand.

The three tests for symmetry in polar coordinates on the preceding page are sufficient to guarantee symmetry, but they are not necessary. That is, a graph may have symmetry even though its equation does not satisfy any of the three tests. For instance, the figure at the right shows the graph of

$r = \theta + 2\pi$. Spiral of Archimedes

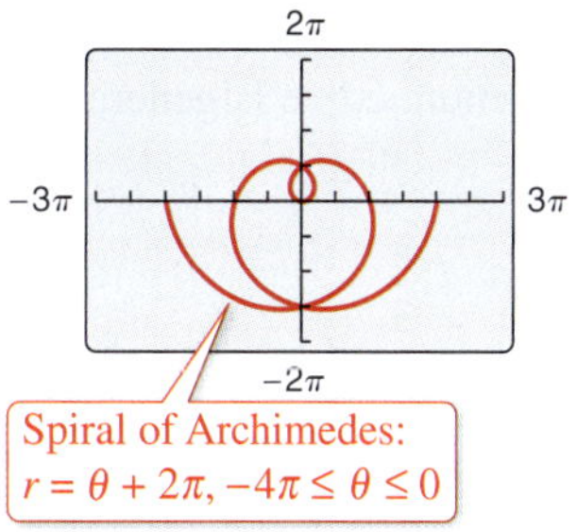

From the figure, you can see that the graph is symmetric with respect to the line $\theta = \pi/2$. Yet the tests on the preceding page fail to indicate symmetry because neither of the following replacements yields an equivalent equation.

Original Equation	*Replacement*	*New Equation*
$r = \theta + 2\pi$	(r, θ) with $(-r, -\theta)$	$-r = -\theta + 2\pi$
$r = \theta + 2\pi$	(r, θ) with $(r, \pi - \theta)$	$r = -\theta + 3\pi$

Note that the equation in Example 1, $r = 4\sin\theta$, has the form $r = f(\sin\theta)$ and its graph is symmetric with respect to the line $\theta = \pi/2$. Also, the equation in Example 2, $r = 3 + 2\cos\theta$, has the form $r = g(\cos\theta)$ and its graph is symmetric with respect to the polar axis. These observations can be generalized to yield the *quick tests for symmetry* listed below.

Quick Tests for Symmetry in Polar Coordinates

1. The graph of $r = f(\sin\theta)$ is symmetric with respect to the line $\theta = \frac{\pi}{2}$.

2. The graph of $r = g(\cos\theta)$ is symmetric with respect to the polar axis.

An additional aid to sketching graphs of polar equations involves knowing the θ-values for which $r = 0$. In Example 1, $r = 0$ when $\theta = 0$. Some curves reach their zeros at more than one point, as shown in Example 3.

EXAMPLE 3 Sketching a Polar Graph

See LarsonPrecalculus.com for an interactive version of this type of example.

Sketch the graph of

$$r = 2\cos 3\theta.$$

Solution

Symmetry: With respect to the polar axis

Zeros of r: $r = 0$ when $3\theta = \frac{\pi}{2}, \frac{3\pi}{2}, \frac{5\pi}{2}$

or $\theta = \frac{\pi}{6}, \frac{\pi}{2}, \frac{5\pi}{6}$

θ	0	$\frac{\pi}{12}$	$\frac{\pi}{6}$	$\frac{\pi}{4}$	$\frac{\pi}{3}$	$\frac{5\pi}{12}$	$\frac{\pi}{2}$
r	2	$\sqrt{2}$	0	$-\sqrt{2}$	-2	$-\sqrt{2}$	0

By plotting these points and using the specified symmetry and zeros, you can obtain the graph shown in Figure 9.62. This graph is called a **rose curve,** and each loop on the graph is called a *petal.* Note how the entire curve is generated as θ increases from 0 to π.

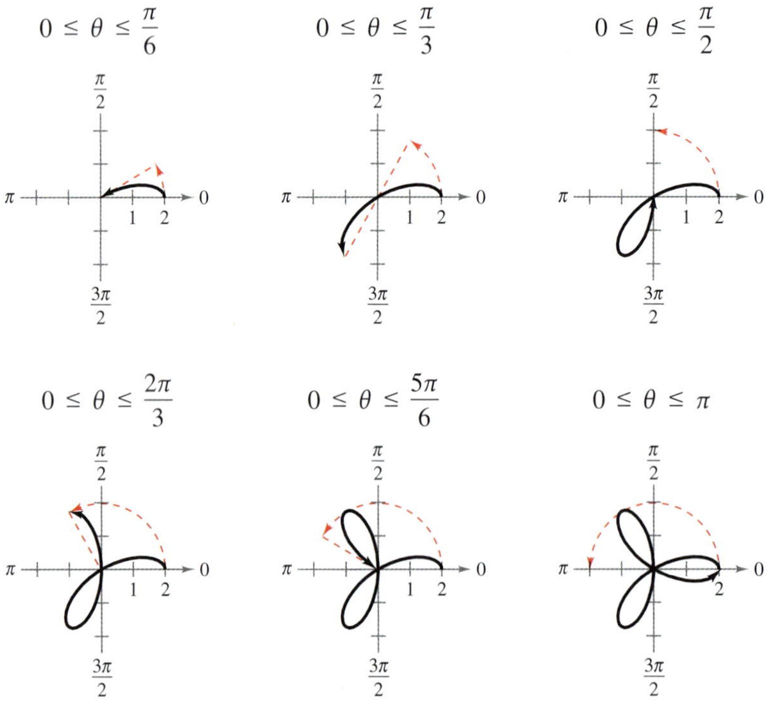

Figure 9.62

Checkpoint Audio-video solution in English & Spanish at LarsonPrecalculus.com

Sketch the graph of $r = 4\sin 2\theta$.

Explore the Concept

Notice that the rose curve in Example 3 has three petals. How many petals do the rose curves $r = 2\cos 4\theta$ and $r = 2\sin 3\theta$ have? Determine the numbers of petals for the curves $r = 2\cos n\theta$ and $r = 2\sin n\theta$, where n is a positive integer.

Special Polar Graphs

Several important types of graphs have equations that are simpler in polar form than in rectangular form. For example, the circle

$$r = 4 \sin \theta$$

in Example 1 has the more complicated rectangular equation

$$x^2 + (y - 2)^2 = 4.$$

Several other types of graphs that have simple polar equations are shown below.

Limaçons

$r = a \pm b \cos \theta,\ r = a \pm b \sin \theta \quad (a > 0, b > 0)$

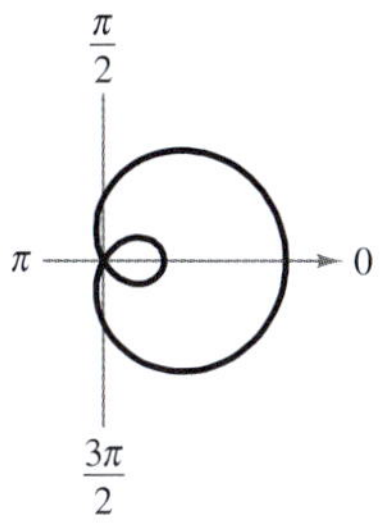

$\frac{a}{b} < 1$

Limaçon with inner loop

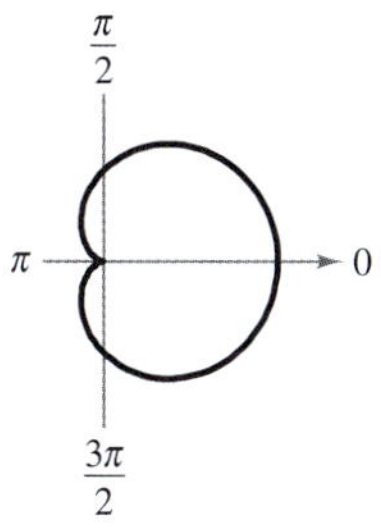

$\frac{a}{b} = 1$

Cardioid (heart-shaped)

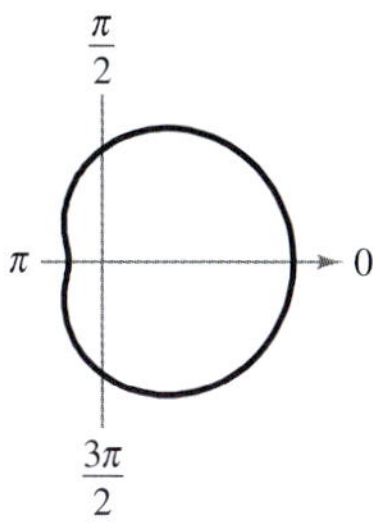

$1 < \frac{a}{b} < 2$

Dimpled limaçon

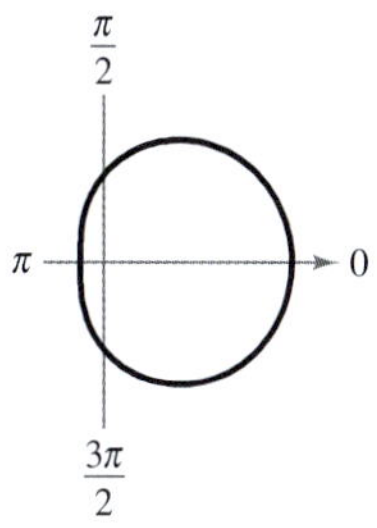

$\frac{a}{b} \geq 2$

Convex limaçon

Rose Curves

n petals when n is odd, $2n$ petals when n is even $(n \geq 2)$

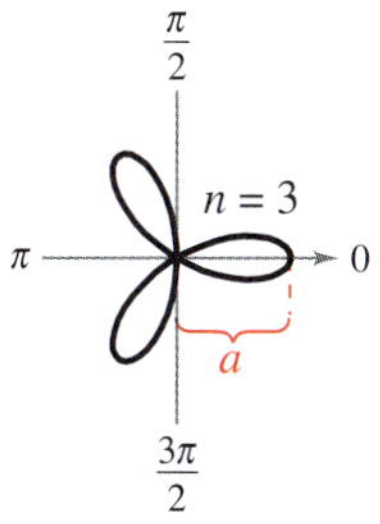

$r = a \cos n\theta$

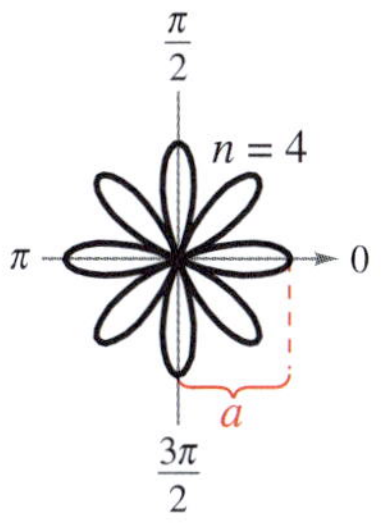

$r = a \cos n\theta$

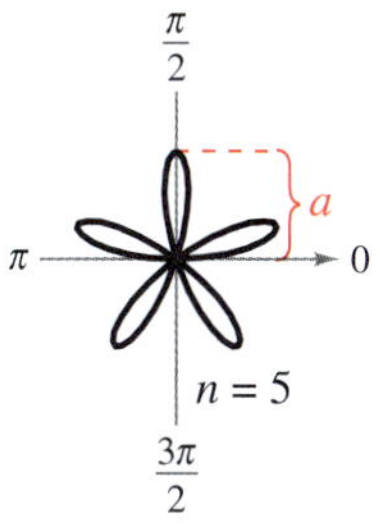

$r = a \sin n\theta$

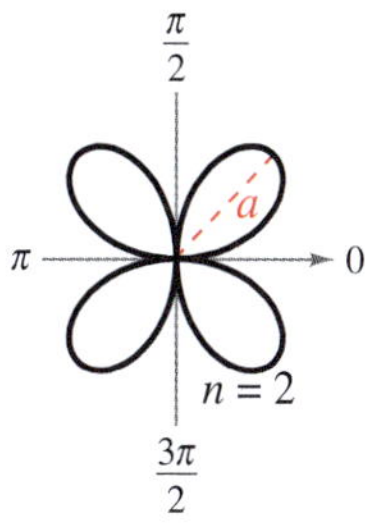

$r = a \sin n\theta$

Circles and Lemniscates

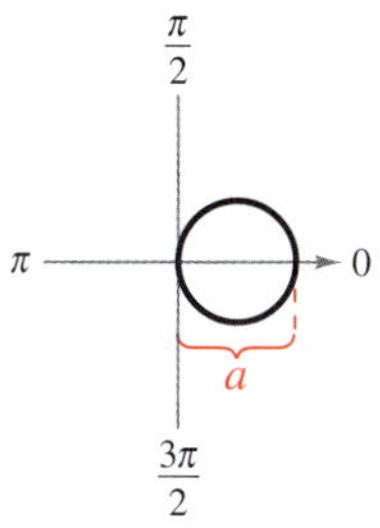

$r = a \cos \theta$

Circle

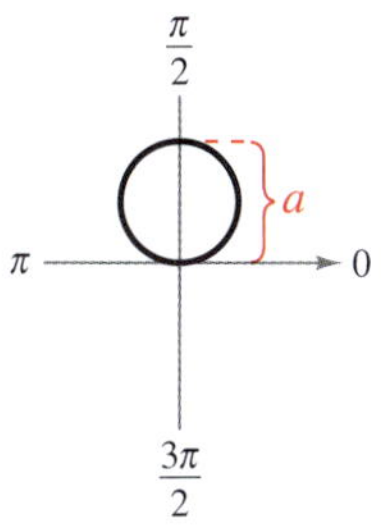

$r = a \sin \theta$

Circle

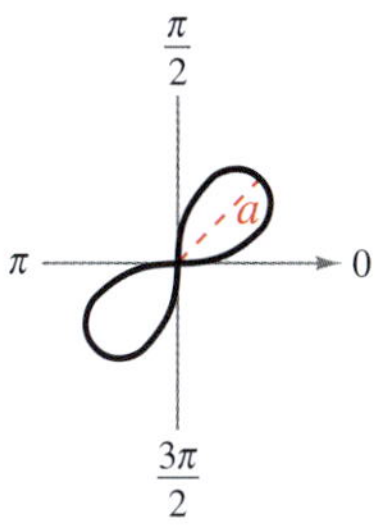

$r^2 = a^2 \sin 2\theta$

Lemniscate

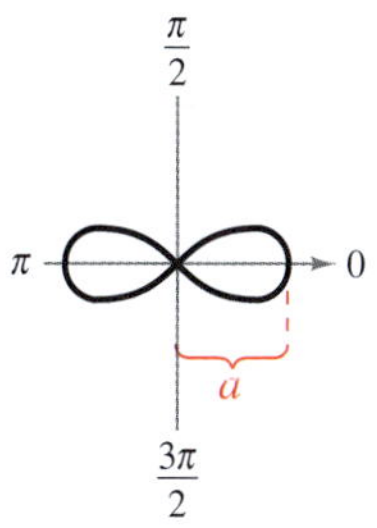

$r^2 = a^2 \cos 2\theta$

Lemniscate

The quick tests for symmetry presented on page 695 are especially useful when graphing rose curves. Because rose curves have the form $r = f(\sin \theta)$ or the form $r = g(\cos \theta)$, you know that a rose curve will be either symmetric with respect to the line $\theta = \pi/2$ or symmetric with respect to the polar axis.

EXAMPLE 4 Graphing a Rose Curve

Use a graphing utility to graph $r = 3 \cos 2\theta$.

Solution

Type of curve: Rose curve with $2n = 4$ petals

Symmetry: With respect to the polar axis, the line $\theta = \frac{\pi}{2}$, and the pole

Zeros of r: $r = 0$ when $\theta = \frac{\pi}{4}, \frac{3\pi}{4}$

Using a graphing utility in *polar* mode, enter the equation, as shown in Figure 9.63 (with $0 \le \theta \le 2\pi$). You should obtain the graph shown in Figure 9.64.

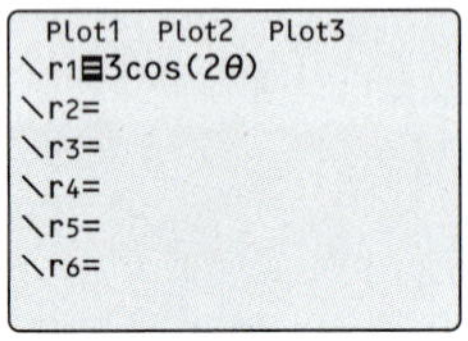

Figure 9.63

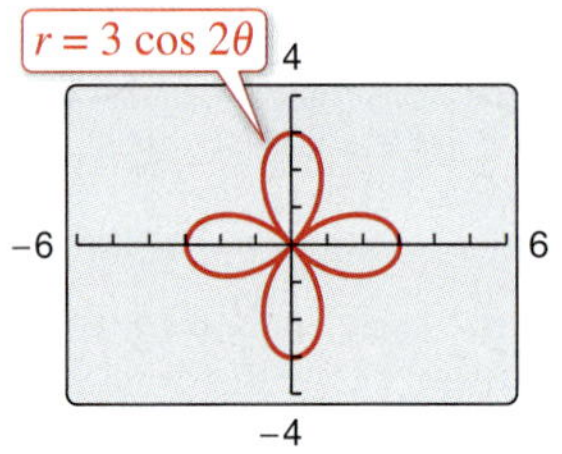

Figure 9.64

Checkpoint *Audio-video solution in English & Spanish at LarsonPrecalculus.com*

Use a graphing utility to graph $r = 3 \sin 3\theta$.

EXAMPLE 5 Graphing a Lemniscate

Use a graphing utility to graph $r^2 = 9 \sin 2\theta$.

Solution

Type of curve: Lemniscate

Symmetry: With respect to the pole

Zeros of r: $r = 0$ when $\theta = 0, \frac{\pi}{2}$

Using a graphing utility in *polar* mode, enter the equation, as shown in Figure 9.65 (with $0 \le \theta \le 2\pi$). You should obtain the graph shown in Figure 9.66.

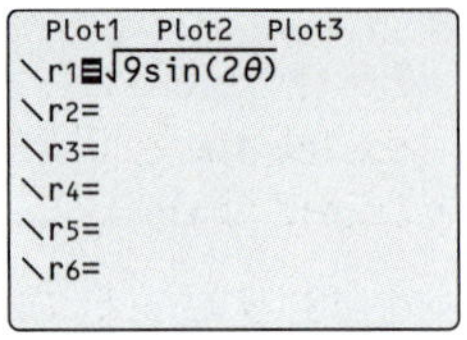

Figure 9.65

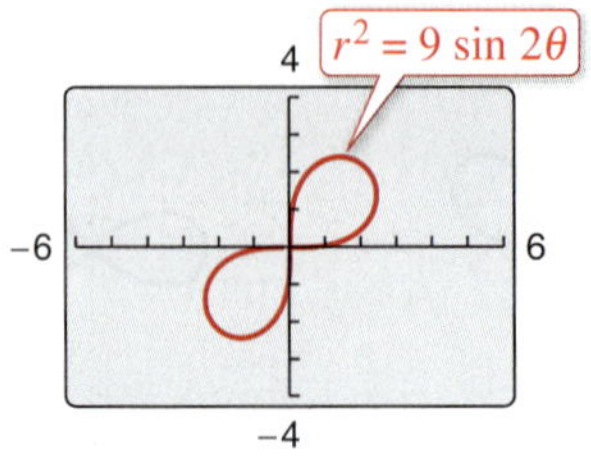

Figure 9.66

Checkpoint *Audio-video solution in English & Spanish at LarsonPrecalculus.com*

Use a graphing utility to graph $r^2 = 4 \cos 2\theta$.

What's Wrong?

You use a graphing utility in *polar* mode to confirm the result in Example 5 and obtain the graph shown below (with $0 \le \theta \le 2\pi$). What's wrong?

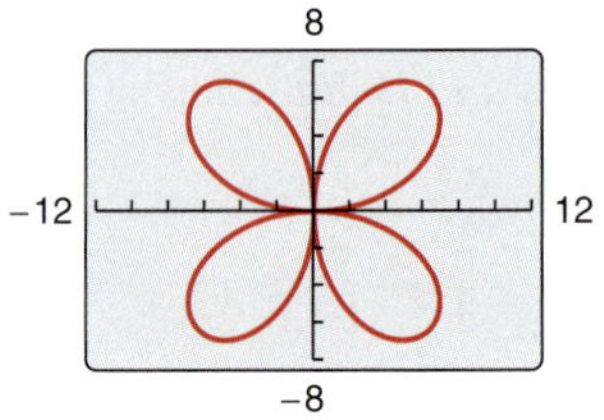

9.6 Exercises

See *CalcChat.com* for tutorial help and worked-out solutions to odd-numbered exercises.
For instructions on how to use a graphing utility, see Appendix A.

Vocabulary and Concept Check

In Exercises 1–4, fill in the blank.

1. The equation $r = 2 + \cos\theta$ represents a _______ .

2. The equation $r = 2\cos\theta$ represents a _______ .

3. The equation $r^2 = 4\sin 2\theta$ represents a _______ .

4. The equation $r = 1 + \sin\theta$ represents a _______ .

5. How can you test whether the graph of a polar equation is symmetric with respect to the line $\theta = \dfrac{\pi}{2}$?

6. Is the graph of $r = 3 + 4\cos\theta$ symmetric with respect to the line $\theta = \dfrac{\pi}{2}$ or to the polar axis?

Procedures and Problem Solving

Identifying Types of Polar Graphs **In Exercises 7–10, identify the type of polar graph.**

7.

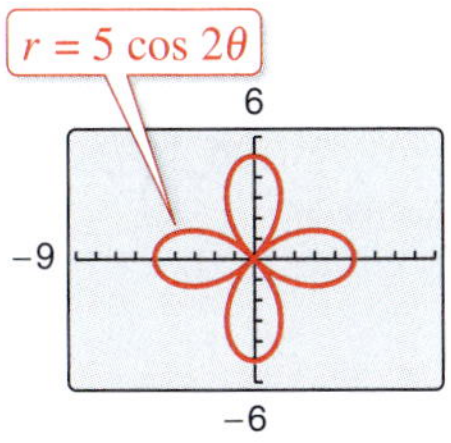

8.

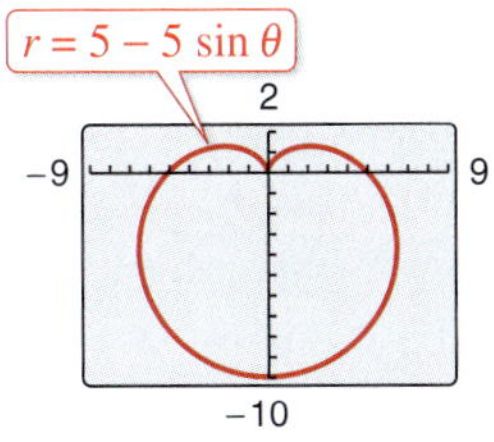

9.

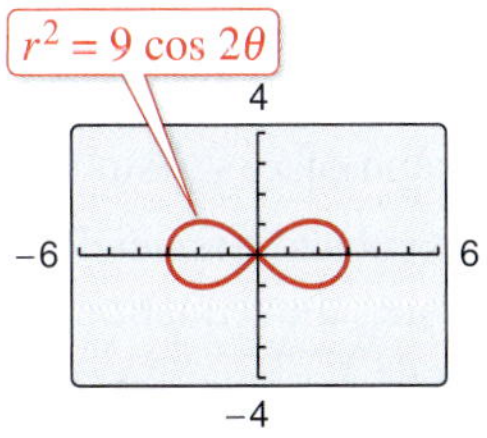

10.

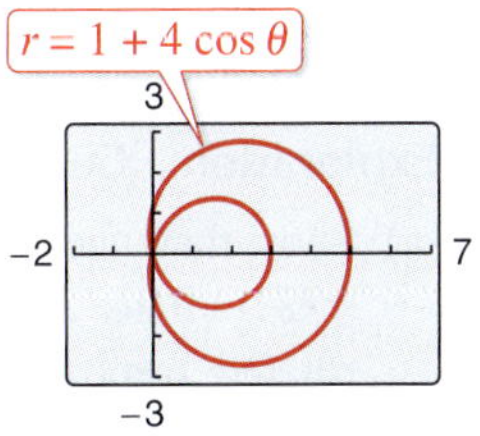

Finding the Equation of a Polar Curve **In Exercises 11–14, determine which equation is represented by the graph of the polar curve.**

11.

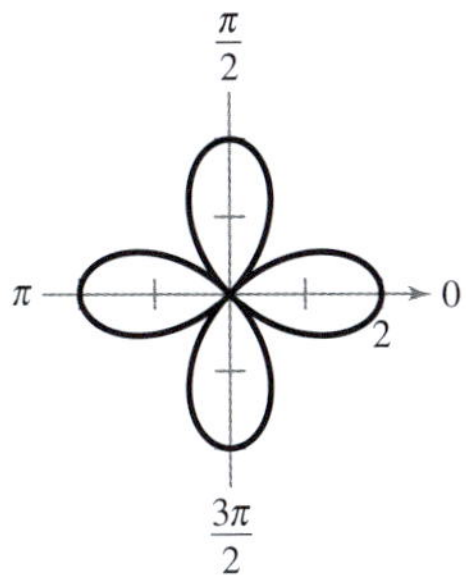

(a) $r = 2\cos 4\theta$

(b) $r = \cos 4\theta$

(c) $r = 2\cos 2\theta$

(d) $r = 2\cos\dfrac{\theta}{2}$

12.

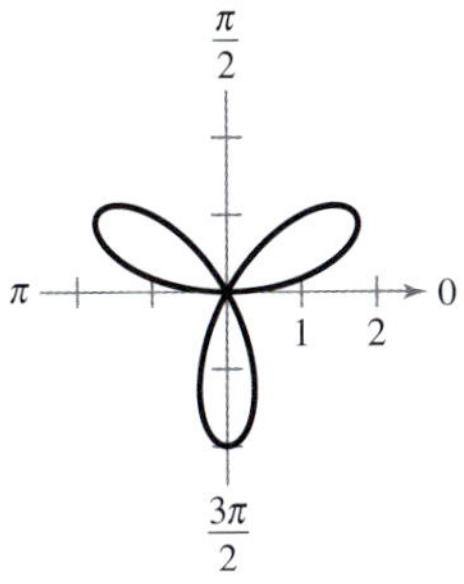

(a) $r = 2\sin 6\theta$

(b) $r = 2\cos\dfrac{3\theta}{2}$

(c) $r = 2\sin\dfrac{3\theta}{2}$

(d) $r = 2\sin 3\theta$

13.

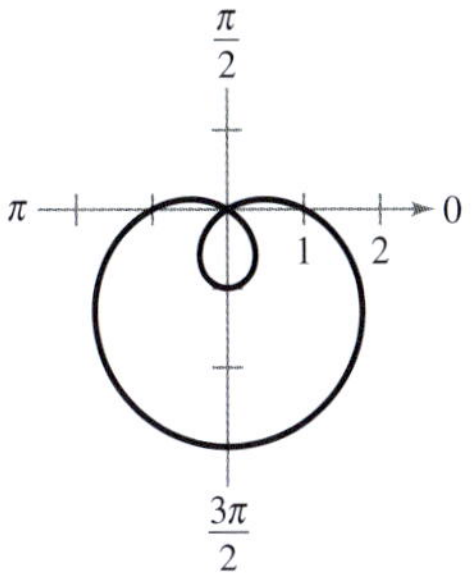

(a) $r = 1 - 2\sin\theta$

(b) $r = 1 + 2\sin\theta$

(c) $r = 1 + 2\cos\theta$

(d) $r = 1 - 2\cos\theta$

14.

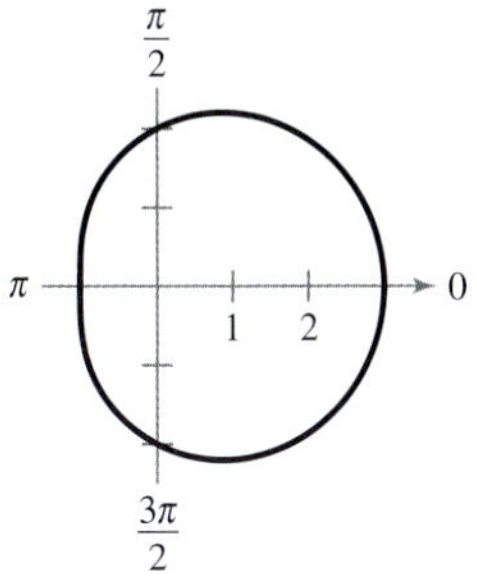

(a) $r = 2 - \cos\theta$

(b) $r = 2 - \sin\theta$

(c) $r = 2 + \cos\theta$

(d) $r = 2 + \sin\theta$

Testing for Symmetry **In Exercises 15–22, test for symmetry with respect to the line $\theta = \pi/2$, the polar axis, and the pole.**

15. $r = 6 + 3\cos\theta$

16. $r = 9\cos 3\theta$

17. $r = \dfrac{4}{1 - \cos\theta}$

18. $r = \dfrac{3}{1 + \sin\theta}$

19. $r = 12\sin\theta$

20. $r = 7\csc\theta\cos\theta$

21. $r^2 = 64\sin 2\theta$

22. $r^2 = 36\sin 2\theta$

Finding Zeros of r **In Exercises 23–26, find any zeros of *r*.**

23. $r = 10 - 10\sin\theta$

24. $r = 6 + 12\cos\theta$

25. $r = 4\cos 3\theta$

26. $r = 3\sin 2\theta$

Using Symmetry to Sketch a Polar Graph **In Exercises 27–34, use symmetry to sketch the graph of the polar equation. Use a graphing utility to verify your graph.**

27. $r = 15$

28. $\theta = -5\pi/3$

29. $r = 3\sin\theta$

30. $r = 2\cos\theta$

31. $r = 3(1 - \cos\theta)$

32. $r = 4(1 + \sin\theta)$

33. $r = 4 + 5\sin\theta$

34. $r = 3 + 6\cos\theta$

Sketching a Polar Graph **In Exercises 35–40, identify and sketch the graph of the polar equation. Identify any symmetry and zeros of r. Use a graphing utility to verify your results.**

35. $r = 5 \cos 3\theta$ **36.** $r = \sin 5\theta$

37. $r = -7 \sin 2\theta$ **38.** $r = 3 \cos 4\theta$

39. $r = 1 + 2 \sin \theta$ **40.** $r = \sqrt{3} - 2 \cos \theta$

Using a Graphing Utility to Graph a Polar Equation **In Exercises 41–54, use a graphing utility to graph the polar equation. Describe your viewing window.**

41. $r = 8 \cos 2\theta$ **42.** $r = -\cos 2\theta$

43. $r = 2(5 - \sin \theta)$ **44.** $r = 6 - 4 \sin \theta$

45. $r = \dfrac{3}{\sin \theta - 2 \cos \theta}$ **46.** $r = \dfrac{6}{2 \sin \theta - 3 \cos \theta}$

47. $r^2 = \sin 2\theta$ **48.** $r^2 = 4 \cos 3\theta$

49. $r = 8 \sin \theta \cos^2 \theta$ **50.** $r = 2 \cos(3\theta - 2)$

51. $r = 2 \csc \theta + 6$ **52.** $r = 4 - \sec \theta$

53. $r = e^{\theta}$ **54.** $r = e^{\theta/2}$

Using a Graphing Utility to Find an Interval **In Exercises 55–60, use a graphing utility to graph the polar equation. Find an interval for θ for which the graph is traced *only once*.**

55. $r = 3 - 4 \cos \theta$ **56.** $r = 2(1 - 2 \sin \theta)$

57. $r = 2 \cos \dfrac{3\theta}{2}$ **58.** $r = 3 \sin \dfrac{5\theta}{2}$

59. $r = 9 \sin \theta$ **60.** $r^2 = 25 \cos 2\theta$

Using a Graphing Utility to Graph a Polar Equation **In Exercises 61–64, use a graphing utility to graph the polar equation and show that the indicated line is an asymptote of the graph.**

	Name of Graph	*Polar Equation*	*Asymptote*
61.	Conchoid	$r = 2 - \sec \theta$	$x = -1$
62.	Conchoid	$r = 2 + \csc \theta$	$y = 1$
63.	Hyperbolic spiral	$r = 3/\theta$	$y = 3$
64.	Strophoid	$r = 2 \cos 2\theta \sec \theta$	$x = -2$

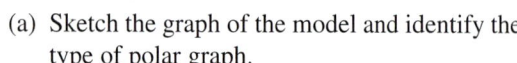

65. ***Why you should learn it*** *(p. 693)* The sound pickup pattern of a microphone is modeled by the polar equation $r = 5 + 5 \cos \theta$, where $|r|$ measures how sensitive the microphone is to sounds coming from the angle θ.

(a) Sketch the graph of the model and identify the type of polar graph.

(b) At what angle is the microphone most sensitive to sound?

66. Geometry The area of the lemniscate $r^2 = a^2 \cos 2\theta$ is a^2. Sketch the graph of $r^2 = 16 \cos 2\theta$. Then find the area of one loop of the graph.

Focusing on Concepts

True or False? **In Exercises 67 and 68, determine whether the statement is true or false. Justify your answer.**

67. The graph of $r = 6 \sin 5\theta$ is a rose curve with five petals.

68. A lemniscate will always have symmetry with respect to the polar axis.

69. Error Analysis Describe the error.

Because $a/b = 4/2 = 2$, the graph of $r = 4 + 2 \cos \theta$ is a dimpled limaçon.

70. HOW DO YOU SEE IT? Determine which graph matches each polar equation.

(a) $r = 5 \sin \theta$ (b) $r = 2 + 5 \sin \theta$

(c) $r = 5 \cos 2\theta$

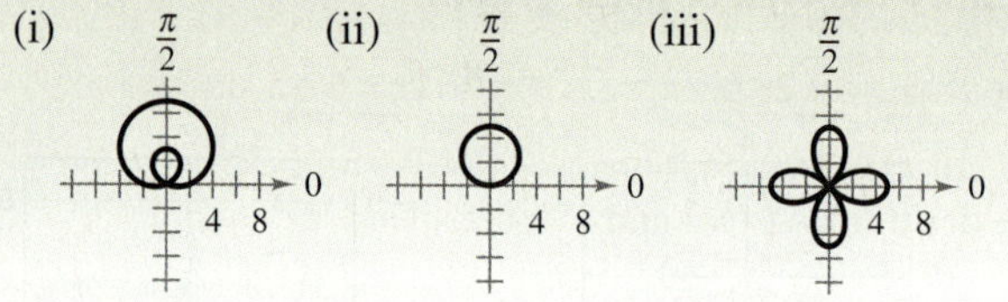

71. Exploration The graph of $r = f(\theta)$ is rotated about the pole through an angle ϕ. Show that the equation of the rotated graph is $r = f(\theta - \phi)$.

72. Exploration Consider the graph of $r = f(\sin \theta)$.

(a) Show that when the graph is rotated counterclockwise $\pi/2$ radians about the pole, the equation of the rotated graph is $r = f(-\cos \theta)$.

(b) Show that when the graph is rotated counterclockwise π radians about the pole, the equation of the rotated graph is $r = f(-\sin \theta)$.

(c) Show that when the graph is rotated counterclockwise $3\pi/2$ radians about the pole, the equation of the rotated graph is $r = f(\cos \theta)$.

Writing an Equation for Special Polar Graphs **In Exercises 73 and 74, use the results of Exercises 71 and 72.**

73. Write an equation for the limaçon $r = 2 - \sin \theta$ after it has been rotated through each angle.

(a) $\pi/4$ (b) $\pi/2$ (c) π (d) $3\pi/2$

74. Write an equation for the rose curve $r = 2 \sin 2\theta$ after it has been rotated through each angle.

(a) $\pi/6$ (b) $\pi/2$ (c) $2\pi/3$ (d) π

Cumulative Mixed Review

Finding the Zeros of a Rational Function **In Exercises 75 and 76, find the zeros (if any) of the rational function.**

75. $f(x) = \dfrac{x^2 - 9}{x + 1}$ **76.** $f(x) = 6 + \dfrac{4}{x^2 + 4}$

9.7 Polar Equations of Conics

Alternative Definition of Conics and Polar Equations

In Sections 9.2 and 9.3, you learned that the rectangular equations of ellipses and hyperbolas take simpler forms when the origin lies at the *center*. As it happens, there are many important applications of conics in which it is more convenient to use a *focus* as the origin. In this section, you will learn that polar equations of conics take simpler forms when a focus lies at the pole.

To begin, consider the following alternative definition of a conic that uses the concept of eccentricity (a measure of the flatness of the conic).

Alternative Definition of a Conic

The locus of a point in the plane that moves such that its distance from a fixed point (focus) is in a constant ratio to its distance from a fixed line (directrix) is a **conic.** The constant ratio is the **eccentricity** of the conic and is denoted by e. Moreover, the conic is an **ellipse** when $0 < e < 1$, a **parabola** when $e = 1$, and a **hyperbola** when $e > 1$. (See Figure 9.67.)

In Figure 9.67, note that for each type of conic, F is the fixed point (focus) from the definition above and is located at the pole.

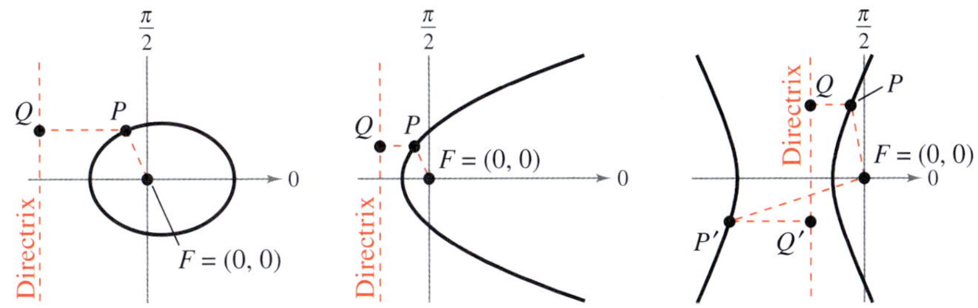

Ellipse: $0 < e < 1$

$\frac{PF}{PQ} < 1$

Parabola: $e = 1$

$\frac{PF}{PQ} = 1$

Hyperbola: $e > 1$

$\frac{PF}{PQ} = \frac{P'F}{P'Q'} > 1$

Figure 9.67

The benefit of locating a focus of a conic at the pole is that the equation of the conic becomes simpler.

Polar Equations of Conics

The graph of a polar equation of the form

$$r = \frac{ep}{1 \pm e \cos \theta}$$

or

$$r = \frac{ep}{1 \pm e \sin \theta}$$

is a conic, where $e > 0$ is the eccentricity and $p > 0$ is the distance between the focus (pole) and the directrix.

(See the proof on page 719.)

What you should learn

- Define conics in terms of eccentricities, and write and graph equations of conics in polar form.
- Use equations of conics in polar form to model real-life problems.

Why you should learn it

The orbits of planets and satellites can be modeled by polar equations. For instance, in Exercise 64 on page 707, you will use a polar equation to model the orbit of a satellite.

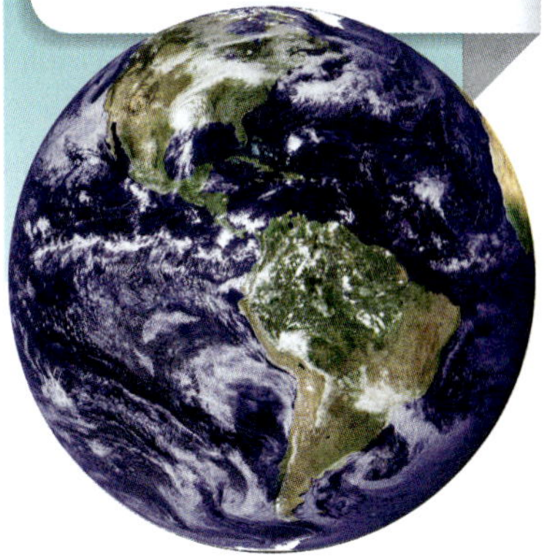

Algebra Help

Notice in the *alternative* definition of a conic that the eccentricity e is a strictly positive constant, as shown in Figure 9.67. So, this definition excludes circles and degenerate conics.

An equation of the form

$$r = \frac{ep}{1 \pm e \cos \theta}$$ Vertical directrix

corresponds to a conic with a vertical directrix and symmetry with respect to the polar axis. An equation of the form

$$r = \frac{ep}{1 \pm e \sin \theta}$$ Horizontal directrix

corresponds to a conic with a horizontal directrix and symmetry with respect to the line $\theta = \pi/2$. Moreover, the converse is also true—that is, any conic with a focus at the pole and having a horizontal or vertical directrix can be represented by one of these equations.

EXAMPLE 1 Identifying a Conic from Its Equation

See LarsonPrecalculus.com for an interactive version of this type of example.

Identify the type of conic represented by the equation $r = \dfrac{15}{3 - 2 \cos \theta}$.

Algebraic Solution

To identify the type of conic, rewrite the equation in the form $r = ep/(1 \pm e \cos \theta)$.

$$r = \frac{15}{3 - 2 \cos \theta}$$

$$= \frac{5}{1 - (2/3) \cos \theta}$$ Divide numerator and denominator by 3.

Because $e = \frac{2}{3} < 1$, you can conclude that the graph is an ellipse.

Graphical Solution

Use a graphing utility in *polar* mode and be sure to use a square setting, as shown in the figure.

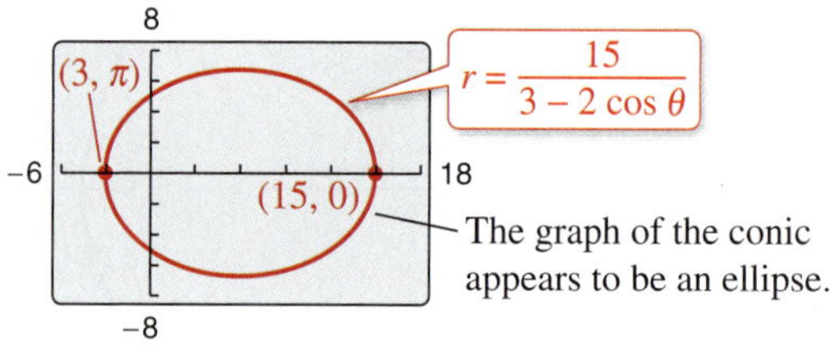
8
(3, π)
$r = \dfrac{15}{3 - 2\cos\theta}$
−6
18
(15, 0)
−8
The graph of the conic appears to be an ellipse.

Checkpoint *Audio-video solution in English & Spanish at LarsonPrecalculus.com*

Identify the type of conic represented by the equation $r = \dfrac{8}{2 - 3 \sin \theta}$.

For the ellipse in Example 1, the major axis is horizontal and the vertices lie at $(r, \theta) = (15, 0)$ and $(r, \theta) = (3, \pi)$. So, the length of the *major* axis is $2a = 18$. To find the length of the *minor* axis, you can use the equations $e = c/a$ and $b^2 = a^2 - c^2$ to conclude that

$$b^2 = a^2 - c^2$$
$$= a^2 - (ea)^2$$
$$= a^2(1 - e^2).$$ Ellipse

Because $e = 2/3$, you have $b^2 = 9^2[1 - (2/3)^2] = 45$, which implies that

$$b = \sqrt{45} = 3\sqrt{5}.$$

So, the length of the minor axis is $2b = 6\sqrt{5}$. A similar analysis for hyperbolas yields

$$b^2 = c^2 - a^2$$
$$= (ea)^2 - a^2$$
$$= a^2(e^2 - 1).$$ Hyperbola

EXAMPLE 2 Analyzing the Graph of a Polar Equation

Analyze the graph of the polar equation $r = \dfrac{32}{3 + 5 \sin \theta}$.

Solution

Dividing the numerator and denominator by 3 produces

$$r = \frac{32/3}{1 + (5/3) \sin \theta}.$$

Because $e = \frac{5}{3} > 1$, the graph is a hyperbola. The transverse axis of the hyperbola lies on the line $\theta = \pi/2$, and the vertices occur at $(r, \theta) = (4, \pi/2)$ and $(r, \theta) = (-16, 3\pi/2)$. Because the length of the transverse axis is 12, you can see that $a = 6$. To find b, write

$$b^2 = a^2(e^2 - 1) = 6^2\left[\left(\frac{5}{3}\right)^2 - 1\right] = 64.$$

So, $b = 8$. You can use a and b to determine that the asymptotes are $y = 10 \pm \frac{3}{4}x$, as shown in Figure 9.68.

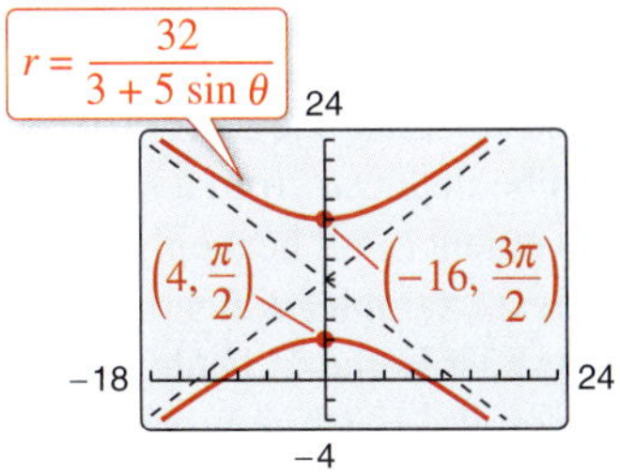

Figure 9.68

Checkpoint Audio-video solution in English & Spanish at LarsonPrecalculus.com

Identify the conic $r = \dfrac{3}{2 - 4 \sin \theta}$ and sketch its graph.

In the next example, you are asked to find a polar equation for a specified conic. To do this, let p be the distance between the pole and the directrix.

1. Horizontal directrix $(y = p)$ above the pole: $r = \dfrac{ep}{1 + e \sin \theta}$

2. Horizontal directrix $(y = -p)$ below the pole: $r = \dfrac{ep}{1 - e \sin \theta}$

3. Vertical directrix $(x = p)$ to the right of the pole: $r = \dfrac{ep}{1 + e \cos \theta}$

4. Vertical directrix $(x = -p)$ to the left of the pole: $r = \dfrac{ep}{1 - e \cos \theta}$

Explore the Concept

Use a graphing utility in *polar* mode to investigate the four orientations shown at the left.

EXAMPLE 3 Finding the Polar Equation of a Conic

Find the polar equation of the parabola whose focus is the pole and whose directrix is the line $y = 3$.

Solution

Because the directrix is horizontal and above the pole, use an equation of the form

$$r = \frac{ep}{1 + e \sin \theta}.$$

Moreover, because the eccentricity of a parabola is $e = 1$ and the distance between the pole and the directrix is $p = 3$, you have the equation

$$r = \frac{3}{1 + \sin \theta}.$$ See Figure 9.69.

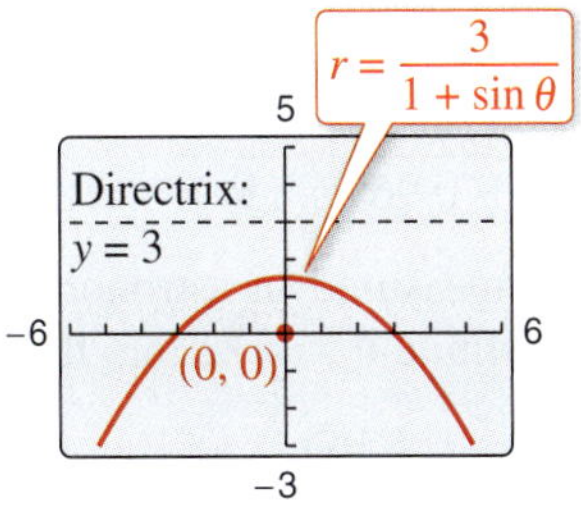

Figure 9.69

Checkpoint Audio-video solution in English & Spanish at LarsonPrecalculus.com

Find a polar equation of the parabola whose focus is the pole and whose directrix is the line $x = -2$.

Application

Kepler's Laws (listed below), named after the German astronomer Johannes Kepler (1571–1630), can be used to describe the orbits of the planets about the sun.

1. Each planet moves in an elliptical orbit with the sun as a focus.
2. A ray from the sun to the planet sweeps out equal areas of the ellipse in equal times.
3. The square of the period (the time it takes for a planet to orbit the sun) is proportional to the cube of the mean distance between the planet and the sun.

Although Kepler simply stated these laws on the basis of observation, they were later validated by Isaac Newton (1642–1727). In fact, Newton was able to show that each law can be deduced from a set of universal laws of motion and gravitation that govern the movement of all heavenly bodies, including comets and satellites. This is illustrated in the next example, which involves the comet named after the English mathematician and physicist Edmund Halley (1656–1742).

If you use Earth as a reference with a period of 1 year and a distance of 1 astronomical unit (an *astronomical unit* is defined as the mean distance between Earth and the sun, or about 93 million miles), then the proportionality constant in Kepler's third law is 1. For example, because Mars has a mean distance to the sun of $d \approx 1.524$ astronomical units, its period P is given by

$$d^3 = P^2.$$

So, the period of Mars is $P \approx 1.88$ years.

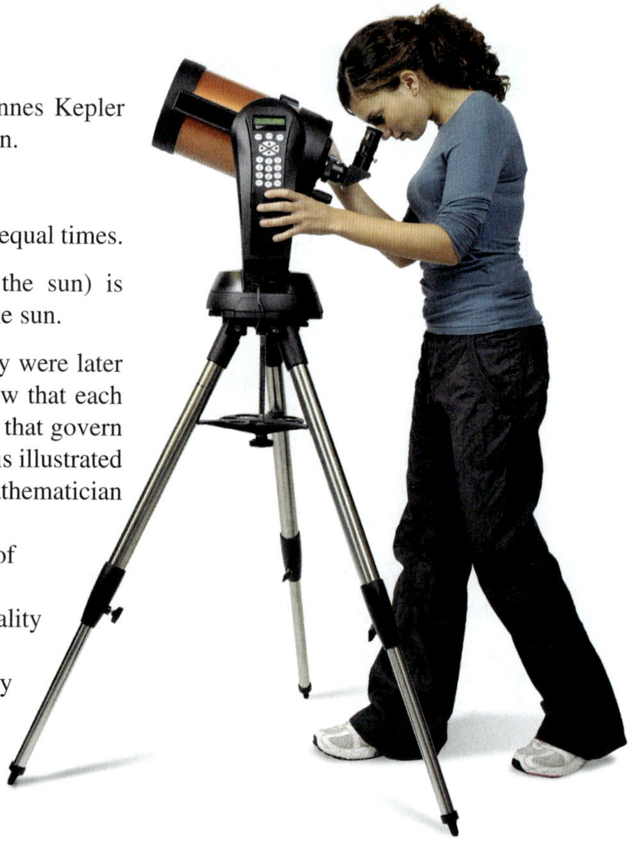

Astronomer

EXAMPLE 4 Halley's Comet

Halley's comet has an elliptical orbit with an eccentricity of $e \approx 0.967$. The length of the major axis of the orbit is approximately 35.88 astronomical units. Find a polar equation for the orbit. How close does Halley's comet come to the sun?

Solution

Using a vertical major axis, as shown in Figure 9.70, choose an equation of the form

$$r = \frac{ep}{1 + e \sin \theta}.$$

Because the vertices of the ellipse occur at $\theta = \pi/2$ and $\theta = 3\pi/2$, you can determine that the length of the major axis is the sum of the r-values of the vertices. That is,

$$2a = \frac{0.967p}{1 + 0.967} + \frac{0.967p}{1 - 0.967} \quad \Rightarrow \quad 35.88 \approx 29.79p.$$

So, $p \approx 1.204$ and $ep \approx (0.967)(1.204) \approx 1.164$. Using this value of ep in the equation, you have

$$r = \frac{1.164}{1 + 0.967 \sin \theta}$$

where r is measured in astronomical units. To find the closest point to the sun (the focus), substitute $\theta = \pi/2$ into this equation to obtain

$$r = \frac{1.164}{1 + 0.967 \sin(\pi/2)} \approx 0.59 \text{ astronomical unit} \approx 55{,}000{,}000 \text{ miles}.$$

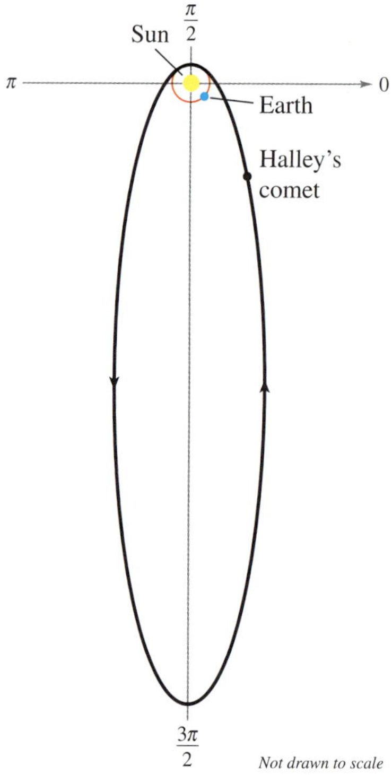

Figure 9.70

✓ Checkpoint *Audio-video solution in English & Spanish at LarsonPrecalculus.com*

Encke's comet has an elliptical orbit with an eccentricity of $e \approx 0.847$. The length of the major axis of the orbit is approximately 4.43 astronomical units. Find a polar equation for the orbit. How close does Encke's comet come to the sun?

9.7 Exercises

See *CalcChat.com* for tutorial help and worked-out solutions to odd-numbered exercises.
For instructions on how to use a graphing utility, see Appendix A.

Vocabulary and Concept Check

1. Fill in the blank: The locus of a point in the plane that moves such that its distance from a fixed point (focus) is in a constant ratio to its distance from a fixed line (directrix) is a _______ .

2. Match the conic with its eccentricity.

(a) $0 < e < 1$ (i) ellipse

(b) $e = 1$ (ii) hyperbola

(c) $e > 1$ (iii) parabola

3. A conic has a polar equation of the form $r = \dfrac{ep}{1 + e\cos\theta}$. Is the directrix vertical or horizontal?

4. A conic with a horizontal directrix has a polar equation of the form $r = \dfrac{ep}{1 - e\sin\theta}$. Is the directrix above or below the pole?

Procedures and Problem Solving

Identifying a Conic **In Exercises 5–8, write the polar equation of the conic for each value of *e* and identify the type of conic represented by the equation. Then use a graphing utility to graph each polar equation.**

(a) $e = 1$ **(b)** $e = 0.5$ **(c)** $e = 1.5$

5. $r = \dfrac{2e}{1 + e\cos\theta}$

6. $r = \dfrac{2e}{1 - e\cos\theta}$

7. $r = \dfrac{2e}{1 - e\sin\theta}$

8. $r = \dfrac{2e}{1 + e\sin\theta}$

Identifying the Polar Equation of a Conic **In Exercises 9–12, match the polar equation with its graph. [The graphs are labeled (a), (b), (c), and (d).]**

(a)

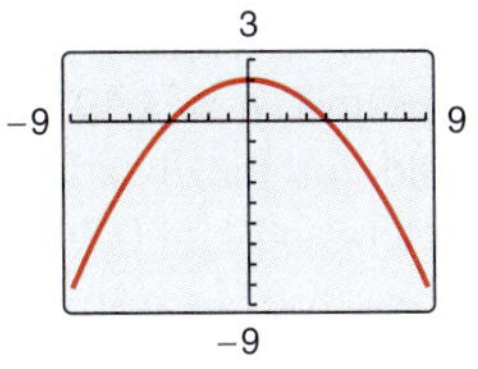

(b)

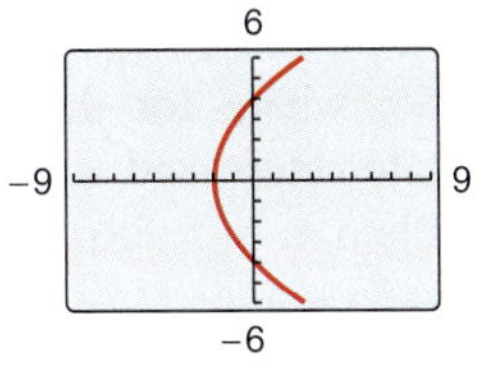

(c)

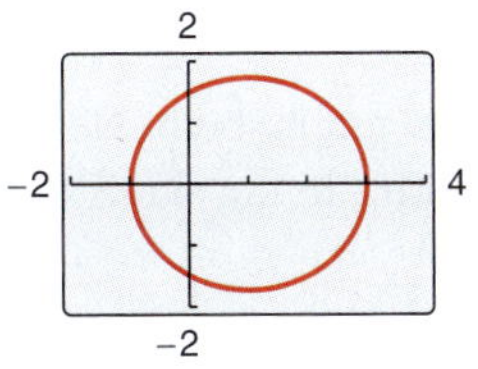

(d)

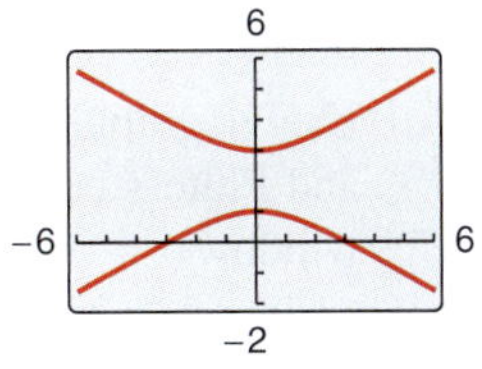

9. $r = \dfrac{4}{1 - \cos\theta}$

10. $r = \dfrac{3}{2 - \cos\theta}$

11. $r = \dfrac{3}{1 + 2\sin\theta}$

12. $r = \dfrac{4}{1 + \sin\theta}$

Identifying a Conic from Its Equation **In Exercises 13–22, identify the type of conic represented by the equation. Use a graphing utility to confirm your result.**

13. $r = \dfrac{3}{1 - \cos\theta}$

14. $r = \dfrac{2}{4 + \sin\theta}$

15. $r = \dfrac{4}{1 - 5\cos\theta}$

16. $r = \dfrac{7}{1 + \sin\theta}$

17. $r = \dfrac{8}{4 + 3\sin\theta}$

18. $r = \dfrac{9}{3 - 2\cos\theta}$

19. $r = \dfrac{6}{2 + \sin\theta}$

20. $r = \dfrac{5}{-1 + \cos\theta}$

21. $r = \dfrac{3}{-4 - 8\cos\theta}$

22. $r = \dfrac{10}{3 + 9\sin\theta}$

Analyzing the Graph of a Polar Equation **In Exercises 23–30, identify the type of conic represented by the polar equation. Then sketch the graph. Use a graphing utility to check your graph.**

23. $r = \dfrac{5}{1 - \sin\theta}$

24. $r = \dfrac{6}{1 + 3\sin\theta}$

25. $r = \dfrac{2}{2 - \cos\theta}$

26. $r = \dfrac{4}{4 + \sin\theta}$

27. $r = \dfrac{30}{10 + 20\sin\theta}$

28. $r = \dfrac{12}{2 - \cos\theta}$

29. $r = \dfrac{-3}{-4 + 2\cos\theta}$

30. $r = \dfrac{-4}{-1 + \cos\theta}$

Graphing a Rotated Conic **In Exercises 31–36, use a graphing utility to graph the rotated conic.**

31. $r = \dfrac{3}{1 - \cos(\theta - \pi/4)}$ (See Exercise 13.)

32. $r = \dfrac{7}{1 + \sin(\theta - \pi/3)}$ (See Exercise 16.)

33. $r = \dfrac{4}{1 - 5\cos(\theta + 3\pi/4)}$ (See Exercise 15.)

34. $r = \dfrac{9}{3 - 2\cos(\theta + \pi/2)}$ (See Exercise 18.)

35. $r = \dfrac{8}{4 + 3\sin(\theta + \pi/6)}$ (See Exercise 17.)

36. $r = \dfrac{10}{3 + 9\sin(\theta + 2\pi/3)}$ (See Exercise 22.)

Finding the Polar Equation of a Conic **In Exercises 37–54, find a polar equation of the conic with its focus at the pole.**

Conic	*Eccentricity*	*Directrix*
37. Parabola	$e = 1$	$x = -1$
38. Parabola	$e = 1$	$y = -4$
39. Ellipse	$e = \frac{1}{2}$	$y = 1$
40. Ellipse	$e = \frac{3}{4}$	$y = -4$
41. Hyperbola	$e = 2$	$x = 1$
42. Hyperbola	$e = \frac{3}{2}$	$x = -1$

Conic	*Vertex or Vertices*
43. Parabola	$\left(1, -\frac{\pi}{2}\right)$
44. Parabola	$(8, 0)$
45. Parabola	$(5, \pi)$
46. Parabola	$\left(10, \frac{\pi}{2}\right)$
47. Ellipse	$(2, 0), (10, \pi)$
48. Ellipse	$\left(2, \frac{\pi}{2}\right), \left(4, \frac{3\pi}{2}\right)$
49. Ellipse	$\left(1, \frac{3\pi}{2}\right), \left(11, \frac{\pi}{2}\right)$
50. Ellipse	$(20, 0), (4, \pi)$
51. Hyperbola	$\left(1, \frac{3\pi}{2}\right), \left(-9, \frac{\pi}{2}\right)$
52. Hyperbola	$(2, 0), (-8, \pi)$
53. Hyperbola	$\left(\frac{3}{5}, \pi\right), (-1, 0)$
54. Hyperbola	$\left(-4, \frac{3\pi}{2}\right), \left(1, \frac{\pi}{2}\right)$

55. Astronomy The planets travel in elliptical orbits with the sun at one focus. Assume that the focus is at the pole, the major axis lies on the polar axis, and the length of the major axis is $2a$ (see figure). Show that the polar equation of the orbit of a planet is

$$r = \frac{(1 - e^2)a}{1 - e\cos\theta}$$

where e is the eccentricity.

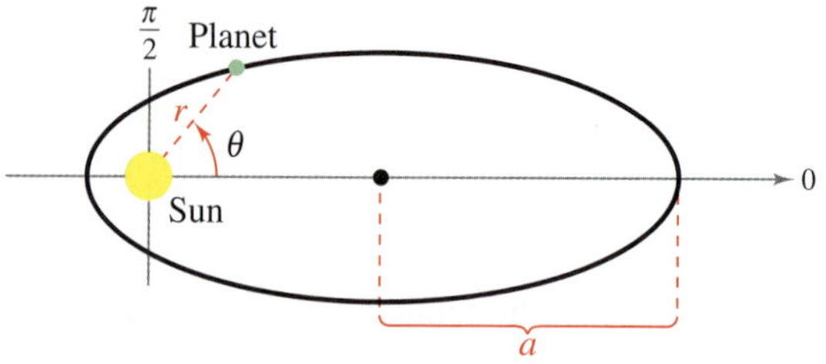

56. Astronomy Use the result of Exercise 55 to show that the minimum distance (*perihelion*) from the sun to a planet is

$$r = a(1 - e)$$

and that the maximum distance (*aphelion*) is

$$r = a(1 + e).$$

Astronomy **In Exercises 57–62, use the results of Exercises 55 and 56 to find the polar equation of the orbit of the planet and the perihelion and aphelion.**

57. Earth $a \approx 9.2957 \times 10^7$ miles, $e \approx 0.0167$

58. Mercury $a \approx 3.5984 \times 10^7$ miles, $e \approx 0.2056$

59. Mars $a \approx 1.4162 \times 10^8$ miles, $e \approx 0.0935$

60. Saturn $a \approx 1.4335 \times 10^9$ kilometers, $e \approx 0.0565$

61. Venus $a \approx 1.0821 \times 10^8$ kilometers, $e \approx 0.0067$

62. Jupiter $a \approx 7.7857 \times 10^8$ kilometers, $e \approx 0.0489$

63. Astronomy Use the results of Exercises 55 and 56, where for the planet Neptune, $a = 4.495 \times 10^9$ kilometers and $e = 0.0113$ and for the dwarf planet Pluto, $a = 5.906 \times 10^9$ kilometers and $e = 0.2488$.

(a) Find the polar equation of the orbit of each planet.

(b) Find the perihelion and aphelion distances for each planet.

(c) Use a graphing utility to graph both Neptune's and Pluto's equations of orbit in the same viewing window.

(d) Is Pluto ever closer to the sun than Neptune? Until recently, Pluto was considered the ninth planet. Why was Pluto called the ninth planet and Neptune the eighth planet?

(e) Do the orbits of Neptune and Pluto intersect? Will Neptune and Pluto ever collide? Why or why not?

64. ***Why you should learn it*** *(p. 701)* On November 27, 1963, the United States launched a satellite named *Explorer 18*. Its low and high points above the surface of Earth were about 119 miles and 122,800 miles, respectively, as shown in the figure. The center of Earth is at one focus of the orbit.

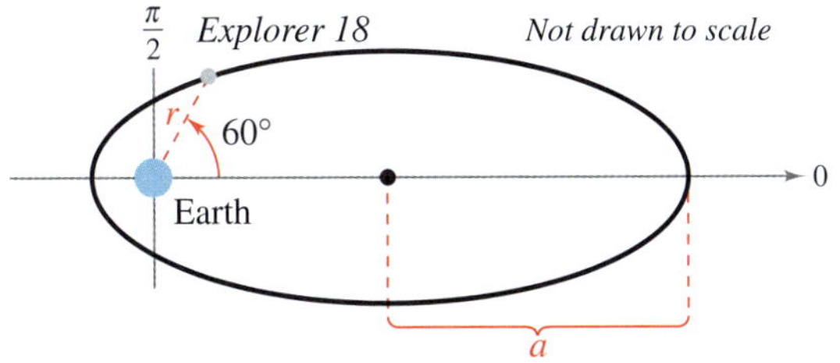

(a) Find the polar equation of the orbit (assume the radius of Earth is 4000 miles).

(b) Find the distance between the surface of Earth and the satellite when $\theta = 60°$.

(c) Find the distance between the surface of Earth and the satellite when $\theta = 30°$.

Focusing on Concepts

True or False? **In Exercises 65–67, determine whether the statement is true or false. Justify your answer.**

65. The graph of $r = 4/(-3 - 3\sin\theta)$ has a horizontal directrix above the pole.

66. The conic represented by the following equation is an ellipse.

$$r^2 = \frac{16}{9 - 4\cos\left(\theta + \frac{\pi}{4}\right)}$$

67. The graph of $r = \dfrac{5}{1 - \sin[\theta - (\pi/4)]}$ can be obtained by rotating the graph of $r = \dfrac{5}{1 + \sin\theta}$ about the pole.

68. **Error Analysis** Describe the error.

For the polar equation $r = \dfrac{3}{2 + \sin\theta}$, $e = 1$.

So, the equation represents a parabola.

69. **Writing** In your own words, define the term *eccentricity* and explain how it can be used to classify conics. Then explain how you can use the values of b and c to determine whether a polar equation of the form

$$r = \frac{a}{b + c\sin\theta}$$

represents an ellipse, a parabola, or a hyperbola.

70. **HOW DO YOU SEE IT?** The graph of $r = \dfrac{e}{1 - e\sin\theta}$ is shown for different values of e. Determine which graph matches each value of e.

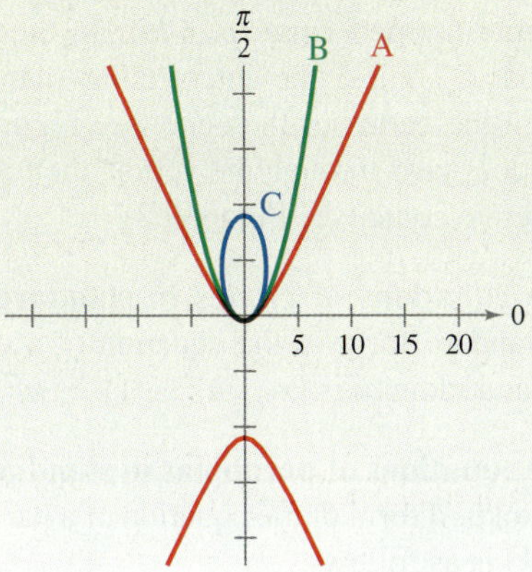

(i) $e = 0.9$ (ii) $e = 1.0$ (iii) $e = 1.1$

71. **Verifying a Polar Equation** Show that the polar equation of the ellipse

$$\frac{x^2}{a^2} + \frac{y^2}{b^2} = 1 \quad \text{is} \quad r^2 = \frac{b^2}{1 - e^2\cos^2\theta}.$$

72. **Verifying a Polar Equation** Show that the polar equation of the hyperbola

$$\frac{x^2}{a^2} - \frac{y^2}{b^2} = 1 \quad \text{is} \quad r^2 = \frac{-b^2}{1 - e^2\cos^2\theta}.$$

Writing a Polar Equation **In Exercises 73–78, use the results of Exercises 71 and 72 to write the polar form of the equation of the conic.**

73. $\dfrac{x^2}{169} + \dfrac{y^2}{144} = 1$

74. $\dfrac{x^2}{9} - \dfrac{y^2}{16} = 1$

75. $\dfrac{x^2}{25} + \dfrac{y^2}{16} = 1$

76. $\dfrac{x^2}{36} - \dfrac{y^2}{4} = 1$

77. Hyperbola One focus: $(5, 0)$
Vertices: $(4, 0), (4, \pi)$

78. Ellipse One focus: $(4, 0)$
Vertices: $(5, 0), (5, \pi)$

79. **Think About It** What conic does the polar equation $r = a\sin\theta + b\cos\theta$ represent?

Cumulative Mixed Review

Evaluating a Trigonometric Expression **In Exercises 80–83, find the exact value of the trigonometric expression when u and v are in Quadrant IV and $\sin u = -\frac{3}{5}$ and $\cos v = 1/\sqrt{2}$.**

80. $\cos(u + v)$

81. $\sin(u + v)$

82. $\sin(u - v)$

83. $\cos(u - v)$

9 Chapter Review

See *CalcChat.com* for tutorial help and worked-out solutions to odd-numbered exercises. For instructions on how to use a graphing utility, see Appendix A.

9.1 *What did you learn?*

Recognize a conic as the intersection of a plane and a double-napped cone *(p. 646)*. In the formation of the four basic conics, the intersecting plane does not pass through the vertex of the cone (see Figure 9.1). When the plane does pass through the vertex, the resulting figure is a degenerate conic (see Figure 9.2).

Write equations of circles in standard form *(p. 647)*. The standard form of the equation of a circle with center (h, k) and radius r is $(x - h)^2 + (y - k)^2 = r^2$.

Write equations of parabolas in standard form *(p. 649)*. The standard form of the equation of a parabola with vertex at (h, k) is as follows.

$(x - h)^2 = 4p(y - k), p \neq 0$ — Vertical axis; directrix: $y = k - p$

$(y - k)^2 = 4p(x - h), p \neq 0$ — Horizontal axis; directrix: $x = h - p$

The focus lies on the axis p units (*directed distance*) from the vertex.

Use the reflective property of parabolas to solve real-life problems *(p. 651)*. The tangent line to a parabola at a point P makes equal angles with (1) the line passing through P and the focus, and (2) the axis of the parabola.

Example

(a) The point $(2, 5)$ is on a circle whose center is $(-4, 6)$. The radius r is the distance between $(2, 5)$ and $(-4, 6)$.

$$r = \sqrt{[2 - (-4)]^2 + (5 - 6)^2} = \sqrt{37}$$

So, the standard form of the equation of the circle is

$$(x + 4)^2 + (y - 6)^2 = 37.$$

(b) To write the standard form of the equation of the parabola with vertex $(3, 1)$ and focus $(1, 1)$, first note that the axis is horizontal and that $p = -2$ (see figure). So, the standard form is

$$(y - 1)^2 = -8(x - 3).$$

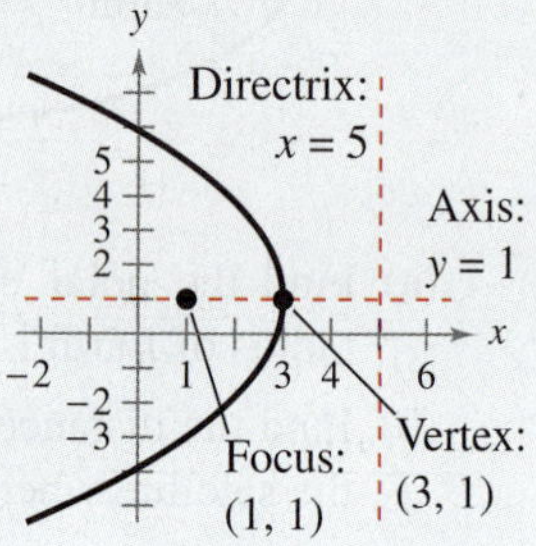

(c) To find the vertex, focus, and directrix of the parabola $y^2 - 2y + 8x - 23 = 0$, write the equation in standard form.

$$(y - 1)^2 = -8(x - 3).$$

Note that the standard form, and consequently the vertex, focus, and directrix, are the same as in part (b).

Forming a Conic Section **In Exercises 1 and 2, state the type of conic formed by the intersection of the plane and the double-napped cone.**

1. 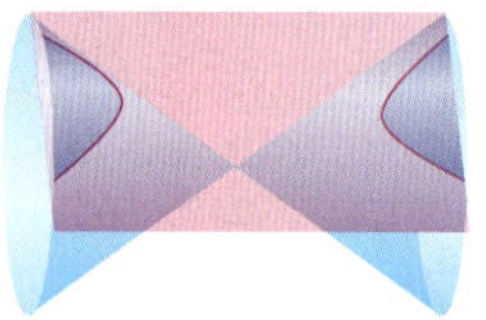

2. 

Finding the Standard Equation of a Circle **In Exercises 3–6, find the standard form of the equation of the circle with the given characteristics.**

3. Center at origin; point on the circle: $(-3, -4)$
4. Center at origin; point on the circle: $(8, -15)$
5. Center: $(2, 4)$; diameter: $2\sqrt{13}$
6. Center: $(-1, 2)$; diameter: $8\sqrt{2}$

Writing the Equation of a Circle in Standard Form **In Exercises 7–10, write the equation of the circle in standard form. Then identify its center and radius.**

7. $\frac{1}{2}x^2 + \frac{1}{2}y^2 = 18$
8. $\frac{3}{4}x^2 + \frac{3}{4}y^2 = 1$
9. $16x^2 + 16y^2 - 16x + 24y - 3 = 0$
10. $4x^2 + 4y^2 + 32x - 24y + 51 = 0$

Sketching a Circle **In Exercises 11 and 12, sketch the circle by hand. Identify its center and radius. Use a graphing utility to verify your graph.**

11. $x^2 + y^2 + 4x + 6y - 3 = 0$
12. $x^2 + y^2 + 8x - 10y - 8 = 0$

Finding the Intercepts of a Circle **In Exercises 13 and 14, find the x- and y-intercepts of the graph of the circle.**

13. $(x - 2)^2 + (y + 4)^2 = 9$
14. $(x + 8)^2 + (y + 5)^2 = 25$

Finding the Vertex, Focus, and Directrix of a Parabola **In Exercises 15–18, find the vertex, focus, and directrix of the parabola, and sketch its graph. Use a graphing utility to verify your graph.**

15. $y = -\frac{1}{8}x^2$
16. $4x - y^2 = 0$
17. $\frac{1}{2}y^2 + 18x = 0$
18. $\frac{1}{4}y - 8x^2 = 0$

Finding the Standard Equation of a Parabola **In Exercises 19–22, find the standard form of the equation of the parabola with the given characteristics.**

19. Vertex: $(0, 0)$
    Focus: $(0, 5)$
20. Vertex: $(-3, 0)$
    Focus: $(0, 0)$

21. 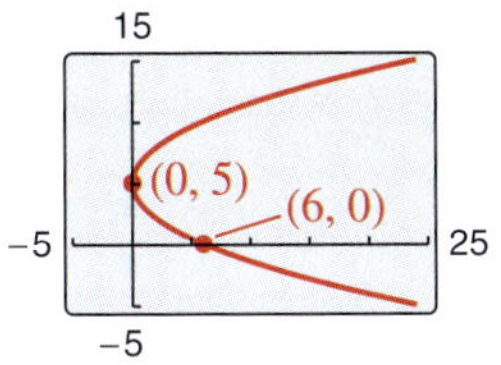

22. 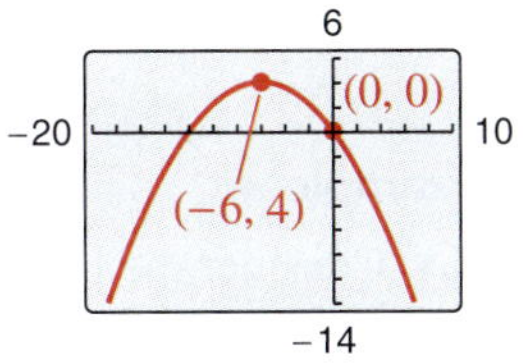

23. **Finding the Tangent Line at a Point on a Parabola** Find an equation of the tangent line to the parabola given by $y^2 = -2x$ at the point $(-8, -4)$.

24. **Architecture** A parabolic archway is 12 meters high at the vertex. At a height of 10 meters, the width of the archway is 8 meters (see figure). How wide is the archway at ground level?

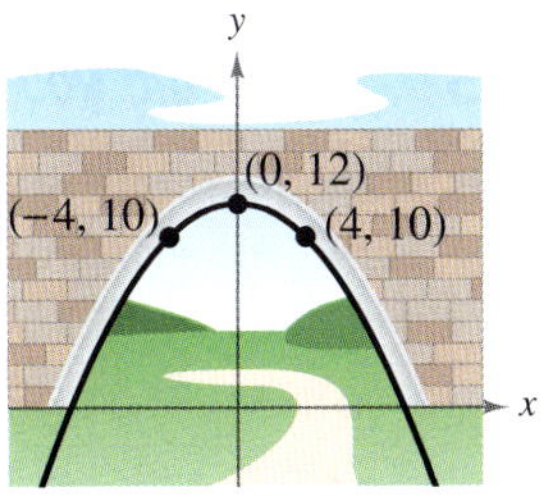

9.2 *What did you learn?*

Write equations of ellipses in standard form *(p. 657)*. The standard form of the equation of an ellipse with center (h, k) and major and minor axes of lengths $2a$ and $2b$, respectively, where $0 < b < a$, is as follows.

$$\frac{(x-h)^2}{a^2} + \frac{(y-k)^2}{b^2} = 1 \quad \text{Major axis is horizontal.}$$

$$\frac{(x-h)^2}{b^2} + \frac{(y-k)^2}{a^2} = 1 \quad \text{Major axis is vertical.}$$

The foci lie on the major axis, c units from the center, with $c^2 = a^2 - b^2$.

Use properties of ellipses to model and solve real-life problems *(p. 661)*. The properties of ellipses can be used to find the greatest and least distances from Earth's center to the moon's center (see Example 5).

Find eccentricities of ellipses *(p. 662)*. The eccentricity e of an ellipse is given by the ratio $e = c/a$.

Example Consider the ellipse

$4x^2 + 25y^2 + 16x - 150y + 141 = 0.$

To find the center, vertices, and foci, first write the equation in standard form.

$$(4x^2 + 16x) + (25y^2 - 150y) = -141$$
$$4(x^2 + 4x) + 25(y^2 - 6y) = -141$$
$$4(x^2 + 4x + 4) + 25(y^2 - 6y + 9) = -141 + 4(4) + 25(9)$$
$$4(x + 2)^2 + 25(y - 3)^2 = 100$$
$$\frac{(x+2)^2}{25} + \frac{(y-3)^2}{4} = 1$$

So, the major axis is horizontal, the center is $(-2, 3)$, $a = 5$, $b = 2$, $c = \sqrt{25 - 4} = \sqrt{21}$, the vertices are $(-2 - 5, 3) = (-7, 3)$ and $(-2 + 5, 3) = (3, 3)$, and the foci are $\left(-2 \pm \sqrt{21}, 3\right)$. (See figure.)

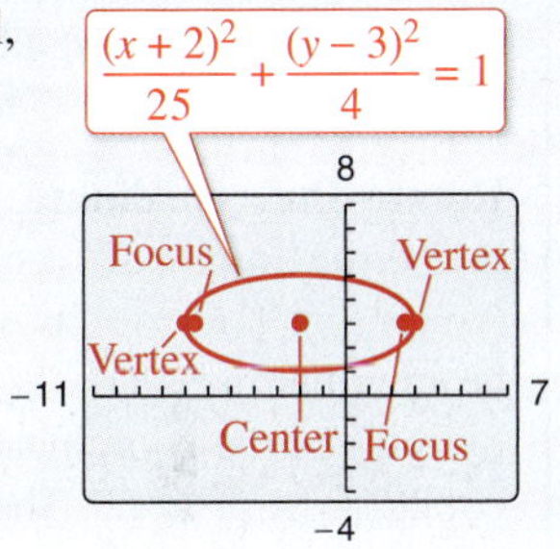

Finding the Standard Equation of an Ellipse **In Exercises 25 and 26, find the standard form of the equation of the ellipse with the given characteristics.**

25. Vertices: $(\pm 5, 0)$; foci: $(\pm 4, 0)$

26. Vertices: $(2, 0)$, $(2, 4)$; foci: $(2, 1)$, $(2, 3)$

Sketching an Ellipse **In Exercises 27–30, find the center, vertices, foci, and eccentricity of the ellipse, and sketch its graph. Use a graphing utility to verify your graph.**

27. $\dfrac{x^2}{25} + \dfrac{y^2}{36} = 1$

28. $\dfrac{(x+5)^2}{1} + \dfrac{(y-1)^2}{9} = 1$

29. $3x^2 + 8y^2 + 12x - 112y + 403 = 0$

30. $16x^2 + 9y^2 - 32x + 72y + 16 = 0$

31. **Architecture** A semielliptical archway is to be formed over the entrance to an estate. The arch is to be set on pillars that are 10 feet apart and is to have a height (atop the pillars) of 4 feet. Where should the foci be placed in order to sketch the arch?

32. **Architecture** You are building a wading pool that is in the shape of an ellipse. Your plans give an equation for the elliptical shape of the pool measured in feet as

$$\frac{x^2}{324} + \frac{y^2}{196} = 1.$$

Find the longest distance across the pool, the shortest distance, and the distance between the foci.

33. **Astronomy** Saturn moves in an elliptical orbit with the sun at one focus. The least distance and the greatest distance of the planet from the sun are 1.3526×10^9 and 1.5145×10^9 kilometers, respectively. Use this information to verify that the eccentricity of the orbit is the value given on page 662.

34. **Astronomy** Mercury moves in an elliptical orbit with the sun at one focus. The eccentricity of Mercury's orbit is $e \approx 0.2056$. The length of the major axis is 72 million miles. Find the standard equation of Mercury's orbit. Place the center of the orbit at the origin and the major axis on the x-axis.

9.3 What did you learn?

Write equations of hyperbolas in standard form *(p. 666)*. The standard form of the equation of a hyperbola with center (h, k) is as follows.

$$\frac{(x-h)^2}{a^2} - \frac{(y-k)^2}{b^2} = 1 \quad \text{Transverse axis is horizontal.}$$

$$\frac{(y-k)^2}{a^2} - \frac{(x-h)^2}{b^2} = 1 \quad \text{Transverse axis is vertical.}$$

The vertices are a units from the center, and the foci are c units from the center. Moreover, $c^2 = a^2 + b^2$.

Find asymptotes of and graph hyperbolas *(p. 668)*.

$y = k \pm (b/a)(x - h)$ Asymptotes for horizontal transverse axis

$y = k \pm (a/b)(x - h)$ Asymptotes for vertical transverse axis

Use properties of hyperbolas to solve real-life problems *(p. 671)*. (See Example 5.)

Classify conics from their general equations *(p. 672)*. The graph of $Ax^2 + Cy^2 + Dx + Ey + F = 0$ is (1) a circle when $A = C$ $(A \neq 0)$, (2) a parabola when $AC = 0$ ($A = 0$ or $C = 0$, but not both), (3) an ellipse when $AC > 0$ (A and C have like signs, $A \neq C$), (4) a hyperbola when $AC < 0$ (A and C have unlike signs).

Rotate the coordinate axes to eliminate the *xy*-term in equations of conics *(p. 673)*. The equation $Ax^2 + Bxy + Cy^2 + Dx + Ey + F = 0$, where $B \neq 0$, can be rewritten as $A'(x')^2 + C'(y')^2 + D'x' + E'y' + F' = 0$ by rotating the coordinate axes through an angle θ, where $\cot 2\theta = (A - C)/B$. The coefficients of the new equation are obtained by substituting $x = x'\cos\theta - y'\sin\theta$ and $y = x'\sin\theta + y'\cos\theta$.

Example

(a) Consider the hyperbola

$$9y^2 - 4x^2 + 18y - 8x - 139 = 0.$$

To find the center, vertices, foci, and asymptotes, first write the equation in standard form (verify this):

$$\frac{(y+1)^2}{16} - \frac{(x+1)^2}{36} = 1.$$

So, the transverse axis is vertical, the center is $(-1, -1)$, $a = 4$, $b = 6$, $c = \sqrt{16 + 36} = 2\sqrt{13}$, the vertices are $(-1, -1 + 4) = (-1, 3)$ and $(-1, -1 - 4) = (-1, -5)$, the foci are $\left(-1, -1 \pm 2\sqrt{13}\right)$, and the equations of the asymptotes are $y = -1 \pm \frac{4}{6}(x + 1)$, or $y = \frac{2}{3}x - \frac{1}{3}$ and $y = -\frac{2}{3}x - \frac{5}{3}$. (See figure.)

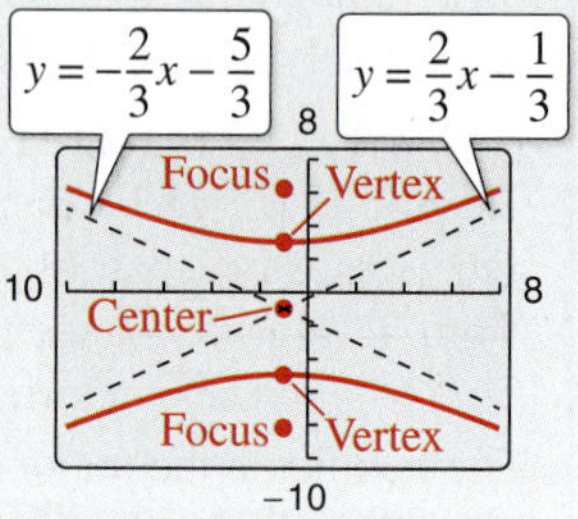

(b) Consider the ellipse $13x^2 + 6\sqrt{3}xy + 7y^2 - 16 = 0$. Because $A = 13$, $B = 6\sqrt{3}$, and $C = 7$,

$$\cot 2\theta = \frac{1}{\sqrt{3}} \Rightarrow 2\theta = \frac{\pi}{3} \Rightarrow \theta = \frac{\pi}{6},$$

$$x = x'\cos\frac{\pi}{6} - y'\sin\frac{\pi}{6} = \frac{\sqrt{3}x' - y'}{2}, \text{ and}$$

$$y = x'\sin\frac{\pi}{6} + y'\cos\frac{\pi}{6} = \frac{x' + \sqrt{3}y'}{2}.$$

Substituting the above expressions for x and y into the equation of the ellipse gives $16(x')^2 + 4(y')^2 = 16$ (verify this), or $(x')^2/1 + (y')^2/4 = 1$.

Finding the Standard Equation of a Hyperbola **In Exercises 35–38, find the standard form of the equation of the hyperbola with the given characteristics.**

35. Vertices: $(\pm 4, 0)$; foci: $(\pm 6, 0)$

36. Vertices: $(0, \pm 1)$; foci: $(0, \pm 2)$

37. Foci: $(3, \pm 2)$; asymptotes: $y = \pm 2(x - 3)$

38. Foci: $(0, 0)$, $(8, 0)$; asymptotes: $y = \pm 2(x - 4)$

Analyzing a Hyperbola **In Exercises 39–44, (a) find the standard form of the equation (b) find the center, vertices, foci, and asymptotes of the hyperbola, and (c) sketch the hyperbola.**

39. $5y^2 - 4x^2 = 20$

40. $x^2 - y^2 = \frac{9}{4}$

41. $9x^2 - 16y^2 - 18x - 32y - 151 = 0$

42. $-4x^2 + 25y^2 - 8x + 150y + 121 = 0$

43. $y^2 - 4x^2 - 2y - 48x + 59 = 0$

44. $9x^2 - y^2 - 72x + 8y + 119 = 0$

45. **Marine Navigation** Radio transmitting station A is located 200 miles east of transmitting station B. A ship is in an area to the north and 40 miles west of station A. Synchronized radio pulses transmitted to the ship at 186,000 miles per second by the two stations are received 0.0005 second sooner from station A than from station B. How far north is the ship?

46. **Physics** Two of your friends live 4 miles apart on the same "east-west" street, and you live halfway between them. You are having a three-way phone conversation when you hear an explosion. Six seconds later, your friend to the east hears the explosion, and your friend to the west hears it 8 seconds after you do. Find equations of two hyperbolas that would locate the explosion. (Assume that the coordinate system is measured in feet and that sound travels at 1100 feet per second.)

Classifying a Conic from a General Equation **In Exercises 47–50, classify the graph of the equation as a circle, a parabola, an ellipse, or a hyperbola.**

47. $6x^2 - x - 2y + 13 = 0$

48. $3x^2 + 5y^2 - 10x - 8y - 90 = 0$

49. $8x^2 + 8y^2 - 7x + 14y + 1 = 0$

50. $-x^2 + 4y^2 + 4x + 3y + 12 = 0$

Rotation of Axes **In Exercises 51–54, rotate the axes to eliminate the xy-term in the equation. Then write the equation in standard form. Sketch the graph of the resulting equation, showing both sets of axes.**

51. $xy - 3 = 0$

52. $x^2 - 4xy + y^2 + 9 = 0$

53. $5x^2 - 2xy + 5y^2 - 12 = 0$

54. $4x^2 + 8xy + 4y^2 + 7\sqrt{2}x + 9\sqrt{2}y = 0$

9.4 *What did you learn?*

Evaluate sets of parametric equations for given values of the parameter *(p. 679)*. If f and g are continuous functions of t on an interval I, then the set of ordered pairs $(f(t), g(t))$ is a plane curve C. $x = f(t)$ and $y = g(t)$ are parametric equations for C, and t is the parameter.

Graph curves that are represented by sets of parametric equations *(p. 680)*. Each set of points (x, y) on a curve represented by a pair of parametric equations is determined from a value chosen for t. Tracing a curve in a specific direction defines the orientation of the curve.

Rewrite sets of parametric equations as single rectangular equations by eliminating the parameter *(p. 682)*. Solve for t in one equation and substitute that value of t into the other equation. The result is the corresponding rectangular equation. You may need to alter the domain of the rectangular equation so that its graph matches the graph of the parametric equations.

Find sets of parametric equations for graphs *(p. 683)*. Parametric representations are not unique (see Example 4).

Example Consider the parametric equations $x = 2\cos t$ and $y = 4\sin t$. Using values of t in the interval $[0, 2\pi]$, the parametric equations yield the points (x, y) in the table.

t	0	$\pi/4$	$\pi/2$	$3\pi/4$	π	$5\pi/4$	$3\pi/2$	$7\pi/4$
x	2	$\sqrt{2}$	0	$-\sqrt{2}$	-2	$-\sqrt{2}$	0	$\sqrt{2}$
y	0	$2\sqrt{2}$	4	$2\sqrt{2}$	0	$-2\sqrt{2}$	-4	$-2\sqrt{2}$

By plotting the points in the order of increasing t, you obtain the ellipse shown. The arrows on the ellipse indicate its orientation as t increases from 0 to 2π. So, when a particle moves on this ellipse, it would start at $(2, 0)$ and then move counterclockwise. Eliminating the parameter, the equation of the ellipse is $4x^2 + y^2 = 16$. (verify this).

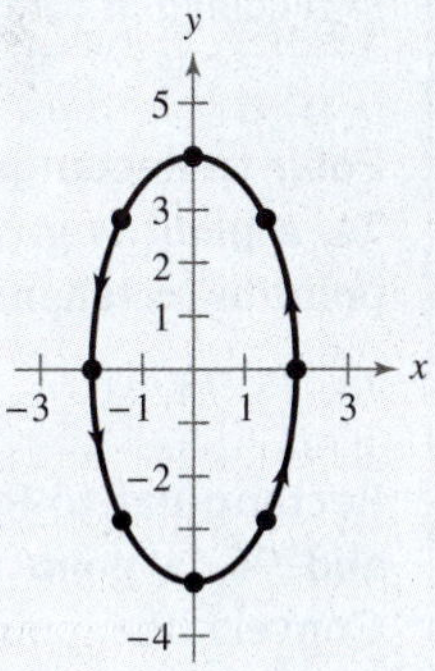

Using Parametric Equations **In Exercises 55 and 56, (a) create a table of x- and y-values for the set of parametric questions using the indicated values of t, and (b) use the table to sketch a graph of the parametric equations.**

55. $x = 3t - 2,\ y = 7 - 4t,\ t = -2, -1, 0, 1,$ and 2

56. $x = \sqrt{t},\ y = 8 - t,\ t = 0, 1, 2, 3,$ and 4

Sketching a Plane Curve and Eliminating the Parameter **In Exercises 57–62, (a) sketch the curve represented by the parametric equations (indicate the orientation of the curve), and (b) eliminate the parameter and write the corresponding rectangular equation whose graph represents the curve. Adjust the domain of the resulting rectangular equation, if necessary.**

57. $x = 2t$
$y = 4t$

58. $x = 4t + 1$
$y = 2 - 3t$

59. $x = t^3$
$y = \frac{1}{2}t^2$

60. $x = \frac{4}{t}$
$y = t^2 - 1$

61. $x = t^2 + 2$
$y = 4t^2 - 3$

62. $x = 5t$
$y = t^2$

Using a Graphing Utility in Parametric Mode **In Exercises 63–68, use a graphing utility to graph the curve represented by the parametric equations. Use the Vertical Line Test to determine whether y is a function of x.**

63. $x = 2t,\ y = 4t$

64. $x = 1 + 4t,\ y = 2 - 3t$

65. $x = t^2,\ y = \sqrt{t}$

66. $x = t + 4,\ y = t^2$

67. $x = 5\cos\theta$
$y = 5\sin\theta$

68. $x = 3 + 3\cos\theta$
$y = 2 + 5\sin\theta$

Finding Parametric Equations for a Graph **In Exercises 69–72, find a set of parametric equations to represent the graph of the given rectangular equation using the parameters (a) $t = x$ and (b) $t = 1 - x$.**

69. $y = 6x + 2$

70. $y = 10 - x$

71. $y = x^2 + 3$

72. $y = 2x^3 + 5x$

9.5 What did you learn?

Plot points and find multiple representations of points in the polar coordinate system *(p. 687)*. Each point P in the plane can be assigned polar coordinates (r, θ), where r is the directed distance from O (the origin or pole) to P, and θ is the directed angle counterclockwise from the polar axis to segment $\overline{OP}$.

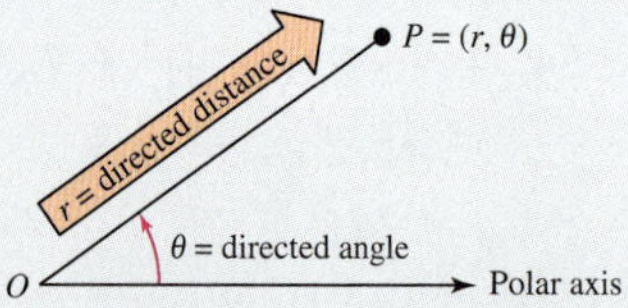

Example

(a) Other representations of the polar coordinates $(1, \pi/4)$ include $(-1, \pi/4 - \pi) = (-1, -3\pi/4)$ and $(-1, \pi/4 + \pi) = (-1, 5\pi/4)$.

(b) To convert the point $(2, 2\pi/3)$ to rectangular coordinates,

$$(x, y) = \left(2 \cos \frac{2\pi}{3}, 2 \sin \frac{2\pi}{3}\right) = \left(-1, \sqrt{3}\right).$$

(c) To convert the third-quadrant point $(-4, -4)$ to polar coordinates,

$\tan \theta = -1/-1 = 1$ ⇨ $\theta = 5\pi/4$ and

$r = \sqrt{(-4)^2 + (-4)^2} = 4\sqrt{2}.$

So, *one* set of polar coordinates is $\left(4\sqrt{2}, 5\pi/4\right)$.

(d) To convert $x^2 + y^2 = 6y$ to polar form,

$$(r \cos \theta)^2 + (r \sin \theta)^2 = 6r \sin \theta$$
$$r^2(\cos^2 \theta + \sin^2 \theta) = 6r \sin \theta$$
$$r = 6 \sin \theta.$$

Convert points *(p. 689)* and equations *(p. 690)* from rectangular to polar form and vice versa. Polar coordinates (r, θ) are related to rectangular coordinates (x, y) as follows.

Polar to rectangular: $x = r \cos \theta$, $y = r \sin \theta$

Rectangular to polar: $\tan \theta = y/x$, $r^2 = x^2 + y^2$

See Example 5 for examples of converting polar equations to rectangular form.

Polar-to-Rectangular Conversion In Exercises 73 and 74, a point is given in polar coordinates. Convert the point to rectangular coordinates.

73. $(5, 7\pi/4)$ **74.** $(-2, 5\pi/3)$

Rectangular-to-Polar Conversion In Exercises 75 and 76, a point is given in rectangular coordinates. Convert the point to polar coordinates. (There are many correct answers.)

75. $(5, -5)$ **76.** $\left(-3, -\sqrt{3}\right)$

Converting a Rectangular Equation to Polar Form In Exercises 77–80, convert the rectangular equation to polar form.

77. $x^2 + y^2 = 81$ **78.** $x^2 + y^2 - 4x = 0$

79. $xy = 10$ **80.** $4x^2 + y^2 = 1$

Converting a Polar Equation to Rectangular Form In Exercises 81–84, convert the polar equation to rectangular form.

81. $r = 14$ **82.** $r = 8 \sin \theta$

83. $r^2 = \sin \theta$ **84.** $\theta = 4\pi/3$

9.6 What did you learn?

Graph polar equations by point plotting *(p. 693)*, and use symmetry and zeros as sketching aids *(p. 694)*. The graph of a polar equation is symmetric with respect to the following when the given substitution yields an equivalent equation.

1. Line $\theta = \pi/2$: Replace (r, θ) with $(r, \pi - \theta)$ or $(-r, -\theta)$.
2. Polar axis: Replace (r, θ) with $(r, -\theta)$ or $(-r, \pi - \theta)$.
3. Pole: Replace (r, θ) with $(r, \pi + \theta)$ or $(-r, \theta)$.

Quick tests for symmetry in polar coordinates: (1) the graph of $r = f(\sin \theta)$ is symmetric with respect to the line $\theta = \pi/2$, and (2) the graph of $r = g(\cos \theta)$ is symmetric with respect to the polar axis.

Recognize special polar graphs *(p. 697)*. Several types of graphs, such as limaçons, rose curves, circles, and lemniscates, are listed on page 697.

Example To graph the polar equation $r = 2 \cos 4\theta$, note that the equation is of the form $r = a \cos n\theta$, where $n = 4$ is even. So, the graph is a rose curve with $2n = 8$ petals. Further, the graph has symmetry with respect to the line $\theta = \pi/2$, the polar axis, and the pole (verify this), and $r = 0$ when

$$4\theta = \frac{(2m + 1)\pi}{2}$$

or where

$$\theta = \frac{(2m + 1)\pi}{8}$$

where m is an integer. You can use the above information, along with point plotting with $0 \le \theta \le 2\pi$, to obtain the graph at the right.

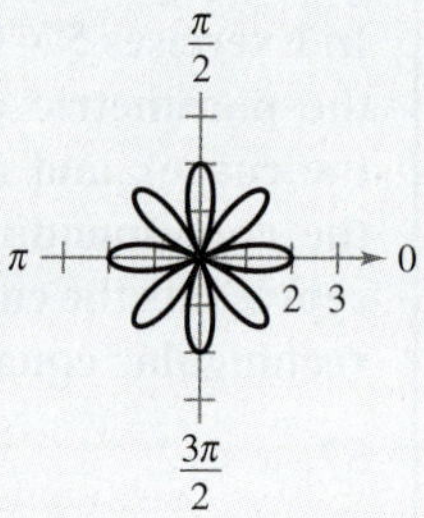

Sketching the Graph of a Polar Equation **In Exercises 85–88, sketch the graph of the polar equation using symmetry, zeros, maximum r-values, and any other additional points.**

85. $r = 5$

86. $\theta = -\dfrac{5\pi}{6}$

87. $r = -2 \cos \theta$

88. $r = 5 \sin \theta$

Sketching a Polar Graph **In Exercises 89–92, identify and sketch the graph of the polar equation. Identify any symmetry and zeros of r. Use a graphing utility to verify your results.**

89. $r = 1 + 3 \sin \theta$

90. $r = 7 - 7 \cos \theta$

91. $r = \cos 5\theta$

92. $r^2 = 6 \sin 2\theta$

9.7 *What did you learn?*

Define conics in terms of eccentricities, and write and graph equations of conics in polar form *(p. 701)*. A conic with eccentricity e is an ellipse when $0 < e < 1$, a parabola when $e = 1$, and a hyperbola when $e > 1$ (see Figure 9.67). Polar equations of conics have the forms

$r = ep/(1 \pm e \cos \theta)$ Vertical directrix, polar axis symmetry

$r = ep/(1 \pm e \sin \theta)$ Horizontal directrix, line $\theta = \pi/2$ symmetry

where $p > 0$ is the distance between the focus (pole) and the directrix, which can be horizontal above or below the pole or vertical to the right or left of the pole (see page 703).

Use equations of conics in polar form to model real-life problems *(p. 704)*. An equation of a conic in polar form can be used to model the orbit of Halley's comet (see Example 4).

Example Use a graphing utility to verify the type of conic, directrix, and symmetry corresponding to each equation.

(a) $r = 1/(1 - 0.4 \cos \theta)$ Ellipse; vertical directrix to the left of the pole; polar axis symmetry

(b) $r = 1/(1 + 0.4 \sin \theta)$ Ellipse; horizontal directrix above the pole; line $\theta = \pi/2$ symmetry

(c) $r = 1/(1 + \cos \theta)$ Parabola; vertical directrix to the right of the pole; polar axis symmetry

(d) $r = 1/(1 - \sin \theta)$ Parabola; horizontal directrix below the pole; line $\theta = \pi/2$ symmetry

(e) $r = 1/(1 - 2 \cos \theta)$ Hyperbola; vertical directrix to the left of the pole; polar axis symmetry

(f) $r = 1/(1 + 2 \sin \theta)$ Hyperbola; horizontal directrix above the pole; line $\theta = \pi/2$ symmetry

Identifying a Conic from Its Equation **In Exercises 93–98, identify the type of conic represented by the equation. Use a graphing utility to confirm your result.**

93. $r = \dfrac{1}{1 + 6 \sin \theta}$

94. $r = \dfrac{2}{1 + \sin \theta}$

95. $r = \dfrac{4}{5 - 3 \cos \theta}$

96. $r = \dfrac{6}{-1 + 4 \cos \theta}$

97. $r = \dfrac{3}{4 - 4 \cos \theta}$

98. $r = \dfrac{5}{6 + 2 \sin \theta}$

Finding the Polar Equation of a Conic **In Exercises 99–102, find a polar equation of the conic with its focus at the pole.**

99. Parabola; vertex: $(2, \pi)$

100. Parabola; vertex: $(3, \pi/2)$

101. Ellipse; vertices: $(5, \pi/2)$, $(1, 3\pi/2)$

102. Hyperbola; vertices: $(1, 0)$, $(-7, \pi)$

103. Astronomy A comet has an elliptical orbit around the sun with an eccentricity of $e \approx 0.9$. The length of the major axis of the orbit is approximately 20 astronomical units. Find a polar equation for the orbit. How close does the comet come to the sun?

104. Astronomy An asteroid takes a parabolic path with Earth as its focus. It is about 6,000,000 miles from Earth at its closest approach. Write the polar equation of the path of the asteroid with its vertex at $\theta = -\pi/2$. Find the distance between the asteroid and Earth when $\theta = -\pi/3$.

Focusing on Concepts

True or False? **In Exercises 105 and 106, determine whether the statement is true or false. Justify your answer.**

105. The graph of $\frac{1}{4}x^2 - y^4 = 1$ is a hyperbola.

106. Only one set of parametric equations can represent the line $y = 4 - 3x$.

107. Think About It The path of a moving object is modeled by the parametric equations $x = 4 \cos t$ and $y = 3 \sin t$, where t is time. How would the path change for each of the following?

(a) $x = 4 \cos 2t$, $y = 3 \sin 2t$

(b) $x = 5 \cos t$, $y = 3 \sin t$

9 Chapter Test

See *CalcChat.com* for tutorial help and worked-out solutions to odd-numbered exercises. For instructions on how to use a graphing utility, see Appendix A.

Take this test as you would take a test in class. After you are finished, check your work against the answers given in the back of the book.

1. Write the equation of the circle $x^2 - 14x + y^2 - 4y + 17 = 0$ in standard form. Identify its center and radius. Use a graphing utility to graph the circle.
2. Find the vertex, focus, and directrix of the parabola represented by $y^2 - 8x = 0$. Use a graphing utility to graph the parabola.
3. Find the standard form of the equation of the ellipse $x^2 + 4y^2 - 24x + 32 = 0$. Find its center, vertices, foci, and eccentricity. Use a graphing utility to graph the ellipse.
4. Find the center, vertices, foci, and asymptotes of the hyperbola represented by $x^2 - 4y^2 - 4x = 0$. Use a graphing utility to graph the hyperbola.
5. Classify the graph of each equation as a circle, a parabola, an ellipse, or a hyperbola.
   (a) $8x^2 + 3x + 4y + 2 = 0$ (b) $-2x^2 + y^2 + 4x + 3 = 0$
6. Rotate the axes to eliminate the xy-term of $xy - x + 3y + 1 = 0$. Then write the equation in standard form. Which conic do the equations represent?

In Exercises 7–9, sketch the curve represented by the parametric equations. Then eliminate the parameter and write the corresponding rectangular equation whose graph represents the curve.

7. $x = 1 - t$, $y = 5t$
8. $x = \sqrt{t^2 + 2}$, $y = t/4$
9. $x = e^{-t}$, $y = 2e^t$

10. Use a graphing utility to graph the curve represented by the parametric equations $x = 5 \cos \theta$ and $y = 2 \sin \theta$.

Collaborative Project

To work a collaborative project involving Topics in Analytic Geometry, visit this textbook's website at *LarsonPrecalculus.com.*

In Exercises 11–13, find a set of parametric equations to represent the graph of the given rectangular equation using the parameters (a) $t = x$ and (b) $t = x + 3$.

11. $y = -3x + 5$
12. $xy = 8$
13. $x^2 + 2y = 4$

14. Convert the polar coordinates $\left(-2, \frac{5\pi}{6}\right)$ to rectangular form.
15. Convert the rectangular coordinates $(2, -2)$ to polar form and find two additional polar representations of this point. (There are many correct answers.)
16. Convert the rectangular equation $x^2 + y^2 - 3x = 0$ to polar form.
17. Convert the polar equation $r = 2 \sin \theta$ to rectangular form.
18. Write a polar equation for graph of the conic at the right.

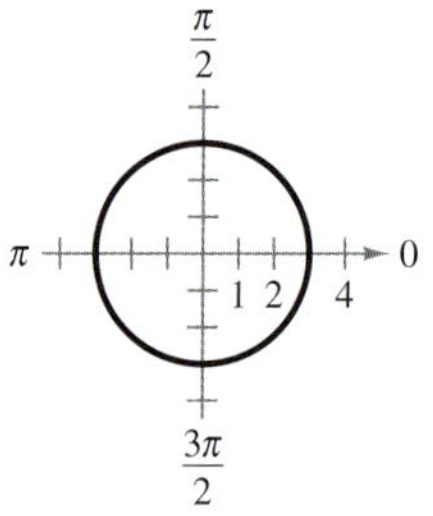

Figure for 18

In Exercises 19–21, identify the conic represented by the polar equation algebraically. Then use a graphing utility to graph the polar equation.

19. $r = 2 + 3 \sin \theta$
20. $r = \dfrac{1}{1 - \cos \theta}$
21. $r = \dfrac{4}{2 + 3 \sin \theta}$

22. Find a polar equation that represents a rose curve that is symmetric with respect to the line $\theta = \pi/2$ and has 7 petals.
23. Sketch a graph of the polar equation $r = 8 \cos 3\theta$. Identify any symmetry and zeros of the graph.
24. Find a polar equation of an ellipse with its focus at the pole, an eccentricity of $e = \frac{1}{4}$, and directrix at $y = 4$.
25. Find a polar equation of a hyperbola with its focus at the pole, an eccentricity of $e = \frac{5}{4}$, and directrix at $y = 2$.

Standardized Test Practice

See *CalcChat.com* for tutorial help and worked-out solutions to odd-numbered exercises.
For instructions on how to use a graphing utility, see Appendix A.

1. Which of the following pairs of functions are inverses?

(A) $f(x) = x + 1, g(x) = x - 1$

(B) $f(x) = x^2, g(x) = -\frac{1}{2}x^2$

(C) $f(x) = 2x^3, g(x) = 8\sqrt[3]{x}$

(D) $f(x) = 3^x, g(x) = -\log_3 x$

2. Which expression is a factor of $x^3 - x^2 + 12$?

(A) $x - 2$ (B) $x + 2$

(C) $x - 12$ (D) $x + 12$

3. What point on the unit circle corresponds to an angle of 150°?

(A) $\left(-\frac{1}{2}, \frac{\sqrt{3}}{2}\right)$ (B) $\left(\frac{1}{2}, -\frac{\sqrt{3}}{2}\right)$

(C) $\left(-\frac{\sqrt{3}}{2}, \frac{1}{2}\right)$ (D) $\left(\frac{\sqrt{3}}{2}, -\frac{1}{2}\right)$

4. $\begin{cases} x - 2y = 8 \\ 2x + 3y = 2 \end{cases}$

What point is the solution of the system above?

(A) $x = 0, y = -4$

(B) $x = 8, y = 0$

(C) $x = 4, y = -2$

(D) $x = 10, y = -1$

5. $\begin{cases} x - y^2 = -1 \\ x - y = 1 \end{cases}$

Which graph represents the system above?

(A)

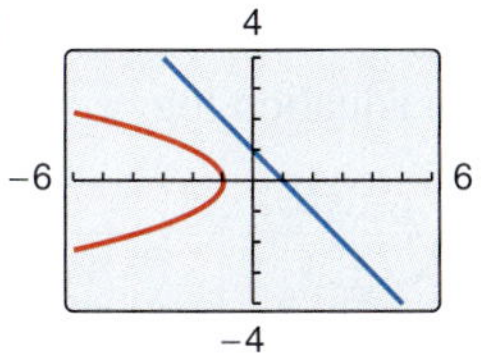

(B)

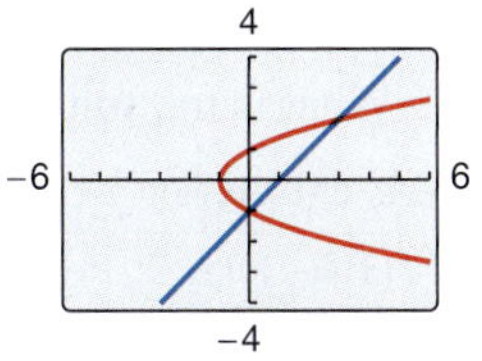

(C)

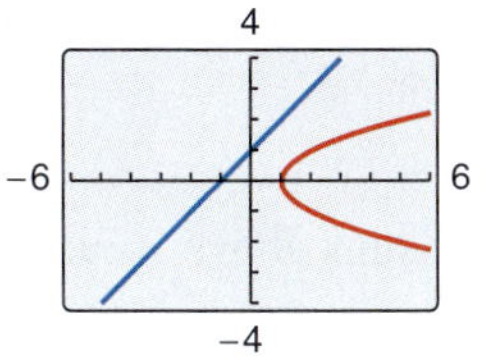

(D)

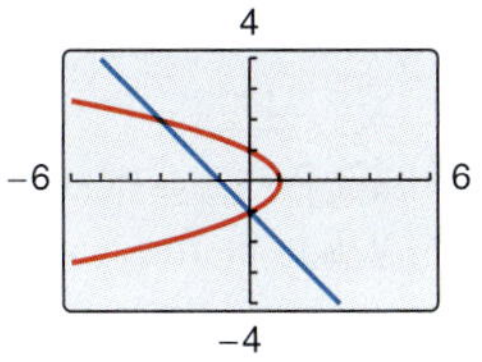

6. A teenager borrows money to pay for a used car. Some of the money is borrowed at 4% and some at 6%. How much is borrowed at 4%, given that the annual interest is \$500 and the amount borrowed at 6% is six times the amount borrowed at 4%?

(A) \$1250

(B) \$6250

(C) \$7500

(D) \$8750

7. $A = \begin{bmatrix} 2 & 4 \\ 0 & -9 \end{bmatrix}$ and $B = \begin{bmatrix} -4 & 0 \\ 2 & 3 \end{bmatrix}$

Use the matrices above to solve for X in the equation $2X - B = A$.

(A) $\begin{bmatrix} -2 & 4 \\ 2 & -6 \end{bmatrix}$ (B) $\begin{bmatrix} 2 & 2 \\ -1 & -6 \end{bmatrix}$

(C) $\begin{bmatrix} -1 & 2 \\ 1 & -3 \end{bmatrix}$ (D) $\begin{bmatrix} -4 & 4 \\ 4 & -12 \end{bmatrix}$

8. Which of the following matrices does *not* have a determinant of 2?

(A) $\begin{bmatrix} 2 & 0 \\ 0 & 1 \end{bmatrix}$ (B) $\begin{bmatrix} 0 & 2 \\ 1 & 1 \end{bmatrix}$

(C) $\begin{bmatrix} 1 & 1 \\ -1 & 1 \end{bmatrix}$ (D) $\begin{bmatrix} -1 & 1 \\ -3 & 1 \end{bmatrix}$

9. A square is defined by the vertices (2, 1), (2, 4), (5, 1), and (5, 4). Which of the following points lies on the perimeter of the square after a reflection in the x-axis?

(A) (2, 3) (B) (−2, 3)

(C) (3, −1) (D) (−3, −1)

10.

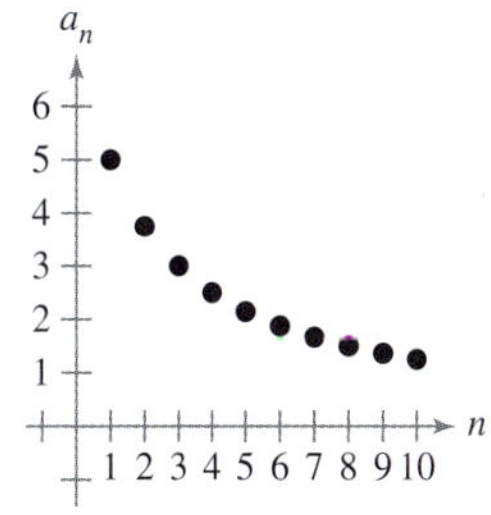

Which of the following sequences is represented by the graph above?

(A) $a_n = \dfrac{15}{n + 2}$ (B) $a_n = \dfrac{15}{2n}$

(C) $a_n = \dfrac{15}{n!}$ (D) $a_n = \dfrac{15}{2n^2}$

11. Which of the following sequences has a 101st term of 2139, where a_1 is the first term of the sequence?

(A) $a_n = 21n + 15$

(B) $a_n = 3250 - 11n$

(C) $a_n = 2450 - 3n$

(D) $a_n = 22n - 19$

12. What is the common ratio of a geometric sequence with a first term of 729 and a sixth term of −3?

(A) −122 (B) $-\dfrac{1}{3}$

(C) $\dfrac{1}{3}$ (D) 122

13. What is the third term in the expansion of $(x - 2y)^6$?

(A) $60x^4y^2$

(B) $240x^2y^4$

(C) $12x^5y$

(D) $15x^4y^2$

14. How many more ways are there to fill the positions of president, vice president, and treasurer from a class of 32 than a class of 16?

(A) 3

(B) 16

(C) 4400

(D) 26,400

15. A six-sided die is rolled twice. Which of the following events has a probability of $\frac{1}{6}$?

(A) rolling 5 or 6 first and 1 second

(B) rolling 2 first and 2 second

(C) rolling 3, 4, or 5 and 3 or 5 second

(D) rolling 1 or 2 first and 3 or 4 second

16.

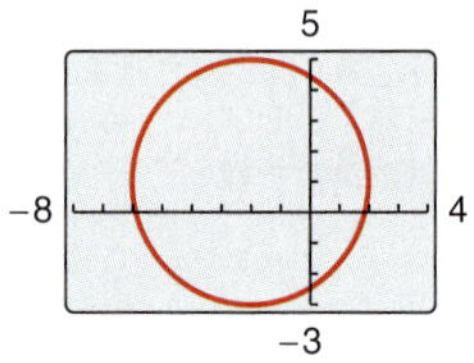

Which of the following equations represents the graph above?

(A) $(x + 2)^2 + (y - 1)^2 = 4$

(B) $(x - 2)^2 + (y + 1)^2 = 4$

(C) $(x + 2)^2 + (y - 1)^2 = 16$

(D) $(x - 2)^2 + (y + 1)^2 = 16$

17. $\dfrac{(x+1)^2}{12} - \dfrac{(y-4)^2}{48} = 1$

The equation above represents a conic. Which of the following is true?

I. The graph of the equation is an ellipse.

II. The point (3, 0) lies on the graph.

III. A focus is $(-1, 5)$.

(A) I only

(B) II only

(C) I and II only

(D) II and III only

18. Which of the following rectangular equations produces the same graph as the parametric equations $x = t - 1$ and $y = t^2$?

(A) $y = (x + 1)^2$

(B) $y = (x - 1)^2$

(C) $y = x^2 + x - 1$

(D) $y = x^2 + x + 1$

19.

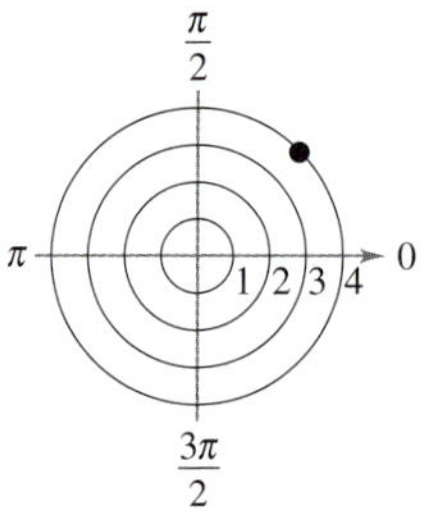

A point in the polar coordinate plane is shown above. Which of the following polar coordinates represents this point?

(A) $\left(-3, -\dfrac{5\pi}{4}\right)$ (B) $\left(3, -\dfrac{\pi}{4}\right)$

(C) $\left(-4, \dfrac{5\pi}{4}\right)$ (D) $\left(-4, \dfrac{\pi}{4}\right)$

20. $r = 2 \sin \dfrac{\theta}{3}$

A polar equation is shown above. On which of the following intervals is the graph traced *only once?*

(A) $0 \le \theta \le \dfrac{\pi}{3}$ (B) $0 \le \theta \le 2\pi$

(C) $0 \le \theta \le 3\pi$ (D) $0 \le \theta \le 6\pi$

21. $r = \dfrac{5}{2 - \sin \theta}$

What is the graph of the polar equation above?

(A) line

(B) ellipse

(C) hyperbola

(D) parabola

22. What is the solution of the equation $\log_3 x = 5$?

23. $\dfrac{4x^2}{(x-1)(x+1)^2} = \dfrac{A}{x-1} + \dfrac{B}{x+1} - \dfrac{2}{(x+1)^2}$

What is the value of $A + B$ in the partial fraction decomposition above?

24. $\displaystyle\sum_{i=1}^{4} \frac{i^2}{5 - i}$

What is the value of the sum above, rounded to the nearest hundredth?

25. ${}_8P_5$

What is value of the expression above?

26. A parabolic archway is 16 meters high at the vertex. At a height of 14 meters, the width of the archway is 12 meters. How wide is the archway at ground level, rounded to the nearest hundredth of a meter?

27. $\dfrac{x^2}{28} + \dfrac{(y+1)^2}{19} = 1$

What is the eccentricity of the ellipse given by the equation above, rounded to the nearest hundredth?

Proofs in Mathematics

Standard Equation of a Parabola (p. 649)

The standard form of the equation of a parabola with vertex at (h, k) is as follows.

$(x - h)^2 = 4p(y - k), \quad p \neq 0$ — Vertical axis; directrix: $y = k - p$

$(y - k)^2 = 4p(x - h), \quad p \neq 0$ — Horizontal axis; directrix: $x = h - p$

The focus lies on the axis p units (*directed distance*) from the vertex. If the vertex is at the origin $(0, 0)$, then the equation takes one of the following forms.

$x^2 = 4py$ — Vertical axis

$y^2 = 4px$ — Horizontal axis

Proof

For the case in which the directrix is parallel to the x-axis and the focus lies above the vertex, as shown in the top figure, if (x, y) is any point on the parabola, then, by definition, it is equidistant from the focus

$(h, k + p)$ — Focus lies above the vertex.

and the directrix

$y = k - p.$ — Directrix is parallel to the x-axis.

Using the Distance Formula, you have

$$\sqrt{(x - h)^2 + [y - (k + p)]^2} = y - (k - p)$$
$$(x - h)^2 + [y - (k + p)]^2 = [y - (k - p)]^2$$
$$(x - h)^2 + y^2 - 2y(k + p) + (k + p)^2 = y^2 - 2y(k - p) + (k - p)^2$$
$$(x - h)^2 + y^2 - 2ky - 2py + k^2 + 2pk + p^2 = y^2 - 2ky + 2py + k^2 - 2pk + p^2$$
$$(x - h)^2 - 2py + 2pk = 2py - 2pk$$
$$(x - h)^2 = 4p(y - k).$$

For the case in which the directrix is parallel to the y-axis and the focus lies to the right of the vertex, as shown in the bottom figure, if (x, y) is any point on the parabola, then, by definition, it is equidistant from the focus

$(h + p, k)$ — Focus lies to the right of the vertex.

and the directrix

$x = h - p.$ — Directrix is parallel to the y-axis.

Using the Distance Formula, you have

$$\sqrt{[x - (h + p)]^2 + (y - k)^2} = x - (h - p)$$
$$[x - (h + p)]^2 + (y - k)^2 = [x - (h - p)]^2$$
$$x^2 - 2x(h + p) + (h + p)^2 + (y - k)^2 = x^2 - 2x(h - p) + (h - p)^2$$
$$x^2 - 2hx - 2px + h^2 + 2ph + p^2 + (y - k)^2 = x^2 - 2hx + 2px + h^2 - 2ph + p^2$$
$$-2px + 2ph + (y - k)^2 = 2px - 2ph$$
$$(y - k)^2 = 4p(x - h).$$

Note that when the vertex of a parabola is at the origin, the two equations above simplify to $x^2 = 4py$ and $y^2 = 4px$, respectively. The proofs of the other cases are similar. ■

Parabolic Paths

There are many natural occurrences of parabolas in real life. For instance, the famous astronomer Galileo Galilei (1564–1642) discovered in the seventeenth century that an object that is projected upward and obliquely to the pull of gravity travels in a parabolic path. Examples of this are the center of gravity of a jumping dolphin and the path of water molecules in a drinking fountain.

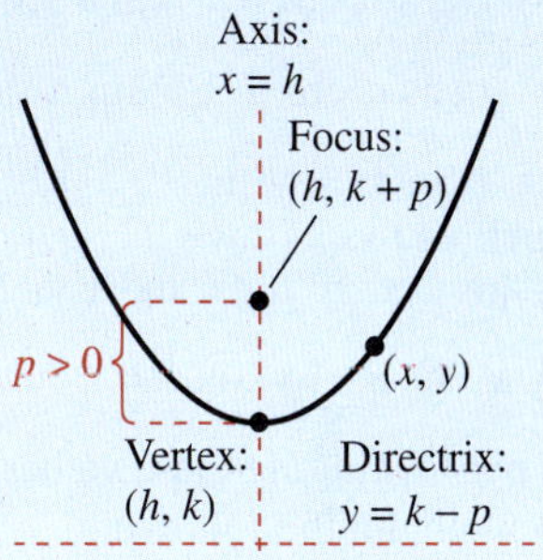

Parabola with vertical axis

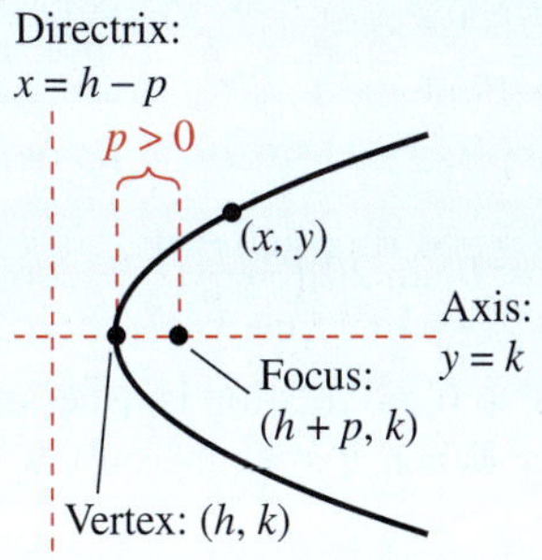

Parabola with horizontal axis

Rotation of Axes to Eliminate an *xy*-Term (p. 673)

The general second-degree equation $Ax^2 + Bxy + Cy^2 + Dx + Ey + F = 0$, where $B \neq 0$, can be rewritten as

$$A'(x')^2 + C'(y')^2 + D'x' + E'y' + F' = 0$$

by rotating the coordinate axes through an angle θ, where

$$\cot 2\theta = \frac{A - C}{B}.$$

The coefficients of the new equation are obtained by making the substitutions $x = x'\cos\theta - y'\sin\theta$ and $y = x'\sin\theta + y'\cos\theta$.

Proof

You need to discover how the coordinates in the xy-system are related to the coordinates in the $x'y'$-system. To do this, choose a point $P(x, y)$ in the original system and attempt to find its coordinates (x', y') in the rotated system. In either system, the distance r between the point P and the origin is the same. So, the equations for x, y, x', and y' are those given in the figures. Using the formulas for the sine and cosine of the difference of two angles, you have the following.

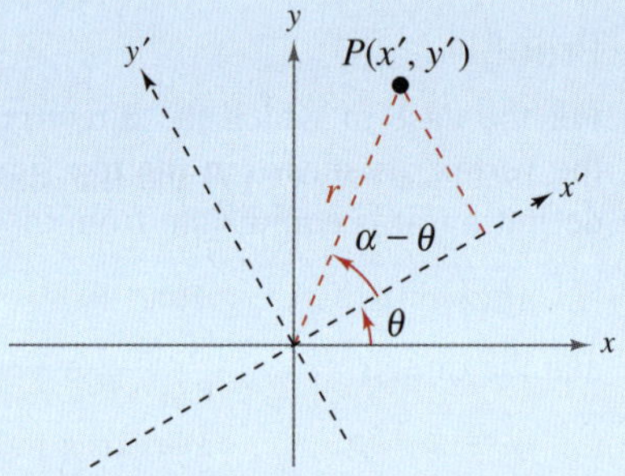

Rotated: $x' = r\cos(\alpha - \theta)$
$y' = r\sin(\alpha - \theta)$

$$\begin{aligned} x' &= r\cos(\alpha - \theta) \\ &= r(\cos\alpha\cos\theta + \sin\alpha\sin\theta) \\ &= r\cos\alpha\cos\theta + r\sin\alpha\sin\theta \\ &= x\cos\theta + y\sin\theta \end{aligned}$$

$$\begin{aligned} y' &= r\sin(\alpha - \theta) \\ &= r(\sin\alpha\cos\theta - \cos\alpha\sin\theta) \\ &= r\sin\alpha\cos\theta - r\cos\alpha\sin\theta \\ &= y\cos\theta - x\sin\theta \end{aligned}$$

Solving this system for x and y yields

$$x = x'\cos\theta - y'\sin\theta \quad \text{and} \quad y = x'\sin\theta + y'\cos\theta.$$

Finally, by substituting these values for x and y into the original equation and collecting terms, you obtain the following.

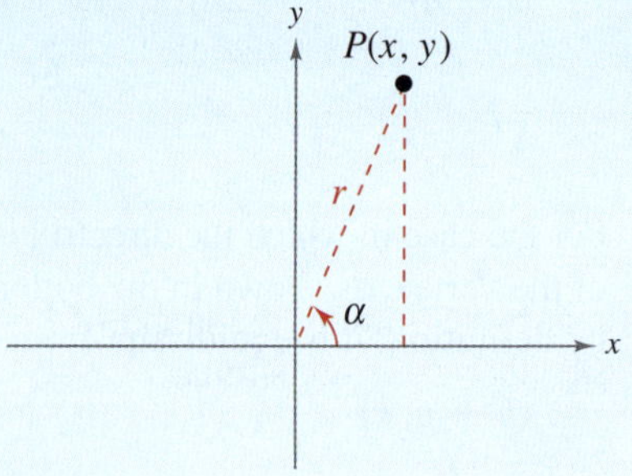

Original: $x = r\cos\alpha$
$y = r\sin\alpha$

$$\begin{aligned} A' &= A\cos^2\theta + B\cos\theta\sin\theta + C\sin^2\theta \\ C' &= A\sin^2\theta - B\cos\theta\sin\theta + C\cos^2\theta \\ D' &= D\cos\theta + E\sin\theta \\ E' &= -D\sin\theta + E\cos\theta \\ F' &= F \end{aligned}$$

To eliminate the $x'y'$-term, you must select θ such that $B' = 0$.

$$\begin{aligned} B' &= 2(C - A)\sin\theta\cos\theta + B(\cos^2\theta - \sin^2\theta) \\ &= (C - A)\sin 2\theta + B\cos 2\theta \\ &= B(\sin 2\theta)\left(\frac{C - A}{B} + \cot 2\theta\right) = 0, \quad \sin 2\theta \neq 0 \end{aligned}$$

When $B = 0$, no rotation is necessary because the xy-term is not present in the original equation. When $B \neq 0$, the only way to make $B' = 0$ is to let

$$\cot 2\theta = \frac{A - C}{B}, \quad B \neq 0.$$

So, you have established the desired results. ■

Polar Equations of Conics (p. 701)

The graph of a polar equation of the form

$$r = \frac{ep}{1 \pm e \cos \theta} \quad \text{or} \quad r = \frac{ep}{1 \pm e \sin \theta}$$

is a conic, where $e > 0$ is the eccentricity and $p > 0$ is the distance between the focus (pole) and the directrix.

Proof

In the figure, consider a vertical directrix, p units to the right of the focus $F(0, 0)$. To show that $r = ed/(1 + e \cos \theta)$ is a conic, let $P(r, \theta)$ be a point on the graph of

$$r = \frac{ep}{1 + e \cos \theta}.$$

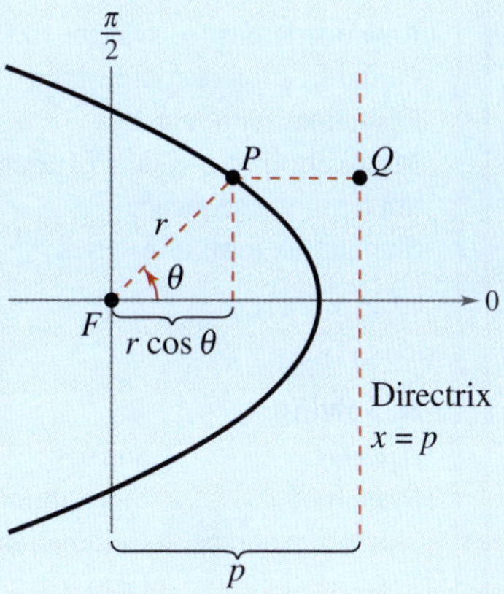

The distance between P and the directrix is

$$\begin{aligned} PQ &= |p - r\cos\theta| \\ &= \left| p - \left(\frac{ep}{1 + e\cos\theta}\right)\cos\theta \right| \\ &= \left| p\left(1 - \frac{e\cos\theta}{1 + e\cos\theta}\right)\right| \\ &= \left|\frac{p}{1 + e\cos\theta}\right| \\ &= \left|\frac{r}{e}\right|. \end{aligned}$$

Moreover, because the distance between P and the pole is simply $PF = |r|$, the ratio of PF to PQ is

$$\frac{PF}{PQ} = \frac{|r|}{|r/e|} = |e| = e$$

and, by definition, the graph of $r = ed/(1 + e \cos \theta)$ must be a conic. The proofs of the other cases are similar. ■

Progressive Summary (Chapters 3–9)

This chart outlines the topics that have been covered so far in this text. Progressive Summary charts appear after Chapters 2, 3, 6, 9 and 11. In each Progressive Summary, new topics encountered for the first time appear in blue.

Transcendental Functions

Exponential, Logarithmic, Trigonometric, Inverse Trigonometric

Rewriting

Exponential form ↔ Logarithmic form
Condense/expand logarithmic expressions
Simplify trigonometric expressions
Prove trigonometric identities
Use conversion formulas
Operations with vectors
Powers and roots of complex numbers

Solving

Equation	*Strategy*
Exponential	Take logarithm of each side
Logarithmic	Exponentiate each side
Trigonometric	Isolate function; Factor, use inverse function
Multiple angle or high powers	Use trigonometric identities

Analyzing

Graphically	*Algebraically*
Intercepts	Domain, Range
Asymptotes	Transformations
Minimum values	Composition
Maximum values	Inverse Properties
	Amplitude, Period
	Reference angles

Numerically
Table of values

Systems and Series

Systems, Sequences, Series

Rewriting

Row operations for systems of equations
Partial fraction decomposition
Operations with matrices
Matrix form of a system of equations
*n*th term of a sequence
Summation form of a series

Solving

Equation	*Strategy*
System of linear equations	Substitution; Elimination; Gaussian; Gauss-Jordan; Inverse matrices; Cramer's Rule

Analyzing

Systems:
Intersecting, parallel, and coincident lines, determinants
Sequences:
Graphing utility in *dot* mode, *n*th term, partial sums, summation formulas

Other Topics

Conics, Parametric and Polar Equations

Rewriting

Standard forms of conics
Eliminate parameters
Rectangular form ↔ Parametric form
Rectangular form ↔ Polar form

Solving

Equation	*Strategy*
Conics	Convert to standard form; Convert to polar form

Analyzing

Conics:
Table of values, vertices, foci, axes, symmetry, asymptotes, translations, eccentricity, rotations
Parametric forms:
Point plotting, eliminate parameters
Polar forms:
Point plotting, special equations, symmetry, zeros, eccentricity, directrix

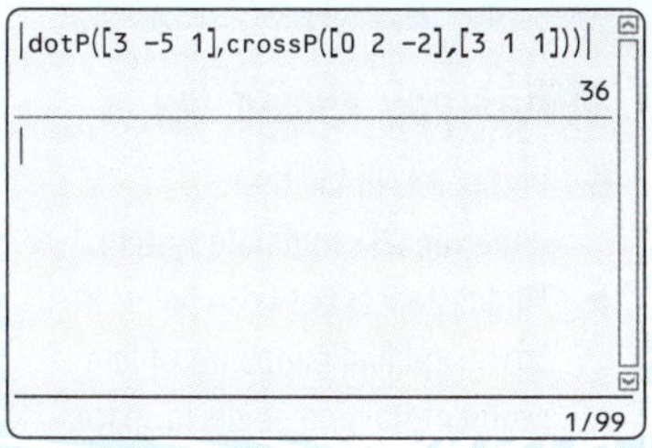

Section 10.3, Example 5
Volume of a Parallelepiped

10 Analytic Geometry in Three Dimensions

Student Resources at LarsonPrecalculus.com

- **Videos** explaining the concepts of precalculus
- **Worked-out solution videos** for all *Checkpoint* exercises
- **Editable spreadsheets** of the data sets in the text
- **Group projects** for each chapter applying concepts to real-life problems

10.1 The Three-Dimensional Coordinate System

The Three-Dimensional Coordinate System

Recall that the Cartesian plane is determined by two perpendicular number lines called the x-axis and the y-axis. These axes, together with their point of intersection (the origin), allow you to develop a two-dimensional coordinate system for identifying points in a plane. To identify a point in space, you must introduce a third dimension to the model. The geometry of this three-dimensional model is called **solid analytic geometry.**

You can construct a **three-dimensional coordinate system** by passing a z-axis perpendicular to both the x- and y-axes at the origin. Figure 10.1 shows the positive portion of each coordinate axis. Taken as pairs, the axes determine three **coordinate planes:** the ***xy*-plane,** the ***xz*-plane,** and the ***yz*-plane.** These three coordinate planes separate the three-dimensional coordinate system into eight **octants.** The first octant is the one in which all three coordinates are positive. In this three-dimensional system, a point P in space is determined by an ordered triple (x, y, z), where x, y, and z are as follows.

$x =$ directed distance from yz-plane to P
$y =$ directed distance from xz-plane to P
$z =$ directed distance from xy-plane to P

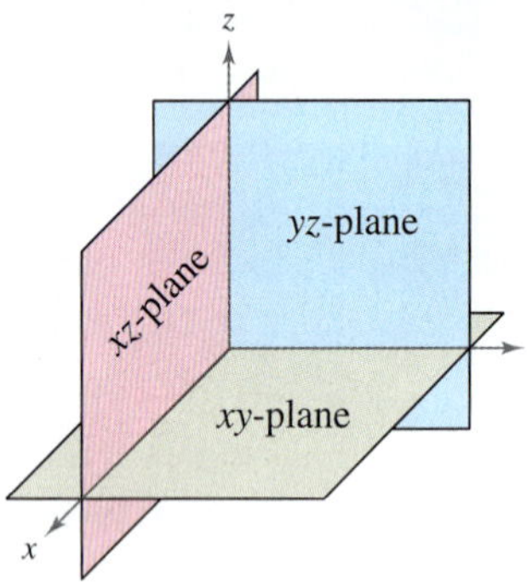

Figure 10.1

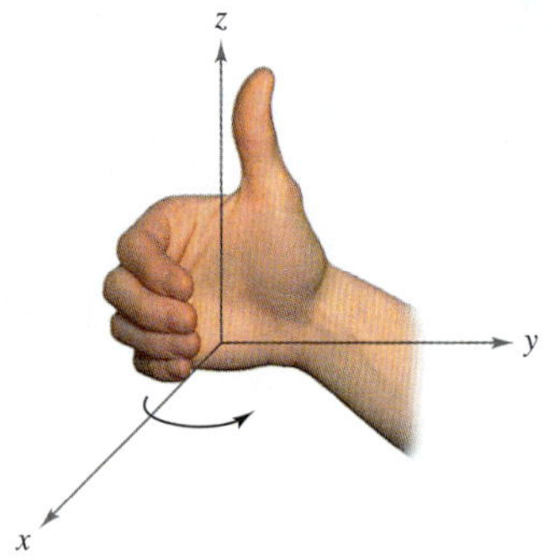

Figure 10.2

A three-dimensional coordinate system can have either a **left-handed** or a **right-handed** orientation. In this text, you will work exclusively with right-handed systems, as illustrated in Figure 10.2. In a right-handed system, Octants II, III, and IV are found by rotating counterclockwise around the positive z-axis. Octant V is below Octant I. Octants VI, VII, and VIII are then found by rotating counterclockwise around the negative z-axis.

What you should learn

- Plot points in the three-dimensional coordinate system.
- Find distances between points in space and find midpoints of line segments joining points in space.
- Write equations of spheres in standard form and find traces of surfaces in space.

Why you should learn it

The three-dimensional coordinate system can be used to graph equations that model materials used in mechanical engineering, such as the spherical shape of a bearing ball, as shown in Exercise 90 on page 728.

EXAMPLE 1 Plotting Points in Space

Plot the points (a) $(2, -3, 3)$, (b) $(-2, 6, 2)$, (c) $(1, 4, 0)$, and (d) $(2, 2, -3)$ in space.

Solution

To plot the point $(2, -3, 3)$, notice that $x = 2$, $y = -3$, and $z = 3$. To help visualize the point, locate the point $(2, -3)$ in the xy-plane (denoted by a cross in the figure). The point $(2, -3, 3)$ lies three units above the cross. You can plot the other points in a similar manner, as shown in the figure.

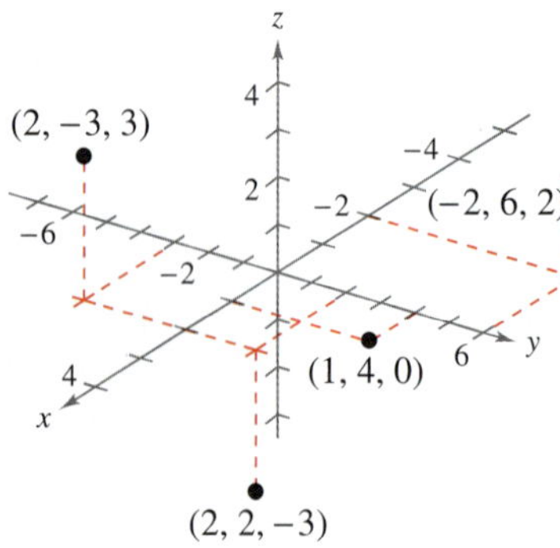

Checkpoint Audio-video solution in English & Spanish at LarsonPrecalculus.com

Plot the points (a) $(-1, 2, -2)$, (b) $(3, -1, -1)$, (c) $(-1, -2, -1)$, and (d) $(-3, -2, 1)$ in space.

The Distance and Midpoint Formulas

Many of the formulas established for the two-dimensional coordinate system can be extended to three dimensions. For example, to find the distance between two points in space, you can use the Pythagorean Theorem twice, as shown in Figure 10.3.

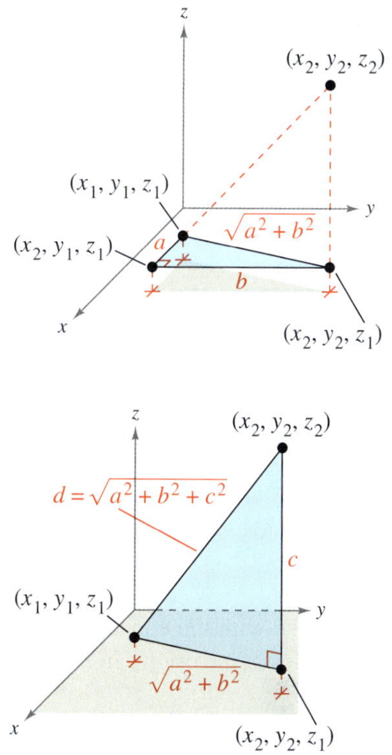

Figure 10.3

Distance Formula in Space

The distance between the points (x_1, y_1, z_1) and (x_2, y_2, z_2) given by the **Distance Formula in Space** is

$$d = \sqrt{(x_2 - x_1)^2 + (y_2 - y_1)^2 + (z_2 - z_1)^2}.$$

EXAMPLE 2 Finding the Distance Between Two Points in Space

Find the distance between

$(1, 0, 2)$ and $(2, 4, -3)$.

Solution

$$\begin{aligned} d &= \sqrt{(x_2 - x_1)^2 + (y_2 - y_1)^2 + (z_2 - z_1)^2} && \text{Distance Formula in Space} \\ &= \sqrt{(2-1)^2 + (4-0)^2 + (-3-2)^2} && \text{Substitute.} \\ &= \sqrt{1 + 16 + 25} && \text{Simplify.} \\ &= \sqrt{42} && \text{Simplify.} \end{aligned}$$

Checkpoint *Audio-video solution in English & Spanish at LarsonPrecalculus.com*

Find the distance between $(0, 1, 3)$ and $(1, 4, -2)$.

Notice the similarity between the Distance Formulas in the plane and in space. The Midpoint Formulas in the plane and in space are also similar.

Midpoint Formula in Space

The midpoint of the line segment joining the points (x_1, y_1, z_1) and (x_2, y_2, z_2) given by the **Midpoint Formula in Space** is

$$\left(\frac{x_1 + x_2}{2}, \frac{y_1 + y_2}{2}, \frac{z_1 + z_2}{2}\right).$$

EXAMPLE 3 Using the Midpoint Formula in Space

Find the midpoint of the line segment joining

$(5, -2, 3)$ and $(0, 4, 4)$.

Solution

Using the Midpoint Formula in Space, the midpoint is

$$\left(\frac{5+0}{2}, \frac{-2+4}{2}, \frac{3+4}{2}\right) = \left(\frac{5}{2}, 1, \frac{7}{2}\right)$$

as shown in Figure 10.4.

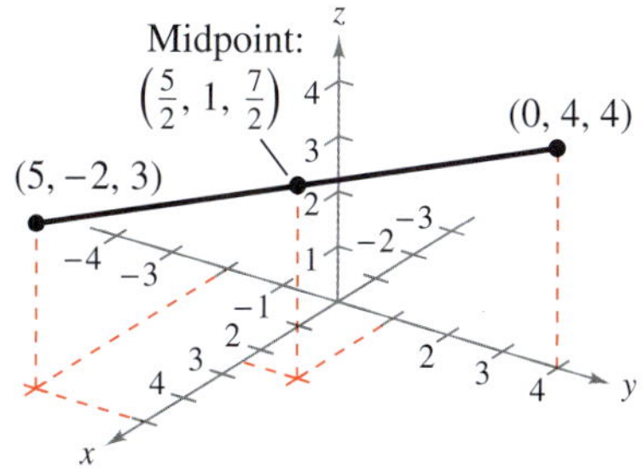

Figure 10.4

Checkpoint *Audio-video solution in English & Spanish at LarsonPrecalculus.com*

Find the midpoint of the line segment joining $(7, -1, -2)$ and $(3, -5, 4)$.

The Equation of a Sphere

A **sphere** with center (h, k, j) and radius r is defined as the set of all points (x, y, z) such that the distance between (x, y, z) and (h, k, j) is r, as shown in Figure 10.5. Using the Distance Formula, this condition can be written as

$$\sqrt{(x - h)^2 + (y - k)^2 + (z - j)^2} = r.$$

By squaring each side of this equation, you obtain the standard equation of a sphere.

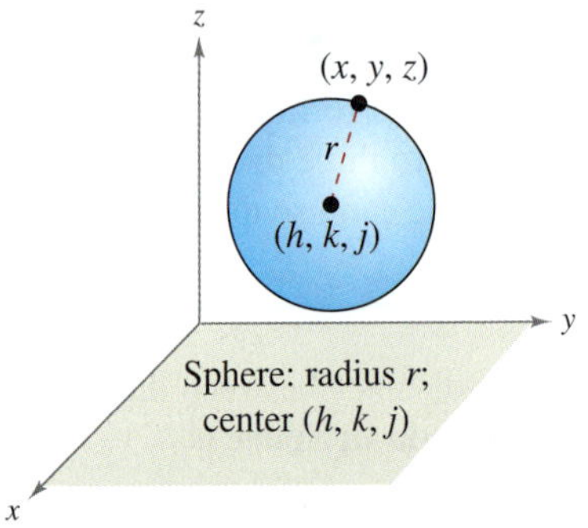

Figure 10.5

Standard Equation of a Sphere

The **standard equation of a sphere** with center (h, k, j) and radius r is given by

$$(x - h)^2 + (y - k)^2 + (z - j)^2 = r^2.$$

Notice the similarity of this formula to the equation of a circle in the plane.

$$(x - h)^2 + (y - k)^2 + (z - j)^2 = r^2 \quad \text{Equation of sphere in space}$$

$$(x - h)^2 + (y - k)^2 = r^2 \quad \text{Equation of circle in the plane}$$

As is true with the equation of a circle, the equation of a sphere is simpler when the center lies at the origin. In this case, the equation is

$$x^2 + y^2 + z^2 = r^2. \quad \text{Sphere with center at origin}$$

Explore the Concept

Find the equation of the sphere that has the points $(3, -2, 6)$ and $(-1, 4, 2)$ as endpoints of a diameter. Explain how this problem gives you a chance to use all three formulas discussed so far in this section: the Distance Formula in Space, the Midpoint Formula in Space, and the standard equation of a sphere.

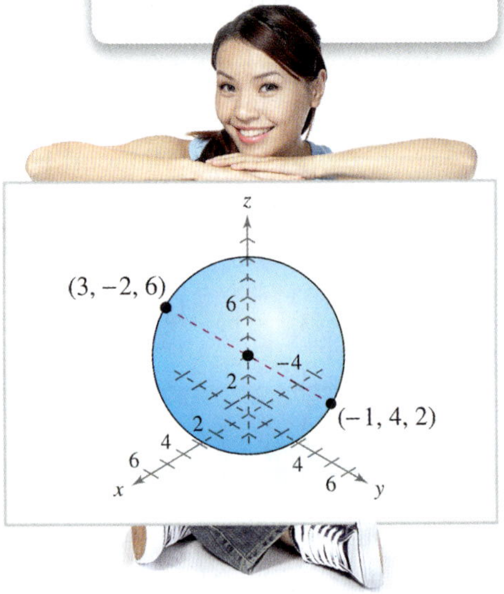

EXAMPLE 4 Finding the Equation of a Sphere

See LarsonPrecalculus.com for an interactive version of this type of example.

Find the standard equation of the sphere with center $(2, 4, 3)$ and radius 3. Does this sphere intersect the xy-plane?

Solution

The equation of the sphere with center $(h, k, j) = (2, 4, 3)$ and radius $r = 3$ is

$$(x - h)^2 + (y - k)^2 + (z - j)^2 = r^2 \quad \text{Standard equation}$$

$$(x - 2)^2 + (y - 4)^2 + (z - 3)^2 = 3^2. \quad \text{Substitute.}$$

From the graph shown in Figure 10.6, you can see that the center of the sphere lies three units above the xy-plane. Because the sphere has a radius of 3, you can conclude that it does intersect the xy-plane—at the point $(2, 4, 0)$.

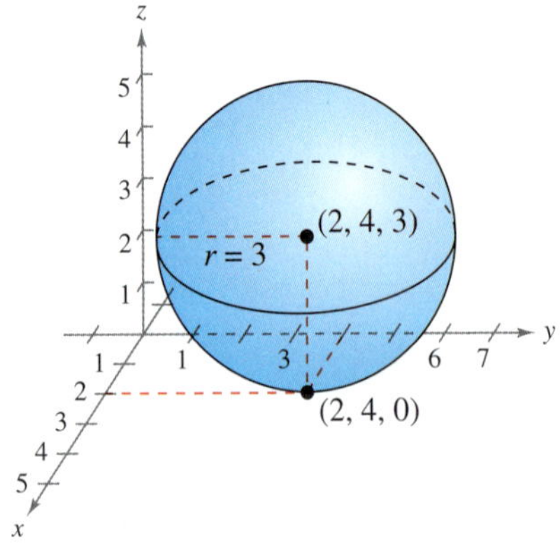

Figure 10.6

Checkpoint *Audio-video solution in English & Spanish at LarsonPrecalculus.com*

Find the standard equation of the sphere with center $(1, 3, 2)$ and radius 4.

EXAMPLE 5 Finding the Center and Radius of a Sphere

Find the center and radius of the sphere $x^2 + y^2 + z^2 - 2x + 4y - 6z + 8 = 0$.

Solution

To obtain the standard equation of this sphere, complete the squares as follows.

$$x^2 + y^2 + z^2 - 2x + 4y - 6z + 8 = 0$$
$$(x^2 - 2x + \quad) + (y^2 + 4y + \quad) + (z^2 - 6z + \quad) = -8$$
$$(x^2 - 2x + 1) + (y^2 + 4y + 4) + (z^2 - 6z + 9) = -8 + 1 + 4 + 9$$
$$(x - 1)^2 + (y + 2)^2 + (z - 3)^2 = (\sqrt{6})^2$$

So, the center of the sphere is $(1, -2, 3)$, and its radius is $\sqrt{6}$. See Figure 10.7.

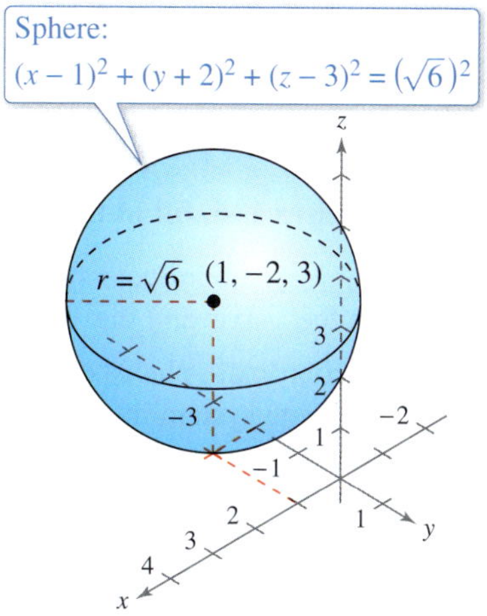

Figure 10.7

Checkpoint Audio-video solution in English & Spanish at LarsonPrecalculus.com

Find the center and radius of the sphere $x^2 + y^2 + z^2 + 6x - 4y + 8z - 7 = 0$.

Note in Example 5 that the points satisfying the equation of the sphere are "surface points," not "interior points." In general, the collection of points satisfying an equation involving x, y, and z is called a **surface in space.**

Finding the intersection of a surface with one of the three coordinate planes (or with a plane parallel to one of the three coordinate planes) helps one visualize the surface. Such an intersection is called a **trace** of the surface. For example, the xy-trace of a surface consists of all points that are common to both the surface *and* the xy-plane.

EXAMPLE 6 Finding a Trace of a Surface

Find the xy-trace of the sphere $(x - 3)^2 + (y - 2)^2 + (z + 4)^2 = 5^2$.

Solution

To find the xy-trace of this surface, use the fact that every point in the xy-plane has a z-coordinate of zero. By substituting $z = 0$ into the original equation, the resulting equation will represent the intersection of the surface with the plane.

$(x - 3)^2 + (y - 2)^2 + (z + 4)^2 = 5^2$ Write original equation.

$(x - 3)^2 + (y - 2)^2 + (0 + 4)^2 = 5^2$ Substitute 0 for z.

$(x - 3)^2 + (y - 2)^2 + 16 = 25$ Simplify.

$(x - 3)^2 + (y - 2)^2 = 9$ Subtract 16 from each side.

$(x - 3)^2 + (y - 2)^2 = 3^2$ Equation of circle

From this form, you can see that the xy-trace is a circle of radius 3, as shown in the figure.

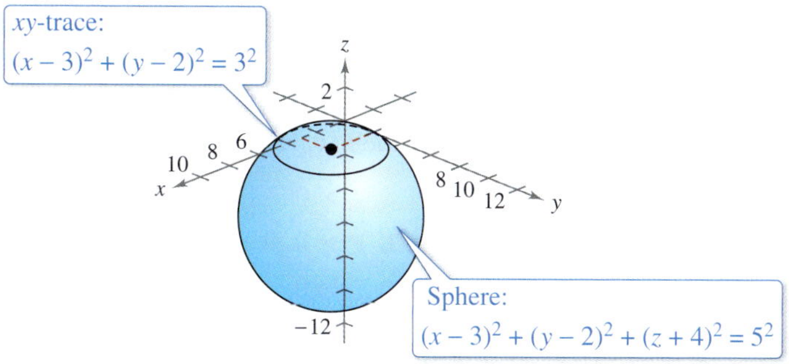

Checkpoint Audio-video solution in English & Spanish at LarsonPrecalculus.com

Find the xz-trace of the sphere $(x - 1)^2 + (y - 3)^2 + (z - 8)^2 = 10^2$.

10.1 Exercises

See *CalcChat.com* for tutorial help and worked-out solutions to odd-numbered exercises. For instructions on how to use a graphing utility, see Appendix A.

Vocabulary and Concept Check

In Exercises 1–6, fill in the blank(s).

1. A _______ coordinate system can be formed by passing a z-axis perpendicular to both the x-axis and the y-axis at the origin.

2. The coordinate planes of a three-dimensional coordinate system separate the coordinate system into eight _______ .

3. The distance between the points (x_1, y_1, z_1) and (x_2, y_2, z_2) can be found using the _______ in Space.

4. A _______ is the set of all points (x, y, z) such that the distance between (x, y, z) and a fixed point (h, k, j) is r.

5. A _______ in _______ is the collection of points satisfying an equation involving x, y, and z.

6. The intersection of a surface with one of the three coordinate planes is called a _______ of the surface.

7. What does the equation $(x - h)^2 + (y - k)^2 + (z - j)^2 = r^2$ represent? What do h, k, j, and r represent?

8. How do you find the yz-trace of the sphere given by $(x - 2)^2 + (y + 1)^2 + (z - 2)^2 = 7^2$?

Procedures and Problem Solving

Approximating the Coordinates of a Point **In Exercises 9–12, approximate the coordinates of the points.**

9.

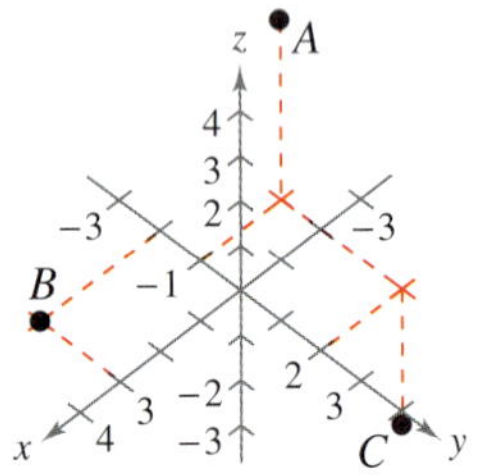

10.

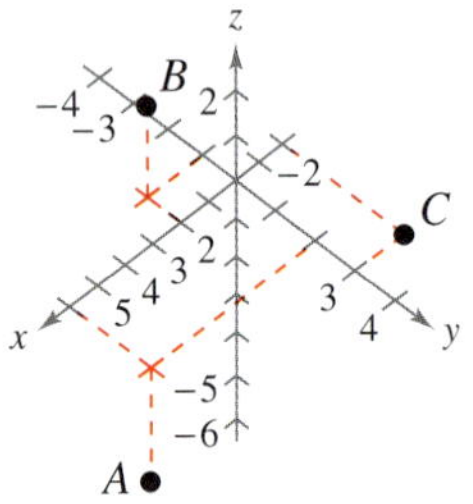

11.

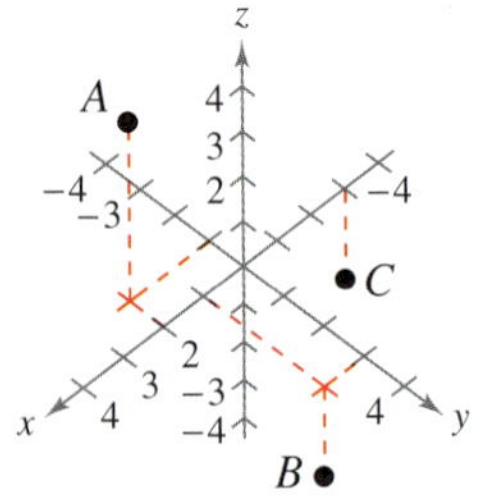

12.

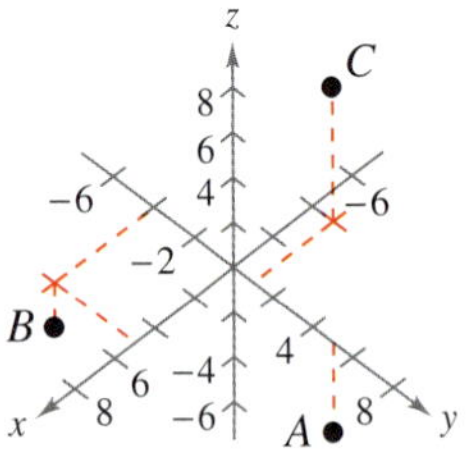

Plotting Points in Space **In Exercises 13–18, plot the points in the same three-dimensional coordinate system.**

13. $(3, 1, 2), (2, -1, -3)$

14. $(3, 0, 1), (-2, -3, -1)$

15. $(1, -3, 0), (-3, 1, 2)$

16. $(0, 3, -4), (2, 1, 2)$

17. $(2, -4, 5), (1, 4, -5)$

18. $(3, -5, 4), (5, -3, -2)$

Finding the Coordinates of a Point **In Exercises 19–24, find the coordinates of the point.**

19. The point is located three units behind the yz-plane, three units to the right of the xz-plane, and five units above the xy-plane.

20. The point is located six units in front of the yz-plane, two units to the left of the xz-plane, and one unit below the xy-plane.

21. The point is located on the x-axis and 11 units in front of the yz-plane.

22. The point is located on the y-axis, nine units to the left of the xz-plane.

23. The point is located in the yz-plane, two units to the right of the xz-plane, and seven units above the xy-plane.

24. The point is located in the xz-plane, one unit behind the yz-plane, and six units above the xy-plane.

Determining the Octant **In Exercises 25–28, determine the octant in which (x, y, z) is located.**

25. $x > 0, y < 0, z > 0$

26. $x < 0, y > 0, z < 0$

27. $x < 0, y > 0, z > 0$

28. $x > 0, y > 0, z > 0$

Determining the Plane **In Exercises 29 and 30, determine the plane in which (x, y, z) is located.**

29. $x = 0, y < 0, z > 0$

30. $x < 0, y < 0, z = 0$

Finding the Distance Between Two Points in Space **In Exercises 31–38, find the distance between the points.**

31.

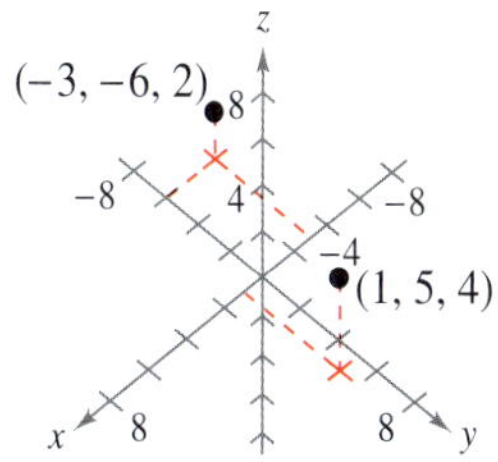

32.

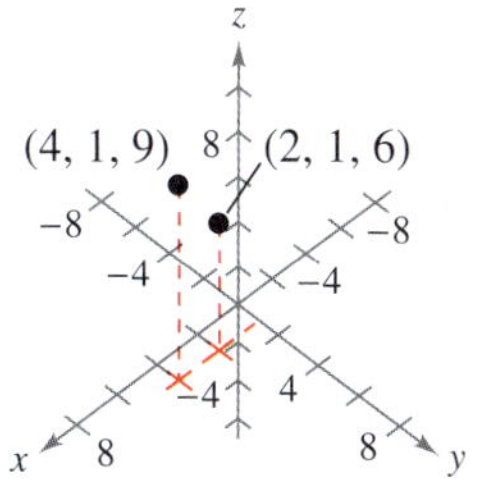

33. (3, 2, 5), (7, 4, 8)

34. (1, 0, 0), (7, 0, 4)

35. (1, 0, −10), (0, −3, 2)

36. (2, −4, 0), (0, 6, −3)

37. (6, −9, 1), (−2, −1, 5)

38. (1, 1, −7), (−2, −3, −7)

Using the Pythagorean Theorem **In Exercises 39–42, find the lengths of the sides of the right triangle. Show that these lengths satisfy the Pythagorean Theorem.**

39.

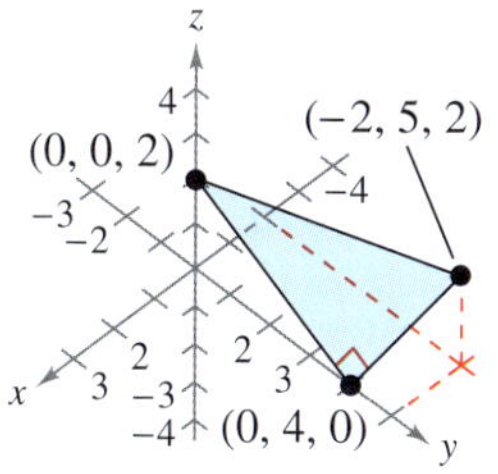

40.

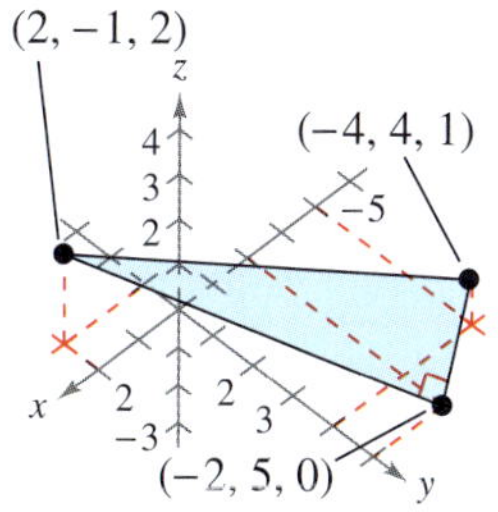

41. (0, 0, 0), (2, 2, 1), (2, −4, 4)

42. (1, 0, 1), (1, 3, 1), (1, 0, 3)

Finding the Side Lengths of a Triangle **In Exercises 43–46, find the lengths of the sides of the triangle with the indicated vertices and determine whether the triangle is a right triangle, an isosceles triangle, or neither.**

43. (1, −3, −2), (5, −1, 2), (−1, 1, 2)

44. (5, 3, 4), (7, 1, 3), (3, 5, 3)

45. (5, 0, 0), (1, 3, −2), (−2, −1, 4)

46. (1, −2, −1), (3, 0, 0), (3, −6, 3)

Finding the Midpoint of a Line Segment in Space **In Exercises 47–52, find the midpoint of the line segment joining the points.**

47. (3, 0, 1), (2, −6, 3)

48. (6, −2, 5), (−4, 2, 7)

49. (1, 5, −1), (2, 2, 2)

50. (−3, 5, 5), (−6, 4, 8)

51. (−3, −9, 0), (−5, −3, −2)

52. (9, −5, 1), (9, −2, −4)

53. Timing a Sky Ride A suspended carriage travels along a cable at 2.25 meters per second directly from (1, 0, 1) to (2, 800, 41), where every measurement is in meters. How long does it take the carriage to travel to the midpoint of the cable?

54. Building a Bridge A bridge is built between two treehouses. The first tree house is at (5, 15, 30) and the second is at (55, 16, 20), where every measurement is in feet. The cost of materials is \$3 per linear foot. What is the cost of materials to replace the bridge?

Finding the Equation of a Sphere **In Exercises 55–64, find the standard equation of the sphere with the given characteristics.**

55.

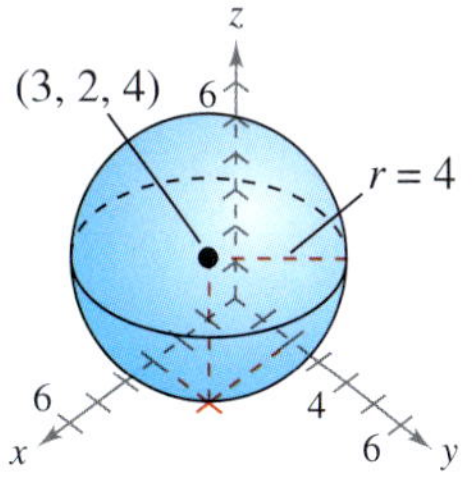

56.

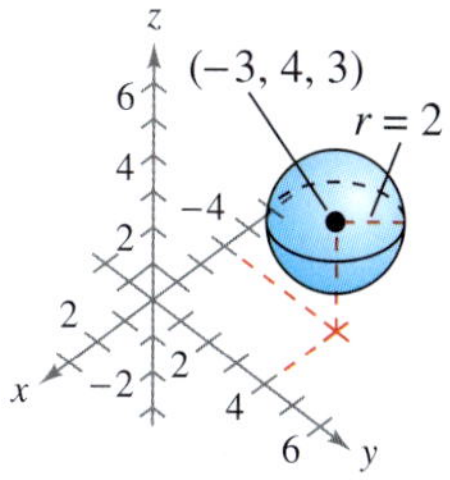

57.

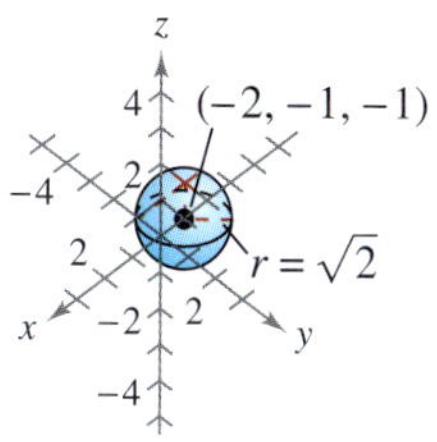

58.

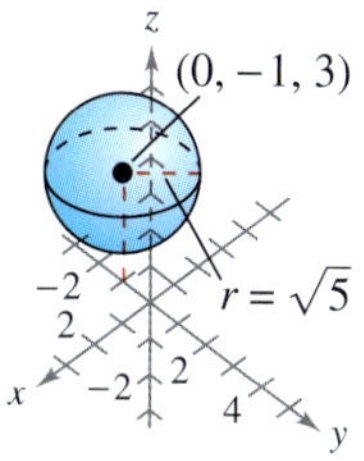

59. Center: (0, 4, 3); radius: 4

60. Center: (2, −1, 8); radius: 7

61. Center: (−3, 7, 5); diameter: 10

62. Center: (0, 5, −9); diameter: 8

63. Endpoints of a diameter: (3, 0, 0), (0, 0, 6)

64. Endpoints of a diameter: (−4, 0, −8), (0, 2, −2)

Finding the Center and Radius of a Sphere **In Exercises 65–76, find the center and radius of the sphere.**

65. $x^2 + y^2 + z^2 - 6x = 0$

66. $x^2 + y^2 + z^2 + 12z = 0$

67. $x^2 + y^2 + z^2 - 4x + 2y = 0$

68. $x^2 + y^2 + z^2 + 2y - 14z = 0$

69. $x^2 + y^2 + z^2 + 4x - 8z + 19 = 0$

70. $x^2 + y^2 + z^2 - 8y - 6z + 13 = 0$

71. $x^2 + y^2 + z^2 - 4x + 2y - 6z + 10 = 0$

72. $x^2 + y^2 + z^2 - 6x + 4y - 8z + 4 = 0$

73. $4x^2 + 4y^2 + 4z^2 - 8x + 16y - 1 = 0$

74. $9x^2 + 9y^2 + 9z^2 - 6x + 18y + 1 = 0$

75. $4x^2 + 4y^2 + 4z^2 - 4x - 32y + 8z + 33 = 0$

76. $9x^2 + 9y^2 + 9z^2 - 18x + 36y + 54z - 126 = 0$

Finding a Trace of a Surface **In Exercises 77–82, sketch the graph of the equation and find the specified trace.**

77. $(x - 1)^2 + y^2 + z^2 = 36$; xz-trace

78. $x^2 + (y + 3)^2 + z^2 = 25$; yz-trace

79. $(x + 2)^2 + (y - 3)^2 + z^2 = 9$; yz-trace

80. $x^2 + (y - 1)^2 + (z + 1)^2 = 4$; xy-trace

81. $x^2 + y^2 + z^2 - 2x - 4z + 1 = 0$; yz-trace

82. $x^2 + y^2 + z^2 - 4y - 6z - 12 = 0$; xz-trace

Graphing a Sphere **In Exercises 83–86, use a three-dimensional graphing utility to graph the sphere.**

83. $x^2 + y^2 + z^2 - 6x - 8y - 10z + 46 = 0$

84. $x^2 + y^2 + z^2 + 6y - 8z + 21 = 0$

85. $4x^2 + 4y^2 + 4z^2 - 8x - 16y + 8z - 25 = 0$

86. $9x^2 + 9y^2 + 9z^2 + 18x - 18y + 36z + 35 = 0$

87. **Crystallography** Crystals are classified according to their symmetry. Crystals shaped like cubes are classified as isometric. The vertices of an isometric crystal mapped onto a three-dimensional coordinate system are shown in the figure. Determine (x, y, z).

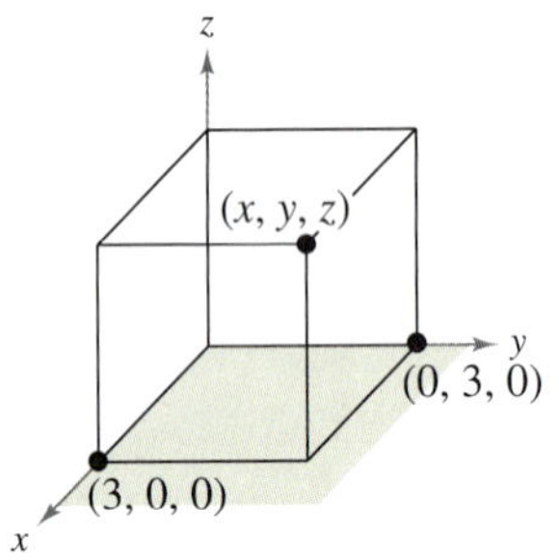

Figure for 87

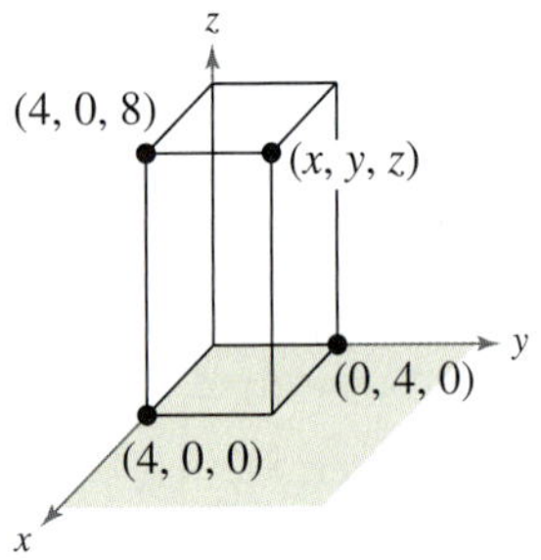

Figure for 88

88. **Crystallography** Crystals shaped like rectangular prisms are classified as tetragonal. The vertices of a tetragonal crystal mapped onto a three-dimensional coordinate system are shown in the figure. Determine (x, y, z).

89. **Architecture** A spherical building has a diameter of 205 feet. The center of the building is placed at the origin of a three-dimensional coordinate system. What is the equation of the sphere?

90. ***Why you should learn it*** *(p. 722)* Ball bearings allow machinery to move at extremely high speeds and carry large loads. A bearing ball is in the shape of a sphere with a volume of 288π cubic millimeters.

(a) Write an equation for the sphere. Let the center of the bearing ball correspond to the origin of the three-dimensional coordinate system.

(b) Use a three-dimensional graphing utility to graph the sphere from part (a).

Error Analysis **In Exercises 91 and 92, describe the error.**

91. The xz-trace of a sphere centered at the origin with a diameter of 8 is $y^2 + z^2 = 16$.

92. Of $(2, 3, 6)$, $(-5, 4, -3)$, and $(0, -8, 0)$, the point $(2, 3, 6)$ is closest to the origin and the point $(-5, 4, -3)$ is farthest from the origin.

Focusing on Concepts

True or False? **In Exercises 93 and 94, determine whether the statement is true or false. Justify your answer.**

93. In the ordered triple (x, y, z) that represents point P in space, x is the directed distance from the xy-plane to P.

94. The surface consisting of all points (x, y, z) in space that are the same distance r from the point (h, k, j) has a circle as its xy-trace.

95. **Think About It** What is the z-coordinate of any point in the xy-plane? What is the y-coordinate of any point in the xz-plane? What is the x-coordinate of any point in the yz-plane?

96. **HOW DO YOU SEE IT?** Approximate the coordinates of each point, and state the octant in which each point lies.

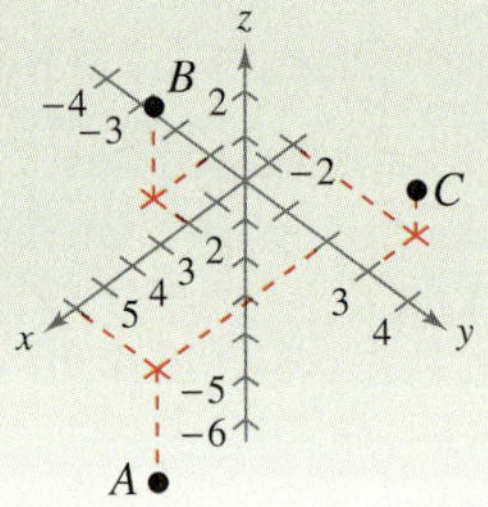

97. **Finding an Endpoint** A line segment has (x_1, y_1, z_1) as one endpoint and (x_m, y_m, z_m) as its midpoint. Find the other endpoint (x_2, y_2, z_2) of the line segment in terms of x_1, y_1, z_1, x_m, y_m, and z_m.

98. **Finding an Endpoint** Use the result of Exercise 97 to find the coordinates of one endpoint of a line segment when the coordinates of the other endpoint and the midpoint are $(3, 0, 2)$ and $(5, 8, 7)$, respectively.

Cumulative Mixed Review

Completing the Square **In Exercises 99–102, solve the quadratic equation by completing the square.**

99. $v^2 + 3v - 2 = 0$

100. $z^2 - 7z - 19 = 0$

101. $x^2 - 5x + 5 = 0$

102. $x^2 + 3x - 1 = 0$

Finding the Dot Product **In Exercises 103 and 104, find the dot product of u and v.**

103. $\mathbf{u} = \langle -4, 1 \rangle$, $\mathbf{v} = \langle 3, 5 \rangle$

104. $\mathbf{u} = \langle -1, 3 \rangle$, $\mathbf{v} = \langle -2, -6 \rangle$

10.2 Vectors in Space

What you should learn

- Find the component forms of, the unit vectors in the same direction of, the magnitudes of, the dot products of, and the angles between vectors in space.
- Determine whether vectors in space are parallel.
- Use vectors in space to solve real-life problems.

Why you should learn it

Vectors in space can be used to represent many physical forces, such as tension in the chains used to support a birdbath, as shown in Exercise 67 on page 735.

Vectors in Space

Similar to vectors in the plane, vectors in space are used to represent quantities, such as force and velocity, that have both magnitude and direction. In space, vectors are denoted by ordered triples

$$\mathbf{v} = \langle v_1, v_2, v_3 \rangle. \qquad \text{Component form}$$

The **zero vector** is denoted by

$$\mathbf{0} = \langle 0, 0, 0 \rangle. \qquad \text{Zero vector}$$

Using the **standard unit vectors** $\mathbf{i} = \langle 1, 0, 0 \rangle$, $\mathbf{j} = \langle 0, 1, 0 \rangle$, and $\mathbf{k} = \langle 0, 0, 1 \rangle$, the **standard unit vector notation** for $\mathbf{v}$ is

$$\mathbf{v} = v_1\mathbf{i} + v_2\mathbf{j} + v_3\mathbf{k} \qquad \text{Unit vector form}$$

as shown in Figure 10.8. If $\mathbf{v}$ is represented by the directed line segment from $P(p_1, p_2, p_3)$ to $Q(q_1, q_2, q_3)$, as shown in Figure 10.9, then the **component form** of $\mathbf{v}$ is produced by subtracting the coordinates of the initial point from the coordinates of the terminal point, as shown below.

$$\mathbf{v} = \langle v_1, v_2, v_3 \rangle = \langle q_1 - p_1, q_2 - p_2, q_3 - p_3 \rangle$$

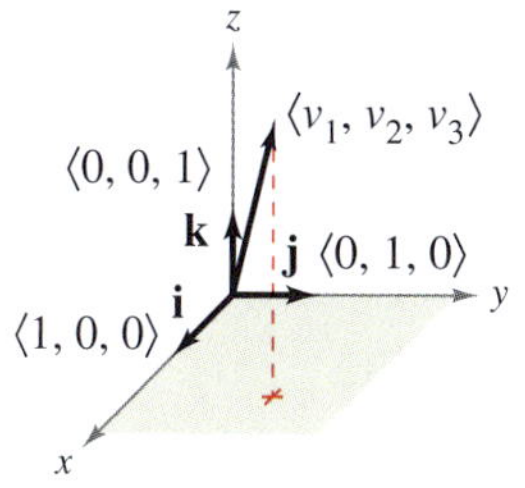

Figure 10.8

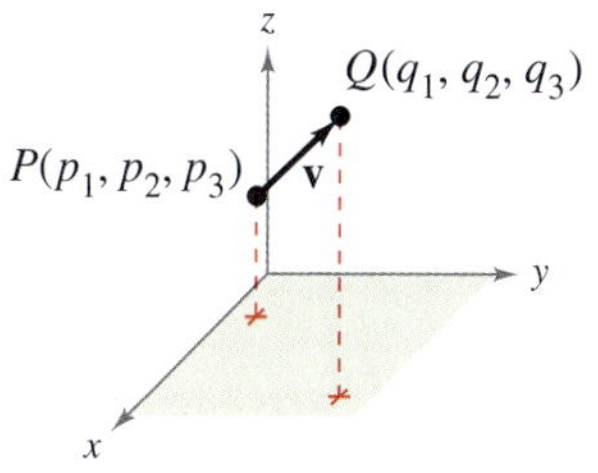

Figure 10.9

Vectors in Space

1. Two vectors are **equal** if and only if their corresponding components are equal.
2. The **magnitude** (or **length**) of $\mathbf{u} = \langle u_1, u_2, u_3 \rangle$ is
$$\|\mathbf{u}\| = \sqrt{u_1^2 + u_2^2 + u_3^2}.$$
3. A **unit vector u** in the direction of **v** is
$$\mathbf{u} = \frac{\mathbf{v}}{\|\mathbf{v}\|}, \ \mathbf{v} \neq 0.$$
4. The **sum** of $\mathbf{u} = \langle u_1, u_2, u_3 \rangle$ and $\mathbf{v} = \langle v_1, v_2, v_3 \rangle$ is
$$\mathbf{u} + \mathbf{v} = \langle u_1 + v_1, u_2 + v_2, u_3 + v_3 \rangle. \qquad \text{Vector addition}$$
5. The **scalar multiple** of the real number c and $\mathbf{u} = \langle u_1, u_2, u_3 \rangle$ is
$$c\mathbf{u} = \langle cu_1, cu_2, cu_3 \rangle. \qquad \text{Scalar multiplication}$$
6. The **dot product** of $\mathbf{u} = \langle u_1, u_2, u_3 \rangle$ and $\mathbf{v} = \langle v_1, v_2, v_3 \rangle$ is
$$\mathbf{u} \cdot \mathbf{v} = u_1v_1 + u_2v_2 + u_3v_3. \qquad \text{Dot product}$$

Note that the properties of vector operations listed in Sections 6.3 and 6.4 are also valid for vectors in space.

EXAMPLE 1 Finding the Component Form of a Vector

See LarsonPrecalculus.com for an interactive version of this type of example.

The vector **v** has initial point (3, 4, 2) and terminal point (3, 6, 4). Find the component form and magnitude of **v**. Then find a unit vector in the direction of **v**.

Solution

The component form of **v** is $\mathbf{v} = \langle 3-3, 6-4, 4-2\rangle = \langle 0, 2, 2\rangle$, which implies that its magnitude is $\|\mathbf{v}\| = \sqrt{0^2+2^2+2^2} = \sqrt{8} = 2\sqrt{2}$. A unit vector in the direction of **v** is

$$\mathbf{u} = \frac{\mathbf{v}}{\|\mathbf{v}\|} = \frac{1}{2\sqrt{2}}\langle 0, 2, 2\rangle = \left\langle 0, \frac{1}{\sqrt{2}}, \frac{1}{\sqrt{2}}\right\rangle = \left\langle 0, \frac{\sqrt{2}}{2}, \frac{\sqrt{2}}{2}\right\rangle.$$

✓ **Checkpoint** *Audio-video solution in English & Spanish at LarsonPrecalculus.com*

The vector **v** has initial point $(1, -4, 3)$ and terminal point $(2, 2, -1)$. Find the component form and magnitude of **v**. Then find a unit vector in the direction of **v**.

EXAMPLE 2 Finding the Dot Product of Two Vectors

The dot product of $\langle 0, 3, -2\rangle$ and $\langle 4, -2, 3\rangle$ is

$$\langle 0, 3, -2\rangle \cdot \langle 4, -2, 3\rangle = 0(4) + 3(-2) + (-2)(3) = 0 - 6 - 6 = -12.$$

Note that the dot product of two vectors is a real number, not a vector.

✓ **Checkpoint** *Audio-video solution in English & Spanish at LarsonPrecalculus.com*

Find the dot product of $\langle 4, 0, 1\rangle$ and $\langle -1, 3, 2\rangle$.

In Section 6.4, you learned that the **angle between two nonzero vectors** is the angle θ between their respective standard position vectors (see Figure 10.10), where $0 \le \theta \le \pi$ or $0° \le \theta \le 180°$. For vectors in the plane or in space, this angle can be found using the dot product. (Note that the angle between the zero vector and another vector is not defined.)

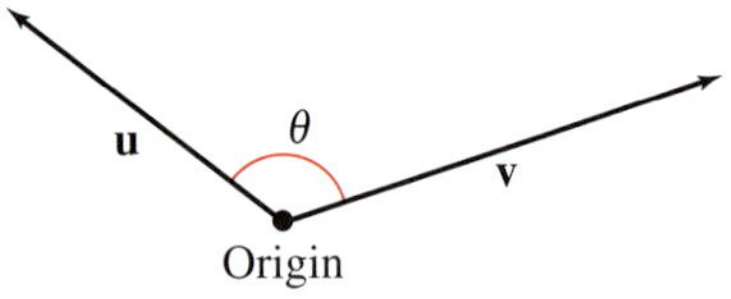

Figure 10.10

Angle Between Two Vectors

If θ is the angle between two nonzero vectors **u** and **v**, then $\cos\theta = \dfrac{\mathbf{u}\cdot\mathbf{v}}{\|\mathbf{u}\|\|\mathbf{v}\|}$.

Recall from Section 6.4 that if the dot product of two nonzero vectors is zero, then the angle between the vectors is 90° and the vectors are **orthogonal**. Note that the standard unit vectors **i**, **j**, and **k** are orthogonal to each other.

EXAMPLE 3 Finding the Angle Between Two Vectors

Find the angle between $\mathbf{u} = \langle 1, 0, 2\rangle$ and $\mathbf{v} = \langle 3, 1, 0\rangle$.

Solution

$$\cos\theta = \frac{\mathbf{u}\cdot\mathbf{v}}{\|\mathbf{u}\|\|\mathbf{v}\|} = \frac{\langle 1, 0, 2\rangle \cdot \langle 3, 1, 0\rangle}{\|\langle 1, 0, 2\rangle\|\|\langle 3, 1, 0\rangle\|} = \frac{3}{\sqrt{50}}$$

So, the angle between **u** and **v** is $\theta = \arccos(3/\sqrt{50}) \approx 64.9°$. (See Figure 10.11.)

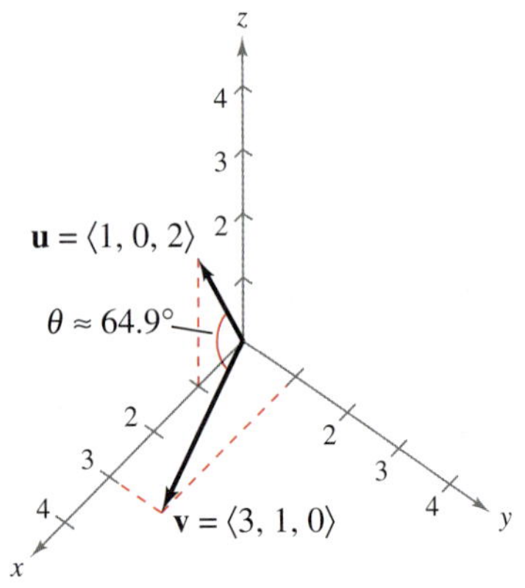

Figure 10.11

✓ **Checkpoint** *Audio-video solution in English & Spanish at LarsonPrecalculus.com*

Find the angle between $\mathbf{u} = \langle 1, -2, 4\rangle$ and $\mathbf{v} = \langle 2, -3, -1\rangle$.

Parallel Vectors

Recall from the definition of scalar multiplication that positive scalar multiples of a nonzero vector **v** have the same direction as **v**, whereas negative multiples have the direction opposite that of **v**. In general, two nonzero vectors **u** and **v** are **parallel** when there is some scalar c such that $\mathbf{u} = c\mathbf{v}$. For instance, in Figure 10.12, the vectors **u**, **v**, and **w** are parallel because $\mathbf{u} = 2\mathbf{v}$ and $\mathbf{w} = -\mathbf{v}$.

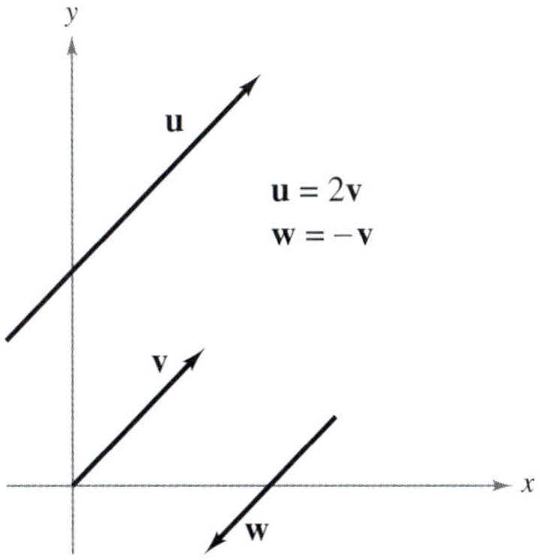

Figure 10.12

Technology Tip

Some graphing utilities have the capability of performing vector operations, such as the dot product. Consult the user's guide for your graphing utility for specific instructions.

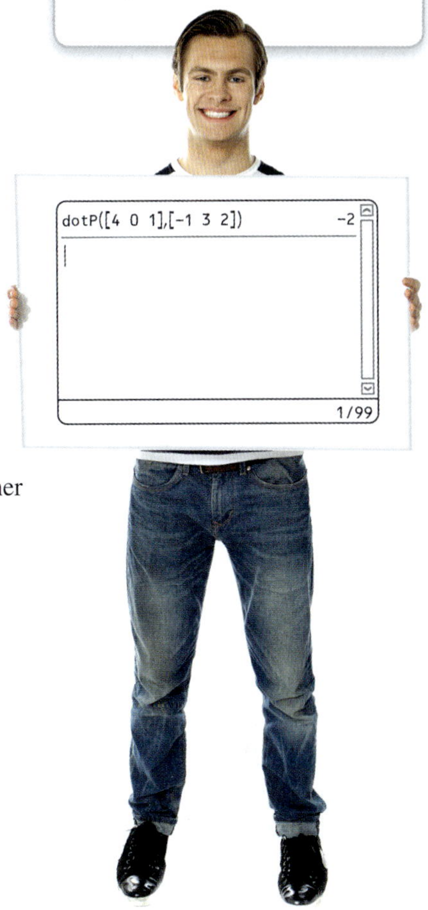

EXAMPLE 4 Parallel Vectors

Vector **w** has initial point $(1, -2, 0)$ and terminal point $(3, 2, 1)$. Determine whether each vector is parallel to **w**.

a. $\mathbf{u} = \langle 4, 8, 2 \rangle$

b. $\mathbf{v} = \left\langle -1, -2, -\frac{1}{2} \right\rangle$

c. $\mathbf{z} = \langle 4, 8, 4 \rangle$

Solution

Begin by writing **w** in component form.

$$\mathbf{w} = \langle 3 - 1, 2 - (-2), 1 - 0 \rangle = \langle 2, 4, 1 \rangle$$

a. Because

$$\mathbf{u} = \langle 4, 8, 2 \rangle = 2\langle 2, 4, 1 \rangle = 2\mathbf{w}$$

you can conclude that **u** *is* parallel to **w**.

b. Because

$$\mathbf{v} = \left\langle -1, -2, -\frac{1}{2} \right\rangle = -\frac{1}{2}\langle 2, 4, 1 \rangle = -\frac{1}{2}\mathbf{w}$$

you can conclude that **v** *is* parallel to **w**.

c. In this case, you need to find a scalar c such that $\langle 4, 8, 4 \rangle = c\langle 2, 4, 1 \rangle$. However, equating corresponding components produces $c = 2$ for the first two components and $c = 4$ for the third. So, the equation has no solution, and the vectors **z** and **w** are *not* parallel.

✓ ***Checkpoint*** *Audio-video solution in English & Spanish at LarsonPrecalculus.com*

Vector **w** has initial point $(2, -2, 1)$ and terminal point $(0, 4, -3)$. Determine whether each vector is parallel to **w**.

a. $\mathbf{u} = \langle -1, 3, -2 \rangle$

b. $\mathbf{v} = \langle -2, 6, -5 \rangle$

c. $\mathbf{z} = \langle 3, -9, 6 \rangle$

You can use vectors to determine whether three points are collinear (lie on the same line). The points P, Q, and R are **collinear** if and only if the vectors $\overrightarrow{PQ}$ and $\overrightarrow{PR}$ are parallel.

EXAMPLE 5 Using Vectors to Determine Collinear Points

Determine whether the points $P(2, -1, 4)$, $Q(5, 4, 6)$, and $R(-4, -11, 0)$ are collinear.

Solution

The component forms of $\overrightarrow{PQ}$ and $\overrightarrow{PR}$ are

$$\overrightarrow{PQ} = \langle 5 - 2, 4 - (-1), 6 - 4 \rangle = \langle 3, 5, 2 \rangle$$

and

$$\overrightarrow{PR} = \langle -4 - 2, -11 - (-1), 0 - 4 \rangle = \langle -6, -10, -4 \rangle.$$

Because $\overrightarrow{PR} = -2\overrightarrow{PQ}$, you can conclude that the vectors are parallel. So, the points P, Q, and R lie on the same line, as shown in the figure.

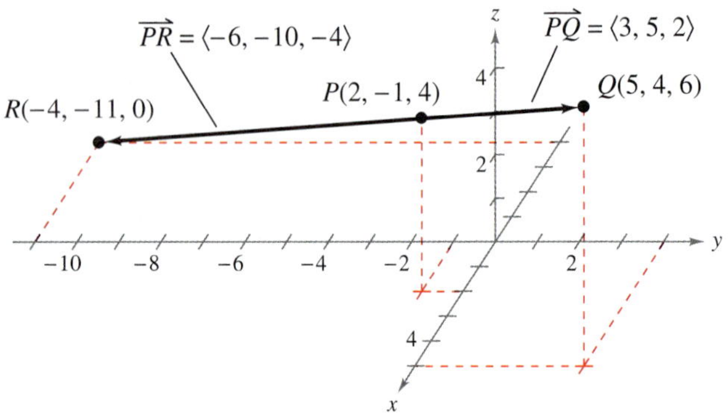

✓ **Checkpoint** *Audio-video solution in English & Spanish at LarsonPrecalculus.com*

Determine whether the points $P(-2, 7, -3)$, $Q(1, 4, 3)$, and $R(3, 2, 7)$ are collinear.

EXAMPLE 6 Finding the Terminal Point of a Vector

The initial point of the vector $\mathbf{v} = \langle 4, 2, -1 \rangle$ is $P(3, -1, 6)$. What is the terminal point of this vector?

Solution

Using the component form of the vector whose initial point is $P(3, -1, 6)$ and whose terminal point is $Q(q_1, q_2, q_3)$, you can write

$$\overrightarrow{PQ} = \langle q_1 - p_1, q_2 - p_2, q_3 - p_3 \rangle = \langle q_1 - 3, q_2 + 1, q_3 - 6 \rangle = \langle 4, 2, -1 \rangle.$$

This implies that

$$q_1 - 3 = 4, \quad q_2 + 1 = 2, \quad \text{and} \quad q_3 - 6 = -1.$$

The solutions of these three equations are

$$q_1 = 7, \quad q_2 = 1, \quad \text{and} \quad q_3 = 5.$$

So, the terminal point is $Q(7, 1, 5)$.

✓ **Checkpoint** *Audio-video solution in English & Spanish at LarsonPrecalculus.com*

The initial point of the vector $\mathbf{v} = \langle 2, -3, 6 \rangle$ is $P(1, -2, -5)$. What is the terminal point of this vector?

Application

The next example shows how to use vectors to solve an equilibrium problem in space.

EXAMPLE 7 Solving an Equilibrium Problem

A weight of 480 pounds is supported by three ropes. As shown in Figure 10.13, the weight is located at $S(0, 2, -1)$. The ropes are tied to the points

$$P(2, 0, 0), \quad Q(0, 4, 0), \quad \text{and} \quad R(-2, 0, 0).$$

Find the force (or tension) on each rope.

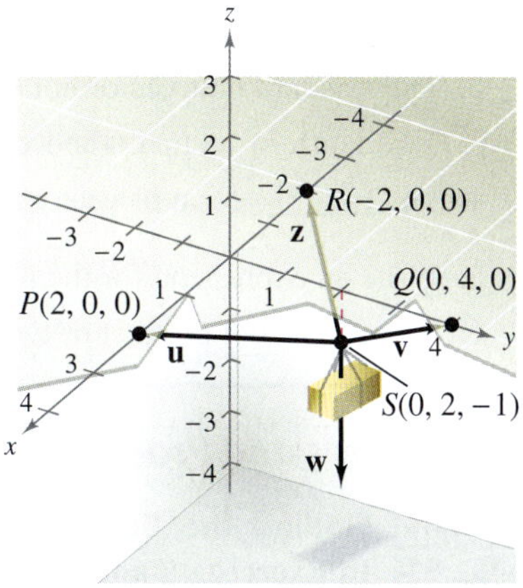

Figure 10.13

Solution

The (downward) force of the weight is represented by the vector

$$\mathbf{w} = \langle 0, 0, -480 \rangle.$$

The force vectors corresponding to the ropes are as follows.

$$\mathbf{u} = \|\mathbf{u}\|\frac{\overrightarrow{SP}}{\|\overrightarrow{SP}\|} = \|\mathbf{u}\|\frac{\langle 2-0, 0-2, 0-(-1)\rangle}{3} = \|\mathbf{u}\|\left\langle \frac{2}{3}, -\frac{2}{3}, \frac{1}{3}\right\rangle$$

$$\mathbf{v} = \|\mathbf{v}\|\frac{\overrightarrow{SQ}}{\|\overrightarrow{SQ}\|} = \|\mathbf{v}\|\frac{\langle 0-0, 4-2, 0-(-1)\rangle}{\sqrt{5}} = \|\mathbf{v}\|\left\langle 0, \frac{2}{\sqrt{5}}, \frac{1}{\sqrt{5}}\right\rangle$$

$$\mathbf{z} = \|\mathbf{z}\|\frac{\overrightarrow{SR}}{\|\overrightarrow{SR}\|} = \|\mathbf{z}\|\frac{\langle -2-0, 0-2, 0-(-1)\rangle}{3} = \|\mathbf{z}\|\left\langle -\frac{2}{3}, -\frac{2}{3}, \frac{1}{3}\right\rangle$$

For the system to be in equilibrium, it must be true that

$$\mathbf{u} + \mathbf{v} + \mathbf{z} + \mathbf{w} = \mathbf{0}$$

or

$$\mathbf{u} + \mathbf{v} + \mathbf{z} = -\mathbf{w}.$$

This yields the following system of linear equations.

$$\begin{cases} \frac{2}{3}\|\mathbf{u}\| \phantom{+ \frac{2}{\sqrt{5}}\|\mathbf{v}\|} - \frac{2}{3}\|\mathbf{z}\| = 0 \\ -\frac{2}{3}\|\mathbf{u}\| + \frac{2}{\sqrt{5}}\|\mathbf{v}\| - \frac{2}{3}\|\mathbf{z}\| = 0 \\ \frac{1}{3}\|\mathbf{u}\| + \frac{1}{\sqrt{5}}\|\mathbf{v}\| + \frac{1}{3}\|\mathbf{z}\| = 480 \end{cases}$$

Using the techniques demonstrated in Chapter 7, you can find the solution of the system to be

$$\|\mathbf{u}\| = 360.0$$

$$\|\mathbf{v}\| \approx 536.7$$

and

$$\|\mathbf{z}\| = 360.0.$$

So, the rope attached at point P has 360 pounds of tension, the rope attached at point Q has about 536.7 pounds of tension, and the rope attached at point R has 360 pounds of tension.

Checkpoint *Audio-video solution in English & Spanish at LarsonPrecalculus.com*

A weight of 240 pounds, located at $S(0, 3, -2)$, is supported by ropes tied to the points $P(3, 0, 0)$, $Q(0, 6, 0)$, and $R(-3, 0, 0)$. Find the force (or tension) on each rope.

10.2 Exercises

See *CalcChat.com* for tutorial help and worked-out solutions to odd-numbered exercises. For instructions on how to use a graphing utility, see Appendix A.

Vocabulary and Concept Check

In Exercises 1–4, fill in the blank.

1. In space, the _______ vector is denoted by $\mathbf{0} = \langle 0, 0, 0 \rangle$.
2. The standard unit vector notation for vector $\mathbf{v} = \langle v_1, v_2, v_3 \rangle$ is _______ .
3. Two nonzero vectors $\mathbf{u}$ and $\mathbf{v}$ are _______ when there is some scalar c such that $\mathbf{u} = c\mathbf{v}$.
4. The _______ form of vector $\mathbf{v} = \overrightarrow{PQ}$ is produced by $\langle q_1 - p_1, q_2 - p_2, q_3 - p_3 \rangle$.
5. Write an expression for the magnitude of vector $\mathbf{v}$, $\|\mathbf{v}\|$, where $\mathbf{v} = \langle v_1, v_2, v_3 \rangle$.
6. What is the dot product of two vectors that are orthogonal?

Procedures and Problem Solving

Finding the Component Form of a Vector **In Exercises 7 and 8, find the component form of the vector v.**

7.

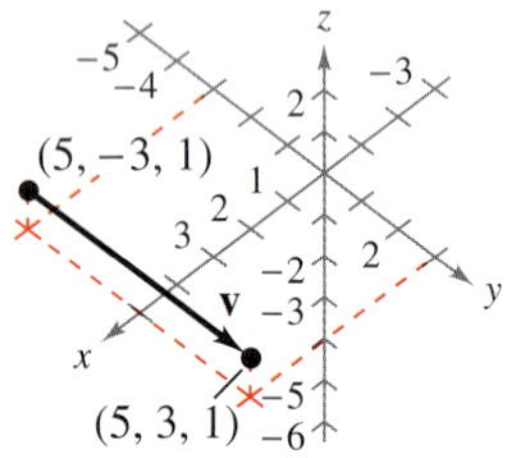

8.

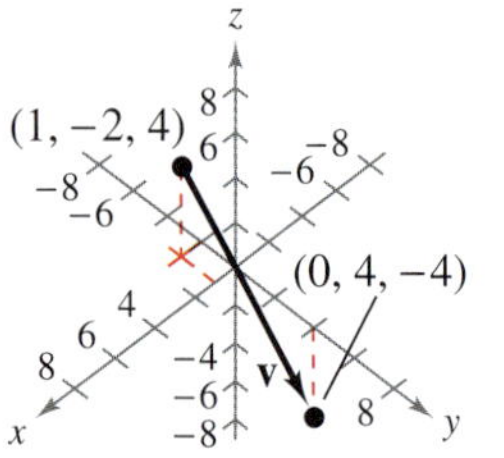

Finding the Component Form of a Vector **In Exercises 9–12, use the initial point and terminal point of the vector v to (a) write the component form of v, (b) find the magnitude of v, and (c) find a unit vector in the direction of v.**

	Initial Point	*Terminal Point*
9.	$(-7, 3, 4)$	$(-3, 0, 0)$
10.	$(0, -1, 0)$	$(0, 2, 1)$
11.	$(-6, 4, -2)$	$(1, -1, 3)$
12.	$(-2, 2, 1)$	$(-5, -2, 6)$

Sketching Vectors **In Exercises 13–16, sketch each scalar multiple of v.**

13. $\mathbf{v} = \langle 1, 1, 3 \rangle$
 (a) $2\mathbf{v}$ (b) $-\mathbf{v}$ (c) $\frac{3}{2}\mathbf{v}$ (d) $0\mathbf{v}$
14. $\mathbf{v} = \langle 1, 2, 3 \rangle$
 (a) $-\mathbf{v}$ (b) $2\mathbf{v}$ (c) $\frac{1}{2}\mathbf{v}$ (d) $\frac{5}{2}\mathbf{v}$
15. $\mathbf{v} = 2\mathbf{i} + 2\mathbf{j} - \mathbf{k}$
 (a) $3\mathbf{v}$ (b) $-2\mathbf{v}$ (c) $\frac{5}{2}\mathbf{v}$ (d) $\frac{3}{2}\mathbf{v}$
16. $\mathbf{v} = \mathbf{i} - 2\mathbf{j} + \mathbf{k}$
 (a) $4\mathbf{v}$ (b) $-2\mathbf{v}$ (c) $\frac{1}{2}\mathbf{v}$ (d) $0\mathbf{v}$

Using Vector Operations **In Exercises 17–22, find the vector z, given $\mathbf{u} = \langle -1, 3, 2 \rangle$, $\mathbf{v} = \langle 1, -2, -2 \rangle$, and $\mathbf{w} = \langle 5, 0, -5 \rangle$. Use a graphing utility to verify your answer.**

17. $\mathbf{z} = \mathbf{u} - 2\mathbf{v}$
18. $\mathbf{z} = 3\mathbf{w} - 2\mathbf{v} + \mathbf{u}$
19. $\mathbf{z} = 2\mathbf{u} - 3\mathbf{v} + \frac{1}{2}\mathbf{w}$
20. $\mathbf{z} = 7\mathbf{u} + \mathbf{v} - \frac{1}{5}\mathbf{w}$
21. $2\mathbf{z} - 4\mathbf{u} = \mathbf{w}$
22. $\mathbf{u} + \mathbf{v} - 2\mathbf{z} = \mathbf{0}$

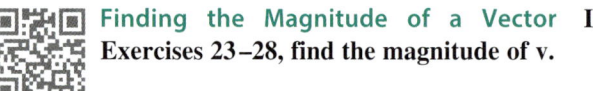

Finding the Magnitude of a Vector **In Exercises 23–28, find the magnitude of v.**

23. $\mathbf{v} = \langle 2, 5, 4 \rangle$
24. $\mathbf{v} = \langle -5, 2, -4 \rangle$
25. $\mathbf{v} = \langle 1, -2, 4 \rangle$
26. $\mathbf{v} = \langle -1, 0, 3 \rangle$
27. $\mathbf{v} = 6\mathbf{i} - \mathbf{j} + 8\mathbf{k}$
28. $\mathbf{v} = 3\mathbf{i} + 7\mathbf{j} - \mathbf{k}$

Finding Unit Vectors **In Exercises 29–32, find a unit vector (a) in the direction of v and (b) in the direction opposite of v.**

29. $\mathbf{v} = 5\mathbf{i} + 12\mathbf{k}$
30. $\mathbf{v} = 3\mathbf{i} + 4\mathbf{k}$
31. $\mathbf{v} = 8\mathbf{i} + 3\mathbf{j} - \mathbf{k}$
32. $\mathbf{v} = -3\mathbf{i} + 5\mathbf{j} + 10\mathbf{k}$

Using a Graphing Utility **In Exercises 33–36, use a graphing utility to determine the specified quantity, where $\mathbf{u} = \langle -3, 2, 1 \rangle$ and $\mathbf{v} = \langle 4, 4.5, -3 \rangle$.**

33. $5\mathbf{u} - 4\mathbf{v}$
34. $4\mathbf{u} + \frac{1}{4}\mathbf{v}$
35. $2\|\mathbf{u} + \mathbf{v}\|$
36. $2\|\mathbf{u}\| + 3\|\mathbf{v}\|$

Finding the Dot Product of Two Vectors **In Exercises 37–40, find the dot product of u and v.**

37. $\mathbf{u} = \langle 4, 4, -1 \rangle$
 $\mathbf{v} = \langle 2, -5, -8 \rangle$
38. $\mathbf{u} = \langle 3, -1, 6 \rangle$
 $\mathbf{v} = \langle 4, -10, 1 \rangle$
39. $\mathbf{u} = \mathbf{i} + 3\mathbf{j} - 5\mathbf{k}$
 $\mathbf{v} = 4\mathbf{i} + 6\mathbf{j} - 7\mathbf{k}$
40. $\mathbf{u} = 6\mathbf{i} + 8\mathbf{j} + 4\mathbf{k}$
 $\mathbf{v} = -\mathbf{i} - 6\mathbf{j} + 3\mathbf{k}$

Finding the Angle Between Two Vectors In Exercises 41–44, find the angle θ between the vectors.

41. $\mathbf{u} = \langle 0, 2, 2 \rangle$, $\mathbf{v} = \langle 3, 0, -4 \rangle$
42. $\mathbf{u} = \langle -1, 3, 0 \rangle$, $\mathbf{v} = \langle 1, 2, -1 \rangle$
43. $\mathbf{u} = 10\mathbf{i} + 40\mathbf{j}$, $\mathbf{v} = -3\mathbf{j} + 8\mathbf{k}$
44. $\mathbf{u} = 8\mathbf{j} - 20\mathbf{k}$, $\mathbf{v} = 10\mathbf{i} - 5\mathbf{k}$

Determining Orthogonal and Parallel Vectors In Exercises 45–50, determine whether u and v are orthogonal, parallel, or neither.

45. $\mathbf{u} = \langle -12, 6, 15 \rangle$, $\mathbf{v} = \langle 8, -4, -10 \rangle$
46. $\mathbf{u} = \langle -8, 16, 4 \rangle$, $\mathbf{v} = \langle 10, -20, -5 \rangle$
47. $\mathbf{u} = \langle 0, 1, 6 \rangle$, $\mathbf{v} = \langle 1, -1, -2 \rangle$
48. $\mathbf{u} = \langle 0, 4, -1 \rangle$, $\mathbf{v} = \langle -3, -5, 9 \rangle$
49. $\mathbf{u} = -2\mathbf{i} + 3\mathbf{j} - \mathbf{k}$, $\mathbf{v} = 2\mathbf{i} + \mathbf{j} - \mathbf{k}$
50. $\mathbf{u} = 2\mathbf{i} - 3\mathbf{j} + \mathbf{k}$, $\mathbf{v} = -\mathbf{i} - \mathbf{j} - \mathbf{k}$

Using Vectors to Determine Collinear Points In Exercises 51–54, use vectors to determine whether the points are collinear.

51. $(5, 4, 1), (7, 3, -1), (4, 5, 3)$
52. $(1, 3, 2), (-1, 2, 5), (3, 4, -1)$
53. $(0, 4, 4), (-1, 5, 6), (-2, 6, 7)$
54. $(-2, 7, 4), (-4, 8, 1), (0, 6, 7)$

Using Vectors to Classify a Triangle In Exercises 55–58, the vertices of a triangle are given. Determine whether the triangle is an acute triangle, an obtuse triangle, or a right triangle. Explain your reasoning.

55. $(1, 2, 0), (0, 0, 0), (-2, 1, 0)$
56. $(-3, 0, 0), (0, 0, 0), (1, 2, 3)$
57. $(2, -3, 4), (0, 1, 2), (-1, 2, 0)$
58. $(1, 1, 3), (-4, 5, 9), (3, -1, 6)$

Finding the Terminal Point of a Vector In Exercises 59–62, the vector v and its initial point are given. Find the terminal point.

59. $\mathbf{v} = \langle 2, -4, 7 \rangle$
Initial point: $(1, 5, 0)$
60. $\mathbf{v} = \langle -2, 15, -8 \rangle$
Initial point: $(5, -6, 0)$
61. $\mathbf{v} = \left\langle 4, \frac{3}{2}, -\frac{1}{4} \right\rangle$
Initial point: $\left(2, 1, -\frac{3}{2}\right)$
62. $\mathbf{v} = \left\langle \frac{5}{2}, -\frac{1}{2}, 4 \right\rangle$
Initial point: $\left(3, 2, -\frac{1}{2}\right)$

Finding a Vector In Exercises 63 and 64, write the component form of v.

63. Vector $\mathbf{v}$ lies in the yz-plane, has magnitude 4, and makes an angle of 45° with the positive y-axis.
64. Vector $\mathbf{v}$ lies in the xz-plane, has magnitude 10, and makes an angle of 60° with the positive z-axis.

Finding a Scalar In Exercises 65 and 66, determine the values of c that satisfy the given conditions.

65. $\|c\mathbf{u}\| = 3$, $\mathbf{u} = \mathbf{i} + 2\mathbf{j} + 3\mathbf{k}$
66. $\|c\mathbf{u}\| = 12$, $\mathbf{u} = -2\mathbf{i} + 2\mathbf{j} - 4\mathbf{k}$

67. ***Why you should learn it*** *(p. 729)* A round birdbath weighs 8 pounds and has a radius of 6 inches. It is supported by three equally-spaced 24-inch chains attached to a tree branch (see figure). Find the tension in each chain.

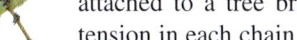

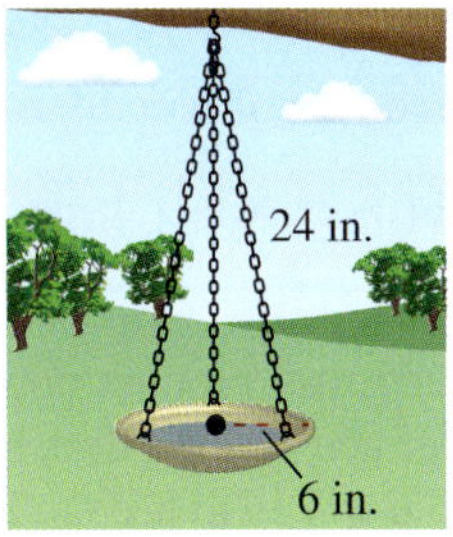

Figure for 67

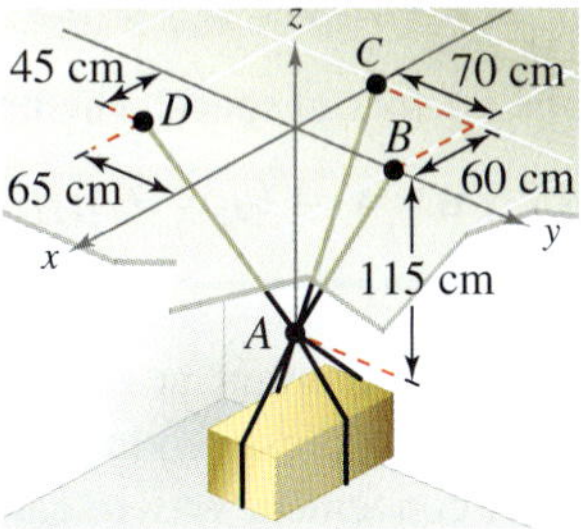

Figure for 68

68. **Forces in Equilibrium** The weight of a crate is 500 newtons. Find the tension in each of the supporting cables shown in the figure.

Focusing on Concepts

True or False? In Exercises 69 and 70, determine whether the statement is true or false. Justify your answer.

69. If the dot product of two nonzero vectors is zero, then the angle between the vectors is a right angle.
70. If $\overrightarrow{AB}$ and $\overrightarrow{AC}$ represent parallel vectors, then points A, B, and C are collinear.

71. **Error Analysis** Describe the error in determining whether $(3, 4, 1)$, $(5, 2, 5)$ and $(9, -2, 13)$ are collinear.

$\mathbf{u} = \langle 5 - 3, 2 - 4, 5 - 1 \rangle = \langle 2, -2, 4 \rangle$
$\mathbf{v} = \langle 3 - 9, 4 - (-2), 1 - 13 \rangle$
$= \langle -6, 6, -12 \rangle$

Because $\mathbf{u} \neq \mathbf{v}$, the points are not collinear.

72. **HOW DO YOU SEE IT?** Use the graph to find $\mathbf{u} \cdot \mathbf{v}$ without calculating. Explain your reasoning.

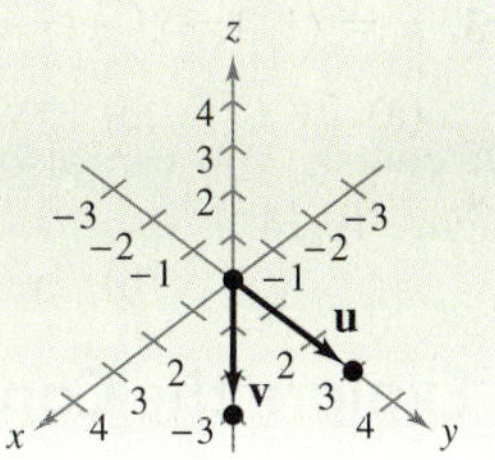

73. **Think About It** The initial and terminal points of $\mathbf{v}$ are (x_1, y_1, z_1) and (x, y, z), respectively. Describe the set of all points (x, y, z) such that $\|\mathbf{v}\| = 4$.

10.3 The Cross Product of Two Vectors

The Cross Product

Many applications in physics, engineering, and geometry involve finding a vector in space that is orthogonal to two given vectors. In this section, you will study a product that yields such a vector. It is called the **cross product,** and it is most conveniently defined and calculated using the standard unit vector form.

Definition of Cross Product of Two Vectors in Space

Let

$\mathbf{u} = u_1\mathbf{i} + u_2\mathbf{j} + u_3\mathbf{k}$ and $\mathbf{v} = v_1\mathbf{i} + v_2\mathbf{j} + v_3\mathbf{k}$

be vectors in space. The **cross product** of **u** and **v** is the vector

$$\mathbf{u} \times \mathbf{v} = (u_2v_3 - u_3v_2)\mathbf{i} - (u_1v_3 - u_3v_1)\mathbf{j} + (u_1v_2 - u_2v_1)\mathbf{k}.$$

What you should learn

- Find cross products of vectors in space.
- Use geometric properties of cross products of vectors in space.
- Use triple scalar products to find volumes of parallelepipeds.

Why you should learn it

The cross product of two vectors in space has many applications in physics and engineering. For instance, in Exercise 65 on page 742, the cross product is used to find the torque on the crank of a bicycle's brake.

It is important to note that the cross product applies only to three-dimensional vectors. The cross product is not defined for two-dimensional vectors.

A convenient way to calculate $\mathbf{u} \times \mathbf{v}$ is to use the *determinant form* with cofactor expansion shown below. (This 3×3 determinant form is used simply to help remember the formula for the cross product—it is technically not a determinant because the entries of the corresponding array are not all real numbers.)

$$\mathbf{u} \times \mathbf{v} = \begin{vmatrix} \mathbf{i} & \mathbf{j} & \mathbf{k} \\ u_1 & u_2 & u_3 \\ v_1 & v_2 & v_3 \end{vmatrix} \begin{matrix} \\ \leftarrow \text{Put "u" in Row 2.} \\ \leftarrow \text{Put "v" in Row 3.} \end{matrix}$$

$$= \begin{vmatrix} \mathbf{i} & \mathbf{j} & \mathbf{k} \\ u_1 & u_2 & u_3 \\ v_1 & v_2 & v_3 \end{vmatrix}\mathbf{i} - \begin{vmatrix} \mathbf{i} & \mathbf{j} & \mathbf{k} \\ u_1 & u_2 & u_3 \\ v_1 & v_2 & v_3 \end{vmatrix}\mathbf{j} + \begin{vmatrix} \mathbf{i} & \mathbf{j} & \mathbf{k} \\ u_1 & u_2 & u_3 \\ v_1 & v_2 & v_3 \end{vmatrix}\mathbf{k}$$

$$= \begin{vmatrix} u_2 & u_3 \\ v_2 & v_3 \end{vmatrix}\mathbf{i} - \begin{vmatrix} u_1 & u_3 \\ v_1 & v_3 \end{vmatrix}\mathbf{j} + \begin{vmatrix} u_1 & u_2 \\ v_1 & v_2 \end{vmatrix}\mathbf{k}$$

$$= (u_2v_3 - u_3v_2)\mathbf{i} - (u_1v_3 - u_3v_1)\mathbf{j} + (u_1v_2 - u_2v_1)\mathbf{k}$$

Note the minus sign in front of the **j**-component. Recall from Section 7.7 that each of the three 2×2 determinants can be evaluated using the diagonal pattern

$$\begin{vmatrix} a_1 & b_1 \\ a_2 & b_2 \end{vmatrix} = a_1b_2 - a_2b_1.$$

Here is an example.

$$\begin{vmatrix} 2 & 4 \\ 3 & -1 \end{vmatrix} = 2(-1) - 4(3) = -2 - 12 = -14$$

Of course, you can also use a graphing utility to find a determinant, as shown in Figure 10.14.

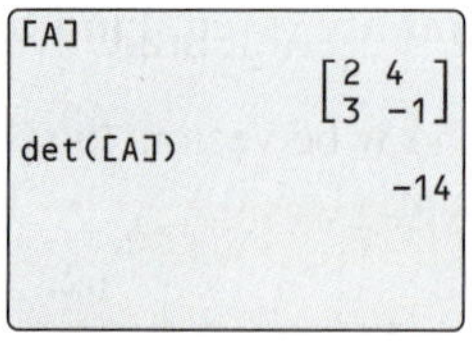

Figure 10.14

Explore the Concept

Find each cross product and sketch the result. What can you conclude?

a. $\mathbf{i} \times \mathbf{j}$ **b.** $\mathbf{i} \times \mathbf{k}$ **c.** $\mathbf{j} \times \mathbf{k}$

EXAMPLE 1 Finding Cross Products

Given $\mathbf{u} = \mathbf{i} + 2\mathbf{j} + \mathbf{k}$ and $\mathbf{v} = 3\mathbf{i} + \mathbf{j} + 2\mathbf{k}$, find each cross product.

a. $\mathbf{u} \times \mathbf{v}$ **b.** $\mathbf{v} \times \mathbf{u}$ **c.** $\mathbf{v} \times \mathbf{v}$

Solution

$$\textbf{a. } \mathbf{u} \times \mathbf{v} = \begin{vmatrix} \mathbf{i} & \mathbf{j} & \mathbf{k} \\ 1 & 2 & 1 \\ 3 & 1 & 2 \end{vmatrix}$$

$$= \begin{vmatrix} 2 & 1 \\ 1 & 2 \end{vmatrix}\mathbf{i} - \begin{vmatrix} 1 & 1 \\ 3 & 2 \end{vmatrix}\mathbf{j} + \begin{vmatrix} 1 & 2 \\ 3 & 1 \end{vmatrix}\mathbf{k}$$

$$= (4 - 1)\mathbf{i} - (2 - 3)\mathbf{j} + (1 - 6)\mathbf{k}$$

$$= 3\mathbf{i} + \mathbf{j} - 5\mathbf{k}$$

$$\textbf{b. } \mathbf{v} \times \mathbf{u} = \begin{vmatrix} \mathbf{i} & \mathbf{j} & \mathbf{k} \\ 3 & 1 & 2 \\ 1 & 2 & 1 \end{vmatrix}$$

$$= \begin{vmatrix} 1 & 2 \\ 2 & 1 \end{vmatrix}\mathbf{i} - \begin{vmatrix} 3 & 2 \\ 1 & 1 \end{vmatrix}\mathbf{j} + \begin{vmatrix} 3 & 1 \\ 1 & 2 \end{vmatrix}\mathbf{k}$$

$$= (1 - 4)\mathbf{i} - (3 - 2)\mathbf{j} + (6 - 1)\mathbf{k}$$

$$= -3\mathbf{i} - \mathbf{j} + 5\mathbf{k}$$

Note that this result is the negative of that in part (a).

$$\textbf{c. } \mathbf{v} \times \mathbf{v} = \begin{vmatrix} \mathbf{i} & \mathbf{j} & \mathbf{k} \\ 3 & 1 & 2 \\ 3 & 1 & 2 \end{vmatrix} = \mathbf{0}$$

✓ **Checkpoint** *Audio-video solution in English & Spanish at LarsonPrecalculus.com*

Given $\mathbf{u} = \mathbf{i} - 3\mathbf{j} + 2\mathbf{k}$ and $\mathbf{v} = 2\mathbf{i} - \mathbf{j} + 2\mathbf{k}$, find each cross product.

a. $\mathbf{u} \times \mathbf{v}$ **b.** $\mathbf{v} \times \mathbf{u}$ **c.** $\mathbf{v} \times \mathbf{v}$

The results obtained in Example 1 suggest some interesting algebraic properties of the cross product. For instance,

$$\mathbf{u} \times \mathbf{v} = -(\mathbf{v} \times \mathbf{u}) \quad \text{and} \quad \mathbf{v} \times \mathbf{v} = \mathbf{0}.$$

These properties, and several others, are summarized in the following list.

Algebraic Properties of the Cross Product

Let **u**, **v**, and **w** be vectors in space, and let c be a scalar.

1. $\mathbf{u} \times \mathbf{v} = -(\mathbf{v} \times \mathbf{u})$
2. $\mathbf{u} \times (\mathbf{v} + \mathbf{w}) = (\mathbf{u} \times \mathbf{v}) + (\mathbf{u} \times \mathbf{w})$
3. $c(\mathbf{u} \times \mathbf{v}) = (c\mathbf{u}) \times \mathbf{v} = \mathbf{u} \times (c\mathbf{v})$
4. $\mathbf{u} \times \mathbf{0} = \mathbf{0} \times \mathbf{u} = \mathbf{0}$
5. $\mathbf{u} \times \mathbf{u} = \mathbf{0}$
6. $\mathbf{u} \cdot (\mathbf{v} \times \mathbf{w}) = (\mathbf{u} \times \mathbf{v}) \cdot \mathbf{w}$

(See the proof on page 757.)

Technology Tip

Some graphing utilities have the capability of performing vector operations, such as the cross product. Consult the user's guide for your graphing utility for specific instructions.

Explore the Concept

Calculate $\mathbf{u} \times \mathbf{v}$ and $-(\mathbf{v} \times \mathbf{u})$ for several values of **u** and **v**. What do your results imply? Interpret your results geometrically.

Geometric Properties of the Cross Product

The first property listed on the preceding page indicates that the cross product is *not commutative*. In particular, this property indicates that the vectors $\mathbf{u} \times \mathbf{v}$ and $\mathbf{v} \times \mathbf{u}$ have equal lengths but opposite directions. The following list gives some other *geometric* properties of the cross product of two vectors.

Geometric Properties of the Cross Product

Let $\mathbf{u}$ and $\mathbf{v}$ be nonzero vectors in space, and let θ be the angle between $\mathbf{u}$ and $\mathbf{v}$.

1. $\mathbf{u} \times \mathbf{v}$ is orthogonal to both $\mathbf{u}$ and $\mathbf{v}$.
2. $\|\mathbf{u} \times \mathbf{v}\| = \|\mathbf{u}\|\|\mathbf{v}\| \sin \theta$
3. $\mathbf{u} \times \mathbf{v} = \mathbf{0}$ if and only if $\mathbf{u}$ and $\mathbf{v}$ are scalar multiples of each other.
4. $\|\mathbf{u} \times \mathbf{v}\| =$ area of parallelogram having $\mathbf{u}$ and $\mathbf{v}$ as adjacent sides.

(See the proof on page 758.)

Both $\mathbf{u} \times \mathbf{v}$ and $\mathbf{v} \times \mathbf{u}$ are perpendicular to the plane determined by $\mathbf{u}$ and $\mathbf{v}$. One way to remember the orientations of the vectors $\mathbf{u}$, $\mathbf{v}$, and $\mathbf{u} \times \mathbf{v}$ is to compare them with the unit vectors $\mathbf{i}$, $\mathbf{j}$, and $\mathbf{k} = \mathbf{i} \times \mathbf{j}$, as shown in Figure 10.15. The three vectors $\mathbf{u}$, $\mathbf{v}$, and $\mathbf{u} \times \mathbf{v}$ form a *right-handed system.*

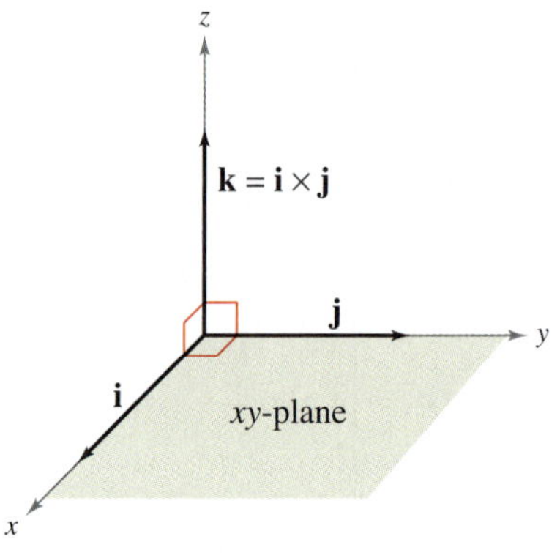

Figure 10.15

EXAMPLE 2 Using the Cross Product

See LarsonPrecalculus.com for an interactive version of this type of example.

Find a unit vector that is orthogonal to both

$\mathbf{u} = 3\mathbf{i} - 4\mathbf{j} + \mathbf{k}$ and $\mathbf{v} = -3\mathbf{i} + 6\mathbf{j}$.

Solution

The cross product $\mathbf{u} \times \mathbf{v}$, as shown in the figure, is orthogonal to both $\mathbf{u}$ and $\mathbf{v}$.

$$\mathbf{u} \times \mathbf{v} = \begin{vmatrix} \mathbf{i} & \mathbf{j} & \mathbf{k} \\ 3 & -4 & 1 \\ -3 & 6 & 0 \end{vmatrix}$$

$$= -6\mathbf{i} - 3\mathbf{j} + 6\mathbf{k}$$

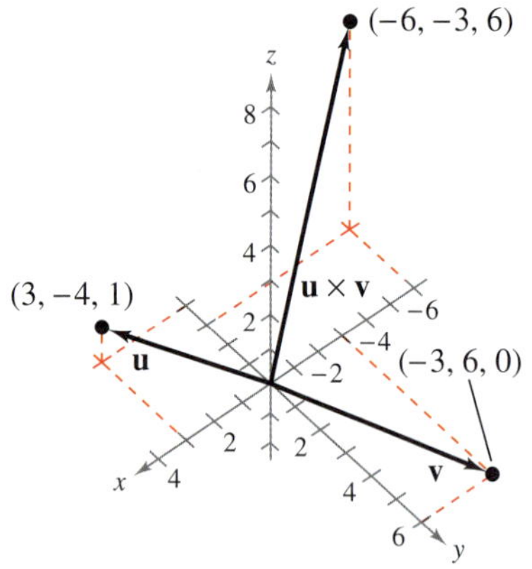

Because

$$\|\mathbf{u} \times \mathbf{v}\| = \sqrt{(-6)^2 + (-3)^2 + 6^2}$$
$$= \sqrt{81}$$
$$= 9$$

a unit vector orthogonal to both $\mathbf{u}$ and $\mathbf{v}$ is

$$\frac{\mathbf{u} \times \mathbf{v}}{\|\mathbf{u} \times \mathbf{v}\|} = -\frac{2}{3}\mathbf{i} - \frac{1}{3}\mathbf{j} + \frac{2}{3}\mathbf{k}.$$

✓ Checkpoint *Audio-video solution in English & Spanish at LarsonPrecalculus.com*

Find a unit vector that is orthogonal to both

$\mathbf{u} = 2\mathbf{i} + 4\mathbf{j} - 3\mathbf{k}$

and

$\mathbf{v} = -2\mathbf{i} - 3\mathbf{j} + 2\mathbf{k}$.

What's Wrong?

You use a graphing utility to find $\mathbf{u} \times \mathbf{v}$ for

$\mathbf{u} = -\mathbf{i} + 4\mathbf{j} + 3\mathbf{k}$

and

$\mathbf{v} = \mathbf{i} - 2\mathbf{k}$

as shown in the figure. You conclude that

$\mathbf{w} = -8\mathbf{i} + \mathbf{j} - 4\mathbf{k}$

is a unit vector orthogonal to both $\mathbf{u}$ and $\mathbf{v}$. What's wrong?

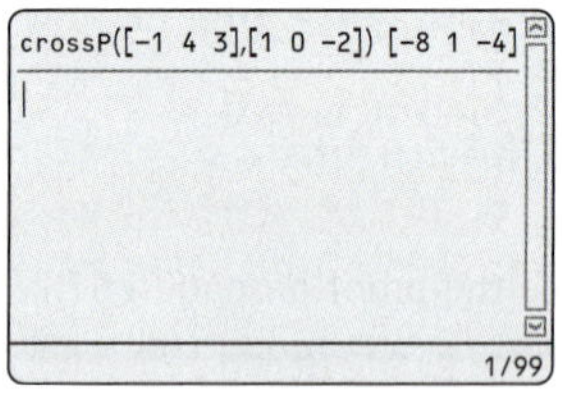

In Example 2, note that you could have used the cross product $\mathbf{v} \times \mathbf{u}$ to form a unit vector that is orthogonal to both $\mathbf{u}$ and $\mathbf{v}$. With that choice, you would have obtained the *negative* of the unit vector found in the example.

The fourth geometric property of the cross product states that $\|\mathbf{u} \times \mathbf{v}\|$ is the area of the parallelogram that has $\mathbf{u}$ and $\mathbf{v}$ as adjacent sides. It follows that the area of a triangle having vectors $\mathbf{u}$ and $\mathbf{v}$ as adjacent sides is $\frac{1}{2}\|\mathbf{u} \times \mathbf{v}\|$.

EXAMPLE 3 Geometric Application of the Cross Product

Show that the quadrilateral with the following vertices is a parallelogram. Then find the area of the parallelogram.

$A(5, 2, 0), \quad B(2, 6, 1), \quad C(2, 4, 7), \quad D(5, 0, 6)$

Solution

From Figure 10.16, you can see that the sides of the quadrilateral correspond to the following four vectors.

$$\overrightarrow{AB} = -3\mathbf{i} + 4\mathbf{j} + \mathbf{k} \qquad \overrightarrow{CD} = 3\mathbf{i} - 4\mathbf{j} - \mathbf{k} = -\overrightarrow{AB}$$

$$\overrightarrow{AD} = 0\mathbf{i} - 2\mathbf{j} + 6\mathbf{k} \qquad \overrightarrow{CB} = 0\mathbf{i} + 2\mathbf{j} - 6\mathbf{k} = -\overrightarrow{AD}$$

Because $\overrightarrow{CD} = -\overrightarrow{AB}$ and $\overrightarrow{CB} = -\overrightarrow{AD}$, you can conclude that $\overrightarrow{AB}$ is parallel to $\overrightarrow{CD}$ and $\overrightarrow{AD}$ is parallel to $\overrightarrow{CB}$. It follows that the quadrilateral is a parallelogram with $\overrightarrow{AB}$ and $\overrightarrow{AD}$ as adjacent sides. Moreover, because

$$\overrightarrow{AB} \times \overrightarrow{AD} = \begin{vmatrix} \mathbf{i} & \mathbf{j} & \mathbf{k} \\ -3 & 4 & 1 \\ 0 & -2 & 6 \end{vmatrix} = 26\mathbf{i} + 18\mathbf{j} + 6\mathbf{k}$$

the area of the parallelogram is

$$\|\overrightarrow{AB} \times \overrightarrow{AD}\| = \sqrt{26^2 + 18^2 + 6^2} = \sqrt{1036} \approx 32.19 \text{ square units.}$$

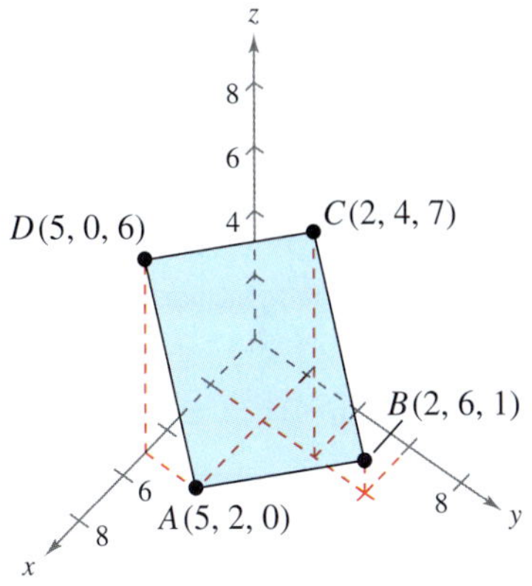

Figure 10.16

Checkpoint *Audio-video solution in English & Spanish at LarsonPrecalculus.com*

Show that the quadrilateral with the following vertices is a parallelogram. Then find the area of the parallelogram.

$A(4, 3, 0), \quad B(3, 4, 2), \quad C(3, 2, 6), \quad D(4, 1, 4)$

EXAMPLE 4 Finding the Area of a Triangle

Find the area of the triangle with vertices $A(4, 3, 0)$, $B(5, 4, 6)$, and $C(4, 1, 5)$.

Solution

The triangle is shown in Figure 10.17. The cross product of $\overrightarrow{AB} = \mathbf{i} + \mathbf{j} + 6\mathbf{k}$ and $\overrightarrow{AC} = 0\mathbf{i} - 2\mathbf{j} + 5\mathbf{k}$ is

$$\overrightarrow{AB} \times \overrightarrow{AC} = \begin{vmatrix} \mathbf{i} & \mathbf{j} & \mathbf{k} \\ 1 & 1 & 6 \\ 0 & -2 & 5 \end{vmatrix} = 17\mathbf{i} - 5\mathbf{j} - 2\mathbf{k}.$$

So, the area of the triangle is

$$A = \tfrac{1}{2}\|\overrightarrow{AB} \times \overrightarrow{AC}\| = \tfrac{1}{2}\sqrt{17^2 + (-5)^2 + (-2)^2} = \tfrac{1}{2}\sqrt{318} \approx 8.9 \text{ square units.}$$

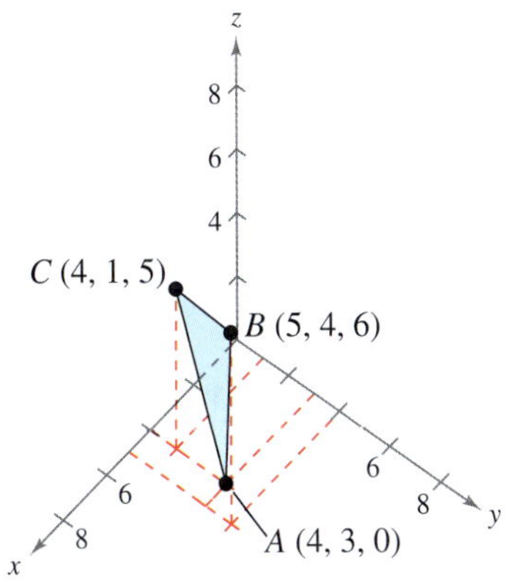

Figure 10.17

Checkpoint *Audio-video solution in English & Spanish at LarsonPrecalculus.com*

Find the area of the triangle with vertices

$A(3, 2, 1)$, $B(3, 1, 7)$, and $C(5, 2, 6)$.

The Triple Scalar Product

For the vectors **u**, **v**, and **w** in space, the dot product of **u** and $\mathbf{v} \times \mathbf{w}$ is called the **triple scalar product** of **u**, **v**, and **w**.

The Triple Scalar Product

For $\mathbf{u} = u_1\mathbf{i} + u_2\mathbf{j} + u_3\mathbf{k}$, $\mathbf{v} = v_1\mathbf{i} + v_2\mathbf{j} + v_3\mathbf{k}$, and $\mathbf{w} = w_1\mathbf{i} + w_2\mathbf{j} + w_3\mathbf{k}$, the triple scalar product is given by

$$\mathbf{u} \cdot (\mathbf{v} \times \mathbf{w}) = \begin{vmatrix} u_1 & u_2 & u_3 \\ v_1 & v_2 & v_3 \\ w_1 & w_2 & w_3 \end{vmatrix}.$$

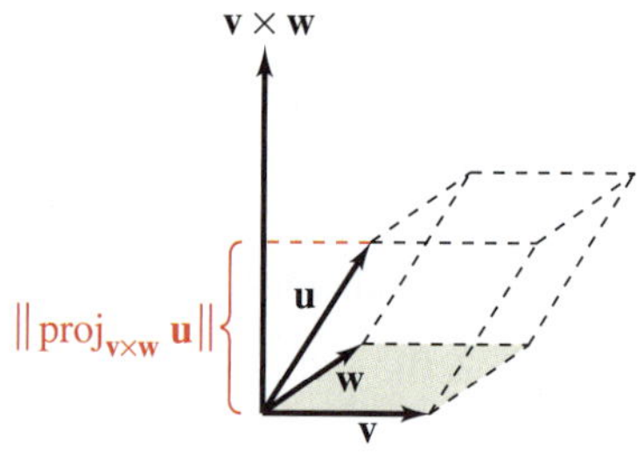

Area of base = $\|\mathbf{v} \times \mathbf{w}\|$
Volume of parallelepiped = $|\mathbf{u} \cdot (\mathbf{v} \times \mathbf{w})|$
Figure 10.18

When the vectors **u**, **v**, and **w** do not lie in the same plane, the triple scalar product $\mathbf{u} \cdot (\mathbf{v} \times \mathbf{w})$ can be used to determine the volume of the parallelepiped (a polyhedron, all of whose faces are parallelograms) with **u**, **v**, and **w** as adjacent edges, as shown in Figure 10.18.

Geometric Property of Triple Scalar Product

The volume V of a parallelepiped with vectors **u**, **v**, and **w** as adjacent edges is given by

$$V = |\mathbf{u} \cdot (\mathbf{v} \times \mathbf{w})|.$$

EXAMPLE 5 Volume by the Triple Scalar Product

Find the volume of the parallelepiped having

$$\mathbf{u} = 3\mathbf{i} - 5\mathbf{j} + \mathbf{k}, \quad \mathbf{v} = 2\mathbf{j} - 2\mathbf{k}, \quad \text{and} \quad \mathbf{w} = 3\mathbf{i} + \mathbf{j} + \mathbf{k}$$

as adjacent edges, as shown in Figure 10.19.

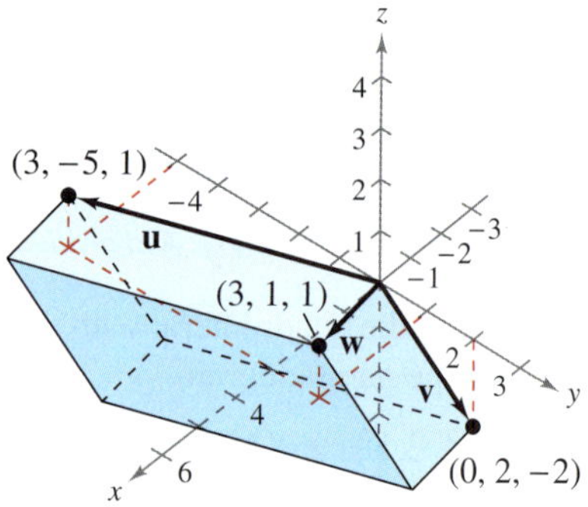

Figure 10.19

Solution

The value of the triple scalar product is

$$\mathbf{u} \cdot (\mathbf{v} \times \mathbf{w}) = \begin{vmatrix} 3 & -5 & 1 \\ 0 & 2 & -2 \\ 3 & 1 & 1 \end{vmatrix}$$

$$= 3\begin{vmatrix} 2 & -2 \\ 1 & 1 \end{vmatrix} - (-5)\begin{vmatrix} 0 & -2 \\ 3 & 1 \end{vmatrix} + 1\begin{vmatrix} 0 & 2 \\ 3 & 1 \end{vmatrix}$$

$$= 3(4) + 5(6) + 1(-6)$$

$$= 36.$$

So, the volume of the parallelepiped is

$$|\mathbf{u} \cdot (\mathbf{v} \times \mathbf{w})| = |36| = 36 \text{ cubic units.}$$

You can confirm this result using a graphing utility, as shown in Figure 10.20.

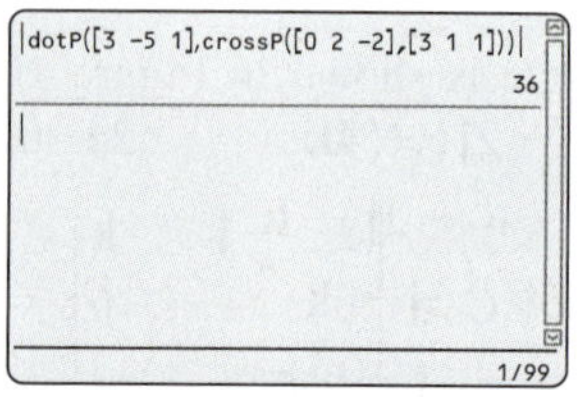

Figure 10.20

✓ ***Checkpoint*** *Audio-video solution in English & Spanish at LarsonPrecalculus.com*

Find the volume of the parallelepiped having

$$\mathbf{u} = 2\mathbf{i} - \mathbf{j} + 4\mathbf{k}, \quad \mathbf{v} = \mathbf{i} + 3\mathbf{j} - 2\mathbf{k}, \quad \text{and} \quad \mathbf{w} = 2\mathbf{i} - 2\mathbf{j} + 3\mathbf{k}$$

as adjacent edges.

10.3 Exercises

See *CalcChat.com* for tutorial help and worked-out solutions to odd-numbered exercises. For instructions on how to use a graphing utility, see Appendix A.

Vocabulary and Concept Check

In Exercises 1–6, fill in the blank.

1. To find a vector in space that is orthogonal to two given vectors, find the _______ of the two vectors.
2. $\mathbf{u} \times \mathbf{u} =$ _______
3. $-(\mathbf{v} \times \mathbf{u}) =$ _______
4. The area of a parallelogram with **u** and **v** as adjacent sides is given by _______ .
5. The dot product of **u** and $\mathbf{v} \times \mathbf{w}$ is called the _______ of **u**, **v**, and **w**.
6. The volume V of a _______ with vectors **u**, **v**, and **w** as adjacent edges is $V = |\mathbf{u} \cdot (\mathbf{v} \times \mathbf{w})|$.

Procedures and Problem Solving

Finding the Cross Product **In Exercises 7–20, find u × v and show that it is orthogonal to both u and v.**

7. $\mathbf{u} = \langle 1, -1, 0 \rangle$, $\mathbf{v} = \langle 0, 1, -1 \rangle$
8. $\mathbf{u} = \langle -1, 1, 0 \rangle$, $\mathbf{v} = \langle 1, 0, -1 \rangle$
9. $\mathbf{u} = \langle -9, 4, 7 \rangle$, $\mathbf{v} = \langle 6, 3, 2 \rangle$
10. $\mathbf{u} = \langle -5, 5, 11 \rangle$, $\mathbf{v} = \langle 2, 2, 3 \rangle$
11. $\mathbf{u} = 6\mathbf{i} + 2\mathbf{j} + \mathbf{k}$, $\mathbf{v} = \mathbf{i} + 3\mathbf{j} - 2\mathbf{k}$
12. $\mathbf{u} = 4\mathbf{i} + 2\mathbf{j} + \mathbf{k}$, $\mathbf{v} = -5\mathbf{i} - \mathbf{j} + 3\mathbf{k}$
13. $\mathbf{u} = 2\mathbf{i} + 4\mathbf{j} + 3\mathbf{k}$, $\mathbf{v} = -\mathbf{i} + 3\mathbf{j} - 2\mathbf{k}$
14. $\mathbf{u} = 3\mathbf{i} - 2\mathbf{j} + \mathbf{k}$, $\mathbf{v} = -2\mathbf{i} + \mathbf{j} + 2\mathbf{k}$
15. $\mathbf{u} = \frac{1}{2}\mathbf{i} - \frac{2}{3}\mathbf{j} + \mathbf{k}$, $\mathbf{v} = -\frac{3}{4}\mathbf{i} + \mathbf{j} + \frac{1}{4}\mathbf{k}$
16. $\mathbf{u} = \frac{2}{5}\mathbf{i} - \frac{1}{4}\mathbf{j} + \frac{1}{2}\mathbf{k}$, $\mathbf{v} = -\frac{3}{5}\mathbf{i} + \mathbf{j} + \frac{1}{5}\mathbf{k}$
17. $\mathbf{u} = \frac{1}{6}\mathbf{k}$, $\mathbf{v} = -6\mathbf{i} - \frac{1}{3}\mathbf{j}$
18. $\mathbf{u} = \frac{2}{3}\mathbf{i}$, $\mathbf{v} = \frac{1}{3}\mathbf{j} - 3\mathbf{k}$
19. $\mathbf{u} = -\mathbf{i} + \mathbf{k}$, $\mathbf{v} = \mathbf{j} - 2\mathbf{k}$
20. $\mathbf{u} = \mathbf{i} - \mathbf{k}$, $\mathbf{v} = -\mathbf{j} + 8\mathbf{k}$

Using a Graphing Utility to Find the Cross Product **In Exercises 21–26, use a graphing utility to find u × v.**

21. $\mathbf{u} = \langle 2, 4, 3 \rangle$, $\mathbf{v} = \langle 0, -2, 1 \rangle$
22. $\mathbf{u} = \langle 4, -2, 6 \rangle$, $\mathbf{v} = \langle -1, 5, 7 \rangle$
23. $\mathbf{u} = \mathbf{i} - 2\mathbf{j} + 4\mathbf{k}$, $\mathbf{v} = -4\mathbf{i} + 2\mathbf{j} - \mathbf{k}$
24. $\mathbf{u} = 2\mathbf{i} - \mathbf{j} + 3\mathbf{k}$, $\mathbf{v} = -\mathbf{i} + \mathbf{j} - 4\mathbf{k}$
25. $\mathbf{u} = 6\mathbf{i} - 5\mathbf{j} + \mathbf{k}$, $\mathbf{v} = \frac{1}{2}\mathbf{i} - \frac{3}{4}\mathbf{j} + \frac{2}{10}\mathbf{k}$
26. $\mathbf{u} = 8\mathbf{i} - 4\mathbf{j} + 2\mathbf{k}$, $\mathbf{v} = \frac{1}{2}\mathbf{i} + \frac{3}{4}\mathbf{j} - \frac{1}{4}\mathbf{k}$

Vector Operations **In Exercises 27–32, evaluate the expression, where $\mathbf{u} = 2\mathbf{i} + \mathbf{j} - 2\mathbf{k}$ and $\mathbf{v} = -\mathbf{i} + 5\mathbf{j} - \mathbf{k}$.**

27. $\mathbf{u} \times (\mathbf{v} \times \mathbf{v})$
28. $\mathbf{v} \times (\mathbf{u} \times \mathbf{u})$
29. $\mathbf{u} \times (-3\mathbf{v})$
30. $(-2\mathbf{u}) \times \mathbf{v}$
31. $\mathbf{u} \cdot (\mathbf{u} \times \mathbf{v})$
32. $\mathbf{v} \cdot (\mathbf{v} \times \mathbf{u})$

Using the Cross Product **In Exercises 33–40, find a unit vector orthogonal to u and v.**

33. $\mathbf{u} = \langle 2, -3, 4 \rangle$, $\mathbf{v} = \langle 0, -1, 1 \rangle$
34. $\mathbf{u} = \langle 1, 1, 0 \rangle$, $\mathbf{v} = \langle 1, 0, -2 \rangle$
35. $\mathbf{u} = 3\mathbf{i} + \mathbf{j}$, $\mathbf{v} = \mathbf{j} + \mathbf{k}$
36. $\mathbf{u} = \mathbf{i} + 2\mathbf{j}$, $\mathbf{v} = \mathbf{i} - 3\mathbf{k}$
37. $\mathbf{u} = \mathbf{i} + 9\mathbf{k} - \mathbf{k}$, $\mathbf{v} = 4\mathbf{i} + \mathbf{j} + \mathbf{k}$
38. $\mathbf{u} = 7\mathbf{i} - \mathbf{j} + \mathbf{k}$, $\mathbf{v} = \mathbf{i} - \mathbf{j} + 6\mathbf{k}$
39. $\mathbf{u} = \mathbf{i} + \mathbf{j} - \mathbf{k}$, $\mathbf{v} = \mathbf{i} + \mathbf{j} + \mathbf{k}$
40. $\mathbf{u} = \mathbf{i} - 2\mathbf{j} + 2\mathbf{k}$, $\mathbf{v} = 2\mathbf{i} - \mathbf{j} - 2\mathbf{k}$

Finding the Area of a Parallelogram **In Exercises 41–46, find the area of the parallelogram that has u and v as adjacent sides.**

41. $\mathbf{u} = \mathbf{k}$, $\mathbf{v} = \mathbf{i} + \mathbf{k}$
42. $\mathbf{u} = 3\mathbf{j}$, $\mathbf{v} = 2\mathbf{i} + \mathbf{k}$
43. $\mathbf{u} = 3\mathbf{i} + 4\mathbf{j} + 6\mathbf{k}$, $\mathbf{v} = 2\mathbf{i} - \mathbf{j} + 5\mathbf{k}$
44. $\mathbf{u} = -2\mathbf{i} + 3\mathbf{j} + 2\mathbf{k}$, $\mathbf{v} = \mathbf{i} + 2\mathbf{j} + 4\mathbf{k}$
45. $\mathbf{u} = \langle 4, 4, -6 \rangle$, $\mathbf{v} = \langle 0, 4, 6 \rangle$
46. $\mathbf{u} = \langle 4, -3, 2 \rangle$, $\mathbf{v} = \langle 5, 0, 1 \rangle$

Geometric Application of the Cross Product **In Exercises 47 and 48, (a) verify that the points are the vertices of a parallelogram and (b) find its area.**

47. $A(2, -1, 4)$, $B(3, 1, 2)$, $C(0, 5, 6)$, $D(-1, 3, 8)$
48. $A(1, 1, 1)$, $B(6, 5, 2)$, $C(7, 7, 5)$, $D(2, 3, 4)$

Finding the Area of a Triangle **In Exercises 49–52, find the area of the triangle with the given vertices.**

49. $(0, 0, 0)$, $(1, 2, 3)$, $(-3, 0, 0)$
50. $(2, 4, 0)$, $(-2, -4, 0)$, $(0, 0, 4)$
51. $(2, 3, -5)$, $(-2, -2, 0)$, $(3, 0, 6)$
52. $(1, -4, 3)$, $(2, 0, 2)$, $(-2, 2, 0)$

Finding the Triple Scalar Product **In Exercises 53–56, find the triple scalar product.**

53. $\mathbf{u} = \langle 2, 3, 3\rangle$, $\mathbf{v} = \langle 4, 4, 0\rangle$, $\mathbf{w} = \langle 0, 0, 4\rangle$

54. $\mathbf{u} = \langle 4, 8, 3\rangle$, $\mathbf{v} = \langle 2, 3, 0\rangle$, $\mathbf{w} = \langle 0, 0, 1\rangle$

55. $\mathbf{u} = 2\mathbf{i} + 3\mathbf{j} + \mathbf{k}$, $\mathbf{v} = \mathbf{i} - \mathbf{j}$, $\mathbf{w} = 4\mathbf{i} + 3\mathbf{j} + \mathbf{k}$

56. $\mathbf{u} = \mathbf{i} + 4\mathbf{j} - 7\mathbf{k}$, $\mathbf{v} = 2\mathbf{i} + 4\mathbf{k}$, $\mathbf{w} = -3\mathbf{j} + 6\mathbf{k}$

Using the Triple Scalar Product to find Volume **In Exercises 57–62, use the triple scalar product to find the volume of the parallelepiped having adjacent edges u, v, and w.**

57.

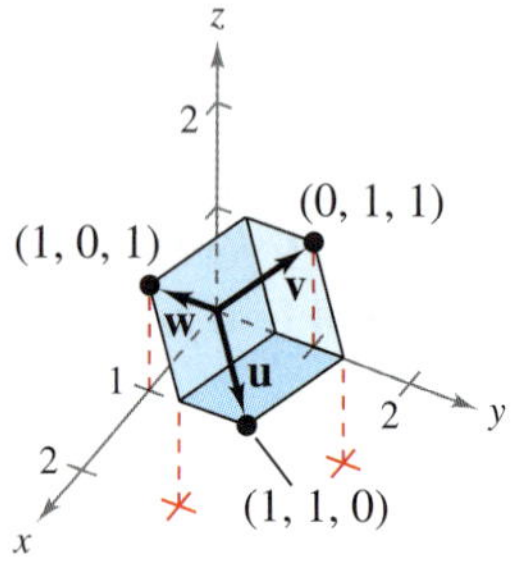

58.

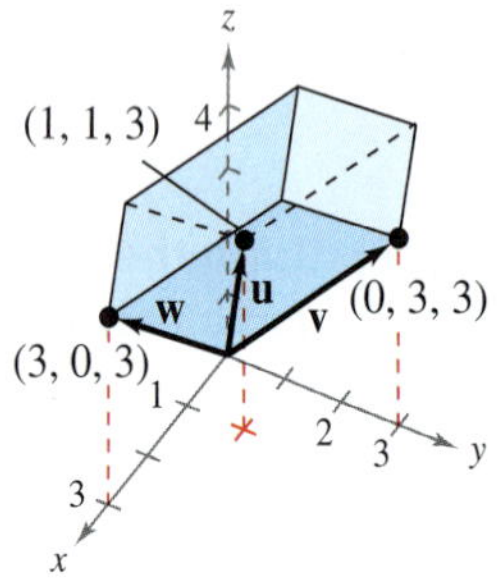

59.

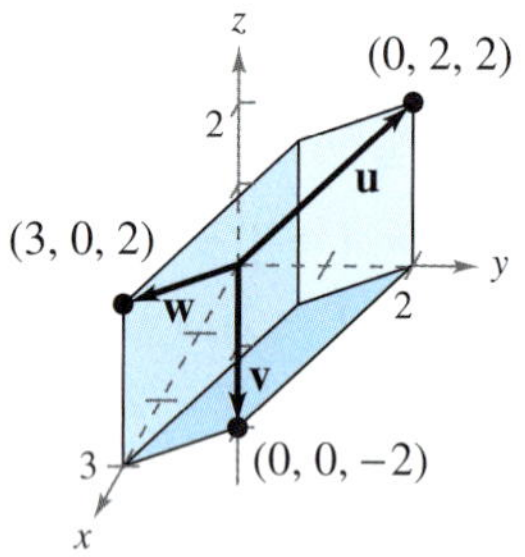

60.

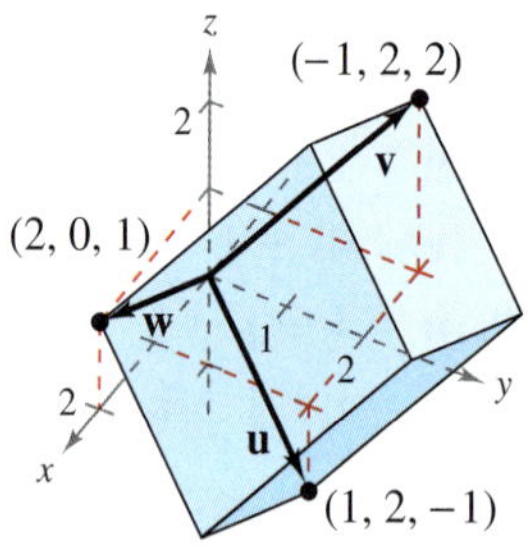

61. $\mathbf{u} = \langle 4, 0, 0\rangle$, $\mathbf{v} = \langle 0, -2, 3\rangle$, $\mathbf{w} = \langle 0, 5, 3\rangle$

62. $\mathbf{u} = \mathbf{i} + \mathbf{j}$, $\mathbf{v} = \mathbf{i} + 2\mathbf{k}$, $\mathbf{w} = \mathbf{j} + \mathbf{k}$

Finding the Volume of a Parallelepiped **In Exercises 63 and 64, find the volume of the parallelepiped with the given vertices.**

63. $A(0, 0, 0)$, $B(3, 0, 0)$, $C(3, -1, 4)$, $D(0, -1, 4)$, $E(3, 4, 2)$, $F(0, 4, 2)$, $G(0, 3, 6)$, $H(3, 3, 6)$

64. $A(0, 0, 1)$, $B(1, 2, 1)$, $C(1, 0, 3)$, $D(0, 1, 2)$, $E(2, 2, 3)$, $F(1, 1, 4)$, $G(1, 3, 2)$, $H(2, 3, 4)$

65. ***Why you should learn it*** *(p. 736)* The brakes on a bicycle are applied by using a downward force of p pounds on the pedal when the six-inch crank makes a 40° angle with the horizontal (see figure). Vectors representing the position of the crank and the force are $\mathbf{V} = \frac{1}{2}(\cos 40°\,\mathbf{j} + \sin 40°\,\mathbf{k})$ and $\mathbf{F} = -p\mathbf{k}$, respectively.

(a) The magnitude of the torque on the crank is given by $\|\mathbf{V} \times \mathbf{F}\|$. Using the given information, write the torque T on the crank as a function of p.

(b) Use the function from part (a) to complete the table.

p	15	20	25	30	35	40	45
T							

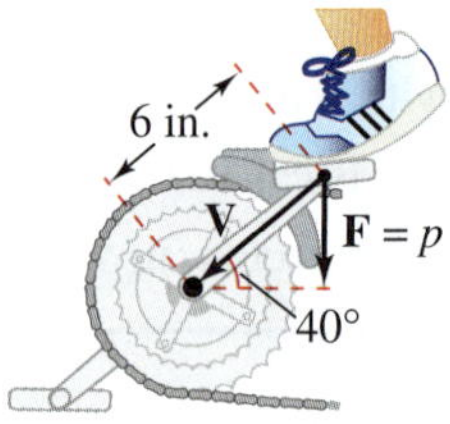

Figure for 65

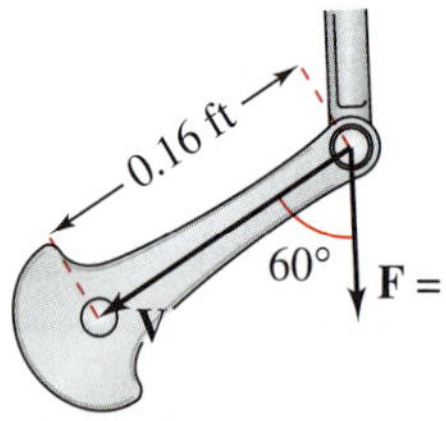

Figure for 66

66. Mechanical Engineering Both the magnitude and direction of the force on a crankshaft rotates. Use the technique given in Exercise 65 to find the magnitude of the torque on the crankshaft shown in the figure.

Focusing on Concepts

True or False? **In Exercises 67 and 68, determine whether the statement is true or false. Justify your answer.**

67. The cross product is not defined for vectors in the plane.

68. If $\mathbf{u}$ and $\mathbf{v}$ are vectors in space that are nonzero and not parallel, then $\mathbf{u} \times \mathbf{v} = \mathbf{v} \times \mathbf{u}$.

69. Proof Prove that $\|\mathbf{u} \times \mathbf{v}\| = \|\mathbf{u}\|\,\|\mathbf{v}\|$ if $\mathbf{u}$ and $\mathbf{v}$ are orthogonal.

70. Think About It When the magnitudes of two vectors are doubled, how will the magnitude of the cross product of the vectors change?

71. Proof Consider the vectors $\mathbf{u} = \langle \cos\alpha, \sin\alpha, 0\rangle$ and $\mathbf{v} = \langle \cos\beta, \sin\beta, 0\rangle$, where $\alpha > \beta$. Find the cross product of the vectors and use the result to prove the identity $\sin(\alpha - \beta) = \sin\alpha\cos\beta - \cos\alpha\sin\beta$.

72. HOW DO YOU SEE IT? The figure shows the two vectors **u** and **v**. Sketch the cross products $\mathbf{u} \times \mathbf{v}$ and $\mathbf{v} \times \mathbf{u}$.

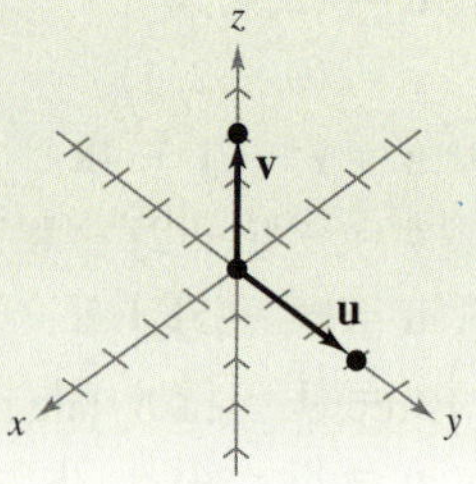

Cumulative Mixed Review

Solving a System **In Exercises 73 and 74, use any method to solve the system.**

73. $\begin{cases} 3x - 7y = 8 \\ 4x + 7y = 6 \end{cases}$

74. $\begin{cases} -2x - 5y = -3 \\ 3x + 4y = 1 \end{cases}$

10.4 Lines and Planes in Space

Lines in Space

In the plane, *slope* is used to determine an equation of a line. In space, it is more convenient to use *vectors* to determine the equation of a line.

In Figure 10.21, consider the line L through the point

$P(x_1, y_1, z_1)$

and parallel to the vector

$\mathbf{v} = \langle a, b, c \rangle.$ Direction vector for L

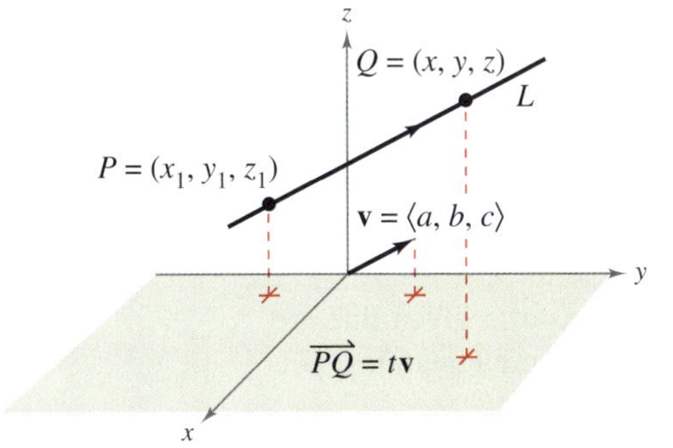

Figure 10.21

The vector $\mathbf{v}$ is the **direction vector** for the line L, and a, b, and c are the **direction numbers.** One way of describing the line L is to say that it consists of all points $Q(x, y, z)$ for which the vector $\overrightarrow{PQ}$ is parallel to $\mathbf{v}$. This means that $\overrightarrow{PQ}$ is a scalar multiple of $\mathbf{v}$, and you can write

$$\overrightarrow{PQ} = t\mathbf{v}$$

where t is a scalar.

$$\begin{aligned}\overrightarrow{PQ} &= \langle x - x_1, y - y_1, z - z_1 \rangle \\ &= \langle at, bt, ct \rangle \\ &= t\mathbf{v}\end{aligned}$$

By equating corresponding components, you can obtain the **parametric equations of a line in space.**

Parametric Equations of a Line in Space

A line L parallel to the nonzero vector

$\mathbf{v} = \langle a, b, c \rangle$

and passing through the point

$P(x_1, y_1, z_1)$

is represented by the parametric equations

$$x = x_1 + at, \quad y = y_1 + bt, \quad \text{and} \quad z = z_1 + ct.$$

When the direction numbers a, b, and c are all nonzero, you can eliminate the parameter t to obtain the **symmetric equations** of a line.

$$\frac{x - x_1}{a} = \frac{y - y_1}{b} = \frac{z - z_1}{c}$$ Symmetric equations

What you should learn

- Find parametric and symmetric equations of lines in space.
- Find equations of planes in space.
- Sketch planes in space.
- Find distances between points and planes in space.

Why you should learn it

Normal vectors to a plane are important in modeling and solving real-life problems. For instance, in Exercise 60 on page 751, normal vectors are used to find the angle between two adjacent sides of a tapered bread pan.

EXAMPLE 1 Finding Parametric and Symmetric Equations

Find parametric and symmetric equations of the line L that passes through the point $(1, -2, 4)$ and is parallel to $\mathbf{v} = \langle 2, 4, -4 \rangle$.

Solution

To find a set of parametric equations of the line, use the coordinates $x_1 = 1$, $y_1 = -2$, and $z_1 = 4$ and direction numbers $a = 2$, $b = 4$, and $c = -4$. (See Figure 10.22.)

$$x = 1 + 2t, \quad y = -2 + 4t, \quad z = 4 - 4t \qquad \text{Parametric equations}$$

Because a, b, and c are all nonzero, a set of symmetric equations is

$$\frac{x - 1}{2} = \frac{y + 2}{4} = \frac{z - 4}{-4}. \qquad \text{Symmetric equations}$$

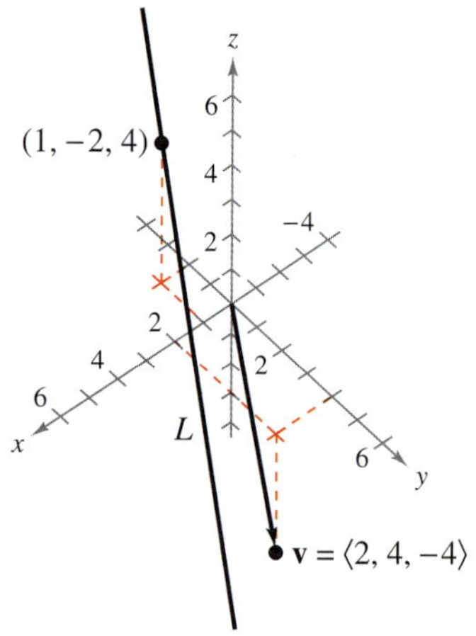

Figure 10.22

✓ ***Checkpoint*** *Audio-video solution in English & Spanish at LarsonPrecalculus.com*

Find parametric and symmetric equations of the line L that passes through the point $(2, 1, -3)$ and is parallel to $\mathbf{v} = \langle 4, -2, 7 \rangle$.

Neither the parametric equations nor the symmetric equations of a given line are unique. For instance, in Example 1, by letting $t = 1$ in the parametric equations, you obtain the point $(3, 2, 0)$. Using this point with the direction numbers

$$a = 2, \quad b = 4, \quad \text{and} \quad c = -4$$

produces the parametric equations

$$x = 3 + 2t, \quad y = 2 + 4t, \quad \text{and} \quad z = -4t.$$

EXAMPLE 2 Finding Equations of a Line Through Two Points

Find parametric and symmetric equations of the line that passes through the points $(-2, 1, 0)$ and $(1, 3, 5)$.

Solution

Begin by using the points $P = (-2, 1, 0)$ and $Q = (1, 3, 5)$. Then a direction vector for the line passing through P and Q is

$$\begin{aligned}\mathbf{v} &= \overrightarrow{PQ}\\ &= \langle 1 - (-2), 3 - 1, 5 - 0 \rangle\\ &= \langle 3, 2, 5 \rangle\\ &= \langle a, b, c \rangle.\end{aligned}$$

Using the direction numbers

$$a = 3, \quad b = 2, \quad \text{and} \quad c = 5$$

with the point $P(-2, 1, 0)$, you can obtain the parametric equations

$$x = -2 + 3t, \quad y = 1 + 2t, \quad \text{and} \quad z = 5t. \qquad \text{Parametric equations}$$

Because a, b, and c are all nonzero, a set of symmetric equations is

$$\frac{x + 2}{3} = \frac{y - 1}{2} = \frac{z}{5}. \qquad \text{Symmetric equations}$$

✓ ***Checkpoint*** *Audio-video solution in English & Spanish at LarsonPrecalculus.com*

Find parametric and symmetric equations of the line that passes through the points $(1, -6, 3)$ and $(3, 5, 8)$.

Planes in Space

You have seen how an equation of a line in space can be obtained from a point on the line and a vector *parallel* to it. You will now see that an equation of a plane in space can be obtained from a point in the plane and a vector *normal* (perpendicular) to the plane.

Consider the plane containing the point $P(x_1, y_1, z_1)$ having a nonzero normal vector $\mathbf{n} = \langle a, b, c \rangle$, as shown in Figure 10.23.

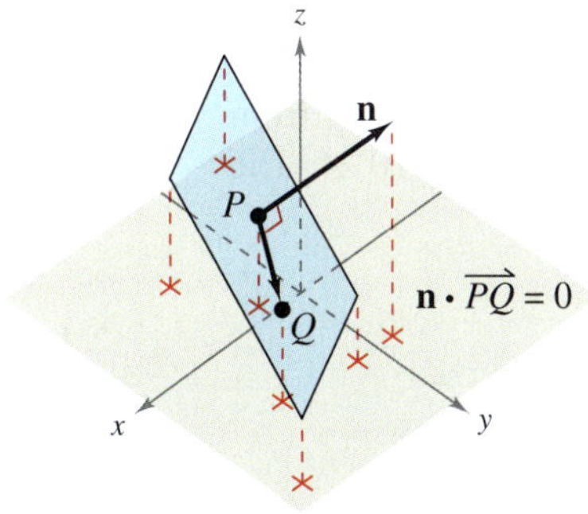

Figure 10.23

This plane consists of all points $Q(x, y, z)$ for which the vector $\overrightarrow{PQ}$ is orthogonal to $\mathbf{n}$. Using the dot product, you can write the following.

$$\mathbf{n} \cdot \overrightarrow{PQ} = 0$$

$$\langle a, b, c \rangle \cdot \langle x - x_1, y - y_1, z - z_1 \rangle = 0$$

$$a(x - x_1) + b(y - y_1) + c(z - z_1) = 0$$

The third equation of the plane is said to be in **standard form.**

Standard Equation of a Plane in Space

The plane containing the point (x_1, y_1, z_1) and having nonzero normal vector $\mathbf{n} = \langle a, b, c \rangle$ can be represented by the **standard form of the equation of a plane**

$$a(x - x_1) + b(y - y_1) + c(z - z_1) = 0.$$ Standard form of equation of plane

By regrouping terms in the standard form of the equation of a plane, you obtain the **general form**

$$ax + by + cz + d = 0.$$ General form of equation of plane

Given the general form of the equation of a plane, it is easy to find a normal vector to the plane. Use the coefficients of x, y, and z to write $\mathbf{n} = \langle a, b, c \rangle$.

Explore the Concept

Consider the four planes listed below.

$$2x + 3y - z = 2$$

$$-2x - 3y + z = -2$$

$$4x + 6y - 2z = 5$$

$$-6x - 9y + 3z = 11$$

What are the normal vectors to each plane? What can you say about the relative positions of these planes in space?

EXAMPLE 3 Finding an Equation of a Plane in Three-Space

See LarsonPrecalculus.com for an interactive version of this type of example.

Find the general equation of the plane passing through the points $(2, 1, 1)$, $(0, 4, 1)$, and $(-2, 1, 4)$.

Solution

To find the equation of the plane, you need a point in the plane and a vector that is normal to the plane. There are three choices for the point, but no normal vector is given. To obtain a normal vector, use the cross product of vectors **u** and **v** extending from the point $(2, 1, 1)$ to the points $(0, 4, 1)$ and $(-2, 1, 4)$, as shown in Figure 10.24. The component forms of **u** and **v** are

$$\mathbf{u} = \langle 0 - 2, 4 - 1, 1 - 1 \rangle = \langle -2, 3, 0 \rangle$$

and

$$\mathbf{v} = \langle -2 - 2, 1 - 1, 4 - 1 \rangle = \langle -4, 0, 3 \rangle.$$

So, a vector normal to the given plane is

$$\mathbf{n} = \mathbf{u} \times \mathbf{v}$$

$$= \begin{vmatrix} \mathbf{i} & \mathbf{j} & \mathbf{k} \\ -2 & 3 & 0 \\ -4 & 0 & 3 \end{vmatrix}$$

$$= 9\mathbf{i} + 6\mathbf{j} + 12\mathbf{k}.$$

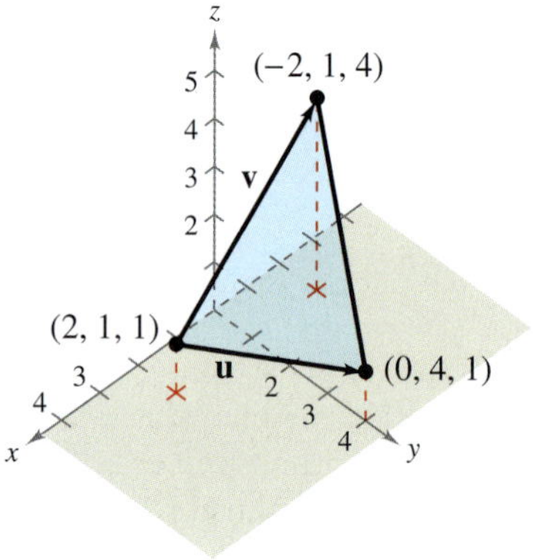

Figure 10.24

So, the direction numbers for **n** are $a = 9$, $b = 6$, and $c = 12$. Using these numbers and the point $(x_1, y_1, z_1) = (2, 1, 1)$, you can determine an equation of the plane to be

$$a(x - x_1) + b(y - y_1) + c(z - z_1) = 0$$

$$9(x - 2) + 6(y - 1) + 12(z - 1) = 0 \quad \text{Standard form}$$

$$9x + 6y + 12z - 36 = 0$$

$$3x + 2y + 4z - 12 = 0. \quad \text{General form}$$

Check that each of the three points satisfies the equation $3x + 2y + 4z - 12 = 0$.

Checkpoint *Audio-video solution in English & Spanish at LarsonPrecalculus.com*

Find the general form of the equation of the plane passing through the points $(1, 0, 5)$, $(-1, -2, 9)$, and $(2, 3, 2)$.

Two distinct planes in three-space either are parallel or intersect in a line. When they intersect, you can determine the angle θ $(0 \le \theta \le 90°)$ between them from the angle between their normal vectors, as shown in the figure. Specifically, if vectors $\mathbf{n}_1$ and $\mathbf{n}_2$ are normal to two intersecting planes, then the angle θ between the normal vectors is equal to the **angle between the two planes** and is given by

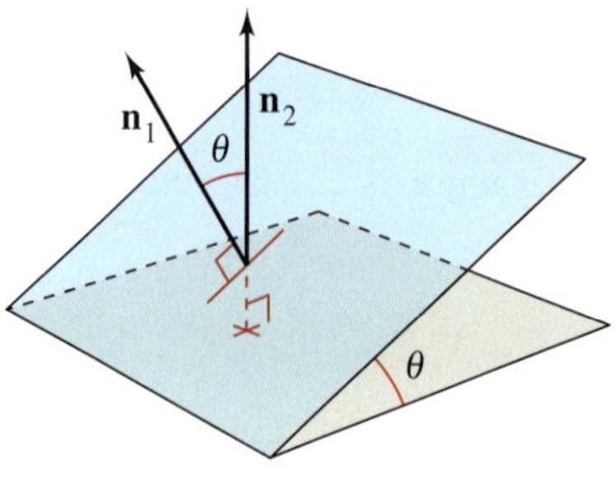

$$\cos\theta = \frac{|\mathbf{n}_1 \cdot \mathbf{n}_2|}{\|\mathbf{n}_1\|\|\mathbf{n}_2\|}. \quad \text{Angle between two planes}$$

Consequently, two planes with normal vectors $\mathbf{n}_1$ and $\mathbf{n}_2$ are

1. *perpendicular* when $\mathbf{n}_1 \cdot \mathbf{n}_2 = 0$.
2. *parallel* when $\mathbf{n}_1$ is a scalar multiple of $\mathbf{n}_2$.

EXAMPLE 4 Finding the Line of Intersection of Two Planes

Find the angle between the two planes given by

$x - 2y + z = 0$ Equation for plane 1

$2x + 3y - 2z = 0$ Equation for plane 2

and find parametric equations of their line of intersection.

Solution

The normal vectors to the planes are $\mathbf{n}_1 = \langle 1, -2, 1 \rangle$ and $\mathbf{n}_2 = \langle 2, 3, -2 \rangle$. Consequently, the angle between the two planes is determined as follows.

$$\cos\theta = \frac{|\mathbf{n}_1 \cdot \mathbf{n}_2|}{\|\mathbf{n}_1\|\|\mathbf{n}_2\|} = \frac{|-6|}{\sqrt{6}\sqrt{17}} = \frac{6}{\sqrt{102}} \approx 0.59409$$

This implies that the angle between the two planes is $\theta \approx 53.55°$, as shown in Figure 10.25. You can find the line of intersection of the two planes by simultaneously solving the two linear equations representing the planes. One way to do this is to multiply the first equation by -2 and add the result to the second equation.

$$\begin{aligned} x - 2y + z &= 0 \quad \Rightarrow \quad -2x + 4y - 2z = 0 \\ 2x + 3y - 2z &= 0 \quad \Rightarrow \quad \underline{2x + 3y - 2z = 0} \end{aligned}$$

$$7y - 4z = 0 \quad \Rightarrow \quad y = \frac{4z}{7}$$

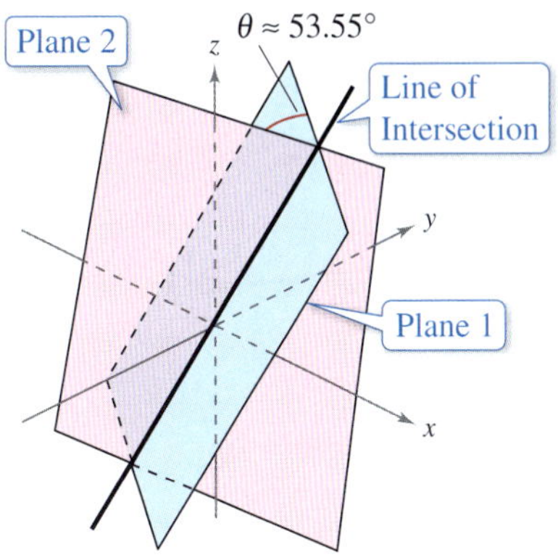

Figure 10.25

Substituting $y = 4z/7$ back into one of the original equations, you can determine that $x = z/7$. Finally, by letting $t = z/7$, you obtain the parametric equations

$x = t = x_1 + at$ Parametric equation for x

$y = 4t = y_1 + bt$ Parametric equation for y

$z = 7t = z_1 + ct.$ Parametric equation for z

Because $(x_1, y_1, z_1) = (0, 0, 0)$ lies in both planes, you can substitute for x_1, y_1, and z_1 in these parametric equations, which indicates that $a = 1$, $b = 4$, and $c = 7$ are direction numbers for the line of intersection.

✓ Checkpoint *Audio-video solution in English & Spanish at LarsonPrecalculus.com*

Find the angle between the two planes given by

$2x + y + 3z = 0$ Equation for plane 1

$4x - 2y - 2z = 0$ Equation for plane 2

and find parametric equations of their line of intersection.

Note that the direction numbers in Example 4 can be obtained from the cross product of the two normal vectors as follows.

$$\begin{aligned} \mathbf{n}_1 \times \mathbf{n}_2 &= \begin{vmatrix} \mathbf{i} & \mathbf{j} & \mathbf{k} \\ 1 & -2 & 1 \\ 2 & 3 & -2 \end{vmatrix} \\ &= \begin{vmatrix} -2 & 1 \\ 3 & -2 \end{vmatrix}\mathbf{i} - \begin{vmatrix} 1 & 1 \\ 2 & -2 \end{vmatrix}\mathbf{j} + \begin{vmatrix} 1 & -2 \\ 2 & 3 \end{vmatrix}\mathbf{k} \\ &= \mathbf{i} + 4\mathbf{j} + 7\mathbf{k} \end{aligned}$$

This means that the *line of intersection of the two planes is parallel to the cross product of their normal vectors.*

Sketching Planes in Space

As discussed in Section 10.1, when a plane in space intersects one of the coordinate planes, the line of intersection is called the *trace* of the given plane in the coordinate plane. To sketch a plane in space, it is helpful to find its points of intersection with the coordinate axes and its traces in the coordinate planes. For example, consider the plane

$$3x + 2y + 4z = 12. \qquad \text{Equation of plane}$$

You can find the xy-trace by letting $z = 0$ and sketching the line

$$3x + 2y = 12 \qquad xy\text{-trace}$$

in the xy-plane. This line intersects the x-axis at $(4, 0, 0)$ and the y-axis at $(0, 6, 0)$. In Figure 10.26, this process is continued by finding the yz-trace and the xz-trace and then shading the triangular region lying in the first octant.

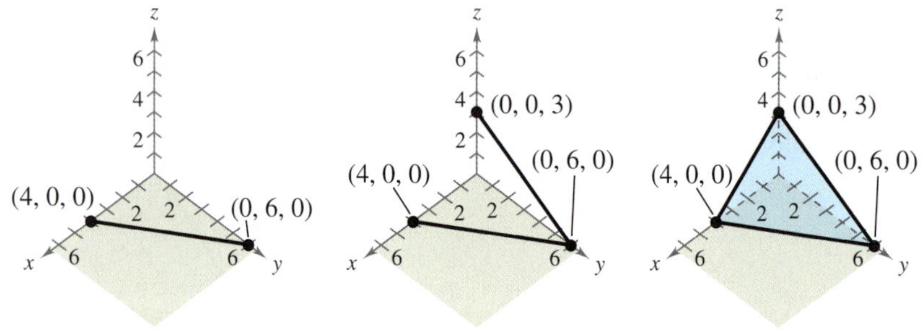

(a) ***xy*-trace** $(z = 0)$: $3x + 2y = 12$ (b) ***yz*-trace** $(x = 0)$: $2y + 4z = 12$ (c) ***xz*-trace** $(y = 0)$: $3x + 4z = 12$

Figure 10.26

Technology Tip

Most three-dimensional graphing utilities and computer algebra systems can graph a plane in space. Consult the user's guide for your utility for specific instructions.

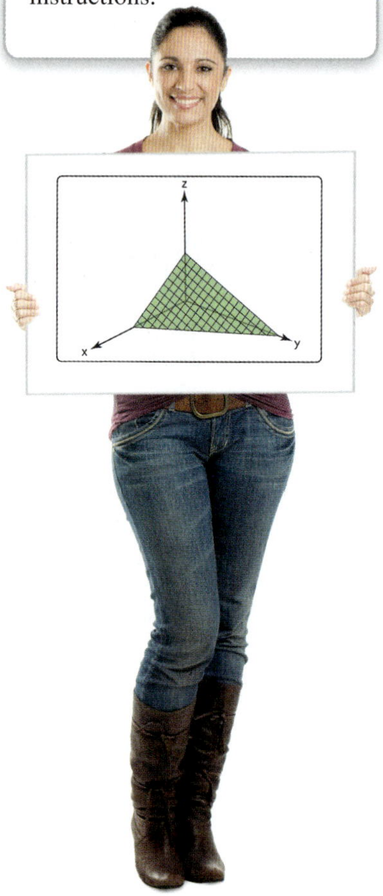

If the equation of a plane has a missing variable, such as

$$2x + z = 1$$

then the plane must be *parallel to the axis* represented by the missing variable, as shown in Figure 10.27.

If two variables are missing from the equation of a plane, then it is *parallel to the coordinate plane* represented by the missing variables, as shown in Figure 10.28. For instance, the graph of

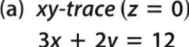

$$z + 2 = 0$$

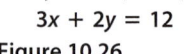

is parallel to the xy-plane. Try confirming this using a three-dimensional graphing utility.

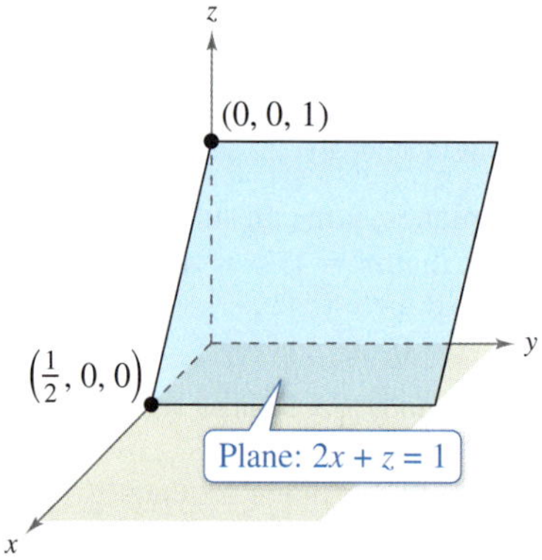

Plane is parallel to y-axis

Figure 10.27

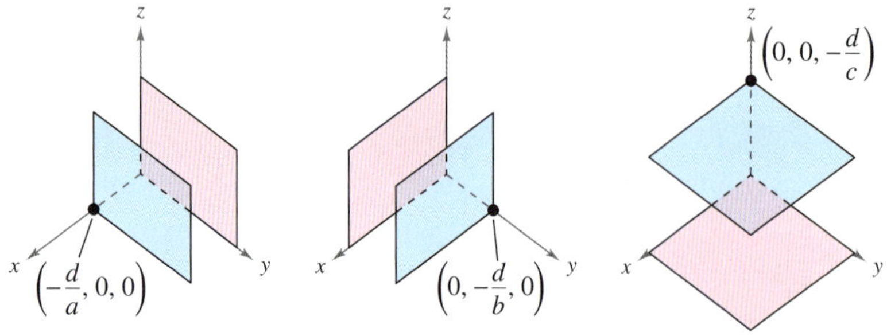

(a) ***Plane*** $ax + d = 0$ ***is parallel to yz-plane.*** (b) ***Plane*** $by + d = 0$ ***is parallel to xz-plane.*** (c) ***Plane*** $cz + d = 0$ ***is parallel to xy-plane.***

Figure 10.28

Distance Between a Point and a Plane

The distance D between a point Q and a plane is the length of the shortest line segment connecting Q to the plane, as shown in Figure 10.29. When P is *any* point in the plane, you can find this distance by projecting the vector

$\overrightarrow{PQ}$

onto the normal vector $\mathbf{n}$. The length of this projection is the desired distance.

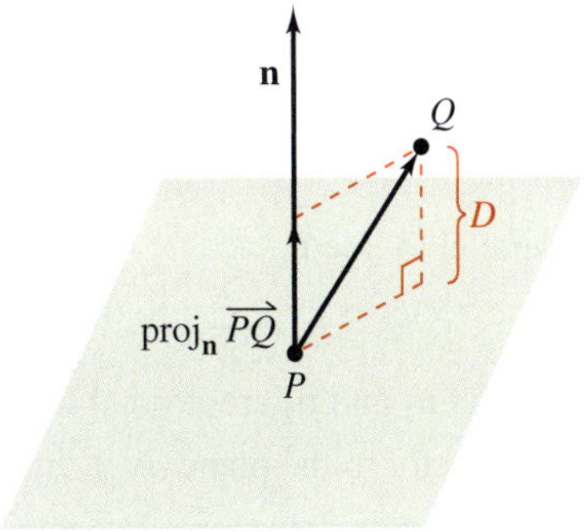

$D = \|\text{proj}_\mathbf{n}\overrightarrow{PQ}\|$
Figure 10.29

Distance Between a Point and a Plane

The **distance between a plane and a point** Q (not in the plane) is

$$D = \|\text{proj}_\mathbf{n}\overrightarrow{PQ}\| = \frac{|\overrightarrow{PQ} \cdot \mathbf{n}|}{\|\mathbf{n}\|}$$

where P is a point in the plane and $\mathbf{n}$ is normal to the plane.

To find a point in the plane given by $ax + by + cz + d = 0$, where $a \neq 0$, let $y = 0$ and $z = 0$. Then, from the equation $ax + d = 0$, you can conclude that the point $(-d/a, 0, 0)$ lies in the plane.

EXAMPLE 5 Finding the Distance Between a Point and a Plane

Find the distance between the point $Q(1, 5, -4)$ and the plane $3x - y + 2z = 6$.

Solution

You know that $\mathbf{n} = \langle 3, -1, 2\rangle$ is normal to the given plane. To find a point in the plane, let $y = 0$ and $z = 0$, and obtain the point $P(2, 0, 0)$. The vector from P to Q is

$$\overrightarrow{PQ} = \langle 1 - 2, 5 - 0, -4 - 0\rangle = \langle -1, 5, -4\rangle.$$

The formula for the distance between a point and a plane produces

$$\begin{aligned} D &= \frac{|\overrightarrow{PQ} \cdot \mathbf{n}|}{\|\mathbf{n}\|} \\ &= \frac{|\langle -1, 5, -4\rangle \cdot \langle 3, -1, 2\rangle|}{\sqrt{9 + 1 + 4}} \\ &= \frac{|-3 - 5 - 8|}{\sqrt{14}} \\ &= \frac{16}{\sqrt{14}}. \end{aligned}$$

✓ Checkpoint *Audio-video solution in English & Spanish at LarsonPrecalculus.com*

Find the distance between the point $Q(-6, 2, 1)$ and the plane $2x + 5y - z = 3$.

The choice of the point P in Example 5 is arbitrary. Try choosing a different point to verify that you obtain the same distance.

10.4 Exercises

See CalcChat.com for tutorial help and worked-out solutions to odd-numbered exercises.
For instructions on how to use a graphing utility, see Appendix A.

Vocabulary and Concept Check

In Exercises 1 and 2, fill in the blank.

1. The _______ vector for a line L is parallel to line L.
2. When a vector is normal to a plane, the vector is _______ to the plane.
3. What is the value of $\mathbf{n}_1 \cdot \mathbf{n}_2$ when vectors $\mathbf{n}_1$ and $\mathbf{n}_2$ are normal to two perpendicular planes?
4. Write the component form of a normal vector to the plane $ax + by + cz + d = 0$.

Procedures and Problem Solving

Finding Parametric and Symmetric Equations **In Exercises 5–10, find (a) a set of parametric equations and (b) if possible, a set of symmetric equations for the line through the point and parallel to the specified vector or line. (There are many correct answers.)**

	Point	*Parallel to*
5.	$(0, 0, 0)$	$\mathbf{v} = \langle 1, 3, 5\rangle$
6.	$(3, -5, 1)$	$\mathbf{v} = \langle 3, -7, -10\rangle$
7.	$(-4, 1, 0)$	$\mathbf{v} = 6\mathbf{j} - 5\mathbf{k}$
8.	$(5, 0, 10)$	$\mathbf{v} = 4\mathbf{i} + 3\mathbf{k}$
9.	$(2, -3, 5)$	$x = 5 + 2t, y = 7 - 3t, z = -2 + t$
10.	$(7, -8, 2)$	$x = 3 + 3t, y = 5 - 2t, z = -7 + t$

A Line Through Two Points **In Exercises 11–18, find (a) a set of parametric equations and (b) if possible, a set of symmetric equations of the line that passes through the two points. (There are many correct answers.)**

11. $(2, 0, 2), (1, 4, -3)$
12. $(2, 5, 0), (4, -3, 6)$
13. $(-3, 8, 15), (1, -2, 16)$
14. $(2, 3, -1), (1, -5, 3)$
15. $(3, 1, 2), (-1, 1, 5)$
16. $(4, 2, -3), (-5, 2, 1)$
17. $\left(-\frac{1}{2}, 2, \frac{1}{2}\right), \left(1, -\frac{1}{2}, 0\right)$
18. $\left(-\frac{3}{2}, \frac{3}{2}, 2\right), \left(0, -5, \frac{3}{2}\right)$

Using Parametric Equations **In Exercises 19 and 20, sketch a graph of the line.**

19. $x = 3 + 2t, y = 2 + t, z = 1 + \frac{1}{2}t$
20. $x = 5 - 2t, y = 1 + t, z = 5 - \frac{1}{2}t$

Finding an Equation of a Plane in Three-Space **In Exercises 21–26, find the general form of the equation of the plane passing through the point and perpendicular to the specified vector or line.**

	Point	*Perpendicular to*
21.	$(2, 1, 2)$	$\mathbf{n} = \mathbf{i}$
22.	$(3, -2, 1)$	$\mathbf{n} = \mathbf{j}$
23.	$(5, 6, 3)$	$\mathbf{n} = -2\mathbf{i} + \mathbf{j} - 2\mathbf{k}$
24.	$(0, 0, 0)$	$\mathbf{n} = 9\mathbf{i} - 3\mathbf{j} + 5\mathbf{k}$
25.	$(2, 1, 3)$	$x = 1 - 2t, y = 3 + t, z = 5 + 2t$
26.	$(0, 0, 6)$	$x = 1 - t, y = 2 + t, z = 4 - 2t$

Finding an Equation of a Plane in Three-Space **In Exercises 27–30, find the general form of the equation of the plane passing through the three points.**

27. $(3, 1, 2), (2, 3, 4), (-3, 4, 2)$
28. $(4, -1, 3), (2, 5, 1), (-1, 2, 1)$
29. $(2, 3, -2), (3, 4, 2), (1, -1, 0)$
30. $(5, -1, 4), (1, -1, 2), (2, 1, -3)$

Finding an Equation of a Plane in Three-Space **In Exercises 31–36, find the general form of the equation of the plane with the given characteristics.**

31. Passes through $(2, 5, 3)$ and is parallel to the xz-plane
32. Passes through $(1, 2, 3)$ and is parallel to the yz-plane
33. Passes through $(0, 2, 4)$ and $(-1, -2, 0)$ and is perpendicular to the yz-plane
34. Passes through $(1, -2, 4)$ and $(4, 0, -1)$ and is perpendicular to the xz-plane
35. Passes through $(2, 2, 1)$ and $(-1, 1, -1)$ and is perpendicular to $2x - 3y + z = 3$
36. Passes through $(1, 2, 0)$ and $(-1, -1, 2)$ and is perpendicular to $2x - 3y + z = 6$

Finding Parametric Equations of a Line **In Exercises 37–42, find a set of parametric equations of the line. (There are many correct answers.)**

37. Passes through $(5, 1, 8)$ and is perpendicular to the plane given by $-x + 3y - 4z = 5$
38. Passes through $(-4, 5, 2)$ and is perpendicular to the plane given by $-x + 2y + z = 5$

39. Passes through $(6, -4, -1)$ and is parallel to $\mathbf{v} = 3\mathbf{i} - 4\mathbf{j}$

40. Passes through $(-1, 4, -3)$ and is parallel to $\mathbf{v} = 5\mathbf{i} - \mathbf{j}$

41. Passes through $(-2, 7, 4)$ and is parallel to $x = -t$, $y = 3 + 4t$, $z = -2 + 5t$

42. Passes through $(6, -3, 8)$ and is parallel to $x = 5 - 2t$, $y = -4 + 9t$, $z = 12t$

Parallel or Perpendicular Planes **In Exercises 43–46, determine whether the planes are parallel, perpendicular, or neither. If they are neither parallel nor perpendicular, find the angle of intersection.**

43. $5x - 3y + z = 4$, $x + 4y + 7z = 1$

44. $2x - 4y - z = 6$, $-8x + 16y + 4z = 5$

45. $x - 5y - z = 1$, $5x - 25y - 5z = -3$

46. $-6x - y + 4z = 1$, $3x - 6y + 3z = -8$

Finding the Line of Intersection of Two Planes **In Exercises 47–50, (a) find the angle between the two planes and (b) find parametric equations of their line of intersection.**

47. $x + y - 2z = 0$, $2x - y + 3z = 0$

48. $x - 3y + 2z = 0$, $3x + 2y - 5z = 0$

49. $x + y - z = 0$, $2x - 5y - z = 1$

50. $3x - 4y + 5z = 6$, $x + y - z = 2$

Sketching a Plane in Space **In Exercises 51–54, plot the intercepts and sketch a graph of the plane.**

51. $x + 2y + 3z = 6$

52. $2x - y + 4z = 4$

53. $x + 2y = 4$

54. $y + z = 5$

Finding the Distance Between a Point and a Plane **In Exercises 55–58, find the distance between the point and the plane.**

55. $(1, 3, 4)$, $4x - 5y + 2z = 6$

56. $(1, 4, 2)$, $x - 2y + z = 3$

57. $(4, -2, -2)$, $2x - y + z = 4$

58. $(3, -4, 7)$, $4x - 3y - 7z = 14$

59. **Mechanical Engineering** A company that produces after-market machine parts is designing a reproduction of a fuel tank with the shape and dimensions shown in the figure. Find the angle between two adjacent sides of the tank.

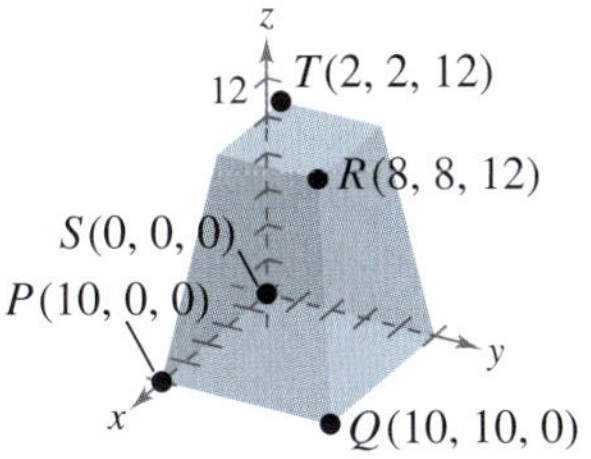

60. ***Why you should learn it*** *(p. 743)* A bread pan is tapered so that a loaf of bread can be easily removed. The figure shows the shape and dimensions of the pan. Find the angle between two adjacent sides of the pan.

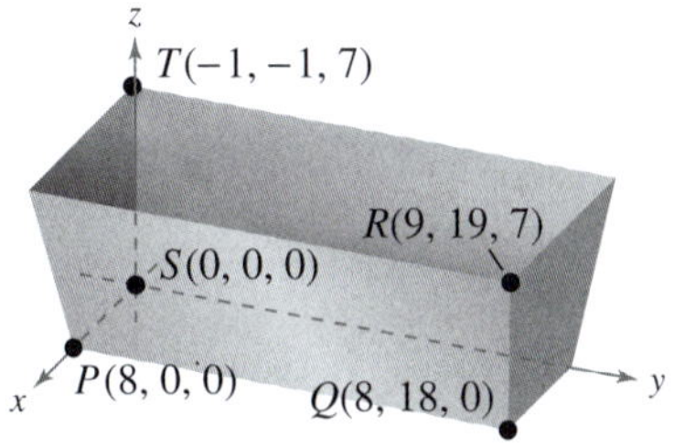

Focusing on Concepts

True or False? **In Exercises 61 and 62, determine whether the statement is true or false. Justify your answer.**

61. Two lines in space are either parallel or they intersect.

62. Two nonparallel planes in space will always intersect.

63. **Error Analysis** Find the error in writing symmetric equations of the line that passes through $(1, 2, 6)$ and is parallel to $\mathbf{v} = \langle 3, 5, 4 \rangle$.

$$\frac{x-3}{1} = \frac{y-5}{2} = \frac{z-4}{6}$$

64. **HOW DO YOU SEE IT?** The figure shows two planes and their normal vectors.

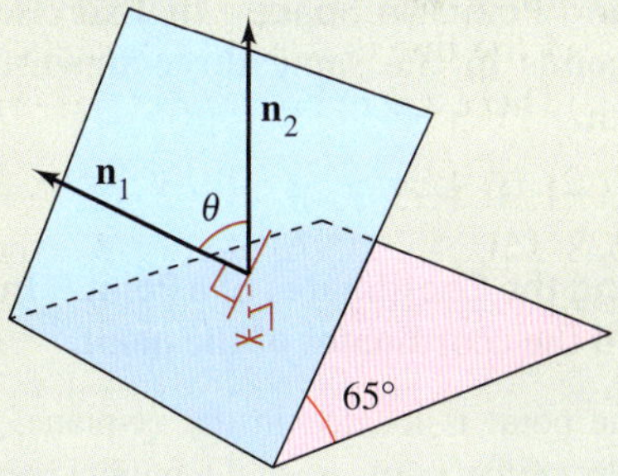

(a) What is the angle between the normal vectors? Explain your reasoning.

(b) A third plane is parallel to $\mathbf{n}_1$. What is the angle between this plane and the blue plane? Explain your reasoning.

(c) A fourth plane is parallel to $\mathbf{n}_2$. Explain how to find the angle between this plane and the blue plane.

65. **Think About It** The direction numbers of two distinct lines in space are $10, -18, 20$, and $-15, 27, -30$. What is the relationship between the lines? Explain.

10 Chapter Review

See CalcChat.com for tutorial help and worked-out solutions to odd-numbered exercises. For instructions on how to use a graphing utility, see Appendix A.

10.1 *What did you learn?*

Plot points in the three-dimensional coordinate system *(p. 722)*. The axes in a three-dimensional coordinate system determine the *xy*-plane, the *xz*-plane, and the *yz*-plane. These three coordinate planes separate the three-dimensional coordinate system into eight octants. A point P in space is determined by an ordered triple (x, y, z).

Find distances between points in space and find midpoints of line segments joining points in space *(p. 723)*. The distance between the points (x_1, y_1, z_1) and (x_2, y_2, z_2) given by the Distance Formula in Space is

$$d = \sqrt{(x_2 - x_1)^2 + (y_2 - y_1)^2 + (z_2 - z_1)^2}.$$

The midpoint of the line segment joining the points (x_1, y_1, z_1) and (x_2, y_2, z_2) given by the Midpoint Formula in Space is

$$\left(\frac{x_1 + x_2}{2}, \frac{y_1 + y_2}{2}, \frac{z_1 + z_2}{2}\right).$$

Write equations of spheres in standard form and find traces of surfaces in space *(p. 724)*. The standard equation of a sphere with center (h, k, j) and radius r is

$$(x - h)^2 + (y - k)^2 + (z - j)^2 = r^2.$$

The intersection of a surface in space with a coordinate plane or a plane parallel to a coordinate plane is called a trace of the surface.

Example

(a) The points $(2, -4, 3)$, $(-2, 4, 2)$, $(1, 5, 0)$, and $(2, 3, -2)$ are shown plotted in space.

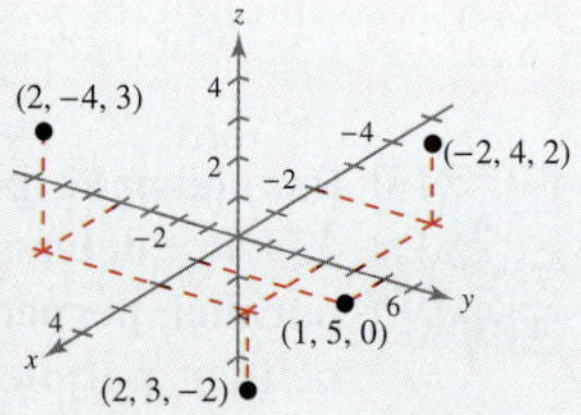

(b) The distance between $(1, 5, 0)$ and $(2, 3, -2)$ is

$$d = \sqrt{(2 - 1)^2 + (3 - 5)^2 + (-2 - 0)^2} = \sqrt{9} = 3.$$

(c) The midpoint of the line segment joining $(1, 5, 0)$ and $(2, 3, -2)$ is

$$\left(\frac{1 + 2}{2}, \frac{5 + 3}{2}, \frac{0 + (-2)}{2}\right) = \left(\frac{3}{2}, 4, -1\right).$$

(d) To find the center and radius of the sphere

$$x^2 + y^2 + z^2 + 2x - 8y + 4z + 16 = 0$$

write its standard equation. So, you have

$$(x + 1)^2 + (y - 4)^2 + (z + 2)^2 = (\sqrt{5})^2.$$

The center is $(-1, 4, -2)$ and the radius is $\sqrt{5}$.

(e) To determine the *yz*-trace of the sphere in part (d), substitute $x = 0$ into the equation and simplify, yielding

$$(y - 4)^2 + (z + 2)^2 = 2^2.$$

So, the *yz*-trace is a circle of radius 2 in the *yz*-plane.

Plotting Points in Space **In Exercises 1 and 2, plot the points in the same three-dimensional coordinate system.**

1. $(5, -1, 3), (-3, 3, 0)$ **2.** $(2, 4, -1), (0, -1, 2)$

Finding the Coordinates of a Point **In Exercises 3 and 4, find the coordinates of the point.**

3. The point is located in the *xy*-plane, nine units to the right of the *xz*-plane, and ten units behind the *yz*-plane.

4. The point is located on the *y*-axis and six units to the left of the *xz*-plane.

Finding the Distance Between Two Points in Space **In Exercises 5 and 6, find the distance between the points.**

5. $(4, 0, -6), (5, -2, 1)$ **6.** $(2, 3, -4), (-1, -2, 0)$

Using the Pythagorean Theorem **In Exercises 7 and 8, find the lengths of the sides of the right triangle with the given vertices. Show that these lengths satisfy the Pythagorean Theorem.**

7. $(3, -2, 0), (0, 3, 2), (0, 5, -3)$

8. $(0, 0, 4), (4, 5, 5), (4, 3, 2)$

Finding the Midpoint of a Line Segment **In Exercises 9 and 10, find the midpoint of the line segment joining the points.**

9. $(-2, 3, 2), (2, -5, 4)$

10. $(-5, -3, 1), (-8, 10, -6)$

Finding the Equation of a Sphere **In Exercises 11 and 12, find the standard equation of the sphere with the given characteristics.**

11. Center: $(3, -2, 4)$; radius: 4

12. Endpoints of a diameter: $(-3, 5, -2), (-1, 7, 4)$

Finding the Center and Radius of a Sphere **In Exercises 13 and 14, find the center and radius of the sphere.**

13. $x^2 + y^2 + z^2 - 4x - 6y + 2z + 6 = 0$

14. $x^2 + y^2 + z^2 - 10x + 6y - 4z + 34 = 0$

Finding a Trace of a Surface **In Exercises 15 and 16, sketch the graph of the equation and find (a) the *xy*-trace, (b) the *yz*-trace, and (c) the *xz*-trace.**

15. $x^2 + (y - 3)^2 + z^2 = 16$

16. $(x + 2)^2 + (y - 1)^2 + z^2 = 9$

10.2 *What did you learn?*

Find the component forms of, the unit vectors in the same direction of, the magnitudes of, the dot products of, and the angles between vectors in space *(p. 729)*.

Component form: $\mathbf{v} = \langle v_1, v_2, v_3 \rangle$

Unit vector form: $\mathbf{v} = v_1\mathbf{i} + v_2\mathbf{j} + v_3\mathbf{k}$, where $\mathbf{i} = \langle 1, 0, 0 \rangle$, $\mathbf{j} = \langle 0, 1, 0 \rangle$, and $\mathbf{k} = \langle 0, 0, 1 \rangle$ are the standard unit vectors

Zero vector: $\mathbf{0} = \langle 0, 0, 0 \rangle$

1. Two vectors are equal if and only if their corresponding components are equal.
2. The magnitude (or length) of $\mathbf{u} = \langle u_1, u_2, u_3 \rangle$ is $\|\mathbf{u}\| = \sqrt{u_1^2 + u_2^2 + u_3^2}$.
3. A unit vector $\mathbf{u}$ in the direction of $\mathbf{v}$ is $\mathbf{u} = \dfrac{\mathbf{v}}{\|\mathbf{v}\|}$, $\mathbf{v} \neq \mathbf{0}$.
4. The sum of $\mathbf{u} = \langle u_1, u_2, u_3 \rangle$ and $\mathbf{v} = \langle v_1, v_2, v_3 \rangle$ is $\mathbf{u} + \mathbf{v} = \langle u_1 + v_1, u_2 + v_2, u_3 + v_3 \rangle$.
5. The scalar multiple of the real number c and $\mathbf{u} = \langle u_1, u_2, u_3 \rangle$ is $c\mathbf{u} = \langle cu_1, cu_2, cu_3 \rangle$.
6. The dot product of $\mathbf{u} = \langle u_1, u_2, u_3 \rangle$ and $\mathbf{v} = \langle v_1, v_2, v_3 \rangle$ is $\mathbf{u} \cdot \mathbf{v} = u_1v_2 + u_2v_2 + u_3v_3$. If the dot product of two nonzero vectors is zero, then the angle between the vectors is 90° and the vectors are orthogonal.

If θ is the angle between two nonzero vectors $\mathbf{u}$ and $\mathbf{v}$, then

$$\cos\theta = \frac{\mathbf{u} \cdot \mathbf{v}}{\|\mathbf{u}\|\|\mathbf{v}\|}.$$

Determine whether vectors in space are parallel *(p. 731)*. Two nonzero vectors $\mathbf{u}$ and $\mathbf{v}$ are parallel when there is some scalar c such that $\mathbf{u} = c\mathbf{v}$.

Use vectors in space to solve real-life problems *(p. 733)*. Vectors can be used to solve equilibrium problems in space (see Example 7).

Example Consider the vector $\mathbf{v}$, whose initial point is $(-4, 1, 6)$ and whose terminal point is $(5, -2, -3)$.

(a) The component form of $\mathbf{v}$ is

$$\langle 5 - (-4), -2 - 1, -3 - 6 \rangle = \langle 9, -3, -9 \rangle.$$

(b) The unit vector form of $\mathbf{v}$ is

$$\mathbf{v} = 9\mathbf{i} - 3\mathbf{j} - 9\mathbf{k}.$$

(c) The magnitude of $\mathbf{v}$ is

$$\|\mathbf{v}\| = \sqrt{9^2 + (-3)^2 + (-9)^2} = \sqrt{171} = 3\sqrt{19}.$$

(d) A unit vector $\mathbf{u}$ in the direction of $\mathbf{v}$ is

$$\mathbf{u} = \frac{1}{3\sqrt{19}}\langle 9, -3, -9 \rangle = \left\langle \frac{3\sqrt{19}}{19}, -\frac{\sqrt{19}}{19}, -\frac{3\sqrt{19}}{19} \right\rangle.$$

(e) The sum of $\mathbf{v}$ and $\mathbf{w} = \langle 6, -4, 12 \rangle$ is

$$\mathbf{v} + \mathbf{w} = \langle 9, -3, -9 \rangle + \langle 6, -4, 12 \rangle = \langle 9 + 6, -3 + (-4), -9 + 12 \rangle = \langle 15, -7, 3 \rangle.$$

(f) The scalar multiple of 5 and $\mathbf{v}$ is

$$5\mathbf{v} = 5\langle 9, -3, -9 \rangle = \langle 45, -15, -45 \rangle.$$

Note that $\mathbf{v}$ and $5\mathbf{v}$ are parallel.

(g) The dot product of $\mathbf{v}$ and $\mathbf{w}$ is

$$\mathbf{v} \cdot \mathbf{w} = \langle 9, -3, -9 \rangle \cdot \langle 6, -4, 12 \rangle = 9(6) + (-3)(-4) + (-9)(12) = -42.$$

(h) To find the angle θ between $\mathbf{v}$ and $\mathbf{w}$, use the results of parts (c) and (g) and the fact that $\|\mathbf{w}\| = 14$ (verify this).

$$\cos\theta = \frac{-42}{(3\sqrt{19})(14)} = -\frac{1}{\sqrt{19}} \quad \Rightarrow \quad \theta \approx 103.3^\circ$$

Finding the Component Form of a Vector **In Exercises 17 and 18, (a) write the component form of the vector v, (b) find the magnitude of v, and (c) find a unit vector in the direction of v.**

	Initial Point	*Terminal Point*
17.	$(4, -2, 1)$	$(-1, 5, 0)$
18.	$(2, -1, 2)$	$(-3, 2, 3)$

Finding the Dot Product of Two Vectors **In Exercises 19–22, find the dot product of u and v.**

19. $\mathbf{u} = \langle -1, 4, 3 \rangle$, $\mathbf{v} = \langle 0, -6, 5 \rangle$

20. $\mathbf{u} = \langle 8, -10, 2 \rangle$, $\mathbf{v} = \langle 5, 5, 5 \rangle$

21. $\mathbf{u} = 2\mathbf{i} + \mathbf{j} - 2\mathbf{k}$, $\mathbf{v} = \mathbf{i} - 3\mathbf{j} + 2\mathbf{k}$

22. $\mathbf{u} = 7\mathbf{i} + 4\mathbf{j} - 2\mathbf{k}$, $\mathbf{v} = -9\mathbf{i} - 5\mathbf{k}$

Finding the Angle Between Two Vectors **In Exercises 23–26, find the angle θ between the vectors.**

23. $\mathbf{u} = \langle 2, -1, 0 \rangle$, $\mathbf{v} = \langle 1, 2, 1 \rangle$

24. $\mathbf{u} = \langle 3, 1, -1 \rangle$, $\mathbf{v} = \langle 4, 5, 2 \rangle$

25. $\mathbf{u} = \langle \sqrt{3}, -1, 3 \rangle$, $\mathbf{v} = \langle -\sqrt{3}, 6, 7 \rangle$

26. $\mathbf{u} = \langle 2\sqrt{2}, -4, 4 \rangle$, $\mathbf{v} = \langle -\sqrt{2}, 1, 2 \rangle$

Determining Orthogonal and Parallel Vectors **In Exercises 27–30, determine whether u and v are orthogonal, parallel, or neither.**

27. $\mathbf{u} = \langle 7, -2, 3 \rangle$, $\mathbf{v} = \langle -1, 4, 5 \rangle$

28. $\mathbf{u} = \langle -4, 3, -6 \rangle$, $\mathbf{v} = \langle 12, -9, 18 \rangle$

29. $\mathbf{u} = 3\mathbf{j} + 2\mathbf{k}$, $\mathbf{v} = 16\mathbf{i} - 12\mathbf{k}$

30. $\mathbf{u} = 6\mathbf{i} + 5\mathbf{j} + 9\mathbf{k}$, $\mathbf{v} = 5\mathbf{i} + 3\mathbf{j} - 5\mathbf{k}$

Using Vectors to Determine Collinear Points In Exercises 31–34, use vectors to determine whether the points are collinear.

31. (5, 2, 0), (2, 6, 1), (2, 4, 7)

32. (6, 3, −1), (5, 8, 3), (7, −2, −5)

33. (3, 4, −1), (−1, 6, 9), (5, 3, −6)

34. (5, −4, 7), (8, −5, 5), (11, 6, 3)

35. Forces in Equilibrium A load of 300 pounds is supported by three cables, as shown in the figure. Find the tension in each of the supporting cables.

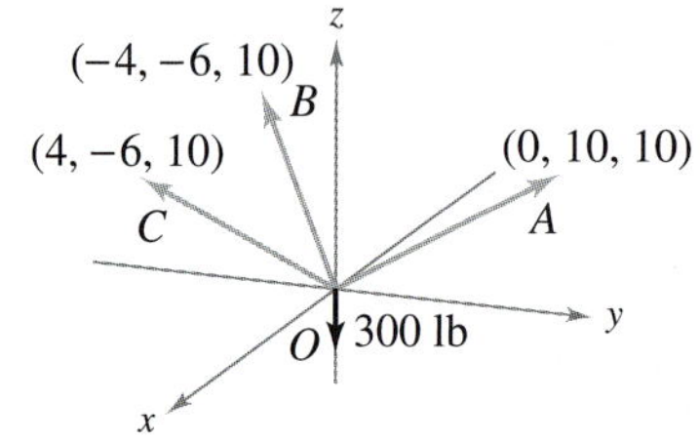

Figure for 35

36. Forces in Equilibrium Determine the tension in each of the supporting cables in Exercise 35 when the load is 200 pounds.

10.3 *What did you learn?*

Find cross products of vectors in space *(p. 736)*, and use geometric properties of cross products of vectors in space *(p. 738)*. Let $\mathbf{u} = u_1\mathbf{i} + u_2\mathbf{j} + u_3\mathbf{k}$ and $\mathbf{v} = v_1\mathbf{i} + v_2\mathbf{j} + v_3\mathbf{k}$ be vectors in space. The cross product of $\mathbf{u}$ and $\mathbf{v}$ is the vector

$$\mathbf{u} \times \mathbf{v} = (u_2v_3 - u_3v_2)\mathbf{i} - (u_1v_3 - u_3v_1)\mathbf{j} + (u_1v_2 - u_2v_1)\mathbf{k}.$$

In *determinant form,*

$$\mathbf{u} \times \mathbf{v} = \begin{vmatrix} \mathbf{i} & \mathbf{j} & \mathbf{k} \\ u_1 & u_2 & u_3 \\ v_1 & v_2 & v_3 \end{vmatrix}.$$

Algebraic and geometric properties of the cross product are listed on pages 737 and 738, respectively

Use triple scalar products to find volumes of parallelepipeds *(p. 740)*. For $\mathbf{u} = u_1\mathbf{i} + u_2\mathbf{j} + u_3\mathbf{k}$, $\mathbf{v} = v_1\mathbf{i} + v_2\mathbf{j} + v_3\mathbf{k}$, and $\mathbf{w} = w_1\mathbf{i} + w_2\mathbf{j} + w_3\mathbf{k}$ the triple scalar product is given by

$$\mathbf{u} \cdot (\mathbf{v} \times \mathbf{w}) = \begin{vmatrix} u_1 & u_2 & u_3 \\ v_1 & v_2 & v_3 \\ w_1 & w_2 & w_3 \end{vmatrix}.$$

The volume V of a parallelepiped with vectors $\mathbf{u}$, $\mathbf{v}$, and $\mathbf{w}$ as adjacent edges is given by $V = |\mathbf{u} \cdot (\mathbf{v} \times \mathbf{w})|$.

Example Consider the vectors $\mathbf{u} = 2\mathbf{i} + 3\mathbf{j} - \mathbf{k}$, $\mathbf{v} = -4\mathbf{i} + \mathbf{j} - 3\mathbf{k}$, and $\mathbf{w} = 6\mathbf{i} - 5\mathbf{j} + 2\mathbf{k}$.

(a) $\mathbf{u} \times \mathbf{v} = (-9 + 1)\mathbf{i} - (-6 - 4)\mathbf{j} + (2 + 12)\mathbf{k}$
$= -8\mathbf{i} + 10\mathbf{j} + 14\mathbf{k}$

(b) $\mathbf{v} \times \mathbf{u} = 8\mathbf{i} - 10\mathbf{j} - 14\mathbf{k}$

(c) $\mathbf{u} \times \mathbf{w} = (6 - 5)\mathbf{i} - (4 + 6)\mathbf{j} + (-10 - 18)\mathbf{k}$
$= \mathbf{i} - 10\mathbf{j} - 28\mathbf{k}$

(d) Using the results of parts (a) and (c),
$\mathbf{u} \times (\mathbf{v} + \mathbf{w}) = \langle -8, 10, 14\rangle + \langle 1, -10, -28\rangle$
$= \langle -7, 0, -14\rangle.$

(e) To find a unit vector that is orthogonal to both $\mathbf{u}$ and $\mathbf{v}$, use the result of part (a) and the fact that $\|\mathbf{u} \times \mathbf{v}\| = 6\sqrt{10}$ (verify this).

$$\frac{\mathbf{u} \times \mathbf{v}}{\|\mathbf{u} \times \mathbf{v}\|} = \frac{1}{6\sqrt{10}}\langle -8, 10, 14\rangle = \left\langle -\frac{2\sqrt{10}}{15}, \frac{\sqrt{10}}{6}, \frac{7\sqrt{10}}{30}\right\rangle$$

(f) To find the volume V of a parallelepiped with $\mathbf{u}$, $\mathbf{v}$, and $\mathbf{w}$ as adjacent edges, use determinant form.

$$\begin{vmatrix} 2 & 3 & -1 \\ -4 & 1 & -3 \\ 6 & -5 & 2 \end{vmatrix} = -70$$

So, the volume is $V = |-70| = 70$ cubic units.

Finding the Cross Product In Exercises 37–40, find u × v and show that it is orthogonal to both u and v.

37. $\mathbf{u} = \langle 0, -4, 6\rangle$
$\mathbf{v} = \langle 3, -1, 5\rangle$

38. $\mathbf{u} = \langle 4, 6, 2\rangle$
$\mathbf{v} = \langle 5, -3, 0\rangle$

39. $\mathbf{u} = \mathbf{i} - 2\mathbf{j} + 12\mathbf{k}$
$\mathbf{v} = 3\mathbf{i} + 5\mathbf{j} - 4\mathbf{k}$

40. $\mathbf{u} = 3\mathbf{i} + 4\mathbf{j} - 8\mathbf{k}$
$\mathbf{v} = 9\mathbf{i} - \mathbf{j} + 6\mathbf{k}$

Using the Cross Product In Exercises 41 and 42, find a unit vector orthogonal to u and v.

41. $\mathbf{u} = -2\mathbf{i} + 3\mathbf{j} - 5\mathbf{k}$
$\mathbf{v} = 10\mathbf{i} - 12\mathbf{j} + 2\mathbf{k}$

42. $\mathbf{u} = 2\mathbf{i} - \mathbf{j} + \mathbf{k}$
$\mathbf{v} = -\mathbf{i} - \mathbf{j} - 2\mathbf{k}$

Geometric Application of the Cross Product In Exercises 43 and 44, (a) verify that the points are the vertices of a parallelogram and (b) find its area.

43. $A(2, -1, 1)$, $B(5, 1, 4)$, $C(0, 1, 1)$, $D(3, 3, 4)$

44. $A(0, 4, 0)$, $B(1, 4, 1)$, $C(0, 6, 0)$, $D(1, 6, 1)$

Parallelepiped Volume In Exercises 45 and 46, find the volume of the parallelepiped with the given vertices.

45. $A(0, 0, 0)$, $B(3, 0, 0)$, $C(0, 5, 1)$, $D(3, 5, 1)$, $E(2, 0, 5)$, $F(5, 0, 5)$, $G(2, 5, 6)$, $H(5, 5, 6)$

46. $A(0, 0, 0)$, $B(2, 0, 0)$, $C(2, 4, 0)$, $D(0, 4, 0)$, $E(0, 0, 6)$, $F(2, 0, 6)$, $G(2, 4, 6)$, $H(0, 4, 6)$

10.4 *What did you learn?*

Find parametric and symmetric equations of lines in space (*p. 743*). A line L parallel to the nonzero (direction) vector $\mathbf{v} = \langle a, b, c \rangle$ and passing through the point $P(x_1, y_1, z_1)$ is represented by the parametric equations $x = x_1 + at$, $y = y_1 + bt$, and $z = z_1 + ct$. When the direction numbers a, b, and c are all nonzero, you can eliminate the parameter to obtain the symmetric equations of a line, $(x - x_1)/a = (y - y_1)/b = (z - z_1)/c$.

Find equations of planes in space (*p. 745*). The plane containing the point (x_1, y_1, z_1) and having nonzero normal vector $\mathbf{n} = \langle a, b, c \rangle$ can be represented by the standard form of the equation of a plane $a(x - x_1) + b(y - y_1) + c(z - z_1) = 0$. If vectors $\mathbf{n}_1$ and $\mathbf{n}_2$ are normal to two intersecting planes, then the angle θ between the normal vectors is equal to the angle between the two planes and is given by $\cos \theta = |\mathbf{n}_1 \cdot \mathbf{n}_2|/(\|\mathbf{n}_1\|\|\mathbf{n}_2\|)$ (see Example 4).

Sketch planes in space (*p. 748*). Examples of how to sketch planes in space are given on page 748.

Find distances between points and planes in space (*p. 749*). The distance between a plane and a point Q (not in the plane) is $D = \|\text{proj}_{\mathbf{n}} \overrightarrow{PQ}\| = |\overrightarrow{PQ} \cdot \mathbf{n}|/\|\mathbf{n}\|$, where P is a point in the plane and $\mathbf{n}$ is normal to the plane.

Example

(a) Consider the line that passes through the points $(1, -4, 2)$ and $(-2, 1, -4)$. A direction vector $\mathbf{v}$ for the line is

$$\mathbf{v} = \langle -2 - 1, 1 - (-4), -4 - 2 \rangle = \langle -3, 5, -6 \rangle.$$

So, the line can be represented by the parametric equations or the symmetric equations below.

$x = 1 - 3t$, $y = -4 + 5t$, and $z = 2 - 6t$ Parametric equations

$$\frac{x - 1}{-3} = \frac{y + 4}{5} = \frac{z - 2}{-6}$$ Symmetric equations

(b) Consider the plane that passes through the points $(-2, 1, -4)$, $(1, -4, 2)$, and $(0, 3, 2)$. (See figure.)

$\mathbf{u} = \langle 0 - 1, 3 - (-4), 2 - 2 \rangle$

$= \langle -1, 7, 0 \rangle$

$\mathbf{v} = \langle -3, 5, -6 \rangle$ See part (a).

$\mathbf{n} = \mathbf{u} \times \mathbf{v}$

$= \langle -42, -6, 16 \rangle$

(Verify this.)

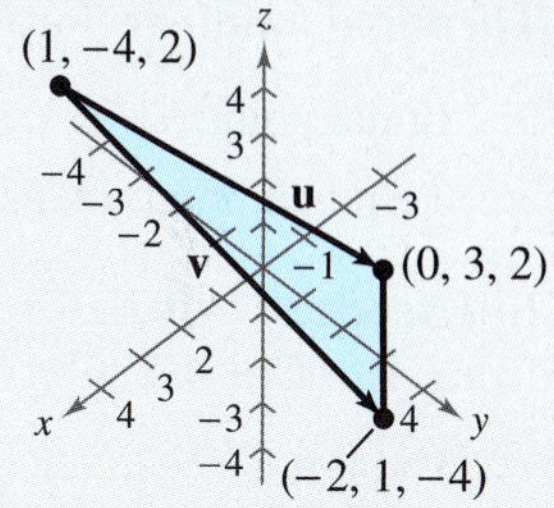

So, an equation of the plane is

$-42(x + 2) - 6(y - 1) + 16(z + 4) = 0$, or

$21x + 3y - 8z + 7 = 0.$

Finding Parametric and Symmetric Equations **In Exercises 47–50, find (a) a set of parametric equations and (b) if possible, a set of symmetric equations for the specified line. (There are many correct answers.)**

47. Passes through $(0, -10, 3)$ and is parallel to $\mathbf{v} = \langle -5, 2, 3 \rangle$

48. Passes through $(0, 1, 7)$ and is parallel to $\mathbf{v} = \langle -2, \frac{5}{2}, 1 \rangle$

49. Passes through $(-1, 3, 5)$ and is parallel to $x = 1 + 2t$, $y = 2 - t$, and $z = 1 + 3t$

50. Passes through $(3, 2, 1)$ and is parallel to $x = y = z$

A Line Through Two Points **In Exercises 51 and 52, find (a) a set of parametric equations and (b) if possible, a set of symmetric equations for the line that passes through the two points. (There are many correct answers.)**

51. $(3, 0, 2)$, $(9, 11, 16)$ **52.** $(-1, 4, 3)$, $(8, 10, 5)$

Finding an Equation of a Plane in Three-Space **In Exercises 53 and 54, find the general form of the equation of the plane with the given characteristics.**

53. Passes through $(0, 0, 4)$, $(5, 0, 2)$, and $(2, 3, 8)$

54. Passes through $(6, -3, 4)$ and is perpendicular to $x = 1 - t$, $y = 2 + t$, and $z = 4 - 2t$

Sketching a Plane in Space **In Exercises 55 and 56, plot the intercepts and sketch a graph of the plane.**

55. $5x - 5y - 2z = 10$

56. $-4x - 3z = 12$

Finding the Distance Between a Point and a Plane **In Exercises 57 and 58, find the distance between the point and the plane.**

57. $(1, 2, 3)$
$2x - y + z = 4$

58. $(0, 0, 0)$
$2x + 3y + z = 12$

Focusing on Concepts

True or False? **In Exercises 59 and 60, determine whether the statement is true or false. Justify your answer.**

59. The cross product is commutative.

60. The triple scalar product of three vectors in space is a scalar.

Verifying Properties **In Exercises 61–64, verify the property.**

61. $\mathbf{u} \cdot \mathbf{u} = \|\mathbf{u}\|^2$ **62.** $\mathbf{u} \times \mathbf{v} = -(\mathbf{v} \times \mathbf{u})$

63. $\mathbf{u} \cdot (\mathbf{v} + \mathbf{w}) = \mathbf{u} \cdot \mathbf{v} + \mathbf{u} \cdot \mathbf{w}$

64. $\mathbf{u} \times (\mathbf{v} + \mathbf{w}) = (\mathbf{u} \times \mathbf{v}) + (\mathbf{u} \times \mathbf{w})$

10 Chapter Test

See *CalcChat.com* for tutorial help and worked-out solutions to odd-numbered exercises.
For instructions on how to use a graphing utility, see Appendix A.

Take this test as you would take a test in class. After you are finished, check your work against the answers given in the back of the book.

1. Plot each point in the same three-dimensional coordinate system.

(a) $(5, -2, 3)$ (b) $(-3, 4, 2)$ (c) $(2, 1, -1)$

In Exercises 2–4, use the points $A(8, -2, 5)$, $B(6, 4, -1)$, and $C(-4, 3, 0)$ to solve the problem.

2. Consider the triangle with vertices A, B, and C. Is it a right triangle? Explain.

3. Find the midpoint of the line segment joining points A and B.

4. Find the standard equation of the sphere for which A and B are the endpoints of a diameter. Sketch the sphere and find its xz-trace.

In Exercises 5 and 6, find the component form and the magnitude of the vector v. Then find a unit vector in the direction of v.

5. Initial point: $(2, -1, 3)$
Terminal point: $(4, 4, -7)$

6. Initial point: $(6, 2, -5)$
Terminal point: $(3, -4, 8)$

In Exercises 7–10, let u and v be the vectors from $A(8, -2, 5)$ to $B(6, 4, -1)$ and from A to $C(-4, 3, 0)$, respectively.

7. Write $\mathbf{u}$ and $\mathbf{v}$ in component form.

8. Find (a) $\|\mathbf{v}\|$, (b) $\mathbf{u} \cdot \mathbf{v}$, and (c) $\mathbf{u} \times \mathbf{v}$.

9. Find the angle θ between $\mathbf{u}$ and $\mathbf{v}$.

10. Find (a) a set of parametric equations and (b) a set of symmetric equations for the line through points A and B.

In Exercises 11–13, determine whether u and v are parallel, perpendicular, or neither.

11. $\mathbf{u} = 2\mathbf{i} - \mathbf{j} + 3\mathbf{k}$
$\mathbf{v} = -8\mathbf{i} + 4\mathbf{j} - 12\mathbf{k}$

12. $\mathbf{u} = \mathbf{i} - 3\mathbf{j}$
$\mathbf{v} = 3\mathbf{i} + 5\mathbf{k}$

13. $\mathbf{u} = \langle 7, -1, 4 \rangle$
$\mathbf{v} = \langle 1, 3, -1 \rangle$

14. Verify that the points $A(2, -3, 1)$, $B(6, 5, -1)$, $C(3, -6, 4)$, and $D(7, 2, 2)$ are the vertices of a parallelogram. Then find the area of the parallelogram.

15. Find the general form of the equation of the plane passing through the points $(-3, -4, 2)$, $(-3, 4, 1)$, and $(1, 1, -2)$.

16. Find the volume of the parallelepiped with the given vertices (see figure).
$A(0, 0, 5)$, $B(0, 10, 5)$, $C(4, 10, 5)$, $D(4, 0, 5)$,
$E(0, 1, 0)$, $F(0, 11, 0)$, $G(4, 11, 0)$, $H(4, 1, 0)$

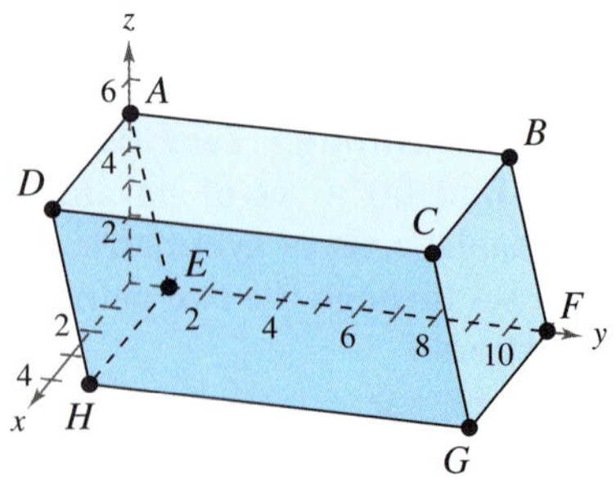

Figure for 16

In Exercises 17 and 18, plot the intercepts and sketch a graph of the plane.

17. $3x + 6y + 2z = 18$

18. $4x - 5z = 20$

19. Find the distance between the point $(2, -1, 6)$ and the plane $3x - 2y + z = 6$.

Collaborative Project

To work a collaborative project involving Analytic Geometry in Three Dimensions, visit this textbook's website at *LarsonPrecalculus.com.*

Proofs in Mathematics

Algebraic Properties of the Cross Product (p. 737)

Let **u**, **v**, and **w** be vectors in space, and let c be a scalar.

1. $\mathbf{u} \times \mathbf{v} = -(\mathbf{v} \times \mathbf{u})$
2. $\mathbf{u} \times (\mathbf{v} + \mathbf{w}) = (\mathbf{u} \times \mathbf{v}) + (\mathbf{u} \times \mathbf{w})$
3. $c(\mathbf{u} \times \mathbf{v}) = (c\mathbf{u}) \times \mathbf{v} = \mathbf{u} \times (c\mathbf{v})$
4. $\mathbf{u} \times \mathbf{0} = \mathbf{0} \times \mathbf{u} = \mathbf{0}$
5. $\mathbf{u} \times \mathbf{u} = \mathbf{0}$
6. $\mathbf{u} \cdot (\mathbf{v} \times \mathbf{w}) = (\mathbf{u} \times \mathbf{v}) \cdot \mathbf{w}$

Notation for Dot and Cross Products

The notation for the dot products and cross products of vectors was first introduced by the American physicist Josiah Willard Gibbs (1839–1903). In the early 1880s, Gibbs built a system to represent physical quantities called *vector analysis*. The system was a departure from William Hamilton's theory of quaternions.

Proof

Let $\mathbf{u} = u_1\mathbf{i} + u_2\mathbf{j} + u_3\mathbf{k}$, $\mathbf{v} = v_1\mathbf{i} + v_2\mathbf{j} + v_3\mathbf{k}$, $\mathbf{w} = w_1\mathbf{i} + w_2\mathbf{j} + w_3\mathbf{k}$, $\mathbf{0} = 0\mathbf{i} + 0\mathbf{j} + 0\mathbf{k}$, and let c be a scalar.

1. $\mathbf{u} \times \mathbf{v} = (u_2v_3 - u_3v_2)\mathbf{i} - (u_1v_3 - u_3v_1)\mathbf{j} + (u_1v_2 - u_2v_1)\mathbf{k}$

$\mathbf{v} \times \mathbf{u} = (v_2u_3 - v_3u_2)\mathbf{i} - (v_1u_3 - v_3u_1)\mathbf{j} + (v_1u_2 - v_2u_1)\mathbf{k}$

So, this implies $\mathbf{u} \times \mathbf{v} = -(\mathbf{v} \times \mathbf{u})$.

2.
$$\begin{aligned}\mathbf{u} \times (\mathbf{v} + \mathbf{w}) &= [u_2(v_3 + w_3) - u_3(v_2 + w_2)]\mathbf{i} - [u_1(v_3 + w_3) - \\ &\qquad u_3(v_1 + w_1)]\mathbf{j} + [u_1(v_2 + w_2) - u_2(v_1 + w_1)]\mathbf{k} \\ &= (u_2v_3 - u_3v_2)\mathbf{i} - (u_1v_3 - u_3v_1)\mathbf{j} + (u_1v_2 - u_2v_1)\mathbf{k} + \\ &\qquad (u_2w_3 - u_3w_2)\mathbf{i} - (u_1w_3 - u_3w_1)\mathbf{j} + (u_1w_2 - u_2w_1)\mathbf{k} \\ &= (\mathbf{u} \times \mathbf{v}) + (\mathbf{u} \times \mathbf{w})\end{aligned}$$

3.
$$\begin{aligned}(c\mathbf{u}) \times \mathbf{v} &= (cu_2v_3 - cu_3v_2)\mathbf{i} - (cu_1v_3 - cu_3v_1)\mathbf{j} + (cu_1v_2 - cu_2v_1)\mathbf{k} \\ &= c[(u_2v_3 - u_3v_2)\mathbf{i} - (u_1v_3 - u_3v_1)\mathbf{j} + (u_1v_2 - u_2v_1)\mathbf{k}] \\ &= c(\mathbf{u} \times \mathbf{v})\end{aligned}$$

4.
$$\begin{aligned}\mathbf{u} \times \mathbf{0} &= (u_2 \cdot 0 - u_3 \cdot 0)\mathbf{i} - (u_1 \cdot 0 - u_3 \cdot 0)\mathbf{j} + (u_1 \cdot 0 - u_2 \cdot 0)\mathbf{k} \\ &= 0\mathbf{i} + 0\mathbf{j} + 0\mathbf{k} = \mathbf{0}\end{aligned}$$

$$\begin{aligned}\mathbf{0} \times \mathbf{u} &= (0 \cdot u_3 - 0 \cdot u_2)\mathbf{i} - (0 \cdot u_3 - 0 \cdot u_1)\mathbf{j} + (0 \cdot u_2 - 0 \cdot u_1)\mathbf{k} \\ &= 0\mathbf{i} + 0\mathbf{j} + 0\mathbf{k} = \mathbf{0}\end{aligned}$$

So, this implies $\mathbf{u} \times \mathbf{0} = \mathbf{0} \times \mathbf{u} = \mathbf{0}$.

5. $\mathbf{u} \times \mathbf{u} = (u_2u_3 - u_3u_2)\mathbf{i} - (u_1u_3 - u_3u_1)\mathbf{j} + (u_1u_2 - u_2u_1)\mathbf{k} = \mathbf{0}$

6. $\mathbf{u} \cdot (\mathbf{v} \times \mathbf{w}) = \begin{vmatrix} u_1 & u_2 & u_3 \\ v_1 & v_2 & v_3 \\ w_1 & w_2 & w_3 \end{vmatrix}$ and $(\mathbf{u} \times \mathbf{v}) \cdot \mathbf{w} = \mathbf{w} \cdot (\mathbf{u} \times \mathbf{v}) = \begin{vmatrix} w_1 & w_2 & w_3 \\ u_1 & u_2 & u_3 \\ v_1 & v_2 & v_3 \end{vmatrix}$

$$\begin{aligned}\mathbf{u} \cdot (\mathbf{v} \times \mathbf{w}) &= u_1(v_2w_3 - w_2v_3) - u_2(v_1w_3 - w_1v_3) + u_3(v_1w_2 - w_1v_2) \\ &= u_1v_2w_3 - u_1w_2v_3 - u_2v_1w_3 + u_2w_1v_3 + u_3v_1w_2 - u_3w_1v_2 \\ &= u_2w_1v_3 - u_3w_1v_2 - u_1w_2v_3 + u_3v_1w_2 + u_1v_2w_3 - u_2v_1w_3 \\ &= w_1(u_2v_3 - v_2u_3) - w_2(u_1v_3 - v_1u_3) + w_3(u_1v_2 - v_1u_2) \\ &= \mathbf{w} \cdot (\mathbf{u} \times \mathbf{v}) \\ &= (\mathbf{u} \times \mathbf{v}) \cdot \mathbf{w}\end{aligned}$$

■

Geometric Properties of the Cross Product (p. 738)

Let $\mathbf{u}$ and $\mathbf{v}$ be nonzero vectors in space, and let θ be the angle between $\mathbf{u}$ and $\mathbf{v}$.

1. $\mathbf{u} \times \mathbf{v}$ is orthogonal to both $\mathbf{u}$ and $\mathbf{v}$.
2. $\|\mathbf{u} \times \mathbf{v}\| = \|\mathbf{u}\|\|\mathbf{v}\| \sin \theta$
3. $\mathbf{u} \times \mathbf{v} = \mathbf{0}$ if and only if $\mathbf{u}$ and $\mathbf{v}$ are scalar multiples of each other.
4. $\|\mathbf{u} \times \mathbf{v}\| =$ area of parallelogram having $\mathbf{u}$ and $\mathbf{v}$ as adjacent sides.

Proof

Let $\mathbf{u} = u_1\mathbf{i} + u_2\mathbf{j} + u_3\mathbf{k}$, $\mathbf{v} = v_1\mathbf{i} + v_2\mathbf{j} + v_3\mathbf{k}$, and $\mathbf{0} = 0\mathbf{i} + 0\mathbf{j} + 0\mathbf{k}$.

1.
$$\mathbf{u} \times \mathbf{v} = (u_2v_3 - u_3v_2)\mathbf{i} - (u_1v_3 - u_3v_1)\mathbf{j} + (u_1v_2 - u_2v_1)\mathbf{k}$$
$$\begin{aligned}(\mathbf{u} \times \mathbf{v}) \cdot \mathbf{u} &= (u_2v_3 - u_3v_2)u_1 - (u_1v_3 - u_3v_1)u_2 + (u_1v_2 - u_2v_1)u_3 \\ &= u_1u_2v_3 - u_1u_3v_2 - u_1u_2v_3 + u_2u_3v_1 + u_1u_3v_2 - u_2u_3v_1 \\ &= 0\end{aligned}$$
$$\begin{aligned}(\mathbf{u} \times \mathbf{v}) \cdot \mathbf{v} &= (u_2v_3 - u_3v_2)v_1 - (u_1v_3 - u_3v_1)v_2 + (u_1v_2 - u_2v_1)v_3 \\ &= u_2v_1v_3 - u_3v_1v_2 - u_1v_2u_3 + u_3v_1v_2 + u_1v_2v_3 - u_2v_1v_3 \\ &= 0\end{aligned}$$

Because two vectors are orthogonal when their dot product is zero, it follows that $\mathbf{u} \times \mathbf{v}$ is orthogonal to both $\mathbf{u}$ and $\mathbf{v}$.

2. Note that $\cos \theta = \dfrac{\mathbf{u} \cdot \mathbf{v}}{\|\mathbf{u}\|\|\mathbf{v}\|}$. Therefore,

$$\begin{aligned}\|\mathbf{u}\|\|\mathbf{v}\| \sin \theta &= \|\mathbf{u}\|\|\mathbf{v}\|\sqrt{1 - \cos^2 \theta} \\ &= \|\mathbf{u}\|\|\mathbf{v}\|\sqrt{1 - \frac{(\mathbf{u} \cdot \mathbf{v})^2}{\|\mathbf{u}\|^2\|\mathbf{v}\|^2}} \\ &= \sqrt{\|\mathbf{u}\|^2\|\mathbf{v}\|^2 - (\mathbf{u} \cdot \mathbf{v})^2} \\ &= \sqrt{(u_1^2 + u_2^2 + u_3^2)(v_1^2 + v_2^2 + v_3^2) - (u_1v_1 + u_2v_2 + u_3v_3)^2} \\ &= \sqrt{(u_2v_3 - u_3v_2)^2 + (u_1v_3 - u_3v_1)^2 + (u_1v_2 - u_2v_1)^2} \\ &= \|\mathbf{u} \times \mathbf{v}\|.\end{aligned}$$

3. If $\mathbf{u}$ and $\mathbf{v}$ are scalar multiples of each other, then $\mathbf{u} = c\mathbf{v}$ for some scalar c.

$$\mathbf{u} \times \mathbf{v} = (c\mathbf{v}) \times \mathbf{v} = c(\mathbf{v} \times \mathbf{v}) = c(\mathbf{0}) = \mathbf{0}$$

If $\mathbf{u} \times \mathbf{v} = \mathbf{0}$, then $\|\mathbf{u}\|\|\mathbf{v}\|\sin \theta = 0$. (Assume $\mathbf{u} \neq \mathbf{0}$ and $\mathbf{v} \neq \mathbf{0}$.) So, $\sin \theta = 0$, and $\theta = 0$ or $\theta = \pi$. In either case, because θ is the angle between the vectors, $\mathbf{u}$ and $\mathbf{v}$ are parallel. Therefore, $\mathbf{u} = c\mathbf{v}$ for some scalar c.

4. The figure at the right is a parallelogram having $\mathbf{v}$ and $\mathbf{u}$ as adjacent sides. Because the height of the parallelogram is $\|\mathbf{v}\| \sin \theta$, the area is

$$\begin{aligned}\text{Area} &= (\text{base})(\text{height}) \\ &= \|\mathbf{u}\|\|\mathbf{v}\| \sin \theta \\ &= \|\mathbf{u} \times \mathbf{v}\|.\end{aligned}$$

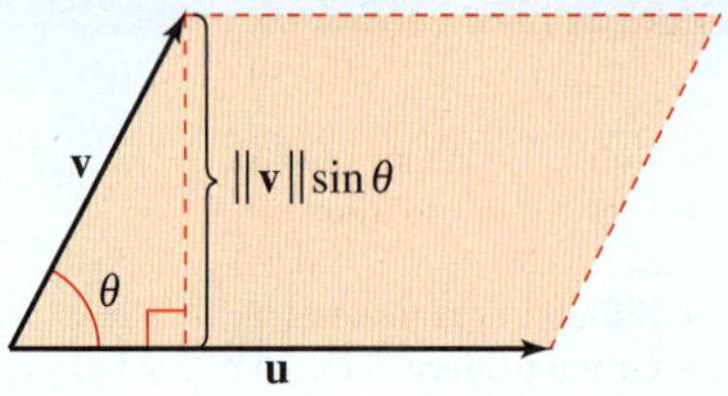

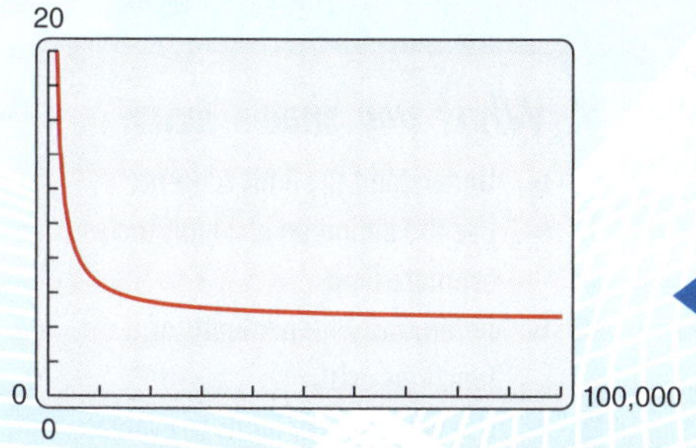

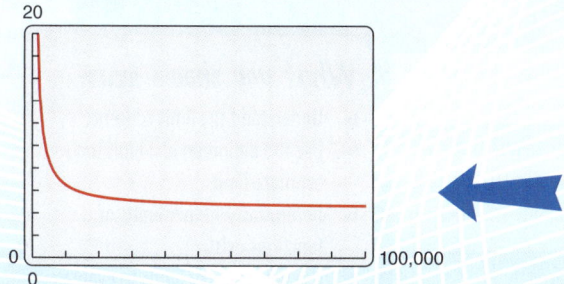

Section 11.4, Example 3
Average Cost

11 Limits and an Introduction to Calculus

Student Resources at LarsonPrecalculus.com

- **Videos** explaining the concepts of precalculus
- **Worked-out solution videos** for all *Checkpoint* exercises
- **Editable spreadsheets** of the data sets in the text
- **Group projects** for each chapter applying concepts to real-life problems

11.1 Introduction to Limits

The Limit Concept

The notion of a limit is a *fundamental* concept of calculus. In this chapter, you will learn how to evaluate limits and how they are used in the two basic problems of calculus: the tangent line problem and the area problem.

What you should learn

- Understand the limit concept.
- Use the definition of a limit to estimate limits.
- Determine whether limits of functions exist.
- Use properties of limits to evaluate limits.

Why you should learn it

The concept of a limit is useful in applications involving maximization. For instance, in Exercise 5 on page 767, the concept of a limit is used to verify the maximum volume of an open box.

EXAMPLE 1 Finding a Rectangle of Maximum Area

Find the dimensions of a rectangle that has a perimeter of 24 inches and a maximum area.

Solution

Let w represent the width of the rectangle and let l represent the length of the rectangle. Because

$$2w + 2l = 24 \qquad \text{Perimeter is 24.}$$

it follows that

$$l = 12 - w$$

as shown in Figure 11.1. So, the area of the rectangle is

$$\begin{aligned} A &= lw && \text{Formula for area} \\ &= (12 - w)w && \text{Substitute } 12 - w \text{ for } l. \\ &= 12w - w^2. && \text{Simplify.} \end{aligned}$$

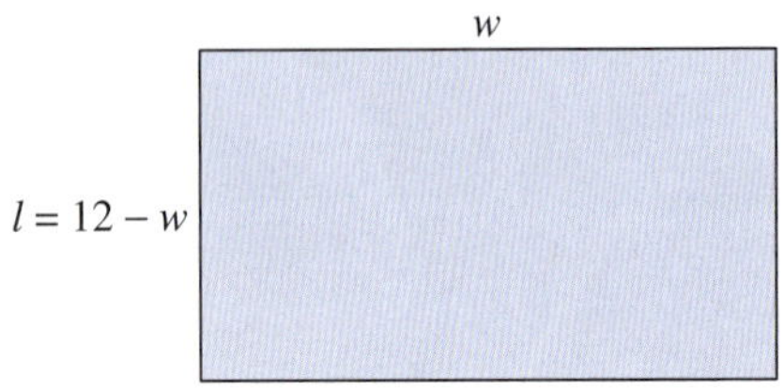

Figure 11.1

Using this model for area, you can experiment with different values of w to see how to obtain the maximum area. After trying several values, it appears that the maximum area occurs when $w = 6$, as shown in the table.

Width, w	5	5.5	5.9	6	6.1	6.5	7
Area, A	35.00	35.75	35.99	36.00	35.99	35.75	35.00

In limit terminology, you can say that "the limit of A as w approaches 6 is 36." This is written as

$$\lim_{w \to 6} A = \lim_{w \to 6} (12w - w^2) = 36.$$

So, the dimensions of a rectangle that has a perimeter of 24 inches and a maximum area are $w = 6$ inches and $l = 12 - 6 = 6$ inches, or 6×6 inches.

Checkpoint *Audio-video solution in English & Spanish at LarsonPrecalculus.com*

Find the dimensions of a rectangle that has a perimeter of 52 inches and a maximum area.

Definition of Limit

Definition of Limit

If $f(x)$ becomes arbitrarily close to a unique number L as x approaches c from either side, then the **limit** of $f(x)$ as x approaches c is L. This is written as

$$\lim_{x\to c} f(x) = L.$$

EXAMPLE 2 Estimating a Limit Numerically

Use a table to estimate the limit: $\lim_{x\to 2} (3x - 2)$.

Solution

Let $f(x) = 3x - 2$. Then construct a table that shows $f(x)$ for two sets of x-values—one set that approaches 2 from the left and one that approaches 2 from the right.

x approaches 2 from the left. → ← x approaches 2 from the right.

x	1.9	1.99	1.999	2	2.001	2.01	2.1
$f(x)$	3.700	3.970	3.997	?	4.003	4.030	4.300

$f(x)$ approaches 4. → ← $f(x)$ approaches 4.

From the table, it appears that the closer x gets to 2, the closer $f(x)$ gets to 4. So, you can estimate the limit to be 4. Figure 11.2 adds further support to this conclusion.

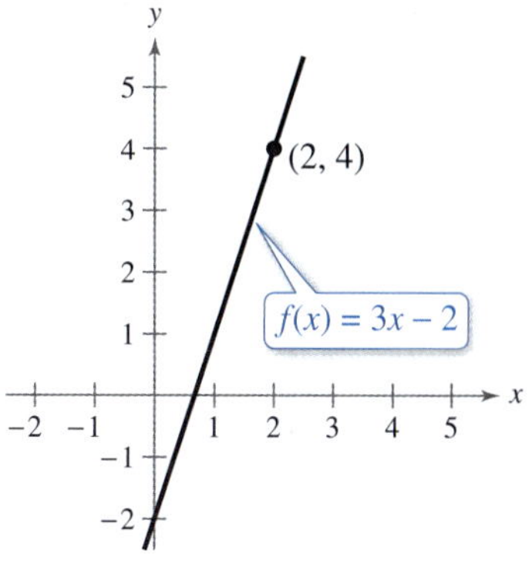

Figure 11.2

Checkpoint Audio-video solution in English & Spanish at LarsonPrecalculus.com

Use a table to estimate the limit: $\lim_{x\to 3} (3 - 2x)$.

In Figure 11.2, note that the graph of $f(x) = 3x - 2$ is continuous. For graphs that are not continuous, finding a limit can be more difficult.

EXAMPLE 3 Estimating a Limit Numerically

Use a table to estimate the limit: $\lim_{x\to 0} \dfrac{x}{\sqrt{x+1} - 1}$.

Solution

Let $f(x) = x/(\sqrt{x+1} - 1)$. Then construct a table that shows $f(x)$ for two sets of x-values—one set that approaches 0 from the left and one that approaches 0 from the right.

x approaches 0 from the left. → ← x approaches 0 from the right.

x	−0.01	−0.001	−0.0001	0	0.0001	0.001	0.01
$f(x)$	1.99499	1.99949	1.99995	?	2.00005	2.00050	2.00499

$f(x)$ approaches 2. → ← $f(x)$ approaches 2.

From the table, it appears that the limit is 2. This limit is reinforced by the graph of f (see Figure 11.3).

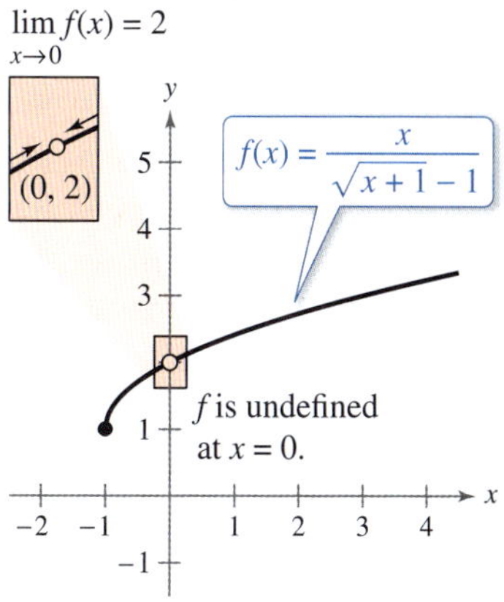

Figure 11.3

Checkpoint Audio-video solution in English & Spanish at LarsonPrecalculus.com

Use a table to estimate the limit: $\lim_{x\to 1} \dfrac{x - 1}{x^2 + 3x - 4}$.

In Example 3, note that $f(x)$ has a limit as x approaches 0 even though the function is not defined at $x = 0$. This often happens, and it is important to realize that *the existence or nonexistence of $f(x)$ at $x = c$ has no bearing on the existence of the limit of $f(x)$ as x approaches c.*

EXAMPLE 4 Using a Graphing Utility to Estimate a Limit

Estimate the limit: $\lim_{x\to 1} \dfrac{x^3 - x^2 + x - 1}{x - 1}$.

Numerical Solution

Let $f(x) = (x^3 - x^2 + x - 1)/(x - 1)$.

Create a table that shows $f(x)$ for several x-values near 1.

X	Y1
.997	1.994
.998	1.996
.999	1.998
1	ERROR
1.001	2.002
1.002	2.004
1.003	2.006
X=1	

From the figure, it appears that the closer x gets to 1, the closer $f(x)$ gets to 2. So, you can estimate the limit to be 2.

Graphical Solution

Use a graphing utility to graph $f(x) = (x^3 - x^2 + x - 1)/(x - 1)$ using a *decimal* setting, as shown in the figure. (Some graphing utilities may use a different viewing window, such as $-6.6 \le x \le 6.6$ and $-1.1 \le y \le 7.1$.)

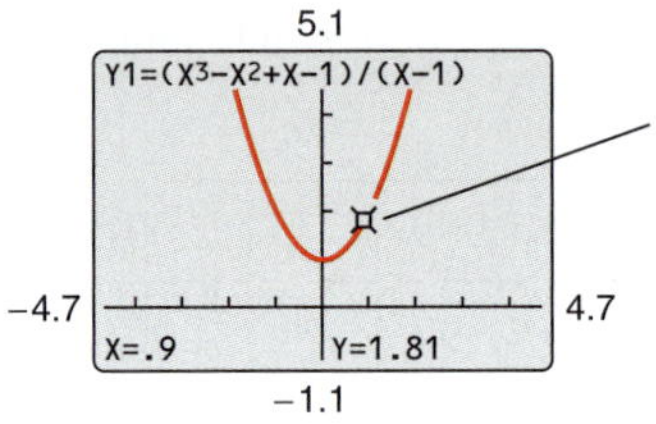

Use the *trace* feature to determine that as x gets closer and closer to 1, $f(x)$ gets closer and closer to 2 from the left and from the right.

From the figure, you can estimate the limit to be 2. As you use the *trace* feature, notice that there is no value given for y when $x = 1$ and that there is a hole or break in the graph at $x = 1$.

Checkpoint *Audio-video solution in English & Spanish at LarsonPrecalculus.com*

Estimate the limit: $\lim_{x\to 2} \dfrac{x^3 - 2x^2 + 3x - 6}{x - 2}$.

EXAMPLE 5 Using a Graph to Find a Limit

Find the limit of $f(x)$ as x approaches 3, where

$$f(x) = \begin{cases} 2, & x \neq 3 \\ 0, & x = 3 \end{cases}.$$

Solution

Because $f(x) = 2$ for all x other than $x = 3$, you can estimate that the limit is 2, as shown in Figure 11.4. So, you can write

$$\lim_{x\to 3} f(x) = 2.$$

Notice that $f(3) = 0$ has no bearing on the existence or value of the limit as x approaches 3. For instance, if the function were defined as

$$f(x) = \begin{cases} 2, & x \neq 3 \\ 4, & x = 3 \end{cases}$$

then the limit as x approaches 3 would be the same.

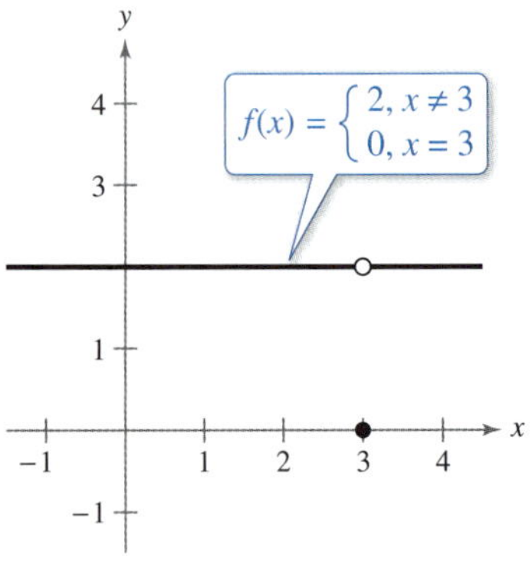

Figure 11.4

Checkpoint *Audio-video solution in English & Spanish at LarsonPrecalculus.com*

Find the limit of $f(x)$ as x approaches 2, where

$$f(x) = \begin{cases} -3, & x \neq 2 \\ 0, & x = 2 \end{cases}.$$

Limits That Fail to Exist

In the next three examples, you will examine some limits that fail to exist.

EXAMPLE 6 Comparing Left and Right Behavior

Show that the limit does not exist.

$$\lim_{x \to 0} \frac{|x|}{x}$$

Solution

Consider the graph of the function given by $f(x) = |x|/x$. In the figure, you can see that for positive x-values

$$\frac{|x|}{x} = 1, \quad x > 0$$

and for negative x-values

$$\frac{|x|}{x} = -1, \quad x < 0.$$

This means that no matter how close x gets to 0, there will be both positive and negative x-values that yield $f(x) = 1$ and $f(x) = -1$. This implies that the limit does not exist.

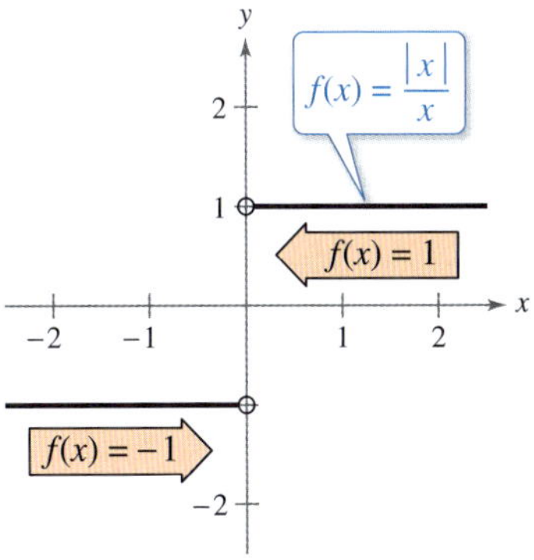

Checkpoint *Audio-video solution in English & Spanish at LarsonPrecalculus.com*

Show that the limit does not exist: $\lim_{x \to 1} \dfrac{2|x-1|}{x-1}$.

What's Wrong?

You use a graphing utility to graph

$$y_1 = \frac{x^3 - 1}{x - 1}$$

using a *decimal* setting, as shown in the figure. (Some graphing utilities may use a different viewing window, such as $-6.6 \le x \le 6.6$ and $-1.1 \le y \le 7.1$.) You use the *trace* feature to conclude that the limit

$$\lim_{x \to 1} \frac{x^3 - 1}{x - 1}$$

does not exist. What's wrong?

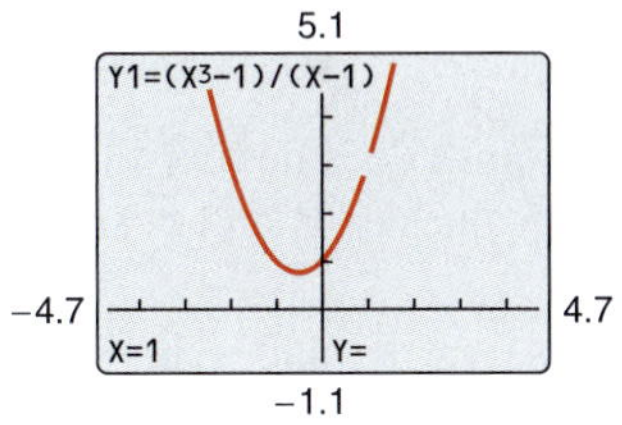

EXAMPLE 7 Unbounded Behavior

Discuss the existence of the limit: $\lim_{x \to 0} \dfrac{1}{x^2}$.

Solution

Let $f(x) = 1/x^2$. In Figure 11.5, note that as x approaches 0 from either the right or the left, $f(x)$ increases without bound. This means that by choosing x close enough to 0, you can force $f(x)$ to be as large as you want. For instance, $f(x)$ will be larger than 100 when you choose x that is within $\frac{1}{10}$ of 0. That is,

$$0 < |x| < \frac{1}{10} \quad \Rightarrow \quad f(x) = \frac{1}{x^2} > 100.$$

Similarly, you can force $f(x)$ to be larger than 1,000,000, when you choose x that is within $\frac{1}{1000}$ of 0.

$$0 < |x| < \frac{1}{1000} \quad \Rightarrow \quad f(x) = \frac{1}{x^2} > 1{,}000{,}000$$

Because $f(x)$ is not approaching a unique real number L as x approaches 0, you can conclude that the limit does not exist.

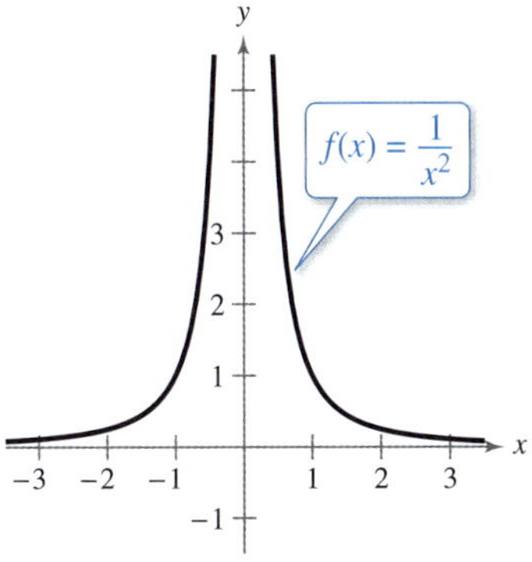

Figure 11.5

Checkpoint *Audio-video solution in English & Spanish at LarsonPrecalculus.com*

Discuss the existence of the limit: $\lim_{x \to 0} \left(-\dfrac{1}{x^2}\right)$.

EXAMPLE 8 Oscillating Behavior

See LarsonPrecalculus.com for an interactive version of this type of example.

Discuss the existence of the limit: $\lim_{x \to 0} \sin \frac{1}{x}$.

Solution

Let $f(x) = \sin(1/x)$. Notice in the figure below that as x approaches 0, $f(x)$ oscillates between -1 and 1.

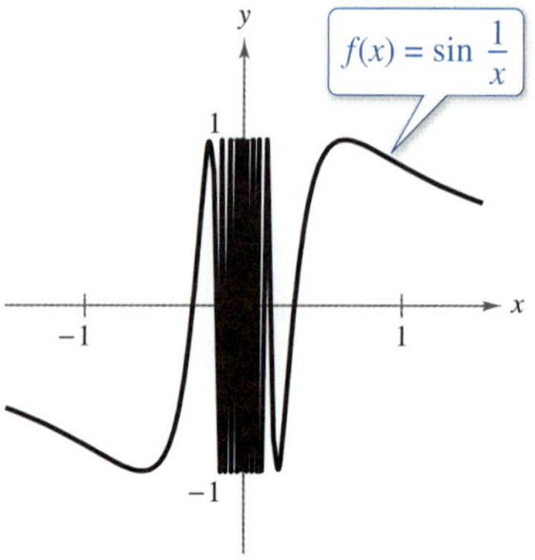

So, the limit does not exist because no matter how close you are to 0, it is possible to choose values of x_1 and x_2 such that

$$\sin \frac{1}{x_1} = 1 \quad \text{and} \quad \sin \frac{1}{x_2} = -1$$

as indicated in the table.

x	$\frac{2}{\pi}$	$\frac{2}{3\pi}$	$\frac{2}{5\pi}$	$\frac{2}{7\pi}$	$\frac{2}{9\pi}$	$\frac{2}{11\pi}$	$x \to 0$
$\sin \frac{1}{x}$	1	-1	1	-1	1	-1	Limit does not exist.

✓ Checkpoint ▶ *Audio-video solution in English & Spanish at LarsonPrecalculus.com*

Discuss the existence of the limit: $\lim_{x \to 1} \cos \frac{1}{x - 1}$.

Examples 6, 7, and 8 show three of the most common types of behavior associated with the *nonexistence* of a limit.

Conditions Under Which Limits Do Not Exist

The limit of $f(x)$ as $x \to c$ does not exist under any of the conditions listed below.

1. $f(x)$ approaches a different number from the right side of c than it approaches from the left side of c. Example 6
2. $f(x)$ increases or decreases without bound as x approaches c. Example 7
3. $f(x)$ oscillates between two fixed values as x approaches c. Example 8

Technology Tip

When using a graphing utility to investigate the behavior of a function near the x-value at which you are trying to evaluate a limit, remember that you cannot always trust the graphs that the graphing utility displays. For instance, consider the incorrect graph shown in Figure 11.6. The graphing utility cannot show the correct graph because $f(x) = \sin(1/x)$ has infinitely many oscillations over any interval that contains 0.

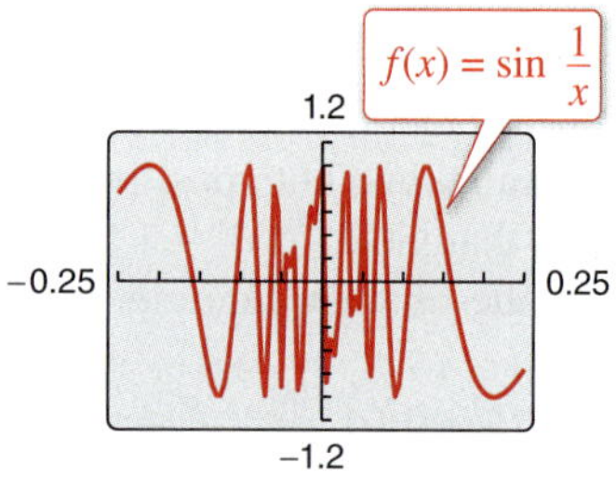

Figure 11.6

Properties of Limits

Sometimes, as in Example 2, the limit of $f(x)$ as $x \to c$ is simply $f(c)$. In such cases, it is said that the limit can be evaluated by **direct substitution.** That is,

$$\lim_{x \to c} f(x) = f(c). \qquad \text{Substitute } c \text{ for } x.$$

There are many "well-behaved" functions, such as polynomial functions, that have this property. Some of the basic ones are included in the list below.

Basic Limits

Let b and c be real numbers and let n be a positive integer.

1. $\lim_{x \to c} b = b$
2. $\lim_{x \to c} x = c$
3. $\lim_{x \to c} x^n = c^n$ (See the proof on page 814.)
4. $\lim_{x \to c} \sqrt[n]{x} = \sqrt[n]{c}$, valid for all c when n is odd and valid for $c > 0$ when n is even

Trigonometric functions can also be included in this list. For instance,

$$\lim_{x \to \pi} \sin x = \sin \pi = 0 \quad \text{and} \quad \lim_{x \to 0} \cos x = \cos 0 = 1.$$

By combining the basic limits with the properties of limits listed below, you can find limits for a wide variety of functions.

Properties of Limits

Let b and c be real numbers, let n be a positive integer, and let f and g be functions with the limits

$$\lim_{x \to c} f(x) = L \quad \text{and} \quad \lim_{x \to c} g(x) = K.$$

1. **Scalar multiple:** $\lim_{x \to c} [bf(x)] = bL$
2. **Sum or difference:** $\lim_{x \to c} [f(x) \pm g(x)] = L \pm K$
3. **Product:** $\lim_{x \to c} [f(x)g(x)] = LK$
4. **Quotient:** $\lim_{x \to c} \dfrac{f(x)}{g(x)} = \dfrac{L}{K}, \quad K \neq 0$
5. **Power:** $\lim_{x \to c} [f(x)]^n = L^n$

Technology Tip

When evaluating limits, remember that there are several ways to solve most problems. Often, a problem can be solved *numerically*, *graphically*, or *algebraically*. You can use a graphing utility to confirm the limits in the examples and in the exercise set numerically using the *table* feature or graphically using the *zoom* and *trace* features.

Explore the Concept

Use a graphing utility to graph the tangent function. What are $\lim_{x \to 0} \tan x$ and $\lim_{x \to \pi/4} \tan x$? What can you say about the existence of the limit $\lim_{x \to \pi/2} \tan x$?

EXAMPLE 9 Basic Limits and Properties of Limits

a. $\lim_{x\to 4} x^2 = (4)^2 = 16$ — Use direct substitution.

b. $\lim_{x\to 4} 5x = 5 \lim_{x\to 4} x = 5(4) = 20$ — Scalar Multiple Property

c. $\lim_{x\to \pi} \dfrac{\tan x}{x} = \dfrac{\lim_{x\to\pi} \tan x}{\lim_{x\to\pi} x} = \dfrac{0}{\pi} = 0$ — Quotient Property

d. $\lim_{x\to 9} \sqrt{x} = \sqrt{9} = 3$ — Use direct substitution.

e. $\lim_{x\to \pi} (x \cos x) = (\lim_{x\to\pi} x)(\lim_{x\to\pi} \cos x) = \pi(\cos \pi) = -\pi$ — Product Property

f. $\lim_{x\to 3} (x + 4)^2 = \left[(\lim_{x\to 3} x) + (\lim_{x\to 3} 4)\right]^2$ — Sum and Power Properties

$= (3 + 4)^2$

$= 49$

✓ Checkpoint Audio-video solution in English & Spanish at LarsonPrecalculus.com

Find (a) $\lim_{x\to 3} \dfrac{1}{4}$, (b) $\lim_{x\to 3} x^3$, and (c) $\lim_{x\to \pi} \dfrac{\cos x}{x}$.

The results of using direct substitution to evaluate limits of polynomials and rational functions are summarized below.

Limits of Polynomial and Rational Functions

1. If p is a polynomial function and c is a real number, then

$\lim_{x\to c} p(x) = p(c)$. (See the proof on page 814).

2. If r is a rational function given by $r(x) = p(x)/q(x)$, and c is a real number such that $q(c) \neq 0$, then

$$\lim_{x\to c} r(x) = r(c) = \frac{p(c)}{q(c)}.$$

EXAMPLE 10 Limits of Polynomial and Rational Functions

Find (a) $\lim_{x\to -1} (x^2 + x - 6)$ and (b) $\lim_{x\to -1} \dfrac{x^2 + x - 6}{x + 3}$.

Solution

a. To evaluate the limit of the polynomial function, use direct substitution.

$$\lim_{x\to -1} (x^2 + x - 6) = (-1)^2 + (-1) - 6 = -6$$

b. The denominator is not 0 when $x = -1$, so you can evaluate the limit of the rational function using direct substitution.

$$\lim_{x\to -1} \frac{x^2 + x - 6}{x + 3} = \frac{(-1)^2 + (-1) - 6}{-1 + 3} = -\frac{6}{2} = -3$$

✓ Checkpoint Audio-video solution in English & Spanish at LarsonPrecalculus.com

Find (a) $\lim_{x\to 3} (x^2 - 3x + 7)$ and (b) $\lim_{x\to 3} \dfrac{x^2 - 3x + 7}{x}$.

Explore the Concept

Sketch the graph of each function. Then find the limits of each function as x approaches 1 and as x approaches 2. What conclusions can you make?

a. $f(x) = x + 1$

b. $g(x) = \dfrac{x^2 - 1}{x - 1}$

c. $h(x) = \dfrac{x^3 - 2x^2 - x + 2}{x^2 - 3x + 2}$

Use a graphing utility to graph each function. Does the graphing utility distinguish among the three graphs? Write a short explanation of your findings.

Explore the Concept

Use a graphing utility to graph the function

$$f(x) = \frac{x^2 - 3x - 10}{x - 5}.$$

Use the *trace* feature to approximate $\lim_{x\to 4} f(x)$. What do you think $\lim_{x\to 5} f(x)$ equals? Is f defined at $x = 5$? Does this affect the existence of the limit as x approaches 5?

11.1 Exercises

See *CalcChat.com* for tutorial help and worked-out solutions to odd-numbered exercises.
For instructions on how to use a graphing utility, see Appendix A.

Vocabulary and Concept Check

In Exercises 1 and 2, fill in the blank.

1. If $f(x)$ becomes arbitrarily close to a unique number L as x approaches c from either side, then the _______ of $f(x)$ as x approaches c is L.

2. The _______ property indicates that $\lim_{x\to c}[bf(x)] = bL$ when b is a real number and f is a function with the limit $\lim_{x\to c}[f(x)] = L$.

3. List three basic limits that result from direct substitution.

4. List three conditions under which limits do not exist.

Procedures and Problem Solving

5. ***Why you should learn it*** *(p. 760)* A packaging machine creates an open box from a square piece of material, 24 centimeters on a side. It cuts equal squares with sides of length x from the corners and turns up the sides.

(a) Verify that the volume V of the box is given by $V = 4x(12 - x)^2$.

(b) The box has a maximum volume when $x = 4$. Use a graphing utility to complete the table and observe the behavior of the function as x approaches 4. Use the table to find $\lim_{x\to 4} V$.

x	3	3.5	3.9	4	4.1	4.5	5
V							

(c) Use the graphing utility to graph the volume function. Verify that the volume is maximum when $x = 4$.

6. Landscape Design A landscaper arranges bricks to cover a region shaped like a right triangle with a hypotenuse of $\sqrt{18}$ meters and an area as large as possible.

(a) Verify that the area A of the triangle is given by $A = \frac{1}{2}x\sqrt{18 - x^2}$.

(b) The triangle has a maximum area when $x = 3$ meters. Use a graphing utility to complete the table and observe the behavior of the function as x approaches 3. Use the table to find $\lim_{x\to 3} A$.

x	2	2.5	2.9	3	3.1	3.5	4
A							

(c) Use the graphing utility to graph the area function. Verify that the area is maximum when $x = 3$ meters.

Estimating a Limit Numerically **In Exercises 7–10, complete the table and use the result to estimate the limit numerically. Determine whether the limit can be reached.**

7. $\lim_{x\to 2}(5x + 4)$

x	1.9	1.99	1.999	2	2.001	2.01	2.1
$f(x)$				?			

8. $\lim_{x\to 0}(2x^2 + x - 4)$

x	-0.1	-0.01	-0.001	0	0.001	0.01	0.1
$f(x)$				?			

9. $\lim_{x\to -2}\dfrac{x + 2}{x^2 - 4}$

x	-2.1	-2.01	-2.001	-2	-1.999	-1.99	-1.9
$f(x)$				?			

10. $\lim_{x\to 1}\dfrac{x^2 + 3x - 4}{x - 1}$

x	0.9	0.99	0.999	1	1.001	1.01	1.1
y				?			

Thomas-Soellner/iStock/Getty Images

Using a Graphing Utility to Estimate a Limit In Exercises 11–26, use the *table* feature of a graphing utility to create a table for the function and use the result to estimate the limit numerically. Use the graphing utility to graph the corresponding function to confirm your result.

11. $\lim_{x \to -1} \frac{x^2 - 1}{x + 1}$

12. $\lim_{x \to 2} \frac{4 - x^2}{x - 2}$

13. $\lim_{x \to 1} \frac{x - 1}{x^2 + 2x - 3}$

14. $\lim_{x \to -3} \frac{x + 3}{x^2 + x - 6}$

15. $\lim_{x \to 6} \frac{\sqrt{x + 3} - 3}{x - 6}$

16. $\lim_{x \to -3} \frac{\sqrt{1 - x} - 2}{x + 3}$

17. $\lim_{x \to -4} \frac{\frac{x}{x + 2} - 2}{x + 4}$

18. $\lim_{x \to 2} \frac{\frac{1}{x + 2} - \frac{1}{4}}{x - 2}$

19. $\lim_{x \to 0} \frac{\sin x}{x}$

20. $\lim_{x \to 0} \frac{\cos x - 1}{x}$

21. $\lim_{x \to 0} \frac{\sin^2 x}{x}$

22. $\lim_{x \to 0} \frac{-x}{\tan x}$

23. $\lim_{x \to 0} \frac{e^{2x} - 1}{2x}$

24. $\lim_{x \to 0} \frac{1 - e^{-4x}}{x}$

25. $\lim_{x \to 2} \frac{\ln(2x - 3)}{x - 2}$

26. $\lim_{x \to 1} \frac{\ln x^2}{x - 1}$

Using a Graph to Find a Limit In Exercises 27–32, graph the function and find the limit (if it exists) as x approaches 2.

27. $f(x) = \begin{cases} 3, & x \neq 2 \\ 1, & x = 2 \end{cases}$

28. $f(x) = \begin{cases} 5, & x \neq 2 \\ -4, & x = 2 \end{cases}$

29. $f(x) = \begin{cases} 2x + 1, & x < 2 \\ x + 3, & x \geq 2 \end{cases}$

30. $f(x) = \begin{cases} 2x - 3, & x < 2 \\ x^2 - 4x + 5, & x \geq 2 \end{cases}$

31. $f(x) = \begin{cases} 2x, & x < 2 \\ x^2, & x \geq 2 \end{cases}$

32. $f(x) = \begin{cases} 2 - x^2, & x < 2 \\ x - 2, & x \geq 2 \end{cases}$

Using a Graph to Find a Limit In Exercises 33–42, use the graph to find the limit (if it exists). If the limit does not exist, explain why.

33. $\lim_{x \to -2} (x^2 - 2)$

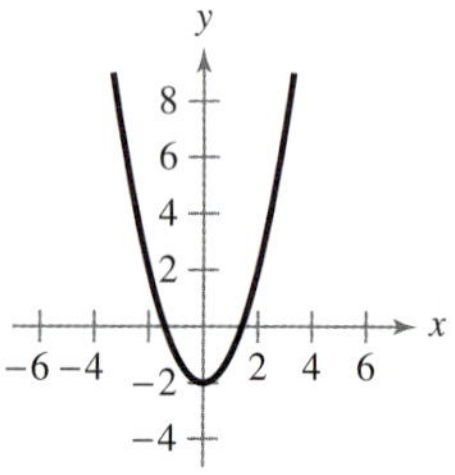

34. $\lim_{x \to -2} \frac{x^2 - 4}{x + 2}$

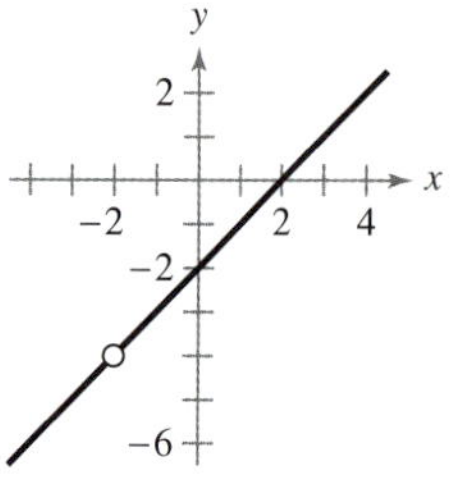

35. $\lim_{x \to -2} \frac{|x + 2|}{x + 2}$

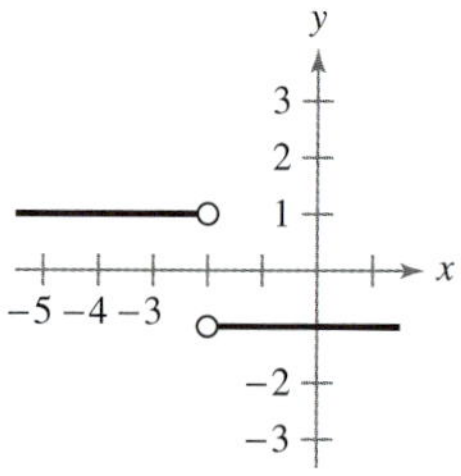

36. $\lim_{x \to 3} \frac{x - 3}{|x - 3|}$

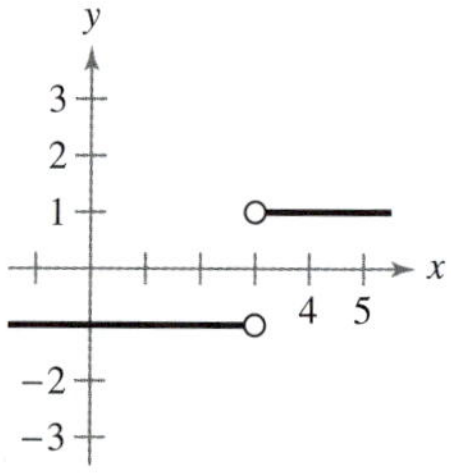

37. $\lim_{x \to 1} \frac{1}{x - 1}$

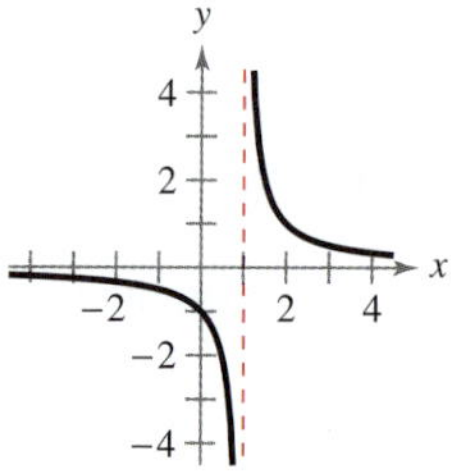

38. $\lim_{x \to \pi/2} \sec x$

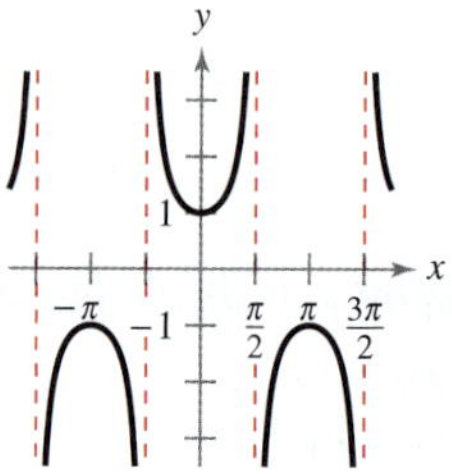

39. $\lim_{x \to -1} \sin \frac{\pi x}{2}$

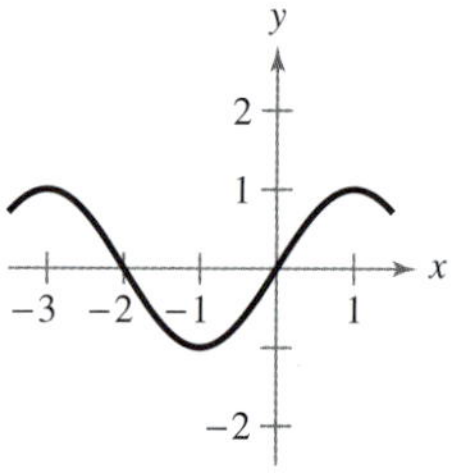

40. $\lim_{x \to \pi/4} -3 \cos x$

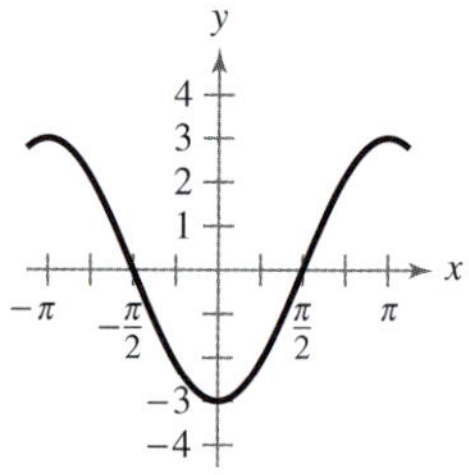

41. $\lim_{x \to 0} 2 \cos \frac{\pi}{x}$

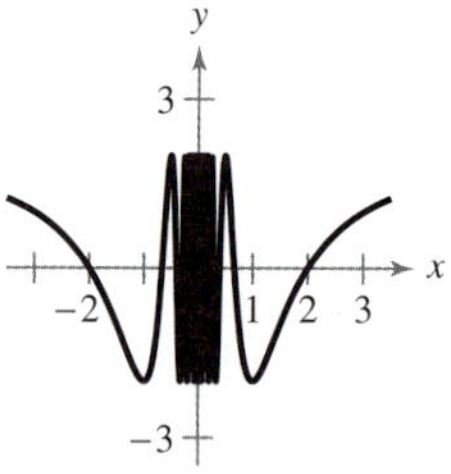

42. $\lim_{x \to 0} \sin \frac{\pi}{x} + 1$

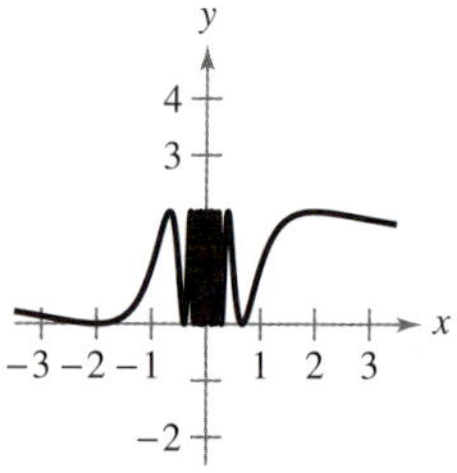

Determining Whether a Limit Exists In Exercises 43–50, use a graphing utility to graph the function and use the graph to determine whether the limit exists. If the limit does not exist, explain why.

43. $f(x) = \frac{5}{2 + e^{1/x}}, \quad \lim_{x \to 0} f(x)$

44. $f(x) = \frac{e^x - 1}{x}, \quad \lim_{x \to 0} f(x)$

45. $f(x) = \sin \pi x, \quad \lim_{x \to -1} f(x)$

46. $f(x) = \cos \frac{1}{x}, \quad \lim_{x \to 0} f(x)$

47. $f(x) = \dfrac{\sqrt{x+3} - 1}{x - 4}, \quad \lim_{x \to 4} f(x)$

48. $f(x) = \dfrac{2 + |x|}{1 - |x|}, \quad \lim_{x \to 1} f(x)$

49. $f(x) = \ln(x + 3), \quad \lim_{x \to 4} f(x)$

50. $f(x) = \ln(7 - x), \quad \lim_{x \to -1} f(x)$

Evaluating Limits **In Exercises 51 and 52, use the given information to evaluate each limit.**

51. $\lim_{x \to c} f(x) = 3, \quad \lim_{x \to c} g(x) = 6$

(a) $\lim_{x \to c} [-2g(x)]$ (b) $\lim_{x \to c} [f(x) + g(x)]$

(c) $\lim_{x \to c} \dfrac{f(x)}{g(x)}$ (d) $\lim_{x \to c} \sqrt{f(x)}$

52. $\lim_{x \to c} f(x) = 5, \quad \lim_{x \to c} g(x) = -2$

(a) $\lim_{x \to c} [f(x) + g(x)]^2$ (b) $\lim_{x \to c} [6f(x)g(x)]$

(c) $\lim_{x \to c} \dfrac{5g(x)}{4f(x)}$ (d) $\lim_{x \to c} \dfrac{1}{\sqrt{f(x)}}$

Evaluating Limits **In Exercises 53 and 54, find (a) $\lim_{x \to 2} f(x)$, (b) $\lim_{x \to 2} g(x)$, (c) $\lim_{x \to 2} [f(x)g(x)]$, and (d) $\lim_{x \to 2} [g(x) - f(x)]$.**

53. $f(x) = x^4, \quad g(x) = \dfrac{\sqrt{x^2 + 5}}{2x^2}$

54. $f(x) = \dfrac{x}{3 - x}, \quad g(x) = \sin \pi x$

Evaluating a Limit by Direct Substitution **In Exercises 55–74, find the limit using direct substitution.**

55. $\lim_{x \to 4} (8 - x^2)$

56. $\lim_{x \to -2} (x^3 - 5x)$

57. $\lim_{x \to -3} (2x^2 + 4x + 1)$

58. $\lim_{x \to -2} (x^3 - 6x + 5)$

59. $\lim_{x \to 4} \left(-\dfrac{8}{x}\right)$

60. $\lim_{x \to -5} \dfrac{6}{x + 4}$

61. $\lim_{x \to -3} \dfrac{3x}{x^2 + 1}$

62. $\lim_{x \to 4} \dfrac{x - 1}{x^2 + 2x + 3}$

63. $\lim_{x \to -2} \dfrac{6 - x^2}{x}$

64. $\lim_{x \to 3} \dfrac{x^2 + 1}{x}$

65. $\lim_{x \to -1} \sqrt{x + 2}$

66. $\lim_{x \to 3} \sqrt[3]{x^2 - 1}$

67. $\lim_{x \to 0} \dfrac{2x + 3}{\sqrt{x + 16}}$

68. $\lim_{x \to 0} \dfrac{\sqrt{x + 1}}{x - 5}$

69. $\lim_{x \to 3} e^{-2x}$

70. $\lim_{x \to e} \ln x$

71. $\lim_{x \to \pi} \sin 2x$

72. $\lim_{x \to \pi} \tan x$

73. $\lim_{x \to 1/2} \arcsin x$

74. $\lim_{x \to 1/4} \arccos 2x$

Focusing on Concepts

True or False? **In Exercises 75 and 76, determine whether the statement is true or false. Justify your answer.**

75. If $f(x) = x^{1/3}$, then the limit of $f(x)$ as x approaches c is $f(c)$.

76. If f is a rational function, then the limit of $f(x)$ as x approaches c is $f(c)$.

77. **Think About It** From Exercises 7 to 10, select a limit that can be reached and one that cannot be reached.

(a) Use a graphing utility to graph the corresponding functions using a standard viewing window. Do the graphs reveal whether the limit can be reached? Explain.

(b) Use the graphing utility to graph the corresponding functions using a *decimal* setting. Do the graphs reveal whether the limit can be reached? Explain.

78. **Think About It** Use the results of Exercise 77 to draw a conclusion as to whether you can use the graph generated by a graphing utility to determine reliably when a limit can be reached.

79. **Think About It**

(a) Given $f(2) = 4$, can you conclude anything about $\lim_{x \to 2} f(x)$? Explain your reasoning.

(b) Given $\lim_{x \to 2} f(x) = 4$, can you conclude anything about $f(2)$? Explain your reasoning.

80. **HOW DO YOU SEE IT?** Use the graph of the function f to decide whether the value of the given quantity exists. If it does, find it. If not, explain why.

(a) $f(0)$

(b) $\lim_{x \to 0} f(x)$

(c) $f(2)$

(d) $\lim_{x \to 2} f(x)$

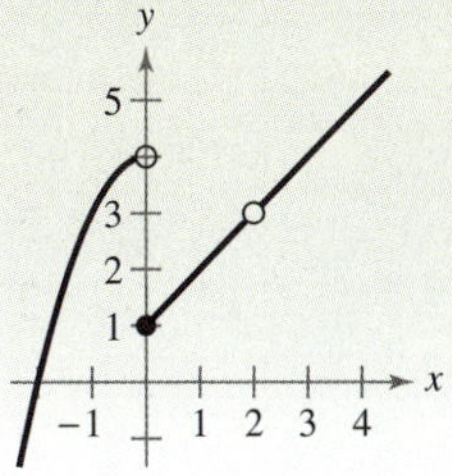

Error Analysis **In Exercises 81 and 82, describe the error.**

81. $\lim_{x \to -16} \sqrt[4]{x} = \sqrt[4]{-16}$ ✗

82. $\lim_{x \to -1} \dfrac{x^2 - 1}{|x + 1|} = 2$ ✗

Cumulative Mixed Review

Simplifying Rational Expressions **In Exercises 83–86, simplify the rational expression.**

83. $\dfrac{5 - x}{3x - 15}$

84. $\dfrac{x^2 - 81}{9 - x}$

85. $\dfrac{x^2 - 12x + 36}{x^2 - 7x + 6}$

86. $\dfrac{x^3 - 8}{x^2 - 4}$

11.2 Techniques for Evaluating Limits

Dividing Out Technique

In Section 11.1, you studied several types of functions whose limits can be evaluated by direct substitution. In this section, you will study several techniques for evaluating limits of functions for which direct substitution fails. For instance, consider using direct substitution to find the limit

$$\lim_{x \to -3} \frac{x^2 + x - 6}{x + 3}.$$

The numerator and denominator are both 0 when $x = -3$, so direct substitution fails. By using a table, however, it appears that the limit of the function as x approaches -3 is -5.

x	-3.01	-3.001	-3.0001	-3	-2.9999	-2.999	-2.99
$\dfrac{x^2 + x - 6}{x + 3}$	-5.01	-5.001	-5.0001	?	-4.9999	-4.999	-4.99

Another way to find the limit of this function is shown in Example 1.

What you should learn

- Use the dividing out technique to evaluate limits of functions.
- Use the rationalizing technique to evaluate limits of functions.
- Use technology to approximate limits of functions graphically and numerically.
- Evaluate one-sided limits of functions.
- Evaluate limits of difference quotients from calculus.

Why you should learn it

Many definitions in calculus involve the limit of a function. For instance, in Exercises 61 and 62 on page 778, the definition of the velocity of a free-falling object at any instant in time involves finding the limit of a position function.

EXAMPLE 1 Dividing Out Technique

See LarsonPrecalculus.com for an interactive version of this type of example.

Find the limit.

$$\lim_{x \to -3} \frac{x^2 + x - 6}{x + 3}$$

Solution

Begin by factoring the numerator and dividing out any common factors.

$$\lim_{x \to -3} \frac{x^2 + x - 6}{x + 3} = \lim_{x \to -3} \frac{(x - 2)(x + 3)}{x + 3} \quad \text{Factor numerator.}$$

$$= \lim_{x \to -3} \frac{(x - 2)\cancel{(x + 3)}}{\cancel{x + 3}} \quad \text{Divide out common factor.}$$

$$= \lim_{x \to -3} (x - 2) \quad \text{Simplify.}$$

$$= -3 - 2 \quad \text{Direct substitution}$$

$$= -5 \quad \text{Simplify.}$$

✓ Checkpoint *Audio-video solution in English & Spanish at LarsonPrecalculus.com*

Find the limit: $\lim_{x \to 4} \dfrac{x^2 - 7x + 12}{x - 4}$.

The procedure for evaluating the limit shown in Example 1 is called the **dividing out technique.** The validity of this technique stems from the fact that when two functions agree at all but a single number c, they must have identical limit behavior at $x = c$. In Example 1, the functions given by

$$f(x) = \frac{x^2 + x - 6}{x + 3} \quad \text{and} \quad g(x) = x - 2$$

agree at all values of x other than $x = -3$. So, you can use $g(x)$ to find the limit of $f(x)$.

The dividing out technique should be applied only when direct substitution produces 0 in both the numerator *and* the denominator. An expression such as $\frac{0}{0}$ has no meaning as a real number. It is called an **indeterminate form** because you cannot, from the form alone, determine the limit. When you try to evaluate a limit of a rational function by direct substitution and encounter this form, you can conclude that the numerator and denominator must have a common factor. After factoring and dividing out, you should try direct substitution again.

EXAMPLE 2 Dividing Out Technique

Find the limit: $\lim_{x \to 1} \frac{x - 1}{x^3 - x^2 + x - 1}$.

Solution

Begin by substituting $x = 1$ into the numerator and denominator.

$$\frac{x - 1}{x^3 - x^2 + x - 1}$$

$1 - 1 = 0$ Numerator is 0 when $x = 1$.

$1^3 - 1^2 + 1 - 1 = 0$ Denominator is 0 when $x = 1$.

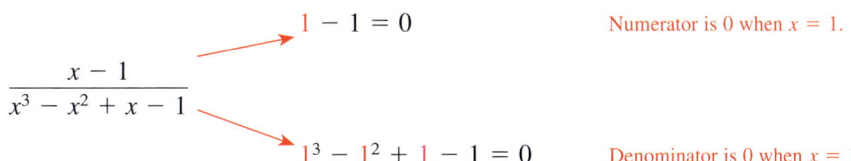

Because both the numerator and denominator are zero when $x = 1$, direct substitution will not yield the limit. To find the limit, you should factor the numerator and denominator, divide out any common factors, and then try direct substitution again.

$$\lim_{x \to 1} \frac{x - 1}{x^3 - x^2 + x - 1} = \lim_{x \to 1} \frac{x - 1}{(x - 1)(x^2 + 1)}$$ Factor denominator.

$$= \lim_{x \to 1} \frac{\cancel{x - 1}}{\cancel{(x - 1)}(x^2 + 1)}$$ Divide out common factor.

$$= \lim_{x \to 1} \frac{1}{x^2 + 1}$$ Simplify.

$$= \frac{1}{1^2 + 1}$$ Direct substitution

$$= \frac{1}{2}$$ Simplify.

This result is reinforced by the figure.

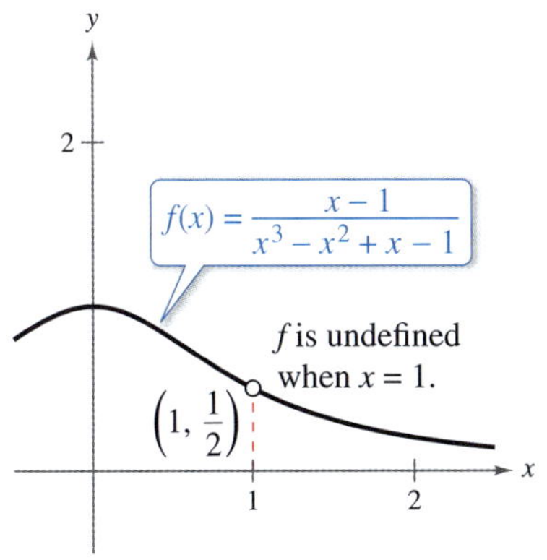

Checkpoint *Audio-video solution in English & Spanish at LarsonPrecalculus.com*

Find the limit: $\lim_{x \to 7} \frac{x - 7}{x^3 - 7x^2 + 7x - 49}$.

Algebra Help

In Example 2, the factorization of the denominator can be obtained by dividing by $(x - 1)$ or by grouping, as shown below.

$$x^3 - x^2 + x - 1$$
$$= x^2(x - 1) + (x - 1)$$
$$= (x - 1)(x^2 + 1)$$

Rationalizing Technique

Another way to find the limits of some functions is first to rationalize the numerator of the function. This is called the **rationalizing technique.** Recall that rationalizing the numerator means multiplying the numerator and denominator by the conjugate of the numerator. For instance, the conjugate of $\sqrt{x} + 4$ is $\sqrt{x} - 4$.

EXAMPLE 3 Rationalizing Technique

Find the limit.

$$\lim_{x \to 0} \frac{\sqrt{x+1} - 1}{x}$$

Solution

By direct substitution, you obtain the indeterminate form $\frac{0}{0}$.

$$\lim_{x \to 0} \frac{\sqrt{x+1} - 1}{x} = \frac{\sqrt{0+1} - 1}{0} = \frac{0}{0}$$ Indeterminate form

In this case, you can rewrite the fraction by rationalizing the numerator.

$$\frac{\sqrt{x+1} - 1}{x} = \left(\frac{\sqrt{x+1} - 1}{x}\right)\left(\frac{\sqrt{x+1} + 1}{\sqrt{x+1} + 1}\right)$$

$$= \frac{(x+1) - 1}{x(\sqrt{x+1} + 1)}$$ Multiply.

$$= \frac{x}{x(\sqrt{x+1} + 1)}$$ Simplify.

$$= \frac{\cancel{x}}{\cancel{x}(\sqrt{x+1} + 1)}$$ Divide out common factor.

$$= \frac{1}{\sqrt{x+1} + 1}, \quad x \neq 0$$ Simplify.

Now you can evaluate the limit by direct substitution.

$$\lim_{x \to 0} \frac{\sqrt{x+1} - 1}{x} = \lim_{x \to 0} \frac{1}{\sqrt{x+1} + 1} = \frac{1}{\sqrt{0+1} + 1} = \frac{1}{1+1} = \frac{1}{2}$$

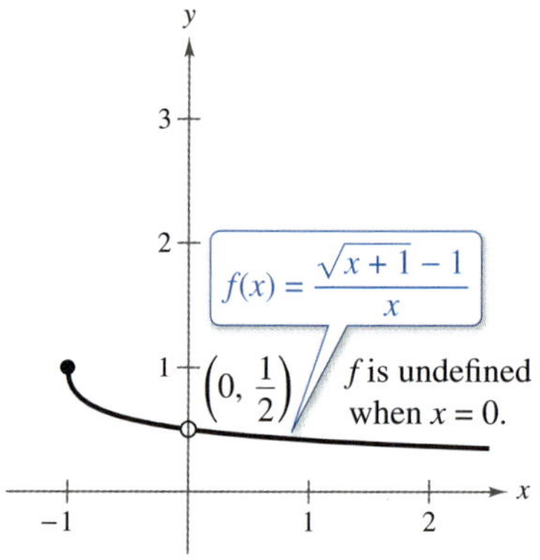

Figure 11.7

You can reinforce your conclusion that the limit is $\frac{1}{2}$ by constructing a table, as shown below, or by sketching a graph, as shown in Figure 11.7.

x	-0.1	-0.01	-0.001	0	0.001	0.01	0.1
$f(x)$	0.5132	0.5013	0.5001	?	0.4999	0.4988	0.4881

Checkpoint *Audio-video solution in English & Spanish at LarsonPrecalculus.com*

Find the limit.

$$\lim_{x \to 0} \frac{1 - \sqrt{1-x}}{x}$$

The rationalizing technique for evaluating limits is based on multiplication by a convenient form of 1. In Example 3, the convenient form is

$$1 = \frac{\sqrt{x+1} + 1}{\sqrt{x+1} + 1}.$$

Using Technology

The dividing out and rationalizing techniques may not work well for finding limits of nonalgebraic functions. You often need to use more sophisticated analytic techniques to find limits of these types of functions.

EXAMPLE 4 Approximating a Limit Numerically

Approximate the limit: $\lim_{x \to 0} (1 + x)^{1/x}$.

Solution

Let $f(x) = (1 + x)^{1/x}$.

Create a table that shows $f(x)$ for several x-values near 0.

X	Y1
−.003	2.7224
−.002	2.721
−.001	2.7196
0	ERROR
.001	2.7169
.002	2.7156
.003	2.7142

X=0

Figure 11.8

Because 0 is halfway between -0.001 and 0.001 (see Figure 11.8), use the average of the values of f at these two x-values to estimate the limit.

$$\lim_{x \to 0} (1 + x)^{1/x} \approx \frac{2.7196 + 2.7169}{2} = 2.71825$$

The actual limit can be found algebraically to be

$$e \approx 2.71828.$$

✓ Checkpoint *Audio-video solution in English & Spanish at LarsonPrecalculus.com*

Approximate the limit: $\lim_{x \to 0} \dfrac{e^x - 1}{x}$.

EXAMPLE 5 Approximating a Limit Graphically

Approximate the limit: $\lim_{x \to 0} \dfrac{\sin x}{x}$.

Solution

Direct substitution produces the indeterminate form $\frac{0}{0}$. To approximate the limit, begin by using a graphing utility to graph $f(x) = (\sin x)/x$, as shown in Figure 11.9. Then use the *zoom* and *trace* features of the graphing utility to choose a point on each side of 0, such as $(-0.0012467, 0.9999997)$ and $(0.0012467, 0.9999997)$. Because 0 is halfway between -0.0012467 and 0.0012467, use the average of the values of f at these two x-coordinates to estimate the limit.

$$\lim_{x \to 0} \frac{\sin x}{x} \approx \frac{0.9999997 + 0.9999997}{2} = 0.9999997$$

It can be shown algebraically that this limit is exactly 1.

✓ Checkpoint *Audio-video solution in English & Spanish at LarsonPrecalculus.com*

Approximate the limit: $\lim_{x \to 0} \dfrac{1 - \cos x}{x}$.

Technology Tip

In Example 4, a graph of $f(x) = (1 + x)^{1/x}$ on a graphing utility would appear continuous at $x = 0$ (see below). But when you try to use the *trace* feature of a graphing utility to determine $f(0)$, no value is given. Some graphing utilities can show breaks or holes in a graph when an appropriate viewing window is used. Because the hole in the graph of f occurs on the y-axis, the hole is not visible.

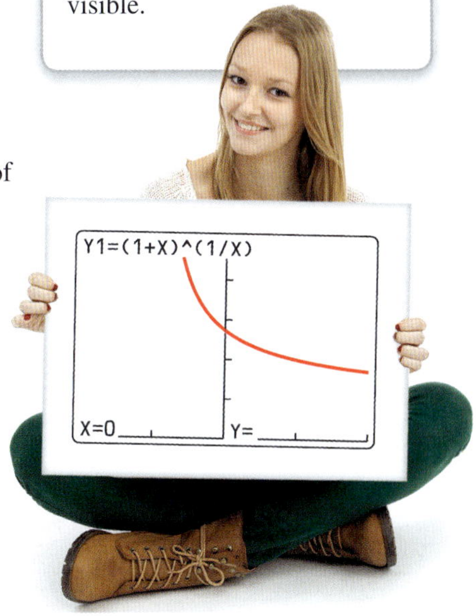

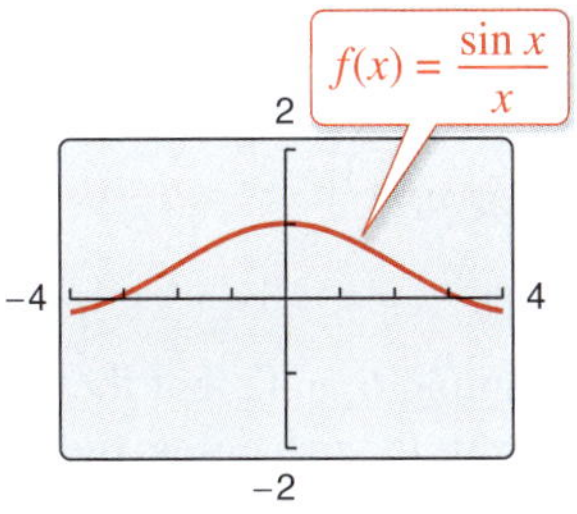

Figure 11.9

One-Sided Limits

In Section 11.1, you saw that one way in which a limit can fail to exist is when a function approaches a different value from the left side of c than it approaches from the right side of c. This type of behavior can be described more concisely with the concept of a **one-sided limit.**

$$\lim_{x \to c^-} f(x) = L_1 \text{ or } f(x) \to L_1 \text{ as } x \to c^- \qquad \text{Limit from the left}$$

$$\lim_{x \to c^+} f(x) = L_2 \text{ or } f(x) \to L_2 \text{ as } x \to c^+ \qquad \text{Limit from the right}$$

EXAMPLE 6 Evaluating One-Sided Limits

Find the limit as $x \to 0$ from the left and the limit as $x \to 0$ from the right for

$$f(x) = \frac{|2x|}{x}.$$

Solution

From the graph of f shown in the figure, you can see that $f(x) = -2$ for all $x < 0$.

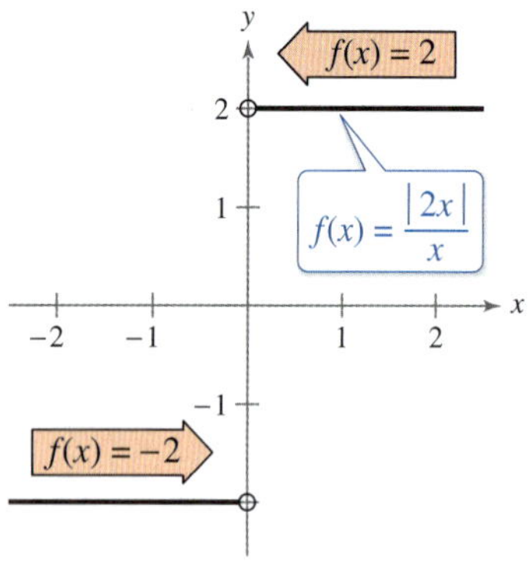

So, the limit from the left is

$$\lim_{x \to 0^-} \frac{|2x|}{x} = -2. \qquad \text{Limit from the left}$$

Because $f(x) = 2$ for all $x > 0$, the limit from the right is

$$\lim_{x \to 0^+} \frac{|2x|}{x} = 2. \qquad \text{Limit from the right}$$

✓ ***Checkpoint*** ▶ *Audio-video solution in English & Spanish at LarsonPrecalculus.com*

Find the limit as $x \to 3$ from the left and the limit as $x \to 3$ from the right for

$$f(x) = \frac{|x - 3|}{x - 3}.$$

In Example 6, note that the function approaches different limits from the left and from the right. In such cases, the limit of $f(x)$ as $x \to c$ does not exist. For the limit of a function to exist as $x \to c$, it must be true that both one-sided limits exist and are equal.

Existence of a Limit

If f is a function and c and L are real numbers, then $\lim_{x \to c} f(x) = L$ if and only if both the left and right limits *exist* and are *equal* to L.

EXAMPLE 7 Finding One-Sided Limits

Find the limit of $f(x)$ as x approaches 1.

$$f(x) = \begin{cases} 4 - x, & x < 1 \\ 4x - x^2, & x > 1 \end{cases}$$

Solution

Remember that you are concerned about the value of f *near* $x = 1$ rather than *at* $x = 1$. So, for $x < 1$, $f(x)$ is given by $4 - x$, and you can use direct substitution to obtain

$$\lim_{x\to 1^-} f(x) = \lim_{x\to 1^-} (4 - x) = 4 - 1 = 3.$$

For $x > 1$, $f(x)$ is given by $4x - x^2$, and you can use direct substitution to obtain

$$\lim_{x\to 1^+} f(x) = \lim_{x\to 1^+} (4x - x^2) = 4(1) - 1^2 = 3.$$

Because the one-sided limits both exist and are equal to 3, it follows that

$$\lim_{x\to 1} f(x) = 3.$$

The graph in Figure 11.10 supports this conclusion.

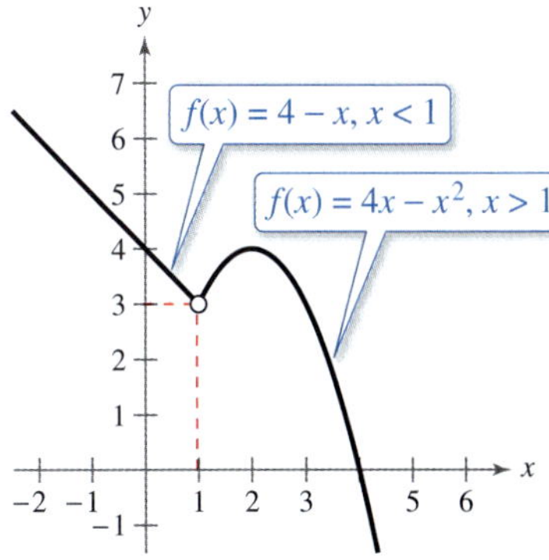

Figure 11.10

Checkpoint *Audio-video solution in English & Spanish at LarsonPrecalculus.com*

Find the limit of $f(x)$ as x approaches -1.

$$f(x) = \begin{cases} -x^2 - 3x, & x < -1 \\ x + 3, & x > -1 \end{cases}$$

EXAMPLE 8 Comparing Limits from the Left and Right

To ship a package overnight, a delivery service charges \$24 for the first pound and \$4 for each additional pound or portion of a pound. Let x represent the weight of a package and let $f(x)$ represent the shipping cost. Show that the limit of $f(x)$ as $x \to 2$ does not exist.

$$f(x) = \begin{cases} 24, & 0 < x \le 1 \\ 28, & 1 < x \le 2 \\ 32, & 2 < x \le 3 \end{cases}$$

Solution

The graph of f is shown in the figure. The limit of $f(x)$ as x approaches 2 from the left is

$$\lim_{x\to 2^-} f(x) = 28$$

whereas the limit of $f(x)$ as x approaches 2 from the right is

$$\lim_{x\to 2^+} f(x) = 32.$$

Because these one-sided limits are not equal, the limit of $f(x)$ as $x \to 2$ does not exist.

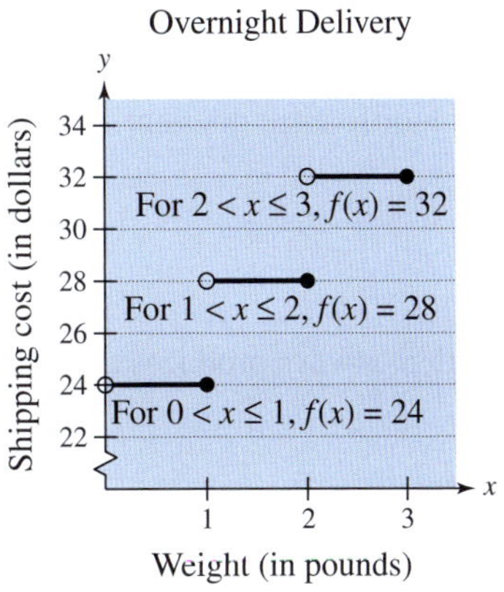

Checkpoint *Audio-video solution in English & Spanish at LarsonPrecalculus.com*

In Example 8, show that the limit of $f(x)$ as $x \to 1$ does not exist.

A Limit from Calculus

In the next section, you will study an important type of limit from calculus—the limit of a *difference quotient.*

EXAMPLE 9 Evaluating a Limit from Calculus

For the function $f(x) = x^2 - 1$, find $\lim_{h \to 0} \frac{f(3 + h) - f(3)}{h}$.

Solution

Direct substitution produces an indeterminate form.

$$\lim_{h \to 0} \frac{f(3 + h) - f(3)}{h} = \lim_{h \to 0} \frac{[(3 + h)^2 - 1] - [(3)^2 - 1]}{h} \quad \text{Direct substitution}$$

$$= \lim_{h \to 0} \frac{9 + 6h + h^2 - 1 - 9 + 1}{h} \quad \text{Expand binomial and evaluate power.}$$

$$= \lim_{h \to 0} \frac{6h + h^2}{h} \quad \text{Combine like terms.}$$

$$= \frac{0}{0} \quad \text{Indeterminate form}$$

By factoring and dividing out, you can find the limit.

$$\lim_{h \to 0} \frac{f(3 + h) - f(3)}{h} = \lim_{h \to 0} \frac{6h + h^2}{h}$$

$$= \lim_{h \to 0} \frac{h(6 + h)}{h} \quad \text{Factor numerator.}$$

$$= \lim_{h \to 0} (6 + h) \quad \text{Divide out common factor and simplify.}$$

$$= 6 + 0 \quad \text{Direct substitution}$$

$$= 6 \quad \text{Simplify.}$$

So, the limit is 6.

✓ Checkpoint ▶ *Audio-video solution in English & Spanish at LarsonPrecalculus.com*

For the function $f(x) = 4x - x^2$, find $\lim_{h \to 0} \frac{f(2 + h) - f(2)}{h}$. ■

Note that for any x-value, the limit of a difference quotient is an expression of the form

$$\lim_{h \to 0} \frac{f(x + h) - f(x)}{h}.$$

Direct substitution into the difference quotient always produces the indeterminate form $\frac{0}{0}$. For instance,

$$\lim_{h \to 0} \frac{f(x + h) - f(x)}{h} = \frac{f(x + 0) - f(x)}{0} \quad \text{Direct substitution}$$

$$= \frac{f(x) - f(x)}{0}$$

$$= \frac{0}{0}. \quad \text{Indeterminate form}$$

11.2 Exercises

See *CalcChat.com* for tutorial help and worked-out solutions to odd-numbered exercises.
For instructions on how to use a graphing utility, see Appendix A.

Vocabulary and Concept Check

In Exercises 1 and 2, fill in the blank.

1. To find a limit of a rational function that has common factors in its numerator and denominator, use the _______ .

2. The expression $\frac{0}{0}$ has no meaning as a real number and is called an _______ because you cannot, from the form alone, determine the limit.

3. Which algebraic technique can you use to find $\lim\limits_{x\to 0} \dfrac{\sqrt{x+4}-2}{x}$?

4. Describe in words what is meant by $\lim\limits_{x\to 0^+} f(x) = -2$.

Procedures and Problem Solving

Using a Graph to Find Limits **In Exercises 5–8, use the graph to find each limit. Then identify another function that agrees with the given function at all but one point.**

5. $h(x) = \dfrac{x^2 + x}{x}$

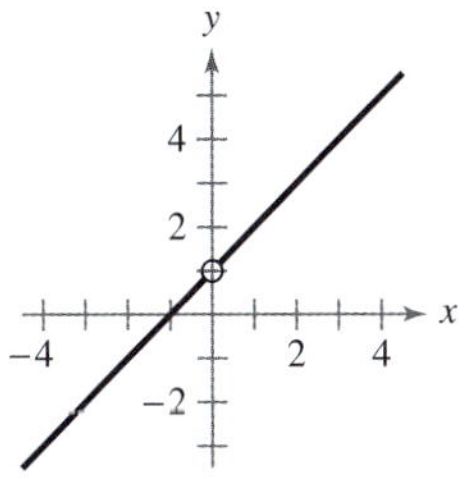

(a) $\lim\limits_{x\to -2} h(x)$

(b) $\lim\limits_{x\to 0} h(x)$

(c) $\lim\limits_{x\to 3} h(x)$

(d) $\lim\limits_{x\to 4} h(x)$

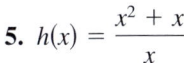

6. $g(x) = \dfrac{-2x^2 + x}{x}$

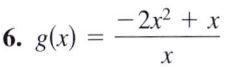

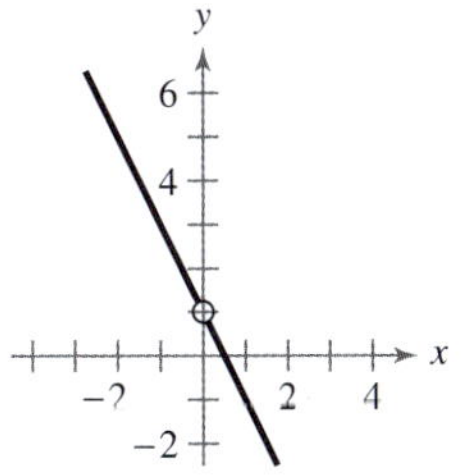

(a) $\lim\limits_{x\to 0} g(x)$

(b) $\lim\limits_{x\to -1} g(x)$

(c) $\lim\limits_{x\to -2} g(x)$

(d) $\lim\limits_{x\to 1} g(x)$

7. $g(x) = \dfrac{x^3 - x}{x - 1}$

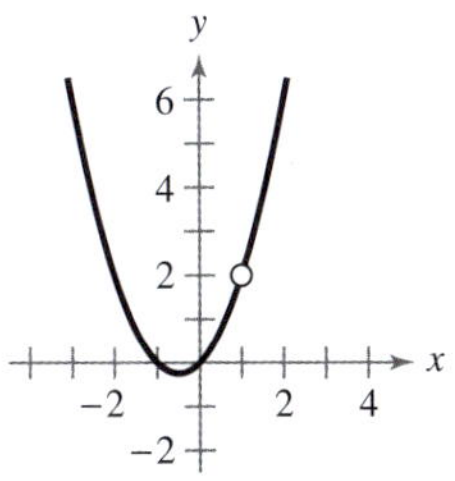

(a) $\lim\limits_{x\to 1} g(x)$

(b) $\lim\limits_{x\to -1} g(x)$

(c) $\lim\limits_{x\to 0} g(x)$

(d) $\lim\limits_{x\to -3} g(x)$

8. $f(x) = \dfrac{4x - x^3}{x + 2}$

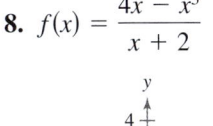

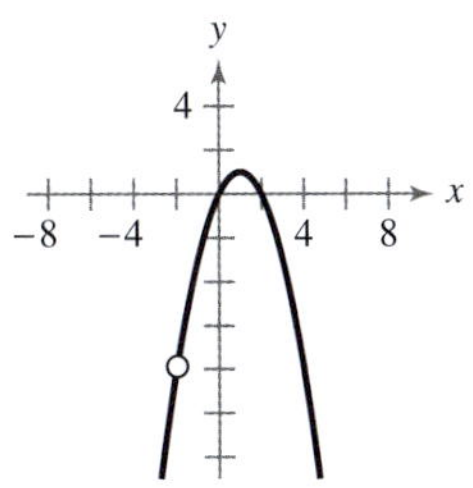

(a) $\lim\limits_{x\to 0} f(x)$

(b) $\lim\limits_{x\to -2} f(x)$

(c) $\lim\limits_{x\to 1} f(x)$

(d) $\lim\limits_{x\to 2} f(x)$

Finding a Limit **In Exercises 9–30, find the limit (if it exists). Use a graphing utility to confirm your result graphically.**

9. $\lim\limits_{x\to 2} \dfrac{x^2 - x - 2}{x - 2}$

10. $\lim\limits_{x\to -1} \dfrac{x^2 + 6x + 5}{x + 1}$

11. $\lim\limits_{x\to -1} \dfrac{1 - 2x - 3x^2}{1 + x}$

12. $\lim\limits_{x\to -3} \dfrac{3 - 5x - 2x^2}{3 + x}$

13. $\lim\limits_{x\to 6} \dfrac{x - 6}{x^2 - 36}$

14. $\lim\limits_{x\to 9} \dfrac{x - 9}{x^2 - 81}$

15. $\lim\limits_{t\to 2} \dfrac{t^3 - 8}{t - 2}$

16. $\lim\limits_{a\to -4} \dfrac{a^3 + 64}{a + 4}$

17. $\lim\limits_{x\to 2} \dfrac{x^5 - 32}{x - 2}$

18. $\lim\limits_{x\to 1} \dfrac{x^5 - 1}{x - 1}$

19. $\lim\limits_{x\to -4} \dfrac{x^2 + x - 12}{x^2 + 6x + 8}$

20. $\lim\limits_{x\to 3} \dfrac{x^2 - 8x + 15}{x^2 - 2x - 3}$

21. $\lim\limits_{x\to -1} \dfrac{x^3 + 2x^2 - x - 2}{x^3 + 4x^2 - x - 4}$

22. $\lim\limits_{x\to -3} \dfrac{x^3 + 2x^2 - 9x - 18}{x^3 + x^2 - 9x - 9}$

23. $\lim\limits_{y\to 0} \dfrac{\sqrt{5+y} - \sqrt{5}}{y}$

24. $\lim\limits_{z\to 0} \dfrac{\sqrt{7-z} - \sqrt{7}}{z}$

25. $\lim\limits_{x\to -3} \dfrac{\sqrt{x+7} - 2}{x + 3}$

26. $\lim\limits_{x\to 2} \dfrac{4 - \sqrt{18 - x}}{x - 2}$

27. $\lim\limits_{x\to 0} \dfrac{\frac{1}{x+1} - 1}{x}$

28. $\lim\limits_{x\to 0} \dfrac{\frac{1}{x-8} + \frac{1}{8}}{x}$

29. $\lim\limits_{x\to 0} \dfrac{\frac{1}{x+4} - \frac{1}{4}}{x}$

30. $\lim\limits_{x\to 0} \dfrac{\frac{1}{2+x} - \frac{1}{2}}{x}$

31. $\lim\limits_{x\to \pi/2} \dfrac{1 - \sin x}{\cos x}$

32. $\lim\limits_{x\to 0} \dfrac{\cos x - 1}{\sin x}$

33. $\lim\limits_{x\to 0} \dfrac{\cos 2x}{\cot 2x}$

34. $\lim\limits_{x\to \pi} \dfrac{\sin x}{\csc x}$

35. $\lim\limits_{x\to \pi/2} \dfrac{\sin x - 1}{x}$

36. $\lim\limits_{x\to \pi} \dfrac{1 + \cos x}{x}$

Approximating a Limit Numerically In Exercises 37–42, use the *table* feature of a graphing utility to create a table for the function and use the result to approximate the limit numerically. Write an approximation that is accurate to three decimal places.

37. $\lim_{x \to 0} \dfrac{e^{2x} - 1}{x}$

38. $\lim_{x \to 0} \dfrac{1 - e^{-x}}{x}$

39. $\lim_{x \to 1} \dfrac{x - 1}{\ln x}$

40. $\lim_{x \to 2} \dfrac{-\ln(x - 1)}{x - 2}$

41. $\lim_{x \to 0}(1 + x)^{3/x}$

42. $\lim_{x \to 0}(1 + 2x)^{1/x}$

Approximating a Limit Graphically In Exercises 43–48, use a graphing utility to graph the function and approximate the limit. Write an approximation that is accurate to three decimal places.

43. $\lim_{x \to 0} \dfrac{\sin 2x}{x}$

44. $\lim_{x \to 0} \dfrac{\sin 3x}{x}$

45. $\lim_{x \to 0} \dfrac{\tan x}{x}$

46. $\lim_{x \to 0} \dfrac{1 - \cos 2x}{x}$

47. $\lim_{x \to 1} \dfrac{1 - \sqrt[3]{x}}{1 - x}$

48. $\lim_{x \to 1} \dfrac{\sqrt[3]{x} - x}{x - 1}$

Evaluating One-Sided Limits In Exercises 49–56, graph the function. Evaluate the corresponding one-sided limits. Then determine the limit (if it exists).

49. $\lim_{x \to 1} \dfrac{|x - 1|}{x - 1}$

50. $\lim_{x \to 2} \dfrac{|x - 2|}{x - 2}$

51. $\lim_{x \to -1} \dfrac{1}{x^3 + 1}$

52. $\lim_{x \to 1} \dfrac{1}{x^2 - 1}$

53. $\lim_{x \to 2} f(x)$ where $f(x) = \begin{cases} x - 1, & x \le 2 \\ 2x - 3, & x > 2 \end{cases}$

54. $\lim_{x \to 1} f(x)$ where $f(x) = \begin{cases} 2x + 1, & x < 1 \\ x + 2, & x \ge 1 \end{cases}$

55. $\lim_{x \to 1} f(x)$ where $f(x) = \begin{cases} 4 - x^2, & x \le 1 \\ 3 - x, & x > 1 \end{cases}$

56. $\lim_{x \to 0} f(x)$ where $f(x) = \begin{cases} x^2 + 3, & x \le 0 \\ x + 5, & x > 0 \end{cases}$

Algebraic-Graphical-Numerical In Exercises 57–60, (a) graphically approximate the limit (if it exists) by using a graphing utility to graph the function, (b) numerically approximate the limit (if it exists) by using the *table* feature of the graphing utility to create a table, and (c) algebraically evaluate the limit (if it exists) by the appropriate technique(s).

57. $\lim_{x \to 2} \dfrac{x^4 - 2x^2 - 8}{x^4 - 6x^2 + 8}$

58. $\lim_{x \to 2} \dfrac{x^4 - 1}{x^4 - 3x^2 - 4}$

59. $\lim_{x \to 16^+} \dfrac{4 - \sqrt{x}}{x - 16}$

60. $\lim_{x \to 0^-} \dfrac{\sqrt{x + 2} - \sqrt{2}}{x}$

Why you should learn it *(p. 770)* In Exercises 61 and 62, use the position function

$$s(t) = -16t^2 + 128$$

which gives the height (in feet) of a free-falling object t seconds after it begins falling. The velocity at time $t = a$ seconds is given by

$$\lim_{t \to a} \frac{s(a) - s(t)}{a - t}.$$

61. Find the velocity when $t = 1$ second.

62. Find the velocity when $t = 3$ seconds.

63. Human Resources A union contract guarantees a 3% salary increase each year. Based on a starting salary of \$50,000, the salaries $f(t)$ (in thousands of dollars) for 3 years are given by

$$f(t) = \begin{cases} 50{,}000, & 0 < t \le 1 \\ 51{,}500, & 1 < t \le 2 \\ 53{,}045, & 2 < t \le 3 \end{cases}$$

where t represents the time in years. Show that the limit of f as $t \to 2$ does not exist.

64. Business The cost of sending a package overnight is \$25.95 for the first pound and \$1.20 for each additional pound or portion of a pound. A plastic mailing bag can hold up to 3 pounds. The cost $f(x)$ of sending a package in a plastic mailing bag is given by

$$f(x) = \begin{cases} 25.95, & 0 < x \le 1 \\ 27.15, & 1 < x \le 2 \\ 28.35, & 2 < x \le 3 \end{cases}$$

where x represents the weight of the package (in pounds). Show that the limit of f as $x \to 1$ does not exist.

65. MODELING DATA

The cost C (in dollars) of making x color photocopies at a copy shop is given by the function

$$C(x) = \begin{cases} 0.42x, & 0 < x \le 99 \\ 0.32x, & 99 < x \le 999 \\ 0.23x, & 999 < x \le 9999 \\ 0.16x, & x > 9999 \end{cases}$$

(a) Sketch a graph of the function.

(b) Find each limit and interpret your result in the context of the situation.

(i) $\lim_{x \to 15} C(x)$ (ii) $\lim_{x \to 450} C(x)$ (iii) $\lim_{x \to 6000} C(x)$

(c) Create a table of values to show numerically that each limit does not exist.

(i) $\lim_{x \to 99} C(x)$ (ii) $\lim_{x \to 999} C(x)$ (iii) $\lim_{x \to 9999} C(x)$

(d) Explain how you can use the graph in part (a) to verify that the limits in part (c) do not exist.

66. MODELING DATA

The cost of hooking up and towing a car is \$80 for the first mile and \$3 for each additional mile or portion of a mile. A model for the cost C (in dollars) is $C(x) = 80 - 3[\![-(x - 1)]\!]$, where x is the distance in miles. (Recall from Section 1.3 that $f(x) = [\![x]\!]$ is defined as the greatest integer less than or equal to x.)

(a) Use a graphing utility to graph C for $0 < x \le 10$.

(b) Complete the table and observe the behavior of C as x approaches 5.5. Use the graph from part (a) and the table to find $\lim_{x \to 5.5} C(x)$.

x	5	5.3	5.4	5.5	5.6	5.7	6
$C(x)$				?			

(c) Complete the table and observe the behavior of C as x approaches 5. Does the limit of $C(x)$ as x approaches 5 exist? Explain.

x	4	4.5	4.9	5	5.1	5.5	6
$C(x)$				?			

Evaluating a Limit from Calculus **In Exercises 67–72, find**

$$\lim_{h \to 0} \frac{f(x + h) - f(x)}{h}.$$

67. $f(x) = 3x - 1$

68. $f(x) = 5 + 7x$

69. $f(x) = \sqrt{x - 2}$

70. $f(x) = \sqrt{x}$

71. $f(x) = x^2 - 3x - 6$

72. $f(x) = 3x^2 - 5x + 1$

Finding Limits **In Exercises 73 and 74, state whether each limit can be evaluated by using direct substitution. Then evaluate or approximate each limit.**

73. (a) $\lim_{x \to 0} x^2 \sin x^2$ (b) $\lim_{x \to 0} \frac{\sin x^2}{x^2}$

74. (a) $\lim_{x \to 0} \frac{x}{\cos x}$ (b) $\lim_{x \to 0} \frac{1 - \cos x}{x}$

Using Graphs to Find a Limit **In Exercises 75–80, use a graphing utility to graph the function and the equations $y = |x|$ and $y = -|x|$ in the same viewing window. Use the graphs to find $\lim_{x \to 0} f(x)$.**

75. $f(x) = x \cos x$

76. $f(x) = |x \sin x|$

77. $f(x) = |x| \sin x$

78. $f(x) = |x| \cos x$

79. $f(x) = x \sin \frac{1}{x}$

80. $f(x) = x \cos \frac{1}{x}$

Focusing on Concepts

True or False? **In Exercises 81 and 82, determine whether the statement is true or false. Justify your answer.**

81. When your attempt to find the limit of a rational function yields the indeterminate form $\frac{0}{0}$, the numerator and denominator of the rational function have a common factor.

82. If $f(c) = L$, then $\lim_{x \to c} f(c) = L$.

83. Error Analysis Describe the error evaluating

$$\lim_{x \to -3} \frac{x^2 - 2x - 15}{x + 3}.$$

The limit does not exist because division by 0 is undefined.

84. Think About It Sketch the graph of a function for which the limit of $f(x)$ as x approaches 1 is 4 but $f(1) \ne 4$.

85. Writing Consider the limit of the rational function $p(x)/q(x)$. What conclusion can you make when direct substitution produces each expression? Write a short paragraph explaining your reasoning.

(a) $\lim_{x \to c} \frac{p(x)}{q(x)} = \frac{0}{1}$ (b) $\lim_{x \to c} \frac{p(x)}{q(x)} = \frac{1}{1}$

(c) $\lim_{x \to c} \frac{p(x)}{q(x)} = \frac{1}{0}$ (d) $\lim_{x \to c} \frac{p(x)}{q(x)} = \frac{0}{0}$

86. HOW DO YOU SEE IT? For the function

$$f(x) = \begin{cases} 2x, & x \le 0 \\ x^2 + 1, & x > 0 \end{cases}$$

shown at the right, find each of the following limits. If the limit does not exist, then explain why.

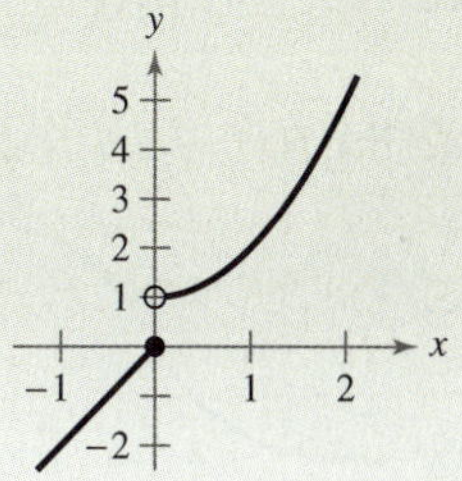

(a) $\lim_{x \to 0^-} f(x)$

(b) $\lim_{x \to 0^+} f(x)$

(c) $\lim_{x \to 0} f(x)$

Cumulative Mixed Review

Identifying a Conic from Its Equation **In Exercises 87 and 88, identify the type of conic represented by the equation. Use a graphing utility to confirm your result.**

87. $r = \frac{3}{1 + \cos \theta}$

88. $r = \frac{12}{3 + 2 \sin \theta}$

11.3 The Tangent Line Problem

Tangent Line to a Graph

Calculus is a branch of mathematics that studies rates of change of functions. If you go on to take a course in calculus, you will learn that rates of change have many applications in real life.

Earlier in the text, you learned that the slope of a line is the rate at which a line rises or falls. This rate (or slope) is the same at every point on the line. For graphs other than lines, the rate at which the graph rises or falls changes from point to point. For instance, in Figure 11.11, the parabola is rising more quickly at the point (x_1, y_1) than it is at the point (x_2, y_2). At the vertex (x_3, y_3), the graph levels off, and at the point (x_4, y_4), the graph is falling.

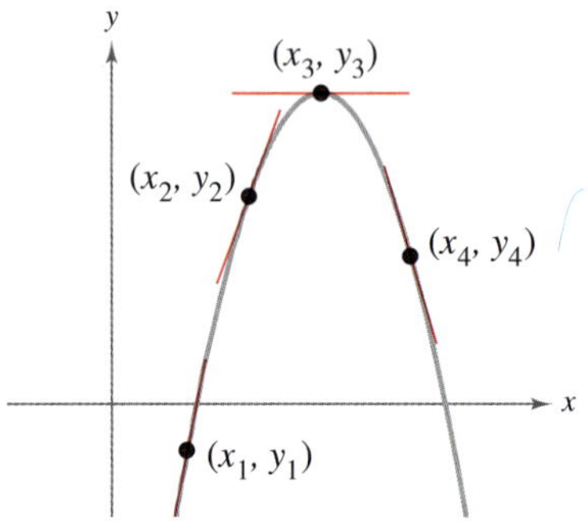

Figure 11.11

To determine the rate at which a graph rises or falls at a *single point*, you can find the slope of the tangent line at that point. In simple terms, the **tangent line** to the graph of a function f at a point

$P(x_1, y_1)$

is the line that best approximates the slope of the graph at the point. Figure 11.12 shows other examples of tangent lines.

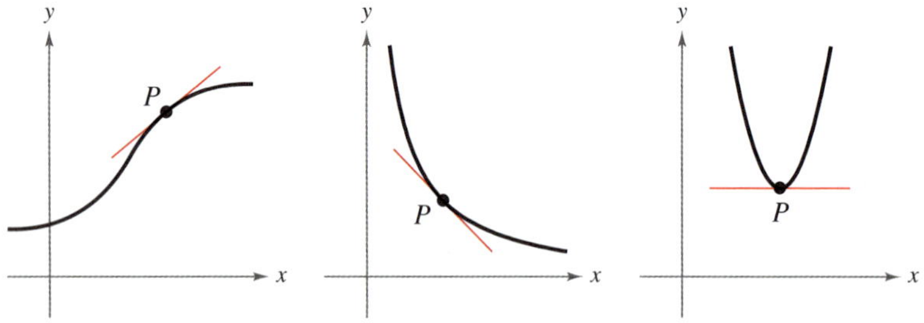

Figure 11.12

From geometry, you know that a line is tangent to a circle when the line intersects the circle at only one point (see Figure 11.13). Tangent lines to noncircular graphs, however, can intersect the graph at more than one point. For instance, in the first graph in Figure 11.12, if the tangent line were extended, it would intersect the graph at a point other than the point of tangency.

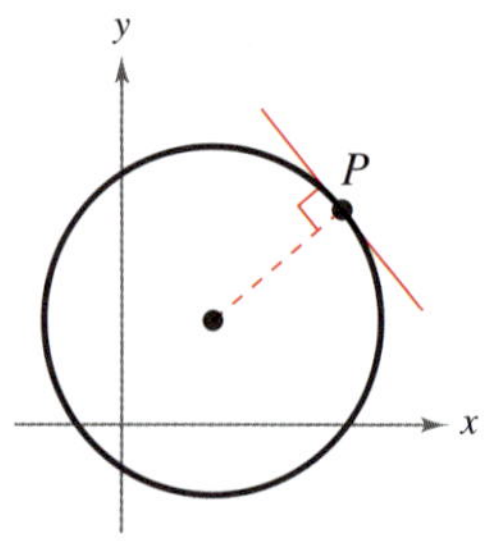

Figure 11.13

What you should learn

- Understand the tangent line problem.
- Use a tangent line to approximate the slope of a graph at a point.
- Use the limit definition of slope to find exact slopes of graphs.
- Find derivatives of functions and use derivatives to find slopes of graphs.

Why you should learn it

The derivative, or the slope of the tangent line to the graph of a function at a point, can be used to analyze rates of change. For instance, in Exercise 69 on page 789, the derivative is used to analyze the rate of change of the volume of a spherical balloon.

Slope of a Graph

Because a tangent line approximates the slope of a graph at a point, the problem of finding the slope of a graph at a point is the same as finding the slope of the tangent line at the point.

EXAMPLE 1 Visually Approximating the Slope of a Graph

Use Figure 11.14 to approximate the slope of the graph of $f(x) = x^2$ at the point (1, 1).

Solution

From the graph of $f(x) = x^2$, you can see that the tangent line at (1, 1) rises approximately two units for each unit change in x. So, you can estimate the slope of the tangent line at (1, 1) to be

$$\text{Slope} = \frac{\text{change in } y}{\text{change in } x} \approx \frac{2}{1} = 2.$$

Because the tangent line at the point (1, 1) has a slope of about 2, you can conclude that the graph of f has a slope of about 2 at the point (1, 1).

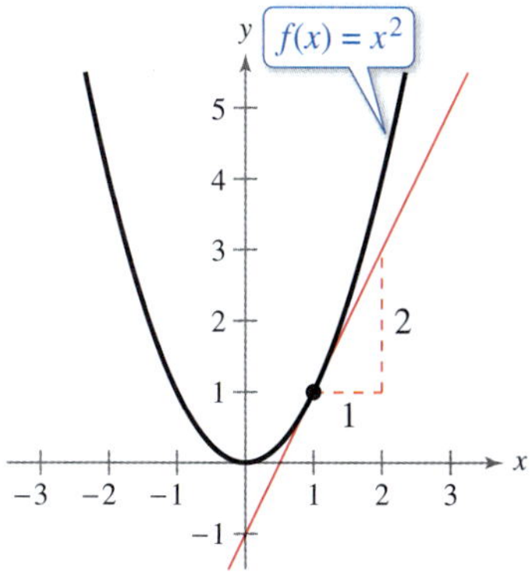

Figure 11.14

Checkpoint *Audio-video solution in English & Spanish at LarsonPrecalculus.com*

Use Figure 11.14 to approximate the slope of the graph of $f(x) = x^2$ at the point (2, 4).

When you are visually approximating the slope of a graph, remember that the scales on the horizontal and vertical axes may differ. When this happens (as it frequently does in applications), the slope of the tangent line is distorted, and you must be careful to account for the difference in the scales.

EXAMPLE 2 Visually Approximating the Slope of a Graph

Figure 11.15 shows the graph of a model for the monthly normal temperatures (in degrees Fahrenheit) for Dallas, Texas. Approximate the slope of this graph at the indicated point and give a physical interpretation of the result. *(Source: National Climatic Data Center)*

Solution

From the graph, you can see that the tangent line at the given point falls approximately 10 units for each one-unit change in x. So, you can estimate the slope at the given point to be

$$\text{Slope} = \frac{\text{change in } y}{\text{change in } x} \approx \frac{-10}{1} = -10 \text{ degrees per month.}$$

This means that the monthly normal temperature in November is about 10 degrees lower than the monthly normal temperature in October.

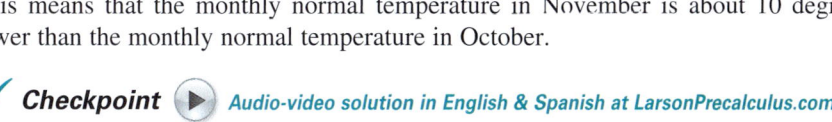

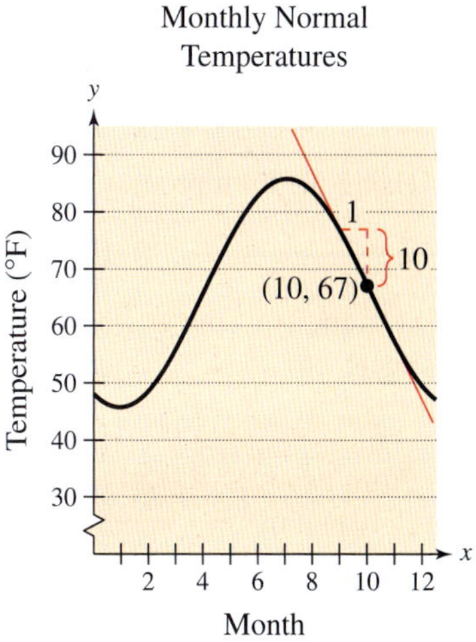

Figure 11.15

Checkpoint *Audio-video solution in English & Spanish at LarsonPrecalculus.com*

In Example 2, approximate the slope of the graph at the point (4, 66) and give a physical interpretation of the result.

Slope and the Limit Process

In Examples 1 and 2, you approximated the slope of a graph at a point by creating a graph and then "eyeballing" the tangent line at the point of tangency. A more systematic method of approximating the slope of a tangent line uses a **secant line** through the point of tangency and a second point on the graph, as shown in Figure 11.16. If $(x, f(x))$ is the point of tangency and

$$(x + h, f(x + h))$$

is a second point on the graph of f, then the slope of the secant line through the two points is given by

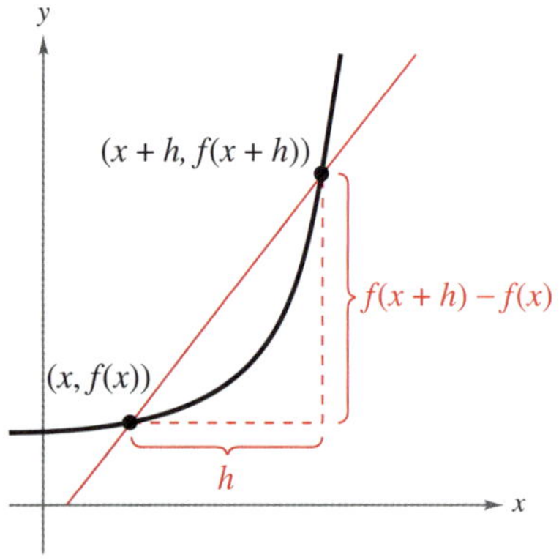

The secant line through $(x, f(x))$ and $(x + h, f(x + h))$
Figure 11.16

Algebra Help

The use of the word *secant* comes from the Latin *secare,* which means to cut and is not a reference to the trigonometric function of the same name.

$$m_{\text{sec}} = \frac{f(x + h) - f(x)}{h}. \qquad \text{Slope of secant line}$$

The right side of this equation is a **difference quotient.** The denominator h is the *change in x*, and the numerator is the *change in y*. The beauty of this procedure is that you obtain more and more accurate approximations of the slope of the tangent line by choosing points closer and closer to the point of tangency, as shown in Figure 11.17.

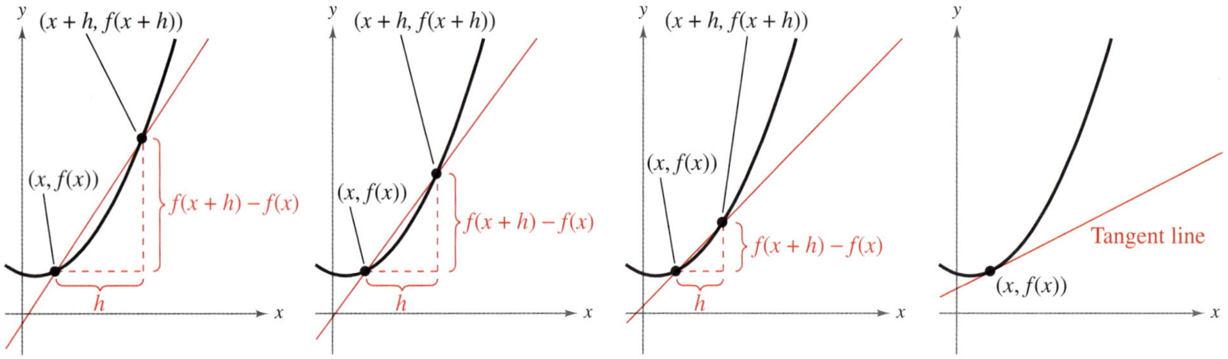

As h approaches 0, the secant line approaches the tangent line.
Figure 11.17

Using the limit process, you can find the *exact* slope of the tangent line at $(x, f(x))$.

Definition of the Slope of a Graph

The **slope** m of the graph of f at the point $(x, f(x))$ is equal to the slope of its tangent line at $(x, f(x))$, and is given by

$$m = \lim_{h \to 0} m_{\text{sec}}$$

$$= \lim_{h \to 0} \frac{f(x + h) - f(x)}{h}$$

provided this limit exists.

From the definition above and from Section 11.2, you can see that the difference quotient is used frequently in calculus. Using the difference quotient to find the slope of a tangent line to a graph is a major concept of calculus.

EXAMPLE 3 Finding the Slope of a Graph

Find the slope of the graph of $f(x) = x^2$ at the point $(-2, 4)$.

Solution

Find an expression that represents the slope of a secant line at $(-2, 4)$.

$$m_{sec} = \frac{f(-2+h) - f(-2)}{h}$$ Set up difference quotient.

$$= \frac{(-2+h)^2 - (-2)^2}{h}$$ Substitute into $f(x) = x^2$.

$$= \frac{4 - 4h + h^2 - 4}{h}$$ Expand terms.

$$= \frac{-4h + h^2}{h}$$ Simplify.

$$= \frac{h(-4+h)}{h}$$ Factor and divide out.

$$= -4 + h, \quad h \neq 0$$ Simplify.

Next, take the limit of m_{sec} as h approaches 0.

$$m = \lim_{h \to 0} m_{sec} = \lim_{h \to 0} (-4 + h) = -4 + 0 = -4$$

The graph has a slope of -4 at the point $(-2, 4)$, as shown in Figure 11.18.

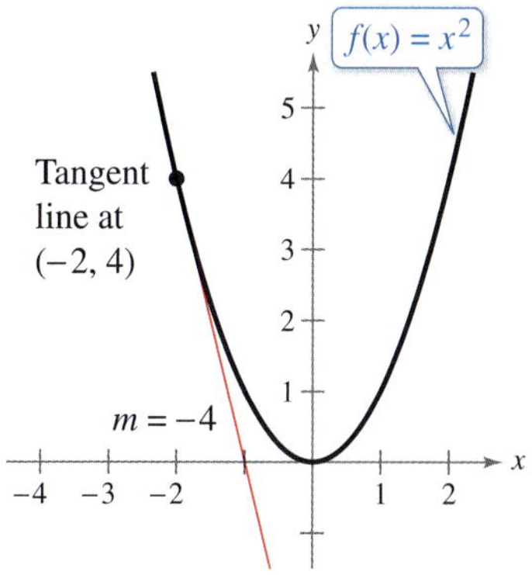

Figure 11.18

Checkpoint *Audio-video solution in English & Spanish at LarsonPrecalculus.com*

Find the slope of the graph of $g(x) = x^2 - 2x$ at the point $(3, 3)$.

EXAMPLE 4 Finding the Slope of a Graph

Find the slope of $f(x) = -2x + 4$.

Solution

$$m = \lim_{h \to 0} \frac{f(x+h) - f(x)}{h}$$ Set up difference quotient.

$$= \lim_{h \to 0} \frac{[-2(x+h) + 4] - (-2x + 4)}{h}$$ Substitute into $f(x) = -2x + 4$.

$$= \lim_{h \to 0} \frac{-2x - 2h + 4 + 2x - 4}{h}$$ Expand terms.

$$= \lim_{h \to 0} \frac{-2h}{h}$$ Divide out.

$$= -2$$ Simplify.

You know from your study of linear functions that the line given by

$$f(x) = -2x + 4$$

has a slope of -2, as shown in Figure 11.19. This conclusion is consistent with that obtained by the limit definition of slope, as shown above.

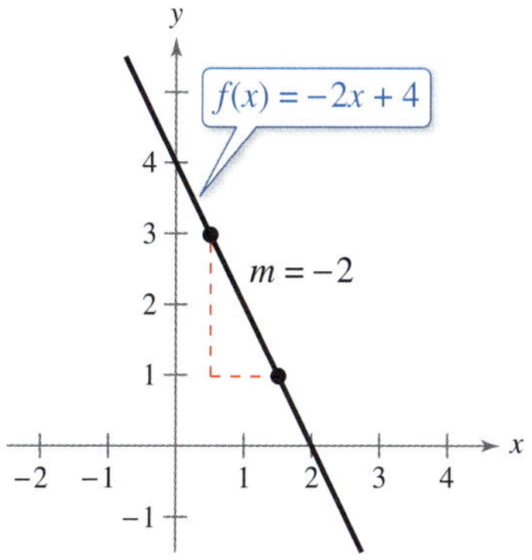

Figure 11.19

Checkpoint *Audio-video solution in English & Spanish at LarsonPrecalculus.com*

Find the slope of $f(x) = -3x + 4$.

It is important that you see the difference between the ways the difference quotients were set up in Examples 3 and 4. In Example 3, you found the slope of a graph at a specific point $(c, f(c))$. To find the slope in such a case, you can use the following form of the difference quotient.

$$m = \lim_{h \to 0} \frac{f(c + h) - f(c)}{h} \qquad \text{Slope at specific point}$$

In Example 4, however, you found a *formula* for the slope at *any* point on the graph. In such cases, you should use x, rather than c, in the difference quotient.

$$m = \lim_{h \to 0} \frac{f(x + h) - f(x)}{h} \qquad \text{Formula for slope}$$

EXAMPLE 5 Finding a Formula for the Slope of a Graph

Find a formula for the slope of the graph of

$$f(x) = x^2 + 1.$$

What are the slopes at the points $(-1, 2)$ and $(2, 5)$?

Solution

$$m_{sec} = \frac{f(x + h) - f(x)}{h} \qquad \text{Set up difference quotient.}$$

$$= \frac{[(x + h)^2 + 1] - (x^2 + 1)}{h} \qquad \text{Substitute into } f(x) = x^2 + 1.$$

$$= \frac{x^2 + 2xh + h^2 + 1 - x^2 - 1}{h} \qquad \text{Expand terms.}$$

$$= \frac{2xh + h^2}{h} \qquad \text{Simplify.}$$

$$= \frac{\cancel{h}(2x + h)}{\cancel{h}} \qquad \text{Factor and divide out.}$$

$$= 2x + h, \quad h \neq 0 \qquad \text{Simplify.}$$

Next, take the limit of m_{sec} as h approaches 0.

$$m = \lim_{h \to 0} m_{sec}$$

$$= \lim_{h \to 0} (2x + h)$$

$$= 2x + 0$$

$$= 2x$$

Using the formula $m = 2x$ for the slope at $(x, f(x))$, you can find the slope at the specified points. At $(-1, 2)$, the slope is

$$m = 2(-1) = -2$$

and at $(2, 5)$, the slope is

$$m = 2(2) = 4.$$

The graph of f is shown in the figure.

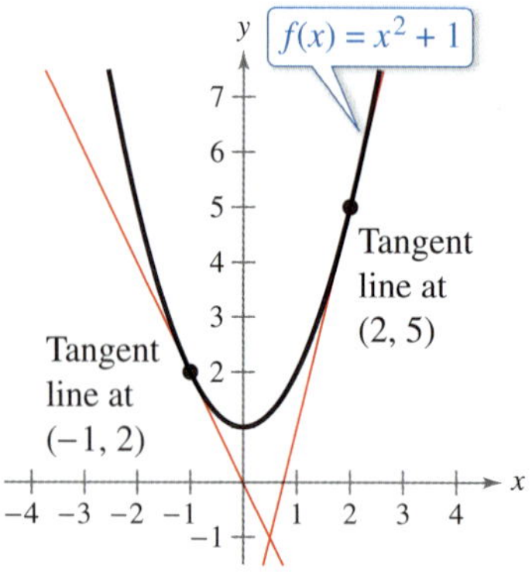

Technology Tip

To check the result in Example 5, first use the point-slope form to find the tangent lines at $(-1, 2)$ and $(2, 5)$, which are $y = -2x$ and $y = 4x - 3$, respectively. Then use a graphing utility to graph the function and the tangent lines

$$y_1 = x^2 + 1$$

$$y_2 = -2x$$

$$y_3 = 4x - 3$$

in the same viewing window. You can also check the result using the *tangent* feature. For instructions on how to use the *tangent* feature, see Appendix A; for specific keystrokes, go to this textbook's *Student Companion Website.*

Checkpoint *Audio-video solution in English & Spanish at LarsonPrecalculus.com*

Find a formula for the slope of the graph of $f(x) = 2x^2 - 3$. What is the slope at the points $(-3, 15)$ and $(2, 5)$?

The Derivative of a Function

In Example 5, you started with the function $f(x) = x^2 + 1$ and used the limit process to derive another function, $m = 2x$, that represents the slope of the graph of f at the point $(x, f(x))$. This derived function is called the **derivative** of f at x. It is denoted by $f'(x)$, which is read as "f prime of x."

Definition of the Derivative

The **derivative** of f at x is given by

$$f'(x) = \lim_{h \to 0} \frac{f(x+h) - f(x)}{h}$$

provided this limit exists.

Remember that the derivative $f'(x)$ is a formula for the slope of the tangent line to the graph of f at the point $(x, f(x))$.

Algebra Help

In Section 1.1, you studied the slope of a line, which represents the *average rate of change* over an interval. The derivative of a function is a formula which represents the *instantaneous rate of change* at a point.

EXAMPLE 6 Finding a Derivative

Find the derivative of $f(x) = 3x^2 - 2x$.

Solution

$$\begin{aligned}
f'(x) &= \lim_{h \to 0} \frac{f(x+h) - f(x)}{h} \\
&= \lim_{h \to 0} \frac{[3(x+h)^2 - 2(x+h)] - (3x^2 - 2x)}{h} \\
&= \lim_{h \to 0} \frac{3x^2 + 6xh + 3h^2 - 2x - 2h - 3x^2 + 2x}{h} \\
&= \lim_{h \to 0} \frac{6xh + 3h^2 - 2h}{h} \\
&= \lim_{h \to 0} \frac{h(6x + 3h - 2)}{h} \\
&= \lim_{h \to 0} (6x + 3h - 2) \\
&= 6x + 3(0) - 2 \\
&= 6x - 2
\end{aligned}$$

So, the derivative of $f(x) = 3x^2 - 2x$ is

$f'(x) = 6x - 2.$ Derivative of f at x

✓ **Checkpoint** *Audio-video solution in English & Spanish at LarsonPrecalculus.com*

Find the derivative of

$f(x) = x^3 + 2x.$

Explore the Concept

Use a graphing utility to graph the function $f(x) = 3x^2 - 2x$. Use the *trace* feature to approximate the coordinates of the vertex of this parabola. Then use the derivative of $f(x) = 3x^2 - 2x$ to find the slope of the tangent line at the vertex. Make a conjecture about the slope of the tangent line at the vertex of an arbitrary parabola that opens upward or downward.

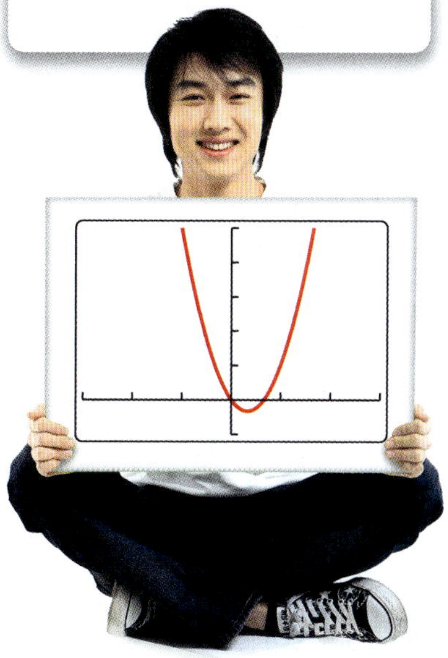

Note that in addition to $f'(x)$, other notations can be used to denote the derivative of $y = f(x)$. The most common are

$$\frac{dy}{dx}, \quad y', \quad \frac{d}{dx}[f(x)], \quad \text{and} \quad D_x[y].$$

wizdata/Shutterstock.com

EXAMPLE 7 Using the Derivative

See LarsonPrecalculus.com for an interactive version of this type of example.

Find $f'(x)$ for $f(x) = \sqrt{x}$. Then find the slopes of the graph of f at the points $(1, 1)$ and $(4, 2)$ and equations of the tangent lines to the graph at the points.

Solution

$$f'(x) = \lim_{h\to0}\frac{f(x+h)-f(x)}{h}$$

$$= \lim_{h\to0}\frac{\sqrt{x+h}-\sqrt{x}}{h}$$

Because direct substitution yields the indeterminate form $\frac{0}{0}$, you should use the rationalizing technique discussed in Section 11.2 to find the limit.

$$f'(x) = \lim_{h\to0}\left(\frac{\sqrt{x+h}-\sqrt{x}}{h}\right)\left(\frac{\sqrt{x+h}+\sqrt{x}}{\sqrt{x+h}+\sqrt{x}}\right)$$

$$= \lim_{h\to0}\frac{(x+h)-x}{h(\sqrt{x+h}+\sqrt{x})}$$

$$= \lim_{h\to0}\frac{\not{h}}{\not{h}(\sqrt{x+h}+\sqrt{x})}$$

$$= \lim_{h\to0}\frac{1}{\sqrt{x+h}+\sqrt{x}}$$

$$= \frac{1}{\sqrt{x+0}+\sqrt{x}}$$

$$= \frac{1}{2\sqrt{x}}$$

Algebra Help

Remember that to rationalize the numerator of an expression, you must multiply the numerator and denominator by the conjugate of the numerator.

At the point $(1, 1)$, the slope is

$$f'(1) = \frac{1}{2\sqrt{1}} = \frac{1}{2}.$$

An equation of the tangent line at the point $(1, 1)$ is

$y - y_1 = m(x - x_1)$ Point-slope form

$y - 1 = \frac{1}{2}(x - 1)$ Substitute $\frac{1}{2}$ for m, 1 for x_1, and 1 for y_1.

$y = \frac{1}{2}x + \frac{1}{2}$. Tangent line

At the point $(4, 2)$, the slope is $f'(4) = \dfrac{1}{2\sqrt{4}} = \dfrac{1}{4}$. An equation of the tangent line at the point $(4, 2)$ is

$y - y_1 = m(x - x_1)$ Point-slope form

$y - 2 = \frac{1}{4}(x - 4)$ Substitute $\frac{1}{4}$ for m, 4 for x_1, and 2 for y_1.

$y = \frac{1}{4}x + 1$. Tangent line

The graphs of f and the tangent lines at the points $(1, 1)$ and $(4, 2)$ are shown in Figure 11.20.

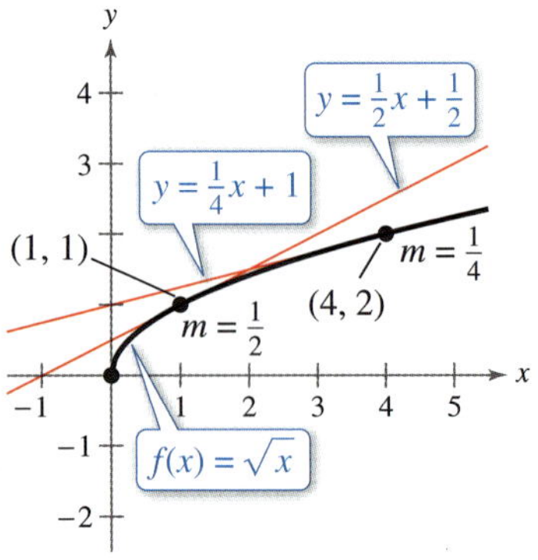

Figure 11.20

Checkpoint Audio-video solution in English & Spanish at LarsonPrecalculus.com

Find the derivative of $f(x) = \sqrt{x - 1}$. Then find the slopes of the graph of f at the points $(2, 1)$ and $(10, 3)$ and equations of the tangent lines to the graph at the points.

11.3 Exercises

See *CalcChat.com* for tutorial help and worked-out solutions to odd-numbered exercises.
For instructions on how to use a graphing utility, see Appendix A.

Vocabulary and Concept Check

In Exercises 1–4, fill in the blank.

1. _______ is the study of the rates of change of functions.
2. The _______ to the graph of a function at a point is the line that best approximates the slope of the graph at the point.
3. To approximate a tangent line to a graph, you can make use of a _______ through the point of tangency and a second point on the graph.
4. The _______ of a function f at x represents the slope of the graph of f at the point $(x, f(x))$.

Procedures and Problem Solving

Visually Approximating the Slope of a Graph In Exercises 5–8, use the figure to approximate the slope of the graph at the point (x, y).

5.

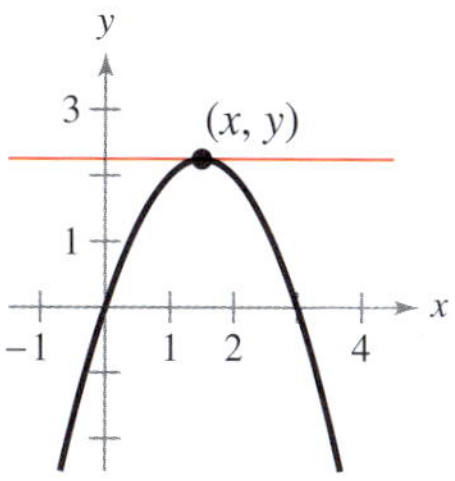

6.

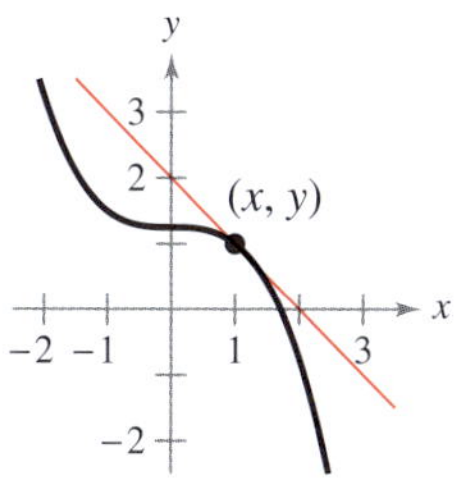

7.

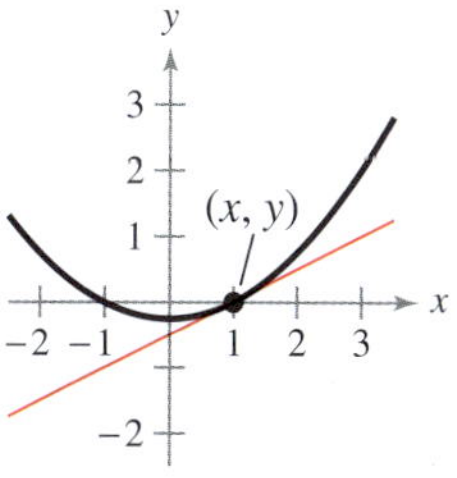

8.

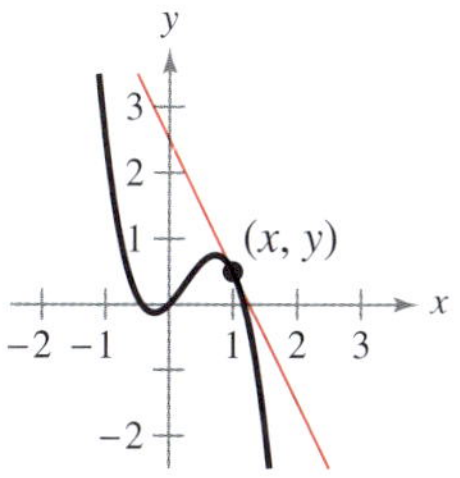

Approximating the Slope of a Tangent Line In Exercises 9–14, use a graphing utility to graph the function and the tangent line at the point $(1, f(1))$. Use the graph to approximate the slope of the tangent line.

9. $f(x) = x^2 - 3$
10. $f(x) = x^2 - 2x + 1$
11. $f(x) = \sqrt{2 - x}$
12. $f(x) = \sqrt{x + 3}$
13. $f(x) = \dfrac{1}{x + 1}$
14. $f(x) = \dfrac{3}{2 - x}$

Finding the Slope of a Graph In Exercises 15–22, use the limit process to find the slope of the graph of the function at the specified point. Use a graphing utility to confirm your result.

15. $f(x) = -2x^2, \quad (-2, -8)$
16. $g(x) = x^2 - 4x, \quad (3, -3)$
17. $g(x) = 11 - 6x, \quad (1, 5)$
18. $h(x) = 5x + 9, \quad (-1, 4)$
19. $g(x) = \dfrac{4}{x}, \quad (2, 2)$
20. $g(x) = \dfrac{1}{x - 4}, \quad \left(0, -\frac{1}{4}\right)$
21. $h(x) = \sqrt{x}, \quad (9, 3)$
22. $h(x) = \sqrt{x + 8}, \quad (-4, 2)$

Finding a Formula for the Slope of a Graph In Exercises 23–28, find a formula for the slope of the graph of f at the point $(x, f(x))$. Then use it to find the slopes at the two specified points.

23. $f(x) = 4 - x^2$
 (a) $(0, 4)$
 (b) $(-1, 3)$
24. $f(x) = x^3 + 1$
 (a) $(-1, 0)$
 (b) $(2, 9)$
25. $f(x) = \dfrac{1}{x + 4}$
 (a) $\left(0, \frac{1}{4}\right)$
 (b) $\left(-2, \frac{1}{2}\right)$
26. $f(x) = \dfrac{1}{x + 2}$
 (a) $\left(0, \frac{1}{2}\right)$
 (b) $(-3, -1)$
27. $f(x) = \sqrt{x - 3}$
 (a) $(4, 1)$
 (b) $(12, 3)$
28. $f(x) = \sqrt{x - 4}$
 (a) $(5, 1)$
 (b) $(8, 2)$

Finding a Derivative In Exercises 29–40, find the derivative of the function.

29. $f(x) = 9 - \frac{1}{3}x$
30. $f(x) = 6 - \frac{1}{2}x$
31. $f(x) = 2x^2 + 3x$
32. $f(x) = x^2 - 3x + 4$
33. $f(x) = 5$
34. $f(x) = -1$
35. $f(x) = \dfrac{1}{x^2}$
36. $f(x) = \dfrac{1}{x^3}$
37. $f(x) = \sqrt{x + 4}$
38. $f(x) = -2\sqrt{x + 1}$
39. $f(x) = \dfrac{1}{\sqrt{x - 9}}$
40. $f(x) = \dfrac{1}{\sqrt{x + 1}}$

Using the Derivative **In Exercises 41–50, (a) find the slope of the graph of f at the given point, (b) find an equation of the tangent line to the graph at the point, and (c) graph the function and the tangent line.**

41. $f(x) = x^2 - 1, (2, 3)$
42. $f(x) = x^2 + 3, (2, 7)$
43. $f(x) = x^3 - 2x, (1, -1)$
44. $f(x) = x^3 - 5x, (2, -2)$
45. $f(x) = x^2 - 7x + 6, (2, -4)$
46. $f(x) = x^2 + 2x + 1, (-3, 4)$
47. $f(x) = \sqrt{6 - x}, (-3, 3)$
48. $f(x) = \sqrt{x - 2}, (3, 1)$
49. $f(x) = \dfrac{1}{x + 5}, (-4, 1)$
50. $f(x) = \dfrac{1}{x - 3}, (4, 1)$

Graphing a Function Over an Interval **In Exercises 51–54, use a graphing utility to graph f over the interval $[-2, 2]$ and complete the table. Compare the value of the derivative with a visual approximation of the slope of the graph.**

x	-2	-1.5	-1	-0.5	0	0.5	1	1.5	2
$f(x)$									
$f'(x)$									

51. $f(x) = \frac{1}{2}x^2$
52. $f(x) = \frac{1}{4}x^3$
53. $f(x) = \sqrt{x + 3}$
54. $f(x) = \dfrac{x^2 - 4}{x + 4}$

Using the Derivative **In Exercises 55–58, find the derivative of f. Use the derivative to determine any points on the graph of f at which the tangent line is horizontal. Use a graphing utility to verify your results.**

55. $f(x) = x^2 - 4x + 3$
56. $f(x) = -x^2 + 4x - 9$
57. $f(x) = 2x^3 - 8x$
58. $f(x) = x^3 + 3x$

Using the Derivative **In Exercises 59–66, use the function and its derivative to determine any points on the graph of f at which the tangent line is horizontal. Use a graphing utility to verify your results.**

59. $f(x) = x^4 - 2x^2, f'(x) = 4x^3 - 4x$
60. $f(x) = 3x^4 + 4x^3, f'(x) = 12x^3 + 12x^2$
61. $f(x) = 2 \cos x + x, f'(x) = -2 \sin x + 1$, over the interval $(0, 2\pi)$
62. $f(x) = x - 2 \sin x, f'(x) = 1 - 2 \cos x$, over the interval $(0, 2\pi)$
63. $f(x) = x^2e^x, f'(x) = x^2e^x + 2xe^x$
64. $f(x) = xe^{-x}, f'(x) = e^{-x} - xe^{-x}$
65. $f(x) = x \ln x, f'(x) = \ln x + 1$
66. $f(x) = \dfrac{\ln x}{x}, f'(x) = \dfrac{1 - \ln x}{x^2}$

67. **MODELING DATA**

The table shows the revenues y (in billions of dollars) for Goodyear Tire & Rubber from 2011 through 2017. *(Source: Goodyear Tire & Rubber)*

DATA

Year	Revenue, y (in billions of dollars)
2011	22.77
2012	20.99
2013	19.54
2014	18.14
2015	16.44
2016	15.16
2017	15.38

Spreadsheet at LarsonPrecalculus.com

(a) Use the *regression* feature of a graphing utility to find a quadratic model for the data. Let t represent the year, with $t = 1$ corresponding to 2011.

(b) Use the graphing utility to graph the model found in part (a). Estimate the slope of the graph when $t = 4$ and interpret the result.

(c) Find the derivative of the model in part (a). Then evaluate the derivative for $t = 4$.

(d) Write a brief statement regarding your results for parts (a) through (c).

68. **MODELING DATA**

The table shows the numbers N (in thousands) of books sold when the price per book is p (in dollars).

DATA

Price, p (in dollars)	Number of books, n (in thousands)
10	900
15	630
20	396
25	227
30	102
35	36

Spreadsheet at LarsonPrecalculus.com

(a) Use the *regression* feature of a graphing utility to find a quadratic model for the data.

(b) Use the graphing utility to graph the model found in part (a). Estimate the slopes of the graph when $p = 15$ and $p = 30$.

(c) Use the graphing utility to graph the tangent lines to the model when $p = 15$ and $p = 30$. Compare the slopes given by the graphing utility with your estimates in part (b).

(d) The slopes of the tangent lines at $p = 15$ and $p = 30$ are not the same. Explain what this means to the company selling the books.

69. ***Why you should learn it*** *(p. 780)* A spherical balloon is inflated. The volume V is approximated by the formula $V(r) = \frac{4}{3}\pi r^3$, where r is the radius.

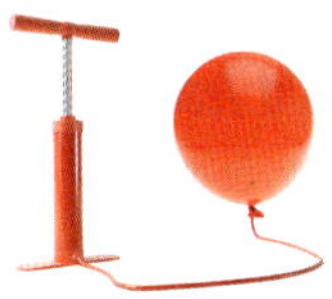

(a) Find the derivative of V with respect to r.

(b) Evaluate the derivative when the radius is 4 inches.

(c) What type of unit would be applied to your answer in part (b)? Explain.

70. Rate of Change An approximately spherical benign tumor is reducing in size. The surface area S is given by the formula $S(r) = 4\pi r^2$, where r is the radius.

(a) Find the derivative of S with respect to r.

(b) Evaluate the derivative when the radius is 3 millimeters.

(c) What type of unit would be applied to your answer in part (b)? Explain.

71. Vertical Motion A water balloon is thrown upward from the top of an 80-foot building with an initial velocity of 64 feet per second. The height or displacement s (in feet) of the balloon can be modeled by the position function $s(t) = -16t^2 + 64t + 80$, where t is the time in seconds since it was thrown.

(a) Find a formula for the instantaneous rate of change of the balloon.

(b) Find the average rate of change of the balloon after the first three seconds of flight. Explain your results.

(c) Find the time at which the balloon reaches its maximum height. Explain your method.

(d) Velocity is given by the derivative of the position function. Find the velocity of the balloon as it impacts the ground.

(e) Use a graphing utility to graph the model and verify your results for parts (a)–(d).

72. Vertical Motion A coin is dropped from the top of a 120-foot building. The height or displacement s (in feet) of the coin can be modeled by the position function $s(t) = -16t^2 + 120$, where t is the time in seconds since it was dropped.

(a) Find a formula for the instantaneous rate of change of the coin.

(b) Find the average rate of change of the coin after the first two seconds of free fall. Explain your results.

(c) Velocity is given by the derivative of the position function. Find the velocity of the coin as it impacts the ground.

(d) Find the time when the velocity of the coin is -70 feet per second.

(e) Use a graphing utility to graph the model and verify your results for parts (a)–(d).

Focusing on Concepts

True or False? **In Exercises 73 and 74, determine whether the statement is true or false. Justify your answer.**

73. The slope of the graph of $y = x^2$ is different at every point on the graph of f.

74. A tangent line to a graph can intersect the graph only at the point of tangency.

75. Writing Consider the graph of a function f.

(a) Explain how you can use a secant line to approximate the tangent line at $(x, f(x))$.

(b) Explain how you can use the limit process to find the exact slope of the tangent line at $(x, f(x))$.

76. HOW DO YOU SEE IT? Match the function with the graph of its *derivative*. Explain your reasoning. [The graphs are labeled (i), (ii), (iii), and (iv).]

(i)

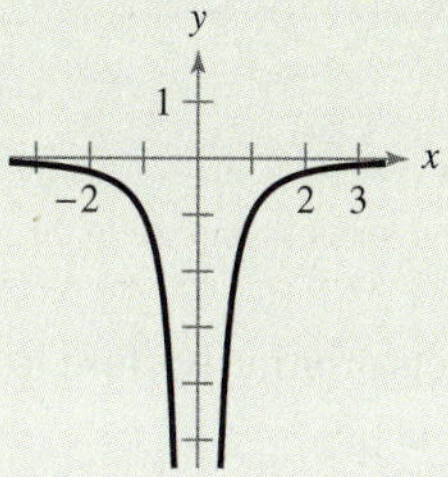

(ii)

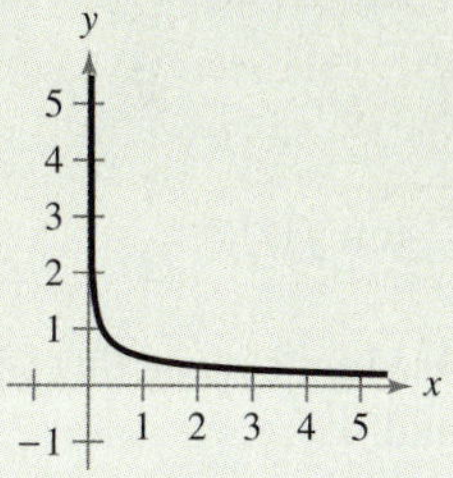

(iii)

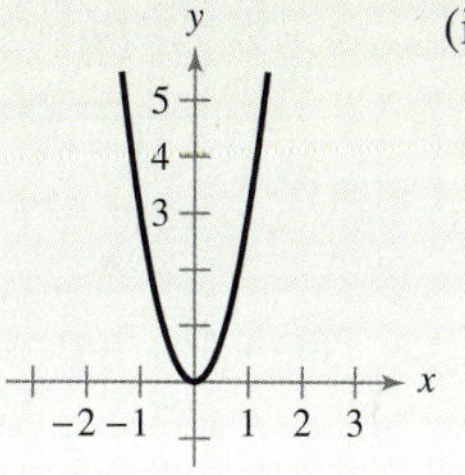

(iv)

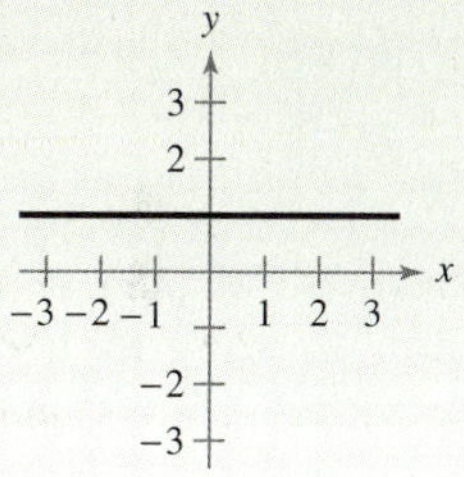

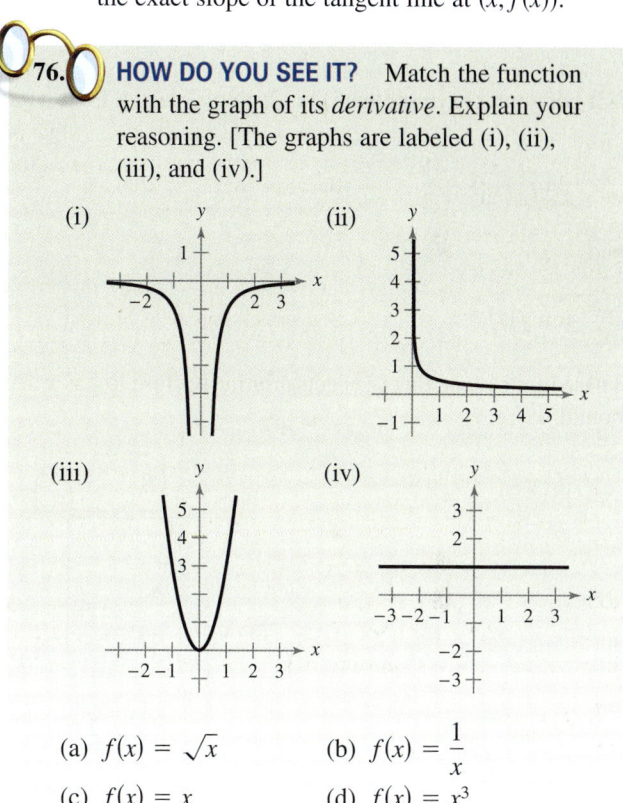

(a) $f(x) = \sqrt{x}$ (b) $f(x) = \dfrac{1}{x}$

(c) $f(x) = x$ (d) $f(x) = x^3$

Cumulative Mixed Review

Using Standard Form to Graph a Parabola **In Exercises 77 and 78, write the quadratic function in standard form and sketch its graph.**

77. $f(x) = x^2 + 5x + 7$

78. $g(x) = -3x^2 + 6x + 5$

79. ***Project: Veterinarians*** To work an extended application analyzing the number of veterinarians employed in the United States, visit the text's website at *LarsonPrecalculus.com.* (*Data Source:* U.S. Bureau of Labor Statistics)

11.4 Limits at Infinity and Limits of Sequences

Limits at Infinity and Horizontal Asymptotes

As pointed out at the beginning of this chapter, there are two basic problems in calculus: finding **tangent lines** and finding the **area** of a region. In Section 11.3, you saw how limits can be used to solve the tangent line problem. In this section and the next, you will see how a different type of limit, a *limit at infinity*, can be used to solve the area problem. To get an idea of what is meant by a limit at infinity, consider the function

$$f(x) = \frac{x+1}{2x}.$$

The graph of f is shown in Figure 11.21. From Section 2.6, you know that

$$y = \frac{1}{2}$$

is a horizontal asymptote of the graph of f. Using limit notation, this can be written as

$$\lim_{x \to -\infty} f(x) = \frac{1}{2}$$ Horizontal asymptote to the left

and

$$\lim_{x \to \infty} f(x) = \frac{1}{2}.$$ Horizontal asymptote to the right

These limits mean that $f(x)$ gets arbitrarily close to $\frac{1}{2}$ as x decreases or increases without bound.

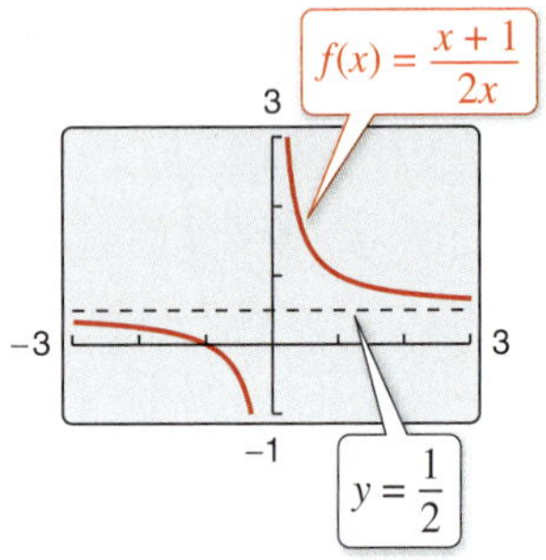

The limit of f(x) as x approaches −∞ or ∞ is ½.
Figure 11.21

Definition of Limits at Infinity

If f is a function and L_1 and L_2 are real numbers, then the statements

$$\lim_{x \to -\infty} f(x) = L_1$$ Limit as x approaches $-\infty$

and

$$\lim_{x \to \infty} f(x) = L_2$$ Limit as x approaches ∞

denote the **limits at infinity.** The first statement is read "*the limit of $f(x)$ as x approaches $-\infty$ is L_1*," and the second is read "*the limit of $f(x)$ as x approaches ∞ is L_2*."

What you should learn

- Evaluate limits of functions at infinity.
- Find limits of sequences.

Why you should learn it

Finding limits at infinity is useful in analyzing functions that model real-life situations. For instance, in Exercise 60 on page 798, you are asked to find a limit at infinity to decide whether you can use a given model to predict the mean salary of a financial manager in the United States.

annebaek/iStock/Getty Images

The next definition is useful when evaluating limits at infinity.

Limits at Infinity

If r is a positive real number, then

$$\lim_{x \to \infty} \frac{1}{x^r} = 0.$$ Limit toward the right

Furthermore, if x^r is defined when $x < 0$, then

$$\lim_{x \to -\infty} \frac{1}{x^r} = 0.$$ Limit toward the left

Limits at infinity share many of the properties of limits listed in Section 11.1. Some of these properties are demonstrated in the next example.

EXAMPLE 1 Evaluating a Limit at Infinity

Find the limit: $\lim_{x \to \infty} \left(4 - \frac{3}{x^2}\right)$.

Algebraic Solution

Use the properties of limits listed in Section 11.1.

$$\begin{aligned}\lim_{x \to \infty} \left(4 - \frac{3}{x^2}\right) &= \lim_{x \to \infty} 4 - \lim_{x \to \infty} \frac{3}{x^2} \\ &= \lim_{x \to \infty} 4 - 3\left(\lim_{x \to \infty} \frac{1}{x^2}\right) \\ &= 4 - 3(0) \\ &= 4\end{aligned}$$

So, the limit of

$$f(x) = 4 - \frac{3}{x^2}$$

as x approaches ∞ is 4.

Graphical Solution

Use a graphing utility to graph

$$y = 4 - \frac{3}{x^2}.$$

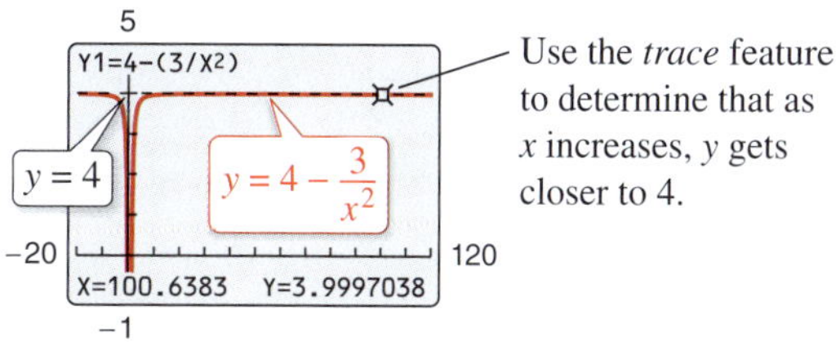
5
Y1=4-(3/X2)
y = 4
y = 4 − 3/x²
−20
120
X=100.6383 Y=3.9997038
−1
Use the *trace* feature to determine that as x increases, y gets closer to 4.

Figure 11.22

From Figure 11.22, you can estimate that the limit is 4. Note in the figure that the line $y = 4$ appears to be a horizontal asymptote to the right.

✓ **Checkpoint** ▶ *Audio-video solution in English & Spanish at LarsonPrecalculus.com*

Find the limit: $\lim_{x \to \infty} \frac{2}{x^2}$.

In Figure 11.22, it appears that the line $y = 4$ is also a horizontal asymptote *to the left*. You can verify this by showing that

$$\lim_{x \to -\infty} \left(4 - \frac{3}{x^2}\right) = 4.$$

The graph of a rational function does not need to have a horizontal asymptote. When it does, however, its left and right asymptotes must be the same.

When evaluating limits at infinity for more complicated rational functions, divide the numerator and denominator by the highest power of x in the denominator. This enables you to evaluate each limit using the limits at infinity at the top of this page.

Explore the Concept

Use a graphing utility to graph the two functions given by

$$y_1 = \frac{1}{\sqrt{x}} \quad \text{and} \quad y_2 = \frac{1}{\sqrt[3]{x}}$$

in the same viewing window. Why doesn't y_1 appear to the left of the y-axis? How does this relate to the statement at the left about the infinite limit

$$\lim_{x \to -\infty} \frac{1}{x^r}?$$

EXAMPLE 2 Comparing Limits at Infinity

See LarsonPrecalculus.com for an interactive version of this type of example.

Find the limit (if it exists) as x approaches ∞ for each function.

a. $f(x) = \dfrac{-2x + 3}{3x^2 + 1}$

b. $f(x) = \dfrac{-2x^2 + 3}{3x^2 + 1}$

c. $f(x) = \dfrac{-2x^3 + 3}{3x^2 + 1}$

Solution

In each case, begin by dividing both the numerator and denominator by x^2, the highest power of x in the denominator.

a. $$\lim_{x\to\infty} \frac{-2x + 3}{3x^2 + 1} = \lim_{x\to\infty} \frac{-\frac{2}{x} + \frac{3}{x^2}}{3 + \frac{1}{x^2}}$$

$$= \frac{-0 + 0}{3 + 0}$$

$$= 0$$

b. $$\lim_{x\to\infty} \frac{-2x^2 + 3}{3x^2 + 1} = \lim_{x\to\infty} \frac{-2 + \frac{3}{x^2}}{3 + \frac{1}{x^2}}$$

$$= \frac{-2 + 0}{3 + 0}$$

$$= -\frac{2}{3}$$

c. $$\lim_{x\to\infty} \frac{-2x^3 + 3}{3x^2 + 1} = \lim_{x\to\infty} \frac{-2x + \frac{3}{x^2}}{3 + \frac{1}{x^2}}$$

In this case, you can conclude that the limit does not exist because the numerator decreases without bound as the denominator approaches 3.

✓ Checkpoint *Audio-video solution in English & Spanish at LarsonPrecalculus.com*

Find the limit (if it exists) as x approaches ∞ for each function.

a. $f(x) = \dfrac{2x}{1 - x^2}$ **b.** $f(x) = \dfrac{2x}{1 - x}$ **c.** $f(x) = \dfrac{2x^2}{1 - x}$

In Example 2, observe that when the degree of the numerator is less than the degree of the denominator, as in part (a), the limit is 0. When the degrees of the numerator and denominator are equal, as in part (b), the limit is the ratio of the leading coefficients. When the degree of the numerator is greater than the degree of the denominator, as in part (c), the limit does not exist. These results seem reasonable when you realize that for large values of x, the highest-powered term of a polynomial is the most "influential" term. That is, a polynomial tends to behave as its highest-powered term behaves as x approaches positive or negative infinity.

Explore the Concept

Use a graphing utility to complete the table below to support the conclusion that

$$\lim_{x\to\infty} \frac{1}{x} = 0.$$

x	10^0	10^1	10^2
$\frac{1}{x}$			

x	10^3	10^4	10^5
$\frac{1}{x}$			

Make a conjecture about

$$\lim_{x\to 0} \frac{1}{x}.$$

Limits at Infinity for Rational Functions

Consider the rational function

$$f(x) = \frac{N(x)}{D(x)}$$

where

$$N(x) = a_n x^n + \cdot \cdot \cdot + a_0 \qquad \text{and} \qquad D(x) = b_m x^m + \cdot \cdot \cdot + b_0.$$

The limit of $f(x)$ as x approaches positive or negative infinity is as follows.

$$\lim_{x \to \pm\infty} f(x) = \begin{cases} 0, & n < m \\ \dfrac{a_n}{b_m}, & n = m \end{cases}$$

If $n > m$, then the limit does not exist.

EXAMPLE 3 Finding the Average Cost

A company manufactures mobile phone protective cases that cost \$4.50 per case to produce. The company's initial investment is \$20,000, which implies that the total cost C of producing x cases is given by $C = 4.50x + 20{,}000$. The average cost $\overline{C}$ per case is given by

$$\overline{C} = \frac{C}{x} = \frac{4.50x + 20{,}000}{x}.$$

Find the average cost per case when (a) $x = 1000$, (b) $x = 10{,}000$, and (c) $x = 100{,}000$. (d) What is the limit of $\overline{C}$ as x approaches infinity?

Solution

a. When $x = 1000$, the average cost per case is

$$\overline{C} = \frac{4.50(1000) + 20{,}000}{1000} \qquad x = 1000$$

$$= \$24.50.$$

b. When $x = 10{,}000$, the average cost per case is

$$\overline{C} = \frac{4.50(10{,}000) + 20{,}000}{10{,}000} \qquad x = 10{,}000$$

$$= \$6.50.$$

c. When $x = 100{,}000$, the average cost per case is

$$\overline{C} = \frac{4.50(100{,}000) + 20{,}000}{100{,}000} \qquad x = 100{,}000$$

$$= \$4.70.$$

d. As x approaches infinity, the limit of $\overline{C}$ is

$$\lim_{x \to \infty} \frac{4.50x + 20{,}000}{x} = \$4.50. \qquad x \to \infty$$

The graph of $\overline{C}$ is shown in Figure 11.23.

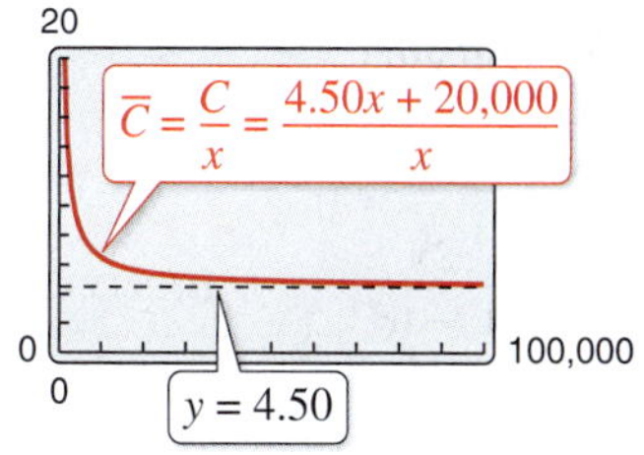

As $x \to \infty$, the average cost per case approaches \$4.50.
Figure 11.23

Checkpoint *Audio-video solution in English & Spanish at LarsonPrecalculus.com*

Repeat Example 3 when the mobile phone protective cases cost \$4.00 per case to produce.

Limits of Sequences

Limits of sequences have many of the same properties as limits of functions. For instance, consider the sequence whose nth term is $a_n = 1/2^n$.

$$\frac{1}{2}, \frac{1}{4}, \frac{1}{8}, \frac{1}{16}, \frac{1}{32}, \ldots$$

As n increases without bound, the terms of this sequence get closer and closer to 0, and the sequence is said to **converge** to 0. Using limit notation, you can write

$$\lim_{n\to\infty} \frac{1}{2^n} = 0.$$

The next definition shows how limits of functions of x can be used to evaluate the limit of a sequence.

Limit of a Sequence

Let L be a real number. Let f be a function of a real variable such that

$$\lim_{x\to\infty} f(x) = L.$$

If $\{a_n\}$ is a sequence such that $f(n) = a_n$ for every positive integer n, then

$$\lim_{n\to\infty} a_n = L.$$

A sequence that does not converge is said to **diverge.** For instance, the terms of the sequence $1, -1, 1, -1, 1, \ldots$ oscillate between 1 and -1. This sequence diverges because it does not approach a unique number.

EXAMPLE 4 Finding the Limit of a Sequence

Find the limit of each sequence. (Assume that n begins with 1.)

a. $a_n = \dfrac{2n+1}{n+4}$

b. $b_n = \dfrac{2n+1}{n^2+4}$

c. $c_n = \dfrac{2n^2+1}{4n^2}$

Solution

a. $\displaystyle\lim_{n\to\infty} \frac{2n+1}{n+4} = 2$ $\quad \frac{3}{5}, \frac{5}{6}, \frac{7}{7}, \frac{9}{8}, \frac{11}{9}, \frac{13}{10}, \ldots \to 2$

b. $\displaystyle\lim_{n\to\infty} \frac{2n+1}{n^2+4} = 0$ $\quad \frac{3}{5}, \frac{5}{8}, \frac{7}{13}, \frac{9}{20}, \frac{11}{29}, \frac{13}{40}, \ldots \to 0$

c. $\displaystyle\lim_{n\to\infty} \frac{2n^2+1}{4n^2} = \frac{1}{2}$ $\quad \frac{3}{4}, \frac{9}{16}, \frac{19}{36}, \frac{33}{64}, \frac{51}{100}, \frac{73}{144}, \ldots \to \frac{1}{2}$

Algebra Help

You can use the definition of limits at infinity for rational functions on page 793 to verify the limits of the sequences in Example 4.

✓ Checkpoint *Audio-video solution in English & Spanish at LarsonPrecalculus.com*

Find the limit of each sequence. (Assume that n begins with 1.)

a. $a_n = \dfrac{n+2}{2n-1}$ **b.** $a_n = \dfrac{n^3+2}{3n^3}$ **c.** $a_n = \dfrac{n+2}{3n^2}$

In the next section, you will encounter limits of sequences, such as that shown in Example 5. A strategy for evaluating such limits is to begin by writing the nth term in rational form. Treating this form as the ratio of two polynomials, determine the limit by comparing the degrees of the numerator and denominator, as shown on page 793.

EXAMPLE 5 Finding the Limit of a Sequence

Find the limit of the sequence whose nth term is

$$a_n = \frac{8}{n^3}\left[\frac{n(n+1)(2n+1)}{6}\right].$$

Algebraic Solution

Begin by writing the nth term in rational form—similar to the ratio of two polynomials.

$$a_n = \frac{8}{n^3}\left[\frac{n(n+1)(2n+1)}{6}\right] \quad \text{Write original } n\text{th term.}$$

$$= \frac{8(n)(n+1)(2n+1)}{6n^3} \quad \text{Multiply fractions.}$$

$$= \frac{8n^3 + 12n^2 + 4n}{3n^3} \quad \text{Write in rational form.}$$

This form shows that the degree of the numerator is equal to the degree of the denominator. So, the limit of the sequence is the ratio of the leading coefficients.

$$\lim_{n\to\infty} \frac{8n^3 + 12n^2 + 4n}{3n^3} = \frac{8}{3}$$

Numerical Solution

Set your graphing utility to *sequence* mode and enter the sequence. Use the *table* feature to create a table that shows the value of a_n as n becomes larger and larger, as shown in Figure 11.24.

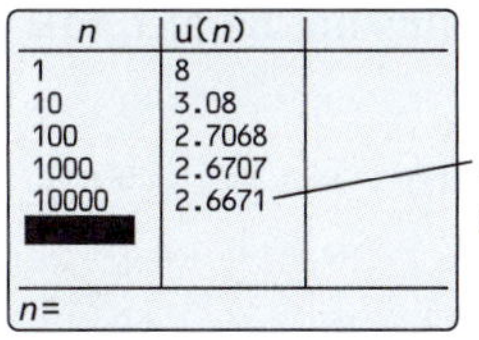

As n increases, a_n approaches 2.667.

Figure 11.24

So, you can estimate that as n approaches ∞, a_n gets closer and closer to $2.667 \approx \frac{8}{3}$.

Checkpoint Audio-video solution in English & Spanish at LarsonPrecalculus.com

Find the limit of each sequence.

a. $a_n = \frac{5}{n^3}\left[\frac{(n+1)(n-1)(2n)}{7}\right]$ **b.** $a_n = \frac{2}{n}\left(n - \frac{2}{n}\left[\frac{n(n-1)}{4}\right]\right)$

The result of Example 5 is supported by the figure, which shows the graph of the sequence a_n and $y = 8/3$.

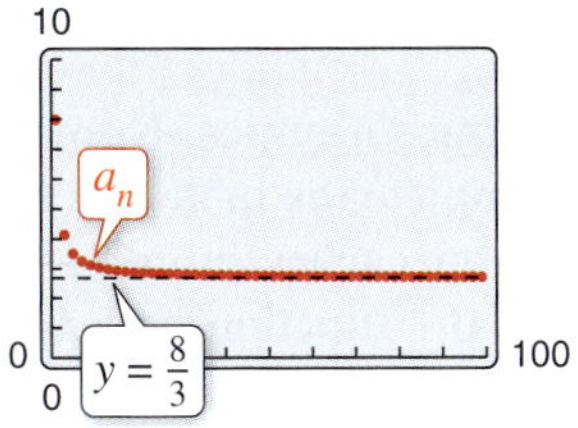

Explore the Concept

In Figure 11.24, the value of a_n approaches its limit of $\frac{8}{3}$ rather slowly. (The first term to be accurate to three decimal places is $a_{4801} \approx 2.667$.) Each sequence below converges to 0. Which converges the quickest? Which converges the slowest? Why? Write a short paragraph discussing your conclusions.

a. $a_n = \frac{1}{n}$ **b.** $b_n = \frac{1}{n^2}$ **c.** $c_n = \frac{1}{2^n}$ **d.** $d_n = \frac{1}{n!}$ **e.** $h_n = \frac{2^n}{n!}$

11.4 Exercises

See CalcChat.com for tutorial help and worked-out solutions to odd-numbered exercises.
For instructions on how to use a graphing utility, see Appendix A.

Vocabulary and Concept Check

In Exercises 1–4, fill in the blank.

1. If r is a positive real number, then $\lim_{x\to\infty} \frac{1}{x^r} = $ _______ .

2. For a rational function, when the degrees of the numerator and denominator are equal, the limit is the _______ of the leading coefficients.

3. A sequence that has a limit is said to _______ .

4. A sequence that does not have a limit is said to _______ .

Procedures and Problem Solving

Identifying the Graph of an Equation **In Exercises 5–10, match the function with its graph, using horizontal asymptotes as aids. [The graphs are labeled (a), (b), (c), (d), (e), and (f).]**

(a)

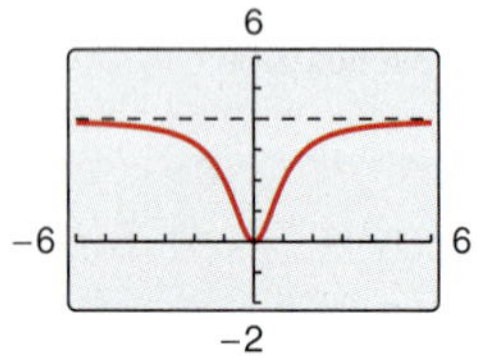

(b)

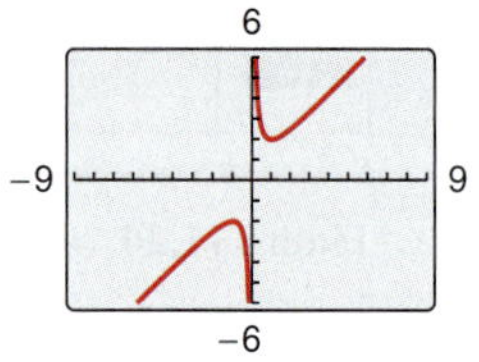

(c)

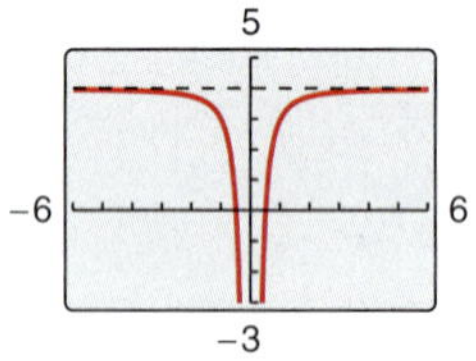

(d)

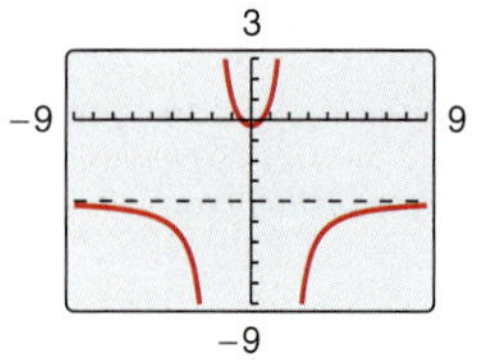

(e)

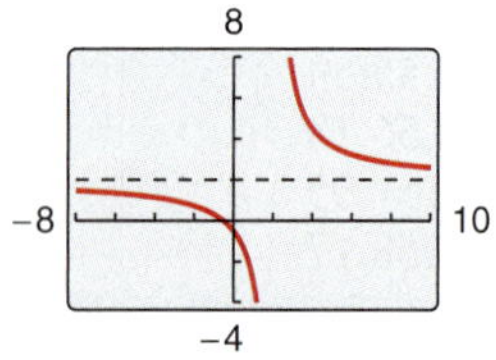

(f)

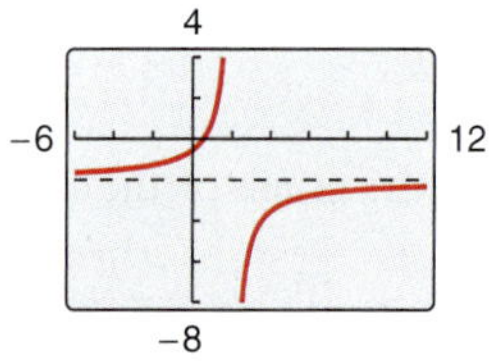

5. $f(x) = \dfrac{2x+1}{x-2}$

6. $f(x) = \dfrac{1-2x}{x-2}$

7. $f(x) = x + \dfrac{1}{x}$

8. $f(x) = 4 - \dfrac{1}{x^2}$

9. $f(x) = \dfrac{4x^2}{x^2+1}$

10. $f(x) = \dfrac{1-4x^2}{x^2-4}$

Evaluating a Limit at Infinity **In Exercises 11–30, find the limit (if it exists). If the limit does not exist, explain why. Use a graphing utility to verify your result graphically.**

11. $\lim_{x\to\infty}\left(\dfrac{4}{x^2}+3\right)$

12. $\lim_{x\to\infty}\left(\dfrac{5}{2x^6}-7\right)$

13. $\lim_{x\to\infty}\dfrac{1+6x}{1-3x}$

14. $\lim_{x\to\infty}\dfrac{2-8x}{2+5x}$

15. $\lim_{x\to-\infty}\dfrac{24x+3}{6x+1}$

16. $\lim_{x\to-\infty}\dfrac{7x+4}{4x+9}$

17. $\lim_{x\to-\infty}\dfrac{4x^2-3}{2-x^2}$

18. $\lim_{x\to-\infty}\dfrac{x^2+3}{5x^2-4}$

19. $\lim_{t\to\infty}\dfrac{t^2}{t+3}$

20. $\lim_{y\to\infty}\dfrac{4y^4}{y^2+3}$

21. $\lim_{t\to\infty}\dfrac{4t^2+3t-1}{3t^2+2t-5}$

22. $\lim_{x\to\infty}\dfrac{5-6x-3x^2}{2x^2+x+4}$

23. $\lim_{y\to-\infty}\dfrac{3+8y-4y^2}{3-y-2y^2}$

24. $\lim_{t\to-\infty}\dfrac{t^2+9t-10}{2+4t-3t^2}$

25. $\lim_{x\to-\infty}\dfrac{-(x^2+3)}{(2-x)^2}$

26. $\lim_{x\to\infty}\dfrac{-(2x^2-6)}{(x-1)^2}$

27. $\lim_{x\to-\infty}\left[\dfrac{6x^2}{(x+1)^2}-4\right]$

28. $\lim_{x\to\infty}\left[7+\dfrac{2x^2}{(x+3)^2}\right]$

29. $\lim_{t\to\infty}\left(\dfrac{1}{3t^2}-\dfrac{5t}{t+2}\right)$

30. $\lim_{x\to\infty}\left[\dfrac{4}{2x+1}+\dfrac{3x^2}{(x-3)^2}\right]$

Algebraic-Graphical-Numerical **In Exercises 31–36, (a) complete the table and numerically estimate the limit as x approaches infinity, (b) use a graphing utility to graph the function and estimate the limit graphically, and (c) find the limit algebraically.**

x	10^0	10^1	10^2	10^3	10^4	10^5	10^6
$f(x)$							

31. $f(x) = \dfrac{2x}{3-x}$

32. $f(x) = \dfrac{x^2}{x^2+4}$

33. $f(x) = \dfrac{7x}{1-x^2}$

34. $f(x) = \dfrac{2x+1}{x^2-1}$

35. $f(x) = 1 - \dfrac{3}{x^2}$

36. $f(x) = \dfrac{1}{x} - 2$

Estimating the Limit at Infinity **In Exercises 37–42, (a) complete the table and numerically estimate the limit as x approaches infinity for the function, and (b) use a graphing utility to graph the function and estimate the limit graphically.**

x	10^0	10^1	10^2	10^3	10^4	10^5	10^6
$f(x)$							

37. $f(x) = x - \sqrt{x^2 + 2}$ **38.** $f(x) = 3x - \sqrt{9x^2 + 1}$

39. $f(x) = 3\left(2x - \sqrt{4x^2 + x}\right)$

40. $f(x) = 4\left(4x - \sqrt{16x^2 - x}\right)$

41. $f(x) = \dfrac{x}{\sqrt{x^2 + 120}}$

42. $f(x) = \dfrac{2\sqrt{x^2 + 9000}}{x}$

Finding the Limit of a Sequence **In Exercises 43–52, write the first five terms of the sequence and find the limit of the sequence (if it exists). If the limit does not exist, explain why. Assume n begins with 1.**

43. $a_n = \dfrac{n + 1}{n^2 + 1}$ **44.** $a_n = \dfrac{n}{n^2 + 1}$

45. $a_n = \dfrac{6n - 2}{2n + 1}$ **46.** $a_n = \dfrac{5n - 1}{2n + 3}$

47. $a_n = \dfrac{n^2}{4n - 2}$ **48.** $a_n = \dfrac{4n^2 + 1}{2n}$

49. $a_n = \dfrac{(n + 1)!}{n!}$ **50.** $a_n = \dfrac{(3n - 2)!}{(3n + 1)!}$

51. $a_n = \dfrac{(-1)^n}{n}$ **52.** $a_n = \dfrac{(-1)^{n+1}}{n^2}$

Finding the Limit of a Sequence **In Exercises 53–56, use a graphing utility to complete the table and estimate the limit of the sequence as n approaches infinity. Then find the limit algebraically.**

n	10^0	10^1	10^2	10^3
a_n				

53. $a_n = \dfrac{1}{n}\left(n + \dfrac{1}{n}\left[\dfrac{n(n + 1)}{2}\right]\right)$

54. $a_n = \dfrac{4}{n}\left(n + \dfrac{4}{n}\left[\dfrac{n(n + 1)}{2}\right]\right)$

55. $a_n = \dfrac{10}{n^3}\left[\dfrac{n(n + 1)(3n + 1)}{6}\right]$

56. $a_n = \dfrac{3n(n + 1)}{n^2} - \dfrac{4}{n^4}\left[\dfrac{n(n + 1)}{2}\right]^2$

57. Average Cost A company's cost function for producing a model of digital binoculars is $C = 34x + 18{,}000$, where C is the cost (in dollars) and x is the number of binoculars produced.

(a) Write a model for the average cost per unit produced.

(b) Find the average costs per unit when $x = 1000$ and $x = 5000$.

(c) Determine the limit of the average cost function as x approaches infinity. Explain the meaning of the limit in the context of the problem.

58. Average Cost The cost function for a new supermarket to recycle x tons of organic material is given by $C = 50x + 2700$, where C is the cost (in dollars).

(a) Write a model for the average cost per ton of organic material recycled.

(b) Find the average costs of recycling 100 tons and 1000 tons of organic material.

(c) Determine the limit of the average cost function as x approaches infinity. Explain the meaning of the limit in the context of the problem.

59. MODELING DATA

The table shows the numbers F (in thousands) of female Selected Reserve personnel in the United States military for the years 2010 through 2016. (*Source:* U.S. Dept. of Defense)

DATA Year	Female personnel, F (in thousands)
2010	153.07
2011	153.92
2012	154.35
2013	155.59
2014	156.18
2015	157.05
2016	158.17

Spreadsheet at LarsonPrecalculus.com

A model for the data is given by

$$F(t) = \frac{153.06 - 23.674t}{1 - 0.1597t + 0.00077t^2}, \quad 0 \le t \le 6$$

where t represents the year, with $t = 0$ corresponding to 2010.

(a) Use a graphing utility to create a scatter plot of the data and graph the model in the same viewing window. How do they compare?

(b) Use the model to predict the numbers of female Selected Reserves in 2017 and 2018.

(c) Find the limit of the model as $t \to \infty$ and interpret its meaning in the context of the situation.

(d) Is this a good model for predicting future numbers of female Selected Reserves? Explain.

60. ***Why you should learn it*** *(p. 790)* The table shows the mean annual salaries S (in thousands of dollars) of a financial manager in the United States for the years 2006 through 2017. *(Source:* U.S. Bureau of Labor Statistics)

DATA

Year	Salary, S (in thousands of dollars)
2006	101.45
2007	106.20
2008	110.64
2009	113.73
2010	116.97
2011	120.45
2012	123.26
2013	126.66
2014	130.23
2015	134.33
2016	139.72
2017	143.53

Spreadsheet at LarsonPrecalculus.com

A model for the data is given by

$$S(t) = \frac{-201.37 + 240.816t - 7.3976t^2}{1 + 1.861t - 0.0707t^2}, \quad 6 \le t \le 17$$

where t represents the year, with $t = 6$ corresponding to 2006.

(a) Use a graphing utility to create a scatter plot of the data and graph the model in the same viewing window. How do they compare?

(b) Use the model to predict the mean salaries in 2018 and 2019.

(c) Find the limit of the model as $t \to \infty$ and interpret its meaning in the context of the situation.

(d) Is this a good model for predicting the mean annual salaries in future years? Explain.

Focusing on Concepts

True or False? **In Exercises 61–64, determine whether the statement is true or false. Justify your answer.**

61. Every rational function has a horizontal asymptote.

62. If a rational function f has a vertical asymptote, then the limit of $f(x)$ as x approaches ∞ exists.

63. If a sequence converges, then it has a limit.

64. When the degrees of the numerator and denominator of a rational function are equal, the limit as x approaches infinity does not exist.

65. Think About It Find functions f and g such that both $f(x)$ and $g(x)$ increase without bound as x approaches ∞, but $\lim_{x\to\infty} [f(x) - g(x)] \neq \infty$.

66. Think About It Use a graphing utility to graph the function

$$f(x) = \frac{x}{\sqrt{x^2 + 1}}.$$

How many horizontal asymptotes does the function appear to have? What are the horizontal asymptotes?

Finding the Limit of a Sequence **In Exercises 67–70, use a graphing utility to create a scatter plot of the terms of the sequence. Determine whether the sequence converges or diverges. If it converges, estimate its limit.**

67. $a_n = 4\left(\frac{2}{3}\right)^n$ **68.** $a_n = 3\left(\frac{3}{2}\right)^n$

69. $a_n = \dfrac{3[1 - (1.5)^n]}{1 - 1.5}$

70. $a_n = \dfrac{3[1 - (0.5)^n]}{1 - 0.5}$

71. Error Analysis Describe the error in finding the limit.

$$\lim_{x\to\infty} \frac{1 - 2x - x^2}{4x^2 + 1} = 0 \quad ✗$$

72. HOW DO YOU SEE IT? Use each graph to estimate $\lim_{x\to\infty} f(x)$, $\lim_{x\to-\infty} f(x)$, and the horizontal asymptote of the graph of f.

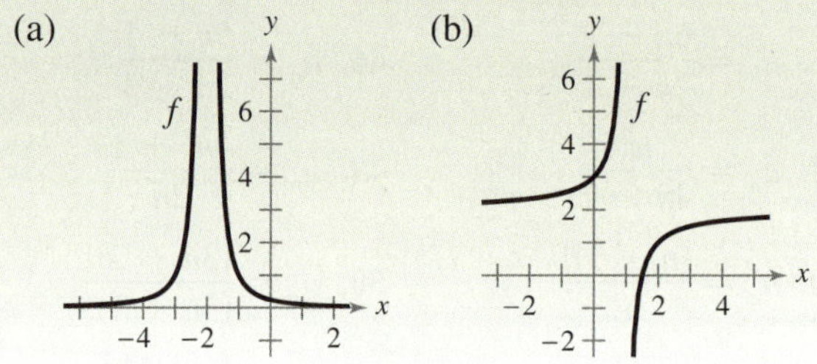

Cumulative Mixed Review

Sketching Transformations **In Exercises 73 and 74, sketch the graphs of y and each transformation on the same rectangular coordinate system.**

73. $y = x^4$

(a) $f(x) = (x + 3)^4$ (b) $f(x) = x^4 - 1$

(c) $f(x) = 2 - 4x^4$ (d) $f(x) = \frac{1}{2}(x - 4)^4$

74. $y = x^3$

(a) $f(x) = (x - 2)^3$ (b) $f(x) = x^3 - 3$

(c) $f(x) = 2 - \frac{1}{4}x^3$ (d) $f(x) = 3(x + 1)^3$

Using Sigma Notation **In Exercises 75 and 76, find the sum.**

75. $\sum_{i=1}^{6} (2i + 3)$ **76.** $\sum_{i=0}^{4} 5i^2$

11.5 The Area Problem

Limits of Summations

Earlier in the text, you used the concept of a limit to obtain a formula for the sum S of an infinite geometric series

$$S = a_1 + a_1 r + a_1 r^2 + \cdot \cdot \cdot = \sum_{i=1}^{\infty} a_1 r^{i-1} = \frac{a_1}{1-r}, \quad |r| < 1.$$

Using limit notation, this sum can be written as

$$S = \lim_{n\to\infty} \sum_{i=1}^{n} a_1 r^{i-1}$$

$$= \lim_{n\to\infty} \frac{a_1(1-r^n)}{1-r} \qquad \sum_{i=1}^{n} a_1 r^{i-1} = a_1\left(\frac{1-r^n}{1-r}\right)$$

$$= \frac{a_1}{1-r}\left[\lim_{n\to\infty} (1-r^n)\right] \qquad \text{Scalar Multiple Property}$$

$$= \frac{a_1}{1-r}. \qquad \lim_{n\to\infty} r^n = 0 \text{ for } |r| < 1$$

The summation formulas and properties listed below are used to evaluate finite and infinite summations. (You studied some of these properties in Section 8.1.)

Summation Formulas and Properties

1. $\sum_{i=1}^{n} c = cn$, c is a constant.
2. $\sum_{i=1}^{n} i = \frac{n(n+1)}{2}$
3. $\sum_{i=1}^{n} i^2 = \frac{n(n+1)(2n+1)}{6}$
4. $\sum_{i=1}^{n} i^3 = \frac{n^2(n+1)^2}{4}$
5. $\sum_{i=1}^{n} (a_i \pm b_i) = \sum_{i=1}^{n} a_i \pm \sum_{i=1}^{n} b_i$
6. $\sum_{i=1}^{n} ca_i = c\sum_{i=1}^{n} a_i$, c is a constant.

EXAMPLE 1 Evaluating a Summation

Evaluate the summation.

$$\sum_{i=1}^{200} i = 1 + 2 + 3 + 4 + \cdot \cdot \cdot + 200$$

Solution

Using Formula 2 with $n = 200$, you can write

$$\sum_{i=1}^{200} i = \frac{200(200+1)}{2} \qquad \sum_{i=1}^{n} i = \frac{n(n+1)}{2}$$

$$= \frac{40{,}200}{2}$$

$$= 20{,}100.$$

✓ **Checkpoint** Audio-video solution in English & Spanish at LarsonPrecalculus.com

Evaluate the summation: $\sum_{i=1}^{10} i^2$.

What you should learn

- Find limits of summations.
- Use rectangles to approximate and limits of summations to find areas of plane regions.

Why you should learn it

Limits of summations are useful in determining areas of plane regions. For instance, in Exercise 40 on page 805, you are asked to find the limit of a summation to determine the area of a parcel of land bounded by a stream and two roads.

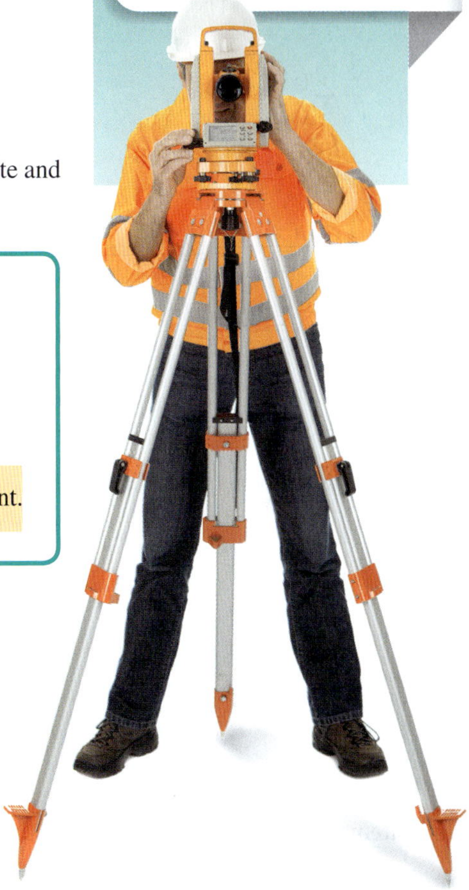

Andrey_Popov/Shutterstock.com

EXAMPLE 2 Evaluating a Summation

Evaluate the summation

$$S = \sum_{i=1}^{n} \frac{i+2}{n^2} = \frac{3}{n^2} + \frac{4}{n^2} + \frac{5}{n^2} + \cdots + \frac{n+2}{n^2}$$

for $n = 10, 100, 1000,$ and $10{,}000$.

Solution

Begin by applying summation formulas and properties to simplify S. In the second line of the solution, note that

$$\frac{1}{n^2}$$

can be factored out of the sum because n is considered to be constant.

$$S = \sum_{i=1}^{n} \frac{i+2}{n^2}$$ Write original form of summation.

$$= \frac{1}{n^2}\sum_{i=1}^{n}(i+2)$$ Factor constant $1/n^2$ out of sum.

$$= \frac{1}{n^2}\left(\sum_{i=1}^{n} i + \sum_{i=1}^{n} 2\right)$$ Write as two sums.

$$= \frac{1}{n^2}\left[\frac{n(n+1)}{2} + 2n\right]$$ Apply Formulas 1 and 2.

$$= \frac{1}{n^2}\left(\frac{n^2+5n}{2}\right)$$ Add fractions.

$$= \frac{n+5}{2n}$$ Simplify.

Algebra Help

Be sure you understand that you cannot factor i out of

$$\sum_{i=1}^{n} i$$

because i is the (variable) index of summation.

Now, evaluate the sum by substituting the appropriate values of n, as shown in the table.

n	10	100	1000	10,000
$\sum_{i=1}^{n} \frac{i+2}{n^2} = \frac{n+5}{2n}$	0.75	0.525	0.5025	0.50025

Checkpoint *Audio-video solution in English & Spanish at LarsonPrecalculus.com*

Evaluate the summation

$$S = \sum_{i=1}^{n} \frac{3i+2}{n^2} = \frac{5}{n^2} + \frac{8}{n^2} + \frac{11}{n^2} + \cdots + \frac{3n+2}{n^2}$$

for $n = 10, 100, 1000,$ and $10{,}000$.

In Example 2, note that the sum appears to approach a limit as n increases. To find the limit of

$$\frac{n+5}{2n}$$

as n approaches infinity, you can use the techniques from Section 11.4 to write

$$\lim_{n\to\infty} \frac{n+5}{2n} = \frac{1}{2}.$$

Be sure you notice the strategy used in Example 2. Rather than separately evaluating the sums

$$\sum_{i=1}^{10} \frac{i+2}{n^2}, \quad \sum_{i=1}^{100} \frac{i+2}{n^2}, \quad \sum_{i=1}^{1000} \frac{i+2}{n^2}, \quad \sum_{i=1}^{10,000} \frac{i+2}{n^2}$$

it was more efficient first to convert to rational form using the summation formulas and properties listed on page 799.

$$S = \underbrace{\sum_{i=1}^{n} \frac{i+2}{n^2}}_{\text{Summation form}} = \underbrace{\frac{n+5}{2n}}_{\text{Rational form}}$$

With this rational form, each sum can be evaluated by simply substituting appropriate values of n.

EXAMPLE 3 Finding the Limit of a Summation

Find the limit of $S(n)$ as $n \to \infty$.

$$S(n) = \sum_{i=1}^{n} \left(1 + \frac{i}{n}\right)^2 \left(\frac{1}{n}\right)$$

Solution

Begin by rewriting the summation in rational form.

$$S(n) = \sum_{i=1}^{n} \left(1 + \frac{i}{n}\right)^2 \left(\frac{1}{n}\right) \qquad \text{Write original form of summation.}$$

$$= \sum_{i=1}^{n} \left(\frac{n^2 + 2ni + i^2}{n^2}\right)\left(\frac{1}{n}\right) \qquad \text{Square } (1 + i/n) \text{ and write as a single fraction.}$$

$$= \frac{1}{n^3} \sum_{i=1}^{n} (n^2 + 2ni + i^2) \qquad \text{Factor constant } 1/n^3 \text{ out of the sum.}$$

$$= \frac{1}{n^3}\left(\sum_{i=1}^{n} n^2 + \sum_{i=1}^{n} 2ni + \sum_{i=1}^{n} i^2\right) \qquad \text{Write as three sums.}$$

$$= \frac{1}{n^3}\left\{n^3 + 2n\left[\frac{n(n+1)}{2}\right] + \frac{n(n+1)(2n+1)}{6}\right\} \qquad \text{Use summation formulas.}$$

$$= \frac{14n^3 + 9n^2 + n}{6n^3} \qquad \text{Simplify.}$$

Algebra Help

As you can see from Example 3, there is a lot of algebra involved in rewriting a summation in rational form. You may want to review simplifying rational expressions if you are having difficulty with this procedure.

In this rational form, you can now find the limit as $n \to \infty$.

$$\lim_{n\to\infty} S(n) = \lim_{n\to\infty} \frac{14n^3 + 9n^2 + n}{6n^3}$$

$$= \frac{14}{6}$$

$$= \frac{7}{3}$$

Checkpoint Audio-video solution in English & Spanish at LarsonPrecalculus.com

Find the limit of $S(n)$ as $n \to \infty$.

$$S(n) = \sum_{i=1}^{n} \left(\frac{2}{n} + \frac{i}{n^2}\right)\left(\frac{1}{n}\right)$$

The Area Problem

You now have the tools needed to solve the second basic problem of calculus: the area problem. The problem is to find the *area* of a plane region bounded by the graph of a nonnegative, continuous function f, the x-axis, and the vertical lines $x = a$ and $x = b$, as shown in Figure 11.25.

When the region is a square, triangle, trapezoid, or semicircle, you can find its area by using a geometric formula. Many other regions, however, require a different approach—one that involves the limit of a summation. The basic strategy is to use a collection of rectangles of equal width that approximates the region, as illustrated in Example 4.

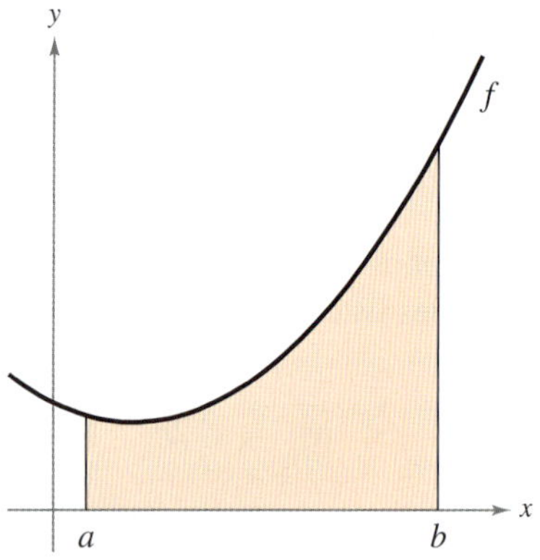

Area under a curve
Figure 11.25

EXAMPLE 4 Approximating the Area of a Region

See LarsonPrecalculus.com for an interactive version of this type of example.

Use the five rectangles in Figure 11.26 to approximate the area of the region bounded by the graph of $f(x) = 6 - x^2$, the x-axis, and the lines $x = 0$ and $x = 2$.

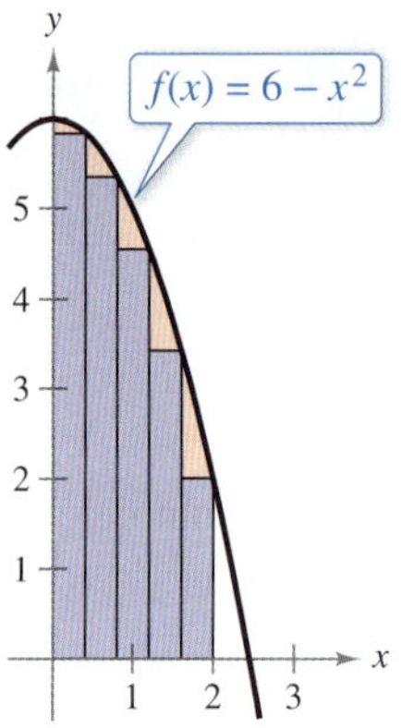

Figure 11.26

Solution

Because the length of the interval along the x-axis is 2 and there are five rectangles, the width of each rectangle is $\frac{2}{5}$. The height of each rectangle can be obtained by evaluating f at the right endpoint of each interval. The five intervals are as follows.

$$\left[0, \frac{2}{5}\right], \quad \left[\frac{2}{5}, \frac{4}{5}\right], \quad \left[\frac{4}{5}, \frac{6}{5}\right], \quad \left[\frac{6}{5}, \frac{8}{5}\right], \quad \left[\frac{8}{5}, \frac{10}{5}\right]$$

Notice that the right endpoint of each interval is $\frac{2}{5}i$ for $i = 1, 2, 3, 4$, and 5. The sum of the areas of the five rectangles is

$$\sum_{i=1}^{5} \underbrace{f\left(\frac{2i}{5}\right)}_{\text{Height}} \underbrace{\left(\frac{2}{5}\right)}_{\text{Width}} = \sum_{i=1}^{5} \left[6 - \left(\frac{2i}{5}\right)^2\right]\left(\frac{2}{5}\right)$$

$$= \frac{2}{5}\left(\sum_{i=1}^{5} 6 - \frac{4}{25}\sum_{i=1}^{5} i^2\right)$$

$$= \frac{2}{5}\left(30 - \frac{44}{5}\right)$$

$$= \frac{212}{25}$$

$$= 8.48.$$

So, the area of the region is approximately 8.48 square units.

✓ ***Checkpoint*** *Audio-video solution in English & Spanish at LarsonPrecalculus.com*

Use the four rectangles in Figure 11.27 to approximate the area of the region bounded by the graph of $f(x) = x + 1$, the x-axis, and the lines $x = 0$ and $x = 2$.

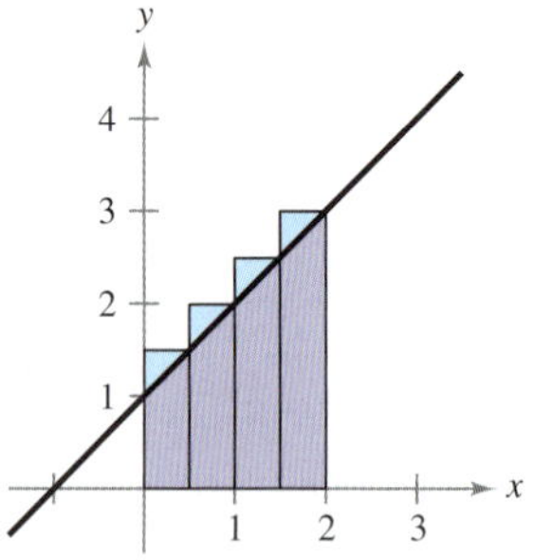

Figure 11.27

By increasing the number of rectangles used in Example 4, you can obtain closer and closer approximations of the area of the region. For instance, using 25 rectangles of width $\frac{2}{25}$ each, you can approximate the area to be $A \approx 9.17$ square units. The table shows even better approximations.

Number of rectangles	5	25	100	1000	5000
Approximate area	8.48	9.17	9.29	9.33	9.33

Based on the procedure illustrated in Example 4, the *exact* **area of a plane region** is given by the limit of the sum of the areas of n rectangles as n approaches ∞, as shown in the next definition.

Area of a Plane Region

Let f be a function that is continuous and nonnegative on the interval $[a, b]$. The **area** A of the region bounded by the graph of f, the x-axis, and the vertical lines $x = a$ and $x = b$ is given by

$$A = \lim_{n\to\infty} \sum_{i=1}^{n} \underbrace{f\left(a + \frac{(b-a)i}{n}\right)}_{\text{Height}} \underbrace{\left(\frac{b-a}{n}\right)}_{\text{Width}}.$$

EXAMPLE 5 Finding the Area of a Region

Find the area of the region bounded by the graph of

$$f(x) = x^2$$

and the x-axis between $x = 0$ and $x = 1$, as shown in Figure 11.28.

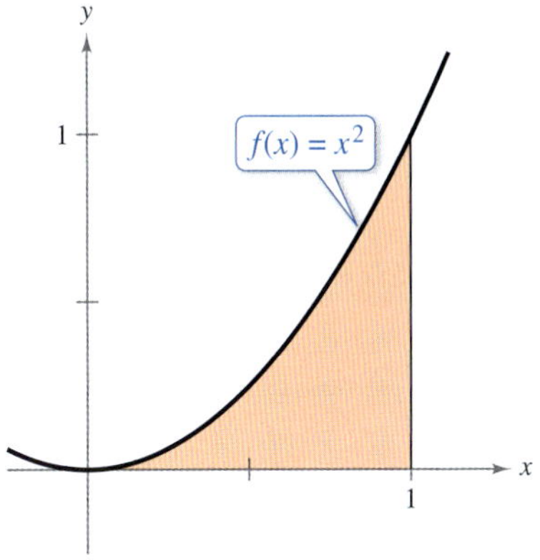

Figure 11.28

Solution

Begin by finding the dimensions of the rectangles.

$$\textit{Width: } \frac{b-a}{n} = \frac{1-0}{n} = \frac{1}{n}$$

$$\textit{Height: } f\left(a + \frac{(b-a)i}{n}\right) = f\left(0 + \frac{(1-0)i}{n}\right) = f\left(\frac{i}{n}\right) = \frac{i^2}{n^2}$$

Next, approximate the area as the sum of the areas of n rectangles.

$$\begin{aligned} A &\approx \sum_{i=1}^{n} f\left(a + \frac{(b-a)i}{n}\right)\left(\frac{b-a}{n}\right) \\ &= \sum_{i=1}^{n} \left(\frac{i^2}{n^2}\right)\left(\frac{1}{n}\right) \\ &= \sum_{i=1}^{n} \frac{i^2}{n^3} \\ &= \frac{1}{n^3}\sum_{i=1}^{n} i^2 \\ &= \frac{1}{n^3}\left[\frac{n(n+1)(2n+1)}{6}\right] \\ &= \frac{2n^3 + 3n^2 + n}{6n^3} \end{aligned}$$

Finally, find the exact area by taking the limit as n approaches ∞.

$$A = \lim_{n\to\infty} \frac{2n^3 + 3n^2 + n}{6n^3} = \frac{1}{3}$$

So, the area of the region is *exactly* $\frac{1}{3}$ square unit.

Checkpoint *Audio-video solution in English & Spanish at LarsonPrecalculus.com*

Find the area of the region bounded by the graph of $f(x) = 3x$ and the x-axis between $x = 0$ and $x = 1$.

11.5 Exercises

See *CalcChat.com* for tutorial help and worked-out solutions to odd-numbered exercises.
For instructions on how to use a graphing utility, see Appendix A.

Vocabulary and Concept Check

In Exercises 1 and 2, fill in the blank.

1. $\sum_{i=1}^{n} i =$ ________

2. $\sum_{i=1}^{n} c =$ ________ , where c is a constant

3. Can you obtain a better approximation of the area of the shaded region shown using 10 rectangles of equal width or 100 rectangles of equal width?

4. Does the limit of the sum of n rectangles as n approaches infinity represent the *exact* area of a plane region or an *approximation* of the area?

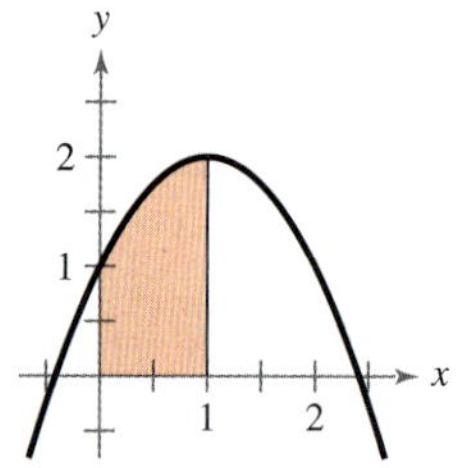

Figure for 3

Procedures and Problem Solving

Evaluating a Summation **In Exercises 5–10, evaluate the sum using the summation formulas and properties.**

5. $\sum_{i=1}^{20} i^3$

6. $\sum_{i=1}^{30} i^2$

7. $\sum_{k=1}^{15} (k^3 + 2)$

8. $\sum_{k=1}^{50} (2k + 1)$

9. $\sum_{j=1}^{25} (j^2 + 2j)$

10. $\sum_{j=1}^{10} (j^3 - 3j^2)$

Finding the Limit of a Summation **In Exercises 11–16, (a) rewrite the sum as a rational function $S(n)$. (b) Use $S(n)$ to complete the table. (c) Find $\lim_{n\to\infty} S(n)$.**

n	10^0	10^1	10^2	10^3	10^4
$S(n)$					

11. $\sum_{i=1}^{n} \frac{i^3}{n^4}$

12. $\sum_{i=1}^{n} \frac{i}{n^2}$

13. $\sum_{i=1}^{n} \frac{3}{n^3}(1 + i^2)$

14. $\sum_{i=1}^{n} \frac{2i + 3}{n^2}$

15. $\sum_{i=1}^{n} \left(\frac{i^2}{n^3} + \frac{2}{n}\right)\left(\frac{1}{n}\right)$

16. $\sum_{i=1}^{n} \left(\frac{5}{n} + \frac{4i^3}{n^3}\right)\left(\frac{1}{n}\right)$

Approximating the Area of a Region **In Exercises 17–20, approximate the area of the region using the indicated number of rectangles of equal width.**

17. $f(x) = x + 4$

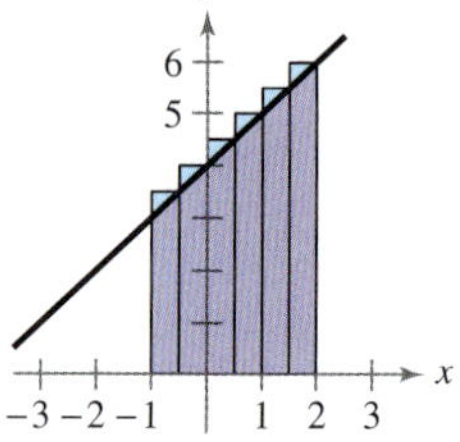

18. $f(x) = 2 - x^2$

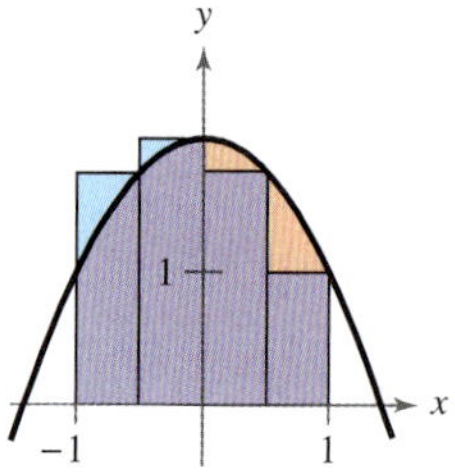

19. $f(x) = \frac{1}{4}x^3$
(8 rectangles)

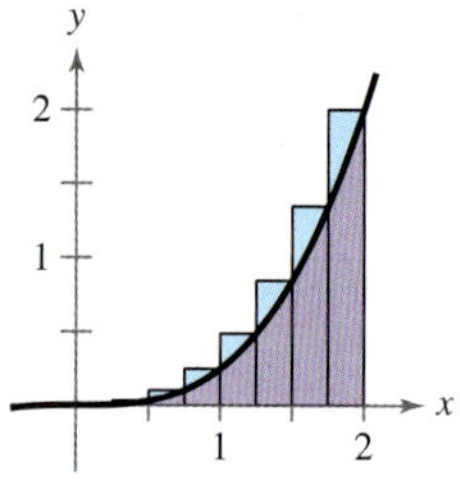

20. $f(x) = \frac{1}{2}(x - 1)^3$
(4 rectangles)

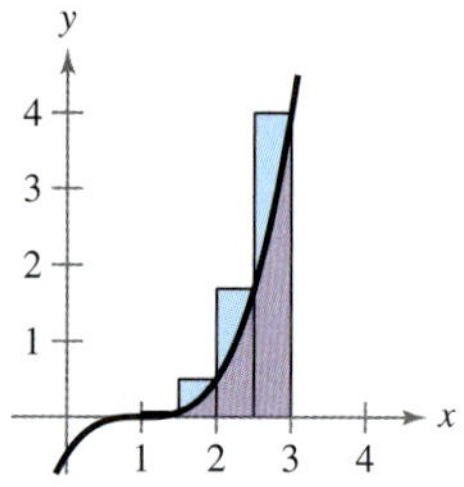

Approximating the Area of a Region **In Exercises 21–24, complete the table to show the approximate area of the region using the indicated numbers n of rectangles of equal width.**

n	4	8	20	50
Approximate area				

21. $f(x) = x^2 + 1$

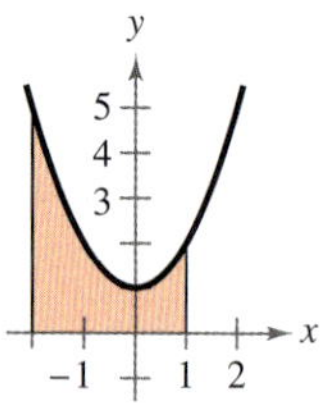

22. $f(x) = -\frac{1}{3}x + 4$

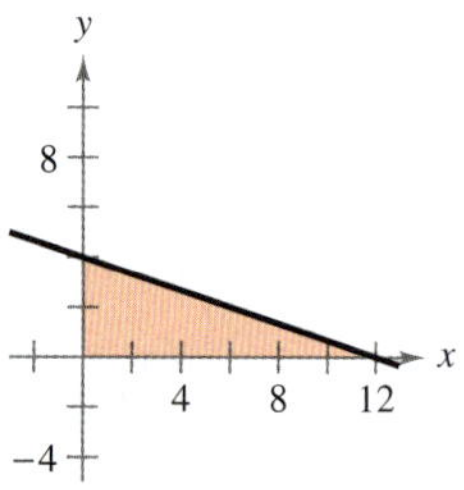

23. $f(x) = \frac{1}{9}x^3$

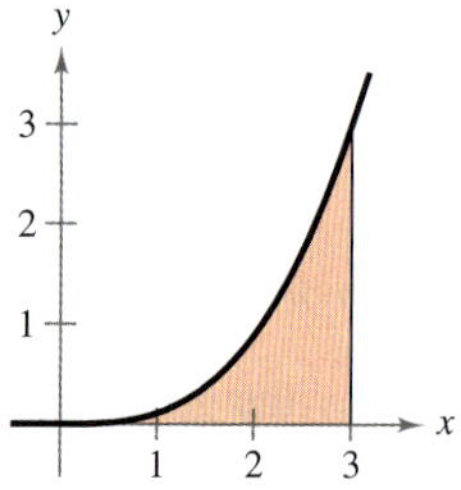

24. $f(x) = 3 - \frac{1}{4}x^3$

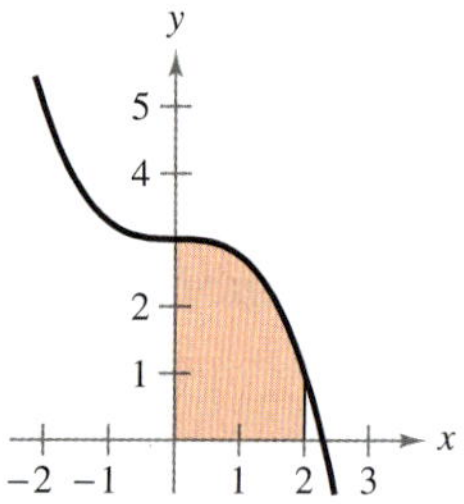

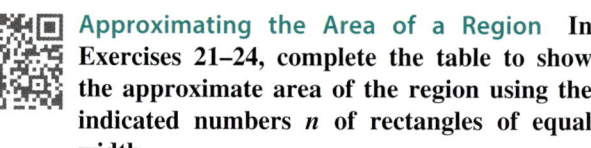

Finding the Area of a Region In Exercises 25–28, complete the table for each finite value of n to approximate the area of the region bounded by the graph of f and the x-axis over the specified interval using the given expression for the sum of the areas of n rectangles. Then find the exact area as $n \to \infty$.

n	4	8	20	50	100	∞
Area						

	Function	*Interval*	*Sum of Areas of n Rectangles*
25.	$f(x) = 2x + 5$	$[0, 4]$	$\dfrac{36n^2 + 16n}{n^2}$
26.	$f(x) = \frac{1}{2}x + 4$	$[-1, 3]$	$\dfrac{18n^2 + 4n}{n^2}$
27.	$f(x) = 9 - x^2$	$[0, 2]$	$\dfrac{46n^3 - 12n^2 - 4n}{3n^3}$
28.	$f(x) = x^2 + 1$	$[4, 6]$	$\dfrac{158n^3 + 60n^2 + 4n}{3n^3}$

Finding the Area of a Region In Exercises 29–36, use the limit process to find the area of the region bounded by the graph of the function and the x-axis over the specified interval.

	Function	*Interval*
29.	$f(x) = x + 1$	$[0, 1]$
30.	$f(x) = 16 - x^2$	$[-2, 1]$
31.	$f(x) = 2 - x^2$	$[-1, 1]$
32.	$f(x) = 3 + x^2$	$[-2, 0]$
33.	$g(x) = 8 - x^3$	$[0, 2]$
34.	$g(x) = 64 - x^3$	$[0, 3]$
35.	$g(x) = 4x - x^3$	$[0, 2]$
36.	$f(x) = x^2 - x^3$	$[0, 1]$

Finding an Area Involving Higher Powers In Exercises 37 and 38, find the area of the function between $x = 0$ and $x = 2$ using the given formula.

37. $f(x) = x^4; \; \sum_{i=1}^{n} i^4 = \dfrac{n(2n + 1)(n + 1)(3n^2 + 3n - 1)}{30}$

38. $f(x) = x^5; \; \sum_{i=1}^{n} i^5 = \dfrac{n^2(2n^2 + 2n - 1)(n + 1)^2}{12}$

39. **Civil Engineering** The boundaries of a parcel of land are two edges modeled by the coordinate axes and a stream modeled by the equation

$$y = (-3.0 \times 10^{-6})x^3 + 0.002x^2 - 1.05x + 400.$$

Use a graphing utility to graph the equation. Find the area of the property. (All distances are measured in feet.)

40. ***Why you should learn it*** *(p. 799)* The table shows the measurements (in feet) of a lot bounded by a stream and two straight roads that meet at right angles (see figure).

x	0	50	100	150
y	450	362	305	268

x	200	250	300
y	245	156	0

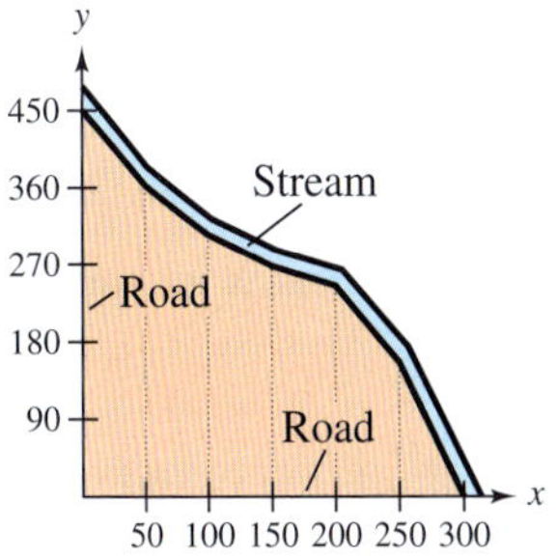

(a) Use the *regression* feature of a graphing utility to find a model of the form $y = ax^3 + bx^2 + cx + d$.

(b) Use the graphing utility to plot the data and graph the model in the same viewing window.

(c) Use the model in part (a) to estimate the area of the lot.

Focusing on Concepts

True or False? In Exercises 41 and 42, determine whether the statement is true or false. Justify your answer.

41. The sum of the first n positive integers is $n(n + 1)/2$.

42. The sum of the squares of the first n even integers is $[2n(n + 1)(2n + 1)]/3$.

43. **Writing** Describe the process of finding the areas of a region bounded by the graph of a nonnegative, continuous function f, the x-axis, and the vertical lines $x = a$ and $x = b$.

44. **HOW DO YOU SEE IT?** Without performing any calculations, choose the value that best approximates the area of the region shown in the graph. [The choices are labeled (a), (b), (c), (d), and (e).]

(a) -2
(b) 1
(c) 4
(d) 6
(e) 9

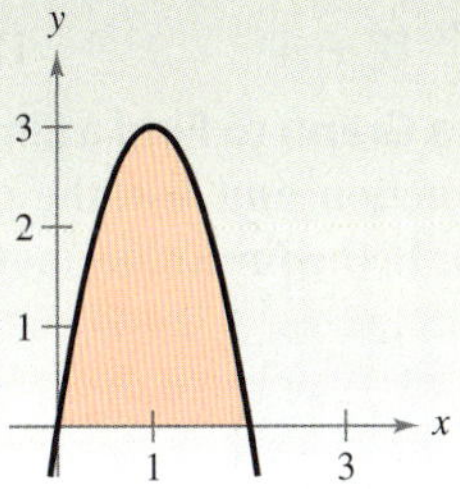

11 Chapter Review

See CalcChat.com for tutorial help and worked-out solutions to odd-numbered exercises. For instructions on how to use a graphing utility, see Appendix A.

11.1 *What did you learn?*

Understand the limit concept *(p. 760)*, and use the definition of a limit to estimate limits *(p. 761)*. If $f(x)$ becomes arbitrarily close to a unique number L as x approaches c from either side, then the limit of $f(x)$ as x approaches c is L. This is written as $\lim_{x \to c} f(x) = L$.

Determine whether limits of functions exist *(p. 763)*. The limit of $f(x)$ as $x \to c$ does not exist under any of the conditions listed below.

1. $f(x)$ approaches a different number from the right side of c than it approaches from the left side of c.
2. $f(x)$ increases or decreases without bound as $x \to c$.
3. $f(x)$ oscillates between two fixed values as $x \to c$.

Use properties of limits to evaluate limits *(p. 765)*. Let b and c be real numbers and let n be a positive integer.

1. $\lim_{x \to c} b = b$ 2. $\lim_{x \to c} x = c$ 3. $\lim_{x \to c} x^n = c^n$
4. $\lim_{x \to c} \sqrt[n]{x} = \sqrt[n]{c}$, valid for all c when n is odd and valid for $c > 0$ when n is even

Let f and g be functions where $\lim_{x \to c} f(x) = L$ and $\lim_{x \to c} g(x) = K$.

1. $\lim_{x \to c} [bf(x)] = bL$ 2. $\lim_{x \to c} [f(x) \pm g(x)] = L \pm K$
3. $\lim_{x \to c} [f(x)g(x)] = LK$ 4. $\lim_{x \to c} \dfrac{f(x)}{g(x)} = \dfrac{L}{K}$, $K \neq 0$
5. $\lim_{x \to c} [f(x)]^n = L^n$

If p is a polynomial function, then $\lim_{x \to c} p(x) = p(c)$.

If r is a rational function given by $r(x) = p(x)/q(x)$ and $q(c) \neq 0$, then $\lim_{x \to c} r(x) = r(c) = p(c)/q(c)$.

Example

(a) Let $f(x) = \dfrac{x}{2 - \sqrt{4 + x}}$. Note that f is undefined at $x = 0$, but from the table, it appears that $\lim_{x \to 0} f(x) = -4$.

x	-0.1	-0.01	-0.001
$f(x)$	-3.9748	-3.9975	-3.9997

x	0.001	0.01	0.1
$f(x)$	-4.0002	-4.0025	-4.0248

Use a graphing utility to graph f and verify this.

(b) Let $f(x) = \dfrac{x - 6}{|x - 6|}$.

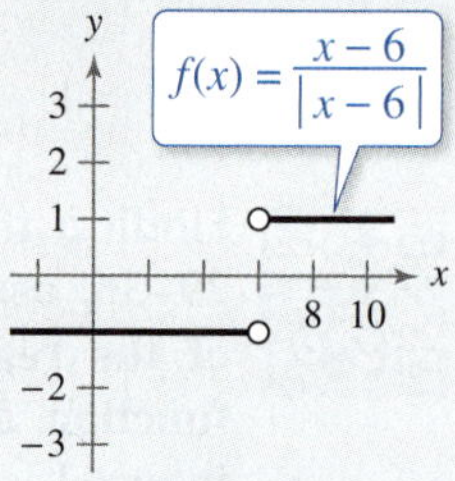

From the graph, $\lim_{x \to 6} f(x)$ does not exist because $f(x) = 1$ when $x > 6$ and $f(x) = -1$ when $x < 6$. So, no matter how close x is to 6, there will be x-values that yield $f(x) = 1$ and $f(x) = -1$.

In parts (c)–(e), let $f(x) = x^2 + 2x + 3$ and let $g(x) = 7 - \sqrt{x}$.

(c) $\lim_{x \to 16} f(x) = 16^2 + 2(16) + 3 = 256 + 32 + 3 = 291$

$\lim_{x \to 16} g(x) = 7 - \sqrt{16} = 7 - 4 = 3$

(d) $\lim_{x \to 16} f(x)g(x) = 291(3) = 873$

(e) $\lim_{x \to 16} \dfrac{f(x)}{g(x)} = \dfrac{291}{3} = 97$

Estimating a Limit Numerically **In Exercises 1–4, use a table to estimate the limit. Determine whether the limit can be reached.**

1. $\lim_{x \to 3} (6x - 1)$
2. $\lim_{x \to 1} (3x^2 - 4x - 4)$
3. $\lim_{x \to 0} \dfrac{\tan x}{2x}$
4. $\lim_{x \to 0} \dfrac{\ln(1 - x)}{x}$

Using a Graph to Find a Limit **In Exercises 5–8, graph the function and use the graph to find the limit (if it exists). If the limit does not exist, explain why.**

5. $\lim_{x \to 1} (3 - x)$
6. $\lim_{x \to 2} \dfrac{1}{x - 2}$
7. $\lim_{x \to -3} \dfrac{|x + 3|}{x + 3}$
8. $\lim_{x \to 1} \dfrac{x^2 - 1}{x - 1}$

Evaluating Limits **In Exercises 9 and 10, use the given information to evaluate each limit.**

9. $\lim_{x \to c} f(x) = 2$, $\lim_{x \to c} g(x) = 5$
   (a) $\lim_{x \to c} [f(x)]^3$ (b) $\lim_{x \to c} [3f(x) - g(x)]$
   (c) $\lim_{x \to c} [f(x)g(x)]$ (d) $\lim_{x \to c} \dfrac{f(x)}{g(x)}$
10. $\lim_{x \to c} f(x) = 8$, $\lim_{x \to c} g(x) = 3$
   (a) $\lim_{x \to c} \sqrt[3]{f(x)}$ (b) $\lim_{x \to c} \dfrac{f(x)}{18}$
   (c) $\lim_{x \to c} [f(x)g(x)]$ (d) $\lim_{x \to c} [f(x) - 2g(x)]$
   (e) $\lim_{x \to c} \dfrac{g(x)}{f(x)}$ (f) $\lim_{x \to c} [f(x)]^{3/2}$

Evaluating a Limit by Direct Substitution **In Exercises 11–22, find the limit using direct substitution.**

11. $\lim_{x\to 4}\left(\frac{1}{2}x + 3\right)$

12. $\lim_{x\to -2}(5 - 2x - x^2)$

13. $\lim_{t\to 3}\frac{t^2+1}{t}$

14. $\lim_{x\to 2}\frac{3x+5}{5x-3}$

15. $\lim_{x\to -2}\sqrt[3]{4x}$

16. $\lim_{x\to -1}\sqrt{5-x}$

17. $\lim_{x\to \pi}\sin 3x$

18. $\lim_{x\to 0}\cos x$

19. $\lim_{x\to -1}e^{-x}$

20. $\lim_{x\to 3}2\ln x$

21. $\lim_{x\to 0}\arccos x$

22. $\lim_{x\to 1}\arctan x$

11.2 *What did you learn?*

Use the dividing out technique to evaluate limits of functions *(p. 770)*. When evaluating a limit of a rational function for which direct substitution produces the indeterminate form 0/0, factor the numerator and denominator, divide out any common factors, and try direct substitution again.

Use the rationalizing technique to evaluate limits of functions *(p. 772)*. This involves rationalizing the numerator of a function by multiplying the numerator and denominator by the conjugate of the numerator.

Use technology to approximate limits graphically and numerically *(p. 773)*. See Examples 4 and 5.

Evaluate one-sided limits of functions *(p. 774)*.

$\lim_{x\to c^-} f(x) = L_1$ or $f(x)\to L_1$ as $x\to c^-$ Limit from the left

$\lim_{x\to c^+} f(x) = L_2$ or $f(x)\to L_2$ as $x\to c^+$ Limit from the right

If f is a function and c and L are real numbers, then $\lim_{x\to c} f(x) = L$ if and only if both the left and right limits exist and are equal to L.

Evaluate limits of difference quotients from calculus *(p. 776)*. For any x-value, the limit of a difference quotient is an expression of the form

$$\lim_{h\to 0}\frac{f(x+h)-f(x)}{h}.$$

Example

(a) $$\lim_{x\to -3}\frac{x+3}{x^2+8x+15} = \lim_{x\to -3}\frac{x+3}{(x+3)(x+5)} = \lim_{x\to -3}\frac{1}{x+5} = \frac{1}{2}$$

(b) $$\lim_{x\to 0}\frac{2-\sqrt{4+x}}{x}$$
$$= \lim_{x\to 0}\left[\left(\frac{2-\sqrt{4+x}}{x}\right)\left(\frac{2+\sqrt{4+x}}{2+\sqrt{4+x}}\right)\right]$$
$$= \lim_{x\to 0}\frac{-x}{x(2+\sqrt{4+x})}$$
$$= \lim_{x\to 0}\frac{-1}{2+\sqrt{4+x}},\quad x\neq 0$$
$$= -\frac{1}{4}$$

(c) Consider $f(x) = \dfrac{x-6}{|x-6|}$ from Example (b) on the preceding page. The limit as $x\to 6$ from the left is $\lim_{x\to 6^-} f(x) = -1$, and the limit as $x\to 6$ from the right is $\lim_{x\to 6^+} f(x) = 1$. Because $\lim_{x\to 6^-} f(x) \neq \lim_{x\to 6^+} f(x)$, the limit as $x\to 6$ does not exist, confirming the result of Example (b) on the preceding page.

Finding a Limit **In Exercises 23–32, find the limit (if it exists). Use a graphing utility to confirm your result graphically.**

23. $\lim_{t\to -2}\frac{t+2}{t^2-4}$

24. $\lim_{t\to 3}\frac{t^2-9}{t-3}$

25. $\lim_{x\to 5}\frac{x-5}{x^2+5x-50}$

26. $\lim_{x\to -1}\frac{x+1}{x^2-5x-6}$

27. $\lim_{x\to 1}\frac{x^2+7x-8}{x^2-3x+2}$

28. $\lim_{x\to -2}\frac{x^2+8x+12}{x^2-3x-10}$

29. $\lim_{x\to -1}\frac{\frac{1}{x+2}-1}{x+1}$

30. $\lim_{x\to 2}\frac{\frac{1}{x-3}+1}{x-2}$

31. $\lim_{u\to 0}\frac{\sqrt{4+u}-2}{u}$

32. $\lim_{v\to 0}\frac{\sqrt{v+9}-3}{v}$

Approximating a Limit **In Exercises 33–40, (a) graphically approximate the limit (if it exists) by using a graphing utility to graph the function and (b) numerically approximate the limit (if it exists) by using the *table* feature of the graphing utility to create a table.**

33. $\lim_{x\to 3}\frac{x-3}{x^2-9}$

34. $\lim_{x\to 4}\frac{4-x}{16-x^2}$

35. $\lim_{x\to 0}e^{-2/x}$

36. $\lim_{x\to 0}\frac{e^{3x}-1}{x}$

37. $\lim_{x\to 0}\frac{\sin 4x}{2x}$

38. $\lim_{x\to 0}\frac{\tan 2x}{x}$

39. $\lim_{x\to 1^+}\frac{\sqrt{2x+1}-\sqrt{3}}{x-1}$

40. $\lim_{x\to 1^+}\frac{1-\sqrt{x}}{x-1}$

Evaluating One-Sided Limits **In Exercises 41–48, graph the function. Evaluate the corresponding one-sided limits. Then determine the limit (if it exists).**

41. $\lim_{x\to0} \frac{|5x|}{x}$

42. $\lim_{x\to0} \frac{|x|}{3x}$

43. $\lim_{x\to4} \frac{4}{x^2 - 16}$

44. $\lim_{x\to8} \frac{1}{x^2 - 64}$

45. $\lim_{x\to5} \frac{|x - 5|}{x - 5}$

46. $\lim_{x\to-2} \frac{|x + 2|}{x + 2}$

47. $\lim_{x\to-2} f(x)$ where $f(x) = \begin{cases} 4 - x, & x \le -2 \\ x^2 - 3, & x > -2 \end{cases}$

48. $\lim_{x\to0} f(x)$ where $f(x) = \begin{cases} x - 6, & x \ge 0 \\ x^2 - 4, & x < 0 \end{cases}$

Evaluating a Limit from Calculus **In Exercises 49 and 50, find** $\lim_{h\to0} \frac{f(x + h) - f(x)}{h}$.

49. $f(x) = 3x - x^2$

50. $f(x) = x^2 - 5x - 2$

11.3 *What did you learn?*

Understand the tangent line problem *(p. 780)*, and use a tangent line to approximate the slope of a graph at a point *(p. 781)*. The tangent line to the graph of a function f at a point $P(x_1, y_1)$ is the line that best approximates the slope of the graph at the point (see figure).

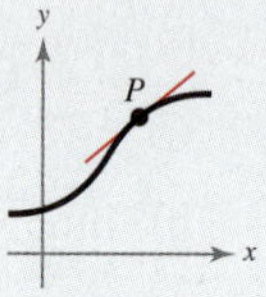

Use the limit definition of slope to find exact slopes of graphs *(p. 782)*. The slope m of the graph of f at the point $(x, f(x))$ is equal to the slope of its tangent line at $(x, f(x))$, and is given by

$$m = \lim_{h\to0} m_{sec} = \lim_{h\to0} \frac{f(x + h) - f(x)}{h}$$

provided this limit exists.

Find derivatives of functions and use derivatives to find slopes of graphs *(p. 785)*. The derivative of f at x is given by

$$f'(x) = \lim_{h\to0} \frac{f(x + h) - f(x)}{h}$$

provided this limit exists. The derivative $f'(x)$ is a formula for the slope of the tangent line to the graph of f at the point $(x, f(x))$.

Example Consider the function $f(x) = x^2 + x + 1$. To find a formula for the slope of the graph, find an expression that represents the slope of the secant line.

$$m_{sec} = \frac{f(x + h) - f(x)}{h}$$

$$= \frac{[(x + h)^2 + (x + h) + 1] - (x^2 + x + 1)}{h}$$

$$= \frac{x^2 + 2xh + h^2 + x + h + 1 - x^2 - x - 1}{h}$$

$$= \frac{2xh + h^2 + h}{h} = 2x + h + 1, \ h \ne 0$$

Taking the limit of m_{sec} as $h \to 0$ gives a formula for the slope of the graph, $m = 2x + 1 = f'(x)$. The figure shows the graphs of f and the tangent line at the point $(-2, 3)$. Verify that the equation of the tangent line is $y = -3x - 3$.

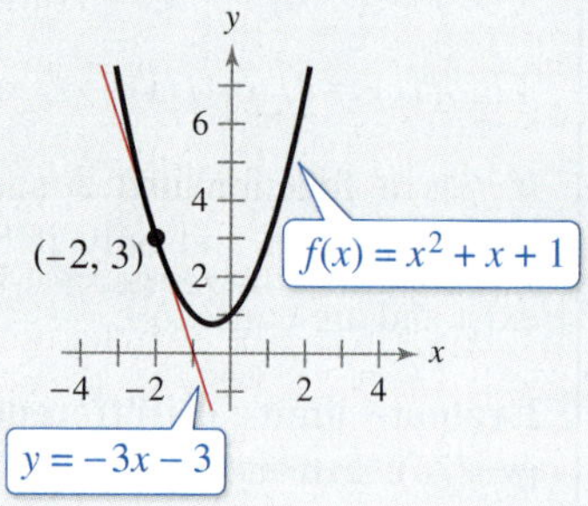

Visually Approximating the Slope of a Graph **In Exercises 51 and 52, use the figure to approximate the slope of the graph at the point (x, y).**

51.

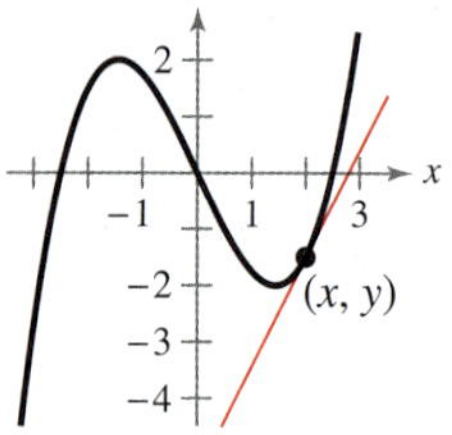

52.

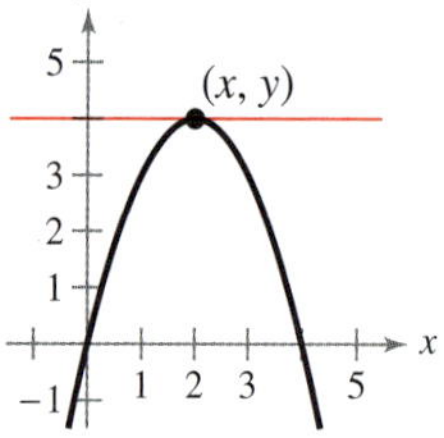

Approximating the Slope of a Tangent Line **In Exercises 53–58, use a graphing utility to graph the function and the tangent line at the point $(2, f(2))$. Use the graph to approximate the slope of the tangent line.**

53. $f(x) = 2x - x^2$

54. $f(x) = 6 - x^2$

55. $f(x) = \sqrt{x + 2}$

56. $f(x) = \sqrt{x^2 + 5}$

57. $f(x) = \frac{6}{x - 4}$

58. $f(x) = \frac{1}{3 - x}$

Finding a Formula for the Slope of a Graph **In Exercises 59–62, find a formula for the slope of the graph of f at the point $(x, f(x))$. Then use it to find the slopes at the two specified points.**

59. $f(x) = x^2 - 4x$
(a) $(4, 0)$
(b) $(-1, 5)$

60. $f(x) = \frac{1}{4}x^4$
(a) $(-2, 4)$
(b) $\left(1, \frac{1}{4}\right)$

61. $f(x) = \frac{4}{x - 6}$
(a) $\left(9, \frac{4}{3}\right)$
(b) $(8, 2)$

62. $f(x) = \sqrt{x}$
(a) $(1, 1)$
(b) $(4, 2)$

Finding a Derivative In Exercises 63–74, find the derivative of the function.

63. $f(x) = 8$
64. $g(x) = -16$
65. $h(x) = 6 - 7x$
66. $f(x) = 9x$
67. $g(x) = 2x^2 + x - 3$
68. $f(x) = -x^3 + 4x$
69. $f(t) = \sqrt{t + 5}$
70. $g(t) = \sqrt{t - 6}$
71. $g(s) = \dfrac{4}{s + 5}$
72. $g(t) = \dfrac{6}{5 - t}$
73. $g(x) = \dfrac{1}{\sqrt{x + 4}}$
74. $f(x) = \dfrac{1}{\sqrt{12 - x}}$

Using the Derivative In Exercises 75–78, (a) find the slope of the graph of f at the given point, (b) find an equation of the tangent line to the graph at the point, and (c) graph the function and the tangent line.

75. $f(x) = 2x^2 - 1, (3, 17)$
76. $f(x) = x^2 + 10, (2, 14)$
77. $f(x) = x^3 + 1, (1, 2)$
78. $f(x) = x^3 - x, (2, 6)$

11.4 *What did you learn?*

Evaluate limits of functions at infinity *(p. 790)*. If f is a function and L_1 and L_2 are real numbers, then the statements $\lim_{x\to-\infty} f(x) = L_1$ and $\lim_{x\to\infty} f(x) = L_2$ denote the limits at infinity.

If r is a positive real number, then $\lim_{x\to\infty} (1/x^r) = 0$. If x^r is defined when $x < 0$, then $\lim_{x\to-\infty} (1/x^r) = 0$.

Consider the rational function $f(x) = N(x)/D(x)$, where $N(x) = a_n x^n + \cdots + a_0$ and $D(x) = b_m x^m + \cdots + b_0$. The limit of $f(x)$ as x approaches positive or negative infinity is as follows.

$$\lim_{x\to\pm\infty} f(x) = \begin{cases} 0, & n < m \\ \dfrac{a_n}{b_m}, & n = m \end{cases}$$

If $n > m$, then the limit does not exist.

Find limits of sequences *(p. 794)*. Let L be a real number. Let f be a function of a real variable such that $\lim_{x\to\infty} f(x) = L$. If $\{a_n\}$ is a sequence such that $f(n) = a_n$ for every positive integer n, then $\lim_{n\to\infty} a_n = L$.

Example

(a) $\lim_{x\to\infty} (5/x^2) = 0$

(b) $\lim_{x\to-\infty} \left(1 - \dfrac{5}{x^3}\right) = \lim_{x\to-\infty} 1 - \lim_{x\to-\infty} \dfrac{5}{x^3} = 1 - 5(0) = 1$

(c) $\lim_{x\to\infty} \dfrac{x + 1}{x^2 + 5x - 3} = 0$

(d) $\lim_{x\to-\infty} \dfrac{2x - 7}{3x^2 - x + 6} = 0$

(e) $\lim_{x\to-\infty} \dfrac{x^2 + 1}{x^2 + 5x - 3} = \dfrac{1}{1} = 1$

(f) $\lim_{x\to\infty} \dfrac{2x^2 - 7}{3x^2 - x + 6} = \dfrac{2}{3}$

(g) $\lim_{x\to\infty} \dfrac{x^3 + 1}{x^2 + 5x - 3}$ does not exist.

(h) $\lim_{x\to-\infty} \dfrac{2x^3 - 7}{3x^2 - x + 6}$ does not exist.

(i) Let $a_n = \dfrac{2n + 1}{n^2}$, $b_n = \dfrac{2n^2 + 1}{n^2}$, and $c_n = \dfrac{2n^3 + 1}{n^2}$.

$\lim_{n\to\infty} a_n = 0$, $\lim_{n\to\infty} b_n = 2$, $\lim_{n\to\infty} c_n$ does not exist.

Evaluating a Limit at Infinity In Exercises 79–90, find the limit (if it exists). If the limit does not exist, explain why. Use a graphing utility to verify your result graphically.

79. $\lim_{x\to\infty} \dfrac{4x}{2x - 3}$
80. $\lim_{x\to\infty} \dfrac{8x}{16x + 4}$
81. $\lim_{x\to-\infty} \dfrac{x + 6}{6 - x}$
82. $\lim_{x\to-\infty} \dfrac{1 - 2x}{x + 2}$
83. $\lim_{x\to\infty} \left(3 - \dfrac{1}{x^2}\right)$
84. $\lim_{x\to\infty} \left(7 + \dfrac{3}{2x^2}\right)$
85. $\lim_{x\to-\infty} \dfrac{2x}{x^2 - 25}$
86. $\lim_{x\to-\infty} \dfrac{3x}{(1 - x)^3}$
87. $\lim_{x\to\infty} \dfrac{x^2}{3x + 4}$
88. $\lim_{y\to\infty} \dfrac{9y^4}{y^2 + 1}$
89. $\lim_{x\to\infty} \left[\dfrac{x}{(x - 2)^2} + 3\right]$
90. $\lim_{x\to\infty} \left[2 - \dfrac{2x^2}{(x + 1)^2}\right]$

Finding the Limit of a Sequence In Exercises 91–100, write the first five terms of the sequence and find the limit of the sequence (if it exists). If the limit does not exist, explain why. Assume n begins with 1.

91. $a_n = \dfrac{n - 1}{n + 1}$
92. $a_n = \dfrac{2n - 3}{5n + 4}$
93. $a_n = \dfrac{2n}{n^2 + 1}$
94. $a_n = \dfrac{3n + 1}{n^2 + 1}$
95. $a_n = \dfrac{n^2}{3n + 2}$
96. $a_n = \dfrac{4n^3 + 1}{2n - 1}$
97. $a_n = \dfrac{(-1)^n}{n^3}$
98. $a_n = \dfrac{(-1)^{n+1}}{n}$
99. $a_n = \dfrac{1}{2n^2}[3 - 2n(n + 1)]$
100. $a_n = \dfrac{3}{n}\left\{n + \dfrac{2}{n}\left[\dfrac{n(n - 1)}{2}\right]\right\}$

11.5 *What did you learn?*

Find limits of summations *(p. 799)*. Summation formulas and properties are below (c is a constant).

1. $\sum_{i=1}^{n} c = cn$ **2.** $\sum_{i=1}^{n} i = \frac{n(n+1)}{2}$

3. $\sum_{i=1}^{n} i^2 = \frac{n(n+1)(2n+1)}{6}$ **4.** $\sum_{i=1}^{n} i^3 = \frac{n^2(n+1)^2}{4}$

5. $\sum_{i=1}^{n} (a_i \pm b_i) = \sum_{i=1}^{n} a_i \pm \sum_{i=1}^{n} b_i$ **6.** $\sum_{i=1}^{n} ca_i = c\sum_{i=1}^{n} a_i$

Use rectangles to approximate and limits of summations to find areas of plane regions *(p. 802)*. A collection of rectangles of equal width can approximate the area of a plane region. Increasing the number of rectangles gives a closer approximation (see Example 4).

Let f be a function that is continuous and nonnegative on $[a, b]$. The area A of the region bounded by the graph of f, the x-axis, and the vertical lines $x = a$ and $x = b$ is

$$A = \lim_{x\to\infty} \sum_{i=1}^{n} \underbrace{f\left(a + \frac{(b-a)i}{n}\right)}_{\text{Height}} \underbrace{\left(\frac{b-a}{n}\right)}_{\text{Width}}.$$

Example Consider the region bounded by the graph of $f(x) = x^2 + 1$ and the x-axis, between $x = 0$ and $x = 3$ (see figure). To approximate the area A of the region with n rectangles, use $(3 - 0)/n = 3/n$ for the width and use

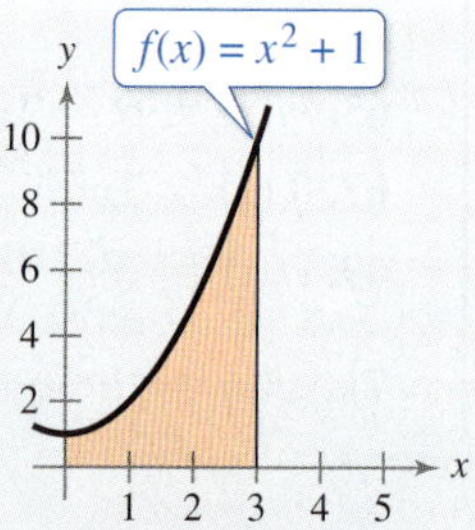

$$f\left(0 + \frac{(3-0)i}{n}\right) = f\left(\frac{3i}{n}\right) = \frac{9i^2}{n^2} + 1$$

where $i = 1, 2, 3, \ldots, n$, for the height of each rectangle. So, an approximation for the area is as follows (verify this).

$$A \approx \sum_{i=1}^{n}\left(\frac{9i^2}{n^2} + 1\right)\left(\frac{3}{n}\right) = \frac{27}{n^3}\sum_{i=1}^{n} i^2 + \frac{1}{n}\sum_{i=1}^{n} 3$$

$$= \frac{24n^2 + 27n + 9}{2n^2}$$

To find the exact area, take the limit as n approaches ∞.

$$A = \lim_{n\to\infty} \frac{24n^2 + 27n + 9}{2n^2} = 12$$

So, the area of the region is *exactly* 12 square units.

Finding the Limit of a Summation **In Exercises 101 and 102, (a) rewrite the sum as a rational function $S(n)$. (b) Use $S(n)$ to complete the table. (c) Find $\lim_{n\to\infty} S(n)$.**

n	10^0	10^1	10^2	10^3	10^4
$S(n)$					

101. $\sum_{i=1}^{n}\left(\frac{4i^2}{n^2} - \frac{i}{n}\right)\left(\frac{1}{n}\right)$ **102.** $\sum_{i=1}^{n}\left[4 - \left(\frac{3i}{n}\right)^2\right]\left(\frac{3i}{n^2}\right)$

Approximating the Area of a Region **In Exercises 103 and 104, approximate the area of the region using the indicated number of rectangles of equal width.**

103. $f(x) = 4 - x$

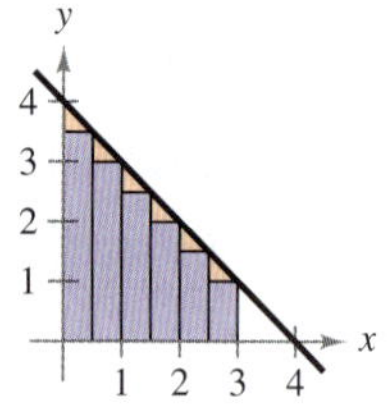

104. $f(x) = 4 - x^2$

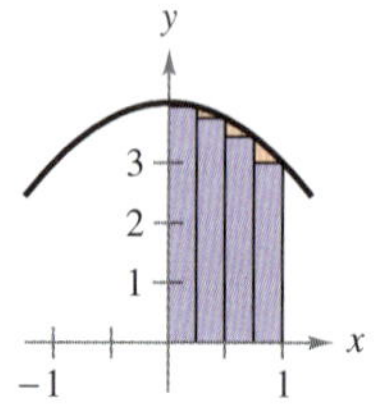

Approximating the Area of a Region **In Exercises 105 and 106, approximate the area of the region using (a) $n = 4$, (b) $n = 8$, (c) $n = 20$, and (d) $n = 50$ rectangles of equal width.**

105. $f(x) = \frac{1}{4}x^2$

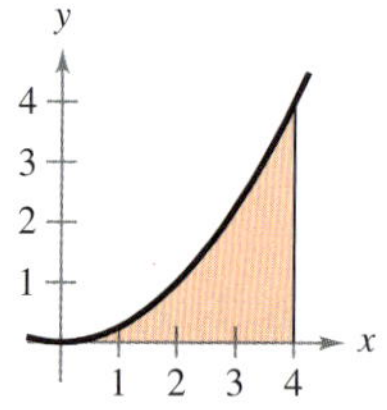

106. $f(x) = 4x - x^2$

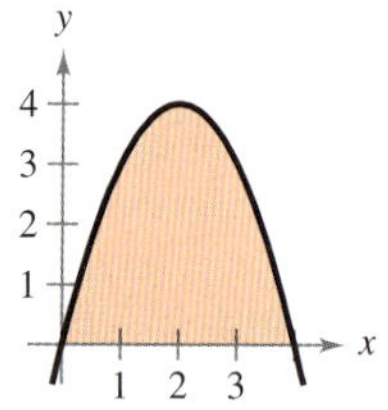

Finding the Area of a Region **In Exercises 107–112, use the limit process to find the area of the region bounded by the graph of the function and the x-axis over the specified interval.**

107. $f(x) = 8 - x$, $[3, 5]$ **108.** $f(x) = 2x - 6$, $[3, 6]$

109. $f(x) = x^2 + 4$, $[0, 3]$ **110.** $f(x) = 6(x - x^2)$, $[0, 1]$

111. $f(x) = x^3 + 1$, $[0, 4]$

112. $f(x) = 8 - x^3$, $[0, 2]$

Focusing on Concepts

True or False? **In Exercises 113 and 114, determine whether the statement is true or false. Justify your answer.**

113. The expression $f'(z)$ gives the slope of the tangent line to the graph of f at the point $(z, f(z))$.

114. If the degree of the numerator $N(x)$ of a rational function $f(x) = N(x)/D(x)$ is greater than the degree of its denominator $D(x)$, then the limit of the rational function as x approaches ∞ is 0.

11 Chapter Test

See *CalcChat.com* for tutorial help and worked-out solutions to odd-numbered exercises. For instructions on how to use a graphing utility, see Appendix A.

Take this test as you would take a test in class. After you are finished, check your work against the answers given in the back of the book.

In Exercises 1–3, use a graphing utility to graph the function and approximate the limit (if it exists). Then find the limit (if it exists) algebraically by using the appropriate technique(s).

1. $\lim_{x \to -2} \dfrac{x^2 - 1}{2x}$ **2.** $\lim_{x \to 1} \dfrac{5 - 4x - x^2}{1 - x}$ **3.** $\lim_{x \to 5} \dfrac{\sqrt{x} - 2}{x - 5}$

In Exercises 4 and 5, use the *table* feature of a graphing utility to estimate the limit numerically. Then use the graphing utility to graph the corresponding function to confirm your result.

4. $\lim_{x \to 0} \dfrac{\sin 2x}{3x}$ **5.** $\lim_{x \to 0} \dfrac{e^{5x} - 1}{x}$

6. Find a formula for the slope of the graph of f at the point $(x, f(x))$. Then use it to find the slope at the specified point.

(a) $f(x) = 3x^2 - 5x - 2$, $(2, 0)$

(b) $f(x) = 2x^3 + 6x$, $(-1, -8)$

In Exercises 7–9, find the derivative of the function.

7. $f(x) = 3 - \dfrac{2}{5}x$ **8.** $f(x) = 2x^2 + 4x - 1$ **9.** $f(x) = \dfrac{1}{x + 1}$

In Exercises 10–12, find the limit (if it exists). If the limit does not exist, explain why. Use a graphing utility to verify your result graphically.

10. $\lim_{x \to \infty} \dfrac{6}{5x - 1}$ **11.** $\lim_{x \to \infty} \dfrac{1 - 4x^2}{8x^2 - 5}$ **12.** $\lim_{x \to \infty} \dfrac{5x^2}{2x - 3}$

In Exercises 13 and 14, write the first five terms of the sequence and find the limit of the sequence (if it exists). If the limit does not exist, explain why. Assume *n* begins with 1.

13. $a_n = \dfrac{n^2 + 3n - 4}{2n^2 + n - 2}$ **14.** $a_n = \dfrac{1 + (-1)^n}{n}$

15. Approximate the area of the region bounded by the graph of $f(x) = 8 - 2x^2$ using the rectangles of equal width shown (see figure).

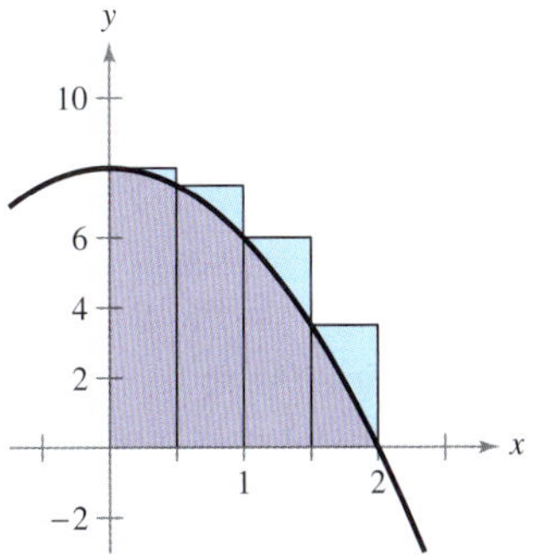

Figure for 15

In Exercises 16 and 17, use the limit process to find the area of the region bounded by the graph of the function and the *x*-axis over the specified interval.

16. $f(x) = x + 3$; interval: $[-2, 2]$

17. $f(x) = 7 - x^2$; interval: $[0, 2]$

18. The table shows the height of a space shuttle during the first 5 seconds of its launch.

(a) Use the *regression* feature of a graphing utility to find a quadratic model $y = ax^2 + bx + c$ for the data.

(b) The value of the derivative of the model at time x is the rate of change of height with respect to time, or the velocity, at that instant. Find the velocity of the shuttle after 5 seconds.

Collaborative Project

To work a collaborative project involving Limits and an Introduction to Calculus, visit this textbook's website at *LarsonPrecalculus.com.*

DATA

Time, *x* (in seconds)	Height, *y* (in feet)
0	0
1	1
2	23
3	60
4	115
5	188

Spreadsheet at LarsonPrecalculus.com

Table for 18

Standardized Test Practice

See CalcChat.com for tutorial help and worked-out solutions to odd-numbered exercises.
For instructions on how to use a graphing utility, see Appendix A.

1.

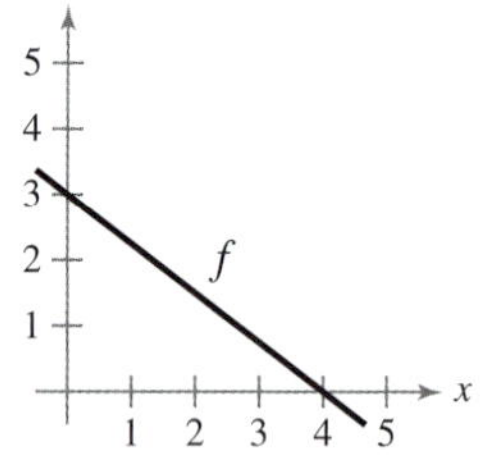

The graph of f is shown above. Which of the following is true?

I. The graph of $g(x) = x - 1$ is perpendicular to the graph of f.

II. The y-intercept of the graph of f is $(0, 3)$.

III. The slope of the graph of f is -4.

(A) I only

(B) II only

(C) I and III only

(D) II and III only

2. For what values of a is $[\![a]\!] = -3$?

(A) $-3 \le a < -2$

(B) $-3 < a \le -2$

(C) $-4 \le a < -3$

(D) $-4 < a \le -3$

3. Which of the following points lies on the graph of $(g \circ f)(x)$, where $f(x) = 2x^2$ and $g(x) = \sqrt{x + 1}$?

(A) $(-1, 2)$ (B) $(0, 2)$

(C) $(1, 2)$ (D) $(2, 3)$

4. $h = -16t^2 + 32t + 49$

The equation above represents the height of a thrown object, where h is the height (in feet) and t is the time (in seconds). Which of the following is the best estimate for when the object hits the ground?

(A) 2 seconds (B) 2.5 seconds

(C) 3 seconds (D) 3.5 seconds

5. Which of the following models predicts the greatest number of hospitalizations y after $t = 5$ days?

(A) $y = 20e^{0.3t}$ (B) $y = 50e^{0.1t}$

(C) $y = 30 \ln t$ (D) $y = 10 \ln 2t$

6. Which of the following angle measures (expressed in radians) is equivalent to an angle measure of 132 degrees?

(A) 2.3 (B) $\dfrac{7\pi}{30}$

(C) $\dfrac{11\pi}{15}$ (D) 4.3

7. Which of the following graphs has an amplitude of 4?

(A)

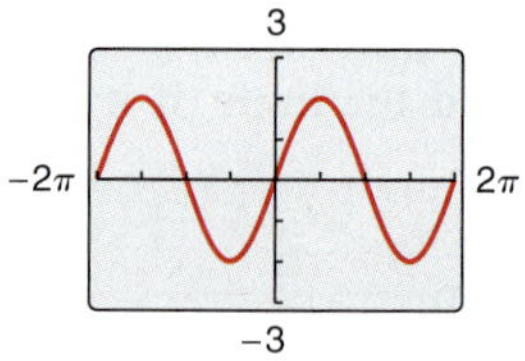

(B)

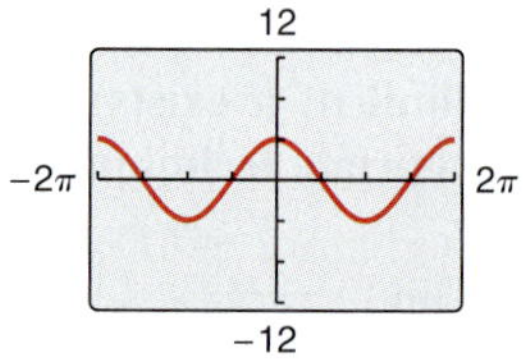

(C)

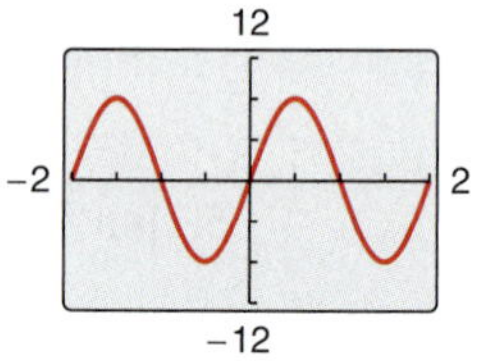

(D)

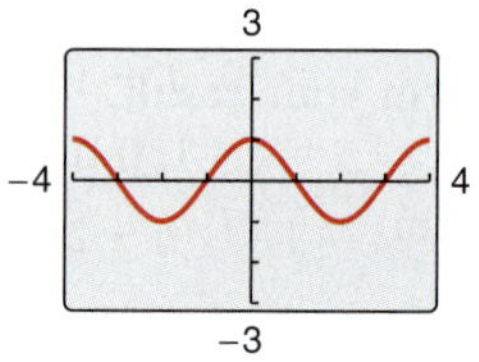

8. Which of the following is *not* a solution of $2 \sin 2x \cos x = -\cos x$?

(A) $\dfrac{\pi}{2}$ (B) $\dfrac{7\pi}{12}$

(C) $\dfrac{11\pi}{12}$ (D) $\dfrac{7\pi}{6}$

9.

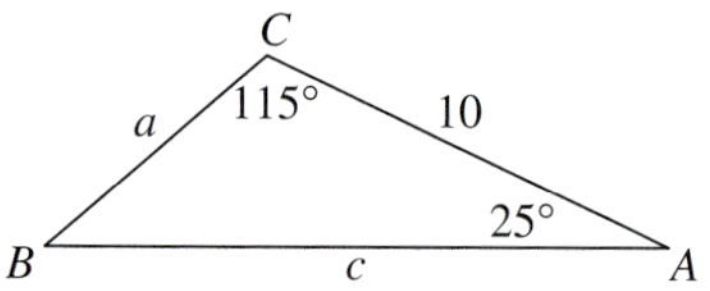

Which of the following equations can be used to solve for c in the triangle above?

(A) $\dfrac{10^2}{\sin 25°} = \dfrac{c}{\sin 115°}$ (B) $\dfrac{6.575}{\sin 115°} = \dfrac{c}{\sin 25°}$

(C) $\dfrac{14.1}{\sin 40°} = \dfrac{c}{\sin 25°}$ (D) $\dfrac{10}{\sin 40°} = \dfrac{c}{\sin 115°}$

10.
$$\begin{aligned} x + 2y - z &= 6 \\ -4x - y + z &= 0 \\ 3x + y - 2z &= 2 \end{aligned}$$

What is the number of solutions of the system above?

(A) zero

(B) one

(C) two

(D) infinitely many

11. The sequence $a_n = 12 + 2n$ predicts the total number of clients for a cosmetologist after n months of advertising. What is the difference between the total after 8 months and the total after 2 months?

(A) 12 (B) 16

(C) 28 (D) 72

12. Which equation represents a circle?

(A) $2x^2 - y = 8$ (B) $2x^2 + y^2 = 8$

(C) $x^2 = 8 - y^2$ (D) $x^2 = 8 + y^2$

13. $x^2 + y^2 + z^2 - 3x - y + 4z - 3 = 0$

What is the center of the sphere represented by the equation above?

(A) $(3, 1, -4)$ (B) $\left(\frac{3}{2}, -\frac{1}{2}, 2\right)$

(C) $\left(\frac{3}{2}, \frac{1}{2}, -2\right)$ (D) $\left(-\frac{3}{2}, \frac{1}{2}, 2\right)$

14. Which of the following vectors have a dot product of -4?

(A) $\mathbf{u} = \langle 0, 4, 0 \rangle$, $\mathbf{v} = \langle 0, 0, -1 \rangle$ (B) $\mathbf{u} = \langle -2, 1, 0 \rangle$, $\mathbf{v} = \langle -2, -1, 0 \rangle$

(C) $\mathbf{u} = \langle -1, 0, 1 \rangle$, $\mathbf{v} = \langle 2, 0, -2 \rangle$ (D) $\mathbf{u} = \langle 1, 5, 0 \rangle$, $\mathbf{v} = \langle -1, 1, 0 \rangle$

15. Which expression gives the area of a triangle with vertices A, B, and C?

(A) $\|\overrightarrow{AB} \times \overrightarrow{AC}\|$ (B) $\frac{1}{2}\|\overrightarrow{AB} \times \overrightarrow{AC}\|$

(C) $\overrightarrow{AB} \cdot \overrightarrow{AC}$ (D) $\frac{1}{2}\overrightarrow{AB} \cdot \overrightarrow{AC}$

16.

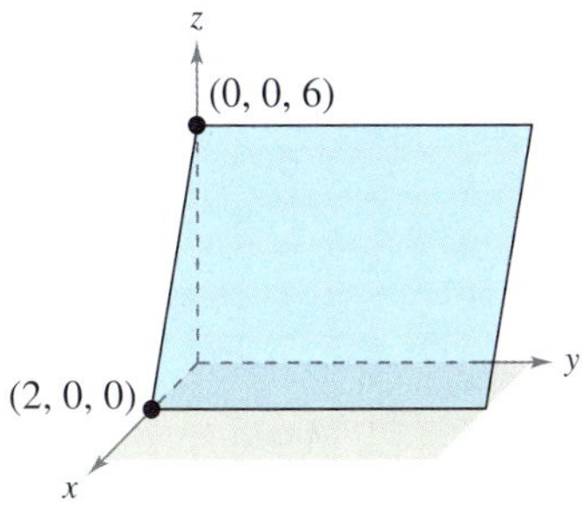

What is the equation of the plane shown above?

(A) $3x + z = 6$

(B) $3y + z = 6$

(C) $6x + y = 0$

(D) $x + 6y = 0$

17. $f(x) = x^3 - 2x^2 + x - 2$

$g(x) = x^2 - 2$

Equations for functions f and g are stated above. Which of the following is true?

I. $\lim_{x\to 0} f(x) < \lim_{x\to 0} g(x)$

II. $\lim_{x\to 2} [f(x)g(x)] = 0$

III. $\lim_{x\to -2} \dfrac{f(x)}{g(x)} = -10$

(A) I only

(B) III only

(C) I and II only

(D) II and III only

18. $f(t) = \begin{cases} 8000, & 0 < t \le 1 \\ 8320, & 1 < t \le 2 \\ 8653, & 2 < t \le 3 \end{cases}$

The storage capacity $f(t)$, in gallons, of a facility is expanded at the beginning of each year. The storage capacities for the next three years are given by the piecewise function above. What is $\lim_{x\to 2} f(t)$?

(A) 0 gallons

(B) 8000 gallons

(C) 8320 gallons

(D) undefined

19. Which of the following functions does *not* have a limit of 3 as x approaches infinity?

(A) $f(x) = \dfrac{x}{3x - 3}$

(B) $g(x) = 3$

(C) $h(x) = \dfrac{3x}{x + 2}$

(D) $j(x) = \dfrac{6x^2 - 6}{2x^2 + 1}$

20. Which of the following formulas gives the exact area under the graph of $f(x) = 3x^2$ on the interval $[-k, k]$?

(A) $2 \lim_{n\to\infty} \sum_{i=1}^{n} 3\left(\frac{ki}{n}\right)^2\left(\frac{k}{n}\right)$

(B) $2 \lim_{n\to\infty} \sum_{i=1}^{n} \left(-k + \frac{2ki}{n}\right)^2\left(\frac{k}{n}\right)$

(C) $\lim_{n\to\infty} \sum_{i=1}^{n} 3\left(\frac{ki}{n}\right)^2\left(\frac{k}{n}\right)$

(D) $\lim_{n\to\infty} \sum_{i=1}^{n} 3\left(-k + \frac{2ki}{n}\right)^2\left(\frac{k}{n}\right)$

21. The function $p(x) = -4x^2 + 12x - 5$ represents the profit earned (in dollars) for an item that costs x dollars. What is the value of x (in dollars) that maximizes the profit earned per item?

22. What is $\left|f\left(\frac{\pi}{4}\right) - g\left(\frac{\pi}{4}\right)\right|$ for

$g(\theta) = \left(\frac{\theta}{\pi}\right)^2$ and $f(\theta) = \frac{1}{4}\tan\theta$?

23. The graph of a quadratic function passes through the points $(-1, 3)$, $(0, 1)$, and $(1, 11)$. What is the leading coefficient of the quadratic function when the function is written in standard form?

24. A fair 12-sided die, labeled from 1 to 12, is tossed twice. What is the probability that the results are an even number followed by a number less than 10?

25. What is the slope of the graph of $y = \sqrt{x}$ at the point $(4, 2)$?

Proofs in Mathematics

Many of the proofs of the definitions and properties presented in this chapter are beyond the scope of this text. Included below are simple proofs for the limit of a power function and the limit of a polynomial function.

Limit of a Power Function (p. 765)

$\lim_{x \to c} x^n = c^n$, where c is a real number and n is a positive integer.

Proof

$$\lim_{x \to c} x^n = \lim_{x \to c}(\underbrace{x \cdot x \cdot x \cdot \cdots \cdot x}_{n \text{ factors}})$$

$$= \underbrace{\lim_{x \to c} x \cdot \lim_{x \to c} x \cdot \lim_{x \to c} x \cdot \cdots \cdot \lim_{x \to c} x}_{n \text{ factors}} \quad \text{Product Property of Limits}$$

$$= \underbrace{c \cdot c \cdot c \cdot \cdots \cdot c}_{n \text{ factors}} \quad \text{Limit of the identity function}$$

$$= c^n \quad \text{Exponential form}$$

Limit of a Polynomial Function (p. 766)

If p is a polynomial function and c is a real number, then

$\lim_{x \to c} p(x) = p(c)$.

Proof

Let p be a polynomial function such that

$$p(x) = a_n x^n + a_{n-1}x^{n-1} + \cdots + a_2 x^2 + a_1 x + a_0.$$

Because a polynomial function is the sum of monomial functions, you can write the following.

$$\lim_{x \to c} p(x) = \lim_{x \to c} (a_n x^n + a_{n-1}x^{n-1} + \cdots + a_2 x^2 + a_1 x + a_0$$

$$= \lim_{x \to c} a_n x^n + \lim_{x \to c} a_{n-1}x^{n-1} + \cdots + \lim_{x \to c} a_2 x^2 + \lim_{x \to c} a_1 x + \lim_{x \to c} a_0$$

$$= a_n c^n + a_{n-1}c^{n-1} + \cdots + a_2 c^2 + a_1 c + a_0 \quad \text{Scalar Multiple Property of Limits and limit of a power function}$$

$$= p(c) \quad p \text{ evaluated at } c$$

Proving Limits

To prove most of the definitions and properties from this chapter, you must use the *formal* definition of a limit. This definition is called the *epsilon-delta definition* and was first introduced by Karl Weierstrass (1815–1897). If you go on to take a course in calculus, you will use this definition of a limit extensively.

Progressive Summary (Chapters 3–11)

This chart outlines the topics that have been covered so far in this text. Progressive Summary charts appear after Chapters 2, 3, 6, 9, and 11. In each Progressive Summary, new topics encountered for the first time appear in blue.

Transcendental Functions

Exponential, Logarithmic, Trigonometric, Inverse Trigonometric

Rewriting

Exponential form ↔ Logarithmic form
Condense/expand logarithmic expressions
Simplify trigonometric expressions
Prove trigonometric identities
Use conversion formulas
Operations with vectors
Powers and roots of complex numbers

Solving

Equation	*Strategy*
Exponential	Take logarithm of each side
Logarithmic	Exponentiate each side
Trigonometric	Isolate function Factor, use inverse function
Multiple angle . . . . or high powers	Use trigonometric identities

Analyzing

Graphically	*Algebraically*
Intercepts	Domain, Range
Asymptotes	Transformations
Minimum values	Composition
Maximum values	Inverse Properties
	Amplitude, Period
	Reference angles

Numerically
Table of values

Systems and Series

Systems, Sequences, Series

Rewriting

Row operations for systems of equations
Partial fraction decomposition
Operations with matrices
Matrix form of a system of equations
*n*th term of a sequence
Summation form of a series

Solving

Equation	*Strategy*
System of . . . . linear equations	Substitution Elimination Gaussian Gauss-Jordan Inverse matrices Cramer's Rule

Analyzing

Systems:
Intersecting, parallel, and coincident lines, determinants
Sequences:
Graphing utility in *dot* mode, *n*th term, partial sums, summation formulas

Other Topics

Conics, Parametric and Polar Equations
3-space, Limits

Rewriting

Standard forms of conics
Eliminate parameters
Rectangular form ↔ Parametric form
Rectangular form ↔ Polar form
Rationalize numerator
Difference quotient

Solving

Equation	*Strategy*
Conics	Convert to standard form Convert to polar form
3-space	Equation of plane Equation of line Cross-product of vectors
Limits	Direct substitution One-sided limits Limits at infinity

Analyzing

Conics:
Table of values, vertices, foci, axes, symmetry, asymptotes, translations, eccentricity
Parametric forms:
Point plotting, eliminate parameters
Polar forms:
Point plotting, special equations, symmetry, zeros, eccentricity, directrix
3-space:
Point plotting, intercepts, traces, vectors

Appendix A: Technology Support

Introduction

Graphing utilities, such as graphing calculators and computers with graphing software, are very valuable tools for visualizing mathematical principles, verifying solutions of equations, exploring mathematical ideas, and developing mathematical models. Although graphing utilities are extremely helpful in learning mathematics, their use does not mean that learning algebra is any less important. In fact, the combination of knowledge of mathematics and the use of graphing utilities enables you to explore mathematics more easily and to a greater depth. It is up to you to learn the capabilities of your graphing utility and to practice using this tool to enhance your mathematical learning.

In this text, there are many opportunities to use a graphing utility, some of which are described below.

Uses of a Graphing Utility

1. Check or validate answers to problems obtained using algebraic methods.
2. Discover and explore algebraic properties, rules, and concepts.
3. Graph functions, and approximate solutions of equations involving functions.
4. Efficiently perform complicated mathematical procedures, such as those found in many real-life applications.
5. Find mathematical models for sets of data.

In this appendix, the features of graphing utilities are discussed from a generic perspective and are listed in alphabetical order. To learn how to use the features of a specific graphing utility, consult your user's manual or go to this textbook's *Student Companion Website*. Additional keystroke guides are available for most graphing utilities, and your college library may have a resource on how to use your graphing utility.

Many graphing utilities are designed to act as "function graphers." In this course, functions and their graphs are studied in detail. You may recall from previous courses that a function can be thought of as a rule that describes the relationship between two variables. These rules are frequently written in terms of x and y. For instance, the equation

$$y = 3x + 5$$

represents y as a function of x.

Many graphing utilities have an *equation editor* feature. To enter an equation, you may have to write the equation in "$y =$ " form, as shown in Figure A.1. (You should note that your *equation editor* screen may not look like the screen shown in Figure A.1.)

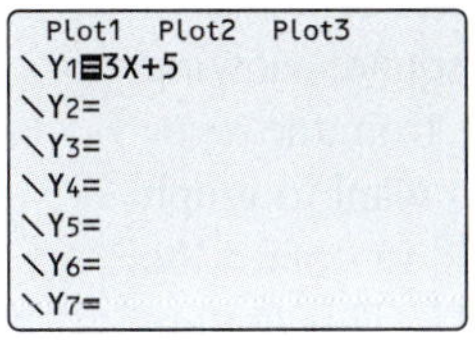

Figure A.1

Cumulative Sum Feature

The *cumulative sum* feature finds partial sums of a series. For instance, to find the first four partial sums of the series

$$\sum_{k=1}^{4} 2(0.1)^k$$

choose the *cumulative sum* feature, which is found in the *operations* menu of the *list* feature (see Figure A.2). To use this feature, you will also have to use the *sequence* feature (see Figure A.2 and page A15). You must enter an expression for the sequence, a variable, the lower limit of summation, and the upper limit of summation, as shown in Figure A.3. After pressing ENTER, you can see that the first four partial sums are 0.2, 0.22, 0.222, and 0.2222. You may have to scroll to the right in order to see all the partial sums.

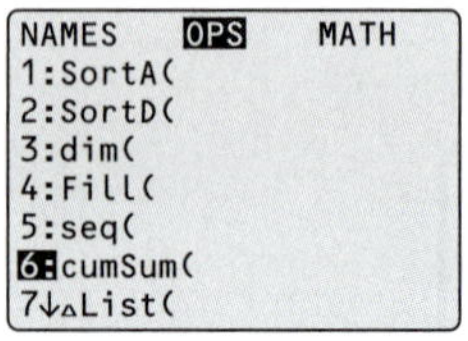

Figure A.2

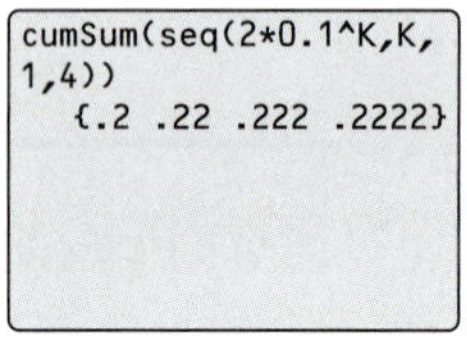

Figure A.3

Technology Tip

As you use your graphing utility, be aware of how parentheses are inserted in an expression. Some graphing utilities automatically insert the left parenthesis when certain calculator buttons are pressed. The placement of parentheses can make a difference between a correct answer and an incorrect answer.

Determinant Feature

The *determinant* feature evaluates the determinant of a square matrix. For instance, to evaluate the determinant of

$$A = \begin{bmatrix} 7 & -1 & 0 \\ 2 & 2 & 3 \\ -6 & 4 & 1 \end{bmatrix}$$

enter the 3×3 matrix in the graphing utility using the *matrix editor*, as shown in Figure A.4. Then choose the *determinant* feature from the *math* menu of the *matrix* feature, as shown in Figure A.5. Once you choose the matrix name, A, press ENTER and you should obtain a determinant of -50, as shown in Figure A.6.

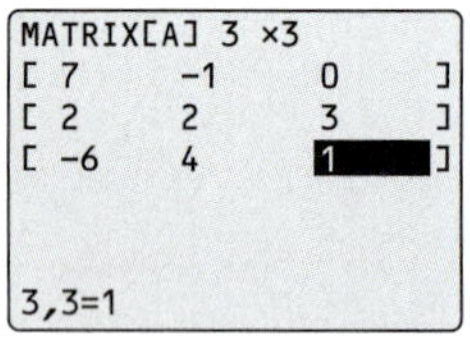

Figure A.4

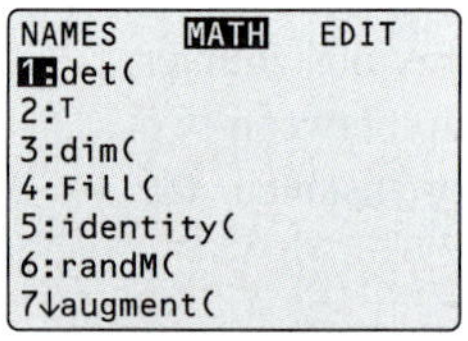

Figure A.5

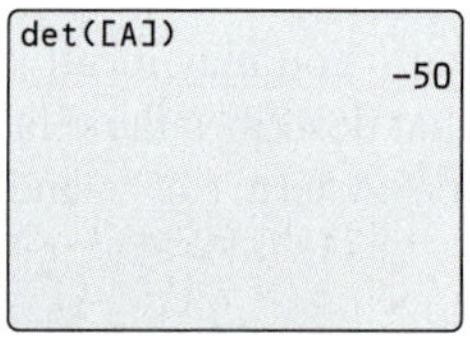

Figure A.6

Draw Inverse Feature

The *draw inverse* feature graphs the inverse function of a *one-to-one* function. For instance, to graph the inverse function of $f(x) = x^3 + 4$, first enter the function in the *equation editor* (see Figure A.7) and graph the function (using a square viewing window), as shown in Figure A.8. Then choose the *draw inverse* feature from the *draw* feature menu, as shown in Figure A.9. You must enter the function you want to graph the inverse function of, as shown in Figure A.10. Finally, press ENTER to obtain the graph of the inverse function of $f(x) = x^3 + 4$, as shown in Figure A.11. This feature can be used only when the graphing utility is in *function* mode.

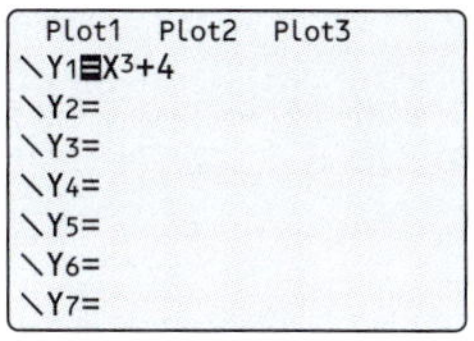

Figure A.7

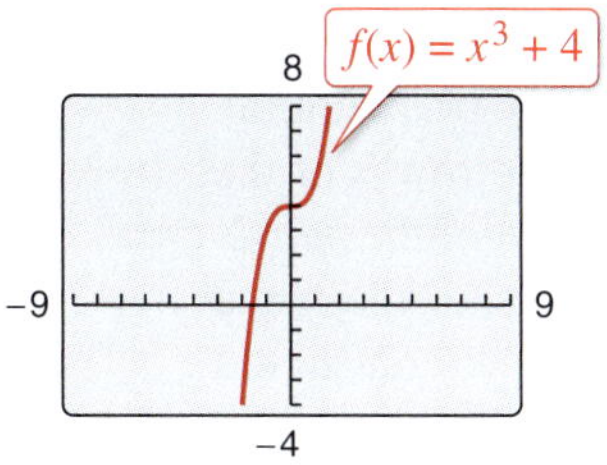

Figure A.8

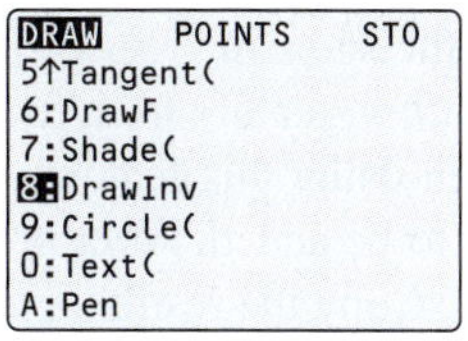

Figure A.9

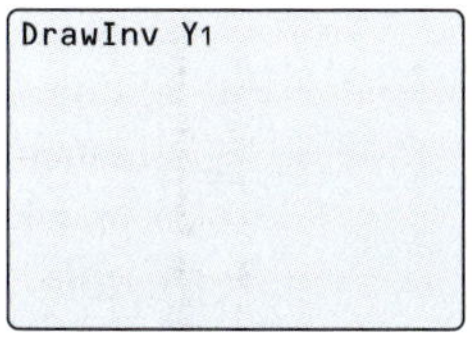

Figure A.10

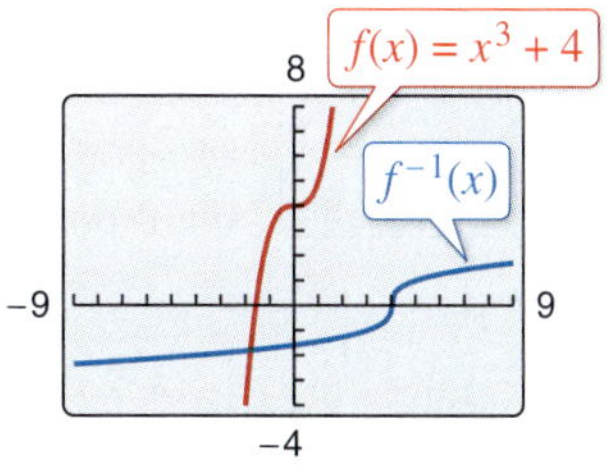

Figure A.11

Elementary Row Operations Features

Most graphing utilities can perform elementary row operations on matrices.

Row Swap Feature

The *row swap* feature interchanges two rows of a matrix. To interchange rows 1 and 3 of the matrix

$$A = \begin{bmatrix} -1 & -2 & 1 & 2 \\ 2 & -4 & 6 & -2 \\ 1 & 3 & -3 & 0 \end{bmatrix}$$

first enter the matrix in the graphing utility using the *matrix editor*, as shown in Figure A.12. Then choose the *row swap* feature from the *math* menu of the *matrix* feature, as shown in Figure A.13. When using this feature, you must enter the name of the matrix and the two rows that are to be interchanged. After pressing ENTER, you should obtain the matrix shown in Figure A.14. Because the resulting matrix will be used to demonstrate the other elementary row operation features, use the *store* feature to copy the resulting matrix to [A], as shown in Figure A.15.

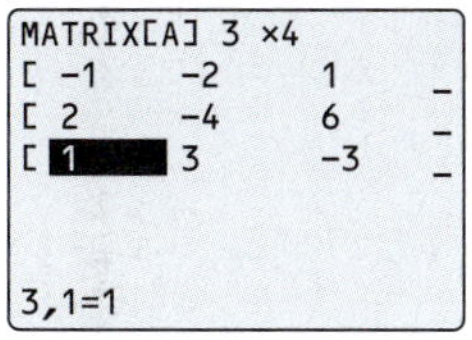

Figure A.12

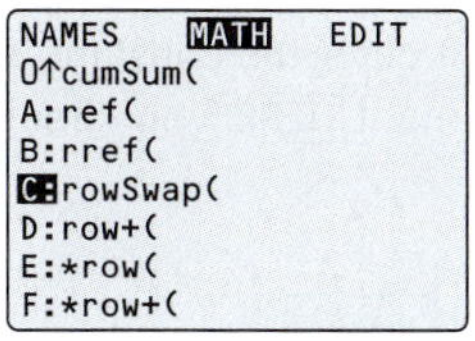

Figure A.13

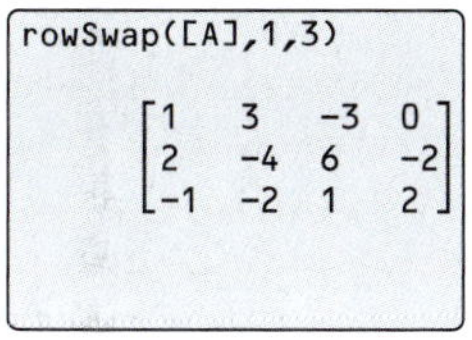

Figure A.14

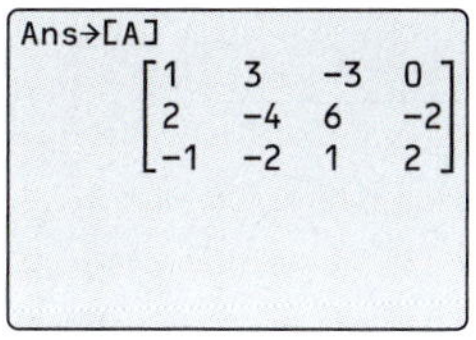

Figure A.15

Technology Tip

The *store* feature of a graphing utility is used to store a value in a variable or to copy one matrix to another matrix. For instance, as shown at the left, after performing a row operation on a matrix, you can copy the answer to another matrix (see Figure A.15). You can then perform another row operation on the copied matrix. To continue performing row operations to obtain a matrix in row-echelon form or reduced row-echelon form, you must copy the resulting matrix to a new matrix before each operation.

Row Addition and Row Multiplication and Addition Features

The *row addition* and *row multiplication and addition* features add a row or a multiple of a row of a matrix to another row of the same matrix. To add row 1 to row 3 of the matrix stored in [A], choose the *row addition* feature from the *math* menu of the *matrix* feature, as shown in Figure A.16. When using this feature, you must enter the name of the matrix and the two rows that are to be added. After pressing ENTER, you should obtain the matrix shown in Figure A.17. Copy the resulting matrix to [A].

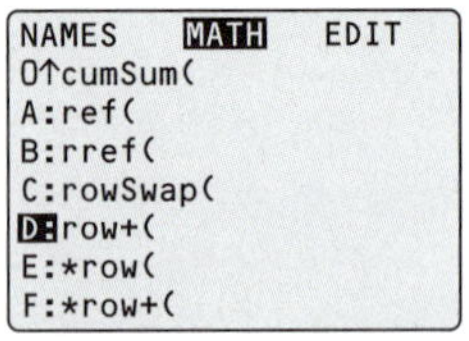

Figure A.16

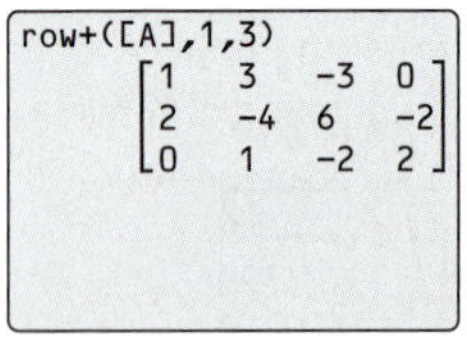

Figure A.17

To add -2 times row 1 to row 2 of the matrix stored in [A], choose the *row multiplication and addition* feature from the *math* menu of the *matrix* feature, as shown in Figure A.18. When using this feature, you must enter the constant, the name of the matrix, the row the constant is multiplied by, and the row to be added to. After pressing ENTER, you should obtain the matrix shown in Figure A.19. Copy the resulting matrix to [A].

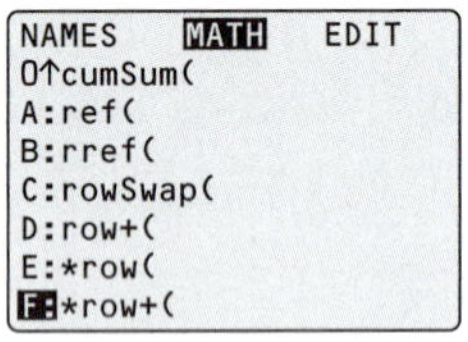

Figure A.18

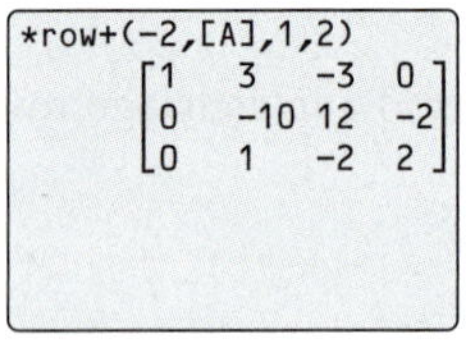

Figure A.19

Row Multiplication Feature

The *row multiplication* feature multiplies a row of a matrix by a nonzero constant. To multiply row 2 of the matrix stored in [A] by

$$-\frac{1}{10}$$

choose the *row multiplication* feature from the *math* menu of the *matrix* feature, as shown in Figure A.20. When using this feature, you must enter the constant, the name of the matrix, and the row to be multiplied. After pressing ENTER, you should obtain the matrix shown in Figure A.21.

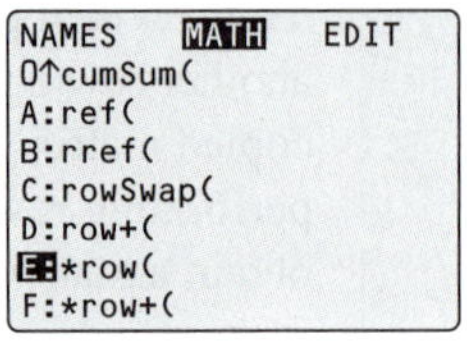

Figure A.20

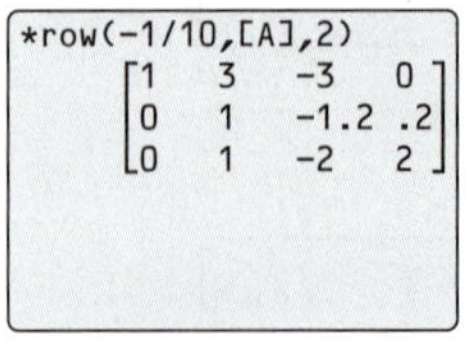

Figure A.21

Intersect Feature

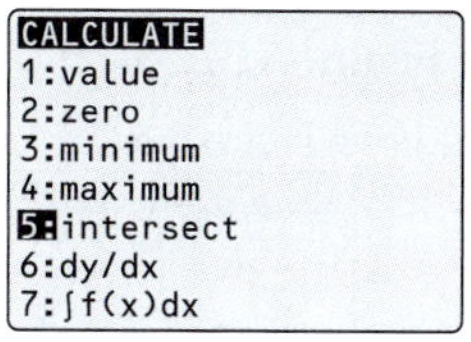

Figure A.22

The *intersect* feature finds a point of intersection of two graphs. The *intersect* feature is found in the *calculate* menu (see Figure A.22).

To find the point of intersection of the graphs of $y_1 = -x + 2$ and $y_2 = x + 4$, first enter the equations in the *equation editor*, as shown in Figure A.23. Then graph the equations, as shown in Figure A.24. Next, use the *intersect* feature to find the point of intersection. Trace the cursor along the graph of y_1 near the intersection and press ENTER (see Figure A.25). Then trace the cursor along the graph of y^2 near the intersection and press ENTER (see Figure A.26). Marks are then placed on the graph at these points (see Figure A.27). Finally, move the cursor near the point of intersection and press ENTER. In Figure A.28, you can see that the coordinates of the point of intersection are displayed at the bottom of the window. So, the point of intersection is $(-1, 3)$.

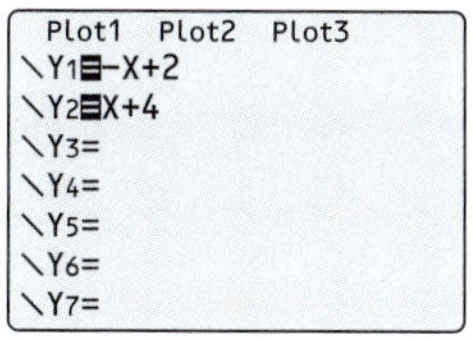

Figure A.23

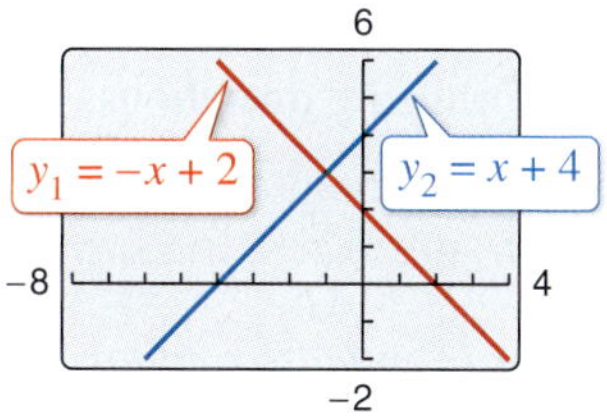

Figure A.24

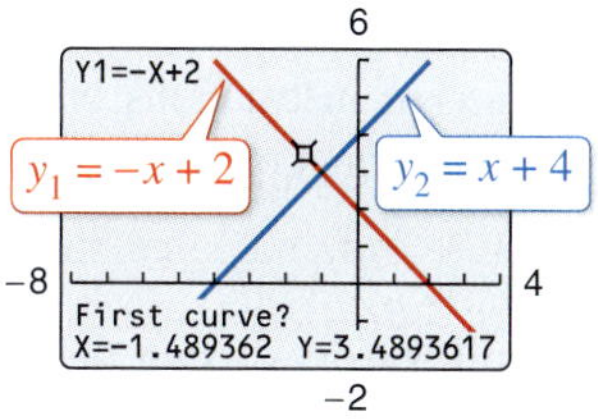

Figure A.25

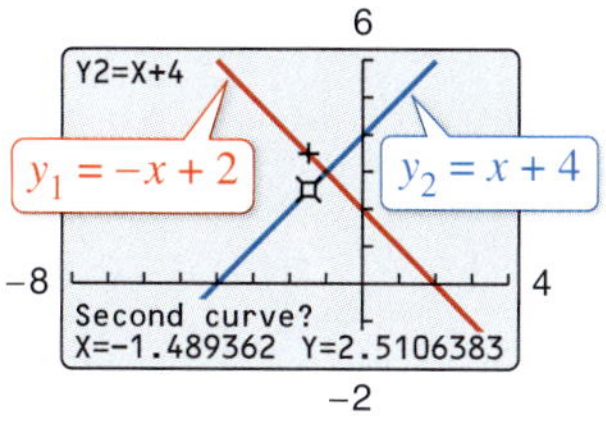

Figure A.26

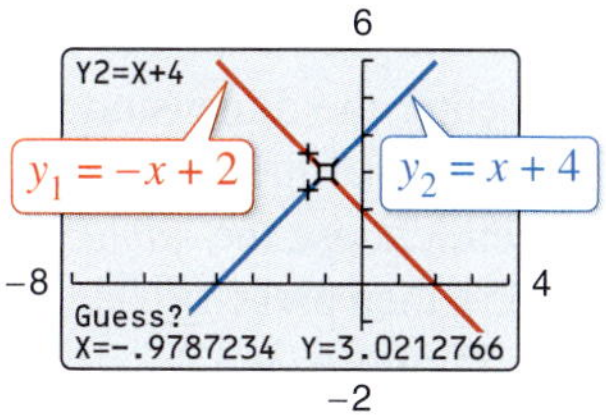

Figure A.27

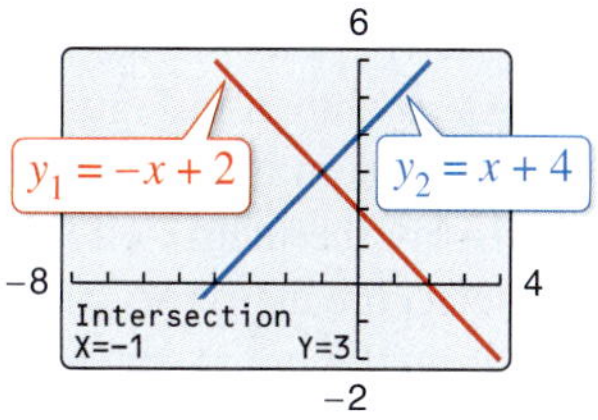

Figure A.28

List Editor

Most graphing utilities can store data in lists. The *list editor* can be used to create tables and to store statistical data. The *list editor* can be found in the *edit* menu of the *statistics* feature, as shown in Figure A.29. To enter the numbers 1 through 10 in a list, first choose a list (L_1) and then begin entering the data into each row, as shown in Figure A.30.

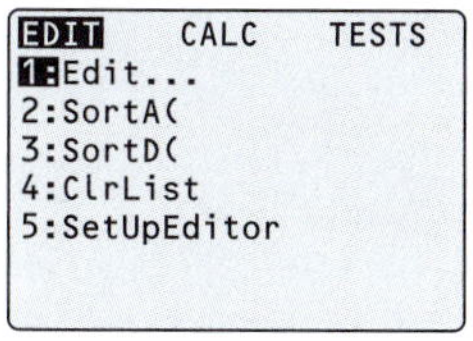

Figure A.29

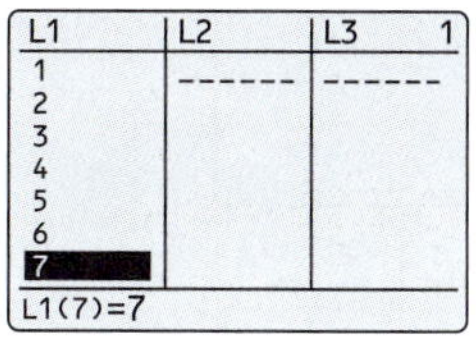

Figure A.30

You can also attach a formula to a list. For instance, you can multiply each of the data values in L_1 by 3. First, display the *list editor* and move the cursor to the top line.

Then move the cursor onto the list to which you want to attach the formula (L_2). Finally, enter the formula $3{*}L_1$ (see Figure A.31) and then press ENTER. You should obtain the list shown in Figure A.32.

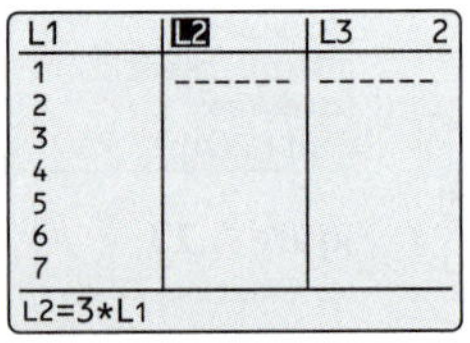

Figure A.31

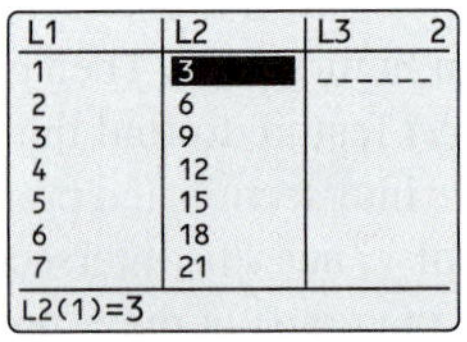

Figure A.32

Matrix Feature

The *matrix* feature of a graphing utility has many uses, such as evaluating a determinant and performing row operations. (For instructions on how to perform row operations, see Elementary Row Operations Features on page A3.)

Matrix Editor

You can define, display, and edit matrices using the *matrix editor*. The *matrix editor* can be found in the *edit* menu of the *matrix* feature. For instance, to enter the matrix

$$A = \begin{bmatrix} 6 & -3 & 4 \\ 9 & 0 & -1 \end{bmatrix}$$

first choose the matrix name [A], as shown in Figure A.33. Then enter the dimension of the matrix (in this case, the dimension is 2×3) and enter the entries of the matrix, as shown in Figure A.34. To display the matrix on the home screen, choose the *name* menu of the *matrix* feature and select the matrix [A] (see Figure A.35), then press ENTER. The matrix A should now appear on the home screen, as shown in Figure A.36.

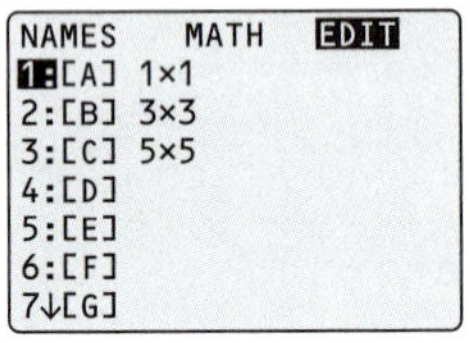

Figure A.33

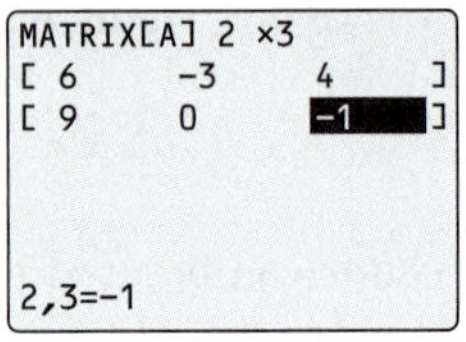

Figure A.34

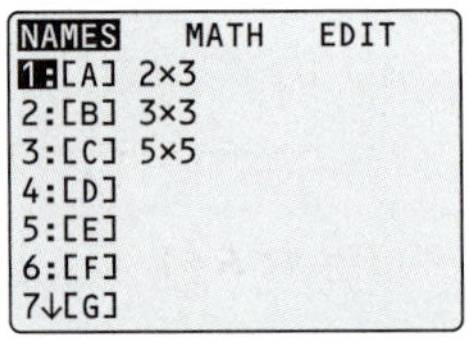

Figure A.35

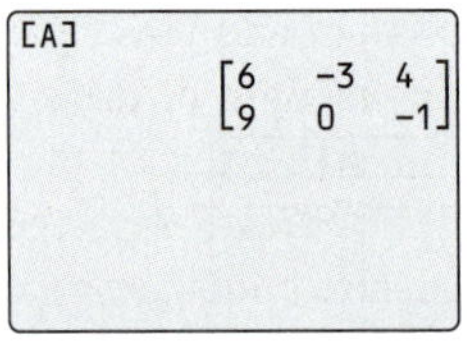

Figure A.36

Matrix Operations

Most graphing utilities can perform matrix operations. To find the sum $A + B$ of the matrices shown at the right, first enter the matrices in the *matrix editor* as [A] and [B]. Then find the sum, as shown in Figure A.37. Scalar multiplication can be performed in a similar manner. For instance, you can evaluate $7A$, where A is the matrix at the right, as shown in Figure A.38. To find the product AB of the matrices A and B at the right, first be sure that the product is defined. Because the number of columns of A (2 columns) equals the number of rows of B (2 rows), you can find the product AB, as shown in Figure A.39.

$$A = \begin{bmatrix} -3 & 5 \\ 0 & 4 \end{bmatrix}$$

$$B = \begin{bmatrix} 7 & -2 \\ -1 & 2 \end{bmatrix}$$

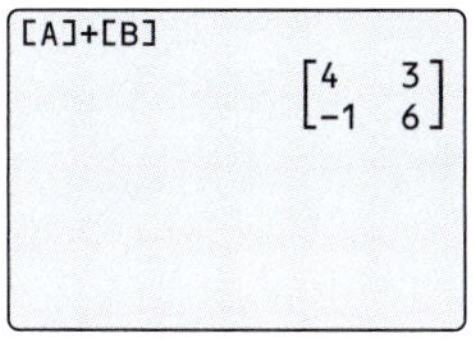

Figure A.37

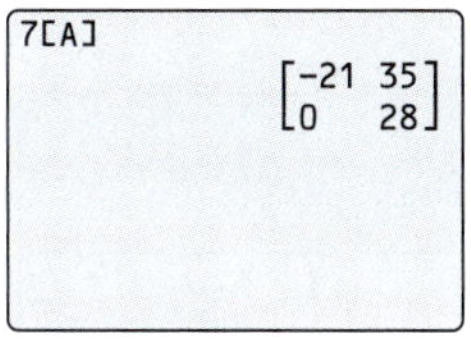

Figure A.38

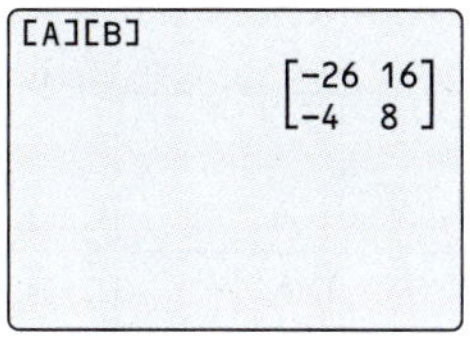

Figure A.39

Inverse Matrix

Some graphing utilities may not have an *inverse matrix* feature. However, you can find the inverse of a square matrix by using the inverse key $\boxed{x^{-1}}$. To find the inverse of the matrix

$$A = \begin{bmatrix} 1 & -2 & 1 \\ -1 & 3 & 0 \\ 2 & 4 & 5 \end{bmatrix}$$

enter the matrix in the *matrix editor* as [A]. Then find the inverse, as shown in Figure A.40.

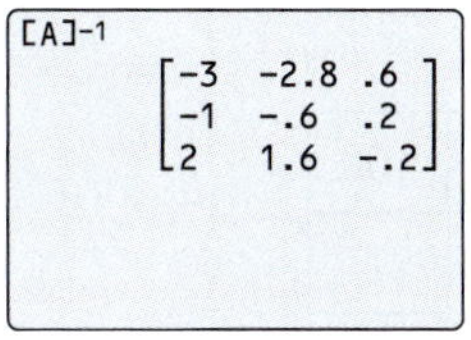

Figure A.40

Maximum and Minimum Features

The *maximum* and *minimum* features find relative extrema of a function. For instance, thc graph of

$$y = x^3 - 3x$$

is shown in Figure A.41. In the figure, the graph appears to have a relative maximum at $x = -1$ and a relative minimum at $x = 1$. To find the exact values of the relative extrema, you can use the *maximum* and *minimum* features found in the *calculate* menu (see Figure A.42). First, to find the relative maximum, choose the *maximum* feature and trace the cursor along the graph to a point left of the maximum and press ENTER (see Figure A.43). Then trace the cursor along the graph to a point right of the maximum and press ENTER (see Figure A.44). Note the two arrows near the top of the display marking the left and right bounds, as shown in Figure A.45. Next, trace the cursor along the graph between the two bounds and as close to the maximum as you can (see Figure A.45) and press ENTER. In Figure A.46, you can see that the coordinates of the maximum point are displayed at the bottom of the window. So, the relative maximum occurs at the point $(-1, 2)$.

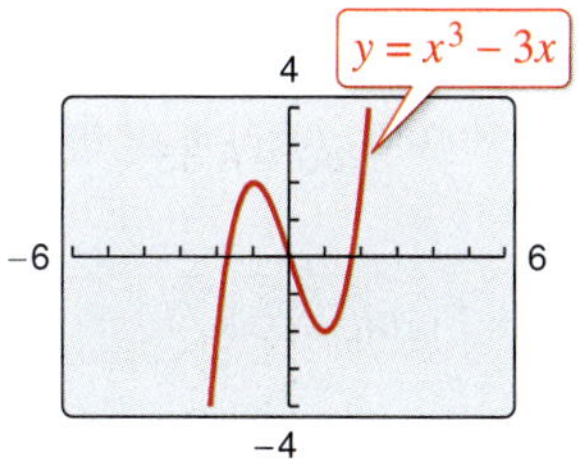

Figure A.41

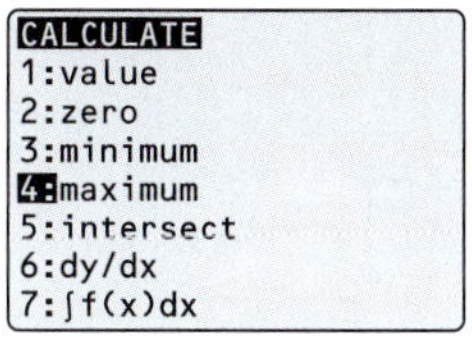

Figure A.42

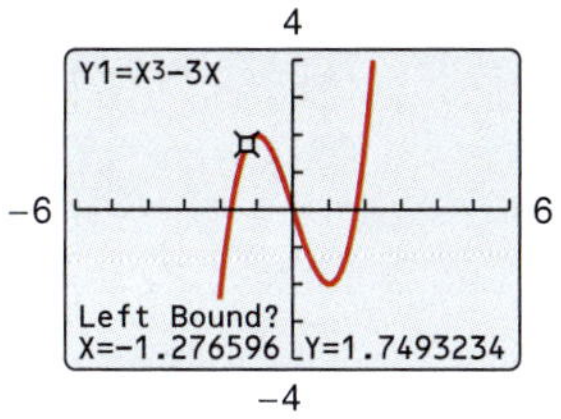

Figure A.43

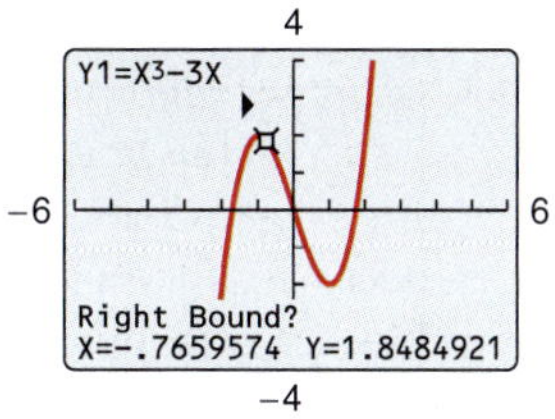

Figure A.44

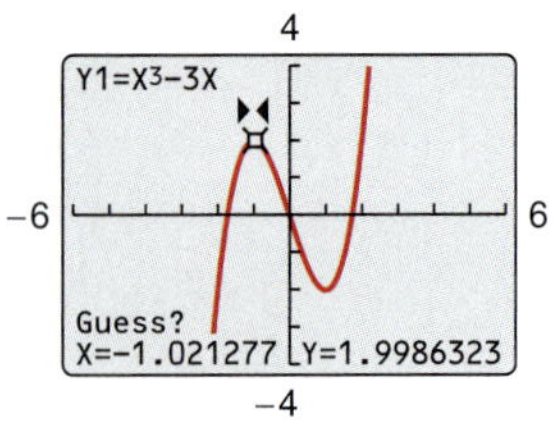

Figure A.45

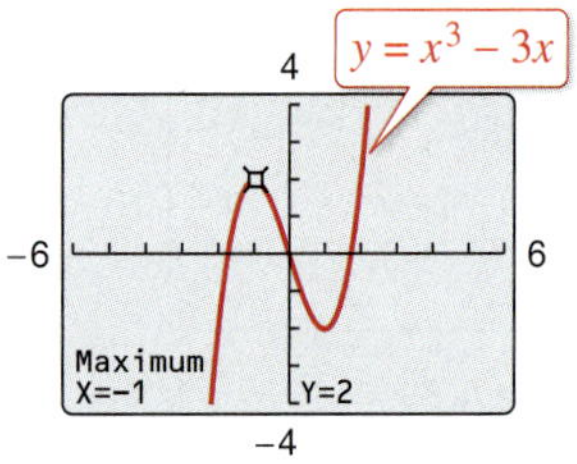

Figure A.46

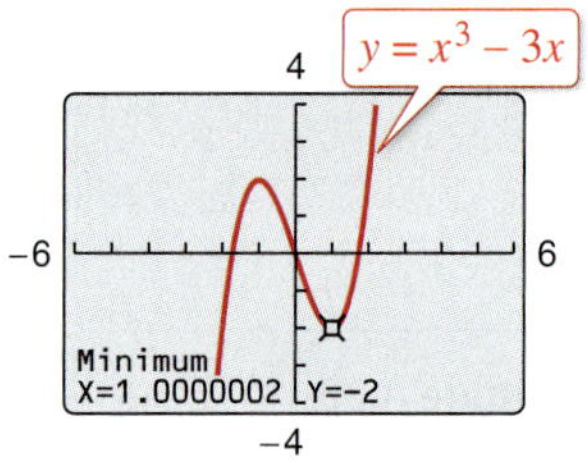

Figure A.47

You can find the relative minimum in a similar manner. In Figure A.47, you can see that the relative minimum occurs at the point $(1, -2)$.

Mean and Median Features

In real-life applications, you often encounter large data sets and want to calculate statistical values. The *mean* and *median* features calculate the mean and median of a data set. For instance, in a survey, 100 people were asked how much money (in dollars) per week they withdraw from an automatic teller machine (ATM). The results are shown in the table below. The frequency represents the number of responses.

Amount	10	20	30	40	50	60	70	80	90	100
Frequency	3	8	10	19	24	13	13	7	2	1

To find the mean and median of the data set, first enter the data in the *list editor*, as shown in Figure A.48. Enter the amount in L_1 and the frequency in L_2. Then choose the *mean* feature from the *math* menu of the *list* feature, as shown in Figure A.49. When using this feature, you must enter a list and a frequency list (if applicable). In this case, the list is L_1 and the frequency list is L_2. After pressing ENTER, you should obtain a mean of

$49.80

as shown in Figure A.50. You can follow the same steps (except choose the *median* feature) to find the median of the data. You should obtain a median of

$50

as shown in Figure A.51.

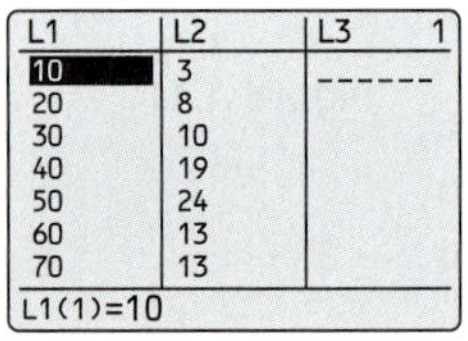

Figure A.48

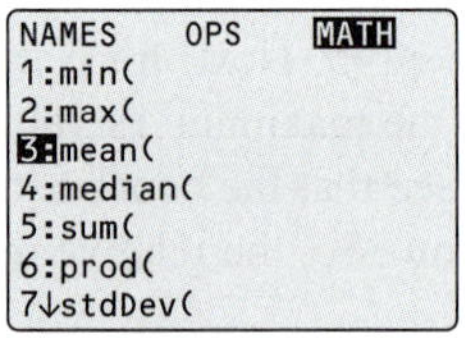

Figure A.49

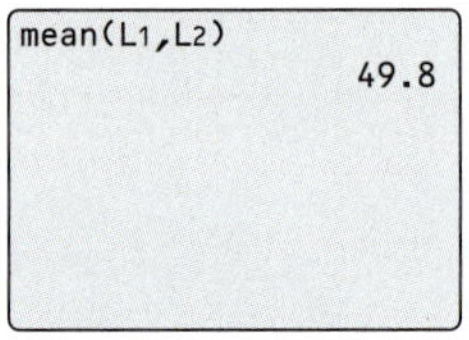

Figure A.50

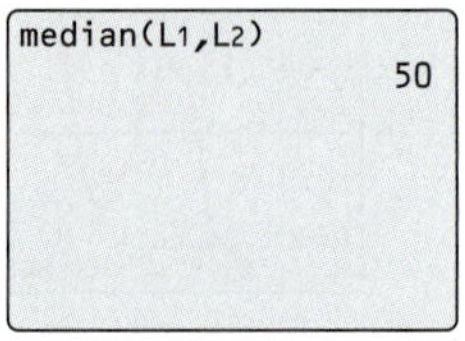

Figure A.51

Mode Settings

Mode settings of a graphing utility control how the utility displays and interprets numbers and graphs. Figure A.52 shows the default mode settings for most graphing utilities.

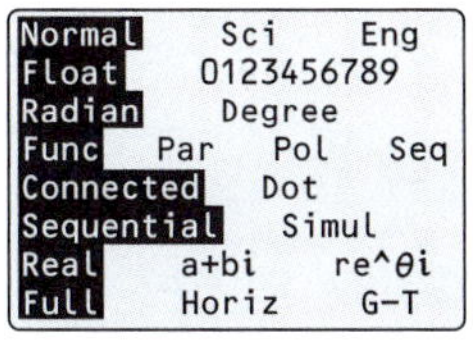

Figure A.52

Radian and Degree Modes

The trigonometric functions can be applied to angles measured in either radians or degrees. When your graphing utility is in *radian* mode, it interprets angle values as radians and displays answers in radians. When your graphing utility is in *degree* mode, it interprets angle values as degrees and displays answers in degrees. For instance, to calculate $\sin(\pi/6)$, make sure the calculator is in *radian* mode. You should obtain an answer of 0.5, as shown in Figure A.53. To calculate $\sin 45°$, make sure the calculator is in *degree* mode, as shown in Figure A.54. You should obtain an approximate answer of 0.7071, as shown in Figure A.55. Without changing the mode of the calculator before evaluating $\sin 45°$, you would obtain an answer of approximately 0.8509, which is the sine of 45 radians.

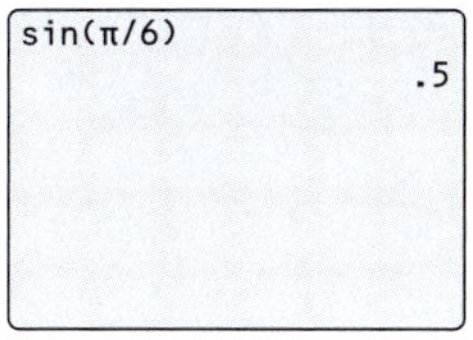

Figure A.53

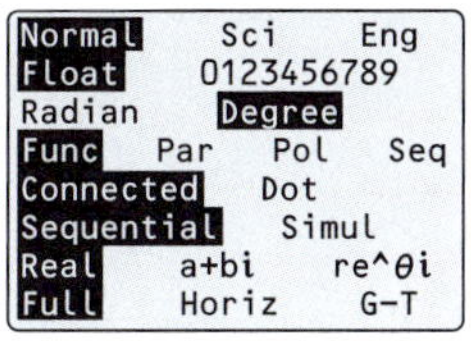

Figure A.54

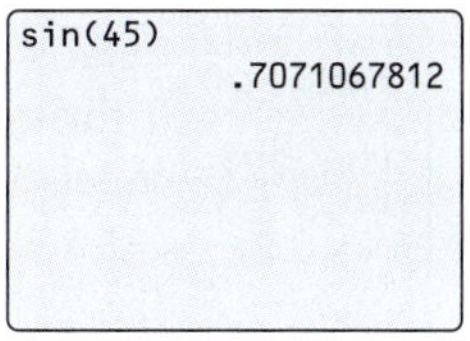

Figure A.55

Function, Parametric, Polar, and Sequence Modes

Most graphing utilities can graph using four different modes.

Function Mode To graph algebraic, transcendental, and nonelementary functions (see page 2), use the *function* mode. For instance, to graph $y = 2x^2$, use the *function* mode, as shown in Figure A.52. Then enter the equation in the *equation editor*, as shown in Figure A.56. Using a standard viewing window (see Figure A.57), you obtain the graph shown in Figure A.58.

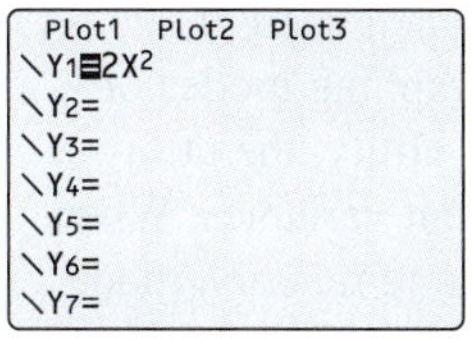

Figure A.56

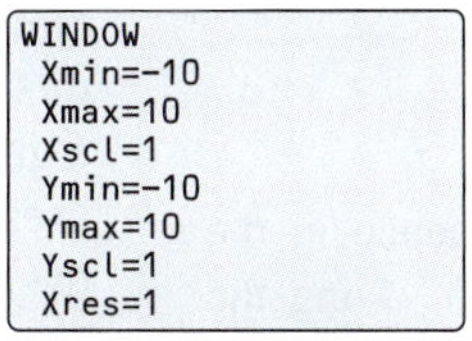

Figure A.57

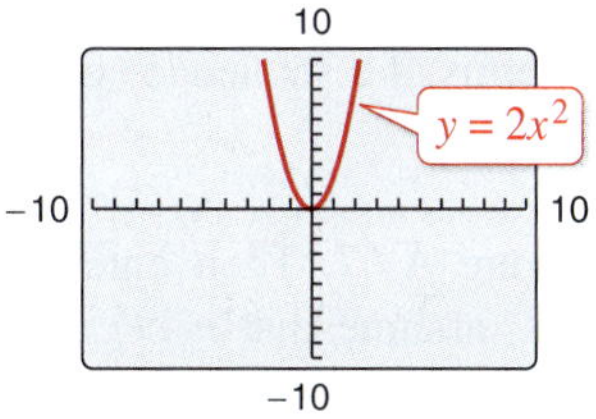

Figure A.58

Parametric Mode To graph parametric equations such as

$$x = t + 1 \quad \text{and} \quad y = t^2$$

use the *parametric* mode, as shown in Figure A.59. Then enter the equations in the *equation editor*, as shown in Figure A.60. Using the viewing window shown in Figure A.61, you obtain the graph shown in Figure A.62.

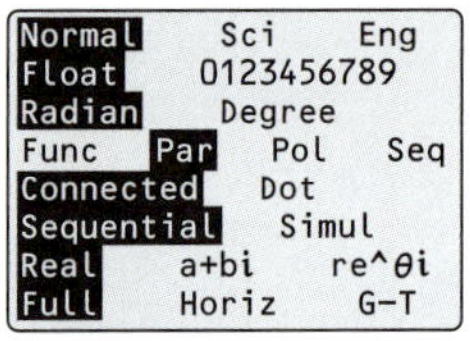

Figure A.59

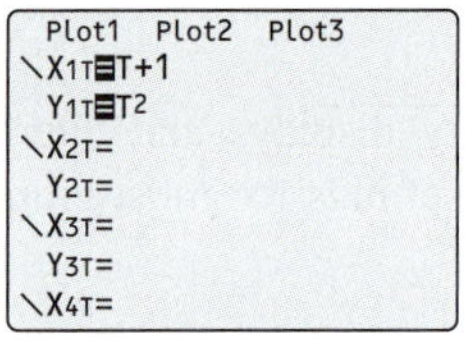

Figure A.60

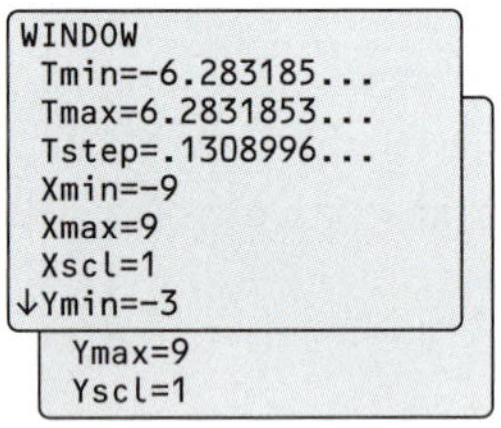

Figure A.61

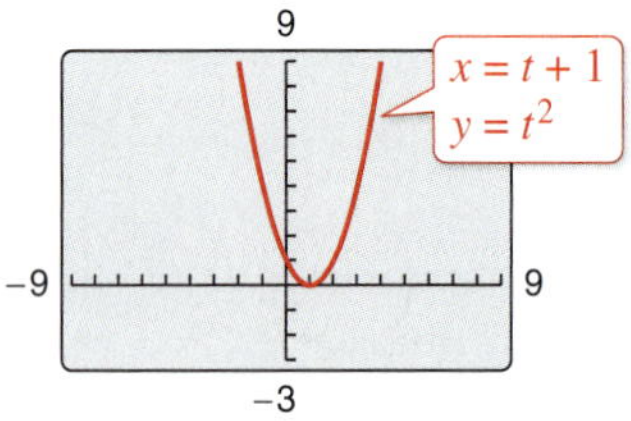

Figure A.62

Polar Mode To graph polar equations of the form $r = f(\theta)$, use the *polar* mode of a graphing utility. For instance, to graph the polar equation

$$r = 2\cos\theta$$

use the *polar* mode (and *radian* mode), as shown in Figure A.63. Then enter the equation in the *equation editor*, as shown in Figure A.64. Using the viewing window shown in Figure A.65, you obtain the graph shown in Figure A.66.

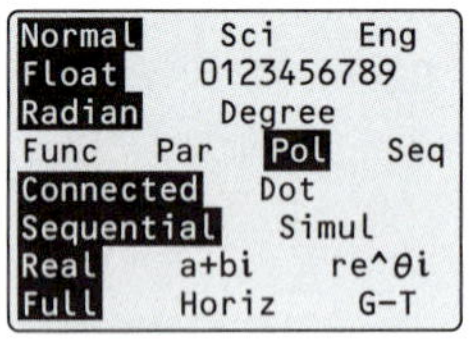

Figure A.63

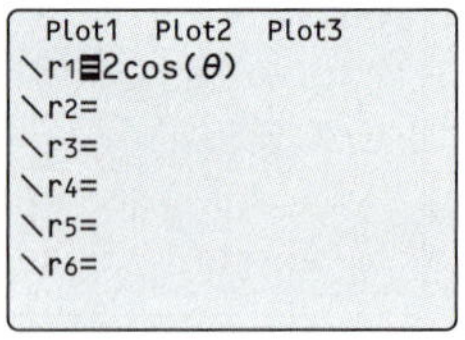

Figure A.64

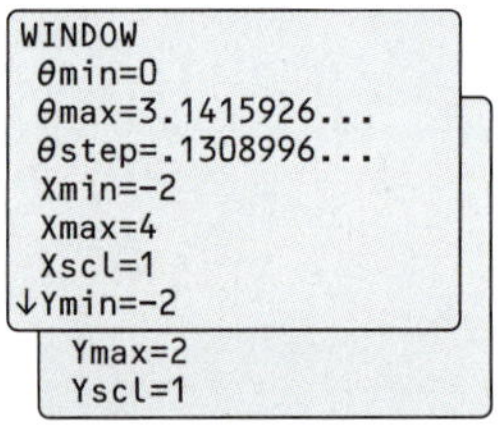

Figure A.65

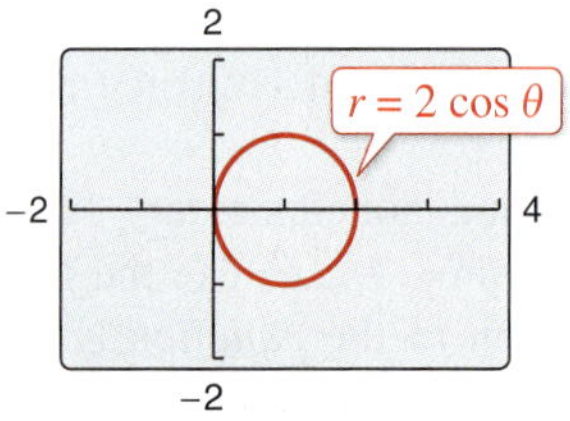

Figure A.66

Sequence Mode To graph the first five terms of a sequence such as

$$a_n = 4n - 5$$

use the *sequence* mode, as shown in Figure A.67. Then enter the sequence in the *equation editor*, as shown in Figure A.68 (assume that n begins with 1). Using the viewing window shown in Figure A.69, you obtain the graph shown in Figure A.70.

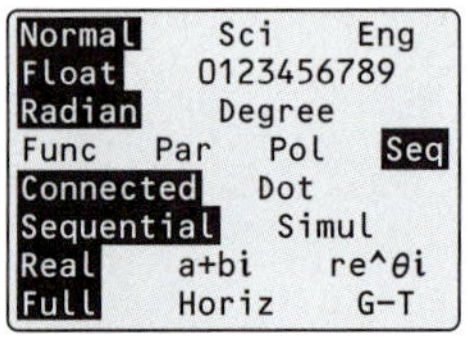

Figure A.67

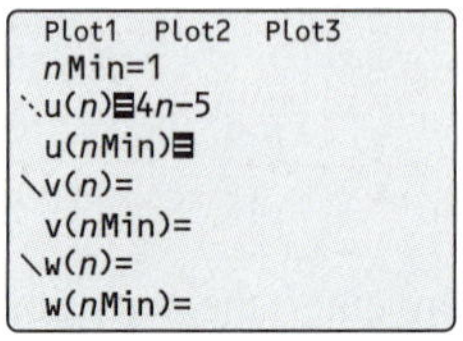

Figure A.68

Technology Tip

Note that when using the different graphing modes of a graphing utility, the utility uses different variables. When the utility is in *function* mode, it uses the variables x and y. In *parametric* mode, the utility uses the variables x, y, and t. In *polar* mode, the utility uses the variables r and θ. In *sequence* mode, the utility uses the variables u (instead of a) and n.

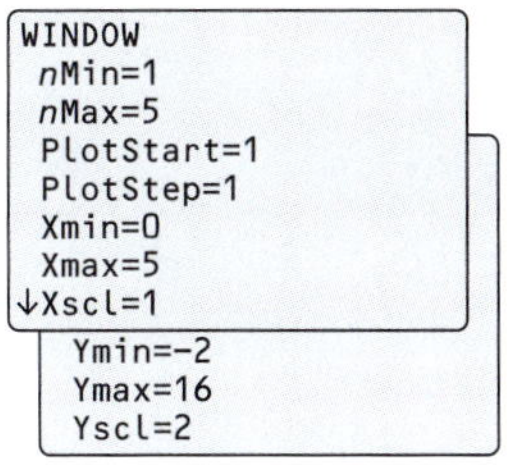

Figure A.69

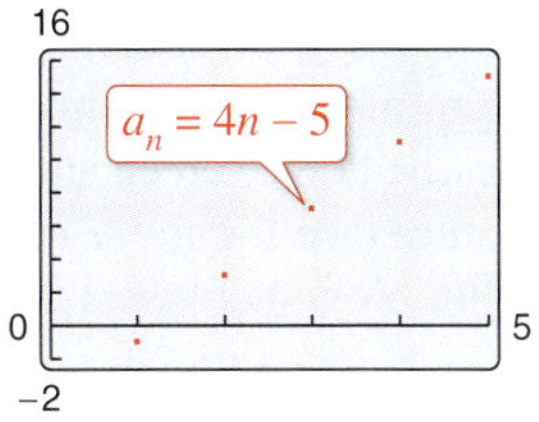

Figure A.70

Connected and Dot Modes

Graphing utilities use the point-plotting method to graph functions. When a graphing utility is in *connected* mode, the utility connects the points that are plotted. When the utility is in *dot* mode, it does not connect the points that are plotted. (On some graphing utilities, *connected* mode is called *thick* or *thin* mode, and *dot* mode is called *dot-thick* or *dot-thin* mode.) For instance, the graph of

$$y = x^3$$

in *connected* mode is shown in Figure A.71. To graph this function using *dot* mode, first change the mode to *dot* mode (see Figure A.72) and then graph the equation, as shown in Figure A.73. As you can see in Figure A.73, the graph is a collection of dots.

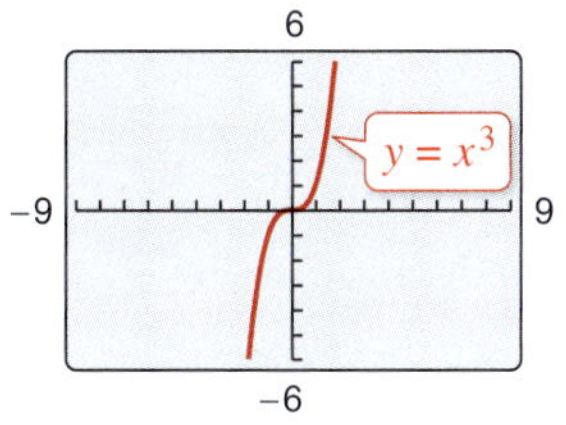

Figure A.71

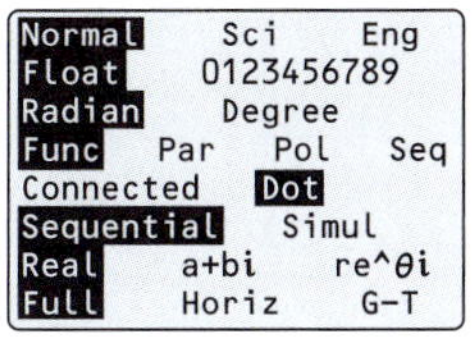

Figure A.72

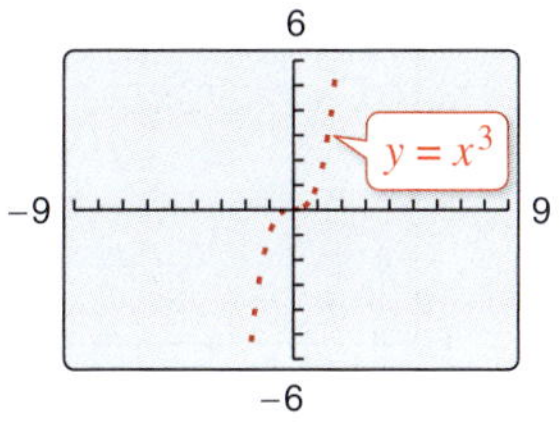

Figure A.73

A problem arises in some graphing utilities when the *connected* mode is used. Graphs with vertical asymptotes, such as rational functions and tangent functions, appear to be connected. For instance, the graph of

$$y = \frac{1}{x + 3}$$

is shown in Figure A.74. Notice how the two portions of the graph appear to be connected with a vertical line at $x = -3$. From your study of rational functions, you know that the graph has a vertical asymptote at $x = -3$ and therefore is undefined when $x = -3$. If your graphing utility draws extraneous vertical lines when graphing functions with vertical asymptotes, you can use the *dot* mode to eliminate the vertical lines, as shown in Figure A.75.

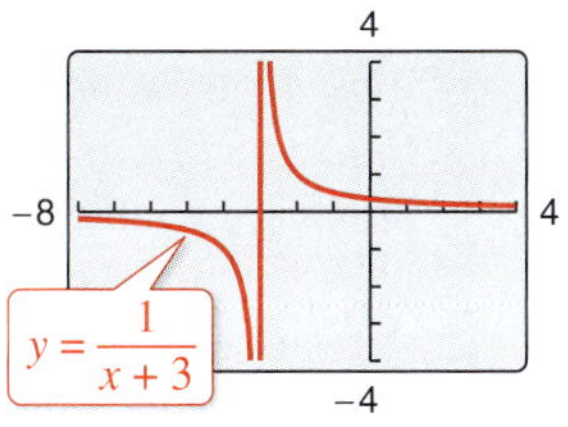

Figure A.74

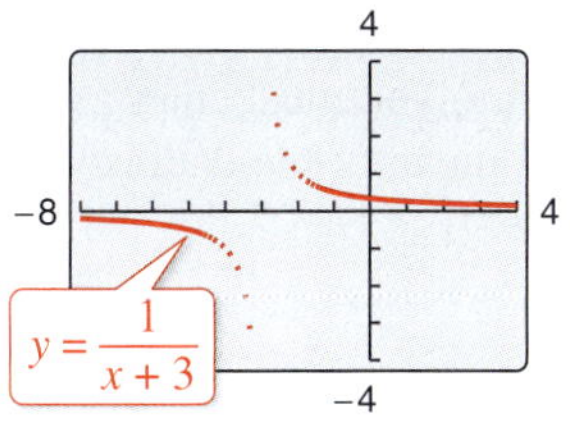

Figure A.75

$_nC_r$ Feature

The $_nC_r$ feature calculates binomial coefficients and the number of combinations of n elements taken r at a time. For instance, to find the number of combinations of eight elements taken five at a time, enter 8 (the n-value) on the home screen and choose the $_nC_r$ feature from the *probability* menu of the *math* feature (see Figure A.76). Next, enter 5 (the r-value) on the home screen and press ENTER. You should obtain 56, as shown in Figure A.77.

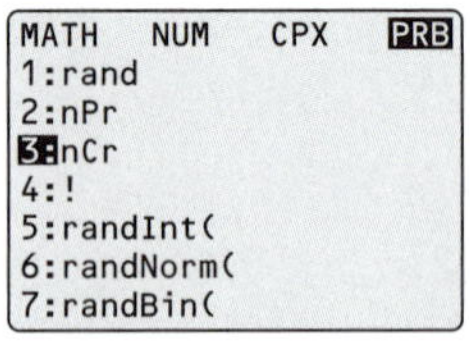

Figure A.76

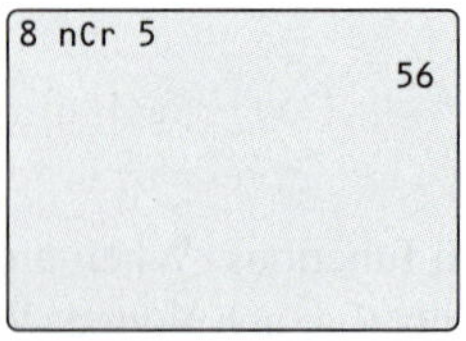

Figure A.77

$_nP_r$ Feature

The $_nP_r$ feature calculates the number of permutations of n elements taken r at a time. For instance, to find the number of permutations of six elements taken four at a time, enter 6 (the n-value) on the home screen and choose the $_nP_r$ feature from the *probability* menu of the *math* feature (see Figure A.78). Next, enter 4 (the r-value) on the home screen and press ENTER. You should obtain 360, as shown in Figure A.79.

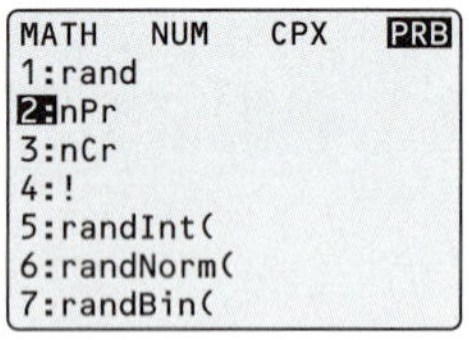

Figure A.78

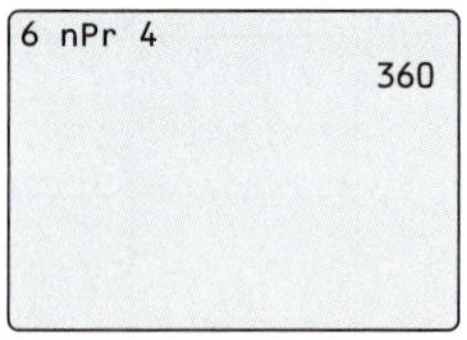

Figure A.79

One-Variable Statistics Feature

Graphing utilities are useful in calculating statistical values for a set of data. The *one-variable statistics* feature analyzes data with one measured variable. This feature outputs the mean of the data, the sum of the data, the sum of the data squared, the sample standard deviation of the data, the population standard deviation of the data, the number of data points, the minimum data value, the maximum data value, the first quartile of the data, the median of the data, and the third quartile of the data. Consider the hourly earnings (in dollars) of 12 retail sales associates.

8.05, 10.25, 8.45, 9.15, 8.90, 8.20,

9.25, 10.30, 8.60, 9.60, 10.05, 11.35

You can use the *one-variable statistics* feature to determine the mean and standard deviation of the data. First, enter the data in the *list editor*, as shown in Figure A.80. Then choose the *one-variable statistics* feature from the *calculate* menu of the *statistics* feature, as shown in Figure A.81. When using this feature, you must enter a list. In this case, the list is L_1. In Figure A.82, you can see that the mean of the data is

$$\bar{x} \approx 9.35$$

and the standard deviation of the data is

$$\sigma x \approx 0.95.$$

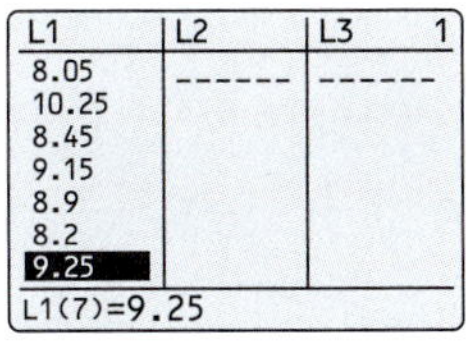

Figure A.80

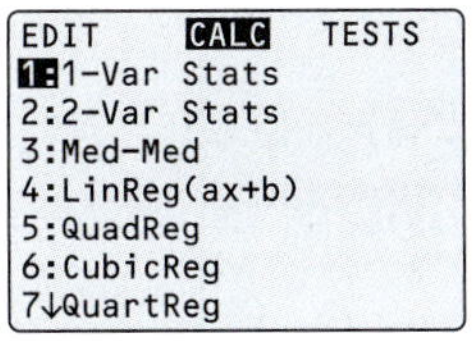

Figure A.81

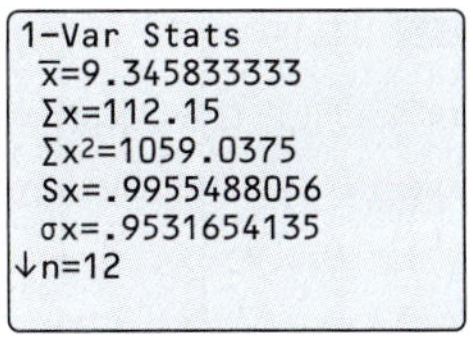

Figure A.82

Regression Feature

Throughout the text, you are asked to use the *regression* feature of a graphing utility to find models for sets of data. Most graphing utilities have built-in regression programs for the models listed below.

Regression	*Form of Model*
Linear	$y = ax + b$ or $y = a + bx$
Quadratic	$y = ax^2 + bx + c$
Cubic	$y = ax^3 + bx^2 + cx + d$
Quartic	$y = ax^4 + bx^3 + cx^2 + dx + e$
Logarithmic	$y = a + b \ln x$
Exponential	$y = ab^x$
Power	$y = ax^b$
Logistic	$y = \dfrac{c}{1 + ae^{-bx}}$
Sine	$y = a \sin(bx + c) + d$

For instance, you can find a linear model for the data in the table.

x	y
6	222.8
7	228.7
8	235.0
9	240.3
10	245.0
11	248.2
12	254.4
13	260.2
14	268.3
15	287.0

Enter the data in the *list editor*, as shown in Figure A.83. Note that L_1 contains the x-values, and L_2 contains the y-values. Now choose the *linear regression* feature from the *calculate* menu of the *statistics* feature, as shown in Figure A.84. In Figure A.85, you can see that a linear model for the data is given by

$$y = 6.22x + 183.7.$$

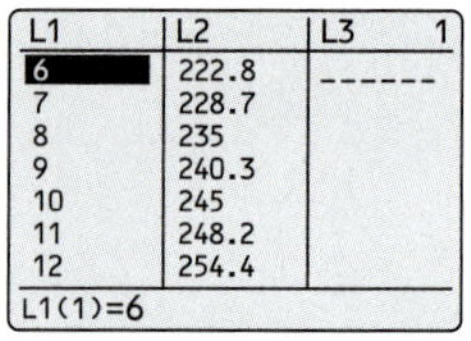

Figure A.83

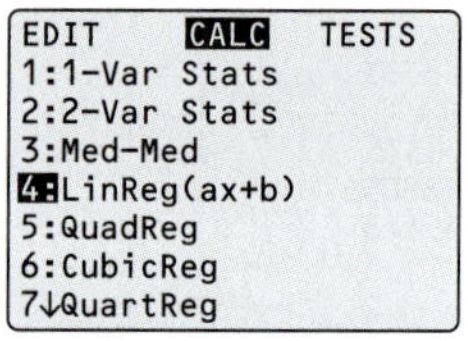

Figure A.84

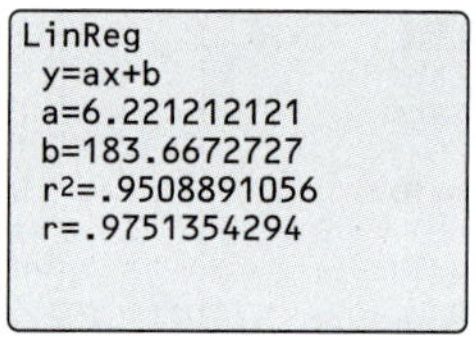

Figure A.85

When you use the *regression* feature of a graphing utility, you will notice that the program may also output an "r-value." (For some calculators, make sure you select the *diagnostics on* feature before you use the *regression* feature. Otherwise, the calculator will not output an r-value.) The r-value, or *correlation coefficient,* measures how well the linear model fits the data. The closer $|r|$ is to 1, the better the fit. For the data above,

$$r \approx 0.97514$$

which implies that the model is a good fit for the data.

Row-Echelon and Reduced Row-Echelon Features

Some graphing utilities have features that can transform a matrix to row-echelon form and reduced row-echelon form. These features can be used to check your solutions to systems of equations.

Row-Echelon Feature

Consider the system of equations and the corresponding augmented matrix shown below.

Linear System

$$\begin{cases} 2x + 5y - 3z = 4 \\ 4x + y = 2 \\ -x + 3y - 2z = -1 \end{cases}$$

Augmented Matrix

$$\begin{bmatrix} 2 & 5 & -3 & \vdots & 4 \\ 4 & 1 & 0 & \vdots & 2 \\ -1 & 3 & -2 & \vdots & -1 \end{bmatrix}$$

You can use the *row-echelon* feature of a graphing utility to write the augmented matrix in row-echelon form. First, enter the matrix in the graphing utility using the *matrix editor*, as shown in Figure A.86. Next, choose the *row-echelon* feature from the *math* menu of the *matrix* feature, as shown in Figure A.87. When using this feature, you must enter the name of the matrix. In this case, the name of the matrix is [A]. You should obtain the matrix shown in Figure A.88. You may have to scroll to the right in order to see all the entries of the matrix.

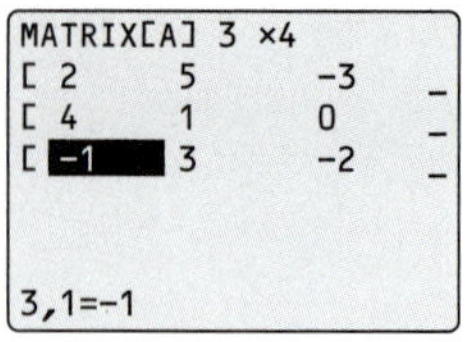

Figure A.86

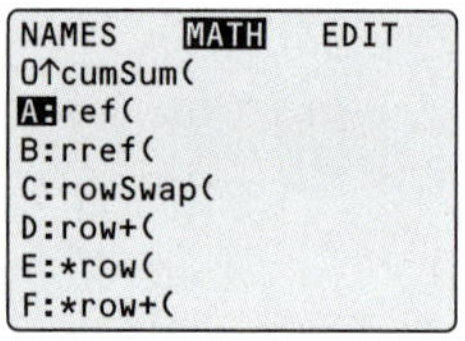

Figure A.87

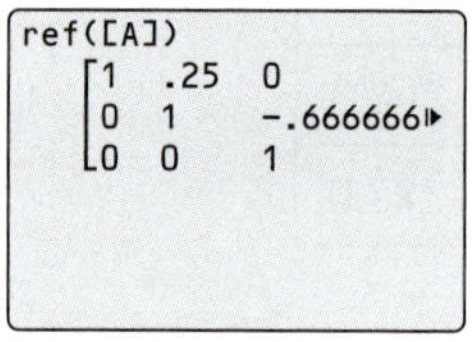

Figure A.88

Reduced Row-Echelon Feature

To write the augmented matrix in reduced row-echelon form, follow the same steps used to write a matrix in row-echelon form except choose the *reduced row-echelon* feature, as shown in Figure A.89. You should obtain the matrix shown in Figure A.90. From Figure A.90, you can conclude that the solution of the system is $x = 3$, $y = -10$, and $z = -16$.

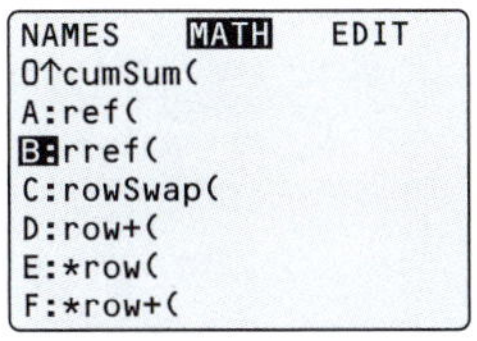

Figure A.89

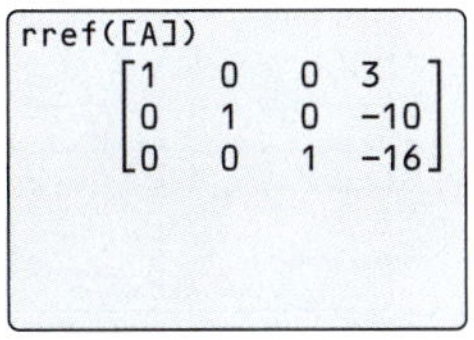

Figure A.90

Sequence Feature

The *sequence* feature is used to display the terms of sequences. For instance, to determine the first five terms of the arithmetic sequence

$a_n = 3n + 5$ Assume n begins with 1.

choose the *sequence* feature from the *operations* menu of the *list* feature, as shown in Figure A.91. When using this feature, you must enter the sequence, the variable (in this case n), the beginning value (in this case 1), and the end value (in this case 5). The first five terms of the sequence are 8, 11, 14, 17, and 20, as shown in Figure A.92. You may have to scroll to the right in order to see all the terms of the sequence.

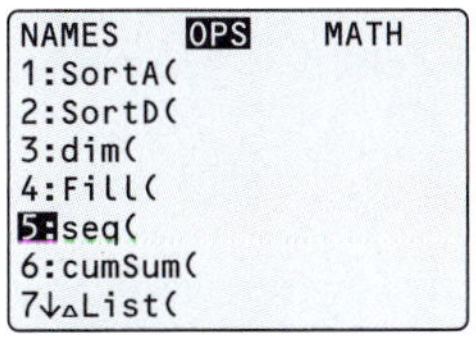

Figure A.91

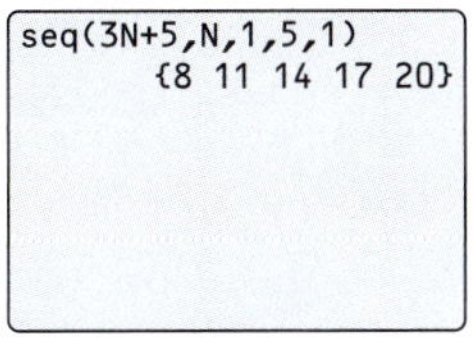

Figure A.92

Shade Feature

Most graphing utilities have a *shade* feature that can be used to graph inequalities. For instance, to graph the inequality

$y \leq 2x - 3$

first enter the equation $y = 2x - 3$ in the *equation editor*, as shown in Figure A.93. Next, using a standard viewing window (see Figure A.94), graph the equation, as shown in Figure A.95.

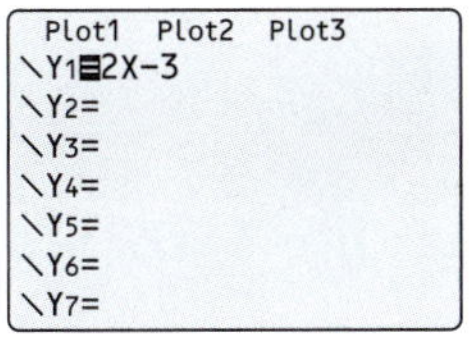

Figure A.93

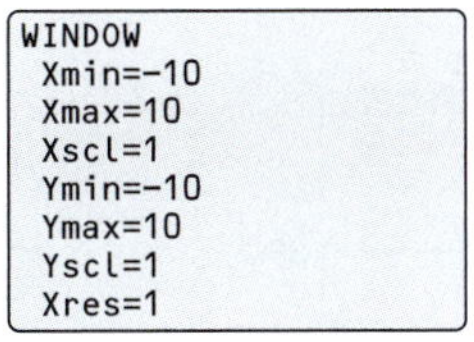

Figure A.94

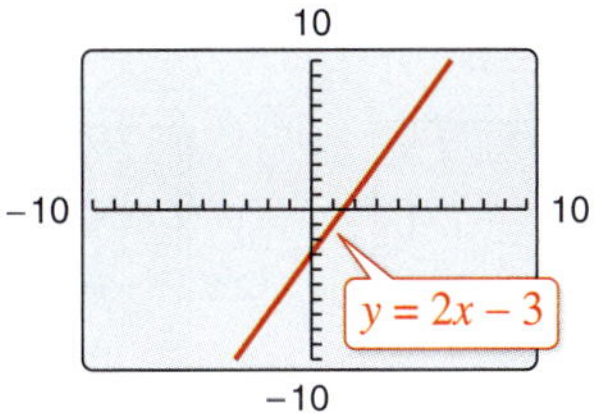

Figure A.95

Because the inequality sign is ≤, you want to shade the region below the line $y = 2x - 3$. Choose the *shade* feature from the *draw* feature menu, as shown in Figure A.96. You must enter a lower function and an upper function. In this case, the lower function is -10 (this is the least y-value in the viewing window) and the upper function is Y_1 $(y = 2x - 3)$, as shown in Figure A.97. Then press ENTER to obtain the graph shown in Figure A.98.

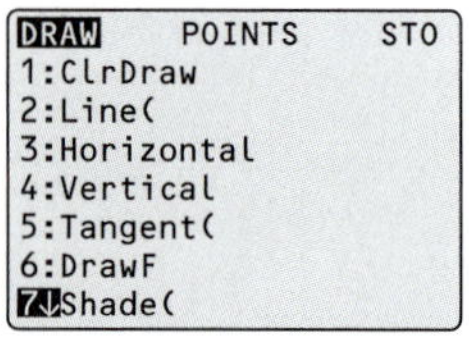

Figure A.96

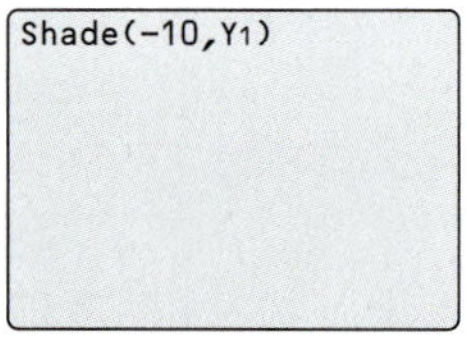

Figure A.97

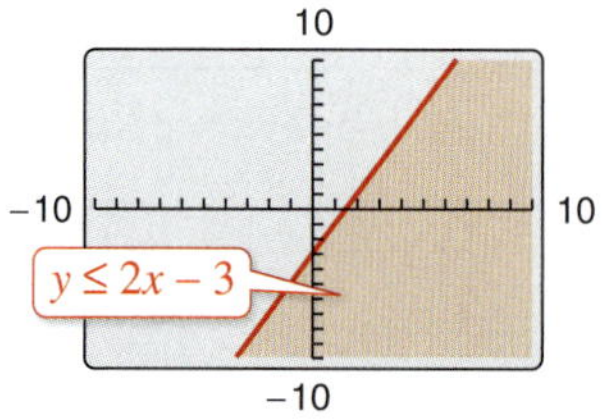

Figure A.98

To graph the inequality

$$y \geq 2x - 3$$

(using a standard viewing window), you would enter the lower function as Y_1 $(y = 2x - 3)$ and the upper function as 10 (the greatest y-value in the viewing window).

Statistical Plotting Feature

The *statistical plotting* feature plots data that are stored in lists. Most graphing utilities can display the following types of plots.

Plot Type	*Variables*
Scatter plot	x-list, y-list
xy line graph	x-list, y-list
Histogram	x-list, frequency
Box-and-whisker plot	x-list, frequency
Normal probability plot	data list, data axis

For instance, use a box-and-whisker plot to represent the data listed below. Then use the graphing utility plot to find the least number, the lower quartile, the median, the upper quartile, and the greatest number.

17, 19, 21, 27, 29, 30, 37, 27, 15, 23, 19, 16

First, use the *list editor* to enter the values in a list, as shown in Figure A.99. Then go to the *statistical plotting editor*. In this editor, you will turn the plot on, select the box-and-whisker plot, select the list you entered in the *list editor*, and enter the frequency of each item in the list, as shown in Figure A.100. Now use the *zoom* feature and choose the *zoom stat* option to set the viewing window and plot the graph, as shown in Figure A.101. Use the *trace* feature to find that the least number is 15, the lower quartile is 18, the median is 22, the upper quartile is 28, and the greatest number is 37.

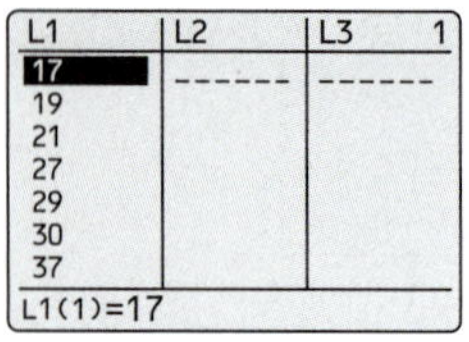

Figure A.99

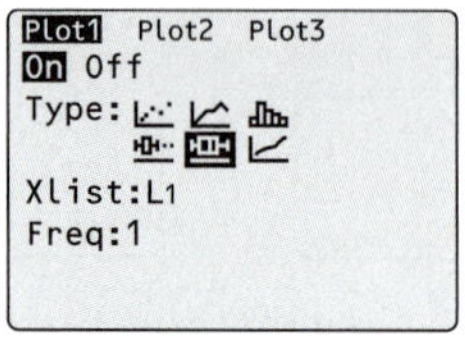

Figure A.100

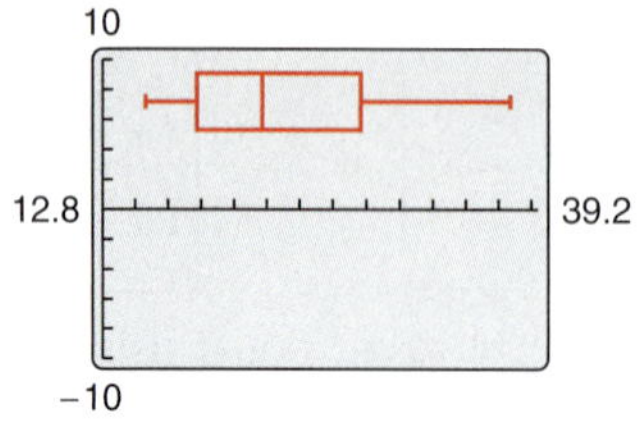

Figure A.101

Sum Feature

The *sum* feature finds the sum of a list of data. For instance, the data below represent a student's quiz scores on 10 quizzes throughout an algebra course.

22, 23, 19, 24, 20, 15, 25, 21, 18, 24

To find the total quiz points the student earned, enter the data in the *list editor*, as shown in Figure A.102. To find the sum, choose the *sum* feature from the *math* menu of the *list* feature, as shown in Figure A.103. You must enter a list. In this case, the list is L_1. You should obtain a sum of 211, as shown in Figure A.104.

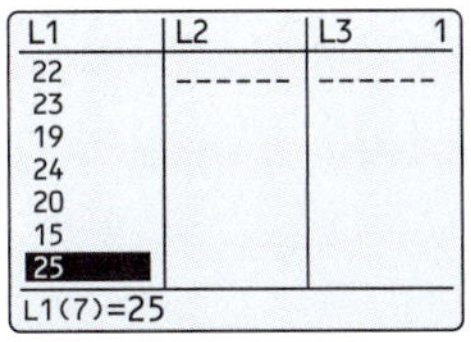

Figure A.102

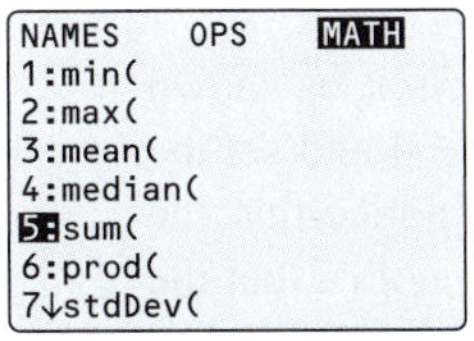

Figure A.103

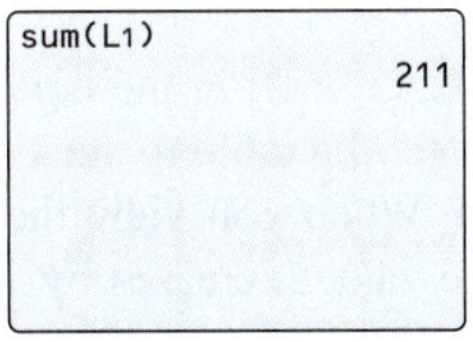

Figure A.104

Sum Sequence Feature

The *sum* feature and the *sequence* feature can be combined to find the sum of a sequence or series. [On some graphing utilities, you can find a sum using the *summation* feature (see below).] For instance, to find the sum

$$\sum_{k=0}^{10} 5^{k+1}$$

first choose the *sum* feature from the *math* menu of the *list* feature, as shown in Figure A.105. Then choose the *sequence* feature from the *operations* menu of the *list* feature, as shown in Figure A.106. You must enter an expression for the sequence, a variable, the lower limit of summation, and the upper limit of summation. After pressing ENTER, you should obtain the sum 61,035,155, as shown in Figure A.107.

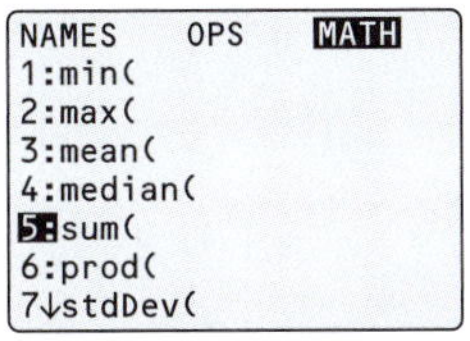

Figure A.105

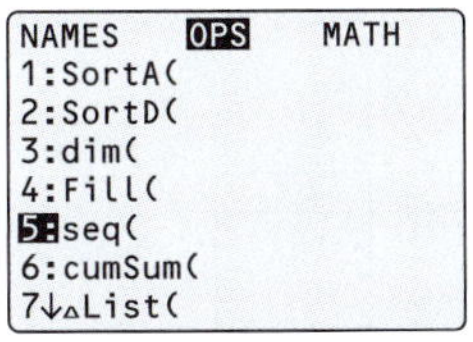

Figure A.106

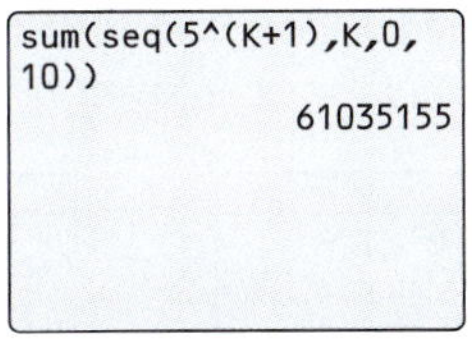

Figure A.107

Summation Feature

On some graphing utilities, you can find a sum using the *summation* feature. For instance, to find the sum

$$\sum_{k=0}^{10} 5^{k+1}$$

choose the *summation* feature from the *math* menu of the *math* feature, as shown in Figure A.108(a). Enter an expression for the sequence, a variable, the lower limit of summation, and the upper limit of summation. You should obtain a sum of 61,035,155, as shown in Figure A.108(b). (This is the same result found using the *sum sequence* feature above.)

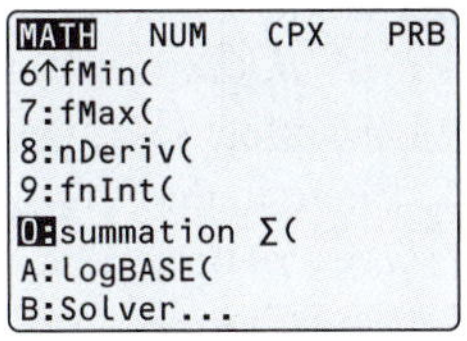

(a)

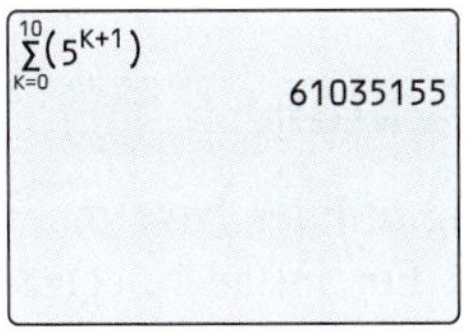

(b)

Figure 108

Table Feature

Most graphing utilities are capable of displaying a table of values with x-values and one or more corresponding y-values. These tables can be used to check solutions of an equation and to generate ordered pairs to assist in graphing an equation by hand.

To use the *table* feature, enter an equation in the *equation editor*. The table may have a setup screen, which allows you to select the starting x-value and the table step or x-increment. You may then have the option of automatically generating values for x and y or building your own table using the *ask* mode (see figure at the right).

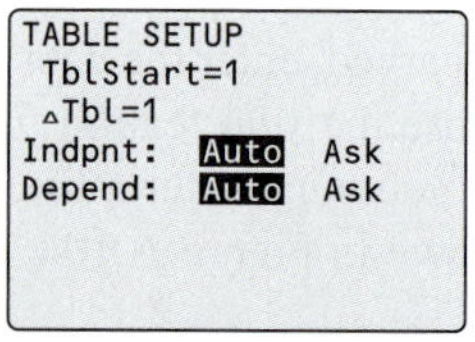

For instance, enter the equation $y = 3x/(x + 2)$ in the *equation editor*, as shown in Figure A.109. In the table setup screen, set the table to start at $x = -4$ and set the table step to 1, as shown in Figure A.110. When you view the table, notice that the first x-value is -4 and that each successive value increases by 1. Also, notice that the Y_1 column gives the resulting y-value for each x-value, as shown in Figure A.111. The table shows that the y-value for $x = -2$ is ERROR. This means that the equation is undefined when $x = -2$.

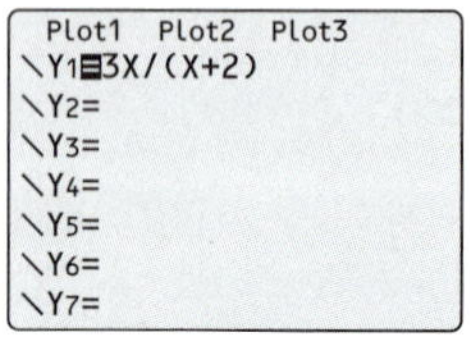

Figure A.109

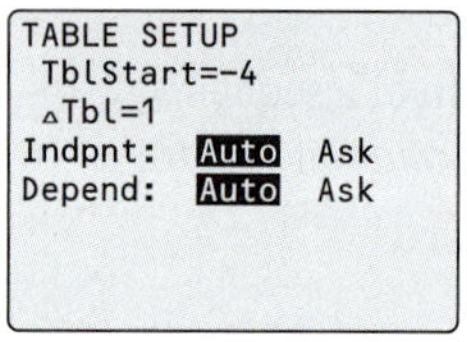

Figure A.110

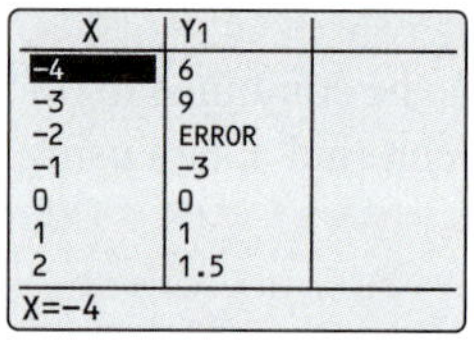

Figure A.111

With the same equation in the *equation editor*, set the independent variable in the table to *ask* mode, as shown in Figure A.112. In this mode, you do not need to set the starting x-value or the table step because you are entering any value you choose for x. You may enter any real number for x. When you enter $x = 1 + \sqrt{3}$, the graphing utility may rewrite the number as a decimal approximation, as shown in Figure A.113. You can continue to build your own table by entering additional x-values to generate y-values, as shown in Figure A.114.

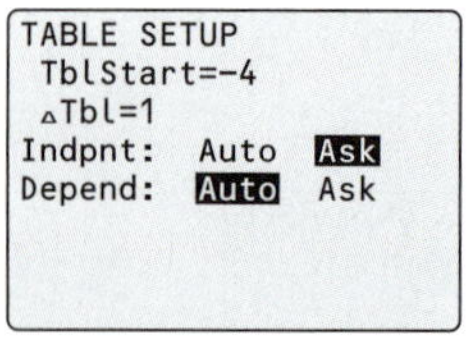

Figure A.112

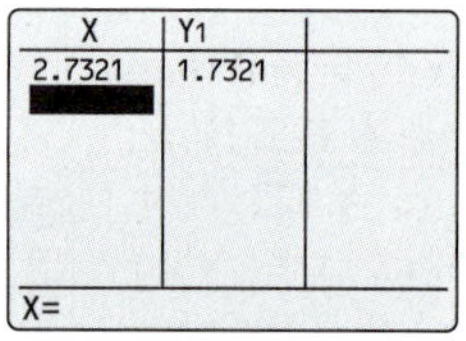

Figure A.113

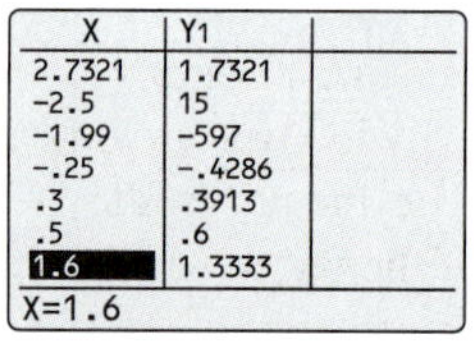

Figure A.114

When you have several equations in the *equation editor*, the table may generate y-values for each equation.

Tangent Feature

Some graphing utilities have the capability of drawing a tangent line to a graph at a given point. For instance, consider the tangent line to the graph of the equation $y = -x^3 + x + 2$ at the point $(1, 2)$. To draw the tangent line, enter the equation for y in the *equation editor*, as shown in Figure A.115. Using the viewing window shown in Figure A.116, graph the equation, as shown in Figure A.117. Next, choose the *tangent* feature from the *draw* feature menu, as shown in Figure A.118. You can either move the cursor to select a point or enter the x-value at which you want the tangent line to be drawn.

Because you want the tangent line to the graph of

$$y = -x^3 + x + 2$$

at the point (1, 2), enter 1 (see Figure A.119) and then press ENTER. The x-value you entered and the equation of the tangent line are displayed at the bottom of the window, as shown in Figure A.120.

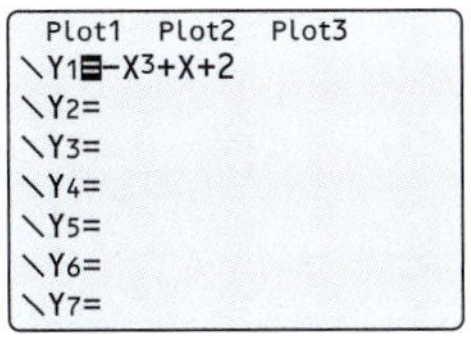

Figure A.115

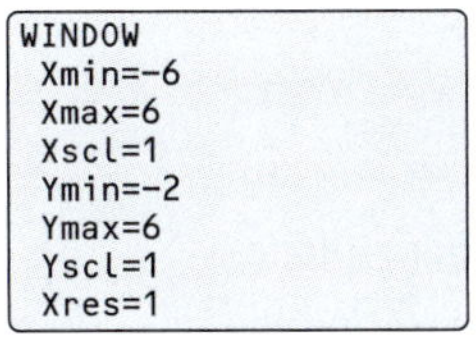

Figure A.116

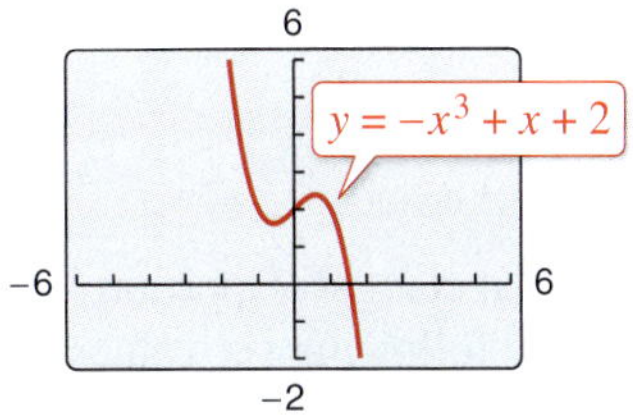

Figure A.117

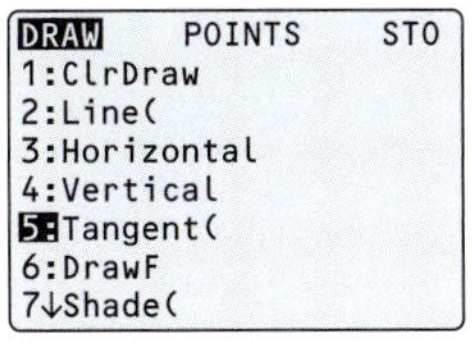

Figure A.118

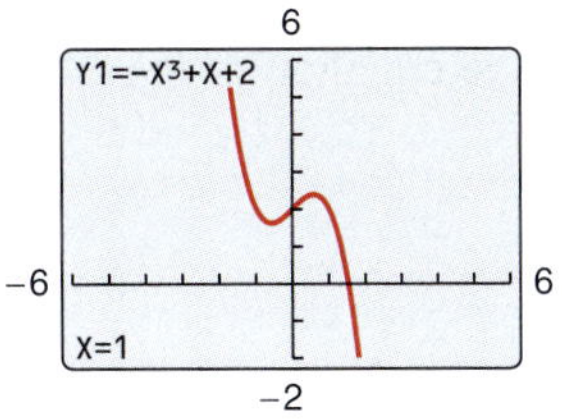

Figure A.119

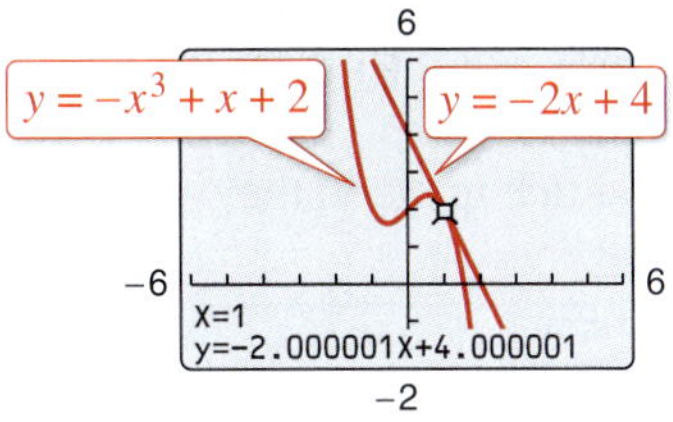

Figure A.120

Trace Feature

For instructions on how to use the *trace* feature, see Zoom and Trace Features on page A23.

Value Feature

The *value* feature finds the value of a function y for a given x-value. To find the value of a function such as

$$f(x) = 0.5x^2 - 1.5x$$

at $x = 1.8$, first enter the function in the *equation editor* (see Figure A.121) and then graph the function (using a standard viewing window), as shown in Figure A.122. Next, choose the *value* feature from the *calculate* menu, as shown in Figure A.123. You will see "X = " displayed at the bottom of the window. Enter the x-value, in this case $x = 1.8$, as shown in Figure A.124. When entering an x-value, be sure it is between the Xmin and Xmax values you entered for the viewing window. Then press ENTER. In Figure A.125, you can see that when $x = 1.8$,

$$y = -1.08.$$

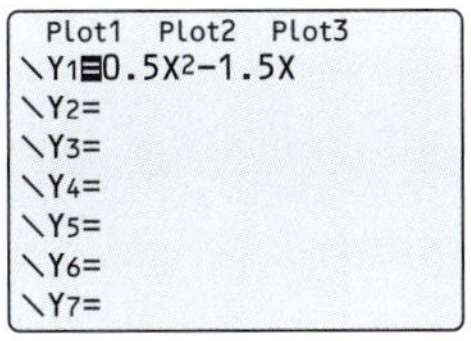

Figure A.121

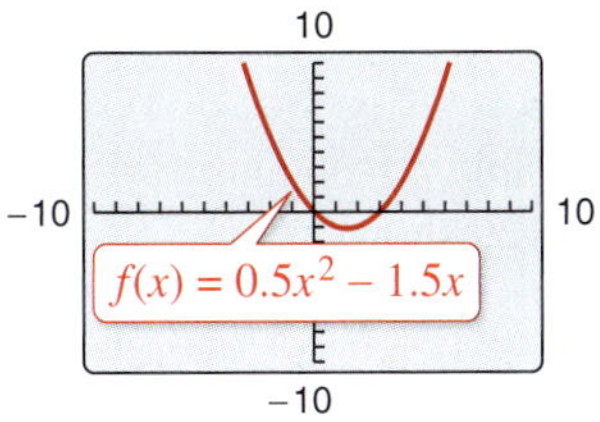

Figure A.122

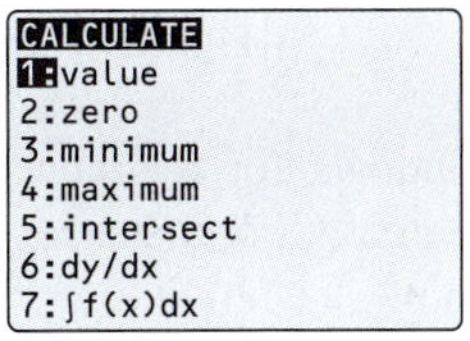

Figure A.123

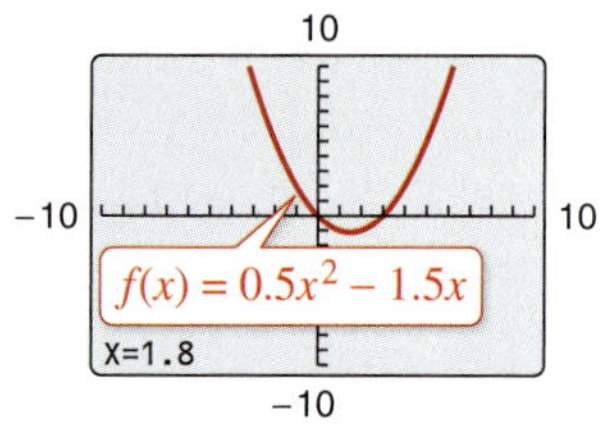

Figure A.124

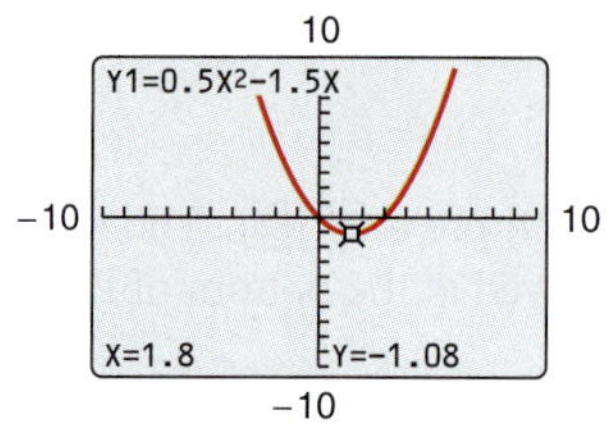

Figure A.125

Viewing Window

A viewing window for a graph is a rectangular portion of the coordinate plane. A viewing window is determined by the following six values (see Figure A.126).

Xmin = the minimum value of x
Xmax = the maximum value of x
Xscl = the number of units per tick mark on the x-axis
Ymin = the minimum value of y
Ymax = the maximum value of y
Yscl = the number of units per tick mark on the y-axis

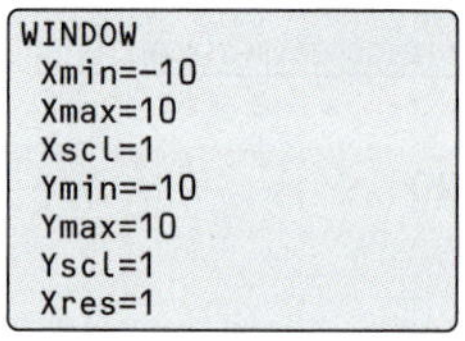

Figure A.126

When you enter these six values in a graphing utility, you are setting the viewing window. On some graphing utilities, there is a seventh value for the viewing window labeled Xres. This sets the pixel resolution (1 through 8). For instance, when

Xres = 1

functions are evaluated and graphed at each pixel on the x-axis. Some graphing utilities have a standard viewing window, as shown in Figure A.127. To initialize the standard viewing window quickly, choose the *standard viewing window* feature from the *zoom* feature menu (see page A23), as shown in Figure A.128.

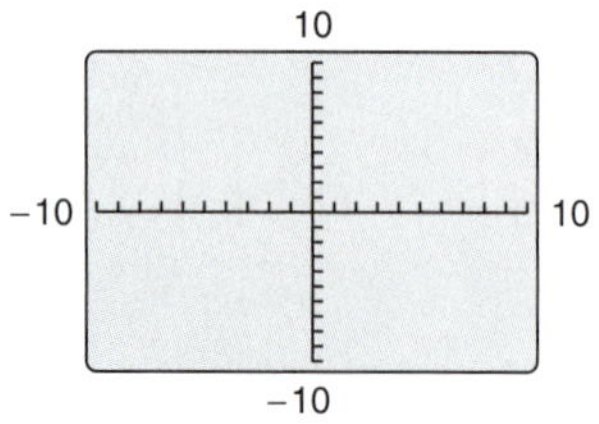

Figure A.127

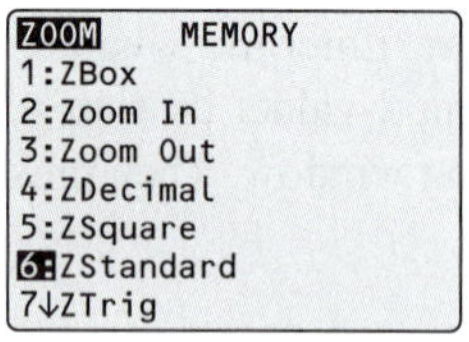

Figure A.128

By choosing different viewing windows for a graph, it is possible to obtain different impressions of the shape of the graph. For instance, Figure A.129 shows four different viewing windows for the graph of

$$y = 0.1x^4 - x^3 + 2x^2.$$

Of these viewing windows, the one shown in part (a) is the most complete.

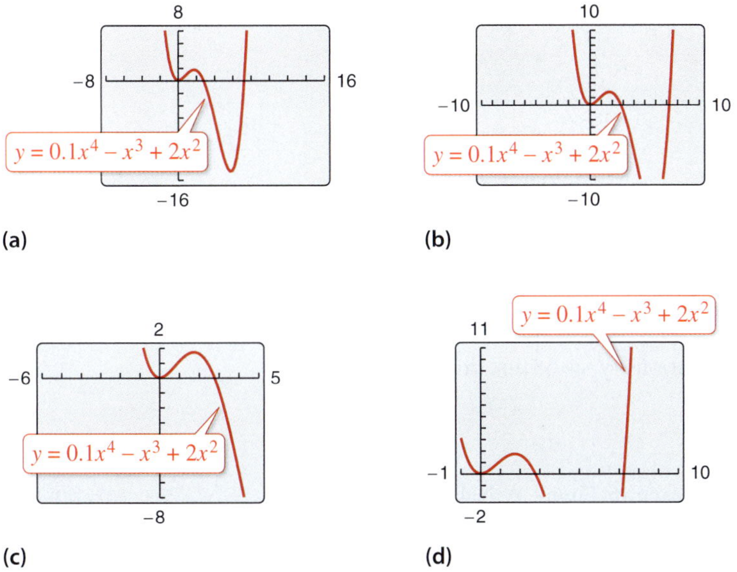

Figure A.129

On most graphing utilities, the display screen is two-thirds as high as it is wide. On such screens, you can obtain a graph with a true geometric perspective by using a square setting—one in which

$$\frac{\text{Ymax} - \text{Ymin}}{\text{Xmax} - \text{Xmin}} = \frac{2}{3}.$$

One such setting is shown in Figure A.130. Notice that the x and y tick marks are equally spaced on a square setting but not on a standard setting (see Figure A.127). To initialize the square viewing window quickly, choose the *square viewing window* feature from the *zoom* feature menu (see page A23), as shown in Figure A.131.

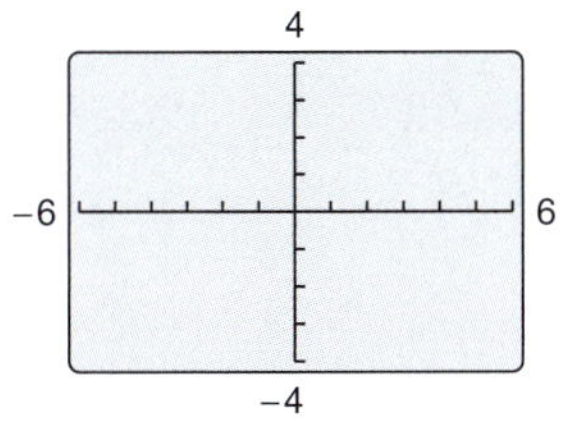

Figure A.130

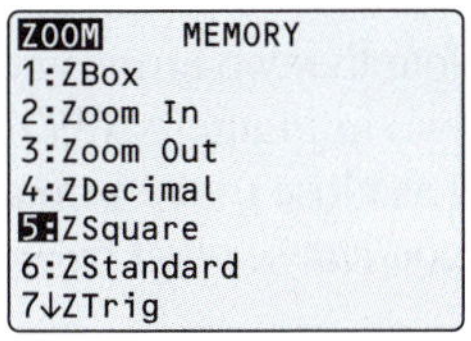

Figure A.131

To see how the viewing window affects the geometric perspective, graph the semicircles

$$y_1 = \sqrt{9 - x^2}$$

and

$$y_2 = -\sqrt{9 - x^2}$$

using a standard viewing window, as shown in Figure A.132. Notice how the circle appears elliptical rather than circular. Now graph y_1 and y_2 using a square viewing window, as shown in Figure A.133. Notice how the circle appears circular. (Note that when you graph the two semicircles, your graphing utility may not connect them. This is because some graphing utilities are limited in their resolution.)

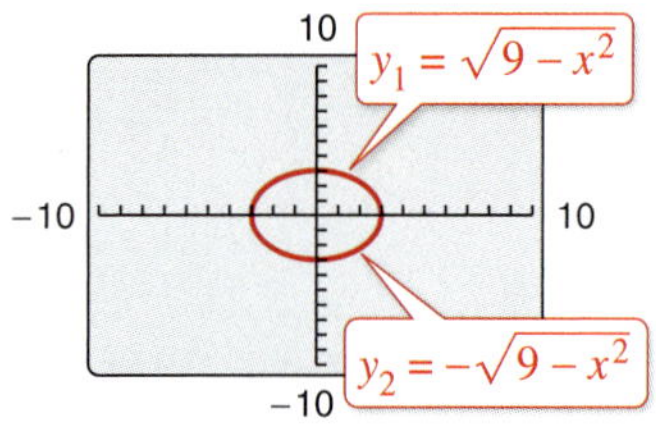

Figure A.132

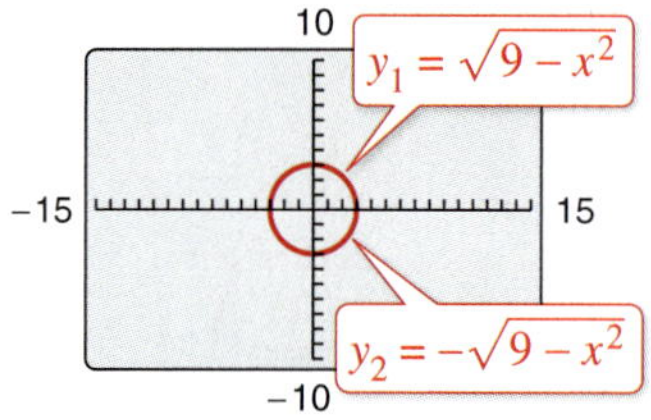

Figure A.133

Zero or Root Feature

The *zero* or *root* feature finds the real zeros of the various types of functions studied in this text. To find the zeros of a function such as

$$f(x) = 2x^3 - 4x$$

first enter the function in the *equation editor*, as shown in Figure A.134. Now graph the equation (using a standard viewing window), as shown in Figure A.135. From the figure, you can see that the graph of the function crosses the x-axis three times, so the function has three real zeros.

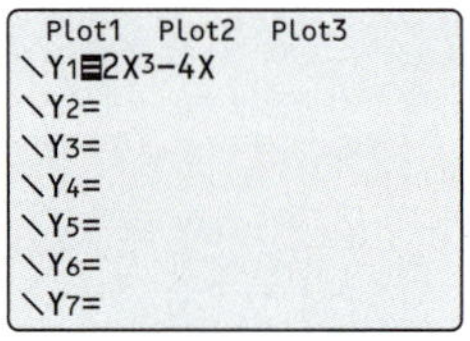

Figure A.134

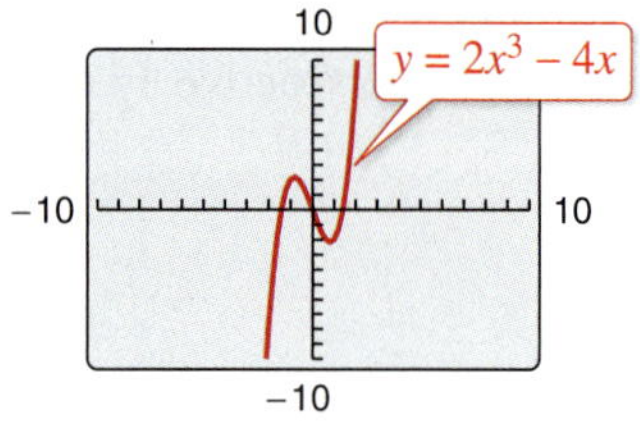

Figure A.135

To find these zeros, choose the *zero* feature found in the *calculate* menu (see Figure A.136). Next, trace the cursor along the graph to a point left of one of the zeros and press ENTER (see Figure A.137). Then trace the cursor along the graph to a point right of the zero and press ENTER (see Figure A.138). Note the two arrows near the top of the display marking the left and right bounds, as shown in Figure A.139. Now trace the cursor along the graph between the two bounds and as close to the zero as you can (see Figure A.140) and press ENTER. In Figure A.141, you can see that one zero of the function is $x \approx -1.414214$.

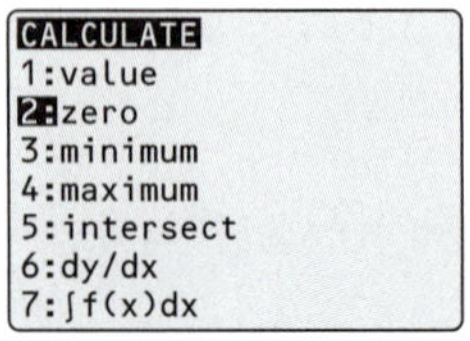

Figure A.136

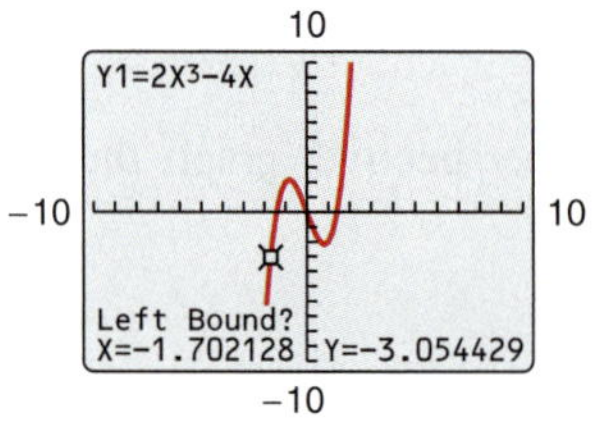

Figure A.137

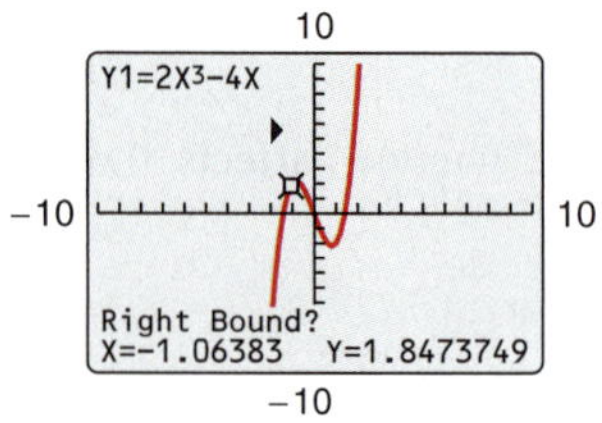

Figure A.138

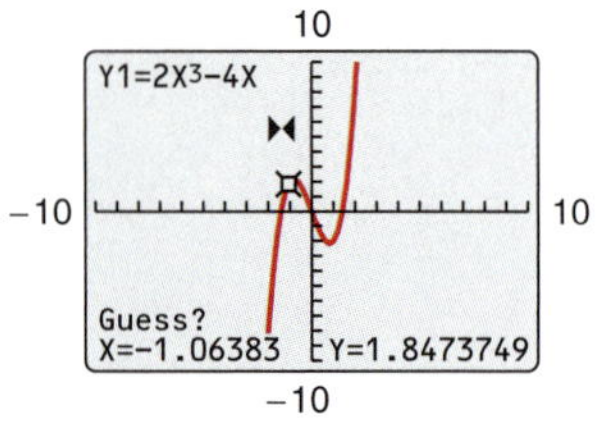

Figure A.139

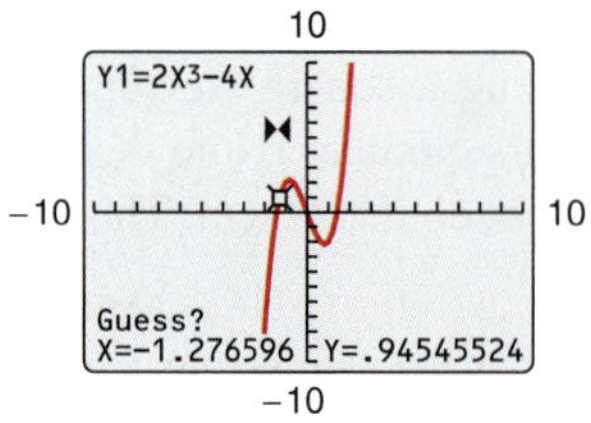

Figure A.140

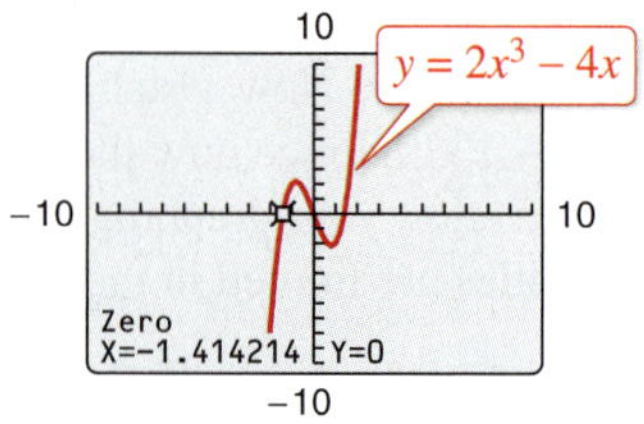

Figure A.141

Repeat this process to determine that the other two zeros of the function are $x = 0$ (see Figure A.142) and $x \approx 1.414214$ (see Figure A.143).

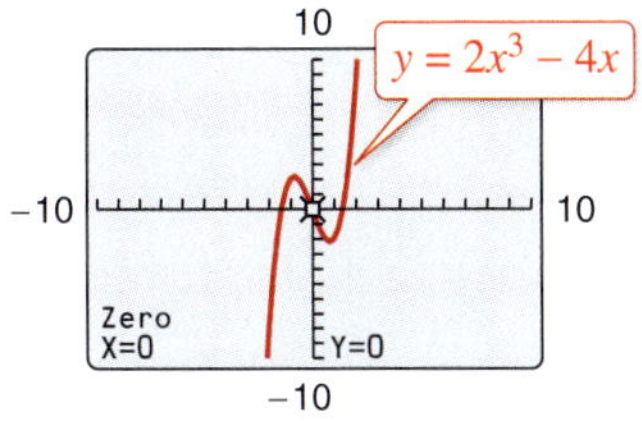

Figure A.142

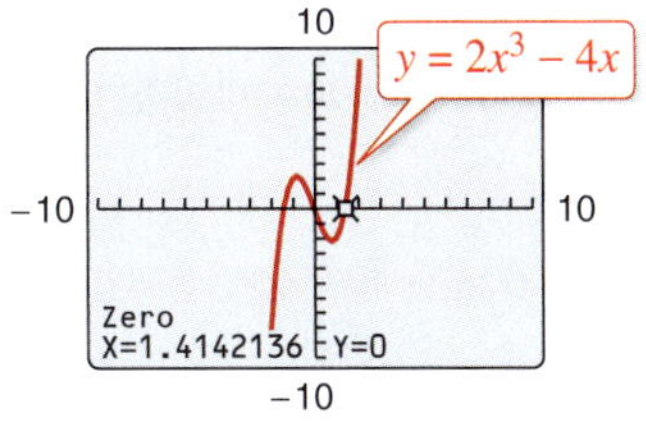

Figure A.143

Zoom and Trace Features

The *zoom* feature enables you to adjust the viewing window of a graph quickly (see Figure A.144). For example, the *zoom box* feature allows you to create a new viewing window by drawing a box around any part of the graph.

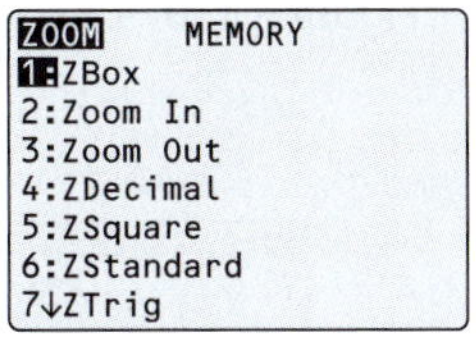

Figure A.144

The *trace* feature moves from point to point along a graph. For instance, enter the equation

$$y = 2x^3 - 3x + 2$$

in the *equation editor* (see Figure A.145) and graph the equation, as shown in Figure A.146. To activate the *trace* feature, press TRACE; then use the arrow keys to move the cursor along the graph. As you trace the graph, the coordinates of each point are displayed, as shown in Figure A.147.

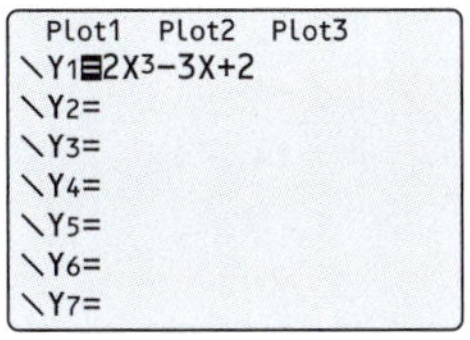

Figure A.145

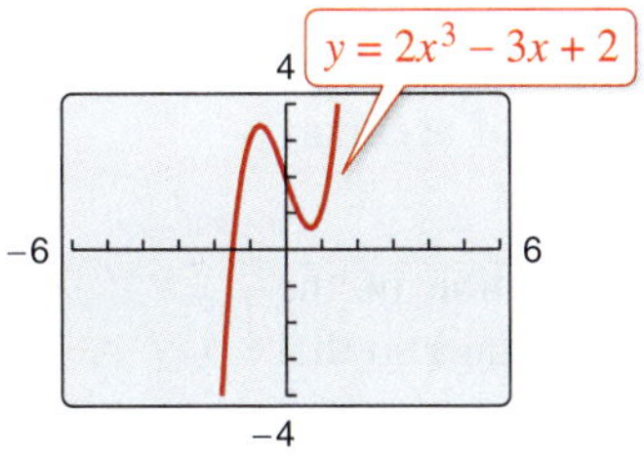

Figure A.146

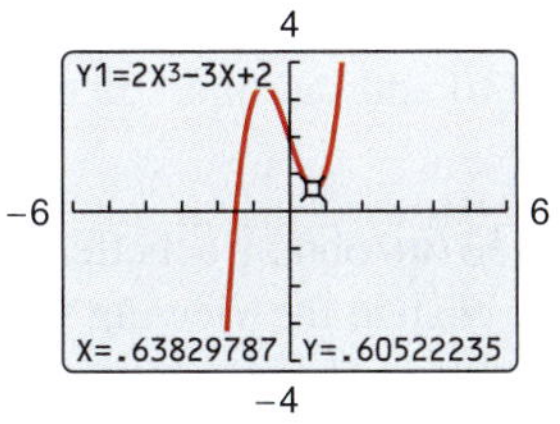

Figure A.147

The *trace* feature combined with the *zoom* feature enables you to obtain better and better approximations of desired points on a graph. For instance, you can use the *zoom* feature to approximate the x-intercept of the graph of

$$y = 2x^3 - 3x + 2.$$

From the viewing window shown in Figure A.146, the graph appears to have only one x-intercept. This intercept lies between -2 and -1. To zoom in on the x-intercept, choose the *zoom-in* feature from the *zoom* feature menu, as shown in Figure A.148. Next, trace the cursor to the point you want to zoom in on, in this case the x-intercept (see Figure A.149). Then press ENTER. You should obtain a graph similar to the one shown in Figure A.150.

Now, using the *trace* feature, you can approximate the x-intercept to be

$x \approx -1.468085$

as shown in Figure A.151. Use the *zoom-in* feature again to obtain a graph similar to the one shown in Figure A.152. Using the *trace* feature, you can approximate the x-intercept to be

$x \approx -1.476064$

as shown in Figure A.153.

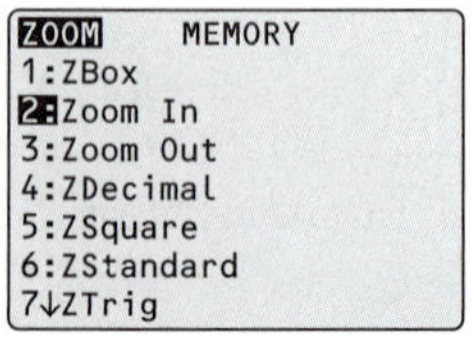

Figure A.148

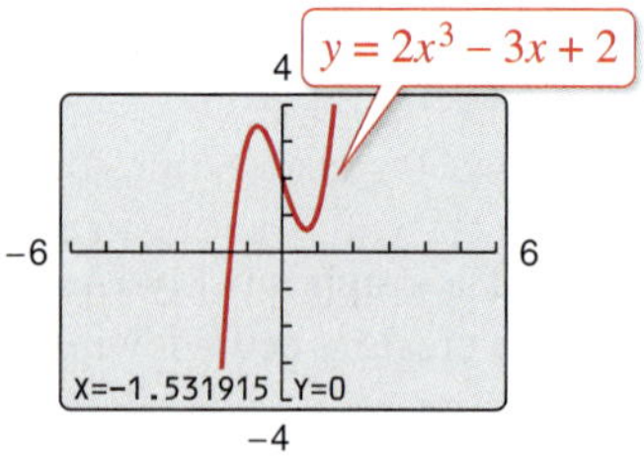

Figure A.149

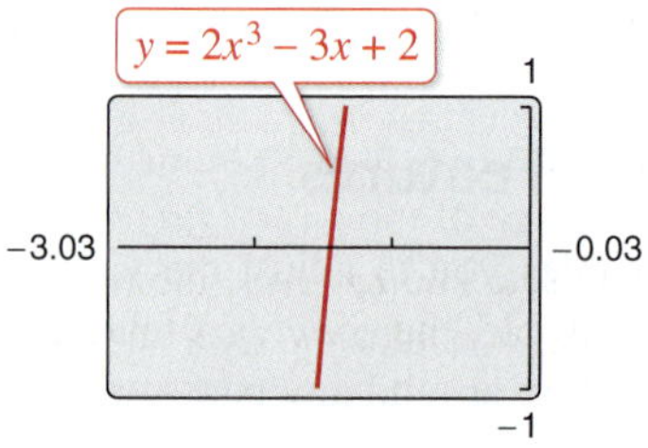

Figure A.150

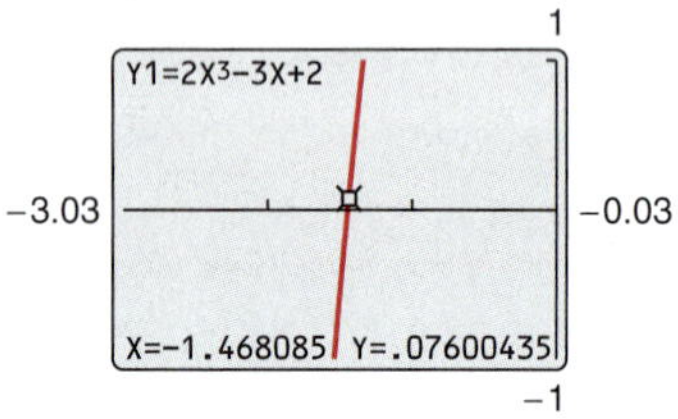

Figure A.151

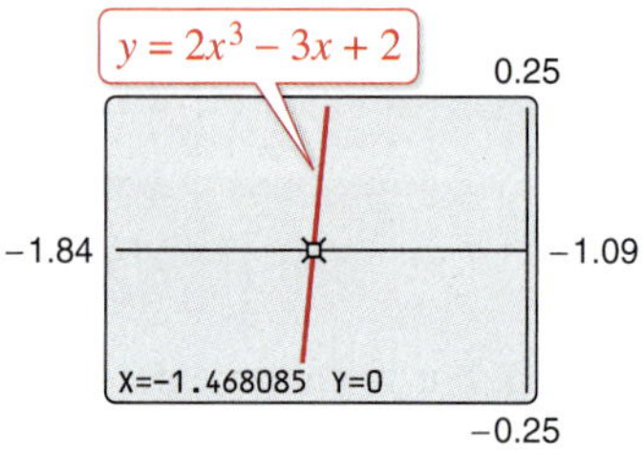

Figure A.152

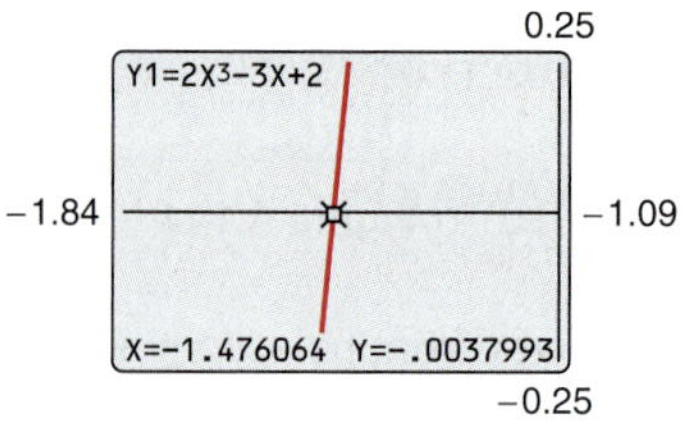

Figure A.153

Here are some suggestions for using the *zoom* feature.

1. With each successive zoom-in, adjust the x-scale so that the viewing window shows at least one tick mark on each side of the x-intercept.
2. The error in your approximation will be less than the distance between two scale marks.
3. The *trace* feature can usually be used to add one more decimal place of accuracy without changing the viewing window.

You can adjust the scale in Figure A.153 to obtain a better approximation of the x-intercept. Using the suggestions above, change the viewing window settings so that the viewing window shows at least one tick mark on each side of the x-intercept, as shown in Figure A.154. From Figure A.154, you can determine that the error in your approximation will be less than 0.001 (the Xscl value). Then, using the *trace* feature, you can improve the approximation, as shown in Figure A.155. To three decimal places, the x-intercept is

$x \approx -1.476.$

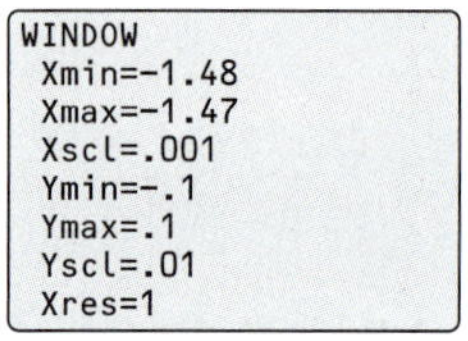

Figure A.154

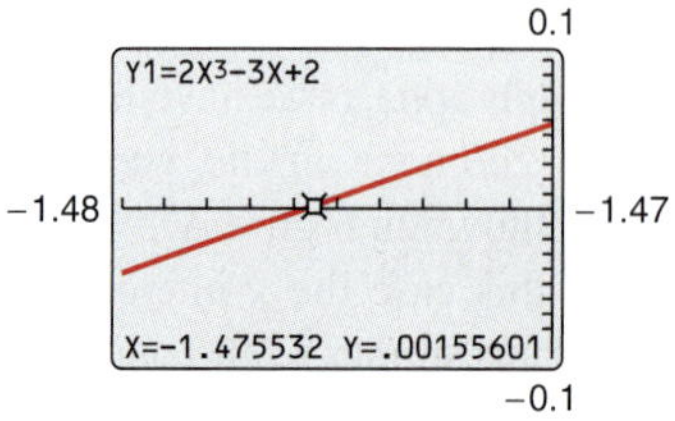

Figure A.155

Appendix B: Concepts in Statistics

B.1 Representing Data Graphically

What you should learn

- Use line plots to order and analyze data.
- Use histograms to represent frequency distributions.
- Use bar graphs to represent and analyze data.
- Use scatter plots to represent and analyze data.
- Use line graphs to represent and analyze data.

Why you should learn it

Double bar graphs allow you to compare visually two sets of data over time. For example, in Exercise 15 on page A32, you will create a double bar graph showing the total college enrollments for women and men.

Line Plots

Statistics is the branch of mathematics that studies techniques for collecting, organizing, and interpreting data. In this section, you will study several ways to organize data. The first is a **line plot,** which uses a portion of a real number line to order numbers. Line plots are especially useful for ordering small sets of numbers (about 50 or less) by hand.

Many statistical measures can be obtained from a line plot. Two such measures are the *frequency* and *range* of the data. The **frequency** measures the number of times a value occurs in a data set. The **range** is the difference between the greatest and least data values. For example, consider the data values

20, 21, 21, 25, 32.

The frequency of 21 in the data set is 2 because 21 occurs twice. The range is 12 because the difference between the greatest and least data values is $32 - 20 = 12$.

EXAMPLE 1 Constructing a Line Plot

The scores from an economics class of 30 students are listed. The scores are for a 100-point exam. *(Spreadsheet at LarsonPrecalculus.com)*

DATA 93, 70, 76, 67, 86, 93, 82, 78, 83, 86, 64, 78, 76, 66, 83, 83, 96, 74, 69, 76, 64, 74, 79, 76, 88, 76, 81, 82, 74, 70

a. Use a line plot to organize the scores.

b. Which score occurs with the greatest frequency?

c. What is the range of the scores?

Solution

a. Begin by scanning the data to find the least and greatest numbers. For the data, the least number is 64 and the greatest is 96. Next, draw a portion of a real number line that includes the interval [64, 96]. To create the line plot, start with the first number, 93, and record a • above 93 on the number line. Continue recording a • for each number in the list until you obtain the line plot shown in the figure.

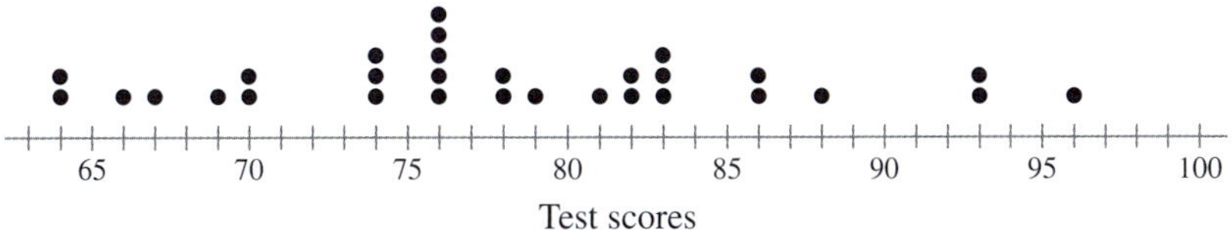

b. From the line plot, you can see that 76 occurs with the greatest frequency.

c. Because the range is the difference between the greatest and least data values, the range of scores is

$96 - 64 = 32.$

✓ **Checkpoint** Audio-video solution in English & Spanish at LarsonPrecalculus.com

Rework Example 1 for the test scores listed. *(Spreadsheet at LarsonPrecalculus.com)*

DATA 68, 73, 67, 95, 71, 82, 85, 74, 82, 61, 87, 92, 78, 74, 64, 71, 74, 82, 71, 83, 92, 82, 78, 72, 82, 64, 85, 67, 71, 62

Histograms and Frequency Distributions

When you want to organize large sets of data, it is useful to group the data into intervals and plot the frequency of the data in each interval. A **frequency distribution** can be used to construct a **histogram.** A histogram uses a portion of a real number line as its horizontal axis. The bars of a histogram are not separated by spaces.

EXAMPLE 2 Constructing a Histogram

The table at the right shows the percent of the resident population of each state and the District of Columbia who were at least 65 years old in 2016. Construct a frequency distribution and a histogram for the data. (*Source:* U.S. Census Bureau)

DATA

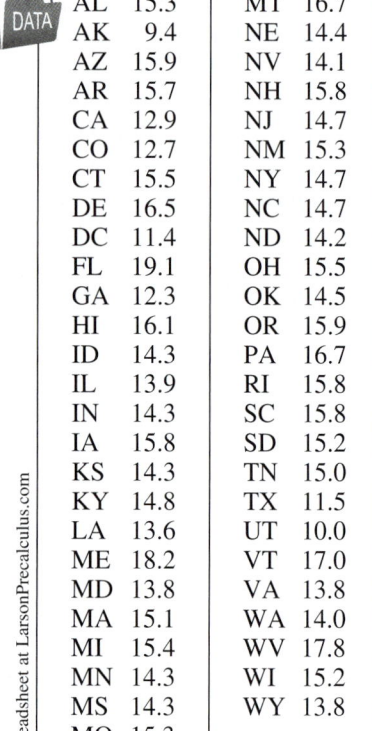

AL	15.3	MT	16.7
AK	9.4	NE	14.4
AZ	15.9	NV	14.1
AR	15.7	NH	15.8
CA	12.9	NJ	14.7
CO	12.7	NM	15.3
CT	15.5	NY	14.7
DE	16.5	NC	14.7
DC	11.4	ND	14.2
FL	19.1	OH	15.5
GA	12.3	OK	14.5
HI	16.1	OR	15.9
ID	14.3	PA	16.7
IL	13.9	RI	15.8
IN	14.3	SC	15.8
IA	15.8	SD	15.2
KS	14.3	TN	15.0
KY	14.8	TX	11.5
LA	13.6	UT	10.0
ME	18.2	VT	17.0
MD	13.8	VA	13.8
MA	15.1	WA	14.0
MI	15.4	WV	17.8
MN	14.3	WI	15.2
MS	14.3	WY	13.8
MO	15.3		

Spreadsheet at LarsonPrecalculus.com

Solution

To begin constructing a frequency distribution, you must first decide on the number of intervals. There are several ways to group the data. However, because the least number is 9.4 and the greatest is 19.1, it seems that six intervals would be appropriate. The first would be the interval

$[9, 11)$ First interval

the second would be

$[11, 13)$ Second interval

and so on. By tallying the data into the six intervals, you obtain the frequency distribution shown below.

Interval	*Tally*
$[9, 11)$	||
$[11, 13)$	𝍸
$[13, 15)$	𝍸 𝍸 𝍸 ||||
$[15, 17)$	𝍸 𝍸 𝍸 𝍸 |
$[17, 19)$	|||
$[19, 21)$	|

You can construct the histogram by drawing a vertical axis to represent the number of states and a horizontal axis to represent the percent of the population 65 and older. Then, for each interval, draw a vertical bar whose height is the total tally, as shown in Figure B.1. Note that the break in the horizontal axis indicates that the numbers 0 through 8 have been omitted.

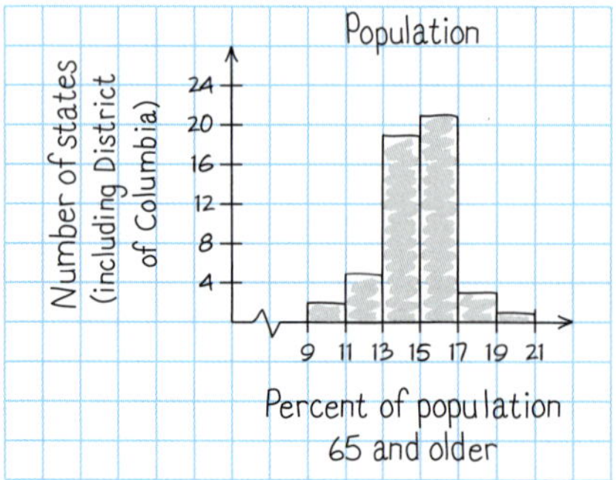

Figure B.1

✓ **Checkpoint** ▶ *Audio-video solution in English & Spanish at LarsonPrecalculus.com*

Construct a frequency distribution and a histogram for the data in Example 1.

Bar Graphs

A **bar graph** is similar to a histogram, except that the bars can be either horizontal or vertical, and the labels of the bars are not necessarily numbers. For instance, the labels of the bars can be the months of a year, names of cities, or types of vehicles. Another difference between a bar graph and a histogram is that the bars in a bar graph are usually separated by spaces.

EXAMPLE 3 Constructing a Bar Graph

The table shows the revenues for Netflix from 2012 through 2017. Construct a bar graph for the data. What can you conclude? (*Source:* Netflix Inc.)

DATA

Year	Revenue (in billions of dollars)
2012	3.609
2013	4.375
2014	5.505
2015	6.780
2016	8.831
2017	11.693

Spreadsheet at LarsonPrecalculus.com

Solution

To create a bar graph, begin by drawing a vertical axis to represent the revenues and a horizontal axis to represent the year. The bar graph is shown below.

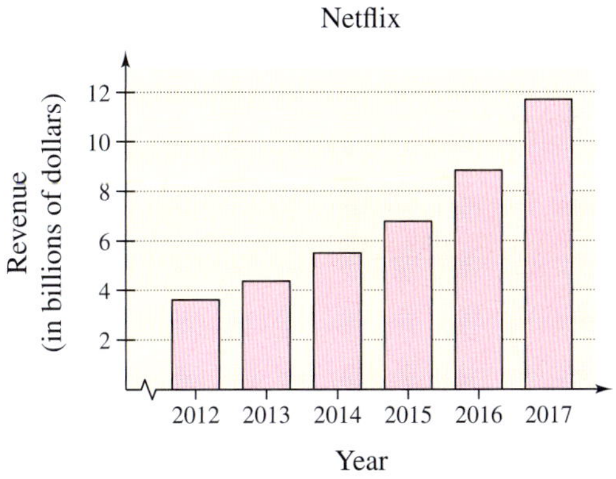

From the bar graph, you can see that the revenues have steadily increased from 2012 through 2017. Also, the revenues in 2017 are more than three times the revenues in 2012.

✓ ***Checkpoint*** *Audio-video solution in English & Spanish at LarsonPrecalculus.com*

The table shows the long-term debts for Netflix from 2012 through 2017. Construct a bar graph for the data. What can you conclude? (*Source:* Netflix Inc.)

DATA

Year	Long-term debt (in billions of dollars)
2012	0.400
2013	0.500
2014	0.900
2015	2.371
2016	3.364
2017	6.499

Spreadsheet at LarsonPrecalculus.com

EXAMPLE 4 Constructing a Double Bar Graph

The table shows the percents of associate degrees awarded to males and females for selected fields of study in the United States in 2016. Construct a double bar graph for the data. (*Source:* U.S. National Center for Education Statistics)

DATA

Field of Study	% Female	% Male
Agriculture and Natural Resources	37.9	62.1
Biological and Biomedical Sciences	65.9	34.1
Accounting	72.3	27.7
Business/Commerce	56.6	43.4
Communication/Journalism	58.7	41.3
Education	88.4	11.6
Engineering	14.6	85.4
Family and Consumer Sciences	95.8	4.2
Emergency Medical Technician	30.5	69.5
Legal Professions	86.0	14.0
Mathematics and Statistics	28.7	71.3
Physical Sciences	41.0	59.0
Precision Production	6.5	93.5
Psychology	75.6	24.4
Social Sciences	65.1	34.9

Spreadsheet at LarsonPrecalculus.com

Solution

For the data, a horizontal bar graph seems to be appropriate. This makes it easier to label and read the bars, as shown below.

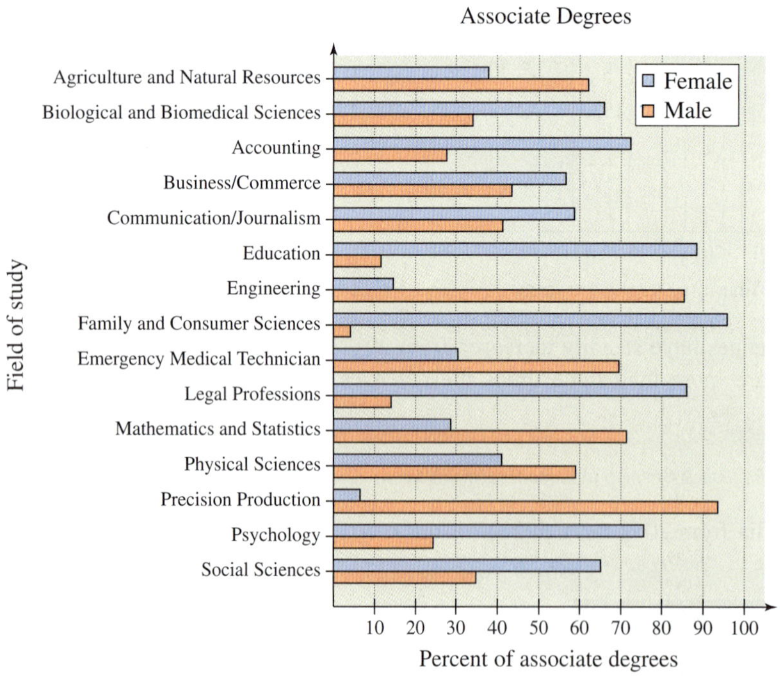

Checkpoint Audio-video solution in English & Spanish at LarsonPrecalculus.com

Construct a double bar graph for the revenues and long-term debts for Netflix in Example 3 and the Checkpoint after Example 3, respectively.

Scatter Plots

Data can be represented graphically by points plotted in a rectangular coordinate system. This type of graph is called a **scatter plot.** (To review the rectangular coordinate system, see Appendix C.1 at this textbook's *Student Companion Website*.)

EXAMPLE 5 Sketching a Scatter Plot

The numbers of employees E (in thousands) in the software publishing industry in the United States from 2008 through 2017 are shown in the table, where t represents the year. Sketch a scatter plot of the data by hand. (*Source:* U.S. Bureau of Labor Statistics)

DATA

Year, t	Employees, E
2008	264
2009	258
2010	261
2011	271
2012	287
2013	300
2014	316
2015	334
2016	359
2017	375

Spreadsheet at LarsonPrecalculus.com

Solution

Before you sketch the scatter plot, it is helpful to represent each pair of values by an ordered pair (t, E) as follows.

(2008, 264), (2009, 258), (2010, 261), (2011, 271), (2012, 287),
(2013, 300), (2014, 316), (2015, 334), (2016, 359), (2017, 375)

To sketch a scatter plot of the data shown in the table, first draw a vertical axis to represent the number of employees (in thousands) and a horizontal axis to represent the year. Then plot the resulting points, as shown in the figure. Note that the break in the t-axis indicates that the numbers 0 through 2007 have been omitted.

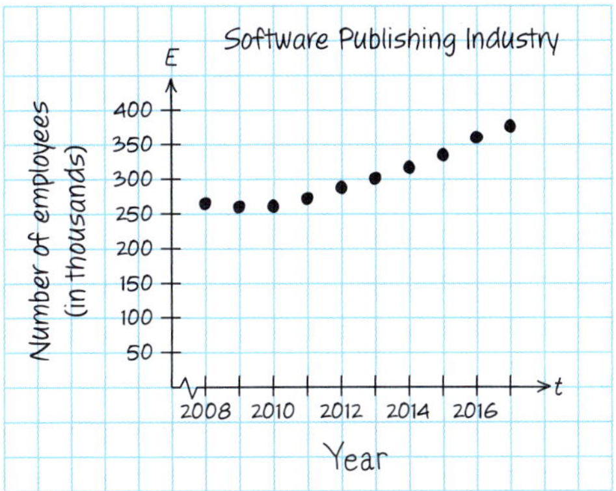

Checkpoint *Audio-video solution in English & Spanish at LarsonPrecalculus.com*

The numbers of employees E (in thousands) in outpatient care centers in the United States from 2008 through 2017 are shown in the table, where t represents the year. Sketch a scatter plot of the data by hand. (*Source:* U.S. Bureau of Labor Statistics)

DATA

Year, t	Employees, E
2008	581
2009	606
2010	649
2011	670
2012	699
2013	731
2014	764
2015	806
2016	853
2017	897

Spreadsheet at LarsonPrecalculus.com

In Example 5, you could have let $t = 1$ represent the year 2008. In that case, the horizontal axis of the graph would not have been broken, and the tick marks would have been labeled 1 through 10 (instead of 2008 through 2017).

Technology Tip

You can use a graphing utility to graph the scatter plot in Example 5. For instructions on how to use the *list editor* and the *statistical plotting* feature, see Appendix A; for specific keystrokes, go to this textbook's *Student Companion Website.*

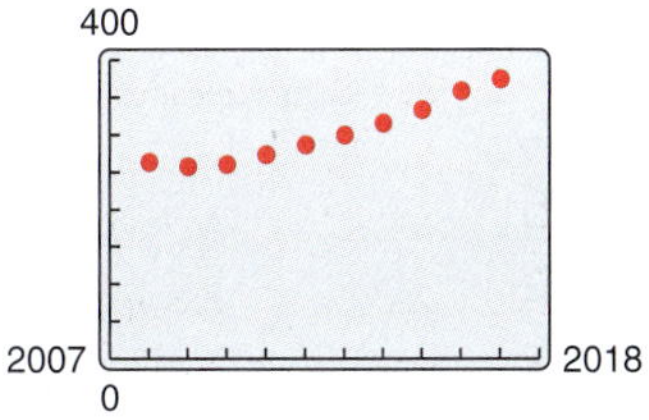

Line Graphs

A **line graph** is similar to a standard coordinate graph. Line graphs are often used to show trends over periods of time.

EXAMPLE 6 Constructing a Line Graph

The table at the right shows the number of immigrants (in thousands) legally entering the United States for each decade from 1901 through 2010. Construct a line graph for the data. What can you conclude? (*Source:* U.S. Department of Homeland Security)

Decade	Number (in thousands)
1901–1910	8,795
1911–1920	5,736
1921–1930	4,107
1931–1940	528
1941–1950	1,035
1951–1960	2,515
1961–1970	3,322
1971–1980	4,399
1981–1990	7,256
1991–2000	9,081
2001–2010	10,501

Solution

Begin by drawing a vertical axis to represent the number of immigrants in thousands. Then label the horizontal axis with decades and plot the points shown in the table. Finally, connect the points with line segments, as shown in Figure B.2. From the line graph, you can see that the number of immigrants hit a low point during the depression of the 1930s. Since then, the number has steadily increased.

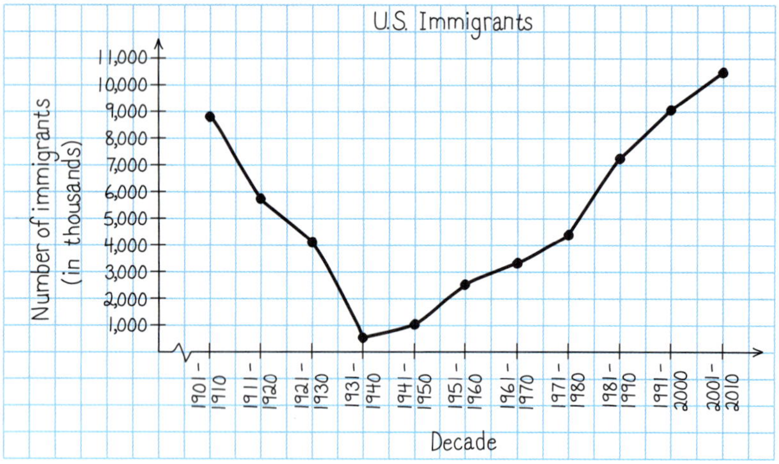

Figure B.2

You can use a graphing utility to check your sketch, as shown in Figure B.3.

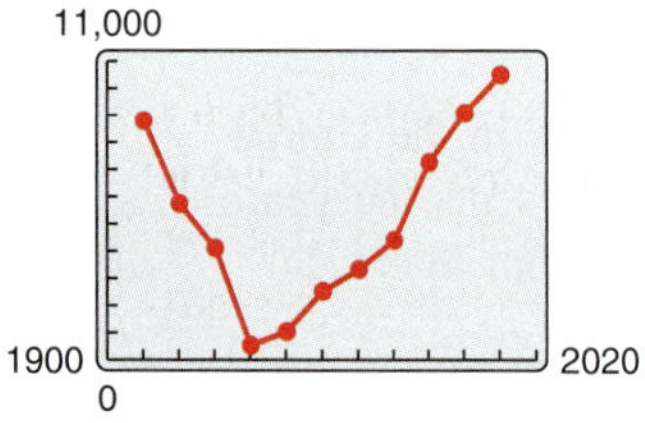

Figure B.3

✓ ***Checkpoint*** *Audio-video solution in English & Spanish at LarsonPrecalculus.com*

Construct a line graph for the data in Example 3.

Technology Tip

You can use the *statistical plotting* feature of a graphing utility to create different types of graphs, such as line graphs. For instructions on how to use the *statistical plotting* feature, see Appendix A; for specific keystrokes, go to this textbook's *Student Companion Website.*

B.1 Exercises

See *CalcChat.com* for tutorial help and worked-out solutions to odd-numbered exercises. For instructions on how to use a graphing utility, see Appendix A.

Vocabulary and Concept Check

In Exercises 1–6, fill in the blank.

1. _______ are useful for ordering small sets of numbers by hand.

2. A _______ uses a portion of a real number line as its horizontal axis, and the bars are not separated by spaces.

3. You can use a _______ to construct a histogram.

4. The bars in a _______ can be either vertical or horizontal.

5. A _______ represents data graphically as points plotted in a rectangular coordinate system.

6. _______ show trends over periods of time.

Procedures and Problem Solving

7. Economics The line plot shows a sample of average prices of diesel fuel at 25 fuel stations throughout the United States.

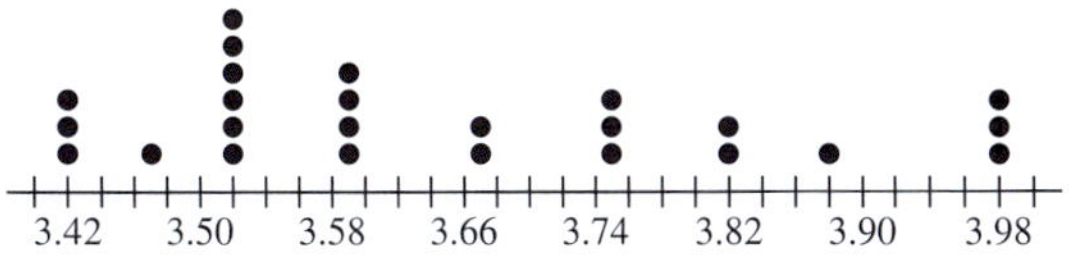

(a) What price occurred with the greatest frequency?

(b) What is the range of prices?

8. Environmental Science The line plot shows the weights (to the nearest hundred pounds) of municipal waste hauled by a garbage truck in 30 trips to a landfill.

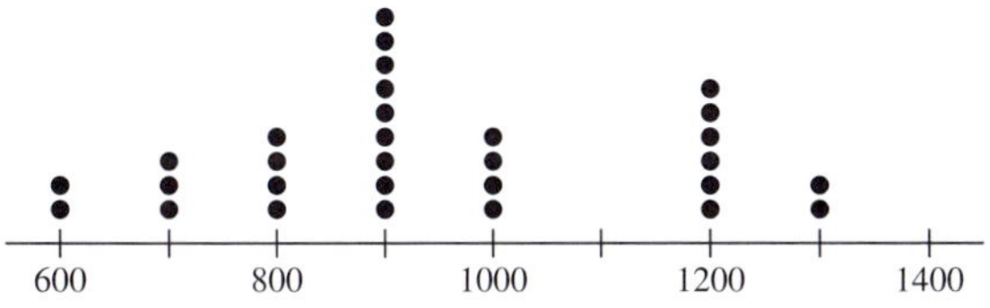

(a) What weight occurred with the greatest frequency?

(b) What is the range of weights?

Education In Exercises 9 and 10, use the following scores from an algebra class of 30 students. The scores are for two 25-point quizzes.

Quiz #1 **20, 15, 14, 20, 16, 19, 10, 21, 24, 15, 15, 14, 15, 21, 19, 15, 20, 18, 18, 22, 18, 16, 18, 19, 21, 19, 16, 20, 14, 12**

Quiz #2 **22, 22, 23, 22, 21, 24, 22, 19, 21, 23, 23, 25, 24, 22, 22, 23, 23, 23, 23, 22, 24, 23, 22, 24, 21, 24, 16, 21, 16, 14**

9. Construct a line plot for each quiz. For each quiz, which score(s) occurred with the greatest frequency?

10. Explain how you can use the line plots to determine which quiz had the greater mean score without calculating the means.

11. Demographics The list shows the percents of individuals living below the poverty level in the 50 states in 2016. Use a frequency distribution and a histogram to organize the data. *(Spreadsheet at LarsonPrecalculus.com)* (*Source:* U.S. Census Bureau)

DATA

AK 10.9	AL 16.3	AR 16.1	AZ 16.6
CA 13.9	CO 9.2	CT 9.5	DE 11.3
FL 14.6	GA 16.8	HI 10.1	IA 10.1
ID 11.7	IL 11.5	IN 12.7	KS 12.7
KY 17.4	LA 19.4	MA 10.6	MD 8.4
ME 12.5	MI 11.9	MN 8.3	MO 11.4
MS 20.1	MT 11.8	NC 14.4	ND 10.9
NE 9.9	NH 6.8	NJ 10.3	NM 18.7
NV 11.5	NY 13.0	OH 13.6	OK 14.4
OR 11.8	PA 11.7	RI 11.6	SC 14.2
SD 14.2	TN 14.8	TX 14.3	UT 8.9
VA 11.1	VT 10.2	WA 11.2	WI 11.1
WV 16.3	WY 10.3		

12. Education The list shows the numbers of operating public school districts in the 50 states in 2017. Use a frequency distribution and a histogram to organize the data. *(Spreadsheet at LarsonPrecalculus.com)* (*Source:* National Education Association)

DATA

AK 54	AL 137	AR 259	AZ 715
CA 1028	CO 178	CT 196	DE 44
FL 75	GA 207	HI 1	IA 333
ID 155	IL 852	IN 402	KS 286
KY 173	LA 147	MA 404	MD 24
ME 214	MI 829	MN 529	MO 556
MS 165	MT 406	NC 115	ND 176
NE 245	NH 165	NJ 702	NM 89
NV 17	NY 691	OH 1026	OK 512
OR 196	PA 796	RI 63	SC 86
SD 150	TN 141	TX 1203	UT 141
VA 132	VT 360	WA 307	WI 422
WV 55	WY 48		

13. Meterology The table shows the normal monthly precipitation (in inches) in Olympia, Washington. Construct a bar graph for the data. Write a brief statement regarding the precipitation in Olympia, Washington, throughout the year. (*Source:* National Oceanic and Atmospheric Administration)

DATA

Month	Precipitation
January	7.84
February	5.27
March	5.29
April	3.54
May	2.33
June	1.76
July	0.63
August	0.94
September	1.71
October	4.60
November	8.63
December	7.46

Spreadsheet at LarsonPrecalculus.com

14. Business The table shows the revenues (in billions of dollars) for Costco Wholesale stores in the years from 1998 through 2017. Construct a bar graph for the data. Write a brief statement regarding the revenue of Costco Wholesale stores over time. (*Source:* Costco Wholesale Corporation)

DATA

Year	Revenue (in billions of dollars)
1998	24.27
1999	27.46
2000	32.16
2001	34.80
2002	38.76
2003	42.55
2004	48.11
2005	52.94
2006	60.15
2007	64.91
2008	72.48
2009	71.45
2010	77.95
2011	88.92
2012	99.14
2013	105.16
2014	112.64
2015	116.20
2016	118.72
2017	129.03

Spreadsheet at LarsonPrecalculus.com

15. *Why you should learn it* *(p. A25)* The table shows the total college enrollments (in thousands) for women and men in the United States from 2008 through 2015. Construct a double bar graph for the data. (*Source:* U.S. National Center for Education Statistics)

DATA

Year	Women (in thousands)	Men (in thousands)
2008	10,914	8,189
2009	11,581	8,733
2010	11,974	9,046
2011	11,976	9,034
2012	11,725	8,919
2013	11,515	8,861
2014	11,410	8,797
2015	11,256	8,721

Spreadsheet at LarsonPrecalculus.com

16. Demographics The table shows the populations (in millions) of the seven most populous cities in the United States in 2010 and 2017. Construct a double bar graph for the data. (*Source:* U.S. Census Bureau)

DATA

City	2010 population (in millions)	2017 population (in millions)
New York, NY	8.19	8.62
Los Angeles, CA	3.80	4.00
Chicago, IL	2.70	2.72
Houston, TX	2.10	2.31
Phoenix, AZ	1.45	1.63
Philadelphia, PA	1.53	1.58
San Antonio, TX	1.33	1.51

Spreadsheet at LarsonPrecalculus.com

17. Accounting The table shows the sales y (in billions of dollars) for Apple Inc. for the years 2007 through 2017. Sketch a scatter plot of the data. (*Source:* Apple Inc.)

DATA

Year	Sales, y (in billions of dollars)
2007	24.0
2008	32.5
2009	36.5
2010	65.2
2011	108.2
2012	156.5
2013	170.9
2014	183.2
2015	231.3
2016	214.2
2017	228.6

Spreadsheet at LarsonPrecalculus.com

18. **Meteorology** The table shows the lowest temperature on record y (in degrees Fahrenheit) in Salt Lake City, Utah, for each month x, where $x = 1$ represents January. Sketch a scatter plot of the data. *(Source:* U.S. National Oceanic and Atmospheric Administration)

DATA

Month, x	Temperature, y
1	−22
2	−14
3	2
4	15
5	25
6	35
7	40
8	37
9	27
10	16
11	−14
12	−15

Spreadsheet at LarsonPrecalculus.com

Economics **In Exercises 19–24, use the line graph, which shows the average prices of a gallon of gasoline (all grades, all formulations) from 2009 through 2017.** *(Source:* U.S. Energy Information Administration)

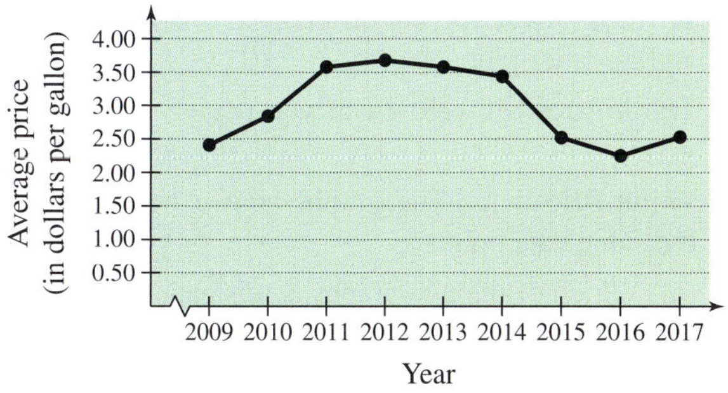

19. Describe the trend in the price of gasoline from 2009 to 2012.

20. Describe the trend in the price of gasoline from 2012 to 2016.

21. Determine when the price of gasoline showed the greatest rate of decrease from one year to the next.

22. Determine when the price of gasoline showed the greatest rate of increase from one year to the next.

23. Approximate the percent decrease in the price per gallon of gasoline from 2012 to 2016.

24. Approximate the percent increase in the price per gallon of gasoline from 2009 to 2013.

25. **Human Resources** The table shows the percents of wives in two-income families in the United States who earned more than their husbands from 2003 through 2015. *(Source:* U.S. Bureau of Labor Statistics)

(a) Construct a line graph for the data.

(b) Write a brief statement describing what the graph reveals.

DATA

Year	Percent
2003	25.2
2004	25.3
2005	25.5
2006	25.7
2007	25.9
2008	26.6
2009	28.9
2010	29.2
2011	28.1
2012	29.0
2013	29.3
2014	28.2
2015	29.3

Spreadsheet at LarsonPrecalculus.com

26. **Economics** The table shows the trade deficits (the differences between imports and exports) of the United States from 2001 through 2017. *(Source:* U.S. Census Bureau)

(a) Construct a line graph for the data.

(b) Write a brief statement describing what the graph reveals.

DATA

Year	Trade deficit (in billions of dollars)
2001	412
2002	468
2003	532
2004	655
2005	772
2006	828
2007	809
2008	816
2009	504
2010	635
2011	725
2012	730
2013	689
2014	734
2015	745
2016	737
2017	796

Spreadsheet at LarsonPrecalculus.com

Education **In Exercises 27–30, refer to the scatter plot, which shows the mathematics entrance test scores x and the final examination scores y in an algebra course for a sample of 10 students.**

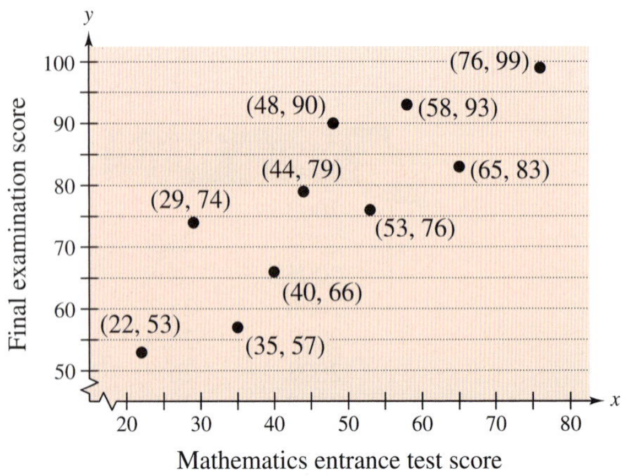

27. Find the entrance test score of any student with a final exam score in the 80s.

28. Find the final exam score of any student with an entrance test score in the 50s.

29. Did the student with the highest entrance test score have the highest final exam score?

30. Does a higher entrance test score necessarily imply a higher final exam score? Explain.

31. **Networking** The list shows the percent of households in each of the 50 states and the District of Columbia with a broadband Internet subscription in 2016. Use a graphing utility to organize the data in the graph of your choice. Explain your choice of graph. *(Spreadsheet at LarsonPrecalculus.com)* (*Source:* U.S. Census Bureau)

DATA

AK 85.7	AL 74.7	AR 70.9
AZ 83.1	CA 85.4	CO 86.9
CT 84.1	DC 79.8	DE 83.3
FL 81.2	GA 80.7	HI 83.2
IA 79.6	ID 79.4	IL 82.0
IN 79.2	KS 80.3	KY 77.3
LA 74.4	MA 85.5	MD 85.8
ME 80.7	MI 80.5	MN 83.5
MO 79.3	MS 70.7	MT 78.9
NC 79.0	ND 81.4	NE 81.6
NH 86.4	NJ 84.2	NM 73.7
NV 80.9	NY 81.7	OH 80.9
OK 77.2	OR 84.9	PA 80.5
RI 82.8	SC 77.0	SD 79.5
TN 76.7	TX 80.5	UT 85.4
VA 83.4	VT 81.1	WA 87.4
WI 81.3	WV 74.2	WY 83.2

32. **Athletics** The table shows the numbers of participants (in thousands) in high school athletic programs in the United States from 2006 through 2017. Organize the data in an appropriate display. Explain your choice of graph. (*Source:* National Federation of State High School Associations)

DATA

Year	Female athletes (in thousands)	Male athletes (in thousands)
2006	2953	4207
2007	3022	4321
2008	3057	4372
2009	3114	4423
2010	3173	4456
2011	3174	4494
2012	3208	4485
2013	3223	4491
2014	3268	4528
2015	3288	4519
2016	3324	4545
2017	3400	4563

Spreadsheet at LarsonPrecalculus.com

Focusing on Concepts

33. **Writing** Describe the differences between a bar graph and a histogram.

34. **Writing** Describe the differences between a line plot and a scatter plot.

35. **Think About It** How can you decide which type of graph to use when you are organizing data?

36. **Think About It** The graphs shown below represent the same data points.
    (a) Which of the two graphs is misleading, and why?
    (b) Discuss other ways in which graphs can be misleading.
    (c) Why would it be beneficial for someone to use a misleading graph?

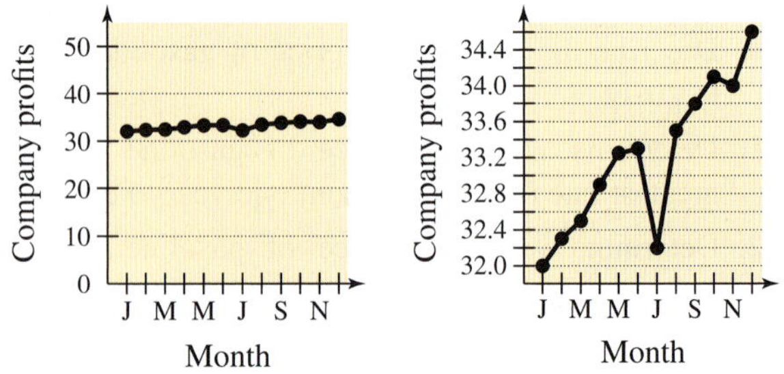

37. **Writing** Name a different type of graph you could use to display the data in the given exercise. Then name a type of graph that you would *not* use to display the data. Explain.
    (a) Exercise 11 (b) Exercise 14

B.2 Measures of Central Tendency and Dispersion

What you should learn

- Find and interpret the mean, median, and mode of a set of data.
- Determine the measure of central tendency that best represents a set of data.
- Find the standard deviation of a set of data.
- Use box-and-whisker plots.

Why you should learn it

Measures of central tendency and dispersion provide a convenient way to describe and compare sets of data. For instance, in Exercise 32 on page A43, the mean and standard deviation are used to analyze the prices of gold for the years 1994 through 2017.

Mean, Median, and Mode

In many real-life situations, it is helpful to describe data by a single number that is most representative of the entire collection of numbers. Such a number is called a **measure of central tendency.** The most commonly used measures are as follows.

1. The **mean,** or **average,** of n numbers is the sum of the numbers divided by n.
2. The **median** of n numbers is the middle number when the numbers are written in numerical order. If n is even, then the median is the average of the two middle numbers.
3. The **mode** of n numbers is the number that occurs most frequently. If two numbers tie for most frequent occurrence, then the collection has two modes and is called **bimodal.**

EXAMPLE 1 Comparing Measures of Central Tendency

An analyst states that the average annual income of a company's 25 employees is \$60,849. The actual annual incomes of the 25 employees are shown below. What are the mean, median, and mode of the incomes? *(Spreadsheet at LarsonPrecalculus.com)*

DATA

\$17,305	\$478,320	\$45,678	\$18,980	\$17,408
\$25,676	\$28,906	\$12,500	\$24,540	\$33,450
\$12,500	\$33,855	\$37,450	\$20,432	\$28,956
\$34,983	\$36,540	\$250,921	\$36,853	\$16,430
\$32,654	\$98,213	\$48,980	\$94,024	\$35,671

Solution

The mean of the incomes is

$$\text{Mean} = \frac{17,305 + 478,320 + 45,678 + 18,980 + \cdots + 35,671}{25}$$

$$= \frac{1,521,225}{25}$$

$$= \$60,849.$$

To find the median, order the incomes as follows.

\$12,500	\$12,500	\$16,430	\$17,305	\$17,408
\$18,980	\$20,432	\$24,540	\$25,676	\$28,906
\$28,956	\$32,654	\$33,450	\$33,855	\$34,983
\$35,671	\$36,540	\$36,853	\$37,450	\$45,678
\$48,980	\$94,024	\$98,213	\$250,921	\$478,320

From this list, you can see that the median income is \$33,450. You can also see that \$12,500 is the only income that occurs more than once. So, the mode is \$12,500.

Checkpoint *Audio-video solution in English & Spanish at LarsonPrecalculus.com*

Find the mean, median, and mode of the following data set.

68, 73, 67, 95, 71, 82, 85, 74, 82, 61

In Example 1, was the analyst telling the truth about the annual incomes? Technically, the person was telling the truth because the average is (generally) defined to be the mean. However, of the three measures of central tendency—*mean:* \$60,849, *median:* \$33,450, *mode:* \$12,500—it seems clear that the median is most representative. The mean is inflated by the two highest salaries.

Choosing a Measure of Central Tendency

Which of the three measures of central tendency is most representative of a particular data set? The answer depends on the distribution of the data *and* the way in which you plan to use the data.

For instance, in Example 1, the mean salary of \$60,849 would not seem very representative to a potential employee (the mean is inflated by the two highest salaries). A better use of the mean is for estimating income tax. For instance, to estimate a 1% income tax for the 25 employees, a tax collector would use 1% of the *mean* annual income of the employees.

EXAMPLE 2 Choosing a Measure of Central Tendency

Which measure of central tendency is most representative of the data given in each frequency distribution?

a.

Number	1	2	3	4	5	6	7	8	9
Frequency	7	20	15	11	8	3	2	0	15

b.

Number	1	2	3	4	5	6	7	8	9
Frequency	9	8	7	6	5	6	7	8	9

c.

Number	1	2	3	4	5	6	7	8	9
Frequency	6	1	2	3	5	5	8	3	0

Solution

a. For this data set, the mean is 4.23, the median is 3, and the mode is 2. Of these, the median or mode is probably the most representative measure.

b. For this data set, the mean and median are each 5, and the modes are 1 and 9 (the distribution is bimodal). Of these, the mean or median is the most representative measure.

c. For this data set, the mean is 4.88, the median is 5, and the mode is 7. Of these, the mean or median is the most representative measure.

✓ ***Checkpoint*** ▶ *Audio-video solution in English & Spanish at LarsonPrecalculus.com*

Which measure of central tendency is the most representative of the data shown in each frequency distribution?

a.

Number	1	2	3	4	5	6	7	8	9
Frequency	1	1	1	1	3	1	1	1	1

b.

Number	1	2	3	4	5	6	7	8	9
Frequency	0	3	4	5	5	3	2	1	6

c.

Number	1	2	3	4	5	6	7	8	9
Frequency	4	3	2	1	0	1	2	3	4

d.

Number	1	2	3	4	5	6	7	8	9
Frequency	11	8	3	2	0	15	7	20	15

Technology Tip

Most graphing utilities have *mean* and *median* features that can be used to find the means and medians of data sets. Enter the data from Example 2(a) in the *list editor* of a graphing utility. Then use the *mean* and *median* features to verify the solution to Example 2(a), as shown below.

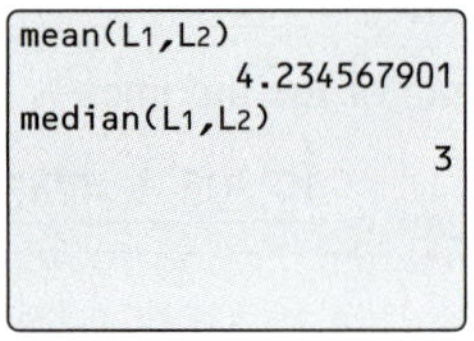

For instructions on how to use the *list*, *mean*, and *median* features, see Appendix A; for specific keystrokes, go to this textbook's *Student Companion Website*.

Variance and Standard Deviation

Very different sets of numbers can have the same mean. You will now study two **measures of dispersion,** which give you an idea of how much the numbers in a set differ from the mean of the set. These two measures are called the *variance* of the set and the *standard deviation* of the set.

Definitions of Variance and Standard Deviation

Consider a set of numbers $\{x_1, x_2, \ldots, x_n\}$ with a mean of $\bar{x}$. The **variance** of the set is

$$v = \frac{(x_1 - \bar{x})^2 + (x_2 - \bar{x})^2 + \cdots + (x_n - \bar{x})^2}{n}$$

and the **standard deviation** of the set is $\sigma = \sqrt{v}$. (The symbol σ is the lowercase Greek letter sigma.)

The standard deviation of a set is a measure of how much a typical number in the set differs from the mean. The greater the standard deviation, the more the numbers in the set vary from the mean. For instance, each of the sets

$$\{5, 5, 5, 5\}, \quad \{4, 4, 6, 6\}, \quad \text{and} \quad \{3, 3, 7, 7\}$$

has a mean of 5. The standard deviations of the sets are 0, 1, and 2, respectively.

$$\sigma_1 = \sqrt{\frac{(5-5)^2 + (5-5)^2 + (5-5)^2 + (5-5)^2}{4}} = 0$$

$$\sigma_2 = \sqrt{\frac{(4-5)^2 + (4-5)^2 + (6-5)^2 + (6-5)^2}{4}} = 1$$

$$\sigma_3 = \sqrt{\frac{(3-5)^2 + (3-5)^2 + (7-5)^2 + (7-5)^2}{4}} = 2$$

EXAMPLE 3 Estimations of Standard Deviation

Consider the three frequency distributions represented by the histograms in Figure B.4. Which data set has the least standard deviation? Which has the greatest?

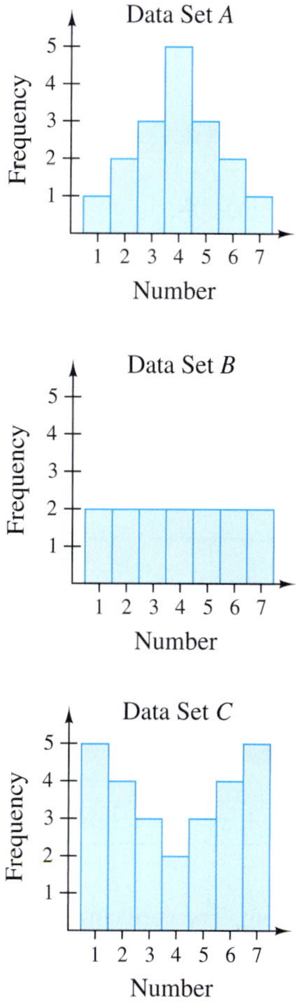

Figure B.4

Solution

Of the three data sets, the numbers in data set A are grouped most closely to the center and the numbers in data set C are the most dispersed. So, data set A has the least standard deviation and data set C has the greatest standard deviation.

Checkpoint *Audio-video solution in English & Spanish at LarsonPrecalculus.com*

Consider the three frequency distributions represented by the histograms shown below. Which data set has the least standard deviation? Which has the greatest?

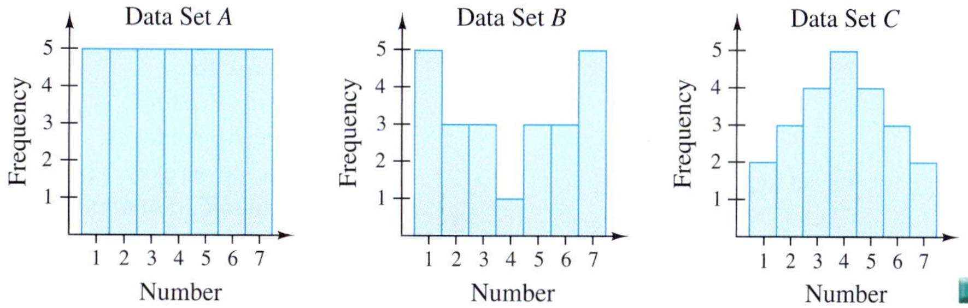

EXAMPLE 4 Finding a Standard Deviation

Find the standard deviation of each data set shown in Example 3.

Solution

Because of the symmetry of each histogram, you can conclude that each has a mean of

$\bar{x} = 4.$

The standard deviation of data set A is

$$\sigma = \sqrt{\frac{(-3)^2 + 2(-2)^2 + 3(-1)^2 + 5(0)^2 + 3(1)^2 + 2(2)^2 + (3)^2}{17}}$$

$$\approx 1.53.$$

The standard deviation of data set B is

$$\sigma = \sqrt{\frac{2(-3)^2 + 2(-2)^2 + 2(-1)^2 + 2(0)^2 + 2(1)^2 + 2(2)^2 + 2(3)^2}{14}}$$

$$= 2.$$

The standard deviation of data set C is

$$\sigma = \sqrt{\frac{5(-3)^2 + 4(-2)^2 + 3(-1)^2 + 2(0)^2 + 3(1)^2 + 4(2)^2 + 5(3)^2}{14}}$$

$$\approx 2.22.$$

These values confirm the results of Example 3. That is, data set A has the least standard deviation and data set C has the greatest.

Checkpoint *Audio-video solution in English & Spanish at LarsonPrecalculus.com*

Find the standard deviation of each data set shown in the Checkpoint with Example 3.

The following alternative formula provides a more efficient way to compute the standard deviation.

Alternative Formula for Standard Deviation

The standard deviation of $\{x_1, x_2, \ldots, x_n\}$ is given by

$$\sigma = \sqrt{\frac{x_1^2 + x_2^2 + \cdots + x_n^2}{n} - \bar{x}^2}.$$

Because of lengthy computations, this formula is difficult to verify. Conceptually, however, the process is straightforward. It consists of showing that the expressions

$$\sqrt{\frac{(x_1 - \bar{x})^2 + (x_2 - \bar{x})^2 + \cdots + (x_n - \bar{x})^2}{n}}$$

and

$$\sqrt{\frac{x_1^2 + x_2^2 + \cdots + x_n^2}{n} - \bar{x}^2}$$

are equivalent. Try verifying this equivalence for the set $\{x_1, x_2, x_3\}$ with

$$\bar{x} = \frac{x_1 + x_2 + x_3}{3}.$$

Technology Tip

Most graphing utilities have *statistical* features that can be used to find different statistical values of data sets, such as the standard deviation. Enter data set A from Example 3 in the *list editor* of a graphing utility. Then use the *one-variable statistics* feature to verify the solution to Example 4, as shown below.

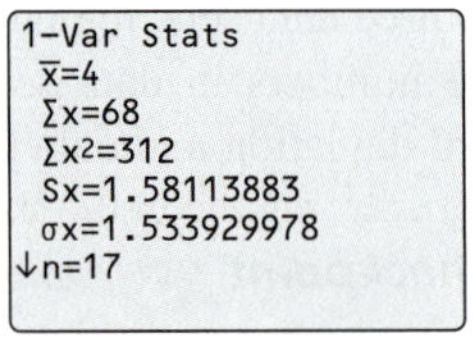

In the figure above, the standard deviation is represented as σx, which is about 1.53. For instructions on how to use the *one-variable statistics* feature, see Appendix A; for specific keystrokes, go to this textbook's *Student Companion Website.*

EXAMPLE 5 Using the Alternative Formula

Use the alternative formula for standard deviation to find the standard deviation of the following set of numbers.

5, 6, 6, 7, 7, 8, 8, 8, 9, 10

Solution

Begin by finding the mean of the set, which is 7.4. So, the standard deviation is

$$\sigma = \sqrt{\frac{5^2 + 2(6^2) + 2(7^2) + 3(8^2) + 9^2 + 10^2}{10} - (7.4)^2}$$

$$= \sqrt{\frac{568}{10} - 54.76}$$

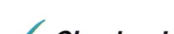

$$\approx 1.43.$$

Checkpoint Audio-video solution in English & Spanish at LarsonPrecalculus.com

Use the alternative formula for standard deviation to find the standard deviation of the following set of numbers.

3, 3, 3, 4, 4, 5, 5, 6, 6, 7

A well-known theorem in statistics, called *Chebychev's Theorem*, states that at least

$$1 - \frac{1}{k^2}$$

of the numbers in a distribution must lie within k standard deviations ($k > 1$) of the mean. So, at least $\frac{3}{4}$ or 75% of the numbers in a collection must lie within two standard deviations of the mean, and at least $\frac{8}{9}$ or $88\frac{8}{9}\%$ of the numbers must lie within three standard deviations of the mean. For most distributions, these percents are low. For instance, in all three distributions shown in Example 3, 100% of the numbers lie within two standard deviations of the mean.

EXAMPLE 6 Describing a Distribution

The table at the right shows the estimated numbers of registered nurses (in thousands) in each state and the District of Columbia in 2017. Find the mean and standard deviation of the numbers. What percent of the numbers lie within two standard deviations of the mean? (*Source:* U.S. Bureau of Labor Statistics)

DATA

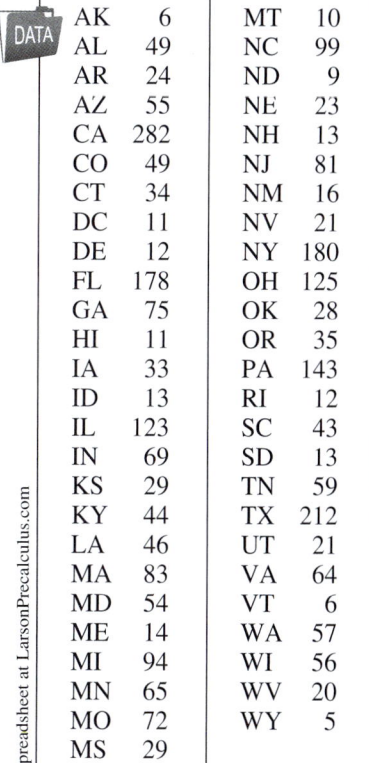

AK	6	MT	10
AL	49	NC	99
AR	24	ND	9
AZ	55	NE	23
CA	282	NH	13
CO	49	NJ	81
CT	34	NM	16
DC	11	NV	21
DE	12	NY	180
FL	178	OH	125
GA	75	OK	28
HI	11	OR	35
IA	33	PA	143
ID	13	RI	12
IL	123	SC	43
IN	69	SD	13
KS	29	TN	59
KY	44	TX	212
LA	46	UT	21
MA	83	VA	64
MD	54	VT	6
ME	14	WA	57
MI	94	WI	56
MN	65	WV	20
MO	72	WY	5
MS	29		

Spreadsheet at LarsonPrecalculus.com

Solution

Begin by entering the numbers in a graphing utility. Then use the *one-variable statistics* feature to obtain

$\bar{x} \approx 57.0$ and $\sigma \approx 57.2$.

The interval that contains all numbers that lie within two standard deviations of the mean is

$$[57.0 - 2(57.2), 57.0 + 2(57.2)] \quad \text{or} \quad [-57.4, 171.4].$$

From the table, you can see that all but four of the numbers (about 92%) lie in this interval—all but the numbers that correspond to the numbers of registered nurses in California, Florida, New York, and Texas.

Checkpoint Audio-video solution in English & Spanish at LarsonPrecalculus.com

In Example 6, what percent of the numbers lie within three standard deviations of the mean?

Box-and-Whisker Plots

Standard deviation is the measure of dispersion that is associated with the mean. **Quartiles** measure dispersion associated with the median.

Definition of Quartiles

Consider an ordered set of numbers whose median is m. The **lower quartile** is the median of the numbers that occur before m. The **upper quartile** is the median of the numbers that occur after m.

EXAMPLE 7 Finding Quartiles of a Set

Find the lower and upper quartiles for the following data set.

34, 14, 24, 16, 12, 18, 20, 24, 16, 26, 13, 27

Solution

Begin by ordering the set.

12, 13, 14,	16, 16, 18,	20, 24, 24,	26, 27, 34
1st 25%	2nd 25%	3rd 25%	4th 25%

This set has 12 numbers. The median of the entire set is $(18 + 20)/2 = 38/2 = 19$. The median of the six numbers that are before 19 is $(14 + 16)/2 = 30/2 = 15$. So, the lower quartile is 15. The median of the six numbers that are after 19 is $(24 + 26)/2 = 50/2 = 25$. So, the upper quartile is 25.

Checkpoint *Audio-video solution in English & Spanish at LarsonPrecalculus.com*

Find the lower and upper quartiles for the following data set.

39, 47, 81, 43, 23, 23, 27, 86, 15, 3, 74, 55

Quartiles are represented graphically by a **box-and-whisker plot,** as shown in Figure B.5. In the plot, notice that five numbers are given: the least number, the lower quartile, the median, the upper quartile, and the greatest number. Also notice that the numbers are spaced proportionally, as though they were on a real number line.

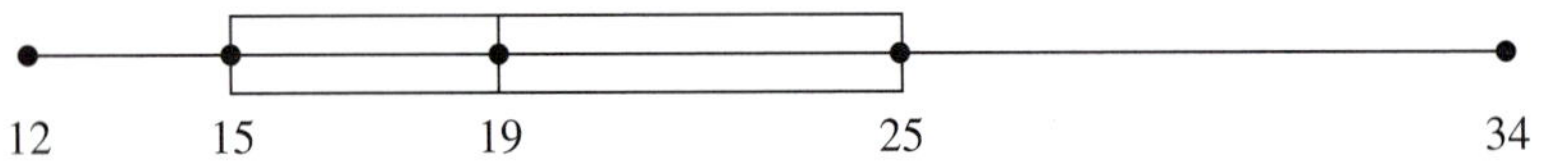

Figure B.5

Technology Tip

You can use a graphing utility to graph the box-and-whisker plot in Figure B.5. After entering the data in the graphing utility's *list editor*, use the *statistical plotting* and *ZoomStat* features to display the box-and-whisker plot, as shown in Figure B.6. For instructions on how to use the *list editor* and *statistical plotting* features, see Appendix A; for specific keystrokes, go to this textbook's *Student Companion Website*.

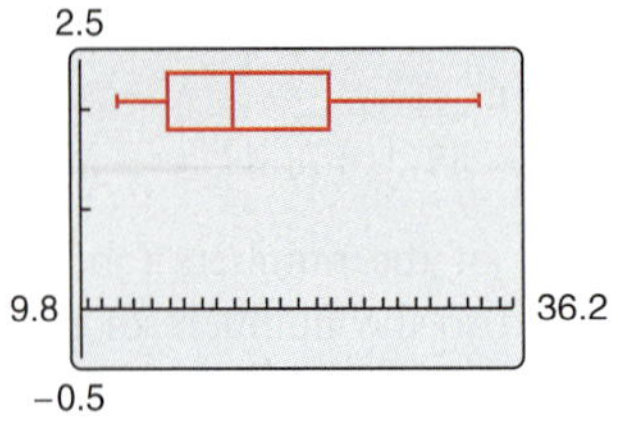

Figure B.6

The next example shows how to find quartiles when the number of elements in a set is not divisible by 4.

EXAMPLE 8 Sketching Box-and-Whisker Plots

Sketch a box-and-whisker plot for each data set.

a. 82, 82, 83, 85, 87, 89, 90, 94, 95, 95, 96, 98, 99

b. 11, 13, 13, 15, 17, 17, 20, 24, 24, 27

Solution

a. This set has 13 numbers. The median is 90 (the seventh number). The lower quartile is 84 (the median of the first six numbers). The upper quartile is 95.5 (the median of the last six numbers). See Figure B.7.

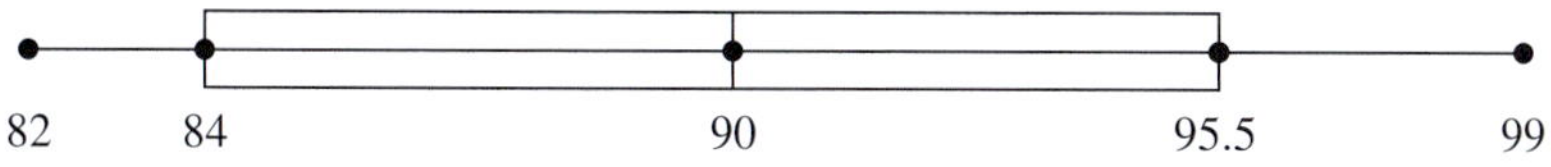

Figure B.7

b. This set has 10 numbers. The median is 17 (the average of the fifth and sixth numbers). The lower quartile is 13 (the median of the first five numbers). The upper quartile is 24 (the median of the last five numbers). See Figure B.8.

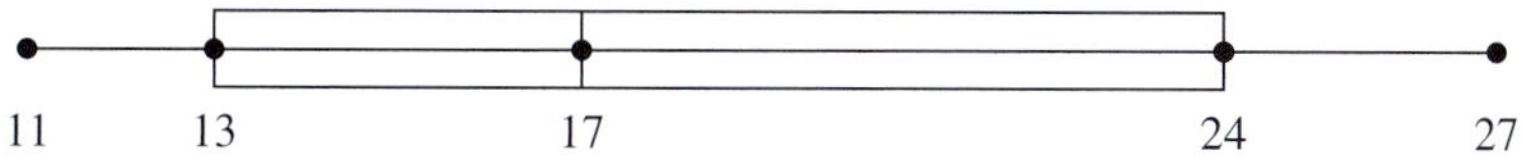

Figure B.8

✓ ***Checkpoint*** *Audio-video solution in English & Spanish at LarsonPrecalculus.com*

Sketch a box-and-whisker plot for the data set: 19, 22, 38, 44, 50, 54, 56, 62, 67, 79, 81, 85, 93.

B.2 Exercises

See *CalcChat.com* for tutorial help and worked-out solutions to odd-numbered exercises. For instructions on how to use a graphing utility, see Appendix A.

Vocabulary and Concept Check

In Exercises 1–4, fill in the blank(s).

1. A single number that is the most representative of a data set is called a _______ of _______ .

2. If two numbers are tied for the most frequent occurrence, then the collection has two _______ and is called _______ .

3. Two measures of dispersion are called the _______ and the _______ of a data set.

4. Quartiles are represented graphically by a _______ .

Procedures and Problem Solving

Comparing Measures of Central Tendency **In Exercises 5–10, find the mean, median, and mode of the data set.**

5. 5, 12, 7, 14, 8, 9, 7, 12

6. 30, 37, 32, 39, 33, 34, 32

7. 5, 12, 7, 24, 8, 9, 7, 12

8. 17, 37, 32, 39, 33, 34, 32

9. 21, 4, 18, 6, 25, 3

10. 2, 10, 19, 3, 7, 5, 3, 10, 31, 11

11. Sales A car rental company kept the following record of the numbers of miles a rental car was driven. What are the mean, median, and mode of the data?

Monday	420	Tuesday	210
Wednesday	310	Thursday	310
Friday	480	Saturday	160

12. Demographics A study was done on families having six children. The table shows the numbers of families in the study with the indicated numbers of girls. Find the mode of the data. Then use the *mean* and *median* features of a graphing utility to find the mean and median of the data.

Number of girls	0	1	2	3	4	5	6
Frequency	1	22	48	54	58	14	3

Spreadsheet at LarsonPrecalculus.com

13. Athletics The table shows the bowling scores of a three-member team for a three-game series.

Team member	Game 1	Game 2	Game 3
Jay	141	202	166
Hank	199	195	205
Buck	182	231	215

Spreadsheet at LarsonPrecalculus.com

(a) Find the mean for each team member.

(b) Find the mean and median of the nine scores.

(c) Which measure of central tendency best describes the nine scores?

14. Sales The selling prices of 12 new homes built in one subdivision are listed.

\$525,000	\$375,000	\$425,000	\$550,000
\$385,000	\$500,000	\$550,000	\$425,000
\$475,000	\$500,000	\$350,000	\$450,000

(a) Find the mean, mode, and median of the prices.

(b) Which measure of central tendency best describes the prices? Explain.

15. Education An English professor records the following scores for a 100-point exam.

99, 64, 80, 77, 59, 72, 87, 79, 92, 88, 90, 42, 20, 89, 42, 100, 98, 84, 78, 91

Which measure of central tendency best describes these test scores?

16. Sales A salesperson sold eight pairs of men's brown dress shoes. The sizes of the eight pairs were

$10\frac{1}{2}$, 8, 12, $10\frac{1}{2}$, 10, $9\frac{1}{2}$, 11, and $10\frac{1}{2}$.

Which measure (or measures) of central tendency best describes (describe) the typical shoe size for these data?

Finding Mean and Standard Deviation **In Exercises 17 and 18, line plots of data sets are given. Determine the mean and standard deviation of each data set.**

17. (a)
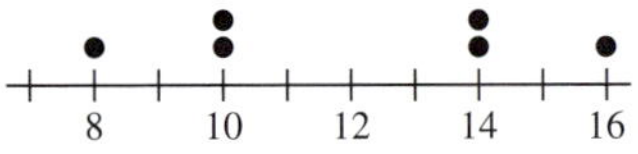

(b)
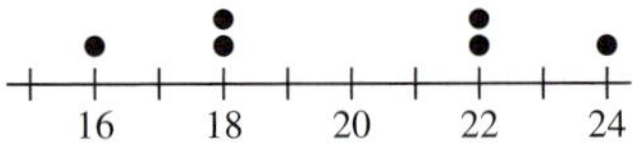

(c)
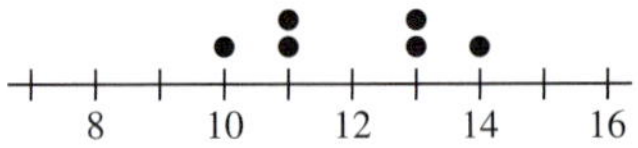

(d)
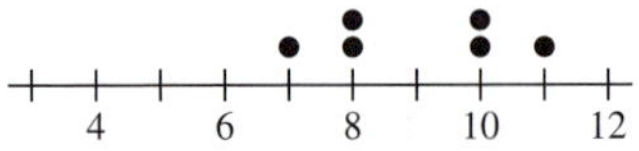

18. (a)
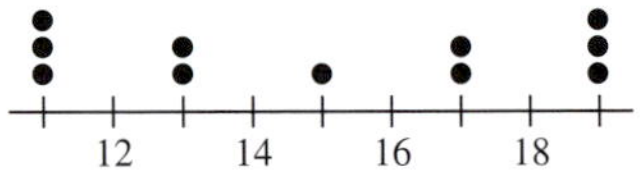

(b)
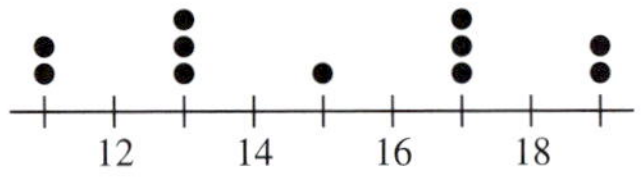

(c)
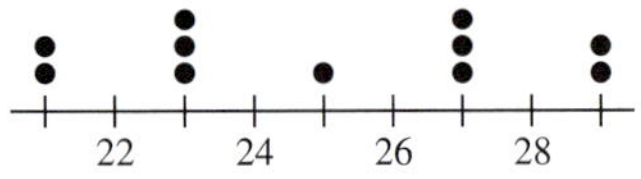

(d)
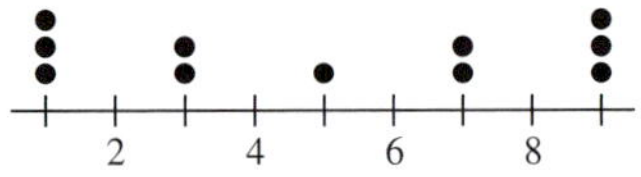

Finding Mean, Variance, and Standard Deviation **In Exercises 19–26, find the mean ($\bar{x}$), variance (v), and standard deviation (σ) of the data set. Use the *one-variable statistics* feature of a graphing utility to verify your answer.**

19. 4, 10, 8, 2

20. 3, 15, 6, 9, 2

21. 1, 1, 1, 5, 5, 5

22. 2, 2, 2, 2, 2, 2

23. 0, 1, 1, 2, 2, 2, 3, 3, 4

24. 1, 2, 3, 4, 5, 6, 7

25. 49, 62, 40, 29, 32, 70

26. 1.5, 0.4, 2.1, 0.7, 0.8

Using the Alternative Formula **In Exercises 27–30, use the alternative formula to find the standard deviation of the data set.**

27. 2, 4, 6, 6, 13, 5

28. 8.1, 6.9, 3.7, 4.2, 6.1

29. 246, 336, 473, 167, 219

30. 9.0, 7.5, 3.3, 7.4, 6.0

31. Athletics The total numbers of points scored during the 2017-2018 NBA regular season by the top 250 point scorers had a mean of 889 and a standard deviation of 383. Use Chebychev's Theorem to determine the intervals containing at least $\frac{3}{4}$ and at least $\frac{8}{9}$ of the point totals. (*Source:* National Basketball Association)

32. ***Why you should learn it*** *(p. A35)* The following data represent the average prices of gold (in dollars per ounce) for the years 1994 through 2017. Use a graphing utility to find the mean, variance, and standard deviation of the data. What percent of the data lie within one standard deviation of the mean? *(Source:* Kitco Precious Metals, Inc.)

384	384	388	331	294	279
279	271	310	363	410	445
603	695	872	972	1225	1572
1669	1411	1266	1160	1251	1257

Sketching a Box-and-Whisker Plot In Exercises 33–36, (a) find the lower and upper quartiles of the data and (b) sketch a box-and-whisker plot for the data without using a graphing utility.

33. 23, 15, 14, 23, 13, 14, 13, 20, 12

34. 11, 10, 11, 14, 17, 16, 14, 11, 8, 14, 20

35. 19, 12, 14, 9, 14, 15, 17, 13, 19, 11, 10, 19

36. 25, 20, 22, 28, 24, 28, 25, 19, 27, 29, 28, 21

Creating a Box-and-Whisker Plot In Exercises 37–40, use a graphing utility to create a box-and-whisker plot for the data.

37. 19, 12, 14, 9, 14, 15, 17, 13, 19, 11, 10, 19

38. 9, 5, 5, 5, 6, 5, 4, 12, 7, 10, 7, 11, 8, 9, 9

39. 20.1, 43.4, 34.9, 23.9, 33.5, 24.1, 22.5, 42.4, 25.7, 17.4, 23.8, 33.3, 17.3, 36.4, 21.8

40. 78.4, 76.3, 107.5, 78.5, 93.2, 90.3, 77.8, 37.1, 97.1, 75.5, 58.8, 65.6

Focusing on Concepts

41. Exploration Compare your answers in Exercise 5 with those in Exercise 7. Which of the measures of central tendency is sensitive to extreme measurements? Explain your reasoning.

42. Exploration

(a) Add 6 to each number in Exercise 8 and find the mean, median, and mode of the revised data set. How are the measures of central tendency changed?

(b) When a constant k is added to each number in a set of data, how will the measures of central tendency change?

43. Think About It Construct a collection of numbers that has a mean of 6, a median of 4, and a mode of 4. If this is not possible, explain why.

44. Think About It Construct a collection of numbers that has a mean of 6, a median of 6, and a mode of 4. If this is not possible, explain why.

45. Think About It Without calculating the standard deviation, explain why the set

$\{4, 4, 20, 20\}$

has a standard deviation of 8.

46. Think About It When the standard deviation of a set of numbers is 0, what does this imply about the set?

47. Education An instructor adds five points to each student's exam score. Will this change the mean and standard deviation of the exam scores? Explain.

48. Exploration

(a) Multiply each number in Exercise 24 by 3 and find the mean, variance, and standard deviation of the revised data set. Compare these measures with what you found in Exercise 24. How did the measures change?

(b) When each number in a data set is multiplied by a constant k, how will the mean, variance, and standard deviation change?

49. Think About It The histograms represent the test scores of two classes of a college course in mathematics. Which histogram has the smaller standard deviation? Explain your reasoning.

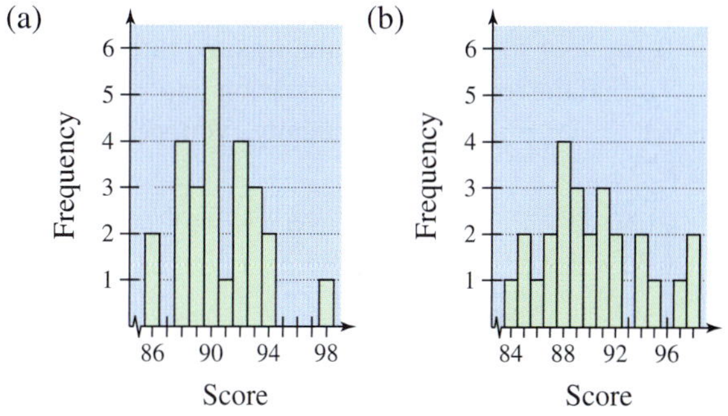

50. Manufacturing A company redesigns a product to last longer. The lifetimes (in months) of 20 units with the original design and 20 units with the new design are listed below. Create a box-and-whisker plot for each set of data, and then interpret your results.

Original Design

15.1	78.3	56.3	68.9
30.6	27.2	12.5	42.7
72.7	20.2	53.0	13.5
11.0	18.4	85.2	10.8
38.3	85.1	10.0	12.6

New Design

55.8	71.5	25.6	19.0
23.1	37.2	60.0	35.3
18.9	80.5	46.7	31.1
67.9	23.5	99.5	54.0
23.2	45.5	24.8	87.8

B.3 Least Squares Regression

What you should learn

- Use the sum of squared differences to determine a least squares regression line.
- Find a least squares regression line for a set of data.
- Find a least squares regression parabola for a set of data.

Why you should learn it

The method of least squares provides a way of creating a mathematical model for a set of data, which can then be analyzed.

In many of the examples and exercises in this text, you have been asked to use the *regression* feature of a graphing utility to find mathematical models for sets of data. The *regression* feature of a graphing utility uses the **method of least squares** to find a mathematical model for a set of data. As a measure of how well a model fits a set of data points

$$\{(x_1, y_1), (x_2, y_2), (x_3, y_3), . . . , (x_n, y_n)\}$$

you can add the squares of the differences between the actual y-values and the values given by the model to obtain the **sum of the squared differences.** For instance, the table shows several values of x and y. The table also shows the values of a linear model $y^* = 0.21x - 2.0$ for each x-value. The sum of the squared differences for the model is 28.0178.

x	70	72	75	76	77	78	80
y	10.9	11.2	11.7	12	12.2	12.3	12.6
y^*	12.70	13.12	13.75	13.96	14.17	14.38	14.80
$(y - y^*)^2$	3.24	3.6864	4.2025	3.8416	3.8809	4.3264	4.84

The model that has the *least* sum of the squared differences is the **least squares regression** line for the data. The least squares regression line for the data in the table is $y = 0.18x - 1.5$. The sum of the squared differences is 0.4032.

To find the least squares regression line $y = ax + b$ for the points $\{(x_1, y_1), (x_2, y_2), (x_3, y_3), . . . , (x_n, y_n)\}$ algebraically, you need to solve the following system for a and b.

$$\begin{cases} nb + \left(\sum_{i=1}^{n} x_i\right)a = \sum_{i=1}^{n} y_i \\ \left(\sum_{i=1}^{n} x_i\right)b + \left(\sum_{i=1}^{n} x_i^2\right)a = \sum_{i=1}^{n} x_i y_i \end{cases}$$

In the system,

$$\sum_{i=1}^{n} x_i = x_1 + x_2 + \cdot \cdot \cdot + x_n$$

$$\sum_{i=1}^{n} y_i = y_1 + y_2 + \cdot \cdot \cdot + y_n$$

$$\sum_{i=1}^{n} x_i^2 = x_1^2 + x_2^2 + \cdot \cdot \cdot + x_n^2$$

$$\sum_{i=1}^{n} x_i y_i = x_1 y_1 + x_2 y_2 + \cdot \cdot \cdot + x_n y_n.$$

Technology Tip

Recall from Section 1.7 that when you use the *regression* feature of a graphing utility to find a linear model, the program may output a correlation coefficient, r. When $|r|$ is close to 1, the linear model is a good fit for the data.

EXAMPLE 1 Finding a Least Squares Regression Line

Find the least squares regression line for $(-3, 0)$, $(-1, 1)$, $(0, 2)$, and $(2, 3)$.

Solution

Begin by constructing a table, as shown below.

x	y	xy	x^2
-3	0	0	9
-1	1	-1	1
0	2	0	0
2	3	6	4
$\sum_{i=1}^{n} x_i = -2$	$\sum_{i=1}^{n} y_i = 6$	$\sum_{i=1}^{n} x_i y_i = 5$	$\sum_{i=1}^{n} x_i^2 = 14$

Applying the system for the least squares regression line with $n = 4$ produces

$$\begin{cases} nb + \left(\sum_{i=1}^{n} x_i\right)a = \sum_{i=1}^{n} y_i \\ \left(\sum_{i=1}^{n} x_i\right)b + \left(\sum_{i=1}^{n} x_i^2\right)a = \sum_{i=1}^{n} x_i y_i \end{cases} \Rightarrow \begin{cases} 4b - 2a = 6 \\ -2b + 14a = 5 \end{cases}.$$

Solving this system of equations produces $a = \frac{8}{13}$ and $b = \frac{47}{26}$. So, the least squares regression line is $y = \frac{8}{13}x + \frac{47}{26}$, as shown in Figure B.9.

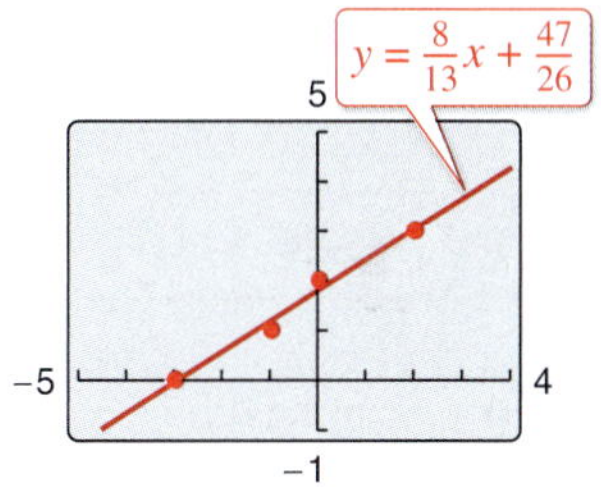

Figure B.9

✓ ***Checkpoint*** *Audio-video solution in English & Spanish at LarsonPrecalculus.com*

Find the least squares regression line for the points $(-1, -1)$, $(0, 0)$, $(1, 2)$, and $(2, 4)$.

The least squares regression parabola $y = ax^2 + bx + c$ for the points

$\{(x_1, y_1), (x_2, y_2), (x_3, y_3), \ldots, (x_n, y_n)\}$

is obtained in a similar manner by solving the following system of three equations in three unknowns for a, b, and c.

$$\begin{cases} nc + \left(\sum_{i=1}^{n} x_i\right)b + \left(\sum_{i=1}^{n} x_i^2\right)a = \sum_{i=1}^{n} y_i \\ \left(\sum_{i=1}^{n} x_i\right)c + \left(\sum_{i=1}^{n} x_i^2\right)b + \left(\sum_{i=1}^{n} x_i^3\right)a = \sum_{i=1}^{n} x_i y_i \\ \left(\sum_{i=1}^{n} x_i^2\right)c + \left(\sum_{i=1}^{n} x_i^3\right)b + \left(\sum_{i=1}^{n} x_i^4\right)a = \sum_{i=1}^{n} x_i^2 y_i \end{cases}$$

B.3 Exercises

See *CalcChat.com* for tutorial help and worked-out solutions to odd-numbered exercises.
For instructions on how to use a graphing utility, see Appendix A.

Finding a Least Squares Regression Line **In Exercises 1–4, find the least squares regression line for the points. Verify your answer with a graphing utility.**

1. $(-4, 1)$, $(-3, 3)$, $(-2, 4)$, $(-1, 6)$

2. $(0, -1)$, $(2, 0)$, $(4, 3)$, $(6, 5)$

3. $(-3, 1)$, $(-1, 2)$, $(1, 2)$, $(4, 3)$

4. $(0, -1)$, $(2, 1)$, $(3, 2)$, $(5, 3)$

Answers to Odd-Numbered Exercises and Tests

Chapter 1

Section 1.1 (page 11)

1. (a) iii (b) i (c) v (d) ii (e) iv **3.** parallel

5. 0 **7.** (a) L_2 (b) L_3 (c) L_1 **9.** $\frac{3}{2}$

11.

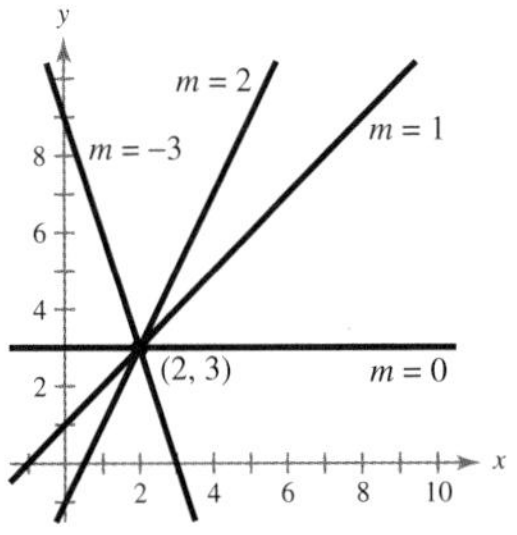

13. $m = -\frac{3}{2}$ **15.** $m = 2$ **17.** $m = 0$

19. m is undefined. **21.** $m = 0.15$

23. $m = -\frac{5}{2}$

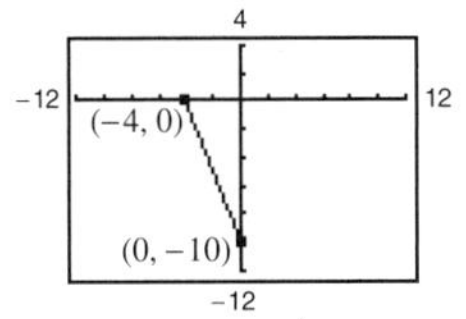

25. m is undefined.

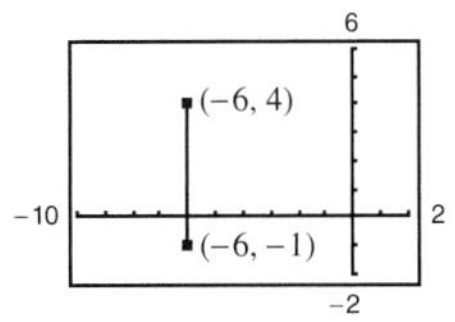

27. *Sample answer:* (0, 1), (3, 1), (−1, 1)

29. *Sample answer:* (−1, −7), (−2, −5), (−5, 1)

31. *Sample answer:* (3, −4), (5, −3), (9, −1)

33. $y = 3x - 2$

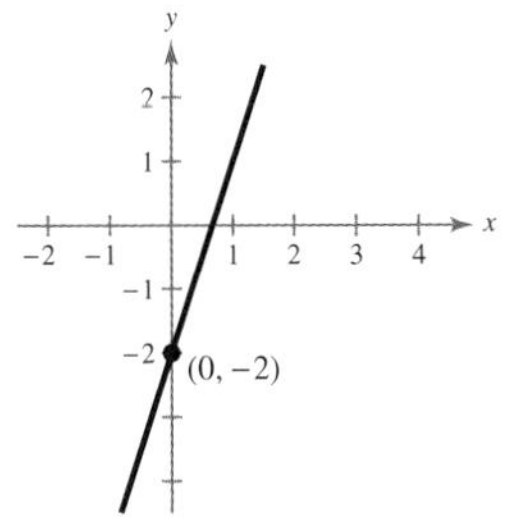

35. $y = -\frac{1}{2}x - 2$

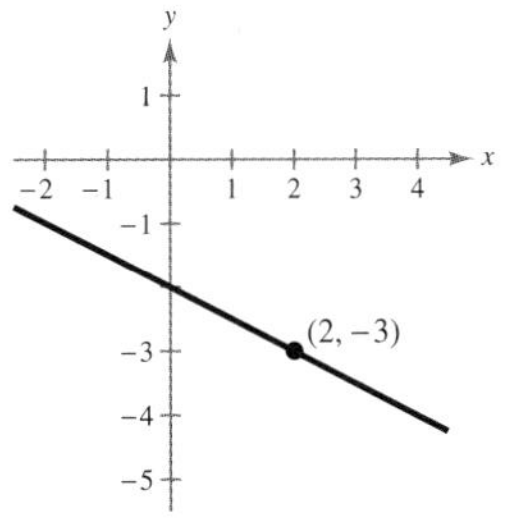

37. $y = \frac{3}{2}$

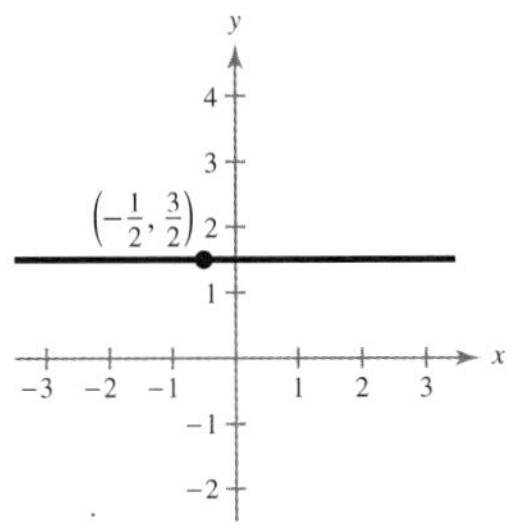

39. $x = 6$

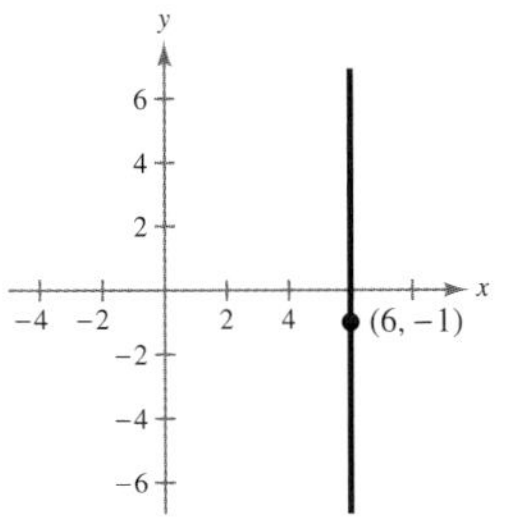

41. $y = 0.09x + 2.94$; $4.83 million

43. $m = \frac{2}{3}$; y-intercept: $(0, -3)$; a line that rises from left to right

45. $m = \frac{2}{5}$; y-intercept: $(0, 2)$; a line that rises from left to right

47. Slope is undefined; no y-intercept; a vertical line at $x = -6$

49. (a) $m = 5$; y-intercept: $(0, 3)$
(b)

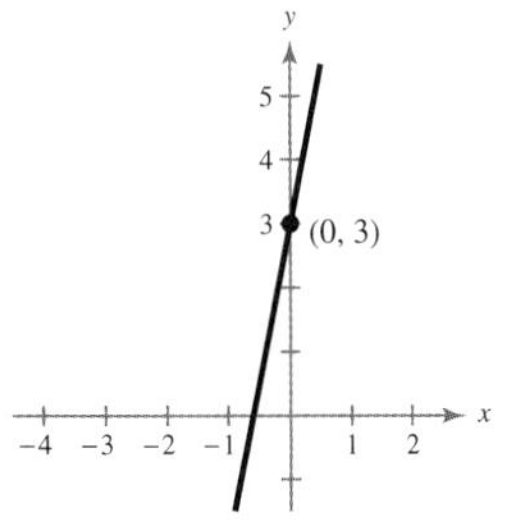

51. (a) Slope is undefined; there is no y-intercept.
(b)

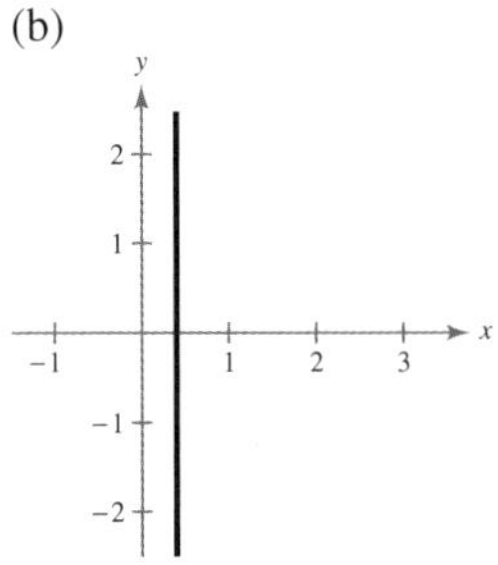

53. (a) $m = 0$; y-intercept: $\left(0, -\frac{5}{3}\right)$
(b)

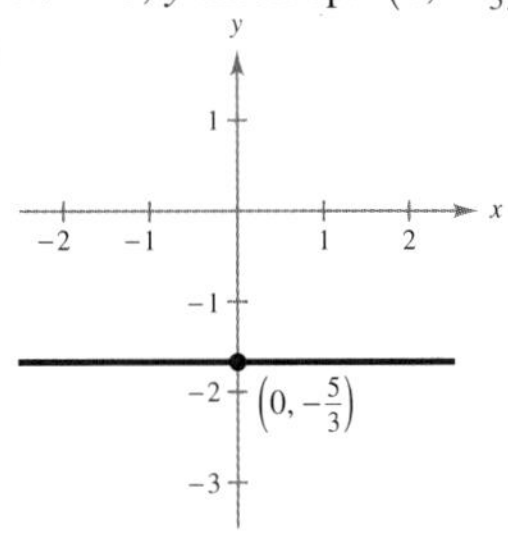

55. $y = 2x - 5$

57. $y = -\frac{3}{5}x + 2$

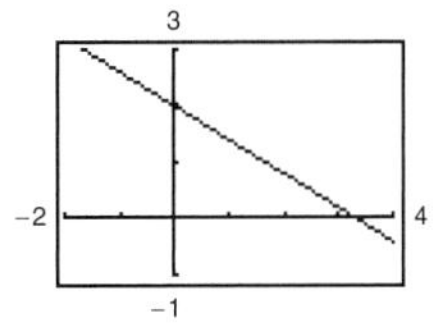

59. Cannot be written in slope-intercept form because the slope is undefined; $x + 8 = 0$

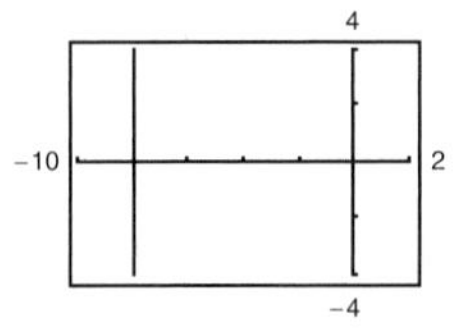

61. $y = -\frac{1}{2}x + \frac{3}{2}$

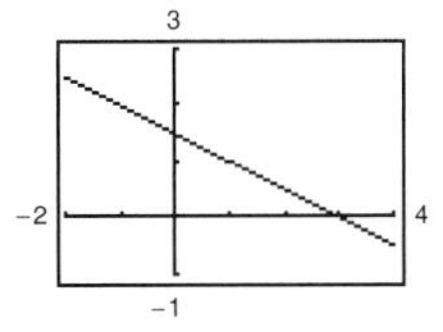

63. $y = \frac{2}{5}x + \frac{1}{5}$

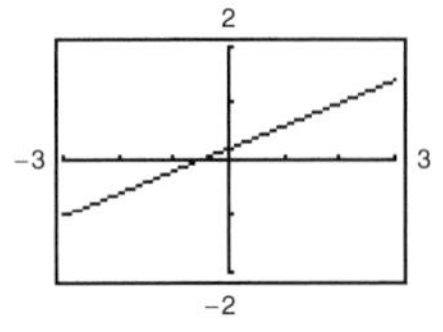

65.

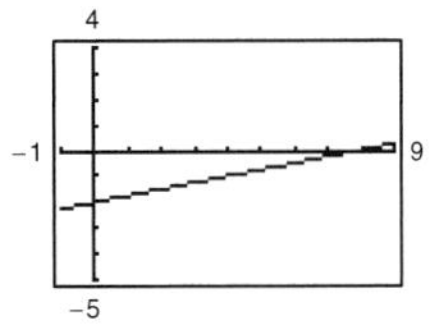

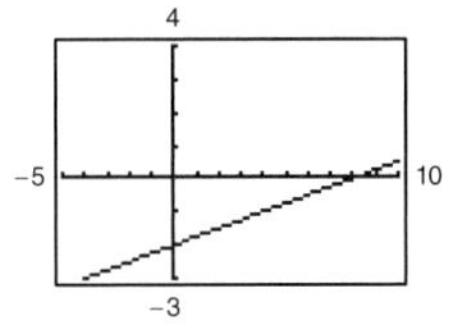

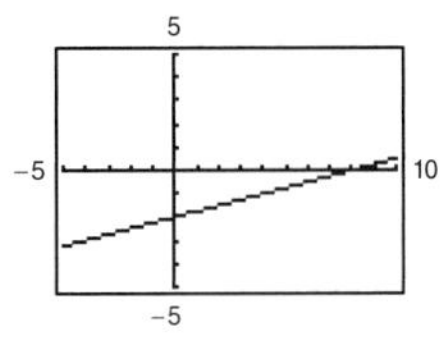

The first and second graphs do not give an accurate view of the slope. The third graph is best because it uses a square setting.

67. Perpendicular **69.** Parallel

71. (a) $y = 2x - 3$ (b) $y = -\frac{1}{2}x + 2$

73. (a) $y = -\frac{3}{4}x + \frac{3}{8}$ (b) $y = \frac{4}{3}x + \frac{127}{72}$

75. (a) $x = 3$ (b) $y = -2$

77. (a) $y = 1$ (b) $x = -5$

79. (a) $y = -\frac{6}{5}x - 6.08$ (b) $y = \frac{5}{6}x + 1.85$

81. $y = 2x + 1$ **83.** $y = -\frac{1}{2}x + 1$

85. The lines $y = \frac{1}{4}x$ and $y = -4x$ are perpendicular.

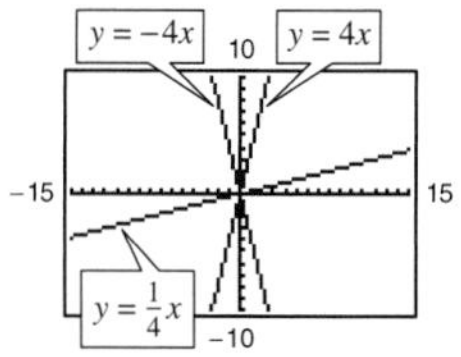

87. The lines $y = -\frac{1}{2}x$ and $y = -\frac{1}{2}x + 3$ are parallel. Both are perpendicular to $y = 2x - 4$.

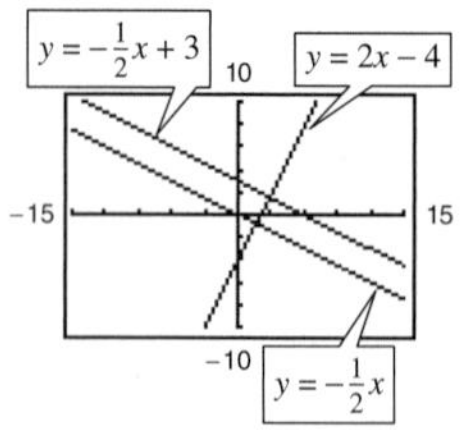

89. 12 ft

91. (a) The greatest increase was from 2011 to 2012, and the greatest decrease was from 2015 to 2016.
(b) $y = 4.91x + 14$
(c) There is an increase of about \$4.91 billion per year.
(d) \$68.01 billion; Answers will vary.

93. $V = 125t + 290$ **95.** $V = -2000t + 56{,}400$

97. (a) $V = 10{,}995 - 1011.5t$
(b)

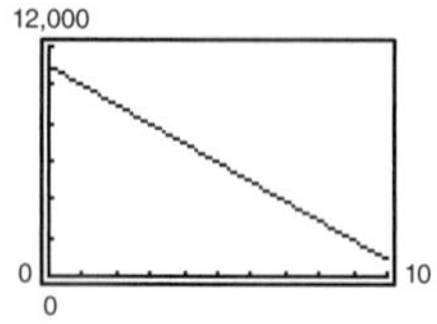

t	0	1	2	3	4
V	10,995	9983.5	8972	7960.5	6949

t	5	6	7	8	9	10
V	5937.5	4926	3914.5	2903	1891.5	880

99. (a) $C = 59t + 84{,}400$ (b) $R = 150t$
(c) $P = 91t - 84{,}400$ (d) $t \approx 927.5$ h

101. (a) Increase of about 398 students per year
(b) About 25,560 students, about 27,550 students, about 29,540 students
(c) $y = 398x + 23{,}570$, where $x = 0$ corresponds to 2000; $m = 398$; The slope determines the average increase in enrollment.

103. False; The slopes $\left(\frac{2}{7}\text{ and } -\frac{11}{7}\right)$ are not equal.

105.

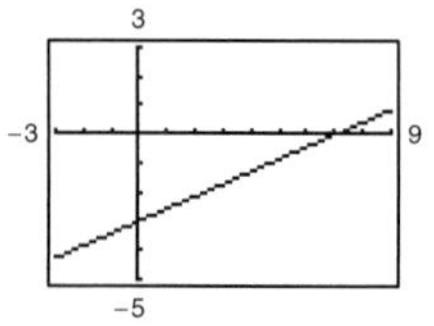

a and *b* represent the *x*- and *y*-intercepts.

107.

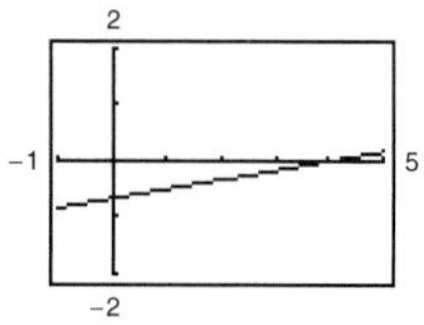

a and *b* represent the *x*- and *y*-intercepts.

109. $9x + 2y - 18 = 0$ **111.** $12x + 3y + 2 = 0$

113. (a) Yes (b) No (c) No (d) No

115. (a) No (b) Yes

117. The slope and y-intercept are switched; The slope is 2, and the y-intercept is $(0, -3)$.

119. No; *Sample answer:* The line $y = 2$ does not have an x-intercept.

121. No **123.** No **125.** $(x - 9)(x + 3)$

127. $(2x - 5)(x + 8)$ **129.** Answers will vary.

Section 1.2 (page 24)

1. domain, range, function **3.** No **5.** No

7. Yes; Each element of the domain is assigned to exactly one element of the range.

9. No; Both elements in the domain are assigned to more than one element in the range. The National Football Conference, is assigned to the Giants, the Saints, and the Seahawks, and the American Football Conference is also assigned to three elements in the range, the Patriots, the Ravens, and the Steelers.

11. No; The input 7 is matched with two outputs.

13. No; The inputs 1, 2, and 3 are each matched with two outputs.

15. (a) Function; Each element in A corresponds to exactly one element in B.
(b) Not a function; The element 1 in A corresponds to two elements, -2 and 1, in B.
(c) Not a function from A to B; It is a function from B to A.

17. Yes; Each input value (year) is matched with exactly one output value (average price).

19. Not a function **21.** Function **23.** Function

25. Not a function **27.** Function **29.** Not a function

31. (a) 7 (b) -11 (c) $3t + 7$

33. (a) 15 (b) $4t^2 - 19t + 27$ (c) $4t^2 - 3t - 10$

35. (a) 1 (b) 2.5 (c) $3 - 2|x|$

37. (a) Undefined (b) $-\dfrac{1}{5}$ (c) $\dfrac{1}{y^2 + 6y}$

39. (a) 1 (b) -1 (c) $\dfrac{|t|}{t}$

41. (a) -1 (b) 2 (c) 6 **43.** (a) 6 (b) 3 (c) 10

45. (a) 0 (b) 4 (c) 5

47. $\{(-2, 9), (-1, 4), (0, 1), (1, 0), (2, 1)\}$

49. $\{(-2, 4), (-1, 3), (0, 2), (1, 3), (2, 4)\}$

51.

t	-5	-4	-3	-2	-1
$h(t)$	1	$\frac{1}{2}$	0	$\frac{1}{2}$	1

53. 5 **55.** $\frac{4}{9}$ **57.** All real numbers x

59. All real numbers t except $t = 0$

61. All real numbers y such that $y \geq -6$

63. All real numbers x except $x = 0, -2$

65. All real numbers y such that $y > 10$

67.

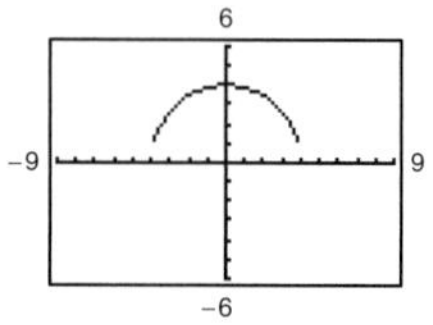

Domain: $-4 \leq x \leq 4$
Range: $0 \leq y \leq 4$

69.

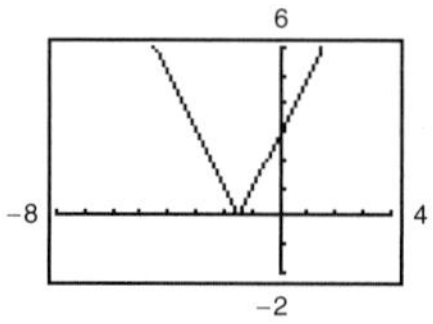

Domain: all real numbers
Range: $y \geq 0$

71.

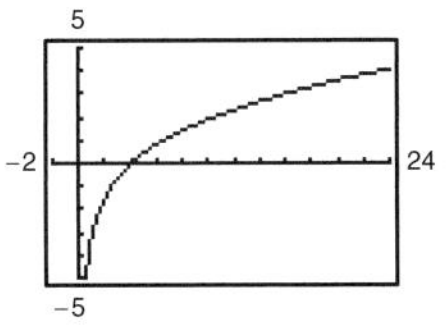

Domain: $x \geq 0$

Range: all real numbers

73. $A = \dfrac{C^2}{4\pi}$

75. (a) $1024\ \text{cm}^3$ (b) Yes

(c) $V = x(24 - 2x)^2$; Yes; $0 < x < 12$

(d)

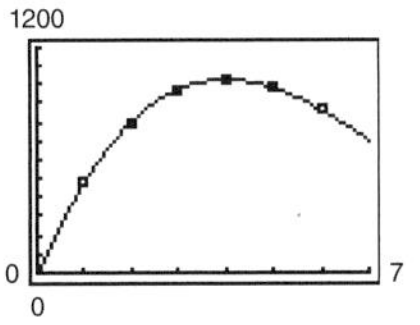

The graph of the equation fits the data points well; Answers will vary.

77. $A = \dfrac{a^2}{2(a-2)},\ a > 2$

79. (a) $C = 68.75x + 248{,}000$ (b) $R = 99.99x$

(c) $P = 31.24x - 248{,}000$

(d) 376,800; \$376,800 profit for 20,000 units

(e) −248,000; \$248,000 loss for 0 units

81. (a) The independent variable is t and represents the year, with $t = 6$ corresponding to 2006. The dependent variable is n and represents the number (in thousands) of structurally deficient bridges.

(b)

t	6	7	8	9	10	11
$n(t)$	75.424	73.961	72.936	72.349	70.9	68.51

t	12	13	14	15	16	17
$n(t)$	66.12	63.73	61.34	58.95	56.56	54.17

(c) The model fits the data well.

(d) *Sample answer:* No; The function may not accurately model other years.

83. Yes; The ball is 6 feet high when it reaches the other child.

85. $2,\ c \neq 0$ **87.** $3x^2 + 3hx + h^2 + 3,\ h \neq 0$

89. False; The range is $[-1, \infty)$.

91. $f(x) = \sqrt{x} + 2$

Domain: $[0, \infty)$

Range: $[2, \infty)$

93. The domain of $f(x)$ includes $x = 1$ and the domain of $g(x)$ does not because you cannot divide by 0. So, the functions do not have the same domain.

95. $\dfrac{12x + 20}{x + 2}$ **97.** $\dfrac{2x^2}{3},\ x \neq -2, 0$

Section 1.3 (page 37)

1. decreasing **3.** $[1, 4]$ **5.** Relative maximum

7. Domain: $(-\infty, \infty)$; range: $(-\infty, 1]$

(a) 0 (b) 1 (c) 0 (d) −3

9. Domain: $(-\infty, \infty)$; range: $(-2, \infty)$

(a) 0 (b) 1 (c) 2 (d) 3

11.

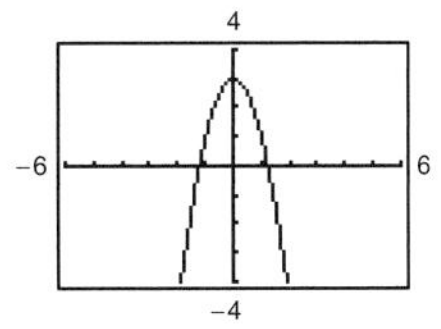

Domain: $(-\infty, \infty)$

Range: $(-\infty, 3]$

13.

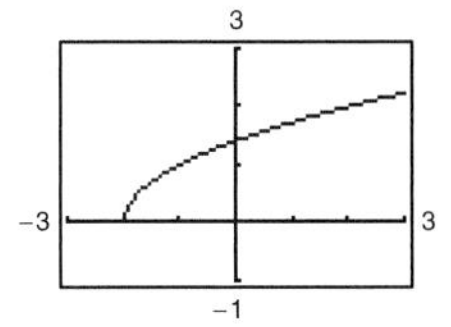

Domain: $[-2, \infty)$

Range: $[0, \infty)$

15.

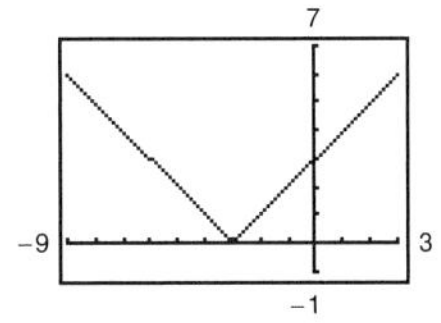

Domain: $(-\infty, \infty)$

Range: $[0, \infty)$

17. (a) $(-\infty, \infty)$ (b) $(-\infty, \infty)$ (c) 2

(d) 2 (e) 1; $(1, 1)$ (f) 5

19. (a) $(-\infty, \infty)$ (b) $[-2, \infty)$ (c) $-1, 3$

(d) −1 (e) −2; $(1, -2)$ (f) 2

21. x-intercepts

23. Function; Graph the given function over the window shown in the figure.

25. Not a function; Solve for y and graph the two resulting functions.

27. Increasing on $(-\infty, \infty)$

29. Decreasing on $(-\infty, -1)$; increasing on $(1, \infty)$

31. (a)

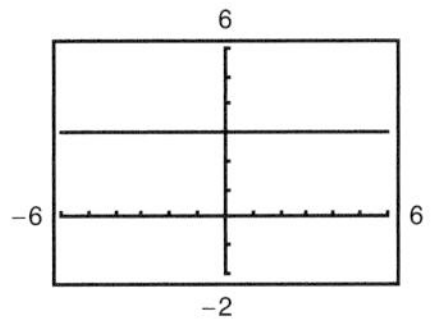

(b) Constant: $(-\infty, \infty)$

33. (a)

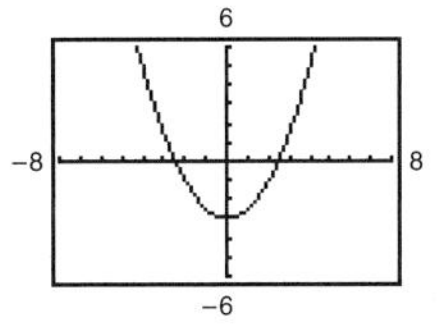

(b) Decreasing on $(-\infty, 0)$

Increasing on $(0, \infty)$

35. (a)

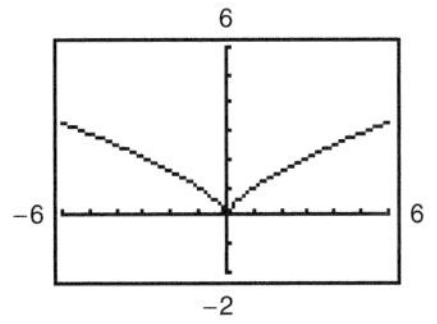

(b) Decreasing on $(-\infty, 0)$

Increasing on $(0, \infty)$

37. (a)

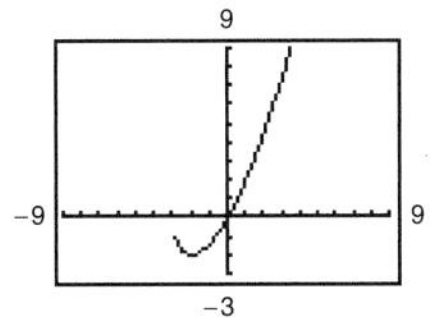

(b) Increasing on $(-2, \infty)$

Decreasing on $(-3, -2)$

39. (a)

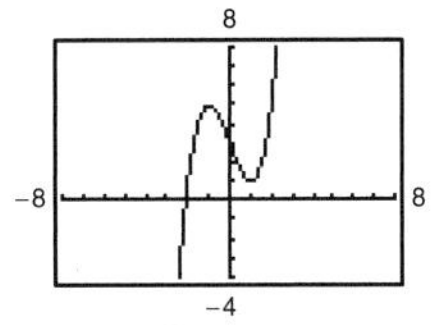

(b) Increasing on $(-\infty, -1)$, $(1, \infty)$

Decreasing on $(-1, 1)$

41. (a)

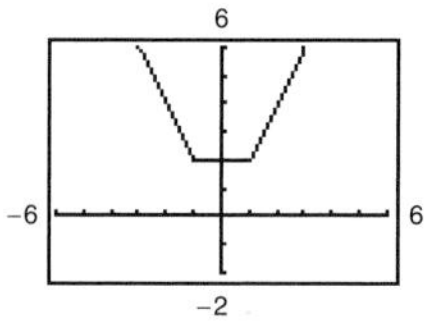

(b) Decreasing on $(-\infty, -1)$
Constant on $(-1, 1)$
Increasing on $(1, \infty)$

43.

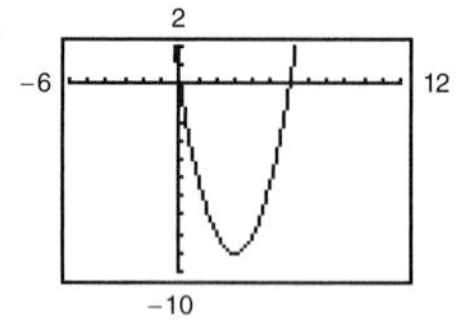

Relative minimum: $(3, -9)$

45.

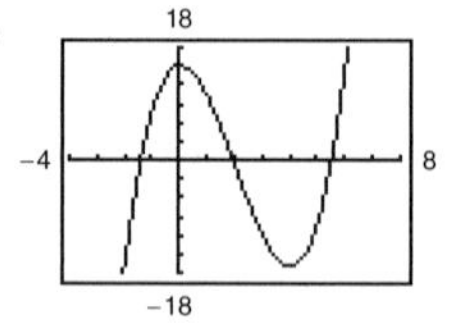

Relative minimum: $(4, -17)$
Relative maximum: $(0, 15)$

47.

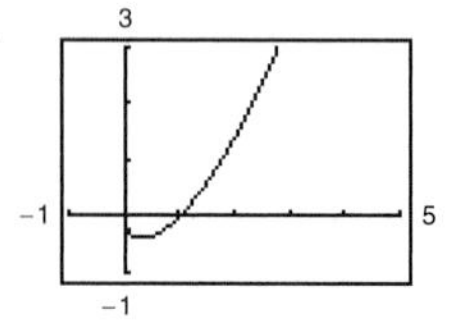

Relative minimum:
about $(0.33, -0.38)$

49.

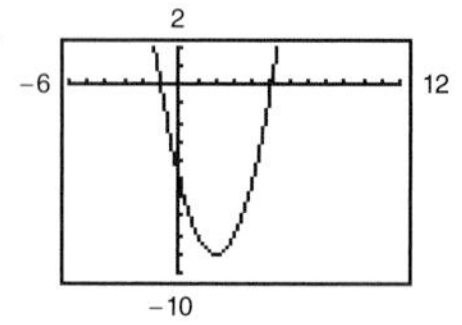

Relative minimum: $(2, -9)$

51.

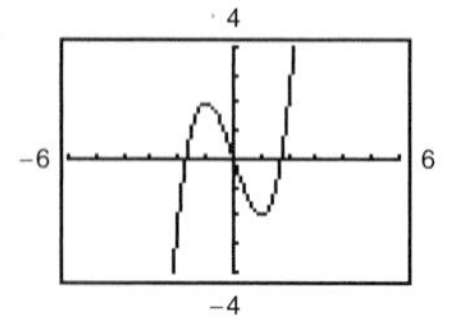

Relative minimum: $(1, -2)$
Relative maximum: $(-1, 2)$

53.

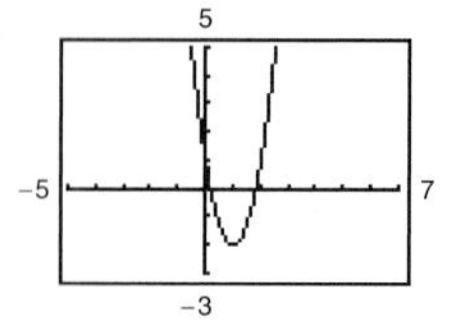

Relative minimum: $(1, -2)$

55.

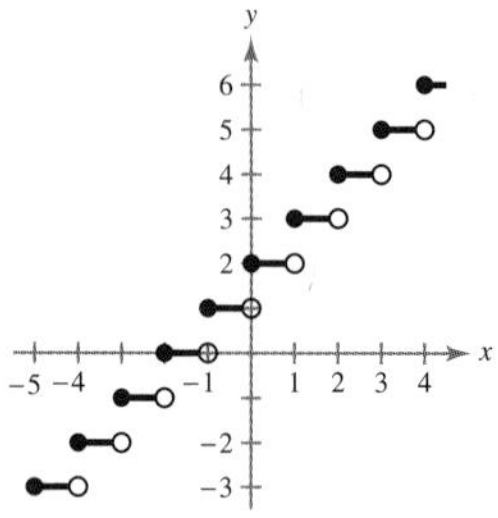

57.

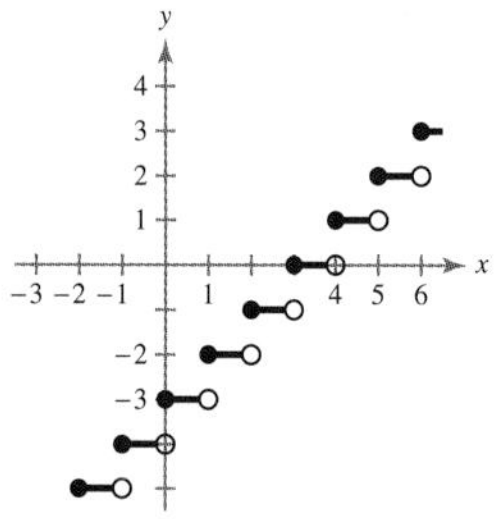

59.

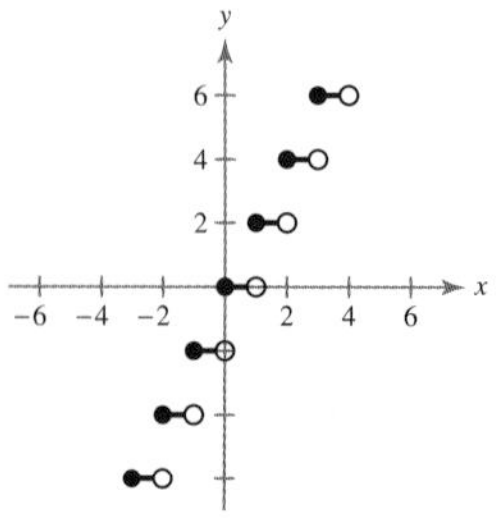

61.

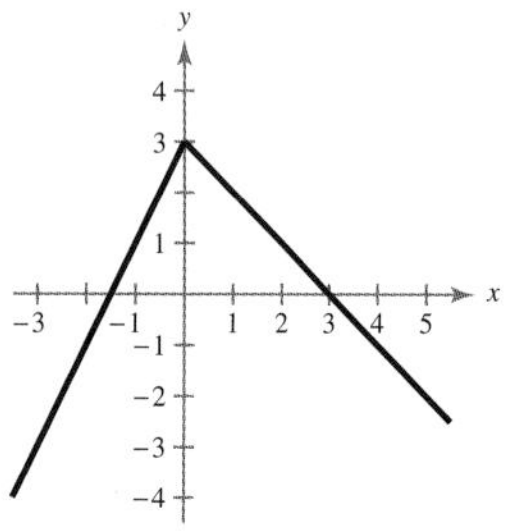

63.

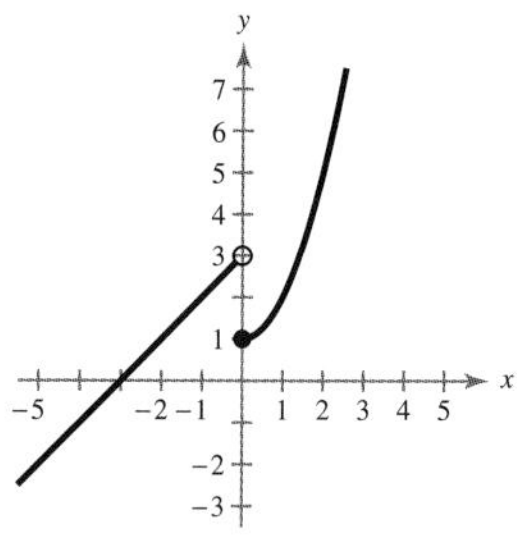

65.

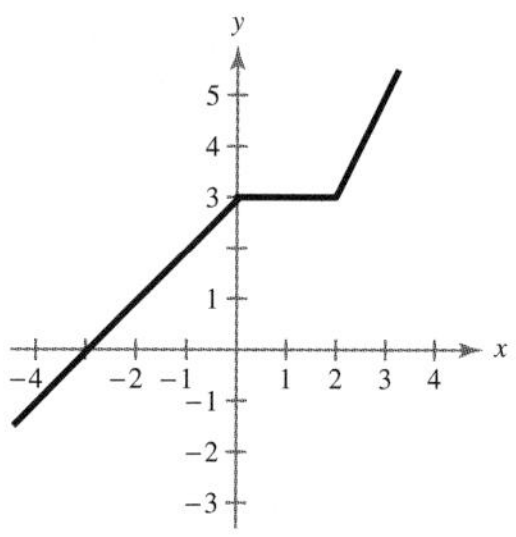

67. Even

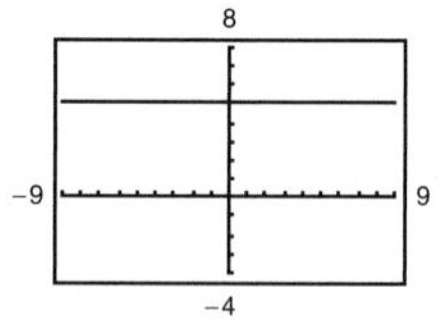

69. Odd

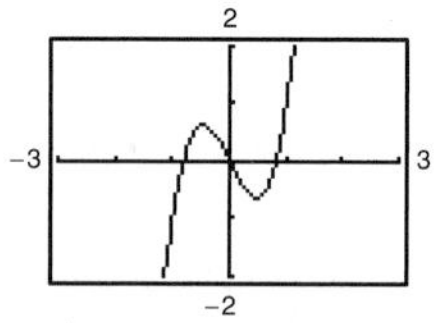

71. Even

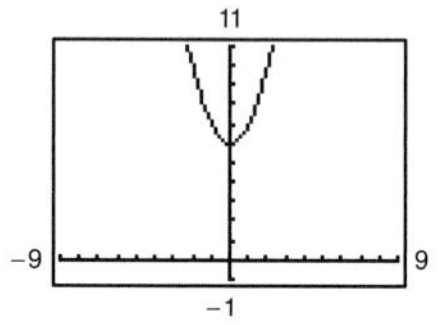

73. Even

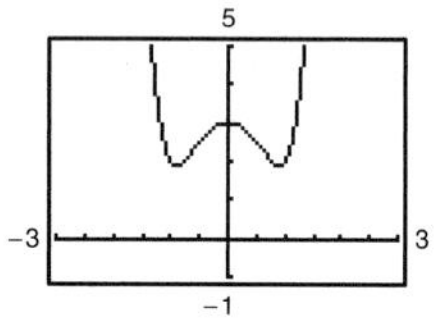

75. Neither

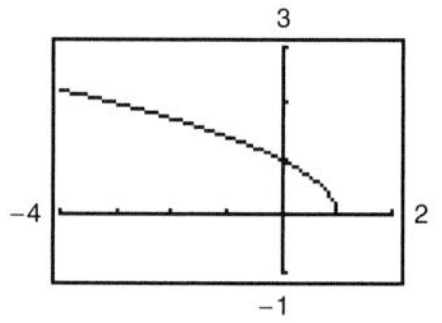

77. Neither

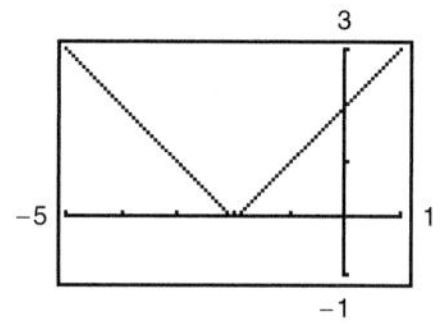

79. (a) $\left(-\frac{3}{2}, 4\right)$ (b) $\left(-\frac{3}{2}, -4\right)$

81. (a) $(2, -9)$ (b) $(2, 9)$

83. (a) $(-x, -y)$ (b) $(-x, y)$ **85.** (a)–(c) Neither

87. (a)–(c) Neither **89.** (a)–(c) Odd

91. (a)–(c) Odd **93.** (a)–(c) Even

95. (a) C_2

(b)

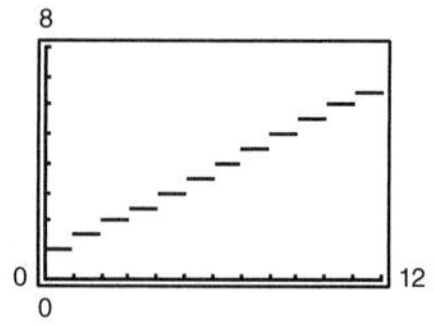

$4.50

97. $h = -x^2 + 4x - 3, 1 \le x \le 3$

99. (a)

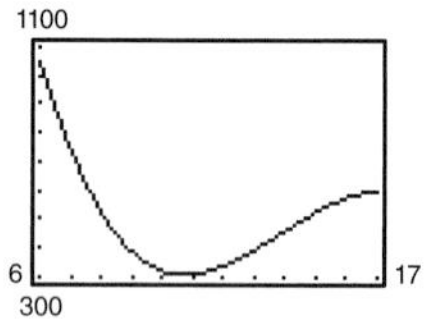

(b) Increasing from 2011 to 2017; Decreasing from 2006 to 2011
(c) About 310,000 in 2011

101. False; Counterexample: $f(x) = \sqrt{1 + x^2}$

103. c **104.** d **105.** b **106.** e **107.** a **108.** f

109. The negative symbol should be divided out of each term, which yields $f(-x) = -(2x^3 + 5)$. So, the function is neither even nor odd.

111. (a) Even. g is a reflection in the x-axis.
(b) Even. g is a reflection in the y-axis.
(c) Even. g is a vertical shift downward.
(d) Neither. g is shifted to the left and reflected in the x-axis.

113. Proof **115.** Terms: $-2x^2, 11x, 3$; coefficients: $-2, 11$

117. Terms: $\frac{x}{3}, -5x^2, x^3$; coefficients: $\frac{1}{3}, -5, 1$

119. (a) -17 (b) -17 (c) $-x^2 + 3x + 1$

121. $h + 4, h \neq 0$

Section 1.4 (page 47)

1. Horizontal shifts, vertical shifts, reflections

3. $-f(x), f(-x)$

5. (a)

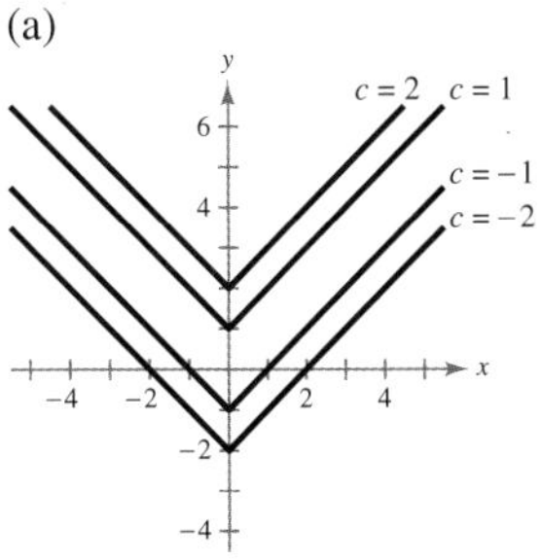

(b)

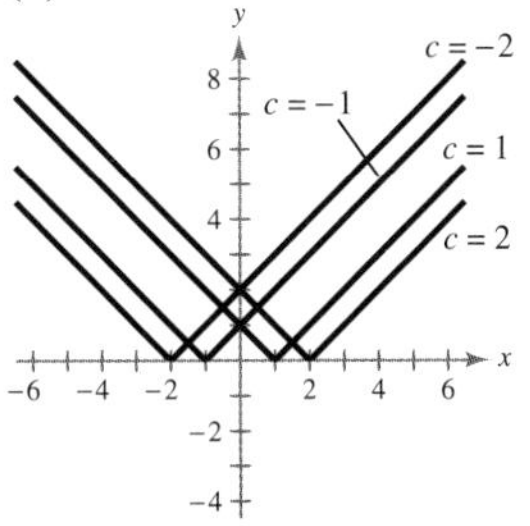

7.

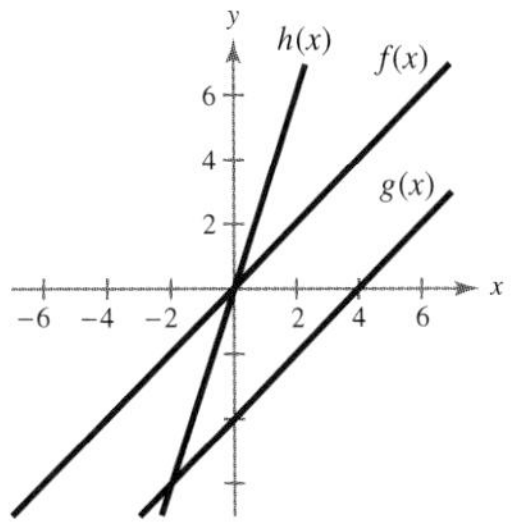

9.

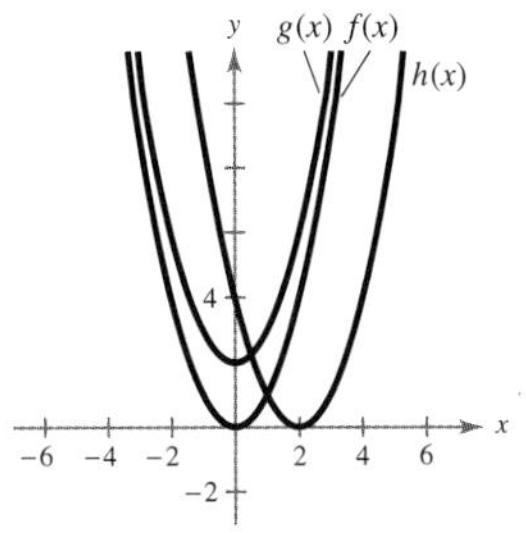

11.

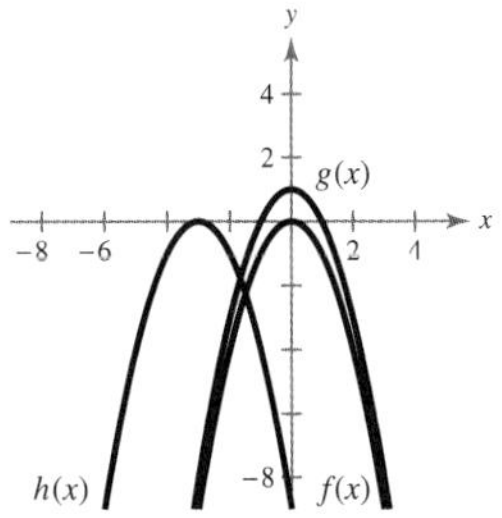

13.

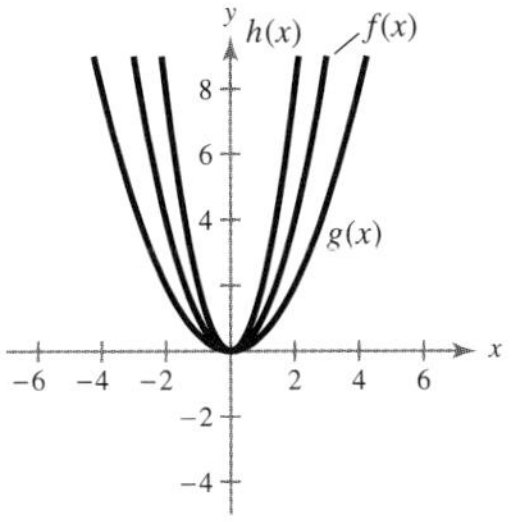

15.

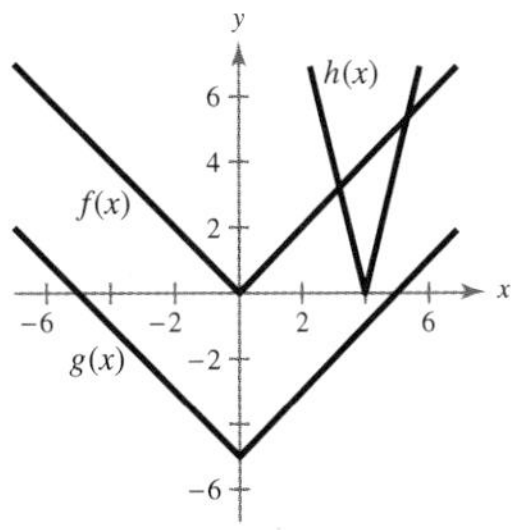

17.

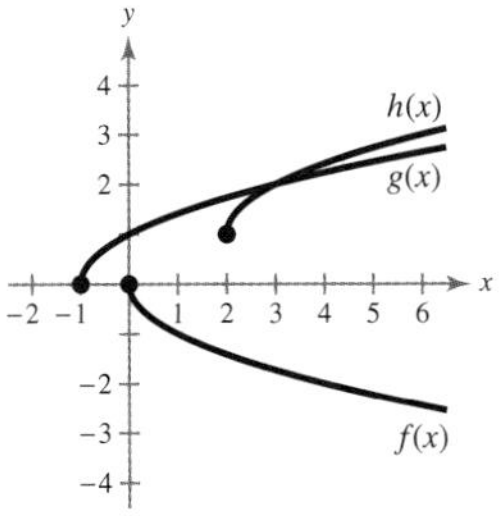

19.

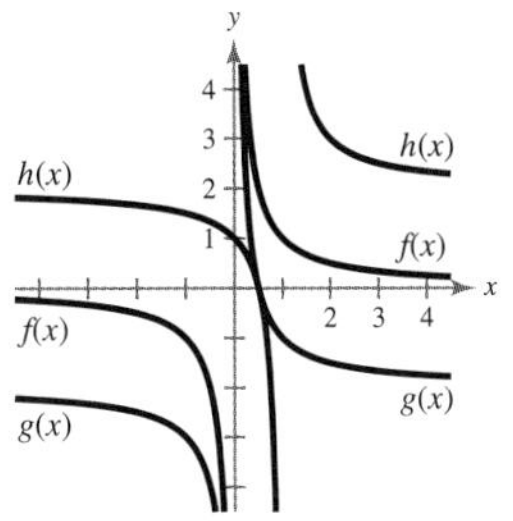

21. (a)

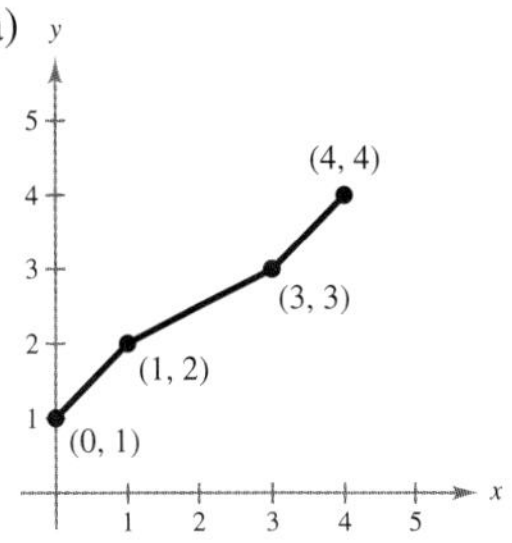

(b)

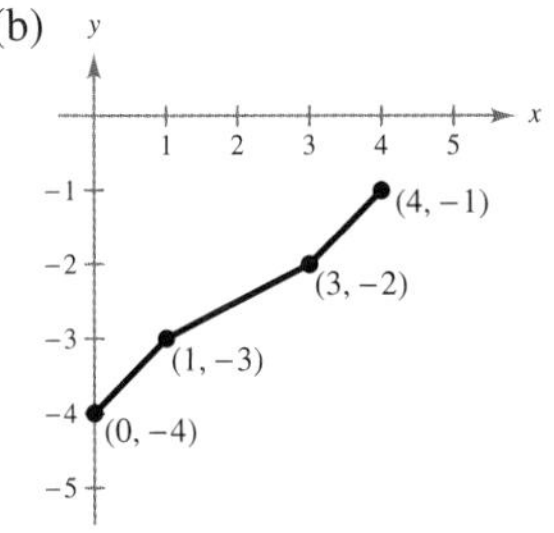

(c)

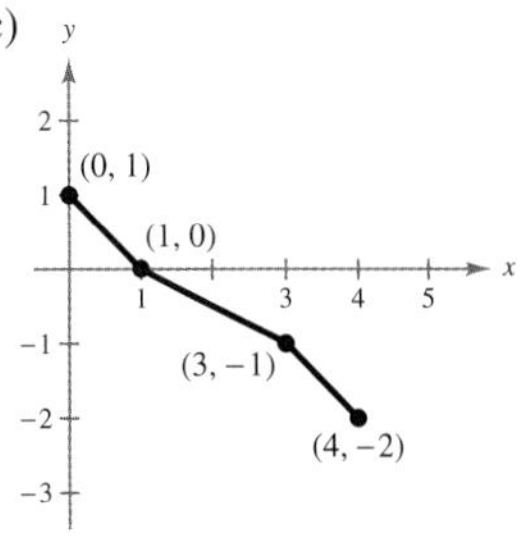

(d)

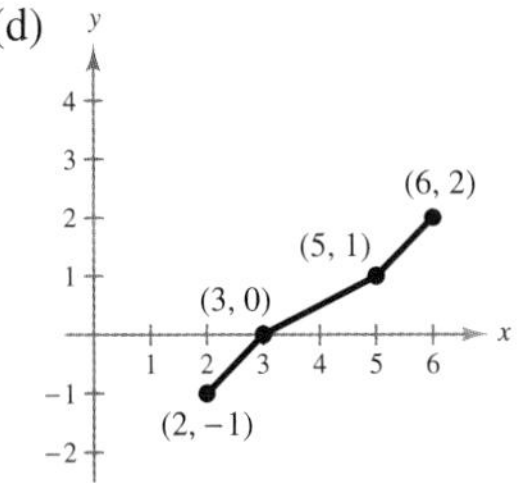

(e)

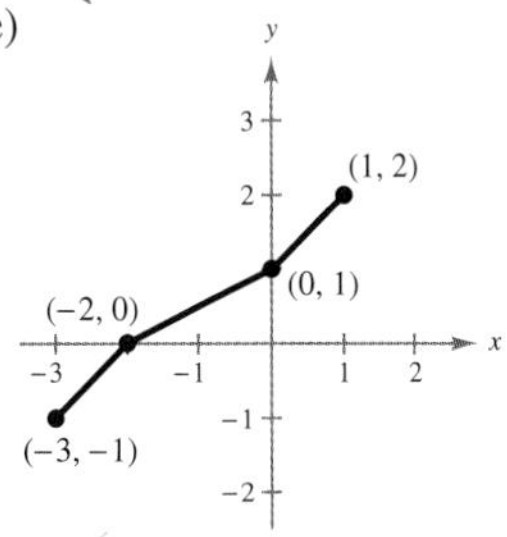

(f)

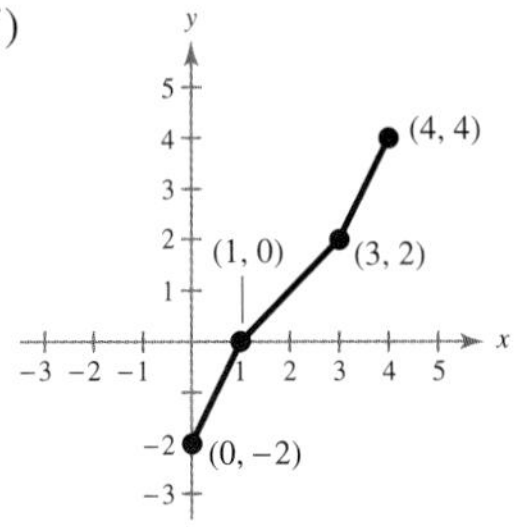

(g)

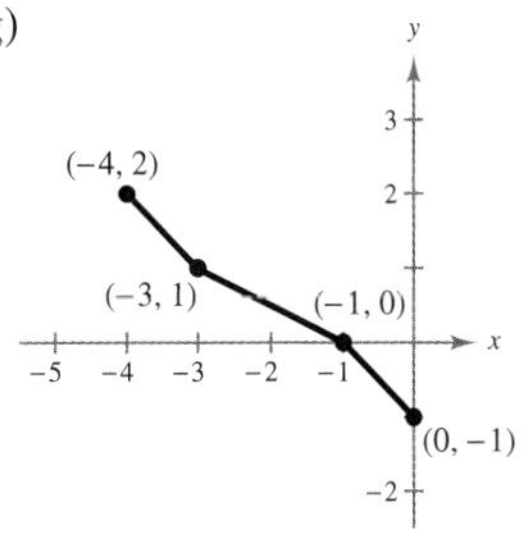

(h)

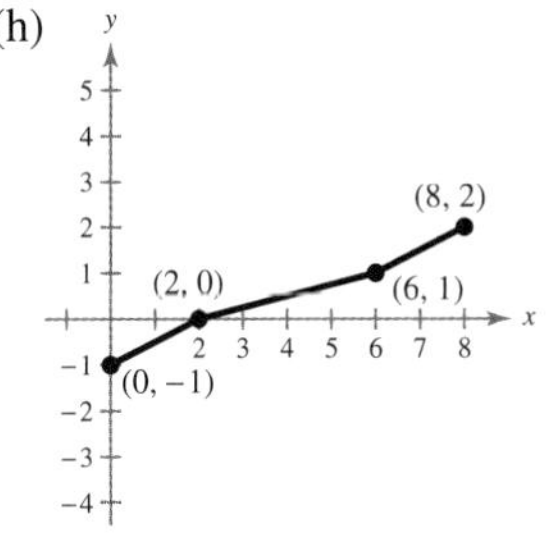

23. Vertical shift four units upward

25. Horizontal shift four units to the right

27. Vertical shift 12 units downward

29. Horizontal shift two units to the right of $y = x^3$; $y = (x - 2)^3$

31. Vertical shift three units downward of $y = x^2$; $y = x^2 - 3$

33. Reflection in the x-axis and a vertical shift one unit upward of $y = \sqrt{x}$; $y = 1 - \sqrt{x}$

35. Reflection in the x-axis

37. Reflection in the y-axis (identical)

39. Reflection in the x-axis or a reflection in the y-axis

41. Vertical shrink **43.** Vertical stretch

45. Horizontal shrink

47.

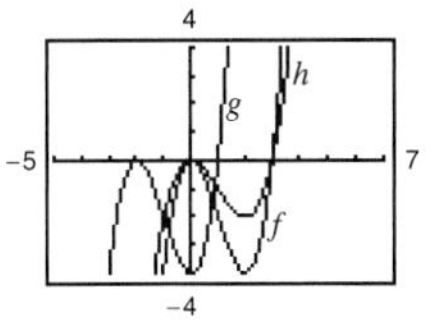

g is a horizontal shift and h is a vertical shrink.

CHAPTER 1

49. (a) $f(x) = x^2$ (b) Upward shift of six units

(c)

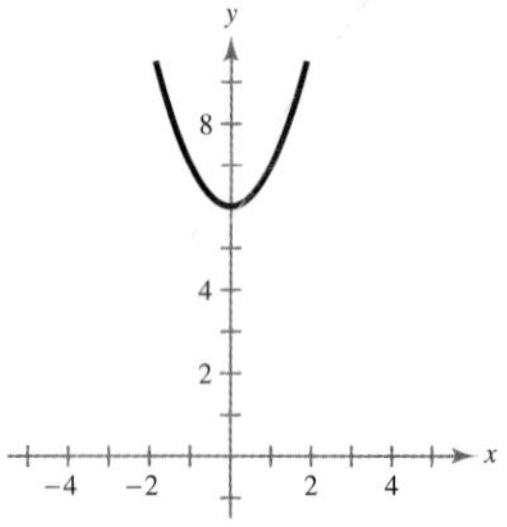

(d) $g(x) = f(x) + 6$

51. (a) $f(x) = x^2$

(b) Horizontal shift five units to the left, reflection in the x-axis, and vertical shift two units upward

(c)

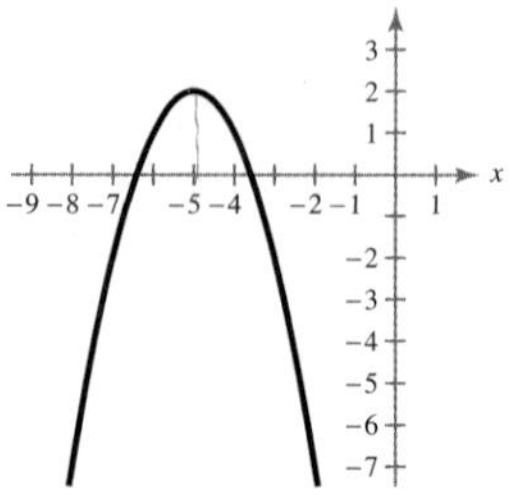

(d) $g(x) = 2 - f(x + 5)$

53. (a) $f(x) = x^3$

(b) Horizontal shift two units to the right and vertical shrink by a factor of $\frac{1}{3}$

(c)

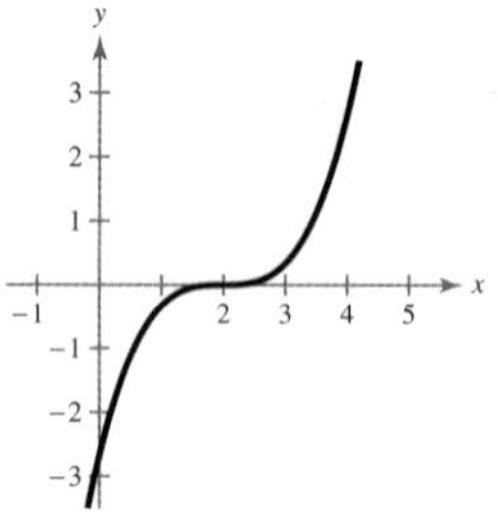

(d) $g(x) = \frac{1}{3}f(x - 2)$

55. (a) $f(x) = \dfrac{1}{x}$

(b) Horizontal shift eight units to the left and vertical shift nine units downward

(c)

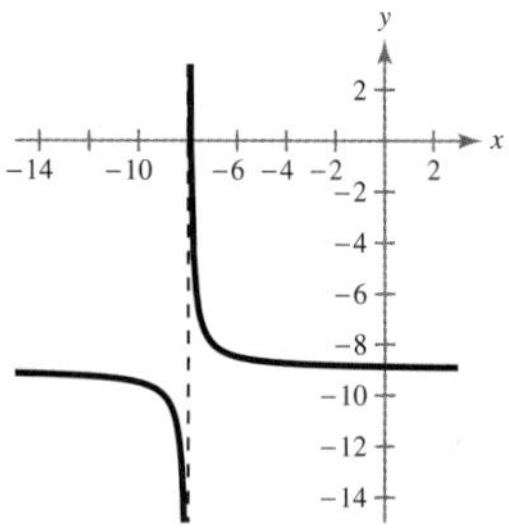

(d) $g(x) = f(x + 8) - 9$

57. (a) $f(x) = |x|$

(b) Horizontal shift one unit to the right, reflection in the x-axis, vertical stretch by a factor of 2, and vertical shift four units downward

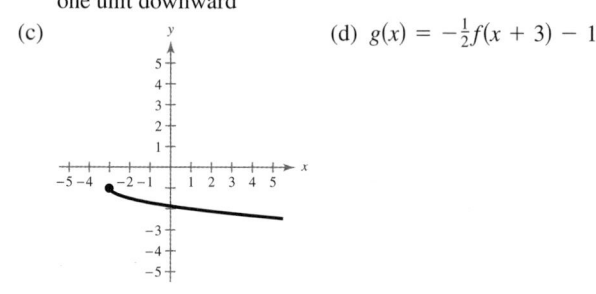

(c)

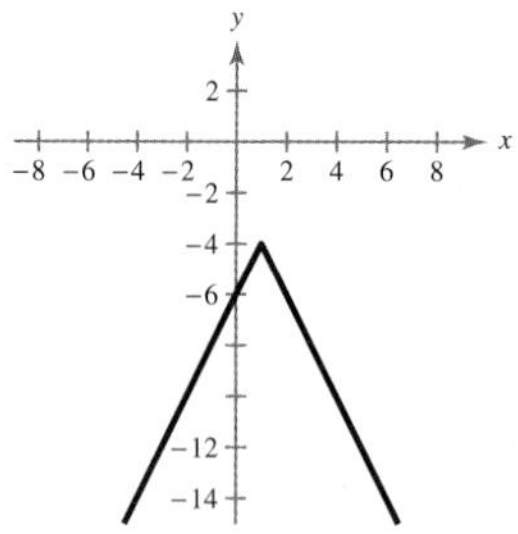

(d) $g(x) = -2f(x - 1) - 4$

59. (a) $f(x) = \sqrt{x}$

(b) Horizontal shift three units to the left, reflection in the x-axis, vertical shrink by a factor of $\frac{1}{2}$, and vertical shift one unit downward

(c)

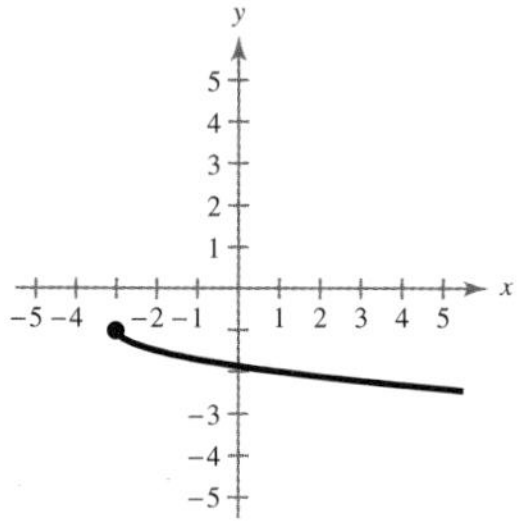

(d) $g(x) = -\frac{1}{2}f(x + 3) - 1$

61. $g(x) = (x - 3)^2 - 7$ **63.** $g(x) = -|x| - 12$

65. (a) Horizontal shift 4.8 units to the right, vertical stretch by a factor of 1.5, and vertical shift 123 units up

(b)

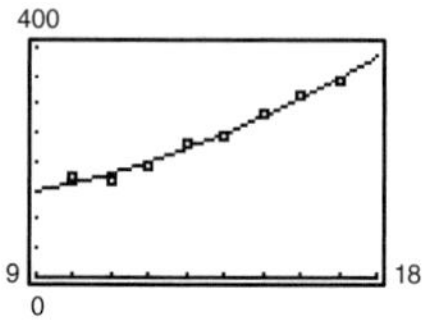

(c) $N(t) = 1.5(t + 5.2)^2 + 123$; Substitute $(t + 10)$ for t in the original model.

67. False; When $f(x) = x^2$, $f(-x) = (-x)^2 = x^2$. Because $f(x) = f(-x)$ in this case, $y = f(-x)$ is not a reflection of $y = f(x)$ across the x-axis in all cases.

69. $x = -2$ and $x = 3$

71. Cannot be determined because it is a vertical shift

73. c **75.** c

77. The graph of $f(x) = x^2$ should have been shifted one unit to the left instead of one unit to the right.

79. Answers will vary. **81.** All real numbers x except $x = 9$

83. All real numbers x such that $-10 \le x \le 10$

Section 1.5 (page 56)

1. addition, subtraction, multiplication, division

3. $g(x)$ **5.** $2x$

7.

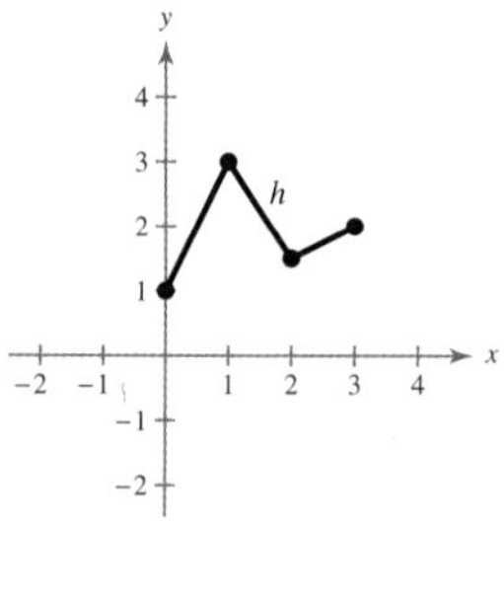

9.

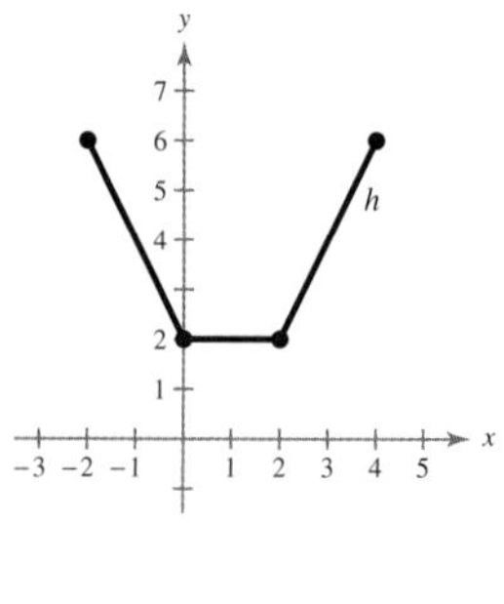

11. (a) $2x$ (b) 6 (c) $x^2 - 9$ (d) $\dfrac{x + 3}{x - 3}$

All real numbers x, except $x = 3$

13. (a) $3x^2 - 5x + 6$ (b) $3x^2 + 5x - 6$

(c) $-15x^3 + 18x^2$ (d) $\dfrac{3x^2}{6 - 5x}$

All real numbers x, except $x = \frac{6}{5}$

15. (a) $x^2 + 5 + \sqrt{1 - x}$ (b) $x^2 + 5 - \sqrt{1 - x}$

(c) $(x^2 + 5)\sqrt{1 - x}$ (d) $\dfrac{x^2 + 5}{\sqrt{1 - x}}, x < 1$

17. (a) $\dfrac{x^4 + x^3 + x}{x + 1}$ (b) $\dfrac{-x^4 - x^3 + x}{x + 1}$ (c) $\dfrac{x^4}{x + 1}$

(d) $\dfrac{1}{x^2(x + 1)}$; all real numbers x except $x = 0, -1$

19. 7 **21.** 5 **23.** 306 **25.** $\frac{8}{23}$ **27.** $-t^2 - t + 5$

29. $-125t^3 + 75t^2 + 10t - 6$ **31.** $\dfrac{t - 1}{t^2 - 8t + 14}$

33.

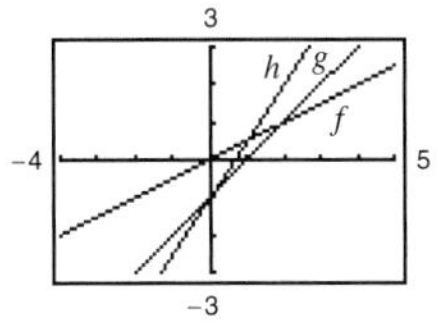

35.

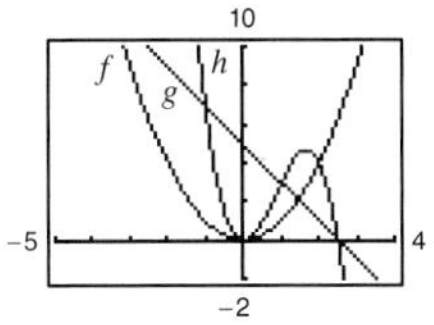

37.

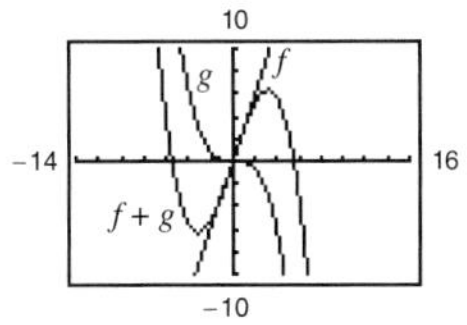

$f(x), 0 \le x \le 2$;
$g(x), x > 6$

39.

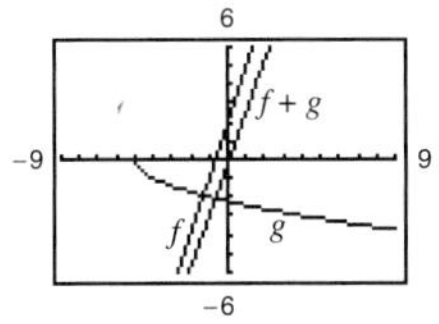

$f(x), 0 \le x \le 2$;
$f(x), x > 6$

41. (a) $2(x + 4)^2$ (b) $2x^2 + 4$ (c) $x + 8$ (d) 32

43. (a) x (b) x (c) $x^9 + 3x^6 + 3x^3 + 2$ (d) 0

45. (a) All real numbers x such that $x \ge 7$

(b) All real numbers x

(c) All real numbers x such that $x \le -\dfrac{\sqrt{7}}{2}$ or $x \ge \dfrac{\sqrt{7}}{2}$

47. (a) All real numbers x

(b) All real numbers x such that $x \ge 0$

(c) All real numbers x such that $x \ge 0$

49. (a) All real numbers x except $x = 0$

(b) All real numbers x except $x = -3$

(c) All real numbers x except $x = -3$

51. (a) All real numbers x (b) All real numbers x

(c) All real numbers x

53. (a) All real numbers x

(b) All real numbers x except $x = \pm 2$

(c) All real numbers x except $x = \pm 2$

55. (a) $(f \circ g)(x) = \sqrt{x^2 + 4}$

$(g \circ f)(x) = x + 4, x \ge -4$;

Domain of $f \circ g$: all real numbers x

(b) $f \circ g \ne g \circ f$

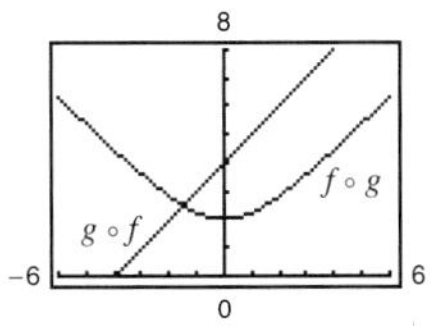

57. (a) $(f \circ g)(x) = x$; $(g \circ f)(x) = x$;

Domain of $f \circ g$: all real numbers x

(b) $f \circ g = g \circ f$

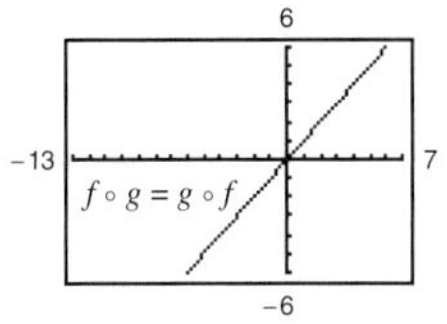

59. (a) $(f \circ g)(x) = x^4$; $(g \circ f)(x) = x^4$;

Domain of $f \circ g$: all real numbers x

(b) $f \circ g = g \circ f$

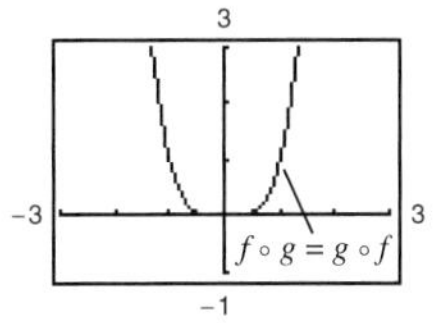

61. (a) $(f \circ g)(x) = x$; $(g \circ f)(x) = x$ (b) $x = x$

(c)

x	0	1	2	3
$g(x)$	$-\frac{4}{5}$	$-\frac{3}{5}$	$-\frac{2}{5}$	$-\frac{1}{5}$
$(f \circ g)(x)$	0	1	2	3

x	0	1	2	3
$f(x)$	4	9	14	19
$(g \circ f)(x)$	0	1	2	3

63. (a) $(f \circ g)(x) = \sqrt{x^2 + 1}$; $(g \circ f)(x) = x + 1, x \ge -6$

(b) $x + 1 \ne \sqrt{x^2 + 1}$

(c)

x	0	1	2	3
$g(x)$	-5	-4	-1	4
$(f \circ g)(x)$	1	$\sqrt{2}$	$\sqrt{5}$	$\sqrt{10}$

x	0	1	2	3
$f(x)$	$\sqrt{6}$	$\sqrt{7}$	$\sqrt{8}$	3
$(g \circ f)(x)$	1	2	3	4

65. (a) $(f \circ g)(x) = |2x^3|$; $(g \circ f)(x) = 2|x|^3$

(b) $|2x^3| = 2|x|^3$

(c)

x	-2	-1	0	1	2
$g(x)$	-16	-2	0	2	16
$(f \circ g)(x)$	16	2	0	2	16

x	-2	-1	0	1	2
$f(x)$	2	1	0	1	2
$(g \circ f)(x)$	16	2	0	2	16

67. (a) 3 (b) 0 **69.** (a) 2 (b) 4

71. $f(x) = x^2, g(x) = 2x + 1$ **73.** $f(x) = \sqrt[3]{x}, g(x) = x^2 - 4$

75. $f(x) = \dfrac{1}{x}, g(x) = x + 2$

77. $f(x) = x^2 + 2x, g(x) = x + 4$

79. (a) $T = \frac{3}{4}x + \frac{1}{15}x^2$
(b)
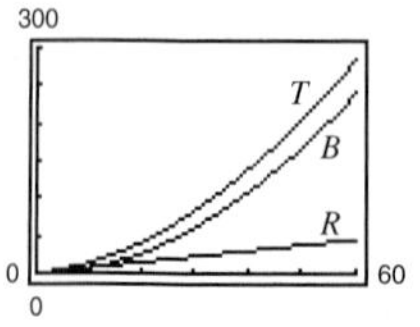

(c) B; For example, $B(60) = 240$, whereas $R(60)$ is only 45.

81. (a) $r(x) = \dfrac{x}{2}$ (b) $A(r) = \pi r^2$

(c) $(A \circ r)(x) = \pi\left(\dfrac{x}{2}\right)^2$; $(A \circ r)(x)$ represents the area of the circular base of the tank with radius $x/2$.

83. (a) $T = 1.3t^2 + 72.4t + 1322$
(b)
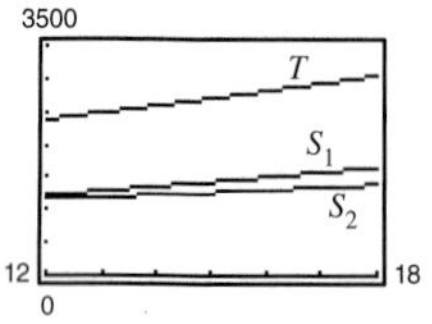

85. (a) $N(T(t))$ or $(N \circ T)(t) = 40t^2 + 590$; $N(T(t))$ or $(N \circ T)(t)$ represents the number of bacteria after t hours outside the refrigerator.
(b) $(N \circ T)(12) = 6350$; There are 6350 bacteria in a refrigerated food product after 12 hours outside the refrigerator.
(c) About 3.9 h

87. $s(t) = \sqrt{(150 - 450t)^2 + (200 - 450t)^2}$
$= 50\sqrt{162t^2 - 126t + 25}$

89. False; $g(x) = x - 3$ **91.** Odd (Proofs will vary.)

93. a, c

95. (a) $O = 12 + Y$; Answers will vary.
(b) Middle child is 8 years old, youngest child is 4 years old.

97. $(f \circ g)(x)$ means you replace every x in $f(x)$ with $g(x)$. So, $(f \circ g)(x) = 2(3x + 2) + 1$ not $(f \circ g)(x) = 3(2x + 1) + 2$.

99. *Sample answer:* $(0, -5), (1, -5), (2, -7)$

101. *Sample answer:* $(0, 7), (1, 4\sqrt{3}), (2, \sqrt{45})$

Section 1.6 (page 67)

1. inverse, f^{-1} **3.** $y = x$ **5.** At most once

7. $f^{-1}(x) = x - 11$ **9.** $f^{-1}(x) = \dfrac{x - 1}{3}$

11. $f^{-1}(x) = x^3$ **13.** c **14.** b **15.** a **16.** d

17. $f(g(x)) = f(\sqrt[3]{x}) = (\sqrt[3]{x})^3 = x$
$g(f(x)) = g(x^3) = \sqrt[3]{x^3} = x$

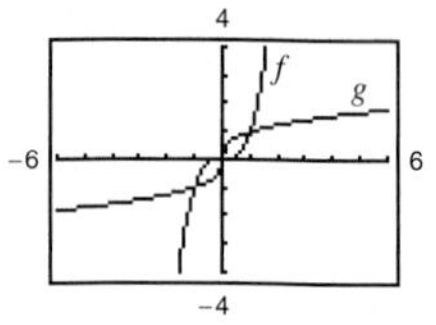

Reflections in the line $y = x$

19. $f(g(x)) = f(x^2 + 4), x \ge 0$
$= \sqrt{(x^2 + 4) - 4} = x$
$g(f(x)) = g(\sqrt{x - 4})$
$= (\sqrt{x - 4})^2 + 4 = x$

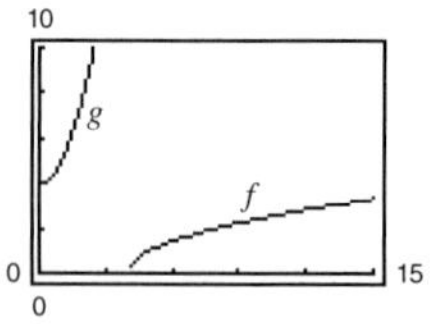

Reflections in the line $y = x$

21. $f(g(x)) = f(\sqrt[3]{1 - x}) = 1 - (\sqrt[3]{1 - x})^3 = x$
$g(f(x)) = g(1 - x^3) = \sqrt[3]{1 - (1 - x^3)} = x$

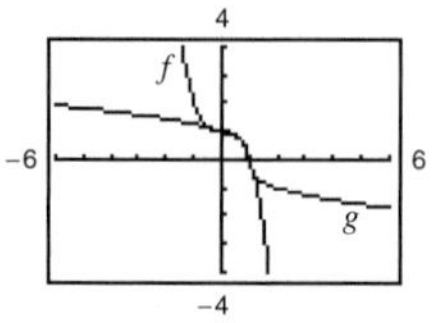

Reflections in the line $y = x$

23. (a) $f(g(x)) = f\left(-\dfrac{2x + 6}{7}\right)$
$= -\dfrac{7}{2}\left(-\dfrac{2x + 6}{7}\right) - 3 = x$
$g(f(x)) = g\left(-\dfrac{7}{2}x - 3\right)$
$= -\dfrac{2\left(-\frac{7}{2}x - 3\right) + 6}{7} = x$

(b)
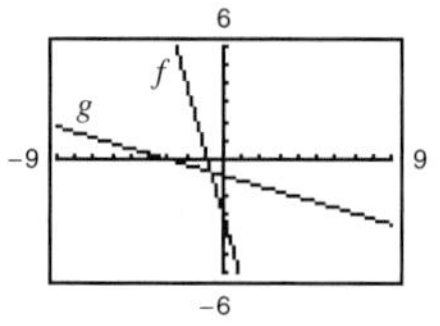

(c)

x	0	2	−2	6
$f(x)$	−3	−10	4	−24

x	−3	−10	4	−24
$g(x)$	0	2	−2	6

25. (a) $f(g(x)) = f(\sqrt[3]{x - 5}) = (\sqrt[3]{x - 5})^3 + 5 = x$
$g(f(x)) = g(x^3 + 5) = \sqrt[3]{(x^3 + 5) - 5} = x$

(b)
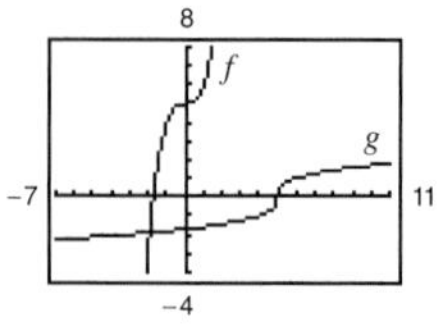

(c)

x	0	1	−1	−2	4
$f(x)$	5	6	4	−3	69

x	5	6	4	−3	69
$g(x)$	0	1	−1	−2	4

27. (a) $f(g(x)) = f(8 + x^2)$
$= -\sqrt{(8 + x^2) - 8}$
$= -\sqrt{x^2} = -(-x) = x, x \le 0$
$g(f(x)) = g\left(-\sqrt{x - 8}\right)$
$= 8 + \left(-\sqrt{x - 8}\right)^2$
$= 8 + (x - 8) = x, x \ge 8$

(b)

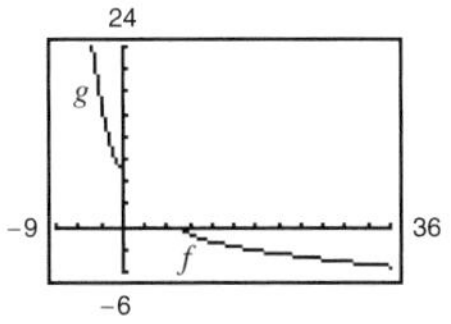

(c)

x	8	9	12	15
$f(x)$	0	-1	-2	$-\sqrt{7}$

x	0	-1	-2	$-\sqrt{7}$
$g(x)$	8	9	12	15

29. (a) $f(g(x)) = f\left(\frac{1}{2}x\right) = 2\left(\frac{1}{2}x\right) = x$
$g(f(x)) = g(2x) = \frac{1}{2}(2x) = x$

(b)

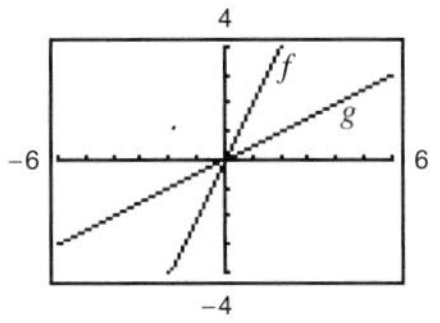

(c)

x	-4	-2	0	2	4
$f(x)$	-8	-4	0	4	8

x	-8	-4	0	4	8
$g(x)$	-4	-2	0	2	4

31. (a) $f(g(x)) = f\left(-\dfrac{5x + 1}{x - 1}\right)$

$$= \frac{\left(-\dfrac{5x + 1}{x - 1}\right) - 1}{\left(-\dfrac{5x + 1}{x - 1}\right) + 5} = \frac{\dfrac{6x}{x - 1}}{\dfrac{6}{x - 1}} = x, x \ne 1$$

$g(f(x)) = g\left(\dfrac{x - 1}{x + 5}\right)$

$$= -\frac{5\left(\dfrac{x - 1}{x + 5}\right) + 1}{\left(\dfrac{x - 1}{x + 5}\right) - 1} = \frac{\dfrac{6x}{x + 5}}{\dfrac{6}{x + 5}} = x, x \ne -5$$

(b)

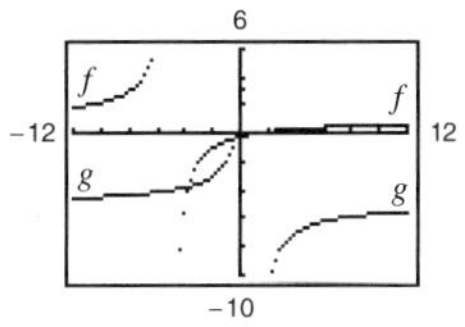

(c)

x	-3	-2	-1	0	2	3	4
$f(x)$	-2	-1	$-\frac{1}{2}$	$-\frac{1}{5}$	$\frac{1}{7}$	$\frac{1}{4}$	$\frac{1}{3}$

x	-2	-1	$-\frac{1}{2}$	$-\frac{1}{5}$	$\frac{1}{7}$	$\frac{1}{4}$	$\frac{1}{3}$
$g(x)$	-3	-2	-1	0	2	3	4

33. Yes; No two elements in the domain of f correspond to the same element in the range of f.

35. No; -3 and 0 both correspond to 6, so f is not one-to-one.

37. Not a function **39.** Function; one-to-one

41. Function; one-to-one

43.

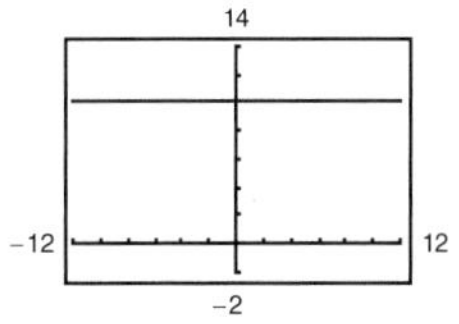

No inverse function

45.

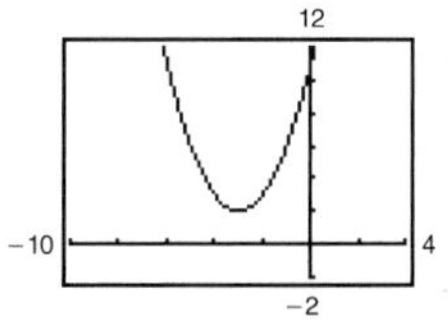

No inverse function

47.

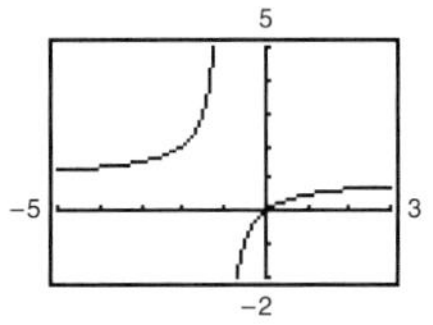

An inverse function exists.

49.

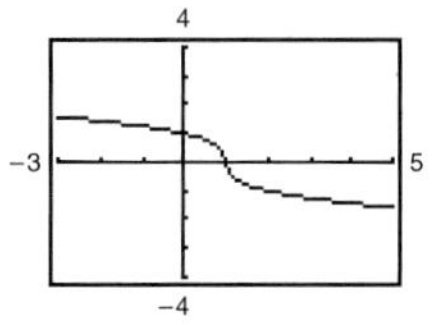

An inverse function exists.

51.

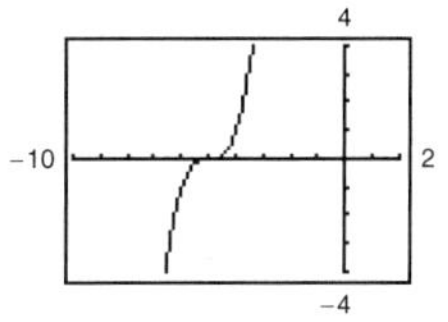

An inverse function exists.

53.

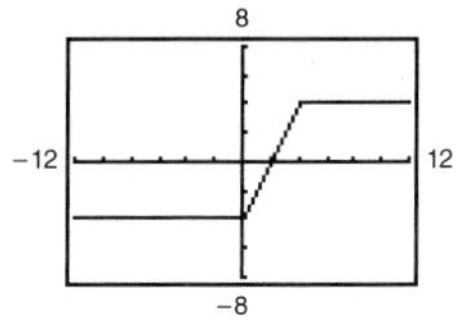

No inverse function

55.

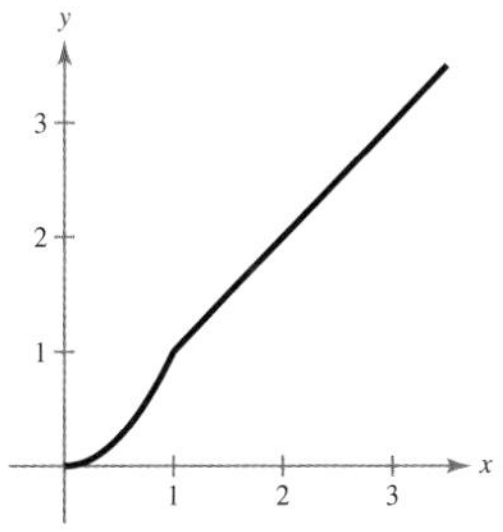

An inverse function exists.

57. No inverse function **59.** $f^{-1}(x) = \dfrac{5x - 4}{3}$

61. No inverse function **63.** $f^{-1}(x) = \sqrt{x} - 3$

65. No inverse function **67.** $f^{-1}(x) = \dfrac{x^2 - 3}{2}, x \ge 0$

69. $f^{-1}(x) = 2 - x, x \ge 0$ **71.** $f^{-1}(x) = \dfrac{5x + 3}{5 - 2x}$

73. (a) $f^{-1}(x) = \dfrac{x + 9}{4}$

(b)

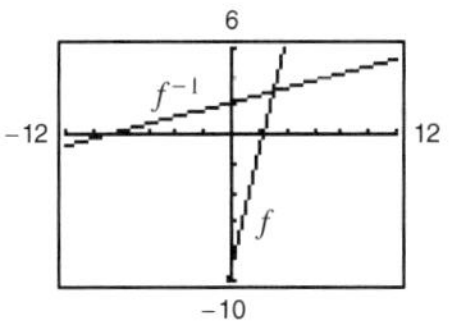

(c) Reflections in the line $y = x$

(d) The domains and ranges of f and f^{-1} are all real numbers x.

75. (a) $f^{-1}(x) = \sqrt[5]{x+2}$

(b)

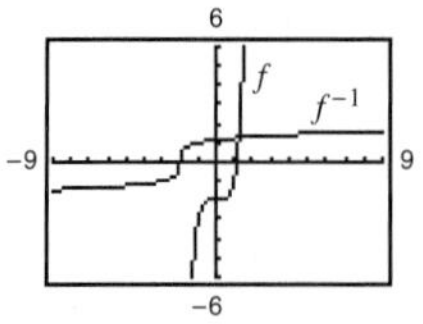

(c) Reflections in the line $y = x$

(d) The domains and ranges of f and f^{-1} are all real numbers x.

77. (a) $f^{-1}(x) = -\sqrt[4]{x}$

(b)

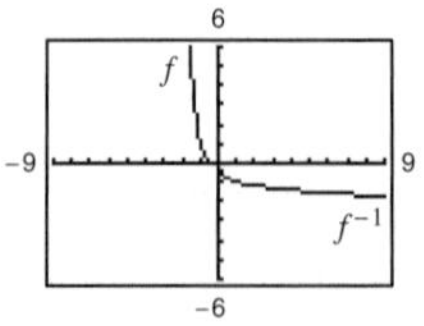

(c) Reflections in the line $y = x$

(d) The domain of f and range of f^{-1} are all real numbers x such that $x \le 0$. The domain of f^{-1} and range of f are all real numbers x such that $x \ge 0$.

79. (a) $f^{-1}(x) = \sqrt{4 - x^2}, 0 \le x \le 2$

(b)

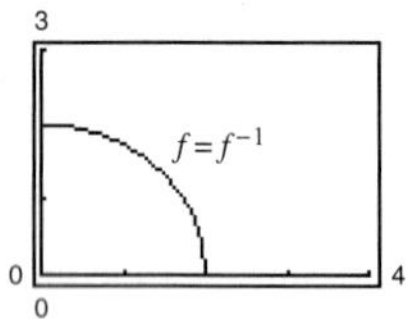

(c) The graphs are the same.

(d) The domains and ranges of f and f^{-1} are all real numbers x such that $0 \le x \le 2$.

81. (a) $f^{-1}(x) = \sqrt[3]{\dfrac{4}{x}}$

(b)

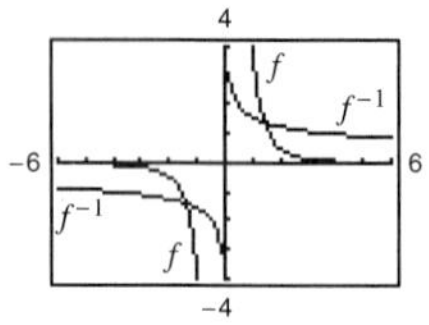

(c) Reflections in the line $y = x$

(d) The domains and ranges of f and f^{-1} are all real numbers x such that $x \ne 0$.

83. $f^{-1}(x) = \sqrt{x} + 2$

Domain of f: all real numbers x such that $x \ge 2$

Range of f: all real numbers y such that $y \ge 0$

Domain of f^{-1}: all real numbers x such that $x \ge 0$

Range of f^{-1}: all real numbers y such that $y \ge 2$

85. $f^{-1}(x) = x - 2, x \ge 0$

Domain of f: all real numbers x such that $x \ge -2$

Range of f: all real numbers y such that $y \ge 0$

Domain of f^{-1}: all real numbers x such that $x \ge 0$

Range of f^{-1}: all real numbers y such that $y \ge -2$

87. $f^{-1}(x) = \sqrt[4]{x} + 1$

Domain of f: all real numbers x such that $x \ge 1$

Range of f: all real numbers y such that $y \ge 0$

Domain of f^{-1}: all real numbers x such that $x \ge 0$

Range of f^{-1}: all real numbers y such that $y \ge 1$

89. $f^{-1}(x) = \dfrac{\sqrt{x} - 3}{4}$

Domain of f: all real numbers x such that $x \ge -\frac{3}{4}$

Range of f: all real numbers y such that $y \ge 0$

Domain of f^{-1}: all real numbers x such that $x \ge 0$

Range of f^{-1}: all real numbers y such that $y \ge -\frac{3}{4}$

91. $f^{-1}(x) = \sqrt{-\dfrac{x+5}{2}}$

Domain of f: all real numbers x such that $x \ge 0$

Range of f: all real numbers y such that $y \le -5$

Domain of f^{-1}: all real numbers x such that $x \le -5$

Range of f^{-1}: all real numbers y such that $y \ge 0$

93. $f^{-1}(x) = \dfrac{x+3}{3}, x \ge 0$

Domain of f: all real numbers x such that $x \ge 1$

Range of f: all real numbers y such that $y \ge 0$

Domain of f^{-1}: all real numbers x such that $x \ge 0$

Range of f^{-1}: all real numbers y such that $y \ge 1$

95. $f^{-1}(x) = x + 3, x \ge 0$

Domain of f: all real numbers x such that $x \ge 4$

Range of f: all real numbers y such that $y \ge 1$

Domain of f^{-1}: all real numbers x such that $x \ge 1$

Range of f^{-1}: all real numbers y such that $y \ge 4$

97.

x	-4	-2	2	3
$f^{-1}(x)$	-2	-1	1	3

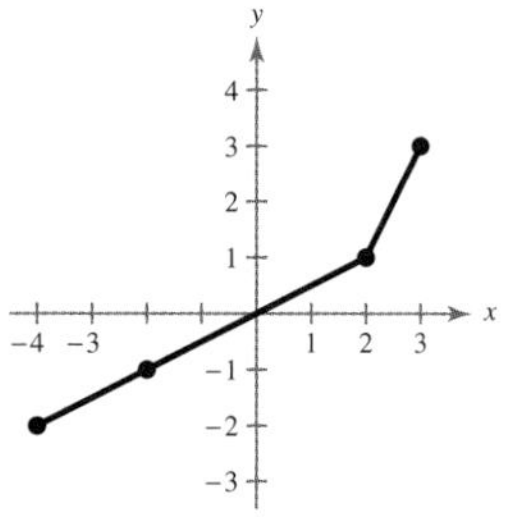

99. 4 **101.** -2 **103.** 0 **105.** 2

107. (a) and (b)

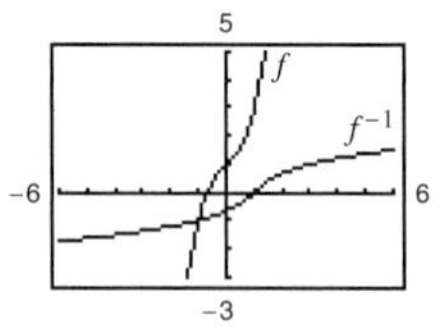

(c) Inverse function because it satisfies the Vertical Line Test

109. (a) and (b)

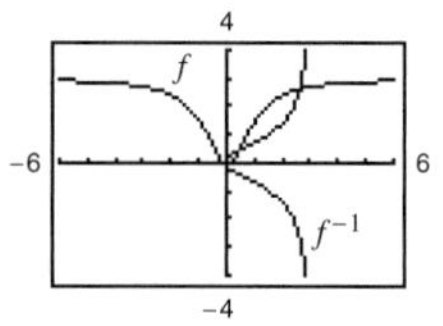

(c) Not an inverse function because it does not satisfy the Vertical Line Test

111. (a) and (b)

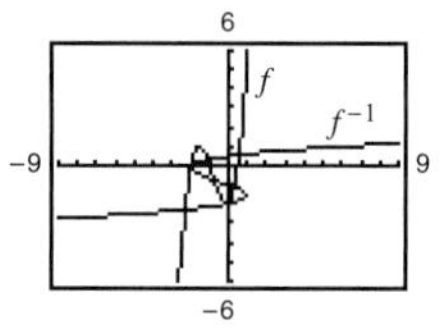

(c) Not an inverse function because it does not satisfy the Vertical Line Test

113. 32 **115.** -168 **117.** $8\sqrt[3]{x} + 24$

119. $\frac{x+1}{2}$ **121.** $\frac{x+1}{2}$

123. (a) f is one-to-one because no two elements in the domain (men's U.S. shoe sizes) correspond to the same element in the range (men's European shoe sizes).
(b) 44 (c) 10 (d) 41 (e) 12

125. (a) 19 9 64 64 4 69 29 4 64 9 104 29 94
(b) $f^{-1}(x) = \frac{x-4}{5}$; What time

127. False; For example, $y = x^2$ is even, but does not have an inverse.

129. This situation could be represented by a one-to-one function. The inverse function would represent the number of miles completed in terms of time in hours.

131. This function could not be represented by a one-to-one function because it oscillates.

133. The graph of f^{-1} is a reflection of the graph of f in the line $y = x$.

135. (a) The function will be one-to-one because no two values of x will produce the same value for $f(x)$.
(b) $f^{-1}(50)$ represents the value of 50 degrees Fahrenheit in degrees Celsius.

137. The domain of g should be restricted so $g(x) = x^4, x \geq 0$.

139. Proof **141.** $5xy, x \neq 0, y \neq -5$ **143.** Function

Section 1.7 (page 76)

1. positive **3.** Negative correlation

5. (a)

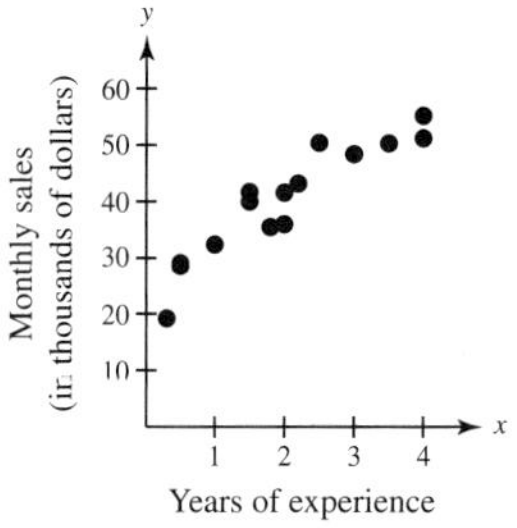

(b) Positively correlated; The more experience the sales representative has, the higher the sales.

7. *Sample answer:*

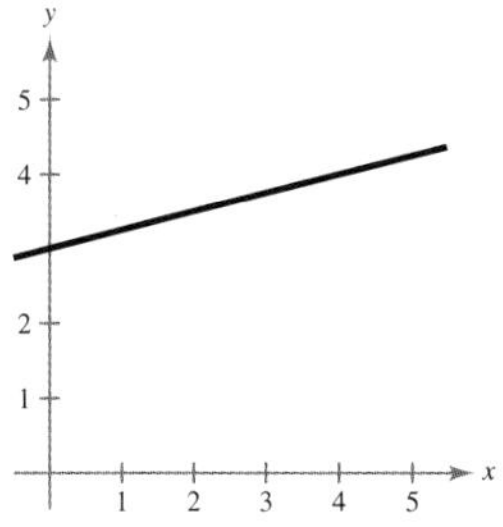

$y = \frac{1}{4}x + 3$

9. *Sample answer:*

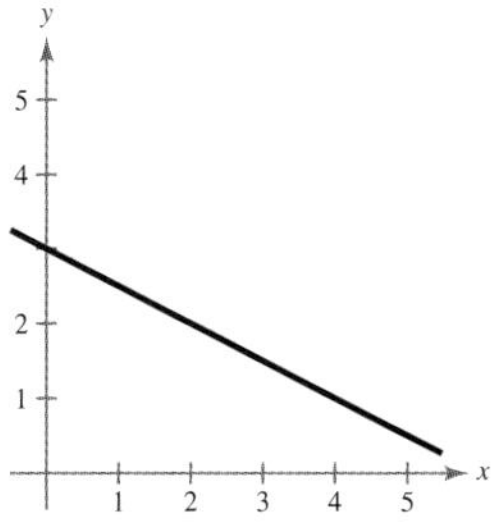

$y = -\frac{1}{2}x + 3$

11. *Sample answer:* (a) and (b)

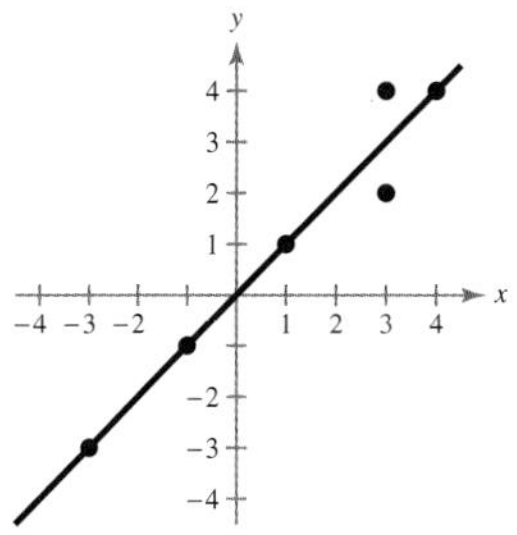

(c) $y = x$

13. *Sample answer:* (a) and (b)

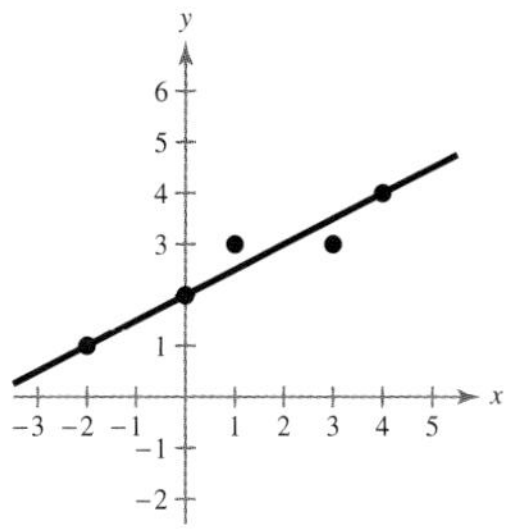

(c) $y = \frac{1}{2}x + 2$

15. *Sample answer:* (a) and (b)

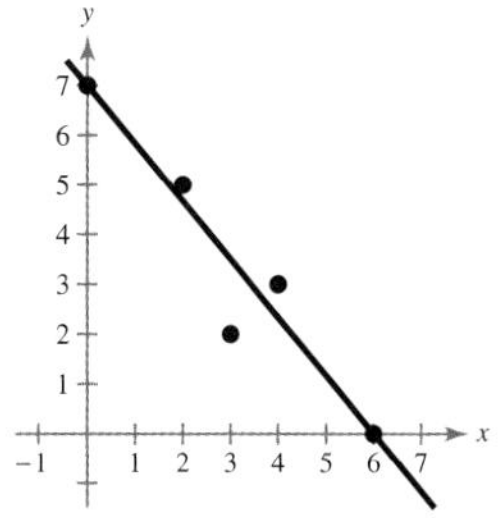

(c) $y = -\frac{7}{6}x + 7$

17. $y = 0.46x + 1.6$
(a)

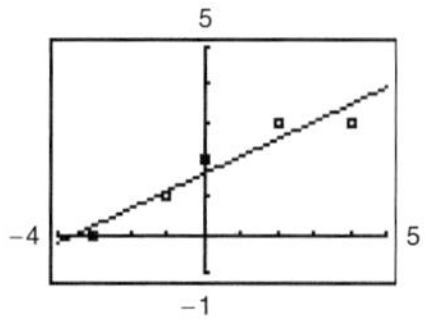

(b)

x	-3	-1	0	2	4
Linear equation	0.22	1.14	1.6	2.52	3.44
Given data	0	1	2	3	3

The model fits the data well.

19. (a)

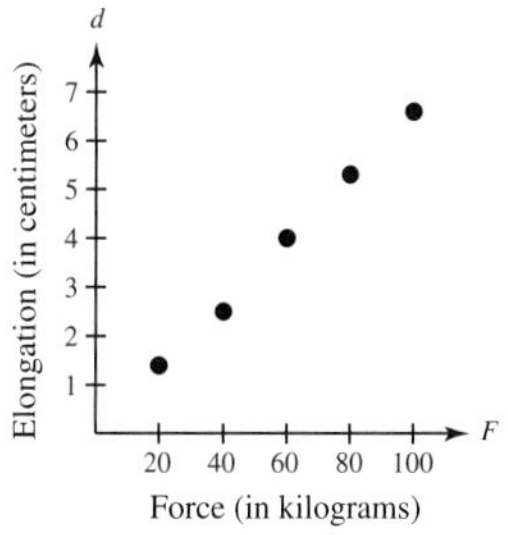

(b) *Sample answer:* $d = 0.07F - 0.3$
(c) $d = 0.066F$ (d) 3.63 cm

21. (a) and (c)

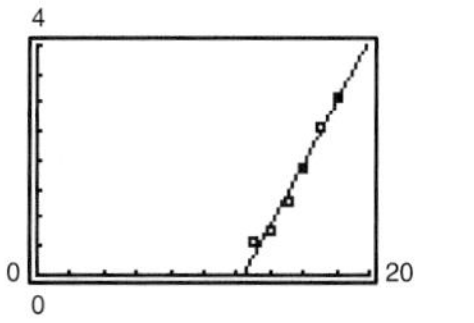

The model fits the data well.
(b) $V = 0.5393t - 6.658$
(d) 2021: \$4.667 billion; 2026: \$7.364 billion; Answers will vary.
(e) 0.5393; The average annual increase in enterprise values is about \$0.5393 billion.

23. (a) and (c)

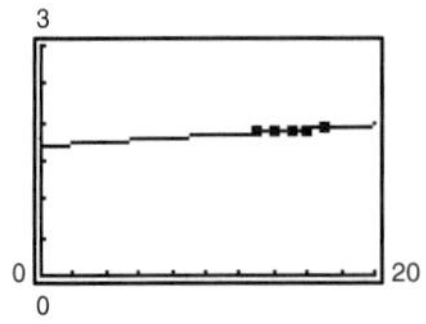

(b) $P = 0.0133t + 1.695$

(d)

Year	2013	2014	2015	2016	2017
Actual	1.867	1.881	1.894	1.908	1.92
Model	1.8679	1.8812	1.8945	1.9078	1.9211

The model fits the data well.

(e) 1.974 million people; Answers will vary.

25. (a) $y = 45.9x + 108$

(b)

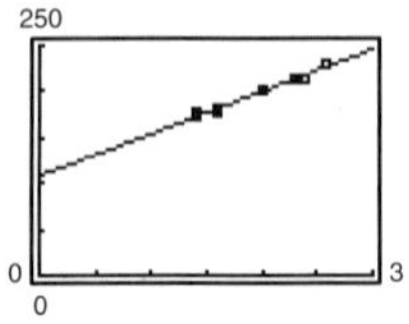

(c) The slope represents the increase in sales due to increased advertising.

(d) $176,850

27. (a) $T = -0.014t + 4.77$; $r \approx -0.8727$

(b) The negative slope means that the winning times are decreasing over time.

(c)

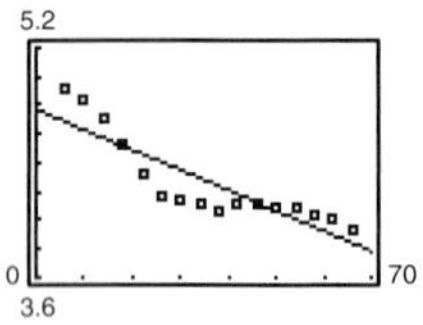

(d)

Year	1956	1960	1964	1968	1972	1976
Actual	4.91	4.84	4.72	4.53	4.32	4.16
Model	4.69	4.63	4.57	4.52	4.46	4.41

Year	1980	1984	1988	1992	1996
Actual	4.15	4.12	4.06	4.12	4.12
Model	4.35	4.29	4.24	4.18	4.13

Year	2000	2004	2008	2012	2016
Actual	4.10	4.09	4.05	4.02	3.94
Model	4.07	4.01	3.96	3.90	3.85

The model does not fit the data well.

(e) The closer $|r|$ is to 1, the better the model fits the data.

(f) No; The winning times have leveled off in recent years, but the model values continue to decrease to unrealistic times.

29. True; To have positive correlation, the y-values tend to increase as x increases.

31. (a) The a-value and b-value were switched. The equation is $y = -1.02x + 9.18$, not $y = 9.18x - 1.02$.

(b) Because r is negative, the data would have a strong negative correlation, not a strong positive correlation.

33. (a) 10 (b) 40 (c) $2w^2 + 5w + 7$

Review Exercises (page 82)

1. $m = \frac{3}{7}$

3. (a) $3x + 2y - 1 = 0$
(b) *Sample answer:* $(-1, 2), (1, -1), (-5, 8)$

5. (a) $x - 10 = 0$
(b) *Sample answer:* $(10, 1), (10, 3), (10, -2)$

7. $y = -1$

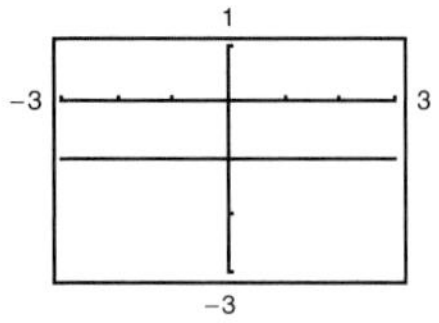

9. $y = \frac{2}{7}x + \frac{2}{7}$

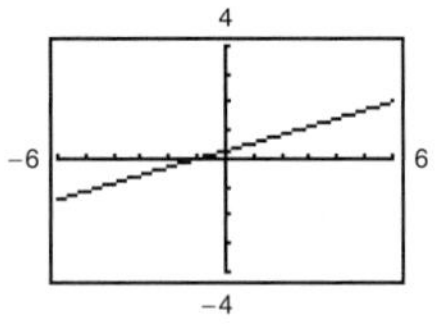

11. (a) $y = \frac{5}{4}x - \frac{23}{4}$ (b) $y = -\frac{4}{5}x + \frac{2}{5}$

13. $V = 1580t - 12{,}640$ **15.** $220,000

17. (a) Not a function; The element 20 in A corresponds to two elements, 4 and 6, in B.
(b) Function; Each element in A corresponds to exactly one element in B.

19. Not a function **21.** Function

23. (a) 10 (b) $b^6 + 1$ (c) $x^2 - 2x + 2$

25. (a) -1 (b) 2 (c) 6

27. All real numbers x except $x = -2$

29. $C = 5.25x + 17{,}500$; $P = 3.18x - 17{,}500$

31. $2h + 4x + 3, h \neq 0$

33.

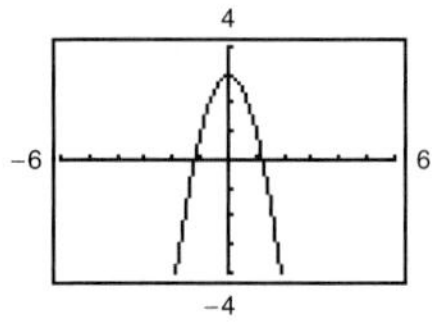

Domain: all real numbers x
Range: $(-\infty, 3]$

35.

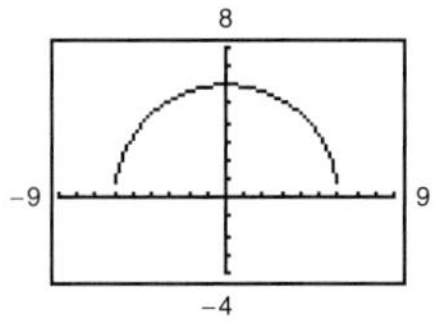

Domain: $[-6, 6]$
Range: $[0, 6]$

37. Not a function; Solve for y and graph the two resulting functions.

39. Increasing on $(-\infty, -1), (1, \infty)$; decreasing on $(-1, 1)$

41.

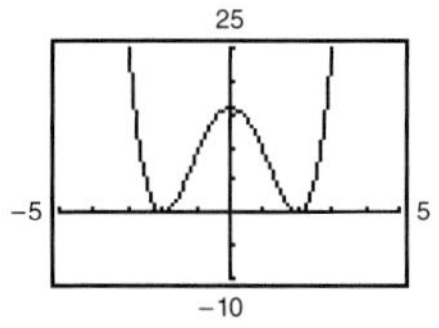

Relative maximum: $(0, 16)$
Relative minima: $(-2, 0), (2, 0)$

43.

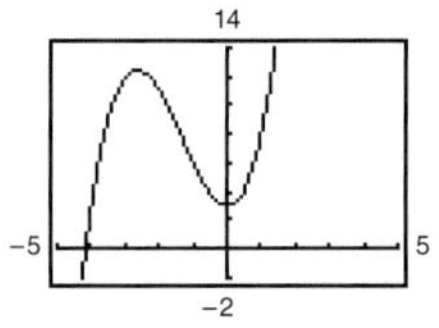

Relative minimum: $(0, 3)$
Relative maximum: about $(-2.67, 12.48)$

45.

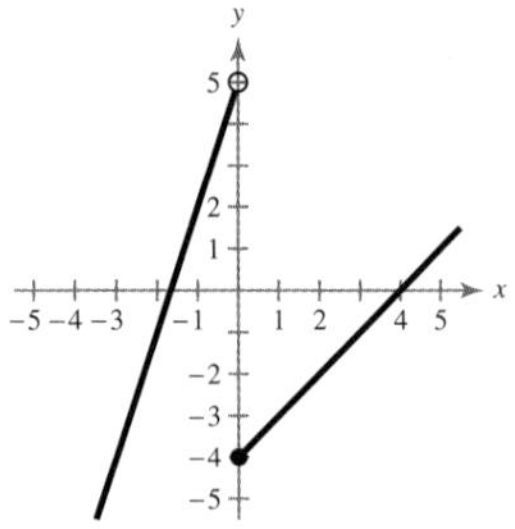

47.

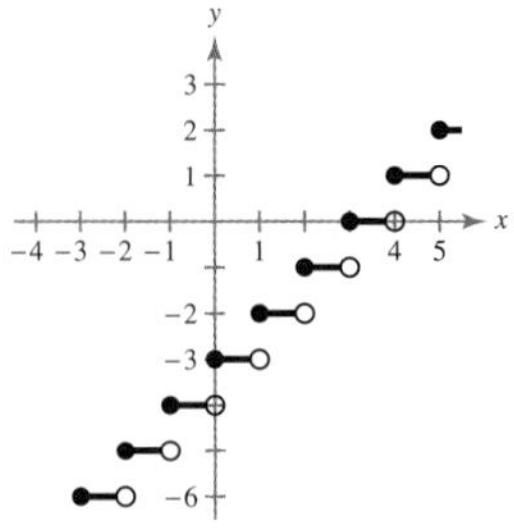

49. Even **51.** Even **53.** Neither **55.** Odd

57. Rational function $f(x) = \frac{1}{x}$
Horizontal shift three units right
$g(x) = \frac{1}{x - 3}$

59. Quadratic function $f(x) = x^2$
Vertical shift one unit upward, horizontal shift two units right
$g(x) = (x - 2)^2 + 1$

61. Absolute value function $f(x) = |x|$
Vertical shift three units upward
$g(x) = |x| + 3$

63. Identity function $f(x) = x$
Horizontal shrink by a factor of $\frac{1}{3}$, vertical shift four units upward
$g(x) = 3x + 4$

65.

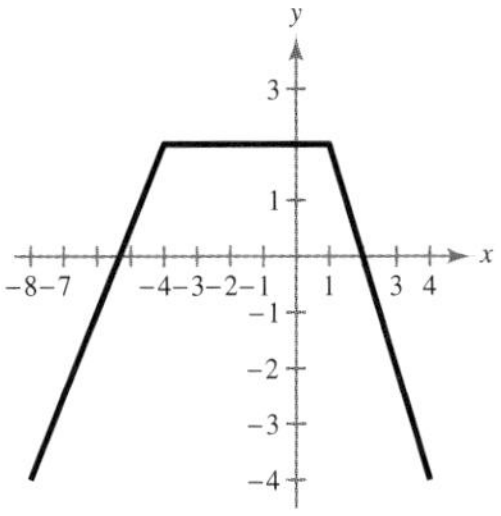

67.

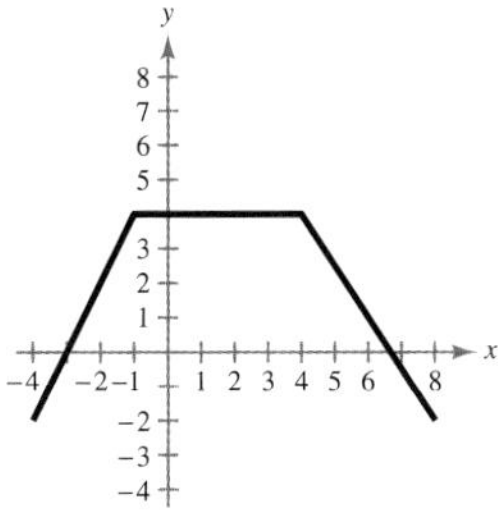

69. (a) Rational function $f(x) = \frac{1}{x}$
(b) Vertical shift six units downward
(c)

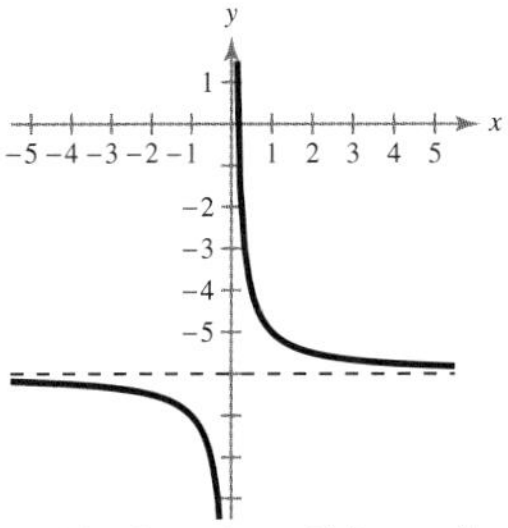

(d) $h(x) = f(x) - 6$

71. (a) Cubic function $f(x) = x^3$
(b) Horizontal shift two units right, vertical shift five units upward
(c)

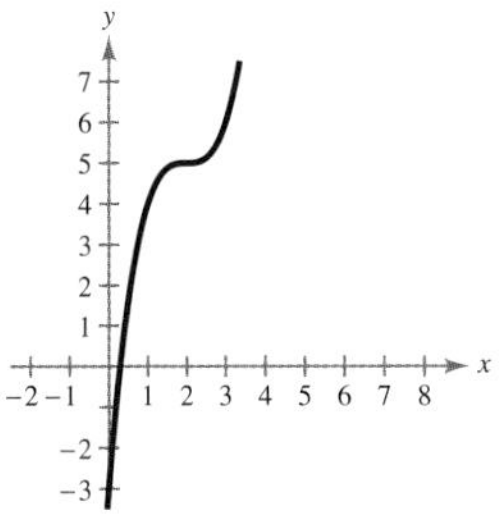

(d) $h(x) = f(x - 2) + 5$

73. (a) Square root function $f(x) = \sqrt{x}$
(b) Reflection in the x-axis, vertical shift six units downward
(c)

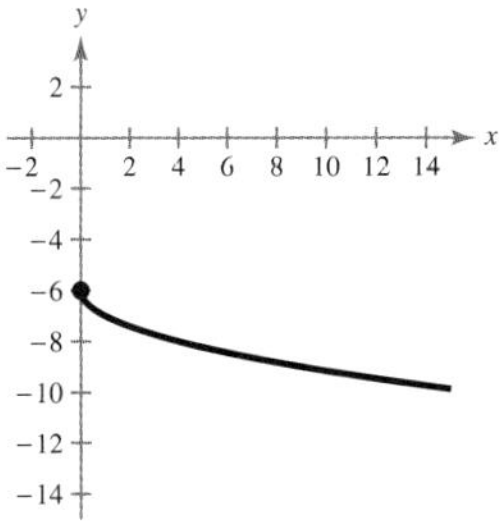

(d) $h(x) = -f(x) - 6$

75. (a) Absolute value function $f(x) = |x|$
(b) Vertical shift nine units upward, horizontal shrink by a factor of $\frac{1}{2}$
(c)

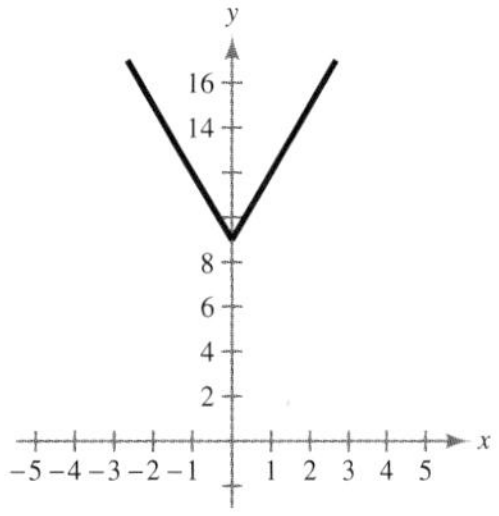

(d) $h(x) = f(3x) + 9$

77. (a) Rational function $f(x) = \frac{1}{x}$
(b) Horizontal shift one unit left, reflection in the x-axis, vertical stretch by a factor of 2, vertical shift three units downward
(c)

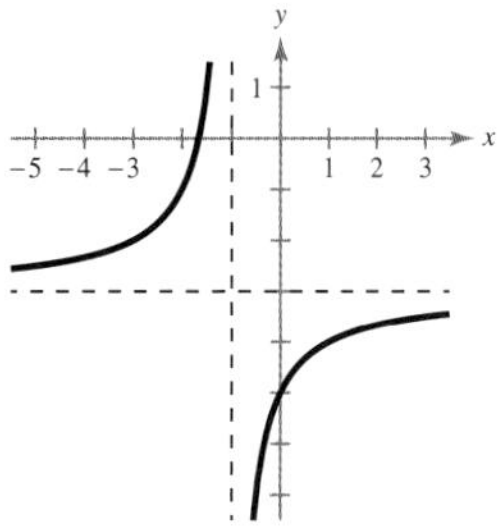

(d) $h(x) = -2f(x + 1) - 3$

79. -7 **81.** 5 **83.** 35 **85.** -97

87. $f(x) = x^2$, $g(x) = x + 3$ **89.** $f(x) = \frac{2}{x}$, $g(x) = x + 4$

91.

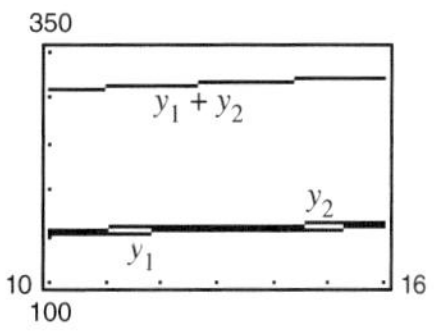

93. $f^{-1}(x) = \frac{x}{6}$ **95.** $f^{-1}(x) = 2x - 6$

97. (a) $f(g(x)) = f\left(\frac{3 - x}{4}\right) = 3 - 4\left(\frac{3 - x}{4}\right) = x$

$g(f(x)) = g(3 - 4x) = \frac{3 - (3 - 4x)}{4} = x$

(b)

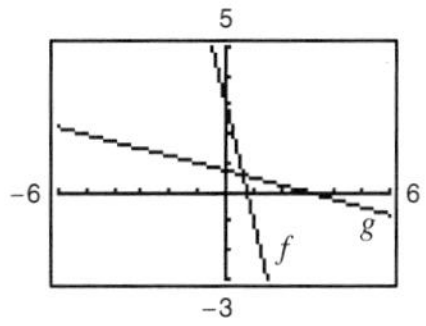

(c)

x	-1	0	1	2	3
$f(x)$	7	3	-1	-5	-9

x	7	3	-1	-5	-9
$g(x)$	-1	0	1	2	3

99.

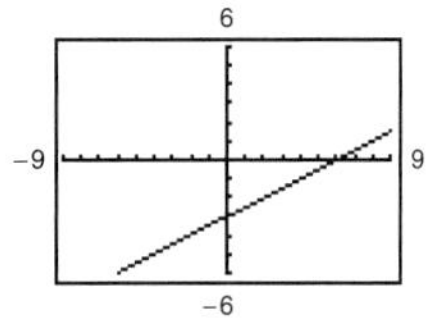

An inverse function exists.

101.

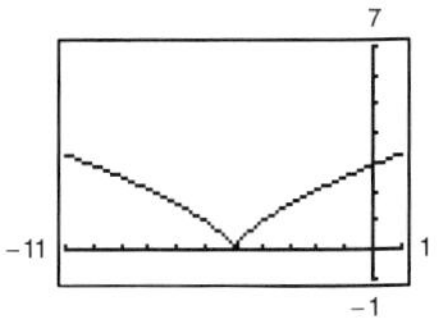

No inverse function

CHAPTER 1

103. $f^{-1}(x) = \dfrac{8x - 3}{7}$ **105.** $f^{-1}(x) = 6 - \dfrac{x^2}{16}, x \geq 0$

107. $y = -0.0043x + 3.779$; No; Answers will vary.

109. False; For example, $f(x) = 4 - x = f^{-1}(x)$.

Chapter Test (page 86)

1. (a) $5x + 2y - 8 = 0$ (b) $2x - 5y + 20 = 0$

2. $y = -x + 1$

3. No. To some x there correspond two values of y.

4. (a) -9 (b) 1 (c) $|t - 4| - 15$ **5.** $(-\infty, 3]$

6. $C = 48.47x + 50{,}000$; $P = 151.52x - 50{,}000$

7. Odd **8.** Even

9. Increasing: $(-2, 0)$, $(2, \infty)$; Decreasing: $(-\infty, -2)$, $(0, 2)$

10. Increasing: $(-2, 2)$; Constant: $(-\infty, -2)$, $(2, \infty)$

11.

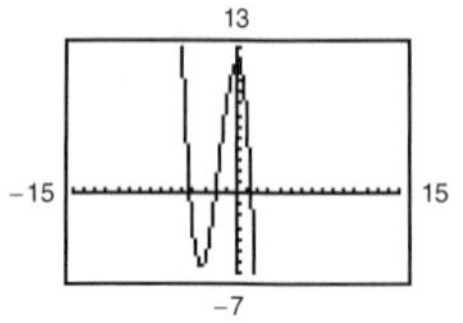

Relative minimum: about $(-3.33, -6.52)$

Relative maximum: $(0, 12)$

12.

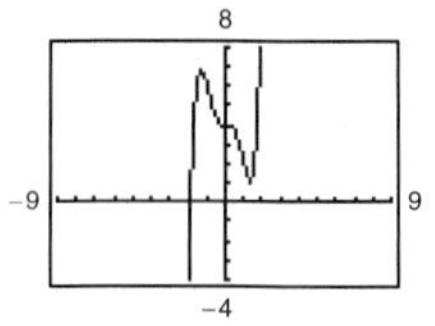

Relative minimum: about $(1.34, 1.10)$

Relative maximum: about $(-1.34, 6.90)$

13. (a) Cubic function $f(x) = x^3$

(b) Horizontal shift five units to the right, reflection in the x-axis, vertical stretch by a factor of 2, and vertical shift three units upward

(c)

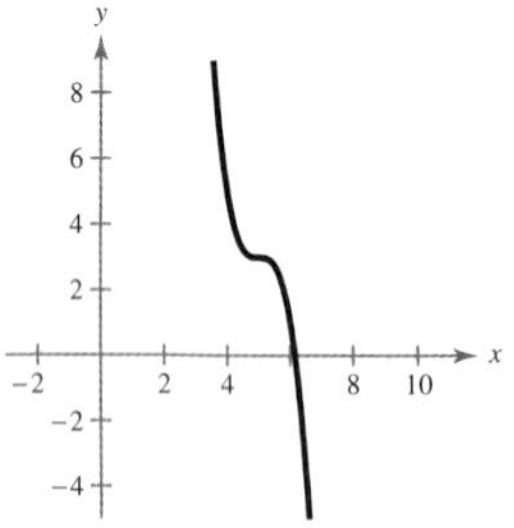

14. (a) Square root function $f(x) = \sqrt{x}$

(b) Reflection in the y-axis, horizontal shift two units to the left, horizontal shrink by a factor of $\frac{1}{9}$

(c)

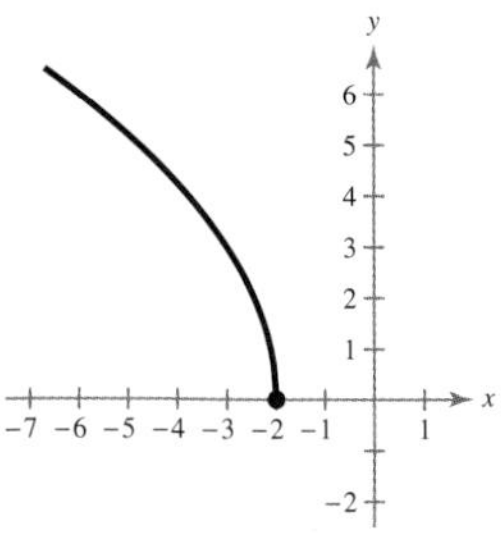

15. (a) Absolute value function $f(x) = |x|$

(b) Reflection in the y-axis (no effect), vertical stretch by a factor of 6, vertical shift eight units downward

(c)

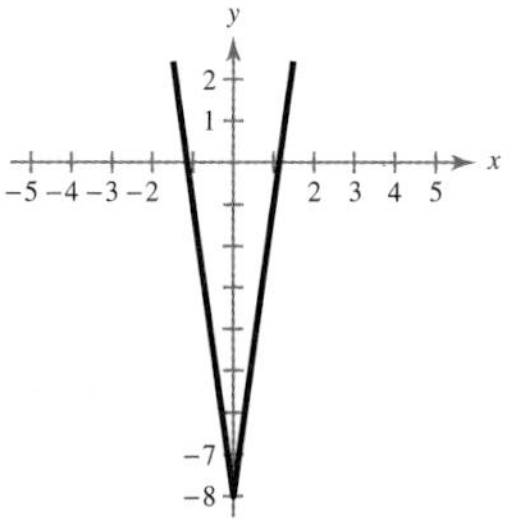

16. (a) $x^2 - \sqrt{2 - x}, (-\infty, 2]$ (b) $\dfrac{x^2}{\sqrt{2 - x}}, (-\infty, 2)$

(c) $2 - x, (-\infty, 2]$ (d) $\sqrt{2 - x^2}, \left[-\sqrt{2}, \sqrt{2}\right]$

17. $f^{-1}(x) = \sqrt[3]{x - 8}$ **18.** No inverse function

19. $f^{-1}(x) = \left(\frac{8}{3}x\right)^{2/3}, x \geq 0$ **20.** $C = 2.8025t + 27.144$; 2021

Chapter 2

Section 2.1 (page 96)

1. nonnegative integer, real **3.** Yes; (2, 3) **5.** c

6. d **7.** b **8.** a

9. (a)

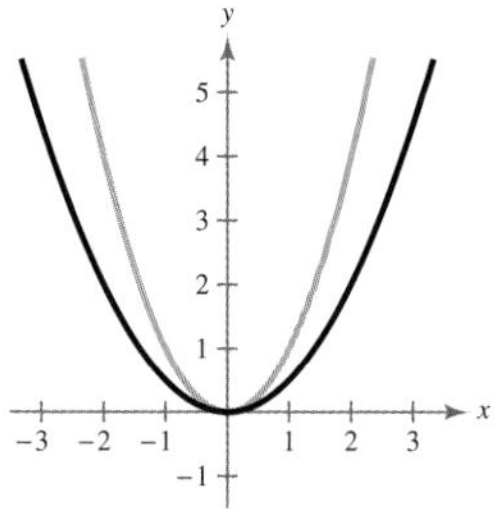

Vertical shrink

(b)

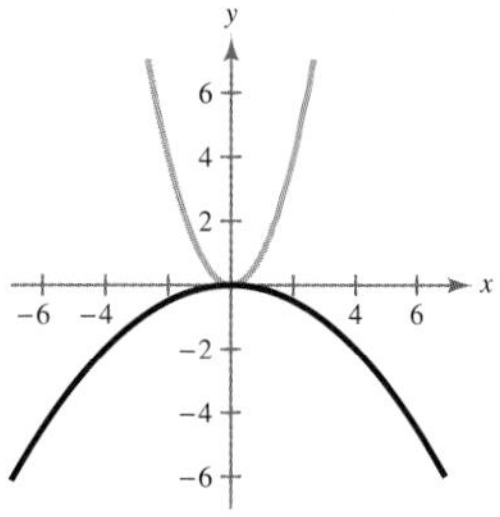

Vertical shrink and a reflection in the x-axis

(c)

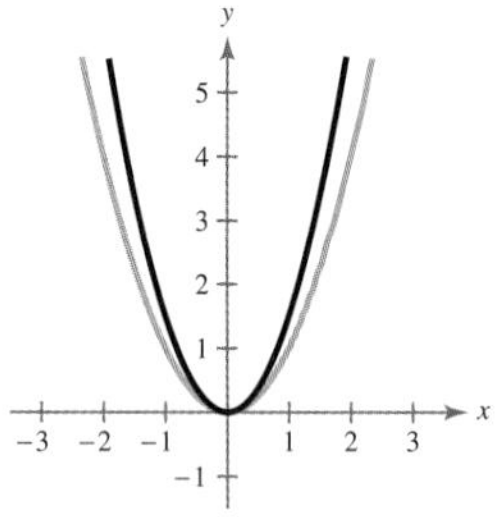

Vertical stretch

(d)

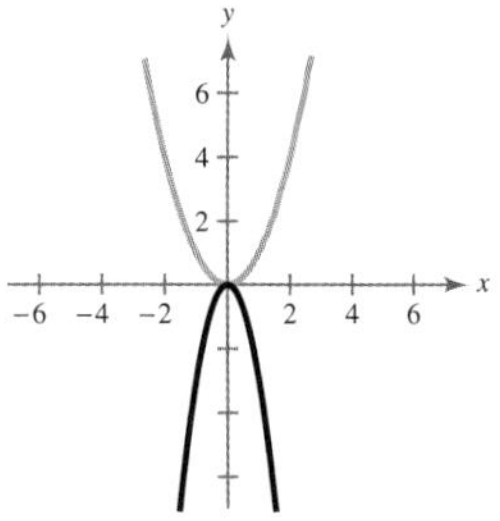

Vertical stretch and a reflection in the x-axis

11. (a)

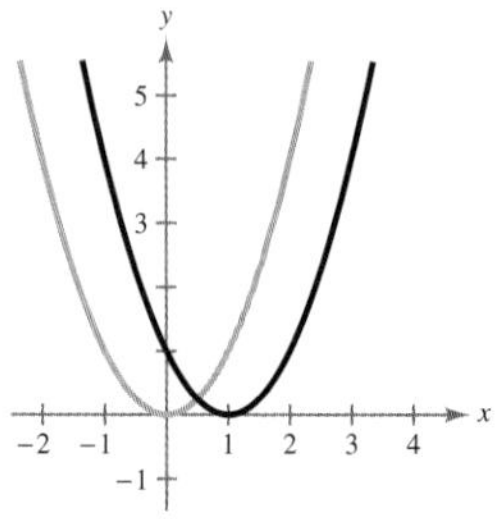

Right shift of one unit

(b)

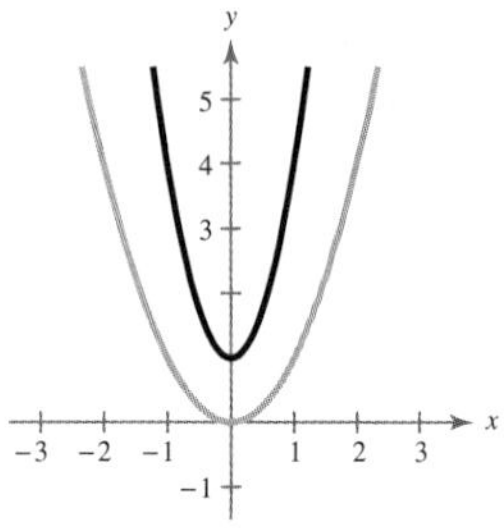

Horizontal shrink and an upward shift of one unit

(c)

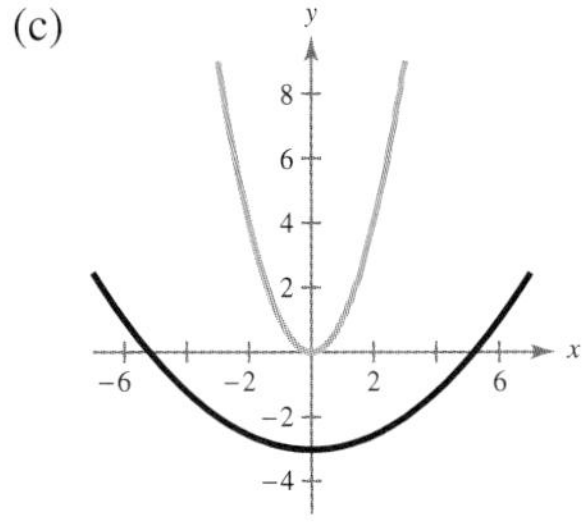

Horizontal stretch and a downward shift of three units

(d)

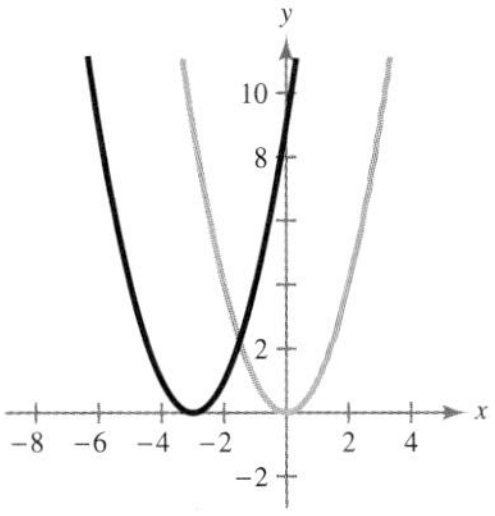

Left shift of three units

13. $f(x) = (x - 3)^2 - 9$

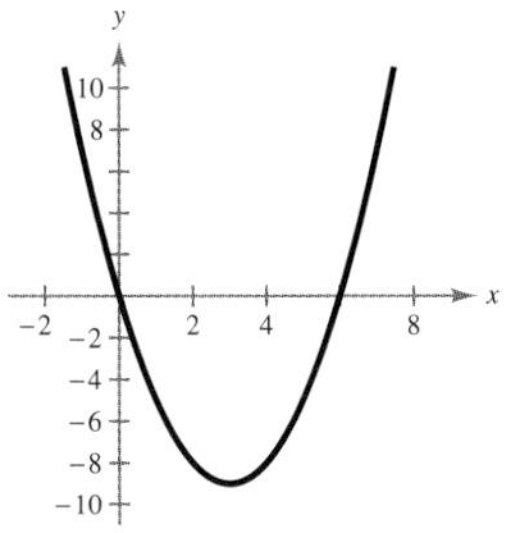

Vertex: $(3, -9)$
Axis of symmetry: $x = 3$
x-intercepts: $(0, 0), (6, 0)$

15. $h(x) = (x - 4)^2$

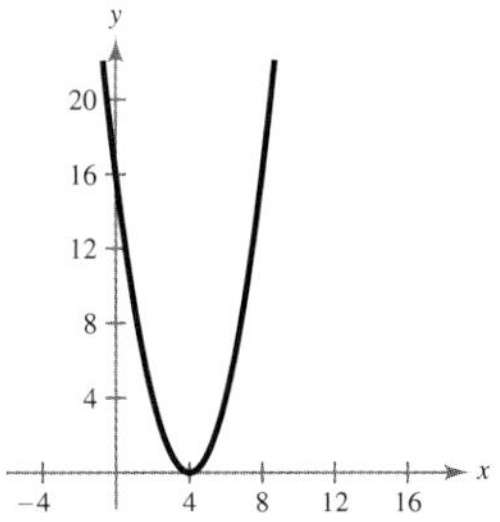

Vertex: $(4, 0)$
Axis of symmetry: $x = 4$
x-intercept: $(4, 0)$

17. $f(x) = (x - 3)^2 - 7$

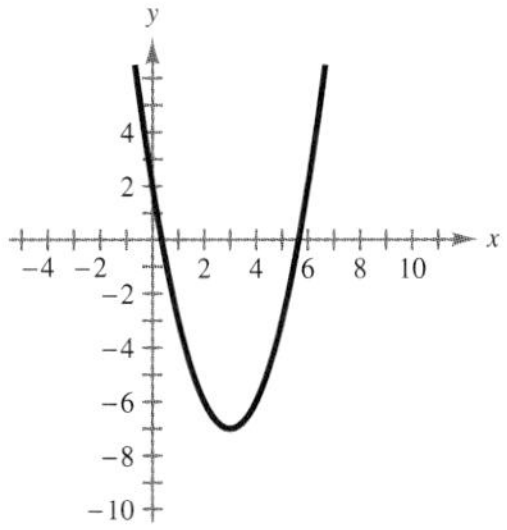

Vertex: $(3, -7)$
Axis of symmetry: $x = 3$
x-intercepts: $(3 \pm \sqrt{7}, 0)$

19. $f(x) = (x - 4)^2 + 5$

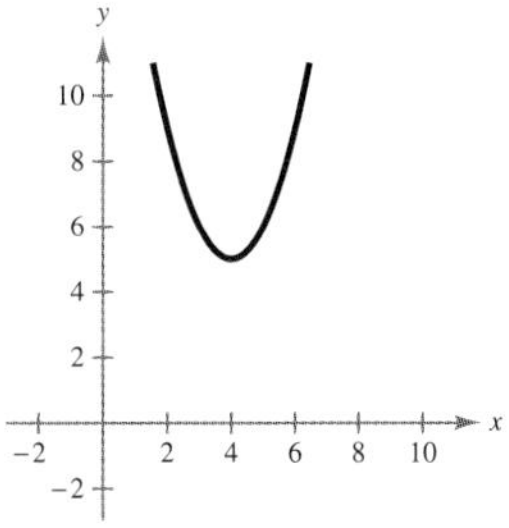

Vertex: $(4, 5)$
Axis of symmetry: $x = 4$
No x-intercept

21. $f(x) = \left(x - \frac{1}{2}\right)^2 + 1$

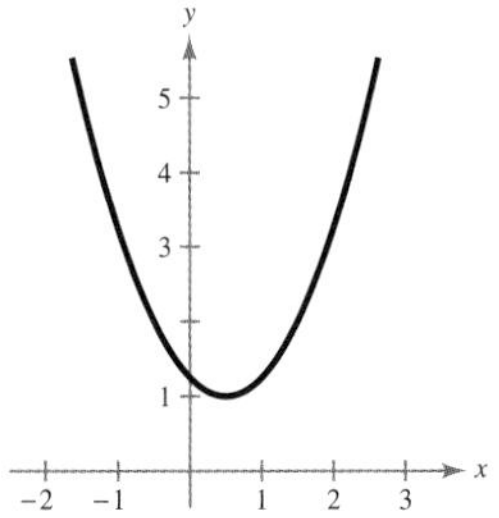

Vertex: $\left(\frac{1}{2}, 1\right)$
Axis of symmetry: $x = \frac{1}{2}$
No x-intercept

23. $f(x) = -(x - 1)^2 + 6$

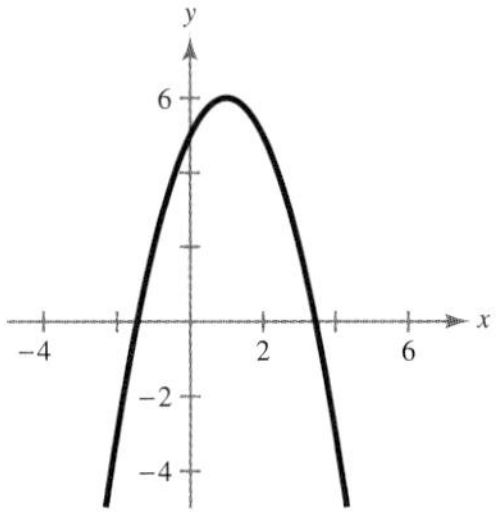

Vertex: $(1, 6)$
Axis of symmetry: $x = 1$
x-intercepts: $(1 \pm \sqrt{6}, 0)$

25. $h(x) = 4\left(x - \frac{1}{2}\right)^2 + 20$

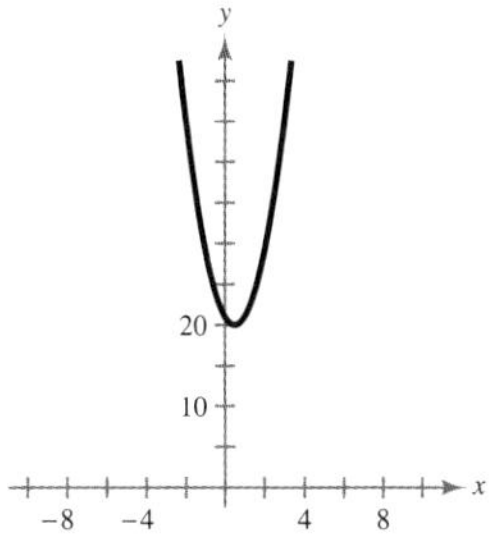

Vertex: $\left(\frac{1}{2}, 20\right)$
Axis of symmetry: $x = \frac{1}{2}$
No x-intercept

27. Parabola opening upward
Vertex: $(-4, -5)$
x-intercepts: $(-4 \pm \sqrt{5}, 0)$

29. Parabola opening downward
Vertex: $(1, 16)$
x-intercepts: $(-3, 0), (5, 0)$

31. Parabola opening downward
Vertex: $(4, 1)$
x-intercepts: $\left(4 \pm \frac{1}{2}\sqrt{2}, 0\right)$

33. $f(x) = -(x + 1)^2 + 4$ **35.** $f(x) = (x + 2)^2 + 5$

37. $f(x) = 4(x - 1)^2 - 2$ **39.** $f(x) = -\frac{104}{125}\left(x - \frac{1}{2}\right)^2 + 1$

41. $(5, 0), (-1, 0)$ **43.** $(-4, 0)$

45.

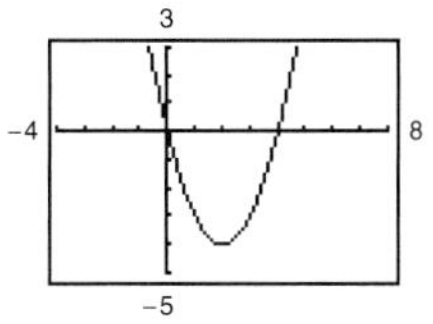

$(0, 0), (4, 0)$

47.

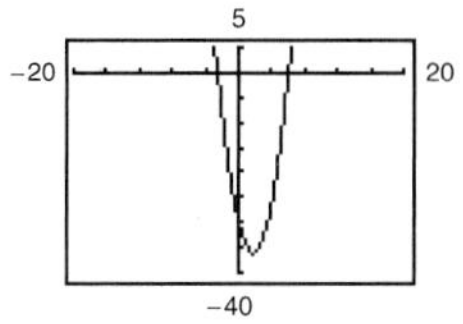

$\left(-\frac{5}{2}, 0\right), (6, 0)$

49.

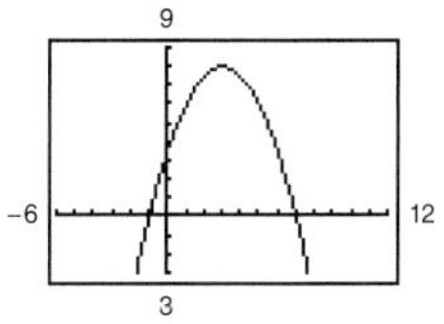

$(7, 0), (-1, 0)$

51. *Sample answer:*
$f(x) = x^2 - 9$
$g(x) = -x^2 + 9$

53. *Sample answer:*
$f(x) = 2x^2 + 7x + 3$
$g(x) = -2x^2 - 7x - 3$

55. 55, 55 **57.** 12, 6

59. (a)

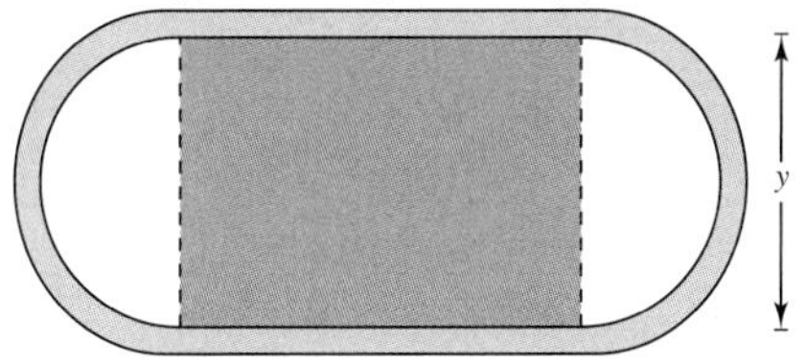

(b) $r = \frac{1}{2}y$; $d = y\pi$ (c) $y = \dfrac{200 - 2x}{\pi}$

(d) $A = x\left(\dfrac{200 - 2x}{\pi}\right)$

(e)

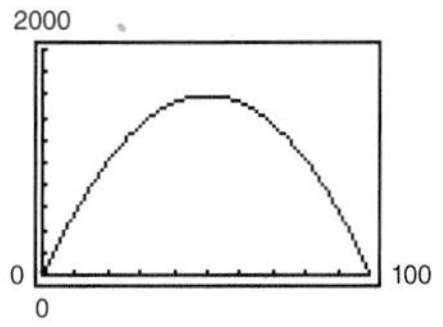

$x = 50$ m, $y = \dfrac{100}{\pi}$ m

61. (a)
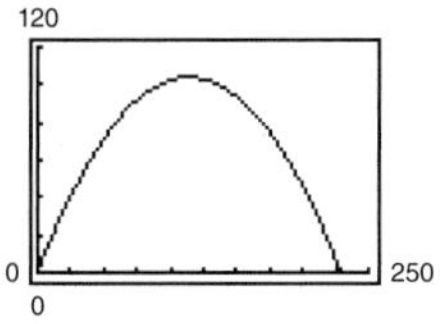

(b) $\frac{3}{2}$ ft

(c) About 104 ft (d) About 228.6 ft

63. (a) $A = -2x^2 + 112x - 600$ (b) $x = 28$ in.

65. (a)
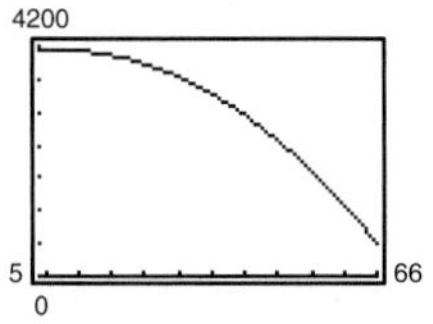

(b) 1958; about 4133 cigarettes

(c) Yes; Consumption began decreasing more.

67. True; The vertex is (5, 0).

69. $2 > 1$, which indicates a vertical stretch.

71. c, d **73.** Horizontal shift z units to the right

75. Vertical stretch $(z > 1)$ or shrink $(0 < z < 1)$ and horizontal shift three units to the right

77. $b = \pm 20$ **79.** $b = \pm 8$ **81.** Proof

83. $y = -x^2 + 5x - 4$; Answers will vary.

85. Model (a); The profits are positive and rising.

87. (1.2, 6.8) **89.** Answers will vary.

Section 2.2 (page 109)

1. continuous **3.** (a) solution (b) $(x - a)$ (c) $(a, 0)$

5. No **7.** $f(x_2) > 0$ **9.** f **10.** h **11.** c

12. a **13.** e **14.** d **15.** g **16.** b

17.
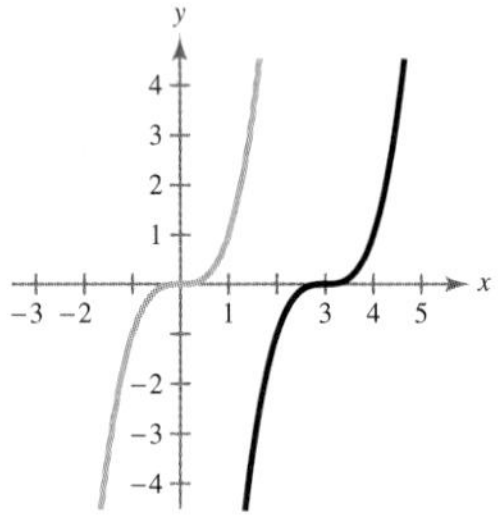

Horizontal shift three units to the right

19.
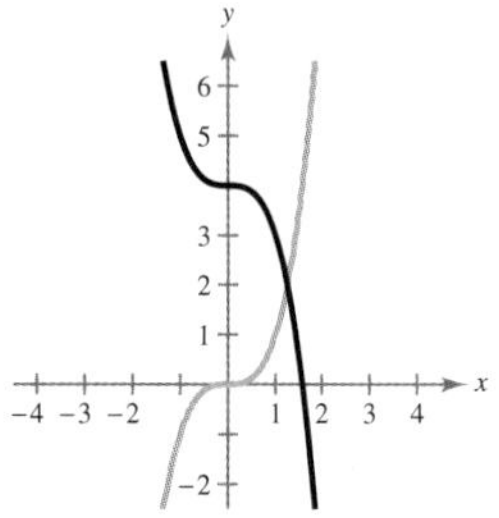

Reflection in the x-axis and vertical shift four units upward

21.
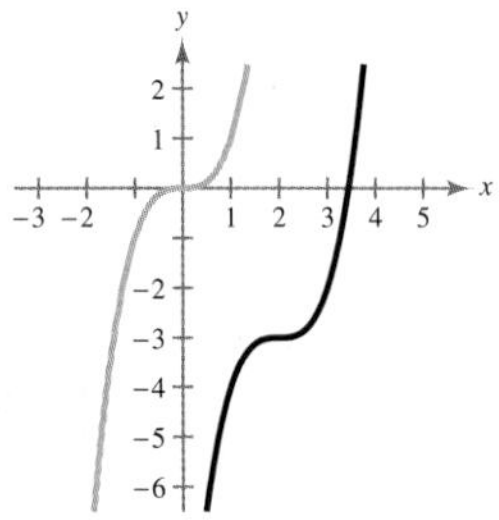

Horizontal shift two units to the right and vertical shift three units downward

23.
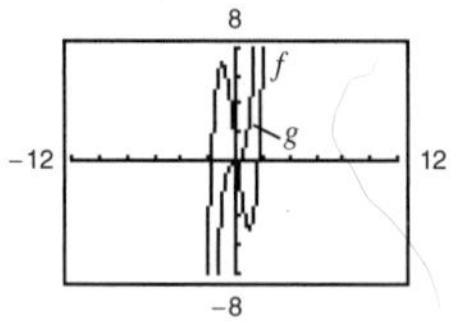

Yes; The leading coefficients are the same.

25.
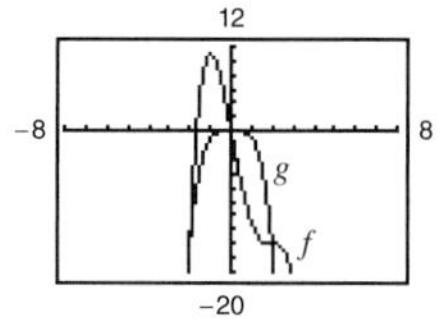

Yes; The leading coefficients are the same.

27.
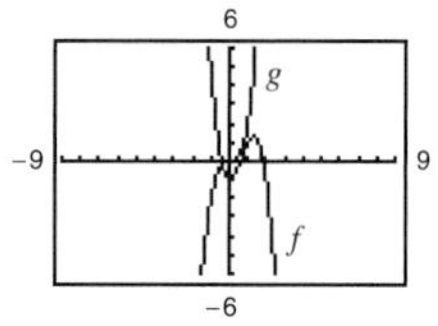

No; The leading coefficients have different signs.

29. Rises to the left, rises to the right

31. Falls to the left, falls to the right

33. Falls to the left, rises to the right

35. Falls to the left, falls to the right

37. (a) $(2 \pm \sqrt{3}, 0)$

(b)
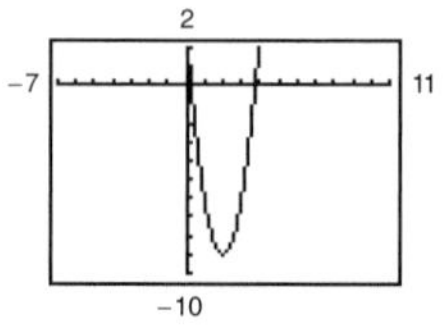

(c) About (0.27, 0), about (3.73, 0); Answers are approximately the same.

39. (a) $(-1, 0)$, $(1, 0)$

(b)
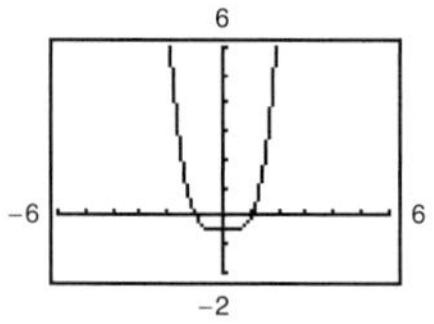

(c) $(-1, 0)$, $(1, 0)$; Answers are the same.

41. (a) $(0, 0)$, $\left(\frac{1}{2}, 0\right)$

(b)
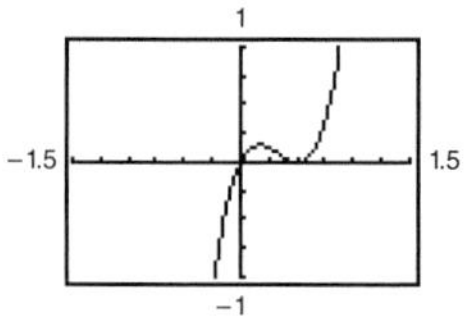

(c) (0, 0), (0.5, 0); Answers are the same.

43. (a) $(0, 0)$, $(\pm\sqrt{2}, 0)$

(b)
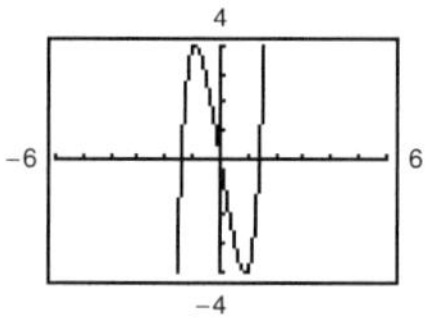

(c) About $(-1.41, 0)$, $(0, 0)$, about $(1.41, 0)$; Answers are approximately the same.

45. (a) $(\pm\sqrt{5}, 0)$

(b)
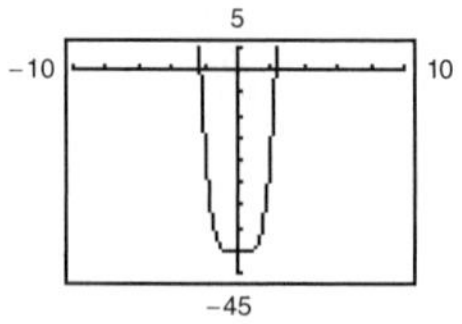

(c) About $(-2.236, 0)$, about $(2.236, 0)$; Answers are approximately the same.

47. (a) $(4, 0)$, $(\pm 5, 0)$

(b)
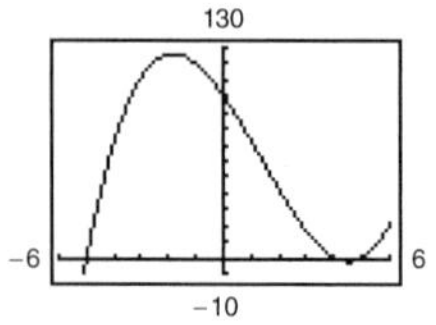

(c) $(-5, 0)$, $(4, 0)$, $(5, 0)$; Answers are the same.

49. ± 5 (multiplicity 1) **51.** 3 (multiplicity 2)

53. $1, -2$ (multiplicity 1)

55. 2 (multiplicity 2), 0 (multiplicity 1)

57. $\dfrac{-5 \pm \sqrt{37}}{2}$ (multiplicity 1)

59.

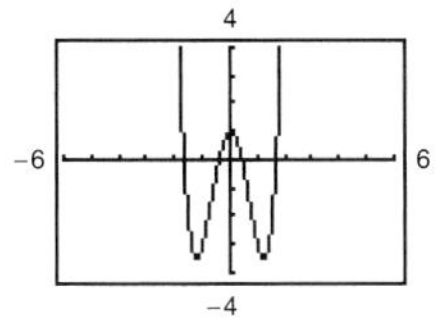

Zeros: $\pm 1.680, \pm 0.421$
Relative minima: $(\pm 1.225, -3.500)$
Relative maximum: $(0, 1)$

61.

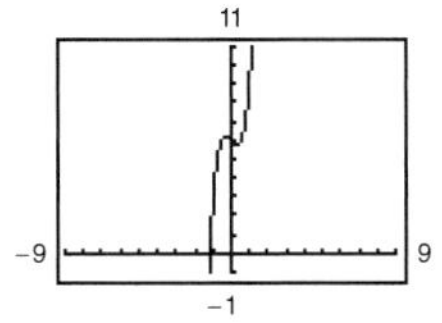

Zero: -1.178
Relative minimum: $(0.324, 5.782)$
Relative maximum: $(-0.324, 6.218)$

63.

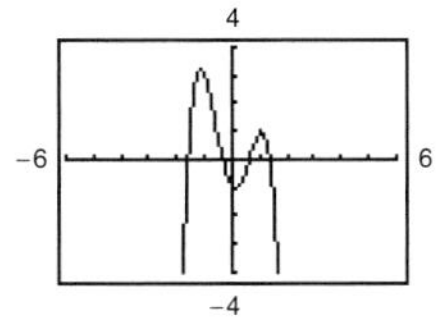

Zeros: $-1.618, -0.366, 0.618, 1.366$
Relative minimum: $(0.101, -1.050)$
Relative maxima: $(-1.165, 3.267), (1.064, 1.033)$

65. *Sample answer:* $f(x) = x^2 - 7x$
67. *Sample answer:* $f(x) = x^3 + 6x^2 + 8x$
69. *Sample answer:* $f(x) = x^4 - 4x^3 - 9x^2 + 36x$
71. *Sample answer:* $f(x) = x^2 - 2x - 1$
73. *Sample answer:* $f(x) = x^3 - 6x^2 + 7x + 2$
75. *Sample answer:* $f(x) = x^4 + 2x^3 - 23x^2 - 24x + 144$
77. *Sample answer:* $f(x) = -x^3 - 4x^2 - 5x - 2$

79. *Sample answer:*

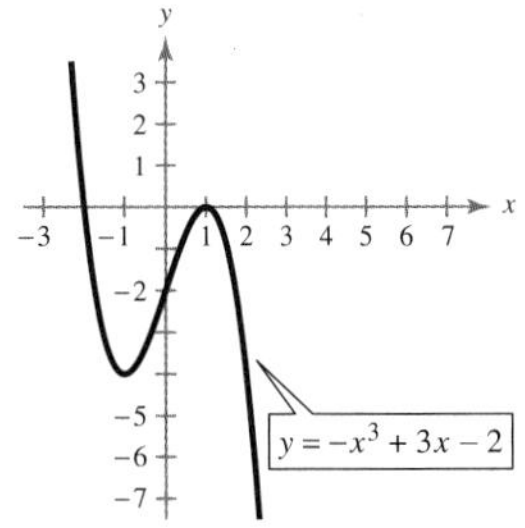

81. *Sample answer:*

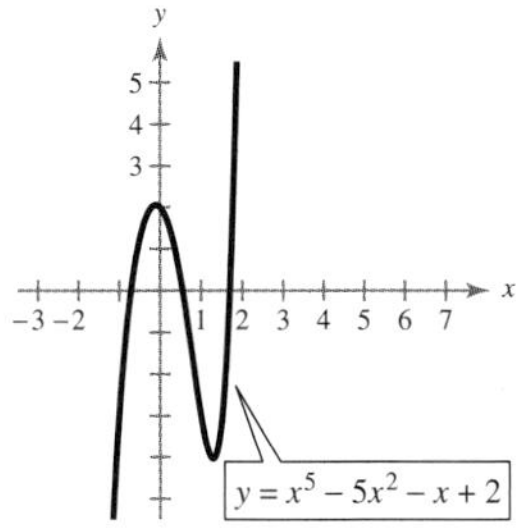

83. (a) Falls to the left, rises to the right
(b) $(0, 0), (\pm 5, 0)$
(c) and (d)

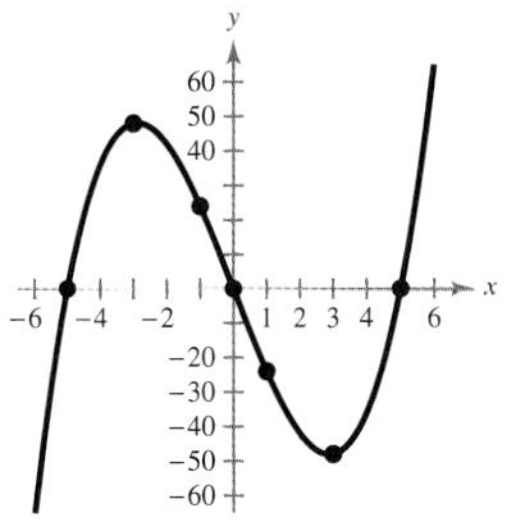

85. (a) Falls to the left, rises to the right
(b) $(0, 0), (3, 0)$
(c) and (d)

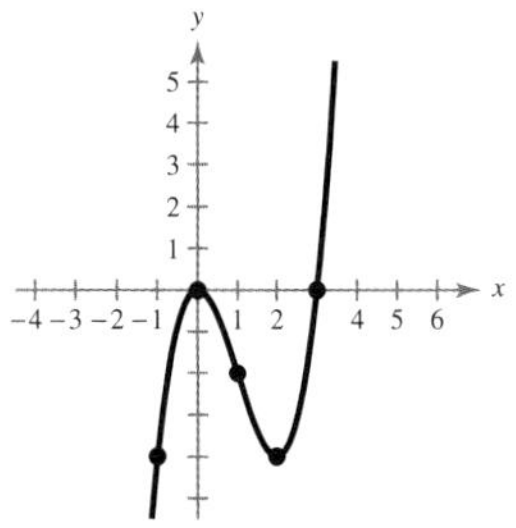

87. (a) Falls to the left, falls to the right
(b) $(\pm 2, 0), (\pm\sqrt{5}, 0)$
(c) and (d)

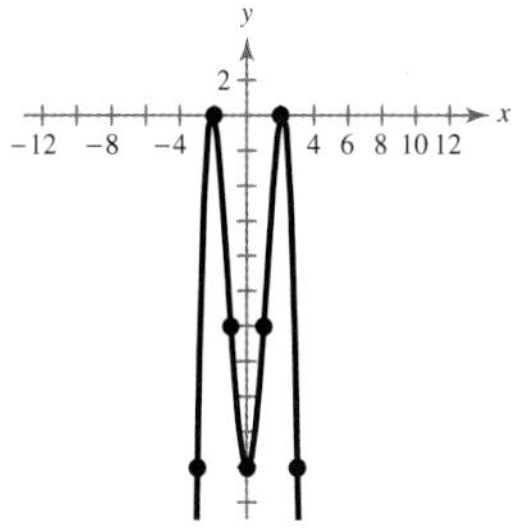

89. (a) Falls to the left, rises to the right
(b) $(\pm 3, 0)$
(c) and (d)

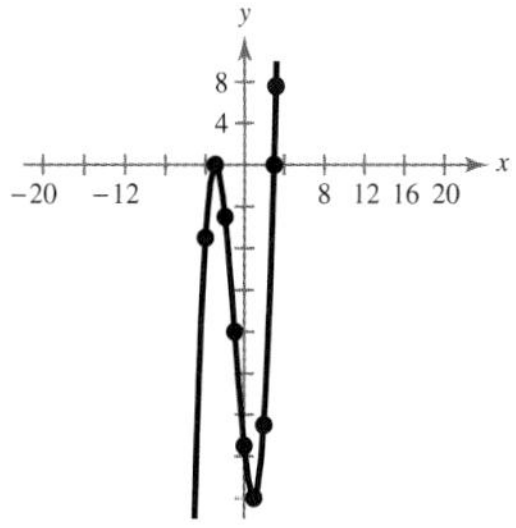

91. (a) Falls to the left, falls to the right
(b) $(-2, 0), (2, 0)$
(c) and (d)

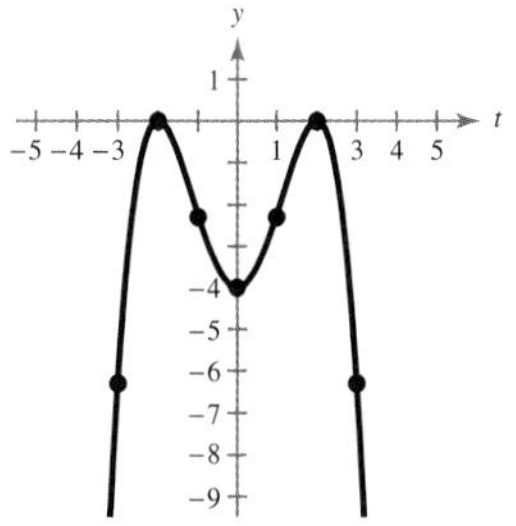

93. (a) $(-1, 0), (1, 2), (2, 3)$ (b) $-0.879, 1.347, 2.532$
95. (a) $(-2, -1), (0, 1)$ (b) $-1.585, 0.779$
97. (a) $(-1, 0), (3, 4)$ (b) $-0.578, 3.418$

99.

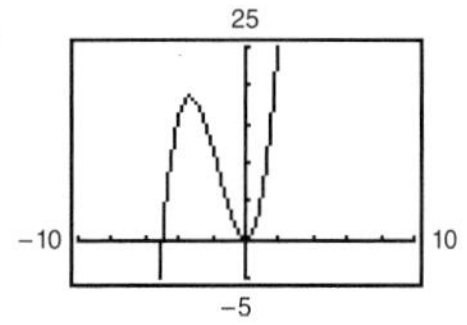

Two x-intercepts

101.

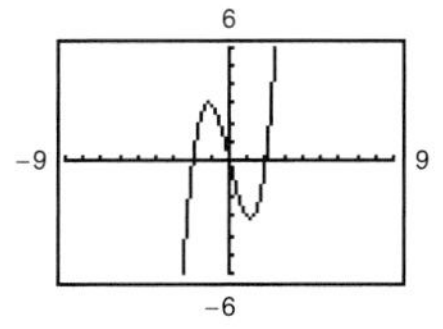

Origin symmetry
Three x-intercepts

103.

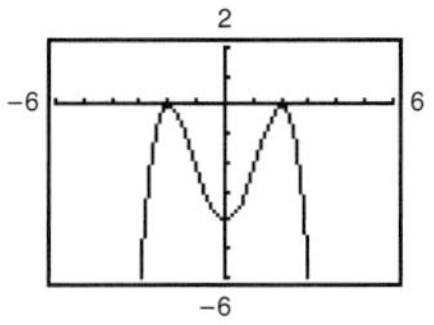

y-axis symmetry
Two *x*-intercepts

105.

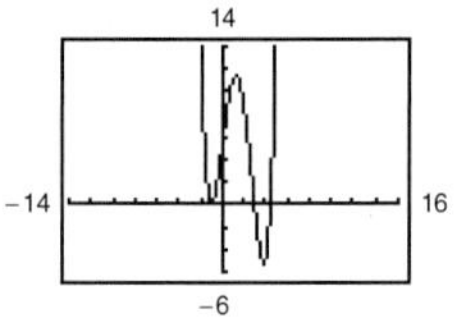

Three *x*-intercepts

107. (a) $V = x(36 - 2x)^2$ (b) $0 < x < 18$

(c)

Height, x	Volume, V
1	1156
2	2048
3	2700
4	3136
5	3380
6	3456
7	3388

24 cm × 24 cm × 6 cm

(d)

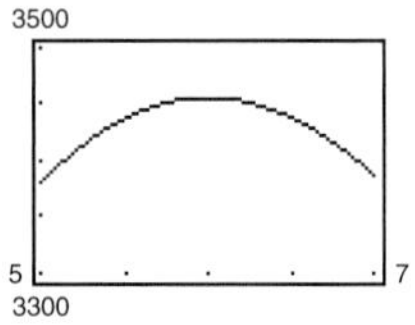

$x = 6$; The results are the same.

109. (200, 160)

111.

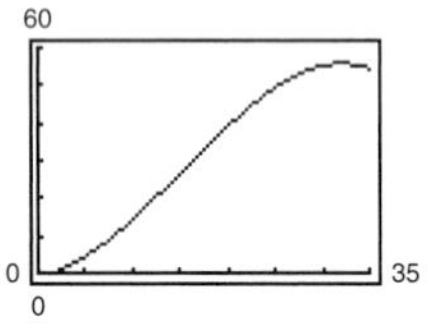

About 15.22 yr

113. True; $f(x) = x^6$ has only one real zero.

115. False; The right-hand and left-hand behaviors of the graph are the same.

117.

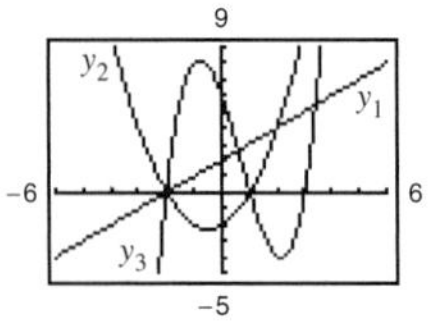

119. The leading coefficient is −1, not 1, so the graph rises to the left and falls to the right.

121. 69 **123.** $-\frac{1408}{49} \approx -28.73$ **125.** 109

127. $x \le -\frac{1}{2}, x \ge 1$

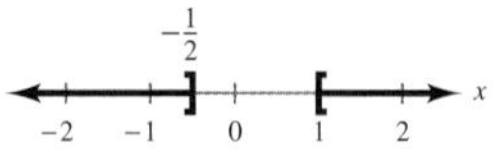

Section 2.3 (page 124)

1. $f(x)$ is the dividend, $d(x)$ is the divisor, $q(x)$ is the quotient, and $r(x)$ is the remainder.

3. constant term, leading coefficient **5.** upper, lower

7. 15 **9.** $(x + 4)(x + 2)$ **11.** $(x + 2)(x + 6)(x - 3)$

13. $x^2 - x - 20 - \dfrac{54}{x - 3}$ **15.** $2x^2 - 6x + 18 - \dfrac{47}{x + 3}$

17. $5x - 1$ **19.** $x - \dfrac{x + 9}{x^2 + 1}$ **21.** $2x - \dfrac{17x - 5}{x^2 - 2x + 1}$

23. $3x^2 - 2x + 5, x \ne 5$ **25.** $6x^2 + 25x + 74 + \dfrac{248}{x - 3}$

27. $9x^2 - 16, x \ne 2$ **29.** $x^2 - 8x + 64, x \ne -8$

31. $4x^2 + 14x - 30, x \ne -\frac{1}{2}$

33.

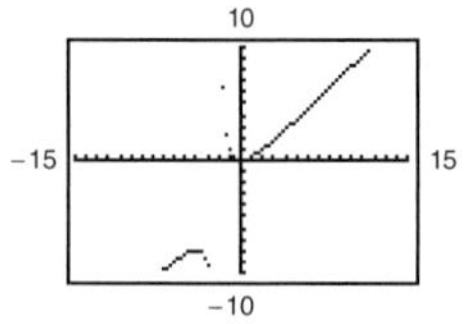

35.

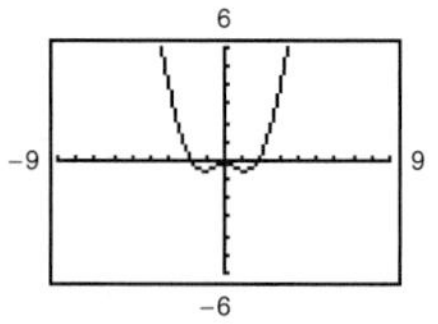

37. $f(x) = (x - 2)(x^2 + x - 16) - 15, f(2) = -15$

39. $f(x) = \left(x - \sqrt{2}\right)\left[x^2 + \left(3 + \sqrt{2}\right)x + 3\sqrt{2}\right] - 8$, $f\left(\sqrt{2}\right) = -8$

41. $f(x) = \left(x - 1 + \sqrt{3}\right)\left[4x^2 - \left(2 + 4\sqrt{3}\right)x - \left(2 + 2\sqrt{3}\right)\right]$, $f\left(1 - \sqrt{3}\right) = 0$

43. (a) −2 (b) 1 (c) 36 (d) 5

45. (a) −35 (b) −22 (c) −10 (d) −211

47. $(x - 5)(x + 4)(x + 1)$; Zeros: 5, −4, −1

49. $(2x + 3)(x - 3)(x - 7)$; Zeros: $-\frac{3}{2}$, 3, 7

51. $\left(x - \sqrt{3}\right)\left(x + \sqrt{3}\right)(x + 2)$; Zeros: $\pm\sqrt{3}, -2$

53. (a) Answers will vary. (b) $(2x - 1)$
(c) $(x + 2)(x - 1)(2x - 1)$ (d) $-2, 1, \frac{1}{2}$
(e)

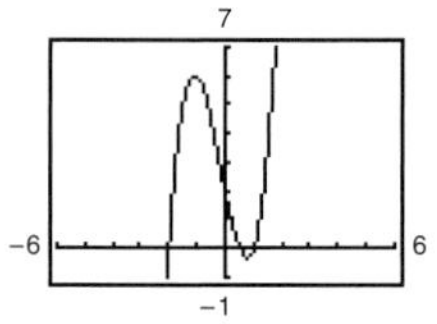

55. (a) Answers will vary. (b) $(4x + 3), (2x - 1)$
(c) $(4x + 3)(2x - 1)(x - 4)(x + 2)$ (d) $-2, -\frac{3}{4}, \frac{1}{2}, 4$
(e)

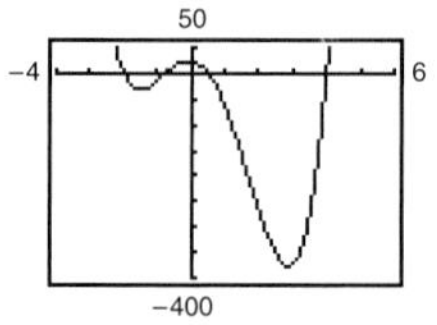

57. (a) Answers will vary. (b) $(x + 7), (3x - 2)$
(c) $(2x + 1)(3x - 2)(x + 7)$ (d) $-7, -\frac{1}{2}, \frac{2}{3}$
(e)

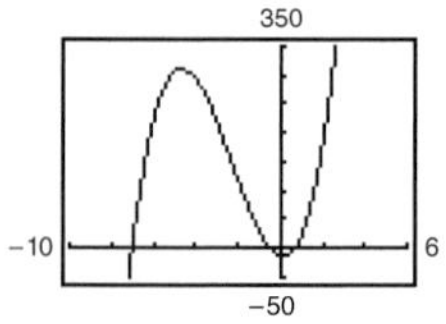

59. $-1, \frac{1}{2}$ **61.** $-1, \frac{3}{2}, 3, 5$

63. 4, 2, or 0 positive real zeros, no negative real zeros

65. 2 or 0 positive real zeros, 2 or 0 negative real zeros

67. (a) 1 positive real zero, 2 or 0 negative real zeros
(b) $\pm 1, \pm 2, \pm 4$
(c)
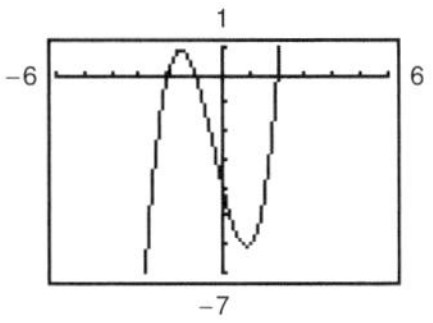

(d) $-2, -1, 2$

69. (a) 3 or 1 positive real zero(s), 1 negative real zero
(b) $\pm 1, \pm 2, \pm 4, \pm 8, \pm\frac{1}{2}$
(c)
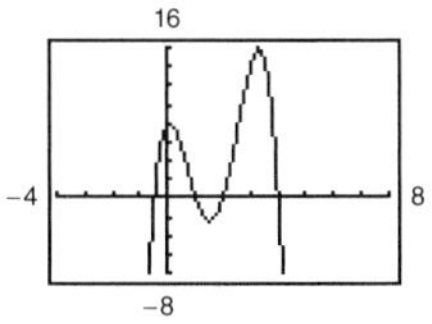

(d) $-\frac{1}{2}, 1, 2, 4$

71. (a) 2 or 0 positive real zeros, 1 negative real zero
(b) $\pm 1, \pm 3, \pm\frac{1}{2}, \pm\frac{3}{2}, \pm\frac{1}{4}, \pm\frac{3}{4}, \pm\frac{1}{8}, \pm\frac{3}{8}, \pm\frac{1}{16}, \pm\frac{3}{16}, \pm\frac{1}{32}, \pm\frac{3}{32}$
(c)
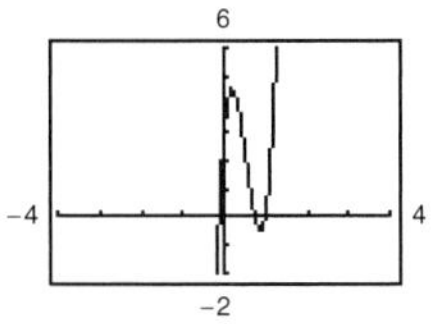

(d) $-\frac{1}{8}, \frac{3}{4}, 1$

73. Answers will vary; about 1.937, about 3.705
75. Answers will vary; about -2.152, about 0.611, about 3.041
77. $\pm\frac{3}{2}, \pm 1$ **79.** $-\frac{1}{3}, \frac{1}{2}, -2$ **81.** d **82.** a **83.** b
84. c **85.** $-\frac{1}{2}, 2 \pm \sqrt{3}, 1$ **87.** $-1, \frac{3}{2}, 4 \pm \sqrt{17}$
89. $-3, 0, \dfrac{3 \pm \sqrt{14}}{5}$ **91.** $-3, 1$ **93.** $-6, \frac{1}{2}, 1$
95. $-3, -\frac{3}{2}, \frac{1}{2}, 4$ **97.** $-3, 0, \dfrac{-1 \pm \sqrt{7}}{4}$
99. $-\frac{3}{2}, -\frac{5}{4}, -1, 1, 2$
101. (a) $-2, 0.268, 3.732$ (b) -2
(c) $h(t) = (t + 2)(t - 2 + \sqrt{3})(t - 2 - \sqrt{3})$
103. (a) $0, 3, 4, -1.414, 1.414$ (b) $0, 3, 4$
(c) $h(x) = x(x - 3)(x - 4)(x + \sqrt{2})(x - \sqrt{2})$
105. (a)
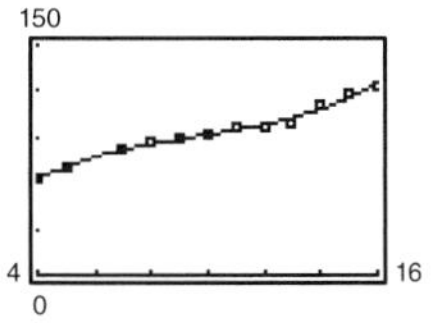

(b) The model fits the data well.
(c) 215.8 subscriptions; *Sample answer:* No, the number of subscriptions is unlikely to increase that rapidly.
107. (a) $V(x) = x^2(120 - 4x)$
(b)
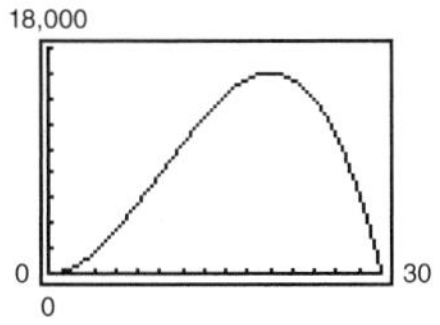

$20 \times 20 \times 40$
(c) $15, \dfrac{15 \pm 15\sqrt{5}}{2}$; $\dfrac{15 - 15\sqrt{5}}{2}$ represents a negative volume.
109. True; $(4x - 7) = 4\left(x - \frac{7}{4}\right)$ **111.** $-2(x - 1)^2(x + 2)$
113. $k = -1$, not 1. **115.** 5 **117.** $\pm\dfrac{\sqrt{17}}{2}$ **119.** $-\dfrac{4}{3}, 5$

Section 2.4 (page 133)

1. (a) ii (b) iii (c) i **3.** 15; $9i$ **5.** $-2 + 4i$
7. $a = -9, b = 4$ **9.** $a = 3, b = 5$ **11.** $2 + 5i$
13. 12 **15.** $1 - 8i$ **17.** -11 **19.** $0.3i$
21. $7 + 3i$ **23.** 1 **25.** $-14 + 20i$ **27.** $\frac{19}{6} + \frac{37}{6}i$
29. $-4.2 + 7.5i$ **31.** $12 + 20i$ **33.** $5 + i$
35. $-20 + 32i$ **37.** 11 **39.** $-13 + 84i$ **41.** $80i$
43. $6 + 2i$; 40 **45.** $-1 - \sqrt{7}i$; 8 **47.** $-\sqrt{29}i$; 29
49. $9 + \sqrt{6}i$; 87 **51.** $-6i$ **53.** $\frac{8}{41} + \frac{10}{41}i$ **55.** $\frac{12}{13} + \frac{5}{13}i$
57. $-\frac{40}{1681} - \frac{9}{1681}i$ **59.** $-\frac{1}{2} - \frac{5}{2}i$ **61.** $3(\sqrt{2} - \sqrt{6})i$
63. $-2\sqrt{3}$ **65.** -10 **67.** $-2 - 4\sqrt{6}i$ **69.** $\pm 5i$
71. $1 \pm i$ **73.** $-2 \pm \dfrac{i}{2}$ **75.** $\dfrac{1}{8} \pm \dfrac{\sqrt{11}}{8}i$ **77.** $2 \pm \sqrt{2}i$
79. $\dfrac{5}{7} \pm \dfrac{5\sqrt{15}}{7}$ **81.** $-1 + 6i$ **83.** $-375\sqrt{3}i$ **85.** i
87. (a) $3.12 - 0.97i$ (b) $12.82 + 5.28i$
89. False; Any real number is equal to its conjugate.
91. $\sqrt{-6}$ should be written in standard form before multiplying.
93. Proof **95.** $16x^2 - 25$ **97.** $3x^2 + \frac{23}{2}x - 2$

Section 2.5 (page 140)

1. Fundamental Theorem, Algebra **3.** n linear factors
5. 3 **7.** 5 **9.** 2 **11 and 13.** Answers will vary.
15. Zeros: $4, -i, i$; One real zero; they are the same.
17. Zeros: $\sqrt{2}i, \sqrt{2}i, -\sqrt{2}i, -\sqrt{2}i$; No real zeros; they are the same.
19. $2 \pm \sqrt{3}$; $(x - 2 - \sqrt{3})(x - 2 + \sqrt{3})$
21. $6 \pm \sqrt{10}$; $(x - 6 - \sqrt{10})(x - 6 + \sqrt{10})$
23. $\pm 5i$; $(x + 5i)(x - 5i)$
25. $\pm\frac{3}{2}, \pm\frac{3}{2}i$; $(2x - 3)(2x + 3)(2x - 3i)(2x + 3i)$
27. $\dfrac{1 \pm \sqrt{223}i}{2}$; $\left(z - \dfrac{1 - \sqrt{223}i}{2}\right)\left(z - \dfrac{1 + \sqrt{223}i}{2}\right)$
29. $\pm i, \pm 3i$; $(x + i)(x - i)(x + 3i)(x - 3i)$
31. $\frac{5}{3}, \pm 4i$; $(3x - 5)(x - 4i)(x + 4i)$
33. $-1, 2 \pm i$; $(x + 1)(x - 2 - i)(x - 2 + i)$
35. $1 \pm \sqrt{5}i, -\frac{1}{5}$; $(5x + 1)(x - 1 + \sqrt{5}i)(x - 1 - \sqrt{5}i)$
37. $2, 2, \pm 2i$; $(x - 2)^2(x + 2i)(x - 2i)$
39. (a) $7 \pm \sqrt{3}$ (b) $(x - 7 - \sqrt{3})(x - 7 + \sqrt{3})$
(c) $(7 \pm \sqrt{3}, 0)$
41. (a) $\frac{3}{2}, \pm 2i$ (b) $(2x - 3)(x - 2i)(x + 2i)$ (c) $\left(\frac{3}{2}, 0\right)$
43. (a) $-6, 3 \pm 4i$ (b) $(x + 6)(x - 3 - 4i)(x - 3 + 4i)$
(c) $(-6, 0)$
45. (a) $\pm 5i, \pm 2i$ (b) $(x + 5i)(x - 5i)(x + 2i)(x - 2i)$
(c) None
47. *Sample answer:* $f(x) = x^3 - 5x^2 + x - 5$
49. *Sample answer:* $f(x) = x^4 - 6x^3 + 14x^2 - 16x + 8$
51. *Sample answer:* $f(x) = x^4 + 3x^3 - 7x^2 + 15x$
53. $f(x) = -2x^4 + 2x^3 + 2x^2 + 2x + 4$
55. $f(x) = x^3 + x^2 - 2x + 12$
57. (a) $(x^2 + 1)(x^2 - 7)$ (b) $(x^2 + 1)(x + \sqrt{7})(x - \sqrt{7})$
(c) $(x + i)(x - i)(x + \sqrt{7})(x - \sqrt{7})$
59. (a) $(x^2 - 6)(x^2 - 2x + 3)$
(b) $(x + \sqrt{6})(x - \sqrt{6})(x^2 - 2x + 3)$
(c) $(x + \sqrt{6})(x - \sqrt{6})(x - 1 - \sqrt{2}i)(x - 1 + \sqrt{2}i)$

61. $1, \pm 2i$ **63.** $-3, 5 \pm 2i$ **65.** $-\frac{2}{3}, 1 \pm \sqrt{3}i$

67. (a) 1.000, 2.000 (b) $-3 \pm \sqrt{2}i$

69. (a) 0.750 (b) $\frac{1}{2} \pm \frac{\sqrt{5}}{2}i$

71. No; Setting $h = 50$ and solving the resulting equation yields imaginary roots.

73. False; A third-degree polynomial must have at least one real zero.

75. The function should be $f(x) = (x + 2)(x - 3.5)(x + i)(x - i)$.

77. Parabola opening downward
Vertex: $\left(\frac{1}{2}, \frac{25}{4}\right)$

Section 2.6 (page 147)

1. rational functions **3.** vertical asymptote

5. (a) Domain: all real numbers x except $x = 1$

(b)

x	$f(x)$
0.5	-2
0.9	-10
0.99	-100
0.999	-1000

x	$f(x)$
1.5	2
1.1	10
1.01	100
1.001	1000

x	$f(x)$
5	0.25
10	$0.\overline{1}$
100	$0.\overline{01}$
1000	$0.\overline{001}$

x	$f(x)$
-5	$-0.1\overline{6}$
-10	$-0.\overline{09}$
-100	$-0.\overline{0099}$
-1000	$-0.\overline{000999}$

(c) f approaches $-\infty$ from the left and ∞ from the right of $x = 1$.

7. (a) Domain: all real numbers x except $x = 1$

(b)

x	$f(x)$
0.5	3
0.9	27
0.99	297
0.999	2997

x	$f(x)$
1.5	9
1.1	33
1.01	303
1.001	3003

x	$f(x)$
5	3.75
10	$3.\overline{3}$
100	$3.\overline{03}$
1000	$3.\overline{003}$

x	$f(x)$
-5	-2.5
-10	-2.727
-100	-2.97
-1000	-2.997

(c) f approaches ∞ from both the left and the right of $x = 1$.

9. (a) Domain: all real numbers x except $x = \pm 1$

(b)

x	$f(x)$
0.5	-1
0.9	-12.79
0.99	-147.8
0.999	-1498

x	$f(x)$
1.5	5.4
1.1	17.29
1.01	152.3
1.001	1502.3

x	$f(x)$
5	3.125
10	$3.\overline{03}$
100	$3.\overline{0003}$
1000	3.000003

x	$f(x)$
-5	3.125
-10	$3.\overline{03}$
-100	$3.\overline{0003}$
-1000	3.000003

(c) f approaches ∞ from the left and $-\infty$ from the right of $x = -1$. f approaches $-\infty$ from the left and ∞ from the right of $x = 1$.

11. a **12.** d **13.** c **14.** e **15.** b **16.** f

17. Vertical asymptote: $x = -1$
Horizontal asymptote: $y = 0$

19. Vertical asymptotes: $x = -3, 2$
Horizontal asymptote: $y = 2$

21. Vertical asymptote: $x = 2$
Horizontal asymptote: $y = -1$
Hole at $x = 0$

23. Vertical asymptote: $x = 0$
Horizontal asymptote: $y = 1$
Hole at $x = -4$

25. (a) Domain: all real numbers x (b) Continuous
(c) Horizontal asymptote: $y = 5$

27. (a) Domain: all real numbers x except $x = 3$
(b) Not continuous
(c) Vertical asymptote: $x = 3$
Horizontal asymptote: $y = 0$

29. (a) Domain of f: all real numbers x except $x = 4$
Domain of g: all real numbers x
(b) Vertical asymptote: none; Hole in f at $x = 4$
(c)

x	1	2	3	4	5	6	7
$f(x)$	5	6	7	Undef.	9	10	11
$g(x)$	5	6	7	8	9	10	11

(d)

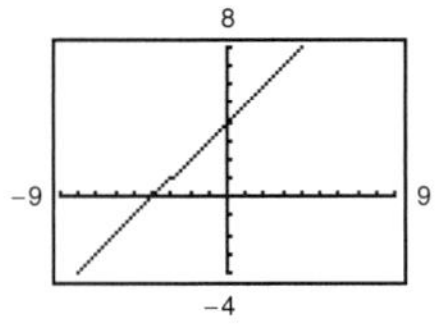

(e) Graphing utilities are limited in their resolution and therefore may not show a hole in a graph.

31. (a) Domain of f: all real numbers x except $x = -1, 3$
Domain of g: all real numbers x except $x = 3$
(b) Vertical asymptote: $x = 3$; Hole in f at $x = -1$
(c)

x	-2	-1	0	1	2	3	4
$f(x)$	$\frac{3}{5}$	Undef.	$\frac{1}{3}$	0	-1	Undef.	3
$g(x)$	$\frac{3}{5}$	$\frac{1}{2}$	$\frac{1}{3}$	0	-1	Undef.	3

(d)
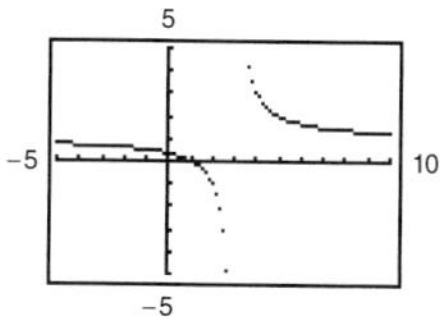

(e) Graphing utilities are limited in their resolution and therefore may not show a hole in a graph.

33. 4; less than; greater than **35.** 0; less than; greater than

37. ± 3 **39.** -3 **41.** $-4, 5$ **43.** $-\frac{1}{3}, 1$

45. (a)
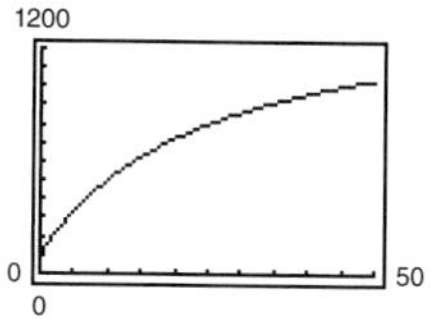

(b) 333 deer, 500 deer, 800 deer

(c) 1500; Because the degrees of the numerator and the denominator are equal, the limiting size is the ratio of the leading coefficients, $60/0.04 = 1500$.

47. (a)

M	200	400	600	800	1000
t	0.472	0.596	0.710	0.817	0.916

M	1200	1400	1600	1800	2000
t	1.009	1.096	1.178	1.255	1.328

The greater the mass, the more time required per oscillation.
(b) $M \approx 1306$ g

49. (a)
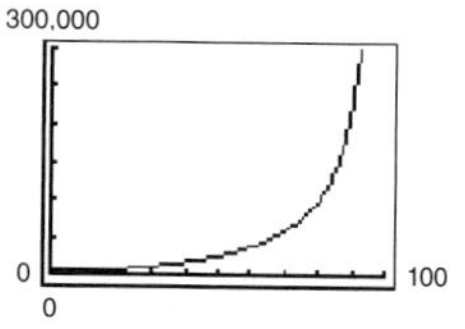

(b) \$4411.76; \$25,000; \$225,000
(c) No; The function is undefined at $p = 100$.

51. False; $\dfrac{1}{x^2 + 1}$ has no vertical asymptote.

53. The graph of f has a vertical asymptote at $x = -2$, not a hole.

55. No; If $x = c$ is also a zero in the denominator, then f is undefined at $x = c$.

57.
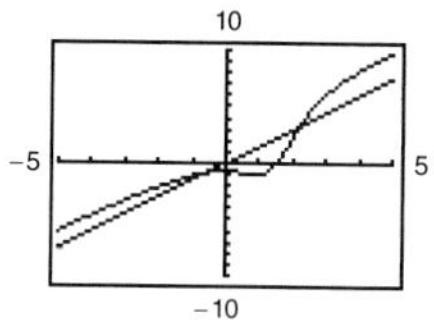

The graphs appear to be about the same as $x \to \pm\infty$.

59. $y = x - 1$
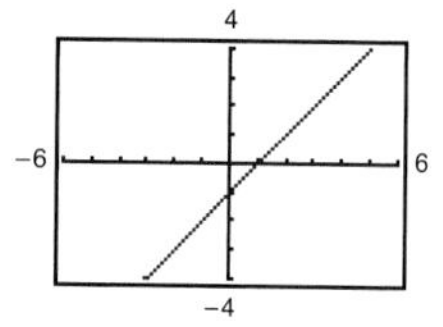

61. Not possible;
The slope is undefined;
$x - 2 = 0$
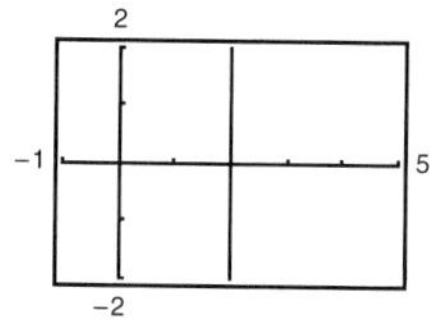

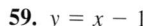

63. $x + 9 + \dfrac{42}{x - 4}$ **65.** $2x^2 - 9 + \dfrac{34}{x^2 + 5}$

Section 2.7 (page 157)

1. slant, asymptote **3.** Yes

5.
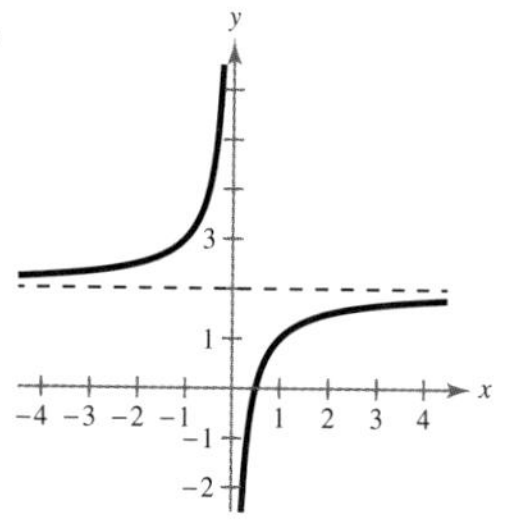

Reflection in the x-axis, vertical shift two units upward

7.
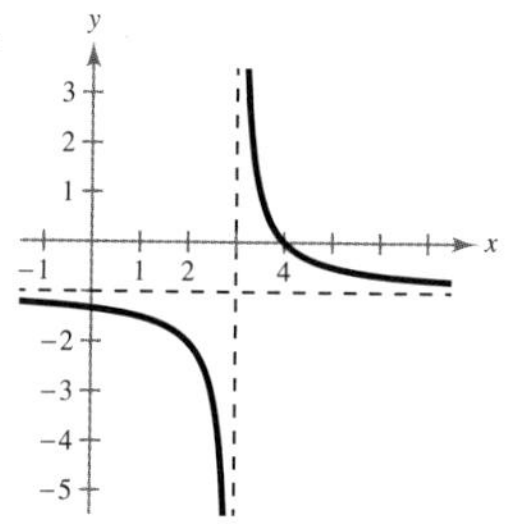

Horizontal shift three units to the right, vertical shift one unit downward

9.
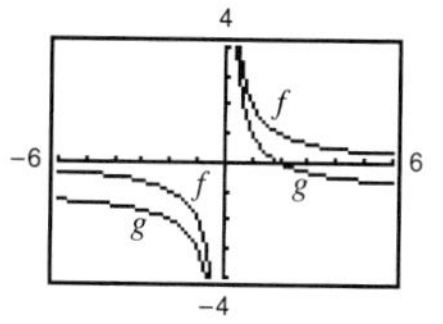

Vertical shift

11.
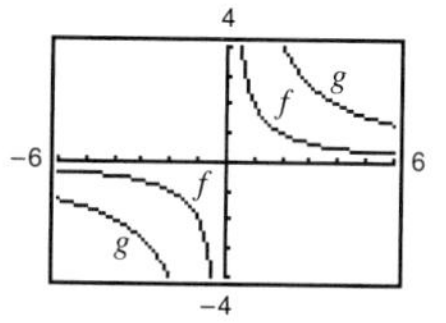

Vertical stretch

13.
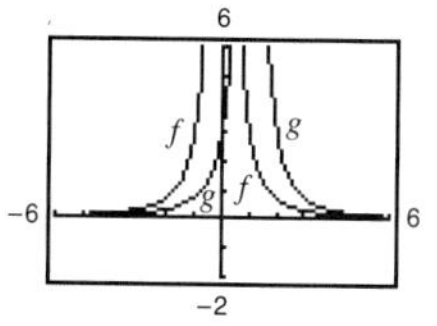

Horizontal shift

15.
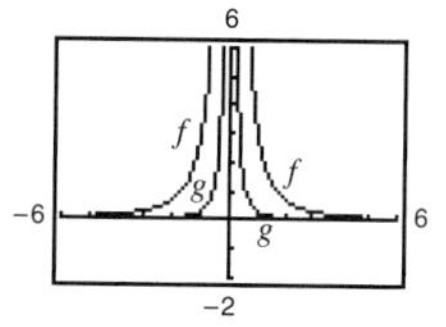

Horizontal shrink

17.
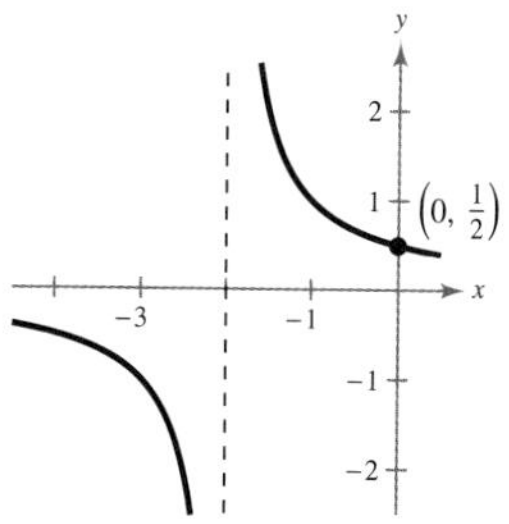

19.
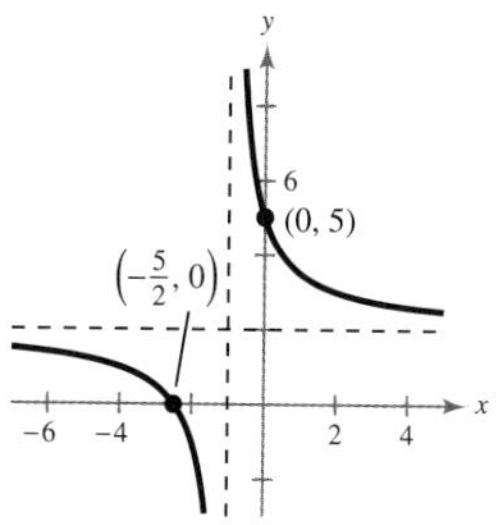

21.
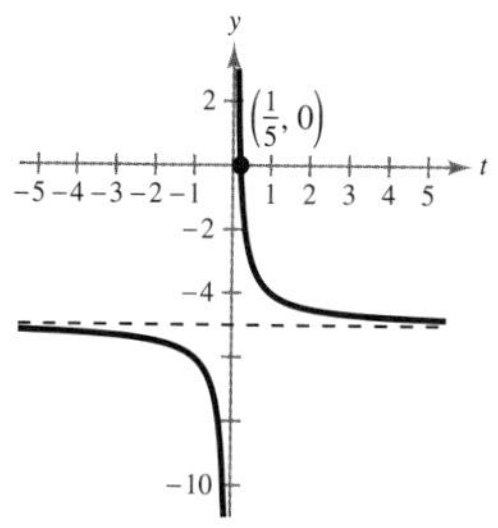

23.
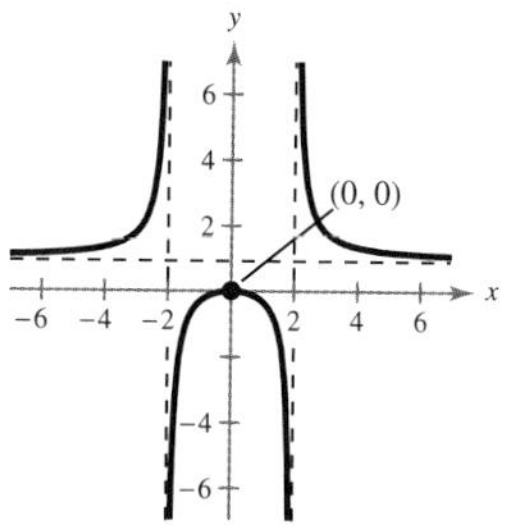

CHAPTER 2

25.

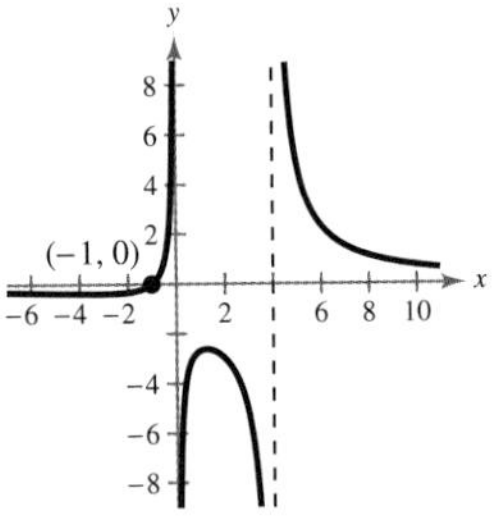

27.

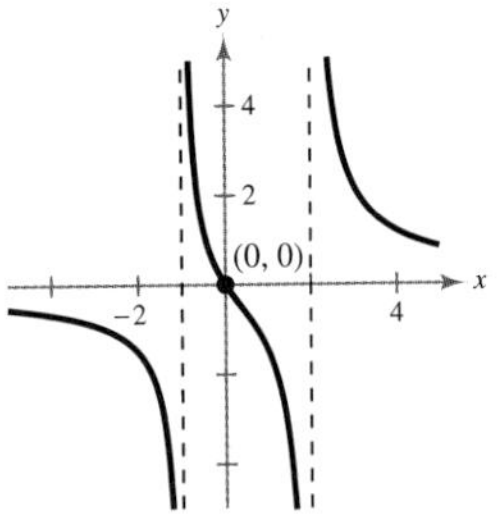

29.

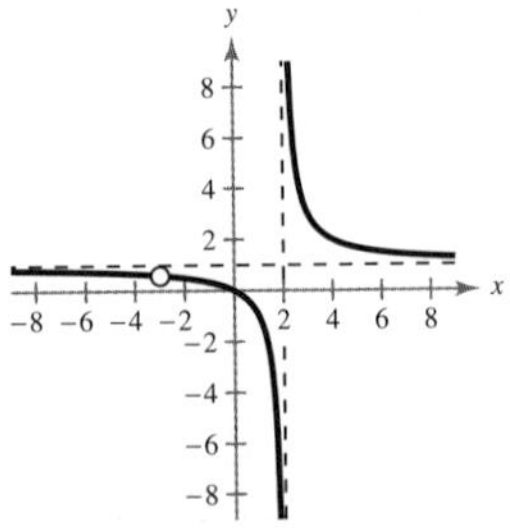

There is a hole at $x = -3$.

31.

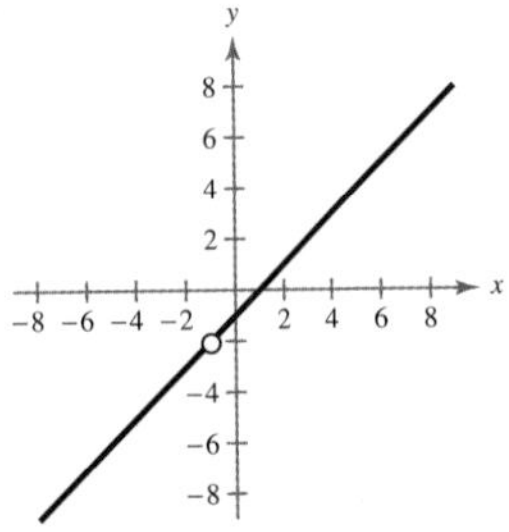

There is a hole at $x = -1$.

33.

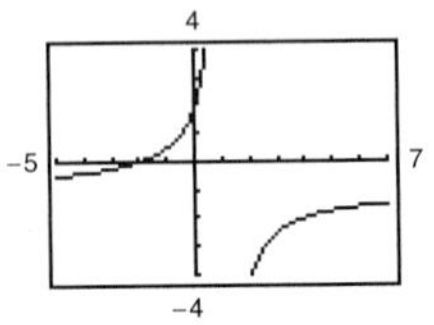

Domain: $(-\infty, 1), (1, \infty)$
Vertical asymptote: $x = 1$
Horizontal asymptote: $y = -1$

35.

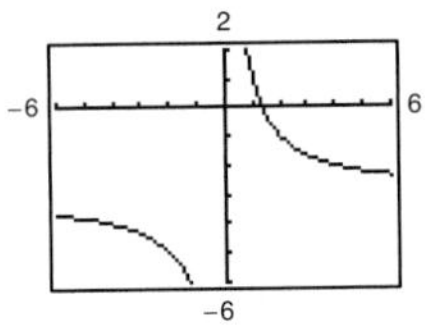

Domain: $(-\infty, 0), (0, \infty)$
Vertical asymptote: $x = 0$
Horizontal asymptote: $y = -3$

37.

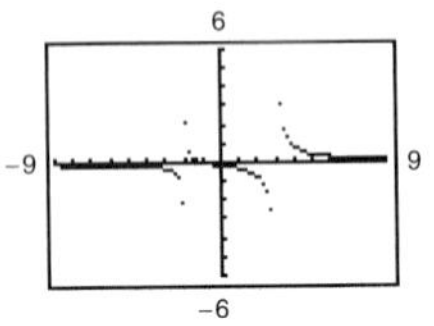

Domain: $(-\infty, -2), (-2, 3), (3, \infty)$
Vertical asymptotes: $x = -2, x = 3$
Horizontal asymptote: $y = 0$

39.

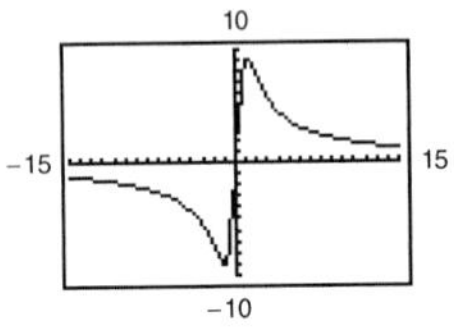

Domain: $(-\infty, 0), (0, \infty)$
Vertical asymptote: $x = 0$
Horizontal asymptote: $y = 0$

41.

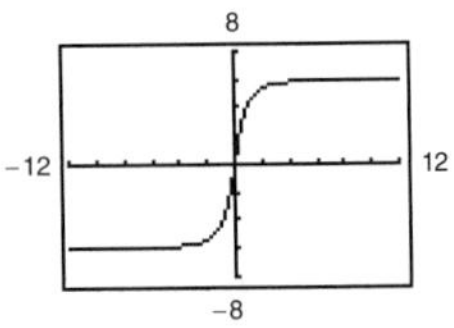

There are two horizontal asymptotes, $y = \pm 6$.

43.

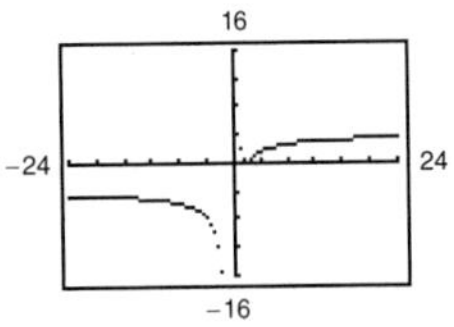

There are two horizontal asymptotes, $y = \pm 4$, and one vertical asymptote, $x = -1$.

45.

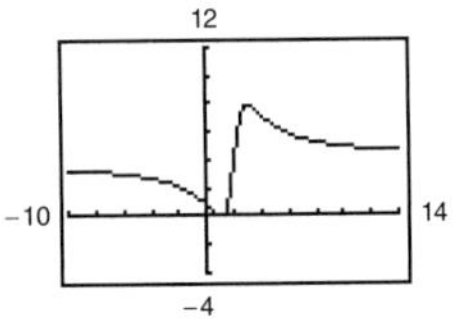

The graph crosses the horizontal asymptote, $y = 4$.

47.

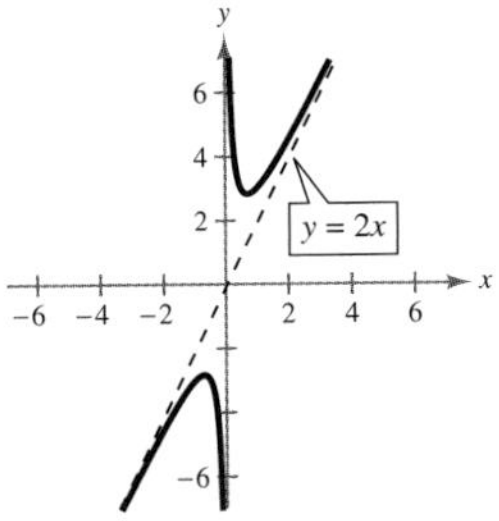

49.

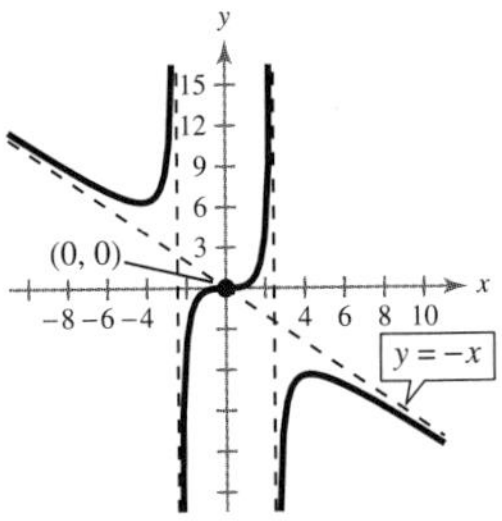

51.

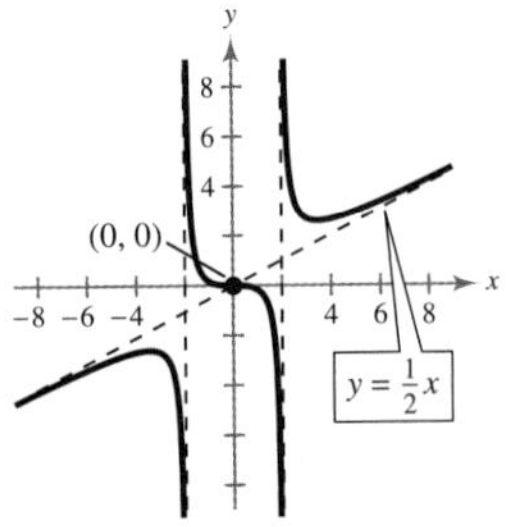

53.

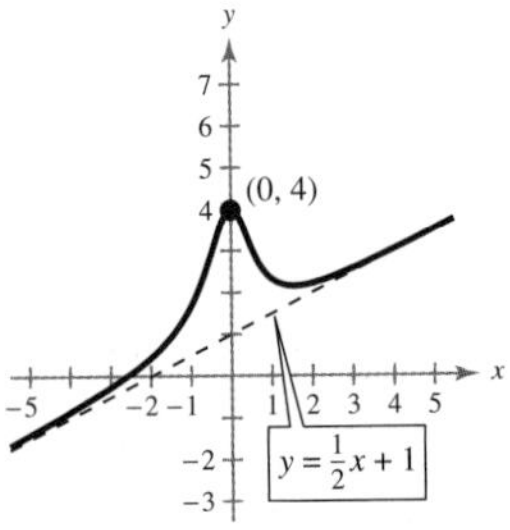

55. $(-1, 0)$ **57.** $(1, 0), (-1, 0)$

59.

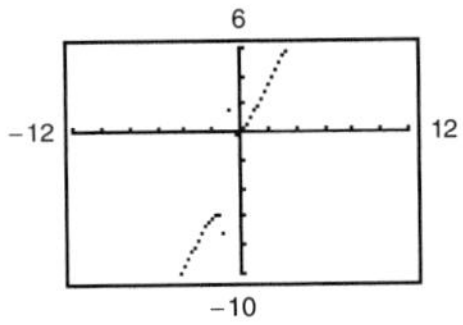

Domain: $(-\infty, -1), (-1, \infty)$
Vertical asymptote: $x = -1$
Slant asymptote: $y = 2x - 1$

61.

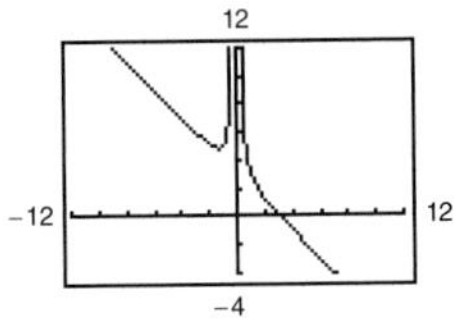

Domain: $(-\infty, 0), (0, \infty)$
Vertical asymptote: $x = 0$
Slant asymptote: $y = -x + 3$

63. Vertical asymptotes: $x = \pm 2$; horizontal asymptote: $y = 1$; slant asymptote: none; holes: none

65. Vertical asymptote: $x = -\frac{3}{2}$; horizontal asymptote: $y = 1$; slant asymptote: none; hole at $x = 2$

67. Vertical asymptote: $x = -2$; horizontal asymptote: none; slant asymptote: $y = 2x - 7$; hole at $x = -1$

69.

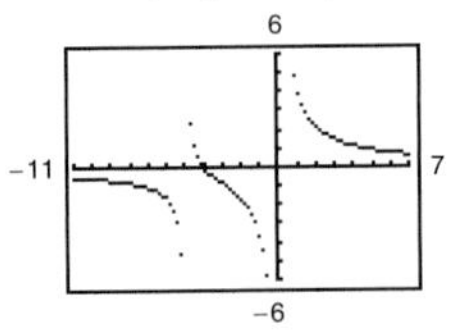

$(-4, 0)$

71.

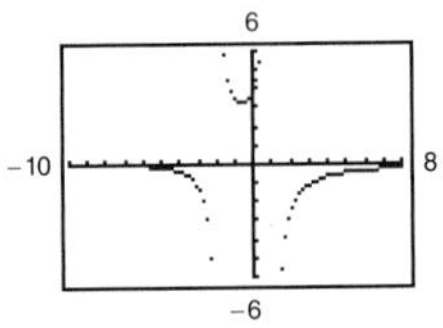

$(-8, 0)$

73.

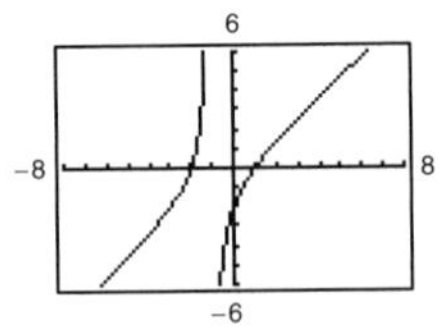

$(-2, 0), (1, 0)$

75.

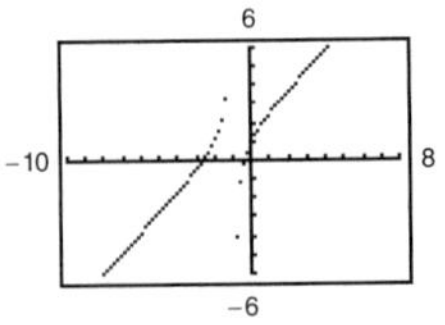

$\left(\frac{-3 \pm \sqrt{5}}{2}, 0\right)$

77.

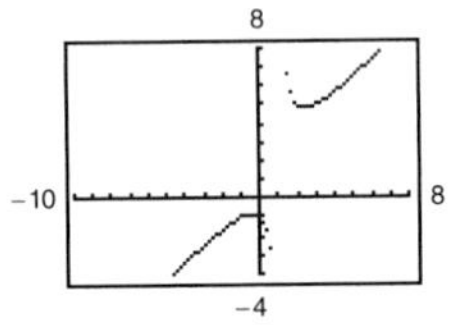

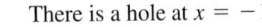

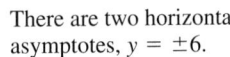

None

79. (a) Answers will vary. (b) $[0, 950]$

(c)

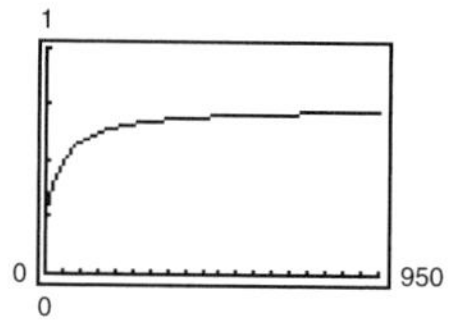

The concentration increases more slowly; 75%

81. (a) Answers will vary. (b) $(2, \infty)$

(c)

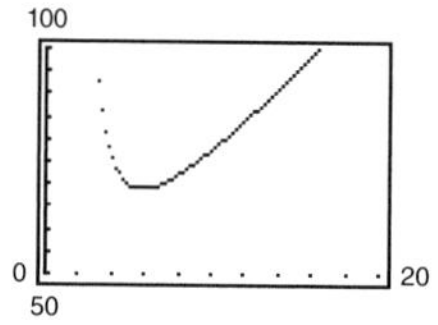

5.9 in $\times$ 11.8 in.

83.

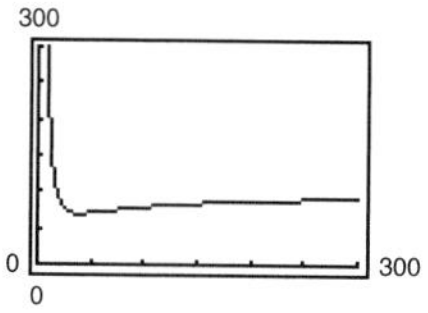

$x \approx 40$

85. (a) $C = 0$. The chemical will eventually dissipate.

(b)

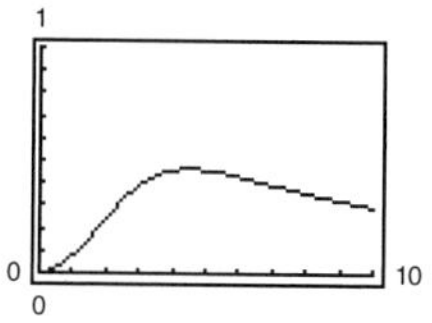

$t \approx 4.5$ h

(c) Before about 2.6 hours and after about 8.3 hours

87. (a) $N = -0.19t + 4.6$

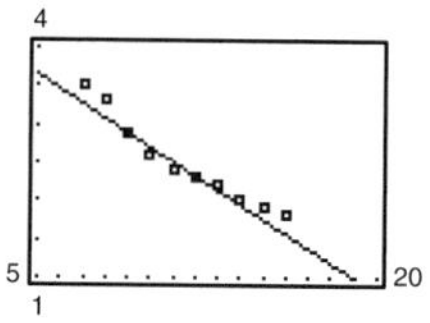

(b) $N = \dfrac{1}{0.030t + 0.072}$

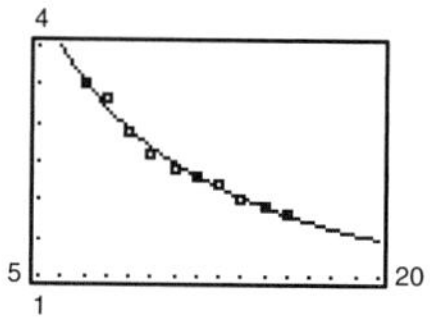

(c) Answers will vary.

89. False; The graph of a rational function is continuous when the polynomial in the denominator has no real zeros.

91.

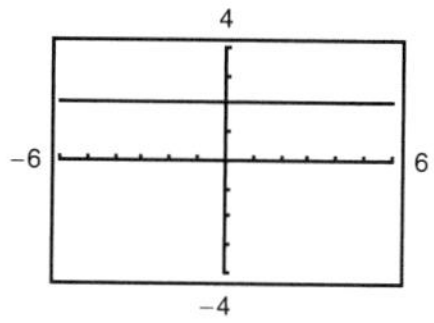

The denominator is a factor of the numerator.

93. *Horizontal asymptotes:*

If the degree of the numerator is greater than the degree of the denominator, then there is no horizontal asymptote.

If the degree of the numerator is less than the degree of the denominator, then there is a horizontal asymptote at $y = 0$.

If the degree of the numerator is equal to the degree of the denominator, then there is a horizontal asymptote at the line given by the ratio of the leading coefficients.

Vertical asymptotes:

Set the denominator equal to zero and solve.

Slant asymptotes:

If there is no horizontal asymptote and the degree of the numerator is exactly one greater than the degree of the denominator, then divide the numerator by the denominator. The slant asymptote is the result, not including the remainder.

95. The parts of the graph should not be connected by the line $x = -3$.

97.

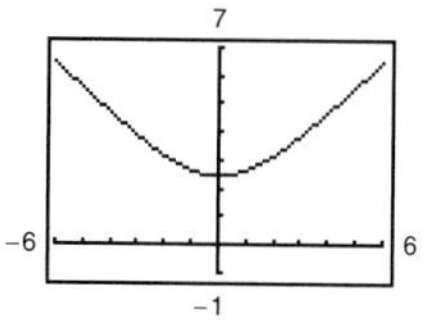

Domain: $(-\infty, \infty)$

Range: $\left[\sqrt{6}, \infty\right)$

99.

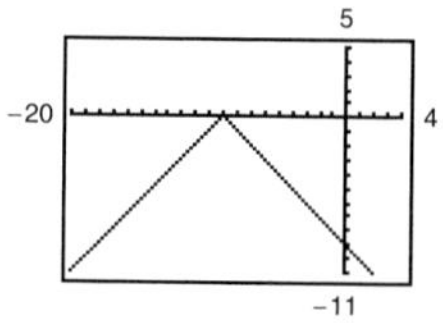

Domain: $(-\infty, \infty)$

Range: $(-\infty, 0]$

101. Answers will vary.

Section 2.8 (page 165)

1. Quadratic **3.** Quadratic **5.** Linear **7.** Neither

9. (a)

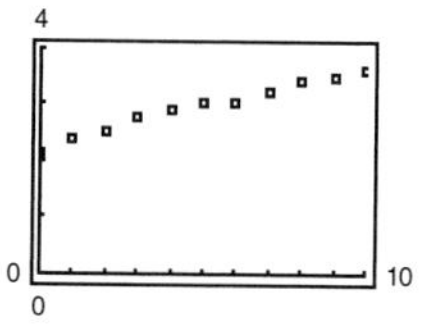

(b) Linear (c) $y = 0.14x + 2.2$

(d)

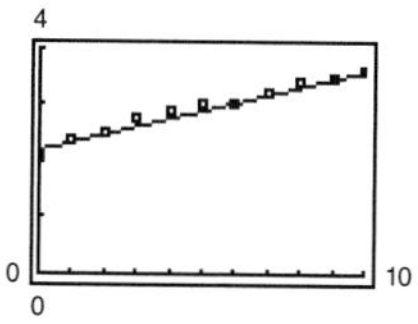

(e)

x	0	1	2	3	4	5
Actual, y	2.1	2.4	2.5	2.8	2.9	3.0
Model, y	2.2	2.3	2.5	2.6	2.8	2.9

x	6	7	8	9	10
Actual, y	3.0	3.2	3.4	3.5	3.6
Model, y	3.0	3.2	3.3	3.5	3.6

11. (a) 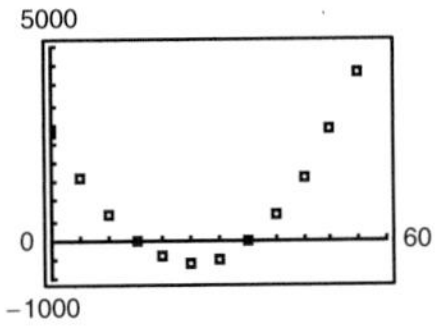

(b) Quadratic

(c) $y = 5.50x^2 - 274.6x + 2822$

(d) 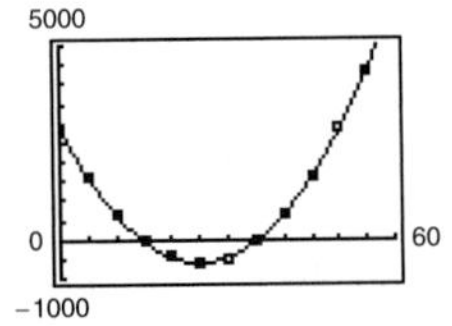

(e)

x	0	5	10	15	20
Actual, y	2795	1590	650	−30	−450
Model, y	2822	1586.5	626	−59.5	−470

x	25	30	35	40
Actual, y	−615	−520	−55	625
Model, y	−605.5	−466	−51.5	638

x	45	50	55
Actual, y	1630	2845	4350
Model, y	1602.5	2842	4356.5

13. (a) 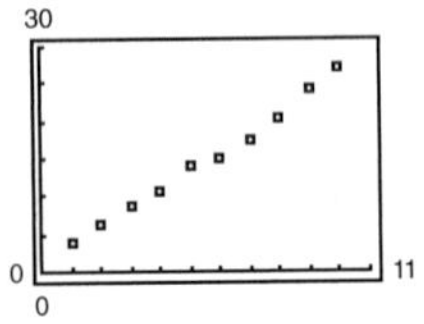

(b) Linear (c) $y = 2.48x + 1.1$

(d) 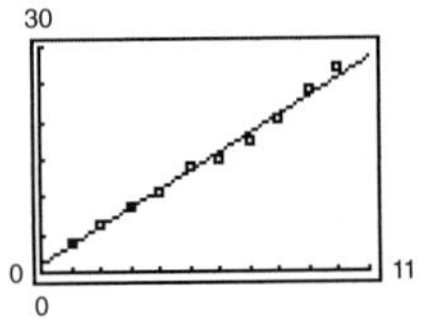

(e)

x	1	2	3	4	5
Actual, y	4.0	6.5	8.8	10.6	13.9
Model, y	3.6	6.1	8.5	11.0	13.5

x	6	7	8	9	10
Actual, y	15.0	17.5	20.1	24.0	27.1
Model, y	16.0	18.5	20.9	23.4	25.9

15. (a) 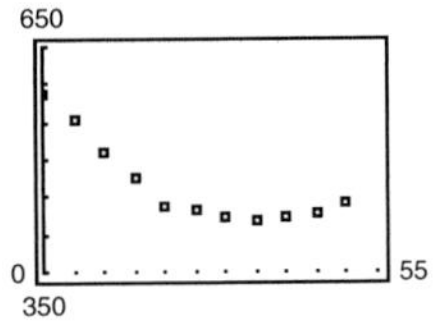

(b) Quadratic

(c) $y = 0.14x^2 - 9.9x + 591$

(d) 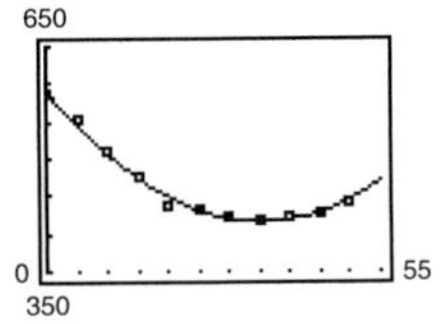

(e)

x	0	5	10	15	20	25
Actual, y	587	551	512	478	436	430
Model, y	591	545	506	474	449	431

x	30	35	40	45	50
Actual, y	424	420	423	429	444
Model, y	420	416	419	429	446

17. (a) and (c) 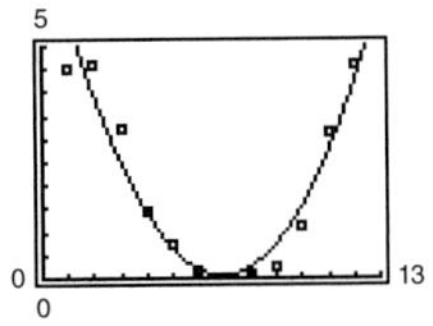

(b) $P = 0.1601t^2 - 2.206t + 7.65$; $r^2 \approx 0.9185$

(d) July (e) July; They are the same.

19. (a) and (c) 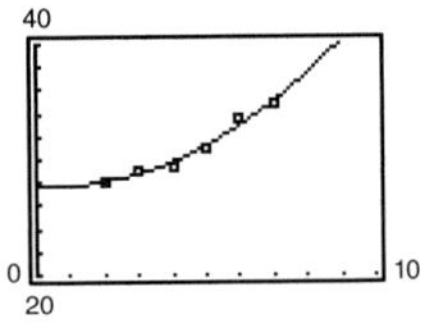

(b) $N = 0.177t^2 - 0.20t + 27.8$

(d) 2021; Answers will vary.

21. $P = 1.16x^2 - 18.8x + 71$; no

23. (a) 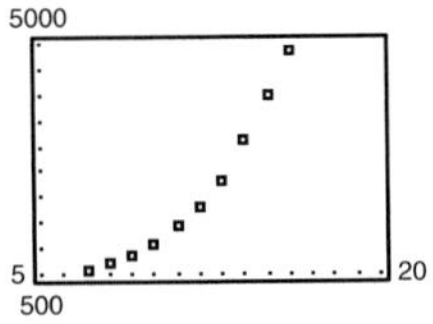

(b) $R = 460.25t - 3216.4$; $r^2 \approx 0.9140$

(c) 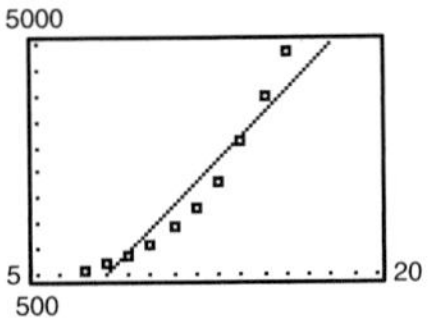

(d) $R = 55.338t^2 - 812.52t + 3645.5$; $r^2 \approx 0.9986$

(e) 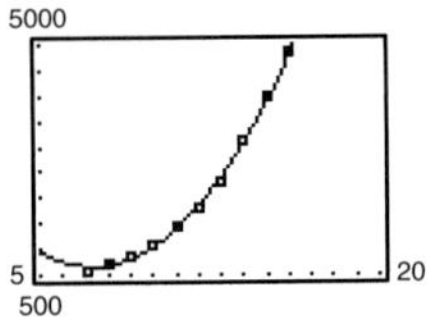

(f) Quadratic; The coefficient of determination is closer to 1.

(g) 2019

25. True; See "Basic Characteristics of Quadratic Functions" on page 91.

27. The model is consistently above the data points.

29. (a) $(f \circ g)(x) = 2x^2 + 5$ (b) $(g \circ f)(x) = 4(x^2 - x + 1)$
(c) $(g \circ g)(x) = x^4 + 6x^2 + 12$

31. (a) $(f \circ g)(x) = x$ (b) $(g \circ f)(x) = x$
(c) $(g \circ g)(x) = \sqrt[3]{\sqrt[3]{x+1} + 1}$

33. $f^{-1}(x) = \dfrac{x-5}{2}$ **35.** $f^{-1}(x) = \sqrt{x-5}$

37. $1 + 3i$; 10 **39.** $5i$; 25

Review Exercises (page 168)

1.

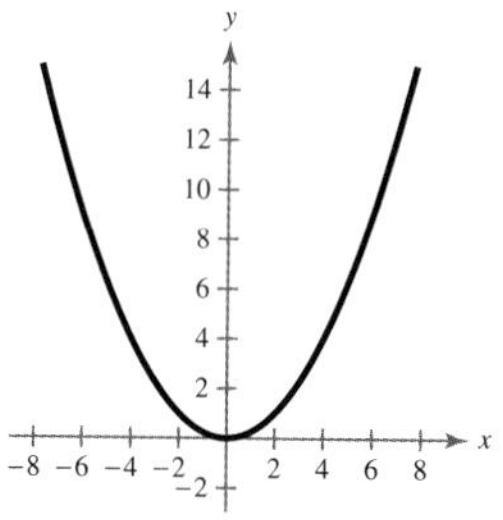

Vertical shrink

3.

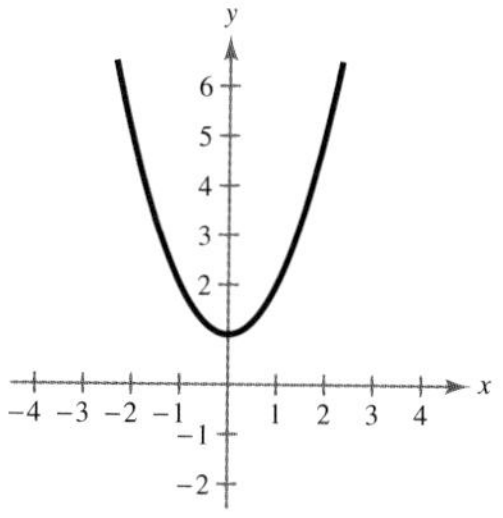

Vertical shift one unit upward

5.

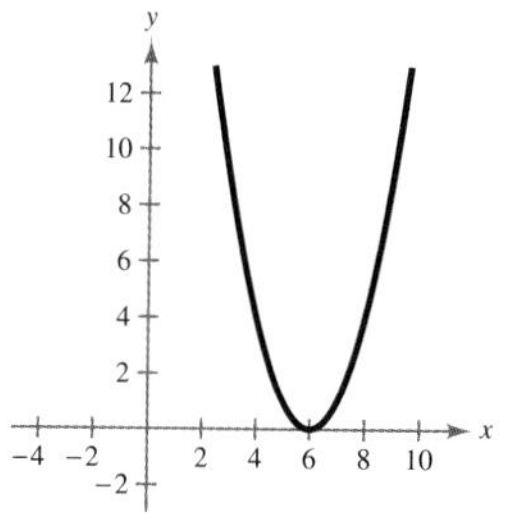

Horizontal shift six units to the right

7. $f(x) = (x+2)^2 - 12$ **9.** $f(x) = -(x+3)^2 + 19$

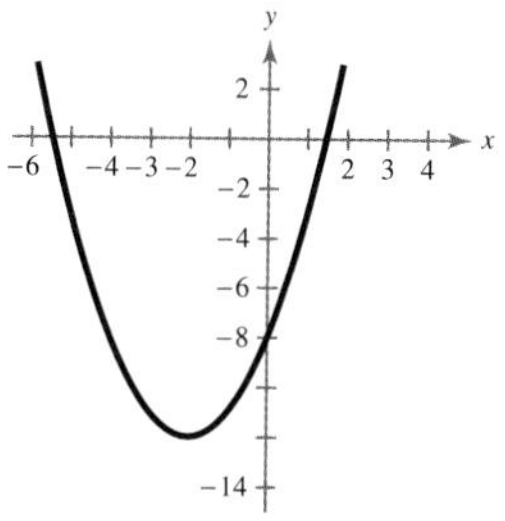

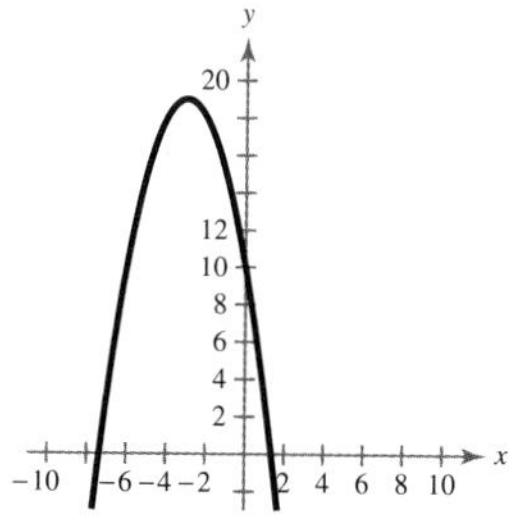

Vertex: $(-2, -12)$
Axis of symmetry: $x = -2$
x-intercepts: $(-2 \pm 2\sqrt{3}, 0)$

Vertex: $(-3, 19)$
Axis of symmetry: $x = -3$
x-intercepts: $(-3 \pm \sqrt{19}, 0)$

11. $f(x) = -(x+3)^2 + 4$

13. (a) $A = x\left(\dfrac{8-x}{2}\right), 0 < x < 8$

(b)

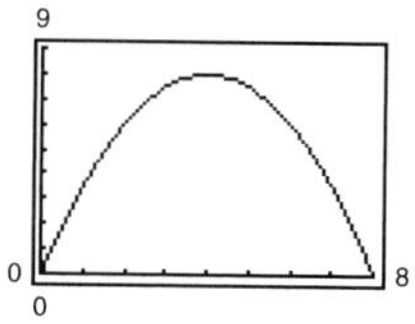

$x = 4, y = 2$

(c) $A = -\frac{1}{2}(x-4)^2 + 8$; $x = 4$, $y = 2$; They are the same.

15.

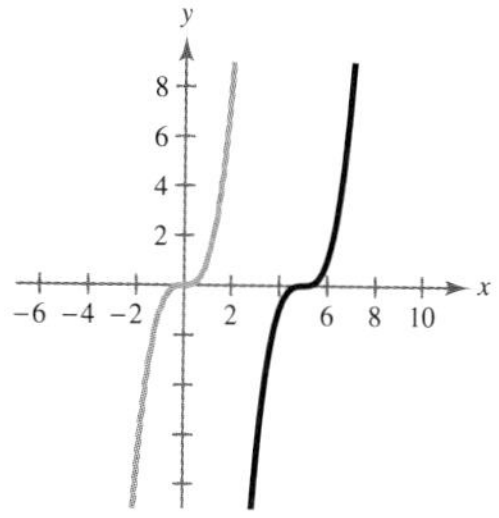

Horizontal shift five units to the right

17.

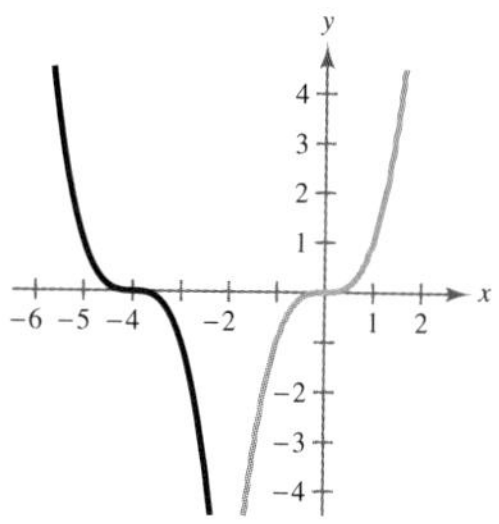

Reflection in the x-axis, horizontal shift four units to the left

19.

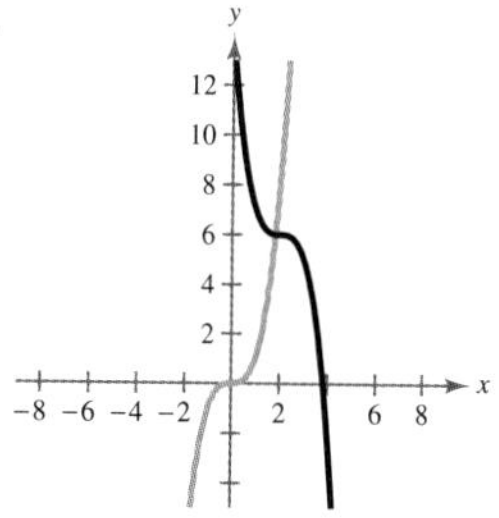

Reflection in the x-axis, horizontal shift two units to the right, vertical shift six units upward

21. Rises to the left, rises to the right

23. Falls to the left, falls to the right

25. (a) $x = -1, 0, 0, 2$
(b)

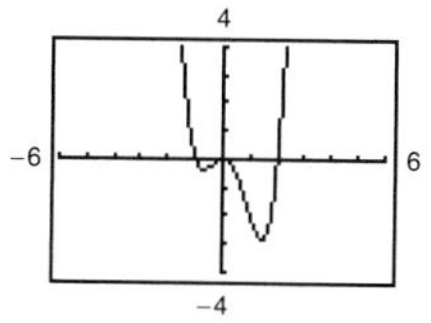

(c) $x = -1, 0, 0, 2$; They are the same.

27. (a) $t = 0, \pm\sqrt{3}$
(b)

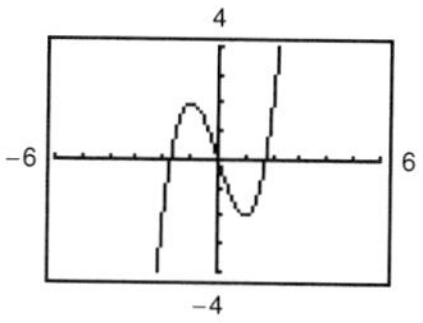

(c) $t = 0, t \approx \pm 1.73$; They are approximately the same.

29. (a) $x = -3, -3, 0$
(b)

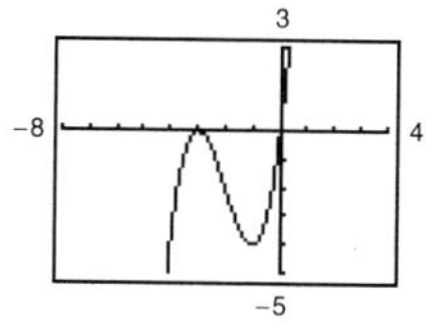

(c) $x = -3, -3, 0$; They are the same.

31. *Sample answer:* $f(x) = x^3 - 12x - 16$

33. (a) Rises to the left, rises to the right (b) $x = -3, -1, 3, 3$
(c) and (d)

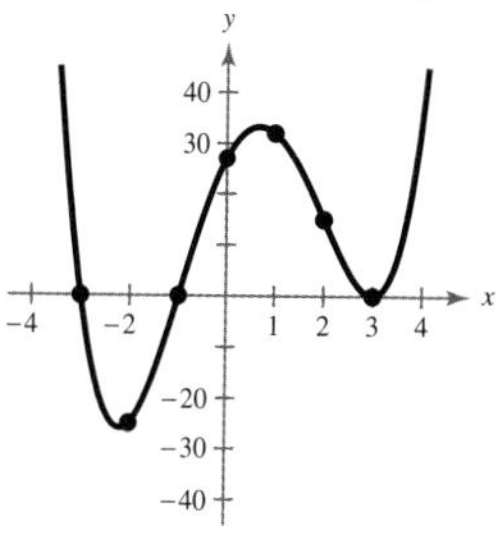

35. (a) $(-3, -2), (-1, 0), (0, 1)$
(b) $x \approx -2.247, -0.555, 0.802$

37. (a) $(-3, -2), (-1, 0), (3, 4)$
(b) $x = -2.979, -0.554, 3.533$

39. $8x + 5 + \dfrac{2}{3x - 2}$ **41.** $x^2 - 2,\ x \neq \pm 1$

43. $5x + 2,\ x \neq \dfrac{3 \pm \sqrt{5}}{2}$ **45.** $3x^2 + 2x + 20 + \dfrac{58}{x - 4}$

47. $6x^3 - 27x,\ x \neq \frac{2}{3}$ **49.** -421

51. (a) Answers will vary. (b) $(2x + 5), (x - 3)$
(c) $f(x) = (x + 6)(2x + 5)(x - 3)$ (d) $x = -6, -\frac{5}{2}, 3$
(e)

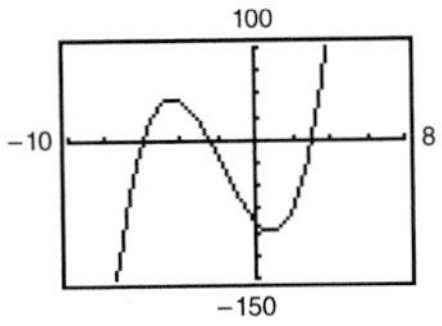

53. $x = -3, -2, 0, 0$ **55.** $x = \frac{3}{4}, 1, 1$

57. 2 or 0 positive real zeros
1 negative real zero

59. 0 positive real zeros
1 negative real zero

61. Answers will vary.
$x = \frac{3}{4}$

63. $x = -1, 2, 3$ **65.** $3 - 4i$ **67.** $-4 - i$

69. $8 + 8i$ **71.** $-1 - 2i$ **73.** $-4 - 46i$ **75.** -80

77. $1 - 6i$ **79.** $\frac{17}{26} + \frac{7}{26}i$ **81.** $x = \pm 4i$

83. $x = -\dfrac{3}{2} \pm \dfrac{\sqrt{15}}{2}i$ **85.** $x = \dfrac{5}{6} \pm \dfrac{\sqrt{47}}{6}i$

87 and 89. Answers will vary.

91. $x = 4, \dfrac{3 \pm \sqrt{15}i}{2}$;
$(x - 4)\left(x - \dfrac{3 + \sqrt{15}i}{2}\right)\left(x - \dfrac{3 - \sqrt{15}i}{2}\right)$

93. $x = 2, -\frac{3}{2}, 1 \pm i$; $(x - 2)(2x + 3)(x - 1 + i)(x - 1 - i)$

95. $x = 0, -1, \pm\sqrt{5}i$; $x^2(x + 1)(x + \sqrt{5}i)(x - \sqrt{5}i)$

97. (a) $x = -3, 4 \pm i$
(b) $(x + 3)(x - 4 - i)(x - 4 + i)$ (c) $(-3, 0)$

99. (a) $\pm 3i, \pm 5i$
(b) $(x - 3i)(x + 3i)(x - 5i)(x + 5i)$ (c) None

101. $f(x) = x^3 - 5x^2 + 9x - 45$

103. $f(x) = x^3 + 6x^2 + 28x + 40$

105. (a) $(x^2 + 9)(x^2 - 2x - 1)$
(b) $(x^2 + 9)\left((x - 1 + \sqrt{2})(x - 1 - \sqrt{2}\right)$
(c) $(x + 3i)(x - 3i)(x - 1 + \sqrt{2})(x - 1 - \sqrt{2})$

107. $x = -3, \pm 2i$ **109.** (a) $x = 1.5$ (b) $x = \pm\frac{1}{2}i$

111. (a) Domain: all real numbers x except $x = -3$
(b) Not continuous
(c) Vertical asymptote: $x = -3$
Horizontal asymptote: $y = -1$

113. (a) Domain: all real numbers x except $x = 6, -3$
(b) Not continuous
(c) Vertical asymptotes: $x = 6, x = -3$
Horizontal asymptote: $y = 0$

115. (a) Domain: all real numbers x except $x = 7$
(b) Not continuous
(c) Vertical asymptote: $x = 7$
Horizontal asymptote: $y = -1$

117. (a) Domain: all real numbers x except $x = \pm\dfrac{\sqrt{6}}{2}$
(b) Not continuous
(c) Vertical asymptotes: $x = \pm\dfrac{\sqrt{6}}{2}$
Horizontal asymptote: $y = 2$

119. (a) Domain: all real numbers x except $x = 5, -3$
(b) Not continuous
(c) Vertical asymptote: $x = -3$
Horizontal asymptote: $y = 0$

121. (a) Domain: all real numbers x except $x = -3, 0, 8$
(b) Not continuous
(c) Vertical asymptotes: $x = -3, x = 0, x = 8$
Horizontal asymptote: $y = 0$

123. -2; greater than; less than

125. (a) \$176 million; \$528 million; \$1584 million
(b)

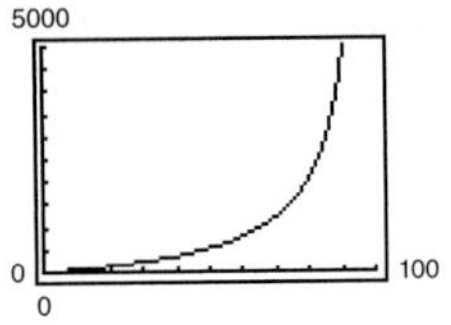

Answers will vary.
(c) No; As $p \to 100$, the cost approaches ∞.

127.

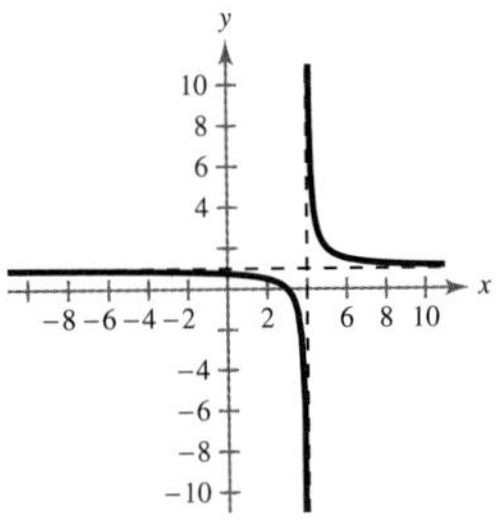

Vertical shift one unit upward, horizontal shift four units to the right

129.

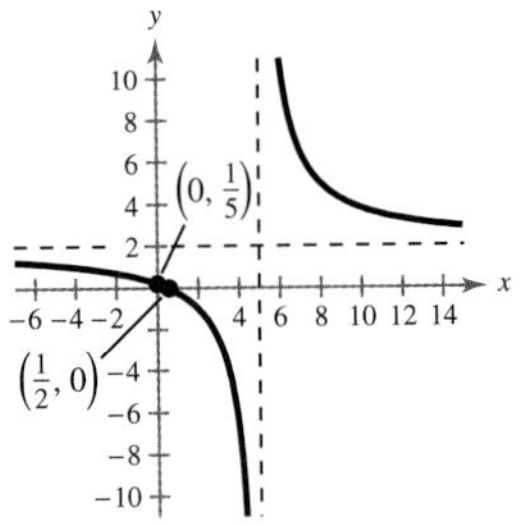

131.

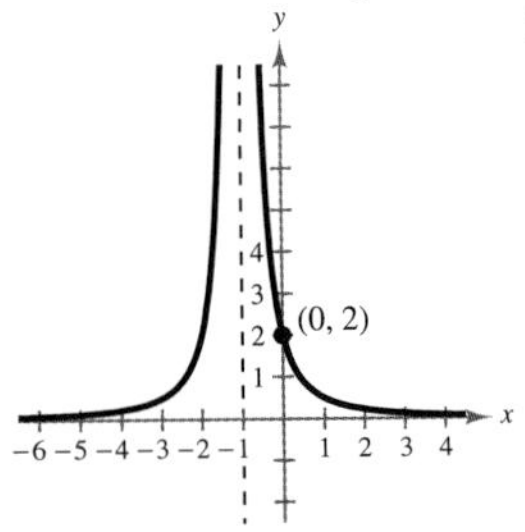

133.

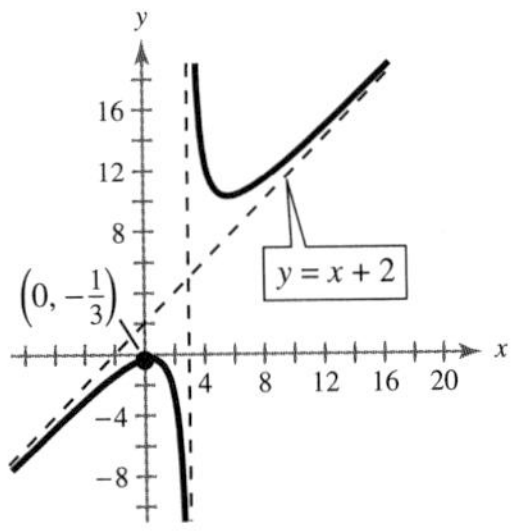

135. Vertical asymptote: $x = -1$
Horizontal asymptote: $y = 1$
Hole at $x = 1$

137. Vertical asymptote: $x = -1$
Slant asymptote: $y = 3x + 2$

139. (a)

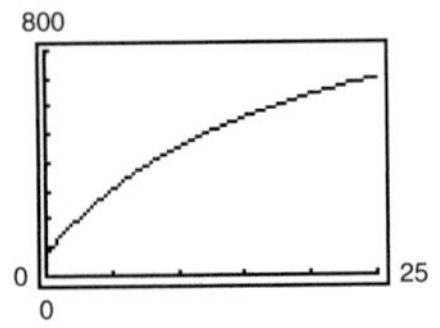

304,000; about 453,333; about 702,222
(b) 1,200,000; N has a horizontal asymptote at $y = 1200$.

141. Quadratic **143.** Linear **145.** Quadratic

147. $y = 0.1312x^2 - 8.329x + 183.22$; yes; $r^2 \approx 0.9941$

149. $y = 10.4602x^2 - 32.484x + 28.38$; no; When graphed, the model does not fit the data well.

151. (a)

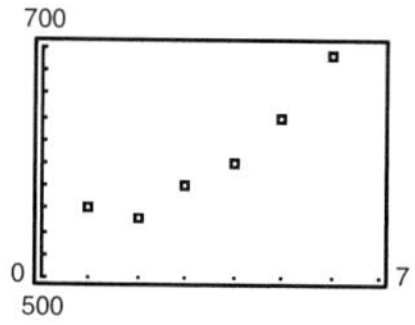

(b) $M = 27.26t + 508.8$; $r^2 \approx 0.8848$

(c)

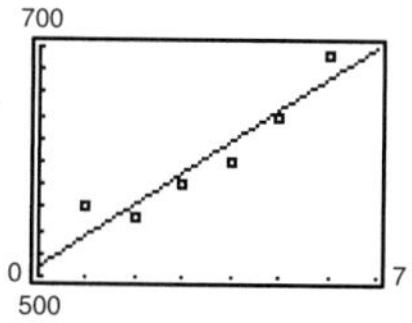

(d) $M = 6.398t^2 - 17.53t + 568.5$; $r^2 \approx 0.9888$

(e)

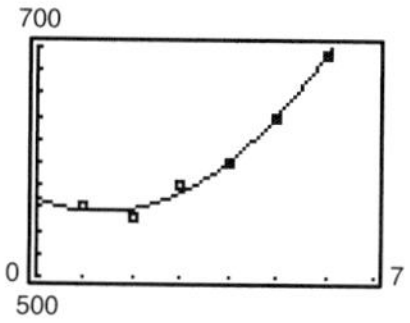

(f) Quadratic; The coefficient of determination is closer to 1.

(g) 2018

153. False; The graph has a minimum value.

155. False; The remainder is -8.

157. False; Because $-8i$ and $4i$ are zeros, and the polynomial has real coefficients, $8i$ and $-4i$ are also zeros. Therefore, the polynomial has at least six zeros, whereas a polynomial of degree 4 can have at most four zeros.

159. False; For the graph of a rational function to have a slant asymptote, the degree of its numerator must be exactly one more than the degree of its denominator.

161. The function and interval values were switched. The function takes on every value between 3 and 9 on the interval $[-1, 2]$.

163. The first step is completed incorrectly: $\sqrt{-4} = 2i \neq 4i$

165. (a) i (b) 1 (c) -1 (d) $-i$

Chapter Test (page 177)

1. Vertex: $\left(-\frac{5}{2}, -\frac{1}{4}\right)$
x-intercepts: $(-3, 0)$, $(-2, 0)$

2. $y = (x - 3)^2 - 6$ **3.** 0, multiplicity 1; $\frac{3}{2}$, multiplicity 2

4.

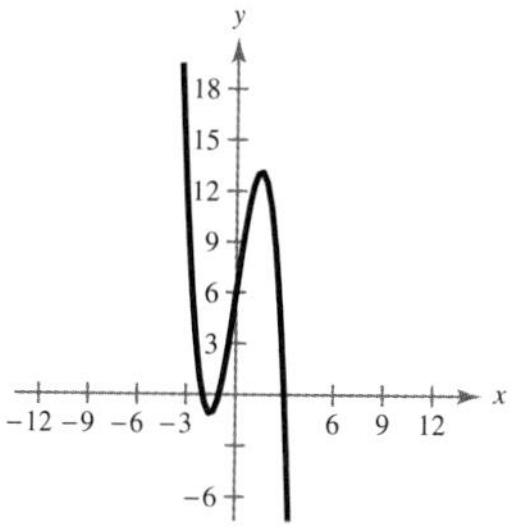

5. $2x + \dfrac{x-3}{x^2+2}$ **6.** $x^3 - x^2 + 5x - 9 + \dfrac{10}{x+5}$ **7.** 13

8. $\pm 1, -\frac{2}{3}$ **9.** $-9 - 18i$ **10.** $6 + \left(2\sqrt{5} + \sqrt{14}\right)i$

11. $13 + 4i$ **12.** $-17 + 14i$ **13.** $5 - 8i$

14. $-1 + 2i$ **15.** $x = \pm 3\sqrt{6}i$ **16.** $x = 1 \pm \sqrt{7}i$

17. $x = -1, 4 \pm \sqrt{3}i$
$(x + 1)\left(x - 4 + \sqrt{3}i\right)\left(x - 4 - \sqrt{3}i\right)$

18. Vertical asymptote: $x = 6$
Horizontal asymptote: $y = 0$
Hole at $x = -6$

19.

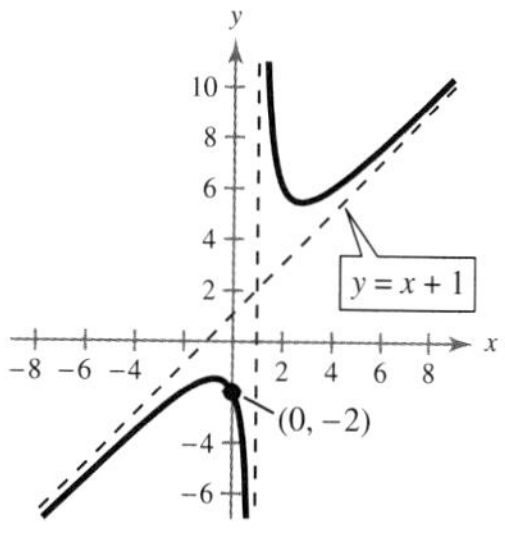

20.

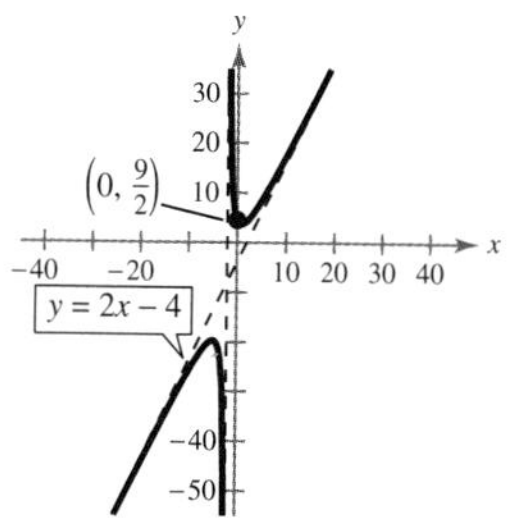

21. (a) and (c)

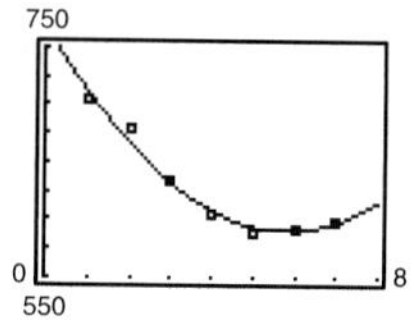

(b) $A = 5.217t^2 - 60.78t + 767.7$

(d) \$712.1 billion, \$667.0 billion, \$632.3 billion, \$608.1 billion, \$594.2 billion, \$590.8 billion, \$597.9 billion

(e) Yes; The estimated results are close to the actual results.

Chapter 3

Section 3.1 (page 191)

1. transcendental **3.** Horizontal shift one unit to the left

5. 0.863 **7.** 0.006

9.

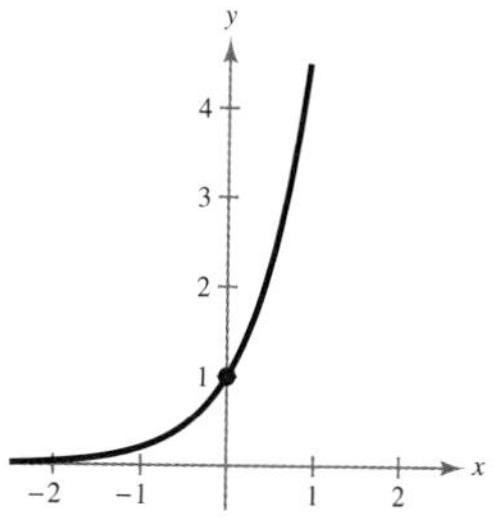

$y = 0$, $(0, 1)$, increasing

11.

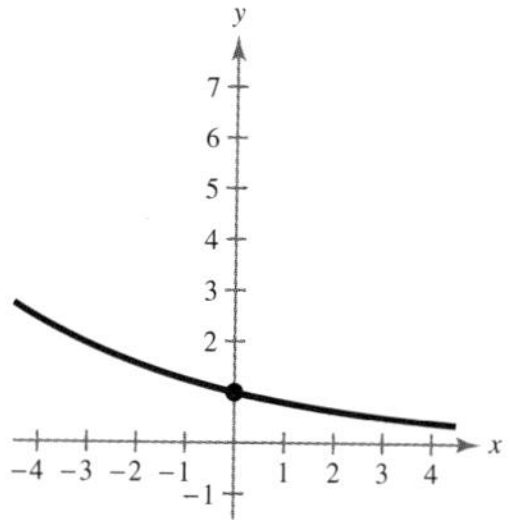

$y = 0$, $(0, 1)$, decreasing

13. d **14.** a **15.** c **16.** b

17. Right shift of five units

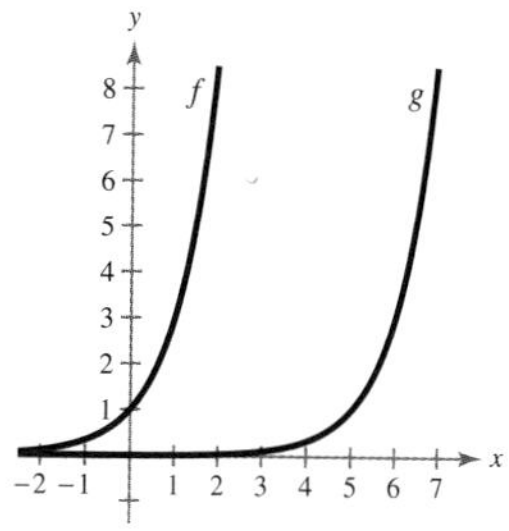

19. Upward shift of five units

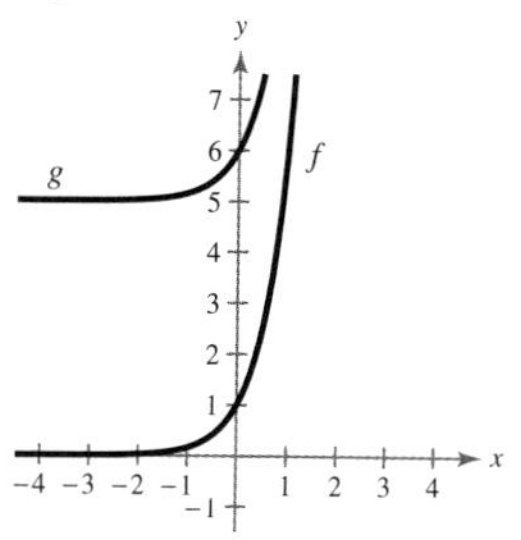

21. Left shift of four units and reflection in the x-axis

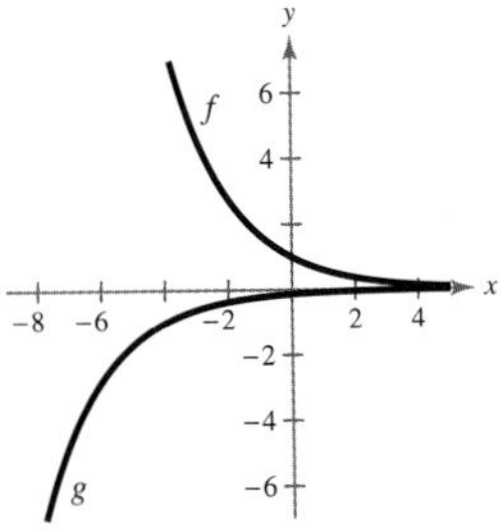

23. Reflection in the x-axis and upward shift of five units

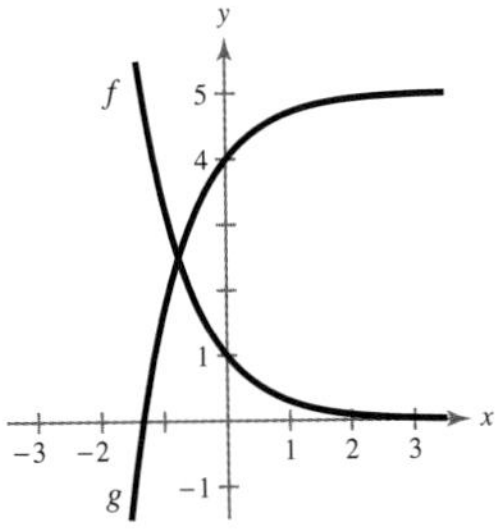

25. Reflection in the y-axis and right shift of three units

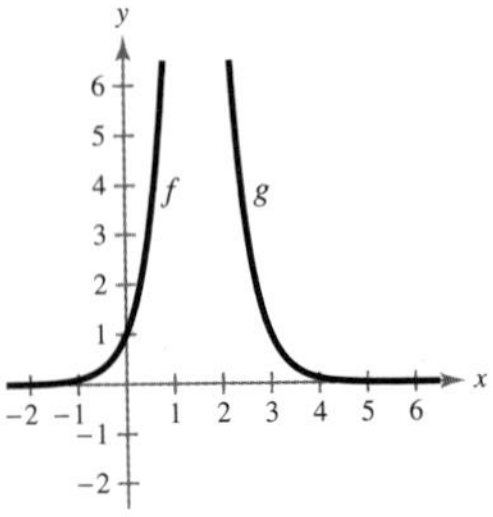

27. Reflection in the y-axis and upward shift of two units

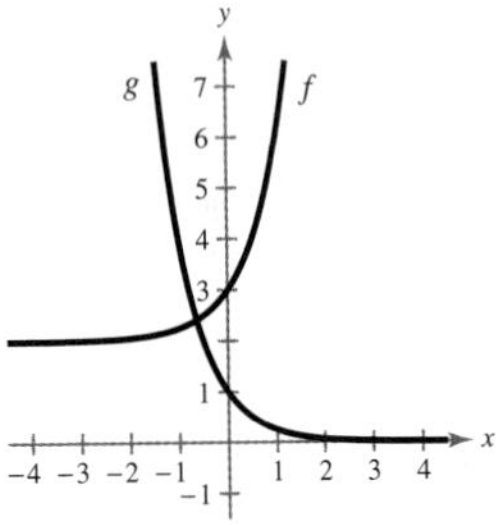

29. (a)

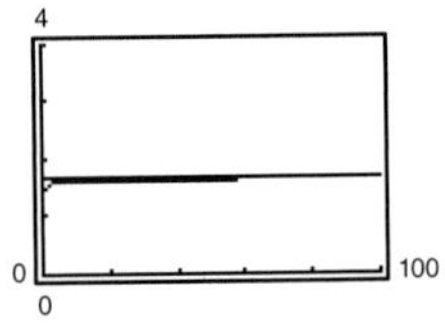

(b) $e^{0.5} \approx 1.6487$

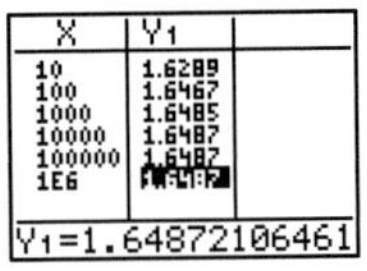

31. 9897.129 **33.** 7166.647

35.

x	-2	-1	0	1	2
$f(x)$	0.44	0.67	1	1.5	2.25

Asymptote: $y = 0$

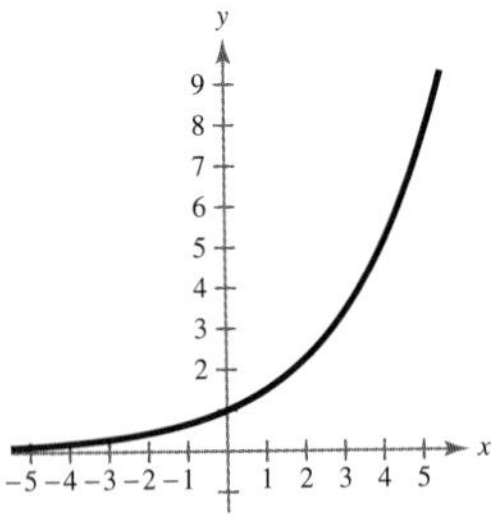

37.

x	-3	-2	-1	0	1
$f(x)$	0.33	1	3	9	27

Asymptote: $y = 0$

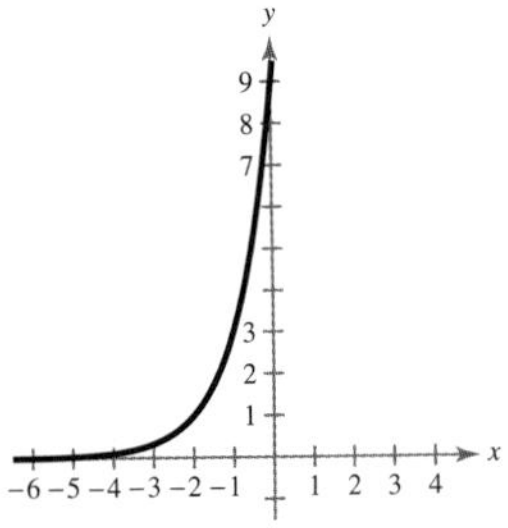

39.

x	-1	0	1	2	3	4
y	1.04	1.11	1.33	2	4	10

Asymptote: $y = 1$

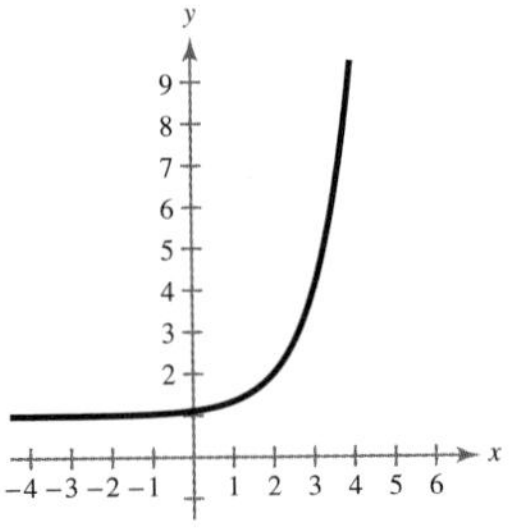

41.

x	-2	-1	0	1	2
$f(x)$	7.39	2.72	1	0.37	0.14

Asymptote: $y = 0$

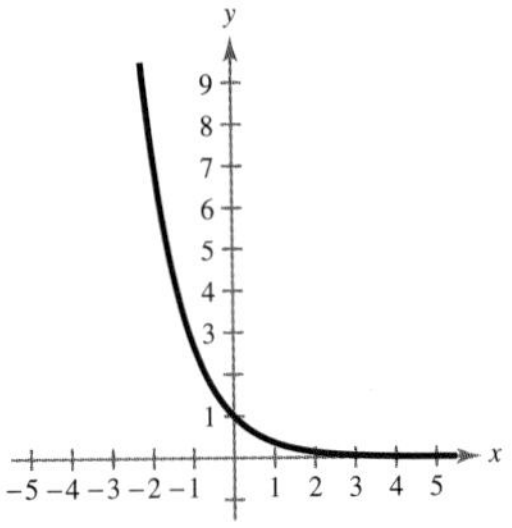

43.

x	-6	-5	-4	-3	-2
$f(x)$	0.41	1.10	3	8.15	22.17

Asymptote: $y = 0$

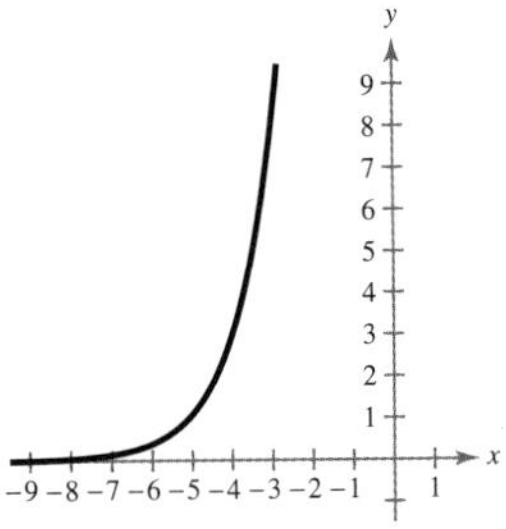

45.

x	3	4	5	6	7
$f(x)$	2.14	2.37	3	4.72	9.39

Asymptote: $y = 2$

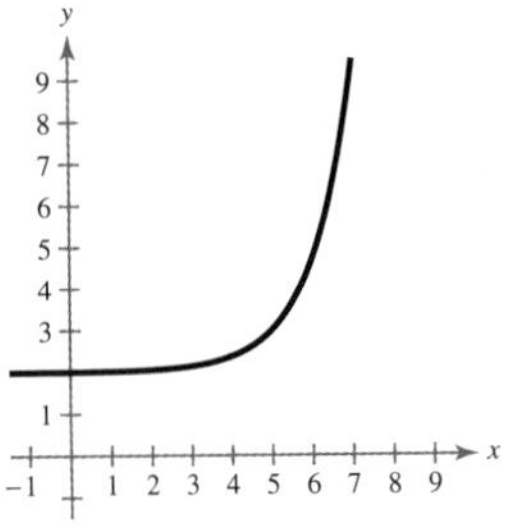

47.

t	-2	-1	0	1	2
$s(t)$	2.54	2.26	2	1.77	1.57

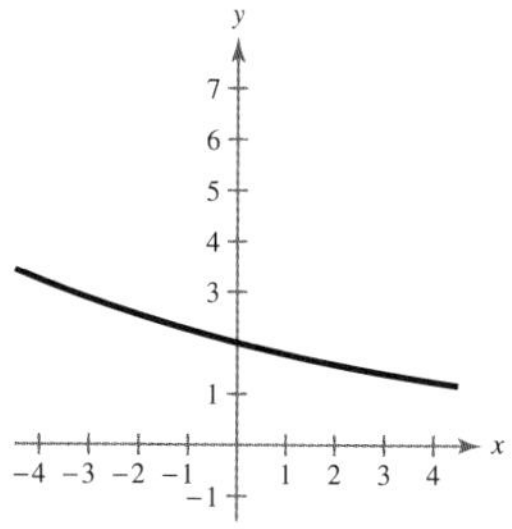

Asymptote: $y = 0$

49. (a)

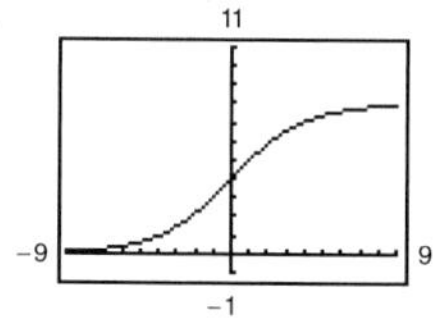

(b) $y = 0, y = 8$

51. (a)

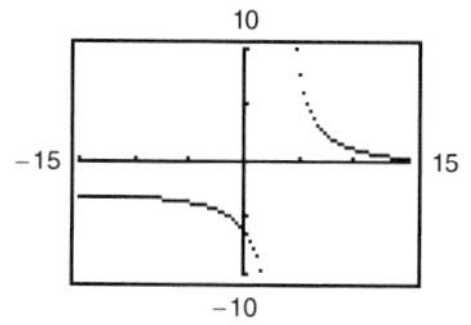

(b) $y = -3, y = 0, x \approx 3.47$

53. (86.350, 1500) **55.** None

57. (a)

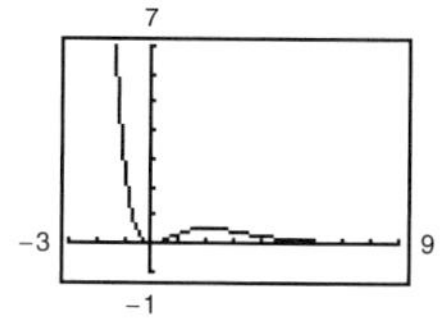

(b) Decreasing on $(-\infty, 0)$, $(2, \infty)$
Increasing on $(0, 2)$

(c) Relative minimum: $(0, 0)$
Relative maximum: about $(2, 0.54)$

59. (a)

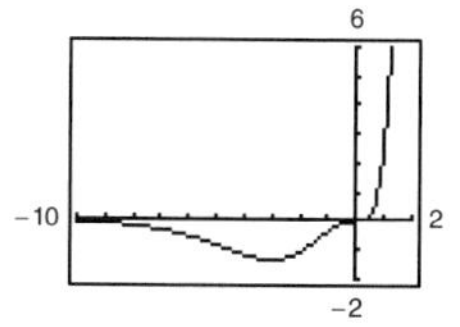

(b) Decreasing on $(-\infty, -3)$
Increasing on $(-3, \infty)$

(c) Relative minimum: about $(-3, -1.344)$

61.

n	1	2	4	12
A	\$1828.49	\$1830.29	\$1831.19	\$1831.80

n	365	Continuous
A	\$1832.09	\$1832.10

63.

n	1	2	4	12
A	\$5477.81	\$5520.10	\$5541.79	\$5556.46

n	365	Continuous
A	\$5563.61	\$5563.85

65.

t	1	10	20
A	\$12,489.73	\$17,901.90	\$26,706.49

t	30	40	50
A	\$39,841.40	\$59,436.39	\$88,668.67

67.

t	1	10	20
A	\$12,181.36	\$13,942.01	\$16,198.31

t	30	40	50
A	\$18,819.75	\$21,865.43	\$25,404.00

69. \$1530.57 **71.** \$18,535.05

73. (a) 10 g (b) About 7.84 g

(c)

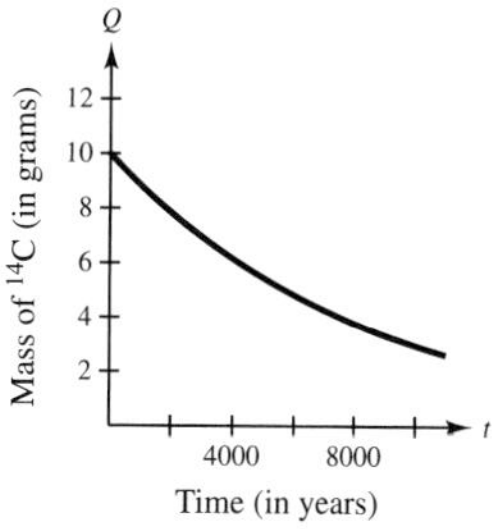

75. (a)

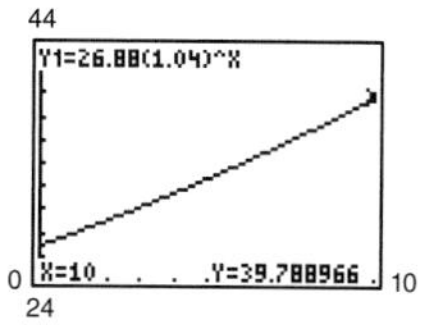

(b)

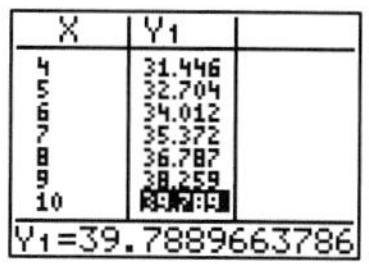

(c) \$39.79

CHAPTER 3

77. (a)

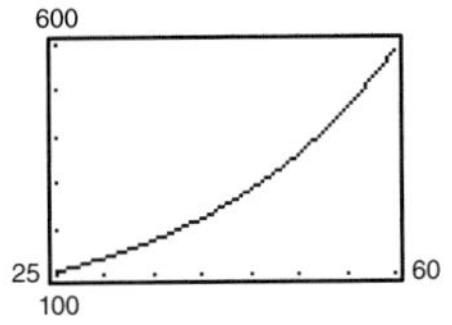

(b)

t	25	26	27	28	29	30
P	107.01	112.34	117.94	123.81	129.98	136.45

t	31	32	33	34	35	36
P	143.24	150.38	157.87	165.73	173.98	182.65

t	37	38	39	40	41	42
P	191.74	201.29	211.32	221.84	232.89	244.49

t	43	44	45	46	47	48
P	256.66	269.44	282.86	296.95	311.74	327.26

t	49	50	51	52	53	54
P	343.56	360.67	378.63	397.48	417.28	438.06

t	55	56	57	58	59	60
P	459.88	482.78	506.82	532.06	558.56	586.37

(c) 2046

79. True; The definition of an exponential function is $f(x) = a^x$, $a > 0, a \neq 1$.

81. d **83.** The range is $(-1, \infty)$, not $(0, \infty)$.

85. $>$ **87.** $>$ **89.** $f^{-1}(x) = \dfrac{x + 7}{5}$

91. $f^{-1}(x) = x^3 - 8$ **93.** Answers will vary.

Section 3.2 (page 201)

1. logarithmic function **3.** $a^{\log_a x} = x$ **5.** $b = a^c$

7. $4^2 = 16$ **9.** $7^{-3} = \frac{1}{343}$ **11.** $32^{2/5} = 4$

13. $2^{1/2} = \sqrt{2}$ **15.** $\log_5 125 = 3$ **17.** $\log_{81} 27 = \frac{3}{4}$

19. $\log_6 \frac{1}{36} = -2$ **21.** $\log_g 4 = a$ **23.** 6 **25.** -3

27. 2.720 **29.** 7.022 **31.** 9 **33.** $\frac{1}{10}$ **35.** 256

37. $3x$ **39.** -3

41.

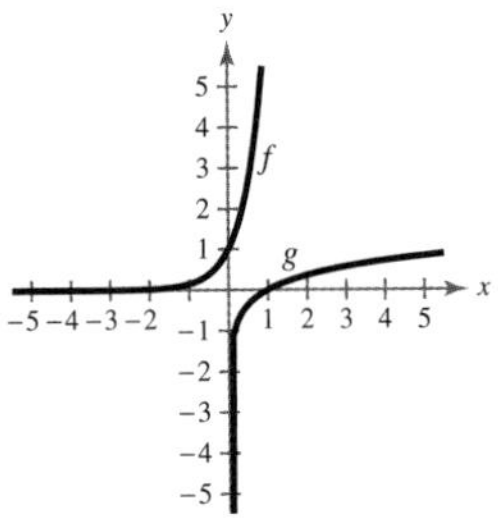

43.

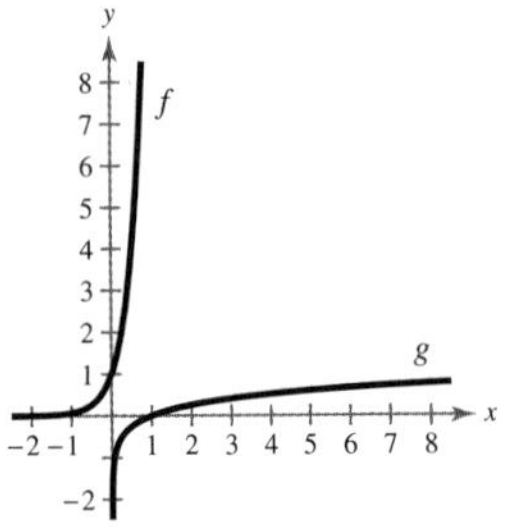

45.

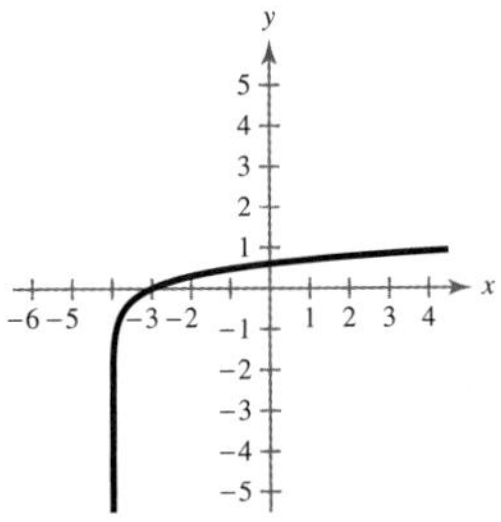

Domain: $(-4, \infty)$
x-intercept: $(-3, 0)$
Vertical asymptote: $x = -4$

47.

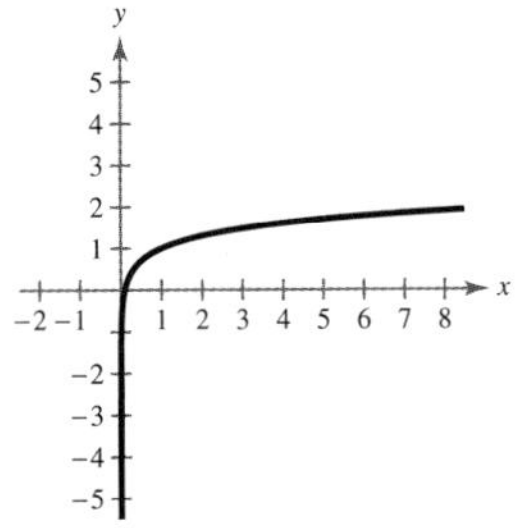

Domain: $(0, \infty)$
x-intercept: $\left(\frac{1}{10}, 0\right)$
Vertical asymptote: $x = 0$

49.

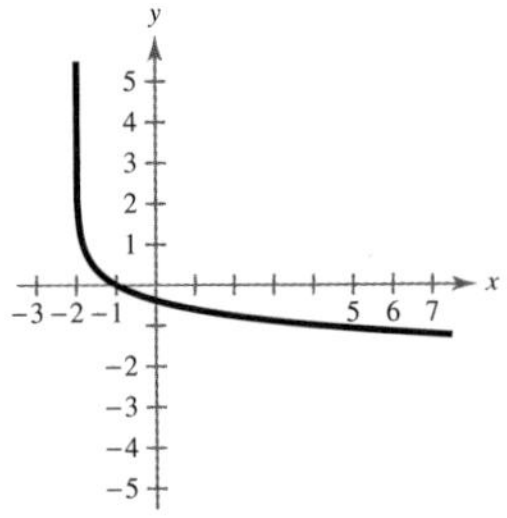

Domain: $(-2, \infty)$
x-intercept: $(-1, 0)$
Vertical asymptote: $x = -2$

51. b **52.** c **53.** d **54.** a

55. Reflection in the y-axis

57. Reflection in the x-axis, vertical shift five units upward

59. Horizontal shift three units to the right and vertical shift two units downward

61. $e^{1.7917\ldots} = 6$ **63.** $e^1 = e$ **65.** $e^{1/2} = \sqrt{e}$

67. $e^{2.1972\ldots} = 9$ **69.** $\ln 20.0855\ldots = 3$

71. $\ln 2.4596\ldots = 0.9$ **73.** $\ln 1.3956\ldots = \frac{1}{3}$

75. $\ln 4.4816\ldots = \frac{3}{2}$ **77.** 2.398 **79.** 6.438

81. 2 **83.** 4 **85.** 0 **87.** 1

89.

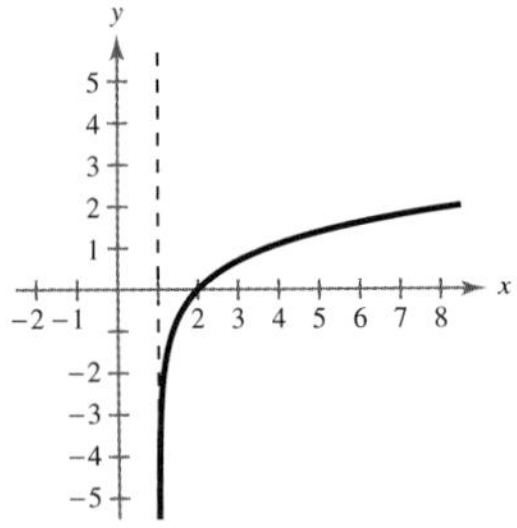

Domain: $(1, \infty)$
Vertical asymptote: $x = 1$
x-intercept: $(2, 0)$

91.

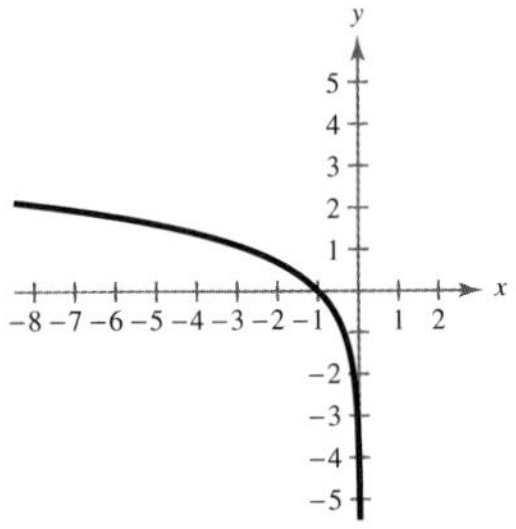

Domain: $(-\infty, 0)$
Vertical asymptote: $x = 0$
x-intercept: $(-1, 0)$

93. Horizontal shift eight units to the left

95. Vertical shift five units downward

97. Horizontal shift one unit to the right and vertical shift two units upward

99. (a)

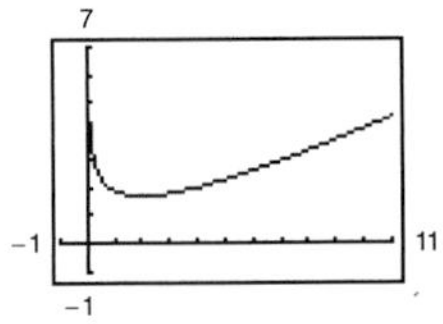

(b) Domain: $(0, \infty)$

(c) Decreasing on $(0, 2)$; increasing on $(2, \infty)$

(d) Relative minimum: $(2, 1.693)$

101. (a)
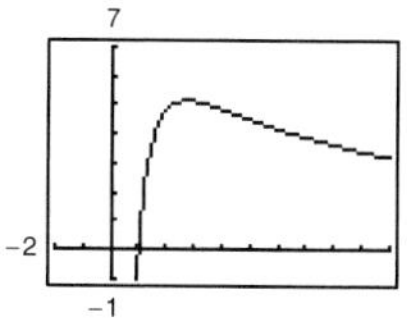

(b) Domain: $(0, \infty)$

(c) Decreasing on $(2.718, \infty)$; increasing on $(0, 2.718)$

(d) Relative maximum: $(2.718, 5.150)$

103. (a)
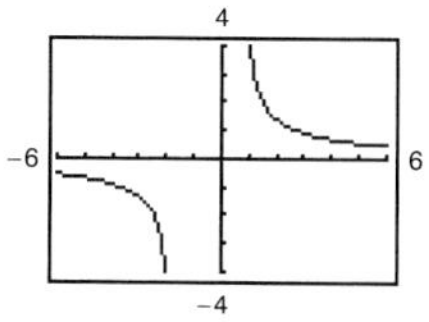

(b) Domain: $(-\infty, -2)$, $(1, \infty)$

(c) Decreasing on $(-\infty, -2)$, $(1, \infty)$

(d) No relative maxima or minima

105. (a)
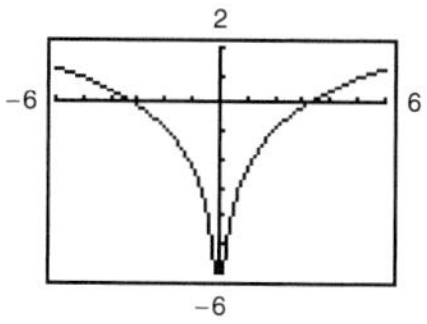

(b) Domain: $(-\infty, 0)$, $(0, \infty)$

(c) Decreasing on $(-\infty, 0)$; increasing on $(0, \infty)$

(d) No relative maxima or minima

107. (a)
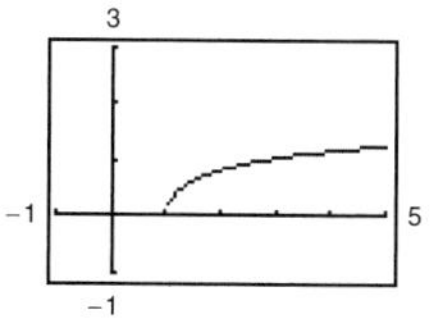

(b) Domain: $[1, \infty)$

(c) Increasing on $(1, \infty)$

(d) Relative minimum: $(1, 0)$

109. 80; about 71.89; about 61.65

111. (a)

K	1	2	4	6	8
t	0	19.80	39.61	51.19	59.41

K	10	12
t	65.79	71.00

It takes about 19.80 years for the principal to double.

(b)
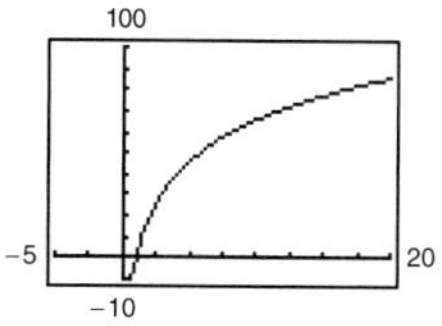

113. (a) 30 yr, 15 yr

(b) Total amount: \$364,813.20; interest: \$164,813.20

Total amount: \$275,398.20; interest: \$75,398.20

115. False; Reflect $g(x)$ about the line $y = x$. **117.** 2

119. $\frac{1}{9}$ **121.** b

123. $\log_a x$ is the inverse of a^x only if $0 < a < 1$ and $a > 1$, so $\log_a x$ is defined only for $0 < a < 1$ and $a > 1$.

125. (a)
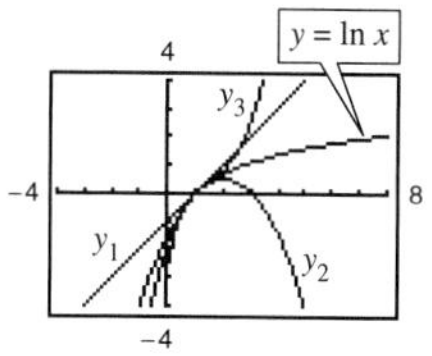

(b)
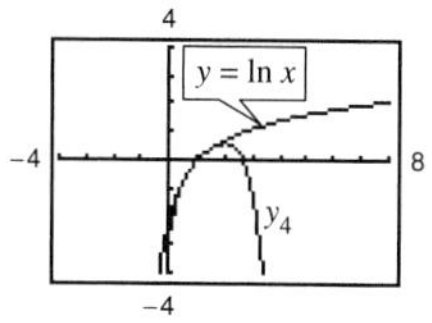

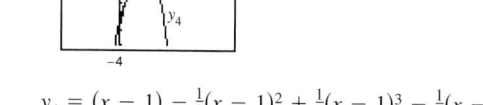

$y_4 = (x - 1) - \frac{1}{2}(x - 1)^2 + \frac{1}{3}(x - 1)^3 - \frac{1}{4}(x - 1)^4$

Answers will vary.

127. (a)

x	1	5	10	10^2
$f(x)$	0	0.32	0.23	0.046

x	10^4	10^6
$f(x)$	0.00092	0.0000138

(b) 0

129. $y = \log_2 x$, so y is a logarithmic function of x.

131. No real solution **133.** $\frac{83}{3}$

Section 3.3 (page 209)

1. change-of-base **3.** $\log_3 24 = \dfrac{\ln 24}{\ln 3}$

5. (a) $\dfrac{\log_{10} x}{\log_{10} 5}$ (b) $\dfrac{\ln x}{\ln 5}$ **7.** (a) $\dfrac{\log_{10} \frac{5}{6}}{\log_{10} a}$ (b) $\dfrac{\ln \frac{5}{6}}{\ln a}$

9. 2.74 **11.** 0.26 **13.** $\ln 5 + \ln 4$ **15.** $2 \ln 5 - \ln 4$

17. 1.0686 **19.** -0.2294

21.
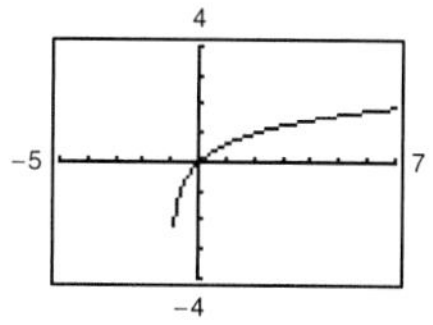

23.
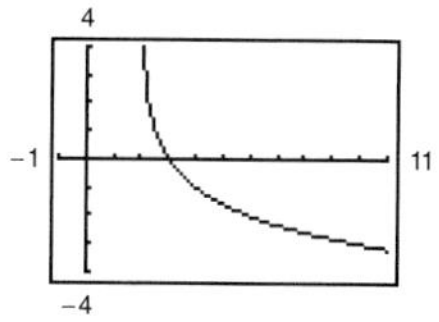

25.
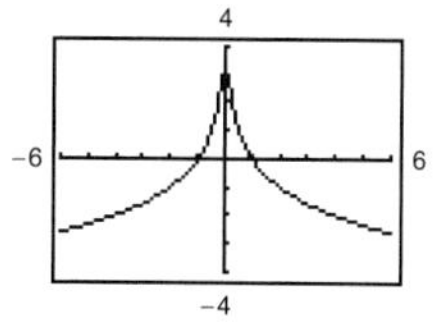

27.
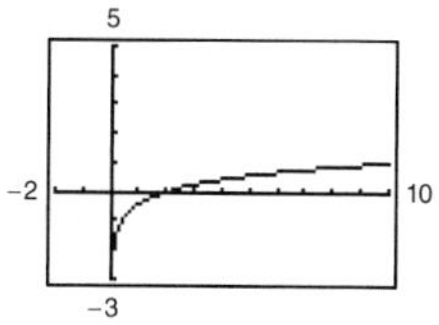

29. $4 + 4 \log_2 3$ **31.** $\ln 6 - 2$

33 and 35. Answers will vary. **37.** $\log_{10} 9 + \log_{10} x$

39. $\log_7 t - \log_7 8$ **41.** $4 \log_8 x$ **43.** $\frac{1}{2} \ln z$

45. $\log_6 x + 3 \log_6 y - 2 \log_6 z$ **47.** $\frac{4}{3} \ln x - \ln y$

49. $4 \log_b x + \frac{1}{2} \log_b y - \frac{1}{2} \log_b(1 + z)$

CHAPTER 3

51. (a)

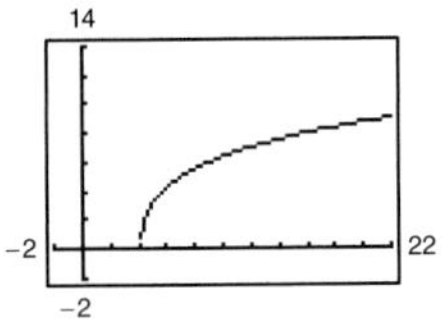

(b)

x	4	5	6	7	8	9	10
y_1	Error	3.22	4.28	4.99	5.55	6.00	6.40
y_2	Error	3.22	4.28	4.99	5.55	6.00	6.40

(c) $y_1 = y_2$

53. (a)

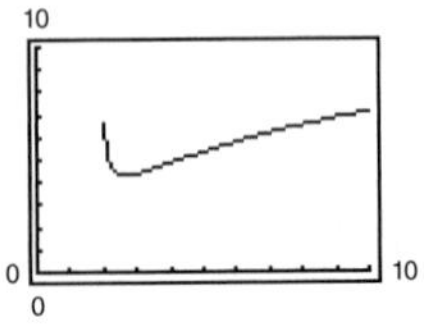

(b)

x	0	3	4	5	6	7	8
y_1	Error	4.39	4.85	5.34	5.78	6.17	6.53
y_2	Error	4.39	4.85	5.34	5.78	6.17	6.53

(c) $y_1 = y_2$

55. $\ln 4x$ **57.** $\log_4 \dfrac{z}{y}$ **59.** $\log_3(x+2)^4$

61. $\ln\left(x\sqrt{x^2+4}\right)$ **63.** $\ln \dfrac{t}{(t-5)^3}$

65. $\ln \dfrac{2(x-2)}{y^3}$ **67.** $\ln \sqrt[3]{\dfrac{x(x+3)^2}{x^2-1}}$

69. (a)

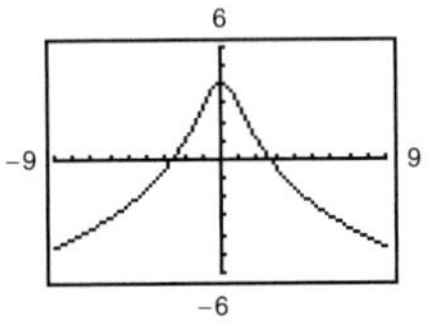

(b)

x	−5	−4	−3	−2	−1
y_1	−2.36	−1.51	−0.45	0.94	2.77
y_2	−2.36	−1.51	−0.45	0.94	2.77

x	0	1	2	3	4	5
y_1	4.16	2.77	0.94	−0.45	−1.51	−2.36
y_2	4.16	2.77	0.94	−0.45	−1.51	−2.36

(c) $y_1 = y_2$

71. (a)

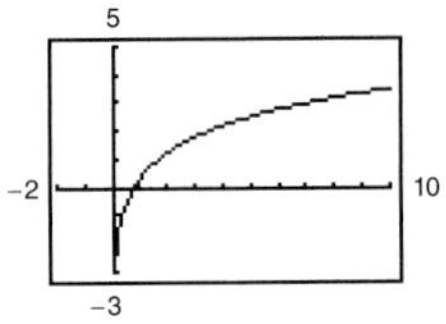

(b)

x	−1	0	1	2
y_1	Error	Error	0.35	1.24
y_2	Error	Error	0.35	1.24

x	3	4	5
y_1	1.79	2.19	2.51
y_2	1.79	2.19	2.51

(c) $y_1 = y_2$

73. 2 **75.** 6.8

77. Not possible; -4 is not in the domain of $\log_2 x$.

79. 3 **81.** -9 **83.** 8 **85.** $\frac{3}{2}$ **87.** $-\frac{1}{2}$

89. (a)

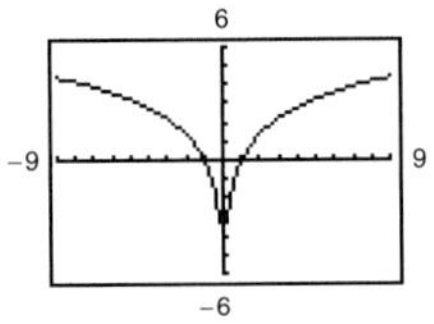

(b)

x	−5	−4	−3	−2	−1
y_1	3.22	2.77	2.20	1.39	0
y_2	Error	Error	Error	Error	Error

x	0	1	2	3	4	5
y_1	Error	0	1.39	2.20	2.77	3.22
y_2	Error	0	1.39	2.20	2.77	3.22

(c) No; The domains differ.

91. (a)

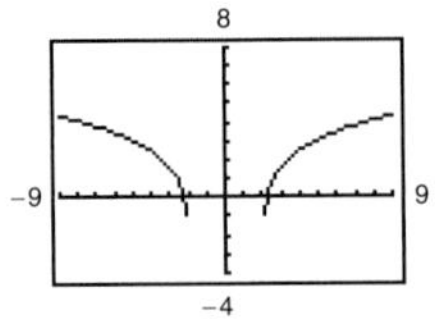

(b)

x	−4	−3	0	3
y_1	Error	Error	Error	1.61
y_2	2.48	1.61	Error	1.61

x	4	5	6
y_1	2.48	3.04	3.47
y_2	2.48	3.04	3.47

(c) No; The domains differ.

93. (a) $\beta = 120 + 10 \log_{10} I$

(b)

I	10^{-4}	10^{-6}	10^{-8}	10^{-10}	10^{-12}	10^{-14}
β	80	60	40	20	0	-20

95. (a)

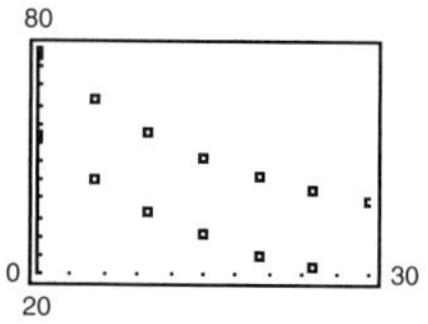

(b) $T = 54.4(0.964)^t + 21$

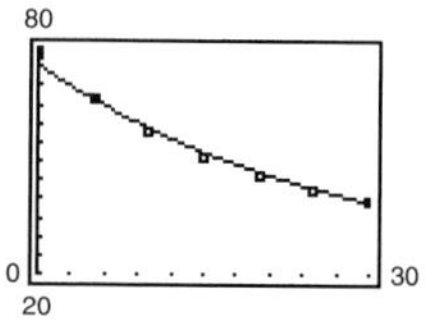

The model fits the data.

(c)

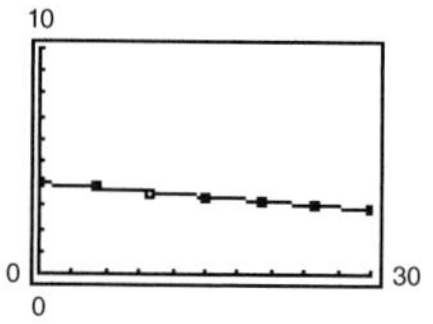

$\ln(T - 21) = -0.037t + 3.997$

$T = e^{(-0.037t+3.997)} + 21$

(d)

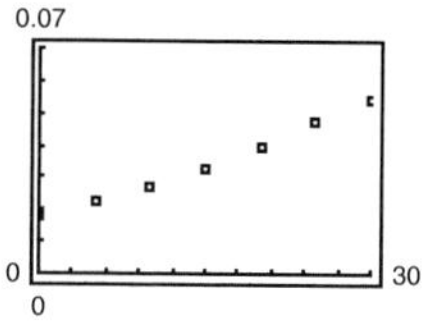

$\dfrac{1}{T - 21} = 0.0012t + 0.0162$

$T = \dfrac{1}{0.0012t + 0.0162} + 21$

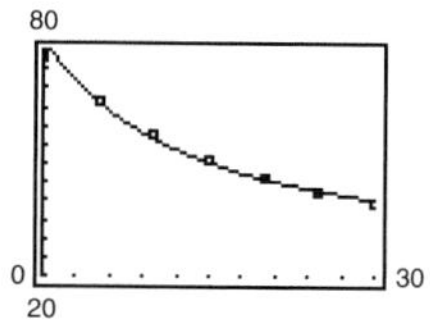

97. True; $\ln(ax) = \ln a + \ln x$ **99.** False; $f(\sqrt{x}) = \frac{1}{2}f(x)$

101. True; $\ln x < 0$ when $0 < x < 1$.

103. The error is an improper use of the Quotient Property of logarithms.

$\ln \dfrac{x^2}{\sqrt{x^2 + 4}} = \ln x^2 - \ln \sqrt{x^2 + 4} = 2 \ln x - \dfrac{1}{2} \ln(x^2 + 4)$

105.

$\ln 1 = 0$	$\ln 9 \approx 2.1972$
$\ln 2 \approx 0.6931$	$\ln 10 \approx 2.3025$
$\ln 3 \approx 1.0986$	$\ln 12 \approx 2.4848$
$\ln 4 \approx 1.3862$	$\ln 15 \approx 2.7080$
$\ln 5 \approx 1.6094$	$\ln 16 \approx 2.7724$
$\ln 6 \approx 1.7917$	$\ln 18 \approx 2.8903$
$\ln 8 \approx 2.0793$	$\ln 20 \approx 2.9956$

107. No; Domain of y_1: $(-\infty, 0)$, $(2, \infty)$; Domain of y_2: $(2, \infty)$

109. $\dfrac{x^3}{64y^4}$, $x \neq 0$ **111.** $\dfrac{3x^4}{2y^3}$, $x \neq 0$

Section 3.4 (page 219)

1. (a) $x = y$ (b) $x = y$ (c) x (d) x **3.** Yes

5. Subtract 3 from both sides. **7.** 2; One-to-One Property

9. 2; One-to-One Property

11. $\ln 2 \approx 0.693$; Inverse Property

13. $e^{-1} \approx 0.368$; Inverse Property

15. 64; Inverse Property **17.** $2, -1$ **19.** $\dfrac{\ln 5}{\ln 3} \approx 1.465$

21. $\ln 39 \approx 3.664$ **23.** $\dfrac{\ln 80}{2 \ln 3} \approx 1.994$

25. $2 - \dfrac{\ln 400}{\ln 3} \approx -3.454$ **27.** $\dfrac{1}{3} \log_{10} \dfrac{3}{2} \approx 0.059$

29. $\dfrac{\ln 12}{3} \approx 0.828$ **31.** $\dfrac{\ln 40}{0.02} \approx 184.444$ **33.** 0

35. $\dfrac{\ln \frac{8}{3}}{3 \ln 2} + \dfrac{1}{3} \approx 0.805$ **37.** $-\dfrac{\ln 2}{\ln 3 - \ln 2} \approx -1.710$

39. $0, \dfrac{\ln 4}{\ln 5} \approx 0.861$ **41.** $\ln \frac{4}{5} \approx -0.223$

43. $\dfrac{\ln 4}{365 \ln\left(1 + \dfrac{0.065}{365}\right)} \approx 21.330$ **45.** 3 **47.** -5.270

49. 138.631 **51.** 2.720 **53.** $\ln 5 \approx 1.609$

55. $\ln \frac{4}{3} \approx 0.288$, $\ln \frac{1}{2} \approx -0.693$

57.

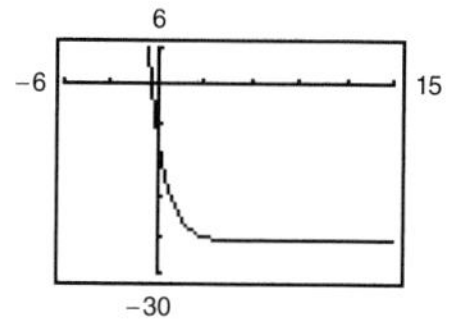

$x \approx -0.427$

59.

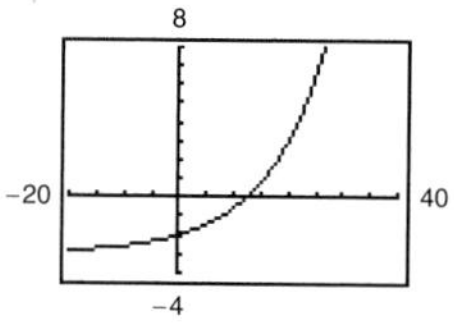

$t \approx 12.207$

61. $\dfrac{e^{2.1}}{6} \approx 1.361$ **63.** $e^{-2} \approx 0.135$

65. $2(3^{11/6}) \approx 14.988$ **67.** $\dfrac{e^5 + 1}{2} \approx 74.707$ **69.** ± 9

71. $e^2 - 2 \approx 5.389$ **73.** $-1 \pm e \approx 1.718, -3.718$

75. $x = 9$ **77.** 1.482 **79.** 0.445 **81.** 0.056

83. ± 0.381

85. No solution; extraneous solution: $\dfrac{e^2}{1 - e^2} \approx -1.157$

87. No solution; extraneous solutions: $-3, -2$

89. No solution; extraneous solution: -7

91. 1; extraneous solution: -3

93. 3; extraneous solution: -2

95. $\dfrac{725 + 125\sqrt{33}}{8} \approx 180.384$;

extraneous solution: $\dfrac{725 - 125\sqrt{33}}{8} \approx 0.866$

97. Answers will vary. **99.** 1.469, 0.001 **101.** 3.398

103. $-1.993, 3.738$ **105.** 0.133, 0.812

107. (a)

x	0.6	0.7	0.8	0.9	1.0
e^{3x}	6.05	8.17	11.02	14.88	20.09

(0.8, 0.9)

(b) 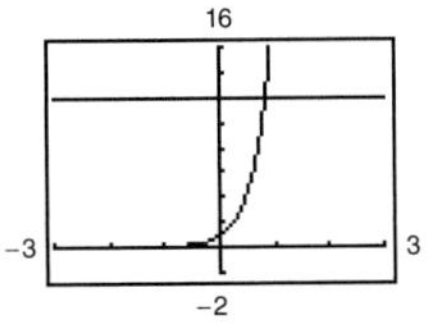

0.828

(c) 0.828

109. (a)

x	5	6	7	8	9
$20(100 - e^{x/2})$	1756	1598	1338	908	200

(8, 9)

(b) 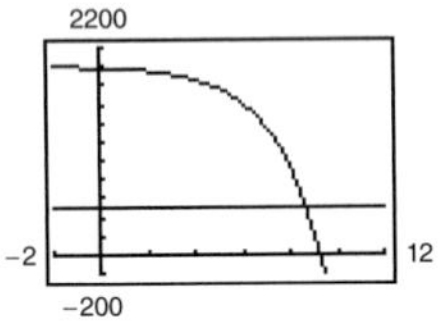

8.635

(c) 8.635

111. (a)

x	2	3	4	5	6
$\ln 2x$	1.39	1.79	2.08	2.30	2.48

(5, 6)

(b) 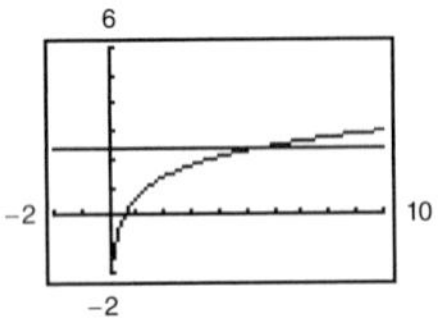

5.512

(c) 5.512

113. (a)

x	12	13	14	15	16
$6 \log_3(0.5x)$	9.79	10.22	10.63	11.00	11.36

(14, 15)

(b) 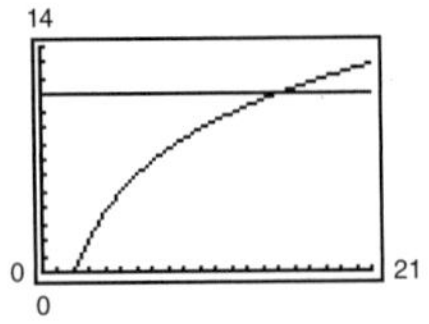

14.988

(c) 14.988

115. $-1, 0$ **117.** 1 **119.** $e^{-1/2} \approx 0.607$

121. $e^{-1} \approx 0.368$ **123.** $\dfrac{\ln y - \ln a}{b}$

125. $b \pm \sqrt{c(\ln a - \ln y)}$ **127.** (a) 9.90 yr (b) 15.69 yr

129. (a) 27.73 yr (b) 43.94 yr

131. (a) 93 weeks (b) 124 weeks **133.** 2010

135. (a) 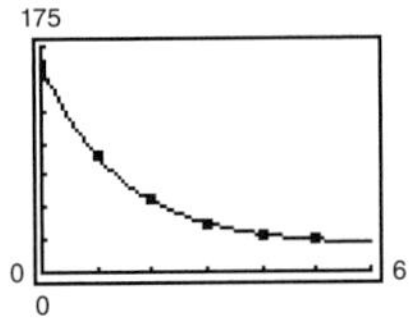

(b) $y = 20$; The object's temperature cannot cool below the room's temperature.

(c) About 1.22 h

137. False; $e^x = 0$ has no solutions.

139. Both sides of the equation should be divided by 2 before taking the natural log of both sides.

141. (a) 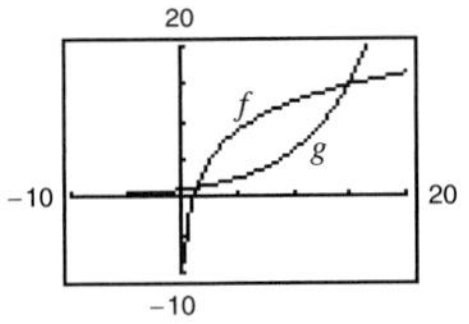

About (1.258, 1.258), about (14.767, 14.767)

(b) $a = e^{(1/e)} \approx 1.445$ (c) $1 < a < e^{(1/e)}$

143. 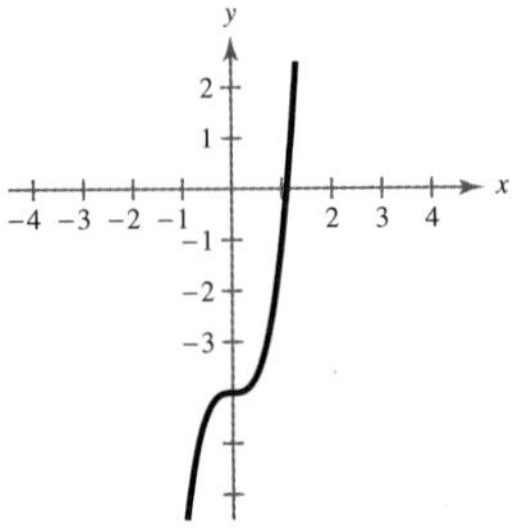

145. 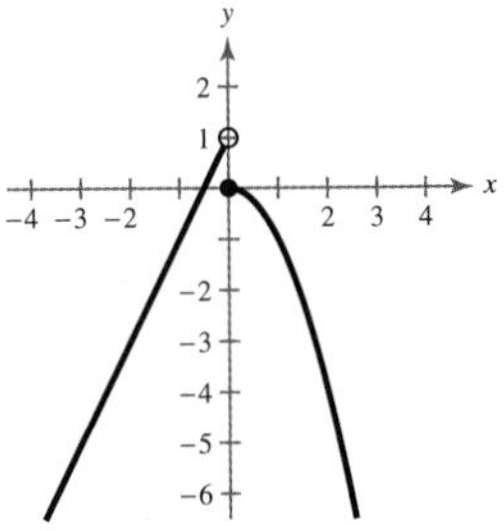

Section 3.5 (page 230)

1. (a) iv (b) i (c) iii (d) vi (e) ii (f) v

3. sigmoidal **5.** Exponential decay

7. c **8.** e **9.** b **10.** a **11.** d **12.** f

13. About 19.8 yr; $1419.07 **15.** About 3.3%; $10,432.26

17. About 1.25%; about 55.45 yr

19. $63,762.82; about 15.40 yr

21.

r	2%	4%	6%	8%	10%	12%
t	54.93	27.47	18.31	13.73	10.99	9.16

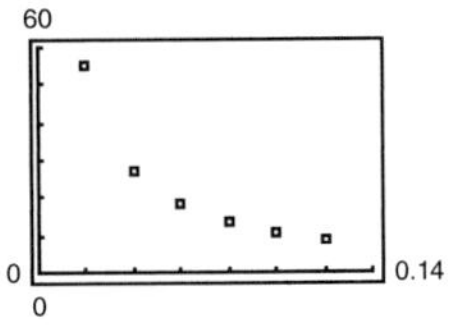

23. 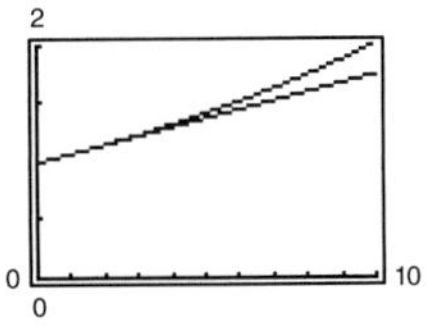

Continuous compounding

25. About 2.26 g **27.** About 1.47 g **29.** About 431.7

31. $y = e^{0.768x}$ **33.** $y = 4e^{-0.2773x}$

35. (a) Increasing; The positive exponent indicates that the model is increasing.

(b) About 7,950,000 people; about 8,141,000 people; about 8,402,000 people

(c) 2026

37. (a) 0.0153 (b) About 6,054,000 people

39. About 15,601 years ago

41. (a) $V = -7527.5t + 41,820$ (b) $V = 41,820e^{-0.2231t}$

(c) 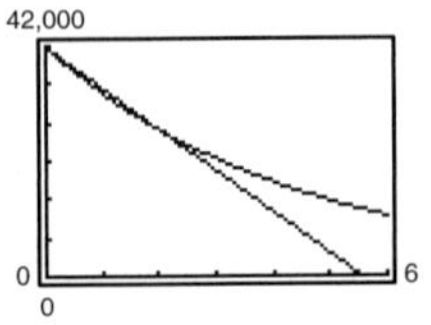

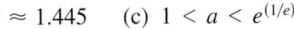

(d) Exponential model (e) $0 < t < 2$; $t \geq 2$

43. (a)
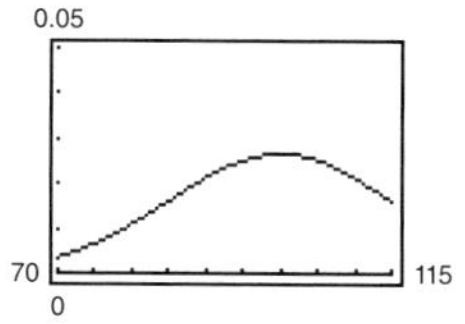

(b) 100

45. (a) About 203 animals (b) About 13 mo
(c)
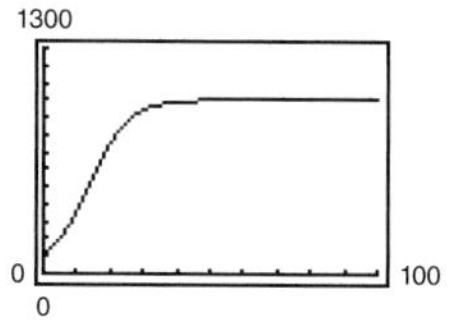

(d) The horizontal asymptotes occur at $p = 1000$ and $p = 0$.
(e) The asymptote at $p = 1000$ means there will not be more than 1000 animals in the preserve.

47. (a) About 7.6 (b) About 7.1 (c) About 5.4 (d) About 8.3

49. (a) 20 dB (b) 70 dB (c) 120 dB (d) 10 dB

51. About 97.49% **53.** About 4.64

55. $10^{-3.2} \approx 6.3 \times 10^{-4}$ moles/L **57.** About 79,433 times

59. (a)
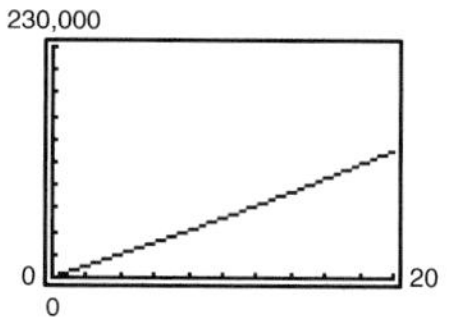

(b) About 33.5 years; yes

61. About 3:00 A.M.

63. False; The domain can be all real numbers.

65. No; Any x-value in the Gaussian model will give a positive y-value.

67. Gaussian model **69.** Exponential growth model

71. The intercept is $(0, a)$, not $(a, 0)$.

73. a; $(0, -3)$, $\left(\frac{9}{4}, 0\right)$ **74.** b; $(0, 2)$, $(5, 0)$

75. d; $(0, 25)$, $\left(\frac{100}{9}, 0\right)$ **76.** c; $(0, 4)$, $(2, 0)$

77. $2x^2 + 3 + \dfrac{3}{x - 4}$ **79.** Answers will vary.

Section 3.6 (page 240)

1. $y = ax^b$ **3.** Scatter plot **5.** Logarithmic model

7. Quadratic model **9.** Exponential model

11.
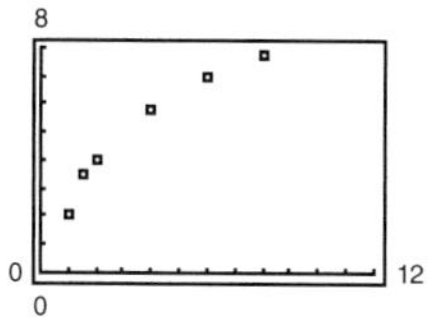

Logarithmic model

13.
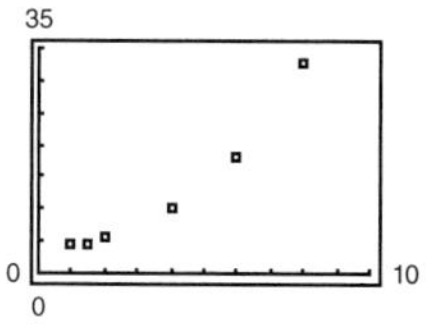

Exponential model

15.
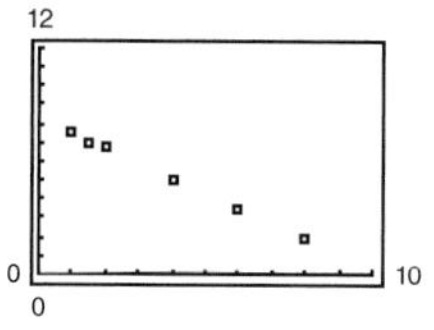

Linear model

17. $y = 4.760(1.2591)^x$; 0.9811
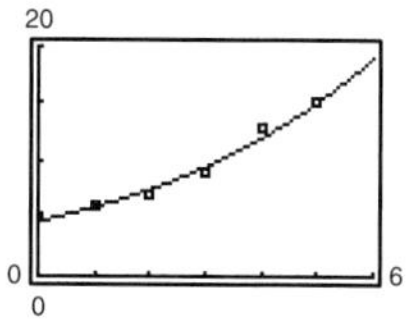

19. $y = 7.707(0.8070)^x$; 0.9448
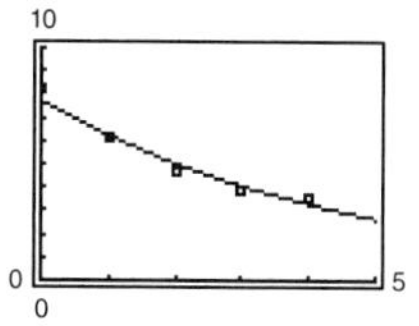

21. $y = 2.083 + 1.257 \ln x$; 0.9867
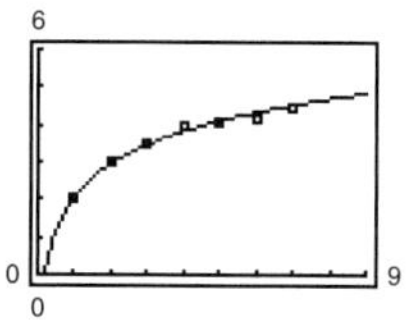

23. $y = -4.730 \ln x + 10.353$; 0.9661
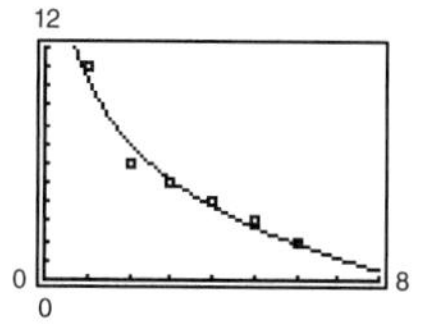

25. $y = 1.987x^{0.760}$; 0.9969
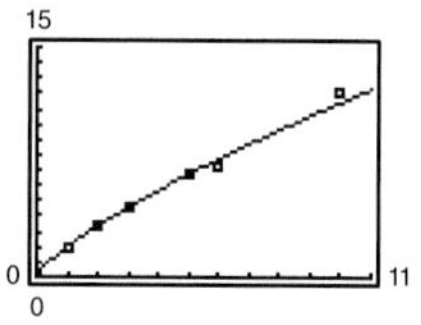

27. $y = 16.103x^{-3.174}$; 0.8816
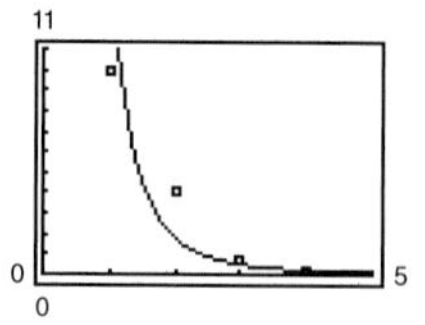

29. (a) Exponential model: $S = 4.114(1.1960)^t$; 0.9986
Power model: $S = 3.153t^{0.766}$; 0.9514
(b) Exponential model:
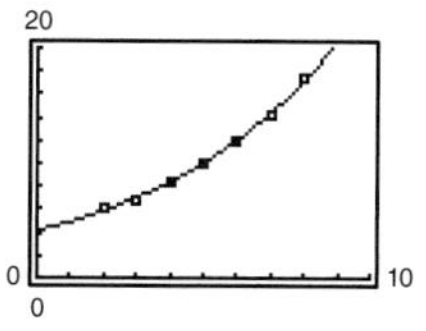

Power model:
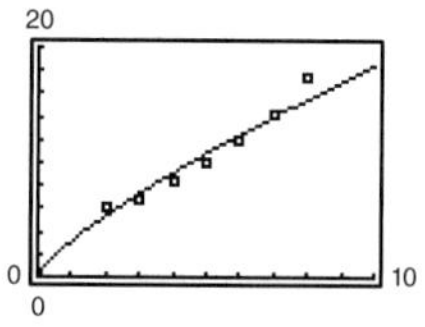

(c) Exponential model; The coefficient of determination is closer to 1.

31. (a) $P = 0.0136t + 2.674$; 0.9911
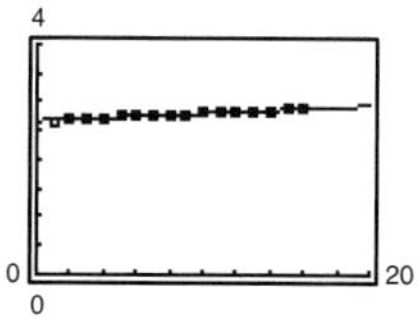

(b) $P = 2.640t^{0.0286}$; 0.9312
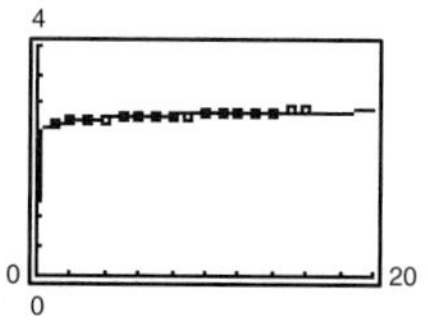

(c) $P = 2.676(1.0049)^t$; 0.9891
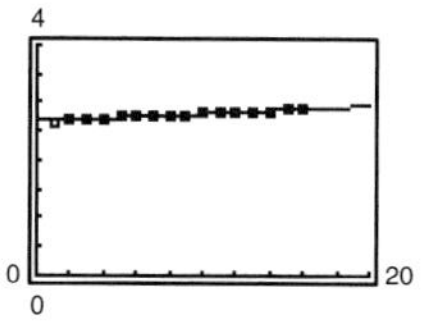

(d) $P = 2.638 + 0.079 \ln t$; 0.9265
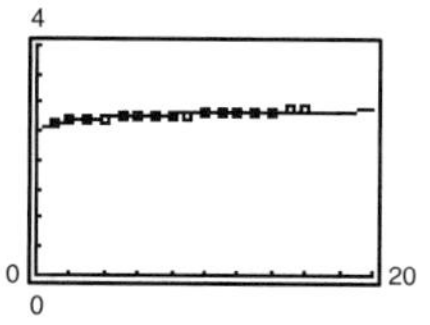

(e) Linear model; The coefficient of determination is closest to 1.
(f) Linear: 3,014,000 people
Power: about 2,895,000 people
Exponential: about 3,024,000 people
Logarithmic: about 2,892,000 people
(g) and (h) Answers will vary.

33. (a) $S = 43.588(1.0668)^t$ (b) $S = 43.588(e^{0.0647t})$
(c) About \$83.24 billion; Answers will vary.

35. (a) $P = \dfrac{86.3934}{1 + 1544.2020e^{-0.2691x}}$

(b)
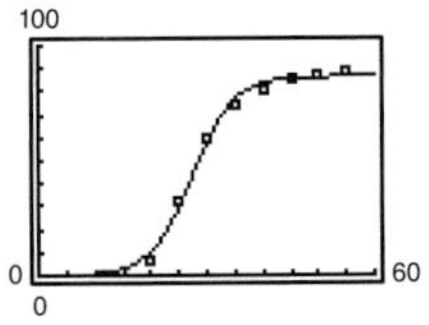

The model fits the data well.

37. True; See page 223. **39.** Answers will vary.

41. The values of x and y were switched. When $y = 2.5$, $x \approx 3$, but when $x = 2.5$, $y \approx 2$.

43. Slope: $-\frac{2}{5}$; y-intercept: $(0, 2)$

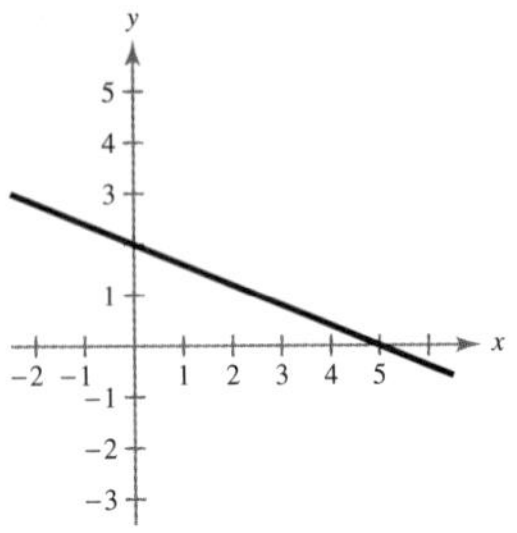

45. Slope: 0.16; y-intercept: $(0, -5)$

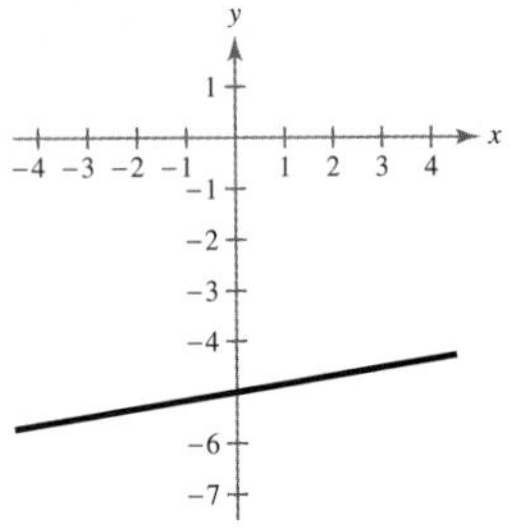

Review Exercises (page 244)

1. 0.033

3.
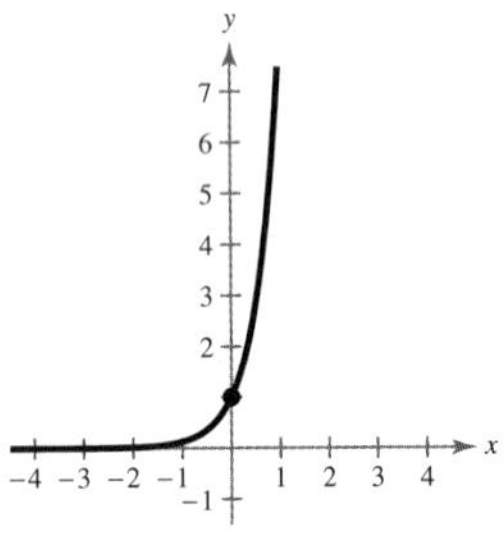

Horizontal asymptote: $y = 0$
y-intercept: $(0, 1)$
Increasing on $(-\infty, \infty)$

5.
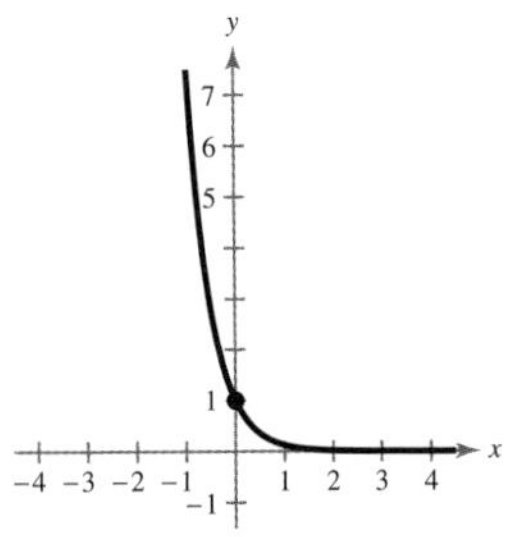

Horizontal asymptote: $y = 0$
y-intercept: $(0, 1)$
Decreasing on $(-\infty, \infty)$

7. c **8.** d **9.** b **10.** a

11.

x	0	1	2	3	4
$h(x)$	0.37	1	2.72	7.39	20.09

Horizontal asymptote: $y = 0$

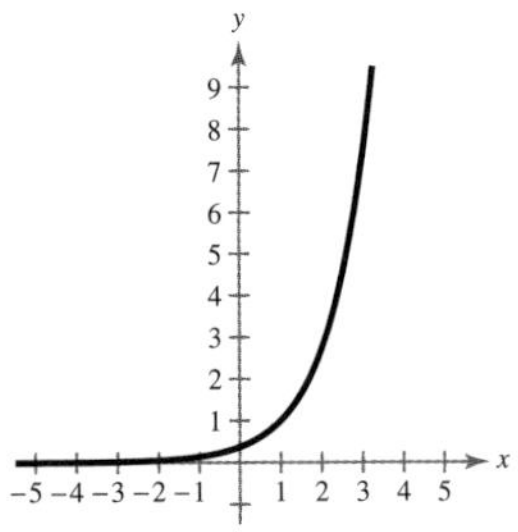

13.

x	-2	-1	0	1	2
$h(x)$	2.86	2.63	2	0.28	-4.39

Horizontal asymptote: $y = 3$

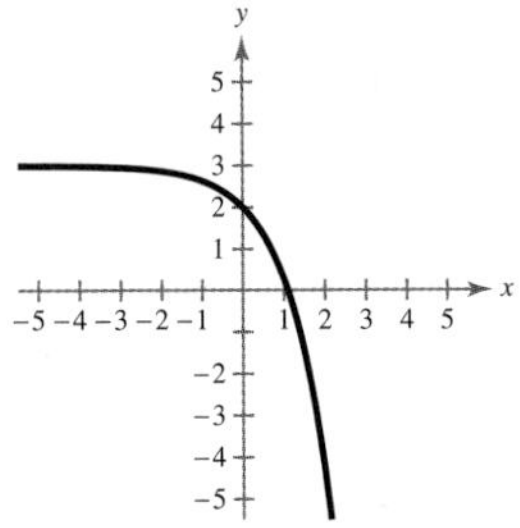

15.

x	-1	0	1	2	3	4
$f(x)$	6.59	4	2.43	1.47	0.89	0.54

Horizontal asymptote: $y = 0$

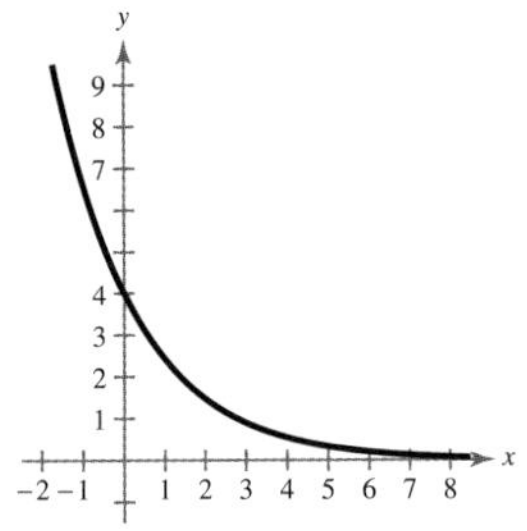

17.

t	1	10	20
A	\$10,832.87	\$22,255.41	\$49,530.32

t	30	40	50
A	\$110,231.76	\$245,325.30	\$545,981.50

19. (a)
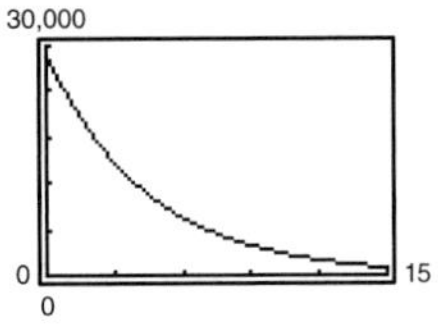

(b) \$14,694.40

(c) When it is first sold; yes; Answers will vary.

21. $5^4 = 625$ **23.** $64^{1/6} = 2$ **25.** $\log_4 64 = 3$

27. $\ln 24.532\ldots = 3.2$ **29.** $\log_{1/2} 8 = -3$

31. 3 **33.** -1

35. Domain: $(0, \infty)$
x-intercept: $(10{,}000, 0)$
Vertical asymptote: $x = 0$

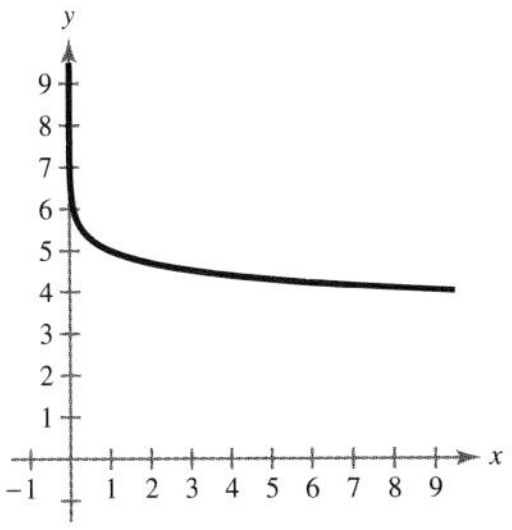

37. Domain: $(1, \infty)$
x-intercept: $(1.000001, 0)$
Vertical asymptote: $x = 1$

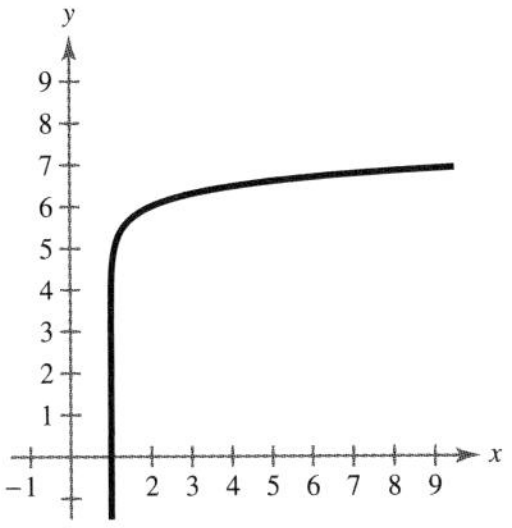

39. 3.953 **41.** 1.282

43.

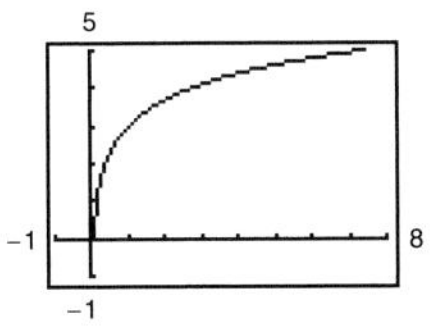

Domain: $(0, \infty)$
x-intercept: $(0.05, 0)$
Vertical asymptote: $x = 0$

45.

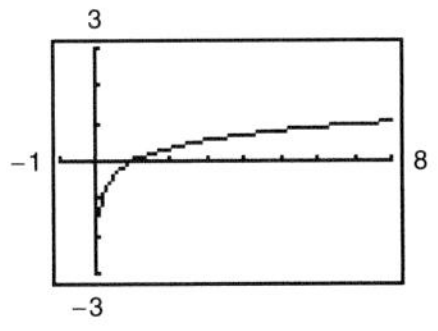

Domain: $(0, \infty)$
x-intercept: $(1, 0)$
Vertical asymptote: $x = 0$

47. (a) $0 \le h < 18{,}000$
(b)

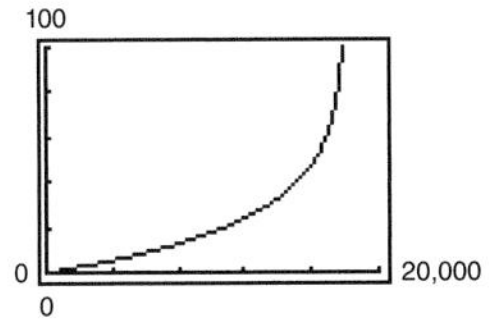

Asymptote: $h = 18{,}000$
(c) The time required to increase its altitude further increases.
(d) About 5.46 min

49. (a) and (b) -0.649 **51.** 1.1292

53.

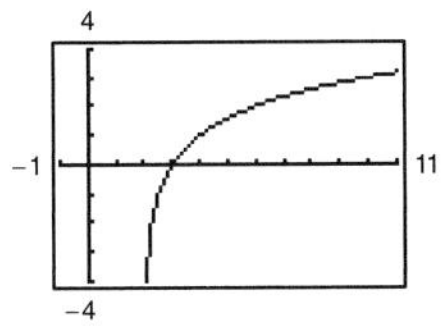

55. $\ln 5 - 2$ **57.** $1 + 2 \log_6 x$

59. $\ln(x + 3) - \ln x - \ln y$ **61.** $\log_2 9x$

63. $\ln \dfrac{\sqrt{2x - 1}}{(x^2 + 1)^2}$ **65.** $\ln \dfrac{3\sqrt[3]{4 - x^2}}{x}$

67. (a)

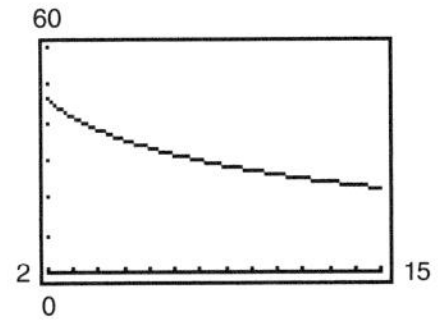

(b) The decrease in productivity starts to level off.

69. 4; One-to-One Property **71.** -3; One-to-One Property

73. -5; One-to-One Property

75. $\ln 4 \approx 1.386$; Inverse Property

77. 4; One-to-One Property **79.** 4096; Inverse Property

81. $e^4 \approx 54.598$; Inverse Property **83.** -0.757

85. 4.459 **87.** 1.760 **89.** 3.916 **91.** 1.609, 0.693

93. 0.9 **95.** 2715.830 **97.** 8.389 **99.** 36.945

101. 53.598 **103.** 4 **105.** -1 **107.** 0.368

109. $t = 9$ **111.** e **112.** b **113.** c **114.** d

115. a **116.** f **117.** 0.0045; about 3,221,000

119. (a) About 11.08 weeks (b) About 24.16 weeks

121. (a) $h = \dfrac{21.253}{1 + 24.284^{-0.131x}}$

(b)

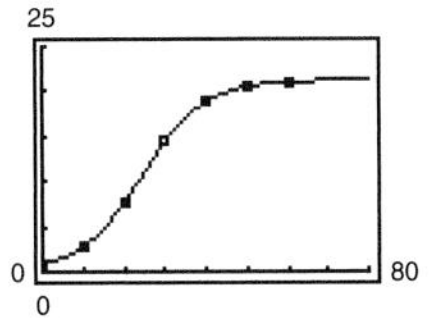

(c) The model fits the data well. (d) About 21.3 m

123. False; $x > 0$

125. $1 < \sqrt{3} < 2$, so $3^1 < 3^{\sqrt{3}} < 3^2$.

Chapter Test (page 250)

1.

x	-2	-1	0	1	2
$f(x)$	100	10	1	0.1	0.01

Horizontal asymptote: $y = 0$

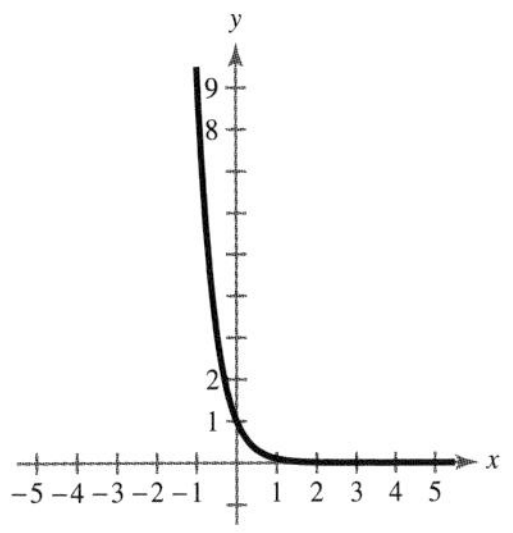

2.

x	0	2	3	4
$f(x)$	-0.03	-1	-6	-36

Horizontal asymptote: $y = 0$

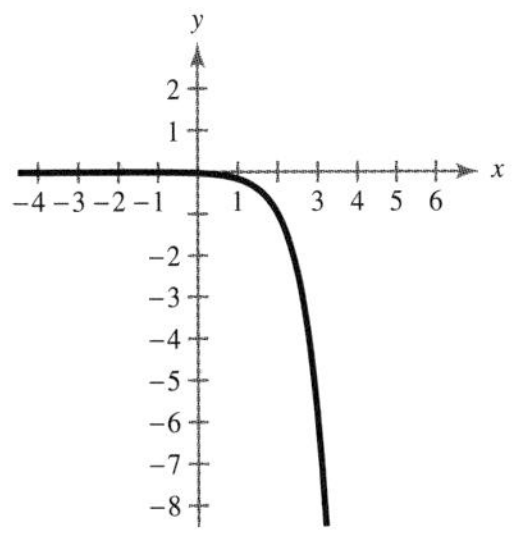

3.

x	-2	-1	0	1	2
$f(x)$	0.9817	0.8647	0	-6.3891	-53.5982

Horizontal asymptote: $y = 1$

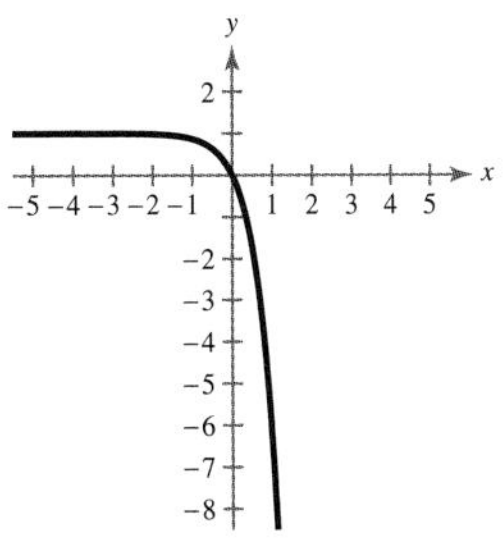

4. -0.89 **5.** 9.2 **6.** 2

7.

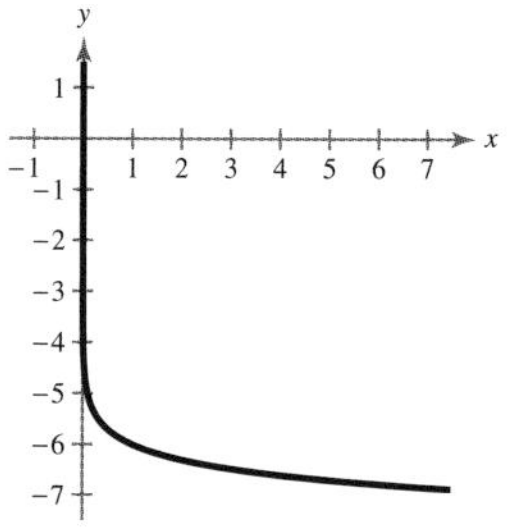

Domain: $(0, \infty)$
x-intercept: $(10^{-6}, 0)$
Vertical asymptote: $x = 0$

8.

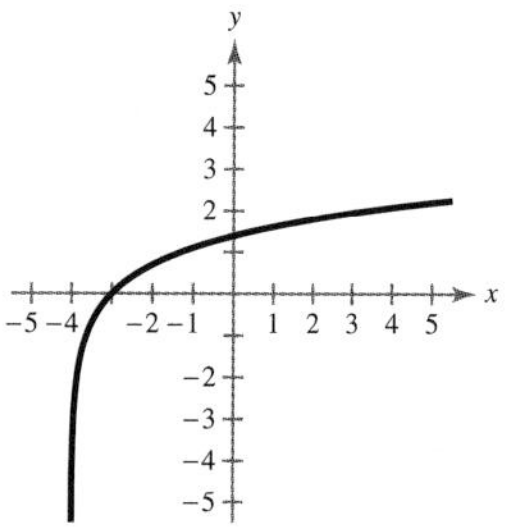

Domain: $(-4, \infty)$
x-intercept: $(-3, 0)$
Vertical asymptote: $x = -4$

9.

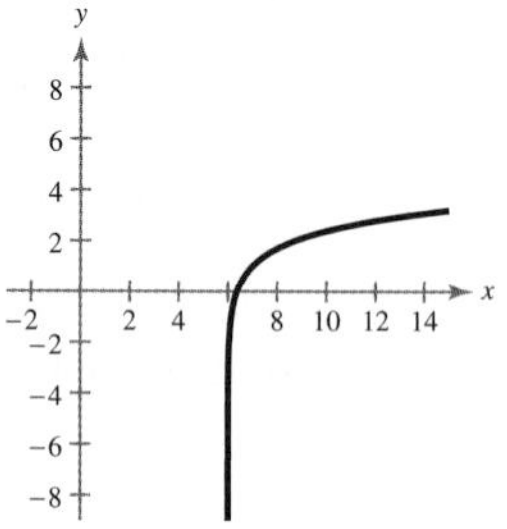

Domain: $(6, \infty)$
x-intercept: $(6.37, 0)$
Vertical asymptote: $x = 6$

10. 1.94 **11.** -0.29 **12.** 1.67
13. $\log_2 3 + 4 \log_2 a$ **14.** $\ln 5 + \frac{1}{2} \ln x - \ln 6$
15. $\ln x + \frac{1}{2} \ln(x + 1) - \ln 2 - 4$ **16.** $\log_3 13y$
17. $\ln \frac{x^4}{y^4}$ **18.** $\log_{10} x^{5/2}$ **19.** $-2, 3$ **20.** $-\frac{1}{3}$
21. $-0.636, 1.750$ **22.** 548.317 **23.** 98
24. 26.667 **25.** About 1.4277, about 0.0001
26. About 0.2049, about 0.6437 **27.** About 0.7336
28. About 0.3645 **29.** About 62.3%
30. (a)

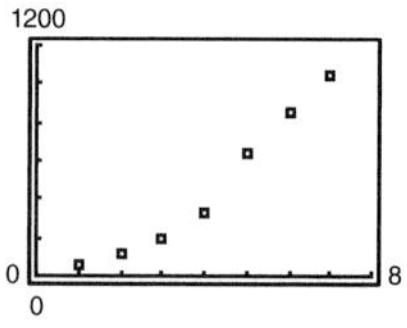

(b) *Sample answer:* Quadratic; $R = 22.432t^2 - 3.08t + 31.4$; $r^2 \approx 0.9892$

(c)

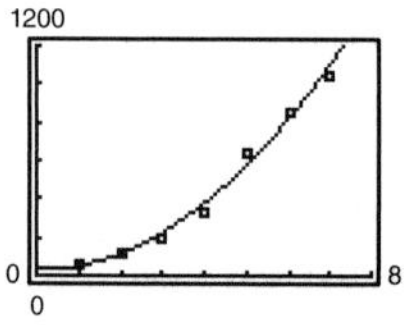

(d) Yes; r^2 is close to 1.

Standardized Test Practice (page 251)

1. C **2.** D **3.** D **4.** B **5.** D **6.** A
7. B **8.** C **9.** D **10.** D **11.** D **12.** C
13. D **14.** C **15.** B **16.** A **17.** C **18.** B
19. C **20.** D **21.** C **22.** 42 **23.** 73 **24.** 5
25. \$2700 **26.** 147 rabbits

Chapter 4

Section 4.1 (page 263)

1. Trigonometry **3.** standard position **5.** $\frac{\pi \text{ rad}}{180°}$
7. 180° **9.** No **11.** 2
13. (a) Quadrant I (b) Quadrant III
15. (a) Quadrant IV (b) Quadrant III
17. (a) Quadrant IV (b) Quadrant III
19. (a)

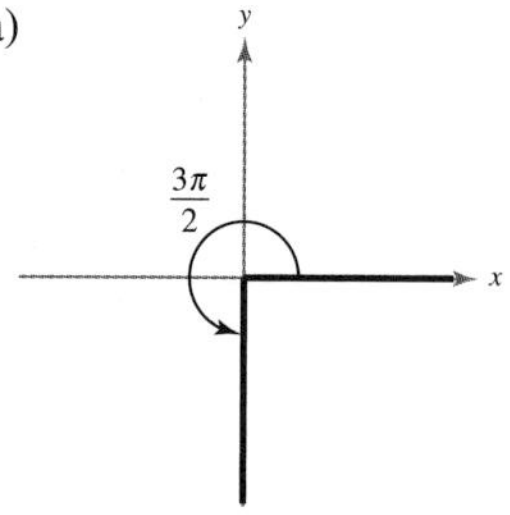

(b)

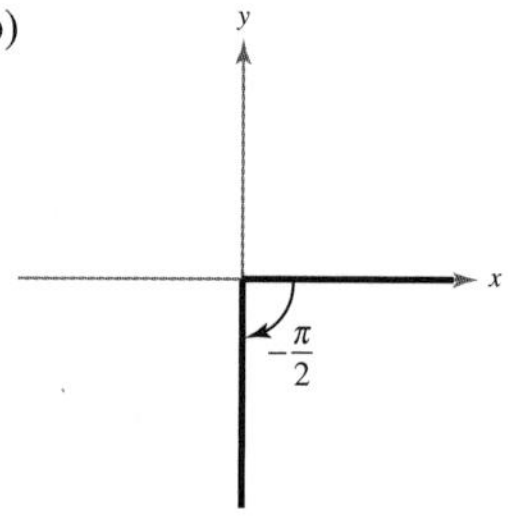

21. (a)

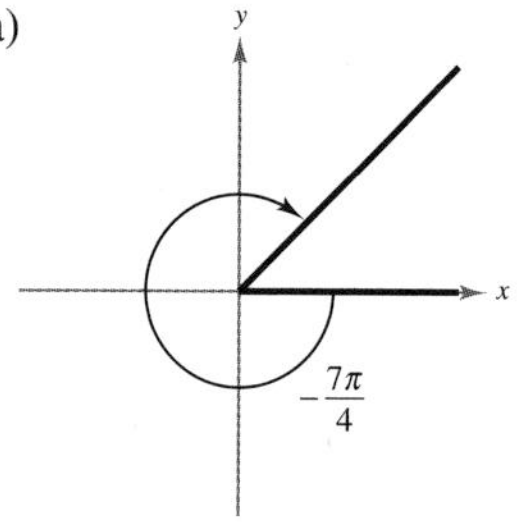

(b)

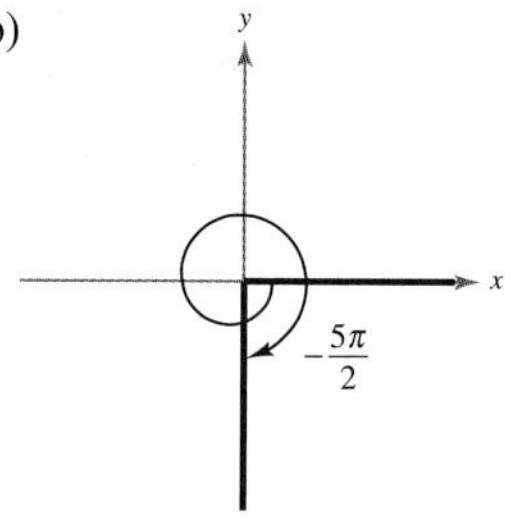

23. (a)

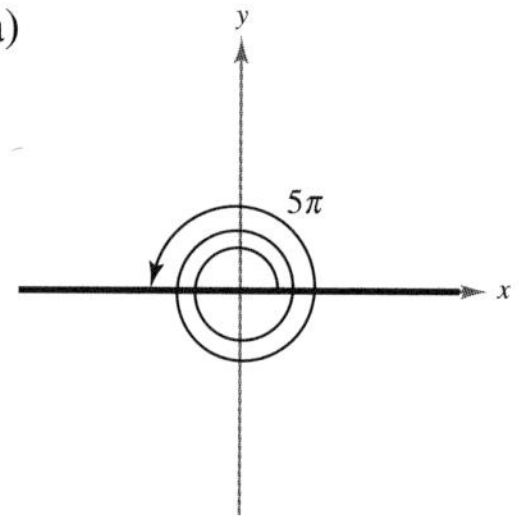

(b)

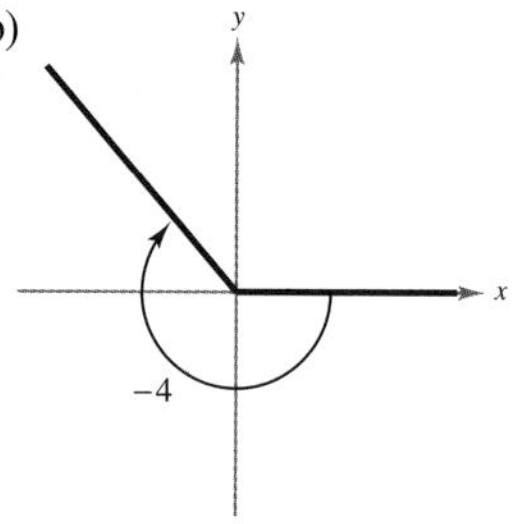

25. *Sample answers:* (a) $\frac{13\pi}{6}, -\frac{11\pi}{6}$ (b) $\frac{7\pi}{6}, -\frac{17\pi}{6}$
27. *Sample answers:* (a) $\frac{\pi}{4}, -\frac{7\pi}{4}$ (b) $\frac{28\pi}{15}, -\frac{32\pi}{15}$
29. 210° **31.** (a) Quadrant I (b) Quadrant III
33. (a) Quadrant III (b) Quadrant IV
35. (a) Quadrant II (b) Quadrant I
37. (a)

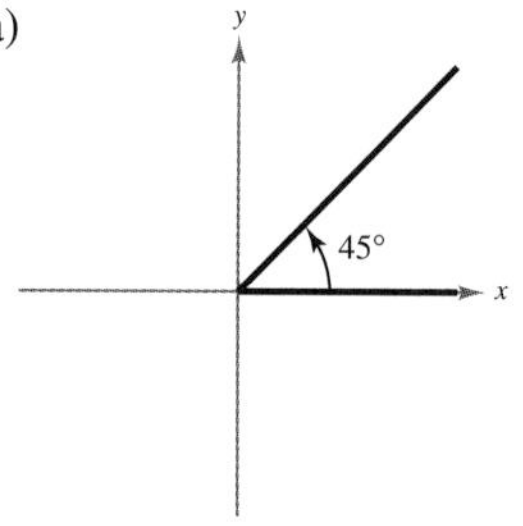

(b)

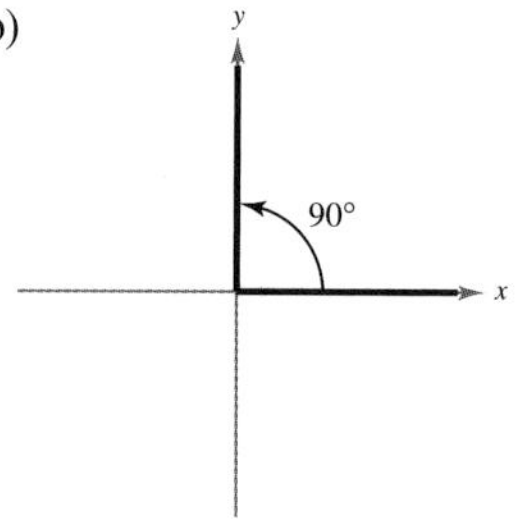

39. (a)
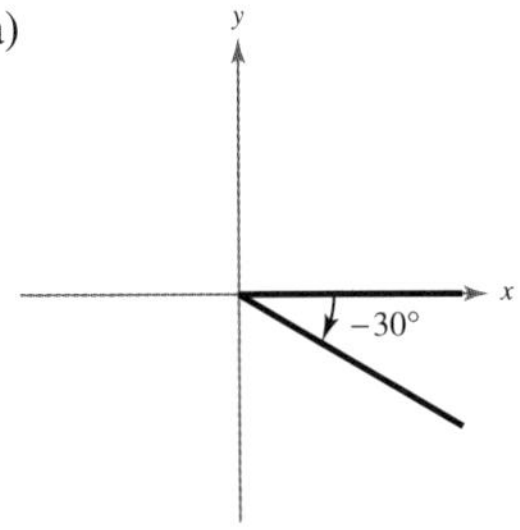

(b)
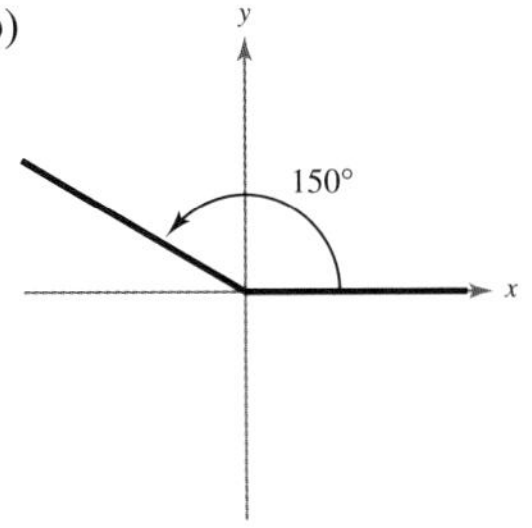

41. (a)
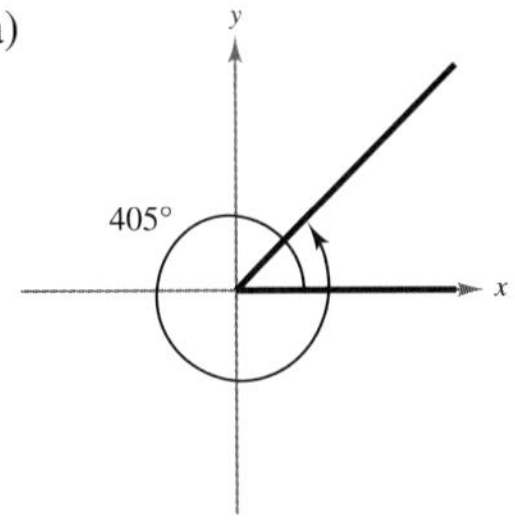

(b)
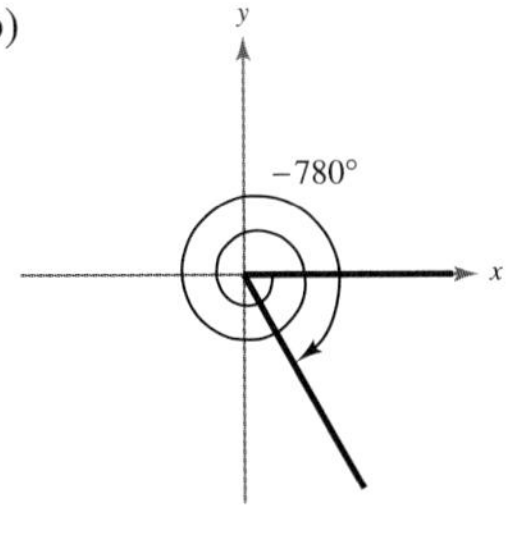

43. *Sample answers:* (a) 480°, −240° (b) 150°, −570°

45. *Sample answers:* (a) 60°, −660° (b) 590°, −130°

47. (a) $\frac{2\pi}{3}$ (b) $-\frac{\pi}{9}$ **49.** (a) $\frac{\pi}{6}$ (b) $-\frac{2\pi}{5}$

51. (a) 270° (b) −210° **53.** (a) −180° (b) 540°

55. 1.986 **57.** −3.776 **59.** −0.014 **61.** 25.714°

63. 1170° **65.** −121.467° **67.** 64.75° **69.** 85.308°

71. −125.01° **73.** 13° 5′15″ **75.** −28° 59′13″

77. 280° 36′ **79.** −345° 7′12″ **81.** −20° 20′24″

83. (a) Complement: 44°; supplement: 134°
(b) Complement: none; supplement: none
(c) Complement: $\frac{\pi}{6}$; supplement: $\frac{2\pi}{3}$
(d) Complement: none; supplement: $\frac{\pi}{4}$

85. (a) Complement: none; supplement: 87°
(b) Complement: 70°; supplement: 160°
(c) Complement: $\frac{3\pi}{8}$; supplement: $\frac{7\pi}{8}$
(d) Complement: none; supplement: none

87. 10π in. ≈ 31.42 in. **89.** 14π in. ≈ 43.98 in. **91.** $\frac{8}{15}$

93. $\frac{70}{29}$ **95.** 22.29 ft **97.** 34.80 mi

99. About 171.57 mi **101.** About 4° 2′33″

103. About 275.02° **105.** About 435.71 km/min

107. (a) 2100π rad/min (b) About 64,324.1 in./min

109. (a) $\frac{1000\pi}{3}$ rad/sec (b) About 62.83 m/sec

111. False; One radian is larger: 1 rad ≈ 57.3°.

113. The sum of the angles should equal 180° = π rad, not 2π.

115. $\frac{50\pi}{3}$ m^2

117. (a) $A = 0.4r^2, r > 0; s = 0.8r, r > 0$

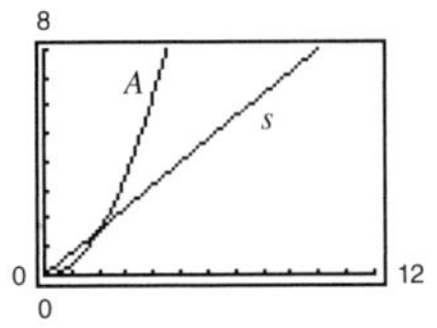

The area function changes more rapidly for $r > 1$ because it is quadratic and the arc length function is linear.

(b) $A = 50\theta, 0 < \theta < 2\pi; s = 10\theta, 0 < \theta < 2\pi$

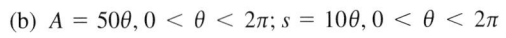

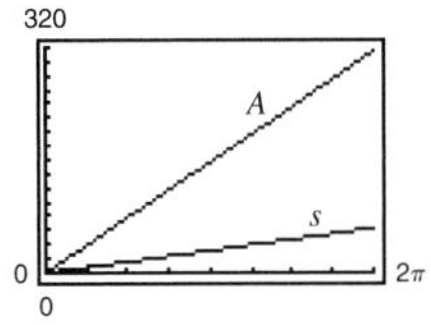

119. Answers will vary.

121.
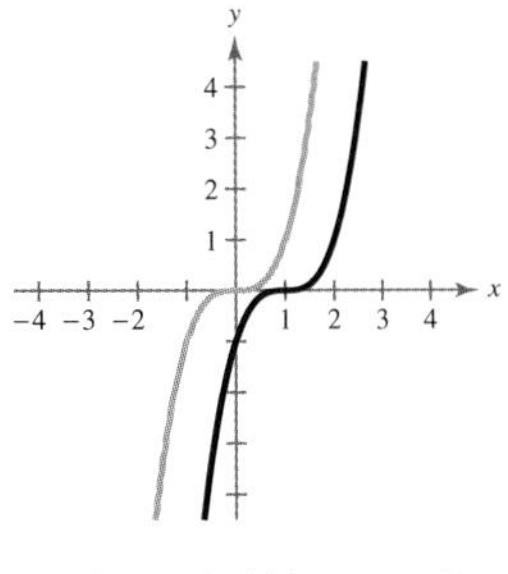

Horizontal shift one unit to the right

123.
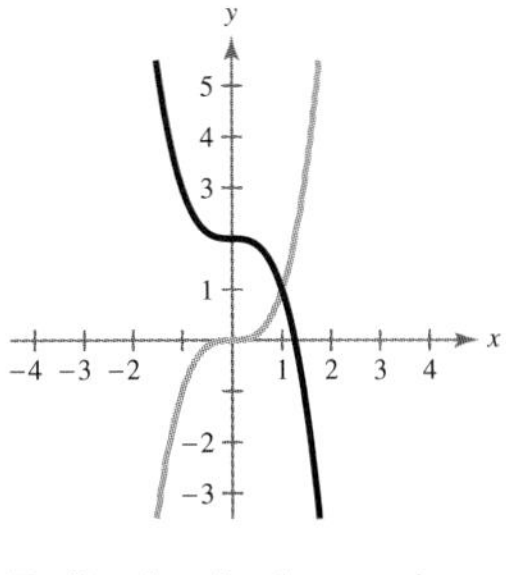

Reflection in the x-axis, vertical shift two units upward

125.
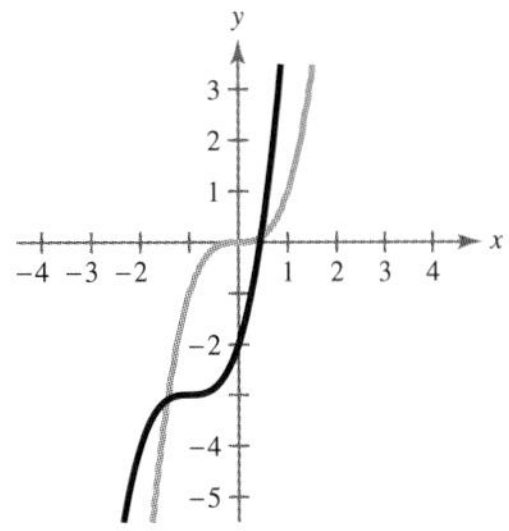

Horizontal shift one unit to the left, vertical shift three units downward

Section 4.2 (page 272)

1. unit circle **3.** odd, even

5. Even: cos, sec; Odd: sin, csc, tan, cot **7.** 8

9. $\sin\theta = \frac{15}{17}$ $\csc\theta = \frac{17}{15}$
$\cos\theta = -\frac{8}{17}$ $\sec\theta = -\frac{17}{8}$
$\tan\theta = -\frac{15}{8}$ $\cot\theta = -\frac{8}{15}$

11. $\sin\theta = -\frac{21}{29}$ $\csc\theta = -\frac{29}{21}$
$\cos\theta = \frac{20}{29}$ $\sec\theta = \frac{29}{20}$
$\tan\theta = -\frac{21}{20}$ $\cot\theta = -\frac{20}{21}$

13. $\left(\frac{\sqrt{3}}{2}, \frac{1}{2}\right)$ **15.** $\left(\frac{\sqrt{2}}{2}, \frac{\sqrt{2}}{2}\right)$ **17.** $(0, -1)$

19. $\left(\frac{1}{2}, -\frac{\sqrt{3}}{2}\right)$ **21.** $\left(-\frac{\sqrt{3}}{2}, -\frac{1}{2}\right)$

23. $\sin\frac{\pi}{4} = \frac{\sqrt{2}}{2}$
$\cos\frac{\pi}{4} = \frac{\sqrt{2}}{2}$
$\tan\frac{\pi}{4} = 1$

25. $\sin\left(-\frac{7\pi}{4}\right) = \frac{\sqrt{2}}{2}$
$\cos\left(-\frac{7\pi}{4}\right) = \frac{\sqrt{2}}{2}$
$\tan\left(-\frac{7\pi}{4}\right) = 1$

27. $\sin\frac{2\pi}{3} = \frac{\sqrt{3}}{2}$
$\cos\frac{2\pi}{3} = -\frac{1}{2}$
$\tan\frac{2\pi}{3} = -\sqrt{3}$

29. $\sin\frac{\pi}{2} = 1$
$\cos\frac{\pi}{2} = 0$
$\tan\frac{\pi}{2}$ is undefined.

CHAPTER 4

31. $\sin\left(-\dfrac{\pi}{6}\right) = -\dfrac{1}{2}$

$\cos\left(-\dfrac{\pi}{6}\right) = \dfrac{\sqrt{3}}{2}$

$\tan\left(-\dfrac{\pi}{6}\right) = -\dfrac{\sqrt{3}}{3}$

33. $\sin\dfrac{2\pi}{3} = \dfrac{\sqrt{3}}{2}$ $\quad \csc\dfrac{2\pi}{3} = \dfrac{2\sqrt{3}}{3}$

$\cos\dfrac{2\pi}{3} = -\dfrac{1}{2}$ $\quad \sec\dfrac{2\pi}{3} = -2$

$\tan\dfrac{2\pi}{3} = -\sqrt{3}$ $\quad \cot\dfrac{2\pi}{3} = -\dfrac{\sqrt{3}}{3}$

35. $\sin\dfrac{4\pi}{3} = -\dfrac{\sqrt{3}}{2}$ $\quad \csc\dfrac{4\pi}{3} = -\dfrac{2\sqrt{3}}{3}$

$\cos\dfrac{4\pi}{3} = -\dfrac{1}{2}$ $\quad \sec\dfrac{4\pi}{3} = -2$

$\tan\dfrac{4\pi}{3} = \sqrt{3}$ $\quad \cot\dfrac{4\pi}{3} = \dfrac{\sqrt{3}}{3}$

37. $\sin\left(-\dfrac{5\pi}{3}\right) = \dfrac{\sqrt{3}}{2}$ $\quad \csc\left(-\dfrac{5\pi}{3}\right) = \dfrac{2\sqrt{3}}{3}$

$\cos\left(-\dfrac{5\pi}{3}\right) = \dfrac{1}{2}$ $\quad \sec\left(-\dfrac{5\pi}{3}\right) = 2$

$\tan\left(-\dfrac{5\pi}{3}\right) = \sqrt{3}$ $\quad \cot\left(-\dfrac{5\pi}{3}\right) = \dfrac{\sqrt{3}}{3}$

39. $\sin\left(-\dfrac{\pi}{2}\right) = -1$ $\quad \csc\left(-\dfrac{\pi}{2}\right) = -1$

$\cos\left(-\dfrac{\pi}{2}\right) = 0$ $\quad \sec\left(-\dfrac{\pi}{2}\right)$ is undefined.

$\tan\left(-\dfrac{\pi}{2}\right)$ is undefined. $\quad \cot\left(-\dfrac{\pi}{2}\right) = 0$

41. 0 **43.** $-\dfrac{1}{2}$ **45.** $\dfrac{\sqrt{3}}{2}$ **47.** $-\dfrac{\sqrt{2}}{2}$

49. (a) $-\frac{1}{3}$ (b) -3 **51.** (a) $-\frac{1}{5}$ (b) -5

53. (a) $-\frac{4}{5}$ (b) $\frac{5}{4}$ **55.** 0.6428 **57.** 0.8090

59. 0.1228 **61.** -0.1288 **63.** 3.2709

65. -1.1547 **67.** 7.1154 **69.** 1.1319

71. (a) About -0.9 (b) About -0.4

73. (a) About 0.25, 2.89 (b) About 1.82, 4.46

75. About 0.79 amp

77. (a) 0.25 ft (b) About 0.02 ft (c) About -0.25 ft

79. False; $\sin(-t) = -\sin t$ means that the function is odd, not that the sine of a negative angle is a negative number.

81. True; $a - 6\pi$ is coterminal with a.

83. It is an odd function.

85. (a) y-axis (b) $\sin t_1 = \sin(\pi - t_1)$ (c) $\cos(\pi - t_1) = -\cos t_1$

87. (a) $\frac{2}{3}$ (b) $-\frac{2}{3}$ **89.** The calculator is in *degree* mode.

91. Answers will vary.

93. Vertical asymptote: $x = 3$; horizontal asymptote: $y = 2$

Section 4.3 (page 282)

1. (a) iii (b) vi (c) ii (d) v (e) i (f) iv

3. elevation, depression **5.** 12

7. $\sin\theta = \frac{3}{5}$ $\quad \csc\theta = \frac{5}{3}$

$\cos\theta = \frac{4}{5}$ $\quad \sec\theta = \frac{5}{4}$

$\tan\theta = \frac{3}{4}$ $\quad \cot\theta = \frac{4}{3}$

9. $\sin\theta = \dfrac{\sqrt{2}}{2}$ $\quad \csc\theta = \sqrt{2}$

$\cos\theta = \dfrac{\sqrt{2}}{2}$ $\quad \sec\theta = \sqrt{2}$

$\tan\theta = 1$ $\quad \cot\theta = 1$

11. $\sin\theta = \frac{8}{17}$ $\quad \csc\theta = \frac{17}{8}$

$\cos\theta = \frac{15}{17}$ $\quad \sec\theta = \frac{17}{15}$

$\tan\theta = \frac{8}{15}$ $\quad \cot\theta = \frac{15}{8}$

The triangles are similar and corresponding sides are proportional.

13. $\cos\theta = \dfrac{\sqrt{11}}{6}$

$\tan\theta = \dfrac{5\sqrt{11}}{11}$

$\csc\theta = \dfrac{6}{5}$

$\sec\theta = \dfrac{6\sqrt{11}}{11}$

$\cot\theta = \dfrac{\sqrt{11}}{5}$

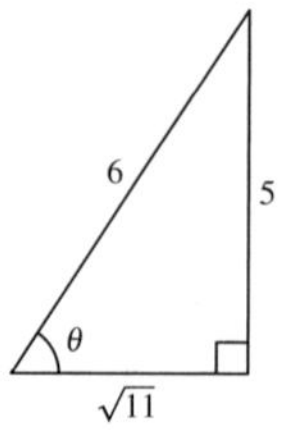

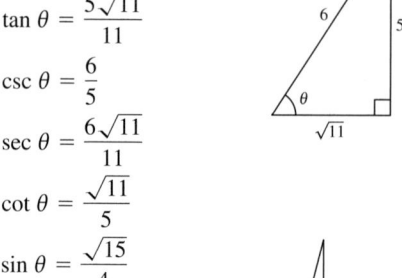

15. $\sin\theta = \dfrac{\sqrt{15}}{4}$

$\cos\theta = \dfrac{1}{4}$

$\tan\theta = \sqrt{15}$

$\csc\theta = \dfrac{4\sqrt{15}}{15}$

$\cot\theta = \dfrac{\sqrt{15}}{15}$

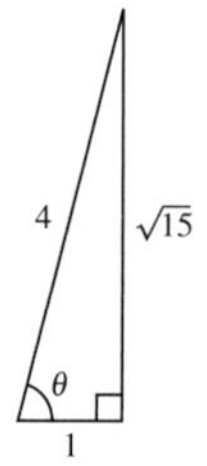

17. $\sin\theta = \dfrac{3\sqrt{10}}{10}$

$\cos\theta = \dfrac{\sqrt{10}}{10}$

$\csc\theta = \dfrac{\sqrt{10}}{3}$

$\sec\theta = \sqrt{10}$

$\cot\theta = \dfrac{1}{3}$

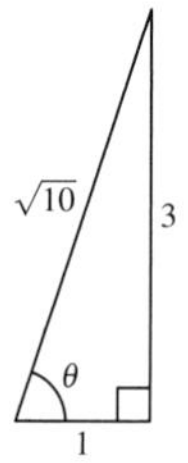

19. $\sin\theta = \dfrac{2\sqrt{13}}{13}$

$\cos\theta = \dfrac{3\sqrt{13}}{13}$

$\tan\theta = \dfrac{2}{3}$

$\sec\theta = \dfrac{\sqrt{13}}{3}$

$\csc\theta = \dfrac{\sqrt{13}}{2}$

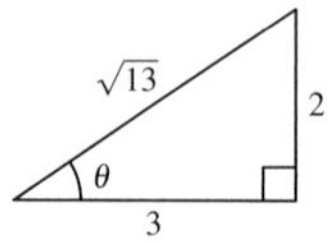

21. $\dfrac{\pi}{6}; \dfrac{\sqrt{3}}{3}$ **23.** $45°; \dfrac{\sqrt{2}}{2}$ **25.** $45°; \sqrt{2}$ **27.** $60°; \dfrac{\pi}{3}$

29. (a) 0.1736 (b) 0.1736 **31.** (a) 1.3499 (b) 1.3430

33. (a) 5.0273 (b) 0.4142 **35.** $\csc\theta$ **37.** $\cot\theta$

39. $\cos\theta$ **41.** $\dfrac{\sin\theta}{\cos\theta}$ **43.** 1 **45.** $\cos\theta$ **47.** $\cot\theta$

49. $\csc\theta$ **51.** (a) $\sqrt{3}$ (b) $\dfrac{1}{2}$ (c) $\dfrac{\sqrt{3}}{2}$ (d) $\dfrac{\sqrt{3}}{3}$

53. (a) $\dfrac{2\sqrt{2}}{3}$ (b) $2\sqrt{2}$ (c) 3 (d) 3

55. (a) $\frac{1}{4}$ (b) $\sqrt{17}$ (c) $\frac{\sqrt{17}}{4}$ (d) 4

57–65. Answers will vary. **67.** (a) $30° = \frac{\pi}{6}$ (b) $30° = \frac{\pi}{6}$

69. (a) $60° = \frac{\pi}{3}$ (b) $45° = \frac{\pi}{4}$

71. (a) $60° = \frac{\pi}{3}$ (b) $45° = \frac{\pi}{4}$

73. $y = 35\sqrt{3}, r = 70\sqrt{3}$ **75.** $x = 8, y = 8\sqrt{3}$

77. (a) About 519.33 ft (b) About 1174.17 ft
(c) About 173.8 ft/min

79. About 160 ft

81. $(x_1, y_1) = (28\sqrt{3}, 28)$
$(x_2, y_2) = (28, 28\sqrt{3})$

83. (a)

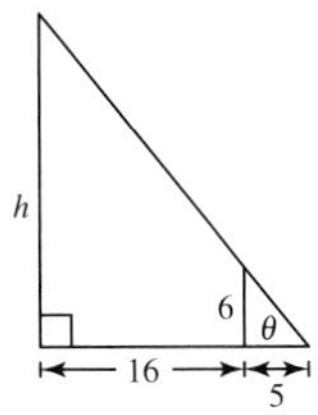

(b) $\tan\theta = \frac{h}{21}$ (c) 25.2 ft

85. True; $\csc x = \frac{1}{\sin x}$ **87.** False; $\frac{\sqrt{2}}{2} + \frac{\sqrt{2}}{2} \neq 1$

89. False; $1.7321 \neq 0.0349$

91. Yes; Use the Pythagorean Theorem. Answers will vary.

93. (a)

θ	0°	20°	40°	60°	80°
$\sin\theta$	0	0.3420	0.6428	0.8660	0.9848
$\cos\theta$	1	0.9397	0.7660	0.5	0.1736
$\tan\theta$	0	0.3640	0.8391	1.7321	5.6713

(b) Sine: increasing
Cosine: decreasing
Tangent: increasing

(c) Answers will vary.

95.

x	−1	0	1	2
$f(x)$	2.05	3	22.09	405.43

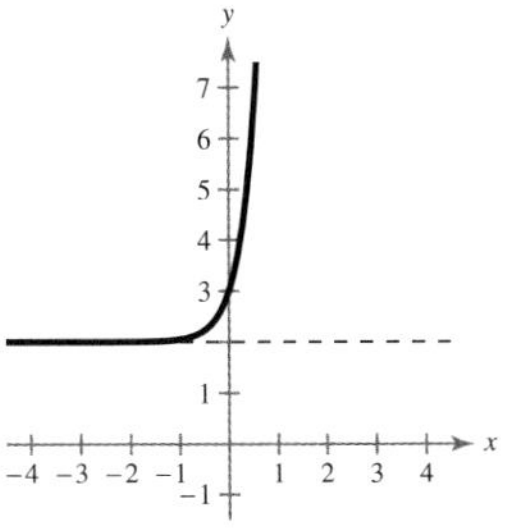

Asymptote: $y = 2$

97.

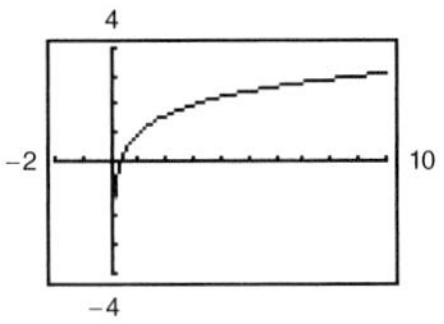

Domain: $(0, \infty)$
Vertical asymptote: $x = 0$
x-intercept: $(\frac{1}{3}, 0)$

Section 4.4 (page 291)

1. $\frac{y}{r}$ **3.** $\frac{y}{x}$ **5.** $\cos\theta$ **7.** reference angle **9.** $0, \pi$

11. (a) $\sin\theta = \frac{3}{5}$ $\csc\theta = \frac{5}{3}$
$\cos\theta = \frac{4}{5}$ $\sec\theta = \frac{5}{4}$
$\tan\theta = \frac{3}{4}$ $\cot\theta = \frac{4}{3}$

(b) $\sin\theta = \frac{\sqrt{2}}{2}$ $\csc\theta = \sqrt{2}$
$\cos\theta = -\frac{\sqrt{2}}{2}$ $\sec\theta = -\sqrt{2}$
$\tan\theta = -1$ $\cot\theta = -1$

13. (a) $\sin\theta = -\frac{1}{2}$ $\csc\theta = -2$
$\cos\theta = -\frac{\sqrt{3}}{2}$ $\sec\theta = -\frac{2\sqrt{3}}{3}$
$\tan\theta = \frac{\sqrt{3}}{3}$ $\cot\theta = \sqrt{3}$

(b) $\sin\theta = -\frac{\sqrt{17}}{17}$ $\csc\theta = -\sqrt{17}$
$\cos\theta = \frac{4\sqrt{17}}{17}$ $\sec\theta = \frac{\sqrt{17}}{4}$
$\tan\theta = -\frac{1}{4}$ $\cot\theta = -4$

15. $\sin\theta = \frac{24}{25}$ $\csc\theta = \frac{25}{24}$
$\cos\theta = \frac{7}{25}$ $\sec\theta = \frac{25}{7}$
$\tan\theta = \frac{24}{7}$ $\cot\theta = \frac{7}{24}$

17. $\sin\theta = -\frac{12}{13}$ $\csc\theta = -\frac{13}{12}$
$\cos\theta = \frac{5}{13}$ $\sec 0 = \frac{13}{5}$
$\tan\theta = -\frac{12}{5}$ $\cot\theta = -\frac{5}{12}$

19. $\sin\theta = \frac{\sqrt{21}}{5}$ $\csc\theta = \frac{5\sqrt{21}}{21}$
$\cos\theta = -\frac{2}{5}$ $\sec\theta = -\frac{5}{2}$
$\tan\theta = -\frac{\sqrt{21}}{2}$ $\cot\theta = -\frac{2\sqrt{21}}{21}$

21. $\sin\theta = -\frac{4\sqrt{41}}{41}$ $\csc\theta = -\frac{\sqrt{41}}{4}$
$\cos\theta = -\frac{5\sqrt{41}}{41}$ $\sec\theta = -\frac{\sqrt{41}}{5}$
$\tan\theta = \frac{4}{5}$ $\cot\theta = \frac{5}{4}$

23. Quadrant III **25.** Quadrant I

27. $\cos\theta = -\frac{4}{5}$
$\tan\theta = -\frac{3}{4}$
$\csc\theta = \frac{5}{3}$
$\sec\theta = -\frac{5}{4}$
$\cot\theta = -\frac{4}{3}$

29. $\sin\theta = -\frac{15}{17}$
$\cos\theta = \frac{8}{17}$
$\csc\theta = -\frac{17}{15}$
$\sec\theta = \frac{17}{8}$
$\cot\theta = -\frac{8}{15}$

CHAPTER 4

31. $\sin\theta = \dfrac{2\sqrt{2}}{3}$

$\cos\theta = -\dfrac{1}{3}$

$\tan\theta = -2\sqrt{2}$

$\csc\theta = \dfrac{3\sqrt{2}}{4}$

$\cot\theta = -\dfrac{\sqrt{2}}{4}$

33. $\sin\theta = 0$ $\csc\theta$ is undefined.

$\cos\theta = -1$ $\sec\theta = -1$

$\tan\theta = 0$ $\cot\theta$ is undefined.

35. $\sin\theta = \dfrac{\sqrt{2}}{2}$ $\csc\theta = \sqrt{2}$

$\cos\theta = -\dfrac{\sqrt{2}}{2}$ $\sec\theta = -\sqrt{2}$

$\tan\theta = -1$ $\cot\theta = -1$

37. $\sin\theta = -\dfrac{3\sqrt{10}}{10}$ $\csc\theta = -\dfrac{\sqrt{10}}{3}$

$\cos\theta = -\dfrac{\sqrt{10}}{10}$ $\sec\theta = -\sqrt{10}$

$\tan\theta = 3$ $\cot\theta = \dfrac{1}{3}$

39. -1 **41.** Undefined   **43.** -1 **45.** Undefined

47. $\theta' = 60°$

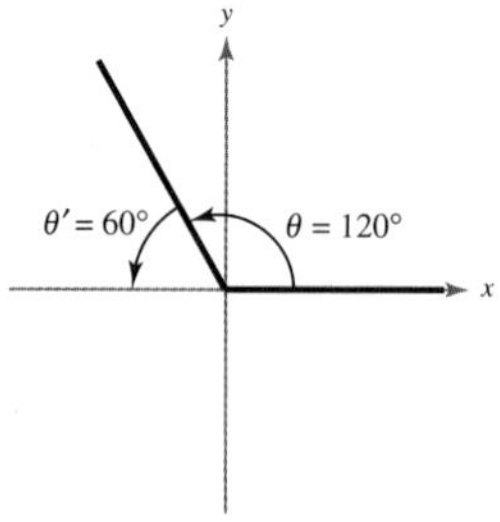

49. $\theta' = 30°$

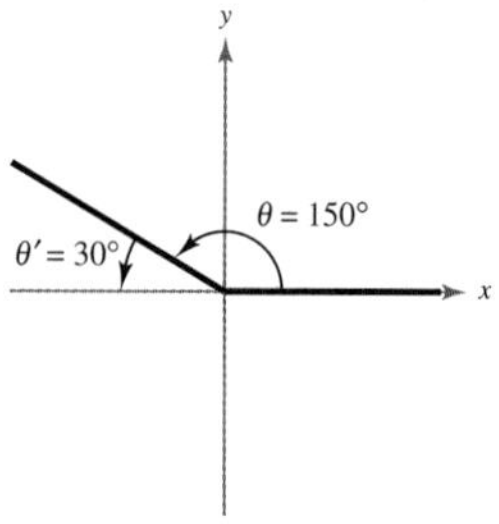

51. $\theta' = 45°$

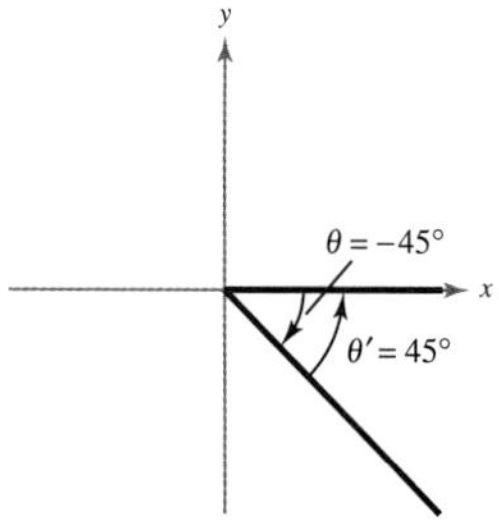

53. $\theta' = \dfrac{\pi}{3}$

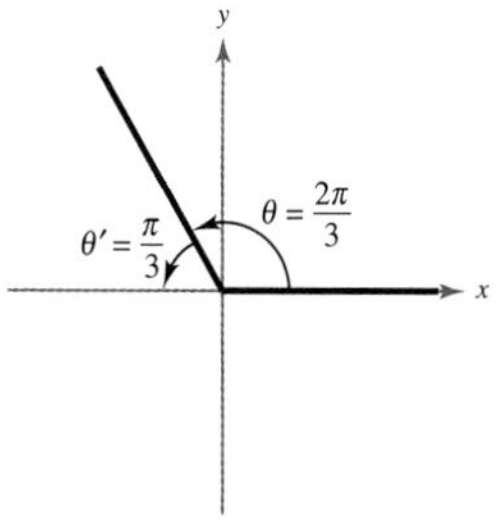

55. $\theta' = \dfrac{\pi}{6}$

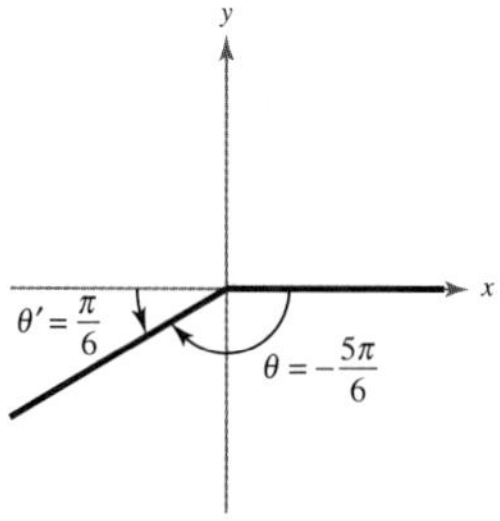

57. $\theta' = \dfrac{\pi}{6}$

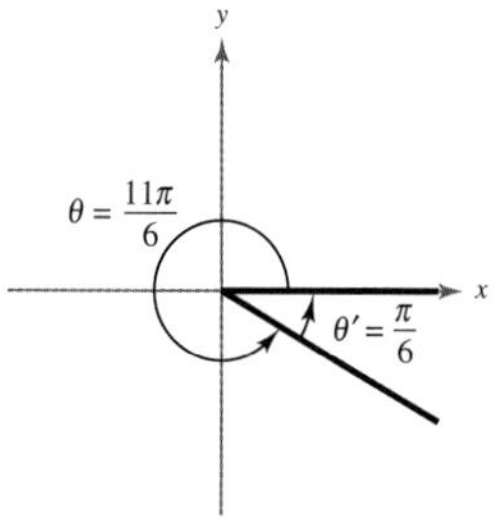

59. $\theta' = 32°$

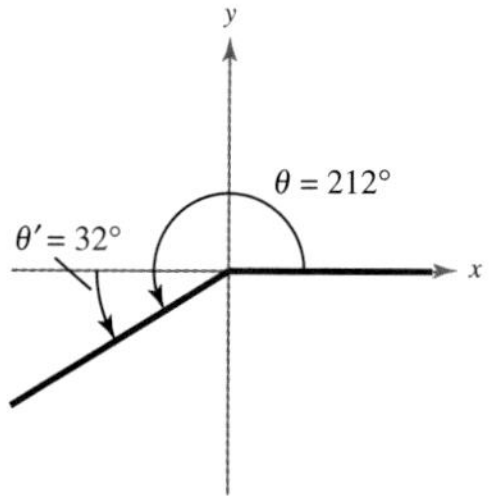

61. $\theta' = 78°$

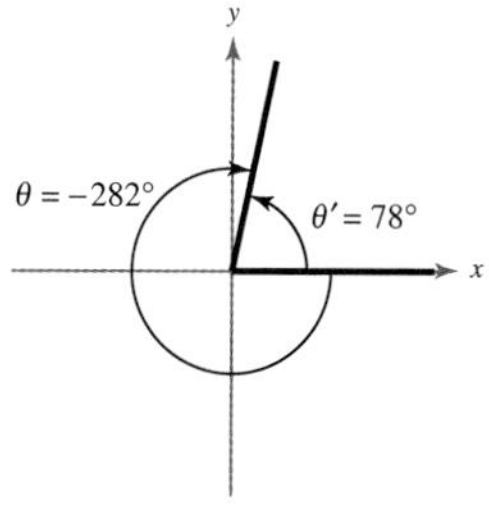

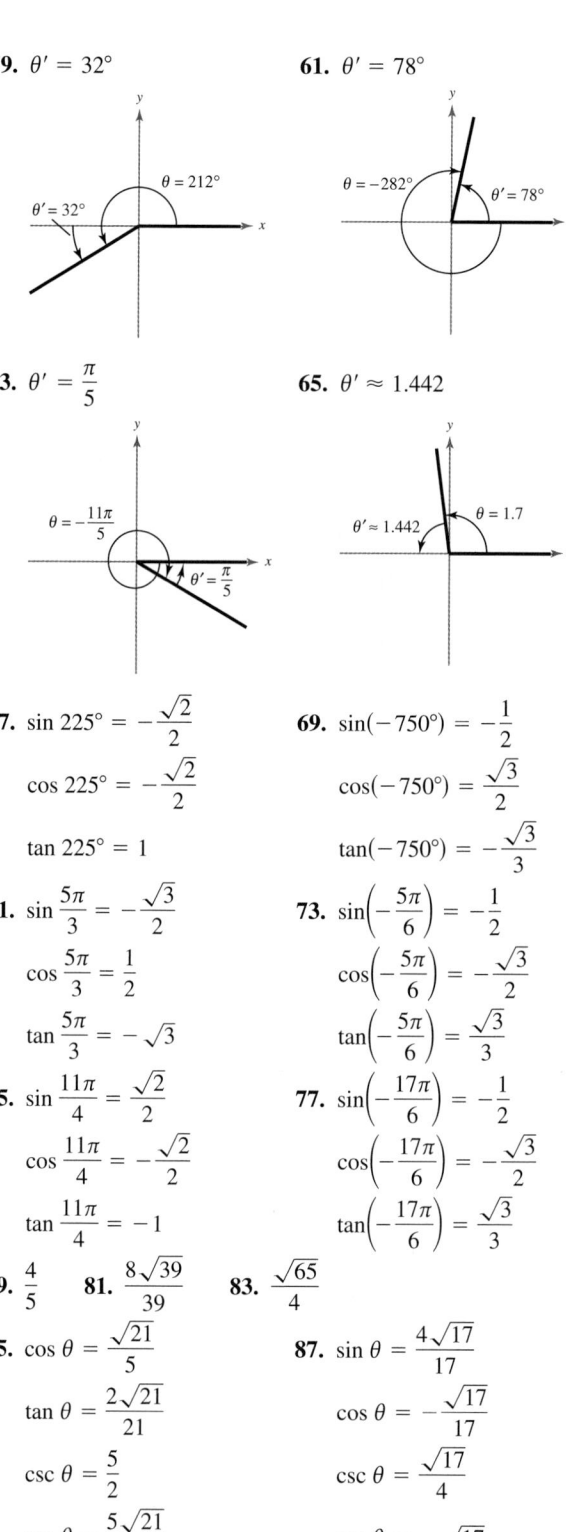

63. $\theta' = \dfrac{\pi}{5}$

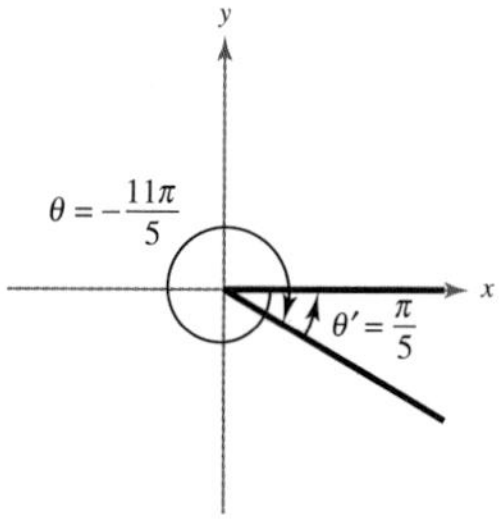

65. $\theta' \approx 1.442$

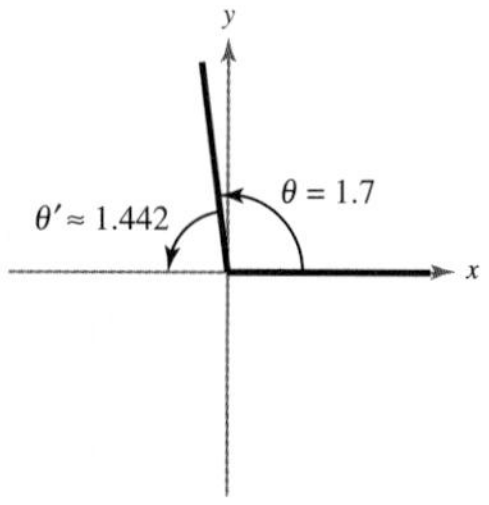

67. $\sin 225° = -\dfrac{\sqrt{2}}{2}$

$\cos 225° = -\dfrac{\sqrt{2}}{2}$

$\tan 225° = 1$

69. $\sin(-750°) = -\dfrac{1}{2}$

$\cos(-750°) = \dfrac{\sqrt{3}}{2}$

$\tan(-750°) = -\dfrac{\sqrt{3}}{3}$

71. $\sin\dfrac{5\pi}{3} = -\dfrac{\sqrt{3}}{2}$

$\cos\dfrac{5\pi}{3} = \dfrac{1}{2}$

$\tan\dfrac{5\pi}{3} = -\sqrt{3}$

73. $\sin\left(-\dfrac{5\pi}{6}\right) = -\dfrac{1}{2}$

$\cos\left(-\dfrac{5\pi}{6}\right) = -\dfrac{\sqrt{3}}{2}$

$\tan\left(-\dfrac{5\pi}{6}\right) = \dfrac{\sqrt{3}}{3}$

75. $\sin\dfrac{11\pi}{4} = \dfrac{\sqrt{2}}{2}$

$\cos\dfrac{11\pi}{4} = -\dfrac{\sqrt{2}}{2}$

$\tan\dfrac{11\pi}{4} = -1$

77. $\sin\left(-\dfrac{17\pi}{6}\right) = -\dfrac{1}{2}$

$\cos\left(-\dfrac{17\pi}{6}\right) = -\dfrac{\sqrt{3}}{2}$

$\tan\left(-\dfrac{17\pi}{6}\right) = \dfrac{\sqrt{3}}{3}$

79. $\dfrac{4}{5}$ **81.** $\dfrac{8\sqrt{39}}{39}$ **83.** $\dfrac{\sqrt{65}}{4}$

85. $\cos\theta = \dfrac{\sqrt{21}}{5}$

$\tan\theta = \dfrac{2\sqrt{21}}{21}$

$\csc\theta = \dfrac{5}{2}$

$\sec\theta = \dfrac{5\sqrt{21}}{21}$

$\cot\theta = \dfrac{\sqrt{21}}{2}$

87. $\sin\theta = \dfrac{4\sqrt{17}}{17}$

$\cos\theta = -\dfrac{\sqrt{17}}{17}$

$\csc\theta = \dfrac{\sqrt{17}}{4}$

$\sec\theta = -\sqrt{17}$

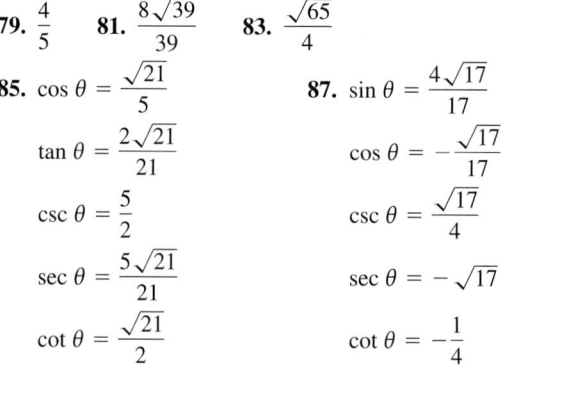

$\cot\theta = -\dfrac{1}{4}$

89. $\sin\theta = -\frac{2}{3}$

$\cos\theta = \frac{\sqrt{5}}{3}$

$\tan\theta = -\frac{2\sqrt{5}}{5}$

$\sec\theta = \frac{3\sqrt{5}}{5}$

$\cot\theta = -\frac{\sqrt{5}}{2}$

91. 0.1736 **93.** 1.4142 **95.** -0.3420 **97.** 1.0247

99. 0.8391 **101.** -3.7321

103. (a) $30° = \frac{\pi}{6}, 150° = \frac{5\pi}{6}$ (b) $210° = \frac{7\pi}{6}, 330° = \frac{11\pi}{6}$

105. (a) $60° = \frac{\pi}{3}, 120° = \frac{2\pi}{3}$ (b) $135° = \frac{3\pi}{4}, 315° = \frac{7\pi}{4}$

107. (a) $150° = \frac{5\pi}{6}, 210° = \frac{7\pi}{6}$ (b) $\frac{11\pi}{6} = 330°, \frac{5\pi}{6} = 150°$

109. (a) $\frac{1+\sqrt{3}}{2}$ (b) $\frac{\sqrt{3}-1}{2}$ (c) $\frac{3}{4}$

(d) $\frac{\sqrt{3}}{4}$ (e) $\frac{\sqrt{3}}{2}$ (f) $\frac{\sqrt{3}}{2}$

111. (a) 0 (b) $\sqrt{2}$ (c) $\frac{1}{2}$ (d) $-\frac{1}{2}$ (e) -1 (f) $\frac{\sqrt{2}}{2}$

113. (a) $\frac{-1-\sqrt{3}}{2}$ (b) $\frac{1-\sqrt{3}}{2}$ (c) $\frac{3}{4}$ (d) $\frac{\sqrt{3}}{4}$

(e) $\frac{\sqrt{3}}{2}$ (f) $-\frac{\sqrt{3}}{2}$

115. (a) $-\frac{1+\sqrt{3}}{2}$ (b) $\frac{1-\sqrt{3}}{2}$ (c) $\frac{3}{4}$ (d) $\frac{\sqrt{3}}{4}$

(e) $\frac{\sqrt{3}}{2}$ (f) $-\frac{\sqrt{3}}{2}$

117. (a) $-\frac{1+\sqrt{3}}{2}$ (b) $\frac{-1+\sqrt{3}}{2}$ (c) $\frac{1}{4}$ (d) $\frac{\sqrt{3}}{4}$

(e) $\frac{\sqrt{3}}{2}$ (f) $-\frac{1}{2}$

119. (a) 1 (b) -1 (c) 0 (d) 0 (e) 0 (f) 0

121. (a) -1 (b) 1 (c) 0 (d) 0 (e) 0 (f) 0

123. (a) 47.8°F (b) 83.8°F (c) 65.8°F

125. (a) 12 mi (b) 6 mi (c) About 6.93 mi

127. True; $0 < \cos\theta < 1$ in Quadrant I, so

$\sin\theta < \frac{\sin\theta}{\cos\theta} = \tan\theta$.

129. False; Sine is positive in Quadrant II.

131.

θ	0°	20°	40°
$\sin\theta$	0	0.3420	0.6428
$\sin(180° - \theta)$	0	0.3420	0.6428

θ	60°	80°
$\sin\theta$	0.8660	0.9848
$\sin(180° - \theta)$	0.8660	0.9848

$\sin\theta = \sin(180° - \theta)$

133. The calculator is in *degree* mode. **135.** 3.449, -1.449

137. 4.908, -5.908

Section 4.5 (page 301)

1. one cycle **3.** amplitude

5. It vertically shifts the graph d units.

7. (a) $x = -2\pi, -\pi, 0, \pi, 2\pi$ (b) $y = 0$

(c) Increasing: $\left(-2\pi, -\frac{3\pi}{2}\right), \left(-\frac{\pi}{2}, \frac{\pi}{2}\right), \left(\frac{3\pi}{2}, 2\pi\right)$

Decreasing: $\left(-\frac{3\pi}{2}, -\frac{\pi}{2}\right), \left(\frac{\pi}{2}, \frac{3\pi}{2}\right)$

(d) Relative maxima: $\left(-\frac{3\pi}{2}, 1\right), \left(\frac{\pi}{2}, 1\right)$

Relative minima: $\left(-\frac{\pi}{2}, -1\right), \left(\frac{3\pi}{2}, -1\right)$

9. Period: 2π; Amplitude: 4 **11.** Period: 8π; Amplitude: 1

13. Period: 4; Amplitude: $\frac{3}{4}$ **15.** Period: 4π; Amplitude: 1

17. Period: 2π; Amplitude: 2

19. Maximum: 3; Minimum: 1

21. Maximum: 7; Minimum: 1

23. *Sample answer:* $[1, 2\pi + 1]$ **25.** *Sample answer:* $[-\pi, \pi]$

27. g is a shift of f π units to the right.

29. g is a reflection of f in the x-axis.

31. g is a shift of f three units upward.

33. g is a reflection of f in the y-axis and has five times the amplitude of f.

35. g has twice the amplitude of f.

37. g is a horizontal shift of f π units to the right.

39.

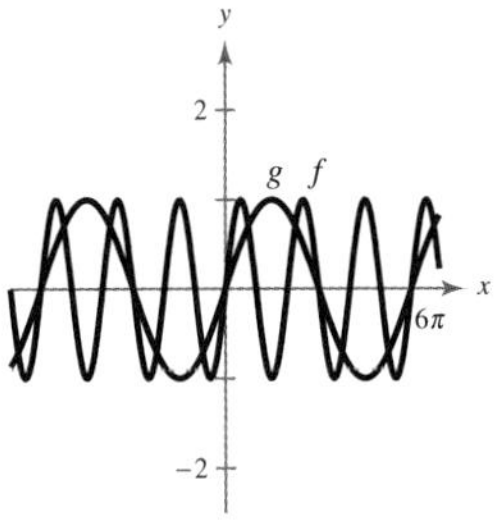

41.

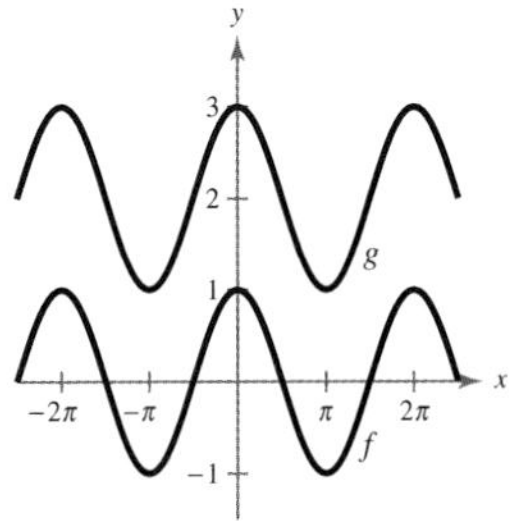

43.

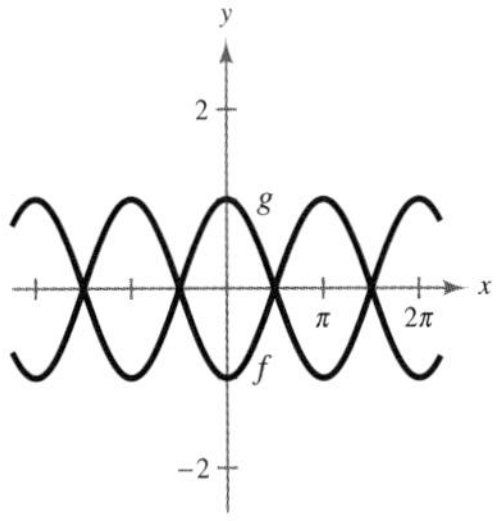

45.

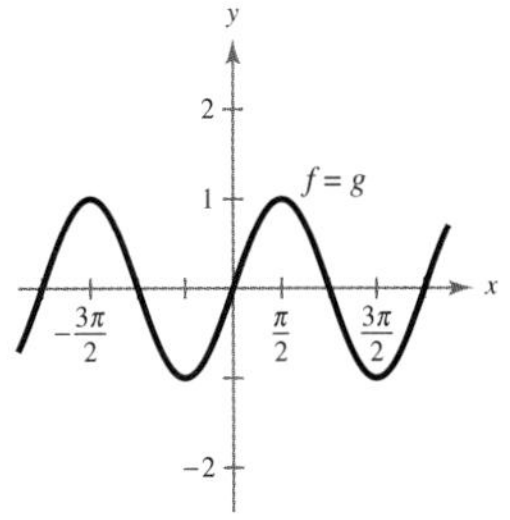

47.

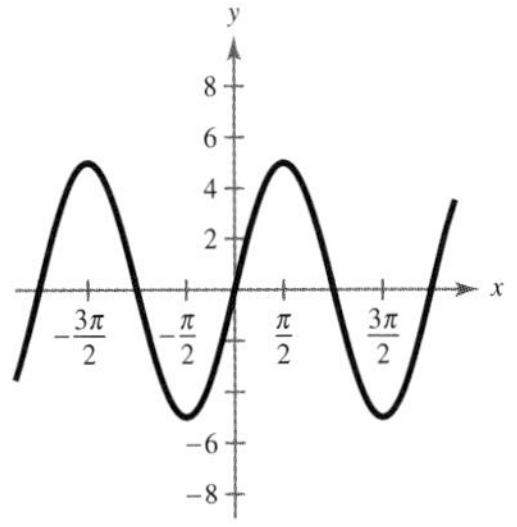

49.

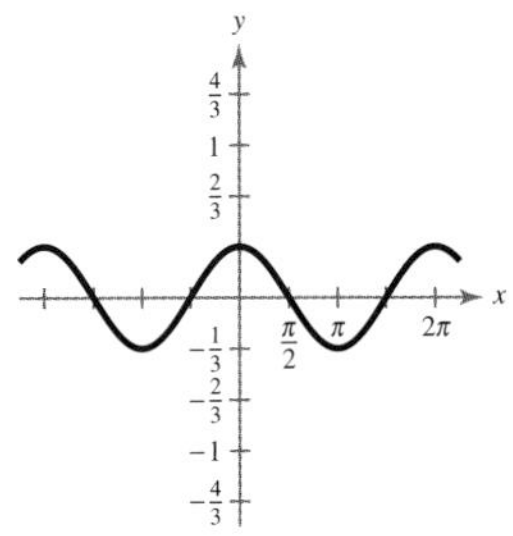

CHAPTER 4

51. 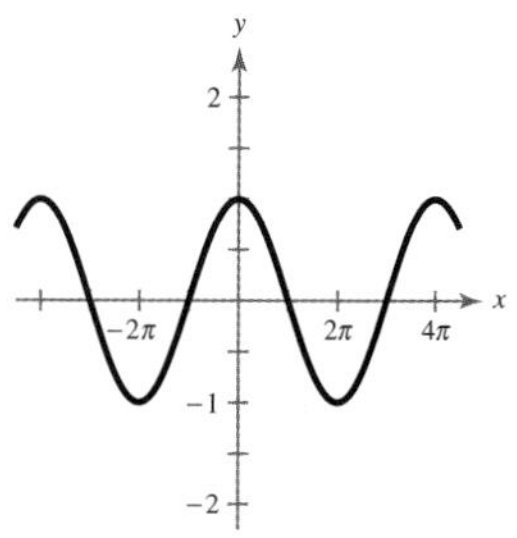

53. 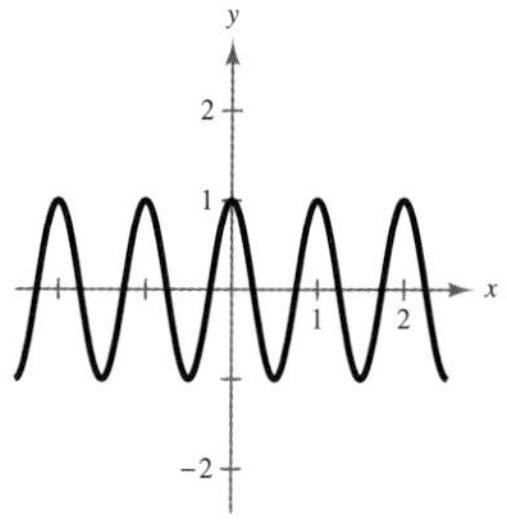

55. 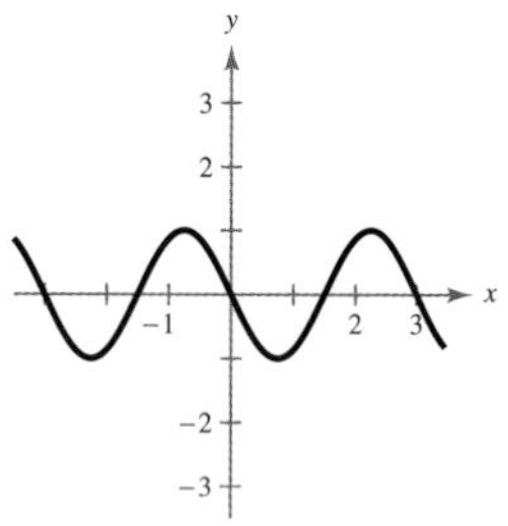

57. 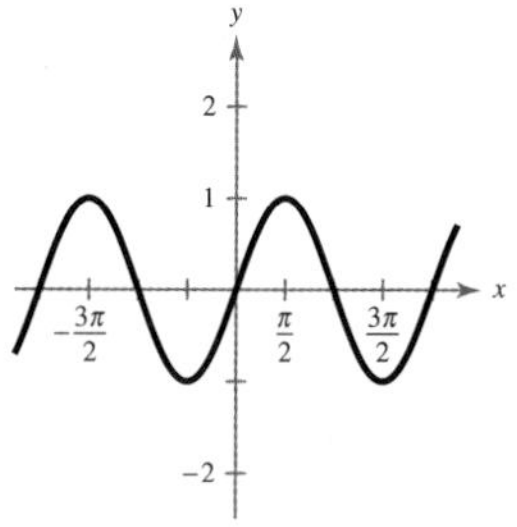

59. 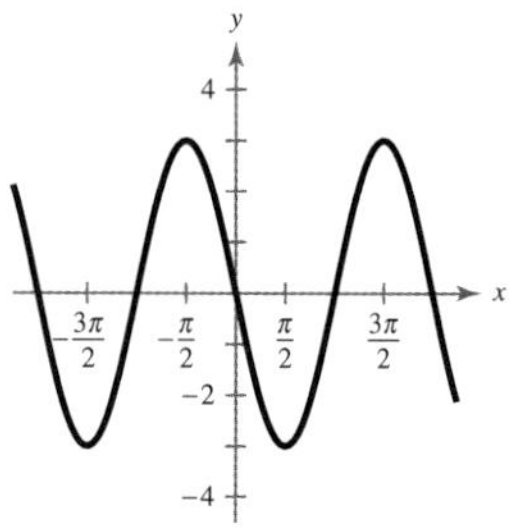

61. 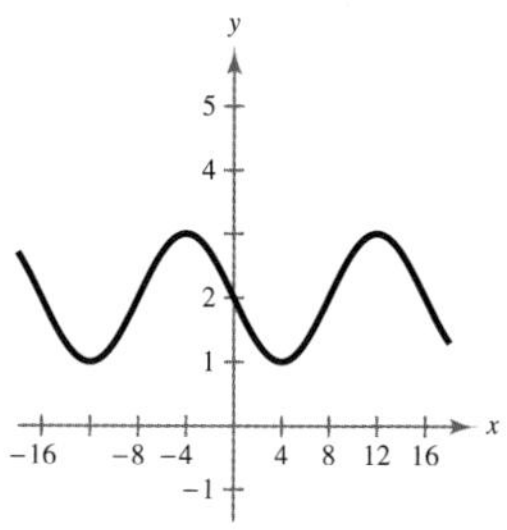

63. 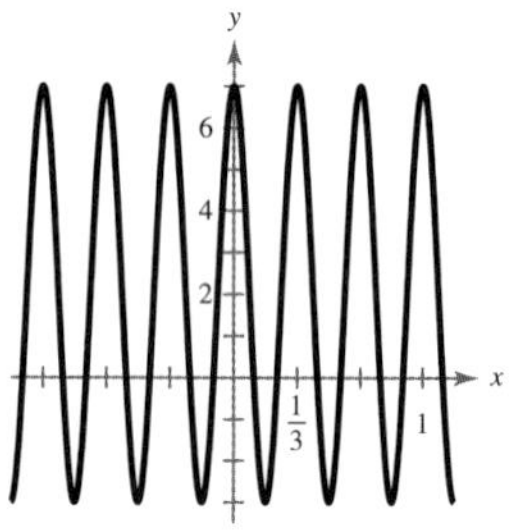

65. 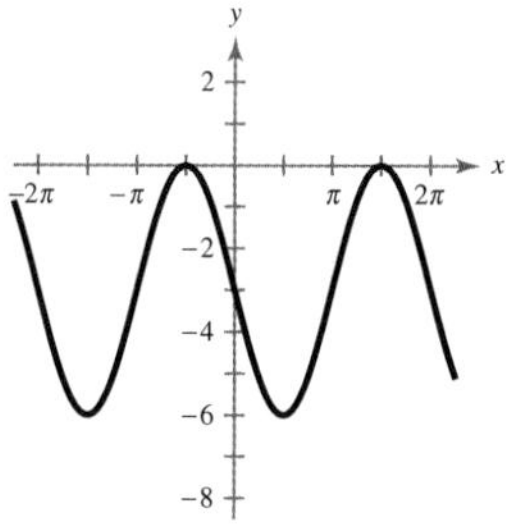

67. 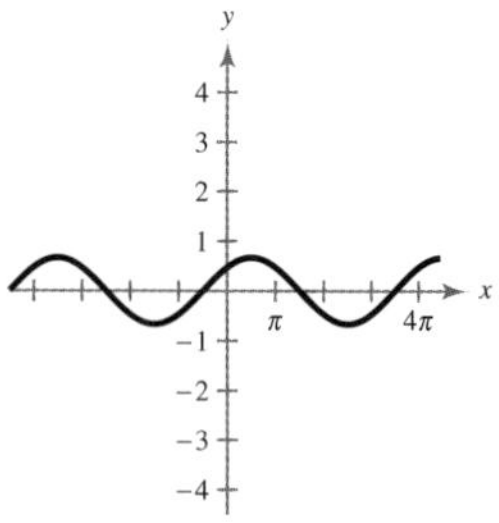

69. 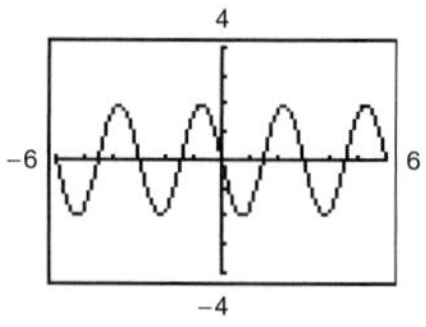

Amplitude: 2
Period: 3

71. 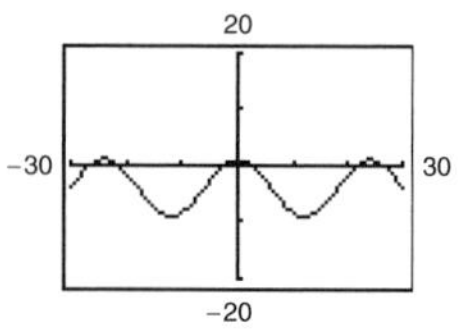

Amplitude: 5
Period: 24

73. 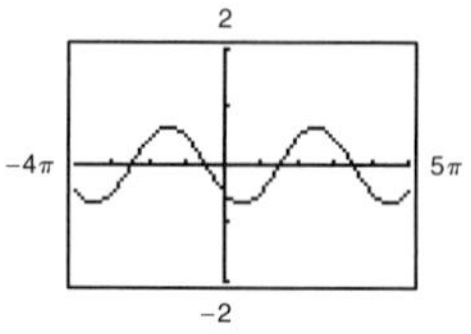

Amplitude: $\frac{2}{3}$
Period: 4π

75. 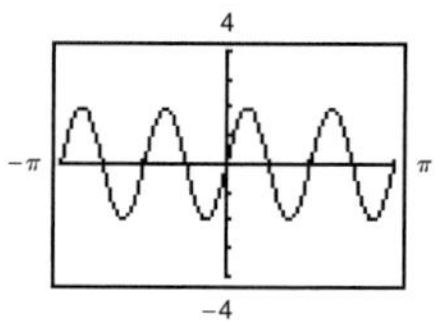

Amplitude: 2
Period: $\frac{\pi}{2}$

77. 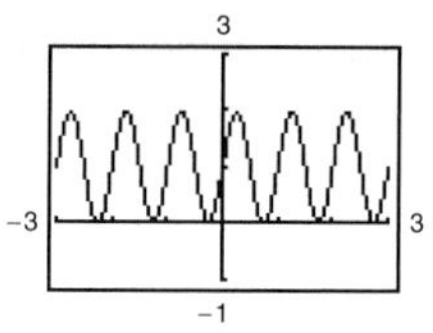

Amplitude: 1
Period: 1

79. 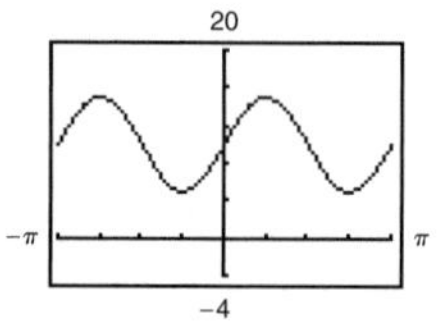

Amplitude: 5
Period: π

81. 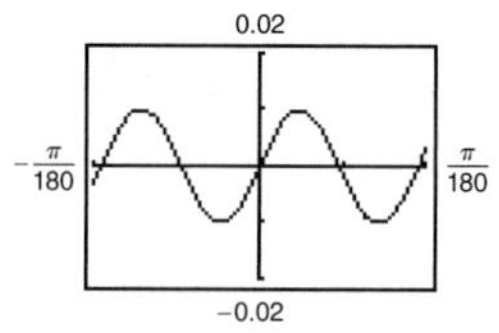

Amplitude: $\frac{1}{100}$
Period: $\frac{1}{60}$

83. $a = -4, d = 4$ **85.** $a = -1, d = -3$

87. $a = -3, b = 2, c = 0$ **89.** $a = 1, b = 1, c = \frac{\pi}{4}$

91. 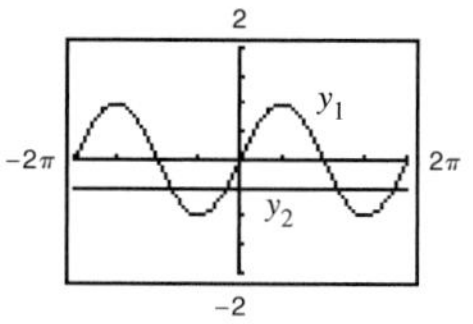

$x = -\frac{5\pi}{6}, -\frac{\pi}{6}, \frac{7\pi}{6}, \frac{11\pi}{6}$

93. 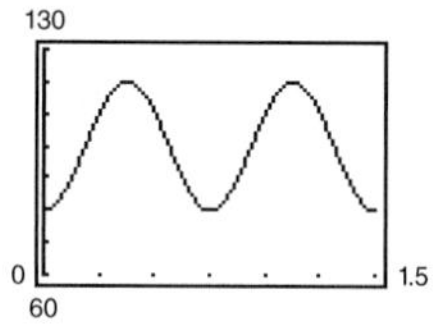

80 beats/min

95. (a) 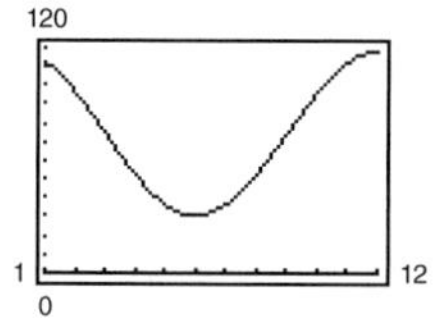

(b) Maximum sales: December
Minimum sales: June

97. (a) 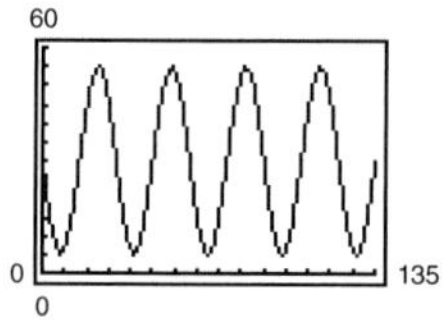

(b) Minimum height: 5 ft
Maximum height: 55 ft

99. (a) and (c)

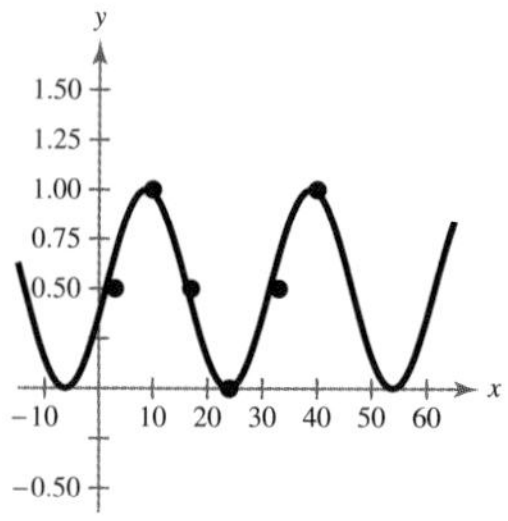

The model fits the data well.

(b) $y = 0.5 \sin(0.209x - 0.628) + 0.5$

(d) About 30 days (e) About 58.9%

101. True; The period of $\sin x$ is 2π. Adding 2π moves the graph one period to the right.

103. False; The function $y = \frac{1}{2}\cos 2x$ has an amplitude that is one-half that of the function $y = \cos x$.

105.

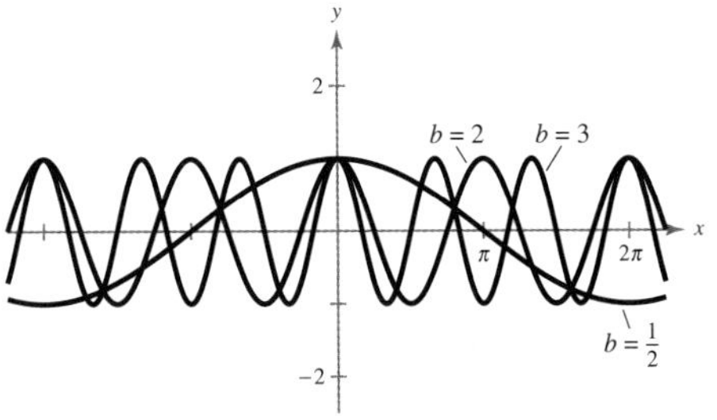

The value of b affects the period of the graph.

$b = \frac{1}{2} \rightarrow \frac{1}{2}$ cycle

$b = 2 \rightarrow 2$ cycles

$b = 3 \rightarrow 3$ cycles

107. a

109. (a)

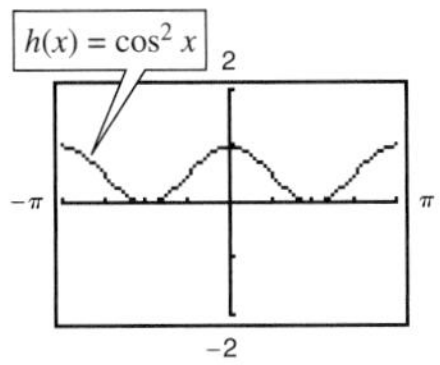

Even

(b)

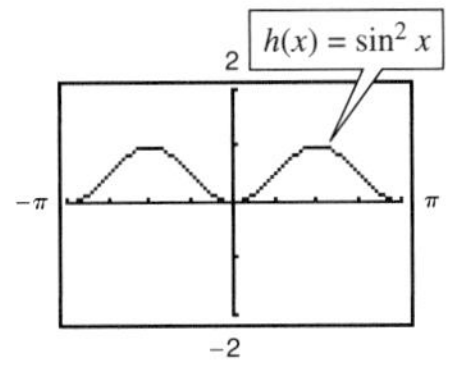

Even

(c)

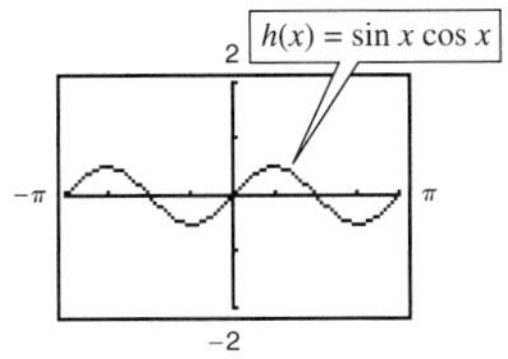

Odd

111. (a)

x	-1	-0.1	-0.01
$\dfrac{\sin x}{x}$	0.8415	0.9983	1.0000

x	-0.001	0	0.001
$\dfrac{\sin x}{x}$	1.0000	Undefined	1.0000

x	0.01	0.1	1
$\dfrac{\sin x}{x}$	1.0000	0.9983	0.8415

(b)

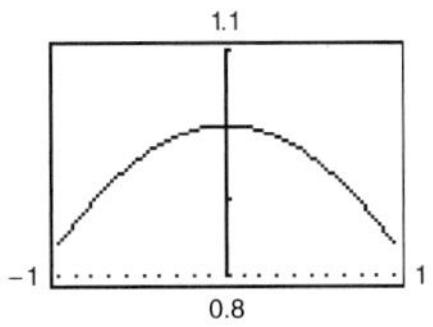

$f \rightarrow 1$ as $x \rightarrow 0$

(c) The ratio approaches 1 as x approaches 0.

113. (a)

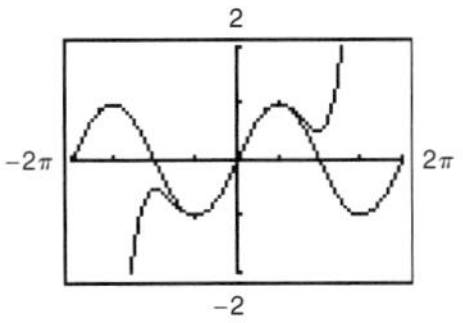

The polynomial function is a good approximation of the sine function when x is close to 0.

(b)

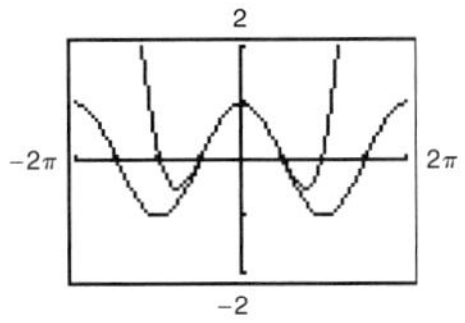

The polynomial function is a good approximation of the cosine function when x is close to 0.

(c) $\sin x \approx x - \dfrac{x^3}{3!} + \dfrac{x^5}{5!} - \dfrac{x^7}{7!}$

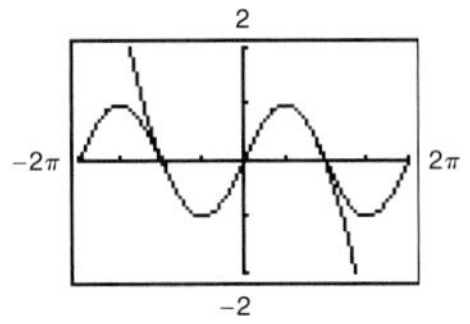

$\cos x \approx 1 - \dfrac{x^2}{2!} + \dfrac{x^4}{4!} - \dfrac{x^6}{6!}$

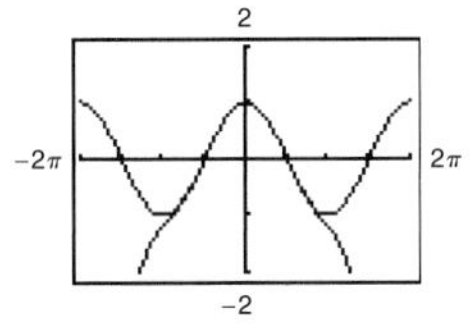

The accuracy increased.

115.

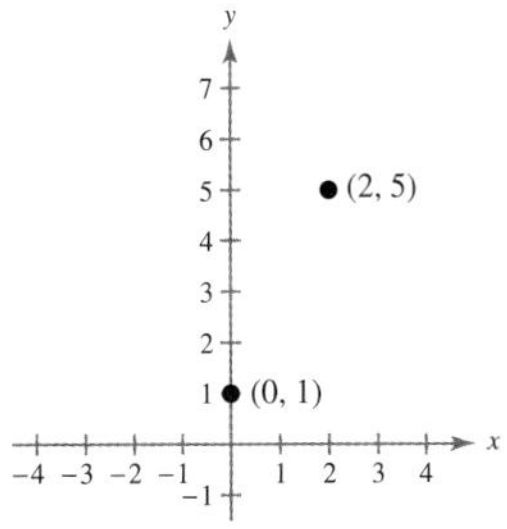

$m = 2$

117. 487.014° **119.** Answers will vary.

Section 4.6 (page 313)

1. vertical **3.** tangent, cotangent

5. (a) $x = -2\pi, -\pi, 0, \pi, 2\pi$ (b) $y = 0$

(c) Increasing on $\left(-2\pi, -\dfrac{3\pi}{2}\right), \left(-\dfrac{3\pi}{2}, -\dfrac{\pi}{2}\right), \left(-\dfrac{\pi}{2}, \dfrac{\pi}{2}\right), \left(\dfrac{\pi}{2}, \dfrac{3\pi}{2}\right), \left(\dfrac{3\pi}{2}, 2\pi\right)$

(d) No relative extrema

(e) $x = -\dfrac{3\pi}{2}, -\dfrac{\pi}{2}, \dfrac{\pi}{2}, \dfrac{3\pi}{2}$

7. (a) No x-intercepts (b) $y = 1$

(c) Increasing on $\left(-2\pi, -\frac{3\pi}{2}\right), \left(-\frac{3\pi}{2}, -\pi\right), \left(0, \frac{\pi}{2}\right), \left(\frac{\pi}{2}, \pi\right)$

Decreasing on $\left(-\pi, -\frac{\pi}{2}\right), \left(-\frac{\pi}{2}, 0\right), \left(\pi, \frac{3\pi}{2}\right), \left(\frac{3\pi}{2}, 2\pi\right)$

(d) Relative minima: $(-2\pi, 1), (0, 1), (2\pi, 1)$

Relative maxima: $(-\pi, -1), (\pi, -1)$

(e) $x = -\frac{3\pi}{2}, -\frac{\pi}{2}, \frac{\pi}{2}, \frac{3\pi}{2}$

9.

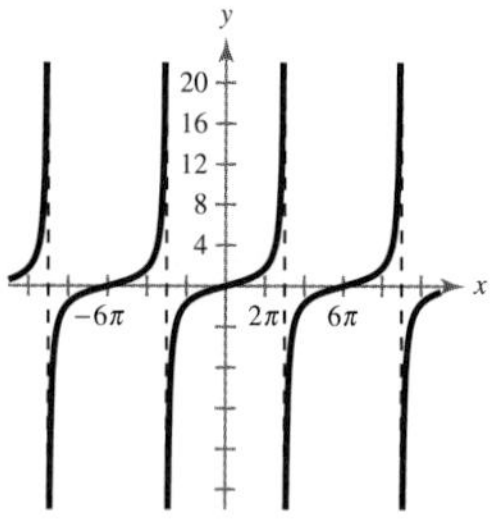

11.

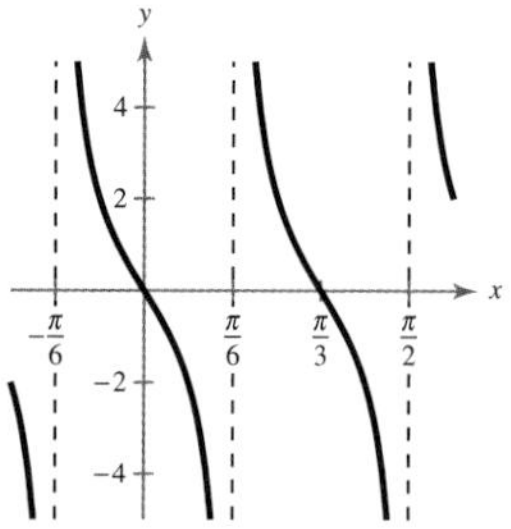

13.

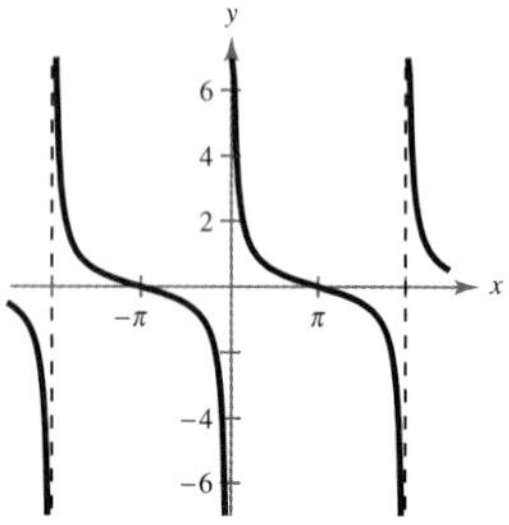

15.

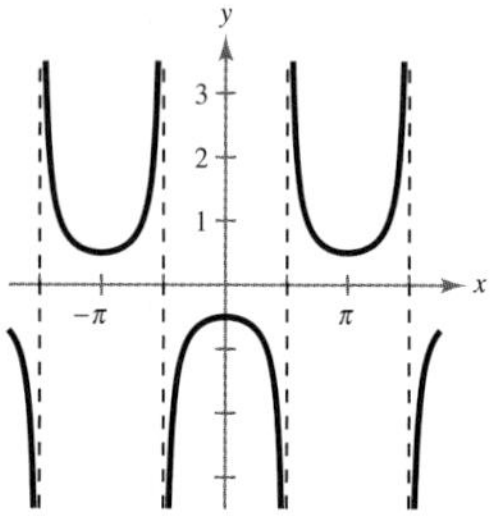

17.

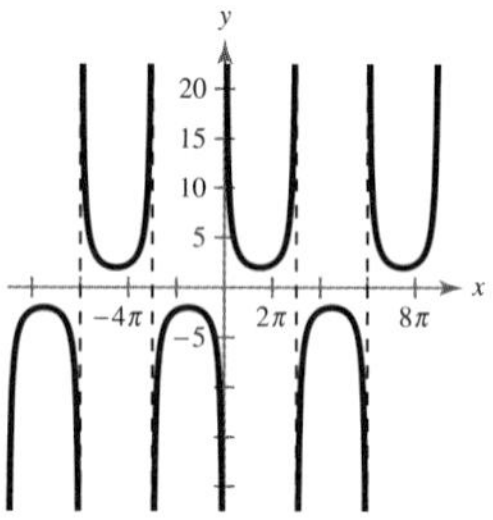

19.

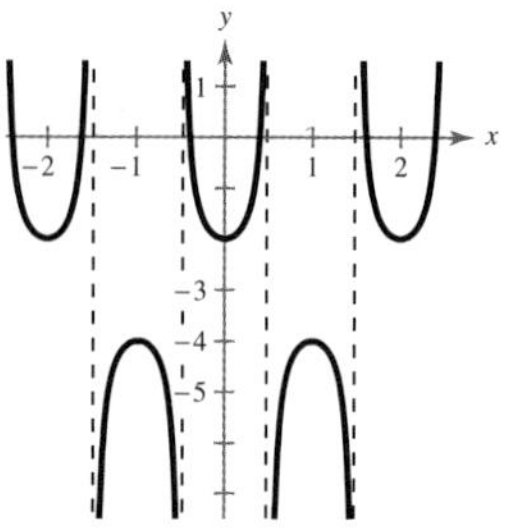

21.

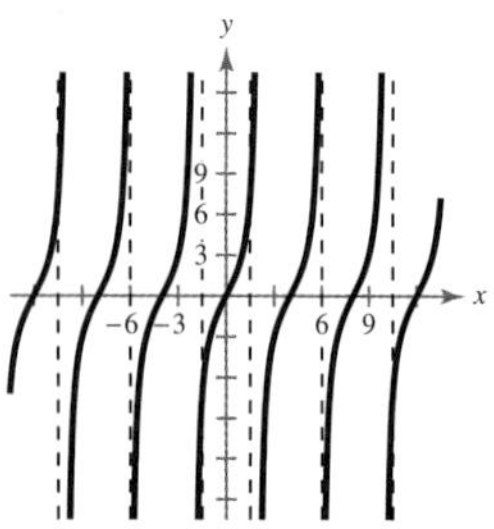

23.

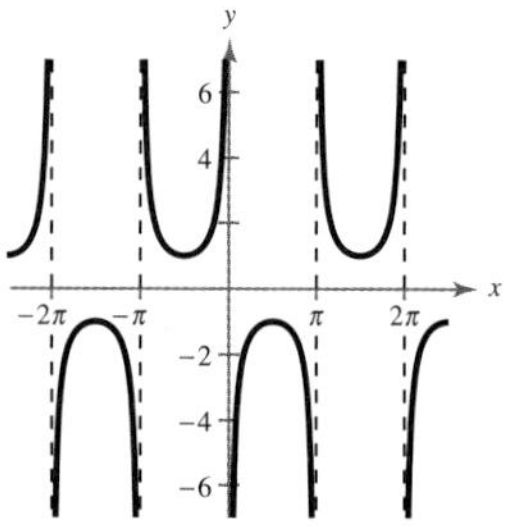

25.

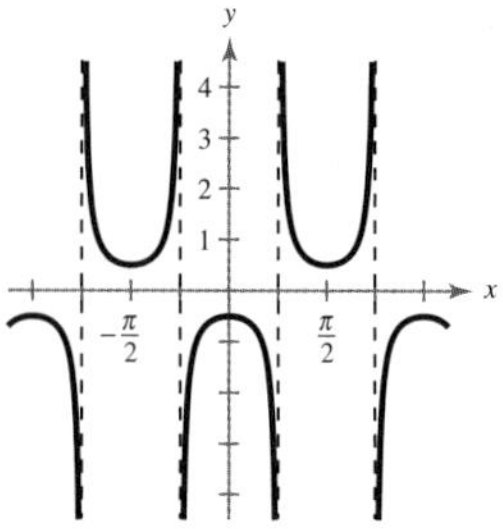

27.

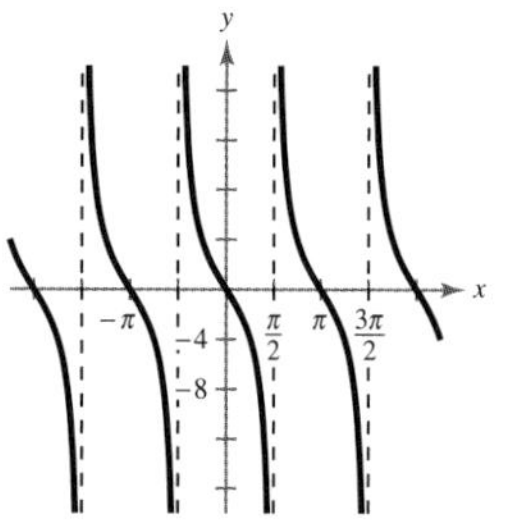

29.

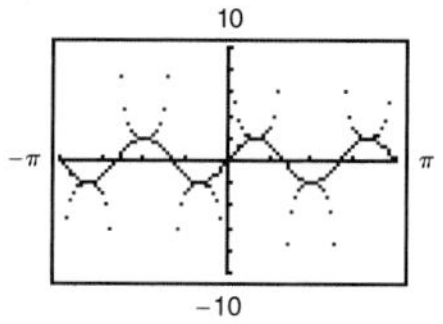

Answers will vary.

31.

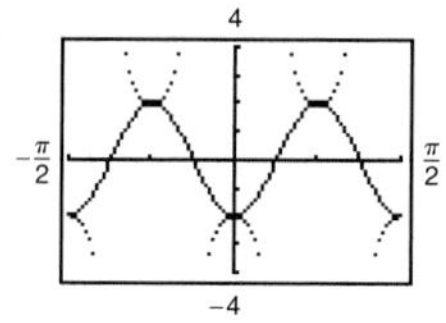

Answers will vary.

33.

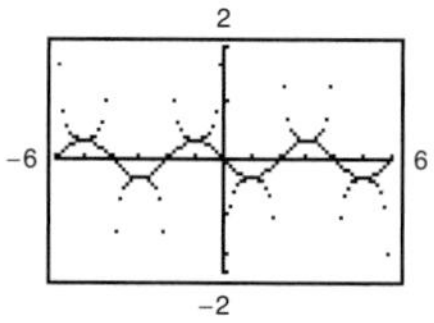

Answers will vary.

35. $-\frac{7\pi}{4}, -\frac{3\pi}{4}, \frac{\pi}{4}, \frac{5\pi}{4}$ **37.** $-\frac{5\pi}{3}, -\frac{\pi}{3}, \frac{\pi}{3}, \frac{5\pi}{3}$

39. $-\frac{7\pi}{6}, -\frac{\pi}{6}, \frac{5\pi}{6}, \frac{11\pi}{6}$ **41.** Even **43.** Odd **45.** Odd

47.

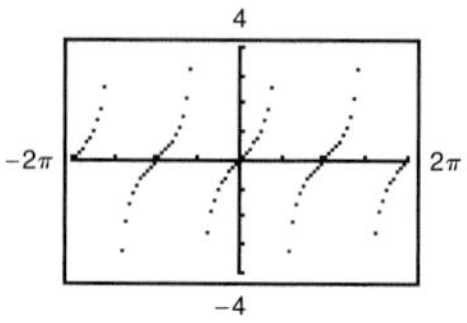

Answers will vary.

49.

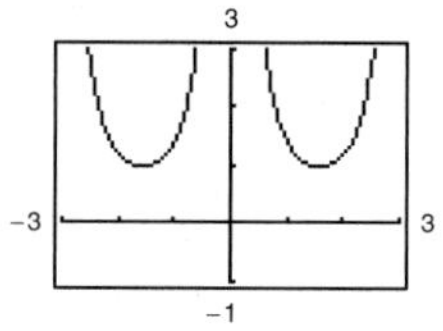

Answers will vary.

51. d; as x approaches 0, $f(x)$ approaches 0.

52. a; as x approaches 0, $f(x)$ approaches 0.

53. b; as x approaches 0, $g(x)$ approaches 0.

54. c; as x approaches 0, $g(x)$ approaches 0.

55.

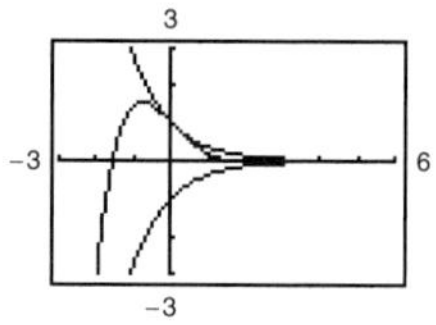

$-e^{-x} \le e^{-x} \cos x \le e^{-x}$

Touches $y = \pm e^{-x}$ at $x = n\pi$

Intercepts at $x = \frac{\pi}{2} + n\pi$

57.

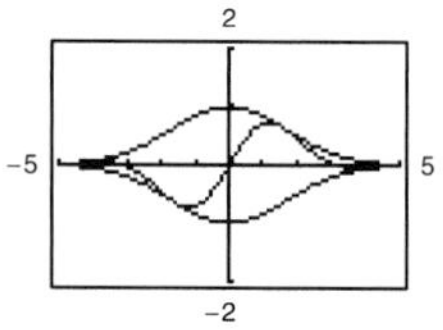

Touches $y = \pm 2^{-x^2/4}$ at $x = \frac{\pi}{2} + n\pi$

Intercepts at $x = n\pi$

59. (a) $f \to -\infty$ (b) $f \to \infty$ (c) $f \to -\infty$ (d) $f \to \infty$

61. (a) $f \to \infty$ (b) $f \to -\infty$ (c) $f \to -\infty$ (d) $f \to \infty$

63. $d = 7 \cot x$

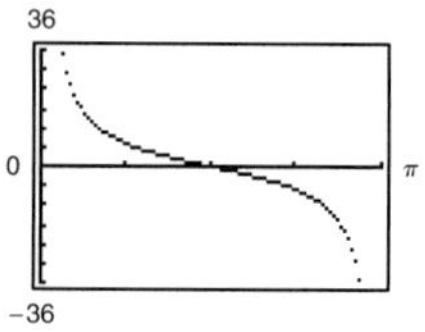

65. (a)

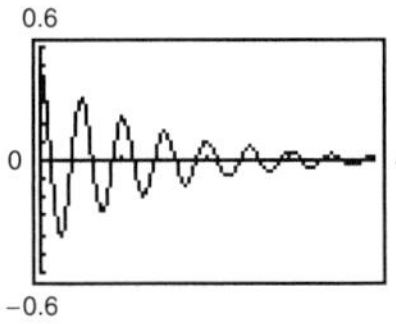

(b) Not periodic and damped; approaches 0 as t increases.

67. (a)

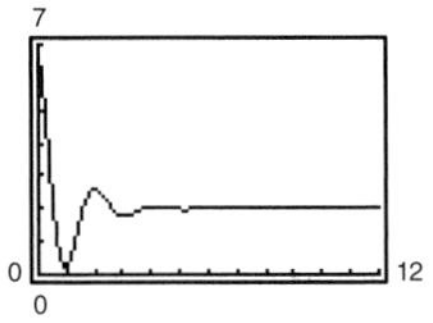

t	0	1	2	3	4
S	6.810	0.248	2.625	1.782	2.074

t	5	6	7	8	9
S	1.975	2.008	1.998	2.001	2.000

t	10	11	12
S	2.000	2.000	2.000

(b) As $t \to 11$, $S \to 2$.

69. True; $\tan\left(-\frac{5\pi}{2}\right)$ is undefined. **71.** True; $\csc x = \dfrac{1}{\sin x}$

73. (a)

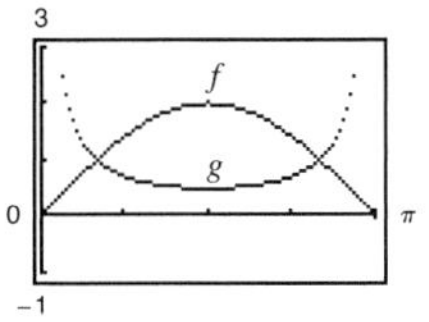

(b) About $0.524 < x < 2.618$

(c) f approaches 0 and g approaches ∞ because g is the reciprocal of f.

75. (a)

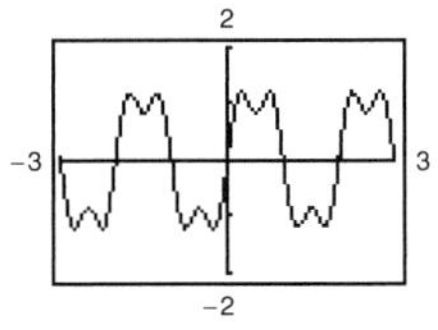

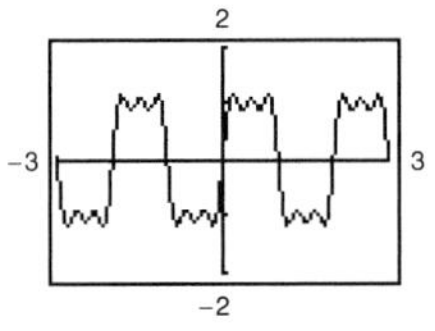

(b) $y_3 = \dfrac{4}{\pi}\left(\sin \pi x + \dfrac{1}{3}\sin 3\pi x + \dfrac{1}{5}\sin 5\pi x + \dfrac{1}{7}\sin 7\pi x\right)$

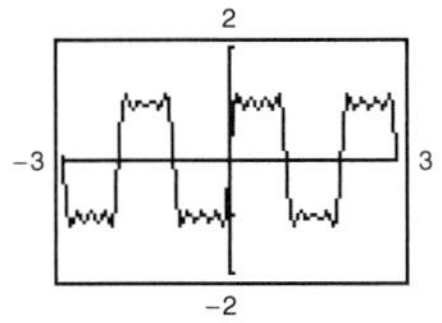

(c) $y_4 = \dfrac{4}{\pi}\left(\sin \pi x + \dfrac{1}{3}\sin 3\pi x + \dfrac{1}{5}\sin 5\pi x + \dfrac{1}{7}\sin 7\pi x + \dfrac{1}{9}\sin 9\pi x\right)$

77. The vertical stretch is by a factor of 3, not 4, and the horizontal shrink is by a factor of $\frac{1}{4}$, not $\frac{1}{3}$.

79. (a)

x	-1	-0.1	-0.01
$\frac{\tan x}{x}$	1.5574	1.0033	1.0000

x	-0.001	0	0.001
$\frac{\tan x}{x}$	1.0000	Undefined	1.0000

x	0.01	0.1	1
$\frac{\tan x}{x}$	1.0000	1.0033	1.5574

(b)

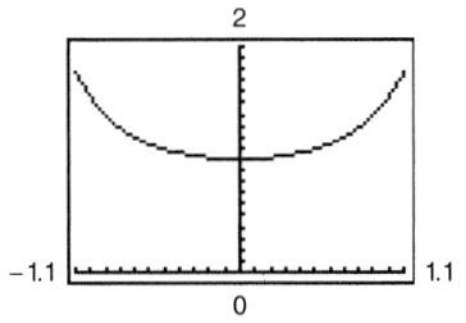

$f \to 1$ as $x \to 0$

(c) The ratio approaches 1 as x approaches 0.

81. Domain: all real numbers x
Intercepts: $(-4, 0)$, $(1, 0)$, $(0, -4)$
No asymptotes

83. Domain: all real numbers x
Intercept: $(0, 5)$
Asymptote: $y = 2$

Section 4.7 (page 324)

1. $y = \sin^{-1} x$, $-1 \le x \le 1$ **3.** $\tan^{-1} x$ or $\arctan x$

5. $\frac{\pi}{6}$ **7.** $\frac{\pi}{6}$ **9.** $\frac{\pi}{6}$ **11.** Not possible **13.** $\frac{2\pi}{3}$

15. $-\frac{\pi}{3}$ **17.** $-\frac{\pi}{3}$

19. (a)

x	-1	-0.8	-0.6	-0.4	-0.2
y	-1.571	-0.927	-0.644	-0.412	-0.201

x	0	0.2	0.4	0.6	0.8	1
y	0	0.201	0.412	0.644	0.927	1.571

(b)

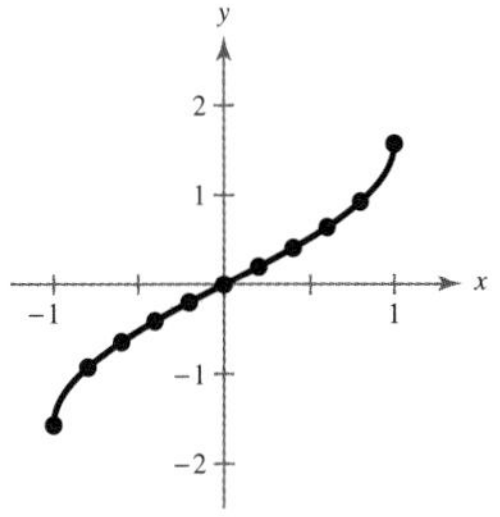

(c)

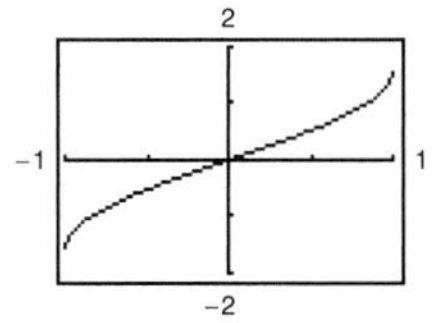

They are the same.

(d) Intercept: $(0, 0)$; symmetric about the origin

21. $\left(-\sqrt{3}, -\frac{\pi}{3}\right), \left(-\frac{\sqrt{3}}{3}, -\frac{\pi}{6}\right), \left(1, \frac{\pi}{4}\right)$ **23.** 1.19

25. -0.85 **27.** -1.25 **29.** Not possible **31.** 1.99

33. 0.74 **35.** 1.07 **37.** -1.50

39.

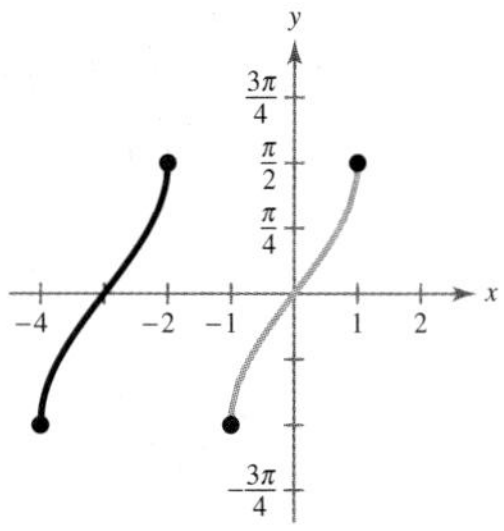

g is a horizontal shift of *f* three units to the left.

41.

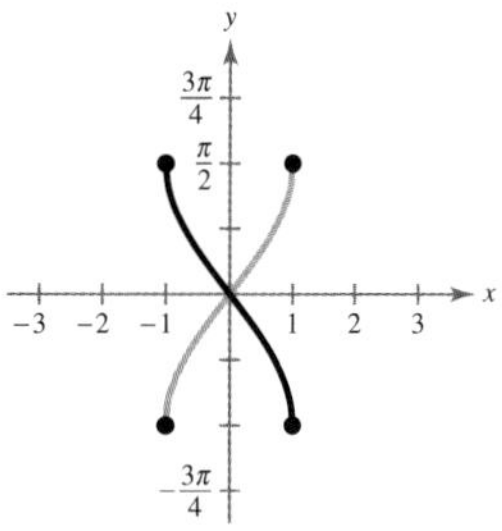

g is a reflection of *f* in the *y*-axis.

43.

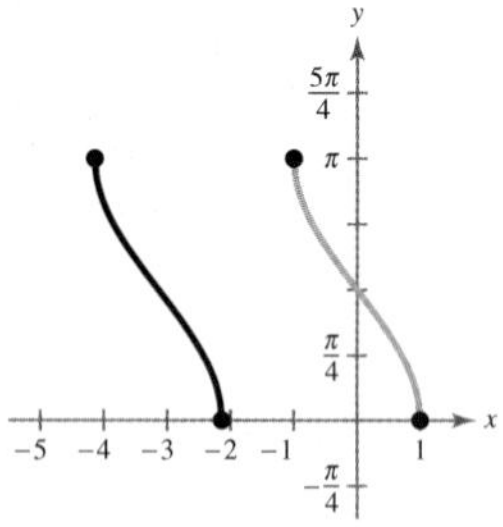

g is a horizontal shift of *f* π units to the left.

45.

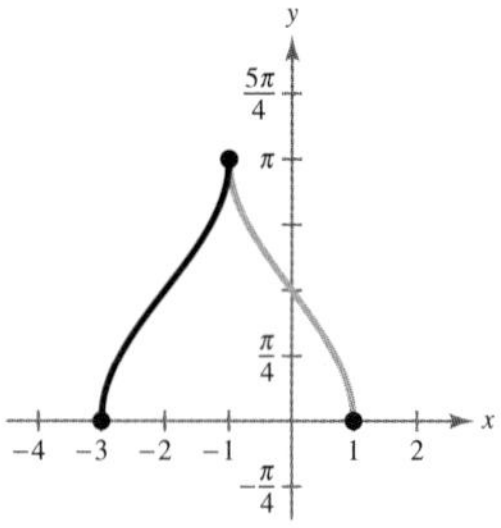

g is a reflection of *f* in the *y*-axis and a horizontal shift of *f* two units to the left.

47.

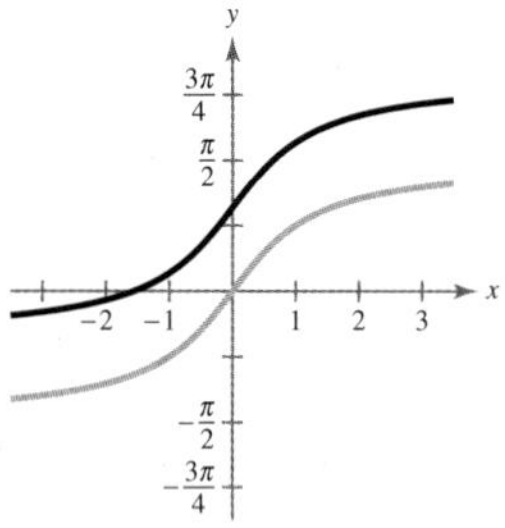

g is a vertical shift of *f* one unit upward.

49.

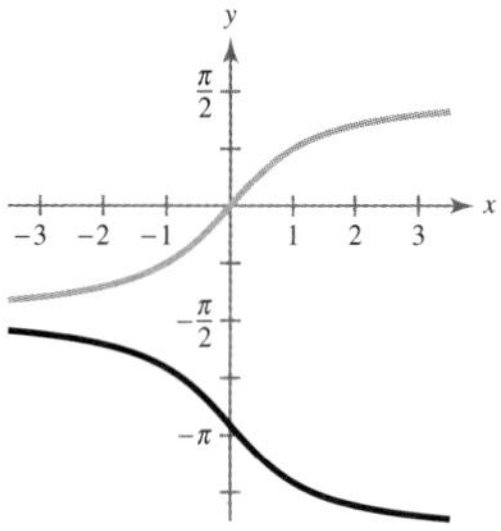

g is a reflection of *f* in the *x*-axis and a vertical shift of *f* three units down.

51. $\theta = \arctan \dfrac{x}{4}$ **53.** $\theta = \arcsin \dfrac{x+2}{5}$

55. $\sqrt{4-x^2}$; $\theta = \arcsin \dfrac{x}{2}$, $\theta = \arccos \dfrac{\sqrt{4-x^2}}{2}$, $\theta = \arctan \dfrac{x}{\sqrt{4-x^2}}$

57. $\sqrt{x^2+2x+5}$; $\theta = \arcsin \dfrac{x+1}{\sqrt{x^2+2x+5}}$, $\theta = \arccos \dfrac{2}{\sqrt{x^2+2x+5}}$, $\theta = \arctan \dfrac{x+1}{2}$

59. 0.3 **61.** Not possible **63.** -0.5 **65.** $-\dfrac{\pi}{6}$

67. $\dfrac{\pi}{2}$ **69.** $\dfrac{3}{5}$ **71.** $\dfrac{\sqrt{5}}{5}$ **73.** $\dfrac{13}{12}$ **75.** $-\dfrac{5}{3}$

77. $-\dfrac{21}{20}$ **79.** 2 **81.** $\sqrt{1-4x^2}$ **83.** $\dfrac{1}{x}$

85. $\sqrt{1-x^2}$ **87.** $\dfrac{\sqrt{9-x^2}}{x}$ **89.** $\dfrac{\sqrt{x^2+a^2}}{x}$

91. $\dfrac{14}{\sqrt{x^2+196}}$ **93.** $\dfrac{|x-1|}{\sqrt{x^2-2x+10}}$

95.

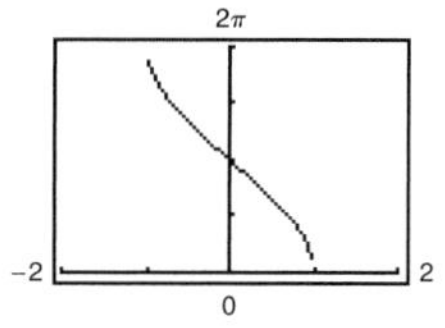

97.

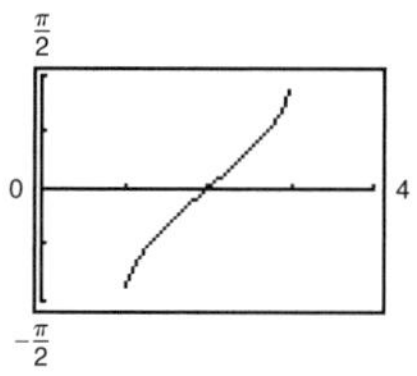

99.

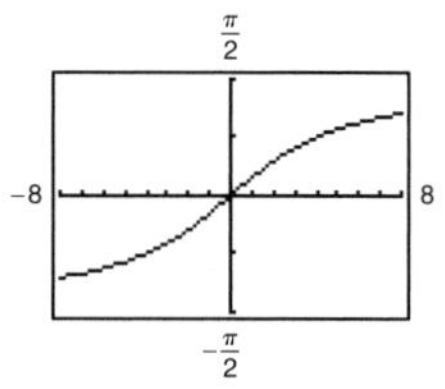

101. $3\sqrt{2}\sin\left(2t + \dfrac{\pi}{4}\right)$

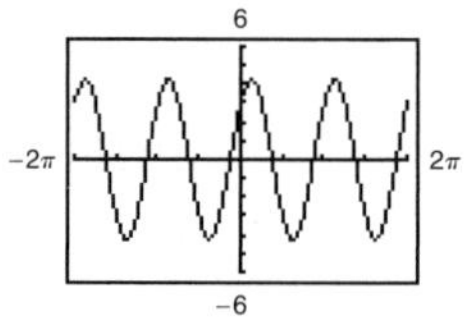

The two forms are equivalent.

103. $\dfrac{\pi}{2}$ **105.** $\dfrac{\pi}{2}$ **107.** π

109. (a)

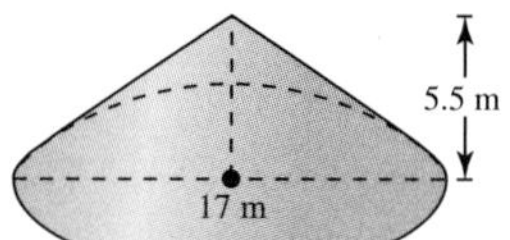

(b) About 0.574 rad
(c) About 8.08 m

111. (a) $\theta = \arctan \dfrac{6}{x}$ (b) About 0.54 rad, about 1.11 rad

113. False; $5\pi/6$ is not in the range of the arcsine function.

115. False; Inverse trigonometric functions do not have the same relationships that trigonometric functions have.

117.

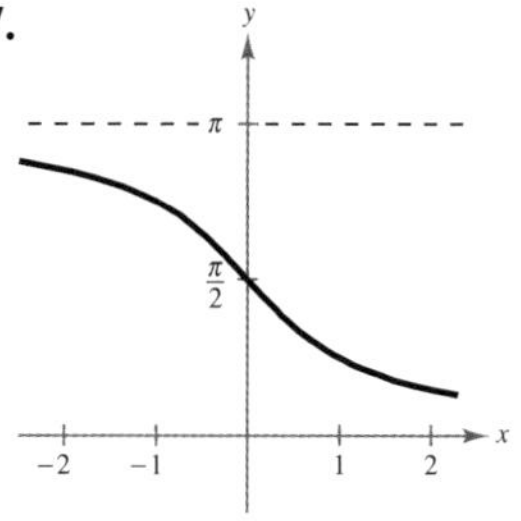

119.

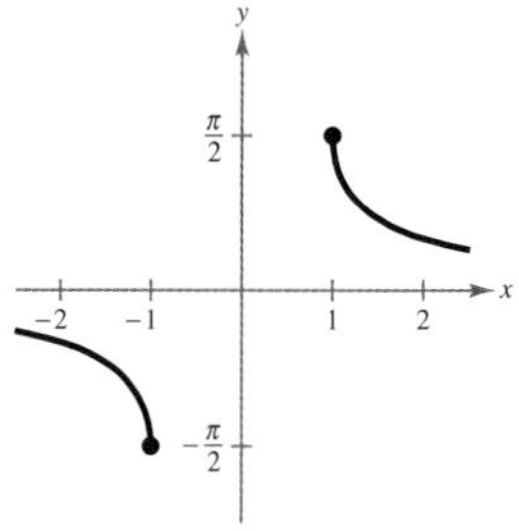

121. $\dfrac{\pi}{4}$ **123.** $\dfrac{5\pi}{6}$ **125 and 127.** Proofs

129. $\dfrac{\sqrt{2}}{2}$ **131.** $\dfrac{\sqrt{3}}{3}$

133. $\cos\theta = \dfrac{\sqrt{11}}{6}$

$\tan\theta = \dfrac{5\sqrt{11}}{11}$

$\csc\theta = \dfrac{6}{5}$

$\sec\theta = \dfrac{6\sqrt{11}}{11}$

$\cot\theta = \dfrac{\sqrt{11}}{5}$

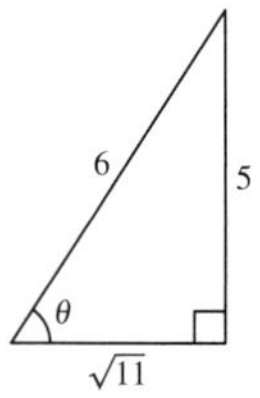

135. $\sin\theta = \dfrac{\sqrt{7}}{4}$

$\tan\theta = \dfrac{\sqrt{7}}{3}$

$\csc\theta = \dfrac{4\sqrt{7}}{7}$

$\sec\theta = \dfrac{4}{3}$

$\cot\theta = \dfrac{3\sqrt{7}}{7}$

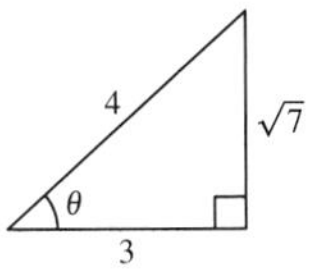

Section 4.8 (page 334)

1. harmonic motion

3.

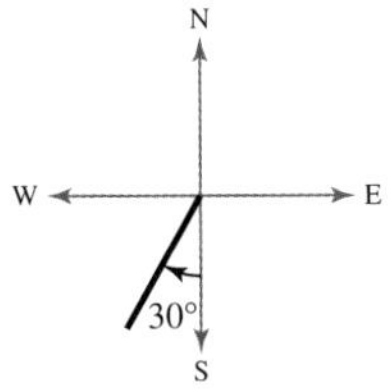

5. $B = 30°$, $a \approx 10.39$, $b = 6$

7. $A = 19°$, $a \approx 4.82$, $c \approx 14.81$

9. $A \approx 53.13°$, $B \approx 36.87°$, $c = 5$

11. $A \approx 17.24°$, $B \approx 72.76°$, $b \approx 51.58$

13. $B = 77° 45'$, $a \approx 91.34$, $b \approx 420.70$

15. 2.50 **17.** 0.50 **19.** 24.10 **21.** About 19.32 ft

23. About 113.63 ft **25.** About 97.14 ft

27. $d = 25(\tan 21.5° - \tan 15°)$; $d \approx 3.15$ m

29. About 1933.32 ft **31.** About 3.23 mi

33. About 75.97° **35.** About 5098.78 ft

37. (a) $h \approx 161.26$ ft

(b)

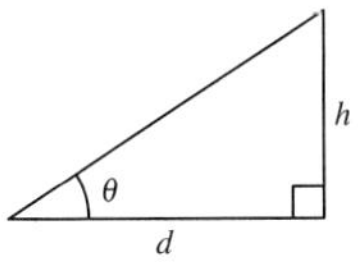

(c) $d = \dfrac{h}{\tan\theta}$

(d) About 914.55; about 443.060; about 279.31; about 192.18; about 135.31

(e) No; Answers will vary.

39. (a) $l = \sqrt{h^2 + 28h + 10{,}196}$ (b) $\theta = \arccos\dfrac{100}{l}$

(c) $h \approx 56$ ft

41. (a) About 61.82°; about 15.64° (b) About 31.10 ft

43. N 76.89° E; about 36.54 nm **45.** N 56.31° W

47. (a) N 58° E (b) About 68.82 m **49.** About 78.69°

51. About 35.26° **53.** $y = \sqrt{3}r$ **55.** $a \approx 12.2$; $b \approx 7.0$

57. $d = 8\sin\pi t$ **59.** $d = 3\cos\dfrac{4\pi t}{3}$

61. (a) 4 (b) 4 (c) 4 (d) $\frac{1}{16}$

63. (a) $\frac{1}{16}$ (b) 70 (c) 0 (d) $\frac{1}{140}$ **65.** $\omega = 528\pi$

67. (a)

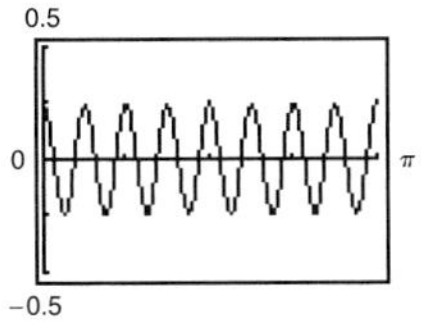

(b) $\dfrac{\pi}{8}$ sec (c) $\dfrac{\pi}{32}$ sec

69. (a)

θ	L_1	L_2	$L_1 + L_2$
0.1	$\dfrac{8}{\sin 0.1}$	$\dfrac{6}{\cos 0.1}$	86.16
0.2	$\dfrac{8}{\sin 0.2}$	$\dfrac{6}{\cos 0.2}$	46.39
0.3	$\dfrac{8}{\sin 0.3}$	$\dfrac{6}{\cos 0.3}$	33.35
0.4	$\dfrac{8}{\sin 0.4}$	$\dfrac{6}{\cos 0.4}$	27.06

(b)

θ	L_1	L_2	$L_1 + L_2$
0.5	$\dfrac{8}{\sin 0.5}$	$\dfrac{6}{\cos 0.5}$	23.52
0.6	$\dfrac{8}{\sin 0.6}$	$\dfrac{6}{\cos 0.6}$	21.44
0.7	$\dfrac{8}{\sin 0.7}$	$\dfrac{6}{\cos 0.7}$	20.26
0.8	$\dfrac{8}{\sin 0.8}$	$\dfrac{6}{\cos 0.8}$	19.76

About 19.76 ft

(c) $L_1 + L_2 = \dfrac{8}{\sin\theta} + \dfrac{6}{\cos\theta}$

(d)

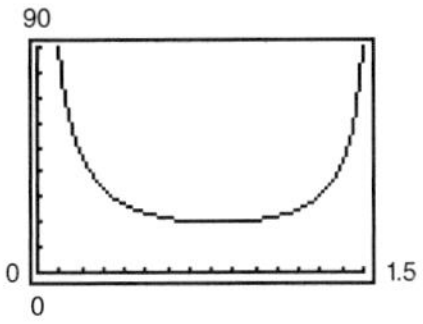

About 19.73 ft; They are almost identical.

71. (a) and (b)

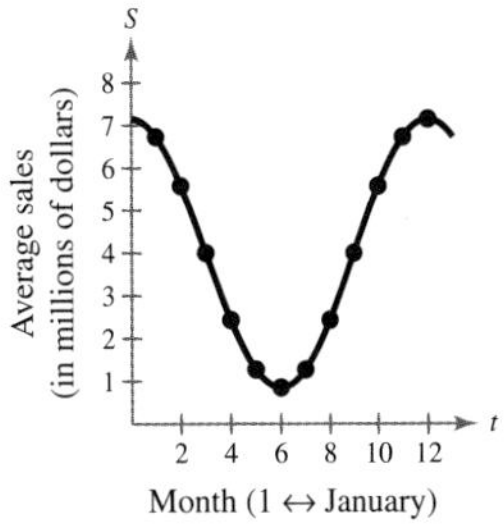

$S = 3.15\sin(0.524t + 1.5686) + 4$

The model fits the data well.

(c) 12 mo; Yes, sales of outerwear are seasonal.

(d) Maximum displacement of 3.15 million dollars from the average sales of 4 million dollars.

73. False; Sine or cosine is used to model harmonic motion.

75. S 58° E, N 58° W; Answers will vary.

77. The bearing is S 40° W, not N 40° W.

79. All real numbers x **81.** All real numbers x

83. All real numbers x such that $x > 6$

Review Exercises (page 340)

1. (a)

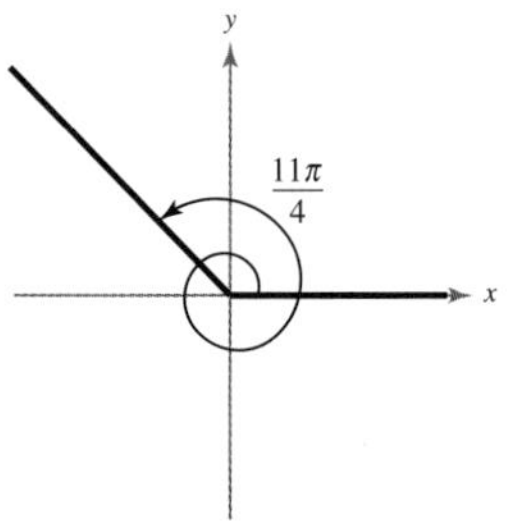

(b) Quadrant II

(c) *Sample answer:* $\frac{3\pi}{4}$, $-\frac{5\pi}{4}$

3. (a)

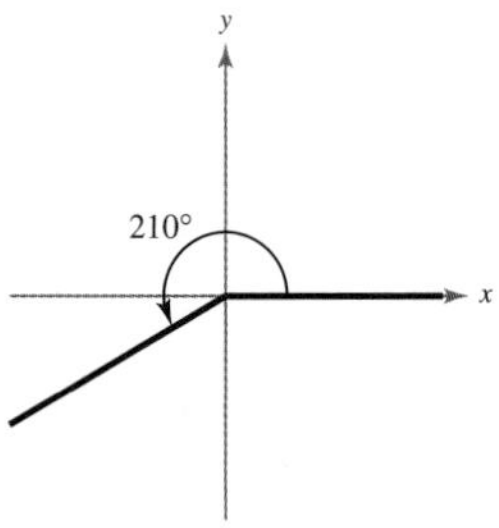

(b) Quadrant III

(c) *Sample answer:* 570°, −150°

5. −1.552 **7.** 128.571 **9.** 153.754 **11.** 126° 55′48″

13. Complement: $\frac{5\pi}{14}$; supplement: $\frac{6\pi}{7}$ **15.** $\frac{6}{5}$ rad

17. (a) $\frac{46\pi}{3}$ m (b) 5π cm **19.** $\left(\frac{\sqrt{2}}{2}, -\frac{\sqrt{2}}{2}\right)$

21. $\left(-\frac{1}{2}, -\frac{\sqrt{3}}{2}\right)$

23. $\sin\frac{5\pi}{4} = -\frac{\sqrt{2}}{2}$ $\csc\frac{5\pi}{4} = -\sqrt{2}$

$\cos\frac{5\pi}{4} = -\frac{\sqrt{2}}{2}$ $\sec\frac{5\pi}{4} = -\sqrt{2}$

$\tan\frac{5\pi}{4} = 1$ $\cot\frac{5\pi}{4} = 1$

25. $\sin 4\pi = 0$ $\csc 4\pi$ is undefined.

$\cos 4\pi = 1$ $\sec 4\pi = 1$

$\tan 4\pi = 0$ $\cot 4\pi$ is undefined.

27. $\sin\left(-\frac{11\pi}{6}\right) = \frac{1}{2}$ $\csc\left(-\frac{11\pi}{6}\right) = 2$

$\cos\left(-\frac{11\pi}{6}\right) = \frac{\sqrt{3}}{2}$ $\sec\left(-\frac{11\pi}{6}\right) = \frac{2\sqrt{3}}{3}$

$\tan\left(-\frac{11\pi}{6}\right) = \frac{\sqrt{3}}{3}$ $\cot\left(-\frac{11\pi}{6}\right) = \sqrt{3}$

29. $\sin\left(-\frac{7\pi}{3}\right) = -\frac{\sqrt{3}}{2}$ $\csc\left(-\frac{7\pi}{3}\right) = -\frac{2\sqrt{3}}{3}$

$\cos\left(-\frac{7\pi}{3}\right) = \frac{1}{2}$ $\sec\left(-\frac{7\pi}{3}\right) = 2$

$\tan\left(-\frac{7\pi}{3}\right) = -\sqrt{3}$ $\cot\left(-\frac{7\pi}{3}\right) = -\frac{\sqrt{3}}{3}$

31. 1 **33.** $-\frac{1}{2}$ **35.** (a) $-\frac{3}{5}$ (b) $-\frac{5}{3}$

37. (a) $\frac{2}{3}$ (b) $\frac{3}{2}$ **39.** −0.3827 **41.** −0.3416

43. $\sin\theta = \frac{\sqrt{65}}{9}$ $\csc\theta = \frac{9\sqrt{65}}{65}$

$\cos\theta = \frac{4}{9}$ $\sec\theta = \frac{9}{4}$

$\tan\theta = \frac{\sqrt{65}}{4}$ $\cot\theta = \frac{4\sqrt{65}}{65}$

45. $\sin\theta = \frac{5\sqrt{61}}{61}$ $\csc\theta = \frac{\sqrt{61}}{5}$

$\cos\theta = \frac{6\sqrt{61}}{61}$ $\sec\theta = \frac{\sqrt{61}}{6}$

$\tan\theta = \frac{5}{6}$ $\cot\theta = \frac{6}{5}$

47. $\cos\theta = \frac{\sqrt{95}}{12}$

$\tan\theta = \frac{7\sqrt{95}}{95}$

$\csc\theta = \frac{12}{7}$

$\sec\theta = \frac{12\sqrt{95}}{95}$

$\cot\theta = \frac{\sqrt{95}}{7}$

49. $\sin\theta = \frac{\sqrt{26}}{26}$

$\cos\theta = \frac{5\sqrt{26}}{26}$

$\csc\theta = \sqrt{26}$

$\sec\theta = \frac{\sqrt{26}}{5}$

$\cot\theta = 5$

51. (a) 0.1045 (b) 1.2656

53. Answers will vary. **55.** About 235 ft

57. $\sin\theta = \frac{4}{5}$ $\csc\theta = \frac{5}{4}$

$\cos\theta = \frac{3}{5}$ $\sec\theta = \frac{5}{3}$

$\tan\theta = \frac{4}{3}$ $\cot\theta = \frac{3}{4}$

59. $\sin\theta = \frac{2\sqrt{53}}{53}$ $\csc\theta = \frac{\sqrt{53}}{2}$

$\cos\theta = -\frac{7\sqrt{53}}{53}$ $\sec\theta = -\frac{\sqrt{53}}{7}$

$\tan\theta = -\frac{2}{7}$ $\cot\theta = -\frac{7}{2}$

61. $\sin\theta = \frac{15\sqrt{481}}{481}$ $\csc\theta = \frac{\sqrt{481}}{15}$

$\cos\theta = \frac{16\sqrt{481}}{481}$ $\sec\theta = \frac{\sqrt{481}}{16}$

$\tan\theta = \frac{15}{16}$ $\cot\theta = \frac{16}{15}$

63. $\sin\theta = -\frac{\sqrt{11}}{6}$

$\cos\theta = \frac{5}{6}$

$\tan\theta = -\frac{\sqrt{11}}{5}$

$\csc\theta = -\frac{6\sqrt{11}}{11}$

$\cot\theta = -\frac{5\sqrt{11}}{11}$

65. $\sin\theta = \frac{\sqrt{21}}{5}$

$\tan\theta = -\frac{\sqrt{21}}{2}$

$\csc\theta = \frac{5\sqrt{21}}{21}$

$\sec\theta = -\frac{5}{2}$

$\cot\theta = -\frac{2\sqrt{21}}{21}$

67. $\theta' = 30°$

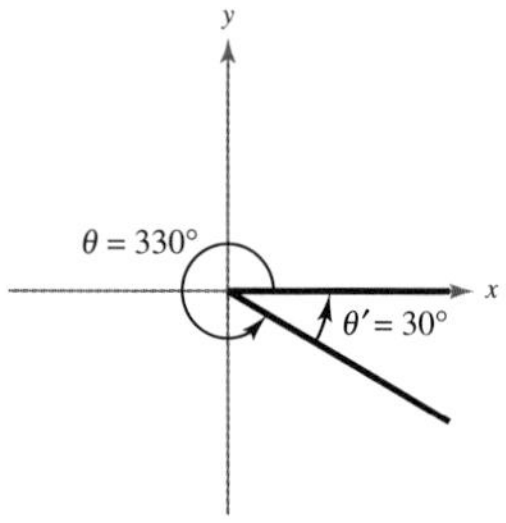

69. $\theta' = \frac{\pi}{4}$

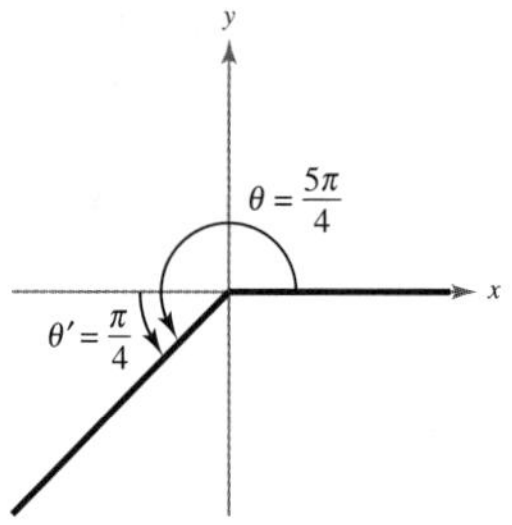

71. $\theta' = 78°$

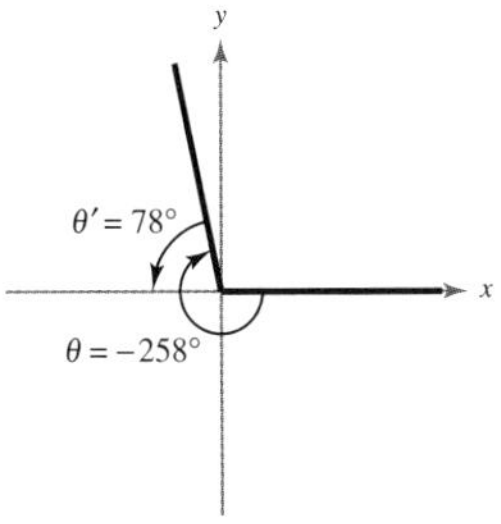

73. $\theta' = \dfrac{\pi}{5}$

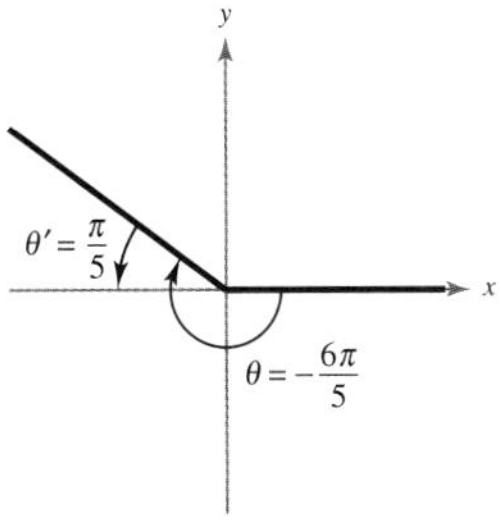

75. $\sin 240° = -\dfrac{\sqrt{3}}{2}$

$\cos 240° = -\dfrac{1}{2}$

$\tan 240° = \sqrt{3}$

77. $\sin(-225°) = \dfrac{\sqrt{2}}{2}$

$\cos(-225°) = -\dfrac{\sqrt{2}}{2}$

$\tan(-225°) = -1$

79. $\sin 4\pi = 0$

$\cos 4\pi = 1$

$\tan 4\pi = 0$

81. $\sin\left(-\dfrac{9\pi}{4}\right) = -\dfrac{\sqrt{2}}{2}$

$\cos\left(-\dfrac{9\pi}{4}\right) = \dfrac{\sqrt{2}}{2}$

$\tan\left(-\dfrac{9\pi}{4}\right) = -1$

83. 0.6494 **85.** 3.2361

87.

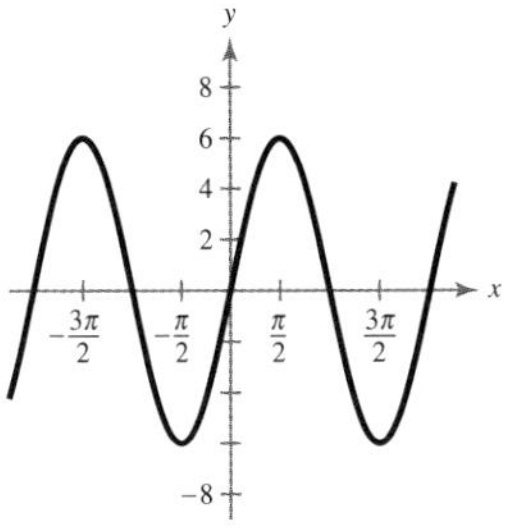

89.

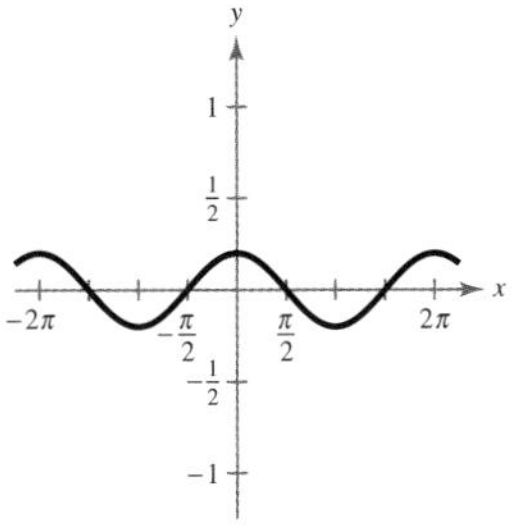

91.

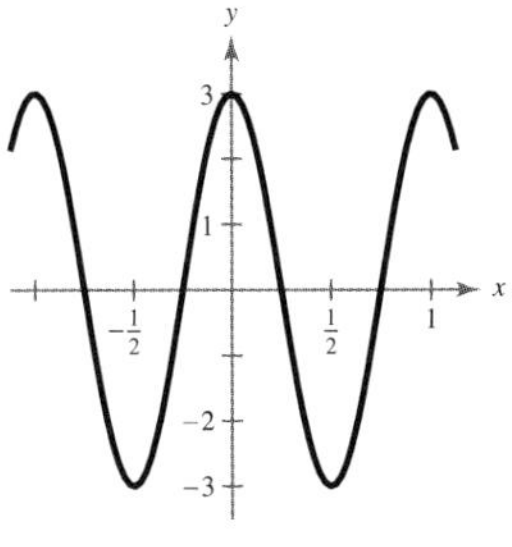

93.

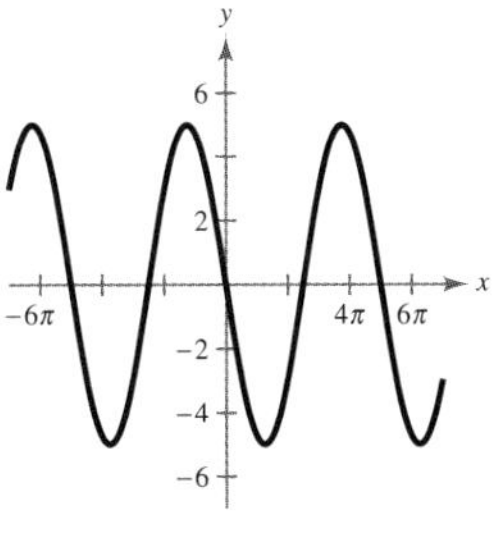

95.

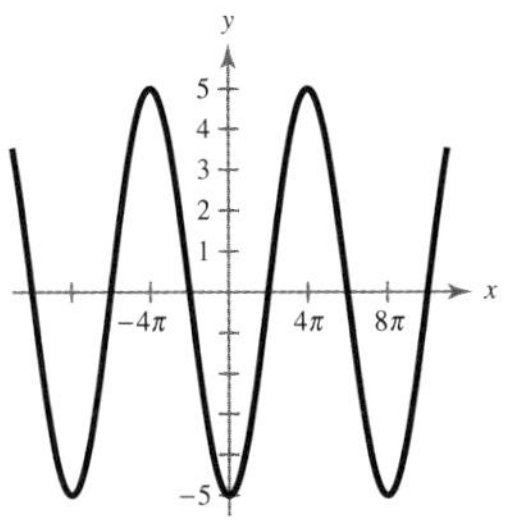

97.

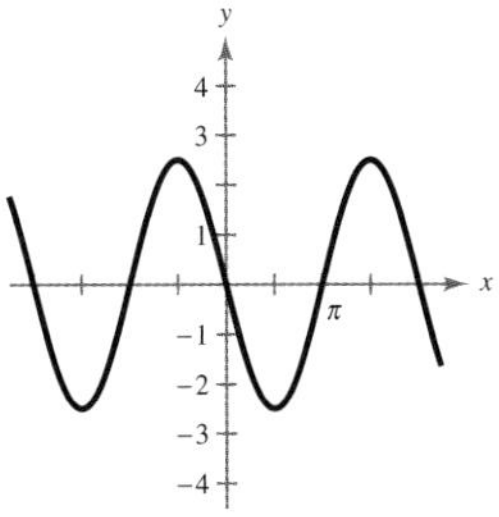

99.

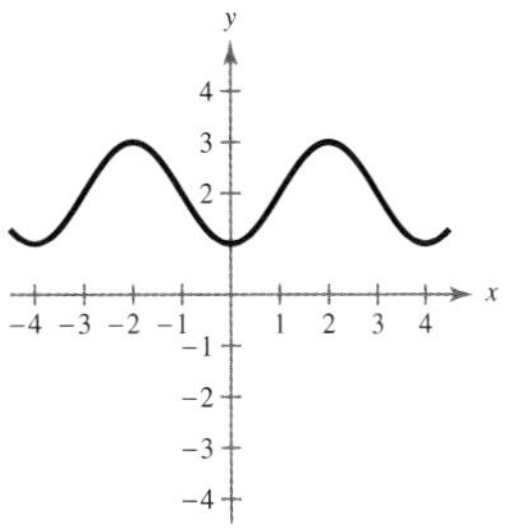

101.

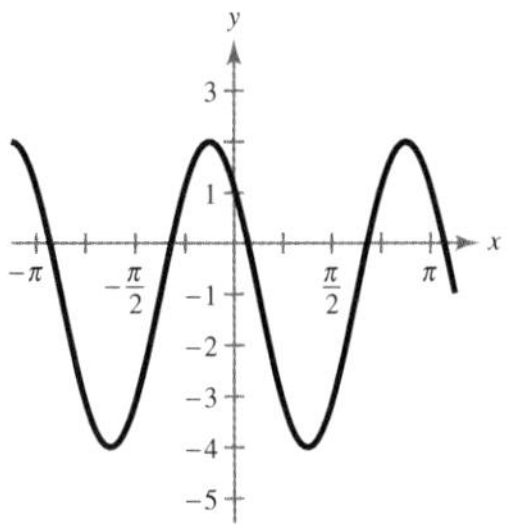

103.

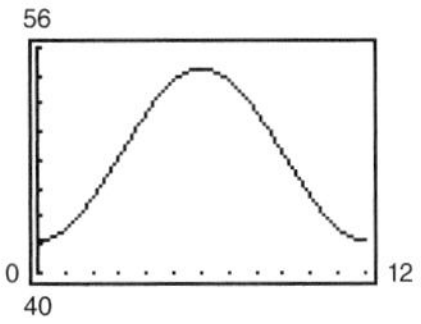

Maximum sales: June
Minimum sales: December

105.

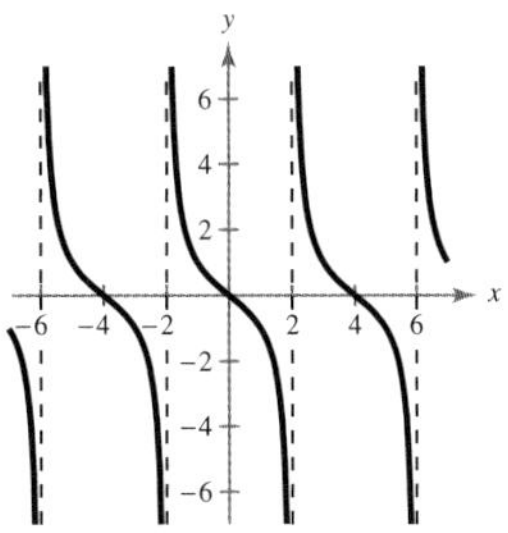

107.

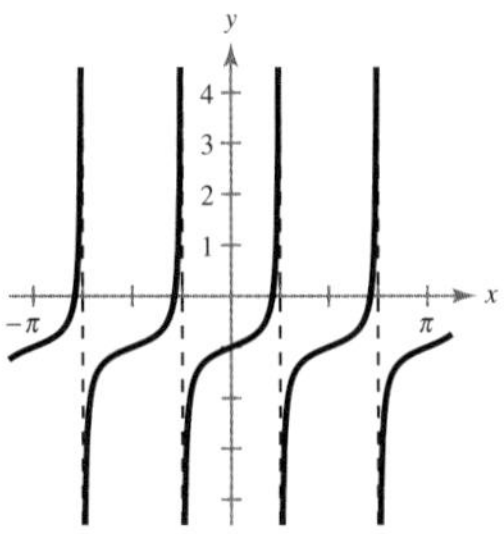

109.

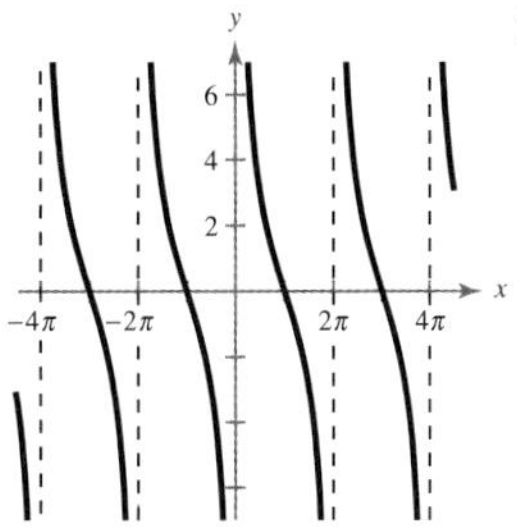

111.

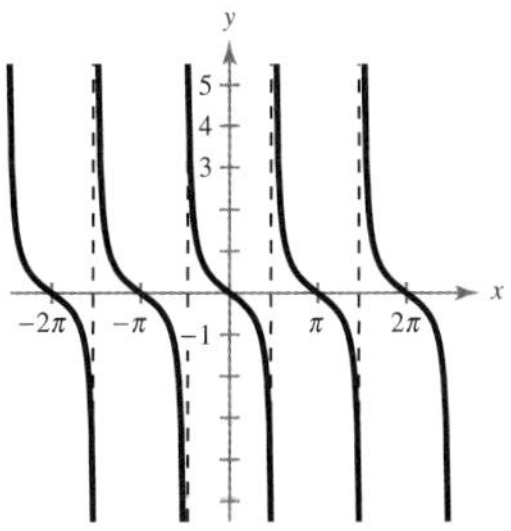

113.

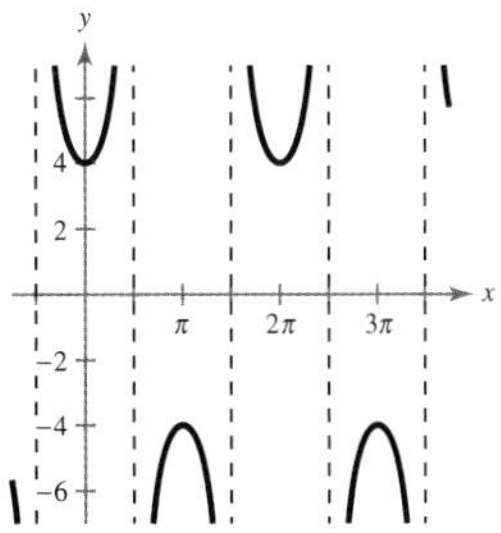

115.

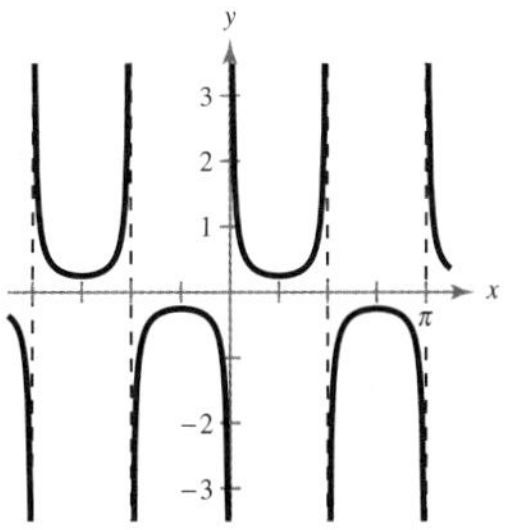

117.

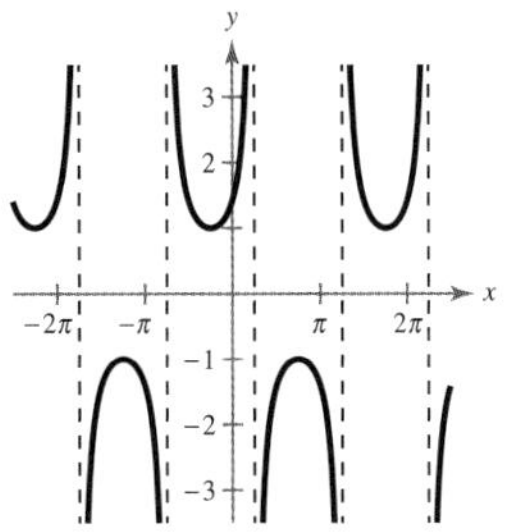

119.

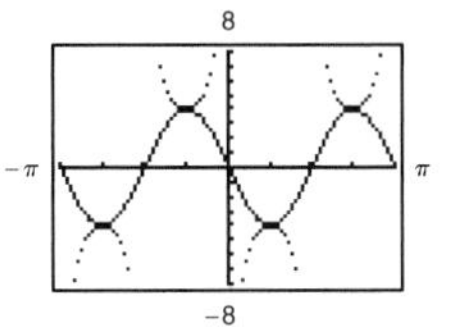

Answers will vary.

121.

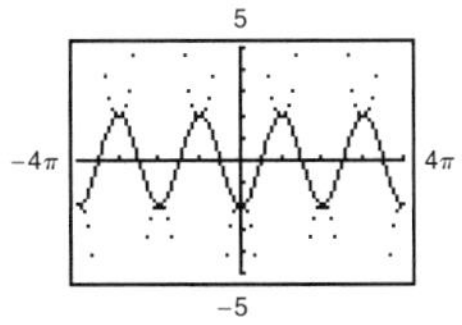

Answers will vary.

123.

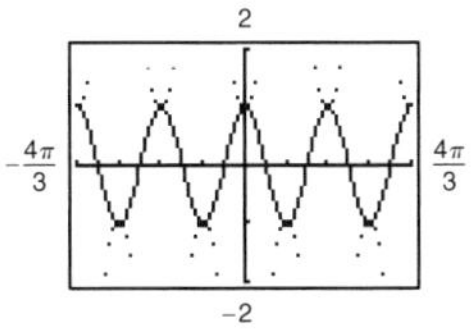

Answers will vary.

CHAPTER 4

125. 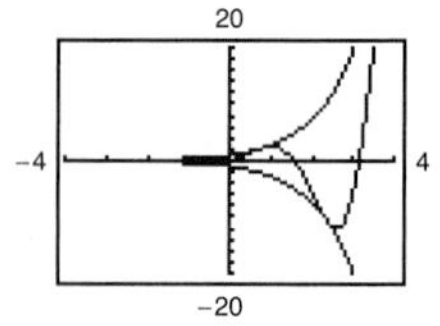

$-e^x \le e^x \sin 2x \le e^x$

Touches $y = \pm e^x$ at $x = \dfrac{\pi}{4} + \dfrac{n\pi}{2}$

Intercepts at $x = \dfrac{n\pi}{2}$

127. 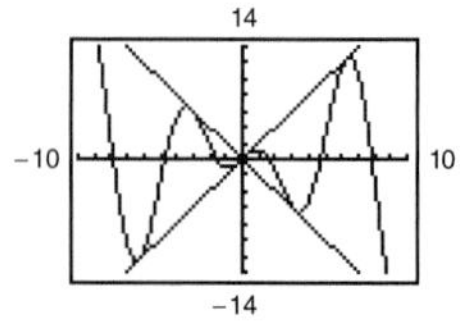

$-2x \le 2x \cos x \le 2x$

Touches $y = \pm 2x$ at $x = n\pi$

Intercepts at $x = \dfrac{\pi}{2} + n\pi$

129. $-\dfrac{\pi}{4}$ **131.** $\dfrac{\pi}{2}$ **133.** 0 **135.** 1.14

137. $\theta = \arcsin \dfrac{x}{16}$ **139.** $\dfrac{1}{\sqrt{1 - x^2}}$ **141.** About 0.071 km

143. About 9.47 mi **145.** $d = 3 \cos \dfrac{2\pi t}{15}$

147. False; y is a function but is not one-to-one on $30° \le \theta \le 150°$.

149. False; Sine or cosine is used to model harmonic motion.

151. (a) 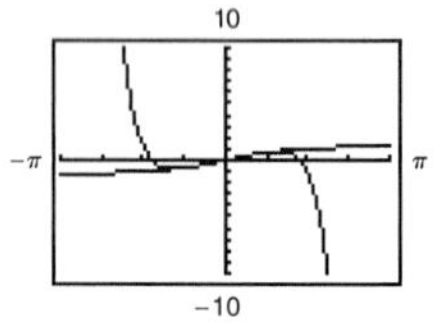

(b) The polynomial function is a good approximation for the arctangent function when x is close to 0.

Chapter Test (page 347)

1. (a) 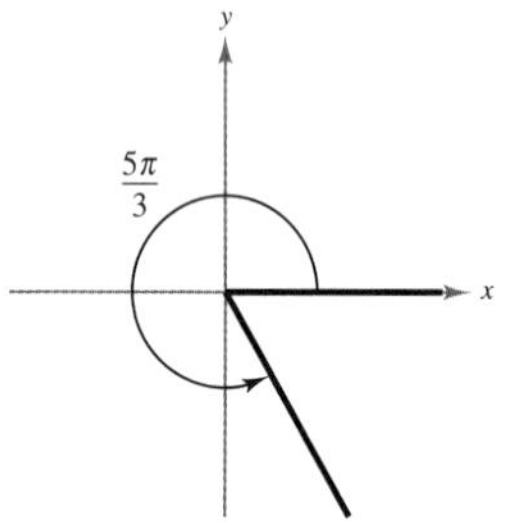

(b) Answers will vary. *Sample answer:* $\dfrac{11\pi}{3}, -\dfrac{\pi}{3}$

(c) 300°

2. Complement: 41°; supplement: 131° **3.** $\dfrac{13\pi}{3}$ yd

4. $44\frac{4}{9} \approx 44.44$ rad/sec

5. $\sin \theta = -\dfrac{24}{25}$

$\cos \theta = \dfrac{7}{25}$

$\tan \theta = -\dfrac{24}{7}$

$\csc \theta = -\dfrac{25}{24}$

$\sec \theta = \dfrac{25}{7}$

$\cot \theta = -\dfrac{7}{24}$

6. $\sin \theta = \dfrac{7\sqrt{53}}{53}$

$\cos \theta = \dfrac{2\sqrt{53}}{53}$

$\csc \theta = \dfrac{\sqrt{53}}{7}$

$\sec \theta = \dfrac{\sqrt{53}}{2}$

$\cot \theta = \dfrac{2}{7}$

7. $\theta' = 75°$ **8.** Quadrant III **9.** 150°, 210°

10. 0.7361

11. 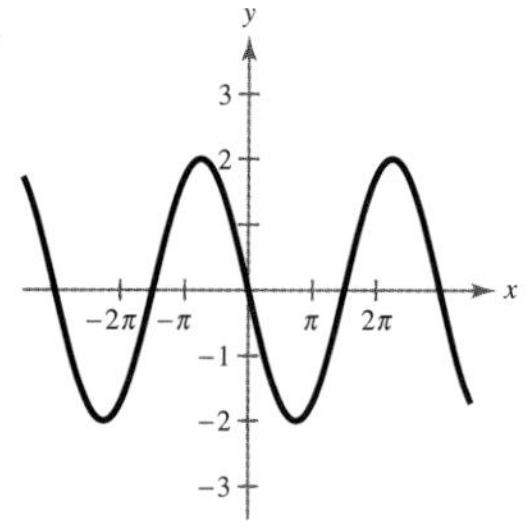

12. 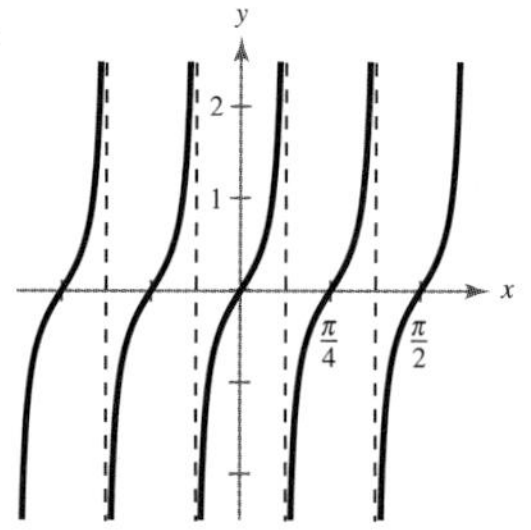

13. 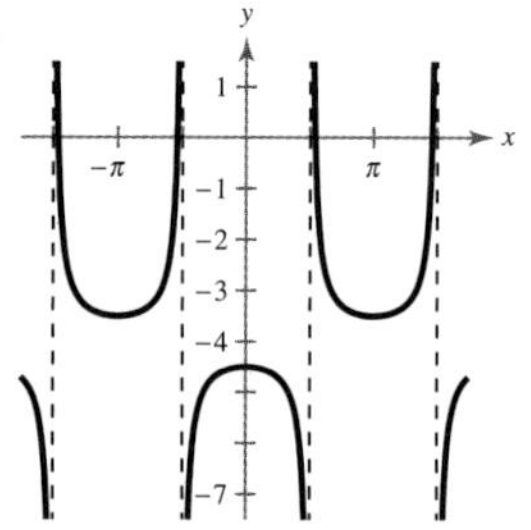

14. 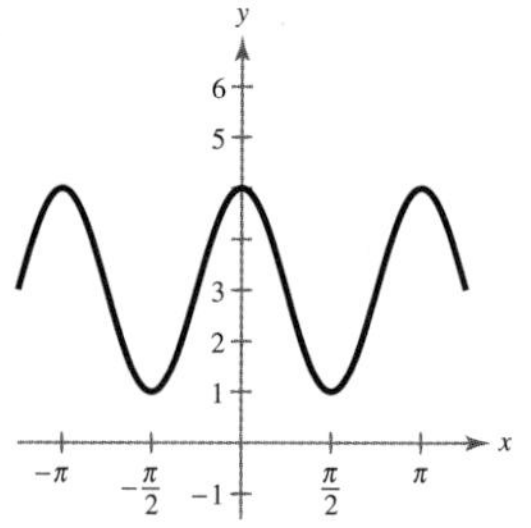

15. 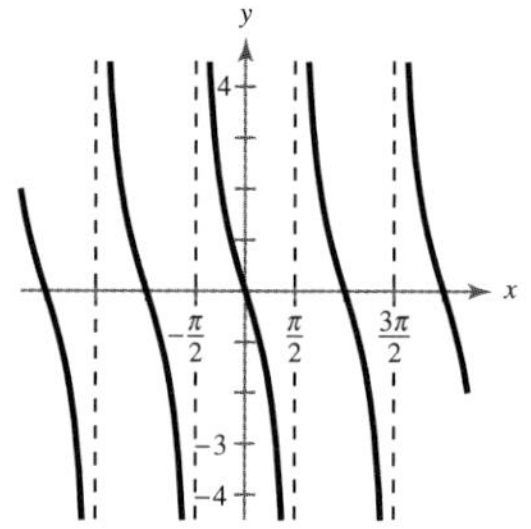

16. 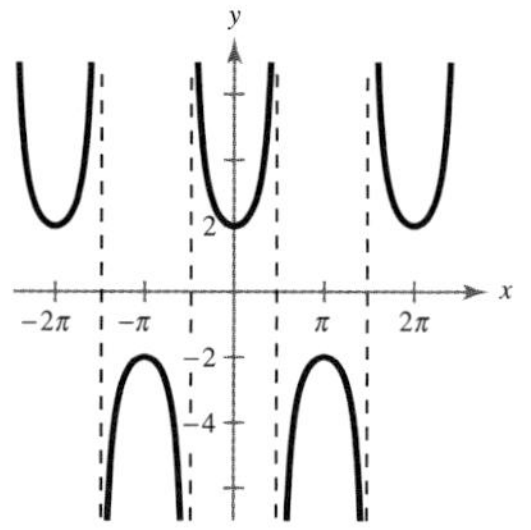

17. 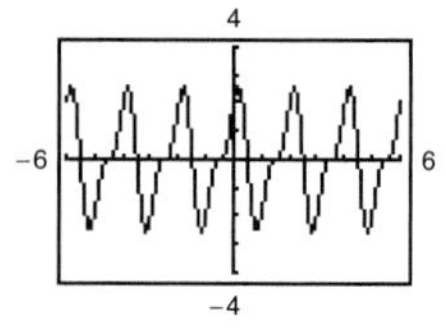

Period: 2

18. 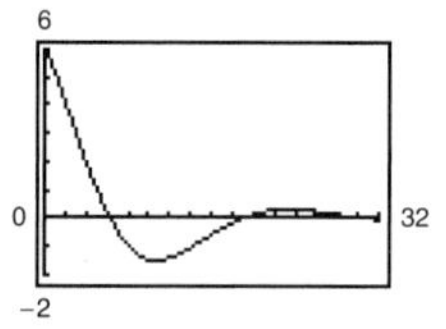

Not periodic

19. $a = 2, b = \dfrac{1}{2}, c = -\dfrac{\pi}{4}$ **20.** $\dfrac{\sqrt{5}}{2}$

21. 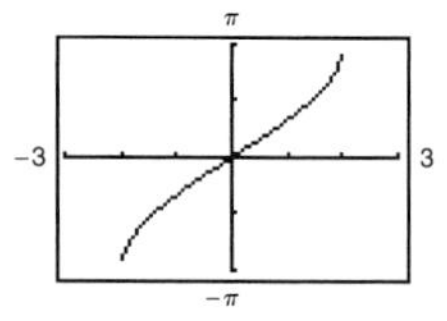

22. 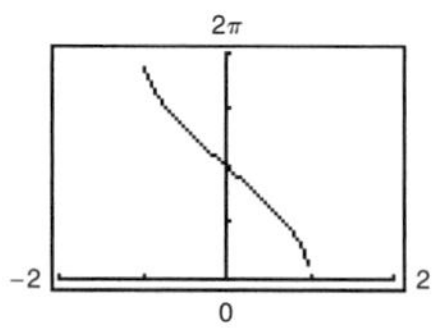

23. 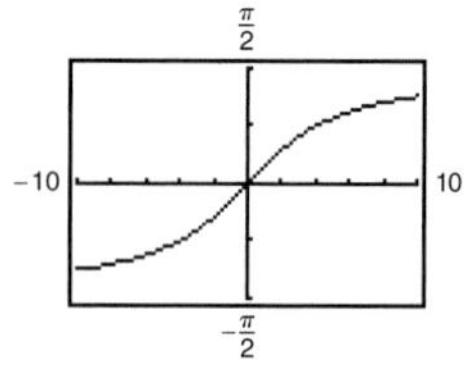

24. S 6.67° E, about 29.90 nm **25.** $d = -6 \cos \pi t$

Chapter 5

Section 5.1 (page 356)

1. (a) iii (b) i (c) ii **3.** $\sin u$ **5.** $\sec u$

7. $\sin x = \frac{\sqrt{21}}{5}$ $\csc x = \frac{5\sqrt{21}}{21}$

$\cos x = -\frac{2}{5}$ $\sec x = -\frac{5}{2}$

$\tan x = -\frac{\sqrt{21}}{2}$ $\cot x = -\frac{2\sqrt{21}}{21}$

9. $\sin \theta = -\frac{3}{4}$ $\csc \theta = -\frac{4}{3}$

$\cos \theta = \frac{\sqrt{7}}{4}$ $\sec \theta = \frac{4\sqrt{7}}{7}$

$\tan \theta = -\frac{3\sqrt{7}}{7}$ $\cot \theta = -\frac{\sqrt{7}}{3}$

11. $\sin \theta = \frac{2\sqrt{6}}{5}$ $\csc \theta = \frac{5\sqrt{6}}{12}$

$\cos \theta = \frac{1}{5}$ $\sec \theta = 5$

$\tan \theta = 2\sqrt{6}$ $\cot \theta = \frac{\sqrt{6}}{12}$

13. $\sin x = \frac{1}{2}$ $\csc x = 2$

$\cos x = \frac{\sqrt{3}}{2}$ $\sec x = \frac{2\sqrt{3}}{3}$

$\tan x = \frac{\sqrt{3}}{3}$ $\cot x = \sqrt{3}$

15. $\sin \phi = \frac{8}{17}$ $\csc \phi = \frac{17}{8}$

$\cos \phi = -\frac{15}{17}$ $\sec \phi = -\frac{17}{15}$

$\tan \phi = -\frac{8}{15}$ $\cot \phi = -\frac{15}{8}$

17. $\sin \theta = 0$ $\csc \theta$ is undefined.

$\cos \theta = -1$ $\sec \theta = -1$

$\tan \theta = 0$ $\cot \theta$ is undefined.

19. d **20.** a **21.** b **22.** f **23.** c **24.** e
25. b **26.** c **27.** f **28.** a **29.** e **30.** d
31. $\cos x$ **33.** $\cos^2 \phi$ **35.** $\sin^2 x$ **37.** $\sec \beta$
39. $\sec x$ **41.** 1 **43.** $-\cot x$ **45.** $1 + \sin y$
47. $\sec x + 1$ **49.** $\sin^4 x$ **51.** $(3 \sin x + 1)(\sin x - 2)$
53. $\csc^2 x(\cot x + 1)$ **55.** $(\csc x - 1)(\csc x + 2)$
57. $1 + 2 \sin x \cos x$ **59.** $\cot^2 x$ **61.** $2 \csc^2 x$
63. $-\cot x$ **65.** $2 \sec x$ **67.** $\cos x$ **69.** $1 + \cos y$
71. $3(\sec x + \tan x)$

73.

x	0.2	0.4	0.6	0.8
y_1	0.1987	0.3894	0.5646	0.7174
y_2	−0.1987	−0.3894	−0.5646	−0.7174

x	1.0	1.2	1.4
y_1	0.8415	0.9320	0.9854
y_2	−0.8415	−0.9320	−0.9854

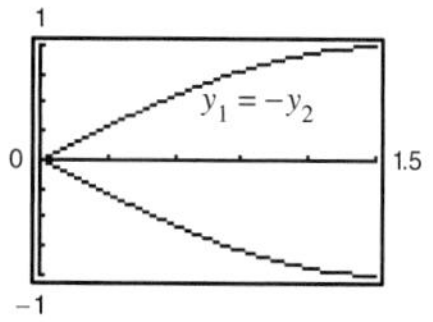

75.

x	0.2	0.4	0.6	0.8
y_1	1.2230	1.5085	1.8958	2.4650
y_2	1.2230	1.5085	1.8958	2.4650

x	1.0	1.2	1.4
y_1	3.4082	5.3319	11.6814
y_2	3.4082	5.3319	11.6814

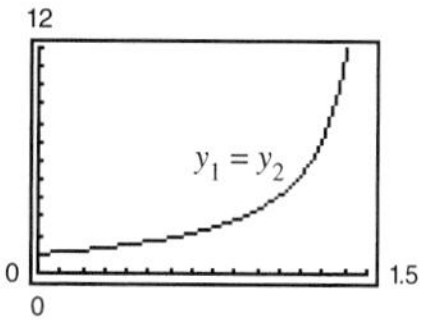

77. $\csc x$ **79.** $\tan x$ **81.** $3 \sin \theta$ **83.** $3 \sec \theta$
85. $7 \cos \theta$ **87.** $3 \sec \theta$ **89.** $\sqrt{2} \cos \theta$ **91.** $\sec^3 \theta$

93. $0 \le \theta \le \pi$ **95.** $0 \le \theta < \frac{\pi}{2}, \frac{3\pi}{2} < \theta < 2\pi$

97. $\ln|\cot \theta|$ **99.** $\ln|\tan x|$ **101.** $\ln|(\cos x)(1 + \sin x)|$
103. (a) and (b) Answers will vary.
105. (a) and (b) Answers will vary. **107.** $\mu = \tan \theta$
109. Answers will vary. **111.** False; $\cos 0 \cdot \sec \frac{\pi}{4} \neq 1$
113. 1, 1 **115.** $-\infty$, 0

117. $\cos \theta = \pm\sqrt{1 - \sin^2 \theta}$

$\tan \theta = \pm\frac{\sin \theta}{\sqrt{1 - \sin^2 \theta}}$

$\csc \theta = \frac{1}{\sin \theta}$

$\sec \theta = \pm\frac{1}{\sqrt{1 - \sin^2 \theta}}$

$\cot \theta = \pm\frac{\sqrt{1 - \sin^2 \theta}}{\sin \theta}$

The sign depends on the choice of θ.

119. $\frac{\sin \theta}{\cos \theta}$ **121.** No inverse function

123. $f^{-1}(x) = \frac{x^2 + 14}{3}, x \ge 0$

125.

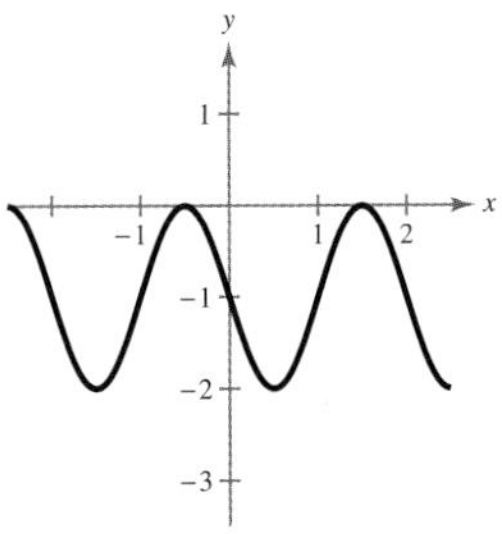

127.

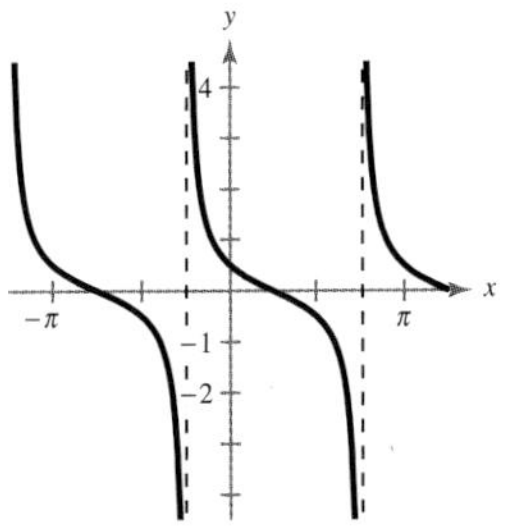

Section 5.2 (page 364)

1. $\cot u$ **3.** $\tan u$ **5.** $\cos^2 u$ **7.** $\cos u$ **9.** No

11–37. Answers will vary.

39.

x	0.2	0.4	0.6	0.8
y_1	1.0203	1.0857	1.2116	1.4353
y_2	1.0203	1.0857	1.2116	1.4353

x	1.0	1.2	1.4
y_1	1.8508	2.7597	5.8835
y_2	1.8508	2.7597	5.8835

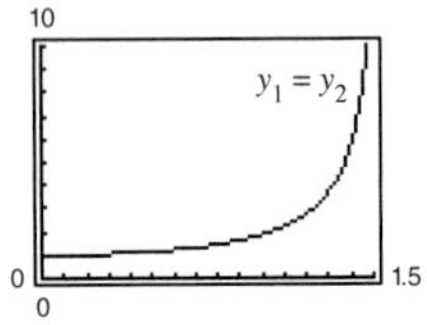

41.

x	0.2	0.4	0.6	0.8
y_1	24.3360	5.5943	2.1366	0.9433
y_2	24.3360	5.5943	2.1366	0.9433

x	1.0	1.2	1.4
y_1	0.4123	0.1511	0.0297
y_2	0.4123	0.1511	0.0297

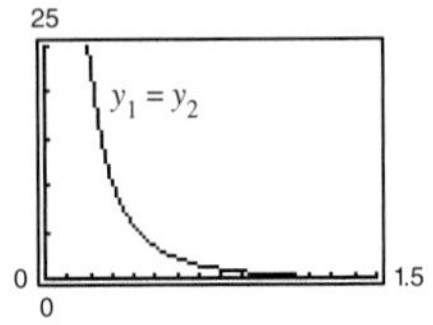

43.

x	0.2	0.4	0.6	0.8
y_1	4.8348	2.1785	1.2064	0.6767
y_2	4.8348	2.1785	1.2064	0.6767

x	1.0	1.2	1.4
y_1	0.3469	0.1409	0.0293
y_2	0.3469	0.1409	0.0293

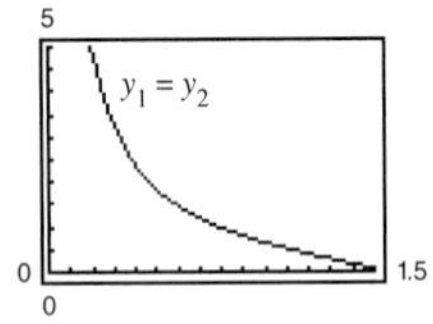

45.

x	0.2	0.4	0.6	0.8
y_1	5.0335	2.5679	1.7710	1.3940
y_2	5.0335	2.5679	1.7710	1.3940

x	1.0	1.2	1.4
y_1	1.1884	1.0729	1.0148
y_2	1.1884	1.0729	1.0148

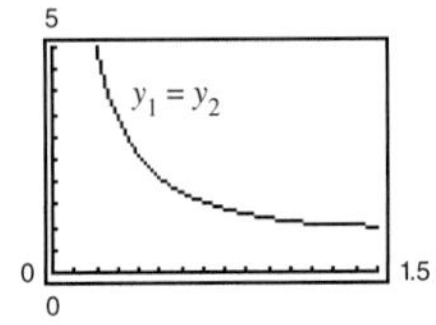

47.

x	0.2	0.4	0.6	0.8
y_1	5.1359	2.7880	2.1458	2.0009
y_2	5.1359	2.7880	2.1458	2.0009

x	1.0	1.2	1.4
y_1	2.1995	2.9609	5.9704
y_2	2.1995	2.9609	5.9704

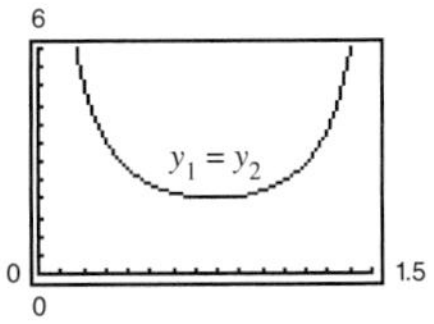

49–65. Answers will vary.

67.

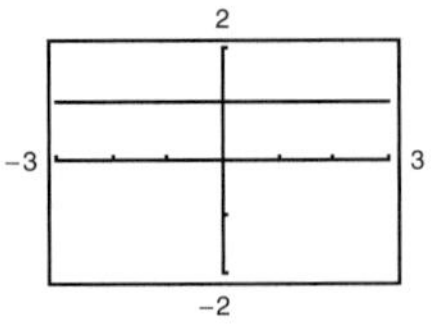

$y = 1$

69.

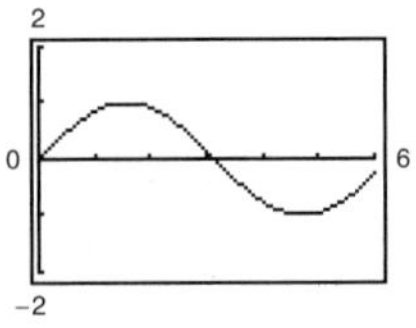

$y = \sin x$

71. Answers will vary. **73.** 1 **75.** 2

77–83. Answers will vary.

85. (a) Answers will vary.

(b)

θ	15°	30°	45°	60°	75°	90°
s	18.66	8.66	5	2.88	1.34	0

(c) Maximum: 15° Minimum: 90° (d) Noon

87. False; $\sin^2\left(\frac{\pi}{4}\right) + \cos^2\left(\frac{\pi}{4}\right) \neq 1 + \tan^2\left(\frac{\pi}{4}\right)$

89. (a) Answers will vary. (b) No. Division by zero.

91. $a \cos \theta$ **93.** $a \sec \theta$ **95.** $\sqrt{\sec^2 x - 1} = |\tan x|; \frac{3\pi}{4}$

97. $1 - \cos^2 \theta = \sin^2 \theta; \pi$ **99.** Answers will vary.

101.

x	−4	−2	0	2	3
y	3.0625	3.25	4	7	11

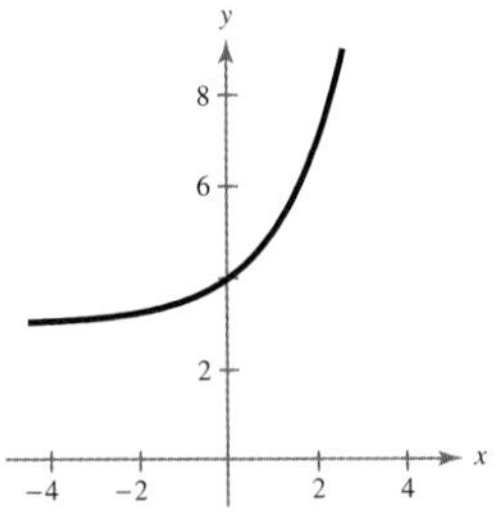

Horizontal asymptote at $y = 3$

103.

x	−3	−2	0	2	4
y	7	3	0	−0.75	−0.9375

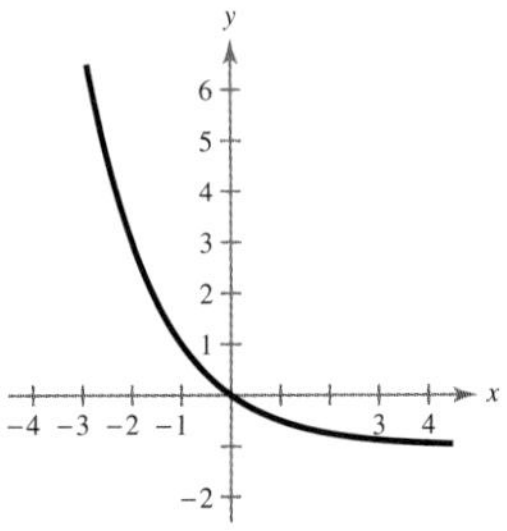

Horizontal asymptote at $y = -1$

Section 5.3 (page 375)

1. general **3.** No **5–9.** Answers will vary.

11. $0°, 180°$ **13.** $150°, 210°$ **15.** $45°, 225°$

17. $\frac{5\pi}{6}, \frac{7\pi}{6}$ **19.** $\frac{\pi}{6}, \frac{7\pi}{6}$ **21.** $\frac{3\pi}{4}, \frac{7\pi}{4}$

23. $\frac{7\pi}{6} + 2n\pi, \frac{11\pi}{6} + 2n\pi$ **25.** $\frac{\pi}{3} + 2n\pi, \frac{2\pi}{3} + 2n\pi$

27. $\frac{2\pi}{3} + 2n\pi, \frac{4\pi}{3} + 2n\pi$ **29.** $n\pi$ **31.** $\frac{\pi}{6} + n\pi, \frac{5\pi}{6} + n\pi$

33. $\frac{\pi}{6} + n\pi, \frac{5\pi}{6} + n\pi$ **35.** $\frac{n\pi}{2}$

37. $n\pi, \frac{\pi}{3} + 2n\pi, \frac{5\pi}{3} + 2n\pi$ **39.** $\frac{\pi}{6}, \frac{5\pi}{6}, \frac{3\pi}{2}$

41. About 1.936, about 2.777, about 5.077, about 5.918

43. $\frac{\pi}{2}, \frac{2\pi}{3}, \frac{4\pi}{3}, \frac{3\pi}{2}$ **45.** $0, \frac{\pi}{2}$

47. About 0.8614, about 5.4218 **49.** About 1.5708

51. About 0.5236, about 2.6180

53. (a)

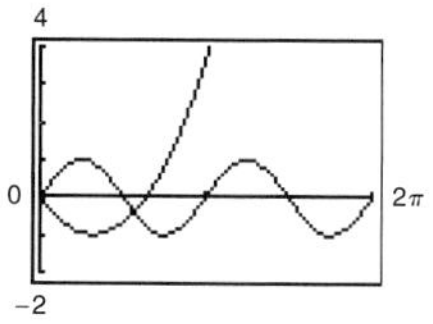

(b) $\sin 2x = x^2 - 2x$
(c) $(0, 0), (1.7757, -0.3984)$

55. (a)

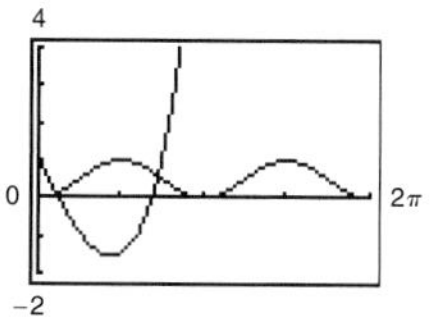

(b) $\sin^2 x = e^x - 4x$
(c) $(0.3194, 0.0986)$, $(2.2680, 0.5878)$

57. $\frac{\pi}{6} + n\pi, \frac{5\pi}{6} + n\pi$ **59.** $\frac{\pi}{12} + \frac{n\pi}{3}$

61. $\frac{\pi}{2} + 4n\pi, \frac{7\pi}{2} + 4n\pi$ **63.** $\frac{\pi}{3} + 2n\pi$ **65.** $-1, 3$

67. ± 2 **69.** About 1.1071, about 4.2487

71. About 0.8603, about 3.4256

73. 0, about 2.6779, about 3.1416, about 5.8195

75. About 0.5236, about 0.7297, about 2.4119, about 2.6180

77. $\arctan 3 + n\pi, \arctan(-4) + n\pi$

79. $\frac{\pi}{4} + n\pi, \arctan 5 + n\pi$ **81.** $\frac{5\pi}{3} + 2n\pi, \frac{\pi}{3} + 2n\pi$

83. $-1.154, 0.534$ **85.** 1.110

87. (a)

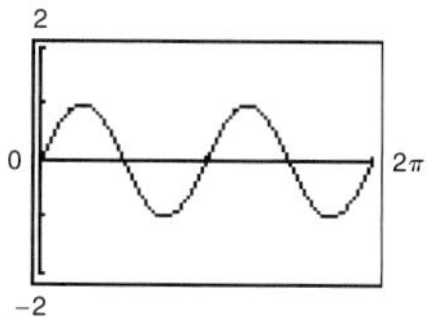

Maxima: (0.7854, 1), (3.9270, 1)
Minima: (2.3562, −1), (5.4978, −1)

(b) $\frac{\pi}{4}, \frac{3\pi}{4}, \frac{5\pi}{4}, \frac{7\pi}{4}$

89. (a)

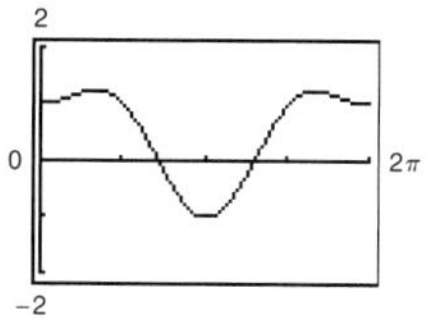

Maxima: (1.0472, 1.25), (5.2360, 1.25)
Minima: (0, 1), (3.1416, −1), (6.2832, 1)

(b) $0, \frac{\pi}{3}, \pi, \frac{5\pi}{3}, 2\pi$

91. (a)

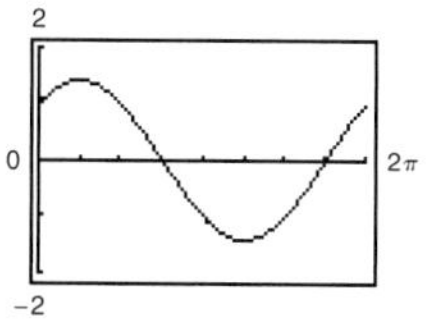

Maximum: (0.7854, 1.4142)
Minimum: (3.9270, −1.4142)

(b) $\frac{\pi}{4}, \frac{5\pi}{4}$

93. 1 **95.** May, June, July

97. (a) All real numbers x except $x = 0$
(b) y-axis symmetry; horizontal asymptote: $y = 1$
(c) Oscillates (d) Infinite number of solutions
(e) Yes; about 0.6366

99. About 0.04 sec, about 0.43 sec, about 0.83 sec

101. About 41.37°, about 48.63°

103. (a)

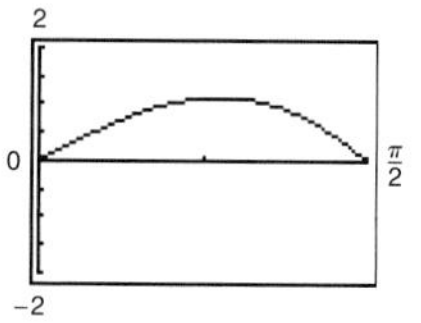

$x \approx 0.86, A \approx 1.12$
(b) About $0.6100 < x < 1.0980$

105. 1

107. False; The range of the sine function does not include 3.4.

109. Answers will vary. **111.** 2.164

113. −3.763 **115.** Answers will vary.

Section 5.4 (page 383)

1. $\sin u \cos v - \cos u \sin v$ **3.** $\frac{\tan u + \tan v}{1 - \tan u \tan v}$

5. $\cos u \cos v + \sin u \sin v$

7. *Sample answer:* $\sin(45° + 30°)$

9. (a) $-\frac{1}{2}$ (b) $-\frac{3}{2}$ **11.** (a) -1 (b) $\frac{1 + \sqrt{3}}{2}$

13. $\sin 105° = \frac{\sqrt{2} + \sqrt{6}}{4}$
$\cos 105° = \frac{\sqrt{2} - \sqrt{6}}{4}$
$\tan 105° = -2 - \sqrt{3}$

15. $\sin 195° = \frac{\sqrt{2} - \sqrt{6}}{4}$
$\cos 195° = -\frac{\sqrt{2} + \sqrt{6}}{4}$
$\tan 195° = 2 - \sqrt{3}$

17. $\sin\left(-\frac{11\pi}{12}\right) = \frac{\sqrt{2} - \sqrt{6}}{4}$
$\cos\left(-\frac{11\pi}{12}\right) = -\frac{\sqrt{2} + \sqrt{6}}{4}$
$\tan\left(-\frac{11\pi}{12}\right) = 2 - \sqrt{3}$

19. $\sin\left(-\frac{5\pi}{12}\right) = -\frac{\sqrt{2} + \sqrt{6}}{4}$
$\cos\left(-\frac{5\pi}{12}\right) = \frac{\sqrt{6} - \sqrt{2}}{4}$
$\tan\left(-\frac{5\pi}{12}\right) = -2 - \sqrt{3}$

21. $\sin \frac{13\pi}{12} = \frac{\sqrt{2} - \sqrt{6}}{4}$

$\cos \frac{13\pi}{12} = -\frac{\sqrt{2} + \sqrt{6}}{4}$

$\tan \frac{13\pi}{12} = 2 - \sqrt{3}$

23. $\sin 285° = -\frac{\sqrt{2} + \sqrt{6}}{4}$

$\cos 285° = \frac{\sqrt{6} - \sqrt{2}}{4}$

$\tan 285° = -2 - \sqrt{3}$

25. $\sin(-105°) = -\frac{\sqrt{2} + \sqrt{6}}{4}$

$\cos(-105°) = \frac{\sqrt{2} - \sqrt{6}}{4}$

$\tan(-105°) = 2 + \sqrt{3}$

27. $\sin 40°$ **29.** $\tan 209°$ **31.** $\cos \frac{13\pi}{36}$

33. $\sin(3.5x + 1.2y)$ **35.** $\frac{\sqrt{3}}{2}$ **37.** $\frac{\sqrt{3}}{2}$ **39.** 0

41.

x	0.2	0.4	0.6	0.8
y_1	0.9605	0.8484	0.6812	0.4854
y_2	0.9605	0.8484	0.6812	0.4854

x	1.0	1.2	1.4
y_1	0.2919	0.1313	0.0289
y_2	0.2919	0.1313	0.0289

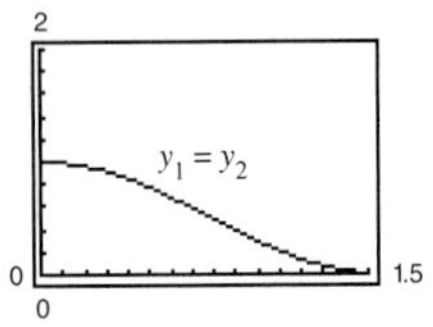

43. $-\frac{13}{85}$ **45.** $-\frac{13}{84}$ **47.** $\frac{4}{3}$ **49.** $\frac{44}{125}$ **51.** 1

53. $\frac{2x^2 - \sqrt{1 - x^2}}{\sqrt{4x^2 + 1}}$ **55.** $\frac{24}{25}$ **57–63.** Proofs

65. $-\sin \theta$ **67.** $-\sec \theta$ **69.** $\frac{\pi}{2}$

71. $0, \frac{\pi}{3}, \pi, \frac{5\pi}{3}$ **73.** $\frac{\pi}{4}, \frac{7\pi}{4}$ **75.** $\frac{\pi}{2}, \pi, \frac{3\pi}{2}$

77. Answers will vary.

79. False; $\cos(u \pm v) = \cos u \cos v \mp \sin u \sin v$

81. (a) All real numbers h except $h = 0$

(b)

h	0.01	0.02	0.05
$f(h)$	0.4957	0.4913	0.4781
$g(h)$	0.4957	0.4913	0.4781

h	0.1	0.2	0.5
$f(h)$	0.4559	0.4104	0.2674
$g(h)$	0.4559	0.4104	0.2674

(c)

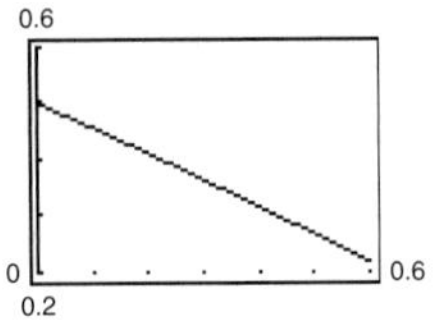

(d) $f \to \frac{1}{2}$ as $h \to 0$

$g \to \frac{1}{2}$ as $h \to 0$

83 and 85. Answers will vary.

87. (a) $\sqrt{2} \sin\left(\theta + \frac{\pi}{4}\right)$ (b) $\sqrt{2} \cos\left(\theta - \frac{\pi}{4}\right)$

89. (a) $13 \sin(3\theta + 0.3948)$ (b) $13 \cos(3\theta - 1.1760)$

91. $2 \cos \theta$ **93.** 15° **95.** $u + v = w$; Answers will vary.

97. (0, 19), (38, 0) **99.** (0, 4), (2, 0), (7, 0)

Section 5.5 (page 392)

1. $\frac{1 + \cos 2u}{2}$ **3.** $-2 \sin\left(\frac{u + v}{2}\right) \sin\left(\frac{u - v}{2}\right)$

5. $\tan \frac{u}{2}$ **7.** (a) ii (b) i (c) iii

9. (a) $\frac{3}{5}$ (b) $\frac{24}{25}$ (c) $\frac{4}{5}$ (d) $\frac{7}{25}$ (e) $\frac{3}{4}$ (f) $\frac{24}{7}$ (g) $\frac{5}{3}$
(h) $\frac{25}{24}$ (i) $\frac{5}{4}$ (j) $\frac{25}{7}$ (k) $\frac{4}{3}$ (l) $\frac{7}{24}$

11. $n\pi, \frac{\pi}{3} + 2n\pi, \frac{5\pi}{3} + 2n\pi$ **13.** $\frac{\pi}{12} + n\pi, \frac{5\pi}{12} + n\pi$

15. $\frac{2n\pi}{3}$ **17.** $\frac{n\pi}{2}$ **19.** $\frac{\pi}{6} + n\pi, \frac{\pi}{2} + n\pi, \frac{5\pi}{6} + n\pi$

21. $\sin 2u = -\frac{24}{25}$

$\cos 2u = \frac{7}{25}$

$\tan 2u = -\frac{24}{7}$

23. $\sin 2u = \frac{4}{5}$

$\cos 2u = \frac{3}{5}$

$\tan 2u = \frac{4}{3}$

25. $\sin 2u = -\frac{\sqrt{3}}{2}$

$\cos 2u = -\frac{1}{2}$

$\tan 2u = \sqrt{3}$

27. $\frac{1}{4 \cos^3 x - 3 \cos x}$ **29.** $3 \sin 2x$ **31.** $\frac{1}{2} \cos 2x$

33. $\frac{1 - \cos 2x}{2}$ **35.** $\frac{\cos x + \cos x \cos 2x}{2}$

37. $\frac{1}{8}(3 + 4 \cos 2x + \cos 4x)$

39. $\frac{1}{32}(2 + \cos 2x - 2 \cos 4x - \cos 6x)$ **41.** $\frac{1}{8}(1 - \cos 4x)$

43. $\frac{1}{2}(1 + \cos x)$ **45.** $\frac{1}{2}(1 - \cos 2x)$

47. (a) $\frac{\sqrt{17}}{17}$ (b) $\sqrt{17}$ (c) $\frac{4\sqrt{17}}{17}$ (d) $\frac{\sqrt{17}}{4}$
(e) $\frac{1}{4}$ (f) 4 (g) $\frac{8}{17}$ (h) $\frac{2\sqrt{17}}{17}$

49. $\sin 75° = \frac{\sqrt{2 + \sqrt{3}}}{2}$

$\cos 75° = \frac{\sqrt{2 - \sqrt{3}}}{2}$

$\tan 75° = \sqrt{3} + 2$

51. $\sin 157° 30' = \frac{\sqrt{2 - \sqrt{2}}}{2}$

$\cos 157° 30' = -\frac{\sqrt{2 + \sqrt{2}}}{2}$

$\tan 157° 30' = 1 - \sqrt{2}$

53. $\sin \frac{\pi}{8} = \frac{\sqrt{2 - \sqrt{2}}}{2}$

$\cos \frac{\pi}{8} = \frac{\sqrt{2 + \sqrt{2}}}{2}$

$\tan \frac{\pi}{8} = \sqrt{2} - 1$

55. $\sin \frac{7\pi}{12} = \frac{\sqrt{2 + \sqrt{3}}}{2}$

$\cos \frac{7\pi}{12} = -\frac{\sqrt{2 - \sqrt{3}}}{2}$

$\tan \frac{7\pi}{12} = -2 - \sqrt{3}$

57. $\sin \frac{u}{2} = \frac{3}{5}$

$\cos \frac{u}{2} = \frac{4}{5}$

$\tan \frac{u}{2} = \frac{3}{4}$

59. $\sin \frac{u}{2} = \frac{\sqrt{50 + 15\sqrt{10}}}{10}$

$\cos \frac{u}{2} = -\frac{\sqrt{50 - 15\sqrt{10}}}{10}$

$\tan \frac{u}{2} = -\sqrt{10} - 3$

61. $\sin \frac{u}{2} = \frac{3\sqrt{10}}{10}$

$\cos \frac{u}{2} = -\frac{\sqrt{10}}{10}$

$\tan \frac{u}{2} = -3$

63. $|\sin 3x|$ **65.** $-|\tan 4x|$ **67.** π

69. $\frac{\pi}{3}, \pi, \frac{5\pi}{3}$ **71.** $\frac{1}{2}(\cos 60° + \cos 270°)$

73. $3\left(\sin \frac{2\pi}{3} + \sin 0\right)$ **75.** $\frac{1}{2}(\cos 2\theta - \cos 8\theta)$

77. $\frac{1}{2}(\sin 2x + \sin 2y)$ **79.** $2 \cos 4\theta \sin \theta$

81. $2 \cos 4x \cos 2x$ **83.** $2 \cos \alpha \sin \beta$ **85.** $-2 \sin \theta \sin \frac{\pi}{2}$

87. $\frac{\sqrt{6}}{2}$ **89.** $-\sqrt{2}$ **91.** $0, \frac{\pi}{4}, \frac{\pi}{2}, \frac{3\pi}{4}, \pi, \frac{5\pi}{4}, \frac{3\pi}{2}, \frac{7\pi}{4}$

93. $\frac{\pi}{6}, \frac{5\pi}{6}$ **95.** $\frac{25}{169}$ **97.** $\frac{4}{13}$

99–111. Answers will vary. **113.** $2x\sqrt{1 - x^2}$

115. $1 - 2x^2$ **117.** $\frac{2x}{x^2 + 1}$

119. (a) $x = 0, \frac{\pi}{3}, \pi, \frac{5\pi}{3}, 2\pi$

(b) $x = \arccos \frac{1 \pm \sqrt{33}}{8}, 2\pi - \arccos \frac{1 \pm \sqrt{33}}{8}$

121. About 23.85° **123.** $x = 2r(1 - \cos \theta)$

125. False; $\sin \frac{x}{2} = -\sqrt{\frac{1 - \cos x}{2}}$ for $\pi \le \frac{x}{2} \le 2\pi$.

127. (a) $f(x) = \frac{3}{4} + \frac{\cos 4x}{4}$

(b) Answers will vary.

Sample answer: $f(x) = \frac{1}{2}(1 + \cos^2 2x)$

(c)–(e) Answers will vary.

129. sin 75° should be positive.

131. (a)

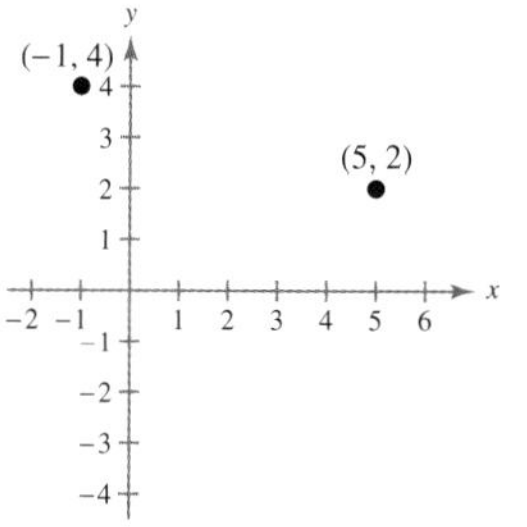

(b) $2\sqrt{10}$ (c) $(2, 3)$

133. About 12.8282 cm

Review Exercises (page 396)

1. $-\cot x$ **3.** $\sin x$ **5.** $|\csc x|$

7. $\sin x = -\frac{4}{5}$
$\cos x = -\frac{3}{5}$
$\csc x = -\frac{5}{4}$
$\sec x = -\frac{5}{3}$
$\cot x = \frac{3}{4}$

9. $\sin \theta = \frac{\sqrt{6}}{3}$
$\cos \theta = -\frac{\sqrt{3}}{3}$
$\tan \theta = -\sqrt{2}$
$\csc \theta = \frac{\sqrt{6}}{2}$
$\cot \theta = -\frac{\sqrt{2}}{2}$

11. $\tan x = \frac{4}{3}$
$\csc x = \frac{5}{4}$
$\sec x = \frac{5}{3}$
$\cot x = \frac{3}{4}$

13. $\tan x = -1$
$\csc x = -\sqrt{2}$
$\sec x = \sqrt{2}$
$\cot x = -1$

15. $\cos^2 x$ **17.** $-\sin \theta$ **19.** 1 **21.** $\csc x$

23. $\sec x + 1$ **25–37.** Answers will vary.

39. $\frac{3\pi}{2} + 2n\pi$ **41.** $\frac{\pi}{3} + 2n\pi, \frac{2\pi}{3} + 2n\pi$ **43.** $\frac{\pi}{3} + n\pi$

45. $\frac{\pi}{3} + n\pi, \frac{2\pi}{3} + n\pi$ **47.** $\frac{\pi}{3} + 2n\pi, \frac{\pi}{2} + n\pi, \frac{5\pi}{3} + 2n\pi$

49. $n\pi$ **51.** $0, \frac{2\pi}{3}, \frac{4\pi}{3}$ **53.** $\frac{3\pi}{2}$ **55.** $\frac{\pi}{8}, \frac{3\pi}{8}, \frac{9\pi}{8}, \frac{11\pi}{8}$

57. $\frac{\pi}{3}, \frac{5\pi}{3}$ **59.** $\frac{7\pi}{12} + n\pi, \frac{11\pi}{12} + n\pi$ **61.** $\frac{\pi}{12} + \frac{n\pi}{6}$

63. $n\pi, \arctan 2 + n\pi$

65. $\frac{3\pi}{2} + 2n\pi, \arcsin \frac{2}{3} + 2n\pi, -\arcsin \frac{2}{3} + (2n + 1)\pi$

67. $\sin 285° = -\frac{\sqrt{2} + \sqrt{6}}{4}$
$\cos 285° = \frac{\sqrt{6} - \sqrt{2}}{4}$
$\tan 285° = -2 - \sqrt{3}$

69. $\sin \frac{31\pi}{12} = \frac{\sqrt{2} + \sqrt{6}}{4}$
$\cos \frac{31\pi}{12} = \frac{\sqrt{2} - \sqrt{6}}{4}$
$\tan \frac{31\pi}{12} = -2 - \sqrt{3}$

71. $\sin(-15°)$ **73.** $\tan 75°$ **75.** $-\frac{4}{5}$ **77.** $-\frac{44}{117}$

79. $\frac{3}{5}$ **81–85.** Proofs **87.** $\frac{\pi}{4}, \frac{7\pi}{4}$

89. $\sin 2u = \frac{20\sqrt{14}}{81}$
$\cos 2u = \frac{31}{81}$
$\tan 2u = \frac{20\sqrt{14}}{31}$

91. $\sin 2u = -\frac{35}{85}$
$\cos 2u = \frac{77}{85}$
$\tan 2u = -\frac{36}{77}$

93 and 95. Answers will vary. **97.** $\frac{1 - \cos 8x}{1 + \cos 8x}$

99. $\frac{1}{128}(3 - 4 \cos 4x + \cos 8x)$

101. $\sin 15° = \frac{\sqrt{2 - \sqrt{3}}}{2}$
$\cos 15° = \frac{\sqrt{2 + \sqrt{3}}}{2}$
$\tan 15° = 2 - \sqrt{3}$

103. $\sin \frac{7\pi}{8} = \frac{\sqrt{2 - \sqrt{2}}}{2}$
$\cos \frac{7\pi}{8} = -\frac{\sqrt{2 + \sqrt{2}}}{2}$
$\tan \frac{7\pi}{8} = 1 - \sqrt{2}$

105. $\sin \frac{u}{2} = \frac{2\sqrt{13}}{13}$
$\cos \frac{u}{2} = \frac{3\sqrt{13}}{13}$
$\tan \frac{u}{2} = \frac{2}{3}$

107. $\sin \frac{u}{2} = \frac{3\sqrt{14}}{14}$
$\cos \frac{u}{2} = \frac{\sqrt{70}}{14}$
$\tan \frac{u}{2} = \frac{3\sqrt{5}}{5}$

109. $-|\cos 4x|$ **111.** $\tan 5x$ **113.** $2\left(\sin \frac{\pi}{2} + \sin 0\right)$

115. $\frac{1}{2}(\cos \alpha - \cos 9\alpha)$ **117.** $2 \cos \frac{5\theta}{2} \cos \frac{\theta}{2}$

119. $2 \cos x \sin \frac{\pi}{4}$ **121.** False; $\cos \frac{\theta}{2} > 0$

123. True; Answers will vary. **125.** $y_3 = y_2 + 1$

127. Answers will vary.

129. No; $\sin \theta = \frac{1}{2}$ has an infinite number of solutions but is not an identity.

131. $y = \frac{1}{2}\sqrt{10} \sin\left(8t - \arctan \frac{1}{2}\right)$

Chapter Test (page 401)

1. $\sin\theta = \dfrac{-3\sqrt{13}}{13}$ **2.** 1 **3.** 1 **4.** $\csc\theta\sec\theta$

$\cos\theta = \dfrac{-2\sqrt{13}}{13}$

$\csc\theta = \dfrac{-\sqrt{13}}{3}$

$\sec\theta = \dfrac{-\sqrt{13}}{2}$

$\cot\theta = \dfrac{2}{3}$

5.

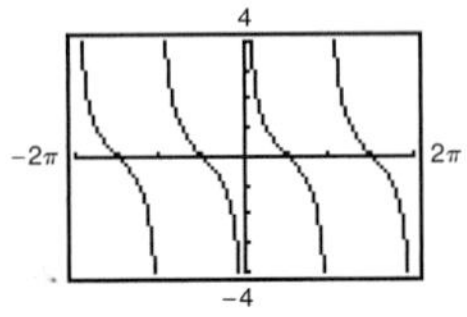

$y_1 = y_2$

6. $6\cos\theta$ **7–12.** Answers will vary. **13.** $\dfrac{\pi}{3}, \dfrac{5\pi}{3}$

14. $\dfrac{\pi}{6}, \dfrac{5\pi}{6}, \dfrac{7\pi}{6}, \dfrac{11\pi}{6}$ **15.** $0, \dfrac{3\pi}{4}, \pi, \dfrac{7\pi}{4}$ **16.** $\dfrac{\pi}{6}, \dfrac{5\pi}{6}, \dfrac{3\pi}{2}$

17. About 0.8922, about 5.3910 **18.** $\dfrac{\pi}{6}, \dfrac{\pi}{2}, \dfrac{5\pi}{6}, \dfrac{3\pi}{2}$

19. $\dfrac{\sqrt{2}-\sqrt{6}}{4}$ **20.** $\dfrac{1}{16}\left(\dfrac{10 - 15\cos 2x + 6\cos 4x - \cos 6x}{1 + \cos 2x}\right)$

21. $\tan 2\theta$ **22.** $2(\sin 6\theta + \sin 2\theta)$ **23.** $-2\sin 2\theta \sin\theta$

24. $-2.596, 0, 2.596$ **25.** About 76.52°

Chapter 6

Section 6.1 (page 412)

1. $\dfrac{b}{\sin B}$ **3.** $\dfrac{1}{2}bc\sin A$; $\dfrac{1}{2}ab\sin C$; $\dfrac{1}{2}ac\sin B$

5. AAS (or ASA) and SSA

7. $C = 95°$, $b \approx 24.59$ in., $c \approx 28.29$ in.

9. $C = 120°$, $a \approx 16.33$ cm, $b \approx 5.98$ cm

11. $C = 74° 15'$, $a \approx 6.41$ km, $c \approx 6.26$ km

13. $B \approx 21.55°$, $C \approx 122.45°$, $c \approx 11.49$

15. $C = 105°$, $a \approx 5.94$, $b \approx 6.65$

17. $B \approx 48.74°$, $C \approx 21.26°$, $c \approx 48.23$

19. $B = 42° 4'$, $a \approx 22.05$, $b \approx 14.88$

21. $A = 48°$, $b \approx 2.29$, $c \approx 4.73$

23. $B \approx 18° 13'$, $C \approx 51° 32'$, $c \approx 40.05$

25. No solution **27.** No solution

29. Two solutions

$B \approx 72.21°$, $C \approx 49.79°$, $c \approx 10.27$

$B \approx 107.79°$, $C \approx 14.21°$, $c \approx 3.30$

31. (a) $b \le 5$ or $b = \dfrac{5}{\sin 36°}$ (b) $5 < b < \dfrac{5}{\sin 36°}$

(c) $b > \dfrac{5}{\sin 36°}$

33. (a) $b \le 10.8$ or $b = \dfrac{10.8}{\sin 10°}$ (b) $10.8 < b < \dfrac{10.8}{\sin 10°}$

(c) $b > \dfrac{10.8}{\sin 10°}$

35. 28.19 square units **37.** 22.12 square units

39. 2888.57 square units **41.** 24.12 m **43.** About 160.39°

45. (a)

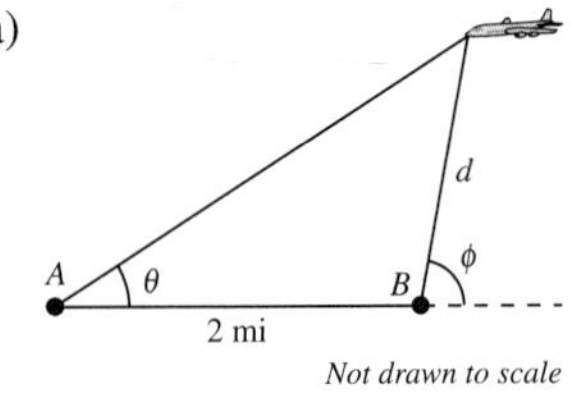

(b) $d = \dfrac{2\sin\theta}{\sin(\phi - \theta)}$ (c) About 3.76 mi

47. $\theta \approx 16.08°$

49. (a) About 1.66 nm (b) About 1.18 h

(c) $d_1 \approx 0.68$ nm; $d_2 \approx 0.78$ nm

51. False; The triangle cannot be solved if only three angles are known.

53. The formula is $A = \frac{1}{2}ab\sin C$, so solve the triangle to find the value of a.

55. The area of the acute triangle is four times the area of the obtuse triangle.

57. 1.40 **59.** −3.46

Section 6.2 (page 419)

1. $c^2 = a^2 + b^2 - 2ab\cos C$ **3.** standard

5. $A \approx 33.56°$, $B \approx 62.18°$, $C \approx 84.26°$

7. $A \approx 49.51°$, $B \approx 55.40°$, $C \approx 75.09°$

9. $B \approx 23.79°$, $C \approx 126.21°$, $a \approx 18.59$

11. $A \approx 30.11°$, $B \approx 43.16°$, $C \approx 106.73°$

13. $A \approx 29.44°$, $C \approx 100.56°$, $a \approx 23.38$

15. $B \approx 27.46°$, $C \approx 32.54°$, $a \approx 11.27$

17. $A \approx 103.52°$, $B \approx 38.24°$, $C \approx 38.24°$

19. $A \approx 154° 14'$, $C \approx 17° 31'$, $b \approx 8.58$

21. $A \approx 33.80°$, $B \approx 103.20°$, $c \approx 0.54$

23. $c \approx 12.07$, $d \approx 5.69$, $\phi = 135°$

25. $d \approx 13.86$, $\theta \approx 68.20°$, $\phi \approx 111.80°$

27. $b \approx 16.96$, $\theta \approx 77.22°$, $\phi \approx 102.78°$

29. Yes; $B \approx 48.57°$, $C \approx 91.43°$, $a \approx 12.86$ m

31. Yes; $A \approx 102.44°$, $C \approx 37.56°$, $b \approx 5.26$

33. No; no solution **35.** No; $C = 103°$, $a \approx 0.82$, $b \approx 0.71$

37. About 23.53 square units **39.** About 19.81 square units

41. About 0.27 square unit **43.** About 0.18 square unit

45. About 483.40 m **47.** (a) N 59.7° E (b) N 72.8° E

49. About 72.28° **51.** About 131.12 ft, about 118.56 ft

53. About 9655.79 ft²

55. (a) $49 = 2.25 + x^2 - 3x\cos\theta$

(b) $x = \frac{1}{2}\left(3\cos\theta + \sqrt{9\cos^2\theta + 187}\right)$

(c)

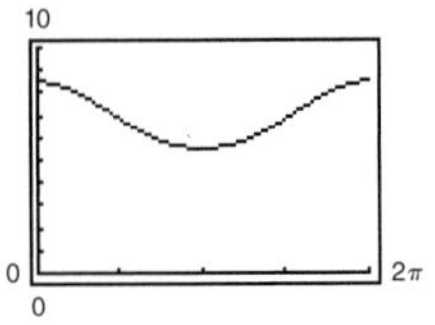

(d) 6 in.

57. True; The third side is found using the Law of Cosines and the other angles are found by the Law of Sines.

59. The values were substituted in the wrong order. It should be $\cos A = \dfrac{4^2 + 6^2 - 5^2}{(2)(4)(6)}$.

61. Proof

63. To solve the triangle using the Law of Cosines, substitute values into $a^2 = b^2 + c^2 - 2bc \cos A$.

Simplify the equation so that you have a quadratic equation in terms of c. Then, find the two values of c, and find the two triangles that model the given information.

Using the Law of Sines will give the same result as using the Law of Cosines.

Sample answer: An advantage of using the Law of Cosines is that it is easier to choose the correct value to avoid the ambiguous case, but its disadvantage is that there are more computations. The opposite is true for the Law of Sines.

65. $-\frac{\pi}{2}$ **67.** $\frac{\pi}{6}$

Section 6.3 (page 431)

1. directed line segment **3.** magnitude
5. standard position **7.** resultant
9. magnitude and direction **11.** scalar **13.** vector
15. scalar
17. Equivalent; **u** and **v** have the same magnitude and direction.
19. Not equivalent; **u** and **v** do not have the same direction.
21. Equivalent; **u** and **v** have the same magnitude and direction.
23. $\langle 1, 3\rangle$, $\|\mathbf{v}\| = \sqrt{10}$ **25.** $\langle 5, 0\rangle$, $\|\mathbf{v}\| = 5$
27. $\langle -8, 6\rangle$, $\|\mathbf{v}\| = 10$ **29.** $\langle -3.6, -3.6\rangle$, $\|\mathbf{v}\| = \frac{18}{5}\sqrt{2}$

31.

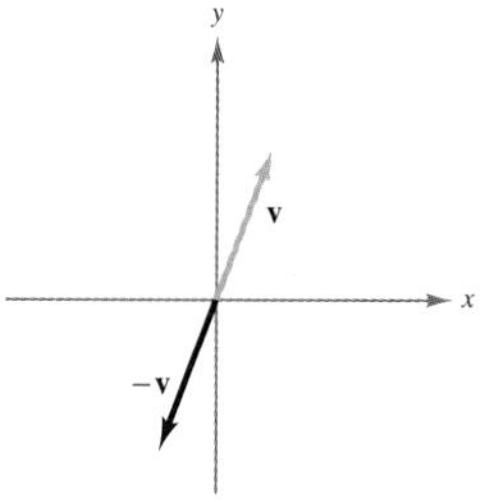

33.

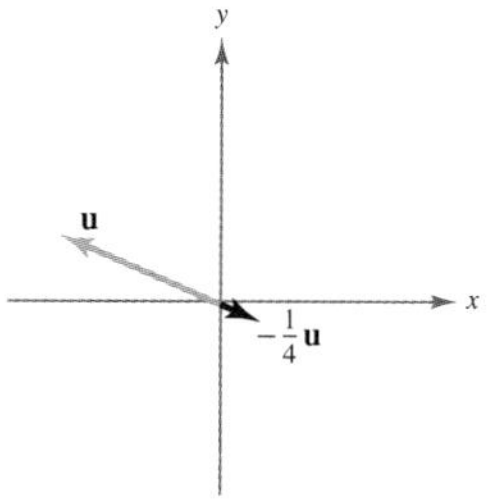

35.

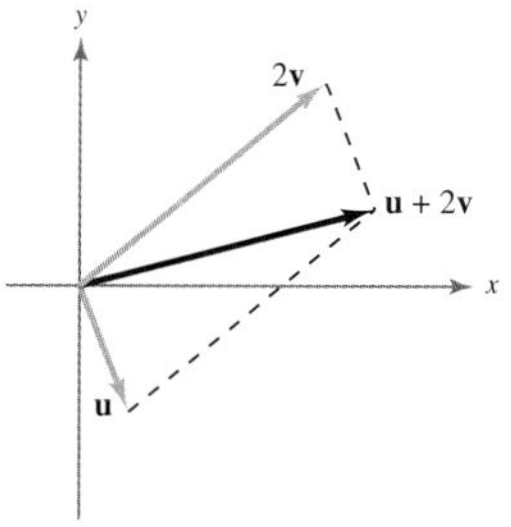

37.

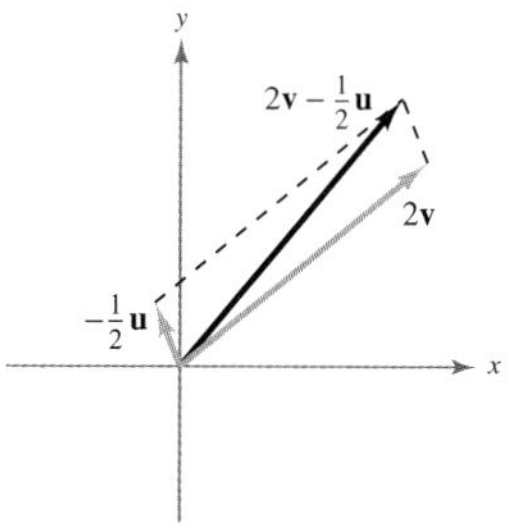

39. (a) $\langle 3, 4\rangle$

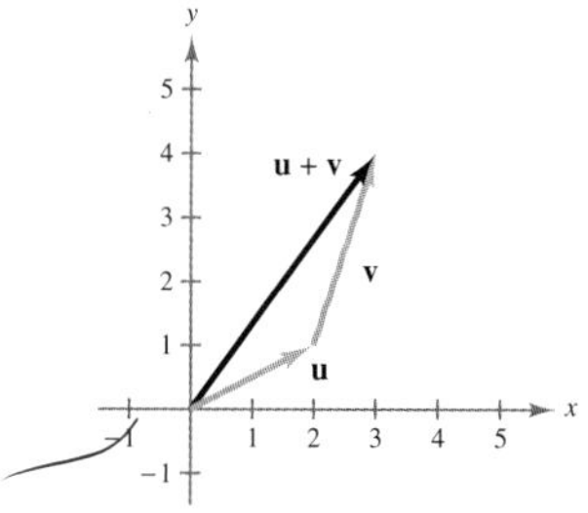

(b) $\langle 1, -2\rangle$

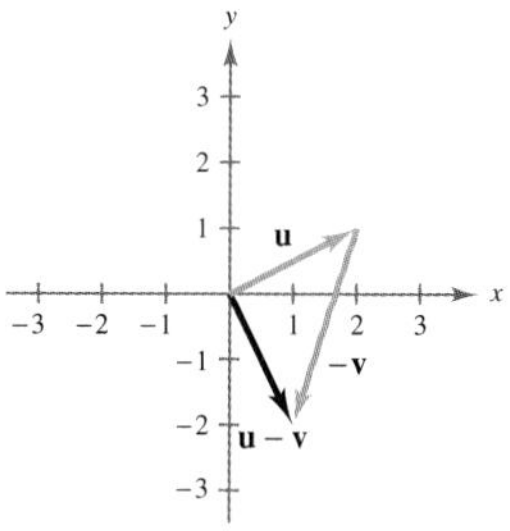

(c) $\langle 1, -7\rangle$

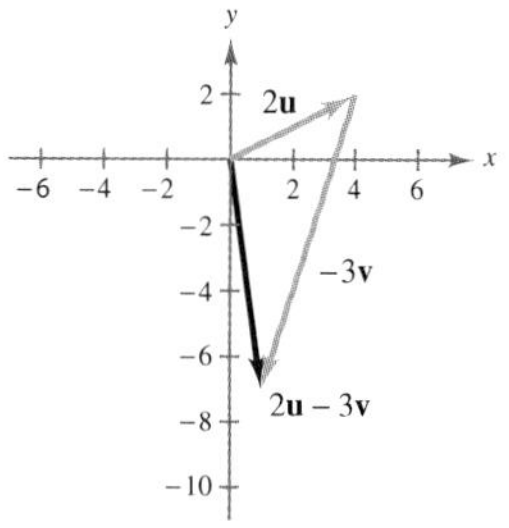

41. (a) $\langle -5, 3\rangle$

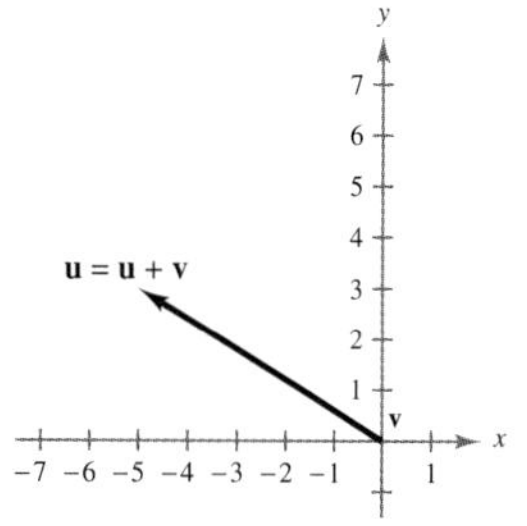

(b) $\langle -5, 3\rangle$

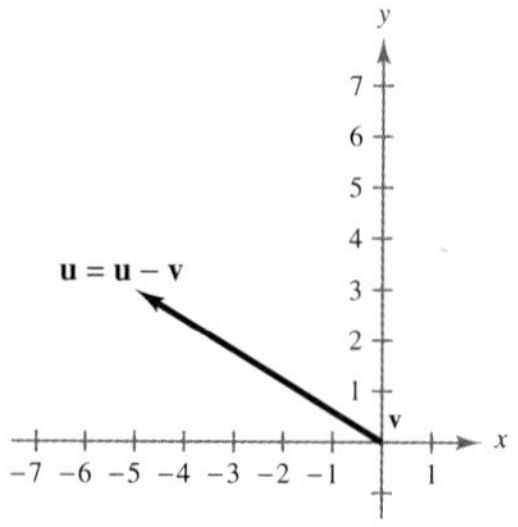

(c) $\langle -10, 6\rangle$

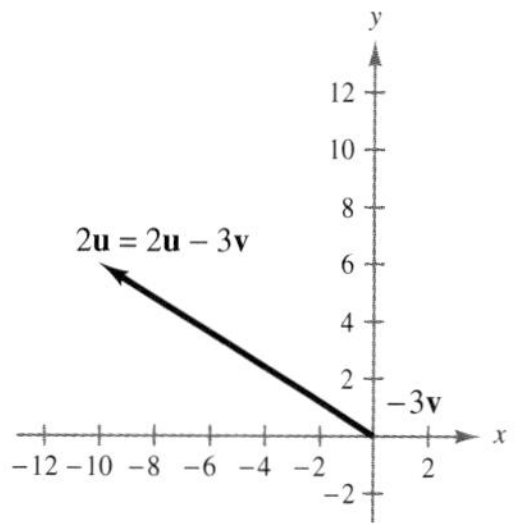

43. (a) $\langle 1, -9\rangle$

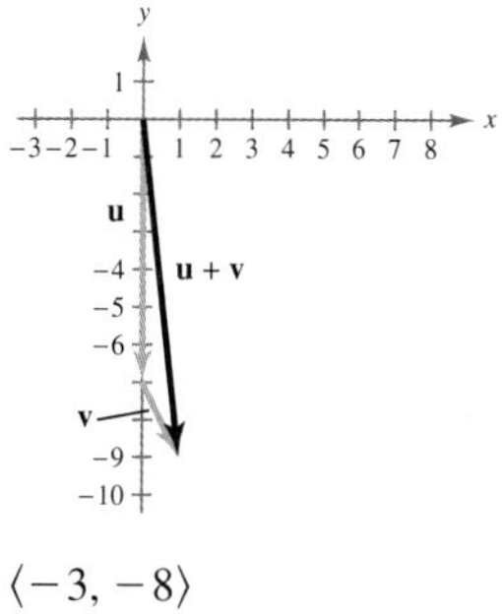

(b) $\langle -1, -5\rangle$

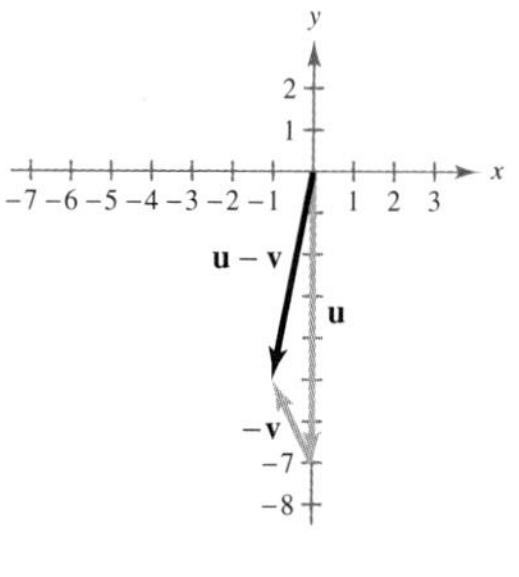

(c) $\langle -3, -8\rangle$

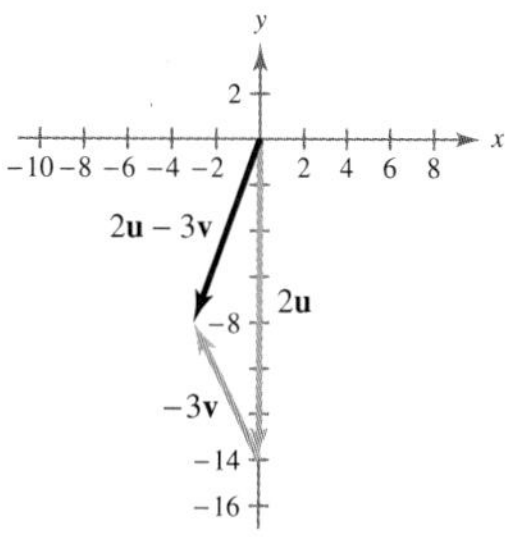

45. $\mathbf{u} + \mathbf{v}$ **47.** $\mathbf{w} - \mathbf{v}$ **49.** 10 **51.** $9\sqrt{5}$
53. $\langle 1, 0\rangle$ **55.** $\left\langle -\frac{\sqrt{2}}{2}, \frac{\sqrt{2}}{2}\right\rangle$ **57.** $\left\langle -\frac{24}{25}, -\frac{7}{25}\right\rangle$
59. $\frac{4}{5}\mathbf{i} - \frac{3}{5}\mathbf{j}$ **61.** $\mathbf{j}$ **63.** $\langle -6, 8\rangle$ **65.** $\frac{21}{5}\mathbf{i} + \frac{28}{5}\mathbf{j}$
67. $5\mathbf{i} - 3\mathbf{j}$ **69.** $6\mathbf{i} - 3\mathbf{j}$

71. $\mathbf{v} = \langle 3, -\frac{3}{2} \rangle$

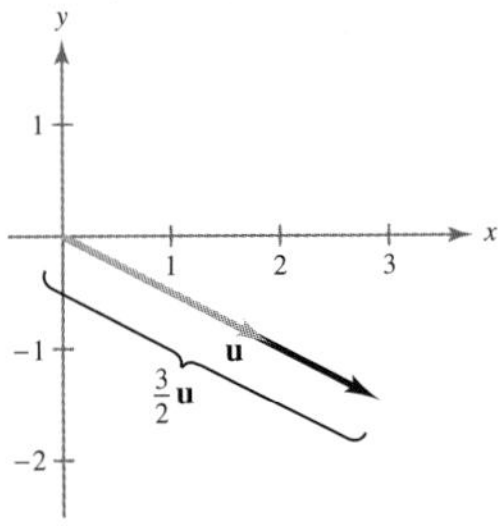

73. $\mathbf{v} = \langle 4, 3 \rangle$

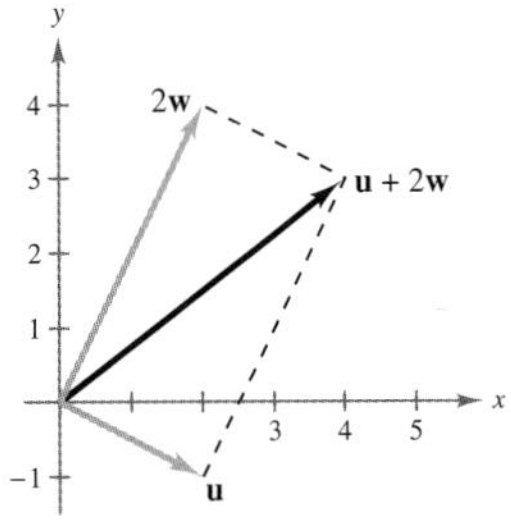

75. $\mathbf{v} = \langle \frac{7}{2}, -\frac{1}{2} \rangle$

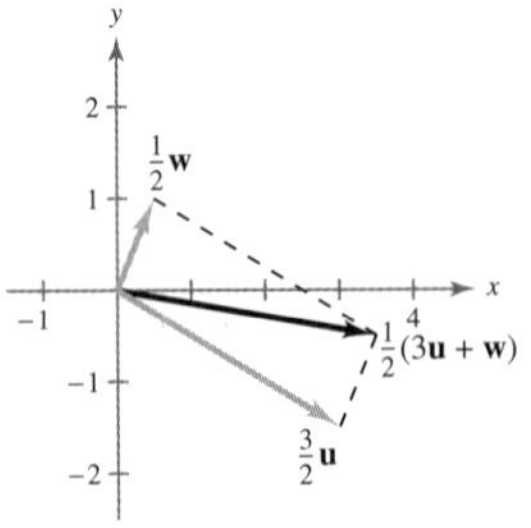

77. $\|\mathbf{v}\| = 5, \theta = 60°$ **79.** $\|\mathbf{v}\| = 6\sqrt{2}, \theta = 315°$

81. $\|\mathbf{v}\| = \sqrt{29}, \theta \approx 111.80°$

83. $\mathbf{v} = \langle 3, 0 \rangle$

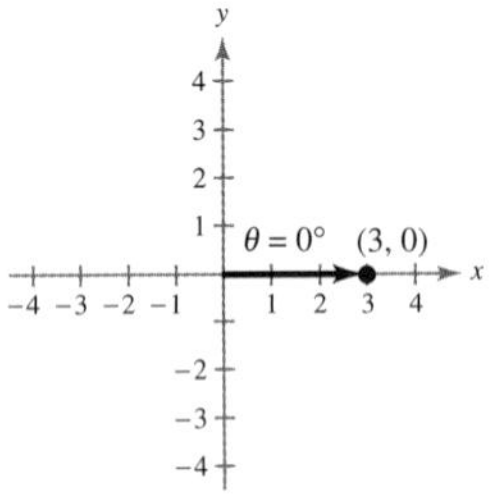

85. $\mathbf{v} = \left\langle -\frac{7}{4}, \frac{7\sqrt{3}}{4} \right\rangle$

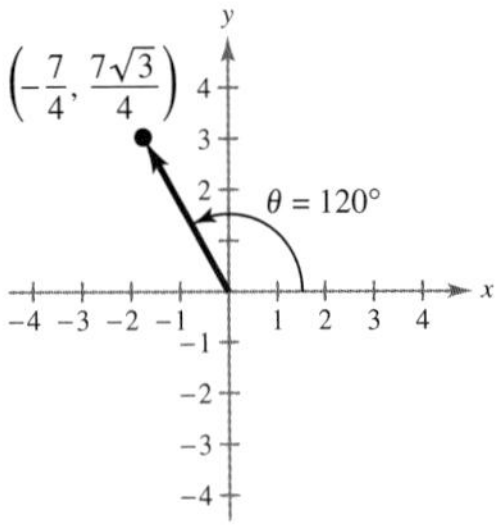

87. $\mathbf{v} = \left\langle -\frac{3\sqrt{6}}{2}, \frac{3\sqrt{2}}{2} \right\rangle$

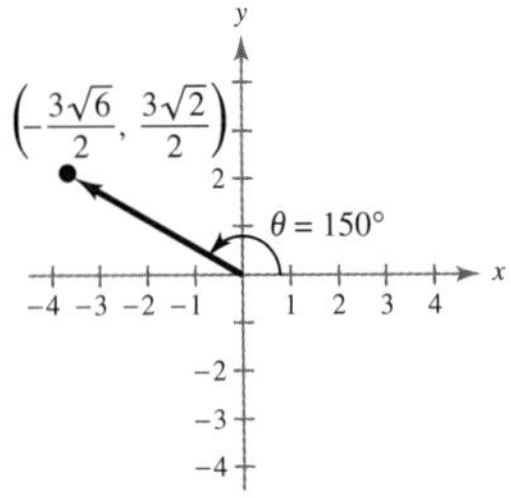

89. $\mathbf{v} = \left\langle \frac{\sqrt{10}}{5}, \frac{3\sqrt{10}}{5} \right\rangle$

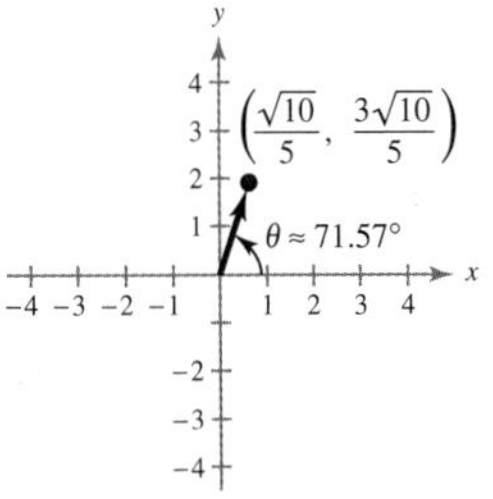

91. $\langle 5, 5 \rangle$ **93.** $\langle -32.84, 31.24 \rangle$ **95.** 90°

97.

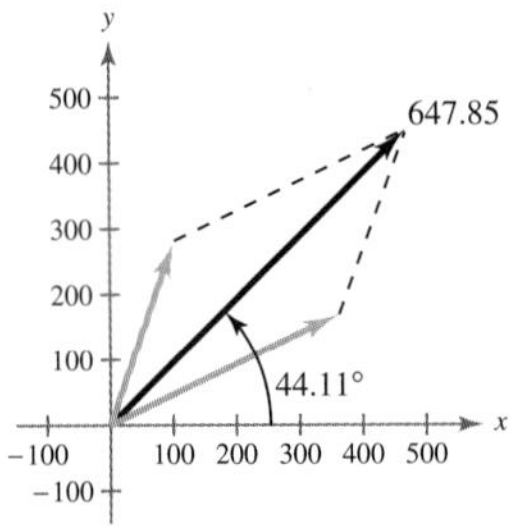

$\|\mathbf{v}\| \approx 647.85, \theta \approx 44.11°$

99. About 62.72°

101. Vertical component: about 45.00 ft/sec
Horizontal component: about 53.62 ft/sec

103. About 19,785.83 lb, about 15,484.09 lb

105. (a) $T = 3000 \sec \theta$; Domain: $0° \le \theta < 90°$

(b)

θ	10°	20°	30°
T	3046.28	3192.53	3464.10

θ	40°	50°	60°
T	3916.22	4667.17	6000

(c)

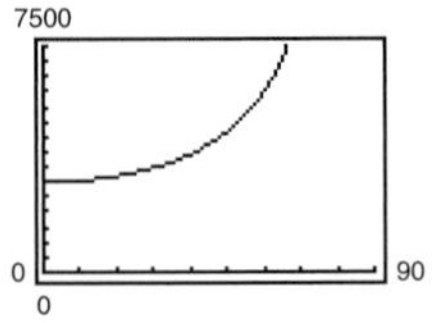

(d) The component in the direction of the motion of the barge decreases.

107. (a) About 12.10°, about 357.85 N

(b) $M = 10\sqrt{660 \cos \theta + 709}$

$\alpha = \arctan \dfrac{15 \sin \theta}{15 \cos \theta + 22}$

(c)

θ	0°	30°	60°	90°
M	370	357.85	322.34	266.27
α	0°	12.10°	23.77°	34.29°

θ	120°	150°	180°
M	194.68	117.23	70
α	41.86°	39.78°	0°

(d)

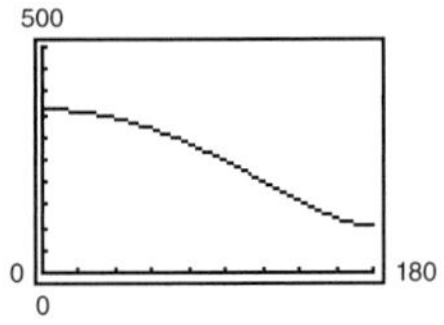

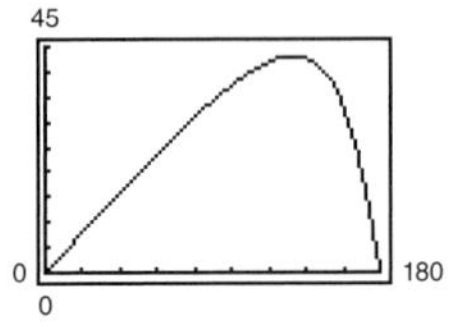

(e) For increasing θ, the two vectors tend to work against each other, resulting in a decrease in the magnitude of the resultant.

109. N 26.67° E, about 130.35 km/h

111. True; The vectors are equivalent. **113.** True; $a = b = 0$

115. (a) 0° (b) 180°

(c) No; The magnitude is equal to the sum when the angle between the vectors is 0°.

117. Answers will vary. **119.** Proof

121. $\langle 1, 3 \rangle$ or $\langle -1, -3 \rangle$

123. $u_1 = 6 - (-3) = 9$ and $u_2 = -1 - 4 = -5$, so $\mathbf{u} = \langle 9, -5 \rangle$

125. $48xy^2, x \ne 0$ **127.** 7.14×10^5

129. $\frac{\pi}{2} + n\pi, \pi + 2n\pi$ **131.** $\frac{\pi}{3} + 2n\pi, \frac{5\pi}{3} + 2n\pi$

133. $10 \csc \theta$

Section 6.4 (page 442)

1. Yes **3.** Scalar **5.** $\left(\frac{\mathbf{u} \cdot \mathbf{v}}{\|\mathbf{v}\|^2}\right)\mathbf{v}$ **7.** -19 **9.** 14

11. 20; scalar **13.** $\langle 24, -12\rangle$; vector **15.** $\mathbf{0}$; vector

17. 60; scalar **19.** $\sqrt{10} - 1$; scalar **21.** -12; scalar

23. $\frac{\pi}{2}$ **25.** About 2.50 **27.** $\frac{\pi}{2}$ **29.** $\frac{5\pi}{12}$

31.

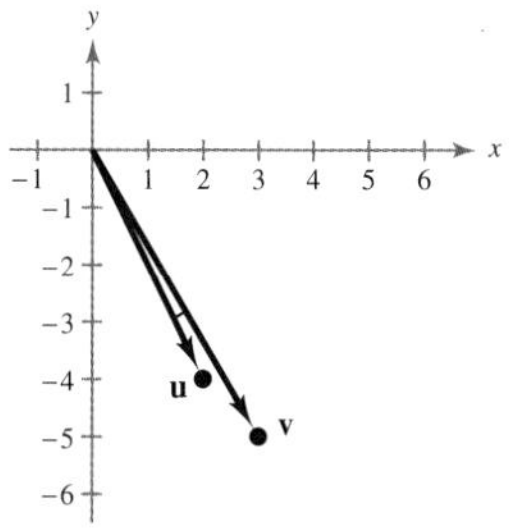

$\theta \approx 4.40°$

33.

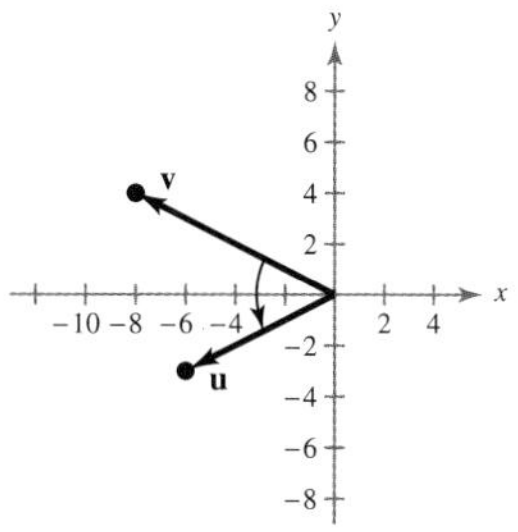

$\theta \approx 53.13°$

35. About 26.57°, about 63.43°, 90°

37. About 41.63°, about 53.13°, about 85.24°

39. $-162\sqrt{2}$ **41.** -20 **43.** Not orthogonal

45. Not orthogonal **47.** Orthogonal **49.** Parallel

51. 3 **53.** $\frac{20}{7}$ **55.** 0 **57.** $\langle 0, 0\rangle$, $\mathbf{u} = \langle 4, 2\rangle + \langle 0, 0\rangle$

59. $\frac{45}{229}\langle 2, 15\rangle$, $\mathbf{u} = \left\langle \frac{90}{229}, \frac{675}{229}\right\rangle + \left\langle -\frac{90}{229}, \frac{12}{229}\right\rangle$ **61.** $\mathbf{u}$

63. $\mathbf{0}$ **65.** $\langle 3, -1\rangle, \langle -3, 1\rangle$ **67.** $-\frac{3}{4}\mathbf{i} - \frac{1}{2}\mathbf{j}, \frac{3}{4}\mathbf{i} + \frac{1}{2}\mathbf{j}$

69. 32

71. (a) 39,761.25; It is the total dollar amount paid to the employees.
(b) Multiply $\mathbf{v}$ by 1.02.

73. (a) $30{,}000 \sin d$

(b)

d	0°	1°	2°	3°
Force	0	523.57	1046.98	1570.08

d	4°	5°	6°	7°
Force	2092.69	2614.67	3135.85	3656.08

d	8°	9°	10°
Force	4175.19	4693.03	5209.45

(c) About 29,885.84 lb

75. (a) $125\sqrt{3}d$

(b)

d	25	50	100
Work	5412.66	10,825.32	21,650.64

77. About 679.73 ft-lb

79. True; The zero vector is orthogonal to every vector.

81. Orthogonal; $\mathbf{u} \cdot \mathbf{v} = 0$ **83.** 1; $\|\mathbf{u}\| = 1$ and $\mathbf{u} \cdot \mathbf{u} = \|\mathbf{u}\|^2$.

85. (a) $\mathbf{u}$ and $\mathbf{v}$ are parallel. (b) $\mathbf{u}$ and $\mathbf{v}$ are orthogonal.

87 and 89. Proofs

91. g is a horizontal shift of f four units to the right.

93. g is a vertical shift of f six units upward.

95. 10 **97.** $\frac{47}{26} - \frac{27}{26}i$

Section 6.5 (page 449)

1. (a, b) **3.** absolute value **5.** c **6.** f **7.** h **8.** a

9. b **10.** g **11.** e **12.** d

13.

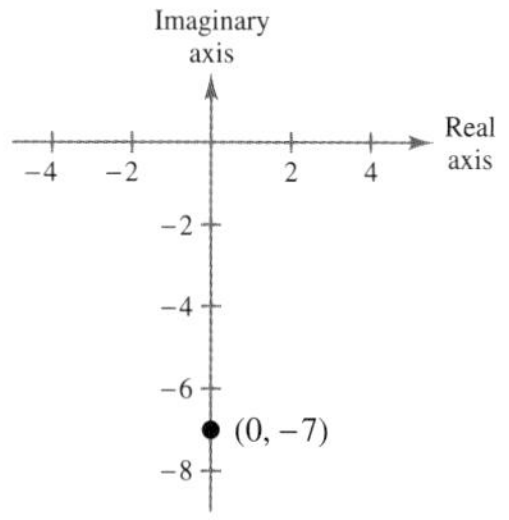

7

15.

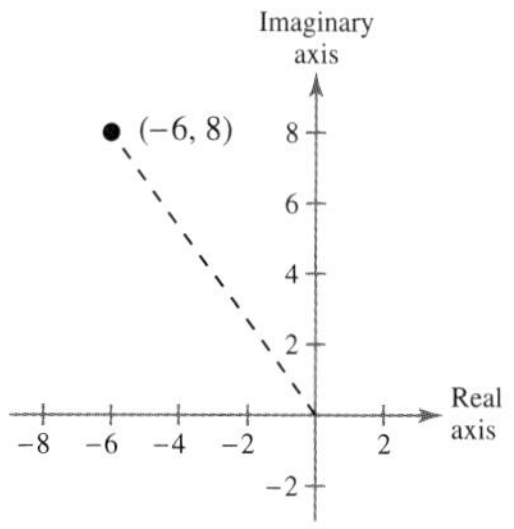

10

17.

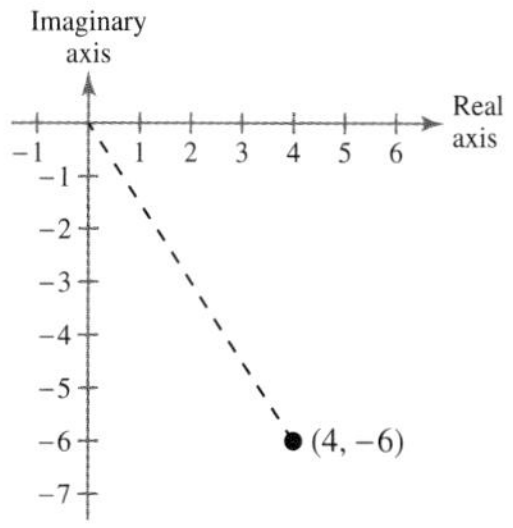

$2\sqrt{13}$

19. z_2 **21.** 4.26 **23.** 1.78 **25.** $5 + 6i$

27. $10 + 4i$ **29.** $5 - 8i$ **31.** $-2 - 2i$

33. $10 - 3i$ **35.** $-6i$

37.

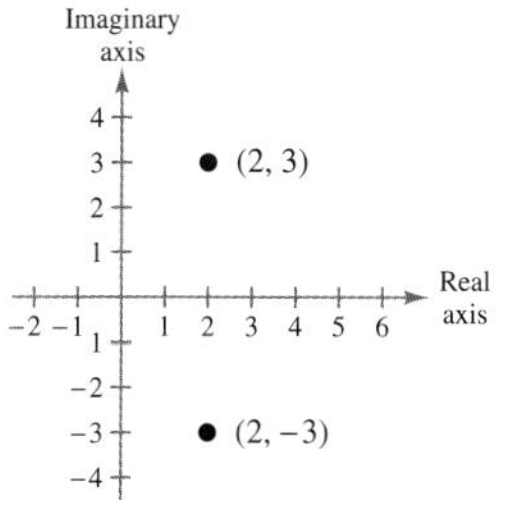

$2 - 3i$

39.

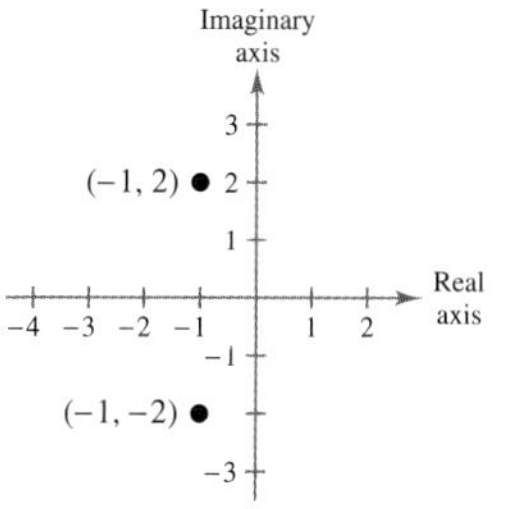

$-1 + 2i$

41. $2\sqrt{2} \approx 2.83$ **43.** $\sqrt{109} \approx 10.44$ **45.** $(4, 3)$

47. $\left(\frac{9}{2}, -\frac{3}{2}\right)$

49.

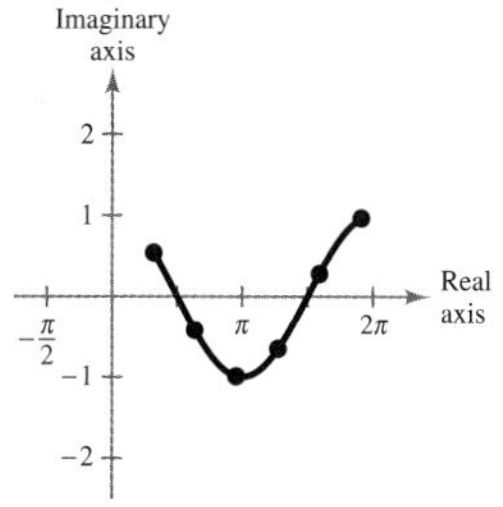

51.

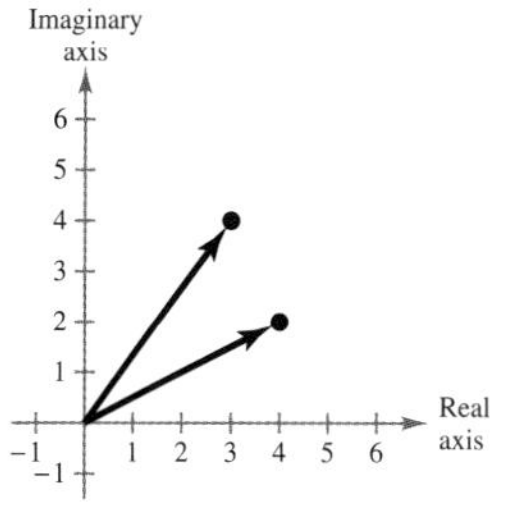

Horizontal: 7 N
Vertical: 6 N

53. False; The modulus is always real.

55. False; $|1 + i| + |1 - i| = 2\sqrt{2}$ and $|(1 + i) + (1 - i)| = 2$

57. Rotation 90° counterclockwise about the origin

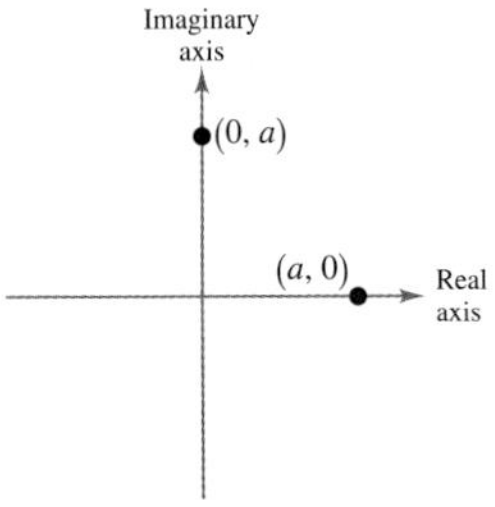

59. Reflection in the vertical axis, or rotation 180° about the origin

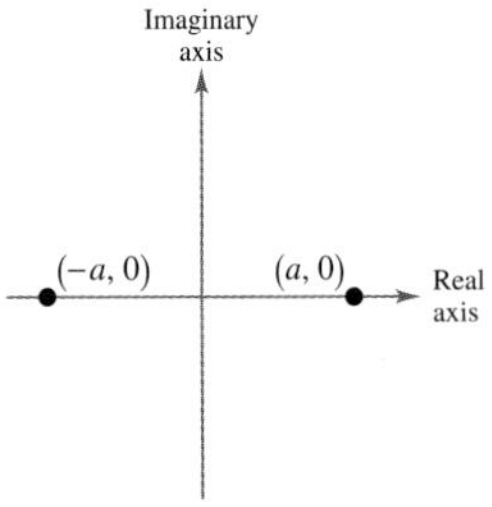

61. Circle **63.** Isosceles **65 and 67.** Answers will vary.

Section 6.6 (page 459)

1. trigonometric form; modulus; argument

3. divide; subtract **5.** *n*th root **7.** $2\left(\cos\frac{\pi}{2} + i\sin\frac{\pi}{2}\right)$

9. $3\sqrt{2}\left(\cos\frac{5\pi}{4} + i\sin\frac{5\pi}{4}\right)$ **11.** $2\left(\cos\frac{11\pi}{6} + i\sin\frac{11\pi}{6}\right)$

13.

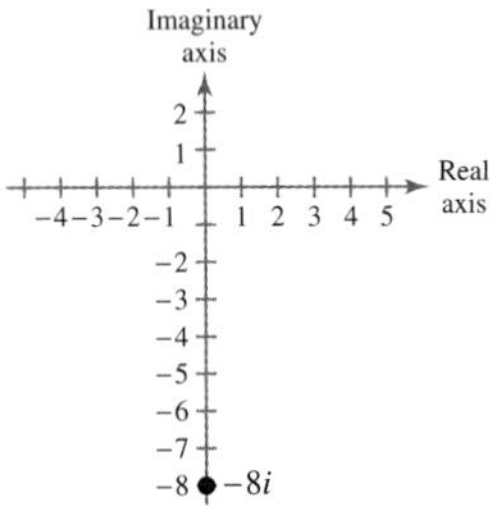

$8\left(\cos\frac{3\pi}{2} + i\sin\frac{3\pi}{2}\right)$

15.

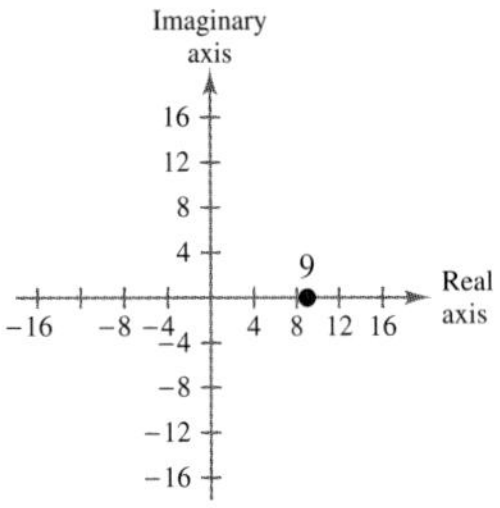

$9(\cos 0 + i\sin 0)$

17.

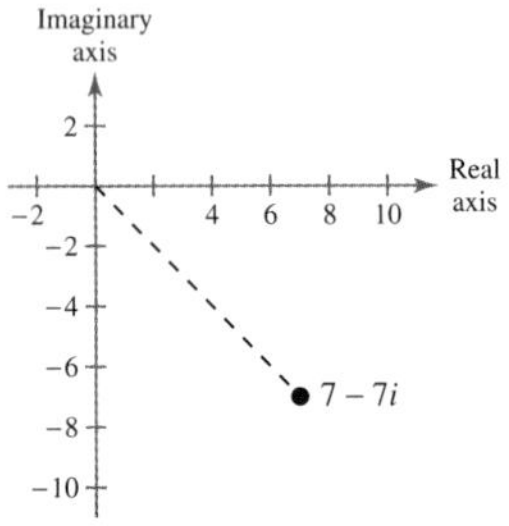

$7\sqrt{2}\left(\cos\frac{7\pi}{4} + i\sin\frac{7\pi}{4}\right)$

19.

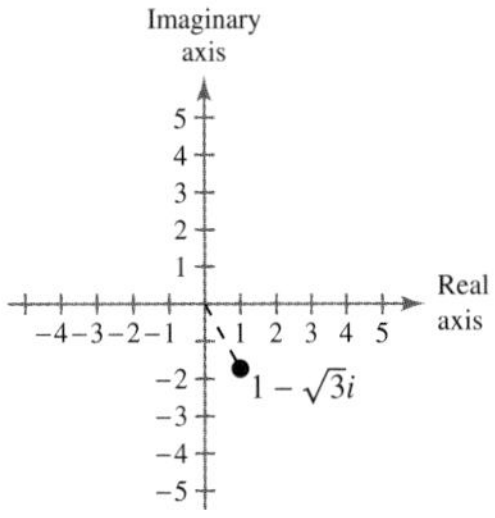

$2\left(\cos\frac{5\pi}{3} + i\sin\frac{5\pi}{3}\right)$

21.

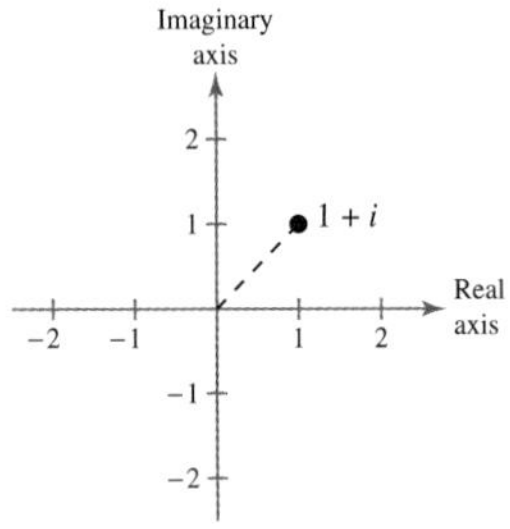

$\sqrt{2}\left(\cos\frac{\pi}{4} + i\sin\frac{\pi}{4}\right)$

23.

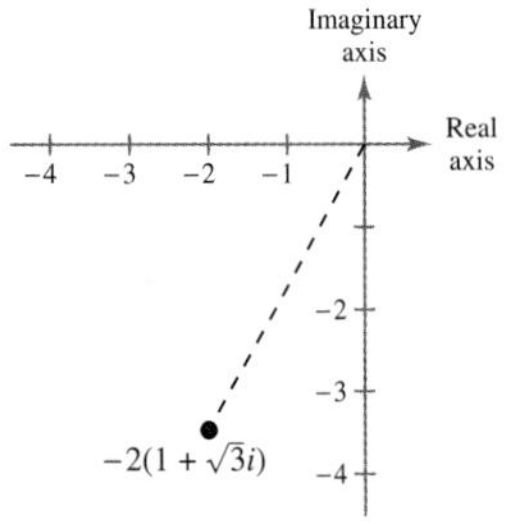

$4\left(\cos\frac{4\pi}{3} + i\sin\frac{4\pi}{3}\right)$

25.

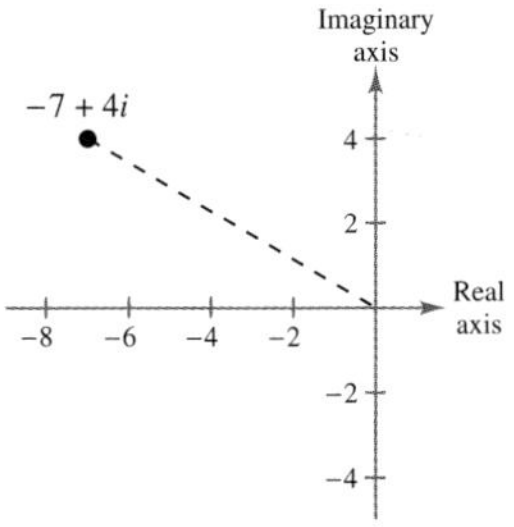

$\sqrt{65}(\cos 2.622 + i\sin 2.622)$

27.

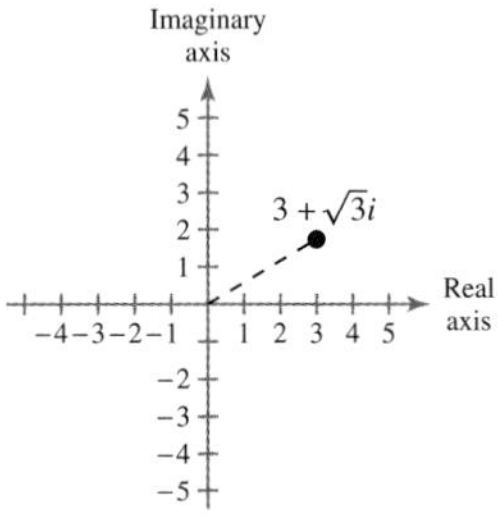

$2\sqrt{3}\left(\cos\frac{\pi}{6} + i\sin\frac{\pi}{6}\right)$

29.

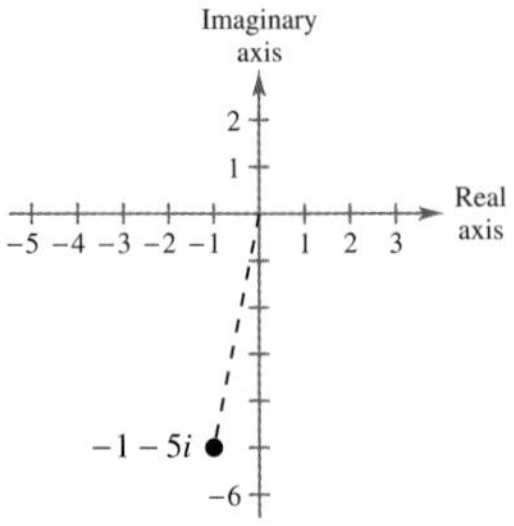

$\sqrt{26}(\cos 4.515 + i\sin 4.515)$

31. $3 + 3\sqrt{3}i$

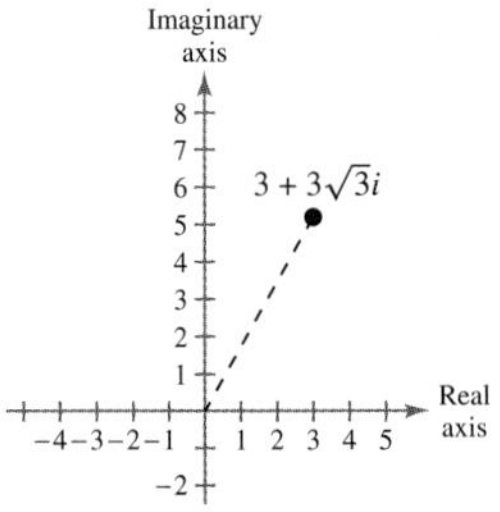

33. $-\frac{21\sqrt{2}}{8} + \frac{21\sqrt{2}}{8}i$

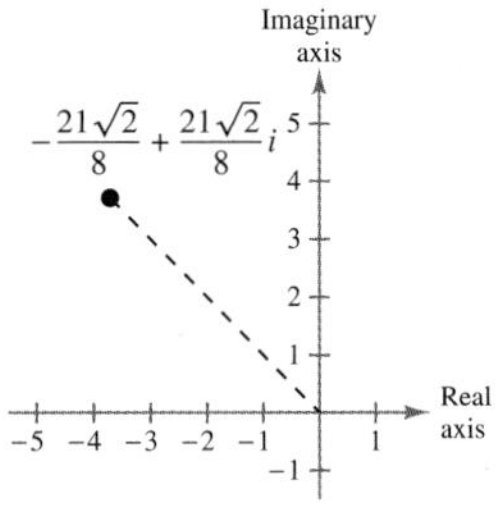

35. $-1 + \sqrt{3}i$

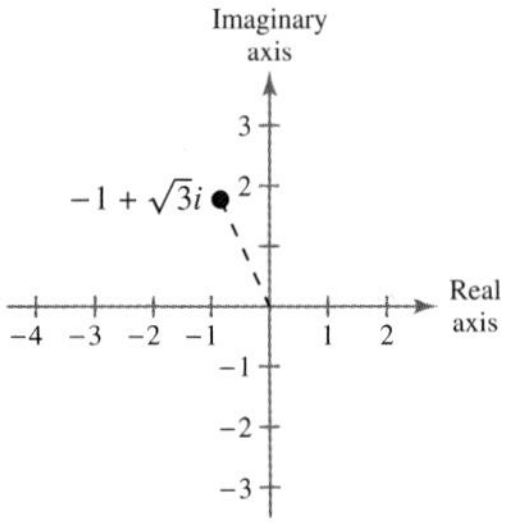

37. $6 - 2\sqrt{3}i$

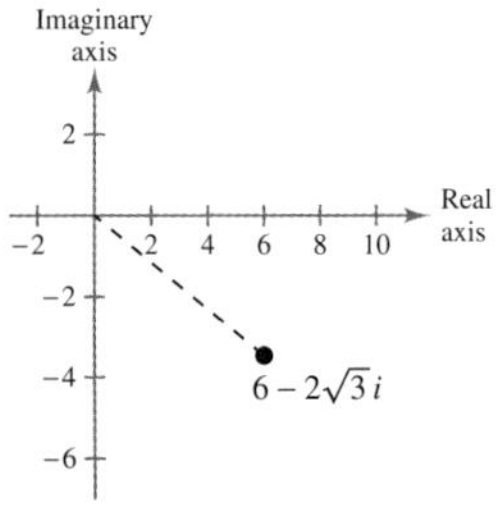

39. $-\frac{5}{2}i$

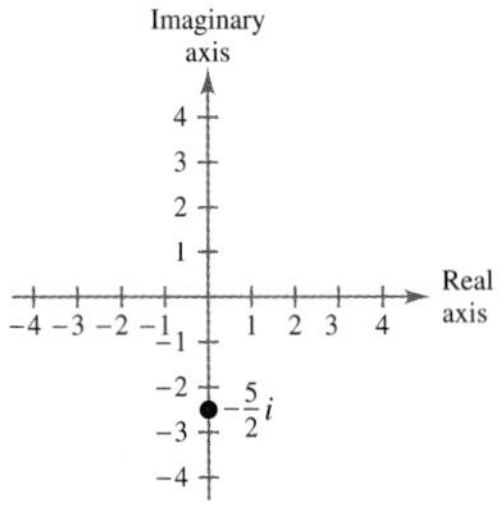

41. $2.8408 + 0.9643i$

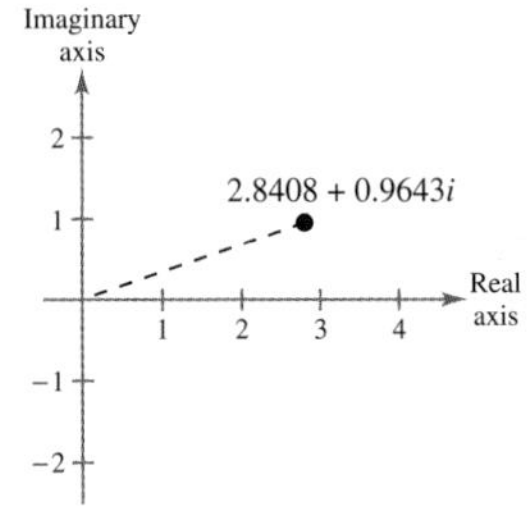

43. $-3.8302 + 3.2139i$ **45.** $4.7693 + 7.6324i$

47. $12\left(\cos\frac{\pi}{3} + i\sin\frac{\pi}{3}\right)$ **49.** $20\left(\cos\frac{2\pi}{3} + i\sin\frac{2\pi}{3}\right)$

51. $\frac{10}{9}(\cos 200° + i\sin 200°)$ **53.** $\frac{1}{3}(\cos 30° + i\sin 30°)$

55. $\frac{9}{8}(\cos \pi + i \sin \pi)$ **57.** $4(\cos 312° + i \sin 312°)$

59. (a) $2\sqrt{2}\left(\cos \frac{7\pi}{4} + i \sin \frac{7\pi}{4}\right)$ (b) and (c) 4

$\sqrt{2}\left(\cos \frac{\pi}{4} + i \sin \frac{\pi}{4}\right)$

61. (a) $2\sqrt{2}\left(\cos \frac{\pi}{4} + i \sin \frac{\pi}{4}\right)$ (b) and (c) 4

$\sqrt{2}\left(\cos \frac{7\pi}{4} + i \sin \frac{7\pi}{4}\right)$

63. (a) $2\left(\cos \frac{3\pi}{2} + i \sin \frac{3\pi}{2}\right)$ (b) and (c) $2 - 2i$

$\sqrt{2}\left(\cos \frac{\pi}{4} + i \sin \frac{\pi}{4}\right)$

65. (a) $2\left(\cos \frac{3\pi}{2} + i \sin \frac{3\pi}{2}\right)$ (b) and (c) $-2 - 2\sqrt{3}i$

$2\left(\cos \frac{11\pi}{6} + i \sin \frac{11\pi}{6}\right)$

67. (a) $2(\cos 0 + i \sin 0)$ (b) and (c) $2 - 2i$

$\sqrt{2}\left(\cos \frac{7\pi}{4} + i \sin \frac{7\pi}{4}\right)$

69. (a) $5(\cos 0.93 + i \sin 0.93)$

$2\left(\cos \frac{5\pi}{3} + i \sin \frac{5\pi}{3}\right)$

(b) and (c) $\left(\frac{3}{4} - \sqrt{3}\right) + \left(\frac{3\sqrt{3}}{4} + 1\right)i$

71. (a) $5(\cos 0 + i \sin 0)$ (b) and (c) $\frac{5}{4} - \frac{5}{4}i$

$2\sqrt{2}\left(\cos \frac{\pi}{4} + i \sin \frac{\pi}{4}\right)$

73. (a) $4\left(\cos \frac{\pi}{2} + i \sin \frac{\pi}{2}\right)$ (b) and (c) $2 - 2i$

$\sqrt{2}\left(\cos \frac{3\pi}{4} + i \sin \frac{3\pi}{4}\right)$

75. -1

77.

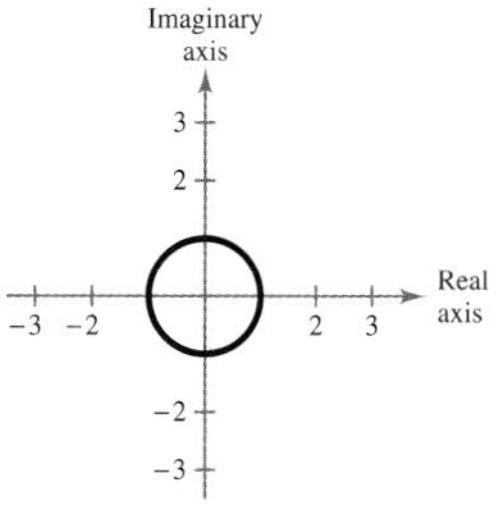

79.

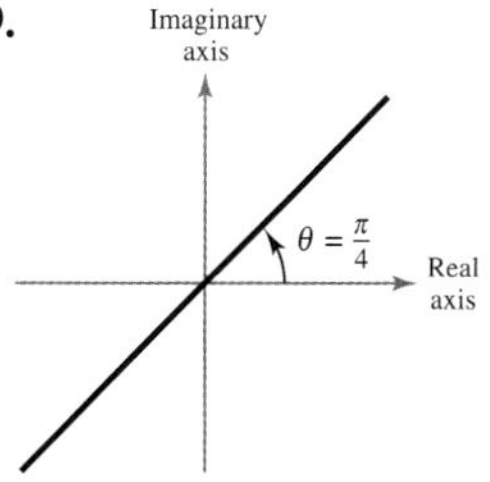

81. $-4 - 4i$ **83.** $2 + 2i$ **85.** $\frac{125}{2} + \frac{125\sqrt{3}}{2}i$ **87.** i

89. 16,384 **91.** $2048 + 2048\sqrt{3}i$ **93.** $\frac{9\sqrt{2}}{2} + \frac{9\sqrt{2}}{2}i$

95. $-597 - 122i$ **97.** $4.5386 - 15.3428i$

99.

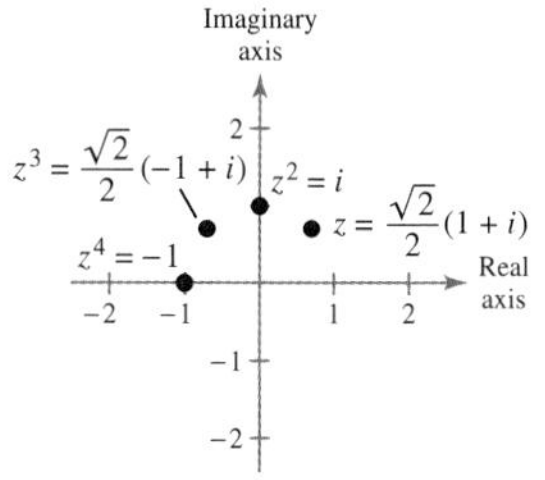

The absolute value of each is 1.

101. Answers will vary. **103.** $\sqrt{2} + \sqrt{2}i, -\sqrt{2} - \sqrt{2}i$

105. $-1.5538 + 0.6436i, 1.5538 - 0.6436i$

107. (a) $\sqrt{5}(\cos 60° + i \sin 60°)$

$\sqrt{5}(\cos 240° + i \sin 240°)$

(b) $\frac{\sqrt{5}}{2} + \frac{\sqrt{15}}{2}i, -\frac{\sqrt{5}}{2} - \frac{\sqrt{15}}{2}i$

(c)

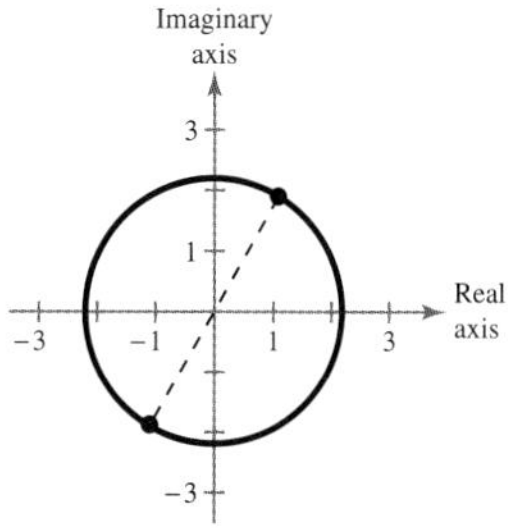

109. (a) $2\left(\cos \frac{2\pi}{9} + i \sin \frac{2\pi}{9}\right)$

$2\left(\cos \frac{8\pi}{9} + i \sin \frac{8\pi}{9}\right)$

$2\left(\cos \frac{14\pi}{9} + i \sin \frac{14\pi}{9}\right)$

(b) $1.5321 + 1.2856i, -1.8794 + 0.6840i,$
$0.3473 - 1.969i$

(c)

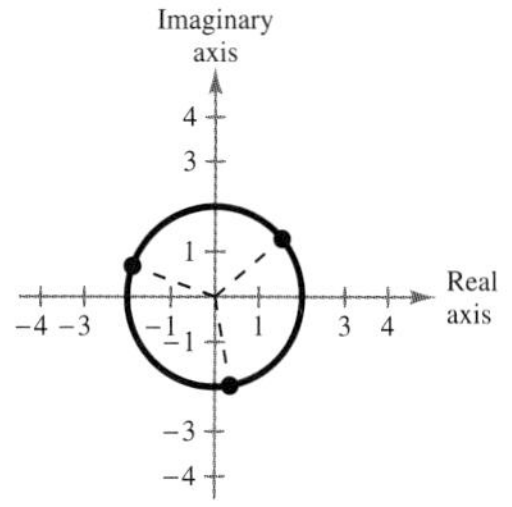

111. (a) $\sqrt[3]{25}\left(\cos \frac{\pi}{2} + i \sin \frac{\pi}{2}\right)$

$\sqrt[3]{25}\left(\cos \frac{7\pi}{6} + i \sin \frac{7\pi}{6}\right)$

$\sqrt[3]{25}\left(\cos \frac{11\pi}{6} + i \sin \frac{11\pi}{6}\right)$

(b) $2.9240i, -2.5323 - 1.4620i, 2.5323 - 1.4620i$

(c)

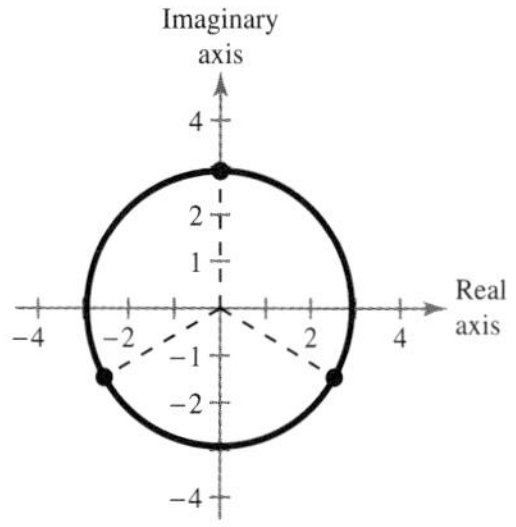

113. (a) $5\left(\cos \frac{4\pi}{9} + i \sin \frac{4\pi}{9}\right)$

$5\left(\cos \frac{10\pi}{9} + i \sin \frac{10\pi}{9}\right)$

$5\left(\cos \frac{16\pi}{9} + i \sin \frac{16\pi}{9}\right)$

(b) $0.8682 + 4.9240i, -4.6985 - 1.7101i,$ $3.8302 - 3.2139i$

(c)

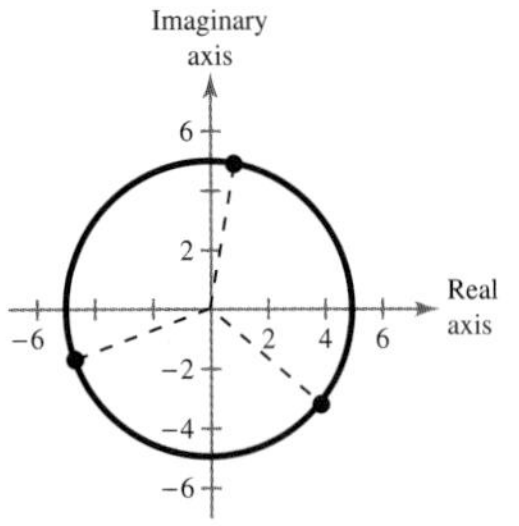

115. (a) $4\left(\cos \frac{\pi}{8} + i \sin \frac{\pi}{8}\right)$

$4\left(\cos \frac{5\pi}{8} + i \sin \frac{5\pi}{8}\right)$

$4\left(\cos \frac{9\pi}{8} + i \sin \frac{9\pi}{8}\right)$

$4\left(\cos \frac{13\pi}{8} + i \sin \frac{13\pi}{8}\right)$

(b) $3.6955 + 1.5307i, -1.5307 + 3.6955i,$ $-3.6955 - 1.5307i, 1.5307 - 3.6955i$

(c)

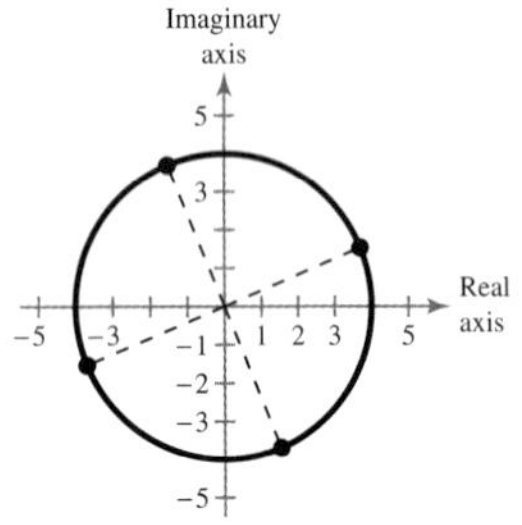

117. (a) $\cos 0 + i \sin 0$

$\cos \frac{2\pi}{5} + i \sin \frac{2\pi}{5}$

$\cos \frac{4\pi}{5} + i \sin \frac{4\pi}{5}$

$\cos \frac{6\pi}{5} + i \sin \frac{6\pi}{5}$

$\cos \frac{8\pi}{5} + i \sin \frac{8\pi}{5}$

(b) $1, 0.3090 + 0.9511i, -0.8090 + 0.5878i,$ $-0.8090 - 0.5878i, 0.3090 - 0.9511i$

(c)

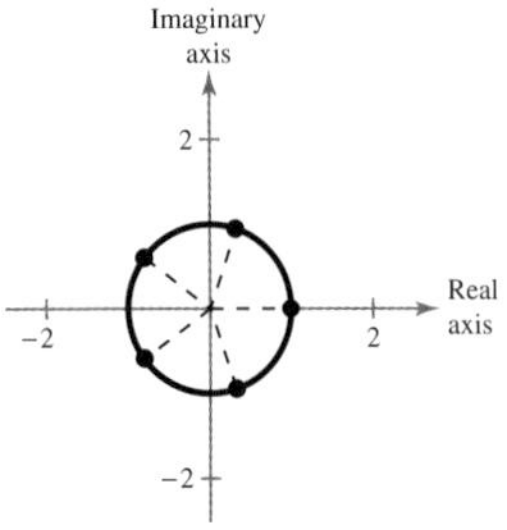

119. (a) $2\left(\cos \frac{\pi}{4} + i \sin \frac{\pi}{4}\right)$

$2\left(\cos \frac{3\pi}{4} + i \sin \frac{3\pi}{4}\right)$

$2\left(\cos \frac{5\pi}{4} + i \sin \frac{5\pi}{4}\right)$

$2\left(\cos \frac{7\pi}{4} + i \sin \frac{7\pi}{4}\right)$

(b) $\sqrt{2} + \sqrt{2}i, -\sqrt{2} + \sqrt{2}i, -\sqrt{2} - \sqrt{2}i, \sqrt{2} - \sqrt{2}i$

(c)

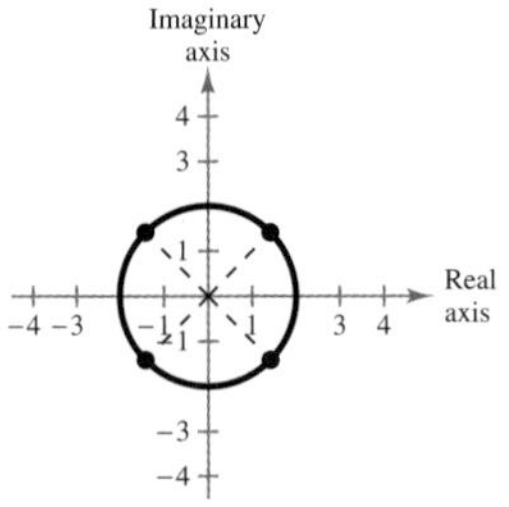

121. (a) $2\sqrt{2}\left(\cos \frac{3\pi}{20} + i \sin \frac{3\pi}{20}\right)$

$2\sqrt{2}\left(\cos \frac{11\pi}{20} + i \sin \frac{11\pi}{20}\right)$

$2\sqrt{2}\left(\cos \frac{19\pi}{20} + i \sin \frac{19\pi}{20}\right)$

$2\sqrt{2}\left(\cos \frac{27\pi}{20} + i \sin \frac{27\pi}{20}\right)$

$2\sqrt{2}\left(\cos \frac{7\pi}{4} + i \sin \frac{7\pi}{4}\right)$

(b) $2.5201 + 1.2841i, -0.4425 + 2.7936i,$ $-2.7936 + 0.4425i, -1.2841 - 2.5201i, 2 - 2i$

(c)

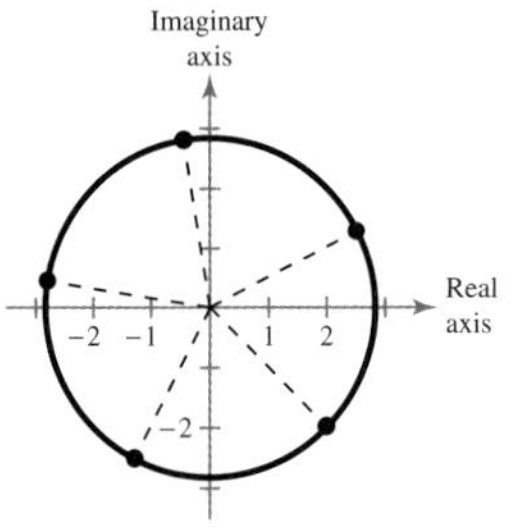

123. $\cos \frac{\pi}{8} + i \sin \frac{\pi}{8}$

$\cos \frac{5\pi}{8} + i \sin \frac{5\pi}{8}$

$\cos \frac{9\pi}{8} + i \sin \frac{9\pi}{8}$

$\cos \frac{13\pi}{8} + i \sin \frac{13\pi}{8}$

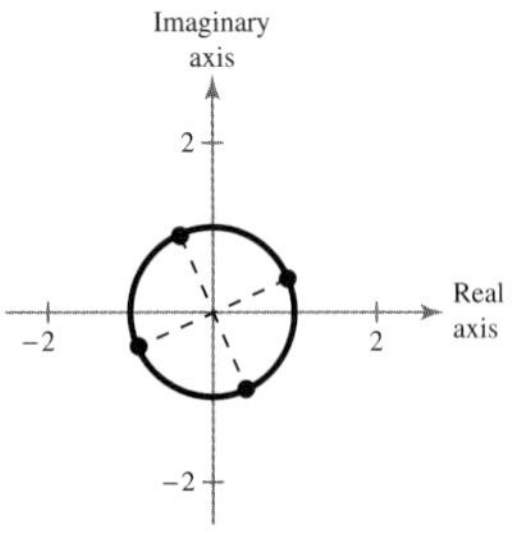

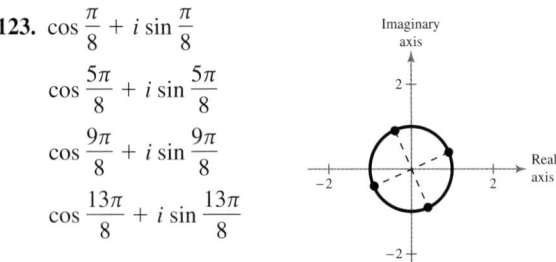

125. $3\left(\cos \frac{\pi}{5} + i \sin \frac{\pi}{5}\right)$

$3\left(\cos \frac{3\pi}{5} + i \sin \frac{3\pi}{5}\right)$

$3(\cos \pi + i \sin \pi)$

$3\left(\cos \frac{7\pi}{5} + i \sin \frac{7\pi}{5}\right)$

$3\left(\cos \frac{9\pi}{5} + i \sin \frac{9\pi}{5}\right)$

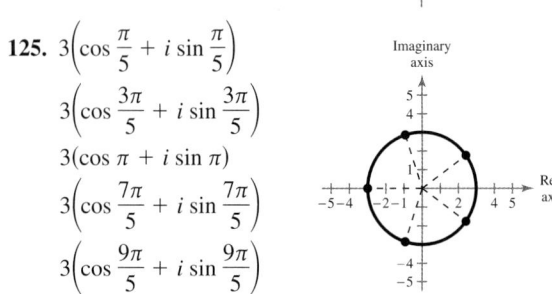

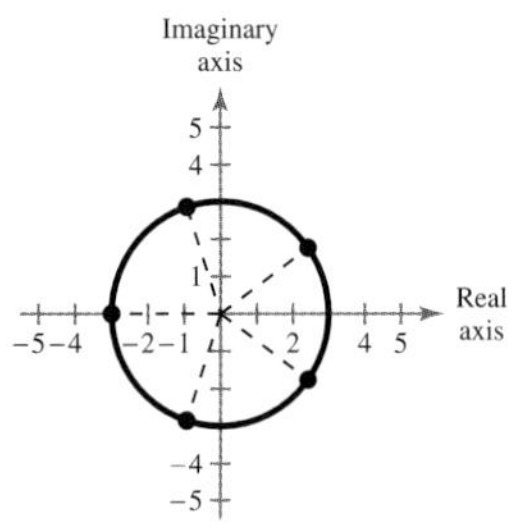

127. $2\left(\cos\frac{3\pi}{8} + i\sin\frac{3\pi}{8}\right)$
$2\left(\cos\frac{7\pi}{8} + i\sin\frac{7\pi}{8}\right)$
$2\left(\cos\frac{11\pi}{8} + i\sin\frac{11\pi}{8}\right)$
$2\left(\cos\frac{15\pi}{8} + i\sin\frac{15\pi}{8}\right)$

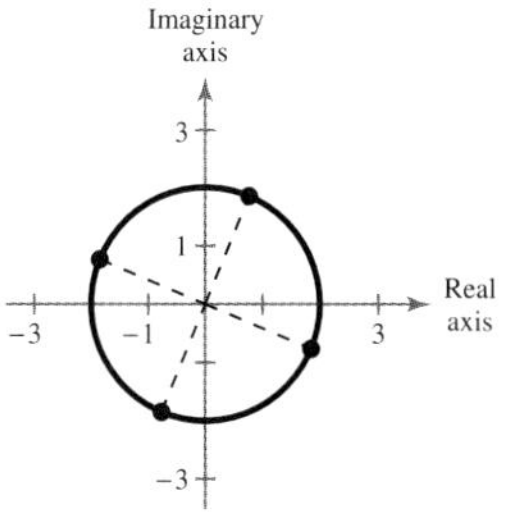

129. $\sqrt[6]{2}\left(\cos\frac{7\pi}{12} + i\sin\frac{7\pi}{12}\right)$
$\sqrt[6]{2}\left(\cos\frac{5\pi}{4} + i\sin\frac{5\pi}{4}\right)$
$\sqrt[6]{2}\left(\cos\frac{23\pi}{12} + i\sin\frac{23\pi}{12}\right)$

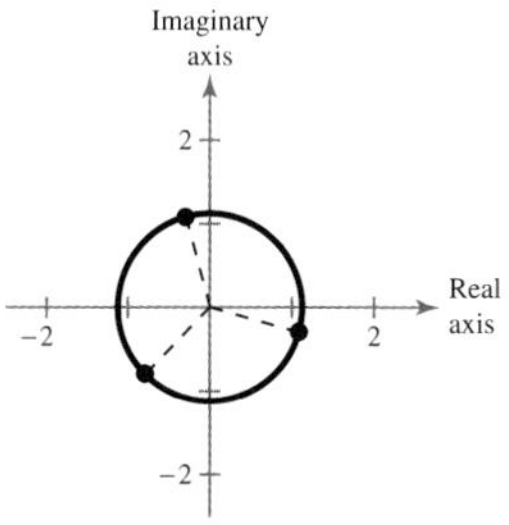

131. $E = 34 + 38i$ **133.** $Z = \frac{3}{2} - \frac{1}{2}i$

135. True; $z_1z_2 = r_1r_2[\cos(\theta_1 + \theta_2) + i\sin(\theta_1 + \theta_2)] = 0$ if and only if $r_1 = 0$ and/or $r_2 = 0$.

137–141. Answers will vary.

143. Maximum displacement: 16; $t = 2$

Review Exercises (page 462)

1. $C = 98°, b \approx 23.13, c \approx 29.90$

3. $A = 66°, b \approx 2.53, c \approx 9.11$

5. $C = 74°\ 15', a \approx 5.84, c \approx 6.48$

7. $B \approx 39.48°, C \approx 65.52°, c \approx 48.24$

9. Two solutions
$A \approx 40.92°, C \approx 114.08°, c \approx 8.64$
$A \approx 139.08°, C \approx 15.92°, c \approx 2.60$

11. About 19.06 **13.** About 221.34 **15.** About 31.01 ft

17. $A \approx 27.82°, B \approx 54.75°, C \approx 97.44°$

19. $A \approx 15.29°, B \approx 20.59°, C \approx 144.11°$

21. $B \approx 9.90°, C \approx 20.10°, a \approx 29.09$

23. $A \approx 86.38°, B \approx 28.62°, c \approx 22.70$

25. $A \approx 29.93°, B \approx 86.18°, C \approx 63.90°$

27. $A \approx 45.76°, B \approx 91.24°, c \approx 21.42$

29. No; $A \approx 77.52°, B \approx 38.48°, a \approx 14.12$

31. Yes; $A \approx 28.62°, B \approx 33.56°, C \approx 117.82°$

33. About 11.62 square units **35.** About 511.71 square units

37. About 0.17 square unit **39.** About 4.29 ft, about 12.63 ft

41. Equivalent; **u** and **v** have the same magnitude and direction.

43. $\langle 7, -7\rangle, \|\mathbf{v}\| = 7\sqrt{2}$

45. (a) $\langle 4, 4\rangle$ (b) $\langle 4, 6\rangle$

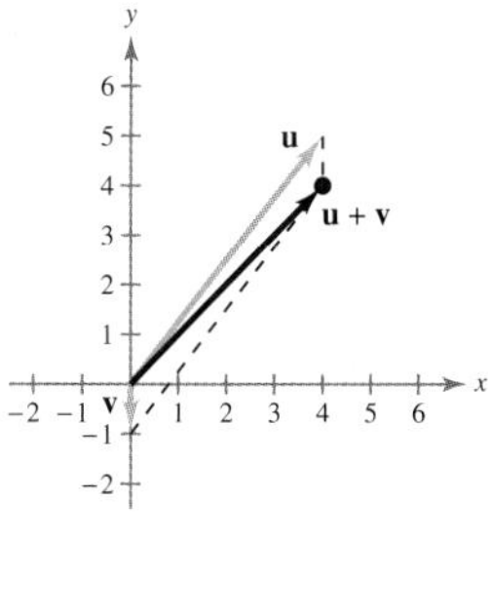

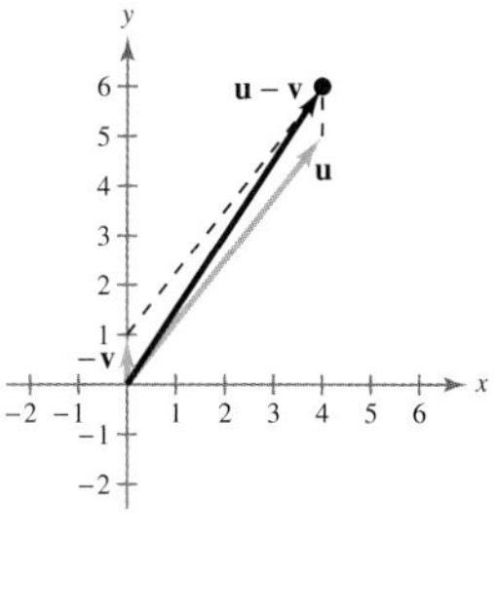

(c) $\langle 12, 15\rangle$ (d) $\langle 20, 23\rangle$

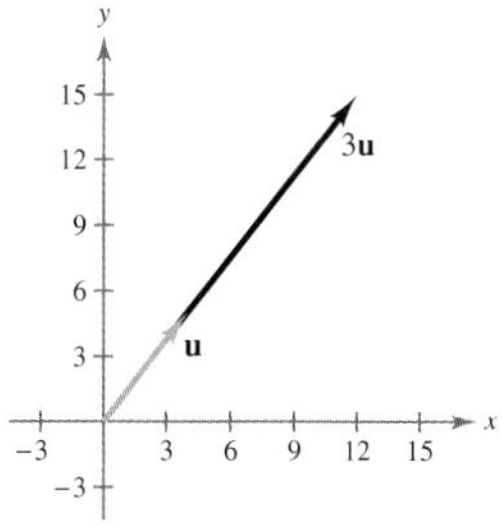

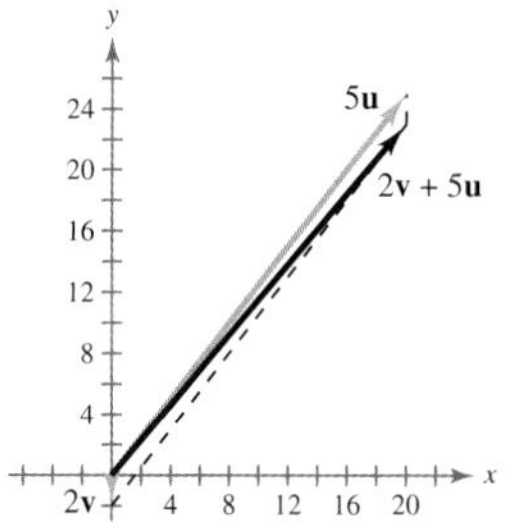

47. $\left\langle \frac{10}{3}, 1\right\rangle$ **49.** $\langle 74, -5\rangle$

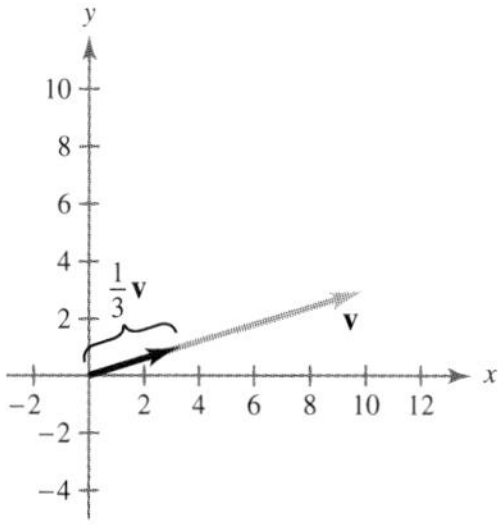

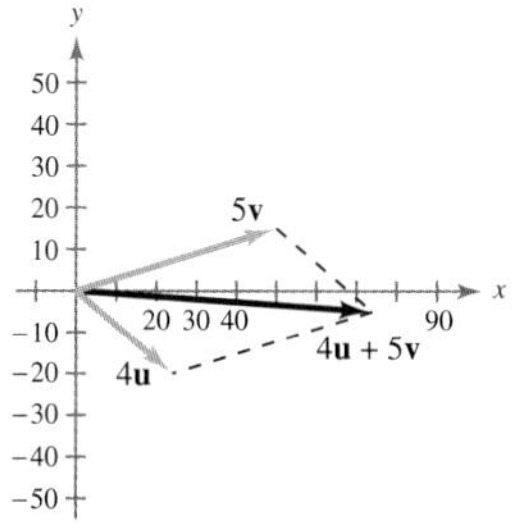

51. $\langle 0, -1\rangle$ **53.** $\frac{8}{17}\mathbf{i} - \frac{15}{17}\mathbf{j}$ **55.** $9\mathbf{i} - 8\mathbf{j}$

57. $\|\mathbf{v}\| = 2\sqrt{2}; \theta = 315°$ **59.** $\|\mathbf{v}\| = \sqrt{41}; \theta \approx 38.7°$

61. $\|\mathbf{v}\| = \sqrt{65}; \theta \approx 119.7°$

63. About 133.92 lb, about 5.55° from the 85-lb force

65. 180 lb **67.** −17 **69.** 52, scalar **71.** 2.802

73.

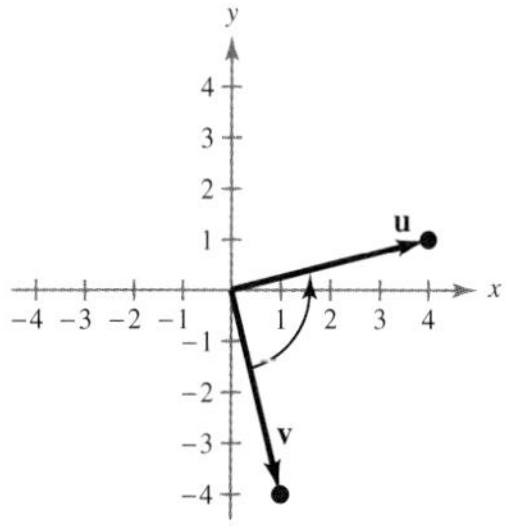

90°

75.

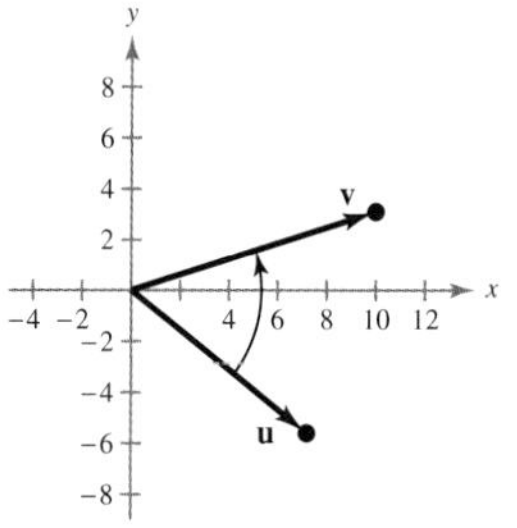

About 55.098°

77. Orthogonal **79.** Neither **81.** $\frac{1}{2}$

83. $\frac{13}{17}\langle -4, -1\rangle, \left\langle -\frac{52}{17}, -\frac{13}{17}\right\rangle + \left\langle -\frac{16}{17}, \frac{64}{17}\right\rangle$

85. $\frac{5}{2}\langle -1, 1\rangle, \left\langle -\frac{5}{2}, \frac{5}{2}\right\rangle + \left\langle \frac{9}{2}, \frac{9}{2}\right\rangle$ **87.** 72,000 ft-lb

89.

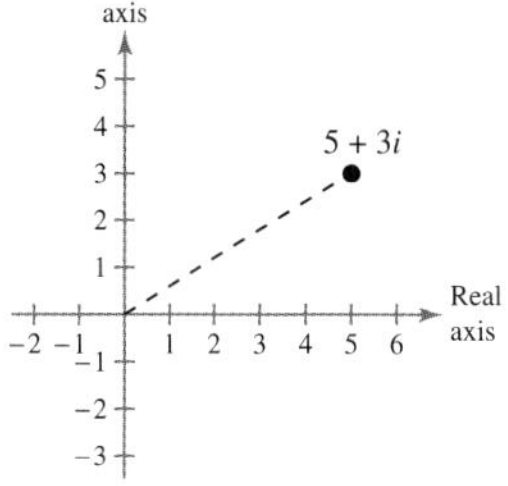

$\sqrt{34}$

91.

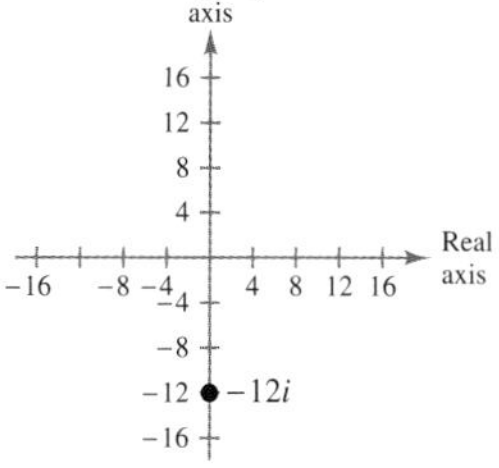

12

93. $3 + i$ **95.** $-2 + i$

97.

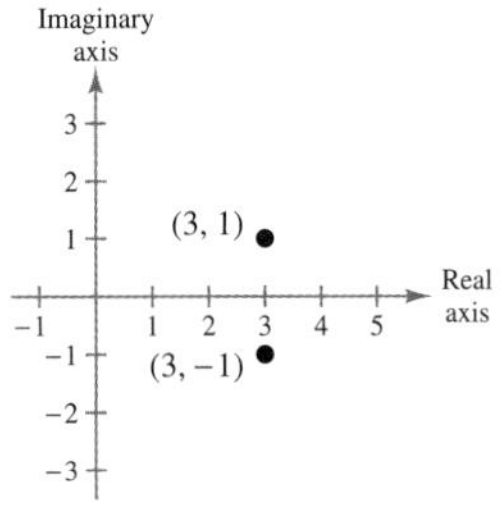

$3 - i$

99. $\sqrt{10}$ **101.** $\left(\frac{5}{2}, 2\right)$

103. $4\left(\cos\frac{\pi}{2} + i\sin\frac{\pi}{2}\right)$ **105.** $2\left(\cos\frac{7\pi}{6} + i\sin\frac{7\pi}{6}\right)$

107. $\sqrt{2} + \sqrt{2}i$

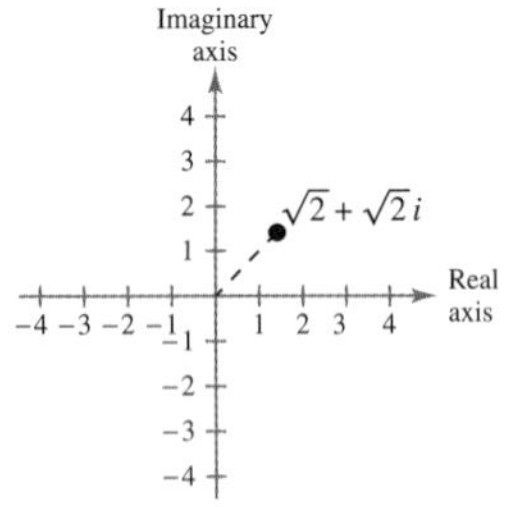

109. $10\left(\cos\frac{3\pi}{4} + i\sin\frac{3\pi}{4}\right)$ **111.** $4(\cos 240° + i\sin 240°)$

113. (a) $2\sqrt{2}\left(\cos\frac{7\pi}{4} + i\sin\frac{7\pi}{4}\right)$ (b) and (c) 12

$3\sqrt{2}\left(\cos\frac{\pi}{4} + i\sin\frac{\pi}{4}\right)$

115. (a) $3\sqrt{2}\left(\cos\frac{7\pi}{4} + i\sin\frac{7\pi}{4}\right)$ (b) and (c) $-\frac{3}{2}i$

$2\sqrt{2}\left(\cos\frac{\pi}{4} + i\sin\frac{\pi}{4}\right)$

117. $-16 - 16\sqrt{3}i$

119. $0.3660 + 1.3660i, -0.3660 - 1.3660i$

121. $\sqrt{2} - \sqrt{2}i, -\sqrt{2} + \sqrt{2}i$

123. (a) $3\left(\cos\frac{\pi}{4} + i\sin\frac{\pi}{4}\right)$

$3\left(\cos\frac{7\pi}{12} + i\sin\frac{7\pi}{12}\right)$

$3\left(\cos\frac{11\pi}{12} + i\sin\frac{11\pi}{12}\right)$

$3\left(\cos\frac{5\pi}{4} + i\sin\frac{5\pi}{4}\right)$

$3\left(\cos\frac{19\pi}{12} + i\sin\frac{19\pi}{12}\right)$

$3\left(\cos\frac{23\pi}{12} + i\sin\frac{23\pi}{12}\right)$

(b) $2.1213 + 2.1213i, -0.7765 + 2.8978i,$
$-2.8978 + 0.7765i, -2.1213 - 2.1213i,$
$0.7765 - 2.8978i, 2.8978 - 0.7765i$

(c)

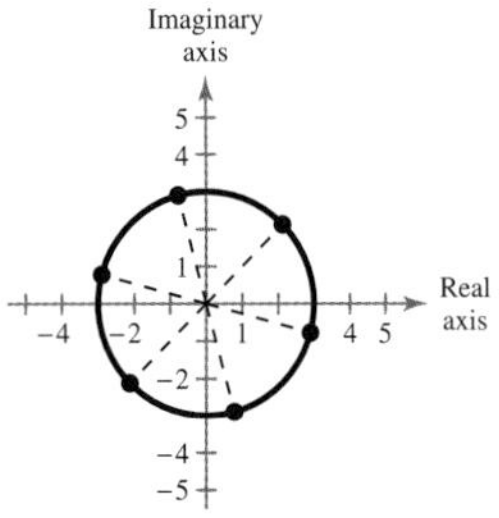

125. (a) $7(\cos 0 + i\sin 0)$

$7\left(\cos\frac{2\pi}{3} + i\sin\frac{2\pi}{3}\right)$

$7\left(\cos\frac{4\pi}{3} + i\sin\frac{4\pi}{3}\right)$

(b) $7, -\frac{7}{2} + \frac{7\sqrt{3}}{2}i, -\frac{7}{2} - \frac{7\sqrt{3}}{2}i$

(c)

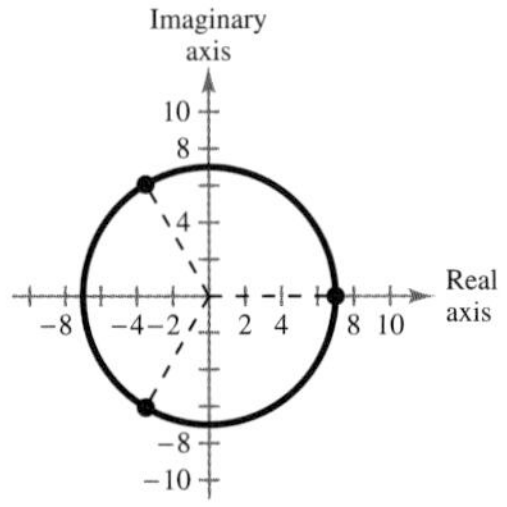

127. $4\left(\cos\frac{\pi}{4} + i\sin\frac{\pi}{4}\right)$

$4\left(\cos\frac{3\pi}{4} + i\sin\frac{3\pi}{4}\right)$

$4\left(\cos\frac{5\pi}{4} + i\sin\frac{5\pi}{4}\right)$

$4\left(\cos\frac{7\pi}{4} + i\sin\frac{7\pi}{4}\right)$

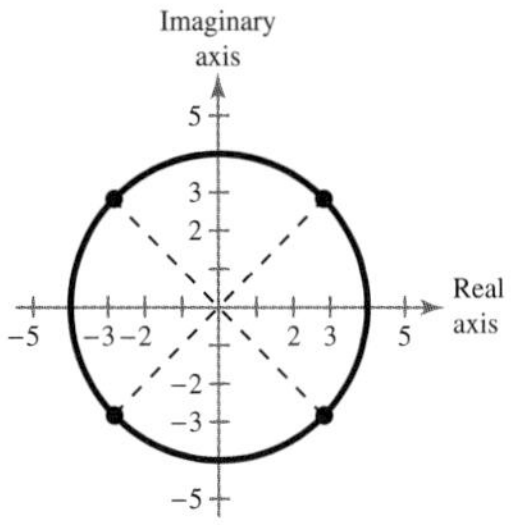

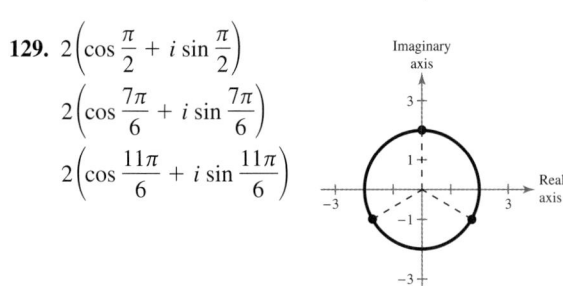

129. $2\left(\cos\frac{\pi}{2} + i\sin\frac{\pi}{2}\right)$

$2\left(\cos\frac{7\pi}{6} + i\sin\frac{7\pi}{6}\right)$

$2\left(\cos\frac{11\pi}{6} + i\sin\frac{11\pi}{6}\right)$

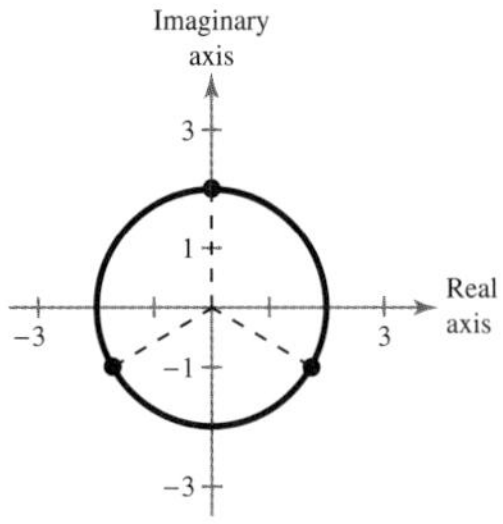

131. True; sin 90° is defined in the Law of Sines.

133. False; AAS and ASA cases have unique solutions.

135. True; $\cos 90° = 0$ **137.** $\frac{a}{\sin A} = \frac{b}{\sin B} = \frac{c}{\sin C}$

139. Direction and magnitude **141.** Answers will vary.

Chapter Test (page 469)

1. $C = 46°, a \approx 13.07, b \approx 22.03$
2. $A \approx 22.33°, B \approx 49.46°, C \approx 108.21°$
3. $B \approx 40.11°, C \approx 104.89°, a \approx 7.12$
4. $A = 88°, a \approx 29.98, c \approx 27.81$
5. Two solutions
   $B \approx 41.10°, C \approx 113.90°, c \approx 38.94$
   $B \approx 138.90°, C \approx 16.10°, c \approx 11.81$
6. No solution **7.** About 675 ft **8.** About 2337 m^2
9. $\mathbf{w} = \langle 12, 13\rangle, \|\mathbf{w}\| \approx \sqrt{313}$

10. (a) $\langle 1, 8 \rangle$

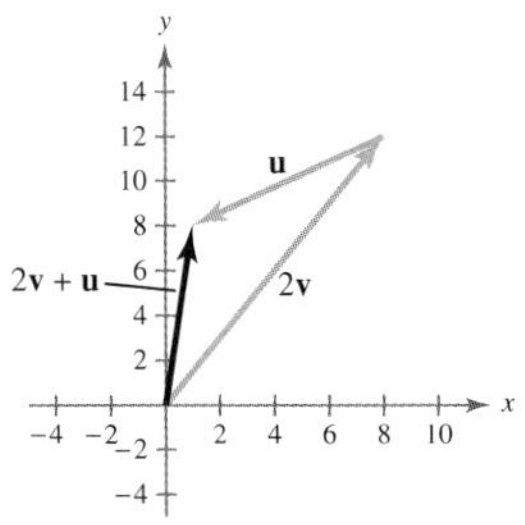

(b) $\langle -19, -22 \rangle$

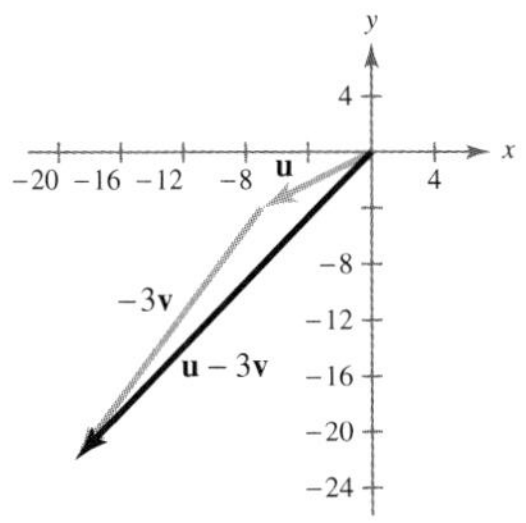

(c) $\langle -51, -44 \rangle$

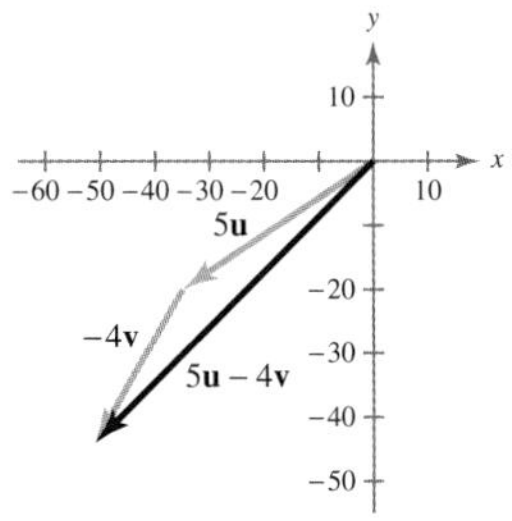

11. (a) $-\mathbf{j}$

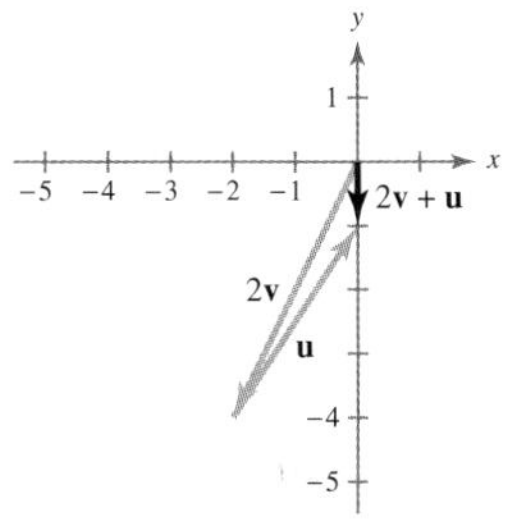

(b) $5\mathbf{i} + 9\mathbf{j}$

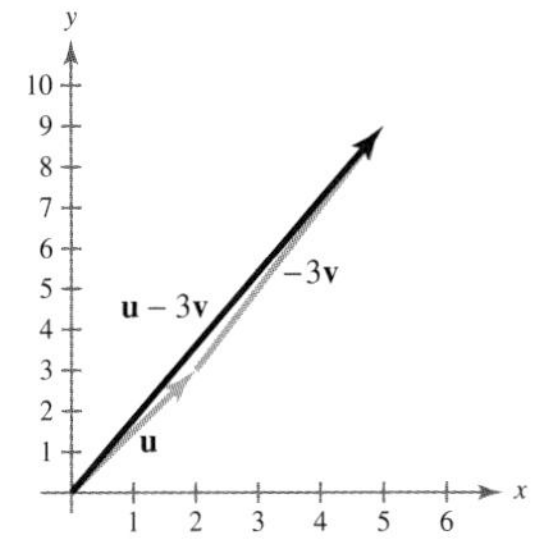

(c) $14\mathbf{i} + 23\mathbf{j}$

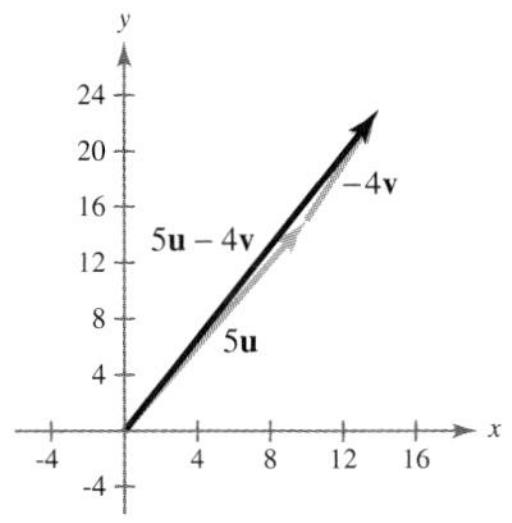

12. $\left\langle \frac{24}{25}, -\frac{7}{25} \right\rangle$ **13.** $\left\langle \frac{18\sqrt{34}}{17}, -\frac{30\sqrt{34}}{17} \right\rangle$

14. $\theta \approx 14.87°$, about 250.15 lb **15.** -1

16. About 105.95° **17.** Yes; $\mathbf{u} \cdot \mathbf{v} = 0$

18. $\left\langle \frac{185}{26}, \frac{37}{26} \right\rangle$; $\mathbf{u} = \left\langle \frac{185}{26}, \frac{37}{26} \right\rangle + \left\langle -\frac{29}{26}, \frac{145}{26} \right\rangle$

19.

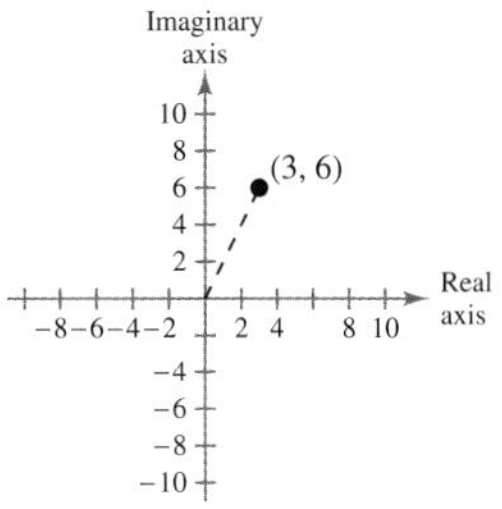

$3\sqrt{5}$

20. $4 - 3i$ **21.** $1 + 11i$ **22.** 5

23.

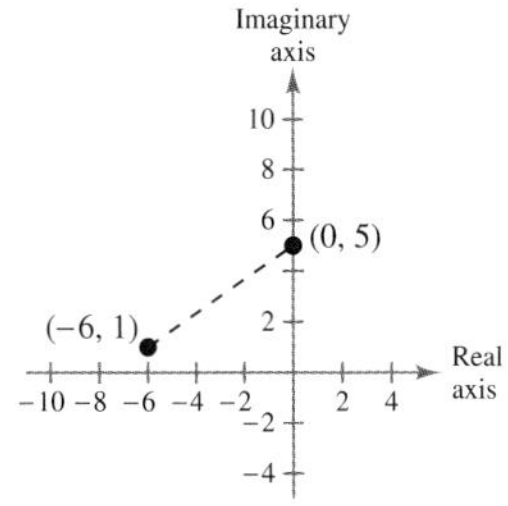

$(-3, 3)$

24. $z = 6\sqrt{2}\left(\cos \frac{3\pi}{4} + i \sin \frac{3\pi}{4}\right)$

25. $-50 - 50\sqrt{3}i$

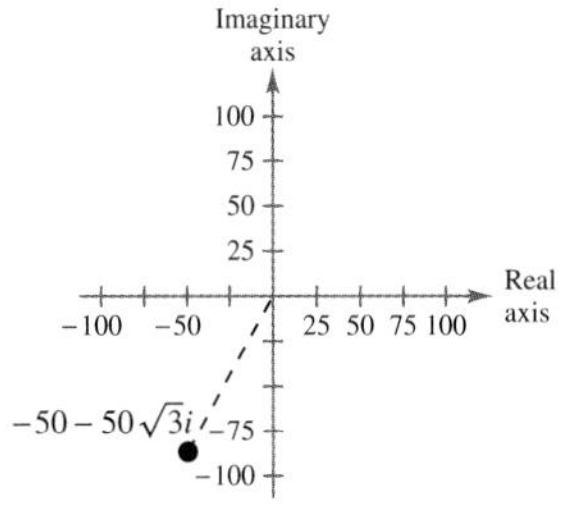

26. $-\frac{6561}{2} + \frac{6561\sqrt{3}}{2}i$ **27.** $5832i$

28. $4\left(\cos \frac{\pi}{12} + i \sin \frac{\pi}{12}\right)$

$4\left(\cos \frac{7\pi}{12} + i \sin \frac{7\pi}{12}\right)$

$4\left(\cos \frac{13\pi}{12} + i \sin \frac{13\pi}{12}\right)$

$4\left(\cos \frac{19\pi}{12} + i \sin \frac{19\pi}{12}\right)$

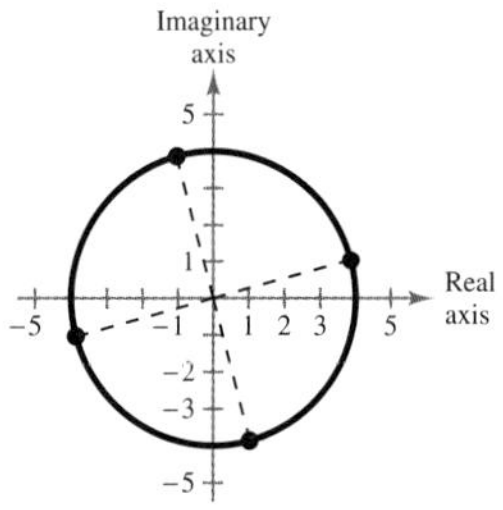

29. $5\left(\cos \frac{\pi}{8} + i \sin \frac{\pi}{8}\right)$

$5\left(\cos \frac{5\pi}{8} + i \sin \frac{5\pi}{8}\right)$

$5\left(\cos \frac{9\pi}{8} + i \sin \frac{9\pi}{8}\right)$

$5\left(\cos \frac{13\pi}{8} + i \sin \frac{13\pi}{8}\right)$

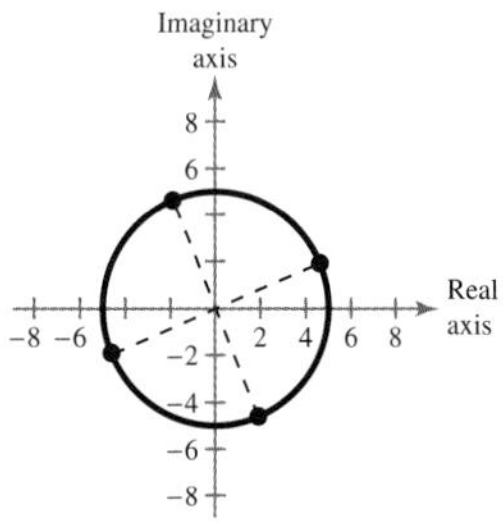

Standardized Test Practice (page 470)

1. D **2.** B **3.** A **4.** A **5.** C **6.** D
7. B **8.** B **9.** C **10.** B **11.** B **12.** C
13. B **14.** D **15.** B **16.** C **17.** A **18.** A
19. B **20.** D **21.** B **22.** −0.81 **23.** 3.2
24. 0.83 **25.** 7.3 mi **26.** 9 **27.** 30.78 lb **28.** 5

Chapter 7

Section 7.1 (page 484)

1. system, equations **3.** substitution **5.** Break-even point
7. (a) No (b) No (c) No (d) Yes
9. (a) No (b) Yes (c) No (d) No
11. $(2, 2)$ **13.** $(2, 6), (-1, 3)$
15. $(0, 2), (\sqrt{3}, 2 - 3\sqrt{3}), (-\sqrt{3}, 2 + 3\sqrt{3})$ **17.** $(4, 4)$
19. $(10, -10)$ **21.** $(\frac{1}{2}, 3)$ **23.** $(1, 1)$ **25.** $(\frac{20}{3}, \frac{40}{3})$
27. No solution **29.** \$5500 at 2%; \$6500 at 6%
31. \$6000 at 2.8%; \$6000 at 3.8% **33.** $(-2, 0), (3, 5)$
35. No real solution **37.** $(0, 0), (1, 1), (-1, -1)$
39. $(-3, 1)$ **41.** $(1, 2)$ **43.** No real solution
45. No real solution **47.** $(3, -0.625)$ **49.** $(8, 3), (3, -2)$
51. $(\pm 1.644, 2.702)$ **53.** $(0, 1)$ **55.** $(3.496, 3.252)$
57. $(2.25, 5.5)$ **59.** $(0, -13), (\pm 12, 5)$ **61.** $(1, 2)$
63. $(-2, 0), (\frac{29}{10}, \frac{21}{10})$ **65.** No real solution **67.** $(0.25, 1.5)$
69. $(0.287, 1.751)$ **71.** $(0, 1), (1, 0)$ **73.** No real solution

75.

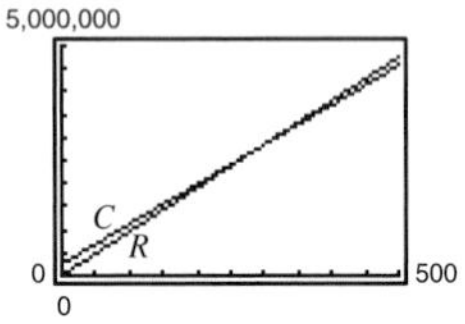

293 units; \$2,784,086

77.

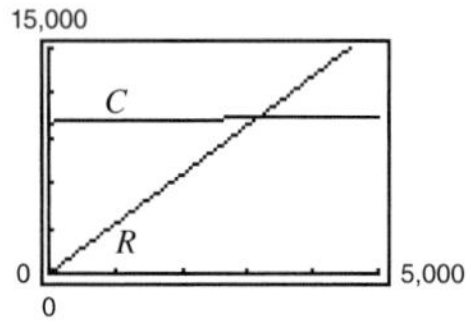

3133 units; \$10,308

79. 6 m × 9 m

81. (a)

Week	Animated	Horror
1	336	42
2	312	60
3	288	78
4	264	96
5	240	114
6	216	132
7	192	150
8	168	168
9	144	186
10	120	204
11	96	222
12	72	240

(b) and (c) $x = 8$

(d) The answers are the same.

(e) During week 8, the same number of animated and horror films were rented.

83. (a) $C = 9.45x + 18{,}000$
$R = 55.95x$

(b)

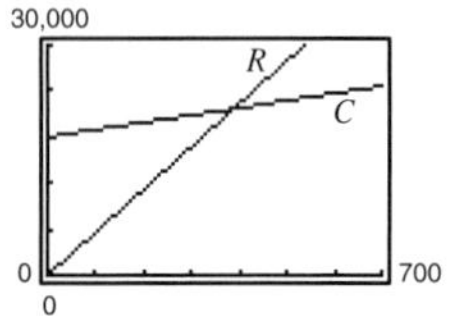

About 387 units

85. 8 mi × 12 mi

87. (a) $\begin{cases} x + y = 20{,}000 \\ 0.055x + 0.075y = 1300 \end{cases}$

(b)

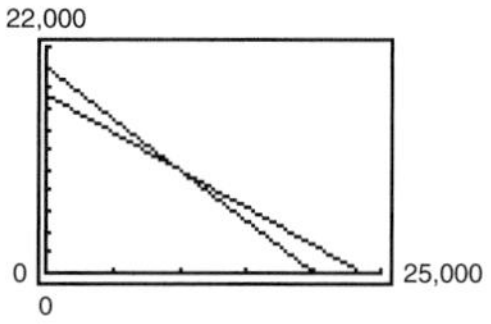

(c) \$10,000; The solution is (10,000, 10,000).

89. (a)

t	Year	Arizona	Massachusetts
0	2010	6377	6574
1	2011	6464.5	6616.5
2	2012	6552	6659
3	2013	6639.5	6701.5
4	2014	6727	6744
5	2015	6814.5	6786.5
6	2016	6902	6829
7	2017	6989.5	6871.5

(b) 2015–2017

(c)

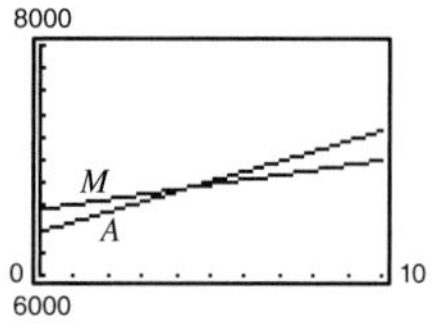

About (4.38, 6760.06)

(d) $\left(\dfrac{197}{45}, \dfrac{121{,}681}{18}\right)$

91. False; You can solve for either variable before back-substituting.

93. (a) The graphs intersect at one point.
(b) The graphs do not intersect.
(c) The graphs are the same line.

95. For a linear system, the result will be a contradictory equation such as $0 = N$, where N is a nonzero real number. For a nonlinear system, there may be an equation with imaginary roots.

97. The graph does not show both intersection points.

99. $y = \frac{2}{7}x + \frac{22}{7}$

101. $x = 4$; A vertical line cannot be written in slope-intercept form.

Section 7.2 (page 493)

1. elimination **3.** Inconsistent **5.** Yes
7. $(2, 1)$ **9.** $(1, -1)$

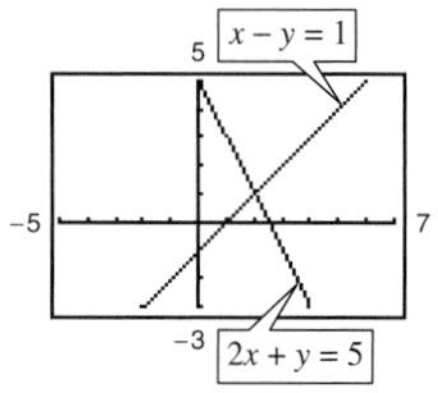

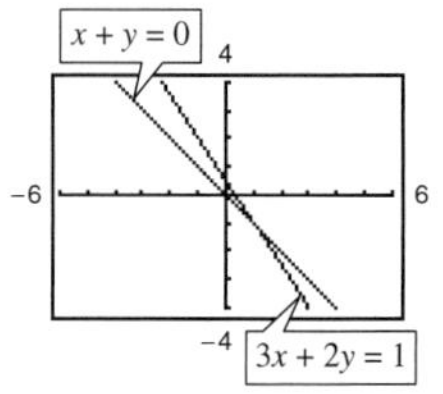

11. Inconsistent

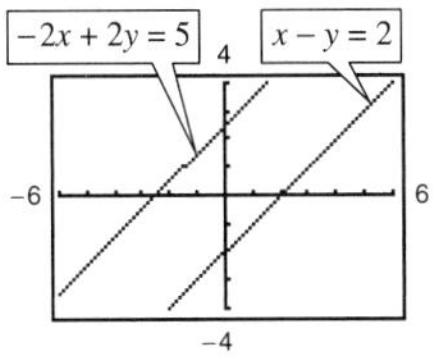

13. $\left(2, \frac{1}{2}\right)$ **15.** $\left(\frac{3}{2}, -\frac{1}{2}\right)$ **17.** $\left(-3, \frac{3}{2}\right)$ **19.** $\left(\frac{5}{2}, 0\right)$

21. Inconsistent **23.** $\left(\frac{2}{9}, \frac{4}{3}\right)$

25. b; One solution, consistent

26. a; Infinitely many solutions, consistent

27. d; No solutions, inconsistent

28. c; One solution, consistent **29.** $\left(\frac{3}{2}, -\frac{1}{2}\right)$

31. Inconsistent **33.** All points on $6x + 8y - 1 = 0$

35. $\left(\frac{7}{2}, -\frac{1}{2}\right)$ **37.** All points on $-5x + 6y = -3$

39. Inconsistent **41.** $(-1, 1)$

43.

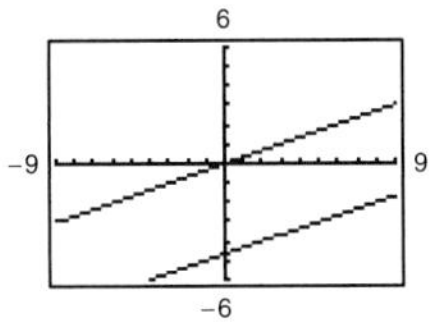

Inconsistent

45.

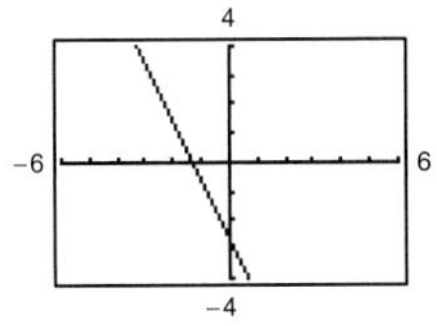

Consistent; all points on $6x + 3y = -8$

47.

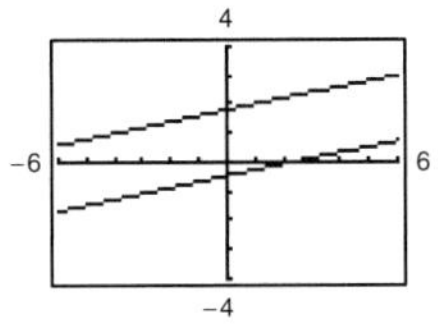

Inconsistent

49.

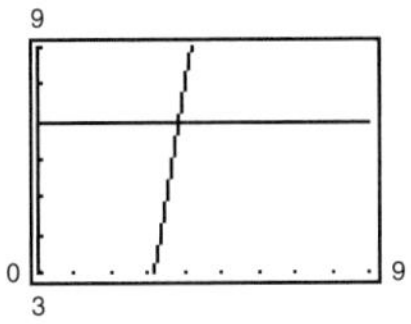

(3.833, 7)

51.

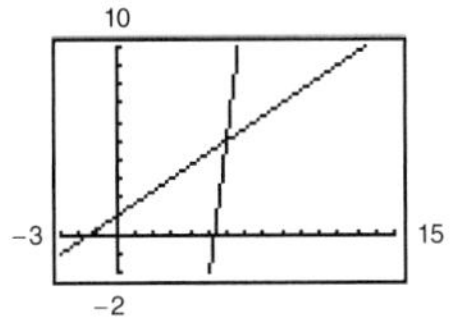

(6, 5)

53.

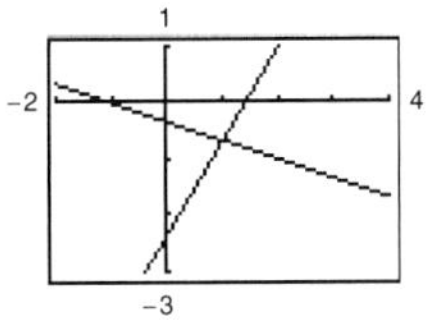

$(1, -0.667)$

55.

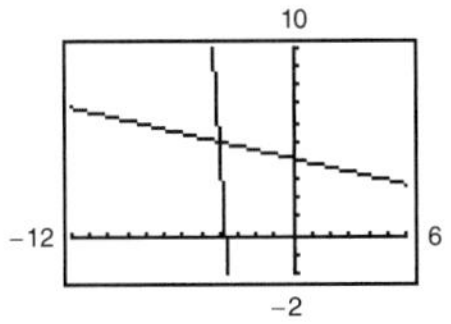

$(-4, 5)$

57. (4, 1); Answers will vary. **59.** $(2, -1)$; Answers will vary.

61. $(6, -3)$; Answers will vary. **63.** $\left(\frac{49}{4}, \frac{33}{4}\right)$; Answers will vary.

65. $\begin{cases} 3 = -a + b \\ -1 = -3a + b \end{cases}$

$a = 2, b = 5$

67. $\begin{cases} -6 = -4a + b \\ -3 = 8a + b \end{cases}$

$a = \frac{1}{4}, b = -5$

69. (240, 404) **71.** (2,000,000, 100)

73. Current: 5 mi/h, boat: 25 mi/h

75. (a) $\begin{cases} 15A + 12C = 15{,}996 \\ A + C = 1175 \end{cases}$

(b) $A = 632, C = 543$; Answers will vary.

(c) $A = 632, C = 543$

77. 9 oranges, 5 grapefruit

79. (a) $S \approx 1017.93, t \approx 4.02$; Answers will vary.

(b) In 2024, both retailers had sales of about $1017.93 million.

(c) The coefficient is the annual increase in sales for each retailer.

(d) The retailers would never have the same amount of sales for any year.

81. $y = 0.97x + 2.1$ **83.** $y = -0.58x + 5.39$

85. (a) and (b) $y = 14x + 19$

(c)

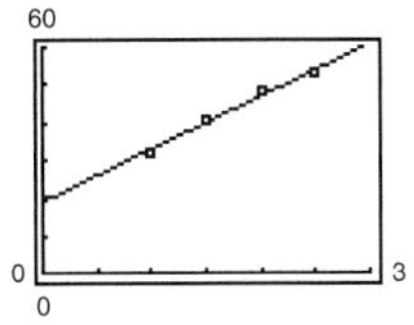

(d) 41.4 bushels per acre

87. True; A linear system can have only one solution, no solution, or infinitely many solutions.

89. False; Sometimes you will be able to get only a close approximation.

91. (a) $k = 18$ (b) $k \neq 18$ **93.** $u = 1; v = -\tan x$

95. $x \leq -\frac{22}{3}$

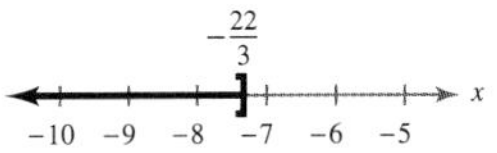

97. $-2 < x < 18$

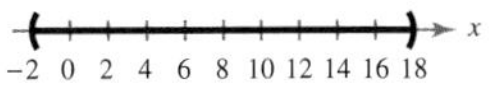

99. $-5 < x < \frac{7}{2}$

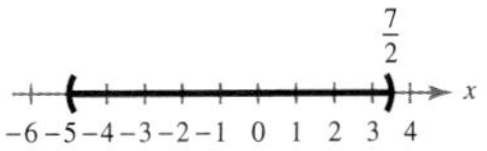

101. $\ln 8x$ **103.** $\log_9 \dfrac{12}{x}$ **105.** $\ln \dfrac{x^2}{x + 2}$

107. Answers will vary.

Section 7.3 (page 507)

1. row-echelon **3.** Gaussian **5.** position

7. partial fraction decomposition **9.** Independent

11. (a) No (b) Yes (c) No (d) No

13. (a) No (b) No (c) Yes (d) No

15. $(-13, -10, 8)$ **17.** (3, 10, 2) **19.** $\left(\frac{11}{4}, 7, 11\right)$

21. $\begin{cases} x - 2y + 3z = 5 \\ y - 2z = 9 \\ 2x - 3z = 0 \end{cases}$

It removed the x-term from Equation 2.

23. $(5, 1, -1)$ **25.** $(2, -3, -2)$ **27.** Inconsistent

29. $(-a + 3, a + 1, a)$ **31.** $\left(\frac{3}{10}, \frac{2}{5}, 0\right)$ **33.** Inconsistent

35. $(2a, 21a - 1, 8a)$ **37.** $\left(-\frac{3}{2}a + \frac{1}{2}, -\frac{2}{3}a + 1, a\right)$

39. *Sample answer:*

$\begin{cases} x + y + z = 1 \\ 2x + y + z = 4 \\ x + y - 3z = -7 \end{cases}$

41. *Sample answer:*

$\begin{cases} x - z = 0 \\ -x + y = 4 \end{cases}$

43.

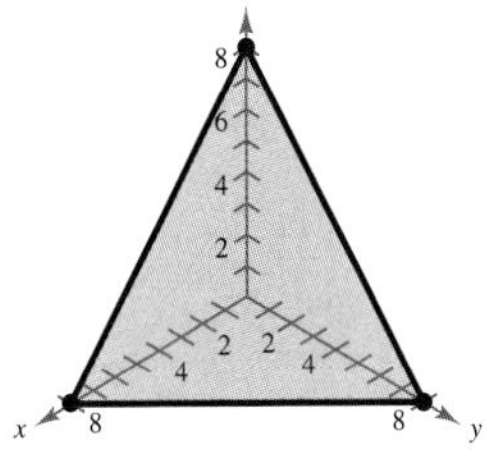

45.

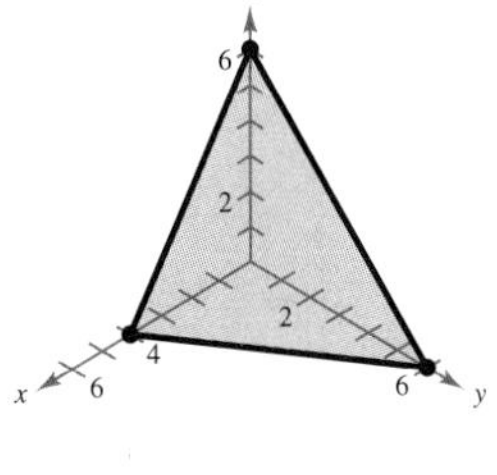

47. Plane **49.** $\frac{A}{x} + \frac{B}{x - 14}$ **51.** $\frac{A}{x} + \frac{B}{x^2} + \frac{C}{x - 10}$

53. $\frac{A}{x - 5} + \frac{B}{(x - 5)^2} + \frac{C}{(x - 5)^3}$

55. $\frac{A}{x} + \frac{B}{x^2} + \frac{Cx + D}{x^2 + 3} + \frac{Ex + F}{(x^2 + 3)^2}$ **57.** $\frac{1}{x} - \frac{1}{x + 1}$

59. $-\frac{3}{x} - \frac{1}{x + 2} + \frac{5}{x - 2}$ **61.** $\frac{2}{x - 1} - \frac{1}{(x - 1)^2}$

63. $\frac{1}{8}\left(\frac{1}{2x + 1} + \frac{1}{2x - 1} - \frac{4x}{4x^2 + 1}\right)$

65. $x + 3 + \frac{6}{x - 1} + \frac{4}{(x - 1)^2} + \frac{1}{(x - 1)^3}$

67. $-\frac{1}{x - 1} + \frac{x + 2}{x^2 - 2}$

69. $\frac{3}{x} - \frac{2}{x - 4}$

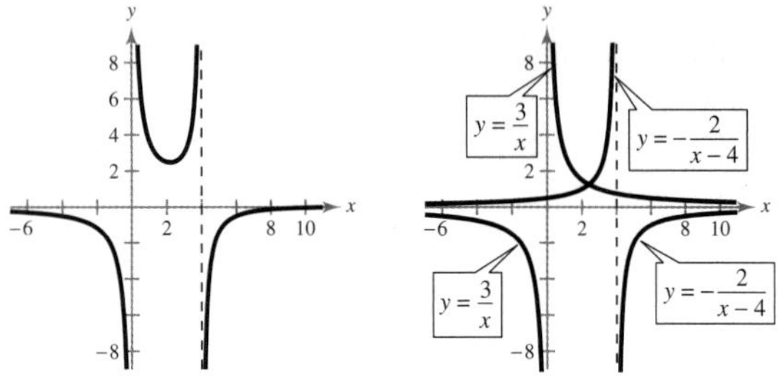

The vertical asymptotes are the same.

71. $s = -16t^2 + 144$ **73.** $s = -16t^2 - 32t + 400$

75. $y = x^2 - 3x$ **77.** $y = -2x^2 - 7x - 4$

79. $x^2 + y^2 - 10x = 0$ **81.** $x^2 + y^2 - 3x + 2y = 0$

83. \$20,000 at 4%, \$2500 at 6%, and \$7500 at 8%

85. $187{,}500 + s$ in certificates of deposit
$187{,}500 - s$ in municipal bonds
$125{,}000 - s$ in blue-chip stocks
s in growth stocks

87. 22 two-point baskets, 13 three-point baskets, 10 free throws

89. $I_1 = 1, I_2 = 2, I_3 = 1$ **91.** $y = -\frac{5}{24}x^2 - \frac{3}{10}x + \frac{41}{6}$

93. $y = x^2 - x$

95. (a) $\begin{cases} 900a + 30b + c = 55 \\ 1600a + 40b + c = 105 \\ 2500a + 50b + c = 188 \end{cases}$

$y = 0.165x^2 - 6.55x + 103$

(b)

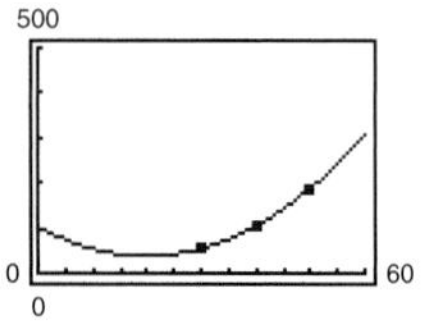

(c) 453 ft

97. $\frac{60}{100 - p} - \frac{60}{100 + p}$

99. True; Three parallel planes have no common intersection point.

101. The leading coefficient of the third equation is not 1.

103. The equation should be $4x = A(x - 2) + B$.

105. No; There are two arithmetic errors. The constant in the second equation should be -11 and the coefficient of z in the third equation should be 2.

107. *Sample answer:* $\begin{cases} 2x + y - 5z = 3 \\ -4x - 2y + 10z = 7 \end{cases}$

109. $x = 5, y = 5, \lambda = -5$ **111.** $\langle 1, 0 \rangle$

113. $\left\langle \frac{\sqrt{26}}{26}, \frac{5\sqrt{26}}{26} \right\rangle$ **115.** $\frac{\pi}{6} + n\pi$

117. Answers will vary.

Section 7.4 (page 521)

1. matrix **3.** Gauss-Jordan elimination **5.** No

7. $a_{46} = -8$ **9.** 1×2 **11.** 3×1 **13.** 2×2

15. $\left[\begin{array}{cc:c} 4 & -3 & -5 \\ -1 & 3 & 12 \end{array}\right]$ **17.** $\left[\begin{array}{ccc:c} 1 & -1 & 2 & 2 \\ 4 & -3 & 1 & -1 \\ 2 & 1 & 0 & 0 \end{array}\right]$

19. $\left[\begin{array}{ccc:c} 7 & -5 & 1 & 13 \\ 19 & 0 & -8 & 10 \end{array}\right]$

21. $\begin{cases} 3x + 4y = 0 \\ x - y = 7 \end{cases}$ **23.** $\begin{cases} 12y + 3z = 0 \\ -2x + 18y = 10 \\ x + 7y - 8z = 43 \end{cases}$

25. Multiply R_1 by $-\frac{1}{4}$.

27. Interchange R_1 and R_2. Add 4 times new R_1 to R_3.

29. $\begin{bmatrix} 1 & 4 & 3 \\ 0 & 2 & -1 \end{bmatrix}$

31. $\begin{bmatrix} 1 & 1 & 4 & -1 \\ 0 & 5 & -2 & 6 \\ 0 & 3 & 20 & 4 \end{bmatrix}, \begin{bmatrix} 1 & 1 & 4 & -1 \\ 0 & 1 & -\frac{2}{5} & \frac{6}{5} \\ 0 & 3 & 20 & 4 \end{bmatrix}$

33. $\left[\begin{array}{ccc:c} 0 & -2 & 0 & 5 \\ 0 & 2 & 1 & 1 \\ 4 & 8 & 4 & -4 \end{array}\right]$ **35.** $\left[\begin{array}{cc:c} -1 & 3 & 9 \\ 4 & -5 & -12 \end{array}\right]$

37. (a) i) $\left[\begin{array}{cc:c} 3 & 0 & -6 \\ 6 & -4 & -28 \end{array}\right]$ ii) $\left[\begin{array}{cc:c} 3 & 0 & -6 \\ 0 & -4 & -16 \end{array}\right]$

iii) $\left[\begin{array}{cc:c} 3 & 0 & -6 \\ 0 & 1 & 4 \end{array}\right]$ iv) $\left[\begin{array}{cc:c} 1 & 0 & -2 \\ 0 & 1 & 4 \end{array}\right]$

(b) $x = -2, y = 4$ (c) Answers will vary.

39. Reduced row-echelon form **41.** Row-echelon form

43. $\begin{bmatrix} 1 & -3 & 3 \\ 0 & 1 & -2 \end{bmatrix}$ **45.** $\begin{bmatrix} 1 & 2 & -1 & 3 \\ 0 & 1 & -2 & 5 \\ 0 & 0 & 1 & -1 \end{bmatrix}$

47. $\begin{bmatrix} 1 & 0 & 0 \\ 0 & 1 & 0 \\ 0 & 0 & 1 \end{bmatrix}$ **49.** $\begin{bmatrix} 1 & 0 & -\frac{3}{7} & -\frac{8}{7} \\ 0 & 1 & -\frac{12}{7} & \frac{10}{7} \end{bmatrix}$

51. $\begin{cases} x - 2y = 4 \\ y = -1 \end{cases}$
$(2, -1)$

53. $\begin{cases} x - y + 2z = 4 \\ y - z = 2 \\ z = -2 \end{cases}$
$(8, 0, -2)$

55. $(7, -5)$ **57.** $(3, 2)$ **59.** Inconsistent

61. $(-4, -3, 6)$ **63.** $(3, -2, 5, 0)$ **65.** $(-1, -4)$

67. $(4, -3, 2)$ **69.** $\left(-\frac{3}{2}a + \frac{3}{2}, \frac{1}{3}a + \frac{1}{3}, a\right)$

71. $(-13, -9, 17)$ **73.** $(0, -6, 2)$ **75.** $(-5a, a, 3)$

77. Yes; $(-1, 1, -3)$ **79.** No **81.** $y = x^2 + 2x + 5$

83. $y = 2x^2 - x + 1$ **85.** $f(x) = -9x^2 - 5x + 11$

87. $f(x) = x^3 - 2x^2 - 4x + 1$

89. (a) $f(x) = -\frac{1}{250}x^2 + \frac{11}{30}x + 5$

(b)

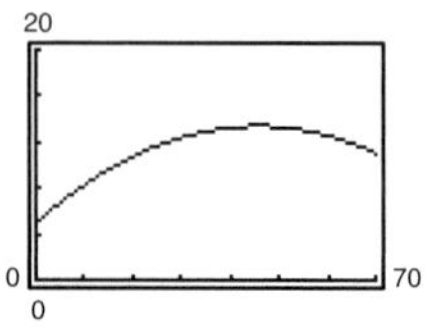

About 13.4 ft

91. $\frac{8x^2}{(x - 1)^2(x + 1)} = \frac{2}{x + 1} + \frac{6}{x - 1} + \frac{4}{(x - 1)^2}$

93. $A = -1, B = 3, C = 2, D = -3$

95. $A_1 = \$200{,}000, A_2 = \$500{,}000, A_3 = \$800{,}000$

97. $\begin{cases} 5.5g + 4.25s + 3.75m = 480 \\ -g + s + m = 0 \\ g + s + m = 100 \end{cases}$

50 lb of glossy, 35 lb of semi-gloss, 15 lb of matte

99. (a) $x_1 = s, x_2 = t, x_3 = 600 - s,$
$x_4 = s - t, x_5 = 500 - t, x_6 = s, x_7 = t$
(b) $x_1 = 0, x_2 = 0, x_3 = 600, x_4 = 0, x_5 = 500,$
$x_6 = 0, x_7 = 0$

101. True; See Example 7.

103. *Sample answer:*
$\begin{cases} x + y + 7z = -1 \\ x + 2y + 11z = 0 \\ 2x + y + 10z = -3 \end{cases}$

105. No; Answers will vary.

107.

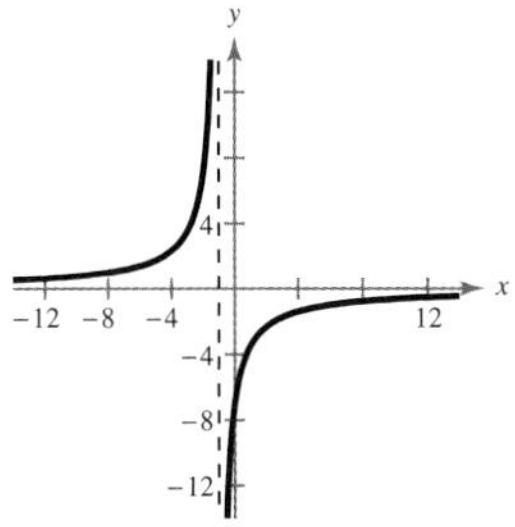

Asymptotes:
$x = -1, y = 0$

109.

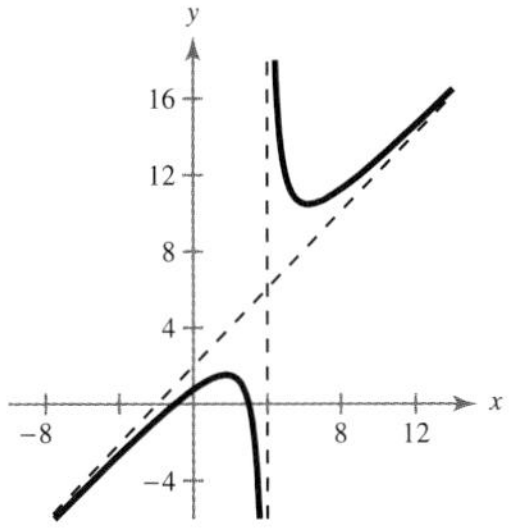

Asymptotes:
$x = 4, y = x + 2$

Section 7.5 (page 536)

1. equal **3.** zero, O

5. (a) iii (b) i (c) iv (d) v (e) ii

7. No **9.** $x = 5, y = -8$ **11.** $x = -1, y = 3$

13. (a) $\begin{bmatrix} 8 & -1 \\ 1 & 7 \end{bmatrix}$ (b) $\begin{bmatrix} 2 & -3 \\ 5 & -5 \end{bmatrix}$
(c) $\begin{bmatrix} 15 & -6 \\ 9 & 3 \end{bmatrix}$ (d) $\begin{bmatrix} 9 & -8 \\ 13 & -9 \end{bmatrix}$

15. (a) $\begin{bmatrix} 9 & 5 \\ 1 & -2 \\ -3 & 15 \end{bmatrix}$ (b) $\begin{bmatrix} 7 & -7 \\ 3 & 8 \\ -5 & -5 \end{bmatrix}$
(c) $\begin{bmatrix} 24 & -3 \\ 6 & 9 \\ -12 & 15 \end{bmatrix}$ (d) $\begin{bmatrix} 22 & -15 \\ 8 & 19 \\ -14 & -5 \end{bmatrix}$

17. (a) $\begin{bmatrix} 5 & 5 & -2 & 4 & 4 \\ -5 & 10 & 0 & -4 & -7 \end{bmatrix}$
(b) $\begin{bmatrix} 3 & 5 & 0 & 2 & 4 \\ 7 & -6 & -4 & 2 & 7 \end{bmatrix}$
(c) $\begin{bmatrix} 12 & 15 & -3 & 9 & 12 \\ 3 & 6 & -6 & -3 & 0 \end{bmatrix}$
(d) $\begin{bmatrix} 10 & 15 & -1 & 7 & 12 \\ 15 & -10 & -10 & 3 & 14 \end{bmatrix}$

19. (a) Not possible (b) Not possible
(c) $\begin{bmatrix} 18 & 0 & 9 \\ -3 & -12 & 0 \end{bmatrix}$ (d) Not possible

21. $\begin{bmatrix} -8 & -7 \\ 15 & -1 \end{bmatrix}$ **23.** $\begin{bmatrix} -3 & -\frac{1}{3} & 1 \\ -4 & \frac{8}{3} & -5 \end{bmatrix}$

25. $\begin{bmatrix} -17.143 & 2.143 \\ 11.571 & 10.286 \end{bmatrix}$ **27.** $\begin{bmatrix} -16.463 & -33.373 \\ 3.706 & -26.654 \\ 2.167 & 18.382 \end{bmatrix}$

29. $\begin{bmatrix} -4 & 6 & -2 \\ 4 & 0 & 10 \end{bmatrix}$ **31.** $\begin{bmatrix} -2 & 0 & 5 \\ -\frac{5}{2} & 0 & \frac{7}{2} \end{bmatrix}$

33. $\begin{bmatrix} 3 & -\frac{1}{2} & -\frac{13}{2} \\ 3 & 0 & -\frac{11}{2} \end{bmatrix}$ **35.** $\begin{bmatrix} 2 & -5 & 5 \\ -5 & 0 & -6 \end{bmatrix}$

37. $\begin{bmatrix} -2 & 51 \\ -8 & 33 \\ 0 & 27 \end{bmatrix}$ **39.** Not possible

41. $\begin{bmatrix} 2 & 0 & 0 \\ 0 & -1 & 0 \\ 0 & 0 & -\frac{1}{3} \end{bmatrix}$ **43.** $\begin{bmatrix} 10 & -40 & 20 \\ -6 & 24 & -12 \\ 8 & -32 & 16 \end{bmatrix}$

45. (a) $\begin{bmatrix} 0 & 15 \\ 6 & 12 \end{bmatrix}$ (b) $\begin{bmatrix} -2 & 2 \\ 31 & 14 \end{bmatrix}$ (c) $\begin{bmatrix} 9 & 6 \\ 12 & 12 \end{bmatrix}$

47. (a) $\begin{bmatrix} 0 & -10 \\ 10 & 0 \end{bmatrix}$ (b) $\begin{bmatrix} 0 & -10 \\ 10 & 0 \end{bmatrix}$ (c) $\begin{bmatrix} 8 & -6 \\ 6 & 8 \end{bmatrix}$

49. $\begin{bmatrix} 124 & -70 \\ 228 & 452 \\ 192 & -72 \end{bmatrix}$ **51.** Not possible **53.** $\begin{bmatrix} 5 & 8 \\ -4 & -16 \end{bmatrix}$

55. $\begin{bmatrix} -4 & 10 \\ 3 & 14 \end{bmatrix}$ **57.** (a) No (b) Yes (c) No (d) No

59. (a) $\langle 4, 7\rangle$ (b) $\langle -2, 3\rangle$ (c) $\langle 8, 1\rangle$

61. (a) $\langle 3, 6\rangle$ (b) $\langle -7, -2\rangle$ (c) $\langle 17, 10\rangle$

63. $\langle 4, -2\rangle$; Reflection in the x-axis

65. $\langle 2, 4\rangle$; Reflection in the line $y = x$

67. $\langle 8, 2\rangle$; Horizontal stretch

69. (a) $\begin{bmatrix} -1 & 1 \\ -2 & 1 \end{bmatrix}\begin{bmatrix} x_1 \\ x_2 \end{bmatrix} = \begin{bmatrix} 4 \\ 0 \end{bmatrix}$ (b) $\begin{bmatrix} 4 \\ 8 \end{bmatrix}$

71. (a) $\begin{bmatrix} 2 & 1 \\ 2 & -3 \end{bmatrix}\begin{bmatrix} x_1 \\ x_2 \end{bmatrix} = \begin{bmatrix} 2 \\ 10 \end{bmatrix}$ (b) $\begin{bmatrix} 2 \\ -2 \end{bmatrix}$

73. (a) $\begin{bmatrix} 1 & -2 & 3 \\ -1 & 3 & -1 \\ 2 & -5 & 5 \end{bmatrix}\begin{bmatrix} x_1 \\ x_2 \\ x_3 \end{bmatrix} = \begin{bmatrix} 9 \\ -6 \\ 17 \end{bmatrix}$ (b) $\begin{bmatrix} 1 \\ -1 \\ 2 \end{bmatrix}$

75. (a) $\begin{bmatrix} 7 & -2 & 5 \\ -6 & 13 & -8 \\ 16 & 11 & -3 \end{bmatrix}$ (b) $\begin{bmatrix} 7 & -2 & 5 \\ -6 & 13 & -8 \\ 16 & 11 & -3 \end{bmatrix}$
They are the same.

77. (a) $\begin{bmatrix} 25 & -34 & 28 \\ -53 & 34 & -7 \\ -76 & 30 & 21 \end{bmatrix}$ (b) $\begin{bmatrix} 25 & -34 & 28 \\ -53 & 34 & -7 \\ -76 & 30 & 21 \end{bmatrix}$
They are the same.

79. (a) and (b) $\begin{bmatrix} 3 & -6 & 0 \\ 3 & 3 & 0 \end{bmatrix}$ **81.** Not possible, undefined

83. Not possible, undefined **85.** (a) and (b) $\begin{bmatrix} 5 & 16 & -13 \\ -8 & 5 & 0 \end{bmatrix}$

87. $\begin{bmatrix} -4 & 0 \\ 8 & 2 \end{bmatrix}$ **89.** $\begin{bmatrix} 85 & 102 & 51 & 34 \\ 119 & 136 & 170 & 68 \end{bmatrix}$

91. $\begin{bmatrix} 84 & 60 & 48 \\ 42 & 120 & 90 \end{bmatrix}$

93. (a) [\$3.50 \$4.00]
(b) [\$837.50 \$1050.00 \$762.50]
The entries represent the total profits made at the three outlets.

95. (a) $\begin{bmatrix} \$23.20 & \$20.50 \\ \$38.20 & \$33.80 \\ \$76.90 & \$68.50 \end{bmatrix}$ The entries represent labor costs at the two plants for the three boat sizes.
(b) \$8.40

97. $\begin{bmatrix} 0.40 & 0.15 & 0.15 \\ 0.28 & 0.53 & 0.17 \\ 0.32 & 0.32 & 0.68 \end{bmatrix}$ P^2 represents the proportion of changes in party affiliations after two elections.

99. True; To add two matrices, you add corresponding entries.

101. Not possible **103.** 3×2 **105.** Not possible

107. 2×3 **109 and 111.** Answers will vary.

113. $AC = BC = \begin{bmatrix} 2 & 3 \\ 2 & 3 \end{bmatrix}, A \neq B$

115. (a) $A^2 = \begin{bmatrix} -1 & 0 \\ 0 & -1 \end{bmatrix}, A^3 = \begin{bmatrix} -i & 0 \\ 0 & -i \end{bmatrix}, A^4 = \begin{bmatrix} 1 & 0 \\ 0 & 1 \end{bmatrix}$
The entries on the main diagonal are i^2 in A^2, i^3 in A^3, and i^4 in A^4.
(b) $B^2 = \begin{bmatrix} 1 & 0 \\ 0 & 1 \end{bmatrix}$
B^2 is the identity matrix.

117. The resulting matrix should be a 1×1 matrix with an entry of $3(1) + 4(2)$.

119. $\ln \dfrac{64}{\sqrt[3]{x^2+3}}$

Section 7.6 (page 547)

1. inverse **3.** No **5–9.** Answers will vary.

11. $\begin{bmatrix} \frac{1}{2} & 0 \\ 0 & \frac{1}{3} \end{bmatrix}$ **13.** $\begin{bmatrix} -3 & 2 \\ -2 & 1 \end{bmatrix}$ **15.** Does not exist

17. Does not exist **19.** $\begin{bmatrix} 1 & 1 & -1 \\ -3 & 2 & -1 \\ 3 & -3 & 2 \end{bmatrix}$

21. $\begin{bmatrix} -\frac{3}{2} & \frac{3}{2} & 1 \\ \frac{9}{2} & -\frac{7}{2} & -3 \\ -1 & 1 & 1 \end{bmatrix}$ **23.** $\begin{bmatrix} -12 & -5 & -9 \\ -4 & -2 & -4 \\ -8 & -4 & -6 \end{bmatrix}$

25. $\dfrac{5}{11}\begin{bmatrix} 0 & -4 & 2 \\ -22 & 11 & 11 \\ 22 & -6 & -8 \end{bmatrix}$ **27.** Does not exist

29. Does not exist **31.** $\dfrac{1}{59}\begin{bmatrix} 16 & 15 \\ -4 & 70 \end{bmatrix}$

33. $\begin{bmatrix} \frac{5}{13} & -\frac{3}{13} \\ \frac{1}{13} & \frac{2}{13} \end{bmatrix}$ **35.** $k = 0$ **37.** $(-23, -18)$

39. $(7, -4, -7)$ **41.** $(2, -2)$ **43.** Not possible

45. $(-4, -8)$ **47.** $(-1, 3, 2)$ **49.** $\left(\frac{11}{8}, -\frac{3}{8}\right)$

51. $\left(-\frac{61}{30}, -\frac{1}{6}, -\frac{3}{10}\right)$ **53.** $(5, 0, -2, 3)$

55. \$4000 in AAA-rated bonds, \$2000 in A-rated bonds, and \$4000 in B-rated bonds

57. \$30,000 in AAA-rated bonds, \$90,000 in A-rated bonds, and \$180,000 in B-rated bonds

59. $I_1 = \frac{1}{2}$ ampere, $I_2 = 3$ amperes, $I_3 = 3.5$ amperes

61. 100 bags for seedlings, 100 bags for general potting, 100 bags for hardwood plants

63. (a) $\begin{cases} 2.5r + 4l + 2i = 300 \\ -r + 2l + 2i = 0 \\ r + l + i = 120 \end{cases}$
(b) $\begin{bmatrix} 2.5 & 4 & 2 \\ -1 & 2 & 2 \\ 1 & 1 & 1 \end{bmatrix}\begin{bmatrix} r \\ l \\ i \end{bmatrix} = \begin{bmatrix} 300 \\ 0 \\ 120 \end{bmatrix}$
(c) 80 roses, 10 lilies, 30 irises

65. True; $AA^{-1} = I = A^{-1}A$.

67. The scalar should be $\frac{1}{3}$, not 3. **69.** Answers will vary.

71. Answers will vary.

$$A^{-1} = \begin{bmatrix} \frac{1}{a_{11}} & 0 & 0 & 0 & \cdots & 0 \\ 0 & \frac{1}{a_{22}} & 0 & 0 & \cdots & 0 \\ 0 & 0 & \frac{1}{a_{33}} & 0 & \cdots & 0 \\ \vdots & \vdots & \vdots & \vdots & \cdots & \vdots \\ 0 & 0 & 0 & 0 & \cdots & \frac{1}{a_{nn}} \end{bmatrix}$$

73. $\ln 3 \approx 1.099$ **75.** $\dfrac{e^{12/7}}{3} \approx 1.851$

Section 7.7 (page 554)

1. determinant **3.** -5 **5.** 6 **7.** 16 **9.** -24

11. -9 **13.** 7.76 **15.** 11.998

17. (a) $M_{11} = -6, M_{12} = 3, M_{21} = 5, M_{22} = 4$
(b) $C_{11} = -6, C_{12} = -3, C_{21} = -5, C_{22} = 4$

19. (a) $M_{11} = 10, M_{12} = -43, M_{13} = 2, M_{21} = -30, M_{22} = 17, M_{23} = -6, M_{31} = 54, M_{32} = -53, M_{33} = -34$
(b) $C_{11} = 10, C_{12} = 43, C_{13} = 2, C_{21} = 30, C_{22} = 17, C_{23} = 6, C_{31} = 54, C_{32} = 53, C_{33} = -34$

21. (a) 96 (b) 96 **23.** (a) -170 (b) -170

25. -9 **27.** 0 **29.** -9 **31.** -168 **33.** 412

35. -336 **37.** 48,834 **39.** 566

41. (a) 1 (b) -1 (c) $\begin{bmatrix} 3 & -5 \\ -5 & 8 \end{bmatrix}$
(d) -1; $|AB| = |A| \cdot |B|$

43. (a) -25 (b) -220 (c) $\begin{bmatrix} -7 & -16 & -1 & -28 \\ -4 & -14 & -11 & 8 \\ 13 & 4 & 4 & -4 \\ -2 & 3 & 2 & 2 \end{bmatrix}$
(d) 5500; $|AB| = |A| \cdot |B|$

45–49. Answers will vary. **51.** $x = \pm 2$ **53.** $x = \pm\frac{5}{2}$

55. $x = -1, 3$ **57.** $x = -4, -1$ **59.** $x = 1, \frac{1}{2}$

61. $x = 3$ **63.** $8uv - 1$ **65.** e^{5x} **67.** $1 - \ln x$

69. r

71. True; Expansion by cofactors on a row of zeros is zero.

73. Answers will vary. *Sample answer:*

$A = \begin{bmatrix} 1 & 0 & -3 \\ 6 & -2 & 7 \\ 9 & 5 & -1 \end{bmatrix}, \quad B = \begin{bmatrix} 3 & 1 & 5 \\ -8 & 1 & 0 \\ -7 & 6 & -2 \end{bmatrix}$

$|A + B| = -328, |A| + |B| = -404$

75. (a) 9 (b) $\begin{bmatrix} \frac{1}{3} & -\frac{1}{3} \\ \frac{2}{9} & \frac{1}{9} \end{bmatrix}$ (c) $\frac{1}{9}$ (d) They are reciprocals.

77. (a) 2 (b) $\begin{bmatrix} -4 & -5 & 1.5 \\ -1 & -1 & 0.5 \\ -1 & -1 & 0 \end{bmatrix}$ (c) $\frac{1}{2}$
(d) They are reciprocals.

79. (a) Columns 2 and 3 are interchanged.
(b) Rows 1 and 3 are interchanged.

81. (a) 3 is factored from the second row.
(b) 2 and 4 are factored from the first and second columns, respectively.

83. (a) 15 (b) -75 (c) -120
The determinant of a triangular matrix is the product of the numbers along the main diagonal.

85. The signs for each coffactors should be $-, +, -$.

87. $(2y - 3)^2$

Section 7.8 (page 566)

1. Cramer's Rule **3.** 7 **5.** $\frac{33}{8}$ **7.** $x = 0, -\frac{16}{5}$

9. Collinear **11.** Not collinear **13.** $x = 3$

15. (0, 0), (0, 3), (6, 0), (6, 3) **17.** $(-4, 3), (-5, 3), (-4, 4), (-5, 4)$

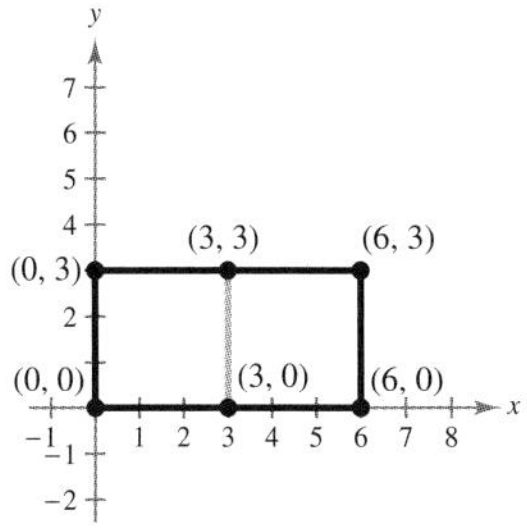

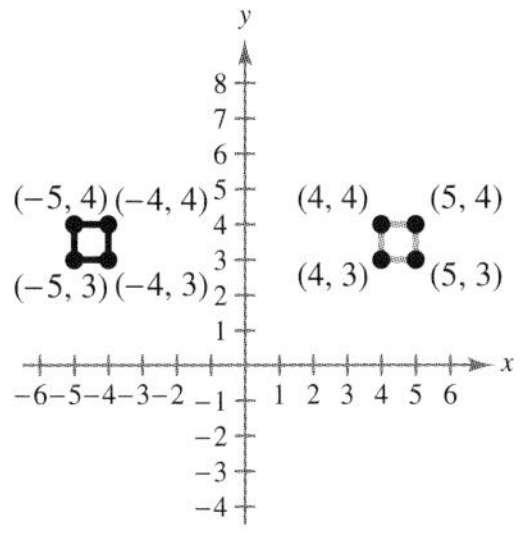

19. 2 square units **21.** 10 square units **23.** $(1, -1)$

25. $(-1, 2)$ **27.** $\left(0, -\frac{1}{2}, \frac{1}{2}\right)$ **29.** Not possible

31. (a) $y = -\frac{1}{7}t^2 + \frac{712}{175}t + \frac{5909}{70}$
(b)

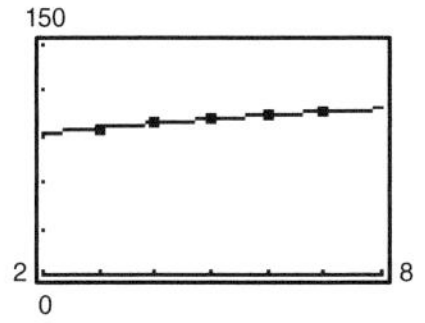

The model fits the data well.

33. (a) [3 15], [13 5], [0 8], [15 13], [5 0], [19 15], [15 14]
(b) 48 81 28 51 24 40 54 95 5 10 64 113 57 100

35. (a) [3 1 12], [12 0 13], [5 0 20], [15 4 1], [25 0 0]
(b) -68 21 35 -66 14 39 -115 35 60 13 -13 -1 25 -25 0

37. HAPPY NEW YEAR **39.** CANCEL ORDERS SUE

41. True; Cramer's Rule divides by the determinant.

43. The matrix should be $\begin{bmatrix} k & 0 \\ 0 & 1 \end{bmatrix}$.

45. Answers will vary. **47.** $x + 4y - 19 = 0$

Review Exercises (page 568)

1. (3, 3) **3.** $(1, -2)$ **5.** (0.25, 0.625) **7.** (5, 4)

9. $(0, 0), (2, 8), (-2, 8)$ **11.** $(2, -0.5)$

13. $(0, 0), (4, -4)$ **15.** (4, 8) **17.** 800 plants

19. 80 m × 200 m **21.** $(-8, 1)$

23. $\left(\frac{5}{2}, 3\right)$ **25.** Infinitely many solutions **27.** (0, 0)

29. No solution

31.

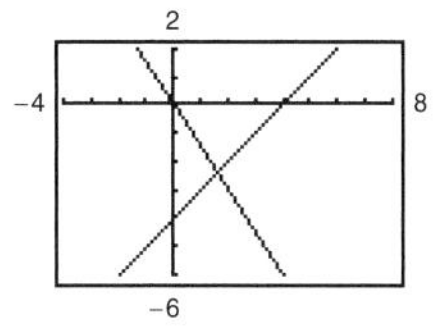

Consistent: $(1.6, -2.4)$

33.

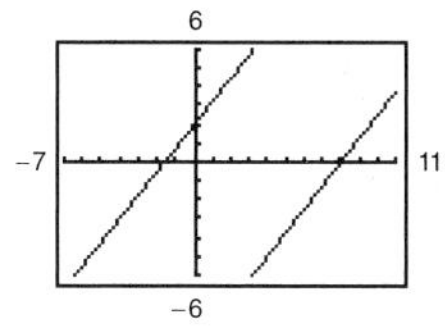

Inconsistent

35.

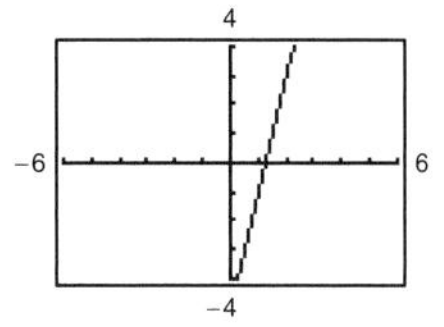

Consistent; All points on the line $8x - 2y = 10$

37. $\left(\frac{500{,}000}{7}, \frac{159}{7}\right)$ **39.** 227 mi/h; 287 mi/h

41. $(0, 4, -2)$ **43.** $\left(-\frac{3}{4}, 0, -\frac{5}{4}\right)$ **45.** $(a - 4, a - 3, a)$

47. $\frac{3}{x + 2} - \frac{4}{x + 4}$ **49.** $\frac{1}{2}\left(\frac{3}{x - 1} - \frac{x - 3}{x^2 + 1}\right)$

51. $y = 2x^2 + x - 5$

53. 4 par-3 holes, 10 par-4 holes, 4 par-5 holes

55. $\left[\begin{array}{cc:c} 6 & -7 & 11 \\ -2 & 5 & -1 \end{array}\right]$ **57.** $\left[\begin{array}{ccc:c} 3 & -5 & 1 & 25 \\ -4 & 0 & -2 & -14 \end{array}\right]$

59. $\begin{bmatrix} 1 & -\frac{5}{3} & 3 \\ 0 & 1 & \frac{9}{7} \end{bmatrix}$ **61.** $\begin{bmatrix} 1 & 1 & 1 \\ 0 & 1 & 2 \\ 0 & 0 & 1 \end{bmatrix}$ **63.** $\begin{bmatrix} 1 & 0 \\ 0 & 1 \end{bmatrix}$

65. $\begin{bmatrix} 1 & 0 & 0 \\ 0 & 1 & 0 \\ 0 & 0 & 1 \end{bmatrix}$ **67.** $(10, -12)$ **69.** $\left(\frac{1}{2}, -\frac{1}{3}, 1\right)$

71. $(a - 3, -2a, a)$ **73.** Inconsistent **75.** $(3, 0, -4)$

77. $x = 12, y = -7$

79. (a) $\begin{bmatrix} 17 & -17 \\ 13 & 2 \end{bmatrix}$ (b) $\begin{bmatrix} -3 & 23 \\ -15 & 8 \end{bmatrix}$
(c) $\begin{bmatrix} 14 & 6 \\ -2 & 10 \end{bmatrix}$ (d) $\begin{bmatrix} 37 & -57 \\ 41 & -4 \end{bmatrix}$

81. $\begin{bmatrix} -13 & -8 & 18 \\ 0 & 11 & -19 \end{bmatrix}$ **83.** $\begin{bmatrix} 52 & 0 \\ -12 & 32 \\ -16 & 32 \end{bmatrix}$

85. $\begin{bmatrix} 21 & -\frac{45}{4} & -\frac{159}{8} \\ \frac{93}{8} & \frac{327}{8} & \frac{291}{8} \end{bmatrix}$ **87.** $\begin{bmatrix} -5 & 1 \\ 0 & -\frac{9}{2} \\ -1 & 4 \end{bmatrix}$

89. $\begin{bmatrix} 14 & -2 & 8 \\ 14 & -10 & 40 \\ 36 & -12 & 48 \end{bmatrix}$ **91.** $\begin{bmatrix} 14 & -22 & 22 \\ 19 & -41 & 80 \\ 42 & -66 & 66 \end{bmatrix}$

93. $\langle -5, -2 \rangle$; Reflection in the line $y = -x$

95. [\$99 \$112]

97. Answers will vary. **99.** $\begin{bmatrix} 4 & -5 \\ 5 & -6 \end{bmatrix}$

101. $\begin{bmatrix} \frac{1}{2} & -1 & -\frac{1}{2} \\ \frac{1}{2} & -\frac{2}{3} & -\frac{5}{6} \\ 0 & \frac{2}{3} & \frac{1}{3} \end{bmatrix}$ **103.** $\begin{bmatrix} \frac{1}{5} & \frac{1}{5} \\ \frac{1}{10} & -\frac{1}{15} \end{bmatrix}$ **105.** $\begin{bmatrix} 1 & -1 \\ 4 & -\frac{7}{2} \end{bmatrix}$

107. (36, 11) **109.** $(2, -1, 0)$ **111.** $(-3, 1)$

113. $(1, 1, -2)$ **115.** -17 **117.** -42

CHAPTER 7

119. (a) $M_{11} = 4, M_{12} = 7, M_{21} = -1, M_{22} = 2$
(b) $C_{11} = 4, C_{12} = -7, C_{21} = 1, C_{22} = 2$

121. (a) $M_{11} = 30, M_{12} = -12, M_{13} = -21, M_{21} = 20,$
$M_{22} = 19, M_{23} = 22, M_{31} = 5, M_{32} = -2, M_{33} = 19$
(b) $C_{11} = 30, C_{12} = 12, C_{13} = -21, C_{21} = -20,$
$C_{22} = 19, C_{23} = -22, C_{31} = 5, C_{32} = 2, C_{33} = 19$

123. 24 **125.** -4 **127.** 279

129. Not collinear; $\frac{13}{2}$ square units **131.** 8 square units

133. $(1, 2)$ **135.** $(0, -3, -3)$

137. $[12\ 15\ 15], [11\ 0\ 15], [21\ 20\ 0], [2\ 5\ 12],$
$[15\ 23\ 0]; -21\ 6\ 0\ -68\ 8\ 45\ 102\ -42$
$-60\ -53\ 20\ 21\ 99\ -30\ -69$

139. False; Only square matrices are nonsingular.

141. An $n \times n$ matrix A has an inverse A^{-1} if $\det(A) \neq 0$.

Chapter Test (page 576)

1. $(4, -2)$ **2.** $(1, 0), (2, 1)$ **3.** $(-2, -3)$ **4.** $\left(\frac{7}{13}, -\frac{85}{26}\right)$

5. $\left(\frac{14}{3}, \frac{8}{3}\right)$; Answers will vary. **6.** $y = -\frac{1}{2}x^2 + x + 6$

7. $\dfrac{5}{x-1} + \dfrac{3}{(x-1)^2}$ **8.** $\dfrac{1}{x} + \dfrac{2}{x^2} - \dfrac{1}{x^2+1}$

9. $(-2a + 1.5, 2a + 1, a)$ **10.** $(5, 2, -6)$

11. (a) $\begin{bmatrix} 1 & 0 & 4 \\ -7 & -6 & -1 \\ 0 & 4 & 0 \end{bmatrix}$ (b) $\begin{bmatrix} 15 & 12 & 12 \\ -12 & -12 & 0 \\ 3 & 6 & 0 \end{bmatrix}$

(c) $\begin{bmatrix} 7 & 6 & 12 \\ -18 & -16 & -2 \\ 1 & 10 & 0 \end{bmatrix}$ (d) $\begin{bmatrix} 36 & 20 & 4 \\ -28 & -24 & -4 \\ 10 & 8 & 2 \end{bmatrix}$

12. $(-2, 3, -1)$ **13.** 67 **14.** -2

15. $\frac{13}{2}$ square units **16.** $\left(1, -\frac{1}{2}\right)$

17. $x_1 = 700 - s - t, x_2 = 300 - s - t,$
$x_3 = s, x_4 = 100 - t, x_5 = t$

Chapter 8

Section 8.1 (page 587)

1. terms **3.** index, upper limit, lower limit

5. (a) Finite sequence (b) Infinite sequence

7. $-3, -1, 1, 3, 5$ **9.** $3, 9, 27, 81, 243$

11. $-2, 4, -8, 16, -32$ **13.** $2, \frac{3}{2}, \frac{4}{3}, \frac{5}{4}, \frac{6}{5}$ **15.** $\frac{1}{2}, \frac{2}{5}, \frac{3}{10}, \frac{4}{17}, \frac{5}{26}$

17. (a) 0, 1, 0, 0.5, 0 (b) $0, 1, 0, \frac{1}{2}, 0$

19. (a) 0.5, 0.75, 0.875, 0.938, 0.969 (b) $\frac{1}{2}, \frac{3}{4}, \frac{7}{8}, \frac{15}{16}, \frac{31}{32}$

21. (a) 1, 0.354, 0.192, 0.125, 0.089 (b) $1, \dfrac{1}{2^{3/2}}, \dfrac{1}{3^{3/2}}, \dfrac{1}{8}, \dfrac{1}{5^{3/2}}$

23. (a) $-1, 0.25, -0.111, 0.063, -0.04$
(b) $-1, \frac{1}{4}, -\frac{1}{9}, \frac{1}{16}, -\frac{1}{25}$

25. (a) and (b) 3, 15, 35, 63, 99

27. 9, 15, 21, 27, 33, 39, 45, 51, 57, 63

29. $3, \frac{5}{2}, \frac{7}{3}, \frac{9}{4}, \frac{11}{5}, \frac{13}{6}, \frac{15}{7}, \frac{17}{8}, \frac{19}{9}, \frac{21}{10}$ **31.** 0, 2, 0, 2, 0, 2, 0, 2, 0, 2

33. $\frac{100}{101}$ **35.** -43 **37.** $\frac{129}{256}$ **39.** $a_n = 5n - 2$

41. $a_n = 6n + 1$ **43.** $a_n = n^2 - 1$ **45.** $a_n = \dfrac{n+1}{n+2}$

47. $a_n = \dfrac{(-1)^{n+1}(n+1)}{2^n}$ **49.** $a_n = 1 + \dfrac{1}{n}$

51. $a_n = (-1)^n + 2(1)^n = (-1)^n + 2$ **53.** 28, 24, 20, 16, 12

55. 3, 4, 6, 10, 18 **57.** 1, 3, 4, 7, 11

59. 5, 8, 11, 14, 17; $a_n = 3n + 2$

61. 81, 27, 9, 3, 1; $a_n = \dfrac{243}{3^n}$

63. (a) 1, 1, 0.5, 0.167, 0.042 (b) $1, 1, \frac{1}{2}, \frac{1}{6}, \frac{1}{24}$

65. (a) 0, 0.5, 0.667, 0.375, 0.133 (b) $0, \frac{1}{2}, \frac{2}{3}, \frac{3}{8}, \frac{2}{15}$

67. (a) and (b) $6, -24, 60, -120, 210$ **69.** $\frac{1}{12}$ **71.** $\frac{33}{2}$

73. $n + 1$ **75.** $\dfrac{1}{2n(2n+1)}$ **77.** c **78.** b

79. d **80.** a

81.

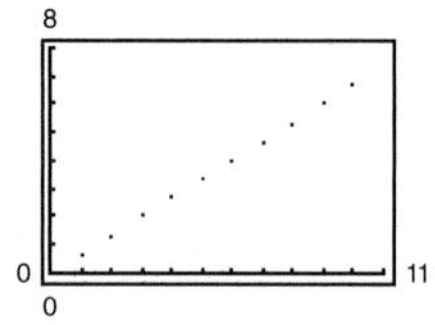

83.

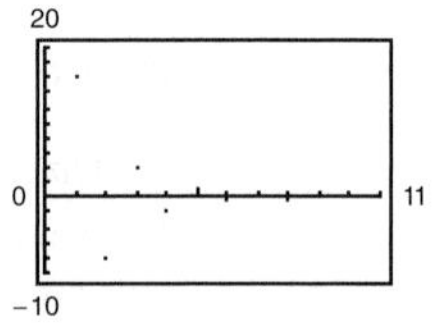

85.

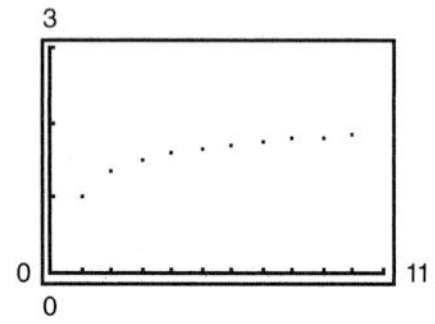

87. 35 **89.** 364 **91.** $\frac{124}{429}$ **93.** 80 **95.** 310

97. 81 **99.** $0.6\overline{3}$ **101.** $\displaystyle\sum_{i=1}^{9} \frac{1}{3i} \approx 0.94299$

103. $\displaystyle\sum_{i=1}^{8} \left[2\left(\frac{i}{8}\right) + 3\right] = 33$ **105.** $\displaystyle\sum_{i=1}^{6} (-1)^i 3^i = 546$

107. $\displaystyle\sum_{i=1}^{5} \frac{2^i - 1}{2^{i+1}} \approx 2.0156$ **109.** (a) $\dfrac{7}{8}$ (b) $\dfrac{15}{16}$ (c) $\dfrac{31}{32}$

111. (a) $-\frac{3}{2}$ (b) $-\frac{5}{4}$ (c) $-\frac{11}{8}$ **113.** $\frac{2}{3}$ **115.** $\frac{7}{9}$

117. $x, \dfrac{x^2}{2}, \dfrac{x^3}{6}, \dfrac{x^4}{24}, \dfrac{x^5}{120}$ **119.** $-\dfrac{x^3}{3}, \dfrac{x^5}{5}, -\dfrac{x^7}{7}, \dfrac{x^9}{9}, -\dfrac{x^{11}}{11}$

121. $-\dfrac{x^2}{2}, \dfrac{x^4}{24}, -\dfrac{x^6}{720}, \dfrac{x^8}{40{,}320}, -\dfrac{x^{10}}{3{,}628{,}800}$

123. $\dfrac{1}{4}, \dfrac{1}{12}, \dfrac{1}{24}, \dfrac{1}{40}, \dfrac{1}{60}; \dfrac{1}{2} - \dfrac{1}{2n+2}$

125. $\dfrac{1}{6}, \dfrac{1}{12}, \dfrac{1}{20}, \dfrac{1}{30}, \dfrac{1}{42}; \dfrac{1}{2} - \dfrac{1}{n+2}$

127. (a) \$5050, \$5100.50, \$5151.50, \$5203.02, \$5255.05, \$5307.60, \$5360.68, \$5414.28
(b) \$7444.32

129. (a) Linear sequence: $R_n = 1573.16n - 280.6, r^2 \approx 0.9455$
Quadratic sequence:
$R_n = 253.007n^2 - 703.90n + 4104.8, r^2 \approx 0.9977$
(b)

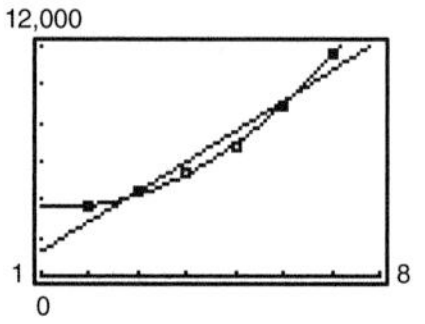

Quadratic sequence; The coefficient of determination is closer to 1.
(c) About \$26,975.7 million (d) 2025
(e) About \$40,791.5 million; \$40,791.6 million

131. True; Use the Properties of Sums.

133. 1, 1, 2, 3, 5, 8, 13, 21, 34, 55, 89, 144;
$1, 2, \frac{3}{2}, \frac{5}{3}, \frac{8}{5}, \frac{13}{8}, \frac{21}{13}, \frac{34}{21}, \frac{55}{34}, \frac{89}{55}$

135. 1, 1, 2, 3, 5

137. $a_{n+1} = \frac{1}{2}a_n + \frac{(1+\sqrt{5})^n + (1-\sqrt{5})^n}{2^{n+1}}$

$a_{n+2} = \frac{3}{2}a_n + \frac{(1+\sqrt{5})^n + (1-\sqrt{5})^n}{2^{n+1}}$

139. $\sum_{k=1}^{4}(3 + 2k^2) = 3(4) + 2\sum_{k=1}^{4}k^2 = 72$ **141.** Proof

143. (a) $\begin{bmatrix} 2 \\ 6 \end{bmatrix}$ (b) $\begin{bmatrix} -10 \\ -15 \end{bmatrix}$ (c) Not possible
(d) Not possible

Section 8.2 (page 596)

1. $a_n = a_1 + (n-1)d$
3. A sequence is arithmetic when the differences between consecutive terms are the same.
5. 8 **7.** -1 **9.** Yes; The common difference is 2.
11. No; There is no common difference.
13. Yes; The common difference is 13.
15. No; There is no common difference.
17. Yes; The common difference is 0.
19. $a_n = 6n - 5$ **21.** $a_n = -7n + 50$ **23.** $a_n = \frac{10}{3}n + \frac{5}{3}$
25. $a_n = -\frac{3}{4}n + \frac{9}{2}$ **27.** $a_n = 103 - 3n$
29. $2, -4, -10, -16, -22$ **31.** $-2, 2, 6, 10, 14$
33. 22.45, 20.725, 19, 17.275, 15.55
35. 15, 19, 23, 27, 31; $d = 4$; $a_n = 11 + 4n$
37. $a_n = 25 - 3n$

39.

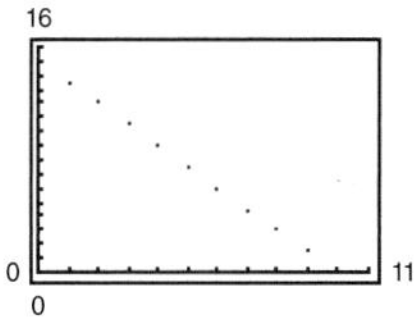

41.

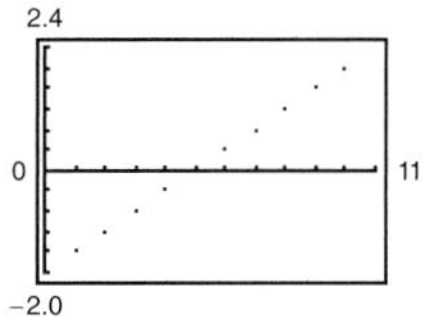

43. -49 **45.** -10.2
47. $-1, 3, 7, 11, 15, 19, 23, 27, 31, 35$
49. 19.25, 18.5, 17.75, 17, 16.25, 15.5, 14.75, 14, 13.25, 12.5
51. 1.55, 1.6, 1.65, 1.7, 1.75, 1.8, 1.85, 1.9, 1.95, 2
53. 110 **55.** -25 **57.** 10,000 **59.** -4585
61. -7020 **63.** 4000 **65.** 129,250 **67.** 10,100
69. 440 **71.** 2777.5 **73.** 14,268 **75.** 405 bricks
77. \$200,000
79. (a) $a_n = 0.070n + 2.03$
(b)

Year	2011	2012	2013	2014
Sales (in thousands of dollars)	2.10	2.17	2.24	2.31

Year	2015	2016	2017	2018
Sales (in thousands of dollars)	2.38	2.45	2.52	2.59

(c) \$18,760 (d) \$20,090; Answers will vary.
81. True; Use the recursion formula, $a_{n+1} = a_n + d$.
83. $x, 3x, 5x, 7x, 9x, 11x, 13x, 15x, 17x, 19x$
85. Given a_1 and a_2, you know that $d = a_2 - a_1$. So, $a_n = a_1 + (n-1)d$.
87. When $n = 50$, $a_n = 2(50) - 1 = 99$.
89. $-7, -4, -1, 2, 5, 8, 11$
$a_1 = -7$, $a_{n+1} = a_n + 3$
Answers will vary.
91. Answers will vary. *Sample answer:* Gauss saw that the sum of the first and last numbers was 101, the sum of the second and second-last numbers was 101, and so on. Seeing that there were 50 such pairs of numbers, Gauss simply multiplied 50 by 101 to get the summation 5050.

$a_n = (n+1)\left(\frac{n}{2}\right)$, where n is the total number of natural numbers.

93. $(1, 5, -1)$ **95.** Answers will vary.

Section 8.3 (page 605)

1. geometric, common **3.** geometric series **5.** $|r| < 1$
7. Yes; $r = 2$ **9.** No **11.** Yes; $r = -\frac{1}{2}$
13. Yes; $r = 0.1$ **15.** No **17.** 6, 18, 54, 162, 486
19. $1, \frac{1}{2}, \frac{1}{4}, \frac{1}{8}, \frac{1}{16}$ **21.** $5, -\frac{1}{2}, \frac{1}{20}, -\frac{1}{200}, \frac{1}{2000}$
23. $1, e, e^2, e^3, e^4$ **25.** 64, 32, 16, 8, 4; $r = \frac{1}{2}$; $a_n = 64\left(\frac{1}{2}\right)^{n-1}$
27. 9, 18, 36, 72, 144; $r = 2$; $a_n = 9(2)^{n-1}$
29. $6, -9, \frac{27}{2}, -\frac{81}{4}, \frac{243}{8}$; $r = -\frac{3}{2}$; $a_n = 6\left(-\frac{3}{2}\right)^{n-1}$
31. (a) and (b) About 43.984
33. (a) About 44.949 (b) $\frac{32{,}768}{729}$ **35.** (a) and (b) -2048
37. (a) About 633.568 (b) $448\sqrt{2}$
39. $a_n = 7(3)^{n-1}$; 45,927 **41.** $a_n = 5(6)^{n-1}$; 50,388,480
43. 9 **45.** $-\frac{2}{9}$

47.

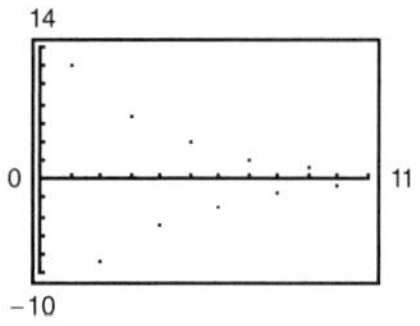

49.

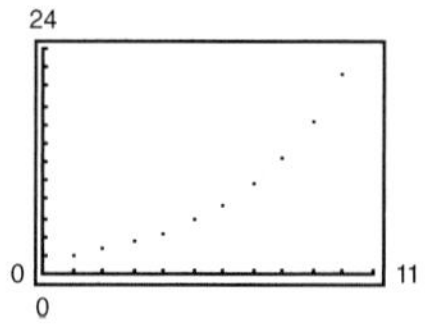

51. $8, 4, 6, 5, \frac{11}{2}$

53.

n	S_n
1	16
2	24
3	28
4	30
5	31
6	31.5
7	31.75
8	31.875
9	31.9375
10	31.96875

55. 511 **57.** 43 **59.** About 29,921.31
61. About 6.40 **63.** About 2092.60
65. $\sum_{n=1}^{6} 5(4)^{n-1}$ **67.** $\sum_{n=1}^{7} 2\left(-\frac{1}{4}\right)^{n-1}$ **69.** 2 **71.** 16
73. $\frac{10}{3}$ **75.** Not possible; $\left|\frac{7}{3}\right| > 1$ **77.** $-\frac{30}{19}$ **79.** 27
81. Not possible; $\left|\frac{5}{2}\right| > 1$ **83.** $-\frac{49}{9}$ **85.** $\frac{4}{11}$ **87.** $\frac{113}{90}$

CHAPTER 8

89. Geometric; $r = 2$; 262,136 **91.** Geometric; $r = \frac{1}{3}$; 135
93. Arithmetic; $d = 6$; 720
95. Geometric; $r = 0.8$; about 28.944 **97.** \$6480.83
99. Answers will vary.
101. (a) \$104,793.08 (b) \$105,055.53
103. (a) \$153,237.86 (b) \$153,657.02 **105.** 126 in.^{2}
107. (a) $T_n = 70(0.8)^n$ (b) About 18.4°F; about 4.8°F
(c)
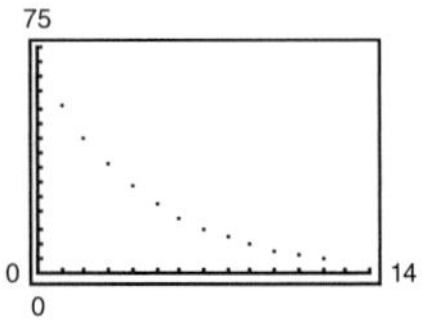

About 3.5 h
109. (a) $S = \sum_{n=0}^{\infty} 4\pi$ (b) $V = \sum_{n=0}^{\infty} \frac{4}{3}\pi\left(\frac{1}{3}\right)^n$
(c) Infinite surface area; Finite volume: 2π cubic units
111. 42 ft
113. False; You can find the nth term by using the formula $a_n = a_1 r^{n-1}$.
115. $3, \frac{3x}{2}, \frac{3x^2}{4}, \frac{3x^3}{8}, \frac{3x^4}{16}$ **117.** $100e^{8x}$
119. (a)
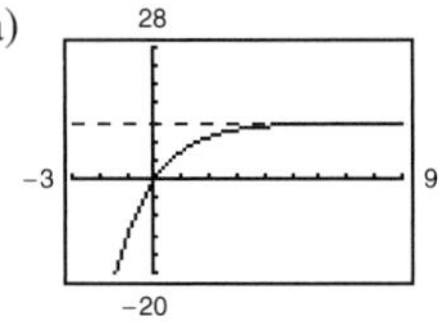

Horizontal asymptote: $y = 12$
Corresponds to the sum of the series

(b)
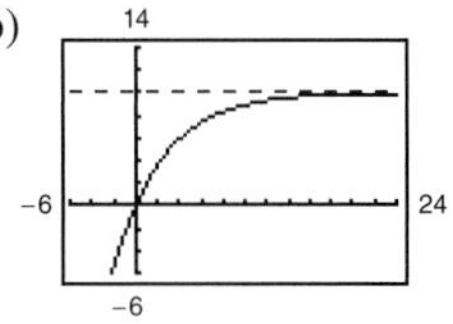

Horizontal asymptote: $y = 10$
Corresponds to the sum of the series
121. Divide the second term by the first to obtain the common ratio. The nth term is the first term times the common ratio raised to the $(n - 1)$th power.
123. -38 **125.** Answers will vary.

Section 8.4 (page 614)

1. ${}_nC_r$ or $\binom{n}{r}$ **3.** Binomial Theorem, Pascal's Triangle
5. 10 **7.** 66 **9.** 165 **11.** 1 **13.** 210
15. 91 **17.** 749,398 **19.** 124,750 **21.** Yes
23. No **25.** $x^6 + 6x^5 + 15x^4 + 20x^3 + 15x^2 + 6x + 1$
27. $y^3 - 12y^2 + 48y - 64$
29. $x^5 + 5x^4y + 10x^3y^2 + 10x^2y^3 + 5xy^4 + y^5$
31. $32x^5 - 80x^4y + 80x^3y^2 - 40x^2y^3 + 10xy^4 - y^5$
33. $64y^3 - 144y^2 + 108y - 27$
35. $64r^6 - 576r^5s + 2160r^4s^2 - 4320r^3s^3 + 4860r^2s^4 - 2916rs^5 + 729s^6$
37. $x^8 + 8x^6 + 24x^4 + 32x^2 + 16$
39. $-x^{10} + 25x^8 - 250x^6 + 1250x^4 - 3125x^2 + 3125$
41. $x^8 + 4x^6y^2 + 6x^4y^4 + 4x^2y^6 + y^8$
43. $729x^{18} - 1458x^{15}y + 1215x^{12}y^2 - 540x^9y^3 + 135x^6y^4 - 18x^3y^5 + y^6$
45. $\frac{1}{x^5} + \frac{5y}{x^4} + \frac{10y^2}{x^3} + \frac{10y^3}{x^2} + \frac{5y^4}{x} + y^5$
47. $\frac{16}{x^4} - \frac{64y}{x^3} + \frac{96y^2}{x^2} - \frac{64y^3}{x} + 16y^4$
49. $-512x^4 + 576x^3 - 240x^2 + 44x - 3$
51. $-4x^6 - 24x^5 - 60x^4 - 83x^3 - 42x^2 - 60x + 20$
53. $27x^{3/2} + 135x + 225\sqrt{x} + 125$
55. $x^2 - 3x^{4/3}y^{1/3} + 3x^{2/3}y^{2/3} - y$ **57.** $61{,}440x^7$
59. $360x^3y^2$ **61.** $1{,}259{,}712x^2y^7$
63. $-4{,}330{,}260{,}000x^3y^9$ **65.** 1,732,104 **67.** 720
69. $-489{,}888$ **71.** 210 **73.** 20 **75.** 6
77. $3125y^5 + 6250y^4 + 5000y^3 + 2000y^2 + 400y + 32$
79. $32x^5 + 240x^4y + 720x^3y^2 + 1080x^2y^3 + 810xy^4 + 243y^5$
81. $81t^4 - 216t^3v + 216t^2v^2 - 96tv^3 + 16v^4$
83. $3x^2 + 3xh + h^2, h \neq 0$
85. $6x^5 + 15x^4h + 20x^3h^2 + 15x^2h^3 + 6xh^4 + h^5, h \neq 0$
87. -4 **89.** $161 + 240i$ **91.** $2035 + 828i$
93. $-115 + 236i$ **95.** $-23 + 208\sqrt{3}i$ **97.** 1
99. $-\frac{1}{8}$ **101.** 1.172 **103.** 510,568.785
105.
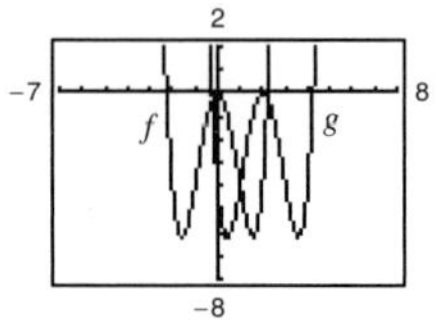

g is shifted two units to the right of f.
$g(x) = x^4 - 8x^3 + 19x^2 - 12x - 4$
107.
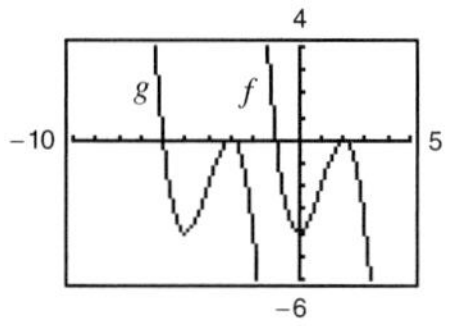

g is shifted five units to the left of f.
$g(x) = -x^3 - 12x^2 - 45x - 54$
109.
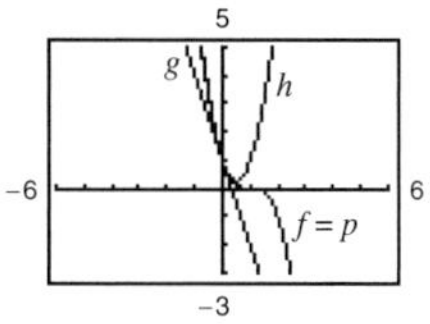

$p(x)$ is the expansion of $f(x)$.
111. About 0.2734 **113.** About 0.1707
115. (a) $g(t) = 0.0055t^2 + 0.303t + 7.6$
(b)
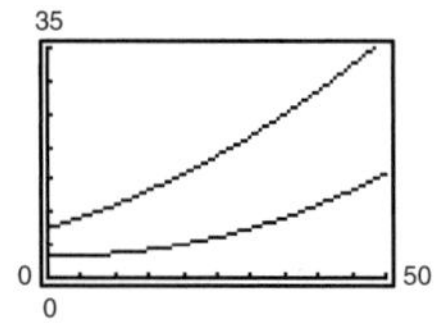

117. True; Pascal's Triangle is made up of binomial coefficients.
119. The first and last numbers in each row are 1. Every other number in each row is formed by adding the two numbers immediately above the number.

1 8 28 56 70 56 28 8 1
1 9 36 84 126 126 84 36 9 1
1 10 45 120 210 252 210 120 45 10 1

121 and 123. Answers will vary.
125. The last factor should be $(-2y)^r$. **127.** $\begin{bmatrix} 1 & 2 \\ -0.5 & -0.5 \end{bmatrix}$

Section 8.5 (page 623)

1. Fundamental Counting Principle **3.** n, r **5.** 8
7. 6 **9.** 11 **11.** 3 **13.** 30 **15.** 1024
17. (a) 900 (b) 648 **19.** 24,000,000
21. (a) 100,000 (b) 20,000 **23.** (a) 10,000 (b) 9000
25. (a) 720 (b) 48 **27.** 120 **29.** 362,880
31. 24 **33.** 20 **35.** 720 **37.** 32,760
39. 197,149,680 **41.** 24,024 **43.** 456,456
45. ABCD, ABDC, ACBD, ACDB, ADBC, ADCB, BACD, BADC, CABD, CADB, DABC, DACB, BCAD, BDAC, CBAD, CDAB, DBAC, DCAB, BCDA, BDCA, CBDA, CDBA, DBCA, DCBA
47. 420 **49.** 20,160 **51.** 15 **53.** 66
55. 1 **57.** 40,920 **59.** 850,668
61. AB, AC, AD, AE, AF, BC, BD, BE, BF, CD, CE, CF, DE, DF, EF
63. 75,287,520 **65.** 5,586,853,480 **67.** 292,201,338
69. 462,000 **71.** 3744 **73.** 5 **75.** 20
77. $n = 5$ or $n = 6$ **79.** $n = 10$ **81.** $n = 3$ **83.** $n = 2$
85. True; The Fundamental Counting Principle determines the number of ways different events can occur by multiplying the number of ways each event occurs.
87. The last factor should be $5 - 3 + 1 = 3$.
89. (a) Permutations; order matters
(b) Combinations; order does not matter
91 and 93. Answers will vary. **95.** $(-2, -8)$

Section 8.6 (page 632)

1. sample space **3.** mutually exclusive **5.** $P(E) = 0$
7. $\{(H, 1), (H, 2), (H, 3), (H, 4), (H, 5), (H, 6), (T, 1), (T, 2), (T, 3), (T, 4), (T, 5), (T, 6)\}$
9. {ABC, ACB, BAC, BCA, CAB, CBA}
11. $\frac{3}{8}$ **13.** $\frac{7}{8}$ **15.** $\frac{3}{13}$ **17.** $\frac{3}{26}$ **19.** $\frac{3}{100}$ **21.** $\frac{9}{25}$
23. $\dfrac{1}{3{,}838{,}380}$ **25.** $\dfrac{1}{8436}$ **27.** $\dfrac{1}{5}$ **29.** 0.25
31. 0.88 **33.** $\frac{4}{13}$ **35.** $\frac{2}{13}$ **37.** No **39.** $\frac{11}{32}$ **41.** $\frac{5}{8}$
43. $\frac{3}{32}$ **45.** No **47.** $\frac{1}{8}$ **49.** $\frac{27}{1000}$ **51.** $\frac{2}{125}$ **53.** $\frac{3}{4}$
55. (a) About 14,638,000 (b) 0.35 (c) 0.90 (d) 0.46
57. (a) $\frac{1}{120}$ (b) $\frac{1}{24}$ **59.** (a) $\frac{14}{55}$ (b) $\frac{12}{55}$ (c) $\frac{54}{55}$
61. (a) 0.1024 (b) 0.5376 (c) 0.4624
63. (a) $\dfrac{1}{15{,}625}$ (b) $\dfrac{4096}{15{,}625}$ (c) $\dfrac{11{,}529}{15{,}625}$
65. (a) $\dfrac{\pi}{4}$ (b) Answers will vary.
67. True; The sum of the probabilities of all outcomes must be 1.
69. The probability is $\left(\frac{3}{15}\right)\left(\frac{3}{15}\right) = \frac{1}{25}$.
71. (a) No; $P(A) + P(B) = 0.76 + 0.58 = 1.34 > 1$. The sum of the probabilities is greater than 1, so A and B cannot be mutually exclusive.
(b) $0.76 \le P(A \cup B) \le 1$
73. 70

Review Exercises (page 636)

1. $\dfrac{2}{3}, \dfrac{4}{5}, \dfrac{8}{9}, \dfrac{16}{17}, \dfrac{32}{33}$ **3.** $a_n = 5n + 2$ **5.** $a_n = \dfrac{2}{2n - 1}$
7. $9, 3, 1, \dfrac{1}{3}, \dfrac{1}{9}$ **9.** $\dfrac{1}{132}$ **11.** 168 **13.** $\displaystyle\sum_{k=1}^{9} 3k^2 = 855$
15. $\frac{8}{9}$
17. \$3015, \$3030.08, \$3045.23, \$3060.45, \$3075.75, \$3091.13, \$3106.59, \$3122.12; the balances in the fund for the first two years
19. No; There is no common difference. **21.** 4, 7, 10, 13, 16
23. 1, 4, 7, 10, 13
25. 35, 32, 29, 26, 23; $d = -3$; $a_n = 38 - 3n$
27. $a_n = 103 - 3n$; 1600 **29.** 110 **31.** \$57,564
33. Yes; $r = 2$ **35.** $4, -1, \frac{1}{4}, -\frac{1}{16}, \frac{1}{64}$
37. $120, 40, \frac{40}{3}, \frac{40}{9}, \frac{40}{27}$; $r = \frac{1}{3}$; $a_n = 120\left(\frac{1}{3}\right)^{n-1}$
39. (a) -0.5 (b) $-\frac{1}{2}$ **41.** 19,531 **43.** 29,493
45. 24 **47.** Not possible; $\left|\frac{3}{2}\right| > 1$
49. (a) $a_n = 130{,}000(0.7)^n$ (b) \$21,849.10
51. 45 **53.** $x^4 + 20x^3 + 150x^2 + 500x + 625$
55. $256x^4 - 768x^3y + 864x^2y^2 - 432xy^3 + 81y^4$
57. $a^5 - 20a^4b + 160a^3b^2 - 640a^2b^3 + 1280ab^4 - 1024b^5$
59. 1 **61.** 210 **63.** (a) 11 (b) 10 **65.** 95,040
67. 5040 **69.** $n = 3$ **71.** 924 **73.** 479,001,600
75. $\frac{1}{9}$ **77.** $\frac{24}{625}$
79. True; $\dfrac{(n + 2)!}{n!} = \dfrac{(n + 2)(n + 1)n!}{n!} = (n + 2)(n + 1)$
81. (a) Each term is obtained by adding the same constant (common difference) to the preceding term.
(b) Each term is obtained by multiplying the same constant (common ratio) by the preceding term.

Chapter Test (page 641)

1. $3, 2, \frac{4}{3}, \frac{8}{9}, \frac{16}{27}$ **2.** 12, 16, 20, 24, 28
3. $-x, \dfrac{x^2}{2}, -\dfrac{x^3}{3}, \dfrac{x^4}{4}, -\dfrac{x^5}{5}$ **4.** 7920 **5.** $\dfrac{1}{n + 1}$
6. $a_n = n^2 + 1$ **7.** $a_n = 5100 - 100n$ **8.** $a_n = 4\left(\frac{1}{2}\right)^{n-1}$
9. $\displaystyle\sum_{n=1}^{12} \frac{2}{3n + 1}$ **10.** $\displaystyle\sum_{n=1}^{\infty} 2\left(\frac{1}{4}\right)^{n-1}$
11. Geometric; $r = \dfrac{1}{7}$; $\dfrac{960{,}800}{16{,}807}$ **12.** Arithmetic; $d = 1$; 45
13. Geometric; $r = -\frac{2}{5}$; $\frac{25}{7}$
14. $81a^4 - 540a^3b + 1350a^2b^2 - 1500ab^3 + 625b^4$
15. $600{,}000x^2$ **16.** 84 **17.** 1140 **18.** 328,440
19. 72 **20.** $n = 3$ **21.** 26,000 **22.** 12,650
23. $\frac{3}{52}$ **24.** $\frac{1}{5005}$ **25.** (a) $\frac{1}{4}$ (b) $\frac{121}{3600}$ (c) $\frac{1}{60}$

Chapter 9

Section 9.1 (page 653)

1. conic section **3.** circle, center **5.** $x^2 + y^2 = 36$
7. $(x - 3)^2 + (y - 7)^2 = 53$ **9.** $(x + 3)^2 + (y + 3)^2 = 61$
11. Center: $(0, 0)$
Radius: 4
13. Center: $(5, 0)$
Radius: 3
15. Center: $(-1, -6)$
Radius: $\sqrt{19}$
17. $x^2 + y^2 = 4$
Center: $(0, 0)$
Radius: 2
19. $x^2 + y^2 = \dfrac{5}{9}$
Center: $(0, 0)$
Radius: $\dfrac{\sqrt{5}}{3}$

21. $(x-1)^2+(y+3)^2=1$
Center: $(1, -3)$
Radius: 1

23. $\left(x+\frac{3}{2}\right)^2+(y-3)^2=1$
Center: $\left(-\frac{3}{2}, 3\right)$
Radius: 1

25. Center: $(0, 0)$
Radius: 1

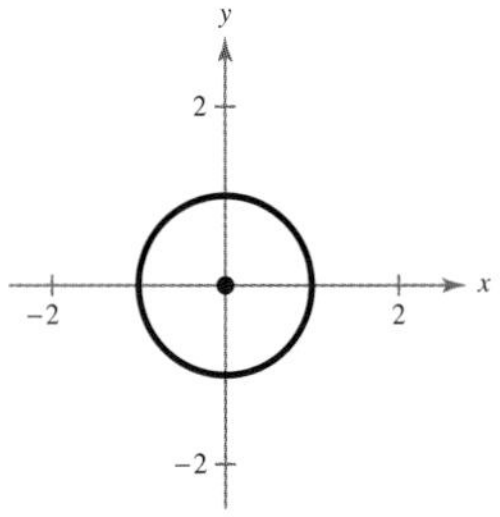

27. Center: $(-4, -1)$
Radius: 3

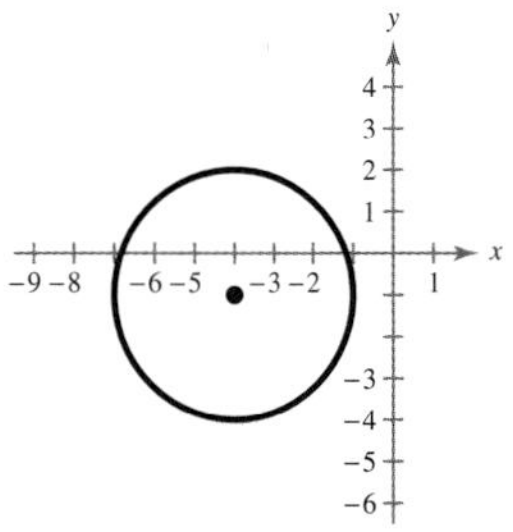

29. Center: $(7, -4)$
Radius: 5

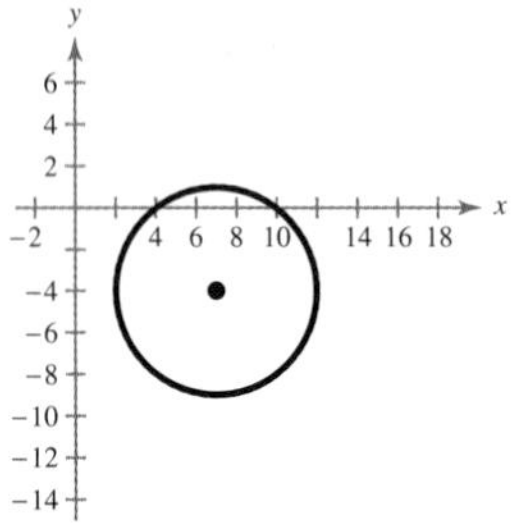

31. Center: $(-1, 0)$
Radius: 6

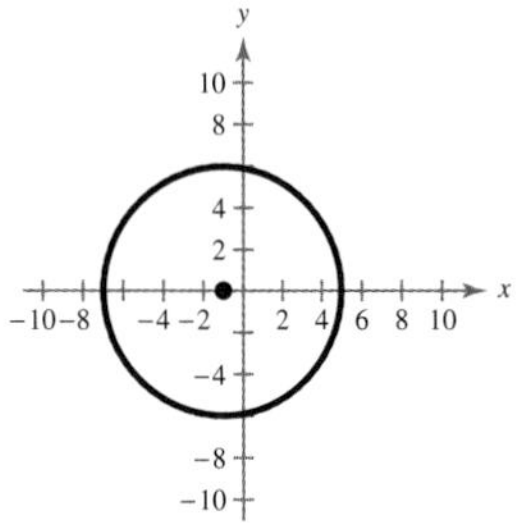

33. x-intercepts: $(-1, 0), (-9, 0)$
y-intercept: $(0, 3)$

35. x-intercepts: $\left(6 \pm \sqrt{7}, 0\right)$
y-intercepts: none

37. x-intercepts: $\left(1 \pm 2\sqrt{7}, 0\right)$
y-intercepts: $(0, 9), (0, -3)$

39. d **40.** b **41.** a **42.** c **43.** $x^2=\frac{3}{2}y$

45. $x^2=8y$ **47.** $y^2=-2x$ **49.** $x^2=-8y$

51. $y^2=8x$ **53.** $y^2=8x$

55. Vertex: $(0, 0)$
Focus: $\left(0, \frac{1}{2}\right)$
Directrix: $y=-\frac{1}{2}$

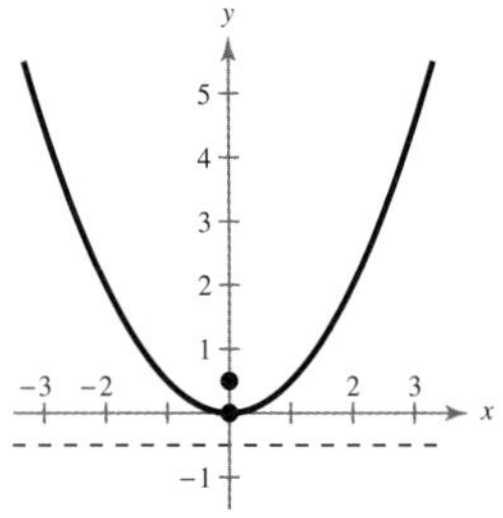

57. Vertex: $(0, 0)$
Focus: $\left(-\frac{3}{2}, 0\right)$
Directrix: $x=\frac{3}{2}$

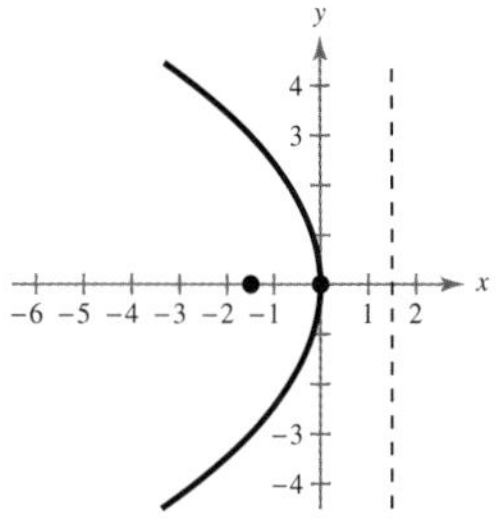

59. Vertex: $(0, 0)$
Focus: $\left(0, -\frac{3}{2}\right)$
Directrix: $y=\frac{3}{2}$

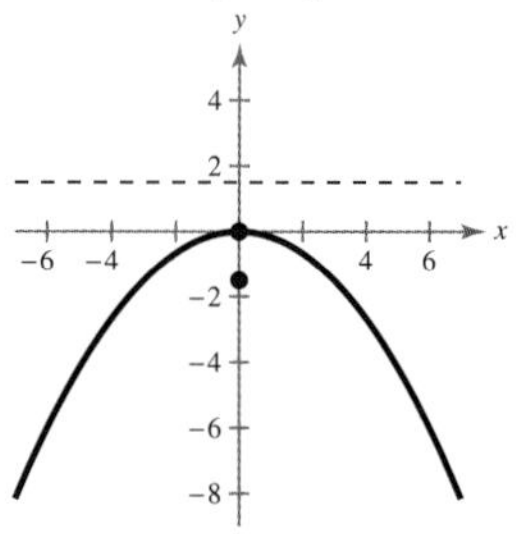

61. Vertex: $\left(-\frac{3}{2}, 2\right)$
Focus: $\left(-\frac{3}{2}, 3\right)$
Directrix: $y=1$

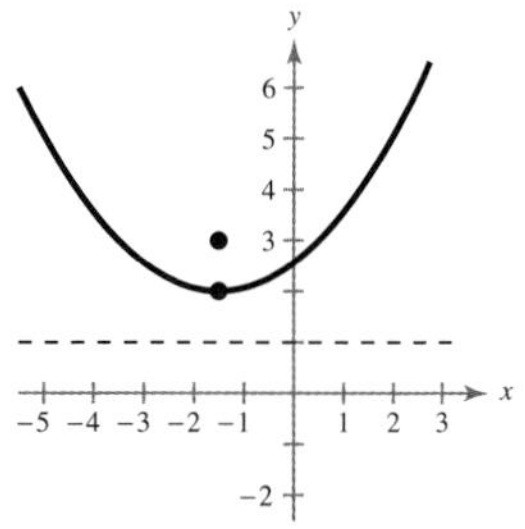

63. Vertex: $\left(\frac{3}{2}, -7\right)$
Focus: $\left(\frac{5}{2}, -7\right)$
Directrix: $x=\frac{1}{2}$

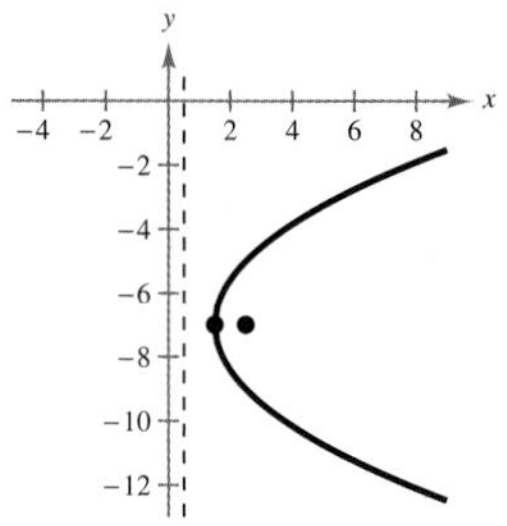

65. Vertex: $(-2, -3)$
Focus: $(-4, -3)$
Directrix: $x=0$

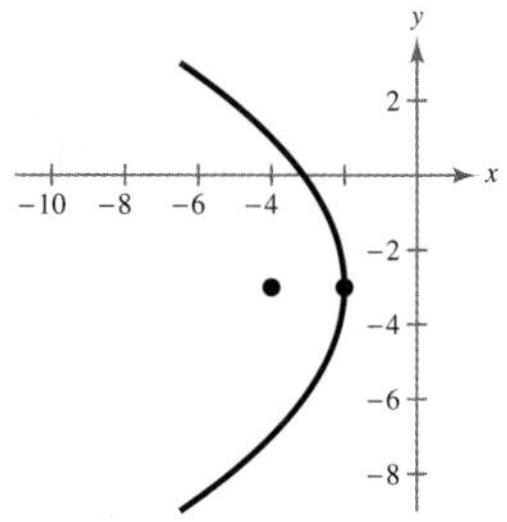

67. Vertex: $(-2, 1)$
Focus: $\left(-2, -\frac{1}{2}\right)$
Directrix: $y=\frac{5}{2}$

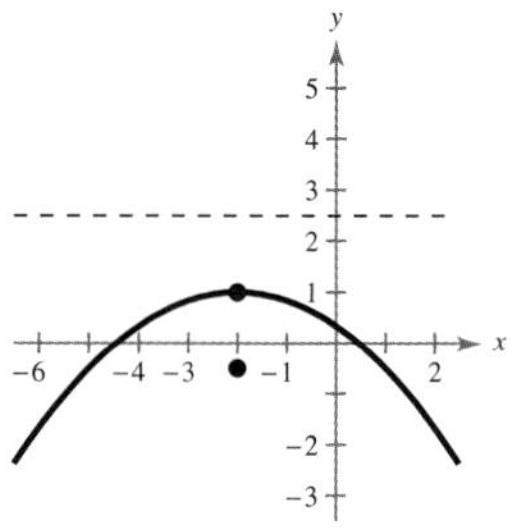

69. Vertex: $\left(\frac{1}{4}, -\frac{1}{2}\right)$
Focus: $\left(0, -\frac{1}{2}\right)$
Directrix: $x=\frac{1}{2}$

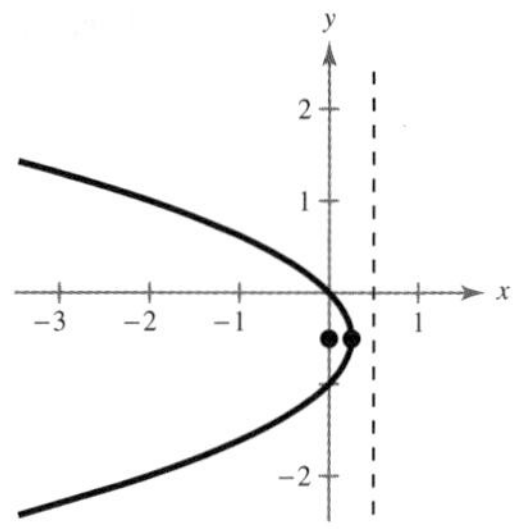

71. Vertex: $(1, 1)$
Focus: $(1, 2)$
Directrix: $y=0$

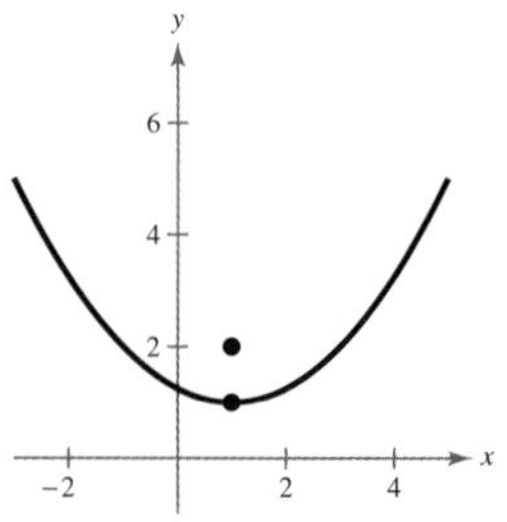

73. $(x-3)^2=-(y-1)$ **75.** $y^2=4(x+4)$

77. $y^2=2(x+2)$ **79.** $(y-2)^2=-8(x-6)$

81. $x^2=8(y-5)$ **83.** $(y-2)^2=8x$

85. $4x-y-8=0$ **87.** $4x-y+2=0$

89.

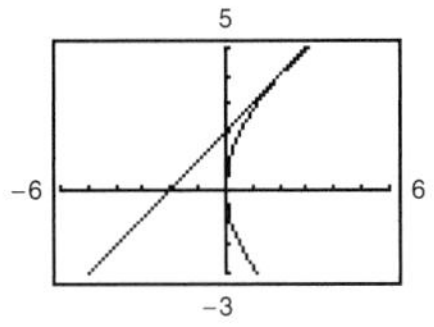

(2, 4)

91. (a) $x^2 + y^2 = 2704$ (b) Yes (c) 2 mi

93. $x^2 = -4(y - 4)$ **95.** $y^2 = 640x$

97. (a)

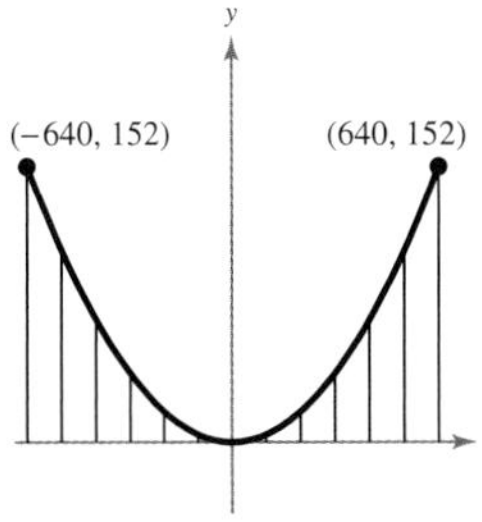

(b) $x^2 = \frac{51{,}200}{19}y$

(c)

x	0	200	400	500	600
y	0	14.844	59.375	92.773	133.59

99. (a) $y^2 = 6x$ (b) About 2.67 in.

101. (a) $x^2 = -49(y - 100)$ (b) 70 ft **103.** $y = \frac{3}{4}x - \frac{25}{4}$

105. $y = \frac{\sqrt{2}}{2}x - 3\sqrt{2}$ **107.** $y = -\frac{3}{4}x + 12$

109. False; $x^2 + (y + 5)^2$ represents a circle with its center at $(0, -5)$ and a radius of 5.

111. True; The vertex is the closest point to the directrix or focus.

113. The diameter is 4, so the radius is 2.

115.

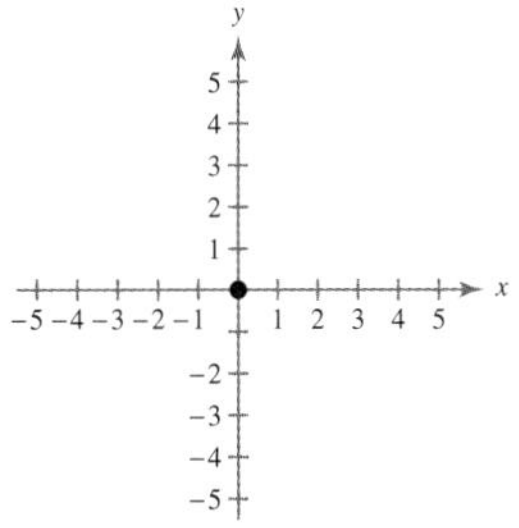

The intersection results in a point.

117. $y = \sqrt{6(x + 1)} + 3$

119. Zero: about -1.35; minimum: about $(0.67, 0.22)$; maximum: about $(-0.67, 3.78)$

121. Zeros: none; minimum: about $(-0.79, 0.81)$

Section 9.2 (page 663)

1. ellipse, foci **3.** minor axis **5.** Vertical **7.** b

8. a **9.** c **10.** d **11.** $\frac{x^2}{4} + \frac{y^2}{16} = 1$

13. $\frac{x^2}{49} + \frac{y^2}{45} = 1$ **15.** $\frac{x^2}{24} + \frac{y^2}{49} = 1$

17. $\frac{5x^2}{48} + \frac{y^2}{16} = 1$ **19.** $\frac{(x + 3)^2}{16} + \frac{(y - 1)^2}{4} = 1$

21. $\frac{(x - 4)^2}{16} + \frac{(y - 2)^2}{1} = 1$ **23.** $\frac{(x - 1)^2}{25} + \frac{(y - 1)^2}{21} = 1$

25. $\frac{(x - 1)^2}{9} + \frac{(y - 3)^2}{4} = 1$ **27.** $\frac{(x - 1)^2}{12} + \frac{(y - 4)^2}{16} = 1$

29. $\frac{(x - 3)^2}{36} + \frac{(y - 2)^2}{32} = 1$

31. Center: $(0, 0)$
Vertices: $(\pm 8, 0)$
Foci: $(\pm\sqrt{55}, 0)$
Eccentricity: $\frac{\sqrt{55}}{8}$

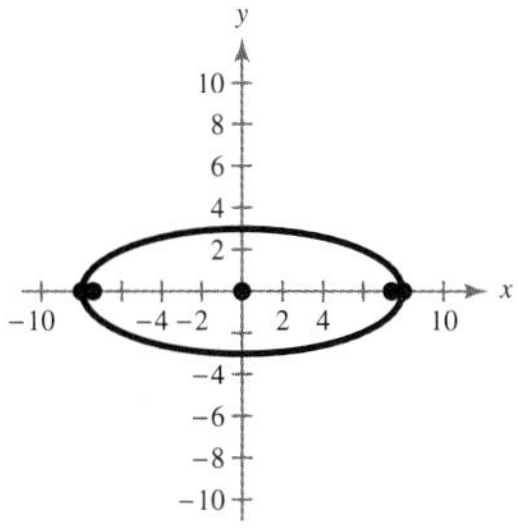

33. Center: $(4, 0)$
Vertices: $(4, \pm 5)$
Foci: $(4, \pm 3)$
Eccentricity: $\frac{3}{5}$

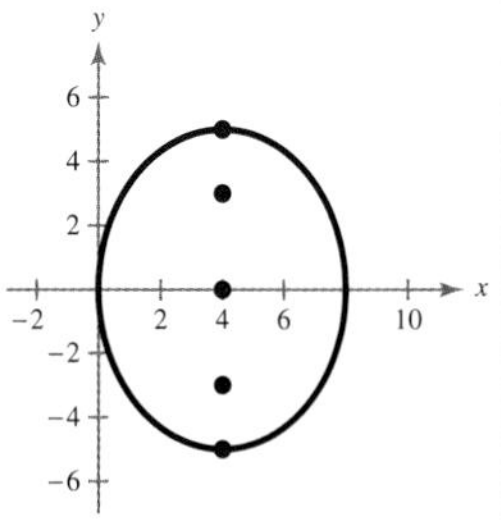

35. Center: $(-5, 1)$
Vertices: $\left(-\frac{7}{2}, 1\right), \left(-\frac{13}{2}, 1\right)$
Foci: $\left(-5 \pm \frac{\sqrt{5}}{2}, 1\right)$
Eccentricity: $\frac{\sqrt{5}}{3}$

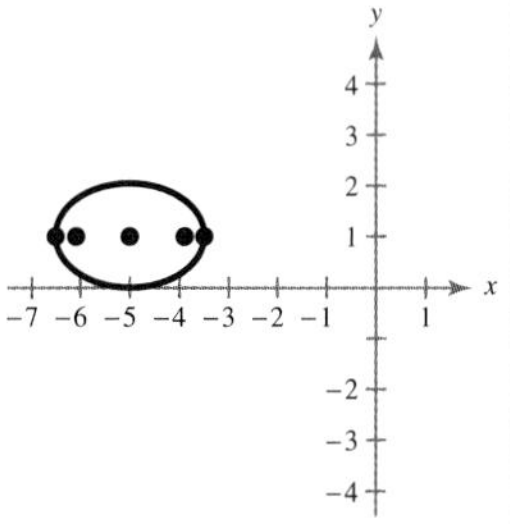

37. Center: $(0, 0)$
Vertices: $(\pm 6, 0)$
Foci: $(\pm 4\sqrt{2}, 0)$
Eccentricity: $\frac{2\sqrt{2}}{3}$

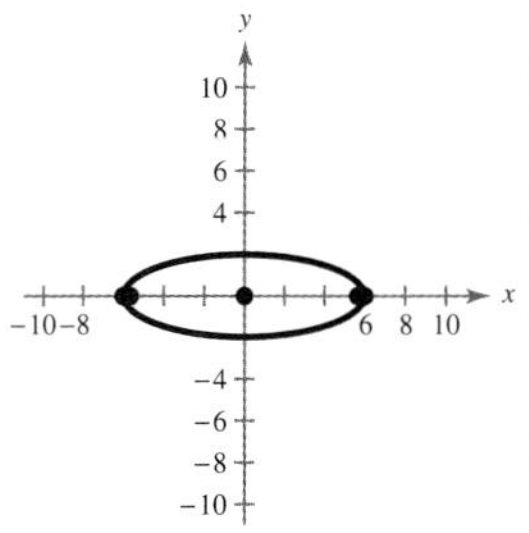

39. Center: $(0, 0)$
Vertices: $(0, \pm 7)$
Foci: $(0, \pm 3\sqrt{5})$
Eccentricity: $\frac{3\sqrt{5}}{7}$

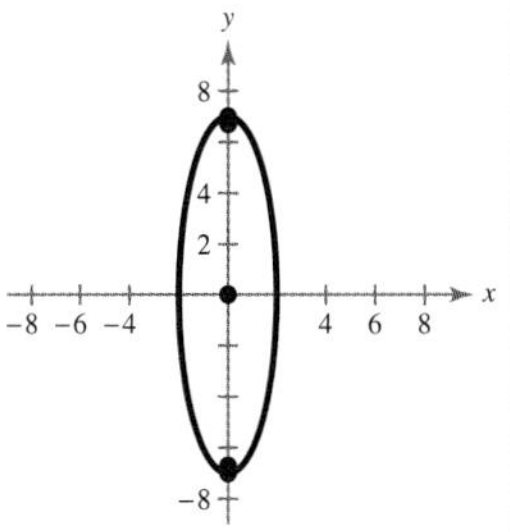

41. Center: $(-2, 3)$
Vertices: $(-2, 6), (-2, 0)$
Foci: $(-2, 3 \pm \sqrt{5})$
Eccentricity: $\frac{\sqrt{5}}{3}$

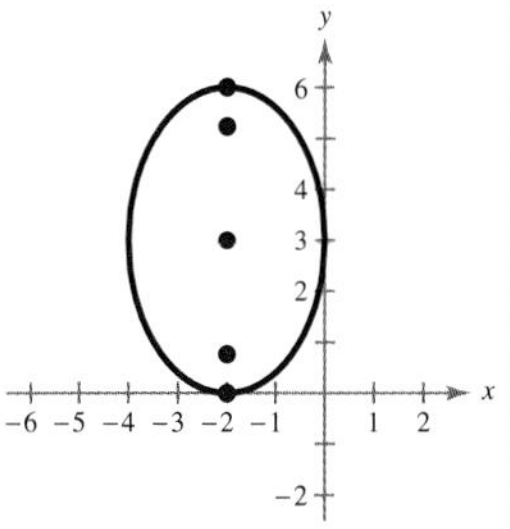

43. Center: $\left(-\frac{3}{2}, \frac{5}{2}\right)$

Vertices: $\left(-\frac{3}{2}, \frac{5 \pm 4\sqrt{3}}{2}\right)$

Foci: $\left(-\frac{3}{2}, \frac{5}{2} \pm 2\sqrt{2}\right)$

Eccentricity: $\frac{\sqrt{6}}{3}$

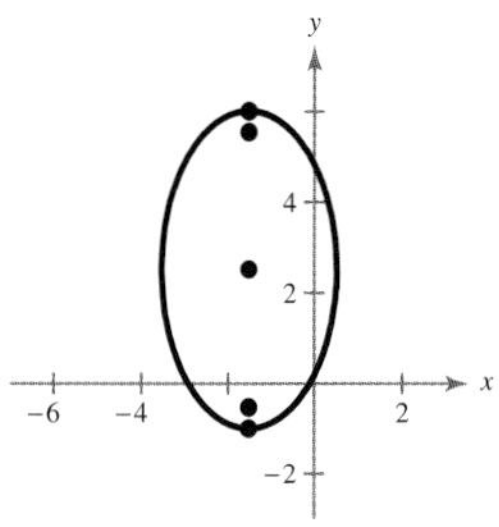

45. Center: $\left(\frac{1}{2}, -1\right)$

Vertices: $\left(\frac{1}{2} \pm \sqrt{5}, -1\right)$

Foci: $\left(\frac{1}{2} \pm \sqrt{2}, -1\right)$

Eccentricity: $\frac{\sqrt{10}}{5}$

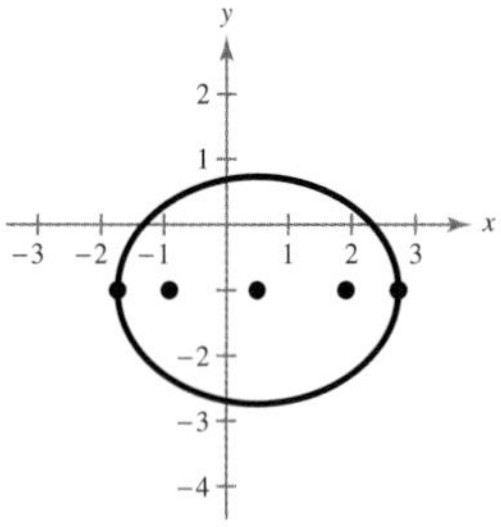

47. $\frac{\sqrt{5}}{3}$ **49.** $\frac{2\sqrt{2}}{3}$ **51.** $\frac{x^2}{25} + \frac{y^2}{16} = 1$

53. $\frac{(x-1)^2}{45} + \frac{(y-7)^2}{81} = 1$ **55.** $\frac{x^2}{318.09} + \frac{y^2}{18.80} = 1$

57. $40\sqrt{33}$ mm ≈ 229.8 mm **59.** 40 units

61.

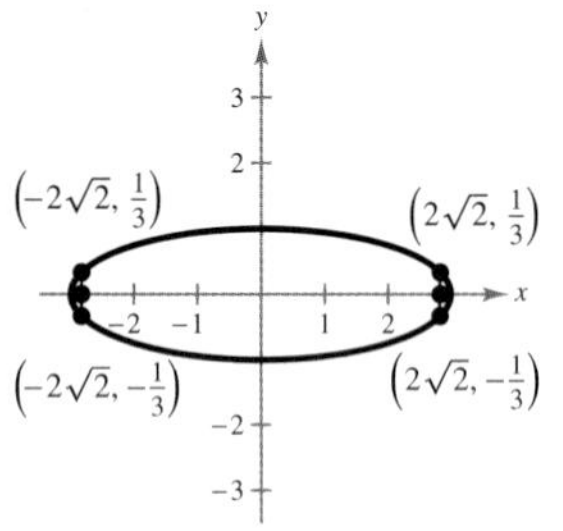

63.

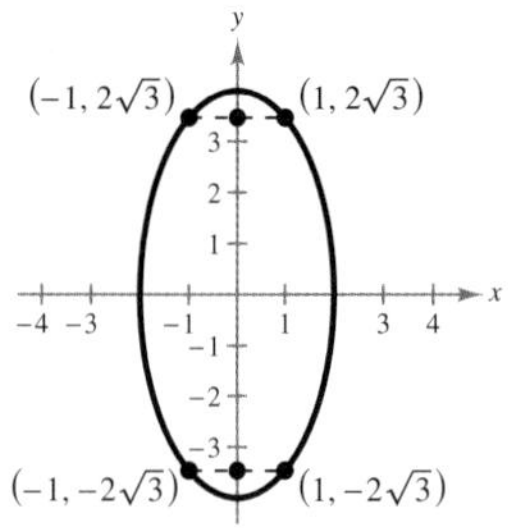

65. True; The ellipse is more elongated when e is close to 1.

67. No; The equation represents an ellipse.

69. The vertices and foci were switched.

71. (a) $2a$

(b) The sum of the distances from the two fixed points is constant.

73. Proof **75.** Arithmetic **77.** 1093

Section 9.3 (page 675)

1. hyperbola **3.** rotation, axes **5.** Two; center

7. a **8.** d **9.** b **10.** c **11.** $\frac{y^2}{4} - \frac{x^2}{12} = 1$

13. $\frac{(x-2)^2}{4} - \frac{y^2}{12} = 1$ **15.** $\frac{(y-5)^2}{16} - \frac{(x-4)^2}{9} = 1$

17. $\frac{y^2}{9} - \frac{4(x-2)^2}{9} = 1$ **19.** $\frac{(x-2)^2}{4} - \frac{(y+3)^2}{8} = 1$

21. Center: $(0, 0)$

Vertices: $(\pm 1, 0)$

Foci: $(\pm\sqrt{2}, 0)$

Asymptotes: $y = \pm x$

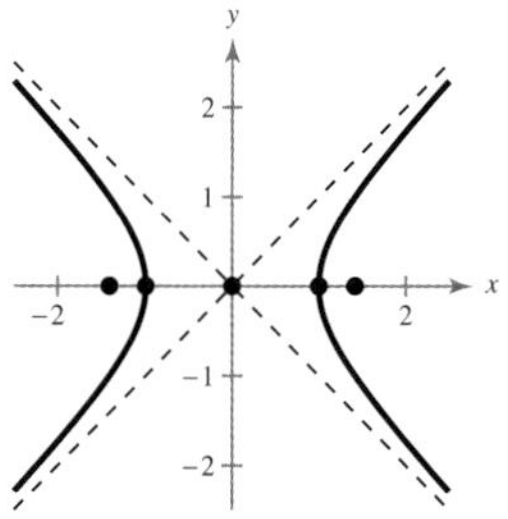

23. Center: $(0, 0)$

Vertices: $(0, \pm 1)$

Foci: $(0, \pm\sqrt{5})$

Asymptotes: $y = \pm\frac{1}{2}x$

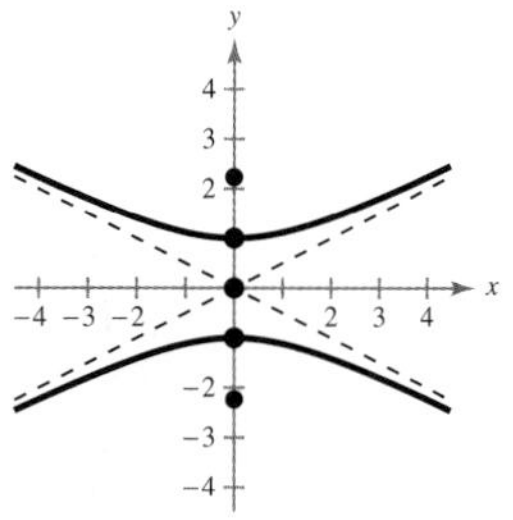

25. Center: $(0, 0)$

Vertices: $(0, \pm 4)$

Foci: $(0, \pm 2\sqrt{5})$

Asymptotes: $y = \pm 2x$

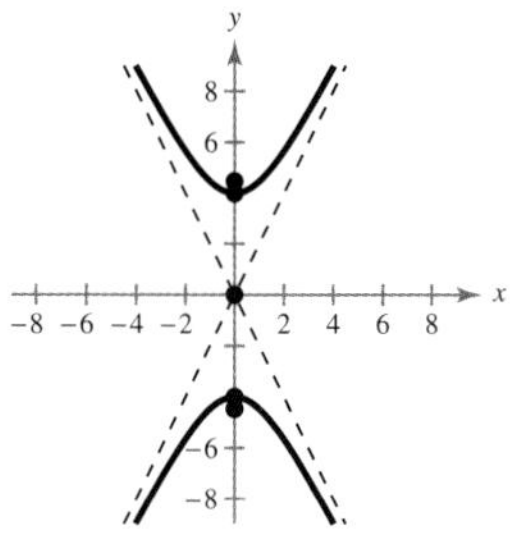

27. Center: $(3, 1)$

Vertices: $(0, 1), (6, 1)$

Foci: $(3 \pm \sqrt{10}, 1)$

Asymptotes:

$y = 1 \pm \frac{1}{3}(x - 3)$

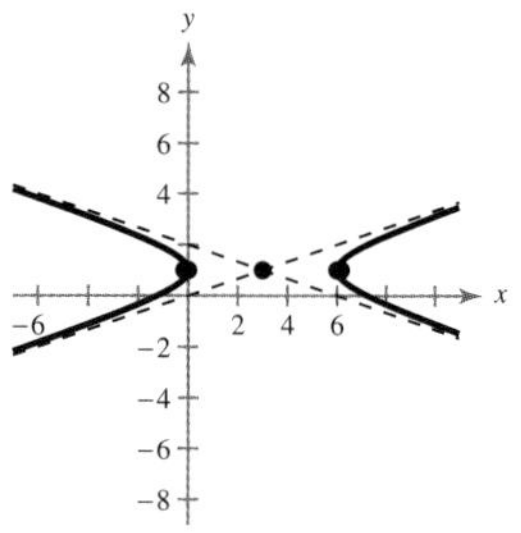

29. Center: $(2, -6)$

Vertices: $\left(2, -\frac{17}{3}\right), \left(2, -\frac{19}{3}\right)$

Foci: $\left(2, -6 \pm \frac{\sqrt{13}}{6}\right)$

Asymptotes: $y = -6 \pm \frac{2}{3}(x - 2)$

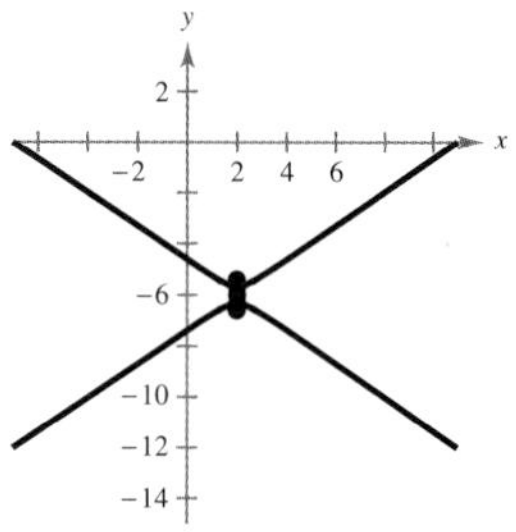

31. (a) $\frac{x^2}{9} - \frac{y^2}{4} = 1$

(b) Center: $(0, 0)$

Vertices: $(\pm 3, 0)$

Foci: $(\pm\sqrt{13}, 0)$

Asymptotes: $y = \pm\frac{2}{3}x$

(c)

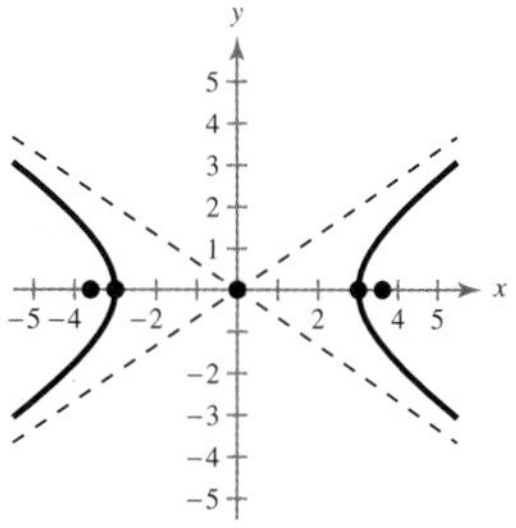

33. (a) $\frac{y^2}{4} - \frac{x^2}{8} = 1$

(b) Center: $(0, 0)$

Vertices: $(0, \pm 2)$

Foci: $(0, \pm 2\sqrt{3})$

Asymptotes: $y = \pm\frac{\sqrt{2}}{2}x$

(c)

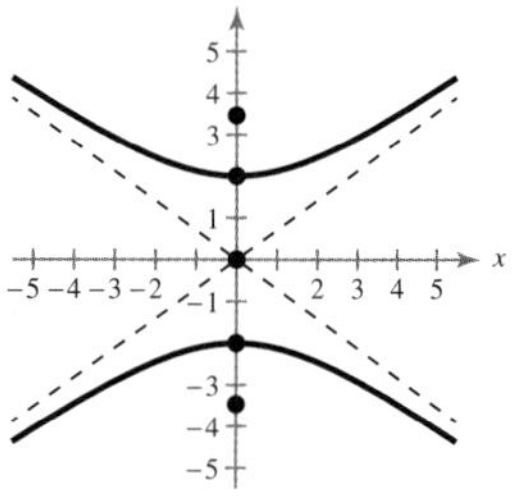

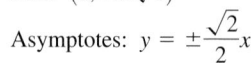

35. (a) $(x-2)^2 - \dfrac{(y+3)^2}{9} = 1$

(b) Center: $(2, -3)$

Vertices: $(3, -3), (1, -3)$

Foci: $\left(2 \pm \sqrt{10}, -3\right)$

Asymptotes: $y = -3 \pm 3(x-2)$

(c)

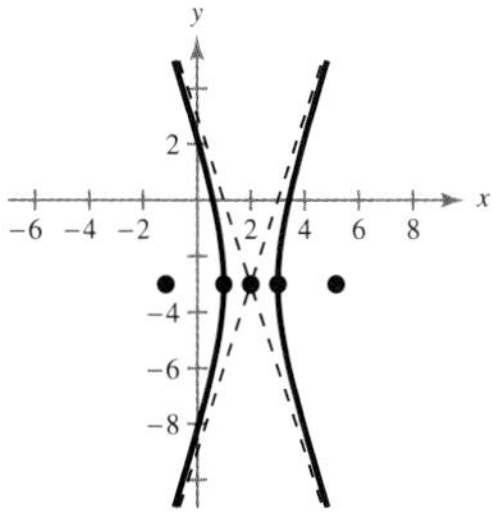

37. (a) $\dfrac{(x+4)^2}{7} - \dfrac{y^2}{2} = 1$

(b) Center: $(-4, 0)$

Vertices: $\left(-4 \pm \sqrt{7}, 0\right)$

Foci: $(-7, 0), (-1, 0)$

Asymptotes: $y = \pm\dfrac{\sqrt{14}}{7}(x+4)$

(c)

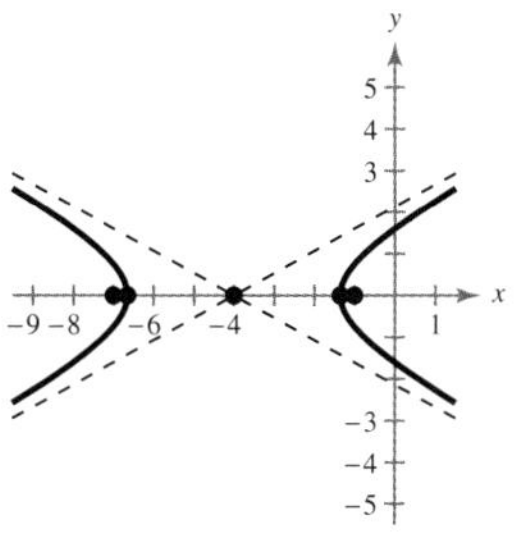

39. (a) $\dfrac{(y+3)^2}{2} - \dfrac{(x-1)^2}{18} = 1$

(b) Center: $(1, -3)$

Vertices: $\left(1, -3 \pm \sqrt{2}\right)$

Foci: $\left(1, -3 \pm 2\sqrt{5}\right)$

Asymptotes: $y = -3 \pm \dfrac{1}{3}(x-1)$

(c)

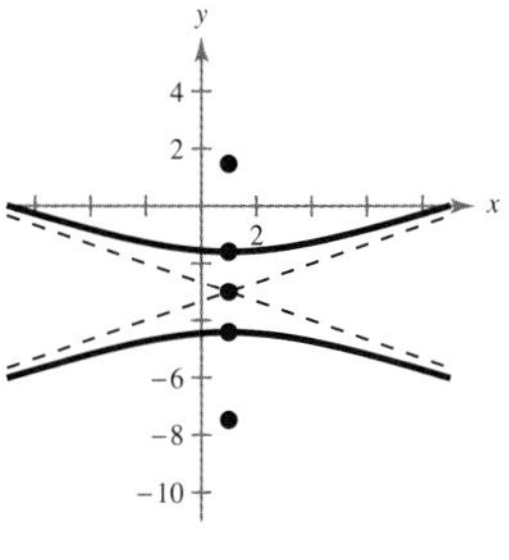

41. $\dfrac{x^2}{1} - \dfrac{y^2}{25} = 1$ **43.** $\dfrac{17y^2}{1024} - \dfrac{17x^2}{64} = 1$

45. $\dfrac{(x-2)^2}{1} - \dfrac{(y-2)^2}{1} = 1$ **47.** $\dfrac{(x-3)^2}{9} - \dfrac{(y-2)^2}{4} = 1$

49. $\dfrac{(x-4)^2}{16} - \dfrac{(y+1)^2}{9} = 1$

51. $\dfrac{x^2}{98{,}010{,}000} - \dfrac{y^2}{13{,}503{,}600} = 1$

53. (a) $x^2 - \dfrac{y^2}{27} = 1$ (b) About 1.89 ft = 22.68 in.

55. $\left(12\sqrt{5} - 12, 0\right) \approx (14.83, 0)$ **57.** Ellipse

59. Hyperbola **61.** Parabola **63.** $(3, 0)$

65. $\left(\dfrac{3+\sqrt{3}}{2}, \dfrac{3\sqrt{3}-1}{2}\right)$ **67.** e **68.** b **69.** f

70. a **71.** d **72.** c

73. $\dfrac{(y')^2}{2} - \dfrac{(x')^2}{2} = 1$

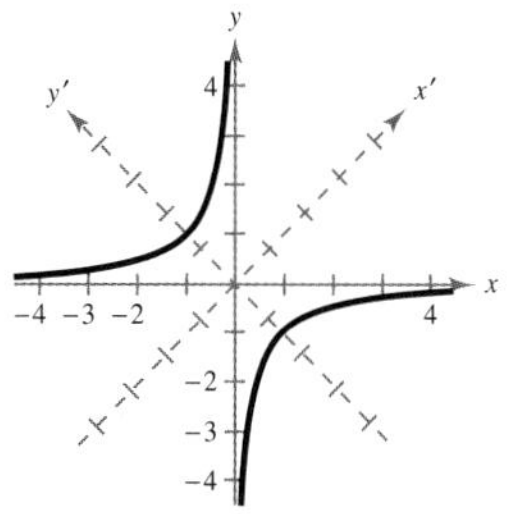

75. $\dfrac{\left(y' + \dfrac{3\sqrt{2}}{2}\right)^2}{12} - \dfrac{\left(x' + \dfrac{\sqrt{2}}{2}\right)^2}{12} = 1$

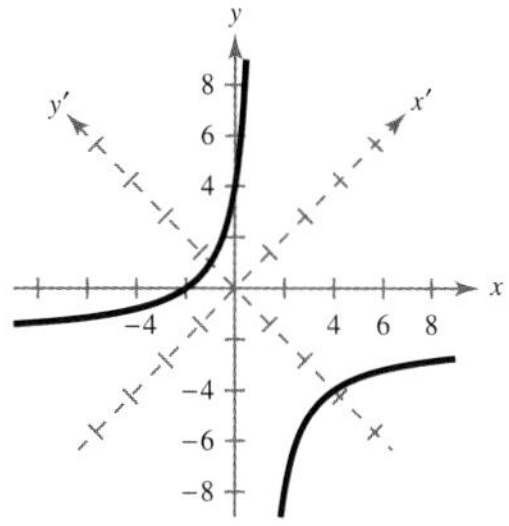

77. $\dfrac{(x')^2}{6} + \dfrac{(y')^2}{3/2} = 1$

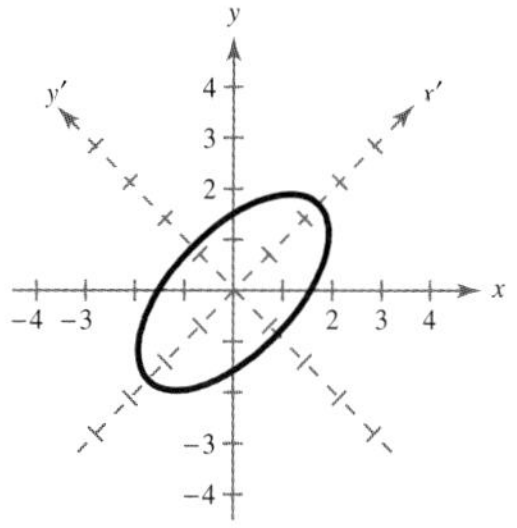

79. $(y')^2 = 4(x'-1)$

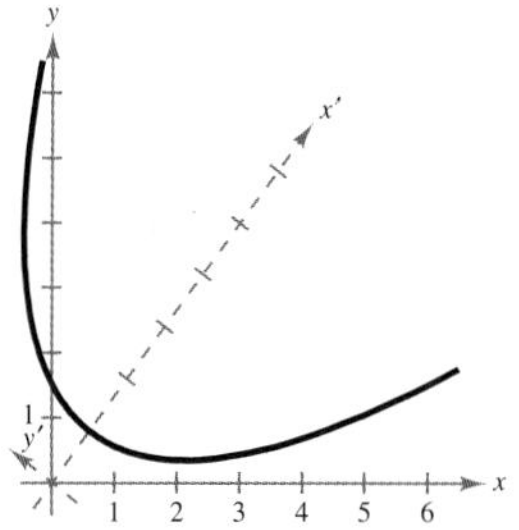

81.

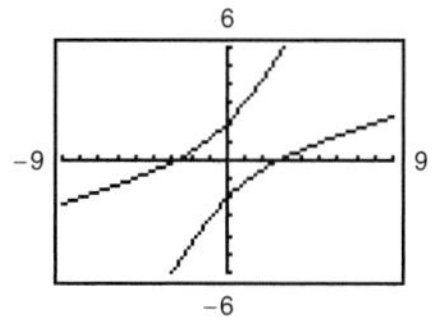

$\theta \approx 37.98°$

Answers will vary.

83.

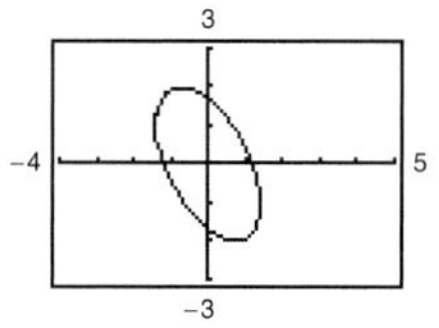

$\theta \approx 28.15°$

Answers will vary.

85.

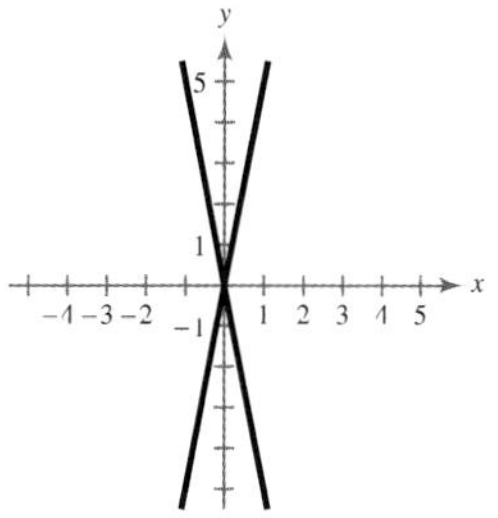

87.

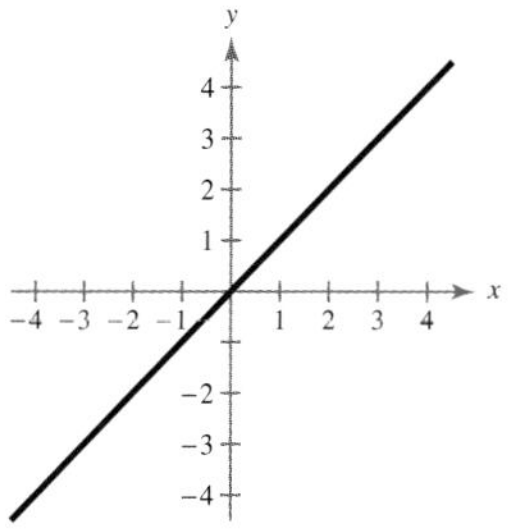

89.

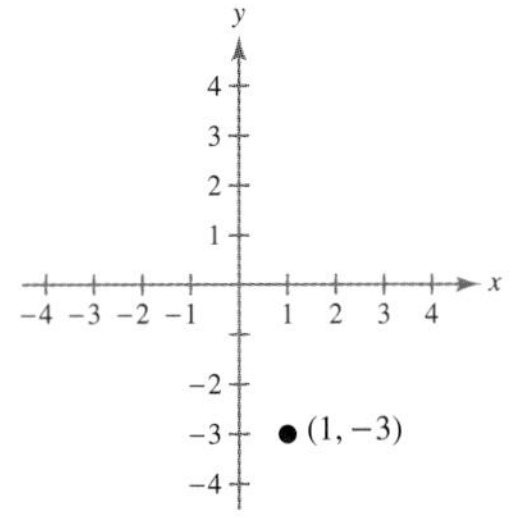

91. True; For a hyperbola, $c^2 = a^2 + b^2$. The larger the ratio of b to a, the larger the eccentricity of the hyperbola, $e = c/a$.

93. True; A hyperbola has perpendicular asymptotes only when $a = b$.

95. Proof **97.** Answers will vary. **99.** Proof

101. $x(x + 4)(x - 4)$ **103.** $2x(x - 6)^2$

105. $2(2x + 3)(4x^2 - 6x + 9)$

107.

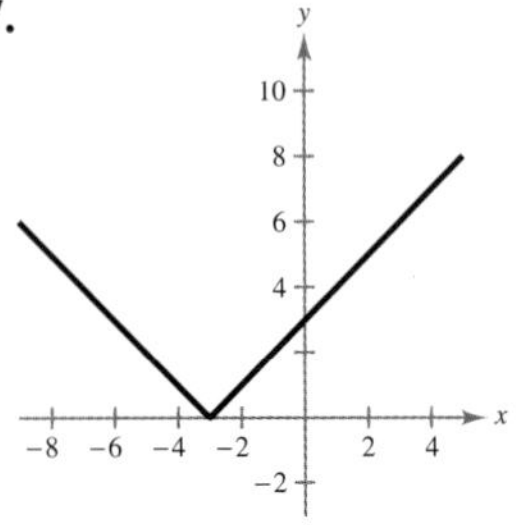

109.

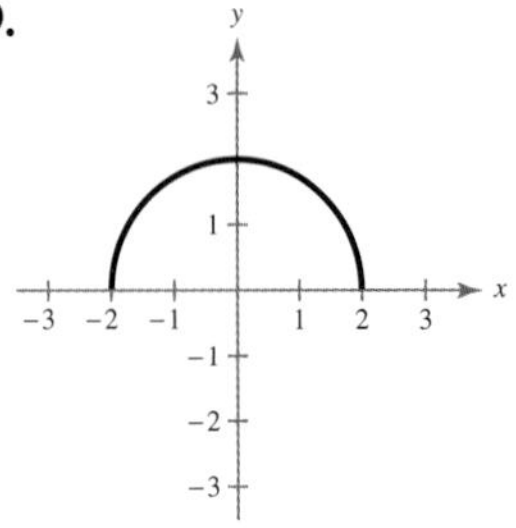

Section 9.4 (page 684)

1. plane curve, parametric equations, parameter

3. Eliminate the parameter.

5. d **6.** a **7.** c **8.** b

9. (a)

t	0	1	2	3	4
x	0	1	$\sqrt{2}$	$\sqrt{3}$	2
y	2	1	0	-1	-2

(b)

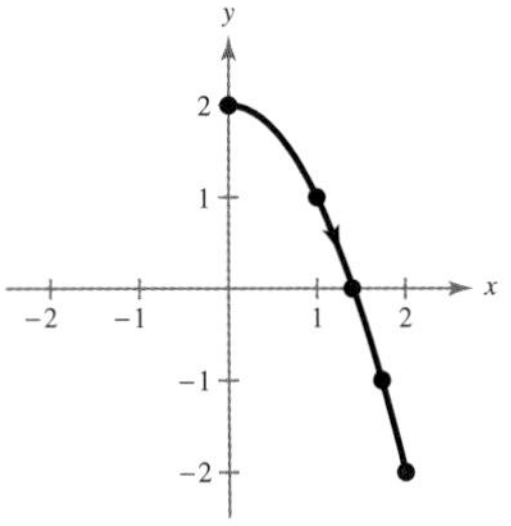

The curve starts at (0, 2) and moves along the right half of the parabola.

(c)

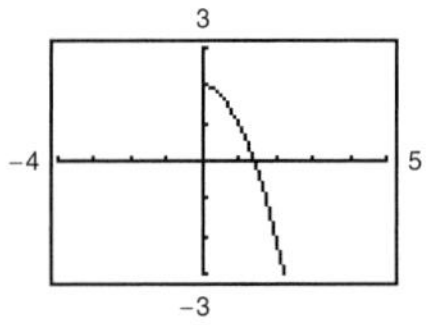

(d) $y = 2 - x^2$

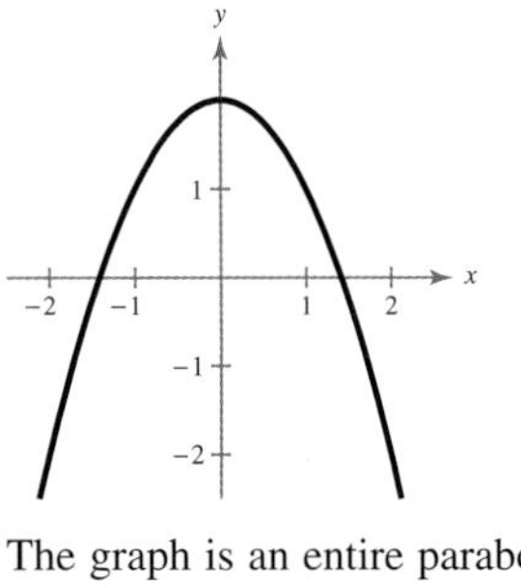

The graph is an entire parabola rather than just the right half.

11. b

13. (a)

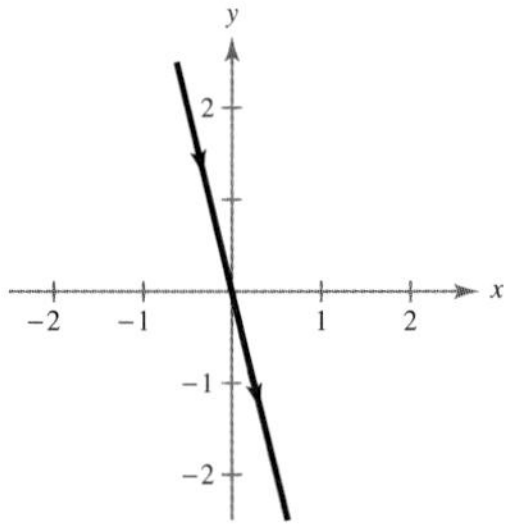

(b) $y = -4x$

15. (a)

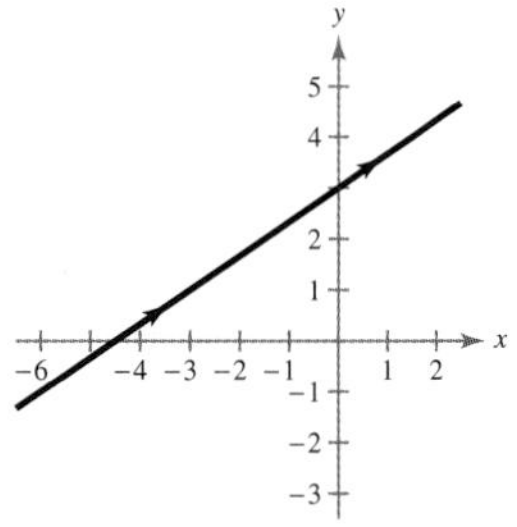

(b) $y = \frac{2}{3}x + 3$

17. (a)

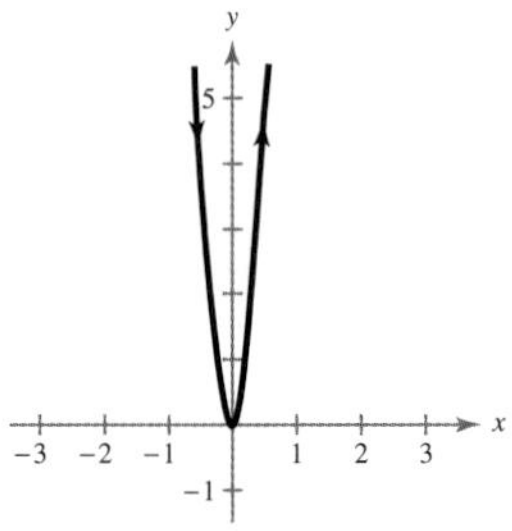

(b) $y = 16x^2$

19. (a)

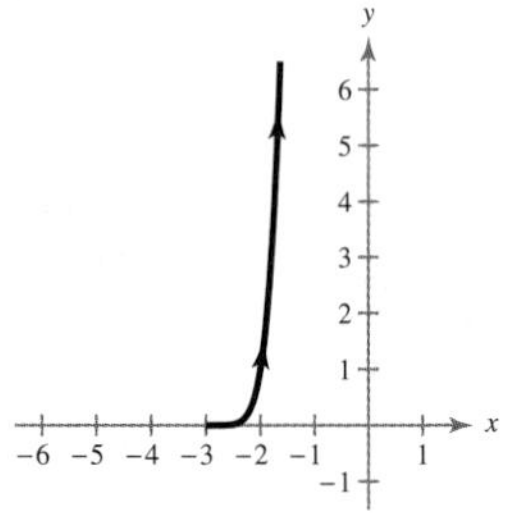

(b) $y = (x + 3)^6, x \geq -3$

21. (a)

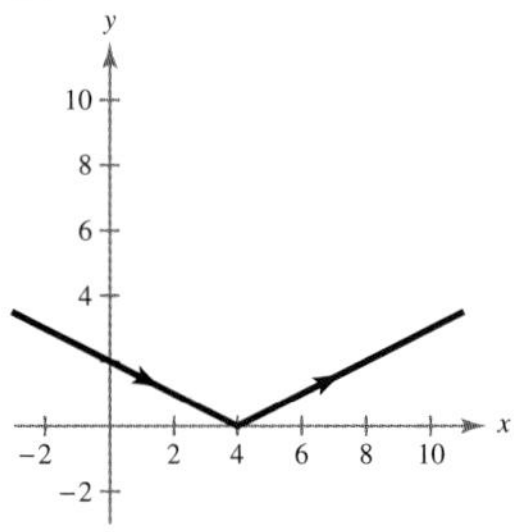

(b) $y = \frac{1}{2}|x - 4|$

23. (a)

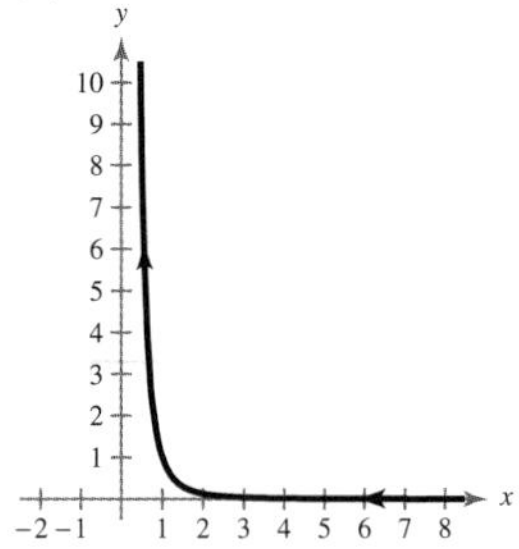

(b) $y = x^{-3}, x > 0$

25. (a)

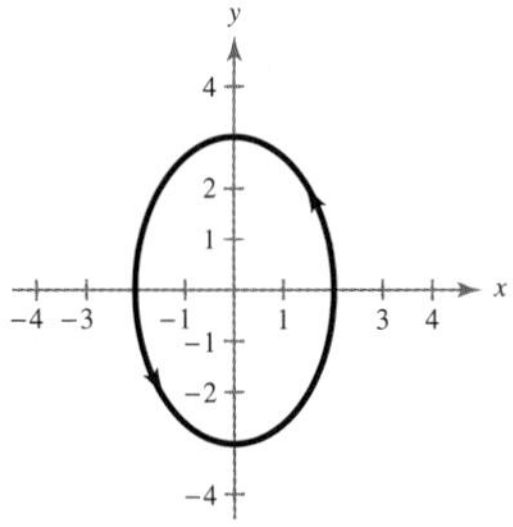

(b) $\dfrac{x^2}{4} + \dfrac{y^2}{9} = 1$

27. (a)

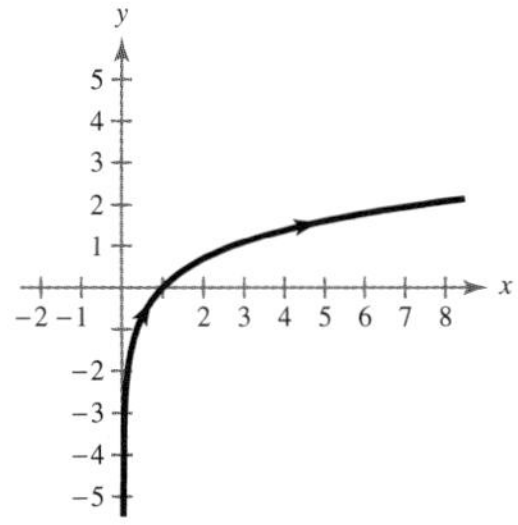

(b) $y = \ln x$

29.

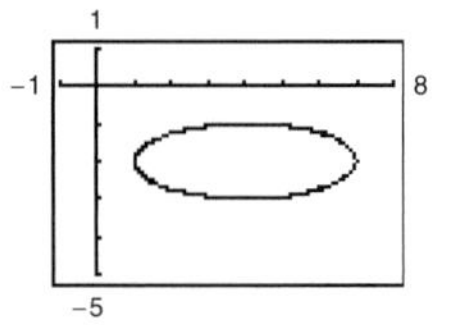

Not a function

31.

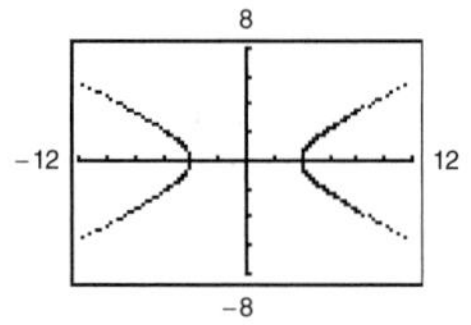

Not a function

33.

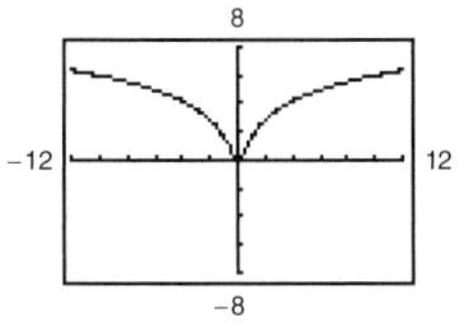

Function

35. Each curve represents a portion of the line $y = 2x + 1$.

Domain	Orientation
(a) $(-\infty, \infty)$	Left to right
(b) $[-1, 1]$	Depends on θ
(c) $(0, \infty)$	Right to left
(d) $(0, \infty)$	Left to right

37. $y - y_1 = \dfrac{y_2 - y_1}{x_2 - x_1}(x - x_1)$ **39.** $\dfrac{(x-h)^2}{a^2} + \dfrac{(y-k)^2}{b^2} = 1$

41. $x = -1 + 7t,\ y = -1 + 10t$

43. $x = 5 \cos \theta,\ y = 3 \sin \theta$

45. (a) $x = t,\ y = -2t + 1$ (b) $x = 2 - t,\ y = 2t - 3$

47. (a) $x = t,\ y = \dfrac{1}{t}$ (b) $x = 2 - t,\ y = \dfrac{1}{2 - t}$

49. (a) $x = t,\ y = 4t^2 + 5$
(b) $x = 2 - t,\ y = 4t^2 - 16t + 21$

51. (a) $x = t,\ y = e^{-t}$ (b) $x = 2 - t,\ y = e^{t-2}$

53.

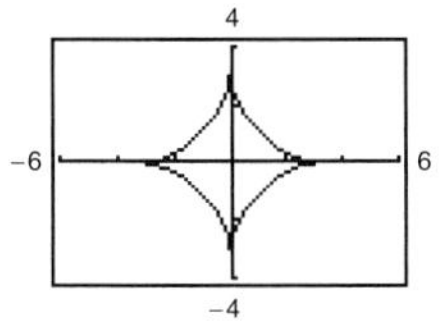

55.

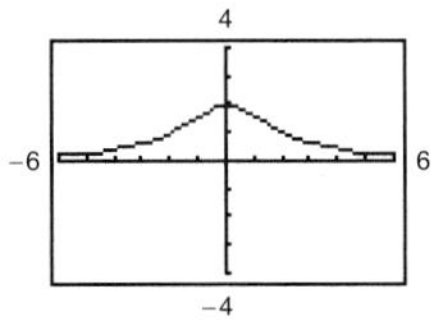

57.

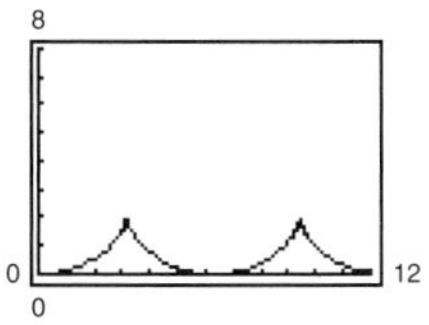

59. b **60.** c **61.** d **62.** a

63. (a) $x = (146.67 \cos \theta)t$
$y = 3 + (146.67 \sin \theta)t - 16t^2$

(b)

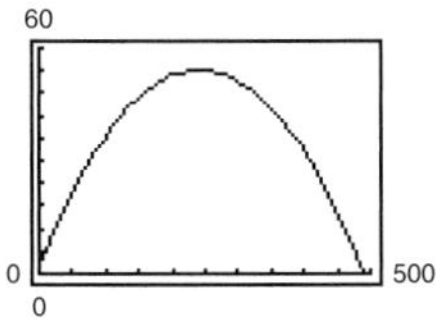

Yes

(c) About 19.4°

65. True; Both sets of parametric equations correspond to $y = x^2 + 1$.

67. False; The set $x = t^2,\ y = t$ does not correspond to y as a function of x.

69. The graph should include points where $t < 0$.

71. (a) Counterclockwise (b) $0 \le \theta < 2\pi$
(c) Answers will vary. *Sample answer:* $x = 6 \cos \theta$, $y = -6 \sin \theta$
(d) The curve begins at the point $(0, 6)$ and is traced in a clockwise direction.

73. $\langle \frac{1}{2}, 0 \rangle$

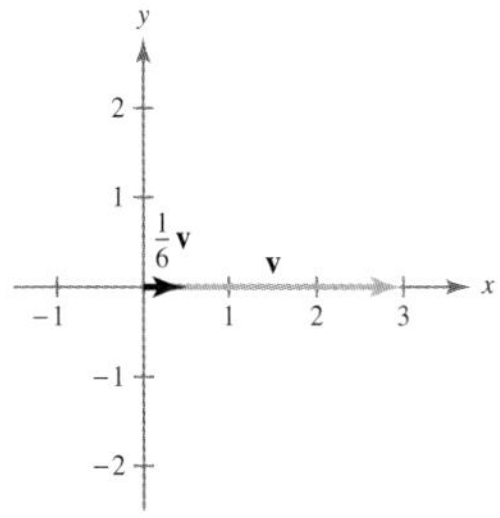

Section 9.5 (page 691)

1. pole

3. $x = r \cos \theta,\ y = r \sin \theta$, and $\tan \theta = \dfrac{y}{x}$, $r^2 = x^2 + y^2$

5. A; $(0, 2)$ **6.** D; $(0, -2)$

7. C; $(0, 0)$ **8.** B; $(\sqrt{2}, \sqrt{2})$

9.

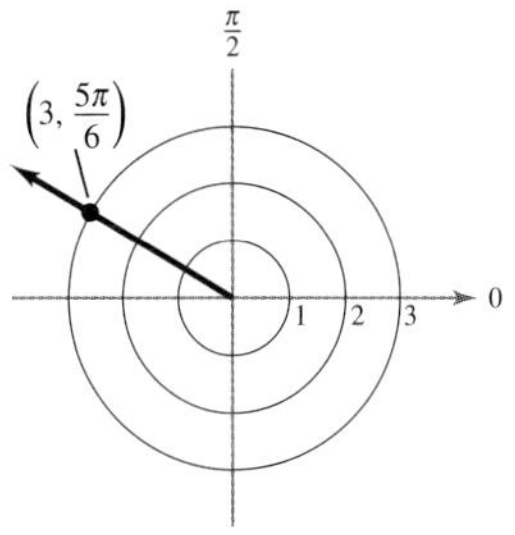

$\left(3, -\dfrac{7\pi}{6}\right), \left(-3, \dfrac{11\pi}{6}\right), \left(-3, -\dfrac{\pi}{6}\right)$

11.

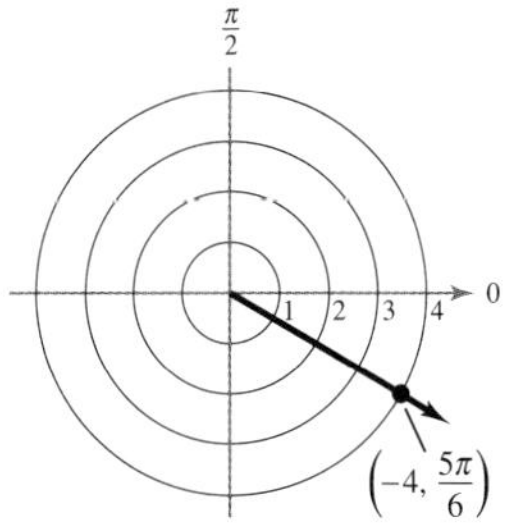

$\left(4, \dfrac{11\pi}{6}\right), \left(-4, -\dfrac{7\pi}{6}\right), \left(4, -\dfrac{\pi}{6}\right)$

13.

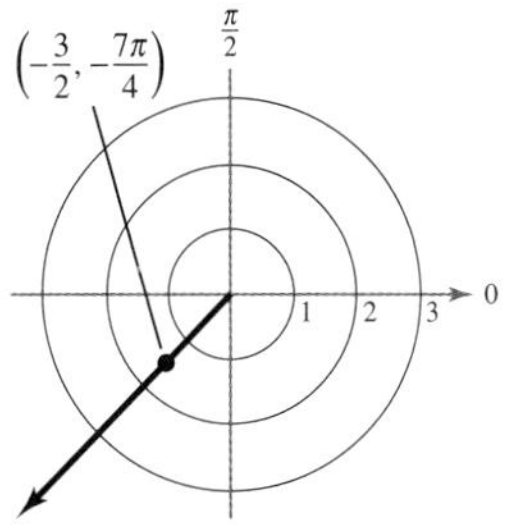

$\left(-\dfrac{3}{2}, \dfrac{\pi}{4}\right), \left(\dfrac{3}{2}, \dfrac{5\pi}{4}\right), \left(\dfrac{3}{2}, -\dfrac{3\pi}{4}\right)$

CHAPTER 9

15.

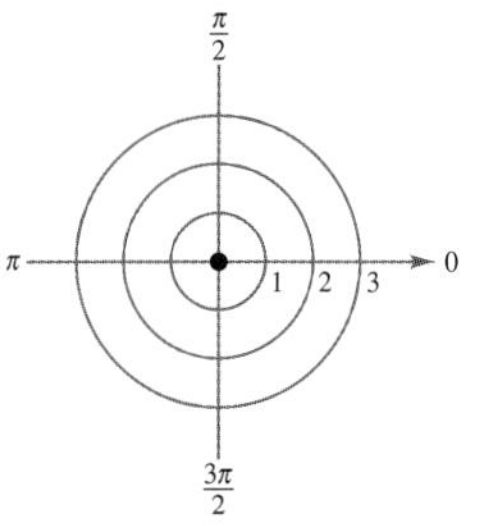

Sample answer: $\left(0, -\frac{11\pi}{6}\right), \left(0, -\frac{5\pi}{6}\right), \left(0, \frac{\pi}{6}\right)$

17. $(-3, 0)$ **19.** $(0, 0)$ **21.** $(\sqrt{3}, 1)$ **23.** $(0, 5)$

25. $(1.53, 1.29)$ **27.** $(-1.42, -0.47)$

29. *Sample answer:* $(7, \pi)$ **31.** *Sample answer:* $\left(\sqrt{2}, \frac{\pi}{4}\right)$

33. *Sample answer:* $\left(\sqrt{6}, \frac{5\pi}{4}\right)$

35. *Sample answer:* $(3.61, -0.59)$

37. *Sample answer:* $(2.65, 2.28)$ **39.** $r = 6$ **41.** $\theta = \frac{\pi}{4}$

43. $r = a \csc \theta$ **45.** $r = -\frac{2}{3 \cos \theta - \sin \theta}$

47. $r^2 = 8 \csc 2\theta$ **49.** $r^2 = 9 \cos 2\theta$

51. $r = 6 \cos \theta$ **53.** $r = 2a \cos \theta$

55. $r = \tan^2 \theta \sec \theta$ **57.** $x^2 + \left(y - \frac{5}{2}\right)^2 = \frac{25}{4}$

59. $y = -\frac{\sqrt{3}}{3}x$ **61.** $x = 0$ **63.** $y = -2$

65. $(x^2 + y^2)^3 = x^2$ **67.** $(x^2 + y^2)^2 = 9x^2y - 3y^3$

69. $y^2 = 4x + 4$ **71.** $4x^2 - 5y^2 = 36y + 36$

73. The graph is a circle centered at the origin with a radius of 7; $x^2 + y^2 = 49$

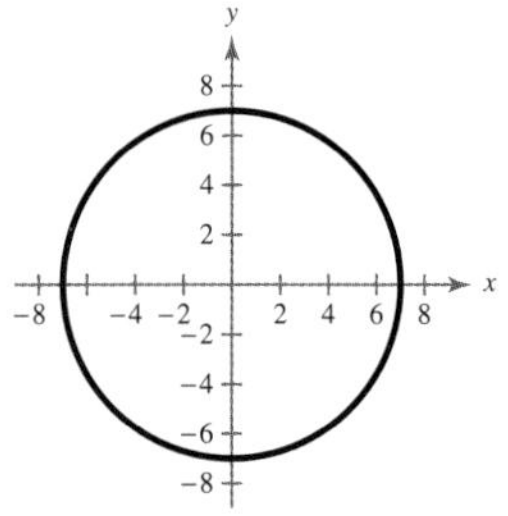

75. The graph consists of all points on the line that makes an angle of $\pi/4$ with the positive x-axis; $x - y = 0$

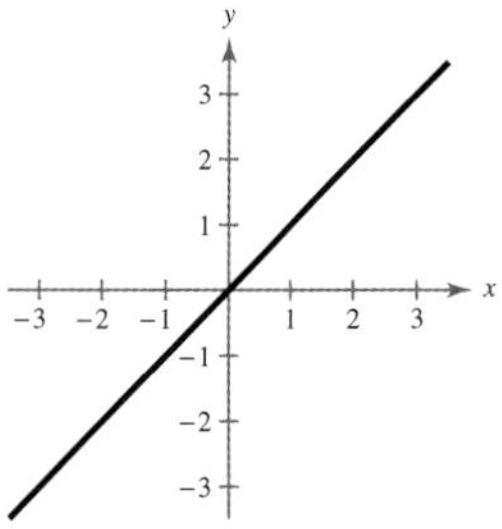

77. The graph is a vertical line through $(5, 0)$; $x = 5$

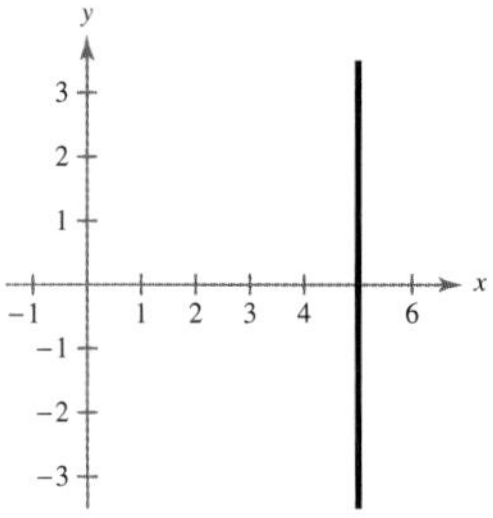

79. (a) $r = 30$

(b) $(30, 5\pi/6)$; 30 represents the distance of the passenger car from the center, and $5\pi/6 = 150°$ represents the angle to which the car has rotated.

(c) $(-25.98, 15)$; The car is about 25.98 feet to the left of the center and 15 feet above the center.

81. True; Because r is a directed distance, (r, θ) can be represented by $(-r, \theta \pm (2n + 1)\pi)$, so $|r| = |-r|$.

83. (a) Answers will vary.

(b) $d = |r_1 - r_2|$; Answers will vary.

(c) $d = \sqrt{r_1^2 + r_2^2}$; Answers will vary.

85. Answers will vary.

87. $(x - h)^2 + (y - k)^2 = h^2 + k^2$

89. $a \approx 16.16$
$b \approx 19.44$
$B = 86°$

Section 9.6 (page 699)

1. convex limaçon **3.** lemniscate

5. Replace (r, θ) with $(r, \pi - \theta)$ or $(-r, -\theta)$ and determine whether the equation is equivalent.

7. Rose curve **9.** Lemniscate **11.** c **12.** d

13. a **14.** c **15.** Polar axis **17.** Polar axis

19. $\theta = \frac{\pi}{2}$ **21.** Pole **23.** $\frac{\pi}{2}$ **25.** $\frac{\pi}{6}, \frac{\pi}{2}, \frac{5\pi}{6}$

27.

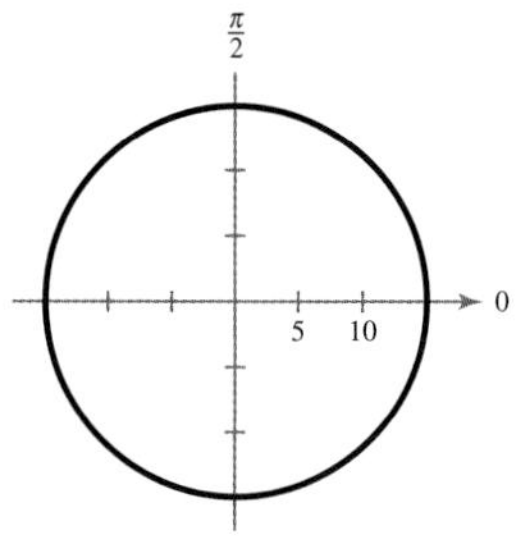

29.

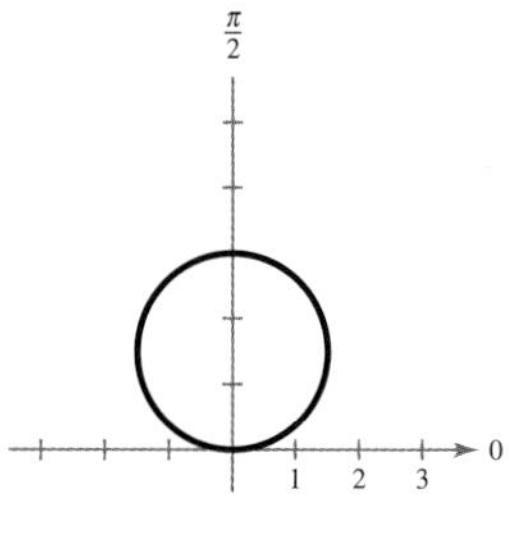

31.

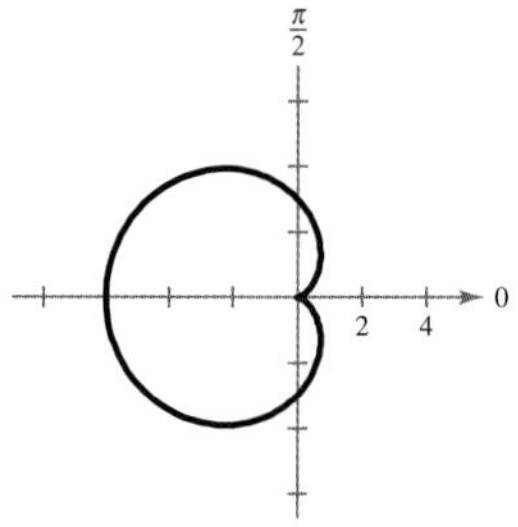

33.

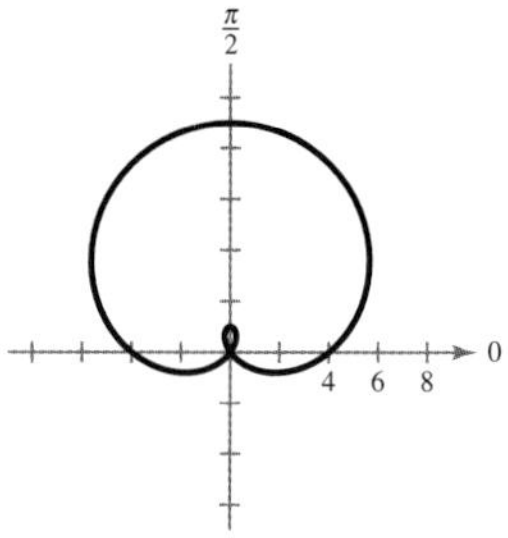

35.

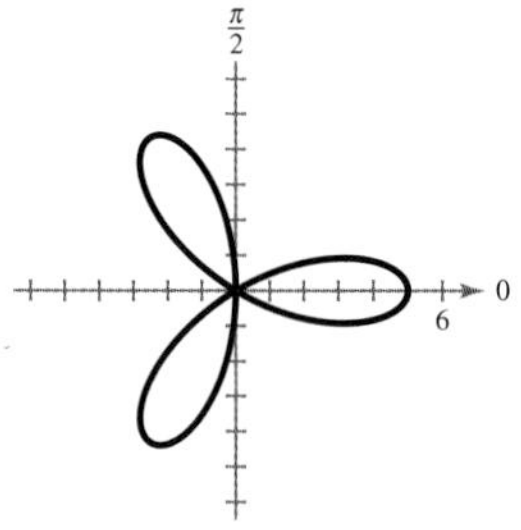

Symmetry: polar axis

Zeros: $\frac{\pi}{6}, \frac{\pi}{2}, \frac{5\pi}{6}$

37.

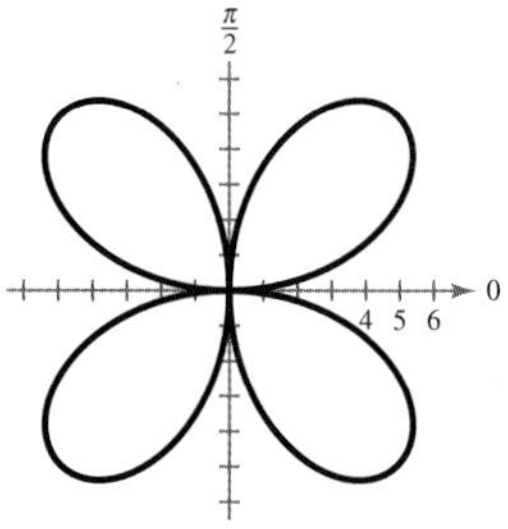

Symmetry: $\theta = \frac{\pi}{2}$, polar axis, pole

Zeros: $0, \frac{\pi}{2}, \pi, \frac{3\pi}{2}, 2\pi$

39.

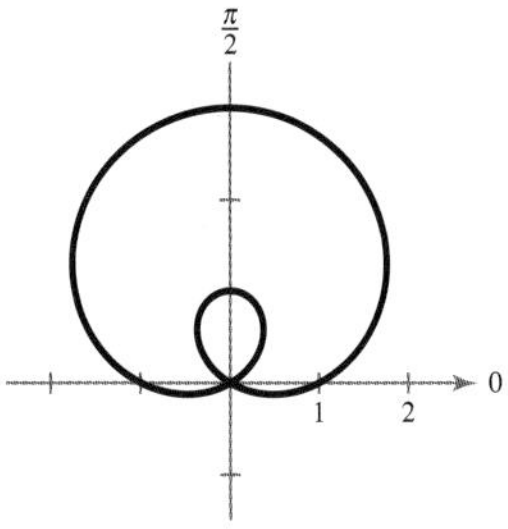

Symmetry: $\theta = \frac{\pi}{2}$

Zeros: $\frac{7\pi}{6}, \frac{11\pi}{6}$

41.

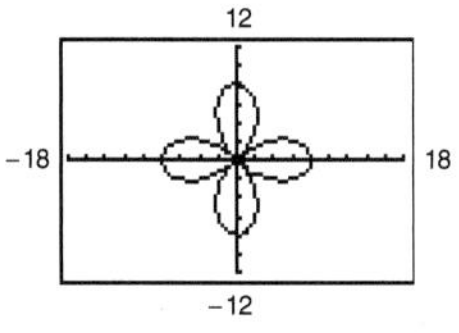

Answers will vary.

43.

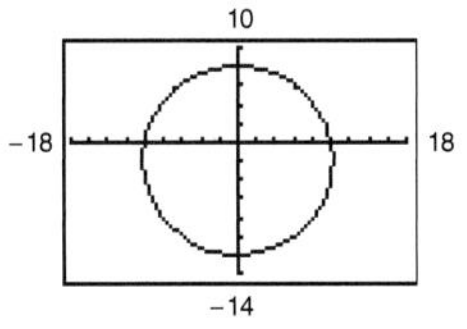

Answers will vary.

45.

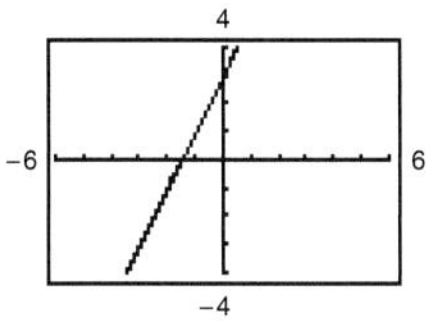

Answers will vary.

47.

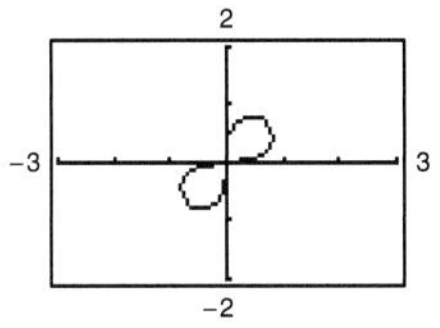

Answers will vary.

49.

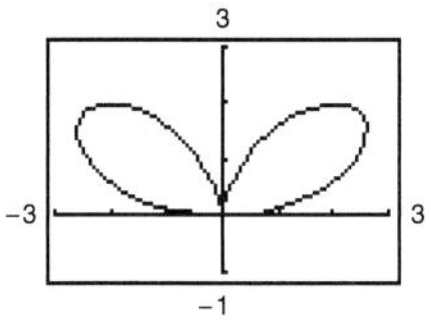

Answers will vary.

51.

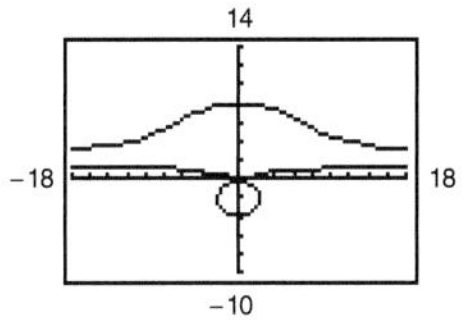

Answers will vary.

53.

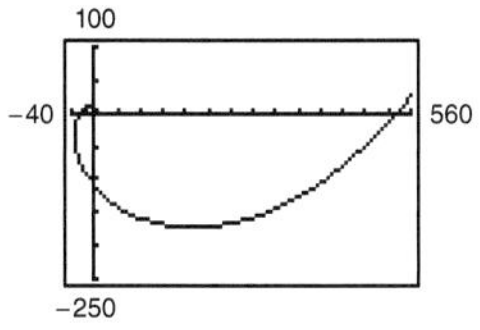

Answers will vary.

55.

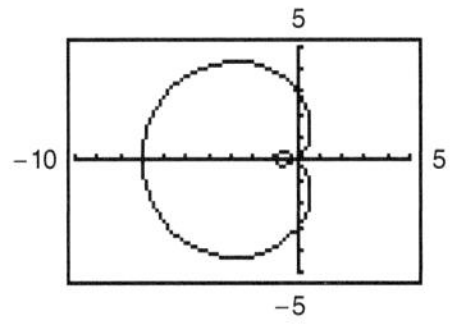

$0 \le \theta < 2\pi$

57.

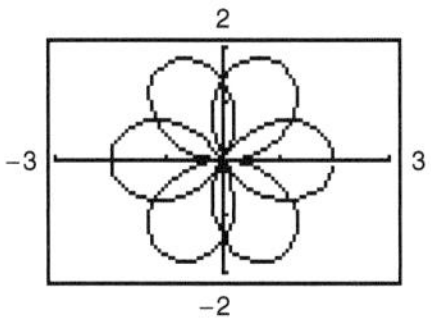

$0 \le \theta < 4\pi$

59.

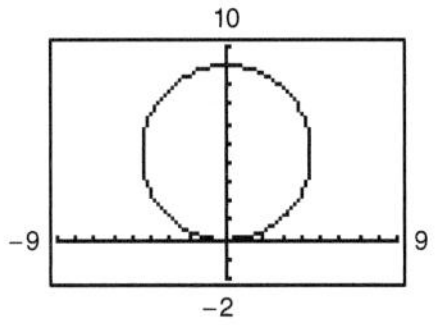

$0 \le \theta < \pi$

61.

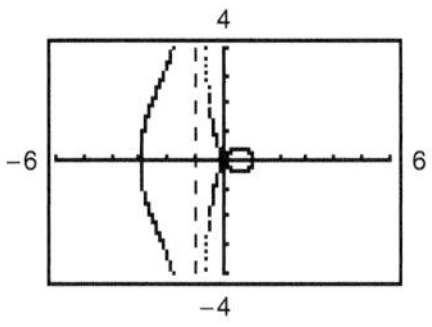

63.

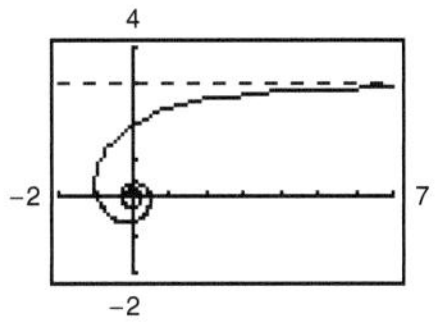

65. (a)

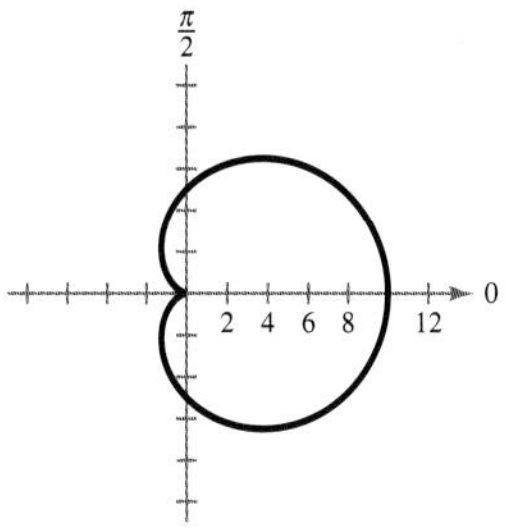

Cardioid

(b) 0

67. True; $n = 5$

69. When $\frac{a}{b} \ge 2$, the graph is a convex limaçon.

71. Answers will vary.

73. (a) $r = 2 - \sin\left(\theta - \frac{\pi}{4}\right)$ (b) $r = 2 + \cos\theta$

(c) $r = 2 + \sin\theta$ (d) $r = 2 - \cos\theta$

75. $x = -3, 3$

Section 9.7 (page 705)

1. conic **3.** vertical

5. (a) $r = \frac{2}{1 + \cos\theta}$, parabola

(b) $r = \frac{3}{1 + 0.5\cos\theta}$, ellipse

(c) $r = \frac{3}{1 + 1.5\cos\theta}$, hyperbola

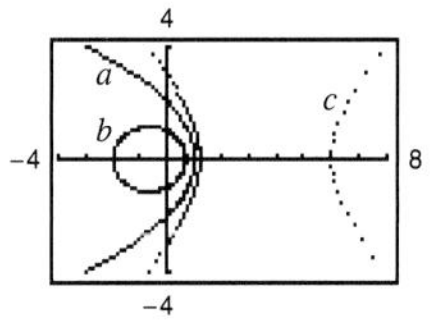

7. (a) $r = \frac{2}{1 - \sin\theta}$, parabola

(b) $r = \frac{1}{1 - 0.5\sin\theta}$, ellipse

(c) $r = \frac{3}{1 - 1.5\sin\theta}$, hyperbola

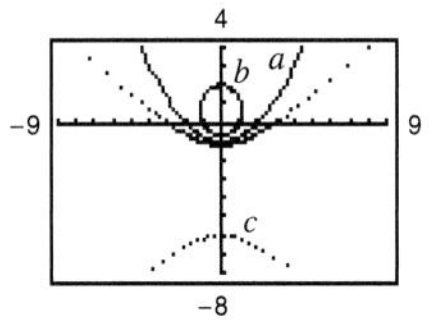

9. b **10.** c **11.** d **12.** a **13.** Parabola

15. Hyperbola **17.** Ellipse **19.** Ellipse **21.** Hyperbola

23. Parabola

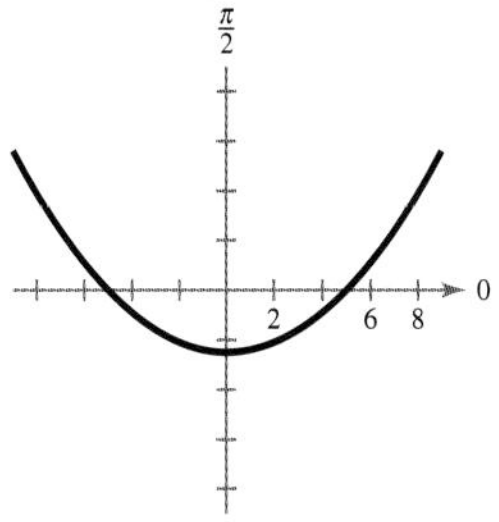

25. Ellipse

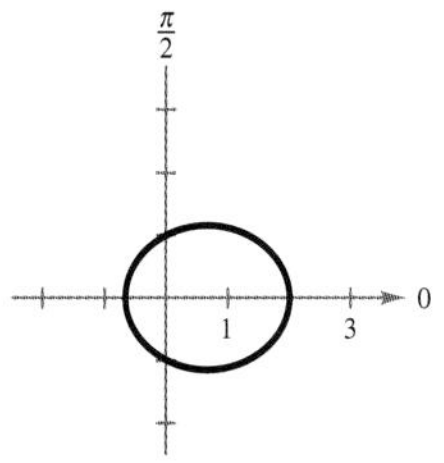

27. Hyperbola

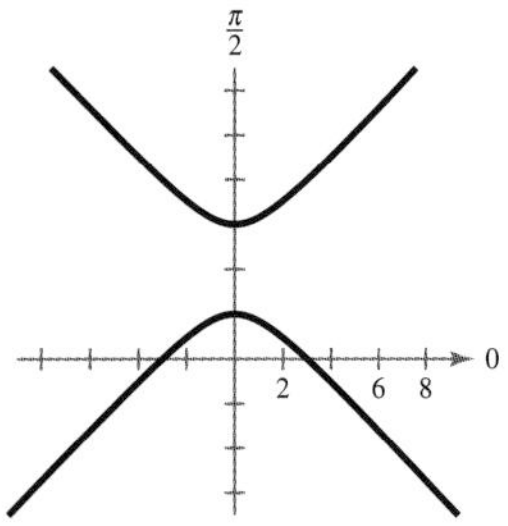

29. Ellipse

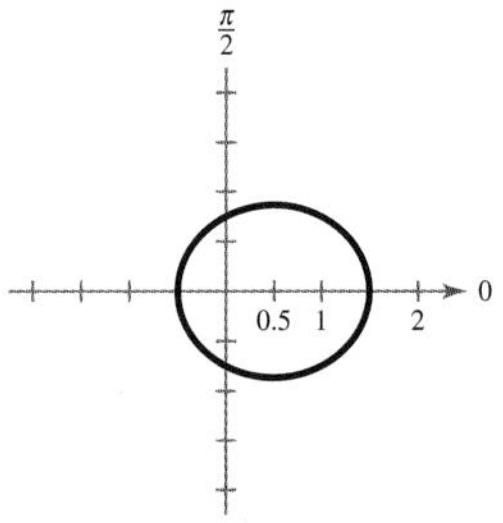

31.

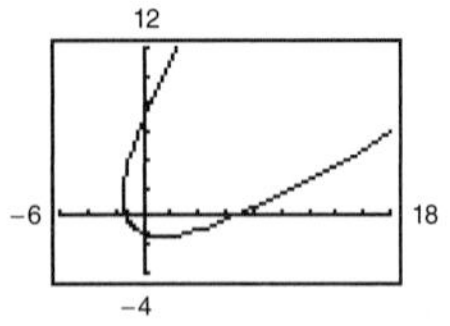

33.

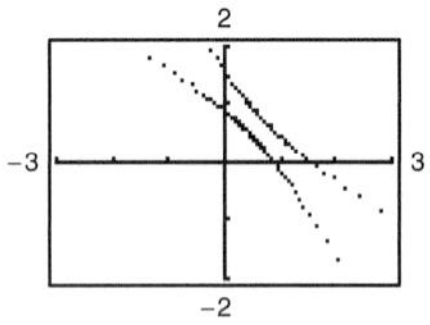

35.

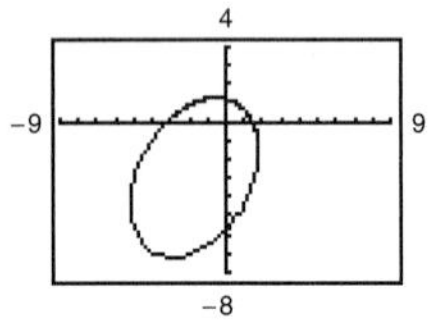

37. $r = \dfrac{1}{1 - \cos\theta}$ **39.** $r = \dfrac{1}{2 + \sin\theta}$

41. $r = \dfrac{2}{1 + 2\cos\theta}$ **43.** $r = \dfrac{2}{1 - \sin\theta}$

45. $r = \dfrac{10}{1 - \cos\theta}$ **47.** $r = \dfrac{10}{3 + 2\cos\theta}$

49. $r = \dfrac{11}{6 - 5\sin\theta}$ **51.** $r = \dfrac{9}{4 - 5\sin\theta}$

53. $r = \dfrac{3}{1 - 4\cos\theta}$ **55.** Answers will vary.

57. $r = \dfrac{9.2931 \times 10^7}{1 - 0.0167\cos\theta}$

Perihelion: about 9.1405×10^7 mi

Aphelion: about 9.4509×10^7 mi

59. $r = \dfrac{1.4038 \times 10^8}{1 - 0.0935\cos\theta}$

Perihelion: about 1.2838×10^8 mi

Aphelion: about 1.5486×10^8 mi

61. $r = \dfrac{1.0821 \times 10^8}{1 - 0.0067\cos\theta}$

Perihelion: about 1.0748×10^8 km

Aphelion: about 1.0894×10^8 km

63. (a) $r_{\text{Neptune}} = \dfrac{4.4944 \times 10^9}{1 - 0.0113\cos\theta}$

$r_{\text{Pluto}} = \dfrac{5.5404 \times 10^9}{1 - 0.2488\cos\theta}$

(b) Neptune: Perihelion: about 4.4442×10^9 km
Aphelion: about 4.5458×10^9 km

Pluto: Perihelion: about 4.4366×10^9 km
Aphelion: about 7.3754×10^9 km

(c)

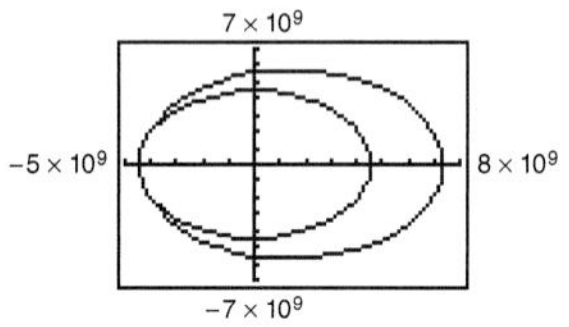

(d) Yes; On average, Pluto is farther from the sun than Neptune.

(e) Using a graphing utility, it would appear that the orbits intersect; No; Pluto and Neptune will never collide because the orbits do not intersect in three-dimensional space.

65. False; The equation can be rewritten as $r = \dfrac{-4/3}{1 + \sin\theta}$.

Because ep is negative, p must be negative, so the directrix has to be below the pole.

67. True; The graph is rotated $\dfrac{5\pi}{4}$ radians clockwise.

69 and 71. Answers will vary.

73. $r^2 = \dfrac{24{,}336}{169 - 25\cos^2\theta}$ **75.** $r^2 = \dfrac{400}{25 - 9\cos^2\theta}$

77. $r^2 = \dfrac{144}{25\cos^2\theta - 16}$ **79.** Circle

81. $-\dfrac{7\sqrt{2}}{10}$ **83.** $\dfrac{7\sqrt{2}}{10}$

Review Exercises (page 708)

1. Hyperbola **3.** $x^2 + y^2 = 25$

5. $(x - 2)^2 + (y - 4)^2 = 13$

7. $x^2 + y^2 = 36$

Center: $(0, 0)$

Radius: 6

9. $\left(x - \frac{1}{2}\right)^2 + \left(y + \frac{3}{4}\right)^2 = 1$

Center: $\left(\frac{1}{2}, -\frac{3}{4}\right)$

Radius: 1

11.

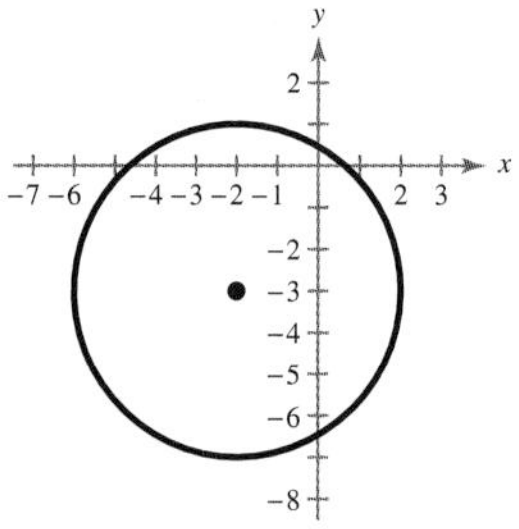

Center: $(-2, -3)$

Radius: 4

13. $\left(0, -4 \pm \sqrt{5}\right)$

15. Vertex: $(0, 0)$

Focus: $(0, -2)$

Directrix: $y = 2$

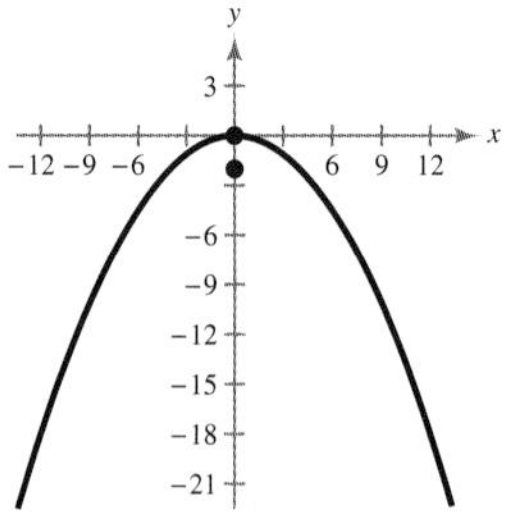

17. Vertex: $(0, 0)$

Focus: $(-9, 0)$

Directrix: $x = 9$

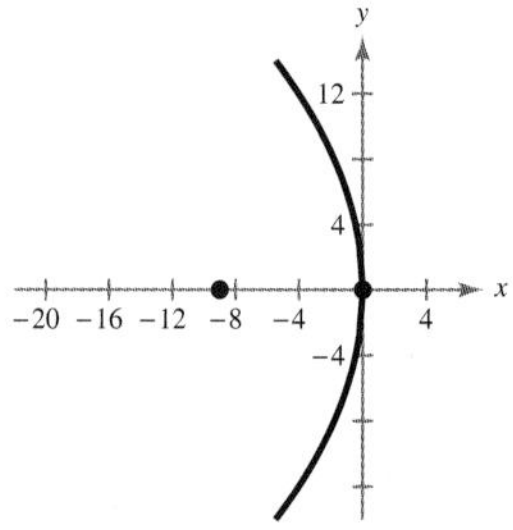

19. $x^2 = 20y$ **21.** $25x = 6(y - 5)^2$ **23.** $y = \frac{1}{4}x - 2$

25. $\dfrac{x^2}{25} + \dfrac{y^2}{9} = 1$

27. Center: $(0, 0)$
Vertices: $(0, \pm 6)$
Foci: $\left(0, \pm \sqrt{11}\right)$
Eccentricity: $\dfrac{\sqrt{11}}{6}$

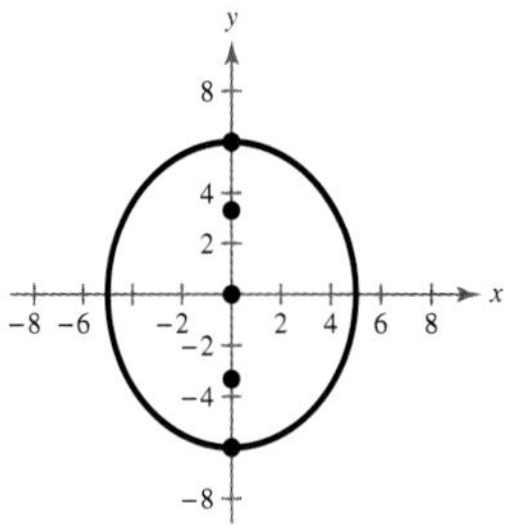

29. Center: $(-2, 7)$
Vertices: $\left(-2 \pm \dfrac{\sqrt{3}}{3}, 7\right)$
Foci: $\left(-2 \pm \dfrac{\sqrt{30}}{12}, 7\right)$
Eccentricity: $\dfrac{\sqrt{10}}{4}$

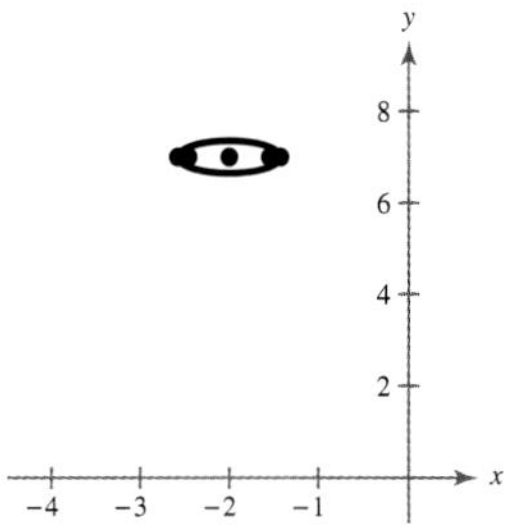

31. The foci should be placed 3 feet on either side of the center at the same height as the pillers.

33. $e \approx 0.0565$ **35.** $\dfrac{x^2}{16} - \dfrac{y^2}{20} = 1$

37. $\dfrac{y^2}{16/5} - \dfrac{(x-3)^2}{4/5} = 1$

39. (a) $\dfrac{y^2}{4} - \dfrac{x^2}{5} = 1$
(b) Center: $(0, 0)$
Vertices: $(0, \pm 2)$
Foci: $(0, \pm 3)$
Asymptotes:
$y = \pm\dfrac{2\sqrt{5}}{5}x$
(c)

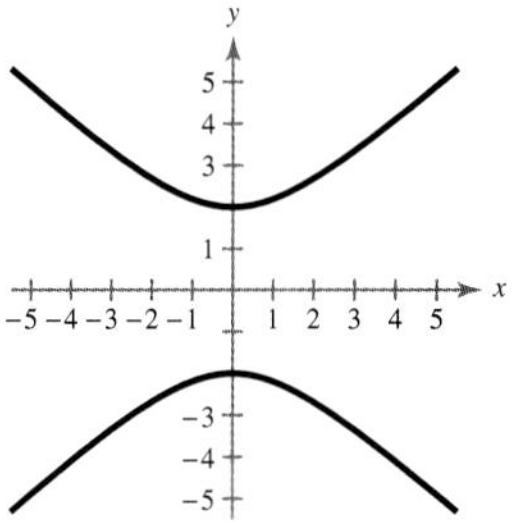

41. (a) $\dfrac{(x-1)^2}{16} - \dfrac{(y+1)^2}{9} = 1$
(b) Center: $(1, -1)$
Vertices: $(5, -1)$, $(-3, -1)$
Foci: $(6, -1)$, $(-4, -1)$
Asymptotes:
$y = -1 \pm \dfrac{3}{4}(x-1)$
(c)

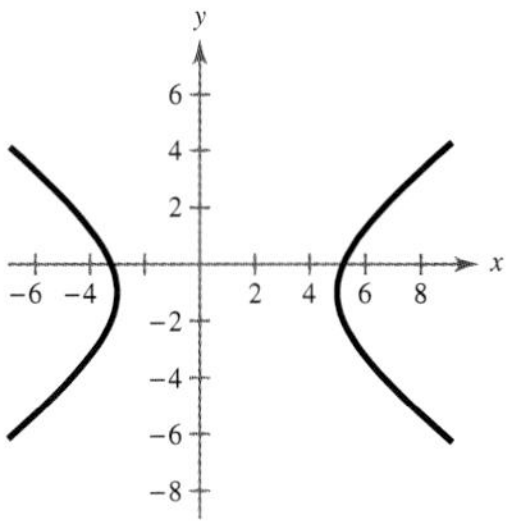

43. (a) $\dfrac{(x+6)^2}{101/2} - \dfrac{(y-1)^2}{202} = 1$
(b) Center: $(-6, 1)$
Vertices:
$\left(-6 \pm \dfrac{\sqrt{202}}{2}, 1\right)$
Foci: $\left(-6 \pm \dfrac{\sqrt{1010}}{2}, 1\right)$
Asymptotes:
$y = 1 \pm 2(x+6)$
(c)

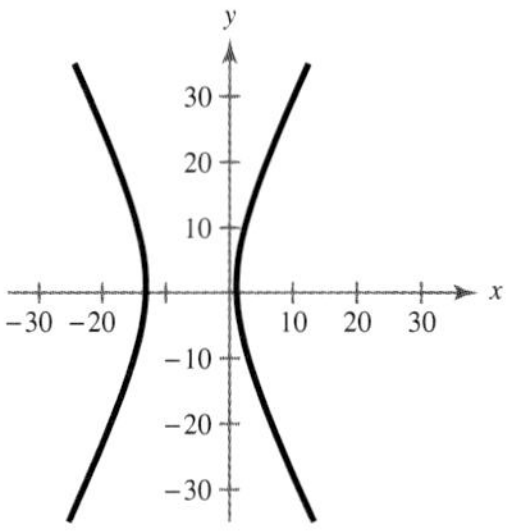

45. About 72 mi **47.** Parabola **49.** Circle

51. $\dfrac{(x')^2}{6} - \dfrac{(y')^2}{6} = 1$

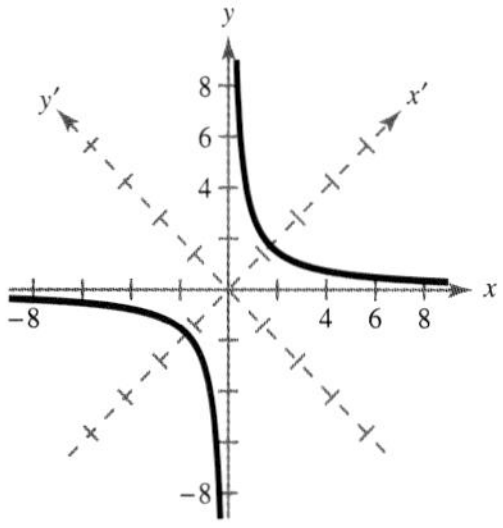

53. $\dfrac{(x')^2}{3} + \dfrac{(y')^2}{2} = 1$

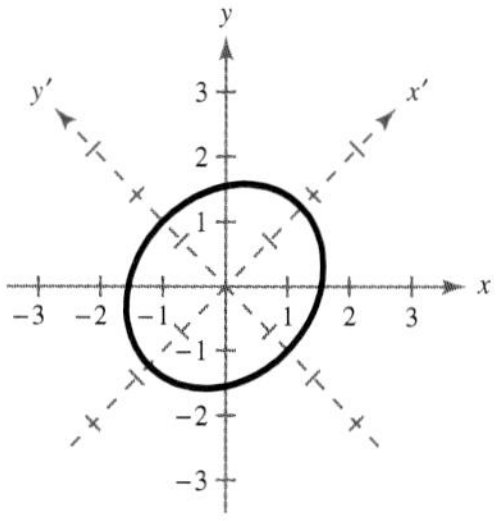

55. (a)

t	-2	-1	0	1	2
x	-8	-5	-2	1	4
y	15	11	7	3	-1

(b)

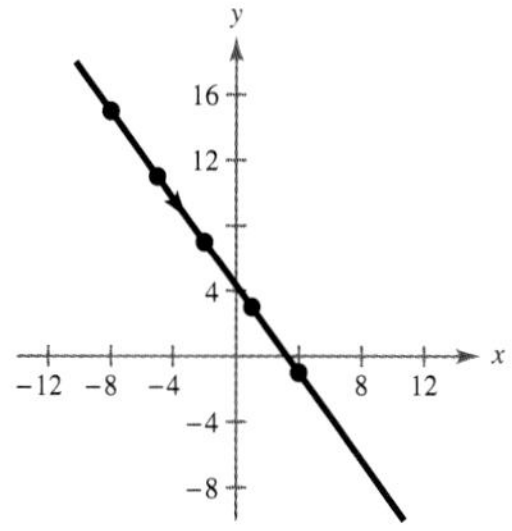

57. (a)

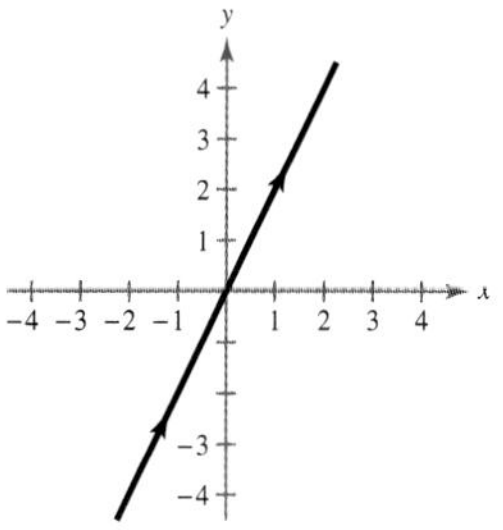

(b) $y = 2x$

59. (a)

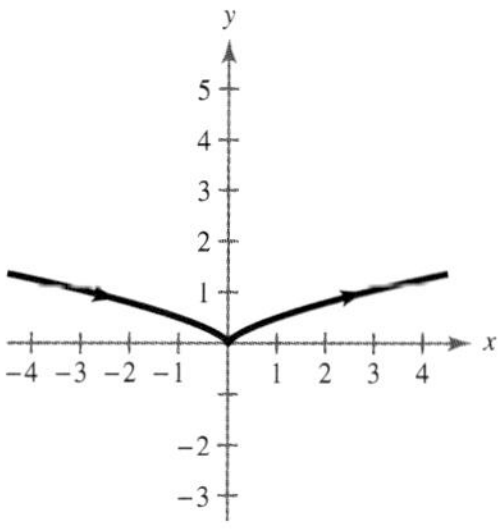

(b) $y = \dfrac{1}{2}x^{2/3}$

61. (a)

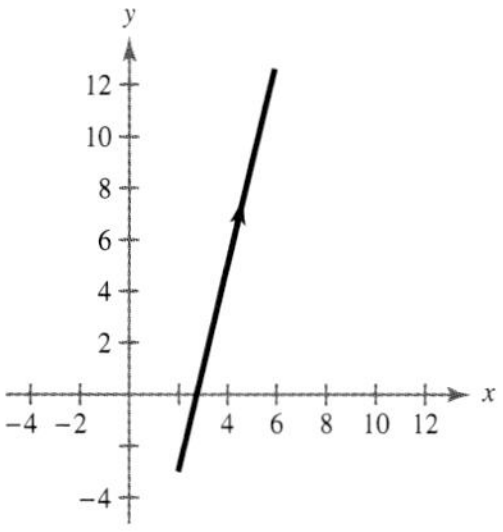

(b) $y = 4x - 11, x \geq 2$

63.

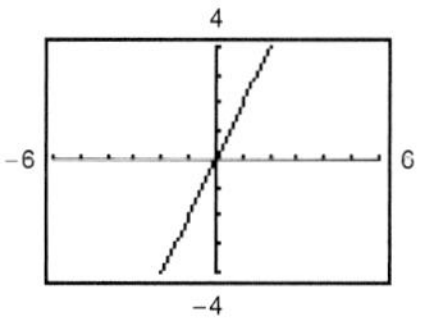

Function

65.

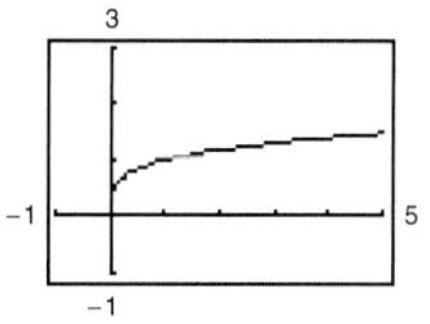

Function

CHAPTER 9

67. 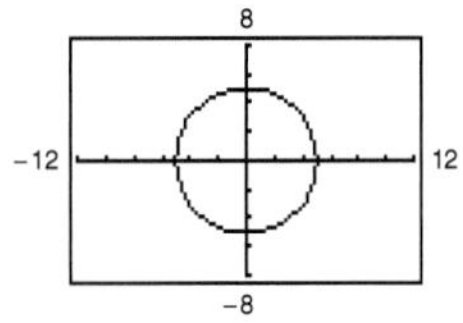

Not a function

69. (a) $x = t, y = 6t + 2$ (b) $x = 1 - t, y = 8 - 6t$

71. (a) $x = t, y = t^2 + 3$ (b) $x = 1 - t, y = t^2 - 2t + 4$

73. $\left(\frac{5\sqrt{2}}{2}, -\frac{5\sqrt{2}}{2}\right)$ **75.** *Sample answer:* $\left(5\sqrt{2}, \frac{7\pi}{4}\right)$

77. $r = 9$ **79.** $r^2 = 10 \sec\theta \csc\theta$

81. $x^2 + y^2 = 196$ **83.** $(x^2 + y^2)^3 = y^2$

85. 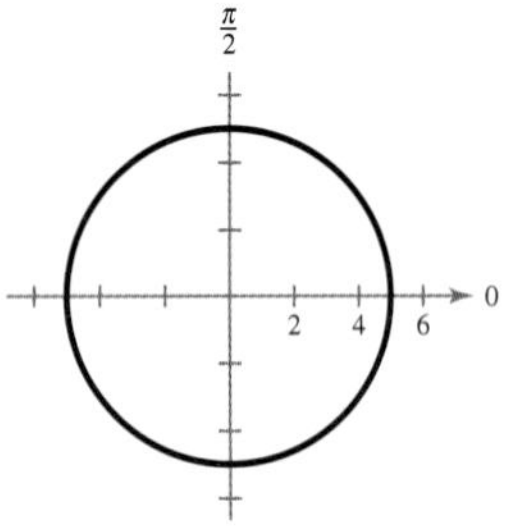

87. 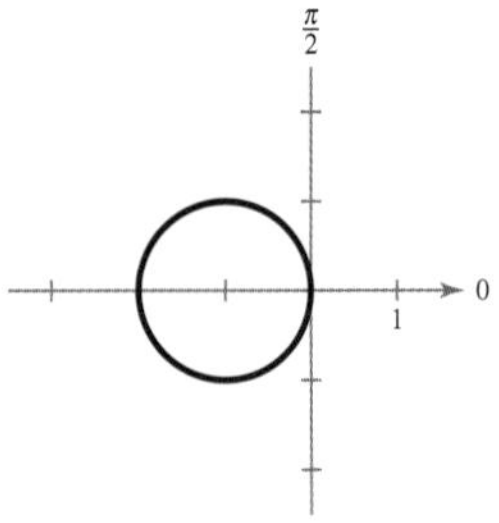

89. Limaçon with inner loop 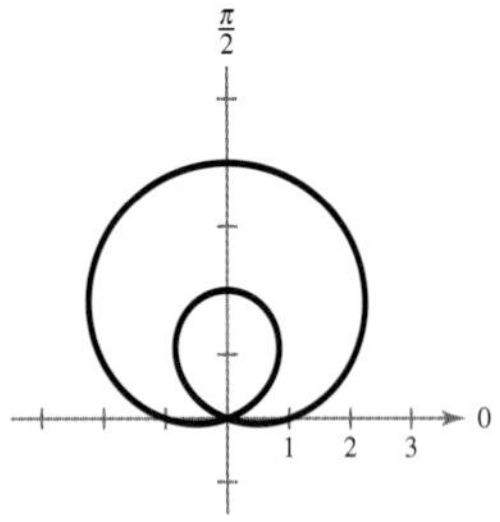

Symmetry: $\theta = \frac{\pi}{2}$

Zeros: $\theta \approx 3.48, 5.94$

91. Rose curve 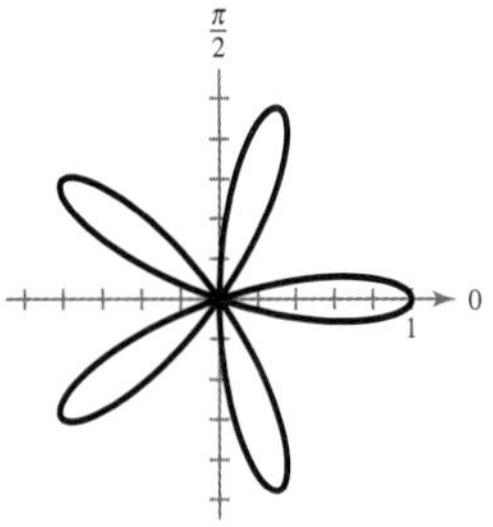

Symmetry: Polar axis

Zeros: $\theta = \frac{\pi}{10}, \frac{3\pi}{10}, \frac{\pi}{2}, \frac{7\pi}{10}, \frac{9\pi}{10}$

93. Hyperbola **95.** Ellipse **97.** Parabola

99. $r = \dfrac{4}{1 - \cos\theta}$ **101.** $r = \dfrac{5}{3 - 2\sin\theta}$

103. *Sample answer:* $r = \dfrac{1.9}{1 - 0.9\cos\theta}$; 1 astronomical unit

105. False; The equation of a hyperbola is a second-degree equation.

107. (a) The time it takes to make one revolution is halved.

(b) The length of the major axis is increased by two units.

Chapter Test (page 714)

1. $(x - 7)^2 + (y - 2)^2 = 36$

Center: $(7, 2)$

Radius: 6

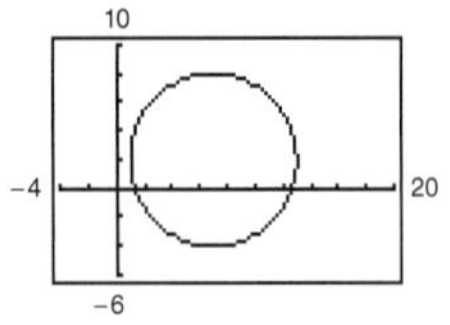

2. Vertex: $(0, 0)$

Focus: $(2, 0)$

Directrix: $x = -2$

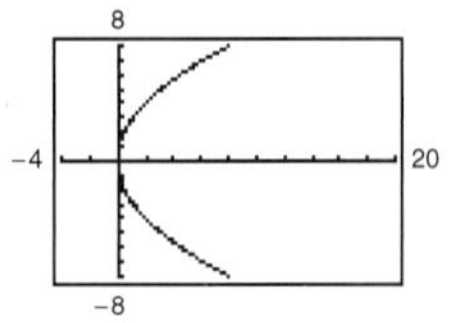

3. $\dfrac{(x - 12)^2}{112} + \dfrac{y^2}{28} = 1$

Center: $(12, 0)$

Vertices: $\left(12 \pm 4\sqrt{7}, 0\right)$

Foci: $\left(12 \pm 2\sqrt{21}, 0\right)$

Eccentricity: $\dfrac{\sqrt{3}}{2}$

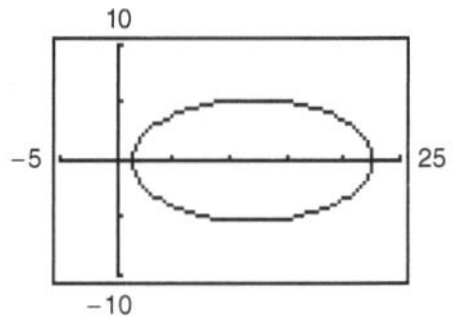

4. Center: $(2, 0)$

Vertices: $(0, 0), (4, 0)$

Foci: $\left(2 \pm \sqrt{5}, 0\right)$

Asymptotes: $y = \pm\frac{1}{2}(x - 2)$

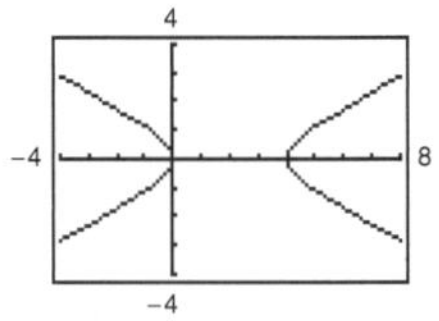

5. (a) Parabola (b) Hyperbola

6. $\dfrac{\left(y' + 2\sqrt{2}\right)^2}{8} - \dfrac{\left(x' + \sqrt{2}\right)^2}{8} = 1$; Hyperbola

7. 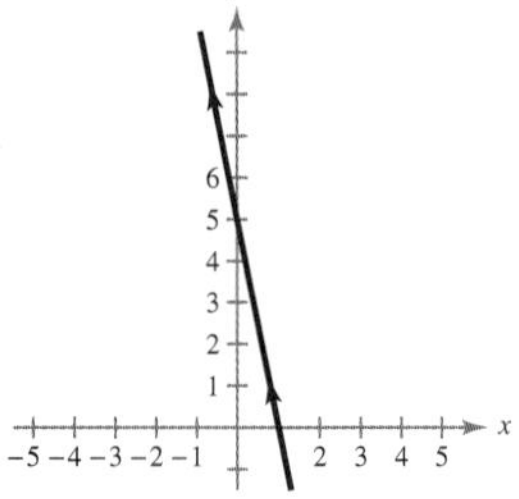

$y = 5 - 5x$

8. 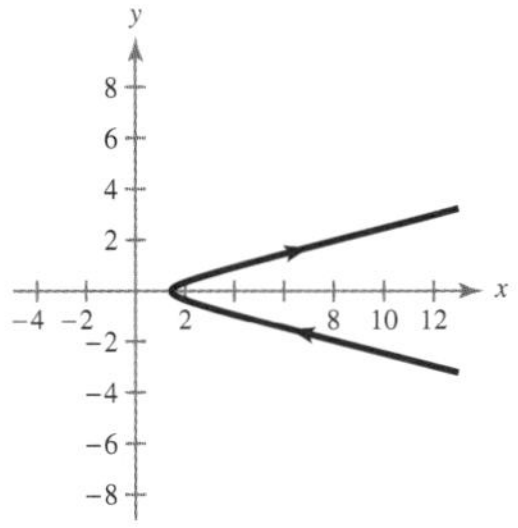

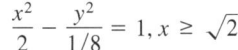

$\dfrac{x^2}{2} - \dfrac{y^2}{1/8} = 1, x \geq \sqrt{2}$

9. 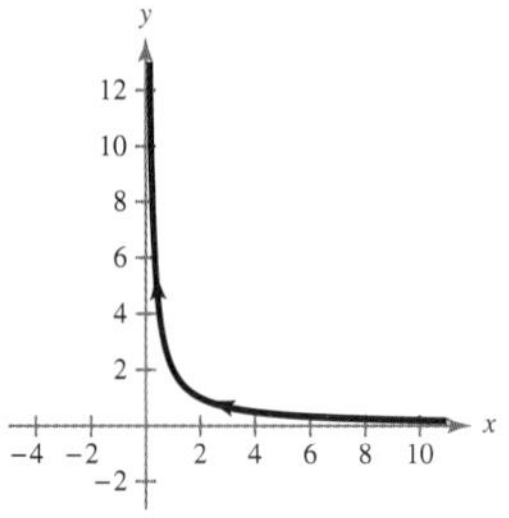

$y = \dfrac{2}{x}, x > 0$

10. 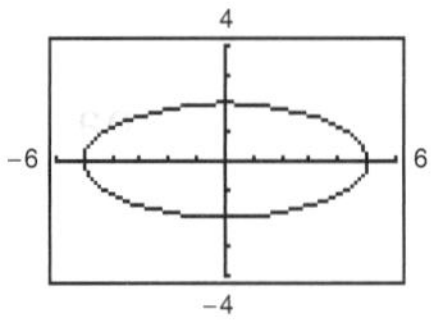

11. (a) $x = t, y = -3t + 5$ (b) $x = t - 3, y = -3t + 14$

12. (a) $x = t, y = \dfrac{8}{t}$ (b) $x = t - 3, y = \dfrac{8}{t - 3}$

13. (a) $x = t, y = \dfrac{4 - t^2}{2}$ (b) $x = t - 3, y = \dfrac{-t^2 + 6t - 5}{2}$

14. $\left(\sqrt{3}, -1\right)$

15. *Sample answer:* $\left(2\sqrt{2}, \frac{7\pi}{4}\right); \left(2\sqrt{2}, -\frac{\pi}{4}\right), \left(-2\sqrt{2}, \frac{3\pi}{4}\right)$

16. $r = 3\cos\theta$ **17.** $x^2 + (y - 1)^2 = 1$ **18.** $r = 3$

19. Limaçon with inner loop

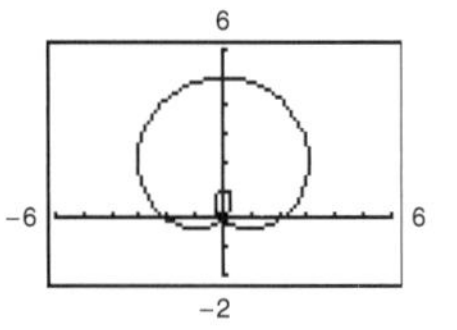

20. Parabola

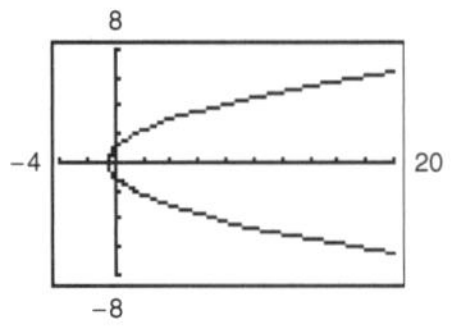

21. Hyperbola

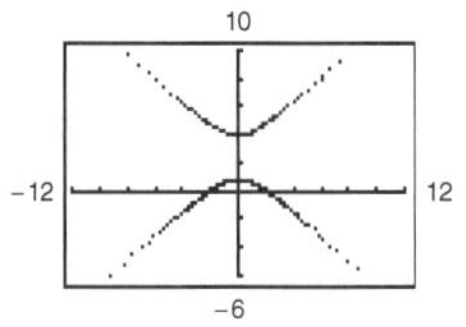

22. $r = a \sin 7\theta, a \neq 0$

23.

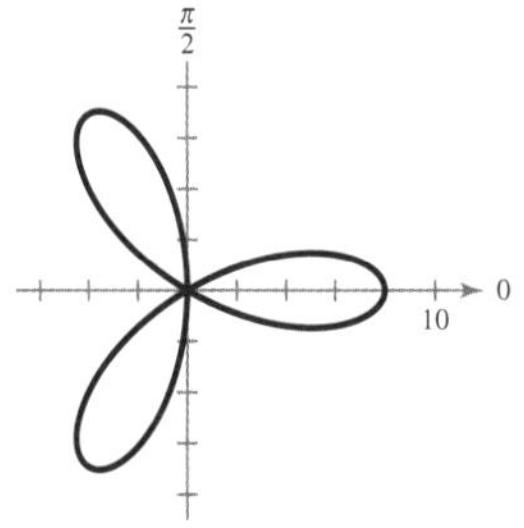

Symmetry: Polar axis

Zeros: $\frac{\pi}{6}, \frac{\pi}{2}, \frac{5\pi}{6}$

24. $r = \dfrac{4}{4 + \sin \theta}$ **25.** $r = \dfrac{10}{4 + 5 \sin \theta}$

Standardized Test Practice (page 715)

1. A **2.** B **3.** C **4.** C **5.** B **6.** A **7.** C
8. B **9.** C **10.** A **11.** B **12.** B **13.** A
14. D **15.** C **16.** C **17.** B **18.** A **19.** C
20. C **21.** B **22.** 243 **23.** 4 **24.** 22.08
25. 6720 **26.** 33.94 **27.** 0.57

Chapter 10

Section 10.1 (page 726)

1. three-dimensional **3.** Distance Formula
5. surface, space
7. A sphere with center (h, k, j) and radius r
9. $A(-2, -1, 4), B(3, -2, 0), C(-2, 2, -3)$
11. $A(2, -1, 4), B(1, 3, -2), C(-3, 0, -2)$

13.

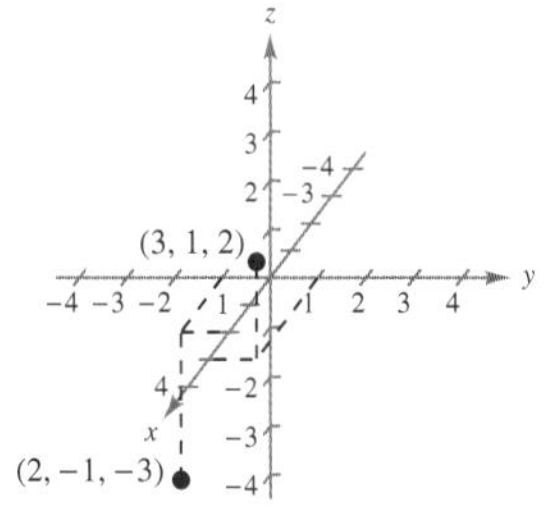

15.

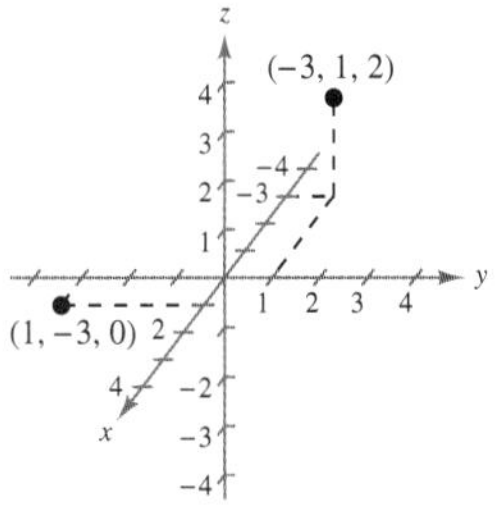

17.

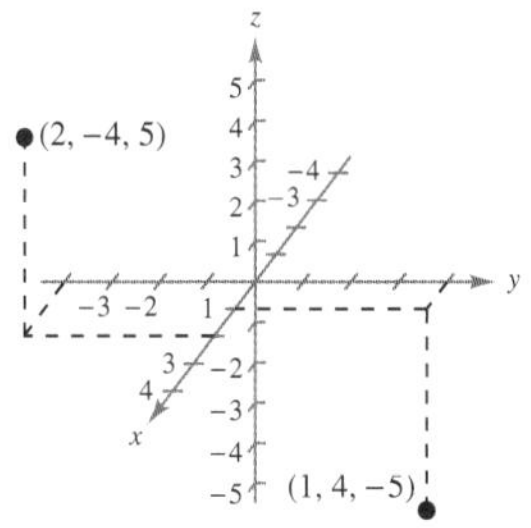

19. $(-3, 3, 5)$ **21.** $(11, 0, 0)$ **23.** $(0, 2, 7)$
25. Octant IV **27.** Octant II **29.** yz-plane
31. $\sqrt{141}$ unitss **33.** $\sqrt{29}$ units **35.** $\sqrt{154}$ units
37. 12 units **39.** $2\sqrt{5}, 3, \sqrt{29}$ **41.** $3, 6, 3\sqrt{5}$
43. $6, 6, 2\sqrt{10}$; isosceles triangle
45. $\sqrt{29}, \sqrt{66}, \sqrt{61}$; neither **47.** $\left(\frac{5}{2}, -3, 2\right)$
49. $\left(\frac{3}{2}, \frac{7}{2}, \frac{1}{2}\right)$ **51.** $(-4, -6, -1)$ **53.** 178 sec
55. $(x - 3)^2 + (y - 2)^2 + (z - 4)^2 = 16$
57. $(x + 2)^2 + (y + 1)^2 + (z + 1)^2 = 2$
59. $x^2 + (y - 4)^2 + (z - 3)^2 = 16$
61. $(x + 3)^2 + (y - 7)^2 + (z - 5)^2 = 25$
63. $\left(x - \frac{3}{2}\right)^2 + y^2 + (z - 3)^2 = \frac{45}{4}$
65. Center: $(3, 0, 0)$; radius: 3
67. Center: $(2, -1, 0)$; radius: $\sqrt{5}$
69. Center: $(-2, 0, 4)$; radius: 1
71. Center: $(2, -1, 3)$; radius: 2
73. Center: $(1, -2, 0)$; radius: $\dfrac{\sqrt{21}}{2}$
75. Center: $\left(\frac{1}{2}, 4, -1\right)$; radius: 3

77.

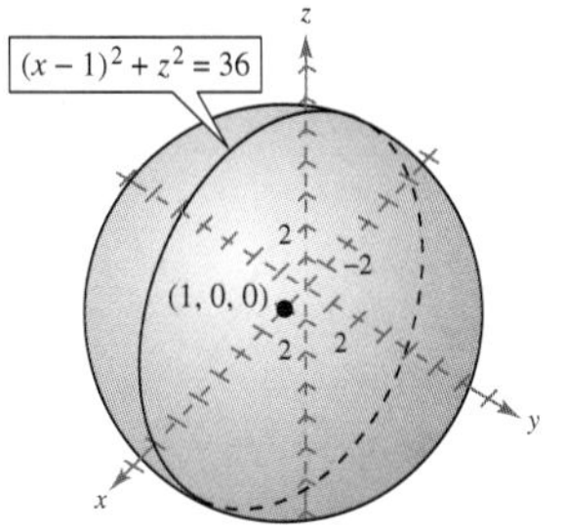

79.

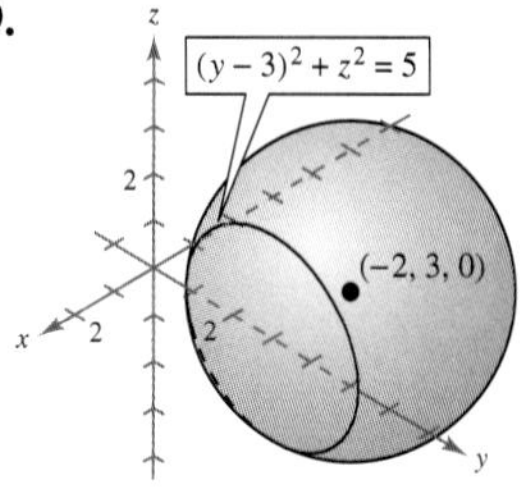

81.

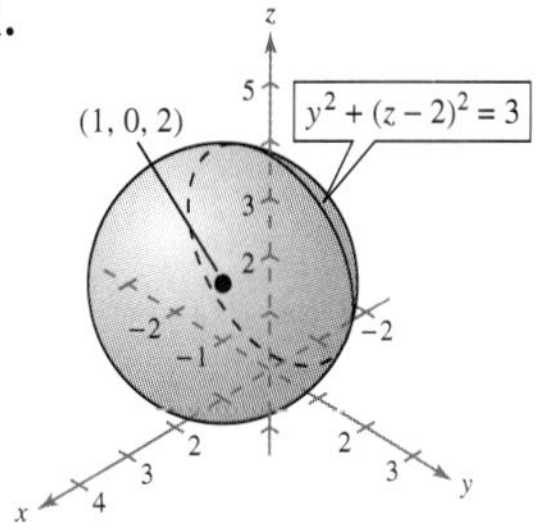

83.

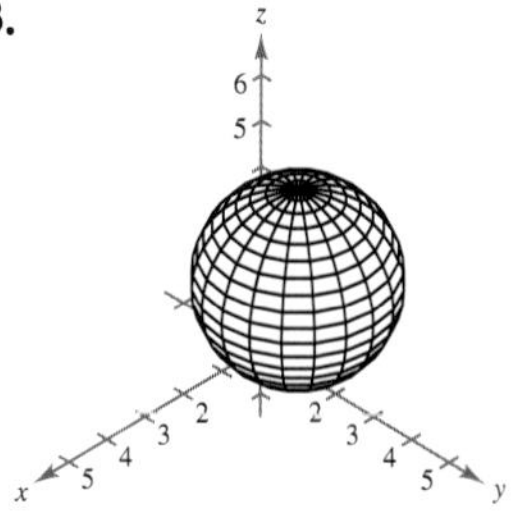

85.

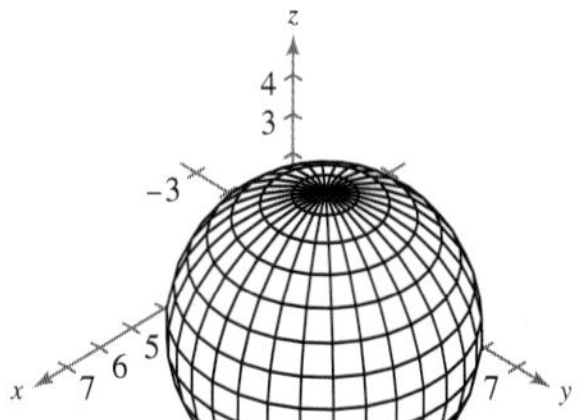

87. $(3, 3, 3)$ **89.** $x^2 + y^2 + z^2 = \dfrac{205^2}{4}$
91. The xz-trace is $x^2 + z^2 = 16$.
93. False; z is the directed distance from the xy-plane to P.
95. 0; 0; 0 **97.** $(x_2, y_2, z_2) = (2x_m - x_1, 2y_m - y_1, 2z_m - z_1)$
99. $y = -\dfrac{3 \pm \sqrt{17}}{2}$ **101.** $x = \dfrac{5 \pm \sqrt{5}}{2}$ **103.** -7

Section 10.2 (page 734)

1. zero **3.** parallel **5.** $\|\mathbf{v}\| = \sqrt{v_1^2 + v_2^2 + v_3^2}$
7. $\langle 0, 6, 0 \rangle$
9. (a) $\langle 4, -3, -4 \rangle$ (b) $\sqrt{41}$ (c) $\dfrac{\sqrt{41}}{41}\langle 4, -3, -4 \rangle$
11. (a) $\langle 7, -5, 5 \rangle$ (b) $3\sqrt{11}$ (c) $\dfrac{\sqrt{11}}{33}\langle 7, -5, 5 \rangle$

13. (a)

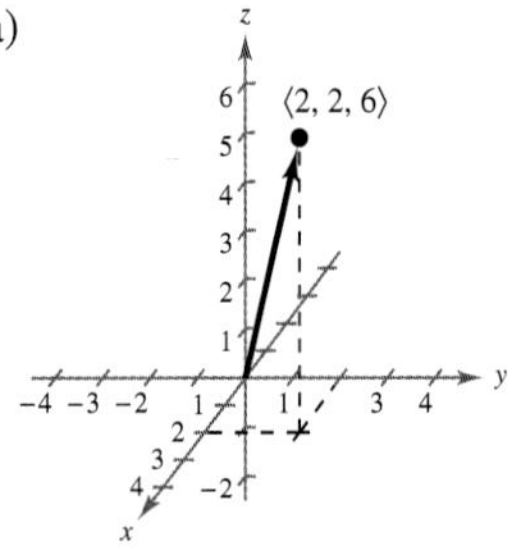

(b)

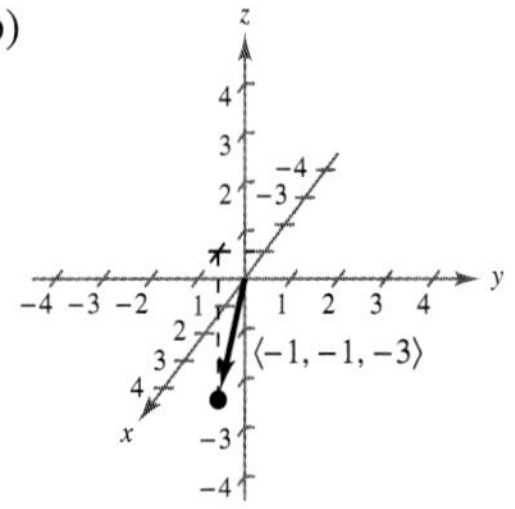

(c)

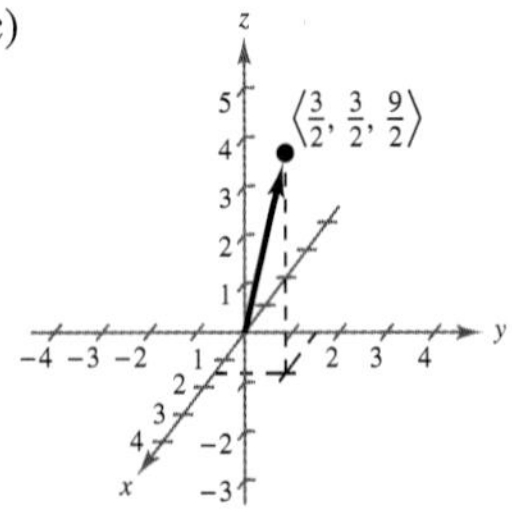

(d)

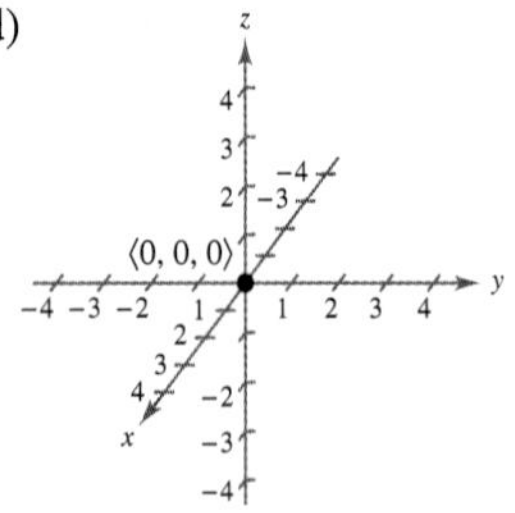

15. (a)

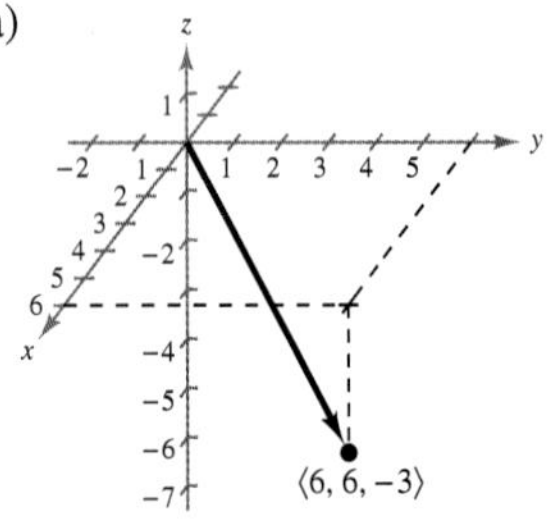

(b)

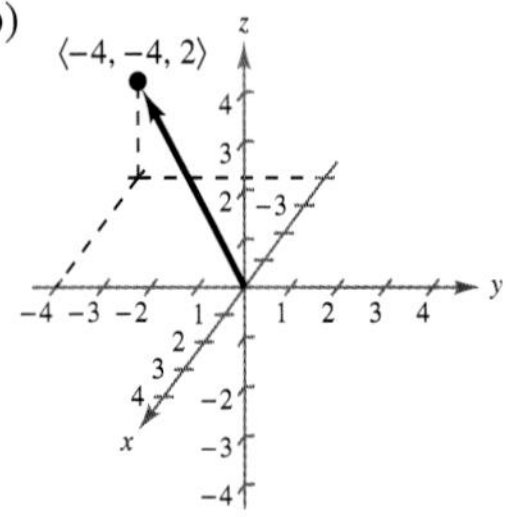

(c)

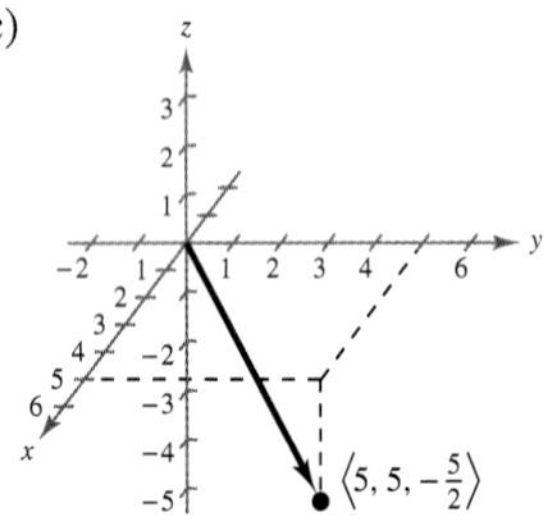

(d)

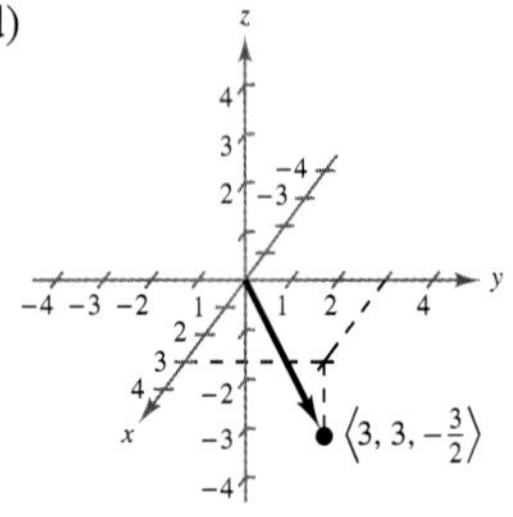

17. $\mathbf{z} = \langle -3, 7, 6 \rangle$ **19.** $\mathbf{z} = \langle -\frac{5}{2}, 12, \frac{15}{2} \rangle$
21. $\mathbf{z} = \langle \frac{1}{2}, 6, \frac{3}{2} \rangle$ **23.** $3\sqrt{5}$ **25.** $\sqrt{21}$ **27.** $\sqrt{101}$
29. (a) $\frac{1}{13}(5\mathbf{i} + 12\mathbf{k})$ (b) $-\frac{1}{13}(5\mathbf{i} + 12\mathbf{k})$
31. (a) $\frac{\sqrt{74}}{74}(8\mathbf{i} + 3\mathbf{j} - \mathbf{k})$ (b) $-\frac{\sqrt{74}}{74}(8\mathbf{i} + 3\mathbf{j} - \mathbf{k})$
33. $\langle -31, -8, 17 \rangle$ **35.** About 13.75 **37.** -4
39. 57 **41.** About 124.45° **43.** About 109.92°
45. Parallel **47.** Neither **49.** Orthogonal
51. Not collinear **53.** Not collinear
55. Right triangle; Answers will vary.
57. Acute triangle; Answers will vary. **59.** (3, 1, 7)
61. $(6, \frac{5}{2}, -\frac{7}{4})$ **63.** $\langle 0, 2\sqrt{2}, 2\sqrt{2} \rangle$ or $\langle 0, 2\sqrt{2}, -2\sqrt{2} \rangle$
65. $\pm\frac{3\sqrt{14}}{14}$ **67.** About 2.75 lb
69. True; $\cos^{-1} 0 = 90°$
71. $\mathbf{u} = -3\mathbf{v}$, so the points are collinear.
73. Sphere of radius 4 centered at (x_1, y_1, z_1)

Section 10.3 (page 741)

1. cross product **3.** $\mathbf{u} \times \mathbf{v}$ **5.** triple scalar product
7. $\langle 1, 1, 1 \rangle$ **9.** $\langle -13, 60, -51 \rangle$ **11.** $-7\mathbf{i} + 13\mathbf{j} + 16\mathbf{k}$
13. $-17\mathbf{i} + \mathbf{j} + 10\mathbf{k}$ **15.** $-\frac{7}{6}\mathbf{i} - \frac{7}{8}\mathbf{j}$ **17.** $\frac{1}{18}\mathbf{i} - \mathbf{j}$
19. $-\mathbf{i} - 2\mathbf{j} - \mathbf{k}$ **21.** $\langle 10, -2, -4 \rangle$
23. $-6\mathbf{i} - 15\mathbf{j} - 6\mathbf{k}$ **25.** $-\frac{1}{4}\mathbf{i} - \frac{7}{10}\mathbf{j} - 2\mathbf{k}$ **27.** $\mathbf{0}$
29. $-27\mathbf{i} - 12\mathbf{j} - 33\mathbf{k}$ **31.** 0 **33.** $\frac{1}{3}\mathbf{i} - \frac{2}{3}\mathbf{j} - \frac{2}{3}\mathbf{k}$
35. $\frac{\sqrt{19}}{19}(\mathbf{i} - 3\mathbf{j} + 3\mathbf{k})$ **37.** $\frac{\sqrt{6}}{90}(10\mathbf{i} - 5\mathbf{j} - 35\mathbf{k})$
39. $\frac{\sqrt{2}}{2}(\mathbf{i} - \mathbf{j})$ **41.** 1 **43.** $\sqrt{806}$ **45.** 56
47. (a) Answers will vary. (b) $6\sqrt{10}$ **49.** $\frac{3\sqrt{13}}{2}$
51. $\frac{1}{2}\sqrt{4290}$ **53.** -16 **55.** 2 **57.** 2 **59.** 12
61. 84 **63.** 54
65. (a) $T(p) = \frac{p}{2}\cos 40°$
(b)

p	15	20	25	30	35	40	45
T	5.75	7.66	9.58	11.49	13.41	15.32	17.24

67. True; The cross product is defined only for three-dimensional vectors.
69 and 71. Proofs **73.** $(2, -\frac{2}{7})$

Section 10.4 (page 750)

1. direction **3.** 0
5. (a) $x = t, y = 3t, z = 5t$ (b) $x = \frac{y}{3} = \frac{z}{5}$
7. (a) $x = -4, y = 1 + 6t, z = -5t$ (b) Not possible
9. (a) $x = 2 + 2t, y = -3 - 3t, z = 5 + t$
(b) $\frac{x-2}{2} = \frac{y+3}{-3} = z - 5$
11. (a) $x = 2 - t, y = 4t, z = 2 - 5t$
(b) $\frac{x-2}{-1} = \frac{y}{4} = \frac{z-2}{-5}$
13. (a) $x = -3 + 4t, y = 8 - 10t, z = 15 + t$
(b) $\frac{x+3}{4} = \frac{y-8}{-10} = z - 15$
15. (a) $x = 3 - 4t, y = 1, z = 2 + 3t$ (b) Not possible
17. (a) $x = -\frac{1}{2} + 3t, y = 2 - 5t, z = \frac{1}{2} - t$
(b) $\frac{x + \frac{1}{2}}{3} = \frac{y-2}{-5} = \frac{z - \frac{1}{2}}{-1}$
19.

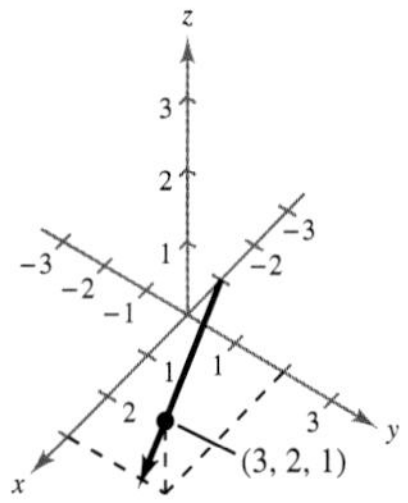

21. $x - 2 = 0$ **23.** $-2x + y - 2z + 10 = 0$
25. $-2x + y + 2z - 3 = 0$ **27.** $-2x - 4y + 3z + 4 = 0$
29. $6x - 2y - z - 8 = 0$ **31.** $y - 5 = 0$
33. $y - z + 2 = 0$ **35.** $7x + y - 11z - 5 = 0$

37. $x = 5 - t, y = 1 + 3t, z = 8 - 4t$
39. $x = 6 + 3t, y = -4 - 4t, z = -1$
41. $x = -2 - t, y = 7 + 4t, z = 4 + 5t$
43. Perpendicular **45.** Parallel
47. (a) About 56.94° (b) $x = t, y = -7t, z = -3t$
49. (a) About 77.83° (b) $x = 1 + 6t, y = t, z = 1 + 7t$
51.

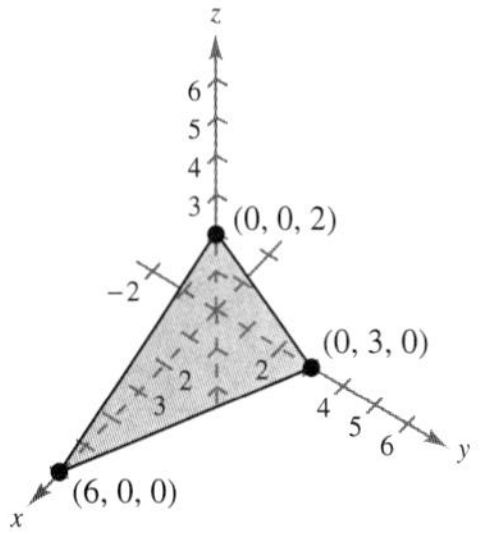

53.

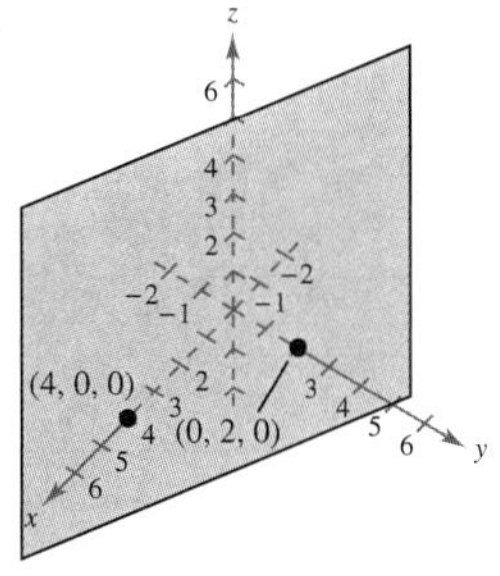

55. $\frac{3\sqrt{5}}{5}$ **57.** $\frac{2\sqrt{6}}{3}$ **59.** About 88.45°
61. False; Lines that do not intersect and are not in the same plane may not be parallel.
63. The point and the vector are reversed. The equations are $\frac{x-1}{3} = \frac{y-2}{5} = \frac{z-6}{4}$.
65. Parallel; $\langle 10, -18, 20\rangle$ is a scalar multiple of $\langle -15, 27, -30\rangle$.

Review Exercises (page 752)

1.

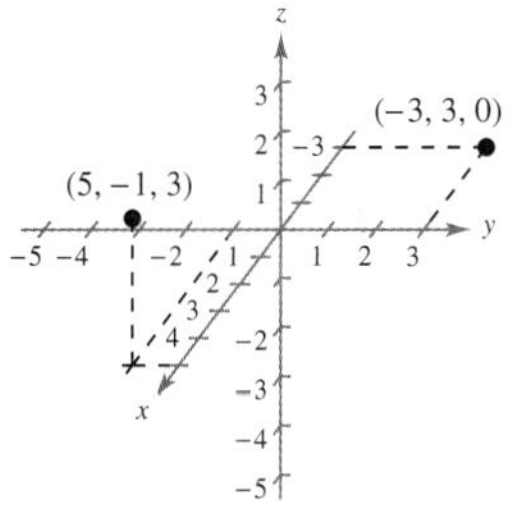

3. $(-10, 9, 0)$ **5.** $3\sqrt{6}$ **7.** $\sqrt{29}, \sqrt{38}, \sqrt{67}$
9. $(0, -1, 3)$ **11.** $(x - 3)^2 + (y + 2)^2 + (z - 4)^2 = 16$
13. Center: $(2, 3, -1)$; radius: $2\sqrt{2}$
15. (a)

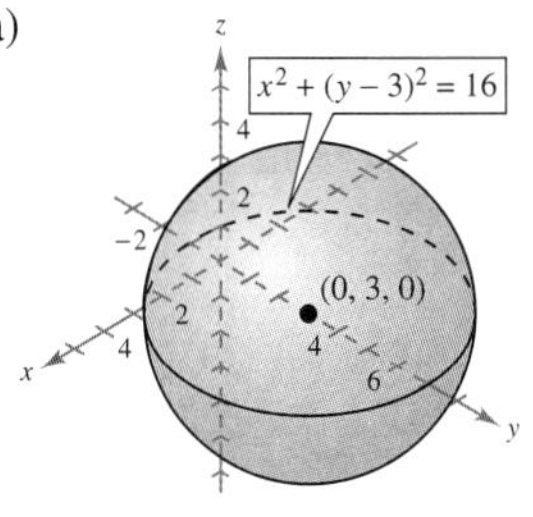

(b)

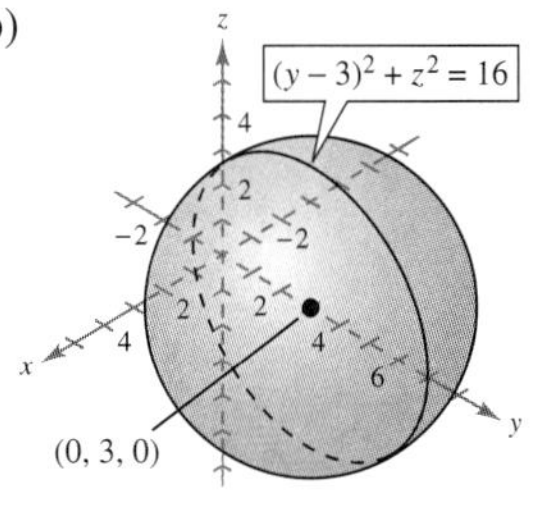

(c)

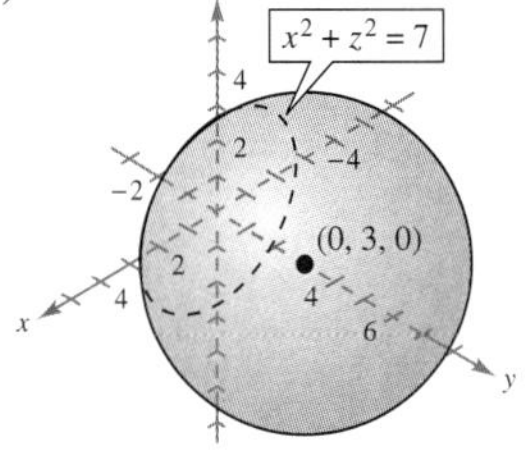

17. (a) $\langle -5, 7, -1\rangle$ (b) $5\sqrt{3}$ (c) $\frac{\sqrt{3}}{15}\langle -5, 7, -1\rangle$
19. -9 **21.** -5 **23.** 90° **25.** About 69.22°
27. Orthogonal **29.** Neither **31.** Not collinear
33. Collinear
35. A: about 159.10 lb of tension
B: about 115.58 lb of tension
C: about 115.58 lb of tension
37. $-14\mathbf{i} + 18\mathbf{j} + 12\mathbf{k}$ **39.** $-52\mathbf{i} + 40\mathbf{j} + 11\mathbf{k}$
41. $-\frac{\sqrt{1267}}{1267}(27\mathbf{i} + 23\mathbf{j} + 3\mathbf{k})$
43. (a) Answers will vary. (b) $2\sqrt{43}$ **45.** 75
47. (a) $x = -5t, y = -10 + 2t, z = 3 + 3t$
(b) $\frac{x}{-5} = \frac{y+10}{2} = \frac{z-3}{3}$
49. (a) $x = -1 + 2t, y = 3 - t, z = 5 + 3t$
(b) $\frac{x+1}{2} = \frac{y-3}{-1} = \frac{z-5}{3}$
51. (a) $x = 3 + 6t, y = 11t, z = 2 + 4t$
(b) $\frac{x-3}{6} = \frac{y}{11} = \frac{z-2}{4}$
53. $2x - 8y + 5z - 20 = 0$
55.

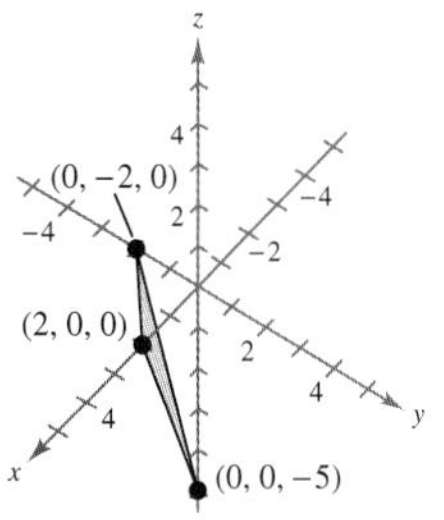

57. $\frac{\sqrt{6}}{6}$ **59.** False; $\mathbf{u} \times \mathbf{v} = -(\mathbf{v} \times \mathbf{u})$
61 and 63. Answers will vary.

Chapter Test (page 756)

1.

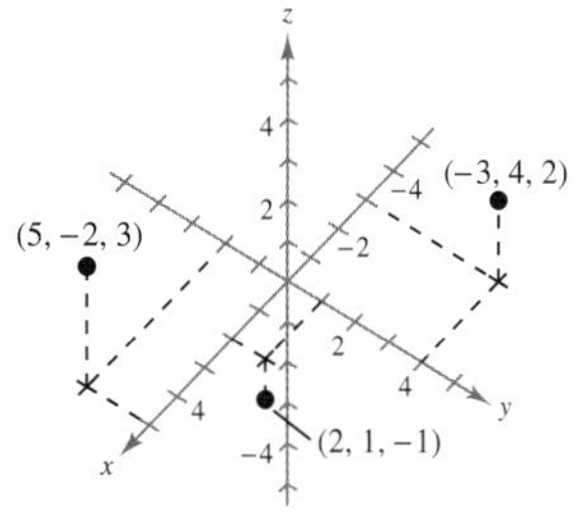

2. No; Answers will vary. **3.** $(7, 1, 2)$
4. $(x - 7)^2 + (y - 1)^2 + (z - 2)^2 = 19$

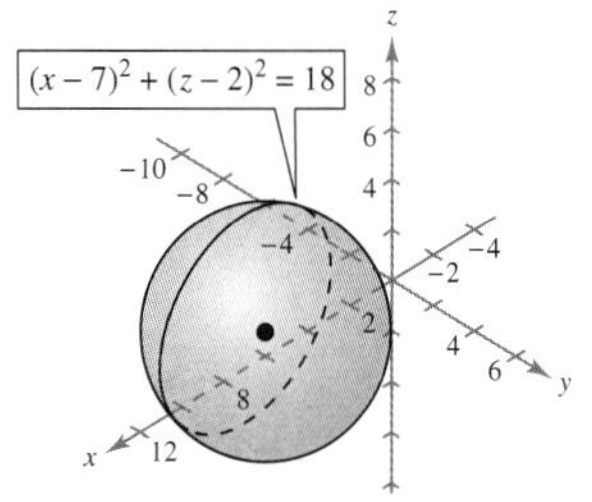

5. $\langle 2, 5, -10\rangle$; $\sqrt{129}$; $\frac{\sqrt{129}}{129}\langle 2, 5, -10\rangle$
6. $\langle -3, -6, 13\rangle$; $\sqrt{214}$; $\frac{\sqrt{214}}{214}\langle -3, -6, 13\rangle$

7. $\mathbf{u} = \langle -2, 6, -6 \rangle$, $\mathbf{v} = \langle -12, 5, -5 \rangle$
8. (a) $\sqrt{194}$ (b) 84 (c) $\langle 0, 62, 62 \rangle$ **9.** About 46.23°
10. (a) $x = 8 - 2t$, $y = -2 + 6t$, $z = 5 - 6t$
(b) $\dfrac{x-8}{-2} = \dfrac{y+2}{6} = \dfrac{z-5}{-6}$
11. Parallel **12.** Neither **13.** Perpendicular
14. Answers will vary; $2\sqrt{230}$
15. $27x + 4y + 32z + 33 = 0$ **16.** 200
17.

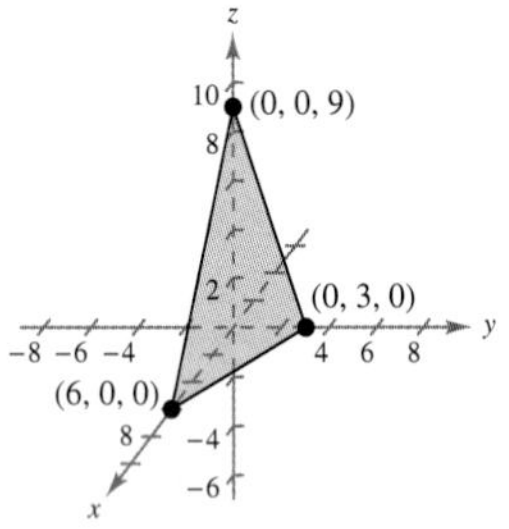

18.

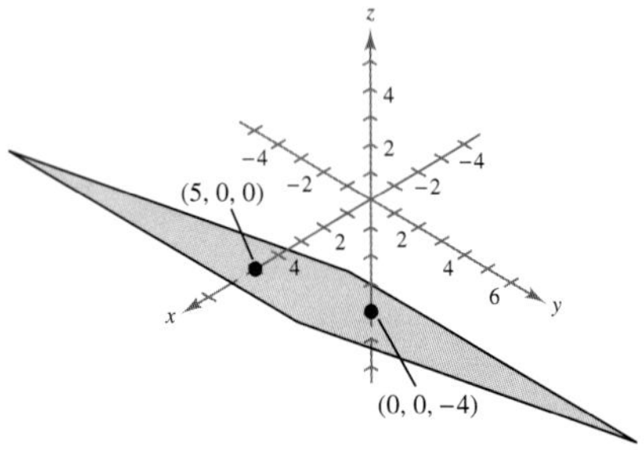

19. $\dfrac{4\sqrt{14}}{7}$

Chapter 11

Section 11.1 (page 767)

1. limit **3.** *Sample answer:* $\lim_{x \to c} b = b$; $\lim_{x \to c} x = c$; $\lim_{x \to c} x^n = c^n$
5. (a) Answers will vary.
(b)

x	3	3.5	3.9	4
V	972	1011.5	1023.5	1024

x	4.1	4.5	5
V	1023.5	1012.5	980

$\lim_{x \to 4} V = 1024$

(c)

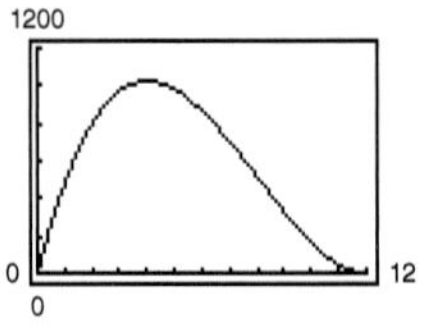

7.

x	1.9	1.99	1.999	2
$f(x)$	13.5	13.95	13.995	14

x	2.001	2.01	2.1
$f(x)$	14.005	14.05	14.5

14; Yes

9.

x	−2.1	−2.01	−2.001	−2
$f(x)$	−0.244	−0.249	−0.250	Error

x	−1.999	−1.99	−1.9
$f(x)$	−0.250	−0.251	−0.256

−0.25; No

11.

x	−1.1	−1.01	−1.001	−1
$f(x)$	−2.1	−2.01	−2.001	Error

x	−0.999	−0.99	−0.9
$f(x)$	−1.999	−1.99	−1.9

−2

13.

x	0.9	0.99	0.999	1
$f(x)$	0.2564	0.2506	0.2501	Error

x	1.001	1.01	1.1
$f(x)$	0.2499	0.2494	0.2439

0.25

15.

x	5.9	5.99	5.999	6
$f(x)$	0.16713	0.16671	0.16667	Error

x	6.001	6.01	6.1
$f(x)$	0.16666	0.16662	0.16621

$\frac{1}{6}$

17.

x	−4.1	−4.01	−4.001	−4
$f(x)$	0.4762	0.4975	0.4998	Error

x	−3.999	−3.99	−3.9
$f(x)$	0.5003	0.5025	0.5263

0.5

19.

x	−0.1	−0.01	−0.001	0
$f(x)$	0.9983	0.99998	0.9999998	Error

x	0.001	0.01	0.1
$f(x)$	0.9999998	0.99998	0.9983

1

21.

x	−0.1	−0.01	−0.001	0
$f(x)$	−0.0997	−0.0100	−0.0010	Error

x	0.001	0.01	0.1
$f(x)$	0.0010	0.0100	0.0997

0

23.

x	-0.1	-0.01	-0.001	0
$f(x)$	0.9063	0.9901	0.9990	Error

x	0.001	0.01	0.1
$f(x)$	1.0010	1.0101	1.1070

1

25.

x	1.9	1.99	1.999	2
$f(x)$	2.2314	2.0203	2.002	Error

x	2.001	2.01	2.1
$f(x)$	1.998	1.9803	1.8232

2

27.

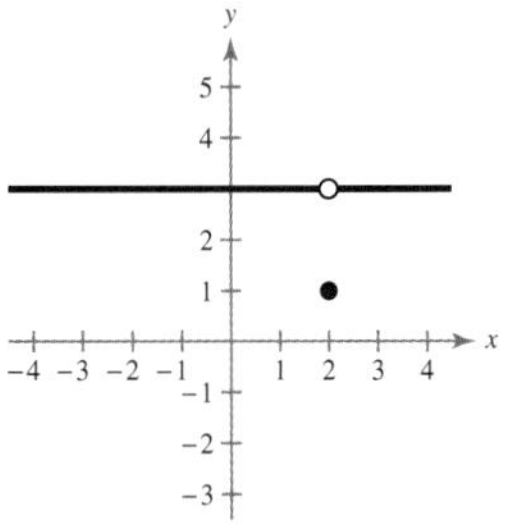

3

29.

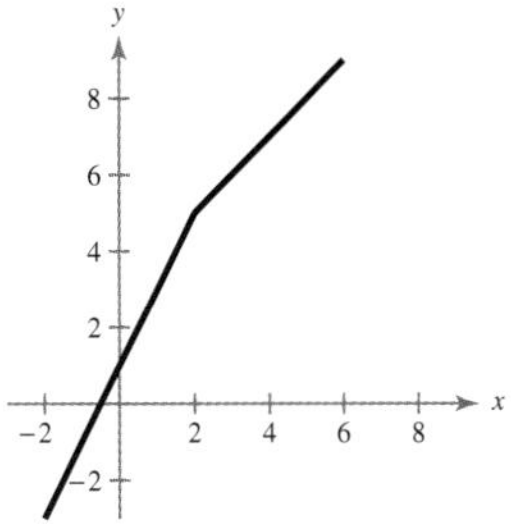

5

31.

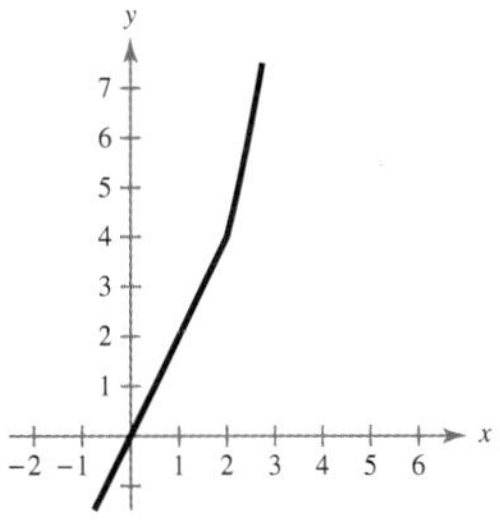

4

33. 2

35. Limit does not exist; The left and right behaviors are different.

37. Limit does not exist; The values are unbounded. **39.** -1

41. Limit does not exist; Function oscillates between -2 and 2.

43.

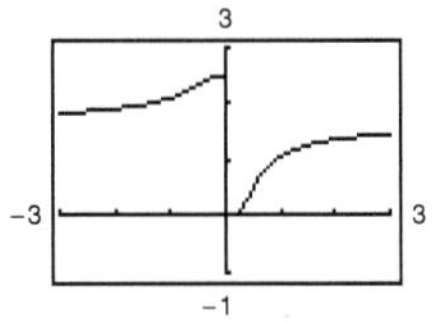

No; The left and right behaviors are different.

45.

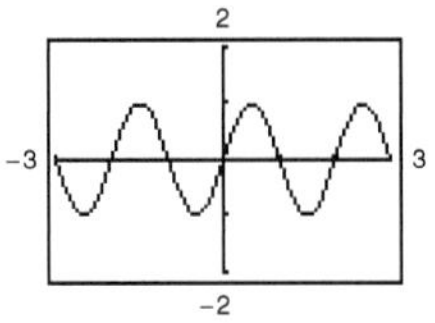

Yes

47.

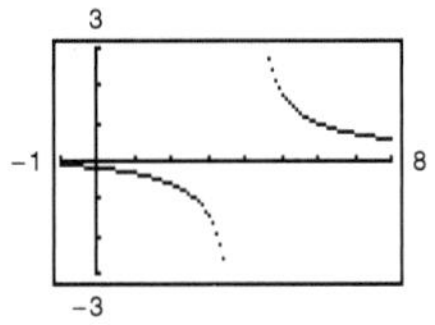

No; The values are unbounded.

49.

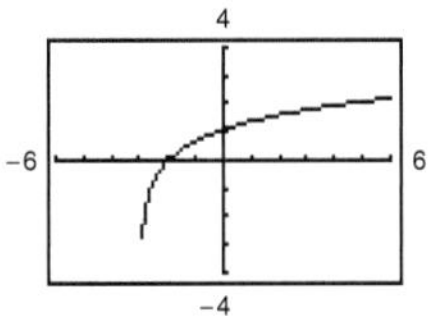

Yes

51. (a) -12 (b) 9 (c) $\frac{1}{2}$ (d) $\sqrt{3}$

53. (a) 16 (b) $\frac{3}{8}$ (c) 6 (d) $-\frac{125}{8}$

55. -8 **57.** 7 **59.** -2 **61.** $-\frac{9}{10}$ **63.** -1

65. 1 **67.** $\frac{3}{4}$ **69.** $\frac{1}{e^6}$ **71.** 0 **73.** $\frac{\pi}{6}$

75. True; $\lim_{x \to c} \sqrt[n]{x} = \sqrt[n]{c}$ when n is odd.

77. (a) and (b) Answers will vary.

79. (a) No; The function may approach different values from the right and left of 4.
(b) No; The function may approach 4 as x approaches 2, but the function could be undefined at $x = 2$.

81. Because n is even and $c < 0$, direct substitution cannot be used.

83. $-\frac{1}{3}, x \neq 5$ **85.** $\frac{x-6}{x-1}, x \neq 6$

Section 11.2 (page 777)

1. dividing out technique **3.** Rationalizing technique

5. (a) -1 (b) 1 (c) 4 (d) 5
$h_2(x) = x + 1$

7. (a) 2 (b) 0 (c) 0 (d) 6
$g_2(x) = x(x + 1)$

9. 3 **11.** 4 **13.** $\frac{1}{12}$ **15.** 12 **17.** 80 **19.** $\frac{7}{2}$

21. $\frac{1}{3}$ **23.** $\frac{\sqrt{5}}{10}$ **25.** $\frac{1}{4}$ **27.** -1 **29.** $-\frac{1}{16}$

31. 0 **33.** 0 **35.** 0

37.

x	-0.1	-0.01	-0.001	0
$f(x)$	1.813	1.980	1.998	Error

x	0.001	0.01	0.1
$f(x)$	2.002	2.020	2.214

2.000

39.

x	0.9	0.99	0.999	1
$f(x)$	0.9491	0.9949	0.9995	Error

x	1.001	1.01	1.1
$f(x)$	1.0005	1.005	1.0492

1.000

41.

x	-0.1	-0.01	-0.001	0
$f(x)$	23.59	20.391	20.116	Error

x	0.001	0.01	0.1
$f(x)$	20.055	19.788	17.449

20.090

43.

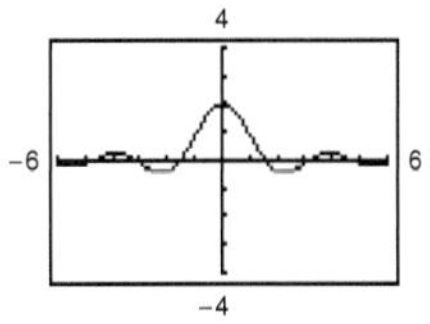

2.000

45.

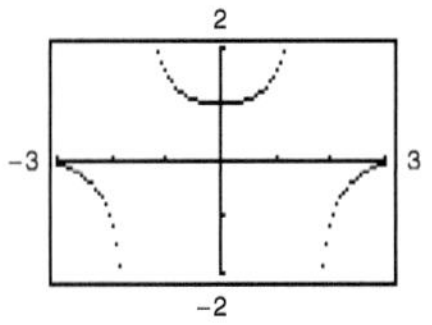

1.000

47.

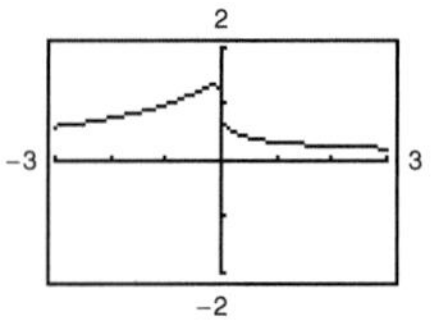

0.333

49.

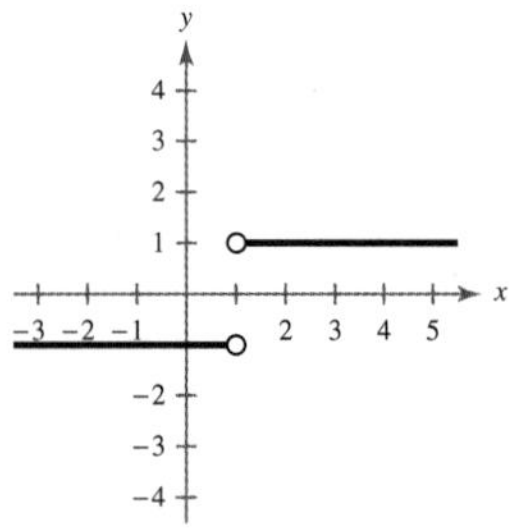

$-1, 1$; Limit does not exist.

51.

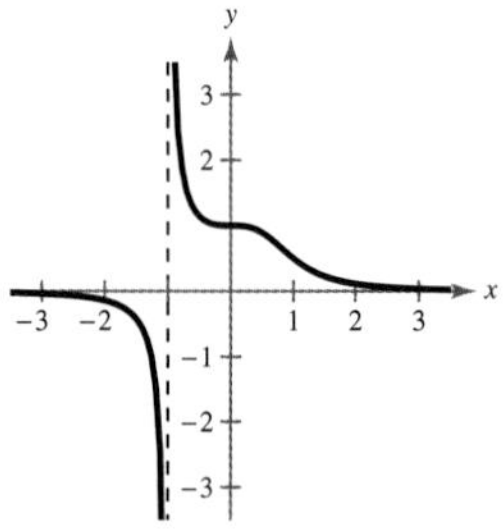

One-sided limits do not exist; Limit does not exist.

53.

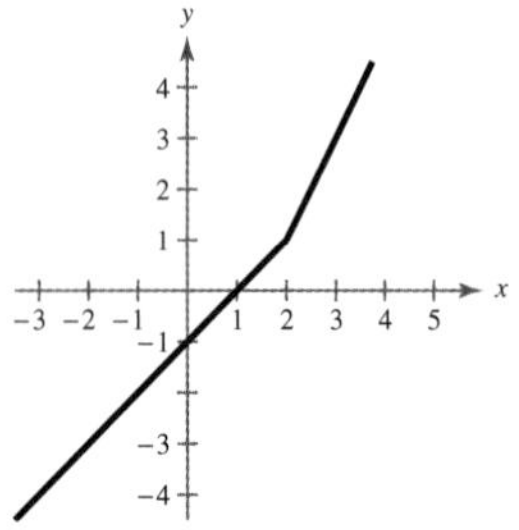

1, 1; 1

55.

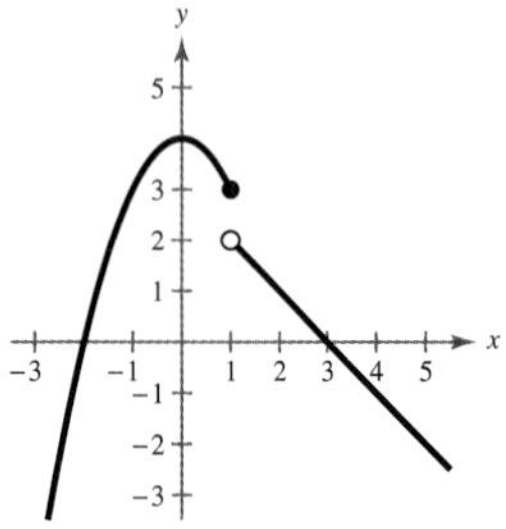

3, 2; Limit does not exist.

57. (a)–(c) 3 **59.** (a)–(c) $-\frac{1}{8}$ **61.** -32 ft/sec

63. Answers will vary.

65. (a)

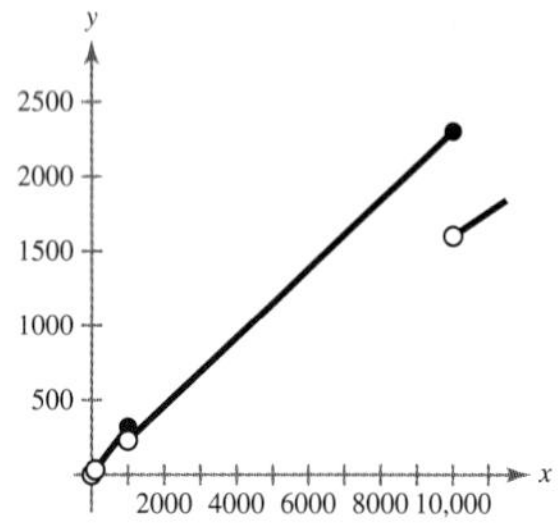

(b) (i) 6.3; 15 color photocopies will cost \$6.30.
(ii) 144; 450 color photocopies will cost \$144.
(iii) 1380; 6000 color photocopies will cost \$1380.

(c) *Sample answers:*

(i)

x	98.999	99	99.001
$C(x)$	41.57958	41.58	31.68032

(ii)

x	998.999	999	999.001
$C(x)$	319.67968	319.68	229.77023

(iii)

x	9998.999	9999	9999.001
$C(x)$	2299.76977	2299.77	1599.84016

(d) The graph shows that one-sided limits are not equal, which can be seen when the graph "jumps" at $x = 99, 999$, and 9999.

67. 3 **69.** $\dfrac{1}{2\sqrt{x-2}}$ **71.** $2x - 3$

73. (a) Yes; 0 (b) No; 1

75.

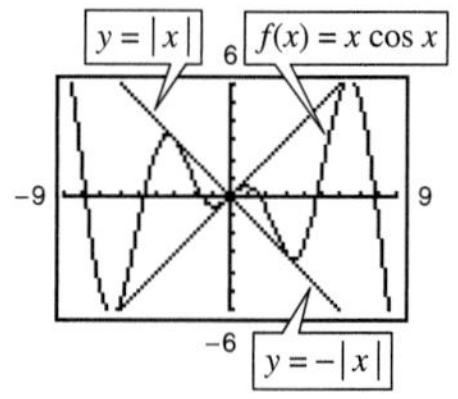

0

77.

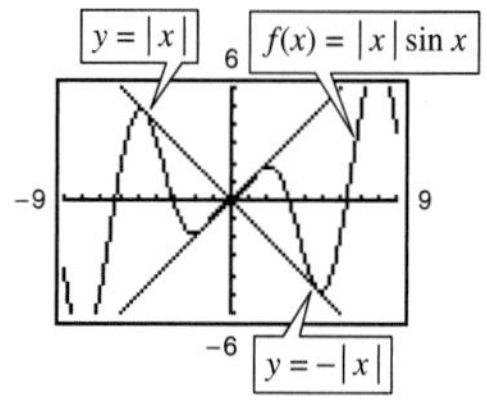

0

79.

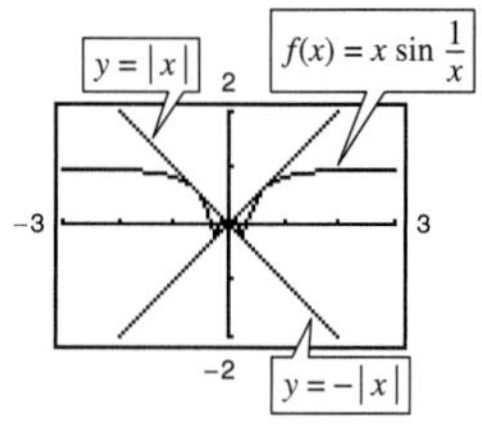

0

81. True; See page 771.

83. Use the dividing out technique to find that the limit is -8.

85. Answers will vary. **87.** Parabola

Section 11.3 (page 787)

1. Calculus **3.** secant line **5.** 0 **7.** $\frac{1}{2}$

9.

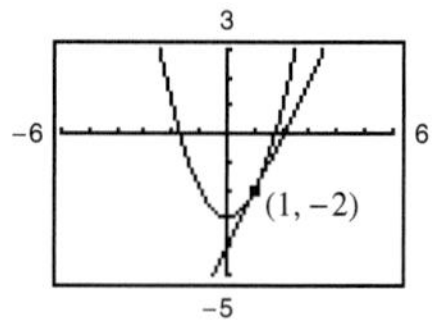

2

11.

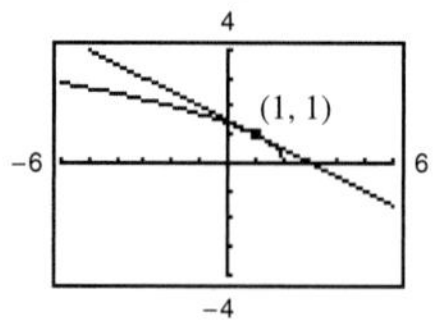

$-\frac{1}{2}$

13.

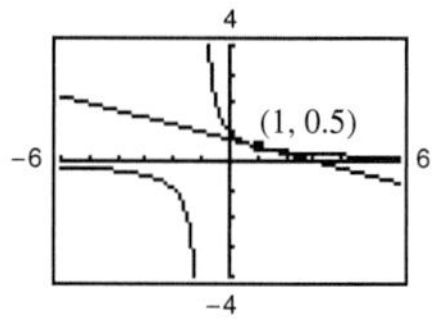

$-\frac{1}{4}$

15. 8 **17.** -6 **19.** -1 **21.** $\frac{1}{6}$

23. $m = -2x$; (a) 0 (b) 2

25. $m = -\dfrac{1}{(x+4)^2}$; (a) $-\dfrac{1}{16}$ (b) $-\dfrac{1}{4}$

27. $m = \dfrac{1}{2\sqrt{x-3}}$; (a) $\dfrac{1}{2}$ (b) $\dfrac{1}{6}$ **29.** $-\dfrac{1}{3}$ **31.** $4x + 3$

33. 0 **35.** $-\dfrac{2}{x^3}$ **37.** $\dfrac{1}{2\sqrt{x+4}}$ **39.** $-\dfrac{1}{2(x-9)^{3/2}}$

41. (a) 4
(b) $y = 4x - 5$
(c)

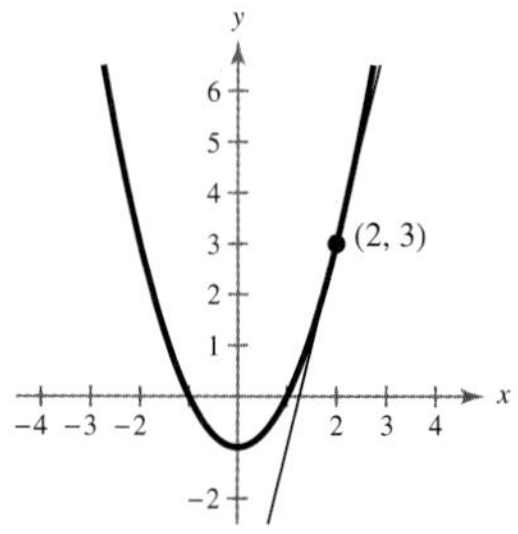

43. (a) 1
(b) $y = x - 2$
(c)

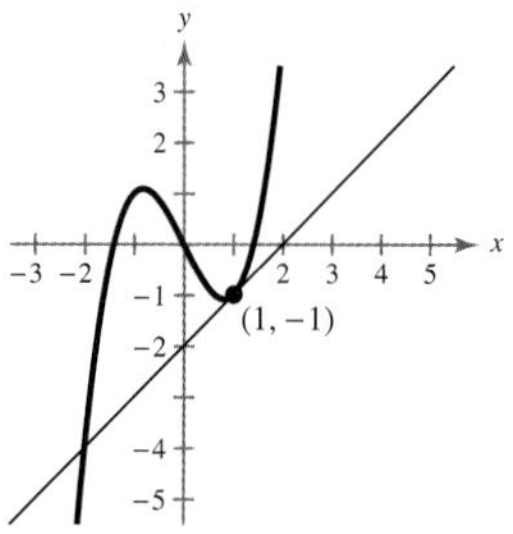

45. (a) -3
(b) $y = -3x + 2$
(c)

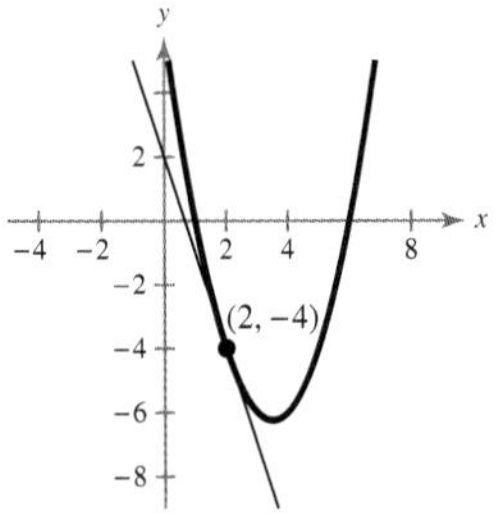

47. (a) $-\frac{1}{6}$
(b) $y = -\frac{1}{6}x + \frac{5}{2}$
(c)

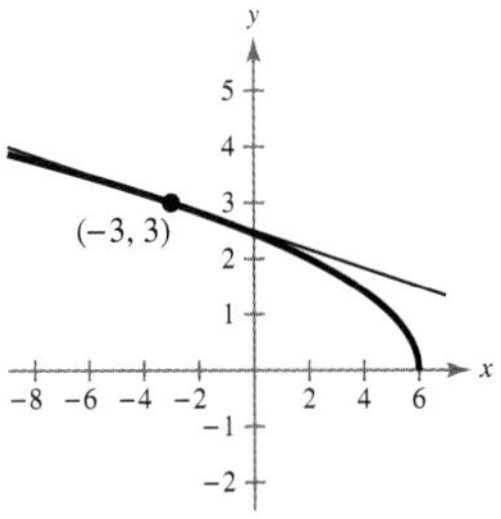

49. (a) -1
(b) $y = -x - 3$
(c)

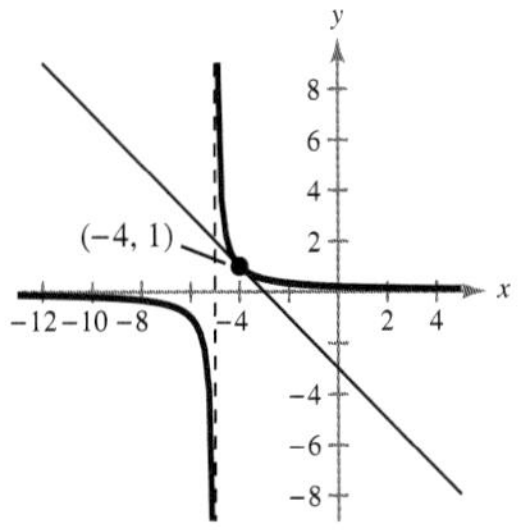

51.

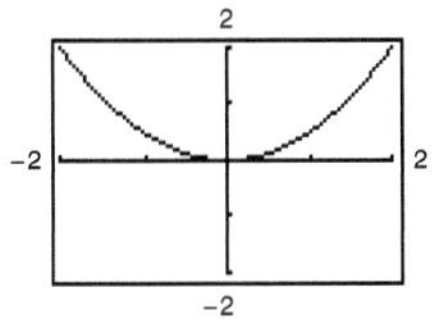

x	-2	-1.5	-1	-0.5	0
$f(x)$	2	1.125	0.5	0.125	0
$f'(x)$	-2	-1.5	-1	-0.5	0

x	0.5	1	1.5	2
$f(x)$	0.125	0.5	1.125	2
$f'(x)$	0.5	1	1.5	2

They appear to be the same.

53.

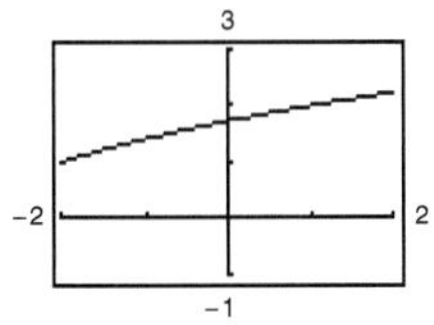

x	-2	-1.5	-1	-0.5	0
$f(x)$	1	1.225	1.414	1.581	1.732
$f'(x)$	0.5	0.408	0.354	0.316	0.289

x	0.5	1	1.5	2
$f(x)$	1.871	2	2.121	2.236
$f'(x)$	0.267	0.25	0.236	0.224

They appear to be the same.

55. $f'(x) = 2x - 4$; $(2, -1)$

57. $f'(x) = 6x^2 - 8$; $\left(-\frac{2\sqrt{3}}{3}, \frac{32\sqrt{3}}{9}\right), \left(\frac{2\sqrt{3}}{3}, -\frac{32\sqrt{3}}{9}\right)$

59. $(-1, -1), (0, 0), (1, -1)$

61. $\left(\frac{\pi}{6}, \sqrt{3} + \frac{\pi}{6}\right), \left(\frac{5\pi}{6}, \frac{5\pi}{6} - \sqrt{3}\right)$ **63.** $(0, 0), (-2, 4e^{-2})$

65. $(e^{-1}, -e^{-1})$

67. (a) $y = 0.1220t^2 - 2.295t + 25.09$
(b)

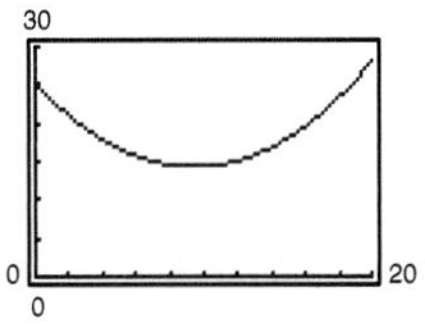

About -1.3; The revenue for Goodyear Tire & Rubber was decreasing by approximately \$1.3 billion per year in 2014.
(c) $y' = 0.2440t - 2.295$; -1.319 (d) Answers will vary.

69. (a) $V'(r) = 4\pi r^2$ (b) About 201.06
(c) Cubic inches per inch; The derivative is a formula for rate of change.

71. (a) $s'(t) = -32t + 64$ (b) 16 ft/sec; Answers will vary.
(c) $t = 2$ sec; Answers will vary. (d) -96 ft/sec
(e)

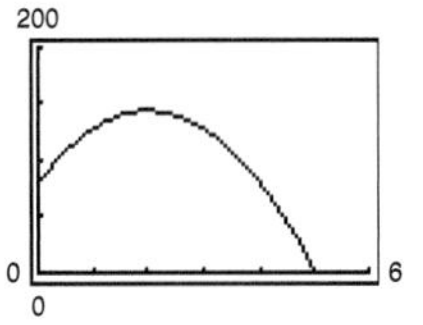

73. True; The graph of the derivative is a line, which is a one-to-one function.

75. (a) Answers will vary.
(b) Use the equation $m = \lim_{h \to 0} \frac{f(x + h) - f(x)}{h}$.

77. $f(x) = (x + 2.5)^2 + 0.75$

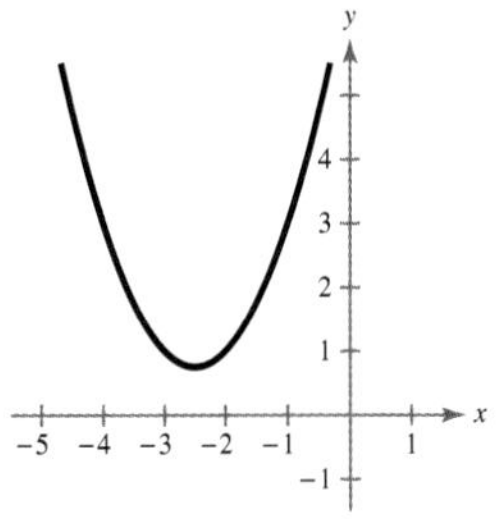

79. Answers will vary.

Section 11.4 (page 796)

1. 0 **3.** converge **5.** e **6.** f **7.** b **8.** c
9. a **10.** d **11.** 3 **13.** -2 **15.** 4 **17.** -4
19. Limit does not exist; The function increases with bound.
21. $\frac{4}{3}$ **23.** 2 **25.** -1 **27.** 2 **29.** -5

31. (a)

x	10^0	10^1	10^2	10^3
$f(x)$	1	−2.857	−2.062	−2.006

x	10^4	10^5
$f(x)$	−2.00060018	−2.000060002

x	10^6
$f(x)$	−2.000006

−2

(b)

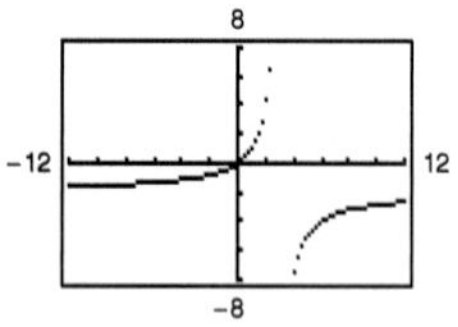

(c) −2

−2

33. (a)

x	10^0	10^1	10^2	10^3
$f(x)$	Error	−0.7071	−0.07	−0.007

x	10^4	10^5	10^6
$f(x)$	−0.0007	−0.00007	−0.000007

0

(b)

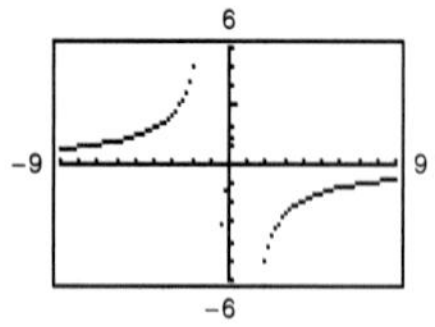

(c) 0

0

35. (a)

x	10^0	10^1	10^2	10^3
$f(x)$	−2	0.97	0.9997	0.999997

x	10^4	10^5	10^6
$f(x)$	0.99999997	0.9999999997	1

1

(b)

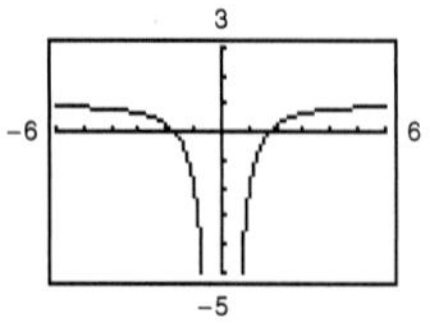

(c) 1

1

37. (a)

x	10^0	10^1	10^2	10^3
$f(x)$	−0.7321	−0.0995	−0.0100	−0.0010

x	10^4	10^5	10^6
$f(x)$	-1.0×10^{-4}	-1.0×10^{-5}	-1.0×10^{-6}

0

(b)

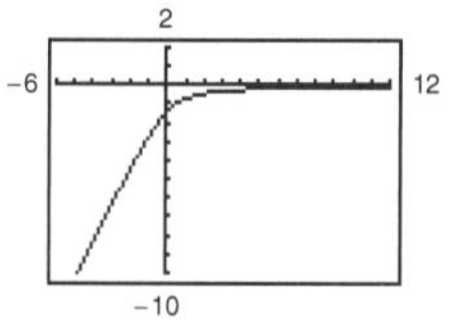

0

39. (a)

x	10^0	10^1	10^2	10^3
$f(x)$	−0.7082	−0.7454	−0.7495	−0.74995

x	10^4	10^5	10^6
$f(x)$	−0.749995	−0.7499995	−0.7500

−0.75

(b)

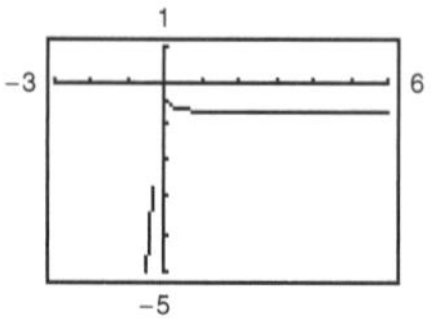

−0.75

41. (a)

x	10^0	10^1	10^2	10^3
$f(x)$	0.0909	0.6742	0.9941	0.9999

x	10^4	10^5	10^6
$f(x)$	1.0000	1.0000	1.0000

1

(b)

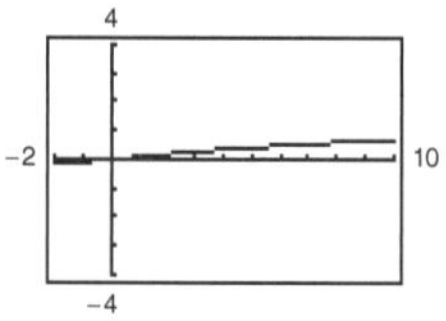

1

43. $1, \frac{3}{5}, \frac{2}{5}, \frac{5}{17}, \frac{3}{13}$; Limit: 0 **45.** $\frac{4}{3}, 2, \frac{16}{7}, \frac{22}{9}, \frac{28}{11}$; Limit: 3

47. $\frac{1}{2}, \frac{2}{3}, \frac{9}{10}, \frac{8}{7}, \frac{25}{18}$; Limit does not exist; The terms increase without bound.

49. 2, 3, 4, 5, 6; Limit does not exist; The terms increase without bound.

51. $-1, \frac{1}{2}, -\frac{1}{3}, \frac{1}{4}, -\frac{1}{5}$; Limit: 0

53.

n	10^0	10^1	10^2	10^3
a_n	2	1.55	1.505	1.5005

1.5

$\lim_{x \to \infty} a_n = \frac{3}{2}$

55.

n	10^0	10^1	10^2	10^3
a_n	13.33	5.683	5.06683	5.0066683

5

$\lim_{x \to \infty} a_n = 5$

57. (a) $\overline{C}(x) = \dfrac{34x + 18{,}000}{x}$ (b) \$52; \$37.60

(c) \$34; As the number of digital binoculars gets very large, the average cost approaches \$34.

59. (a)

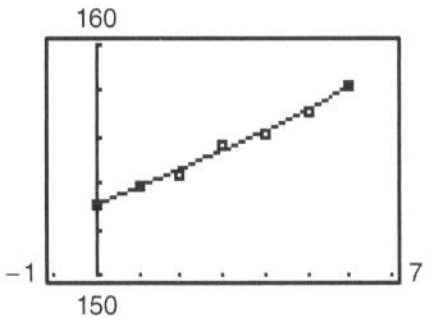

The model fits the data well.

(b) 2017: about 157,889
2018: about 159,128

(c) 0; As time passes, the number of female Selected Reserves personnel approaches 0.

(d) Answers will vary.

61. False; $y = \dfrac{x^2}{x+1}$ does not have a horizontal asymptote.

63. True; See page 794.

65. *Sample answer:* $f(x) = x^2,\ g(x) = x^2$

67.

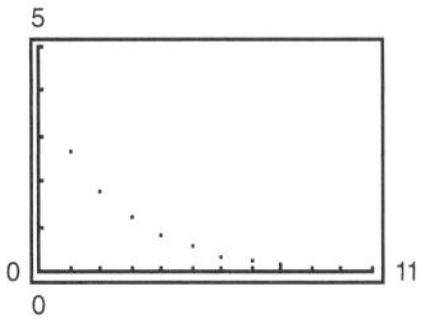

Converges; 0

69.

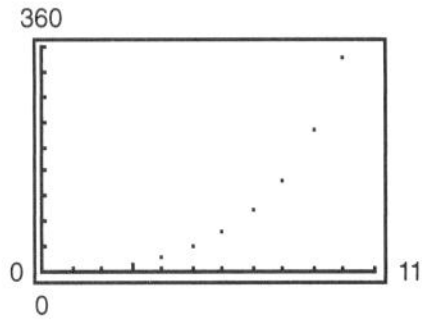

Diverges

71. The limit is $-\frac{1}{4}$.

73.

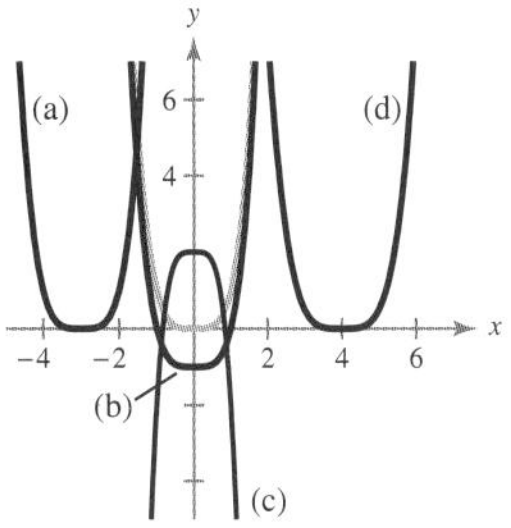

75. 60

Section 11.5 (page 804)

1. $\dfrac{n(n+1)}{2}$ **3.** 100 rectangles **5.** 44,100 **7.** 14,430

9. 6175

11. (a) $S(n) = \dfrac{n^2 + 2n + 1}{4n^2}$

(b)

n	10^0	10^1	10^2	10^3	10^4
$S(n)$	1	0.3025	0.2550	0.2505	0.2501

(c) $\dfrac{1}{4}$

13. (a) $S(n) = \dfrac{2n^2 + 3n - 7}{2n^2}$

(b)

n	10^0	10^1	10^2	10^3	10^4
$S(n)$	6	1.185	1.0154	1.0015	1.00015

(c) 1

15. (a) $S(n) = \dfrac{14n^2 + 3n + 1}{6n^3}$

(b)

n	10^0	10^1	10^2	10^3	10^4
$S(n)$	3	0.2385	0.0234	0.0023	0.0002

(c) 0

17. 14.25 **19.** 1.27

21.

n	4	8	20	50
Approximate area	5.156	5.508	5.786	5.912

23.

n	4	8	20	50
Approximate area	3.5156	2.8477	2.4806	2.3409

25.

n	4	8	20	50	100	∞
Area	40	38	36.8	36.32	36.16	36

27.

n	4	8	20	50	100	∞
Area	14.25	14.81	15.13	15.25	15.29	$\frac{46}{3}$

29. $\frac{3}{2}$ **31.** $\frac{10}{3}$ **33.** 12 **35.** 4 **37.** $\frac{32}{5}$

39.

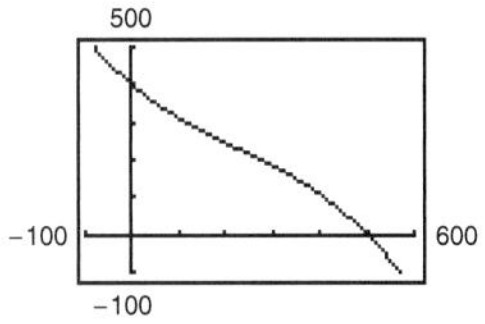

105,208.33 ft^2

41. True; See Formula 2 on page 799. **43.** Answers will vary.

Review Exercises (page 806)

1. 17; Yes **3.** 0.5; No

5.

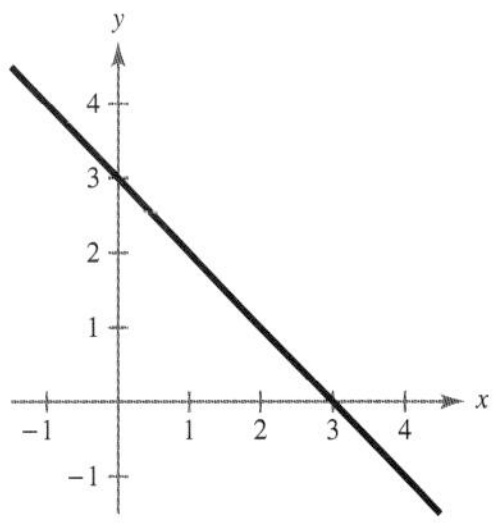

2

7.

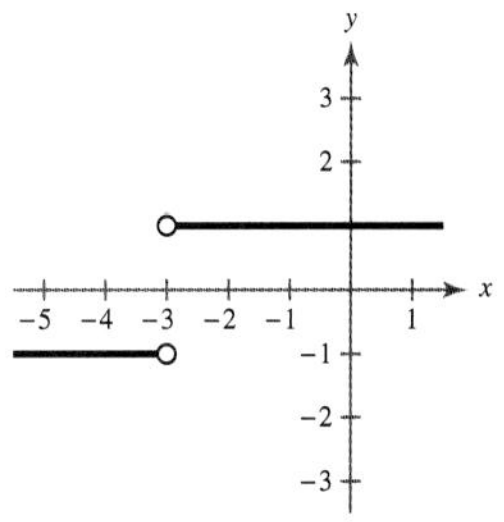

Limit does not exist; The left and right behaviors are different.

9. (a) 8 (b) 1 (c) 10 (d) $\frac{2}{5}$ **11.** 5 **13.** $\frac{10}{3}$

15. -2 **17.** 0 **19.** e **21.** $\dfrac{\pi}{2}$ **23.** $-\dfrac{1}{4}$ **25.** $\dfrac{1}{15}$

27. -9 **29.** -1 **31.** $\frac{1}{4}$ **33.** (a) and (b) About 0.17

35. Limit does not exist. **37.** (a) and (b) 2

39. (a) and (b) About 0.577

41.

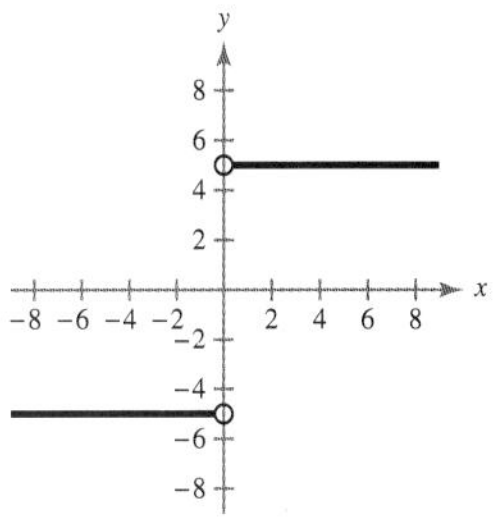

-5, 5; Limit does not exist.

43.

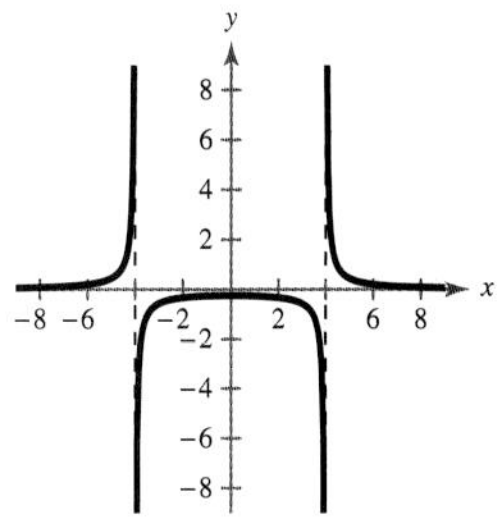

One-sided limits do not exist; Limit does not exist.

45.

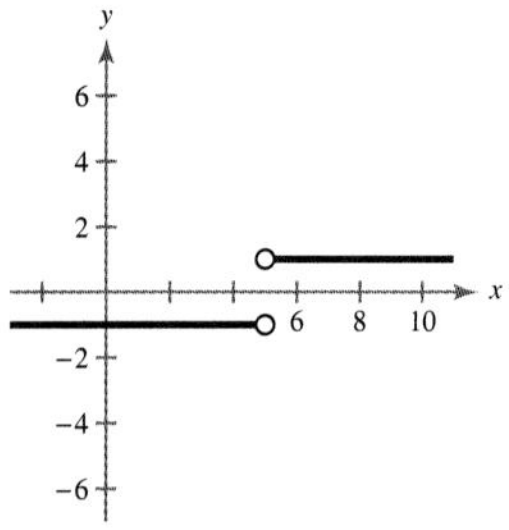

$-1, 1$; Limit does not exist.

47.

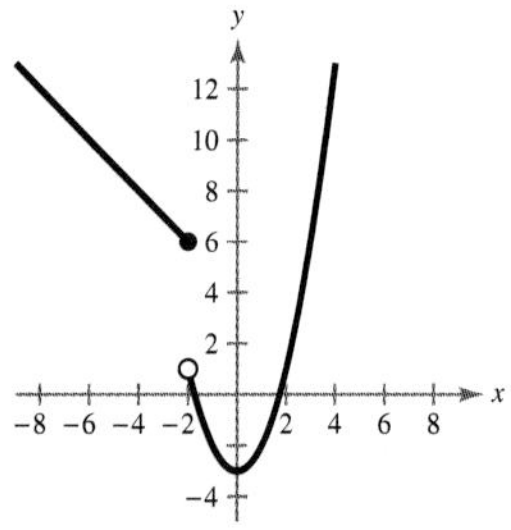

6, 1; Limit does not exist.

49. $3 - 2x$ **51.** 2

53.

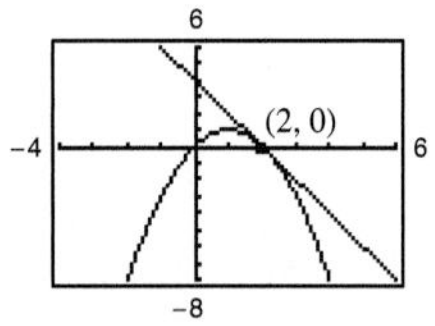

-2

55.

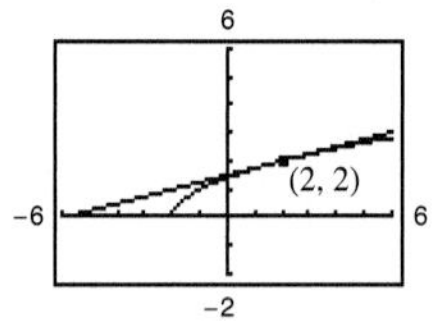

$\frac{1}{4}$

57.

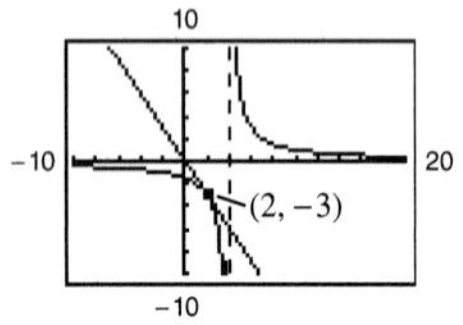

$-\frac{3}{2}$

59. $m = 2x - 4$; (a) 4 (b) -6

61. $m = -\dfrac{4}{(x-6)^2}$; (a) $-\dfrac{4}{9}$ (b) -1 **63.** $f'(x) = 0$

65. $h'(x) = -7$ **67.** $g'(x) = 4x + 1$ **69.** $f'(t) = \dfrac{1}{2\sqrt{t+5}}$

71. $g'(s) = -\dfrac{4}{(s+5)^2}$ **73.** $g'(x) = -\dfrac{1}{2(x+4)^{3/2}}$

75. (a) 12 (b) $12x - 19$
(c)

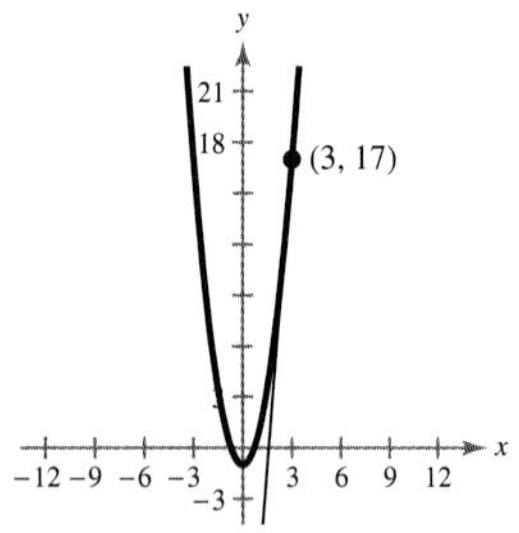

77. (a) 3 (b) $y = 3x - 1$
(c)

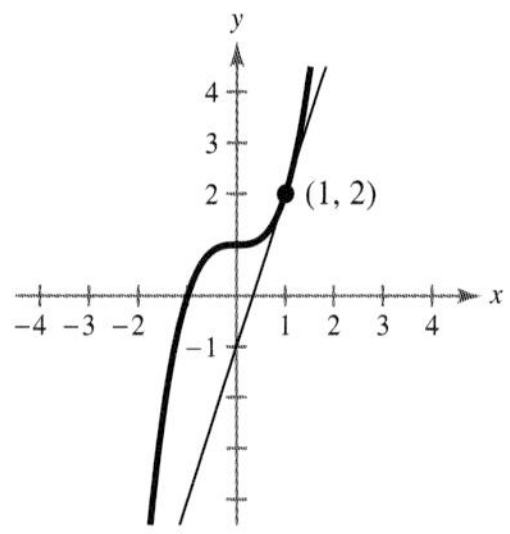

79. 2 **81.** -1 **83.** 3 **85.** 0

87. Limit does not exist; The function increases without bound.

89. 3 **91.** $0, \frac{1}{3}, \frac{1}{2}, \frac{3}{5}, \frac{2}{3}$; Limit: 1 **93.** $1, \frac{4}{5}, \frac{3}{5}, \frac{8}{17}, \frac{5}{13}$; Limit: 0

95. $\frac{1}{5}, \frac{1}{2}, \frac{9}{11}, \frac{8}{7}, \frac{25}{17}$; Limit does not exist; the terms increase without bound.

97. $-1, \frac{1}{8}, -\frac{1}{27}, \frac{1}{64}, -\frac{1}{125}$; Limit: 0

99. $-\frac{1}{2}, -\frac{9}{8}, -\frac{7}{6}, -\frac{37}{32}, -\frac{57}{50}$; Limit: -1

101. (a) $S(n) = \dfrac{5n^2 + 9n + 4}{6n^2}$

(b)

n	10^0	10^1	10^2	10^3	10^4
$S(n)$	3	0.99	0.8484	0.8348	0.8335

(c) $\dfrac{5}{6}$

103. 6.75

105. (a) 7.5 (b) 6.375 (c) 5.74 (d) 5.4944 **107.** 8

109. 21 **111.** 68 **113.** True; See page 785.

Chapter Test (page 811)

1.

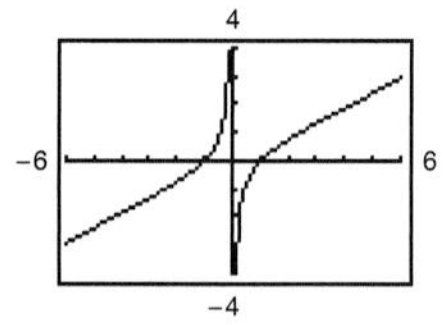

-0.75

$\lim_{x \to -2} f(x) = -\frac{3}{4}$

2.

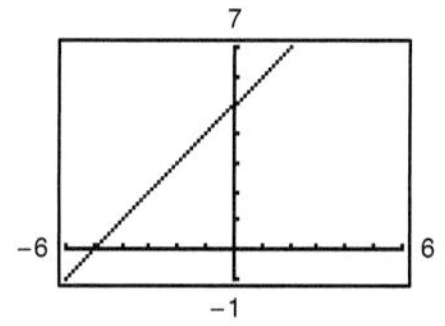

6

$\lim_{x \to 1} f(x) = 6$

3.

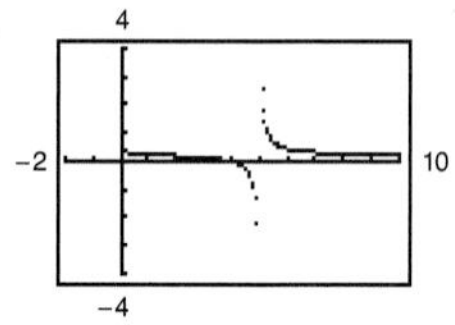

Limit does not exist.

4. About 0.6667 **5.** 5

6. (a) $m = 6x - 5$; 7 (b) $m = 6x^2 + 6$; 12

7. $f'(x) = -\frac{2}{5}$ **8.** $f'(x) = 4x + 4$

9. $f'(x) = -\dfrac{1}{(x+1)^2}$ **10.** 0 **11.** $-\dfrac{1}{2}$

12. Limit does not exist; The function increases without bound.

13. $0, \frac{3}{4}, \frac{14}{19}, \frac{12}{17}, \frac{36}{53}$; Limit: $\frac{1}{2}$ **14.** $0, 1, 0, \frac{1}{2}, 0$; Limit: 0

15. 12.5 **16.** 12 **17.** $\frac{34}{3}$

18. (a) $y = 8.79x^2 - 6.2x - 0.4$ (b) 81.7 ft/sec

Standardized Practice Test (page 812)

1. B **2.** A **3.** D **4.** C **5.** A **6.** C
7. B **8.** D **9.** D **10.** B **11.** A **12.** C
13. C **14.** C **15.** B **16.** A **17.** D **18.** D
19. A **20.** A **21.** \$1.50 **22.** $\frac{3}{16}$ **23.** 6
24. 0.375 **25.** $\frac{1}{4}$

Appendix B

Appendix B.1 (page A31)

1. Line plots **3.** frequency distribution **5.** scatterplot

7. (a) \$3.52 (b) 0.56

9. Quiz 1:

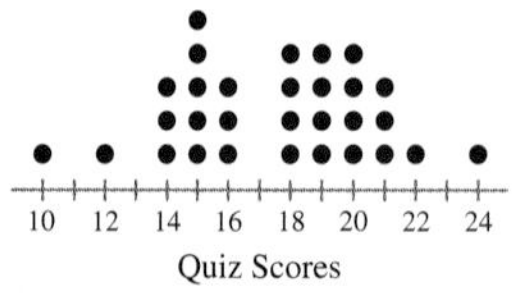

15

Quiz 2:

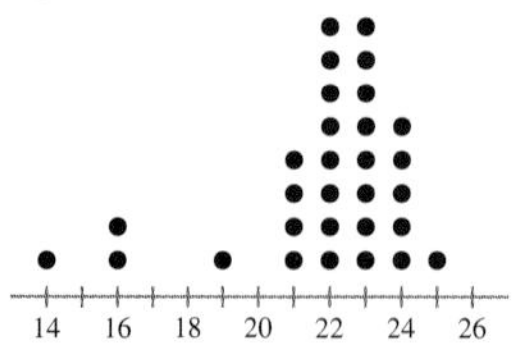

22 and 23

11. *Sample answer:*

Interval	Tally
[6, 9)	||||
[9, 12)	~~||||~~ ~~||||~~ ~~||||~~ ~~||||~~ ||||
[12, 15)	~~||||~~ ~~||||~~ |||
[15, 18)	~~||||~~ |
[18, 21)	|||

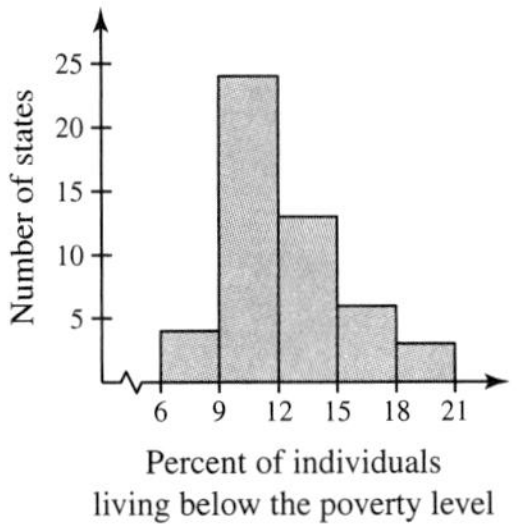

13.

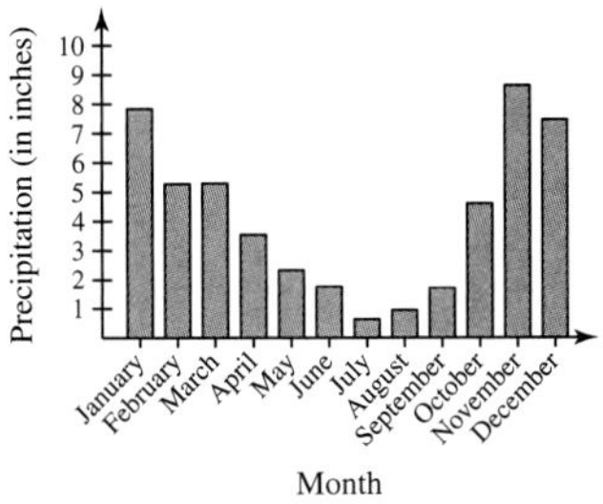

Answers will vary.

15.

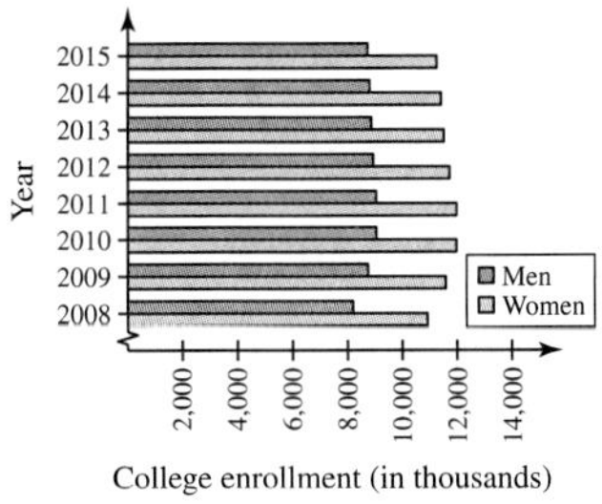

17.

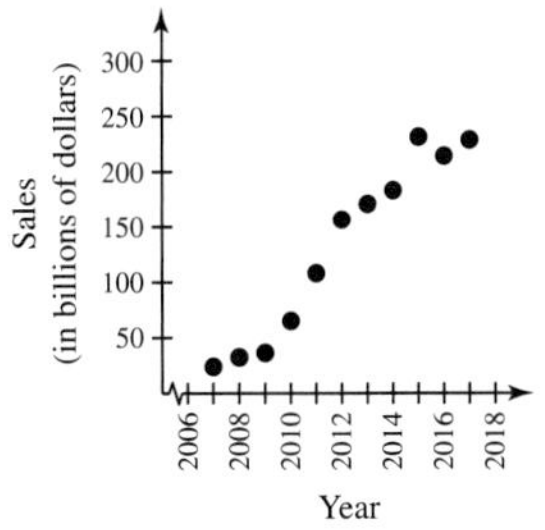

19. The price increased from 2009 to 2012.

21. 2014 to 2015 **23.** About 40%

25. (a)

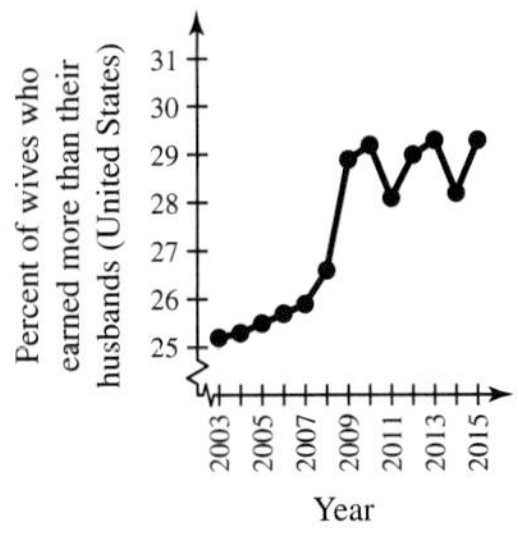

(b) Answers will vary.

27. 65 **29.** Yes

31. *Sample answer:*

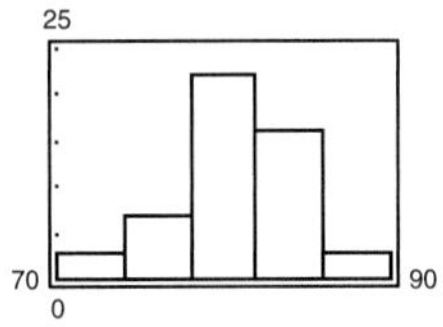

A histogram is best because the data are percents within a year that do not relate to increasing or decreasing behavior.

33. A bar graph is similar to a histogram, except that the bars can be either horizontal or vertical and the labels of the bars are not necessarily numbers. Another difference between a bar graph and a histogram is that the bars in a bar graph are usually separated by spaces.

35. Answers will vary.

37. (a) *Sample answer:* Because the data are from a single year and do not relate to increasing or decreasing behavior, a bar graph could be used. Because the data do not need to be shown as trends over time, a line graph would not be used.

(b) *Sample answer:* Because the data relate to increasing and decreasing behavior over time, a line graph could be used. Because the data are from separate years, a histogram would not be used.

Appendix B.2 (page A41)

1. measure, central tendency **3.** variance, standard deviation

5. Mean: 9.25; median: 8.5; modes: 7, 12

7. Mean: 10.5; median: 8.5; modes: 7, 12

9. Mean: about 12.83; median: 12; mode: none

11. Mean: 315; median: 310; mode: 310

13. (a) Jay: $169\frac{2}{3}$; Hank: $199\frac{2}{3}$; Buck: $209\frac{1}{3}$

(b) $192\frac{8}{9}$; 199 (c) *Sample answer:* Median

15. *Sample answer:* Median

17. (a) $\bar{x} = 12; \sigma \approx 2.83$ (b) $\bar{x} = 20; \sigma \approx 2.83$

(c) $\bar{x} = 12; \sigma \approx 1.41$ (d) $\bar{x} = 9; \sigma \approx 1.41$

19. $\bar{x} = 6, v = 10, \sigma \approx 3.16$ **21.** $\bar{x} = 3, v = 4, \sigma = 2$

23. $\bar{x} = 2, v = \frac{4}{3}, \sigma \approx 1.15$ **25.** $\bar{x} = 47, v = 226, \sigma \approx 15.03$

27. About 3.42 **29.** About 107.42

31. $[123, 1655]$; $[-260, 2038]$

33. (a) Upper quartile: 21.5
Lower quartile: 13

(b)

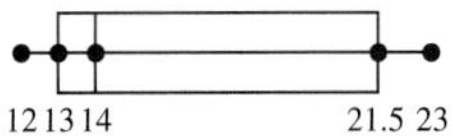

35. (a) Upper quartile: 11.5
Lower quartile: 18

(b)

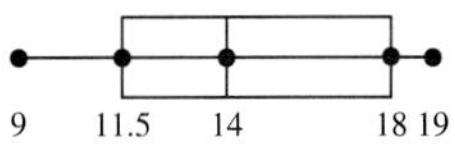

37.

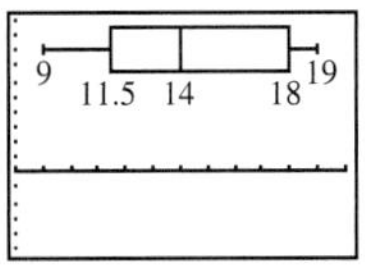

39.

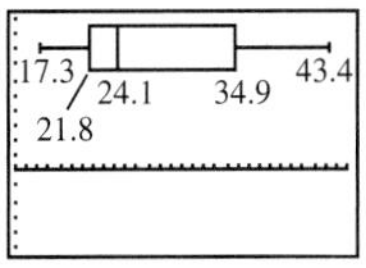

41. Mean; Answers will vary. **43.** *Sample answer:* {4, 4, 10}

45. $\bar{x} = 12$ and $|x_i - 12| = 8$ for all x_i.

47. The mean will increase by 5, but the standard deviation will not change.

49. a; *Sample answer:* The scores in (b) are more evenly distributed.

Appendix B.3 (page A45)

1. $y = 1.6x + 7.5$ **3.** $y = 0.262x + 1.93$

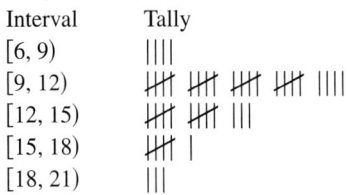

APPENDIX B

Explore the Concepts

Chapter 1

Section 1.1 (page 4)

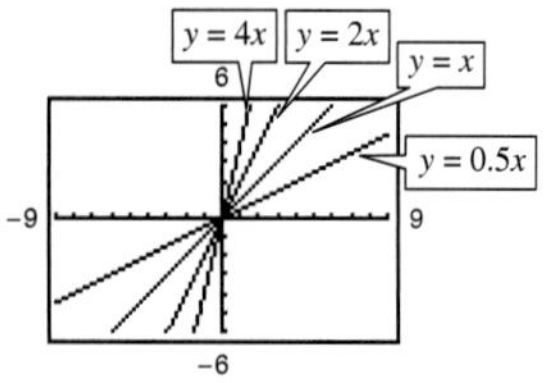

The lines become steeper.

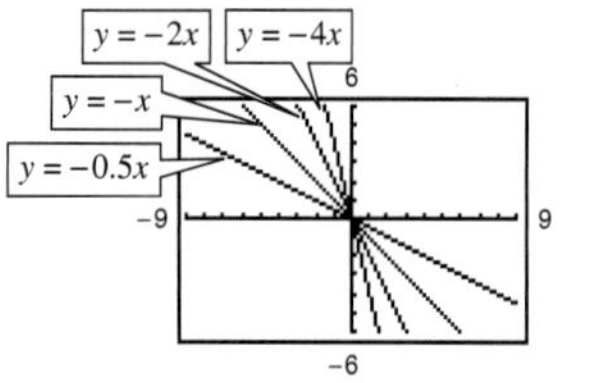

The slopes fall from left to right.

Section 1.1 (page 8)

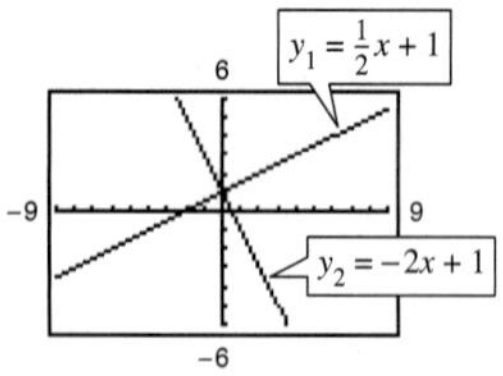

The lines are perpendicular.

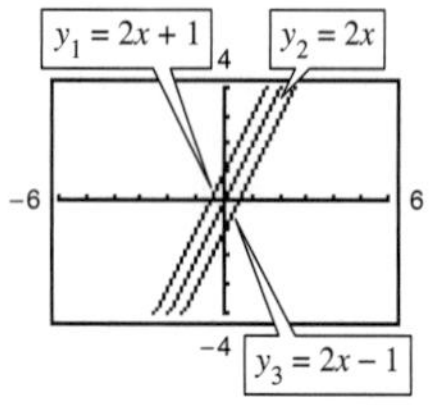

The lines are parallel.

Section 1.2 (page 17)

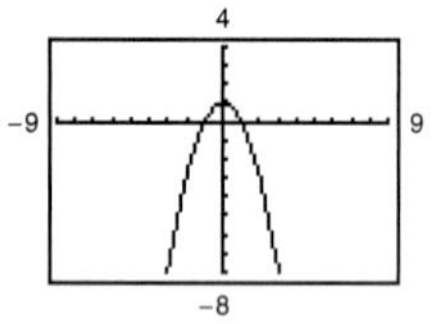

Answers will vary. *Sample answer:* The graph passes the Vertical Line Test, so each x-value has at most one y-value.

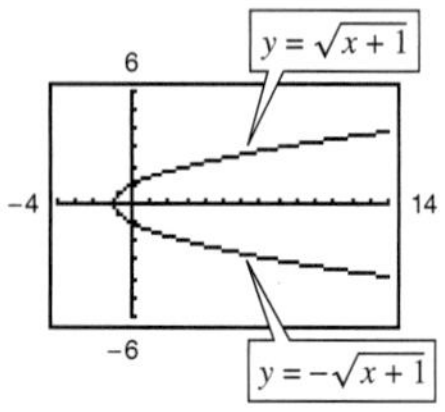

Answers will vary. *Sample answer:* y is not a function of x because there are two values of y for each value of x that is greater than -1.

Section 1.2 (page 20)

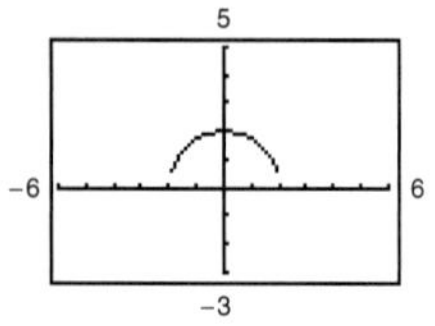

Domain: $[-2, 2]$

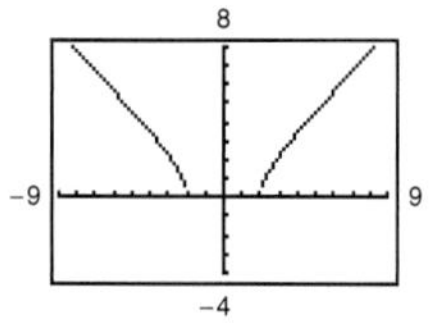

Domain: $(-\infty, -2] \cup [2, \infty)$
Yes; -2 and 2

Section 1.3 (page 35)

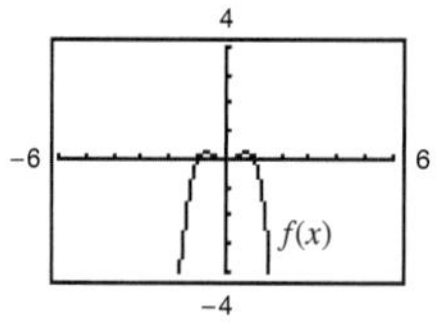

Even

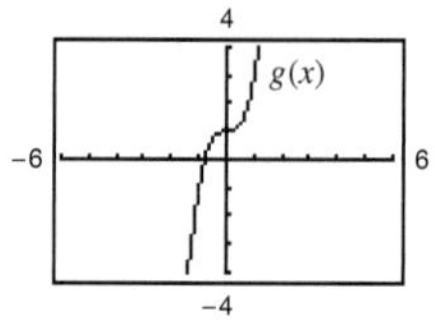

Neither

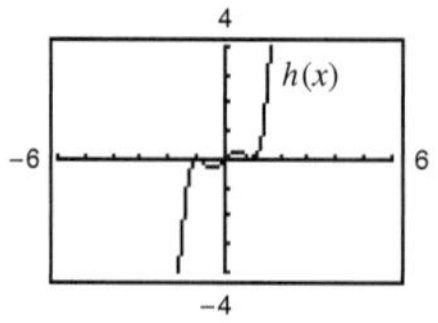

Odd

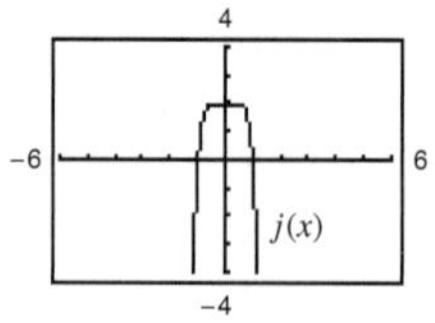

Even

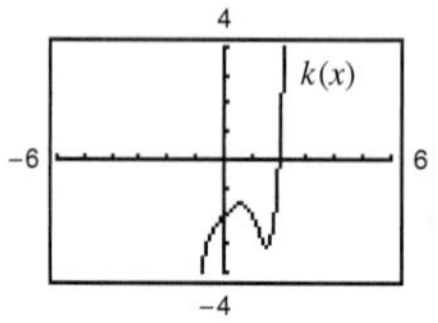

Neither

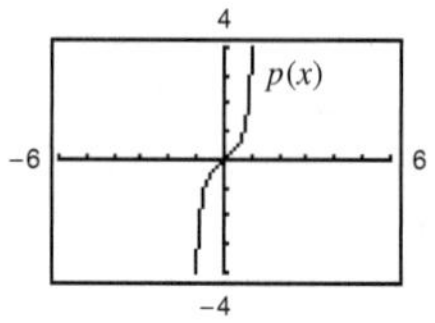

Odd

(a) The exponents are all odd.
(b) The exponents are all even.
(c) The equations of odd functions contain terms with all odd exponents.
(d) The equations of even functions contain terms with all even exponents.
(e) The equations of functions that are neither even nor odd include terms with both even and odd exponents.

Section 1.4 (page 42)

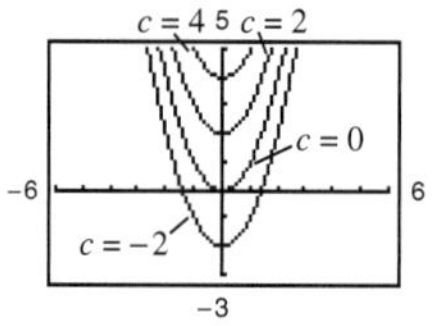

c shifts the graph vertically.

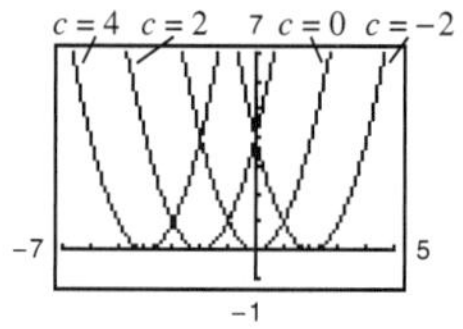

c shifts the graph horizontally.

Section 1.4 (page 44)

a.

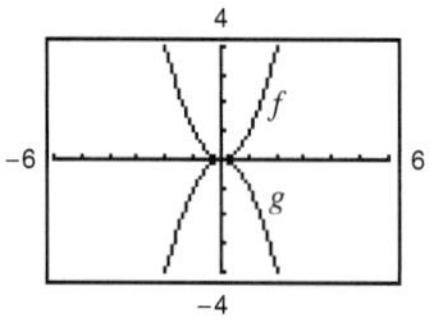

Reflection in the x-axis

b.

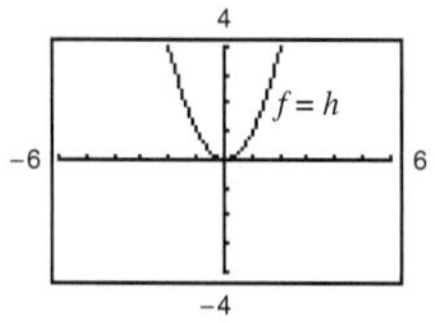

Reflection in the y-axis

Section 1.5 (page 52)

No.

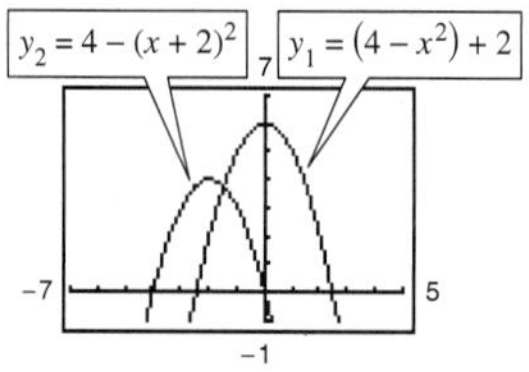

There are two different graphs. y_1 represents $f \circ g$ and y_2 represents $g \circ f$.

Section 1.5 (page 54)

a. Answers will vary. *Sample answer:* Let $g(x) = x^3 - 2$ and $f(x) = |x|$. $h(x) = f(g(x))$

b. Answers will vary. *Sample answer:* Let $q(x) = |x|$ and $p(x) = x^3 - 2$. $r(x) = p(q(x))$

Chapter 2

Section 2.1 (page 91)

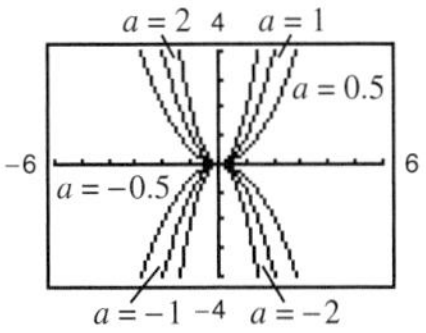

The larger the value of $|a|$, the narrower the graph.

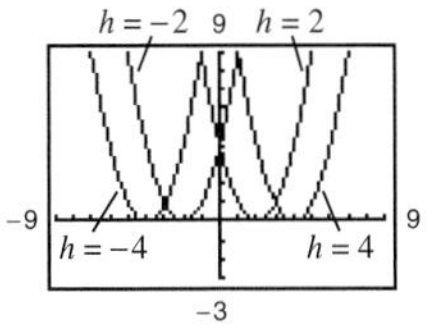

A negative h shifts the graph h units to the left. A positive h shifts the graph h units to the right.

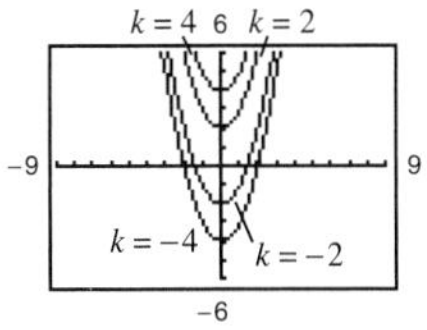

A negative k shifts the graph k units downward. A positive k shifts the graph k units upward.

Section 2.2 (page 101)

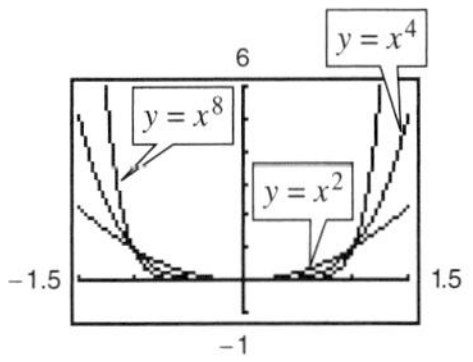

In the interval $(-1, 1)$, $y = x^8$ is on the bottom. Outside the interval $(-1, 1)$, $y = x^2$ is on the bottom.

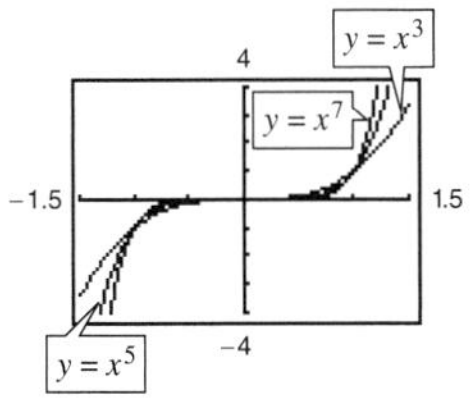

In the intervals $(-\infty, -1)$ and $(0, 1)$, $y = x^7$ is on the bottom. In the intervals $(-1, 0)$ and $(1, \infty)$, $y = x^3$ is on the bottom.

Section 2.2 (page 102)

a. Degree 3 (odd); +1

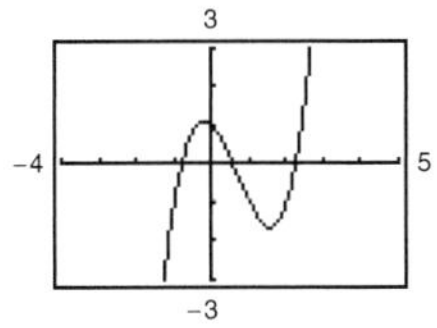

b. Degree 5 (odd); +2

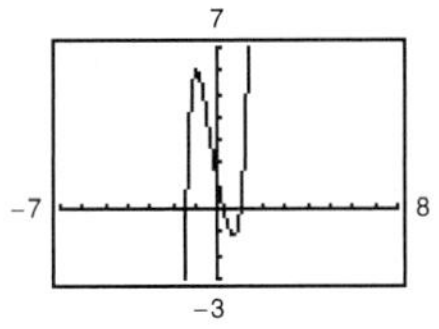

c. Degree 5 (odd); −2

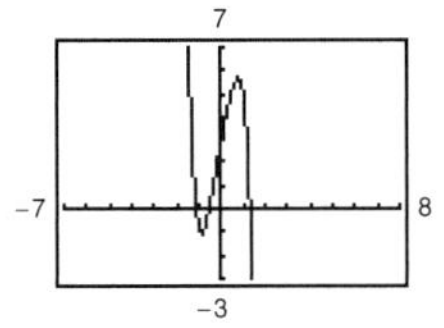

d. Degree 3 (odd); −1

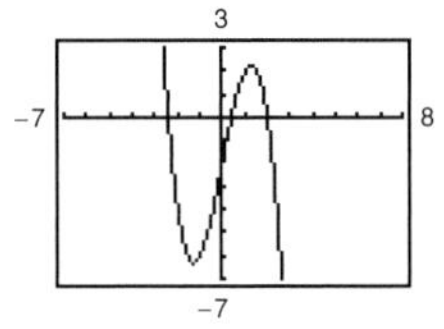

e. Degree 2 (even); +2

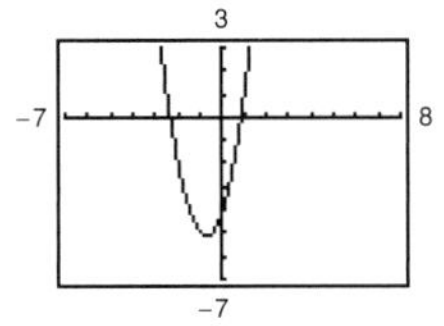

f. Degree 4 (even); +1

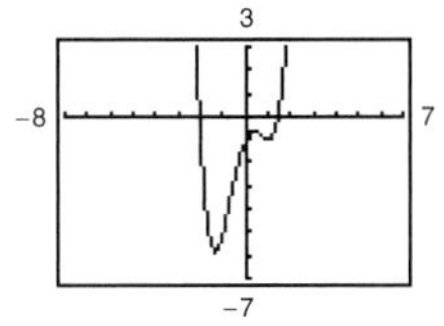

g. Degree 2 (even); −1

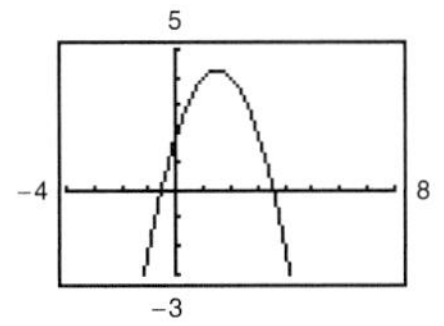

h. Degree 6 (even); −1

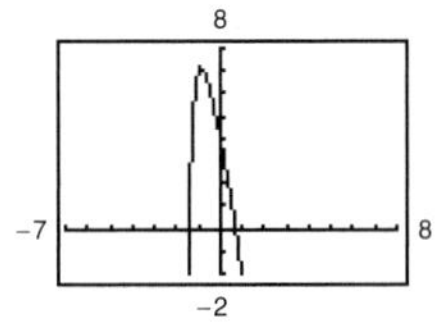

A function of odd degree falls to the left and rises to the right when the leading coefficient is positive. It rises to the left and falls to the right when the leading coefficient is negative.

A function of even degree rises to the left and to the right when the leading coefficient is positive. It falls to the left and to the right when the leading coefficient is negative.

Section 2.2 (page 103)

The degree of the polynomial is the same as the number of zeros and there is one more zero than there are relative minima and relative maxima.

Section 2.3 (page 116)

1; This is the remainder after division by $x + 3$.

Section 2.3 (page 122)

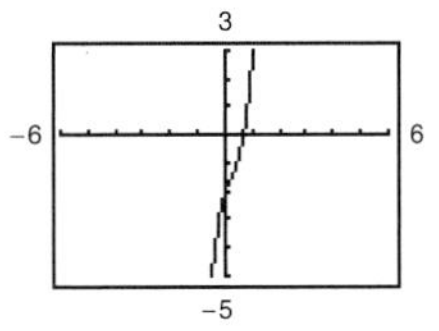

The real zero corresponds to the point at which the graph intersects the x-axis.

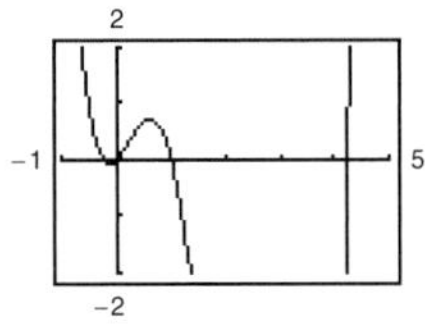

The graph intersects the x-axis four times. There are four real zeros.

Section 2.3 (page 123)

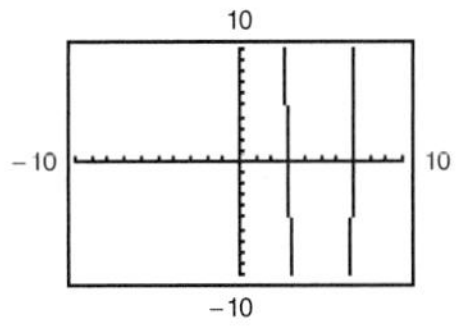

Answers will vary. There appear to be two zeros: (3, 0) and (7, 0). The real zeros are about −14.80, about 2.97, and about 7.02.

Section 2.4 (page 130)

$i, -1, -i, 1, i, -1, -i, 1$; The pattern repeats after every fourth power. Divide the power by 4. When the remainder is 1, the answer is i. When the remainder is 2, the answer is -1. When the remainder is 3, the answer is $-i$. When the remainder is 0, the answer is 1.

Section 2.6 (page 142)

Using the *table* feature of a graphing utility, as x approaches 0 from the left, $f(x)$ decreases. As x approaches 0 from the right, $f(x)$ increases. Using the *trace* feature of a graphing utility, $f(x)$ is undefined when $x = 0$.

Section 2.6 (page 143)

Using a table of values, the graph of $f(x) = \dfrac{2x+1}{x+1}$ is not continuous, the graph of $f(x) = \dfrac{4}{x^2+1}$ is continuous, and the graph of $f(x) = \dfrac{2}{(x-1)^2}$ is not continuous. When the graph of a function has an asymptote, the function may still be continuous. For example, the graph of the function $f(x) = \dfrac{4}{x^2+1}$ has a horizontal asymptote at $y = 0$, but the function is continuous.

Section 2.6 (page 146)

x	-1	2	4	8
$f(x)$	Error	1.3333	1.6	1.7778

x	20	100	1000
$f(x)$	1.9048	1.9802	1.998

The vertical asymptote occurs at $x = -1$.
The horizontal asymptote occurs at $y = 2$.
Answers will vary.

Section 2.7 (page 152)

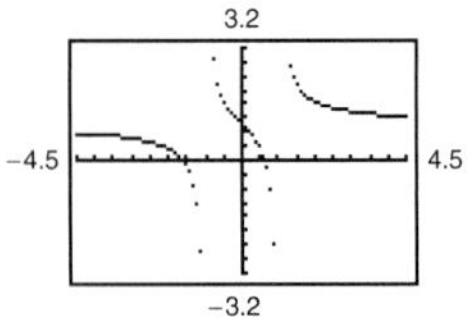

There are breaks at $x = -1$ and $x = 1$, and a hole at $x = 0$;
No; At $x = 0$, the denominator is undefined.

Section 2.7 (page 155)

It is possible for the graph of a rational function to cross its horizontal asymptote or its slant asymptote. (Explanations will vary.)

Chapter 3

Section 3.1 (page 184)

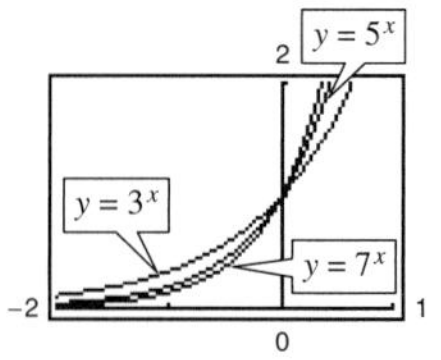

$y = 3^x$ is on top in the interval $(-\infty, 0)$.
$y = 7^x$ is on the bottom in the interval $(-\infty, 0)$.
$y = 7^x$ is on top in the interval $(0, \infty)$.
$y = 3^x$ is on the bottom in the interval $(0, \infty)$.

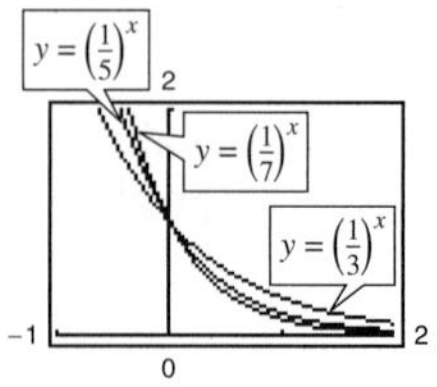

$y = \left(\frac{1}{7}\right)^x$ is on top in the interval $(-\infty, 0)$.
$y = \left(\frac{1}{3}\right)^x$ is on the bottom in the interval $(-\infty, 0)$.
$y = \left(\frac{1}{3}\right)^x$ is on top in the interval $(0, \infty)$.
$y = \left(\frac{1}{7}\right)^x$ is on the bottom in the interval $(0, \infty)$.

For b between 0 and 1, the graph of $y = b^x$ rises to the left and falls to the right. For b greater than 1, the graph of $y = b^x$ falls to the left and rises to the right.

Section 3.1 (page 185)

$Y_1 = \frac{1}{3}Y_2$

Section 3.1 (page 186)

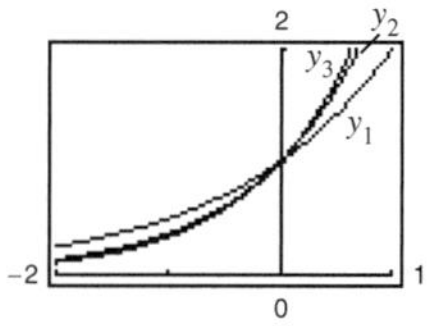

Answers will vary. $e \approx 2.71828$

Section 3.1 (page 187)

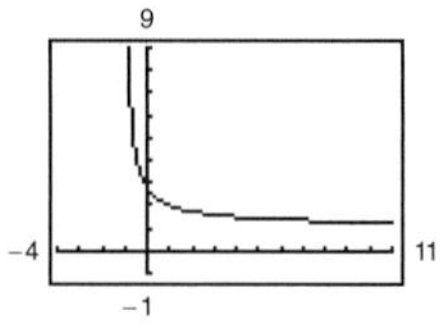

Near $x = 0$, the graph approaches e. There is no y-intercept. In Example 5, as $x \to \infty$, the function approaches e.

x	-0.1	-0.01	-0.001	-0.0001
y_1	2.868	2.732	2.7196	2.7184

x	0	0.0001	0.001
y_1	Error	2.7181	2.7169

Section 3.1 (page 188)

(a) \$5466.09 (b) \$5466.35 (c) \$5466.36 (d) \$5466.38
No; Answers will vary.

Section 3.2 (page 197)

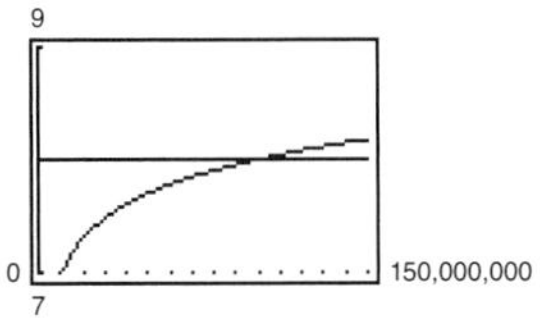

(100,000,000, 8)
$\log_{10} 100{,}000{,}000 = 8$

Section 3.3 (page 207)

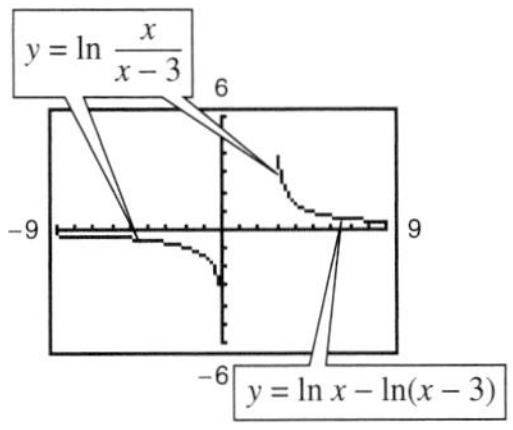

No; Answers will vary.

Section 3.6 (page 237)

$y = -0.365x^2 + 5.57x - 3.1$
Answers will vary.

Chapter 4

Section 4.2 (page 269)

1. 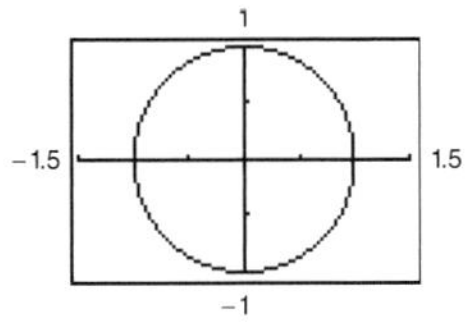

   The graph is a circle.
2. The t-values represent the central angle in radians; the x- and y-values represent the location on the coordinate plane.
3. $-1 \le x \le 1, -1 \le y \le 1$

Section 4.3 (page 278)

The answer is always 1, because this is a trigonometric identity. The result is the same in radian and degree modes.

Section 4.5 (page 295)

This program shows the "unwrapping" of the sine function.

Section 4.6 (page 309)

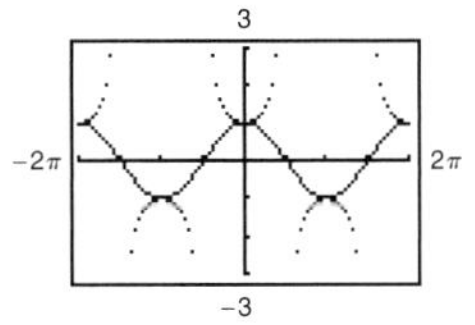

The relative minimum of the cosine function is a relative maximum of the secant function. The relative maximum of the cosine function is a relative minimum of the secant function. As x approaches the zeros of the cosine function, the graph of the secant function approaches infinity.

Section 4.7 (page 322)

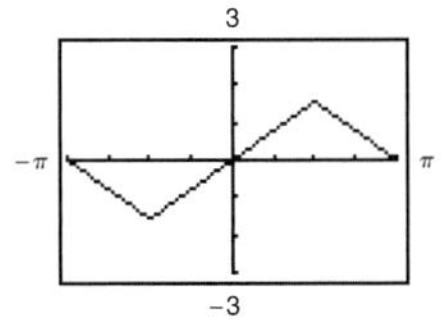

Domain: all real numbers; range: $-\pi/2 \le y \le \pi/2$; 4 is not in the range of the arcsin function.

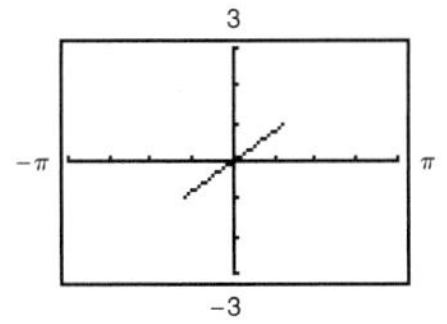

Domain: $-1 \le x \le 1$; range: $-1 \le y \le 1$; 4 is not in the domain of the arcsine function.

Chapter 5

Section 5.3 (page 369)

When you divide both sides of the equation by $\cot x$, you lose the solution $x = \pi/2$. Therefore this is not a valid method to use when solving equations.

Section 5.3 (page 371)

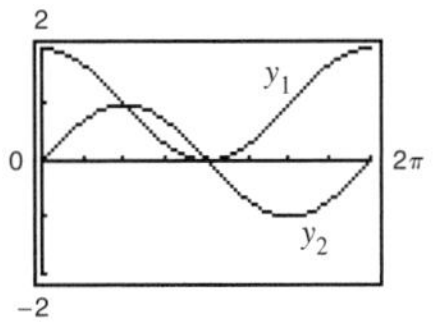

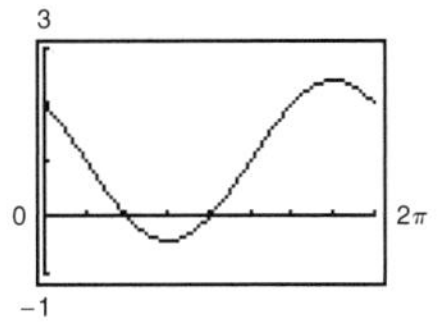

Both methods produce the same solutions, $\pi/2$ and π.
Preferences will vary.

Section 5.4 (page 380)

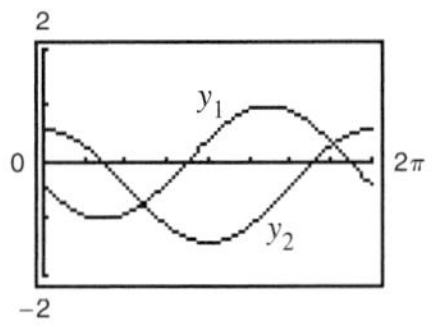

The graphs are not the same. The expressions are not equivalent.

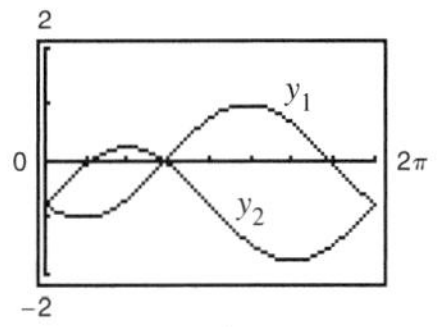

The graphs are not the same. The expressions are not equivalent.

Chapter 6

Section 6.2 (page 415)

Pythagorean Theorem; It is just a special case of the more general Law of Cosines.

Section 6.2 (page 418)

Yes; Each formula has advantages and disadvantages, depending on the angles or sides given.

Section 6.6 (page 455)

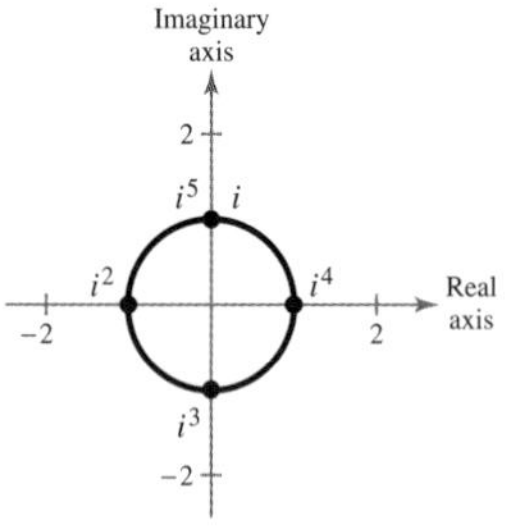

$\cos(\pi/2) + i\sin(\pi/2)$, $\cos\pi + i\sin\pi$,
$\cos(3\pi/2) + i\sin(3\pi/2)$, $\cos 2\pi + i\sin 2\pi$,
$\cos(5\pi/2) + i\sin(5\pi/2)$; θ increases by $\pi/2$.

Section 6.6 (page 456)

The given equation can be written as

$$x^4 = -16 = 16(\cos\pi + i\sin\pi)$$

which means that you can solve the equation by finding the four fourth roots of -16. Each of these roots has the form

$$\sqrt[4]{16}\left(\cos\frac{\pi + 2\pi k}{4} + i\sin\frac{\pi + 2\pi k}{4}\right).$$

Using $k = 0, 1, 2$, and 3, you obtain the roots $\sqrt{2} + \sqrt{2}i$, $-\sqrt{2} + \sqrt{2}i$, $-\sqrt{2} - \sqrt{2}i$, and $\sqrt{2} - \sqrt{2}i$.

Section 6.6 (page 458)

For $n \geq 3$, the vertices of the polygon in the viewing window are the nth roots of unity.

Chapter 7

Section 7.1 (page 481)

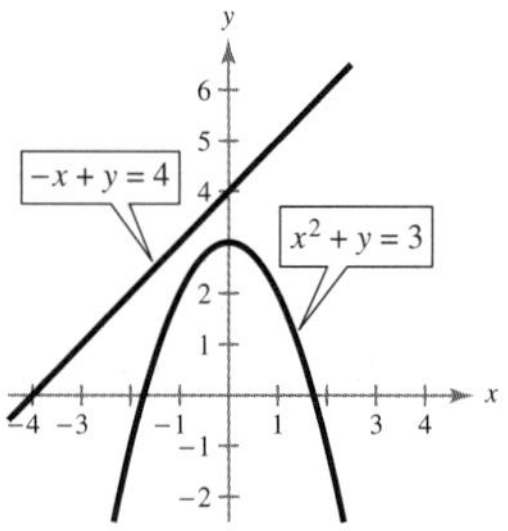

No; Both solutions are imaginary.

Section 7.2 (page 488)

$\left(\frac{3}{2}, -\frac{1}{4}\right)$; elimination

Section 7.2 (page 490)

a.

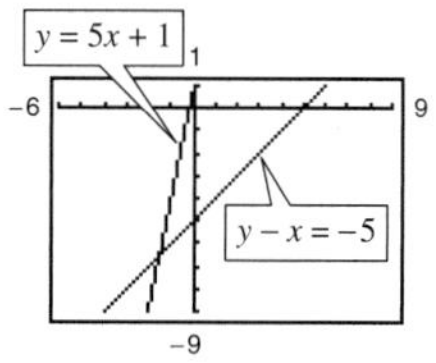

The lines intersect at one point and have different slopes.

b.

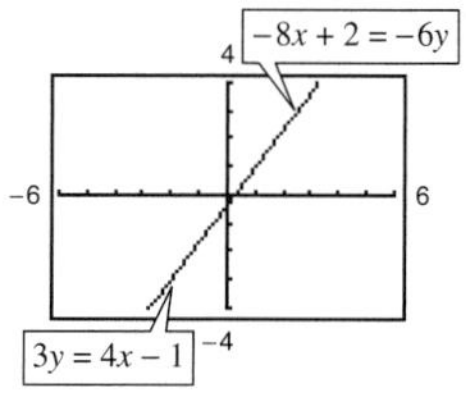

The lines are identical and have the same slope.

c.

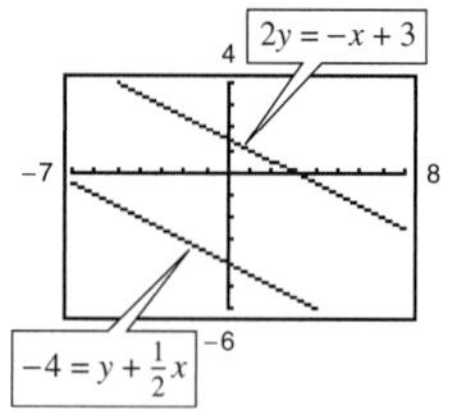

The lines are parallel. The lines have the same slope and are distinct.

Section 7.3 (page 505)

Answers will vary.

Section 7.5 (page 528)

Corresponding entries of A and B are opposites.

$$A + B = \begin{bmatrix} 0 & 0 \\ 0 & 0 \end{bmatrix}$$

Answers will vary.

Section 7.5 (page 531)

$$AB = \begin{bmatrix} 4 & 7 \\ 8 & 15 \end{bmatrix}, BA = \begin{bmatrix} 3 & 4 \\ 11 & 16 \end{bmatrix}, (AB)C = \begin{bmatrix} 12 & 7 \\ 24 & 15 \end{bmatrix},$$

$$A(BC) = \begin{bmatrix} 12 & 7 \\ 24 & 15 \end{bmatrix}$$

Matrix multiplication is associative but not commutative.

Section 7.6 (page 542)

$$A^{-1} = \begin{bmatrix} -2 & -3 & -1 \\ -3 & -4 & -1 \\ -4 & -6 & -1 \end{bmatrix}; AB = BA = \begin{bmatrix} 1 & 0 & 0 \\ 0 & 1 & 0 \\ 0 & 0 & 1 \end{bmatrix}$$

Multiplication of a matrix and its inverse is commutative and gives the identity matrix.

Section 7.6 (page 543)

$(AB)^{-1} = B^{-1}A^{-1}$

Section 7.6 (page 545)

Singular Mat; Row 2 is a multiple of Row 1.

Section 7.7 (page 551)

Invalid Dim; The matrix is not square.

Chapter 8

Section 8.1 (page 583)

$3 \cdot 5! = 360$; $(3 \cdot 5)! \approx 1.31 \times 10^{12}$; Answers will vary.

Section 8.1 (page 586)

Cube \ Number of blue faces	0	1	2	3
$3 \times 3 \times 3$	1	6	12	8
$4 \times 4 \times 4$	8	24	24	8
$5 \times 5 \times 5$	27	54	36	8
$6 \times 6 \times 6$	64	96	48	8
$n \times n \times n$	$(n-2)^3$	$6(n-2)^2$	$12(n-2)$	8

Section 8.2 (page 592)

The differences between consecutive terms are the same in each sequence.

Section 8.3 (page 603)

When $r = -1$, the series is $a_1 - a_1 + a_1 - a_1 + \cdot \cdot \cdot$, which does not have a finite sum. Similarly, when $r = 1$, the series $a_1 + a_1 + \cdot \cdot \cdot$ does not have a finite sum. Finally, a series with $r > 1$ or $r < -1$, such as $\sum_{n=0}^{\infty} \left(\frac{3}{2}\right)^n$, does not have a finite sum.

Section 8.4 (page 610)

a. 1, 1 **b.** 1, 1 **c.** 1, 1 **d.** 7, 7 **e.** 8, 8 **f.** 10, 10

The coefficients of the first and last terms of a binomial expansion are always 1. The coefficients of the second and next-to-last terms of a binomial expansion to the nth power are always n. Answers will vary.

Section 8.4 (page 613)

126, 126; 7, 7; 495, 495; 1, 1; 120, 120

Symmetry between the right and left sides

Section 8.6 (page 627)

Answers will vary; 250

Chapter 9

Section 9.2 (page 658)

To obtain ellipses that are almost circular, place the thumbtacks close together. To obtain long and narrow ellipses, move the thumbtacks farther apart.

Section 9.4 (page 681)

a.

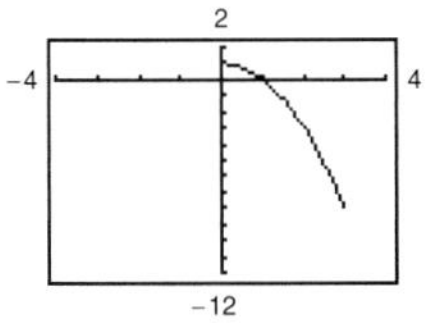

b.

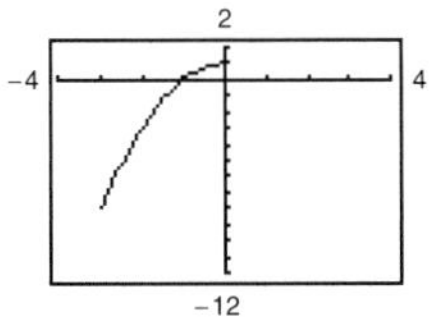

c.

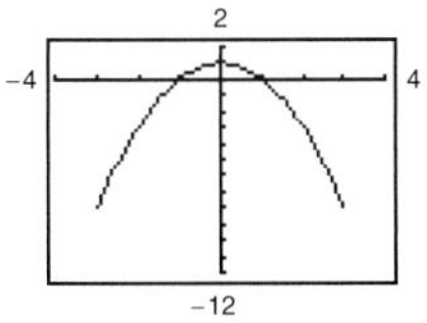

The graphs are portions of a parabola that graph from left to right. When $x = -t$, the graphs are reflections in the x-axis of the previous graphs from left to right.

Section 9.5 (page 688)

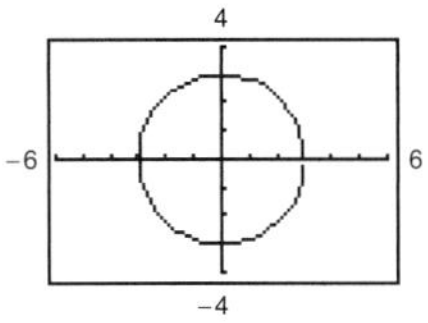

a. Yes

b. Yes; Trace until $\theta = 5\pi/4 + 2n\pi$.

Section 9.6 (page 696)

Eight petals; three petals; When n is odd, there are n petals. When n is even, there are $2n$ petals.

Section 9.7 (page 703)

Answers will vary.

Chapter 10

Section 10.1 (page 724)

$(x-1)^2 + (y-1)^2 + (z-4)^2 = 17$

Midpoint Formula locates center; Distance Formula finds radius; standard equation finds equation.

Section 10.3 (page 736)

a. $\mathbf{k}$

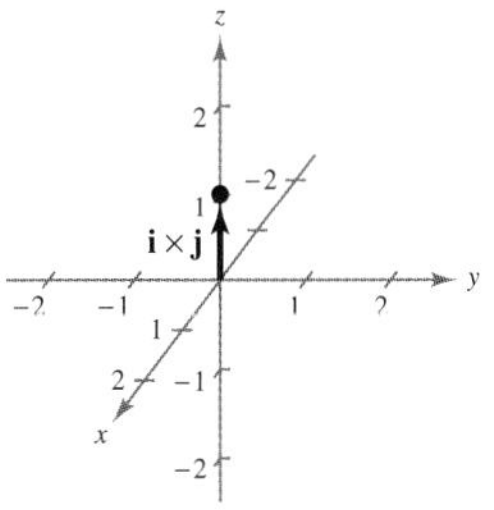

b. $-\mathbf{j}$

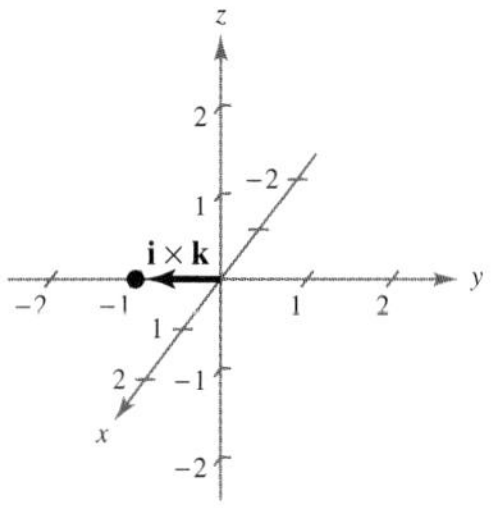

c. $\mathbf{i}$

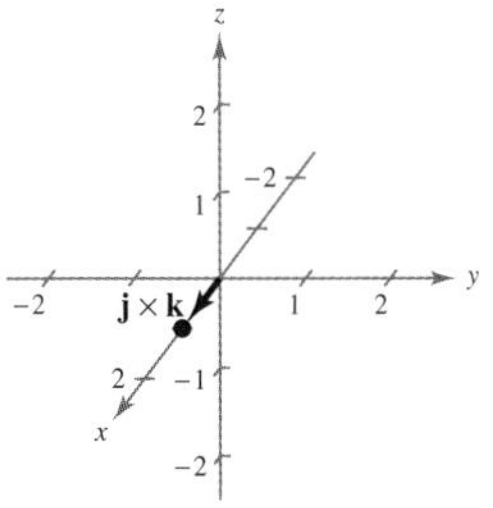

The cross product of two unit vectors results in the third unit vector, and the x-, y-, and z-axes in space are all orthogonal to each other.

Section 10.3 (page 737)

$\mathbf{u} \times \mathbf{v} = -(\mathbf{v} \times \mathbf{u})$

Sample answer: The vectors $\mathbf{u} \times \mathbf{v}$ and $\mathbf{v} \times \mathbf{u}$ have equal lengths but opposite directions.

Section 10.4 (page 745)

$\mathbf{n} = \langle 2, 3, -1 \rangle$ $\mathbf{n} = \langle 4, 6, -2 \rangle$

$\mathbf{n} = \langle -2, -3, 1 \rangle$ $\mathbf{n} = \langle -6, -9, 3 \rangle$

The planes are parallel to each other.

Chapter 11

Section 11.1 (page 765)

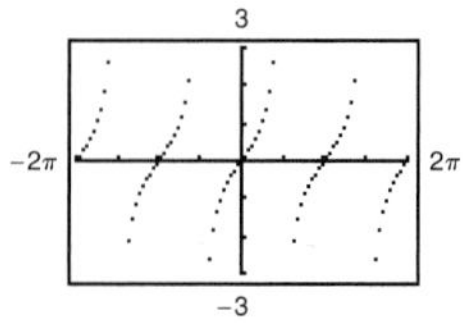

$\lim_{x\to 0} \tan x = 0$; $\lim_{x\to \pi/4} \tan x = 1$; $\lim_{x\to \pi/2} \tan x$ does not exist.

Section 11.1 (page 766)

a.

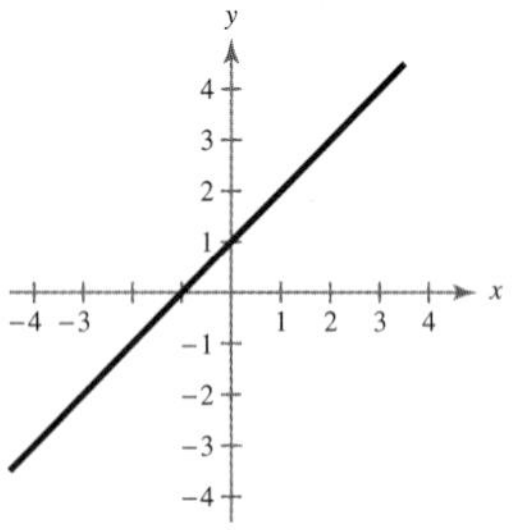

2; 3; Answers will vary.

b.

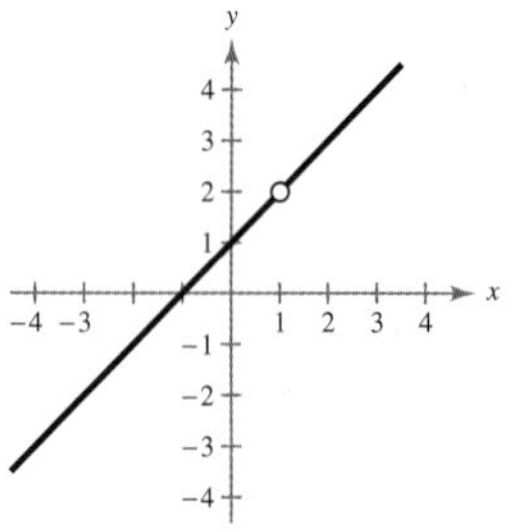

2; 3; Answers will vary.

c.

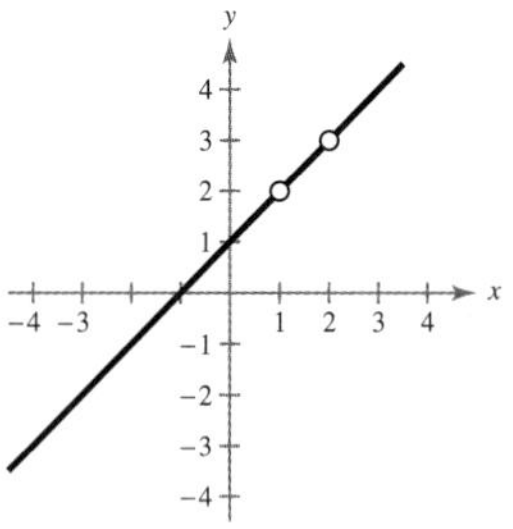

2; 3; Answers will vary.

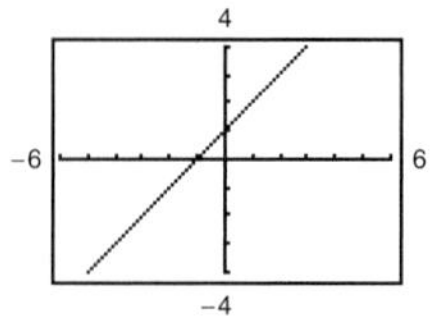

No; The graphs of all three functions appear to be the same line. $f(x) = x + 1$.

Section 11.1 (page 766)

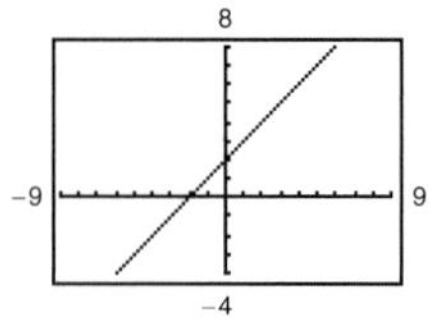

6; 7; no; no

Section 11.3 (page 785)

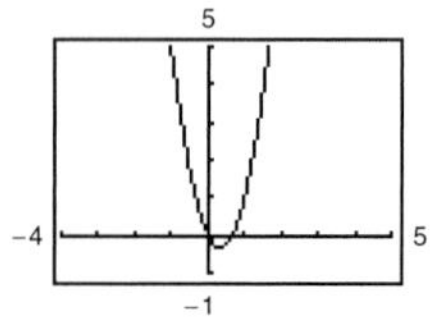

About (0.33, −0.33); 0; 0

Section 11.4 (page 791)

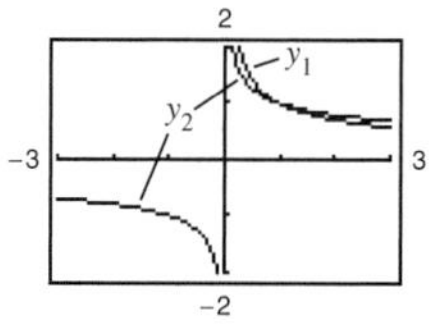

$\sqrt{x}$ is undefined for $x < 0$; Answers will vary.

Section 11.4 (page 792)

x	1	10	100	1000	10,000	100,000
$\frac{1}{x}$	1	0.1	0.01	0.001	1E − 4	1E − 5

$\lim_{x\to 0} \frac{1}{x}$ does not exist.

Section 11.4 (page 795)

d; a; Answers will vary.

What's Wrong?

Chapter 1

Section 1.1 (page 10)

The viewing window on your calculator must be changed to a square setting. This will show that the lines are not perpendicular.

Section 1.3 (page 34)

The domain of y_1 should be all x-values less than or equal to zero and the domain of y_2 should be all x-values greater than zero in order to define $f(x)$ properly.

Section 1.6 (page 65)

$f(x) = x^2$ does not have an inverse function because it does not pass the Horizontal Line Test.

Chapter 2

Section 2.6 (page 144)

The horizontal asymptote is $y = 2$. It appears that $y = 1$ is the horizontal asymptote because the viewing window is not large enough.

Chapter 3

Section 3.1 (page 185)

The function $g(x)$ was entered as 3^x + 2 rather than 3^(x + 2).

Chapter 4

Section 4.3 (page 281)

The calculator is in *radian* mode instead of *degree* mode.

Section 4.5 (page 296)

The calculator is in *degree* mode instead of *radian* mode.

Chapter 5

Section 5.2 (page 363)

The graphs appear to coincide because the viewing window is too large. From the table on the graphing utility, for different values of x, y_1 and y_2 differ slightly.

Chapter 6

Section 6.6 (page 452)

The answer $1.047 \approx \frac{\pi}{3}$ lies in the first quadrant and $z = -2 - 2\sqrt{3}i$ lies in the third quadrant.

Chapter 7

Section 7.2 (page 492)

The system does have a solution when $x = 39{,}600$, but the viewing window on the graphing utility is too small for this to be seen in the graph.

Chapter 8

Section 8.3 (page 602)

a_1 is 7(0.5), not 7. So it should be 7 * 0.5 * (1 − 0.5^6)/(1 − 0.5).

Chapter 9

Section 9.4 (page 683)

The graphing utility uses a minimum value of 0 for t.

Section 9.6 (page 698)

The graph shows $r = 9 \sin 2\theta$ when it should be $r = \sqrt{9 \sin 2\theta}$ and $r = -\sqrt{9 \sin 2\theta}$.

Chapter 10

Section 10.3 (page 738)

$\mathbf{w} = -8\mathbf{i} + \mathbf{j} - 4\mathbf{k}$ is a vector orthogonal to both **u** and **v** but is not a *unit* vector.

Chapter 11

Section 11.1 (page 763)

In using the arrow keys to trace along the graph in a *decimal* setting, you move at increments of 0.1 along the x-axis. These increments are too large to show that the value of the function approaches 3 as you approach $x = 1$.

Checkpoints

Chapter 1

Section 1.1

1. (a) 2 (b) $-\frac{3}{2}$ (c) 0 **2.** $y = 2x - 13$

3. $y = 3.559x + 98.404$; \$119.758 billion

4. $m = \frac{1}{2}$; y-intercept: $(0, -2)$; a line that rises from left to right

5. (a)

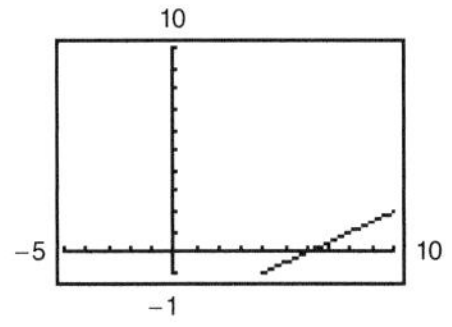

(b)

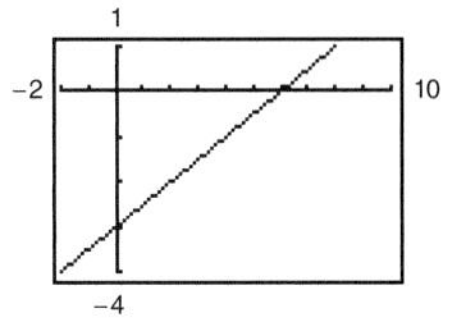

(c)

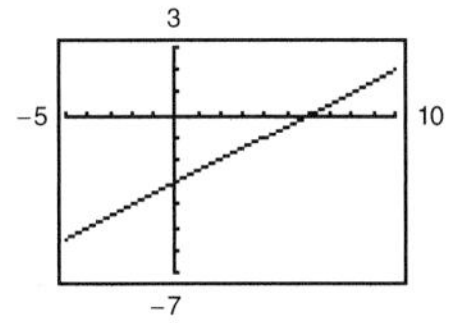

The first graph does not show both intercepts. The third graph is best because it shows both intercepts and gives the most accurate view of the slope by using a square setting.

6. $y = \frac{5}{3}x + \frac{23}{3}$ **7.** $y = -\frac{3}{5}x - \frac{7}{5}$

8. The lines $y = \frac{1}{2}x$ and $y = -2x$ are perpendicular.

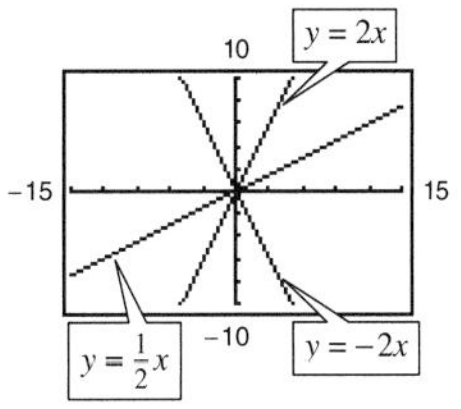

Section 1.2

1. Function **2.** (a) Not a function (b) Function

3. (a) -2 (b) -38 (c) $-3x^2 + 6x + 7$

4. $f(-2) = 5$, $f(2) = 1$, $f(3) = 2$

5. (a) $\{-2, -1, 0, 1, 2\}$ (b) All real numbers x except $x = 3$

6. (a) All real numbers r such that $r > 0$
(b) All real numbers x such that $x \geq 16$

7. Domain: $[-2, 2]$
Range: $[0, 2]$

8. 2013: 897 thousand registered vessels **9.** No
2014: 900 thousand registered vessels

10. $2x + h + 2$, $h \neq 0$

Section 1.3

1. (a) All real numbers x except $x = -3$
(b) $f(0) = 3$; $f(3) = -6$ (c) $(-\infty, 3]$

2. Domain: $[1, \infty)$
Range: $[0, \infty)$

3. Not a function

4.

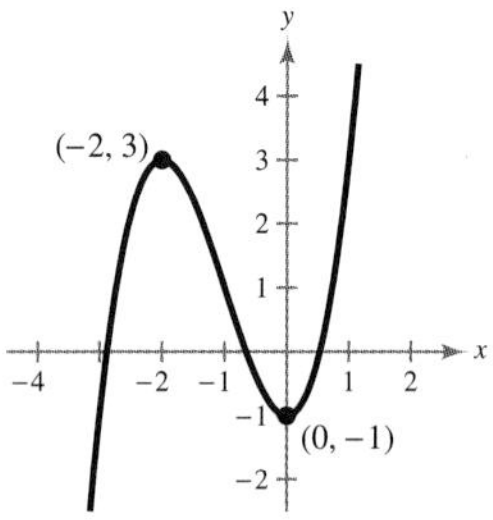

Increasing on $(-\infty, -2)$ and $(0, \infty)$
Decreasing on $(-2, 0)$

5. $(-0.875, 6.0625)$

6. Relative minimum: $(1, -7)$
Relative maximum: $(-2, 20)$

7. About 34° F

8.

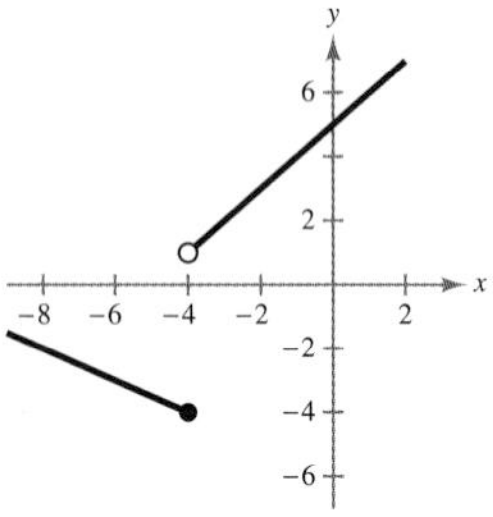

9.

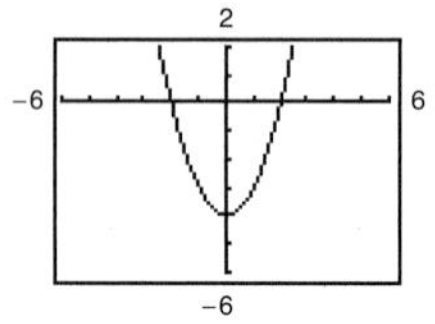

Even

10. (a) Neither; No symmetry (b) Even; y-axis symmetry
(c) Odd; Origin symmetry

Section 1.4

1. (a)

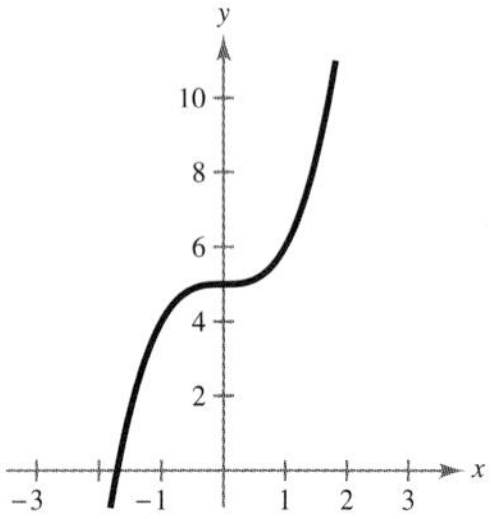

(b)

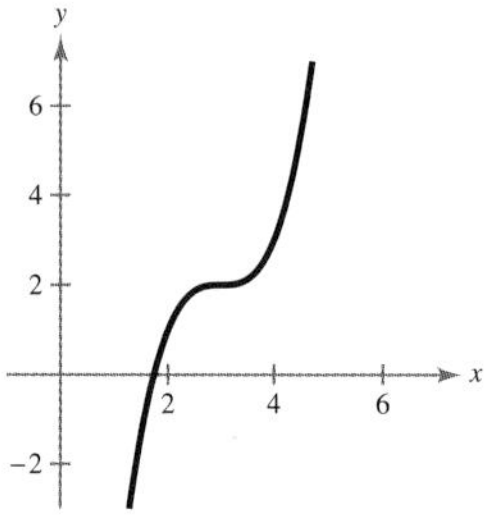

2. $k(x) = x^2 - 1$ **3.** $k(x) = -(x + 2)^2$

4. Reflection in the x-axis

5. (a) Vertical stretch (b) Vertical shrink

6. (a) Horizontal shrink (b) Horizontal stretch

Section 1.5

1. $x^2 - x + 1; 3$ **2.** $x^2 + x - 1; 11$ **3.** $-x^3 + x^2; -18$

4. $\left(\frac{f}{g}\right)(x) = \frac{\sqrt{x-3}}{\sqrt{16-x^2}}$; Domain: $[3, 4)$
$\left(\frac{g}{f}\right)(x) = \frac{\sqrt{16-x^2}}{\sqrt{x-3}}$; Domain: $(3, 4]$

5. $(x - 1)^2; 1$

6. (a) $8x^2 + 7$ (b) $16x^2 + 80x + 101$ (c) 9

7. All real numbers x **8.** (a) x^2 (b) x^2

9. $f(x) = \frac{1}{5}\sqrt[3]{x},\ g(x) = 8 - x$

10. (a) $32t^2 + 36t + 204$ (b) About 4.5 h

Section 1.6

1. $f^{-1}(x) = 5x;\ f(f^{-1}(x)) = \frac{1}{5}(5x) = x,\ f^{-1}(f(x)) = 5\left(\frac{1}{5}x\right) = x$

2. $f^{-1}(x) = x - 7;\ f(f^{-1}(x)) = (x - 7) + 7 = x,$
$f^{-1}(f(x)) = (x + 7) - 7 = x$

3. $f(g(x)) = \left(\sqrt[5]{x}\right)^5 = x \quad g(f(x)) = \sqrt[5]{x^5} = x$

4. $g(x) = 7x + 4$

5.

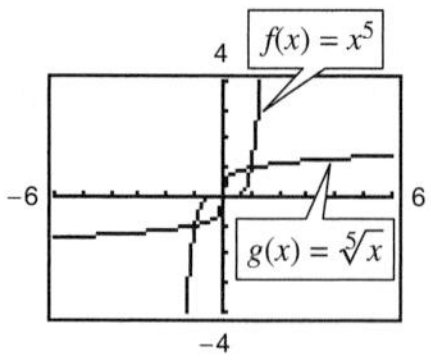

6.

x	-2	-1	0	1	2	3	4
$f(x)$	-32	-1	0	1	32	243	1024

x	-32	-1	0	1	32	243	1024
$g(x)$	-2	-1	0	1	2	3	4

7. Yes **8.** No **9.** $f^{-1}(x) = \frac{x + 3}{2}$

10. $f^{-1}(x) = x^3 - 10$ **11.** $f^{-1}(x) = \sqrt{x} + 1$

Section 1.7

1.

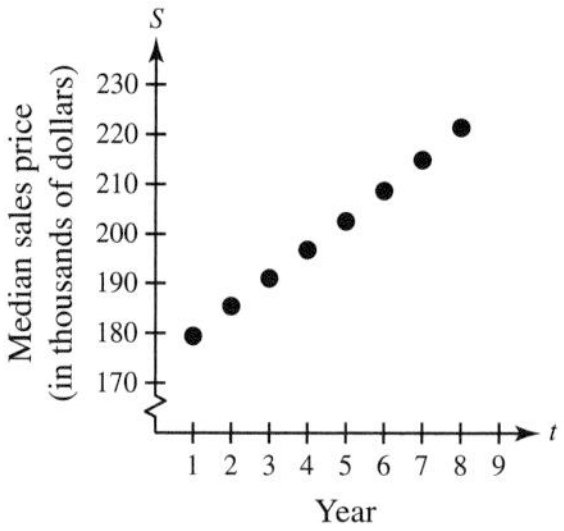

2. (a)

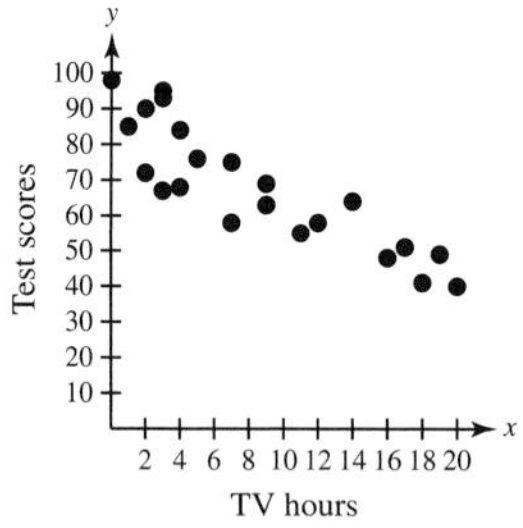

(b) Negative correlation; The more time a student spent watching TV, the lower his or her test score tended to be.

3. *Sample answer:* $S = 5.9t + 173.1$

4. (a) $N = 7375.0t + 4988$
(b) The model is a good fit for the data.

5. $e = 399.9g - 653$

Chapter 2

Section 2.1

1. (a)

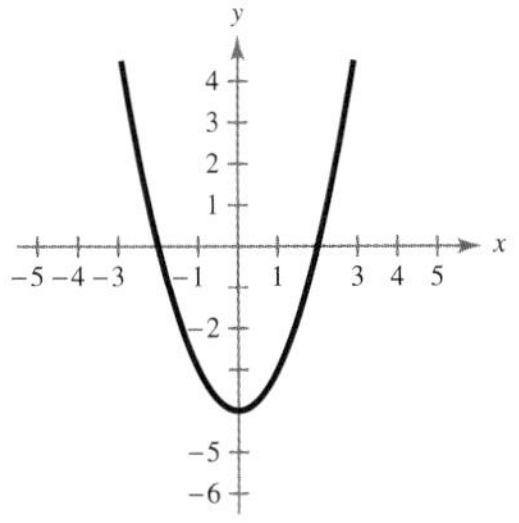

Vertical shift 4 units downward

(b)

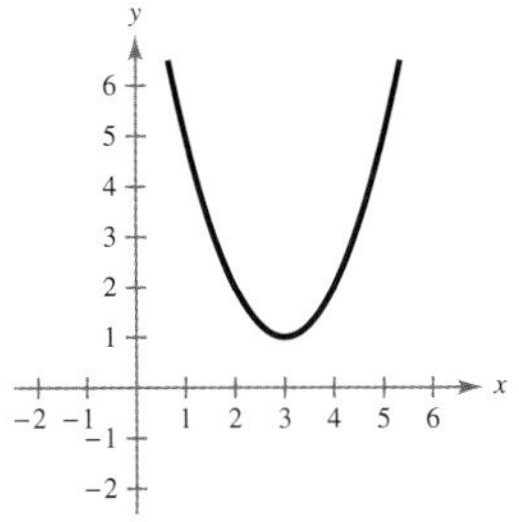

Horizontal shift 3 units right, vertical shift one unit upward

2. Parabola opening upward with vertex $(1, 1)$

3. Parabola opening upward with vertex $(2, -1)$
x-intercepts: $(1, 0)$ and $(3, 0)$

4. $f(x) = (x + 4)^2 + 11$ **5.** About 39.7 ft

Section 2.2

1. (a)

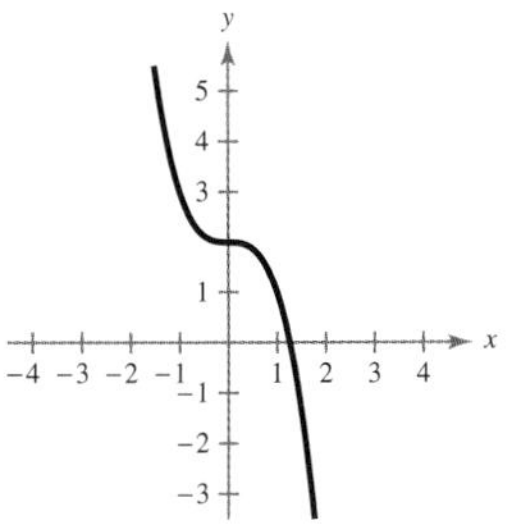

(b)

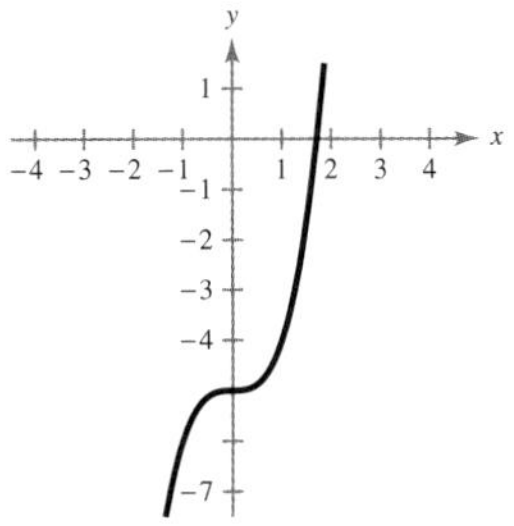

(c)

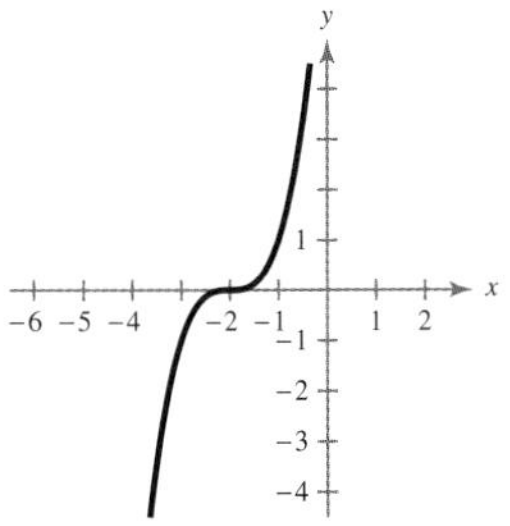

2. Falls to the left, rises to the right

3. (a) Falls to the left and to the right
(b) Rises to the left, falls to the right

4. $x = -3, 0, 2$

5. Zeros: $0, 1, -3$
Relative maximum: about $(-1.87, 6.06)$
Relative minimum: about $(0.54, -0.88)$

6. $x = 0, 2$ **7.** *Sample answer:* $f(x) = x^4 - 5x^2 + 4$

8.

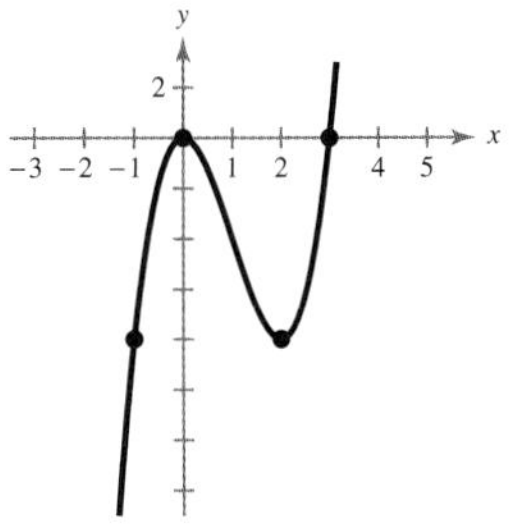

9.

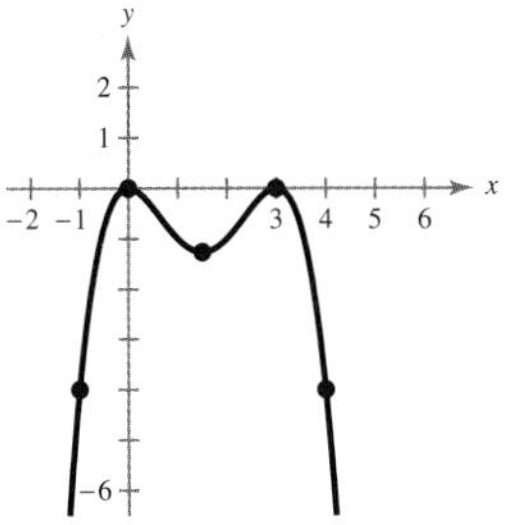

10. $(-1, 0), (0, 1), (3, 4)$

Section 2.3

1. $(x + 4)(3x + 7)$ **2.** $x^2 + x + 3, x \neq 3$

3. $2x^3 - x^2 + 3 - \dfrac{6}{3x + 1}$ **4.** $5x^2 - 2x + 3, x \neq -2$

5. (a) 1 (b) 396 (c) $-\frac{13}{2}$ (d) -17

6. $f(-3) = 0$; $(x - 5), (x + 2)$ **7.** $-1, 2, 4$ **8.** $-3, \frac{1}{2}, 2, 5$

9. 3 or 1 positive real zero(s), no negative real zeros

10. $\dfrac{1}{2}$ **11.** $-2, 0, 1 \pm \dfrac{\sqrt{15}}{3}$

Section 2.4

1. (a) $12 - i$ (b) $-2 + 7i$ (c) i (d) 0

2. (a) $18 - 6i$ (b) 41 (c) $12 + 16i$

3. (a) 45 (b) 29 **4.** $\frac{3}{5} + \frac{4}{5}i$ **5.** $-2\sqrt{7}$

6. $-\dfrac{7}{8} \pm \dfrac{\sqrt{23}}{8}i$

Section 2.5

1. 4 **2.** Answers will vary.

3. $(x + 4)(x - 4)(x + 4i)(x - 4i)$
$x = \pm 4, \pm 4i$

4. $f(x) = x^4 + 45x^2 - 196$

5. $f(x) = -(x^4 + x^3 + 2x^2 + 4x - 8)$

6. (a) $f(x) = (x^2 + 1)(x^2 - 3)$
(b) $f(x) = (x^2 + 1)(x + \sqrt{3})(x - \sqrt{3})$
(c) $f(x) = (x + i)(x - i)(x + \sqrt{3})(x - \sqrt{3})$

7. $\frac{2}{3}, \pm 4i$

Section 2.6

1. All real numbers x except $x = 1$;
f approaches $-\infty$ from the left and ∞ from the right of $x = 1$

2. Vertical asymptotes: $x = -1, x = 1$
Horizontal asymptote: $y = 5$

3. Vertical asymptote: $x = 0$
Horizontal asymptote: $y = 1$
Hole at $x = -5$

4. (a) All real numbers x (b) None (c) $y = 3$

5. (a) \$63.75 million; about \$208.64 million; \$1020 million
(b) No. The function is undefined at $p = 100$.

Section 2.7

1. (a)

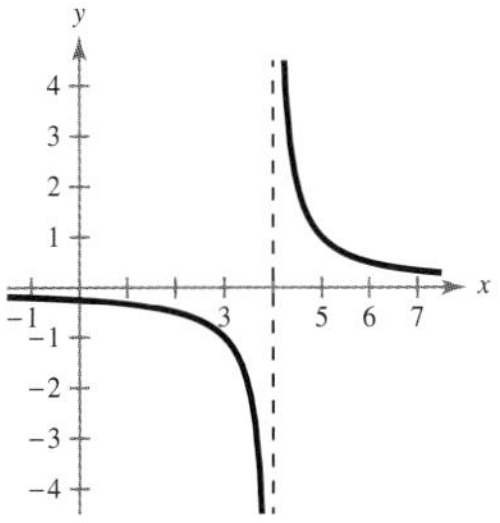

Horizontal shift 4 units right

(b)

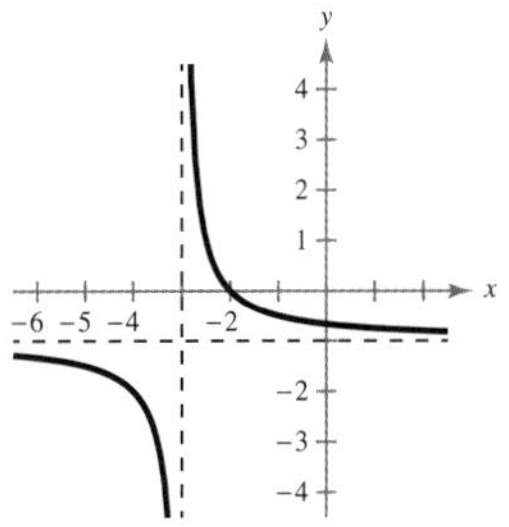

Horizontal shift 3 units left, vertical shift 1 unit downward

2.

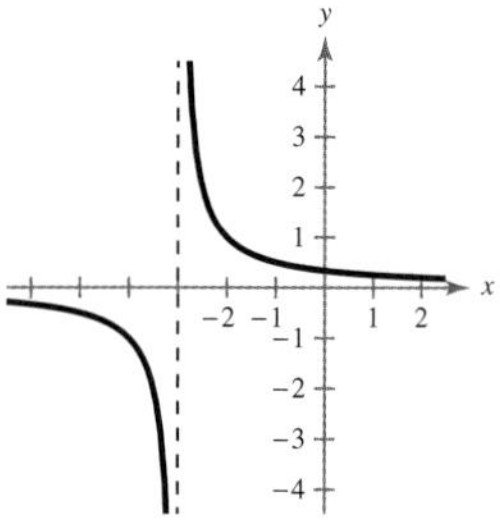

3.

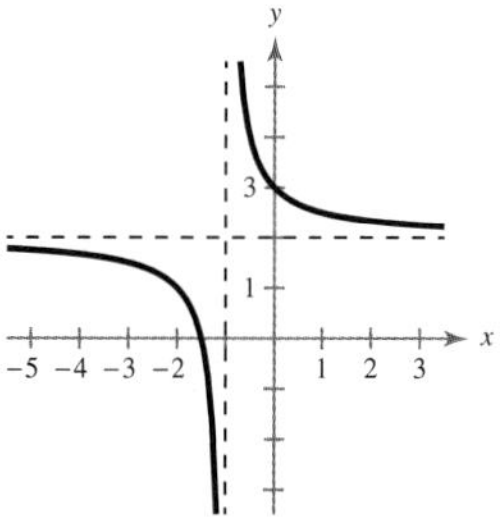

4.

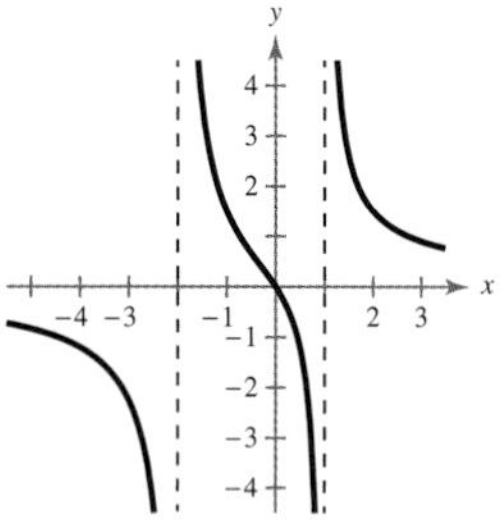

5.

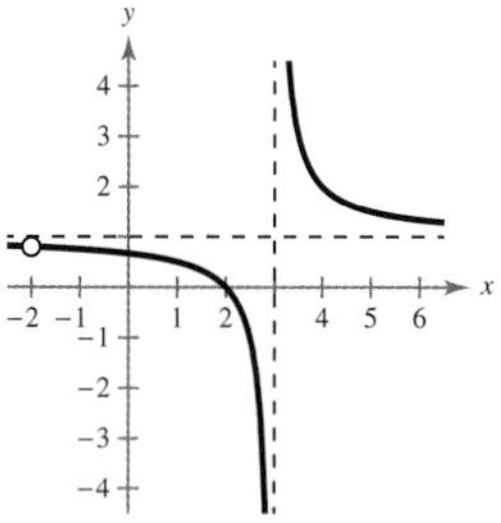

6.

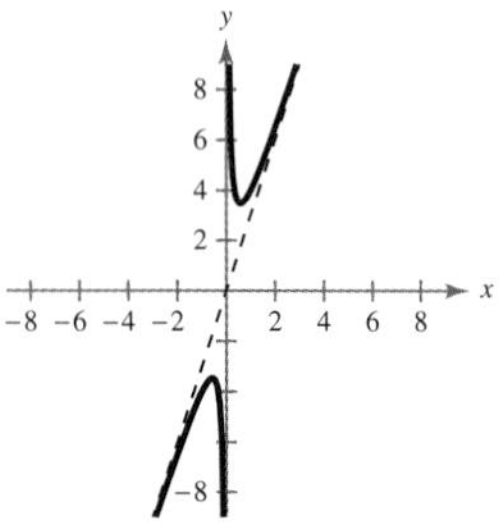

7. About 12.9 in. by about 6.5 in.

Section 2.8

1. Quadratic

2. (a)

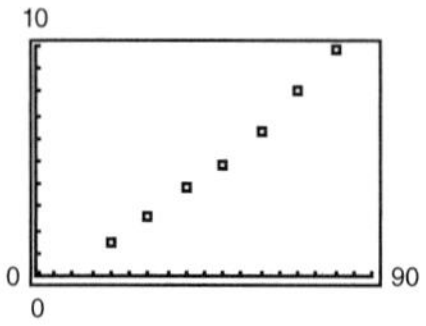

(b) $y = 0.0008x^2 + 0.06x$ (c) About 5.7 seconds

3. $y = -0.61x^2 + 2.3x + 219$; year 10

4. $y = 1.72x + 67.1$; $y = -0.128x^2 + 5.05x + 46.3$; quadratic

Chapter 3

Section 3.1

1. 0.0528248

2.

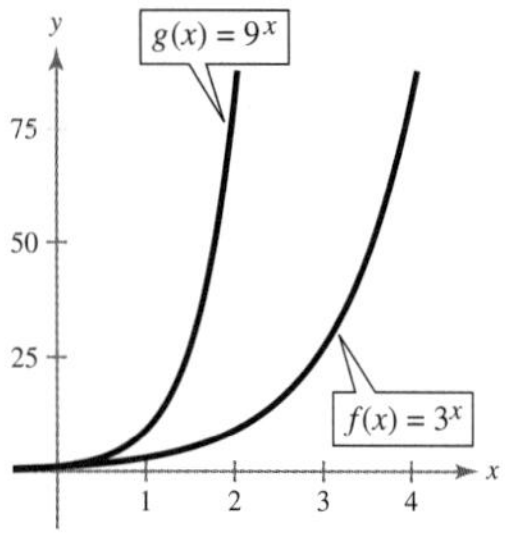

3.

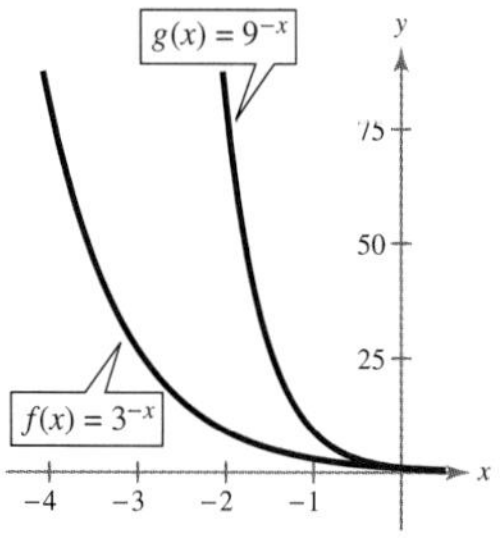

4. (a) Shift the graph of f two units to the right.
(b) Shift the graph of f three units up.
(c) Reflect the graph of f in the y-axis and shift three units down.

5.

x	10	100	1000
$(1 + 2/x)^x$	6.1917	7.2446	7.3743

x	10,000	100,000	1,000,000
$(1 + 2/x)^x$	7.3876	7.3889	7.3890

6. (a) 1.3498588 (b) 0.3011942 (c) 492.7490411

7.

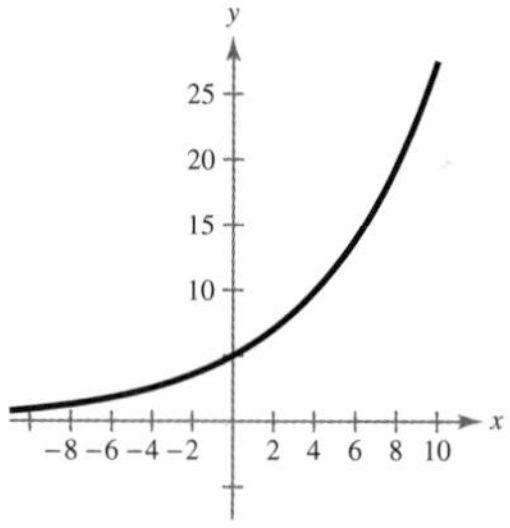

8. \$11,520.76

9. (a) \$7927.75 (b) \$7935.08 (c) \$7938.78

10. About 0.22 g

11. (a) 10 (b) About 42
(c)

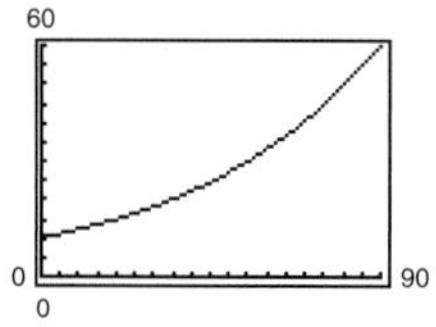

Section 3.2

1. (a) 0 (b) -3 (c) 4

2. (a) 2.4393327 (b) -0.5606673 (c) Error
(d) -0.3010300

3. (a) 2 (b) 3

4.

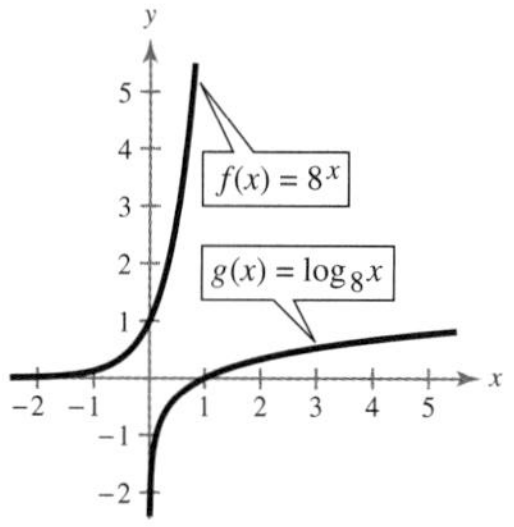

5.

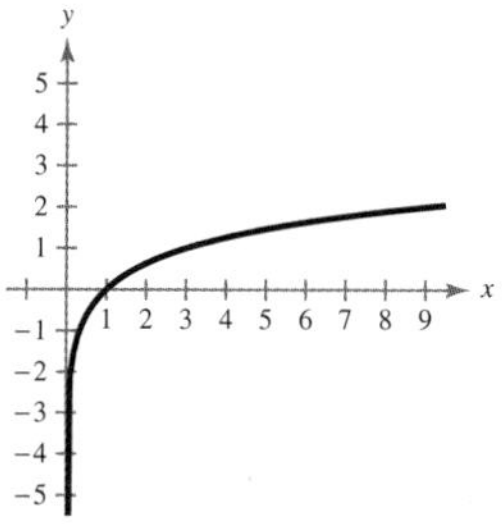

6. (a)

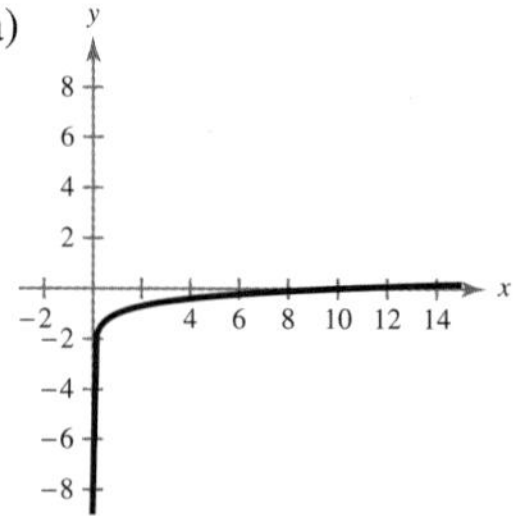

(b)

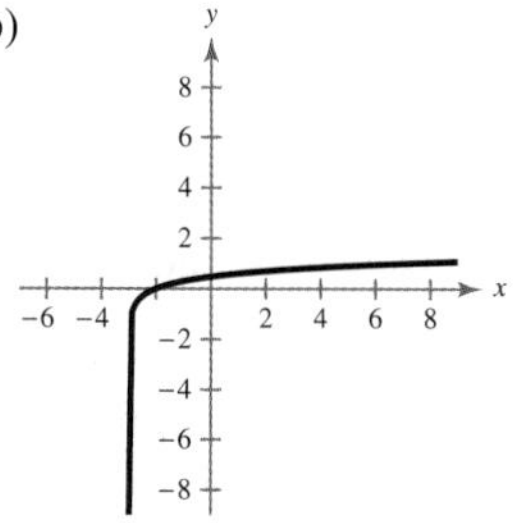

7. (a) -4.6051702 (b) 1.3862944 (c) 1.3169579
(d) Error

8. (a) $\frac{1}{3}$ (b) 0 (c) $\frac{3}{4}$ (d) 7 **9.** $(-3, \infty)$

10. (a) About 70.84 (b) About 61.18 (c) About 59.61

Section 3.3

1. 3.5850 **2.** 3.5850

3. (a) $\log_{10} 3 + 2\log_{10} 5$ (b) $2\log_{10} 3 - 3\log_{10} 5$

4. Answers will vary. **5.** $\log_3 4 + 2\log_3 x - \frac{1}{2}\log_3 y$

6. $\log_{10} \dfrac{(x+3)^2}{(x-2)^4}$ **7.** $\ln y = \dfrac{2}{3}\ln x$

Section 3.4

1. (a) 9 (b) 216 (c) ln 5 (d) $-\frac{1}{2}$

2. (a) ln 10 (b) $\log_5 16$

3. $\frac{1}{3}\ln 12$ **4.** $\log_2 \frac{7}{6} - 5 \approx -4.78$ **5.** ln 3, ln 4

6. (a) $e^{2/3}$ (b) 7 **7.** $e^{-2/3}$ **8.** $\frac{32}{3}$ **9.** 10

10. $\log_a \dfrac{1}{x} = \log_a 1 - \log_a x = -\log_a x$ **11.** 4.079

12. About 13.2 years; It takes longer for your money to double at a lower interest rate.

13. 2017

Section 3.5

1. 2021 **2.** 400 bacteria **3.** About 38,000 yr

4. 533 **5.** About 7 days **6.** 30 dB

Section 3.6

1. Exponential

2. $y = 16.639(0.500)^x$, $r^2 \approx 0.9986$;
$y = 9.578x^{-1.957}$, $r^2 \approx 0.8739$;
exponential

3. $y = 2.460 + 1.090 \ln x$, $r^2 \approx 0.9826$;
$y = 0.346x + 2.323$, $r^2 \approx 0.9023$;
logarithmic

4. About 5.4 days; This is the half-life of the substance.

5. $y = 0$, $y = 100$; As the number of egg masses increases, the percent of defoliation approaches 100%. Because the domain is $x \geq 0$, there is no context for the asymptote $y = 0$.

Chapter 4

Section 4.1

1. (a) $\dfrac{\pi}{4}, -\dfrac{7\pi}{4}$ (b) $\dfrac{5\pi}{3}, -\dfrac{7\pi}{3}$ **2.** (a) $\dfrac{5\pi}{12}$ (b) $\dfrac{16\pi}{9}$

3. (a) 30° (b) 300°

4. (a) Complement: 54°
Supplement: 144°
(b) Complement: none
Supplement: 89°
(c) Complement: $\dfrac{\pi}{3}$
Supplement: $\dfrac{5\pi}{6}$
(d) Complement: none
Supplement: $\dfrac{\pi}{6}$

5. 24π in. $\approx$ 75.40 in. **6.** About 0.84 cm/sec

7. (a) 4800π rad/min (b) About 60,319 in./min

Section 4.2

1. (a) $\sin \dfrac{\pi}{2} = 1$ $\csc \dfrac{\pi}{2} = 1$
$\cos \dfrac{\pi}{2} = 0$ $\sec \dfrac{\pi}{2}$ is undefined.
$\tan \dfrac{\pi}{2}$ is undefined. $\cot \dfrac{\pi}{2} = 0$

(b) $\sin 0 = 0$ $\csc 0$ is undefined.
$\cos 0 = 1$ $\sec 0 = 1$
$\tan 0 = 0$ $\cot 0$ is undefined.

(c) $\sin\left(-\dfrac{5\pi}{6}\right) = -\dfrac{1}{2}$ $\csc\left(-\dfrac{5\pi}{6}\right) = -2$
$\cos\left(-\dfrac{5\pi}{6}\right) = -\dfrac{\sqrt{3}}{2}$ $\sec\left(-\dfrac{5\pi}{6}\right) = -\dfrac{2\sqrt{3}}{3}$
$\tan\left(-\dfrac{5\pi}{6}\right) = \dfrac{\sqrt{3}}{3}$ $\cot\left(-\dfrac{5\pi}{6}\right) = \sqrt{3}$

(d) $\sin\left(-\dfrac{3\pi}{4}\right) = -\dfrac{\sqrt{2}}{2}$ $\csc\left(-\dfrac{3\pi}{4}\right) = -\sqrt{2}$
$\cos\left(-\dfrac{3\pi}{4}\right) = -\dfrac{\sqrt{2}}{2}$ $\sec\left(-\dfrac{3\pi}{4}\right) = -\sqrt{2}$
$\tan\left(-\dfrac{3\pi}{4}\right) = 1$ $\cot\left(-\dfrac{3\pi}{4}\right) = 1$

2. (a) 0 (b) $-\dfrac{\sqrt{3}}{2}$ (c) 0.3

3. (a) 0.7818315 (b) 1.0997502

Section 4.3

1. $\sin\theta = \dfrac{1}{2}$ $\csc\theta = 2$
$\cos\theta = \dfrac{\sqrt{3}}{2}$ $\sec\theta = \dfrac{2\sqrt{3}}{3}$
$\tan\theta = \dfrac{\sqrt{3}}{3}$ $\cot\theta = \sqrt{3}$

2. 1; $\sqrt{2}$; $\sqrt{2}$ **3.** $\tan 60° = \sqrt{3}$, $\tan 30° = \dfrac{\sqrt{3}}{3}$

4. 0.5665501 **5.** (a) 0.96 (b) $\frac{7}{24}$ **6.** (a) $\frac{1}{2}$ (b) $\sqrt{5}$

7. Answers will vary. **8.** 40 ft **9.** 60°

10. About 17.6 ft; about 17.2 ft

Section 4.4

1. $\sin\theta = \dfrac{3\sqrt{13}}{13}$, $\cos\theta = -\dfrac{2\sqrt{13}}{13}$, $\tan\theta = -\dfrac{3}{2}$

2. $\cos\theta = -\frac{3}{5}$, $\tan\theta = -\frac{4}{3}$ **3.** Undefined; 0

4. (a) 33° (b) $\dfrac{4\pi}{9}$ (c) $\dfrac{\pi}{5}$

5. (a) $-\dfrac{\sqrt{2}}{2}$ (b) $-\dfrac{1}{2}$ (c) $-\dfrac{\sqrt{3}}{3}$ **6.** $-\dfrac{3}{5}$

Section 4.5

1.

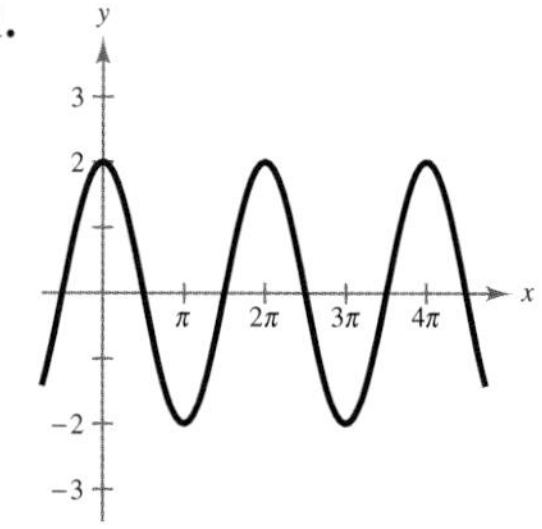

2.

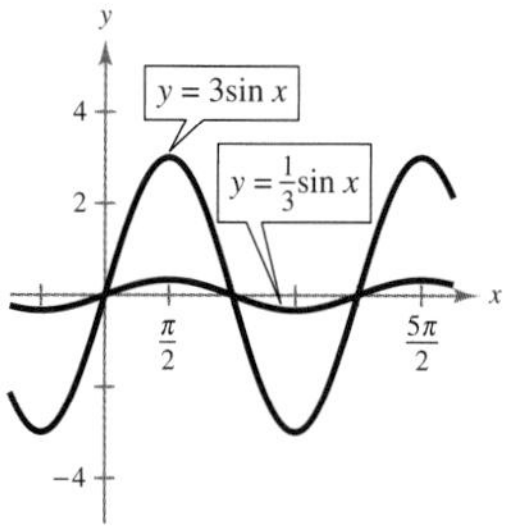

3.

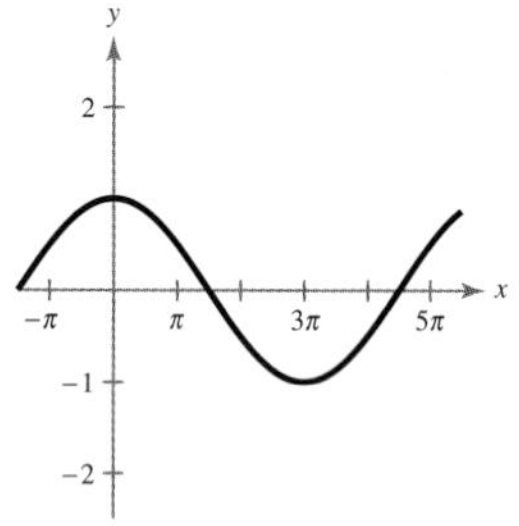

4. Period: 2π
Amplitude: 2
$\left[\frac{\pi}{2}, \frac{5\pi}{2}\right]$ corresponds to one cycle.
Key points: $\left(\frac{\pi}{2}, 2\right), (\pi, 0), \left(\frac{3\pi}{2}, -2\right), (2\pi, 0), \left(\frac{5\pi}{2}, 2\right)$

5. Period: 2
Amplitude: $\frac{1}{2}$
$[-1, 1]$ corresponds to one cycle.
Key points: $(-1, 0), \left(-\frac{1}{2}, -\frac{1}{2}\right), (0, 0), \left(\frac{1}{2}, \frac{1}{2}\right), (1, 0)$

6. Period: 2π
Amplitude: 2
$[0, 2\pi]$ corresponds to one cycle.
Key points: $(0, -3), \left(\frac{\pi}{2}, -5\right), (\pi, -7), \left(\frac{3\pi}{2}, -5\right), (2\pi, -3)$
Shifted 5 units down from $y = 2\cos x$

7. *Sample answer:* $y = 2\cos(x - \pi)$

8. $y = 5.6\sin(0.524t - 0.525) + 5.7$

Section 4.6

1.

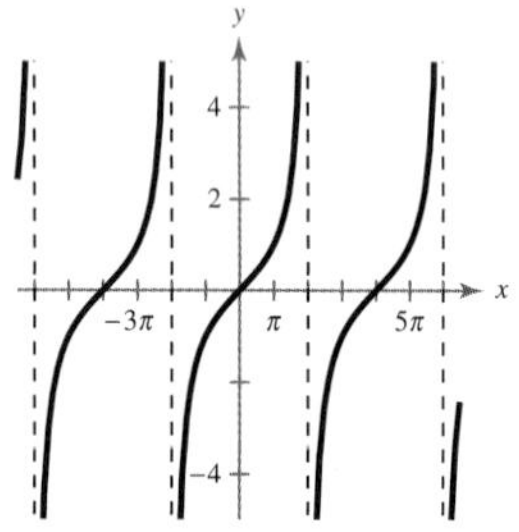

2.

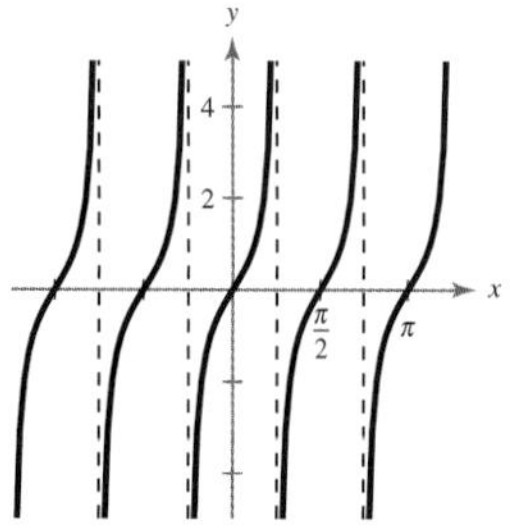

3.

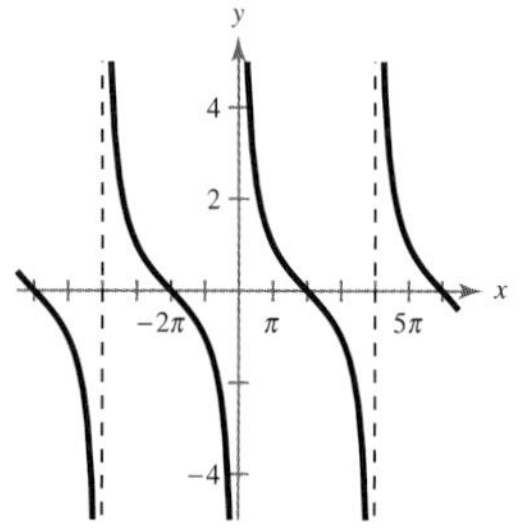

4.

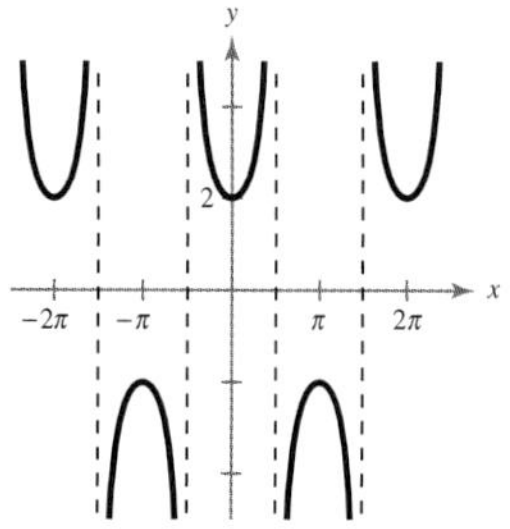

5.

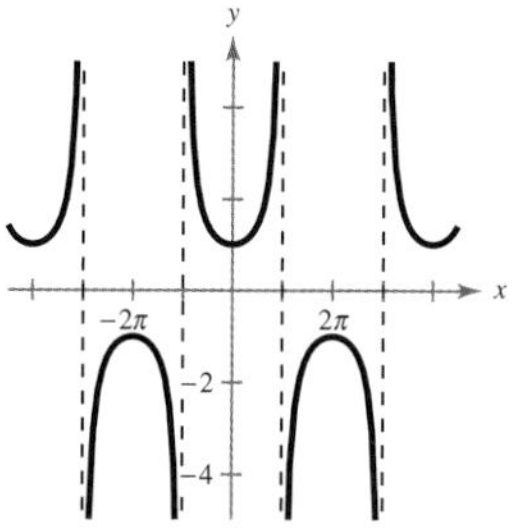

6. The graph touches the curves $y = -e^x$ and $y = e^x$ at $x = \frac{\pi}{8} + \frac{n\pi}{4}$ and has intercepts at $x = \frac{n\pi}{4}$.

Section 4.7

1. (a) $\frac{\pi}{2}$ (b) Not possible

2.

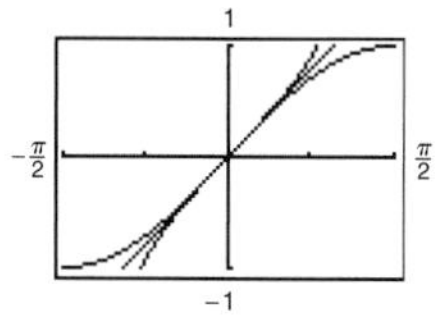

3. (a) $\frac{3\pi}{4}$ (b) $\frac{\pi}{3}$

4. (a) 1.3670516 (b) Not possible (c) 1.9273001

5. g is shifted one unit to the left of f.

6. (a) -14 (b) $-\frac{\pi}{4}$ (c) 0.54 **7.** $\frac{4}{5}$ **8.** $\sqrt{x^2 + 1}$

Section 4.8

1. $B = 70°, a \approx 5.46, c \approx 15.96$

2. About 15.8 ft **3.** About 15.1 ft **4.** About 3.58°

5. Bearing: N 53° W, Distance: about 3.2 nm

6 $d = 6\sin\frac{2\pi}{3}t$; Frequency: $\frac{1}{3}$ cycle/sec

7. (a) 4 (b) 3 cycles per unit of time (c) 4 (d) $\frac{1}{12}$

Chapter 5

Section 5.1

1. $\sin x = -\frac{\sqrt{10}}{10}, \cos x = -\frac{3\sqrt{10}}{10}, \tan x = \frac{1}{3}, \csc x = -\sqrt{10},$
$\sec x = -\frac{\sqrt{10}}{3}, \cot x = 3$

2. $-\sin x$

3. (a) $(1 + \cos\theta)(1 - \cos\theta)$ (b) $(2\csc\theta - 3)(\csc\theta - 2)$

4. $(\tan x + 1)(\tan x + 2)$ **5.** $\sin x$ **6.** $1 + \sin\theta$

7. $3\cos\theta$

Section 5.2

1–7. Answers will vary.

Section 5.3

1. $\frac{3\pi}{2} + 2n\pi$ **2.** $\frac{\pi}{4}, \frac{3\pi}{4}$ **3.** $\frac{\pi}{3} + n\pi, \frac{2\pi}{3} + n\pi$

4. $n\pi$ **5.** $\frac{\pi}{6}, \frac{\pi}{2}, \frac{5\pi}{6}$ **6.** $\frac{\pi}{4} + n\pi, \frac{3\pi}{4} + n\pi$

7. $0, \frac{3\pi}{2}$ **8.** $\frac{\pi}{6} + n\pi, \frac{\pi}{3} + n\pi$ **9.** $\frac{\pi}{2} + 2n\pi$

10. $\arctan \frac{3}{4} + n\pi$, $\arctan(-2) + n\pi$
11. $-2.7607, -1.8049, 0, 1.8049, 2.7607$
12. (a) $32.3°, 85.5°$ (b) $54.7°$

Section 5.4

1. (a) $\dfrac{\sqrt{2} + \sqrt{6}}{4}$ (b) $\dfrac{\sqrt{2} + \sqrt{6}}{4}$ **2.** $-\dfrac{63}{65}$

3. $\dfrac{x + \sqrt{1 - x^2}}{\sqrt{2}}$ **4.** Answers will vary.

5. (a) $-\cos\theta$ (b) $-\tan\theta$ **6.** $\dfrac{\pi}{3}, \dfrac{5\pi}{3}$

7. Answers will vary.

Section 5.5

1. $\dfrac{\pi}{3} + 2n\pi, \pi + 2n\pi, \dfrac{5\pi}{3} + 2n\pi$
2. $\sin 2\theta = \frac{24}{25}$, $\cos 2\theta = \frac{7}{25}$, $\tan 2\theta = \frac{24}{7}$
3. $4\cos^3 x - 3\cos x$ **4.** $\dfrac{\cos 4x - 4\cos 2x + 3}{\cos 4x + 4\cos 2x + 3}$
5. $-\dfrac{\sqrt{2 - \sqrt{3}}}{2}$ **6.** $\dfrac{\pi}{3}, \pi, \dfrac{5\pi}{3}$ **7.** $\dfrac{1}{2}\sin 8x + \dfrac{1}{2}\sin 2x$
8. $\dfrac{\sqrt{2}}{2}$ **9.** $0, \dfrac{\pi}{6}, \dfrac{\pi}{2}, \dfrac{5\pi}{6}, \pi, \dfrac{7\pi}{6}, \dfrac{3\pi}{2}, \dfrac{11\pi}{6}$

Chapter 6

Section 6.1

1. $C = 105°, b \approx 45.25, c \approx 61.82$ **2.** About 18.32 ft
3. $B \approx 12.39°, C \approx 136.61°, c \approx 16.01$
4. $b\left(\dfrac{\sin A}{a}\right) \approx 3.0311 > 1$
5. Two solutions:
$B \approx 70.4°, C \approx 51.6°, c \approx 4.16$ ft
$B \approx 109.6°, C \approx 12.4°, c \approx 1.14$ ft
6. About 212.72 yd^2
7.

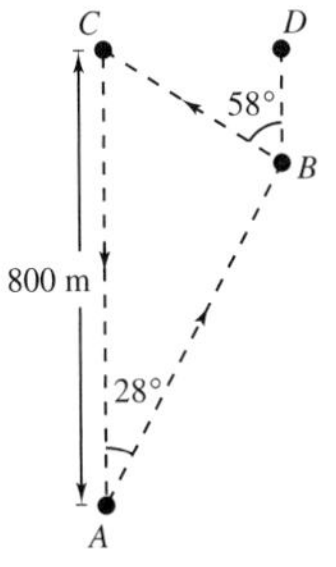

1856.59 m

Section 6.2

1. $A \approx 26.38°, B \approx 36.34°, C \approx 117.28°$
2. $B \approx 59.7°, C \approx 40.3°, a \approx 18.26$
3. About 202 ft **4.** N 15.37° E **5.** About 19.90 ft^2

Section 6.3

1. $\|\overrightarrow{PQ}\| = \|\overrightarrow{RS}\| = \sqrt{10}$, $\text{slope}_{\overrightarrow{PQ}} = \text{slope}_{\overrightarrow{RS}} = \frac{1}{3}$
$\overrightarrow{PQ}$ and $\overrightarrow{RS}$ have the same magnitude and direction, so they are equal.

2. $\mathbf{v} = \langle -5, 6\rangle, \|\mathbf{v}\| = \sqrt{61}$
3. (a) $\langle 4, 6\rangle$ (b) $\langle -2, 2\rangle$ (c) $\langle -7, 2\rangle$
4. (a) $3\sqrt{17}$ (b) $2\sqrt{13}$ (c) $5\sqrt{13}$
5. $\left\langle \dfrac{6\sqrt{37}}{37}, -\dfrac{\sqrt{37}}{37}\right\rangle$ **6.** $-6\mathbf{i} - 3\mathbf{j}$ **7.** $11\mathbf{i} - 14\mathbf{j}$
8. (a) 135° (b) About 209.74° **9.** $\langle -50\sqrt{2}, -50\sqrt{2}\rangle$
10. About 2405 lb **11.** About 451.8 mi/h, $\theta \approx 305.1°$

Section 6.4

1. -6 **2.** (a) $\langle -36, 108\rangle$ (b) 43 (c) $2\sqrt{10}$
3. 45° **4.** Yes **5.** $\frac{1}{17}\langle 64, 16\rangle$; $\frac{1}{17}\langle -13, 52\rangle$
6. About 38.8 lb **7.** About 1212 ft-lb

Section 6.5

1.

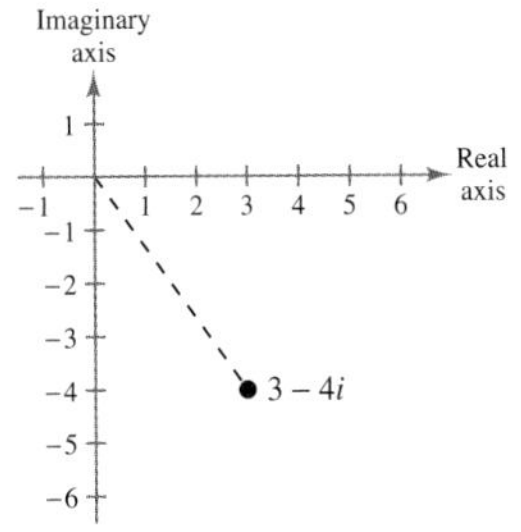

5

2. $4 + 3i$ **3.** $1 - 5i$ **4.** $2 + 3i$

5.

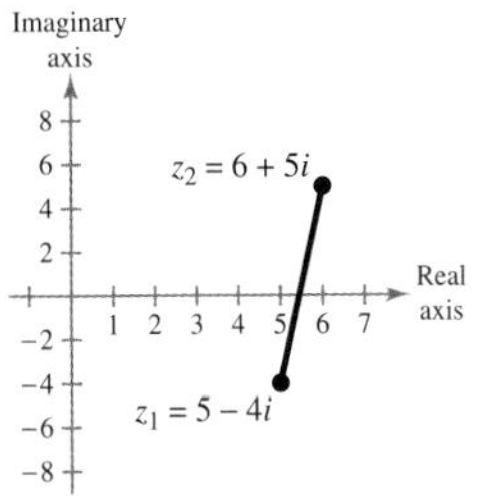

$\sqrt{82} \approx 9.06$ units

6.

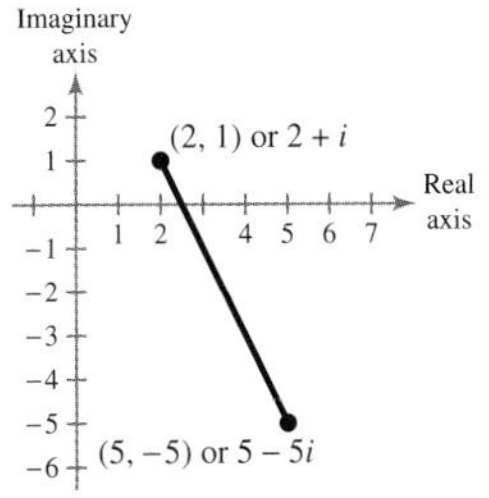

$\left(\frac{7}{2}, -2\right)$

Section 6.6

1. $z = 3\left(\cos\dfrac{\pi}{2} + i\sin\dfrac{\pi}{2}\right)$ **2.** $6\sqrt{2}\left(\cos\dfrac{7\pi}{4} + i\sin\dfrac{7\pi}{4}\right)$
3. $-4 + 4\sqrt{3}i$ **4.** 10 **5.** $12i$ **6.** $\dfrac{\sqrt{3}}{2} + \dfrac{1}{2}i$
7. -4 **8.** $1, i, -1, -i$
9. $\sqrt[3]{3} + \sqrt[3]{3}i, -1.9701 + 0.5279i, 0.5279 - 1.9701i$

Chapter 7

Section 7.1

1. (3, 3) **2.** \$6250 at 6.5%, \$18,750 at 8.5%
3. $(-3, -1), (2, 9)$ **4.** No real solution **5.** (1, 3)
6. About 14,286 bottles

Section 7.2

1. $\left(\frac{3}{4}, \frac{5}{2}\right)$ **2.** $(3, -1)$

3.

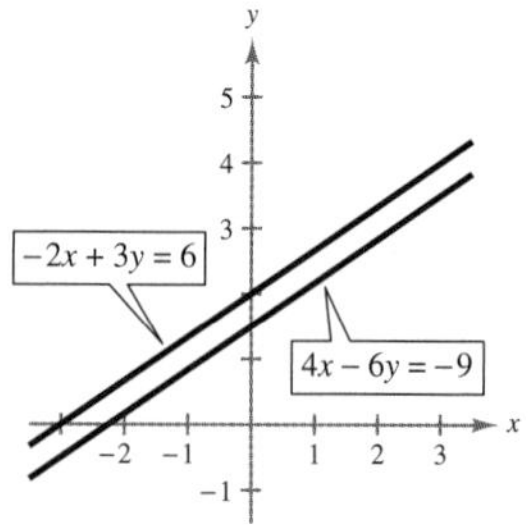

No solution, inconsistent

4. No solution

5. Infinitely many solutions

6. About 471.18 mi/h; about 16.63 mi/h

Section 7.3

1. $(2, -3, 3)$ **2.** $(1, 2, 3)$ **3.** No solution

4. Infinitely many solutions: $(-23a + 22, 15a - 13, a)$

5. Infinitely many solutions: $\left(\frac{1}{4}a, \frac{17}{4}a - 3, a\right)$

6. $-\frac{3}{2x+1} + \frac{2}{x-1}$ **7.** $-\frac{3}{x} + \frac{4}{x^2} + \frac{3}{x+1}$

8. $s = -16t^2 + 20t + 100$; The object was thrown upward at a velocity of 20 feet per second from a height of 100 feet.

9. $y = \frac{1}{3}x^2 - 2x$

Section 7.4

1. 2×3 **2.** $\begin{bmatrix} 1 & 1 & 1 & \vdots & 2 \\ 2 & -1 & 3 & \vdots & -1 \\ -1 & 2 & -1 & \vdots & 4 \end{bmatrix}$; 3×4

3. 6 **4.** Answers will vary. $(-1, 0, 1)$

5. Reduced row-echelon form

6. $(4, -2, 1)$ **7.** No solution **8.** $(7, 4, -3)$

9. $(3a + 8, 2a - 5, a)$

Section 7.5

1. $a_{11} = 6, a_{12} = 3, a_{21} = -2, a_{22} = 4$

2. $\begin{bmatrix} 6 & -2 \\ 2 & 3 \end{bmatrix}$ **3.** $\begin{bmatrix} 12 & -3 \\ 0 & 12 \\ -9 & 24 \end{bmatrix}$ **4.** $\begin{bmatrix} 12 & -11 \\ 2 & 6 \\ -11 & 10 \end{bmatrix}$

5. $\begin{bmatrix} -6 & 6 \\ -10 & 6 \end{bmatrix}$ **6.** $\begin{bmatrix} 5 & 0 \\ -1 & 4 \end{bmatrix}$ **7.** $\begin{bmatrix} -1 & 30 \\ 2 & -4 \\ 1 & 12 \end{bmatrix}$

8. $\begin{bmatrix} -3 & -22 \\ 3 & 10 \\ -5 & 10 \end{bmatrix}$ **9.** $\begin{bmatrix} 70 & -17 & 73 \\ 32 & 11 & 6 \\ 16 & -38 & 70 \end{bmatrix}$

10. (a) $\begin{bmatrix} -5 \\ 1 \end{bmatrix}$ (b) $\begin{bmatrix} 17 \\ 23 \end{bmatrix}$

11. $\begin{bmatrix} -3 \\ 1 \end{bmatrix}$; Reflection in the y-axis

12. (a) $\begin{bmatrix} -2 & -3 \\ 6 & 1 \end{bmatrix}\begin{bmatrix} x_1 \\ x_2 \end{bmatrix} = \begin{bmatrix} -4 \\ -36 \end{bmatrix}$ (b) $\begin{bmatrix} -7 \\ 6 \end{bmatrix}$

13. Total cost for women's team: \$2490
Total cost for men's team: \$2871

Section 7.6

1. $AB = I$ and $BA = I$

2. $\begin{bmatrix} 3 & 2 \\ 1 & 1 \end{bmatrix}$ **3.** $\begin{bmatrix} -4 & -2 & 5 \\ -2 & -1 & 2 \\ -1 & 0 & 1 \end{bmatrix}$ **4.** $\begin{bmatrix} \frac{4}{23} & \frac{1}{23} \\ -\frac{3}{23} & \frac{5}{23} \end{bmatrix}$

5. $(-2, 1, 7)$

Section 7.7

1. (a) -7 (b) 10 (c) 0 **2.** 0.64

3. $M_{11} = -9, M_{12} = -10, M_{13} = 2, M_{21} = 5, M_{22} = -2,$
$M_{23} = -3, M_{31} = 13, M_{32} = 5, M_{33} = -1$
$C_{11} = -9, C_{12} = 10, C_{13} = 2, C_{21} = -5, C_{22} = -2,$
$C_{23} = 3, C_{31} = 13, C_{32} = -5, C_{33} = -1$

4. -31

Section 7.8

1. 9 square units **2.** Collinear

3. $(0, 0), (2, 0), (0, 4), (2, 4)$ **4.** 10 square units

5. $(3, -2)$ **6.** $(2, -3, 1)$

7. $[15 \quad 23 \quad 12][19 \quad 0 \quad 1][18 \quad 5 \quad 0]$
$[14 \quad 15 \quad 3][20 \quad 21 \quad 18][14 \quad 1 \quad 12]$

8. 110 -39 -59 25 -21 -3 23 -18 -5 47
-20 -24 149 -56 -75 87 -38 -37

9. OWLS ARE NOCTURNAL

Chapter 8

Section 8.1

1. 3, 5, 7, 9 **2.** $1, \frac{3}{2}, \frac{1}{3}, \frac{3}{4}$

3. (a) $a_n = 4n - 3$ (b) $a_n = 2n + 1$

4. 6, 7, 8, 9, 10 **5.** 1, 3, 4, 7, 11 **6.** $2, 4, 5, \frac{14}{3}, \frac{41}{12}$

7. $4(n + 1)$ **8.** 44 **9.** (a) 0.5555 (b) $\frac{5}{9}$

10. (a) \$1000, \$1002.50, \$1005.01 (b) \$1127.33

Section 8.2

1. 2, 5, 8, 11; $d = 3$ **2.** $a_n = 5n - 6$

3. -3, 1, 5, 9, 13, 17, 21, 25, 29, 33, 37 **4.** 79

5. 217 **6.** 43,560 **7.** \$2,500,000

Section 8.3

1. $-12, 24, -48, 96$; $r = -2$ **2.** 2, 8, 32, 128, 512

3. 104.02 **4.** $a_n = 4(5)^{n-1}$; 195,312,500 **5.** $\frac{2187}{32}$

6. About 2.667 **7.** (a) 10 (b) 6.25 **8.** \$3500.85

Section 8.4

1. (a) 462 (b) 36 (c) 1 (d) 1

2. (a) 21 (b) 21 (c) 14 (d) 14

3. $x^4 + 4x^3 + 6x^2 + 4x + 1$

4. $x^4 - 4x^3 + 6x^2 - 4x + 1$

5. (a) $81y^4 - 108y^3 + 54y^2 - 12y + 1$
(b) $32x^5 + 80x^4y + 80x^3y^2 - 40x^2y^3 + 10xy^4 - y^5$

6. $y^6 + 15y^4 + 75y^2 + 125$

7. (a) $1120a^4b^4$ (b) $-412,500,000$

8. 1, 9, 36, 84, 126, 126, 84, 36, 9, 1

Section 8.5

1. 3 ways **2.** 36 ways **3.** 27,000 combinations
4. 2,600,000 numbers **5.** 24 permutations **6.** 20 ways
7. 1260 ways **8.** 21 ways

Section 8.6

1. $\{HH1, HH2, HH3, HH4, HH5, HH6, HT1, HT2, HT3, HT4, HT5, HT6, TH1, TH2, TH3, TH4, TH5, TH6, TT1, TT2, TT3, TT4, TT5, TT6\}$

2. (a) $\frac{1}{8}$ (b) $\frac{1}{4}$ **3.** $\frac{320}{2353} \approx 0.136$ **4.** $\frac{1}{962{,}598}$

5. $\frac{4}{13} \approx 0.308$ **6.** $\frac{66}{529} \approx 0.125$ **7.** $\frac{121}{900} \approx 0.134$

8. About 0.0000853

Chapter 9

Section 9.1

1. $(x - 1)^2 + (y + 1)^2 = 8$

2.

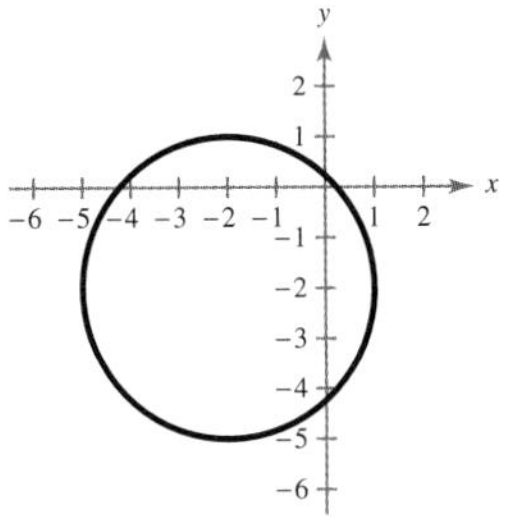

Center: $(-2, -2)$; radius: 3

3. x-intercept: $(2, 0)$
y-intercepts: $\left(0, -3 + \sqrt{5}\right), \left(0, -3 - \sqrt{5}\right)$

4. $x^2 = \frac{3}{2}y$ **5.** $(2, -3)$ **6.** $(y + 3)^2 = 8(x - 2)$

7. $y = 6x - 3$

Section 9.2

1. $\frac{(x - 2)^2}{7} + \frac{(y - 3)^2}{16} = 1$

2.

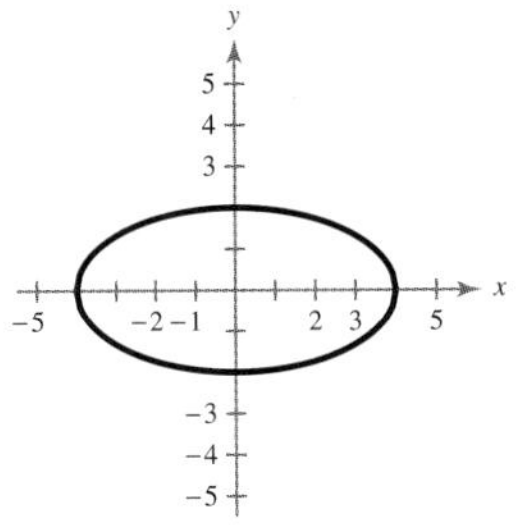

3.

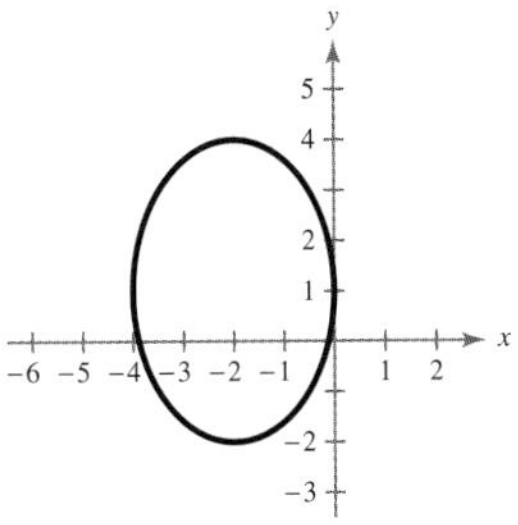

Center: $(0, 0)$
Vertices: $(-4, 0), (4, 0)$

4. Center: $(-1, 3)$; vertices: $(-4, 3), (2, 3)$; foci: $(-3, 3), (1, 3)$

5. Greatest distance: about 4.094 astronomical units
Least distance: about 0.336 astronomical units

Section 9.3

1. $\frac{(y + 1)^2}{9} - \frac{(x - 2)^2}{7} = 1$

2.

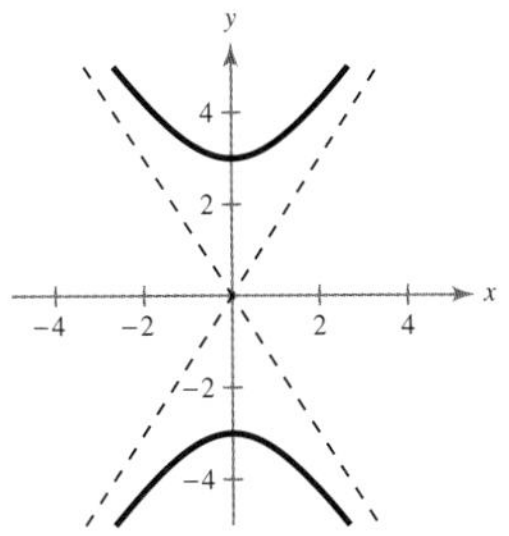

3. Foci: $\left(\pm\sqrt{13}, 1\right)$
Asymptotes: $y = 1 \pm \frac{3}{2}x$

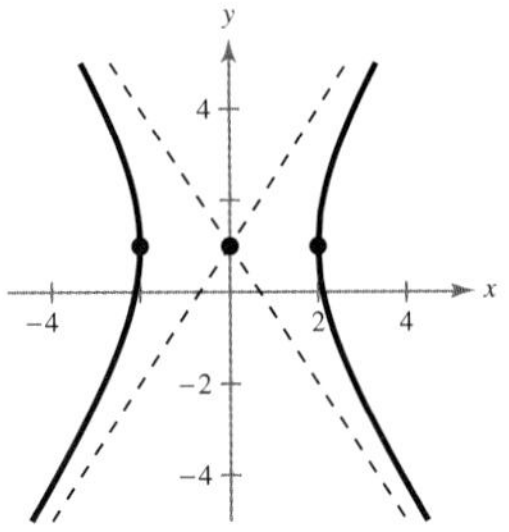

4. $\frac{(x - 6)^2}{9} - \frac{(y - 2)^2}{4} = 1$

5. The explosion occurred somewhere on the right branch of the hyperbola $\frac{x^2}{4{,}840{,}000} - \frac{y^2}{2{,}129{,}600} = 1.$

6. (a) Circle (b) Hyperbola (c) Ellipse (d) Parabola

7. $\frac{(y')^2}{12} - \frac{(x')^2}{12} = 1$

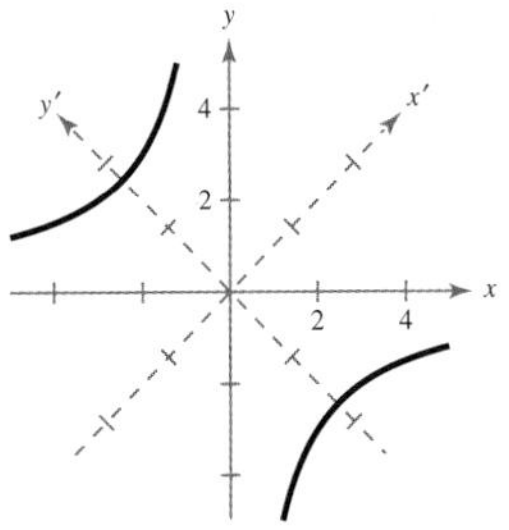

Section 9.4

1.

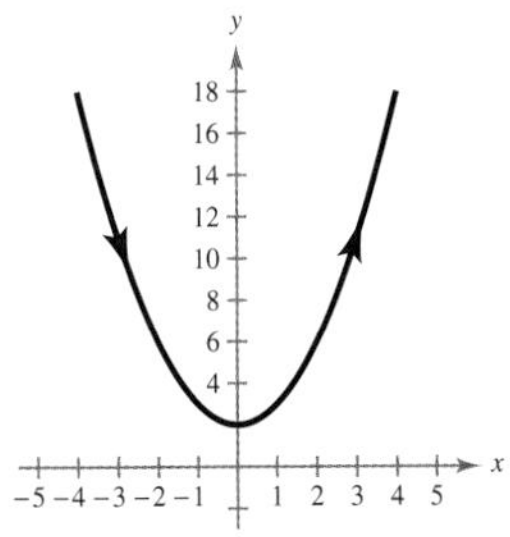

2.

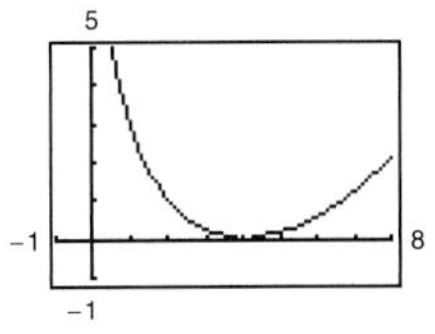

y is a function of x.

3. $y = 1 + 2x^2, x > 0$

4. (a) $x = t, y = t^2 + 2$ (b) $x = 1 - t, y = t^2 - 2t + 3$

Section 9.5

1. (a)

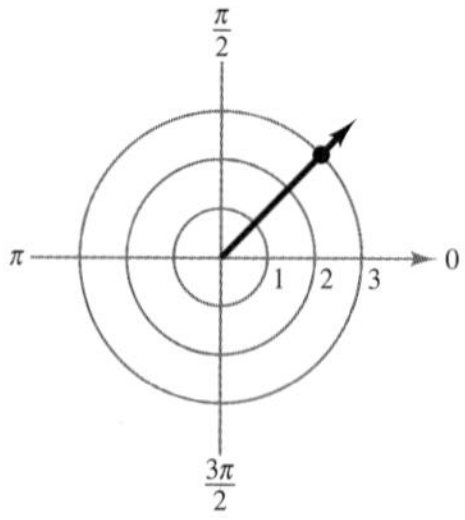

(b)

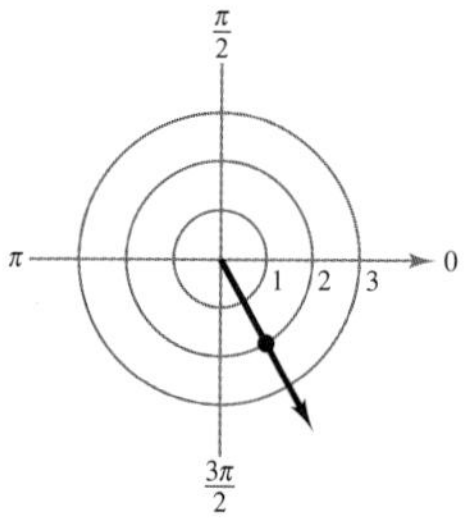

(c)

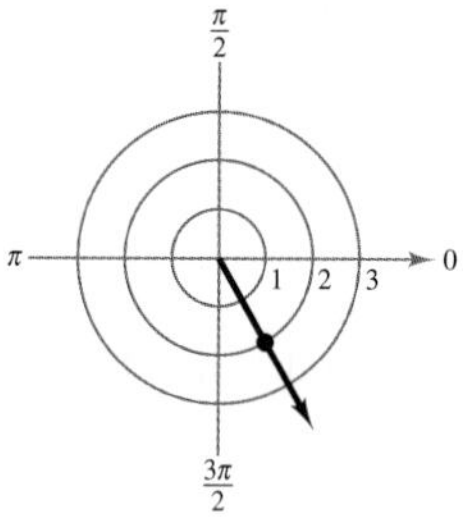

2. (a)

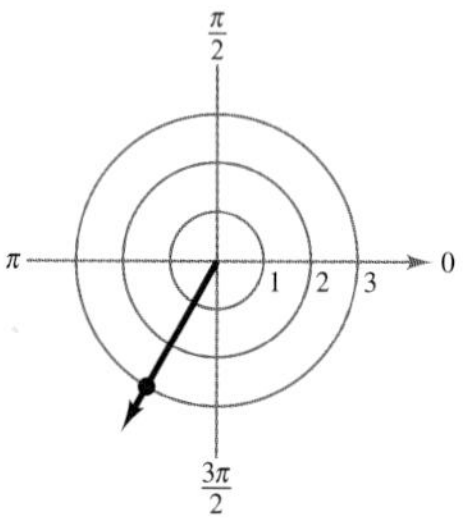

$\left(3, -\frac{2\pi}{3}\right), \left(-3, \frac{\pi}{3}\right), \left(-3, -\frac{5\pi}{3}\right)$

(b)

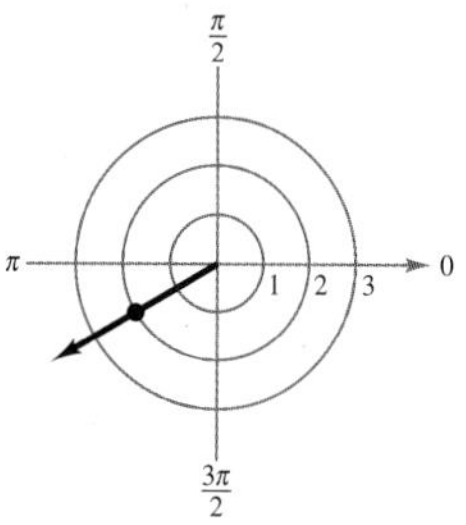

$\left(2, \frac{7\pi}{6}\right), \left(-2, -\frac{11\pi}{6}\right), \left(-2, \frac{\pi}{6}\right)$

(c)

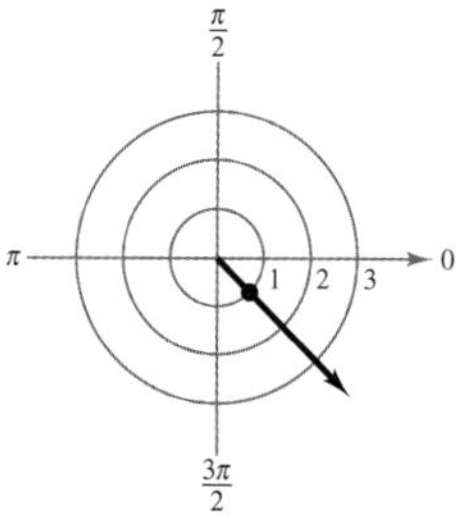

$\left(-1, -\frac{5\pi}{4}\right), \left(1, \frac{7\pi}{4}\right), \left(1, -\frac{\pi}{4}\right)$

3. $(2, 2\sqrt{3})$ **4.** $\left(2, \frac{\pi}{2}\right)$

5. The graph consists of all points that are 3 units away from $(0, 3)$; $x^2 + (y - 3)^2 = 9$

Section 9.6

1.

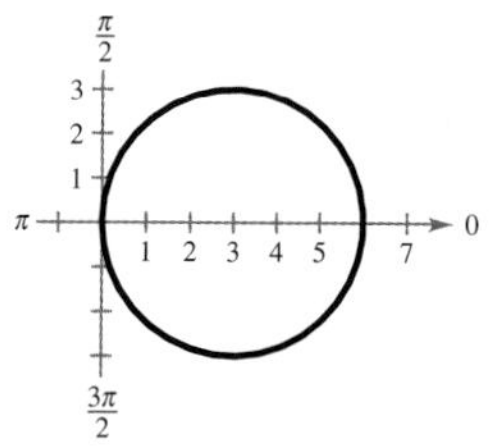

2.

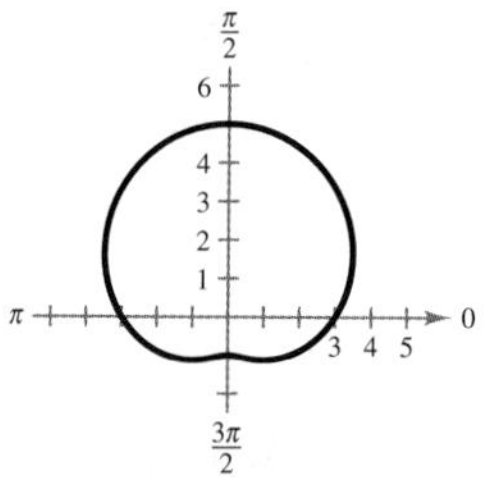

3.

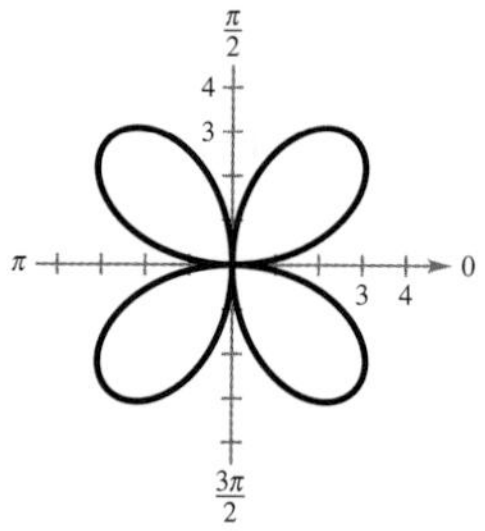

4.

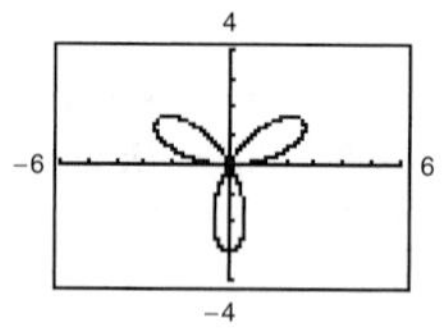

5.

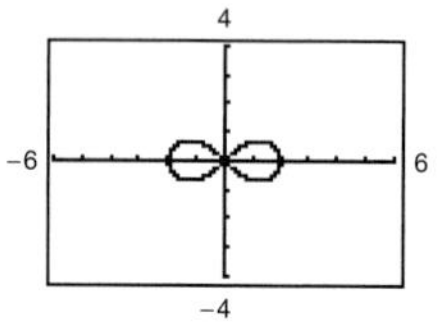

Section 9.7

1. Hyperbola

2. Hyperbola

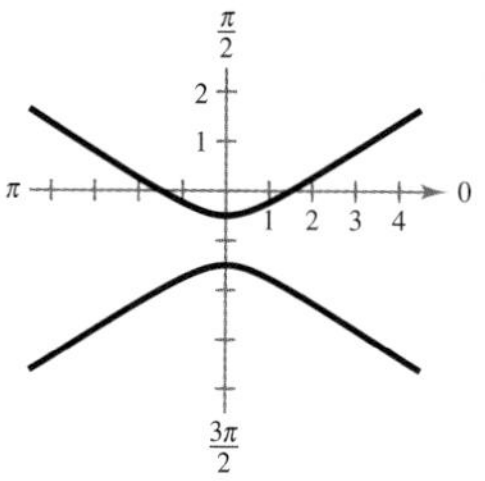

3. $r = \dfrac{2}{1 - \cos\theta}$

4. $r = \dfrac{0.626}{1 + 0.847 \sin\theta}$; about 0.340 astronomical unit

Chapter 10

Section 10.1

1.

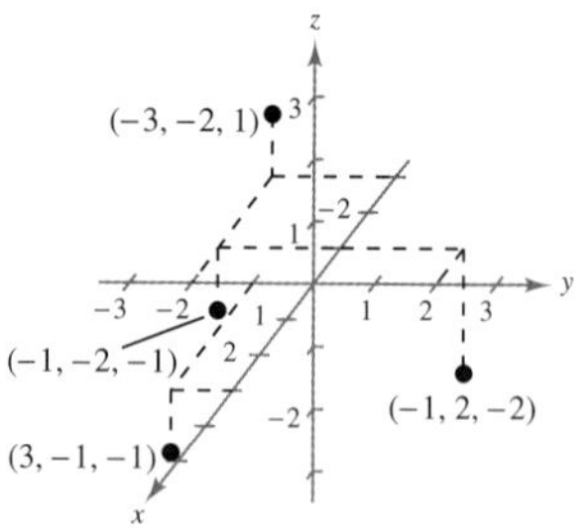

2. $\sqrt{35}$ **3.** $(5, -3, 1)$

4. $(x - 1)^2 + (y - 3)^2 + (z - 2)^2 = 16$

5. Center: $(-3, 2, -4)$; radius: 6

6. $(x - 1)^2 + (z - 8)^2 = 91$

Section 10.2

1. $\langle 1, 6, -4\rangle$; $\sqrt{53}$; $\left\langle \frac{\sqrt{53}}{53}, \frac{6\sqrt{53}}{53}, -\frac{4\sqrt{53}}{53} \right\rangle$ **2.** -2
3. About 76.5° **4.** (a) Yes (b) No (c) Yes
5. Collinear **6.** $Q(3, -5, 1)$
7. About 140.7 lb, about 216.3 lb, about 140.7 lb

Section 10.3

1. (a) $-4\mathbf{i} + 2\mathbf{j} + 5\mathbf{k}$ (b) $4\mathbf{i} - 2\mathbf{j} - 5\mathbf{k}$ (c) $\mathbf{0}$
2. $-\frac{1}{3}\mathbf{i} + \frac{2}{3}\mathbf{j} + \frac{2}{3}\mathbf{k}$
3. $\overrightarrow{AB} = -\mathbf{i} + \mathbf{j} + 2\mathbf{k}$ is parallel to $\overrightarrow{DC} = -\mathbf{i} + \mathbf{j} + 2\mathbf{k}$.
$\overrightarrow{AD} = 0\mathbf{i} - 2\mathbf{j} + 4\mathbf{k}$ is parallel to $\overrightarrow{BC} = 0\mathbf{i} - 2\mathbf{j} + 4\mathbf{k}$.
Area $= 2\sqrt{21} \approx 9.17$
4. 6.6 **5.** 15

Section 10.4

1. Parametric equations: $x = 2 + 4t, y = 1 - 2t, z = -3 + 7t$
Symmetric equations: $\frac{x-2}{4} = \frac{y-1}{-2} = \frac{z+3}{7}$
2. Parametric equations: $x = 1 + 2t, y = -6 + 11t, z = 3 + 5t$
Symmetric equations: $\frac{x-1}{2} = \frac{y+6}{11} = \frac{z-3}{5}$
3. $3x + y + 2z - 13 = 0$
4. $90°$; $x = t, y = 4t, z = -2t$ **5.** $\frac{\sqrt{30}}{5}$

Chapter 11

Section 11.1

1. 13 in. × 13 in.

2.

x	2.9	2.99	2.999	3	3.001	3.01	3.1
$f(x)$	-2.8	-2.98	-2.998	?	-3.002	-3.02	-3.2

-3

3.

x	0.9	0.99	0.999	1
$f(x)$	0.20408	0.200401	0.200040	?

x	1.001	1.01	1.1
$f(x)$	0.199960	0.199601	0.196078

0.2

4. 7 **5.** -3 **6.** Answers will vary.
7. The limit does not exist. **8.** The limit does not exist.
9. (a) $\frac{1}{4}$ (b) 27 (c) $-\frac{1}{\pi}$ **10.** (a) 7 (b) $\frac{7}{3}$

Section 11.2

1. 1 **2.** $\frac{1}{56}$ **3.** $\frac{1}{2}$ **4.** 1 **5.** 0 **6.** -1; 1
7. 2 **8.** Answers will vary. **9.** 0

Section 11.3

1. 4
2. 10 degrees per month; This means that you can expect the monthly normal temperature in May to be about 10 degrees higher than the normal temperature in April.
3. 4 **4.** -3 **5.** $m = 4x$; -12; 8 **6.** $f'(x) = 3x^2 + 2$
7. $f'(x) = \frac{1}{2\sqrt{x-1}}$; $\frac{1}{2}$; $\frac{1}{6}$; $y = \frac{1}{2}x$; $y = \frac{1}{6}x + \frac{4}{3}$

Section 11.4

1. 0 **2.** (a) 0 (b) -2 (c) Limit does not exist.
3. (a) \$24.00 (b) \$6.00 (c) \$4.20 (d) \$4.00
4. (a) $\frac{1}{2}$ (b) $\frac{1}{3}$ (c) 0 **5.** (a) $\frac{10}{7}$ (b) 1

Section 11.5

1. 385 **2.** 1.85; 1.535; 1.5035; 1.50035 **3.** 0
4. 4.5 square units **5.** $\frac{3}{2}$ square units

Appendix B

Appendix B.1

1. (a)

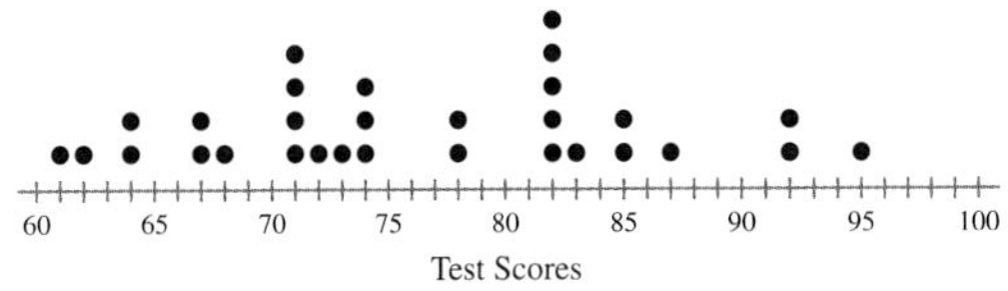

(b) 82 (c) 34

2. *Sample answer:*

Interval — Tally
[60, 65) ||||
[65, 70) |||
[70, 75) 卌 ||||
[75, 80) ||
[80, 85) 卌 |
[85, 90) |||
[90, 95) ||
[95, 100) |

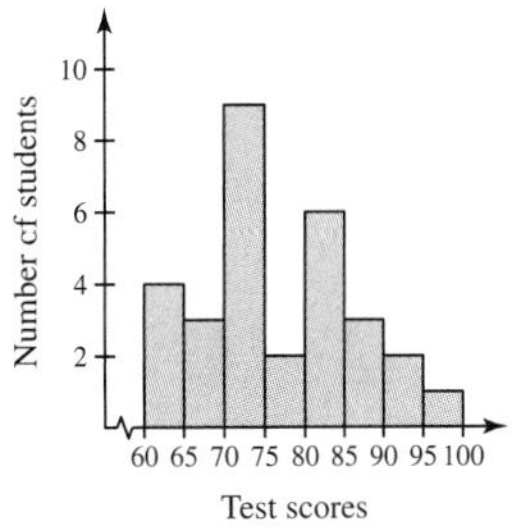

3.

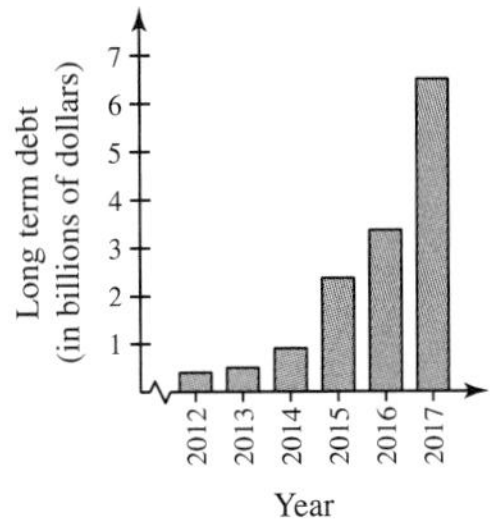

4.

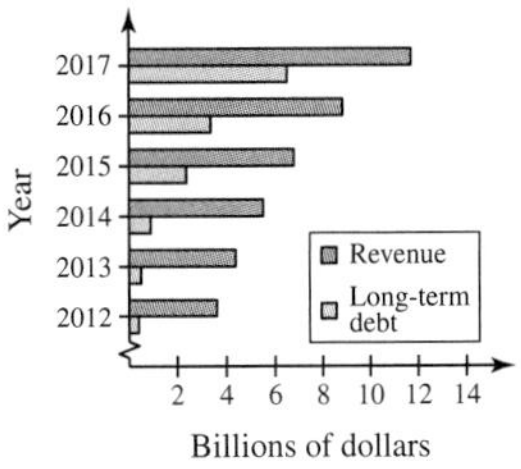

Answers will vary.

5.

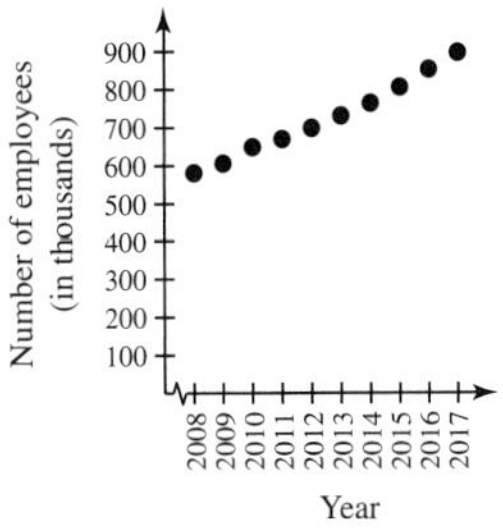

6.

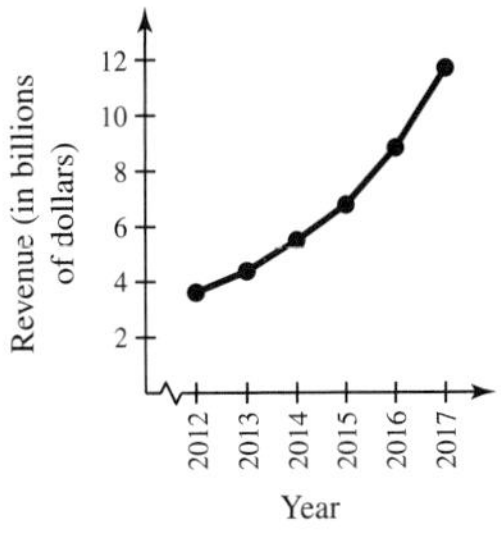

Appendix B.2

1. Mean: 75.8; median: 73.5; mode: 82

2. (a) *Sample answer:* Mean or median
(b) *Sample answer:* Median
(c) *Sample answer:* Mean
(d) *Sample answer:* Median

3. C; B **4.** A: 5; B: about 2.28; C: about 1.72

5. About 1.36 **6.** About 98%

7. Lower quartile: 23; Upper quartile: 64.5

8.

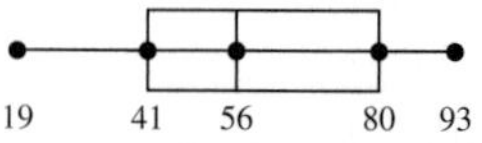

Appendix B.3

1. $y = \frac{17}{10}x + \frac{2}{5}$

Library of Parent Functions Review

Chapter 4 (page 348)

1. $|x|$ **2.** x^2 **3.** $\dfrac{1}{x}$ **4.** $[\![x]\!]$ **5.** x^3 **6.** $\tan x$

7. $\log_a x, a > 1$ **8.** $\sqrt{x}$ **9.** $a^x, a > 1$ **10.** $\sin x$

11. $\csc x$ **12.** $\arcsin x$

13. $f(x) = x^3$; vertical shift of f two units downward; $g(x) = x^3 - 2$

14. $f(x) = x^2$; reflection in the x-axis, vertical shift of f two units upward; $g(x) = -x^2 + 2$

15. $f(x) = |x|$; horizontal shift of f two units right; $g(x) = |x - 2|$

16. $f(x) = \sqrt{x}$; vertical shift of f three units downward; $g(x) = \sqrt{x} - 3$

17. $f(x) = [\![x]\!]$; horizontal shift of f one unit left; $g(x) = [\![x + 1]\!]$

18. $f(x) = \dfrac{1}{x}$; horizontal shift of f one unit left; $g(x) = \dfrac{1}{x + 1}$

19. a **20.** c

21.

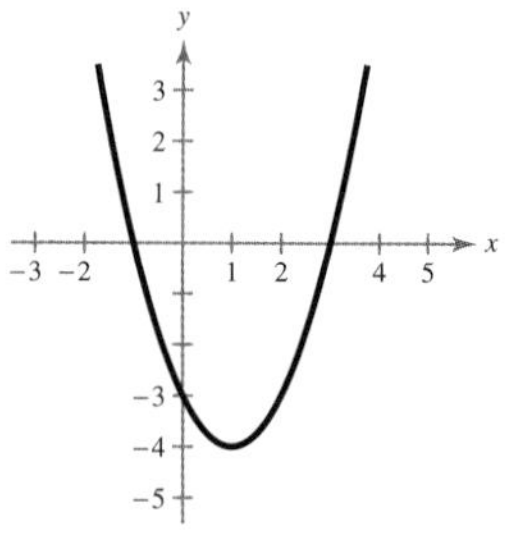

22.

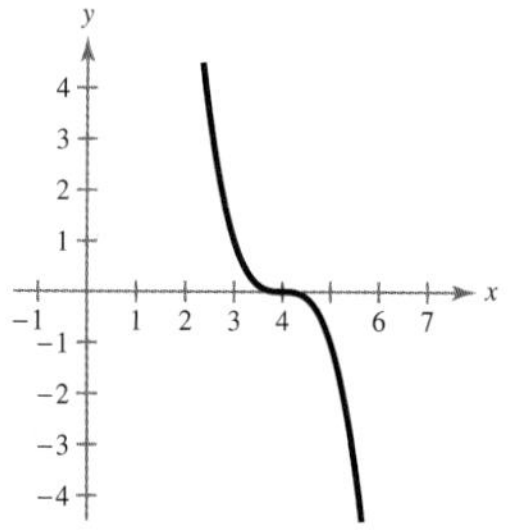

23.

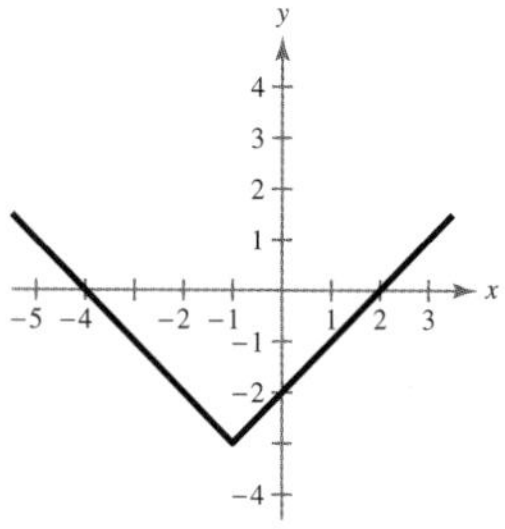

24.

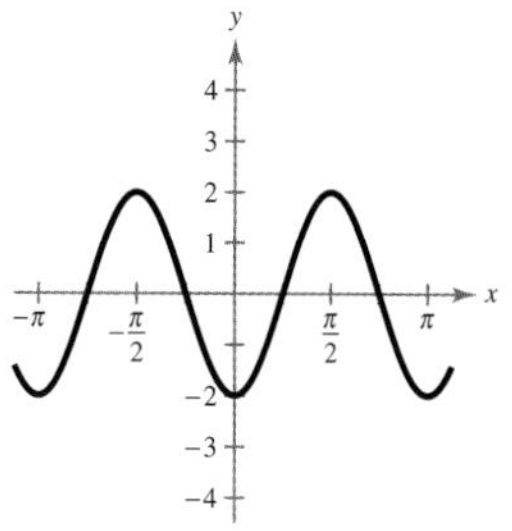

Index of Selected Applications

Index

Q

R

T

Library of Parent Functions Summary

Linear Function ***(p. 6)***

$f(x) = x$

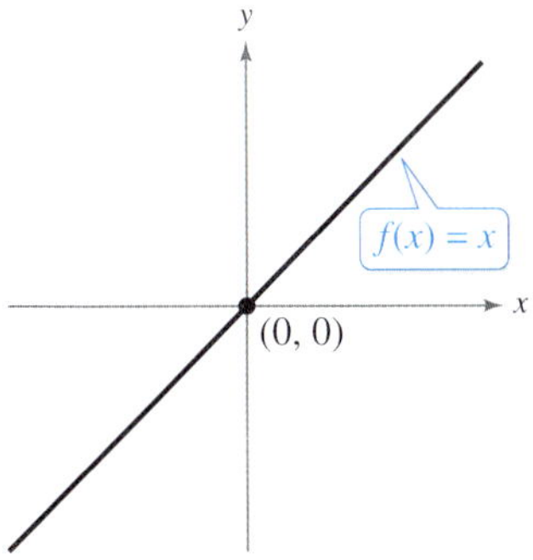

Domain: $(-\infty, \infty)$
Range: $(-\infty, \infty)$
Intercept: $(0, 0)$
Increasing

Absolute Value Function ***(p. 19)***

$f(x) = |x| = \begin{cases} x, & x \geq 0 \\ -x, & x < 0 \end{cases}$

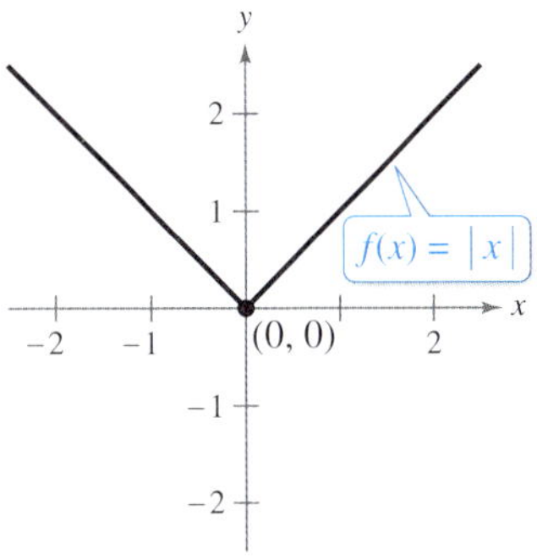

Domain: $(-\infty, \infty)$
Range: $[0, \infty)$
Intercept: $(0, 0)$
Decreasing on $(-\infty, 0)$
Increasing on $(0, \infty)$
Even function
y-axis symmetry

Square Root Function ***(p. 20)***

$f(x) = \sqrt{x}$

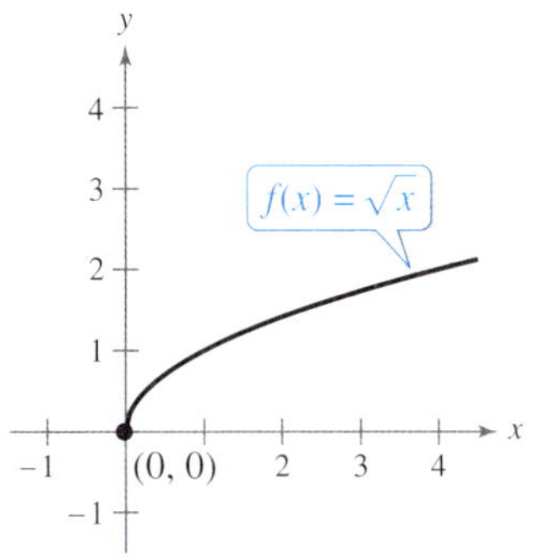

Domain: $[0, \infty)$
Range: $[0, \infty)$
Intercept: $(0, 0)$
Increasing on $(0, \infty)$

Greatest Integer Function ***(p. 34)***

$f(x) = [\![x]\!]$

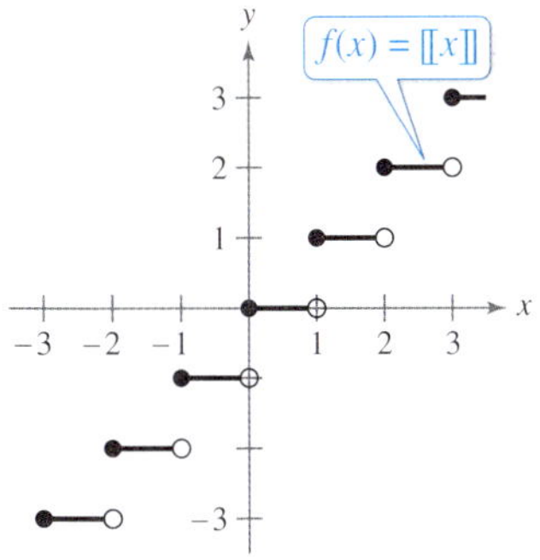

Domain: $(-\infty, \infty)$
Range: the set of integers
x-intercepts: in the interval $[0, 1)$
y-intercept: $(0, 0)$
Constant between each pair of consecutive integers
Jumps vertically one unit at each integer value

Quadratic Function ***(p. 92)***

$f(x) = x^2$

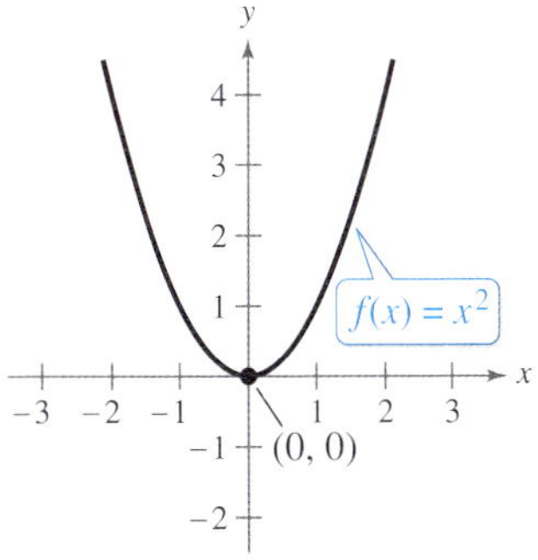

Domain: $(-\infty, \infty)$
Range: $[0, \infty)$
Intercept: $(0, 0)$
Decreasing on $(-\infty, 0)$
Increasing on $(0, \infty)$
Even function
Axis of symmetry: $x = 0$
Relative minimum or vertex: $(0, 0)$

Cubic Function ***(p. 101)***

$f(x) = x^3$

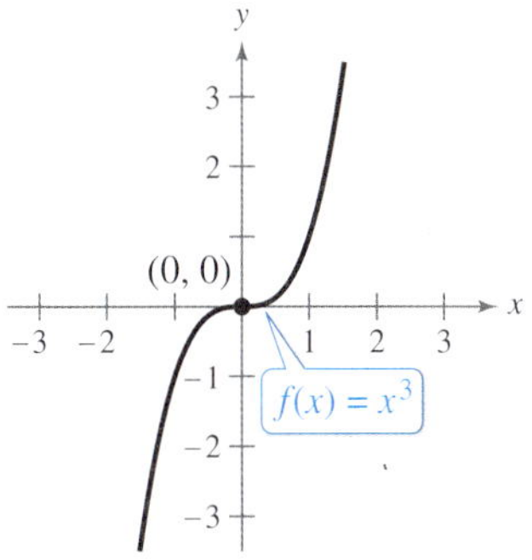

Domain: $(-\infty, \infty)$
Range: $(-\infty, \infty)$
Intercept: $(0, 0)$
Increasing on $(-\infty, \infty)$
Odd function
Origin symmetry

Rational Function ***(p. 152)***

$f(x) = \dfrac{1}{x}$

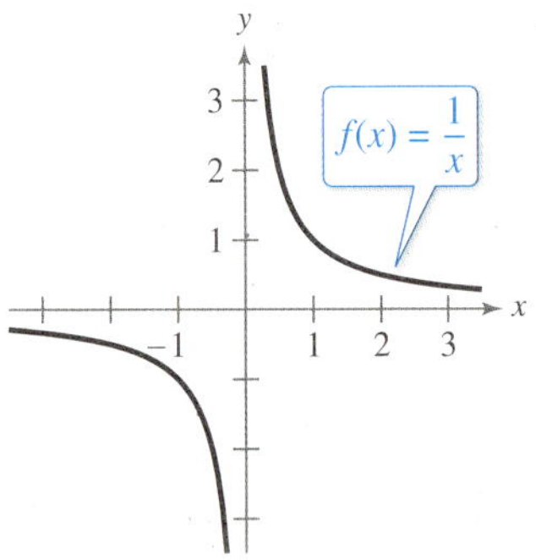

Domain: $(-\infty, 0) \cup (0, \infty)$
Range: $(-\infty, 0) \cup (0, \infty)$
No intercepts
Decreasing on $(-\infty, 0)$ and $(0, \infty)$
Odd function
Origin symmetry
Vertical asymptote: y-axis
Horizontal asymptote: x-axis

Exponential Function ***(p. 184)***

$f(x) = a^x, a > 1$

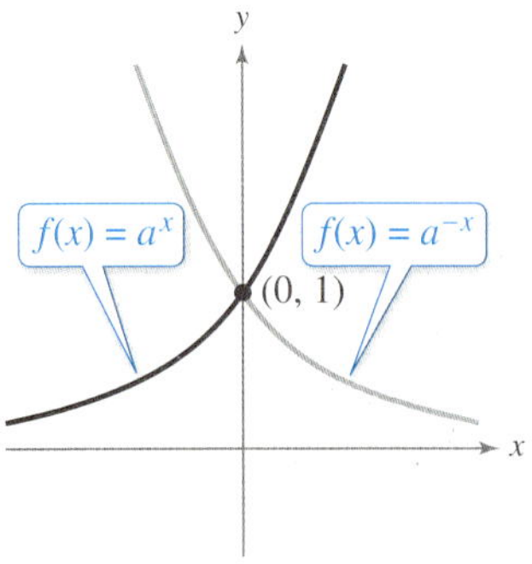

Domain: $(-\infty, \infty)$
Range: $(0, \infty)$
Intercept: $(0, 1)$
Increasing on $(-\infty, \infty)$
for $f(x) = a^x$
Decreasing on $(-\infty, \infty)$
for $f(x) = a^{-x}$
x-axis is a horizontal asymptote
Continuous

Logarithmic Function ***(p. 197)***

$f(x) = \log_a x, a > 1$

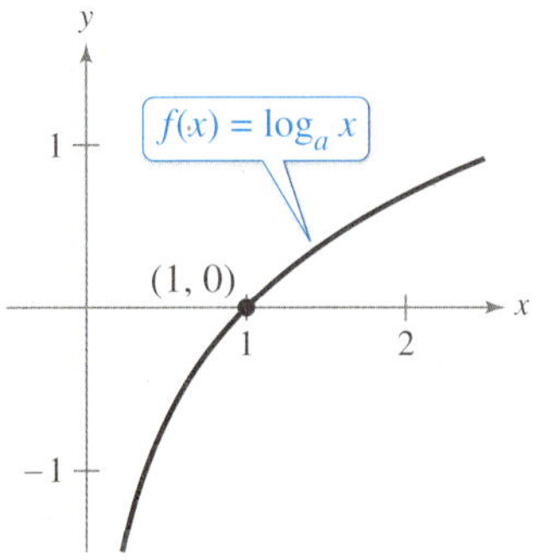

Domain: $(0, \infty)$
Range: $(-\infty, \infty)$
Intercept: $(1, 0)$
Increasing on $(0, \infty)$
y-axis is a vertical asymptote
Continuous
Reflection of graph of $f(x) = a^x$
in the line $y = x$

Sine Function ***(p. 295)***

$f(x) = \sin x$

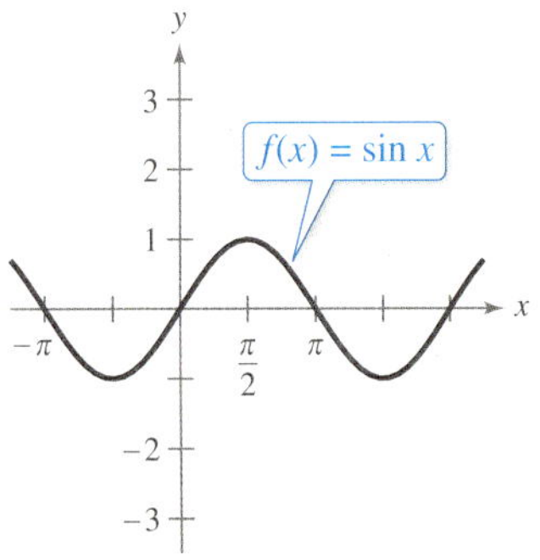

Domain: $(-\infty, \infty)$
Range: $[-1, 1]$
Period: 2π
x-intercepts: $(n\pi, 0)$
y-intercept: $(0, 0)$
Odd function
Origin symmetry

Cosine Function ***(p. 295)***

$f(x) = \cos x$

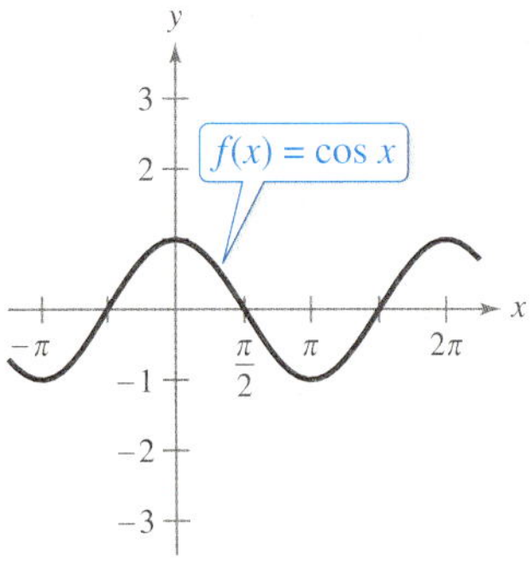

Domain: $(-\infty, \infty)$
Range: $[-1, 1]$
Period: 2π
x-intercepts: $\left(\dfrac{\pi}{2} + n\pi, 0\right)$
y-intercept: $(0, 1)$
Even function
y-axis symmetry

Tangent Function ***(p. 306)***

$f(x) = \tan x$

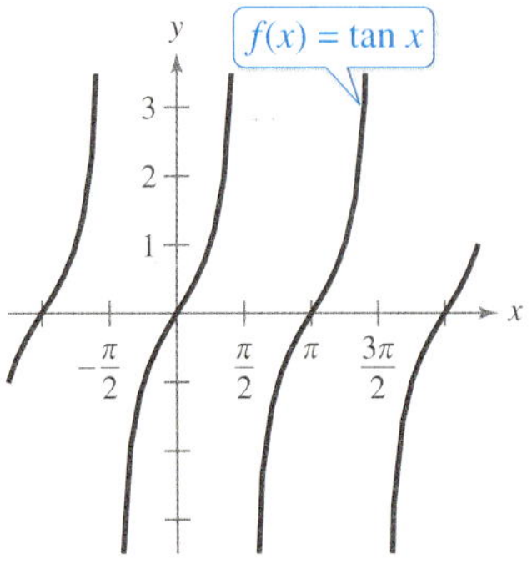

Domain: $x \neq \dfrac{\pi}{2} + n\pi$
Range: $(-\infty, \infty)$
Period: π
x-intercepts: $(n\pi, 0)$
y-intercept: $(0, 0)$
Vertical asymptotes: $x = \dfrac{\pi}{2} + n\pi$
Odd function
Origin symmetry